R. V. Krstić

Illustrated Encyclopedia of Human Histology

With 1576 Figures

Springer-Verlag
Berlin Heidelberg New York Tokyo 1984

Dr. Radivoj V. Krstić
Professeur associé
Université de Lausanne
Institut d'Histologie et d'Embryologie
CH-1011 Lausanne

ISBN 978-3-642-51598-9 ISBN 978-3-642-51596-5 (eBook)
DOI 10.1007/978-3-642-51596-5

Library of Congress Cataloging in Publication Data. Krstić, Radivoj V., 1935–. Illustrated encyclopedia of human histology. 1. Histology-Dictionaries. I. Title. QM550.2.K77 1984 611'.003'21 84-10569

Softcover reprint of the hardcover 1st edition 1984

Reproduction of the figures: Gustav Dreher GmbH, Stuttgart

Typesetting, printing and bookbinding: Brühlsche Universitätsdruckerei, Giessen
2121/3130-543210

Dedicated to Mrs. Mira Ferić
and to the illustrious memory
of Mr. Jerko Ferić, C.E.

Preface

During the past three decades, histology has seen enormous progress, thanks to new techniques and new investigation instruments. Numerous discoveries of important structures and morphofunctional phenomena have been described in a wealth of papers of ever increasing size and complexity. These publications have become difficult to follow, not only because of their number, but also because of a disparity of terminology and the multitude of synonyms employed by different authors. All of this makes reading and comprehension of the progress that has been achieved laborious, even for histologists, but especially for students, researchers in other basic branches of medicine, or clinicians who have to consult histological texts during their studies or investigations.

In order to facilitate the orientation of all those interested in histology, a concise and practical volume in encyclopedic style, defining and, at the same time, illustrating fundamental histological terms, enumerating synonyms, and describing morphofunctional phenomena has become necessary, both because a work of this genre does not exist and because the list of Latin terms elaborated by the International Anatomical Nomenclature Committee in 1977 does not give illustrations or definitions of histological terms. The present work attempts to redress this deficiency.

With regard to the entries, the reader will find the name of the item followed by, in parentheses, any synonyms and the Latin term, if one exists, according to Nomina Anatomica (4th edn., Amsterdam, Oxford: Excerpta Medica, 1977) marked by an asterisk. Then comes a definition of the item, a description which is as elementary as possible, and in many cases one or more illustrations. Cross-references between related items or phenomena are indicated by arrows in the text. After the entry, the reader will find additional cross-references pertaining to the item described, as well as a list of suggested further reading.

It is clear that the elaboration of a practical work which has the goal of giving up-to-date direct visual information as rapidly as possible must necessarily be a technical compromise in terms of the number of items, the number and format of illustrations, the size of characters, and the length of texts. For this reason, texts are laconic, literature appearing in periodicals is cited without titles, and illustrations are small, demonstrating only the structures related to the item in question.

Even though the major part of the halftone illustrations, pen drawings, and diagrams are originals which I have drawn for this book, a certain number of them have been reproduced, thanks to the kidness of Springer-Verlag from my books *Ultrastructure of the Mammalian Cell, Springer-Verlag 1979,* and *Die Gewebe des Menschen und der Säugetiere, Springer-Verlag 1984.* Here, I would like to express my deep gratitude to Mr. J. Wilewski, who helped me in the execution of a number of pen drawings and diagrams for the present book.

During the several years it took to elaborate this work, many people have kindly aided me in different ways, and I thank them very cordially. Without their help, this work would have remained only an idea.

For technical assistance in light and electron microscopy, I am grateful to Miss G. Borel, Mrs. C. Thommen-Wäckerlin, Mrs. A. Lochman, Mr. P.-A. Milliquet, and particularly to Mrs. D. Richard who, among other duties, kindly acted as my secretary.

For corrections of the English, I thank warmly my former students, Miss A. Dorward, Mr. M. Seftel, and Mr. D. Rappaport. I am particularly grateful to Mrs. E. Warwood, who had the courage to reread and recorrect all the texts. I am obliged to Mrs. M. Devolz for her very efficient typing, and to Mrs. A. Ghelber for her help in researching scientific literature. Also, I am most grateful to Springer-Verlag for the acceptance of my ideas concerning the layout of this book. Finally, many thanks go to my wife and to my son for the great patience and sympathy they have shown during the several years of preparation of the present work.

I am fully aware that a work such as this must necessarily contain imperfections, although I have attempted to reduce them to a minimum. For those which have escaped me, I am, naturally, fully responsible; here, I simply ask the benevolent reader for understanding; at the same time I am open and thankful for all constructive criticisms and suggestions.

Le Mont-sur-Lausanne, Summer 1984 Radivoj V. Krstić

Most frequently used symbols and abbreviations

*	Latin term according to Nomina Anatomica, 4th edn., Amsterdam, Oxford: Excerpta Medica, 1977
↑	see under
ADP	adenosine diphosphate
ATP	adenosine triphosphate
cAMP	cyclic 3′ 5′-adenosine monophosphate
CNS	central nervous system
DNA	deoxyribonucleic acid
EM	electron microscope(y)
ER	endoplasmic reticulum
LM	light microscope(y)
mRNA	messenger ribonucleic acid
PNS	peripheral nervous system
rER	rough endoplasmic reticulum
RNA	ribonucleic acid(s)
SEM	scanning electron microscope(y)
sER	smooth endoplasmic reticulum
TEM	transmission electron microscope(y)
tRNA	transfer ribonucleic acid

Items are always abbreviated by capitalization of the initial letter, followed by a period, irrespective of plurals and genitives.

A

Å (Ångström): A unit of wavelength = 10^{-8} cm; in the ↑ SI system = 0.1 nm. Å is now out of use in ↑ histology.

↑ Nanometer

A-bands, of myofibrils (stria A*): 1) Light microscope = dark stained, doubly refractile and, in ↑ polarizing microscope, anisotropic cylindrical segments (A) along ↑ myofibrils (Mf) of ↑ cardiac muscle cells and ↑ skeletal muscle fibers. In exceptional preparations, a paler ↑ H-band traversing the center of A. can be seen. Length of A. varies according to muscle examined (approx. 1.5–2 μm). 2) Transmission electron microscope, longitudinal section = principal constituents of A. are parallel arrays of ↑ myosin myofilaments (My), slightly thicker parts of which are bound together by slender cross connections (CC), giving rise to a transversal density, ↑ M-line. Extending to ↑ Z-lines of ↑ I-bands, ↑ actin myofilaments (Ac) interdigitate with sets of myosin myofilaments of adjacent A. In relaxed muscle, actin myofilaments do not meet within A., leaving a clear zone between, ↑ H-band. 3) Transmission electron microscope, cross section = a hexagonal disposition of myosin myofilaments at levels of M-line and H-band; in remaining part of A., six actin myofilaments are arranged regularly around myosin myofilaments. The 10 to 20 nm-large interspaces between two myosin myofilaments are traversed by fine cross bridges corresponding to ↑ heavy meromyosin (HMM) which extends radially from each myofilament toward neighboring actin myofilaments. Length of A. is determined by length of myosin myofilaments and remains constant in all phases of contraction. There are no structural differences between A. of cardiac and skeletal muscle.

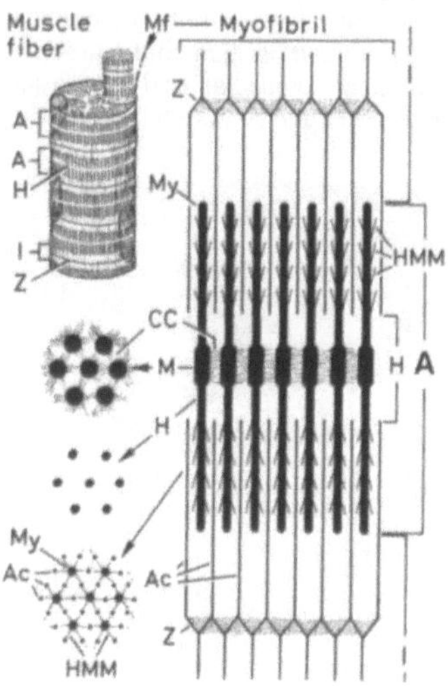

↑ Bands, of striated muscles

ABP: ↑ Actin-binding protein; ↑ Androgen-binding protein

Absorption cavity (erosion tunnel, resorption cavity, lacuna ossea*): A tubular cavity (AC) formed by osteolytic activity of ↑ osteoclasts (Oc) in compact ↑ bone. Osteoclasts are followed by blood vessels (BV) and young connective tissue cells (CC), which differentiate into ↑ osteoblasts (Ob). The latter deposit new bone ↑ lamellae (BL) around inner surface of A., thus forming new ↑ osteons. A. are frequent during secondary ↑ bone formation.

↑ Howship's lacunae

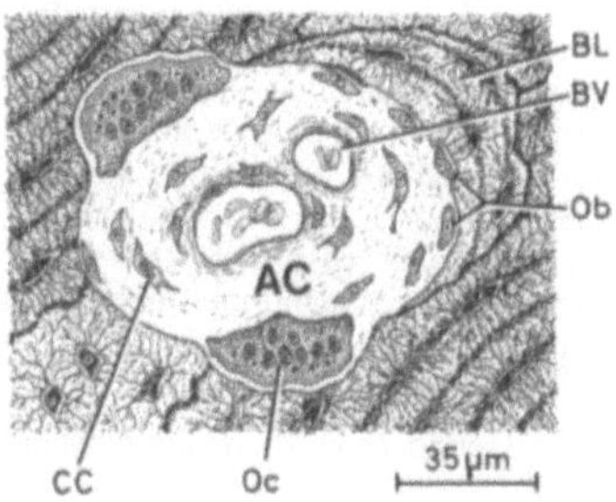

Absorption intestinal, of lipids: A process of passage of lipids through intestinal epithelium. Sequence: Triglycerides (Tr) of dietary fats are hydrolyzed by pancreatic lipase (L) to free fatty acids (FA) and monoglycerides (Mo). These combine in gut lumen with bile salts (BS) to form micelles (Mi) of about 2–5 nm. In contact with ↑ microvilli (Mv), fatty acids and monoglycerides diffuse, without requiring energy, across the plasmalemma (P) of ↑ absorptive cell (AC). Membranes of smooth ↑ endoplasmic reticulum (sER) resynthesize triglycerides from absorbed fatty acids and monoglycerides using their own enzymatic system; lipid droplets (D) are visible first in lumina of apical smooth endoplasmic reticulum. They are then transported to the rough endoplasmic reticulum (rER), where a protein component is incorporated under expenditure of energy. Resulting ↑ lipoprotein droplets migrate to ↑ Golgi apparatus (G) where a carbohydrate moiety and membranous investment are added. Such glycolipoprotein droplets coalesce with lateral plasmalemma of absorptive cells and are released by ↑ exocytosis. Particles of newly synthesized fat, fat ↑ chylomicrons (Ch), are transported through lateral intercellular clefts (IC), cross basal lamina (BL), and enter ↑ lacteals (La). (Fig., left modified from Cardell et al. 1967)

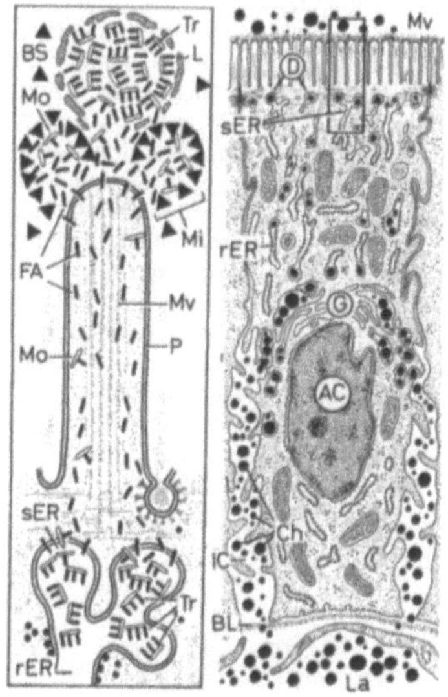

Cardell, R.R., Badenhausen, S., Porter, R.K.: J. Cell Biol. *34*, 123 (1967); Marenus, K.D., Sjöstrand, F.S.: J. Ultrastr. Res. *79*, 92 and 110 (1982); Ockner, R.K., Isslbacher, K.J.: Rev. Physiol. Biochem. Pharmacol. *71*, 107 (1974); Ohsima, Y.: Arch. Histol. Jpn. *40*, 153 (1977); Schonfeld, G., Bell, E., Alpers, D.H.: J. Clin. Invest. *61*, 1539 (1978)

Absorption intestinal, of polysaccharides: These substances are first broken down by ptyalin and pancreatic β-amylase to di- and monosaccharides, then absorbed without significant morphological change through ↑ absorptive cells into blood ↑ capillaries.

Absorption intestinal, of proteins: Prior to absorption, ingested proteins are hydrolyzed to amino acids. Only in suckling period of newborns can intact proteins and antibodies penetrate ↑ absorptive cells by ↑ micropinocytosis and reach circulation.

Yamamoto, T.: Arch. Histol. Jpn. *45*, 1 (1982)

Absorptive cells (enterocyte, cellula columnaris*): Tall columnar epithelial cells (height 18–25 μm) lining most of luminal surface of small intestine. Elliptical frequently invaginated nucleus

(N) with prominent nucleolus (Nu) is located in lower part of cell body. Cytoplasm contains numerous filamentous mitochondria (M), a well-developed supranuclear ↑ Golgi apparatus (G), and ramified profiles of smooth ↑ endoplasmic reticulum (sER) immediately below ↑ terminal web (TW). Rough ↑ endoplasmic reticulum (rER) is frequently continuous with the smooth. Presence of some lysosomes (Ly) and an appreciable number of free ribosomes (R). Microvilli (Mv) and their ↑ glycocalyx (Gc) form a ↑ striated border (SB). ↑ Micropinocytotic vesicles (Mp) are scattered in terminal web area and on lateral and basal cell surfaces; many lateral interdigitations (Int) strengthen contact with neighboring cells. Well-developed ↑ junctional complex (JC) surrounds ↑ apical pole; a ↑ basal lamina (BL) underlies basal cell pole. A. originate from stem cells of ↑ Lieberkühn's crypts; life span of A. is about 1.5–3 days. Function: absorption of digestive end products.

↑ Absorption intestinal, of lipids, polysaccharides, proteins

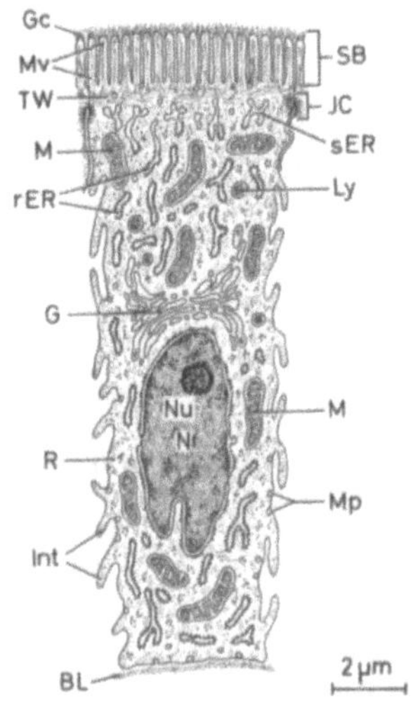

Böck, P., Tillmann, B., Osterkamp, U.: Z. mikrosk.-anat. Forsch. *94*, 1077 (1980); Cheng, H., Bjerknes, M.: Anat. Rec. *203*, 251 (1982); Colony Moxey, P., Trier, J.S.: Anat. Rec. *195*, 463 (1979); Pavelka, M., Ellinger, R.: J. Ultrastruct. Res. *77*, 210 (1981); Stenling, R., Helander, H.F.: Cell Tissue Res. *217*, 11 (1981)

Acanthocyte: A term synonymously used for a crenated ↑ erythrocyte.

Acanthosomes: ↑ Coated vesicles

Accessory glands, of male reproductive system: ↑ Bulbourethral glands; ↑ Prostate; ↑ Seminal vesicles

Accessory lacrimal glands: ↑ Lacrimal glands, accessory

Accidental involution, of thymus: ↑ Thymus, accidental involution of

Accidents, of cell division: Abnormal events that may occur during ↑ meiosis and ↑ mitosis resulting in numerical and morphological anomalies of ↑ chromosomes.

↑ Chromosomes, anomalies of

Accommodation, of eye: ↑ Eye, accommodation of

A-cells, of gastrointestinal tract (glucagon-producing cells): Round or oval pale cells scattered in ↑ gastrointestinal tract of dogs and cats, possibly also in other mammals. Morphologically, A. are very similar to ↑ A-cells of ↑ pancreatic islets. Their ↑ secretory granules, measuring about 430 nm with a round dense core surrounded by a clear halo, are located in basal cell pole and are indistinguishable from granules of pancreatic A-cells. On the basis of immunohistochemical analyses, it is believed that A. produce ↑ enteroglucagon.

↑ Endocrine cells, of gastrointestinal tract

Baetens, D., Rufener, C., Srikant, B.C., Dobbs, R., Unger, R., Orci, L.: J. Cell Biol. *69*, 455 (1976)

A-cells, of pancreatic islets (alpha cell, glucagon cell, cellula alpha*): Argyrophilic cells located predominantly at periphery of ↑ pancreatic islets making up about 20% of their volume. Polygonal in shape, A. contain a deeply invaginated nucleus (N), with a conspicuous nucleolus (Nu), a well-developed Golgi apparatus (G), some short cisternae of rough ↑ endoplasmic reticulum (rER), elongated mitochondria (M), few lysosomes ↑ lipofuscin pigment, and round strongly electron-dense ↑ unit membrane-bound secretory granules (S) that are insoluble in alcohol. Granules are uniform in size (300 nm) and are synthesized within Golgi apparatus. A halo of low density separates spherical osmiophilic core from granule's membrane. Contents of secretory granules are discharged from cell body by ↑ exocytosis (inset). A. originate from neural crest and migrate into pancreatic buds. Those of ventral pancreatic bud produce ↑ pancreatic polypeptide, and those of dorsal pancreatic bud produce ↑ glucagon. BL = ↑ basal lamina, Cap = capillary, P = ↑ plasmalemma

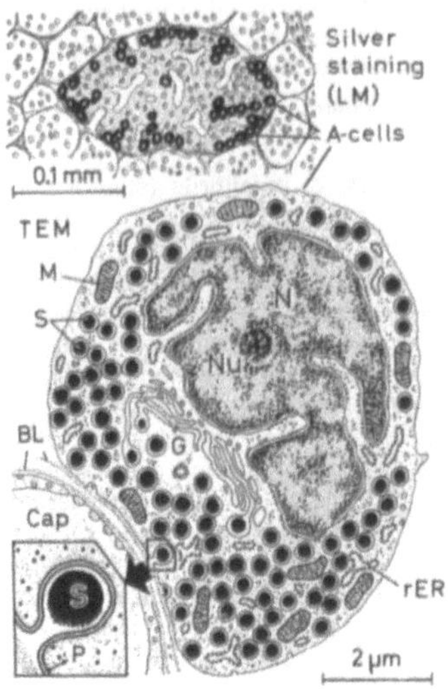

Unger, R.H., Orci, L. (eds.): Glucagon. Current Endocrinology Series. New York, Amsterdam: Elsevier 1981

A-cells, of synovial membrane: ↑ Synovial cells

Acellular cementum: ↑ Cementum

Acervuli (brain sand, concretion, psammoma body, corpus arenaceum*): Mulberry-shaped calcifications, 400 µm–5 mm or more in diameter, scattered in ↑ parenchyma of ↑ pineal organ and ↑ choroid plexus. Chemically A. are composed of ↑ hydroxyapatite with a small quantity of ↑ strontium. A. in pineal organ have various numbers of spherical lobes (L), whereas those of choroid plexus have a smooth surface but an identical chemical composition. In the tissue, a basket of ↑ reticular and ↑ collagen microfibrils surrounds each extracellu-

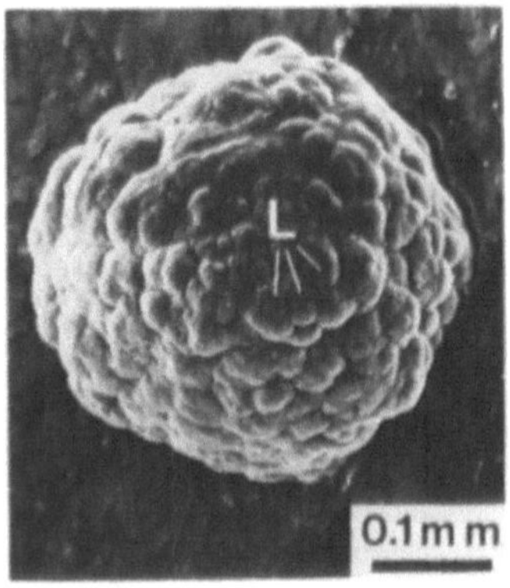

lar A. Since number of A. increases with age, it is possible that they represent a product of cell degeneration. (Fig. = human)

Allen, D.J., Allen, J.S., DiDio, L.J., McGrath, J.A.: J. Submicrosc. Cytol. *13*, 675 (1981)

Acervuli, origin and formation of: It is believed that ↑ acervuli in ↑ pineal organ may occur independently both in vacuoles of ↑ pinealocytes and in intercellular spaces in the form of crystallization globuli (G), 2–5 μm in diameter. After death of pinealocyte, globuli reach extracellular spaces where they increase in volume by additional crystallization and/or fusion with other globuli and form a lobe (L). Several lobes unite to constitute an acervulus. Therefore, internal structure of an acervulus may have uni- or pluricentric concentric organization of successive layers of ↑ hydroxypatite (H). Data concerning A. of choroid plexus are lacking.

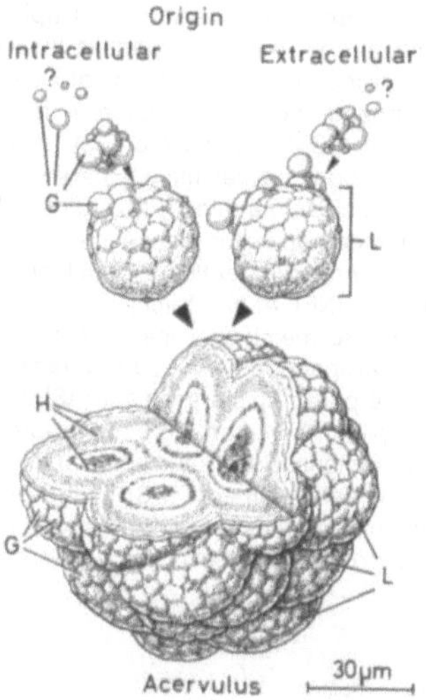

Krstić, R.: Cell Tissue Res. *174*, 129 (1976)

Acetylcholine: The ↑ neurotransmitter substance of some ↑ neurons of ↑ central nervous system, of all ↑ synapses between pre- and postganglionic neurons of ↑ autonomic nervous system, of all parasympathetic postganglionic neurons, of all ↑ motor end plates, and of some postganglionic sympathetic axons innervating ↑ sweat glands. Bound to a lipoprotein carrier, A. is stored in ↑ synaptic vesicles and released into ↑ synaptic cleft when an action potential reaches axonal arborization. A. activates receptor sites of postsynaptic membrane, increasing its permeability and generating a wave of depolarization. (See physiology texts for further information)

↑ Synapse, cholinergic

Acetylcholinesterase: The enzyme responsible for inactivation of ↑ acetylcholine. Inhibition of A. provokes an increase of acetylcholine within ↑ synaptic clefts, thus interrupting transmission from one ↑ neuron to another.

Gautron, J.: Histochemistry *76*, 469 (1982)

Achromaffin paraganglia: ↑ Paraganglia, parasympathetic

Achromatic apparatus (apparatus mitoticus*): A term designating ↑ astrosphere and ↑ spindle apparatus because of their lack of affinity for ordinary dyes.

↑ Chromatic apparatus

Achromatic objective: ↑ Objective

Achromatin: A former term for nonstainable portion of ↑ chromatin, i.e., ↑ euchromatin.

Acid dyes (anionic dyes): Stains capable of forming salts with a positively charged (basic) biological substance. Aniline blue, eosin, light green, orange G, phloxin, etc. are A.

Acid hydrolases: A collective term for a number of lysosomal hydrolytic enzymes active at acid pH (acid phosphatase, acid ribonuclease, catepsin, β-glucuronidase, β-galactosidase, aryl-sulfatase, and peroxidase).

↑ Lysosomes

Acid phosphatase: ↑ Acid hydrolases

Acidophilia (eosinophilia): The property of basic protein groups contained in ↑ nucleolus, ↑ cytoplasm and extracellular material (↑ collagen fibrils, etc.) of binding with an ↑ acid dye and appearing pink or red.

Acidophilic cells, of hypophysis: ↑ Hypophysis, acidophilic cells of

Acidophilic erythroblast: ↑ Erythroblast, orthochromatic

Acidophils: ↑ Hypophysis, acidophilic cells of

Acinar glands: ↑ Tubuloacinar glands

Acinus, of glands (acinus glandularis*): A mulberry-shaped secretory unit

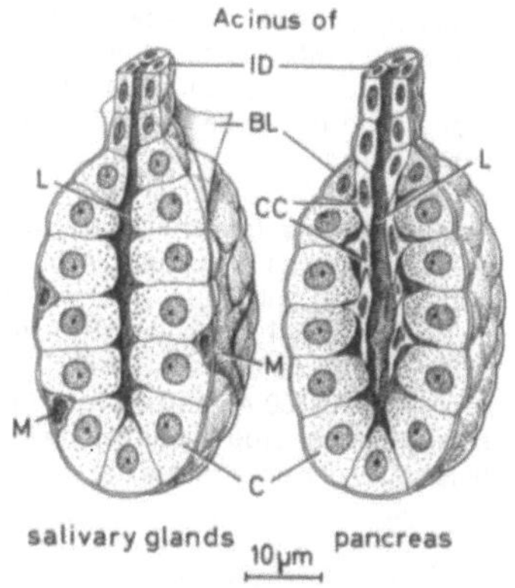

formed by a group of ↑ exocrine ↑ serous glandular cells (C). Forming a single layer, the cells converge toward a minute central lumen (L); they overlie a ↑ basal lamina (BL). Each A. is connected to a narrow ↑ intercalated duct (ID), which in ↑ pancreatic A. begins with ↑ centroacinar cells (CC). M = ↑ myoepithelial cells.

Acinus, of liver: The area of liver ↑ parenchyma involving segments of two neighboring portal lobules supplied by ↑ inlet venule and terminal branches of ↑ interlobular artery (IA), and drained by branches of interlobular ↑ bile duct (BD). Liver A. is subdivided into three zones along a decreasing gradient of oxygen and metabolite concentration, cells of zone I being closest, those of zone III farthest from vascular supply. The concept of liver A. is not ubiquitously accepted. (Fig. modified from Rappaport et al., 1954) CV = ↑ central veins, of liver; IV = ↑ interlobular vein

↑ Liver lobule, classic; ↑ Liver lobule, portal

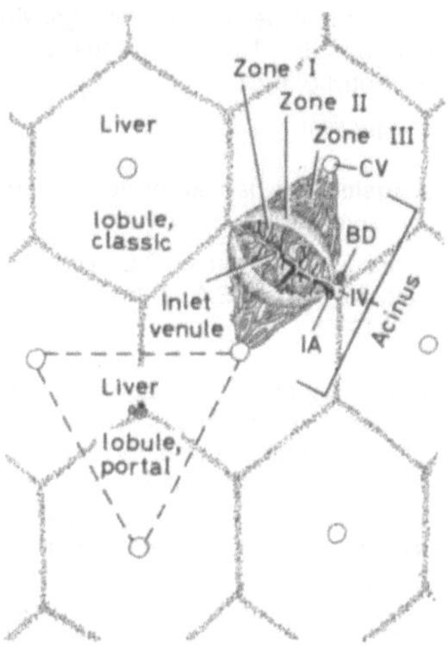

Rappaport, A.M., Borowy, A.J., Lougheed, W.M., Lolto, W.N.: Anat. Rec. *119*, 11 (1954); Schultz, M., Hildebrand, R.: Cell Tissue Res. *231*, 643 (1983)

Acinus, of lung (primary lobule, lobulus pulmonaris primarius*): A morphofunctional pulmonary unit served by a respiratory ↑ bronchiole (RB), including ↑ alveolar ducts (AD) with ↑ alveoli (A). 200–300 A. form a ↑ lung lobule. (In the fig., proportions of lung lobule have not been adhered to)

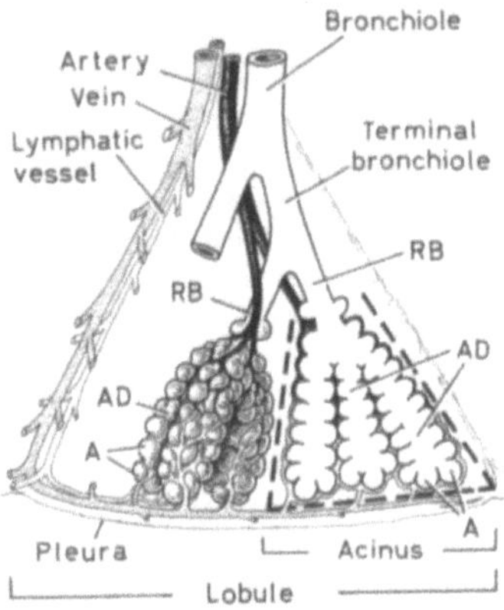

Boyden, E.A.: Anat. Rec. *169*, 282 (1971)

Acoustic fibers: ↑ Corti's organ, innervation of

Acoustic ganglion: ↑ Spiral ganglion

Acridine orange: ↑ Fluorochromes

Acromegaly: Unusual growth of bones of skull and extremities accompanied by progressive enlargement of head, face, hands, feet, ↑ penis, etc. as a result of hypersecretion of ↑ growth hormone by ↑ somatotropes of ↑ adenohypophysis.

↑ Gigantism

Acrosin: A trypsinlike proteolytic enzyme present in ↑ acrosome.

Johnson, L.A., Garner, D.L., Truitt-Gibert, A.J., Lessly, B.A.: J. Andrology *4*, 222 (1983)

Acrosomal cap (head cap): A smooth surfaced cisterna (AC), about 70 nm wide, covering the anterior pole of ↑ spermatid nucleus (SN). A. is delimited by outer (Ao) and inner (Ai) acrosomal membranes and filled with homogenous, moderately osmiophilic material. In the area of A., ↑ nuclear envelope (NE) loses its ↑ pores and becomes

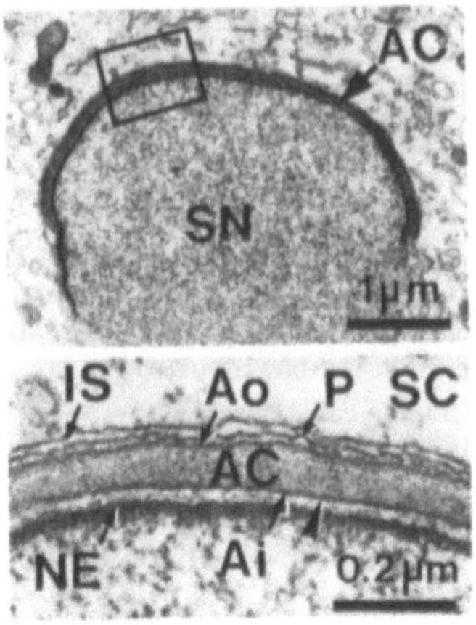

denser from nuclear side because of chromatin condensation. A roughly 25-nm gap (arrowhead) between inner acrosomal membrane and nuclear envelope contains granulofilamentous material. A. is separated from intercellular space (IS) and ↑ Sertoli's cell (SC) by plasmalemma (P) of spermatid. A. develops by spreading of ↑ acrosomal vesicle over spermatid nucleus; it marks anterior tip of the future ↑ spermatozoon nucleus. (Figs. = human)

↑ Spermiogenesis

Acrosomal granule (granulum acrosomale*): A round or elliptical mass of moderate to strongly osmiophilic material (AG) contained in ↑ acrosomal vesicle (AV). A. is rich in ↑ glycoproteins and therefore ↑ PAS-positive. In the course of ↑ spermiogenesis, A. increases in size through fusion of new ↑ proacrosomal granules (arrowhead), becoming flattened and spreading to fill entire acrosomal vesicle. SN = spermatid nucleus, G = ↑ Golgi apparatus. (Fig. = human)

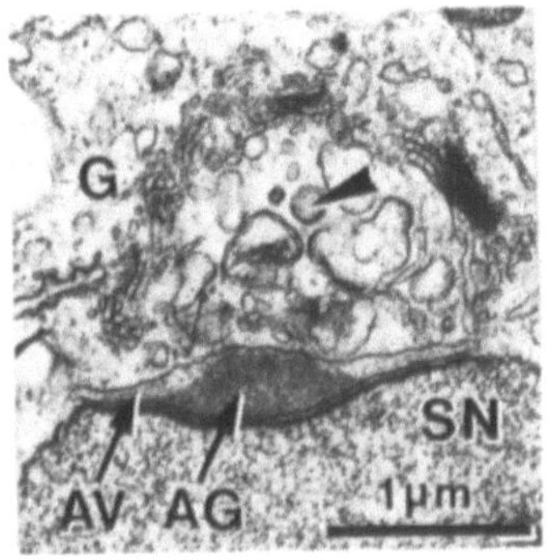

Acrosomal phase, of spermiogenesis: ↑ Spermiogenesis

Acrosomal vesicle (acrosomal vacuole, vesicula acrosomalis*): A

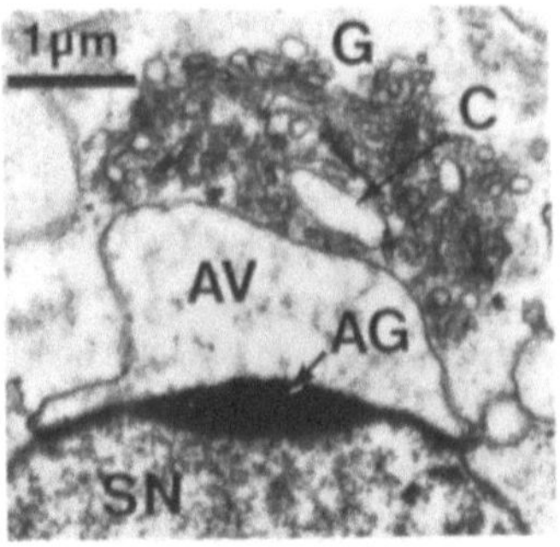

spherical, membrane-enclosed structure (AV) at anterior pole of a ↑ spermatid nucleus (SN). A. contains ↑ acrosomal granule (AG) and an electron-lucent flocculent material. A. is formed by fusion of membranes of ↑ proacrosomal granules and cisternae (C) of Golgi apparatus (G). A. is the future ↑ acrosome. (Fig. = human)

↑ Spermiogenesis

Acrosome (acrosoma*): Caplike, membrane-limited ↑ organelle closely overlying anterior two-thirds of ↑ spermatozoon nucleus (N). Outer A. membrane (OM) is covered by spermatozoon plasmalemma (P). A layer of perinuclear material (PM), about 25 nm wide, separates inner A. membrane (IM) from outer leaflet of ↑ nuclear envelope (NE). At tip of nucleus a fold of inner A. membrane leaves subacrosomal space (SS). A. is subdivided into three parts. Apical segment (AS) is of simple form in man (fig.), but assumes characteristic shapes in some animals. Principal segment (PS) represents main portion of A. and corresponds to ↑ acrosomal cap of some authors. In

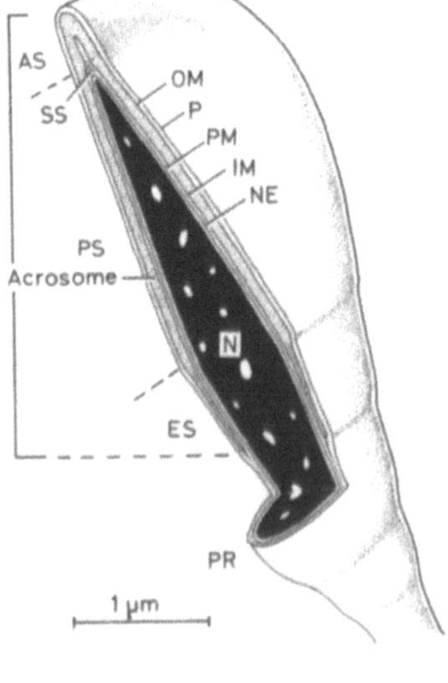

this region, A. membranes are about 50 nm apart. In equatorial segment (ES), A. membranes are closer (25 nm) and A. content is condensed. A ↑ postacrosomal region (PR) follows A. A. contains carbohydrates, ↑ hyaluronidase, some lysosomal ↑ acid hydrolases and ↑ acrosin. All enzymes are active during penetration of spermatozoon though ↑ corona radiata and ↑ zona pellucida of secondary ↑ oocyte.

↑ Acrosome reaction; ↑ Spermiogenesis

Acrosome, formation of: 1) Golgi phase = in ↑ Golgi apparatus (G) of a ↑ spermatid appear small ↑ proacrosomal granules (PG) rich in carbohydrates, which coalesce to form an ↑ acrosomal granule (AG). Its limiting membrane fuses with outer membrane (OM) of ↑ nuclear envelope. Newly synthesized granules and cisternae (C) of Golgi apparatus form an ↑ acrosomal vesicle (AV) around the acrosomal granule, contributing to its enlargement. 2) Cap phase = limiting membrane of acrosomal vesicle expands (arrows) over anterior half of nucleus as a thin head cap (HC). 3) Acrosomal phase = substance of acrosomal vesicle is gradually distributed throughout cap, which finally constitutes ↑ acrosome.

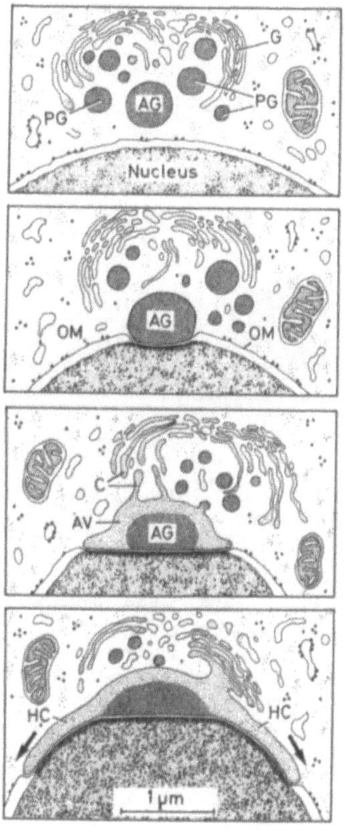

↑ Spermiogenesis

Hermo, L., Clermont, Y., Rambourg, A.: Anat. Rec. *193*, 243 (1979); Sinowatz, F., Wrobel, K.-H.: Cell Tissue Res. *219*, 511 (1981)

Acrosome reaction: A sequence of structural changes of capacitated ↑ spermatozoon head in vicinity of a secondary ↑ oocyte. A. in mammals: Outer acrosomal membrane (OAM) fuses with overlying plasmalemma (P), thus creating numerous openings (Op) through which acrosomal enzymes are liberated (arrows) to dissolve ↑ zona pellucida (ZP). At end of A., only inner acrosomal (IAM) membrane covers anterior part of spermatozoon head. Plasmalemma of ↑ postacrosomal region (PR) subsequently fuses with microvilli (Mv) of oocyte, the process of fusion now involving inner acrosomal membrane. Contact of two ↑ germ cells induces bulges (B) from oocyte surface, which quickly envelop the spermatozoon head.

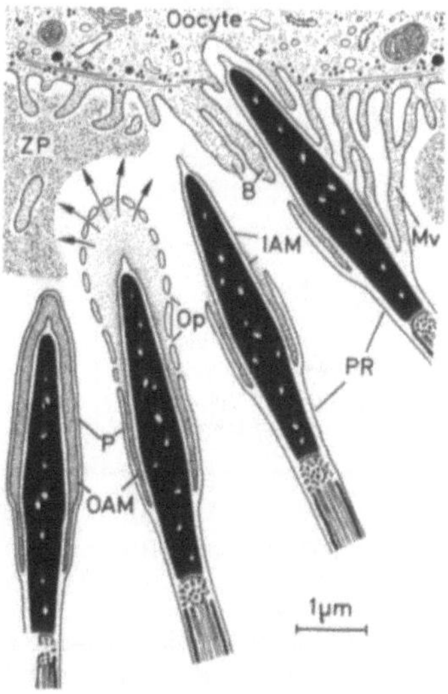

Katz, D.F., Yanagimachi R.: Biol. Reprod. *25*, 785 (1981); Shalgi, R., Phillips, D.: J. Ultrastruct. Res. *71*, 154 (1980); Shapiro, B.M., Eddy, E.M.: Int. Rev. Cytol. *66*, 257 (1980)

Acrylic resins: Plastic materials used for ↑ embedding in transmission ↑ electron microscopy.

ACTH: ↑ Adrenocorticotropic hormone

ACTH-cells: ↑ Corticotropes

ACTH-RF: ↑ Adrenocorticotropic hormone-releasing factor; ↑ Releasing factors

Actin-binding protein (ABP): A proteinaceous substance that cross-links ↑ actin microfilaments of ↑ macrophages in vitro. Together with ↑ myosin, A. can change consistency of ↑ ectoplasm during spreading of ↑ pseudopodia and ↑ phagocytosis.

Stendahl, O.I., Hartwig, J.H., Brotschi, E.A., Stossel, T.P.: J. Cell Biol. *84*, 215 (1980)

Actin microfilaments: ↑ Actin myofilaments

Actin myofilaments (actin microfilament, myofilamentum tenue*): Protein microfilaments, 5–7 nm diameter, and more than 1 µm long, present in a large variety of cells. A. of ↑ myofibrils consist of **g**lobular subunits of G-actin, 5.6 nm in diameter (m.w. about 42000 daltons), joined into longitudinal polymers of **f**ibrillar F-actin. Two rows of F-actin entwine helicoidally to form an A. Between two F-strands lie narrow ↑ tropomyosin (Tr) molecules, along which ↑ troponin molecules (Tn) attach at regular intervals of 40 nm. A. are regularly oriented in ↑ cardiac muscle cells and ↑ skeletal muscle fibers where they are anchored to ↑ Z-lines, form ↑ I-bands, and interdigitate between ↑ myosin filaments of ↑ A-bands. In ↑ smooth muscle cells, A. are attached to ↑ dense bodies. A. are found also in nonmuscle cells with contractile capacity (↑ endothelial cells, ↑ macrophages, ↑ myofibroblasts, etc.), and in some cell structures (↑ microvilli, ↑ terminal web, etc.).

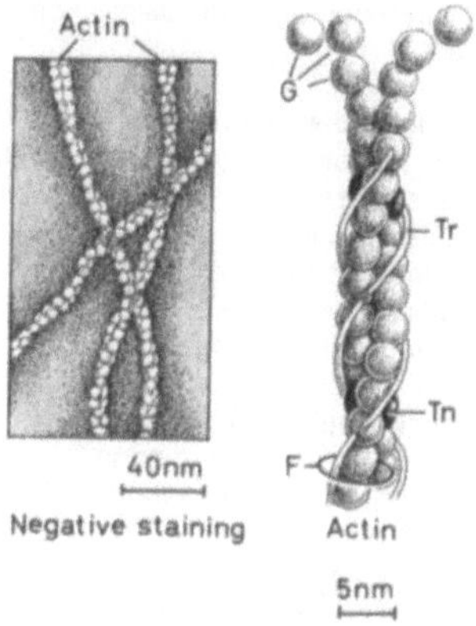

DeRosier, D.J., Tilney, L.G.: How Actin Filaments Pack into Bundles. Cold Spring Harbor Symposia on Quantitative Biology, Vol. 46, Part 2. New York: Cold Spring Harbor Laboratory, 1982;

Gröschel-Stewart, U.: Int. Rev. Cytol. *65*, 194 (1980); Oosawa, F.: Macromolecular Assembly of Actin. In: Stracher, A. (ed.): Muscle and Nonmuscle Motility, Vol. 1. New York: Academic Press 1983; Pollard, T.D.: J. Cell Biol. *91*, 156s (1981)

Actinin (α-actinin): A contractile protein found in ↑ Z-lines of striated muscles and in ↑ dense bodies of ↑ smooth muscle cells.

Jockusch, H., Jockusch, B.M.: Dev. Biol. *75*, 231 (1980)

Actinlike microfilaments: Thread-shaped structures (arrows) about 6 nm thick and 0.5 μm or more in length found in ↑ cytoplasmic matrix of a great variety of cells, mostly attached to cell ↑ organelles and plasmalemma. It is believed that A. are contractile and involved in intracellular displacement of organelles, cell movements, and ↑ exocytosis. (Fig. = ↑ interstitial cell of testis, human)

↑ Capsular epithelium, of Bowman's capsule

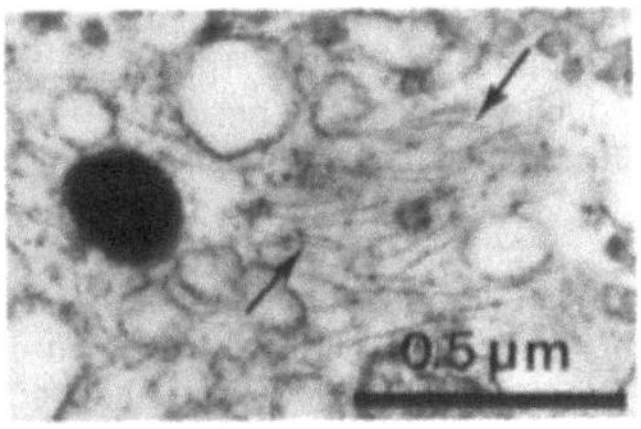

Aguas, A.P.: J. Ultrastruct. Res. *74*, 175 (1981)

Actomyosin: A result of mixing of ↑ actin and ↑ myosin in vitro. A. contracts in the presence of ↑ adenosine triphosphate.

Addison's disease (hypoadrenocorticism): Primary A.: a chronically insufficient synthesis of adrenocortical hormones especially ↑ mineralocorticoids (↑ aldosterone). Secondary A. = an insufficient synthesis of hypophyseal ↑ adrenocorticotropic hormone.

Adenohypophyseal hormones: ↑ Adrenocorticotropic hormone (ACTH); ↑ Follicle-stimulating hormone (FSH); ↑ Growth hormone (GH or STH); ↑ Lipotropins (LPH); ↑ Luteinizing hormone/ ↑ interstitial cell-stimulating hormone (LH/ICSH); ↑ Thyrotropic hormone (TSH); ↑ Prolactin (LTH); ↑ Melanocyte-stimulating hormone (MSH)

Adenohypophysis (anterior lobe, lobus anterior*): Anterior part of ↑ hypophysis composed of pars distalis, pars tuberalis, and pars intermedia.

Bhatnagar, A.S. (ed.): The Anterior Pituitary Gland. New York: Raven Press 1983; Tixier-Vidal., Farquhar, M.G. (eds.): The Anterior Pituitary. New York: Academic Press 1975

Adenomyosis: ↑ Endometriosis

Adenosine triphosphate (ATP): The most important energy-rich compound of cell energy-transporting system. Together with adenosine diphosphate (ADP), ATP is synthesized in ↑ mitochondria (M) by oxidative phosphorylation during cell respiration.

ATP $\rightleftarrows$ ADP + P + 7300 cal or 30543 J (↑ SI).

Within mitochondria ADP is energetically recharged by addition of a phosphate group (P) to produce an ATP molecule. (See biochemistry texts for further information.)

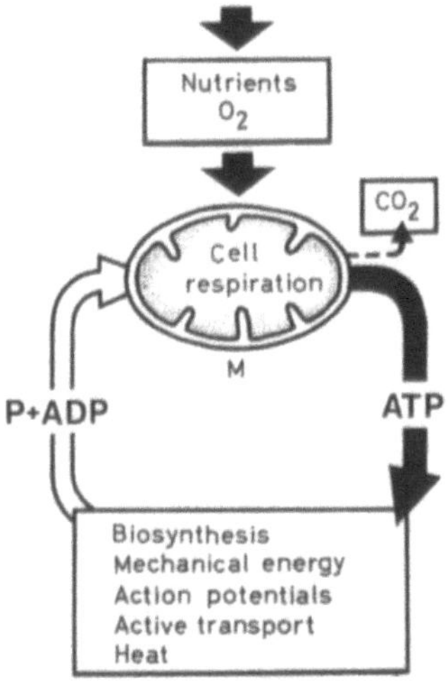

Hinkle, P.C., McCarty, R.E.: Sci. Am. 238/3, 104 (1978)

Adenyl cyclase: A membrane-bound enzyme responsible, together with ↑ cyclic AMP, for transmission and amplification of information carried by polypeptide hormones.

↑ Polypeptide hormones, action on the cell of

Frowein, J.: Dtsch. Med. Wochenschr. *97*, 1918 (1972); Schulze, W.: Histochemistry *75*, 133 (1982); Stefanini, S., Farrace, M.G., Ciofi Luzatto, A.: Cell Mol. Biol. *29*, 291 (1983); Sutherland, E.W.: JAMA *214*, 1281 (1970)

ADH: ↑ Antidiuretic hormone

Adhering junction: ↑ Zonula adherens

Adipokinetic hormones: ↑ Lipotropins

Adipose cells, brown (brown fat cell, brown lipocyte, multilocular adipose cell, plurivacuolar adipose cell, lipocytus multivesicularis*): Polygonal cells measuring 15–25 μm with an eccentrically located nucleus (N) and one or two nucleoli (Nu). Cytoplasm contains numerous large round mitochondria (M) with long transversal cristae, small Golgi apparatus (G), poorly developed cisternae of smooth (sER) and rough (rER) endoplasmic reticulum, some ↑ lysosomes, free ribosomes, ↑ glycogen particles, ↑ micropinocytotic vesicles, and abundant ↑ lipid droplets (L). Cytochrome of mitochondria and lipochrome contained within lipid droplets give a brown tinge to A. Each A. is enclosed by a ↑ basal lamina (BL) and a basketry of ↑ reticular and ↑ collagen microfibrils. Capillaries of nonfenestrated type (Cap) and ↑ adrenergic nerve endings (NE) come in contact with nearly every A.

↑ Lipid droplets, microfilaments of

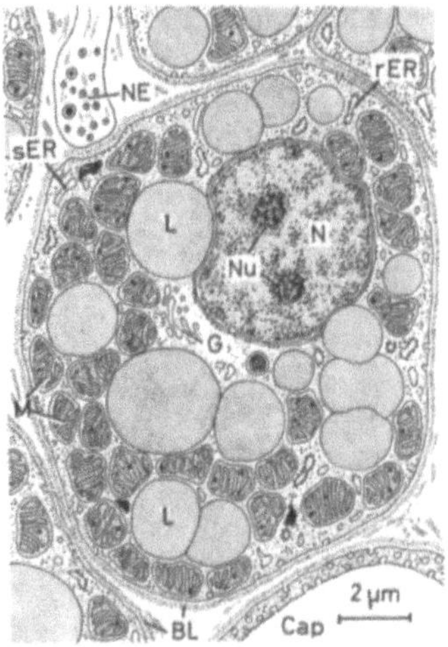

Suter, E.: Lab. Invest. *21*, 246 (1969); Nedergaard, J., Lindberg, O.: Int. Rev. Cytol. *74*, 188 (1982)

Adipose cells, white (lipocytus*): Large spherical or polyhedral cells (25–100 μm) containing a single central lipid droplet (L) devoid of boundary membrane. Only a single 5-nm thick interface condensation (C), reinforced by parallel microfilaments (Mf) 5 nm thick, separates lipid droplet from surrounding cytoplasm. With nucleus (N) pressed to cell periphery and a thin rim of cytoplasm (Cy), A. has a signet-ring appearance. Some filamentous mitochondria (M), a few cisternae of rough (rER) and tubules of smooth ↑ endoplasmic reticulum (sER), poorly developed Golgi apparatus, and moderate number of free ribosomes lie in cytoplasmic sheet. Many ↑ micropinocytotic vesicles (MV) and occasional lysosomes are found. In cytoplasmic girdle synthesized ↑ lipid droplets (LD) are discharged into main lipid mass (arrows). Each A. is enveloped by a ↑ basal lamina (BL) and a basketry of ↑ reticular fibers (RF). Capillaries (Cap) come in contact with every A., while nerve endings do not.

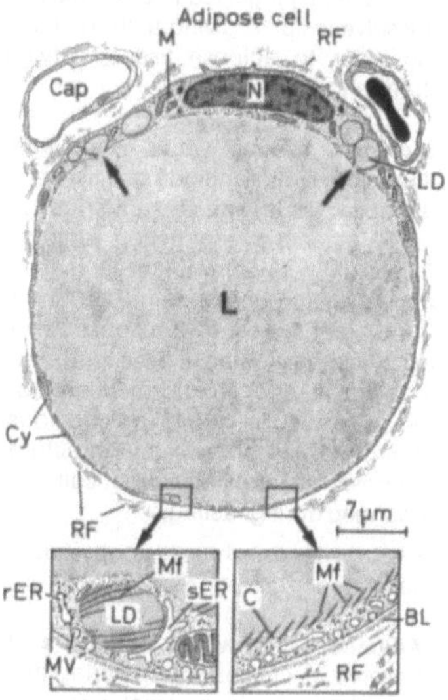

Adipose cells, white, histophysiological changes in fasting: After 24 h: Increase in number of ↑ micropinocytotic vesicles only. After 48 h: Decrease in size of ↑ lipid droplets (L); many adipose cells become multivacuolated. After 72 h: Adipose cells decrease in volume but contain many lipid droplets; proliferation of smooth endoplasmic reticulum; cell surface becomes irregular but no change in mitochondrial size and shape. Continued fasting: Adipose cells appear fusiform ("serous cells"), harboring few or no lipid droplets. The described process is reversible.

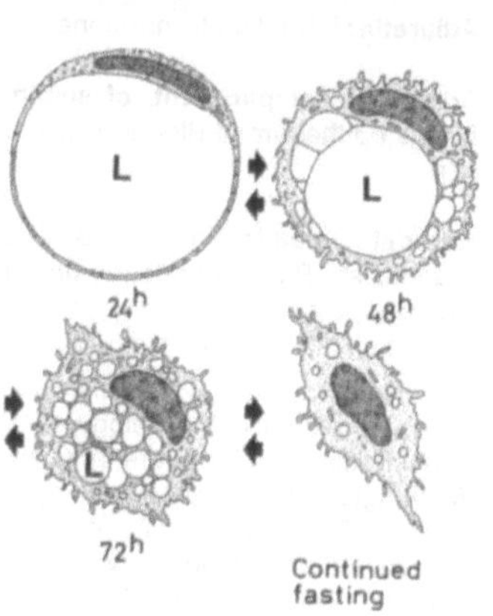

Slavin, B.G.: Int. Rev. Cytol. *33*, 297 (1972)

Adipose cells, white, histophysiology of: Uptake, synthesis, storage and mobilization of fat as calorie-rich material. All these events are regulated by action of lipogenetic (i.e. ↑ insulin, ↑ prostaglandin E_1) and lipolytic hormones and ↑ neurotransmitters (↑ catecholamines, ↑ glucagon, ↑ ACTH, ↑ lipotropins, ↑ TSH, ↑ LH, ↑ STH, ↑ thyroxine, ↑ serotonin, ↑ secretin) under control of ↑ neurosecretory system. However, morphofunctional mechanisms of lipid deposition and mobilization are highly complex and not fully known. Simplified functional schema: Capillary lipoprotein lipase breaks down triglycerides of blood ↑ chylomicrons to fatty acids (FA). These are absorbed by fat cells and resynthesized to triglycerides. Both glucose (Glu) and amino acids (AA) can be converted into fatty acids to yield triglycerides. During caloric demand, lipolytic hormones or neurotransmitters bound to corresponding receptors on plasmalemma of adipose cell activate, through ↑ adenyl cyclase-cAMP system, dormant tissue lipase, which then breaks down triglycerides to glycerol and fatty acids. The latter pass across cell membrane and enter capillaries. Here they are fixed to a carrier albumin and transported to other cells to be used as fuel. Inhibition of adenyl cyclase-cAMP system by lipogenetic hormones leads to deposition of lipids. Cap = capillaries

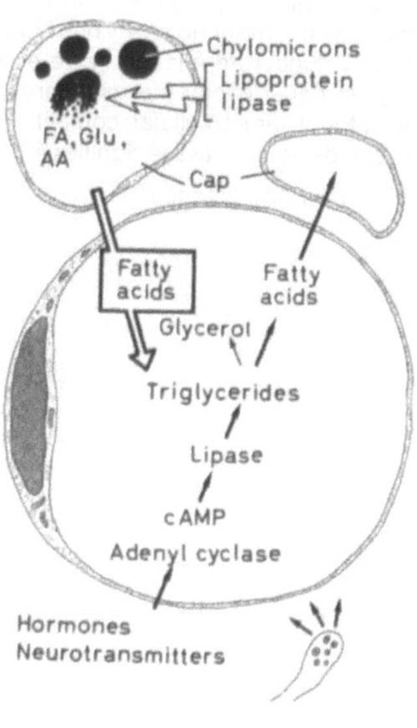

Slavin, B.G.: Int. Rev. Cytol. *33*, 297 (1972)

Adipose tissue (textus adiposus*): A kind of ↑ connective tissue characterized by large, densely grouped ↑ adipose cells. Depending on whether they contain several ↑ lipid droplets or only one large lipid droplet, A. is subdivided into brown ↑ A. and white ↑ A., respectively.

Adipose tissue, brown (textus adiposus fuscus*): A close cell union consisting of brown ↑ adipose cells (BAC) resembling an epithelium, especially an endocrine gland (adrenal cortex). Lobular organization of A. is pronounced, with some white ↑ adipose cells (WAC) within lobules. A. is very richly vascularized and innervated.

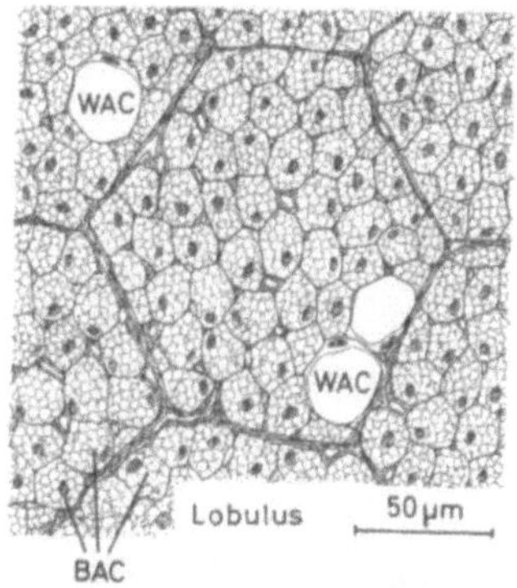

Bargmann, W., Hehn, G.V., Lindner, E.: Z. Zellforsch. *85*, 601 (1968); Schmidt, F.G., Donat, K., Budras, K.-D.: Z. mikrosk.-anat. Forsch. *96*, 885 (1982) Suter, E.: Lab. Invest. *21*, 246 (1969)

Adipose tissue, brown, distribution of: In hibernating species and laboratory rodents, brown adipose tissue is found in and around thorax (anterior mediastinum, along great vessels of neck and thoracic aorta, costovertebral angle, axillae and as an interscapular fat body). Brown adipose tissue occurs also near ↑ thyroid gland and renal ↑ hilus. In human newborn, brown adipose tissue makes up 2%–5% of body weight and is located in axillae, at nape, in posterior triangle of neck, in vicinity of thyroid gland, carotid arteries, and renal hilus. Through coalescence of ↑ lipid droplets, brown adipose tissue diminishes postnatally in humans, coming to resemble white ↑ adipose tissue. Under some conditions (old age, chronic wasting diseases, starvation), brown adipose tissue reappears in described areas.

Adipose tissue, brown, functions of: 1) Brown adipose tissue is a heat generator ("chemical furnace") of greatest importance for hibernators. In nonhibernating mammals and human newborns, this tissue facilitates adaptation to ambient temperatures. ↑ Norepinephrine released from ↑ adrenergic nerve endings (NE) acts through corresponding membrane receptors on ↑ adenyl cyclase-cAMP system to activate lipase that breaks down triglyceride of ↑ lipid droplets (L) to glycerol and fatty acids. Mitochondrial oxidation of the latter with subsequent heat production warms circulating blood and maintains body temperature during winter dormancy. 2) Brown adipose tissue is also a calorie-rich reserve material.

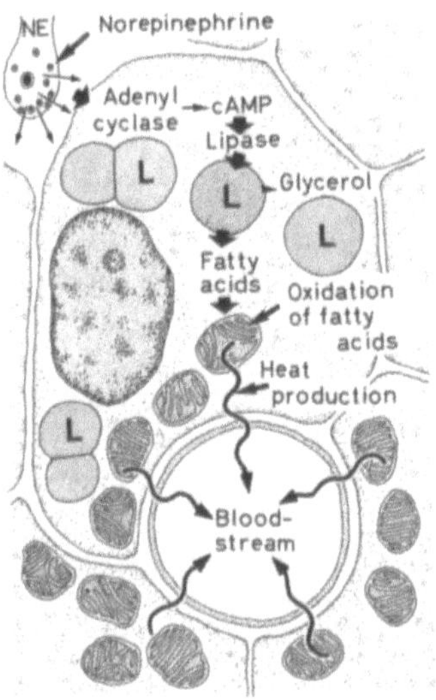

Smith, R.E., Horwitz, B.A.: Physiol. Rev. *49*, 330, 1969; Suter, E.R., Staubli, W.: J. Histochem. Cytochem. *18*, 100 (1970); Tanuma, Y., Ohata, M., Ito, T., Yokochi, C.: Arch. Histol. Jpn. *39*, 117 (1976)

Adipose tissue, histogenesis of: ↑ Mesenchymal cells differentiate into ↑ reticular cells which form a lobular glandlike mass of ↑ epithelioid cells, the primitive fat organ (4th fetal month). In cells, ↑ lipid droplets accumulate, which do not coalesce or fuse to form a single large droplet; reticular cells stop dividing, round off, and primitive fat organ becomes a fat lobulus. Thus, as a kind of ↑ reticular tissue, both brown and white adipose tissue originates from primitive fat organ, however, A. is not completely understood.

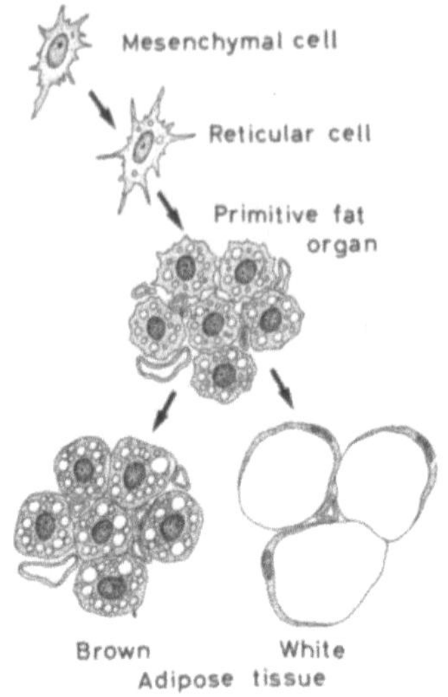

Slavin, B.G.: Anat. Rec. *195*, 63 (1979); Hahn, P., Novak, M.: J. Lipid Res. *16*, 79 (1975); Van, R.L.R., Roncari, D.A.K.: Cell Tissue Res. *195*, 317 (1978)

Adipose tissue, white (ordinary adipose tissue, textus adiposus*): A union of densely packed white ↑ adipose cells (AC). A. is subdivided into small lobules by connective tissue septa (S) with blood, lymphatic vessels (V), and nerve fibers. In histological sections fat is usually dissolved, so that A. appears as a network with round or spherical meshes. Between the cells run capillaries (Cap), assuring a very rich blood supply. Fine ↑ adrenergic nerve endings of periarteriolar plexus innervate A. without direct cellular contact. (In this book adipose tissue signifies A.)

↑ Synapse, "by distance"

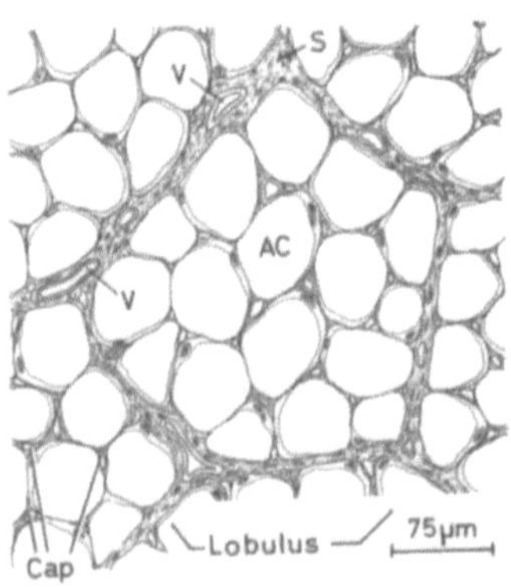

Cahill, G.F., Renold, A.E.: Adipose Tissue – a Brief History. In: Angel, A., Hollenberg, C.H., Roncari, D.A.K. (eds.): Adipocyte and Obesity. New York: Raven Press 1983; Motta, P.: J. Microsc. *22*, 15 (1975); Slavin, B.G.: Int. Rev. Cytol. *33*, 297 (1972)

Adipose tissue, white, amount in terms of body weight of: At 25 years of age, 15% in men and 26% in women; at 55 years, 25% in men and 38% in women. Mean values for all ages = 16% in men and 20% in women.

Adipose tissue, white, color of: Exogenous liposoluble ↑ pigments (↑ carotenoids) ingested in a normal diet dissolve in lipids, giving a characteristic color to adipose tissue, varying from white to deep yellow.

Adipose tissue, white, functional kinds of: 1) Storage adipose tissue = adipose tissue with fat readily available for energy production (adipose tissue of ↑ hypodermis, ↑ mesenteries, ↑ omenta, retroperitoneum). 2) Structural adipose tissue = adipose tissue with the role of an elastic pad, mechanical support, and protection (adipose tissue in orbit, articulations, palms, soles, ↑ cheek, etc.). Structural adipose tissue remains practically unchanged during fasting.

↑ Adipose cells, white, histophysiological changes in fasting in

Adipose tissue, white, functions of: Reserve of calorie-rich material, thermoisolation, binding of water, replacement of involuted organs (↑ thymus, red ↑ bone marrow), elastic pads, mechanical support, space reservation for undeveloped organs (↑ mammary gland).

Adipose type, of synovial membrane: ↑ Synovial membrane

Adiuretin: ↑ Antidiuretic hormone

Adluminal compartment, of seminiferous epithelium: ↑ Blood-testis barrier

Adrenal glands (glandula suprarenalis*): Paired endocrine organs overlying cranial pole of ↑ kidneys, consisting of cortex and medulla.

Blaschko, H., Sayers, G., Smith, A.D. (eds.): Handbook of Physiology, sec. 7, Endocrinology, Vol. 6. Adrenal Gland. Washington: American Physiological Society 1975; Christy, N.P. (ed.): The Human Adrenal Cortex. New York: Harper & Row, Publishers 1971

Adrenal glands, cortex, control of secretion of: The hypothalamic ↑ adrenocorticotropin-releasing factor (ACTH-RF) secreted under various influences, mainly stress, acts on ↑ corticotropes of the anterior lobe of ↑ hypophysis to produce ↑ adrenocorticotrophic hormone (ACTH), which stimulates zona fasciculata to synthesize corticosteroids. Among their peripheral effects is inhibition of synthesis of both ACTH-RF and ACTH through a negative ↑ feedback mechanism (interrupted lines). Decrease in renal arterial blood pressure or blood volume provokes discharge of ↑ renin from ↑ juxtaglomerular cells (JGC). This enzyme converts plasmal ↑ angiotensinogen to ↑ angiotensin I. In ↑ lung, ↑ subfornical organ (SFO) and, possibly ↑ organum vasculosum laminae terminalis (OVLT), the latter is further modified to ↑ angiotensin II, which stimulates cells of zona glomerulosa to produce ↑ aldosterone. Through reabsorption of sodium and water in ↑ distal tubules, aldosterone brings about an increase in intravascular fluid volume and a concomitant rise in blood pressure, inhibiting juxtaglomerular cells to secrete renin (interrupted lines).

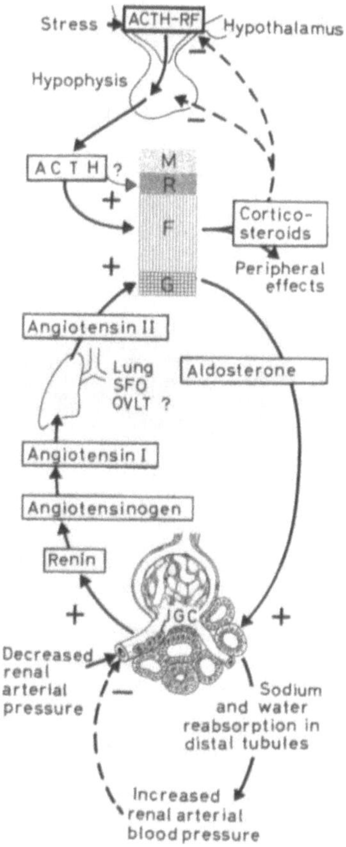

Gill, G.N.: Pharmacol. Ther. B *2*, 313 (1976)

Adrenal glands, cortex, histophysiology of: Of all endocrine glands, structure of adrenal cortex is the most variable. Following hypophysectomy (Hect), both zona fasciculata (F) and reticularis (R) atrophy, while zona glomerulosa (G) remains well preserved. In a hypophysectomized and ACTH-treated animal (Hect + ACTH), adrenal cortex has a similar structure to that in normal animal. Following ACTH injection or exposure to stress (ACTH/Stress), animal with an intact ↑ hypophysis will show a considerable hypertrophy of both zona fasciculata and reticularis but no reaction of zona glomerulosa. When animal is maintained on a sodium-deficient diet (Sodium def.), zona glomerulosa hypertrophies leading to an increase in ↑ aldosterone production. M = medulla

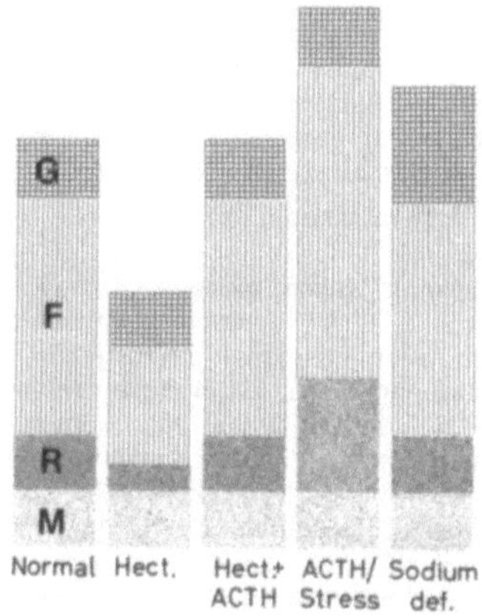

Neville, A.M., O'Hare, M.J.: The Human Adrenal Cortex. Berlin, Heidelberg, New York: Springer-Verlag 1982; Nussdorfer, G.G.: Int. Rev. Cytol. *64*, 307 (1980)

Adrenal glands, cortex of (cortex glandulae suprarenalis*): A three-layered glandular ↑ parenchyma surrounding medulla. Structure: 1) Capsule (C) = ↑ dense connective tissue; 2) zona glomerulosa (ZG) = round or horse-shoe-shaped nests of clear cells; 3) zona fasciculata (ZF) = a long cord of large polygonal cells with foamy cytoplasm; 4) zona reticularis (ZR) = anastomosing cords of small well-stained cells containing ↑ lipofuscin. Zona glomerulosa produces ↑ mineralocorticoid hormones, mainly under action of ↑ renin-angiotensin system; zona fasciculata is site of secretion of ↑ glucocorticoid hormones under action of ↑ ACTH; zona reticularis is site of production of weak ↑ androgenic hormones.

↑ Adrenal glands, cortex, control of secretion; ↑ Cells, of zona glomerulosa; ↑ Cells, of zona fasciculata; ↑ Cells, of zona reticularis

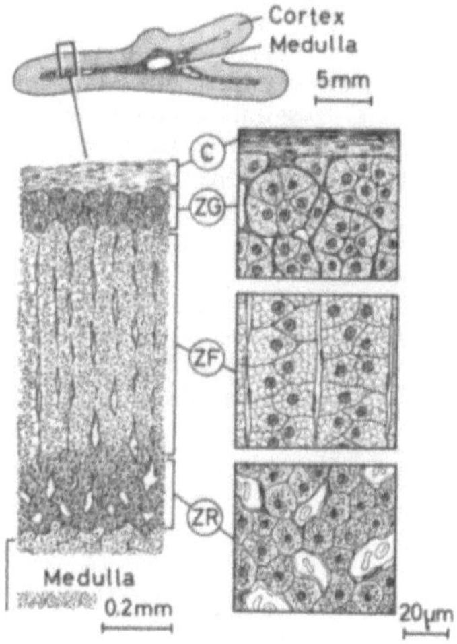

Jones, I., Chester, M., Henderson, I.W. (eds.): General Comparative and Clinical Endocrinology of the Adrenal Cortex. Vol. 2. New York: Academic Press 1978; Nussdorfer, G.G., Mazzocchi G., Meneghelli, V.: Int. Rev. Cytol. *55*, 291 (1978); Zwierzina, W.D.: Acta Anat. *103*, 409 (1979)

Adrenal glands, cortex, of newborn: The cells of peritoneal epithelium proliferate during 5th and 6th weeks of development, forming primitive or fetal cortex (FC). The latter is enveloped by a thin layer of closely packed cells that give rise to permanent or definitive cortex (DC). Two weeks after birth, fetal cortex undergoes ↑ involution, while definitive cortex differentiates over a period of about 3 years to form zones of adrenal cortex. G = zona glomerulosa, F = zona fasciculata, R = zona reticularis. (See embryology texts for further information)

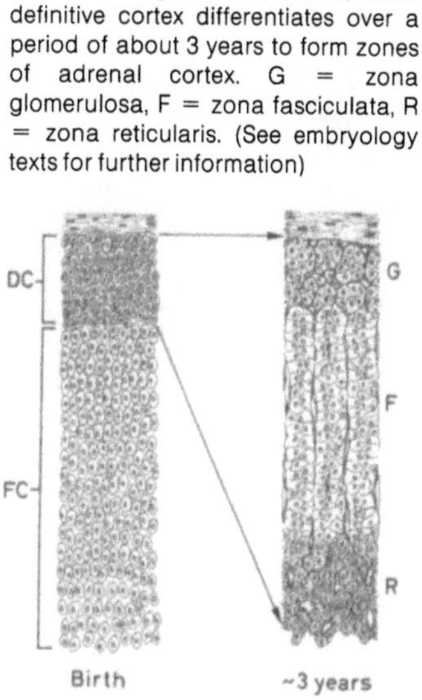

Adrenal glands, cortex, regeneration of: Not definitely clarified. It is likely that zona glomerulosa serves as a germinative zone, since autologous capsule grafts (C) containing some glomerulosa cells (G) regenerate a new cortex. Assumption is that cells of zona glomerulosa migrate toward medulla (arrows) and complete their life cycle in zona reticularis. F = zona fasciculata, R = zona reticularis

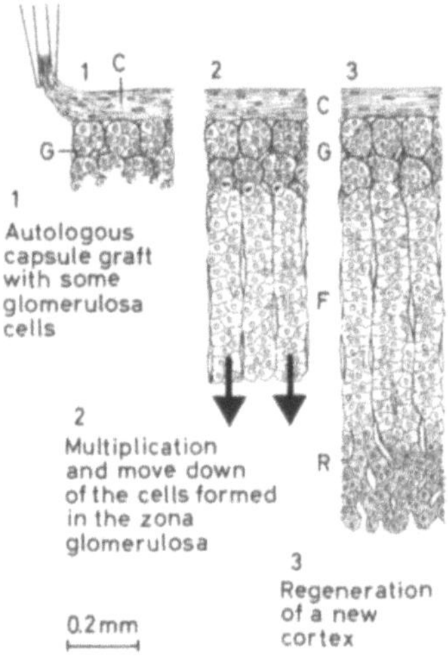

Belloni, A.S., Vassanelli, P., Robba, C., Rebuffat, P., Mazzocchi, G., Nussdorfer, G.G.: J. Anat. *135*, 245 (1982)

Adrenal glands, innervation of: Nerve impulses leaving ↑ hypothalamus reach sympathetic neurons situated in intermediolateral columns (IC) of ↑ spinal cord at level of Th_8–L_1. From here, sympathetic axons pass through ↑ sympathetic trunk, leaving it as splanchnic nerves (SN). Majority of these ↑ preganglionic fibers enter ↑ neuroglandular synaptic contacts with ↑ chromaffin cells (CC) of medulla; part of preganglionic nerve fibers serve to innervate cortex. Some preganglionic fibers terminate in celiac ganglion, its postganglionic axons (interrupted lines) innervating blood vessels (V) of adrenal glands. Thus, stimulation of sympathetic system provokes a rapid release of ↑ catecholamines into bloodstream (arrows).

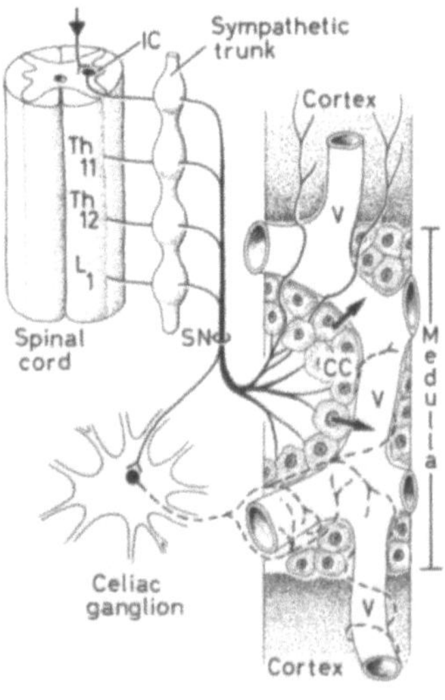

Migally, N.: Anat. Rec. *194*, 105 (1979)

Adrenal glands, medulla, control of secretion of: ↑ Adrenal glands, innervation of

Adrenal glands, medulla, histophysiology of : Physical or emotional stress causes rapid depletion of medullary hormones, which, together with sympathetic activity, adapt organism to immediate situation ("fight or flight" syndrome). Prolonged stressogenic influences may induce a hypertrophy of adrenal cortex via ↑ hypothalamo-hypophyseal complex.

↑ Adrenal glands, innervation of

Adrenal glands, medulla of (medulla*): ↑ Epithelioid anastomosing cords of round polygonal or columnar ↑ chromaffin cells (C) in close contact with capillaries (Cap). Special methods reveal two kinds of endocrine cells giving positive ↑ chromaffin reaction. ↑ Epinephrine producing cells (majority) produce, take up, and release ↑ epinephrine, and ↑ norepinephrine producing cells do likewise with ↑ norepinephrine. Occasional sympathetic ganglion cells (GC), probably related to vascular innervation, may also be scattered among chromaffin cells.

↑ Autonomic neurons, of sympathetic ganglia

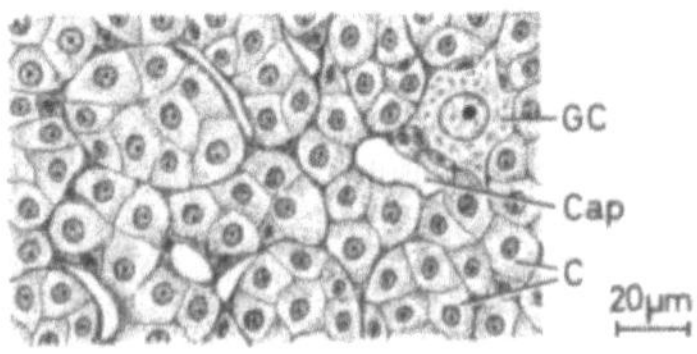

Carmichael, S.W., Ulrich, R.G.: Mikroskopie *40*, 53 (1983); Westhead, E.W.: TINS *6*, 254 (1983)

Adrenal glands, vascularization of: Small suprarenal arteries (SA), arising from ↑ aorta, inferior phrenic artery, and renal artery pierce adrenal capsule (C) and form capsular plexus (CP), which gives rise to a network of cortical sinusoid ↑ capillaries (Cap). These continue through cortex (arrow), collecting all secreted cortical hormones, to medulla (M), where they end in venous radicles (V) disposed around medullary cords. Medullary arterioles, or arteriolae perforantes (AP), extend without collaterals from capsular plexus through cortex, ending in medullary capillaries. These fenestrated capillaries and venous radicles join to form medullary veins (MV), which finally compose single ↑ suprarenal vein (SV), its wall contains numerous longitudinally oriented ↑ intimal cushions (IC). These muscle pads help regulate blood flow through adrenal cortex (concentration of hormones?). Thus, medulla is supplied with fresh blood by arteriolae perforantes, and blood that has passed along cortical cells by way of venous radicles. In this way, ↑ corticosteroids in venous capillaries influence synthesis of ↑ catecholamines by ↑ chromaffin cells. G = zona glomerulosa, F = zona fasciculata, R = zona reticularis

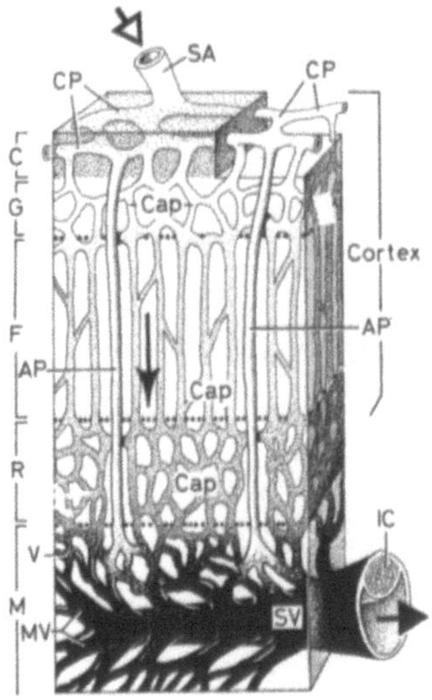

Adrenaline: ↑ Epinephrine

Adrenaline storage granules: ↑ Epinephrine storage granules

Adrenergic nerve endings: The terminations of nearly all ↑ postganglionic sympathetic axons with ↑ synaptic vesicles containing ↑ norepinephrine as ↑ neurotransmitter. Its release induces a response of ↑ smooth muscle and/or glandular cells. Under light microscope, A. can be demonstrated by silver impregnation or by ↑ fluorescence techniques (yellow-green fluorescence after tissue exposition to ↑ formalin vapors). Ultrastructurally, A. are characterized by their lack of ↑ Schwann's sheath, frequent varicosities (V), mutual synaptic contacts, and occasional contacts with ↑ cholinergic fibers. A. contain clusters of 40–60-nm ↑ dense core ↑ synaptic vesicles (SV) giving a positive ↑ chromaffin reaction. Besides mitochondria (M), smooth cisternae (sER), ↑ neurotubules (Nt), and ↑ neurofilaments (Nf), A. enclose 50–60-nm "empty" vesicles and occasionally 80–120-nm catecholamine-negative vesicles (LV), probably containing ↑ serotonin. A. are enveloped in a basal lamina (BL), lying generally free in an intercellular or ↑ pericapillary space, although they can make contact with smooth muscle and/or glandular cells. A. of CNS have no basal lamina. C = dense core

↑ Autonomic nervous system; ↑ Sympathetic nervous system; ↑ Synapse, "by distance"

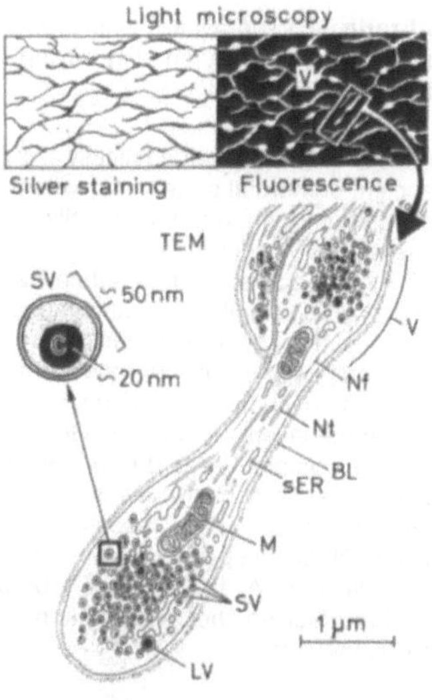

Avery, J.K., Cox, C.F., Chiego, D.J., Jr.: Anat. Rec. *198*, 59 (1980); Archakova, L.I.: Neirofiziologiya *12*, 86 (1980)

Adrenergic nerve fibers (adrenergic system): Efferent ↑ nonmyelinated nerve fibers releasing ↑ norepinephrine from their endings. A. include all postganglionic nerve fibers of ↑ sympathetic nervous system.

↑ Adrenergic nerve endings; ↑ Cholinergic nerve endings

Adrenergic synapse: ↑ Synapse, adrenergic

Adrenochromes: Brown-colored water-insoluble polymers of ↑ catecholamines, formed as a result of ↑ chromaffin reaction. A. originate from ↑ epinephrine, noradrenochromes from ↑ norepinephrine. A. are chemically related to precursor of ↑ melanin.

Adrenocorticotropes: ↑ Corticotropes

Adrenocorticotropic hormone (ACTH, adrenocorticotrophin, corticotropin): A polypeptide ↑ hormone secreted by adenohypophyseal ↑ corticotropes stimulating release of ↑ corticosteroids from zona fasciculata of ↑ adrenal glands. Together with corticosteroids, A. induces lipolysis, massive lymphocyte destruction, thymic ↑ involution, and has a general anti-inflammatory effect. A. is also found in ↑ hypothalamus.

Pelletrei, G., Désy, L.: Cell Tissue Res. *196*, 525 (1979)

Adrenocorticotropic hormone-releasing factor(ACTH-RF, corticotropin-releasing factor, CRF, corticotropin-releasing hormone, CRH): A short-lived polypeptide synthesized and released under various influences (mainly stress) from hypothalamic ↑ neurosecretory neurons. Via hypophyseoportal circulation, it reaches ↑ corticotropes stimulating them to produce ↑ adrenocorticotropic hormone.

↑ Adrenal glands, cortex, control of secretion of; ↑ Hypophysis, vascularization of

Kawata, M., Hashimoto, K., Takahara, J., Sano, Y.: Cell Tissue Res. *230*, 239 (1983) and Arch. Histol. Jpn. *46*, 183 (1983); Liposits, Z., Görcs, T., Sétalo, G., Lengvari, I., Flerko, B., Vigh, S., Schally, A.V.: Cell Tissue Res. *229*, 191 (1983)

Ad-spermatogonia: ↑ Spermatogonia

Adventitia: ↑ Tunica adventitia

Adventitial cells: A kind of ↑ reticular cells separated from vascular sinus of ↑ bone marrow by a thin ↑ basal lamina (BL) of ↑ littoral cells (Li). A. are stellate cells with branched cytoplasmic processes that penetrate between hematopoetic elements (H). In a clear cytoplasm, A. contain an elliptical nucleus (N), a few large mitochondria (M), a small Golgi apparatus (G), some short cisterns of rough endoplasmic reticulum (rER), scattered ↑ residual bodies (RB), and a few free ribosomes. A. can accumulate ↑ lipid droplets, and are believed to have a phagocytic function.

↑ Bone marrow, vascular sinuses of

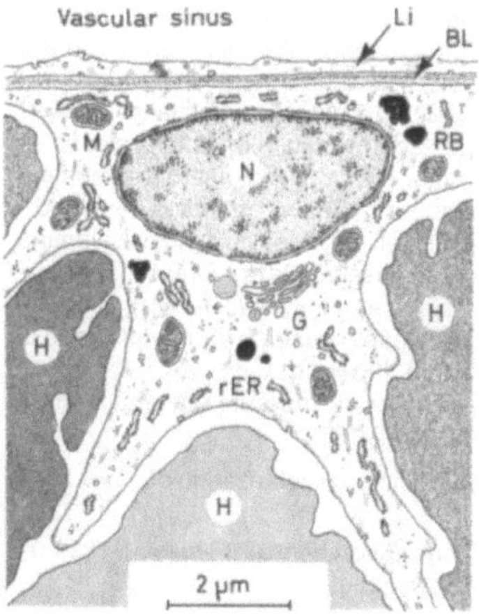

A-face, of plasmalemma: A former designation for ↑ PF-face of freeze-cleaved cell membrane.

↑ Cell membrane, freeze-cleaving of

Afferent: Incoming

Afferent I and II fibers: ↑ Neuromuscular spindles

Afferent arterioles (arteriola glomerularis afferens*): Lateral branches (Aff), 0.1–0.6 mm long, of an ↑ interlobular artery (IA) supplying blood to ↑ renal corpuscles (RC). Before entering corpuscle, A. give off some capillaries (Cap) for vascularization of neighboring convoluted tubules. In corpuscle, A. divide into ↑ glomerular capillaries (GC). Structure: 1) Tunica intima = a continuous endothelial layer (E) and fenestrated internal ↑ elastic lamina (IEL). 2) Tunica media = one to two

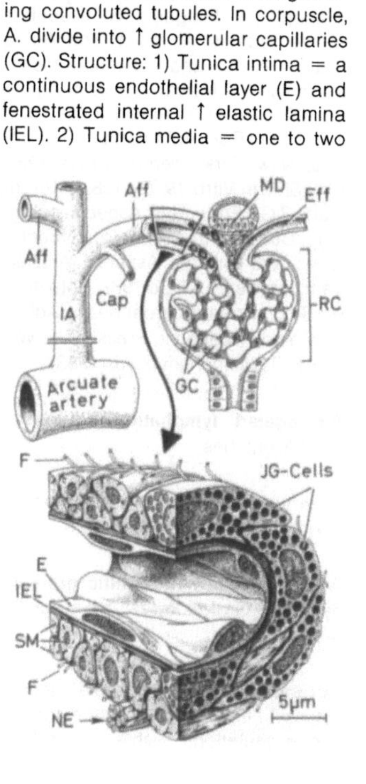

layers of circularly arranged ↑ smooth muscle cells (SM), between which are situated ↑ adrenergic nerve endings (NE). 3) Tunica adventitia = a few ↑ collagen, ↑ reticular, and ↑ elastic fibers (F). In preglomerular segment (about 25 µm proximal to renal corpuscle), A. loose their elastic lamina and muscle cells are replaced by ↑ juxtaglomerular cells (JG-Cells), lying immediately beneath ↑ macula densa (MD). Blood in A. has a higher colloidal-osmotic pressure than in ↑ efferent arteriole (Eff).

↑ Kidney, vascularization of

Age involution of thymus: ↑ Thymus, age involution of

Ageing (senescence): The gradual irreversible development of structural and functional changes in the cell and its environment not due to diseases or trauma. Morphologically, A. of a cell is characterized by vacuolization of ↑ nucleus, cytoplasm, and rough endoplasmic reticulum, reduction in number of mitochondria, loss of ↑ metaplasm, decrease in secretory activity, and accumulation of ↑ lipid droplets and ↑ lipofuscin. Functionally, A. is expressed by a greater mechanical fragility and a lesser enzymatic activity. In A., ↑ ground substance, water, and ↑ proteoglycans diminish. Dead cells may disappear by ↑ lysis, ↑ phagocytosis, or ↑ desquamation.

Giacobini, E., Filogamo, G., Giacobini G., Vernadakis, A. (eds.): The Aging Brain: Cellular and Molecular Mechanisms in the Nervous System. Aging, Vol. 20. New York: Raven Press 1982; Johnson, J.E.Jr.: Aging and Cell Structure. New York: Plenum Press 1981; Kim, S.U.: In Vitro *19*, 73 (1983); Korenman, S.G.: Endocrine Aspects of Aging. Current Endocrinology Series. Basic and Clinical Aspects. Amsterdam: Elsevier Publ. Co 1982; Rothstein, M.: Biochemical Approaches to Aging. New York: Academic Press 1982; Wilson, P.D.: Histochem. J. *15*, (1983)

Aggregated lymphatic nodules: ↑ Peyer's patches

Agranular endoplasmic reticulum: ↑ Endoplasmic reticulum, smooth

A-granules, of neutrophilic granulocytes: ↑ Granulocytes, neutrophilic, granules of

Agranulocytes (nongranular leukocytes): ↑ Leukocytes without cytoplasmic granulations visible under light microscope. ↑ Lymphocytes and ↑ monocytes are A.

Air-blood barrier: ↑ Blood-air barrier

Airways, of lung: ↑ Pulmonary airways

Albinism: The general lack of pigmentation. Two causes: 1) Absence of ↑ melanoblast-melanocyte system; 2) congenital lack of ↑ tyrosonase in epidermal ↑ melanocytes and, consequently, inability to convert colorless ↑ premelanosomes to black ↑ melanin granules.

↑ Melanogenesis

Albuginea: ↑ Tunica albuginea

Albuminoid: A stronlgy ↑ acidophilic substance chemically similar to ↑ keratin, partially responsible for loss of ↑ basophilia in deep ↑ cartilage areas.

AL-cells: ↑ A-like cells

Alcian blue: A copper-containing stain for ↑ light- and ↑ transmission electron microscopic demonstration of ↑ glycosaminoglycans.

Aldehyde-containing fixatives: Substances having free aldehyde groups which react with amino, carboxyl, and indole groups of proteins, thereby precipitating proteins.

↑ Formaldehyde; ↑ Glutaraldehyde; ↑ Paraformaldehyde

Aldehyde-fuchsin: A stain for demonstration of ↑ elastic fibers and lamellae, as well as for ↑ neurosecretory granules.

Aldehydes, histochemical detection of: 1) Free aldehyde groups existing normally in tissues are demonstrated by plasmal reaction with ↑ Schiff's reagent. 2) Selective oxydation of ↑ glycogen and ↑ glycoproteins induces formation of free aldehyde groups detected with Schiff's reagent (↑ PAS-reaction). 3) Selective hydrolysis of ↑ nucleic acids liberates aldehyde groups of ↑ deoxyribonucleic acid, which also react with Schiff's reagent (↑ Feulgen reaction).

Aldosterone: The main ↑ mineralocorticoid hormone, secreted by ↑ cells of zona glomerulosa of ↑ adrenal glands. A. increases reabsorption of sodium and water and excretion of potassium by ↑ distal tubules of kidney. Secretion of A. is controlled by ↑ juxtaglomerular apparatus, and ↑ renin-↑ angiotensin system.

↑ Adrenal glands cortex, control of secretion of

Fanestil, D.D., Kipnowski, J.: Klin. Wochenschr. *60*, 1180 (1982)

Aldosterone-stimulating factor: A hypothetical hypothalamic substance acting on production of ↑ aldosterone.

A-like cells (AL-cells): A category of ↑ endocrine cells, not definitively classified, found in intestinal mucosa of the cat and man. A. are characterized by a resemblance to ↑ A-cells of ↑ pancreatic islets. Whereas in the cat ↑ secretory granules have a uniform size (300 nm), they vary in man from 100–400 nm; in both species granules are predominantly infranculear.

↑ Endocrine cells, of gastrointestinal tract

Osaka, M., Sasagawa, T., Fujita, T.: Arch. Histol. Jpn. *35*, 235 (1973); Vassallo, G., Solcia, E., Capella, C.: Z. Zellforsch. *98*, 333 (1969)

Alizarin: A ↑ supravital ↑ acid dye for selective demonstration of calcifying ↑ bone matrix.

Allocortex: A phylogenetically very ancient primordial cortex, embryologically, structurally, and functionally different from ↑ isocortex. A. is composed of: A) Archicortex = area subcallosa (AS), gyrus cinguli (GC), and hippocampus (Hip), characterized by one layer only of ↑ neurons. B) Paleocortex = ↑ olfactory bulb (OB) and uncus (U) with three layers of neurons. General structure of three-layered A.: I) Lamina zonalis = associative elements; II) lamina granularis = receptive elements; III) lamina pyramidalis = effectory elements. A. extends over 1/12 of brain surface and belongs functionally to ↑ limbic system.

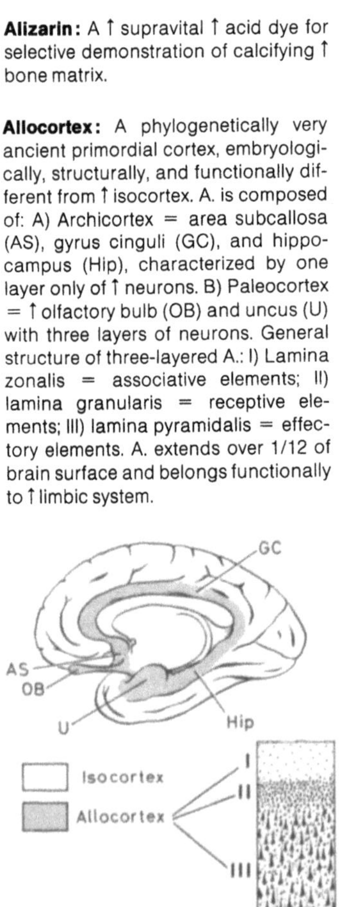

↑ Brain cortex, heterotypical isocortex of; ↑ Brain cortex, homotypical isocortex of; ↑ Brain cortex, phylogenetic and structural subdivision of

Stephan, H.: Allocortex. In: Bargmann, W. (ed.): Handbuch der mikroskopischen Anatomie des Menschen. Vol. 4, Part 9. Berlin, Heidelberg, New York: Springer-Verlag 1975

Allogen hair: A term for a growing ↑ hair.

Allograft: ↑ Transplantation

Allograft rejection: The destructive action of ↑ killer cells on allograft by secretion of ↑ lymphokines. A. can be prevented by ↑ cortisol or ↑ cytochalasin B administration.

Alloxan: A drug which selectively destroys ↑ B-cells of ↑ pancreatic islets, inducing in this way a permanent diabetes mellitus in experimental animals.

Alpha cells, of hypophysis: ↑ Hypophysis, acidophilic cells of

Alpha cells, of pancreatic islets: ↑ A-cells, of pancreatic islets

Alpha granules, of platelets: Round and/or elongated structures, enclosed by a ↑ unit membrane, averaging 0.2 µm in diameter with finely granulated osmiophilic content (α). It is believed that platelet factor 3, some enzymes (↑ acid phosphatase, β-glucuronidase), and ↑ fibrinogen are contained in A. (Fig. = rat)

↑ Platelets

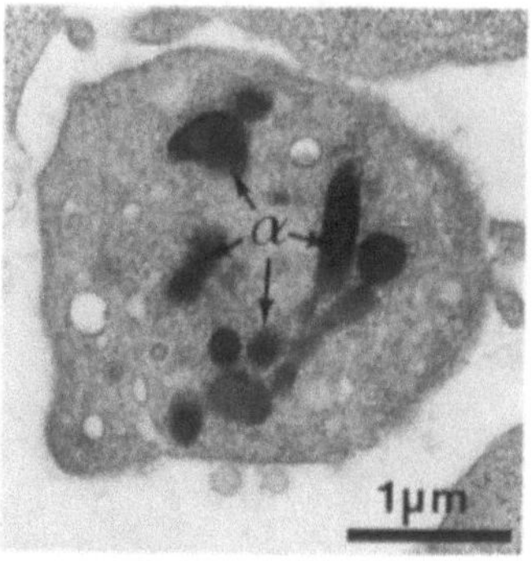

Alpha motor fiber: A large ↑ myelinated nerve fiber (12–20 µm in diameter) innervating ↑ extrafusal muscle fibers.

Alpha particles, of glycogen: ↑ Glycogen

Alpha tubulin: A protein which, with beta ↑ tubulin, constitutes ↑ microtubules.

Alum: ↑ Mordant

Alveolar bone: A lamellar ↑ bone forming ↑ alveoli around roots of ↑ teeth. A. represents a store of rapidly mobilizable calcium for organism. ↑ Periosteum of A. forms ↑ periodontal ligament.

↑ Bone, metabolic; ↑ Lamina dura

Alveolar cells, of mammary gland: ↑ Mammary gland, epithelium of

Alveolar cells, type I (membranous pneumonocyte, pneumonocyte type I, small alveolar cell, pulmonary epithelial cell, squamous alveolar cell, cellula respiratoria s. squamosa*): Very attenuated squamous epithelial cells in contact with alveolar air. Together with flattened nucleus, ↑ perikaryon (P) contains a small Golgi apparatus, a few small mitochondria, some short rough endoplasmic cisternae, numerous ↑ micropinocytotic vesicles, and free ribosomes. Remaining cytoplasm forms an extremely thin uninterrupted sheet (S) with a surface area of about 4000 µm². A. join together to form a continuous alveolar lining; they lie on a thin basal lamina (BL). By ↑ micropinocytosis, A. can transport a small amount of inhaled particulate material to underlying tissue. A. participate in formation and function of ↑ blood-air barrier through their permeability to gases. Cap = capillaries

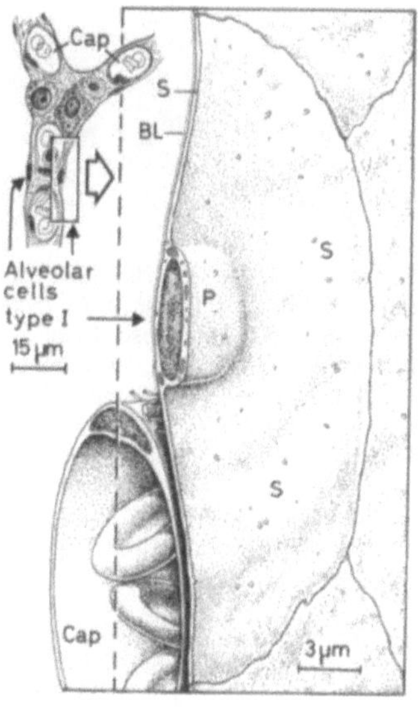

Alveolar cells, type II (granular pneumonocyte, great alveolar cell, niche cell, pneumonocyte type II, septal cell, cellula magna s. granularis*): Round or cuboidal secretory alveolar cells, 10–12 µm in diameter, devoid of lateral cytoplasmic processes. Each of these relatively numerous cells is located in a slight depression of ↑ alveolar wall (AW). Its round nucleus (N) has a central position. All ↑ organelles, particularly rough endoplasmic cisternae (rER), are well developed. Apical cytoplasm contains a various number of ↑ multiversicular bodies (MvB) and ↑ cytosomes (Cyt). Excreted from cell (inset), lamellar content of the latter furnishes a ↑ surfactant, coating whole alveolar surface. Laterally, A. enter into contact with cytoplasmic sheets of ↑ alveolar cells type I. Free cell surface is studded with short irregular microvilli (Mv). BL = basal lamina, Cap = capillaries.

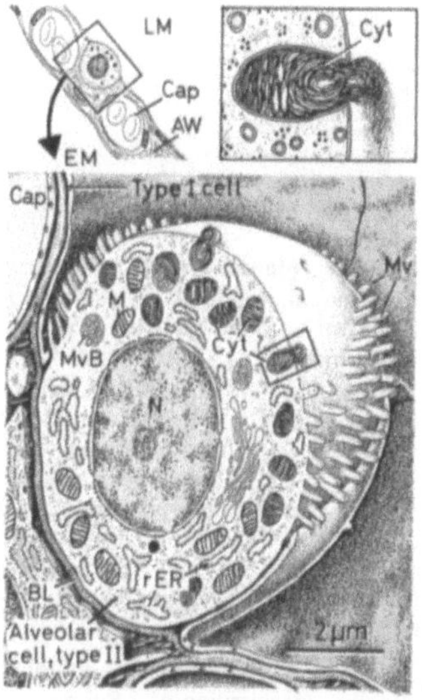

Douglas, W.H.J., McAteer, J.A., Smith, J.R., Braunschweiger, W.R.: Int. Rev. Cytol. Suppl. *10*, 45 (1979); Miller, M.L., Andringa, A., Vinegar, A.: Ultrastruct. Res. *79*, 85 (1982)

Alveolar crest fibers: Some of the ↑ cementoalveolar fibers extending from acellular ↑ cementum to border of an ↑ alveolus of a tooth.

Alveolar ducts (ductus alveolaris*): Branches of a respiratory ↑ bronchiole (RB). A. have essentially no walls, only rings of ↑ smooth muscle cells situated within knoblike enlargments of ↑ alveolar septa (arrowheads). Numerous ↑ alveoli (A) open into A. Each A. ends in an ↑ alveolar sac (asterisk). TB = terminal ↑ bronchiole

↑ Acinus, of lung

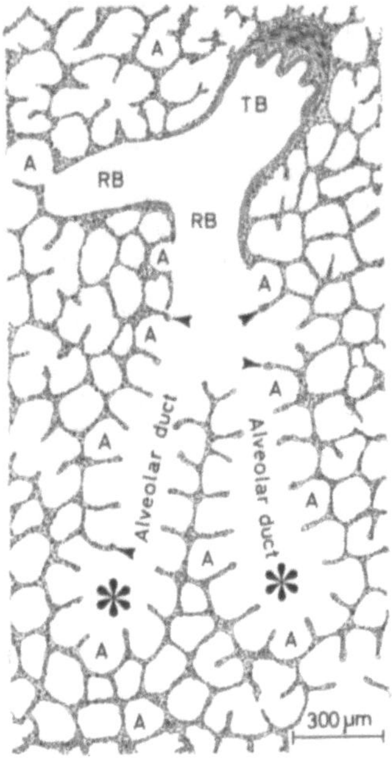

Alveolar glands (saccular gland, glandula alveolaris*): ↑ Glands consisting of a varied number of alveoli. Two varieties: 1) Unistratified A. (in man, only alveoli of lactating ↑ mammary gland with ↑ apocrine mode of secretion). 2) Pluristratified A. = a) simple A. (small pear-shaped ↑ sebaceous glands of ↑ skin); b) branched A. (large sebaceous glands of skin, ↑ tarsal glands). All pluristratified A. have ↑ holocrine mode of secretion.

↑ Alveolus, of mammary gland; ↑ Glands, classification of; ↑ Sebaceous follicles

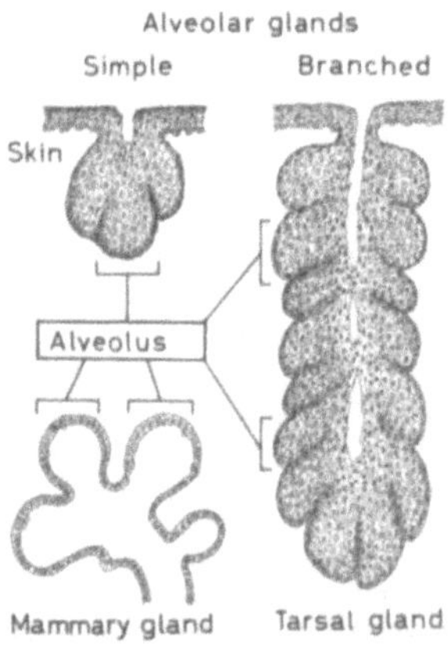

Alveolar lining: ↑ Surfactant

Alveolar macrophages (alveolar phagocyte, dust cell, phagocytus alveolaris*): Large intra-alveolar cells with ameboid properties moving on alveolar surface. Irregular in shape, A. have a deeply invaginated nucleus (N), well-developed ↑ organelles, numerous ↑ lysosomes (Ly), and ↑ phagolysosomes (PhLy). Free surface of A. is covered with ↑ pseudopodia (Ps) and irregularly shaped ↑ microvilli (Mv). A. phagocytize inhaled particles (dust, carbon, bacteria). A. can then either reenter ↑ alveolar wall and migrate via pulmonary lymphatics to regional ↑ lymph nodes, or wander through air conduits to ↑ larynx to be swallowed or expectorated with bronchial ↑ mucus. The question whether A. represent an autochthonous intra-alveolar cell population or transformed ↑ monocytes that have left the bloodstream and entered alveolar cavities is still not elucidated. Cap = capillaries, G = Golgi apparatus, M = mitochondria

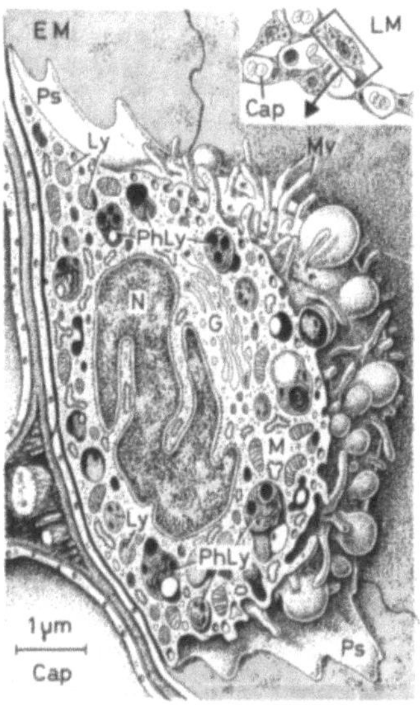

Ferin, J.: Anat. Rec. *203*, 265 (1982); Migally, N., Murthy R.C., Doye, A., Zambernard, J.: J. Submicrosc. Cytol. *14*, 621 (1982); Radzun, J.H., Parwaresch, M.R., Kreipe, H.J.: Histochem. Cytochem. *31*, 318 (1983); Sannes, P.L., Eguchi, M., Spicer, S.S.: Lab. Invest. *41*, 135 (1979), Sorokin, S.P.: Anat. Rec. *206*, 117 and 145 (1983); Tarling, J.D., Coggle, J.E.: Cell Tissue Kinet. *15*, 577 (1982)

Alveolar pores (pore of Kohn, porus septi*): Physiological openings (arrowheads) in ↑ alveolar septum permitting air circulation between ↑ alveoli (A). (Fig. = rat)

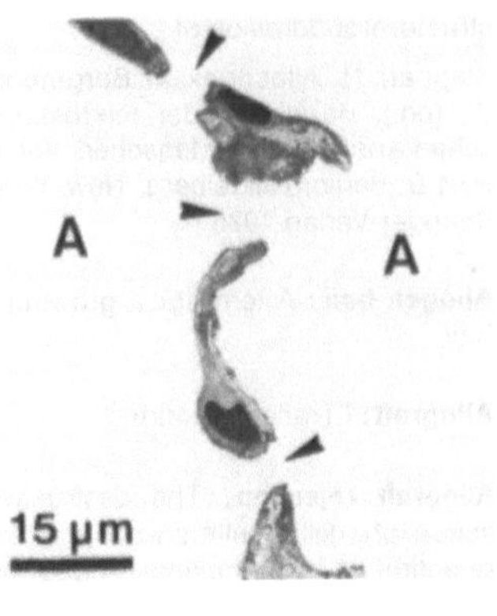

Parra, S.C., Gaddy, L.R., Takaro, T.: Lab. Invest. *38*, 8 (1978)

Alveolar sac (sacculus alveolaris*): A cluster of ↑ alveoli at end of an ↑ alveolar duct.

Alveolar septum (alveolar wall, interalveolar septum, septum interalveolare*): A wall common to two adjacent ↑ alveoli. Structure: 1) Inner alveolar lining = ↑ alveolar cells, type I (I); ↑ alveolar cells, type II (II); their ↑ basal lamina (BL); and ↑ alveolar macrophages (AM). 2) Pulmonary capil-

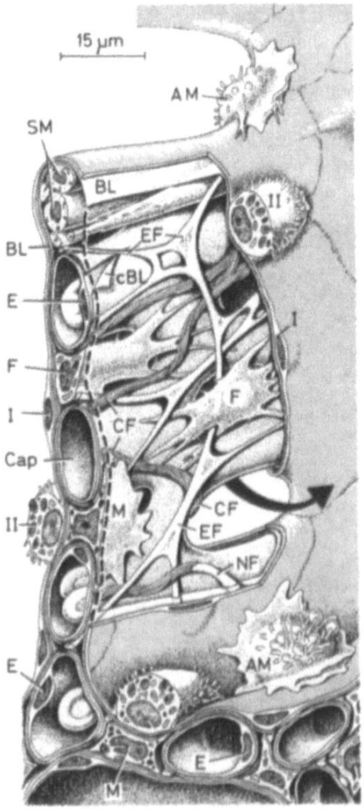

laries (Cap) = ↑ endothelial cells (E) and capillary basal lamina (cBL). 3) Interstitial connective tissue space = ↑ fibroblasts and ↑ fibrocytes (F); ↑ macrophages (M); occasional ↑ mast cells; ↑ lymphocytes and eosinophilic ↑ granulocytes; ↑ smooth muscle cells (SM) around alveolar opening; basket of ↑ elastic fibers (EF); bundles of ↑ collagen fibrils (CF); and ↑ nonmyelinated nerve fibers (NF). An arrow passes through ↑ alveolar pore.

↑ Alveolus of lung, elastic basket of

Fox, B., Bull, T.B., Guz, A.: J. Anat. *131*, 683 (1980); Rosenquist, T.H.: Anat. Rec. *200*, 447 (1981)

Alveolar surface area, of lung: The total surface of all lung ↑ alveoli. In man, it has been calculated to be 143 m^2.

Gehr, P., Bakofen, M., Weibel, E.R.: Respir. Physiol. *32*, 121 (1978)

Alveolar type, of synovial membrane: ↑ Synovial membrane

Alveolar wall: ↑ Alveolar septum

Alveolocapillary membrane: ↑ Blood-air barrier

Alveolus (pl. alveoli): A small saclike cavity.

Alveolus, of glands (alveolus glandularis*): A saccular secretory end piece of an ↑ alveolar gland.

Alveolus, of lung (alveolus pulmonaris*): A minute sphere (A) (mean

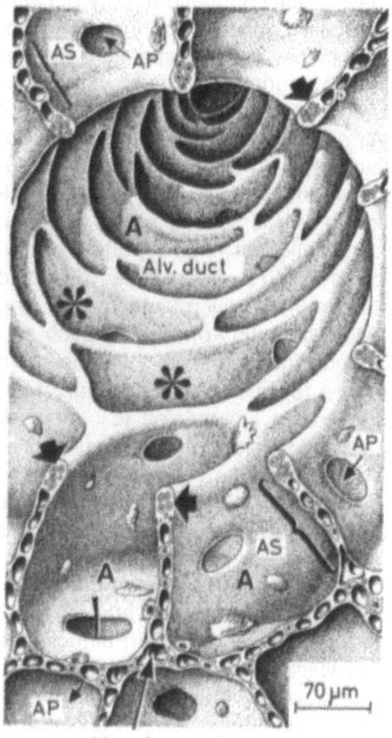

Alveolar septum

diameter about 300 µm in man) connected by an aperture (*) to an ↑ alveolar duct (Alv. duct), ↑ alveolar sac, or respiratory ↑ bronchiole. Adjacent alveoli are separated by very complex common walls, ↑ alveolar septa (AS). Opening of each A. is ringed by ↑ smooth muscle cells located in knoblike swellings at summit of alveolar septa (arrows). Neighboring alveoli communicate through ↑ alveolar pores (AP). In man, there are approximately 150×10^6 A.

↑ Alveolar surface area, of lung

Scheuermann, D.W., De Groodt-Lassel, M.H.A.: Verh. Anat. Ges. *75*, 307 (1981)

Alveolus of lung, elastic basket of: A complex network of branched ↑ elastic fibers (EF) surrounding lung ↑ alveoli (A) with the function of aiding expiration. (Fig. = 40-µm-thick ↑ section, human, ↑ orcein staining)

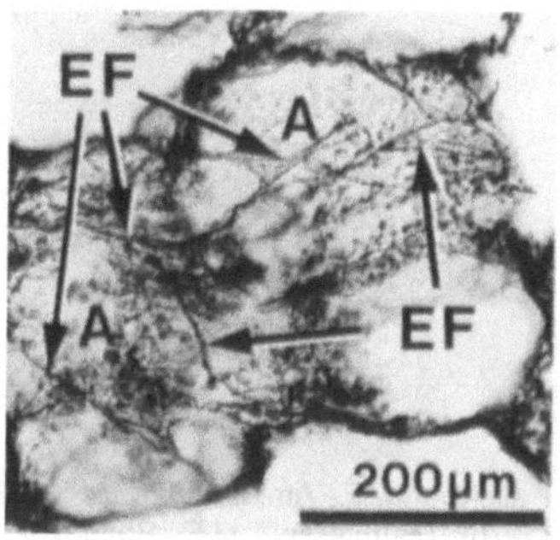

Alveolus, of mammary gland (alveolus glandulae mammae*): The secretory saclike glandular portion (A) of a lactating ↑ mammary gland. (Fig. = human)

↑ Alveolar glands

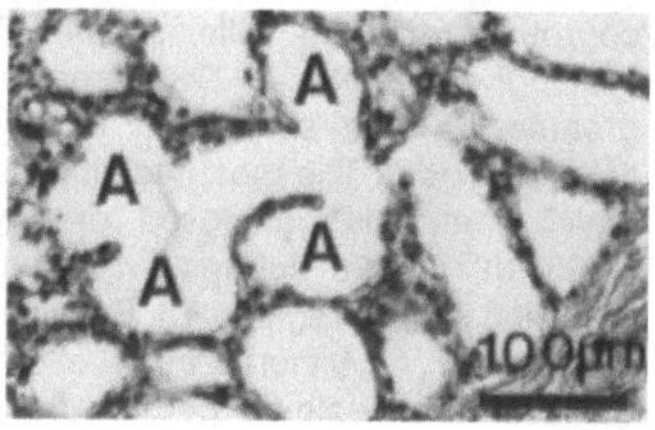

Alveolus, of tooth (alveolus dentalis*): A conical bony socket (arrows) in jaw in which root of tooth (T) is attached to ↑ alveolar bone by ↑ periodontal ligament (PL) and its ↑ cementoalveolar fibers.

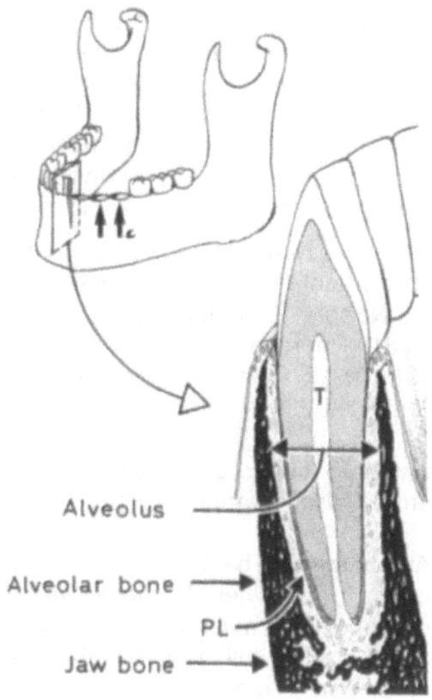

Amacrine cells (neurocytus amacrinus*): Axonless retinal ↑ neurons (A) located at vitreal side of inner granular layer. All ↑ dendrites (D), some with functional properties of an ↑ axon, branch in inner plexiform layer (IPL). Two kinds of A. in primates: 1) Diffuse A., with dendrites spreading throughout thickness of inner plexiform layer; 2) stratified A., with den-

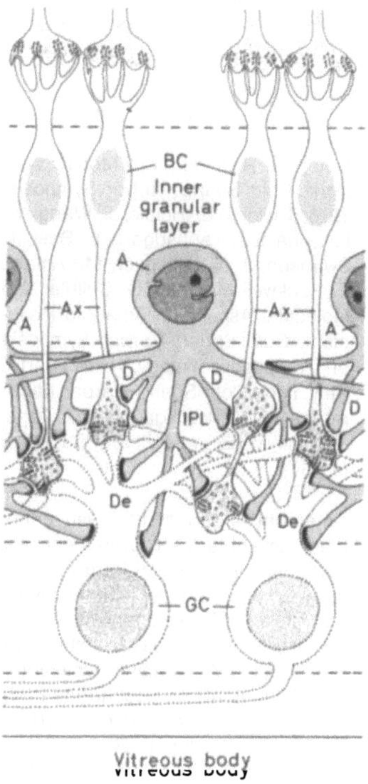

drites ramifying in one or two horizontal levels in inner plexiform layer. Processes of A. form numerous ↑ synaptic contacts with axons (Ax) of ↑ bipolar cells (BC) and with dendrites (De) of ↑ ganglion cells (GC). A. are considered intraretinal association neurons.

Cohen, J., Hadjiconstantinou, M., Neff, N.H.: Brain Res. *260*, 125 (1983); Kolb, H., Nelson, T.: Vision Res. *21*, 1625 (1981); Pourcho, R.G.: Brain Res. *252*, 101 (1982); Sakai, H., Hashimoto, Y.: Brain Res. *270*, 345 (1983)

Ameboid cells: Cells capable of ↑ ameboidism; ↑ wandering cells.

Ameboidism: The property of ↑ histiocytes, ↑ monocytes, ↑ granulocytes, ↑ lymphocytes, ↑ mast cells, ↑ microglia, and ↑ plasma cells to move about as an ameba. The cells protrude and retract ↑ filipodia (F), ↑ pseudopodia (P), and ↑ lamellipodia (L) with which they move within tissues (e.g., granulocytes 0.5–1 µm/s). A. depends on temperature, pH, and certain chemical substances. A. is a condition necessary for ↑ phagocytosis. A. is provoked by displacement of attaching points of ↑ microfilaments and ↑ microtubules on cytoplasmic side of ↑ cell membrane.

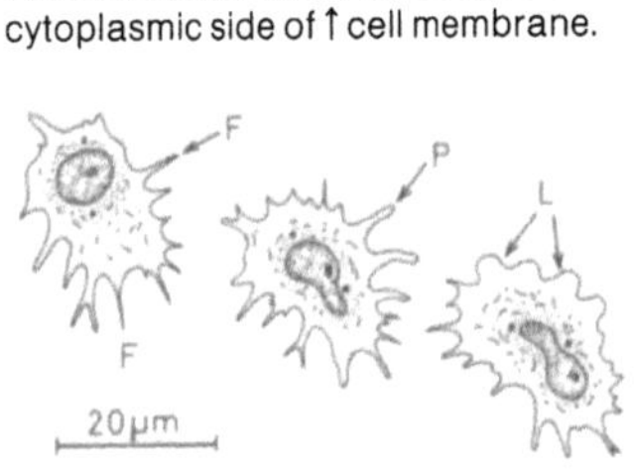

Goldman, R.D. (ed.): Cell Motility. New York, Cold Spring Harbor Symposia, 1976; Taylor D.L., Heiple, J., Wang, Y.-L., Luna, E.J., Tanasugarn, L., Brier, J., Swanson, J., Fechheimer, M., Amato, P., Rockwell, M., Daley, G.: Cellular and Molecular Aspects of Ameboid Movement. Cold Spring Harbor Symposia on Quantitative Biology, Vol. 46, Part 1. New York: Cold Spring Harbor Laboratory 1982; Tokunaga, M., Tokunaga, J., Niimi, M.: Biomed. Res. *2*, Suppl. 13 (1981)

Ameloblasts (adamantoblast, ganoblast, ameloblastus*): Cylindrical cells (A), a number of which form a single-layered inner enamel epithelium responsible for ↑ amelogenesis. As epithelial cells, A. lie on basal lamina (BL). Basal half of A. contains nucleus (N) and the majority of mitochondria (M). Very numerous cisternae of rough endoplasmic reticulum (rER) and a considerable quantity of free ribosomes characterize A. as protein-synthetizing cells. From well-developed supranuclear Golgi apparatus (G), small ↑ secretory granules (S) arise reaching ↑ Tomes' processes (TP). Here, granules are discharged into extracellular space to form organic matrix (Mtr) of ↑ enamel (E). Once amelogenesis is achieved, A. disappear. D = ↑ dentin, Pd = ↑ predentin, O = ↑ odontoblasts, SR = ↑ stellate reticulum (See embryology texts for further information)

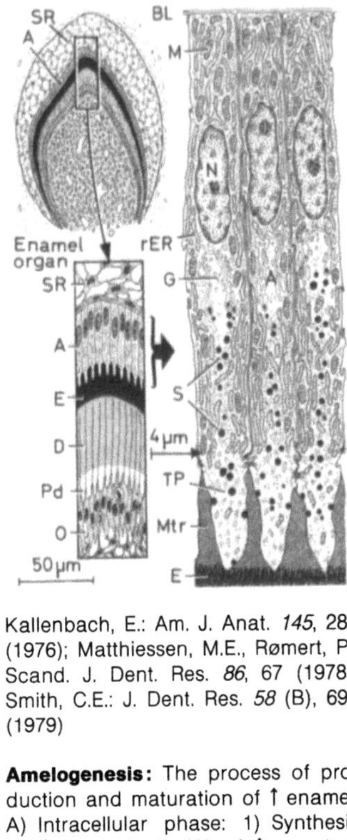

Kallenbach, E.: Am. J. Anat. *145*, 283 (1976); Matthiessen, M.E., Rømert, P.: Scand. J. Dent. Res. *86*, 67 (1978); Smith, C.E.: J. Dent. Res. *58* (B), 695 (1979)

Amelogenesis: The process of production and maturation of ↑ enamel. A) Intracellular phase: 1) Synthesis by ↑ ameloblasts (A) of ↑ secretory granules (S) containing enamel organic matrix (Mtr) composed of ↑ glycoproteins, mineral salts, sulfated ↑ proteoglycans, etc. 2) Discharge of granules at free surface of ameloblasts (arrows). The most recently secreted matrix is composed of tiny (~25 nm) tubular subunits visible in so-called growth regions. B) Extracellular phase: 3) Mineralization of organic matrix: needle-shaped ↑ hydroxy apatite crystals (H) appear in matrix. 4) With its ↑ Tomes' process, each ameloblast forms one ↑ enamel prism bound by a thin layer of noncalcified matrix, ↑ prism sheath (PS), as well as by a calcified ↑ interprismatic substance (Int). 5) Maturation of enamel = a loss of organic material and water, increase of mineralization. As enamel thickens, ameloblasts migrate outward (hollow arrow) from tooth papilla.

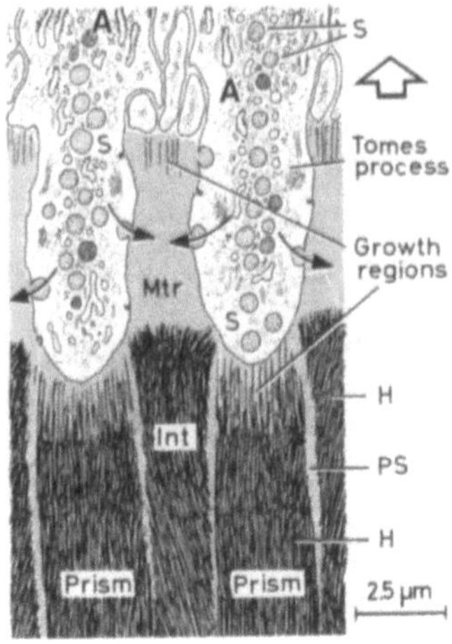

Warshawsky, H.: J. Biol. Buccale *7*, 105 (1979); Weiss, M.P., Voegel, J.C., Frank, R.M.: J. Ultrastruct Res. *76*, 286 (1981)

Amiantine degeneration: ↑ Asbestos cartilage

Amicrons: Particles with dimensions of ions and molecules taken up by cells in the course of ↑ athrocytosis.

Amitosis (direct nuclear division): A special form of nuclear division observed in highly differentiated cells having limited mitotic activity but with frequently ↑ polyploid nuclei (↑ liver and kidney parenchymal cells, ↑ autonomic neurons, ↑ pinealocytes, ↑ cardiac muscle cells, etc.). In the course of A., no ↑ chromosomes appear. A. begins (1) with an elongation of ↑ nucleus and division of ↑ nucleolus, followed, in the second phase (2), by a constriction of middle portion of nucleus. A furrow is produced by a 1 to 1.5-µm-thick microfilamentous ring (MfR), amitotic apparatus, which surrounds nuclear neck. In the third phase (3), a microtubular loop (MtL) formed by ↑ centrioles (C), replaces microfilamentous ring. Microtubules shorten, making nuclear constriction deeper until nucleus is divided into two approximately equal parts. Since A. is obviously not accompanied by cytoplasmic division, genetic material remains in the same cell. Thus, A. results in formation of a single ↑ binucleated cell which may become multinucleated after successive amitotic divisions (giant inflammatory cells). It seems that one of starting factors for A. is an unfavorable relation between nuclear volume and nuclear surface of polyploid nuclei in cells mentioned above. Normally, these cells cannot multiply their number through mitosis but through A. can multiply number of their nuclei in answer to high functional demand. However, A. is generally insufficiently understood, especially the distribution of genetic material to the two daughter cells in cases where A. ends in ↑ cytokinesis. (Modified from Pehlemann 1973)

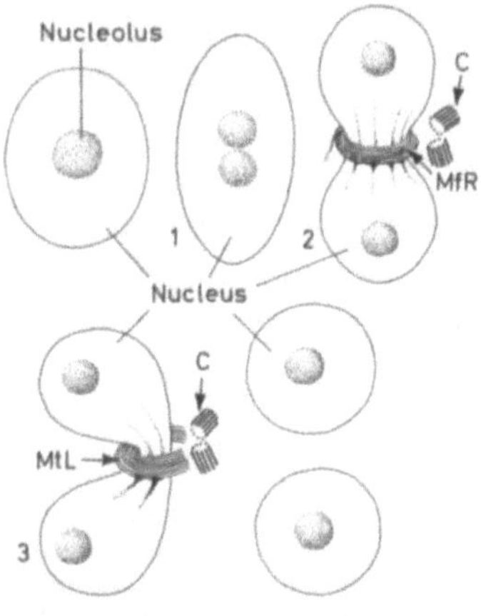

Bucher, O.: Zum Problem der Amitose. In: Handbuch der allgemeinen Pathologie. Vol. 2, Part 2. Der Zellkern I. Berlin, Heidelberg, New York: Springer-Verlag 1971; Pehlemann, F.-W.: Verh. Anat. Ges. *67*, 619 (1973); Pfitzer, P.: Pathol. Res. Pract. *167*, 292 (1980)

Amitotic apparatus: ↑ Amitosis

Amniotic epithelium: A ↑ simple cuboidal to thin ↑ squamous stratified epithelium lining amniotic cavity (AC) and ↑ umbilical cord (UC). In the case of simple cuboidal A., the cells are characterized by an oval or elliptical nucleus, a few mitochondria, a well-developed ↑ Golgi apparatus, sometimes distended rough endoplasmic cisternae, membrane-bound vacuoles (V), ↑ tonofilaments, and ↑ glycogen (Gly) accumulations. Lateral cell surfaces are highly interdigitated; apical cell surface bears numerous microvilli (Mv); basal cell pole shows a moderately developed ↑ basal labyrinth (BLt) and basal processes (BP) of various lengths, affixed with ↑ hemidesmosomes to ↑ basal lamina (BL). Numerous branched ↑ intercellular canaliculi (IC) are present between cells. Near insertion of umbilical cord, A. becomes squamous stratified with occasional ↑ keratinization; its structure corresponds in general to that of a thin ↑ epidermis. A., which produces amniotic fluid, is an example of secretory surface epithelium. (See embryology texts for further information)

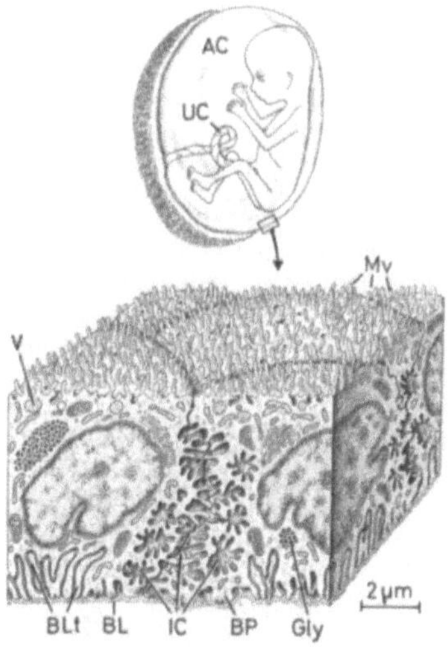

Hoyes, A.D.: J. Anat. *105*, 145 (1969); King, B.F.: Anat. Rec. *190*, 113 (1978); Tiedemann, K.: Anat. Embryol. *158*, 75 (1979)

Amphicrine glands: ↑ Glands simultaneously ↑ exocrine for one product and ↑ endocrine for another. In some glands, one epithelium is responsible for exocrine secretion (↑ pancreatic acinar cells, ↑ seminiferous epithelium), and another epithelium or tissue for hormonal production (↑ pancreatic islets, ↑ interstitial cells, of testis). ↑ Liver parenchymal cells have both functions (exocrine secretion of bile, endocrine secretion of glucose and proteins).

Amphicytes: ↑ Satellite cells, of peripheral neurons

Ampulla hepatopancreatica* (ampulla of Vater): A flasklike enlargment formed by fusion of end parts of ↑ ductus choledochus with ↑ pancreatic duct(s) within submucosa of ↑ duodenum and surrounded by ↑ duodenal glands (DG). Structure: 1) Tunica mucosa (TM): a) epithelium = a ↑ simple tall columnar epithelium; b) lamina propria (LP) = a ↑ loose connective tissue with pear-shaped mucoid glands (G); tunica mucosa forms short valvules (V). 2) Tunica mucularis (TMu) = circularly arranged smooth muscle bundles forming ↑ sphincter of Oddi (SO). 3) Tunica adventitia (TA) = a loose connective tissue separating A. from duodenal glands and duodenal tunica muscularis (DMu). P = ↑ pancreas

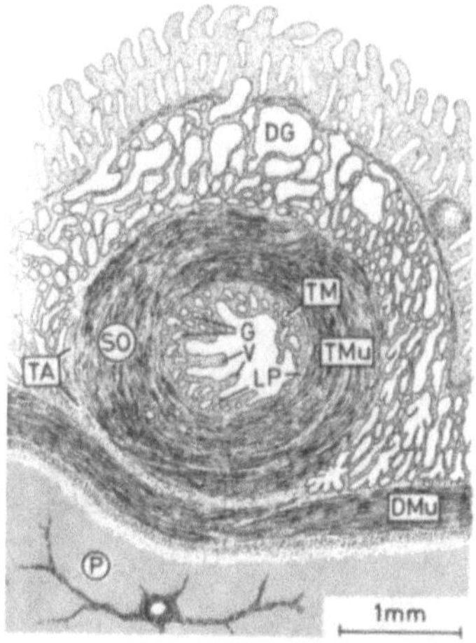

Ampulla, of ductus deferens (ampulla ductus deferentis*): A terminal, about 2–3 cm long, enlarged segment of ↑ ductus deferens situated just prior to beginning of ↑ ejaculatory duct. Structure: 1) Tunica mucosa (TM) is very irregularly folded and consists of a ↑ simple cuboidal epithelium and a loose lamina propria. 2) Tunica muscularis (TMu) = ↑ smooth muscle cells arranged in an inner circular (IC) and an outer longitudinal layer (OL). 3) Tunica adventitia (TA) = a ↑ loose connective tissue.

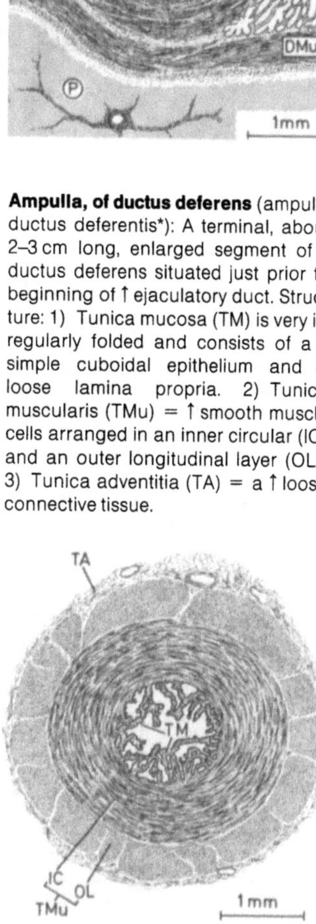

Ampulla, of oviduct: ↑ Oviduct

Ampulla, of semicircular ducts (ampulla membranacea*): An enlargement (A) of each ↑ semicircular duct (SD) at transition point with utricle (U). Each A. contains a ↑ crista ampullaris (CA) with a ↑ cupula (C). In the living state, cupula spans entire height of A. Structure of wall of A. corresponds to that of wall of membranous ↑ labyrinth.

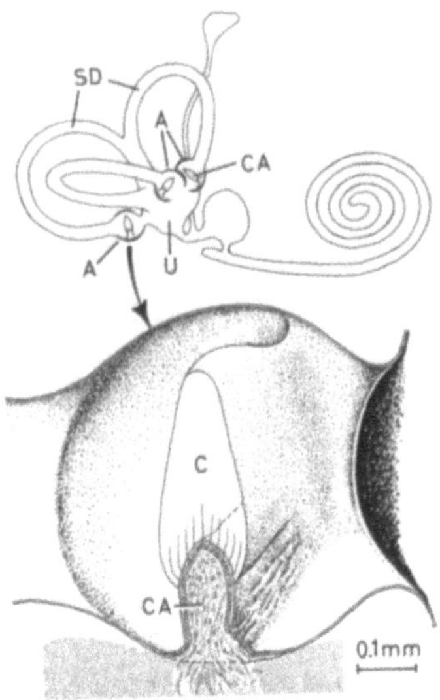

Ampulla, of Vater: ↑ Ampulla hepatopancreatica

Amylases: Enzymes secreted by ↑ salivary glands and by exocrine ↑ pancreas which split starch and/or ↑ glycogen into water-soluble carbohydrates.

Anal canal (canalis analis*): The 2 to 3-cm-long terminal part of ↑ rectum extending from upper to lower border of internal anal sphincter (IS). Structure: A) Zona columnaris or hemorrhoidalis. Upper limit of this zone is indicated by the line marking transition from ↑ simple columnar epithelium of ↑ rectum to ↑ stratified squamous epithelium of A. (arrow), visible as a toothed pectinate line (PL). Terminal part of rectal mucosa makes five to ten longitudinal folds, anal or rectal columns (ANC), which delimitate pocketlike anal sinuses (AS). Lower ends of columns are united to form transverse semilunar folds, anal valves (AV). Stratified squamous epithelium extends over lower parts of anal columns; ↑ Lieberkühn's crypts are not present. Lamina muscularis mucosae (LMM) of rectum breaks up into bundles, which continue only as far as region of anal columns as dilator muscle of Rudinger. A network of submucosal veins with small subendothelial cushions is situated within columns; it anastomoses with internal hemorrhoidal venous plexus (IHP) forming a ↑ corpus cavernosum recti. Pathological dilatation of these veins leads to internal hemorrhoids. Circular layer (CL) of rectal tunica muscularis becomes gradually thicker in columnar zone. B) Zona intermedia. Approximately 1-cm-long segment of A., from lower limit of anal valves to the line where stratified squamous epithelium merges into ↑ epidermis (linea ano-cutane, LAC). Subepithelial layer contains many ↑ elastic fibers, large veins of internal hemorrhoidal plexus, ↑ lymphocytes, ↑ mast cells, tactile ↑ corpuscules, and ↑ sebaceous glands (SG) not associated with hair follicles. Circular layer of rectal musculature ends in this zone as smooth internal sphincter of anus (IS).

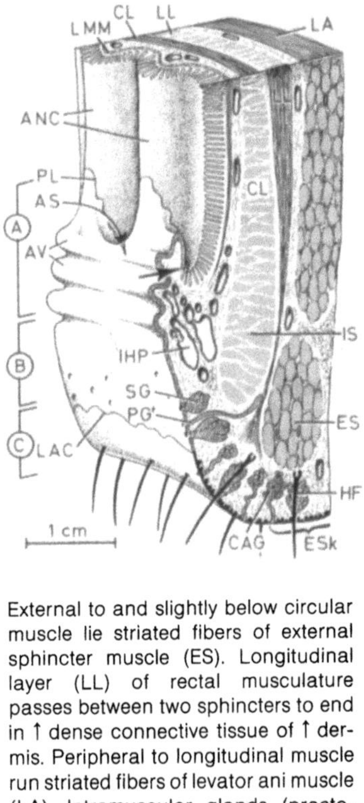

External to and slightly below circular muscle lie striated fibers of external sphincter muscle (ES). Longitudinal layer (LL) of rectal musculature passes between two sphincters to end in ↑ dense connective tissue of ↑ dermis. Peripheral to longitudinal muscle run striated fibers of levator ani muscle (LA). Intramuscular glands (proctodeal glands, PG), lined by simple columnar or stratified squamous epithelium, are located between sphincter muscles. C) Zona cutanea. Terminal segment of A. below linea anocutanea (LAC), covered by a moderately keratinized and pigmented epidermis, is continuous with external ↑ skin (ESk). In subepidermal layer, there are eccrine ↑ sweat glands and sebaceous glands; the latter are associated with ↑ hair follicles (HF). Apocrine ↑ sweat glands, or circumanal glands (CAG), also present here, begin their function after puberty and remain under control of ↑ sex hormones.

Fenger, C., Lyon, H.: Histochem. J. *14*, 631 (1982)

Anal columns: ↑ Anal canal

Anal valves: ↑ Anal canal

Analyzer: A Nicol prism or sheet of Polaroid film mounted above the ↑ objective of a ↑ polarizing microscope. A. serves to analyze deviation of plane of polarization of light provoked by the object.

Anaphase: The stage of cell division in which daughter ↑ chromosomes move from ↑ equatorial plate to opposing poles of the cell.

↑ Meiosis; ↑ Mitosis

Anaplasia (dedifferentiation): The return of differentiated cells to state of immature, embryonic ones with accelerated reproduction.

Anastomosis, arteriovenous (anastomosis arteriovenosa*): A coiled shunt (A) directly connecting an ↑ arteriole (Art) with a ↑ venule (V) often found in distal parts of extremities (↑ microvascular bed in ↑ skin of fingertips and toes, ↑ nail beds, ↑ lips, nose, ↑ auricle). In comparison with its caliber, A. has a thick wall (12–15 μm), a variable length (30–100 μm), and rich vasomotor innervation. When A. is contracted under stimulation of sympathetic nerves, an increased amount of blood passes through ↑ capillaries and warms periphery; when A. dilates, blood can bypass terminal network and flow directly into venule. A. are also found in ↑ gastrointestinal tract, ↑

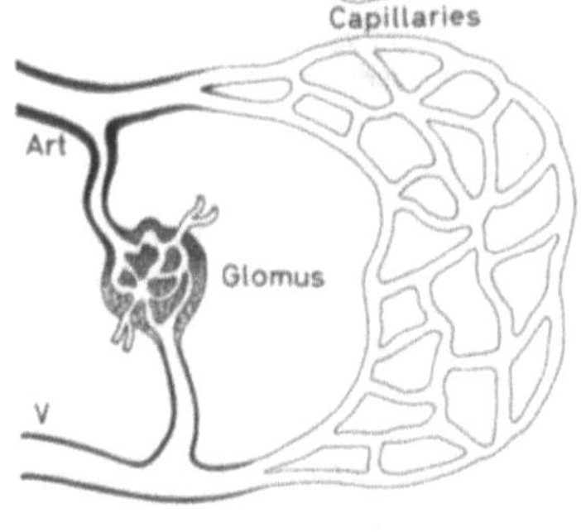

thyroid, and ↑ erectile tissue. Specially structured A., called ↑ glomi, represent an essential part of ↑ aortic, ↑ carotid, and ↑ coccygeal bodies.

Clara, M.: Die arteriovenösen Anastomosen. 2nd edn. Wien: Springer-Verlag 1956; Staubesand, J.: Morphologie des arterio-venösen Anastomosen. In: Von Bertelheimer, H. and Küchmeister, H. (eds.): Kapillaren und Interstitium. Stuttgart: Thieme 1955

Anchoring filaments: ↑ Lymphatic capillaries

Anchoring villi (villus ancoralis*): ↑ Placental villi anchored in ↑ basal plate of ↑ decidua basalis.

↑ Deciduotrophoblastic complex; ↑ Placenta, full-term, structure of

Enders, A.C.: Am. J. Anat. *122*, 419 (1968)

Androgen-binding protein (ABP): A specific protein secreted by ↑ Sertoli's cells under influence of ↑ follicle-stimulating hormone (FSH). A. is found within ↑ semiferous tubules and excurrent system of testis; it may bind considerable amounts of ↑ testosterone necessary for ↑ spermatogenesis.

French, F.S., Ritzén, E.M.: J. Reprod. Fertil. *32*, 479 (1973)

Androgenic hormones: ↑ Androgens

Androgens (androgenic hormones): Hormones stimulating activity of male sex organs and development of male sex characteristics (masculinization). The most potent of A., ↑ testosterone, is produced by ↑ interstitial cells of ↑ testis; weak A., dehydroepiandrosterone and its sulfate, are produced by zona reticularis of ↑ adrenal cortex. A small quantity of A. is secreted by ↑ hilus cells of ↑ ovary.

Androspermatozoa: ↑ Spermatozoa carying 22 ↑ autosomes and the ↑ Y-chromosome.

↑ Gynospermatozoa; ↑ Sex, determination of

Portsmann, T., Portsmann, B., Rohde, W., Reich, W., Wass, R., Dörner, G.: Dermatol. Monatsschr. *165*, 514 (1979)

Androsterone: ↑ Testosterone

Anemia: Any condition characterized by: 1) A decreased number of ↑ erythrocytes per liter, 2) a diminution of ↑ hemoglobin in 100 ml blood, and 3) a diminution of volume of packed blood cells.

Aneuploidy: The state of cells whose ↑ diploid chromosomal number is not an exact multiple of 2 × 23 (in man), i.e., cells with incomplete or more than complete chromosomal set (...44, ...45, ...47, ...48, etc.).

Bond, D.J., Chandley, A.C.: Aneuploidy. Oxford Monographs on Medical Genetics, Vol. 11. Oxford: Oxford University Press 1983

Angioarchitectonics (angioarchitecture): The arrangement of blood vessels in a tissue or organ.

Angioblasts: ↑ Blood islands

Angiotensin I: An inactive decapeptide converted in lung to ↑ angiotensin II.

Schweisfurth, H.: Dtsch. Med. Wochenschr. *107*, 1815 (1982)

Angiotensin II: An octapeptide formed by removal of two amino acids from ↑ angiotensin I under influence of a lung converting enzyme. A. is the most powerful stimulus for production and release of ↑ aldosterone and is the most potent vasoconstrictor known.

↑ Adrenal glands cortex, control of secretion in

Angiotensinogen: A plasma α_2-globulin synthesized by ↑ liver parenchymal cells and converted by ↑ renin to ↑ angiotensin I

↑ Adrenal glands cortex, control of secretion

Richoux, J.P., Cordonnier, J.L., Bouhnik, J., Clauser, E., Corvol, P., Menard, J., Grignon, G.: Cell Tissue Res. *233*, 439 (1983)

Ångström unit: ↑ Å

Anionic dyes: ↑Acid dyes

Anisocytosis: The abnormal variation in size of ↑ erythrocytes.

Anisotropy: A property of some inorganic and organic structures having more than one ↑ refractive index of polarized light along different optical axes with respect to nonrandom orientation of their molecules. A. is studied by means of ↑ polarizing microscope

↑ Birefringence

Annular gap junction: ↑ Nexus, annular

Annular nexus: ↑ Nexus, annular

Annular ridge, of optic papilla: ↑ Optic papilla

Annulate lamellae (lamellae annulatae*): Cytoplasmic ↑ organelles predominantly observed in rapidly dividing and proliferating cells (embryonic, ↑ germ, neoplastic). A. consist of a various number (3–20) of parallel smooth cisternae (SC) each containing a large number of regularly arranged fenestrations (F) that exactly resemble ↑ nuclear pore complexes (NP) in their dimensions and structure. These pores are closed with a fine diaphragm (D) and lined by ↑ annuli. In the area of adjacent fenestrations lies an osmiophilic substance. Because of similarity with nuclear pores, it has been suggested that A. arise by delamination from ↑ nuclear envelope (NE). It is also believed that A. are a special type of ↑ endoplasmic reticulum since fenestrated cisternae may be continuous with rough endoplasmic reticulum (rER). Function of A. is still unknown.

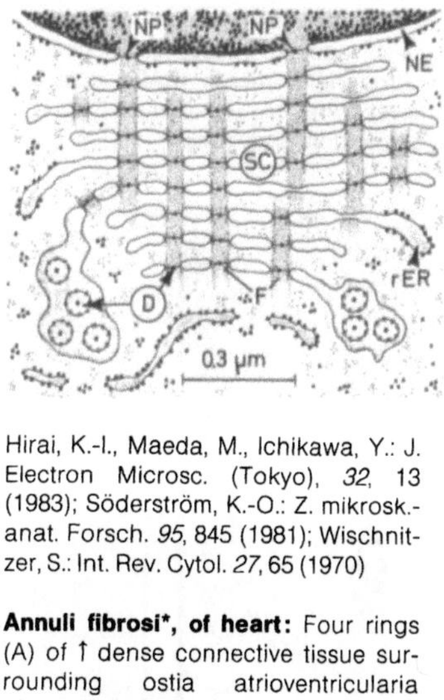

Hirai, K.-I., Maeda, M., Ichikawa, Y.: J. Electron Microsc. (Tokyo), *32*, 13 (1983); Söderström, K.-O.: Z. mikrosk.-anat. Forsch. *95*, 845 (1981); Wischnitzer, S.: Int. Rev. Cytol. *27*, 65 (1970)

Annuli fibrosi*, of heart: Four rings (A) of ↑ dense connective tissue surrounding ostia atrioventricularia (OAD, OAS) and roots of ↑ aorta (Ao) and truncus pulmonaris (TP). Annulus dexter (AD), annulus sinister (AS), and annulus of aorta come into contact, forming trigona fibrosa (T), also com-

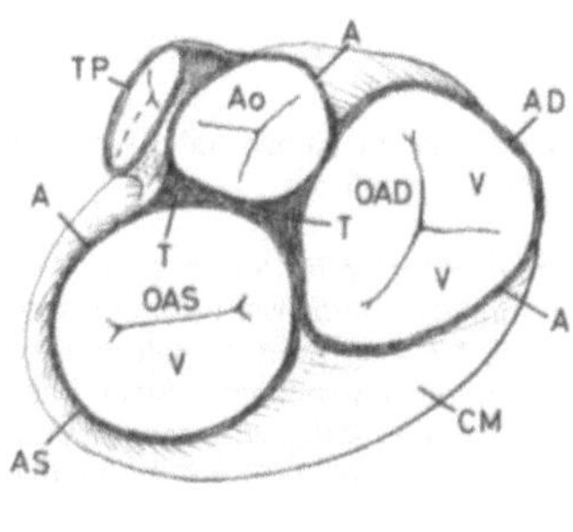

posed of dense connective tissue. A. serve as attaching lines for roots of heart ↑ valves (V) and cardiac muscle fibers (CM). (View from above)

Annuli fibrosi, of intervertebral disc: ↑ Intervertebral disc

Annulospiral nerve endings: ↑ Neuromuscular spindle

Annulus fibrocartilagineus: ↑ Tympanic membrane

Annulus, of nuclear pores: ↑ Nuclear pores

Annulus, of spermatozoon: A ring of dense material (A) at junction between middle piece (MP) and principal piece (PP) of ↑ spermatozoon tail fused with spermatozoon plasmalemma (P). In ↑ spermatid, A. is a small dense ring fixed to flagellar membrane; in the course of ↑ spermiogenesis, it moves down flagellum (F) probably to prevent caudal sliding of mitochondria (M) during movements of tail. In some species (rodents), A. delimits a circular retroannular recess (RAR). It is believed that A. is derived from ↑ chromatoid body (CB).

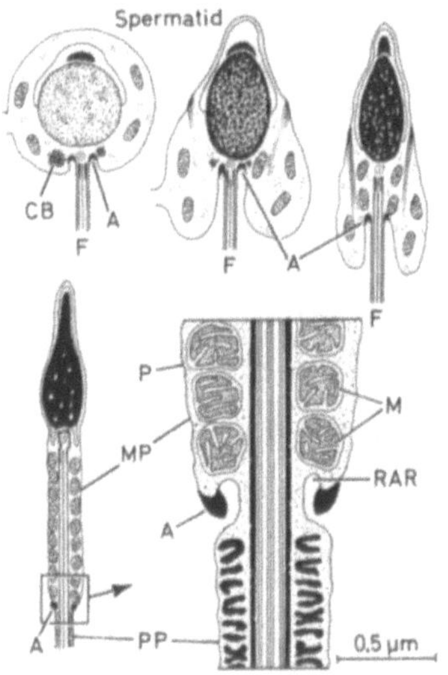

Anomalies, chromosomal: ↑ Chromosomes, anomalies of

Anovulatory cycle: An ↑ ovarian cycle of normal duration without ↑ ovulation (no ripe follicle) and, therefore, minimal endometrial changes. Since ↑ corpus luteum is lacking, endometrial epithelium remains in proliferative phase until menstrual bleeding occurs. It is believed that a healthy woman has three to four A. a year.

Antennulae microvillares: Microfilaments of ↑ glycocalyx (Glyc) present at tips of ↑ microvilli (Mv) of ↑ absorption cells of intestine, ↑ gallbladder epithelial cells, etc.

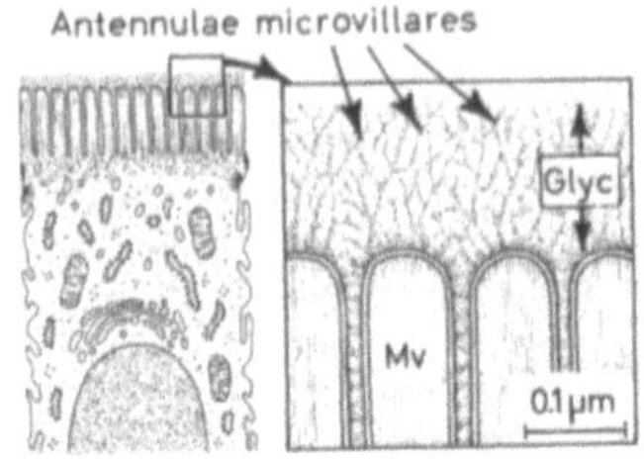

Yamada, K.: J. Morphol. *124*, 1 (1968)

Anterior chamber, of eye (camera anterior bulbi*): A space (AC) delimited by posterior aspect of ↑ cornea (C), anterior surface of ↑ iris (I), iridocorneal angle (A), and central portion of anterior aspect of ↑ lens (L). A. contains ↑ aqueous humor. A. communicates with ↑ posterior chamber through a fine slit between iris and lens.

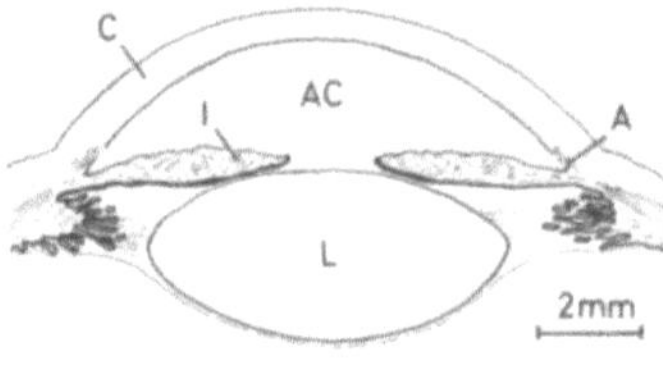

Anterior column cells: ↑ Motor neurons

Anterior epithelium, of cornea: ↑ Cornea

Anterior lingual glands (glandula lingualis anterior, glandula apicis linguae, Blandin's gland, Nuhn's gland, glandula anterior linguae*): Small paired ↑ salivary glands situated near tip of ↑ tongue on each side of frenulum. A. are composed of mixed secretory tubules with serous ↑ demilunes and, in some portions, of tubules containing only ↑ mucous cells. Excretory canals open on inferior lingual surface.

Anterior lobe, of hypophysis: ↑ Hypophysis, pars distalis of

Anterograde transneuronal degeneration: ↑ Transneuronal degeneration

Antibodies: ↑ Immunoglobulins; ↑ Plasma cells

Anticodon: The sequence of three bases in ↑ transfer RNA molecule which is complementary to ↑ codon of ↑ messenger RNA. A. is situated in part of molecule directly opposite amino acid acceptor end; A. is only segment of transfer RNA molecule capable of reading message written on the codon. (See molecular biology texts for further information)

↑ Protein synthesis

Antidiuretic hormone (ADH, adiuretin, vasopressin): A hypothalamic hormone synthesized predominantly in ↑ neurosecretory cells of ↑ nucleus supraopticus. Linked to appropriate ↑ neurophysin, A. travels in ↑ neurosecretory granula with ↑ axoplasmic flow along ↑ hypothalamohypophyseal tract to be stored in ↑ neurohypophysis. Following an increase of plasma osmolatity and/or decrease in blood volume, ADH-filled granula are discharged by ↑ exocytosis through basal lamina into ↑ pericapillary space where they escape oberservation. Here, A. and neurophysin enter circulation in molecular dispersion. A. raises blood pressure by stimulating constriction of small blood vessels and, by activating cAMP, provokes an increase in water reabsorption into ↑ distal convoluted tubules, with consequent decrease in urine output. Insufficient production of A. leads to ↑ diabetes insipidus.

Baertschi, A.J., Dreifuss, J.J. (eds.): Neuroendocrinology of Vasopressin, Corticoliberin and Opiomelanocortins. New York: Academic Press 1982; Kawata, M., Sano, Y.: Anat. Embryol. *165*, 151 (1982); Krisch, B., Becker, K., Bargmann, W.: Z. Zellforsch. *123*, 47 (1972)

Antiphlogistic hormones: A former term for ↑ glucocorticoids because of their pronounced anti-inflammatory effect.

Antisphering substance: A factor present in plasma that maintains regular biconcave form of ↑ erythrocytes and prevents occurrence of crenated ↑ erythrocytes in isotonic solution.

Antral follicle: ↑ Ovarian follicle, secondary

Antrum, of ovarian follicles (antrum folliculare*): A cavity among ↑ follicular cells of secondary vesicular and mature ↑ ovarian follicles filled with ↑ liquor folliculi.

Anulus: ↑Annulus

Aorta*: The largest ↑ elastic artery of the body. Structure: 1) Tunica intima (TI) (about 150 µm thick) = a) ↑ endothelium (E); b) subendothelial layer (SL) rich in interlacing ↑ collagen and elastic ↑ fibers, small longitudinally disposed bundles of ↑ smooth muscle cells, and some fibroblastlike cells. There is no distinct internal ↑ elastic lamina between intima and media. 2) Tunica media (about 2 mm thick) = 50–75 concentrically arranged 2 to 3-µm-thick fenestrated ↑ elastic laminae (EL) interconnected by elastic fibers and circumferentially disposed smooth muscle cells. In interspaces between membranes there are a few ↑ fibroblasts, ↑ fibrocytes, occasional ↑ wandering cells, and a ↑ ground substance containing ↑ chondroitin sulfate. A. and other elastic arteries have no distinct external ↑ elastic lamina. 3) Tunica adventitia (TA) = a very thin, ↑ loose connective tissue with ↑ vasa vasorum (VV), lymphatic vessels, and ↑ myelinated and ↑ nonmyelinated nerve fibers. Intima and inner part of media are nourished by diffusion from circulating blood; ↑ vasa vasorum vascularize external part of A. wall and adventitia.

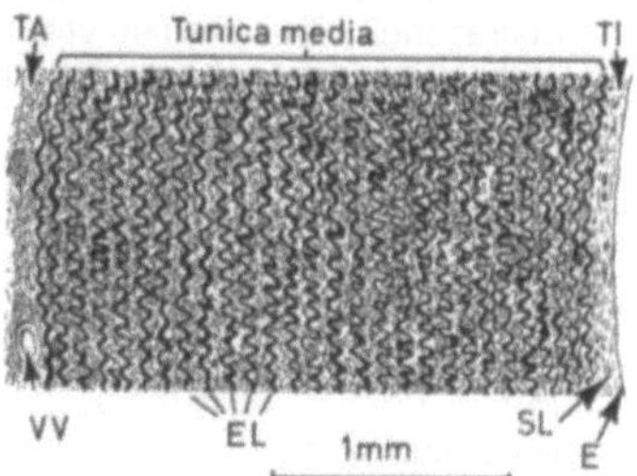

Aortic body (glomus aorticum*): One of several small vascular organs situated near arch of ↑ aorta, with structure and function identical to ↑ carotid body.

Abbott, C.P., Howe, A.: Acta Anat. *81*, 609 (1972); Easton, J., Howe, A.: Cell Tissue Res. *232*, 349 (1983); Hansen, J.T.: Cell Tissue Res. *196*, 511 (1979)

Aortic valves: ↑Valves, aortic

Apatite crystals: ↑ Hydroxyapatite crystals

Aperture, numerical: ↑ Numerical aperture

Apex, of tooth (apex dentis*): End of tooth root perforated by a small hole, ↑apical foramen.

↑Teeth

Apical differentiations: A collective term for specialized structures on free surface of epithelial cells (↑ cilia, ↑ cytofila, ↑ microplicae, ↑ microvilli, ↑ stereocilia).

Apical fibers: Certain ↑ cementoalveolar fibers extending from ↑ cementum of tip of a tooth root to its ↑ alveolus.

Apical foramen (foramen apicale dentis*): A small opening at ↑ apex of tooth root allowing communication between pulp chamber and ↑ periodontal ligament. Blood and lymphatic vessels and nerve fibers pass through A.

Apical pits (apical canaliculus, apical invagination, apical tubules): ↑ Micropinocytotic invaginations (Ap) arising at base of certain ↑ microvilli (Mv) and extending into cytoplasm. A. may elongate, giving rise to tubular membrane-bound bodies (TMB), which pinch off to form apical vesicles (AV). All these structures are very frequent in ↑ proximal tubule cell of kidney but are also found in ↑ absorptive intestinal cells, nonciliated cells of ↑ ductuli efferentes, and ↑ marginal cells of ↑ stria vascularis. A. and apical vesicles are believed to be active in uptake of proteins and membrane recycling. (Fig. = proximal tubule cell, rat, ↑ ruthenium red staining)

↑Clathrin

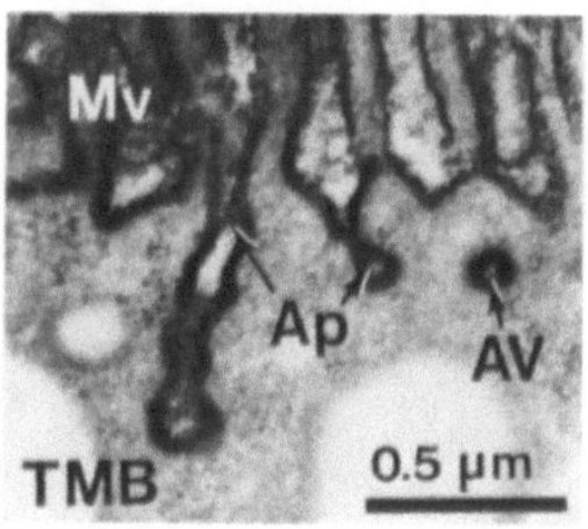

Christensen, E.I.: Eur. J. Cell Biol. *29*, 43 (1982)

Apical pole (apex cellularis*): The part of a cell forming free cell surface; a term used mainly in describing ↑ simple, ↑ pseudostratified, and glandular epithelia, as well as some other cells (↑ odontoblasts). Lateral cell surface A. bears ↑ junctional complex (JC).

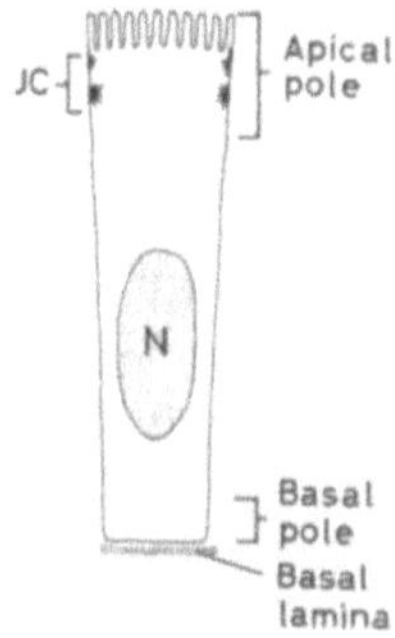

Apical vesicles: ↑Apical pits

Apical vesicles, of thyroid follicular cells: ↑ Unit membrane-bound structures (AV) (50–200 nm in diameter) filled with fine-granular, moderately osmiophilic material. A. are concentrated in ↑ apical pole of cell, immediately below apical plasmalemma with which they fuse before discharging their contents into follicular lumen (FL). A. arise from ↑ Golgi apparatus (G) and appear to transport ↑ thyroglobulin from this ↑ organelle to follicular lumen. Ly = ↑ lysosomes (Fig. = rat)

↑Thyroid follicular cells

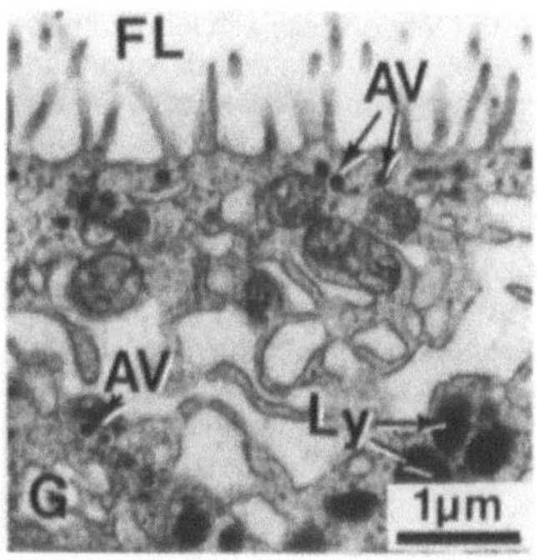

Apochromatic objective: ↑Objective

Apocrine glands (glandula apocrina*): ↑ Exocrine glands employing ↑ apocrine mechanism of ↑ secretion.

Apocrine secretion: ↑ A form of release of secretory product in which protruding ↑ apical pole with secretory material separates from cell body, enters glandular lumen, and becomes a product of secretion. During its life cycle, secretory cell restores lost cytoplasm and resynthesizes secretory product (e.g., A. of ↑ lipid droplets in ↑ lactation). (Modified after Bargmann and Knoop 1959)

↑ Eccrine secretion; ↑ Secretion

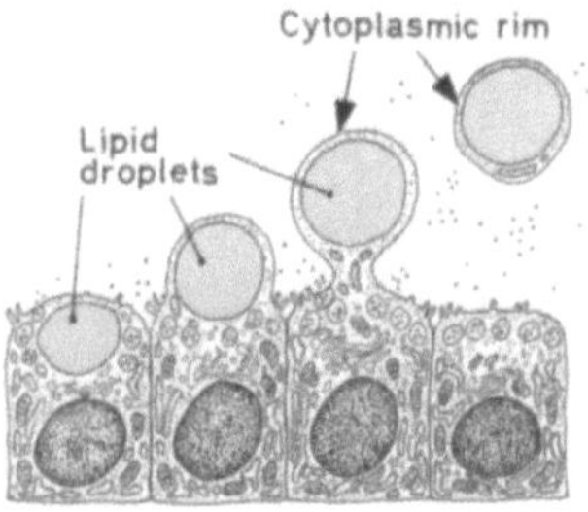

Bargmann, W., Knoop, A.: Z. Zellforsch. *49*, 344 (1959)

Apocrine sweat glands: ↑ Sweat glands, apocrine

Apolar neurons: Nerve cells in earliest stage of development without any processes. A. cease to exist when ↑ differentiation begins.

Aponeurosis*: A ↑ dense regular connective tissue in which parallel, wavy bundles of collagen fibers are arranged in multiple sheets or lamellae. In general, bundles of adjacent lamellae cross each other at approximately 90 °C; between lamellae there is an interchange of fibrils. Cells of A. are similar to ↑ tendon cells with long, ramified processes that follow direction and crossing of fibers. Therefore, A. can be considered a flat tendon. Corneal stroma is an example of specially differentiated A.

↑ Cornea; ↑ Fascia

Aponeurosis linguae* (lingual aponeurosis, fascia linguae*): A layer of ↑ dense connective tissue belonging to deep zone of ↑ lamina propria of dorsal lingual mucosa serving as insertion for lingual muscles.

↑ Tongue

Apparato reticolare interno*: ↑ Golgi apparatus

Apparatus, juxtaglomerular: ↑ Juxtaglomerular apparatus

Apparatus, of Golgi: ↑ Golgi apparatus

Appendages, of skin: ↑ Skin appendages

Appendices epiploicae* (epiploic appendages): Pendulous accumulations of white ↑ adipose tissue in ↑ tela subserosa of ↑ colon covered by ↑ peritoneum.

Appendix (processus vermiformis, appendix vermiformis*): A wormlike evagination of ↑ cecum, about 10–15 cm long and up to 8 mm wide. As A. belongs to ↑ large intestine, it has no ↑ intestinal villi. Structure: 1) Tunica mucosa (TM): a) epithelium (E) = a ↑ simple columnar epithelium with ↑ striated border and numerous ↑ goblet cells, forming crypts of ↑ Lieberkühn (CL), of irregular shape and varying lengths. Occasional ↑ Paneth's cells and an appreciable quantity of ↑ enterochromaffin cells are located deep in crypts; b) lamina propria (LP) = ↑ lymphatic tissue with numerous secondary ↑ lymphatic nodules (LN); c) lamina muscularis mucosae (LMM) = a very thin layer of ↑ smooth muscle cells often perforated or masked by lymphatic tissue. 2) Tela submucosa (TS) = a ↑ loose connective tissue with blood and lymphatic vessels and nerve fibers. Internal layer of submucosa may be occupied by lymphatic tissue. 3) Tunica muscularis (TMu) = very thin; both internal circular and external longitudinal layers are clearly delimited. 4) Tela subserosa (TSs) = a loose connective tissue with vessels and nerve fibers. 5) Tunica serosa (TSe) = ↑ mesothelial cells of visceral ↑ peritoneum. Subserosa and serosa form mesoappendix (Mes). Because of its richness in intestinal endocrine cells, A. should not be considered a rudimentary organ. Under stress, its thin wall can be perforated as a consequence of destructive action of ↑ corticosteroids on lymphatic tissue.

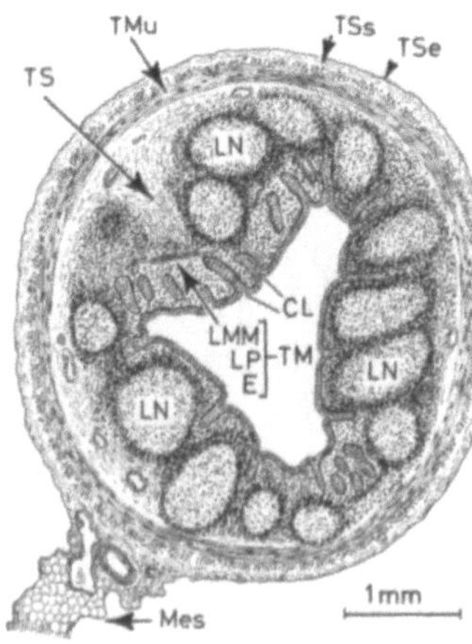

Bockman, D.E.: Arch. Histol. Jpn. *46*, 271 (1983); Gorgollon, P.: J. Anat. *126*, 87 (1978)

Appositional growth, of bone: ↑ Bone, appositional growth of

Appositional growth, of cartilage: ↑ Cartilage, appositional growth of

Ap-spermatogonia: ↑ Spermatogonia

APUD-cells (amine precursor uptake and decarboxylation): A concept grouping certain apparently unrelated endocrine cells, some in ↑ endocrine glands, others in nonendocrine tissues, having numerous morphofunctional similarities. Ultrastructural characteristics: a well-developed Golgi apparatus, few cisternae of rough endoplasmic reticulum, round and fixation-labile mitochondria, a considerable number of round, 10 to 30-nm membrane-bound ↑ secretory granules with a dense core. Functional properties: Production of a polypeptide hormone of low molecular weight, synthesis of biogenic amines and/or ↑ serotonin from their precursors with aid of l-amino acid decarboxylase, storage of biogenic amines predominantly within secretory granules. A. include ↑ endocrine cells of ↑ gastrointestinal tract, ↑ chromaffin cells of ↑ adrenal medulla, pancreatic ↑ A-, ↑ B-, and ↑ D-cells, ↑ C-cells of thyroid gland, hypophyseal ACTH-, STH-, and MSH-cells, type I cells of ↑ cartoid and ↑ aortic bodies, ↑ NEB-cells, urethral and tracheal endocrine cells, and type I cells of chromaffin ↑ paraganglia. The common origin of A. from neural crest has not yet been confirmed.

Ayer-Le Lievre, C., Fontaine-Perus, J.: Arch. Histol. Jpn. *45*, 409 (1982); Pearse, A.G.E.: Mikroskopie *36*, 257 (1980); Winckler, J.: Klin. Wochenschr. *54*, 49 (1976)

Aqueous humor (humor aquosus*): A clear, thin, watery, slightly alkaline fluid (index of refraction 1.33) secreted by epithelium of ↑ ciliary body (CB). A. flows from ↑ posterior chamber (PCh) through pupil into ↑ anterior chamber (ACh), where it is reabsorbed through ↑ trabecular meshwork (TM) and canal of ↑ Schlemm (SC). A. contains less protein (0.02%), ↑ urea, and glucose than blood plasma, but more ascorbate, pyruvate, lactate, and depolymerized ↑ hyaluronic acid; it has no ↑ fibrinogen. A., which also penetrates ↑

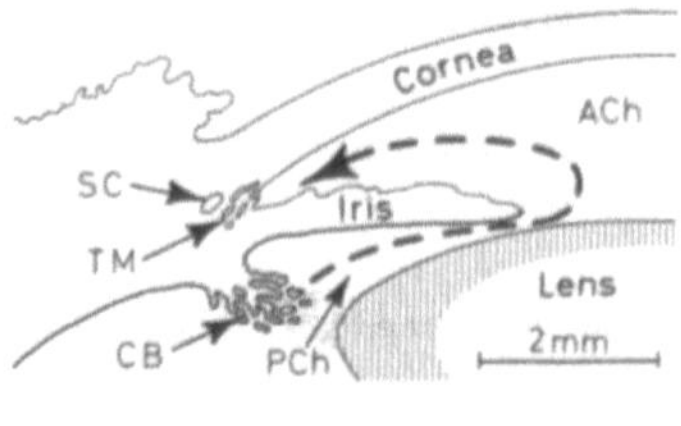

vitreous body, is responsible for maintenance of intraocular pressure and for metabolic supply of ↑ lens. If the very accurate balance of secretion and reabsorption of A. is disturbed, intraocular pressure increases to over 25 mm Hg leading to ↑ glaucoma.

↑ Ciliary epithelium

Arachnoid (arachnoidea encephali*): The middle, avascular meninx separated from ↑ dura mater (DM) by ↑ subdural space (SDS). Structure: 1) Arachnoid membrane (AM): five to eight layers of flattened, densely packed ↑ fibroblasts (F) held together by numerous ↑ nexus. Surface of arachnoid membrane in contact with subdural space is covered with a lining of thin squamous mesotheliumlike epithelial cells (EC). Local proliferation of both types of cells forms ↑ arachnoid villi (AV). At the level of ↑ subarachnoid angle, arachnoid membrane continues as ↑ perineural endothelium of peripheral ↑ nerves. 2) Arachnoid trabeculae (AT) = ribbon- or pillar-shaped strands forming a three-dimensional spider's, web pattern between arachnoid membrane and ↑ pia mater (PM). Trabeculae, consisting of thin cytoplasmic processes of fibroblasts, are attached to pia mater. 3) Subarachnoid space (SAS) = a space between arachnoid membrane and pia mater traversed by trabeculae; it is filled with ↑ cerebrospinal fluid, contains numerous blood vessels (V), many ↑ macrophages, ↑ lympohocytes, perivascular ↑ mast cells, and many embryonic mesenchymal cells. It is likely that cerebrospinal fluid seeps through arachnoid villi (i.e. ↑ Pacchionian granulations) to be eliminated in bloodstream (i.e., sinuses of dura mater). Subarachnoid space communicates with ↑ endoneurial space.

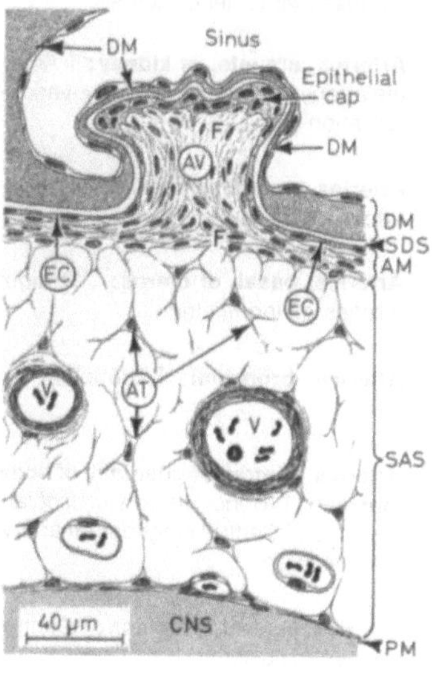

↑ Meninges

Merchant, R.E., Kelleher, J.J., Low, F.N.: J. Submicrosc. Cytol. *11*, 293 (1979); Zaki, W.: Bull. Assoc. Anat. (Nancy) *61*, 283 (1977)

Arachnoid granulations: ↑ Pacchionian granulations

Arachnoid membrane: ↑ Arachnoid

Arachnoid villi (Pacchionian corpuscles): Avascular, compact, buttonlike projections of arachnoid membrane predominantly penetrating dural sinuses and, more rarely, bone. A. are covered by a uni- or pluristratified epithelial cap and a thin layer of ↑ dura mater. A. are considered to be precursors of ↑ Pacchionian granulations.

↑ Arachnoid

Arantius' nodule (fibrous nodule, nodulus valvulae semilunaris*): A rounded accumulation of ↑ dense connective tissue at middle of free margin of ↑ aortic and ↑ pulmonic valves.

Arched collecting tubules (tubulus renalis arcuatus*): The intracortical initial segment of ↑ collecting tubules of kidney.

Archicortex: ↑ Allocortex

Archipallium: ↑ Allocortex

Archiplasm: ↑ Hof, of nucleus

Arcuate arteries, of kidney (arteriae arcuatae*): ↑ Muscular arteries, branches of interlobar arteries running across bases ↑ of medullary ↑ pyramids between renal cortex and medulla. Each A. gives off numerous ↑ interlobular arteries.

↑ Kidney, vascularization of

Arcuate arteries, of uterus: ↑ Muscular arteries, branches of uterine arteries. A. lie in stratum vasculare of ↑ myometrium and anastomose with corresponding vessels from the other side. Each A. gives off several ↑ coiled arteries.

↑ Endometrium, vascularization of

Arcus senilis (arcus lipoides, gerontoxon): The opaque, grayish ring near corneal ↑ limbus, frequent in the aged. A. results from lipid deposits (mixture of fat and cholesterol) in intercellular spaces of corneal stroma and ↑ Descemet's membrane.

Area cribrosa, of renal papilla (area cribrosa*): The area on tip of ↑ renal papilla where ↑ papillary ducts open.

↑ Collecting tubules, of kidney

Area postrema*: A ↑ circumventricular organ (AP) situated in caudal angle of IV ventricle, at beginning of central canal (C) of ↑ spinal cord. A. consists of modified neurons ("parenchymal cells") separated by large intercellular spaces and connected by nerve fibers with tractus solitarius. A. is covered by specialized ↑ ependymal cells, richly vascularized by fenestrated ↑ capillaries surrounded by large ↑ perivascular spaces, thus having no ↑ blood-brain barrier. A. seems to be involved in chemoreception, osmoreception, ↑ neurosecretion, and regulation of some autonomic functions (vomiting?), but its exact function remains unknown. A. in the human is a paired organ. (Fig. = mongolian gerbil, courtesy Biomedical Research)

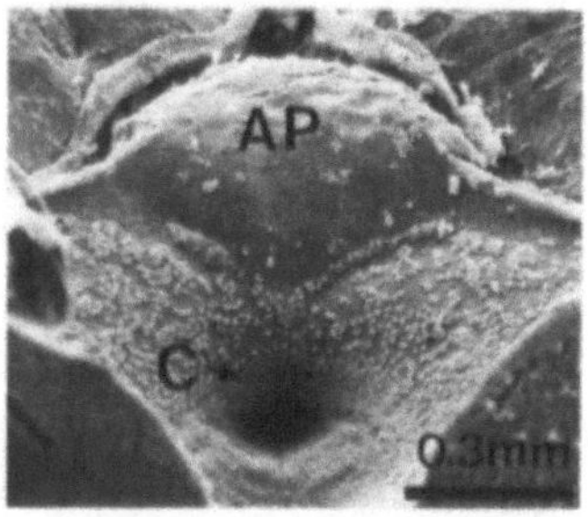

Bird, E., Cardone, C., Contreras, J.: Brain Res. *270*, 193 (1983); Gotow, T., Hashimoto, P.H.: Cell Tissue Res. *201*, 207 (1979); Karasawa, N., Yoshida, M., Kondo, Y., Nagatsu, T., Nagatsu, I.: Acta Histochem. Cytochem. *16*, 138 (1983); Krstić, R.: Biomed. Res. *2*, Suppl. 129 (1981); Miller, A.D., Wilson, V.J.: Brain Res. *270*, 154 (1983)

Area striata: ↑ Striate area

Areae gastricae: ↑ Gastric areas

Areas, of brain cortex: A more or less sharply delimited parcel of ↑ brain cortex with its own characteristic histological structure and function. At present, Brodmann's mapping and numbering (47 A.) is used for reference.

Truex, R.C., Carpenter, M.B.: Human Neuroanatomy. 6th edn. Baltimore: Williams and Wilkins 1969

Areola, of mammary gland (areola mammae*): The circular area of hairless ↑ skin surrounding ↑ nipple. Structure: 1) ↑ Epidermis (E) = a keratinized ↑ stratified squamous epithelium becoming pigmented at puberty and particularly during pregnancy. Epidermis forms very deep irregular ↑ epidermal ridges (R). 2) ↑ Dermis (D): a) papillary layer (PL) = unusually long, branched ↑ dermal papillae with many capillaries and Meissner's ↑ corpuscles; b) reticular layer (RL) = a ↑ dense connective tissue, rich in ↑ elastic fibers and smooth muscle bundles (MB), both attached to epidermis, rare ↑ sebaceous glands, ↑ areolar glands (AG), and a few nerve fibers. 3) ↑ Hypodermis (Hyp) = a ↑ loose connective tissue disposed in layers parallel to epidermis containing some adipose lobules. Under the influence of cold or psychical stimuli, muscular elements contract and skin of A. wrinkles.

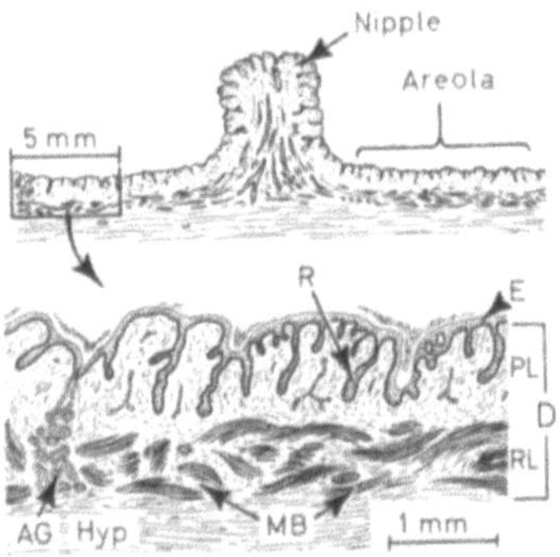

Areolar glands (Montgomery's glands, glandulae areolares*): Small cutaneous ↑ tubuloalveolar glands, considered to transitional between ↑ sweat and ↑ mammary glands. There are 10–12 A. per ↑ areola, predominantly located at the margin in the form of small round elevations. A. moisten ↑ nipple of the breast and ↑ lips of the nursing child.

Areolar tissue: ↑ Loose connective tissue

Argentaffin cells: ↑ Enterochromaffin cells

Argentaffin reaction: ↑ Argentaffinity

Argentaffinity: The capacity of some cellular and tissue components to reduce silver from its salts without any special pretreatment.

Argyrophilia: The ability of tissue elements to reduce siver from silver salts after pretreatment with a chemical reducing agent (formol, hydrochinon, etc.).

Gallyas, F.: Histochemistry *74*, 393, 409 and 423 (1982)

Argyrophilic cells: A group of endocrine cells with ↑ secretory granules, reducing silver salts only after exposition to a reducing agent. Based on immunocytochemical studies, ↑ G-cells, ↑ enterochromaffinlike cells of stomachal ↑ fundus, ↑ pancreas, fetal lung, etc. are classified among A.

↑ APUD-cells

Argyrophilic fibers: ↑ Reticular fibers

Argyrophilic reaction: ↑ Argyrophilia

Arneth's formula: Normal approximative ratio of neutrophilic ↑ granulocytes based on number of nuclear lobes. Frequencies: one lobe 5%, (↑ band neutrophils), two lobes 35%, three lobes 41%, four lobes 17%, five lobes 2%. A shift to the left (L) signifies that many young cells with one or two segments have entered blood; a shift to the right (R) signifies an incapacity of ↑ bone marrow to produce new neutrophilic granulocytes since the number of old elements with four or more lobes has increased.

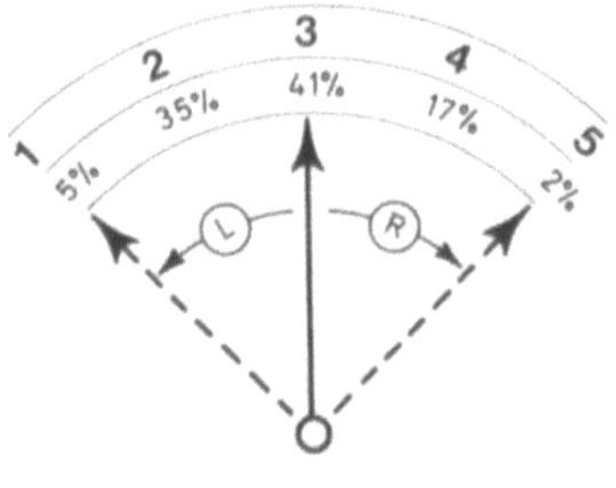

Arrector pili muscle (musculus arrector pili*): A small, flat bundle of ↑ smooth muscle cells inserted between papillary layer (PL) of ↑ dermis and connective tissue sheath of ↑ hair follicle (HF), passing below ↑ sebaceous gland (SG). Contraction of A. (fig. on

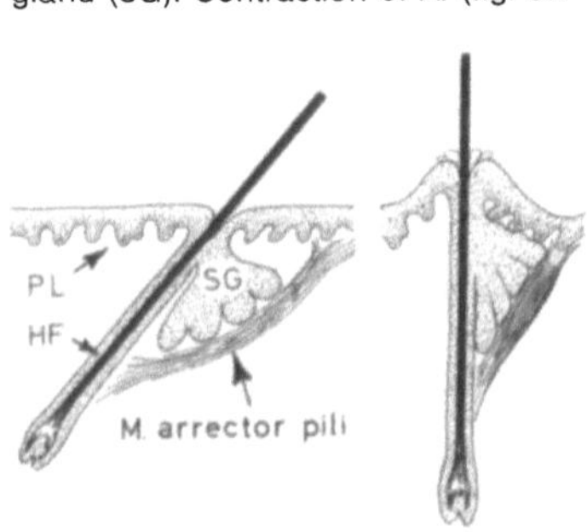

the right), provoked by ↑ sympathetic nervous system under influence of cold, fear, anger, etc., erects ↑ hair, depressing ↑ skin at insertion point of muscle, and elevating zone immediately surrounding hair (cutis anserina = "goose flesh"). Simultaneously, contraction of A. presses sebaceous gland, aiding in expulsion of ↑ sebum. A. are lacking in ↑ eyelashes, hairs of eyebrows, ↑ tragi, and ↑ vibrissae.

Arteria profunda penis: ↑ Deep artery, of penis

Arteries (arteriae*): Blood vessels conveying blood away from heart. With the exception of ↑ pulmonary and umbilical arteries, all arteries carry oxygenated blood. General structure: 1) Innermost layer = tunica intima; 2) middle layer = tunica media; 3) external layer = tunica adventitia or externa. With regard to the structure of tunica media, A. are divided into ↑ elastic A., ↑ muscular A., ↑ hybrid A., and ↑ mixed A.

Arteries, age-related changes of: Starting in early decades of life, following physiological changes gradually occur in arterial wall: Irregular thickening of intima, frequent splitting of inner ↑ elastic lamina, ↑ hypoplasia of ↑ smooth muscle cells, granular deposition of calcium salts and lipids in media, relative increase in ↑ elastic fibers with decrease in their elasticity, increase in collagen fibers and ↑ glycosaminoglycans. Result: Stiffness of arterial wall. A. affect first ↑ aorta and coronary and brain arteries; small muscular ↑ arteries and ↑ arterioles remain practically unaffected. Not to be confused with arteriosclerosis.

Arteries, arcuate, of kidney: ↑ Arcuate arteries, of kidney; ↑ Kidney, vascularization of

Arteries, arcuate, of uterus: ↑ Arcuate arteries, of uterus

Arteries, basal, of uterus: ↑ Straight arteries, of endometrium

Arteries, bronchial: ↑ Bronchial arteries

Arteries, central longitudinal, of bone marrow: ↑ Bone marrow, vascularization of; ↑ Central longitudinal artery, of bone marrow

Arteries, central, of spleen: ↑ Central arteries, of spleen

Arteries, coiled, of endometrium: ↑ Coiled arteries, of endometrium

Arteries, coiled, of penis: ↑ Helicine arteries, of penis

Arteries, coronary: ↑ Coronary arteries

Arteries, deep, of penis: ↑ Deep arteries, of penis

Arteries, elastic: ↑ Elastic arteries

Arteries, helicine, of penis: ↑ Helicine arteries, of penis

Arteries, hybrid: ↑ Hybrid arteries

Arteries, interlobular, of kidney: ↑ Interlobular arteries, of kidney

Arteries, interlobular, of liver: ↑ Interlobular arteries, of liver

Arteries, mixed: ↑ Mixed arteries

Arteries, muscular: ↑ Muscular arteries

Arteries, muscular, innervation of: ↑ Muscular arteries, innervation of

Arteries, muscular, small: ↑ Muscular arteries, small

Arteries, penicillar: ↑ Penicillar arteries

Arteries, pulp: ↑ Penicillar arteries

Arteries, spiral, of uterus: ↑ Coiled arteries, of endometrium

Arteries, terminal: ↑ Terminal arteries

Arteries, uteroplacental: ↑ Uteroplacental arteries

Arterioles (arteriola*): Smallest of ↑ muscular arteries with a diameter of less than 100 µm. Structure: 1) Tunica intima: a) ↑ endothelium (E) = a simple nonfenestrated squamous epithelium; b) a thin internal ↑ elastic lamina (IEL) in A. larger than 50 µm; in small A., replaced by ↑ basal lamina. Processes of endothelial cells penetrate through openings in elastic lamina forming ↑ myoendothelial junctions. 2) Tunica media (TM) = one or two layers of helicoidally arranged spindle-shaped ↑ smooth muscle cells surrounded by basal lamina, ↑ collagen, and ↑ reticular fibers. 3) Tunica adventitia (TA, with same thickness as media) = a ↑ loose connective tissue containing ↑ macrophages, ↑ mast cells, and ↑ nonmyelinated nerve fibers.

↑ Microvasculature

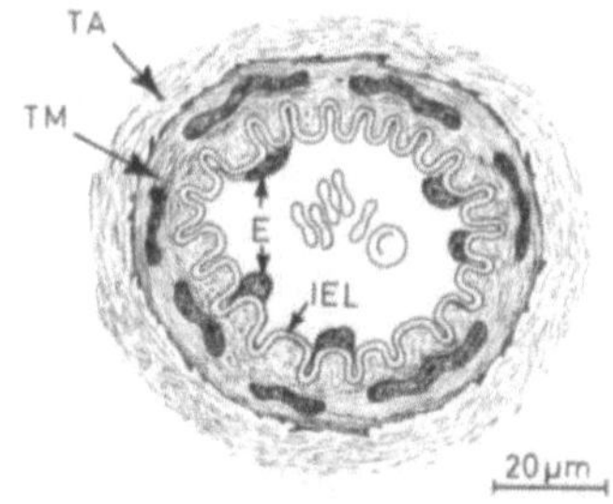

Hyde, B.J., Braekevelt, C.R.: Anat. Anz. *153*, 45 (1983)

Arterioles, afferent: ↑ Afferent arterioles

Arterioles, efferent: ↑ Efferent arterioles

Arterioles, sheathed, of spleen: ↑ Capillaries, sheathed, of spleen

Arteriovenous anastomoses: ↑ Anastomoses, arteriovenous

Articular cartilage (cartilago articularis*): A layer of hyaline ↑ cartilage, about 0.2–6 mm thick, covering articular surfaces of bone. Structure: 1) Zone of hyaline cartilage (HC) = a) tangential zone or gliding layer (GL) of flattened ↑ chondrocytes; b) transitional zone (TZ) of randomly distributed, round chondrocytes; c) radial zone (RZ) of columns of chondrocytes. 2) Thin, wavy, basophilic line of calcified cartilage indicates functional limit (FL) between plastic hyaline cartilage and hard calcified cartilage underneath. 3) Zone of calcified cartilage (CC). Collagen fibers (CF) are tangential in gliding layer, then turn and run between column up to bone without penetrating it. The very irregular surface between A. and bone assures a solid contact between these two tissues and marks histological limit (HL). A. is almost exclusively nourished from ↑ synovial fluid. A. is not covered by ↑ perichondrium and thus cannot regenerate. A. of temporomandibular and sternoclavicular articulations consists of fibrous ↑ cartilage since mandibula and clavicula are mesenchymal ↑ bones.

Articular cartilage

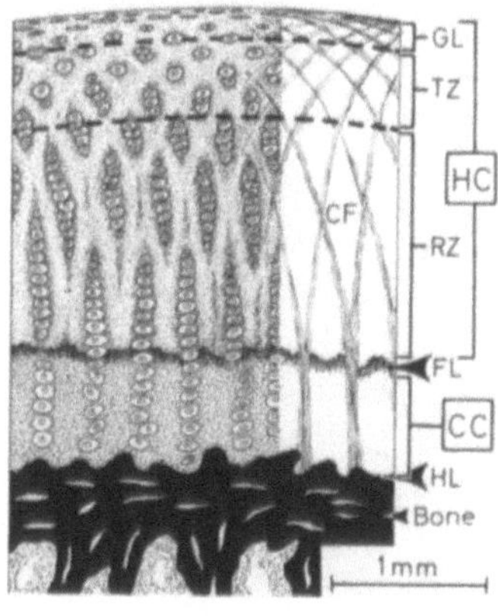

Bloebaum, R.D., Wilson, A.S.: J. Anat. *131*, 333 (1980); Draenert, Y., Draenert, K.: Verh. Anat. Ges. *73*, 909 (1979); Ghadially, R.N., Young, N.K., Lalonde, J.-M.A.: J. Anat. *135*, 685 (1982); Minns, R.J., Stevens, F.S.: J. Anat. *123*, 437 (1977); Poole, A.R., Pidoux, I., Reiner, A., Rosenberg, L.: J. Cell Biol. *93*, 921 (1982)

Articular disc (joint cartilage, intra-articular cartilage, discus articularis*): A plate (D) of fibrous ↑ cartilage situated within an articular cavity. A. is attached at its periphery to articular capsule (C) and subdivides joint into two cavities. A. is very poorly vascularized and devoid of nerve fibers; it permits correct fitting together of incongruent articular surfaces and increases articular contact surfaces, thereby distributing pressure on the ↑ joint. A. is incompletely covered by ↑ synovial membrane and bathed by ↑ synovial fluid. The A. are present only in articulations where one of the bones develops according to mechanism of direct ↑ bone formation (articulatio sternoclavicularis, articulatio temporomandibularis).

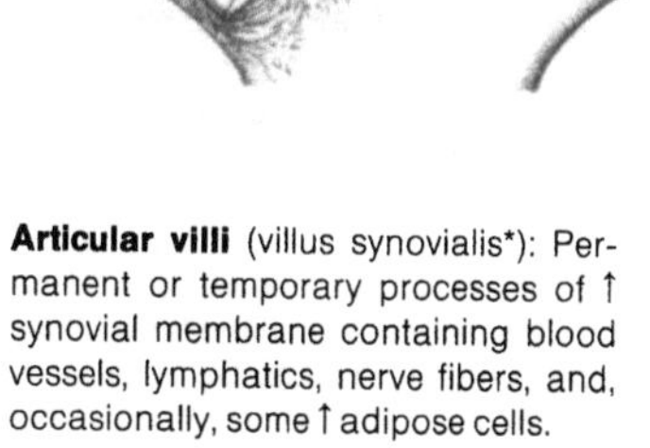

Articular villi (villus synovialis*): Permanent or temporary processes of ↑ synovial membrane containing blood vessels, lymphatics, nerve fibers, and, occasionally, some ↑ adipose cells.

Articulations: ↑ Joints

Artifact: Any alteration of cells, tissues, or organs caused by manipulations or technical procedures during preparation for microscopic analyses.

Aryepiglottic folds (plicae aryepiglotticae*): Folds streched between the of ↑ arytenoid cartilages and ↑ epiglottis at laryngeal entrance. A. are covered with nonkeratinized ↑ stratified squamous epithelium.

Arytenoid cartilages, of larynx (cartilago arythenoidea*): Pieces of predominantly hyaline ↑ cartilage; only the vocal processes consist of elastic ↑ cartilage.

Arytenoid glands (glandulae arytenoideae*): A group of ↑ mixed ↑ tubuloalveolar glands located along rim of ↑ aryepiglottic folds.

↑ Laryngeal glands

Asbestos cartilage (amiantine degeneration): A degenerative change observed principally in large pieces of aged hyaline ↑ cartilage. With decrease in water content in the ↑ ground substance, collagen fibers become visible under the ordinary microscope. Macroscopically, A. appears shiny and glossy, resembling asbestos.

↑ Cartilage, degeneration of; ↑ Degeneration

Hough, A.J., Mottram, F.C., Sokoloff, L.: Am. J. Pathol. *73*, 201 (1973)

Ascendent degeneration, of peripheral nerve fibers: ↑ Nerve fibers, of peripheral nervous system, degeneration and regeneration of

Ascending colon: ↑ Colon

A-spermatogonia: ↑ Spermatogonia

Association neurons: ↑ Neurons located between ↑ sensory and ↑ motor neurons (↑ association neurons of ↑ spinal cord, etc.), or between ↑ sensory neurons (↑ amacrine cells, ↑ horizontal cells, etc.).

Association neurons, of retina: ↑ Amacrine cells; ↑ Horizontal cells

Association neurons, of spinal cord: Small multipolar ↑ neurons (A) situated in zona spongiosa. The axons of A. leave the gray matter and participate in formation of fasciculi proprii (FP) of the white matter. Within the fasciculus, axons divide into ascending (As) and descending (De) branches; the ascending and descending branches have three and two collaterals (C), respectively. All of them make synaptic contacts with ↑ motor neurons (M) and connect segments of ↑ spinal cord. A. are in synaptic contact with short collaterals (c) of the ascending branch (AB) of a ↑ pseudounipolar cell (PC) axon (Ax); the descending branch (DB) synapses directly with ↑ motor neurons.

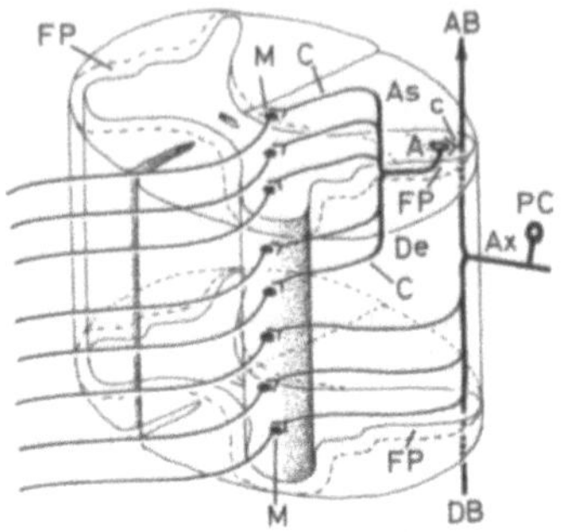

Aster: ↑ Astrosphere

Astral rays (radiatio polaris*): Microtubular bundles radiating from ↑ astrosphere of a cell undergoing division.

Astroblast: The precursor cell of an ↑ astrocyte

Astrocytes (macroglia, astrocytus*): Kind of large ↑ neuroglial cells of CNS. With ↑ perivascular (PF) and ↑ subpial feet (SF), A. constitute the ↑ membrana limitans gliae perivascularis (MLGP) and superficialis (MLGS), which delimit CNS from surrounding tissues and establish a microenvironment for ↑ neurons (Ne). Other processes of A. cover surfaces of neurons and participate in an exchange of metabolites between nerve cells and blood and/or ↑ cerebrospinal fluid. A. also form a supporting framework for neurons and ↑ synapses. After trauma to ↑ nervous tissue, A. form a glial scar (gliosis). A. are divided into: ↑ A., fibrous, A., plasmofibrillar, ↑ A., protoplasmic, and ↑ A., velate. ↑ Bergmann's and ↑ Müller's cells are modified A. Cap = capillary

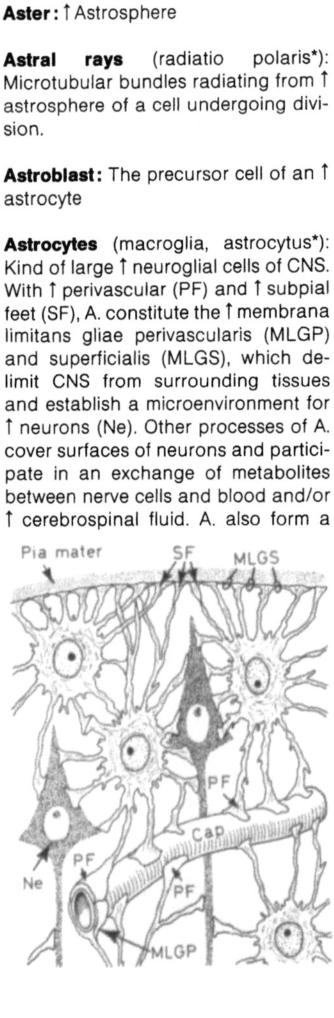

Allt, G.: Trends Neurosci. *3*, 72 (1980)

Astrocytes, fibrous (astrocytus fibrosus*): Kind of ↑ astrocytes occurring in the ↑ white matter of CNS. The cell body, about 10–12 µm in diameter, sens out long, thin, infrequently branched processes (P) predominantly parallel to ↑ nerve fibers (NF). The nucleus (N) is poor in ↑ heterochromatin; the cytoplasm contains a few mitochondria, a small Golgi apparatus, some cisternae of the rough endoplasmic reticulum, free ribosomes, and ↑ glycogen particles (Gly). The perikaryon and processes enclose a varying number of ↑ gliofibrils (Gf). A. supply nerve fibers of CNS with metabolites; with their ↑ perivascular feet (arrow), A. contact capillaries (Cap) and participate in formation of ↑ membrana limitans gliae perivascularis. Also, with their ↑ subpial feet, A. can form ↑ membrana limitans gliae superficialis. The processes of A. serve as support for tracts of nerve fibers.

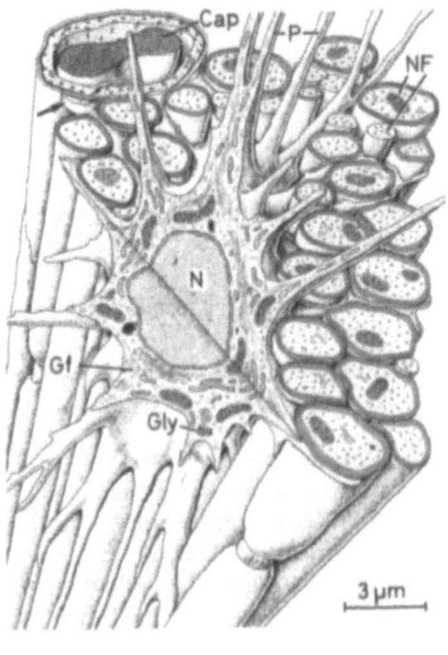

Astrocytes, mixed: ↑ Astrocytes, plasmofibrillar

Astrocytes, plasmofibrillar (mixed astrocytes): ↑ Astrocytes found at the boundary between gray and ↑ white matter; their processes that spread into the ↑ gray matter have a protoplasmatic character, those penetrating the white matter are fibrous.

Astrocytes, protoplasmic (astrocytus protoplasmaticus*): Variety of ↑ astrocytes (A) with a large polygonal body (15–25 μm), located within the gray mass of CNS. Variously long processes (P) emerge from the cell body; these are applied to capillaries (Cap) on one side and to nerve cells (NC) on the other. The enlarged terminations of these processes, ↑ perivascular feet (PF), form both ↑ membrana limitans gliae perivascularis and superficialis. The lateral surfaces of A. are very irregular and serve as a support for ↑ nerve fibers (NF) and ↑ synapses (Sy). The nucleus of A. is poor in ↑ heterochromatin; in the cytoplasm are scattered some mitochondria, short cisternae of the rough endplasmic reticulum, a small Golgi apparatus, the ↑ centriole, and free ribosomes. Through the cytoplasm and cell expansions run numerous ↑ gliofibrils (Gf).

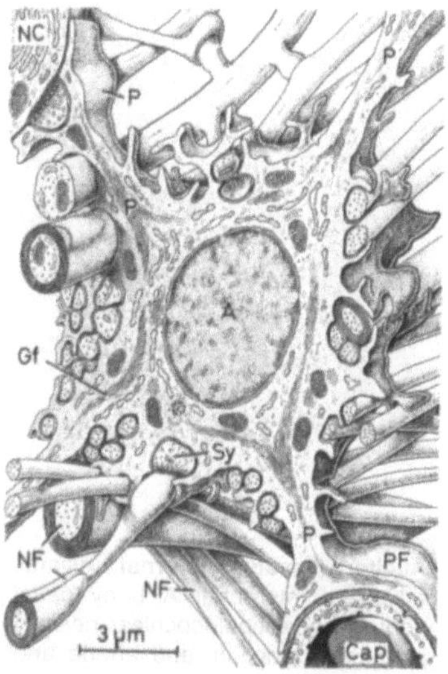

Ling, E.A., Patterson, J.A., Privat, A., Mori, J.S.: J. Comp. Neurol. *149*, 543 (1973); Philipps, D.E.: Z. Zellforsch. *140*, 145 (1973)

Astrocytes, velate: Variety (AV) of ↑ astrocytes found in granular layer of ↑ cerebellum, characterized by a number of extremely thin, sheetlike extensions (arrows) surrounding ↑ cerebellar glomeruli (CG), ↑ granule cells (GC), and capillaries (Cap). Internal structure and function are identical to those in other astrocytes.

Chan-Palay, V., Palay, L.S.: Z. Anat. Entwick. Gesch. *138*, 1 (1972)

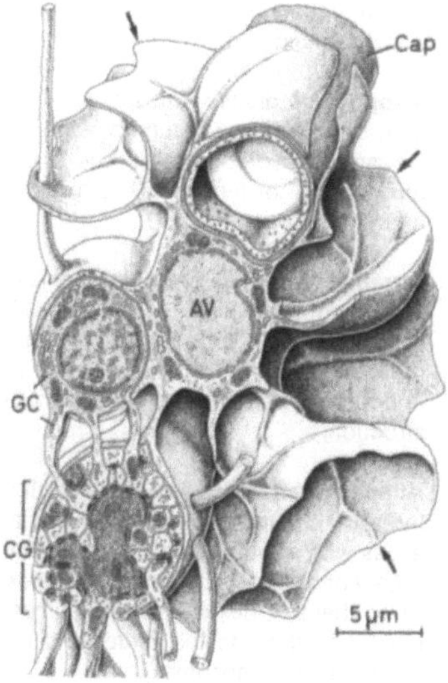

Astroglia: The collective term for all varieties of ↑ astrocytes.

Astroglial end feet: ↑ Perivascular feet; ↑ Subpial feet

Astrosphere (aster): A starlike cytoplasmic differentiation in the form of radiating ↑ microtubules (astral rays) extending outward from ↑ centrosphere (C) of a dividing cell. A. is resistant to ordinary staining methods.

↑ Diaster

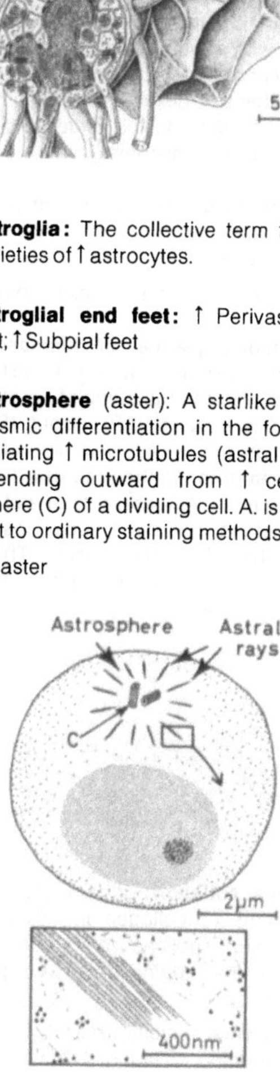

Bajer, A.S., Mole-Bajer, J.: Asters, Poles, and Transport Propreties Within Spindlelike Microtubule Arrays. Cold Spring Harbor Symposia on Quantitative Biology, Vol. 46, Part 1. New York: Cold Spring Harbor Laboratory 1982

Athrocytosis: The capacity of cells to absorb and retain solid particles of colloidal dimensions as shown by ↑ vital staining. Absorbed particles which have the diameter of ions and molecules are called amicrons; when the particles (1–100 nm) are visible with the ↑ transmission electron microscope they are termed ultramicrons (U). Whereas amicrons pass the cell membrane without any visible morphological changes, the principle of ↑ micropinocytosis plays an important role in A. of ultramicrons. These particles attach themselves to the ↑ glycocalyx (1), the plasmalemma (P) invaginates (2), and they are transported into the cell body within ↑ micropinocytotic vesicles (3, 4).

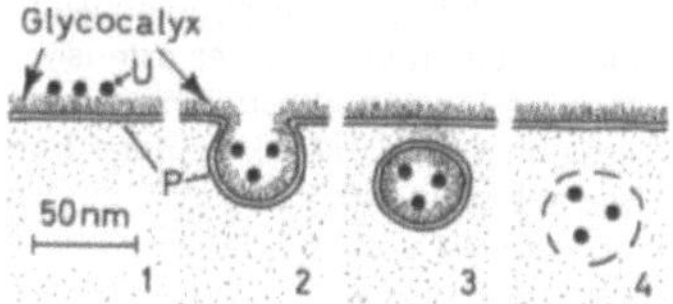

ATP: ↑ Adenosine triphosphate

Atresia, of ovarian follicles (follicular atresia): The process of growth interruption, ↑ degeneration, and removal of ↑ ovarian follicles, which can begin at any stage of follicular development. Starting at birth, A. reaches a maximum between birth and puberty, becoming slower after puberty. Of about 400000 primordial follicles, only 400–500 reach full maturation. The reason(s) for A. are not known.

↑ Ovarian follicles, atretic

Atretic follicles: ↑ Ovarian follicles, atretic

Atrial granules: Membrane-bound granules (AG), 0.3–0.4 μm in diameter, with electron-dense content, found in ↑ endoplasm of atrial ↑ cardiac muscle cells. A. give neither ↑ chromaffin nor catecholamine reactions. In man, A. appear at about the 7th week of fetal development. Function still unknown. (Fig. = rat)

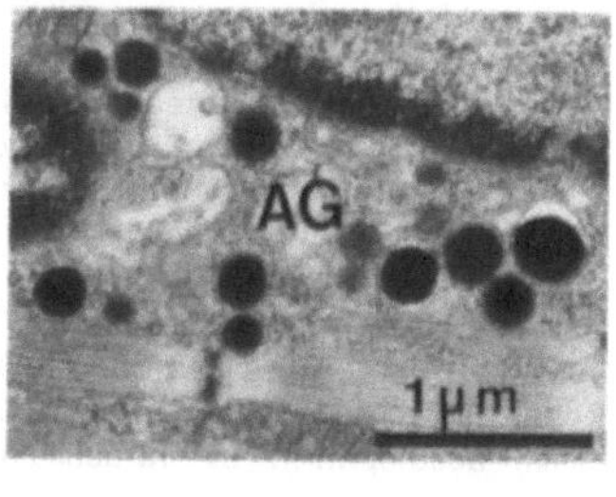

Saetersdal, T., Rotevatn, S., Myklebust, R., Ødegården, S.: Anat. Embryol. *160*, 1 (1980); Saetersdal, T., Ødegården, S., Rotevatn, S., Engedal, H.: Cell Tissue Res. *290*, 345 (1980); Seiden, D.: Anat. Rec. *194*, 587 (1979); Yunge, L., Benchimol, S., Cantin, M.: Cell Tissue Res. *207*, 1 (1980)

Atrioventricular bundle (A-V bundle, bundle of His, truncus fasciculi atrioventricularis*): A part, about 15 mm long and 2–3 mm wide, of ↑ impulse-conducting system of heart, beginning in ↑ atrioventricular node and penetrating interventricular septum. Here, A. divides into two branches (crus dextrum* and crus sinistrum*) destined for the corresponding ventricles. Each branch ramifies extensively forming ↑ Purkinje fibers, which establish contacts, via ↑ Purkinje and ↑ transitional cells, with ordinary ↑ cardiac muscle cells.

Marino, T.A.: Cell Tissue Res. *206*, 271 (1980); Mochet, M., Moranec, L., Guillemot, H., Hatt, R.Y.: J. Mol. Cell Cardiol. *7*, 869 (1975)

Atrioventricular node (A-V node, node of Aschoff-Tawara, nodus atrioventricularis*): Part of ↑ impulse-conducting system of heart, formed by cordlike groups (C) of ↑ nodal and ↑ transitional cells. The cords unite to form the ↑ atrioventricular bundle.

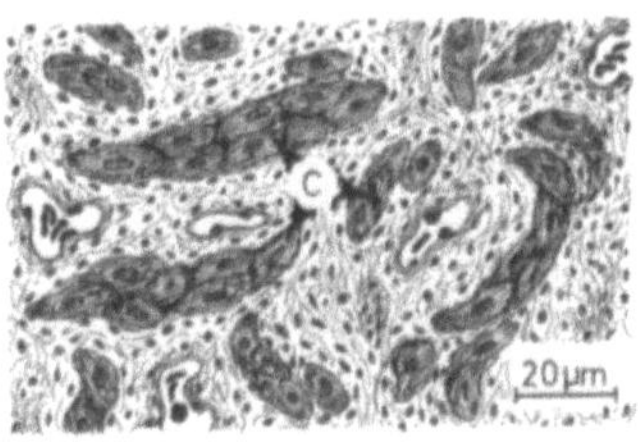

Bhatnagar, K.P., Spoonamore, B.A.: Acta Anat. *105*, 157 (1979); Moravec, M., Moravec, J.: J. Ultrastruct. Res. *81*, 47 (1982); Roberts, N.K., Castelman, K.R.: Anat. Rec. *195*, 699 (1979); Weihe, E., Kalmbach, P.: Cell Tissue Res. *192*, 77 (1978)

Atrioventricular valves: ↑ Valves, atrioventricular

Atrophy (atrophia*): Decrease in volume of a cell, tissue or organ. A. can be physiological (↑ A., brown, ↑ A., of disuse, ↑ A., senile) or pathological.

Atrophy, brown (atrophia fusca): ↑ Atrophy of certain tissues or organs (heart muscle, ↑ liver parenchymal cells) accompanied by intracellular accumulation of ↑ lipofuscin. A. occurs with advancing age.

Atrophy, of disuse (atrophia ex inactivitate): Atrophy resulting from decreased function of a cell, tissue, or organ (e.g., A. of female sexual organs after ↑ menopause).

Atrophy, senile (atrophia senilis): Atrophy of a cell, tissue, or organ with advancing age, probably from a decrease in the anabolic process due to endocrine changes and vascular obstruction.

Attachment devices (junctiones cellulares*): Specialized plasmalemmal areas with a function to hold together adjacent cells (↑ junctional complex, of epithelia, ↑ nexus, ↑ desmosomes) and/or fix them to the underlying tissue (↑ hemidesmosomes).

Attachment plaque: A term sometimes erroneously used as a synonym for ↑ dense bodies of ↑ smooth muscle cells.

↑ Attachment plaque, of desmosome

Attachment plaque, of desmosome (lamina desmosomatica*): A well-delimited, flattened accumulation of osmiophilic material (AP), about 15 nm thick, on the cytoplasmic side of the cell membrane in the area of a ↑ desmosome (D). A. represents one of two symmetrical halves of a desmosome into which ↑ tonofilmaments (Tf) are inserted in the form of hairpin loops. (Fig. = human)

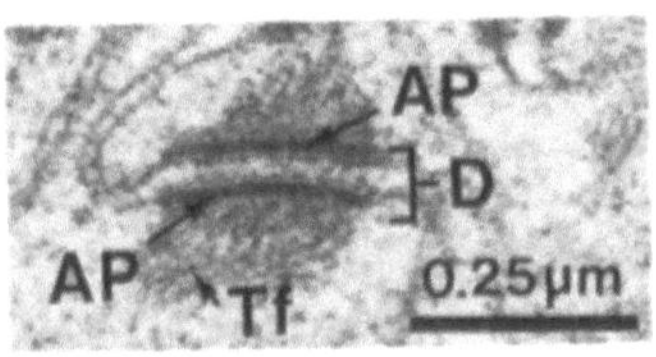

Atypical collagen: Certain rare, natural forms of collagen in which the molecules polymerize side-to-side without overlapping or in hexagonal patterns.

↑ Descemet's membrane; ↑ Fibrous long-spacing collagen, ↑ Long-spacing collagen

Atypical epithelium: An ↑ epithelium in which the cells do not form a close cell union, but rather are separated from one another either by liquid (↑ stellate reticulum) or by other cells (thymic epithelial ↑ cytoreticulum).

Auditory cells: ↑ Hair cells, of Corti's organ

Auditory meatus, external (meatus acusticus externus*): A canal about 25 mm long and 8 mm wide lined with ↑ skin and situated between ↑ auricle and ↑ tympanic membrane. Structure: 1) Cartilaginous part (outer half, toward auricle) = a thin ↑ epidermis (E) without papillae; numerous ↑ hairs (H), ↑ sebaceous (SG), and ↑ ceruminous (CG) glands in ↑ dermis, the latter is firmly attached to the ↑ perichondrium (P), rendering the skin of A. completely immobile. Elastic ↑ cartilage (C) of A. is continous with cartilage of ↑ auricle. 2) Osseous part (inner half, toward tympanic membrane) = very thin ↑ skin continous with cutaneous layer of tympanic membrane. The skin gradually loses its hairs and glands and lies fixed to the ↑ periosteum.

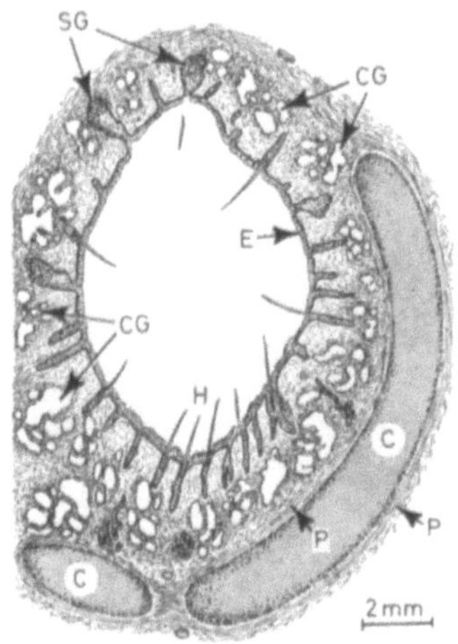

Auditory meatus, internal (meatus acusticus internus*): A bony canal containing vestibulocochlear nerve, ↑ vestibular ganglion, and arteria and vena labyrinthi. The end of A. is base of ↑ modiolus.

Auditory ossicles (ossicula auditus*): Three small mature ↑ bones (incus, malleus, stapes) without epiphyses, interconnected by diarthrodial ↑ joints. The periosteal surface of A. is covered with simple cuboidal epithelium of ↑ tympanic cavity. Malleus and incus have small marrow cavities; patches of hyaline ↑ cartilage may be found on manubrium mallei, which is attached to ↑ tympanic membrane, and on footplate of stapes (fixed by a ↑ syndesmosis to border of ↑ oval window). A. transmit vibrations, sent through tympanic membrane as sound waves, to the ↑ oval window.

Auditory strings: Compact bundles (AS) of about 8 to 10-µm-thick, parallel collagenlike microfibrils stretched between ↑ spiral ligament and ↑ lamina spiralis ossea. A. are embedded in an amorphous mass belonging to ↑ basilar membrane (BM) of ↑ cochlear duct; in zona pectinata, A. form two layers. There are about 20000 A. in man; A. have a length of 0.04 mm in basal cochlear turn and 0.5 mm in ↑ helicotrema. Each frequency of a sound transmitted to ↑ cochlear fluids provokes the vibration of the corresponding A. and consecutive excitation of ↑ hair cells located above A. (Fig. = rat, transverse section)

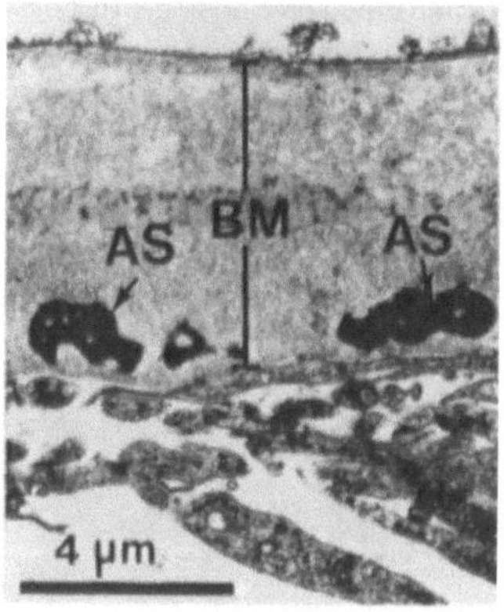

Auditory teeth, of Huschka (dentes acustici*): Regularly spaced radial ridges (AT) at upper surface of vestibular lip (VL) of ↑ limbus spiralis (LS). A. consist of vertically arranged ↑ collagen fibers, extending slightly beyond the border of the lip to form about 2500 toothlike projections. ↑ Interdental cells (IC) lie between A. C = ↑ Corti's organ, RM = ↑ Reissner's membrane, TM = ↑ tectorial membrane.

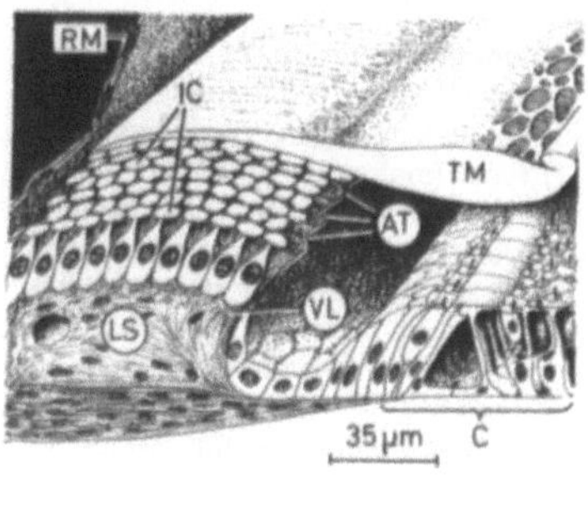

Auditory tubes (Eustachian tube, tuba pharyngotympanica, tuba auditiva*): Paired canals, each about 4 cm long, connecting ↑ tympanic cavity with ↑ pharynx. Structure: A) Osseous part (the third toward the tympanic cavity). Tunica mucosa (TM) = a simple ciliated columnar epithelium overlying a thin lamina propria, devoid of glands, in close contact with bone. B) Cartilaginous part (the two-thirds toward the pharynx). Tunica mucosa is thicker than in osseous part, containing compound ↑ tubuloalveolar glands (Gl), as well as abundant ↑ lymphocytes grouped in nodules (LN), which form tubal ↑ tonsils near pharyngeal opening. Epithelium becomes ↑ pseudostratified with many ciliated and ↑ goblet cells (inset) as it approaches the pharynx (Ph). Toward the pharynx, elastic ↑ cartilage (C) becomes hyaline ↑ cartilage.

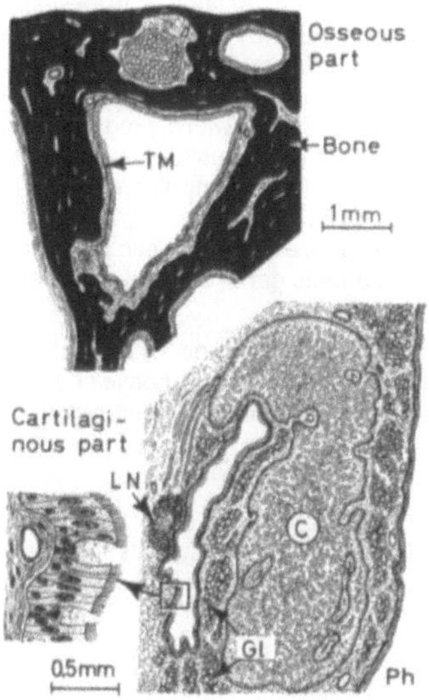

Auerbach's plexus: ↑ Myenteric plexus

Auricle (pinna, auricula*): Part of external ear covered by delicate ↑ skin (S) with a distinct ↑ hypodermis (Hy) only on posterior surface (PS). Skin of anterior surface (AS) is immobile, being attached directly to ↑ perichondrium (P). ↑ Sebaceous glands surrounding small ↑ hairs (H) may be found on posterior surface of A. as well as rare ↑ sweat glands. Skeleton of A. is formed by a plate of elastic ↑ cartilage (EC) with flexible perichondrium containing abundant ↑ elastic fibers. Lobule is covered by thin skin similar to that of cartilaginous part of A. Hypodermis of lobule is very rich in ↑ adipose tissue and blood vessels.

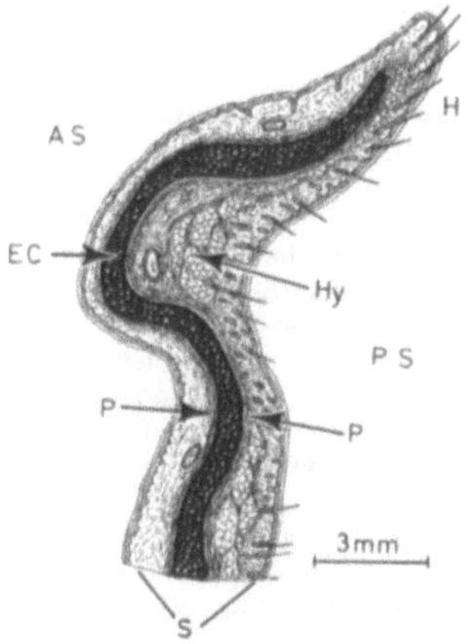

Auricular cartilage (cartilago auricularis*): ↑ Cartilage, elastic

Autochromosomes: ↑ Autosomes

Autodesmosomes: ↑ Desmosomes (A) formed between apposing processes of the same cell. A. are found in ↑ chorion laeve and ↑ amniotic epithelium; they can be involved in reduction of cell surface. (Fig. after Petry 1980)

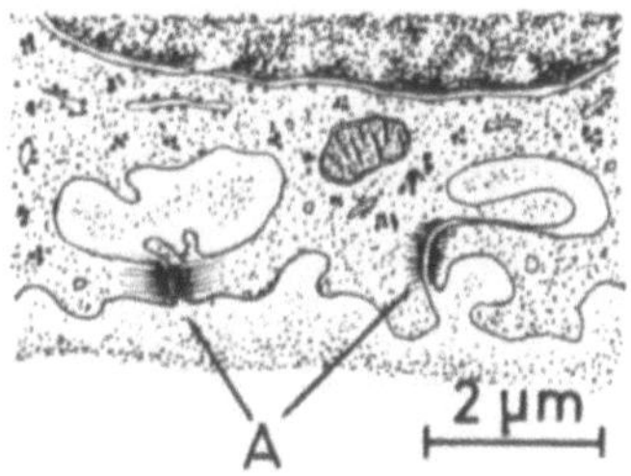

Petry, G.: Eur. J. Cell Biol. *23*, 129 (1980)

Autofluorescence: The capacity of some natural substances (↑ porphyrins, vitamin A) or ↑ inclusions (↑ lipofuscin) to become fluorescent when exposed to ultraviolet light.

↑ Fluorescence

Autograft (autografting, autoplastic grafting, autotransplantation): The transfer of tissues or organs to a new location in the body of the same individual.

↑ Transplantation

Autolysis: Self digestion of cells in absence of oxygen by their own enzymes liberated after disruption of the lysosomal membrane.

Autolysosomes: ↑ Autophagosomes

Autolytic vacuoles: ↑ Autophagosomes

Automatic cell recognition: A computer-assisted procedure for instrumental diagnosis and sorting of cells. In a ↑ cytophotometer, the ↑ condenser (Co) focuses the field diaphragm (FD) to produce a 0.5–2 μm spot of light in plane of cell to be analyzed (arrow). The cell's image is projected by microscope (M) onto measuring diaphragm (D) of photomultiplier (Ph). A fast-scanning device (SD) moves microscope stage (S) in direction of X- and Y-axes so that the spot scans entire cell surface. Absorptions and/or extinctions of light are correspondingly coded in an analog-digital converter (CU) as a series of numbers, which are then transmitted on line to a computer (C). The described system is able to distinguish, using a multitude of parameters, cells belonging to two cytomorphologically similar cell populations which cannot be positively distinguished visually (error less than 5%). Another possibility for A. is punctual decomposition of a projected image by a television tube (TV), coding (CU) the nuances obtained, and treating then in a computer. This system is somewhat faster than the first one but less precise. L = light source, Mo = monochromator

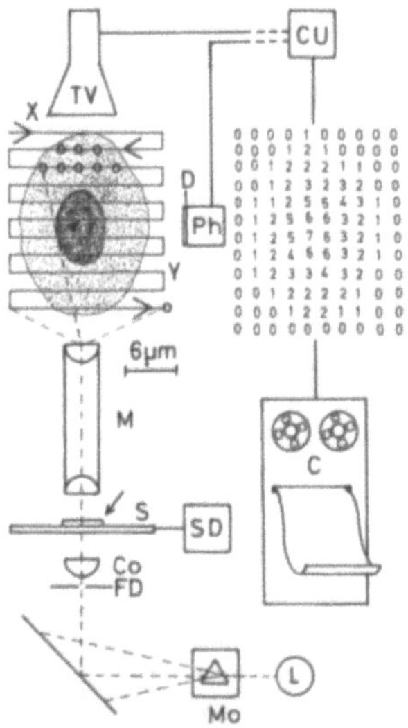

Barz, H., Kunze, K.D., Voss, K., Simon, H.: Exp. Pathol. *14*, 55 (1977); Bradbury, S.: J. Micros. *115*, 137 (1979); Preston, K.Jr.: Clinical Use of Automated Microscopes for Cell Analysis. In: Preston, K., Onoe, M. (eds.): Digital Processing of Biomedical Image. New York: Plenum Press 1976; Rosenfeld, A. (ed.): Digital Picture Analysis. Berlin, Heidelberg, New York: Springer-Verlag 1976; Tanaka, N., Ueno, T., Ishikawa, A., Konoike, K., Shimaoka, Y., Yamauchi, K., Hosoi, S., Okamoto, Y., Tsunekawa, S.: Anal. Quant. Cytol. *4*, 279 (1982); Dytch, H.E., Bartels, P.H., Bibbo, M., Pishotta, F.T., Wied, G.L.: Anal. Quant. Cytol. *4*, 263 (1982)

Autonomic ganglia (vegetative ganglia, ganglion autonomicum*): All ganglia belonging to ↑ sympathetic and ↑ parasympathetic nervous systems. Structure: 1) Capsule (C): a) external layer = a ↑ dense connective tissue continuous with ↑ epineurium of related pre- and postganglionic trunks; b) inner layer = ↑ perineurial epithelium and perineurial connective tissue continuous with ↑ perineurium of the already mentioned trunks. 2) ↑ Endoneurium (E) fills interior of A. = a ↑ loose connective tissue containing ↑ multipolar neurons (N) with short dendrites synapsing with ↑ preganglionic fibers and great numbers of ↑ nonmyelinated and rare ↑ myelinated nerve fibers (NF). No morphological differences between sympathetic and parasympathetic ganglia. RC = rami communicantes.

↑ Autonomic neurons, of sympathetic nervous system; ↑ Autonomic neurons, of parasympathetic nervous system

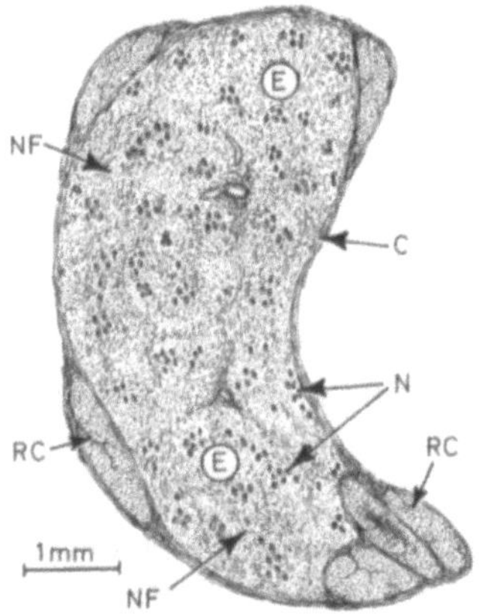

Autonomic nerve endings: ↑ Adrenergic nerve endings; ↑ Cholinergic nerve endings

Autonomic nerve fibers: ↑ Adrenergic nerve fibers; ↑ Cholinergic nerve fibers

Autonomic nervous system (vegetative nervous system, systema nervosum autonomicum*): The efferent (motor) system responsible for automatic innervation of ↑ smooth musculature, ↑ myocardium, and ↑ glands. Composition (general schema): 1) An appropriate part of ↑ central nervous system (CNS). 2) Preganglionic nerve fibers connecting CNS with autonomic ganglia. 3) ↑ Autonomic ganglia (sympathetic and parasympathetic). 4) ↑ Postganglionic nerve fibers connecting autonomic ganglia with effector organ. A. is divided into ↑ sympathetic and ↑ parasympathetic nervous systems. Neurons of afferent or viscerosensitive fibers belong to A. and are located in ↑ spinal ganglion without distinction between sympathetic and parasympathetic elements.

↑ Adrenergic nerve endings: ↑ Cholinergic nerve endings

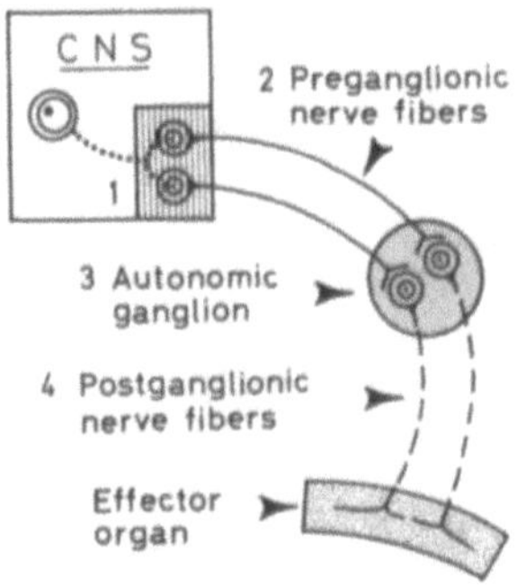

Brooks, C.,Mc.: J. Auton. Nerv. Syst. *7*, 199 (1983); Guérin, J., Bioulac, B., Henry, P., Loiseau, P.: Système Nerveux Végétatif, Basle, Sandoz-Edition 1979; Nilsson, S.: Autonomic Nerve Function in the Vertebrates. Zoophysiology, Vol. 13. Berlin, Heidelberg, New York: Springer Verlag 1983

Autonomic neurons, of parasympathetic ganglia: ↑ Neurons structurally very similar to ↑ ganglion cells of ↑ spinal ganglia.

Autonomic neurons, of sympathetic ganglia: Nerve cells situated in the sympathetic ganglia. Two types of A.: 1) Type I (majority of ganglion cells) = multipolar, frequently ↑ binucleated neurons, 20–35 μm in diameter, of very variable form, with short ↑ dendrites (D) and long ↑ axon (A). Neurons have a large ovoid or spherical nucleus, an abundance of ↑ Nissl bodies, a well-developed Golgi apparatus, frequent ↑ lipofuscin granules, and numerous small ↑ dense-core vesicles (50 μm) containing ↑ norepinephrine. These vesicles are scattered in dendrites, ↑ perikaryon, and, particularly, in the axons. By ↑ axoplasmic flow, they reach axon endings and are discharged into extracellular space. A. are surrounded by ↑ satellite cells (SC). In outlying sympathetic ganglia, satellite cells may be partially absent but ↑ Schwann's cells accompany the axons everywhere. 2) Type II (SGC cells, small granule-containing cells, SIF cells, small intensely fluorescent cells) = small, grouped neurons (10–20 µm) giving strong fluorescence after treatment with formaldehyde vapor. Cells have a small elliptical nucleus, more condensed ↑ heterochromatin than neurons of type I, less developed ↑ organelles, few Nissl bodies, and numerous large dense-core vesicles (0.1–0.3 µm) containing dopamine, which can be released into ↑ pericapillary spaces. The dendrites (D), ↑ soma, and axon (A) are incompletely surrounded by satellite cells (SC).

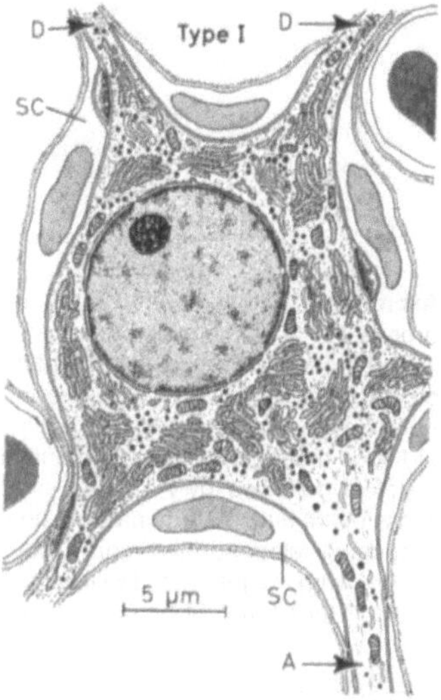

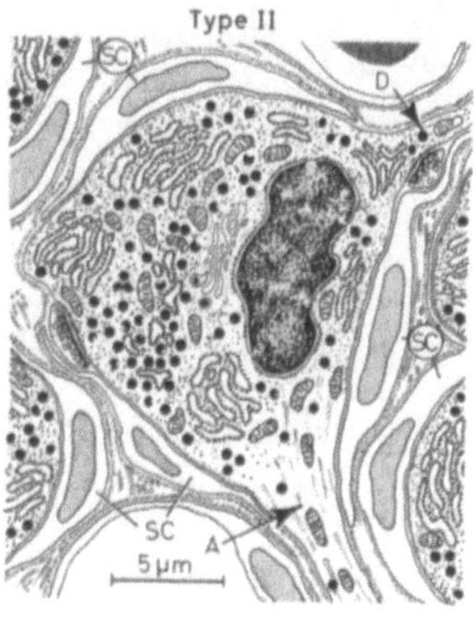

Anderson, M.J., Waxman, G., Laufer, M.: Anat. Rec. *205*, 73 (1983); Autillo-Touati, A.: Histochemistry *60*, 189 (1979); Eränkö, O., Soinila, S., Päivärinta, H. (eds.): Histochemistry and Cell Biology of Autonomic Neurons, SIF Cells, and Paraneurons. Adv. in Biochem. Psychopharmacology, Vol. 25. New York: Raven Press 1980; Partanen, M., Hervonen, A., Santer, R.M.: Histochemistry *66*, 99 (1980); Tay, S.S.W., Wong, W.C., Ling, E.A.: J. Anat. *136*, 35 (1983)

Autonomic neurons, parasympathetic, of spinal cord: ↑ Neurons located in ↑ nucleus intermediomedialis (NIM) and intermediolateralis (NI) of gray substance of sacral part of ↑ spinal cord (S_2–S_4); morphologically, very similar to sympathetic nerve cells, but less frequent. Their axons reach parasympathetic ganglia of corresponding organs via ventral roots.

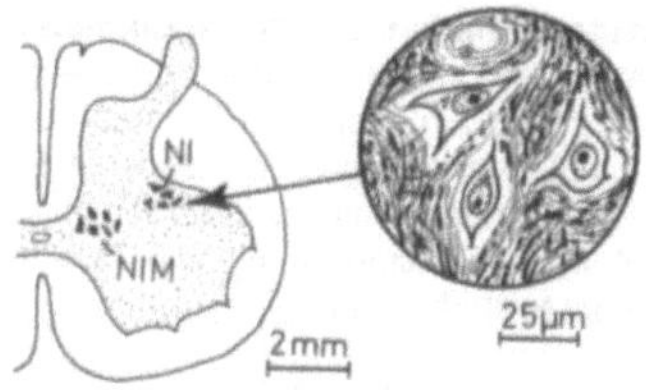

Autonomic neurons, sympathetic, of spinal cord: Fusiform ↑ neurons, about 25 µm in diameter, located in ↑ nucleus intermediolateralis (Ni) of intermediolateral gray columns (IGC) of thoracic and upper lumbal ↑ spinal cord (C_8–L_3). A. have all the characteristics of a neuron. ↑ Axons of A. leave spinal cord and reach sympathetic ganglia (SG) via ventral roots (VR). These ↑ cholinergic preganglionic fibers (PF) are ↑ myelinated and form the white rami communicantes (WRC).

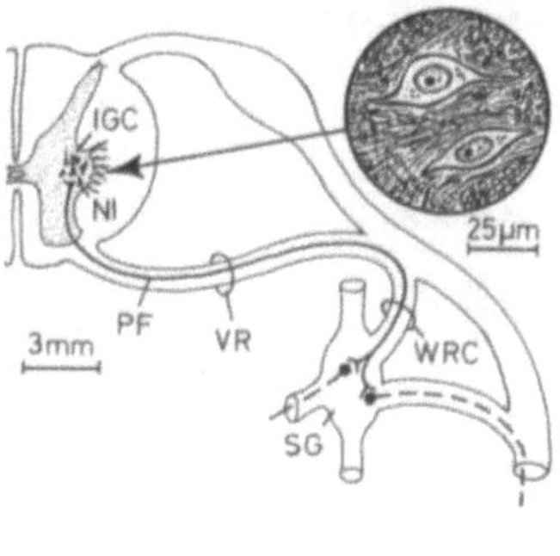

Autophagic vacuoles: ↑ Autophagosomes

Autophagocytosis: ↑ Autophagy

Autophagosomes (autolysome, autolytic vacuole, autophagic vacuole, autosomal body, cytolysosome, cytosegresome, vacuola autophagica*): Kind of secondary ↑ lysosome in which segregated cytoplasmic areas with ↑ organelles [mitochondrium (M), ribosomes (R)] or ↑ inclusions are digested. At the start of formation of A., segregated cytoplasmic area is surrounded by an ↑ isolation membrane (arrowheads). Later, its inner sheet disappears. (Fig. = ↑ mesenchymal cell, rat)

↑ Autophagy: ↑ Crinophagy

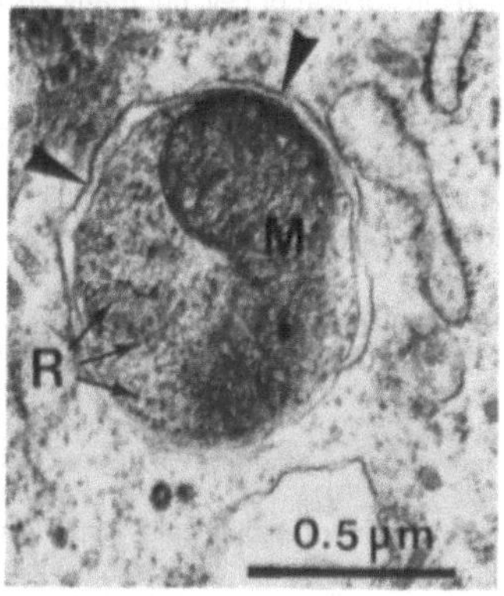

Sakai, M., Ogawa, K.: Histochemistry *76*, 479 (1982)

Autophagy (autophagocytosis): The process of digestion of segregated parts of the cell's own cytoplasm by ↑ lysosomes. Sequence: 1) Appearance of a semilunar ↑ isolation membrane (arrowheads) in cytoplasm. 2) This membrane then envelops and segre-

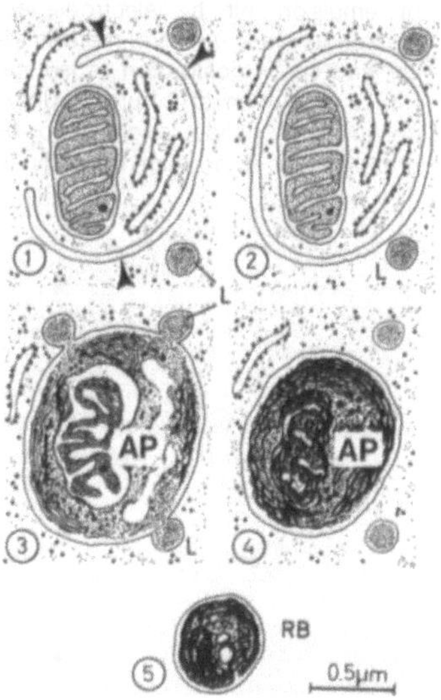

gates a cytoplasmic area containing ↑ organelles and/or ↑ inclusions. 3) Primary lysosomes (L) approach membrane, fuse with it, and discharge their enzymes into what is now termed an ↑ autophagosome (AP). 4) Intravacuolar digestion and destruction of vacuolar content with appearance of dark lamellar structures and amorphous material within autophagosome. 5) Formation of ↑ residual body (RB). A. is involved in normal turnover of organelles, remodeling of cytoplasm, and elimination of excess secretory product (↑ crinophagy). Stress, injuries, toxins, etc., induce A., which aids in survival of cell, but many also lead to its death.

Ishikawa, T., Furuno, K., Kato., K.: Exp. Cell. Res. *144*, 15 (1983); Ogawa, K.: Acta Histochem. Cytochem. *14*, 362 (1981); Thyberg, J., Hedin, U., Stenseth, K.: Eur. J. Cell. Biol. *29*, 24 (1982)

Autoradiography (radioautography): A cytochemical technique for localization of biologically important molecules involved in cell metabolism by means of radioactive isotopes (^{3}H, ^{14}C, ^{35}S, ^{45}Ca, ^{125}I, etc.). Procedure: 1) Injection of biologically important molecule labeled with radioactive isotope into a living animal. 2) After some minutes, or longer, the organs are fixed and prepared for light and/or ↑ electron microscopy. 3) ↑ Sectioning of embedded organs. 4) Covering of sections with a photographic emulsion. 5) Exposure of emulsion, for several weeks, to the action of radioactive emission of isotope contained in sections. The isotope emits β-radiation, i.e., electrons, acting on the silver halide (AgBr). 6) Development – sites of emulsion hit by electrons are marked as black grains of metallic silver and show localization of isotope with a resolution of 0.1 µm. 7) Unexposed silver halide is dissolved by photographic fixer. Aim of A. is to analyze various cell activities, principally the process of synthesis. For DNA, tritiated (^{3}H) thymidine is used, for RNA, ^{3}H-cytidine, and for ↑ glycogen, ^{3}H-glucose. (Modified after Caro 1964)

Autoradiography: Advances in Methods and Applications. J. Histochem. Cytochem. *29*, 107 (1981); Caro, L.G.: High Resolution Autoradiography. In: Prescott, D.M. (ed.): Methods in Cell Physiology, Vol. 1. New York: Academic Press (1964); Droz, B., Sandoz, D. (eds.): Techniques in Autoradiography. J. Microsc. Biol. Cell *27*, 71 (1976); Glauert, A.M. (ed.): Practical Methods in Electron Microscopy. Vol. 6, Part 1, Autoradiography and Immunocytochemistry. Amsterdam: Elsevier/North Holland (1980); Hirosawa, K., Hama, K.: Electron Microsc. (Tokyo) *31*, 405 (1982)

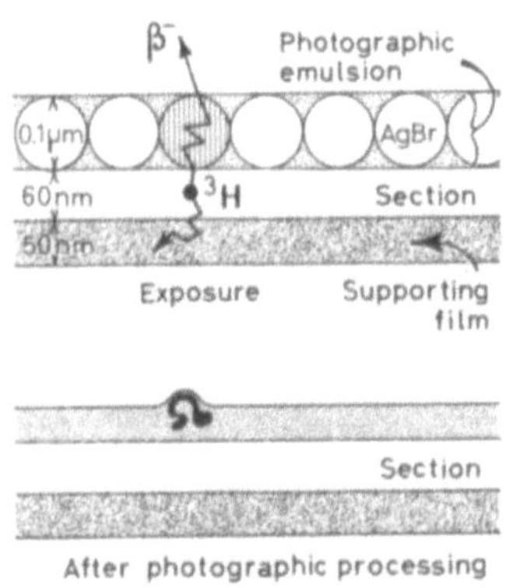

Autoregulation, of arterial diameter: Ability of ↑ muscular artery to change its caliber in response to metabolic stimuli expressing local needs of the tissue. Arterial A. is based on sensibility of arterial ↑ smooth muscle cells to react to such stimuli.

Autosomal bodies: ↑ Autophagosome

Autosomes (autochromosomes): Any ↑ chromosome other than ↑ heterosomes, i.e., sex chromosomes. In human ↑ diploid cells, A. occur as 22 homologous pairs of paternal (P) and maternal (M) chromosomes. A. occur singly in ↑ germ cells. C = ↑ centromere

↑ Karyotype

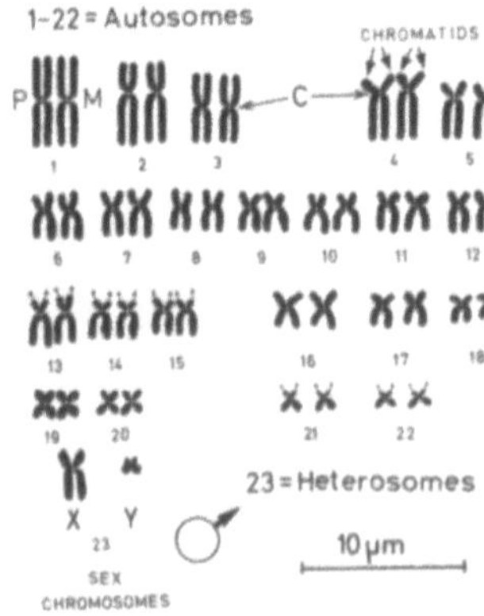

Autotransplantation: ↑ Autograft; ↑ Transplantation

Axial "filament" complex: ↑ Axoneme, of cilium

Axillary skin: Thin haired ↑ skin of axillar region characterized by presence of apocrine ↑ sweat glands.

Axis cylinder: ↑ Axon

Axoaxonal synapse: ↑ Synapse, axoaxonal

Axodendritic synapse: ↑ Synapse, axodendritic

Axolemma*: The plasmalemma (Al) of an ↑ axon (A), continuing cell membrane of ↑ perikaryon. In ↑ axon hillock of axon, A. shows an electron-dense undercoating, about 20 nm thick, missing only at sites of axoaxonal ↑ synapses. (Fig. = optic nerve fibers, rat)

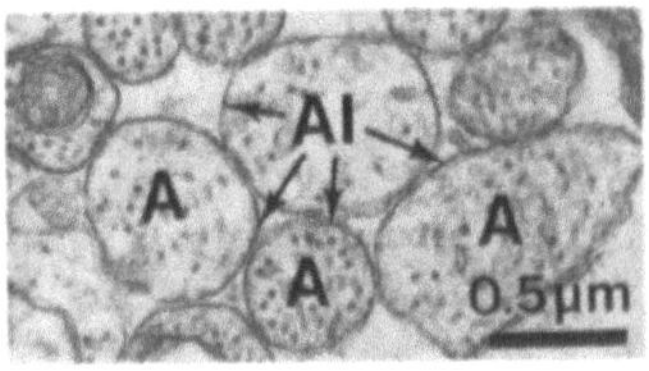

Axon* (axis cylinder, neurite): A unique process of a nerve cell, of variable length but with a relatively constant diameter (1–20 µm), carrying nerve impulses away from cell body. A. begins at ↑ axon hillock continuous with initial segment, followed by A. in the strict sense of the term. A. terminates in fine, branching telodendria or end arborization, with slight end swellings, ↑ boutons terminaux. Structure: 1) ↑ Ax-

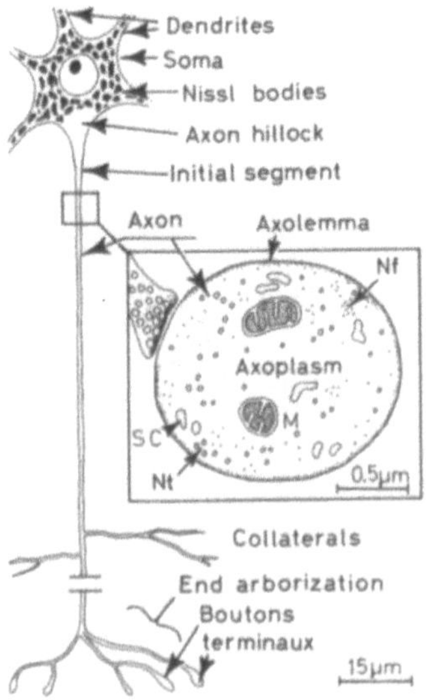

olemma; 2) ↑ axoplasm with long, thin mitochondria (M), abundant ↑ neurotubuli (Nt) and ↑ neurofilaments (Nf), some smooth cisternae (SC), and occasional ↑ multivesicular bodies. Axoplasm is devoid of ↑ Nissl bodies and free ribosomes. Some A. give off collaterals or side branches. An A. can be myelinated or nonmyelinated.

Bray, G.M., Rasminsky, M., Aguayo, A.J.: Ann. Rev. Neurosci. *4*, 127 (1981); Lasek, R.J., Brady, S.T.: The Axon: a Prototype for Studying Expressional Cytoplasm. Cold Spring Harbor Symposia on Quantitative Biology, Vol. 46, Part. 1. New York: Cold Spring Harbor Laboratory 1982; Quatacker, J.: Histochem. J. *13*, 109 (1981); Schnapp, B.J., Reese, T.S.: J. Cell Biol. *94*, 667 (1982); Waxman, S.G. (ed.): Physiology and Phatology of Axons. New York: Raven Press 1978

Axon collaterals (ramus collateralis*): ↑ Axon side branches.

Axon hillock (colliculus axonis*): The conical elevation of ↑ soma of nerve cell from which ↑ axon arises. A. contains mitochondria, ↑ neurofilaments, ↑ neurotubuli, a few smooth cisternae, very rarely free ribosomes, and, in ↑ neurosecretory cells, many ↑ neurosecretory granules. ↑ Nissl bodies are practically absent in A.

Palay, S.L., Sotello, C., Peters, A., Orkand, P.M.: J. Cell Biol. *38*, 193 (1968)

Axonal flow: ↑ Axoplasmic flow

Axonal reaction: ↑ Nerve fibers of peripheral nervous system, degeneration and regeneration of

Axonal spheroids: Swellings along ↑ myelinated and ↑ nonmyelinated axons containing ↑ mitochondria, dense bodies, vesicles, and ↑ tubules of smooth endoplasmic reticulum. Existence of A. is probably an age-dependent phenomenon.

Leonhardt, H.: Cell Tissue Res. *174*, 9 (1976)

Axonal transport: ↑ Axoplasmic flow

Axoneme, of cilium (axial "filament" complex, ciliary shaft, filamentum axiale*): The microtubular central cylinder common to all ↑ cilia and ↑ spermatozoon tail consisting of an inner core composed of two single ↑ microtubules and nine external microtubular pairs (doublets). A. of a cilium is formed from ↑ basal body by polymerization of microtubular proteins.

↑ Axoneme, of spermatozoon; ↑ Cilia, cross section of

Axoneme, of spermatozoon (axonema, filamentum axiale*): The microtubular core (A) of ↑ spermatozoon tail composed of a central microtubular pair (CM) surrounded by nine external microtubular pairs (EM). A. is the product of the distal centriole. (Fig. = rat)

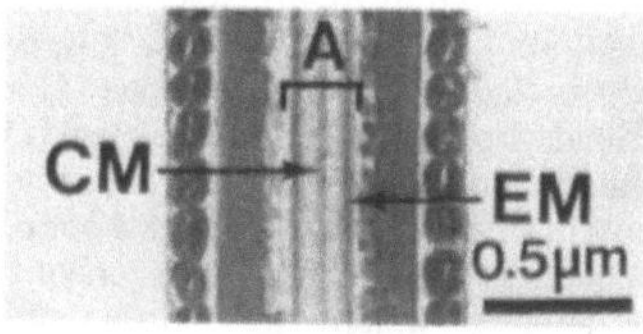

Axons, myelinated: Axons enveloped by a ↑ myelin sheath.

↑ Myelinated nerve fibers

Axons, naked: Axons (A) devoid of any sheath, present in abundance in ↑ gray matter of ↑ brain, ↑ spinal cord, and ↑ retina. ↑ Autonomic nerve fibers are frequently A. (Fig. = ↑ optic fibers, rat)

↑ Free nerve endings

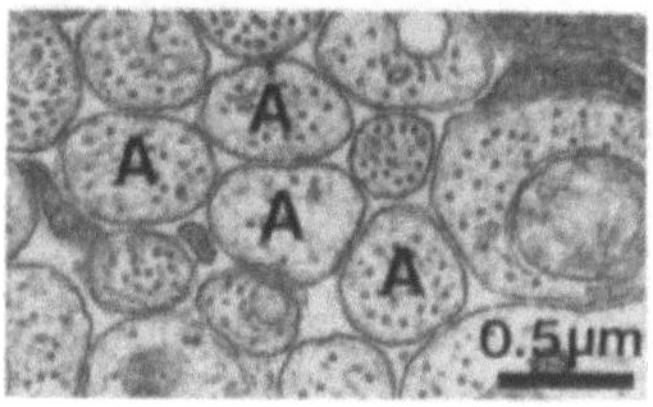

Axons, neurosecretory: ↑ Neurosecretory axons

Axons, nonmyelinated: ↑ Axons devoid of a ↑ myelin sheath.

↑ Nonmyelinated nerve fibers

Axons, primary degeneration of: Changes in proximal segment of a sectioned axon extending one to two ↑ internodal segments in a clean cut and 2–3 cm in lacerated wounds, with identical features to those found in Wallerian degeneration.

↑ Nerve fibers of peripheral nervous system, degeneration and regeneration of

Axons, supraependymal: ↑ Supraependymal axons

Axoplasm (axoplasma*): The ↑ cytoplasm of the ↑ axon.

↑ Axoplasmic flow

Axoplasmic flow (axonal flow, axonal transport, axoplasmic transport): The continuous circulation of the ↑ axoplasm away from (anterograde flow) and toward (retrograde flow) ↑ perikaryon. 1) Anterograde flow: a) Slow A. (1–3 mm/day), transporting newly synthesized axoplasm down axon to replace catabolized axoplasm (↑ axon cannot synthesize axoplasm since it is almost completely devoid of free ribosomes and rough endoplasmic reticulum); this flow is probably effected by peristaltic contractions of glial sheath. b) Fast A. (100–500 mm/day) carries glycopolysaccharides, phospholipids, ↑ neurotransmitter substances, vesicles, smooth endoplasmic cisternae, ↑ neurosecretory granules, and ↑ mitochondria. It is believed that fast A. is probably accomplished by ↑ neurofilaments and by ↑ neurotubuli with macromolecular arms moving material in both cellulifugal and cellulipetal directions. 2) Retrograde A. = cellulipetal movement of axoplasm, transporting proteins and other substances taken up by nerve endings.

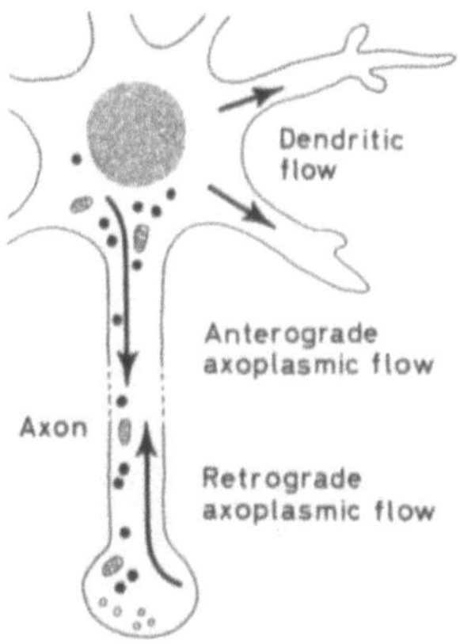

Hammerschlag, R., Stone, G., Bolen, F.A., Lindsey, J.D., Ellisman, M.H.: J. Cell Biol. *93*, 568 (1982); Ochs, S.: Axoplasmic Transport and its Relation to Other Nerve Functions, New York: John Wiley and Sons 1982; Schwartz, J.H.: Ann. Rev. Neurosci. *2*, 467 (1979); Tsukita, S., Ishikawa, H.: J. Cell Biol. *81*; 513 (1980); Weiss, D.G.; Axoplasmic Transport. Berlin, Heidelberg, New York: Springer-Verlag 1982

Axosomatic synapse: ↑ Synapse, axosomatic

Axospinous synapse: ↑ Synapse, axospinous

Axotomy: The cutting of an ↑ axon.

Azan staining: A combined staining with **az**ocarmine G and **an**iline blue. Results: ↑ collagen and ↑ reticular fibers, blue; nuclei, red; cytoplasm, pink; ↑ erythrocytes, reddish-orange; muscle fibers, reddish-violet; ↑ neuroglial cells, red; ↑ mucus, blue; cell granulations; blue, red, or yellow.

Azurophilic granules (primary granules, granulum azurophilicum*): Unit membrane-bound granules of about 0.25–0.5 μm in diameter found in eosinophilic and neutrophilic ↑ granulocytes, ↑ lymphocytes, ↑ monocytes, and some of their precursors. Specifically stained with methylene azure, A. appear reddish purple. A. enclose a fine granular, moderately to strongly osmiophilic material. Since A. contain several lytic enzymes, they are considered to be primary ↑ lysosomes involved in ↑ phagocytosis.

↑ Granulocytes, neutrophilic, granules of

Azygos vein (vena azygos*): A large propulsive ↑ vein with identical structure to ↑ vena cava inferior.

B

Bacillary layer, of ↑ retina: A layer composed of inner and outer segments of ↑ photoreceptors; B. is situated between ↑ pigment epithelium and ↑ outer limiting membrane.

Balbiani's vitelline body, of oocytes: An accumulation of rough ↑ endoplasmic reticulum (rER), a ↑ Golgi apparatus (G), ↑ mitochondria (M), ↑ lysosomes (Ly), ↑ annulatae lamellae (AL), ↑ lipid droplets (L), and vesicles (V) situated next to the nucleus (N). B. acts possibly as initial center for formation, multiplication, and accumulation of ↑ organelles, which are finally distributed throughout ↑ ooplasm before ↑ yolk deposition starts.

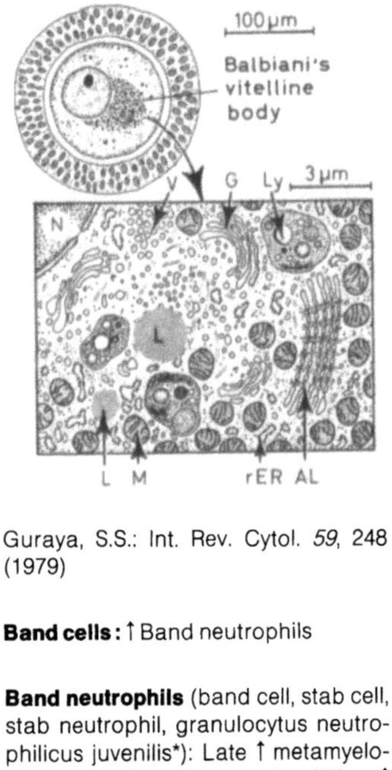

Guraya, S.S.: Int. Rev. Cytol. *59*, 248 (1979)

Band cells: ↑ Band neutrophils

Band neutrophils (band cell, stab cell, stab neutrophil, granulocytus neutrophilicus juvenilis*): Late ↑ metamyelocytes or juvenile neutrophilic ↑ granulocytes with typically curved, crescent-shaped nucleus (N) in the form of a band, found in 1%–5% of neutrophilic granulocytes.

↑ Arneth's formula

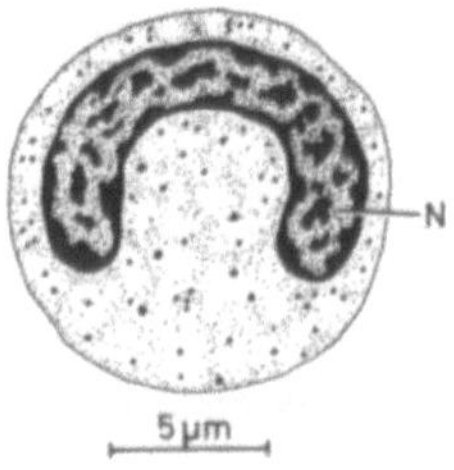

Band, of Baillarger, inner (stria laminae pyramidalis internae*): A strip of ↑ myelinated nerve fibers parallel to surface of brain isocortex and situated in its ganglionic layer.

↑ Brain cortex, homotypical isocortex of

Band, of Baillarger, outer (stria laminae granularis internae*): A strip of ↑ myelinated nerve fibers parallel to surface of brain isocortex and situated in its inner granular layer.

↑ Brain cortex, homotypical isocortex of

Band, of Bechterew (stria laminae granularis externae*): A strip of ↑ myelinated intracortical associative fibers located in outer granular layer of brain.

↑ Brain cortex, homotypical isocortex of

Banding patterns, of chromosomes: ↑ Chromosomes, banding patterns of

Bands, of Büngner: ↑ Büngner's band; ↑ Nerve fibers of peripheral nervous system, degeneration and regeneration of

Bands, of striated muscles: A repetitive, transverse banding pattern visible along ↑ myofibrils (Mf) of longitudinally sectioned ↑ cardiac and ↑ skeletal muscle fibers. The double refractile **a**nisotropic, therefore, ↑ A-bands appear bright in polarized light; the **i**sotropic, therefore, ↑ I-bands appear dark. Under the normal light microscope after usual staining or under the transmission electron microscope, the contrast is reversed – the A-bands appear dark, the I-bands, clear. Each I-band is bisected by a dark ↑ Z-line. The segment between two successive Z-lines is termed a ↑ sarcomere (S). (Fig. = ↑ skeletal muscle fiber, rat)

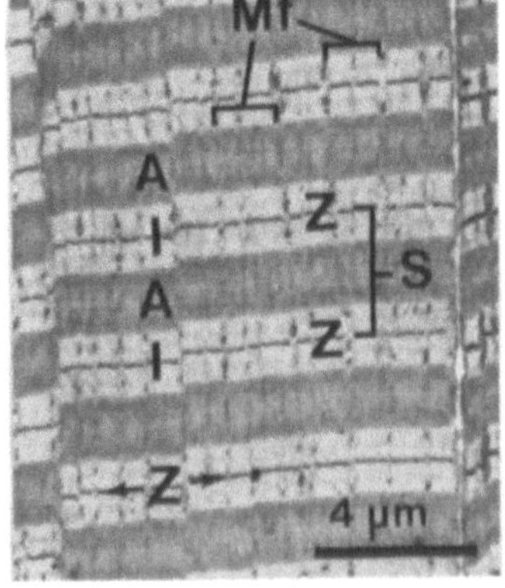

Bar, synaptic: ↑ Synaptic bars

Bar, terminal: ↑ Terminal bars

Baroreceptor field: ↑ Carotid sinus

Baroreceptors: Specialized ↑ mechanoreceptor areas in walls of heart and some blood vessels sensitive to changes in blood pressure. ↑ Carotid sinus and arcus aortae are B.

Taha, A.A.M., Abdel-Magied, E.M., King, A.S.: J. Anat. *137*, 197 (1983)

Barr body (chromatin body, corpusculum chromatini sexualis*): A small, lentricular mass, about 1 µm in size, of intensely staining facultative ↑ heterochromatin (BB) apposing inner aspect of ↑ nuclear envelope or associated with ↑ nucleolus in female ↑ interphase nuclei. B. represents inactivated long arm of one of the two ↑ X-chromosomes remaining clumped through interphase; it is visible under the light microscope and helps to determine the genetic sex of an individual. About 90% of female, and 10% of male, interphase nuclei contain a B. or B.-like structure.

↑ "Drumstick;" ↑ Neurons, structure of; ↑ Sex chromatin

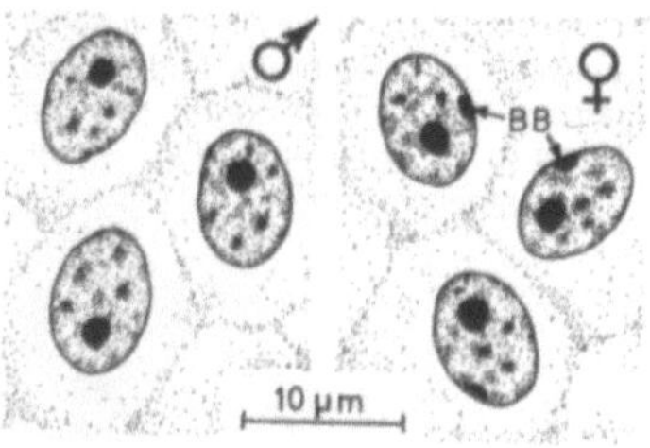

Schellens, J.P.M., James, J., Hoeben, K.A.: Biol. Cell *35*, 11 (1979)

Barrier, blood-air: ↑ Blood-air barrier

Barrier, blood-aqueous: ↑ Blood-aqueous barrier

Barrier, blood-brain: ↑ Blood-brain barrier

Barrier, blood-ganglion: ↑ Blood-ganglion barrier

Barrier, blood-liquor: ↑ Blood-liquor barrier

Barrier, blood-nerve: ↑ Blood-nerve barrier

Barrier, blood-retinal: ↑ Blood-retinal barrier

Barrier, blood-testis: ↑ Blood-testis barrier

Barrier, blood-thymus: ↑ Blood-thymus barrier

Barrier, blood-urine: ↑ Glomerular filtration membrane

Barrier, chorioallantoic: ↑ Placental barrier

Barrier, osmotic: ↑ Osmotic barrier

Barrier, permeability: ↑ Permeability barrier

Barrier, placental: ↑ Placental barrier

Barrier, protective permeability, of stomach: ↑ Protective permeability barrier, of stomach

Bartholin's glands: ↑ Vestibular glands

Basal bodies, of cilia: ↑ Cilia, basal bodies of

Basal cells, of epidermis (keratinocyte, epidermocytus basalis*): Cuboidal or columnar cells forming deepest layer of ↑ epidermis. B. have a relatively large ↑ heterochromatin-rich nucleus (N), a small ↑ Golgi apparatus, scarce mitochondria and cisternae of rough endoplasmic reticulum, very numerous free ribosomes, scattered ↑ tonofilaments bound into ↑ tonofibrils (Tf), and some mainly supranuclear ↑ melanin granules (MG). Cell borders are very irregular; basal plasmalemma forms several ↑ basal processes (BP) penetrating between ↑ reticular and ↑ collagen microfibrils of ↑ dermal papillae. Lateral cell processes contact similar processes of neighboring cells. B. lie on ↑ basal lamina (BL); they are

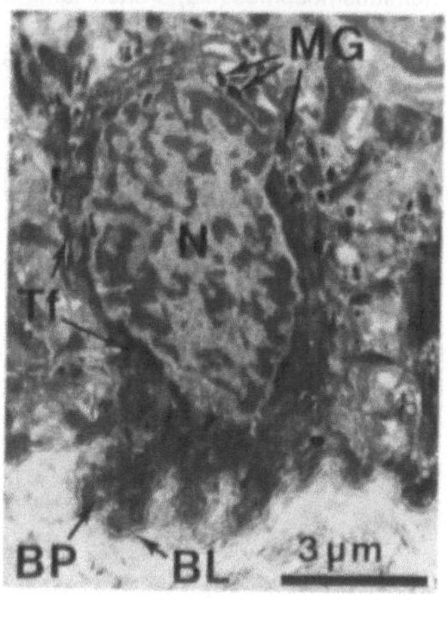

considered stem cells of epidermis since they have a high mitotic activity. (Fig. = human)

↑ Epidermis, ultrastructure of

Lavker, R.M., Sun, T.T.: Science *215*, 1239 (1982)

Basal cells, of sebaceous glands: ↑ Sebaceous cells

Basal cells, of various epithelia: Small, unspecialized, round or pyramidal cells with poorly developed ↑ organelles. B. lie on ↑ basal lamina, but do not reach free epithelial surface. By division and further ↑ differentiation, B. replace lost epithelial cells.

↑ Ductus epididymidis, epithelium of; ↑ Epidermis; ↑ Olfactory epithelium; ↑ Prostate, epithelium of; ↑ Stria vascularis; ↑ Taste buds, structure of; ↑ Trachea, epithelium of

Basal compartment, of seminiferous epithelium: ↑ Blood-testis barrier

Basal foot: ↑ Cilia, basal foot of

Basal granular cells: ↑ Enterochromaffin cells

Basal infoldings: Invaginations (BI) of basal plasmalemma into cell body with function of increasing contact surface with extracellular milieu. B. are about 20 nm wide and of varying depths; B. comprise part of ↑ basal labyrinth. (Fig. = ↑ distal tubule cell, rat)

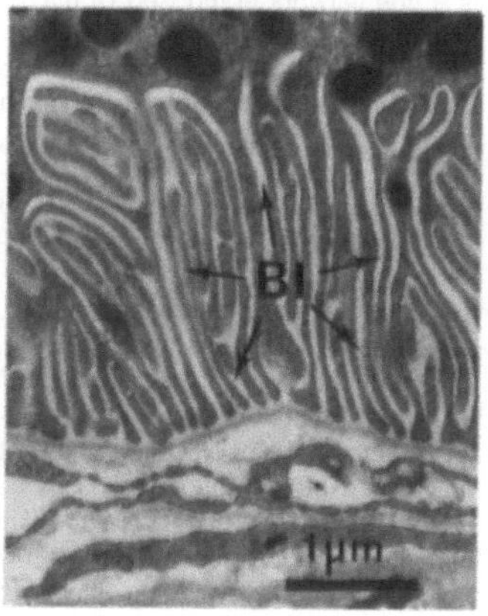

Basal labyrinth: A system (BLt) of ↑ basal infoldings (BI) and cytoplasmic compartments between them. Compartments are always open toward cytoplasm. B. is found in cells engaged in active transport of fluid and ions, such as ↑ clear cells of ↑ sweat glands, cell of ↑ proximal and ↑ distal tubules of nephron, ↑ striated duct cells of salivary glands, ↑ ciliary epithelial cells, ↑ marginal cells of ↑ stria vascularis, cells of ↑ choroid plexus. In general, compartments of B. are too small to contain ↑ mitochondria; however, in ↑ distal tubule cells of ↑ nephron (fig.) and in striated duct cells, compartments frequently contain mitochondria (M) oriented perpendicularly to ↑ basal lamina (BL), giving a pattern of basal striation. N = nucleus

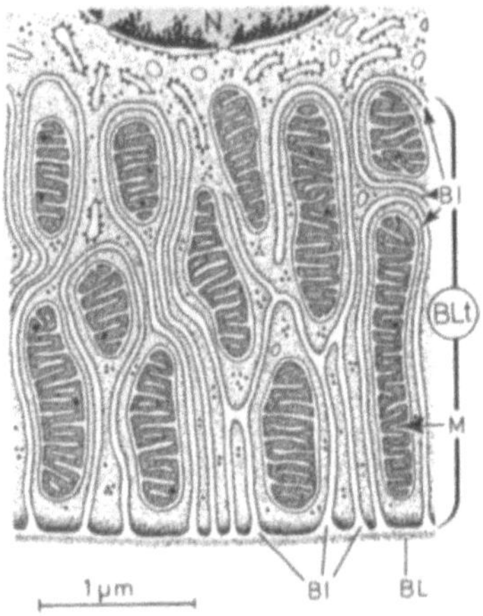

Basal lamina (basement lamina, lamina basalis*): A ↑ PAS-positive, 20 to 100-nm-thick, moderately osmiophilic layer (BL) underlying all epithelia and surrounding several classes of cells. B. is a fine feltwork composed of microfilaments 3–4 nm in diameter embedded in a moderately electron-dense matrix. In general, B. is separated from cell by a clear gap, lamina lucida* (LL), through which rune fine microfilaments (arrowheads) attaching B. to cell membrane (CM). ↑ Reticular microfibrils, belonging to sheaths underlying or surrounding the respective cell, penetrate into B. from opposite side. B. is a mixture of ↑ proteoglycans (↑ heparan sulfate, ↑ laminin, ↑ fibronectin), collagen of type IV, which does not polymerize into microfibrils, plus 10%–20% carbohydrates. B. is produced by the cell supporting it and probably serves as a semipermeable filter and agent directing some cell activities (↑ secretion). B. is part of ↑ basement membrane with which it is frequently confused. (Fig. = corneal epithelium, rat)

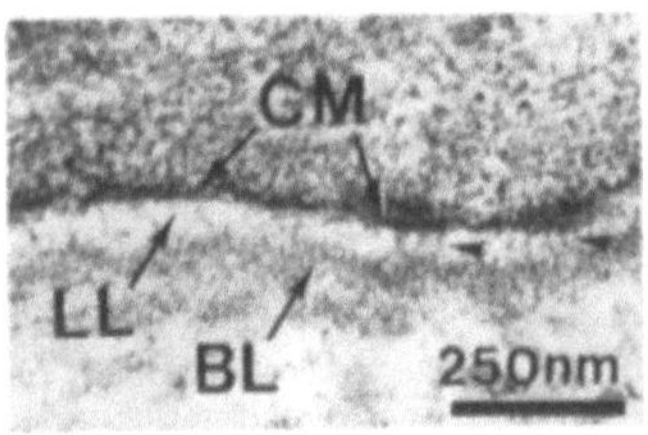

Csato, W., Merker, H.-J.: Cell Tissue Res. *228*, 85 (1983); Garbi, C., Wollman, S.H.: J. Cell Biol. *94*, 489 (1982); Junqueira, L.C.U., Montes, G.S., Toledo, O.M.S., Bexiga, S.R.R., Gordilho, M.A., Brentani, R.R.: Histochem. J. *15*, 785 (1983); Laurie, G.W., Leblond, C.P.: J. Histochem. Cytochem. *31*, 159 (1983); Sawada, H.: Biomed. Res. *2*, Suppl. 125 (1981)

Basal lamina, of glomerular capillaries: ↑ Glomerular basal lamina

Basal myoepithelial cells: ↑ Basket cells, of salivary glands; ↑ Myoepithelial cells

Basal plate, of placenta (decidual plate): Portion of ↑ decidua basalis between trophoblastic shell and zona spongiosa.

Basal plate, of spermatozoon: A layer of dense material accumulated in ↑ implantation fossa of ↑ spermatozoal head. B. is fused by means of fine perpendicular filaments with capitulum, forming a kind of articulation between spermatozoal head and neck.

↑ Spermatozoon, neck of

Basal pole: Zone of cell body in epithelial cells situated between ↑ nucleus and ↑ basal lamina.

↑ Apical pole

Basal processes: Cytoplasmic protrusions (BP) of ↑ basal cells (BC) of ↑ epidermis, very irregular in shape and lengt, often branched, penetrating into underlying tissue. By means of ↑ hemidesmosomes (Hd), ↑ basal lamina (BL) is firmly attached in B. From opposite side, ↑ reticular microfibrils (RMf) are anchored in basal lamina, fixing it to ↑ collagen fibrils (C). B. are cellular differentiations which contribute to maintenance and reinforcement of contact between epidermis and ↑ dermis; they are particularly developed in mechanically exposed skin (palms and soles).

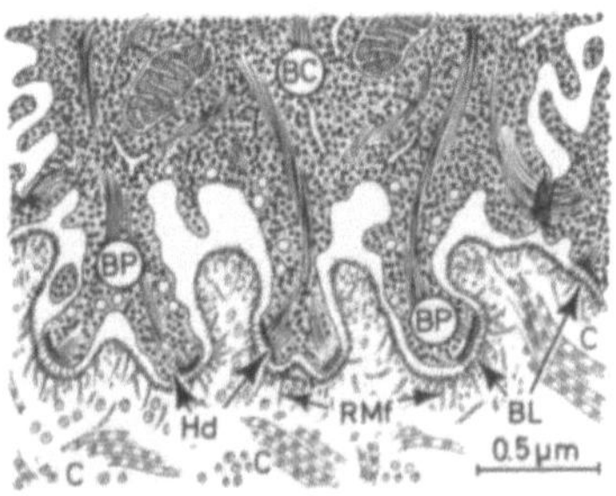

↑ Dermoepidermal junction

Basal striation (striatura basalis*): A pattern of infranuclear lines due to presence of long ↑ mitochondria oriented perpendicularly to ↑ basal lamina (BL). B. are particularly visible unter the light microscope and after appropriate staining in ↑ distal convoluted tubule of ↑ nephron (1) and in ↑ striated ducts of salivary glands (2).

↑ Basal infoldings; ↑ Basal labyrinth

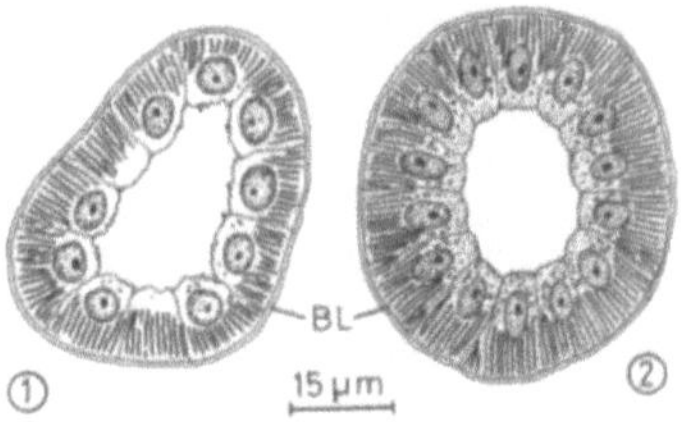

Basal web: A network of densely packed ↑ microfilaments (BW) attached to basal plasmalemma of some epithelial cells (mostly in ↑ simple epithelia). B. reinforces the connection between cell membrane and ↑ basal lamina (BL); it may contract. B. belongs to ↑ cytoskeleton. (Fig. = corneal epithelium, rat)

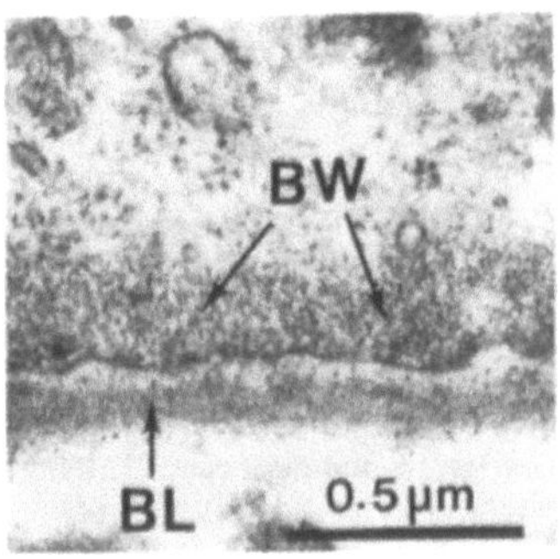

Basalis (stratum basale endometrii*): The thin, deepest layer of ↑ endometrium in contact with ↑ myometrium. During proliferative phase of every ↑ menstrual cycle, B. regenerates ↑ functionalis.

Basement membrane (membrana basalis*): An extracellular supporting layer (BM) found under basal surface of all epithelia, as well as around all cells surrounded by ↑ connective tissue but not belonging to it (↑ cardiac and ↑ smooth muscle cells, ↑ skeletal muscle fibers, peripheral ↑ neuroglial cells). Structure: 1) ↑ Basal lamina (BL). 2) Reticular lamina (RL) or lamina fibroreticularis* = a condensed ↑ ground substance with ↑ reticular microfibrils making loops within basal lamina and intermingling with neighboring ↑ collagen microfibrils (CMf). In this way, the reticular lamina fixes cells to connective tissue. It is ↑ argyrophilic, but does not always have to be present. Hd = ↑ hemidesmosomes. (Fig. = corneal epithelium, rat)

↑ Glomerular basal lamina

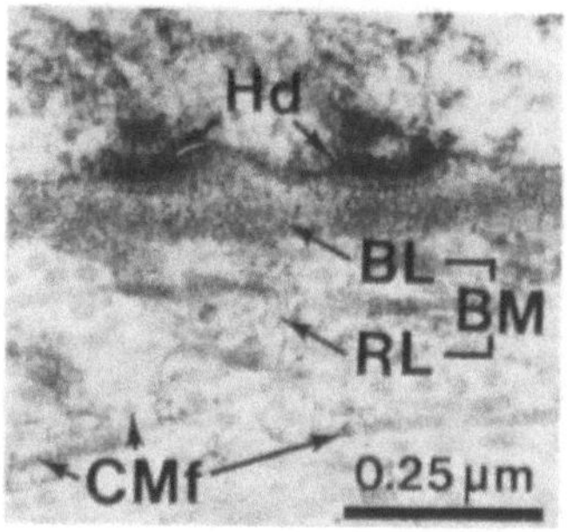

Alquier, C., Fayet, G., Hovsepian, S., Michel-Béchet, M.: Cell Tissue Res. *200*, 69 (1979); Kefalides, N.A., Alper, R., Clark, C.C.: Int. Rev. Cytol. *61*, 167 (1979); Merker, H.-J., Barrach, H.-J.: Eur. J. Cell Biol. *26*, 111 (1982); Wicha, M.S., Liotta, L.A., Garbisa, S., Kidwell, R.W.: Exp. Cell Res. *124*, 181 (1979)

Basic dyes (cationic dyes): Stains such as basic fuchsin, carmin, ↑ hematoxylin, methylene blue, etc. capable of forming a salt linkage with a negatively charged (acidic) biological substance.

↑ Basophilia

Basic protein: The major constituent of ↑ myelin present in major dense lines. B. is responsible for formation and maintenance of compact structure of myelin.

Omlin, F.X., Webster, deF.H., Palkovits, C.G., Cohen, S.R.: J. Cell Biol. *95*, 242 (1982)

Basilar membrane (membrana spiralis*): Floor (BM) of ↑ cochlear duct stretched between ↑ spiral ligament (SL) and ↑ lamina spiralis ossea (LSO). B. delimits cochlear duct (CD) toward scala tympani (ST); ↑ Corti's organ (CO) rests on B. Width of B. is about 0.16 mm in basal coil and 0.52 mm in ↑ helicotrema; its length is about 30 mm. Parts: A) Zona arcuata = inner third of B., attached to lamina spiralis ossea, contains vas spirale (VS); B) zona pectinata = external two-thirds, attached to spiral ligament. Structure (beginning from scala tympani): 1) A layer of flattened ↑ mesothelial cells (M), which probably produce 2) an amorphous substance (A) containing parallel ↑ auditory strings (AS), about 8–10 nm thick. In zona pectinata, the strings form two layers – one immediately beneath Corti's organ and another, thicker, situated farther down the scala tympani. Vibrations of B. are transmitted to ↑ hair cells (HC) and converted into adequate bioelectrical impulses.

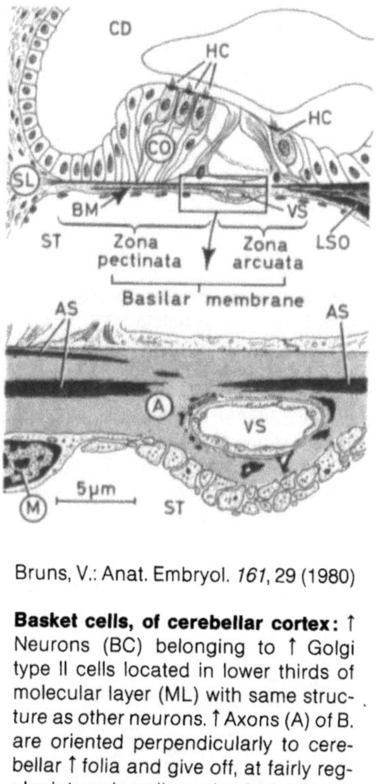

Bruns, V.: Anat. Embryol. *161*, 29 (1980)

Basket cells, of cerebellar cortex: ↑ Neurons (BC) belonging to ↑ Golgi type II cells located in lower thirds of molecular layer (ML) with same structure as other neurons. ↑ Axons (A) of B. are oriented perpendicularly to cerebellar ↑ folia and give off, at fairly regular intervals, collaterals which branch and surround, like baskets (B), bodies of the ↑ Purkinje cells (PC). The axons also participate in formation of supraganglionic nervous plexus (sg). The ↑ dendrites (D) make synaptic contacts with ↑ parallel fibers (PF) in outer half of molecular layer. Axonal endings of B. form inhibitory ↑ synapses on bodies of Purkinje cells.

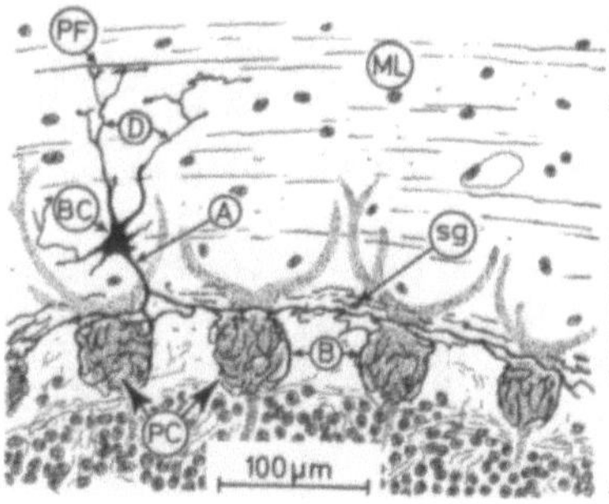

↑ Cerebellum, cyto- and myeloarchitectonics of; ↑ Interneurons, of cerebellar cortex

Léranth, C., Hamori, J.: Z. mikrosk.-anat. Forsch. *95*, 1 (1981)

Basket cells, of mammary gland (myoepitheliocytus stellatus*): Ramified ↑ myoepithelial cells surrounding ↑ alveoli of ↑ mammary gland like baskets. Contraction of B. diminishes diameter of alveoli and expulses ↑ milk into excretory canalicular system.

↑ Myoepithelial cells, of sweat glands

Basket cells, of salivary glands (basal myoepithelial cells, myoepitheliocytus stellatus*): Stellate ↑ myoepithelial cells located between ↑ glandular cells and sharing a common ↑ basal lamina with them. Ramified processes of B. surround ↑ acini, and ↑ mixed and ↑ mucous tubules like baskets. Contraction of B. facilitates expulsion of secretory product into excretory canalicular system.

↑ Myoepithelial cells, of sweat glands

Basket, elastic: ↑ Alveolus of lung, elastic basket of

Baskets, of axons: ↑ Basket cells, of cerebellum

Basophilia: A property of acid proteins in nucleus, cytoplasm, and extracellular material (↑ mucus) to bind ↑ basic dyes and appear blue. Nuclei (DNA content), rough ↑ endoplasmic reticulum, free ↑ ribosomes, and ↑ polyribosomes (RNA content) exhibit strong B.

↑ Ergastoplasm; ↑ Erythropoiesis; ↑ Nissl bodies

Basophilic bodies, of liver: The light-microscopic aspect of clumps (BB) of rough ↑ endoplasmic reticulum after staining with methylene blue.

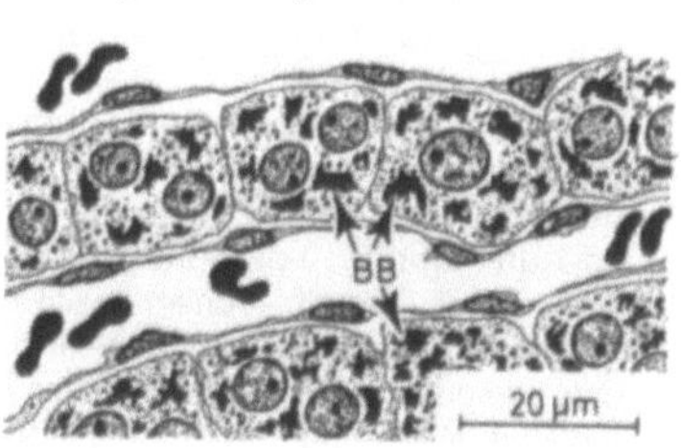

Basophilic cells, of hypophysis: ↑ Hypophysis, basophilic cells, of

Basophilic erythroblast: ↑ Erythroblast, basophilic

Basophilic granules (granulum basophilicum*): Specific granules of basophilic ↑ granulocytes.

Basophilic granulocytes: ↑ Granulocytes, basophilic

Basophilic infiltration: An occasional penetration of ↑ basophils (BC) of pars intermedia (PI) into ↑ neurohypophysis (N). (Fig. = human)

↑ Hypophysis, pars intermedia of

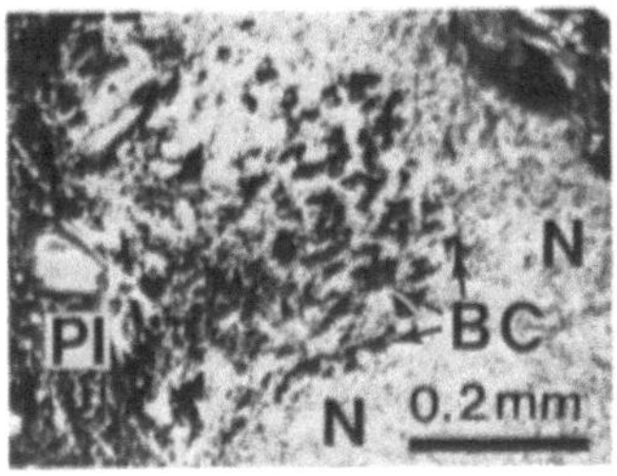

Basophilic metamyelocyte: ↑ Metamyeolocyte, basophilic

Basophilic myelocyte: ↑ Myelocyte, basophilic

Basophilic normoblast: ↑ Erythroblast, basophilic

Basophils, of adenohypophysis: ↑ Hypophysis, basophilic cells of

Basophils, of blood: ↑ Granulocytes, basophilic

Basophils, of pars intermedia ((MSH-cells, cellula melanotrophica*): Polygonal or prismatic cells, grouped in strands, groups, and follicles. B. are characterized by an elliptical, eccentric, heterochromatin-poor nucleus, numerous small mitochondria, a well-developed Golgi apparatus, and short, flattened rough endoplasmic cisternae predominantly located in periphery of cell. Cytoplasm is filled with secretory granules (S), 200–300 nm in diameter, arising from Golgi apparatus (G); the granules show a various degree of ↑ osmiophilia, contain ↑ glycoproteins, and are strongly ↑ basophilic and ↑ PAS- and resorcin–fuchsin-positive. The secretory product of B., ↑ melanocyte-stimulating hormone, stimulates ↑ melanin production in the body. Morphologically, B. are similar to ↑ corticotropes of pars distalis.

↑ Basophilic infiltration; ↑ Hypophysis, pars intermedia of

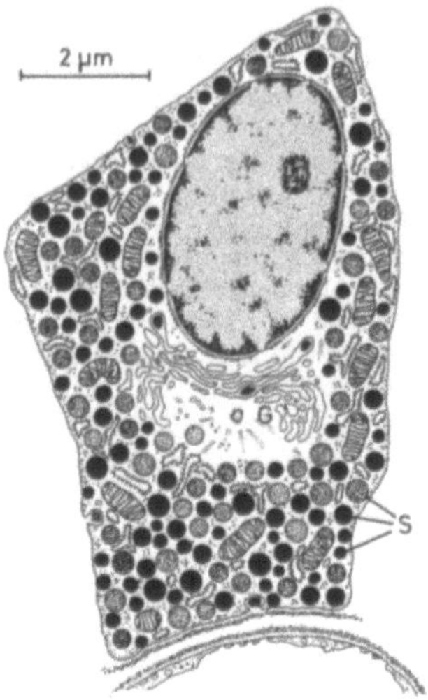

B-cells, of pancreatic islets (beta cells, insulin cells, cellula beta*): Polygonal endocrine cells making up about 75% of cell population of ↑ pancreatic islets, predominantly located in interior of latter. B. have an oval and frequently invaginated nucleus with a moderate amount of ↑ heterochromatin and a conspicuous nucleolus. Cytoplasm encloses a well-developed Golgi apparatus (G), large and numerous mitochondria, short, flattened, and not particularly numerous rough endoplasmic cisternae, some free ribosomes, rarely lysosomes, and ↑ lipofuscin granules. From Golgi apparatus originate numerous alcohol-soluble secretory granules (S), 200 nm in diameter which are delimited by ↑ unit membrane and display a marked clear halo around the central osmiophilic core. In some species (bat, cat, dog, man), one or several rectangular crystalloids may form the core (inset). By ↑ exocytosis, content of granules, insulin, is discharged into ↑ pericapillary space (PS) from which it reaches the capillaries (Cap).

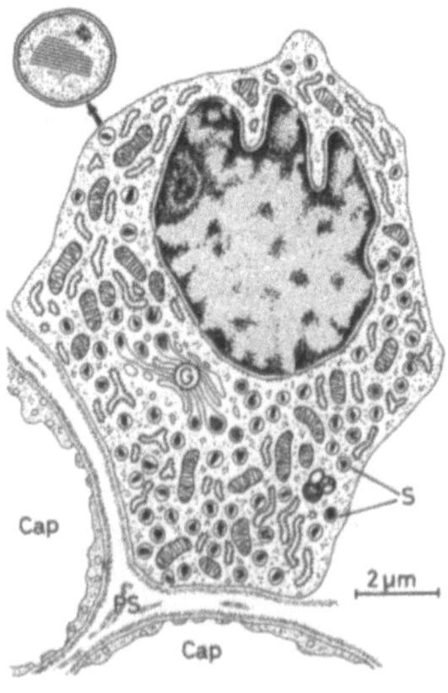

Orci, L.: Morphologic Events Underlying the Secretion of Peptide Hormones. In: James, V.H.T., (ed.): Excerpta Medica International Congress Series No. 143. Amsterdam: Excerpta Medica 1976; Orci, L., Perrelet, A., Freind, D.S.: J. Cell Biol. *75*, 23 (1977); Ravazzola, M., Perrelet, A., Roth, J., Orci, L.: Proc. Natl. Acad. Sci. USA *78*, 5661 (1981)

B-cells, of synovial membrane: ↑ Synovial cells

Bellini's ducts: ↑ Papillary ducts

Bergmann's neuroglial cells: Type of modified ↑ astrocyte (BC) of cerebellar cortex with an ultrastructure corresponding to that of other ↑ astrocytes. Situated at level of ↑ Purkinje cells (PC), B. send out several main processes perpendicularly to the surface, the branched endings of which participate in formation of ↑ membrana limitans gliae superficialis. Short lateral branches of main processes envelope dendrites of Purkinje cells. It is believed that B. play an important role in guidance of immature ↑ granule cells during postnatal development of ↑ cerebellum.

↑ Cerebellum, glioarchitectonics of

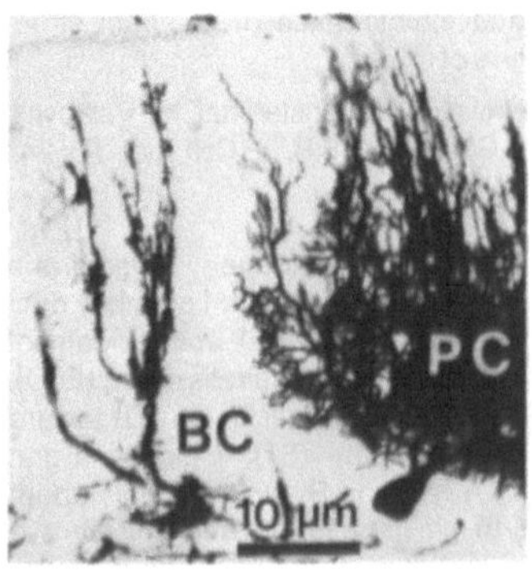

Shiga, T, Ichikawa, M., Hirata, Y.: Anat. Embryol. *167*, 203 (1983); Sommer, T., Lagenauer, C., Schachner, M.: J. Cell Biol. *90*, 448 (1981)

Berlin blue reaction (Prussian blue reaction): A histochemical method for demonstration of ferric salts (Fe^{3+}) using potassium ferrocyanide.

↑ Turnbull reaction

Bertin's columns: ↑ Renal columns

Best's carmine staining: A histochemical method for demonstration of ↑ glycogen, not fully specific since it also stains ↑ mucus, ↑ fibrin, and ↑ mast cell granules; now replaced by ↑ PAS-reaction.

Beta$_2$-basophils, of adenohypophysis: ↑ Thyrotropes

Beta cells, of adenohypophysis: ↑ Thyrotropes

Beta cells, of pancreatic islets: ↑ B-cells, of pancreatic islets

Beta globulins: Blood proteins involved in transport of ↑ hormones, metal ions, and lipid.

↑ Transferrin

Beta particles, of glycogen: ↑ Glycogen

Beta tubulin: A protein which, together with alpha ↑ tubulin, constitutes the ↑ microtubules.

Betz's cells: Giant pyramidal ↑ neurons (B), up to 80 µm in width and 120 µm in height, situated in ganglionic layer of Brodmann's areas 4 and 6 (area gigantopyramidalis) of ↑ brain cortex. Their ↑ dendrites branch in molecular layer; some of their ↑ axons form corticospinal tract and make synaptic contacts with ↑ motor neurons of ↑ spinal cord; other axons run through corpus callosum to opposite cortex. Structurally, B. correspond to other neurons. (Fig. = human, silver staining)

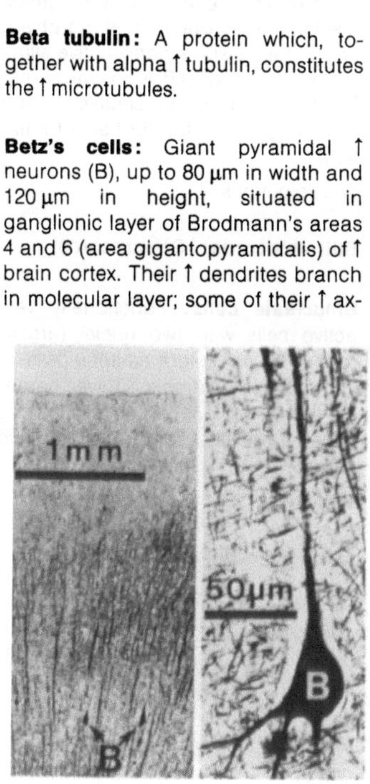

↑ Brain cortex, phylogenetic and structural division of

BF: ↑ Blastogenic factor

B-face, of plasmalemma: A former designation for ↑ EF-face of freeze-cleaved cell membrane.

↑ Freeze-cleaving, terminology of

BFU-E: ↑ Burst-forming units-erythropoietic

B-granules, of neutrophilic granulocytes: ↑ Granulocytes, neutrophilic, granules of

Bichat's fat pad (corpus adiposum buccae*): The accumulation of somewhat incapsulated white ↑ adipose tissue in the ↑ cheek; particularly well-developed in the newborn.

Bielschowsky's silver staining: Silver nitrate staining for demonstration of fibrillar structures – ↑ collagen fibrils (brown), ↑ neurofibrils, and ↑ reticular fibrils (black).

↑ Argyrophilia

Bile: A yellow-brown very viscous fluid filling ↑ gallbladder. B. is an exocrine product of ↑ liver parenchymal cells and consists of ↑ bile pigments, bile acids, ↑ mucus, and water. (See biochemistry and physiology texts for further information)

↑ Bile pathways

Geigy Scientific Tables. Body Fluids. 8th edn. Basle: Ciba-Geigy 1983; Jones, A.L., Schmucker, D.L., Renston, R.H., Murakami, T.: Dig. Dis. Sci. *25*, 609 (1980)

Bile canaliculi (bile capillaries, canaliculi biliferi*): Roughly 0.5 to 1.5-µm-wide channels (BC) formed by apposing ↑ liver parenchymal cells (L). B., therefore, have no wall of their own. The narrow zone of cytoplasm adjacent to B. is devoid of ↑ organelles, but contains numerous microfilaments, which insert into ↑ attaching plaques of ↑ desmosomes (D). This zone shows staining properties similar to those of ↑ terminal web of ↑ absorptive intestinal cells, indicating a comparable chemical composition. Plasmalemma bordering B. forms ↑ microvilli (Mv), which

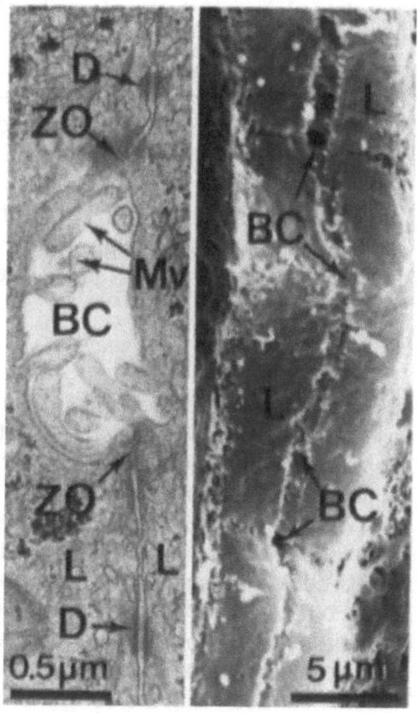

protrude into lumen of B. Zonulae occludents (ZO), discontinuous ↑ nexus, and desmosomes separate B. from other intercellular spaces. B. convey ↑ bile from central parts of classic ↑ liver lobule to terminal ↑ bile ductules. There are no anastomoses of B. from lobule to lobule. (Figs. = rat)

Biava, C.G.: Lab. Invest. *13*, 840 (1964); Gebhardt, R., Jung, W., Robenek, H.: Eur. J. Cell Biol. *29*, 68 (1982); Marinozzi, G., Muto, M., Correr, S., Motta, P.: J. Submicrosc. Cytol. *9*, 127 (1977)

Bile ducts, interlobular (ductus interlobularis bilifer*): A richly anastomosing network of bile channels (BD) into which empty ↑ bile ductules (BDI). Located within ↑ portal canals (PC), B. accompany branches of portal vein. Toward hepatic ↑ hilus B. enlarge gradually, passing from simple cuboidal to simple columnar epithelium, with occasional mucus-secreting areas. Cells lining B. have a round or elliptical nucleus, better developed mitochondria than in those of bile ductules, and prominent Golgi apparatus. Cytoplasm contains numerous ↑ micropinocytotic vesicles and sometimes cholesterol crystals. Cells lie on

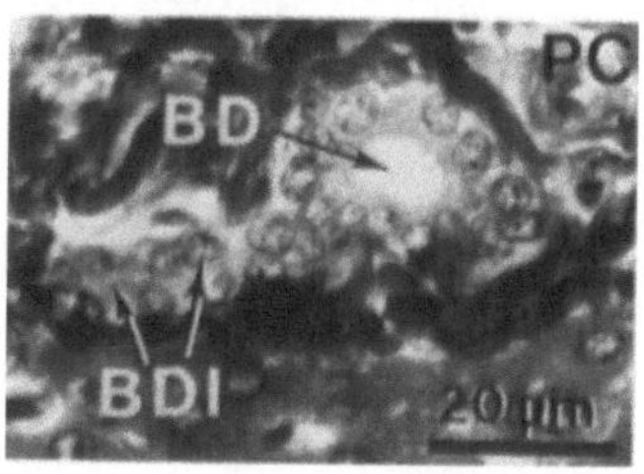

basal lamina and bear microvilli at their ↑ apical poles. ↑ Junctional complexes and numerous interdigitations seal the intercellular contacts. With increase in caliber of B., surrounding layers of ↑ dense connective tissue become thicker and contain ↑ elastic fibers as well as occasional ↑ smooth muscle cells. In hepatic hilus, main B. fuse to form left and right lobar bile ducts. (Fig. = human)

↑ Bile pathways

Bile ductules (cholangiole, perilobular bile ductule, ductulus bilifer*): Short, narrow bile channels situated in ↑ portal canals (PC), connecting terminal ↑ bile ductules with interlobular ↑ bile ducts (IBD). Ductule cells have flattened nuclei, inconspicuous ↑ organelles, but well-developed cytoplasmic interdigitations, apical ↑ junctional complexes, and numerous microvilli at the free surface. All B. are surrounded by a ↑ basal lamina (BL).

↑ Bile pathways

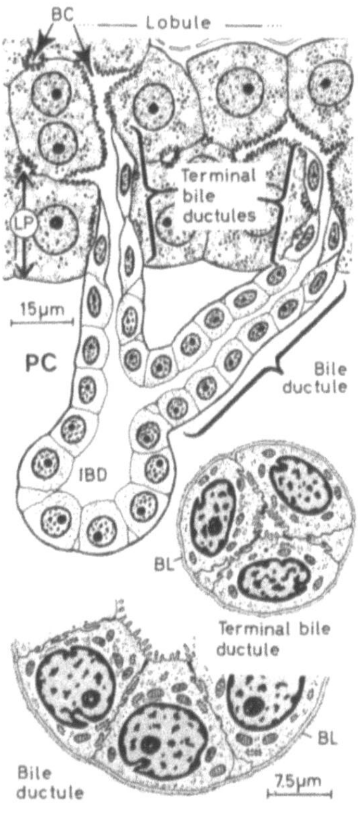

Bile ductules, intralobular: ↑ Bile ductules, terminal

Bile ductules, perilobular: ↑ Bile ductules

Bile ductules, terminal (canal of Hering, intralobular bile ductule): Short, narrow channels located at periphery of classic ↑ liver lobule, connecting ↑ bile canaliculi (BC) to ↑ bile ductules. At first, B. are lined by one or two flattened cells, which become gradually cuboidal after penetration of ↑ limiting plate (LP). These cells are small, pale, with an elliptical nucleus, scarce mitochondria, a moderately developed Golgi apparatus, short, occasional, rough endoplasmic cisternae, and some irregular ↑ microvilli at ↑ apical pole. Cells are firmly held together by numerous interdigitations and ↑ junctional complexes. A ↑ basal lamina (BL) surrounds B., except at areas of contact between ↑ hepatocytes and ductule cells. (See fig. under ↑ Bile ductules)

↑ Bile pathways

Bile pathways (biliary spaces): A system of channels conveying ↑ bile from ↑ liver (L) to ↑ duodenum (D). Parts: A) Intrahepatic bile ducts: I) Intralobular channels = 1) ↑ bile canaliculi; 2) ↑ bile ductules, terminal. II) Interlobular channels = 3) ↑ bile ductules; 4) ↑ bile ducts, interlobular. B) Extrahepatic bile ducts: 5) Left and right lobar bile ducts; 6) common ↑ hepatic duct; 7) ↑ cystic duct; 8) ↑ ductus choledochus.

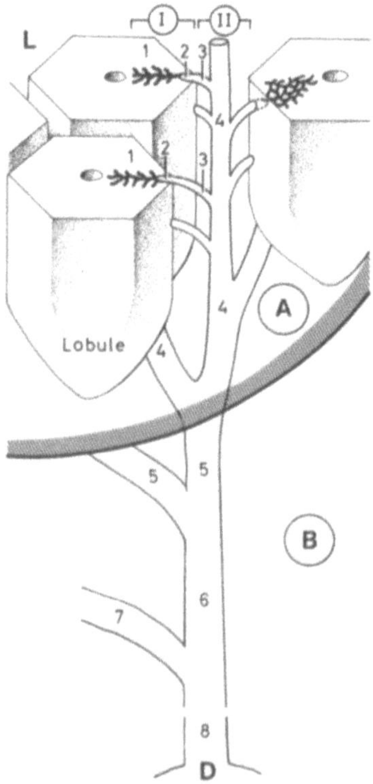

Vonnahme, F.J.: Cell Tissue Res. *215*, 207 (1981)

Bile pigments: A group of iron-free ↑ pigments formed by opening of porphyrin ring of heme during destruction of ↑ hemoglobin. The first B. is ↑ biliverdin, almost completely converted into ↑ bilirubin, which is in turn converted to another B. under influence of intestinal flora. (See biochemistry and physiology texts for further information)

Bile preductules: A notion, not generally accepted, of small short bypasses breaking off at an angle from terminal ↑ bile ductules and draining ↑ bile into ↑ bile ductules. B. are not surrounded by ↑ liver parenchymal cells in their course.

Biliary spaces: ↑ Biliary pathways

Bilirubin: An orange to red, iron-free endogenous ↑ pigment formed from ↑ hemoglobin by macrophages of ↑ mononuclear phagocyte system after destruction of aged ↑ erythrocytes. Released into circulation, B. is taken up by ↑ liver parenchymal cells, conjugated in rough endoplasmic reticulum with glucuronide to form bilirubin glucuronide, and excreted into ↑ bile canaliculi. (See biochemistry and physiology texts for further information)

Biliverdin: An endogenous ↑ pigment resulting from opening of porphyrin ring of heme with a consecutive elimination of iron. Most B. is converted into ↑ bilirubin. (See biochemistry and physiology texts for further information)

↑ Bile pigments

Billroth's cords: ↑ Splenic cords

Binucleate cells: Functionally very active cells with two nuclei (arrow) formed as a result of amitotic division of ↑ nucleus. Examples: ↑ liver parenchymal cells (fig.), ↑ facet cells, gastric ↑ parietal cells, some ↑ autonomic neurons. (Fig. = human)

↑ Amitosis

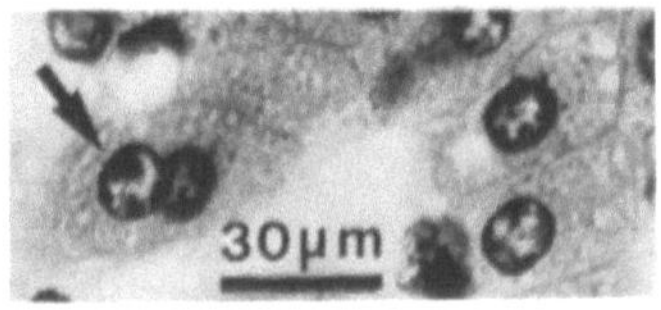

Weatley, D.: Exp. Cell Res. *74*, 455 (1972)

Bioblast: A concept referring to intracellular parasites which have established a felicitous, symbiotic relationship with their eucaryotic host. Since ↑ mitochondria contain DNA and RNA of certain types of ↑ procaryotes and are able to multiply independently of nuclear genetic program, it is believed they represent ancestral procaryotes which penetrated ↑ eucaryotes and established with them a successful functional cooperation, resulting in considerable evolution of the eucaryotes. Some not fully understood observations suggest that ↑ centrioles could also belong to B.

Küntzel, H., Kochel, H.G.: Nature *293*, 751 (1981); Nass, S.: Int. Rev. Cytol. *25*, 55 (1969)

Biomembranes: ↑ Cytomembranes

Biomicroscope: An optical instrument for observation of living tissues. The slitlamp microscope is a kind of B.

↑ Eye, in the study of living tissues; ↑ Transparent chamber

Bipolar cells: ↑ Bipolar neurons

Bipolar layer, of retina (inner nuclear layer, stratum nucleare internum*): A middle retinal layer containing ↑ perikarya of ↑ bipolar neurons (by far most neurons), ↑ Müller's cells, ↑ amacrine, and ↑ horizontal cells.

↑ Retina

Bipolar neurons (neuronum bipolare*): Nerve cells with two processes emerging from opposite points of cell body (CB). One process is peripheral, ↑ dendrite (D), conducting impulses toward the cell body (white arrow), and the other is central, ↑ axon (A), conducting impulses away from the cell body (black arrow). Both processes may branch. B. are relatively rare in vertebrates; they are present in ↑ vestibular and ↑ spiral ganglia of inner ear, in ↑ bipolar cell layer of ↑ retina, and in ↑ olfactory epithelium (↑ olfactory cells). During development of nervous system, most ↑ neurons are bipolar with internal structure corresponding to that in other ↑ neurons.

↑ Bipolar neurons, of retina; ↑ Bipolar neurons, of spiral ganglion; ↑ Bipolar neurons, of vestibular ganglion

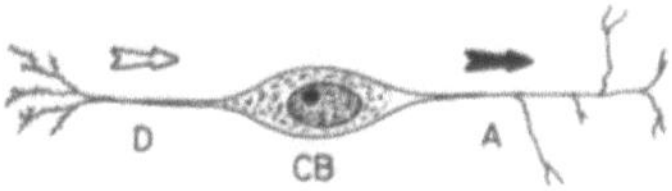

Bipolar neurons, of retina (bipolar cell, neurocytus bipolaris*): Retinal neurons (BC) extending from outer (OPL) to inner plexiform layer (IPL). With their dendrites (D), B. make ↑ synapses with ↑ photoreceptor axons (a); with their axons (Ax), B. synapse with dendrites of ↑ ganglion (GC) and ↑ amacrine (A) cells. The relatively small ↑ perikaryon, located in inner granular layer (IGL), contains an oval nucleus, numerous mitochondria, a well-developed Golgi apparatus, several small ↑ Nissl bodies, and some ↑ neurotubuli. Three types of B.: 1) Rod B. (RBC) connecting several rod cells (RC) with a single ganglion cell; 2) midget B. (MBC) connecting a single cone cell (CC) with a single ganglion cell (area of macula lutea); and 3) flat or diffuse B. (FBC) synapsing with many cone axons. The totality of B. comprises the retinal ganglion.

↑ Triad, of retinal neurons

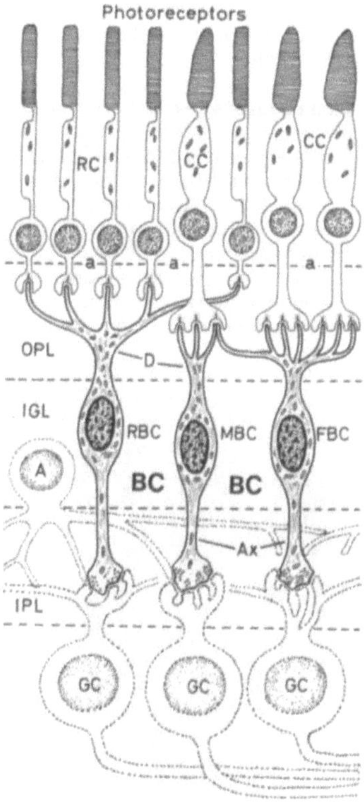

Kaneko, A.: TINS *6*, 219 (1983)

Bipolar neurons, of spiral ganglion: Fusiform ↑ neurons whose unique ↑ dendrite (D) reaches ↑ hair cells and whose axon (A) participates in formation of cochlear nerve. B. display a round, frequently invaginated nucleus (N) poor in ↑ heterochromatin; there are a voluminous nucleolus (Nu), numerous mitochondria and ↑ Nissl bodies (NB), a well-developed Golgi apparatus, some lysosomes, ↑ neurofibrils, and ↑ neurotubuli. ↑ Schwann's cells (SC) closely adhere to B; the former may synthesize ↑ myelin lamellae (ML) surrounding the bodies and processes of some, but not all, B. A ↑ basal lamina (BL) separates Schwann's cells from ↑ endoneurium (E). Function of myelinated B. is still unknown.

↑ Spiral ganglion

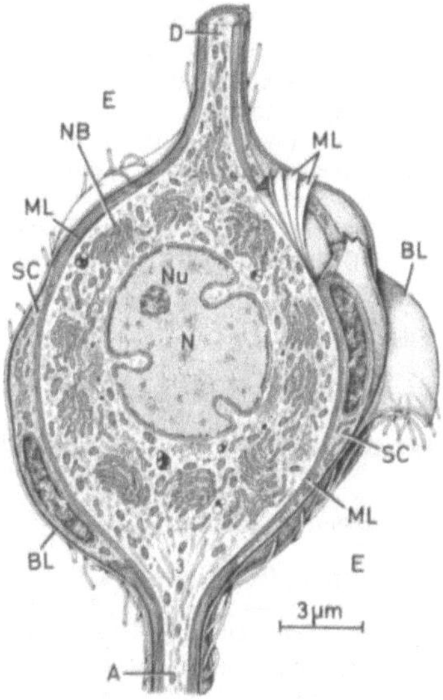

Bipolar neurons, of vestibular ganglion: Spindle-shaped nerve cells whose unique ↑ dendrite reaches one of ↑ vestibular cells of either ↑ maculae

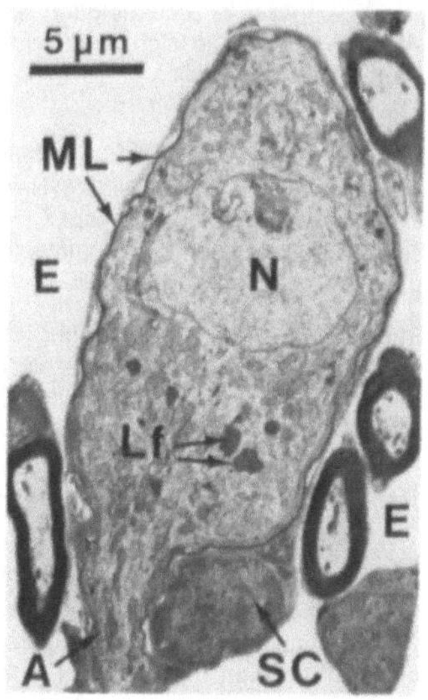

or ↑ cristae ampullares. Their ↑ axons (A) form vestibular nerve. Besides possessing general structural characteristics of ↑ neurons, B. are surrounded by ↑ Schwann's cells (SC), which furnish them a system of ↑ myelin lamellae (ML) with an unknown functional significance. E = ↑ endoneurium, Lf = ↑ lipofuscin granules, N = nucleus. (Fig. = mongolian gerbil)

↑ Vestibular ganglion

Birbeck's granules: ↑ Langerhans' bodies

Birefringence: The property of some inorganic and organic structures (↑ collagen fibers, ↑ striated muscle fibers, ↑ cell membrane, ↑ myelin sheath, etc.) to split a ray of polarized light into two rays vibrating in planes perpendicular to one another with different velocities and different ↑ refraction indices. B. is studied with ↑ polarizing microscope.

Bivalent: ↑ Tetrad

Bladder, urinary: ↑ Urinary bladder

Blandin's glands: ↑ Anterior lingual glands

Blast cells: ↑ Immunoblasts

Blastogenic factor (BF): A ↑ lymphokine secreted by activated ↑ T-lymphocytes stimulating transformation and clonal multiplication of nonsensitized lymphocytes into ↑ immunoblasts.

Blind spot: ↑ Optic papilla

Blisters: ↑ Membrane blisters

Blood (sanguis*): A fluid connective tissue composed of circulating formed elements and a liquid intercellular substance, the blood plasma.

↑ Blood, composition of

Bessis, M.: Living Blood Cells and their Ultrastructure. Berlin, Heidelberg, New York: Springer Verlag 1973; Kapf, C.T., Jandl, J.H.: Le Sang. Atlas commenté d'hématologie. Paris: MEDSI 1982

Blood-air barrier (air-blood barrier, alveolocapillary membrane): A very thin (2.2±0.2 µm in the human) multilayered biological membrane between air and blood. Structure: 1) An extremely thin cytoplasmic sheet of ↑ alveolar cells type I (AC I) covered with ↑ surfactant (S); 2) ↑ basal lamina (BL) of alveolar cells; 3) capillary basal lamina (cBL); and 4) very flattened, nonfenestrated cytoplasm of capillary ↑ endothelial cells (E). Both basal laminae fuse where alveolar and endothelial cells are opposed. Gas exchange is effected by passive diffusion (arrows). Cap = capillaries.

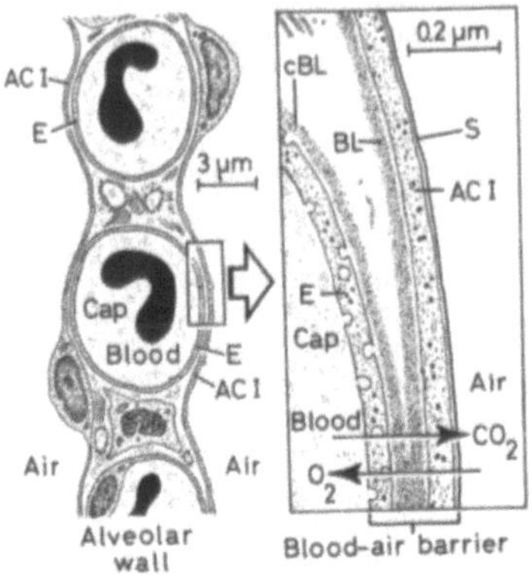

Bartels, H.: Cell Tissue Res. *198*, 269 (1979); Gehr, P., Bachofen, M., Weibel, E.R.: Respir. Physiol. *32*, 121 (1978); Meban, C.: J. Anat. *131*, 299 (1980); Vaccaro, C.A., Brody, J.S.: J. Cell Biol. *91*, 427 (1981); Weibel, E.R.: AJR *133*, 1021 (1979)

Blood-aqueous barrier: A functional barrier preventing free diffusion of fluids between blood (B) and ↑ posterior eye chamber (C). B. is formed by ↑ junctional complexes (arrowheads) connecting ↑ apical poles of nonpigmented cells (NC) of ↑ ciliary body (CB). B. is responsible for maintaining special chemical composition of ↑ aqueous humor. PC = pigment cells

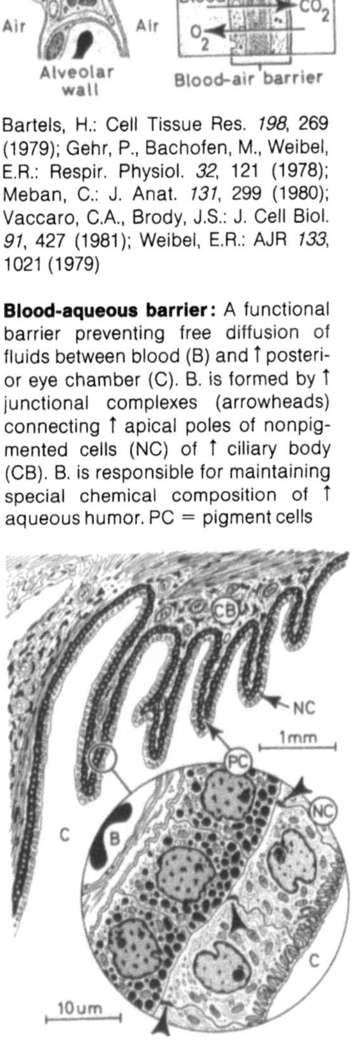

Vegge, T.: Z. Zellforsch. *114*, 309 (1971)

Blood-brain barrier (hematoencephalic barrier): A selective physiological barrier between blood (B) and tissue of ↑ central nervous system. Structure: 1) Nonfenestrated capillary endothelial cells (E) with extensive ↑ zonulae occludentes: 2) capillary ↑ basal lamina (BL); and 3) ↑ perivascular end feet (PF) of ↑ astrocytes (= ↑ membrana limitans gliae perivascularis) closely apposed to basal lamina. No ↑ pericapillary space. B. assures a constant and optimal environmental composition for ↑ neurons. Whereas gases and components of simple solutions diffuse easily through B., it is able to block passage of certain substances (e.g., medicaments, bile pigments) into central nervous system. It is believed that this process is the result of openings and closings of zonula occludentes; astrocyte processes seem to play a less important role in barrier effect. (Fig. = rat)

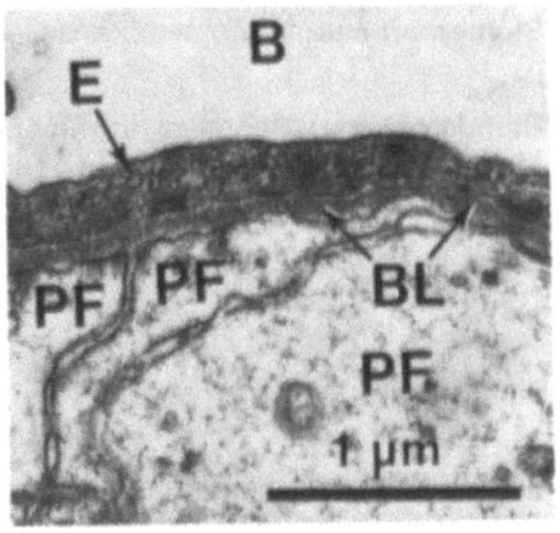

Brightman, M.W.: Eye Res. Suppl. 1 (1977); Van Deurs, B.: Int. Rev. Cytol. *65*, 117 (1980)

Blood-brain barrier, barrier-free areas of: Well-delimited areas of brain, generally vascularized by fenestrated ↑ capillaries surrounded by large ↑ pericapillary spaces: ↑ area postrema, ↑ choroid plexus, ↑ infundibulum, ↑ median eminence, ↑ neurohypophysis, ↑ organum vasculosum laminae terminalis, ↑ pineal organ, ↑ subcommissural organ, and ↑ subfornical organ.

Blood, buffy coat of: A thin, gray-white layer between column of packed ↑

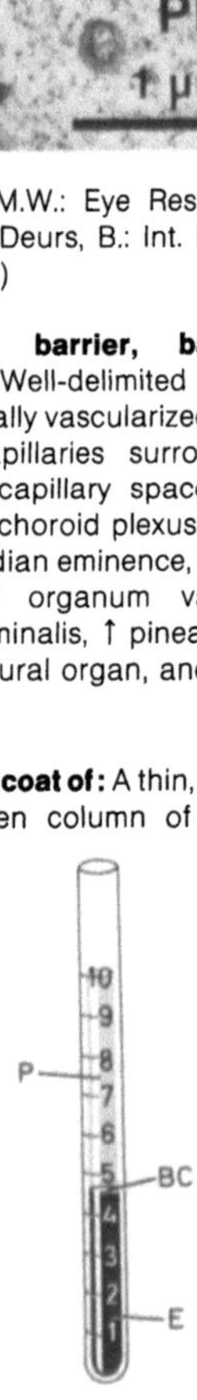

erythrocytes (E) and plasma (P), occurring after centrifugation or sedimentation of blood when clotting is prevented. B. forms about 1% of total blood volume and consists of ↑ leukocytes and ↑ platelets.

Blood cell formation: ↑ Erythropoiesis; ↑ Granulopoieses; ↑ Lymphopoieses; ↑ Monopoiesis; ↑ Thrombopoiesis

Blood cell formation, theories of: ↑ Hematopoiesis, theories of

Blood circulatory system: ↑ Cardiovascular system

Blood clotting (coagulatio sanguis): A process transforming fluid blood into a semisolid gel in order to limit hemorrhage. Sequence: 1) Clotting is initiated as blood begins to flow out through injured vessel wall (W) and as ↑ platelets (P) adhere, in presence of von Willebrandt's factor, to ↑ basal lamina and ↑ collagen fibrils (Col) of underlying tissue. 2) Platelets then release ↑ serotonin, adenosine diphosphate, and ↑ thromboplastin (platelet factor 3). 3) This process is also stimulated by tissue thromboplastin, Ca^{2+}, and other clotting factors released from injured area. 4) Both tissue and platelet thromboplastin convert ↑ prothrombin to ↑ thrombin. The latter transforms ↑ fibrinogen to ↑ fibrin, which polymerizes as a feltwork of fine, cross-striated microfibrils (F) trapping platelets and blood cells to form a clot (C). 5) Serotonin provokes contraction of vascular ↑ smooth muscle cells and adenosine diphosphate is necessary for shortening of thrombosthenin released from aggregated and disintegrated platelets. Through its polymerization and shortening, the clot contracts and stops the bleeding. The described process is still far from being unterstood; the depicted schema gives only a broad idea. (See physiology texts for further information)

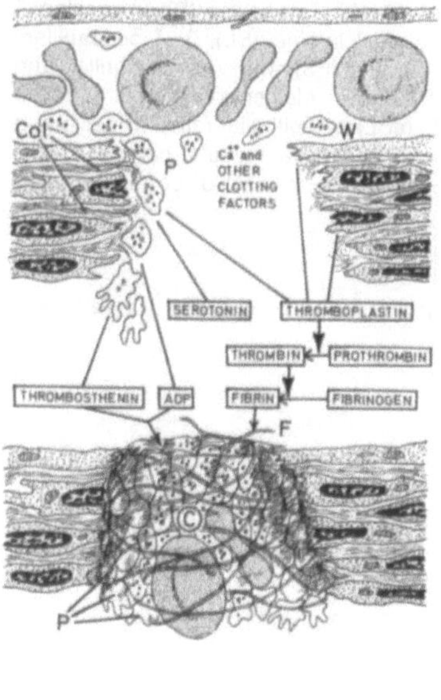

Zucker, M.B.: Sci. Am. *242/6*, 70 (1980)

Blood, composition of: The blood is composed of a liquid phase – blood plasma – and formed elements, i.e., cells and platelets. (Asterisks indicate the letters under which the corresponding terms are described)

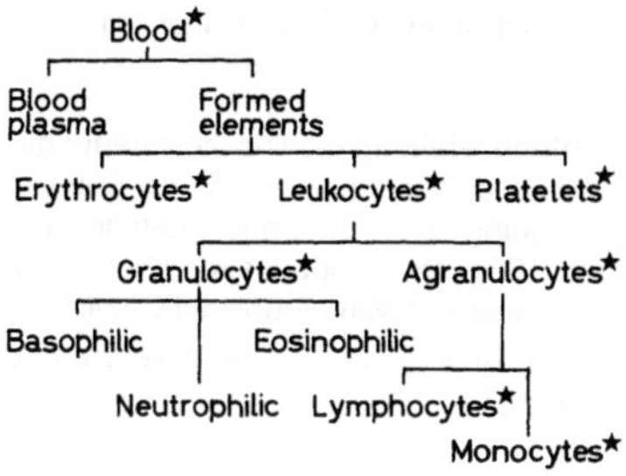

Blood count: Calculation of number of ↑ erythrocytes, ↑ platelets, and ↑ leukocytes in a given volume of blood, as well as determination of percentages of various categories of leukocytes in stained blood smears. The normal values in the adult are:

	mm^3	l (SI)
Erythrocytes	♀ $4.5 \pm 0.5 \times 10^6$ ♂ $5.0 \pm 0.5 \times 10^6$	$4.0\text{–}5.5 \times 10^{12}$/l
Platelets	250000–500000	$250\text{–}500 \times 10^9$/l
Leukocytes	5000–10000	$5\text{–}10 \times 10^9$/l
Granulocytes		Range
basophilic		0–2%
eosinophilic		1–4%
neutrophilic		40–75%
band neutrophils		5–10%
Lymphocytes		20–45%
Monocytes		2–8%

Blood dust: ↑ Hemoconia

Blood, formed elements of: Morphologically and functionally characteristic cells and corpuscles circulating in normal blood: 1) Red blood cells (↑ erythrocytes and ↑ reticulocytes); 2) ↑ leukocytes = a) granulocytes (basophilic, eosinophilic, and neutrophilic ↑ granulocytes), b) agranulocytes (↑ lymphocytes and ↑ monocytes); 3) ↑ platelets.

↑ Blood, composition of

Blood-ganglion barrier: A selectively permeable boundary between blood and ↑ ganglion cells of ↑ peripheral nervous system ↑ ganglia. B. is similar to ↑ blood-nerve barrier and consists of continuous ↑ endothelium of blood ↑ capillaries and capsule of ganglion.

Depace, D.M.: Anat. Rec. *204*, 357 (1982)

Blood islands (insulae sanguineae*): Aggregates of ↑ mesenchymal cells in wall of yolk sac. B. are first sites of ↑ hematopoiesis. The central cells of the aggregates differentiate into primitive ↑ erythroblasts; the peripheral cells, angioblasts, become ↑ endothelial cells of primitive blood vessels. (See embryology texts for further information)

↑ Hematopoiesis, prenatal

Hamilton, W.J., Boyd, J.D., Mossman, H.W.: Human Embryology, 4th ed Baltimore: Williams & Wilkins 1972

Blood-liquor, barrier: A functional barrier morphologically represented by ↑ junctional complexes (arrows) between adjacent epithelial cells of ↑ choroid plexus. B. keeps blood solutes from reaching cerebral ventricles and assures a constant specific composition of ↑ cerebrospinal fluid (CF). Cap = capillary

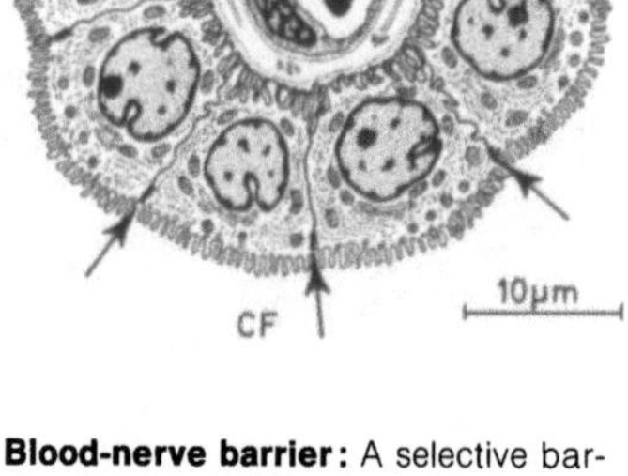

Blood-nerve barrier: A selective barrier (interrupted lines) between blood (B) and ↑ endoneurial spaces (ES) surrounding ↑ nerve fibers (NF). Structure: 1) Cell membranes, basal lamina, and ↑ zonulae occludentes of innermost cell layer of ↑ perineurial epithelium (PE); 2) basal lamina enveloping endoneurial precapillaries (PCap) and capillaries; and 3) luminal and basal cell membranes and zonulae occludentes of ↑ endothelial cells of endoneurial capillaries (ECap). B. regulates movements of small and large molecules into endoneurial spaces,

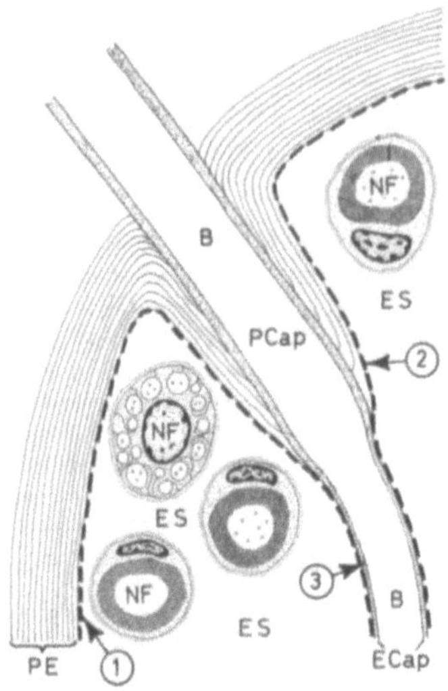

thereby assuring stability of endoneurial environment. B. inhibits entry of some substances into peripheral ↑ nerves.

Aker, D.F.: Anat. Rec. *174*, 21 (1972)

Blood, origin of erythrocytes: ↑ Erythropoiesis

Blood, origin of granulocytes: ↑ Granulopoiesis

Blood, origin of lymphocytes: ↑ Lymphopoiesis

Blood, origin of monocytes: ↑ Monopoiesis

Blood, origin of platelets: ↑ Thrombopoiesis

Blood-retinal barrier: A common term for several structures functioning

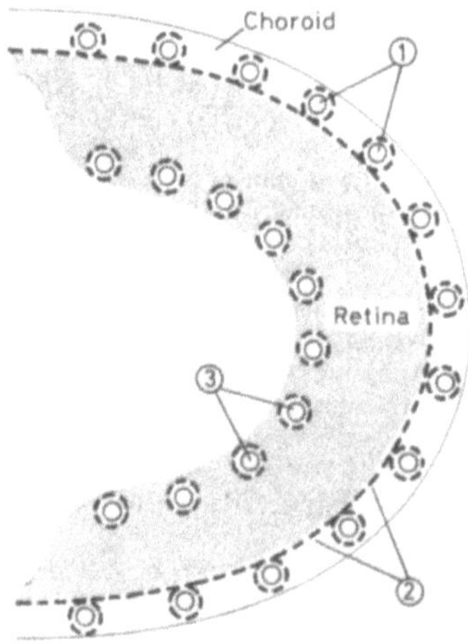

as a selective barrier between blood and ↑ retina preventing free diffusion of blood solutes into retinal tissue. Structure and elements (interrupted lines): 1) ↑ Endothelial cells of choroid vessels, especially capillaries, with their basal laminae; 2) ↑ pigment epithelium with its ↑ zonulae occludentes; and 3) endothelial cells of retinal capillaries, their basal laminae, and ↑ perivascular feet of ↑ retinal neuroglial cells. This last element (3) is structurally very similar to ↑ blood-brain barrier.

Blood sinuses: ↑ Capillaries, sinusoidal

Blood, staining of: The most frequently used techniques for B. are Giemsa staining and panoptic staining according to Pappenheim, combining Giemsa with May-Grünwald staining.

Wittekind, D.: Verh. Anat. Ges. *74*, 183 (1980)

Blood supply: Vascularization: see under organ concerned.

Blood, supravital staining of: A method for ↑ staining ↑ reticulocytes by mixing a freshly obtained blood droplet with ↑ brillant cresyl blue or 9-aminoacridin.

↑ Supravital staining

Blood-testis barrier: A selective functional barrier formed by complicated Sertoli-Sertoli (S) intercellular junctions (IJ), which delimit, at the same time, the adluminal compartment (AC) from the basal compartment (BC) of ↑ seminiferous epithelium. Whereas sub-

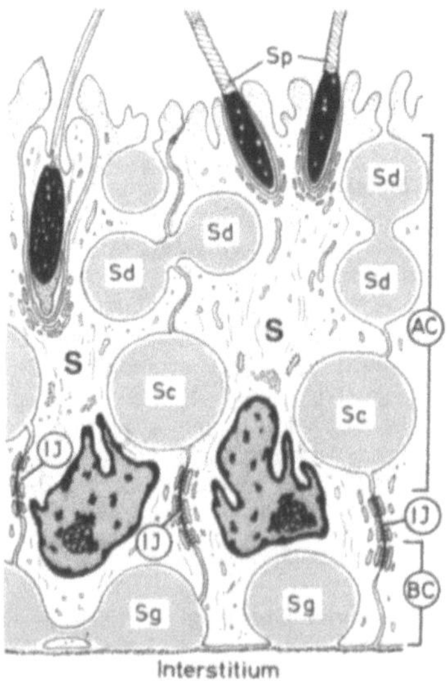

stances in extracellular space of interstitium may penetrate relatively freely into basal compartment containing ↑ spermatogonia (Sg), B. hinders further diffusion of these substances into adluminal compartment and maintains special microenvironment created by ↑ Sertoli's cells for ↑ differentiation of ↑ germ cells lodged here (↑ spermatocytes, Sc; ↑ spermatids, Sd; ↑ spermatozoa, Sp). B. permits accumulation of high concentrations of ↑ androgen-binding protein in adluminal compartment and prevents "foreign" proteins produced by germ cells from reaching bloodstream and provoking an autoimmune body reaction.

Cavicchia, J.C., Moviglia, G.A.: Anat. Rec. *205*, 387 (1983); Gravis, C.J.: Z. mikrosk.-anat. Forsch. *93*, 321 (1979); Marcaillou, C., Szöllösi, A.: J. Ultrastruct. Res. *70*, 128 (1980); Neaves, B.W.: The Blood-Testis Barrier. In: Johnson, A.D., Gomes, W.R. (eds.): The Testis. Vol. 4. New York, San Francisco, London: Academic Press 1977; Russel, L.D.: Am. J. Anat. *155*, 259 (1979)

Blood vascular system: ↑ Cardiovascular system

Blood vessels (vasa sanguinea*): A system of continuous tubes, part of ↑ cardiovascular system, distributing blood to and collecting it from tissues and organs of body.

↑ Arteries; ↑ Capillaries; ↑ Veins

Blood-thymus barrier: A functional and selective barrier separating ↑ T-lymphocytes (T) from blood of cortical capillaries (CCap) of ↑ thymus. Structure: 1) Continuous ↑ endothelial cells (E) with ↑ pericytes (P); 2) capillary ↑ basal lamina (BL); 3) ↑ pericapillary space (PS) with various fibrillar and cellular elements; 4) basal lamina (eBL) of epithelial-reticular cells; and 5) ↑ epithelial-reticular cells (EC). B. is complete in most of cortex; in juxta-

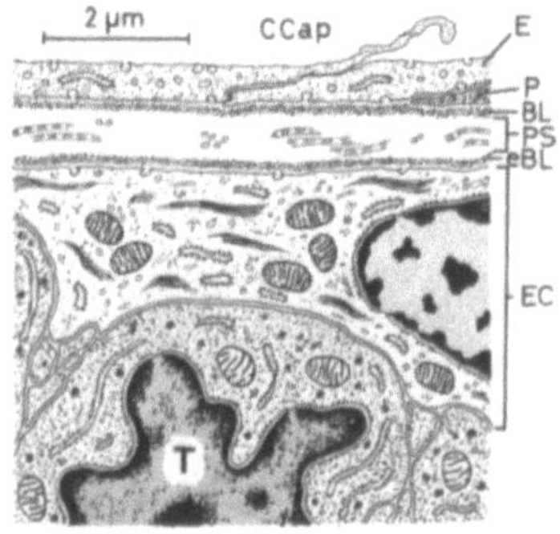

medullary cortex, around cortical venules, and in medulla, B. is not complete, allowing macromolecules to penetrate from blood into thymic ↑ parenchyma. B. prevents antigens circulating in bloodstream from reaching thymic cortex where T-lymphocytes are formed.

Janossy, G., Thomas, J.A.; Bollum, F.J. et al.: J. Immunol. *125*, 202 (1980); Porter, R., Whelan, J. (eds.): Microenvironments in Hematopoietic and Lymphoid Differentiation. Ciba Foundation Symposium 84. London: Pitman (1981); Raviola, E., Karnovsky, M.J.: J. Exp. Med. *136*, 466 (1972)

B-lymphocytes (bursa-dependent): A functional class of ↑ lymphocytes (about 65% of circulating lymphocytes) responsible for humoral immunity. Surface of B. is covered with about 150 000 molecules of ↑ immunoglobulin of IgM type, which can form complexes with some antigens. After interiorization of these complexes into cell by ↑ micropinocytosis, B. become activated under influence of ↑ T-lymphocytes (↑ helper cells) and initiate ↑ differentiation into ↑ immunoblasts. The latter divide every 6 h and generate more B., which differentiate mainly into ↑ plasma cells, secreting ↑ immunoglobulin G (IgG). The rest of B. become ↑ memory cells, which react more rapidly in contact with the antigen with which they had already been in contact. B. are produced in ↑ bone marrow and pass into ↑ bursa analogue where they mature and differentiate. From bursa analogue, they wander into ↑ lymphatic tissue. Nonactivated B. have a life span of about 12 days, the memory cells several years.

↑ Immune response; ↑ Lymphatic organs, thymus independent zones of

Bastian, F.O., Middleditch, P.R., Bossart, M.I.: J. Histochem. Cytochem. 2 *27*, 1343 (1979); Boesen, A.M., Hokland, P.: Virchows's Arch (Cell Pathol.) *41*, 107 (1982); Cho, Y., De Bruyn, P.P.H.: J. Ultrastruct. Res. *74*, 259 (1981); Kataoka, K., Minowada, J.: Arch. Histol. Jpn. *42*, 355 (1979); Roath, S., Newell, D., Polliack, A., Alexander, E., Peck-Sun, L.: Nature *273*, 15 (1978)

BMP: ↑ Bone inductor substance

Bodian silver staining: An impregnation technique with silver protein complex (Protargol).

Bodies, cavernous, of clitoris: ↑ Cavernous bodies, of clitoris

Bodies, cavernous, of penis: ↑ Cavernous bodies, of penis

Bodies, Malpighian, of spleen: ↑ Splenic nodules

Bodies, multilamellar, of alveolar cells type II: ↑ Multilamellar bodies

Bodies, multivesicular: ↑ Multivesicular bodies

Bodies, myeloid: ↑ Myeloid bodies

Bodies, ν: ↑ Nucleosomes

Bodies, of Call-Exner: ↑ Call-Exner bodies

Bodies, of Glügge: ↑ Compound granular corpuscles

Bodies, of Hassall: ↑ Hassall's bodies

Bodies, of Hassall-Henle: ↑ Hassall-Henle bodies

Bodies, of Hensen: ↑ Hensen's bodies

Bodies, of Herring: ↑ Herring bodies

Bodies, of Langerhans: ↑ Langerhans' bodies

Bodies, of Nissl: ↑ Nissl bodies

Bodies, of Russell: ↑ Russell's fuchsinophilic bodies

Bodies, polar: ↑ Polar bodies

Bodies, residual: ↑ Residual bodies

Bodies, residual, of Regnaud: ↑ Residual bodies, of Regnaud

Bodies, wormlike: ↑ Wormlike bodies

Body, aortic: ↑ Aortic body

Body, basal: ↑ Cilia, basal body of

Body, carotid: ↑ Carotid body

Body, chromatoid: ↑ Chromatoid body

Body, intermediate: ↑ Mid-body

Body, of Barr: ↑ Barr body

Body, of Flemming: ↑ Mid-body

Body, pineal: ↑ Pineal organ

Body, spongious: ↑ Spongious body, of male urethra

Body, vitreous: ↑ Vitreous body

Body, ultimobranchial: ↑ Ultimobranchial body

Bombesin: ↑ Neurotransmitter substances

Bone (textus osseus*): A specialized ↑ supporting tissue composed of specific ↑ bone cells and intercellular substance. Intercellular substance consists of organic ↑ bone matrix and ↑ bone mineral.

Bourne, G.H. (ed.): The Biochemistry and Physiology of Bone. 2nd edn. Vols. 1–4. New York: Academic Press 1972; Holtrop, M.E.: Lab. Sci. *5*, 264 (1975); Knese K.-H.: Stützgewebe und Skelettsystem. In: Möllendorf, W. v., Bargmann, W. (eds.): Handbuch der mikroskopischen Anatomie des Menschen. Vol. 2. Die Gewebe. Part 5. Berlin, Heidelberg, New York: Springer-Verlag 1979; Parfitt, A.M.: The Physiologic and Clinical Significance of Bone Histomorphometric Data. In: Recker, R.R (ed.): Bone Histomorphometry Techniques and Interpretation. Boca Raton: CRC Press 1982; Vaughan, J.M.: The Physiology of Bone. Oxford: Clarendon Press 1981

Bone, absorption cavity of: ↑ Absorption cavity

Bone, alveolar: ↑ Alveolar bone

Bone, appositional growth of: A progressive deposition by ↑ osteoblasts (Ob) of additional ↑ osteoid layers (OL) which will be calcified onto bone (B) surface during bone formation. B. is generally accompanied by ↑ osteoclastic (Oc) destruction of innermost layers. In this way, skull increases its volume, and long bones their diameter.

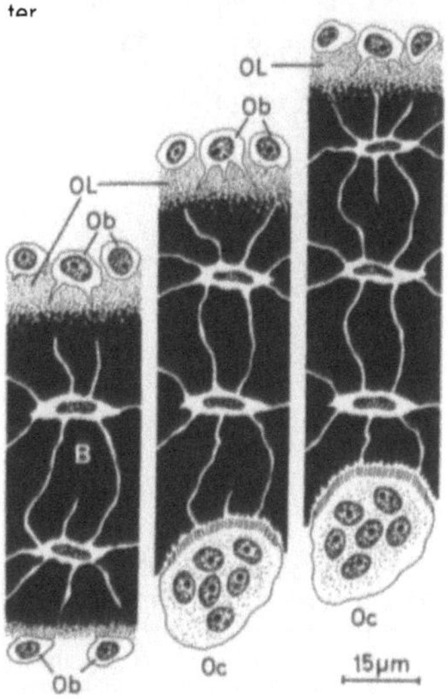

Bone, calcification of: Principally two parallel, very complex, and not fully understood processes. Working hypotheses: 1) Heterogeneous nucleation. Formation of ↑ hydroxyapatite crystals (HA) is initiated from metastable solution of calcium salts by ↑ collagen microfibrils (CM) with aid of ↑ chondroitin sulfates and other ↑ proteoglycans of ground substance (GS). First crystallization nuclei (CN) are lodged within "holes" between tails (T) and heads (H) of successive ↑ tropocollagen molecules (Tr); by deposition of new material, hydroxyapatite crystals grow and spread out into ground substance. 2) Matrix vesicles. Calcium (Ca) and phosphate (P) from extracellular fluid (EF) enter ↑ osteoblast (Ob) and unite within ↑ inner chamber of ↑ mitochondria to form calcium phosphate (CaP). Calcium phosphate leaves ↑ organelle, moves toward peripheral cytoplasm, and accumulates in an unknown manner within ↑ matrix vesicles (MV). Here amorphous calcium phosphate initiates formation of crystalline needles of hydroxyapatite (HA), which grow by deposition of new crystals, perforate limiting membrane of vesicle, and finally spread out into ground substance. This process is aided by alkaline phosphatase contained within matrix vesicles and that secreted by osteoblasts.

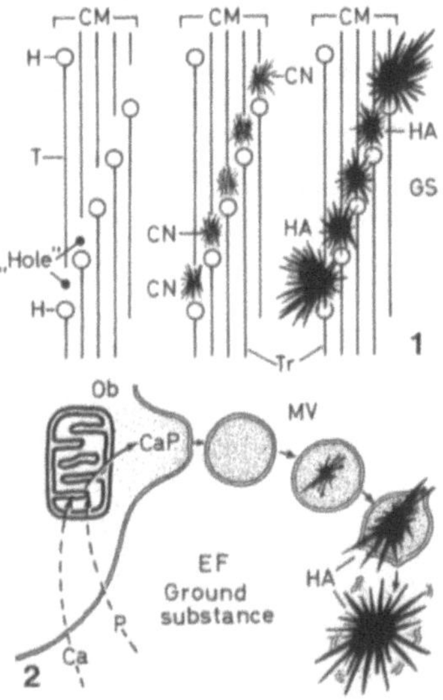

Anderson, H.C., Reynolds, J.: Dev. Biol. *34*, 211 (1973); Slavkin, H.C., Croissant, R.D., Bringas, P., Matosian, P., Wilson, P., Mino, W., Guenther, H.: Fed. Proc. *35*, 127 (1976); Urist, M.R.: Biochemistry of Calcification. In: Bourne, G.H. (ed.): The Biochemistry and Physiology of Bone, 2nd edn. Vol. 4. New York, London: Academic Press 1976

Bone canaliculi: ↑ Canaliculi, of bone

Bone, cancellous (spongy bone substantia spongiosa, trabecular bone, os spongiosum*): A three-dimensional meshwork (CB) of branching bone ↑ trabeculae (T) delimiting a honeycomb system of intercommunicating spaces occupied by ↑ bone marrow (BM). B. forms ↑ epiphyses of long bones and interior of short bones. Compact (C) and B. are continuous with each other. B. is a kind of lamellar ↑ bone devoid of ↑ osteons. Its ↑ osteocytes are supplied with metabolites by diffusion from bone marrow. All cavities in B. are lined with ↑ endosteum.

↑ Diploë

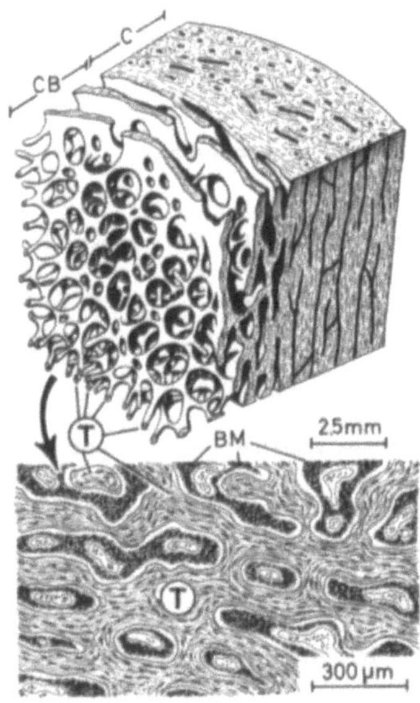

Singh, I.: J. Anat. *127*, 305 (1978); Whitehouse, W.J.: J. Pathol. *116*, 213 (1975)

Bone cells: ↑ Osteoprogenitor cells; ↑ Osteoblasts; ↑ Osteocytes, ↑ Osteoclasts

Bone, cement lines of: ↑ Cement lines

Bone, coarsely bundled: ↑ Bone, immature

Bone collagen: One organic component of ↑ bone matrix consisting of cross-striated microfibrils 50–70 nm in diameter with a periodicity of 64–68 nm. B. does not swell in dilute acids and possesses more intermolecular bonds than ordinary ↑ collagen fibrils. The holes between heads and tails of ↑ tropocollagen molecules can start calcification of ↑ osteoid, since first ↑ hydroxyapatite crystals occur at this level. Fibrils in mature lamellar bone are very regularly arranged: They lie parallel within each ↑ lamella and run helicoidally around ↑ Haversian canal; fibrils of two adjoining lamellae cross under various angles producing a pattern of alternating dark and light rings in polarized light. B. makes up about 25%–30% of dry bone mass. B. belongs to type I ↑ collagen.

↑ Bone, calcification of; ↑ Bone, compact

Hohling, H.J., Kreilos, R., Neubauer, G., Boyde, A.: Z. Zellforsch. *122*, 36 (1971); Katz, E.P., Li, S.-T.: J. Mol. Biol. *80*, 1 (1973); Knese, K.-H.: Stützgewebe und Skelettsystem. In: Möllendorff, W. v., Bargmann, W. (eds.): Handbuch der mikroskopischen Anatomie des Menschen. Vol. 2. Die Gewebe. Part 5. Berlin, Heidelberg, New York: Springer-Verlag 1979

Bone, compact (cortical bone, dense bone, substantia compacta, os compactum*): A solid continuous mass (C) of densely packed bony ↑ lamellae (Lm) arranged in following patterns: 1) Outer circumferential lamellae (OCL) or periosteal lamellae = several lamellae extending around entire periphery of shaft immediately beneath ↑ periosteum (P). 2) ↑ Osteons (O) = concentrically arranged lamellae around ↑ Haversian canals (HC). 3) ↑ Interstitial system of lamellae (IS) =

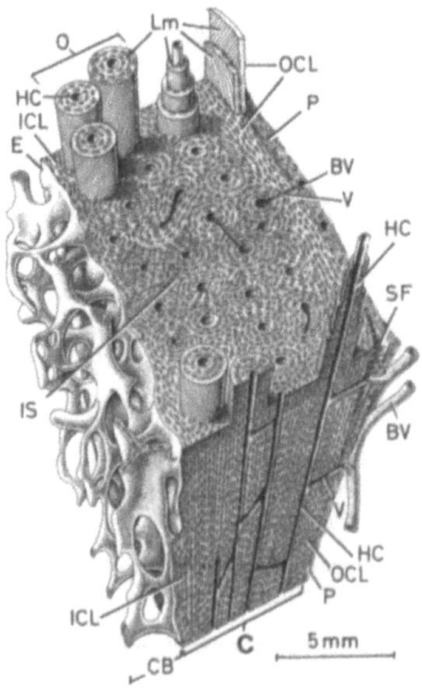

lamellae situated between osteons (2+3=Haversian bone). 4) Inner circumferential (ICL) or endosteal lamellae = several inconstant lamellae extending around medullary cavity immediately beneath ↑ endosteum (E). ↑ Collagen fibrils in adjoining lamellae cross at various angles. B. forms shafts of long bones and surrounds all other bones as a layer of varying thickness. Outer surface of B. is enveloped with periosteum, which is anchored to both outer circumferential lamellae and to interstitial system of lamellae by ↑ Sharpey's fibers (SF). Between cancellous ↑ bone (CB) and B. there is a gradual transition. V = ↑ Volkmann's canal, BV = blood vessels

↑ Bone, flat, of skull

Knese, K.-H.: Stützgewebe und Skelettsystem. In: Möllendorff, W. v., Bargmann, W. (eds.): Handbuch der mikroskopischen Anatomie des Menschen. Vol. 2. Die Gewebe. Part 5. Berlin, Heidelberg, New York: Springer-Verlag 1979

Bone, composition of: Variable during life. B. in adult: Water 20%, collagen 25%–30% of dry mass, highly polymerized sulfated ↑ proteoglycans and citrate 1.25%, bone mineral 65%–70%. Water content decreases with age.

Bone, dense: ↑ Bone, compact

Bone, effects of hormones on: ↑ Calcitonin increases rate of new ↑ bone formation by stimulating ↑ osteoblasts and deposition of calcium salts within ↑ bone matrix; calcitonin can inhibit ↑ osteoclasts. ↑ Growth hormone controls bone growth by stimulating interstitial growth and metabolism within ↑ epiphyseal plates: Lack of growth hormone results in dwarfism, excess before puberty in ↑ gigantism, and after puberty in ↑ acromegaly. ↑ Parathormone controls bone resorption by acting on ↑ osteocytic osteolysis and on osteoclasts, it also provokes appearance of new osteoclasts; excess leads to large bone resorption (= osteitis fibrosa). ↑ Sex hormones accelerate closure of ↑ epiphyseal plates; prepubertal castration provokes eunuchoid gigantism. ↑ Thyroxin influences maturation and metabolism of bone. (See physiology texts for further information)

Baud, C.A.: Schweiz. Med. Wochenschr. *28*, 717 (1968); Parfitt, A.M.: Metabolism *25*, 809, 909, 1033, 1157 (1976)

Bone, effects of vitamins on: Vitamin A deficiency disturbs indirect ↑ bone formation. Vitamin C deficiency provokes disordered synthesis of collagen and ↑ ground substance. Vitamin D deficiency is responsible for deranged maturation of collagen fibrils and irregular calcium salt deposition in ↑ bone matrix (= rickets). (See physiology texts for further information)

Rassmussen, H., Bordier, P.: Metab. Bone Dis. Relat Res. *1*, 7 (1978)

Bone, epiphyseal plate of: ↑ Epiphyseal plate

Bone, flat, of skull: A lamellar bone composed of a middle spongy layer of cancellous ↑ bone (diploë, D) situated between an outer (OT) and an inner table (IT) or plate of compact ↑ bone. ↑ Periosteum covering outer surface of B. is termed pericranium (PC); inner surface is lined by ↑ dura mater (DM). Almost all B. are membranous ↑ bones.

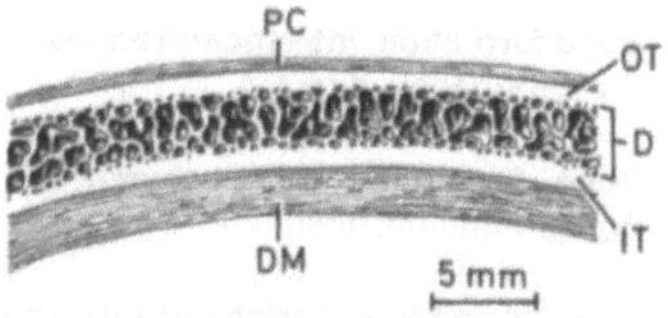

Bone formation (ossification, osteogenesis*): ↑ Histogenesis of bones involving direct and indirect mechanisms.

↑ Bone formation, direct; ↑ Bone formation, indirect

Bone formation, direct (intramembranous bone formation, mesenchymal bone formation, osteogenesis membranacea*): A process of direct transformation of ↑ mesenchymal tissue into bone. Sequence: 1) Condensation of richly vascularized mesenchymal tissue and ↑ differentiation of ↑ mesenchymal cells (MC) into ↑ osteoblasts (Ob), which produce first irregularly oriented ↑ collagen fibrils (F). 2) Osteoblasts form clumps, produce ↑ osteoid (O), and surround themselves with this material. 3) Each osteoblast becomes an ↑ osteocyte (Oc), sequestered in its own ↑ lacuna; osteoid calcifies and first bony islands, ↑ spicules (S), appear. Fusion of spicules produces ↑ trabeculae of immature ↑ bone which will be replaced in course of secondary ↑ bone formation by mature ↑ bone. Bones developing according to this mechanism – mandibula, skull vault, flat bones of face, clavicle, terminal phalanges, and ↑ periosteal bony band around shaft of long bones – are collectively referred to as membranous ↑ bones.

↑ Bone formation, perichondral

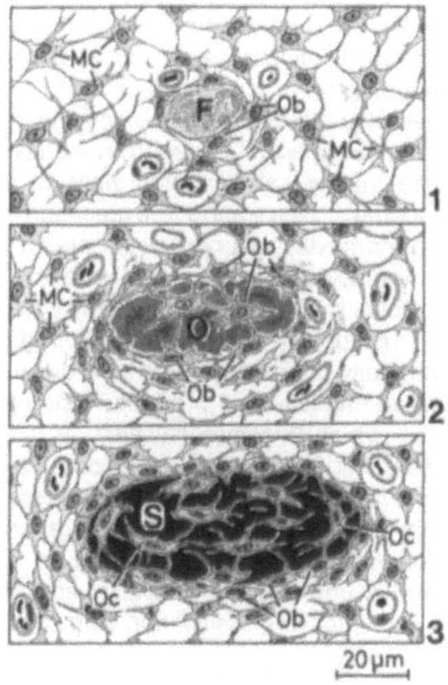

Bernard, G.W., Pease, D.C.: Am. J. Anat. *125*, 271 (1969); Canalis, E.: Endocr. Rev. *4*, 62 (1983)

Bone formation, ectopic: A spontaneous appearance of ↑ bone from ↑ connective tissue in organs without direct structural or functional relation to skeleton (↑ arteries, ↑ eye, ↑ kidney, ↑ muscles, ↑ tendons, etc.).

Bone formation, endochondral: ↑ Bone formation, indirect

Bone formation, indirect (chondral ossification, osteogenesis cartilaginea*): A process of ↑ histogenesis of majority of bones, starting with formation of a ↑ periosteal bony band around ↑ diaphysis of a hyaline cartilaginous piece, according to mechanism of direct ↑ bone formation, and continuing with endochondral ossification, during which hyaline ↑ cartilage is destroyed and replaced by ↑ bone. Sequence: 1) A periosteal bony band (1, 2, 3, 4; B) appears around diaphysis (D). 2) Inside diaphysis, cartilage cells hypertrophy and matrix calcifies. 3) From ↑ periosteum (P), a vascular ↑ periosteal bud (PB) invades interior of diaphysis; in this way, primary or diaphyseal ossification center and primitive marrow cavity (PC) are formed. 4) Ossification zone progresses from diaphysis toward ↑ epiphyses (E) as two fronts in which may be seen following zones: 5a) Reserve zone = hyaline ↑ cartilage with quiescent ↑ chondrocytes (Ch); 5 b) multiplication, proliferation, or maturing zone = chondrocytes divide (arrow) and arrange themselves in parallel columns of flattened cells (isogenic groups); 5 c) hypertrophic zone = ↑ hypertrophy of chondrocytes; 5 d) calcification zone = almost entire matrix around chondrocytes calcifies, cells degenerate; 5 e) ↑ erosion zone = ↑ chondroclasts (Cc) destroy noncalcified transversally oriented walls of ↑ lacunae separating hypertrophic chondrocytes; chondroclasts are followed by capillary loops (CL) and ↑ osteoblasts (Ob); 5 f) ossification zone = osteoblasts congregate on longitudinally oriented remants (CR) of calcified cartilage and produce ↑ osteoid (O), which becomes calcified. In this way, first ↑ spicules (S) of endochondral bone appear. Secondary or epiphyseal centers (EC) of ossification appear, as described in points 2, 3, and 4, and enlarge radially. At end of each epiphysis, ossification stops and

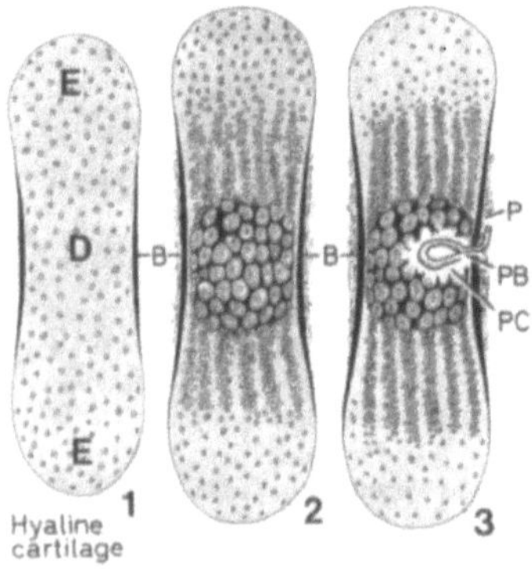

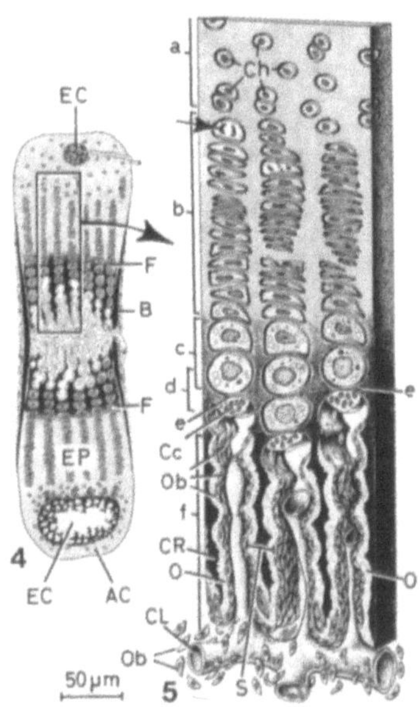

persisting cartilage becomes ↑ articular cartilage (AC). On the opposite side, diaphyseal front of ossification progresses toward epiphyseal front; they are separated by transversal cartilaginous ↑ epiphyseal plate (EP). B. is mechanism permitting long bones to grow in length. B. results in formation of immature ↑ bone, which will be replaced in course of secondary ↑ bone formation by mature ↑ bone.

Bone formation, perichondral: ↑ Cartilage canals

Bagnall, K.M., Harris P.F., Jones, P.R.M.: Anat. Rec. *203*, 293 (1982); Knese, K.-H.:Stützgewebe und Skelettsystem. In: Mollendorff, W.v., Bargmann, W. (eds.): Handbuch der mikroskopischen Anatomie des Menschen. Vol. 2. Die Gewebe. Part 5. Berlin, Heidelberg, New York: Springer-Verlag 1979

Bone formation, intracartilaginous: ↑ Bone formation, indirect

Bone formation, intramembranous: ↑ Bone formation, direct

Bone formation, mesenchymal: ↑ Bone formation, direct

Bone formation, perichondrial: The direct ↑ bone formation around diaphysis of cartilage model by activated ↑ osteoprogenitor cells of ↑ perichondrium. After having initiated bone formation, perichondrium becomes the ↑ periosteum, which continues to produce, by direct ossification, ↑ periosteal bony band. P. represents initial phase of indirect ↑ bone formation.

Bone formation, secondary (Haversian ossification, secondary ossification): A process of destruction of immature ↑ bone (IB) containing some primary ↑ osteons (PO) and its replacement by bone composed of secondary ↑ osteons (SO). Sequence: 1) ↑ Osteoclasts (Oc) erode immature bone, forming ↑ absorption cavities (AC) into which penetrate blood vessels (BV) and ↑ osteoprogenitor cells, which differentiate into ↑ osteoblasts (Ob). At a certain moment, enlargement of an absorption cavity ceases and osteoblasts synthesize concentric ↑ lamellae of a secondary osteon (SO). 2) Simultaneously, new absorption cavities appear here and there, followed by formation of a third generation of osteons (TO). 3) This is then succeeded by fourth, fifth, and higher orders of osteons (NO) until all immature bone is replaced by mature ↑ bone. Remnants of earlier gener-

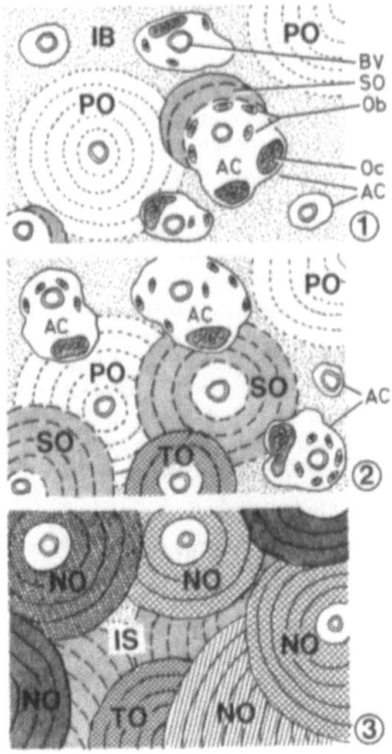

ations of osteons, now situated among new generations of osteons, form ↑ interstitial lamellae (IS).

↑ Bone, internal remodeling of; ↑ Bone repair

Bone, functions of: 1) Formation of skeleton. 2) Attachment points for skeletal muscles. 3) Protection by formation of bony cages (skull, chest, pelvis). 4) Hematopoietic organ by harboring and protecting ↑ bone marrow. 5) Mineral homeostasis of body. (See physiology texts for further information)

Bone, ground substance of (substantia fundamentalis ossea*): An amorphous, slightly ↑ PAS-positive, and weakly metachromatic extracellular substance forming about 5% of organic ↑ bone matrix. Composition: sulfated ↑ proteoglycans containing ↑ chondroitin sulfate, ↑ keratan sulfate, and ↑ hyaluronic acid. B. is mainly produced by ↑ osteoblasts.

↑ Metachromasia

Bourne, G.H. (ed.): The Biochemistry and Physiology of Bone. 2nd edn. Vols 1–3. New York: Academic Press 1971/72; Frasca, P., R.A. Katz, J.L.: Scan. Electron Microsc. 1981/III, 339 (1981)

Bone growth: ↑ Bone, appositional growth of; ↑ Bone formation, indirect; ↑ Bone, long, growth in diameter of; ↑ Epiphyseal plate

Bone, Haversian system, primitive: ↑ Osteon, primary

Bone, Haversian system, secondary: ↑ Osteon, secondary

Bone healing, primary: ↑ Bone repair

Bone healing, secondary: ↑ Bone repair

Bone, histophysiology of: ↑ Bone, effects of hormones on; ↑ Bone, effects of vitamins on

Bone immature (coarsely bundled bone, nonlamellar bone, primary bone, reticulated bone, woven bone, os membranaceum primarium*): The early result of direct and indirect ↑ bone formation (IB). ↑ Collagen fibrils of B. run in various directions; spaces between ↑ trabeculae are relatively large and tortuous; they contain blood vessels and connective tissue. ↑ Osteocytes (Oc) are uniformly distributed, but without any special orientation; in rare cases they have a disposition that resembles a primary ↑ osteon. B. will be replaced in course of secondary ↑ bone formation by mature ↑ bone (MB). In adult organism, only some small parts of bony ↑ labyrinth, external ↑ auditory meatus, and insertion areas of large ↑ tendons are made of B.

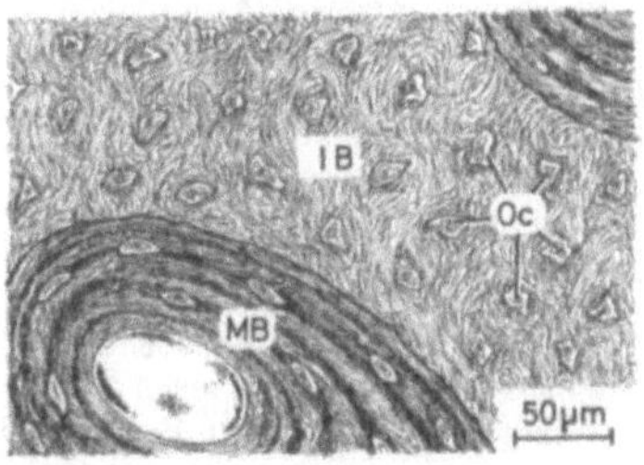

Bone inductor substance: A protein capable of inducing and/or activating ↑ bone formation. B. is present as ↑ bone morphogenetic protein (BMP) in calcified ↑ cartilage, ↑ dentin, and ↑ bone.

Ostrowski, K., Wlodarski, K.: Induction of Heterotopic Bone Formation. In: Bourne, G.H. (ed.): The Biochemistry and Physiology of Bone. Vol. 3. New York and London: Academic Press 1971; Urist, M.R.: Biologic Initiators of Calcification. In: Zippin, I. (ed.): Biologic Meneralization. New York: John Wiley & Sons, Inc. 1973; Urist, M.R.: Mikulski, A., Boyd, S.D.: Arch. Surg. *110*, 416 (1975)

Bone, innervation of: From ↑ periosteum, fine ↑ nonmyelinated nerve fibers penetrate into ↑ perivascular spaces of ↑ Haversian canals.

Thurston, T.J.: J. Anat. *134*, 719 (1982)

Bone, internal remodeling of (bone turnover): A process of destruction of former generations of ↑ osteons and their replacement by a new generation according to mechanism of secondary ↑ bone formation. B. is an active process starting after secondary bone formation and continues throughout life as a result of equilibrated action of ↑ osteoclasts and ↑ osteoblasts. In young children, the remodeling rate is about 200%, in adults only 1%.

Kimmel, D.B., Jee, W.S.S.: Anat. Rec. *203*, 31 (1982); Tran Van, P., Vignery, A., Baron, R.: Anat. Rec. *202*, 445 (1982)

Bone, interstitial substance of: A mass composed of ↑ bone matrix and inorganic salts surrounding ↑ osteocytes

Bone, interstitial system of lamellae: ↑ Interstitial lamellae, of bone

Bone, intramembranous: ↑ Bone, membranous

Bone, lamellae of: ↑ Lamellae, of bone

Bone, lamellar: ↑ Bone, mature

Bone, long, growth in diameter of: A progressive deposition of additional bony layers (ABL) on outer surface of ↑ periosteal bony band (PB) around diaphysis of long bones. Simultaneous with this appositional ↑ bone growth, ↑ osteoclasts erode interior of periosteal bony band (arrowheads) and enlarge medullary cavity (MC).

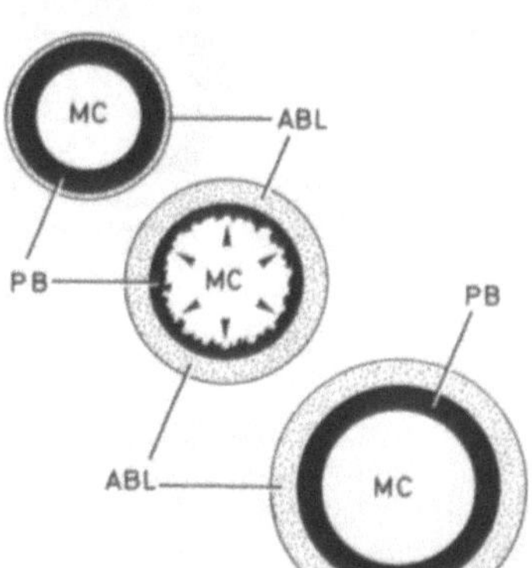

Bone, long, growth in length of: ↑ Bone formation, indirect; ↑ Epiphyseal plate

Bone, lymph circulation in: ↑ Lymphatic capillaries are present in about 3% of ↑ haversian canals. In general, an insufficiently studied subject.

Bone marrow (myeloid tissue, medulla ossium*): A richly vascularized ↑ connective tissue specialized in production of all blood elements. As the principal ↑ hematopoietic tissue, B. is located within medullary cavities of long bones and intertrabecular spaces of flat bones. B. is separated from bone by ↑ endosteum. Structure: 1) ↑ Stroma (S) = a network of ↑ reticular fibers (RF) and ↑ reticular cells (RC) with phagocytotic properties. 2) Vascular compartment (VC) = incompletely lined vascular sinuses (VS) accompanied by ↑ adventitial cells, ↑ fibrocytes, ↑ mast cells, and ↑ plasma cells. 3) Hematopoietic compartment, extravascular tissue, or ↑ parenchyma (HC) = cords of ↑ hematopoietic stem cells, developing and mature blood cells, ↑ megakaryocytes (M), ↑ macrophages, potential osteoblasts, rarely ↑ osteoclasts, and ↑ adipose cells (AC). In the adult, B. makes up about 3.5%–6% of total body weight. Morphologically and functionally, one can distinguish red and yellow ↑ B. B. has no lymphatics; its innervation is vasomotor.

↑ Bone marrow, vascular sinuses of; ↑ Granulopoiesis; ↑ Erythropoiesis; ↑ Lymphopoiesis; ↑ Monopoiesis; ↑ Thrombopoiesis

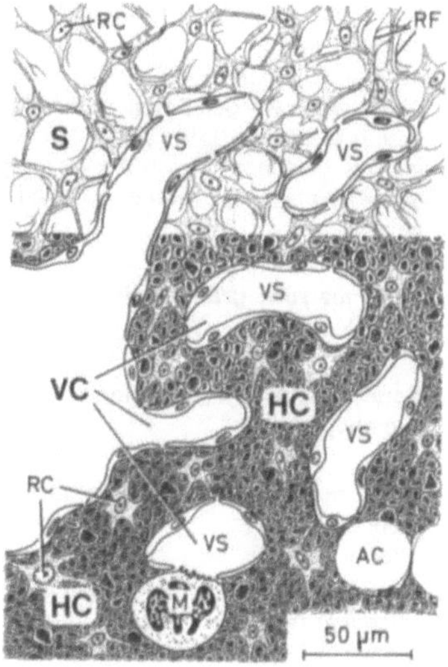

Biermann, A., Graf von Keyserlingk, D.: Acta Anat. *100*, 34 (1978); Riedler, G.F.,

Zingg, R.: Tabulae haemetologicae. Basel: Rocom, Hoffmann-La Roche 1977; Weiss, L., Chen, L.T.: Blood Cell *1*, 617 (1975); Weiss, L.: Anat. Rec. *186*, 161 (1976); Wickramasinghe, S.N.: Human Bone Marrow. Oxford: Blackwell Scientific Publications 1975

Bone marrow, differential count of: The determination of percentages of various nucleated cells in stained bone marrow smears. Normal values in adult are:

	Mean value%	Range%
Erythropoiesis	25.6	18.4–33.8
Proerythroblasts	0.6	0.2–1.3
Erythroblasts:		
basophilic	1.4	0.5–2.4
orthochromatic	2	0.4–4.6
polychromatophilic	21.6	17.9–29.2
Granulopoiesis		
Myeloblasts	0.9	0.2–1.5
Promyelocytes	3.3	2.1–4.1
Myelocytes:		
basophilic	0.3	0–0.5
eosinophilic	1.5	0.5–3
neutrophilic	12	5–19
Metamyelocytes	15.9	9.6–24.6
Granulocytes:		
basophilic	0.2	0–0.7
eosinophilic	2	0.5–4
neutrophilic	20	7–30
Thrombopoiesis		
Megakaryocytes	0.1	0–0.4
Lymphocytes	16.2	11.1–23.2
Monocytes	0.3	0–0.8
Plasma cells	1.3	0.4–3.9
Reticular cells	0.3	0–0.9
Myeloid / erythoid ratio		3:1–4:1

(After Wintrobe 1968, 1975)

Wintrobe, M.M.: Clinical Hematology, 6th and 7th edns Philadelphia: Lea and Febiger 1968 and 1975

Bone marrow, erythropoiesis in: ↑ Erythropoiesis

Bone marrow, fatty: ↑ Bone marrow, yellow

Bone marrow, gelatinous (medulla ossium gelatinosa*): The yellow ↑ bone marrow; gelatinous consistency is due to swelling by water uptake of ↑ reticular cells.

Bone marrow, granulopoiesis in: ↑ Granulopoiesis

Bone marrow, lymphopoiesis in: ↑ Lymphopoiesis

Bone marrow, monopoiesis in: ↑ Monopoiesis

Bone marrow, primary (primitive bone marrow): The first bone marrow formed by ↑ differentiation from vascular mesenchyme, which has invaded primary ossification center in course of indirect ↑ bone formation. B. is a richly vascularized tissue composed of primitive hematopoietic elements, ↑ osteoblasts, ↑ osteoclasts, and ↑ chondroclasts. In course of further development, B. gradually becomes red ↑ bone marrow

Bone marrow, red (medulla ossium rubra*): A hematopoietically active ↑ bone marrow; color is derived from large amount of ↑ hemoglobin in developing and mature ↑ erythrocytes. B. is contained in all bones at birth; in the course of ↑ ageing, it is gradually transformed into yellow ↑ bone marrow by accumulation of adipose cells. In the adult, B. persists in vertebral bodies, ribs, sternum, and pelvis.

Bone marrow, thrombopoiesis in: ↑ Thrombopoiesis

Bone marrow, vascular sinuses of (vas sinusoideum*): About 5 to 25-µm-wide, triple-layered vascular spaces. Structure: 1) Flattened, nonfenestrated, ↑ littoral cells (Li) connected by ↑ zonulae adherentes; 2) basal lamina (not visible); 3) ↑ adventitial cells (AC) covering most of outer surface of sinuses. Littoral cells may separate from one another or be perforated (arrowheads) to permit passage of newly formed blood elements (asterisk) into circulation; simultaneously, basal lamina and adventitial cells are displaced, probably under influence of ↑ erythropoietin. (Fig. = rat)

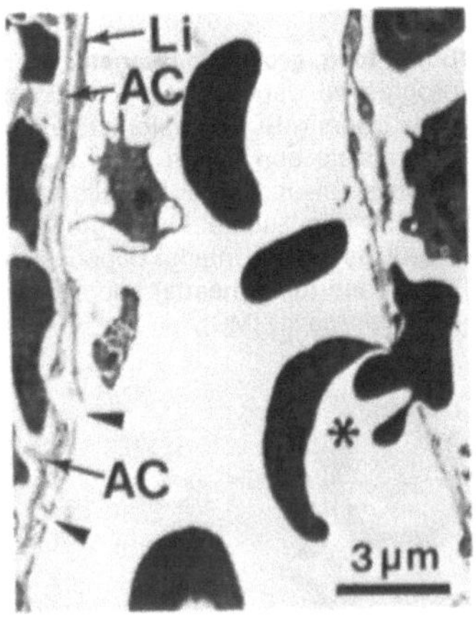

Cambell, F.: Am. J. Anat. *135*, 521 (1972); Weiss, L.: Blood *36*, 189 (1970)

Bone marrow, vascularization, of: From nutrient artery of long bone ↑ central longitudinal arteries arise which give off predominantly radial ↑ arterioles emptying into vascular sinuses. From here, blood reaches ↑ venules, then ↑ central longitudinal vein, and, finally, nutrient vein.

↑ Bone marrow, vascular sinuses of

Draenert, K., Draenert, Y.: Scan. Electron Microsc. 1980/4, 113 (1980)

Bone marrow, yellow (fatty bone marrow, medulla ossium flava*): ↑ Bone marrow with a large number of ↑ adipose cells (AC) giving it a yellowish color. B. has, in general, little hematopoietic activity but may transform itself, according, to functional demand, into red ↑ bone marrow.

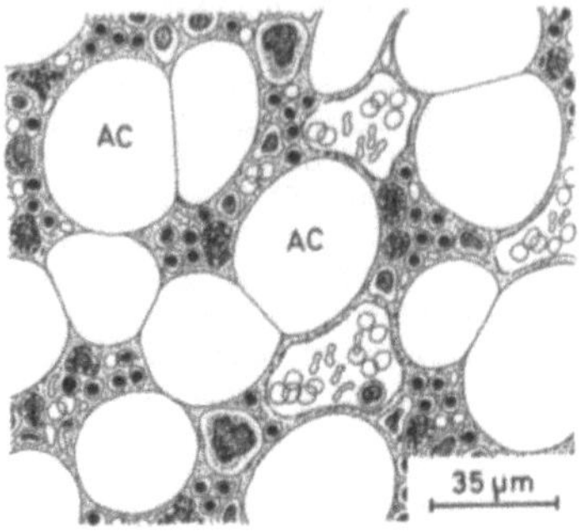

Bone matrix: Extracellular organic components of ↑ bone (↑ collagen fibrils and ↑ ground substance). A term often used to denote the complex of organic and inorganic material (↑ hydroxyapatite crystals) surrounding ↑ osteocytes.

Bone, mature (lamellar bone): A kind of ↑ bone composed of parallel and concentrically oriented ↑ collagen fibrils disposed in layers (↑ lamellae), separated by regularly arranged ↑ osteocytes. B. is formed after destruction of immature ↑ bone as a result of secondary ↑ bone formation. B. is statically and dynamically perfectly adapted to physical forces acting on skeleton. Cancellous ↑ bone and compact ↑ bone are B.

Boyde, A.: Scanning Electron Microscope Studies of Bone. In: G.H. Bourne (ed.): The Biochemistry and Physiology of Bone, Vol. 1. New York: Academic Press 1972; Holtrop, M.E.: Ann. Clin. Leb. Sci. *5*, 264 (1975); Born, V., Dvorak, M.Y.: Z. mikrosk.-anat. Forsch. *88*, 836 (1974)

Bone, membranous (bone mesenchymal, os membranaceum*): ↑ Bone formed according to mechanism of direct ↑ bone formation.

Bone, mesenchymal: ↑ Bone, membranous

Bone, metabolic: The functional category of ↑ bone composed of young, incompletely calcified ↑ osteons, which represent a rapidly mobilizable store of calcium.

↑ Alveoli, of teeth; ↑ Bone, alveolar; ↑ Bone, structural

Bone mineral: Mainly calcium phosphate in form of ↑ hydroxyapatite, calcium bicarbonate, calcium fluoride, calcium citrate, and magnesium chloride. Certain other elements, such as Pb or Ra, as well as ↑ bone-seeking isotopes, may also compose B. if ingested.

Bone morphogenetic protein (BMP): ↑ Bone inductor substance

Bone, nonlamellar: ↑ Bone, immature

Bone, nutritional effects on: ↑ Bone, effects of hormones on; ↑ Bone, effects of vitamins on

Bone, primary: ↑ Bone, immature

Bone, primary osteon of: ↑ Osteon, primary

Bone regeneration (bone repair): The process of bone reconstruction and healing after fracturation. A) Primary healing (under excellent immobilization = ↑ osteosynthesis). 1) Immediate death of ↑ bone on both sides of fracture line. 2) ↑ Osteoprogenitor cells of ↑ Haversian canals belonging to nearest living zone proliferate and differentiate into ↑ osteoclasts, which resorb dead bone. Other osteoprogenitor cells transform themselves into ↑ osteoblasts, which reconstruct new ↑ osteons in a manner similar to secondary ↑ bone formation. There is no ↑ callus. B) Secondary healing = callus formation (immobilization not optimal). 1) After fracture, superficial periosteal osteoprogenitor cell differentiate under relatively avascular conditions into ↑ chondroblasts and ↑ chondrocytes, thereby producing a fibrocartilaginous callus (FC) bridging gap between fragments. 2) Osteoprogenitor cells of inner periosteal layer and of ↑ endosteum proliferate and differentiate under conditions of good vascularization into osteoblasts. 3) Osteoblasts form new bony ↑ trabeculae (Tr) around fibrocartilaginous callus on periosteal side (= external callus, EC) and on endosteal side (= internal callus, IC). 4) Fibrocartilaginous callus is gradually eroded by osteoclasts and entirely replaced by cancellous ↑ bone (CB) as in indirect ↑ bone formation (= bony callus). 5) Cancellous bone is then remodeled as in internal ↑ bone remodeling and finally replaced by compact ↑ bone.

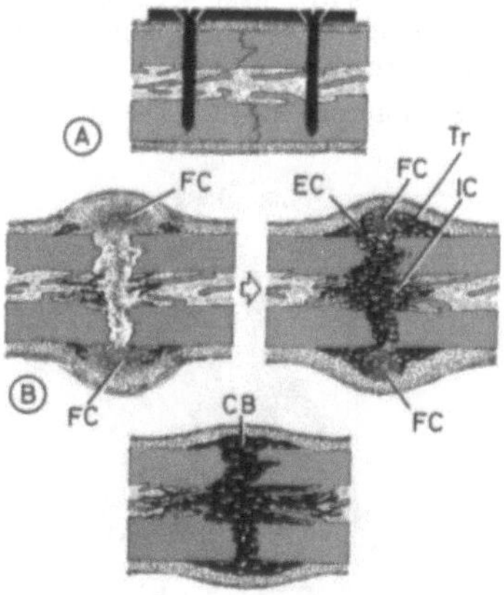

Ham, A.W., Harris, W.R.: Repair and Transplantation of Bone. In: Bourne, G.H. (ed.): The Biochemistry and Physiology of Bone. 2nd edn, Vol. 3. New York: Academic Press 1975

Bone remodeling: ↑ Bone formation, secondary

Bone repair: ↑ Bone regeneration

Bone, reticulated: ↑ Bone, immature

Bone, secondary osteon of: ↑ Osteon, secondary

Bone-seeking isotopes: Very dangerous radioactive isotopes (^{90}Sr, ^{259}Pu, and others) which occur after fission of uranium and become part of ↑ bone mineral, causing severe damage ↑ bone marrow.

Bone, sesamoid: A bone formed within a ↑ tendon (patella, os pisiforme, etc.).

Bone spicules: ↑ Spicules, of bone

Bone, spongy: ↑ Bone, cancellous

Bone, structural: Functional category of mature ↑ bone composed of old, heavily calcified ↑ osteons with a primarily mechanical function; less available store of mobilizable calcium than metabolic ↑ bone.

Bone tissue: ↑ Bone

Bone trabeculae: ↑ Trabeculae, of bone

Bone, trabecular: ↑ Bone, cancellous

Bone transplantation: Only possible when ↑ periosteum (P) is present. 1) Autografts (A) survive and even proliferate, inducing production of new bone ↑ trabeculae (Tr); autografts are finally resorbed. 2) Homografts stimulate regenerative potentiality of ↑ bone but are ultimately rejected in course of ↑ immune response of organism. 3) Heterografts, having lost a part of their antigenicity by refrigeration, may favorize new bone production. It is thought that stimulative action of grafts on ↑ bone regeneration is based on presence of a specific ↑ bone inductor substance within transplant itself.

↑ Transplantation

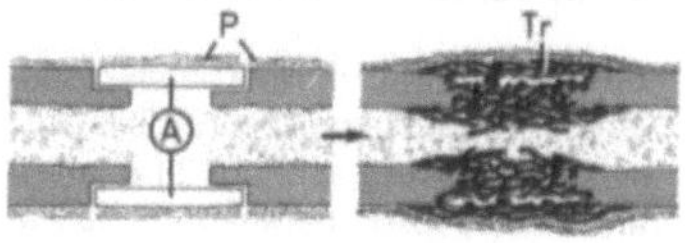

Feik, S.A., Storey, E.: J. Anat. *136*, 1 (1983); Ham, A.W., Harris, W.R.: Repair and Transplantation of Bone. In: Bourne, G.H. (ed.): The Biochemistry and Physiology of Bone. 2nd edn. Vol. 3. New York: Academic Press 1976

Bone turnover: ↑ Bone, internal remodeling of

Bone, vascularization of: Most of a compact ↑ bone is vascularized from ↑ periosteum by periosteal vessels conveying blood via ↑ Volkmann's canals into ↑ Haversian canals. A small part of compact ↑ bone is supplied from medullary cavity by branches of medullary artery, whose finest extensions penetrate into Volkmann's canals to reach innermost Haversian canals. Cancellous ↑ bone is supplied by diffusion from medullary cavity.

Knese, K.-H.: Stützgewebe und Skelettsystem. In : Möllendorff W. v., Bargmann, W. (eds.): Handbuch der mikroskopischen Anatomie des Menschen. Vol. 2. Die Gewebe. Part 5. Berlin, Heidelberg, New York: Springer-Verlag 1979

Bone, Wormian: ↑ Wormian Bone

Bone, woven: ↑ Bone, immature

Bony spiral lamina: ↑ Lamina spiralis ossea

Border, brush: ↑ Brush border

Border cells, of Corti's organ (cellula limitans interna*): Slender cells (BC) adjacent to inner ↑ phalangeal cells (IPC). B. delimit inner boundary of ↑ Corti's organ; their height gradually diminishes as they become squamous cells lining inner spiral sulcus (ISS).

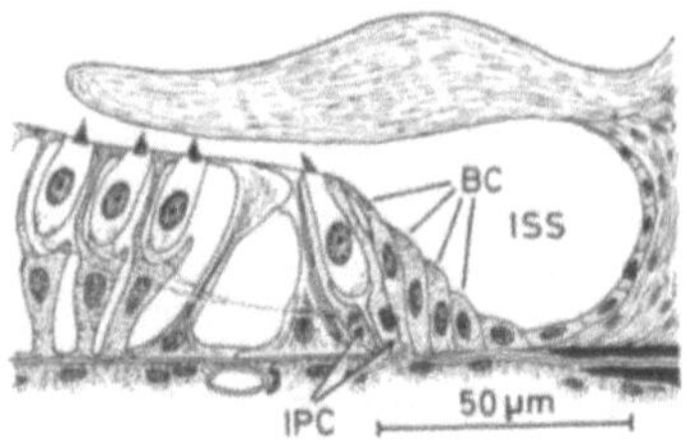

Border, ruffled: ↑ Ruffled border

Border, striated: ↑ Striated border

Böttcher's cells (cellula sustentacularis*): A group of polyhedral cells (BC) situated only in basal coil of cochlea and interposed between ↑ basilar membrane (BM) and cells of ↑ Claudius (CC). B. have a large round nucleus and are connected by numerous interdigitations to neighboring cells. B. are thought to have a secretory or absorptive function.

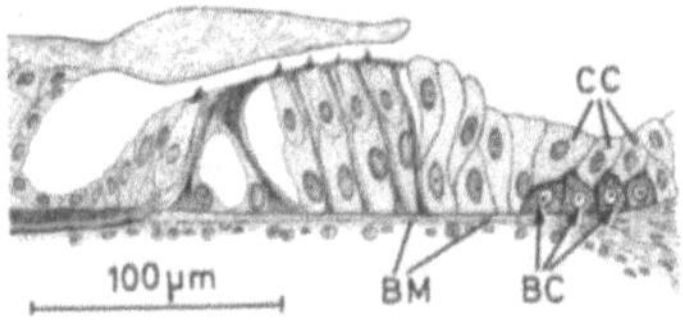

Brodmann, G., Giebel, W.: Arch. Otorhinolaryngol *220*, 105 (1978); Ishiyama, E., Cutt, R.A., Keels, E.W.: Ann Otol. Rhin. Laryngol. *79*, 54 (1970)

Böttcher's crystals: Crystals of varying form appearing in ↑ semen as it begins to cool and dry. It is thought that B. consist of phosphate of spermine, a polyamine substance originating in ↑ prostate.

Bouin's fixative: A fixative mixture, widely used in light microscopy, containing acetic acid, formol, and picric acid.

Bouton terminal (bulba terminalis*): A tiny nonmyelinated end swelling of an ↑ axon (A) adhering either to soma of a nerve cell (axosomatic ↑ synapse), to its dendrite (axodendritic ↑ synapse), or to another axon (axoaxonal ↑ synapse). B. is enclosed by ↑ axolemma (Al), whose portion facing other nerve cell is referred to as ↑ presynaptic membrane (PM). ↑ Presynaptic density, arranged in a hexagonal lattice (HL), adheres to presynaptic membrane, which is perforated by ↑ diffusion channels (DC). ↑ Axoplasm of B. contains some tubules of smooth ↑ endoplasmic reticulum (sER), ↑ mitochondria (M), ↑ neurofilaments (Nf), ↑ neurotubules (Nt), and ↑ synaptic vesicles (SV). PsM = postsynaptic membrane, SC = ↑ synaptic cleft. (Modified after Akert 1971)

↑ Synapse, chemical

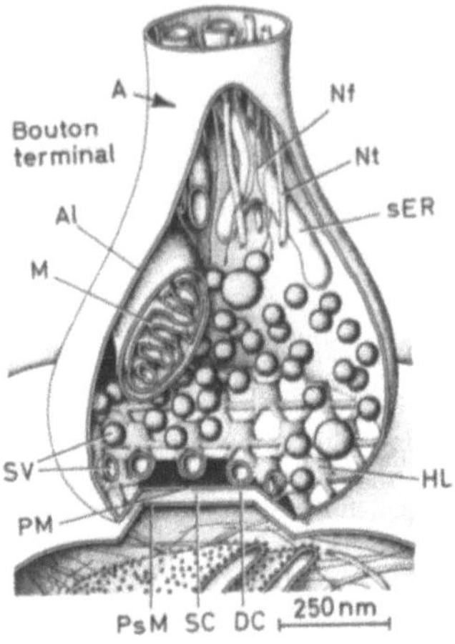

Akert, K.: Klin. Wochenschr. *49*, 509 (1971)

Bowman's capsule (capsula glomeruli*): A double-walled envelope (BC) surrounding ↑ glomerulus (G) of ↑ renal corpuscle. Structure: 1) Visceral wall = ↑ podocytes (P), which surround ↑ glomerular capillaries (GC); 2) parietal wall or ↑ capsular epithelium (CE) = ↑ simple squamous epithelium. At ↑ vascular pole (VP) of renal corpuscle, podocytes are continuous with capsular epithelium; at ↑ urinary pole (UP), capsular epithelium is continuous with cuboidal epithelium of ↑ neck region of ↑ nephron (PT). The two walls of B. delimitate ↑ Bowman's space (BS).

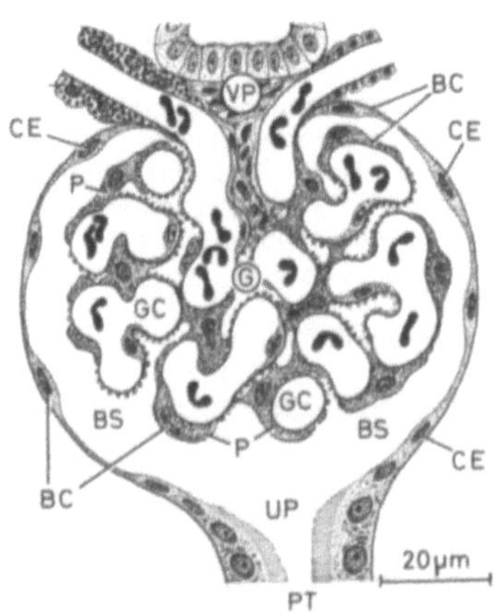

Bowman's glands: ↑ Olfactory glands

Bowman's membrane (lamina limitans anterior*): A condensed outer layer (BM), about 6–10 µm thick, of corneal stroma (S). B. is separated from epithelium (E) by a very distinct ↑ basement membrane. B. consists of randomly oriented 18-nm-thick ↑ collagen fibrils; neither ↑ glycosaminoglycans nor ↑ elastic fibers are present. At corneal ↑ limbus, B. ends abruptly; some mammals have no B.

↑ Cornea

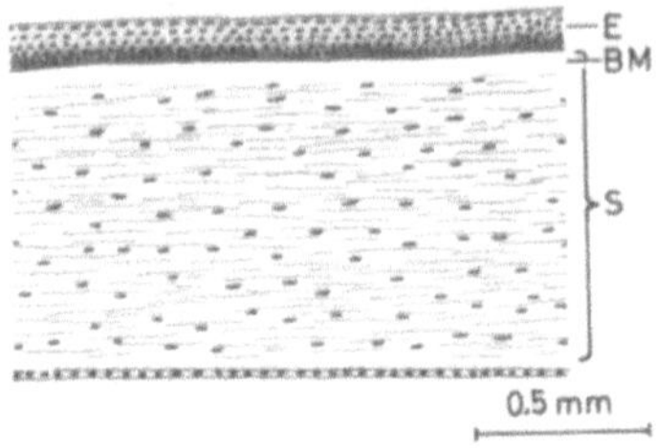

Bowman's space (capsular space, lumen capsulae*): A space between the two walls of ↑ Bowman's capsule. B. is continuous with lumen of rest of ↑ nephron and contains primary urine (glomerular filtrate); hydrostatic pression within B. is 15 mm Hg.

Bradykinin: A nonapeptide secreted by ↑ sweat glands, ↑ salivary glands, exocrine ↑ pancreas, and, partially, by neutrophilic ↑ granulocytes. Not found in blood under normal conditions, B. acts to increase capillary permeability, provoking relaxation of vascular ↑ smooth muscle cells and contraction of visceral smooth muscle cells; also related to process of ↑ blood clotting. Most potent vasodilator active in man yet discovered. (See physiology texts for further information)

Brain (encephalon*): Nervous tissue within cranium comprising cerebral hemispheres, basal ganglia, brain stem, and ↑ cerebellum. Macroscopic structure of cerebral hemispheres and cerebellum: 1) Cortex (cerebral and cerebellar) = ↑ gray matter of CNS; 2) medulla = ↑ white matter. In the present book, the term ↑ brain cortex is used to signify the cerebral cortex.

Braak, H.: Architectonics of the Human Telencephalic Cortex. Berlin, Heidelberg, New York: Springer-Verlag 1980; Schmitt, F.O., Worden, F.G., Adelman, G., Dennis, S.G. (eds.): The Organization of the Cerebral Cortex. Cambridge, Mass: MIT Press 1981

Brain cisternae (cisternae subarachnoidales*): Particularly large ↑ subarachnoid spaces in some areas along ↑ cerebrum and ↑ cerebellum. B. contain practically no trabeculae; they are filled with ↑ cerebrospinal fluid and communicate, through foramen of Magendi (apertura mediana ventriculi quarti*) and foramina of Luschka (aperturae laterales ventriculi quarti*) with ↑ brain ventricles.

Brain, connective tissue sheets of: ↑ Meninges

Brain cortex, heterotypical isocortex of: A type of neocortex in which some cellular layers are more developed than others; a basic six-layered structure is present. 1) Agranular type or motor cortex = prominent pyramidal (III) and ganglionic layers (V), the latter with ↑ Betz's cells (B). Poorly developed inner granular layer (IV). Localization: gyrus precentralis (areas 4–6). 2) Granular type or sensory cortex (coniocortex) = a well-developed inner granular layer, poorly developed pyramidal and ganglionic layers. Localization: somatic sensory cortex (areas 1–3), auditory cortex (areas 41 and 42), visual cortex (areas 17–19).

↑ Brain cortex, homotypical isocortex of; ↑ Brain cortex, phylogenetic and structural subdivision of; ↑ Striate area

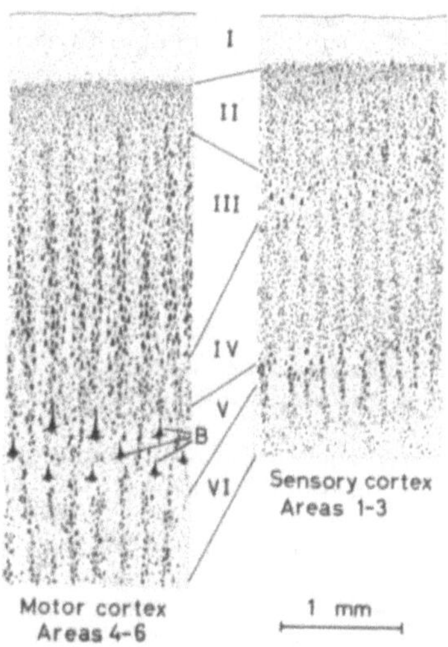

Brazier, M.A.B., Petsche, H. (eds.): Architectonics of the Cerebral Cortex. IBRO Monographs Series. Vol. 3. New York: Raven Press 1978

Brain cortex, homotypical isocortex of: The basic type of isocortex with the usual six layers; makes up major part of frontal, parietal, and temporal lobes. Structure: A) Cytoarchitectonics. 1) Molecular layer (lamina molecularis* or plexiformis*) = a few ↑ neurons, among which horizontal cells of ↑ Cajal are found; a layer with associative function. II) Outer granular layer (lamina granularis externa*) = very numerous, small neurons about 10 μm in diameter; their ↑ dendrites ramify in lamina molecularis, whereas ↑ axons take either direction of ↑ white matter or turn toward molecular layer; presence of cells with short axon (↑ Golgi type III neurons). A layer with receptory function. III) Pyramidal layer (lamina pyramidalis externa*) = neurons with pyramid-shaped ↑ perikarya about 40 μm in diameter; presence of Golgi type II neurons and ↑ Martinoti's cells. Dendrites of pyramidal cells branch in molecular layer; their axons make contact with subcortical nuclei. A layer with associative function. IV) Inner granular layer (lamina granularis interna*) = many small, irregularly shaped neurons of Golgi type II: main receptory layer, doubled in ↑ striata area. V) Ganglionic layer (internal pyramidal layer, lamina pyramidalis interna*) = voluminous pyramidal cells with dendrites branching in molecular layer; axons enter white mass as centrifugal or commissural fibers; in gyrus precentralis (areas 4–6), presence of cells of ↑ Betz; main effectory and associative layer. VI) Layer of polymorphic cells (lamina multiformis*) = irregular fusiform and angular nerve cells. VII) Lamina infima = inconstant, thin layer of small fusiform cells. Functionally, neurons in whole B. are organized into vertical ↑ columns. B) Myeloarchitectonics. 1) Tangential layer of fibers (lamina tangentialis); 2) ↑ band of Bechterew; 3) outer ↑ band of Baillarger; 4) inner ↑ band of Baillarger; 5) lamina substriata; 6) ↑ radial columns.

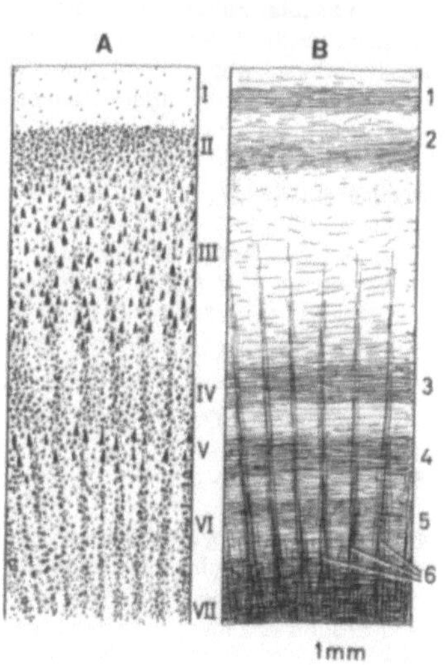

↑ Brain cortex, heterotypical isocortex of

Carpenter, M.B.: Human Neuroanatomy. 7th edn. Baltimore: Williams & Wilkins Co. 1976; Hyvärinen, J.: The Parietal Cortex of Monkey and Man. Berlin, Heidelberg, New York: Springer-Verlag 1982

Brain cortex, neuronal circuits of: B. are only partially understood. An afferent impulse (Aff) is transmitted to ↑ Golgi type II cells (GC) of layer IV; from here, information is conveyed via their axons (a) to: 1) Apical dendrites (ad) of pyramidal cells (PC), resulting in limited excitation of a small column (SC); and 2) basal dendrites (bd) of pyramidal cells, provoking excitation of an entire ↑ column. Some inhibitory ↑ interneurons (not represented), such as basket cells, modulate received information, whereas others, such as stellate cells, probably have an excitatory action. This basic neuronal circuit can also be influenced by afferent impulses of cortical origin passing through layer I and by direct connections of afferent fibers with dendrites of pyramidal cells of layers III and V. (Modified after Szentagothai 1975)

↑ Columns, of brain

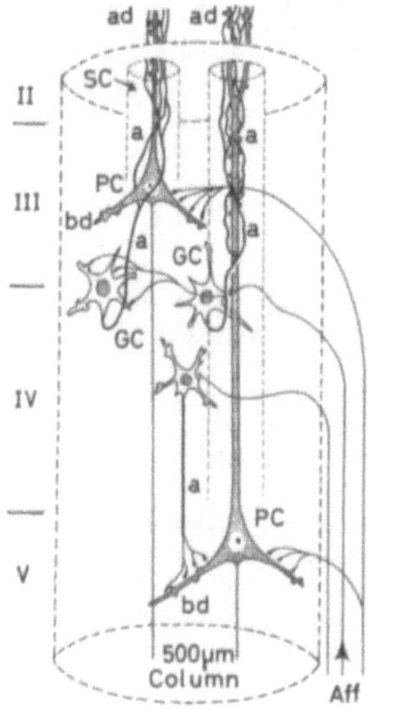

Creutzfeldt, O.D.: Cortex Cerebri. Berlin, Heidelberg, New York: Springer-Verlag 1983; Gallatz, K., Palkovits, M., Szentagothai, J.: Folia Morphol. *30*, 133 (1982); McGeer, P.L., Eccles, J.C., McGeer, E.G.: Molecular Neurobiology of the Mammalian Brain. New York: Plenum Publ. Co. 1978

Brain cortex, penetration of blood vessels into: Turning into brain from ↑ subarachnoid space (SAS), an arterial vessel (AV) is accompanied by an extension of ↑ pia mater (PM) and surrounded by a conical ↑ perivascular space (PS), also called space of ↑ Virchow-Robin. It becomes gradually narrower and disappears totally in area where ↑ perivascular feet (PF) of ↑ astrocytes (A) form ↑ membrana limitans gliae perivascularis (MGP), which is in direct contact with outer aspect of ↑ capillaries (Cap). Veins leave brain in the same manner. Blood vessels enter and leave ↑ spinal cord in an identical way. (In fig., ↑ neurons were omitted.) Ar = arachnoid, DM = ↑ dura mater, MGS = ↑ membrana gliae superficialis, SDS = ↑ subdural space, SF = ↑ subpial feet

↑ Barrier, blood-brain; ↑ Membrana limitans gliae superficialis

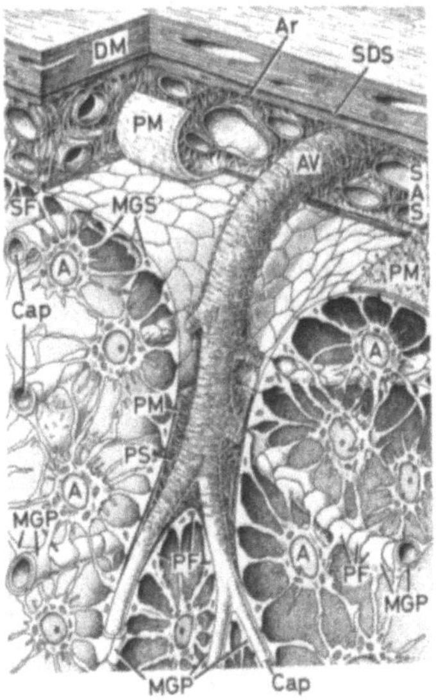

Frederickson, R.G., Low, F.N.: Am. J. Anat. *125*, 123 (1969)

Brain cortex, phylogenetic and structural subdivision of: A) Allocortex or paleocortex (AL) = phylogenetically very ancient, primordial cortex with only one to three cell layers (laminae); extends over 1/12 of brain surface (↑ limbic system, olfactory cortex). B) Isocortex or neocortex (IS) = phylogenetically younger cortex identical development from mantle zone of neural tube, covering 11/12 of cerebral surface; characterized by six cell layers. 1) Homotypical isocortex (HOM) = commonest type of brain cortex with usually six laminae. 2) Heterotypical isocortex (HET) = some laminae more developed than others: a) Agranular (AGR) or motor cortex characterized by well-developed pyramidal cell layer (III) and ganglionic layer (V) with ↑ Betz's cells (B); poorly developed inner graunular layer. b) Granular (GR) or sensory cortex (coniocortex) characterized by prominent outer (II) and inner granular layers (IV) with poorly developed pyramidal and ganglionic cell layers.

↑ Brain cortex, heterotypical isocortex of; ↑ Brain cortex, homotypical isocortex of; ↑ Striate area

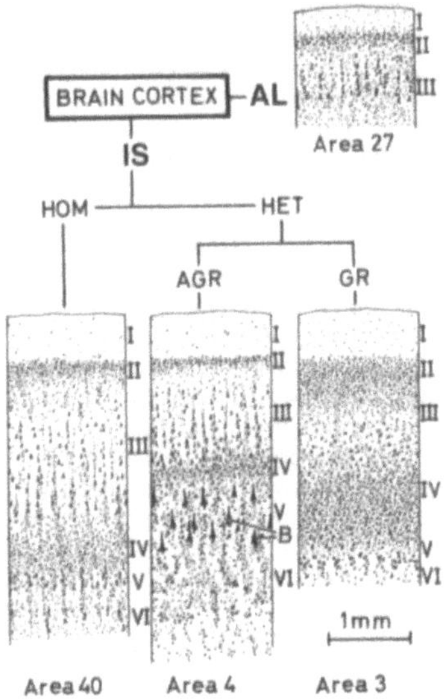

Carpenter, M.B.: Human Neuroanatomy. 7th edn. Baltimore: Williams & Wilkins Co. 1976

Brain sand: ↑ Acervuli

Brain, vascularization of: The brain is supplied with blood by carotids, basilar arteries, and circle of Willis. All arteries are ↑ muscular arteries; almost all ↑ capillaries are continuous.

Brain ventricles (ventriculi*): Local dilatations of intracerebral system of anastomosing channels filled with ↑ cerebrospinal fluid (lateral ventricles, third and fourth ventricles). B. are lined with ↑ ependyma; ↑ choroid plexus penetrates into B.

↑ Circumventricular organs

Breast: ↑ Mammary gland

Bridges, along myosin filament: Six short ↑ heavy meromyosin (HMM-S 1) processes arranged in a helical pattern around a ↑ myosin filament and repeated every 43 nm. B. are absent from center of ↑ H-band (bridge-free ↑ pseudo-H band).

Bridges, intercellular: ↑ Intercellular bridges

Bright-field microscopy: The most frequent technique for diascopic light microscopic ↑ illumination in which observation field is homogeneously illuminated by the ↑ condenser (C) and appears bright. The light in B. passes through the preparation (P); the image is formed almost completely as a result of different absorptions of light at level of a stained histological ↑ section

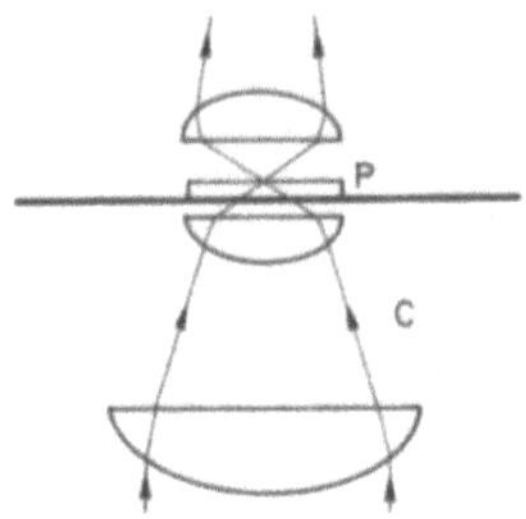

Brilliant cresyl blue: An oxasin dye for ↑ supravital staining of ↑ reticulocytes by precipitating their persisting rRNA into an irregular web or reticulum.

Broad ligament, of uterus (ligamentum latum): A mass of ↑ loose connective tissue at each lateral side of ↑ uterus. B. contains many collagen fibers, a few elastic ones, blood and lymphatic vessels, and nerve fibers. Outer surface of B. is covered by ↑ peritoneum continuous with ↑ perimetrium.

Brodmann's areas: ↑ Areas, of brain cortex

Brodmann's areas 1–3: ↑ Brain cortex, heterotypical isocortex of

Brodmann's areas 4–6: ↑ Brain cortex, heterotypical isocortex of

Brodmann's area 17: ↑ Striate area

Bronchi (bronchus*): Tubular organs whose purpose is to conduct, moisten, and warm inspired air and to trap airborne particles and transport them, via ↑ ciliary motion, to ↑ trachea. B. are subdivided into primary or extrapulmonary ↑ B. and secondary or intrapulmonary ↑ B. Branching of B. forms part of ↑ pulmonary airways.

Bronchi, extrapulmonary: ↑ Bronchi, primary

Bronchi, lobar: Secondary ↑ bronchi conducting air to pulmonary lobes.

Bronchi, primary (bronchi extrapulmonary, bronchi majores*): Two branches arising at tracheal bifurcation and extending to pulmonary ↑ hilus. Histological structure of B. is identical to that of ↑ trachea

Bronchi, secondary (bronchi intrapulmonary, bronchi*): Branching segments of intrapulmonary airways between primary ↑ bronchi and ↑ bronchioles. Structure: 1) Tunica mucosa (TM): a) ↑ bronchial epithelium (E); b) lamina propria (LP) = a ↑ loose connective tissue with strong, longitudinal, elastic fibers and some scattered ↑ lymphatic nodules. 2) Tunica muscularis (TMu) = a spiral layer of ↑ smooth muscle cells. 3) Tela submucosa (TSm) = a loose connective tissue with a venous plexus. 4) Tunica conjunctivo-cartilaginea (TCC) = irregular plates of hyaline ↑ cartilage (HC) connected with ↑ dense connective tissue; in small B., cartilage becomes elastic. ↑ Mixed ↑ bronchial glands (BG) are situated mainly between cartilaginous pieces and in submucosa. 5) Tunica adventitia (TA) = a dense connective tissue containing ↑ bronchial arteries (BA) and veins (BV), lymphatic vessels, and nerve fibers. Topographically, B. are subdivided into lobar and segmental B.; their structure, but not their caliber, is the same as described above for B.

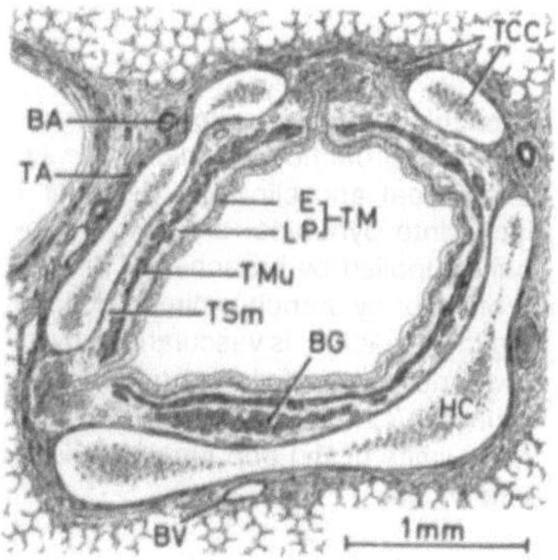

Bronchi, segmental: ↑ Bronchi, secondary

Bronchial arteries (rami bronchiales*): ↑ Muscular arteries (AB), originating from ↑ aorta or intercostal arteries. B. accompany division of bronchial tree up to respiratory ↑ bronchioles. At level of their ↑ alveoli, B. anastomose with branches of ↑ pulmo-

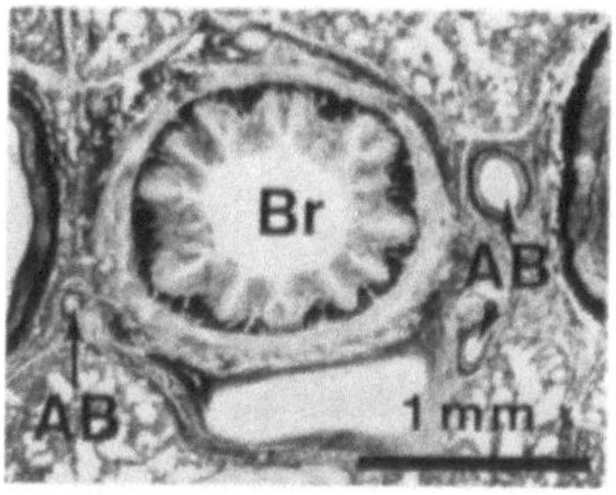

nary artery. B. are located within tunica adventitia of ↑ bronchi (Br) and ↑ bronchioles; blood passing through a capillary network collects in submucosal venous plexus before reaching bronchial veins. B. provide blood supply to ↑ bronchi, bronchioles, pulmonary interstitium, and ↑ pleura as well as walls of pulmonary arteries and veins. B. are nutrient pulmonary vessels. (Fig. = calf, orcein staining)

↑ Lungs, vascularization of

Bronchial cartilages (cartilagines bronchiales*): Hyaline ↑ cartilage pieces in primary large, and medium secondary ↑ bronchi; elastic ↑ cartilage pieces in small secondary bronchi

Bronchial epithelium: A ↑ pseudostratified ciliated epithelium very similar to that in ↑ trachea; endocrine cells in B. are grouped to form ↑ neuroepithelial bodies

↑ Trachea, epithelium of

Mariassy, A.T., Plopper, C.G.: Anat. Rec. *205*, 263 (1983); Wasano, K., Yamamoto, T.: Cell Tissue Res. *201*, 197 (1979)

Bronchial glands (glandulae bronchiales*): ↑ Mixed, ↑ tubuloacinous glands (BG) located in tela submucosa (TSm) of primary and secondary ↑ bronchi. Structure: 1) Serous or hydrotic cells (SC) = prismatic cells with an elliptical nucleus in a central position, a few mitochondria and an inconspicuous Golgi apparatus, rare cisternae of rough endoplasmic reticulum, apical smooth endoplasmic ↑ tubules, rich lateral interdigitations, deep intercellular ↑ canaliculi, and a ↑ basal labyrinth. Cytoplasm encloses a large amount of ↑ glycogen and small translucent granules measuring 50 nm. 2) Mucous cells (MC) = columnar cells with a dense nucleus in ↑ basal pole, a well-developed supranuclear Golgi apparatus, many voluminous secretory granules in ↑ apical pole. Both types of cells produce several types of mucins (sialomucins, sulfomucins) whose expulsion is aided by 3) ↑ Myoepithelial cells (arrowhead). 4) Nonspecialized cuboidal cells are present in large excretory ducts (ED). 5) Occasional ↑ oncocytes may be also found. B. are richly vascularized by fenestrated ↑ capillaries (Cap) and abundantly innervated by ↑ nonmyelinated nerve fibers (NF). The secretion of B. is a protective and hydrating product for ↑ bronchial tree.

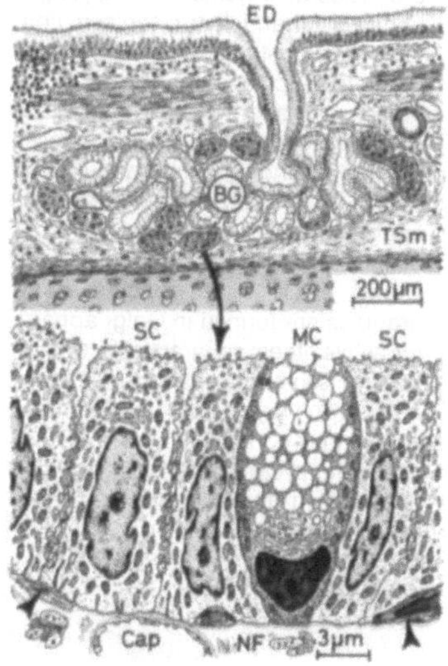

Sorokin, S.: Am. J. Anat. *117*, 311 (1965)

Bronchial tree (arbor bronchialis*): A branching system of channels conducting air to ↑ respiratory zone of lungs. B. makes up part of ↑ pulmonary airways and consists of primary and secondary ↑ bronchi, ↑ bronchioles, and terminal ↑ bronchioles.

Bronchial veins (venae bronchiales*): Small, thin-walled ↑ veins receiving blood from bronchial venous plexus. B. are situated within tunica adventitia of secondary ↑ bronchi and drain their blood into pulmonary veins

Bronchiolar cells: ↑ Bronchiolar epithelium

Bronchiolar epithelium: A ↑ simple columnar ciliated epithelium lining bronchiolar tree up to respiratory ↑ bronchiole; from here B. loses its ↑ cilia and becomes cuboidal; presence of ↑ neuroepithelial bodies. Structure: 1) Ciliated bronchial cells (CC) (majority) = prismatic cells with an elliptical nucleus, a small nucleolus, voluminous mitochondria, a small Golgi apparatus, some rough endoplasmic cisternae, ↑ lysosomes, and ↑ residual bodies. ↑ Apical pole bears long ↑ microvilli and cilia (C). Numerous ↑

micropinocytotic vesicles at apical plasmalemma are possibly related to uptake of airborne material. Through ciliary motion, dust particles are transported toward ↑ bronchi. 2) Nonciliated bronchial cells or Clara cells (NC) = prismatic cells with bulbous apex, an elongated nucleus, many voluminous mitochondria, a well-developed apical or lateral Golgi apparatus, and a considerable number of ↑ glycogen particles. Infranuclear cytoplasm contains ↑ ergastoplasm (rER); in supranuclear cytoplasm are concentrated an appreciable number of smooth endoplasmic ↑ tubules (sER) surrounding membrane-limited, round, and rod-shaped electron-opaque granules (SG), about 0.5 µm in diameter. Cytochemical studies have shown their dual origin: Some granules are formed in Golgi apparatus, while others separate directly from sER. In some species, granules may have a fine-crystalline or filamentous matrix; their content, a mixture of ↑ glycosaminoglycans and cholesterol, is released on free epithelial surface to serve as a protective lining.

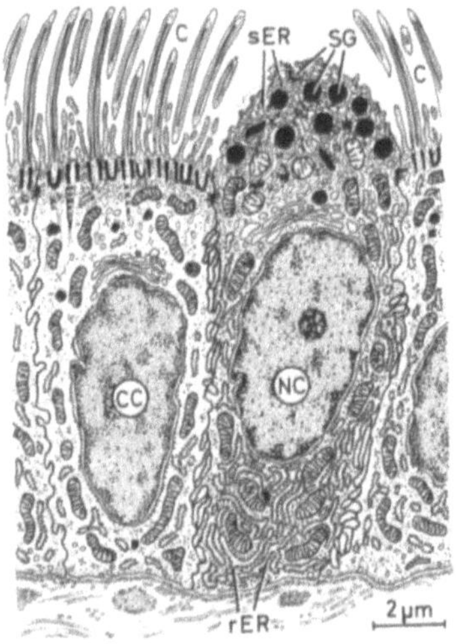

Andrews, P.: Biomed. Res. *2*, Suppl. 281 (1981); Baert, J., Vanderberghe, M.P.: Anat. Rec. *201*, 283 (1981); Devereux, T.R., Fouts, J.R.: In Vitro *16*, 958 (1980); Etherton, J.E., Purchase, F.H., Corrin, B.: J. Anat. *129*, 305 (1979); Taira, K., Shibaskai, S.: Arch. Histol. Jpn. *41*, 351 (1978)

Bronchioles (bronchioli*): Segments of ↑ pulmonary airways situated between secondary ↑ bronchi (B) and ↑ alveolar ducts (AD). Entering ↑ lung lobule (LL), a preterminal ↑ bronchiole (PB) branches into four to eight terminal ↑ bronchioles (TB), each giving rise to two or more respiratory ↑ bronchioles (RB). Each respiratory bronchiole ramifies into ↑ two to ten alveolar ducts.

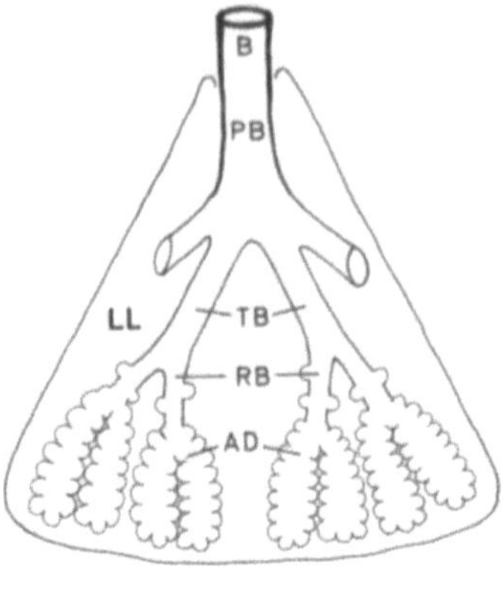

Bronchioles, preterminal: One of ↑ bronchioles as it enters a ↑ lung lobule before its division into terminal ↑ bronchioles. Structurally identical to terminal bronchiole. The term B. is not generally accepted.

Bronchioles, respiratory (bronchiole alveolar, bronchiolus respiratorius*): Branches (RB) of a terminal ↑ bronchiole (TB); the first ↑ alveoli (A) occur in wall of B. Structure: 1) Tunica mucosa (TM): a) epithelium = a ↑ simple cuboidal nonciliated epithelium interrupted in areas of alveoli; b) lamina propria = a very thin layer of ↑ loose connective tissue with longitudinal ↑ elastic fibers. Where there are alveoli, a capillary network replaces connective tissue. 2) Tunica muscularis (TMu) = a thin and incomplete layer of ↑ smooth muscle cells; alveoli lie outside muscular layer. 3) Tunica adventitia (TA) = a very reduced loose connective tissue containing finest branches of pulmonary artery, anastomosing at level of alveoli with branches of ↑ bronchial arteries; presence of fine lymphatics and autonomic nerve fibers. There are three orders of B., all having same structure. B. open into ↑ alveolar ducts (AD).

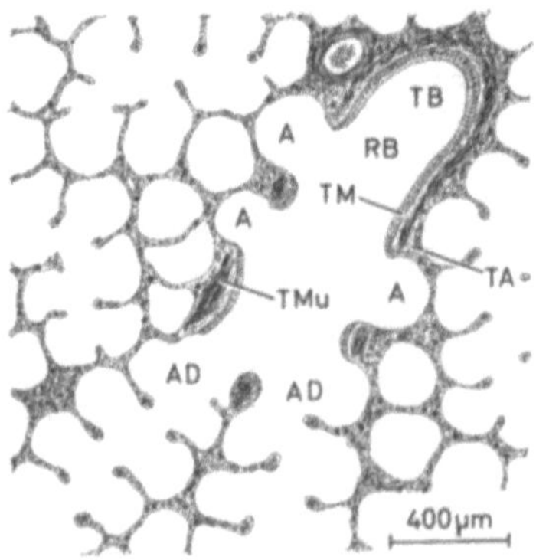

Young, C.D., Moore, W.G., Hutchins, G.M.: Anat. Rec. *198*, 245 (1980)

Bronchioles, terminal (bronchiolus terminalis*): Branches of a preterminal ↑ bronchiole. Structure: 1) Tunica mucosa (TM): a) epithelium (E) = ↑ bronchiolar epithelium; b) lamina propria (LP) = a thin layer of ↑ loose connective tissue with subepithelial, longitudinal, strong, ↑ elastic fibers. 2) Tunica muscularis (TMu) = a layer of spirally arranged ↑ smooth muscle cells. 3) Tunica adventitia (TA) = a loose connective tissue containing small branches of ↑ bronchial arteries (BA), ↑ bronchial veins (BV), lymphatic vessels, and autonomic nerve fibers. Each B. ramifies into two or more respiratory ↑ bronchioles. Mucosal folds are provoked by postmortem muscular shortening.

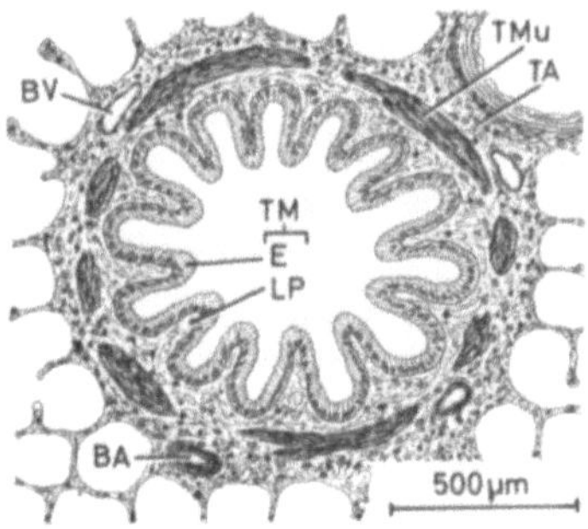

Bronchopulmonary segments (segmenta bronchopulmonalia*): An anatomical and clinical division of ↑ lungs into pyramidal or conical subunits supplied by branches of lobar ↑ bronchi or by branches directly following them; each B. is vascularized by its own artery which follows branching of bronchi. Veins lie intersegmentally and mark limits of B. Each lung is divided into ten B.

Brown adipose tissue: ↑ Adipose tissue, brown

Bruch's membrane (complexus basalis*): An elastic membrane (BM) located between ↑ pigment epithelium (PE) of ↑ retina and ↑ choriocapillaris (C) of ↑ choroid. B. is 1–3 µm thick and consists of: 1) Basal lamina of capillaries (CBL); 2) external layer of delicate ↑ collagen microfibrils (ECF); 3)network of ↑ elastic fibers (EF); 4) inner layer of collagen microfibrils (ICF);

5) basal lamina of pigment epithelium (PBL). (Fig. = rat)

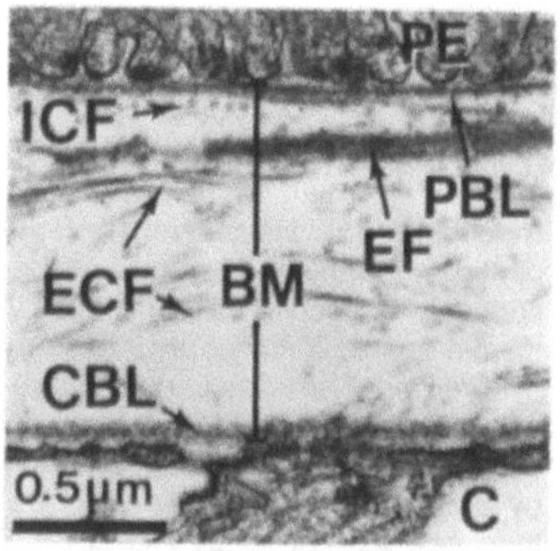

Braekevelt, R.C.: Anat. Embryol. *166*, 415 (1983); Essner, E., Pino, R.M.: Eur. J. Cell Biol. *27*, 251 (1982); Pino, R.M., Essner, E., Pino, L.C.: J. Histochem. Cytochem. *30*, 245 (1982)

Brücke's muscle (fibrae meridionales*): Meridionally arranged ↑ smooth muscle cells of ↑ ciliary muscle of ↑ ciliary body.

Brunner's glands: ↑ Duodenal glands

Brush border (limbus penicillatus*): Closely packed 3 to 6-µm-long ↑ microvilli (BB) found at apical surface of ↑ proximal tubule cells of kidney (fig.), at resorptive surface or ↑ osteoclasts, at ↑ syncitiotrophoblast, etc. B. enormously increases active cell surface; also visible under a light microscope. (Fig. = rat)

↑ Straited border

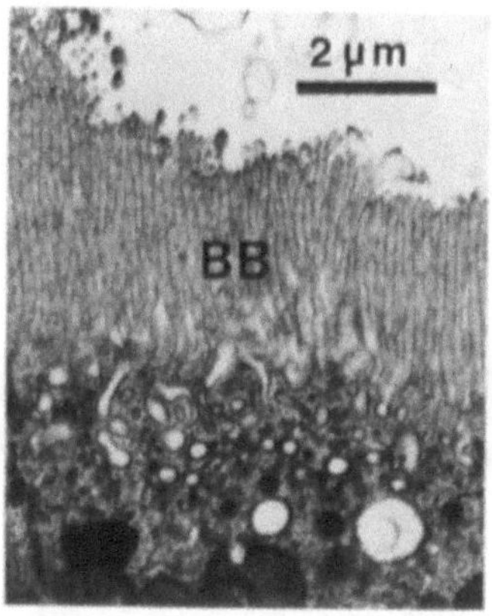

Craig, S., Lancashire, C.L.: J. Cell Biol. *84*, 655 (1980); Ljoda, Z.: Histochemistry *64*, 205 (1979); Scherberich, J.E., Gauhl, C., Mondorf, W.: Curr. Probl. Clin. Biochem. *8*, 85 (1977)

Brush cells: A collective term for all cells of ↑ tracheal epithelium bearing ↑ microvilli at apical surface.

B-spermatogonia: ↑ Spermatogonia

Bubbling, during cell division: ↑ Macropinocytosis of nutrient material from cell environment into cell body during anaphase.

Buccal glands (glandulae buccales*): ↑ Mixed, ↑ tubuloalveolar glands scattered in tunica mucosa of ↑ cheeks. ↑ Mucous tubules frequently open directly into ↑ striated ducts. B. moisten inner surface of cheeks.

Bud, periosteal: ↑ Periosteal bud

Buds, taste: ↑ Taste buds

Bulb, olfactory: ↑ Olfactory bulb

Bulbar conjunctiva (tunica conjunctiva bulbi*): The portion of ↑ conjunctiva covering anterior surface of ↑ sclera

Bulbourethral glands (Cowper's glands, glandulae bulbo-urethrales*): Pea-sized, paired, ↑ tubuloalveolar glands (BG) disposed on each side of urethral bulb (B) and connected with cavernous urethra (U) by fairly long excretory ducts. Structure: 1) Connective capsule (C) = a ↑ dense connective tissue with some ↑ skeletal muscle fibers (MF) belonging to diaphragma urogenitale. 2) ↑ Stroma = connective septa dividing parenchyma into small

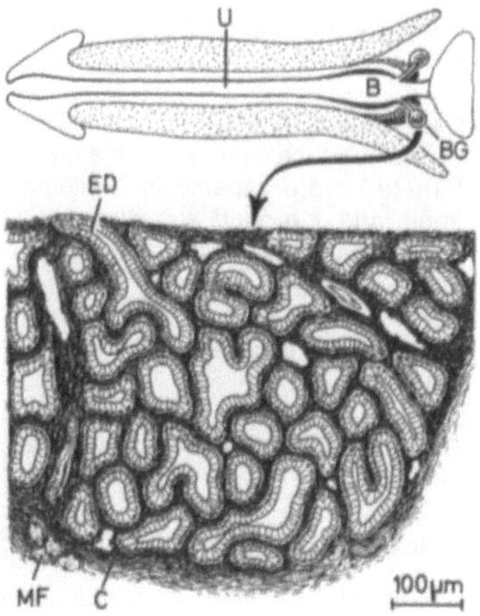

lobules. 3) ↑ Parenchyma = a ↑ simple cuboidal or columnar epithelium, secretory cells of which are of mucous type with a flattened basal nucleus and clear supranuclear cytoplasm filled with secretory granules. Depending on functional stimulation, morphology of epithelium undergoes great changes. Around ↑ tubuloalveoli there are ↑ myoepithelial cells. Intralobular exretory ducts (ED) are lined with simple columnar epithelium, which becomes ↑ pseudostratified at urethral orifices. B. produce transparent, viscous, ↑ mucoid substance rich in ↑ glycosaminoglycans (galactose, galactosamine, galacturonic acid, ↑ sialic acid, methylpentose) acting as lubricant during erotic stimulation and coitus.

Hellgren, L., Mylius, E., Vincent, J.: J. Submicrosc. Cytol. *14*, 683 (1982); Latalski, M., Obuchowska, D.: Z. mikrosk.-anat. Forsch. *93*, 391 (1979); Sikorski, A.: Folia Morphol. *30*, 182 (1982)

Bulbs, terminal, of Krause: ↑ Krause's terminal bulbs

Bulbus pili: ↑ Hair bulb

Bundle, atrioventricular: ↑ Atrioventricular bundle

Bundle, of His: ↑ Atrioventricular bundle

Büngner's bands: Proliferated ↑ Schwann's cells (BB) arranged as tubes (Schwann's tubes) and solid cords uniting proximal and distal stumps of a sectioned peripheral ↑ nerve fascicle. Regenerating axons (A) grow out from proximal stump (PS) toward distal stump (DS) following direction of B. Thus, B. not only give support and direction to ↑ axons, but also give rise to new ↑ Schwann's sheath, permitting axons to grow down distal stump and regenerate sectioned nerve fascicle. Interposition of connective tissue between stumps prevents formation of B.

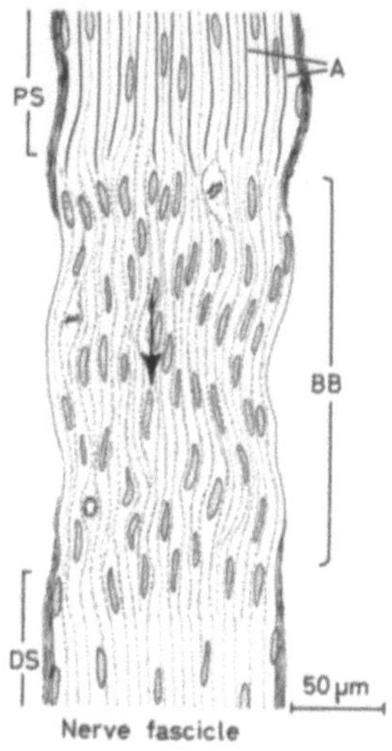

↑ Nerve fibers of peripheral nervous system, degeneration and regeneration of

Bursa analogue (bursa equivalent): Unknown organ(s) or their parts which, in mammals, would have a function in ↑ differentiation of ↑ B-lymphocytes, analogous to ↑ bursa of Fabricius in birds. It is thought that ↑ lymphatic nodules scattered along mammalian gut and ↑ Peyer's patches can partially assume this role.

↑ Lymphocyte circulation

Bursa, of Fabricius: A ↑ lymphoepithelial organ of birds, in form of a cloacal diverticle, involved in production of ↑ B-lymphocytes to respond to antigen by ↑ differentiation into antibody-secreting lymphocytes and ↑ plasma cells. It seems that B. receives stem cells from ↑ bone marrow and induces their differentiation into B-lymphocytes. B., which gave its name to B-lymphocytes, does not exist as a definite organ in mammals.

↑ Bursa analogue

Behrens v. Rautenfeld, D., Budras, K.-D.: Folia Morphol. *28*, 168 (1980); Glick, B.: Int. Rev. Cytol. *48*, 354 (1977); Glick, B., Olah, I.: Anat. Rec. *204*, 341 (1982)

Burst-forming units-erythropoietic (BFU-E, BFU-erythroid): A category of erythropoietic stem cells considered by some to be earliest of cell types committed to ↑ erythropoiesis. B. are practically ↑ erythropoietin insensitive; ↑ colony-forming units-erythropoietic are considered a later stage than B.

Button terminal: ↑ Bouton terminal

C

C3 receptor: The site on cell membrane of ↑ phagocytes which interacts with ↑ complement.

Cabot rings: Remnants of nonextruded nuclei from orthochromatic ↑ erythroblasts.

Cage, of filaments: A detergent-stable network (C) of 7 to 11-nm-thick polypeptide microfilaments surrounding mammalian ↑ spindle apparatus (S). It is thought that C. forms as cells enter division and serves to maintain structural continuity of cytoplasm during this process. (After Zieve et al. 1980)

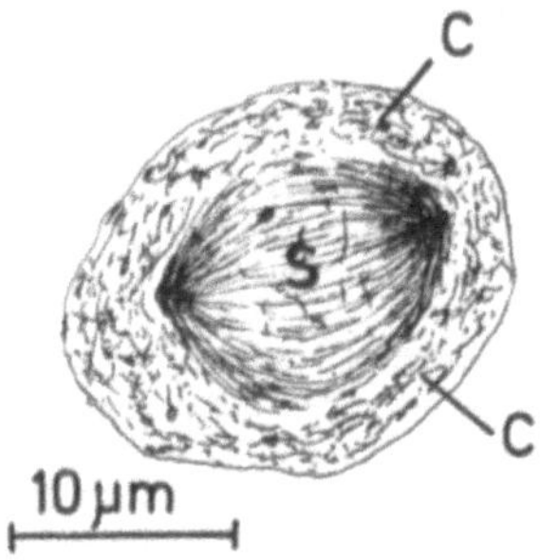

Zieve, G., Heidemann, S.R., McIntosh, R.J.: J. Cell Biol. *87*, 160 (1980)

Cajal's horizontal cells: ↑ Horizontal cells, of Cajal

Cajal's interstitial cells: ↑ Interstitial cells, of Cajal

Calcification: A process of deposition of insoluble calcium salts, mainly in form of ↑ hydroxyapatite, in ↑ collagen fibrils and ↑ ground substance of ↑ bone, ↑ cartilage, ↑ cementum, ↑ dentin, and ↑ enamel.

Calcification, of cartilage: ↑ Cartilage, calcification of

Calcification, of dentin: ↑ Dentin, calcification of

Calcification, of enamel: ↑ Amelogenesis

Calcification vesicles: ↑ Matrix vesicles

Calcification zone, in bone formation (cartilago calcificata*): A zone of calcified matrix of hyaline ↑ cartilage almost completely surrounding hypertrophied ↑ chondrocytes and their ↑ lacunae during indirect ↑ bone formation. Transversally oriented portions separating lacunae remain noncalcified, enabling rapid chondrolysis by ↑ chondroclasts.

Calcifying globules: ↑ Matrix vesicles

Calcitonin (thyrocalcitonin): A polypeptide hormone of 32 amino acids secreted by ↑ C-cells of mammalian ↑ thyroid and by ↑ ultimobranchial body of lower vertebrates. As partial antagonist to ↑ parathyroid hormone, C. lowers blood calcium level, probably by inhibiting bone resorption by ↑ osteoclasts, as well as by increasing ↑ calcification of ↑ bone matrix. (See biochemistry and physiology textbooks for further information)

Zabel, M.: Histochemistry *77*, 269 (1983)

Calcium-accumulating vesicles: ↑ Matrix vesicles

Calcium phosphate, of bone: The main mineral of ↑ bone deposited in form of ↑ hydroxyapatite.

Call-Exner bodies: Small cavities (CEB), 3–10 µm in diameter, found among ↑ granulosa cells (GC) of growing and mature ↑ ovarian follicles and filled with ↑ PAS-positive material. This stainability is based on presence of a filligreelike excess of ↑ basal lamina (BL), lining each cavity, as well as a flocculent proteinaceous material. Functional significance of C. is unknown. It is thought that C. could represent channels for circulation of ↑ liquor folliculi.

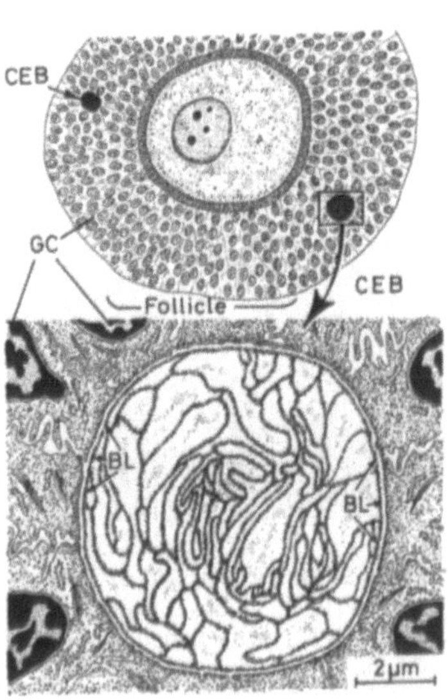

Korfmeister, K.-H.: Verh. Anat. Ges. *77*, 373 (1983); Stolić, E., Sprumont, P.: Acta Anat. *110*, 91 (1981)

Callus: A fusiform mass of bone formed around two fragments of fractured bone during ↑ bone regeneration.

Callus, bony: ↑ Bone regeneration

Callus, external: ↑ Bone regeneration

Callus, fibrocartilaginous: ↑ Bone regeneration

Callus formation: ↑ Bone regeneration

Callus, internal: ↑ Bone regeneration

Calmodulin: A specific Ca^{2+}-binding polypeptide (m.w. about 16500 daltons) present in all ↑ eucaryotic cells, acting as an intracellular receptor for this ion. C. can mediate many Ca^{2+}-dependent processes in synaptic transmission. Biochemical analyses have suggested possible association of C. with ↑ clathrin-coated vesicles, ↑ nexus, ↑ spindle apparatus, ↑ cilia, and ↑ microvilli.

Blum, J.J., Hayes, A., Jamieson, G.A., Vanaman, C.T.: J. Cell Biol. *87*, 386 (1980); Roufogalis, B.D.: Trends in Neurosciences, October 1980, 238; Wai Yiu Cheung: Sci. Am. *246/6*, 48 (1982); Wallace, R.W., Tallant, E.A., Cheung, W.Y.: Multifunctional Role of Calmodulin in Biologic Processes. Cold Spring Harbor Symposia on Quantitative Biology, Vol. 46, Part 2. New York: Cold Spring Harbor Laboratory, 1982; Willingham, M.C, Wehland, J., Klee, C.B., Richert, N.D., Ruthe-ford, A.V., Pastan, I.H.: J. Histochem. Cytochem. *31*, 445 (1983)

Calyceal synapse: ↑ Synapse, calyceal

Calyces, major (calices renales majores*): Two or three outpocketings (CM) of ↑ renal pelvis (P). Each C. branches into several minor calyces (CMi). Structure: 1) Tunica mucosa (TM): a) epithelium (E) = ↑ transitional epithelium; b) lamina propria (LP) = a well-vascularized and innervated ↑ loose connective tissue. 2) Tunica muscularis (TMu) = a thin layer of longitudinal and circular ↑ smooth muscle cells. 3) Tunica adventitia (TA) = a very thin layer of loose connective tissue continuous with adipose tissue of ↑ renal sinus (S). C. are part of extrarenal collecting system.

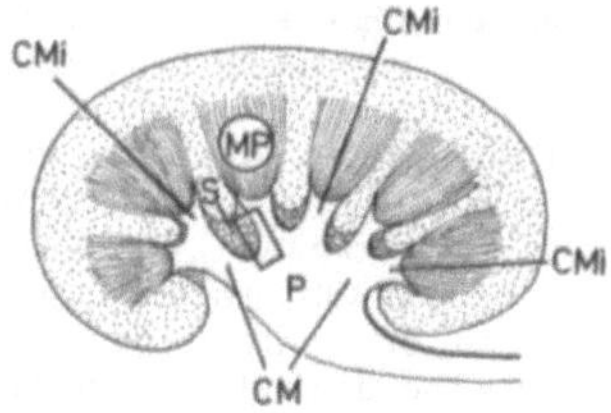

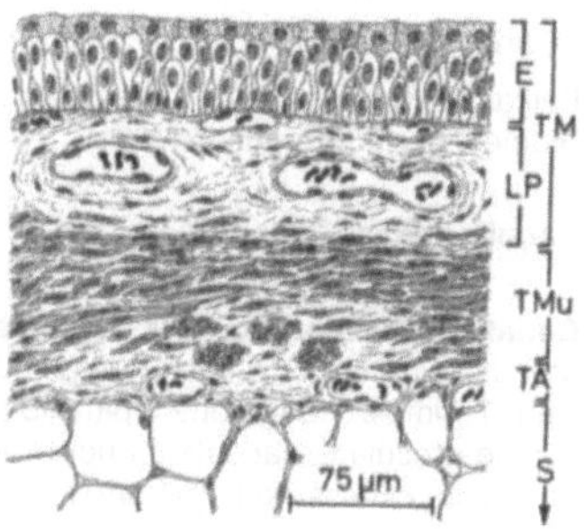

Dixon, J.S., Golsing, J.A.: J. Anat. *135*, 129 (1982)

Calyces, minor (calices renales minores*): Two to four branches (CMi) of each major ↑ calyx (CM), each capping a medullary pyramid (MP). Structure almost identical to major calyces. Circular layer of musculature forms a ring around pyramid; contraction of these fibers moves ↑ urine toward ↑ renal pelvis (P). Therefore, C. are considered primary pacemaker site for ureteral peristalsis. C. are part of extrarenal collecting system. (See fig. under ↑ Calyces, major)

Longrigg, N.: Lancet *1*, 253 (1975); Van den Bulcke, C., Keen, E.N., Fine, H.: J. Urol. *103*, 783 (1970)

Calyces, of axons: ↑ Basket cells, of cerebellum

Cambium, of periosteum (stratum osteogenicum*): The inner osteogenic layer (C) of ↑ periosteum (P) in growing and young bones; does not exist in adult bones. F = fibrous layer, ML = middle layer (Fig. = human fetus)

↑ Osteoprogenitor cells

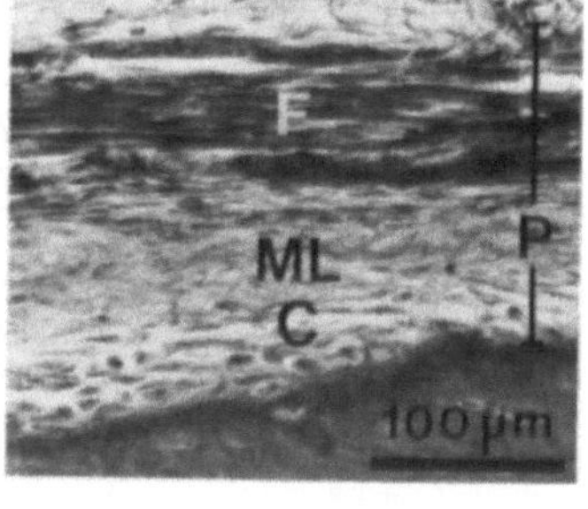

Kukletova, M., Oudran, L.: Z. mikrosk.-anat. Forsch. *97*, 158 (1983)

cAMP: ↑ Cyclic AMP

Canal, of Cloquet: ↑ Hyaloid canal

Canal, of Schlemm: ↑ Schlemm's canal

Canaliculi, apical: ↑ Apical pits

Canaliculi, bile: ↑ Bile canaliculi

Canaliculi, intercellular (secretory capillaries): Very fine and often branched intercellular passages (arrows) formed by apposition of two adjacent ↑ glandular cells. C. are generally studded with many ↑ microvilli (Mv) and do not reach basal lamina (BL). C. increase secretory surface. ↑ Bile canaliculi are also C.

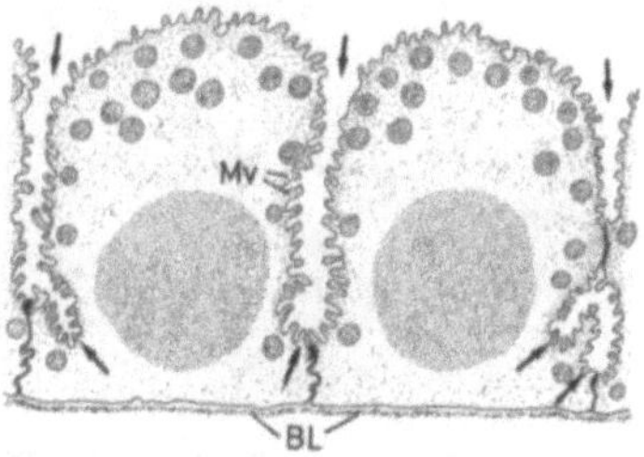

Fujita, T., Kobayashi, S., Serizawa, Y.: Biomed. Res. *2*, Suppl. 115 (1981); Zaitsu, A.: Arch. Histol. Jpn. *39*, 339 (1976)

Canaliculi, intracellular (intracellular channels): Variously deep, branched, invaginations of cell membrane into cell body. C. are particularly well developed in ↑ parietal cells of ↑ gastric glands proper (fig.), where they are studded with many microvilli (Mv); C. open at ↑ apical pole (arrow), thereby increasing cell's secretory surface. C. are less developed in cells of ↑ cytrophoblast and ↑ lutein cells. In white ↑ adipose cells, ↑ liver parenchymal cells, and brown ↑ adipose cells, C. have a form of short intracellular tubular invagination.

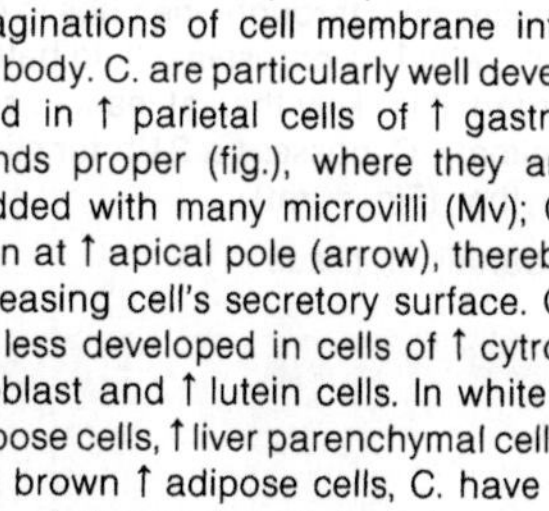

Blanchette-Mackie, J.E., Scow, R.O.: Anat. Rec. *203*, 205 (1982)

Canaliculi, lacrimal: ↑ Lacrimal ducts

Canaliculi, of bone (canaliculus osseus*): Very slender, branching tunnels (C), 0.1–0.2 µm in diameter, radiating in all directions from each ↑ lacuna (L) of ↑ osteocytes (O). Wall of C. consists of calcified ↑ bone matrix (CM) and calcified ↑ collagen microfibrils. Within C. a very delicate process (P) of an osteocyte is located, joining here, by ↑ nexus and/or ↑ zonula occludens (J), the similar process of an adjacent osteocyte. C. of ↑ osteon's innermost lacunae reach ↑ Haversian canal (HC). Through C., ↑ tissue fluid and metabolites circulate to and from Haversian canal. In general, C. do not penetrate ↑ cement lines.

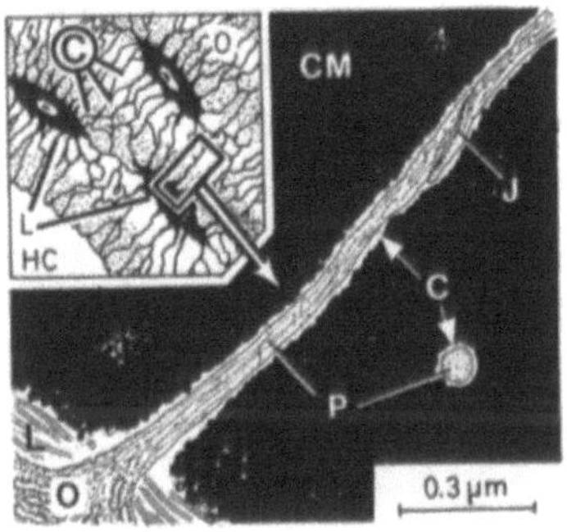

Furseth, R.: Scand. J. Dent. Res. *81*, 339 (1973)

Canaliculi, pineal: A tubular or cleft-like system of channels (C), 1–20 µm wide, between cells of ↑ pineal organ. C. communicate with ↑ perivascular spaces and, through openings in capsule, with ↑ cerebrospinal fluid bathing organ. It is likely that, at least in some animals, C. possess a 24-h functional rhythm. (Fig. = rat)

↑ Canaliculi, tissue

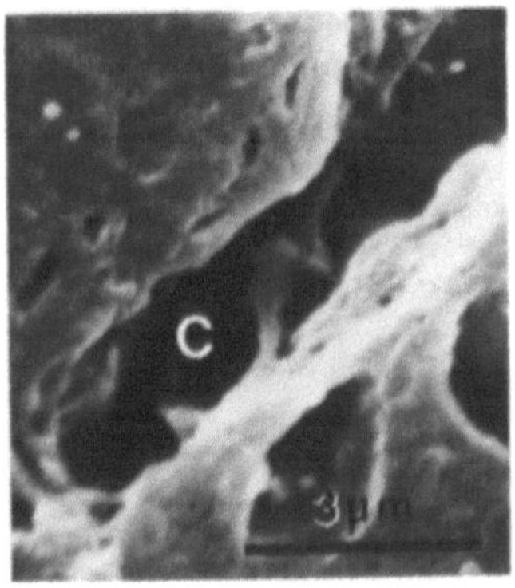

Krstić, R.: Biomed. Res. *2*, Suppl. 129 (1981); Quay, W.B.: Am. J. Anat. *139*, 81 (1974)

Canaliculi, tissue (parenchymatous channels, intercellular canaliculi): Irregular, tubular, or cleft-like 2 to 15-µm-wide spaces of various length lined only by parenchymal epithelial cells of some organs. C. communicate on one side with intercellular spaces and on other with ↑ perivascular spaces and are involved in circulation of ↑ tissue fluid.

↑ Canaliculi, pineal

Fujita, K., Kobayashi, S., Serizawa, Y.: Biomed. Res. *2*, Suppl. 115 (1981)

Canalis analis: ↑ Anal canal

Canalis utriculosaccularis (ductus utriculosaccularis*): A fine tubule lined with ↑ simple squamous epithelium joining utricule to saccule. ↑ Endolymphatic duct arises medially from C.

↑ Labyrinth, membranous

Canals, of Hering: ↑ Bile ductules, terminal

Canals, of Volkmann: ↑ Volkmann's canals

Canals, semicircular: ↑ Semicircular canals

Cancellous bone: ↑ Bone, cancellous

Candidate stem cells (transitional hematopoietic cells, presumptive hematopoietic cells, pluripotential stem cells, lymphocytelike cells, Q-cells, X-cells): Primitively structured, undifferentiated, lymphocytelike cells present in ↑ bone marrow. Small C. are 7–9 µm in diameter, large ones about 12 µm. Nucleus is large, making ↑ nucleoplasmic index favorable to nucleus; nucleus is more irregularly shaped than that of a ↑ lymphocyte. In addition to one to two large nucleoli, nucleus contains a considerably smaller amount of ↑ heterochromatin than that of a lymphocyte. In thin cytoplasmic rim, which may be pale and/or ↑ basophilic, are scattered a few small mitochondria and a small to moderate number of free ribosomes. Golgi apparatus and rough endoplasmic reticulum are minimally developed or absent. C. are capable of maintaining themselves by mitotic division and differentiating into precursors of any other blood cell lineage. C. concept is compatible with concept of ↑ colony-forming units for existence of which there is as yet only indirect evidence; in absence of direct proof, some authors prefer the term C.

↑ Hemocytoblast

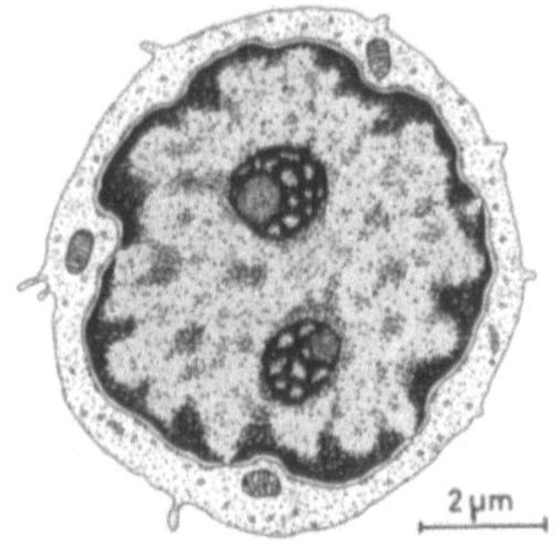

Hematopoietic Stem Cells: Ciba Foundation Symposium. New Series. London: Elsevier 1973; Yoffey, J.M.: Int. Rev. Cytol. *62*, 311 (1980)

Cap phase, of acrosome formation: ↑ Acrosome formation

Cap phase, of spermiogenesis: ↑ Spermiogenesis

Capacitance vessels: Veins comprising a variable blood reservoir because of their distensibility.

Capacitation: The process during which ↑ spermatozoa acquire ability to fertilize secondary ↑ oocytes. C. is characterized by progressive weakening, vesiculation and disruption of ↑ acrosome. C. takes place in female genital tract and seems to be dependent upon its integrity. According to species, C. requires 1–6 h.

↑ Acrosome reaction

Bedford, J.M.: Biol. Reprod. Suppl. *2*, 128 (1970); Koehler, J.K., Kinsey, W.H.: Scan. Electron Microsc. 1977/II, 325 (1977); Shapiro, B.M., Eddy, E.M.: Int. Rev. Cytol. *66*, 257 (1980)

Capillaries (vasa capillaria*): Terminal ramifications of blood vessels in form of endothelial tubes (C) devoid of ↑ smooth muscle cells. C. measure 3–

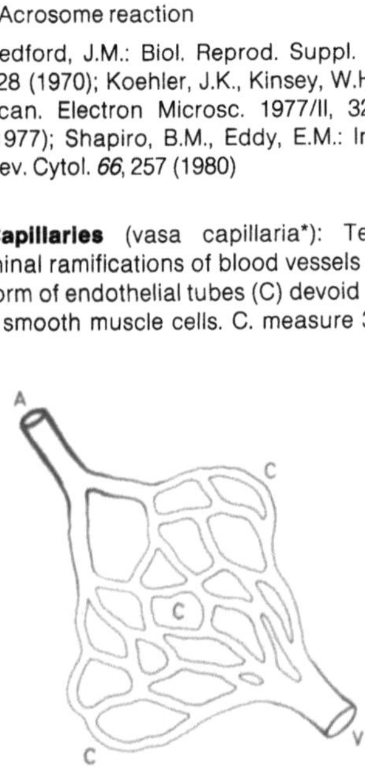

10 µm in diameter, 200–1000 µm in length, and form a richly branched network between smallest ↑ arteriole (A) and ↑ postcapillary venule (V). Wall of C. consists of ↑ endothelium, its ↑ basal lamina, and a few ↑ pericytes. C. are surrounded by a thin adventitia composed of a framework of ↑ reticular and ↑ collagen fibrils.

↑ Capillaries, types of

Capillaries, arterial (vas capillare arteriale*): Blood vessels arising directly from a ↑ precapillary arteriole and situated between them and true ↑ capillaries; structurally identical to corresponding true capillaries.

Capillaries, blood: ↑ Capillaries

Capillaries, continuous (capillaries, nonfenestrated): ↑ Capillaries with uninterrupted ↑ endothelium containing a varying number of ↑ micropinocytotic vesicles. Depending on thickness of endothelial layer, one can distinguish two kinds of C.: 1) Continuous thick endothelium = 0.3 to 0.8-µm-thick endothelium (E), characterized by very numerous ↑ micropinocytotic vesicles (Mv), 50–70 nm in diameter, which open at both basal and luminal surfaces; sometimes these vesicles fuse to form ↑ transendothelial channels (TC). ↑ Endothelial cells are held together by ↑ zonulae occludentes, ↑ nexus (J), and rare ↑ desmosomes. ↑ Basal lamina (BL) is 20–50 nm thick and continuous; it splits and encloses occasional ↑ pericytes and their processes (P). This kind of C. is found in brown ↑ adipose tissue (fig.), ↑ muscular tissues, ↑ testis, and ↑ ovary. 2) Continuous thin endothelium = 0.1 to 0.2-µm-thick endothelium (E) provided with a small number of micropinocytotic vesicles; basal lamina (BL) is almost always continuous; ↑ pericytes (P) are scarce along this kind of C., which is found in CNS, ↑ lung, ↑ connective tissue, ↑ vasa recta of kidney, ↑ spleen, ↑ thymus, ↑ lymph nodes, ↑ bone marrow, and in ↑ Haversian canals. (Fig. = brain, rat)

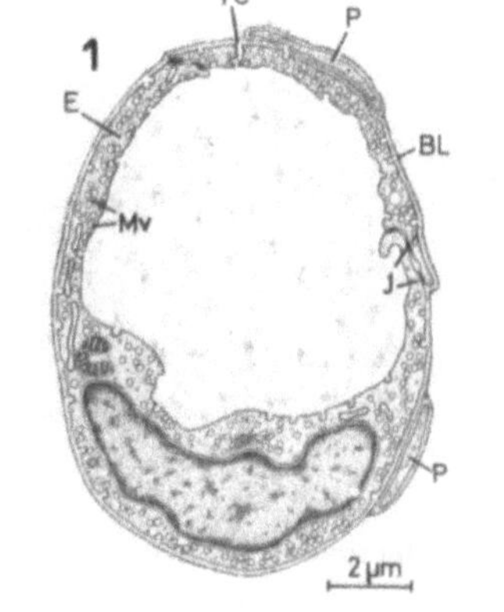

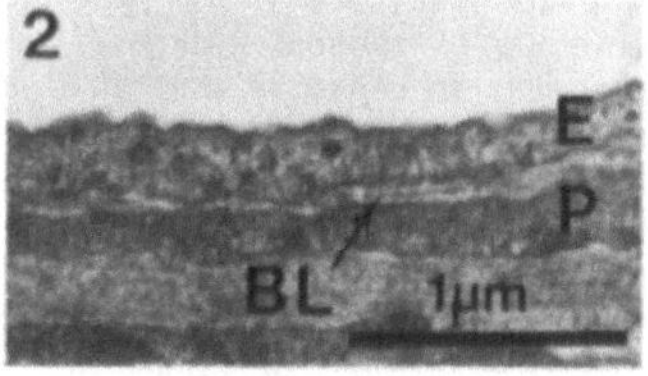

Ludatscher, R.M., Stehbens, W.E.: Z. Zellforsch. *97*, 167 (1969); Simionescu, N., Simionescu, M., Palade, G.E.: J. Cell Biol. *64*, 586 (1975); Simionescu, N., Simionescu, M., Palade, G.E.: Thromb. Res. Suppl. II, *8*, 257 (1976)

Capillaries, discontinuous: ↑ Capillaries, sinusoidal; ↑ Capillaries, of adrenal cortex

Capillaries, ectatic: ↑ Endometrium, vascularization of; ↑ Lacunae, of endometrium

Capillaries, fenestrated (capillaries porous): A type of capillary with very thin ↑ endothelium averaging 80 nm and perforated by ↑ capillary pores (P). Endothelial cells (EC), are held together by ↑ zonulae occuludentes and ↑ nexus (J), occurring in thicker endothelial areas which are devoid of pores, but provided instead with ↑ micropinocytotic vesicles (Mv). Basal lamina (BL) is continuous; ↑ pericytes (Pe) are rare. C. are found in ↑ endocrine glands, ↑ kidney, ↑ choriocapillaris, etc.

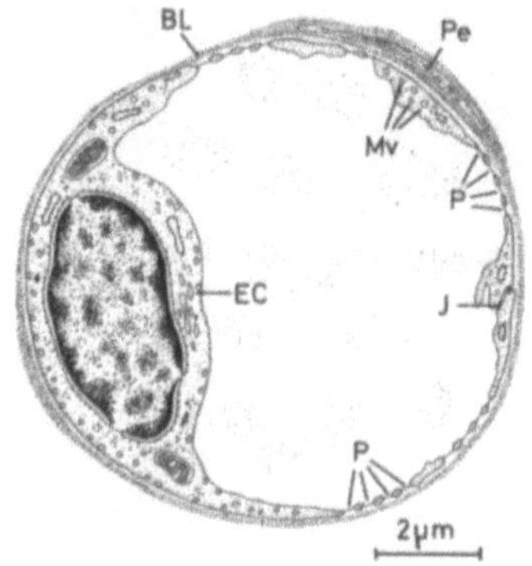

Clementi, F., Palade, G.E.: J. Cell Biol. *41*, 33 (1969); Maul, G.G.: J. Ultrastruct. Res. *36*, 768 (1971); Simionescu, M., Simionescu, N., Palade, G.E.: J. Cell Biol. *60*, 128 (1974); Wolff, J., Merker, H.J.: Z. Zellforsch. *73*, 174 (1966)

Capillaries, glomerular: ↑ Glomerular capillaries

Capillaries, lymphatic: ↑ Lymphatic capillaries

Capillaries, marginal fold of: ↑ Marginal fold

Capillaries, nonfenestrated: ↑ Capillaries, continuous

Capillaries, of adrenal cortex: ↑ Capillaries belonging to sinusoidal type with numerous large holes (H) and few ↑ capillary pores (P) in their ↑ endothelial cells. (Fig. = rat)

↑ Adrenal gland, vascularization of; ↑ Capillaries, sinusoidal

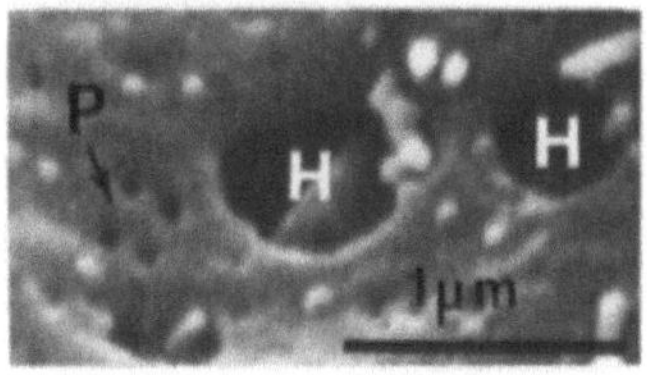

Capillaries, permeability of: ↑ An active and passive exchange of substances across ↑ capillary wall. Permeability is result of intracapillary hydrostatic pressure, oncotic pressure of plasma, hydrostatic and oncotic pressures of ↑ tissue fluid, concentration gradients of substances inside and outside capillaries, passive diffusion of substances across endothelial cell (lipid-soluble molecules), and structure of capillary wall (↑ capillary pores, ↑ caveolae of endothelial cells, ↑ transendothelial channels, intercellular ↑ junctions, etc.) (See physiology texts for further information)

Palade, G.E., Simionescu, M., Simionescu, N.: Acta Physiol. Scand. Suppl. *463*, 11 (1979); Simionescu, N., Simionescu, M., Palade, G.E.: Thromb. Res. Suppl. II, *8*, 257 (1976)

Capillaries, secretory: ↑ Canaliculi, intercellular

Capillaries, sheathed of spleen (sheathed arterioles, arteriola vaginata*): Branches of ↑ penicillar arteries with a luminal diameter of about 15 µm. Some segments of C. are enveloped by ↑ ellipsoids of Schweigger-Seidel. Wall of C. contains few undifferentiated ↑ smooth muscle cells. C. are continuous with terminal arterial ↑ capillaries.

↑ Spleen, vascularization of

Capillaries, sinusoidal (blood sinuses, discontinuous capillaries, sinusoids, vascular sinuses, vas capillare sinusoideum*): A type of ↑ capillary with 0.5 to 3-µm-wide intercellular and transcellular holes (H). ↑ Endothelial cells (E) are held together by ↑ nexus and ↑ zonulae occludentes (J) and contain, besides other ↑ organelles, an appreciable number of ↑ lysosomes (Ly) and ↑ micropinocytotic vesicles. Basal lamina (BL) is almost entirely absent; its traces may be seen in some places. ↑ Pericytes are very rare. This type of C. is found in ↑ liver (fig.), ↑ spleen, ↑ bone marrow, ↑ adrenal cortex, and ↑ adenohypophysis.

↑ Capillaries, of adrenal cortex

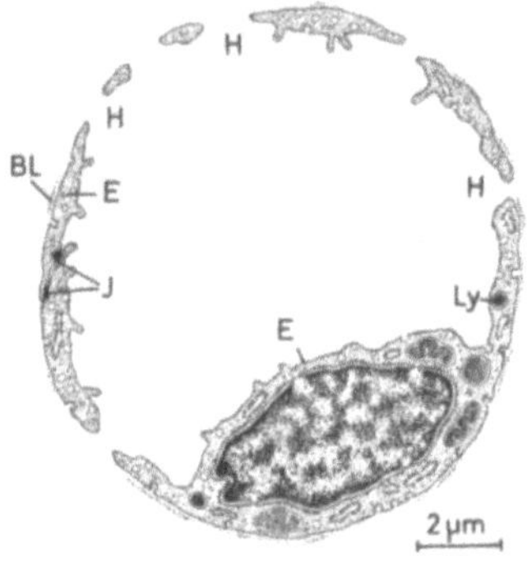

Motta, P.: Cell Tissue Res. *164*, 327 (1975); Muto, M.: Arch. Histol. Jpn. *37*, 368 (1975); Wisse, E.: J. Ultrastruct. Res. *31*, 125 (1970)

Capillaries, true: ↑ Capillaries

Capillaries, types of: Depending on position of ↑ capillaries between arterial and venous circulations, as well as continuity of their walls, they are classified: 1) ↑ Capillaries, arterial; 2) capillaries, true = a) ↑ capillaries, continuous, b) ↑ capillaries, fenestrated; c) ↑ capillaries, sinusoidal; 3) ↑ capillaries, venous.

Capillaries, venous (vas capillare venosum*): Blood vessels, 3–10 µm in diameter, formed by confluence of two to three true ↑ capillaries and continuous with ↑ postcapillary venule. Structurally identical to corresponding true ↑ capillaries.

Anderson, B.G., Anderson, W.D.: Biomed. Res. Suppl. *2*, 209 (1981)

Capillary bed: ↑ Total anastomosing capillary ramifications in a tissue or organ. In C., blood flows slowly and under low pressure.

Capillary pores, of fenestrated capillaries (fenestrae): About 50 to 80-nm-wide round openings (P) perforating thin ↑ endothelium (E) of fenestrated ↑ capillaries. C. are usually closed by a single-layered membrane or diaphragm (D), 4–6 nm thick, having central thickening (arrowhead) of about 12 nm in diameter. There are about 20–60 C./µm². Morphological evidence of existence of a class of smaller C. (9 nm in diameter) is still lacking. Large water-soluble molecules pass through C. BL = ↑ basal lamina (Fig. = ↑ parathyroid gland, rat)

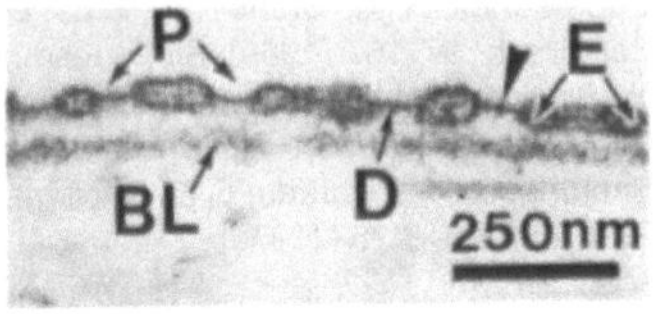

Capillary pores, of glomerular capillaries: Transcellular circular openings (P) in extremely attenuated areas of ↑ endothelial cells. Pores average 100 nm in diameter and frequently lack a diaphragm. There are about 40 pores/µm² in the rat, covering 30% of entire surface of endothelial cell. (Figs. = rat)

↑ Glomerular capillaries

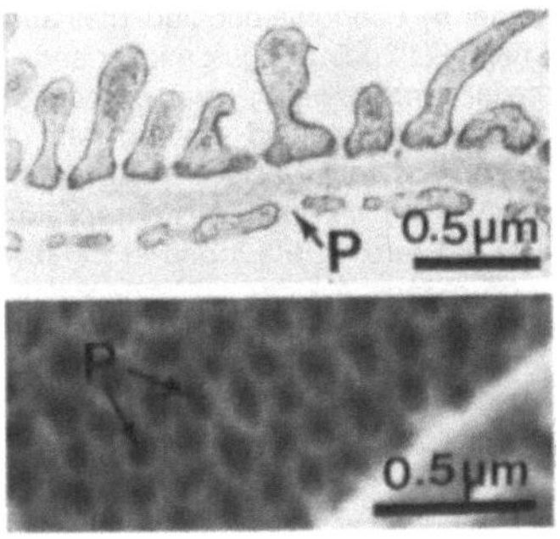

Capillary wall: A layer of cells and noncellular structures separating blood (B) from surrounding tissue (T). Structure: 1) Tunica intima = a) ↑ endothelium (E); b) its ↑ basal lamina (BL); c) a few ↑ pericytes (P). 2) Tunica media = virtually absent. 3) Tunica adventia (TA) = a pericapillary sheath of ↑ loose connective tissue rich in ↑ reticular fibers.

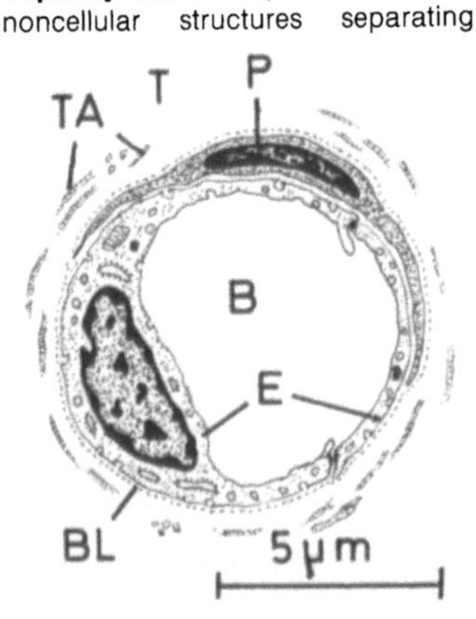

Capitulum: ↑ Spermatozoon, connecting piece of

Capsula: ↑ Capsule

Capsula articularis: ↑ Capsule, of joint

Capsula glomeruli: ↑ Bowman's capsule

Capsular cells: ↑ Satellite cells, of peripheral neurons

Capsular epithelium, of Bowman's capsule (parietal wall, pars externa capsulae glomeruli*): A layer of polygonal, flattened ↑ epithelial cells (E) with ↑ perikarya bulging slightly toward ↑ Bowman's space (BS). ↑ Epithelium rests upon a basal lamina (BL), which is frequently multiple. Mitochondria and rough endoplasmic reticulum are relatively well developed; through cytoplasm run bundles of contractile ↑ actinlike microfilaments (asterisk), whose contraction may influence glomerular filtration rate through an increase in pressure within Bowman's space. (Fig. = rat)

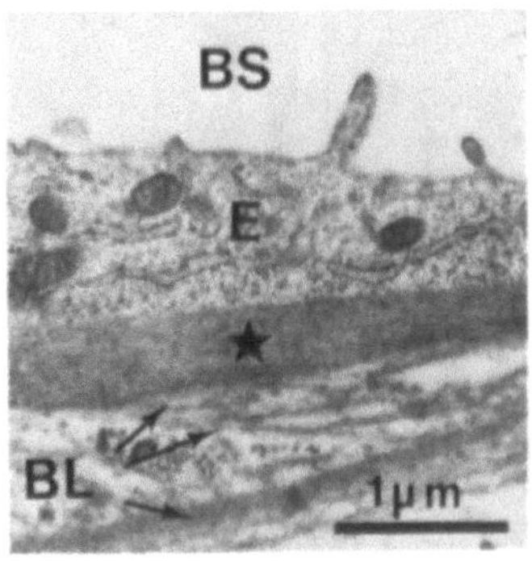

Capsular space: ↑ Bowman's space

Capsule (capsula*): A well-delimited envelope of variable thickness surrounding an organ or part of it. In general, C. consists of ↑ dense connective tissue, except for C. of ↑ lens, ↑ vitreous body, and ↑ renal glomerulus.

Capsule, of chondrocyte (territorial matrix): A deeply ↑ basophilic and ↑

metachromatic zone (C) of ↑ cartilage matrix, 1–3 μm wide, surrounding ↑ chondrocytes (Ch) and ↑ isogenic groups (IG). Stainability of C. is based on presence of concentrated ↑ chondromucoprotein. Under light microscope, there is, in general, a narrow space (arrowheads) between C. and chondrocyte as a result of shrinkage of latter during ↑ fixation. Under transmission electron microscope, C. appears as a clear zone around chondrocyte and contains delicate ↑ collagen microfibrils (CF) in close contact with ↑ matrix vesicles (MV). Amount of chondromucoprotein is considerably lower in interterritorial matrix (ITM).

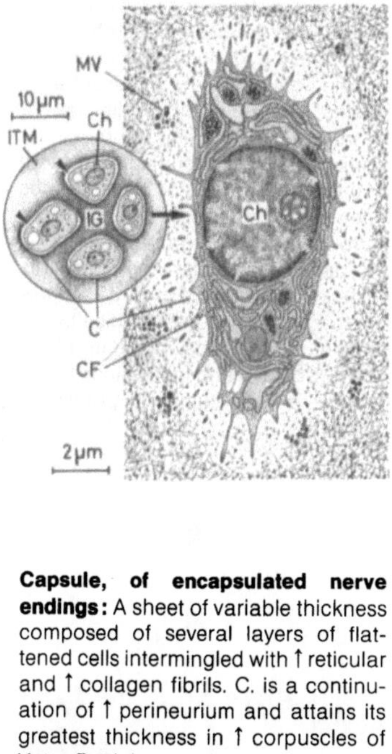

Capsule, of encapsulated nerve endings: A sheet of variable thickness composed of several layers of flattened cells intermingled with ↑ reticular and ↑ collagen fibrils. C. is a continuation of ↑ perineurium and attains its greatest thickness in ↑ corpuscles of Vater-Pacini.

Capsule, of Glisson: ↑ Capsule, of liver

Capsule, of joint (capsula articularis*): A thick layer of ↑ dense connective tissue (Cp) enveloping a ↑ diarthrosis. Structure: 1) ↑ Synovial membrane (SM) = inner layer of C. containing numerous blood (BV) and lymphatic vessels, as well as sensitive nerve endings. 2) Fibrous layer or fibrous capsule (FL) = a poorly vascularized and innervated dense connective tissue sheet continuous with ↑ periosteum (P); this layer is subdivided into an inner portion with longitudinally (L) and an outer portion with circularly (C) oriented ↑ collagen fibers. Depending on anatomical localization of diarthrosis, collagen bundles may have other directions. Thickenings of C. form intra-articular ligaments.

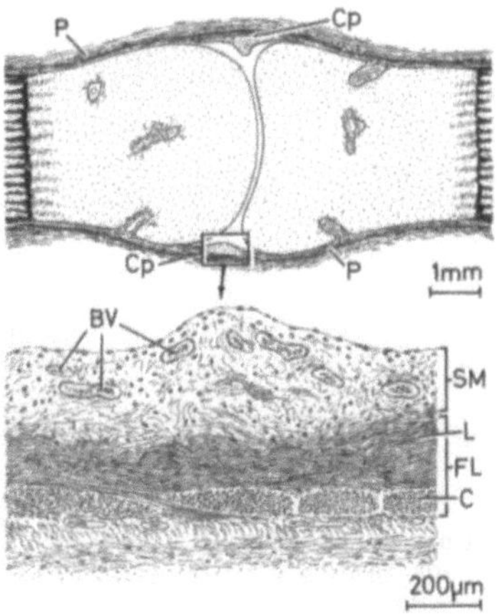

Capsule, of lens (capsula lentis*): A thick basal lamina (CL) of lens epithelium (LE). On anterior surface of ↑ lens, C. measures 10–18 μm, whereas on posterior it is 5–9 μm thick. C. is composed of ↑ glycoproteins synthesized in thin layers (L) by lens epithelium. Material of C. is an elastic network formed of short, interwoven 3 to 5-nm-thick microfilaments (F). C. isolates lens from rest of body; it is impermeable to ↑ macrophages and antigens, but plays an important role in lens metabolism. C. serves as insertion site for ↑ zonular fibers. Elasticity of C. is essential for eye accomodation.

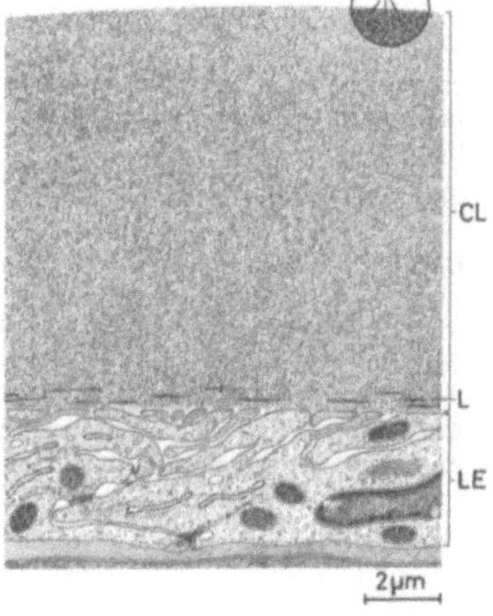

Rafferty, N., Esson, E.A.: J. Ultrastruct. Res. *46*, 239 (1974)

Capsule, of liver (capsule of Glisson): A thin layer of moderately ↑ dense connective tissue (G) enclosing ↑ liver and covered by peritoneal ↑ mesothelium (P). From C., connective tissue extends into peripheral zone of liver ↑ parenchyma separating ↑ lobules and surrounding vessels and ↑ bile pathways. (Fig. = human)

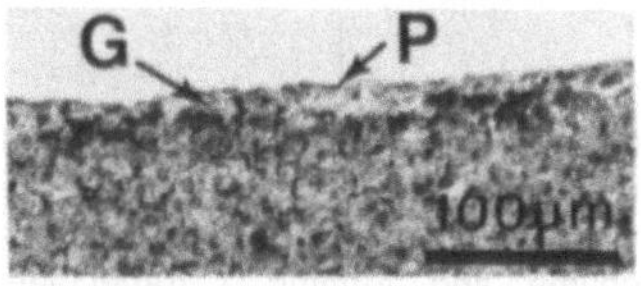

Capsule, of lymph node (capsula*): A thin layer of ↑ dense connective tissue (C) enclosing ↑ lymph nodes. C. contains some ↑ fibroblasts, occasional ↑ smooth muscle cells, and a network of fine ↑ elastic fibers. To outer surface of C. adheres some ↑ adipose (A) and ↑ loose connective tissue; inner aspect of C. is well delimited and lined by ↑ littoral cells (LC) of ↑ marginal sinus (MS), which at convex side of lymph node separates C. from cortical ↑ parenchyma (CP), except in areas of the ↑ hilus, where C. is thickened. C. is traversed by afferent lymphatic vessels (LV), whose ↑ endothelium (E) is continuous with littoral cells. C. sends numerous trabeculae into ↑ parenchyma.

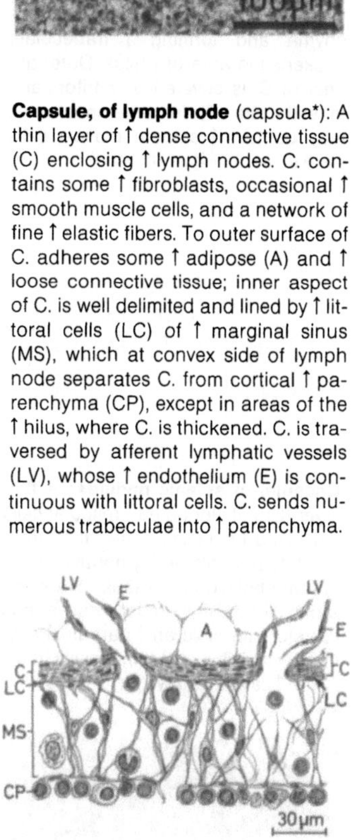

Capsule, of prostate (capsula prostatica*): A thick ↑ dense connective tissue (C) surrounding ↑ prostate and

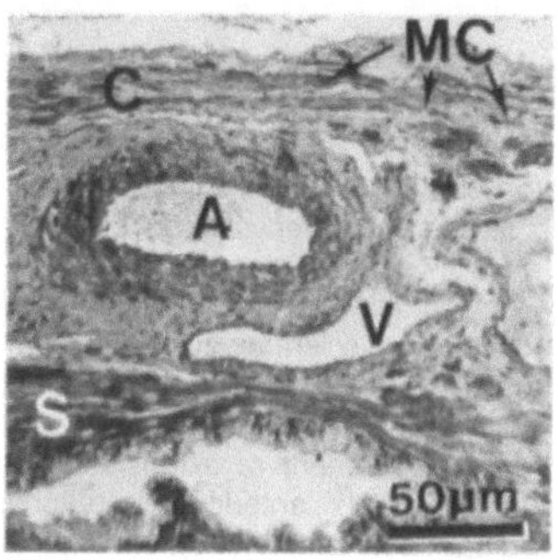

continuous with prostatic ↑ stroma (S). Together with many ↑ elastic fibers, ↑ muscular arteries (A), ↑ veins (V), and strands of ↑ smooth muscle cells (MC) continuous with those scattered in ↑ stroma, C. contains, at periphery, occasional ↑ skeletal muscle fibers belonging to diaphragma urogenitale. (Fig. = human)

Capsule, of spleen (tunica fibrosa* et tunica serosa*): A ↑ dense connective tissue (C) enveloping splenic ↑ parenchyma and forming ↑ trabeculae; thickened in area of ↑ hilus. Outer aspect of C. is covered by peritoneal ↑ mesothelium (P). In some species, C. may contain ↑ smooth muscle cells (rare in man). (Fig. = human)

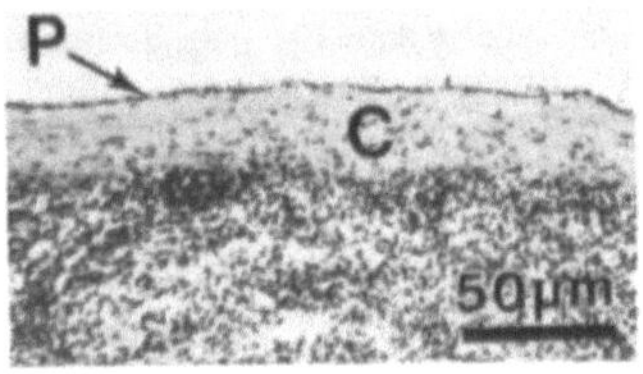

Capsule, of splenic nodules: A thin layer (Cp) of elongated ↑ reticular cells delimiting ↑ splenic nodules (SN) from ↑ red pulp or diffuse ↑ lymphatic tissue in ↑ lymphatic nodules. C. is covered by a mass of small ↑ lymphocytes, and is therefore invisible on ↑ paraffin ↑ sections; it is visible on ↑ semithin sections. CA = ↑ central artery (Fig. = dog)

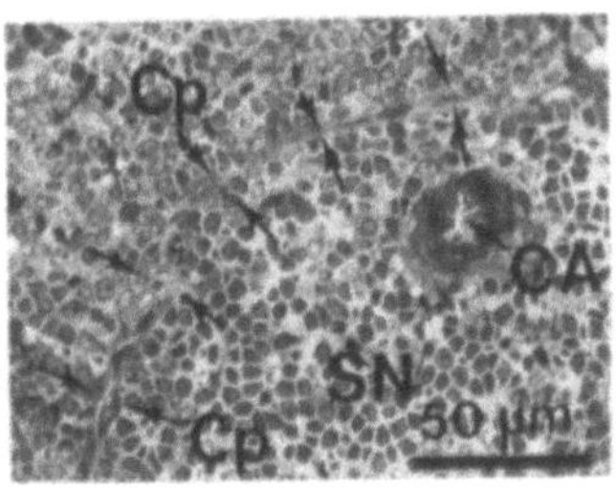

Capsule, of Tenon (fascia oculi, vagina bulbi*): A ↑ dense connective tissue sheath (C) enveloping outer aspect of ↑ sclera (S) and setting it apart from orbital ↑ adipose tissue (AT). A narrow ↑ episcleral space (ES) separates C. from sclera. C. begins at ↑ optic nerve (ON) and terminates just anterior to attachment of extrinsic ocular muscles (EM); anteriorly, it reaches bulbar ↑ conjunctiva (BC). C. serves as spherical articular surface for movements of ↑ eyeball.

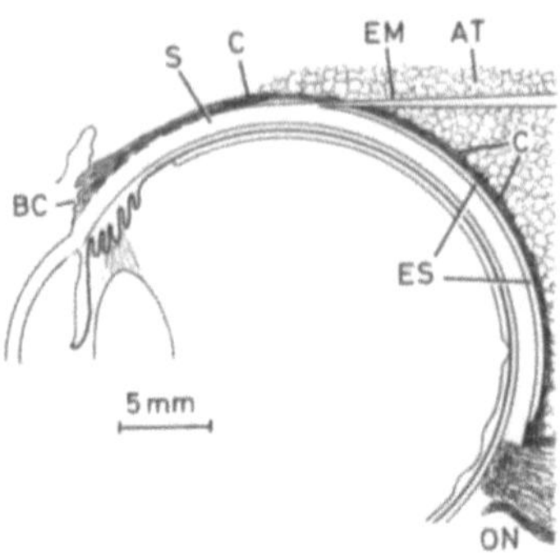

Capsule, of testis: ↑ Tunica albuginea, of testis

Capsule, of thymus: A thin layer of moderately ↑ dense connective tissue enclosing the ↑ thymus. C. contains a variable number of ↑ macrophages, ↑ mast cells, ↑ plasma cells, eosinophilic ↑ granulocytes, and ↑ adipose cells. During ↑ involution of organ, adipose cells increase in number. Septa arise from C. separating thymic lobules.

Capsule, of vitreous body: ↑ Vitreous body

Caput gallinaginis: ↑ Colliculus seminalis

Carbohydrates: A group of chemical compounds including small molecules, such as glucose and sucrose (= mono- and disaccharides), and macromolecular substances, such as starch and ↑ glycogen (= polysaccharides). C. are among most important energy-supplying materials used by cells.

Sharon, N.: Sci. Am. *243/5*, 80 (1980)

Cardia (cardiac orifice, ostium cardiacum*): Opening of ↑ esophagus into ↑ stomach, characterized by ↑ esophagogastric junction.

Cardiac area (cardiac portion, pars cardiaca*): A ring-shaped zone (CA) of ↑ stomach, about 5–30 mm wide, surrounding ↑ cardia (C) and characterized by presence of ↑ cardiac glands. Layers as in stomach. F = ↑ fundus, of stomach

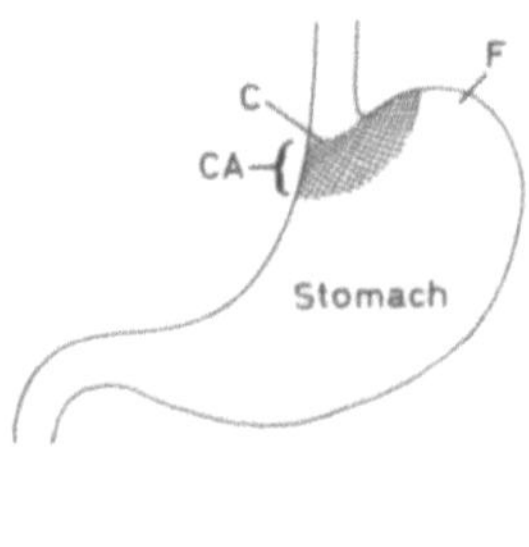

Cardiac glands, esophageal (glandulae cardiacae esophagi*): Inconstantly branched and coiled ↑ mucous glands in lower part of ↑ esophagus, near ↑ cardia. C. are similar to ↑ cardiac glands of stomach, are lined with cuboidal and columnar cells, and have frequent ampullar dilatations.

↑ Esophageal glands proper

Cardiac glands, of stomach (glandula cardiaca*): Richly ramified, slightly coiled, ↑ tubular glands (CG) located in ↑ cardiac area of ↑ stomach. C. are lined solely by columnar, mucus-producing cells beginning at ↑ esophagogastric junction (EGJ). Cells have a basal flattened nucleus; supranuclear portion of cell is occupied by a large quantity of ↑ mucous droplets. Glands open directly into ↑ foveolae gastricae (arrows). ↑ Enterochromaffin cells and occasional ampullar enlargements (asterisk) are present in C. C. contribute to formation of gastric juice by secretion of ↑ mucus, calcium phosphate, sodium and potassium bicarbonate, and sodium and potassium chloride.

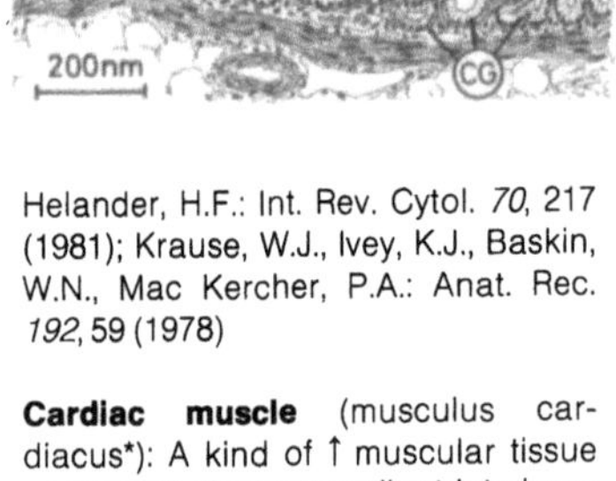

Helander, H.F.: Int. Rev. Cytol. *70*, 217 (1981); Krause, W.J., Ivey, K.J., Baskin, W.N., Mac Kercher, P.A.: Anat. Rec. *192*, 59 (1978)

Cardiac muscle (musculus cardiacus*): A kind of ↑ muscular tissue composed of transversally striated very ramified ↑ cardiac muscle fibers (MF), which anastomose to form a complex three-dimensional network constituting ↑ myocardium. ↑ Intercalated discs (D) divide fibers into ↑ cardiac muscle cells (arrows). Structure of transverse ↑ bands is identical to that of ↑ skeletal muscle fibers. Between fibers, there is an abundant ↑ endomysium (E), rich in blood and lymphatic capillaries (Cap) and nerve fibers.

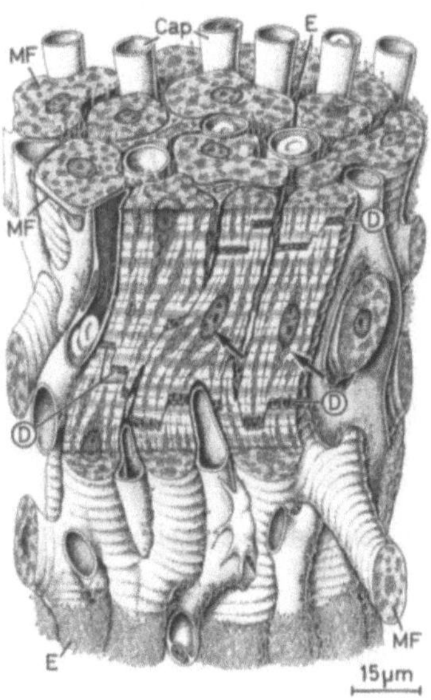

Cardiac muscle cells (myocytus cardiacus*): Simple or branched segments of a ↑ cardiac muscle fiber (100–150 µm long and 10–20 µm thick), limited by two or more ↑ intercalated discs (D). C. contain one to two nuclei (N) in

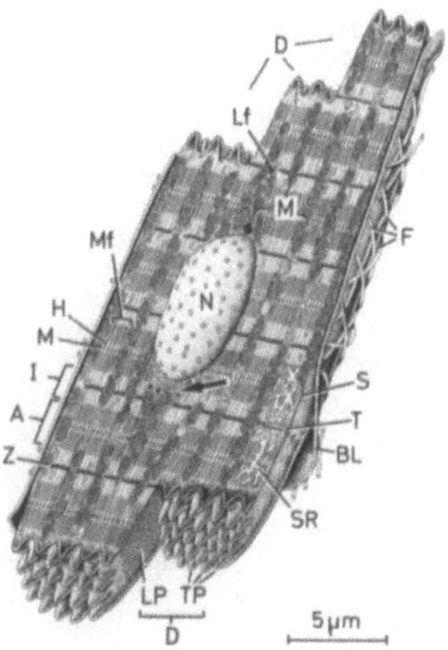

a central position. ↑ Myofibrils (Mf) are longitudinally oriented, diverging as they approach nuclear poles and leaving a conical sarcoplasmic zone, ↑ endoplasm (arrow), in which are located mitochondria, Golgi apparatus, rough endoplasmic cisternae, part of ↑ glycogen contained in cell, a few ↑ lipid droplets, occasional ↑ lysosomes, ↑ lipofuscin granules (Lf) (in the aged), and ↑ atrial granules. Mitochondria (M) are large and abundant; they are disposed in rows along myofibrils and possess many densely packed cristae; between ends of mitochondria, lipid droplets can be found. Arrangement in ↑ sarcoplasm of ↑ actin and ↑ myosin myofilaments forms the ↑ bands characteristic of ↑ striated muscles (↑ A, ↑ I, ↑ H, ↑ M, ↑ Z). Actin myofilaments anchor to transversal portion (TP) of intercalated disc. Myofilaments are not attached to longitudinal portion (LP) of intercalated disc, which is covered by hexagonal arrays of ↑ membrane-associated particles. Sarcoplasmic reticulum (SR) and ↑ T-tubules are similar, but not identical, to those of ↑ skeletal muscle; they form ↑ dyads. Outer aspect of ↑ sarcolemma (S) is covered with a basal lamina (BL) reinforced by a network of ↑ reticular and delicate ↑ collagen fibrils (F).

↑ Striated muscle fibers, molecular biology of contraction of; ↑ Striated muscle fibers, morphology of contraction of

Segretain, D., Rambourg, A., Clermont, Y.: Anat. Rec. *200*, 139 (1981)

Cardiac muscle cells, in pulmonary veins: A layer of ↑ cardiac muscle cells exists in adventitia of ↑ pulmonary veins near junction with left atrium. In some animals (bat, rat, shrew), C. contribute to formation of media and extend deeply into intrapulmonary veins. Contraction of C. acts as a dynamic valve to facilitate venous return.

Cardiac muscle cells, mechanism of contraction in: ↑ Striated muscle fibers, morphology of contraction of

Cardiac muscle cells, sarcoplasmic reticulum of (reticulum sarcoplasmaticum*): A system of relatively simple, longitudinal, smooth channels (SR) extending over full length of ↑ sarcomere (Sm); ↑ sarcoplasmic reticulum occupies longitudinal clefts between ↑ myofibrils (Mf). Instead of ↑ terminal cisternae, small terminal expansions (E) contact T-tubules and form ↑ dyads (Dy) rather than ↑ triads. Some expansions of sarcoplasmic reticulum are applied to inner aspect of ↑ sarcolemma (S) forming ↑ subsarcolemmal cisternae (SC).

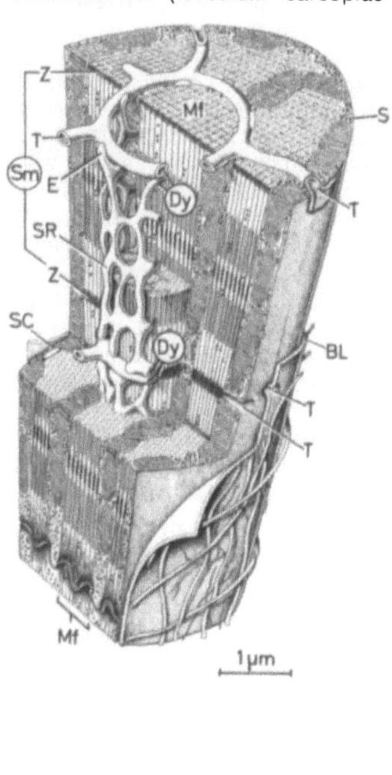

↑ Cardiac muscle cells, T-tubules of

Scales, D.J.: J. Ultrastruct. Res. *83*, 1 (1983)

Cardiac muscle cells, T-tubules of (tubulus transversus*): Deep transversal invaginations of ↑ sarcolemma (S) into ↑ sarcoplasm at level of ↑ Z-line. T-tubules in ↑ cardiac muscle cells are much larger than in ↑ skeletal muscle, permitting ↑ basal lamina (BL) to extend into T-tubules as a thin intratubular coating. C. contribute to exchange of metabolites between cardiac muscle cells and extracellular space and aid in transmission of electric impulses. (See fig. under ↑ Cardiac muscle cells, sarcoplasmic reticulum of)

Cardiac muscle fibers (myofibra cardiaca*): Ramified cords of ↑ cardiac muscle cells held together by ↑ intercalated discs and anchoring to ↑ cardiac skeleton.

↑ Cardiac muscle

Cardiac muscle, regeneration of: Repair of cardiac muscle is impossible

Cardiac skeleton: A ↑ dense connective tissue of heart separating auricular from ventricular musculature. C. is composed of four ↑ annuli fibrosi, two ↑ trigona fibrosa, and ↑ septum membranaceum. C. is supporting tissue of heart and insertion zone for ↑ cardiac muscle fibers.

Cardiovascular system (blood circulatory system, circulatory system, systema cardiovasculare*): A specialized circuit of continuous, closed branching tubes (↑ blood vessels) carrying ↑ blood pumped by ↑ heart. Arrows indicate direction of blood flow; ↑ microvascular bed is framed. (Terms marked with an asterisk are explained under corresponding letters)

↑ Microvasculature

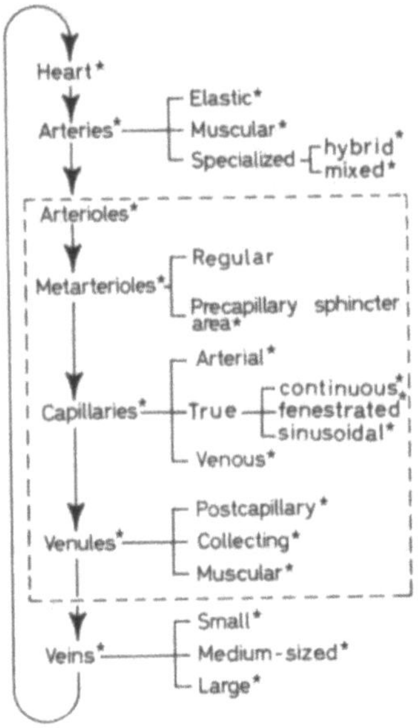

Carina, of trachea (carina tracheae*): A cartilaginous sagittal ridge situated at lower end of windpipe where it divides into two primary ↑ bronchi. Here, C. receives additional support from last ↑ cartilage of ↑ trachea, which extends below bifurcation. Structure of C. corresponds to that of ↑ trachea.

Carotenoids: A group of yellowish to orange liposoluble exogenous ↑ pigments formed in several kinds of vegetables (carrots, tomatoes). Because of their solubility in fat, C. are classed as ↑ lipochromes. Ingested with food, C. are responsible for color of ↑ adipose tissue, reticular zone of ↑ adrenal gland, ↑ yolk, and ↑ corpus luteum. Some C. are provitamins and may be converted into vitamin A in body.

Carotid body (intercarotid body, glomus caroticum*): An inconspicuous highly vascular ↑ paraganglion at bifurcation of each common carotid artery near ↑ carotid sinus (CS). Structure: 1) Fine connective capsule (C). 2) ↑ Parenchyma = clusters and strands of granular ↑ chromaffin ↑ glomus cells type I surrounded by nongranular ↑ glomus cells type II. 3) Sinusoidal ↑ capillaries (Cap) in close contact with both kinds of cells. 4) Efferent nerve fibers of sinus branch of glossopharyngeal nerve and postganglionic fibers of sympathetic superior cervical ganglia form a dense network and contact type I cells. 5) Some ↑ free nerve endings are present in ↑ pericapillary spaces. Excitatory impulses generated in these fibers (suspected to be true chemoreceptor nerve endings) are transmitted to vasomotor center of medulla oblongata. C. is an endocrine organ and a ↑ chemoreceptor reacting to a fall in oxygen concentration, an increase in carbon dioxide, and a decrease in blood pH. C. gives a positive ↑ chromaffin reaction.

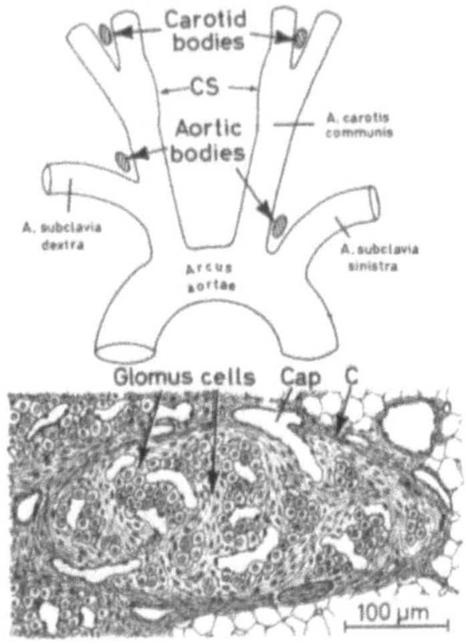

Abraham, A.: Z. mikrosk.-anat. Forsch. *95*, 33 (1981); Biscoe, T.J.: Physiol. Rev. *51*, 437 (1971); Böck, P.: The Paraganglia. In: Handbuch der mikroskopischen Anatomie. Vol. 6, Part 8. Berlin, Heidelberg, New York: Springer-Verlag 1982; Verna, A.: Int. Rev. Cytol. *60*, 271 (1979)

Carotid sinus (baroreceptor field, sinus caroticum*): A slight dilatation of initial segment of internal carotid artery at bifurcation of common carotid artery opposite ↑ carotid body. Media of C. is characterized by an increase in thickness of ↑ elastic laminae compared with other parts of arterial wall; adventitia is relatively thick. Both are very rich in nerve endings from carotid branch of glossopharyngeal nerve. Nerve terminals are located in inner third of adventitia; they show local thickenings filled with numerous mitochondria and ↑ glycogen particles. Nerve endings react to changes in blood pressure, they are, therefore, considered ↑ baroreceptors. Groups of ↑ glomus cells type I have been found outside adventitia.

Clarke, J.A., Burgh Daly de, M.: Cell Tissue Res. *216*, 603 (1981); Gorgas, K., Reinecke, M., Weihe, E., Forssmann, W.G.: Anat. Embryol. *167*, 347 (1983); Yates, R.D., Chen, I.: Cell Tissue Res. *205*, 473 (1980)

Carrier proteins: Substances bound to certain hormones (↑ steroid hormones, ↑ thyroxin) for their transport in blood.

Cartilage (textus cartilagineus*): An avascular ↑ supporting tissue composed of cells, ↑ chondrocytes, and a solid intercellular matrix. Latter consists of fibers (↑ collagen, ↑ elastic) and ↑ cartilage ground substance. With a few exceptions, each piece of C. is surrounded by ↑ perichondrium. There are three kinds of C.: elastic, fibrous, and hyaline ↑ C.

Anderson, D.R.: Am. J. Anat. *114*, 403 (1964); Hall, B.K. (ed.): Cartilage. Vols. 1–3. New York: Academic Press 1983; Thyberg, J.: Histochem. J. *9*, 259 (1977)

Cartilage, appositional growth of: ↑ Cartilage, growth of

Cartilage, articular: ↑ Articular cartilage

Cartilage bones (endochondral bones, osteochondral complexes): ↑ Bones formed in place of cartilaginous model after its almost total destruction in course of indirect ↑ bone formation. C. include base of skull, vertebrae, pelvis, and bones of extremities.

Cartilage, calcification of: In hyaline and fibrous ↑ cartilage, mechanism of C. is principally the same as in calcification of ↑ bone. However, spreading of ↑ hydroxyapatite crystals from ↑ matrix vesicles to surrounding ↑ cartilage ground substance is impaired, either because of too low a concentration of calcium ions or because chemical composition of ↑ cartilage ground substance is not suitable for hydroxyapatite deposition.

Ali, S.Y.: Calcification of Cartilage. In: Hall, B.K.: Cartilage. Vol. 1. New York: Academic Press 1983; Bonucci, E., Dearden, L.C.: Fed. Proc. *35*, 163 (1976); Davis, W.L., Jones, R.G., Knight, J.P., Hagler, K.H.: J. Histochem. Cytochem. *30*, 221 (1982)

Cartilage canals (canalis cartilaginis*): Channels (CC) running through large masses of hyaline ↑ cartilage [costal cartilages, ↑ epiphyses (E) of long bones, etc.]. C. are lined by a thin ↑ perichondrium (P) composed of primitive connective tissue cells (TC), which may fuse to form ↑ chondroclasts (Ch) or differentiate into ↑ osteoblasts. C. contain large blood vessels, capillaries, and ↑ fibroblasts. C. not only play an important role in nourishment of large cartilaginous pieces, but also, because of osteoblasts, in establishment of secondary ossification centers.

↑ Bone formation, indirect

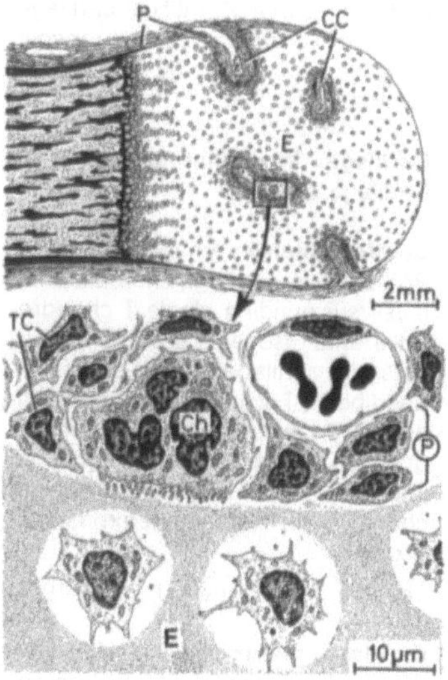

Kugler, J.H., Tomlinson, A., Wagstaff, A., Ward, S.: J. Anat. *129*, 493 (1979)

Cartilage cells: ↑ Chondrocytes

Cartilage, collagen in: Present in all kinds of ↑ cartilage. Because of its concentration, visible under ordinary light microscope in fibrous ↑ cartilage, but masked by matrix ↑ chondromucoprotein in elastic and hyaline cartilage; here, ↑ collagen fibers are visible in polarized light or with transmission electron microscope. In all kinds of cartilage, collagen fibers display a normal cross striation. In general, fibers are randomly distributed, but in fibrous cartilage and in some pieces of hyaline cartilage they are organized in bundles whose direction corresponds to mechanical forces acting on these cartilaginous pieces. C. belongs to type II collagen.

↑ Collagen, types of

Fahmy, A., Hillman, W., Talley, P., Long, V.: Am. J. Bone Joint Surg. *51A*, 802 (1969); Mayne, R., Vondermark, K.: Collagens of Cartilage. In. Hall, B.K. (ed.): Cartilage. Vol. 1. New York: Academic Press 1983

Cartilage, degeneration of: Gradual reduction in capacity of ↑ chondrocytes to produce ↑ chondromucoprotein induces decrease of water in ↑ cartilage matrix and ↑ collagen fibers become unmasked; elastic ↑ cartilage hardly degenerates.

↑ Asbestos cartilage

Cartilage, development of: ↑ Cartilage, histogenesis of

Cartilage, effects of hormones on: ↑ Cartilage, histophysiology of

Cartilage, elastic (cartilago elastica*): A kind of ↑ cartilage characterized by a network of branching ↑ elastic fibers (E) up to 2 µm thick and occasional 1 to 2-µm-thick fenestrated ↑ elastic lamellae (L) in intercellular substance. In peripheral zone, C. also contains ↑ collagen fibers (CF), which in contrast to elastic fibers are masked by ↑ chondromucoprotein; both fibers are continuous with ↑ perichondrium (P). ↑ Chondrocytes (Ch) of C. are round or oval in shape with an ultrastructure corresponding to that of chondrocytes of other cartilages. In deep zone of C., chondrocytes form small ↑ isogenic groups of only two to four cells. C. does not calcify with age. Cap = capillary

↑ Epiglottic cartilage

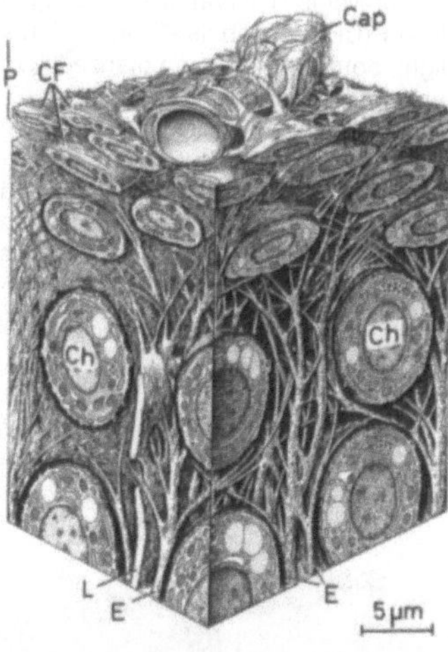

Cox, R.W.: J. Anat. *123*, 283 (1977); Kostović-Knežević, Lj., Bradamante, Ž., Švajger, A.: Cell Tissue Res. *218*, 149 (1981); Nielsen, E.H., Bytzer, P.: J. Anat. *129*, 823 (1979)

Cartilage, elastic, of the body: The ↑ cartilage of ↑ auricle, external ↑ auditory meatus, ↑ auditory tube, ↑ epiglottis, vocal process of ↑ arytenoid cartilage, corniculate cartilage, and bronchial cartilage in small secondary ↑ bronchi.

Cartilage, fibrous (fibrocartilage, cartilago fibrosa*): Very solid ↑ cartilage characterized by a large amount of strong collagen fibers (CF). Only in immediate vicinity of ↑ chondrocytes (Ch) is ↑ chondromucoprotein of ↑ cartilage ground substance in sufficient concentration to mask fibers; elsewhere they are visible under ordinary microscope. Chondrocytes are single or in small ↑ isogenic groups of two cells; their inner structure corresponds to that of other ↑ chondrocytes. In some cases, C. is continuous with ↑ dense connective tissue of ↑ ligaments of ↑ joints; in ↑ intervertebral disks and ↑ symphysis pubis, C. is continuous with hyaline ↑ cartilage. (Fig. = human)

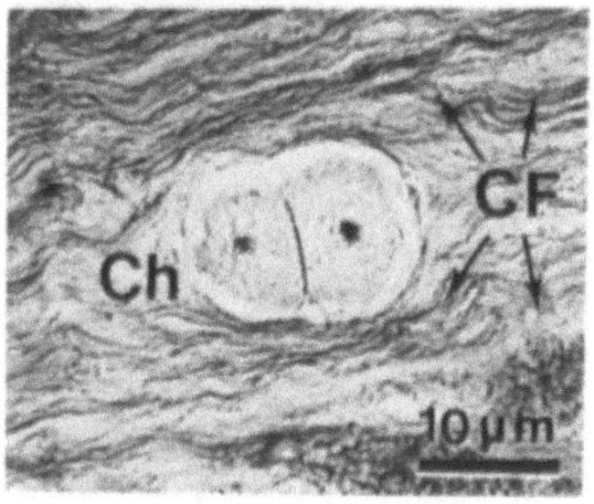

Cartilage, fibrous, of the body: ↑ The cartilage of symphises (↑ symphisis pubis, ↑ intervertebral disc), ↑ articular discs of bones which develop by direct ↑ bone formation (mandibula, clavicula), borders of glenoid fossa of shoulder joint, cotyloid ligament of acetabulum, and at sites of junctions of some ↑ tendons with ↑ bones.

Cartilage, ground substance of: A firm homogenous mass rich in ↑ chondromucoprotein and therefore ↑ PAS-positive, ↑ basophilic, and ↑ metachromatic. C. is produced by ↑ chondrocytes partially in diffuse form and partially in form of ↑ matrix granules. ↑ Collagen and ↑ elastic fibers are scattered throughout C. In hyaline and elastic ↑ cartilage, C. prevents collagen fibers from being observed in ordinary light.

Hall, B.K. (ed.): Cartilage. Vol. 1. New York: Academic Press 1983; Kuettner, B.K., Pauli, B.U., Gall, G., Memoli, V.A., Schenk, R.K.: J. Cell Biol. *93*, 743 (1982); Takagi, M., Parmley, R., Toda, Y., Austin, R.L.: J. Histochem. Cytochem. *30*, 1179 (1982)

Cartilage, growth of: 1) Appositional growth = a progressive deposition by ↑ chondroblasts (Cb) of inner chondrogenic layer of ↑ perichondrium (P) of additional layers of intercellular substance (IS) on surface of cartilage causing its growth in width. 2) Interstitial growth = ↑ mitoses (M) of ↑ chondrocytes (Cc) giving rise to formation of ↑ isogenic groups (IG) and a consequent increase in production of ↑ cartilage matrix (CM). In this way, cartilaginous piece expands from within. This manner of growth is limited to moderately young cartilage.

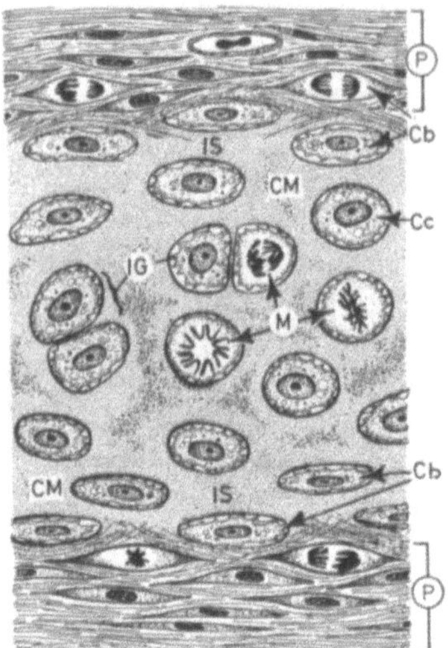

Cartilage, histogenesis of: ↑ Mesenchymal cells (MC) aggregate to form ↑ protochondral tissue (PT) in which they differentiate into ↑ chondroblasts (Ch), which surround themselves with ↑ cartilage matrix (CM). Chondroblasts finally differentiate into ↑ chondrocytes (Cc) situated within their own ↑ lacunae. Continuing growth of cartilage is assured by appositional and interstitial growth.

↑ Cartilage, growth of

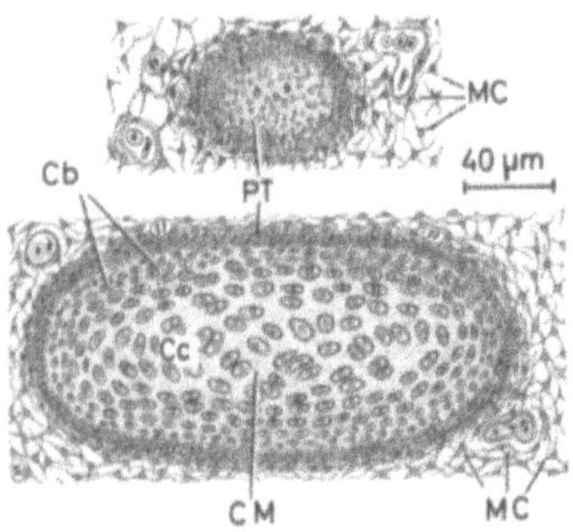

Hall, B.K. (ed.): Cartilage, Vol. 2. Development, Differentiation, and Growth. New York: Academic Press 1983; Knese, K.-H.: Gegenbaurs morphol. Jahrb. *125*, 758 (1979)

Cartilage, histophysiology of: 1) Effects of vitamins: Vitamin A stimulates maturation of ↑ epiphyseal plates; vitamin C is needed for synthesis and maintenance of ↑ collagen fibers and ↑ cartilage ground substance; vitamin D stimulates calcification of cartilage. 2) Effects of hormones: ↑ Growth hormone stimulates secretory activity of ↑ chondrocytes and mitotic activity in epiphyseal plates; ↑ thyroxin stimulates metabolism of chondrocytes; ↑ estrogens and male sex hormones encourage growth of cartilage and fibrillogenesis; ↑ ACTH and ↑ cortisol retard maturation of cartilage and its replacement by bone at level of epiphyseal plate.

↑ Somatomedins; ↑ Somatostatin

Cartilage, hyaline (cartilago hyalina*): The most frequent kind of ↑ cartilage in body, with a solid glasslike ↑ cartilage matrix (CM) and ↑ chondrocytes (Ch) lodged in their own ↑ lacunae (L). ↑ Isogenic groups are frequent in C. and composed of many cells, particularly in deep zone of C. ↑ Mitoses (arrow) within C. are an expression of interstitial growth. ↑ Chondromucoprotein prevents observation of ↑ collagen fibers (CF) and fibrils (F) in ordinary light; both are visible in polarized light and with transmission electron microscope. In most C. pieces, fibers are randomly disposed but always continuous with ↑ perichondrium (P). With age, C. many calcify and be replaced by bone; in some cases, C. undergoes degeneration forming ↑ asbestos cartilage. Like other cartilages, C. is devoid of blood and lymphatic vessels, nerve fibers, and ↑ wandering cells.

↑ Cartilage, growth of

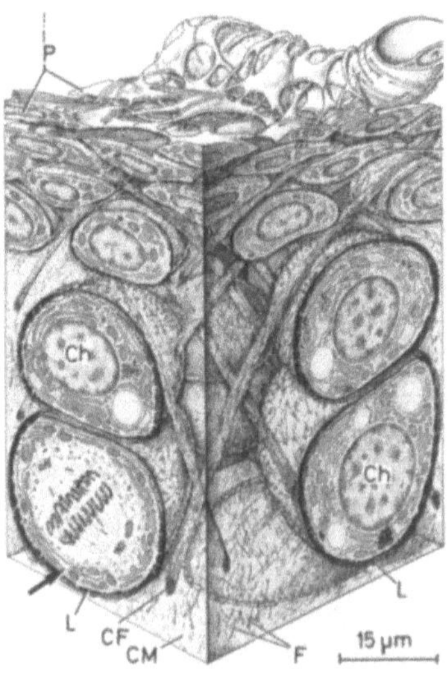

Stofft, E., Graf, J.: Acta Anat. *116*, 114 (1983)

Cartilage, hyaline, of the body: The cartilage of embryonal model, ↑ epiphyseal plates, ↑ synchondroses, ↑ articular cartilage, costal cartilage, nasal cartilage, thyroid cartilage, ↑ arytenoid cartilage (partially), triticeal cartilage, tracheal ↑ cartilage, and bronchial cartilage in primary large and medium secondary ↑ bronchi.

Cartilage, interstitial growth of: ↑ Cartilage, growth of

Cartilage matrix: A metachromatic, ↑ PAS-positive, strongly ↑ basophilic mass surrounding single ↑ chondrocytes and ↑ isogenic groups. C. consists of ↑ cartilage ground substance and fibers (↑ collagen and ↑ elastic fibers). C. also contains ↑ matrix granules; topographically, it is divided into territorial matrix (= ↑ capsule, of chondrocyte) and ↑ interterritorial matrix.

Thyberg, J.: Histochem. J. *9*, 259 (1977)

Cartilage, matrix vesicles of: ↑ Cartilage, calcification of; ↑ Matrix vesicles

Cartilage, nutrition of: ↑ As avascular tissue, ↑ cartilage is nourished by diffusion through ↑ cartilage matrix from ↑ perichondrium. Since diffusion rate is low, C. is classified among ↑ bradytrophic tissues.

↑ Cartilage canals

Cartilage, of auditory tube (cartilago tubae auditivae*): ↑ Auditory tube; ↑ Cartilage, elastic

Cartilage, of external auditory meatus (cartilago meatus acustici*): ↑ Auditory meatus, external; ↑ Cartilage, elastic

Cartilage, regeneration of: Possible only in growing cartilage due to chondrogenic activity of ↑ perichondrium. In damaged adult cartilage, ↑ regeneration produces only ↑ dense connective tissue. In cases of experimentally damaged ↑ cartilage matrix, ↑ chondrocytes are able to compensate its loss.

Cartilage, transplantation of: Possible with success in autografts; allografts die but induce a chondrogenic or osteogenic reaction; ↑ perichondrium must be present.

↑ Transplantation

Cartilage zones, in bone growth: ↑ Bone formation, indirect

Cartilagelike tissue: ↑ Chondroid tissue

Cartilages, of larynx (cartilagines laryngis*): Pieces of hyaline and elastic ↑ cartilage forming skeleton of ↑ larynx.

Cartilages, of trachea (cartilagines tracheales*): About 15–20 horseshoe-shaped hyaline rings (C) with an outer ↑ perichondrium considerably thicker than their inner perichondrium (P). A system of crossing collagen fibers (F) running from external to internal perichondrium is visible in polarized light (inset). C. are bound together by ↑ fibroelastic tissue to a vertical column forming tracheal skeleton.

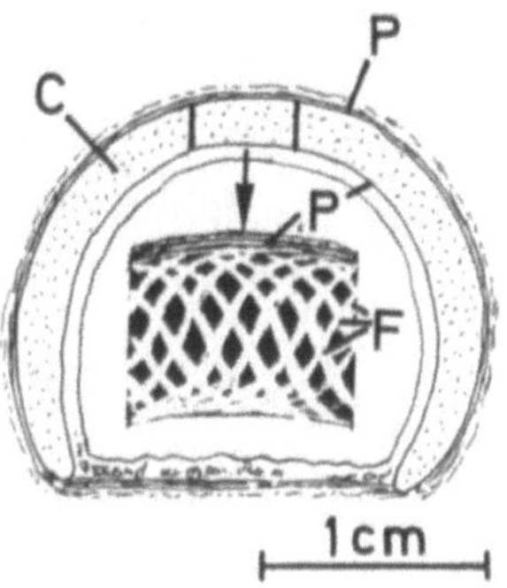

Caruncle, lacrimal: ↑ Lacrimal caruncle

Casein: A nutritive ↑ milk protein produced by epithelial cells of lactating ↑ mammary gland. C. is released together with dissolved milk proteins in form of ↑ casein granules.

Casein granules (granulum proteini*): Very dense, spherical membraneless structures (P), measuring 70 nm – 0.5 µm in diameter, found in alveolar lumen of lactating ↑ mammary gland and in secretory vacuoles (SV) of its epithelial cells. C. are composed of a micellar agglomeration, 2.5 nm in diameter, arranged in a crystalline lattice. C. are synthesized on ribosomes of rough endoplasmic reticulum and

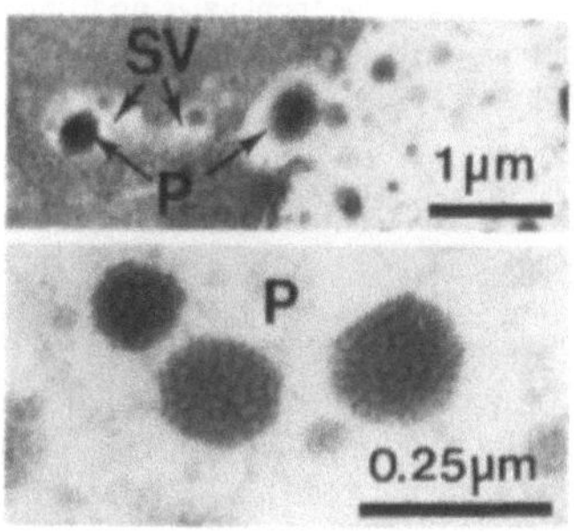

then pass into its cisternae to be transferred to Golgi apparatus, where proteins condense within secretory vacuoles. By fusion of membranes of these vacuoles with plasmalemma, C. are released from cells by an ↑ eccrine secretion. (Fig. = rat)

↑ Lactation

Castration cells: Hypertrophic and strongly vacuolized ↑ gonadotropes occurring in ↑ adenohypophysis after castration.

Kurosumi, K,. Fujita, H.: Functional Morphology of Endocrine Glands. Stuttgart: Georg Thieme 1975

Catalase: An enzyme present in matrix of ↑ peroxisomes and responsible for positive peroxidase reaction of these ↑ organelles. (See biochemistry texts for further information)

Fahimi, H.D.: J. Cell Biol. *43*, 275 (1969)

Catecholamine storage granules (chromaffin granules): Cytoplasmic granules containing ↑ catecholamines.

↑ Epinephrine storage granules; ↑ Norepinephrine storage granules

Winkler, H., Smith, A.D.: The Chromaffin Granule and the Storage of Catecholamines. In: Blaschko, H., Sayers, G., Smith, A.D. (eds.): Handbook of Physiology, Vol. VI. Adrenal Gland. Baltimore: Waverly Press 1975

Catecholamines: Chemical substances belonging to group of pyrocatechols with an alkylamine lateral chain. In presence of an aqueous solution of metallic salts, or other oxidizing agents, C. give a positive ↑ chromaffin reaction. Most important C. are ↑ epinephrine and ↑ norepinephrine, produced and released by corresponding cells of adrenal medulla, where they are contained inside ↑ epinephrine and ↑ norepinephrine storage granules. ↑ Enterochromaffin cells of ↑ gastrointestinal tract, sympathetic ↑ paraganglia, and type I ↑ glomus cells also produce C. Norpinephrine is contained in ↑ synaptic vesicles of ↑ adrenergic nerve endings.

Cationic dyes: ↑ Basic dyes

Caveolae, of endothelial cells (plasmalemmal vesicles): ↑ Micropinocytotic vesicles (C), about 60 nm in diameter, formed by plasmalemma (P) of ↑ endothelial cells (E). Fusion of several C. produces ↑ transendothelial channels. C. are found in particularly great numbers in continuous ↑ capillaries. It is thought that C. are responsible in part for transendothelial transport. C. are structurally very similar to ↑ caveolae of ↑ smooth muscle cells. (Fig. = rat)

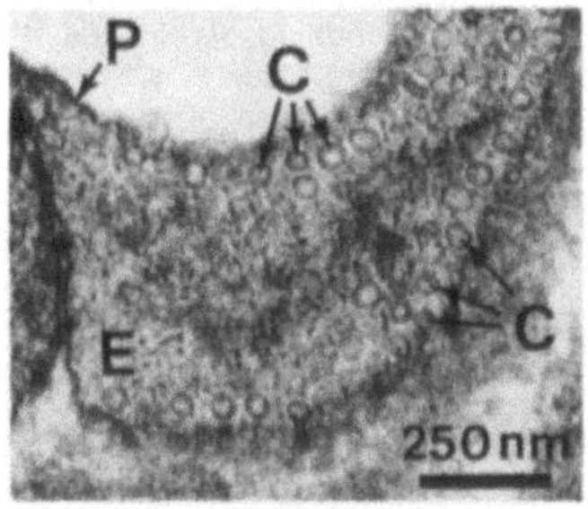

Karnovsky, M.J.; Shea, S.M.: Microvasc. Res. *2*, 353 (1970)

Caveolae, of smooth muscle cells (caveolae intracellulares): ↑ Micropinocytotic vesicles (C), about 50 nm in diameter, with unit membrane continuous with ↑ sarcolemma (S) of ↑ smooth muscle cell. C. may occur during process of ↑ vesiculation, either as single C., more complex tubules (T), or branched chains (Ch) of fused C. It is thought that C. are both analogous to and homologous with ↑ T-system of striated muscle fibers.

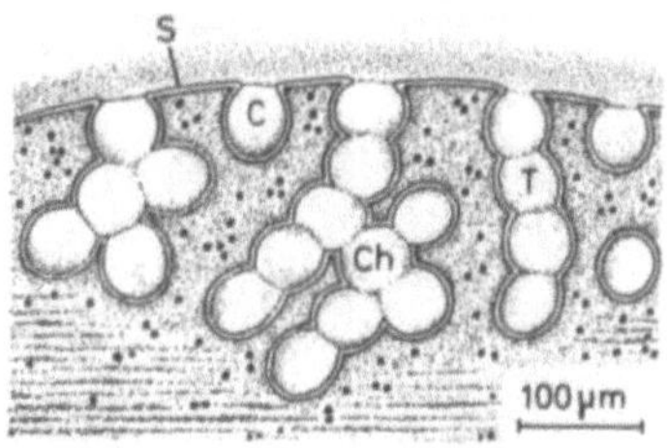

Forbes, M.S., Rennels, M.L., Nelson, E.: J. Ultrastruc. Res. *67*, 325 (1979); Sawada, H.: Biomed. Res. *2*, Suppl. 153 (1981)

Cavernous bodies, of clitoris (corpora cavernosa clitoridis*): Short, paired, cylindrical structures composed of ↑ erectile tissue and enveloped by a ↑ tunica albuginea. C. are homologous to ↑ cavernous bodies of ↑ penis and have an identical structure.

↑ Clitoris

Cavernous bodies, of penis (corpora cavernosa penis*): Two parallel apposed cylinders (CC) forming dorsal part of ↑ penis; posteriorly, they separate and affix to ascending rami of corresponding pubic bones. Structure: 1) ↑ Tunica albuginea (TA) = a dense fibrous capsule, about 1 mm thick, in which ↑ collagen fibers are mainly arranged in an outer longitudinal and inner circular layer; both are intermingled with an elastic network. In anterior portion of penis, albugineae fuse incompletely forming septum penis* (SP). 2) ↑ Erectile tissue = a sponge-like system composed of trabeculae and ↑ cavernous sinuses. Through center of each C. runs ↑ deep artery of penis (DA) from which arise ↑ helicine arteries (HA). a) Trabeculae (T) = ↑ dense connective tissue septa consisting of ↑ fibrocytes (F), occasional ↑ fibroblasts (Fb), bundles of ↑ smooth muscle cells (MC), ↑ collagen (Cf) and ↑ elastic fibers (Ef), and helicine arteries and capillaries. Trabeculae are lined by an ↑ endothelium (E) resting on basal lamina. b) ↑ Cavernous sinuses (CS) = vascular spaces between trabeculae.

↑ Erection

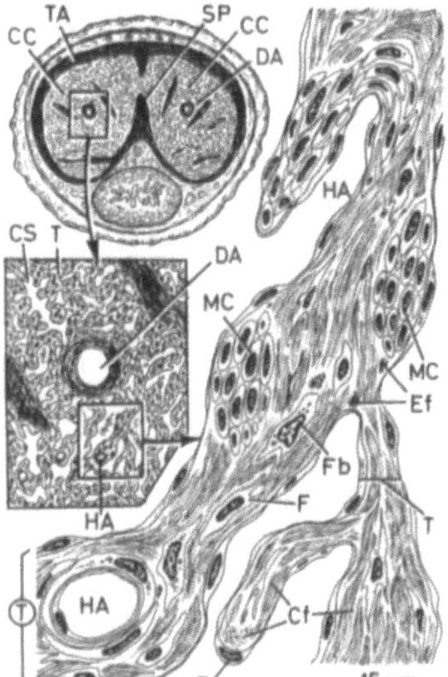

Mariani, G., Battaglia, G.: Arch. Ital. Anat. Embriol. *82*, 285 (1977)

Cavernous sinuses, of cavernous body of penis (cavernae corporum cavernosorum*): Irregular honeycomb spaces (S) separated by trabeculae (T) into which terminal branches of ↑ helicine arteries (H) empty directly; blood is drained by efferent veins. Arrows = bundles of ↑ smooth muscle cells. (Fig. = human)

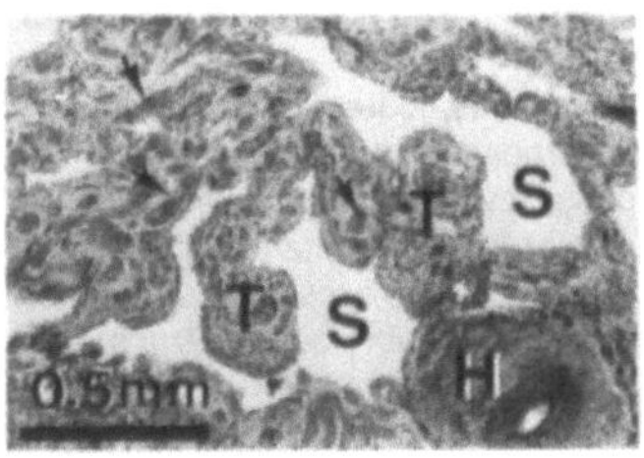

↑ Cavernous bodies, of penis

Cavity, tympanic: ↑ Tympanic cavity

C-cells, of pancreatic islets: A category of rare, pale, nongranulated cells found in ↑ pancreatic islands of man, rat, and guinea pig. C. have an elliptical nucleus, round mitochondria, a prominent Golgi apparatus, a limited number of dense ↑ secretory granules (SG), a few flattened and elongated rough endoplasmic cisternae, occasional lysosomes (Ly), and a small number of free ribosomes. C. rest on a basal lamina (BL). Exact function of C. is not known; it is thought they may represent a stage in secretory cycle or a degenerative form of both ↑ A- and ↑ B-cells of ↑ pancreatic islets.

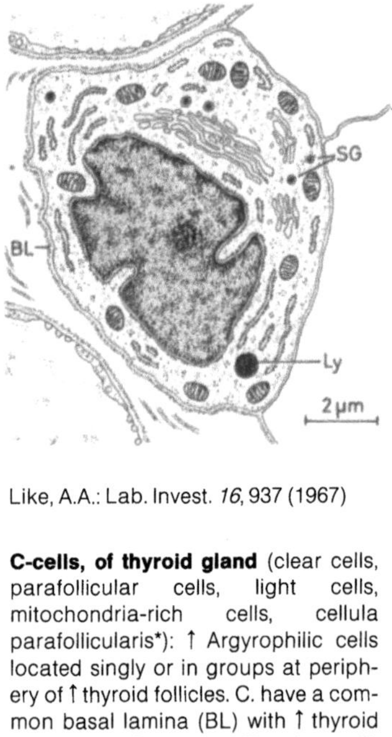

Like, A.A.: Lab. Invest. *16*, 937 (1967)

C-cells, of thyroid gland (clear cells, parafollicular cells, light cells, mitochondria-rich cells, cellula parafollicularis*): ↑ Argyrophilic cells located singly or in groups at periphery of ↑ thyroid follicles. C. have a common basal lamina (BL) with ↑ thyroid follicular cells, but unlike them, C. hardly ever reach follicular cavity. C. are round or polygonal with a globular nucleus (N). Cytoplasm contains numerous oval mitochondria, scattered cisternae of rough endoplasmic reticulum, a few lysosomes, and a moderate amount of free ribosomes. From well-developed Golgi apparatus (G) arise osmiophilic ↑ secretory granules (SG), 25 nm in diameter, surrounded by a ↑ unit membrane. Central core of granules is often separated from membrane by a clear zone (inset). Granules contain the hormone ↑ calcitonin and also, in some animals, ↑ serotonin and ↑ somatostatin. Release of calcitonin is regulated by a positive ↑ feedback by blood calcium level. Hypercalcemia provokes discharge of calcitonin from C.; hypocalcemia has an opposite effect. C. are classed among the ↑ APUD-cells. Cap = capillary

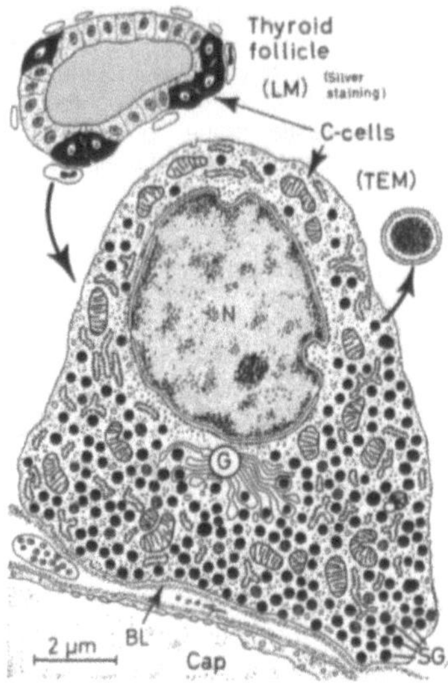

Kameda, Y.: Arch. Histol. Jpn. *46*, 221 (1983); Kameda, Y.: Cell Tissue Res. *225*, 693 (1982); Kameda, Y., Oyama, H., Endoh, M., Horino, M.: Anat. Rec. *204*, 161 (1982); Nunez, E.A., Gershon, M.D.: Int. Rev. Cytol. *52*, 1 (1978); Zabel, M.: Histochemistry *75*, 419 (1982)

C-cells, of ultimobranchial body: Endocrine cells morphologically identical to those of thyroid ↑ C-cells. Nonexistent in man, C. produce ↑ calcitonin in some species (birds, reptiles).

↑ Ultimobranchial body

CCK-PZ: ↑ Cholecystokinin-pancreozymin

Cecum (caecum*): A blind pouch located at proximal end of large ↑ intestine between ileocecal valve and ↑ appendix. Structure identical to that of ↑ colon.

Snipes, R.L.: Anat. Embryol. *157*, 329 (1979)

Cell (cellula*): The smallest unit of a living organism capable of independent existence.

Alberts, B., Bray, D., Lewis, J., Raff, M., Roberts, K., Watson, J.D.: Molecular Biology of the Cell. New York: Garland Publishing 1983; Fawcett, D.W.: The

Cell. 2nd edn. Philadelphia London Toronto: W.B. Saunders Company 1981; Ude, J., Koch, M.: Die Zelle. Stuttgart: Gustav Fischer 1983

Cell axis: An imaginary line passing through ↑ centrosphere and center of ↑ nucleus; in general, this line is perpendicular to ↑ basal lamina if present.

Cell biology: ↑ Cytology

Cell center: ↑ Centrosphere

Cell coat: ↑ Glycocalyx

Cell compartment: Different cell or ↑ organelle media separted from one another by ↑ cytomembranes (e.g., ↑ reticuloplasm separated from cytoplasmic matrix; mitochondrial membrane space from ↑ mitochondrial matrix; ↑ nuclear membrane from reticuloplasm).

Cell component: Any organized region of ↑ cytoplasm composed of macromolecular arrays (e.g., mitochondrion, cell membrane, microtubule, ribosome, Golgi apparatus). C. is formed by ↑ cell constituents.

De Duve, C.: General Principles. In: Roodyn, B.B. (ed.): Enzyme Cytology. New York: Academic Press 1967

Cell constituents: Individual molecules forming ↑ cell components (e.g., lipids, proteins, sugars, ↑ nucleic acids).

Cell cortex: ↑ Ectoplasm

Cell culture: Removal of living cells from body and transfer to a natural or artificial nutrient medium in vitro where they can be kept for various periods of time under well-defined conditions.

Cell cycle (cyclus cellularis*): Sum of morphological and biochemical events occurring in life of a cell between its formation and end of its division into two daughter cells. Four phases: 1) G_1-phase = postmitotic phase; each ↑ chromosome consists of one DNA molecule. Duration is extremely variable and depends upon type of cell and mitotic turnover – from 7 h in rapidly dividing cells to whole life of an organism in nondividing or postmitotic cells (e.g., ↑ neurons). In the latter case, cells abandon C. and remain until cell death in so-called G_0-phase. However, some cells in this phase preserve capacity of reacting to stimuli and reintegrating C. 2) S-phase = phase of synthesis; gradual replication of DNA; each chromosome consists of two DNA molecules, duration about 7 h. 3) G_2-phase = postsynthetic or premitotic phase; duration about 4–5 h. Cells destined to be ↑ polyploid abandon C. in this phase; they can become ↑ binucleate through ↑ amitosis (A). Some of them become mononuclear by nuclear fusion (F) and can reintegrate C. in G_2-phase. 4) M-phase = ↑ mitosis; duration about 1 h.

↑ Generation time; ↑ Interphase; ↑ Somatomedins

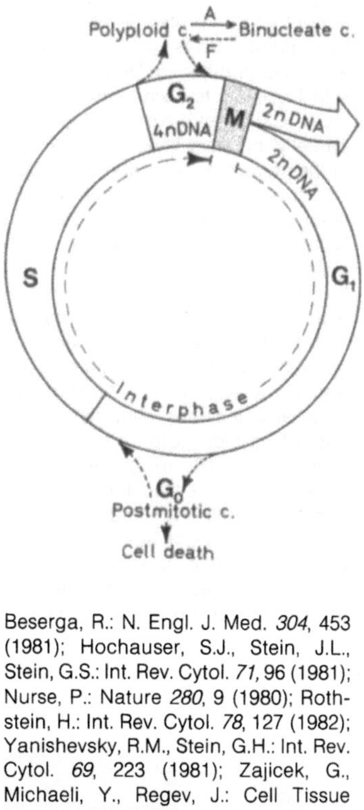

Beserga, R.: N. Engl. J. Med. *304*, 453 (1981); Hochauser, S.J., Stein, J.L., Stein, G.S.: Int. Rev. Cytol. *71*, 96 (1981); Nurse, P.: Nature *280*, 9 (1980); Rothstein, H.: Int. Rev. Cytol. *78*, 127 (1982); Yanishevsky, R.M., Stein, G.H.: Int. Rev. Cytol. *69*, 223 (1981); Zajicek, G., Michaeli, Y., Regev, J.: Cell Tissue Kinet. *12*, 229 (1979)

Cell death: A definitive arrest of all vital phenomena in a cell. C. can represent a genetically controlled event in course of embryonal development and in the healthy organism or it can be provoked by physical and chemical injuries. Dying cells lose water, their cytoplasm becomes denser and strongly stained. Cell volume and intensity of enzymatic activity decrease; waste products (↑ residual bodies, ↑ lipofuscin) and/or, in some cases, ↑ lipid droplets accumulate in cytoplasm. Nucleus undergoes either ↑ karyolysis, ↑ karyorrhexis, or ↑ karyopyknosis.

↑ Nucleus, degeneration and death of

Bowen, I.D., Lockshin, R.A. (eds.): Cell Death in Biology and Pathology. London: Chapman & Hall 1981; Wyllie, A.H., Kerr, J.F., Currie, A.R.: Int. Rev. Cytol. *68*, 251 (1980)

Cell differentiation: ↑ Differentiation, of cells

Cell division: ↑ Meiosis; ↑ Mitosis

Cell fractionation: A method of isolation, from mechanically disrupted cells, of nuclei, ↑ mitochondria, ↑ lysosomes, ↑ secretory granules, and other ↑ cell components by means of density gradient ↑ centrifugation.

Cell inclusions: ↑ Inclusions

Cell junctions: ↑ Attachment devices; ↑ Junctional complex, of epithelia

Cell membrane (cytolemma, plasmolemma, plasmalemma, plasma membrane, cytolemma*): A fine, metabolically active sheath (CM) representing outer boundary of a living cell; it separates internal cell milieu from extracellular environment. C. is visible under light microscope, due to dye deposits upon it. Under transmission electron microscope, C. appears as a 7.5 to 11-nm-thick trilaminar sheath (also called ↑ unit membrane) composed of two external dark protein layers (PL) and a central, electron-transparent, bimolecular phospholipid layer (LL). Under very high power, fine transversal bridges (B) across phospholipid layer and globular structures (G) in protein layers are visible. Free surface of C. is generally covered by a ↑ glycocalyx (Gc) of variable thickness; C. underlies ↑ cytoskeleton (C) of cell membrane on cytoplasmic side.

↑ Cell membrane, cytoskeleton of: ↑ Cell membrane, models of

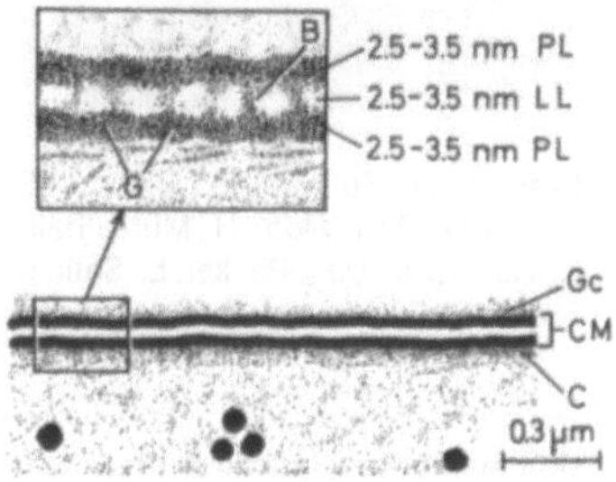

Jamieson, G.A., Robinson, D.M. (eds.): Mammalian Cell Membranes. Vol. 2. The Diversity of Membranes. London. Butterworths 1977; Lodish, H.F., Rothman, J.E.: Sci. Am. *240*, 38 (1979); Quinn, P.J.: The Molecular Biology of Cell Membranes. London: Macmillan 1976; Singer, S.J.: Ann. Rev. Biochem. *43*, 805 (1975)

Cell membrane, chemical composition of: Proteins make up about 50%–60% of its dry weight; lipids 20%–30% (mainly phospholipids, but also cholesterol, lipoproteins, and liposaccharides); and carbohydrates 10%, in form of free oligosaccharides or bound to fat and proteins.

Cell membrane, cytoskeleton of: A very thin layer of proteins (C) attached to cytoplasmic face of plasmalemma (P) giving it rigidity and controlling protein diffusion. C. is continuous with ↑ ectoplasm (E).

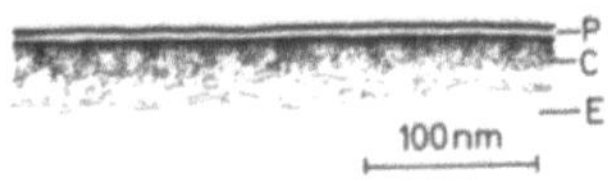

Kirkpatrick, F.H.: Bio Systems *11*, 93 (1979); Nermut, M.V.: Eur. J. Cell. Biol. *25*, 265 (1981)

Cell membrane, freeze-cleaving of: ↑ Free-cleaving technique splits C. at level of phopholipid layer; resulting image has two faces: 1) EF-face with only a few 9-nm ↑ membrane-associated particles (arrow) and some shallow pits (arrowheads); 2) PF-face characterized by numerous 9-nm membrane-associated particles (arrows). Asterisk = fractured intercellular space (Fig. = ↑ pinealocyte, rat)

↑ Freeze-cleaving, terminology of

Branton, D., Bullivant, S., Gilula, N.S., Karnovsky, M.J., Moor, H., Mühlethaler, K., Northcote, D.H., Packer, L., Satir, B., Speth, V., Staehelin, L.A., Streere, R.L., Weinstein, R.S.: Science *190*, 54 (1975)

Cell membrane, functions of: Attachment to other cells and to ↑ basal lamina, selective permeability, specific enzymatic activity through enzymes bound to it, support for hormonal and immunological receptors, transmembranous transport (↑ macro- and ↑ micropinocytosis, ↑ endocytosis, ↑ exocytosis), movement (↑ pseudopodia, ↑ filopodia, ↑ lamellipodia), formation of ↑ myelin sheath, exchange of information between cells, etc. (See physiology texts for further information)

Kaplan, J.: Science *212*, 14 (1981)

Cell membrane, models of: Partially hypothetical concepts intending to explain structure of cell membrane. Among many others, following are the most well known. 1) Classic concept of Davson and Daniell: a) Between two protein layers (PL) a bimolecular phospholipid layer (LL) is situated. Hydrophilic ends (Hf) of its molecules are in contact with protein layer, whereas hydrophobic ends (Hb) intermingle with one another. b) During fixation with ↑ osmium tetroxide, atoms of osmium (Os) penetrate protein layer and partially into lipid layer, which results in their blackening (c) and characteristic structure as ↑ unit membrane. 2) Fluid mosaic model of Singer and Nicolson: This modern concept postulates that phospholipid layer (LL) is fluid; it consists, as in classic model, of a biomolecular layer of molecules with hydrophilic ends (Hf) in contact with both watery cytoplasmic matrix (CM) and extracellular milieu. Phospholipid layer also includes cholesterol (Ch) and glycolipids (Glp). Phospholipid layer proteins (P) and protein aggregates swim in this fluid, forming a mosaic. Some proteins extend throughout entire phospholipid layer, others are incorporated into it, whereas a third group lie over and under it. Some proteins send out, above the surface, ↑ glycoprotein chains of various lengths (Gpc), which constitute, together with ends of the glycolipid molecules, the ↑ glycocalyx (Glc). This model explains existence of bridges across phospholipid layer, existence of ↑ membrane-associated particles within cell membrane, and the latter's asymmetry

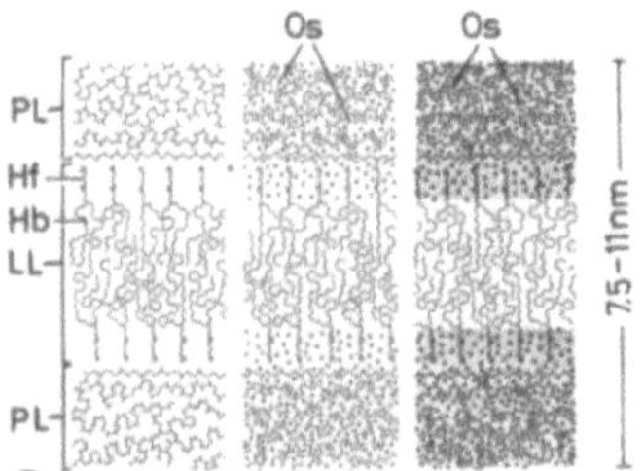

①

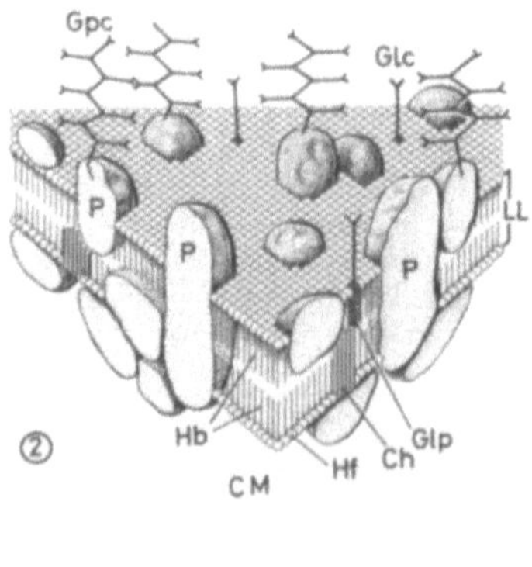

②

Gomperts, B.: The Plasma Membrane. Models for Structure and Function. New York: Academic Press 1977; Robertson, I.D.: J. Cell Biol. *91*, 189s (1981); Singer, J.S., Nicolson, G.L.: Science *175*, 720 (1972)

Cell membrane, specializations of: Structural differentiations of plasmalemma comprising ↑ attachment devices and devices for increasing cell surface (↑ brush border, ↑ stereocilia, ↑ striated border, ↑ microvilli).

Cell Surface: Surface Organization. In: Organization of the Cytoplasm. Vol. 46, Part 2. New York: Cold Spring Harbor Symposia on Quantitative Biology 1982

Cell number: The adult human organism consists of about 10^{13}–10^{14} cells.

Cell organelles: ↑ Organelles

Cell sap: ↑ Cytoplasmic matrix

Cell shape: Depending on function, cells may be globular, fusiform, stellate, flattened, cuboidal, columnar, pyramidal, etc. Cells having an unusual C. are ↑ podocytes, ↑ phalangeal cells, ↑ pillars, etc.

Cell size: C. varies widely, ranging from 5–6 µm (↑ granule cells, of cerebellar cortex; head of ↑ spermatozoon) to 150–200 µm (secondary ↑ oocyte). Some ↑ neurons exceed 1 m in length.

Cell structure: A mammalian cell consists of ↑ protoplasm (= ↑ nucleus and ↑ cytoplasm) enclosed by a ↑ cell membrane. Cytoplasm is composed of ↑ organelles, ↑ euplasm (not represented), ↑ inclusions, and ↑ cytoplasmic matrix.

↑ Metaplasm

Carr, K.E., Toner, P.G. Cell Structure. An Introduction to Biomedical Electron Microscopy. 3rd edn. Edinburgh, London, Melbourne, New York: Churchill Livingstone 1982

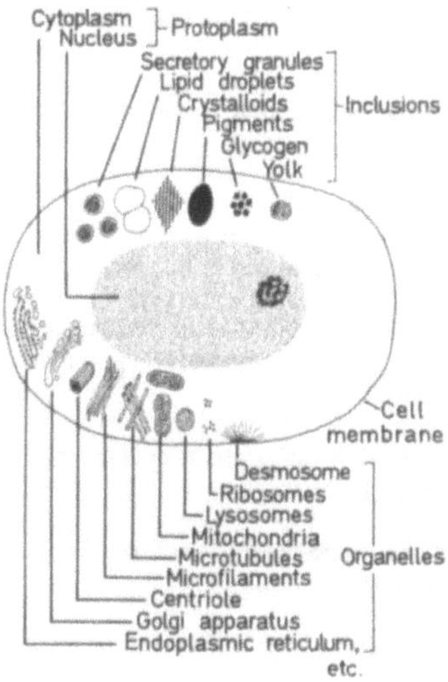

Cell surface protein: ↑ Fibronectin

Cell wall: A cellulose layer of variable thickness covering outer surface of plant cell membrane.

Celloidin: An ↑ embedding material used in light microscopy especially for preparation of ↑ bone, sensory organs, and CNS. C. embedding makes it possible to obtain 15 to 100 µm-thick sections of very large surfaces. C. is very inflammable and explosive (nitrocellulose).

Cells, A, of gastrointestinal tract: ↑ A-cells, of gastrointestinal tract

Cells, A, of pancreatic islets: ↑ A-cells, of pancreatic islets

Cells, A, of synovial membrane: ↑ Synovial cells

Cells, absorptive: ↑ Absorptive cells

Cells, acidophilic, of hypophysis: ↑ Hypophysis, acidophilic cells of

Cells, ACTH: ↑ Corticotropes

Cells, action of polypeptide hormones on: ↑ Polypeptide hormones, action on the cell of

Cells, action of steroid hormones on: ↑ Steroid hormones, action on the cell of

Cells, adipose, brown: ↑ Adipose cells, brown

Cells, adipose, white: ↑ Adipose cells, white

Cells, adrenocortical: ↑ Cells, of zona fasciculata; ↑ Cells, of zona glomerulosa; ↑ Cells, of zona reticularis

Cells, adrenocorticotropic: ↑ Corticotropes

Cells, adventitial: ↑ Adventitial cells, of bone marrow

Cells, AL: ↑ A-like cells

Cells, alpha, of hypophysis: ↑ Hypophysis, acidophilic cells of

Cells, alpha, of pancreatic islets: ↑ A-cells, of pancreatic islets

Cells, alveolar, of mammary gland: ↑ Mammary gland, epithelium of

Cells, alveolar, type I and II, of lung: ↑ Alveolar cells type I and II

Cells, amacrine: ↑ Amacrine cells

Cells, ameboid: ↑ Ameboid cells

Cells, annulate lamellae of: ↑ Annulate lamellae

Cells, APUD: ↑ APUD-cells

Cells, argentaffin: ↑ Enterochromaffin cells

Cells, argyrophilic: ↑ Argyrophilic cells

Cells, attachment devices of: ↑ Attachment devices

Cells, auditory: ↑ Hair cells, of Corti's organ

Cells, automatic recognition of: ↑ Automatic cell recognition

Cells, B: ↑ B-lymphocytes

Cells, B, of pancreatic islets: ↑ B-cells, of pancreatic islets

Cells, B, of synovial membrane: ↑ Synovial cells

Cells, band: ↑ Band neutrophils

Cells, basal granular: ↑ Enterochromaffin cells

Cells, basal, of epidermis: ↑ Basal cells, of epidermis

Cells, basal, of sebaceous glands: ↑ Sebaceous cells

Cells, basal, of various epithelia: ↑ Basal cells, of various epithelia

Cells, basket, of mammary gland: ↑ Basket cells, of mammary gland

Cells, basket, of salivary glands: ↑ Basket cells, of salivary glands

Cells, basophilic, of hypophysis: ↑ Hypophysis, basophilic cells of

Cells, beta, of adenohypophysis: ↑ Thyrotropes

Cells, beta, of pancreatic islets: ↑ B-cells, of pancreatic islets

Cells, binucleate: ↑ Binucleate cells

Cells, biological membranes of: ↑ Cytomembranes

Cells, bipolar: ↑ Bipolar neurons

Cells, border, of Corti's organ: ↑ Border cells, of Corti's organ

Cells, bronchiolar: ↑ Bronchiolar epithelium

Cells, brush: ↑ Brush cells

Cells, C, of pancreatic islets: ↑ C-cells, of pancreatic islets

Cells, C, of thyroid gland: ↑ C-cells, of thyroid gland

Cells, C, of ultimobranchial body: ↑ C-cells, of ultimobranchial body

Cells, candidate stem: ↑ Candidate stem cells

Cells, capsular: ↑ Satellite cells, of peripheral neurons

Cells, cardiac muscle: ↑ Cardiac muscle cells

Cells, castration: ↑ Castration cells

Cells, centrifugation of: ↑ Cell fractionation

Cells, centroacinar: ↑ Centroacinar cells

Cells, chief, of carotid body: ↑ Glomus cells, of carotid body

Cells, chief, of chromaffin paraganglia: ↑ Glomus cells, of carotid body, type I; ↑ Paraganglia, sympathetic

Cells, chief, of gastric glands proper: ↑ Chief cells, of gastric glands proper

Cells, chief, of parasympathetic paraganglia: ↑ Paraganglia, parasympathetic

Cells, chief, of parathyroid glands: ↑ Chief cells, of parathyroid glands

Cells, chief, of pineal organ: ↑ Pinealocytes

Cells, choristoma: ↑ Choristoma cells

Cells, chromacity horizontal: ↑ Horizontal cells

Cells, chromaffin: ↑ Chromaffin cells

Cells, chromophobic, of hypophysis: ↑ Hypophysis, chromophobic cells of

Cells, chromophobic, of pars intermedia: ↑ Hypophysis pars intermedia, chromophobic cells of

Cells, ciliated, of bronchiolar epithelium: ↑ Bronchiolar epithelium

Cells, ciliated, of oviducts: ↑ Oviducts, epithelium of

Cells, ciliated, of tracheal epithelium: ↑ Trachea, epithelium of

Cells, ciliated, of uterus: ↑ Uterus, epithelium of

Cells, classes of: ↑ Eucaryotes; ↑ Procaryotes

Cells, clear, of eccrine sweat glands: ↑ Clear cells, of eccrine sweat glands

Cells, clear, of parathyroid glands: ↑ Chief cells, of parathyroid glands

Cells, clear, of pineal organ: ↑ Pinealocytes

Cells, clear, of taste buds: ↑ Taste buds, structure of

Cells, clear, of thyroid gland: ↑ C-cells, of thyroid gland

Cells, clump: ↑ Clump cells

Cells, collagen-synthesizing: ↑ Collagen-synthesizing cells

Cells, colloid, of Langendorff: ↑ Colloid cells, of Langendorff

Cells, cone: ↑ Photoreceptors

Cells, cone, bipolar: ↑ Bipolar cells, of retina

Cells, cornified: ↑ Horny cells

Cells, cortical, of hair shaft: ↑ Cortical cells, of hair shaft

Cells, cuticular: ↑ Cuticle, of hair shaft; ↑ Cuticle, of the inner root sheath; ↑ Matrix cells, of hair

Cells, D, of gastrointestinal tract: ↑ D-cells, of gastrointestinal tract

Cells, D_1, of gastrointestinal tract: ↑ D_1-cells, of gastrointestinal tract

Cells, D, of pancreatic islets: ↑ D-cells, of pancreatic islets

Cells, dark chief, of parathyroid glands: ↑ Chief cells, of parathyroid glands

Cells, dark, of arched collecting tubules: ↑ Collecting tubules, of kidney

Cells, dark, of eccrine sweat glands: ↑ Mucoid cells, of eccrine sweat glands

Cells, dark, of pineal organ: ↑ Interstitial cells, of pineal organ

Cells, dark of taste buds: ↑ Taste buds, structure of

Cells, dark, problem of: ↑ Dark cells, problem of

Cells, decidual, of ovary: ↑ Decidual cells, of ovary

Cells, decidual, of uterus: ↑ Decidual cells, of uterus

Cells, delta, of hypophysis: ↑ Delta cells, of hypophysis

Cells, delta, of pancreatic islets: ↑ D-cells, of pancreatic islets

Cells, dendritic: ↑ Dendritic cells

Cells, differentiation of: ↑ Differentiation, of cells

Cells, dust: ↑ Alveolar macrophages

Cells, E: ↑ E-cells

Cells, E, of pancreas: ↑ E-cells, of pancreas

Cells, EC: ↑ Enterochromaffin cells

Cells, EC_1: ↑ Enterochromaffinlike cells

Cells, EC_2: ↑ Motilin cells

Cells, ECL: ↑ Enterochromaffinlike cells

Cells, ectoplasmic zone of: ↑ Ectoplasm

Cells, effector: ↑ Effector cells

Cells, endocrine: ↑ Endocrine cells

Cells, endocrine, of gastrointestinal tract: ↑ Endocrine cells, of gastrointestinal tract

Cells, endoplasmic reticulum of: ↑ Endoplasmic reticulum

Cells, endoplasmic zone of: ↑ Endoplasm

Cells, endothelial: ↑ Endothelial cells

Cells, enterochromaffin: ↑ Enterochromaffin cells

Cells, enterochromaffinlike: ↑ Enterochromaffinlike cells

Cells, enteroglucagon-producing: ↑ A-cells, of gastrointestinal tract

Cells, ependymal: ↑ Ependymal cells

Cells, epinephrine-producing: ↑ Epinephrine-producing cells

Cells, epithelial: ↑ Epithelial cells

Cells, epithelial, of lung: ↑ Alveolar cells type I and II

Cells, epithelial-reticular: ↑ Epithelial-reticular cells, of thymus

Cells, epithelioid: ↑ Epithelioid cells

Cells, erythropoietin-responsive: ↑ Erythropoietin-responsive cells

Cells, F: ↑ Pancreatic polypeptide cells; ↑ Synovial cells

Cells, facet: ↑ Facet cells

Cells, fat: ↑ Adipose cells, brown and white

Cells, fat-storing: ↑ Perisinusoidal cells

Cells, fixed, of connective tissue proper: ↑ Fixed cells, of connective tissue proper

Cells, follicle-associated: ↑ M-cells

Cells, follicular, of adenohypophysis: ↑ Follicular cells, of adenohypophysis

Cells, follicular, of ovarian follicles: ↑ Follicular cells, of ovarian follicles

Cells, follicular, of thyroid gland: ↑ Thyroid follicular cells

Cells, G: ↑ G-cells

Cells, gamma: ↑ Hypophysis, chromophobic cells of

Cells, ganglion, of adrenal medulla: ↑ Ganglion cells, of adrenal medulla

Cells, ganglion, of retina: ↑ Ganglion cells, of retina

Cells, ganglion, of spinal ganglion: ↑ Ganglion cells, of spinal ganglion

Cells, gastrin: ↑ G-cells

Cells, germ: ↑ Germ cell

Cells, giant, of trophoblast: ↑ Giant cells, of trophoblast

Cells, glial: ↑ Neuroglia

Cells, glomus, of aortic body: ↑ Glomus cells, of carotid body

Cells, glucagon: ↑ A-cells, of pancreatic islets

Cells, goblet: ↑ Goblet cells

Cells, Golgi apparatus of: ↑ Golgi apparatus

Cells, graft-rejection: ↑ Killer cells

Cells, granule: ↑ Granule cells

Cells, granule, of cerebellar cortex: ↑ Granule cells, of cerebellar cortex

Cells, granule, of cerebral cortex: ↑ Granule cells, of cerebral cortex

Cells, granule, of olfactory bulb: ↑ Granule cells, of olfactory bulb

Cells, granulosa lutein: ↑ Granulosa lutein cells

Cells, granulosa, of ovarian follicles: ↑ Granulosa cells, of ovarian follicles

Cells, ground substance of: ↑ Cytoplasmic matrix

Cells, hair: ↑ Hair cells, inner and outer, of Corti's organ

Cells, hair, of crista ampullaris: ↑ Vestibular cells

Cells, halo: ↑ Halo cells

Cells, helper: ↑ Helper cells

Cells, hematopoietic stem: ↑ Hematopoietic stem cells

Cells, hepatic: ↑ Liver parenchymal cells

Cells, hilus: ↑ Hilus cells

Cells, "homing": ↑ "Homing"

Cells, horizontal, of Cajal: ↑ Cajal's horizontal cells

Cells, horny: ↑ Horny cells

Cells, hyaloid: ↑ Hyalocytes

Cells, hyaloplasm of: ↑ Cytoplasmic matrix

Cells, I: ↑ I-cells

Cells, interdental: ↑ Interdental cells

Cells, intermediate: ↑ Trachea, epithelium of

Cells, intermediate, neuroglial: ↑ Intermediate neuroglial cells

Cells, intermediate, of stria vascularis: ↑ Stria vascularis

Cells, interstitial, of Cajal: ↑ Cajal's interstitial cells

Cells, interstitial, of liver: ↑ Perisinusoidal cells

Cells, interstitial, of ovary: ↑ Interstitial cells, of ovary

Cells, interstitial, of pineal organ: ↑ Interstitial cells, of pineal organ

Cells, interstitial, of testis: ↑ Interstitial cells, of testis

Cells, isolation of: ↑ Centrifugation, density gradient

Cells, juvenile: ↑ Juvenile cells

Cells, juxtaglomerular: ↑ Juxtaglomerular cells

Cells, K: ↑ K-cells

Cells, lacis: ↑ Mesangium, extraglomerular

Cells, lining: ↑ Littoral cells

Cells, littoral, of bone marrow: ↑ Littoral cells, of bone marrow

Cells, littoral, of lymph nodes: ↑ Littoral cells, of lymph nodes

Cells, littoral, of splenic sinusoids: ↑ Endothelial cells, of splenic sinusoids

Cells, liver parenchymal: ↑ Liver parenchymal cells

Cells, living, study of: ↑ Cytology, methods of

Cells, M: ↑ M-cells

Cells, macroglia: ↑ Astrocytes; ↑ Oligodendrocytes

Cells, marginal: ↑ Marginal cells, of stria vascularis

Cells, mast: ↑ Mast cells

Cells, matrix: ↑ Matrix cells, of hair

Cells, medullary, of hair shaft: ↑ Medullary cells, of hair shaft

Cells, mesangial: ↑ Mesangial cells

Cells, mesenchymal: ↑ Mesenchymal cells

Cells, mesothelial: ↑ Mesothelial cells

Cells, microglia: ↑ Microglia

Cells midget bipolar: ↑ Midget bipolar cells

Cells, mononuclear wandering: ↑ Mononuclear wandering cells

Cells, motilin: ↑ Motilin cells

Cells, mucoid, of sweat glands: ↑ Mucoid cells, of sweat glands

Cells, mucous, of bronchial glands: ↑ Bronchial glands

Cells, mucous, of gastric glands proper: ↑ Neck mucous cells

Cells, mucous, of gastric surface: ↑ Surface mucous cells, of gastric mucosa

Cells, mucous, of mucous and mixed glands: ↑ Mucous cells, of mucous and mixed glands

Cells, mural: ↑ Mural cells

Cells, myoepithelial: ↑ Myoepithelial cells

Cells, myoepithelioid: ↑ Juxtaglomerular cells

Cells, N: ↑ N-cells

Cells, NEB: ↑ NEB-cells

Cells, neck mucous, of gastric glands proper: ↑ Neck mucous cells, of gastric glands proper

Cells, nerve: ↑ Neurons

Cells, neurilemmal: ↑ Schwann's cells

Cells, neuroglial: ↑ Neuroglia

Cells, neurosecretory: ↑ Neurosecretory neurons

Cells, nodal: ↑ Nodal cells

Cells, nonpigmented, of ciliary body: ↑ Ciliary body

Cells, norepinephrine-producing: ↑ Norepinephrine-producing cells

"Cells, nurse": ↑ Sertoli's cells

Cells, of anterior column: ↑ Motor neurons

Cells, of Betz: ↑ Betz's cells

Cells, of bone: ↑ Osteoblasts; ↑ Osteoclasts; ↑ Osteocytes

Cells, of Böttcher: ↑ Böttcher's cells

Cells, of Cajal: ↑ Interstitial cells, of Cajal; ↑ Horizontal cells, of Cajal

Cells, of cartilage: ↑ Chondroblasts; ↑ Chondroclasts; ↑ Chondrocytes

Cells, of cervical glands: ↑ Cervical glands, cells of

Cells, of Clara: ↑ Bronchiolar epithelium

Cells, of Claudius: ↑ Claudius' cells

Cells, of Deiters: ↑ Phalangeal cells, outer

Cells, of distal tubule: ↑ Distal tubule of nephron, cells of

Cells, of ductus epididymidis: ↑ Ductus epididymidis, epithelium of

Cells, of Fananas: ↑ Cerebellum, glioarchitectonics of; ↑ Fananas' cells

Cells, of Feyerter: ↑ E-cells

Cells, of Golgi: ↑ Granule cells, of cerebellar cortex

Cells, of Hensen: ↑ Hensen's cells

Cells, of Hofbauer: ↑ Hofbauer's cells

Cells, of Hortega: ↑ Microglia

Cells, of intercalated ducts: ↑ Intercalated duct cells

Cells, of Ito: ↑ Perisinusoidal cells

Cells, of Kolmer: ↑ Kolmer's cells

Cells, of Kupffer: ↑ Kupffer's cells

Cells, of Kurloff: ↑ Kurloff's cells

Cells, of Langendorff: ↑ Colloid cells, of Langendorff

Cells, of Langerhans: ↑ Langerhans' cells, of epidermis

Cells, of Langhans: ↑ Trophoblast

"Cells," of mastoid process (cellulae mastoideae*): Irregular air-filled cavities in temporal bone continuous with tympanic cavity and lined with a ↑ simple squamous epithelium.

Kawabata, I.: Biomed. Res. *2*, Suppl. 433 (1981)

Cells, of Merkel: ↑ Merkel's cells

Cells, of Meynert: ↑ Striate area

Cells, of Müller: ↑ Müller's cells

Cells, of Paneth: ↑ Paneth's cells

Cells, of proximal convoluted tubule: ↑ Proximal convoluted tubule, cells of

Cells, of Purkinje: ↑ Purkinje cells, of cerebellum; ↑ Purkinje cells, of impulse-conducting system

Cells, of Renshaw: ↑ Renshaw's cells

Cells, of Rouget: ↑ Pericytes

Cells, of Schwann: ↑ Schwann's cells

Cells, of seminiferous epithelium: ↑ Spermatogonia; ↑ Spermatocyte, primary; ↑ Spermatocyte, secondary; ↑ Spermatid; ↑ Spermatozoa; ↑ Sertoli's cells

Cells, of Sertoli: ↑ Sertoli's cells

Cells, of straight portion of proximal tubule of nephron: ↑ Straight portion of proximal tubule, cells of

Cells, of striated ducts: ↑ Striated ducts, cells of

Cells, of uterine cervical glands: ↑ Cervical glands, cells of

Cells, of zona fasciculata (spongiocyte, spongiocytus*): Large polyhedral cells with a central spherical nucleus and a voluminous nucleolus. In abundant cytoplasm, a great number of spherical ↑ mitochondria with tubules (M) are scattered. Golgi apparatus is moderately developed, cisternae of rough endoplasmic reticulum (rER) are few, while tubules of smooth endoplasmic reticulum (sER) are numerous. An abundance of ↑ lipid droplets (LD) fills cytoplasm, giving C. a foamy appearance. ↑ Glycogen particles (Gly), free ribosomes, and ↑ microperoxisomes are present in considerable numbers. Cell surface sends many irregular microvilli in direction of capillaries (Cap). Secretion mechanism of C. is not understood: Occasional ↑ exocytosis of lipid droplets (arrows) cannot be considered hormonal secretion. C. produce ↑ glucocorticoids.

↑ Adrenal glands, cortex of

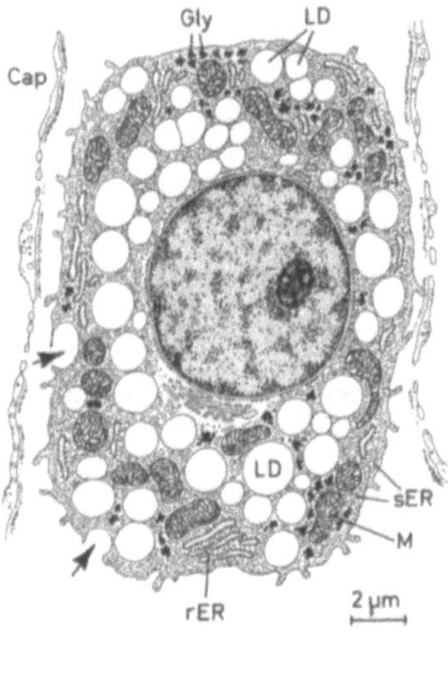

Miao, P., Black, V.H.: J. Cell Biol. *94*, 241 (1982)

Cells, of zona glomerulosa: Large columnar cells arranged in spherical nests, round groups, or arcades. Nucleus is spherical with a conspicuous nucleolus. Cytoplasm contains a few mitochondria (M) with cristae, a small Golgi apparatus (G), some short cisternae of rough endoplasmic reticulum (rER), a moderate number of smooth endoplasmic ↑ tubules (sER), scattered ↑ lipid droplets (LD), and a few ↑ lipofuscin granules (Lf). From surface, irregular microvilli protrude into large intercellular spaces. C. produce ↑ aldosterone; however, mechanism of its secretion is not clear, probably by diffusion through plasmalemma. Cap = capillary

↑ Adrenal glands, cortex of

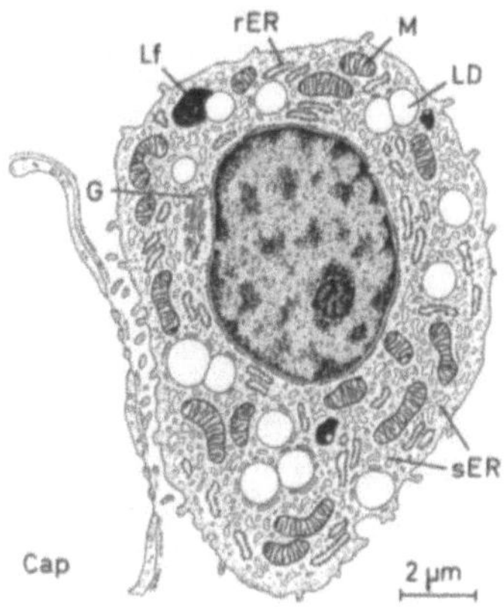

Cells, of zona reticularis: Small polyhedral cells with a central nucleus and an inconspicuous nucleolus. The electron-opaque cytoplasm contains numerous small spherical ↑ mitochondria with tubules (M), a small Golgi apparatus, rare rough endoplasmic cisternae (rER), but an abundance of smooth endoplasmic tubules (sER). ↑ Lipid droplets (LD) are few; ↑ lipofuscin granules (Lf) are numerous. Cell surface is very irregular and studded with short irregular microvilli. ↑ Exocytosis of lipid droplets (arrow) can be observed, however, it is doubtful whether this phenomenon signifies release of hormone(s). Cap = capillary

↑ Adrenal glands, cortex of

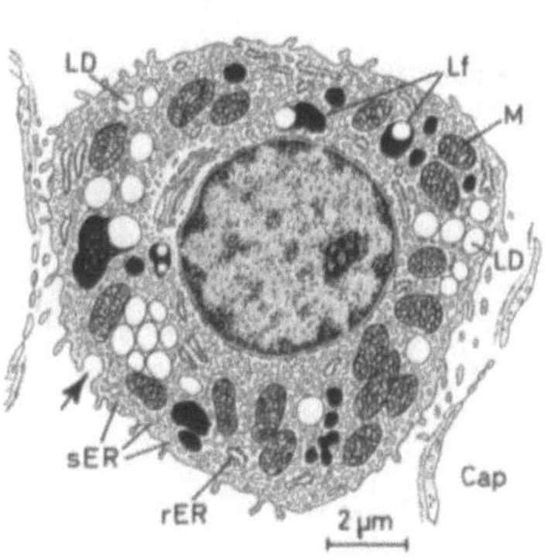

Cells, olfactory: ↑ Olfactory cells; ↑ Olfactory epithelium

Cells, oligodendroglia: ↑ Oligodendrocytes

Cells, oligomucous: ↑ Oligomucous cells

Cells, osteoprogenitor: ↑ Osteoprogenitor cells

Cells, oxyntic: ↑ Parietal cells, of gastric glands proper

Cells, oxyphilic: ↑ Oxyphilic cells, of parathyroid glands

Cells, pacemaker: ↑ Nodal cells

Cells, pancreatic acinar: ↑ Pancreatic acinar cells

Cells, pancreatic polypeptide: ↑ Pancreatic polypeptide cells

Cells, parietal, of gastric glands proper: ↑ Parietal cells, of gastric glands proper

Cells, peripheral, of sweat glands: ↑ Sweat glands, eccrine, excretory duct cells of

Cells, perisinusoidal: ↑ Perisinusoidal cells

Cells, peritubular contractile: ↑ Myofibroblasts

Cells, phalangeal, inner and outer: ↑ Phalangeal cells, inner and outer

Cells, pheochrome: ↑ Chromaffin cells

Cells, pigment epithelial: ↑ Pigment epithelial cells

Cells, plasma: ↑ Plasma cells

Cells, polypeptide hormone-producing: ↑ Endocrine cells

Cells, prickle: ↑ Prickle cells

Cells, principal, of ductus epididymidis: ↑ Ductus epididymidis, epithelium of

Cells, principal, of gastric glands proper: ↑ Parietal cells, of gastric glands proper

Cells, principal, of parathyroid glands: ↑ Chief cells, of parathyroid glands

Cells, protein-secreting: ↑ Basophilic cells with an extensively developed rough ↑ endoplasmic reticulum, mostly in form of ↑ ergastoplasm with very numerous free ribosomes and ↑ mitochondria with cristae (↑ chief cells of ↑ gastric glands proper, ↑ pancreatic acinar cells, ↑ plasma cells, some ↑ neurons, etc.).

Cells, pseudounipolar: ↑ Pseudounipolar cells

Cells, pyroninophilic: ↑ Immunoblasts

Cells, Q: ↑ Candidate stem cells

Cells, red blood: ↑ Erythrocytes

Cells, renewal of: ↑ Renewal, of cells

Cells, reserve, of adenohypophysis: A term designating ↑ chromophobes.

Cells, reticular: ↑ Reticular cells

Cells, rhagiocrine: ↑ Rhagiocrine cells

Cells, rod: ↑ Photoreceptors

Cells, rod bipolar: ↑ Bipolar neurons, of retina

Cells, satellite, of peripheral neurons: ↑ Satellite cells, of peripheral neurons

Cells, satellite, of skeletal muscle fibers: ↑ Satellite cells, of skeletal muscle fibers

Cells, sebaceous: ↑ Sebaceous cells

Cells, secretory, of lacrimal glands: ↑ Lacrimal glands, secretory cells of

Cells, secretory, of mammary gland: ↑ Mammary gland, epithelium of

Cells, senescence of: ↑ Ageing

Cells, septal: ↑ Alveolar cells type II

Cells, serous, of bronchial glands: ↑ Bronchial glands

Cells, serous, of salivary glands: ↑ Serous cells, of salivary glands

Cells, sheath: ↑ Schwann's cells

Cells, SIF: ↑ Autonomic neurons, of sympathetic ganglia

Cells, smooth muscle: ↑ Smooth muscle cells

Cells, smudge: ↑ Smudge cells

Cells, somatic: ↑ Somatic cells

Cells, somatotropic: ↑ Somatotropes, of hypophysis

Cells, spinous: ↑ Epidermis; ↑ Prickle cells

Cells, stab: ↑ Band neutrophils

Cells, stellate, of adenohypophysis: ↑ Stellate cells, of adenohypophysis

Cells, stellate, of cerebellum: ↑ Stellate cells, of cerebellar cortex

Cells, stellate, of liver: ↑ Perisinusoidal cells

Cells, stellate, of pineal organ: ↑ Stellate cells, of pineal organ

Cells, stem, of hematopoiesis: ↑ Stem cells, of hematopoiesis

Cells, stem, of spermatogenic epithelium: ↑ Spermatogonia

Cells, steroid hormone-producing: ↑ Endocrine cells

Cells, superficial, of transitional epithelium: ↑ Facet cells

Cells, superficial, of sweat glands: ↑ Sweat glands, eccrine, excretory duct cells of

Cells, supporting, of carotid body: ↑ Glomus cells, of carotid body

Cells, supporting, of Corti's organ: ↑ Claudius' cells; ↑ Hensen's cells; ↑ Phalangeal cells; ↑ Pillars

Cells, supporting, of cristae ampullares and maculae sacculi and utriculi: ↑ Supporting cells of cristae ampullares and maculae sacculi and utriculi

Cells, supporting, of olfactory epithelium: ↑ Olfactory epithelium

Cells, supporting, of parasympathetic paraganglia: ↑ Paraganglia, parasympathetic

Cells, supporting, of pineal organ: ↑ Interstitial cells, of pineal organ

Cells, supporting, of sympathetic paraganglia: ↑ Carotid body; ↑ Paraganglia, parasympathetic

Cells, supporting, of taste buds: ↑ Taste buds, structure of

Cells, suppressor: ↑ Suppressor cells

Cells, supraependymal: ↑ Supraependymal cells

Cells, synovial: ↑ Synovial cells

Cells, T: ↑ T-lymphocytes; ↑ Transitional cells

Cells, tactile: ↑ Tactile cells

Cells, taste: ↑ Taste buds, structure of

Cells, tendon: ↑ Tendon cells

Cells, theca lutein: ↑ Theca lutein cells

Cells, thecal: ↑ Thecal cells

Cells, thyroid follicular: ↑ Thyroid follicular cells

Cells, thyroidectomy: ↑ Thyroidectomy cells

Cells, tracheal endocrine: ↑ Tracheal endocrine cells

Cells, transitional: ↑ Transitional cells

Cells, types of: ↑ An organism consists of: 1) ↑ Somatic cells making up structure of body; 2) ↑ germ cells for continuation of species.

Cells, U: ↑ Ultimobranchial follicles

Cells, ultimobranchial: ↑ C-cells, of ultimobranchial body

Cells, urethral chromaffin: ↑ Urethral chromaffin cells

Cells, vestibular, type I and II: ↑ Vestibular cells, type I and II

Cells, visual, of retina: ↑ Photoreceptors

Cells, vitreous: ↑ Hyalocytes

Cells, wandering, of connective tissue proper: ↑ Wandering cells, of connective tissue proper

Cells, X: ↑ Candidate stem cells

Cells, zymogenic, of gastric glands: ↑ Chief cells, of gastric glands proper

Cellular cementum: ↑ Cementum

"Cellular" connective tissue: An inappropriate term for a kind of ↑ loose connective tissue with a considerable prevalence of fixed, cellular elements (↑ fibroblasts and ↑ fibrocytes) over fibrillar ones. "C." composes major part of ovarian stroma, where its spindle-shaped cells form characteristic, irregular whorls (W). During pregnancy, some fixed cells can become ↑ interstitial cells. ↑ Wandering cells are rare. "C." also constitutes lamina propria of ↑ endometrium; here fixed cells (mostly fibroblasts) are stellate and wandering cells, especially ↑ macrophages, are more frequent. Fixed cells differentiate into ↑ decidual cells in course of pregnancy.

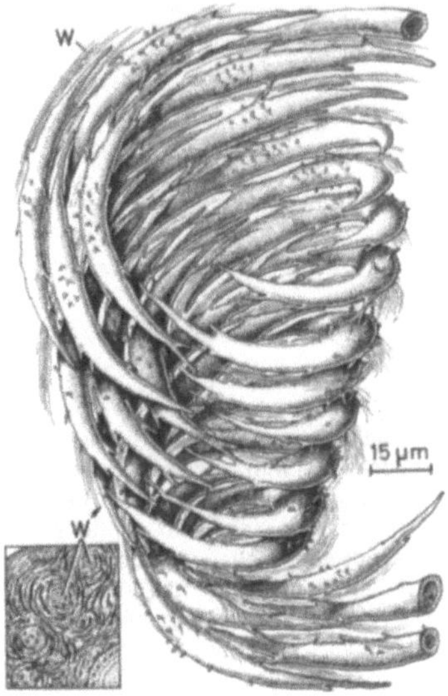

Cellular immunity: A part of ↑ immune response mediated by ↑ T-lymphocytes and ↑ macrophages by means of direct destruction of antigens.

↑ Killer cells

Cement, intercellular: ↑ Intercellular cement

Cement lines (cementing line, Ebner's line, incremental line, linea ce-

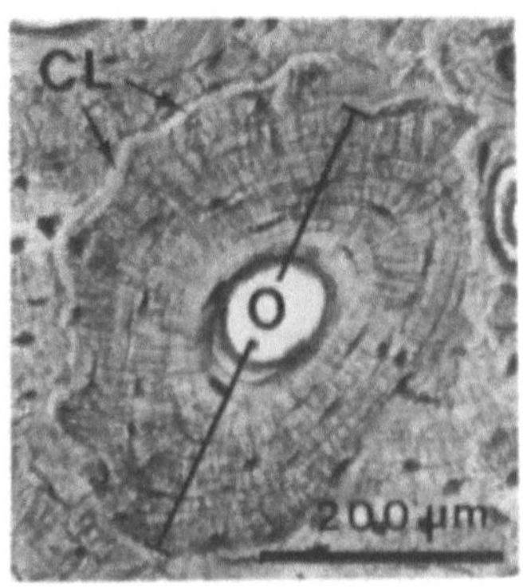

mentalis*): Distinct, refractile borders of collagen-poor ↑ bone matrix (CL) forming outer limit of secondary ↑ osteons (O). C. have staining properties different from other osteon lamellae, a different orientation of their collagen fibrils, and are, in general, not traversed by ↑ canaliculi of bone. (Fig. = human)

↑ Interstitial lamellae, of bone

Cementicles (Malassez' rests, relictum epitheliale*): Calcified remains of Hertwig's epithelial sheath found in ↑ periodontal ligament near surface of ↑ cementum. (See embryology texts for further information)

Cementing lines: ↑ Cement lines

Cementoalveolar fibers: ↑ Sharpey's fibers

Cementoblasts: Stellate cells of mesenchymal origin, precursors of ↑ cementocytes.

↑ Mesenchyme

Cementocytes (cementocytus*): Stellate cells (C) situated in corresponding lacunae within ↑ cementum. C. resemble ↑ osteocytes, but their processes are less numerous, shorter, and more irregular. Cell processes lie within canaliculi, which interconnect C.; processes are directed away from ↑ dentin (D) toward ↑ periodontal ligament, which assures nutrition for C.

↑ Canaliculi, of osteocytes

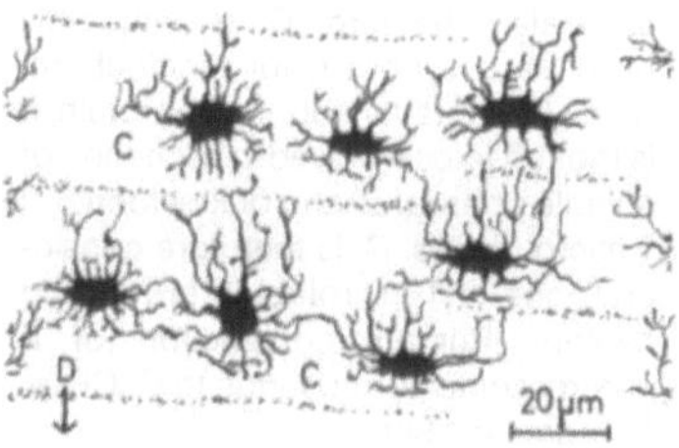

Cementosomes: ↑ Membrane-coating granules

Cementum* (secondary cementum, substantia ossea dentis): A special, avascular bonelike tissue (C) covering ↑ dentin (D) of neck and root of tooth. Two kinds of C.: 1) Cellular C. (CC) or cementum cellulare* = a system of lamellae (L), cells (↑ *cementocytes,* Cc), and matrix surrounding deep part of root. 2) Acellular cementum (AC) or cementum noncellulare* = a thin layer of C. composed only of matrix surrounding neck of tooth. ↑ Sharpey's fibers (ShF) irradiate from ↑ periodontal ligament (PL) into C. With age, C. increases in thickness by apposition of new lamellae at its surface; this C. may develop ↑ Haversian canals and become vascularized. Destruction of periodontal ligament leads to necrosis of entire C. Some epithelial cells, left over from dental development, may be incorporated into C.

↑ Cementicles

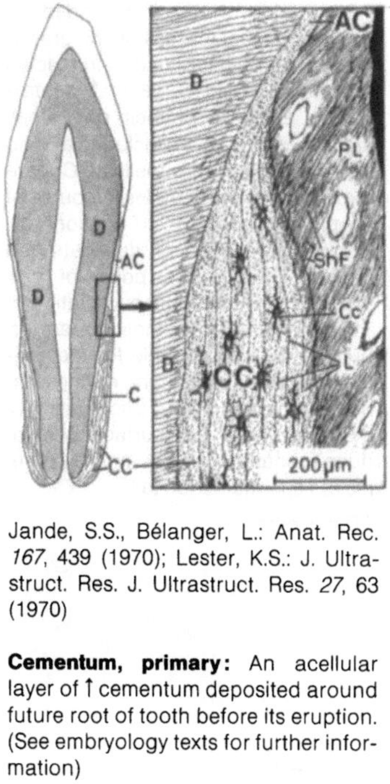

Jande, S.S., Bélanger, L.: Anat. Rec. *167*, 439 (1970); Lester, K.S.: J. Ultrastruct. Res. J. Ultrastruct. Res. *27*, 63 (1970)

Cementum, primary: An acellular layer of ↑ cementum deposited around future root of tooth before its eruption. (See embryology texts for further information)

Cementum, secondary: ↑ Cementum

Center, of chondrification: ↑ Protochondral tissue

Central arteries, of spleen (arteria folicularis, follicular arteries, arteria centralis*): ↑ Muscular arteries, branches of ↑ trabecular arteries, each surrounded in its course through splenic ↑ parenchyma by a ↑ periarterial lymphatic sheath. Germinal center (GC) of ↑ splenic nodules displaces C. in an excentric position in ↑ white pulp. Larger C. have distinct inner ↑ elastic lamina, which is not found in smaller C., the structure of which is similar to that of ↑ arterioles. In such cases, arterial wall is composed of one or two layers of ↑ smooth muscle cells. Tunica adventitia is replaced by ↑ lymphatic tissue. From C. arise numerous radial branches for vascularization of white pulp.

↑ Spleen, vascularization of

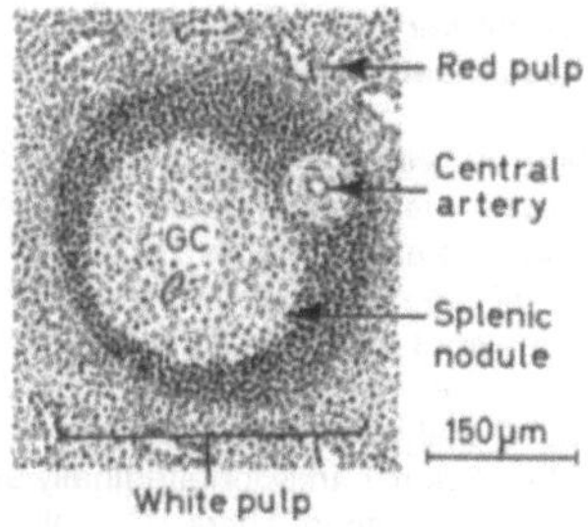

Central fovea: ↑ Fovea centralis

Central longitudinal arteries, of bone marrow: Branches of nutrient artery running parallel to shaft of long bones. From C., slender radial branches arise which empty into vascular sinuses. Occasional branches of C. enter ↑ haversian canals.

↑ Bone marrow, vascular sinuses of

Central longitudinal vein, of bone marrow: A thin-walled vein (V) receiving blood from vascular sinuses (VS) and ending in nutrient vein. (Fig. = rat)

↑ Bone marrow, vascularization of

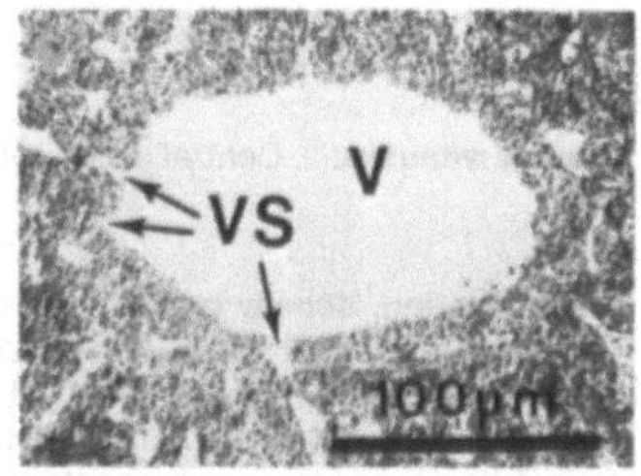

Central nervous system (CNS, systema nervosum centrale*): A mass of ↑ nervous tissue within cranium and canal of vertebral column comprising ↑ brain and ↑ spinal cord. Macroscopic structure: 1) Gray matter or substantia grisea* = ↑ perikarya of ↑ neurons, their ↑ dendrites, proximal portion of long ↑ axons, all of short axons, and ↑ neuroglia cells of CNS. 2) White matter or substantia alba* = largely prevailing ↑ myelinated axons of neurons located in gray matter or in ganglia outside CNS and neuroglial cells of CNS. In cerebral hemispheres and cerebellum, gray matter is peripheral; in spinal cord it is central.

Nieuwenhuys, R., Voogd, J., van Huijzen, C.: The Human Central Nervous System. 2nd edn. Berlin, Heidelberg, New York: Springer-Verlag 1981

Central nervous system, connective tissue sheaths of: ↑ Meninges

Central veins, of liver (central venules, terminal hepatic venules, vena centralis*): Venous vessels (CV), 50–150 µm in diameter, each situated in center of classic ↑ liver lobule. Wall of C. is very thin, consisting only of ↑ endothelial cells surrounded by delicate, predominantly longitudinally oriented ↑ collagen fibers. C. collects blood from ↑ liver sinusoids (S) through numerous openings in its wall (arrows). C. continues into intercalated ↑ veins. (Fig. = human)

↑ Liver, vascularization of

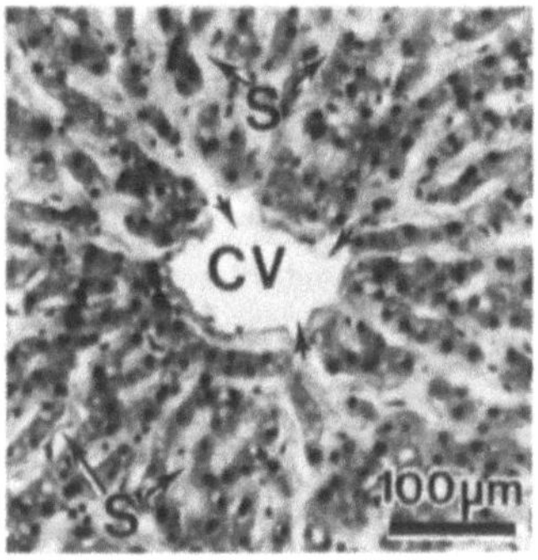

Central venules: ↑ Central veins, of liver

Centrifugation, density gradient (differential centrifugation): A method permitting isolation of various subcellular fractions and/or ↑ organelles from ↑ homogenates by ultracentrifugation. Homogenate is put in a centrifuge tube on top of a stratified solution (sucrose, cesium salts, etc.) that increases in density toward the bottom; after centrifugation, each subcellular fraction or organelle has settled on top of the layer whose density corresponds to its own.

De Duve, C.: Science *189*, 186 (1975); De Duve, C., Beaufay, H.: J. Cell Biol. *91*, 293s (1981); Price, C.A.: Centrifugation in Density Gradients. New York: Academic Press 1982

Centrifugation, differential: ↑ Centrifugation, density gradient

Centriole, distal, of spermatozoon: ↑ Spermatozoon, neck of

Centriole, proximal, of spermatozoon: ↑ Spermatozoon, neck of

Centrioles (centriolum*): Roughly 0.15–0.2 µm wide and 0.3–0.5 µm long, membraneless ↑ organelles composed of nine microtubular triplets (T) arranged as a minute cylinder. Only microtubule A of each triplet has a round profile; those of microtubules B and C are crescent-shaped. Microtubule A sends out two short arms: One (1) is connected to C microtubule of next triplet, and the other (2) is directed toward center of C., where it reaches a longitudinal microfilament (Mf), about 7.5 nm thick. C. are surrounded by dense spherical masses of fine granular material, ↑ satellites (S), connected with each triplet. C. lie in ↑ centrosphere. Because of their own small amount of DNA, C are included among self-duplicating organelles. C. are formed from precentrioles in course of G_2 phase of cell cycle. Functions: 1) During cell division, C. duplicate and migrate to opposite poles of cell. There, C. not only induce formation of ↑ microtubules of ↑ spindle apparatus, but are also responsible for their depolymerization shortening during anaphase. 2) In course of ↑ ciliogenesis, C. migrate toward cell surface and induce formation of ↑ cilia and ↑ axoneme of ↑ spermatozoon tail.

↑ Diplosome

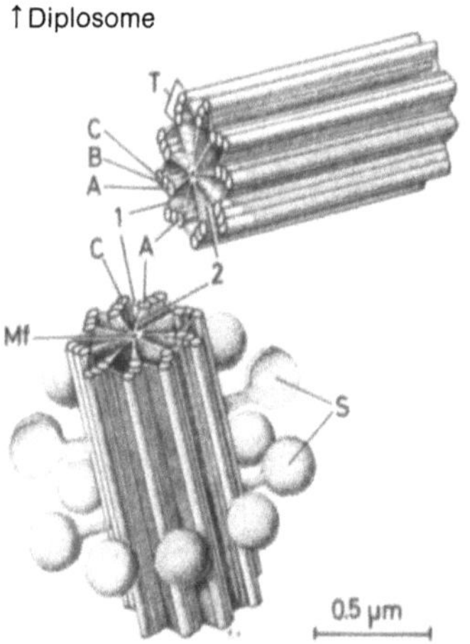

Peterson, S.P., Berns, M.W.: Int. Rev. Cytol. *64*, 81 (1980); Vorobjev, I.A., Chentsov, Yu.S.: J. Cell Biol. *93*, 938 (1982); Wheatley, D.N.: The Centriole. A Central Enigma of Cell Biology. Amsterdam: Elsevier 1982

Centriologenesis: The process of formation of ↑ centrioles.

↑ Ciliogenesis

Centroacinar cells (cellula centroacinosa*): Small, flattened cells (Ca) incompletely bordering lumen (L) of pancreatic ↑ acini. Direct continuation of C. outside acinus forms ↑ intercalated duct. C. have a relatively voluminous nucleus, few mitochondria, a small Golgi apparatus, and only rare rough endoplasmic cisternae. AC = ↑ pancreatic acinar cells. (Fig. = rat)

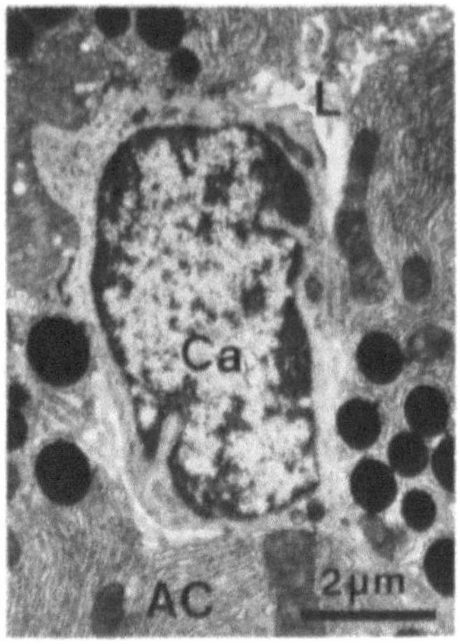

Williams, D.W., Kendall, M.D.: J. Anat. *135*, 173 (1982)

Centromere (centromerus*): A small clear connection of nonchromatin material between two sister ↑ chromatids at point where arms (A) of ↑ chromosome (Chr) meet. During ↑ mitosis and ↑ meiosis, ↑ microtubules (Mt) of ↑ spindle apparatus anchor in centrometric region by means of a special structure, ↑ kinetochore. At onset of anaphase, C. divides and chromatids separate. The term C. is frequently used as a synonym for kinetochore. Since C. contains ribonucleoprotein, it is though to be involved in formation of spindle apparatus and of kinetochoral ↑ microtubules. C. is therefore considered one of ↑ microtubule organizing centers. Primary constriction of a chromosome corresponds to C. CF = ↑ chromatin fiber

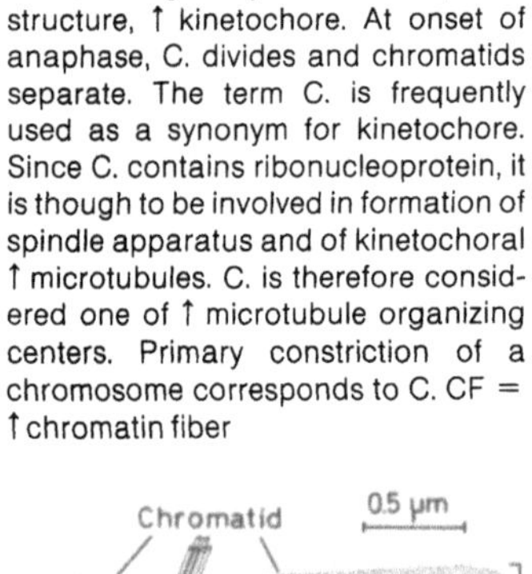

Rieder, C.: J. Cell Biol. *80*, 1 (1979); Schwarzacher, H.G.: Chromosomes in Mitosis and Interphase. In: Bargmann, W. (ed.): Handbuch der mikroskopischen Anatomie des Menschen. Vol. 1, Part 3. Berlin, Heidelberg, New York: Springer-Verlag 1976; Vig, B.K.: Experientia *37*, 566 (1981)

Centrosome: ↑ Centrosphere

Centrosphere (cell center, centrosome, cytocentrum*): A specialized mass of cytoplasm (C), resistant to ordinary staining methods, surrounding ↑ centriole(s) (arrow). Apart from some ↑ microtubuli and smooth vesicles, ↑ organelles and free ribosomes are lacking in C. C. is situated near nucleus (N) or in a concavity of Golgi apparatus (G). (Fig. = ↑ mesenchymal cell, chick embryo)

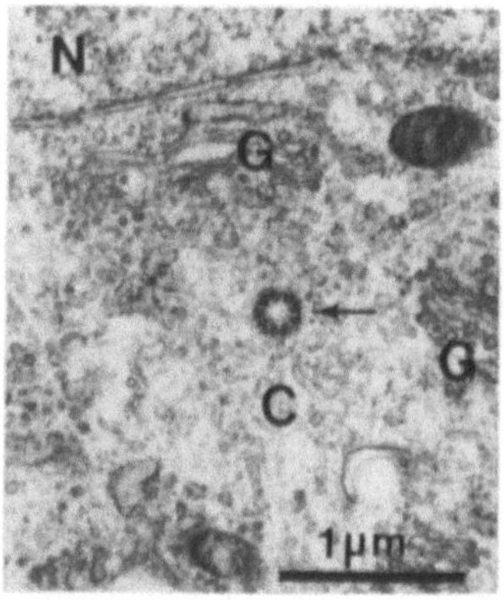

Maunoury, R.: Biol. Cell *36*, 91 (1979)

Cerebellum*: Part of ↑ brain composed of ↑ gray matter, cerebellar cortex, and ↑ white matter, the cerebellar medulla. Cortex forms long, narrow folds, ↑ folia, separated by sulci. Structure: 1) Cortex: a) molecular layer (ML) or stratum moleculare* is covered by ↑ pia mater (PM) and contains only a few ↑ neurons; b) ganglionic layer (GL) or stratum ganglionare* = ↑ Purkinje cells (PC) enclosed by baskets (B) formed by axons of ↑ basket cells, which also participate in formation of supraganglionic nerve plexus (sg). Infraganglionic nerve plexus (ig) is composed of fibers irradiating from medulla; c) granular layer (GrL) or stratum granulare* = very numerous ↑ granule cells (GC), less numerous large ↑ granule cells, and ↑ cerebellar glomeruli (CG). 2) Medulla (M) = nerve fibers.

↑ Cerebellum, cyto- and myeloarchitectonics of; ↑ Cerebellum, glioarchitectonics of; ↑ Cerebellum, neuronal circuits of

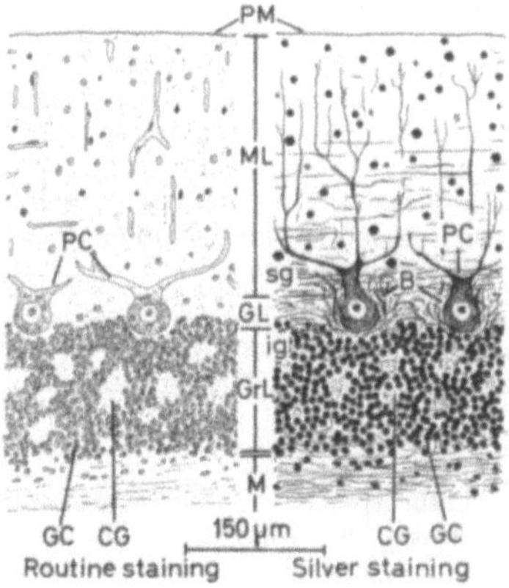

Ghez, C., Fahn, S.: The Cerebellum. In: Kandel, E.R. and Schwartz, J.H. (eds.): Principles of Neural Science. New York Amsterdam Oxford: Elsevier North Holland 1981; Palay, S.L., Chan-Palay, V. (eds.): The Cerebellum-New Vistas. Berlin, Heidelberg, New York: Springer-Verlag 1982

Cerebellum, cyto- and myeloarchitectonics of: 1) Cortex: a) molecular layer (ML) = ↑ stellate cells (SC), ↑ basket cells (BC), all dendritic (d) ramifications of ↑ Purkinje cells (PC), ↑ climbing fibers (CF), ↑ dendrites of large ↑ granule cells (LGC), and ↑ axons of ↑ granule cells (GC), i.e., parallel fibers (PF); b) ganglionic layer (GL) = ↑ perikarya of Purkinje cells, baskets (B) of basket cells, as well as the supra- (sg) and infraganglionar (ig) nerve plexuses; c) granular layer (GrL) = granular cells (GC) with their dendrites, large granule cells (LGC) with their axons, ↑ cerebellar glomeruli (CG), ↑ mossy (MF) and climbing fibers (CF). 2) Medulla (M): a) afferent fibers = climbing and mossy fibers; b) efferent fibers = axons of Purkinje cells. a = axons, d = dendrites

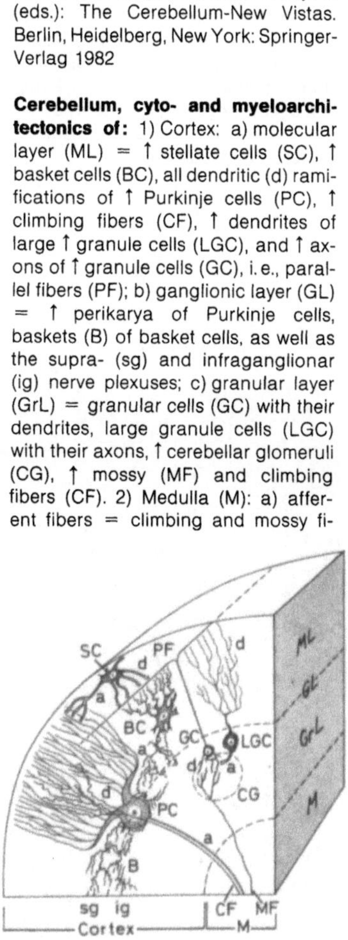

Braak, E., Braak, H.: Anat. Embryol. *166*, 67 (1983); Lange, W.: Z. Zellforsch. *134*, 129 (1972)

Cerebellum, glioarchitectonics of: 1) Cortex: a) molecular layer (ML) = ↑ oligodendrocytes (O), ↑ Fananas' cells (FC), extensions of ↑ Bergmann's cells (BC); b) ganglionic layer (GL) = Fananas' cells, perikarya of Bergmann's cells, oligodendrocytes; c) granular layer (GrL) = protoplasmic ↑ astrocytes (PA), velate ↑ astrocytes (VA), oligodendrocytes. 2) Medulla (M) = fibrous ↑ astrocytes (FA), oligodendrocytes. PC = ↑ Purkinje cells

↑ Glioarchitectonics

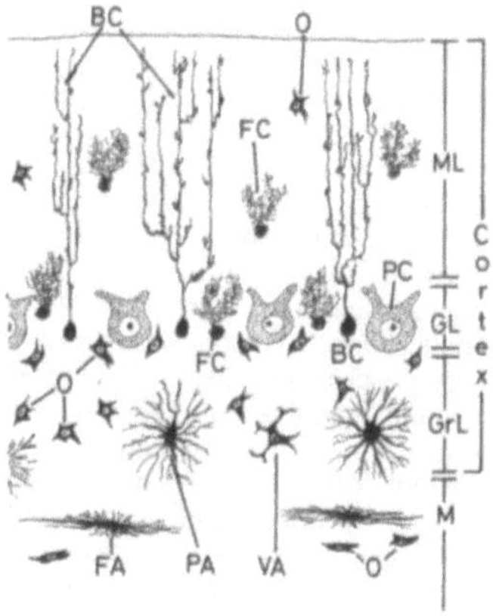

Cerebellar glomeruli (glomerula cerebellaria*): Small anuclear areas (G) between ↑ neurons of granular layer of cerebellar cortex. C. consist of extremely complex synaptic connections between ↑ mossy fibers, dendrites of small ↑ granule cells, and axons of large ↑ granule cells. C. are incompletely surrounded by platelike extensions of velate ↑ astrocytes. (Fig. = human)

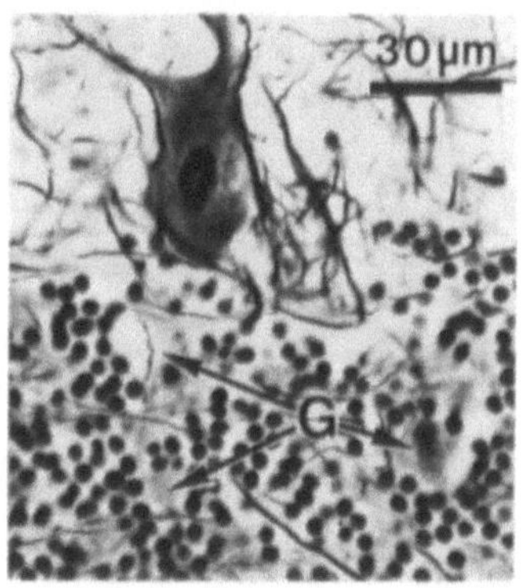

Landis, D.D., Weinstein, L.A., Halperin, J.J.: Develop. Brain Res. *8*, 231 (1983)

Cerebellum, neuronal circuits of: 1) Inhibitory circuit of Purkinje cells: ↑ Mossy fibers (MF) transfer impulses via dendrites of ↑ granule cells (GC) to their ↑ parallel fibers (PF), which form excitatory ↑ synapses with dendrites of many ↑ Purkinje cells (PC) and dendrites of both ↑ basket (BC) and ↑ stellate cells (SC). On the other hand, axons (a) of basket cells form inhibitory synapses with perikarya of Purkinje cells, and axons (a) of stellate cells form inhibitory synapses with dendrites (d) of Purkinje cells. 2) Inhibitory circuit of granule cells: Mossy fibers transfer impulses to granule cells, i.e., parallel fibers which form ↑ synapses with dendrites (d) of large ↑ granule cells (LGC). Their axons and dendrites of granule cells form inhibitory axodendritic ↑ synapses in ↑ cerebellar glomeruli (CG). 3) Excitatory fibers of Purkinje cells = ↑ climbing fibers (CF). Collaterals (C) of Purkinje cells form synapses with axons of neighboring Purkinje cells and large granule cells. Afferent fibers of cerebellum: a) Mossy fibers (form synapses with granule cells), b) climbing fibers (form synapses with dendrites of Purkinje cells, large granule cells, and stellate cells). Efferent fibers of cerebellum: Axons of Purkinje cells (AP) ending in central cerebellar nuclei. ML = molecular layer, GL = ganglionic layer, GrL = granular layer, M = medulla

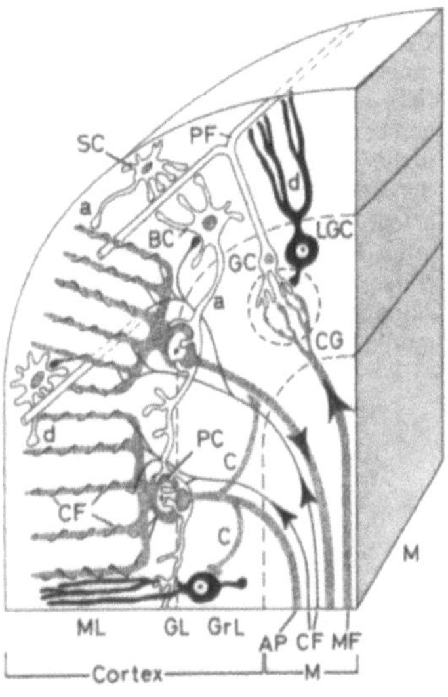

Cerebral cortex: ↑ Brain cortex

Cerebrospinal fluid (liquor cerebrospinalis): A transparent liquid of low specific gravity (1004–1008), chiefly secreted by ↑ choroid plexus but also by ↑ pia mater and brain substance. C. contains practically no protein, some inorganic salts, dextrose, a few ↑ lymphocytes, and ↑ macrophages. C. circulates from lateral ventricles toward third and fourth ventricles, from which it passes into ↑ subarachnoid space to be eliminated into cranial venous sinuses, very probably through ↑ Pacchionian granulations.

Wood, J.H. (ed.): The Neurobiology of Cerebrospinal Fluid. Vol. I and II. New York and London: Plenum Press 1980 and 1983

Cerebrospinal fluid-contacting neurons (CSF-contacting neurons, liquor-contacting neurons): Small ↑ neurons (CF) with ↑ perikarya located among ↑ ependymal cells (EC) or in ↑ subependymal layer of third ventricle. Whereas axon (A) of C. penetrates ↑ neuropil (N), some branched ↑ dendrites (D) pass between ependymal cells and enter, partly as ↑ supraependymal dendrites, ventricular lumen (VL), where they are bathed by ↑ cerebrospinal fluid. Structure of C. corresponds to that of other neurons. Function of C. is not clear; it is believed they represent a kind of cerebral ↑ osmoreceptor. Cap = capillary

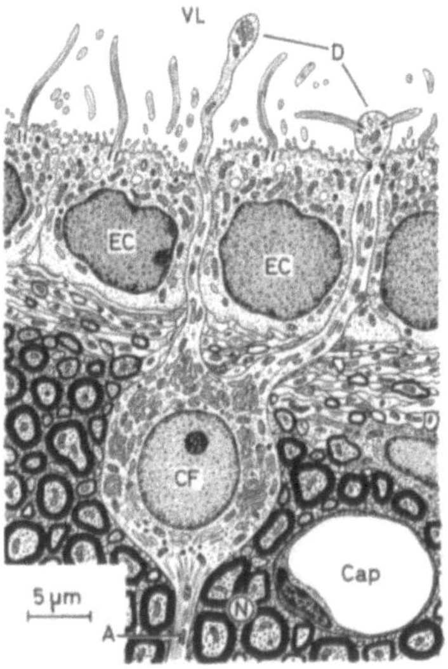

Korf, H.W.: Viglietti-Panzica, C., Panzica, G.C.: Cell Tissue Res. *228*, 149 (1983); Vigh, B., Vigh-Teichmann, I.: The CSF Contacting Neurosecretory Cell: A Protoneuron. In: Farner, D.S., Lederis, K. (eds.): Proc. 8th Int. Symp. on Neurosecretion. New York: Plenum Press 1981; Vigh-Teichmann, I., Vigh, B., Aros, B., Jenner, L., Sikora, K., Kovacs, J.: Z. mikrosk.-anat. Forsch. *93*, 609 (1979)

Ceruloplasmin: A blue-colored α-globulin of blood that binds, transports, and releases nearly all the copper of the organism, thereby contributing to control of its utilization. A molecule of C. contains eight atoms of copper and has a molecular weight of about 132000 daltons. C. plays a role in ↑ erythropoiesis.

Mather, J.P.: In Vitro *18*, 990 (1982)

Cerumen: A brownish, bitter, waxy substance present in external ↑ auditory meatus. C. is a mixture of ↑ ceruminous and ↑ sebaceous gland secretions mixed with desquamated epithelial cells. C. protects external auditory meatus from desiccation and from penetration by insects through its insecticidal action.

Ceruminous glands (glandulae ceruminosae*): Branched, ↑ tubuloalveolar glands (CG) of external ↑ auditory meatus, considered to be modified apocrine ↑ sweat glands. Coils of C. have a large lumen and are lined by a ↑ simple cuboidal epithelium rich in ↑ lipid droplets and pigment granules. Very numerous ↑ myoepithelial cells (My) surround tubuloalveoli. Excretory ducts (D) are relatively narrow; they are lined by a stratified epithelium and open either directly (arrow) onto skin surface or, together with ↑ sebaceous glands (SG), into a ↑ hair follicle (HF). C. produce yellowish secretion which constitutes a major part of ↑ cerumen.

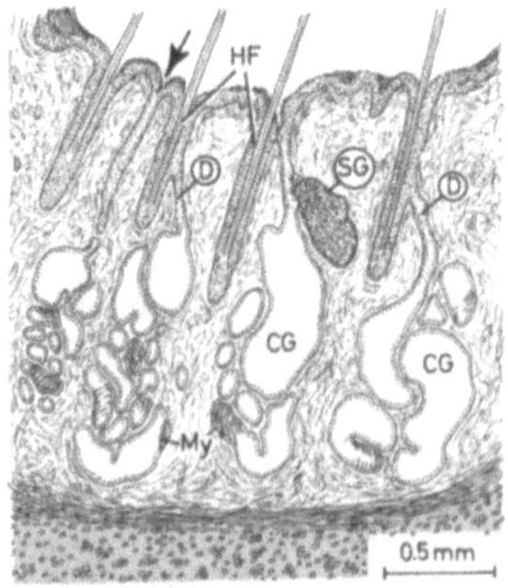

Bende, M.: J. Laryngol. Otol. *95*, 11 (1981); Testa-Riva, F., Puxeddu, P.: Anat. Rec. *196*, 363 (1980)

Cervical glands (glandulae cervicales uteri*): Large, long, branched ↑ tubular glands (CG) of ↑ cervix of ↑ uterus. C. are lined with one layer of columnar, mucus-secreting cells. During proliferative and secretory phases of ↑ menstrual cycle, C. produce highly viscous ↑ mucus; cell nuclei are basal (1). At mid-cycle, when cells secrete actively under estrogen stimulation, nuclei rise to center of cells (2); mucus becomes watery and facilitates spermatozoal migration. In pregnancy, C. enlarge and interglandular tissue diminishes. After ↑ menopause, mucosa and glands gradually atrophy; secretion ceases. Glandular openings sometimes become obstructed, leading to an accumulation of mucus in glands (= ↑ Nabothian cysts).

↑ Cervical glands, cells of

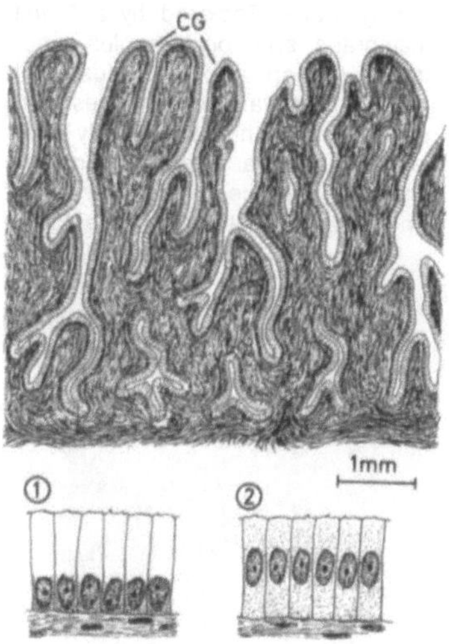

Cervical glands, cells of (cellulae columnares*): Tall columnar cells, about 20 µm in height lining ↑ cervical glands. During most of ↑ menstrual cycle, nucleus is in a basal position; it is elliptical and contains a prominent nucleolus. Cytoplasm encloses small mitochondria, a large supranuclear Golgi apparatus (G), a ↑ centriole, and a well-developed, predominantly basal rough endoplasmic reticulum. Numerous, 0.3 to 1-µm-large, membrane-bound secretory granules (S) with homogenous, highly osmiophilic material arise from Golgi apparatus and fill upper two-thirds of cell body, from which they are discharged by ↑ exocytosis. ↑ Apical pole of cell bears short ↑ microvilli (Mv). Adjacent cells are attached to each other by well-developed ↑ junctional complexes (J), and lateral cell surfaces are interdigitated with those of other C. Presence of dark cells among clear cells is probably a consequence of different functional states. Cells lie on ↑ basal lamina (BL).

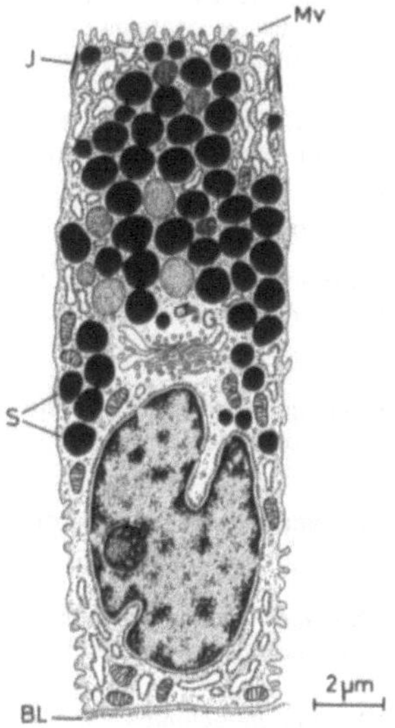

Cervix, of teeth: ↑ Neck, of teeth

Cervix, of uterus (cervix uteri*): Tube-like lower part of ↑ uterus. Its central cavity, cervical canal (CC), begins near isthmus uteri with internal os (IO) and terminates with external os (EO) at lower end of C., which protrudes into upper ↑ vagina (V); this part of C. is called portio vaginalis* (PV). Structure: A) Cervical canal: 1) Endocervix (E) or tunica mucosa = a mucous membrane, about 3–5 mm thick, displaying longitudinal and transversal folds, plicae palmatae* (PP): a) epithelium = ↑ simple columnar epithelium predominantly composed of mucous cells with some ciliated cells; this epithelium forms ↑ cervical glands (CG). Near external os, epithelium becomes abruptly ↑ stratified squamous nonkeratinized (arrow) lining portio vaginalis; b) lamina propria (LP) = a predominantly ↑ dense connective tissue. 2) ↑ Myometrium (My) = ↑ smooth muscle cells arranged in spiral bundles; presence of numerous ↑ elastic fibers. 3) Perimetrium (not represented) = a ↑ loose connective tissue at vesical surface; ↑ peritoneum at lateral and posterior surfaces. B) Portio vaginalis: 1) Exocervix (Ex), ectocervix, or tunica mucosa: a) epithelium (Ep) = stratified squamous nonkeratinized epithelium continuous with that of vagina; b) lamina propria (LP) = a ↑ dense connective tissue without glands. Apart from in cervical glands, there are no cyclic changes in endo- and exocervix; these are not, therefore, sloughed off. NC = ↑ Nabothian cyst

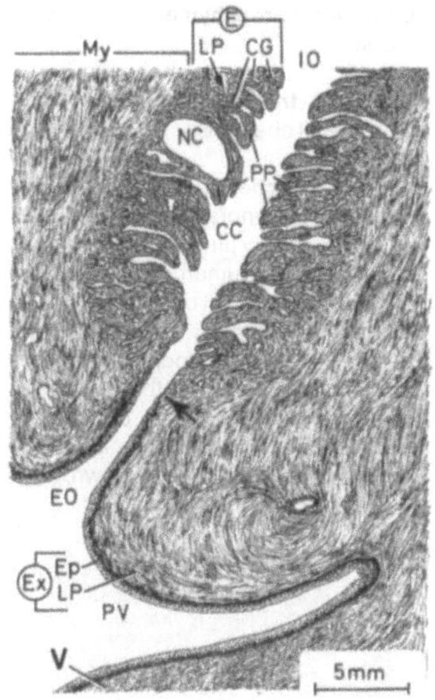

Davina, J.H.M., Stadhouders, A.M., Lamers, G.E.M., van Haelst, U.J.G.M., Kenemans, P.: Scanning Electron Microscopy 1981/III, 37 (1981), Hafez, E.S.E.: Reproduction, *5*, 243 (1981); Vickery, B.H., Bennett, J.P.: Physiol. Rev. *48*, 135 (1968)

CFU: ↑ Colony-forming unit

CFU-C: ↑ Colony-forming unit-culture

CFU-E: ↑ Colony-forming unit-erythropoietic

CFU-G: ↑ Colony-forming unit-granulopoietic

CFU-M: ↑ Colony-forming unit-megakaryocytic

CFU-S: ↑ Colony-forming unit-splenic

Chalone: A specific mitotic inhibitory substance elaborated by a tissue and active only within that tissue; not specific to a given species. C. are constantly produced to control a specific cell population by a negative ↑ feedback mechanism, probably acting on G_1 and G_2 phases of ↑ cell cycle.

Cairnie, A.B., Lala, P.K., Osmond, D.G. (eds.): Stem Cells of Renewing Cell Populations. New York, San Francisco, London: Academic Press 1976; Leith, J.T.: Cell Tissue Kinet. *11*, 433 (1978); Rytömaa, T., Toivonen, H.: Mech. Ageing and Development *9*, 471 (1979)

Chamber, transparent: ↑ Transparent chamber

Channels, parenchymal: ↑ Canaliculi, tissue

Channels, transendothelial: ↑ Transendothelial channels

Charcot-Böttcher crystals: ↑ Peculiar, needlelike, membraneless ↑ inclusion (CBC), up to 25 μm long and 1–3 μm thick, found in human ↑ Sertoli's cells. Structure of C. is not compact; in some areas, they display defects (arrowheads) filled by ↑ cytoplasmic matrix. C. consist of 15-nm-thick parallel microfilaments (arrows) converging toward ends of crystal; chemical composition and functions unknown. (Fig. = human)

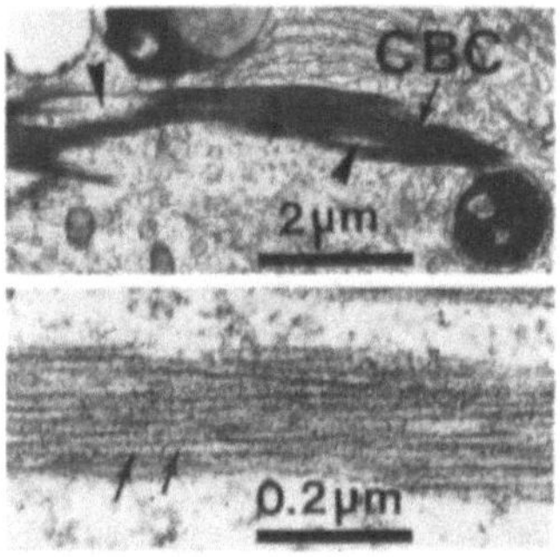

Sohval, A.R., Suzuki, Y., Gabrilova, J.L., Churg, J.: J. Ultrastruct. Res. *34*, 83 (1971)

Cheek (bucca*): Lateral wall of mouth, about 10 mm thick. Structure. Outer surface (pars cutanea) = a thin ↑ skin with hairs, ↑ sebaceous and ↑ sweat glands; ↑ hypodermis (H) is rich in ↑ adipose tissue, particularly in newborns (= ↑ Bichat's fat pad) and adheres to musculus buccinator (MB). Inner surface (pars mucosa): 1) Tunica mucosa (TM) = a) epithelium (E) = ↑ stratified squamous nonkeratinized epithelium; b) lamina propria (LP) = a ↑ loose connective tissue. 2) Tela submucosa (TS) = a ↑ loose connective tissue containing ↑ mixed ↑ buccal glands (BG). 3) Tunica muscularis (TMu) = ↑ skeletal muscle fibers (MF) of buccinator muscle. With exception of epithelium, all layers of C. contain many ↑ elastic fibers.

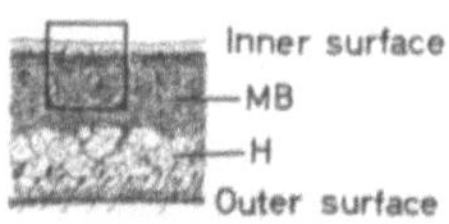

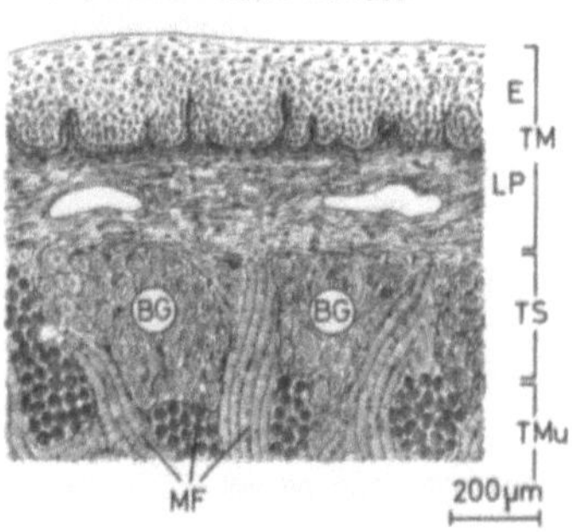

↑ Juxtaoral organ

Schroeder, H.E., Dörig-Schwarzenbach, A.: Cell Tissue Res. *224*, 89 (1982)

Chemical mediators: ↑ Neurotransmitter substances

Chemical synapse: ↑ Synapse, chemical

Chemical synapse, types of: ↑ Synapse, chemical, types of

Chemoarchitectonics: The arrangement of histochemical properties of a ↑ tissue or ↑ organ.

Chemoreceptors: Categories of cells and nerve endings sensitive to various chemical substances (↑ olfactory cells, ↑ taste cells), as well as to changes in oxygen and carbon dioxide tension in blood, including pH. Best-defined of second group are ↑ aortic and ↑ carotid bodies; some of nervi vasorum are also C.

Chemotaxis (chemotropism): A property of ameboid cells (↑ histiocytes, ↑ granulocytes, ↑ microglia, etc.) to be attracted and move toward certain substances, such as the split product of complement, certain bacterial products, denatured proteins (positive C.). Acids and alkalies have an opposite, repellent effect (negative C.).

Lackie, J.M., Wilkinson, P.C. (eds): Biology of the Chemotactic Response. Cambridge: Cambridge University Press 1981

Chemotropism: ↑ Chemotaxis

Chiasma (decussatio*): The point(s) where ↑ chromatids of a ↑ tetrad cross during prophase I of ↑ meiosis. C. is a morphological expression of phenomenon of ↑ crossing over. C. contains a piece of ↑ synaptonemal complex that will ultimately disappear.

Comings, D.E., Okada, T.A.: Adv. Cell Mol. Biol. *2*, 310 (1972)

Chief cells, of carotid body: ↑ Glomus cells, of carotid body

Chief cells, of chromaffin paraganglia: ↑ Glomus cells, type I; ↑ Paraganglia, sympathetic

Chief cells, of gastric glands proper (peptic cells, zymogenic cells, cellulae principales*): Basophilic cuboidal or low columnar cells located in lower third or lower half of ↑ gastric glands proper. Nucleus is spherical and in a basal position; nucleolus is prominent. Apical plasmalemma, covered by a distinct ↑ glycocalyx, forms short and irregular microvilli (Mv). Cytoplasm contains relatively few mitochondria, an extensive rough endoplasmic reticulum (rER), and a well-developed supranuclear Golgi apparatus (G) from which voluminous ↑ secretory granules (SG) arise in form of ↑ zymogen granules. Enclosed by a ↑ unit membrane, secretory granules accumulate in ↑ apical pole. By fusion of limiting membrane of granules with plasmalemma, their low-density content is released (arrows) into glandular lumen. C. produce and release ↑ pepsinogen and ↑ gastric intrinsic factor.

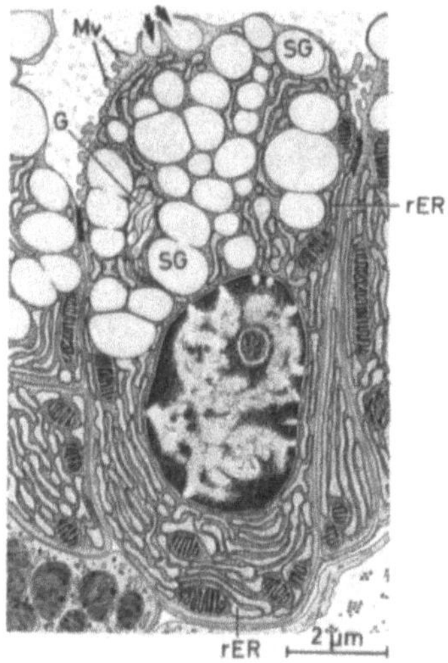

Helander, H.F.: Int. Rev. Cytol. *70*, 217 (1981)

Chief cells, of parasympathetic paraganglia: ↑ Paraganglia, parasympathetic

Chief cells, of parathyroid glands (principal cell, cellula principalis*): Small, polygonal, pale, slightly ↑ acidophilic cells with a voluminous central nucleus and prominent nucleoli forming majority of cells in ↑ parathyroid gland. Cytoplasm contains small scattered mitochondria, an inconspicuous

Golgi apparatus in a juxtanuclear position and short, flattened rough endoplasmic cisternae. A few ↑ lysosomes may be present, together with ↑ lipofuscin granules, ↑ glycogen particles, ↑ lipid droplets, and a small number of free ↑ ribosomes. ↑ Secretory granules (S) are associated with Golgi apparatus and measure, depending upon species, 100–500 nm (inset, bar = 200 nm); they contain a dense, fine granular core surrounded by a clear halo. Granules are enclosed by a ↑ unit membrane; they are released into intercellular and pericapillary spaces, but this phenomenon is difficult to observe. C. are generally in contact with ↑ pericapillary space (asterisk); a basal lamina covers this cell surface. C. are held together by means of ↑ junctional complexes and ↑ desmosomes; in some species (rat), C. display extensive interdigitations and numerous microvillar processes. C. are found in two varieties representing their two functional phases: Light C. (LC) or cellulae principales lucidae* have a translucent cytoplasmic matrix, a few secretory granules and a large amount of glycogen (resting phase). Dark C. (DC) or cellulae principales densae* have an electron-opaque cytoplasmic matrix, a relatively large number of secretory granules, but little glycogen (secretory phase). It is thought that C. synthesize and release ↑ parathormone. (Fig. = monkey, inset = rat)

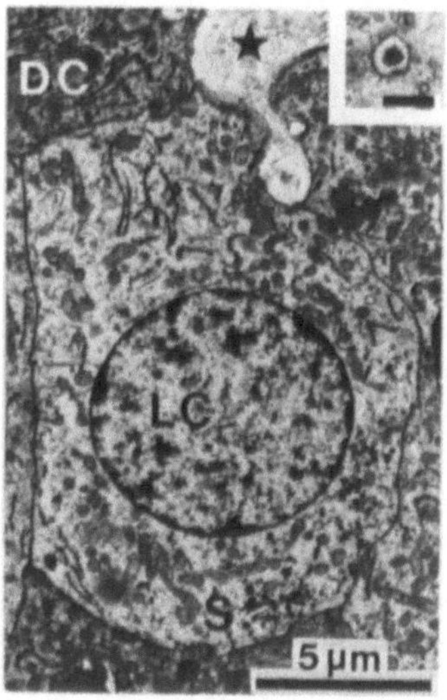

Nunez, A.E., Whalen, J.P., Krook, L.: Am. J. Anat. *134*, 459 (1972); Setoguti, T., Takagi, M.: Okajimas Fol. Anat. Jpn. *56*, 337 (1980); Wild, P.: Acta Anat. *108*, 340 (1980)

Chief cells, of pineal organ: ↑ Pinealocytes

Cholangioles: ↑ Bile ductules

Cholecystokinin-pancreozymin (CCK-PZ): A polypeptide ↑ tissue hormone probably produced by ↑ I-cells of ↑ duodenum, ↑ jejunum, and ↑ ileum, provoking contractions of ↑ gallbladder and secretion of ↑ pancreatic acinar cells.

Choledochoduodenal junction: The connection between ↑ ductus choledochus, ↑ pancreatic duct, and ↑ duodenum, characterized by presence of ↑ sphincter of Oddi and ↑ ampulla hepatopancreatica.

Boyden, E.A.: Surg. Gynecol. Obstet. *104*, 641 (1957)

Cholinergic nerve endings: Terminations of nerve fibers that release, ↑ acetylcholine upon stimulation. Morphologically, C. are characterized by presence of translucid, spherical ↑ synaptic vesicles (SV). (Fig. = ↑ bouton terminal of a. C. in ↑ myenteric plexus, rat)

↑ Adrenergic nerve fibers; ↑ Cholinergic nerve fibers

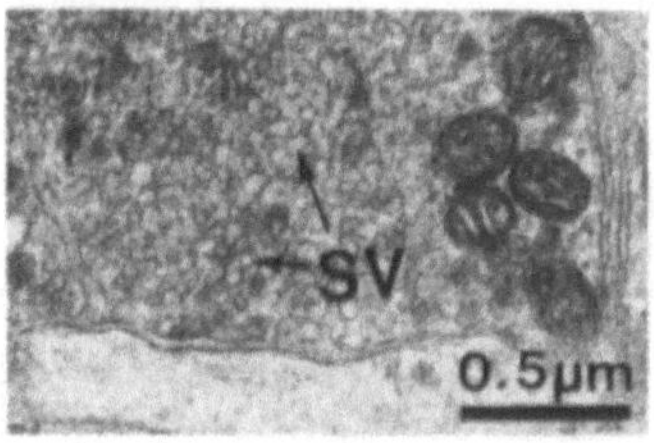

Cholinergic nerve fibers (cholinergic system): Efferent ↑ nonmyelinated nerve fibers releasing ↑ acetylcholine from their terminations. C. include all fibers of ↑ parasympathetic nervous system, preganglionic fibers of ↑ sympathetic nervous system, and postganglionic sympathetic fibers innervating ↑ skin and ↑ sweat glands. C. are also present in CNS.

↑ Adrenergic nerve endings; ↑ Cholinergic nerve endings

Cholinergic synapse: ↑ Synapse, cholinergic

Chondroblasts (chondroblastus*): Young cartilage cells with a small, flattened to oval body, a relatively large nucleus, and poorly developed ↑ organelles. As they differentiate into ↑ chondrocytes, C. enlarge, becoming spherical; their organelles (rough endoplasmic reticulum, Golgi apparatus, etc.) develop and the cells begin to synthesize and secrete ↑ cartilage matrix.

Chondroclasia: Process of destruction of hyaline ↑ cartilage by ↑ chondroclasts.

Laster, K.S., Ash, M.M. Jr.: J. Ultrastruct. Res. *74*, 46 (1981)

Chondroclasts: Giant, multinucleated cells resorbing calcified cartilage during indirect ↑ bone formation; morphologically identical to ↑ osteoclasts

Knese, K.H.: Acta Anat. *83*, 275 (1972)

Chondrocytes (cartilage cell, chondrocytus*): Flattened to spherical cells located in corresponding cavities, lacunae, of ↑ cartilage matrix (CM). Nucleus (N) is predominantly round with one to two nucleoli (Nu). Cytoplasm contains scarce, small mitochondria, a large Golgi apparatus (G), and well-developed rough endoplasmic cisternae (rER). Very numerous free ribosomes, ↑ coated vesicles, and microfibrillar bundles are also present, as well as ↑ glycogen particles (Gly) and ↑ lipid droplets (L), which are frequently larger than nucleus. Surface of C. is provided with numerous microvilli (Mv) and microridges, which penetrate pericellular matrix, referred to as the ↑ capsule (Cp). C. synthesize and secrete material for both ↑ collagen fibrils (Col) and ↑ elastic fibers (E), as well as for ↑ ground substance of ↑ cartilage matrix.

Stockwell, R.A.: Biology of Cartilage Cells. Cambridge: Cambridge University Press 1979

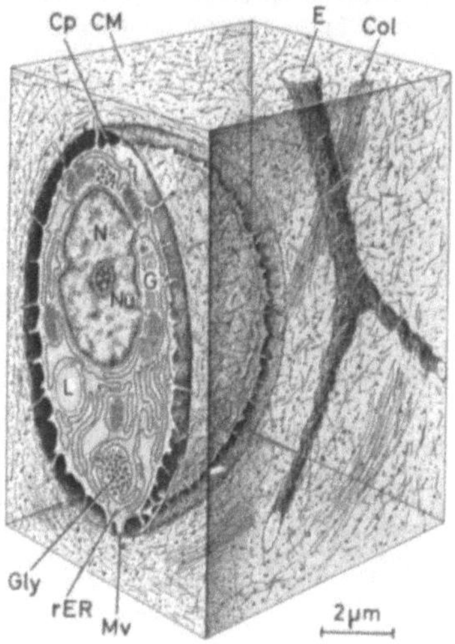

Chondrocytes, synthesis of intercellular components by: 1) ↑ Cartilage ground substance. Polypeptide chains are formed from amino acids by ↑ ribosomes of rough endoplasmic reticulum, accumulated in its cisternae, and transferred to Golgi apparatus, where ↑ chondroitin sulfate is incorporated into protein molecule; this ↑ chondromucoprotein is then discharged from Golgi vacuoles into extracellular space. 2) Fibrous components = ↑ collagen fibrils and ↑ elastic fibers. Pathway is similar to that for ground substance, but perhaps bypassing Golgi apparatus.

Chondroid tissue (cartilagelike tissue, pseudocartilage, fibrohyaline tissue, vesicular supporting tissue): A union of closely apposed, vesicular cells surrounded by a thin capsule and resembling ↑ chondrocytes. Numerous collagen fibrils and only a small amount of ↑ glycosaminoglycans are present in intercellular substance. C. is very rarely present in higher vertebrates under normal conditions (sometimes in trigona fibrosa), but if present, may persist throughout life. C. is considered to be a transitory step in histogenesis of hyaline ↑ cartilage.

↑ Annuli fibrosi; ↑ Capsule, of chondrocyte

Chondroitin sulfates: ↑ Glycosaminoglycans composed of repeating units of N-acetylgalactosamine, glucuronic acid, and sulfated hexosamine. The strongly acidic sulfate group in C. is responsible for ↑ basophilia, solidity, and water content of ↑ cartilage matrix. Also present in ↑ bone, ↑ cornea, and ↑ skin.

Chondromucoprotein (chondroprotein): A ↑ proteoglycan, principal constituent of ↑ cartilage ground substance produced by ↑ chondrocytes. C. is a copolymer of a protein, ↑ chondroitin-4-sulfate (chondroitin sulfate A), and chondroitin-6-sulfate (chondroitin sulfate C). It is strongly acidic and partially responsible for ↑ basophilia, ↑ metachromasia, and ↑ PAS-positivity of ↑ cartilage ground substance.

↑ Glycosaminoglycans

Chondrones: ↑ Isogenic groups

Chorda dorsalis: ↑ Notochord

Chordae tendineae* (tendinous cords): Collageno-elastic strands (CT), enveloped by ↑ endocardium, running from papillary muscles to free margins of atrioventricular ↑ valves. (Fig. = papillary muscle, human)

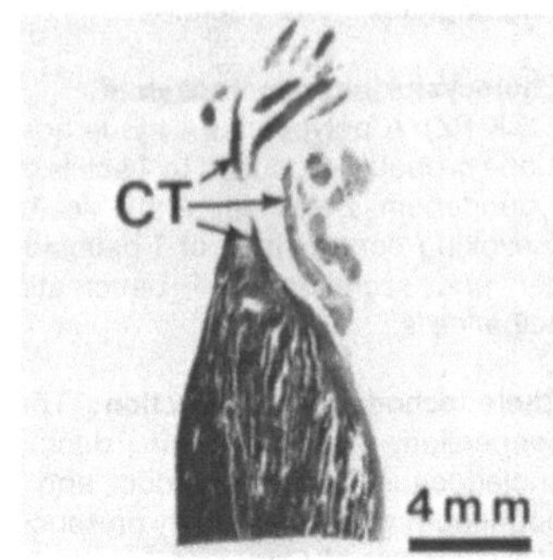

Lim, K.O.: Jpn. J. Physiol. *30*, 455 (1980; Noble, C.W., Hamlett, W.C., McCann, P., Morse, D.E.: Micron *14*, 97 (1983)

Chordoid tissue: ↑ Notochordal tissue

Choriocapillaris (choriocapillary layer, choroidocapillaris, lamina choroidocapillaris*): A dense capillary network (C) of ↑ choroid in intimate contact with ↑ pigment epithelium (PE), separated from it by ↑ Bruch's membrane (arrows). Cytoplasm of endothelial cells facing pigment epithelium is extremely thin and fenestrated; opposite capillary side is thick and devoid of fenestrations. ↑ Pericytes and nuclei of endothelial cells are situated almost exclusively on this capillary side. Capillaries are disposed in one

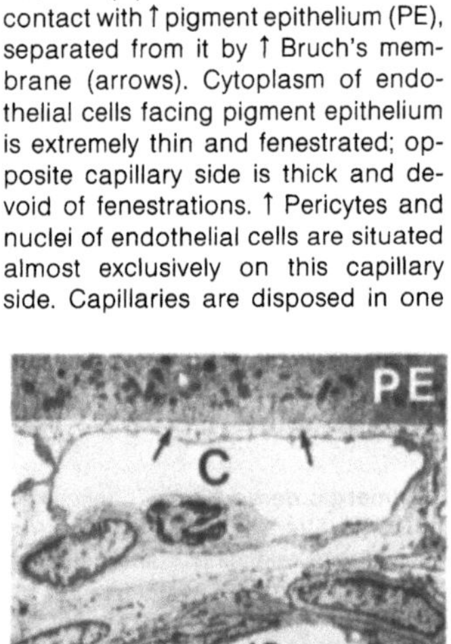

plane and cover between 85% and 95% of external retinal surface; they are arranged in more or less distinctly delimited, polygonal segments. A ↑ metarteriole (A) reaches center of a segment and spreads into capillary network, while veins (V) begin at periphery of segments. (Fig. = rat)

Braekevelt, C.R.: Anat. Embryol. *166*, 415 (1983); Essner, E., Gordon, S.R.: Cell Tissue Res. *231*, 571 (1983); Krstić, R.: Albrecht v. Graefes Arch. Clin. Exp. Ophthalmol. *205*, 245 (1978); Matsusaka, T.: Jpn. J. Opthalmol. *20*, 330 (1976)

Chorion*: Outermost of extraembryonic membranes, consisting of ↑ chorionic mesoderm and ↑ trophoblast. (See embryology texts for further information)

↑ Chorion frondosum; ↑ Chorion laeve; ↑ Chorionic villi

Chorion frondosum*: Part of ↑ chorion (CF) covered with ↑ chorionic villi (CV) and bathed by maternal blood; this part of chorion forms fetal placenta. C. is divided by decidual septa into ↑ cotyledons. CL = ↑ chorion laeve

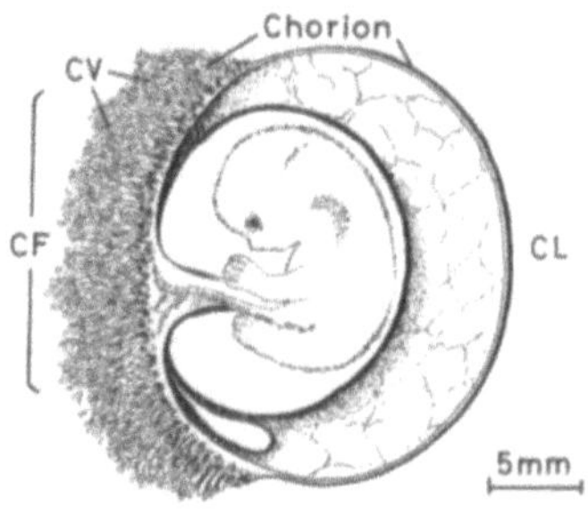

Chorion laeve*: Portion of ↑ chorion (CL) without ↑ chorionic villi (CV) located opposite ↑ chorion frondosum (CF). (See fig. under ↑ Chorion frondosum)

Chorionic epithelium: ↑ Trophoblast

Chorionic gonadotropin: ↑ Human chorionic gonadotropin

Chorionic mesoderm (mesoderma chorionicum*): Part of extraembryonic ↑ mesoderm forming principal mass of ↑ chorion, ↑ chorionic plate, and stroma of secondary and tertiary ↑ chorionic villi. C. is composed of: 1) Loosely arranged ↑ fibroblasts and ↑ fibrocytes, which produce precursors of collagen fibrils and ground substance;

2) ↑ Hofbauer cells. (See embryology texts for further information)

↑ Placenta, full-term, structure of

Chorionic plate: Part of fetal placenta.

↑ Placenta, full-term, structure of

Chorionic somatomammotropin: ↑ Human chorionic somatomammotropin

Chorionic thyrotropin: ↑ Human chorionic thyrotropin

Chorionic villi (villi choriales*): Early processes of ↑ chorion covered with the ↑ trophoblast. Three stages in formation of C.: 1) Primary C. (I) = cords of trophoblast proliferating from chorion into ↑ endometrium. Primary C. are lined by ↑ syncytiotrophoblast (S); the cores are formed by ↑ cytotrophoblast (C). After 15 days of pregnancy, proliferation of ↑ chorionic mesoderm transforms primary C. into secondary C. 2) Secondary C. (II) = processes of chorion with a mesodermal core (M) lined by cytotrophoblast (C) and syncytiotrophoblast (S). Through differentiation of blood vessels within chorionic mesoderm, secondary C. become tertiary C. (days 21–23 of pregnancy). 3) Tertiary C. (III) = processes of chorion containing fetal blood vessels (V) with nucleated or anucleated erythrocytes (E). After formation of ↑ placenta, C. are referred to as ↑ placental villi. H = ↑ Hofbauer cells (See embryology texts for further information)

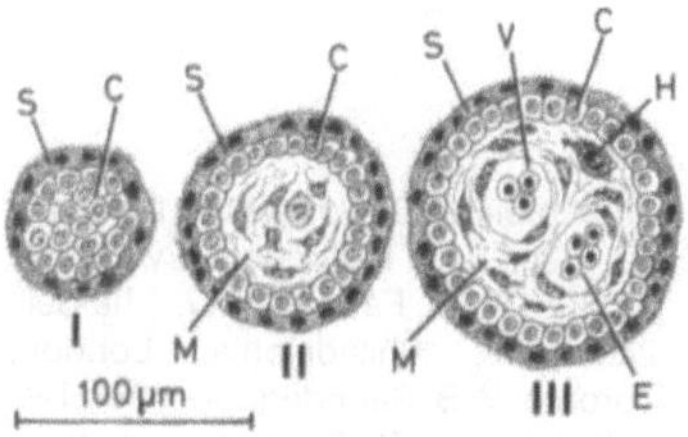

Sala, M.A., Matheus, M., Valeri, V.: Microscopica Acta *82*, 339 (1980)

Choristoma cells: Large, round cells with coarse, ↑ PAS-positive granules, occasionally found in various tissues and organs. C. are result of a maldevelopment of a type of tissue not normally found at the site.

Choroid (choroidea*): A portion of ↑ uvea extending posteriorly from ↑ ora serrata and situated between ↑ pigment epithelium (PE) and ↑ sclera (Sc). Structure: 1) ↑ Bruch's membrane (BM). 2) ↑ Choriocapillaris (C). 3) Vessel layer or lamina vasculosa* (VL) = large ↑ muscular arteries (A), originating from ciliary arteries and veins (V) which drain into vortex veins. 4) Lamina suprachoroidea* (LS) = several layers of flattened, fibroblastlike cells, some ↑ macrophages, and ↑ melanophores (M) contained in meshes of a network of ↑ collagen and ↑ elastic fibers. 5) Stroma, or substantia propria* (S) = ↑ pigment connective tissue containing ↑ fibrocytes, ↑ fibroblasts, macrophages, rare ↑ melanocytes, melanophores, and some ↑ mast cells; it contains collagen and elastic fibers, as well as predominantly ↑ nonmyelinated nerve fibers with occasional ↑ bipolar nerve cells where the nerve fibers cross. A potential cleft, spatium perichoroideale* (SP), separates C. from sclera. It contains a fluid, resembling ↑ cerebrospinal fluid, which is drained through sclera into orbital lymphatic vessels. C. supplies external retinal layers with oxygen and nutrients and contributes to maintenance of intraocular pressure.

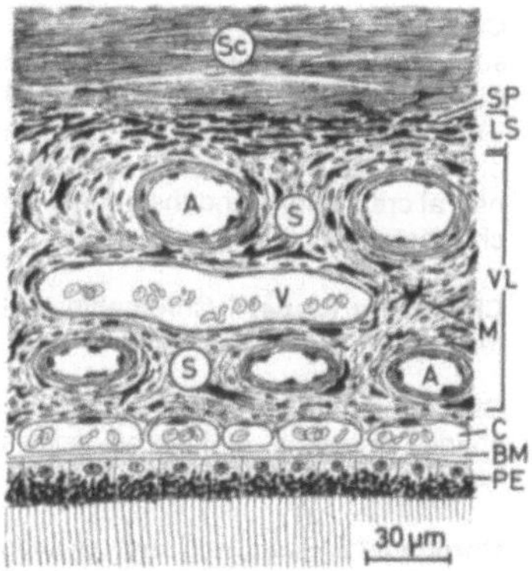

Choroid plexus (plexus chorioideus*): An organ in form of very ramified, villouslike processes (P) projecting from ↑ pia mater (PM) into brain ventricles (V). Localization: roofs of third and fourth ventricles, parts of walls of lateral ventricles. Structure: 1) Epithelium (E) = a single layer of cuboidal cells with convex ↑ apical pole and a round central nucleus. Cytoplasm contains numerous, rod-shaped mitochondria, a small Golgi apparatus, short rough endoplasmic cisternae, free ribosomes, ↑ glycogen particles, and rare lysosomes. Apical plasmalemma forms a great number of long ↑ microvilli with expanded tips. Lateral cell surfaces are moderately interdigitated with neighboring cells; all cells are held together by ↑ junctional complexes. Basal plasmalemma forms infoldings resembling a ↑ basal labyrinth (arrow). Cells rest on a basal lamina. Macrophagelike cells (↑ Kolmer's cells) have been found on epithelium. 2) Lamina propria (LP) or tela choroidea* = a richly vascularized ↑ loose connective tissue of pia mater. ↑ Capillaries (Cap) are of fenestrated type; large ↑ pericapillary spaces contain ↑ macrophages. ↑ Acervuli (Ac) may be found in older individuals. C. produces ↑ cerebrospinal fluid by selective ultrafiltration.

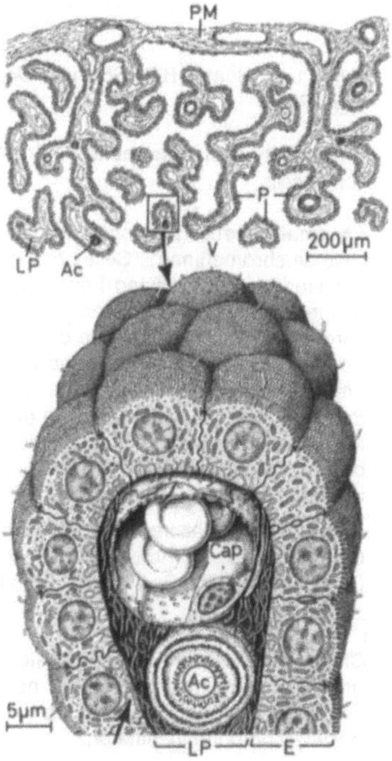

Agnew, W.F., Alvarez, R.B., Yuen, T.G.H., Crews, A.K.: Cell Tissue Res. *208*, 261 (1980); Peters, A., Swan, R.C.: Anat. Rec. *194*, 325 (1979); Van Deurs, B.: Anat. Rec. *195*, 73 (1979)

Choroidal tela: ↑ Tela choroidea

"Christmas tree": A figurative term for ribosomal DNA (rDNA) ↑ cistrons of ↑ nucleolar organizer region (having appearance of a tree trunk) and the lateral preribosomal 45 S ribosomal RNA (rRNA) molecules of variable length that have just been synthesized (appearing as branches of tree). At beginning (B) of cistron, molecules are shortest; they are longest at end (E) of cistron. Granules on cistron are thought to be RNA polymerase (P). Inactive DNA segments (S) separate "C." An arrow indicates direction of synthesis. (See molecular biology texts for further information) (Modified after Miller and Beatty 1969)

↑ Fibrillar centers; ↑ Nucleolus, functions of

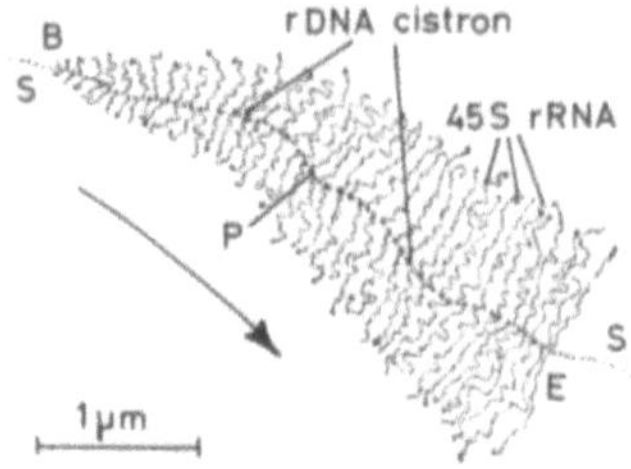

Miller, L.O., Beatty, B.R.: Science *164*, 955 (1969)

Chromacity horizontal cells: ↑ Horizontal cells

Chromaffin cells (pheochrome cells, cellulae chromaffinae*): Cells producing, storing, and releasing ↑ catecholamines and giving a positive ↑ chromaffin reaction. They are derived from neural crest and innervated by ↑ preganglionic sympathetic fibers (endocrine ↑ parenchyma of medulla of ↑ adrenal glands, cells of sympathetic ↑ paraganglia, type I ↑ glomus cells, ↑ urethral chromaffin cells, etc.)

Pollard, H.B., Pazoles, C.J., Creutz, C.E., Zinder, O.: Int. Rev. Cytol. *58*, 160 (1979); Taxi, J.: Int. Rev. Cytol. *57*, 283 (1979)

Chromaffin granules: Secretory granules of ↑ epinephrine- and ↑ norepinephrine-producing cells giving positive ↑ chromaffin reaction.

↑ Epinephrine storage granules; ↑ Norepinephrine storage granules

Schmidt, W., Winkler, H.: Eur. J. Cell Biol. *27*, 96 (1982)

Chromaffin paraganglion: ↑ Paraganglion, sympathetic

Chromaffin reaction (pheochrome reaction): A capacity of ↑ catecholamine storage granules to become oxidized in presence of an aqueous solution of metallic salts (potassium bichromate, ferric chloride) or other oxidizing agents. Resulting brown color is consequence of formation of ↑ adrenochrome. Adrenal medulla (M) gives a highly positive C. Under transmission electron microscope (bottom), reaction product is represented by very electron-dense granules. C = cortex of ↑ adrenal gland, E = ↑ epinephrine-producing cell, NE = ↑ norepinephrine-producing cell (Fig., top = adrenal gland, guinea pig; fig., bottom = rat)

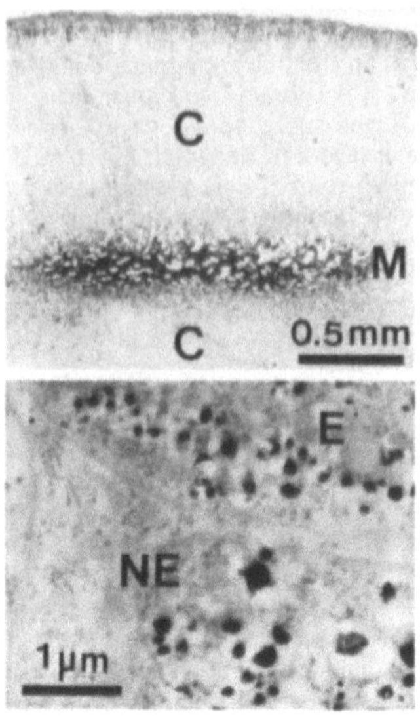

Singh, I.: Anat. Anz. *115*, 81 (1964); Wood, J.G.: Barrnett, R.J.: Anat. Rec. *145*, 301 (1963)

Chromaffin system (chromaffin tissue): A common term for adrenal medulla, sympathetic ↑ paraganglia, and ↑ carotid body, characterized by morphological similarity of their cells, their neural crest origin, and their positive ↑ chromaffin reaction.

Chromaffin tissue: ↑ Chromaffin system

Chromatic apparatus: A deeply stained ensemble of ↑ chromosomes during ↑ mitosis or ↑ meiosis.

Chromatid: Each of two strands formed by longitudinal division of a ↑ chromosome. C. consists of only one ↑ chromatin fiber, i.e., of only one DNA molecule.

Chromatin (chromatinium*): A deeply stainable substance visible under light microscope as irregular clumps of ↑ basophilic material in fixed and colored interphasic nuclei. C. represents DNA-protein complex of ↑ chromosomes; it consists of ↑ euchromatin and ↑ heterochromatin.

Cameron, J., Pavlat, W.A., Jeter, J.R.: Anat. Rec. *194*, 547 (1979); Iino, A., Nagai, S.: Biomed. Res. *2*, Suppl. 91 (1981); Kornberg, R.D.: Ann. Rev. Biochem. *49*, 931 (1977); Oda, T.: Acta Histochem. Cytochem. *15*, 543 (1982); Valencia, R.: J. Physiol. (Paris) *78*, 653 (1982/83)

Chromatin body: ↑ Barr body

Chromatin fiber: An irregularly coiled chain of variable length composed of ↑ nucleosomes interconnected by internucleosomal DNA (iDNA). C. forms ↑ chromosomes during ↑ interphase; each ↑ chromatid consists of one C. In G_1 phase of ↑ cell cycle, C. is loosely and irregularly winded; densely packed coils of C. correspond to ↑ heterochromatin, and uncoiled or poorly coiled segments to ↑ euchromatin. With onset of cell division, C. becomes more densely, but still irregularly coiled and forms a prophase chromosome. Cs. are attached to internal face of ↑ nuclear membrane and to circumference of ↑ nuclear pores. Chromosomes of most ↑ eucaryotic cells are composed of one C. H = ↑ histone octamer, nDNA = nucleosomal DNA, H 1 = histone H 1

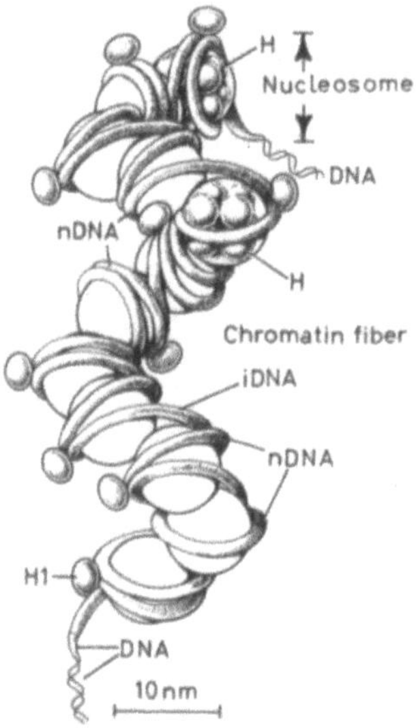

Ehrenpreiss, J.G., Zirne, R.A., Denidenko, O.Je.: Folia Histochem. Cytochem. *19*, 217 (1981); Fawcett, D.W.: The Cell. 2nd edn. Philadelphia, London, Toronto: W.B. Saunders Co. 1981; Ris, J., Kornberg, R.D.: Chromosome Structure and Levels of Chromosome Organization. In: Cell Biology, Vol. 2. New York, San Francisco, London: Academic Press 1979; Schwarzacher, H.G.: Bull. Schweiz. Akad. Med. Wiss. *34*, 337 (1978)

Chromatin, in interphase nuclei: ↑ Euchromatin; ↑ Heterochromatin

Chromatin, intranucleolar (nucleolar chromatin): Thin strands of ↑ chromatin located within pars fibrosa of ↑ nucleolus and termed ↑ fibrillar centers. DNA of C. controls synthesis of RNA in nucleolus. (See molecular biology texts for further information)

↑ Ribonucleic acids, synthesis of

Raška, I., Rychter, Z., Smetana, K.: Z. mikrosk.-anat. Forsch. *97*, 15 (1983)

Chromatin, nucleolar: ↑ Chromatin, intranucleolar

Chromatin, nucleolus-associated: ↑ Chromatin, perinucleolar

Chromatin particles: ↑ Karyosomes

Chromatin, perinucleolar (nucleolus-associated chromatin): Clumps of ↑ heterochromatin (H) associated with ↑ nucleolus (Nu). Fine strands of C. penetrate, as intranucleolar ↑ chromatin, between meshes of ↑ nucleolonema. (Fig. = ↑ pinealocyte, rat)

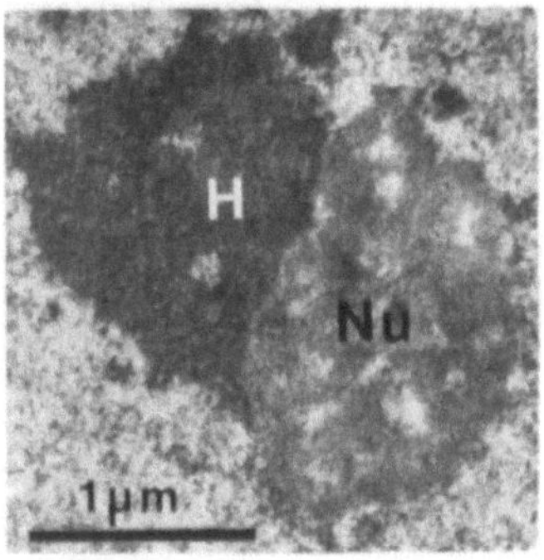

Chromatoid body (corpus chromatoideum*): An electron-dense, ↑ basophilic, fibrogranular structure (CB) found in cytoplasm of developing male ↑ germ cells, generally situated opposite ↑ acrosome (A). C. can be associated with ↑ nuage. C. is presumably composed of basic proteins and RNA. Chromatoid material occurs first in nucleus of pachytene ↑ spermatocyte from which it passes through ↑ nuclear pores into cytoplasm. C. is believed to furnish ring around distal

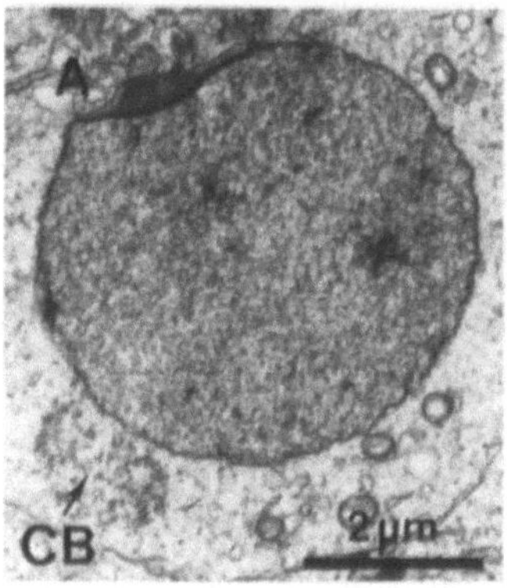

centriole and to participate, through condensation, in formation of ↑ annulus. (Fig. = ↑ spermatid, human)

Parvinen, M., Parvinen, L.-M.: J. Cell Biol. *80*, 621 (1979); Söderström, K.-O.: Z. mikrosk.-anat. Forsch. *92*, 417 (1978)

Chromatolysis, retrograde: Progressive disappearance of ↑ Nissl bodies from ↑ perikaryon of a nerve cell whose ↑ axon has been sectioned.

↑ Nerve fibers of peripheral nervous system, degeneration and regeneration of

Cragg, B.G.: Brain Res. *23*, 1 (1970)

Chrome alum-hematoxylin phloxine stain: A method for demonstrating ↑ A-cells of ↑ pancreatic islets and ↑ neurosecretory granules.

Chromidial substance: The former term for ↑ ergastoplasm.

Chromium shadowing: ↑ Shadowing

Chromocenters: ↑ Karyosomes

Chromogranin: A specific protein contained in ↑ catecholamine storage granules bound to ↑ epinephrine or ↑ noradrenaline.

Chromomembrin-A: A chemical substance present in membranes of ↑ epinephrine and ↑ norepinephrine storage granules.

↑ Catecholamines

Chromomeres (chromomerus*): Darkly stained transversal bands (Chm) in giant polytenic ↑ chromosomes (Chr) of certain insects consisting of densely arranged coils of ↑ chromatin fiber.

↑ Chromosomes, banding patterns of; ↑ Chromosomes, polytenic

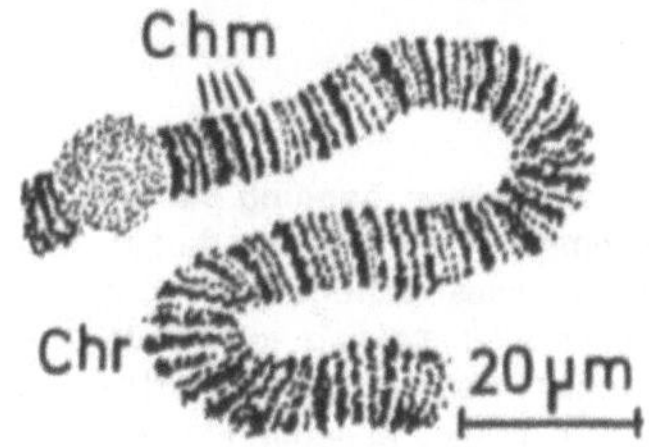

Sass, H.: J. Cell Biol. *45*, 269 (1980)

Chromomere, of platelets: ↑ Platelets

Chromonema*: The thinnest longitudinally oriented ↑ chromatin thread of a ↑ chromatid still visible with light microscope. The term C. is rarely used.

Zatsepina, O.V., Poliakov, V.Yu., Chentsov, Yu.S.: Tsitologiia *25*, 123 (1983)

Chromophilia: Affinity of a cell, ↑ cell component, or tissue to dyes.

Chromophilic substance, of neurons: Synonym for ↑ Nissl bodies.

Chromophobia: Resistance of a cell, ↑ cell component, or tissue to stains.

Chromophobic cells, of hypophysis: ↑ Hypophysis, chromophobic cells of

Chromophobic cells, of pars intermedia: ↑ Hypophysis, pars intermedia of, chromophobic cells of

Chromosomal „fibers": ↑ Microtubules, kinetochoral

Chromosomal microtubules: ↑ Microtubules, kinetochoral

Chromosome female-determining: Since all secondary ↑ oocytes carry one ↑ X-chromosome, X-chromosome of ↑ gynospermatozoa determines female sex of future individual giving, after ↑ fertilization, a ↑ zygote with 44 ↑ autosomes and XX ↑ sex chromosomes.

Chromosome, male-determining: The male sex is determined by ↑ Y-chromosome of ↑ androspermatozoa giving, after ↑ fertilization, a ↑ zygote with 44 ↑ autosomes and XY ↑ sex chromosomes.

Chromosome, X-: ↑ X-chromosome

Chromosome, Y-: ↑ Y-chromosome

Chromosomes (chromosoma*): Membraneless, intensely stainable, permanent organelles of cell nucleus composed of ↑ chromatin fiber(s) (CF). C. are visible under light microscope as small threads about 1–2 µm thick and several micrometers in length, but only during M-phase of ↑ cell cycle. Structure: G_1-phase = only one irregularly and loosely coiled chromatin fiber forms C., which shows two constrictions: a) primary constriction (PC) is present in all C. and corresponds to ↑ centromere (Ce); b) secondary constriction (SC) is present only in five C. in man; it corresponds to ↑ nucleolus organizer region. The same five C. have ↑ satellites (S). G_2-phase = C. shortens because of further coiling of chromatin fiber; after its doubling, during S-phase, C. consists of two sister ↑ chromatids, each formed of only one chromatin fiber. Chromatids are held together by centromere; they lie so closely to one another that they are not discernible as separate structures under light microscope. M-phase = during metaphase (M), chromatids separate from one another, but are still held together in centromeric region by an unreplicated chromatin fiber segment of DNA. Chromatin fiber continues to coil and condense; thus, the whole C. shortens and the major coils (MC) and bands (B) appear. In anaphase, centromere divides, and diffuse decondensation takes place in anaphase and telophase.

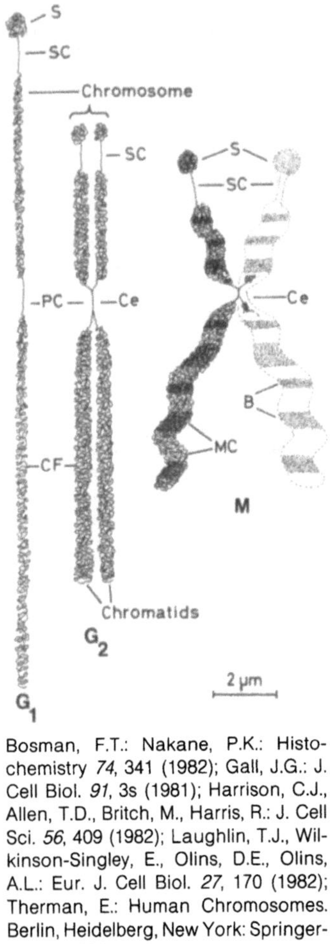

Bosman, F.T.: Nakane, P.K.: Histochemistry *74*, 341 (1982); Gall, J.G.: J. Cell Biol. *91*, 3s (1981); Harrison, C.J., Allen, T.D., Britch, M., Harris, R.: J. Cell Sci. *56*, 409 (1982); Laughlin, T.J., Wilkinson-Singley, E., Olins, D.E., Olins, A.L.: Eur. J. Cell Biol. *27*, 170 (1982); Therman, E.: Human Chromosomes. Berlin, Heidelberg, New York: Springer-Verlag 1980

Chromosomes, acrocentric (telocentric chromosomes, chromosoma acrocentricum*): ↑ Chromosomes with a ↑ centromere (Ce) very close to one end, making one of their arms very short or even imperceptible. Chr = ↑ chromatid

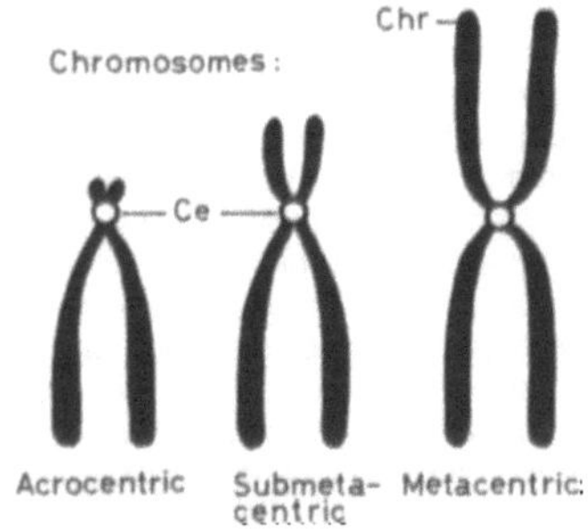

Chromosomes, anomalies of: Any numerical and/or morphological deviation from standard pattern. 1) Anomaly of number of chromosomes as a consequence of ↑ chromosome nondisjunction: a) ↑ autosomes = trisomy 21 = an extra chromosome 21 producing ↑ Down's syndrome (45 + XY or 45 + XX); b) ↑ sex chromosomes = Klinefelter's syndrome (44 + XXY), Turner's syndrome (44 + X), "super female" (44 + XXX), etc. 2) Anomaly of form of chromosomes = a) deletion = disconnection and loss of a whole or part of a chromosome arm; b) translocation = abnormal transfer of a segment of one chromosome to a nonhomologous chromosome during prophase I of ↑ meiosis; c) isochromosomes = transversal, instead of longitudinal, division of ↑ centromere (Ce), resulting in two metacentric isochromosomes with two identical arms of neighboring ↑ chromatids (Chr). (See genetics texts for further information)

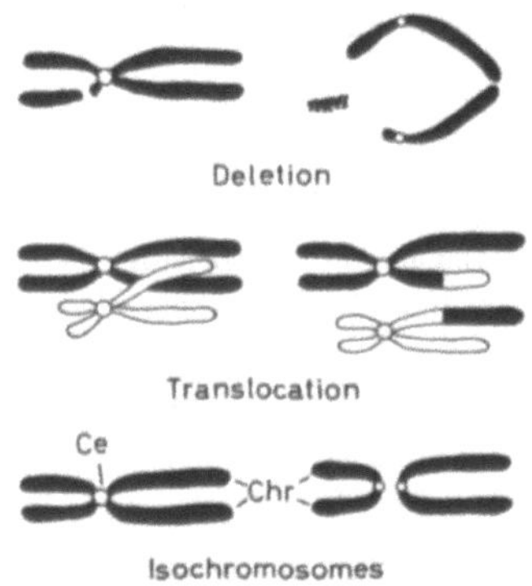

Chromosomes, banding patterns of: Symmetric, transversal, darker and lighter lines occurring in chromosomes after staining with dilute acid Giemsa stain and certain ↑ fluorochromes (quinacrine mustards). Although significance of these patterns is not clear, they help to distinguish between morphologically very similar chromosomes of human ↑ karyotype, e.g., chromosomes 6, 7, 8, 9, 10, 11, 12, ↑ X-chromosome, and certain chromosomal anomalies (not to be confused with banding patterns of polytenic ↑ chromosomes).

↑ Chromomere; ↑ Chromosomes, anomalies of

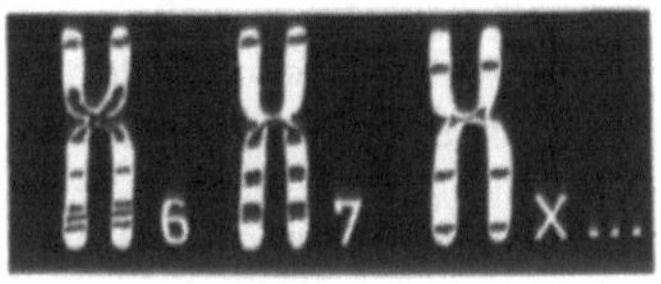

Curtis, D., Horobin, R.W.: Histochem J. *14*, 911 (1982); Drets, M.E., Shaw, M.W.: Proc. Natl. Acad. Sci. USA *68*, 2073 (1971); Shiraishi, N., Fujihara, T., Maeda, S., Kazama, T., Takahashi, R., Sugiyama, T.: Acta Histochem. Cytochem. *15*, 176 (1982); Zheng, H.-Z., Burkholder, G.D.: Exp. Cell Res. *141*, 117 (1982)

Chromosomes, crossing-over: ↑ Crossing over

Chromosomes, giant: Enormous polytenic ↑ chromosomes, about 2 mm long and 10 µm thick, found in salivary glands, tracheae, gut, etc. of some insects. These C. are always in prophase.

Chromosomes, homologous: Morphologically identical chromosomes forming pairs in a ↑ diploid cell. One member of pair originates from father (paternal chromosome), the other one from mother (maternal chromosome). There are 23 homologous chromosome pairs in human femal and 22 in the male, because ↑ sex chromosomes X and Y are morphologically different. Although morphologically identical, C. do not necessarily carry identical genes.

↑ Autosomes; ↑ Karyotype

Chromosomes, metacentric (chromosoma metacentricum*): ↑ Chromosomes divided by ↑ centromere into two segments of almost equal length.

↑ Chromosomes, acrocentric

Chromosomes, nondisjunction of: The failure of homologous ↑ chromosomes to reach opposite cell poles during meiotic anaphase I, so that one cell receives one chromosome more, and the other one chromosome less. In the case of ↑ autosomes, most well-known example is ↑ Down's syndrome, where a person has one chromosome 21 more = trisomy 21. In the case of ↑ sex chromosomes, a nondisjunction results in the appearance of Klinefelter's syndrome (44 autosomes + XXY), Turner's syndrome (44 + X), "super female" (44 + XXX), etc. (See genetics texts for further information)

↑ Chromosomes, anomalies of

Chromosomes, nucleolar organizer regions of: ↑ Nucleolus organizer region

Chromosomes, number of: C. varies greatly from organism to organism. There are, for example, 20 chromosomes in corn, 24 in the tomato, 26 in the frog, 32 in the bee, 46 in man, 48 in the potato, 60 in the ox, 78 in the dog, 500–520 in the fern.

Chromosomes, number of, in meiosis: ↑ Meiosis

Chromosomes, number of, in mitosis: ↑ Mitosis

Chromosomes, polytenic: A phenomenon comprising multiple divisions of ↑ chromosomes of insects and some other animals, in which ↑ chromatids do not separate, but rather arrange themselves parallel to one another. This results in formation of giant ↑ chromosomes, about a thousand times larger than somatic chromosomes. A C. represents a bundle of hundreds of chromatids, i.e., ↑ chromatin fibers (CF), whose densely coiled segments form bands or ↑ chromomeres (Chm); their longitudinal uncoiled portions represent interbands (Ib). C. do not exist in man. (Modified after Sass 1980)

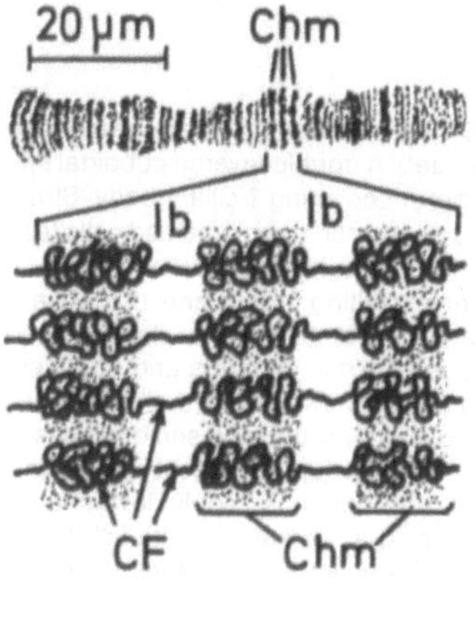

Berendes, H.D.: Int. Rev. Cytol. *35*, 61 (1973); Grond, C.J., Derksen, J.: Eur. J. Cell Biol. *30*, 144 (1983); Mott, M.R., Burnett, E.J., Hill, J.R.: J. Cell Sci. *45*, 15 (1980); Sass, H.: J. Cell Sci. *45*, 269 (1980); Sorsa, V.: Cell Differentiation *12*, 137 (1983)

Chromosomes, satellites of (SAT-chromosome, chromosoma satelliti-ferum*): Small, round, or elongated bodies (S) connected to rest of ↑ chromosome by a delicate, threadlike chromatin filament corresponding to ↑ nucleolus organizer region (NOR). Five human chromosomes carry satellites, which seem to consist of constitutive ↑ heterochromatin.

↑ Karyotype

John, B., Miklos, G.L.G.: Int. Rev. Cytol. *58*, 1 (1979)

Chromosomes, sex: ↑ Autosomes; ↑ Karyotype; ↑ Sex chromosomes

Chromosomes, submedian: ↑ Chromosomes, submetacentric

Chromosomes, submetacentric (submedian chromosome, chromosoma submetacentricum*): ↑ Chromosomes divided by ↑ centromere into arms of unequal length.

↑ Chromosomes, acrocentric

Chromosomes, telocentric: ↑ Chromosomes, acrocentric

Chromosomes, telomeres of (telomerus*): The polarized extremities (T) of ↑ chromosomes, preventing other chromosomes or their segments from fusing with these chromosomes. ↑ Chromatin fiber (CF) begins or ends in telomere.

Chyle: A turbid, milky, white or yellowish fluid occurring in ↑ lacteals during digestion of fats. C. consists of ↑ lymph and ↑ chylomicrons; it is drained through intestinal lymphatic vessels to ↑ thoracic duct.

Chylomicrons: Highly refractile, membraneless fatty bodies, 1–3 µm in diameter, occurring in ↑ lymph and ↑ blood after a fatty meal.

↑ Absorptive cells; ↑ Chyle; ↑ Lipoproteins

Chyme: A semifluid mass of partly digested food formed in distal region of ↑ stomach; C. then passes in small portions into ↑ duodenum.

Chymotrypsins: Proteolytic enzymes produced by ↑ pancreatic acinar cells.

CIF: ↑ Cloning-inhibiting factor

Cilia* (kinocilia, kinetocilia*): Motile processes (C), about 5–10 µm long and 0.2 µm wide, projecting from free surface of numerous kinds of cells. Structure (longitudinal section): 1) Ciliary membrane (CM) is continuous with plasmalemma (P). 2) ↑ Axoneme or ciliary shaft (A) = nine pairs or doublets of longitudinal ↑ microtubules (Mt) continuous with microtubular triplets (Tr) of basal body (BB). Central sheath (CS) surrounding central microtubular pair begins only at level of free cell surface. All elements of axoneme are embedded in ciliary matrix (M). 3) Basal body (BB) with basal foot (BF). 4) Rootlet (R) = a conical, transversally banded structure with repeated periods of 55–70 nm, composed of parallel, 3 to 7-nm-thick microfilaments. Rootlet penetrates at times as far as nucleus; it may be absent in some cells.

↑ Cilia, basal body of; ↑ Cilia, basal foot of; ↑ Cilia, cross section of

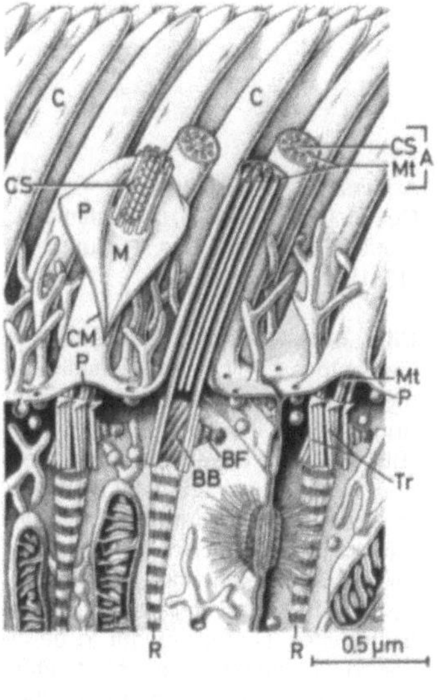

Gibbons, R.I.: J. Cell Biol. *91*, 107s (1981); Dentler, W.L.: Int. Rev. Cytol. *72*,

1 (1983); Kuhn, C., III, Engelman, W.: Cell Tissue Res. *186*, 491 (1978), Warner, F.D.: J. Cell Sci. *20*, 101 (1976)

Cilia, basal body of (kinetosome, corpusculum basale*): A small cylinder, about 0.5 µm long and 0.15 µm wide, at base of each ↑ cilium, considered to be a modified ↑ centriole. C. consists of nine microtubular triplets (Tr); inner microtubules of each triplet grow out and form nine microtubular pairs of ciliary shaft. Central microtubular pair is absent in C. From C., conical, striated rootlet emerges; basal foot (BF) with ↑ microtubule organizing center (MTOC) is attached to C. After numerous duplications (over 100), centriole gives rise to C., which migrates to ↑ apical cell pole; here it forms ↑ cilia.

↑ Ciliogenesis

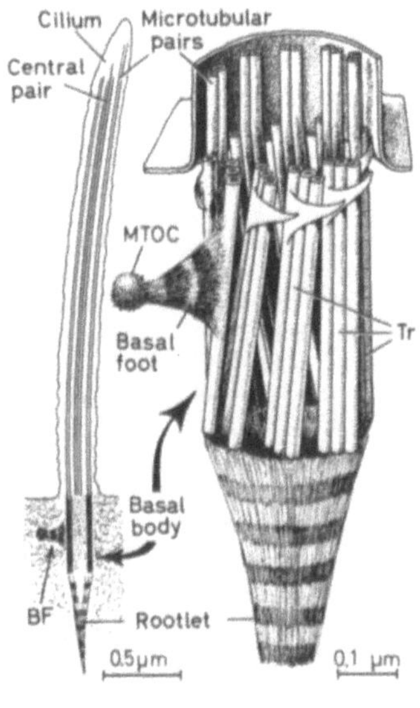

Anderson, R.W.G., Brenner, R.M.: J. Cell Biol. *50*, 10 (1971); Mitsushima, A., Inoué, T.: Biomed. Res. *2*, Suppl. 87 (1981)

Cilia, basal foot of (pes basalis*): A filamentous, conical, transversally striated structure (B) attached at a right angle to lateral surface of basal body (BB). All C. are oriented in same direction toward basal bodies of next cilia. In vicinity of, or in contact with C. is located a ↑ microtubular organizing center, which probably initiates polymerization of axonemal microtubules (Mt) in this area. C. are thought to be involved in transmission of beating rhythm from one cilium to another. (Fig. = ciliated oviduct cell, rat)

↑ Cilia, basal body of

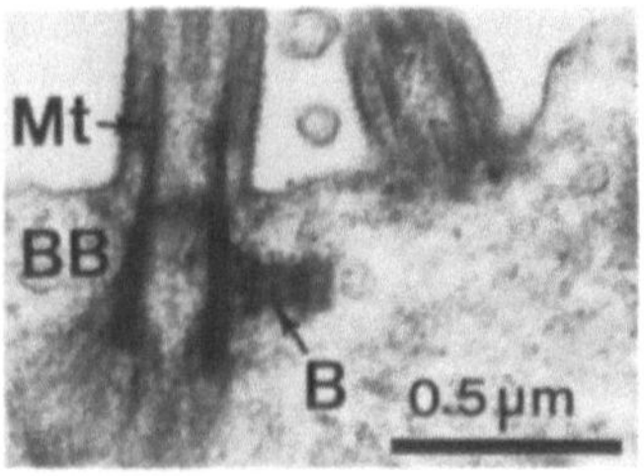

Tilney, L.G., Goddard, J.: J. Cell Biol. *46*, 564 (1970)

Cilia, cross section of: I) Ciliary membrane = plasmalemma. II) Axoneme or ciliary shaft = two single central ↑ microtubules surrounded by a ring of nine regularly disposed microtubular pairs or doublets (formula 9+2). Each doublet is composed of microtubules ("subfibrils") A and B; the latter is incomplete. Microtubule A sends two rows of short, slightly divergent, ↑ dynein arms toward next doublet. Microtubules A are linked to each other by nexin bridges. The doublets are continuous with two inner subunits of basal body. Central microtubules, embedded in central sheath, begin at base of cilium. In tip of cilium, doublets become singlets (only one microtubule).

↑ Cilia, basal body of; ↑ Ciliary motion

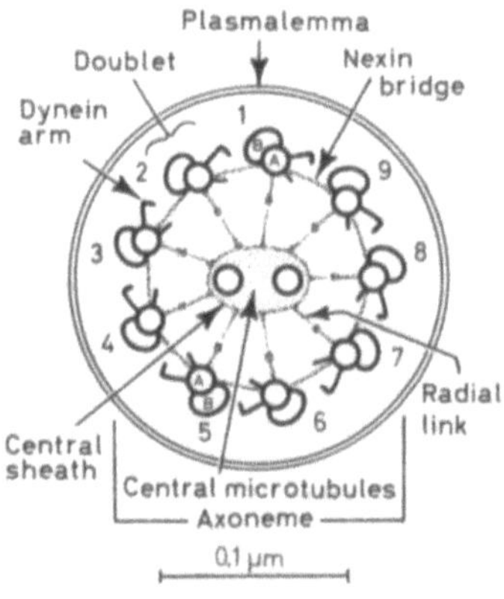

Haimo, L.T., Rosenbaum, J.L.: J. Cell Biol. *91*, 125s (1981); Warner, F.D.: J. Cell Sci. *20*, 101 (1976)

Cilia, derivates of: Modified cilia are ↑ olfactory cilia and outer segments of ↑ photoreceptors. ↑ Spermatozoal tail is also a specially differentiated cilium.

Wen, G.Y., Soifer, D., Wisniewski, H.M.: Anat. Embryol. *165*, 315 (1982)

Cilia, motion of: ↑ Ciliary motion

Cilia, of eyelids: ↑ Eyelashes

Ciliary body (corpus ciliare*): A fibromuscular ring located between ↑ ora serrata (OS) and root of ↑ iris (Ir), representing direct anterior continuation of ↑ choroid. Parts: A) Anterior portion or pars plicata, consists of 70–80 ↑ ciliary processes (CP) to which ↑ zonular fibers (ZF) of ↑ ciliary zonule (CZ) are attached. B) Posterior portion or pars plana (orbiculus ciliaris) is devoid of processes. Structure: I) ↑ Ciliary epithelium (CE) = a double-layered cuboidal epithelium composed of pigmented (PC) and nonpigmented cells (NC). II) Ciliary stroma (CS) = a highly vascularized, ↑ loose connective tissue containing ↑ fibroblasts, ↑ mast cells, and ↑ melanocytes; capillaries are fenestrated. III) Ciliary muscles (smooth): 1) Circular band or Müller's muscle (MM), innervated by ↑ parasympathetic nervous system; 2) Meridional-radial band or Brücke's muscle (BM), probably innervated by ↑ sympathetic nervous system. Ciliary muscles have been shown, according to recent research, to be largely interwoven, forming a single muscle attached to ↑ scleral spur (SS). C. is involved in accomodation of eye and is responsible for production of ↑ aqueous humor. L = ↑ lens

↑ Eye, accomodation of

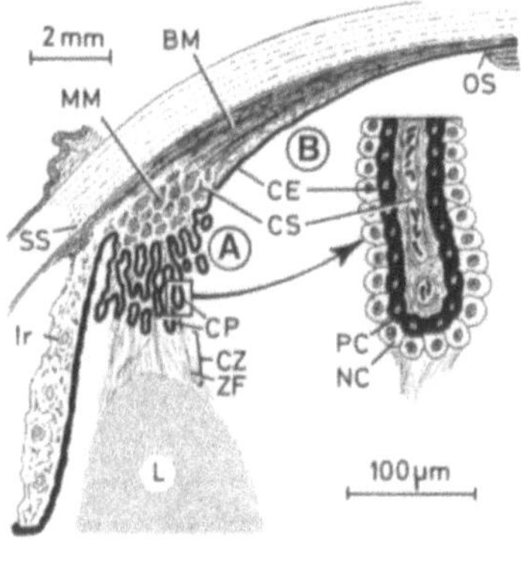

Ciliary epithelium (pars ciliaris retinae*): A double-layered cuboidal epithelium covering ↑ ciliary body. Structure: 1) Inner, nonpigmented cells (epithelium nonpigmentosum*), lying on inner limiting membrane (ILM), have an elliptical or round nucleus, numerous large mitochondria and rough endoplasmic cisternae, a well-developed Golgi apparatus, and some tubules of smooth endoplasmic reticulum. Basal plasmalemma is highly invaginated, forming a ↑ basal labyrinth (BLt). Apical ↑ junctional complexes (JC) hold cells together and constitute ↑ blood-aqueous barrier. Nonpigmented cells have a phagocytotic activity and produce ↑ zonular fibers (ZF). 2) Outer, pigmented cells (epithelium pigmentosum*) are separated from stroma (S) by basal lamina (BL). Cells contain a more or less irregular nucleus, some mitochondria, a few rough endoplasmic cisternae, numerous profiles of smooth endoplasmic reticulum, a small Golgi apparatus, and a number of ↑ melanosomes and ↑ melanin granules (MG). C. produces ↑ aqueous humor.

↑ Ciliary processes

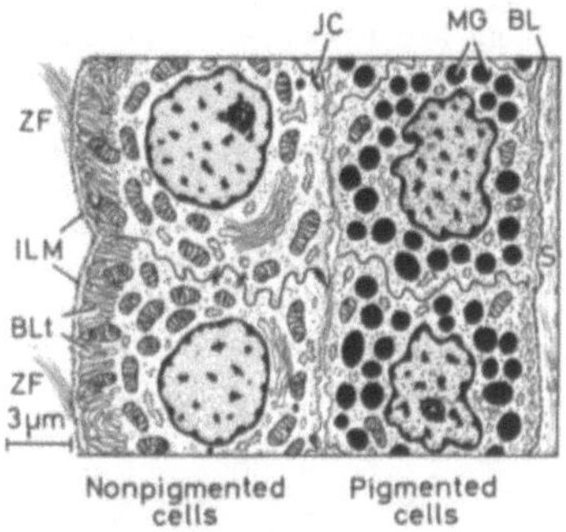

Kondo, K., Coca-Prados, M., Sears, M.: Cell Tissue Res. *233*, 629 (1983)

Ciliary glands, of Moll (glandulae ciliares*): Modified apocrine ↑ sweat glands (CG) situated at edges of ↑ eyelids, where their excretory ducts (e) open onto free surface or into ↑ hair follicles of ↑ eyelashes (E). C. are characterized by enlarged terminal portions. (Fig. = human)

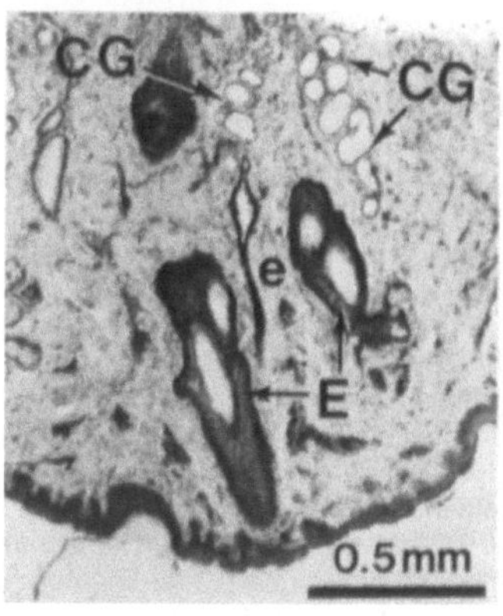

Ciliary motion: In living cells, ↑ cilia beat with a frequency of 10–25 beats/s and always in a genetically programed direction. 1) Metachronal rhythm (M) = when observed in a plane parallel to direction of motion (arrow), movement of cilia forms successive waves which transport ↑ mucus and particulate material. 2) Isochronal rhythm (I) = when observed in a plane perpendicular to direction of motion, all cilia are in same phase of movement. Motion of each cilium is composed of a rapid, stiff stroke forward (= effective stroke, ES), followed by a slower, more flexible stroke backward (= recovery stroke, RS). At beginning of motion, all nine doublets are connected together by dynein arms (D) at same level. Movement is initiated by displacement toward ciliary tip of insertion points of dynein arms on B-microtubule with a consecutive sliding of doublets in relation to one another. Thus, there is no shortening of microtubules. During this movement, the effective stroke is always directed perpendicularly to central sheath (CS) in a plane passing between doublets 5 and 6, whose arms are probably immobile. The central sheath seems to conduct impulses for sliding of doublets, while the necessary energy is provided by ↑ dynein in presence of ATP and Ca^{2+} and Mg^{2+} ions. Synchronization of the ciliary movement spreads through the basal feet. A = A-microtubule

↑ Cilia, basal foot of

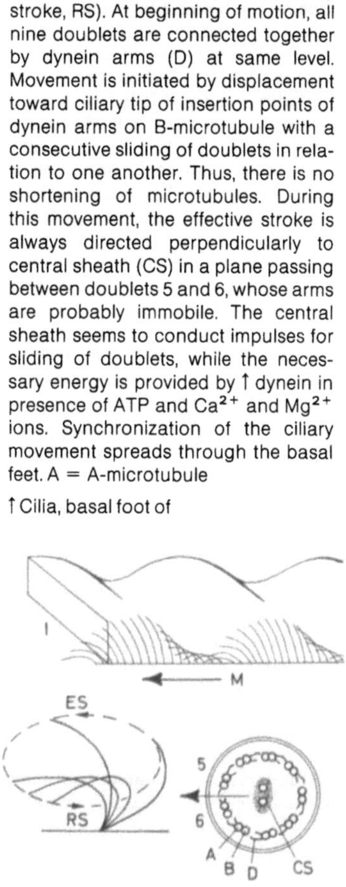

Ciliary muscle, of Riolan (musculus ciliaris*): A small circular bundle of palpebral portion of musculus orbicularis oculi situated between ↑ hair follicles of ↑ eyelashes and ↑ tarsus of ↑ eyelids. Contraction of C. presses margins of eyelids against ↑ cornea and ↑ sclera.

Blake, J.R., Sleigh, M.A.: Biol. Rev. *49*, 85 (1974); Goodenough, U.W., Heuser, J.E.: J. Cell Biol. *95*, 798 (1982); Satir, P.: Sci. Am *231/4*, 44 (1974)

Ciliary muscles: ↑ Ciliary body

Ciliary processes (processus ciliaris*): About 80 radiating folds (CP) of ↑ ciliary body projected into ↑ posterior chamber of eye. C. are lined with ↑ ciliary epithelium. Apart from their role in secretion of ↑ aqueous humor, C. serve as one of insertion sites for ↑ zonular fibers. (Fig. = mongolian gerbil)

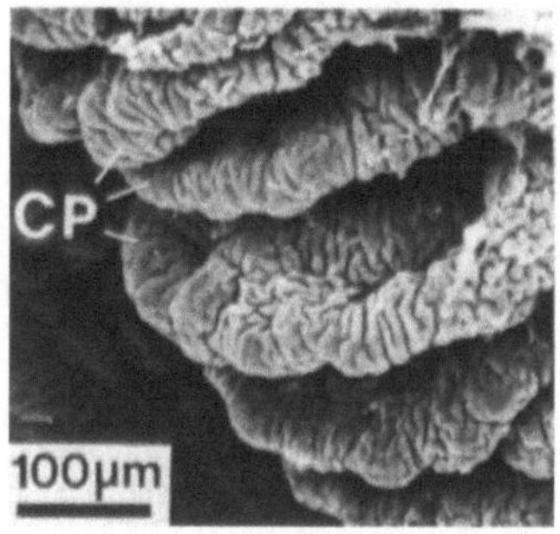

Ciliary shaft: ↑ Axoneme; ↑ Cilia, cross section of

Ciliary stroma: ↑ Ciliary body

Ciliary zonule (suspensory ligament, zonula Zinnii, zonula ciliaris*): A system of fine, radial ↑ zonular fibers (ZF) representing circular suspensory apparatus (CZ) of ↑ lens (L). C. is formed by two bands of zonular fibers: 1) Longer fibers arise from region of ↑ ora serrata (OS), ending on anterior surface of lens ↑ capsule (= anterior zonular sheath, AZS). 2) Shorter fibers arise from ↑ ciliary epithelium covering ↑ ciliary processes (CP), attaching themselves to ↑ capsule of posterior lens surface (= posterior zonular sheath, PZS). This sheath also forms a part of anterior boundary of ↑ vitreous body (VB). The two bands are separated by a triangular channel, canal of Petit (P).

↑ Ciliary body

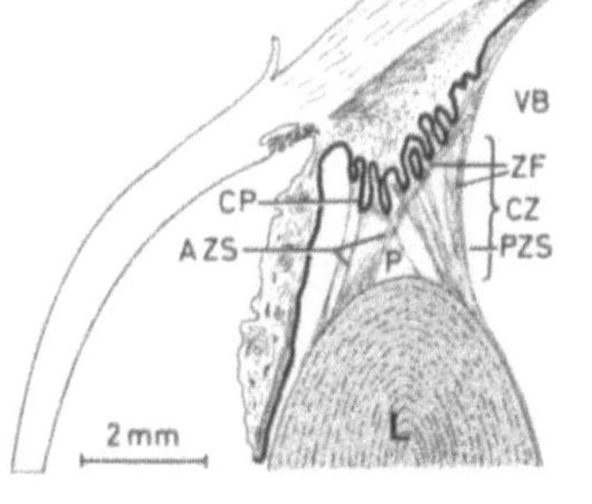

Ciliated cells, of bronchiolar epithelium: ↑ Bronchiolar epithelium

Ciliated cells, of oviducts: ↑ Oviducts, epithelium of

Ciliated cells, of tracheal epithelium: ↑ Trachea, epithelium of

Ciliated cells, of uterus: ↑ Uterus, epithelium of

Ciliogenesis: Process of formation of ↑ cilia. Two pathways: A) Acentriolar C. (most frequent): 1) ↑ Fibrosomes coalesce around ↑ deuterosome; 2) multiple procentrioles develop around deuterosome; 3) under influence of ↑ microtubule organizing center (MTOC), new ↑ centrioles are formed, by polymerization of microtubular subunit, and migrate to ↑ apical pole; 4) here, they become basal bodies (BB) and initiate, together with MTOC, formation of ↑ axonemes of cilia. B) Centriolar C.: 1) Duplication of preexisting centrioles at end of S-phase of ↑ cell cycle produces procentrioles, which transform into centrioles; 2) these newly formed centrioles move toward cell surface, become basal bodies (BB), and start growth of ciliary shaft (thick arrow). Central microtubular doublet is last to be formed. Material for microtubular polymerization comes from granulofilamentous substance (GFS) at tip of cilium; this substance disappears when cilium reaches full development. (Modified after Fawcett 1972)

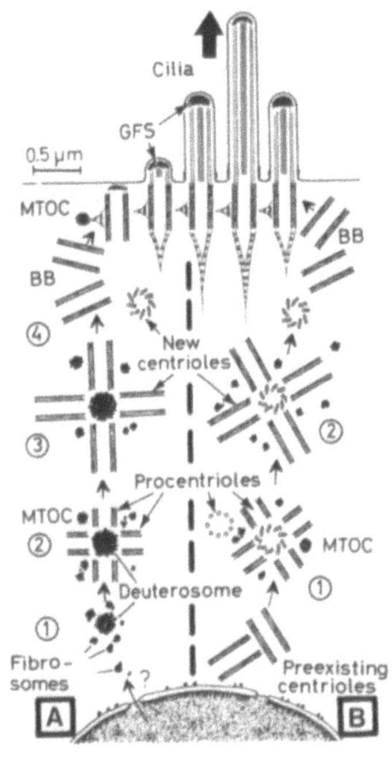

↑ Cilium, basal body of

Chailley, B., Boisvieux-Ulrich, E., Sandoz, D.: Biol. Cell *46*, 51 (1982); Dirksen, R.E.: J. Cell Biol. *51*, 286 (1971); Fawcett, D.W.: Genetics of the Spermatozoon. Edinburgh: Beatty R.A. and Glueckson-Waelsch 1972; Vacek, Z.: Folia Morphol. *21*, 247 (1973)

Cinephotomicrography (cinemicroscopy): A method of taking pictures on cine film in order to demonstrate activity of living cells in ↑ phase contrast or ↑ interference microscope. A special device arranges lapse of time in which frames are to be taken. Rapid cellular movements, such as ↑ ciliary motion and contraction of ↑ striated muscle fibers, need many frames per second; projection of film at a normal speed of about 18 frames/s makes filmed movements appear slower. For slow motion (↑ ameboidism, cell division, ↑ phagocytosis), the frames are taken with a lapse of several seconds; projected at speed of 18 frames/s, these events appear greatly accelerated.

Rose, G.G. (ed.): Cinemicrography in Cell Biology. New York, London: Academic Press 1963

Circulating pool, or granulocytes: ↑ Granulocytes, pools of

Circulatory system: Specialized circuit of continuous, closed branching tubes (vessels) carrying blood pumped by heart. ↑ Arteries; ↑ Capillaries; ↑ Heart; ↑ Veins

Circumanal glands: ↑ Anal canal

Circumferential lamellae, of bone (lamellae circumferentiales externae et internae*): Outer C. (outer basic lamellae, OL) = a system of several uninterrupted parallel bony ↑ lamellae located immediately beneath ↑ periosteum (P) and surrounding ↑ diaphysis. Inner C. (inner basic lamellae, IL) = an inconstant system of bony lamellae adjacent to ↑ endosteum (E).

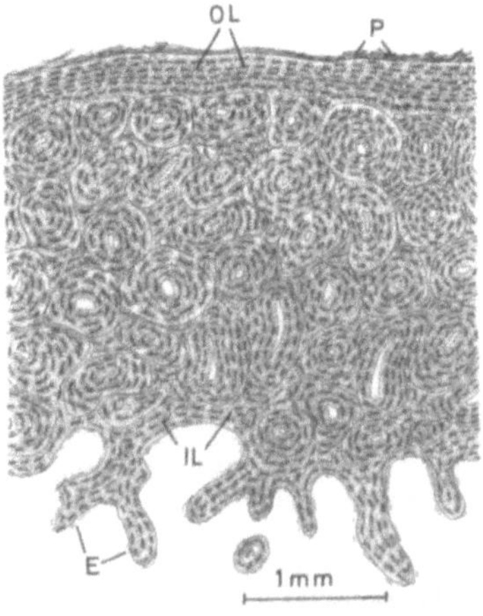

Circumpulpar dentin: ↑ Dentin, circumpulpar

Circumvallate papillae (papilla vallata*): A type of broad tongue papillae (CP), 1–3 mm wide and 1–3 mm tall, arranged in a row anterior to and parallel with sulcus terminalis. The 10–12 C. in man are at level of epithelium of ↑ tongue and surrounded by a circular furrow or papillary crypt (F). Upper epithelium, exposed to mechanical influences, is considerably interdigitated with numerous secondary ↑ connective papillae (SP) not present in area of furrow. Instead, there are about 250 ↑ taste buds (TB) present in this area. ↑ Serous glands (G) (↑ Ebner's glands) open into depths of furrow.

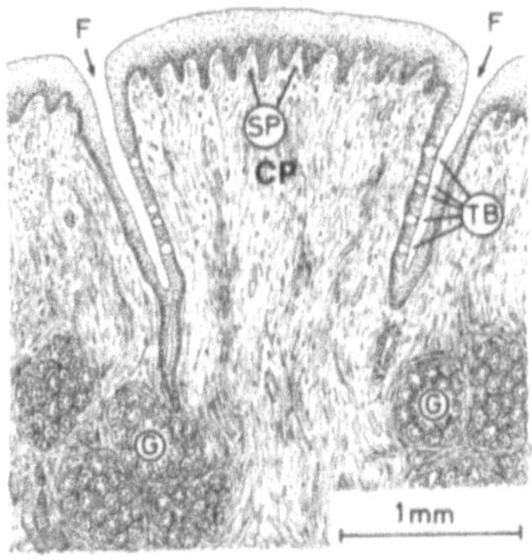

Circumventricular organs: Well-delimited and richly vascularized agglomerations of specialized ↑ ependymal and subependymal cells located on surfaces bordering ↑ brain ventricles; all C. are in direct contact with ↑ cerebrospinal fluid inside and outside ↑ brain ventricles. These organs comprise: ↑ Area postrema (AP), ↑ median eminence (ME), ↑ neurohypophysis (NH), ↑ organum vasculosum laminae terminalis (OVLT), ↑ choroid plexus (CP), ↑ pineal organ (PO), ↑ subcommissural organ (SCO), and ↑ subfornical organ (SFO). C. have secretory and receptive functions; some are involved in detoxication of cerebrospinal fluid and some C. may possess an as yet insufficiently examined endocrine activity. All C. are devoid of ↑ blood-brain barrier.

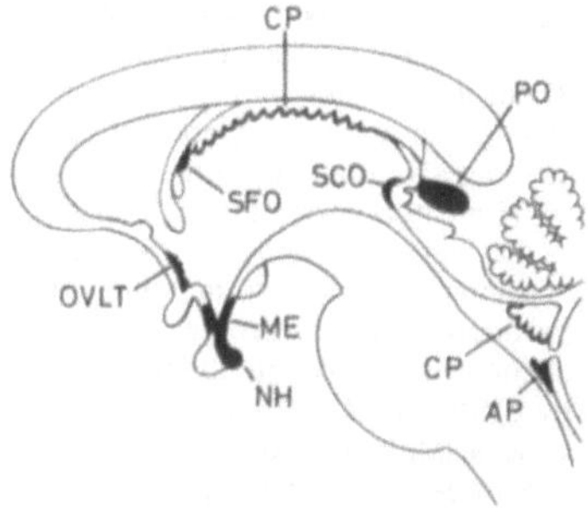

Krstić, R.: Biomed. Res. *2*, Suppl., 129 (1981); Leonhardt, H.: Ependym und circumventriculäre Organe. In: Oksche, A. and Vollrath, L. (eds.): Handbuch der mikroskopischen Anatomie des Menschen. Vol. 4. Nervensystem. Part 10: Neuroglia. Berlin, Heidelberg, New York: Springer-Verlag 1980; Mestres, P.: Scan. Electron Microsc. *2*, 137 (1978); Thompson, S.A.: Cell Tissue Res. *225*, 79 (1982)

Cis face: ↑ Forming face, of Golgi apparatus

Cisterna, perinuclear: ↑ Perinuclear cisterna

Cisternae, of brain: ↑ Brain cisternae

Cisternae, of endoplasmic reticulum (cisternae*): Spaces (C) of rough ↑ endoplasmic reticulum enclosed by nearly parallel ↑ unit membranes. C. may be expanded in some cells. (Fig. top = ↑ follicular cell, of ovarian follicle, rat; fig. bottom = ↑ thyroid follicular cell, rat)

↑ Reticuloplasm

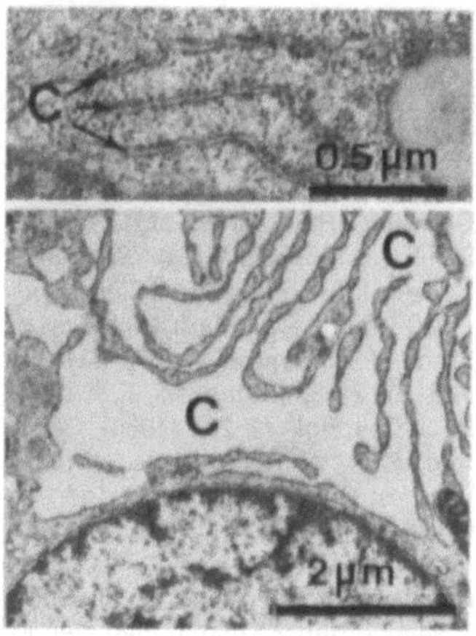

Cisternae, of Golgi apparatus: ↑ Lamellae, of Golgi apparatus

Cistron: A DNA sequence of a ↑ genome involved in coding of genetic information and responsible for synthesis of a definite product (i.e., ↑ messenger RNA, ↑ ribosomal RNA, ↑ transfer RNA). C. is considered smallest functional unit of heredity, equivalent to the ↑ gene.

Clara cells: ↑ Bronchiolar epithelium

Clasmatocytes: The former term for ↑ macrophages of ↑ loose connective tissue.

Clathrin: A 180000-dalton polypeptide composing major part of ↑ coated vesicles and involved in ↑ synaptic vesicle membrane recycling, membrane transport processes, protein uptake by ↑ apical pits, etc.

Harrison, S.C., Kirchhausen, T.: Cell *33*, 650 (1983); Lin, C.-T., Garbern, J., Wu, J.-Y.: J. Histochem. Cytochem. *30*, 853 (1982); Pearse, B.M.F., Crowther, R.A.: Packing of Clathrin Into Coats. Cold Spring Harbor Symposia on Quantitative Biology, Vol.46, Part 2. New York: Cold Spring Harbor Laboratory 1982; Puszkin, S., Schook, W.J.: Clathrin: A Study of its Properties and Mechanochemical Role in Cell Functions. In: Stracher, A. (ed.): Muscle and Nonmuscle Motility, Vol.2. New York: Academic Press 1983

Claudius' cells (cellula sustentacularis*): Clear, cuboidal cells (CC) lining sulcus spiralis externus (SSE); outwardly, C. are adjacent to ↑ stria vascularis (SV), and inwardly to ↑ Hensen's cells (HC). C. have a pale cytoplasm and a predominantly apical nucleus; their ↑ organelles are poorly developed. C. are a kind of ↑ supporting cells for ↑ Corti's organ (CO).

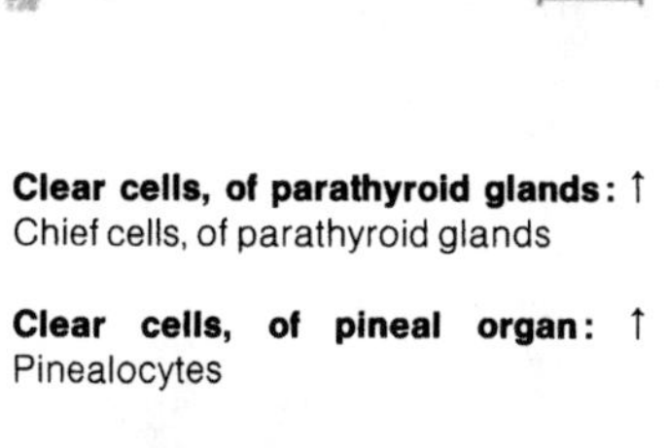

Clear cells, of parathyroid glands: ↑ Chief cells, of parathyroid glands

Clear cells, of pineal organ: ↑ Pinealocytes

Clear cells, of sweat glands (cellula lucida*): Cuboidal or polygonal cells with pale cytoplasm and round nucleus predominantly in a central position. Cytoplasm contains abundant mitochondria, some short rough endoplasmic cisternae, a small Golgi apparatus associated with a few dense granules, several ↑ lipid droplets (L), and an appreciable amount of particulate ↑ glycogen (Gly). C. are interdigitated with numerous ↑ microvilli (Mv), except on surface toward ↑ myoepithelial cells (MC). Short microvilli cover apical surface. Basal plasmalemma forms a moderately developed ↑ basal labyrinth (BLt). C. are believed to be involved in transepithelial fluid transport and in salt excretion. BL = basal lamina

↑ Sweet glands, apocrine; ↑ Sweat glands eccrine

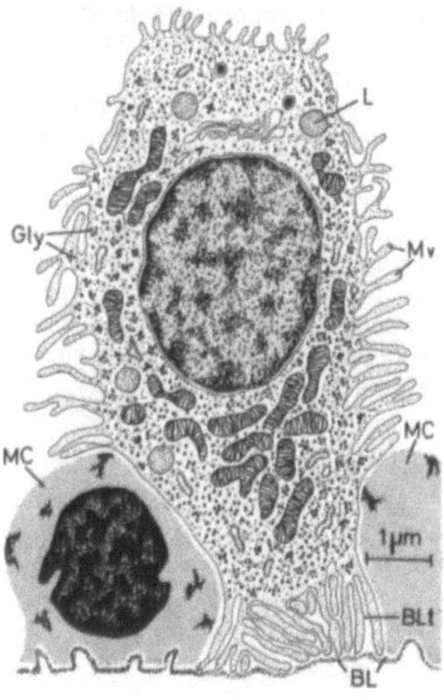

Ellis, R.A.: Eccrine Sweat Glands: Electron Microscopy, Cytochemistry and Anatomy. In: Gans, O. and Steigleder, G.K. (eds.): Normale und pathologische Anatomie der Haut. Berlin, Heidelberg, New York: Springer-Verlag 1969; Hashimoto, K.: J. Ultrastruct Res. *37*, 504 (1971)

Clear cells, of taste buds: ↑ Taste buds, structure of

Clear cells, of thyroid gland: ↑ C-cells, of thyroid gland

Cleft, of Rathke: ↑ Rathke's cleft

Cleft, synaptic, of motor end plate: ↑ Synaptic cleft, of motor end plate

Cleft, synaptic, of neuronal synapse: ↑ Synaptic cleft, of neuronal synapse

Clefts, of Schmidt-Lanterman: ↑ Schmidt-Lanterman clefts

Climbing fibers: One of the two kinds of afferent fibers entering ↑ cerebellar cortex. C. are not branched; they climb along ↑ dendrites of ↑ Purkinje cells establishing excitatory ↑ synapses with them.

↑ Mossy fibers

Blank, N.K., Seil, F.J., Leiman, A.L.: Brain Res. *271*, 135 (1983)

Clitoris*: Part of ↑ external female genital organs consisting of two small corpora cavernosa (CC), which, as crura clitoridis, are affixed to os pubis (OP). The crura unite and end in a short corpus clitoridis (Co). A rudimentary glans clitoridis (GC) underlies corpus; it is related to bulbi vestibuli (BV). Corpora cavernosa are lined by a dense tunica albuginea (TA); structurally, they are identical to ↑ corpora cavernosa penis. C. is covered with thin nonkeratinized ↑ stratified squamous epithelium (E) overlying vascular stroma with high papillae and numerous free nerve endings. Glans is very rich in Meissner's, Ruffini's, Krause's, and Pacini's ↑ corpuscles, as well as intraepithelial nerve endings. Some ↑ mucous glands are located in area of frenula (F).

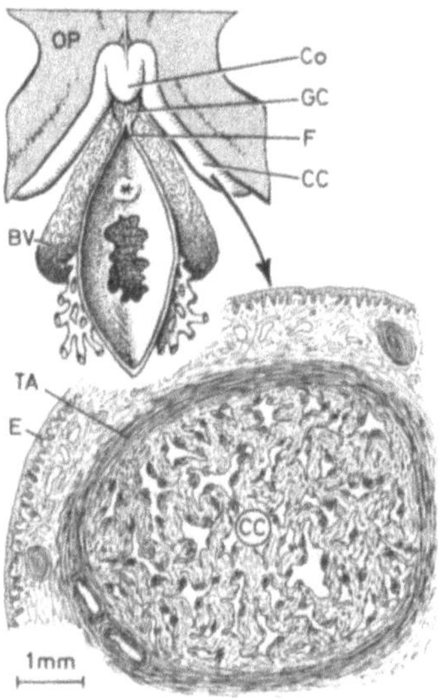

Clone: A population of biological units (cell ↑ organelles, cells, or organisms) deriving from a single ancestor as a result of asexual reproduction. Examples: Reproduction of ↑ chromosomes or cell organelles having autoreproductive power (↑ centrioles, ↑ mitochondria), reproduction of cells by ↑ mitosis. Members of a C. are genetically identical, however, mutation may change them.

Cloning-inhibiting factor (CIF): A ↑ lymphokine that inhibits ↑ mitosis in tissue culture cells.

Cloquet's canal: ↑ Vitreous body

"Closed circulation," of spleen: ↑ Spleen, vascularization of

Clotting, blood: ↑ Blood clotting

Club hair (resting hair, quiescent hair, telogen hair): A ↑ hair with a short ↑ hair follicle (HF) lacking firm attachment to ↑ dermis (D). Follicle of C. is formed by ↑ matrix cells which have ceased to divide; they keratinize in situ and form a thickening whose structure resembles surface ↑ epidermis. C. can be easily plucked from follicle; in normal replacement of hair, about 5% are C.

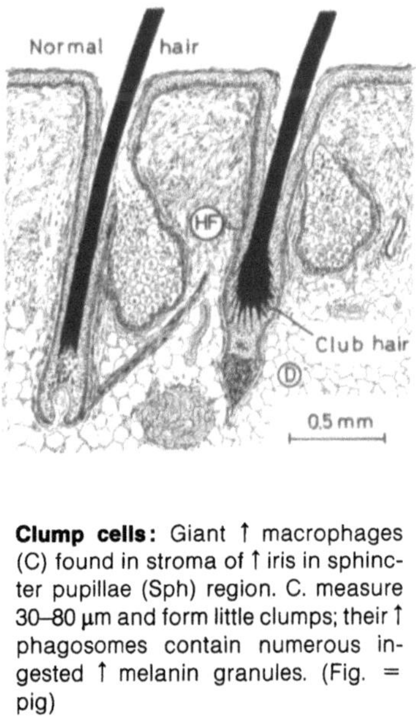

Clump cells: Giant ↑ macrophages (C) found in stroma of ↑ iris in sphincter pupillae (Sph) region. C. measure 30–80 µm and form little clumps; their ↑ phagosomes contain numerous ingested ↑ melanin granules. (Fig. = pig)

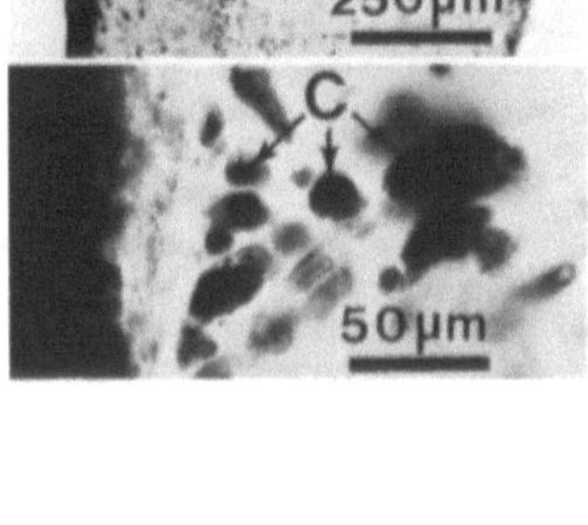

Coated vesicles (acanthosomes, spiny bodies): Vesicles (V) about 50–80 nm in diameter with a moderately osmiophilic amorphous content. Delicate radial bristles (Br) 17 nm long and set 15 nm apart extend from outer leaflet of ↑ unit membrane (M). The bristles and their extensions form penta- and hexagonal patterns around vesicle = clathrin basket. C. may be formed: 1) By pinching off of ↑ micropinocytotic vesicles (Mp); 2) budding off from ribosome-free areas of rough endoplasmic reticular (rER) cisternae; 3) budding off from Golgi apparatus (G). In the first case, C. are involved in ↑ micropinocytosis, in the second, C. transport the polypeptide product to Golgi apparatus, and in the third, they transport lytic enzymes to ↑ lysosomes (Ly). C. associated with rER and Golgi apparatus having a transport role are called ↑ transport vesicles.

↑ Clathrin; ↑ GERL

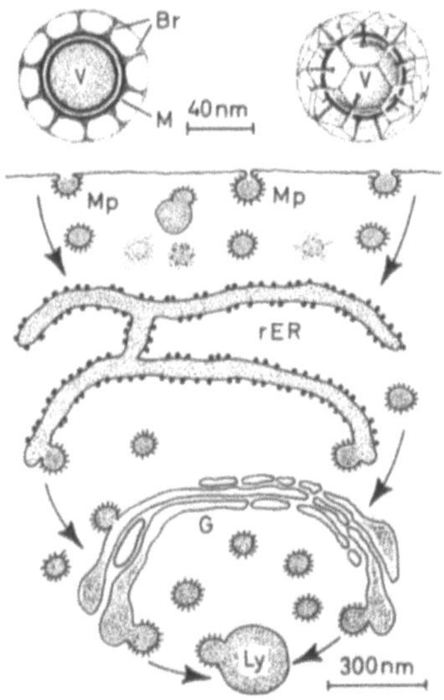

Crowther, R.A., Finch, J.T., Pearse, B.M.F.: J. Mol. Biol. *103*, 785 (1976), Heuser, J., Evans, L.: J. Cell Biol. *84*, 560 (1980); Imhof, B.A., Marti, U., Boller, K., Frank, H., Birchmeier, W.: Exp. Cell Res. *145*, 199 (1983); Kanaseki, T., Kadota, K.: J. Cell Biol. *42*, 202 (1969); Ockleford, C.D., Whyte, A. (eds.): Coated Vesicles. Cambridge: Cambridge University Press 1980

Coccygeal body (corpus coccigeum*): A small organ 2.5 mm in diameter, located in front of tip of coccyx. C. consists of numerous arteriovenous ↑ anastomoses embedded in a ↑ dense connective tissue. ↑ Smooth muscle cells have an epithelioid appearance, but no endocrine function.

Cochlea*: Part of bony ↑ labyrinth consisting of a spiral canal, about 35 mm long and 3–9 mm in diameter, turning two and a half times around axis, or modiolus (M), and ending finally in ↑ helicotrema (H). C. also contains ↑ lamina spiralis ossea (LS), which, with ↑ basilar membrane (BM) and ↑ Reissner's membrane (RM) of ↑ cochlear duct (CD), subdivides C. canal into scala vestibuli (SV) and scala tympani (ST). ↑ Spiral ganglion (SG) is located at base of lamina spiralis ossea. (Fig. = human)

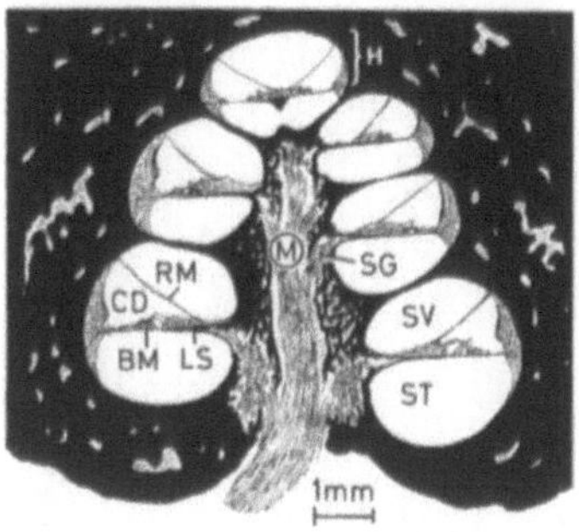

Cochlear duct (scala media, ductus cochlearis*): Part (CD) of membranous ↑ labyrinth located in ↑ cochlea (C). C. emerges as a tubular diverticulum from saccule (S) and turns two and a half times to end blindly in cochlear cupula (Cu). C. is separated from scala tympani (ST) by ↑ basilar membrane (BM) and ↑ lamina spiralis ossea (LS), from scala vestibuli (SV) by ↑ Reissner's membrane (RM), and from bone (B) by ↑ spiral ligament (SL). C. contains ↑ Corti's organ (CO) and ↑ endolymph. ↑ Perilymph circulates in both scalae.

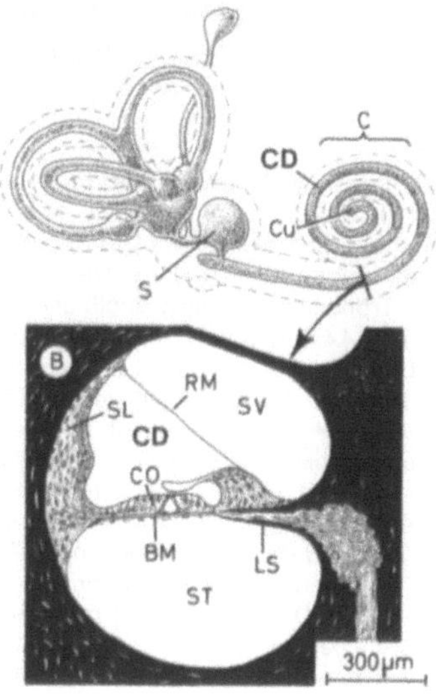

Cochlear fluids: A collective term for both ↑ endolymph and ↑ perilymph.

Cochlear ganglion: ↑ Spiral ganglion

Codon: A sequence of three bases along ↑ messenger RNA molecule that determines, via complementary ↑ anticodon of ↑ transfer RNA, the specific amino acid of a polypeptide chain or the end of ↑ protein synthesis. (See molecular biology texts for further information)

Cohnheim's fields: Small polygonal groups (C) of bundled ↑ myofibrils (Mf) separated by myofibril-free ↑ sarcoplasm, visible with light microscope in transversally sectioned ↑ skeletal (Sk) and ↑ cardiac (Ca) muscle fibers. Some authors consider C. to be a shrinkage ↑ artifact. In longitudinal section, C. correspond to ↑ Leydig-Kölliker columns.

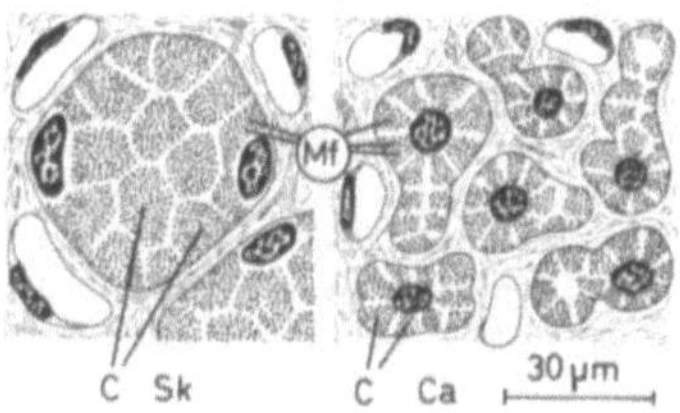

Coiled arteries, of endometrium (spiral artery, of endometrium, arteria spiralis*): ↑ Muscular arteries (C), branches of uterine ↑ arcuate arteries, having a very tortuous course. Ends of C., lying in functionalis (F), are sensitive to sex hormones and degenerate and regenerate during each ↑ menstrual cycle. C. also constrict and dilate rhythmically during the cycle, playing an important role in menstrual bleeding. B = basalis, G = ↑ uterine glands. (Fig. = ↑ decidua parietalis, human)

↑ Endometrium, vascularization of; ↑ Uteroplacental arteries

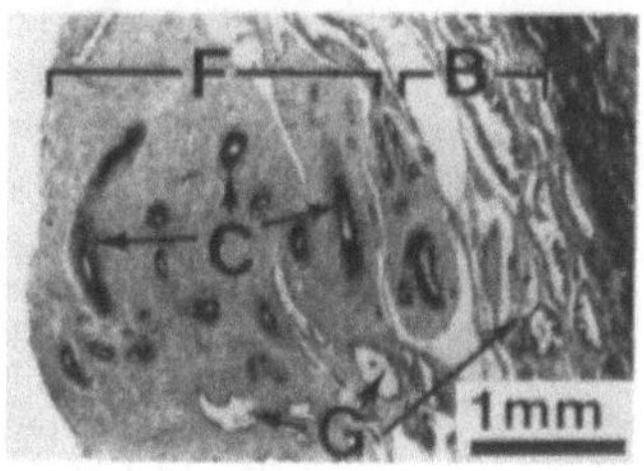

Colchicine: A plant alkaloid blocking polymerization of kinetochoral ↑ microtubules of ↑ spindle apparatus, thereby arresting cell division in metaphase.

Cold insoluble globulin: ↑ Fibronectin

Collagen: The principal protein of white fibers of ↑ connective and ↑ supporting tissues. C. is insoluble in water, swells in dilute acids, can be diluted in strong acids or alkalies, is digested by ↑ collagenases and proteases, yields ↑ gelatin upon boiling, is ↑ birefringent, and stainable with most ↑ acid dyes. C. has a high content of glycine, alanine, hydroxyproline, and proline, but is poor in sulfur and contains no tryptophan. C. represents 30%–35% of total protein content of human organism.

↑ Collagen, types of

Hay, E.: J. Cell Biol. *91*, 205s (1981); Minor, R.R.: Am. J. Pathol. *98*, 225 (1980); Ramachandran, G.N., Reddi, A.H. (eds.): Biochemistry of Collagen. New York, London: Plenum Press 1976

Collagen, arrangement of molecules in: ↑ Collagen fibrillogenesis; ↑ Tropocollagen

Collagen, biosynthesis of: Following transcription (Tc) and translation (Tl) of genetic information, ↑ polyribosomes (Pr) synthesize, using amino acids (alanin, hydroxyproline, proline, hydroxylysine, etc.), polypeptide chains, so called pro-α chains (α), which begin and end with nonhelical

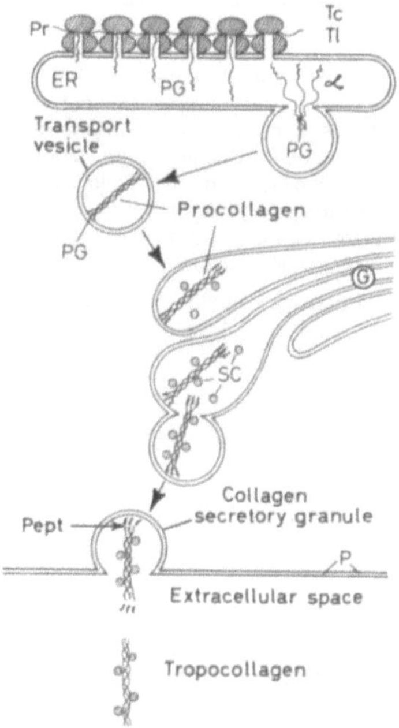

peptide groups (PG), referred to as propeptides or ↑ registration peptides. The chains accumulate within ↑ cisternae of endoplasmic reticulum (ER); three such chains assemble to form a helical ↑ procollagen molecule. These molecules are transferred inside ↑ transport vesicles to Golgi apparatus (G), where a sugar component (SC) is incorporated with them. ↑ Collagen secretory granules then carry these molecules to plasmalemma (P), where plasmalemmal procollagen peptidase (Pept) cleaves the peptide groups, thereby forming ↑ tropocollagen molecules secreted into extracellular space. By their polymerization, these molecules form ↑ collagen fibers. Some steps of C. are poorly understood; this schematic description must, therefore, be considered provisional.

↑ Collagen-synthesizing cells, ↑ Fibroblast, functions of

Kleinman, H.K., Klebe, R.J., Martin, G.R.: J. Cell Biol. *88*, 473 (1981); Mooon-Il Cho, Garant, P.R.: Anat. Rec. *201*, 577 (1981); Nemetschek, Th.: Verh. Anat. Ges. *75*, 47 (1981); Prockop, D.J., Kivirikko, K.I., Tuderman, L., Gutzman, N.A.: N. Engl. J. Med. *301*, 12 (1979)

Collagen, calcification of: The impregnation of collagen structures with calcium salts. In some tissues (↑ bone, ↑ cementum, ↑ dentin, hyaline and fibrous ↑ cartilage), collagen may act as initiator for deposit of ↑ hydroxyapatite crystals (C). The precise interrelationship between hydroxyapatite and ↑ tropocollagen (Tr) is not known; polypeptide chains apparently wrap themselves around the crystals. There is evidence that the first crystallization nuclei occur in "holes" (h) between heads and tails of successive tropocollagen molecules where there is no chemical bond. Calcification spreads along microfibrils from these points. However, collagen is not solely responsible for this calcification.

↑ Bone, calcification of; ↑ Cartilage, calcification of; ↑ Collagen, fibrillogenesis of

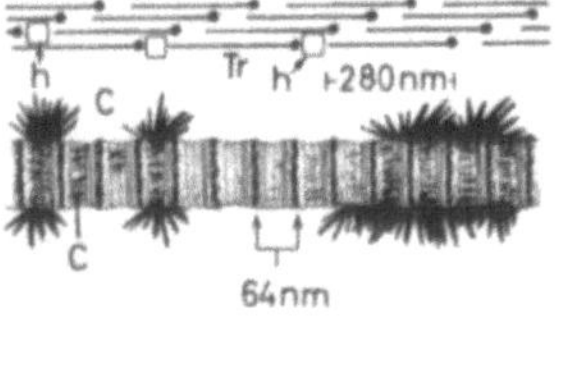

Collagen, degradation of: A process of destruction of collagen carried out by ↑ collagenases and other proteases during growth, remodeling, ↑ involution, ↑ regeneration, and injury; it is well balanced by a simultaneous ↑ collagen biosynthesis. Collagenases initiate degradation by breaking down ↑ tropocollagen molecule at a specific site about 75 nm from its carboxyterminus. Resulting asymmetric fragments are destroyed by proteases in extracellular space and/or phagocytized by ↑ macrophages and broken down in their secondary ↑ lysosomes. In some cases, ↑ fibroblasts and ↑ macrophages may phagocytize entire ↑ collagen microfibrils.

Deporter, D.A., Ten Cate, A.R.: J. Periodontol. *51*, 429 (1980); Svoboda, E.L.A., Shiga, A., Deporter, D.A.: Anat. Rec. *199*, 473 (1981)

Collagen fibers (fibra collagenosa*): Units of white fibers (Fb) of ↑ connective and supporting tissues, each composed of a various number of ↑ collagen fibrils (Fi). Several C. form bundles (fasciculi), 5–15 µm thick, of indefinite length, running in all directions. In connective tissues not under tension, C. have irregular undulating courses; interchanges of fibrils take place between them. C. are very resistant to traction, their extensibility being only about 5%. The term C. is also used generally to describe collagen material. (Fig. top = ↑ dermis, human; fig. bottom = dermis, rat)

↑ Collagen fibers, submicroscopic and microscopic organization of

Collagen fibers, submicroscopic and microscopic organization of: In course of ↑ collagen biosynthesis, ↑ fibroblasts (F) and other collagen-producing cells synthesize ↑ procollagen from amino acids (AA); it is rapidly converted to ↑ tropocollagen molecules (Tr) 1.5 nm thick. In extracellular space, they polymerize, producing ↑ protofibrils (Pr) about 4.5 nm thick. Protofibrils unite to form a ↑ collagen microfibril (Mf) 5–20 nm in diameter. Several microfibrils make up a ↑ collagen fibril (Fi) 0.3–2 µm across, several fibrils associate into a ↑ collagen fiber (Fb), and several fibers may join together into a bundle of fibers (not shown). The elements of C. are held together by ↑ proteoglycans.

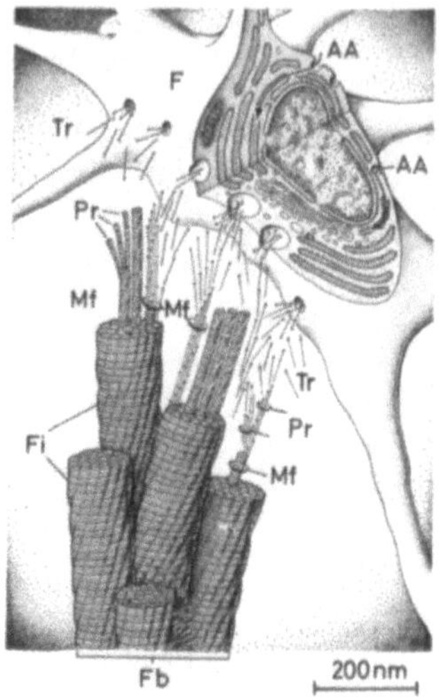

Collagen fibril (fibrilla collagenosa*): A bundle (Fi) of ↑ collagen microfibrils (Mf) 0.3–2 µm in diameter. Several C. constitute a ↑ collagen fiber. (Fig. = ↑ dermis, rat)

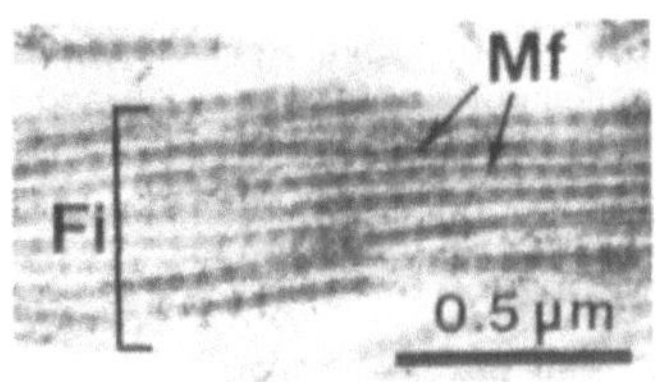

Collagen, fibrillogenesis of: Process of formation of ↑ collagen fibers through polymerization of ↑ tropocollagen molecules (Tr). Current model of C.: Tropocollagen has a length of 280 nm = 4.4 D; D equals 64 nm. In protofibrils (P), heads (H) and tails (T) of successive tropocollagen molecules overlap. Axial displacement of adjacent protofibrils by a distance D results in a microfibril (Mf) with a repeated cross banding of D. Each D band is subdivided into an overlap zone (OZ) of 0.4 D and a hole zone (HZ) or "hole" (h) of 0.6 D. In the case of positively contrasted microfibrils, there are 12 dark bands between major repeated periods, corresponding to lateral chains (LC) of tropocollagen molecule. In the case of negative contrast, D is of same length, but with only two bands: dark bands into which "stain" penetrates (corresponding to "holes" between molecules) and impenetrable clear zones occupied by collagen material. Lateral linkage of tropocollagen molecules is brought about by lysyl and hydroxylysyl residues, which become deaminated by lysyl oxidase secreted by ↑ fibroblasts. According to collagen type, arrangement of tropocollagen molecules, i.e., protofibrils, can be straight or helicoidal.

↑ Contrasting, in transmission electron microscopy

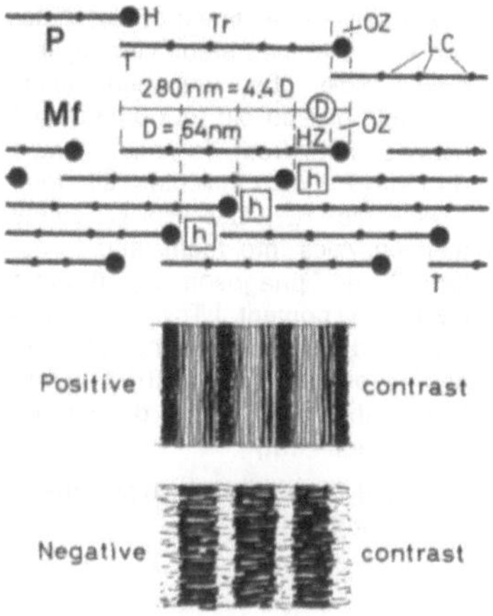

Reale, E., Benazzo, F., Ruggeri, A.: J. Submicrosc. Cytol. *13*, 135 (1981)

Collagen formation: ↑ Collagen, biosynthesis of

Collagen microfibril (microfibrilla*): Fundamental element of ↑ collagen fibrils and fibers. Diameter of C. varies between 20 and 300 nm, depending on tissue and species; length is variable. Periodicity of C. is 64 nm. Each C. consists of three or more ↑ protofibrils. (Fig. = ↑ tendon, rat)

↑ Collagen fibers, submicroscopic and microscopic organization of; ↑ Collagen, fibrillogenesis of

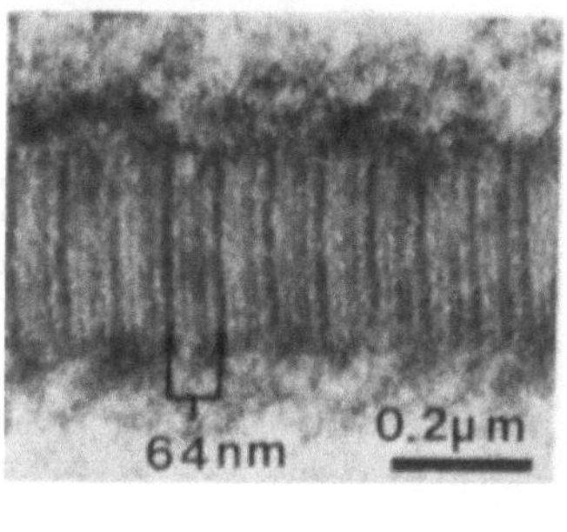

Bouteille, M., Pearse, D.C.: J. Ultrastruct Res. *35*, 314 and 339 (1971); Lillie, J.H., Mac Callus, D.K., Scaletta, L.J., Occhino, J.C.: J. Ultrastruct Res. *58*, 134 (1977)

Collagen molecule: ↑ Tropocollagen

Collagen, "native": A form of ↑ collagen displaying usual banding pattern of 64 nm. C. may also form in vitro if a neutral salt solution of collagen is warmed to 37 °C. For some authors, C. is collagen obtained without any technical procedure; in this case it has a periodicity of 70 nm.

Collagen secretory granules: Unit membrane-bound tubular structures (CG, transversal section) up to 0.4 µm long and 0.1 µm wide, filled with densely packed ↑ procollagen molecules. C. arise from Golgi ↑ cisternae (GC) as rectangular vesicles (RV) whose content gradually densifies into a fine fibrillar material. C. are then transported along cytoskeletal microfilaments to plasmalemma, where they expel their contents by ↑ exocytosis into extracellular space. (Fig. = ↑ fibroblast, rat)

↑ Collagen, biosynthesis of

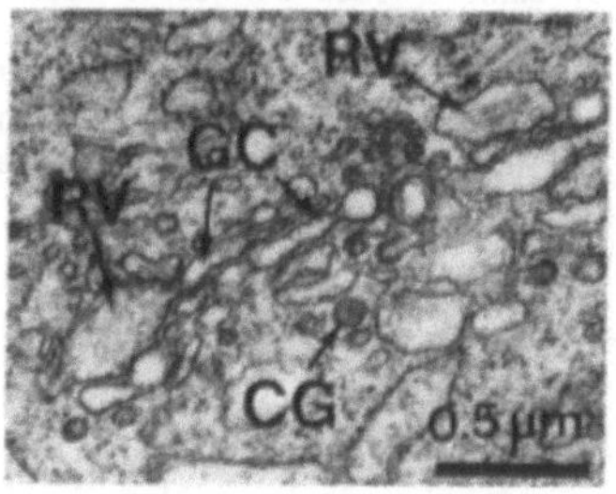

Cho, M.-I., Garant, P.R.: Anat. Rec. *199*, 309 and 459 (1981)

Collagen-synthesizing cells: All cells able to produce ↑ collagen. Major part of collagen of developing organism is synthesized by ↑ fibroblasts, ↑ chondroblasts, ↑ odontoblasts, and ↑ osteoblasts; in adult organism, collagen turnover is assured by ↑ fibrocytes, ↑ chondrocytes, ↑ osteoblasts, and ↑ osteocytes. Epithelial cells produce type IV collagen, constituting ↑ basal laminae. There is some evidence that ↑ smooth muscle cells may also synthesize collagen.

Collagen, types of: Four types of genetically different C. have been isolated, varying in amino acid composition, polymerization sequence, and quantity of sugar. Type I $[\alpha 1(I)]_2\alpha 2$ is found in ↑ bone, ↑ dermis, ↑ dentin, ↑ ligaments, and ↑ tendons. Type II $[\alpha 1(II)]_3$ has been isolated from hyaline ↑ cartilage; it has a higher content of glycosylated hydroxylysine than type I. Type III $[\alpha 1(III)]_3$ originates in ↑ cardiovascular system, intestine, and ↑ uterus; it contains more hydroxyproline than type II and some half-cystine residues. Type IV $[\alpha 1(IV)]_3$ is characteristic of ↑ basal laminae; it is richer in glycosylated hydroxyproline than type III and also includes 3- and 4-hydroxylysine with more half-cystine than type III. The molecules of type IV do not polymerize; they are covalently linked to ↑ proteoglycans.

Bornstein, P., Sage, H.: Ann. Rev. Biochem. *49*, 957 (1980); Junqueira, L.C.U., Cossermelli, W., Brentani, R.; Arch. Histol. Jpn. *41*, 267 (1978); Laurie, G.W., Leblond, C.P., Martin, G.R.: J. Histochem. Cytochem. *30*, 983 (1982); Oberbäumer, I., Wiedemann, H., Timpl. R., Kühn, K.: The EMBO Journal *1*, 805 (1982); Von der Mark, K.: Int. Rev. Connect. Tissue Res. *9*, 265 (1981)

Collagenases: Proteases catalyzing hydrolysis of ↑ collagen and playing an important role in ↑ collagen degradation. C. is produced by ↑ fibroblasts, ↑ granulocytes, ↑ macrophages, epidermal, and ↑ synovial cells. Its concentration is increased in ↑ uterus during postgravidal involution.

Harris, E.D., Jr., Krane, S.M.: N. Engl. J. Med. *291*, 605 and 652 (1974)

Collateral ganglia: ↑ Prevertebral ganglia

Collaterals, of axon: ↑ Side branches of an ↑ axon

Collecting system, extrarenal: A system of tubes conducting ↑ urine from ↑ kidney to exterior. C. comprises: ↑ calyces majores and minores, ↑ renal pelvis, ↑ ureter, ↑ urinary bladder, and ↑ urethra.

Collecting system, intrarenal: ↑ Collecting tubules, of kidney

Collecting tubules, of kidney: An intrarenal canalicular system serving as a conduit for ↑ urine. Components: 1) Arched collecting tubules (ACT) (situated in cortex), descending into ↑ medullary ray (MR) and merging with other arched tubules to form 2) straight or medullary collecting tubules (SCT) running down medulla. Only in inner zone of medulla (IZ) do they join, at acute angles, five to seven similar tubules to form 3) ↑ papillary duct (PD). Diameter of tubules and height of lining epithelium gradually increase. Structure: 1) ↑ Arched collecting tubules = a ↑ simple cuboidal epithelium with dark, intercalated cells (DC) (cellula densa*) rich in mitochondria, some apical microvilli and basal infoldings, and light or principal cells (LC) (cellula lucida*) with a small Golgi apparatus, a few mitochondria, and, rarely, basal infoldings. 2) Straight collecting tubules (tubulus colligens rectus*) = simple columnar epithelial cells with an ellipsoidal, heterochromatin-poor nucleus, a small Golgi apparatus, some mitochondria, very short rough endoplasmic profiles, and shallow basal infoldings. ↑ Apical pole is convex and smooth; lateral cell surfaces are without interdigitations. 3) ↑ Papillary ducts = a ↑ simple columnar epithelium continuous with that of ↑ renal papilla (P). All nuclei are at same level; cytoplasm contains scattered mitochondria, a small Golgi apparatus, some flattened rough endoplasmic cisternae, and an abundance of ribosomes. Apical pole bulges into lumen; lateral cell surfaces form many microvillous processes penetrating into wide intercellular spaces. Papillary ducts measure 200–300 µm in diameter and open onto an area cribriformis (AC) at tip of each papilla. C. are also involved in final reabsorption of water under influence of ↑ antidiuretic hormone acting on ↑ junctional complexes (JC) between the cells. CP = connecting portions of ↑ distal convoluted tubule

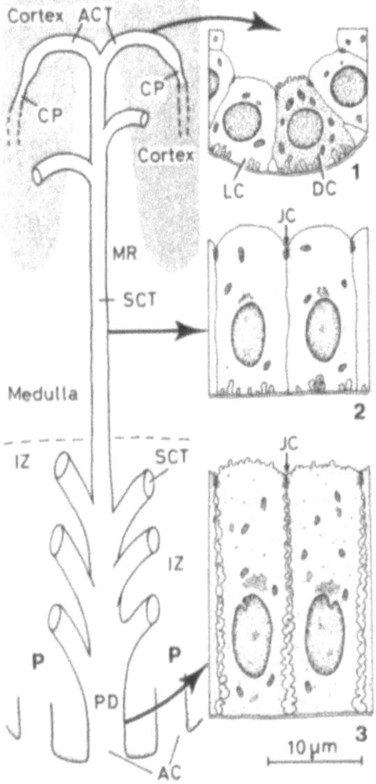

↑ Cortical collecting tubules; ↑ Nephron

Brown, D., Roth, J., Kumpulainen, T., Orci, L.: Histochemistry *75*, 209 (1982); Kaissling, B.: Curr. Probl. Clin. Biochem. *8*, 435 (1977)

Collecting veins, of liver: Thin-walled veins, each formed of several ↑ intercalated veins. Several C. join to form hepatic veins.

↑ Liver, vascularization of

Colliculus prostaticus: ↑ Colliculus seminalis

Colliculus seminalis* (caput gallinaginis, colliculus prostaticus, Müllerian hillock, verumontanum): An elevated segment (CS) of ↑ urethral crest, about 5 mm high and 3–5 mm wide. C. is covered by ↑ stratified columnar epithelium (E); interior is occupied by several glandlike ramifications (R) of ↑ utriculus prostaticus (UP) surrounded by some ↑ loose connective tissue. Through C. run ↑ ejaculatory ducts (ED), which emerge, together with duct of utriculus prostaticus, at surface of C. ↑ Prostatic ducts (PD) open at base of the C.

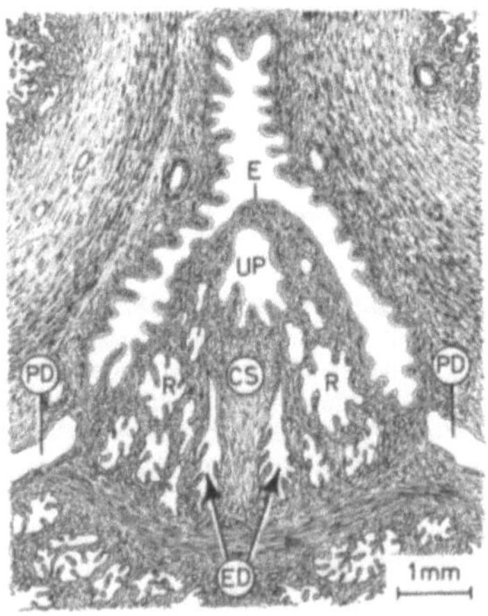

Colloid cells, of Langendorff: Heavily stained, mostly flattened ↑ thyroid follicular cells with a pyknotic nucleus, probably representing degenerative forms.

Colloid droplets: ↑ Colloid vacuoles

Colloid, of thyroid follicles (colloidum*): An amorphous, gelatinous, ↑ PAS-positive substance filling ↑ thyroid follicles. C. is storage product of secretory activity of ↑ thyroid follicular cells; it consists mainly of ↑ thyroglobulin and some ↑ thyroid hormones (↑ thyroxin and ↑ triiodothyronine). Under transmission electron microscope, C. may appear as very pale to moderately dense fine granular material. When thyroid hormones are needed, follicular cells incorporate C. in the form of ↑ colloid vacuoles.

Colloid vacuoles, of thyroid follicular cells (colloid droplets): ↑ Unit membrane-bound ↑ PAS-positive globules (C), 0.5–3 µm in diameter, filled with moderately dense, fine granular material. C. are located predominantly in ↑ apical pole; they represent small amounts of ↑ colloid introduced by ↑ pseudopodia from follicular lumen (FL) into ↑ thyroid follicular cell. Here, ↑ lysosomes (Ly) fuse with C., empty their enzymes into them, and transform C. into ↑ phagosomes (Ph) with a low density content. ↑ Thyroglobulin is split inside phagosome into globulin and thyroid hormones (↑ thyroxin and ↑ triiodothyronine), which diffuse into blood. (Fig. = rat)

↑ Thyroid follicular cells, biosynthesis of hormones in

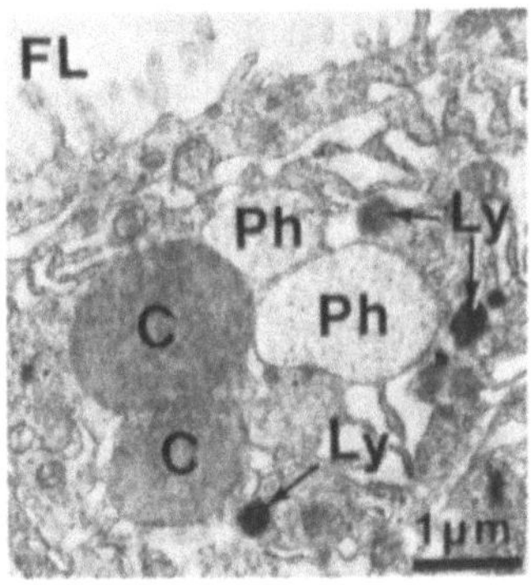

Colon*: Part of ↑ large intestine consisting of ascending (AC), transverse (TC), descending (DC), and sigmoid segments (SC). Structure: 1) Tunica mucosa (TM): a) epithelium (E) = a ↑ simple columnar epithelium forming ↑ Lieberkühn's crypts (LC), in which ↑ goblet cells predominate over ↑ absorptive cells; b) lamina propria (LP) = a well-vascularized and innervated ↑ loose connective tissue with numerous ↑ wandering cells; c) lamina muscularis mucosae (LMM) = a thin layer of ↑ smooth muscle cells. 2) Tela submucosa (TSm) = a loose connective tissue with large blood and lymphatic vessels, many ↑ adipose cells, and ↑ submucosal plexus; presence of solitary ↑ lymphatic nodules (LN). 3) Tunica muscularis (TMu) = smooth muscle cells disposed in an inner circular (IC) and an outer longitudinal (OL) layer. The latter forms thickenings, taeniae (T), whose tonus provokes sacculations, the haustrae (H). Between two haustrae, the wall is thrown into ↑ plicae semilunares (PS). 4) Tela subserosa (TSs) and/or tunica adventitia (TA) = a loose connective tissue connecting tunica muscularis to tunica serosa or to abdominal wall. Accumulations of adipose tissue in this layer form ↑ appendices epiploicae (AE). 5) Tunica serosa (TSer) = peritoneal ↑ mesothelium covering intraperitoneal areas of C. Functions: Concentration and solidification of unabsorbed contents of small intestine by absorption of water; secretion of protective mucous substances.

↑ Rectum

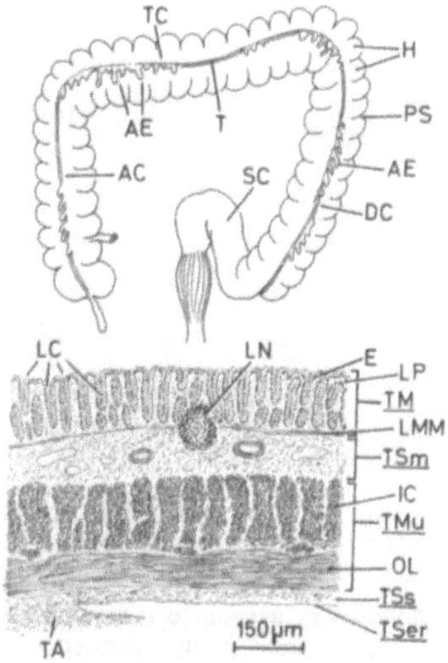

Bustos-Fernandez, L. (ed.): Colon. Structure and Function. New York: Plenum Publ. Corp. 1983

Colony-forming units (CFU, colony-forming cells, CFC): Presumptive pluripotent, small, lymphocytelike ↑ hematopoietic stem cells capable of forming, by clonal reproduction, colonies of proliferating hematopoietic cells in spleen of a lethally irradiated mouse. C. seem to be precursor cells of at least three kinds of morphologically recognizable cells: CFU-E, which give rise to **e**rythropoietic series under influence of ↑ erythropoietin (ERC = erythropoietin responsive cells); CFU-G, which give rise to **g**ranulopoietic series; CFU-M, from which arise **m**egakaryocytes. Some authors use C. specifically to denote stem cell of all blood cells, although direct evidence for such a cell is still lacking. According to some authors, symbol CFU should replace the term ↑ hemocytoblast.

↑ Candidate stem cell; ↑ Clone; ↑ Hematopoiesis, current view of; ↑ Totipotent hematopoietic stem cells

Frassoni, F., Testa, N.G., Lord, B.I.: Cell Tissue Kinet. *15*, 447 (1982); Hara, H., Ogawa, M.: Am. J. Hematol. *4*, 23 (1978); Valeriote, F., Tolen, S.: Cell Tissue Kinet. *16*, 1 (1983), Van Bekkum, D.W.: The Appearance of the Multipotential Stem Cell. In: Baum, J.S., Ledney, G.D. (eds.): Experimental Hematology Today. New York, Springer-Verlag 1977; Vos, O., Wilschut, J.C.I.: Cell Tissue Kinet. *12*, 257 (1979)

Colony-forming units-culture (CFU-C): A category of ↑ colony-forming units developing and differentiating in a tissue **c**ulture.

Colony-forming units-erythropoietic (CFU-E, erythropoietin responsive cells, ERC): A category of ↑ colony-forming units sensitized to ↑ erythropoietin and capable of transforming themselves into the erythropoietic lineage. According to some authors, C. arise from ↑ burst-forming units-erythropoietic.

↑ Erythropoiesis; ↑ Hematopoiesis, current view of

Nijhof, W., Wierenga, P.K.: J. Cell Biol. *96*, 386 (1983)

Colony-forming units-granulopoietic (CFU-G): A category of ↑ colony-forming units differentiating into **g**ranulopoietic lineage.

↑ Granulopoiesis; ↑ Hematopoiesis, current view of

Colony-forming units-megakaryocytic (CFU-M): A class of ↑ colony-forming units differentiating into megakaryocytopoietic lineage.

↑ Hematopoiesis, current view of; ↑ Megakaryocytopoiesis

Colony-forming units-splenic (CFU-S): The designation for ↑ colony-forming units developing in **s**pleen of irradiated animals.

Eaglesom, C.C., Riches, E.G., Wright, E.G.: J. Anat. *135*, 811 (1982)

Colony-stimulating activators: ↑ Granulopoiesis, control of

Colony-stimulating factor (CSF): A proteinaceous substance (m.w. about 45000 daltons) secreted by ↑ histiocytes and ↑ monocytes increasing number of hematopoietic colonies in ↑ bone marrow and stimulating ↑ granulopoiesis.

↑ Granulopoiesis, control of

Okabe, T., Ohsawa, N.: Cell. Mol. Biol. *27*, 579 (1981)

Color, of eye: C. depends on number and pigmentation of ↑ melanophores and ↑ melanocytes in stroma of ↑ iris. When they are few or absent, heavily pigmented cells of posterior epithelium seen through colorless stroma give iris a gray or blue tone. With an increase in number of melanocytes and melanophores, iris becomes greenish to dark brown. In ↑ albinism, iris takes pink color from presence of its own blood vessels and those of ↑ retina seen by transparency.

Colostrum (foremilk): A small amount of white opalescent liquid secreted from ↑ mammary gland at termination of pregnancy. In histological preparation, C. appears as an ↑ acidophilic, amorphous mass containing ↑ colostrum bodies, and cell fragments. C. is rich in lactoproteins and antibodies but very poor in lipids. C. has laxative properties; its secretion continues only 1–3 days after parturition.

↑ Witch's milk

Colostrum bodies: Large, round cells (C) filled with ↑ lipid droplets present in ↑ colostrum and ↑ milk. C. are believed to be ↑ macrophages which have phagocytized lipid droplets and penetrated into lumen of ↑ tubuloalveoli. (Fig. = human, end of pregnancy)

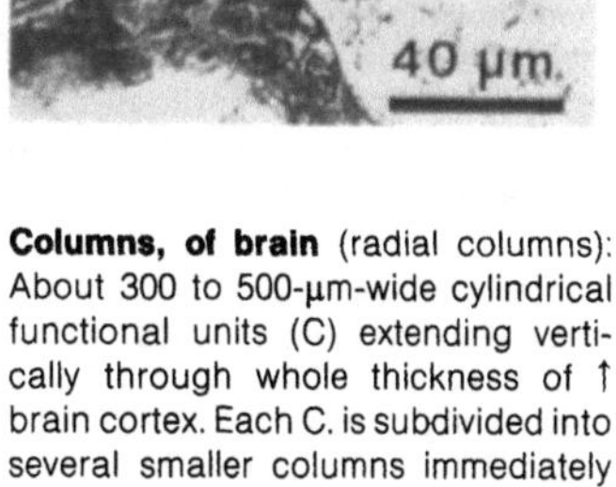

Columns, of brain (radial columns): About 300 to 500-µm-wide cylindrical functional units (C) extending vertically through whole thickness of ↑ brain cortex. Each C. is subdivided into several smaller columns immediately surrounding apical dendrites of pyramidal cells, together with correspond-

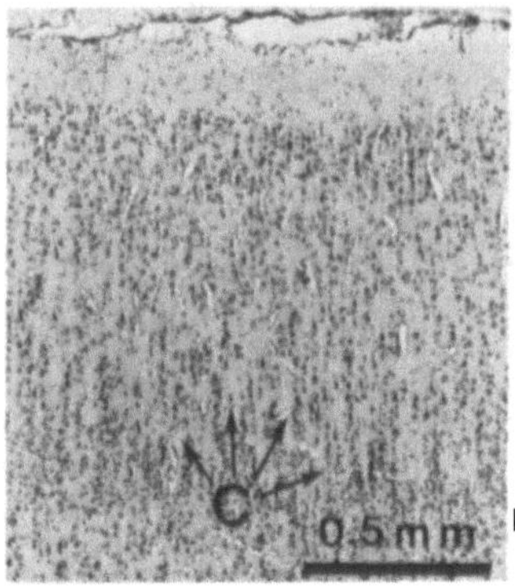

↑ Brain cortex, neuronal circuits of

Compact bone: ↑ Bone, compact

Compartment: ↑ Cell compartment

Complement: A system of proteins in serum that interact with each other and with antibodies to kill bacteria or cells by ↑ lysis. Antibodies and C. provide a finer degree of selectivity for ↑ phagocytosis than the nonspecific form of phagocytosis.

↑ C 3 receptor

Component: ↑ Cell component

Compound glands: ↑ Exocrine glands, each with a ramified excretory duct, unlike ↑ simple glands.

↑ Glands, classification of

Compound granular corpuscles ("Gitterzellen," granulo-adipose bodies of Glügge): Aggregates of ↑ microglia containing large ↑ lipid droplets, ↑ phagosomes, and ↑ residual bodies found around foci of neuronal ↑ degeneration within ↑ central nervous system.

Conarium: Synonym for ↑ pineal organ

Concanavalin A (Con A): Either of two crystalline globulins extracted from jack bean (Concavalia ensiformis) agglutinating mammalian blood and stimulating ↑ T-lymphocytes.

Conchae nasales* (turbinate bones): Three flaplike mucous folds with a central bony lamella projecting into nasal cavity. Structure: 1) Tunica mucosa (TM): a) epithelium (E) = a ↑ pseudostratified ciliated epithelium; b) lamina propria (LP) = a ↑ loose connective tissue with ↑ nasal glands (NG), solitary ↑ lymphatic nodules (LN), extensive venous plexus (VP) with ↑ swell bodies (SB) (present in middle and inferior C.). Lamina propria is firmly attached to ↑ periosteum (P) of bony platelet (B). Engorgement results from constriction of deep, thick walled veins (V) and dilatation of arteries (A) filling venous plexus via capillaries. (Fig. = middle C.)

↑ Respiratory mucosa, vascularization of

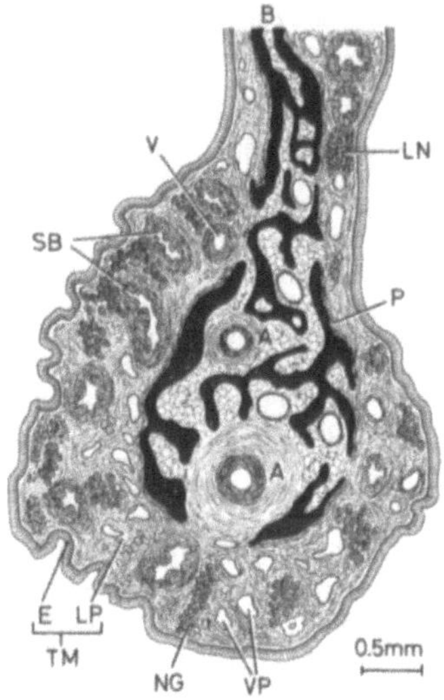

Concretions, of choroid plexus: ↑ Acervuli

Concretions, pineal: ↑ Acervuli

Concretions, prostatic: ↑ Prostatic concretions

Condenser: A glass or ↑ electromagnetic lens serving to focus light rays or electron beam onto the specimen.

Condensing vacuoles (condensing granules, presecretory granules): Oval or spherical ↑ inclusions, 0.1–0.8 µm in diameter (CV), arising from Golgi cisternae by accumulation of a material of low electron density. C. are enclosed in a ↑ unit membrane and occur predominantly near ↑ maturing face of ↑ Golgi apparatus (G). C. contain an immature secretory product, which is condensed up to 25-fold, densified, and transformed into a final secretory product; C. then become mature ↑ secretory granules (SG). (Fig. = ↑ pancreatic acinar cell, rat)

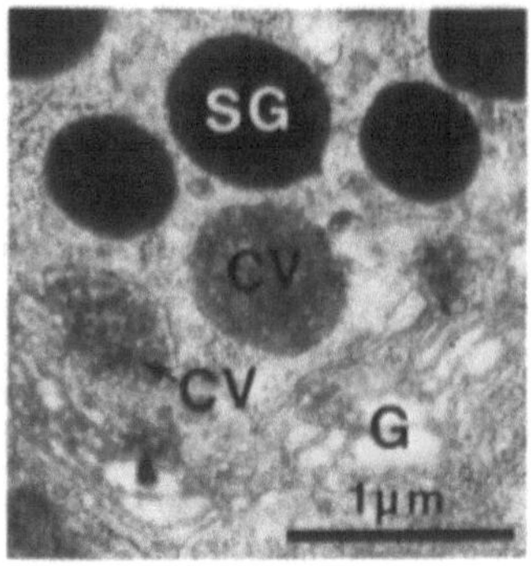

Condensing vesicles: ↑ Unit membrane-delimited spherical structures (CV), 0.5–2 µm in diameter, grouped in supranuclear cytoplasm of proximal tubule cells of ↑ nephron. C. represent concentration sites of reabsorbed proteins from tubular lumen and are considered a kind of lysosome. (Fig. = rat)

↑ Apical pits; ↑ Proximal tubule of nephron, cells of

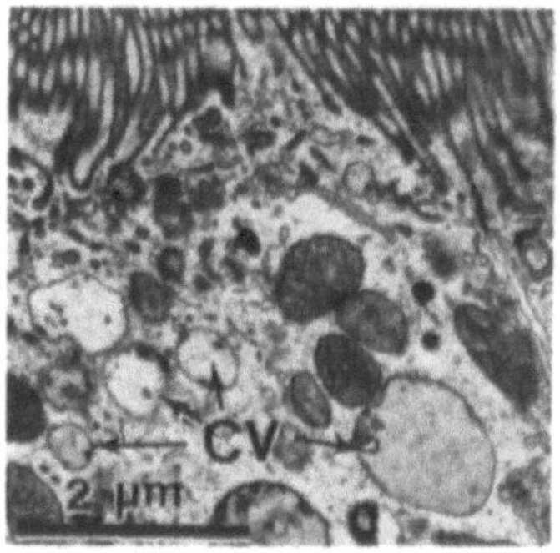

Neiss, W.F.: Histochemistry *77*, 63 (1983)

Conducting system, of heart: ↑ Impulse-conducting system, of heart

Conductive tissue, of heart: ↑ Impulse-conducting system, of heart

Conductivity, of nerve fibers: Ability to transmit excitations along ↑ nerve fibers from one site to another.

Cone cells: ↑ Photoreceptors

Cone cells, bipolar: ↑ Bipolar cells, of retina

Cone pedicle (pes terminalis coni*): A terminal flattened enlargement (CP) of ↑ axon of a cone cell. C. contains a few large mitochondria (M), ↑ synaptic ribbons (arrowheads), and a large number of ↑ synaptic vesicles. ↑ Dendrites of ↑ bipolar and ↑ horizontal cells, as well as axons of the latter, establish synaptic contacts with C. Similar to C. is the ↑ rod spherule. (Fig. = monkey)

↑ Photoreceptors

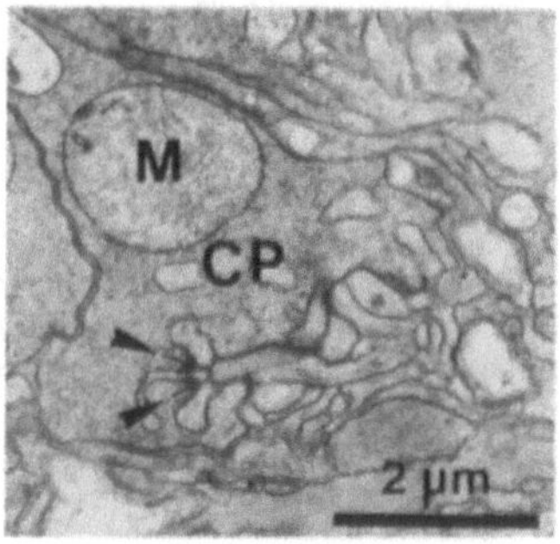

Coni vasculosi (lobuli epididymidis*): About five to ten conical lobules (CV), 10 mm in length, with tips emerging from ↑ mediastinum testis (MT) and bases reaching free surface of epididymal head (H). Each C. is formed by numerous coils of a ↑ ductulus efferens (DE). C. are separated from one another by delicate, vascular connective septa (S). RT = ↑ rete testis

↑ Epididymis

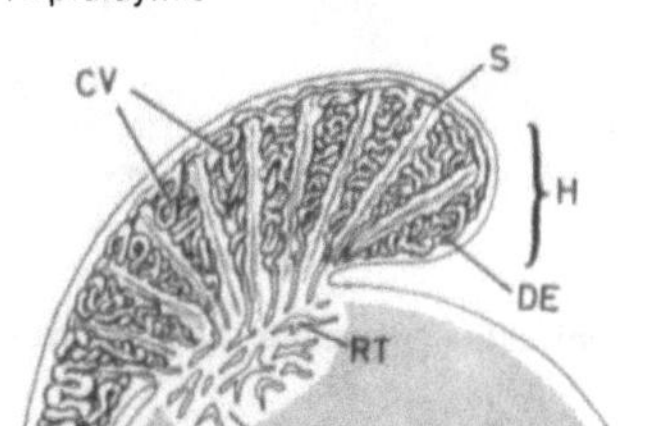

Coniocortex: ↑ Brain cortex, phylogenetic and structural division of

Conjunctiva (tunica conjunctiva*): A thin, transparent mucous membrane covering anterior surface of ↑ sclera (= bulbar conjunctiva, BC) and posterior surface of ↑ eyelids (= palpebral conjunctiva, PC). The two parts are continuous in ↑ fornix (F). 1) Bulbar conjunctiva (tunica conjunctiva bulbi*): a) epithelium (E) = a nonkeratinized, ↑ stratified squamous epithelium, four to five cells thick; in basal cells, ↑ melanin granules may be found; occasionally, presence of ↑ goblet cells. At ↑ limbus (Li), epithelium of C. is continuous with epithelium of ↑ cornea (EC). b) Lamina propria (LP) = a ↑ loose connective tissue with ↑ elastic fibers, ↑ wandering cells, occasional ↑ melanocytes, and arterial, venous, and lymphatic networks ending in closed loops at limbus; very numerous, naked, sensory nerve endings, Krause's ↑ corpuscles, and touch receptors. 2) Palpebral conjunctiva (tunica conjunctiva palpebrarum*): a) Epithelium (E) = stratified columnar epithelium, two to four cells thick, with numerous, very voluminous goblet cells (G) and ↑ lymphocytes (L), which pass between epithelial cells. At lid margins (LM), epithelium is continuous with ↑ epidermis (Ep). b) Lamina propria (LP) = a well-vascularized, delicate loose connective tissue containing numerous lymphocytes, ↑ plasma cells (P), ↑ macrophages, scattered ↑ lymphatic nodules, and diffuse lymphatic infiltrations. Some blood capillaries form glomerular structures of unknown nature.

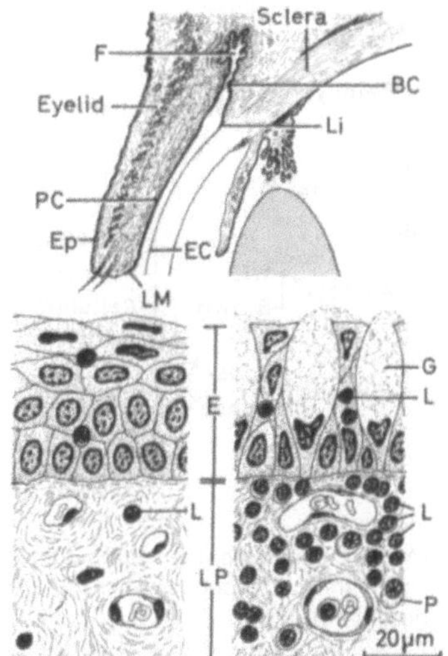

Connective and supporting tissues: A group of ↑ tissues of mesenchymal origin, belonging to one of four fundamental tissues of body. All C. represent loose cell unions with large interspaces containing various fibers and a ground substance. According to consistency of the ground substance, one can distinguish: 1) Fluid ground substance = ↑ connective tissue proper; 2) firm and resilient ground substance = ↑ cartilage; 3) calcified ground substance = ↑ cementum, ↑ bone, and ↑ dentin. Tissues mentioned under 2) and 3) are classified as ↑ supporting tissues.

↑ Connective tissue proper, ground substance of

Connective papillae: Conical protrusions (P) of ↑ loose connective tissue of ↑ lamina propria (LP) between ↑ epithelial ridges (ER) of ↑ stratified squamous nonkeratinized epithelium (E). C. strengthen contact of connective tissue with epithelium. Some C. are ↑ vascular papillae. (Fig. = ↑ esophagus, human)

↑ Dermal papillae

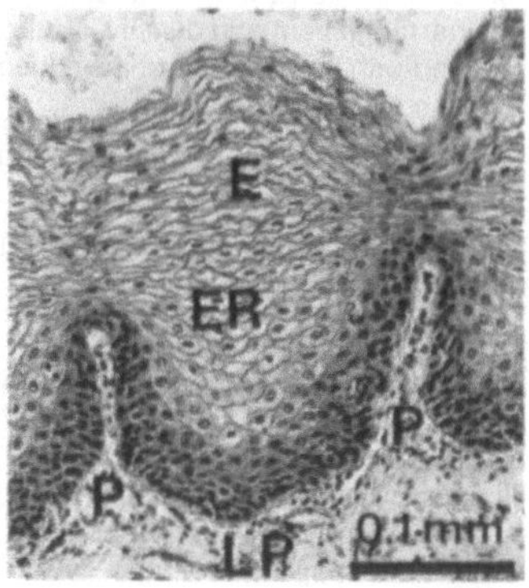

Connective tissue proper (textus connectivus*): A vascularized and innervated, loose, jellylike, cell union with large interspaces containing various fibers and a fluid ground substance.

↑ Connective tissue proper, composition of; ↑ Connective tissue proper, ground substance of

Connective tissue proper, composition of: In general, each type of connective tissue (C) is composed as follows:

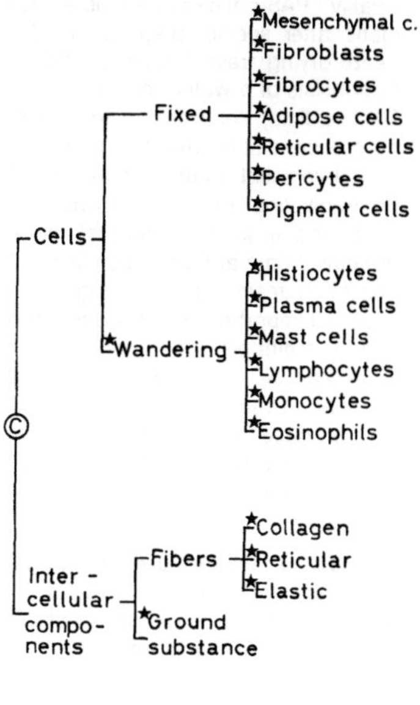

Some authors consider mesenchymal cells and sessile histiocytes to be constituents of mature connective tissue, including neutrophilic ↑ granulocytes in the category of wandering cells. Only certain pigment cells are of ectodermic origin; whereas all other cellular elements originate from ↑ mesenchymal tissue by ↑ differentiation. Different kinds of connective tissue, are result of quantitative variations of cellular or fibrillar elements, as well as variations in their arrangement and density. (Terms marked with an asterisk are described under corresponding letters)

↑ Connective tissue proper, ground substance of

Aterman, K.: Histochem. J. *13*, 341 (1981)

Connective tissue proper, effect of hormones on: ↑ Growth hormone, ↑ testosterone, ↑ estrogens, ↑ TSH, and ↑ thyroxin stimulate development and proliferation of C. ↑ ACTH and ↑ corticoids have opposite effect.

Connective tissue proper, functions of: Mechanical support (fibers), exchange of metabolites between blood and tissue (ground substance), storage of reserve energy material (↑ adipose cells), protection against infection (↑ wandering cells), regeneration after injury (↑ fibroblasts).

Connective tissue proper, ground substance of: An optically transparent, homogenous fluid of variable viscosity in which all components of connective tissue proper bathe. Invisible on histological sections, it appears weakly ↑ PAS-positive and ↑ metachromatic after special preparations (↑ freeze drying, fresh ↑ frozen sections). C. consists of a water-rich phase (= ↑ tissue fluid), relatively limited under normal conditions, and has a colloid-rich, interstitial matrix containing ↑ proteoglycans (mostly ↑ hyaluronic acid) and a small amount of blood proteins. Some authors also classify ↑ basement membranes among C. C. plays an important role in transportation of metabolites between ↑ capillaries and cells; with age, amount of water and hyaluronic acid diminishes. C. is produced by ↑ fibroblasts, ↑ mast cells, and partially by ↑ macrophages and ↑ smooth muscle cells.

Connective tissue proper, metaplasia of: Under certain conditions, ↑ reticular tissue can transform itself into ↑ adipose tissue (transformation of red ↑ bone marrow into yellow ↑ bone marrow).

↑ Metaplasia

Connective tissue proper, regeneration of: After injury, connective tissue proper generally regenerates well, however, this depends on age and vascularization of the tissue in question. ↑ Fibroblasts are stimulated, producing mostly ↑ collagen fibers (very few ↑ elastic fibers); during remodeling, fibroblasts can also remove fibers, since they probably also synthesize ↑ collagenases. Overproduction of collagen fibers leads to formation of fibrous scars.

Connective tissue proper, subdivision of: Depending upon predominance of cell and fiber types, as well as arrangement and density of latter, connective tissue proper (C) can be subdivided as follows:

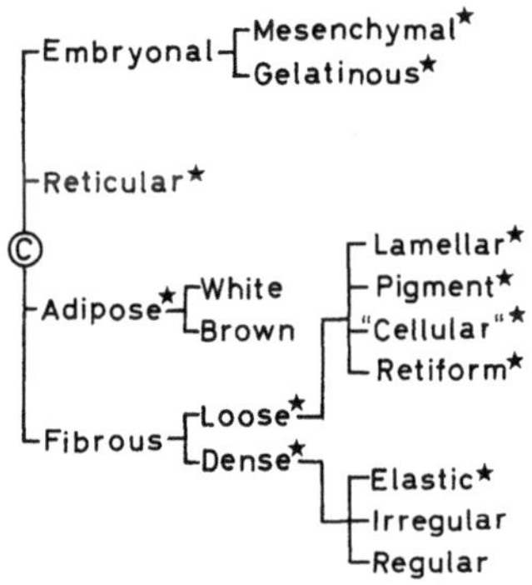

(Asterisks indicate the letters under which the corresponding tissues are described)

Connective tissue proper, undifferentiated cells of: A population of spindle-shaped cells with an elongated nucleus and poorly developed ↑ organelles, found along ↑ capillaries. C. are believed to correspond to ↑ mesenchymal cells persisting to a certain degree even in mature connective tissue proper.

Connexin (gap junction protein): A protein isolated from ↑ nexus.

Finbow, M.E., Shuttleworth, J., Hamilton, A.E., Pitts, J.D.: The EMBO Journal *2*, 1479 (1983); Janssen-Timmen, U., Dermitzel, R., Frixen, U., Leibstein, A., Traub, O., Willecke, K.: The EMBO Journal *2*, 295 (1983); Revel, J.-P., Nicholson, B.J., Yancey, S.B.: Partial Sequence and Turnover of Rat Liver Gap Junction Protein. Cold Spring Harbor Symposia on Quantitative Biology, Vol. 46, Part 2. New York: Cold Spring Harbor Laboratory 1982

Connexons: Hexagonally arranged 9-nm subunits of a ↑ nexus. A fine canal passes through each C., thereby establishing direct communication between two adjacent cells.

Goodenough, D.A., Revel, J.-P.: J. Cell Biol. *45*, 272 (1970)

Constituent: ↑ Cell constituent

Contact inhibition: Capacity of normal cells to cease growth and to aggregate when they meet other normal cells. C. depends upon presence of ↑ sialic acid linked to proteins of ↑ cell membrane.

Contour lines, of Owen (lineae incrementales dentini*): Concentric layers (arrows) of nonhomogenously calcified ↑ dentin (D) occurring as a result of appositional growth. P = pulp cavity. (Fig. = human)

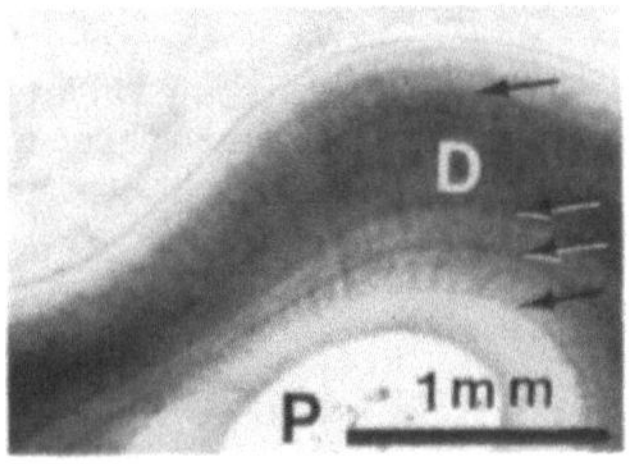

Contractile interstitial cell: ↑ Myofibroblast

Contractile ring: A circle (r) of subplasmalemmal ↑ actin microfilaments (A) mixed with ↑ filamin; C. is oriented transversally to long axis of dividing cell. C. is clearly visible at bottom of cleavage furrow during anaphase and telophase; its contraction leads to ↑ cytokinesis. Mt = microtubules of ↑ spindle apparatus

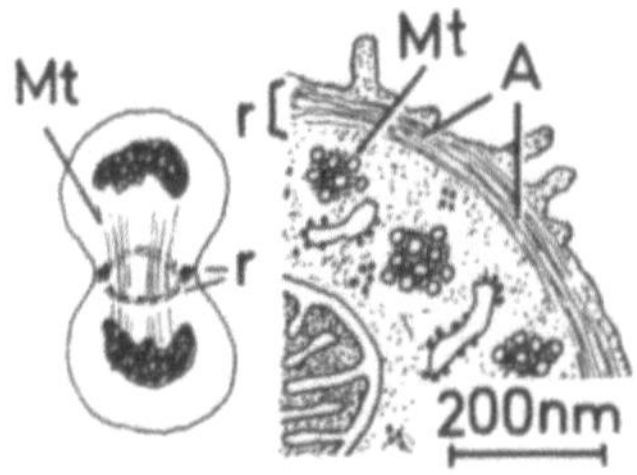

Shroeder, T.E.: Proc. Natl. Acad. Sci. USA *70*, 1688 (1973)

Contraction: A process of shortening of ↑ muscular tissues, various cells, and cell expansions with consumption of energy.

Contraction of smooth muscle cells: ↑ Smooth muscle cells, mechanism of contraction of

Contraction of striated muscle fibers, molecular biology of: ↑ Striated muscle fibers, molecular biology of contraction of

Contraction of striated muscle fibers, morphology of: ↑ Striated muscle fibers, morphology of contraction of

Contrast enhancement: ↑ Contrasting, in transmission electron microscopy; ↑ Sputtering

Contrasting, in transmission electron microscopy: A treatment of ↑ ultrathin sections or particulate material (isolated ↑ collagen microfibrils, viruses, etc.) with salts of heavy metals, such as U, Pb, W, in order to increase scattering of electrons on structures binding these heavy atoms and to enhance contrast of image. 1) Positive C. = specimen binds atoms of heavy metal and appears "stained" on photographic print. 2) Negative C. = specimen does not bind with atoms of heavy metal, which forms deposits around substratum; thus, on photographic print, it occurs unstained, i.e., as a negative.

↑ Sputtering; ↑ "Staining," in transmission electron microscopy

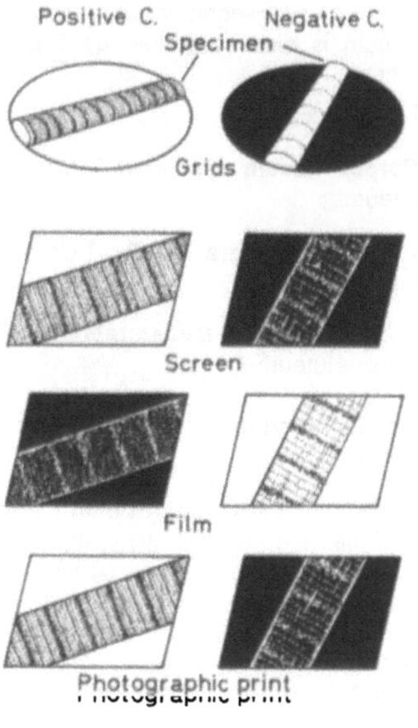

Glauert, A.M. (ed.): Practical Methods in Electron Microscopy. Staining Methods for Sectioned Material. Vol. 5, Part 1. Amsterdam: Elsevier/North Holland 1980

Conus elasticus (membrana fibroelastica infraglottica*): A short tube (CE) beginning with a round opening at inner and superior surface of cricoid cartilage (CC) and ending as a sagittally oriented cleft under tunica mucosa of ↑ vocal cords (VC). C. is composed of predominantly longitudinally oriented ↑ elastic fibers intermingled with ↑ collagen fibers. C. is part of ↑ membrana fibroelastica laryngis. MS = membrana fibroelastica supraglottica; TC = ↑ thyroid cartilage

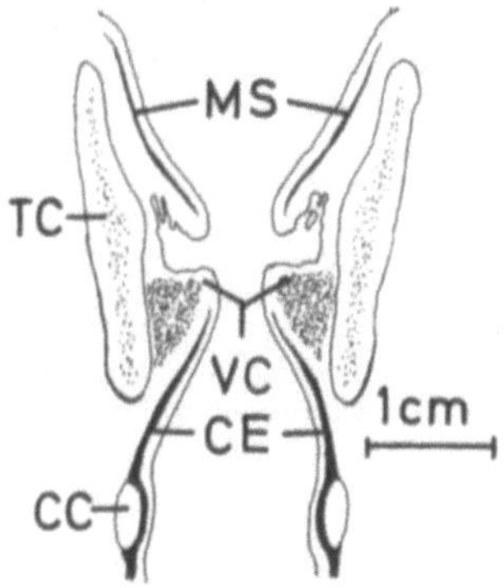

Cords, medullary: ↑ Lymph node, medullary cords of

Cords, of Billroth: ↑ Billroth's cords

Cords, spermatic: ↑ Spermatic cords

Cords, vocal: ↑ Vocal cords

Corium: ↑ Dermis

Cornea*: A convex, transparent part of outer eye coat belonging to ↑ dioptric media of eye (refraction index 1.367), with a thickness of 0.8–0.9 mm in the center and 1.1 mm at the periphery. Structure: 1) Epithelium (anterior corneal epithelium, epithelium anterius*) (Ep) = a ↑ stratified squamous epithelium, about 50 μm thick, composed of five to eight cell layers. Basal cells are connected to basal lamina by numerous ↑ hemidesmosomes. Other cells largely interdigitate by interlacing undulations reinforced by numerous ↑ desmosomes. All the cells contain a large number of ↑ tonofilaments but few ↑ organelles. Superficial cells are fairly flattened and bear ↑ microplicae on their free surface. Epithelium is very rich in sensitive nerve endings; it regenerates well. 2) Bowman's membrane (B) = a condensed outer layer of corneal stroma. 3) Stroma (substantia propria*) (S) = a transparent, specially differentiated ↑ aponeurosis, composed of parallel fibers arranged in many lamellae (L). Fibers of a lamella cross at various angles (right angle in center of C.). Stroma's ground substance contains ↑ chondroitin sulfate and ↑ keratan sulfate, responsible for its transparency. Between lamellae lie stroma's cells, ↑ keratocytes. Numerous nerve endings, but no blood vessels or lymphatics, are present in stroma. 4) Descemet's membrane (lamina limitans posterior*) = a specialized basal lamina (D) of endothelium (En). 5) Endothelium (posterior corneal epithelium, epithelium posterius*, endothelium corneae*) = a ↑ simple squamous epithelium with richly interdigitating, metabolically active cells. Endothelium protects corneal stroma against swelling.

↑ Arcus senilis; ↑ Hassall-Henle bodies

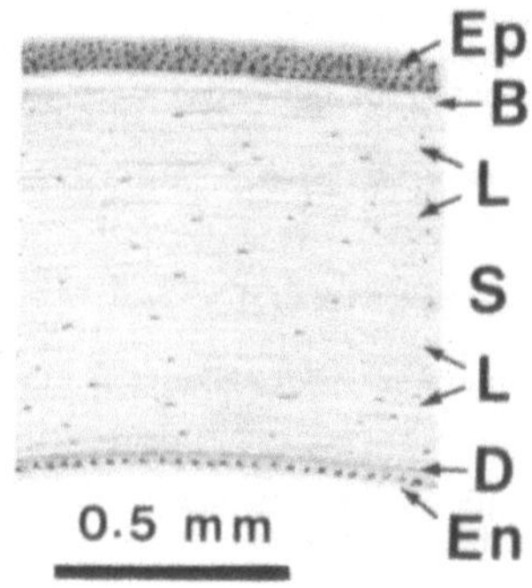

Craig, S.S., Parry, D.A.D.: J. Ultrastruct. Res. *74*, 232 (1981); McDevitt, D.S. (ed.): Cell Biology of the Eye. New York: Academic Press 1982

Cornealocytes: ↑ Keratocytes

Corneoscleral coat: Synonym for ↑ fibrous tunic of ↑ eyeball.

Corniculate cartilage (cartilago corniculatae*): ↑ Elastic cartilage

Cornified cells: ↑ Horny cells

Corona radiata*: Several layers (CR) of radially arranged ↑ follicular cells (FC) nearest oocyte (O) of mature and preovulatory ↑ ovarian follicle. Cells of C. have same morphology as ↑ granulosa cells; those lying on ↑ zona pellucida (ZP) extend long microvilli, which make contact with similar projections of oocyte, predominantly with-

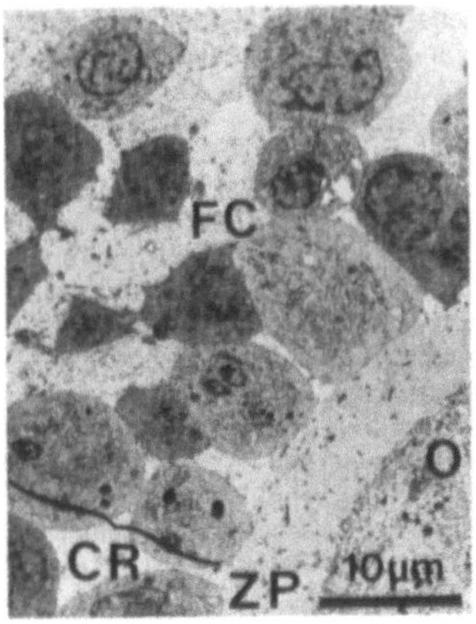

in zona pellucida; some microvilli may, however, directly appose surface of oocyte. After ↑ ovulation, C. accompanies oocyte.

Coronary arteries (arteriae coronariae cordis*): ↑ Muscular arteries arising from ↑ aorta. C. are located in deeper layers of ↑ epicardium and supply ↑ myocardium with blood. Branches of C. are ↑ terminal arteries.

Corpora amylacea: ↑ Prostatic concretions

Corpora arenacea: ↑ Acervuli

Corpora cavernosa clitoridis: ↑ Cavernous bodies, of clitoris

Corpora cavernosa penis: ↑ Cavernous bodies, of penis

Corpora lutea atretica: An inappropriate term given to large degenerating ↑ ovarian follicles.

↑ Ovarian follicles, atretic

Corpus albicans* (corpus fibrosum): A well-delimited, oval white scar (CA) present in deep stroma of ↑ ovary after puberty. C. results from gradual ↑ involution of a ↑ corpus luteum whose epithelium is progressively replaced by ↑ dense connective tissue proliferated from theca externa. At beginning of its formation, C. measures 6–8 mm, but diminishes in course of several weeks to microscopic dimensions. In center of C., a small accumulation of ↑ hemosiderin can sometimes be found.

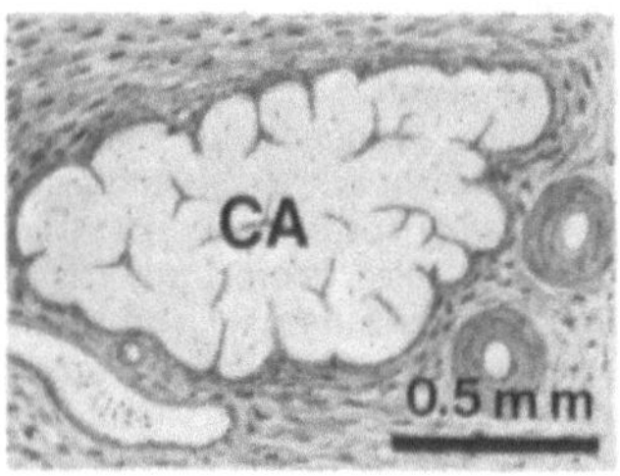

↑ Theca folliculi

Corpus cavernosum recti*: Sum of glomerular arteriovenous ↑ anastomoses within tela submucosa of ↑ anal canal.

Staubesand, J.: Phlebol. Proctol. *1*, 55 (1972)

Corpus ciliare: ↑ Ciliary body

Corpus fibrosum: ↑ Corpus albicans

Corpus luteum* (luteal gland): A yellowish endocrine organ (CL) formed at site of a ruptured mature ↑ ovarian follicle. C. arises after ↑ ovulation, through ↑ differentiation of ↑ granulosa and theca interna cells into ↑ granulosa lutein and ↑ theca lutein cells; simultaneously, connective tissue of theca externa and blood vessels proliferate, forming highly vascular stroma of C. Former antrum (A) of follicle becomes invaded by ↑ loose connective tissue with ↑ macrophages sometimes containing ↑ hemosiderin. Epithelial cells form strands, separated from one another by connective tissue septa (S); granulosa lutein cells (G) are clear; theca lutein cells (T) have a peripheral position and a dark cytoplasm. Many capillaries (C) run between the cells. Two kinds of C.: 1) C. spurium (of menstruation, of ovulation, corpus luteum cyclicum*) = ↑ implantation does not occur; C. undergoes regression in course of 14 days. 2) C. verum (of pregnancy, corpus luteum graviditatis*) = occurs in event of pregnancy. Under influence of ↑ HGC, C. becomes larger (up to 2.5 cm) and persists until 5th–6th month of pregnancy before beginning to involute. Result of regression of both kinds of C. is a ↑ corpus albicans (arrowhead). C. secretes a small amount of ↑ estrogens, but a considerable amount of ↑ progesterone.

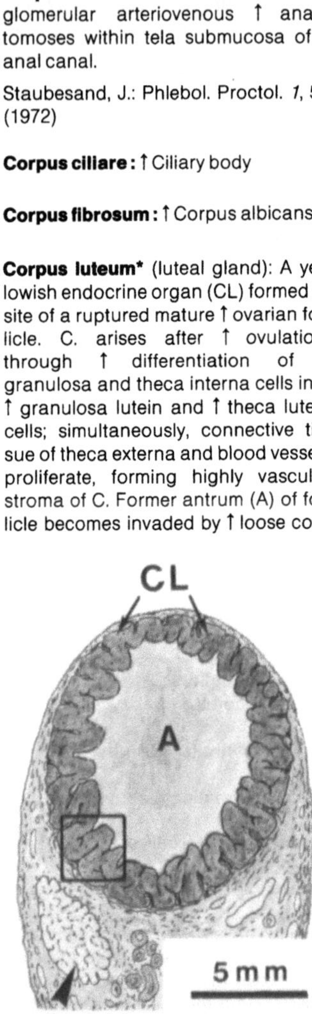

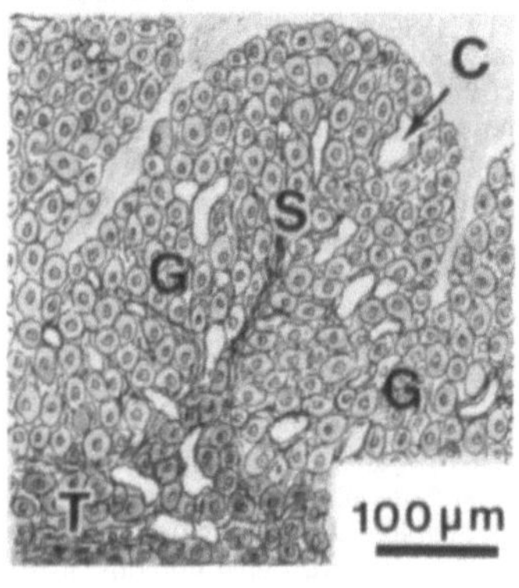

↑ Thecal cells

Booher, C., Enders, A.C., Hendrickx, A.G., Hess, D.L.: Am. J. Anat. *160*, 17 (1981); Channing, C.P., Marsh, I., Sadler, W.A. (eds.): Ovarian Follicular and Corpus Luteum Function. In: Advances in Experimental Medicine and Biology, Vol. 12. New York: Plenum Press (1979); Crisp, T.M., Dessouky, D.A., Denys, F.R.: Am. J. Anat. *127*, 37 (1970); Gemmel, R., Stacy, B.D.: Am. J. Anat. *155*, 1 (1979); Pendergrass, P.B., Principato, R., Reber, M.: J. Submicrosc. Cytol. *13*, 527 (1981)

Corpus luteum, control of formation and secretion of: The ↑ differentiation of ↑ granulosa and theca interna cells into ↑ lutein cells is influenced by ↑ luteinizing hormone. In some animals (rat, mouse), secretion of ↑ corpus luteum is also controlled by ↑ prolactin.

↑ Thecal cells

Corpus luteum cyclicum: ↑ Corpus luteum

Corpus luteum graviditatis: ↑ Corpus luteum

Corpus luteum menstruationis: ↑ Corpus luteum

Corpus, of stomach: ↑ Stomach, corpus of

Corpus, of uterus (corpus uteri*): Part of ↑ uterus above isthmus comprising about two-thirds of nonpregnant organ. Structure: 1) ↑ Endometrium (E) with numerous ↑ uterine glands (UG) surrounded by very cellular endometrial stroma (ES); 2) ↑ myometrium (M)

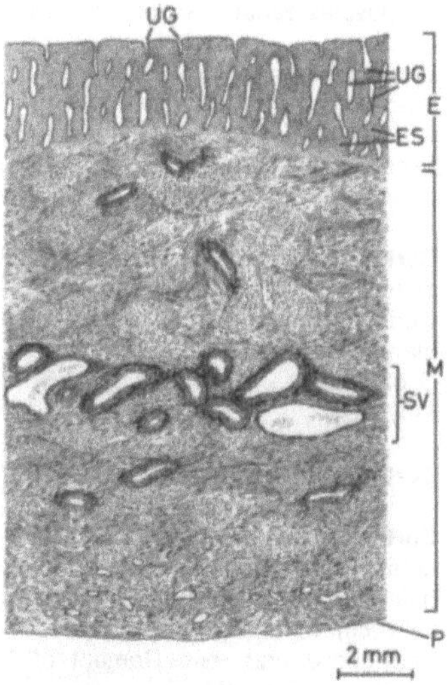

with large blood vessels in stratum vasculare (SV); 3) ↑ perimetrium (P) continuous with ↑ peritoneum lining ↑ broad ligaments; 4) ↑ parametrium only at lateral sides of C. where broad ligaments insert.

Corpus spongiosum: ↑ Spongious body, of male urethra

Corpus ultimobranchiale: ↑ Ultimobranchial bodies

Corpus uteri: ↑ Corpus, of uterus

Corpus vitreum: ↑ Vitreous body

Corpuscles: 1): A collective term for ↑ erythrocytes and ↑ leukocytes. 2) (encapsulated endings, corpusculum nervosum capsulatum*): Well-delimited round or oval structures formed by a lamellated connective tissue capsule around one or more peripheral ↑ nerve endings in contact with either a group of specialized connective cells, ↑ intrafusal muscle fibers, or ↑ tendon fibers. To C. belong sensory receptors in ↑ skin (↑ C. genital, ↑ C., of Golgi-Mazzoni, ↑ C., of Krause, ↑ C., of Meissner, ↑ C., of Merkel, ↑ C., of Ruffini, ↑ C., of Vater-Pacini) and in ↑ skeletal muscles and ↑ tendons (↑ Golgi tendon organ, ↑ neuromuscular spindle)

Watanabe, I., Yamada, E.: Arch. Histol. Jpn. *46*, 173 (1983)

Corpuscles, genital (corpuscle of Dogiel, corpusculum genitale): Small round or spindle-shaped encapsulated organs (CG) located below ↑ epidermis (E) of external genital organs, ↑ nipples, and within and near ↑ cavernous bodies of ↑ penis and ↑ clitoris. C. are surrounded by a lamellar capsule (C); nerve fibers (NF) penetrating C. lose their ↑ myelin sheath and ramify within in form of interlacing glomeruluslike coils. Tactile stimulation of C. provokes vasodilation and filling of cavernous bodies, secretion of ↑ bulbourethral and ↑ Bartholin's glands, and sexual and motor phenomena accompanying ↑ ejaculation and orgasm. (Fig. = silver staining)

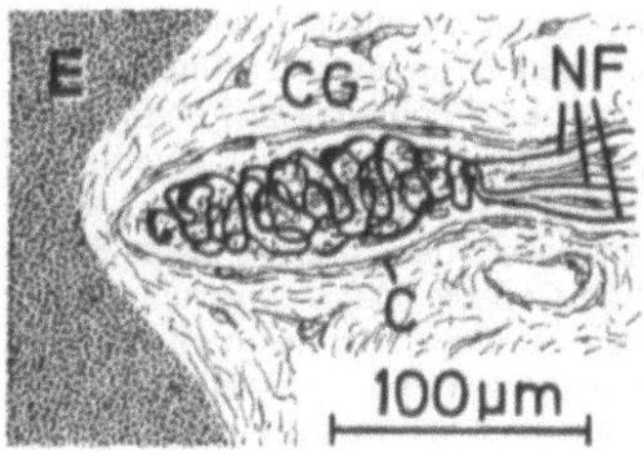

Corpuscles, lingual: A variety of ↑ corpuscles of Krause located within tunica mucosa of ↑ tongue. A lamellated capsule (C) encloses glomeruluslike coils of a branched, single, nonmyelinated axon (A). C. are believed to represent a receptor for cold.

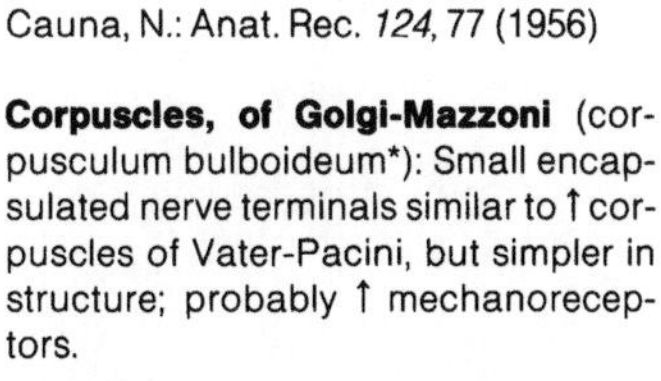

Cauna, N.: Anat. Rec. *124*, 77 (1956)

Corpuscles, of Golgi-Mazzoni (corpusculum bulboideum*): Small encapsulated nerve terminals similar to ↑ corpuscles of Vater-Pacini, but simpler in structure; probably ↑ mechanoreceptors.

Corpuscles, of Krause (Krause's terminal bulbs): A category of small, encapsulated, sensory endings (40–150 µm) situated in ↑ connective papillae just below epithelium (E); frequent in ↑ conjunctiva, ↑ epiglottis, and oral cavity. Structure: After entering bulb, ↑ myelinated fiber (MF) loses its ↑ myelin sheath and branches repeatedly as nonmyelinated axon (A) between specially preformed lamellae (L) of ↑ Schwann's cells (SC). C. is surrounded by a capsule (Ca) composed of flattened capsule cells (CC), representing terminal extension of ↑ perineurium. Capillaries have been observed in some C., which are thought to be re-

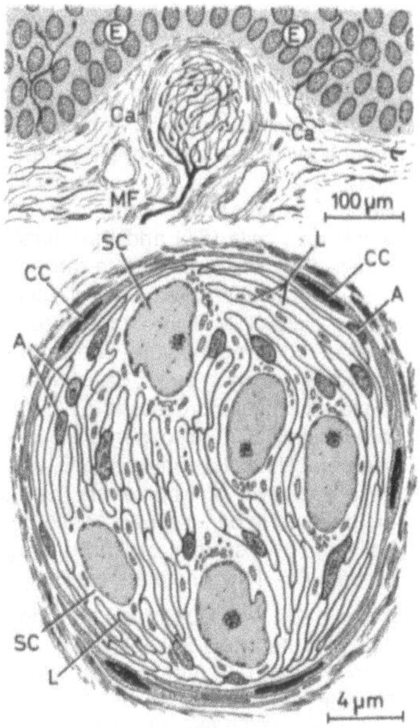

ceptors for cold, possibly also ↑ mechanoreceptors. Very similar to C., but larger, are genital and lingual ↑ corpuscles.

Chouchkov, Ch.: Adv. Anat. Embryol. Cell Biol. *54*, (1978); Halata, Z.: Adv. Anat. Embryol. Cell Biol. *50*, 1 (1975); Kellner, G.: Z. mikrosk.-anat. Forsch. *75*, 130 (1966)

Corpuscles, of Meissner (Meissner's touch corpuscle, corpusculum tactus*): Encapsulated nerve terminals roughly 120 µm long and 70 µm wide, situated at tip of ↑ dermal papillae; par-

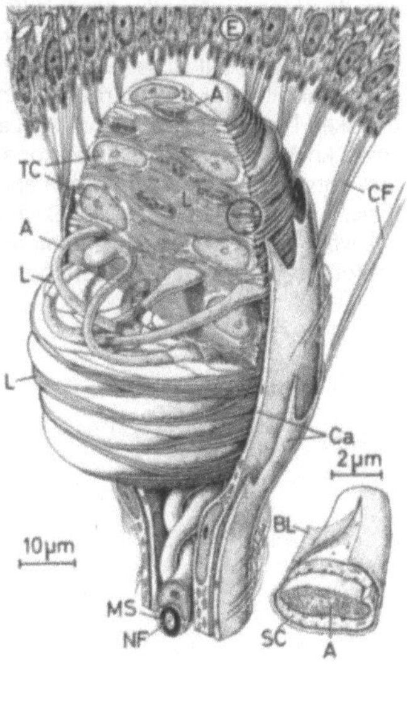

ticularly frequent on finger pads, toes, ↑ lips, edges of ↑ eyelid, and genital ↑ skin. Structure: A lamellar connective tissue capsule (Ca) continuous with ↑ perineurium, incompletely encloses C. After having lost their myelin sheath (MS), several axons (A) penetrate C. Here, accompanied only by ↑ Schwann's cells (SC) and their basal lamina (BL), they constitute a ramified plexus between stacked, pear-shaped, tactile cells (TC). These tactile cells, considered to be modified Schwann's cells, form very numerous, deeply interdigitated, lamellar processes (L). ↑ Collagen fibrils (CF) also penetrate between tactile cells, connecting C. with basal cell layer of ↑ epidermis (E). C. are believed to be ↑ mechanoreceptors activated by distortion of surrounding tissue. NF = ↑ myelinated nerve fiber

Andres, K.H., von Düring, M.: Morphology of Cutaneous Receptors. In: Autrum, H., Jung, R., Loewenstein, W.R., MacKay, D.M., Teuber, H.L. (eds.): Handbook of Sensory Physiology. Vol. 2, Berlin, Heidelberg, New York: Springer-Verlag 1973; Castano, P., Ventura, R.G.: J. Submicrosc. Cytol. *11*, 185 (1979)

Corpuscles, of Merkel (meniscus tactus*): Morphofunctional units composed of a ↑ Merkel cell and the adhering nerve ending.

Saxod, R.: Biol. Cell *37*, 61 (1980)

Corpuscles, of Ruffini: Flattened, encapsulated nerve terminals, 0.25–1.5 mm long, located in ↑ dermis of fingers, toes, soles of feet, capsules of ↑ joints, ↑ ciliary body, ↑ dura mater, and along blood vessels. Structure: A relatively well-developed connective capsule (C) surrounds organ; on entering C., several nerve fibers (NF) lose their ↑ myelin sheath and form a dense arborization of naked ↑ axons (A), each ending in a small, knoblike expansion. Fine axonal branches can leave C. and reach epithelium and/or blood vessels. C. are believed to be principally thermoreceptors (cold?), but also ↑ mechanoreceptors involved in proprioception. (Fig. = silver staining)

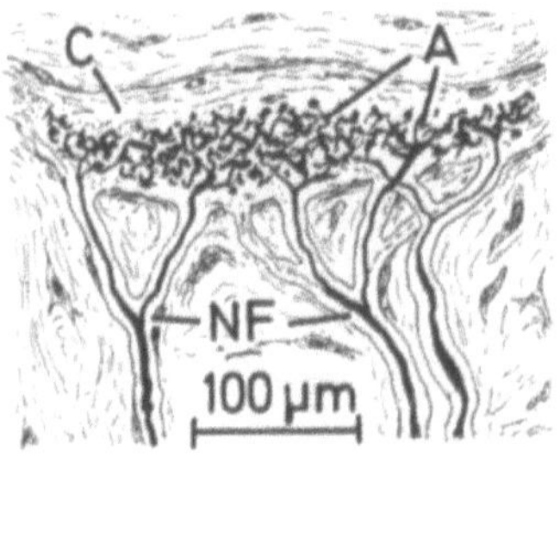

↑ Proprioceptors

Halata, Z., Munger, B.L.: Cell Tissue Res. *219*, 437 (1981)

Corpuscles, of thymus: ↑ Hassall's bodies

Corpuscles, of Vater-Pacini (corpusculum lamellosum*): Firm oval bodies up to 5 mm in diameter, located in ↑ dermis, under tunicae mucosae, in ↑ cornea, ↑ conjunctiva, ↑ heart, ↑ mesenteries, ↑ pancreas, ↑ peritoneum, near large blood vessels, and along arteriovenous ↑ anastomoses. Structure: Capsule (Ca) is continuous with ↑ perineurium (P); outer core (OC), or bulbus externus, consists of 10–60 concentric lamellae (L) formed by flattened ↑ fibrocytes (F). Between lamellae are a lymphlike fluid, ↑ collagen microfibrils (Mf), and some capillaries (Cap). Upon entering C., a single nerve fiber (NF) loses its ↑ myelin sheath (MS), continuing as an axon (A) surrounded only by enormously ramified and interdigitating ↑ Schwann's cells (SC), which form inner core (IC), or bulbus internus. Axon gives off some side branches and ends as a bulbous expansion (B). Each mechanical displacement of lamellae provokes a depolarization wave in axon; C. are thought to represent receptors for pressure and vibration.

↑ Mechanoreceptors

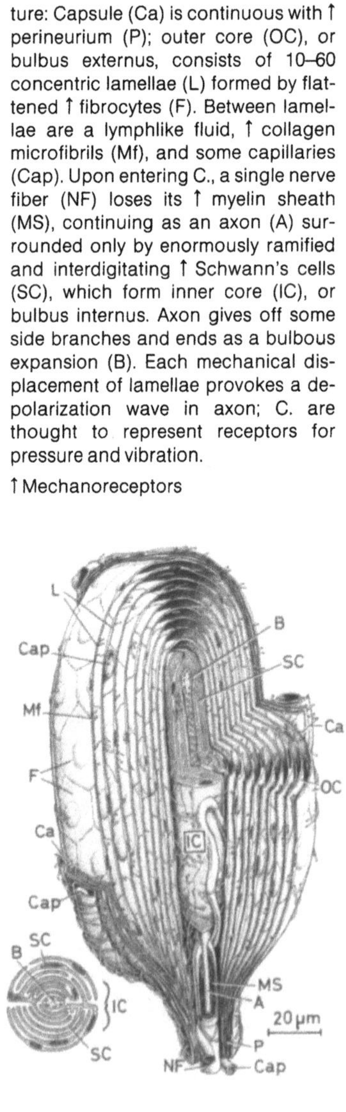

Ide, C., Saito, T.:Acta Histochem. Cytochem. *13*, 298 (1980)

Corpuscles, renal: ↑ Renal corpuscle

Corpuscles, salivary: ↑ Salivary corpuscles

Corpuscula thymica: ↑ Hassall's bodies

Cortex: The outer portion of an organ, or its part, as distinct from the inner or medullary portion (e.g., C. of ↑ adrenal gland; C. of ↑ brain; ↑ C. of hair shaft; C. of ↑ kidney; C. of ↑ lymph node; C. of ↑ thymus).

Cortex, of brain: ↑ Brain cortex

Cortex, of hair shaft (cortex pili*): A keratinized central structure (C) of ↑ hair shaft. C. is formed by ↑ cortical cells (cc), which become fully keratinized above hair root. Content of ↑ melanin granules (M) in C. determines color of hair. Presence of air in C. gives hair a gray to white color. C. is always enveloped by ↑ cuticle (Cu); presence of ↑ medulla in C. is not constant. (Fig. = mongolian gerbil)

↑ Hair bulb; ↑ Matrix cells, of hair

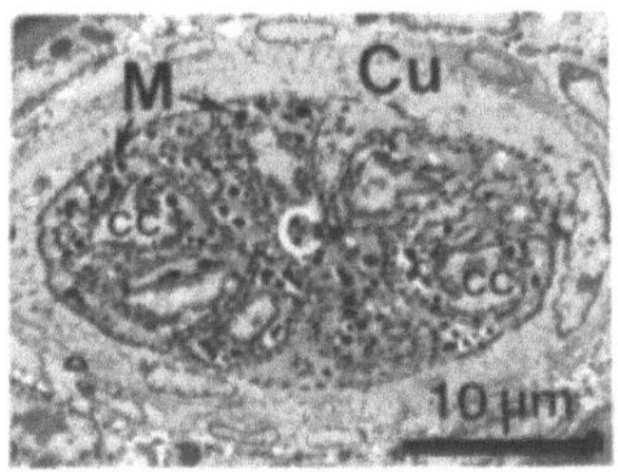

Riggott, J., M., Wyatt, E.H.: J. Anat. *130*, 121 (1980)

Cortex, of kidney: ↑ Kidney; ↑ Renal cortex

Cortex, of lymph node (cortex*): Peripheral, strongly stained ↑ lymphatic tissue (C) of a ↑ lymph node. 1) Outer cortex (OC) = ↑ lymphatic nodules (LN) and internodular diffuse lymphatic tissue (I). 2) Inner cortex (IC) (deep cortex, paracortical area, paracortical zone, tertiary cortex, paracortex*) = a diffuse lymphatic tissue situated between lymphatic nodules and medulla (M). Inner cortex continues without clear demarcation into ↑ medullary cords (MC). Whereas lymphatic nodules consist predominantly of ↑ B-lymphocytes (bone marrow-dependent zone), inner cortex consists chiefly of ↑ T-lymphocytes (thymus-dependent zone). Occasional agglomerations of T-lymphocytes in

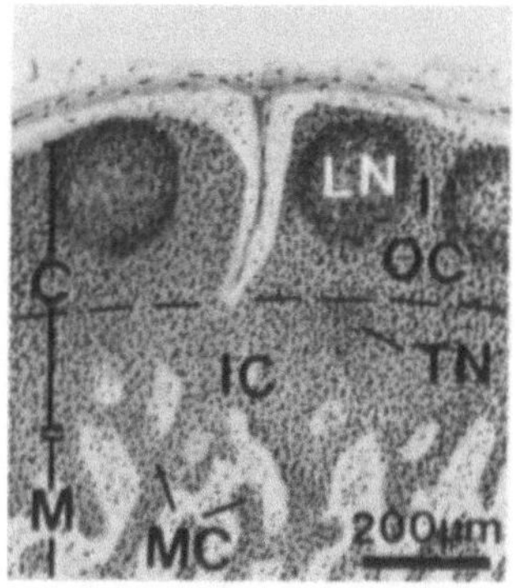

inner cortex, devoid of germinal centers, are termed tertiary nodules (TN). Numerous ↑ postcapillary venules run through inner cortex.

Bélisle, C., Sainte-Marie, G.: Anat. Rec. *201*, 553 (1981)

Cortical cells, of hair shaft: Concentrically arranged fusiform cells (CC) forming ↑ cortex of ↑ hair shaft. Cytoplasm of younger cells (top) contains many peripheral ↑ tonofibrils (Tf) and scattered ↑ melanin granules (M). Older cells (bottom) are almost completely filled with hard ↑ keratin (K). Between keratin masses, free ribosomes (r) are present. Nucleus disappears in cells situated above root of hair. C. are held together by an amorphous intercellular substance. C. are formed from ↑ matrix cells of hair. Cu = ↑ cuticle of hair shaft. (Figs. = mongolian gerbil)

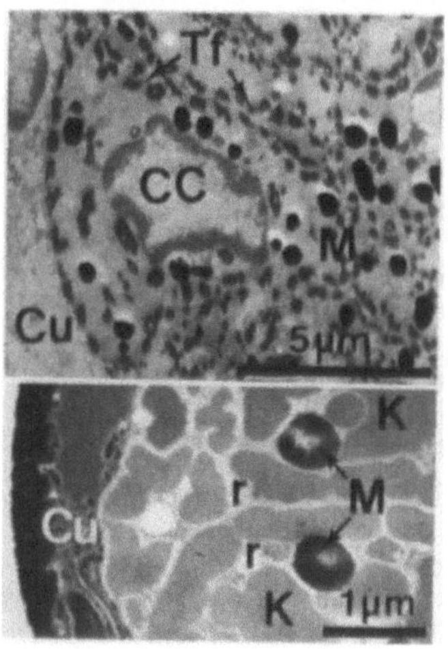

Muto, H., Ozeki, N., Yoshioka, I.: Acta Anat. *109*, 13 (1981), and Acta Dermatol. *76*, 101 (1981)

Cortical collecting tubules, of kidney: All arched ↑ collecting tubules and portion of straight collecting tubules situated within ↑ medullary rays.

↑ Collecting tubules, of kidney

Cortical cytoplasm: ↑ Ectoplasm

Cortical granules, of oocytes (granula corticalia*): Spherical, ↑ unit membrane-bound ↑ inclusions, 0.1–0.3 µm in diameter, with electron-dense, fine granular, homogenous content (CG). Synthesis of C. is initiated in cisternae of rough endoplasmic reticulum and terminated in Golgi apparatus (G). C. contain proteins and ↑ proteoglycans. Function of C. is not known; it is believed that they are released from egg during ↑ acrosome reaction to prevent ↑ polyspermy. (Fig. = rat)

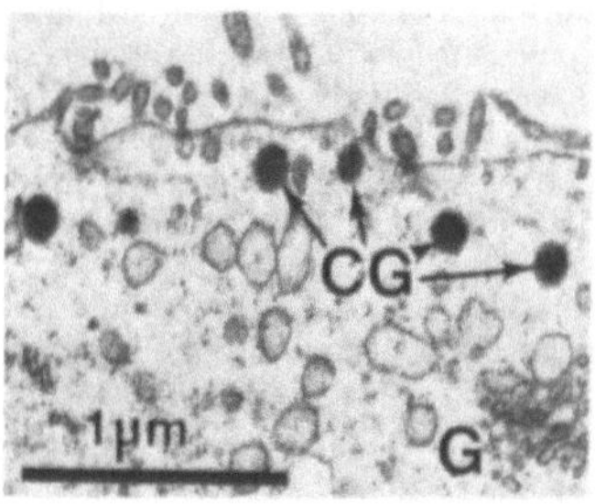

Gulyas, B.J.: Int. Rev. Cytol. *63*, 357 (1980); Guraya, S.S.: Int. Rev. Cytol. *78*, 257 (1982)

Corticoids: A collective term for numerous hormones produced and released by adrenal cortex: ↑ Mineralocorticoids (zona glomerulosa), ↑ glucocorticoids (zona fasciculata), and weak ↑ androgenic hormones (mainly zona reticularis).

↑ Adrenal glands, cortex of

Corticoliberin: ↑ Adrenocorticotropic hormone-releasing factor; ↑ Releasing factors

Corticosterone: ↑ Corticoids; ↑ Glucocorticoids

Corticotropes (adrenocorticotropes, ACTH-cells, ACTH/LPH cells, β_1-cells, corticotropic cells, cellula corticotropica*): Polygonal or irregularly stellate, PAS-negative cells of ↑ adenohypophysis. Nucleus is eccentric, with some deep invaginations of ↑ nuclear envelope and a small nucleolus. Abundant and electron-lucent cytoplasm contains spherical, elongated mitochondria, well-developed, multiple ↑ Golgi fields (G) associated with ↑ secretory

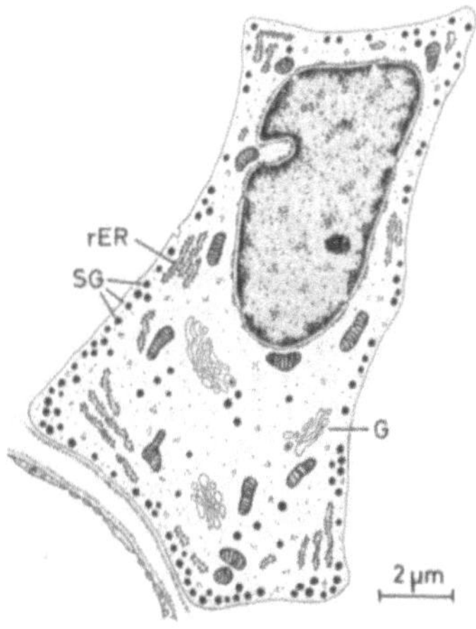

granules (SG), sparse, short, and flattened rough endoplasmic cisternae (rER), a few free ribosomes, and occasional lysosomes. Relatively less numerous ↑ secretory granules (about 200–250 nm in diameter) mainly occupy peripheral cytoplasm. Since they cannot be resolved with the light microscope, C. are classified with ↑ chromophobes. According to other analyses, C. belong to ↑ basophils. It is believed that they produce ↑ ACTH, ↑ lipotropins, ↑ melanocyte-stimulating hormone (βMSH), and a substance similar to ↑ endorphins.

↑ Precursor cells, of adenohypophysis

Girod, G.: Bull. Assoc. Anat. (Nancy) *62*, 417 (1977); Larsson, L.-I.: J. Histochem. Cytochem. *28*, 133 (1980); Li, J.Y., Dubois, M.P., Dubois, P.M.: Cell Tissue Res. *204*, 37 (1979); Yoshimura, F., Nogami, H.: Cell Tissue Res. *219*, 221 (1981)

Corticotropin: ↑ Adrenocorticotropic hormone

Corticotropin-releasing factor: ↑ Adrenocorticotropic hormone-releasing factor; ↑ Releasing factors

Corti's organ (organum spirale*): An epithelial ridge of highly specialized cells situated on ↑ basilar membrane (BM) of ↑ cochlear duct (CD). Structure: 1) Supporting cells = inner (IP) and outer ↑ pillars (OP), inner (IPh), and outer ↑ phalangeal cells (OPh), ↑ Hensen's cells (HC), ↑ Böttcher's cells (BC), ↑ Claudius' cells (CC). Corti's tunnel (CT) is delimited by pillars; ↑ Nuel's spaces (NS) are situated between outer pillars and innermost of outer phalangeal cells. 2) Sensory cells = inner (IH) and outer ↑ hair cells (OH) connected with nerve fibers (NF)

= dendrites of ↑ spiral ganglion ↑ bipolar cells. 3) ↑ Spiral limbus (LS), a part of C. lying on ↑ lamina spiralis ossea (SL); with its vestibular (VL) and tympanic lips (TL), limbus delimits internal spiral tunnel (IST). Outer spiral tunnel, or sulcus (OSS), is located between ↑ spiral prominence and C. 4) ↑ Tectorial membrane (MT), as a product of ↑ interdental cells (IC), emerges from vestibular lip. 5) ↑ Basilar membrane (BM). C. is the specific auditory receptor. RM = ↑ Reissner's membrane

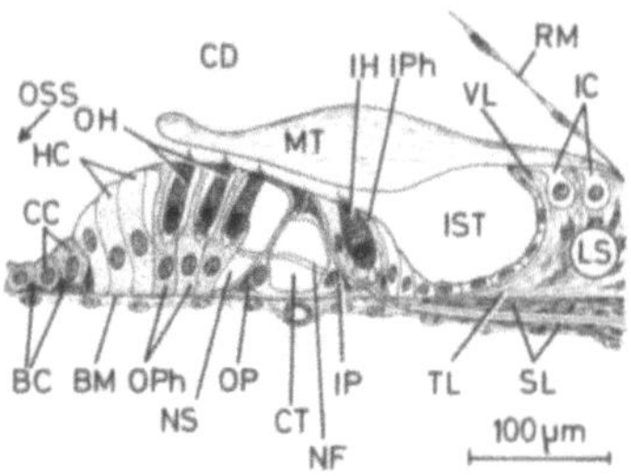

Fujimoto, S., Yamamoto, K., Hayabuchi, I., Yoshizuka, M.: Arch. Histol. Jpn *44*, 223 (1981); Harada, Y.: Biomed. Res. *2*, Suppl. 391 (1981); Saito, K.: J. Ultrastruct. Res. *71*, 222 (1980)

Corti's organ, innervation of: Very complicated and not fully explored. 1) Afferent fibers: a) Acoustic fibers (AF) = ↑ dendrites of ↑ bipolar cells (BC) of ↑ spiral ganglion, radiating in parallel bundles to nearest inner (IHC) and outer ↑ hair cells (OHC). b) Spiral fibers (SF) = dendrites of bipolar cells less numerous and thicker than those of acoustic fibers, probably making ↑ synapses only with outer hair cells; these fibers turn sharply and follow a spiral course, establishing contacts with other outer hair cells. 2) Efferent fibers (EF) = ↑ axons of neurons situated in superior olivary nucleus (SON) and belonging to olivo-cochlear tract. These fibers probably inhibit activity of hair cells. 3) Sympathetic fibers (SyF) = axons of neurons of superior cervical ganglion (SGC) for innervation of blood vessels (BV) in Corti's organ, spiral ganglion, and ↑ stria vascularis. (Fig. modified after Spoendlin 1966)

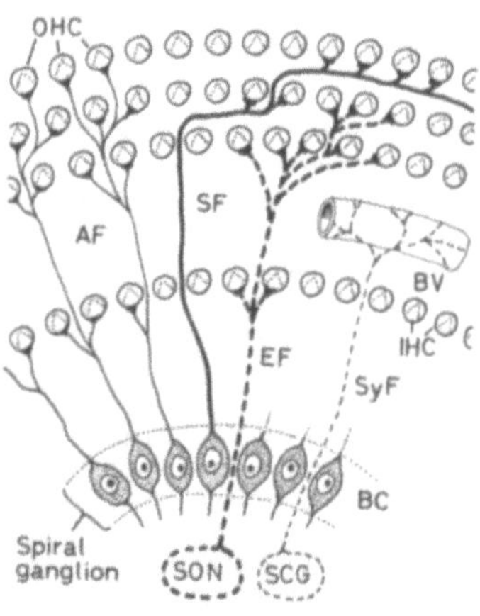

Bredberg, G.: Biomed. Res. *2*, Suppl. 403 (1981); Jones, D.G., Eslami, H.: Cell Tissue Res. *231*, 533 (1983); Liberman, M.C.: Hear Res. *3*, 189 (1980); Omata, T., Schaetzle, W.: Arch. Otorhinolaryngol. *229*, 175 (1980); Speondlin, H.: The Organization of the Cochlear Receptor. Adv Otorhinolaryngol. Vol. 13. Basle: Karger 1966

Corti's tunnel (inner tunnel, cuniculum internum*): A triangular endolymphatic space (C) delimited by ↑ pillars (P) and ↑ basilar membrane (BM). C. extends along length of ↑ ductus cochlearis and communicates through clefts between pillars with ↑ Nuel's spaces (N). Through C. run nerve fibers in contact with outer ↑ hair cells. TM = ↑ tectorial membrane. (Fig. = mongolian gerbil)

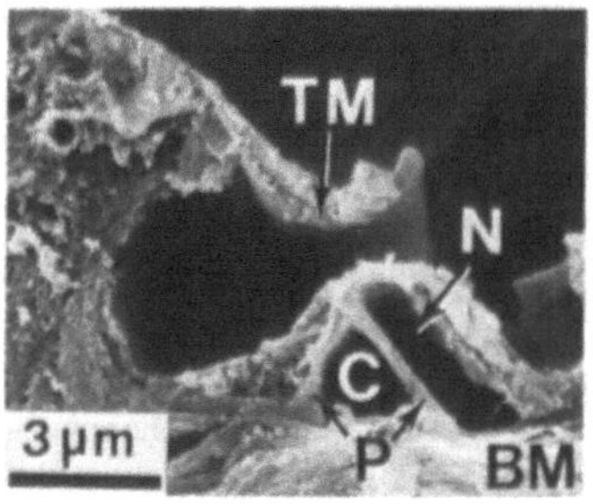

Cortisol: ↑ Corticoids; ↑ Glucocorticoids

Costal cartilage: ↑ Cartilage, hyaline

Costameres: Transverse, riblike bands at ↑ sarcolemma (S) encircling ↑ cardiac and ↑ skeletal muscle fibers perpendicular to their longitudinal axes. C. contain ↑ vinculin and represent regions of physical connection between sarcolemma and underlying ↑ I-bands of ↑ myofibrils (Mf). (Fig. modified after Pardo et al. 1983)

Pardo, J.V., D'Angelo Siciliano, J., Craig, S.: J. Cell Biol. *97*, 1081 (1983)

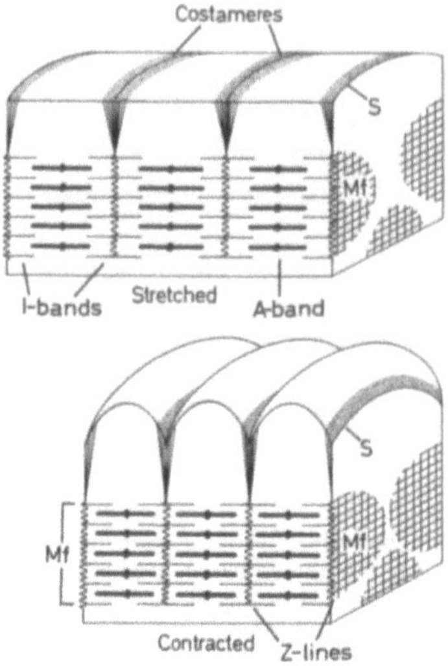

Cotyledon (fetal lobule, cotyledo*): A cup-shaped compartment of fetal placenta formed by one stem villus and all its branches; each C. is incompletely surrounded by decidual septa and provided with a separate blood supply.

↑ Placenta, full-term, structure of

Cover dentin: ↑ Dentin, cover

Coverslip (cover glass): A very thin piece of glass (0.17–0.18 mm) for covering stained sections.

Cowper's glands: ↑ Bulbourethral glands

CPD: ↑ Critical point drying

Craniosacral division (craniosacral nerves): Synonym for ↑ parasympathetic nervous system.

Crenation, of erythrocyte: ↑ Erythrocyte, crenated

Cresyl violet: A dye used for ↑ supravital staining of ↑ reticulocytes and ↑ Nissl bodies on histological sections.

↑ Brilliant cresyl blue

Cretinism: Infantile hypothyroidism which persists into adulthood. C. may originate from thyroid agenesis or inadequate maternal intake of iodine during gestation; it results in physical and mental underdevelopment.

Crevice, gingival: ↑ Gingival crevice

CRF: ↑ Adrenocorticotropic hormone-releasing factor; ↑ Releasing factors

CRH: ↑ Adrenocorticotropic hormone-releasing factor; ↑ Releasing factors

Cricoid cartilage (cartilago cricoidea*): A hyaline ↑ cartilage of ↑ larynx.

"Cri-du-chat" syndrome: ↑ Deletion

Crinophagic vacuole: ↑ Crinophagy

Crinophagy: A kind of ↑ autophagy during which an excess of ↑ secretory granules (S) is inactivated and degraded within an ↑ autophagosome (A) (crinophagic vacuole). (Fig. = ↑ neurosecretory axon, rat)

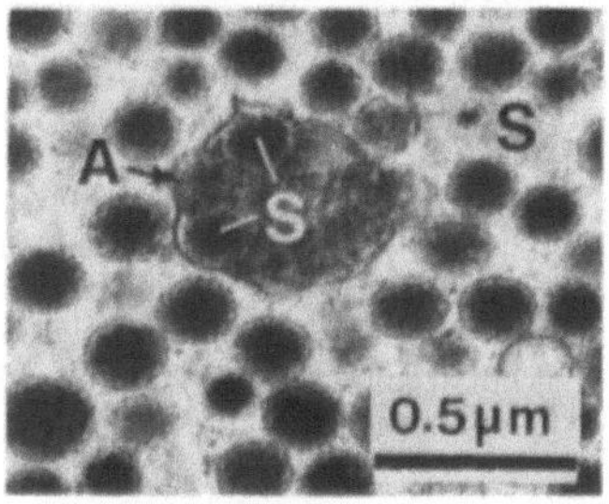

Crista ampullaris*: A small, ridgelike sensory organ (CA) located in ↑ ampulla (A) of each ↑ semicircular duct, perpendicular to its axis. Structure: 1) ↑ Neuroepithelium (E) = ↑ vestibular cells type I and II and ↑ supporting cells. Sensory hairs of vestibular cells penetrate ↑ cupula (C) lying above C. 2) Lamina propria (LP) = a well-vascularized, ↑ loose connective tissue attached to bone (B). Numerous ↑ myelinated nerve fibers belonging to vestibular nerve traverse this layer. C. responds to angular acceleration in plane of canals when cupula with its sensory hairs is displaced by ↑ endolymph. (Fig. = guinea pig)

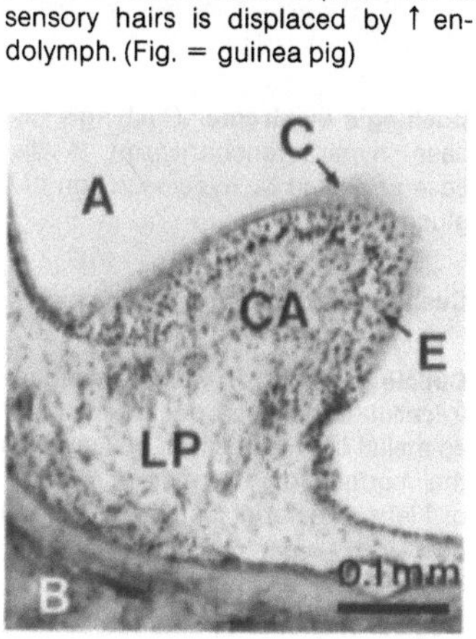

Harada, Y.: Biomed. Res. *2*, Suppl. 391 (1981); Kessel, R.G., Kardon, R.H.: Scan. Electron Microsc. *3*, 967 (1979); Parker, D.E.:Sci. Am. *243/5*, 98 (1980)

Cristae cutis: ↑ Epidermal ridges

Cristae, mitochondrial (crista mitochondrialis*): Folds (C) of inner ↑ mitochondrial membrane (IM), of various length and about 20 nm wide, projecting in form of incomplete and frequently perforated plates into ↑ mitochondrial matrix (MM). C. delimit an intracristal space (IS) continuous with ↑ outer chamber (OC). C. contain ↑ elementary particles involved in oxidation and phosphorylation. Thus, aim of C. is to amplify contact surface between matrix and inner mitochondrial membrane in order to increase energy production. In highly active mitochondria, C. are numerous and densely packed (↑ adipose cells, brown; ↑ cardiac muscle cells, etc.).

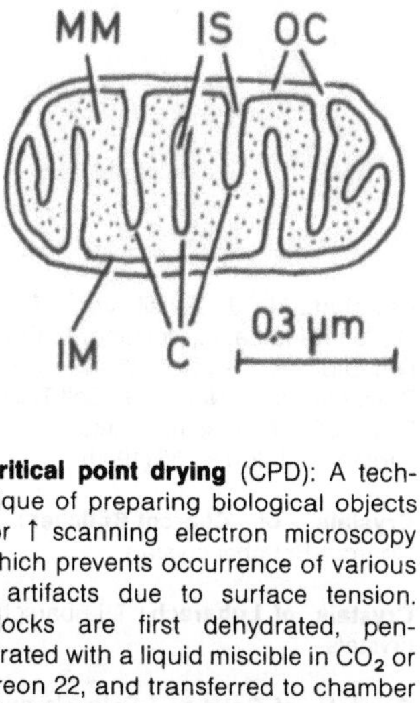

Critical point drying (CPD): A technique of preparing biological objects for ↑ scanning electron microscopy which prevents occurrence of various ↑ artifacts due to surface tension. Blocks are first dehydrated, penetrated with a liquid miscible in CO_2 or Freon 22, and transferred to chamber of critical point apparatus. Chamber is then heated to a critical temperature, causing pressure in chamber to increase. Critical point is reached at 73 atm and 31 °C (for CO_2), and both gaseous and liquid phases exist without an interface meniscus, i.e., without surface tension.

↑ Dehydration

Tanaka, K., Iino, A.: Stain Technol. *49*, 203 (1974); Turner, R.H., Green, C.D.: J. Microsc. *97*, 357 (1973)

Crossing over (crossover, decussatio*): The exchange of corresponding segments with their ↑ genes between ↑ chromatids (Chr) of a ↑ tetrad during prophase I of ↑ meiosis. Ch = ↑ chiasma, C = ↑ centromere

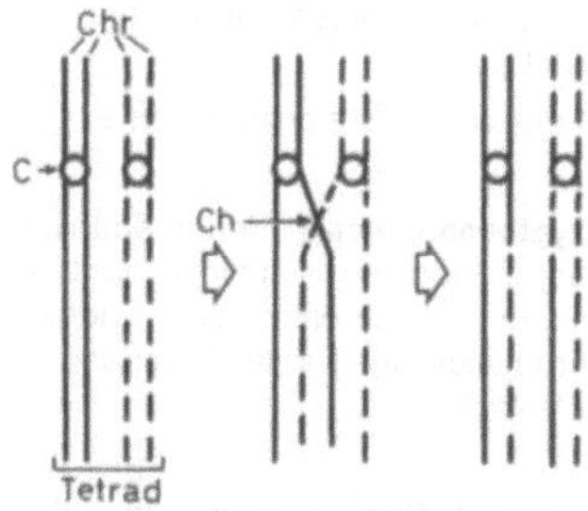

Crown, of tooth (corona dentis*): Part of ↑ tooth covered with ↑ enamel (E); also called anatomical C. Clinical C. is part projecting above ↑ gingiva (G).

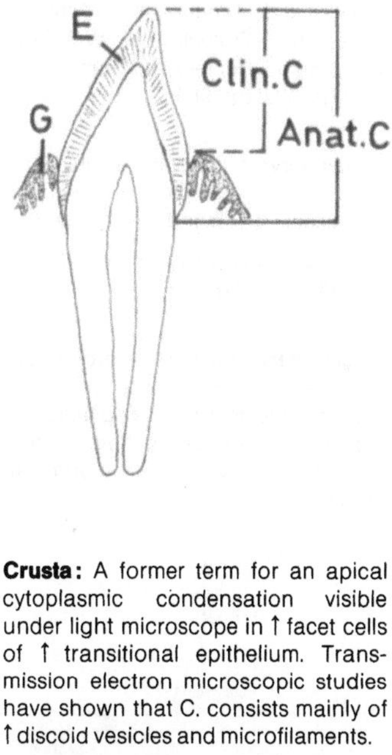

Crusta: A former term for an apical cytoplasmic condensation visible under light microscope in ↑ facet cells of ↑ transitional epithelium. Transmission electron microscopic studies have shown that C. consists mainly of ↑ discoid vesicles and microfilaments.

Teutsch, H.F.: Verh. Anat. Ges. *71*, 1223 (1977)

Cryostat: A refrigeratorlike apparatus containing in its chamber (Ch) a mechanical ↑ microtome (M) and kept, together with frozen tissue blocks (B), at low temperature (−18° to −30 °C). ↑ Sectioning is carried out at same low

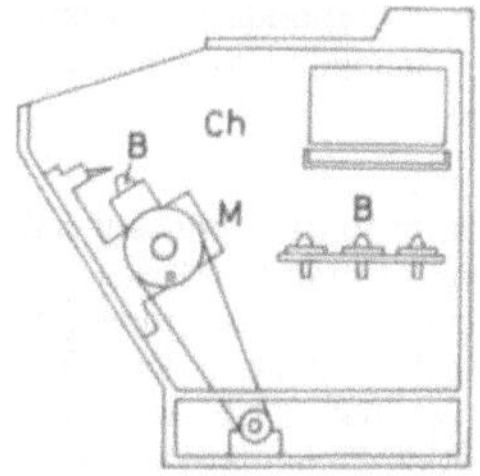

temperature, thereby giving the best results.

↑ Freezing; ↑ Freezing microtome

Cryptoendomitosis (endoreduplication): ↑ Endomitosis at molecular level; neither ↑ chromosomes nor ↑ chromatids are visible. C. leads to ↑ polyploidy.

Cryptorchidism (cryptorchism): Failure of a ↑ testis to descend into ↑ scrotum.

Hadžiselimović, F.: Adv. Anat. Embryol. Cell Biol. *53*, 1 (1977); Hadžiselimović, F.: Cryptorchidism. Management and Implications. With Contributions by Cromie, W.J., Hinman, F., Höcht, B., Kogan, S.J., Trulock, T.S., Woodard, J.R. Berlin, Heidelberg, New York: Springer-Verlag 1983

Crypts (crypta*): A pitlike, obviously nonglandular, infolding of a surface epithelium into an underlying tissue

↑ Crypts, of iris; ↑ Crypts, of tonsils, ↑ Lieberkühn's crypts

Crypts, of iris: Fusiform and/or irregular excavations (C) of anterior surface of ↑ iris, sometimes extending deep into iridic stroma. ↑ Aqueous humor circulates through C. (Fig. = mongolian gerbil)

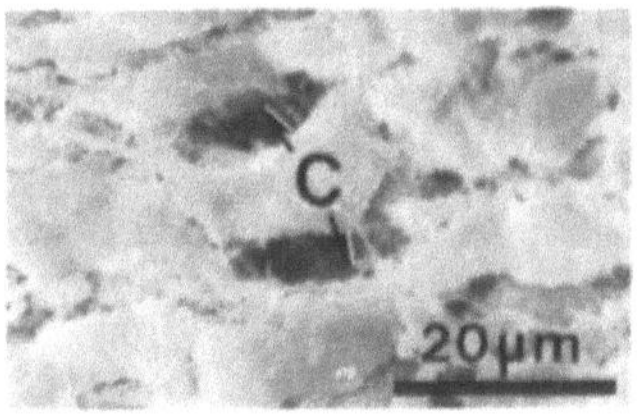

Dietrich, D.E., Witmer, R., Franz, H.E.: Albrecht v. Graefes Arch. Klin. Exp. Ophthalmol. *182*, 321 (1971)

Crypts of Lieberkühn, of colon and rectum: ↑ Lieberkühn's crypts, of colon and rectum

Crypts of Lieberkühn, of small intestine: ↑ Lieberkühn's crypts, of small intestine

Crypts, of tonsils, primary and secondary (cryptae tonsillares primariae et secondariae*): Deep invaginations (C) of nonkeratinized ↑ stratified squamous epithelium (E) extending into underlying ↑ lymphatic tissue (LT) (= primary C.). In the case of lingual ↑ tonsil (fig.), C. are rarely branched, but those of palatine ↑ tonsils always are (= secondary C.). C. are surrounded by lymphatic tissue containing ↑ lymphatic nodules (LF); along C., epithelium loses its typical structure at many sites because of an intense ↑ diapedesis (D) of ↑ lymphocytes. C. contain detritus (asterisk) composed of degenerating lymphocytes, desquamated epithelial cells, and bacteria. In lingual tonsil, ducts (arrowhead) of ↑ mucous glands (MG) open into bottom of C.; this is not the case in palatine tonsil, where a long retention of detritus may provoke infections and calcification of detritus.

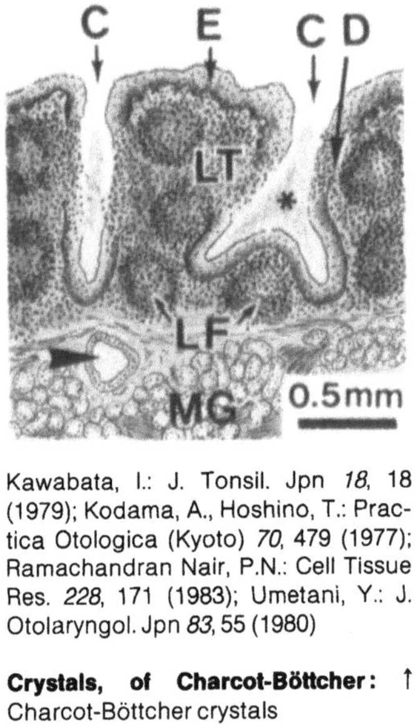

Kawabata, I.: J. Tonsil. Jpn *18*, 18 (1979); Kodama, A., Hoshino, T.: Practica Otologica (Kyoto) *70*, 479 (1977); Ramachandran Nair, P.N.: Cell Tissue Res. *228*, 171 (1983); Umetani, Y.: J. Otolaryngol. Jpn *83*, 55 (1980)

Crystals, of Charcot-Böttcher: ↑ Charcot-Böttcher crystals

Crystals, of Lubarsch: ↑ Lubarsch's crystals

Crystals, of Reinke: ↑ Reinke's crystals

CSF: ↑ Colony-stimulating factor

CSF-contacting neurons: ↑ Cerebrospinal fluid-contacting neurons

Cumulus oophorus (cumulus ovifetus, discus proligerus*): A half-spherical accumulation of ↑ granulosa cells

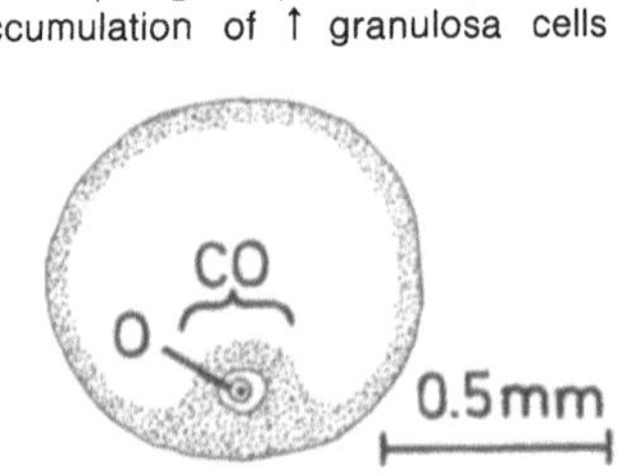

(CO) around primary ↑ oocyte (O) in a mature ovarian ↑ follicle. Some cells of C. persist around oocyte after ↑ ovulation to form ↑ corona radiata.

Tesarik, J., Dvorak, M.: J. Ultrastruct. Res. *78*, 60 (1982)

Cuneiform cartilage (cartilago cuneiformis): An elastic ↑ cartilage of ↑ larynx.

Cupping, of disc: ↑ Optic papilla

Cupula, of cristae ampullares (cupula*): A gelatinous noncellular flap (C) located above each ↑ crista ampullaris (CA) and perpendicular to ↑ ampulla of a ↑ semicircular duct. Hairs (H) of ↑ vestibular cells penetrate C., which is of ↑ glycoprotein nature, more viscous than surrounding ↑ endolymph but of unknown origin. During rotatory movements of head, the endolymph displaces C. (arrows), which excites ↑ vestibular cells. In the living state, tip of C. (asterisk) extends to opposite wall of ampulla, but shrinkage often occurs during preparation, so that the tip does not extend as high as normal. PS = ↑ planum semilunatum

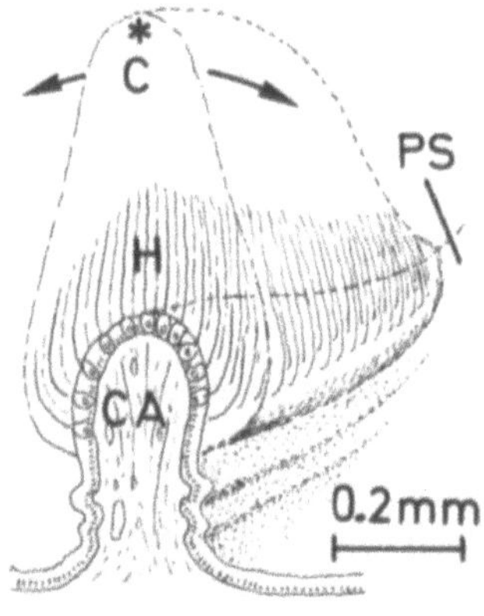

Cushing's syndrome (Cushing's disease, hyperadrenocorticism): A disease provoked by hypersecretion of ↑ glucocorticoids.

Cushions, intimal: ↑ Intimal cushions

Cuticle (cuticula*): A well-delimited extracellular mass produced by some epithelial cells in the form of a layer on their surface (e.g., ↑ enamel; ↑ capsule, of ↑ lens; ↑ tectorial membrane; ↑ zona pellucida).

Cuticle, of enamel: ↑ Enamel cuticle

Cuticle, of hair shaft (epidermicula, cuticula pili*): A thin layer of flattened cells (Cu) located between ↑ cuticle of inner root sheath (C) and ↑ cortex (Co) of ↑ hair shaft (HS). Near ↑ hair papilla, cells are nucleated; more superficially, they become tightly compact, keratinized, nonnucleated scales (S) with edges facing upward. Together with a considerable amount of amorphous ↑ keratin (AK), cells contain some free ribosomes (R). Cells are held together by a dense intercellular substance (IS). In ↑ hair follicle (HF), cells are imbricated with downward-oriented edges of cuticle cells of inner root sheath. C. develop from ↑ matrix cells of hair. HL = ↑ Henle's layer; M = ↑ medulla, of hair shaft

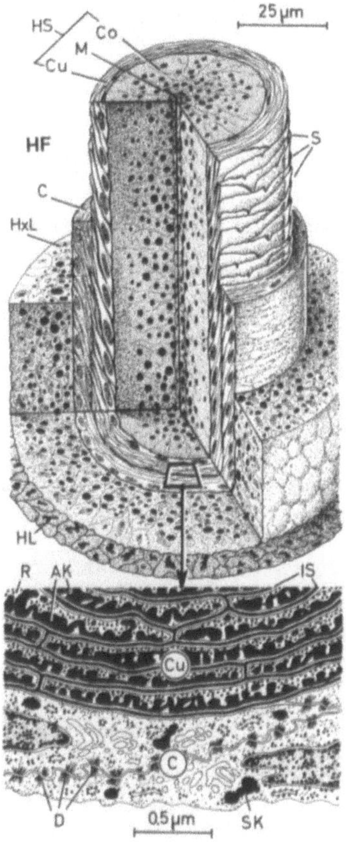

Woods, J.L., Orwin, F.G.: J. Ultrastruct Res. *80*, 230 (1982)

Cuticle, of inner root sheath (cuticula vaginae*): The single layer of flattened, overlapping cells (C) between hair shaft ↑ cuticle (Cu) and ↑ Huxley's layer (HxL) of ↑ hair follicle (HF). With their edges directed toward bottom of follicle, cells of C. interdigitate with cells of hair shaft cuticle. Cells contain soft ↑ keratin (SK); those situated more deeply in follicle contain both ↑ keratohyalin and ↑ tonofibrils. All cells are held together by numerous ↑ desmosomes (D). C. disappears at level of ↑ sebaceous glands. C. derives from hair ↑ matrix cells. (See fig. under ↑ Cuticle, of hair shaft)

Cuticula: ↑ Cuticle

Cuticular cells, of hair: ↑ Cuticle, of hair shaft; ↑ Cuticle, of inner root sheath

Cuticular plate: A particularly thick (0.5–2.5 µm) and dense ↑ terminal web (TW) at ↑ apical pole of ↑ hair and ↑ vestibular cells. C. of hair cells consists of ↑ actin myofilaments connected by thin, 3-nm-cross-linking filaments giving free surface of cells a certain rigidity and capacity to squeeze surface and push ↑ hairs (H) upward to ↑ tectorial membrane, thus increasing their sensitivity. Composition of C. in vestibular cells is probably similar to that of hair cells. C = central core of the hairs. (Fig. = rat)

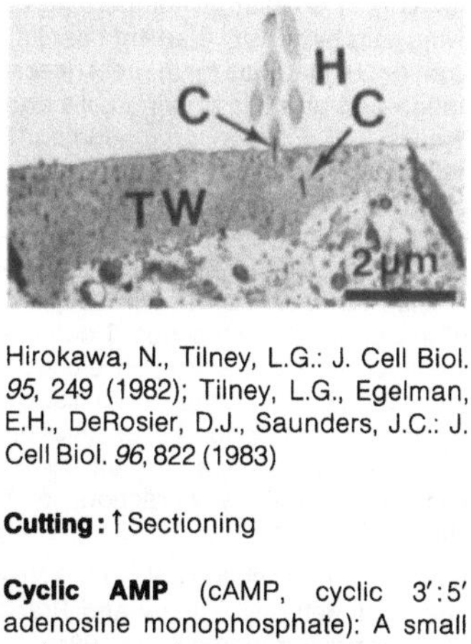

Hirokawa, N., Tilney, L.G.: J. Cell Biol. *95*, 249 (1982); Tilney, L.G., Egelman, E.H., DeRosier, D.J., Saunders, J.C.: J. Cell Biol. *96*, 822 (1983)

Cutting: ↑ Sectioning

Cyclic AMP (cAMP, cyclic 3′:5′ adenosine monophosphate): A small molecular substance regulating as "second messenger" specific enzymatic reactions of cell by activation of its ↑ kinases. C. is also involved in control of ↑ gene activity.

↑ Polypeptide hormone, action on the cell of

Pastan, I.: Sci. Am. *227*, 97 (1972); Robison, G., Butcher, R.W., Sutherland, E.W.: Cyclic AMP. New York, London: Academic Press 1971; Trakht, I., N., Grozdova, I.D., Vasiliev, V.Yu., Severin, E.S.: Bio Systems *12*, 305 (1980)

Cycloheximide: An antibiotic inhibiting protein synthesis on ↑ ribosomes of ↑ eucaryotes.

Cyst, of Rathke: ↑ Rathke's cyst

Cystic duct (ductus cysticus*): A roughly 4-cm-long segment of ↑ bile pathway connecting ↑ gallbladder with ↑ ductus choledochus. Structure: 1) Tunica mucosa (TM): a) epithelium (E) = a ↑ simple columnar epithelium; b) lamina propria (LP) = a subepithelial ↑ loose connective tissue with numerous ↑ lymphocytes. Tunica mucosa can be thrown into crescentic folds, ↑ spiral valve. 2) Tunica fibromuscularis (TF) = occasional ↑ smooth muscle cells intermingled with numerous ↑ collagen and ↑ elastic fibers; presence of glandulae mucosae biliosae (GM) and numerous nerve terminals. 3) Tunica adventitia (TA) = a loose connective tissue containing blood and lymphatic vessels. Delimitation of layers in C. is not definite; description given here is theoretical.

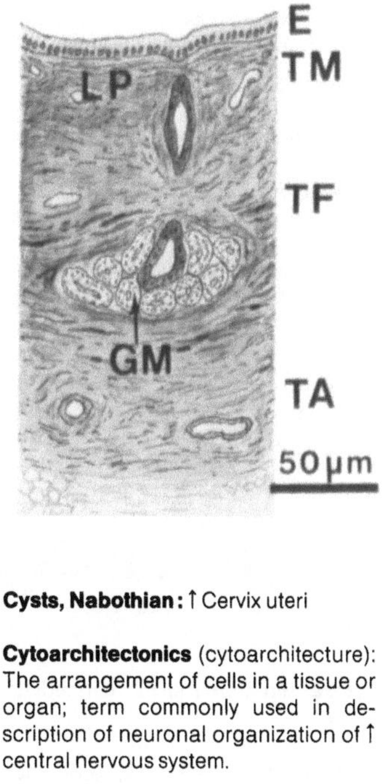

Cysts, Nabothian: ↑ Cervix uteri

Cytoarchitectonics (cytoarchitecture): The arrangement of cells in a tissue or organ; term commonly used in description of neuronal organization of ↑ central nervous system.

Cytocentrum: ↑ Centrosphere

Cytochalasins: Antibiotics of fungal origin that provoke inhibition of membrane and protoplasmic movements through disintegration of ↑ actin and ↑ actinlike microfilaments. Thus, C. block contraction of ↑ smooth muscle cells, beating of ↑ cardiac muscle cells, migration of cells, ↑ cytokinesis, ↑ endocytosis, ↑ exocytosis, etc.

Bray, D.: Nature *282*, 671 (1979); Brown, S.S., Spudich, J.A.: J. Cell Biol. *88*, 487 (1981); Tanenbaum, S.W.:

Cytochalasins. Amsterdam: Elsevier/North Holland 1978

Cytochemistry: A branch of ↑ cytology investigating chemical composition of substances in cell and their localization and dynamics by means of microscopic methods (↑ autoradiography, staining reactions, etc.).

Cytochemistry, methods of: ↑ Autoradiography, ↑ cytospectrophotometry, ↑ electron probe microanalysis, ↑ immunocytochemistry, staining reactions (↑ Feulgen reaction, ↑ periodic acid-Schiff reaction, etc.).

Cytocrinia (cytocrine secretion): The transfer of ↑ melanin granules from ↑ melanocytes to adjacent ↑ basal and ↑ prickle cells of ↑ epidermis. Process of C. is not sufficiently understood; it is thought that melanin granules are transported to adjacent cells in melanocyte processes that may penetrate neighboring cells.

↑ Epidermal-melanin unit

Cytodieresis: ↑ Cytokinesis

Cytofila: An ↑ apical differentiation in form of microvillous, sometimes branched processes (Cf) emerging between ↑ cilia (C) of various ciliated epithelial cells (CC). BB = ↑ basal bodies

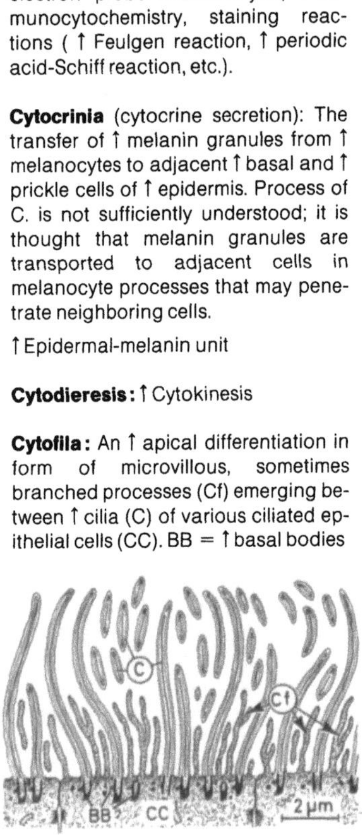

Cytokeratin filaments: ↑ Intermediate filaments

Cytokinesis* (cytodieresis): A process of cleavage of cell body and its definitive separation into two daughter cells following ↑ karyokinesis. C. begins in telophase by equatorial constriction of a ↑ contractile ring (r) of ↑ actin microfilaments. During C., ↑ mid-body (arrowhead) comes into being.

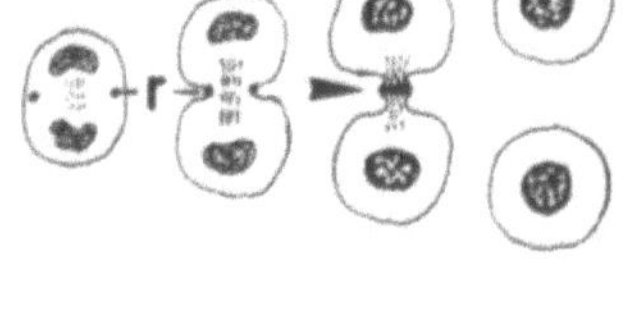

↑ Mitosis

Asnes, C.F., Schroeder, T.E.: Exp. Cell Res. *122*, 327 (1979); Beams, W.H., Kessel, R.G.: Sci. Am. *64*, 279 (1976)

Cytolemma: ↑ Cell membrane

Cytology (cell biology, cytologia*): A science analyzing normal morphology, chemistry, and physiology of cells. Main branches of C. are ↑ cytochemistry and ↑ cytophysiology.

Fawcett, D.W.: The Cell. 2nd edn. Philadelphia, London, Toronto: W.B. Saunders Co. 1981; Sheeler, P., Bianchi, D.E.: Cell Biology: Structure, Biochemistry, and Function. 2nd edn. Chichester: J. Wiley & Sons Ltd. 1983

Cytology, methods of: All procedures and instrumental techniques used in exploration of cells. A) Methods for direct observation of living cells and tissues: 1) ↑ Exteriorization and ↑ transillumination of organs; 2) ↑ transparent chamber method; 3) ↑ cell and ↑ organ culture; 4) mechanical ↑ micromanipulation; 5) ↑ laser, ultraviolet, and proton microbeams; 6) ↑ vital and ↑ supravital staining; 7) isolation of components of living cells by density gradient ↑ centrifugation. B) Methods for indirect observation and analysis of living cells and tissues: 1) ↑ Fixation, ↑ embedding, ↑ sectioning, and ↑ staining; 2) ↑ cytochemistry; 3) ↑ autoradiography; 4) microscopic analysis (↑ fluorescence microscope, ↑ interference microscope, ordinary ↑ light microscope, ↑ phase contrast microscope, ↑ polarizing microscope, ↑ scanning electron microscope, ↑ scanning transmission electron microscope, ↑ transmission electron microscope, ↑ ultraviolet microscope, ↑ X-ray diffraction); 5) ↑ stereology.

Kiernan, J.A.: Histological and Histochemical Methods: Theory and Practice. Oxford: Pergamon Press 1981

Cytolysosomes: ↑ Autophagosomes

Cytomembranes (biomembranes): All metabolically active boundaries separating the ↑ cell, its ↑ organelles, and ↑ inclusions from their surroundings. These include ↑ cell membrane, membrane of ↑ endoplasmic reticulum, ↑ mitochondrial membranes, Golgi membranes, etc. Although all C. have a similar thickness and ultrastructure, they differ in chemical composition and function.

Pelttari, A., Helminen, H.J.: Biol. Cell *47*, 343 (1983); Sato, R., Ohnishi, S.I. (eds.): Structure, Dynamics and Biogenesis of Biomembranes. Tokyo: Japan Scientific Societies Press; New York: Plenum Publishing Corp. 1982

Cytomorphosis: The sequence of gradual changes taking place over a period of 15–30 days, during which cells of stratum germinativum of ↑ epidermis transform themselves into lifeless keratinized plates after having lost their ↑ organelles.

Cytopempsis: ↑ Transcytosis

Cytophotometer (histophotometer, microphotometer): A device for semiquantitative analysis of substances in cells or products of cytochemical reactions, based on absorption of visible light. C. consists of a light source (L), ↑ condenser (C), specimen stage (S), microscope (M), photometer (P), and a registration instrument (R). Principle: Image of cell or tissue formed in microscope is projected onto a measuring diaphragm (D) of photometer, which measures only absorption of light passing through a small opening (Op) in diaphragm. Thus, C. can furnish only relative values obtained after comparison of stained sections, or sections after cytochemical reaction with untreated control sections. At present, ↑ cytospectrophotometer and ↑ microdensitometer are preferred to C.

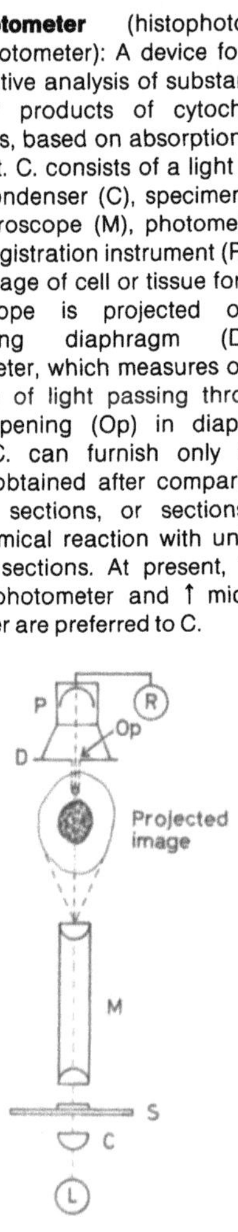

Pett, D., Wimmer, M.: Microphotometric Determination of Enzyme Activities in Cryostat Sections by the Gel Film Technique. In: Trends in Enzyme Histochemistry and Cytochemistry. Ciba Foundation Symposium. Amsterdam: Excerpta Medica 1980; Sandritter, W.: Acta Histochem. Suppl. *26*, 15 (1982)

Cytophotometry (histophotometry, microphotometry): A technique for semiquantitative analysis of concentration of substances in cells or products of cytochemical reactions using ↑ cytophotometer.

James, J.: Histochem. J. *15*, 95 (1983); Krug, H.: Histo- und Zytophotometrie. Einführung in die Absorptionsmethoden. Unter Mitarbeit von G.-R. Voss. Jena: VEB Gustav Fischer 1980; Krug, H., Fritsch, S. (eds.): Quantitative Mikroskopie. Acta Histochem. Suppl. *26*. Jena: Gustav Fischer 1982

Cytophysiology: The branch of ↑ cytology studying morphology of physiological changes in cells.

Cytoplasm (cytoplasma*): Part of ↑ protoplasm surrounding ↑ nucleus. C. represents soluble cell phase and consists of 75% water, 20% proteins, 3% lipids, 1% carbohydrates, 1% salts and ↑ nucleic acids. Constituents of C. are: 1) ↑ Cytoplasmic matrix; 2) ↑ organelles; 3) ↑ inclusions; 4) ↑ euplasm.

↑ Cell, structure of

Penman, S., Fulton, A., Capco, D., Ben Ze'ev, A., Wittelsberger, S., Tse, C.F.: Cytoplasmic and Nuclear Architecture in Cells and Tissue: Form, Functions, and Mode of Assembly. Cold Spring Harbor Symposia on Quantitative Biology, Vol. 46, Part 2. New York: Cold Spring Harbor Laboratory 1982; Schliva, M., Van Blerkom, J., Pryzwansky, K.B.: Structural Organization of the Cytoplasm. Cold Spring Harbor Symposia on Quantitative Biology, Vol. 46, Part 1. New York: Cold Spring Harbor Laboratory 1982

Cytoplasm, zones of: ↑ Zones, of cytoplasm

Cytoplasmic bridge: ↑ Intercellular bridge, true

Cytoplasmic inclusions: ↑ Inclusions

Cytoplasmic matrix (cell sap, cytosol, ground cytoplasm, ground substance of cytoplasm, hyaloplasma*): A transparent, liquid, structureless substance in which cell ↑ organelles and ↑ inclusions are suspended. C. contains a lattice of ↑ microtrabeculae, whose interspaces assure a rapid diffusion of water and soluble metabolites. C. undergoes rapid reversible sol-gel transformations; most metabolic processes and synthetic activities take place in C.

Brinkley, B.R.: Summary: Organization of the Cytoplasm. Cold Spring Harbor Symposia on Quantitative Biology, Vol. 46, Part 2. New York: Cold Spring Harbor Laboratory 1982; Porter, K.R., Anderson, K.: Eur. J. Cell Biol. *29*, 83 (1982); Porter, K.R., Beckerle, M., McNiven, M.: The Cytoplasmic Matrix. In: McIntosh, J.R. (ed.): Modern Cell Biology, Vol. 2. New York: Alan R. Liss Inc. 1983

Cytoplasmic organelles: ↑ Organelles

Cytoreticulum, epithelial: Thymic ↑ epithelial-reticular cells forming three-dimensional parenchymal framework of ↑ thymus. ↑ Lymphocytes lodge in meshes of C.; it represents an ↑ atypical epithelium.

Papiernik, M., Nabarra, B.: Thymus *3*, 345 (1981)

Cytosegresomes: ↑ Autophagosomes

Cytoskeleton: System of ↑ microfilaments (Mf), ↑ microtubules (Mt) ↑ intermediate filaments (If), and ↑ microtrabeculae (not represented) responsible for dynamic organization of cytoplasm and transport of information through cell body. Microfilaments run through whole cell body and anchor into ↑ desmosomes (D) and ↑ cell membrane (CM). With microtubules they participate in formation of ↑ basal (BW) and ↑ terminal web (TW). Microtrabeculae connect microfilaments and microtubules with cell ↑ organelles in a very dynamic network, which is constantly being synthesized and broken down. N = ↑ nucleus

↑ Intermediate microfilaments

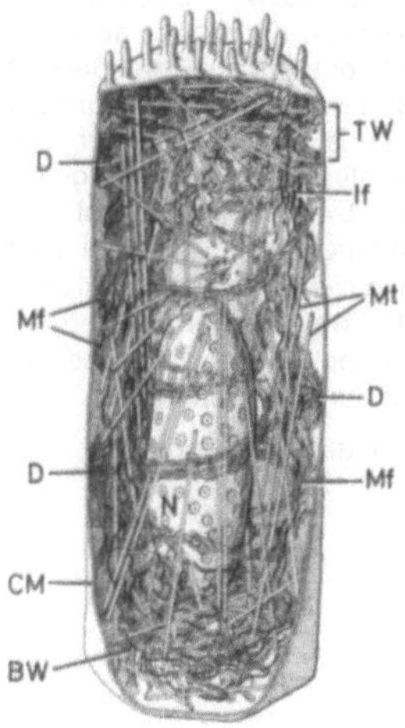

Bell, P.B.: Scan. Electron Microsc. 1981/II, 139 (1981); Brabander, M.J. de: La Recherche *145*, 810 (1983); Trump, B.F., Berezesky, I.K., Phelps, P.C.: Scan. Electron Microsc., 1981/II, *435* (1981); Wilson, L. (ed.): Methods in Cell Biology, Vol. 24. The Cytoskeleton. Part A and Part B. New York: Academic Press 1982

Cytoskeleton, of cell membrane: ↑ Cell membrane, cytoskeleton of

Cystosol: ↑ Cytoplasmic matrix

Cytosomes: ↑ Multilamellar bodies

Cytospectrophotometer (microspectrophotometer): An instrument for optical analysis of concentration and identification of substances in cells. Principle: C. consists of a mercury or xenon light source (L), a monochromator (Mo), producing different wavelengths to obtain absorption spectrum, a ↑ condenser (C), which focuses field diaphragm (FD) in the form of a 2-μm spot of light (P) in plane of cell to be analyzed, a microscope (M), a photomultiplier (Ph), and a recording instrument (R). Concentration of a given substance at a selected point (P) of cell's projected image is measured on basis of absorption maximum of substance at a defined wavelength (260–700 nm). Combined with a scanning device (SD) that moves specimen stage (SS) and connected on-line with a computer (Co), a C. can be used in ↑ automatic cell recognition.

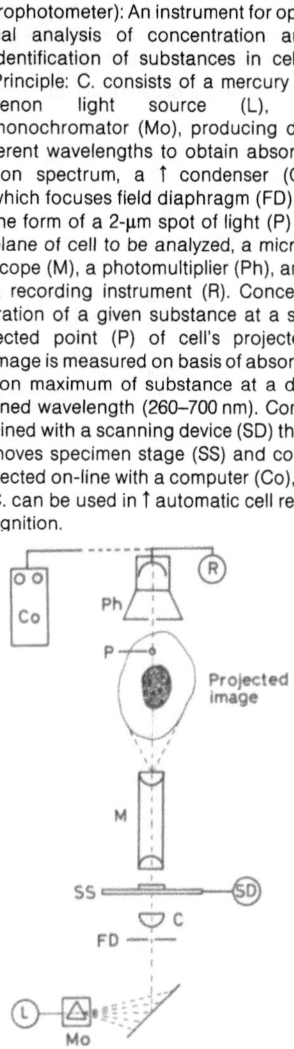

Dubois, J.-P., Hemet, J., Metayer, J., Ducastelle, T., Raoult, J.-P., Doublet, D.: Biol. Cell *39*, 55 (1980); Trapp, L.: Instrumentation for Recording Microspectrophotometry. In: Wied, G.L. (ed.): Introduction to Quantitative Cytochemistry. New York, London: Academic Press 1966; Zeiss, C.: Pamphlet No 40-811/L-e (1964)

Cytospectrophotometry (microspectrophotometry): A technique of identification and quantification of chemical substances in a cell based on analysis of their absorption spectra with a ↑ cytospectrophotometer.

Cytotoxic lymphocytes: ↑ Killer cells

Cytotrophoblast (cytotrophoblastus*): The inner layer of well-delimited cells (C) of ↑ trophoblast. Number of cells in C. gradually decreases from the 4th month of pregnancy. S = ↑ syncytiotrophoblast. (Fig. = human) (See embryology texts for further information)

↑ Placental villi, of early placenta

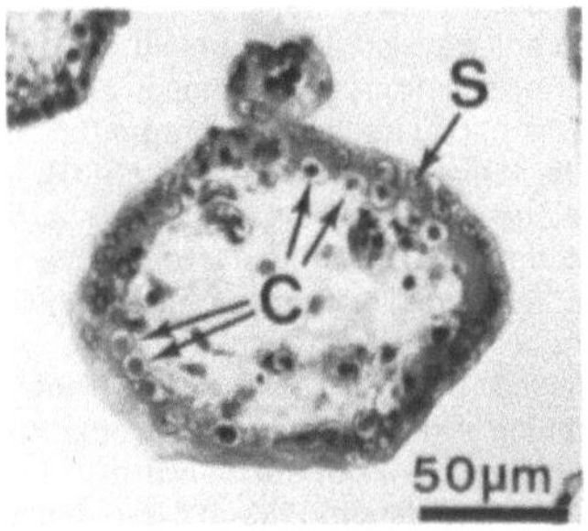

D

DAB: ↑ Diaminobenzidine

Dark cells, of arched collecting tubule: ↑ Collecting tubules, of kidney

Dark cells, of eccrine sweat glands: ↑ Mucoid cells, of eccrine sweat glands

Dark cells, of pineal organ: ↑ Interstitial cells, of pineal organ

Dark cells, of taste buds: ↑ Taste buds, structure of

Dark chief cells, of parathyroid glands: ↑ Chief cells, of parathyroid glands

Dark cells, problem of: Cells with more osmiophilic ↑ cytoplasmic matrix than adjacent cells of same class, but with same ultrastructure. Occurrence and significance of D. are poorly understood; it is believed that they may represent a special cytophysiological state and/or a fixation ↑ artifact, depending upon kind of ↑ fixation.

Bucher, O., Krstić, R.: Z. Anat. Entwickl. Gesch. *141*, 319 (1973)

Dark-field microscope: A kind of diascopic light-microscopic ↑ illumination used in study of living cells and bacteria. No direct light enters ↑ objective (Obj); only peripheral light rays formed either by a diaphragm (D) or a special dark-field ↑ condenser (DFC) are obliquely projected onto preparation (P). Object appears with bright contours against a dark observation field. However, little or no information is obtained concerning object's internal structure. Image in D. is formed by diffraction of light between boundaries having different ↑ refractive indices. D. permits visualization of smaller objects than in ↑ bright-field microscopy; it is now superseded by ↑ phase contrast microscope.

↑ Illumination, diascopic

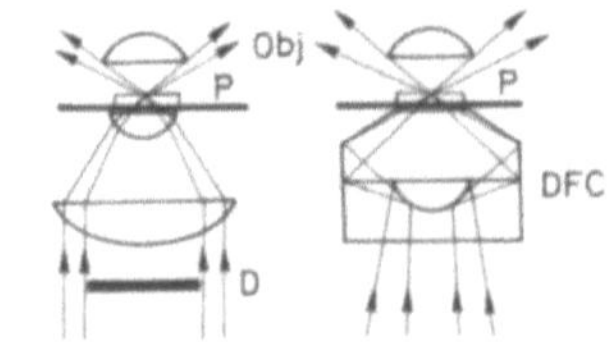

Dark reaction: ↑ Terminal degeneration

Dartos: ↑ Tunica dartos

D-cells, of gastrointestinal tract: ↑ Endocrine cells scattered in epithelium of ↑ fundus of ↑ stomach and in ↑ duodenum; structurally very similar to ↑ D-cells of ↑ pancreatic islets. On basis of immunocytochemical studies, it is believed that D. produce ↑ somatostatin, D. belong to ↑ APUD-cells.

Kusumoto, Y., Iwanaga, T., Ito, S., Fujita, T.: Arch. Histol. Jpn *42*, 459 (1979)

D_1-cells, of gastrointestinal tract: A category of ↑ endocrine cells of ↑ gastrointestinal tract particularly concentrated in ↑ colon, probably secreting ↑ vasoactive intestinal peptide. D_1 belong to ↑ APUD-cells.

Welbourn, R.B.: Hexagone "Roche" *5*, 8 (1981)

D-cells, of pancreatic islets (delta cells): Oval or polygonal endocrine cells making up about 5% of cell population of ↑ pancreatic islets. Nucleus is round; cytoplasm contains elongated mitochondria, a well-developed Golgi apparatus, groups of flattened rough endoplasmic cisternae, a few ↑ lysosomes, some ↑ microtubules, and a moderate amount of free ribosomes. From Golgi apparatus arise 220 to 350-nm, ↑ unit membrane-bound ↑ secretory granules (SG) filled with fine granular, moderately osmiophilic material, which is excreted by ↑ exocytosis (E). Although some analyses indicate D. to be altered ↑ A-cells, immunocytochemical studies suggest that these cells produce ↑ gastrin and ↑ somatostatin. D. are included among ↑ APUD-cells. Cap = capillary

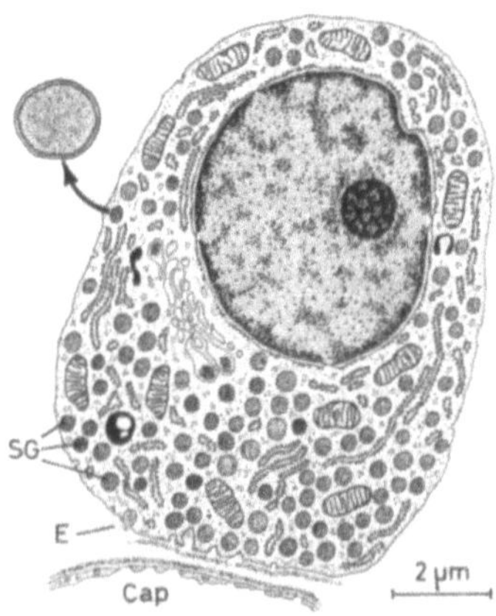

Cavallero, C., Spagnoli, L.G., Cavallero, M.: Arch. Histol. Jpn *36*, 307 (1974); Grube, D., Bohn, R.: Arch.

Histol. Jpn. *46,* 327 (1983); Schusdziarra, V., Ipp, E., Harris, V., Dobbs, R.E., Raskin, P., Orci, L., Unger, R.H.: Metabolism *27,* Suppl. 1, 1227 (1978); Wilander, E., Westermark, P.: Cell Tissue Res. *168,* 33 (1976)

Decalcification (demineralization): A process of dissolution of inorganic salts from different mineralized tissues (↑ bone, calcified ↑ cartilage, ↑ dentin, ↑ cementum, ↑ enamel) by solutions of nitric acid, formic acid, trichloracetic acid, ↑ EDTA, etc.

Clark, P.: Am. J. Clin. Pathol. *24,* 1113 (1954)

Decidua basalis*: Portion of ↑ endometrium (DB) underlying implantation site (IS). DP = decidua parietalis, DC = decidua capsularis, L = lumen of uterus, IO = internal os. (See embroyology texts for further information)

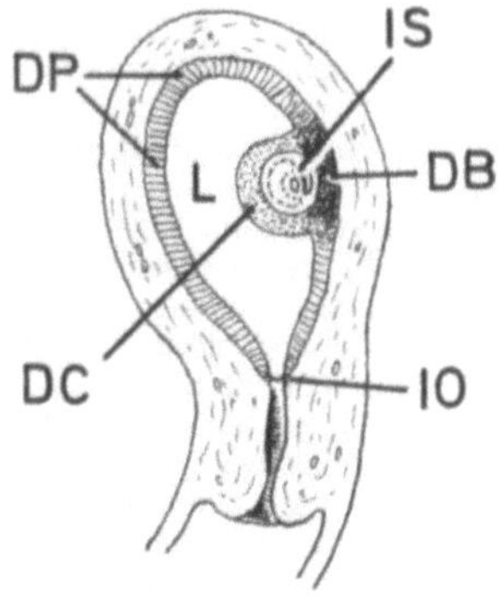

Decidua basalis*, structure of: Portion (DB) of ↑ endometrium of a pregnant ↑ uterus between implantation site and ↑ myometrium (M). Structure (beginning from trophoblastic shell, TS): 1) Zona compacta (ZC) = intermingled ↑ decidual (DC) and ↑ cytotrophoblastic cells (Tr) (= ↑ deciduotrophoblastic complex); 2) zona spongiosa (ZSp) = flattened and elongated ↑ uterine glands (UG) parallel to surface; 3) zona basalis (ZB) = narrow uterine glands. Zona compacta and zona spongiosa constitute ↑ basal or decidual plate (BP), which is composed of: a) ↑ deciduotrophoblastic complex; b) ↑ Nitabuch's fibrinoid; c) part of zona spongiosa. To basal plate are anchored ↑ anchoring villi (AV); in second half of pregnancy, basal plate thins out. During delivery, placenta separates at level of Nitabuch's fibrinoid; endometrium regenerates from zona basalis.

↑ Placenta, full-term, structure of

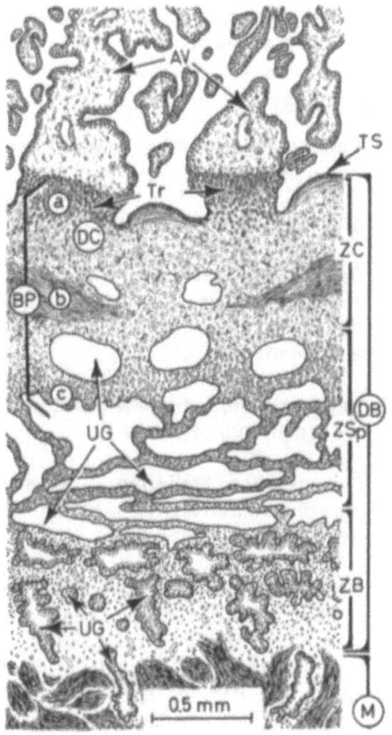

Decidua capsularis*: A thin layer of ↑ endometrium (DC) between implantation site (IS) and lumen (L) of ↑ uterus. (See fig. under ↑ Decidua basalis)

Decidua parietalis*: The ↑ endometrium (DP) lining entire cavity of ↑ uterus down to internal os (IO), except sites of ↑ decidua basalis and ↑ decidua capsularis. Around 4th month of pregnancy, D. and decidua capsularis fuse, obliterating lumen (L) of uterus. (See fig. under ↑ Decidua basalis)

Decidual cells, of ovary: Groups of round or oval cells with pale cytoplasm originating from stromal cells in cases of ovarian pregnancy.

Decidual cells, of uterus (cellulae deciduales*): Large, round, or polygonal, pale cells with an oval central nucleus and a prominent nucleolus. Cytoplasm contains scattered mitochondria, an inconspicuous Golgi apparatus, many elongated and frequently enlarged rough endoplasmic cisternae (rER) with moderately dense ↑ reticuloplasm, a few ↑ lysosomes, numerous free ribosomes, a considerable amount of ↑ glycogen (Gly), and aggregated ↑ lipid droplets (L). Microfilamentous bundles (Mf) and ↑ coated vesicles (CV) are mainly concentrated in ↑ ectoplasm (E). Surface of D. displays irregular microvilli and coarse processes; D. are connected by ↑ nexus with adjacent D., forming an ↑ epithelioid cell union. D. arise in course of ↑ decidualization from fibroblastlike cells of endometrial stroma. It is believed that D. may synthesize proteins.

↑ Endometrium

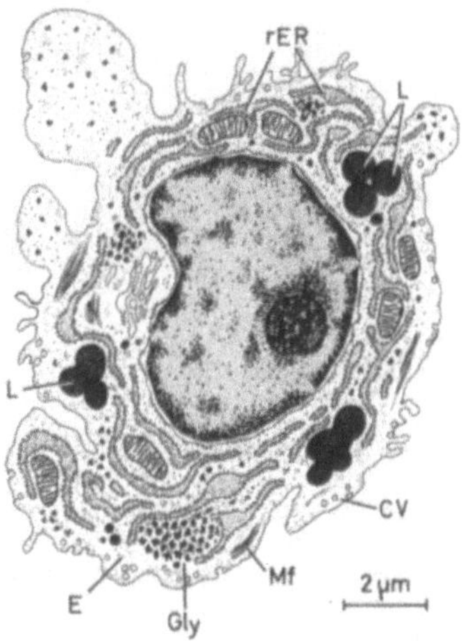

Finn, C.A.: Adv. Reprod. Physiol. *5,* 1 (1971)

Decidual plate: ↑ Basal plate; ↑ Decidua basalis

Decidual septa: ↑ Placenta, full-term, structure of

Decidualization: The process of transformation of fibroblastlike cells of endometrial stroma into ↑ decidual cells under influence of implanted blastocyst, its ↑ human chorionic gonadotropin, and ↑ progesterone of ↑ corpus luteum of pregnancy. (See embryology texts for further information)

Lundkvist, O., Nilsson, B.O.: Cell Tissue Res. *225,* 355 (1982)

Deciduate placenta: The type of ↑ placenta in which ↑ endometrium is transformed into ↑ decidua. Separation of D. at birth is accompanied by tearing of maternal tissue and a considerable loss of maternal blood. Human placenta is a D.

Deciduotrophoblastic complex: The zone of intimate intermingling of ↑ cytotrophoblastic cells (C) of ↑ anchoring villi (AV) with ↑ decidual cells (DC) of ↑ decidua basalis. D. represents a unique example of direct contact between cells with different ↑ genotypes without ↑ graft rejection reaction. (Fig. = human)

↑ Placenta, full-term, structure of

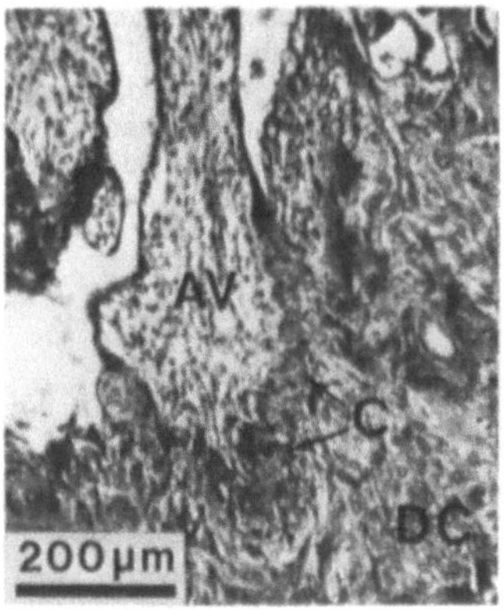

Dedifferentiation: ↑ Anaplasia

Deep arteries, of penis (arteria profunda penis*): Paired ↑ muscular arteries (A) each running through the corresponding ↑ cavernous body (CB) and giving off ↑ helicine arteries (HA). D. are possibly provided with ↑ intimal cushions; they are responsible for nutrition and ↑ erection of ↑ penis (Fig. = human newborn)

↑ Penis, vascularization of

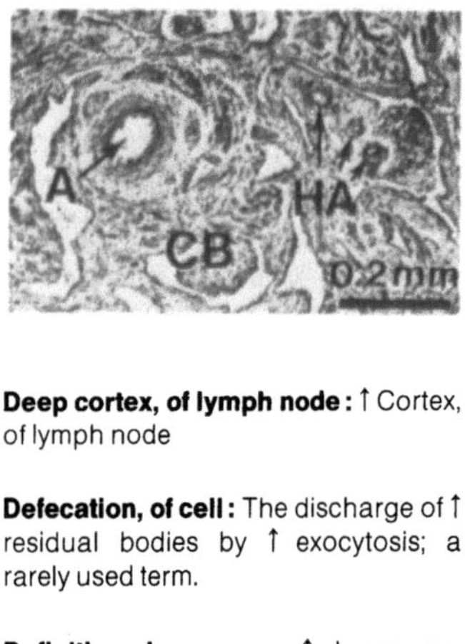

Deep cortex, of lymph node: ↑ Cortex, of lymph node

Defecation, of cell: The discharge of ↑ residual bodies by ↑ exocytosis; a rarely used term.

Definitive lysosome: ↑ Lysosome, secondary

Degeneration (degeneratio): The process of gradual qualitative worsening of metabolic functions, accompanied by retrogressive morphological changes of cells, tissues, and organs. D. terminates in ↑ necrosis, but in the beginning it is reversible.

Degeneration, ascendent, of peripheral nerve fibers: ↑ Nerve fibers of peripheral nervous system, degeneration and regeneration of

Degeneration, descendent, of peripheral nerve fibers: ↑ Nerve fibers of peripheral nervous system, degeneration and regeneration of

Degeneration, primary, of peripheral nerve fibers: ↑ Nerve fibers of peripheral nervous system, degeneration and regeneration of

Degeneration, secondary, of peripheral nerve fibers: ↑ Nerve fibers of peripheral nervous system, degeneration and regeneration of

Degeneration, transneuronal: ↑ Transneuronal degeneration

Degeneration, Wallerian: ↑ Nerve fibers of peripheral nervous system, degeneration and regeneration of

Dehydration: The process of gradual removal of water from: 1) Fixed and washed tissue blocks to prepare them for ↑ embedding; 2) stained ↑ sections to prepare them for mounting in resinous mounting media. D. is accomplished by transferring sections into progressively increasing concentrations of alcohol, acetone, etc.

Deiters' cells: ↑ Phalangeal cells, outer

Delayed hypersensitivity: The cell-mediated ↑ immune response to antigen taking a longer time to become clinically detectable than the humoral immune response.

Deletion: A ↑ chromosome anomaly consisting of breaking off of a ↑ chromosome arm or its part. During anaphase, portion without a ↑ centromere may fail to move to pole and may be lost. "Cri-du-chat" syndrome is provoked by D. of shorter arm of chromosome 5.

Delta cells, of hypophysis (cellulae gonadotrophicae*): Two types of large ↑ basophilic cells present in pars distalis of ↑ hypophysis. D. produce both ↑ follicle-stimulating hormone (↑ gonadotrope type I) and ↑ luteinizing hormone (↑ gonadotrope type II). In man, there are probably neither immunocytochemical nor morphological differences between these two cell types.

Delta cells, of pancreatic islets: ↑ D-cells, of pancreatic islets

Demilunes (Gianuzzi's demilunes, semiluna serosa*): Darkly stained crescentlike groups (D) of ↑ serous cells at end of ↑ mixed tubules (MT) of ↑ mixed glands. Through narrow intercellular channels, serous cells discharge their product into lumen of tubules. (Fig. = ↑ submandibular gland, human)

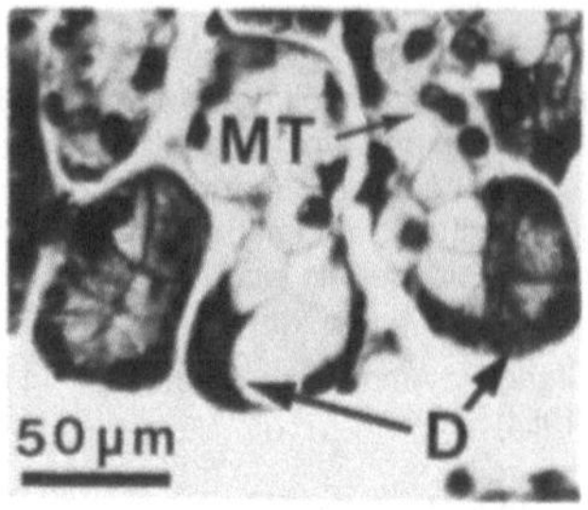

Uddin, M., Tyler, D.W.: Experientia *34*, 609 (1978)

Dendrites (dendritum*): Widely branched processes (D) extending from ↑ soma (S) of a nerve cell. Large stem D. contain all types of cell ↑ organelles, free ribosomes, ↑ neurotubules (Nt), and ↑ neurofilaments (Nf); with reduction of diameter of D., Golgi apparatus (G) disappears, but ↑ Nissl bodies (NB) are still present. In small branches, the latter are reduced to short rough endoplasmic cisternae and occasional free ribosomes; both these structures are lacking in the smallest D.; number of mitochondria may, however, be relatively increased. Neurofilaments and neurotubules are always numerous. Both are thought to be responsible for ↑ dendritic transport. ↑ Axons (A) of other ↑ neurons end on D., forming axo-dendritic ↑ syn-

apses (ADS). To increase synaptic surface area, most dendrites form ↑ dendritic spines (DS). D. are the principal receptive projections receiving and conducting nerve impulses toward cell body.

Dendrites, supraependymal: ↑ Supraependymal dendrites

Dendritic cells: ↑ Argyrophilic cells with numerous radiating processes forming a network in dark region of germinal centers of ↑ lymphatic nodules. D. are a category of ↑ reticular cells.

Dendritic field: ↑ The space occupied by all ↑ dendrites of a ↑ neuron. D. is generally round or oval; it may be fusiform or flattened in one dimension (↑ Purkinje cells).

Dendritic flow (dendritic transport): The cellulifugal movement of various substances and cell ↑ organelles along ↑ dendrites at a rate of about 3 mm/h. ↑ Neurofilaments and ↑ neurotubules probably play an essential role in D.

↑ Axoplasmic flow

Dendritic gemmules: ↑ Dendritic spines

Dendritic macrophages: The term occasionally used to designate ↑ reticular cells.

↑ Dendritic cells

Dendritic spines (dendritic gemmules, dendritic thorns, gemmula dendritica*): Peduncular projections of a ↑ dendrite (D) having a 0.25 to 0.5 µm-wide stalk and an ovoid tip (S) about 1–2 µm in diameter. Middle portion of a dendrite is richest in D. Apart from rare ↑ neurotubules and occasional mitochondria, D. contain ↑ spine apparatus (asterisk). A considerable amount of dense, amorphous material lies on ↑ presynaptic membrane facing ↑ axon (A). D. form principal synaptic surfaces of many, but not all, dendrites. They are, however, inconstant structures, since they diminish and become rare with increasing age, as well as after sensory deprivation or absence of afferent stimuli. (Figs. = rat)

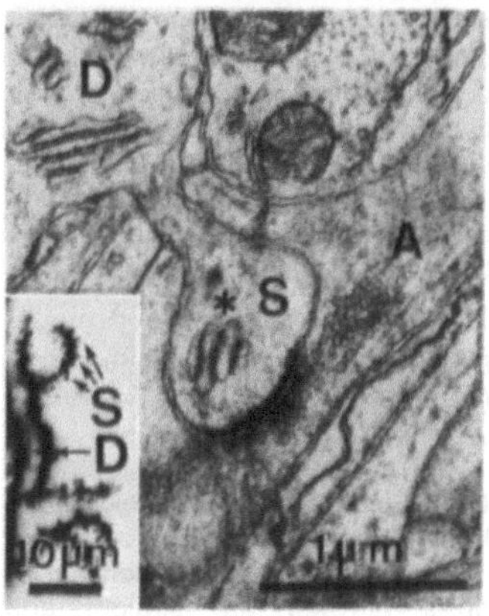

Jones, E.G., Powell, T.P.S.: J. Cell Sci. *5*, 509 (1969); Koch, C., Poggio, T.: TINS *6*, 80 (1983); Špaček, J.: J. Anat. *131*, 723 (1980); Špaček, J., Hartmann, M.: Anat. Embryol. *167*, 289 (1983)

Dendritic thorns: ↑ Dendritic spines

Dendrodendritic synapse: ↑ Synapse, dendrodendritic

Dense bodies: ↑ Membrane-coating granules

Dense bodies, of smooth muscle cells (dense patches, fusiform density, area densa*): Elongated, highly electron-dense structures (DB) lying free in ↑ sarcoplasm or in contact with ↑ sarcolemma (S). D. represent attachment points for ↑ actin myofilaments and are interconnected with ↑ intermediate microfilaments. D. may have capacity to shorten, since they contain the contractile protein ↑ actinin. D. are homologous to ↑ Z-lines of ↑ skeletal muscle fibers. (Fig. = rat)

↑ Desmin microfilaments

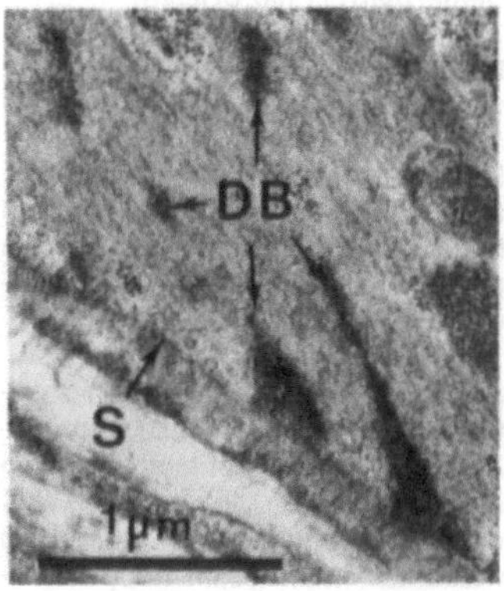

Bond, M., Somlyo, A.V.: J. Cell Biol. *95*, 403 (1982); Tsukita, S., Tsukita, S., Ishikawa, H.: Cell Tissue Res. *229*, 233 (1983)

Dense connective tissue (textus connectivus fibrosus compactus*): A kind of ↑ fibrous connective tissue with densely packed ↑ elastic and ↑ collagen fibers. D. include ↑ elastic tissue, irregular ↑ D., and regular ↑ D.

Dense connective tissue, irregular (textus connectivus fibrosus compactus irregularis*): A kind of ↑ fibrous connective tissue in which ↑ collagen fibers largely predominate over cellular elements. D. has neither constant form nor regular arrangement of its fibers, which are organized in strong bundles (CB) running in various directions. Between bundles lie sporadic flattened cells, mostly ↑ fibrocytes (F) and some ↑ elastic fibers (E). D. forms ↑ capsules of organs, reticular layer of ↑ dermis, ↑ fasciae, ↑ perichondrium, and ↑ periosteum. Vascularization and innervation of D. are poor.

Dense connective tissue, regular (textus connectivus fibrosus compactus regularis*): A kind of ↑ fibrous connective tissue with abundant parallel ↑ collagen fibers, which give it a constant form. Cellular elements of D. are scarce. D. includes ↑ aponeuroses, ↑ ligaments, ↑ meniscal tissue, and ↑ tendons.

Dense-core vesicles (G-vesicles, vesicula densa s. granularis*): ↑ Unit membrane-enclosed ↑ synaptic vesicles (arrows), 40–60 nm in diameter, with a generally eccentric highly electron-dense granulum measuring about 25 nm. D. are characteristic of ↑ adrenergic nerve endings and fibers (ANF); D. contain neurotransmitter ↑ norepinephrine, ↑ dopamine-β-hydroxylase, and a lipid carrier substance, reacting with ↑ zinc iodide-osmium tetroxide. D. are present in G- ↑ synapses as well as in ↑ synapses "by distance." (Fig. = ↑ pineal organ, rat)

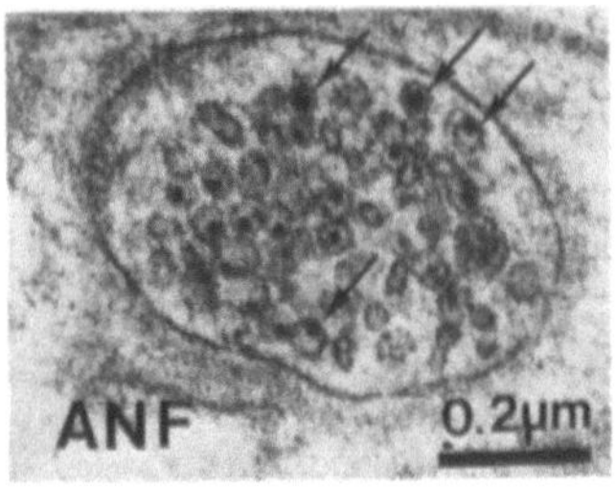

Dense fibers, outer, of spermatozoon tail (fibrae densae externae*): Set of nine proteinaceous pillars (DF) surrounding ↑ axoneme (A) along most of ↑ spermatozoon tail length. Each D. is continuous with one of nine striated columns (SC) of ↑ spermatozoon neck and progressively tapers along tail, ending at a point on principal piece. Here it is attached to one doublet of axoneme. Exact function of D. is unknown; they are interpreted as passive stiffening elements of spermatozoon tail. M = ↑ mitochondrial sheath. (Fig. = rat)

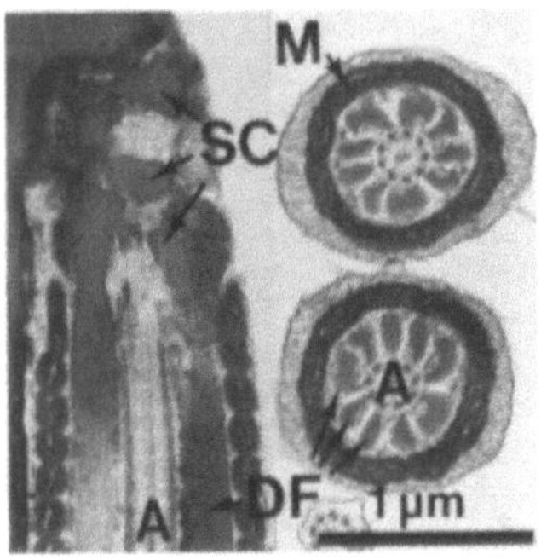

Irons, J.M., Clermont, Y.: Anat. Rec. *202*, 463 (1982)

Dense granules, of mitochondria: ↑ Intramitochondrial granules

Dense lamina: ↑ Fibrous lamina

Dense patches, of smooth muscle cells: ↑ Dense bodies, of smooth muscle cells

Density gradient centrifugation: ↑ Centrifugation, density gradient

Density, presynaptic: ↑ Presynaptic density

Density, postsynaptic: ↑ Postsynaptic density

Dental lamina (lamina dentalis*): A continuous epithelial ridge sinking into underlying ↑ mesenchyme. D. forms several budlike thickenings and precedes tooth formation. (See embryology texts for further information)

Dental papilla (papilla dentalis*): ↑ Mesenchymal tissue (DP) partially surrounded by enamel organ (EO). D. represents precursor of dental ↑ pulp; ↑ odontoblasts (O) producing ↑ dentin (D) differentiate at contact surface between D. and ↑ ameloblasts (A). DS = ↑ dental sac

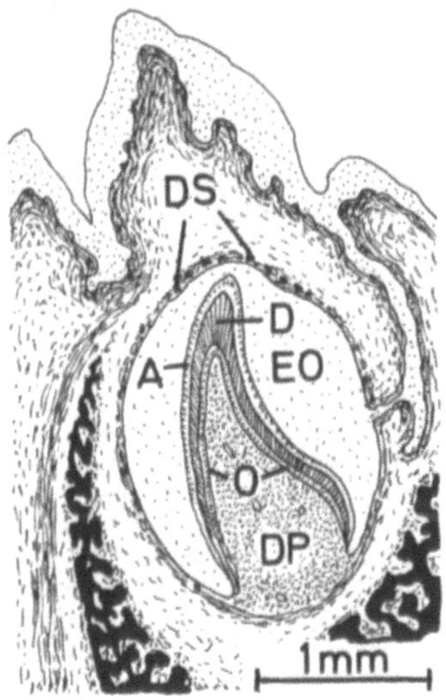

Dental pulp: ↑ Pulp, of teeth

Dental sac (saccus dentalis*): A well-vascularized, thin layer of ↑ mesenchymal tissue surrounding enamel organ. D. assures nutrition of enamel organ and includes formation of root of tooth. (See fig. under ↑ Dental papilla and embryology texts for further information)

Dentin (dentinum*): A yellowish, semitransparent, hard tissue (D) surrounding ↑ pulp cavity (P) of tooth. In region of ↑ crown (Cr), D. is covered with ↑ enamel (E), and in region of root (R) with ↑ cementum (C). Through calcified ground substance run innumerable ↑ dentinal tubules (DT) to which ↑ collagen fibers are in general perpendicularly oriented. Some fibers, extending in a direction parallel to free surface of crown, also surround the tubules. D. consists of two layers: 1) Circumpulpar ↑ D. (cd) (the larger part) with straight tubules; 2) cover or mantle ↑ D. (md), with branched tubules. Incomplete calcification of ↑ ground substance provokes formation of ↑ interglobular spaces (IS) in superficial coronal D., which continue as ↑ Tomes granular layer (TL) beneath dentinocemental junction. The coarser ↑ contour lines of Owen (OL) are a consequence of appositional growth of D. ↑ Odontoblasts (O) produce D.; it consists of 20% organic material and 80% inorganic salts. About 90% of organic material is ↑ collagen; the remainder is ↑ proteoglycans of ground substance.

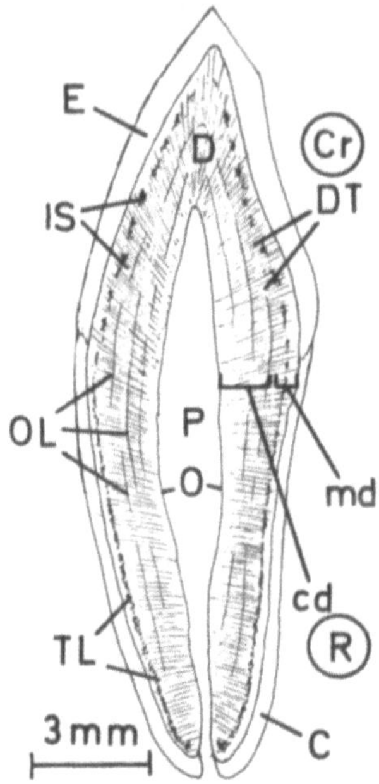

Gunji, T.: Arch. Histol. Jpn *45*, 45 (1982)

Dentin, calcification of: The process of deposition of calcium phosphate in form of ↑ hydroxyapatite in organic matrix (↑ predentin) secreted by ↑ odontoblasts. Mineralization is apparently started by dentinal matrix vesicles, following same ↑ calcification mechanism as that of bone. The first calcification points (nuclei) enlarge concentrically by deposition of new crystals and fuse together, sometimes leaving crystal-free ↑ interglobular spaces filled only with predentin. Calcification process is probably started by ↑ collagen fibers.

↑ Collagen, calcification of

Bernard, G.W.: J. Ultrastruct Res. *41*, 1 (1972); Nylen, M.U.: J. Dent. Res. *58* (B), 922 (1979)

Dentin, circumpulpar: A thick inner layer of ↑ dentin (D) directly surrounding ↑ pulp cavity (P); in D. ↑ dentinal tubules (arrows) are straight and without ramifications. O = ↑ odontoblasts. (Fig. = human)

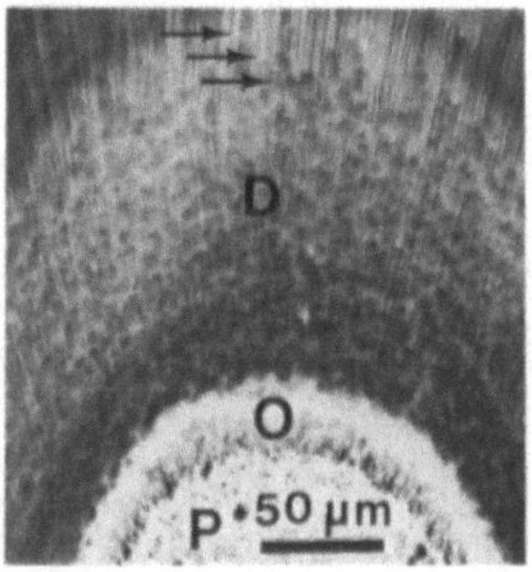

Dentin, cover (mantle dentin): A thin, peripheral layer of ↑ dentin in which ↑ dentinal tubules are branched.

Dentinal globules (globuli dentinales*): Spherical structures of calcified ↑ dentin (DG) delimiting ↑ interglobular spaces (IS). (Fig. = human)

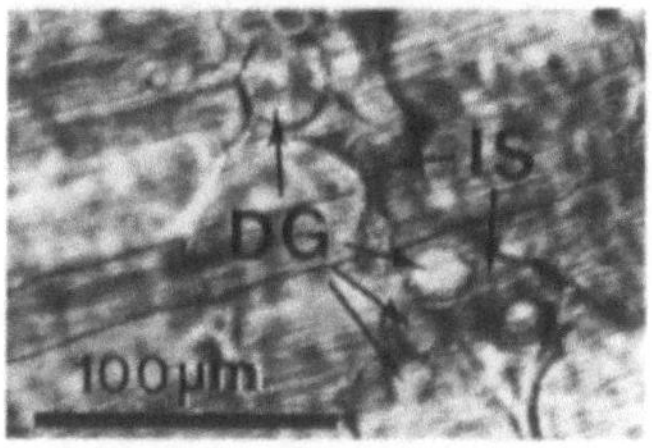

Dentinal matrix vesicles: ↑ Unit membrane-bound granules (D) about 0.2–0.5 µm in diameter found in ↑ predentin (P). D. contain a moderately to highly electron-dense material in which first appear needlelike ↑ hydroxyapatite crystals. Starting within D., crystallization process spreads outside, leading to gradual calcification of ↑ dentin. D. are believed to be secreted by ↑ odontoblasts. Op = odontoblastic process. (Fig. = embryo, rat)

↑ Matrix vesicles; ↑ Matrix vesicles, biogenesis of

Almuddaris, M.F., Dougherty, W.J.: Am. J. Anat. *155*, 223 (1979); Katchburian, E., Severs, N.J.: J. Anat. *134*, 615 (1982); Larsson, Å.: Z. Anat. Entwickl. Gesch. *142*, 103 (1973); Weinstock, M.: J. Ultrastruct. Res. *61*, 218 (1977)

Dentinal tubules (tubuli dentinales*): Fine canaliculi (DT), 0.5–4 µm in diameter, extending radially from ↑ odontoblasts (O) up to ↑ dentinoenamel and dentinocemental junction (C). In circumpulpar ↑ dentin, (cd) D. are straight and unramified; in cover ↑ dentin (m) they are branched and anastomose with one another. Each D. contains a ↑ Tomes' fiber (TF); nerve fibers run through some D. Layer of dentin immediately surrounding D. forms the highly calcified ↑ Neumann's sheath (NS). D = ↑ dentin

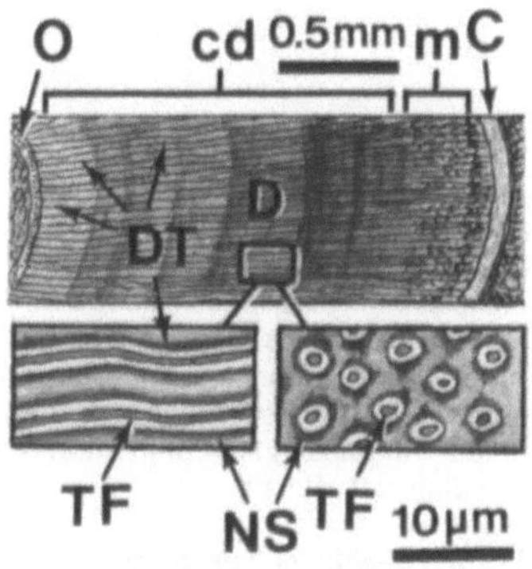

Byers, M.R., Dong, W.K.: Anat. Rec. *205*, 441 (1983); Holland, G.R.: Anat. Rec. *200*, 437 (1981); Hattyasy, D.: Z. mikrosk.-anat. Forsch. *96*, 679 (1982)

Dentinoenamel junction (junctio dentinoenameli*): The irregularly shaped contact surface (S) between ↑ dentin (D) and ↑ enamel (E) characterized by presence of numerous needlelike fusiform processes of dentin with ↑ dentinal tubules (DT) penetrating a short distance into enamel.

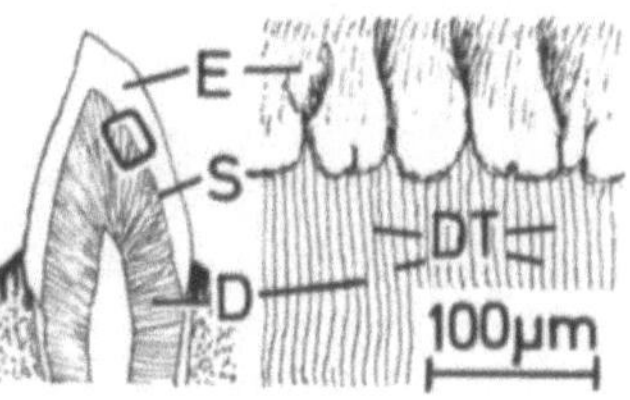

Whittaker, D.K.: J. Anat. *125*, 323 (1978)

Dentinogenesis: The process of formation and maturation of ↑ dentin.

↑ Dentin, calcification of; ↑ Odontoblasts

Deoxycorticosterone: A ↑ mineralocorticoid hormone secreted by ↑ cells of zona glomerulosa of adrenal cortex.

↑ Adrenal glands, cortex of

Deoxyribonuclease: An enzyme secreted by exocrine ↑ pancreas that degrades ↑ deoxyribonucleoprotein.

Deoxyribonucleic acid (DNA): ↑ Nucleic acid carrying almost all genetic information of cell in specific sequence of bases of its ↑ nucleotides. DNA is mainly located in ↑ nucleus, only about 2% is concentrated in ↑ mitochondria and ↑ centriole.

Hancock, R., Hughes, M.E.: Biology of the Cell *44*, 201 (1982); Novi, A.M.: Klin. Wochenschr. *54*, 961 (1976)

Deoxyribonucleic acid, amount in meiosis of: ↑ Meiosis

Deoxyribonucleic acid, amount in mitosis of: ↑ Mitosis

Deoxyribonucleic acid, detection of: ↑ Nucleic acids, detection of

Deoxyribonucleic acid, extrachromosomal: All DNA contained outside ↑ nucleus, i.e., DNA in ↑ mitochondria and ↑ centriole.

↑ Mitochondrial DNA

Cummings, D.J., Borst, P., Dawid, I.B., Weissman, S.M. (eds.): Extrachromosomal DNA. New York: Academic Press 1979

Deoxyribonucleic acid, in cell division: ↑ Meiosis; ↑ Mitosis

Deoxyribonucleic acid, mitochondrial: ↑ Mitochondrial DNA

Deoxyribonucleic acid, staining for: ↑ Nucleic acids, detection of

Deoxyribonucleic acid, structure of: A roughly 2-nm thick molecule of indefinite length composed of two right-handed helical nucleotide chains (Ch) forming a double helix around the same central axis. A nucleotide consists of a phosphoric acid molecule (P), the pentose sugar deoxyribose (D), and one base (B), which can be either adenine (A), thymine (T), cytosine (C), or guanine (G). Nucleotides of a chain are linked together by means of phosphate diester bridges with deoxyribose. The two chains are held together by two hydrogen bonds between adenine and thymine and by three hydrogen bonds between cytosine and guanine. Sequence of bases determines structure of ↑ gene. DNA is linked with proteins to form ↑ deoxyribonucleoprotein. (See biochemistry and molecular biology texts for further information (Modified after De Robertis et al. 1975)

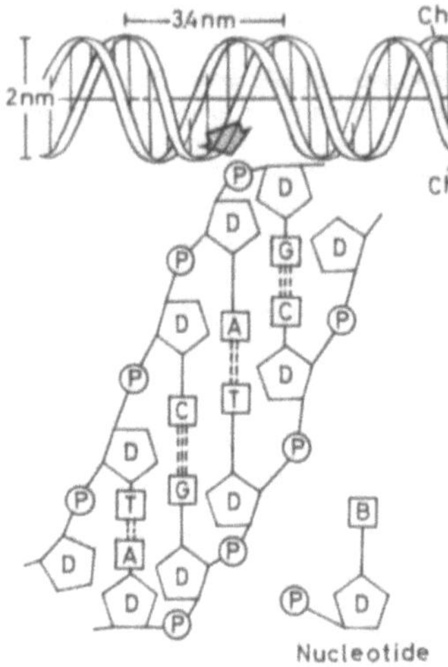

Alberts, B., Bray, D., Lewis, J., Raff, M., Roberts, K., Watson, J.: Molecular Biology of the Cell. New York: Garland Publishers 1983; De Robertis, E.D.P., Saez, A.F., De Robertis, E.M.F.Jr.: Cell Biology, 6th edn. Philadelphia, London, Toronto: Saunders Co. (1975); Hancock, R., Huges, M.E.: Biol. Cell *44*, 201 (1982); Structures of DNA. Cold Spring Harbor Symposia on Quantitative Biology, Vol. 47. New York: Cold Spring Harbor Laboratory 1983

Deoxyribonucleic acid, synthesis of: During S-phase of ↑ cell cycle, double helix chain (DNA) unwinds and separates into two single polynucleotide strands; each of these acts as a matrix to which single nucleotides (N) bind with their complementary bases – adenine (A) with thymine (T) and cytosine (C) with guanine (G). Thus, a new DNA molecule consists of half an original molecule and half newly synthesized material (= semiconservative duplication). It seems that duplication begins at attachment point of ↑ chromatin fiber, i. e., ↑ chromosome, to nuclear membrane. This process is controlled by DNA-polymerases and polynucleotide ligase. P = phosphate, D = deoxyribose. (See molecular biology texts for further information)

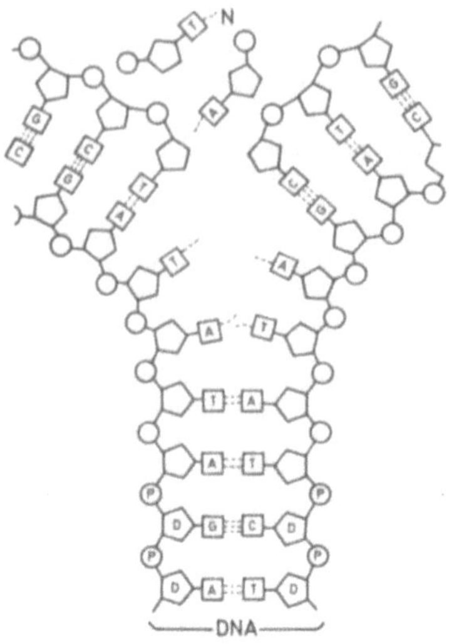

Kornberg, A.: DNA Replication. Oxford: W.H. Freeman & Co. 1980; Kornberg, A.: 1982 Supplement to DNA Replication. Oxford: W.H. Freeman & Co. 1982

Deoxyribonucleoprotein (DNP): A complex formed by ↑ deoxyribonucleic acid with ↑ histones. D. has a diameter of about 5–7 nm and coils into an irregularly convoluted strand, which is in fact ↑ chromatin fiber, with an average diameter of 25 nm.

Dermal papilla, of hair follicle: ↑ Hair papilla

Dermal papillae (papillae dermales*): Conical protrusions (P) of stratum papillare of ↑ dermis interdigitating with ↑ epidermal ridges (ER). D. are formed of ↑ loose connective tissue and contain ↑ lymphatic capillaries, nerve fibers, and capillary loops supplying nutrients to ↑ epidermis. At tip of some D., a ↑ corpuscle of Meissner (M) may be present. D., together with epidermal ridges, contribute to stronger fixation of epidermis to underlying tissue. (Fig. = human)

↑ Dermoepidermal junction; ↑ Vascular papillae

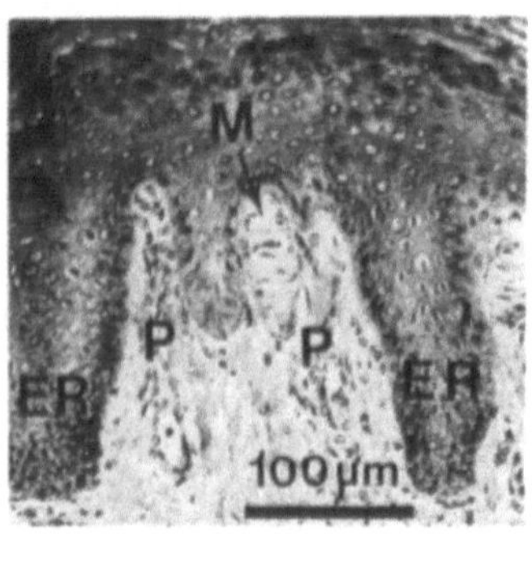

Dermatan sulfate: A ↑ glycosaminoglycan of ground substance of ↑ connective tissue proper. D. has a molecular weight of about 30 000 daltons; it is found in ↑ skin, ↑ tendons, heart ↑ valves, blood vessels, and connective tissue of ↑ lung.

↑ Connective tissue proper, ground substance of

Dermatoglyphics (fingerprints): A genetically determined configuration of horny ↑ epidermal ridges (E) and grooves in tips of fingers, toes, and volar surfaces of hands and feet. D. appear as loops, whorls, arches, and combinations of these. Surface ridges are caused by ↑ epidermis following contours of underlying ↑ dermal papillae (P) (= papillary lines). Being absolutely individual and permanent, D. are used for personal identification. (Fig., top = finger of newborn; fig., bottom = obliquely sectioned skin of adult human finger)

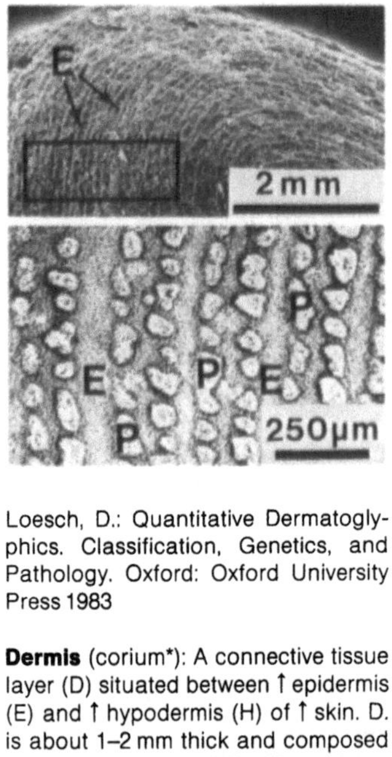

Loesch, D.: Quantitative Dermatoglyphics. Classification, Genetics, and Pathology. Oxford: Oxford University Press 1983

Dermis (corium*): A connective tissue layer (D) situated between ↑ epidermis (E) and ↑ hypodermis (H) of ↑ skin. D. is about 1–2 mm thick and composed of two layers: 1) Papillary layer (stratum papillare*, Sp) = a ↑ loose connective tissue forming ↑ dermal papillae, which interdigitate with ↑ epidermal ridges; 2) reticular layer (stratum reticulare*, Sr) = a ↑ dense connective tissue with bundles of ↑ collagen fibers running in different directions. Both layers are rich in ↑ elastic fibers; there is a gradual transition between the two layers. Within D. of ↑ areolae, ↑ penis, and ↑ scrotum there are some ↑ smooth muscle cells. Cross-striated muscles of facial expression end in D.

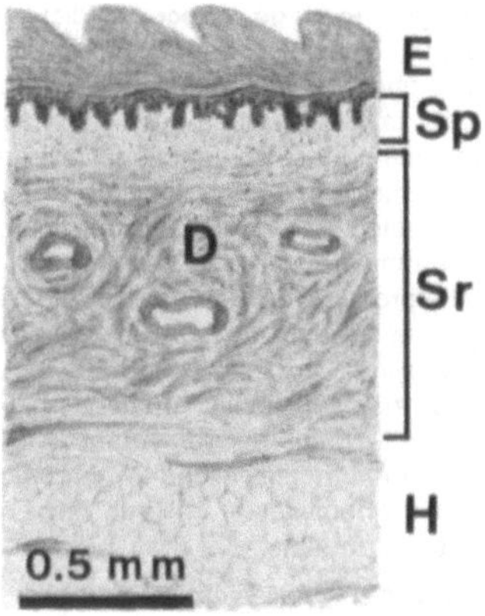

↑ Arrector pili muscle; ↑ Tunica dartos

Montagna, W., Bentley, J.P., Robson, R.L. (eds.): The Dermis. Adv. Biol. Skin *10*, (1970)

Dermoepidermal junction: Firm contact between ↑ epidermis (E) and underlying ↑ dermis (D). D. comprises: 1) An interlocking between ↑ dermal papillae (DP) and ↑ epidermal ridges (ER); 2) ↑ basal processes (P) of epidermal ↑ basal cells (BC). In mechanically exposed areas of ↑ skin (finger pads, soles, palms), D. is particularly strong. BL = ↑ basal lamina, HD = ↑ hemidesmosomes. (Fig., top = human; other figs. = mongolian gerbil and rat)

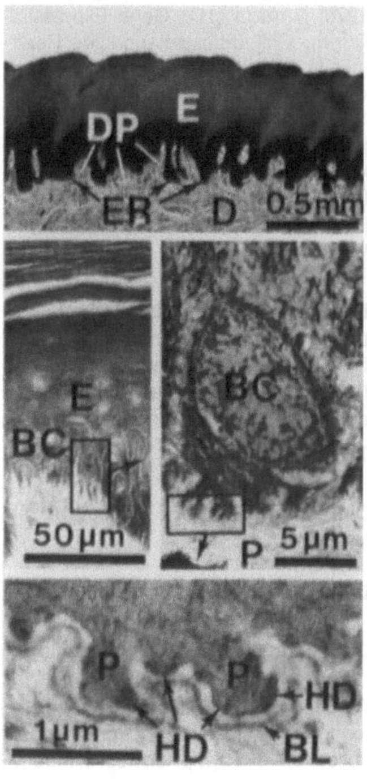

Briggman, R.A., Wheeler, C.E.: J. Invest. Dermatol. *65*, 71 (1975); Briggman, R.A.: J. Invest. Dermatol. *78*, 1 (1982)

Descemet's membrane (lamina limitans posterior*): The very thick (5–10 µm) and strongly ↑ PAS-positive ↑ basal lamina (DM) of ↑ endothelium (E) of ↑ cornea. In young individuals, D. is homogenous; in older individuals its anterior layer (AL) presents cross bands (arrowheads) about 100 µm apart and connected with one another by 10-µm-thick microfilaments. Tangentially sectioned (TS), the bands appear as numerous stacks of nodes (N) connected at same level by the microfilaments (arrows) to form a hexagonal pattern. This material is an atypical kind of collagen. D. ends at ↑ limbus as a circular thickening (↑ Schwalbe's line) considered to be one of insertion points for the ciliary muscles. C = corneal stroma. (Fig. of collagen after Jakus 1956)

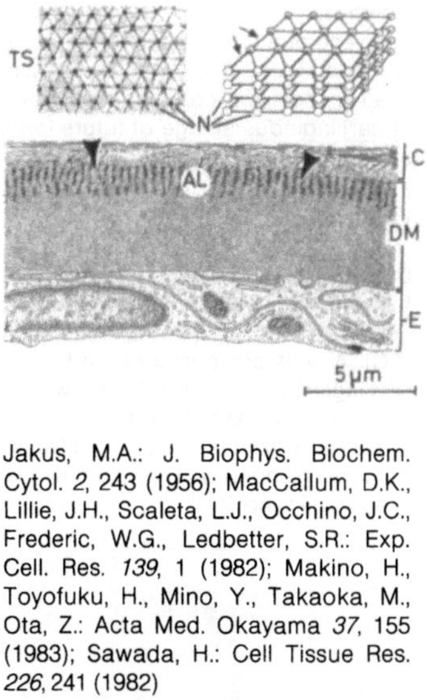

Jakus, M.A.: J. Biophys. Biochem. Cytol. *2*, 243 (1956); MacCallum, D.K., Lillie, J.H., Scaleta, L.J., Occhino, J.C., Frederic, W.G., Ledbetter, S.R.: Exp. Cell. Res. *139*, 1 (1982); Makino, H., Toyofuku, H., Mino, Y., Takaoka, M., Ota, Z.: Acta Med. Okayama *37*, 155 (1983); Sawada, H.: Cell Tissue Res. *226*, 241 (1982)

Descendent degeneration, of peripheral nerve fibers: ↑ Nerve fibers of peripheral nervous system, degeneration and regeneration of

Descending colon: ↑ Colon

Desmin microfilaments (intermediate-sized myofilaments): Class of ↑ intermediate microfilaments found in ↑ M- and ↑ Z-lines of ↑ striated muscle fibers and linking ↑ dense bodies of ↑ smooth muscle cells with one another and with ↑ sarcolemma.

Desmoglycogen: The only indirectly stainable, water-insoluble fraction of ↑ glycogen bound to different structures (about 80% of body's total glycogen content).

↑ Lyoglycogen

Desmosine: An amino acid, a constituent of ↑ elastic fibers; it acts, together with the iso-desmosine, on polymerization of elastic fibers.

Desmosomes (desmosoma s. macula adherens*): Disclike intercellular ↑ attachment devices scattered on cell surface. Each D. consists of two ↑ attachment plaques (AP), one on each opposing ↑ cell membrane. Plaques are separated by a 25-nm-wide space filled with low-density material often bisected by an electron-dense central or intermediate line (CL) representing condensation of outer borders of ↑ glycocalyx of both cells. Intercellular material in area of D. frequently presents a fine, perpendicular striation corresponding to transmembrane linker filaments (TLF); it contains ↑ sialic acid, ↑ glycosaminoglycans, and proteins acting together as a glue with stability depending upon presence of calcium ions. Mitochondria are present in vicinity of D., since D require energy for their existence. Tf = ↑ tonofilaments

↑ Organelles

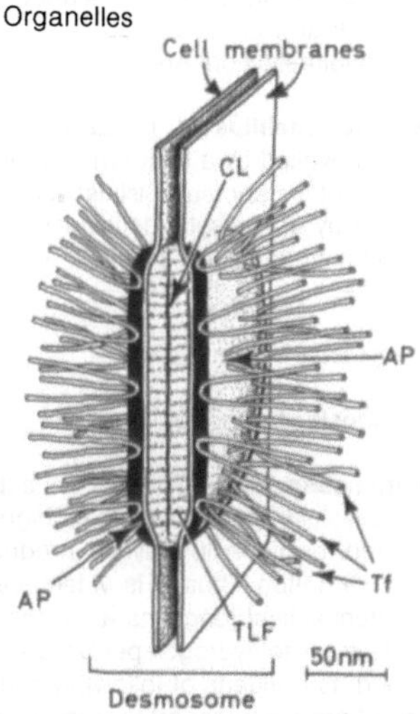

Allen, T.D., Potten, C.S.: J. Ultrastruct. Res. *51*, 94 (1975), Caputo, R., Peluchetti, D.: J. Ultrastruct. Res. *61*, 44 (1977); Gorbsky, G., Steinberg, M.S.: J. Cell Biol. *90*, 243 (1981); Leloup, R., Laurent, L., Ronveaux, M.-F., Drochmans, P., Wanson, J.-C.: Biol. Cell *34*, 137 (1979); Staehelin, L.A., Hull, B.E.: Sci. Am. *238/5*, 141 (1978)

Desquamation: The shedding off of the most superficial layer of any surface, e.g., of ↑ epidermis in form of dead keratinized scales.

Determination, of sex: ↑ Sex, determination of

Detumescence: The returning of an ↑ erectile tissue to its flaccid state after ↑ erection.

Deuterosomes (dense body, generative complex, procentriolar organizer): Membraneless clumps (D) of highly osmiophilic, finely filamentous substance, about 0.2 µm in diameter, situated in supranuclear cytoplasm and surrounded by ↑ fibrosomes (F). D. are believed to act as procentriolar organizers during acentriolar ↑ ciliogenesis. P = procentriole. (Fig. = ciliated oviduct cell, rat)

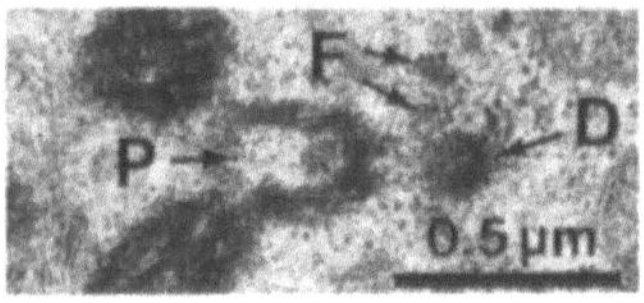

DFP: ↑ Diisopropyl fluorophosphate

Diabetes insipidus: A disease characterized by inability of ↑ kidney to concentrate ↑ urine, caused by a lack of ↑ antidiuretic hormone.

Diabetes mellitus: A metabolic disease provoked by a hypofunction of ↑ B-cells of ↑ pancreatic islets accompanied by an ↑ insulin deficiency and impairment of cell's ability to utilize glucose.

Diacytosis: ↑ Transcytosis

Diakinesis: A phase of ↑ meiosis.

Diaminobenzidine (DAB): A substance yielding an insoluble, highly colored, osmiophilic polymer product upon oxidation. Thus, D. is widely used in cytochemical reactions as a donor of electrons to hydrogen peroxide, employed as a means of following pathway of protein transport in a variety of cells. D. is important for locating cytochrome c, cytochrome oxidase, ↑ hemoglobin, ↑ myoglobin and for immunocytochemical identification of exogenous and endogenous proteins under light and transmission electron microscopes.

↑ Horseradish peroxidase

Barrnett, R.J. (ed.): Symposium in Electron Cytochemistry: Electron Microscopy Society of America, 32nd Annual Meeting. New York: J. Wiley & Sons, Inc. 1976

Diapedesis: 1) Active passage of ameboid blood cells and passive passage of ↑ erythrocytes through intact wall of blood vessels. 2) Active migration of ↑ wandering cells into an epithelium (E), e.g., ↑ lymphocytes (L) from ↑ lymphatic nodules (LN), ↑ macrophages and neutrophilic ↑ granulocytes from underlying connective tissue. D. of lymphocytes is frequent in ↑ tonsils, ↑ appendix, and ↑ ileum. (Fig. = lingual ↑ tonsil, human)

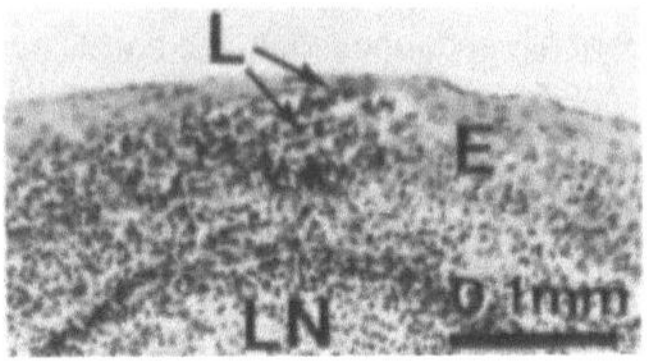

Diaphragm, of nuclear pores: ↑ Nuclear pores

Diaphysis* (shaft): The cylindrical midpiece of a long ↑ bone formed of ↑ compact bone. The term D. is also used to describe corresponding part of cartilaginous anlage of future long bone.

↑ Bone formation, indirect; ↑ Epiphysis; ↑ Metaphysis

Diarthrosis (articulatio synovialis*): A fully movable joint between two bones (B), separated by an articular cavity (AC). A D. is composed of ↑ articular cartilages (HC), ↑ capsule (C) with ↑ synovial membrane (SM), and, in some cases, an articular ↑ disc, ↑ meniscus, or ↑ labia glenoidalia. Articular cavity is filled with ↑ synovial fluid.

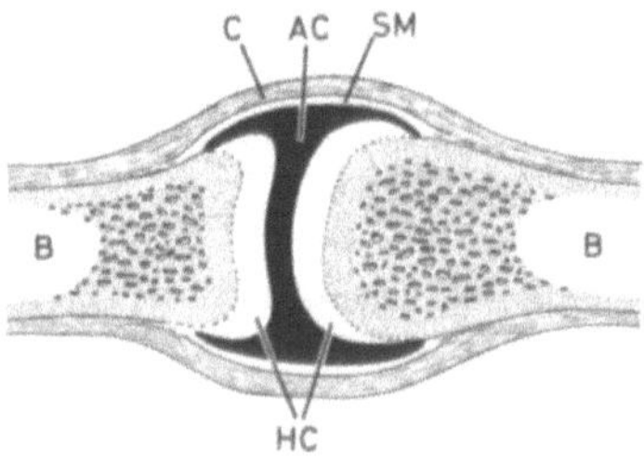

Diascopic illumination: ↑ Illumination, diascopic

Diaster* (aster filialis*): A double star figure of ↑ chromosomes occurring in anaphase of ↑ mitosis.

Dictyate stage, of ovocyte: ↑ Dictyotene stage

Dictyosomes: The former term for ↑ Golgi fields; still used in botany.

Dictyotene stage, of oocyte (dictyate stage): The specific "resting" period for ↑ chromosomes in all primary ↑ oocytes taking place between prophase and metaphase of reductional division of ↑ meiosis. Morphologically, D. is characterized by a very fine network of ↑ chromatin. D. may persist throughout all reproductive life or until oocyte begins its preovulatory growth. It is believed that very long D. (15–30 years) does not sufficiently protect genetic material against environmental factors and could therefore lead to ↑ chromosomal anomalies in children born by aged parturients.

↑ Down's syndrome; ↑ Oogenesis; ↑ Ovarian follicles, growth of

Differential centrifugation: ↑ Centrifugation, density gradient

Differentiation, of cells: The process of gradual specialization of cells to functions definitively different from those of original cell type. D. is accompanied by new activities of ↑ genes and consecutively different specific proteins and morphology. Under normal conditions, once a cell has differentiated, it cannot revert to undifferentiated state.

↑ Anaplasia; ↑ Regeneration

Flickinger, R.A.: Int. Rev. Cytol. *75*, 229 (1982); Gehring, J. (ed.): Genetic Mosaics and Cell Differentiation. Berlin, Heidelberg, New York: Springer-Verlag 1978; Gurdon, J.B.: Gene Expression During Cell Differentiation. In: Readings in Genetics and Evolution. Chap. 4. New York: Oxford University Press 1973; Monroy, A. (ed.): Cell Differentiation. Amsterdam: Elsevier/North Holland: 1978; Nover, L., Luckner, M., Parthier, B. (eds.): Cell Differentiation. Berlin, Heidelberg, New York: Springer-Verlag 1982

Differentiation, of histological sections: The partial removal of dye(s) from overstained histological sections in order to accentuate staining differences of tissue components.

↑ Staining, regressive

Diffraction, electron: ↑ Electron diffraction

Diffraction, optic: ↑ Optic diffraction

Diffraction, X-ray: ↑ X-ray diffraction

Diffuse lymphatic tissue, of gastrointestinal tract: A noncircumscribed collection of ↑ lymphocytes, ↑ macrophages, ↑ plasma cells, and eosinophilic ↑ granulocytes found in ↑ laminae propriae and ↑ telae submucosae. Number and predominance of type of cited cells depends upon bacterial flora and presence or absence of intestinal parasites. Together with overlying epithelium, D. produces IgA, which lines luminal surface and acts as an "antiseptic paint."

↑ Immunoglobulins

Walker, W.A., Hong, R.: J. Pediatr. *83*, 517 (1973)

Diffusion channels: Minute passages (about 1 nm in diameter) traversing presynaptic membrane and connecting ↑ synaptic vesicles with ↑ synaptic cleft. Synaptic vesicles discharge their contents through D. into synaptic cleft; D. are inconstant features since they occur only at moment of synaptic transmission.

↑ Bouton terminal; ↑ Synapse, chemical

Akert, K.: Klin. Wochenschr. *49*, 509 (1971)

Digestion, intracellular: Process of breakdown, by lysosomal enzymes, of either foreign substances introduced by ↑ phagocytosis into cell body (↑ heterophagy) or of cell's own parts (↑ autophagy).

Digestive system: ↑ Gastrointestinal tract

Digestive tract: ↑ Gastrointestinal tract

Digital arteriovenous anastomosis: ↑ Glomus

Dihydroxyphenylalanine (DOPA): Product of conversion of amino acid tyrosine under influence of ↑ tyrosinase. D. is used for cytochemical demonstration of ↑ melanocytes.

Diiodotyrosine (DIT): Result of iodination of tyrosine with oxidized iodide ion. One molecule of D. coupled with one molecule of monoiodotyrosine forms ↑ triiodothyronine; two molecules of D. form ↑ thyroxine.

↑ Thyroid follicular cells, biosynthesis of hormones in

Diisopropyl fluorophosphate (DFP): A substance irreversibly blocking ↑ acetylcholinesterase.

Dilator muscle (musculus dilatator pupillae*): A thin layer of radially disposed, pigmented ↑ myoepithelial cells (D) between stroma (S) and posterior pigment epithelium (PE) of ↑ iris. Anterior half of D. is free of ↑ melanin and therefore visible as a clear band. D. is of ectodermal origin and innervated by sympathetic fibers; its contraction enlarges pupillary diameter. (Fig. = pig)

Dilator muscle, of Rudinger: ↑ Anal canal

Dioptric media: Transparent and refractive structures of ↑ eye comprising ↑ aqueous humor, ↑ cornea, ↑ lens, and ↑ vitreous body.

Diphyletic theory, of hematopoiesis: ↑ Hematopoiesis, theories of

Diploë*: A layer of cancellous ↑ bone (D) between outer and inner compact bone tables (T) of flat skull bones.

↑ Bone, flat, of skull

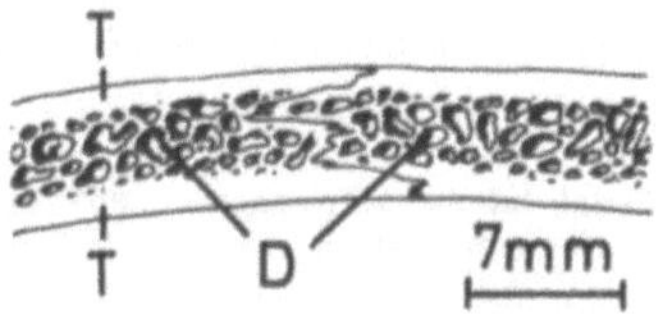

Diploidy (cellula diploidea*): The state of most ↑ somatic cell nuclei, containing twice ↑ haploid chromosomal number (2 × 23 = 46 in man), whereby one member of each ↑ chromosome pair is derived from the father and one from the mother.

↑ Polyploidy

Diplosome (diplosoma*): A pair of ↑ centrioles (Ce) oriented at right angles to one another. (Fig. = ↑ chief cell of parathyroid gland, rat)

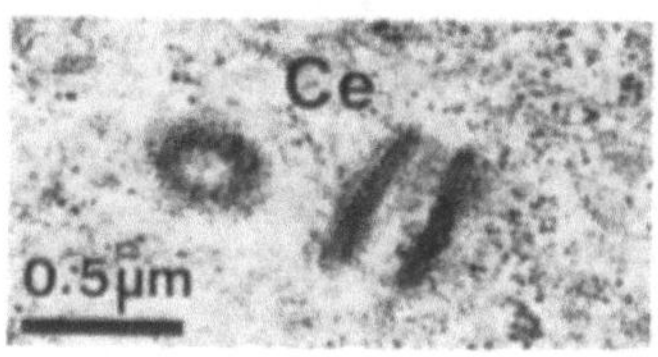

Diplotene: ↑ Meiosis

Direct nuclear division: ↑ Amitosis

Disc, articular: ↑ Articular disc

Disc, intercalated: ↑ Intercalated disc

Disc, intervertebral: ↑ Intervertebral disc

Discochondria: Special kind of ↑ mitochondria (D) in form of asymmetric discs. Thickened part of D. contains tubules (T), whereas flattened part is characterized by bulbous expansions (B) of inner mitochondrial membrane. D. are classified among ↑ mitochondria with tubules; they have been found in cells of adrenal cortex of hedgehog. Function unknown.

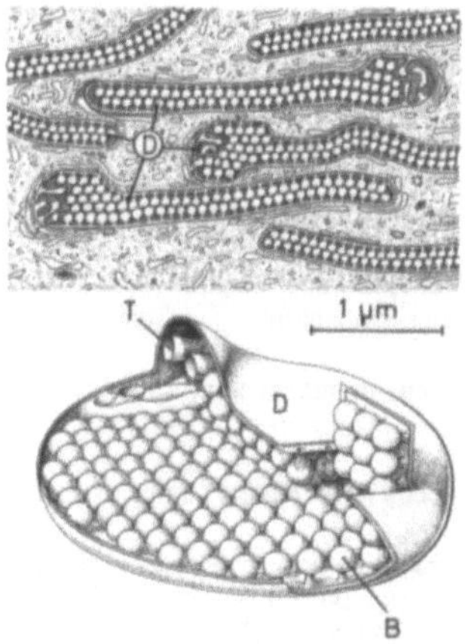

Lindner, E.: Z. Zellforsch. *72*, 212 (1966)

Discoid vesicles (fusiform vesicles): Lenticular structures (DV), 0.3–0.8 µm long and 0.12–0.15 µm wide, found in apical cytoplasm of ↑ facet cells of ↑ transitional epithelium; particularly abundant in empty ↑ urinary bladder. D. are lined by a trilaminar asymmetric membrane of same thickness (about 12 nm) and structure as luminal plasmalemma. Inner leaflet (inset: arrowhead) is thicker (4.3 nm) than cytoplasmic leaflet (2.8 nm). D. are synthesized by Golgi apparatus, grouped in stacks, and transported to cell's luminal surface where they fuse with apical plasmalemma. This gives a very irregular outline to free cell surface. Due to stiffness of membrane segments of D., earlier D. can be recognized (asterisks) as shallow concavities, so-called ↑ plaques, of the cell's luminal surface (c). D. are believed to represent a membrane reserve in case of epithelial distension (filling of urinary

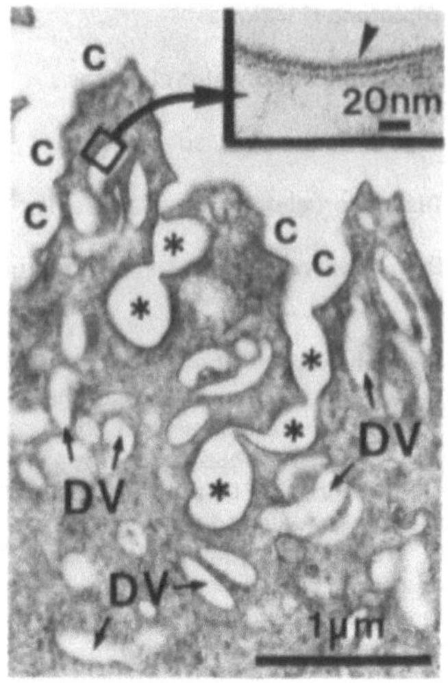

bladder) and for replacement of plaques. (Figs. = monkey)

Alroy, J., Merk, F.B., Morré, J.D., Weinstein, R.S.: Anat. Rec. *203*, 429 (1982); Severs, N.J., Warren, R.C., Barnes, S.H.: J. Ultrastruct. Res. *77*, 160 (1981)

Discontinuous capillaries: ↑ Capillaries, sinusoidal

Discs, membranous: ↑ Membranous discs, of photoreceptors

Discus nervi optici: ↑ Optic papilla

Disse's space (perisinusoidal space, spatium perisinusoideum*): The narrow perivascular space (D) between wall of ↑ liver sinusoid (S) and ↑ liver parenchymal cells (H). D. contains some bundles of ↑ reticular microfibrils and occasional ↑ perisinusoidal cells. Through openings in sinusoidal wall (arrowhead), blood plasma fills D. and comes into contact with surface and microvilli (Mv) of liver parenchymal cells. Thus, D. are of great importance in liver function and in formation of liver lymph. (Fig. = rat)

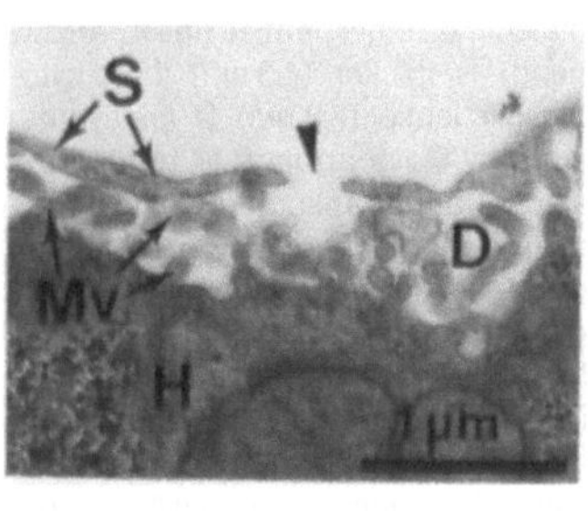

Distal convoluted tubule (pars convoluta partis distalis tubuli nephroni*): Part of ↑ nephron situated between straight portion of ↑ distal tubule and arched ↑ collecting tubule. D. (asterisks) is lined by ↑ simple cuboidal epithelium; it is about 5 mm long and 20–50 μm wide. Nuclei of cells lie in ↑ apical pole close to lumen. D. is situated exclusively in cortex; a short segment of D. contacts ↑ renal corpuscle between arteriola ↑ afferens and efferens; this segment is the ↑ macula densa. (Fig. = human)

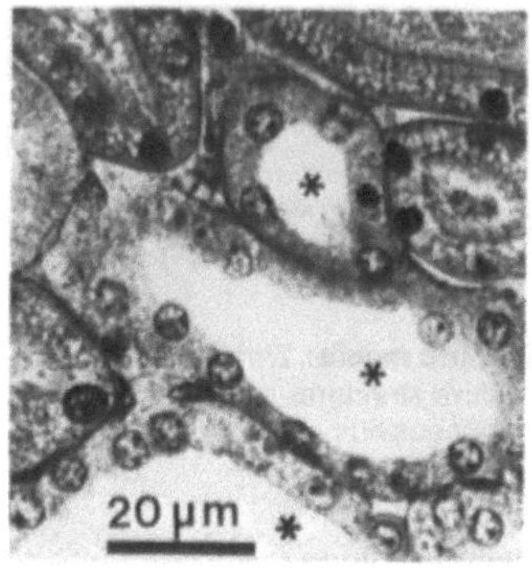

Distal tubule, of nephron (pars distalis tubuli nephroni*): Segment of ↑ nephron situated between straight part of ↑ proximal tubule and arched ↑ collecting tubule. D. is composed of a straight portion and a ↑ distal convoluted tubule with ↑ macula densa.

↑ Distal tubule of nephron, cells of

Distal tubule of nephron, cells of (nephrocyte, cellula cuboidea*): Cuboidal cells with a round nucleus in apical position and richly infolded basal plasmalemma forming a well-developed ↑ basal labyrinth (BLt) with long mitochondria (M) in its compartments. Cytoplasm also contains a few cisternae of rough endoplasmic reticulum, a supranuclear Golgi apparatus (G), ↑ centrioles, occasional lysosomes, and a moderate amount of free ribosomes. Apical plasmalemma forms some short irregular microvilli. There are only a few ↑ micropinocytotic vesicles in apical cytoplasm. D. are connected with neighboring cells by ↑ junctional complexes (J) and numerous lateral interdigitations, also participating in formation of basal labyrinth. D. lie on a basal lamina (BL).

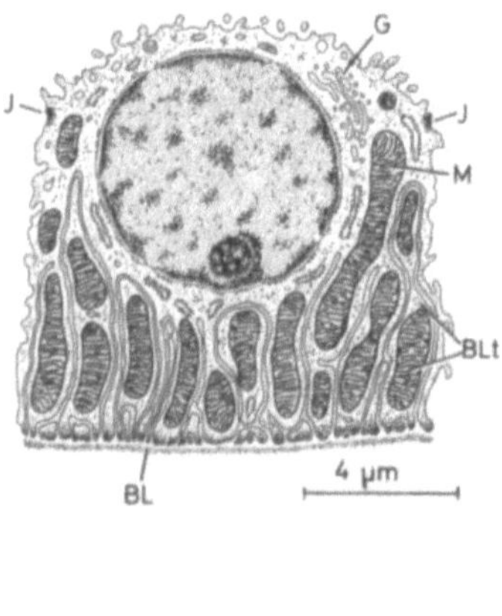

↑ Proximal tubule of nephron, cells of

Distributing arteries: ↑ Muscular arteries

DIT: ↑ Diiodotyrosine

Diverticula, of gallbladder: ↑ Gallbladder

Division, of cells: ↑ Meiosis; ↑ Mitosis

DNA: ↑ Deoxyribonucleic acid

DNP: ↑ Deoxyribonucleoprotein

DOPA: ↑ Dihydroxyphenylalanine

Dopamine: An intermediate in tyrosine metabolism, precursor of ↑ norepinephrine and ↑ epinephrine. Present in adrenal medulla and also in cells of cerebral basal ganglia where it acts as a ↑ neurotransmitter substance.

↑ Synapse, aminergic

Lackovic, Z., Neff, H.N.: Life Sci. *32*, 1665 (1983)

Dopamine β-hydroxylase: An enzyme present in ↑ catecholamine storage granules responsible for converting ↑ dopamine to ↑ norepinephrine.

↑ Dense-core vesicles

Down's syndrome (mongolism, trisomy 21): A ↑ chromosomal anomaly characterized by three number 21 chromosomes; D. is a consequence of ↑ nondisjunction.

Draining veins: Large veins of supracardiac region of body characterized by relatively poor development of smooth muscle tissue in both tunica media and tunica adventitia since these veins return blood to heart by gravitation. ↑ Vena cava superior and external jugular vein are D.

"Drumstick": A small elongated appendix (D) of nucleus of neutrophilic ↑ granulocyte of females, contains one clumped arm of one of ↑ X-chromosomes.

↑ Barr body; ↑ Heterochromatin, facultative; ↑ Sex chromatin

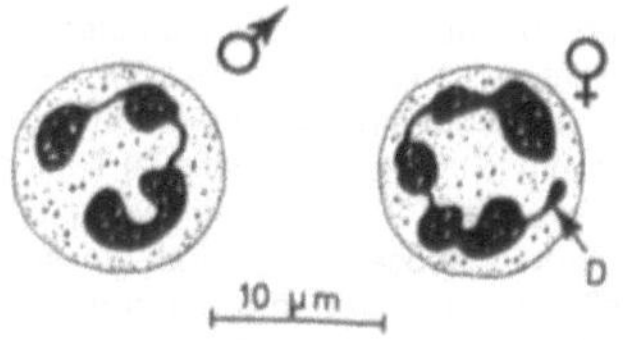

Duct (ductus*, ductulus*): A cellular tube serving as passage for air, secretory and excretory products, ↑ lymph, ↑ spermatozoa, and as a site for location of some sensory organs.

Duct, cochlear: ↑ Cochlear duct

Duct, cystic: ↑ Cystic duct

Duct, of Santorini: ↑ Pancreatic ducts

Duct, of Wirsrung: ↑ Pancreatic ducts

Duct, thoracic: ↑ Thoracic duct

Ducts, alveolar, of lung: ↑ Alveolar ducts

Ducts, ejaculatory: ↑ Ejaculatory ducts

Ducts, excretory, of testis: A system of channels conducting ↑ spermatozoa from ↑ seminiferous tubules (ST) to ↑ urethra (U). D. are: ↑ tubuli recti (TR), ↑ rete testis (RT), ↑ ductuli efferentes (DE), ↑ ductus epididymidis (DEp), ↑ ductus deferens (DD), and ↑ ejaculatory duct (ED).

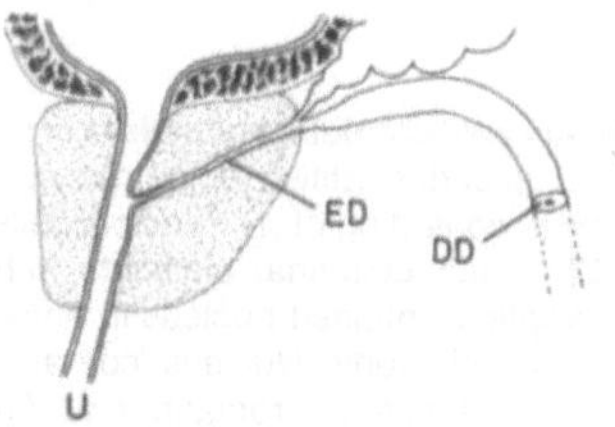

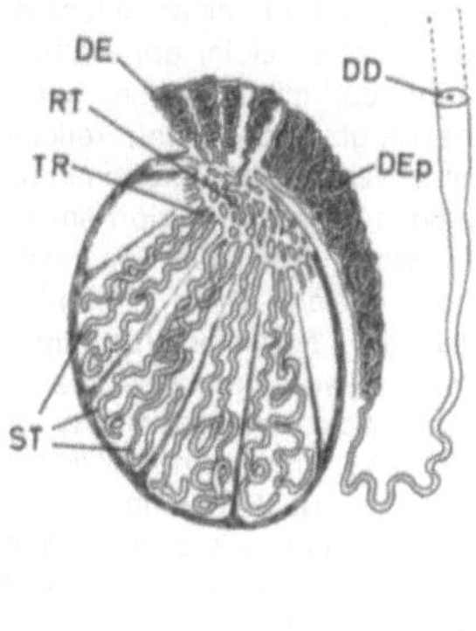

Ducts, intercalated: ↑ Intercalated ducts

Ducts, lactiferous: ↑ Lactiferous ducts

Ducts, of Bellini: ↑ Collecting tubules, of kidney

Ducts, of Luschka: Peculiar, blind, tubular structures situated in tunica adventitia of hepatic surface and neck of ↑ gallbladder. D. do not open onto gallbladder surface; they are thought to represent remnants of fetal bile ducts.

Ducts, of mammary gland: ↑ Lactiferous ducts

Ducts, of salivary glands proper (excretory ducts): A system of channels of variable length conducting ↑ saliva from ↑ acini (A) and ↑ mucous tubules (MT) to vestibule or floor of mouth. To D. belong ↑ intercalated ducts (ID), ↑ striated ducts (SD), ↑ interlobular ducts (ILD), and the main ducts (MD), i.e., ↑ parotid duct, ↑ submandibular duct, and sublingual duct.

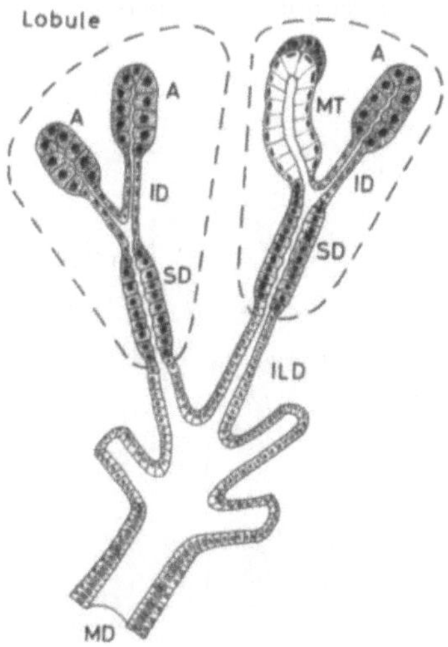

Ducts, pancreatic: ↑ Pancreatic ducts

Ducts, papillary: ↑ Collecting tubules, of kidney

Ducts, striated, of salivary glands: ↑ Striated ducts

Ductuli biliferi: ↑ Bile ductules

Ductuli efferentes (ductuli efferentes testis*): 8–20 canaliculi connecting ↑ rete testis with ↑ ductus epididymidis. Each D. is about 18 cm long but coiled into a ↑ conus vasculosus; these come together, forming head of ↑ epididymis. Structure: 1) Tunica mucosa = alternating patches of ↑ simple and ↑ pseudostratified columnar epithelium (E), giving it a festooned outline. 2) Tunica muscularis (TM) = a thin layer of ↑ smooth muscle cells; their peristaltic contraction pushes ↑ spermatozoa toward ductus epididymidis. 3) Tunica adventitia (TA) = a richly vascularized and innervated ↑ loose connective tissue. S = ↑ spermatozoa (Fig. = human)

↑ Ductuli efferentes, epithelium of

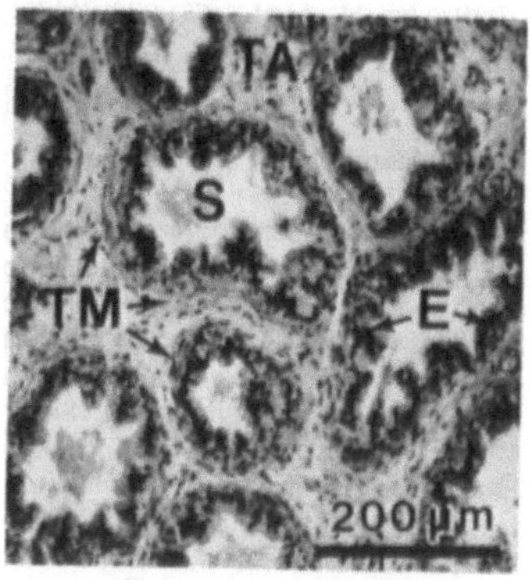

Ductuli efferentes, epithelium of: Alternating patches of a ↑ simple columnar nonciliated epithelium (NCE) and patches of ↑ pseudostratified epithelium (PE) composed of ciliated (CC) and basal cells (BC). 1) Nonciliated cells have a predominantly basal nucleus, a well-developed Golgi apparatus, and a variable number of ↑ secretory granules (S). ↑ Apical pole is studded with microvilli (Mv); in some cases it may bulge into lumen. It is thought that nonciliated cells are of absorptive

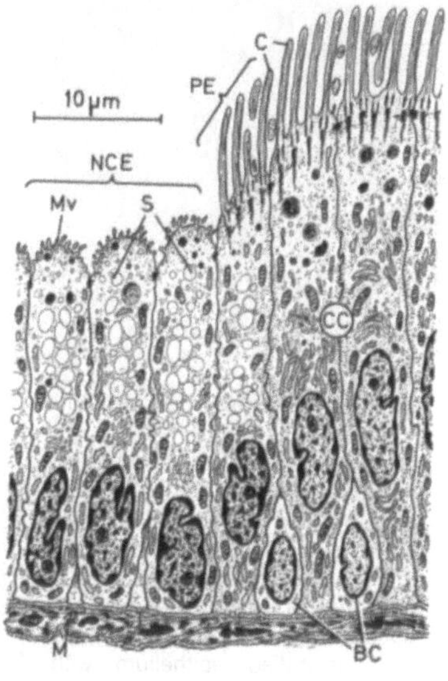

and secretory nature. 2) Ciliated cells are longer than nonciliated cells; nuclei of ciliated cells, occupy middle portion of cell bodies. All cell ↑ organelles are well developed; ↑ cilia (C) beat toward ↑ ductus epididymidis and transport ↑ spermatozoa. 3) Basal cells have a conical shape, an oval or round nucleus, and poorly developed organelles. They probably serve as replacements for ciliated cells. M = tunica muscularis

Jones, R., Hamilton, D.W., Fawcett, D.W.: Am. J. Anat. *156*, 373 (1979); Morita, I.: Arch. Histol. Jpn *26*, 341 (1966)

Ductuli lacrimales: ↑ Lacrimal ducts

Ductus choledochus* (common bile duct): A canal, roughly 6–8 cm long and 3–5 mm wide, part of extrahepatic ↑ bile pathways. D. begins at point of junction with ↑ cystic duct and ends in ↑ ampulla hepatopancreatica. Structure: 1) Tunica mucosa (TM): a) epithelium = a ↑ simple tall columnar epithelium; b) lamina propria = a ↑ loose connective tissue rich in ↑ elastic fibers and ↑ mucous glands (G). 2) Tunica muscularis (TMu) = an incomplete layer of spirally and circularly arranged ↑ smooth muscle cells; in duodenal wall, this layer becomes complete and forms sphincter muscle (musculus sphincter ductus choledochus). 3) Tunica adventitia (TA) = a loose connective tissue.

↑ Sphincter, of Oddi

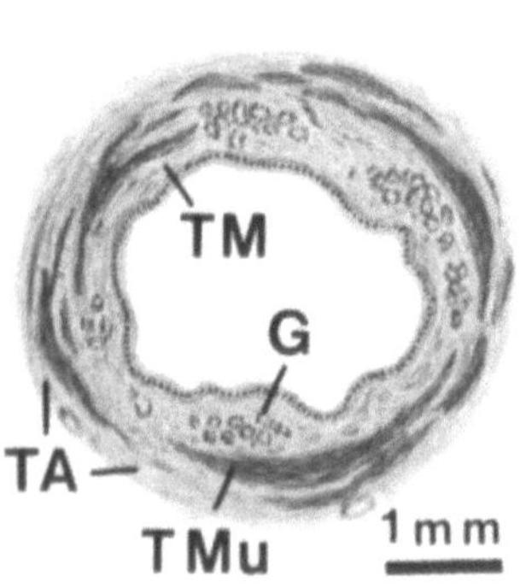

Ductus cochlearis: ↑ Cochlear duct

Ductus deferens*: A musculoepithelial tube, about 30 cm long and 3–5 mm thick, which begins in tail of ↑ epididymis and extends to ↑ ejaculatory ducts. Structure: 1) Tunica mucosa (TM) is longitudinally folded and consists of: a) epithelium (E) = a ↑ pseudostratified epithelium with ↑ stereocilia; b) lamina propria (LP) = a very thin layer of ↑ dense connective tissue filling epithelial folds. 2) Tunica muscularis (TMu) = a very thick layer of helicoidally arranged ↑ smooth muscle cells disposed in inner longitudinal (IL), middle circular (MC), and outer longitudinal layers (OL). On ↑ sarcolemma of almost every smooth muscle cell end autonomic motor nerve endings, which explains fast contractions of D. during ↑ ejaculation. 3) Tunica adventitia (TA) = a highly vascularized and innervated ↑ loose connective tissue of ↑ spermatic cord. Analysis of complicated arrangement of smooth muscle bundles of tunica muscularis has shown that at start of ejaculation it begins to contract, resulting in a shortening of the whole D. As a consequence, epididymal contents pass into D. During subsequent contractions at end of ejaculation, epididymal contents are injected under high pressure into ejaculatory ducts.

↑ Ampulla, of ductus deferens

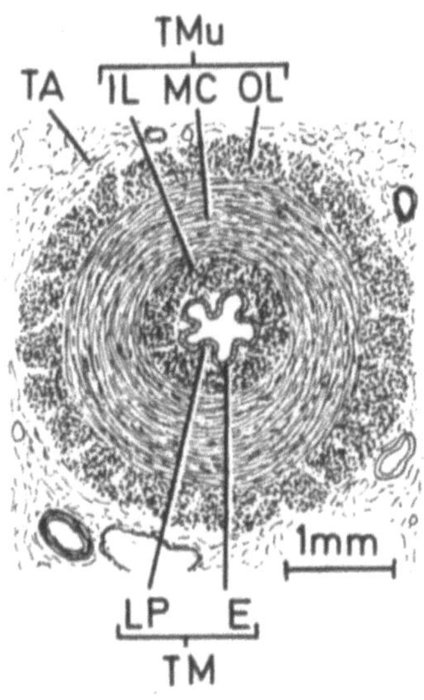

Goerttler, K.: Gegenbaurs morphol. Jahrb. *74*, 550 (1934); Sjöstrand, N.O.: Smooth Muscle of Vas Deferens and Other Organs in the Male Reproductive Tract. In: Bülbring, E., Brading, A., Jones, A.W., Tomita, T. (eds.): Smooth Muscle. London: Edward Arnold 1981

Ductus endolymphaticus: ↑ Endolymphatic duct

Ductus epididymidis*: A highly convoluted single canal forming body and tail of ↑ epididymis. D. is formed by gradual fusion of ↑ ductuli efferentes and continues into ↑ ductus deferens. Structure: 1) Tunica mucosa (TM) = a tall ↑ pseudostratified columnar epithelium (E) with ↑ stereocilia. 2) Tunica muscularis (TMu) = a thin layer of circularly disposed ↑ smooth muscle cells; toward tail, musculature becomes thicker and double-layered; it is finally continuous with that of ductus deferens with which it contracts simultaneously during ↑ ejaculation. 3) Tunica adventitia (TA) = a ↑ loose connective tissue constituting interstitium of epididymis. D. is richly innervated; it is a reservoir for ↑ spermatozoa. Here, they are nonmotile because of acid reaction of epididymal secretion. Total length of straightened D. is about 4 m. S = spermatozoa. (Fig. = human)

↑ Ductus epididymidis, epithelium of

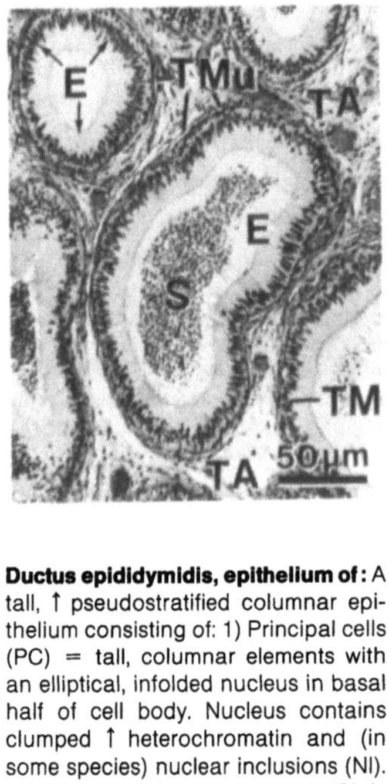

Ductus epididymidis, epithelium of: A tall, ↑ pseudostratified columnar epithelium consisting of: 1) Principal cells (PC) = tall, columnar elements with an elliptical, infolded nucleus in basal half of cell body. Nucleus contains clumped ↑ heterochromatin and (in some species) nuclear inclusions (NI). Supranuclear Golgi apparatus (G) is enormous, mitochondria numerous. The rough endoplasmic reticulum is well developed and predominantly located in basal cell portion; smooth endoplasmic reticulum is extensive above nucleus. Cell's ↑ apical pole is filled with ↑ lysosomes (L), ↑ lipofuscin granules, and ↑ multivesicular bodies (MVB). Apical plasmalemma forms a myriad of ↑ micropinocytotic vesicles and a bouquet of long ↑ stereocilia (St). Principal cells secrete many enzymes necessary for metabolism and maturation of ↑ spermatozoa. Principal cells are also involved in resorption of fluid leaving ↑ testis; they lie on basal lamina (BL). 2) Basal cells (BC) = small, round, or pyramidal cells adjacent to basal lamina lodged between bases of principal cells. ↑ Organelles are few and relatively simple in structure. Cytoplasm may contain some ↑ lipid droplets. Basal cells are believed to divide and differentiate into principal cells. Among epithelial cells are intraepithelial lymphocytes and ↑ macrophages (about 2% of cell population), formerly called "halo cells."

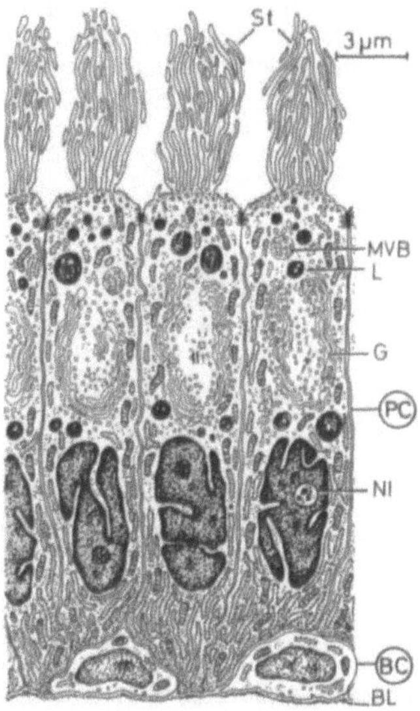

Abe, K., Takano, H., Ito, T.: Arch. Histol. Jpn *46*, 51 and 69 (1983); Cavicchia, J.C.: Cell Tissue Res. *201*, 451 (1979); Wang, Y.F., Holstein, A.F.: Cell Tissue Res. *233*, 517 (1983); Yeung, C.H., Cooper, T.G.: Cell Tissue Res. *226*, 407 (1982)

Ductus pancreaticus: ↑ Pancreatic ducts

Ductus perilymphaticus: ↑ Perilymphatic duct

Ductus reuniens: A fine channel, part of membranous ↑ labyrinth, uniting saccule and ↑ cochlear duct.

Ductus semicirculares: ↑ Semicircular ducts

Ductus thoracicus: ↑ Thoracic duct

Ductus utriculosaccularis: A minute canal connecting ↑ utricle with ↑ saccule of ↑ labyrinth.

Duodenal glands (Brunner's glands, glandulae duodenales*): Richly branched tubular glands (DG), located in ↑ tela submucosa (TS) of ↑ duodenum. They also penetrate ↑ plicae circulares and are sometimes found above lamina muscularis mucosae (LMM). D. may extend into pylorus for a short distance; as a rule, they disappear in distal two-thirds of duodenum. Excretory ducts (E) open into crypts of ↑ Lieberkühn (CL). Secretory cells are cuboidal or low-columnar; their nucleus is flattened and basal with a large nucleolus; cytoplasm contains many rod-shaped mitochondria, an expanded supranuclear Golgi apparatus (G), ↑ ergastoplasmlike rough endoplasmic reticulum, and, in ↑ apical pole, numerous ↑ secretory granules of variable density (SG) with a glycoproteinaceous content. Occasional ↑ Paneth's cells, ↑ oxyntic cells, ↑ goblet cells, and ↑ enterochromaffin cells may also be found in D. D. produce a clear ↑ mucoid secretion (pH 8.2–9.3) for neutralizing gastric juice, thus protecting duodenal mucosa. Secretion also contains a proteolytic enzyme activated by hydrochloric acid, an enterokinase which transforms trypsinogen to ↑ trypsin, and ↑ urogastrone. TM = tunica mucosa

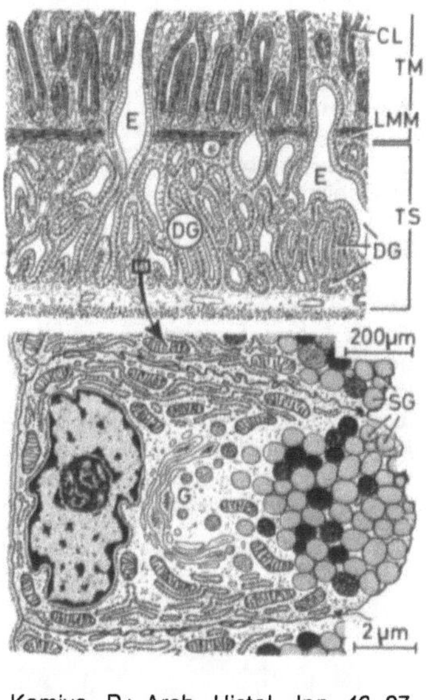

Kamiya, R.: Arch. Histol. Jpn *46*, 87 (1983); Krause, W.J.: Am. J. Anat. *162*, 167 (1981); Poddar, S., Jacob, S.: Histochemistry *65*, 67 (1979); Scott, C.A., Flickinger, C.J.: Anat. Rec. *206*, 267 (1983); Treasure, T.: J. Anat. *127*, 299 (1978)

Duodenum*: The initial segment of small intestine, about 25 cm long, situated between ↑ pylorus and flexura duodenojejunalis. Structure: 1) Tunica mucosa (TM), thrown into ↑ plicae circulares (PC) and consisting of: a) epithelium (E) = a ↑ simple columnar epithelium composed of ↑ absorptive and ↑ goblet cells; epithelium is continuous with that of crypts of ↑ Lieberkühn (LC); b) lamina propria (LP) = a ↑ loose connective tissue forming ↑ intestinal villi (V); c) lamina muscularis mucosae (LMM) = a thin layer of ↑ smooth muscle cells. 2) Tela submucosa (TS) = a richly vascularized and innervated loose connective tissue containing numerous ↑ duodenal glands (DG). 3) Tunica muscularis (TMu) = smooth muscle tissue organized in an inner circular layer, stratum circulare (SC), and an outer longitudinal layer, stratum longitudinale (SL). 4) Tela subserosa (TSs) = a loose connective tissue. 5) Tunica serosa (TSe) = peritoneal ↑ mesothelium.

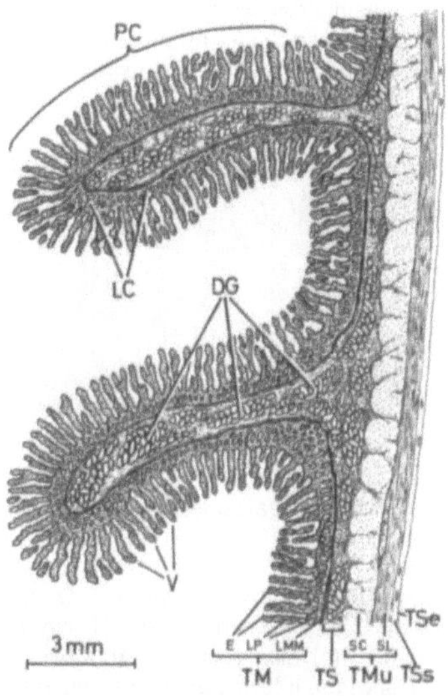

Dura mater* (pachymeninx): The thickest and hardest of ↑ meninges (D), made up of ↑ dense connective tissue indistinctly subdivided into two layers. Outer layer is thick, representing ↑ periosteum of skull ↑ bones and contains blood vessels (arrows), nerve fibers, and lymphatics for itself and for

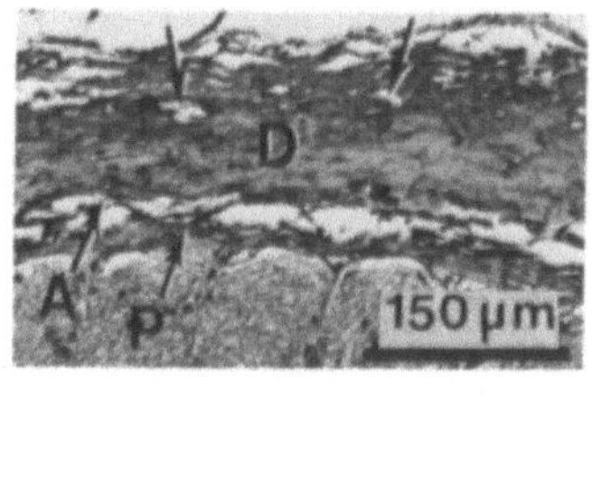

bone. Venous sinuses are also enclosed in this layer. Inner layer is thin, and its aspect facing ↑ subdural space is lined by a layer of ↑ mesothelium. In vertebral canal, D. is completely separated from ↑ periosteum of vertebrae by ↑ epidural space. D. envelops ↑ brain, ↑ spinal cord, and ↑ optic nerve. A = ↑ arachnoid, P = ↑ pia mater (Fig. = ↑ optic nerve, dog)

Dust cells: ↑ Alveolar macrophages

Dyad: 1) (chromosoma bivalens*): Two sister ↑ chromatids held together by ↑ centromere during pachytene stage of meiotic prophase I. A pair of D. form a ↑ tetrad.

↑ Meiosis

2) A term sometimes used to designate a characteristic asymmetric image (D) composed of a small expansion of discontinuous ↑ terminal cisterna of ↑ sarcoplasmic reticulum (SR) of ↑ cardiac muscle cells closely apposed to ↑ T-tubule (T). Z = ↑ Z-line (Fig. = rat)

↑ Triad

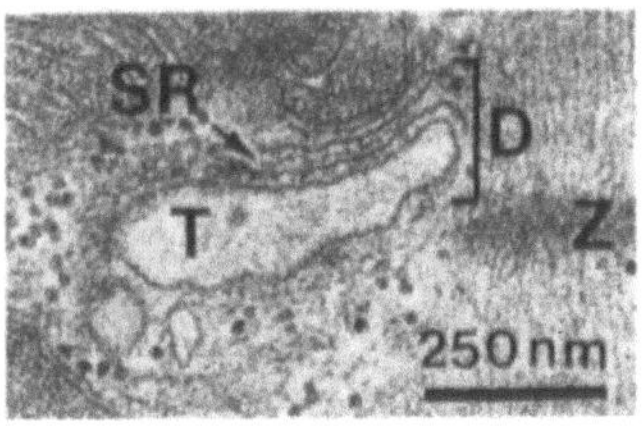

Dyes: Chemical substances used for ↑ staining.

Dynein: An ATP-splitting enzyme, similar to ↑ myosin found in ↑ axonemata of ↑ cilia and ↑ spermatozoon tails where it forms arms of A-microtubules. D. plays an important role in movements of cilia and flagella.

Baccetti, B., Burrini, A.G., Pallini, V., Renieri, T.: J. Cell Biol. *88*, 102 (1980); Goodenough, U.W., Heuser, J.E.: J. Cell Biol. *95*, 798 (1982); Johnson, K.A., Wall, J.S.: J. Cell Biol. *96*, 669 (1982); Warner, F.D., Mitchell, D.R.: Int. Rev. Cytol. *66*, 1 (1980)

E

Ear (auris*): The organ of hearing consisting of three parts: 1) External E. or auris externa* = ↑ auricle (A) and ↑ external auditory meatus (EAM); 2) middle E. or auris media* = ↑ tympanic membrane (TM), ↑ auditory ossicles (AO), ↑ tympanic cavity (TC), and ↑ auditory tube (AT); 3) internal E. or auris interna* = bony and membranous ↑ labyrinth (BML).

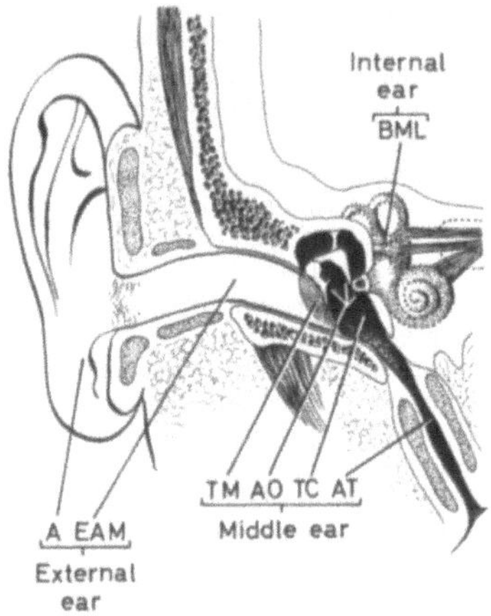

Ear drum: ↑ Tympanic membrane

Early erythroblast: ↑ Erythroblast, basophilic

EBI: ↑ Erythroblastic islands

Ebner's glands (glandula gustatoria*). Purely ↑ serous glands (E) situated among skeletal muscle fibers underlying vallum (V) of ↑ circumvallate papillae (CP). Excretory ducts of E. open into furrow. Function of E. is rinsing of taste-provoking molecules from trench. (Fig. = human)

↑ Taste buds, structure of

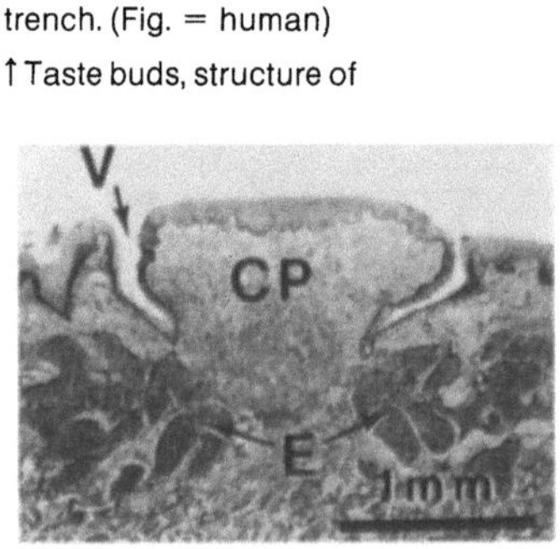

Ebner's lines: ↑ Cement lines

EC-cells: ↑ Enterochromaffin cells

EC_1-cells: ↑ Enterochromaffinlike cells

EC_2-cells: ↑ Motilin cells

Eccrine glands (glandula eccrina*): All glands employing ↑ eccrine mechanism of ↑ secretion.

Eccrine secretion (merocrine secretion): A mode of release of secretory product by ↑ exocytosis without any loss of cytoplasm. Membrane of ↑ secretory granules (S) fuses with apical plasmalemma (P) and contents of granules are discharged onto epithelial surface or into glandular lumen. The great majority of secretory cells function in this manner.

↑ Secretion

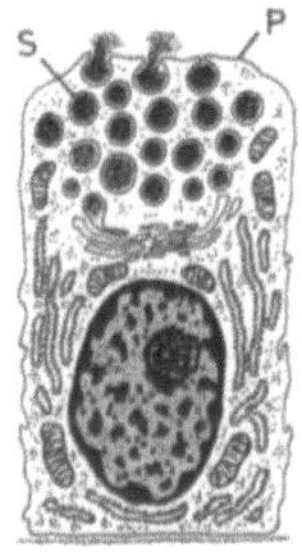

Eccrine sweat glands: ↑ Sweat glands, eccrine

E-cells (Feyerter's cells, clear cells): A concept of disseminated intraepithelial ↑ endocrine cells with several common morphological and functional characteristics. 1) E. lie scattered in various epithelia, have a predominantly clear cytoplasm, and contain basal ↑ secretory granules, which may be ↑ chromaffin, ↑ argyrophilic, ↑ argentaffin, or, rarely, unreactive. 2) E. are in close relation to ↑ adrenergic nerve fibers. 3) E. produce various hormones which enter circulation, but also act locally. 4) E. may transform themselves into carcinoids and carcinomas. To E. belong ↑ endocrine cells of ↑ gastrointestinal tract, cells of ↑ neuroepithelial bodies of lung, and urethral E. Concept of ↑ APUD-cells largely overlaps that of E.

Edmondson, N.A., Lewis, D.J.: Thorax *35*, 371 (1980); Hernandez-Vasquez, A., Will, J., Quay, W.B.: Cell Tissue Res. *189*, 179 (1978); Ratzenhofer, M.: Wien Klin. Wochenschr. *83*, 22 (1971)

E-cells, of pancreas: Polygonal cells with moderately osmiophilic secretory granules, 400–500 nm in diameter, found in ↑ pancreatic islets of dog and opossum; function unknown.

Munger, B.L., Caramia, F., Lacy, P.E.: Z. Zellforsch. *67*, 776 (1965)

ECF-A: ↑ Eosinophilic chemotactic factor-A

Echinocyte: ↑ Erythrocyte, crenated

ECL-cells: ↑ Enterochromaffinlike cells

Ectatic capillaries: ↑ Endometrium, vascularization of; ↑ Lacunae, of endometrium

Ectocervix: ↑ Cervix, of uterus

Ectoderm (ectoderma*): Outermost of primitive germ layers giving rise to ↑ epidermis, ↑ epidermal appendages, ↑ ameloblasts, sensory epithelia, ↑ lens, ↑ retina, epithelium of membranous ↑ labyrinth, all ↑ neurons, etc. (See embryology texts for further information)

Ectoplasm (cell cortex, cortical cytoplasm, exoplasm, exoplasma*): 1) E. of plasmalemma = a zone forming a rather viscous, peripheral, clear, cytoplasmic rim (E), about 50 nm thick, between ↑ cytoskeleton (C) of cell membrane (CM) and deeper cytoplasm (Cy). E. contains ↑ micropinocytotic vesicles (MV) and microfilamentous bundles (Mf) whose contraction may control exo- and endocytotic processes; E. is also capable of rapid sol-gel transformation. ↑ Microtrabeculae (Mt) anchor in E. 2) E. of cell = in some cells, especially motile ones (MC), E. is a particularly large (up to 2 µm) peripheral cytoplasm free of ↑ organelles

↑ Actin-binding protein; ↑ Osteoclast; ↑ Zones, of cytoplasm

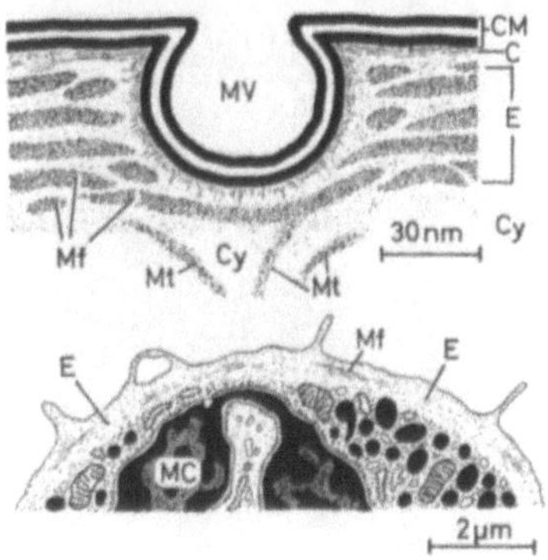

Edema: A local or generalized abnormal accumulation of water in ↑ ground substance of ↑ connective tissues. Most local E. is consequence of increased permeability of ↑ capillaries and ↑ venules provoked by ↑ histamine.

↑ Mast cells

EDTA (**e**thylene**d**iamine**t**etraacetic **a**cid): A chelating agent used for ↑ decalcification.

EF-face, in freeze-cleaving (**e**xtraplasmatic fracture **f**ace): The split half (EF) of cell membrane (CM) facing ↑ cytoplasm (C). E. shows only a few 9-nm particles (arrowheads) and some shallow pits (arrow). Complementary to E. is ↑ PF-face, facing away from cytoplasm. F = fracturing line in middle of lipid bilayer

↑ Cell membrane, freeze-cleaving of; ↑ Freeze-cleaving, terminology of

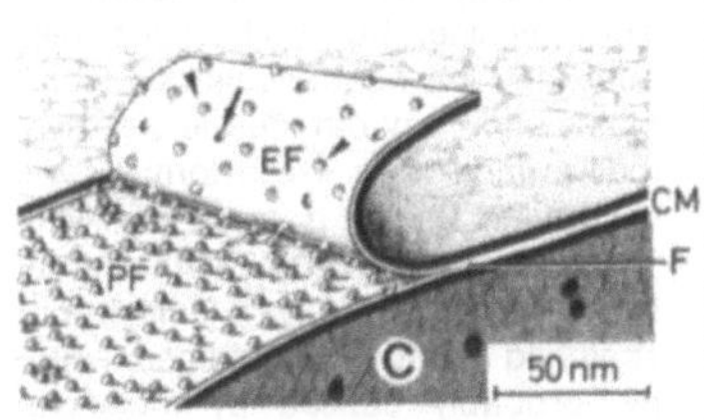

Effector cells: A collective term for both activated ↑ B- and ↑ T-lymphocytes and ↑ plasma cells.

↑ Immune response

Efferent: Outgoing

Efferent arteriole (arteriola efferens*): A thin-walled blood vessel (E) formed by fusion of ↑ glomerular capillaries (G). E. leaves ↑ renal corpuscle (RC) at its ↑ vascular pole, carrying blood away from corpuscle. E. of juxtamedullary ↑ nephrons turn toward medullary pyramids as arterial ↑ vasa recta; E. of cortical nephrons break up into a ↑ peritubular capillary network. A = ↑ afferent arteriole. (Fig. = monkey)

↑ Kidney, vascularization of

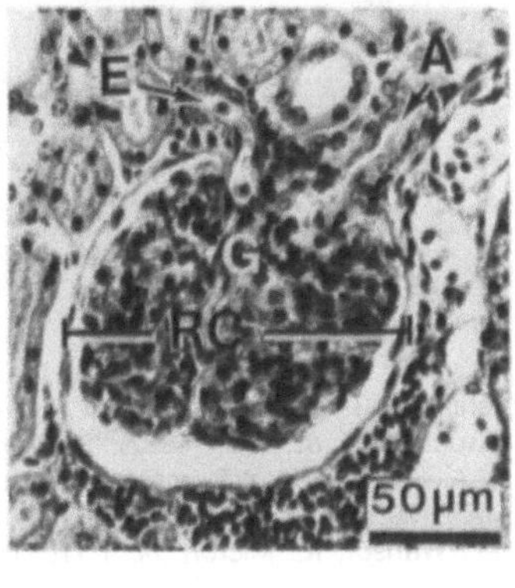

Efferent I, II, and III fibers: ↑ Neuromuscular spindle

Efferent nerve fibers, myelinated, terminations of: ↑ Motor end plate

Efferent nerve fibers, nonmyelinated, terminations of: ↑ Adrenergic nerve endings; ↑ Cholinergic nerve endings

Egg: ↑ Ovum

Ejaculate: Quantity of ↑ semen emitted during one ↑ ejaculation. Average volume of an E. is about 3.5 ml, containing 200–300 million ↑ spermatozoa.

Ejaculation: The emission of ↑ semen.

↑ Ductus deferens

Ejaculatory ducts (ductus ejaculatorius*): About 1 cm long and 0.1–0.3 mm wide, paired tubes (ED) connecting ↑ ampulla (A) of ↑ ductus deferens (DD) with urethral lumen (U) at tip of ↑ colliculus seminalis (CS). At their beginning, E. are lined with ↑ simple cuboidal epithelium, which becomes low ↑ pseudostratified, and, in region of colliculus, ↑ stratified columnar. A moderately dense lamina propria (LP), rich in elastic networks, surrounds E. Dispersed or bundled ↑ smooth muscle cells (M) originating from ampulla may be found within lamina propria. During ↑ ejaculation, ↑ spermatozoa, together with product of ↑ seminal vesicles (SV), are injected into urethral lumen.

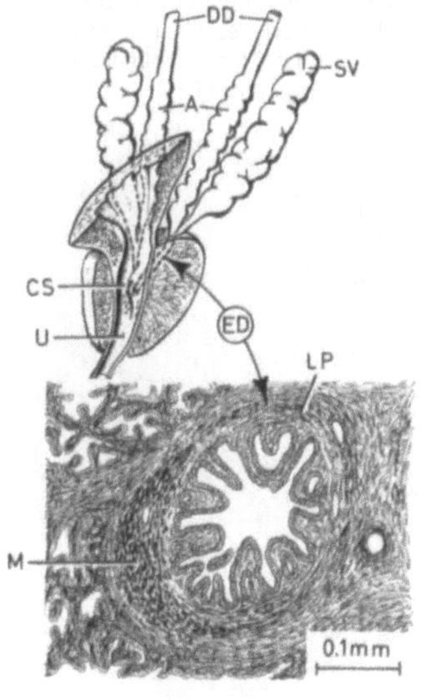

Aumüller, G.: Prostate Gland and Seminal Vesicles. In: Oksche, A., Vollrath, L. (eds.): Handbuch der mikroskopischen Anatomie des Menschen. Vol. 7.

Harn- und Geschlechtsapparat, Part 6. Berlin, Heidelberg, New York: Springer-Verlag 1979; Cossu, M., Marcello, M.F., Esai, E., Testa Riva, F., Riva, A.: Acta. Anat. *116*, 225 (1983); Schlager, F.: Z. mikrosk.-anat. Forsch. *76*, 268 (1967)

Elastase: A pancreatic and macrophagic enzyme digesting ↑ elastin.

Elastic arteries (conducting arteries, arteria elastotypica*): Any of several large vessels near heart conducting blood to ↑ muscular arteries. General characteristics include high content of ↑ elastic fibers and fenestrated ↑ elastic laminae in tunica media. ↑ Aorta, innominate (brachiocephalic) artery, common carotid and subclavian arteries are E.

Elastic baskets: ↑ Alveolus of lung, elastic basket of

Elastic cartilage: ↑ Cartilage, elastic

Elastic connective tissue (textus connectivus elasticus*): A kind of irregular ↑ dense connective tissue composed predominantly of strong ↑ elastic fibers (E) several microns thick. Slender ↑ fibrocytes (Fc) and delicate ↑ collagen fibrils (C) occupy spaces between elastic fibers. In cross section, elastic fibers appear as polygons without internal structure. E. forms ligamentum nuchae, ligamenta flava (LF), ↑ vocal ligaments (V), ligamentum stylohyoideum, ligamentum suspensorium penis, etc. A variety of E. is present in wall of blood vessels (↑ elastic laminae) and in ↑ lung (↑ elastic baskets surrounding ↑ alveoli).

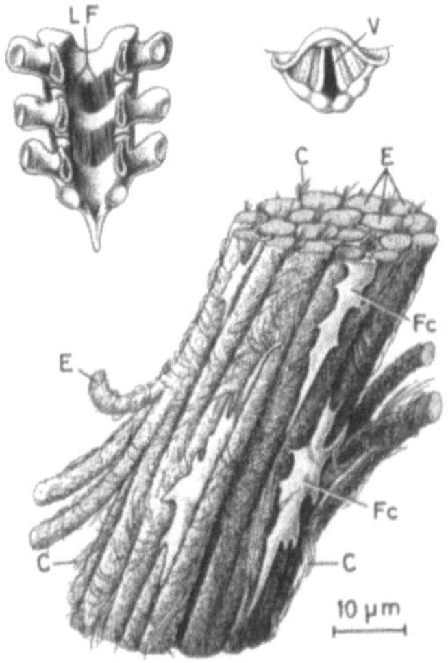

Kadar, A.: The Elastic Fiber. Normal and Pathological Conditions in the Arteries. Experimentelle Pathologie, Supplementbände, Vol. 5. Jena: Fischer Verlag 1979; Robert, A.M., Robert, D.: Biology and Pathology of Elastic Tissue. Frontiers of Matrix Biology, Vol. 8. Basle: Karger 1980

Elastic fibers (fibra elastica*): Long, branched, and anastomosed yellowish fibers, 0.1–10 µm thick (E), occurring in variable amounts in different ↑ connective and ↑ supporting tissues. Adult E. consists of a clear central amorphous mass of ↑ elastin (pars amorpha*) surrounded by a mantle of numerous 11 to 13-nm-wide microfibrils (Mf) (pars filamentosa*), with indistinct periodicity and a light central core. Some microfibrils (arrow) are embedded within mass of elastin. Microfibrils contain hydrophilic amino acids, no hydroxyproline, no desmosine, no isodesmosine, but half-cystine residues, and 5% neutral sugars. ↑ Fibroblasts and ↑ smooth muscle cells synthesize molecules which form E. and microfibrils. E. have a strong affinity for ↑ resorcin fuchsin and ↑ orcein. Unstretched E. are ↑ anisotropic; stretched E. show pronounced isotropy, probably caused by molecular structure of elastin. (Fig., top = human ↑ loose connective tissue, resorcin fuchsin staining; fig., bottom = ↑ dermis, rat)

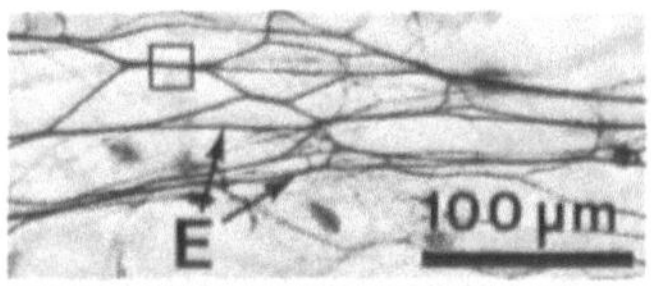

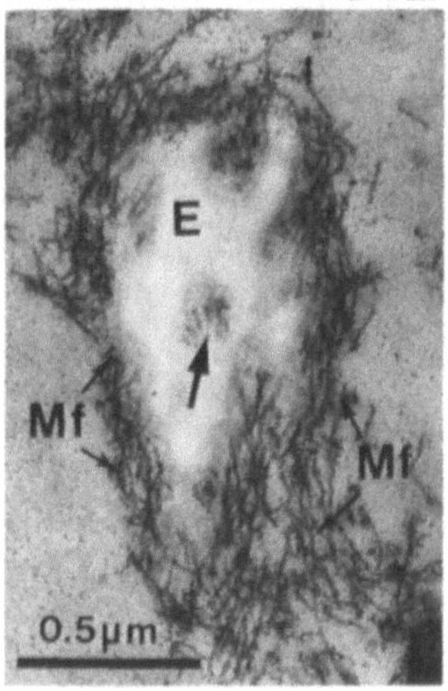

Ichimura, T., Hashimoto, P.H.: J. Ultrastruct. Res. *81*, 172 (1982); Kewley, M.A., Steven, E.S., Williams, G.: J. Anat. *123*, 129 (1977); Pasons, D.F., Marko, M., Wansor, K.: Micron *14*, 1 (1983); Romhanyi, G.: Histochemistry *77*, 133 (1983); Serafini-Fracassini, A., Field, J.M., Hinnie, J.: J. Ultrastruct. Res. *65*, 190 (1978)

Elastic fibers, staining of: ↑ Orcein; ↑ Resorcin fuchsin

Elastic lamina, external (elastica externa, membrana elastica externa*): The thin and frequently discontinuous elastic sheath (EE) situated between tunica media (TM) and tunica adventitia (TA) of most ↑ muscular arteries. (Fig. = human, ↑ orcein staining)

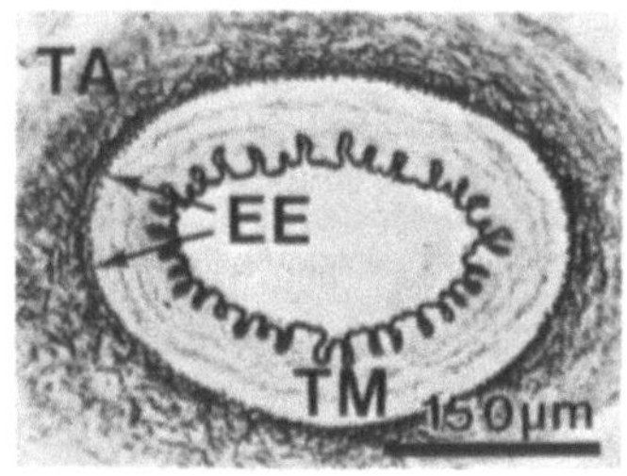

Elastic lamina, internal (elastica interna, membrana elastica interna*): The elastic sheath (EI) forming limit between tunica intima (TI) and tunica media (TM) of ↑ muscular arteries, some veins, and ↑ thoracic duct. E. appears on histological sections as an undulated line, resulting from postmortem arterial collapse. Under transmission electron microscope, E. appears as a translucent mass underlying ↑ endothelial cells (E) of tunica intima. In older individuals, E. may be split into two or more laminae. (Fig., top = man, ↑ orcein staining; fig., bottom = rat)

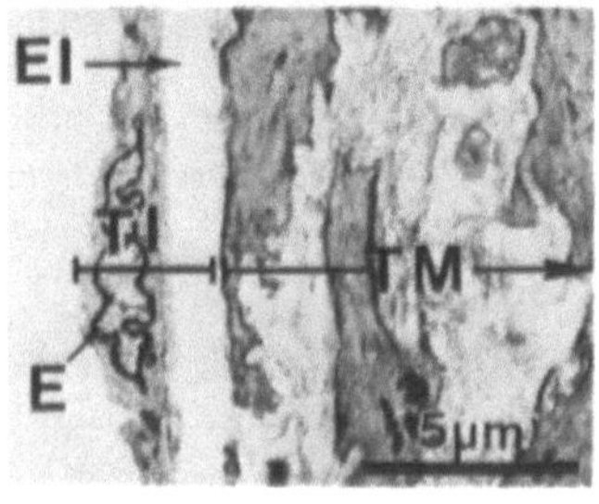

Elastic laminae, fenestrated (membranae fenestratae elasticae*): About

30–75 incomplete concentric elastic sheaths (EL) making up tunica media of ↑ elastic arteries. E. are approximately 2–3 µm thick and 5–15 µm apart; they are interconnected with obliquely arranged ↑ smooth muscle cells (SM) and bundles of ↑ collagen microfibrils (c). ↑ Fibroblasts are also found between E.; ↑ wandering cells pass through openings in E. E = ↑ endothelium. (Fig. = elastic artery, rat)

↑ Aorta

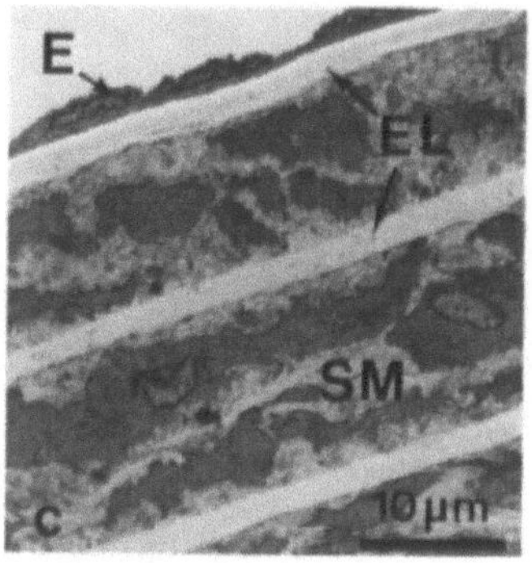

Wasano, K., Yamamoto, T.: J. Electron Microsc. *32*, 33 (1983)

Elastica interna: ↑ Elastic lamina, internal

Elastica externa: ↑ Elastic lamina, external

Elastin: A highly refractile, alkali- and acid-resistant, weakly stainable, insoluble protein constituting amorphous component of ↑ elastic fibers. Glycine represents about one-third of E., and proline about 10%. The rare amino acids desmosine and isodesmosine are probably involved in cross-linking in and between polypeptide chains. E. is produced by ↑ fibroblasts, ↑ smooth muscle cells, and, possibly, by other mesenchymal derivatives and secreted into extracellular space in form of proelastin or tropoelastin molecules. Extracellular polymerization of modified proelastin molecules provokes formation of insoluble elastic fibers surrounded by microfibrils. It has been postulated that molecules of E. form a three-dimensional network of random chains joined by covalent cross-links or a mass of easily deformable globular corpuscles. E. is also capable of triggering first crystallization nuclei of ↑ hydroxyapatite.

Gotte, L., Giro, M.G., Volpin, D., Horne, W.R.: J. Ultrastruct. Res. *46*, 23 (1974); Sandberg, L.B., Soskel, N.T., Leslie, J.G.: N. Engl. J. Med. *304*, 566 (1981)

Electrical synapse: ↑ Synapse, electrical

Electromagnetic lens: A very precise coil of isolated copper wire around a nonmagnetic iron cylinder. Electric current induces formation of an electromagnetic field, which refracts electron beam passing through axis of cylinder. By changing intensity of current in E., one can increase or decrease electron refraction, i.e., one can change magnification as well as focus the object. In ↑ scanning electron microscope, E. are used to reduce electron stream to a very fine beam.

↑ Transmission electron microscope

Hawkes, P.W. (ed.): Magnetic Electron Lenses. Topics in Current Physics, Vol. 18. Berlin, Heidelberg, New York: Springer-Verlag 1982; Ruska, E.: The Early Development of Electron Lenses and Electron Microscopy. Stuttgart: Hirzel Verlag 1980

Electron diffraction: A technique for investigation of regular biological structures (mostly organic crystals) using diffraction of an electron beam (EB). Principle: Electron beam produced by electron gun (EG) is concentrated by ↑ electromagnetic ↑ condenser lens (C) upon object (O) to be analyzed; here it is diffracted, and diffraction image is projected by a ↑ projective (P) onto screen or film. E. of polycrystalline specimens produces a series of concentric rings (R), each of which corresponds to one lattice plane. Diameters of rings are important for identification of analyzed material. Diffraction image of a monocrystal gives a series of spots (S) whose distance from center is used for calculating distances of lattice planes. Compared with ↑ X-ray diffraction, E. requires a much shorter exposition time and permits analyses of 10^{-12} g of material. Since wavelength of electrons at 80 kV is 0.004 nm, E. permits analysis of atomic structures.

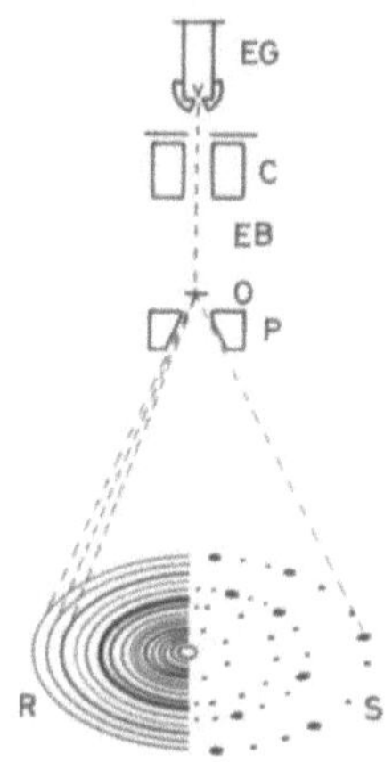

Glauert, A.M. (ed.): Practical Methods in Electron Microscopy, Vol. 1, Part 2. Electron Diffraction and Optical Diffraction Techniques. Amsterdam: Elsevier/North Holland 1974

Electron microscope microanalyzer (EMMA): ↑ Electron probe microanalyzer

Electron microscopes: ↑ A large family of ↑ microscopes using electrons for image formation.

↑ Scanning electron microscope; ↑ Scanning transmission electron microscope; ↑ Transmission electron microscope; ↑ Transmission electron microscope, high-voltage

Electron microscopy: All preparative methods and instrumental techniques permitting an analysis of microscopic preparations with an ↑ electron microscope.

↑ Embedding; ↑ Fixation; ↑ Ultramicrotomy

Griffith, J.D. (ed.): Electron Microscopy in Biology. Vols. 1 and 2. New York: John Wiley & Sons 1981 and 1982; Johannessen J.V. (ed.): Electron Microscopy in Human Medicine, Vols. 1 and 2. New York: McGraw Hill 1978; Rease, D.C., Porter, K.R.: J. Cell Biol. *91*, 287s (1981); Schimmel, G., Vogell, W. (eds.): Methods of Electron Microscopy with English Summaries. Stuttgart: Wissenschaftliche Verlagsgesellschaft 1981; Weakley, B.S.: Technique for Electron Microscopy. Edinburgh: Churchill Livingstone 1982

Electron probe microanalysis (X-ray microanalysis): A technique for immediate in situ identification of chemical elements and their semiquantitative or quantitative measurement using an ↑ electron probe microanalyzer.

de Bruijn, W.C.: Scanning Electron Microscopy 1981/II, 357 (1981); Goldstein, J.I., Newbury, D.E., Echlin, P., Joy, D.C., Fiori, C., Lifshin, E.: Scanning Electron Microscopy and X-Ray Microanalysis. New York: Plenum Press 1981; Hutchinson, Th. E., Somlyo, A.P. (eds.): Microprobe Analysis of Biological Systems. New York, London: Academic Press 1981; Roomans, G.M., Shelburne, J.D. (eds.): Basic Methods in Biological X-Ray Microanalysis. Chicago: Scanning Electron Microscopy Inc. 1983; Sumner, A.T.: Histochemical J. *15*, 501 (1983)

Electron probe microanalyzer (EPMA, X-ray microanalyzer, microsonde): A complex electro-optic instrument for in situ identification and quantitative measurement of chemical elements heavier than sodium. Principle: An electron stream, produced in high vacuum by a cathode (C), is reduced by ↑ electromagnetic lenses (L) to a fine beam (b), 0.1–1 µm in diameter, and projected onto an object (O) (histological section, ultrathin ↑ section, or scanning electron microscope specimen). From impact point, secondary electrons (se) and X-rays (X) emerge. A special X-ray detector (D) measures emitted X-ray spectrum, which is characteristic for a given element. Data obtained are sent, via an amplifier (A), as a series of counts to a multichannel analyzer (MCA), which immediately displays and records results in form of a curve. By changing setting of E. to a new X-ray spectrum, topographic distribution of another element can be obtained from same object. E. is often combined with ↑ scanning electron microscope (SEM) whose scanning generator (SG) allows electron beam to scan entire surface of specimen. Combined with ↑ transmission electron microscope, electron beam transmits object and forms, simultaneously with chemical analysis of a given element, an image of the object in a process called electron microscope microanalysis or EMMA. Accuracy of E. can reach 10^{-15} g.

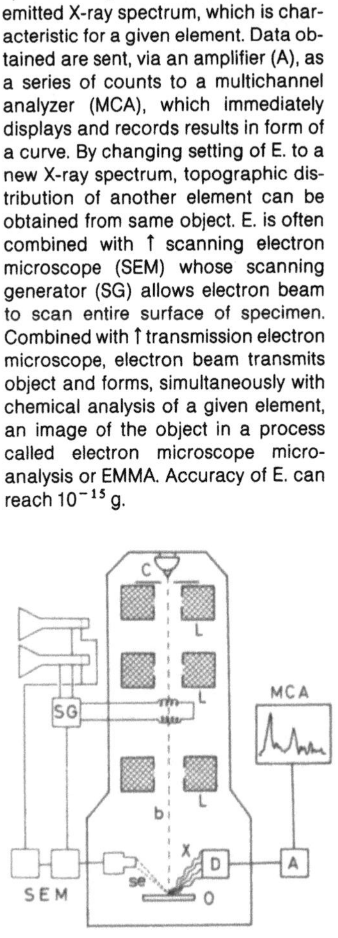

Chandler, J.A.: X-ray Microanalysis in the Electron Microscope. In: Practical Methods in Electron Microscopy, A.M. Glauert (ed.), Amsterdam, New York, Oxford: North Holland 1977; Fiori, C.E.: J. Histochem. Cytochem. *29*, 1029 (1981)

Electronic coupling area: ↑ Nexus; ↑ Synapse, electrical

Electrotonic synapse: ↑ Nexus; ↑ Synapse, electrical

Eleidin: A strongly refractive acidophilic substance related to ↑ keratin and present in stratum lucidum of ↑ epidermis.

Elementary particles, of mitochondria (ATPosomes, electron transport particles, F_1-particles, inner membrane subunits, oxisomes, particula elementaria*): Small corpuscles (EP) incorporated into inner ↑ mitochondrial membrane (IM). After treating mitochondrion with a hypotonic solution, E. become visible on negatively contrasted preparations (insert). Each E. consists of a globular head (H) connected by a slender stalk (S) to base plate (B). Latter contains several coupling factors (CF) incorporated into lipid layers (L) of internal membrane and, in protein leaflet (Pr), enzymes of respiratory chain (RC) and cytochrome-C molecules (CC). There are about 10^4–10^5 E./mitochondrion. Synthesis of ATP from adenosine diphosphate (ADP) and phosphate (P) takes place within head of E. EM = external mitochondrial membrane

↑ Contrasting, in transmission electron microscopy

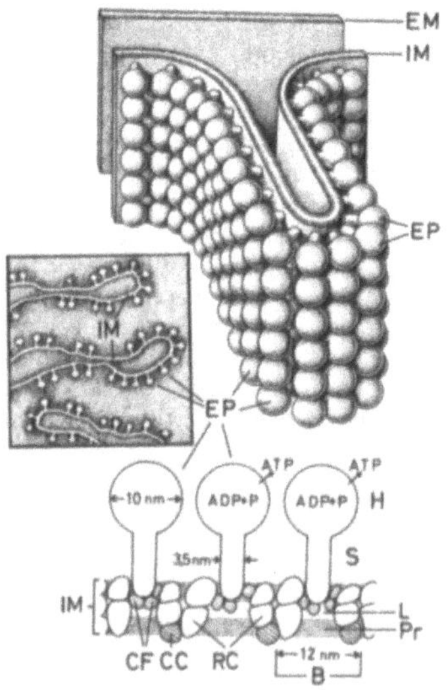

Lehninger, A.L.: The Molecular Organization of Mitochondrial Membranes. In: Advances in Cytopharmacology Vol. I, New York: Raven Press 1971; Sjöstrand, F.S., Cassel, R.Z.: J. Ultrastruct. Res. *63*, 111 (1978); Telford, J.N., Racker, E.: J. Cell Biol. *57*, 580 (1973)

Ellipsoids, of cones and rods: ↑ Photoreceptors

Ellipsoids, of Schweigger-Seidel (sheath of Schweigger-Seidel, arteriola ellipsoidea*): A spindle-shaped accumulation of ↑ white pulp enveloping some segments of sheathed ↑ capillaries (C) of ↑ spleen. E. consist of ↑ lymphocytes (L), ↑ plasma cells (P), ↑ reticular cells (R), and ↑ reticular fibrils connected with reticular fibrils of ↑ red pulp. In man, E. are poorly developed and average 50–100 µm in length and 20–60 µm in width.

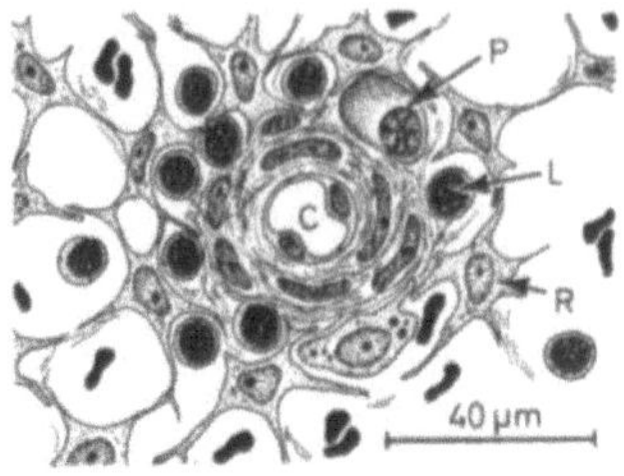

Blue, J., Weiss, L.: Am. J. Anat. *161*, 115 and 189 (1981); Hatae, T.: Arch. Histol. Jpn *41*, 177 (1978)

Embedding: A procedure consisting of gradual introduction of melted ↑ paraffin, ↑ epoxy resins, or a plastic material into a dehydrated block in order to aid ↑ sectioning for ↑ light or transmission ↑ electron microscopy.

Bancroft, J.D., Stevens, A. (eds.): Theory and Practice of Histological Technique, 2nd edn. London: Churchill Livingstone 1982; Glauert, A.M.: Fixation, Dehydration, and Embedding of Biological Specimens. Practical Methods in Electron Microscopy. Amsterdam, New York, Oxford: North Holland 1980

Embryology (embryologia*): The science studying normal development of the organism from ↑ fertilization of secondary ↑ oocyte to birth.

Embryonal tissues: Transitory tissues belonging to ↑ connective tissue proper and characterized by a relatively small number of cells and very large interspaces filled with a jellylike ↑ ground substance. To E. belong ↑ gelatinous and ↑ mesenchymal tissues.

↑ Connective tissue proper, ground substance of

Emiocytosis: ↑ Exocytosis

EMMA: ↑ Electron probe microanalyzer

Emperipolesis: The occasional penetration of a blood cell (mostly ↑ lymphocytes) into cytoplasm of a ↑ megakaryocyte or other cell.

Burns, E.R., Zucker-Franklin, D., Valentine, F.: Lab. Invest. *47*, 99 (1982); Djaldetti, M., Strauss, Z.: J. Submicrosc. Cytol. *14*, 407 (1982)

"En passant" synapse: ↑ Synapse, "en passant"

Enamel (substantia adamantina, enamelum*): A bluish-white to yellowish substance (E) covering ↑ dentin (D) in region of tooth ↑ crown. E. is composed of ↑ enamel prisms (EP), ↑ interprismatic substance, and ↑ enamel sheaths, all running through whole thickness of E. In nondecalcified preparations, one can see, approximately perpendicular to surface of E., ↑ lines of Hunter-Schreger (S) and a system of concentric ↑ lines of Retzius (R). E. is hardest substance of body and consists of 90% ↑ hydroxyapatite and 6%–8% calcium carbonate, calcium fluoride, and magnesium carbonate. The remainder is water and an organic matrix. During ↑ decalcification, E. is dissolved. ↑ Adamantoblasts synthesize E. in course of ↑ amelogenesis.

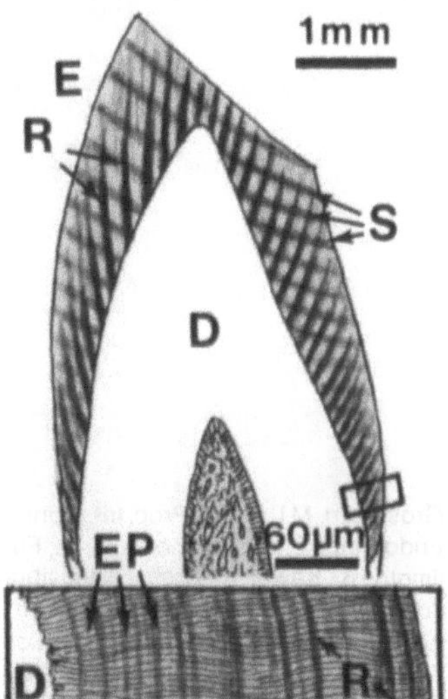

Daculsi, G., Kerebel, B.: J. Ultrastruct. Res. *65*, 163 (1978); Driessens, F.C.M., Heyligers, H.M.J., Wöltgens, J.H.M., Verbeeck, R.M.H.: J. Biol. Buccale *10*, 199 (1982); Frank, R.M.: J. Dent. Res. *58* (B) 684 (1979); Munholz, C.O.G., Leblond, C.P.: Calcif. Tissue Res. *15*, 221 (1974); Whittaker, D.K.: Arch. Oral Biol. *27*, 383 (1982)

Enamel cuticle (cuticula dentis, Nasmyth's membrane, cuticula enameli*): Two extremely thin layers, about 1 µm thick, covering free surface of recently erupted teeth. Outer layer is keratinized, acellular, and probably a remnant of perforated gingival epithelium. Inner layer is amorphous and probably represents final product of ↑ ameloblasts before they vanish. E. disappears rapidly.

Newman, H.N.: Arch. Oral Biol. *25*, 49 (1980)

Enamel lamellae (lamella enameli*): Ribbonlike accumulations (EL) of organic material, roughly 1–5 µm wide, predominantly of salivary origin, extending from enamel surface (E) along cracks between ↑ enamel prisms (EP) and sometimes continuing up to ↑ dentinoenamel junction (DJ).

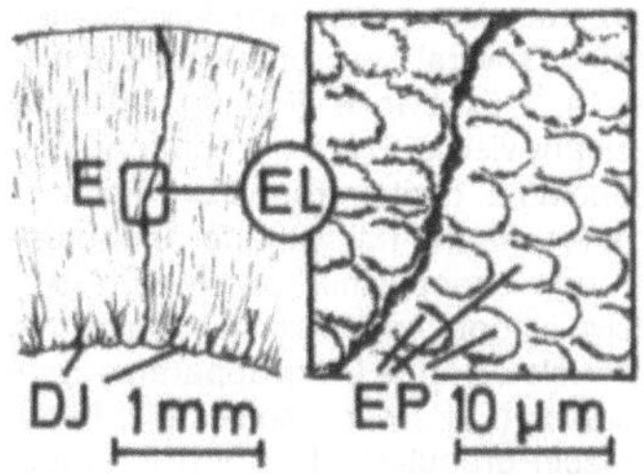

Enamel needles: ↑ Hydroxyapatite crystals

Enamel organ (organum enamelare*): A bell-shaped epithelial formation responsible for development of ↑ enamel. (See embryology texts for further information)

Matthiessen, M.E., Rømert, P.: Cell Tissue Res. *205*, 361 (1980)

Enamel prisms (enamel rod, prisma enameli*): Basic constitutive elements (EP) of ↑ enamel. Each E. has an arcade-shaped, oval, or polygonal profile measuring 3–5 µm in diameter. E. run a highly complex course through whole thickness of enamel, perpendicular to surface of ↑ dentin. In an E., densely packed ↑ hydroxyapatite crystals (H) are all oriented in same direction, whereas those belonging to ↑ interprismatic substance (IS), between E., have different directions. Each E. is surrounded by an ↑ enamel sheath (ES).

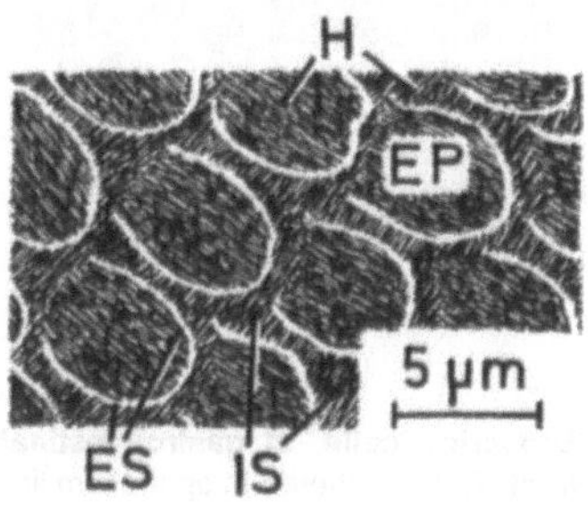

↑ Tomes' process

Boyde, A.: Bull. Group Int. Rech. Sci. Stomatol Odontol *12*, 151 (1969); Boyde, A., Martin, L.: Anat. Embryol. *165*, 193 (1982); Risnes, S.: Am. J. Anat. *154*, 419 and *155*, 245 (1979)

Enamel pulp: ↑ Stellate reticulum

Enamel sheath (prismatic rod sheath, prism sheath, membrana prismatis*): A layer, about 0.1–0.2 µm wide, of nonmineralized organic ground substance between ↑ enamel prisms.

Enamel spindles (fusus enameli*): Fusiform processes of dentinal matrix (ES) penetrating a short distance into ↑ enamel (E) at ↑ dentinoenamel junction (DJ) level. (Fig. = human)

Enamel tufts (fasciculus enameli*): Groups of poorly calcified, twisted ↑ enamel prisms (ET) embedded in an abundant matrix, extending from ↑ dentinoenamel junction (DJ) into ↑ enamel (E).

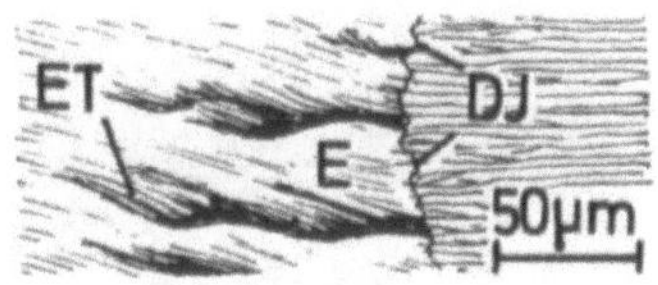

Encapsulated endings: ↑ Corpuscles

Encephalon: ↑ Brain

End bulb, of Krause: ↑ Corpuscle, of Krause

End plate: ↑ Motor end plate

Endocapillary layer (endoepithelial layer): The ↑ glycocalyx (EL) of ↑ endothelial cells covering their luminal surface. Exact chemical nature of E. unknown; the fact that it may be visualized after staining with ↑ rutenium red or ↑ alcian blue (fig.) indicates a ↑ glycosaminoglycan nature. E. is believed to stabilize circulating molecules and to facilitate their contact with plasmalemma of endothelial cells. (Fig. = rat)

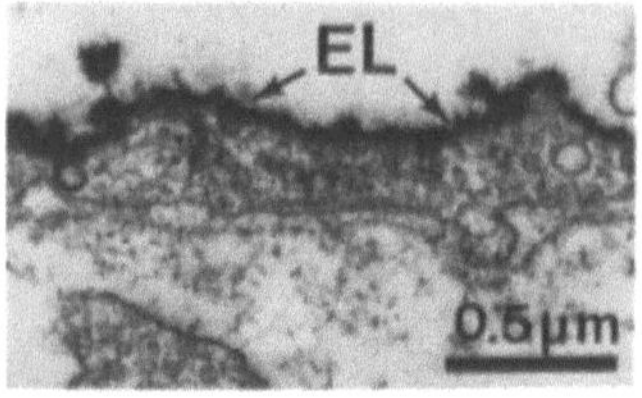

Copley, A.L., Scheinthal, B.M.: Exp. Cell Res. *59*, 491 (1970)

Endocardium*: The internal layer of ↑ heart. Structure: 1) ↑ Endothelium* (E). 2) Subendothelial layer or stratum subendotheliale* (SS) = a thin layer of ↑ loose connective tissue with few ↑ fibroblasts and delicate ↑ collagen fibrils. 3) Myoelastic layer or stratum myoelasticum* (SM) = a relatively ↑ dense connective tissue with thick ↑ collagen and ↑ elastic fibers and scattered ↑ smooth muscle cells (c). 4) Subendocardial layer or tela subendocardialis* (TS) = a loose connective tissue continuous with ↑ endomysium (Em) of ↑ myocardium (U), missing from papillary muscles and ↑ chordae tendineae. E. has no blood vessels; it is vascularized from heart cavities and by subendocardial blood vessels. E. contains ↑ myelinated and ↑ nonmyelinated nerve fibers, as well as lymphatic capillaries. ↑ Purkinje fibers run through subendocardial layer of ventricles. (Fig. = sheep)

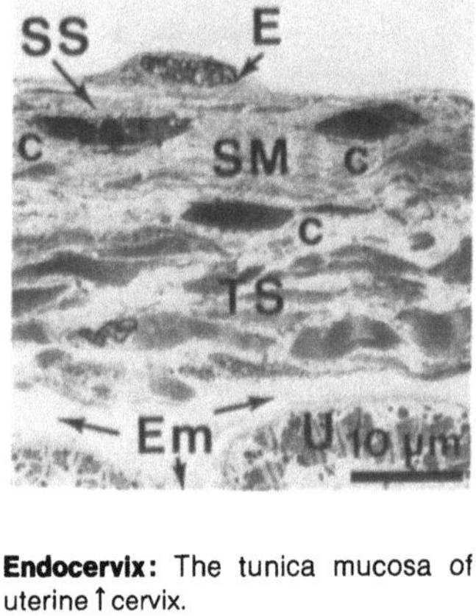

Endocervix: The tunica mucosa of uterine ↑ cervix.

Endochondral bones: ↑ Cartilage bones

Endochondral ossification: A process consisting of destruction of hyaline ↑ cartilage of cartilaginous model and its replacement by ↑ bone in course of indirect ↑ bone formation

Endocrine cells (endocrinocyti*): Cells releasing their secretory product directly into blood, ↑ cerebrospinal fluid, and/or ↑ lymph. Structurally, there are two types of E.: 1) Protein, ↑ glycoprotein, and polypeptide hormone-producing E. have elongated ↑ mitochondria with cristae (M), a well-developed rough endoplasmic reticulum (rER) in the form of stalks of flattened cisternae, occasional smooth endoplasmic tubules (sER), and a prominent Golgi apparatus (G), which is associated with various numbers of 200 to 350-nm ↑ unit membrane-enclosed ↑ secretory granules (SG) with a more or less osmiophilic content, a moderate amount of free ribosomes, rare ↑ lysosomes, and several microtubular bundles (Mt). ↑ Exocytosis (Ex) is in general clearly visible. 2) Steroid hormone-producing E. contain round or elliptical ↑ mitochondria (M) with tubules, sparse rough endoplasmic cisternae, extensive smooth endoplasmic tubules and vesicles, a large Golgi apparatus, lysosomes (Ly), ↑ peroxisomes (P), ↑ lipofuscin granules (Lf), a few free ribosomes, and frequent large ↑ lipid droplets (L) whose cholesterol serves as a hormonal precursor. Exocytosis is not visible.

↑ Cells, of zona fasciculata, glomerulosa, and reticularis; ↑ Follicular cells, of ovarian follicles; ↑ Interstitial cells, of testis; ↑ Lutein cells

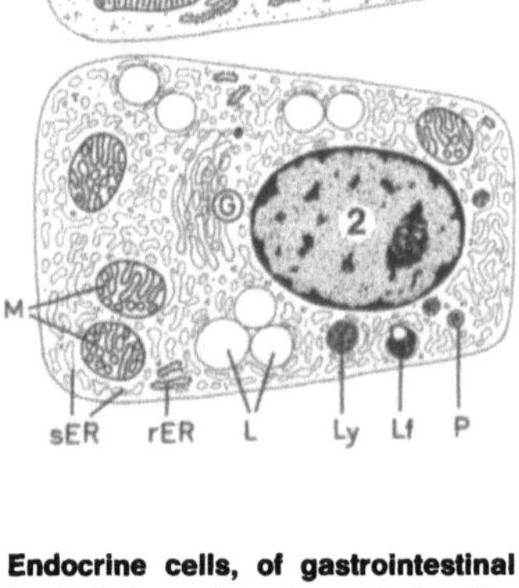

Endocrine cells, of gastrointestinal tract: Cells scattered in epithelium lining ↑ gastrointestinal tract. E. are characterized by clear cytoplasm, elliptical nucleus, small mitochondria and Golgi apparatus, short rough endoplasmic cisternae, and a moderate number of ↑ unit membrane-bound ↑ secretory granules with an osmiophilic content. E. are very similar to one another; only by immunocytochemical methods is it possible to differentiate categories of these cells. Fig. displays types of E. with their probable corresponding hormones. All E. belong to the group of ↑ APUD cells. (Abbreviations and terms marked with an asterisk are explained under corresponding letters) F = ↑ fundus of stomach, D = ↑ duodenum, J = ↑ jejunum, I = ↑ ileum, C = ↑ colon (Modified after Grossman 1976 and Junqueira and Carneiro 1980)

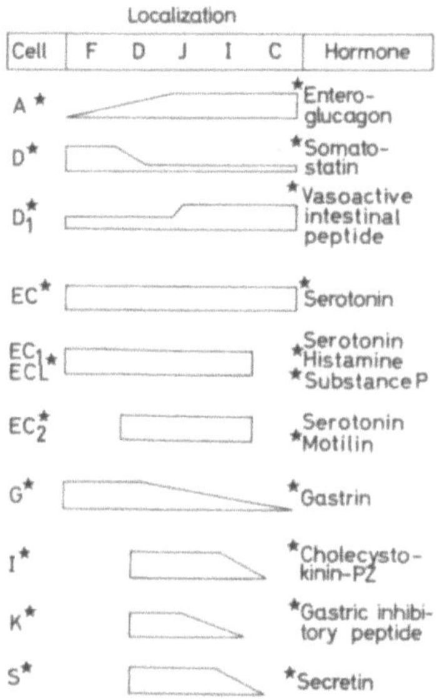

Grossman, M.I. (1976) Proc. Int. Congr. Endocrinol. 2:1, 1976; Inokuchi, H., Fujimoto, S., Kawai, K.: Arch. Histol. Jpn. *46*, 137 (1983); Junqueira, L.C., Carneiro, J.: Basic Histology, 3rd edn. Los Altos: Lange 1980; Lefranc, G., Richard, E., Pradal, G.: Biol. Cell. *46*, 189 (1982); Lehy, T., Peranzi, G., Cristina, M.L.: Histochemistry *71*, 67 (1981); Rizzotti, M., Domenghini, C., Castaldo, L.: Basic Appl. Histochem. *24*, 79 (1980); Welbourn, R.B.: Hexagone "Roche" *9/5*, 8 (1981)

Endocrine glands (glandula endocrina*): Organs of variable volume and structure composed of ↑ endocrine cells. E. have no excretory canals since they discharge their products directly into ↑ blood (= hemocrinia), ↑ lymph (= lymphocrinia), or ↑ cerebrospinal fluid (= hydrencephalocrinia). (See under name of specific E.)

Jones, T.C., Mohr, U., Hunt, R.D., Capen, C.C. (eds): Endocrine System. Monographs on Pathology of Laboratory Animals. Berlin, Heidelberg, New York, Tokyo: Springer-Verlag 1983; Ketelbant-Balasse, P.: The Endocrine System. In: Hodges, G.M. and Hallowes, R.C. (eds.): Biomedical Research Applications of Scanning Electron Microscopy, Vol. 2. London, New York: Academic Press 1980

Endocrine glands, control of secretion in: The activity of endocrine glands can be controlled by simple or complex ↑ feedback mechanisms, or without such mechanisms. An example of control in absence of any feedback mechanism is ↑ suckling reflex.

↑ Releasing factors

Endocrine secretion: The discharge of ↑ hormones into an internal milieu of body (↑ blood, ↑ lymph, ↑ cerebrospinal fluid). Release of glucose by liver may be also considered E.

Endocytosis: The interiorization of fluid and/or solid materials into a cell.

↑ Athrocytosis; ↑ Phagocytosis; ↑ Pinocytosis

Brown, M.S., Anderson, R.G.W., Goldstein, J.L.: Cell *32*, 663 (1983), Herzog, V.: Verh. Anat. Ges. *76*, 41 (1982); Salisbury, J.L., Condeelis, J.S., Maihle, N.J., Satir, P.: Receptor-Mediated Endocytosis by Clathrin-Coated Vesicles: Evidence for a Dynamic Pathway. Cold Spring Harbor Symposia on Quantitative Biology, Vol. 46, Part 2. New York: Cold Spring Harbor Laboratory 1982; Silverstein, S.C., Steinman, R.M., Cohn, Z. A.: Ann. Rev. Biochem. *46*, 669 (1977); Steinman, R.M., Mellman, I.S., Muller, W.A., Cohn, Z. A.: J. Cell Biol. *96*, 1 (1983)

Endoderm (endoderma embryonicum*): The innermost of primitive germ layers, giving rise to epithelia of ↑ gastrointestinal tract, ↑ parenchyma of ↑ liver, ↑ pancreas, etc. (See embryology texts for further information)

Endoepithelial glands: An isolated secretory cell or group (EG) of exocrine glandular cells (GC) situated within a surface epithelium (SE). ↑ Goblet cells are considered unicellular E. Pluricellular E. are present in ↑ respiratory (fig.) and ↑ urethral epithelia.

↑ Glands, classification of; ↑ Urethral glands, endoepithelial; ↑ Urethral glands, of male urethra

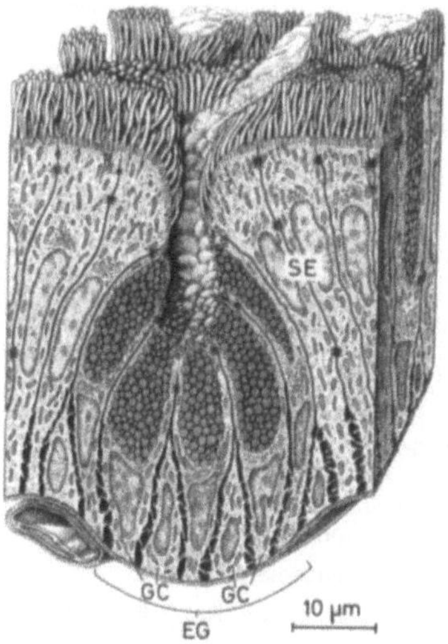

Endogenous pigments: A large group of ↑ pigments synthesized in organism. E. include ↑ hemoglobin, ↑ melanin, ↑ myoglobin, ↑ lipofuscin, etc.

Endolymph (endolympha*): A watery fluid filling membranous ↑ labyrinth. E. is produced by ↑ stria vascularis and eliminated by cells of ↑ endolymphatic sac into ↑ subarachnoid space. In ↑ cochlear duct, E. transmits vibrations to ↑ hair cells; motion of E. displaces ↑ hairs of sensory cells of ↑ cristae ampullares and excites them. E. is very poor in protein and sodium, but rich in potassium.

Ormerod, F.C.: J. Laryngol. Otol. *74*, 659 (1960)

Endolymphatic duct (ductus endolymphaticus*): A fine canal running from ↑ ductus utriculosaccularis to ↑ endolymphatic sac. E. is lined by a ↑ simple squamous to cuboidal epithelium and conducts ↑ endolymph to endolymphatic sac.

↑ Labyrinth, membranous

Endolymphatic sac (sacculus endolymphaticus*): An enlarged, saclike end of ↑ endolymphatic duct located in ↑ subarachnoid space on posterior surface of petrous portion of temporal bone. ↑ Simple squamous or cuboidal epithelium of E. absorbs ↑ endolymph and phagocytizes cellular debris reaching E. E. may be involved in equilibration of intralabyrinthal pressure.

↑ Labyrinth, membranous

Harada, Y.: Biomed. Res. *2*, Suppl. 391 (1981)

Endometriosis: A pathological penetration of ↑ uterine glands into ↑ myometrium.

Endometrium*: The 4 to 6-mm-thick tunica mucosa (E) of ↑ uterus. Structure: 1) Epithelium (Ep) = a single layer of columnar secretory and ciliated cells; ↑ cilia beat toward ↑ vagina. Epithelium forms simple, slightly branched ↑ uterine glands (UG) reaching ↑ myometrium (My). 2) Lamina propria or stroma endometrialis* (LP) = a richly vascularized, ↑ "cellular" connective tissue with abundant ground substance, delicate collagen and elastic fibers, numerous lymphatic vessels, and fairly sporadic nerve fibers. E. can be subdivided on the basis of its morphofunctional behavior during period of fertility into: a) functionalis or stratum functionale* (F) = a layer shed during every ↑ menstrual cycle, consisting of a superficial stratum compactum* (C), rich in cells, and an underlying stratum spongiosum* (S), rich in blood vessels; b) basalis or stratum basale* (B) = a relatively thin, deep layer in contact with myometrium; this layer never desquamates during menstrual cycle and regenerates functionalis during proliferative phase.

↑ Uterus, epithelium of

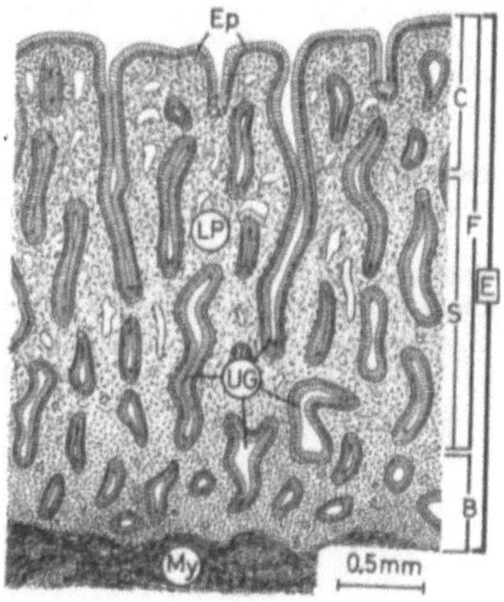

Dallenbach-Hellweg, G.: Endometrium. Berlin, Heidelberg, New York: Springer-Verlag 1981; Schmidt-Matthiesen, H.: The Normal Human Endometrium. New York: McGraw Hill 1963

Endometrium, gravid cycle of: The sequence of morphological and functional changes following ↑ fertilization (Fe) of secondary ↑ oocyte (O) and ↑ implantation (Im) of blastocyst (Bc). About 5 days after ↑ ovulation (Ov),

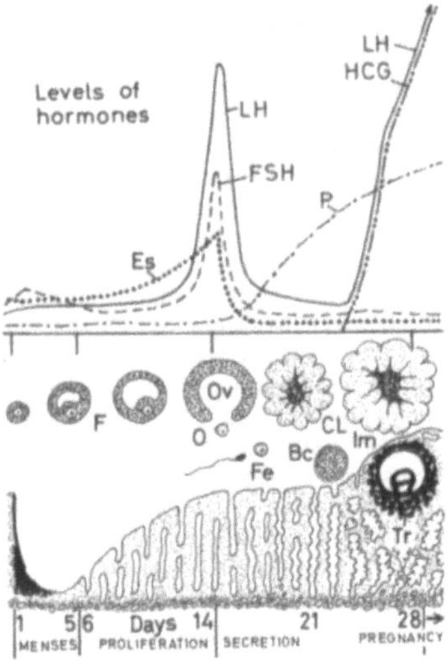

↑ ovum reaches ↑ uterus at blastocyst stage and penetrates ↑ endometrium. Simultaneously, ↑ trophoblast produces ↑ human chorionic gonadotrophin (HCG), which together with ↑ luteinizing hormone (LH) sustains life span of ↑ corpus luteum (CL) and prolongs its secretion of ↑ progesterone (P) until 4th month of pregnancy. Stromal fibroblasts become ↑ decidual cells; endometrium transforms itself gradually into maternal ↑ placenta. After interruption of lactation, normal ↑ menstrual cycle is restored. Es = estrogen level, F = ↑ ovarian follicles, FSH = ↑ follicle-stimulating hormone level

Endometrium, menstrual cycle of: ↑ Menstrual cycle

Endometrium, vascularization of: Uterine artery gives off six to ten ↑ arcuate arteries whose radial branches

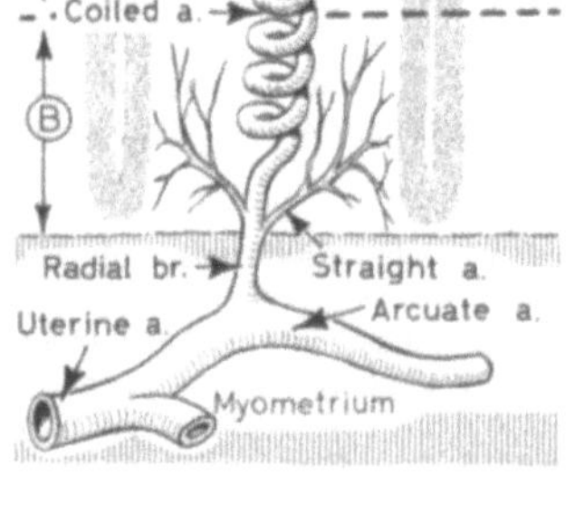

pass through inner muscular layer of ↑ myometrium and basalis (B) and become ↑ coiled arteries. They run parallel to ↑ uterine glands (UG) and terminate in a subepithelial capillary network (CN), which supplies functionalis (F) with blood. Coiled arteries give off a few straight or basal arteries for vascularization of basalis. From capillary network, blood is collected in venous ↑ lacunae (L), which empty into endometrial veins (V). Some authors consider lacunae to be arterial vessels and refer to them as ectatic capillaries. Distal segment of coiled arteries degenerates and regenerates with each ↑ menstrual cycle; straight arteries do not react to hormones. (Modified after Weiss and Greep 1977)

↑ Uteroplacental arteries

Weiss, L., Greep, R.O.: Histology 4th edn. New York: McGraw-Hill Book Comp. 1977

Endomitosis (endopolyploidy): The process of DNA replication in which ↑ chromosomes and ↑ chromatids are formed as in ↑ mitosis, however, ↑ nuclear membrane fails to break down and ↑ spindle apparatus is absent. E. leads to ↑ polyploidy.

↑ Cryptoendomitosis

Nagl, W.: Endopolyploidy and Polyteny in Differentiation and Evolution. Amsterdam: Elsevier/North-Holland 1978

Endomysium*: A highly vascularized and innervated thin sheath of ↑ loose connective tissue (E) surrounding each muscle fiber (F) in cardiac and skeletal muscles. E. consists of a network of ↑ reticular and ↑ collagen fibrils with some ↑ fibroblasts and occasional ↑ wandering cells. Cap = capillaries. (Fig. = cardiac muscle, human)

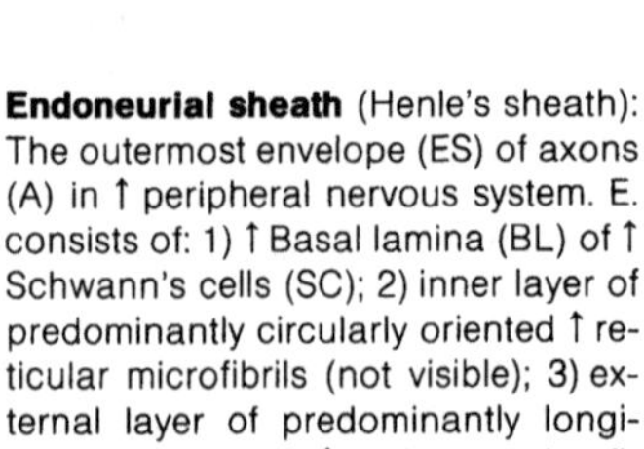

Endoneurial sheath (Henle's sheath): The outermost envelope (ES) of axons (A) in ↑ peripheral nervous system. E. consists of: 1) ↑ Basal lamina (BL) of ↑ Schwann's cells (SC); 2) inner layer of predominantly circularly oriented ↑ reticular microfibrils (not visible); 3) external layer of predominantly longitudinally oriented ↑ collagen microfi-

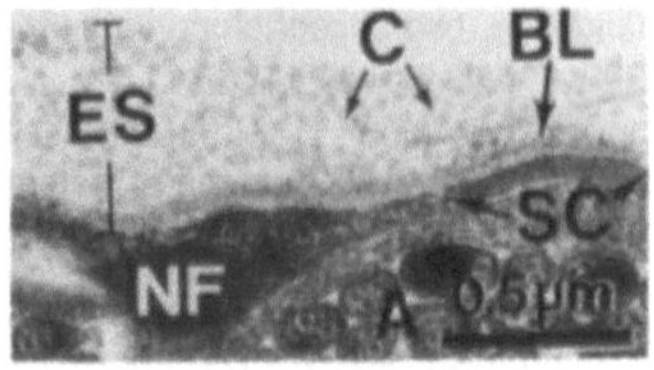

brils (C). NF = ↑ nonmyelinated nerve fiber (Fig. = mongolian gerbil)

Endoneurial space: A space (ES) surrounding ↑ nerve cells (N) and ↑ nerve fibers (NF) of ↑ peripheral nervous system. E. is filled with ↑ endoneurium and communicates with ↑ subarachnoid space. (Fig. = ↑ spiral ganglion, rat)

↑ Arachnoid

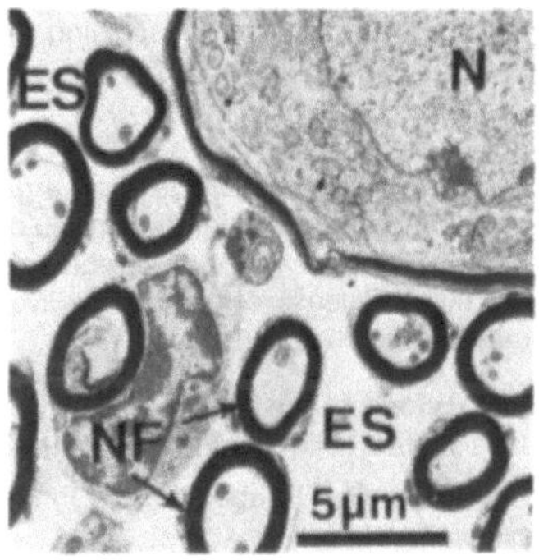

Endoneurium*: A delicate ↑ loose connective tissue (E) surrounding ↑ nerve cells and ↑ nerve fibers (NF) of ↑ peripheral nervous system. In ↑ ganglia, E. connects ganglion cells to capsule; in ↑ nerve fascicles, E. forms thin septa (S) that group nerve fibers into primary bundles and connect them to ↑ perineurium (P). ↑ Collagen

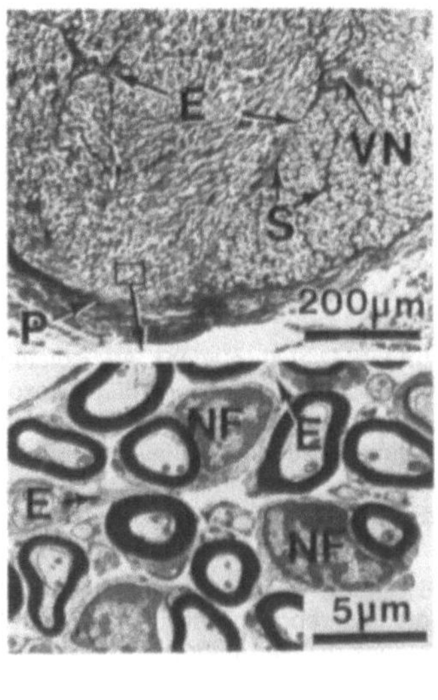

fibrils of E. are longitudinally oriented. E. carries a variable number of ↑ vasa nervorum (VN), lymphatics, and nerve fibers. (Fig., top = human; fig., bottom = rat)

↑ Endoneurial space

Endopeptidases: A group of proteolytic enzymes splitting central peptide bonds of proteins. E. are produced by ↑ pancreatic acinar cells and comprise ↑ trypsin, chymotrypsin, and ↑ elastase.

Endoplasm, of cardiac and smooth muscle cells: The conical sarcoplasmic zone (E) at nuclear poles (NP) of ↑ cardiac and ↑ smooth muscle cells. E. is free of ↑ myofilaments (Mf), but contains cell ↑ organelles, most of the ↑ atrial granules, ↑ glycogen particles, and ↑ lipofuscin granules. (Fig. = smooth muscle cell, rat)

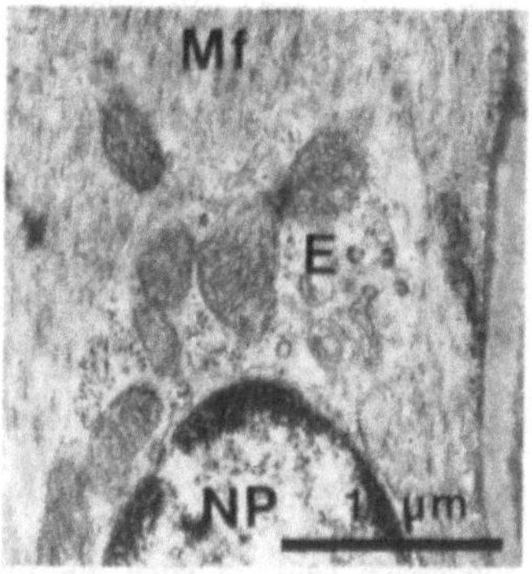

Endoplasm, of cytoplasm (endoplasma*): Part of ↑ cytoplasm with exclusion of ↑ centrosphere and ↑ ectoplasm. E. contains ↑ nucleus, cell ↑ organelles (with exception of ↑ centriole), and ↑ inclusions.

↑ Zones, of cytoplasm

Endoplasmic reticulum (reticulum endoplasmicum*): An ↑ organelle in the form of an irregular ↑ unit membrane-limited network of anastomosing ↑ cisternae (C), ↑ tubules (T), ↑ lamellae (L), and ↑ saccules (S). Depending on presence or absence of ↑ ribosomes attached to cytoplasmic aspect of E., one can distinguish rough and smooth ↑ E. Interior of E. always communicates with ↑ perinuclear cisterna (PC).

Löwe, H.: Acta Histochem. Suppl.-B. *17*, 13 (1976)

Endoplasmic reticulum, rough (rER, granular ER., reticulum endoplasmaticum granulosum*): A kind of ↑ endoplasmic reticulum whose surface facing cytoplasmic matrix is studded with ↑ ribosomes (R) and ↑ polyribosomes (P) involved in ↑ protein synthesis. Newly formed polypeptide chains penetrate cavities of E. in form of a flocculent ↑ reticuloplasm. In some cells, tubules of smooth ↑ endoplasmic reticulum, as well as ↑ transport vesicles bulge from E. Cavity of E. communicates with ↑ perinuclear cisterna (PC) and furnishes external leaflet (Ext) of ↑ nuclear envelope. A form of E. with parallel flattened cisternae (C) or lamellae is termed ↑ ergastoplasm. E. is ↑ basophilic because of its content of ribosomal RNA. E. is particularly developed in protein-producing cells (↑ plasma cells, parietal cells of ↑ gastric glands proper, ↑ pancreatic acinar cells, etc.). NP = ↑ nuclear pores, N = ↑ nucleus

↑ Nissl bodies

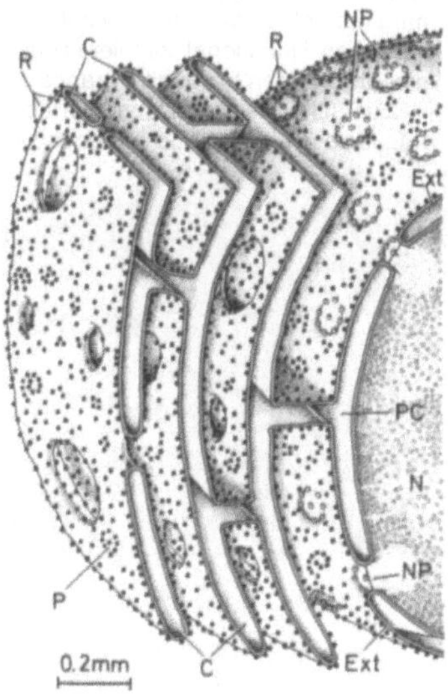

Bergeron, M., Thiéry, G.: Biol. Cell *42*, 43 (1981); Broadwell, R.D., Cataldo, A.M.: J. Histochem. Cytochem. *31*, 1077 (1983); Thiéry, G., Gaffiero, P., Bergeron, M.: Am. J. Anat. *167*, 479 (1983)

Endoplasmic reticulum, smooth (agranular ER, sER, reticulum endoplasmicum non-granulosum*): A ribosome-free, three-dimensional, close-meshed dynamic system of 20–60-nm anastomosing ↑ tubules (T) and ↑ cisternae lined by ↑ unit membranes. E. is well developed in ↑ steroid hormone-producing cells; it is involved in synthesis of ↑ glycogen and lipids from fatty acids. E. also plays a role in cholesterol metabolism, in detoxication of some endogenous and exogenous substances (e.g., phenobarbital), and in Ca^{2+} accumulation. In ↑ megakaryocytes, E. delimits future ↑ platelets; during telophase, it reconstitutes nuclear membrane. Role of E. in ↑ pigment epithelial cells of ↑ retina is not clear. Rough ↑ endoplasmic reticulum synthesizes membranes of E.; both organelles are in continuity; bulging of tubules of E. from rough endoplasmic cisternae is frequent. Exact mechanism of E's function is still poorly understood. A kind of E. is represented by ↑ sarcoplasmic reticulum of ↑ cardiac and skeletal muscle fibers. (Fig. = ↑ follicular cell of an ovarian follicle, rat)

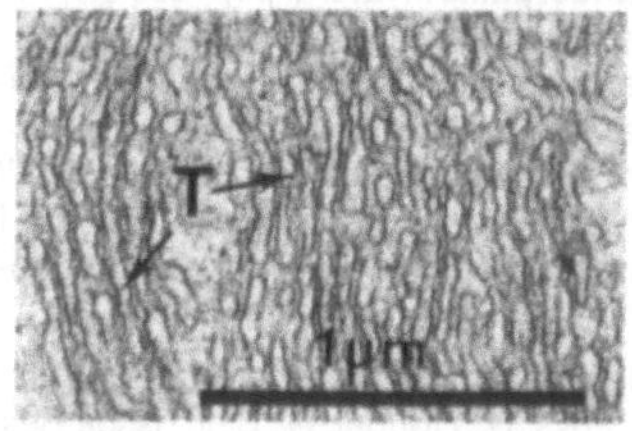

Cardell, J.J., Jr.: Int. Rev. Cytol. *48*, 221 (1979)

Endoplasmic reticulum, smooth, origin of: It is generally assumed that smooth endoplasmic reticulum (sER) originates by bulging from rough endoplasmic reticulum (rER). ↑ Ribosomes (R) of rough endoplasmic reticulum synthesize membranes of smooth endoplasmic reticulum, which bulge (arrows) from edges of lamellae (L) and form ↑ tubules (T) of smooth endoplasmic reticulum. G = ↑ glycogen. (Fig. = ↑ liver parenchymal cell, rat)

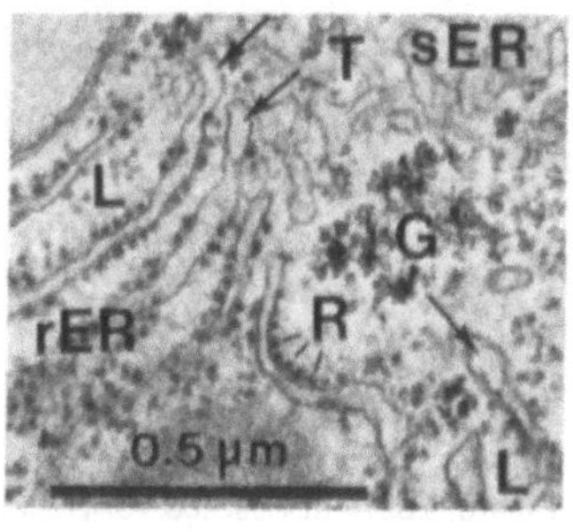

Higgins, J.A., Barrnet, R.J.: J. Cell Biol. *55*, 282 (1972)

Endoplasmic reticulum, transitional: A kind of smooth ↑ endoplasmic reticulum (TR) emerging from rough endoplasmic reticulum near ↑ forming face of ↑ Golgi apparatus (G). Fragmentation of E. tubules produces ↑ transport vesicles (TV).

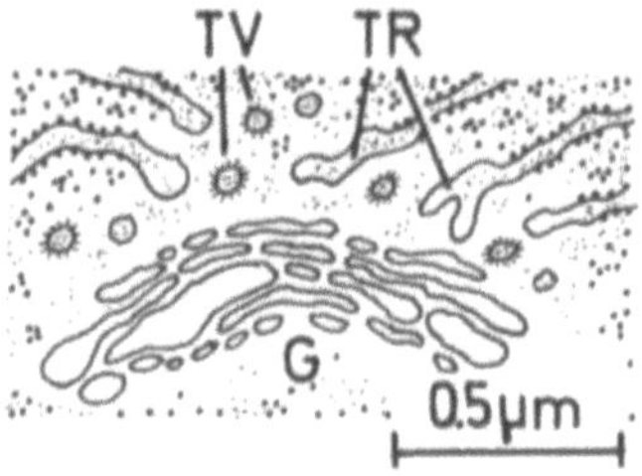

Endopolyploidy: ↑ Endomitosis

Endoreduplication: ↑ Cryptendomitosis

Endorphins: Natural peptides synthesized by some hypothalamic ↑ neurons and able to link with endogenous opiate receptors; their biological action resembles that of opiate alkaloid. It has been suggested that E. act as ↑ neurotransmitters and neuromodulators.

Bloch, B., Bugnon, C., Fellmann, D., Lenys, D., Gouget, A.: Cell Tissue Res. *204*, 1 (1979); Li, C.H.: Cell *31*, 504 (1982); Li, J.Y., Dubois, M.P., Dubois P.M.: Cell Tissue Res. *204*, 37 (1979); Malick, B., Bell, R.M.S. (eds.): Endorphins. Basle: Marcel Dekker (1982); Watkins, W.B.: Cell Tissue Res. *207*, 65 (1980)

Endosteum*: A thin cellular layer (E), composed mainly of ↑ osteoprogenitor cells (OC), lining marrow cavity of ↑ diaphysis and all cavities of cancellous ↑ bone (B). Cells of E. may differentiate into ↑ osteoblasts and/or fuse into ↑ osteoclasts. (Fig. = rat)

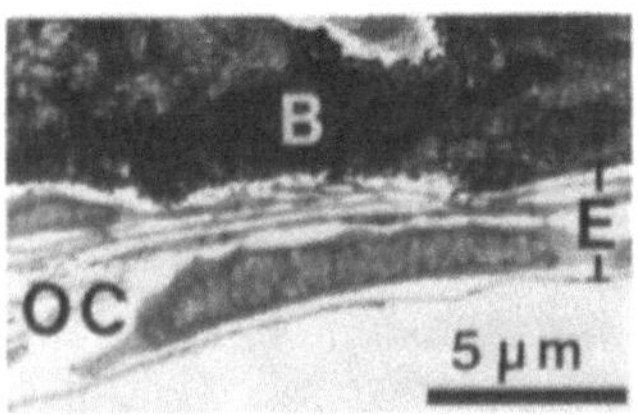

Menton, D.N., Simmons, D.J., Orr, B.Y., Pulard, S.B.: Anat. Rec. *203*, 157 (1982)

Endotendineum (peritendineum internum*): A ↑ loose connective tissue (E) with nerve fibers, blood, and lymphatic vessels surrounding primary tendinous bundles (TB). E. is continuous with ↑ epitendineum. (Fig. = calf)

↑ Tendon

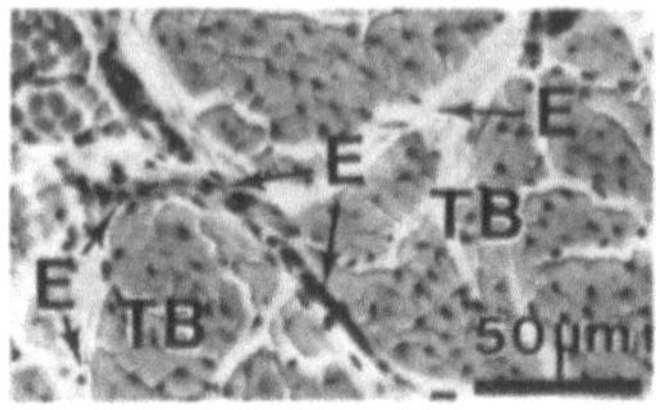

Endothelial cells (endotheliocytus fenestratus et nonfenestratus*): Flattened epithelial cells, 0.1–1 µm thick, 10–15 µm wide, and 25–50 µm long with polygonal or irregular wavy outlines forming ↑ endothelium. Nucleus (N) is higher than surrounding cytoplasm; it is elongated parallel to axis of the vessel. In addition to some mitochondria (M), a ↑ centriole, a small Golgi apparatus (G), a few lysosomes, and a moderate amount of free ribosomes, E. contain microfibrillar bundles, ↑ micropinocytotic vesicles (↑ caveolae), and scattered ↑ specific endothelial organelles (SEO). E. are held together at their edges by ↑ zonulae adherentes and occludentes (J); the cells frequently overlap and form ↑ marginal folds (MF). In ↑ lymphatic capillaries, ↑ junctional complexes may temporarily open to permit entrance of ↑ tissue fluid into lumen. Free surface of E. is covered by a fine ↑ endocapillary layer; luminal plasmalemma forms ↑ microvilli (Mv) mostly grouped over nucleus. Number of microvilli diminishes with increasing speed of bloodstream. A distinct ↑ basal lamina (BL) underlies E. E. containing fenestrations in their cytoplasm form fenestrated ↑ capillaries and those without such fenestrations form continuous ↑ capillaries. E. are very rich in different enzymes (ATPase, nucleoside phosphatase, adenylcyclase, etc.); their surface bears receptor sites for ↑ angiotensin. Antihemophilic factor VIII, thromboplastin, and blood group substances, A, B, and H are also present in E. E. have a low mitotic capacity (life span about 100 days) and regenerate slowly; lost E. seem to be replaced by circulating blood cells, ↑ fibroblasts, ↑ smooth muscle cells (SMC), adjacent E., or undifferentiated subendothelial cells.

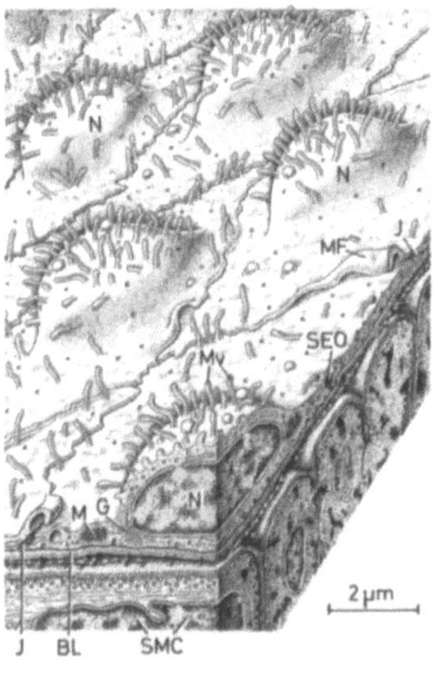

Altura, B.M. (ed.): Vascular Endothelium and Basement Membranes. Basel: Karger 1980; Handenschild, C.C., Schwartz, S.M.: Lab. Invest. *41*, 407 (1979); Hüttner, I., Gabbiani, G.: J. Lab. Invest. *47*, 409 (1982); Messmer, K., Hammersen, F. (eds.): Structure and Function of Endothelial Cells. Progress in Applied Microcirculation, Vol. I. Basle: Karger (1983); Schwartz, S.M., Standeart, D.M., Chi, E.Y.: Lab. Invest. *42*, 507 (1980); Simionescu, M., Palade, G.E.: J. Cell Biol. *68*, 705 (1976)

Endothelial cells, of bone marrow sinuses: ↑ Littoral cells

Endothelial cells, of liver sinusoids (endotheliocytus*): Very flattened squamous cells (E) with an elongated small nucleus, poorly developed ↑ organelles, and a great many ↑ micropinocytotic vesicles. Cytoplasm of E. is extremely thin and perforated by irregular holes and fenestrations, frequently grouped to form ↑ sieve plates (arrows). These openings allow blood plasma, but not blood cells, to gain direct access to ↑ liver parenchymal cells (LC). E. lack basal lamina; they do not phagocytize. Together with ↑ Kupffer's cells, E. form lining of ↑ liver sinusoids

(S) and internal limit of ↑ Disse's space (D).

Wright, P.L., Smith, K.F., Day, W.A., Frazer, R.: Anat. Rec. *206*, 385 (1983)

Endothelial cells, of splenic sinusoids (littoral cells, endotheliocytus bacilliformis*): Spindle-shaped cells (E) lining ↑ splenic sinusoids (S) in a fencelike pattern. Nucleus is elongated and displays a conspicuous nucleolus. Cytoplasm contains a moderate number of mitochondria, a small Golgi apparatus, some short rough endoplasmic cisternae, and a few lysosomes. In certain areas, E. are interconnected by ↑ nexus; elsewhere E. are separated from one another by interendothelial clefts (arrow) measuring 1–7 µm, thus allowing passage of ↑ erythrocytes (Er) and ↑ wandering cells. E. overlie an incomplete basal lamina, which is lacking in the area of interendothelial clefts. Reticular ring fibers surrounding sinus anchor in basal lamina. ↑ Adventitial cells (A) are present over outer aspect of basal lamina. E. have only limited phagocytic capacity. (Fig. = rat)

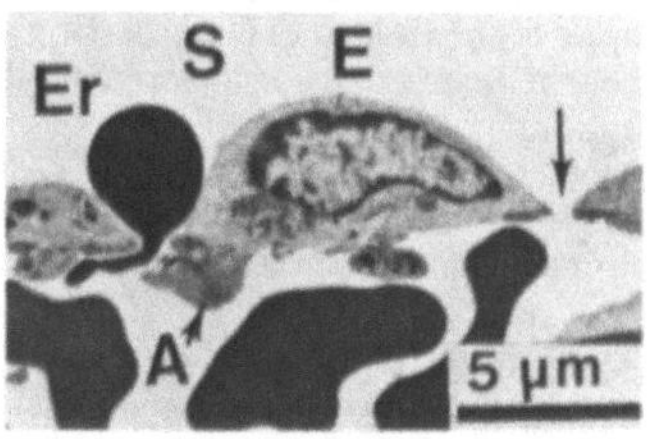

Endothelial organelles: ↑ Specific endothelial organelles

Endothelium* (vasis capillaris*): The internal lining layer of ↑ cardiovascular system and ↑ lymphatic vessels, composed of ↑ endothelial cells.

Fishman, A.P. (ed.): Endothelium. New York: New York Academy of Sciences, Vol. 401, 1982; Freudenberg, N., Freudenberg, K.-H.: The Vascular Endothelial System. Stuttgart: Gustav Fischer 1982; McGuire, P.G., Twietmeyer, T.A.: Artery *11*, 252 (1983)

Endothelium, continuous: ↑ Capillaries, continuous

Endothelium, discontinuous: ↑ Capillaries, sinusoidal

Endothelium, fenestrated: ↑ Capillaries, fenestrated

Endothelium, of cornea: ↑ Cornea

Endothelium, perineural: ↑ Perineural endothelium

Enkephalins: Two pentapeptides widely distributed in CNS. In ↑ neurons of caudate nucleus, E. are contained within ↑ synaptic vesicles. Pharmacological properties of E. are similar to those of narcotic drugs.

Miller, R.J., Pickel, V.M.: J. Histochem. Cytochem. *28*, 903 (1980); Rossier, J., Chapouthier, G.: La Recherche *13*, 1296 (1982)

Enterochromaffin cells (argentaffin cells, basal granular cells, EC-cells, Kulchitsky's cells, pheochrome cells, argentaffinocytus*): Small ↑ endocrine cells located at bases of ↑ gastric glands, proper, as well as in crypts of ↑ Lieberkühn. E. are also found in exocrine and endocrine pancreas. Cell body is triangular or polygonal with its larger side on basal lamina (BL), which also underlies neighboring epithelial cells. As a rule, E. do not extend to surface of epithelium; sometimes, their narrow processes with irregular microvilli may penetrate the lumen. Nucleus is voluminous and spherical; cytoplasm contains some small mitochondria, a small infranuclear Golgi apparatus (G), and only short flattened rough endoplasmic cisternae (rER). In basal cell pole are accumulated numerous ↑ unit membrane-bound, dense ↑ secretory granules (SG), 150–330 nm in diameter, produced by Golgi apparatus. Granules give simultaneously ↑ argentaffin and ↑ chromaffin reactions; they seem to contain ↑ serotonin and ↑ motilin. E. are classified among ↑ APUD cells.

↑ Motilin cells

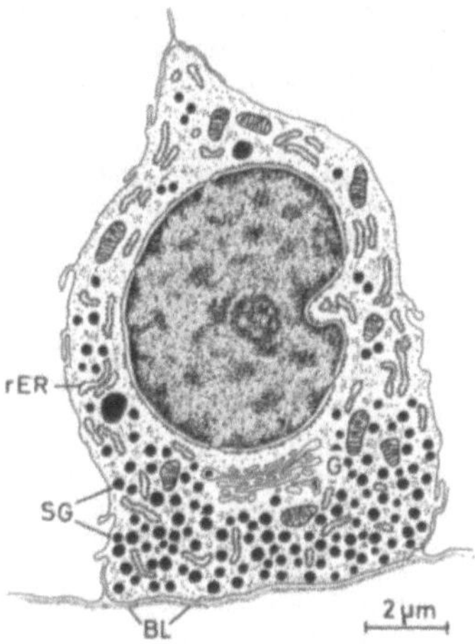

Coupland, R.E., Fujita, T. (eds.): Chromaffin, Enterochromaffin and Related Cells. Amsterdam, Oxford, New York: 1976; Forssmann, W.G., Grube, D.: Z. Zellforsch. *140*, 535 (1973); Inokuchi, H., Kawai, K., Takeuchi, Y., Sano, Y.: Histochemistry *74*, 453 (1982); Kobayashi, S., Iwanaga, T., Fujita, T., Yanaihara, N.: Arch. Histol. Jpn *43*, 85 (1980)

Enterochromaffinlike cells (ECL-cells, EC_1-cells): Irregular, oval, or polygonal cells found in lower two-thirds of ↑ gastric glands proper (fig.), ↑ pancreatic islets, and fetal lung. In addition to a round or elliptical nucleus, E. contain some small mitochondria, a quite well-developed Golgi apparatus, and some short flattened rough endoplasmic cisternae. Their ↑ secretory granules (inset) measure 300–350 nm across and contain an eccentric highly electron-dense core, about 150 nm in diameter. E. rest on basal lamina and do not reach glandular lumen. E. probably produce and store ↑ serotonin, ↑ histamine, and ↑ substance P. It has been suggested that E. are involved in control of acidopeptic secretion. (Fig. = rat)

↑ Endocrine cells, of gastrointestinal tract.

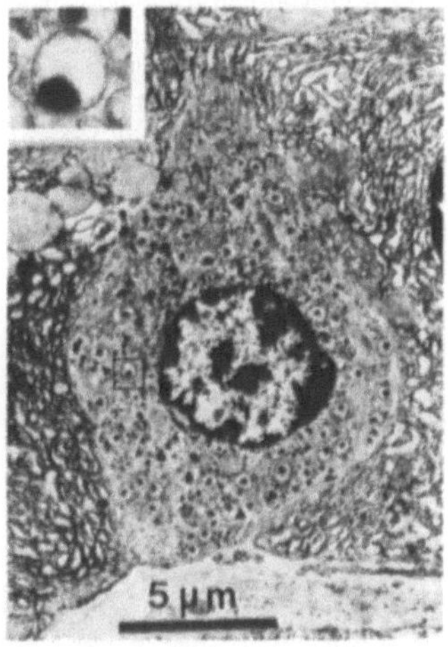

Capella, C., Hage, E., Solcia, E., Usellini, L.: Cell Tissue Res. *186*, 25 (1978); Osaka, M., Sasagawa, T., Fujita, T.: Arch Histol. Jpn *37*, 73 (1974)

Enterocytes: ↑ Absorptive cells

Enteroglucagon: A hormone presumably secreted by ↑ A-cells of ↑ gastrointestinal tract.

↑ Endocrine cells, of gastrointestinal tract

Enteroglucagon-producing cells, of gastrointestinal mucosa: ↑ A-cells, of gastrointestinal tract

Envelope, nuclear: ↑ Nuclear envelope

Eosin: A pink ↑ acid dye widely used in combination with ↑ hemalum for routine ↑ hematoxylin and eosin staining.

Eosinophilia: ↑ Acidophilia

Eosinophilic chemotactic factor-A (ECF-A): An acidic peptide of low molecular weight of about 500 daltons. Produced and released by human basophilic ↑ granulocytes and ↑ mast cells, E. provokes an accumulation of eosinophilic ↑ granulocytes at sites of allergic reaction.

Eosinophilic granules (granulum eosinophilicum*): Membrane-bound specific (secondary) granules (EG), about 0.5–1.5 µm across, of eosinophilic ↑ granulocytes containing a fine, osmiophilic matrix and one or more angular, tightly lamellate crystalloids (C), with ultrastructure depending upon the species. E. contain most lysosomal enzymes found in ↑ azurophilic granules of neutrophilic ↑ granulocytes. E. are very rich in arginine whose cationic guanidinium group is responsible for their ↑ acidophilia. It is believed that a considerable amount of stable ↑ myeloperoxidase is concentrated within crystalloids. E. can be first observed at stage of late ↑ myelocyte. (Fig. = mongolian gerbil)

↑ Granulopoiesis

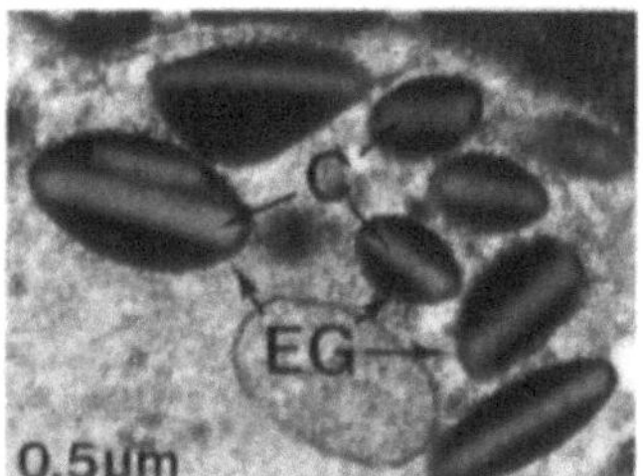

Lewis, D.M., Lewis, J.C., Loegering, A., Gleich, J.G.: J. Cell Biol. *77*, 702 (1978); Okuda, M., Takenaka, T., Kawabori, S., Ogami, Y.: J. Submicrosc. Cytol. *13*, 465 (1981)

Eosinophilic granulocytes: ↑ Granulocytes, eosinophilic

Eosinophilic metamyelocyte: ↑ Metamyelocyte, eosinophilic

Eosinophilic myelocyte: ↑ Myelocyte, eosinophilic

Eosinophils: ↑ Granulocytes, eosinophilic

Ependyma*: A ↑ simple cuboidal to columnar epithelium (E), composed mostly of ↑ ependymal cells with or without ↑ cilia (C) and a few ↑ tanycytes, lining ↑ brain ventricles and central canal of ↑ spinal cord. Ciliated ependymal cells are involved in circulation of ↑ cerebrospinal fluid and transport of some metabolites and ↑ neurotransmitter substances. (Figs. = rat)

↑ Circumventricular organs; ↑ Ependymal differentiations

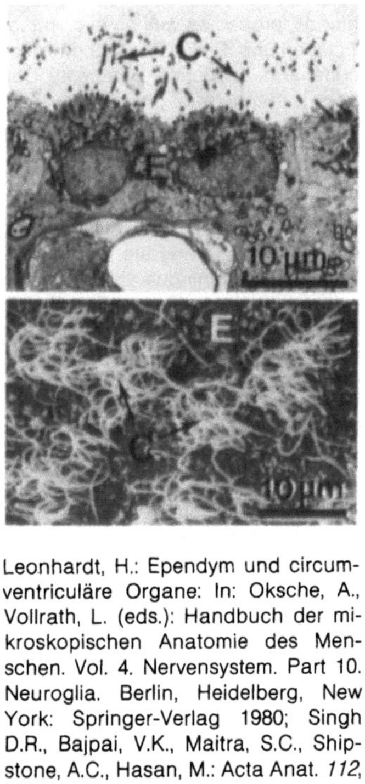

Leonhardt, H.: Ependym und circumventriculäre Organe: In: Oksche, A., Vollrath, L. (eds.): Handbuch der mikroskopischen Anatomie des Menschen. Vol. 4. Nervensystem. Part 10. Neuroglia. Berlin, Heidelberg, New York: Springer-Verlag 1980; Singh D.R., Bajpai, V.K., Maitra, S.C., Shipstone, A.C., Hasan, M.: Acta Anat. *112*, 365 (1982)

Ependymal cells (ependymocytus*): Cuboidal to columnar epithelial cells (E) representing majority of cells of ↑ ependyma. Nucleus is elliptical with a prominent nucleolus; cytoplasm contains numerous, rod-shaped mitochondria, a well-developed Golgi apparatus, ↑ centrioles, short rough endoplasmic cisternae, some smooth endoplasmic tubules, a few free ribosomes, ↑ lysosomes, ↑ glycogen particles, and bundles of 6 to 8-nm-thick microfilaments. Almost every E. bears a tuft of ↑ cilia (C) and ↑ junctional complexes hold adjacent E. together.

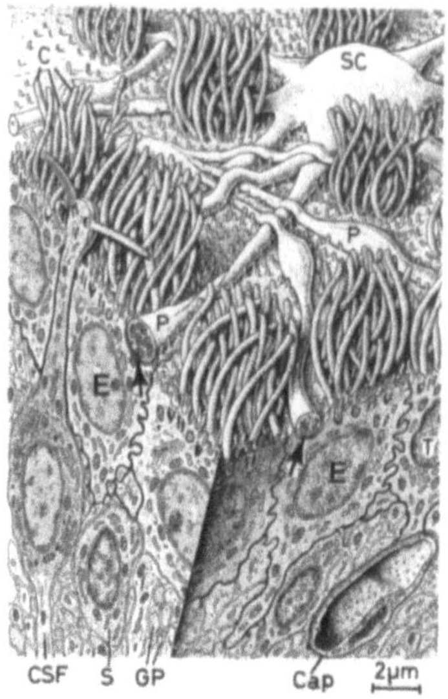

There is no basal lamina between E. and underlying ↑ subependymal cells (S) and glial processes (GP). In some areas of ventricles, processes of ↑ cerebrospinal fluid-contacting neurons (CSF) penetrate between E. ↑ Tanycytes (T) are a kind of E. possessing a coarse basal extension in contact with capillaries (Cap). In some regions of ventricles, ↑ supraependymal cells (SC) overlie E. Extensions (P) of the former, which in some cases are ↑ axons, establish synaptic contacts (arrows) with E. E. belong to ↑ neuroglia and are involved in selective transfer of substances from ↑ cerebrospinal fluid to brain parenchyma, partially via intercellular clefts and partially through E. themselves.

Araki, M., Sato, F., Saito, T.: Arch. Histol. Jpn. *46*, 191 (1983); Millhouse, E.O.: Z. Zellforsch. 149 (1972)

Ependymal differentiations: Agglomerations of specially differentiated ↑ ependymal cells forming, together with underlying nerve tissue, some ↑ circumventricular organs (↑ area postrema, ↑ organum vasculosum laminae terminalis, ↑ subcommissural organ, ↑ subfornical organ).

EPF: ↑ Exophthalmus-producing factor

Epicardium*: Visceral layer of ↑ pericardium covering outer surface of heart in the form of a serous membrane. Structure: 1) ↑ Mesothelium (M) = a single layer of squamous epithelial cells; 2) subepicardial layer or tela subepicardiaca* (S) = a ↑ loose connective tissue rich in ↑ elastic fibers, blood and lymphatic vessels (V), nerve fibers (N), and containing an individually variable amount of ↑ adipose tissue (asterisk), predominantly along coronary vessels. (Fig. = human)

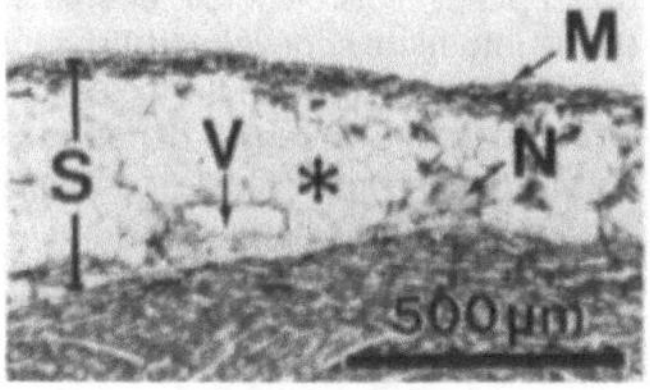

Epidermal appendages: Organs developing from ↑ epidermis in course of ↑ organogenesis – ↑ hairs, ↑ nails, and epithelia of ↑ mammary, ↑ sebaceous, and ↑ sweat glands.

Epidermal growth factor (EGF, urogastrone): A simple polypeptide chain of low molecular weight containing three disulfide bonds. E. has been isolated from mouse submandibullary gland and human urine. Main actions of E. are stimulation of ↑ mitosis in ↑ epidermis and its ↑ keratinization, acceleration of proliferation and ↑ differentiation of various epithelia, hepatic hypertrophy, and inhibition of gastric secretion.

↑ Growth factors

Carpenter, G., Cohen, S.: Ann. Rev. Biochem. *48*, 193 (1979); Das, M.: Int. Rev. Cytol. *78*, 233 (1982); Moore, G.P.M., Panaretto, B.A., Robertson, D.: Anat. Rec. *205*, 47 (1983); Steidler, N.E., Reade, P.C.: J. Anat. *135*, 413 (1982)

Epidermal melanin unit: The ratio between one ↑ melanocyte and adjacent cells of stratum germinativum of ↑ epidermis. Under normal conditions, E. is 1 melanocyte to 36 epidermal cells.

Epidermal pegs: ↑ Epidermal ridges

Epidermal proliferative unit (EPU): A roughly hexagonal epidermal column (E) of nine to ten cells representing source of cell replacement in ↑ epidermis. In center of the base, one Langerhans' (L) and one to two stem cells (S) are located, surrounded by six to seven ↑ basal cells (B). Basal cells flatten and migrate (arrows) into the column, either from center or periphery of base. Each migration is accompanied by mitosis (M) of adjacent cell. Thus, columns consist of former basal cells in course of ↑ differentiation and transformation into keratinized plates (P).

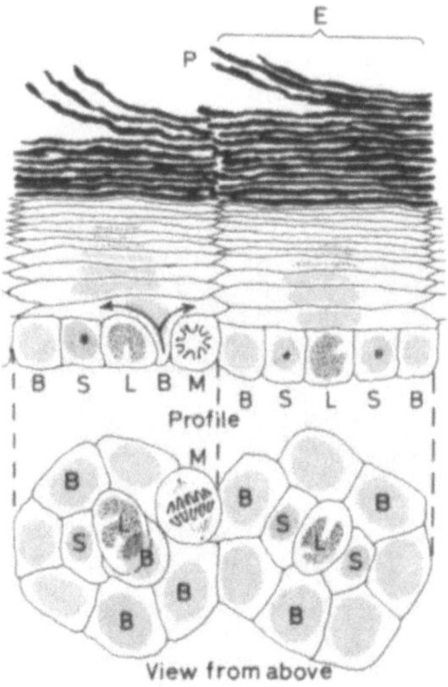

Cairnie, A.B., Lala, P.K., Osmond, D.G. (eds.): Stem Cells of Renewing Cell Populations. New York, San Francisco, London: Academic Press (1976); Potten, C.S.: Int. Rev. Cytol. *69*, 271 (1981)

Epidermal ridges (epidermal pegs): Epidermal projections (R) interdigitating with ↑ dermal papillae (p) of dermal papillary layer. Depending on mechanical forces to which skin area concerned is exposed, E. can be simple (e.g., abdominal skin, A) or very long and ramified (e.g., sole, S). Like all ↑ epithelial ridges, E. assure firm contact between ↑ epidermis and underlying tissue. hR = horny ↑ epidermal ridges

↑ Dermoepidermal junction

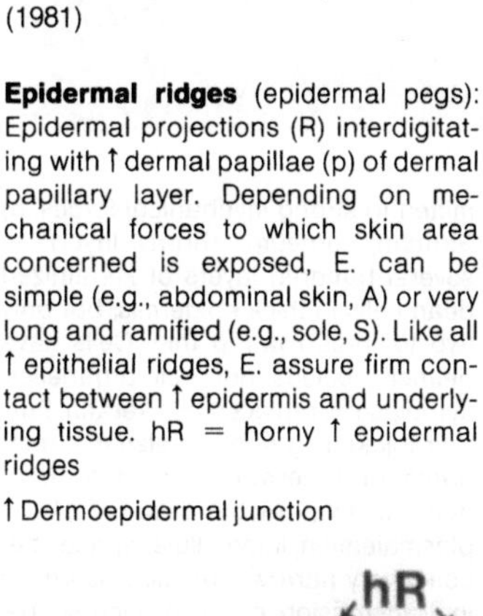

Epidermal ridges, horny (surface epidermal ridges, cristae cutis*): Characteristic regular raised lines (hR) of stratum corneum of ↑ epidermis on volar surface of hands and plantar surface of feet arranged in spiral and concentric patterns. (See fig. under ↑ epidermal ridges)

↑ Dermatoglyphics; ↑ Epidermis, of palms and soles

Epidermis: ↑ Epidermis, light microscopy of; ↑ Epidermis, ultrastructure of

Epidermis, cell replacement of: ↑ Epidermal proliferative units; ↑ Keratopoiesis

Epidermis, functions of: The renewal by mitotic activity of cell layers from ↑ epidermal proliferative units (mainly at night), protection of body against UV radiation, protection against bacterial invasion and slight trauma, protection against water penetration, prevention of evaporation.

Epidermis*, light microscopy of: The superficial layer of ↑ skin, consisting of ↑ stratified squamous keratinized epithelium. Structure (beginning from basal lamina, BL): 1) Stratum basale*; 2) stratum spinosum*; 3) stratum granulosum*; 4) stratum lucidum*; 5) stratm corneum*; 6) stratum disjunctum*. Epidermal downgrowth forms ↑ epidermal ridges (ER), extensively indented with ↑ dermal papillae (DP) of ↑ dermis; in thick hairless skin (volar surface of hands and plantar surface of feet), E. displays horny ↑ epidermal ridges (hR).

↑ Dermoepidermal junction; ↑ Epidermis, ultrastructure of

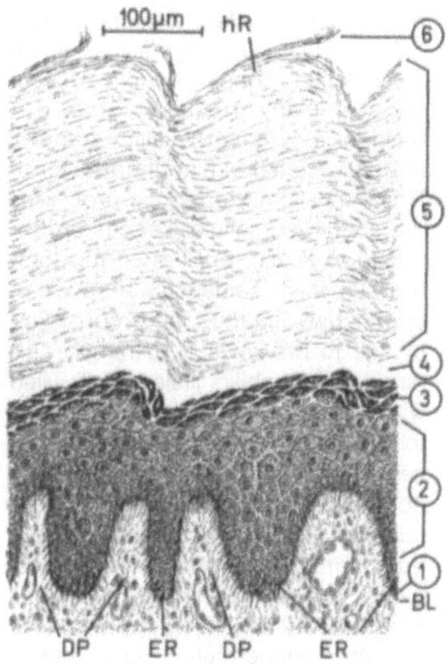

Epidermis, of palms and soles: Areas of particularly thick ↑ epidermis characterized by presence of all epidermal layers with particularly numerous and

deep ↑ epidermal ridges (ER), a very thick stratum corneum (SC), and horny ↑ epidermal ridges (hR).

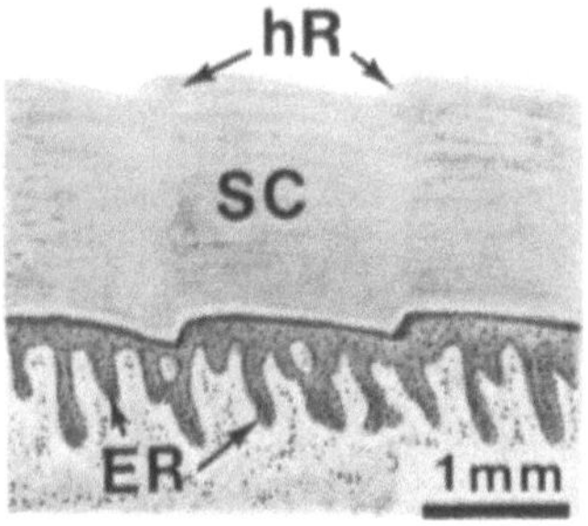

Epidermis, specialization of: ↑ Epidermal appendages

Epidermis*, ultrastructure of: 1) Stratum basale* = a single layer of ↑ basal cells (BC) lying on ↑ basal lamina (BL). 2) Stratum spinosum* = two to ten layers of voluminous ↑ prickle cells (PC). Since these two layers are responsible for renewal of lost surface cells, they are termed stratum germinativum. ↑ Melanocytes, ↑ Langerhans', ↑ and ↑ Merkel's cells (not shown) are scattered among these two layers. 3) Stratum granulosum* = three to five rows of spindle-shaped cells in thick epidermis (soles, palms, and palmar surface of fingers); only one row in thin epidermis. Long cell axes are parallel to skin surface. Cells have an oval nucleus with condensed ↑ chromatin; cell ↑ organelles are almost distinguishable because of presence of many irregularly shaped ↑ basophilic, ↑ keratohyalin granules (KG) associated with abundant ↑ tonofilaments (Tf). Cytoplasmic processes are short; intercellular spaces are reduced to 25 nm. 4) Stratum lucidum* (only in thick epidermis) = a clear, bright, homogenous layer consisting of one to two rows of extremely flattened ↑ acidophilic, lifeless cells filled with a large amount of densely packed tonofilaments covering nucleus and cell organelles. Keratohyalin granules apparently disappear; ↑ eleidin contained in this layer is believed to be a transformation product of keratohyalin. Plasmalemma becomes thickened (12 nm). Intercellular spaces fill with a moderately electron-dense material originating probably from contents of discharged ↑ membrane-coating granules. Stratum lucidum is also considered to be a sliding layer inserted between stratum germinativum and stratum corneum in skin sub-

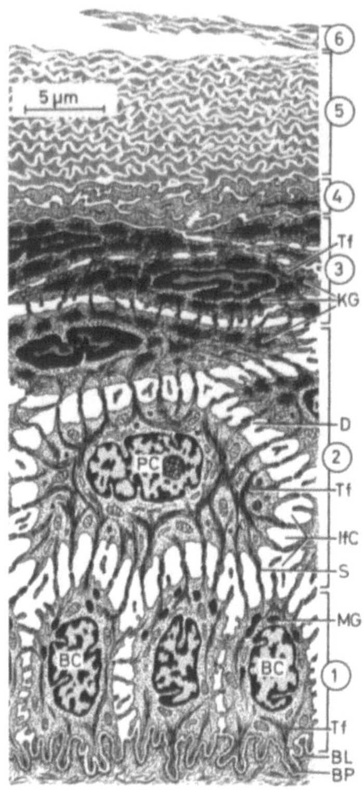

mitted to strong mechanical forces. 5) Stratum corneum* (horny layer) = several hundred layers of keratinized dead cells in thick epidermis, but only two to three rows in thin. Cells have neither nucleus nor cell organelles; cytoplasm is replaced by ↑ keratin. Further thickening of plasmalemma; remnants of ↑ desmosomes in form of dense osmiophilic bands are visible on plasmalemma; intercellular spaces become very narrow, but still discernible in transmission electron microscope. 6) Stratum disjunctum* = most superficial layer of flattened, keratinized scales, which are continuously being loosened and desquamated in the form of single squama cornea* or as sheets. BP = ↑ basal processes, D = desmosomes, IfC = ↑ interfacial canals, MG = ↑ melanin granules, S = spines.

↑ Intercellular bridges, false, ↑ Epidermal proliferative units; ↑ Epidermis, light microscopy of

Breatnach, A.S.: J. Invest. Dermatol. *65*, 2 (1975); Green, H., Fuchs, E., Watt, F.: Differentiated Structural Components of the Keratinocyte. Cold Spring Harbor Symposia on Quantitative Biology, Vol. 46, Part 1. New York: Cold Spring Harbor Laboratory 1982; Lavker, R.M., Sun, T.T.: Science *215*, 1239 (1982)

Epididymis*: A crescent-shaped encapsulated organ (E) situated at posterior and superior surfaces of ↑ testis (Te). E. consists of a caput or head (H), composed of 8–20 ↑ ductuli efferentes grouped in ↑ coni vasculosi (CV) or lobuli epididymidis. Ductuli efferentes empty into highly convoluted ↑ ductus epididymidis (DE), which forms corpus or body (B) of E. In region of cauda or tail (T), ductus epididymidis is continuous with ↑ ductus deferens (DD).

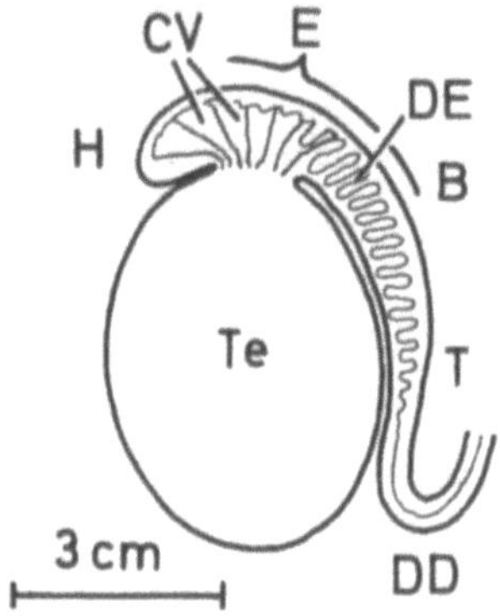

Francavilla, S., De Martino, C., Scorza Barcellona, P., Natali, P.G.: Cell Tissue Res. *233*, 523 (1983); Holstein, A.F.: Morphologische Studien am Nebenhoden des Menschen. In: Bargmann, W., Doerr, W. (eds.): Zwanglose Abhandlung aus dem Gebiet der normalen und pathologischen Anatomie. Stuttgart: G. Thieme 1969; Ogebin-Crist, M.C.: Biol. Reprod. Suppl. *1*, 155 (1969)

Epidural space (spatium epidurale*): A space between ↑ periosteum of vertebral canal and outer surface of ↑ dura mater. E. contains adipose tissue, blood, and lymphatic vessels and represents an elastic envelope for dural sac.

Epiglottic cartilage (cartilago epiglottica*): A sheetlike piece of elastic ↑ cartilage forming skeleton of ↑ epiglottis. Between small ↑ isogenic groups (IG),

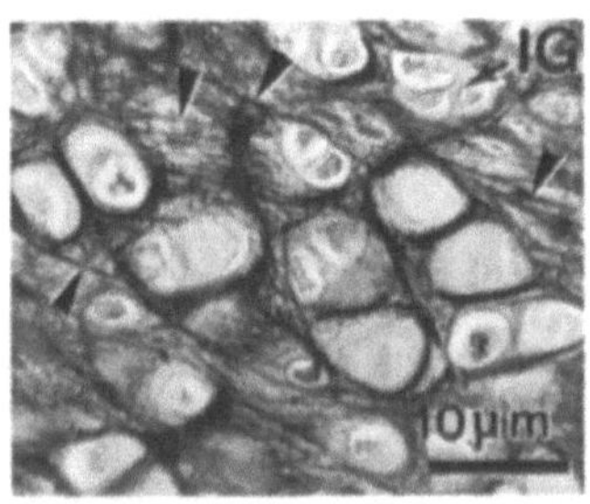

intercellular substance contains many ↑ elastic fibers (arrowheads). (Fig. = human, ↑ resorcin fuchsin staining)

Epiglottic glands (glandulae epiglotticae*): Small ↑ mixed glands (EG) situated in tunica mucosa (TM) and in corresponding holes of ↑ epiglottic cartilage (EC). (Fig. = human, ↑ orcein staining)

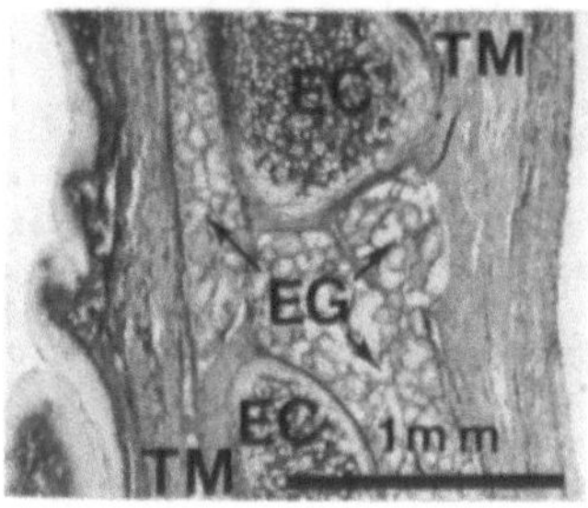

Epiglottis*: An unpaired, spoon-shaped organ representing cover of ↑ larynx. Structure: 1) Tunica mucosa (TM): a) epithelium (E) = a nonkeratinized ↑ stratified squamous epithelium at oral (O) and laryngeal sides (L); transition (arrowhead) to ↑ pseudostratified ciliated epithelium takes place at a variable depth in ventriculus laryngis (VL); b) lamina propria (LP) = a ↑ loose connective tissue with blood and lymphatic vessels and nerve fibers; ↑ epiglottic glands (EG) are predominantly located on laryngeal side in corresponding holes of 2) ↑ epiglottic cartilage (EC) which forms skeleton of E.

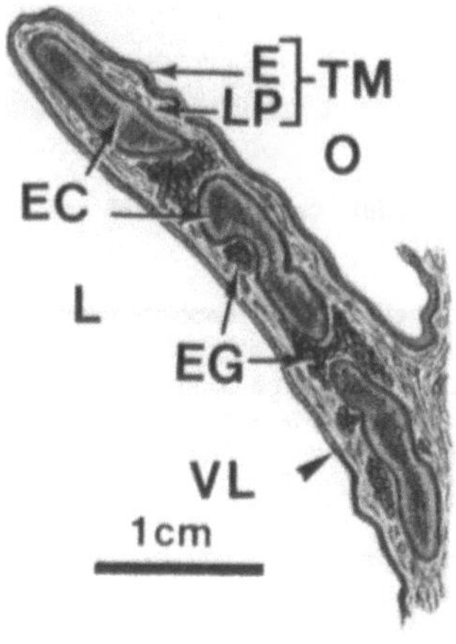

Epimysium*: A ↑ loose connective tissue sheath (E) enveloping an entire skeletal muscle; exteriorly, E. is coninuous with ↑ fascia (F), and interiorly with ↑ perimysium externum (PE). S = ↑ skeletal muscle fibers. (Fig. = human)

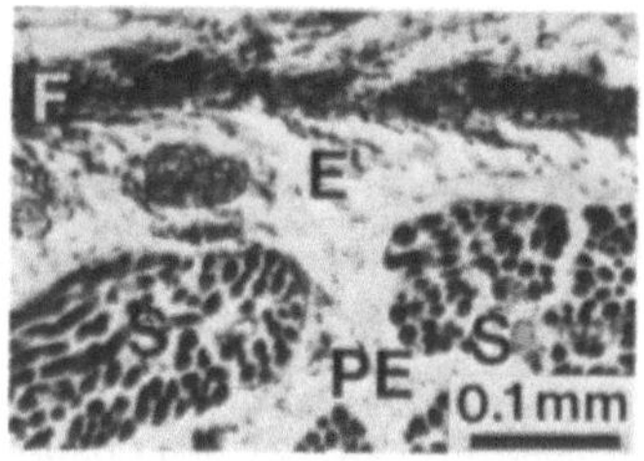

Epinephrine (adrenaline): One of two hormones of medulla of ↑ adrenal gland. Secreted by ↑ epinephrine-producing cells, E. plays a role as ↑ neurotransmitter substance in some postganglionic nerve fibers and in some ↑ neurons of ↑ central nervous system. Effects: Increase of heart rate and cardiac output without increase of blood pressure, increase of oxygen consumption, increase of blood sugar, release of ↑ ACTH from the hypophysis, etc. E. acts as first messenger by activation of ↑ cyclic AMP.

↑ Epinephrine storage granules

Epinephrine-producing cells (epinephrocytus*): ↑ Chromaffin cells forming part of medullary ↑ parenchyma of ↑ adrenal gland. Cell body is polyhedral with an eccentric nucleus and a conspicuous nucleolus. Cytoplasm encloses a moderate number of voluminous mitochondria, short cisternae of rough endoplasmic reticulum, a few lysosomes, rare ↑ multivesicular bodies, and some free ribosomes. A large number of ↑ epinephrine storage granules (SG) separate themselves from the large Golgi apparatus (G). ↑ Cholinergic preganglionic fibers (F) may form neuroglandular ↑ synapses with E. PS = ↑ pericapillary space

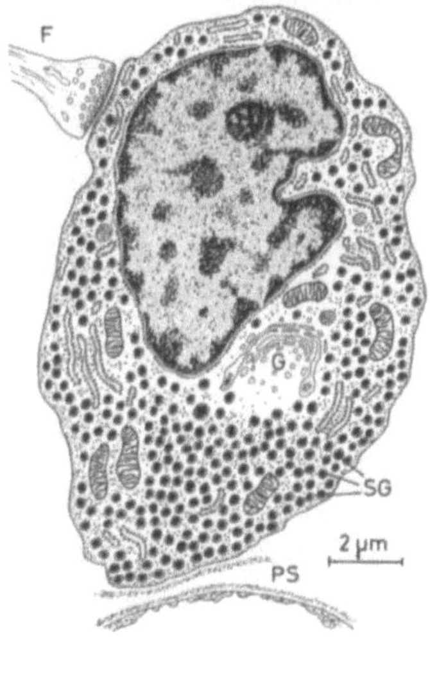

↑ Norepinephrine-producing cells

Epinephrine storage granules (adrenaline storage granules): Unit membrane-limited structures (EG), 15–30 nm in diameter, located in cytoplasm of ↑ epinephrine-producing cells of adrenal medulla (in man, E. and ↑ norepinephrine-storage granules are present in a single type of cell). E. contain fine granular, moderately osmiophilic material representing ↑ epinephrine bound to ↑ chromogranin. In E. is also concentrated dopamine β-oxidase, and in the limiting membranes lysolecithin and chromomembrin – A. E. are produced in Golgi apparatus (G, arrows), which also furnishes their surrounding membranes. It is thought that entire contents of E. are released by ↑ exocytosis into extracellular space from which the hormone enters the circulation. (Fig. = rat)

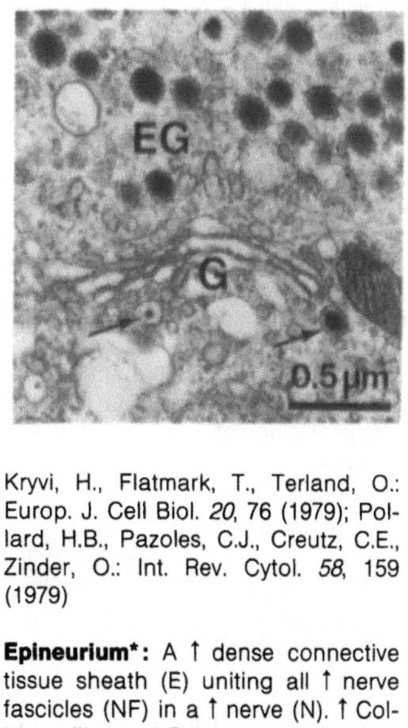

Kryvi, H., Flatmark, T., Terland, O.: Europ. J. Cell Biol. *20*, 76 (1979); Pollard, H.B., Pazoles, C.J., Creutz, C.E., Zinder, O.: Int. Rev. Cytol. *58*, 159 (1979)

Epineurium*: A ↑ dense connective tissue sheath (E) uniting all ↑ nerve fascicles (NF) in a ↑ nerve (N). ↑ Collagen fibers of E. have a longitudinal slightly undulating course. E. also penetrates between nerve fascicles, where

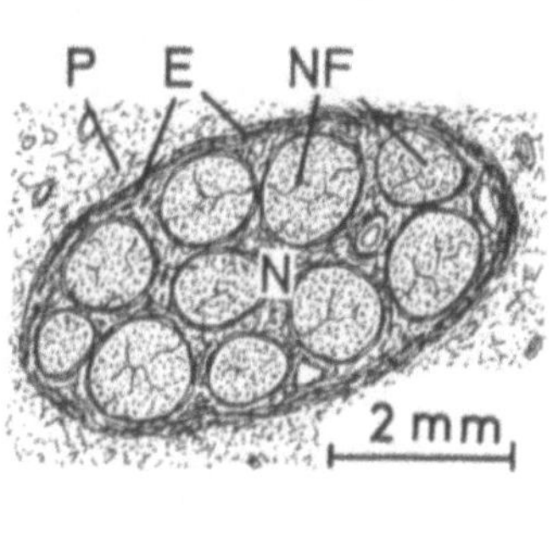

it contains blood and lymphatic vessels, ↑ nervi nervorum, and some ↑ adipose tissue. External to E. is ↑ paraneurium (P).

Epipharynx: A portion of ↑ pharynx.

Epiphyseal cartilage: ↑ Epiphyseal plate

Epiphyseal disc: ↑ Epiphyseal plate

Epiphyseal ossification center: ↑ Bone formation, indirect

Epiphyseal plate (epiphyseal disc, epiphyseal cartilage, linea epiphysialis*): A disc of hyaline ↑ cartilage (EP) between the fronts of diaphyseal (D) and epiphyseal (E) ossification centers. Structure: 1) Relatively small zone of reserve cartilage immediately adjacent to epiphyseal bone (EB); 2) considerable proliferation zone with formation of columns predominantly toward ↑ diaphysis; 3) hypertrophic zone; 4) calcification zone. Through multiplication of ↑ chondrocytes, E. compensates cartilage destruction at border of diaphysis; in this way, ossification front is continually moving away from diaphysis. As far as E. can balance loss of cartilage, bone grows in length. At end of growing period, interstitial growth within E. ceases; E. is completely removed, epiphyseal and diaphyseal marrow cavities become continuous (closure of epiphysis): This signifies end of bone's longitudinal growth. ↑ Growth hormone, ↑ sex hormones, ↑ somatostatin, etc. control histophysiology of E. AC = ↑ articular cartilage

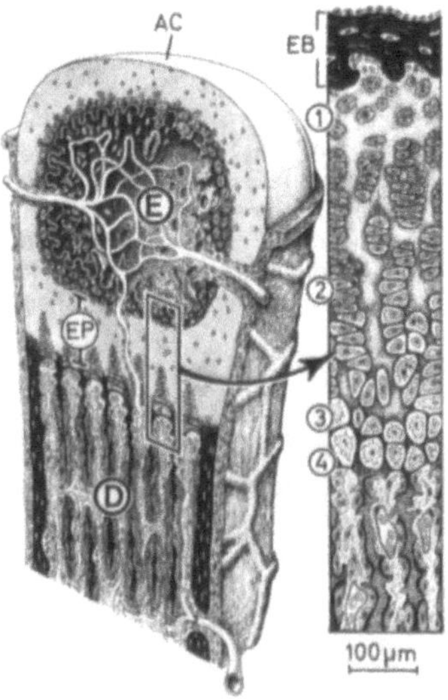

↑ Bone formation, indirect; ↑ Cartilage, growth of; ↑ Synchondrosis

Althoff, J., Quint, P., Krefting, E.-R., Höhling, H.J.: Histochemistry *74*, 541 (1982); Barckhaus, R.H., Höhling, H.J.: Cell Tissue Res. *186*, 541 (1978); Dixon, A.D., Sarnat, B.G. (eds.): Factors and Mechanisms Influencing Bone Growth. Progress in Clinical and Biological Research, 101. New York: Alan R. Liss 1982

Epiphysis cerebri: ↑ Pineal organ

Epiphysis, closure of: ↑ Epiphyseal plate

Epiphysis, of bone (epiphysis*): One of two extremities (proximal and distal) of a long bone.

↑ Diaphysis

Epiplexus cells: ↑ Kolmer's cells

Epiploic appendages: ↑ Appendices epiploicae

Episcleral space (space of Tenon, spatium intervaginale, spatium episclerale*): A narrow slit between ↑ sclera and ↑ capsule of Tenon filled with extremely ↑ loose connective tissue. E. is traversed by tendons of extrinsic eye muscles.

Epitendineum (peritendineum externum): A ↑ dense connective tissue (E), with blood and lymphatic vessels and nerve fibers surrounding a ↑ tendon. E. is continuous toward interior of tendon with ↑ endotendineum (En), and toward exterior with ↑ paratendineum. In tendons provided with a ↑ synovial vagina, E. is reduced and connects visceral ↑ mesothelium (M) to tendon. (Fig. = calf)

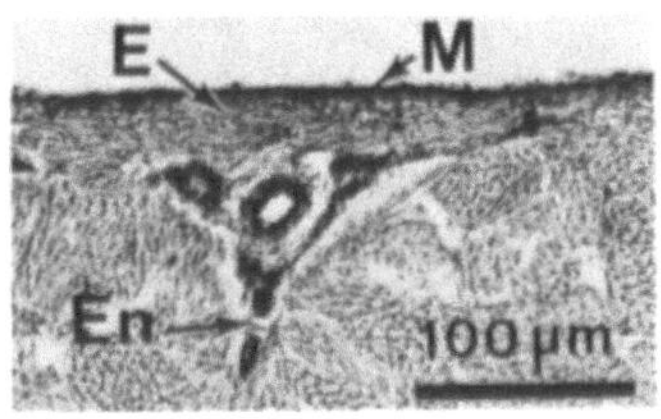

Epithalamo-epiphyseal complex: The anatomical and functional unit of neurosecretory system comprising habenular ganglia, ↑ pineal organ, and ↑ subcommissural and ↑ subfornical organs. E. is thought to represent an antagonist of ↑ hypothalamohypophyseal complex particularly concerning control of gonadal, thyroid, and adrenal activities.

Lahoz Gimeno, M.: An. Anat. *31*, 377 (1982)

Epithelial cells (cellulae epitheliales*): A collective term for all cells belonging to a surface or a glandular ↑ epithelium.

Epithelial cells, brush border of: ↑ Brush border

Epithelial cells, cilia of: ↑ Cilia

Epithelial cells, junctional complex of: ↑ Junctional complex

Epithelial cells, lateral surfaces of: At their lateral surfaces, most epithelial cells form microvillous ↑ interdigitations, short flaplike processes (P), and projections bearing halves of ↑ desmosomes at their tips. All mentioned lateral specializations strengthen cell union. (Fig. = ↑ surface mucous cells, dog)

↑ Epidermis; ↑ Intercellular bridges, false

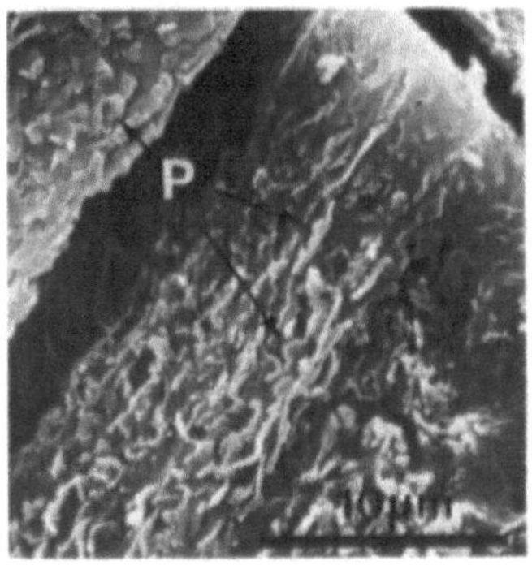

Epithelial cells, microvilli of: ↑ Microvilli

Epithelial cells, specializations of cell surface in: Differentiations of apical plasmalemma comprising ↑ brush border, ↑ cilia, ↑ microvilli, ↑ microridges, ↑ stereocilia, and ↑ striated border.

Epithelial cells, stereocilia of: ↑ Stereocilia

Epithelial-reticular cells, of thymus (epithelioreticulocytus thymi*): Stellate or irregularly shaped cells with elongated cytoplasmic extensions contacting adjacent E. by means of ↑ desmosomes (D). Nucleus is oval and measures 7–10 µm at its greatest diameter; it contains finely dispersed ↑ chromatin; nucleolus is prominent. Cytoplasm encloses sparse mitochondria, a moderately developed Golgi apparatus, and short rough endoplasmic cisternae. Number of free ribosomes can be considerable. Through cytoplasm run numerous ↑ tonofibrils (Tf) ending in ↑ attachment plaques of ↑ desmosomes. Some E. contain small (secretory?) granules, 0.1 µm in diameter (Gr), ↑ multivesicular bodies (MB), ↑ keratohyalin granules (K), a few ↑ lysosomes, ↑ glycogen particles (Gly), and spherical vacuoles (V) with microvillous projections and amorphous contents. E. form epithelial ↑ cytoreticulum of ↑ thymus with meshes filled with ↑ lymphocytes (L). E. provide instructions to ↑ T-lymphocytes and probably synthesize a lymphopoietic hormone. E. of thymic medulla tend to come into close contact with one another, giving rise to ↑ Hassall's bodies.

↑ Epithelium, atypical

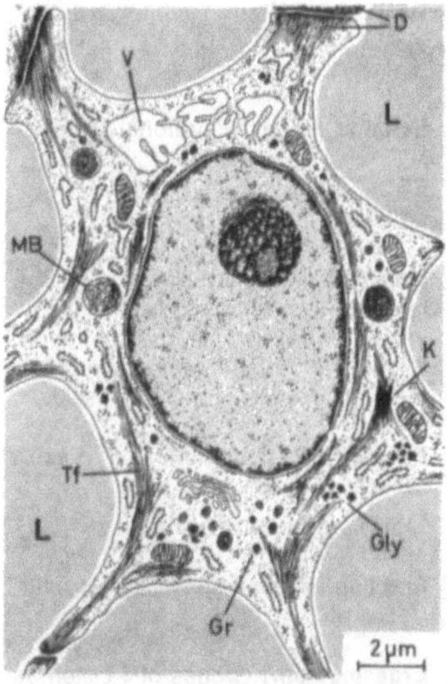

Savino, W., Santa-Rosa, G.L.: Arch. Histol. Jpn *45*, 139 (1982); Wijngaert, F.P. van de et al.: Cell Tissue Res. *237*, 227 (1984)

Epithelial ridges (cristae epitheliales*): Projections of various depth of nonkeratinized ↑ stratified squamous epithelium into lamina propria. E. are well developed at dorsal surface of ↑ tongue, ↑ gingiva, ↑ esophagus, and ↑ vagina.

↑ Epidermal ridges

Epithelial tissue: ↑ Epithelium

Epithelioid cells: Tightly apposed, nonepithelial cells resembling an ↑ epithelium. Examples: Closely packed, specialized ↑ smooth muscle cells of ↑ glomi, some arteriovenous ↑ anastomoses, and ↑ afferent arterioles (↑ juxtaglomerular cells), characterized by large round bodies, a small amount of ↑ myofilaments, and numerous ↑ secretory granules; closely packed ↑ macrophages in chronic inflammations; closely packed ↑ chromaffin cells of adrenal medulla.

Epithelium (epithelial tissue, textus epithelialis*): A close, avascular cell union with very narrow intercellular clefts containing little intercellular substance. E. covers outer surface of body and lines spaces and tubes within body; it forms ↑ glands and other structures derived therefrom. E. are divided into: 1) Surface ↑ E.; 2) glandular ↑ E.; 3) sensory ↑ E.

↑ Nervous tissue

Epithelium, atypical: ↑ Atypical epithelium

Epithelium, basal lamina of: ↑ Basal lamina

Epithelium, basal surface of: ↑ Basement membrane

Epithelium, brush border of: ↑ Brush border

Epithelium, collagen synthesis of: ↑ An epithelium is capable of synthesizing ↑ collagen type IV of ↑ basal laminae.

Epithelium, extraneous cells of: Presence of ↑ wandering cells, primarily ↑ lymphocytes and ↑ phagocytes, in an ↑ epithelium.

↑ Diapedesis

Epithelium, false: Former term, no longer employed, for epithelia originating from ↑ mesenchyme (epithelia of blood and lymphatic vessels, epithelia of pleural, pericardial, and peritoneal cavities).

↑ Mesothelium

Epithelium, functions of: 1) Protection of underlying tissue; 2) ↑ absorption of nutritive substances; 3) ↑ secretion of various products; 4) ↑ excretion of toxic substances; 5) diffusion of gases; 6) filtration of light; 7) surface transport of ↑ mucus by ciliary beating; 8) sensory reception; 9) support of sensory and ↑ germ cells.

Epithelium, "fuzzy" coat of: ↑ Glycocalyx

Epithelium, glands of: ↑ Goblet cells; ↑ Endoepithelial glands

Epithelium, glandular (epithelium glandulare*): An ↑ epithelium whose main function is ↑ secretion.

↑ Glands

Epithelium, glandular, classification of: ↑ According to presence or absence of an excretory canalicular system, E. is divided into ↑ exocrine and ↑ endocrine glands.

↑ Glands, classification of

Epithelium, innervation of: Nerve fibers penetrating ↑ epithelium lose their ↑ myelin sheath and ramify among epithelial cells, accompanied only by ↑ Schwann's cells. In some cases, nerve fibers form neuroglandular ↑ synapses with epithelial cells.

Epithelium, junctional complex of: ↑ Junctional complex

Epithelium, lateral cell compartment of: In general, ↑ epithelial cells are surrounded by a space only 20–30 nm wide. In stratum spinosum of ↑ epidermis, these spaces are very large (↑ interfacial canals) and act as capillaries. ↑ Intercellular canaliculi are formed by two adjacent cells and open onto cell surface. In absorptive and water-transporting epithelia (small intestine, ↑ gallbladder), intercellular spaces are frequently dilated as a result of active transport of sodium and water from lumen into capillaries.

Kaye, G.I., Wheeler, H.O., Whitlock, R.T., Lane, N.: J. Cell Biol. *30*, 237 (1966)

Epithelium, loss of form: The transformation of an ↑ epithelium into another structure, e.g., transformation of epithelium of lens vesicle, under influence of optic vesicle, into ↑ lens fibers. (See embryology texts for further information)

Epithelium, metaplasia of: ↑ Metaplasia

Epithelium, of organs: See under name of specific epithelium or organ.

Epithelium, origin of: An epithelium can originate from ↑ ectoderm, ↑ endoderm, or ↑ mesoderm. (See embryology texts for further information)

Epithelium, regeneration of: Under normal conditions, ↑ regeneration of surface epithelia is complete: Lost cells of simple epithelia are regenerated from special areas of undifferentiated cells (gastric ↑ foveolae, ↑ Lieberkühn's crypts) or by division of neighboring cells; shed cells of ↑ pseudostratified and ↑ stratified epithelia regenerate by mitotic activity and further ↑ differentiation of ↑ basal cells. In ↑ uterus, epithelium is regenerated during proliferative phase of ↑ menstrual cycle. Life span of epithelial cells varies from 3 days (intestinal ↑ absorptive cells) to 90 days (↑ epidermis). Glandular epithelia do not regenerate so well as surface epithelia.

Epithelium, seminiferous: ↑ Seminiferous epithelium

Epithelium, sensory: An ↑ epithelium in which some cells are specialized to react to specific external impulses or substances.

↑ Corti's organ; ↑ Macula of saccule and utricle; ↑ Olfactory epithelium; ↑ Taste buds

Epithelium, sensory, of inner ear: ↑ Corti's organ; ↑ Crista ampullaris; ↑ Macula of saccule and utricle

Epithelium, specializations for attachment and communication of: ↑ Junctional complex

Epithelium, surface (epithelium superficiale*): An ↑ epithelium lining an external or internal surface of body.

Epithelium, surface, classification of: According to arrangement and shape of cells, epithelia are classified as follows:

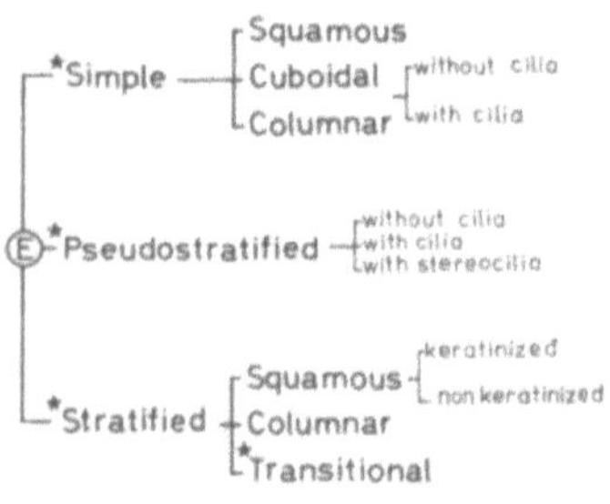

In ↑ simple and ↑ pseudostratified epithelia, all cells are in contact with ↑ basal lamina, whereas in ↑ stratified epithelia only ↑ basal cells lie on basal lamina. (Asterisks indicate letters under which the corresponding epithelia are described)

Epithelium, surface coat of: ↑ Glycocalyx

Epithelium, transporting: A functional class of surface ↑ epithelia (mostly simple) acting as a selective barrier between an interstitium and a lumen to maintain their physiological differences. To do so, E. may: a) Permit a free diffusion of substances through the cells or intercellular spaces; b) aid this diffusion; c) actively transport the substances under energy consumption. Epithelia of intestine, ↑ gallbladder, ↑ nephron, etc. are E.

Giebisch, G. (ed.): Transport Organs: In: Membrane Transport in Biology. Vol. 4. Heidelberg, Berlin, New York: Springer-Verlag 1979; Kriz, W., Kaissling, B., Schiller, A., Taugner, R.: Klin. Wochenschr. *57*, 967 (1979)

Epithelium, vascularization of: As an avascular tissue, ↑ epithelium is supplied with nutrients from capillaries situated within ↑ vascular papillae that interdigitate with ↑ epithelial ridges. (Exception: epithelium of ↑ stria vascularis)

Epon: An epoxy resin used in ↑ transmission electron microscopy as an ↑ embedding medium.

Eponychium*: A fold of keratinized epidermal layer (E) spreading onto free surface of ↑ nail plate (NP). H = ↑ hyponychium

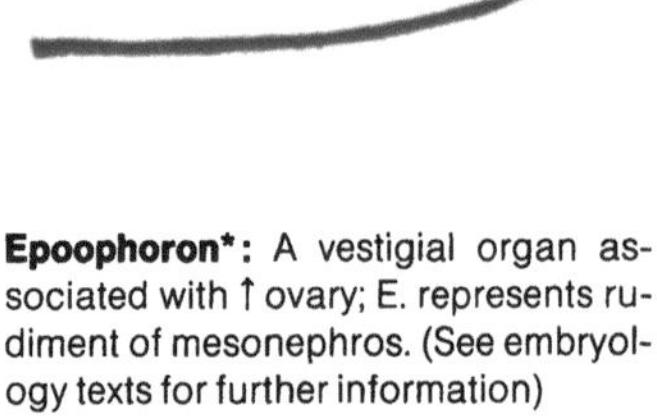

Epoophoron*: A vestigial organ associated with ↑ ovary; E. represents rudiment of mesonephros. (See embryology texts for further information)

Epoxy resins: Polymerizing ↑ embedding media for light and ↑ transmission electron microscopy (Epon, Araldite, Durcupan, etc.).

EPU: ↑ Epidermal proliferative units

Equational division: Any division of a cell in which each daughter cell receives an identical number of ↑ chromatids. Every division of ↑ mitosis and second division of ↑ meiosis are E.

↑ Reductional division

Equatorial plate (metaphase plate, lamina equatorialis*): The arrangement (E) of ↑ chromosomes (Chr) in a plane perpendicular to ↑ spindle apparatus (S) and equidistant from each ↑ diplosome (D) at metaphase of ↑ mitosis and ↑ meiosis. Viewed from profile, E. is somewhat irregular; viewed from one of poles, E. occurs as a ring of chromosomes (equatorial ring).

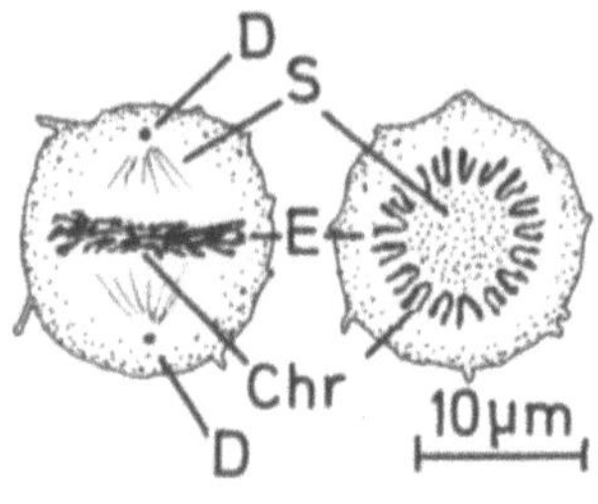

Equatorial ring: ↑ Equatorial plate

ERC: ↑ Colony-forming units-erythropoietic

Erectile tissue: A system of ↑ trabeculae and ↑ cavernous sinuses in ↑ cavernous bodies of ↑ clitoris and ↑ penis. A kind of E. forms ↑ swell bodies of inferior and middle ↑ conchae nasales

Erection: The turgid state of ↑ erectile tissues of ↑ clitoris and ↑ penis.

↑ Penis, vascularization of

Erection, of penis: ↑ Penis, vascularization of

Ergastoplasm: Groups of ↑ cisternae (C) of rough ↑ endoplasmic reticulum arranged parallel to each other, pres-

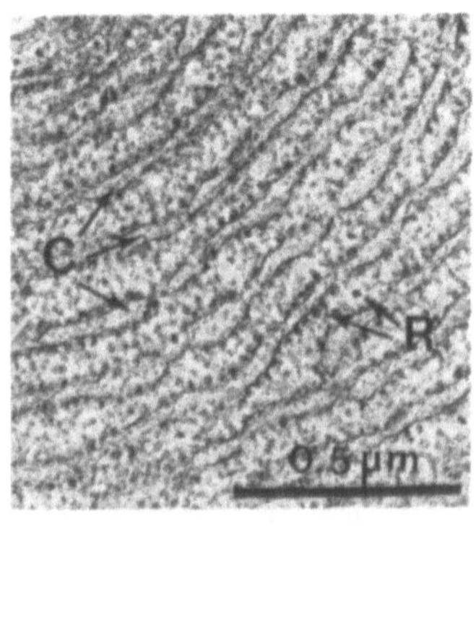

ent in protein-producing cells, such as ↑ fibroblasts, ↑ plasma cells, ↑ liver parenchymal cells, ↑ pancreatic acinar cells (fig.), ↑ neurons, etc. RNA of bound and free ↑ ribosomes (R) and ↑ polyribosomes is responsible for strong ↑ basophilia of E. (Fig. = rat)

↑ Nissl bodies; ↑ Protein synthesis

Tanaka, K.: Int. Rev. Cytol. *68*, 97 (1980)

Erosion tunnel: ↑ Absorption cavity

Erosion zone, of hyaline cartilage (zone resorbens cartilaginea*): Destruction front of calcified hyaline ↑ cartilage (HC) caused by ↑ chondroclasts (Cc) in the course of indirect ↑ bone formation. Chondroclasts destroy noncalcified transversal partitions (arrow) between lacunae of hypertrophied ↑ chondrocytes (Ch). Chondroclasts are followed by capillaries (Cap) and ↑ osteoblasts (Ob), which anchor to undestroyed longitudinal remnants of calcified cartilage (C), where they produce ↑ osteoid (O)

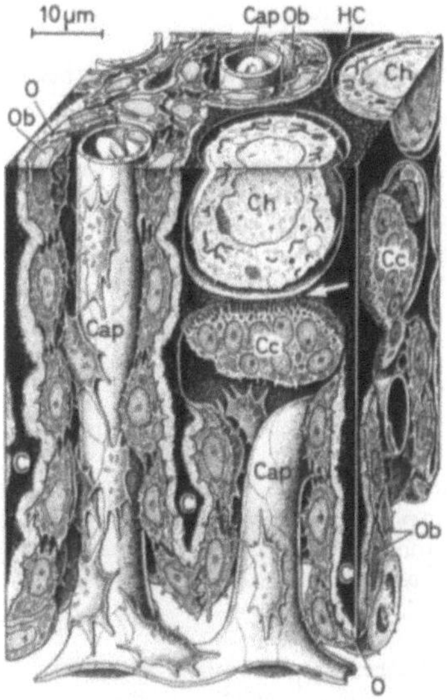

Erythroblast, acidophilic: ↑ Erythroblast, orthochromatic

Erythroblast, basophilic (basophilic normoblast, early erythroblast, prorubricyte, erythroblastus basophilicus*): The first cell of erythropoietic lineage, 10–12 µm in diameter, with a round ↑ heterochromatin-rich nucleus containing nucleolar fragments. Besides sporadic small mitochondria and a moderate number of short rough endoplasmic cisternae, cytoplasm contains an abundance of ↑ ribosomes (R) and ↑ polyribosomes (Pr) responsible for ↑ basophilia of E. Remainder of Golgi apparatus may be present; other ↑ organelles are lacking. In cytoplasm, there is only a little ↑ hemoglobin (Hb) (↑ MCH = 12 pg), which appears as fine granular masses between ribosomes. A marginal ring of ↑ microtubules (Mt) maintains cell shape. Some areas (inset) of plasmalemma are covered by ↑ glycocalyx (Gc); ↑ ferritin (Fe) particles, necessary for hemoglobin synthesis, attach themselves to this coat to be incorporated into cell within ↑ micropinocytotic vesicles (MV). E. are often seen in contact with neighboring ↑ reticular cells and/or ↑ macrophages, from which they may receive ferritin particles by ↑ rhopheocytosis. E. undergo numerous mitotic divisions, producing a large number of polychromatophilic ↑ erythroblasts.

↑ Erythropoiesis

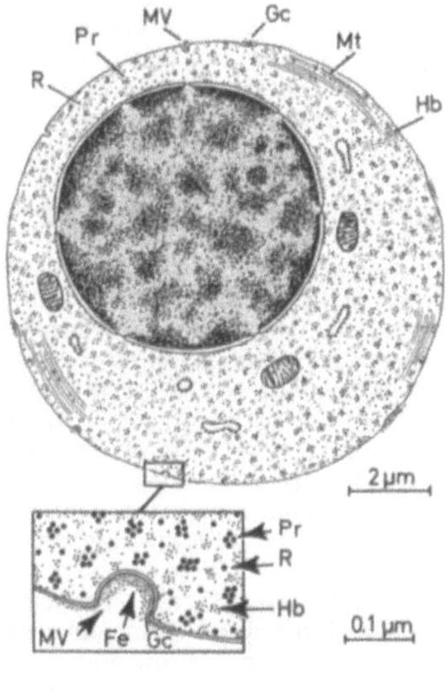

Heynen, M.J., Verwilghen, R.L.: Cell Tissue Res. *224*, 397 (1982)

Erythroblast, definitive: The ↑ erythroblasts formed in liver and spleen during hepatosplenic phase of prenatal ↑ hematopoiesis. Structurally, E. correspond to erythroblasts found in ↑ bone marrow during postnatal ↑ hematopoiesis.

Erythroblast, orthochromatic (acidophilic erythroblast, late erythroblast, metarubricyte, normoblast, erythroblastus acidophilicus*): An ovoid cell, 7–14 µm in diameter. Nucleus is eccentric, round, and almost completely filled with coarse ↑ heterochromatin

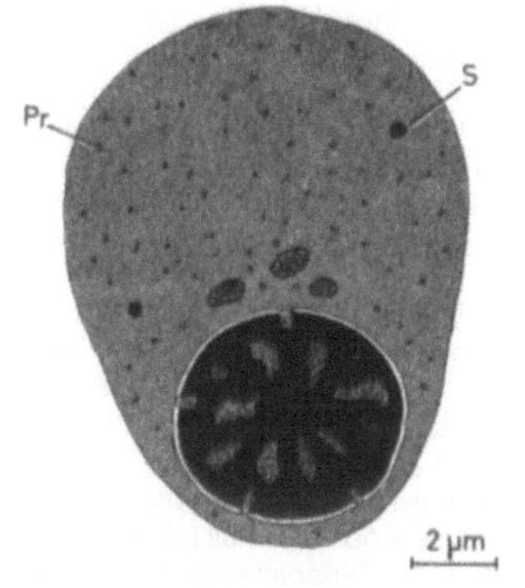

masses reminiscent of spokes of a wheel; it generally has the pattern of a ↑ pyknotic nucleus. Only a few small mitochondria are present within cytoplasm, which is acidophilic under light microscope, and very dense in transmission electron microscope because of a great concentration of ↑ hemoglobin (↑ MCH = 24 pg). Small ↑ siderosomes (S) and some polyribosomes (Pr) may still be present in cytoplasm. E. have no mitotic activity and transform, after nuclear extrusion into ↑ reticulocytes.

↑ Erythroblast, orthochromatic, extrusion of nucleus from; ↑ Erythropoiesis

Erythroblast, orthochromatic, extrusion of nucleus from: Through contraction of cell body (shortening of ↑ microtubuli?), nucleus (N) of orthochromatic ↑ erythroblast (E), surrounded by a thin rim of cytoplasm (C), is pinched off. This is accompanied by formation of several microvilli (Mv) around narrow cytoplasmic area through which nucleus leaves cell body. After expulsion, cell is termed the ↑ reticulocyte (R); extruded nuclei, still lined by thin cytoplasmic film, are phagocytized by ↑ macrophages of ↑ bone marrow; this material is used in nucleotide metabolism.

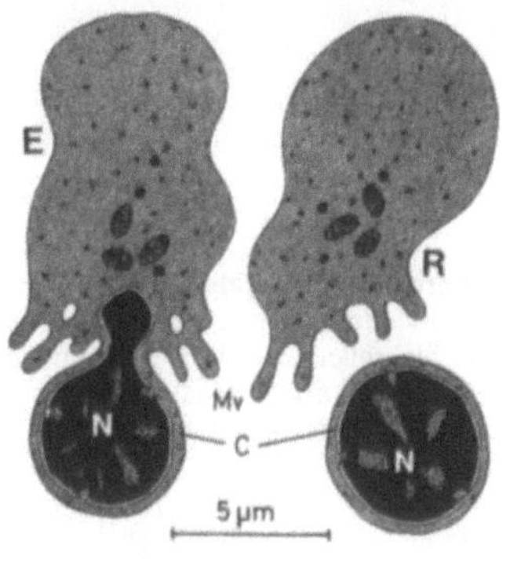

↑ Cabot rings; ↑ Howell-Jolly bodies; ↑ Erythropoiesis

Erythroblast, polychromatophilic (intermediate erythroblast, macroblast, polychromatic erythroblast, polychromatic normoblast, rubricyte, erythroblastus polychromatophilicus*): A cell of erythropoietic series measuring 10–15 μm in diameter. In round or polygonal cell body, spherical nucleus occupies a considerable part of cell volume; it contains a large mass of very condensed ↑ heterochromatin and possesses no nucleolus (= no neoformation of ↑ ribosomes). Other than some scattered mitochondria, small ↑ siderosomes (S), a centriole, marginal ↑ microtubules (Mt), very occasional remnants of rough endoplasmic reticulum (rER), and sporadic ↑ lysosomes (Ly), there are no other cell ↑ organelles. Cytoplasm contains a moderate and decreasing number of (basophilic) ↑ polyribosomes (Pr) involved in synthesis of (acidophilic) ↑ hemoglobin, whose concentration increases throughout life of cell (↑ MCH = 21 pg). Depending upon proportion of polyribosomes and hemoglobin, cytoplasm of E. may appear purplish-blue, blue-green, or green-orange under light microscope; hence the term polychromatophilic. Under transmission electron microscope, it is moderately osmiophilic. E. divides and differentiates into orthochromatic ↑ erythroblast.

↑ Erythropoiesis

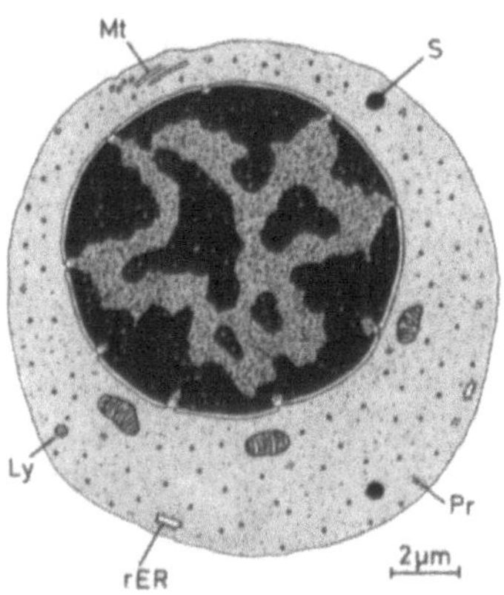

Erythroblastic islands (EBI): A morphofunctional unit composed of a central ↑ reticular cell (R) surrounded by a ring of ↑ erythroblasts (E). Reticular cell is needed for maturation of erythroblasts; a variety of substances, ↑ ferritin, etc., are exchanged between reticular cell and erythroblasts.

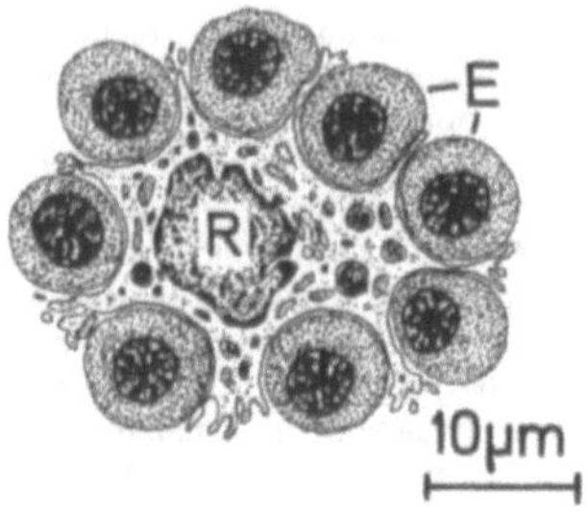

↑ Rhopheocytosis

Prenant, M.: Biol. Cell *38*, 9 (1980)

Erythroblasts, primitive: ↑ Erythroblasts present in wall of yolk sac during mesoblastic phase of prenatal ↑ hematopoiesis.

Erythrocyte (red blood cell, RBC, red corpuscle, rubricyte, erythrocytus*): An anucleate biconcave disc (E), 7.2–7.5 μm in diameter, 1.9 μm thick with a surface area of 140 μm². Total surface area of all E. in circulating blood is about 3800 m². Surrounded by a very elastic ↑ unit membrane, E. is filled with fine granular, osmiophilic material, ↑ hemoglobin (↑ MCH = 30 pg). ↑ Organelles and other cytoplasmic structures are lacking. In stagnant blood or in thick smears, E. form ↑ rouleaux. Energy for various functions of E. (preservation of biconcave shape, maintenance of hemoglobin in its reduced state, maintenance of electrolyte equilibrium) is furnished in the form of ATP by utilization of glucose via Embden-Meyerhof glycolytic pathway and by oxidative pentose phosphate pathway. As a final stage of erythrocyte differentiation. E. have a life of about 120 days, after which they are phagocytized and destroyed by ↑ macrophages of ↑ spleen, ↑ bone marrow, and ↑ liver. There are 5×10^{12} E./l blood in a man and 4.5×10^{12} E./l blood in a woman. (Fig. = dog)

↑ MCV; ↑ Reticulocytes; ↑ SI units

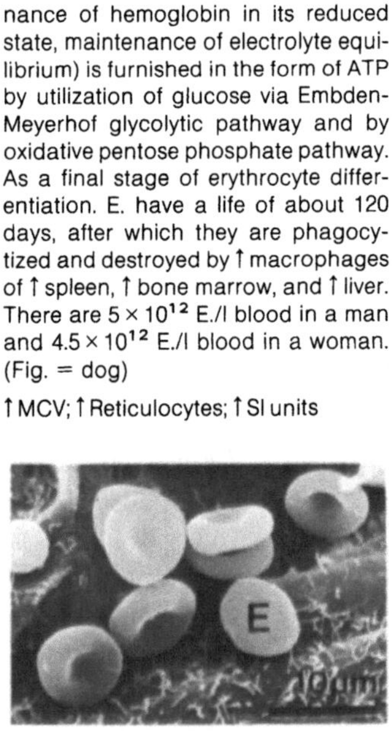

Branton, D.: Membrane Cytoskeletal Interactions in the Human Erythrocyte. Cold Spring Harbor Symposia on Quantitative Biology; Vol. 46, Part 1. New York: Cold Spring Harbor Laboratory 1982; Deuticke, R., Haest, C.W.M., Fischer, T.M.: Verh. Anat. Ges. *74*, 203 (1980); Timme, A.H.: J. Ultrastruct. Res. *77*, 199 (1981); Hattori, A.: Biomed. Res. *2*, Supp. 199 (1981)

Erythrocyte, crenated (echinocyte): An ↑ erythrocyte after loss of water, provoked by hypertonic media with consecutive folding of its plasmalemma and a burrlike appearance. In absence of ↑ antisphering substance, E. occur in isotonic solution, but disappear by addition of the former. (Fig. = rat)

Erythrocyte, development in bone marrow of: ↑ Erythropoiesis

Erythrocyte, extrusion of nucleus from: ↑ Erythroblast, orthochromatic, extrusion of nucleus from

Erythrocyte, formation of: ↑ Erythropoiesis

Erythrocyte ghost (umbra erythrocytica*): An empty membrane of an ↑ erythrocyte remaining after ↑ hemolysis.

Erythrocyte, hypochromic: An erythrocyte with an abnormally small quantity of ↑ hemoglobin (↑ MCH less than 27 pg).

Erythrocyte, normochromic: An ↑ erythrocyte containing a physiological quantity of ↑ hemoglobin (↑ MCH between 27 and 34 pg).

Erythrocyte, origin of: ↑ Erythropoiesis

Erythrocyte, polychromatophilic: ↑ Reticulocyte

Erythrocyte, rouleaux formation of: ↑ Erythrocyte

Erythropoiesis (erythrocytopoiesis*): The formation and maturation of ↑ erythrocytes. Sequence: By mytosis, ↑ colony-forming units and ↑ CFU-E multiply in ↑ bone marrow; some of them lose their pluripotentiality and transform into ↑ proerythroblasts. After a few divisions, some proerythroblasts differentiate into basophilic ↑ erythroblasts, which undergo multiplication and give rise to slightly larger polychromatophilic ↑ erythroblasts. There are two generations of polychromatophilic erythroblasts from which orthochromatic ↑ erythroblasts arise by division; these stop dividing. After extrusion of nucleus (N), these cells become ↑ reticulocytes, which mature into erythrocytes and enter, together with some reticulocytes, circulation. From proerythroblast to erythrocyte, both cell and nuclear volumes gradually diminish; hemoglobin content increases and the ↑ basophilia becomes ↑ acidophilia. Duration of sequence – about 3–7 days. E. in adult takes place exclusively in red bone marrow; in intrauterine life (2nd–9th months), it also occurs in liver and spleen. It is likely that ↑ erythropoietin represents main factor controlling E.

↑ Burst-forming units-erythropoietic; ↑ Erythropoiesis, cell division during; ↑ Hematopoiesis, current view of

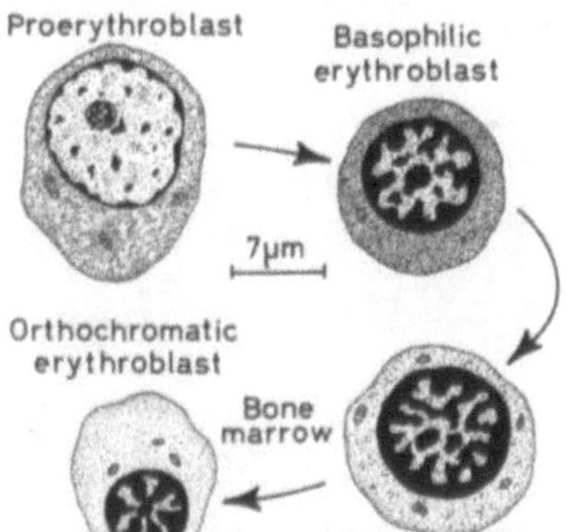

Golde, D.W., Cline, M.J., Metcalf, D., Fox, F.C. (eds.): Hematopoietic Cell Differentiation. New York: Academic Press (1978); Harrison, P.R.: Nature *262,* 353 (1976); Prenant, M.: Biol. Cell *38,* 9 (1980); Riedler, G.F., Zingg, R.: Tabulae haematologicae. Basel; Rocom, Hoffmann-La Roche (1977); Rifkind, R.A., Marks, P.A.: Blood Cells *1,* 417 (1975)

Erythropoiesis, cell divisions during: A ↑ proerythroblast differentiates from a ↑ totipotent hematopoietic stem cell (THCS), ↑ colony-forming unit, and ↑ CFU-E under the influence of ↑ erythropoietin, and then divides, giving two generations of basophilic ↑ erythroblasts (BE) from which arise two generations of polychromatophilic ↑ erythroblasts (PE). They stop dividing and differentiate by way of orthochromatic ↑ erythroblasts (OE) and ↑ reticulocytes (R) to ↑ erythrocytes (E). Thus, from a proerythroblast 16 erythrocytes could be produced; however, the number is lower, since 10% of erythroblasts die during ↑ differentiation. BFU-E = ↑ burst-forming units-erythropoietic, N = expulsed nuclei

↑ Hematopoiesis, current view of

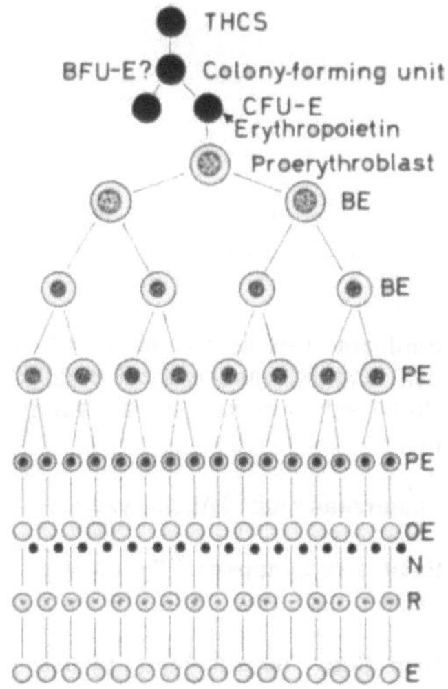

Erythropoiesis-stimulating factor: ↑ Erythropoietin

Erythropoietin (erythropoiesis-stimulating factor): A ↑ glycoprotein tissue hormone (m.w. about 70000 daltons) whose precursor, erythrogenin, is synthesized predominantly in ↑ kidney. E. acts on erythropoietin-responsive cells (ERC), which differentiate via ↑ CFU-E into ↑ proerythroblasts. Through stimulation of hemoglobin synthesis, E. is involved in an unknown manner in control of ↑ erythropoiesis. Main stimulating factor liberating E. is hypoxia.

Fischer, J.W. (ed.): Kidney Hormones, Vol. 2. Erythropoietin. New York: Academic Pres (1977)

Erythropoietin-responsive cells: ↑ Colony-forming unit-erythropoietic

ES-face, of cell membrane: ↑ Freeze-cleaving, terminology of

Esophageal glands proper (glandulae esophageae propriae*): Small, compound ↑ tubuloalveolar glands (EG) located in tela submucosa of upper half of ↑ esophagus. The secretory portion, consisting of ↑ mucous cells, unites into a cystically dilated main duct (D), which passes through lamina muscularis mucosae (LMM) and opens onto surface. At their beginning, ducts are lined with ↑ simple columnar epithelium, which becomes ↑ pseudostratified and finally ↑ stratified squamous nonkeratinized epithelium. (Fig. = human)

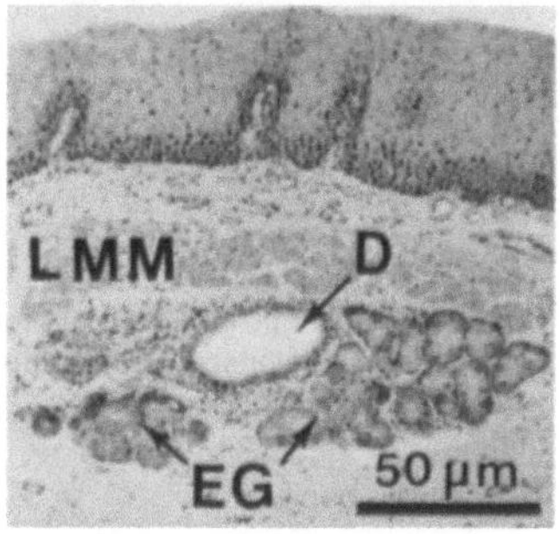

Esophagogastric junction: The transitional line of ↑ stratified squamous epithelium of ↑ esophagus into ↑ simple columnar epithelium of ↑ cardiac area.

↑ Cardiac glands, of stomach

Esophagus*: A tubular organ, roughly 25 cm long, extending from ↑ pharynx to ↑ stomach. Structure: 1) Tunica mucosa (TM): a) epithelium (E) = a ↑ stratified squamous nonkeratinized epithelium; b) lamina propria (LP) = a ↑ loose connective tissue forming numerous papillae (P); c) lamina muscularis mucosae (LM) = predominantly longitudinally oriented ↑ smooth muscle cells. 2) Tela submucosa (TS) = a richly vascularized and innervated loose connective tissue, also containing ↑ esophageal glands proper and a venous plexus near ↑ cardia. Both tunica mucosa and tela submucosa form longitudinal folds (F). 3) Tunica muscularis (TMu) = inner circular (IC) and outer longitudinal (OL) layers. In upper third, musculature is skeletal; in middle segment, ↑ skeletal muscle fibers and smooth muscle cells are intermingled; in inferior third (fig.), musculature is smooth. Circular layer in the upper extremity of E. forms superior esophageal sphincter and in lower extremity it forms inferior esophageal sphincter. 4) Tunica

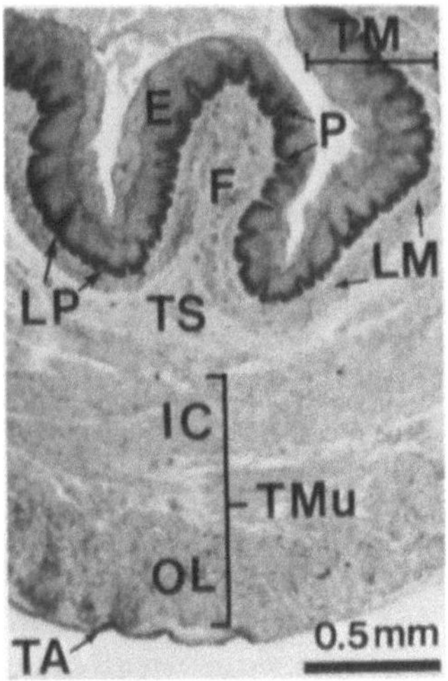

adventitia (TA) = a loose connective tissue. A short intra-abdominal segment of E. is enveloped by ↑ tela subserosa and ↑ tunica serosa. (Fig. = human)

↑ Connective papillae

Hopwood, D., Logan, K.R., Bouchier, I.A.D.: Virchows Arch. B. Cell Pathol. *26*, 345 (1978); Siew, S., Goldstein, M.L.: Scan. Electron Microsc. 1981/4, 173 (1981)

Estradiol (folliculin): The female sex hormone produced under influence of ↑ FSH by ↑ granulosa of ↑ ovarian follicles and theca interna cells. E. is derived from cholesterol; it stimulates development of secondary sexual characteristics in women and development of ↑ uterus. In ↑ menstrual cycle, E. stimulates ↑ regeneration of ↑ endometrium. Among other actions, E. accelerates closure of ↑ epiphyseal plates.

↑ Thecal cells

Mizuhira, V., Shiihashi, M., Sugiura, Y.: Acta Histochem. Cytochem. *15*, 572 (1982)

Estrogens: All hormones or substances with effects similar to those of ↑ estradiol (i.e., estrone, stilbestrol = synthetic E. with a different chemical structure from ↑ estradiol, etc.).

Ethylenediaminetetraacetic acid: ↑ EDTA

Eucaryotes: Highly evolved cells characterized by a ↑ nuclear envelope around nuclear material, i.e., all animal and plant cells, including fungi.

↑ Procaryotes

McQuade, A.B.: BioSystems *16*, 39 (1983)

Euchromatin (euchromatinum*): The dispersed and weakly stainable ↑ chromatin (E) among dense masses of ↑ heterochromatin (H). E. represents uncoiled segments of ↑ chromatin fibers. E. is deeply involved in control of specific activities of cell. (Fig. = ↑ follicular cell of ↑ ovarian follicles, rat)

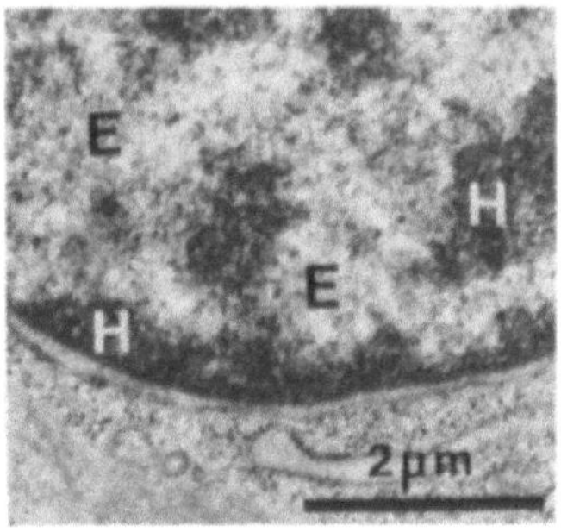

Back, F.: Int. Rev. Cytol. *45*, 25 (1976); Mello M.L.S.: Histochem. J. *15*, 739 (1983)

Euplasm: Cell components that appear and disappear in a cell during its life (↑ astrosphere, ↑ spindle apparatus).

Eustachian tube: ↑ Auditory tube

Excitatory synapse: ↑ Synapse, excitatory

Excretion: The elimination by cells of chemically simple useless noxious waste products of metabolism and some drugs, mostly without energy consumption (urea, etc.).

↑ Secretion

Excretory ducts, of salivary glands: ↑ Ducts, of salivary glands proper

Excretory ducts, of testis: ↑ Ducts, excretory, of testis

Excretosecretory ducts: ↑ Striated ducts

Exocervix: ↑ Cervix, of uterus

Exocrine glands (glandula exocrina*): ↑ Glands having an excretory canalicular system that conducts secretory product(s) onto an internal or external surface of body. (See under name of particular gland)

Exocrine secretion: The release of secretory products via a more or less complicated canalicular system (ducts) onto internal or external surface of body.

↑ Apocrine secretion; ↑ Eccrine secretion; ↑ Holocrine secretion; ↑ Secretion

Exocytosis (emiocytosis): The releasing process of membrane-bound nondiffusible material [↑ secretory granules (S); ↑ residual bodies] from cell. During E., boundary membrane fuses with plasmalemma (P) by a fine stalk (arrows); here plasmalemma opens and contents of a secretory granule and/or residual body are discharged. In ↑ exocrine cells (fig., top) and ↑ thyroid follicular cells, contents of secretory granules are gradually discharged, whereas in ↑ endocrine cells (fig., bottom), E. occurs between basal lamina (BL) and basal plasmalemma (P), where contents of granules are for a short time morphologically identifiable (asterisk). Cytoskeletal ↑ microfilaments and ↑ microtubules of ↑ ectoplasm, together with calcium ions, seem to play important, but still unknown, role in control of E. (Fig., top = ↑ pancreatic acinar cell, rat; fig., bottom = ↑ somatotrope, rat)

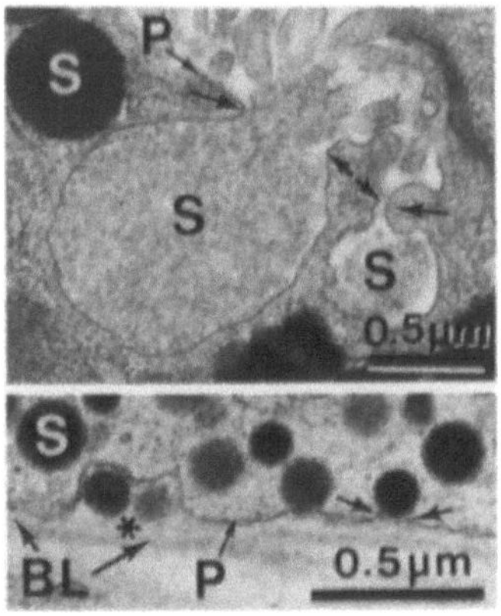

Allison, A.C., Davies, P.: Symp. Soc. Exp. Biol. *28*, 419 (1974); Cell Surface: Synthesis, Exocytosis, Endocytosis. In: Organization of the Cytoplasm. Vol. 46, Part 2. New York: Cold Spring Harbor Symposia on Quantitative Biology 1982; Tanaka, P., De Camilli, P., Meldolesi, J.: J. Cell Biol. *84*, 438 (1980); Vilmart, J., Plattner, H.: J. Histochem. Cytochem. *31*, 626 (1983)

Exoepithelial glands: A group of exocrine glandular cells situated below a surface ↑ epithelium and connected to surface by an excretory duct. Apart from ↑ endoepithelial glands, all glands are E.

↑ Glands, classification of

Exopeptidases: A group of proteolytic enzymes attacking terminal bonds of peptides or proteins. E. are synthesized by ↑ pancreatic acinar cells and comprise carboxypeptidase A and B.

Exophthalmus-producing factor (EPF): A substance provoking swelling of retrobulbar white ↑ adipose tissue. It is thought that E. is a precursor of ↑ thyrotropic hormone which under normal conditions does not leave ↑ hypophysis.

Exoplasm: ↑ Ectoplasm

Expanded tip nerve endings: A group of peripheral ↑ nerve endings with branches terminating in a series of bulbous or discoidal enlargements that make contact with specialized epithelial cells (e.g., ↑ Merkel's cells).

Explantation: The transfer of living cells or organs from an organism to an artificial nutrient medium for culture.

Exteriorization: A technique for microscopic observation of living organs outside the body, possible because of their long and mobile vascular pedicle. E., combined with dia- and/or epidiascopic ↑ illumination, permits study of dynamic processes (↑ absorption, blood and lymph circulation, ↑ secretion, ↑ ovulation, etc.).

External auditory meatus: ↑ Auditory meatus, external

External ear (auris externa*): Parts of organ of hearing receiving sound waves and transmitting them to ↑ tympanic membrane.

↑ Auditory meatus, external; ↑ Auricle; ↑ Ear

External elastic lamina: ↑ Elastic lamina, external

External genital organs, female: ↑ Clitoris; ↑ Labia majora; ↑ Labia minora; ↑ Vestibular glands

External genital organs, male: ↑ Penis; ↑ Scrotum

External glial limiting membrane: ↑ Membrana gliae limitans superficialis

External iliac vein (vena iliaca externa*): A large ↑ propulsive vein with structure identical to ↑ vena cava inferior.

External sphincter, of anus: ↑ Anal canal

External tongue process: ↑ Tongue processes, of oligodendrocytes

Exteroceptors: A group of sensory receptors receiving stimuli from body surface (free nerve endings in ↑ epidermis, ↑ corpuscles).

Extracellular fluid: ↑ Tissue fluid

Extracellular matrix: ↑ Ground substance

Extrafusal muscle fibers: All ordinary ↑ skeletal muscle fibers situated outside capsule of ↑ neuromuscular spindle.

Extrahepatic bile ducts: ↑ Bile pathways

Extrapulmonary bronchi: ↑ Bronchi, primary

Extrusion zone: ↑ The site at apex of ↑ intestinal villus (V) marked by a distinct cleft in ↑ epithelium (arrows) where ↑ absorptive and ↑ goblet cells are shed at end of their migratory cycle from ↑ Lieberkühn's crypts to tip of villus. (Fig. = rat)

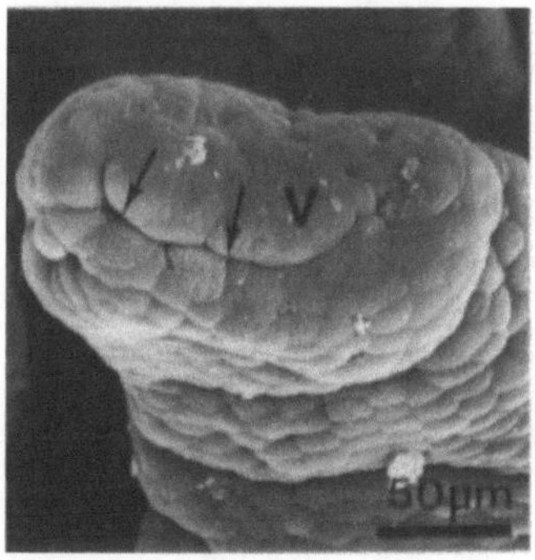

Exudate, serous: ↑ Serous exudate

Eye (oculus*): The organ of vision composed of ↑ eyeball and accessory organs of E.

↑ Eye, accessory organs of

Fine, B.S., Yanoff, M.: Ocular Histology. New York: Harper & Row Publ. 1972; Hansson, H.-A.: Biomed. Res. *2*, Suppl. 465 (1981); McDevitt, D.S. (ed.): Cell Biology of the Eye. New York: Academic Press 1982

Eye, accessory organs of (adnexa): A group of organs assuring protection, movement, and lubrication of ↑ eyeball. To E. belong: ↑ eyelids, ↑ extraocular muscles, ↑ lacrimal glands, ↑ lacrimal ducts, ↑ lacrimal sacs, and ↑ nasolacrimal canals.

Eye, accomodation of: The increase in convexity of surface of ↑ lens. Contraction of circular muscle fibers (CM) of ↑ ciliary body (CB) relaxes traction of ↑ zonular fibers of ↑ ciliary zonules (CZ) on lens (L); its surface becomes more convex, and eye accomodates to near vision. Contraction of meridional-radial muscle fibers (MM) has opposite effect.

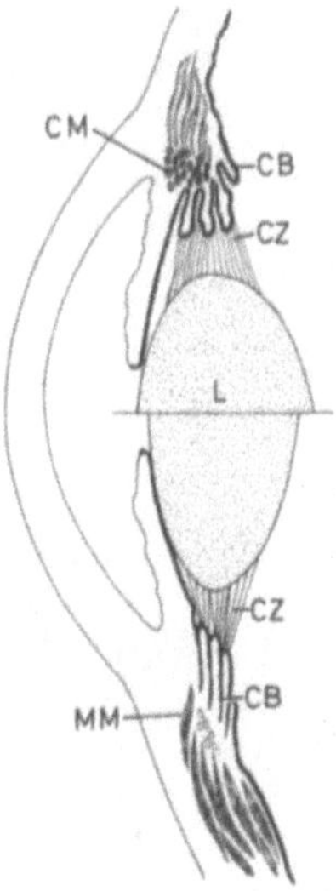

Eye, adnexa of: ↑ Eye, accessory organs of

Eye, anterior chamber of: ↑ Anterior chamber, of eye

Eye, aqueous humor of: ↑ Aqueous humor

Eye, Bowman's membrane of: ↑ Bowman's membrane; ↑ Cornea

Eye, choroid of: ↑ Choroid

Eye, ciliary body of: ↑ Ciliary body

Eye, ciliary muscles of: ↑ Ciliary body

Eye conjunctiva, bulbar and palpebral: ↑ Conjunctiva

Eye, cornea of: ↑ Cornea

Eye, Descemet's membrane of: ↑ Cornea; ↑ Descemet's membrane

Eye, dioptric media of (refractive media): Tissues and fluids contained within ↑ eyeball whose function is to refract light rays and form, at level of ↑ photoreceptors, an image free of distortion. To E. belong ↑ cornea, ↑ aqueous humor, ↑ lens, and ↑ vitreous body.

Eye, fibrous tunic of ↑ (tunica fibrosa oculi*): Outermost coat of ↑ eyeball composed of ↑ cornea and ↑ sclera.

Eye, focusing of: ↑ Ciliary body; ↑ Eye, accomodation of

Eye, in the study of living tissues: ↑ Anterior chamber (A) of ↑ eye represents a natural ↑ transparent chamber convenient for study of denervated autologous transplants (T). Initially, the transplant is supplied with nutrients from ↑ aqueous humor, later it becomes vascularized from corneoiridial angle (arrow). Changes in transplanted tissue or organ are observed through ↑ cornea (C) with a ↑ biomicroscope.

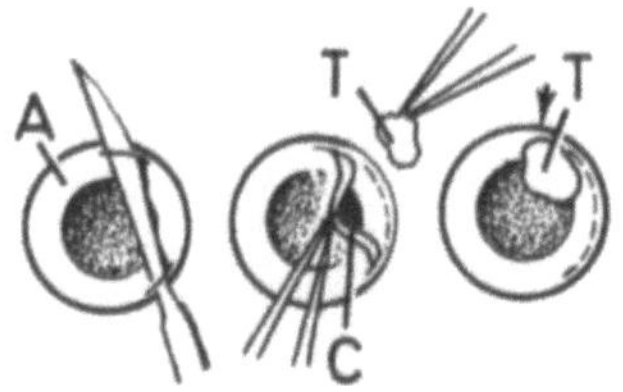

Fawcett, D.W., Wislocki, G.B., Waldo, C.M.: Am. J. Anat. *81*, 413 (1947)

Eye, lymphatics of: Present only in bulbar ↑ conjuncitva; absent in ↑ eyeball.

Eye, muscles of: 1) Extraocular muscles = six skeletal muscles for movements of ↑ eyeball. 2) Intraocular muscles = a) smooth muscles of ↑ ciliary body responsible for accomodation of ↑ eye; b) ↑ dilator and ↑ sphincter muscles of ↑ iris.

Eye, pigment, epithelium of: ↑ Pigment epithelium; ↑ Pigment epithelial cells

Eye, posterior chamber of: ↑ Posterior chamber, of eye

Eye, Schlemm's canal of: ↑ Schlemm's canal

Eye, vascularization of: The arteria ophthalmica branches into two almost independent systems of vessels: 1) Retinal system = arteria and vena centralis retinae* run below inner limiting membrane and vascularize ↑ retina with their branches up to outer plexiform layer and ↑ optic nerve. 2) Ciliary system = long and short posterior ciliary arteries (arteriae ciliares posteriores longae et breves*) for blood supply of entire middle coat of ↑ eyeball and retina up to inner nuclear layer. Blood is predominantly drained into four venae vorticosae*. Both systems anastomose in area of ↑ optic papilla. (See gross anatomy texts for further information)

Eye, vitreous body of: ↑ Vitreous body

Eye, zonules, of: ↑ Ciliary zonules

Eyeball (bulbus oculi*): A very complex spherical organ whose wall is composed of three coats or tunics: 1) Outer coat (tunica fibrosa bulbi*); 2) middle coat (tunica vasculosa bulbi*); 3) internal coat (tunica interna s. sensoria bulbi*). Subdivision of each of these tunics is given on fig. In its interior, E. contains ↑ aqueous humor (AH), ↑ lens, and ↑ viterous body. (Asterisks indicate letters under which particular structures are described)

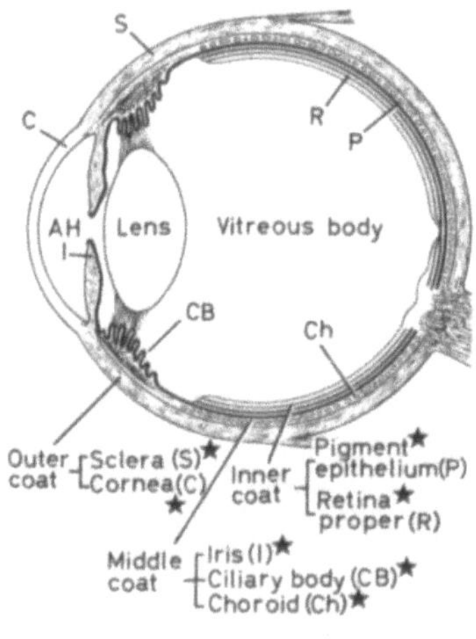

Eyelashes (cilia*): Coarse ↑ hairs inserted in three to four rows along margin of ↑ eyelids. E. are inserted obliquely and extend deep into subepithelial tissue. In ↑ hair follicle of an E., ↑ glands of Zeiss and part of ↑ ciliary glands of Moll open; there is no ↑ arrector pili muscle. An E. is replaced after about 100–150 days.

Eyelids (lids, palpebrae*): Actively mobile conjunctivo-cutaneous folds (E) covering anterior surface of ↑ eyeball. Structure (superior E. beginning from

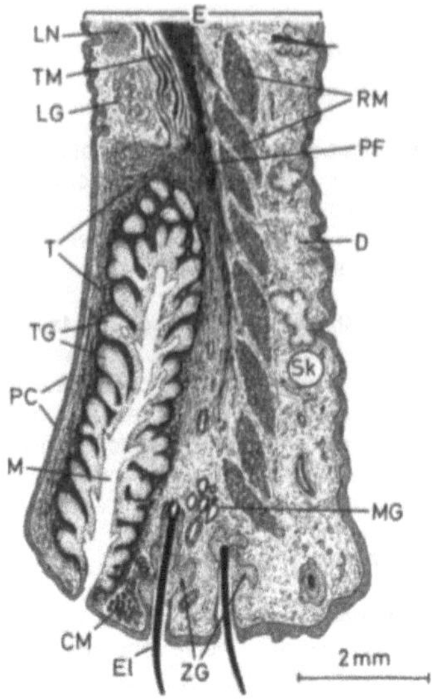

anterior surface): 1) Thin skin (Sk) with scattered ↑ hairs, ↑ sebaceous and ↑ sweat glands embedded in a very loose ↑ dermis (D) containing some ↑ melanocytes, but practically devoid of ↑ adipose tissue. 2) Palpebral portion of ring muscle (RM) of eye (musculus orbicularis oculi*); part of this muscle between tarsus (T) and ↑ eyelashes (El) is ↑ ciliary muscle of Riolan (CM). 3) Palpebral fascia (PF) = ↑ tendon of palpebral levator (musculus levator palpebrae superioris*) attached to tarsus and dermis. 4) Tarsal plate of tarsus (T) = a ↑ dense connective tissue plate enclosing ↑ tarsal glands (TG). 5) ↑ Tarsal muscle of musculus tarsalis (TM) = a smooth muscle attached to tarsus and fascia; its tonus keeps rima oculi open. 6) ↑ Eyelashes (El) associated with ↑ glands of Zeiss (ZG) and ↑ ciliary glands of Moll (MG). 7) Palpebral ↑ conjunctiva (PC). Between tarsus and conjunctival fornix of upper E., accessory ↑ lacrimal glands (LG) are sometimes present. ↑ Lymphatic nodules (LN) are scattered on conjunctival side of E. M = common lumen of tarsal glands

Halata, Z., Munger, B.L.: Anat. Rec. *198*, 657 (1980)

Eyepiece: ↑ Ocular

F

Fabricius, bursa of: ↑ Bursa, of Fabricius

Facet cells (superficial cell, epitheliocytus superficialis*): Large polyhedral or very flattened cells composing superficial layer of ↑ transitional epithelium. F. are frequently binucleate with conspicuous nucleoli. Cytoplasm contains a moderate number of small mitochondria, a well-developed and often multiple Golgi apparatus, short cisternae of rough endoplasmic reticulum, some ↑ lysosomes, ↑ multivesicular bodies, particulate ↑ glycogen, and numerous microfilaments (Mf). Free cell surface in relaxed epithelium is deeply folded; each fold displays numerous, small, V-shaped indentations, a consequence of fusion of ↑ discoid vesicles (DV) with apical plasmalemma, which contains cerebroside as a major component of its polar lipid fraction. Viewed from above, apical plasmalemma displays well-delimited, rigid, polygonal ↑ plaques (P) in a cobblestone arrangement. Lateral cell borders interdigitate to a large extent with adjacent cells; presence of ↑ junctional complexes (JC) and small ↑ desmosomes. Basal cell processes of various lengths penetrate between subjacent cells; several authors believe that these processes can reach basal lamina (BL). When epithelium is stretched, lateral interdigitations, V-shaped indentations, and apical folds disappear, permitting an increase in luminal surface of up to tenfold without disruption. In such cases, apical plasmalemma becomes smooth and F. very flattened. Increase in free cell surface seems to be supported by fusion of numerous, newly synthesized ↑ discoid vesicles that are added to apical cell membrane. F. are responsible for formation and maintenance of osmotic ↑ barrier between urine and ↑ tissue fluid.

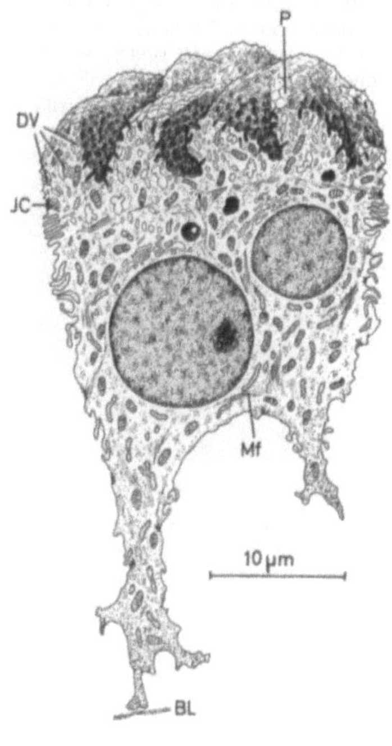

↑ Crusta

Knutton, S., Robertson, J.D.: J. Cell Sci. *22*, 355 (1976); Scheidegger, G.: Acta Anat. *107*, 268 (1980); Severs, N.J., Warren, R.C.: J. Ultrastruct. Res. *64*, 124 (1978); Warren, R.C., Hicks, R.M.: J. Ultrastruct. Res. *64* 327 (1978)

F-actin: A fibrillar polymer of ↑ actin myofilament.

Factor, blastogenic: ↑ Blastogenic factor

Factor, cloning-inhibiting: ↑ Cloning-inhibiting factor

Factor, lymphocyte-transforming: ↑ Lymphocyte-transforming factor

Factor, migration-inhibiting: ↑ Migration-inhibiting factor

Factor, proliferation-inhibiting: ↑ Proliferation-inhibiting factor

Factors, releasing: ↑ Releasing factors

Fallopian tubes: ↑ Oviducts

Fañanas cells: Small ↑ neuroglial cells present in molecular layer and between ↑ Purkinje cells of ↑ cerebellum.

↑ Cerebellum, glioarchitectonics of

Fascia*: The outermost ↑ dense connective tissue sheath of a skeletal muscle (F). In general, internal aspect of F. contacts ↑ epimysium (E); in some cases, an F. represents insertion site for muscle fibers. Such F. contain fewer ↑ elastic fibers than those solely enveloping a muscle. Collagen bundles of an F. are arranged in multiple parallel sheaths or lamellae. In each lamella, collagen fibers have a wavy course and cross at various angles. Structurally, F. corresponds to an ↑ aponeurosis.

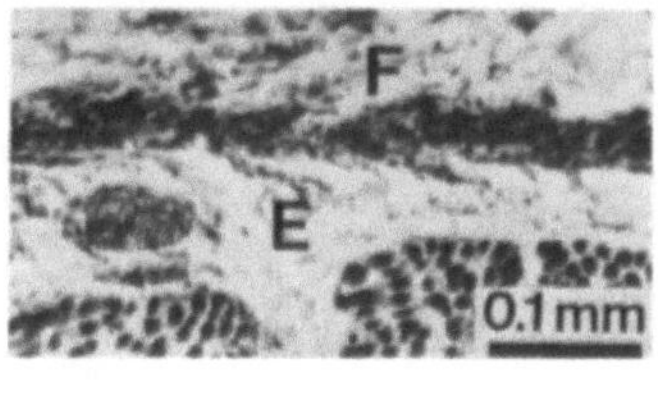

Fascia adherens (intermediate junction): A junctional specialization (FA) in form of discontinuous irregularly outlined areas making up greater part of an ↑ intercalated disc. Cell membrane of F. is unchanged; immediately subjacent to it lies some filamentous electron-dense material to which ↑ actin myofilaments of nearest ↑ I-band (I) anchor. In areas of F., adjacent ↑ cardiac muscle cells are separated by a 15 to 20-nm-wide intercellular cleft filled with moderately dense, amorphous, glycoprotein material with an occasional central line of higher electron density. F. contribute to cohesion of ↑ cardiac muscle cells. (Fig. = rat)

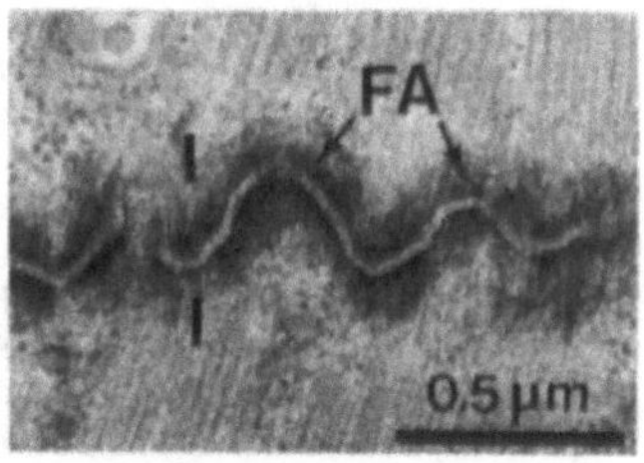

Fascia linguae: ↑ Aponeurosis linguae

Fascia occludens: A term sometimes used to designate ↑ nexus.

Fascia penis superficialis and profunda*: Elastic connective tissue sheaths enveloping all three erectile bodies of ↑ penis.

Fascicle, of nerve: ↑ Nerve fascicle

Fascicle, of skeletal muscle: ↑ Muscle fascicle

Fasciculi longitudinales: ↑ Sphincter, of Oddi

Fasting: ↑ Adipose cell, white, histophysiological changes in fasting

Fat: The principal calorie-rich reserve material of the body occurring in a large variety of cells in the form of ↑ lipid droplets or lipid vacuoles. At body temperature, F. is in the form of oil. Lipid comprises 60%–85% of its weight, 5%–30% is water, and 2%–3% protein; 90%–95% of lipid is in form of triglycerides. Small quantities of free fatty acids, cholesterol and phospholipids are also present.

↑Inclusions

Fat-storing cells: ↑ Perisinusoidal cells

Fatty bone marrow: ↑ Bone marrow, yellow

F-cells: ↑Pancreatic polypeptide cells; ↑Synovial cells

Fc receptor: The site on plasmalemma of ↑ phagocytes which interacts with Fc fragment of ↑ immunoglobulin G molecule during ↑ phagocytosis.

Mellman, I.S.: Endocytosis, Membrane Recycling and Fc Receptor Function. In: Ciba Foundation Symposia, Vol. 92. London: Pitman 1982; Unkeless, J.C., Fleit, H., Mellman, I.S.: Adv. Immunol. *31*, 247 (1981)

Feedback: A kind of control mechanism of activity of an ↑ endocrine gland in which production and output of a hormone is increased (positive F.) or reduced (negative F.) as result of its own action. 1) Examples of negative F.: ↑ Parathyroid hormone mobilizes, via ↑ osteoclasts, calcium from bone; resulting increase in blood calcium reduces further parathyroid hormone release. ↑ Insulin induces entry of glucose into cells; resulting decrease in blood sugar suppresses output of insulin. Both examples represent simple F. not influenced by nervous system. A more complex F. controls activity of zona fasciculata of adrenal cortex: A stressogenic factor provokes, via CNS, release of ↑ ACTH-RF from hypothalamus, followed by a consecutive release of ↑ ACTH from hypophysis and its action on production and release of ↑ corticosteroids. These then act by negative F. upon hypothalamus to reduce discharge of ACTH-RF. 2) Example of positive F.: Increase in blood calcium stimulates ↑ C-cell to synthesize more ↑ calcitonin.

↑Releasing factors

Female-determining chromosome: ↑ Chromosome, female-determining

Female reproductive system: A group of organs destined for production of female ↑ germ cells (↑ ovaries) and transport (↑ oviducts) to ↑ uterus, where ↑ ovum develops during gestational period due to a newly formed organ, ↑ placenta. A canal, ↑ vagina, connects inner reproductive organs with exterior. F. also comprises ↑ mammary glands for nourishment of newborn with ↑ milk.

Ferenczy, A.: The Female Reproductive System. In: Hodges, G.M. and Hallowes, R.C. (eds.): Biomedical Research Applications of Scanning Electron Microscopy, Vol. 2, London, New York: Academic Press 1980

Femtoliter: ↑ fl

Fenestra ovalis: ↑ Oval window

Fenestra vestibuli: ↑ Oval window

Fenestrae, of Golgi apparatus: Small openings in ↑ lamellae of ↑ Golgi apparatus.

Fenestrated capillaries: ↑ Capillaries, fenestrated

Fenestrated laminae: ↑ Elastic laminae

Ferritin (granulum ferritini*): An iron protein complex in form of electron-dense, 8–9-nm molecules (F). An octahedral molecule of F. consists of 24% iron hydroxide bound to protein, apoferritin, capable of holding as many as 2500 iron atoms. Main mass of F. is stored in ↑ macrophages, ↑ reticular cells, and ↑ Kupffer's cells as a result of destruction of ↑ erythrocytes. From here, F. is returned to circulation by ↑ exocytosis, bound to plasmalemmal receptors (R), and taken up by ↑ erythroblasts (E) by ↑ micropinocytosis for synthesis of ↑ hemoglobin. F. can also penetrate erythroblasts by ↑ rhopheocytosis. Agglomerations of F. in erythroblasts form ↑ siderosomes. When it circulates, F. is linked to a plasma iron-binding globulin, transferrin. F. is storage form of iron; some authors classify it among endogenous ↑ pigments. There is about 0.45 g F. in a 70-kg body; simple or immunolabeled F. is frequently used as a tracer in various experiments. (Fig. = rat)

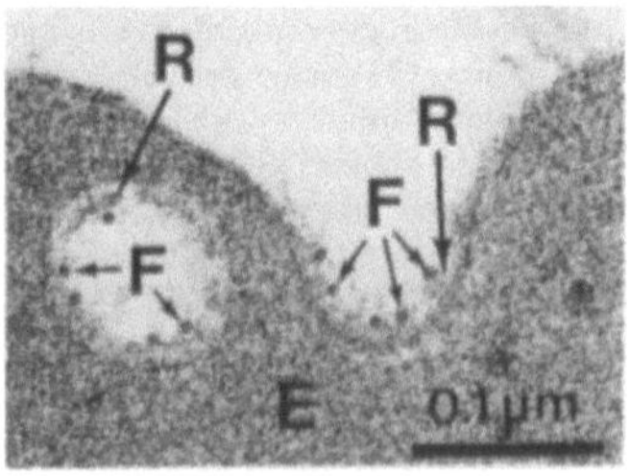

Chrichton, R.R.: N. Engl. J. Med. *284*, 1413 (1971); London, M.F., Michel, C.C., White, I.F.: J. Physiol. (Lond.) *296*, 97 (1979); Munro, H.N., Linder, M.C.: Physiol. Rev. *58*, 317 (1978)

Fertilization (fecundatio*): Penetration of ↑ spermatozoon into secondary ↑ oocyte followed by: 1) Reestablishment of ↑ diploid chromosomal set after fusion of oocyte and spermatozoon nuclei; 2) triggering of segmentation process; 3) sex determination. (See embryology texts for further information)

↑Acrosome reaction; ↑Yolk

Austin, C.R., Short, R.V. (eds.): Reproduction in Mammals, Book. 1. Germ Cells and Fertilization. 2nd edn. Cambridge: Cambridge University Press 1982; Meyer, N.L., Longo, F.J.: Anat. Rec. *195*, 357 (1979)

Fetal lobule: ↑ Cotyledon

Fetal-maternal junction: ↑ Deciduotrophoblastic complex

Fetal placenta: ↑ Placenta, full-term, structure of

Feulgen reaction: A highly specific staining method for cytochemical localization of ↑ deoxyribonucleic acid (DNA). Mechanism of F.: 1) Mild acid hydrolysis with HCl removes purine bases adenine (A) and guanine (G) from deoxyribose (D); 2) liberates aldehyde groups (AG) at these points; 3) free aldehyde groups react, in a manner not completely understood, with ↑ Schiff's reagent and produce an insoluble purple product. It seems that specificity of F. is based on the very similar distance between aldehyde groups and two sulfur atoms of Schiff's reagent. For specificity control of F., sections are pretreated with deoxyribonuclease and an F. subsequently effected. Combined with ↑

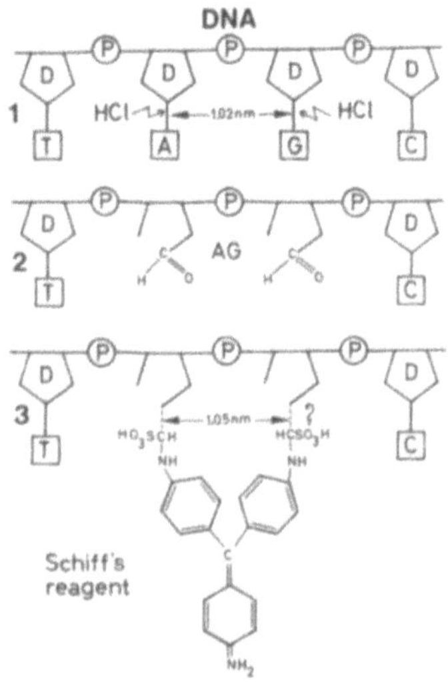

cytospectrophotometry, F. permits calculation of DNA content of a single nucleus. C = cytosine, P = phosphoric acid, T = thymine

Gabe, M.: Techniques histologiques, Paris: Masson & Cie 1968; Hale, J.A.: Feulgen Microspectrophotometry and its Correlation with Other Cytochemical Methods. In: G.L. Wied (ed.): Introduction to Quantitative Cytochemistry. New York and London: Academic Press 1966; Lessler, A.M.: Int. Rev. Cytol. *2*, 231 (1953); Pearse, A.G.E.: Histochemistry, 3rd edn. London: Churchill 1968

Feyerter's cells: ↑ E-cells

Fiber spectrum, of peripheral nerve: The range of nerve fiber diameters in a ↑ nerve. Since diameter of a fiber is related to its conduction velocity, F. indicates range of conduction velocities of a nerve. ↑ Nerve fibers are classified into: Group A (↑ myelinated fibers, diameters, 5–20 µm, conduction velocity 30–120 m/s); group B (↑ autonomic preganglionic fibers, diameters 1.5–4 µm, velocity 5–30 m/s); group C (↑ nonmyelinated fibers, diameters 0.1–1.5 µm, velocity 0.5–2 m/s). (See physiology texts for further information)

Fibers, cementoalveolar: ↑ Cementoalveolar fibers

Fibers, collagen: ↑ Collagen fibers

Fibers, elastic: ↑ Elastic fibers

Fibers, extrafusal: ↑ Extrafusual fibers; ↑ Neuromuscular spindle

Fibers, gamma: ↑ Neuromuscular spindle

Fibers, intrafusual: ↑ Intrafusal fibers; ↑ Neuromuscular spindle

Fibers, muscle: ↑ Muscle fibers

Fibers, nerve: ↑ Nerve fibers

Fibers, of connective and supporting tissues (fibrae textus connectivi*): ↑ Collagen fibers; ↑ Elastic fibers; ↑ Reticular fibers

Fibers, of Korff: ↑ Korff's fibers

Fibers, of Sharpey: ↑ Sharpey's fibers

Fibers, of Tomes: ↑ Tomes' fibers

Fibers, perforating: ↑ Sharpey's fibers

Fibers, Purkinje, of heart: ↑ Purkinje fibers

Fibers, reticular: ↑ Reticular fibers

Fibrillar centers (nucleolar fibrillar centers): Relatively well-delimited, rounded clear areas (FC) within strands of pars fibrosa (Fi) of ↑ nucleolus. Each F. consists of electron-dense granules (gr) 8–28 nm in diameter and filaments about 5 nm thick (not shown). Granules correspond to sections of ↑ chromatin fiber of ↑ nucleolus organizer region in ↑ chromosomes, i.e., to trunks of ↑ "Christmas trees." Filaments represent branches of "Christmas trees" and form dark fibrillar material of pars fibrosa. F. constitute intranucleolar ↑ chromatin. (Fig. = ↑ pancreatic acinar, cells, rat)

↑ Nucleolus, functions of

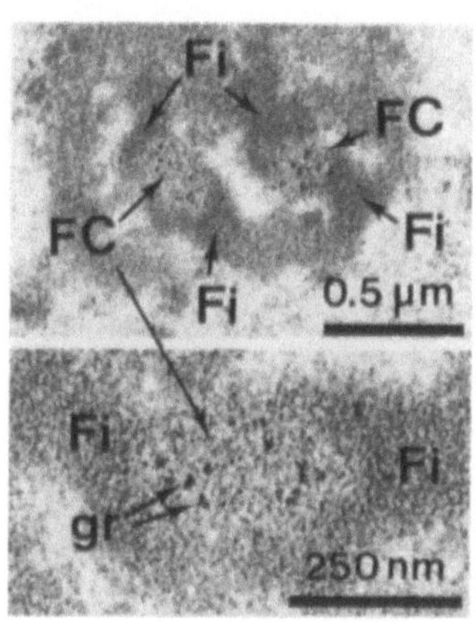

Raška, I., Rychter, Z., Smetana, K.: Z. mikrosk.-anat. Forsch. *97*, 15 (1983)

Fibrillogenesis, collagen: ↑ Collagen fibrillogenesis

Fibrils, collagen: ↑ Collagen fibrils

Fibrils, reticular: ↑ Reticular fibers

Fibrin: An acidophilic feltwork of cross-striated protein fibrils (F) (periodicity 18–24 nm) formed as result of polymerization of plasma ↑ fibrinogen under influence of ↑ thrombin. F. plays an important role in ↑ blood clotting. (Fig. = rat)

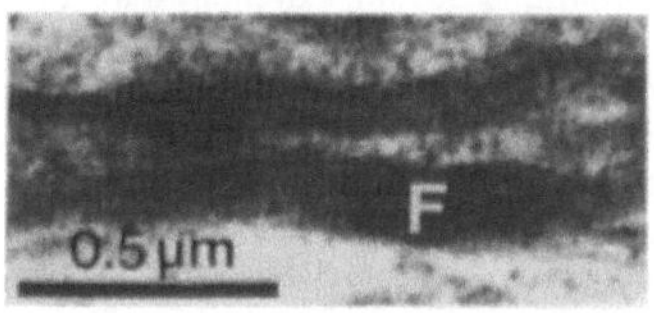

Mosesson, M.W., Doolittle, R.F. (eds.): Molecular Biology of Fibrinogen and Fibrin. Annals of the New York Academy of Sciences, Vol. 408. New York: New York Academy of Sciences 1983

Fibrinogen: A plasma protein synthesized by ↑ liver; under influence of ↑ thrombin, F. polymerizes and becomes ↑ fibrin. Recently, F. has been found in ↑ α-granules of ↑ platelets

Henschen, A., Graeff, H., Lottspeich, F. (eds.): Fibrinogen. Recent Biochemical and Medical Aspects. Hawthorne: Walter De Gruyter 1982; Haverkatem F., Henschen, A., Nieuwenhuizen, W., Straub, P.W. (eds.): Fibrinogen. Structure, Functional Aspects, Metabolism. Hawthorne: De Gruyter 1983

Fibrinoid: A group of extracellular acidophilic substances related to ↑ fibrin. F. occurs in various layers of ↑ placenta as fibrillar or homogenous masses, in which some dead cells and occasional calcifications are scattered. F. increases progressively during second half of pregnancy (↑ Langhans' subchorionic F., ↑ Nitabuch's F., and ↑ Rohr's F.) Gradual accumulation of F. diminishes placental circulation and its metabolism. It is thought that F. is formed as a consequence of hyaline degeneration of ↑ trophoblast.

Fibroarchitectonics: The spacial arrangement of ↑ nerve fibers in ↑ central nervous system. A branch of F. is ↑ myeloarchitectonics.

Fibroblastlike cells: ↑ Synovial cells

Fibroblasts (fibroblastus*): Plump, spindle-shaped, or stellate cells (Fb) present in ↑ connective tissues proper. With their thin processes (P), they contact adjacent F. and/or ↑ fibrocytes (Fc). F. are clearly disposed along ↑ collagen fibers (CF). Nucleus is elliptical with one to two nucleoli; cytoplasm contains filiform mitochondria, a prominent Golgi apparatus (G), ↑ centrioles, well-developed labyrinthine rough endoplasmic (rER) cisternae with flocculent ↑ reticuloplasm, many free ribosomes, occasional ↑ lipid droplets, ↑ coated vesicles and occasional lysosomes, and bundles of ↑ actin microfilaments involved in release of secretory product and in cell motility (↑ myofibroblasts). Near Golgi apparatus, there are small membrane-enclosed ↑ collagen secretory vesicles, (V) with flocculent, moderately osmiophilic material representing precursor of secretory product (↑ procollagen). When stimulated (wound healing), F. rapidly proliferate, become basophilic,

Golgi apparatus and rER enlarge, and number of collagen secretory vesicles increases. F. are limited motile cells responsible for synthesis of ↑ collagen, ↑ reticular and ↑ elastic (EF) fibers, and ↑ proteoglycans of ground substance. Although F. are considered to be well-differentiated cells, they may transform, under some conditions, into ↑ adipose cells. F. are derived from ↑ mesenchymal cells.

↑ Connective tissue proper, ground substance of

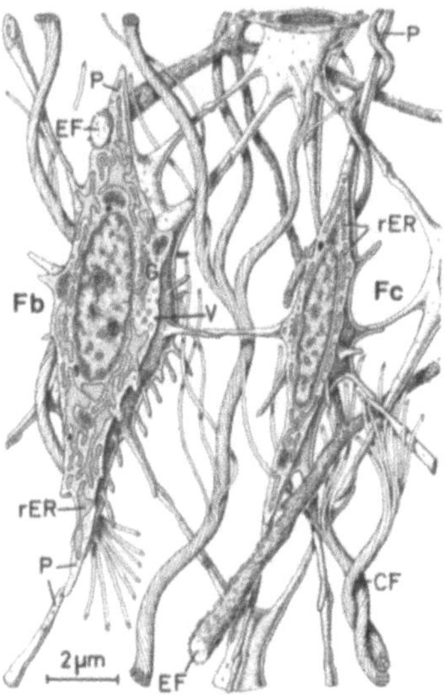

Couchman, J.R., Rees, D.A.: Eur. J. Cell Biol. *27*, 47 (1982); Jackson, D.S., Bentley, J.P.: Connective Tissue Cell. In: F. Beck, B. Lloyd (eds.): The Cell in Medical Science, Vol. III. London, New York, San Francisco: Academic Press 1975; Kischer, C.W.: Tex. Rep. Biol. Med. *32*, 699 (1974); Ten Cate, A.R., Deporter, D.A.: Anat. Rec. *182*, 1 (1975)

Fibroblasts, functions of: The synthesis of material for fibers and ground substance. From amino acids (hydroxyproline, hydroxylysine, desmosine, isodesmosine, etc.), polypeptides are assembled on ↑ ribosomes of rough endoplasmic reticulum (rER) and transported in ↑ transport vesicles to ↑ Golgi apparatus (G), where a sugar component is incorporated. Vesicles (V) loaded with precursor material for all four types of macromolecules arise from both Golgi apparatus and rER and empty their contents into extracellular space. ↑ Fibroblasts produce: 1) ↑ Procollagen (Pc), which is converted into ↑ tropocollagen (Tc), latter forms both ↑ collagen (C) and ↑ reticular (R) microfibrils; 2) microfibrils or microfibrillar protein (Mf) of ↑ elastic fibers and ↑ proelastin (Pe), which polymerizes to form ↑ elastin of ↑ elastic fibers (E); 3) proteoglycans (Pg) of ground substance, composed of ↑ hyaluronic acid (HA), and proteoglycan subunits (PS); 4) structure glycoproteins (SG) (↑ fibronectin and ↑ laminin). Fibroblasts may also phagocytize and degrade collagen microfibrils in the course of their remodeling and turnover.

↑ Connective tissue proper, ground substance of; ↑ Fibrillogenesis

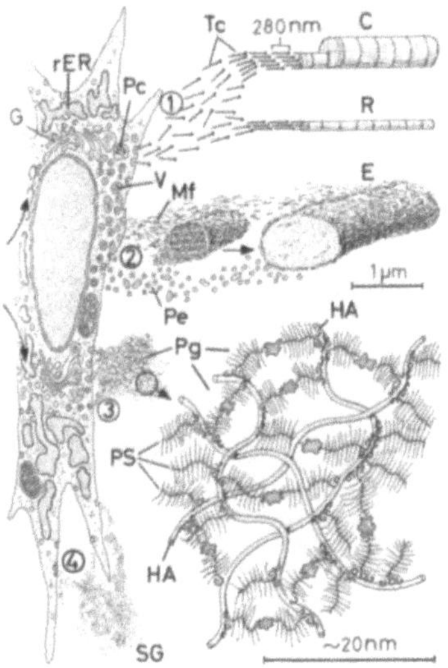

Hay, E.D. (ed.): Cell Biology of Extracellular Matrix. New York: Plenum Press 1981; Melcher, A.H., Chan, J.: J. Ultrastruct. Res. *77*, 1 (1981); Robert, L.: Biol. Med. (Paris) *4*, 1 (1975); Ross, R.: Philos. Trans. R. Soc. Lond. (Biol.) *271*, 247 (1975); Svoboda, E.L.A., Shiga, A., Deporter, M.: Anat. Rec. *199*, 473 (1981)

Fibroblasts, pericryptal: ↑ Pericryptal fibroblasts

Fibrocartilage: ↑ Cartilage, fibrous

Fibrocytes (fibrocytus*): Elongated, fusiform, or stellate cells (Fc) with very long (up to 50 µm), extremely thin, and often ramified processes (P) contacting, by ↑ nexus and ↑ zonulae occludentes, adjacent F. and/or neighboring ↑ fibroblasts. F. are adult form of ↑ fibroblasts; they are common elements of ↑ connective tissue proper and always situated among ↑ collagen fibrils (CF). Nucleus is elongated with condensed chromatin; there is only a small amount of cytoplasm with a few small mitochondria, an inconspicuous Golgi apparatus, some short and flattened rough endoplasmic cisternae, occasional ↑ lysosomes and a moderate number of free ribosomes. Secretory activity of F. is limited; however, under some conditions (wound healing), they may activate and transform again into fibroblasts. F. do not move in connective tissue. However, in ↑ chorionic mesoderm, F. may differentiate into ↑ macrophages (= ↑ Hofbauer cells). (Fig. = rat)

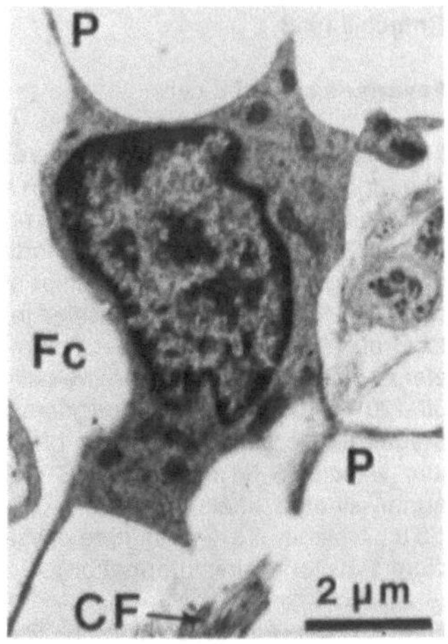

Schmidt, W., Pfaller, K.: Z. mikrosk.-anat. Forsch. *93*, 813 (1979)

Fibroelastic tissue (fibroelastic membrane): A ↑ dense connective tissue with an abundance of ↑ elastic fibers (e.g., F. connecting ↑ tracheal cartilages). The term F. is not generally accepted.

Fibrohyaline tissue: ↑ Chondroid tissue

Fibronectin (cold insoluble globulin, LETS glycoprotein, cell surface protein, CSF): A multimeric, adhesive ↑ glycoprotein present on cell surface, ↑ in basal laminae, connective tissue ↑ ground substance, and circulating in blood. F. plays a role in interactions between cell and extracellular ground substance. Because of its affinity for collagen, F. assures adhesion of ↑ fibroblasts and other cells to their collagen fibrillar support.

↑ Fibroblasts, functions of

Gulati, A.K., Reddi, A.H., Zalewski, A.A.: Anat. Rec. *204*, 175 (1982); Hedman, K.: J. Histochem. Cytochem. *28*, 1233 (1980); Hsieh, P., Segal, R., Chen, B.: J. Cell Biol. *87*, 14 (1980); Hynes, R.O., Yamada, K.: J. Cell Biol. *95*, 369 (1982);

Ruoslahti, E., Engvall, E., Hayman, E.G.: Coll. Relat. Res. *1*, 95 (1981)

Fibrosomes (dark fibrous granules, fibrous granules): Membraneless osmiophilic structures, 20–60 nm in diameter, involved, together with ↑ deuterosome, in acentriolar ↑ ciliogenesis.

Fibrous cartilage: ↑ Cartilage, fibrous

Fibrous connective tissue (textus connectivus fibrosus*): A kind of ↑ connective tissue proper rich in ↑ collagen, ↑ reticular, and ↑ elastic fibers. Based upon their amount and arrangement, one can distinguish ↑ loose and ↑ dense F.

Fibrous granules: ↑ Ciliogenesis, ↑ Fibrosomes

Fibrous lamina (dense lamina, nuclear lamina, nuclear limiting zone, zonula nucleum limitans): An electron-lucent to moderately dense layer (FL), about 12–30 nm thick, of fine 3 to 5-nm filaments interposed between inner ↑ nuclear membrane (M) and ↑ chromatin (C). F. is lacking in areas of ↑ nuclear pores. It is thought that F. consists of three polypeptides (lamins A, B, and C) with function of maintaining form of ↑ nucleus. (Fig = ovarian ↑ follicular cells)

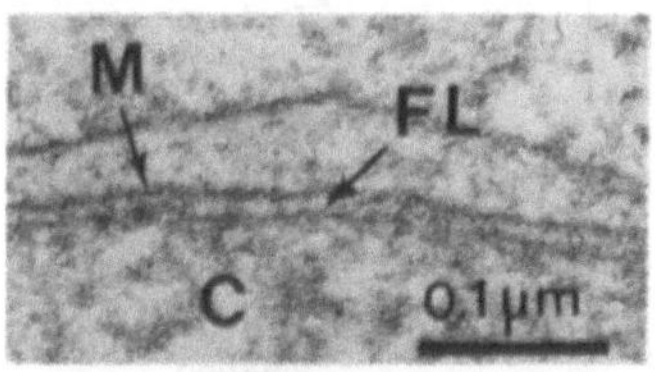

Cohen, A.H., Sundeen, J.R.: Anat. Rec. *186*, 471 (1976); Gerace, L., Blobel, G.: Nuclear Lamina and the Structural Organization of the Nuclear Envelope. Cold Spring Harbor Symposia on Quantitative Biology, Vol. 46, Part 2, New York: Cold Spring Harbor Laboratory 1982; Kirschner, R.H., Rusli, M., Martin, T.E.: J. Cell Biol. *72*, 118 (1977)

Fibrous long-spacing collagen (fibrous long-spacing fibers, FLS): Collagen fibers (FLS) in which ↑ tropocollagen molecules polymerize without overlapping. Therefore, F. is characterized by a cross-banding with a periodicity of 240–300 nm or alternating light and dark bands spaced 80–120 nm. Although rare, F. can be found in connective tissue of various organs. Function and significance of F. are un-

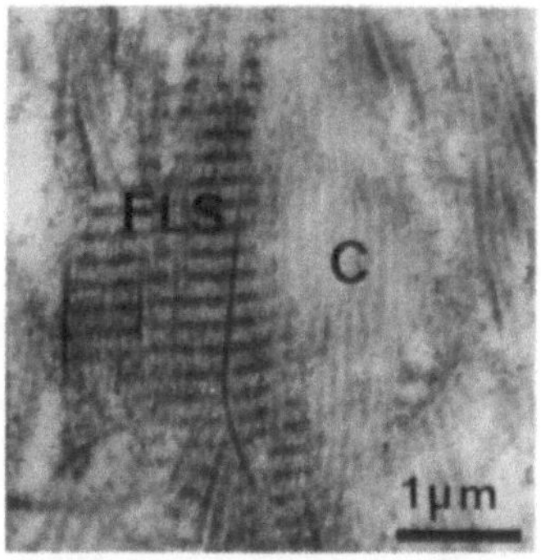

known. C = ordinary ↑ collagen microfibrils. (Fig. = ↑ dermis, mongolian gerbil)

↑ Long-spacing collagen; ↑ Segment long-spacing collagen

Kajikawa, K., Nakanishi, I., Yamamura, T.: Lab. Invest. *43*, 410 (1980)

Fibrous nodule: ↑ Arantius' nodule

Fibrous sheath, of spermatozoon (vagina fibrosa*): The main envelope (F) of principal piece of ↑ spermatozoon tail located between plasmalemma (P) and outer ↑ dense fibers (DF). F. consists of longitudinal continuous ventral (V) and dorsal (D) columns connected by regularly spaced circumferential ribs (R). Septumlike inward extensions of these columns occupy place of outer dense fibers 3 and 8 that have terminated a short distance beyond ↑ annulus. Columns divide principal piece asymmetrically into a major compartment (M) with four dense fibers and a minor compartment (Mi) with only three dense fibers. F. is made up of a fine granular dense material.

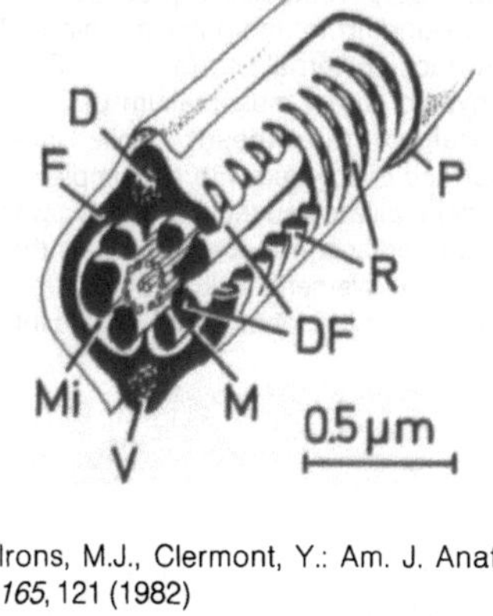

Irons, M.J., Clermont, Y.: Am. J. Anat. *165*, 121 (1982)

Fibrous vagina, of tendon (vagina fibrosa tendinis*): A tube of ↑ dense connective tissue with an interior lined with outer leaf of ↑ synovial vagina of ↑ tendon. F. is firmly connected to ↑ periosteum, representing stable sliding tube for some tendons.

Fibrous veins (vena fibrotypica*): A term sometimes used to designate ↑ sinuses of ↑ dura mater.

Fibrous type, of synovial membrane: ↑ Synovial membrane

Fields, of Cohnheim: ↑ Cohnheim's fields

Fila olfactoria*: Bundles (F) of ↑ nonmyelinated nerve fibers composed of ↑ Schwann's cells and axons of olfactory cells. F. are situated in lamina propria of ↑ olfactory mucosa, penetrating cribriform plate of ethmoid bone to reach ↑ olfactory bulbs. F. are lined by thin ↑ perineurium (P); in interior of the bundles, ↑ endoneurium (E) delimits voluminous Schwann's cells (S), each containing 50–150 axons in its cytoplasm. ↑ Subarachnoid space communicates with ↑ endoneural space of F. (propagation of infection). (Fig. = dog)

↑ Olfactory epithelium

Filaments, actin: ↑ Actin myofilaments

Filaments, intermediate: ↑ Intermediate filaments

Filaments, intrasynaptic: ↑ Intrasynaptic filaments

Filaments, myosin: ↑ Myosin myofilaments

Filaments, of lipid droplets: ↑ Lipid droplets, filaments of

Filamin: A high-molecular weight, actin-binding phosphoprotein, found in mammalian ↑ smooth muscle cells, ↑ platelets, ↑ fibroblasts, ↑ macrophages, kidney and ↑ liver parenchymal cells, and in ↑ terminal web of intestinal epithelial cells. During ↑ cytokinesis, F. is concentrated in cleavage furrow, where it probably contributes to formation of ↑ contractile ring.

Nunnally, M.H., D'Angelo, J.M., Craig, S.W.: J. Cell Biol. *87*, 219 (1980)

Filiform papillae, of tongue (papillae filiformes*): The predominant kind of 0.5 to 3-mm-long papillae covered by a thin keratinized layer (K). Pointed tips of F. are oriented toward ↑ pharynx. Lamina propria (LP) consists of relatively ↑ dense connective tissue; it frequently forms one or two secondary papillae (SP). F. have an abrasive function during mastication.

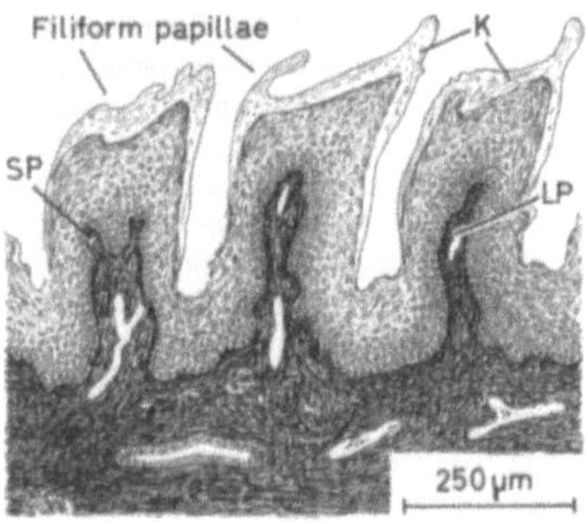

Boshell, J.L., Wilborn, W.H., Singh, B.B.: Acta Anat. *114*, 97 (1982); Fernandez, B., Suarez, T., Zapata, A.: J. Anat. *126*, 487 (1978); Stelfik, D.E., Singh, B.B., McKinney, R.V.Jr., Boshell, J.L.: Acta Anat. *117*, 21 (1983)

Filipin: A polyene antibiotic used in biophysical study of various cell membranes (plasmalemma, Golgi membranes, membranes of ↑ secretory granules) because of its property of binding specifically and stoichiometrically to certain membrane sterols. Resulting filipin-sterol complexes aggregate in clusters, which are particularly visible on ↑ freeze-cleaving ↑ replicas and in transmission electron microscope as lacelike series of local membrane enlargements, 20–30 nm in diameter

Gotow, T., Hashimoto, H.P.: Cell Tissue Res. *230*, 689 (1983); Matsuda, H., Fujita, H., Ishimura, K.: Acta Histochem. Cytochem. *16*, 112 (1983); Montesano, R., Ravazzola, M., Orci, L.: Cell Biol. Int. Rep. *7*, 194 (1983); Orci, L, Perrelet, A., Montesano, R.: J. Histochem. Cytochem. *31*, 952 (1983)

Filopodia: Slender cytoplasmic processes (F), 3–10 µm long and 0.2 µm wide, extending from ↑ perikaryon (P) of a ↑ macrophage. Exact significance of F., which can be branched, is unknown. L = ↑ lamellipodium, Ps = ↑ pseudopodia

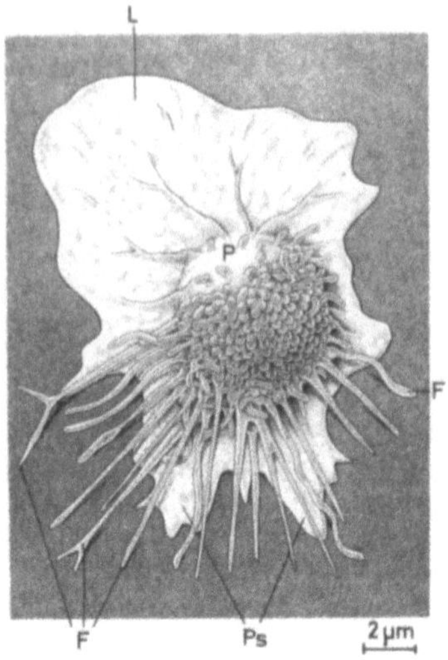

Filtration barrier, of glomerulus: ↑ Glomerular filtration membrane

Filtration pressure (glomerular filtration pressure): The difference between hydrostatic pressure in ↑ glomerular capillaries (90 mm Hg) and the sum of opposing hydrostatic pressure in ↑ Bowman's space (15 mm Hg) and oncotic pressure of blood (30 mm Hg) = 90 − (15+30) = 45 mm Hg. Some authors estimate F. to be 18 mm Hg. (See physiology texts for further information)

↑ Glomerular filtration membrane

Filtration slit, of glomerulus: ↑ Glomerular slit membrane

Fimbriae, of oviduct (fimbriae tubae uterinae*): Tentaclelike processes (F) surrounding infundibular margin of ↑ oviduct. External surface of F. is covered with ↑ mesothelium of ↑ mesosalpinx (arrowhead), while inner surface is covered with a ↑ simple columnar ciliated epithelium (arrow) with scattered secretory cells. It seems that ciliary movement, oriented toward ↑ uterus, moves secondary ↑ oocyte into ampulla. O = ↑ ovary. (Fig. = monkey)

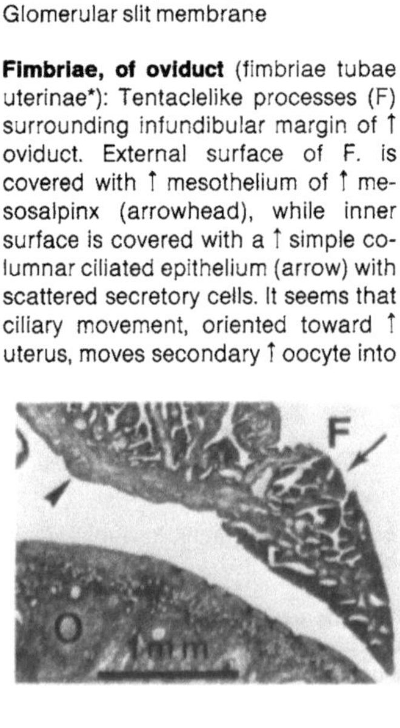

↑ Oviduct, epithelium of

Fingerprint figures: ↑ Tubular myelin

Fingerprints: ↑ Dermatoglyphics

"Fish eye cells": ↑ Nerve fibers of peripheral nervous system, degeneration and regeneration of

Fixation: The procedure of preservation, stabilization, and maintenance of structure of cells, tissues, and organs, for microscopic observations and histochemical reactions. Ideal F. would not provoke any morphological or chemical ↑ artifacts, and thus reflect a true living state. Since such an F. is evidently impossible, many F. methods have been developed with aim of approaching this ideal. The principle of majority of F. is to render cell components insoluble by chemical or physical precipitations or cross-linking cell's protein components. An overview of F. methods is shown in following diagram, whereby ↑ freeze-substitution and ↑ freeze-drying are methods between chemical and physical F. Asterisks designate letters under which the subject is described.

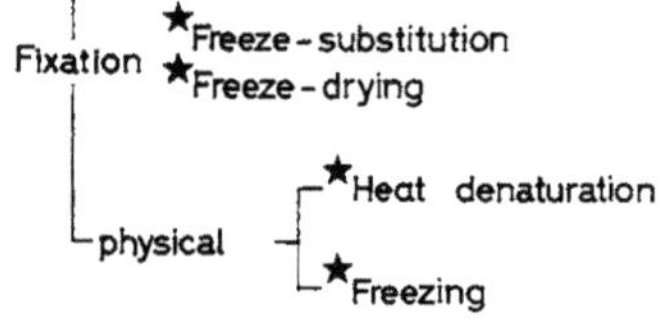

F. prevents ↑ autolysis, postmortem decomposition and prepares material for subsequent ↑ staining.

Fixation, chemical: The procedure of stabilization of cell components with different chemical substances and/or their mixtures. Four types of F.: 1) F. through dehydration (alcohol, acetone); 2) F. through formation of salts (mercuric chloride, potassium bichromate); 3) F. through changes in colloid swelling (acetic acid, trichloracetic acid, sulfosalicylic acid, chromium acid); 4) F. through cross-linkage of proteins (↑ formaldehyde, ↑ glutaraldeyhde, ↑ osmium tetroxide).

↑ Immersion; ↑ Perfusion

Glauert, A.M.: Practical Methods in Electron Microscopy. Vol. 3, Part 1. Fixation, Dehydration and Embedding of Biological Specimens. Amsterdam: Elsevier/North Holland 1980; Hayat, M.A.: Fixation for Electron Microscopy. New York: Academic Press 1981

Fixation, physical: The preservation of morphological and/or chemical characteristics of a biological object by means of increase or decrease in temperature. Only ↑ freezing gives satisfactory results; heat denaturation can be considered a theoretical possibility.

↑ Freeze-cleaving; ↑ Ultracryotomy

Fixed cells, of connective tissue proper: A practically nonmotile population of cells responsible for: 1) Production and general maintenance of ↑ ground substance and fibers (↑ fibroblasts, ↑ fibrocytes, ↑ reticular cells); 2) storage of reserve caloric material (↑ adipose cells, white). Some authors also include fixed macrophages (↑ histiocytes) and ↑ mesenchymal cells in this group.

↑ Ameboidism

fl (femtoliter): 10^{-15} of a liter

Flagellum*: A very long, motile cell process with a complicated inner structure. In mammals, only F. is ↑ spermatozoon tail.

↑ Cilia

Flat diffuse bipolar cells: ↑ Bipolar cells, of retina

Flat midget bipolar cells: ↑ Bipolar cells, of retina

Flemming body: ↑ Mid-body

Flower spray endings (termination nervi racemosa*): Afferent sensory nerve fibers branching around nonnucleated segment of ↑ nuclear chain fibers of a ↑ neuromuscular spindle.

FLS: ↑ Fibrous long-spacing collagen

Fluffy border: ↑ Ruffled border

Fluid, cerebrospinal: ↑ Cerebrospinal fluid

Fluid mosaic model: ↑ Cell membrane, models of

Fluorescein: A fluorescent dye widely employed in ↑ fluorescence microscopy. F. is most frequently used conjugated with antibodies for immunocytochemical localization of an antigen-antibody complex. In ↑ biomicroscopy of eye, F. is used for localization of superficial defects of ↑ cornea.

Fluorescence: A capacity of some natural substances or dyes to absorb light of one frequency and emit light of another. In practice, a specimen is exposed to invisible ultraviolet spectrum, which excites fluorescent substances to emit light within visible range. Wavelength of emitted light is longer than UV light, lower in energy, and depends upon chemical nature of illuminated substance.

↑ Autofluorescence; ↑ Fluorescence microscopy; ↑ Immunofluorescence

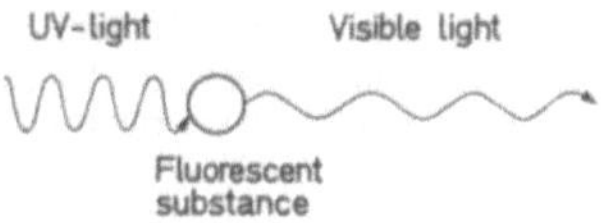

Fluorescence microscope: A light microscope in which object (Ob) is illuminated with blue or UV light, thus becoming fluorescent. Principle: in F. with blue light (BL), an excitatory filter (EF) absorbs all wavelengths except that of blue light. After provoking ↑ fluorescence in the object (Ob), a yellow filter (YF) arrests only blue light, allowing all other wavelengths (i.e., colors) emitted by the object to pass. In F. with UV light, light passes through anticaloric filters (AF), then excitatory filters (EF), and finally a filter absorbing all other wavelengths except that of UV light. Fluorescent spectra formed by object (Ob) are filtered with UV filter (UVF), which arrests UV radiations, protects eyes, and permits observation of all colors emitted by the object. F. in blue light is very simple, but doses not permit observation of blue fluorescence. New F. have epidiascopic ↑ illumination.

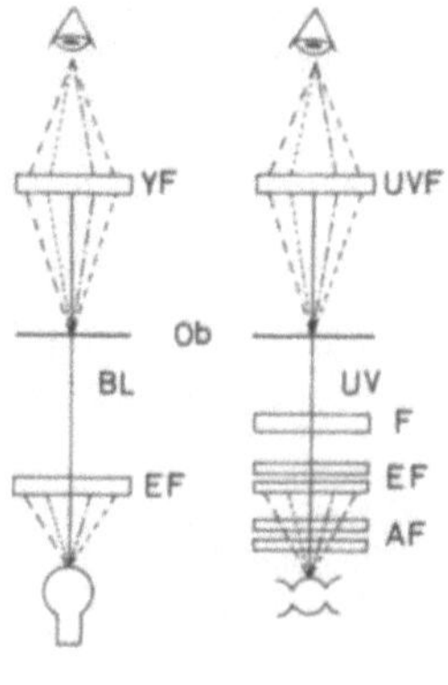

Fluorescent antibody staining: An immunocytochemical technique of labeling antibodies with a ↑ fluorochrome in order to visualize them as a result of an antigen-antibody reaction or in phagocytized form.

Fluorescent staining: Any staining with dye giving ↑ fluorescence. Among F. one can also consider intravital introduction into the body of ↑ fluorochromes, which label different tissues (e.g., labeling of bone, dentin and enamel with antibiotic cyclines, corneal stroma with ↑ fluorescein in cases of corneal injury, etc.).

Fluorochromes: Fluorescent dyes used to stain cells and tissues for analysis by ↑ fluorescence microscope (acridine orange, ↑ fluorescein, auramine, ↑ quinacrine mustards, rhodamine, etc.).

Folia, of cerebellum (folium cerebelli*): Long, narrow folds (F) of ↑ gray matter (gm) with white matter (wm) in interior. F. are separated by furrows, sulci (S). Parallel to axis of F. are ↑ parallel fibers; perpendicular to axis of F. are dendritic arborization of ↑ Purkinje cells and axons of ↑ basket cells. A group of F. forms a lobule (L) of cerebellum. (Fig. = monkey)

↑ Cerebellum, neuronal circuits of

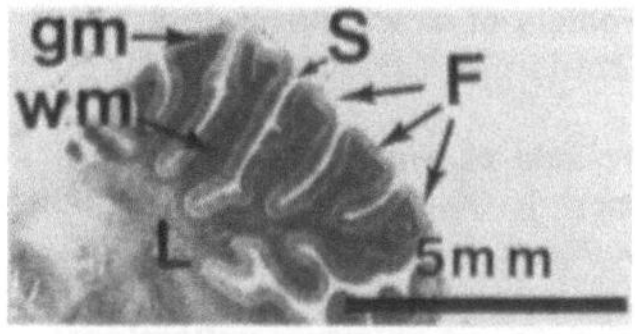

Foliate papillae, of tongue (papillae foliatae*): Leaflike, parallel folds (FP) of tunica mucosa of ↑ tongue situated at its dorsolateral sides. Folds are covered by ↑ stratified squamous epithelium (E); lateral aspects of folds contain many ↑ taste buds (TB). Lamina propria (LP) of F. forms three sec-

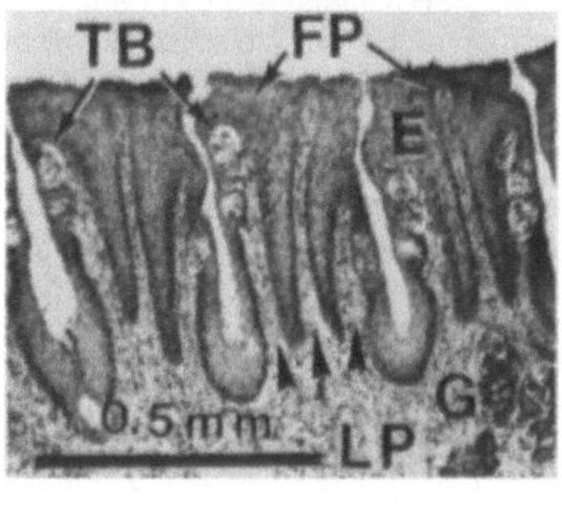

ondary papillae (arrowheads). Small ↑ serous glands (G) open into bottom of furrows separating the folds. F. occur only in infants and some animals; in adult humans they are rudimentary. (Fig. = rabbit)

Liu, H.C., Lee, J.C.: Acta Anat. *112*, 310 (1982)

Follicle-associated epithelial cells: ↑ M-cells

Follicle of ovary, antral: ↑ Ovarian follicle, secondary

Follicle of ovary, atretic: ↑ Ovarian follicle, atretic

Follicle of ovary, control of growth of: ↑ Ovarian follicle, control of growth of

Follicle of ovary, growth of: ↑ Ovarian follicle, growth of

Follicle of ovary, mature: ↑ Ovarian follicle, mature

Follicle of ovary, preovulatory: ↑ Ovarian follicle, preovulatory

Follicle of ovary, primary: ↑ Ovarian follicle, primary

Follicle of ovary, primordial: ↑ Ovarian follicle, primordial

Follicle of ovary, secondary solid: ↑ Ovarian follicle, secondary solid

Follicle of ovary, secondary vesicular: ↑ Ovarian follicle, secondary vesicular

Follicle of ovary, unilaminar: ↑ Ovarian follicle, primordial

Follicle of thyroid gland: ↑ Thyroid follicles

Follicle-stimulating hormone (FSH): A ↑ glycoprotein hormone with a molecular weight of about 30000 daltons secreted under influence of ↑ FSH/LH – releasing factor by ↑ gonadotropes type I of ↑ adenohypophysis. In the female, FSH stimulates growth and maturation of ovarian ↑ follicles, as well as secretion of ↑ estrogens by ↑ follicular cells. In the male, FSH stimulates growth of ↑ seminiferous tubules and ↑ spermatogenesis. FSH also acts, probably by way of ↑ cyclic AMP, on ↑ Sertoli's cells to produce ↑ androgen-binding protein.

↑ Releasing factors

Chappel, S.C., Ulloaaguirre, A., Coutifaris, C.: Endocrine Rev. *4*, 179 (1983)

Follicle-stimulating hormone gonadotropes: ↑ Gonadotropes type I

Follicle-stimulating hormone-releasing factor: ↑ Gonadotropin-releasing factor

Follicle, ultimobranchial: ↑ Ultimobranchial follicles

Follicles, lingual: ↑ Lingual follicles

Follicles, of hair: ↑ Hair follicle

Follicles, of hypophysis: ↑ Hypophyseal follicles

Follicles, of ovary: ↑ Ovarian follicles

Follicular cells, of hypophysis: Cuboidal cells lining occasional ↑ follicles (F) of pars distalis of ↑ hypophysis. ↑ Apical pole of certain F. bears some ↑ cilia penetrating into colloid (Co). The term F. is descriptive, since ↑ acidophilic (A), ↑ basophilic (B), and ↑ chromophobic cells (C) can participate in composition of follicular wall. (Fig. = human)

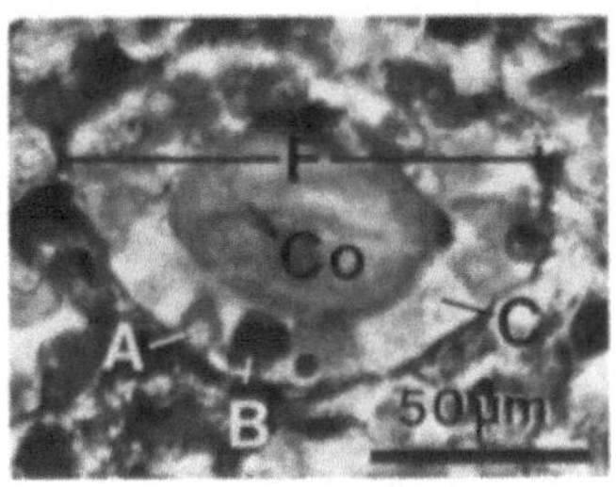

Follicular cells, of ovarian follicles: A single layer of epithelial cells (F) surrounding ↑ oocyte (O) of primordial (PF) and primary ↑ ovarian follicles (PrF). In primordial follicles, F. are flat; as they grow, F. become cuboidal and columnar. By rapid proliferation under influence of ↑ follicle-stimulating hormone, F. constitute in secondary and mature ovarian follicles a stratified epithelium stratum granulosum, composed of ↑ granulosa cells. Low F. have a flattened nucleus and poorly developed ↑ organelles. During follicular development, F. become tall columnar, their nuclei elliptical and more voluminous, cytoplasm fills with free ribosomes, mitochondria and rough endoplasmic cisternae become more numerous, Golgi apparatus enlarges, ↑ lipid droplets increase in number. F. are held together predominantly by ↑ nexus. On one side, F. are in contact with oocyte or ↑ zona pellucida (ZP); they are separated from ovarian stroma (S) by a ↑ basal lamina (BL).

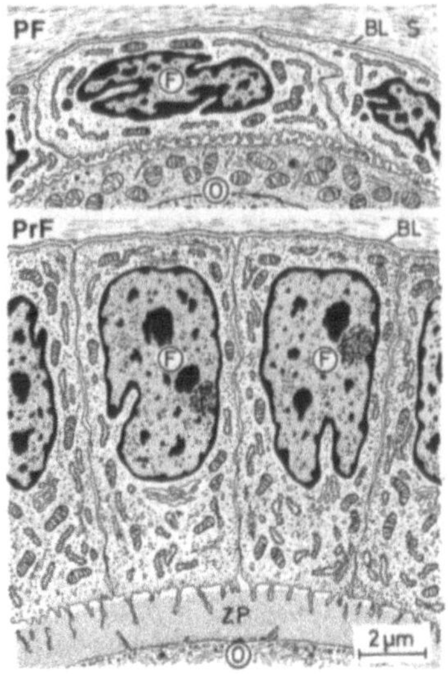

Follicular cells, of thyroid gland: ↑ Thyroid follicular cells

Follicular fluid: ↑ Liquor folliculi

Follicular fluid, secondary: ↑ A very thin fluid secreted at an increased rate into ↑ antrum folliculi a few hours before ↑ ovulation. Production of F. plays an important role in ovulation.

Folliculi lymphatici lienales: ↑ Splenic nodules

Folliculin: ↑ Estradiol

Fontana's spaces: Spaces (F) between trabeculae (T) of ↑ trabecular meshwork; on the one side, F. communicate with ↑ anterior chamber (AC) of eye, on the other side with ↑ Schlemm's canal (SC). Through F. flows ↑ aqueous humor. C = ↑ cornea, I = ↑ iris

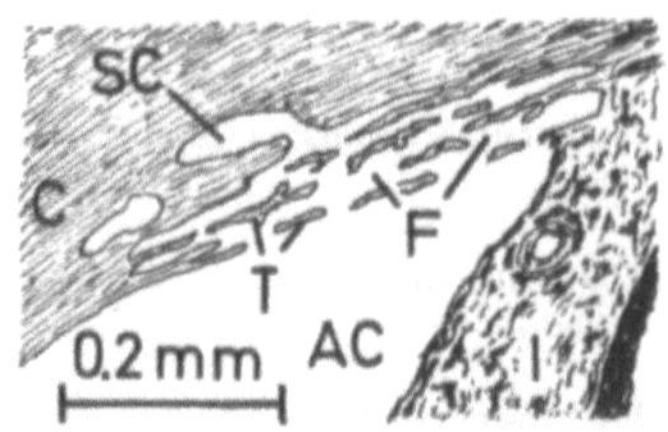

Foot plate: ↑ Perivascular feet

Foramen, apical: ↑ Apical foramen

Foramen caecum linguae*: A small invagination at apex of V-shaped line of tongue's ↑ circumvallate papillae; F. is rudiment of thyroglossal duct. (See embryology texts for further information)

Foramina nervosa, of lamina spiralis: Fine openings at free edge of ↑ lamina spiralis ossea through which ↑ nerve fibers enter organ of ↑ Corti. When nerve fibers (i.e., ↑ dendrites of ↑ bipolar cells of ↑ spiral ganglion) leave F., they become nonmyelinated.

Foreign-body giant cells: ↑ Giant cells

Formalin (formol): A 38%–40% aqueous solution of formaldehyde. Concentrations of 4% or 10% are most frequently used for ↑ fixation.

Formazan: An insoluble reddish violet substance, product of reduction of a soluble ↑ tetrazolium salt. F. indicates, after appropriate histochemical reaction, light- and transmission electron microscopic localization of various enzymes.

Altman, F.P.: Prog. Histochem. Cytochem. *9*, 1 (1976)

Forming face, of Golgi apparatus (cis face, immature face, proximal face): The aspect (F) of ↑ Golgi apparatus oriented toward ↑ nucleus. F. is usually convex; with F. coalesce ↑ transport vesicles (T) originating from rough ↑ endoplasmic reticulum. First two lamellae of F. are highly osmiophilic and responsible for impregnation of Golgi apparatus after Golgi's method for light microscopy. F. contains a high concentration of thiamine pyrophosphatase and nucleotide diphosphatases. Surface of Golgi apparatus opposite F. is ↑ maturing face (M).

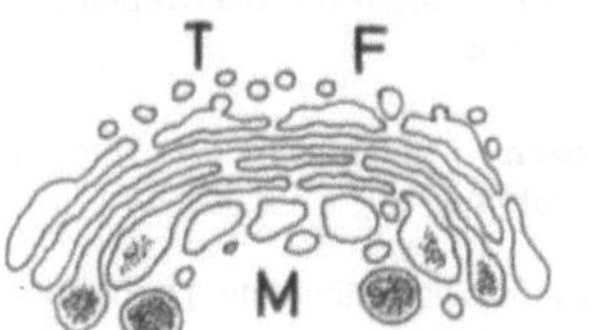

Fornices, of eye (fornix conjunctivae superior et inferior*): The area of transition (F) between bulbar (BC) and palpebral ↑ conjunctiva (PC). Structure: 1) Epithelium (E) = a folded ↑ stratified columnar epithelium, two to four cells thick, with some large ↑ goblet cells; 2) lamina propria (LP) = a very ↑ loose connective tissue with large tissue spaces and ↑ wandering cells. (Fig. = human)

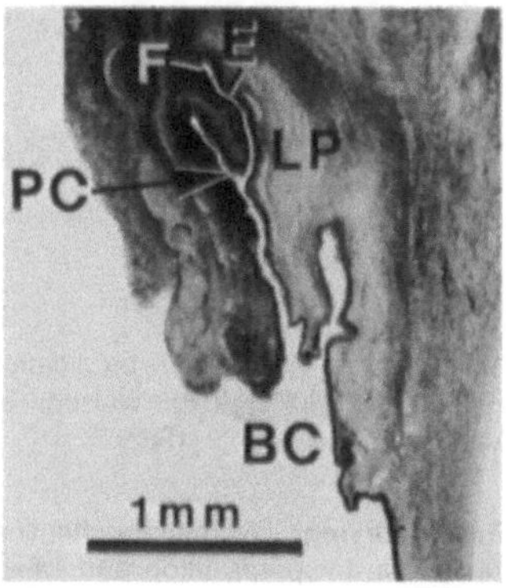

Fossa, implantation, of spermatozoon: ↑ Implantation fossa

Fossa navicularis (fossa navicularis urethrae masculinae*): The terminal, enlarged part (F) of pars spongiosa of male ↑ urethra. F. is lined by ↑ stratified squamous epithelium (E) continuous with that covering ↑ glans penis. Occasional ↑ goblet cells and intraepithelial cystae with colloidlike contents may be found. Lamina propria (LP), with distinct papillae and some smooth muscle cells, separates F. from adjacent veins of ↑ corpus spongiosum (CS). (Fig. = human newborn)

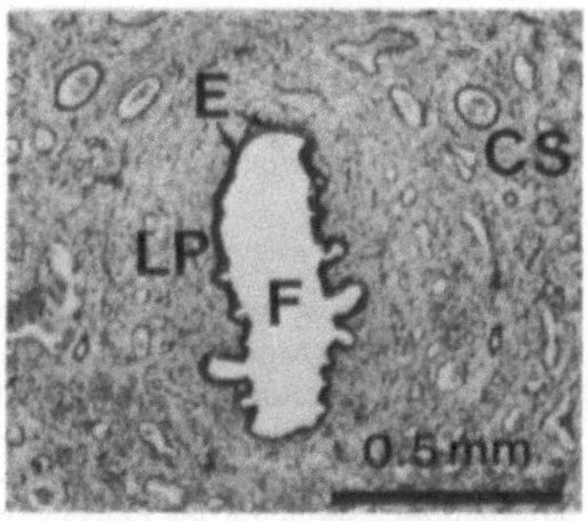

Fovea centralis* (central fovea, "macula"): The deepest central zone (FC) of ↑ macula lutea (ML) about 0.5 mm in diameter. F. is characterized by presence of particularly long macula cones (MC) and complete absence of rods. Except for cones, all other retinal layers are displaced laterally, allowing light to act directly on ↑ photoreceptors. Each of about 30 000 cones makes synaptic contact with only one ↑ bipolar cell; F. is, therefore, point of greatest visual acuity. Central point of F. is devoid of blood vessels (= avascular central territory) and supplied only from ↑ choroid (Ch); nearest retinal capillaries are at a distance of about 275 µm from central point.

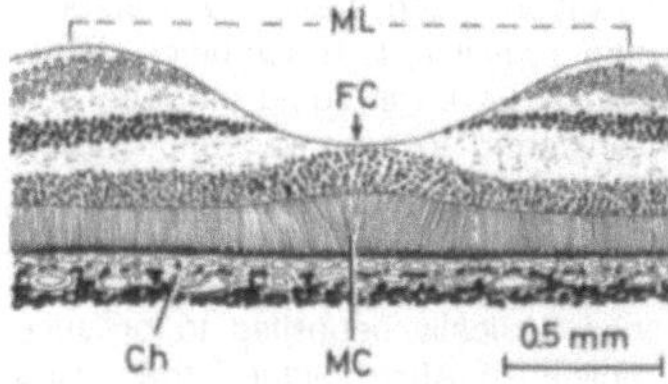

Foveolae gastricae* (gastric pits): Furrows (FG), about 100–200 µm deep, in surface epithelium of gastric mucosa lined by ↑ surface mucous cells (arrowheads). There are about 3.5 million F.; two to five ↑ gastric glands proper (G) open into bottom of each F. M = ↑ mucus. (Fig. = dog)

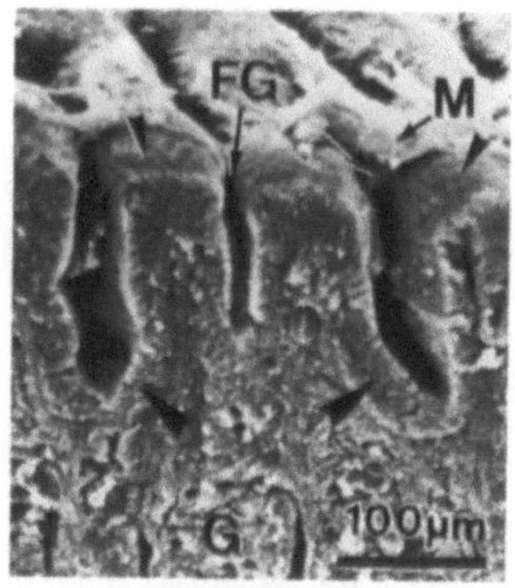

Siew, S., Goldstein, M.L.: Scan. Electron Microsc. 1981/4, 173 (1981)

Foveolae, of stomach: ↑ Foveolae gastricae

Fractionation, of cell components: ↑ Centrifugation, density gradient

Free macrophage: ↑ Histiocyte, wandering

Free nerve endings: A group of terminations of ↑ nerve fibers that end, after having lost all their coverings (including their ↑ Schwann's cell investment) and without any structural specialization, among cells of epithelial, connective, and muscular tissues and in numerous organs.

↑ Axons, naked; ↑ Nerve endings; ↑ Nociceptors

Cauna, N.: Anat. Rec. *198*, 643 (1980)

Free ribosomes: ↑ Ribosomes, free

Freeze-cleaving (freeze-fracturing): A method for preparing a very accurate platinum-carbon ↑ replica of fracture face through a frozen cell without any ↑ fixation, ↑ dehydration, or ↑ embedding. Principle: 1) Tissue block (TB) is quickly frozen in liquid nitrogen and transferred to high vaccum; 2) here, specimen is fractured with a microtome ↑ knife (K); 3) at fractured surface, a thin layer of carbon (C) is then vertically deposited to produce replica (R). After oblique ↑ shadowing of replica with platinum-carbon (PtC), specimen is 4) warmed; replica is cleaned with a strong acid and mounted on a grid for observation in transmission electron microscope. F. is especially valuable for study of ↑ cytomembranes. (Fig. modified after Niedermayer 1978)

↑ Freeze-cleaving, terminology of; ↑ Freeze-etching

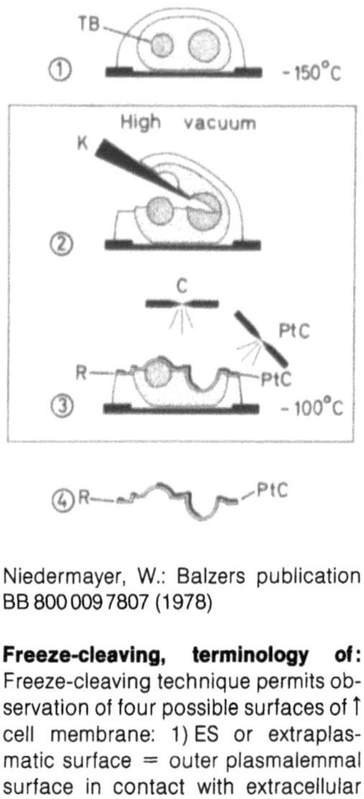

Niedermayer, W.: Balzers publication BB 8000097807 (1978)

Freeze-cleaving, terminology of: Freeze-cleaving technique permits observation of four possible surfaces of ↑ cell membrane: 1) ES or extraplasmatic surface = outer plasmalemmal surface in contact with extracellular milieu (E); 2) PS or plasmatic surface = plasmalemmal aspect in contact with cytoplasm (C); 3) ↑ EF or extraplasmatic fracture face = split half of plasmalemma facing toward cytoplasm; 4) ↑ PF or plasmatic fracture face = split half of plasmalemma facing away from cytoplasm. Arrowheads indicate ↑ membrane-associated particles, which probably result from cleaving of protein molecules (P) traversing membrane. (Fig. modified after Niedermeyer 1978)

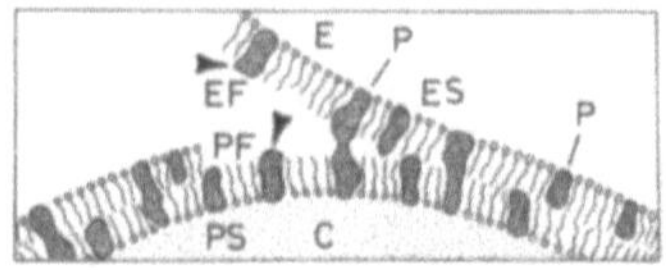

EF = extraplasmatic fracture face
ES = extraplasmatic surface
PF = plasmatic fracture face
PS = plasmatic surface

Niedermeyer, W.: Balzers publication BB 8000097807 (1978); Welsch, U.: Verh. Anat. Ges. *76*, 29 (1982)

Freeze-drying: A physicochemical technique for preparation and ↑ fixation of biological objects consisting of rapid freezing of blocks in liquid nitrogen or air at −160° to −195 °C and dehydration in a vacuum at low temperature (−30° to −40 °C). Very rapid freezing (quenching) prevents formation of tissue-disrupting ice crystals and drying at a low temperature directly transforms ice into a gas. After drying, tissue is fixed in formaldehyde vapors at 50°–80 °C, embedded in ↑ paraffin in vacuum, and cut with an ordinary ↑ microtome. Advantages: Homogenous fixation throughout piece; no shrinking; preservation of enzymes, soluble substances, and lipids. Disadvantage: Complicated and expensive F. apparatus.

↑ Freeze-substitution

Pearse, A.G.E.: Histochemistry. Theoretical and Applied. 3rd edn. London: Churchill Ltd. 1968; Sjöstrand, F.S., Kretzer, F.: J. Ultrastruct. Res. *53*, 1 (1975)

Freeze-etching: A technique in which specimen is kept, after ↑ freeze-cleaving, in high vacuum at −100 °C in vicinity of a cold trap for a short while before replication. Under such environmental conditions, ice crystals of fractured face sublimate, so that ice level (L) is lowered, revealing new details (D) not visible after simple freeze-cleaving.

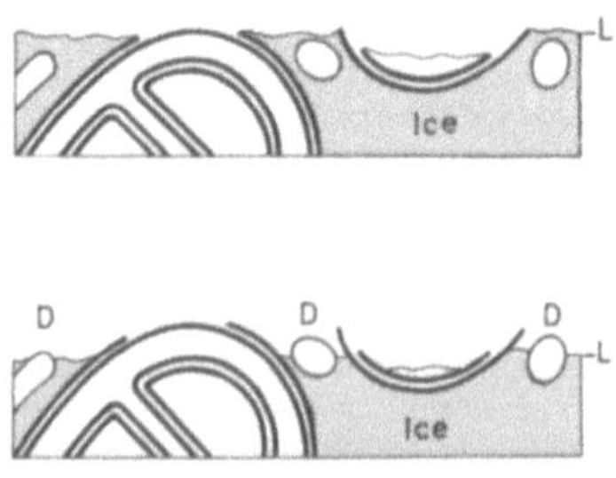

Glauert, A.M. (ed.): Practical Methods in Electron Microscopy. Vol. 8. Amsterdam: Elsevier/North Holland 1980; Orci, L., Perrelet, A.: Freeze-Etch Histology. Berlin, Heidelberg, New York: Springer-Verlag 1975; Freeze-etching. Acta Histochem. Suppl. XXIII, 1 (1981)

Freeze-fracture-etching: A combined technique comprising ↑ freeze-cleaving followed by ↑ freeze-etching of same specimen. F. is the most modern method for microscopic study of ↑ cytomembranes.

Freeze-fracturing: ↑ Freeze-cleaving

Freeze-substitution: A modification of ↑ freeze-drying consisting of rapid freezing of a biological object in liquid nitrogen or helium and replacement of ice by alcohol or other solvents at low temperature (−70 °C). F. is considered to be the best technique for preservation of enzyme activity.

Freezing: A method of physical ↑ fixation and hardening of tissues permitting very rapid production of sections. A fixed or native piece is frozen with carbon dioxide gas under pressure and cut with a ↑ freezing microtome or in a ↑ cryostat. Advantages: Great speed; ↑ embedding not necessary, thus maintenance of enzyme activities and ↑ lipids. Disadvantages: Sections extremely fragile, no possibility of performing many staining methods, dissolution of ↑ glycogen, etc.

↑ Freeze-cleaving

Freezing microtome: A simple mechanical ↑ microtome with ↑ knife and object-holder cooled with carbon dioxide gas under pressure in order to permit ↑ sectioning of frozen tissue blocks. Today, F. has been replaced by ↑ cryostat.

Frozen sections: ↑ Freezing microtome

FSH: ↑ Follicle-stimulating hormone

FSH-gonadotropes: ↑ Gonadotropes type I

FSH/LH-RF: ↑ Gonadotropin-releasing factor

FSH-RF (follicle-stimulating hormone-releasing factor): ↑ Gonadotropin-releasing factor

F – synaptic vesicles: ↑ Synaptic vesicles

Fuchsin: A group of dyes widely used in histological staining methods (acid F., basic F., carbol F., leukofuchsin).

Fuchsin-sulfurous acid: ↑ Schiff's reagent

Functionalis (stratum functionale*): The superficial three-quarters of ↑ endometrium which is shed every ↑ menstrual cycle.

Fundic glands: ↑ Gastric glands proper

Fundus, of stomach: A dome-shaped bulge above level of ↑ cardiac area. Structurally, F. corresponds to corpus of ↑ stomach.

Fundus, of uterus (fundus uteri*): Dome-shaped upper portion of body of ↑ uterus from which ↑ oviducts extend. Histological structure identical to that of uterine ↑ corpus.

Fungiform papillae, of tongue (papillae fungiformes*): A kind of papillae (F) with clublike upper part and narrowed stalk. F. are higher than neighboring ↑ filiform papillae (Fi) and located predominantly on dorsal surface of lingual apex. F. are covered by nonkeratinized stratified squamous epithelium associated with secondary papillae (SP); epithelium may contain occasional ↑ taste buds. Underlying lamina propria (LP) is well vascularized; F. therefore appear red.

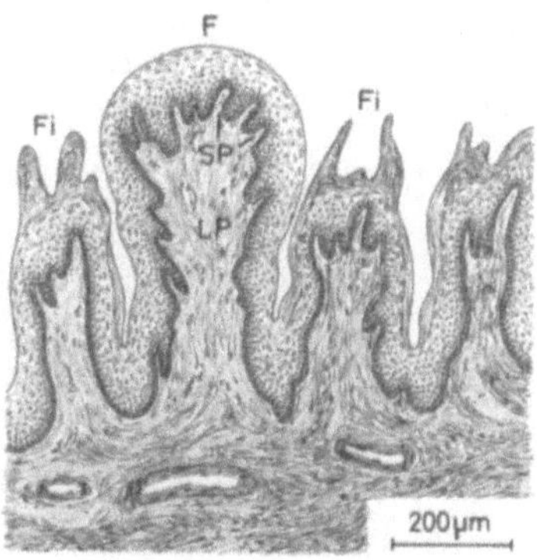

Arvidson, K.: Acta Otolaryngol. *81*, 496 (1976); Beckers, H.W., Eisenacher, W.: Morphologie der Papillae fungiformes. Adv. Anat. Embryol. Cell Biol. *50*/6, 1 (1975)

Fusiform density: ↑ Dense bodies, of smooth muscle cells

Fusiform vesicles: ↑ Discoid vesicles

Fusimotor fibers: ↑ Nerve fibers, fusimotor

Fusimotor neurons: ↑ Gamma motor neurons

Fuzzy coat: ↑ Glycocalyx

G

GABA: ↑ Gamma-aminobutyric acid

G-actin: Globular subunits of ↑ actin myofilaments measuring approximately 5.6 nm in diameter.

Galactophores: ↑ Lactiferous ducts

Galactopoiesis: The maintenance of ↑ lactation.

Gallbladder (vesica biliaris*): A musculoepithelial sac for storage and condensation of ↑ bile. Structure: 1) Tunica mucosa (TM) is thrown into many primary and secondary folds (F): a) epithelium (E) = a ↑ simple tall columnar epithelium without goblet cells; deep epithelial invaginations penetrating into muscular layer are termed diverticula (D) or Rokitansky-Aschoff crypts, b) lamina propria (LP) = a ↑ loose connective tissue containing mucous ↑ tubuloalveolar glands (G) in neck region only. 2) Tunica fibromuscularis (TMu) = irregularly arranged smooth muscle bundles intermingled with a considerable amount of ↑ collagen and ↑ elastic fibers. 3) Tela subserosa (TSs) = a loose connective tissue with many blood and lymphatic vessels and sensory nerve fibers. 4) Tunica serosa (TSe) = ↑ peritoneum or tunica adventitia (TA) in areas not covered by peritoneum = richly innervated and vascularized loose connective tissue. A tenfold condensation of bile takes place in G.

↑ Cholecystokinin; ↑ Gallbladder, epithelium of

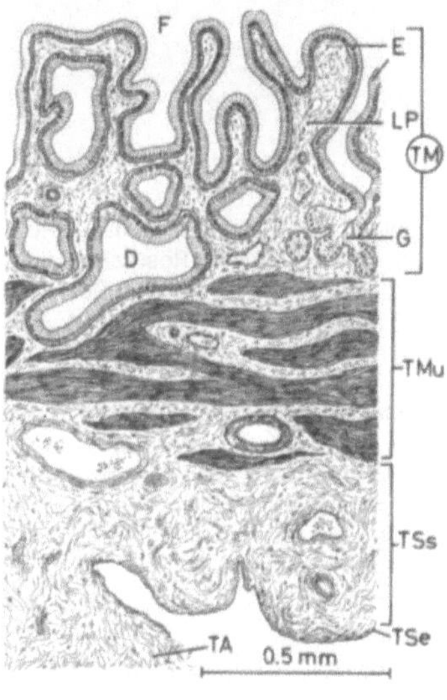

MacPherson, B.R., Scott, G.W., Lennon, F.: Cell Tissue Res. *233*, 161 (1983)

Gallbladder, epithelium of: Tall columnar epithelial cells with an oval nucleus and a conspicuous nucleolus. Cytoplasm contains a moderate number of mitochondria, a small supranuclear Golgi apparatus, short cisternae of rough endoplasmic reticulum, some lysosomes, and free ribosomes. Membrane-limited secretory granules, measuring 0.2–0.3 µm with moderately dense and fine granular material, are scattered in ↑ apical pole. Contents of granules are emptied at cell surface and form a protective layer against bile salts. Free cell surface bears numerous short microvilli; lateral cell surfaces are studded with fingerlike processes interdigitating with adjacent cells. Well-developed ↑ junctional complexes (arrowheads) seal neighboring cells. G. is a secretory surface epithelium producing a small amount of protective ↑ mucus. This epithelium is also involved in concentration of ↑ bile by pumping of sodium and water into intercellular spaces, causing an osmotic gradient that drives water from epithelium through basal lamina (BL) into vessels of lamina propria. (Fig. = mongolian gerbil)

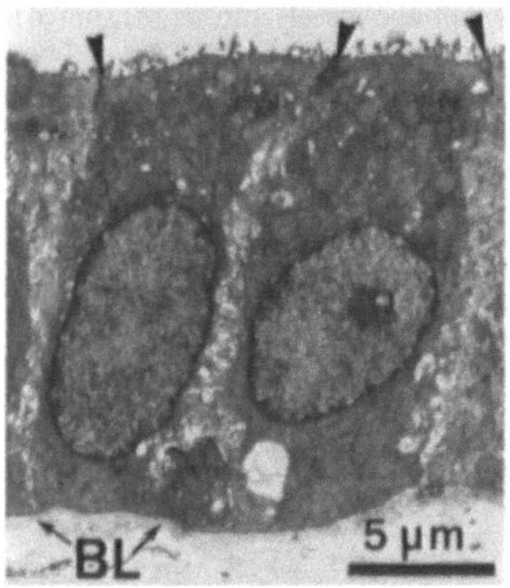

Luciano, L., Reale, E.: Cell Tissue Res. *201*, 37 (1979); Rostgaard, J., Fredericksen, O.: Cell Tissue Res. *215*, 223 (1981); Wahlin, T.: Acta Anat. *103*, 468 (1979)

Gallocyanin-chrome alum: A chrome lake of oxasine dye gallocyanin; stain forms bonds between chromium atoms and phosphate groups of DNA. G. is used for staining and ↑ cytophotometric measurements of DNA content in nuclei.

Sandritter, W., Kiefer, G., Rick, W.: Gallocyanin Chrome Alum. In: G.L. Wied (ed.): Introduction to Quantitative Cytochemistry, New York, London: Academic Press 1966

Gamete: ↑ Germ cell

Gametocyte: ↑ Germ cell

Gamma-aminobutyric acid (GABA): A possible inhibitory ↑ neurotransmitter of ↑ central nervous system, present in high concentrations in ↑ Purkinje cells and possibly in ↑ synaptic vesicles of ↑ synaptic bars and ribbons.

Itoga, E., Kito, S., Nakamura, Y., Togo, M., Kishida, T.: Acta Histochem. Cytochem. *15*, 608 (1982); Okada, Y., Roberts, E. (eds.): GABA- a Global View. Amsterdam: Excerpta Medica 1982

Gamma cells, of adenohypophysis: ↑ Hypophysis, chromophobic cells of

Gamma fibers: ↑ Fusimotor fibers; ↑ Neuromuscular spindle

Gamma globulins: Blood proteins including antibodies (↑ immunoglobulins).

↑ Plasma cells

Gamma motor neurons: Small nerve cells of CNS that innervate, via their fusimotor fibers, ↑ intrafusal muscle fibers of ↑ neuromuscular spindle.

Ganglion: A group of ↑ perikarya of ↑ neurons occurring outside and inside ↑ central nervous system. Structurally and functionally, G. can be divided into: 1) G. of peripheral ↑ nervous system = a) sensory G. with ↑ pseudounipolar or ↑ bipolar cells; b) ↑ autonomic G. (sympathetic and parasympathetic) with multipolar ↑ autonomic nerve cells. 2) G. of central nervous system = clusters of multipolar neurons surrounded by ↑ neuropil.

Ganglion cell layer, of retina (stratum ganglionare*): A single row of ↑ ganglion cells situated between outer plexiform layer and layer of optic nerve fibers. At temporal side of ↑ optic papilla there are two rows of ganglion cells, and in area of ↑ macula lutea eight to ten rows.

Ganglion cells, of adrenal medulla (neurocytus*): Multipolar sympathetic neurons (G) occurring singly or in small groups among cells of adrenal medulla. G. are incompletely surrounded by ↑ satellite cells and are identical in structure to type I ↑ autonomic neurons of ↑ autonomic ganglia. Axons of G. synapse with

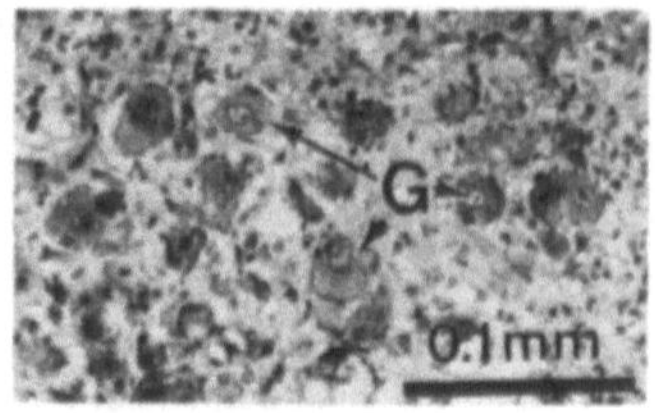

other G. and medullar cells; G. innervate blood vessels of adrenal gland. ↑ Preganglionic sympathetic nerve fibers establish synaptic contacts with G. (Fig. = human)

Ganglion cells, of retina (neurocytus ganglionaris*): Multipolar nerve cells representing third ↑ neuron of optic tract; with their ↑ perikarya, measuring 10–30 µm, G. form ↑ ganglion cell layer (GCL). G. have an eccentric globular nucleus with finely distributed chromatin and a prominent nucleolus. Abundant ↑ neuroplasm contains numerous rod-shaped mitochondria, a well-developed multiple Golgi apparatus (G), ↑ Nissl bodies (NB), ↑ lysosomes (Ly), bundles of ↑ neurofibrils (Nf), ↑ neurotubules, and an abundance of ↑ polysomes and free ↑ ribosomes. ↑ Dendrites (D) of G. make synaptic contacts with ↑ axons (A) of ↑ bipolar cells (BC) and extensions of ↑ amacrine cells (AC). Axons (Ax) of G. (= ↑ optic fibers) form nerve fiber layer (NFL) and extend to corpus geniculatum laterale; they become myelinated behind lamina cribriformis. Types of G: 1) Midget G. (MG), located in central parts of ↑ retina with a single dendrite ramifying around axon terminal of a bipolar cell; 2) diffuse G. (DG), with dendrites spreading

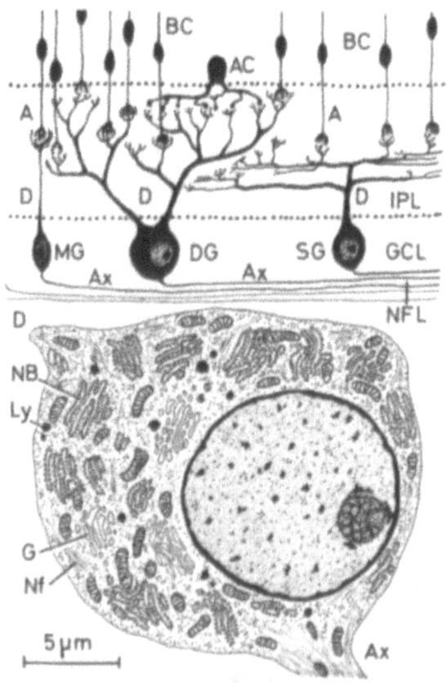

throughout entire inner plexiform layer (IPL); 3) stratified G. (SG), with dendrites branching in strata. G. are in frequent contact with ↑ astrocytes; there are about 400 000 G. in human retina.

Fine, B.S.: Arch. Ophthalmol (Chicago) *69,* 83 (1963); Hebel, R.: Z. mikrosk.-anat. Forsch. *94,* 521 (1980); Perry, V.H.: The Ganglion Cell Layer of the Mammalian Retina. In: Osborne, N.N., Chader, J. (eds.): Progress in Retinal Research, Vol. 1. Oxford: Pergamon Press 1982

Ganglion cells, of spinal ganglion (neurocytus pseudounipolaris*): A ↑ pseudounipolar nerve cell (GC) with a large spherical ↑ perikaryon intimately surrounded by ↑ satellite cells (S). Nucleus is round and displays finely dispersed ↑ chromatin and a prominent nucleolus. Cytoplasm contains numerous mitochondria, ↑ Nissl bodies (NB), a multiple Golgi apparatus (G), ↑ lysosomes, ↑ lipofuscin, ↑ neurofilaments, and ↑ neurotubules. From ↑ axon hillock (AH) extends a unique cell process of variable length that divides in the form of a T into an efferent (Eff) and an afferent (Aff) branch; afferent branch plays the role of a ↑ dendrite, and efferent branch the role of an ↑ axon. However, there is no structural difference between the two branches, since both are myelinated (My) by ↑ Schwann's cells (SC). There is no basal lamina interposed between G. and satellite cells; this is present (BL) only at their exterior aspect and delimitates them from ↑ endoneurium (E), which is rich in ↑ myelinated (My) and ↑ nonmyelinated nerve fibers and capillaries (Cap).

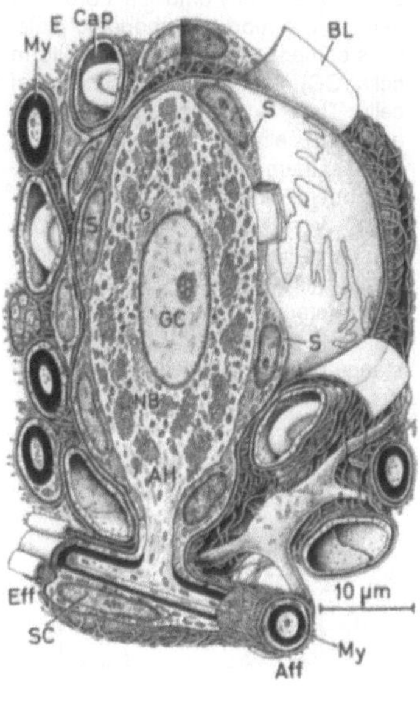

Ganglion, cervical inferior: A sympathetic ganglion with same structure as an ↑ autonomic ganglion.

Ganglion, cochlear: ↑ Spiral ganglion

Ganglion, collateral: ↑ Prevertebral ganglion

Ganglion fasciculi optici*: Total ↑ ganglion cells of ↑ retina.

Ganglion, of autonomic nervous system: ↑ Autonomic ganglion

Ganglion, of myenteric plexus (ganglion plexus myenterici*): Small groups of ↑ multipolar nerve cells (N) surrounded by an incomplete layer of ↑ satellite cells, scattered between circular (CL) and longitudinal (LL) layers of tunica muscularis of ↑ gastrointestinal tract. (Fig. = human)

↑ Intestine, innervation of; ↑ Myenteric plexus

Ganglion, of Scarpa: ↑ Vestibular ganglion

Ganglion, paravertebral: ↑ Vertebral ganglion

Ganglion, pseudounipolar cells of: ↑ Ganglion cells, of spinal ganglion

Ganglion retinae*: All ↑ bipolar cells of ↑ retina.

Ganglion, satellite cells of: ↑ Satellite cells, of peripheral neurons

Ganglion, spinal: ↑ Spinal ganglion

Ganglion, spiral: ↑ Spiral ganglion

Ganglion, terminal: ↑ Terminal ganglion

Ganglion, vertebral: ↑ Vertebral ganglion

Ganglion, vestibular: ↑ Vestibular ganglion

Ganglionic layer, of brain cortex: ↑ Brain cortex, homotypical isocortex of

Gap junction: ↑ Nexus

Gap junction protein: ↑ Connexin

Gärtner's duct: A paired vestigial canal occasionally found in ↑ mesosalpinx between ↑ ovary and ↑ oviduct, extending caudally through lateral wall of ↑ uterus, cervix, and ↑ vagina. G. corresponds to Wolff's canal. (See embryology texts for further information)

Gastric antipernicious anemia factor: ↑ Gastric intrinsic factor

Gastric areas (areae gastricae*): Convex mammilated patterns (GA) forming microrelief of tunica mucosa of ↑ stomach. ↑ Foveolae gastricae (FG) open onto surface of G.

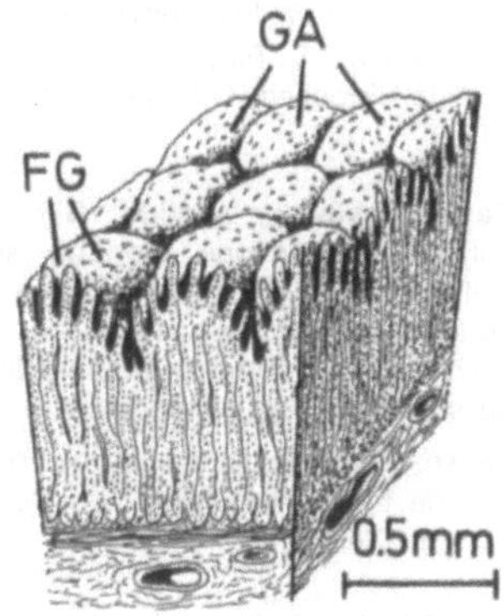

Gastric glands (glandulae gastricae*): All glands present in tunica mucosa of ↑ stomach: 1) ↑ Cardiac glands; 2) ↑ gastric glands proper; 3) ↑ intermediate glands; 4) ↑ pyloric glands.

Gastric glands proper (fundic glands, glandula gastrica propria*): Closely packed, tubular ↑ heterocrine glands, 30–50 µm in diameter, perpendicular to lamina muscularis mucosae (arrowhead). Base (F) of G. can be slightly enlarged and occasionally divided into two branches. Three to five G. open into a ↑ foveola gastrica (FG). Structure: 1) ↑ Neck mucous cells grouped in neck (N) of G.; 2) ↑ parietal cells (P) predominate in body (B); 3) ↑ chief cells (C) prevail in lower half and base of G.; 4) scattered ↑ enterochromaffin cells (E), present only in lower third and bottom of G. A basal lamina separates G. from lamina propria. There are about 15 million G.; they produce gastric juice. (Fig. = rat)

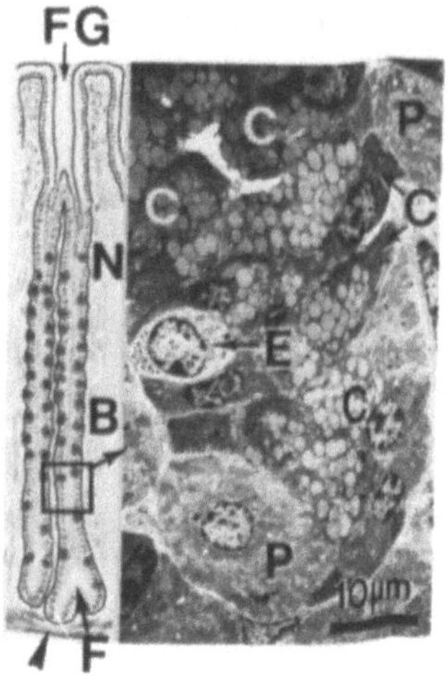

Gannon, B., Browning, J., O'Brien, P.: J. Anat. *135*, 667 (1983); Helander, H.F.: Int. Rev. Cytol. *70*, 217 (1981); Suganuma, T., Suzuki, S., Tsuyama, S., Murata, F.: Acta Histochem. Cytochem. *14*, 534 (1981)

Gastric inhibitory peptide (GIP): A polypeptide hormone probably secreted by ↑ K-cells in tunica mucosa of ↑ duodenum, ↑ jejunum, and ↑ ileum. G. inhibits gastric secretion and motility, but stimulates ↑ insulin secretion.

Brown, J.C.: Gastric Inhibitory Polypeptide. Berlin, Heidelberg, New York: Springer-Verlag 1982

Gastric intrinsic factor (gastric antipernicious anemia factor): A ↑ glycoprotein (m.w. over 53000 daltons) synthesized and released by ↑ parietal cells of ↑ gastric glands proper. In lumen of ↑ stomach, G. binds with vitamin B_{12} and facilitates the latter's absorption by ↑ micropinocytosis in small intestine (mainly ↑ ileum). Failure to produce G. leads to impairment of ↑ erythroblast maturation, consequently to pernicious anemia.

Levine, J.S., Nakane, P.K., Allen, R.H.: J Cell Biol. *90*, 644 (1981)

Gastric pits: ↑ Foveolae gastricae

Gastrin: A polypeptide hormone synthesized by ↑ G-cells of ↑ pyloric and duodenal mucosa. G. stimulates HCl production in ↑ parietal cells of ↑ gastric glands proper.

Gastrin cells: ↑ G-cells

Gastrointestinal hormones: ↑ Tissue hormones produced by ↑ endocrine cells of ↑ gastrointestinal tract (↑ cholecystokinin-pancreozymin, ↑ gastrin, ↑ gastric inhibitory peptide, ↑ glucagon, ↑ somatostatin, etc.).

Gastrointestinal tract (digestive tract, apparatus digestorius*): A continuous series of tubular organs beginning with mouth and ending with anus. General structure: 1) Tunica mucosa (TM) = a) epithelium (Ep); b) lamina propria (LP); c) lamina muscularis mucosae (LMM). 2) Tela submucosa (TS). 3) Tunica muscularis (TMu) = a) inner circular layer (ic); b) outer longitudinal layer (ol). 4) Tela subserosa (TSs). 5) Tunica serosa (TSe) and, in some segments, tunica adventitia (TA). Specialized glands (SG) are associated with G. E = ↑ esophagus, S = ↑ stomach, D = ↑ duodenum, J = ↑ jejunum, I = ↑ ileum, C = ↑ colon, M = ↑ mesentery

↑ Intestine, innervation of

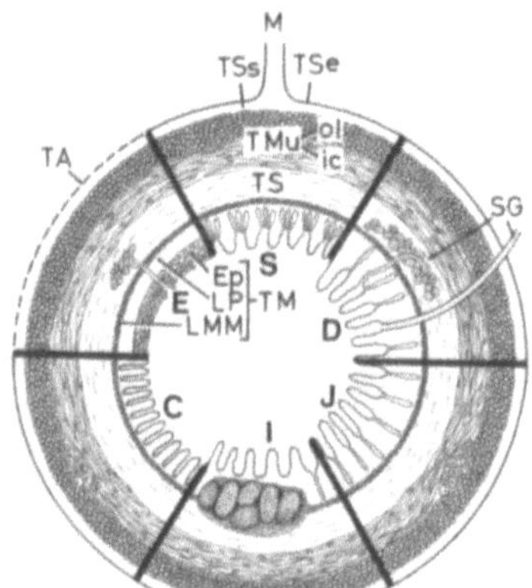

Gabella, G.: Int. Rev. Cytol. *59*, 129 (1976); Toner, P.G., Carr, K.E.: The Digestive System. In: Hodges, G.M. and Hallowes, R.C. (eds.): Biomedical Research Applications of Scanning Electron Microscopy. Vol. 1. London, New York: Academic Press 1979

Gastrointestinal tract, endocrine cells of: ↑ Endocrine cells, of gastrointestinal tract

G-cells (gastrin cells): Pyramidal endocrine cells scattered in epithelium of ↑ pyloric antrum, pylorus, and ↑ duodenum. G. have an oval or elliptical nucleus with a prominent nucleolus. Cytoplasm contains small mitochondria, a well-developed Golgi apparatus, short cisternae of rough endoplasmic reticulum, a few lysosomes, and a considerable number of free ribosomes. Almost the entire cytoplasm is occupied by ↑ unit membrane-bound ↑ secretory granules (SG), about 0.3 μm, with fine granular moderately osmio-

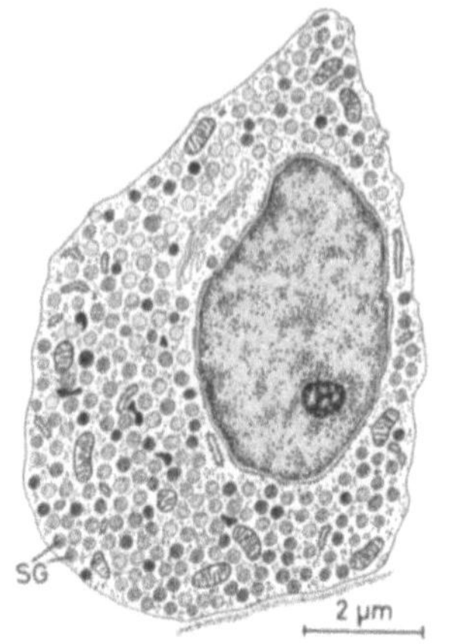

philic contents. Immunocytochemical studies have shown that G. synthesize and release ↑ gastrin. G. are classified as ↑ APUD-cells.

↑ Endocrine cells, of gastrointestinal tract

Fujita, T.: Gastro-Entero-Pancreatic Endocrine System. Stuttgart: Thieme 1974; Garaud, J.-C., Doffoel, M., Stock, C., Grenier, J.F.: Biol. Cell *44*, 165 (1982)

Gelatin: A denaturated protein formed from ↑ collagen by boiling in water or by chemical treatment.

Gelatinous bone marrow: ↑ Bone marrow, gelatinous

Gelatinous tissue (mucous connective tissue, Wharton's jelly, textus mucoideus connexens*): An embryonal ↑ connective tissue located around blood vessels of ↑ umbilical cord, and near its insertion to ↑ chorionic plate. G. is composed of stellate or fusiform cells (GC) resembling ↑ mesenchymal cells. Their extensions often contact adjacent cells. In large intercellular spaces, there are delicate collagen fibrils but no blood and lymphatic capillaries or nerve fibers. Ground substance is very rich in polymerized ↑ hyaluronic acid, which attracts water and confers a certain rigidity to G. In histological sections, ground substance is precipitated and appears fi-

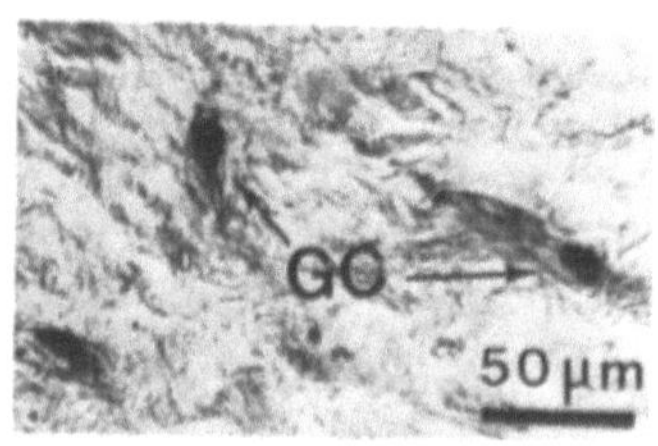

brillar. Although morphologically very similar to ↑ mesenchymal tissue, G. is a definitive product incapable of further ↑ differentiation. (Fig. = human)

↑ Connective tissue proper, ground substance of

Gemmules: ↑ Dendritic spines

Gene: A functional unit of heredity, carrier of a hereditary characteristic. G. is made up of a specific base sequence along a molecule of ↑ deoxyribonucleic acid. Three types of G.: 1) Structural G. = a long nucleotide sequence coding ↑ messenger RNA for synthesis of proteins carried by ↑ euchromatin; 2) G. coding ↑ transfer RNA = a short nucleotide sequence carried by ↑ heterochromatin; 3) G. coding ↑ ribosomal RNA = a short nucleotide sequence carried by ↑ heterochromatin. (See genetics texts for further information)

↑ Cistron; ↑ Ribonucleic acids, synthesis of

Mange, A., Mange, E.: Genetics: Human Aspects. Philadelphia: Saunders College 1980; Rieger, R., Michaelis, A., Green, M.M.: Glossary of Genetics and Cytogenetics. 4th edn. Berlin, Heidelberg, New York: Springer-Verlag 1983; Russel, P.J.: Genetics. Berlin, Heidelberg, New York: Springer Verlag 1983; Sutton, E.H.: An Introduction to Human Genetics. 3rd edn. Philadelphia: Saunders College 1980; Watson, D.D.: Molecular Biology of the Gene. 3rd edn. New York: Benjamin 1976

Generation time: The time needed for a cell to complete a full ↑ cell cycle. In rapidly dividing cells, G. is about 12 h; in human gastric epithelium, G. is 1–2 days.

Generative complex, of cilia: ↑ Deuterosome

Genital corpuscle: ↑ Corpuscle, genital

Genitalia, female/male: All reproductive organs of male and female.

Genitourinary system: A collective term for both reproductive and excretory organs.

Gennari's line: ↑ Striate area

Genome: All ↑ genes of an organism contained in a ↑ haploid set of ↑ chromosomes.

Genophore: ↑ Mitochondrial nucleus

Genotype: Total genetic information contained in ↑ chromosomes of an individual, i.e., its genetic contitution. (See genetics texts for further information)

GERL: A very complex spatial association of innermost cisterna of ↑ maturing face of Golgi apparatus (G) with a hydrolase-rich cisterna (GERL) of smooth ↑ **e**ndoplasmic **r**eticulum giving rise to ↑ **l**ysosomes (Ly) and ↑ coated vesicles (CV). GERL cisterna originates from rough ↑ endoplasmic reticulum (rER) and communicates with it through tubules (T). From GERL emerge digitiform processes (P) that penetrate fenestrae of Golgi cisternae. Coated vesicles and lysosomes bulge and pinch off from edges of GERL, which plays a very important role in ↑ secretion. TV = ↑ transport vesicles, TR = transitional ↑ endoplasmic reticulum.

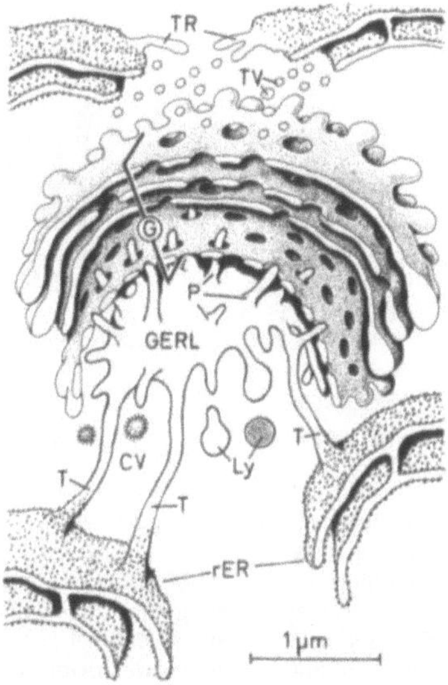

Broadwell, R.D., Cataldo, A.M.: J. Histochem. Cytochem. *31*, 1077 (1983); Fujita, H., Sawano, F.: J. Histochem. Cytochem. *31*, 227 (1983); Noivikoff, P.M., Yam, A.: J. Histochem. Cytochem. *26*, 1 (1978); Olivier, C., Hand, A.R.: J. Histochem. Cytochem. *31*, 1041 (1983)

Gerlach's tonsil: ↑ Tonsil, tubar

Germ cell (gamete, gametocyte, gametus*): A mature ↑ haploid reproductive cell capable of uniting with corresponding cell of another individual of opposite sex to form a ↑ zygote. Ovulated secondary ↑ oocyte is female G.; ↑ capacitated ↑ spermatozoon is male G. G. contains 3.4×10^{-12} g ↑ deoxyribonucleic acid per nucleus.

↑ Cells, types of; ↑ Oogenesis; ↑ Spermatogenesis

Austin, C.R., Short, R.V. (eds.): Reproduction in Mammals. Book 1. Germ Cells and Fertilization. 2nd edn. Cambridge: Cambridge University Press 1982; Kuwana, T., Fujimoto, T.: Anat. Rec. *205*, 21 (1983)

Germinal center (centrum germinale*): A central pale area (GC) of a secondary ↑ lymphatic nodule or a ↑ splenic nodule (fig.). G. are production sites for ↑ B-lymphocytes that move from G. toward ↑ mantle (M) or toward overlying epithelium. Also present in G., besides B-immunoblasts, are ↑ plasma cells, ↑ macrophages, and some ↑ T-lymphocytes.

↑ Immunoblast

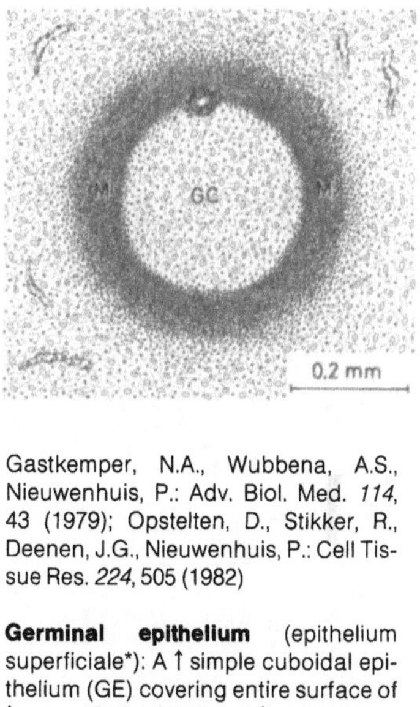

Gastkemper, N.A., Wubbena, A.S., Nieuwenhuis, P.: Adv. Biol. Med. *114*, 43 (1979); Opstelten, D., Stikker, R., Deenen, J.G., Nieuwenhuis, P.: Cell Tissue Res. *224*, 505 (1982)

Germinal epithelium (epithelium superficiale*): A ↑ simple cuboidal epithelium (GE) covering entire surface of ↑ ovary. Cells of G. lie on ↑ tunica albuginea (TA) and bear many microvillous processes at their ↑ apical poles (arrows). The term G. is inaccurate since G. has nothing to do with origin of primordial ↑ oocytes; the term "superficial

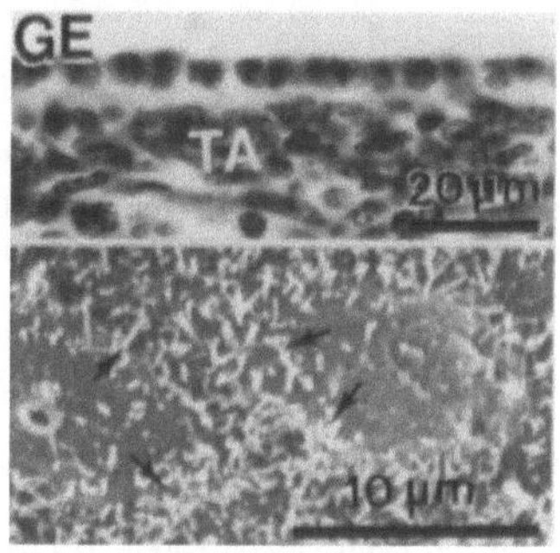

epithelium" is more appropriate. (Figs. = rat)

Motta, P.M., Makabe, S.: Biomed. Res. *2*, 325 (1981)

GFAP: ↑ Glial fibrillary acidic protein

GH: ↑ Growth hormone

GH-cells, of adenohypophysis: ↑ Somatotropes

Ghost erythrocyte: ↑ Erythrocyte ghost

Giant cells: Very large multinucleated cells found around foreign objects resistant to intracellular digestion or around foci of certain bacterial infections (tuberculosis, leprosy). G. are in fact fused ↑ macrophages.

Giant cells, of trophoblast (cellular gigantica trophoblastica*): Very large multinucleated cells (GC) scattered among ↑ decidual cells (DC) of ↑ decidua basalis. G. are considered remainders of outer blastocyst wall which have a role of destroying ↑ endometrium during ↑ implantation. Some G. are also separated ↑ trophoblastic islands that have invaded decidua basalis of early ↑ placenta (fig.). (Fig. = human)

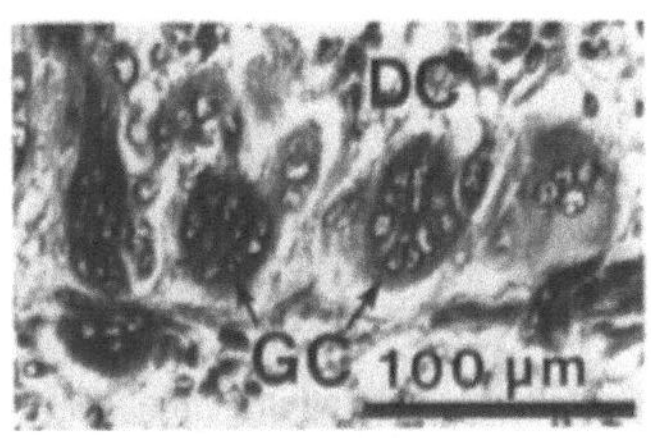

Giant chromosome: ↑ Chromosome, giant

Gianuzzi's demilunes: ↑ Demilunes

Giemsa staining: A technique for staining ↑ blood and ↑ bone marrow.

GIF: ↑ Somatostatin

Gigantism: Giant growth provoked by pathological hyperfunction of ↑ somatotropes before epiphyseal closure, inducing continued growth in length of bones

↑ Acromegaly; ↑ Epiphyseal plate; ↑ Growth hormone

Gingiva* (gum): A tunica mucosa (G) covering ↑ alveolar bone (AB) and surrounding teeth (T). Structure: 1) Epithelium (E) = a slightly keratinized ↑ stratified squamous epithelium with deep ↑ epithelial ridges. Epithelium of small furrow around teeth or ↑ gingival crevice (GC) is devoid of ridges and keratinized layer. 2) Lamina propria (LP) = a poorly vascularized ↑ dense connective tissue firmly attached to ↑ periosteum (P) of alveolar bone. Long ↑ connective papillae interdigitating with epithelial ridges consist of loose connective tissue; this interdigitation and attachment of lamina propria to periosteum assures complete immobility of G. on bone. In G. are also present ↑ elastic fibers and numerous ↑ corpuscles of Krause. C = ↑ cementum, D = ↑ dentin, En = ↑ enamel, Pe = ↑ periodontal ligament

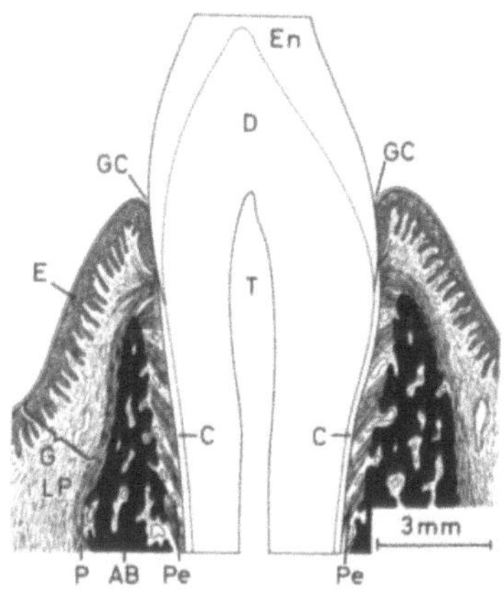

Listgarten, M.A.: Am. J. Anat. *114*, 49 (1964)

Gingival crevice (sulcus gingivalis*): A small furrow (GC) of ↑ gingiva surrounding ↑ tooth. ↑ Stratified squamous nonkeratinized epithelium (E) of G. adheres intimately to enamel surface (En) and becomes lower toward ↑ alveolar bone (AB); epithelium becomes bistratified at bottom of G. There are no ↑ connective papillae in G., only occasional agglomerations of ↑ lymphocytes (AL) and ↑ plasma cells can occur. C = ↑ cementum, D = ↑ dentin

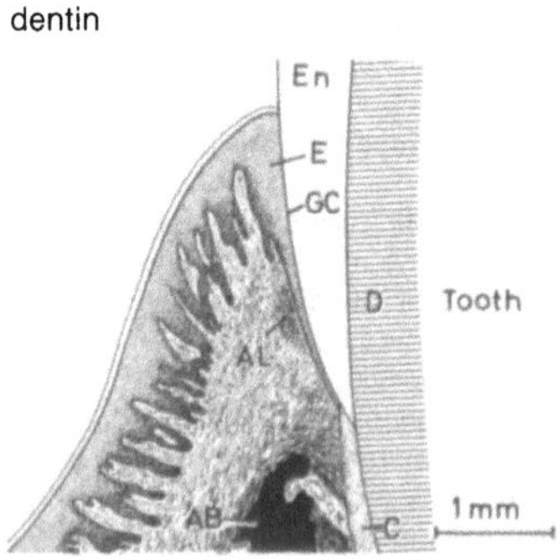

Gingival epithelium: A ↑ stratified squamous epithelium (E) with a thin keratinized layer (K) and deep ↑ epithelial ridges (R) interdigitating with ↑ connective papillae (P) of lamina propria (LP). ↑ Keratinization and papillae are absent in ↑ gingival crevice. (Fig. = human)

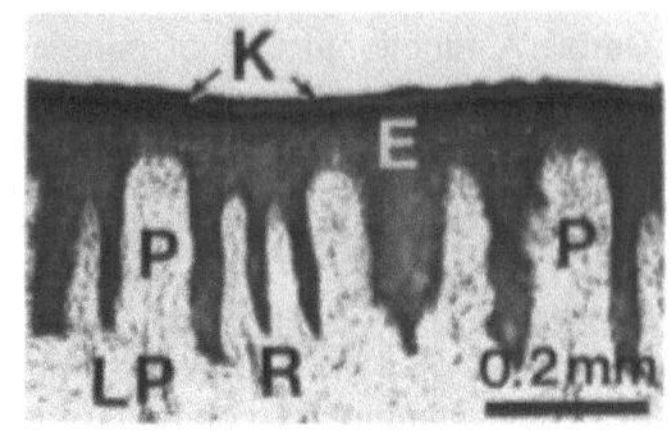

GIP: ↑ Gastric inhibitory peptide

"Gitterzellen": ↑ Compound granular corpuscles

Glabrous skin: A hairless ↑ skin like that of palms, soles, ↑ glans penis, ↑ labia minora.

Gland (glandula*): A secreting organ composed of ↑ glandular cells.

↑ Glands, classification of

Gland, interstitial, of ovary: ↑ Interstitial gland, of ovary

Gland, mammary: ↑ Mammary gland

Gland, of Blandin: ↑ Anterior lingual gland

Gland, thyroid: ↑ Thyroid gland

Gland, trabecular: ↑ Trabecular gland

Glands, accessory, of male reproductive tract: ↑ Accessory glands, of male reproductive tract

Glands, acinar: ↑ Glands, classification of; ↑ Tubuloacinar glands

Glands, adrenal: ↑ Adrenal glands

Glands, alveolar: ↑ Alveolar glands; ↑ Glands, classification of

Glands, amphicrine: ↑ Amphicrine glands, ↑ Glands, classification of

Glands, apocrine: ↑ Apocrine glands; Glands, ↑ classification of

Glands, areolar: ↑ Areolar glands

Glands, bronchial: ↑ Bronchial glands

Glands, buccal: ↑ Buccal glands

Glands, bulbourethral: ↑ Bulbourethral glands

Glands, cardiac, of esophagus: ↑ Cardiac glands, of esophagus

Glands, cardiac, of stomach: ↑ Cardiac glands, of stomach

Glands, ceruminous: ↑ Ceruminous glands

Glands, classification of: Glands are classified according to following criteria: 1) Presence of excretory ducts = a) ↑ exocrine glands; b) ↑ endocrine glands; c) ↑ amphicrine glands. 2) Relationship to surface epithelium = a) ↑ endoepithelial glands; b) ↑ exoepithelial glands. 3) Number of layers of glandular cells = a) ↑ unistratified glands; b) ↑ pluristratified glands. 4) Shape of secretory end pieces = a) ↑ tubular glands; b) ↑ tubuloacinar glands; c) ↑ tubuloalveolar glands; d) ↑ alveolar glands. 5) Branching of excretory ducts = a) ↑ simple glands; b) ↑ compound glands. 6) Mode of secretion = a) ↑ apocrine glands; b) ↑ eccrine glands; c) ↑ holocrine glands. 7) Composition of secretory product = a) ↑ homocrine glands; b) ↑ heterocrine glands. 8) Viscosity of secretory product = a) ↑ serous glands; b) ↑ mucous glands; c) ↑ mixed glands. The present schema intends to summarize G. only partially. Asterisks indicate letters under which the particular glands are described.

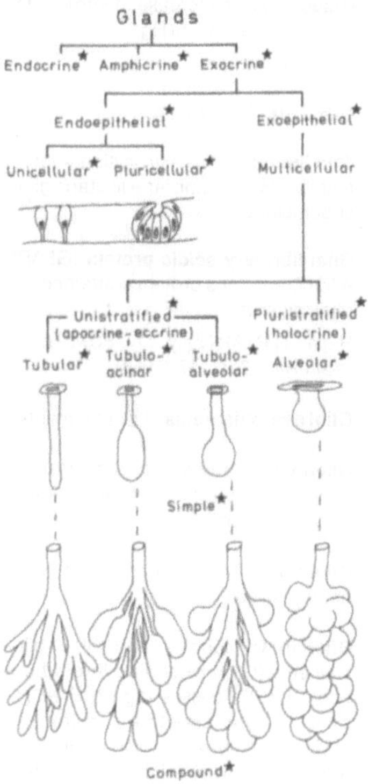

Glands, compound: ↑ Compound glands; ↑ Glands, classification of

Glands, duodenal: ↑ Duodenal glands

Glands, eccrine: ↑ Eccrine glands; ↑ Glands, classification of

Glands, endocrine: ↑ Endocrine glands; ↑ Glands, classification of

Glands, endoepithelial: ↑ Endoepithelial glands; ↑ Glands, classification of

Glands, epiglottic: ↑ Epiglottic glands

Glands, esophageal: ↑ Esophageal glands proper

Glands, exocrine: ↑ Exocrine glands; ↑ Glands, classification of

Glands, exoepithelial: ↑ Exoepithelial glands; ↑ Glands, classification of

Glands, formation of: In primitive surface epithelium (SE) groups of secretory cells (SC) differentiate and proliferate into underlying connective tissue. Cell group destined to become an exocrine gland (EG), however, remains in contact with free surface by a future excretory duct (ED), other groups separate from surface epithelium and transform into ↑ endocrine glands (EnG), with or without follicles (F). Arrows indicate direction of secretion. Cap = capillaries

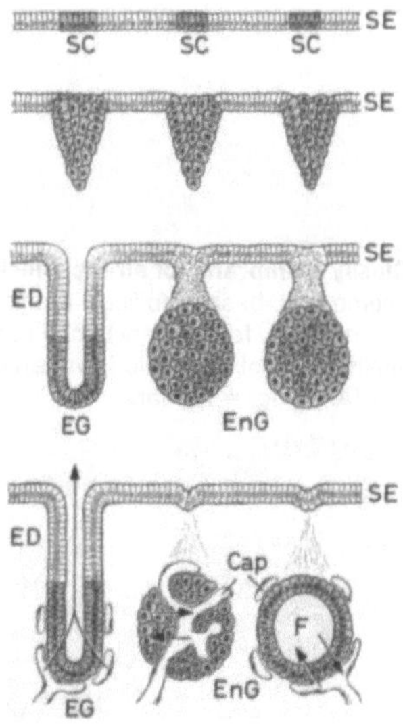

Glands, fundic, of stomach: ↑ Gastric glands proper

Glands, gastric: ↑ Gastric glands; ↑ Gastric glands proper

Glands, glossopalatine: ↑ Glossopalatine glands

Glands, heterocrine: ↑ Heterocrine glands; ↑ Glands, classification of

Glands, holocrine: ↑ Holocrine glands; ↑ Glands, classification of

Glands, homocrine: ↑ Homocrine glands; ↑ Glands, classification of

Glands, intermediate, of stomach: ↑ Intermediate glands, of stomach

Glands, intraepithelial: ↑ Endoepithelial glands; ↑ Glands, classification of

Glands, labial: ↑ Labial glands

Glands, laryngeal: ↑ Laryngeal glands

Glands, Meibomian: ↑ Tarsal glands

Glands, mixed: ↑ Mixed glands; ↑ Glands, classification of

Glands, mucous: ↑ Mucous glands; ↑ Glands, classification of

Glands, multicellular: ↑ Pluricellular glands; ↑ Glands, classification of

Glands, of Bartholin: ↑ Vestibular glands

Glands, of Bowman: ↑ Olfactory glands

Glands, of Brunner: ↑ Duodenal glands

Glands, of Ebner: ↑ Ebner's glands; ↑ Circumvallate papillae

Glands, of Lieberkühn: ↑ Lieberkühn's crypts

Glands, of Littré: ↑ Urethral glands, of male urethra

Glands, of Moll: ↑ Ciliary glands

Glands, of Montgomery: ↑ Areolar glands

Glands, of Nuhn: ↑ Anterior lingual glands

Glands, of Skene: ↑ Urethral glands, of female urethra

Glands, of Zeis (glandulae sebaceae*): Modified ↑ sebaceous glands opening into ↑ hair follicles of ↑ eyelashes. Product of G. lubricates border of ↑ eyelids, preventing flow of ↑ tears outside border.

Glands, olfactory: ↑ Olfactory glands

Glands, palatine: ↑ Palatine glands

Glands, pluricellular: ↑ Pluricellular glands; ↑ Glands, classification of

Glands, pluristratified: ↑ Pluristratified glands; ↑ Glands, classification of

Glands, posterior, of tongue: ↑ Posterior glands, of tongue

Glands, preputial: ↑ Preputial glands

Glands, pyloric: ↑ Pyloric glands

Glands, saccular: ↑ Glands, alveolar

Glands, salivary, major: ↑ Salivary glands, major

Glands, salivary, minor: ↑ Salivary glands, minor

Glands, serous: ↑ Serous glands; ↑ Glands, classification of

Glands, simple: ↑ Simple glands; ↑ Glands, classification of

Glands, sublingual: ↑ Sublingual glands

Glands, submandibular: ↑ Submandibular glands

Glands, sweat: ↑ Sweat glands

Glands, tracheal: ↑ Tracheal glands

Glands, tubular: ↑ Tubular glands; ↑ Glands, classification of

Glands, tubuloacinar: ↑ Tubuloacinar glands; ↑ Glands, classification of

Glands, tubuloalveolar: ↑ Tubuloalveolar glands; ↑ Glands, classification of

Glands, unicellular: ↑ Unicellular glands; ↑ Glands, classification of

Glands, urethral: ↑ Urethral glands

Glandulae vestibulares nasi: Coiled apocrine ↑ sweat glands that empty into ↑ hair follicle of ↑ vibrissae.

Glandular cells (cellula secretoria*): Cells specialized in secreting substances of definite chemical composition and physiological significance.

↑ Epithelium, classification of; ↑ Secretion

Glans clitoridis: ↑ Clitoris

Glans penis*: Conical enlargment of ↑ corpus spongiosum at end of ↑ penis. Posterior surface of G. is concave and covers ends of ↑ corpora cavernosa like a cap. In fetus and newborn (fig.), G. is covered with a ↑ stratified squamous nonkeratinized epithelium (E) fused with that of ↑ prepuce (P) until age of 1–2 years. After separation of prepuce, G. is covered by a very thin, hairless ↑ skin, firmly attached to underlying tissue. Near corona of G., there are large ↑ sebaceous glands producing part of ↑ smegma. Corpus spongiosum of G. consists of dense connective tissue containing large irregularly shaped anastomosing veins (V) with ↑ intimal cushions. ↑ Fossa navicularis (FN) is situated in center of G. On inferior surface of G., a connective strand, frenulum (F), fixes prepuce to G. G. is very rich in free nerve endings and genital ↑ corpuscles.

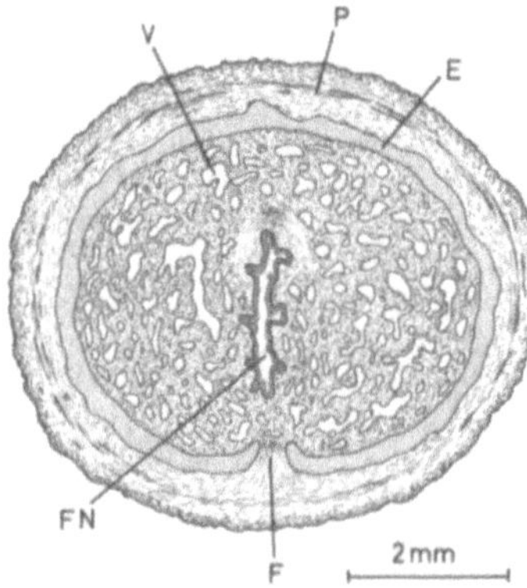

Glassy membrane, of atretic follicles (membrana basalis folliculi atretici*): An irregularly folded, very thick ↑ basal lamina (GM) of an atretic ↑ ovarian follicle (AF). (Fig. = human)

Glassy membrane, of eye: ↑ Bruch's membrane

Glassy membrane, of hair follicle: A thick ↑ basement membrane (G) separating outer root sheath (RS) from fibrous tissue sheath (F). (Fig. = mongolian gerbil)

↑ Hair follicle

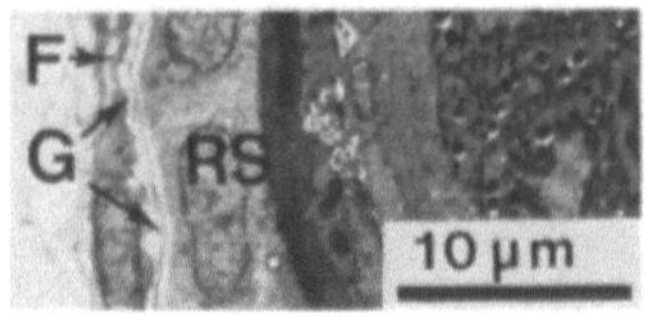

Glassy membrane, of ovarian follicles (membrana basalis folliculi*): A ↑ basal lamina (arrowheads) separating ↑ follicular cells (F) from ↑ theca interna cells (T). (Fig. = rabbit)

↑ Glassy membrane, of atretic follicle; ↑ Ovarian follicles

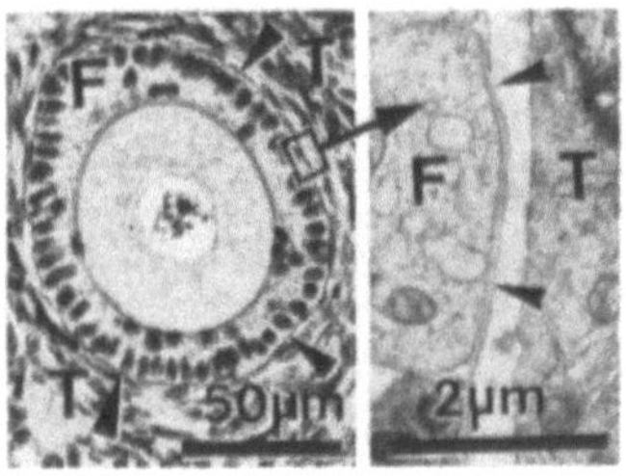

Glaucoma: An increase in intraocular pressure over 25 mm Hg.

↑ Aqueous humor

Glia cells: ↑ Neuroglia

Glia limitans: The astrocytic ↑ subpial feet forming ↑ membrane limitans gliae superficialis.

Glial fibrillary acidic protein (GFAP): A 50000-daltons protein contained in ↑ gliofilaments.

Higley, H.R., McNulty, J.A., Rowden, G.: Brain Res. *304*, 117 (1984)

Glial microfilaments: ↑ Gliofilaments

Glioarchitectonics: A spatial arrangement of ↑ neuroglial cells in ↑ central nervous system.

Glioblast: The precursor of a ↑ neuroglial cell.

Gliofibrils: Bundles (Gf) of ↑ gliofilaments (Gfm) extending throughout body and processes (fig.) of ↑ astrocytes. G. form skeleton of astrocytes and give rigidity to their extensions. Amount of G. is greater in fibrous than in protoplasmic astrocytes. (Fig. = mongolian gerbil)

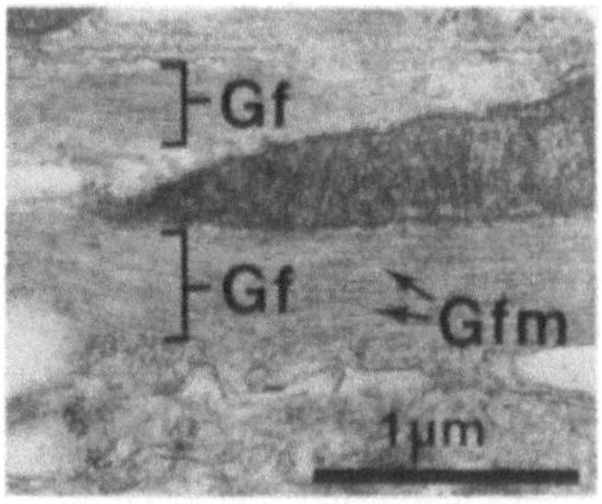

Gliofilaments (glial microfilaments, gliofilamentum*): ↑ Microfilaments (Gfm) present in cytoplasm, processes, and end feet (fig.) of ↑ astrocytes. Diameter of G. averages 5–10 nm; their length is difficult to determine (probably a few micrometers). G. are predominantly composed of ↑ glial fibrillary acidic protein. Bundles of G. constitute ↑ gliofibrils. (Fig. = mongolian gerbil)

Goldman, J.E., Schaumburg, H.H., Norton, W.T.: J. Cell Biol. *78*, 426 (1978)

Gliosis: ↑ Astrocytes

Gliosomes: A rarely employed term for lysosomes of ↑ astrocytes. Some authors designate mitochondria of astrocytes as G.

Glisson's capsule: ↑ Capsule, of liver

Glomerular axis: ↑ Mesangium

Glomerular basal lamina (glomerular basement membrane, membrana basalis*): A continuous (approx. 250–450 nm thick) three-layered feltwork of fine filamentous ↑ ground substance inserted between ↑ pedicels (Pe) of ↑ podocytes and endothelial cells (E) of ↑ glomerular capillaries. Beginning at endothelial cells, G. is composed of lamina rara interna (LRI), lamina densa (LD), and lamina rara externa (LRE). G. consists of nonaggregated macromolecules of type IV collagen, approx. 1.5–2 nm thick, embedded in a ↑ proteoglycan matrix rich in ↑ sialic acid, therefore ↑ PAS-positive. Average distance between filaments at level of lamina densa is 2–3 nm; it becomes greater toward ↑ pedicels. G. comprises part of ↑ glomerular filtration membrane. G. is continuously renewed by endothelial cells and removed by podocytes and ↑ mesangial cells. (Fig. = rat)

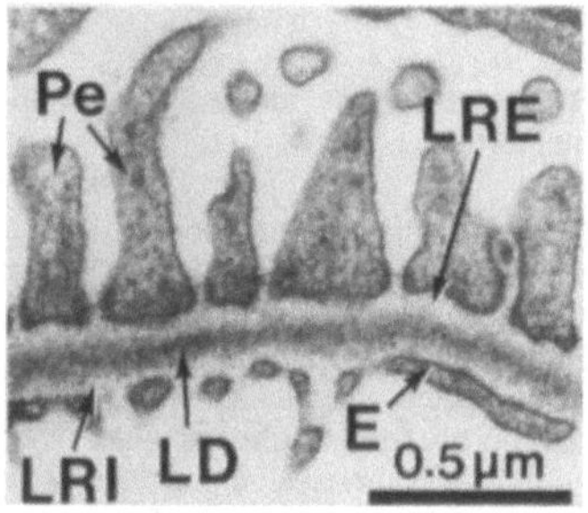

Carlsson, E.C., Meezan, E., Brendel, K., Kenney, M.C.: Anat. Rec. *200*, 421 (1981); Huang, T.W.: Renal Physiol. *3*, 312 (1980); Lubec, G. (ed.): Glomerular Basement Membrane. Parts 1 and 2. Basle: Karger 1980 and 1981; Martinez-Hernandez, A., Amenta, P.S.: Lab. Invest. *48*, 657 (1983); Ota, Z., Makino, H., Takaya, Y., Ofuji, T.: Renal Physiol. *3*, 317 (1980); Reale, E., Luciano, L., Kühn, K.-W.: J. Histochem. Cytochem. *31*, 662 (1983)

Glomerular basement membrane: ↑ Glomerular basal lamina

Glomerular capillaries (rete capillare glomerulare*): Terminal branches of ↑ afferent arteriole forming ↑ glomerulus. G. are of fenestrated type, with larger pores (arrowheads) than in other capillaries; nuclei of ↑ endothelial cells (E) are rich in ↑ heterochromatin, may bulge into lumen, and are generally oriented toward ↑ mesangium. Internal structure of endothelial cells is similar to that of other endothelial cells. A thick ↑ glomerular basal lamina (BL) surrounds endothelial cells; outer aspect of basal lamina is covered by pedicels (Pe) of ↑ podocytes (P). All G. reunite to form ↑ efferent arteriole. Total surface area of all G. ist about 1.5 m².

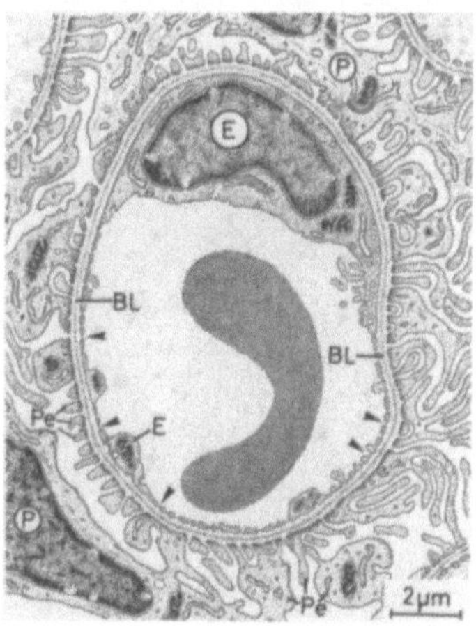

Farquhar, M.G.: Structure and Function in Glomerular Capillaries. In: Kefalides, N. (ed.): Biology and Chemistry of Basement Membrane. New York: Academic Press 1978

Glomerular capillary wall: ↑ Glomerular filtration membrane

Glomerular filtrate: An ultrafiltrate of blood plasma formed by ↑ glomerular filtration membrane under ↑ filtration pressure. Besides water, G. contains phosphates, creatinine, uric acid, ↑ urea, and small traces of albumin; thus, G. is very similar to ↑ interstitial fluid. Almost all G. is reabsorbed in various segments of ↑ nephron.

Glomerular filtration barrier: ↑ Glomerular filtration membrane

Glomerular filtration membrane (glomerular filtration barrier, blood-urine barrier): A three-layered wall separating blood capillaries (Cap) from ↑ Bowman's space (BS). Structure: 1) Fenestrated endothelium (E) of ↑ glomerular capillaries; 2) ↑ glomerular basal lamina (BL) with lamina rara ex-

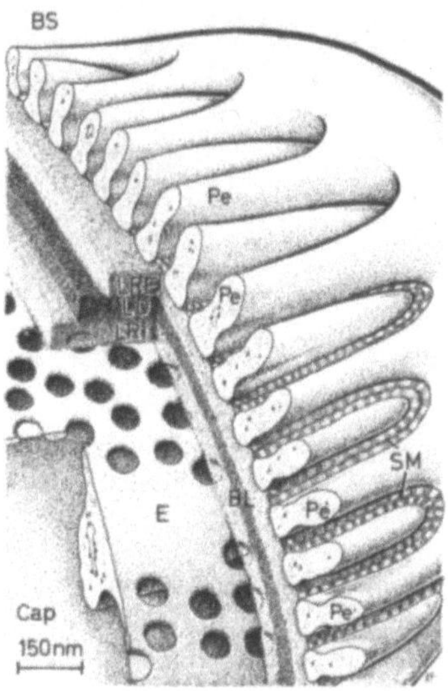

terna (LRE), lamina densa (LD), and lamina rara interna (LRI); 3) ↑ pedicels (Pe) of ↑ podocytes with a ↑ glomerular slit membrane (SM) stretched between them. Due to ↑ filtration pressure, ↑ glomerular filtrate is forced through G. Small molecules, phosphates, urea, etc. pass through G. Protein molecules up to 40000 daltons (albumins) also pass through this membrane; larger protein molecules are held back by glomerular basal lamina.

Glomerular lobule: ↑ Renal glomerulus

Glomerular slit membrane (filtration slit, slit membrane): A highly ordered diaphragm (SM), about 5 nm thick, stretched across slit between podocyte ↑ pedicels (Pe). G. appears as a zipperlike structure with alternating cross bridges (B) extending from ↑ podocyte plasma membrane to a central filament (F) running parallel and equidistant from pedicels. Total area of G. accounts for approximately 2%–3% of total ↑ glomerular filtration membrane area. It is believed that G. represents principal filtration barrier to plasma macromolecules. G. is visible only after special fixation. BL = ↑ glomerular basal lamina. (Inset after Ryan et al. 1974)

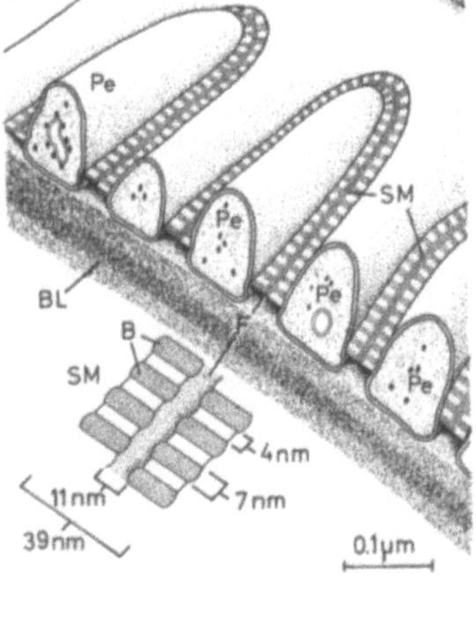

Karnowsky, M.J., Ryan, G.B.: J. Cell Biol. *65*, 233 (1975); Ryan, G.B., Rodewald, R., Karnowsky, M.J.: Lab. Invest. *33*, 461 (1975)

Glomerular stalk: ↑ Mesangium

Glomeruli, cerebellar: ↑ Cerebellar glomeruli

Glomeruli, of central nervous system: Aggregation of synaptic contacts in some areas of CNS, ensheathed almost completely by flattened astrocytic processes.

↑ Cerebellar glomeruli; ↑ Olfactory glomeruli; ↑ Vellate astrocytes

Glomeruli, olfactory: ↑ Olfactory glomeruli

Glomerulus, of kidney: ↑ Renal glomerulus

Glomi (digital arteriovenous anastomosis, glomus organ, Hoyer-Grosser's organ, anastomosis arteriovenosa glomeriformis*): Small skin organs in the form of coiled groups of highly organized arteriovenous ↑ anastomoses surrounded by a fine connective capsule (C). G. may be 2.5 mm in diameter and are found in ↑ nail bed, pads of fingers (fig.) and toes, ↑ auricle, hands, feet, and some other organs of body. Afferent arteriole (A) enters capsule, loses its internal elastic lamina, so that ↑ endothelium, rich in ↑ specific endothelial organelles, lies directly on a muscular layer. Arterial lumen is small because of a relatively thick tunica media in which ↑ smooth muscle cells resemble epithelial cells [= ↑ epithelioid cells (EC)]. There are two types of these cells: Type I cells, more frequent, are rich in mitochondria and contain few myofilaments, whereas type II cells appear dense because of an abundance of myofilaments. Adventitia contains a dense network of both ↑ myelinated and ↑ nonmyelinated nerve fibers. Blood flows from G. into short thin-walled veins that drain into periglomic veins (P). It finally reaches ordinary skin veins. G. as specialized bypasses contribute to regulation of blood flow, pressure, temperature, and conservation of heat. (Fig. = human)

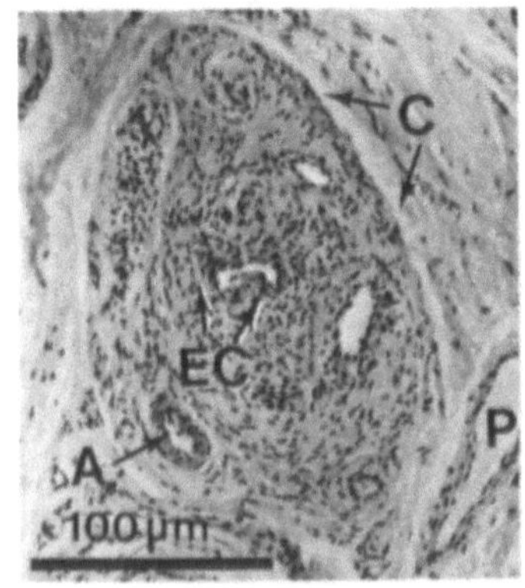

Gorgas, K., Böck, P., Tischendorf, F., Curri, S.: Anat. Embryol. *150*, 269 (1977)

Glomus aorticum: ↑ Aortic body

Glomus cells, of carotid body: Two types of cells constituting ↑ carotid body. 1) Glomus cells, type I (chief cell, granulocytus glomeris principalis*) = polygonal cells with a light, round nucleus, abundant mitochondria and free ribosomes, a well-developed ↑ ergastoplasm, and a ↑ Golgi apparatus. Cytoplasm is filled with numerous small ↑ unit membrane-bound, dense core granules (G) containing ↑ norepinephrine, ↑ dopamine, and ↑ serotonin. Efferent nerve fibers (E), rich in mitochondria and ↑ synaptic vesicles, enter into contact with type I cells. Upon arrival of nerve impulses, cells may discharge ↑ catecholamines via ↑ pericapillary spaces (PS) into capillaries (Cap). Here they serve as secondary ↑ neurotransmitters of efferent

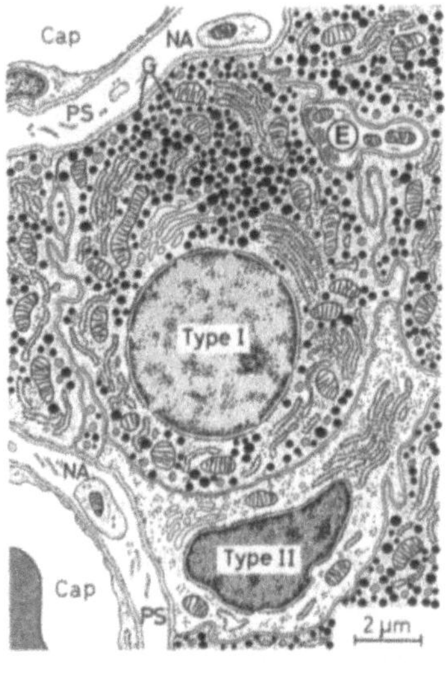

system. Type I cells are not ↑ chemoreceptors, but rather links of an efferent chain which controls sensitivity of sensor [probably naked ↑ axons (NA) situated within pericapillary spaces]. Giving positive ↑ chromaffin reaction, type I cells resemble cells of adrenal medulla. Type I cells are classed as ↑ APUD cells. 2) Glomus cells, type II (supporting cell, nongranulocytus glomeris sustentans*) = fusiform nongranulated cells surrounding type I glomus cells. Type II cells have a nucleus rich in ↑ heterochromatin, containing in their cytoplasm some mitochondria and cisternae of rough endoplasmic reticulum, a small Golgi apparatus, and moderate amounts of free ribosomes. It is believed that type II cells derive from ↑ Schwann's cells.

Böck, P.: The Paraganglia. In: Oksche, A., Vollrath, L. (eds.): Handbuch der mikroskopischen Anatomie des Menschen, Vol. 6, Part 8. Berlin, Heidelberg, New York: Springer-Verlag 1982; Clarke, J.A., de Burgh Daly, M.: Anat. Embryol. *166*, 169 (1983); Grönblad, M.: Cell Tissue Res. *229*, 627 (1983); Hensen, J.T.: J. Ultrastruct. Res. *77*, 189 (1981)

Glossopalatine glands (glandula glossopalatina): Small ↑ mixed salivary glands opening at posterior end of plica sublingualis.

Glottis*: The part of laryngeal wall limiting its median fissure, rima glottidis (RG). In anterior two-thirds, it is delimited by ↑ vocal cords (VC), and in its posterior third by ↑ arytenoid cartilage. G. is vocal apparatus of ↑ larynx. CC = cricoid cartilage, TC = ↑ thyroid cartilage, VeF = ↑ ventricular folds

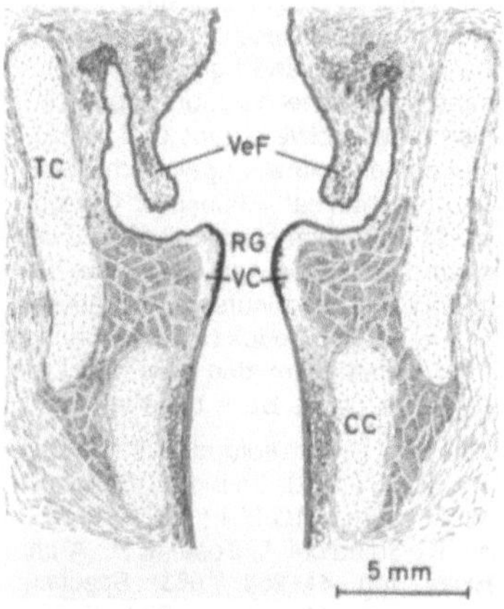

Glucagon: A polypeptide hormone (m.w. about 3200 daltons) secreted by ↑ A-cells of ↑ pancreatic islets. G. stimulates liver ↑ glycogenesis and raises blood glucose by activation of ↑ adenyl cyclase-cAMP system.

Lefebvre, P.: Glucagon I and II. Berlin, Heidelberg, New York: Springer-Verlag 1983

Glucagon cells: ↑ A-cells, of pancreatic islets

Glucocorticoids: ↑ Steroid hormones (G) produced predominantly in zona fasciculata, partially in zona reticularis of adrenal cortex. G. provoke intracellular storage of ↑ glycogen, decreased rate of protein synthesis, increase in protein breakdown, and lipid accumulation in ↑ adipose cells. G. have an anti-inflammatory and antiallergic action; they also prevent rejection reaction reducing activity of ↑ killer cells. G. include cortisol (hydrocortisone), cortisone, corticosterone, etc. Secretion of G. is controlled by ↑ ACTH-RF and ↑ ACTH through a negative ↑ feedback mechanism (interrupted lines). (See biochemistry and physiology texts for further information)

↑ Cells, of zona fasciculata

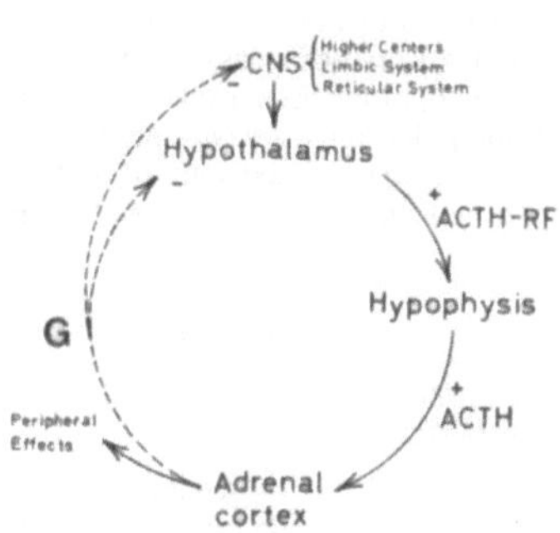

Baxter, J.D., Rousseau, G.G.: Glucocorticoid Hormone Action. Monographs on Endocrinology *12*, 1 (1979)

Glutaraldehyde: An ↑ aldehyde containing fixative widely employed in ↑ electron microscopy.

↑ Fixation

Glutaurine: A parathyroid hormone acting on metabolism of vitamin A, on ↑ connective tissues enhancing synthesis of ↑ proteoglycans, and involved in control of activity of ↑ lysosomes.

Fener, L., Török, O., Csaba, G.: Acta Morphol Acad. Sci. Hung. *26*, 87 (1978)

Glycocalyx* (cell coat, fuzzy coat, glycolemma, surface coat): A variously thick (10 nm–5 µm), moderately osmiophilic layer (G) covering external surface of cell membrane, of which it is considered to be a part. G. consists of an internal amorphous substance and external fine filaments, ↑ antennulae microvillares (arrows). G. is ↑ PAS-positive; in ↑ transmission electron microscope, it can be demonstrated with ↑ alcian blue, colloidal iron or thorium, ↑ lanthanum, ↑ ruthenium red, periodic acid-thiocarbohydrazide, etc. G. is composed of ↑ glycolipids, ↑ glycoproteins containing high amounts of ↑ sialic acid, and ↑ proteoglycans; it is synthesized by coordinated activity of rough endoplasmic reticulum and Golgi apparatus and transported in vesicles to cell surface. Functions: 1)

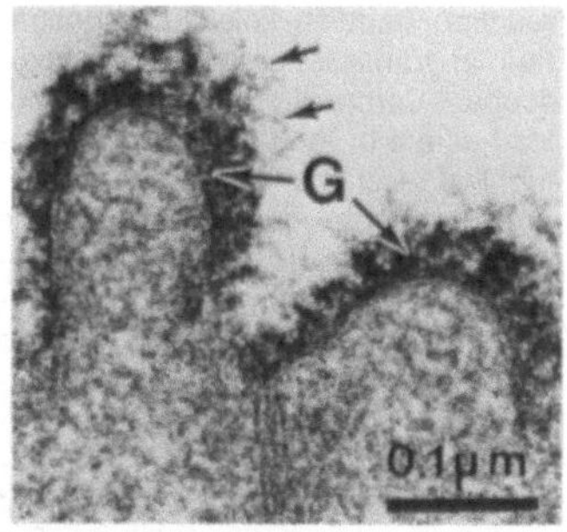

Genetically programed specific cell recognition at molecular level (G. carries antigens of histocompatibility and many others); 2) maintaining cell microenvironment through filtration of substances and generation of negative charge of cell surface; 3) absorption of nutrients and terminal digestion of carbohydrates and proteins by surface enzymes; 4) protection of cell against both proteolytic and mucolytic enzymes. (Fig. = palpebral ↑ conjunctiva, mongolian gerbil)

↑ Apical pits; ↑ Endocapillary layer; ↑ Neurobiotaxis

Brunser, O., Luft, J.H.: J. Ultrastruct. Res. *31*, 291 (1970); Ito, S.: Philos. Trans. R. Soc. Lond. (Biol.) *268*, 55 (1974); Luft, J.H.: Int. Rev. Cytol. *45*, 291 (1976); Singer, S.J., Rothfield, L.I.: Neurosci. Res. Program Bull. *11*, 181 (1973)

Glycoconjugates: Ubiquitous components of animal cells associated with a wide range of biological functions, including membrane interactions inside and outside cell. G. comprise ↑ glycolipids, ↑ proteoglycans, ↑ glycoproteins, and ↑ glycosaminoglycans.

Corfield, A.P., Schauer, R.: Biol. Cell *36*, 213 (1979); Horowitz, M.I. (ed.): The Glycoconjugates. Vols. 1–4. New York: Academic Press 1980–1982; Monsigny, M., Schrével, J. (eds.): Membrane Glycoconjugates. Ivry sur Seine: Soc. Fr. Micr. Electronique 1982

Glycogen (granulum glycogeni*): The energy-rich polysaccharide ↑ inclusion readily convertible into glucose. G. is produced by intimate involvement of smooth endoplasmic reticulum (sER); pathways of its synthesis are still insufficiently understood. G. is abundant in ↑ liver parenchymal cells (fig.), ↑ skeletal muscle fibers, ↑ cardiac muscle cells, cells of adrenal cortex, etc. Histochemical demonstration under light microscope: Iodine (result = brown), Best's carmin (red), ↑ PAS-reaction (reddish-violet). Demonstration in transmission electron microscope: ↑ Contrasting of ↑ sections with lead citrate or lead hydroxide; "staining" of sections with thiocarbohydrazide after oxidation with periodic acid. Two kinds of G. particles are visible in transmission electron microscope: β-particles are round with an average diameter of 15–30 nm; complexes of β-particles form rosettes or α-particles of about 50–100 nm. G. particles are situated outside the sER and have no surrounding membrane. (Fig. = rat)

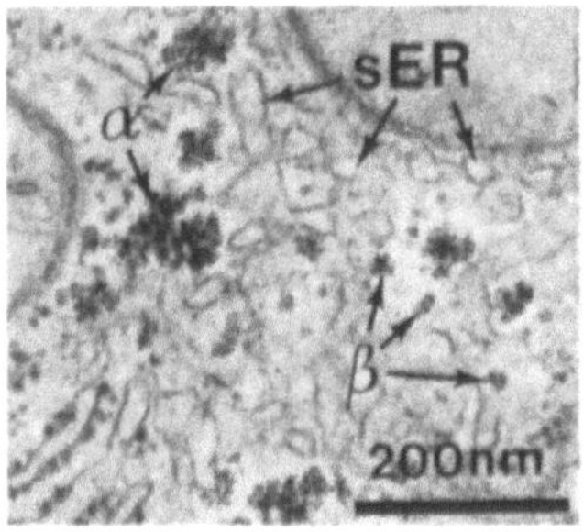

Babcock, M.B., Cardell, R.R.: Am. J. Anat. *140*, 299 (1974); De Bruijn, W.C.: J. Ultrastruct. Res. *42*, 29 (1973); Geyer, G.: Ultrahistochemie. 2nd edn. Stuttgart: Gustav Fischer Verlag 1973; Pearse, A.G.E.: Histochemistry. Theoretical and Applied. 3rd edn. Edinburgh, London: Churchill Livingstone 1972; Stiffler, J.S., Cardell, E.L., Cardell, R.R.Jr.: Am. J. Anat. *160*, 363 (1981)

Glycogenesis: The process of polymerization of glucose to ↑ glycogen by means of a series of enzymatic reactions. G. is closely associated with membranes of smooth endoplasmic reticulum (known to be rich in glucose-6-phosphatase); however, morphofunctional relationships between G. and smooth endoplasmic reticulum are not understood. (See biochemistry and physiology texts for further information)

Glycolemma: ↑ Glycocalyx

Glycolipids: Lipids containing a hexose sugar.

Glycoproteins: Protein macromolecules containing several different monosaccharides not arranged in the form of repeating disaccharide groups as in ↑ glycosaminoglycans. ↑ Thyroglobulin, ↑ gonadotropic hormones (except ↑ prolactin), ↑ mucus, ↑ mucoid, ↑ zona pellucida, etc. are G.

Gottschalk, A. (ed.): Glycoproteins: Their Composition, Structure and Function. Amsterdam: Elsevier 1966; Lennarz, W.J. (ed.): The Biochemistry of Glycoproteins and Proteoglycans. New York, London: Plenum Press 1980

Glycosaminoglycans (mucopolysaccharides): Polysaccharide molecules made up of repeating disaccharides, each of which consists of a hexuronic acid linked with an amino sugar (hexosamine). Molecules of G. are very hydrophilic and combine with many molecules of water. Through this water, diffusion processes of ↑ connective and supporting tissues take place. Following diagram displays classification of G. (Terms marked with an asterisk are described under corresponding letters)

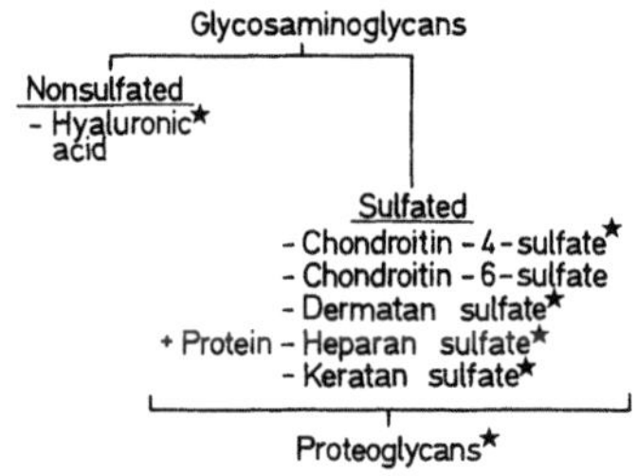

Varma, R.S., Varma, R. (eds.): Glycosaminoglycans and Proteoglycans. Basel, München: Karger 1982

Glycosaminolipids: Lipids containing amino sugars.

Gn-RF: ↑ Gonadotropin-releasing factor

Goblet cells (mucous cells, cellula caliciformis*): Columnar basophilic cells scattered in epithelia of ↑ conjunctiva, small and large ↑ intestines, ↑ nasopharynx, ↑ larynx, ↑ trachea, and ↑ bronchi. In active G. (A), nucleus (N) is flattened or cuplike and located in basal cell pole. Cytoplasm contains mitochondria, a well-developed Golgi apparatus (G), rough endoplasmic cisternae in parallel arrangement, and many free ribosomes; these two latter structures are responsible for ↑ basophilia of G. Arising from Golgi apparatus are numerous, round membrane-bound ↑ mucous droplets (M) with variously dense fine granular content. Droplets fill entire supranuclear cytoplasm and determine goblet-shaped appearance of cells. Droplets are discharged from cells by fusion of their membranes with apical plasmalemma. In some cases, G. expel al-

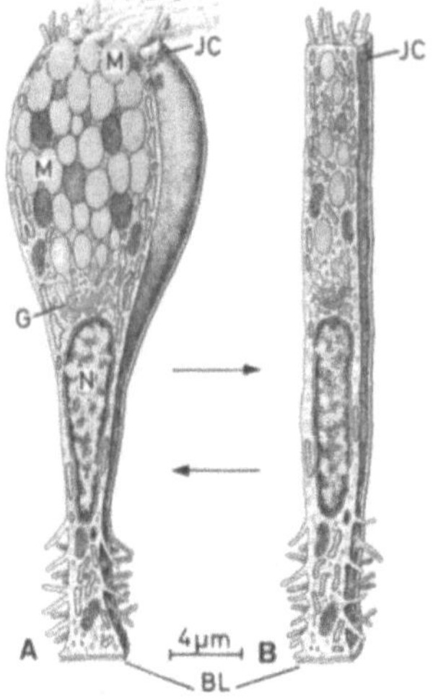

most all droplets and become very slender (B); they refill their cytoplasm during next secretory cycle. G. are joined to adjacent cells by ↑ junctional complexes (JC). Product of G. is basophilic, ↑ PAS-positive, and metachromatic since it is composed of ↑ glycoproteins and ↑ glycosaminoglycans; it is destined for lubrication of lumen of respective organs and protection of their epithelia against chemical and mechanical influences. Considered ↑ unicellular glands, G. have a life span of 2–4 days with one or two secretory cycles. Precursors of intestinal G. are ↑ oligomucous cells originating from ↑ undifferentiated cells of ↑ Lieberkühns' crypts. BL = basal lamina

Altmann, G.G., Leblond, C.P.: J. Cell Sci. *56*, 83 (1982); Cheng, H., Bjerknes, M.: Anat. Rec. *203*, 251 (1982); Kurosumi, K., Shibuichi, I., Tosaka, H.: Arch. Histol. Jpn. *44*, 263 (1983); Specian, R.D., Neutra, M.R.: J. Cell Biol. *85*, 626 (1980)

Goiter (struma): A pathological enlargement of ↑ thyroid gland due to an excess production of ↑ colloid with insufficient concentration of iodine in blood.

Gold chloride staining: Post-treatment of silver-stained sections with gold chloride in order to convert brownish color of impregnated structures into bluish-black and to eliminate weakly impregnated background.

Gold shadowing: ↑ Shadowing

Golgi apparatus (Golgi complex, Golgi region, Golgi zone, dictyosome, internal reticular apparatus, apparatus reticularis internus, complexus golgiensis*): A heterogenous cell ↑ organelle (G) present in all cells except ↑ erythrocytes. Structure: 1) Light microscope = a reticular network of anastomosing canals (G) near or around nucleus (N). 2) Transmission electron microscope: a) ↑ lamellae (L) or cisternae = flattened fenestrated disclike sacs, 0.5–5 µm in diameter, arranged in a slightly curved stack with convexity generally oriented toward nucleus, ↑ forming face (FF); concave side of stack is termed ↑ maturing face (MF). Lamellae are 15–30 nm apart and ↑ hyaloplasm between them is free of ribosomes. Number of lamellae varies from 3 to 30 or more. Vesicles and vacuoles pinch off from their margins;

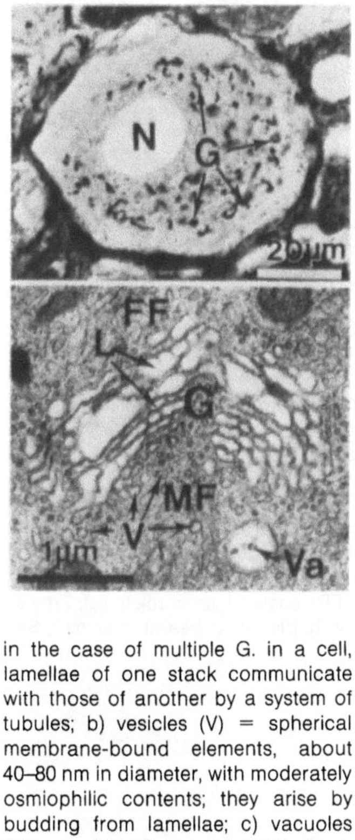

in the case of multiple G. in a cell, lamellae of one stack communicate with those of another by a system of tubules; b) vesicles (V) = spherical membrane-bound elements, about 40–80 nm in diameter, with moderately osmiophilic contents; they arise by budding from lamellae; c) vacuoles (Va) = 0.1–1.0-µm spherical structures that pinch off from periphery of lamellae on maturing face in some ↑ exocrine cells; they contain a moderately osmiophilic secretory product in the course of condensation (= ↑ condensing vacuoles). (Fig., top = ↑ pseudounipolar cell, human, ↑ Golgi staining; fig., bottom = ↑ epididymis, rat)

Farquhar, M.G., Palade, G.E.: J. Cell Biol. *91*, 77s (1981); Miyamoto, T., Tanaka, K.: Arch. Histol. Jpn. *45*, 23 (1982); Tartakoff, A.M.: Trends Bio. Sci. *7*, 174 (1982); Slot, J.W., Geuze, H.J.: J. Histochem. Cytochem. *31*, 1049 (1983); Whaley, G.M.: The Golgi Apparatus. Vienna, New York: Springer-Verlag 1975

Golgi apparatus, functions of: 1) Synthesis of polysaccharides and ↑ glycoproteins (↑ glycocalyx, ↑ mucus); 2) incorporation of sugar component in glycoproteins transported from the rough ↑ endoplasmic reticulum; 3) concentration of secretory product (↑ condensing vacuoles) and formation of ↑ secretory granules; 4) supplying of secretory granules with newly synthesized limiting membrane; 5) synthesis of specialized apical membranes (↑ plaques). Functions of G. are controlled by nucleus.

Farquhar, M.G.: Membrane Recycling in Secretory Cells – Pathway to the Golgi Complex. In: Ciba Foundation Symposia, Vol. 92. London: Pitman 1982; Fleischer, B.: J. Histochem. Cytochem. *31*, 1033 (1983); Leblond, C.P., Bennett, G.: Role of the Golgi Apparatus in Terminal Glycosylation. In: Brinkley, R.B., Porter, K.R. (eds.): International Cell Biology, 1976–1977. New York: The Rockefeller University Press 1977; Morré, J.D., Ovtracht, L.: J. Ultrastruct. Res. *74*, 284 (1981); Whaley, G.W., Dauwalder, M.: Int. Rev. Cytol. *58*, 199 (1979)

Golgi apparatus, membranes of: Golgi lamellae or cisternae (L) are formed by a ↑ unit membrane about 6 nm thick at ↑ forming face (FF) and up to 8 nm at ↑ maturing face (MF). Concentration of ↑ thiamine pyrophosphatase (TPP) decreases from forming face toward maturing face, whereas concentration of acid phosphatase (AP) is highest in lamellae of maturing face.

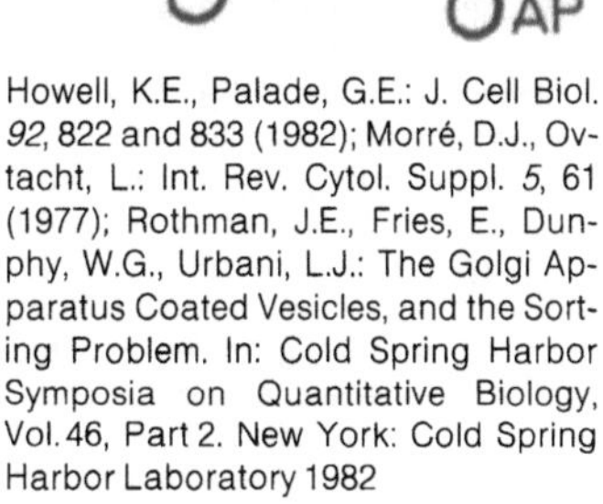

Howell, K.E., Palade, G.E.: J. Cell Biol. *92*, 822 and 833 (1982); Morré, D.J., Ovtacht, L.: Int. Rev. Cytol. Suppl. *5*, 61 (1977); Rothman, J.E., Fries, E., Dunphy, W.G., Urbani, L.J.: The Golgi Apparatus Coated Vesicles, and the Sorting Problem. In: Cold Spring Harbor Symposia on Quantitative Biology, Vol. 46, Part 2. New York: Cold Spring Harbor Laboratory 1982

Golgi apparatus, multiple: A term designating several ↑ Golgi fields in a cell. Although not completely correct, this term is widely used in describing cells.

Golgi apparatus, origin of: It seems likely that ↑ Golgi apparatus (G) originates either from rough ↑ endoplasmic reticulum (ER) or external leaflet of ↑ nuclear envelope (NE), from which smooth ↑ transport vesicles (TV) bulge, separate, and fuse to form a lamella (L). Several lamellae associate into a definitive Golgi apparatus. In this way, Golgi apparatus is constantly renewed.

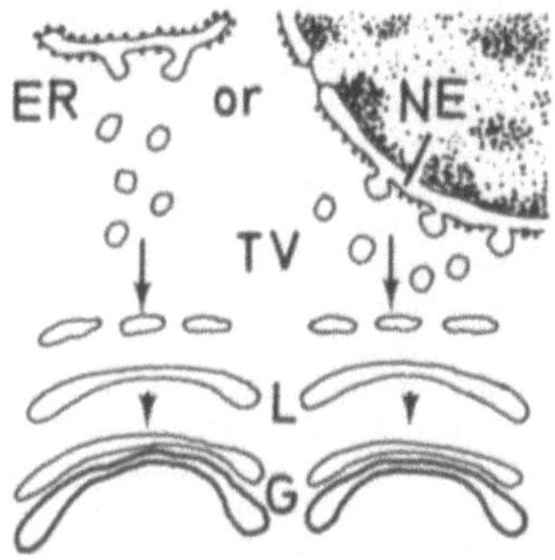

Golgi cells: A term sometimes used to designate large ↑ granule cells of cerebellar cortex.

Golgi cisternae: Synonym for lamellae of ↑ Golgi apparatus.

Golgi complex: ↑ Golgi apparatus

Golgi field (dictyosome): Stacks (GF) of lamellae (L) of ↑ Golgi apparatus (G) surrounding nucleus (not shown) and communicating with one another through a system of anastomosing and branching tubules (T).

↑ GERL

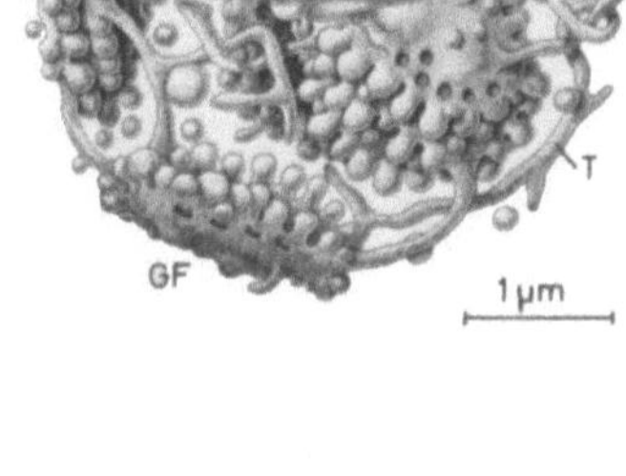

Palay, S.L., Chan-Palay, V.: J. Microsc. *97*, 41 (1973); Rambourg, A., Clermont, Y., Hermo, L.: Three Dimensional Structure of the Golgi Apparatus. In: Hand, A.R. and Oliver, C. (eds.): Methods in Cell Biology 23. New York: Academic Press 1981

Golgi-Mazzoni corpuscle: ↑ Corpuscle, of Golgi-Mazzoni

Golgi phase, of spermiogenesis: ↑ Spermiogenesis

Golgi region: ↑ Golgi apparatus

Golgi staining: A method of combined fixation and impregnation of entire tissue blocks with potassium bichromate, ↑ osmium tetroxide, and silver nitrate without staining of sections. G. is a useful technique for study of CNS.

Golgi tendon organ (tendon organ, neurotendinous spindle, fusus neurotendineus*): A fusiform structure, about 1 mm long, situated at transition between ↑ skeletal muscle fibers (MF) and ↑ tendon fibrils (TF). Structure: 1) Capsule (C) = flattened ↑ fibrocytes (F) surrounded by numerous ↑ collagen microfibrils (CF). Capsule is continuous with ↑ perineurium (P). 2) Interior of G. contains a varying number of tendon fibrils. After entering capsule, one or two ↑ myelinated sensory nerve fibers (NF) lose their ↑ myelin sheath and give rise to numerous ↑ nonmyelinated branches (B) entwined among tendon fibrils. Axons (A) are accompanied by ↑ Schwann's cells (SC) almost to their endings. During

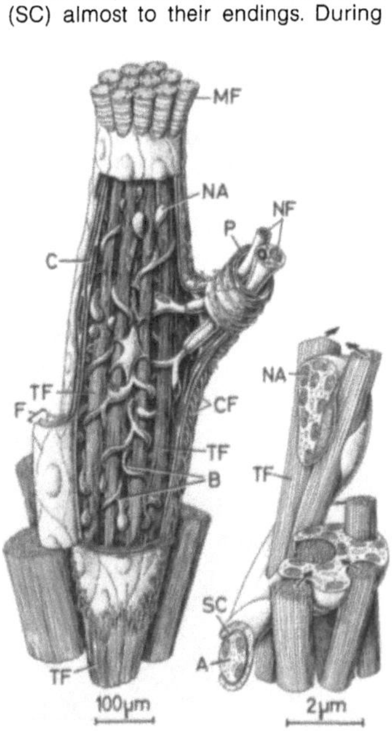

muscle contraction, stretching of tendon fibrils compresses naked ↑ axons (NA) in direction of arrows (inset) and excites them. G. are ↑ proprioceptors.

Bridgemen, C.F.: J. Comp. Neurol. *138*, 369 (1970); Schoultz, T.W., Swett, J.E.: J. Neurocytol. *1*, 1 (1972)

Golgi type I neurons (neuronum multipolare longiaxonicum*): A category of large nerve cells with long ↑ axons leaving surrounding ↑ gray matter to terminate either in another part of ↑ central nervous system or in some other tissue or organ (muscles, glands, skin, etc.). G. include ↑ neurons that contribute to formation of association, commissural, and projection fibers, as well as peripheral ↑ nerves (↑ Purkinje cells of cerebellum, ↑ pyramidal neurons of ↑ brain cortex, ↑ motor neurons of ↑ spinal cord, etc.).

Golgi type II neurons (neuronum multipolare breviaxonicum*): A category of small nerve cells with short ↑ axons remaining in neighboring ↑ gray matter. G. include all ↑ granule cells, association and commissural neurons of ↑ spinal cord, ↑ stellate cells, ↑ basket cells of ↑ cerebellum, ↑ Martinotti's cells, ↑ Renshaw's cells, etc.

↑ Interneurons

Golgi zone: ↑ Golgi apparatus

Golgiosome: The fragmented ↑ Golgi apparatus in the form of small smooth vesicles during ↑ mitosis and ↑ meiosis.

Gomphosis: A ↑ synarthrosis between root of tooth and ↑ alveolar bone. G. is a type of ↑ syndesmosis.

↑ Periodontal ligament

Gonadoliberin: ↑ Gonadotropin-releasing factor

Gonadotropes type I (FSH-cell, delta cell, cellula gonadotrophica*): Large oval, basophilic cells situated along blood capillaries (Cap) in pars distalis of ↑ hypophysis. Nucleus is round and eccentric; cytoplasm encloses scattered mitochondria, enlarged irregularly shaped cisternae of ↑ rough endoplasmic reticulum (rER) filled with moderately osmiophilic flocculent ↑ reticuloplasm, a large Golgi apparatus associated with numerous immature secretory granules, some lysosomes, and free ribosomes. G. contain two kinds of ↑ secretory granules (SG): the smaller ones measure about 200 nm, the larger ones about 500 nm; all granules are surrounded by a ↑ unit membrane and appear basophilic in light-microscopic preparations. In transmission electron microscope, smaller granules are more electron dense than larger ones. ↑ Exocytosis (E) of granule contents is frequent. G. produce ↑ follicle-stimulating hormone.

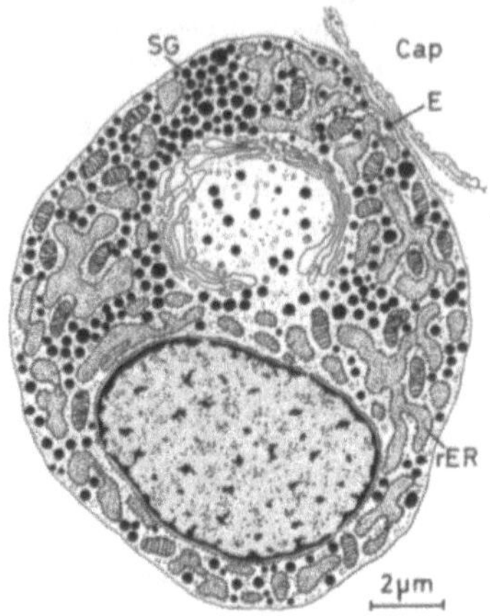

Kurosumi, K., Fujita, H.: Functional Morphology of Endocrine Glands. Stuttgart: Georg Thieme 1975

Gonadotropes type II (delta cell, ICSH-cell, LH-cell luteinizing hormone gonadotrope, cellula gonadotrophica*): Polygonal basophilic cells located along blood capillaries (Cap) in pars distalis of ↑ hypophysis. Nucleus is elongated, eccentric, and slightly infolded; cytoplasm contains small mitochondria, an inconspicuous Golgi apparatus, sporadic flattened cisternae of rough ↑ endoplasmic reticulum (rER), some ↑ lipid droplets (L), ↑ residual bodies, and free ribosomes. ↑ Se-

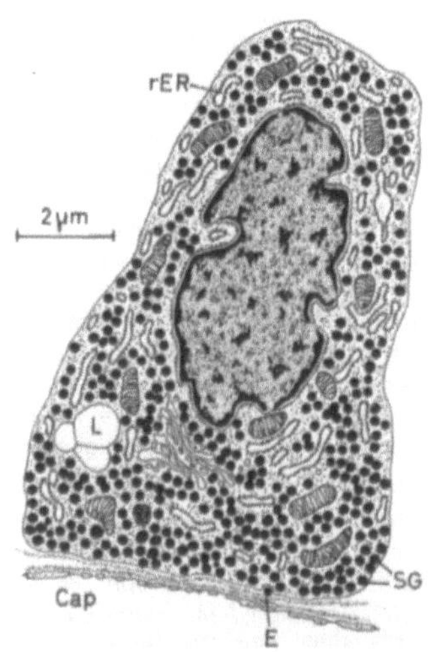

cretory granules (SG) are ↑ unit membrane-enclosed; they contain an electron-dense material and measure about 250 nm in diameter. G. produce ↑ LH in female and ↑ ICSH in male. Secretory product is released by ↑ exocytosis (E). The fig. is valid only for some animals since in man there are no morphological and immunocytochemical differences between G. and ↑ gonadotropes type I.

Girod, G.: Introduction à l'étude des glandes endocrines. 2nd edn. Villeurbanne: SIMEP S.A. 1980; Soji, T.: Arch. Histol. Jpn *45*, 335 (1982)

Gonadotropic hormones (gonadotropins): Hypophyseal hormones controlling function of female and male reproductive systems. G. include ↑ follicle-stimulating hormone (FSH), ↑ luteinizing hormone (LH), ↑ interstitial cell-stimulating hormone (ICSH), identical to LH, and ↑ prolactin. Release of G. is controlled by ↑ gonadotropin-releasing factor and ↑ luteotropic hormone-releasing factor.

Flamigin, C., Givens, J.R. (eds.): The Gonadotropins: Basic Science and Clinical Aspects in Female. Serono Symposia, Vol. 42. London: Academic Press 1982

Gonadotropin-releasing factor (GnRF, FSH/LH-RF): A decapeptide secreted by ↑ neurosecretory neurons in mediobasal nuclei of ↑ hypothalamus and discharged into ↑ portal network of ↑ adenohypophysis. G. has a potent action on type I and II ↑ gonadotropes, inducing production of ↑ FSH and ↑ LH/ICSH. Small amounts of G. are also produced in ↑ organum vasculosum laminae terminalis and ↑ subfornical organ.

↑ Releasing factors

Bennet-Clarke, C., Joseph, S.A.: Cell Tissue Res. *221*, 493 (1982); Dees, W.I., McArthur, N.H.: Anat. Rec. *200*, 281 (1981); McCann, S.M.: J. Endocrinol. Invest. *6*, 243 (1983); Scott, P.M., Knigge, K.M.: Cell Tissue Res. *219*, 393 (1981); Schaison, G.: Ann. Endocrinol. *43*, 247 (1982)

Gonosomes: ↑ Sex chromosomes

Goormaghtigh's cells: ↑ Mesangium, extraglomerular

G_0 phase: The prolonged G_1 phase of ↑ cell cycle in nonrenewing tissues. ↑ Neurons are in G_0 phase.

G_1 phase: The postmitotic phase of ↑ cell cycle, duration of which varies greatly depending on cell type. In nonrenewing tissues, G_1 is designated ↑ G_0 phase.

G_2 phase: The relatively short premitotic phase of ↑ cell cycle. In cells destined to become ↑ polyploid, G_2 may last indefinitely.

Graafian follicle: ↑ Ovarian follicle, mature

Graft-rejection cells: ↑ Killer cells

Graft-rejection reaction: All humoral and cellular events ocurring during ↑ immune response in an attempt to eliminate a homo- or heterograft as a source of foreign antigens.

↑ Killer cells; ↑ Transplantation

Grafting: ↑ Transplantation

Granular cytoplasmic body: ↑ Nucleoluslike body

Granular fibrillar body: ↑ Nucleoluslike body

Granular pneumonocyte: ↑ Alveolar cells type II

Granule (granulum): A ↑ unit membrane-bound spherical structure, 0.1–2 µm in diameter, with more or less osmiophilic contents.

↑ Secretory granules

Granule, acrosomal: ↑ Acrosomal granule

Granule cells: Small multipolar ↑ Golgi type II neurons (G). G. have a few short ↑ dendrites and a short ↑ axon resting in zone immediately surrounding ↑ perikaryon. Nucleus is relatively large and round, filling almost entire cytoplasm. Thin cytoplasmic rim encloses small mitochondria, an inconspicuous Golgi apparatus, a few rough endoplasmic cisternae, some ↑ neurofibrils and ↑ neurotubules, and a small number of free ribosomes. There are three categories of G.: 1) ↑ G. of cerebellar cortex (fig.); 2) ↑ G. of cerebral cortex; 3) ↑ G. of olfactory bulb. O = ↑ oligodendrocyte. (Fig. = rat)

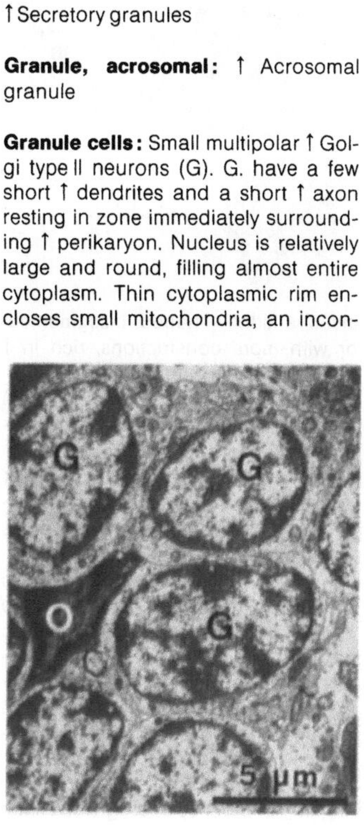

Granule cells, of cerebellar cortex: A category of multipolar ↑ interneurons, about 5–10 µm in diameter, forming major part of ↑ cerebellar granular layer (GL). Two kinds of G.: 1) Granule cells or small granule cells (gr) = very small ↑ neurons, 5–6 µm in diameter, with a relatively voluminous nucleus and only three to five short dendrites (D), whose clawlike endings make synaptic contacts with ↑ mossy fibers (MF) within cerebellar ↑ glomeruli (CG). ↑ Axon (A) of each small G. enters molecular layer (ML) and divides like a "T" into two branches parallel to long axis of ↑ folium (F) [= ↑ parallel fibers (PF)]. Each branch establishes synaptic contact with dendrites of many ↑ Purkinje cells (PC), ↑ basket cells (BC), ↑ stellate cells (SC), and large granule cells (g). 2) Large granule cells or Golgi cells (g) = neurons measuring about 8–10 µm, whose dendrites establish synaptic contact with parallel fibers (PF), whereas their axons enter into synaptic contact with mossy fibers. Large granule cells have an inhibitory action on activity of granule cells (gr). G. are neurons belonging to ↑ Golgi type II neurons; they play a very important role as modulators of synaptic transmission in cerebellum. Internal structure of G. is as in all ↑ granule cells.

↑ Cerebellum, neuronal circuits of

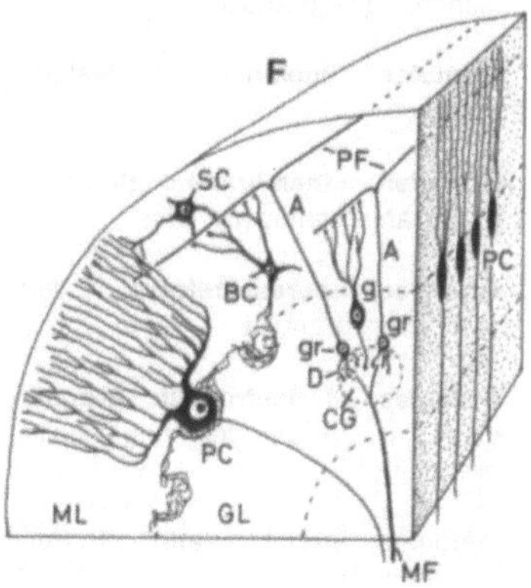

Castejon, O.J.: Scan. Electron Microsc. 1981/IV, 105 (1981)

Granule cells, of cerebral cortex: A category of small ↑ multipolar neurons with few ↑ dendrites and an ↑ axon with collaterals branching locally, participating in formation of cerebral cortex layers. Ultrastructure of G. corresponds to that other ↑ granule cells. As ↑ interneurons belonging to ↑ Golgi type II, G. have an important integrative function.

Granule cells, of olfactory bulb: Small ↑ neurons located in inner granular layer of ↑ olfactory bulb. G. seem to lack an axon; however, their dendrites have an axonal function. G. make special dendrodendritic ↑ synapses with ↑ mitral cell dendrites in the form of reciprocal ↑ synapses. Structurally, G. correspond to other ↑ granule cells.

Granule, proacrosomal: ↑ Proacrosomal granule

Granules, alpha, of platelets: ↑ Platelets

Granules, atrial: ↑ Atrial granules

Granules, azurophilic: ↑ Azurophilic granules

Granules, beta, of pancreatic cells: ↑ B-cells, of pancreatic islets

Granules, delta, of pancreatic cells: ↑ D-cells, of pancreatic islets

Granules, eosinophilic: ↑ Eosinophilic granules

Granules, keratohyalin: ↑ Keratohyalin granules

Granules, lipofuscin: ↑ Lipofuscin

Granules, matrix, of mitochondria: ↑ Mitochondrial granules

Granules, melanin: ↑ Melanin granules

Granules, membrane-coating: ↑ Membrane-coating granules

Granules, neurosecretory: ↑ Neurosecretory granules

Granules, of neutrophilic granulocytes: ↑ Granulocytes, neutrophilic, granules of

Granules, perichromatin: ↑ Perichromatin granules

Granules, presecretory: ↑ Zymogen granules

Granules, secretory: ↑ Secretory granules

Granules, trichohyalin: ↑ Trichohyalin granules

Granules, zymogen: ↑ Zymogen granules

Granulo-adipose bodies: ↑ Compound granular corpuscles

Granulocytes: A collective term for polymorphonuclear ↑ leukocytes because of their ↑ azurophilic (primary) and specific (secondary) granules.

↑ Granulopoiesis; ↑ Granulocyte, basophilic; ↑ Granulocyte, eosinophilic; ↑ Granulocyte, neutrophilic

Ryder, M.I., Niederman, R., Taggart, E.J.: Anat. Rec. *203*, 317 (1982)

Granulocytes, basophilic (basophils, granulocytus basophilicus*): The rarest category of polymorphonuclear ↑ leukocytes (0–2% of all circulating leukocytes), 10–12 μm in diameter. Nucleus is kidney-shaped, bisegmented, or with more constrictions, rich in ↑ heterochromatin; no evident nucleoli. Cytoplasm encloses a small ↑ Golgi apparatus, small sporadic mitochondria, flattened and short rough endoplasmic cisternae, a variable number of free ribosomes, numerous β-glycogen particles, and some lysosomes (Ly), giving a positive peroxidase reaction, referred to as ↑ azurophilic granules. Present in cytoplasm are large, ↑ unit membrane-limited specific granules (BG), about 1.5 μm in diameter, which vary in size and shape; they are stainable with ↑ basic and metachromatic dyes. Granules are water-soluble and filled, in humans (fig.), with closely packed, dense particles (15 nm); in other species, granules are filled with crystalloids or myelin figures. Granules contain ↑ histamine, ↑ heparin, and ↑ leukotrienes; they may be discharged from cell body by ↑ exocytosis, causing a degranulation of G. observed in hypotonic shock, under influence of histamine liberators, immunoglobulin E, and in a number of allergic and inflammatory reactions. Although G. display pronounced ↑ ameboidism, their surface is generally smooth with occasional short cytoplasmic protrusions. G. are ↑ phagocytes, although feeble ones. Despite some similarities with ↑ mast cells, G. probably do not belong to same class of cells since mast cells have no peroxidase-positive granules.

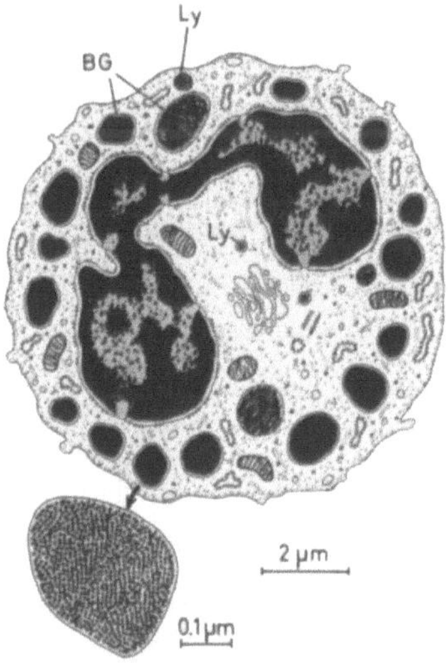

↑ Metachromasia

Austen, K.F.: Harvey Lect. *73*, 93 (1979); Behrendt, H.: Verh. Anat. Ges. *74*, 261 (1980); Dvorak, A., Osage, J.E., Dvorak, H.F., Galli, J.S.: Lab. Invest. *45*, 58 (1981); Herbert, H., Lindberg, M.: J. Ultrastruct. Res. *78*, 215 (1982)

Granulocytes, eosinophilic (eosinophils, granulocytus azurophilicus*): Kind of polymorphonuclear ↑ leukocyte (1%–4% of all circulating ↑ leukocytes) about 12 μm in diameter. G. have a bilobed nucleus (N) without an apparent nucleolus, some mitochondria, a small Golgi apparatus with associated ↑ diplosome, several short rough endoplasmic cisternae, and occasional lysosomes (↑ azurophilic granules, AG). Cytoplasm is filled with numerous ↑ eosinophilic granules (EG). G. present a weak ↑ ameboidism and phagocytize only in presence of antibacterial antibodies. ↑ Histamine release or presence of other eosinotactic substances formed as a consequence of antigen-antibody reaction (allergy, parasites) increases number of circulating G. (eosinophilia); ↑ ACTH and ↑ cortisol have an op-

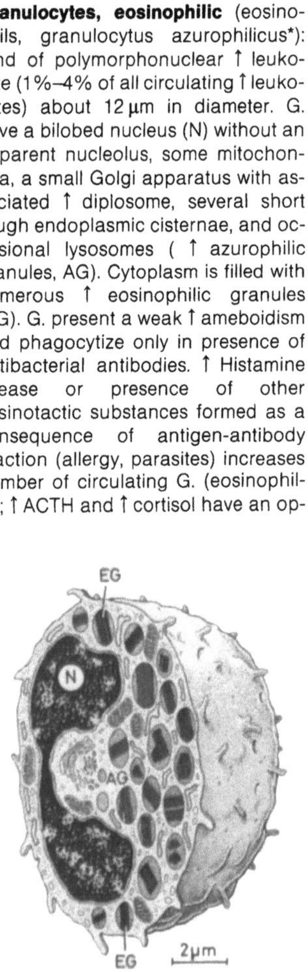

posite effect (eosinopenia). G. are constant ↑ wandering cells of ↑ loose connective tissue.

↑ Eosinophilic chemotactic factor-A

Beeson, P.B., Bass, D.A.: The Eosinophil. Philadelphia: W.B. Saunders 1977; Pimenta, P.F.P., De Souza, W.: J. Submicrosc. Cytol. *14*, 227 (1982); Wulfhekel, U.: Verh. Anat. Ges. *74*, 223 (1980)

Granulocytes, neutrophilic (neutrophils, polys, granulocytus neutrophilicus segmentonuclearis*): The most frequent kind of polymorphonuclear ↑ leukocyte (40%–75% of all circulating leukocytes), 10–12 μm in diameter. In young G., nucleus consists of two segments, and in older ones of three to five segments connected by thin chromatin strands. ↑ Chromatin is coarse and very condensed. Cytoplasm contains a few mitochondria, a small Golgi apparatus (G) with associated ↑ centrioles, some short rough endoplasmic cisternae, reddish-purple ↑ azurophilic granules (AG), and a large number of specific granules (SG), stainable with acid and basic dyes; hence the neutrophilic character. G. are very motile cells and function as ↑ microphages, particularly in presence of ↑ immunoglobulin G; because of their pronounced ↑ chemotaxis, G. may leave blood vessels and enter tissues. About 50% of G. are marginated to capillary ↑ endothelium. Life span – about 8 days.

↑ Arneth's formula; ↑ Band neutrophils; ↑ "Drumstick;" ↑ Granulocyte, neutrophilic, granules of; ↑ Pus

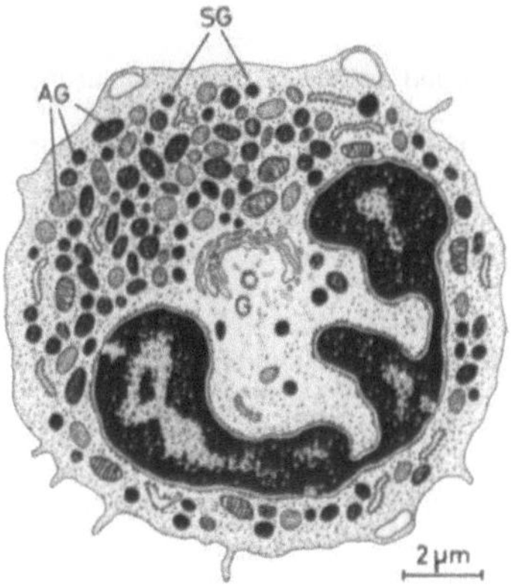

Hoffstein, S.T., Friedman, R.S., Weissmann, G.: J. Cell Biol. *95*, 234 (1982); Klebanoff, S.J., Clarck, R.A.: The Neutrophil. Function and Clinical Disorders. Amsterdam: Elsevier/North Holland 1978; Lisiewicz, J.: Human Neutrophils. Bowe, Md.: The Charles Press Publ. 1980; Przywansky, K.B., Schliwa, M., Porter, K.R.: Eur. J. Cell Biol. *30*, 112 (1983)

Granulocytes, neutrophilic, granules of: About 200 granules per cell; two types: 1) Azurophilic granules [A-granules, primary granules (AG)] = ↑ unit membrane-bound granules, measuring about 500 nm, with strongly osmiophilic matrix. After corresponding staining with methylene azure, azurophilic granules appear reddish-purple. These granules make up 20% of granule population, and are considered primary ↑ lysosomes, since they contain acid phosphatase, aryl-sulfatase, β-galactosidase, β-glucuronidase, 5-nucleotidase, and D-amino acid oxidase and peroxidase. During ↑ phagocytosis, azurophilic granules discharge their contents into ↑ phagolytic vacuoles. They are first granules to appear in Golgi area of neutrophils (and also basophilic and eosinophilic ↑ granulocytes) at stage of ↑ promyelocyte. 2) Specific granules [β-granules, secondary granules (SG)] = unit membrane-limited, spherical or rod-shaped granules, measuring about 200 nm, without lysosomic properties, displaying a weakly osmiophilic interior with occasional crystalloids. They contain ↑ lysozyme, ↑ phagocytin, and alkaline phosphatase, and are also discharged into phagolytic vacuoles. (Fig. = rat)

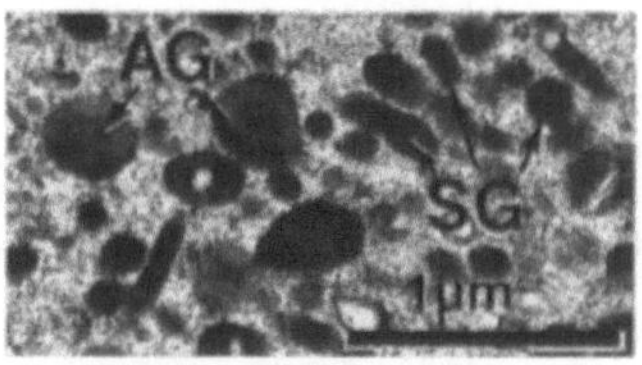

Brederoo, P., Meulen, van der, J.: Cell Tissue Res. *228*, 433 (1983); Brown, W.J., Shannon, W.A. Jr., Snell, W.J.: J. Cell Biol. *96*, 1030 (1983); Calamai, E.G., Spitznagel, J.K.: Lab. Invest. *46*, 597 (1982); Shannon, A.W., Zellmer, D.M.: Histochem. J. *14*, 847 (1982); Wittekind, D., Kretschmer, V.: Acta Histochem. Suppl. *21*, 39–45 (1980)

Granulocytes, pools of: All ↑ granulocytes in blood vessels. 1) Circulating pool = granulocytes transported with blood; 2) marginating pool = granulocytes, marginated against walls of vessels. Together with granulocytes of ↑ bone marrow, marginating pool represents one of two major sources of ready-reserve ↑ granulocytes.

Granulocytes, ready-reserve of: Quickly mobilizable ↑ granulocytes grouped into two pools: 1) Storage pool in ↑ bone marrow; 2) marginating pool in blood vessels (about 50% of circulating granulocytes).

↑ Granulocytes, pools of

Granulomere (granulomerus*): The inner granule containing zone of ↑ platelets.

Granulopoiesis (myelopoiesis, granulocytopoiesis*): The formation and ↑ differentiation of polymorphonuclear leukocytes. Sequence: By mitotic division, pluripotent ↑ colony-forming units (CFU) generate new stem cells in ↑ bone marrow, some of which lose their pluripotentiality and differentiate into ↑ myeloblasts. They develop, with formation of ↑ azurophilic granules (AG), into ↑ promyelocytes, which divide to give rise to other promyelocytes. A number of these differentiate during following 7.5 days into ↑ myelocytes. At this stage, specific granules (SG) occur which allow distinction between basophilic (B), eosinophilic (E), and neutrophilic (N) myelocytes. Fur-

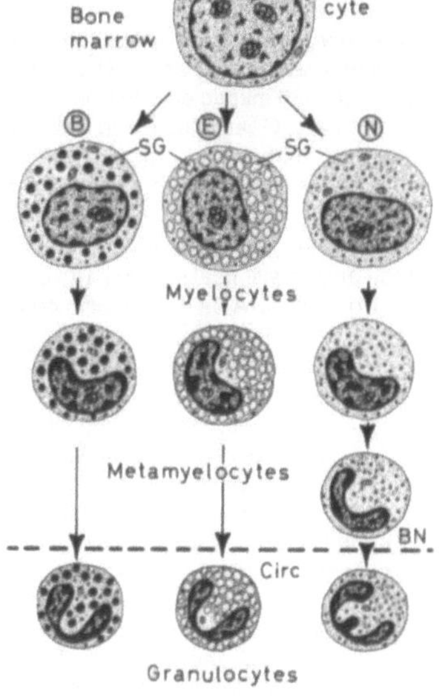

ther mitoses lead to generation of ↑ metamyelocytes, which are able to divide only in early phase of their development. In later phase divisions cease; after 6.5 days, metamyelocytes become mature polymorphonuclear ↑ granulocytes, which enter into the circulation (Circ). In neutrophilic series, transformation passes through stage of ↑ band neutrophil (BN). Myeloblasts and promyelocytes have, in general, round nuclei, myelocytes ellipsoidal nuclei, metamyelocytes kidney-shaped nuclei, band neutrophils crescent-shaped nuclei, and mature granulocytes segmented nuclei. Cell size decreases, beginning with promyelocyte stage. In adult, G. is carried out in red bone marrow; in intrauterine life (2nd–9th month), it also takes place in ↑ liver and ↑ spleen.

↑ Hematopoietic stem cells; ↑ Hematopoiesis, current view of

Golde, D.W., Cline, M.J., Metcalf, D., Fox, F.C. (eds.): Hematopoietic Cell Differentiation. New York: Academic Press 1978

Granulopoiesis, control of: Differentiation and growth of a stem cell into eosinophilic and neutrophilic ↑ granulocytes depends mainly on a group of substances collectively called colony-stimulating activators, among which principal is ↑ colony-stimulating factor. Release of granulocytes from bone marrow is stimulated by a protein, neutrophilic-releasing factor. ↑ Granulopoiesis is inhibited by ↑ chalones and possibly by a negative ↑ feedback system based on number of neutrophilic granulocytes; G. is far from completely understood.

Granulosa cells, of ovarian follicles (cellula granulosa*): Cells (G) derived through high mitotic activity from ↑ follicular cells. G. are applied in form of stratified epithelium to ↑ glassy membrane (arrowheads) of secondary vesicular and mature ↑ ovarian follicles. Initially, G. have an elongated form and an elliptical nucleus with one or two nucleoli. A pale cytoplasm contains only a few small mitochondria, an inconspicuous Golgi apparatus, moderately developed rough endoplasmic cisternae, small ↑ lipid droplets, and some free ribosomes. With further growth of follicle, mitoses increase in number, cells become rounded, their ↑ organelles more prominent, ↑ lipid droplets more frequent, and free ribosomes more numerous. Intercellular spaces (IS) enlarge considerably and coalesce, forming ↑ antrum folliculi. G. produce,

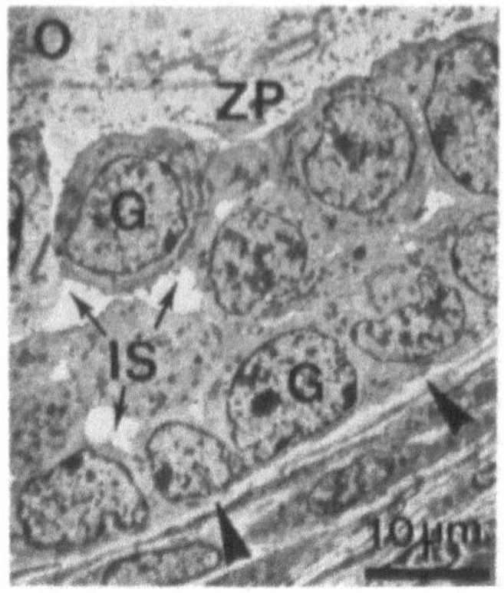

under influence of ↑ follicle-stimulating hormone, ↑ estrogens until ↑ ovulation; afterward, they differentiate into ↑ granulosa lutein cells. O = secondary ↑ oocyte, ZP = ↑ zona pellucida. (Fig. = rat)

↑ Call-Exner bodies, ↑ Corona radiata

Apkarian, R., Curtis, J.C.: Scan. Electron Microsc. 1981/IV, 165 (1981)

Granulosa lutein cells (follicular lutein cells, granulosoluteocytus*): Large polygonal cells (up to 30 µm) composing major part of ↑ corpus luteum. G. have an elliptical nucleus with a prominent nucleolus; cytoplasm contains many spherical mitochondria (M) with tubules, a well-developed and multiple Golgi apparatus (G), scattered stacks of short rough endoplasmic (rER) cisternae, extensive smooth endoplasmic reticulum (sER) composed of networks of branching and anastomosing tubules, ↑ lipid droplets (LD), small lysosomes (Ly), ↑ multivesicular bodies (MvB), ↑ lipofuscin (Lf), bundles of microfilaments, moderate amounts of free ribosomes, and occasional intracellular ↑ canaliculi. G. differentiate, in the course of ↑ luteinization, from ↑ granulosa cells of ↑ ovarian follicles. G. produce ↑ progesterone and small amounts of ↑ estrogens under influence of ↑ LH. During involution of corpus luteum, G. become filled with lipid droplets, nucleus becomes pyknotic, and cells degenerate. Recent studies have shown that G. in pregnancy produce ↑ relaxin.

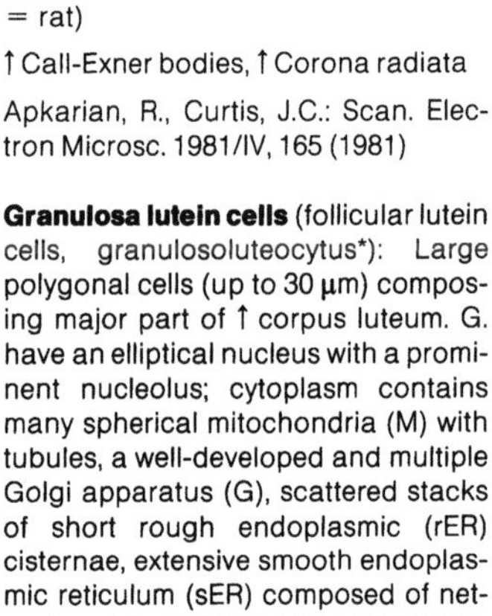

↑ Theca lutein cells

Crisp, T.M., Derousky, D.A., Denys, F.R.: Am. J. Anat. *127*, 37 (1970); Gemmel, R.T., Stacy, B.D.: Cell Tissue Res. *197*, 413 (1979); Meswerdt, W., Mueller, O., Brandau, H.: Arch. Gynäk. *222*, 115 (1977)

Gravid cycle: ↑ Endometrium, gravid cycle of

Gray matter, of CNS (substantia grisea*): Accumulation in form of layers, clusters (nuclei), and ganglia of bodies of ↑ neurons, their ↑ dendrites, and segments or all of their ↑ axons.

Gray's synapse type I: ↑ Synapse, chemical, types of

Gray's synapse type II: ↑ Synapse, chemical, types of

Great alveolar cells: ↑ Alveolar cells type II

GRF: ↑ Growth hormone-releasing factor

Ground cytoplasm: ↑ Cytoplasmic matrix

Ground substance (extracellular matrix, substantia fundamentalis*): One of the two constituents of intercellular components of ↑ connective and ↑ supporting tissue in form of a variously viscous mass surrounding cells and fibers. G. consists of water and an interstitial matrix composed of ↑ proteoglycans; it is, therefore, slightly ↑ PAS-positive and weakly ↑ metachromatic. In connective tissues, G. can be sol or gel; in ↑ cartilage, it is firm because of presence of ↑ chondromucoprotein; in ↑ bone, ↑ cementum, and ↑ dentin, it is calcified. G. is an intermediate transport milieu between cells and capillaries. Some authors also classify ↑ basement membrane with G.

↑ Edema

Hawkes, S., Wang, J.L. (eds.): Extracellular Matrix. New York: Academic Press 1982; Hay, E.D. (ed.): Cell Biology of Extracellular Matrix. New York: Plenum Press 1981

Ground substance, of blood vessels: Extracellular matrix present in wall of blood vessels and around ↑ capillaries. G. consists of ↑ proteoglycans. In arteries, both ↑ chondroitin and ↑ heparan sulfate predominate; in veins, ↑ dermatan sulfate is prevalent. G. plays an important role in permeability of vasculature.

Wight, T.N., Ross, R.: J. Cell Biol. *67*, 660 (1975)

Ground substance, of cartilage: ↑ Cartilage, ground substance of

Ground substance, of cells: ↑ Cytoplasmic matrix

Ground substance, of connective tissue: ↑ Connective tissue proper, ground substance of

Ground substance, of cytoplasm: ↑ Cytoplasmic matrix

Growth: The capacity of cells to constitute and increase, by assimilation of raw materials, a living mass in its entire complexity, and to reproduce its structure (excluding tumor formation). There are two kinds of G.: 1) Numerical G., or ↑ hyperplasia; 2) volumetric G., or ↑ hypertrophy. Increase in size of a young organism is based on numerical G. by mitotic cell division from ovum. Volumetric G. is rare and found in highly differentiated cells with no or very weak mitotic activity. G. is influenced by genetic factors, hormones (↑ growth hormone, ↑ thyroxin, ↑ corticoids, ↑ sexual hormones, ↑ somatostatin), diurnal and seasonal rhythms, age, nutrition, and functional demand. Exact manner in which G. is controlled is not known.

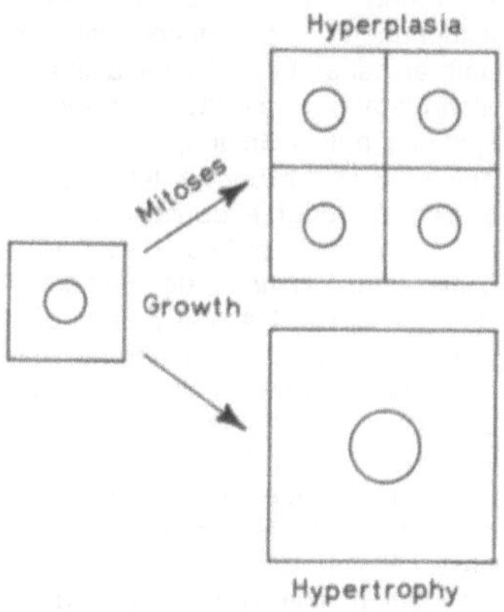

Nicolini, C. (ed.): Cell Growth. New York: Plenum Press 1982

Growth cone: ↑ Retraction bulb

Growth factors: Polypeptide substances of various origins able to control proliferation of cells by acting more or less specifically on a given cell type. G. are considered messengers between target cells and ↑ growth hormone. G. include ↑ colony-stimulating factor, ↑ epidermal growth factor, ↑ monocyte-macrophage growth factor, ↑ nerve growth factor, and ↑ somatomedins.

Adolphe, M.: Biol. Cell *38*, 1 (1980)

Growth hormone (GH, somatotropic hormone, somatotropin, STH): A polypeptide hormone (20000 daltons) produced by ↑ somatotropes of ↑ adenohypophysis. G. stimulates cell multiplication in ↑ epiphyseal plates, provokes rapid uptake of amino acids by cells, and consecutive increase in protein mass of organism. Secretion of G. is controlled by ↑ growth hormone-releasing factor and ↑ somatostatin. Action of G. on target cells is mediated by ↑ somatomedins.

↑ Acromegaly; ↑ Gigantism

Growth hormone-inhibiting factor: ↑ Somatostatin

Growth hormone-releasing factor (GRF, STH-RF, somatoliberin, somatotropic hormone-releasing factor): A hypophyseotropic peptide of unknown structure stimulating ↑ somatotrope cells to produce ↑ growth hormone.

↑ Releasing factors

Grumose bodies: ↑ Unit membrane-enclosed spherical or elongated structures (GB) (0.1–1.0 μm in length) filled with a fine granular and moderately osmiophilic material associated with Golgi apparatus and extensions of ↑ pinealocytes. G. are considered presecretory granules of pinealocytes; distinction between primary ↑ lysosomes and G. is difficult. (Fig. = rat)

Ariens Kappers, J., Schadé, J.P. (eds.): Structure and Function of the Epiphysis Cerebri. Progress in Brain Research, Vol. 10. Amsterdam, London, New York: Elsevier Publ. Co 1965

G-synapse: ↑ Synapse, G-

Gum: ↑ Gingiva

Gustatory cells: ↑ Taste buds, structure of

Gut (intestinum*): A segment of ↑ gastrointestinal tract between ↑ pylorus and ↑ anal canal. G. consists of ↑ appendix, ↑ colon, ↑ duodenum, ↑ ileum, ↑ jejunum, and ↑ rectum.

Gut-associated lymphoepithelial tissue (GALT): A ↑ lymphoepithelial tissue of ↑ appendix and ↑ Peyer's patches.

Gut endocrine cells: ↑ Endocrine cells, of gastrointestinal tract

Gut hormones: ↑ Gastrointestinal hormones

Gynospermatozoa: ↑ Spermatozoa carrying 22 ↑ autosomes and ↑ X-chromosome. G. have about 6%–7% larger nuclear surface than ↑ androspermatozoa.

↑ Sex, determination of

Portsmann, T., Portsmann, B., Rohde, W., Reich, W., Wass, R., Dörner, G.: Dermatol. Monatsschr. *165*, 514 (1979)

H

H and E staining: ↑ Hematoxylin and eosin staining

Hair (pilus*): A keratinized filiform cutaneous organ, 0.05–0.6 mm thick and from several millimeters to over a meter in length, originating from a tubular invagination of ↑ epidermis (E). H. are distributed over nearly entire surface of body. They are absent on volar surfaces and sides of hand and fingers, and plantar surfaces and sides of foot and toes, ↑ glans penis, clitoris, and ↑ labia minora. Structure: 1) ↑ Hair root = ↑ hair follicle, ↑ hair bulb, and ↑ hair papilla; 2) ↑ hair shaft; 3) ↑ arrector pili muscle (AP).

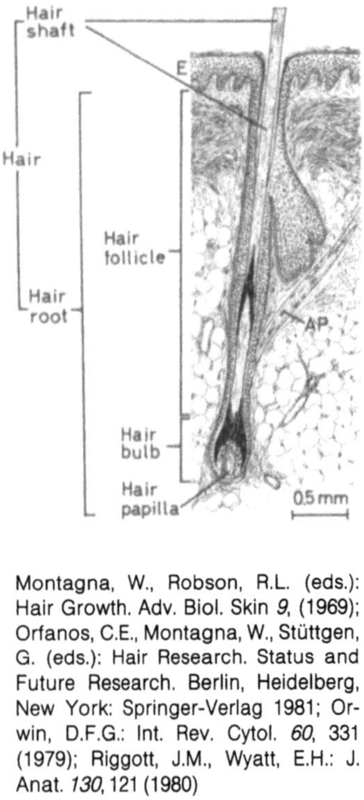

Montagna, W., Robson, R.L. (eds.): Hair Growth. Adv. Biol. Skin *9*, (1969); Orfanos, C.E., Montagna, W., Stüttgen, G. (eds.): Hair Research. Status and Future Research. Berlin, Heidelberg, New York: Springer-Verlag 1981; Orwin, D.F.G.: Int. Rev. Cytol. *60*, 331 (1979); Riggott, J.M., Wyatt, E.H.: J. Anat. *130*, 121 (1980)

Hair, allogen: ↑ Allogen hair

Hair bulb (bulbus pili*): The thickened end of ↑ hair root. Structure: 1) ↑ Matrix cells (MC) surround ↑ hair papilla (HP), divide mitotically, and form ↑ medulla (M) of ↑ hair shaft, ↑ cortex (C) of hair shaft, ↑ cuticle (Cu) of hair shaft, internal root sheath (IRS) composed of ↑ cuticle (Ct) of internal root sheath, ↑ Huxley's layer (Hx), and ↑ Henle's layer (H). 2) Among matrix cells are scattered ↑ melanocytes (Me), which transfer their ↑ melanin granules to cells of hair matrix and cortex. 3) Outer root sheath (ORS). 4) ↑ Glassy membrane (GM), which separates H. from connective tissue sheath (CTS). 5) ↑ Hair papilla (HP) = richly vascularized conical ↑ loose connective tissue with inductive and nutritive functions for hair development and growth.

↑ Hair follicle

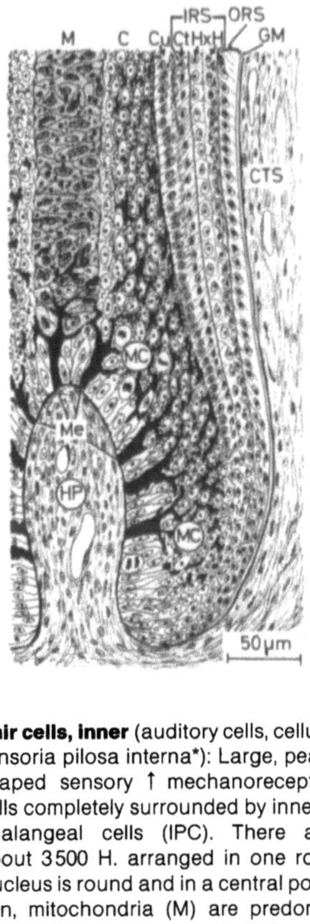

Hair cells, inner (auditory cells, cellula sensoria pilosa interna*): Large, pear-shaped sensory ↑ mechanoreceptor cells completely surrounded by inner ↑ phalangeal cells (IPC). There are about 3500 H. arranged in one row. Nucleus is round and in a central position, mitochondria (M) are predominantly infranuclear and below ↑ cuticular plate (CP). Golgi apparatus (G) is perinuclear and inconspicuous, rough endoplasmic cisternae are rare, as are ↑ lysosomes; ↑ subsurface cisternae (SC) are poorly developed. At ↑ apical pole, H. bear about 60 hairs (H) disposed in a shallow "U" shape with its convexity directed toward inner ↑ pillars (IP) and ↑ centriole (C). Toward ↑ modiolus, H. are in contact with ↑ border cells (BC). Some ↑ synaptic bars (SB) contact basal plasmalemma facing afferent nerve endings (AF); efferent nerve endings (EF) also contact base of H. Considerable morphological differences between inner and outer H. suggest that their functions may be different.

↑ Corti's organ

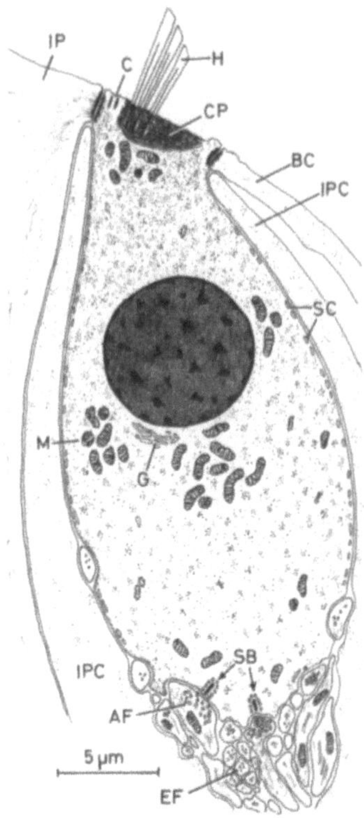

Babel, J., Bischoff, A., Spoendlin, H.: Ultrastructure of the Peripheral Nervous System and Sense Organs. Stuttgart: Georg Thieme 1970; Bruns, V., Goldbach, M.: Anat. Embryol. *161*, 51 (1980); Liberman, M.C.: Hear Res. *3*, 189 (1980)

Hair cells, of crista ampullaris: ↑ Vestibular cells

Hair cells, of maculae sacculi and utriculi: ↑ Vestibular cells

Hair cells, outer (auditory cells, cellula sensoria pilosa externa*): Tall, columnar, sensory ↑ mechanoreceptor cells each lying in cuplike depression of an outer ↑ phalangeal cell (OPC). H. have a round basal nucleus, small mitochondria (M) concentrated in both apical and basal cytoplasm and along lateral cell surfaces, an inconspicuous apical Golgi apparatus (G), numerous ↑ subsurface cisternae (SC), a few short rough endoplasmic cisternae, ↑ Hensen's bodies (HB), ↑ lysosomes (Ly), and moderate numbers of free ribosomes. ↑ Apical pole is surrounded by ↑ reticular membrane (RM) and reinforced by a well-developed ↑ cuticular plate (CP); apical pole bears 100–200 ↑ hairs (H), of which the tallest, oriented toward ↑ centriole (C), contact ↑ membrana tectoria (MT). In basal cell pole, H. have ↑ synaptic bars (SB) facing small knoblike endings of

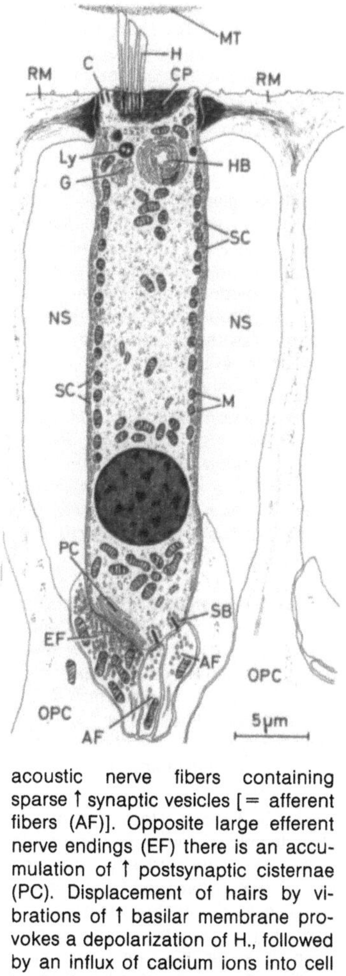

acoustic nerve fibers containing sparse ↑ synaptic vesicles [= afferent fibers (AF)]. Opposite large efferent nerve endings (EF) there is an accumulation of ↑ postsynaptic cisternae (PC). Displacement of hairs by vibrations of ↑ basilar membrane provokes a depolarization of H., followed by an influx of calcium ions into cell body. Calcium ions cause ↑ synaptic vesicles surrounding synaptic bars to discharge ↑ neurotransmitter they contain into ↑ synaptic cleft, which excites afferent nerve fibers. Thus, outer and inner H. and ↑ vestibular cells are extremely sensitive mechanoelectric transducers, converting mechanical forces into an electric signal. There are about 20000 H. arranged in three to five rows. NS = ↑ Nuel's tunnel

↑ Corti's organ

Bodian, D.: Anat. Rec. *197*, 379 (1980); Hudspeth, J.A.: Sci. Am. *248*/1, 42 (1983); Omata, T., Schaetzle, W.: Arch. Otorhinolaryngol. *229*, 175 (1980); Saito, K.: J. Ultrastruct. Res. *71*, 222 (1980); Saito, K.: Cell Tissue Res. *229*, 467 (1983)

Hair, club: ↑ Club hair

Hair, color of: H. depends on amount of ↑ melanin granules in ↑ cortex of ↑ hair shaft. Penetration of air between ↑ cortical and medullary cells or directly into latter, combined with a decrease of pigment production in aged, cause hairs to become gray or white.

Hair, connective tissue sheath of (fibrous sheath, bursa pili*): Three indistinctly separated layers (CTS) of thick collagen fibers with some ↑ fibroblasts and ↑ fibrocytes surrounding ↑ hair follicle. ↑ Glassy membrane (arrowhead) separates H. from outer root sheath (ORS). H. is derived from ↑ dermis; ↑ smooth muscle cells of ↑ arrector pili muscle are attached to it. HS = ↑ hair shaft, IRS = inner root sheath (Fig. = human)

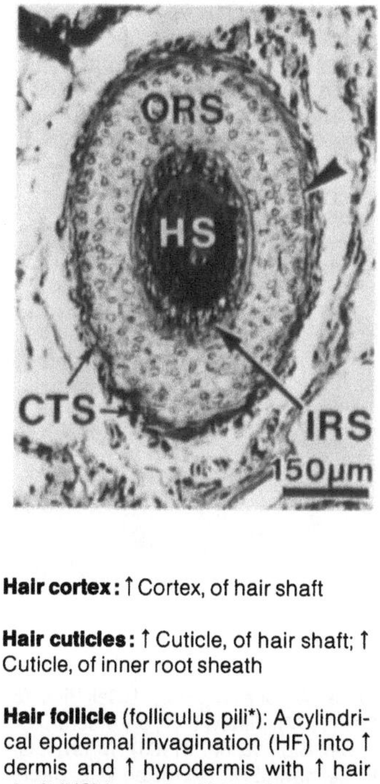

Hair cortex: ↑ Cortex, of hair shaft

Hair cuticles: ↑ Cuticle, of hair shaft; ↑ Cuticle, of inner root sheath

Hair follicle (folliculus pili*): A cylindrical epidermal invagination (HF) into ↑ dermis and ↑ hypodermis with ↑ hair shaft (HS) in center. H. terminates as a bulbous expansion, ↑ hair bulb (B), with a concavity into which a well-vascularized connective tissue, ↑ hair papilla (P) penetrates. Structure: 1) Inner root sheath (IRS) [present from papilla to level of ↑ sebaceous gland openings (SG) into H.] = a) ↑ cuticle (C) of inner root sheath; b) ↑ Huxley's layer (HxL); c) ↑ Henle's layer (HL). 2) Outer root sheath (ORS) = a direct continuation of invaginated ↑ epidermis. Its stratum corneum extends to level of sebaceous glands; stratum granulosum (G) penetrates somewhat deeper; stratum spinosum (SS) and stratum basale (SB) intermingle with ↑ matrix cells of hair. Basal lamina of epidermis is continuous with ↑ glassy membrane (GM). 2) Connective tissue

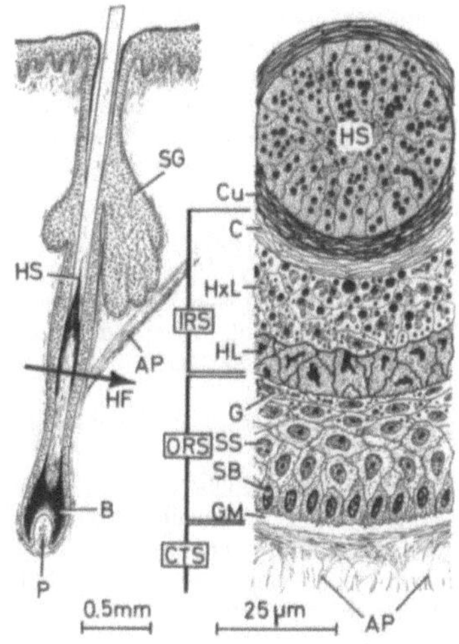

sheath (CTS). AP = ↑ arrector pili muscle, Cu = ↑ cuticle, of hair shaft

↑ Hair, connective tissue sheath of

Orwin, D.F.G.: Int. Rev. Cytol. *60*, 331 (1979); Pinkus, H., Iwasaki, T., Mishima, Y.: J. Anat. *133*, 19 (1981); Rigott, J.M., Wyatt, E.H.: J. Anat. *130*, 121 (1980)

Hair, innervation of: Two kinds of nerve fibers end in and around ↑ hairs: 1) Sensory fibers (F) run parallel to ↑ hair shaft (HS) and end in outer root sheath (ORS) where they become ↑ nonmyelinated. 2) Lanceolate sensory axon terminals (LT) are present in connective sheath (CS) of some hairs. Axons (A) are sandwiched between short ↑ Schwann's cell (SC); all three struc-

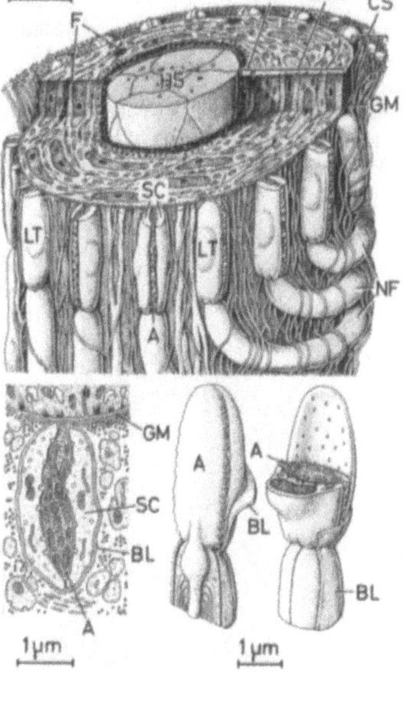

tures are oriented perpendicularly to ↑ glassy membrane (GM). Nerve fibers (NF) are ↑ myelinated up to two last Schwann's cells. Every displacement of hairs stimulates axons. IRS = inner root sheath, BL = basal lamina

Andres, K.H., Düring, M.v.: Morphology of Cutaneous Receptors. In: Handbook of Sensory Physiology, Vol.2. Berlin, Heidelberg, New York: Springer-Verlag 1973; Halata, Z., Munger, B.L.: Anat. Rec. *198*, 657 (1980)

Hair medulla: ↑ Medulla, of hair shaft

Hair papilla (papilla pili*): A cone-shaped loose ↑ connective tissue (P) with blood and ↑ lymphatic capillaries and ↑ nerve fibers occupying ↑ hair bulb (HB). H. is surrounded by ↑ matrix cells (arrowheads). (Fig. = human)

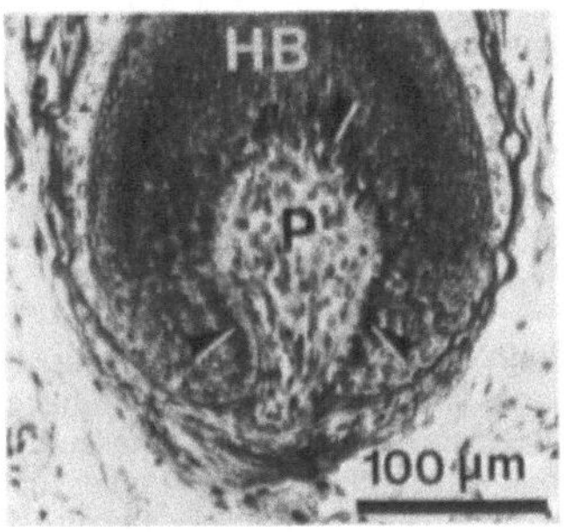

Hair, resting: ↑ Club hair

Hair root (radix pili*): Portion of ↑ hair situated below surface of ↑ epidermis. H. consists of ↑ hair follicle, ↑ hair bulb, and ↑ hair papilla. Some authors designate a group of ↑ matrix cells situated on dome of hair papilla as H.

Hair shaft (scapus pili*): The portion of ↑ hair situated in center of ↑ hair follicle and emerging above surface of ↑ epidermis. Structure: 1) ↑ Medulla (M); 2) ↑ cortex (C); 3) ↑ cuticle (Cu) of H. (Fig. = human)

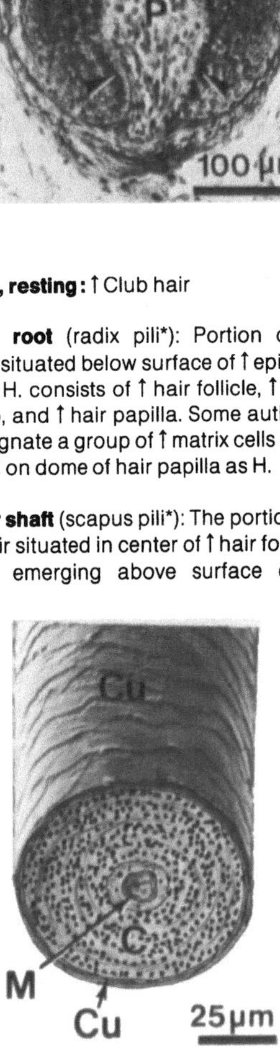

Hair, telogen: ↑ Club hair

Hairs of hair cells, of Corti's organ: About 60–200 rigid ↑ stereocilia (H), 2–12 µm long and 0.1–0.25 µm thick, arranged in two to six rows in a "W" or "U" pattern. Tallest H. are always directed toward ↑ stria vascularis; H. of following rows decrease in height. Each H. has a narrowed base; through its center a central filamentous core (CF) extends for a few micrometers and penetrates into and sometimes through ↑ cuticular plate (CP). Core is surrounded by many 12 to 15-nm-thick microfilaments (Mf), probably ↑ actin. Tallest H. contact ↑ tectorial membrane. Displacement of H. provokes a depolarization wave of ↑ hair cells (HC). OPC = outer ↑ phalangeal cell; RM = ↑ reticular membrane

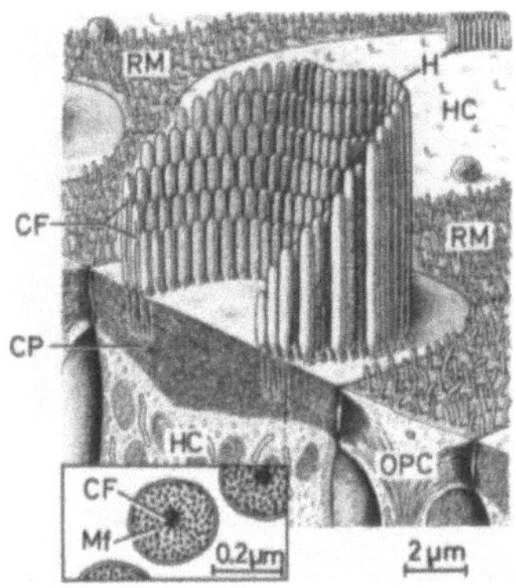

Flock, Å., Cheung, H.C.: J. Cell Biol. *75*, 339 (1977); Tilney, L.G., Saunders, J.C.: J. Cell Biol. *96*, 807 (1983); Tilney, L.G., Engelman, E.H., DeRosier, D.J., Saunders, J.C.: J. Cell Biol. *96*, 822 (1983)

Half desmosomes: ↑ Hemidesmosomes

Halo cells: A former term for intraepithelial ↑ lymphocytes and ↑ macrophages in epithelium of ductus epididymis.

↑ Ductus epididymis, epithelium of

Wang, Y.F., Holstein, A.F.: Cell Tissue Res. *233*, 517 (1983)

Hammar's myoid cells: ↑ Myoid cells, of thymus

Haploidy (cellula haploidea*): State of ↑ nucleus of ↑ spermatozoa and secondary ↑ oocytes containing only half the ↑ diploid number of ↑ chromosomes (46:2 = 23 in human) as a consequence of ↑ meiosis. Through ↑ fertilization, diploid number of chromosomes is restored.

↑ Oogenesis; ↑ Polyploidy; ↑ Spermatogenesis

Harderian glands: Accessory ↑ lacrimal glands present in a large number of vertebrates, but absent in humans. Located in orbit, H. produce a secretion for lubrication of ↑ cornea. Presence of a considerable number of ↑ lymphocytes and ↑ plasma cells in subepithelial layer suggests that H. are involved in ↑ immune response.

Nadakavukaren, M.J., Lin, W.L.: Cell Tissue Res. *233*, 209 (1983); Payne, A.P., McGadey, J., Johnston, H.S., Moore, M.R.: J. Anat. *135*, 451 (1982); Sakai, T.: Arch. Histol. Jpn. *44*, 299 (1981); Schramm, U.: Cell Tissue Res. *205*, 85 (1980)

Hassall-Henle bodies: Dome-shaped protrusions from periphery of ↑ Descemet's membrane into ↑ anterior chamber. H. are composed of atypical, randomly oriented, ↑ collagen microfibrils with a periodicity of about 100 nm. H. increase in frequency after the age of 20 years; function unknown.

Hassall's bodies (Hassall's corpuscles, thymic corpuscles, corpuscula thymica*): Concentrically arranged groups (HB) of flattened ↑ epithelial-reticular cells of ↑ thymus medulla (M). Formation of H. begins with degeneration of an epithelial-reticular cell – swelling of its nucleus and cytoplasm. Soon, this cell is surrounded by one or more other epithelial-reticular cells (EC), which are organized circumferentially and connected closely to one another by numerous ↑ desmosomes. New epithelial-reticular cells from ↑ cytoreticulum wind themselves in a concentric pattern around those already present, leaving very narrow intercellular spaces. In central cells, ↑ keratohyalin granules (KG) and numerous ↑ tonofilaments appear, the amount and size increasing with increase in H. diameter. As innermost cells become distant from blood capillaries (Cap), they swell, degenerate, and transform into keratinized (K) and/or necrotic (N) material, which may also calcify. H. frequently measure over 100 µm in diameter, increasing in number and size with age, as well as in some stressogenic situations. Function of H. is poorly understood; they may be involved in immunological reactions. Some authors consider them to be groups of exhausted epithelial-reticular cells. C = cortex, L = ↑ lymphocytes

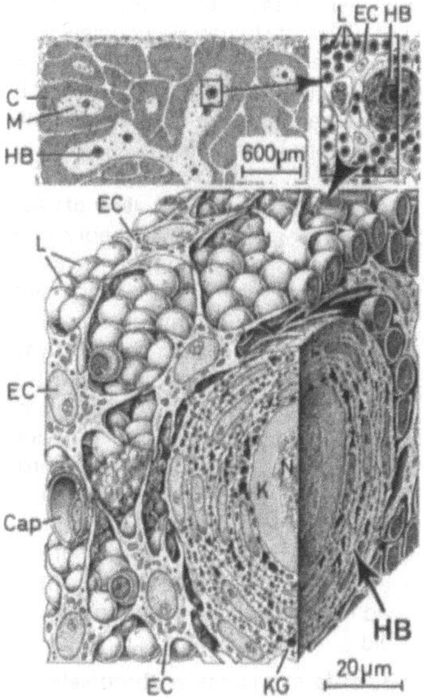

Gaudecker, B.v.: Cell Tissue Res. *186*, 507 (1978); Itoh, T., Kasahara, S., Aizu, S., Kato, K., Takeuchi, M. Mori, T.: Cell Tissue Res. *226*, 469 (1982)

Hassall's corpuscles: ↑ Hassall's bodies

Haustra*: A sacculation of colonic wall delimited by two ↑ plicae semilunares.

↑ Colon

Haversian bone: The central part of thick ↑ compact bone occupied by ↑ osteons and ↑ interstitial system of ↑ lamellae.

↑ Bone formation, secondary

Haversian canals (canalis centralis*): Vascular channels, 20–120 µm in diameter, located in center of ↑ osteons. H. are parallel to each other and to axis of long bones. H. contain one or two small blood vessels [capillaries (Cap); ↑ postcapillary venules and/or ↑ arterioles]. In ↑ perivascular space (PVS) are situated a few ↑ osteoprogenitor cells (Op), some resting ↑ osteoblasts (Ob), inactive ↑ osteoclasts (Ocl), macrophages (M), nerve fibers (NF), and ↑ lymphatic capillaries. ↑ Canaliculi of nearest ↑ osteocytes (Oc), situated between concentric ↑ lamellae (L), reach perivascular space of H. H. communicate with one another, with bone surface, and with marrow cavity through ↑ Volkmann's canals.

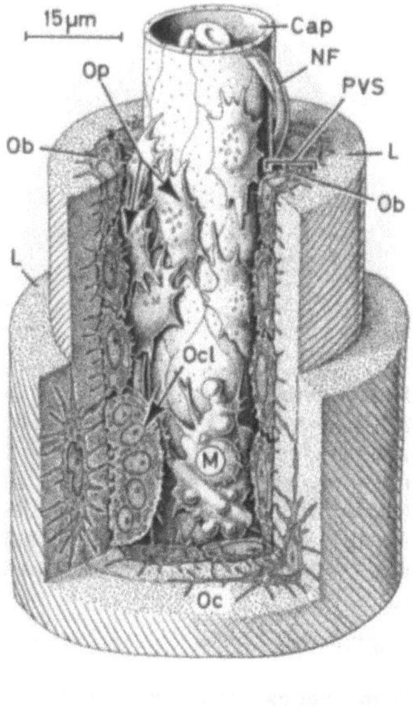

Haversian ossification: ↑ Bone formation, secondary

Haversian system, "definitive": A partially incorrect term for result of secondary ↑ bone formation, during which primary ↑ osteons are replaced by "definitive" ones. Secondary bone formation is a constant process which continues throughout life; one cannot, therefore, speak of a "definitive" Haversian system, since in the course of internal ↑ bone remodeling, second, third, fourth, fifth, and higher orders of secondary ↑ osteons occur.

Haversian system, primitive: ↑ Osteon, primary

Haversian system, secondary: ↑ Osteon, secondary

Haversian system, stages in formation of: ↑ Bone formation, secondary

H-band (stria H., zona lucida*): A clear zone (H) transversing center of ↑ A-band of cardiac and skeletal ↑ myofibrils. In middle of H. is located ↑ M-line. H. is devoid of ↑ actin myofilaments. Cross-bridges between ↑ myosin filaments are absent in narrow clear

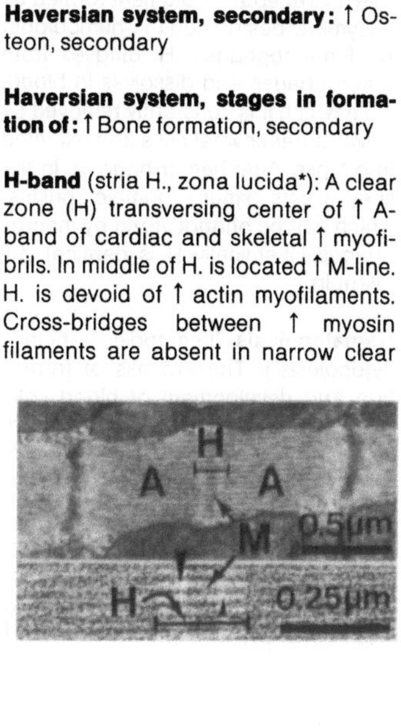

zones immediately adjacent to M-line: this zone is called ↑ pseudo H-band (arrowheads). In contracted myofibrils, H. is narrow; in relaxed myofibrils, it becomes wider. (Fig. = rat)

HCG: ↑ Human chorionic gonadotropin

HCS: ↑ Human chorionic somatomammotropin

HCT: ↑ Human chorionic thyrotropin

Head cap, of spermatozoon: ↑ Acrosomal cap

Heart (cor*): A hollow muscular organ pumping blood from ↑ veins and propelling it into ↑ arteries. Wall of H. consists of three layers: 1) ↑ Endocardium (E) = inner layer; 2) ↑ myocardium (M) = middle, contractile layer; 3) ↑ epicardium (Ep) = external layer. (Fig. = wall of human left ventricle)

↑ Cardiac skeleton

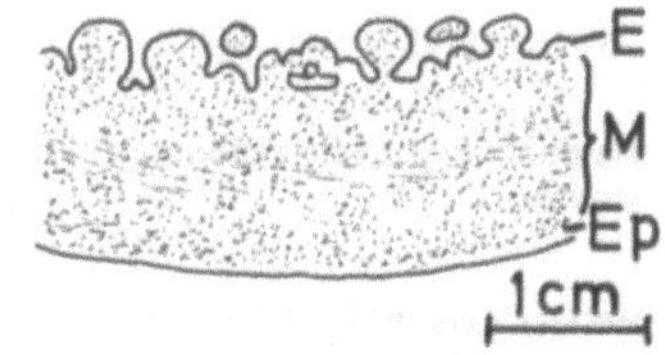

Morgan, H.E. (ed.): Cellular Biology of the Heart. New York: Academic Press 1982

Heart, annuli fibrosi of: ↑ Annuli fibrosi

Heart, chondroid tissue of: ↑ Annuli fibrosi; ↑ Chondroid tissue

Heart, epicardium of: ↑ Epicardium

Heart, impulse-conducting system of: ↑ Impulse-conducting system, of heart

Heart, innervation of: 1) Intrinsic nerves = ↑ myelinated and ↑ nonmyelinated nerve fibers, branches of vagus nerve and ↑ sympathetic trunk. 2) Sensory fibers emerge from vagus nerve and ↑ spinal ganglia; some of them end in encapsulated ↑ baroreceptors of atrial ↑ myocardium.

Heart muscle: ↑ Cardiac muscle

Heart, Purkinje fibers of: ↑ Purkinje fibers

Heart skeleton: ↑ Cardiac skeleton

Heart, valves of: ↑ Valves atrioventricular

Heart, vascularization of: ↑ The heart is supplied by two ↑ coronary arteries with branches ending as ↑ terminal arteries; venous blood is collected by cardiac veins.

Tillmanns, H., Kübler, W., Zebe. H. (eds.): Microcirculation of the Heart. Berlin, Heidelberg, New York: Springer-Verlag 1982

Heat denaturation: A physical ↑ fixation method consisting of coagulation of cell proteins at high temperature. H. is no longer used in ↑ histology, but still is in microbiology.

Heavy meromyosin (HMM): Subunit of ↑ myosin molecule subdivided into two parts: 1) H. S-1 with m.w. about 120000 daltons and 2) H. S-2 with m.w. of 60000 daltons. H. is mobile part of myosin molecule.

↑ Myosin myofilament

Heidenhein's azan staining: ↑ Azan staining

Heister's valve: ↑ Spiral valve

Helicine arteries, of penis (coiled arteries, of penis; arteria helicina*): ↑ Muscular arteries, branches of ↑ deep artery of ↑ penis. H. run through ↑ trabeculae and empty directly into cavernous sinuses of ↑ cavernous bodies. Structure: 1) Tunica intima seems to have ↑ intimal cushions; 2) very thick tunica media. When penis is in its flaccid state, H. are highly contracted and tortuous under influence of ↑ sympathetic nervous system, letting only a small amount of blood enter sinuses. During ↑ erection, under influence of parasympathetic vasodilatator nerves, H. become relaxed and straightened, allowing blood to fill sinuses.

↑ Penis, vascularization of

Böck, P., Gorgas, K.: Arch. Histol. Jpn. *40*, 265 (1977)

Helicoidal polyribosome: ↑ Polyribosome, helicoidal

Helicotrema*: The apex (H) of ↑ cochlea where scala vestibuli (SV) and scala tympani (ST) communicate. CD = ↑ cochlear duct (Fig. = mongolian gerbil)

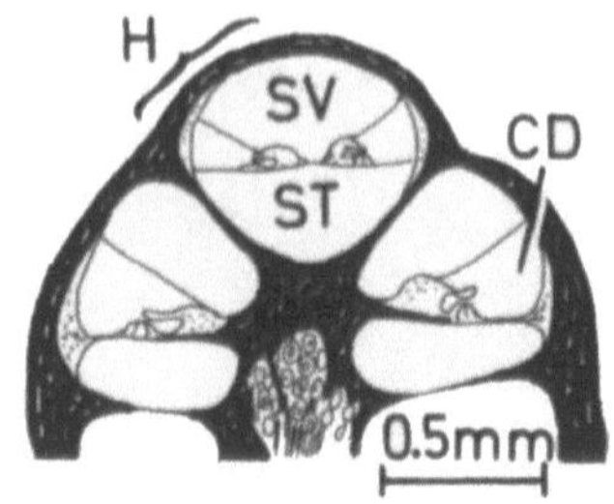

Helper cells: A subpopulation of ↑ T-lymphocytes which bring a specific antigen directly, or via surface of ↑ macrophage, to preconditioned ↑ B-lymphocytes, thus amplifying production of antibodies. H. may also interact with ↑ effector cells during ↑ immune response.

↑ Lymphatic nodules

Hemal nodes: ↑ Hemolymph nodes

Hemalum: A nuclear stain obtained by combining ↑ hematein with a ↑ mordant.

Hematein: Oxidation product of ↑ hematoxylin with very weak tinctorial capacity. Combined only with a ↑ mordant, H. can be used as nuclear stain.

↑ Hemalum

Hematocrit: The volume of ↑ erythrocytes in a sample of blood.

Hematoencephalic barrier: ↑ Blood-brain barrier

Hematoidin (granulum hematoidini*): A yellow-brown, endogenous, iron-free, extrahepatic ↑ pigment formed by enzymatic destruction of ↑ hemoglobin by ↑ macrophages. H. diffuses from macrophages and dissolves in blood, from which it is constantly removed by ↑ liver parenchymal cells and excreted into ↑ bile. After hemorrhage, H. in tissues forms small rhombic prisms, which can be phagocytized by macrophages. Chemically, H. is identical to ↑ bilirubin.

Hematopoiesis (hemopoiesis, hemocytopoiesis*): The process of formation and development of blood cells and other formed elements of blood. H. is not fully understood and, therefore, is subject of numerous controversies.

↑ Candidate stem cell; ↑ Colony-forming units; ↑ Erythropoiesis; ↑ Granulopoiesis; ↑ Hematopoiesis, current view of; ↑ Hematopoieisis, theories of; ↑ Lymphopoiesis; ↑ Megakaryopoiesis; ↑ Monopoiesis; ↑ Thrombopoiesis; ↑ Totipotent hematopoietic stem cells

Kelemen, E.M., Calvo, W., Fliedner, T.M.: Atlas of Human Hemopoietic Development. Berlin, Heidelberg, New York: Springer-Verlag 1979; Theml, H.: Taschenatlas der Hämatologie. Stuttgart: Georg Thieme 1983

Hematopoiesis, current view of: According to monophyletic theory of ↑ hematopoiesis, ↑ totipotent hematopoietic stem cell (THCS) originates from primitive reticulum cell. From former, there arises a stem cell for ↑ lymphopoiesis (↑ immunoblast) and a ↑ colony-forming unit (CFU) capable of transforming into cells of granulopoietic (CFU-G), monopoietic, erythropoietic (CFU-E), or megakaryopoietic (CFU-M) series. In the light of numerous controversies in the field of hematopoiesis, the fig. should be considered only conditional and temporarily valid.

↑ Burst-forming units-erythropoietic

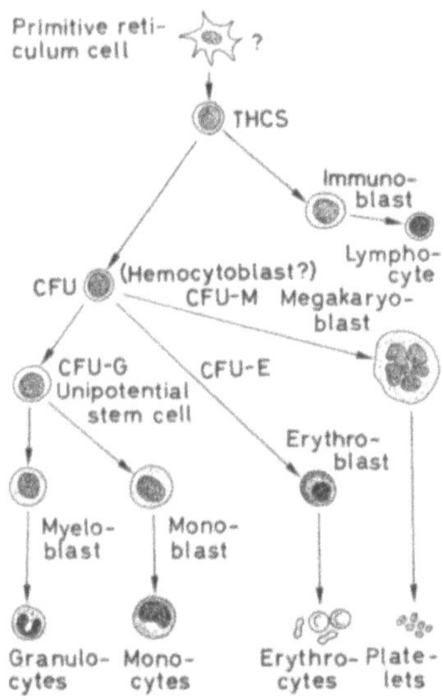

Fleischmann, R.A., Custer, P.R., Mintz, B.: Cell *30*, 351 (1982)

Hematopoiesis, extramedullary: ↑ Hematopoiesis taking place in wall of yolk sac, ↑ liver, and ↑ spleen. In prenatal life, H. is physiological.

↑ Hematopoiesis, prenatal

Hematopoiesis, fetal: ↑ Hematopoiesis, prenatal

Hematopoiesis, medullary: ↑ Hematopoiesis taking place within ↑ bone marrow.

Hematopoiesis, postnatal: ↑ Hematopoiesis throughout postnatal life. Under normal conditions, all blood cells are produced in ↑ bone marrow.

Hematopoiesis, prenatal: ↑ Hematopoiesis during intrauterine life (IL). Phases: 1) Mesoblastic phase (M), 1st–3rd embryonal months = hematopoiesis within ↑ blood islands in wall of yolk sac; nearly all cells are nucleated primitive erythroblasts. 2) Hepatosplenic phase (HS), 2nd–9th months = hematopoiesis within ↑ mesenchyme of ↑ liver and ↑ spleen (S) anlage; formation of definitive ↑ erythroblasts, which lose their nuclei about 3rd–4th embryonal months; occurrence of ↑ granulocytes. 3) Myeloid (My) or medullary phase, after 5th embryonal month = hematopoiesis within ↑ bone marrow; all hematopoietic cell lineages are present.

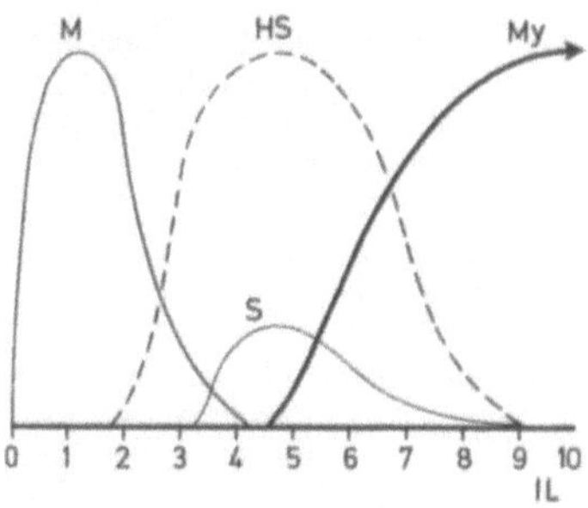

Nessi, A.C., Bozzini, C.E.: J. Embryol. Exp. Morphol. *52*, 13 (1979)

Hematopoiesis, schools of: ↑ Hematopoiesis, theories of

Hematopoiesis, theories of: ↑ Very controversial points of view intending to explain origin of formed elements of blood with regard to hematopoietic potential of earliest hematopoietic cell. 1) Monophyletic theory: All formed elements of blood derive from one kind of ↑ totipotent hematopoietic stem cells and ↑ colony-forming units by ↑ differentiation. 2) Diphyletic theory: All formed elements of blood originate from two kinds of stem cells, i.e., one stem cell (↑ myeloblast) for ↑ granulopoiesis, ↑ lymphopoiesis, and ↑ monopoiesis and another stem cell (endothelial cell of bone marrow sinuses or intersinusoidal capillaries) for ↑ erythropoiesis and ↑ megakaryopoiesis. 3) Polyphyletic theory: Each lineage of formed elements originates from a special stem cell. With discovery of colony-forming units and totipotent hematopoietic stem cells, monophyletic theory has gained in importance.

↑ Hematopoiesis, current view of

Metcalf, D., Moore, M.A.: Hematopoietic Cells. Amsterdam: North Holland Publishing Co. 1971

Hematopoietic cell lineages: ↑ Erythropoiesis; ↑ Granulopoiesis; ↑ Lymphopoiesis; ↑ Monopoiesis; ↑ Thrombopoiesis

Hematopoietic compartment, of bone marrow: All cords (C) composed of ↑ reticular, hematopoietic, and blood cells separating vascular sinuses (S). H. is occupied by ↑ adipose tissue in yellow ↑ bone marrow.

↑ Bone marrow, vascular sinuses of; ↑ Vascular compartment, of bone marrow

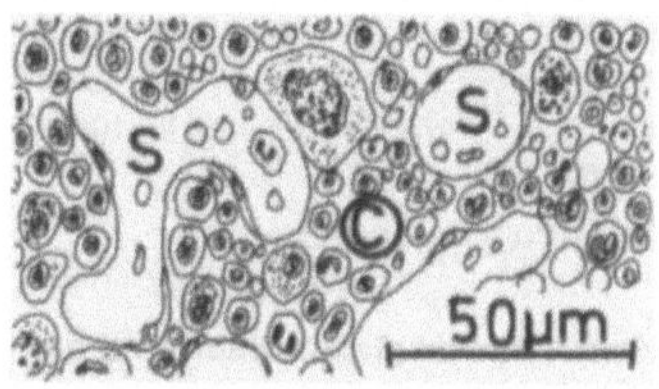

Hematopoietic stem cells: Cells capable both of ↑ differentiation into any blood cells and of maintaining themselves by mitotic division. Concept of H. is partially hypothetical, since only indirect evidence, given by ↑ colony-forming units and ↑ totipotent hematopoietic stem cells, exists about their presence. For this reason, many investigators prefer the term ↑ candidate stem cells. A calculation has shown that in adult mouse there is one H. per 10000 nucleated marrow cells.

↑ Hemocytoblast

Cairnie, A.B., Lala, P.K., Osmond, D.G. (eds.): Stem Cells of Renewing Cell Populations. New York, San Francisco, London: Academic Press 1976; Emura, I., Sekiya, M., Ohnishi, Y.: Arch. Histol. Jpn. *46*, 229 (1983); Fleischman, R.A., Custer, R.P., Mintz, B.: Cell *30*, 351 (1982); Killmann, S.V.A., Cronkite, E.P., Mullerberat, C.N. (eds.): Hematopoietic Stem Cells. Copenhagen: Munksgaard 1983; Ogawa, M., Porter, P.N., Nakahata, T.: Blood *61*, 823 (1983); Quesenberry, P., Levitt, L.: N. Engl. J. Med. *201*, 755 (1979)

Hematopoietic tissues (textus hemopoeticus*): Tissues specialized in production of blood cells. In prenatal life, H. are located in extraembryonal mesoblast of yolk sac, ↑ liver, ↑ spleen, and ↑ bone marrow. In postnatal life, only red ↑ bone marrow remains as functional H.

↑ Hematopoiesis, prenatal

Hematoxylin: A natural plant dye obtained by extraction from logwood and widely used for nuclear staining. H. is colorless: only its oxidation product, ↑ hematein, is a true dye.

Hematoxylin and eosin staining (H. and E. staining): A somewhat incorrect term for the most commonly used ↑ staining method in ↑ histology. Since hemalum and not hematoxylin stains the nuclei, the term hemalum and eosin staining would be more correct. Result: Nuclei are grayish blue, cytoplasm and intercellular substances (fibers, colloid, etc.) are various shades of pink.

↑ Eosin; ↑ Hematein

Hemidesmosomes (half desmosomes, hemidesmosoma*): A basal attachment device (Hd) of some epithelial cells responsible for firm fixation of these cells to underlying ↑ connective tissue. H. consist of a dense attaching plate (arrowhead), to which ↑ tonofilaments (Tf) anchor. From outer side of H., very fine filaments radiate into basal lamina (BL); gap between H. and the latter is about 25 nm; it contains one or several laminar densities (LD). H. are found where epithelium is exposed to considerable mechanical forces – ↑ cornea, ↑ epidermis, ↑ mouth, ↑ esophagus, ↑ vagina, etc. (Fig. = ↑ epidermis, mongolian gerbil)

↑ Basal processes; ↑ Dermoepidermal junction

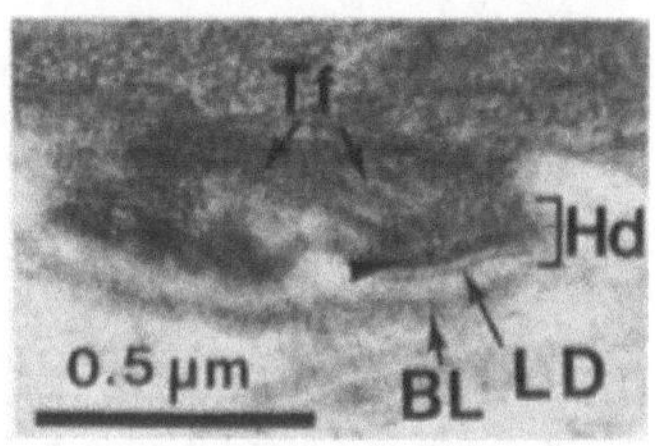

Buck, R.C.: Virchows Arch. B Cell Pathol. *41*, 1 (1982); Gipson, I.K., Grill, S.M., Spurr, S.J., Brennan, S.J.: J. Cell Biol. *97*, 849 (1983); Kelly, D.E.: Cell Tissue Res. *172*, 289 (1976); Kelly, D.E., Kuda, A.M.: Anat. Rec. *199*, 1 (1981)

Hemochorial placenta (placenta haemochorialis, placenta chorioallantoica*): A type of placenta in which ↑ placental villi are bathed by maternal blood. Human ↑ placenta belongs to this type.

Hemoconia (blood dust): Fine particles of diverse origin (fragments of blood and endothelial cells, particles of absorbed material) visible in fresh blood as bright points in ↑ dark-field microscope due to the Tyndall phenomenon.

Hemocrinia: The discharge of a hormone from ↑ endocrine cells or an ↑ endocrine gland directly into blood.

Hemocyanin: The copper-containing respiratory pigment of many invertebrates. Conjugated with certain antibodies, H. is used as a marker in ultrastructural ↑ immunochemistry.

Hemocytoblast (hemocytoblastus*): A multipotential ↑ hematopoietic stem cell, capable, according to monophyletic theory of ↑ hematopoiesis, of differentiating into any blood cell type. H. is a relatively large cell with slightly basophilic cytoplasm and a spherical nucleus with two to four nucleoli. Apart from an appreciable number of free ribosomes, cell ↑ organelles are poorly developed. According to some authors, the term H. should be replaced by the term ↑ colony-forming units, although the latter have not yet been identified morphologically.

↑ Candidate stem cell: ↑ Hematopoiesis, current view of

Ham, A.W., Cormack, D.H.: Histology. 8th edn. Philadelphia Toronto: Lippincott Co. 1979

Hemoglobin (Hb): The red, endogenous, iron-containing, respiratory pigment of ↑ erythrocytes (m.w. 64500 daltons). H. contains 600 amino acids arranged in four polypeptide chains with amino acid sequence controlled by two structural ↑ genes α and β. About 96% of human H. belongs to H. A, about 2% to H. A_2, and 2% fetal H. or H. F that predominates during intrauterine life. H. is the first of all ↑ pigments produced by developing organism; it occurs in ↑ blood islands. In ↑ bone marrow, synthesis starts at ↑ basophilic erythroblast stage. H. transports oxygen from lungs to tissue and carbon dioxide to lungs. Destruction of H. results in formation of several pigments. (See biochemistry and physiology texts for further information)

↑ Hematoidin; ↑ Hemosiderin; ↑ Ferritin

Dickerson, R.E., Geis, I.: Hemoglobin: Structure, Function, Evolution, and Pathology. Menlo Park, California: Benjamin Cummings 1983; Weatherall, D.J.: Cell *33*, 644 (1983)

Hemolymph nodes (hemal nodes, hemolymphonodus*): ↑ Lymph nodes having a high content of ↑ erythrocytes in meshes of ↑ lymphatic tissue. H. are located along large blood vessels and near ↑ kidneys and ↑ spleen; they are present in rodents and ruminants, but not in man. H. are enclosed by a dense capsule, but have no afferent lymphatic vessels; at ↑ hilus, an artery and a vein enter and leave. Function of H. consists, like in the spleen, in destruction of erythrocytes.

Hogg, C.M., Reid, O., Scothorne, R.J.: Anat. *135*, 291 (1982); Nopajaroonsri, C., Luk, S., Simon, G.T.: J. Ultrastruct. Res. *48*, 325 (1974)

Hemolysis: The swelling and disruption of ↑ erythrocytes provoked by hypotonic solutions, toxins, chemical agents, etc. accompanied by liberation of ↑ hemoglobin into surrounding medium.

↑ Erythrocyte ghost

Hemorrhoidal plexus, internal: ↑ Anal canal

Hemosiderin (granulum hemosiderini*): A water-insoluble, brownish, iron-containing endogenous ↑ pigment in form of membrane-bound granules or small dense masses in cytoplasm of ↑ macrophages. Interior of granules consists of densely packed ↑ ferritin particles containing 24%–36% iron. H. is formed during enzymatic destruction of ↑ hemoglobin by macrophages. It is normally present in very small amounts in splenic macrophages, ↑ Kupffer's cells, and ↑ reticular cells of ↑ bone marrow. It is believed to serve as a storage site for ↑ iron. H. yields positive histochemical reaction for iron.

Ghadially, F.N.: J. Submicrosc. Cytol. *11*, 271 (1979)

Henle's fiber layer, of retina: The radiating pattern of inner plexiform layer in ↑ macula lutea.

Henle's layer of hair follicle (stratum epitheliale pallidum*): A single layer (H) of flattered to cuboidal cells situated between outer root sheat (ORS) and ↑ Huxley's layer (Hx) of ↑ hair follicle. During hair growth, cells of H. gradually fill with soft ↑ keratin and flatten out while nuclei become ↑ pyknotic. Through its clear appearance, H. resembles stratum lucidum of ↑ epidermis. H. seems to be involved in

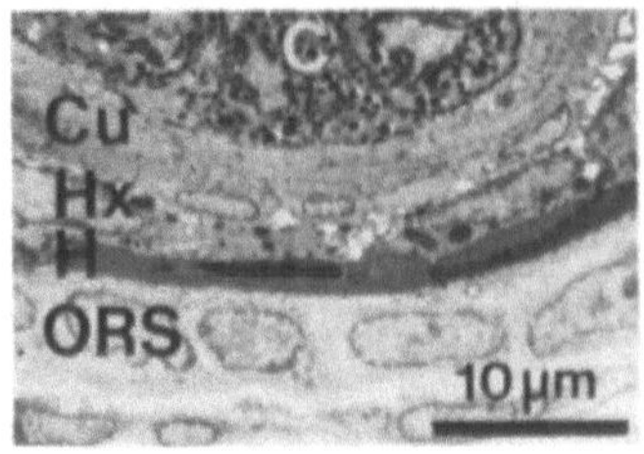

↑ keratinization and maturation of ↑ cortex (C) of hair since it desquamates at level of ↑ sebaceous glands, just where keratinization of cortex is accomplished. Cu = ↑ cuticle, of hair shaft (Fig. = mongolian gerbil)

Henle's loop: ↑ Loop, of Henle

Henle's sheath: ↑ Endoneurial sheath

Hensen's bodies: Whorllike aggregations of smooth ↑ endoplasmic cisternae, 1–2 µm in diameter, found in outer ↑ hair cells of ↑ Corti's organ; functional significance unknown.

Hensen's cells (cellulae limitantes externae*): Supporting cells (HC) of ↑ Corti's organ situated beyond outer ↑ phalangeal cells (OPC). Initially, H. are same height as phalangeal cells, but they rapidly become lower and pass into ↑ Claudius' cells (CC). H. have a round nucleus and an abundance of pale cytoplasm with poorly developed ↑ organelles. In preparations, H. are frequently very swollen. In some species, they contain prominent ↑ lipid droplets, which seem to be discharged into ↑ endolymph. It is believed that H. modulate transmission of vibrations.

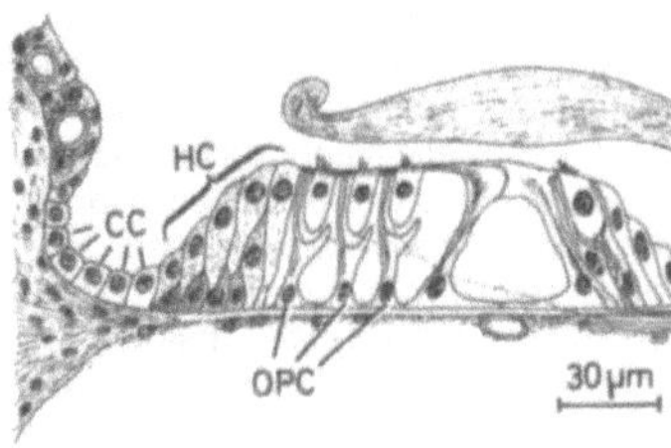

Merchan, M.A., Merchan, J.A., Ludena, M.D.: J. Anat. *131*, 519 (1980)

Hensen's stripe: ↑ Tectorial membrane

Heparan sulfate: A ↑ glycosaminoglycan present in ground substance of ↑ aorta, ↑ liver, and ↑ lung.

↑ Ground substance, of blood vessels

Heparin: A ↑ proteoglycan (m.w. 17000 daltons) contained in specific granules of basophilic ↑ granulocytes and ↑ mast cells. H. is responsible for strong ↑ basophilia and ↑ metachromasia of these granules. As a powerful blood anticoagulant and antilipemic factor, H. forms part of ↑ ground substance of ↑ connective tissue proper.

Hepatic artery (arteria hepatica*): A ↑ muscular artery with terminal branches, ↑ interlobular arteries, assuring nutrient circulation of liver stroma, extrahepatic ↑ bile ducts, and ↑ gallbladder. Only a small volume of blood enters directly into ↑ liver sinuses.

↑ Liver, vascularization of

Hepatic cells: ↑ Liver parenchymal cells

Hepatic duct, common (ductus hepaticus communis*): A canal, roughly 3 mm wide, formed by fusion of left and right hepatic ducts. Structure: 1) Tunica mucosa (TM): a) epithelium = a ↑ simple columnar epithelium; b) lamina propria = a ↑ loose connective tissue rich in ↑ elastic fibers, containing some ↑ lymphocytes and eosinophilic ↑ granulocytes. 2) Tunica muscularis (TMu) = an incomplete layer of longitudinally and obliquely oriented ↑ smooth muscle cells. 3) Tunica adventitia (TA) = a loose connective tissue with thin branches of ↑ hepatic artery.

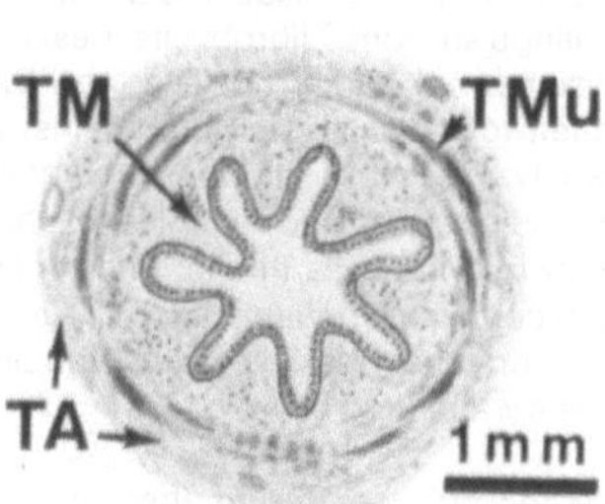

Hepatic plates (liver plates, laminae hepaticae*): Continuous, fenestrated, one-cell-thick laminae (HP) separated by ↑ liver sinusoids (S) and converging toward ↑ central vein (CV). H. are lined by ↑ endothelial (EC) and ↑ Kupffer's cells (KC). At periphery of classic ↑ lobule, H. are continuous with ↑ limiting plate (LP). IV = ↑ interlobular vein

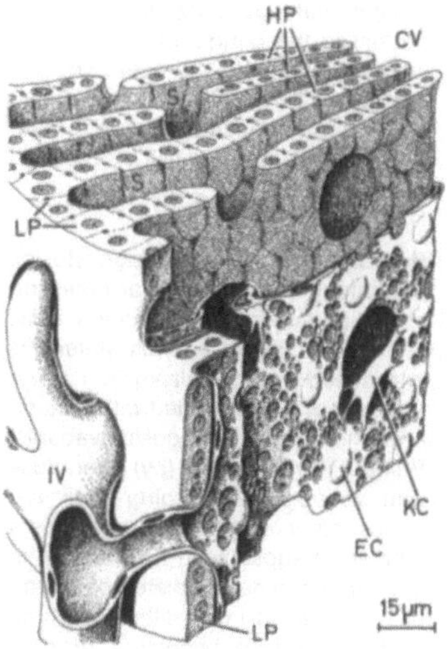

Hepatic sinusoids: ↑ Liver sinusoids

Hepatic vascularization: ↑ Liver, vascularization of

Hepatic veins (venae hepaticae*): ↑ Veins collecting blood from ↑ liver. Branches of H. are ↑ central veins, ↑ intercalated veins, and ↑ collecting veins. H. have no valves; they are surrounded by an abundant ↑ loose connective tissue.

↑ Liver, vascularization of

Hepatocytes: ↑ Liver parenchymal cells

Hepatosplenic phase, of hematopoiesis: ↑ Hematopoiesis, prenatal

Hering's canal: ↑ Bile ductules, terminal

Herring bodies (corpuscula neurosecretoria accumulata*): Spherical thickenings (HB) along ↑ neurosecretory axons (NA) of ↑ hypothalamohypophyseal tract, particularly abundant in ↑ infundibulum and ↑ neurohypophysis. H. are in fact accumulations of very numerous ↑ neurosecretory granules (NG) within ↑ axoplasm (Ax) of neurosecretory axons. H. stain brilliant blue or purple with chrome ↑ alum hematoxylin phloxin, aldehyde fuchsin, or aldehyde thionin after permanganate oxidation.

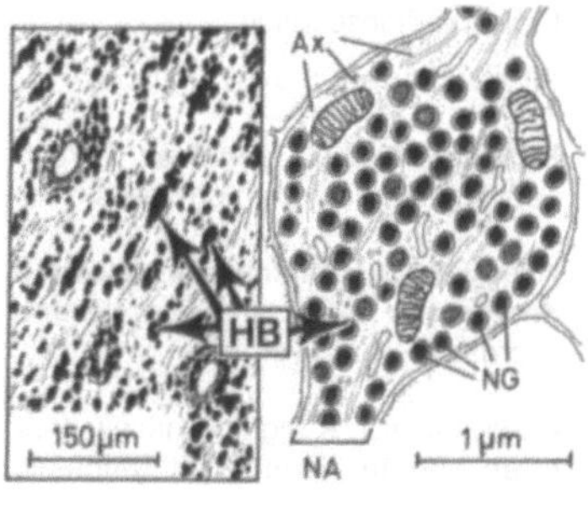

↑ Hypophysis, lobus nervosus of

Heterochromatin (heterochromatinum*): Dense masses of ↑ chromatin (H) visible in nucleus during ↑ interphase. Usually, H. is distributed along ↑ nuclear envelope (NE) and often interrupted at ↑ nucleopores (arrowheads); it is scattered as irregular clumps in ↑ nucleoplasm and partly attached to ↑ nucleolus (perinucleolar ↑ chromatin). In some cells, H. has a characteristic arrangement (↑ erythroblasts, ↑ plasma cells) and represents persisting coiling of ↑ chromatin fibers. Nuclei with large masses of H. are relatively inactive. ↑ Genes coding ↑ transfer and ↑ ribosomal RNA are located in H. Two kinds of H.: 1) ↑ H., constitutive and 2) ↑ H., facultative. ↑ Sex chromatin is also H. E = ↑ euchromatin (Fig. = ↑ endothelial cell, rat)

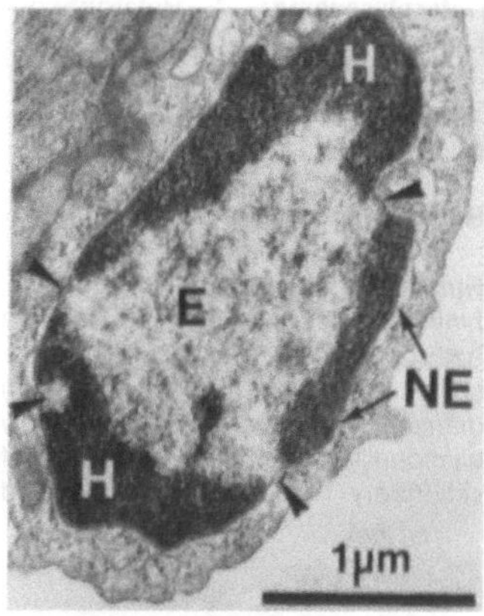

Back F.: Int. Rev. Cytol. *45*, 25 (1976); Mello, M.L.S.: Histochem. J. *15*, 739 (1983); Schwarzacher, H.G.: Chromosomes in Mitosis and Interphase. In: Bargmann, W. (ed.): Handbuch der mikroskopischen Anatomie des Menschen, Vol. 1, Part 3. Berlin, Heidelberg, New York: Springer-Verlag 1976; Valencia, R.: J. Physiol. (Paris) *78*, 653 (1982/83)

Heterochromatin, constitutive: A kind of constantly condensed ↑ heterochromatin with identical localization in homologous ↑ chromosomes. It seems that H. is genetically almost inactive;

however, its role in coding ↑ ribosomal RNA cannot be completely excluded.

De Robertis, E.D., Saez, F.A., De Robertis, E.M.F. Jr.: Cell Biology. 6th edn. Philadelphia, London, Toronto: W.B. Saunders Co. 1975

Heterochromatin, facultative: A kind of ↑ heterochromatin observed only in certain regions of one homologous ↑ chromosome, whereas corresponding regions of other homologous chromosome remain euchromatic. In regions of H., ↑ genes are inactivated. Two categories of H.: 1) Gonosomic H. = inactivated long arm of ↑ X-chromosome in female interphase nuclei, occurring as ↑ Barr body or ↑ "drumstick;" 2) autosomic H. = not sufficiently understood.

Heterochromosomes: ↑ Sex chromosomes

Heterocrine glands: ↑ Glands composed of two or more kinds of cell and secreting simultaneously two or more products (e.g., ↑ gastric glands proper).

Heterograft: ↑ Transplantation

Heterolysosome: ↑ Heterophagosome

Heterophagic vacuole: ↑ Heterophagosome

Heterophagocytosis: ↑ Heterophagy

Heterophagosome (heterolysosome, heterophagic vacuole, vacuola heterophagica*): A ↑ unit membrane-enclosed structure (Ph), 0.5–8.0 μm in diameter, with heterogenous, strongly osmiophilic contents. H. is a kind of secondary ↑ lysosome with ingested foreign material (bacteria, cells, cell debris) in the course of digestion. In later stages of this process, H. becomes morphologically identical with an ↑ autophagosome. (Fig. = ↑ macrophage with ingested cell, mouse)

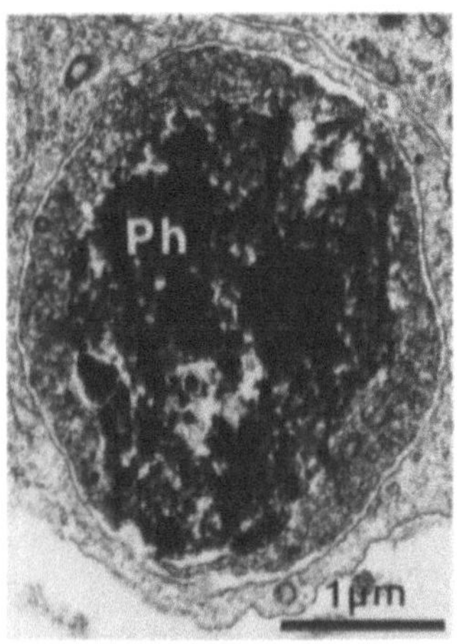

↑ Heterophagy

Heterophagy (heterophagocytosis): The intake and intracellular digestion of foreign material (M) (bacteria, cells, cell debris) by ↑ phagocytes. Material is introduced into cytoplasm by ↑ pseudopodia and surrounded with a membrane to form a ↑ phagolytic vacuole. Primary ↑ lysosomes (Ly) then fuse with vacuole, discharging their enzymes into it, and vacuole becomes a ↑ heterophagosome (Ph). Material in its interior can be completely or incompletely digested; in the latter case, indigestible residues become ↑ residual bodies (RB). Some cells are able to eliminate these by ↑ exocytosis (E); others accumulate them as ↑ lipofuscin pigment.

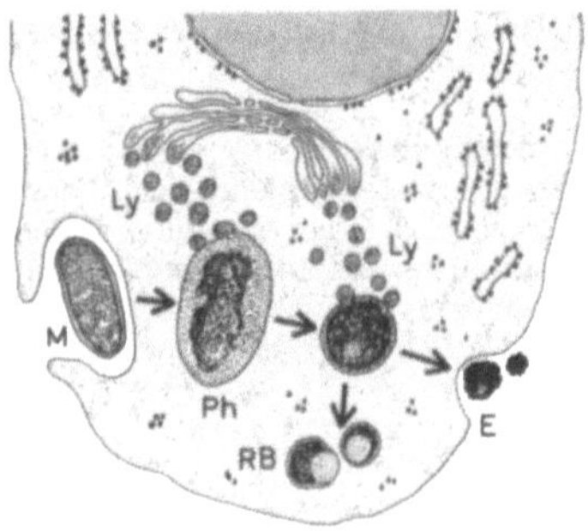

Heterophils: ↑ Granulocyte, neutrophilic

Heterosomes: ↑ Sex chromosomes

Heterotypical isocortex: ↑ Brain cortex, heterotypical isocortex of

Hibernation: Winter sleep. In animals known as hibernators all vital processes are greatly reduced during winter; body temperature, although considerably reduced, is maintained by heating of blood in interscapular mass of brown ↑ adipose tissue.

↑ Adipose tissue, brown, function of

High-endothelium venules: ↑ Postcapillary venules, of lymph nodes

High-voltage transmission electron microscope: ↑ Transmission electron microscope, high-voltage

Hilus (hilum*): A slightly shallow area of an ↑ organ where blood, lymphatic vessels, and nerve fibers enter and leave. (See fig. under ↑ Stroma)

Hilus cells: Large clustered cells located around ↑ hilus of ↑ ovary and near insertion line of ↑ mesovarium. H. contain an appreciable amount of ↑ lipochrome pigment, many ↑ lipid droplets, smooth endoplasmic tubules, and sometimes small crystalloids. Cells are closely adjacent to capillaries and ↑ nonmyelinated nerve fibers. During pregnancy and ↑ menopause, frequency of H. increases. They are believed to be endocrine cells secreting ↑ androgenic hormones.

Hilus, of kidney (hilus renalis*): A medially oriented vertical slit of kidney surface through which pass branches of renal artery, renal vein, lymphatics, nerves, and ↑ renal pelvis. H. communicates with ↑ renal sinus.

HIOMT: ↑ Hydroxyindole-O-methyltransferase

His, bundle of: ↑ Atrioventricular bundle

Histamine: A biogenic amine synthesized from histidine by decarboxylation in basophilic ↑ granulocytes, ↑ mast cells, and some ↑ endocrine cells of ↑ gastrointestinal tract. Liberated from these cells as result of an allergic or inflammatory reaction, H. provokes capillary and arteriolar vasodilatation; it stimulates ↑ phagocytosis of dead cells and their debris.

Histiocytes, fixed (macrophagocytus stabilis*): Stellate or fusiform cells stretched along ↑ collagen fibrils of ↑ loose connective tissue and difficult to distinguish from ↑ fibroblasts. Besides a somewhat darker nucleus than the latter, cytoplasm of H. contains a variety of small, ↑ unit membrane-bound granules and vacuoles. H. phagocytize vital dyes in situ; under certain conditions (inflammation), H. may abandon their sessile position and become wandering ↑ H.

Histiocytes, wandering (free macrophage, tissue macrophage, macrophagocytus nomadicus*): Ameboid and highly phagocytic cells of varying shape present in meshes of ↑ loose connective tissue, notably along capillaries. H. measure about 15 μm in diameter, nucleus is round or oval, nucleolus is conspicuous. Cytoplasm contains moderately developed ↑ organelles and a variable number of primary ↑ lysosomes (PLy), secondary ↑ lysosomes (SLy), ↑ phagolysosomes

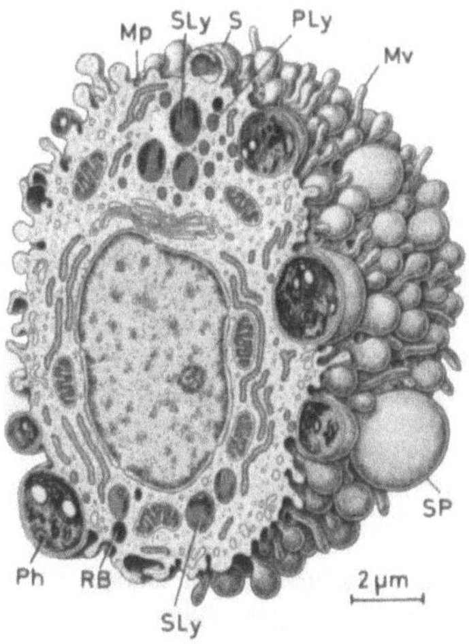

(Ph), and ↑ residual bodies (RB). Cell surface is studded with many irregular microvilli (Mv), ↑ pseudopodia of various lengths, spherical projections (SP), and short, winglike cytoplasmic sheaths (S), representing ↑ macropinocytosis. ↑ Ectoplasm encloses many micropinocytotic vesicles (Mp). H. are endowed with ↑ chemotaxis, ↑ endocytosis, phagocytic recognition, ↑ phagocytosis, and ↑ exocytosis of lysosomal enzymes and semidigested molecules in extracellular space. H. secrete certain proteins of ↑ complement system, pyrogen proteins, ↑ interferon, ↑ lysozyme, ↑ collagenase, ↑ elastase,and ↑ colony-stimulating factor. H. are actively involved in ↑ immune response and activated not only by products of microorganisms but also by ↑ lymphokines released by ↑ T-lymphocytes. Through ↑ amitosis, H. may become multinucleated cells. H. originate from fixed ↑ H. and ↑ monocytes which have left circulation. Life span. 2–3 months.

Histochemistry: The branch of ↑ histology analyzing chemical nature of tissue components by means of microscopic methods.

↑ Cytochemistry

Horobin, R.W.: Histochemistry. Stuttgart, New York: Gustav Fischer 1982; Pearse, A.G.E.: Histochemistry, Theoretical and Applied. 3rd edn. Edinburgh and London: Churchill Livingstone 1972; Stoward, P.J., Polak, J.M. (eds.): Histochemistry: The Widening Horizons of its Applications in the Biomedical Sciences. Chichester, New York: John Wiley & Sons 1981

Histocompatibility: All immunological conditions permitting a successful homograft ↑ transplantation. H. depends on degree of genetic parentage between donor and host.

Klein, J., Rammensee, H.-G., Nagy, Z.A.: Naturwissenschaften *70,* 265 (1983)

Histogenesis: The branch of ↑ embryology studying normal development of ↑ tissues.

Histological preparation: ↑ Microscopic preparation, in electron microscopy; ↑ Microscopic preparation, in light microscopy

Histological sections: ↑ Sections

Histology (histologia*): The science studying normal morphology, chemistry, and physiology of ↑ tissues. Main branches of H. are ↑ histochemistry and ↑ histophysiology. The term H. includes general ↑ H. and special ↑ H.

↑ Cytology

Fujita, T., Tanaka, K., Tokunaga, J.: SEM Atlas of Cells and Tissues. Tokyo, New York: Igaku-Shoin 1981; Kaiser, H.E.: Gebenbaurs Morphol. Jahrb. *129,* 137 (1983); Rogers, A.W.: Cell and Tissues. An Introduction to Histology and Cell Biology. London: Academic Press 1983

Histology, general (histologia generalis*): The branch of ↑ histology that includes ↑ cytology and the study of the four fundamental ↑ tissues, i.e., histology in the strict sense of the term.

Histology, methods of: ↑ Cytology, methods of

Histology, special (microscopic anatomy, histologia specialis*): The branch of ↑ histology studying structure of ↑ organs with ↑ light and ↑ electron microscopes.

Histometry: ↑ Stereology

Histones: Globular basic proteins representing a core around which DNA molecule winds to form a ↑ nucleosome. There are several classes of H. (H1, H2A, H2B, etc.); in nucleosome, core consists of four pairs of H. particles (= H. octamer). It is likely that H1 "seals off" DNA (= nucleosomal DNA) wound around H. core.

↑ Chromatin fiber

Biard-Roche, J., Gorka, C., Lawrence, J.-J.: The EMBO Journal *1,* 1487 (1982); Kornberg, R.D., Thomas, O.J.: Science *184,* 865 (1974)

Histophotometer: ↑ Cytophotometer

Histophysiology: The branch of ↑ histology studying morphological changes of ↑ tissues and ↑ organs in relation to their normal function(s). The term H. is frequently used as a synonym for function.

Historadiography: ↑ Autoradiography

↑ **Hof, of nucleus** (archiplasm): A clear cytoplasmic zone (H) adjacent to one side of nucleus, mostly in its shallow depression. Transmission electron microscopic investigations have shown that H. contains ↑ centrioles and a ↑ Golgi apparatus. H. is particularly visible in ↑ monocytes (M) and ↑ plasma cells (P).

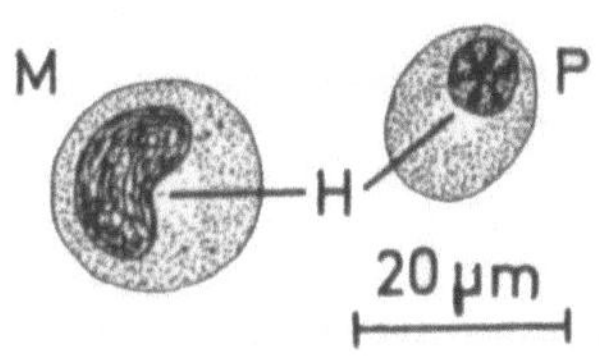

Hofbauer cells: Large, elliptical to round mobile cells present in mesodermal core of ↑ chorionic villi. Nucleus is oval or elongated with a prominent nucleolus. Cytoplasm contains numerous mitochondria, a well-developed Golgi apparatus (G), numerous but short cisternae of rough endoplasmic reticulum, a considerable number of small and large ↑ lysosomes (Ly), scattered ↑ glycogen particles, and free ribosomes. Large numbers of

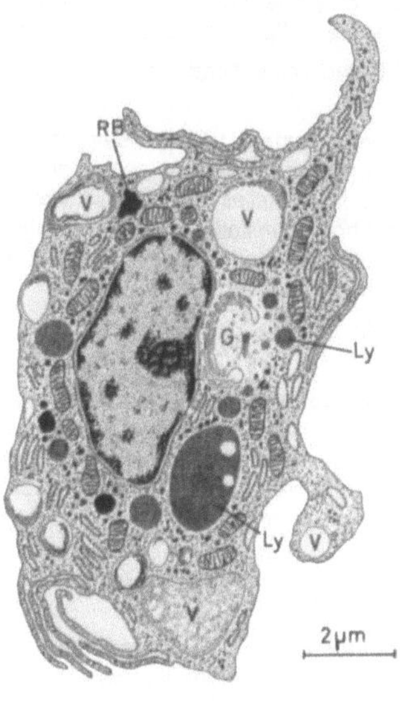

vacuoles (V), without visible contents or fine granular material, are located predominantly at cell periphery; presence of ↑ residual bodies (RB). From cell surface emerge thick cytoplasmic processes and microvillous extensions of various lengths. Since number of H. increases toward end of pregnancy, it is thought that they are involved in remodeling of chorionic stroma in relation to lengthening and branching of chorionic villi. H. are, therefore, believed to have a macrophagic role, although their exact function is unknown.

Castellucci, M., Zaccheo, D., Pescetto, G.: Cell Tissue Res. *210*, 235 (1980) and Verh. Anat. Ges. *74*, 357 (1980); Enders, A.C., King, B.F.: Anat. Rec. *167*, 231 (1970)

Holocrine glands (glandula holocrina*): All ↑ glands functioning according to mechanism of ↑ holocrine secretion.

Holocrine secretion: The gradual transformation of a whole cell into a secretory product. All ↑ sebaceous glands produce their product according to H. Production of ↑ spermatozoa is a kind of H.

"Homing": The capacity of cells to interact with one another to form ↑ tissues; "H". is controlled by carbohydrate receptors (mainly ↑ sialic acid) on ↑ cell membrane.

↑ Glycocalyx

Homocrine glands: All glands composed of only one cell type and secreting only one product (e.g., exocrine ↑ pancreas, ↑ pyloric glands).

↑ Glands, classification of

Homogenate: A highly heterogenous suspension of various cell fragments and ↑ organelles obtained as a result of ↑ homogenization. Analysis of different fractions of H. is possible through density gradient ↑ centrifugation.

↑ Microsomes

Homogenization: The procedure of mechanical disruption of cells in a special apparatus, the homogenizer. H. produces artificial structures, such as ↑ microsomes.

Homograft: ↑ Transplantation

Homologous chromosomes: ↑ Chromosomes, homologous

Homotypical isocortex: ↑ Brain cortex, homotypical isocortex of

Horizontal cells (neurocytus horizontalis*): Retinal ↑ interneurons (HC) with perikarya situated in outer part of inner nuclear layer (INL). Nucleus is large with finely structured chromatin and a prominent nucleolus; cytoplasm contains elongated mitochondria, ↑ Nissl bodies (NB), a large Golgi apparatus, some cisternae of smooth endoplasmic reticulum, a few lysosomes, a moderate number of free ribosomes, ↑ neurofilaments, and at times a rod-shaped crystalloid [= Kolmer's inclusion (KI)]. ↑ Dendrites (D) are directed exclusively toward outer plexiform layer (OPL), where they establish synaptic contact with ↑ cone pedicles (CP), ↑ rod spherules (RS), and dendrites of ↑ bipolar cells (BC). ↑ Axon (A) runs horizontally and contacts ↑ photoreceptors and bipolar cells. H. are, therefore, believed to be able to conduct impulses in both directions along their entire length. There are two functional categories of H.: 1) Luminosity H., which respond to illumination of photoreceptors by hyperpolarization; 2) chromaticity H., which may hyperpolarize and depolarize depending on wavelength of light.

↑ Retina

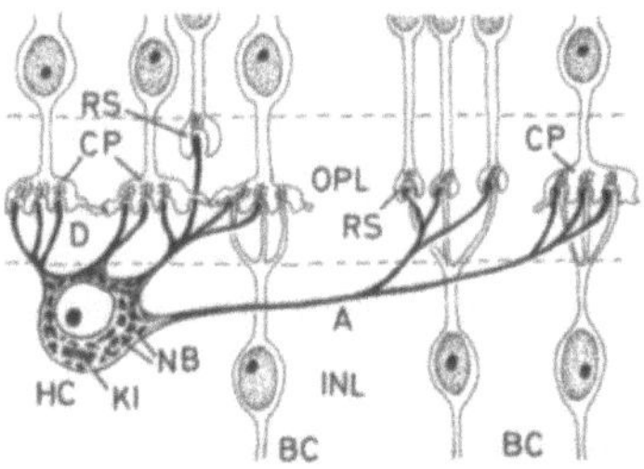

Reichenbach, A., Wohlrab, F.: Z. mikrosk.-anat. Forsch. *97*, 240 (1983); Stephan, P., Weiler, R.: Cell Tissue Res. *221*, 443 (1981); Wassle, H., Boycott, B.B., Peichl, L.: Proc. R. Soc. London [Biol.] *203*, 1152 (1978)

Horizontal cells, of Cajal: Small ↑ neurons present only in molecular layer of ↑ brain cortex. Long branching ↑ dendrites emerge horizontally from fusiform cell body; their ↑ axons form tangential fibers in molecular layer.

Hormone: A chemical substance produced in minute quantities by an ↑ endocrine cell, tissue, or organ and released either into blood, ↑ cerebrospinal fluid, ↑ lymph, or ↑ nervous tissue. An H. has a highly specific regulatory action on ↑ histophysiology of particular target cells, tissues, or organs elsewhere in the body due to specific plasmalemmal receptors for corresponding H.

↑ Polypeptide hormones, action on the cell of: ↑ Steroid hormones, action on the cell of

Marks, F. (ed.): Molekulare Biologie der Hormone. Stuttgart: Gustav Fischer 1979

Hormone, adrenocorticotropic: ↑ Adrenocorticotropic hormone

Hormone, antidiuretic: ↑ Antidiuretic hormone

Hormone, follicle-stimulating: ↑ Follicle-stimulating hormone

Hormone, growth: ↑ Growth hormone

Hormone, interstitial cell-stimulating: ↑ Interstitial cell-stimulating hormone

Hormone, lactogenic: ↑ Prolactin

Hormone, parathyroid: ↑ Parathyroid hormone

Hormone, somatotropic: ↑ Growth hormone

Hormone, thyroid-stimulating: ↑ Thyrotropic hormone

Hormones, adenohypophyseal: ↑ Adenohypophyseal hormones

Hormones, adipokinetic: ↑ Lipotropins

Hormones, adrenal cortical: ↑ Aldosterone; ↑ Corticosteroids

Hormones, gastrointestinal: ↑ Gastrointestinal hormones

Hormones, gonadotropic: ↑ Gonadotropic hormones

Hormones, hypophyseotropic: ↑ Releasing factors

Hormones, neurohypophyseal: ↑ Neurohypophyseal hormones

Hormones, of kidney: ↑ Kidney hormones

Hormones, placental: ↑ Placental hormones

Hormones, polypeptide, action on the cell of: ↑ Polypeptide hormones, action on the cell of

Hormones, releasing: ↑ Releasing factors

Hormones, sex: ↑ Sex hormones

Hormones, steroid, action on the cell of: ↑ Steroid hormones, action on the cell of

Hormones, thyroid: ↑ Calcitonin; ↑ Thyroxin; ↑ Triiodothyronine

Hormones, tissue: ↑ Tissue hormones

Horny cells (cornified cells, keratinized cells, squama cornea*): Anuclear plaques (HC) forming keratinized layer of ↑ epidermis. Interior of H. is completely filled with soft ↑ keratin (K), which gives cells physical and chemical resistance. Plasmalemma of H. is very thick; its inner leaflet (arrows) measures 15 nm and constitutes most resistant component of H. This plasmalemmal layer is formed by an amorphous protein stabilized by disulfide and undefined chemical bonds; ↑ keratin filaments anchor in this layer. Between H., remants of ↑ desmosomes and an intercellular material (arrowhead) are present. H. have a very low water content. G = granular layer (Fig., top = mongolian gerbil; fig., bottom = human)

↑ Keratinization

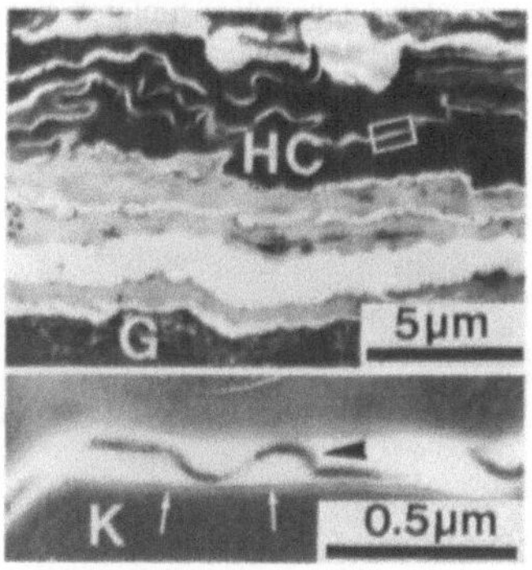

Dale, B.A., Holbrook, K.A., Steinert, P.M.: Nature *276*, 729 (1978)

Horny layer: ↑ Epidermis

Horny matrix: ↑ Keratin, soft

Horseradish peroxidase (HRP): An exogenous hemoprotein enzyme for light and transmission electron microscopic identification of protein transport pathway. Intravenously injected, localization of H. is visualized under light microscope after incubation of ↑ sections with hydrogen peroxide and benzidine. For demonstration of H. in transmission electron microscope, blocks are treated with 3,3'- ↑ diaminobenzidine (DAB), which in presence of H. forms an insoluble osmiophilic polymer, osmium black, at site of peroxidase activity. H. is used for localization of numerous enzymes; it is also very useful in ultrastructural ↑ immunocytochemistry.

Bundgaard, M., Møller, M.: J. Histochem. Cytochem. *29*, 331 (1981); Hanker, J.S.: In: Barrnett, J.R. (ed.): Symposium in Electron Cytochemistry. New York: John Wiley & Sons, Inc. 1976

Hortega's cell: ↑ Microglia

Howell-Jolly bodies: Remnants of nonextruded nucleus from orthochromatic ↑ erythroblast.

Howship's lacunae (resorption lacunae, lacunae erosionis*): Shallow concavities (HL) on surface of ↑ bone (B) and ↑ dentin, generally occupied by ↑ osteoclasts (Oc) and probably formed by their erosive action.

↑ Absorption cavity; ↑ Teeth, replacement of

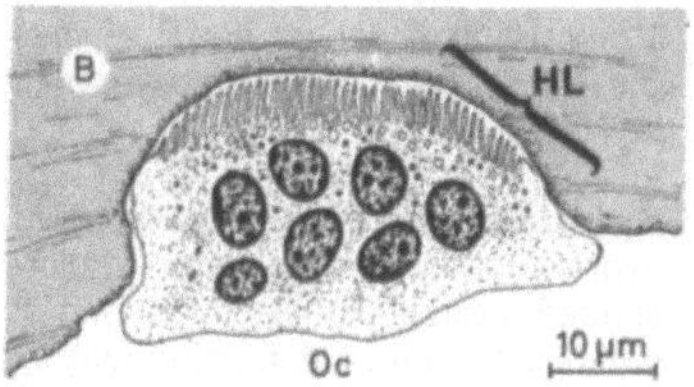

Loutit, J.F., Nisbet, N.W.: Lancet *2*, 26 (1979)

Hoyer-Grosser's organs: ↑ Glomi

HPL: ↑ Human chorionic somatomammotropin

Human chorionic gonadotropin (HCG, hCG): A placental ↑ glycoprotein hormone probably produced by ↑ cytotrophoblast converting ↑ corpus luteum menstruationis into corpus luteum graviditatis and maintaining function of the latter until 4th–5th month of pregnancy.

Dreskin, R.B., Spicer, S.S., Greene, W.B.: J. Histochem. Cytochem. *18*, 862 (1979); Kasai, K., Yoshida, Y.: Acta Histochem. Cytochem. *12*, 283 (1979)

Human chorionic somatomammotropin (HCS, human placental lactogen, HPL): A placental polypeptide hormone probably produced by ↑ syncytiotrophoblast with lactogenic and growth-promoting effect; considered the growth hormone of pregnancy.

Human chorionic thyrotropin (HCT): A placental protein hormone probably synthesized by ↑ syncytiotrophoblast stimulating development and secretion of ↑ thyroid gland during pregnancy.

Human placental lactogen: ↑ Human chorionic somatomammotropin

Humor aquosus*: ↑ Aqueous humor

Humor vitreus: ↑ Vitreous body

Humoral immunity: Part of ↑ immune response based on ability of ↑ B-lymphocytes to transform themselves into ↑ plasma cells, which then secrete ↑ immunoglobulins into blood.

Hunter-Schreger lines: ↑ Lines, of Hunter-Schreger

Huschke's auditory teeth: ↑ Auditory teeth

Huxley's layer, of ↑ hair follicle (stratum epitheliale granuliferum*): One to three strata of flattened to cuboidal cells (Hx) situated between ↑ Henle's layer (H) and ↑ cuticle of inner root sheath (C). Cells of H. contain ↑ trichohyalin granules (arrowheads) and thick bundles of ↑ tonofibrils. During ↑ hair growth, cells of H. become completely filled with soft ↑ keratin and desquamate at level of ↑ sebaceous gland openings. ORS = outer root sheath. (Fig. = mongolian gerbil)

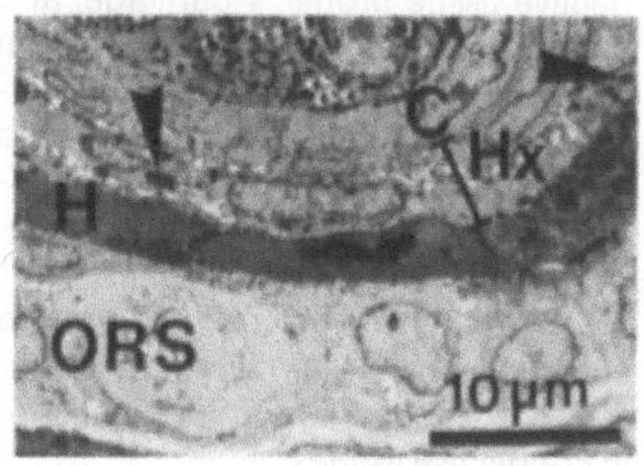

Hyaline cartilage: ↑ Cartilage, hyaline

Hyalocytes (hyaloid cells, vitreous cells): Cells of ↑ vitreous body. H. are stellate elements with irregular processes, an oval or elliptical nucleus, a prominent Golgi apparatus, and some ↑ lysosomes. In addition to phagocytic activity, H. are involved in synthesis of microfibrils and hydrophilic ↑ glycosaminoglycans (predominantly ↑ hyaluronic acid).

Hyaloid canal (canal of Cloquet, canalis hyaloideus*): A remnant (C) of hyaloid artery of ↑ vitreous body extending between ↑ lens and ↑ papilla of optic nerve (P). H. is about 1 mm wide and very inconspicuous in postnatal life. R = ↑ retina. (Fig. = mongolian gerbil)

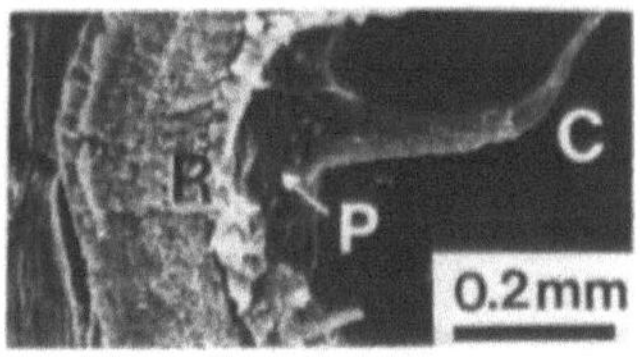

Hyaloid cells: ↑ Hyalocytes

Hyaloid membrane: The slightly thickened anterior part of ↑ vitreous body capsule.

Hyaloideocapsular ligament: Fine microfibrillar connections between ↑ lens capsule and ↑ vitreous body.

Hyalomere (hyalomerus*): The peripheral cytoplasmic rim of ↑ platelets without granules and ↑ organelles.

Hyaloplasm: ↑ Cytoplasmic matrix

Hyaluronic acid: A nonsulfated ↑ glycosaminoglycan with about 2500 repetitive units of N-acetylglycosamine and D-glucuronic acid. Bound to a protein and polymerized, H. gives viscosity to ↑ ground substance of ↑ connective tissue proper, ↑ cartilage, and blood vessels; it is found in ↑ synovial fluid and in ↑ vitreous body. In ↑ gelatinous tissue, H. is highly polymerized and hydrophilic, which explains a certain turgescence of this tissue. Combined with sulfate ions, H. is responsible for increased viscosity of ground substance. H. acts as lubricating medium in articulations and as plasticizer in connective tissues.

Fraser, J.E.R., Appelgren, L.-E., Laurent, C.T.: Cell Tissue Res. *233*, 285 (1983); Singley, C.T., Solursh, M.: Histochemistry *65*, 93 (1980)

Hyaluronidase ("spreading factor"): An enzyme capable of depolymerizing ↑ hyaluronic acid of ↑ ground substance of ↑ connective tissues. H. of ↑ spermatozoon dissolves ↑ zona pellucida of secondary ↑ oocyte. Bacterial H. spreads infections in connective tissues.

Hybrid arteries: A short segment presenting an abrupt transition between ↑ elastic and ↑ muscular arteries. Tunica media is composed of a distinct inner muscular layer and an external layer with ↑ elastic laminae. Visceral branches of abdominal ↑ aorta are H.

Hydatid, of Morgagni (appendix vesiculosa*): A lateral, cystlike enlargement of ↑ epoophoron.

Hydration: The process of gradual introduction of water into histological ↑ sections after removal of ↑ paraffin. H. is accomplished by transferring sections into progressively decreasing concentrations of alcohol, ending with water, in which sections are stained.

Hydrencephalocrinia: The discharge of ↑ hormone(s) from ↑ endocrine cells into ↑ cerebrospinal fluid.

Hydrocortisone: ↑ Glucocorticoids

Hydrotic cells, of bronchial glands: Cells with osmotic function in ↑ bronchial glands in some animals.

Hydroxyapatite [$Ca_{10}(PO_4)_6(OH)_2$]: The main mineral of ↑ acervuli, ↑ bone, ↑ cementum, ↑ dentin, and ↑ enamel present in form of ↑ hydroxyapatite crystals.

Hydroxyapatite crystals (crystallum hydroxyapatiti*): Needle- or leaflet-shaped crystals (H) 40–60 nm long, 2 nm thick, 30 nm wide, forming bulk of ↑ acervuli (A), ↑ bone, ↑ cementum, ↑ dentin, and ↑ enamel (E). In bone, cementum, and dentin, H. are bound mainly to ↑ collagen microfibrils and lie parallel with them; H. may also appear within microfibrils, in ↑ ground substance, and in ↑ matrix vesicles. In enamel, H. are more densely packed and much longer than in bone and dentin (300–1000 nm long, 40–120 nm thick). In acervuli, H. are considerably less numerous, but have same form and chemical composition as in bone. Total surface area of 1 g H. is enormous and varies from 60 to 250 m^2. H. amass Na^+, K^+, Mg^{2+} ions, and water molecules on their surfaces; this layer, the "hydration shell" is involved in exchange of ions between, H. and body fluids and acts as a buffer. (Fig., top = pineal acervulus, human; fig., bottom = enamel, newborn rat)

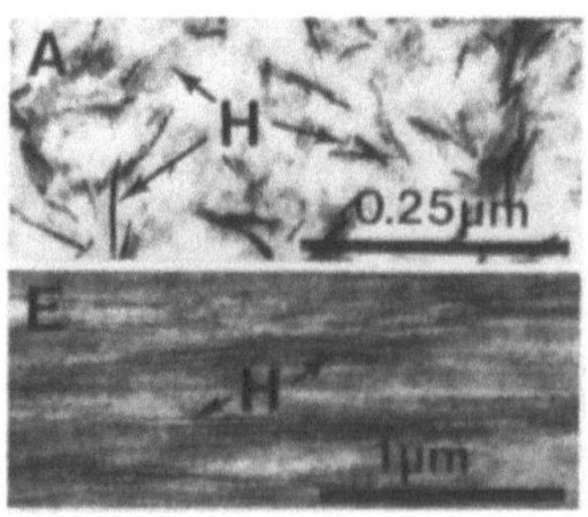

↑ Collagen, calcification of

Arends, J., Jongloed, W.L.: J. Biol. Buccale *6*, 161 (1978); Kerebel, B., Daculsi, G., Kerebel, L.M.: J. Dent. Res. *58*(B), 844 (1979)

Hydroxyindole-O-methyltransferase (HIOMT): An enzyme present in ↑ pinealocytes catalyzing N-acetylserotonin to ↑ melatonin. Activity of H. is inhibited by light, melatonin synthesis takes place, therefore, during the night.

↑ Pineal organ, histophysiology of

5-hydroxytryptamine: ↑ Serotonin

Hymen*: A transversal semilunal fold or fenestrated membrane, 1–3 mm thick, situated at lower end of ↑ vagina in the virgin. H. is covered at vulvar and vaginal sides by ↑ stratified squamous nonkeratinized epithelium (E) firmly interdigitated with underlying ↑ connective papillae (P). Skeleton of H. forms moderately vascularized ↑ loose connective tissue (CT) rich in ↑ elastic fibers. After defloration, ruptures cicatrize in 2–3 days, but H. never regenerates. VF = vaginal fold

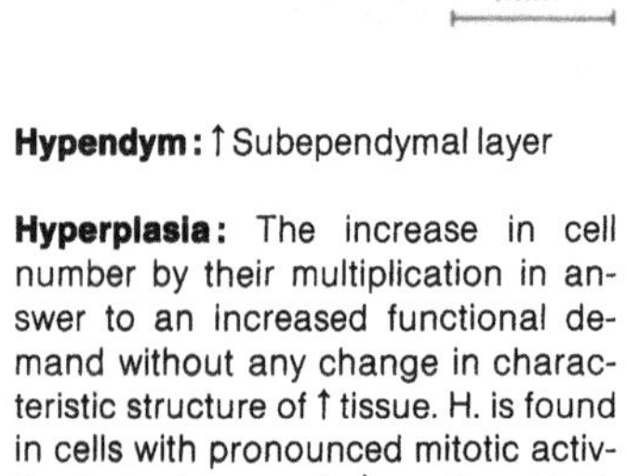

Hypendym: ↑ Subependymal layer

Hyperplasia: The increase in cell number by their multiplication in answer to an increased functional demand without any change in characteristic structure of ↑ tissue. H. is found in cells with pronounced mitotic activity (e.g., increased ↑ erythropoiesis during prolonged journeys at high altitudes).

↑ Growth; ↑ Hypertrophy; ↑ Involution

Hypertrophic zone, of hyaline cartilage: ↑ Bone formation, indirect

Hypertrophy: The increase in cell volume resulting from an increased functional demand. H. is found in highly differentiated cells with very weak mitotic activity (e.g., H. of myometrial ↑ smooth muscle cells in pregnancy, H. of ↑ cardiac muscle cells in sportsmen).

↑ Growth; ↑ Hyperplasia

Anderson, W.A., Sadler, W. (eds.): Perspective in Differentiation and Hypertrophy. New York: Elsevier/North Holland 1982

Hypodermis (subcutis, tela subcutanea*): The deepest layer of ↑ skin, consisting of ↑ loose connective tissue with numerous blood and lymphatic vessels, ↑ nerve fibers, scattered ↑ glomi, ↑ corpuscles of Vater-Pacini, ↑ hair bulbs, and a variable amount of ↑ adipose tissue (panniculus adiposus).

Spearman, R.I.C.: Structure and Function of Subcutaneous Tissue. In: Jarret, A. (ed.). Physiology and Pathophysiology of the Skin. Vol. 7. London: Academic Press 1982

Hyponychium*: The epidermal thickening under free edge of ↑ nail plate.

↑ Eponychium

Hypophyseal follicles: Saclike cell groups found in pars distalis and pars intermedia of ↑ hypophysis. Whereas H. of pars distalis are composed of various hypophyseal cells, those in pars intermedia consist only of basophilic and chromophobic cells. H. of pars intermedia are remnants of hypophyseal cleft. A colloid of unknown chemical nature fills H.

↑ Follicular cells, of hypophysis

Hypophyseal stalk (pedunculus infundibularis*): A narrow tissue strand (HS) connecting ↑ hypophysis with diencephalon (D). H. consists of pars tuberalis (PT) of ↑ adenohypophysis (A) incompletely surrounding ↑ infundibulum (I).

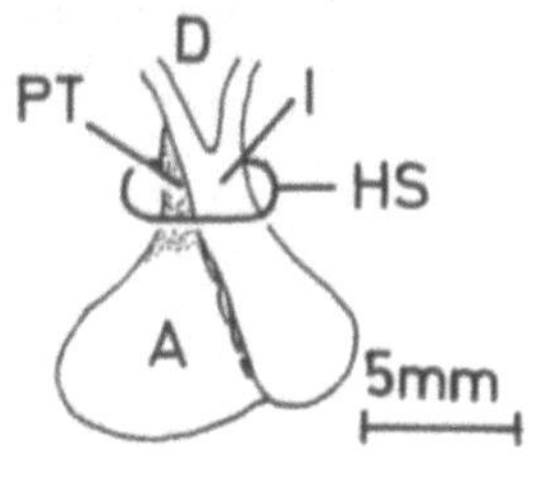

Hypophyseoportal system: ↑ Venules connecting capillaries of ↑ median eminence and ↑ infundibulum with capillaries of pars distalis of ↑ hypophysis.

↑ Hypophysis, vascularization of

Hypophyseotropic hormones: ↑ Releasing factors

Hypophysis* (glandula pituitaria*): An endocrine gland weighing about 0.5 g, located at base of brain in sella turcica and covered by diaphragma sellae. Structure: 1) Adenohypophysis*: a) pars tuberalis; b) pars intermedia; c) pars distalis (all originating from ↑ Rathke's pouch). 2) Neurohypophysis*): a) ↑ median eminence; b) ↑ infundibulum*; c) lobus nervosus* (all of nervous origin). H. is enveloped by a thick capsule (C) of ↑ dense connective tissue continuous with ↑ dura mater.

↑ Hypophysis, lobus nervosus of; ↑ Hypophysis, pars distalis of; ↑ Hypophysis, pars intermedia of: ↑ Hypophysis, pars tuberalis of

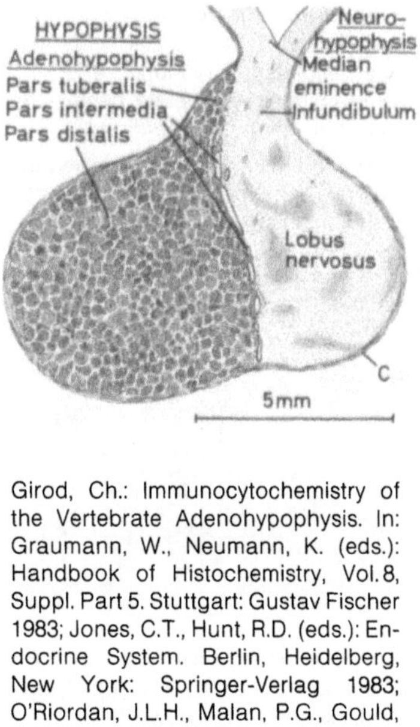

Girod, Ch.: Immunocytochemistry of the Vertebrate Adenohypophysis. In: Graumann, W., Neumann, K. (eds.): Handbook of Histochemistry, Vol. 8, Suppl. Part 5. Stuttgart: Gustav Fischer 1983; Jones, C.T., Hunt, R.D. (eds.): Endocrine System. Berlin, Heidelberg, New York: Springer-Verlag 1983; O'Riordan, J.L.H., Malan, P.G., Gould, R.P.: Essentials of Endocrinology. Oxford: Blackwell Scientific Publications 1982; Osamura, R.Y., Nakane, P.K.: Acta Histochem. Cytochem. *15*, 294 (1982)

Hypophysis, acidophilic cells of (acidophils, α-cells, cellula acidophilica*): About 40% of cells in ↑ adenohypophysis whose ↑ secretory granules have affinity for acid dyes (↑ eosin, orange G, azocarmine) and appear pink or red in sections stained with ↑ azan or ↑ hematoxylin and eosin. ↑ Somatotropes and ↑ mammotropes are H.

Hypophysis, basophilic cells of (basophils, β-cells, cellula basophilica*): About 10% of cells of ↑ adenohypophysis whose secretory granules have a common affinity for basic dyes (aniline blue, methylene blue, resorcin fuchsin), as well as giving a ↑ PAS-positive reaction. Three kinds of H. – ↑ gonadotropes type I and II and ↑ thyrotropes.

Hypophysis, chromophobic cells of (γ-cells, reserve cells, cellula chromophobica*): A feebly stainable group of cells of ↑ adenohypophysis. Under light microscope cells appear pale and devoid of secretory granules. On the basis of immunocytochemical and transmission electron microscopic studies, it has been found that this group of cells consists of relatively large ↑ corticotropes and small ↑ precursor cells; both kinds together make up about 50% of cells of pars distalis.

↑ Precursor cells, of adenohypophysis

Hypophysis, lobus nervosus* of (neurohypophysis, pars nervosa): Part of ↑ hypophysis posterior to pars tuberalis and pars intermedia; H. is a direct continuation of bottom of diencephalon. Structure: Main mass of H. forms termination of fibers (F) of ↑ hypothalamohypophyseal tract. After special staining (fig., bottom), numerous ↑ Herring bodies (HB) and accumulations of neurosecretory material around capillaries (C) become visible. The genuine cells of H. are ↑ pituicytes (P); at times, some ↑ choristoma cells

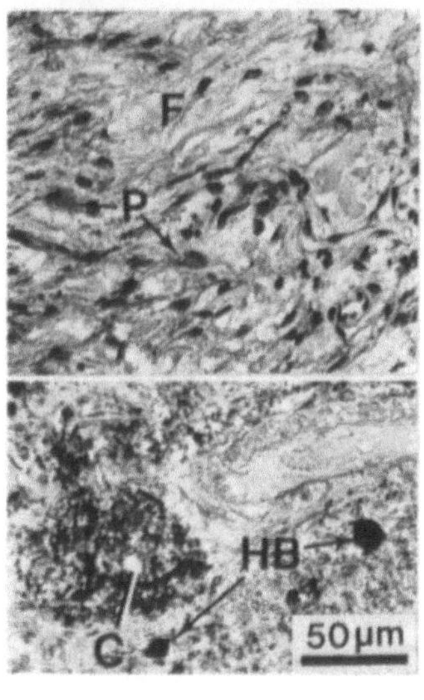

can be found. In H. are stocked hypothalamic hormones ↑ oxytocin and ↑ antidiuretic hormone, which are transported by ↑ axoplasmic flow from ↑ nucleus supraopticus and paraventricularis. H. is abundantly vascularized, mainly by inferior hypophyseal arteries. (Fig., top = human; fig., bottom = dog, chrome alum hematoxylin phloxin staining)

Gainer, H., Sarne, Y., Brownstein, M.J.: J. Cell Biol. *73*, 366 (1977); Moses, A.M., Share, L. (eds.): Neurohypophysis. Basel: Karger 1977

Hypophysis, pars distalis* of: The largest part of ↑ hypophysis delimited by a segment of its capsule, pars tuberalis and pars intermedia. Structure: Irregular clumps and anastomosing cords of epithelial cells surrounded by basal lamina and a dense network of ↑ reticular fibers. Stroma consists of rare, mostly perivascular collagen fibers. Cell cords are separated by an extensive number of fenestrated ↑ capillaries, i.e., ↑ sinusoids (S). According to their affinity to acid and basic dyes, cells of H. are subdivided into acidophilic (A), basophilic (B), and chromophobic (C) cells. Immunohistochemical studies have shown that H. is composed of ↑ corticotropes, ↑ gonadotropes, ↑ mammotropes, ↑ thyrotropes, and ↑ precursor cells. H. controls, via ↑ hypothalamus or directly, adrenal cortex, ↑ thyroid, gonads, ↑ growth, etc.; it is indispensable for life. Disappearance of target organs results in a considerable change in composition of cell population in H. with occurence of new cell types (↑ castration cells, ↑ thyroidectomy cells). (Figs. = human)

↑ Hypophysis, acidophilic cells of; ↑ Hypophysis, basophilic cells of; ↑ Hypophysis, chromophobic cells of

Herbert, D.C., Silverman, A.Y.: Cell Tissue Res. 230, 233 (1983)

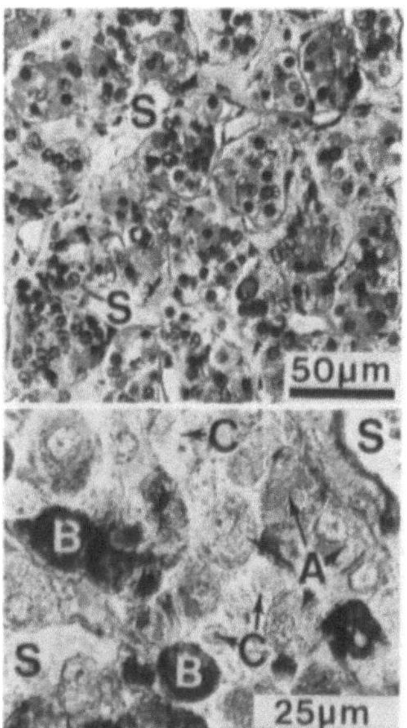

Hypophysis, pars intermedia* of: The part (PI) of ↑ hypophysis located between pars distalis (PD) and lobus nervosus (LN). Structure: Irregular strands, groups, and follicles (F) of basophilic and chromophobic cells. Basophilic cells can invade pars nervosa (↑ basophilic infiltration). Tubular glands of mucous type surrounded by a small amount of ↑ lymphatic tissue can be found in pars intermedia. H. is relatively poorly vascularized by short capillary loops originating from superior hypophyseal artery and connected by short portal venules with capillary network of pars distalis. Some ↑ neurosecretory hypothalamic nerve fibers penetrate H. from pars nervosa. H. produces ↑ melanocyte-stimulating hormone and, in some animals, the opiate peptide β-endorphin. (Fig. = human)

↑ Hypophysis, pars intermedia of, chromophobic cells of; ↑ Hypophysis, vascularization of

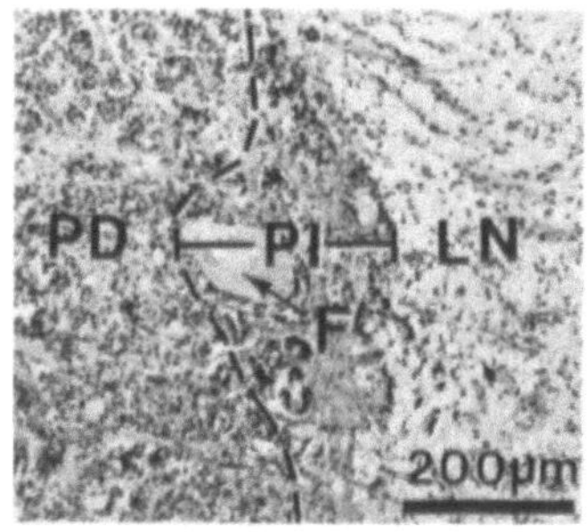

Perry, R.A., Robinson, P.M., Ryan, G.B.: Cell Tissue Res. *217*, 211 (1981); Saland, L.C.: Anat. Rec. *200*, 315 (1981)

Hypophysis, pars intermedia of, chromophobic cells of: Weakly stained polygonal cells (arrows) scattered among basophilic cells (BC). It is not clear whether chromophobic cells represent a functional stage of basophilic cells or a different kind of cell producing a specific hormone. (Fig. = human)

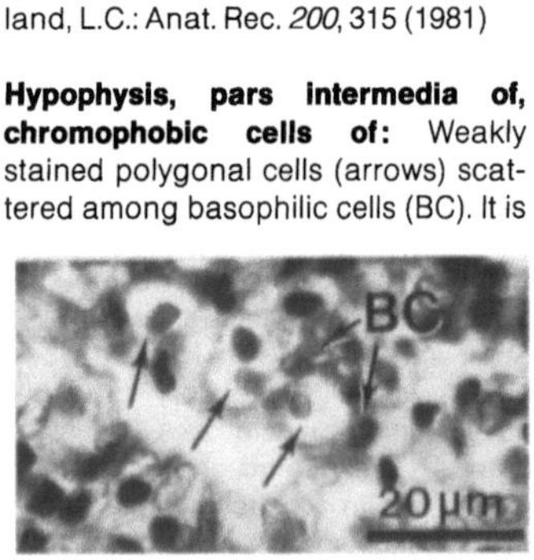

Hypophysis, pars nervosa of: ↑ Hypophysis, lobus nervosus of

Hypophysis, pars tuberalis* of (pars infundibularis): A small portion (PT) of ↑ adenohypophysis extending upward from pars distalis (PD) and incompletely surrounding ↑ infundibulum (Inf). A small segment of H. can be present on posterior aspect of infundibulum. Anterior surface of H. is covered by ↑ dura mater (D); H. is separated from infundibulum by ↑ pia mater (arrowheads). Predominantly basophilic cells of H. form longitudinally arranged cords and follicles (F). Through H. pass numerous capillaries and some portal vessels. There is no special hormone secreted in H. (Figs. = human)

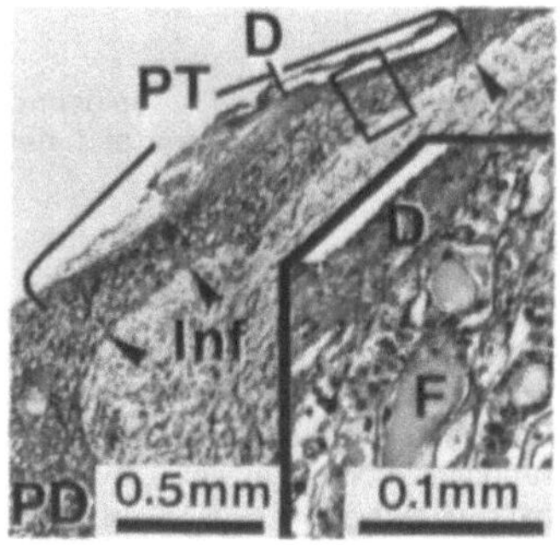

Hypophysis, reserve cells of: ↑ Hypophysis, chromophobic cells of

Hypophysis, stellate cells of: ↑ Stellate cells, of adenohypophysis

Hypophysis, vascularization of: Hypophysis is supplied with blood from two pairs of superior and inferior hypophyseal arteries, branches of internal carotids. 1) Superior hypophyseal arteries (SHA) form, in pars tuberalis (PT), special capillary loops (CL) that penetrate ↑ median eminence (ME) and infundibulum (I). These loops are drained by portal veins (PV) connecting capillary plexus of pars tuberalis with that of pars distalis (PD). A branch of superior hypophyseal arteries, loral artery (LA), bypasses portal circulation and forms similar, but considerably smaller, capillary loops in pars intermedia (PI), whose short portal veins (PV) anastomose with capillaries of pars distalis. 2) Inferior hypophyseal arteries (IHA) form a capillary network within pars nervosa (PN), anastomosis of which with capillaries of pars distalis is rare. Direct anastomosis between superior and inferior hypophyseal arteries can exist. Capillary networks empty into hypophyseal veins (HV) that drain into cavernous sinus (CS). Neurosecretory fibers of ↑ tuberoinfundibular tract (TIT) end on capillary loops of pars tuberalis, whereas fibers of ↑ supraoptico- and paraventriculohypophyseal tract (SHT) terminate on capillary loops of pars distalis. Venous blood, carrying ↑ releasing factors and hypophyseal hormones, follows direction of the arrows.

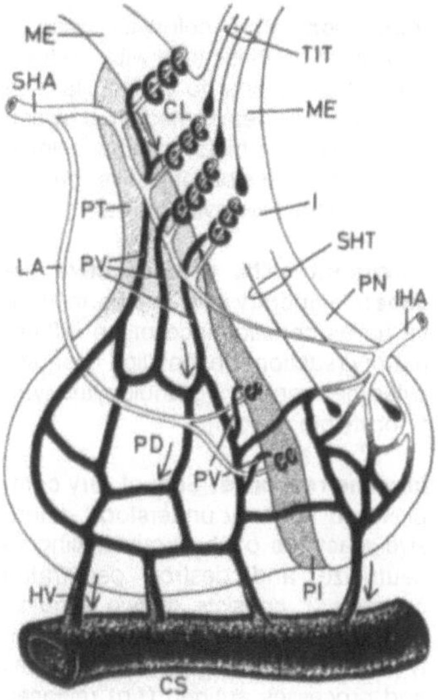

Hypoplasia*: Underdevelopment of a ↑ tissue or ↑ organ due to decrease in number of cells.

Hypothalamohypophyseal complex: Intimate morphofunctional relationship between ↑ hypothalamus and ↑ hypophysis. H. is considered to be the highest level of control of endocrine functions of the organism.

↑ Hypothalamohypophyseal tract

Hypothalamohypophyseal tract (tractus hypothalamohypophysialis*): A densely packed bundle (HHT) of 50000–100000 nonmyelinated ↑ neurosecretory axons arising from various hypothalamic nuclei and converging on infundibular stem to end in ↑ neurohypophysis and ↑ infundibulum. Composition: 1) Supraoptico- and paraventriculohypophyseal tract (SHT) = axons arising from cell bodies of ↑ supraoptic (NSO) and ↑ paraventricular nucleus (NPV) and ending in ↑ pericapillary spaces of pars nervosa (PN). 2) Tuberoinfundibular tract (TIT) = axons originating from nucleus ventromedialis (NVM), dorsomedialis (NDM), and infundibularis (NI) ending around capillary loops (CL) in infundibular stem (I). H. represents a way of transport of hypothalamic hormones by ↑ axoplasmic flow: Supraopticohypophyseal tract carries ↑ antidiuretic hormone and ↑ oxytocin, whereas tuberoinfundibular tract transports ↑ releasing factors.

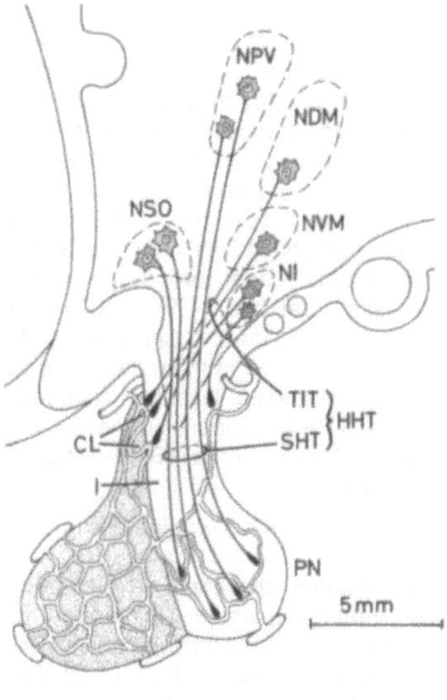

Hypothalamus*: Basal part of diencephalon delimited rostrally by lamina terminalis (LT), basally by chiasma opticum (CO), ↑ median eminence (ME), and tuber cinereum (TC), caudally by corpora mammilaria (CM), and laterally by nucleus subthalamicus (not shown). ↑ Neurons of H. form more or less distinctly demarcated groups, the nuclei, which are subdivided into two groups: 1) Myelin-poor H. nuclei = nucleus preopticus (NP), ↑ nucleus supraopticus (NSO), ↑ nucleus paraventricularis (NPV), nucleus anterius (NA), nucleus ventromedialis (NVM), area dorsalis (AD), nucleus dorsomedialis (NDM), nucleus posterius (NPo), nucleus infundibularis (NI), nuclei tuberales (NT) and lateral H. zone (LHZ). 2) Myelin-rich H. nucleus = nucleus corporis mammilaris (CM). H. and ↑ hypophysis are connected by ↑ hypothalamohypophyseal tract passing through infundibulum (I) and hypothalamic capillaries, which join portal veins of ↑ adenohypophysis. In this way, a large morphofunctional unit arises, the ↑ hypothalamohypophyseal complex. Functions: 1) H. acts as a neuroendocrine gland, since its nuclei (NSO and NPV) synthesize ↑ oxytocin and ↑ antidiuretic hormone, which reach ↑ neurohypophysis (NH) by ↑ axoplasmic flow. 2) H. is the highest center for control and coordination of numerous endocrine functions based on production of ↑ releasing factors in its nucleus ventromedialis and nucleus infundibularis. 3) H. is central region for control of important vegetative functions. (Modified after Schiebler et al. 1977)

↑ Hypophysis, vascularization of

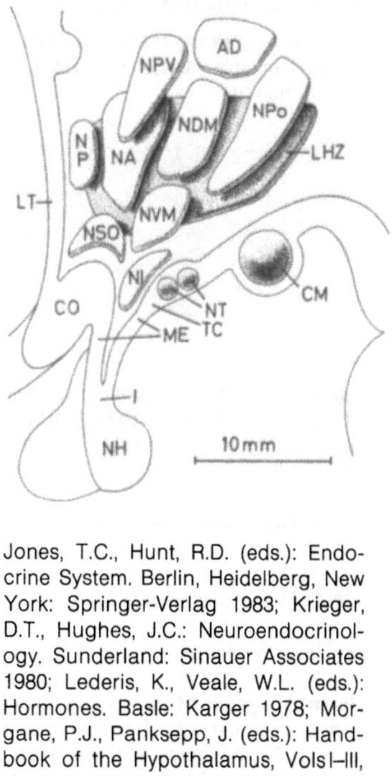

Jones, T.C., Hunt, R.D. (eds.): Endocrine System. Berlin, Heidelberg, New York: Springer-Verlag 1983; Krieger, D.T., Hughes, J.C.: Neuroendocrinology. Sunderland: Sinauer Associates 1980; Lederis, K., Veale, W.L. (eds.): Hormones. Basle: Karger 1978; Morgane, P.J., Panksepp, J. (eds.): Handbook of the Hypothalamus, Vols I–III, New York and Basle: Marcel Dekker 1980–1982; Schiebler, I.H. et al.: Lehrbuch der gesamten Anatomie des Menschen. Berlin, Heidelberg, New York: Springer-Verlag 1977

I

I-bands (stria I*): Lightly stained isotropic segments (I) along ↑ myofibrils of ↑ cardiac muscle cells and ↑ skeletal muscle fibers. I. consist only of hexagonally arranged ↑ actin myofilaments (a) anchored in ↑ Z-line, which is located precisely in middle of I. During contraction, I. shorten due to deeper penetration of actin myofilaments between ↑ myosin myofilaments of contiguous ↑ A-bands. (Fig. = rat)

↑ Sarcomere, transversal sections of: ↑ Striated muscle fiber, mechanism of contraction of

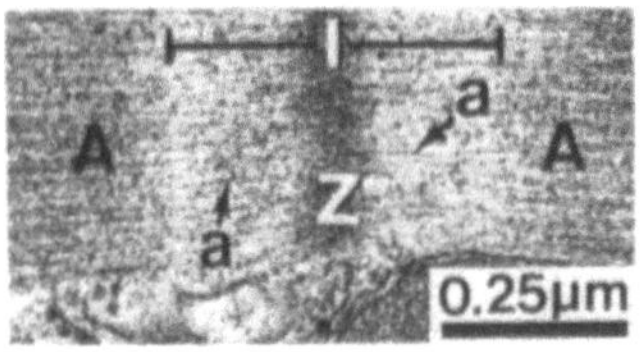

I-cells: A category of ↑ endocrine cells of ↑ gastrointestinal tract which probably secrete ↑ cholecystokinin-pancreozymin.

ICSH: ↑ Interstitial cell-stimulating hormone

iDNA: ↑ Internucleosomal DNA

Ig: ↑ Immunoglobulins

Ileum*: Segment of small intestine between ↑ jejunum and ↑ cecum. Structure: 1) Tunica mucosa (TM): a) epithelium (E) = a ↑ simple columnar epithelium covering ↑ intestinal villi (V) and crypts of ↑ Lieberkühn (C); b) lamina propria (LP) = a ↑ loose connective tissue forming core of each villus and lying between crypts; c) lamina muscularis mucosae (LMM) = a layer of ↑ smooth muscle cells. Villi of I. are less numerous and smaller than those of jejunum; ↑ plicae circulares (PC) are few or absent. 2) Tela submucosa (TS) = a loose connective tissue with ↑ Peyer's patches (PP). 3) Tunica muscularis ((TMu) = smooth muscle tissue organized in inner circular and outer longitudinal layers. 4) Tela subserosa (TSs) = a loose connective tissue. 5) Tunica serosa (TSe) = visceral ↑ peritoneum.

↑ Absorptive cells; ↑ Goblet cells

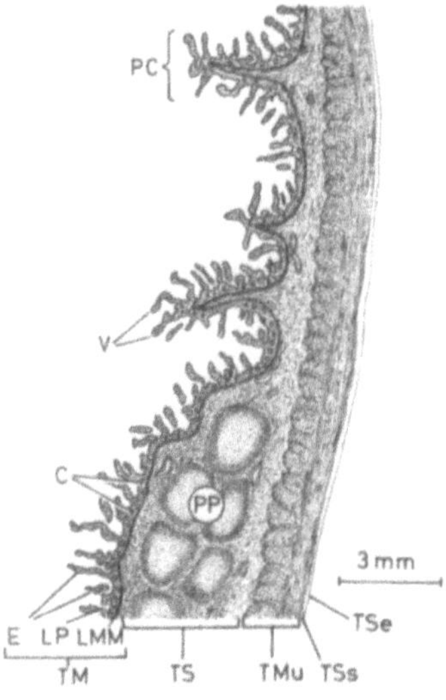

Illumination, diascopic: A technique of light-microscopic illumination in which light passes through object. Depending on wavelength (λ) or on kind of light, I. is used in the following branches of ↑ light microscopy: 1) Ultraviolet light (λ < 400 nm) = a) ↑ fluorescence microscopy; b) ↑ histophotometry. 2) Shortwave light (λ 400–500 nm) = fluorescence microscopy in blue light. 3) Ordinary light (λ about 550 nm): a) ↑ bright-field microscopy; b) ↑ dark-field microscopy; c) ↑ phase contrast microscopy; d) ↑ interference microscopy. 4) Monochromatic light = histophotometry. 5) Infrared light = histophotometry. 6) Polarized light = ↑ polarizing microscopy.

Illumination, epidiascopic: A combined technique for illumination in light microscope using simultaneously transmitted and reflected light.

↑ Illumination, diascopic; ↑ Illumination, episcopic

Illumination, episcopic: A technique of light-microscopic illumination in which only reflected light (RL) from specimen (S) is used for formation of image. Ultraviolet light, shortwave light (blue fluorescence), and ordinary light may be employed for I. O = ↑ objective

↑ Fluorescence microscope

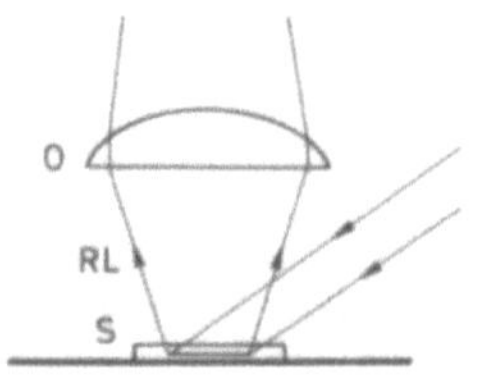

Immediate sensitivity: The antibody-mediated ↑ immune response to antigen accompanied by an instantly visible reaction (edema, pain, shock, fever, etc.).

Immersion fixation: The simplest method of chemical ↑ fixation consisting simply of dropping an object into fixation solution. Advantage – simplicity; disadvantages – requirement of small tissue blocks, ↑ artifacts due to progressive penetration of fixation solution (↑ autolysis of central cells, etc.). I. is indicated for fixation of avascular or weakly vascularized tissues of organs (↑ epidermis, ↑ cartilages, ↑ lens, etc.).

Immersion objective: ↑ Oil-immersion objective

Immersion oil: A colorless, viscous, natural or synthetic fluid with ↑ refractive index identical to crown glass (n = 1.515). I. serves as medium between ↑ coverslip of a histological specimen and ↑ oil-immersion objective, permitting maximal ↑ resolving power.

Immigrant cells, of connective tissues: Temporary cells in a connective tissue as consequence of an inflammatory reaction (basophilic ↑ granulocytes, neutrophilic ↑ granulocytes, lymphocyte precursors).

Immune response: Sum of very complex and not fully understood defensive reactions of the body aiming to neutralize and destroy penetrated antigens. I. consists of two simultaneous processes: 1) Humoral immunity. ↑ Macrophage (M) phagocytizes and processes antigen (Ag) released from an antigen-producing agent (bacteria, endogenous antigens, grafts, parasites, viruses, etc.), rendering it immunogenic. Macrophage interacts directly with a ↑ T-lymphocyte by means of specialized complementary structures on its surface. T-lymphocyte becomes activated and produces microhumoral substances and/ or directly cooperates with a ↑ B-lymphocyte, which in turn differentiates via a B-immunoblast into ↑ plasma cell (P). The latter secretes ↑ immunoglobulin G (IgG) into blood, neutralizing the antigen. A certain number of ↑ B-lymphocytes remain as ↑ memory cells (Mem) with a special surface ↑ immunoglobulin M. In subsequent contacts with the antigen (Ag), the immunoglobulin M links with it; antigen-antibody complex is then interiorized by ↑ micropinocytosis into the B-lymphocyte, inducing its ↑ differentiation into a plasma cell. Mediated by ↑ immunoglobulins, I. is rapid and referred to as ↑ immediate sensitivity. Memory cells have

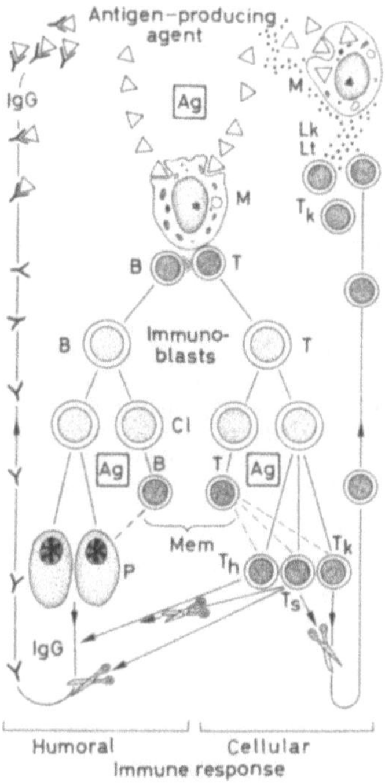

a life span of several years. 2) Cellular immunity. Stimulated by the macrophage and its processed antigen, T-lymphocyte proliferates and forms, probably via T-immunoblasts, a ↑ clone (Cl) from which several subpopulations of T-lymphocytes differentiate: a) Helper cells (T_h), which interact with B-lymphocytes and plasma cells (P) to amplify production of IgG; b) killer cells (T_k), which release cytotoxic substances such as ↑ lymphotoxin (Lt) and ↑ lymphokines (Lk), activating phagocytic activity of local and passing macrophages against the antigen and antigen-producing agent (= inflammation); c) suppressor cells (T_s), which control activity of B-lymphocytes, helper and killer cells. Since cellular I. takes longer to manifest itself clinically, it is referred to as delayed hypersensitivity. Part of the T-lymphocytes remain as memory cells in circulation and react to repeated penetration of antigen by transformation into above-cited lymphocyte subpopulations.

↑ Immunoblast

Bier, O.G., Dias Da Silva, W., Götze, D., Mota, I.: Fundamentals of Immunology. Berlin, Heidelberg, New York: Springer-Verlag 1981; Feldmann, M., Rosenthal, A., Erb, P.: Int. Rev. Cytol. *60*, 149 (1979); Hood, L.E., Wood, W.B.: Immunology. Amsterdam: Addison Wesley 1978

Immune response, primary: A series of morphological and biochemical events resulting in proliferation and ↑ differentiation of ↑ B-lymphocytes into ↑ immunoblasts, ↑ memory cells, and ↑ plasma cells following primary contact with an appropriate antigen. I. may extend over a period of several weeks.

Immune response, secondary: A rapid reaction of ↑ memory cells accompanied by high increase in circulating antibodies (IgG) after secondary contact with antigen that provoked primary ↑ immune response.

Immune system: All cells, tissues, and organs protecting the body from external and internal antigens. I. comprises ↑ lymphatic and ↑ lymphoepithelial organs (↑ lymph nodes, ↑ thymus, ↑ tonsils, and ↑ spleen), all aggregated and solitary ↑ lymphatic nodules scattered in various organs, and all ↑ lymphocytes, ↑ plasma cells, and ↑ macrophages of organism.

Davies, G.E.: Histochem. J. *13*, 879 (1981); Kimball, J.W.: Introduction to Immunology. Berne: Hans Huber 1983; McConnel, I., Munro, A., Waldman, A.: The Immune System. London: Blackwell Scientific 1981; Schwartz L.M. (ed.): Compendium of Immunology. Vols. I–III. Florence: S & AE Scientific and Academic Editions 1980–1983

Immunoblast (blast cell, centroblast, germinoblast, pyrononphilic cell, lymphoblast, lymphoblastus*): A large mononuclear cell, up to 25 µm in diameter, arising from both small ↑ B- and ↑ T-lymphocytes following antigen, mediator, or mitogen stimulation. Thus, two kinds of morphologically identical B- and T-I. exist, each of which generates B-lymphocytes or T-lymphocytes; B-I. also differentiate into ↑ plasma cells. I. has a large round or ovoid nucleus, numerous peripheral clumps of ↑ heterochromatin, and one to three prominent nucleoli. Cytoplasmic rim, 1–4 µm wide, contains several small spherical mitochondria, a moderately developed Golgi apparatus (G), a prominent ↑ centriole (C), some short rough endoplasmic cisternae, and a large amount of RNA in the form of myriads of free ↑ ribosomes and ↑ polyribosomes (R) stainable in light microscopy with pyronin. B-I. are concentrated in ↑ germinal centers of ↑ lymph nodules, whereas T-I. are scattered in thymodependent zones of ↑ lymph nodes and ↑ spleen.

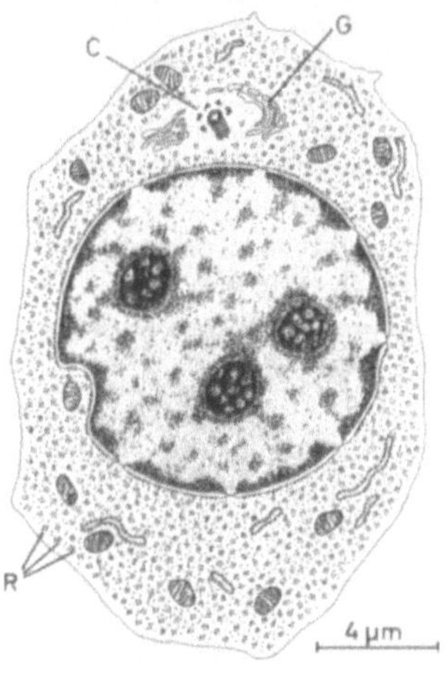

Immunocytochemistry: A branch of ↑ cytochemistry dealing with microscopic localization of endogenous proteins on the basis of antigen-antibody reaction. I. uses technique of ↑ immunofluorescence under light microscope and labeling of antibodies with ↑ ferritin, ↑ hemocyanin, ↑ horseradish peroxidase, etc. in transmission electron microscope.

Bosman, F.T.: Histochemical J. *15*, 189 (1983); Bullock, G.R., Petrusz, P. (eds.): Techniques in Immunocytochemistry. Vol. 1. New York: Academic Press 1982; Childs, G.V.: J. Histochem. Cytochem. *31*, 168 (1983); Heitz, Ph.U.: Acta Histochemica, Suppl. *25*, 17 (1982); Pilgrim, Chr. (ed.): Immunzytochemie. Acta Histochem. Suppl. XXV. Jena: VEB Gustav Fischer 1982

Immunofluorescence: A technique for visualization of antigens (Ag) using antibodies (Ab) labeled with a ↑ fluorochrome (F). Resulting antigen-antibody complex (Ag–Ab) within cells and tissues is localized with high precision under ↑ fluorescence microscope.

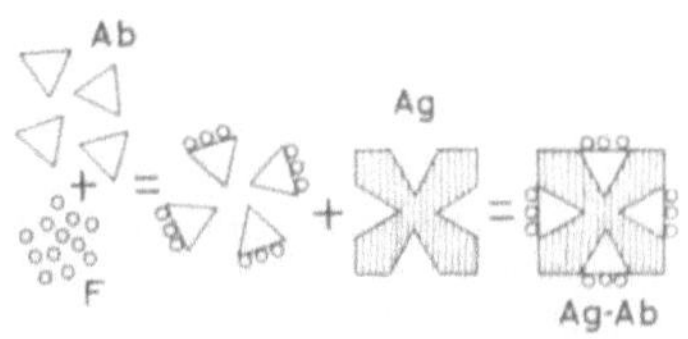

Immunoglobulin A (IgA): A local antibody (m.w. about 170000 daltons) produced by ↑ plasma cells and ↑ lymphocytes of lamina propria immediately underlying epithelia of ↑ respiratory system, ↑ gastrointestinal tract, ↑ salivary glands, and genitourinary system. I. passes upward from lamina propria through epithelium; during this passage, two molecules of I. are assembled into a dimer by a glycoprotein

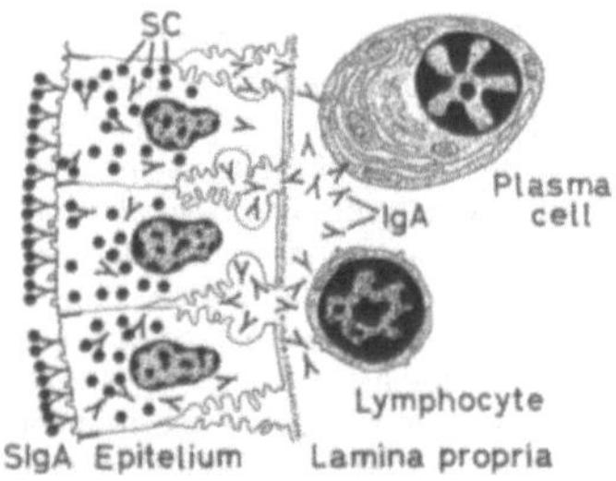

secretory component (SC) produced by epithelial cells. Dimers are relased at free epithelial surface as secretory I. (SIgA), resistant to proteolytic enzymes. I. protects epithelia against viral and bacterial invasions. (Modified after Tomasi 1972)

Brandtzaeg, P., Baklein, K.: Immunology of the Gut. Ciba Foundation Symposium 46. Amsterdam: Elsevier 1977; Brown, W.R., Isobe, Y., Nakane, P.K.: Gastroenterology *71*, 985 (1976); Tomasi, T.B.: N. Engl. J. Med. *287*, 500 (1972)

Immunoglobulin D (IgD): ↑ Immunoglobulins

Immunoglobulin E (IgE): ↑ Immunoglobulins

Immunoglobulin G (IgG): A category of antibodies produced by ↑ plasma cells, representing bulk of plasma ↑ immunoglobulins. Molecule of I. measures 3.5 nm in width and 20 nm in length and consists of two symmetrical halves, each composed of two polypeptide chains arranged in parallel,

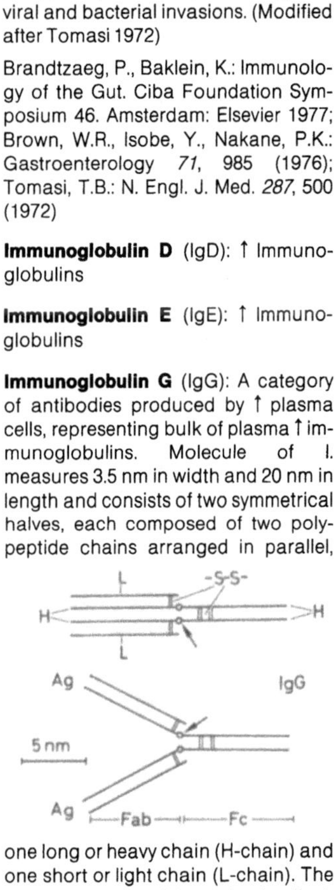

one long or heavy chain (H-chain) and one short or light chain (L-chain). The four chains are held together by disulfide bonds (-S-S-). I. may acquire a "T" or "Y" shape due to hinge-points (arrows) on H-chain. Fab segment of I. combines with antigens (Ag), but Fc segment does not. I. has a molecular weight of about 150000 daltons and encloses 1400 amino acids. I. is responsible for ↑ humoral immunity. (See immunology texts for further information)

Silverton, W.E., Navia, M.A., Davies, D.R.: Proc. Natl. Acad. Sci. USA *74*, 5142 (1977)

Immunoglobulin M (IgM): ↑ Immunoglobulins

Immunoglobulins (Ig): A group of structurally similar plasma proteins, mostly globulins, acting as antibodies against antigens. Kinds of I. = 1) ↑ Immunglobulin A. 2) IgE = a class of surface Ig fixed to plasmalemma of ↑ mast cells and ↑ leukocytes in contact with antigens; IgE induces release of ↑ histamine by ↑ mast cells, ↑ leukotrienes by leukocytes, and provokes some allergic reactions. 3) ↑ Immunoglobulin G. 4) IgM = a surface I. (m. w. about 1000000 daltons), first I. produced by ↑ B-lymphocytes during ↑ immune response and progressively replaced by immunoglobulin G. 5) IgD = a class of poorly understood I.

Nisonoff, A.: Introduction to Molecular Immunology. Oxford: Blackwell Scientific Publications 1982

Implantation: 1) The process of penetration of the blastocyst into pars compacta of ↑ endometrium occurring 6–7 days after ↑ fertilization. (See embryology texts for further information). 2) Insertion of inert materials or small devices into organism with function of replacing lost organs or tissues (artificial heart, heart valves, head of femur, blood vessels, lens, pacemaker, etc.).

1) Enders, A.C., Hendrickx, A.G., Schlafke, S.: Am. J. Anat. *167*, 275 (1983); 1) Glass, R.H., Aggeler, J., Spindle, A., Pedersen, R.A., Werb, Z.: J. Cell Biol. *96*, 1108 (1983)

Implantation fossa, of spermatozoon (articular fossa): A slightly concave caudal pole (IF) of spermatozoon nucleus lined by a poreless ↑ nuclear envelope (NE) whose interspace, about 10 nm wide, is bridged by regular densities (D) between its leaves, probably contributing to reinforcement of this region of envelope. I. contains ↑ basal plate (BP), fused with capitulum (C) of ↑ spermatozoon tail. I. represents articular surface between spermatozoon head and its tail. Ce = proximal centriole.

↑ Spermatozoon, connecting piece of; ↑ Spermatozoon, tail of

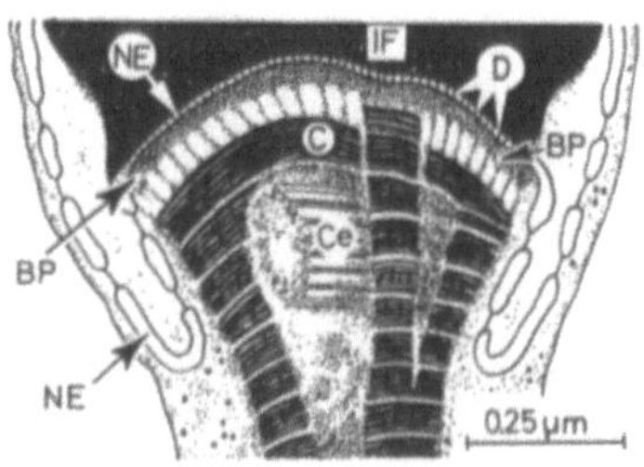

Impulse-conducting system, of heart (conductive tissue, systema conducens cardiacum*): A system of specially modified ↑ cardiac muscle cells with function of generating and conducting impulses of heart contraction to various parts of ↑ myocardium, as well as to assure proper succession of beat of atria and ventricles. Composition: 1) ↑ Sinuatrial node or node of Keith and Flack (SA); 2) ↑ atrioventricular node or node of Aschoff and Tawara (AVN); 3) ↑ atrioventricular bundle or bundle of His (AVB), with its left and right branches; 4) ↑ Purkinje fibers (PF). I. is isolated from surrounding myocardium by a connective tissue sheath and consists of ↑ nodal cells, ↑ Purkinje cells, and ↑ transitional cells.

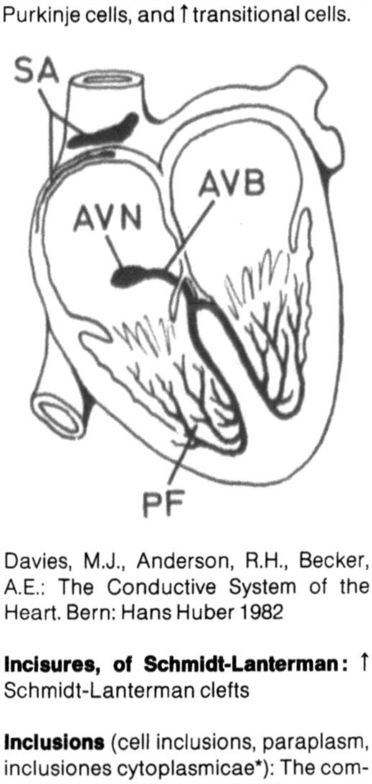

Davies, M.J., Anderson, R.H., Becker, A.E.: The Conductive System of the Heart. Bern: Hans Huber 1982

Incisures, of Schmidt-Lanterman: ↑ Schmidt-Lanterman clefts

Inclusions (cell inclusions, paraplasm, inclusiones cytoplasmicae*): The common name for chemically inert substances that appear and disappear in cell body as a result of cell's activity. I. are not essential for maintenance of cell's life. Large amounts of I. can be temporarily stored in the cell. I. include ↑ crystals, ↑ glycogen, ↑ lipids, ↑ secretory granules, ↑ pigments, ↑ keratohyalin, ↑ trichohyalin, and ↑ yolk.

Incremental lines: ↑ Cement lines

Incus: ↑ Auditory ossicles

Indirect cellular division: ↑ Mitosis

Indoleamines: ↑ Biogenic amines derived from tryptophan; considered ↑ neurotransmitters. ↑ Serotonin and ↑ melatonin are I.

Informosomes: Free cytoplasmic ribonucleoprotein particles (sedimentation rate below 80 S or more than 90 S) associated with ↑ messenger RNA, but not associated with ↑ ribosomes. I. seem to keep information for variously long periods of time before definitive genetic expression. I. may also be involved in modulation of ↑ protein synthesis.

Infundibulum*, of hypophysis (neural stalk): Narrow part of ↑ neurohypophysis connecting ↑ median eminence with lobus nervosus. I. consists of nonmyelinated ↑ neurosecretory axons of ↑ hypothalamohypophyseal tract.

↑ Hypophysis

Infundibulum, of oviduct (infundibulum tubae uterinae*): A funnel-shaped abdominal opening of ↑ oviduct surrounded by ↑ fimbriae.

Inhibiting factors (inhibiting hormones, statins): Substances produced and released from some ↑ neurosecretory neurons and/or endocrine cells inhibiting action of ↑ adenohypophyseal hormones. Best known I. are ↑ melanocyte-stimulating hormone-inhibiting factor, ↑ somatostatin, and prolactin-inhibiting factor.

↑ Releasing factors

Inhibitory interneurons: ↑ Interneurons inhibiting synaptic transmission between two neuronal circuits. They are predominantly ↑ Golgi cells type II, but some ↑ Golgi cells type I (↑ Purkinje cells) can inhibit other neurons.

Inhibitory synapse: ↑ Synapse, inhibitory

Injection, vascular: The introduction of colored mixtures (gelatin + India ink, lithium carmin) or plastic mass into blood or lymphatic vessels for study of vascular bed of a tissue or organ under light and transmission electron microscope.

↑ Perfusion

Inlet venules (interlobular venules, terminal portal venules): Fine venous vessels (IV), branches of ↑ interlobular veins (IIV), running for short distances between classic ↑ liver lobules (LL) and feeding blood into ↑ liver sinusoids. CV = ↑ central vein (Fig. = human)

↑ Liver, vascularization of

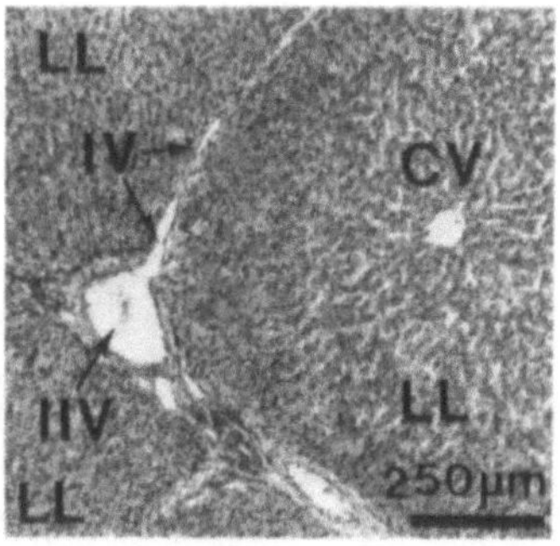

Inner band, of Baillarger: ↑ Brain cortex, homotypical isocortex of

Inner chamber, of mitochondria: A large intramitochondrial compartment (IC) delimited by inner mitochondrial membrane (IM). I. contains ↑ mitochondrial matrix.

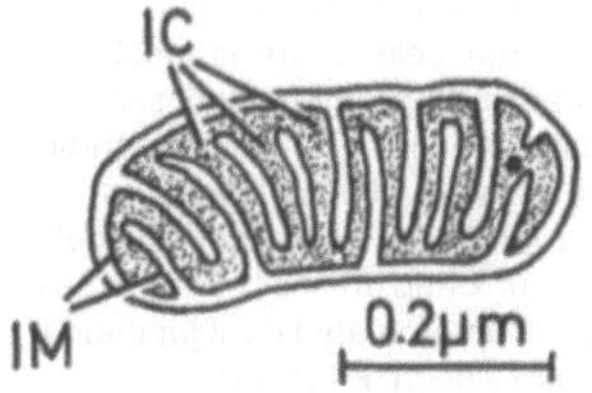

Inner circumferential lamellae: ↑ Bone, compact

Inner cortex, of lymph node: ↑ Lymph node, cortex of

Inner ear (labyrinth, auris interna*): A complex system of bony cavities forming bony ↑ labyrinth, in which membranous ↑ labyrinth is suspended. In I. are situated organs of equilibrium (↑ macula of saccule and utricle, ↑ cristae ampullares) and organ of hearing (↑ Corti's organ).

Harada, Y.: Biomed. Res. *2*, Suppl. 391 (1981); Lundquist, P.-G., Wersäll, J.: Biomed. Res. *2*, Suppl. 379 (1981)

Inner ear, cochlea of: ↑ Cochlea

Inner ear, cochlear duct of: ↑ Cochlear duct

Inner ear, cristae ampullares of: ↑ Cristae ampullares

Inner ear, endolymphatic duct of: ↑ Endolymphatic duct

Inner ear, endolymphatic sac of: ↑ Endolymphatic sac

Inner ear, macula of saccule and utricle of: ↑ Macula of saccule and utricle of

Inner ear, membranous labyrinth of: ↑ Labyrinth, membranous

Inner ear, semicircular canals of: ↑ Semicircular canals and ducts

Inner limiting membrane, of retina (membrana limitans interna*): All conical feet of ↑ Müller's cells and their basal lamina separating ↑ retina from ↑ vitreous body.

Inner mitochondrial membrane: ↑ Mitochondrial membranes

Inner nuclear layer, of retina (stratum granulosum internum*): A layer of nuclei of ↑ amacrine, ↑ bipolar, ↑ horizontal, and ↑ Müller's cells.

Inner nuclear membrane: ↑ Nuclear membrane, inner

Inner plexiform layer, of retina (stratum reticulare internum*): A synaptic contact zone of ↑ axons of ↑ bipolar cells with ↑ dendrites of ↑ ganglion cells, as well as dendrites of ↑ amacrine cells with axons of bipolar cells and dendrites of ↑ ganglion cells.

Inner root sheath, of hair (vagina radiculáris interna*): Three epithelial layers between outer epithelial sheath and ↑ cuticle of hair shaft of a ↑ hair follicle. I. consists of ↑ cuticle, ↑ Huxley's layer, and ↑ Henle's layer, and originates from ↑ matrix cells of hair.

Inner segment, of photoreceptors (segmentum internum*): Part of ↑ photoreceptors protruding outside ↑ outer limiting membrane and containing practically all the cell ↑ organelles. I. is subdivided into ellipsoid and myoid.

Innervation: The supply of a ↑ tissue or ↑ organ with ↑ nerve fibers (motor, sensitive, autonomic).

Insulin: A polypeptide hormone (m.w. about 6000 daltons) produced by ↑ B-cells of ↑ pancreatic islets. Action of I. is ubiquitous, consisting of stimulation of uptake of glucose by various cells (particularly ↑ liver parenchymal cells, ↑ adipose cells, ↑ skeletal muscle fibers) and subsequent lowering of glucose level in blood. Through stimulation of corresponding enzymes (hexosephosphate, etc.), intracellular glucose is metabolized to yield energy or polymerized to ↑ glycogen. In adipose cells, I. stimulates conversion of glucose to fatty acids and triglycerides; I. increases uptake of amino acids and synthesis of proteins. (See physiology texts for further information)

Insulin cells: ↑ B-cells, of pancreatic islets

Intercalated cells, of arched collecting tubules: ↑ Collecting tubules, of kidney

Intercalated disc (discus intercalatus*): A complicated steplike intercellular junction between two adjacent ↑ cardiac muscle cells (CC). I. always occurs at ↑ Z-line level and consists of two portions: 1) Portion transverse to ↑ myofibrils (Mf) shows a complex pattern of short blunt interdigitations (Int) reinforced by ↑ desmosomes (D) and ↑ fasciae adherentes (FA); ↑ actin myofilaments anchor in both. This portion assures cohesion of cardiac cells. 2) Longitudinal portion consists of a variously long ↑ nexus (N) through which contraction impulses pass from one cell to another. T = openings of ↑ T-tubules

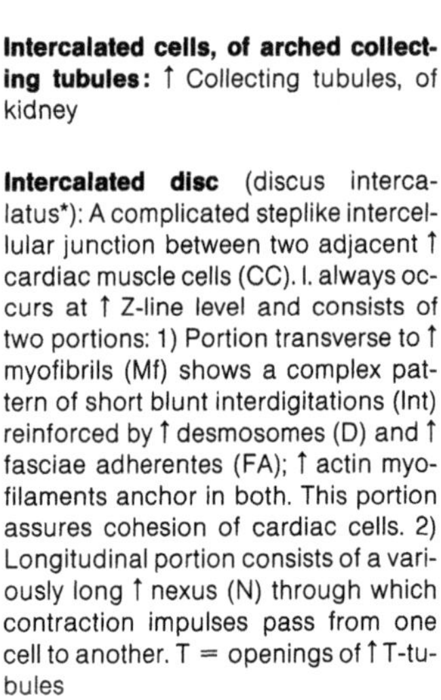

Robinson, T.F.: Cell Tissue Res. *211*, 353 (1980)

Intercalated duct cells: Small squamous to cuboidal cells (IDC) with a round or deeply invaginated nucleus occupying almost entire cytoplasm. In addition to some ↑ organelles and free ↑ ribosomes, I. can contain small serous or mucous secretory granules. (Fig. = ↑ pancreas, rat)

Auger, D.W., Harrison, J.D.: Arch. Oral Biol. *27*, 79 (1982)

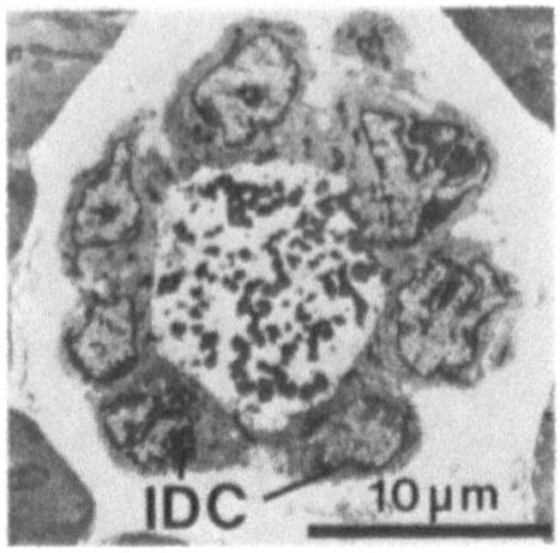

Intercalated ducts (ductus intercalatus*): Tubules (ID) about 20 µm wide and of variable length connecting ↑ acini (A) with ↑ striated ducts (SD) in ↑ salivary glands (fig.) and with interlobular ducts in ↑ pancreas. I. are lined by ↑ simple squamous to cuboidal epithelium, which begins in salivary glands at level of acinar neck and in interior of acini in pancreas as ↑ centroacinar cells. I. are longest in pancreas and ↑ parotid gland, shorter and less frequent in ↑ submandibular gland, and almost absent in ↑ sublingual gland. ↑ Myoepithelial cells often surround them. Granular cells at acinar-intercalated duct junction have recently been described.

↑ Intercalated duct cells

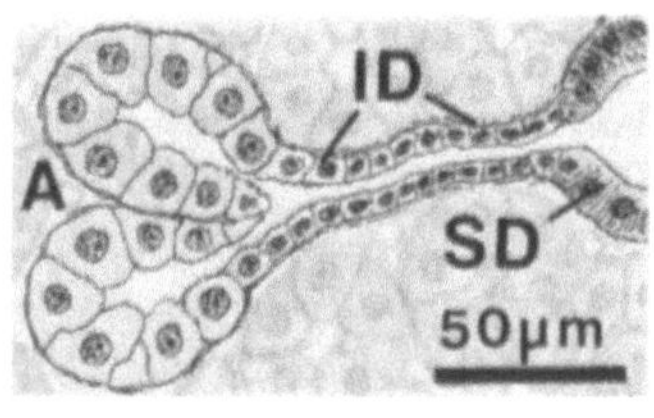

Qwarnström, E.E., Hand, A.R.: Anat. Rec. *206*, 181 (1983)

Intercalated neurons: ↑ Spinal cord, neurons of

Intercalated veins (sublobular vein, vena sublobularis*): Venous vessels formed by two or more ↑ central veins of classic ↑ liver lobules. Several I. unite to form a ↑ collecting vein.

↑ Liver, vascularization of

Intercellular bridge, false: A term designating apparent continuity under light microscope of cytoplasmic spines (S) connecting adjacent ↑ prickle cells. Transmission electron microscope has shown that no such continuity exists, since the spines are firmly attached at

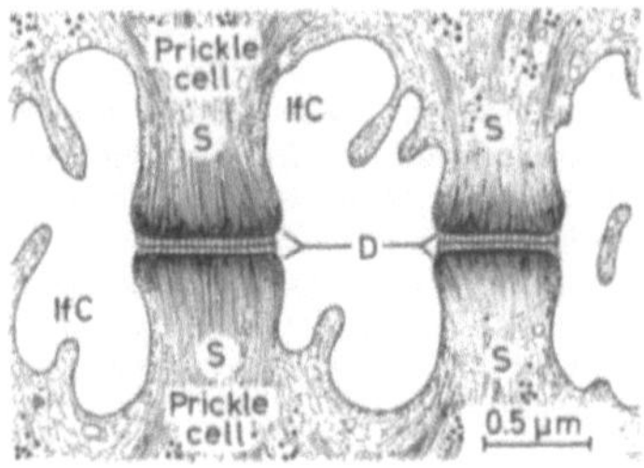

their ends by strong ↑ desmosomes (D). IfC = ↑ interfacial canals

↑ Epidermis, ultrastructure of

Lempert, T.E., Elias, P.M.: Anat. Rec. *193*, 927 (1979)

Intercellular bridge, true (cytoplasmic bridges): The communication (I), 1–3 µm in width, between cytoplasm of two daughter cells formed as a result of their incomplete or arrested ↑ cytokinesis. Presence of I. is a regular feature during ↑ spermatogenesis: Beginning with ↑ spermatogonia type B, I. persist until late ↑ spermatids (S). On cytoplasmic side of plasmalemma, I. are reinforced by a layer of dense material (arrows). (Fig. = human)

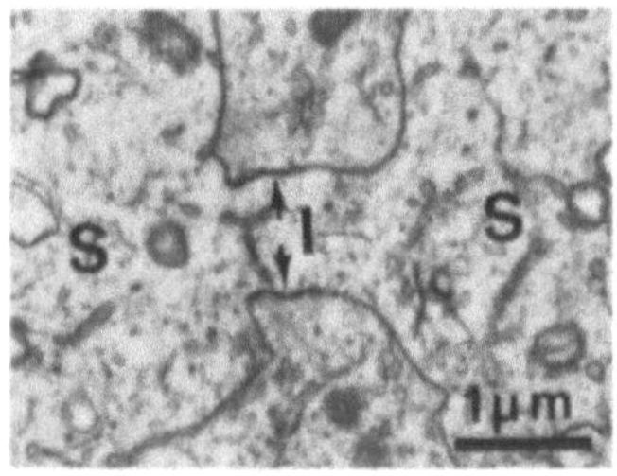

Linares, C.: Z. mikrosk.-anat. Forsch. *97*, 245 (1983); Teixeira, C.S.R.: Folia Anat. Univ. Conimbrigensis *47*, 55 (1982)

Intercellular canaliculi: ↑ Canaliculi, intercellular

Intercellular cement: A ↑ PAS-positive, ↑ argentaffin ↑ glycosaminoglycan-containing substance present between epithelial cells, probably acting as a glue, increasing adherence of cells.

↑ Lanthanum

Intercellular junctions: ↑ Junctions, intercellular

Intercellular substance, of connective and supporting tissues (substantia intercellularis*): All structures found between cells of these tissues – ↑ ground substance (sometimes calcified = ↑ bone, ↑ cementum, ↑ dentin) and various fibers (↑ collagen, ↑ elastic, ↑ reticular).

Interchromatin granules: Granules of about 20 nm (Int) occurring in clusters between ↑ heterochromatin (H) clumps. It is thought that I. are ribonucleoprotein structures. (Fig. = ↑ pancreatic acinar cell, rat; heterochromatin partially bleached with ↑ EDTA)

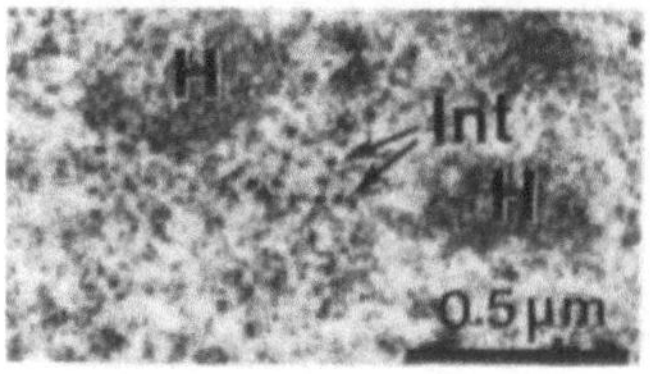

Krzyzowska-Gruca, S., Zborek, A., Gruca, S.: Cell Tissue Res. *231*, 427 (1983); Wassef, M.: J. Ultrastruct. Res. *69*, 121 (1979)

Intercristal space: Portion of ↑ inner chamber of ↑ mitochondria situated between two ↑ cristae mitochondriales.

Interdental cells (cellula interdentalis*): Tall amphoralike cells (IC) lying in rows between ↑ auditory teeth (AT) at upper surface of ↑ spiral limbus (SL). I. have clear cytoplasm and a central nucleus with finely dispersed chromatin. Cytoplasm contains rod-shaped mitochondria, a well-developed supranuclear Golgi apparatus, sparse rough endoplasmic cisternae, and some lysosomes. Neck (N) of the cell is very narrow; thin apical cytoplasmic sheaths (CS) cover auditory teeth, maintaining contacts with adjacent cells. I. produce a mucopolysaccharide substance which transforms gradually into ↑ tectorial membrane (MT).

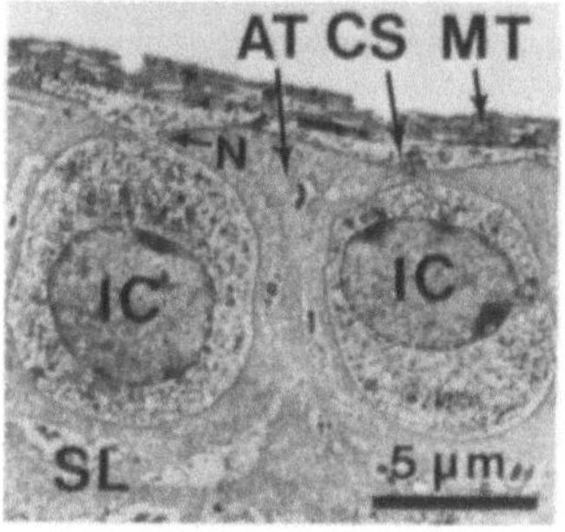

Thorn, L., Arnold, W., Schinko, I.: Verh. Anat. Ges. *73*, 673 (1979)

Interdigitations: Cell processes (I) of various dimensions extending between similar processes of neighboring cells to reinforce cell union. I. are most frequent in ↑ epithelial tissues and ↑ smooth muscle tissue. (Fig. = parathyroid gland, rat)

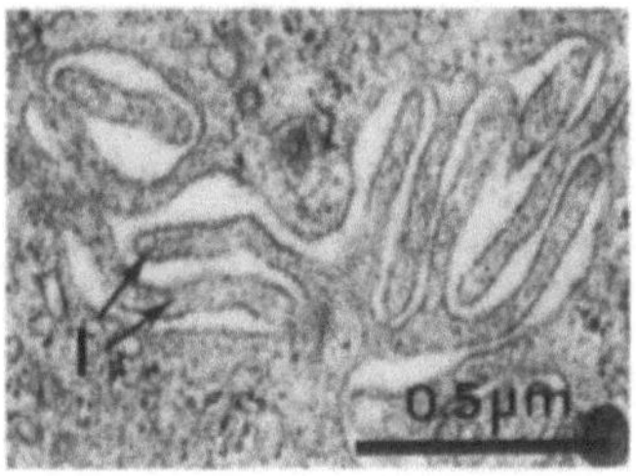

Interfacial canals: Large intercellular spaces (IfC) between adjacent ↑ prickle cells of ↑ epidermis which act as capillaries in this nonvascularized tissue. D = ↑ desmosome

↑ Intercellular bridge, false

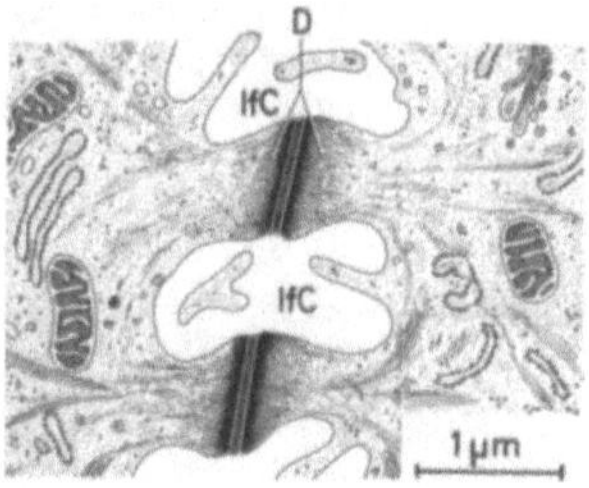

Interfascicular oligodendrocytes: ↑ Oligodendrocytes

Interference microscope: A light bright-field ↑ microscope for optical quantitative examination of unstained and living cells. Principle: A coherent light beam (L) is split by a complex prism (P_1) into two parallel beams (B_1, B_2). One of them (B_1) passes through specimen (S), the other (B_2) does not. After refraction in two separate ↑ objectives (O_1, O_2), the two beams are then recombined in another prism (P_2) and interfere with one another. Measurement of phase shift, i.e., manner in which they interfere, permits precise calculation of ↑ refractive index, thickness, and mass of cells or some of their components. C_1, C_2 = ↑ condensers, Po = polarizers (Fig. = Leitz I.; modified after Richards 1966)

Krug, H., Fritsch, R.S. (eds.): Quantitative Mikroskopie. Acta Histochem. Suppl. 26. Jena: Gustav Fischer 1982; Richards, O.W.: An Introduction to the Theory of Interference Microscopy. In: Wied, G.L. (ed.): Introduction to Quantitative Cytochemistry. London and New York: Academic Press 1966; Tauber, G.: Acta Histochemica, Suppl. XXVI, 81–94 (1982)

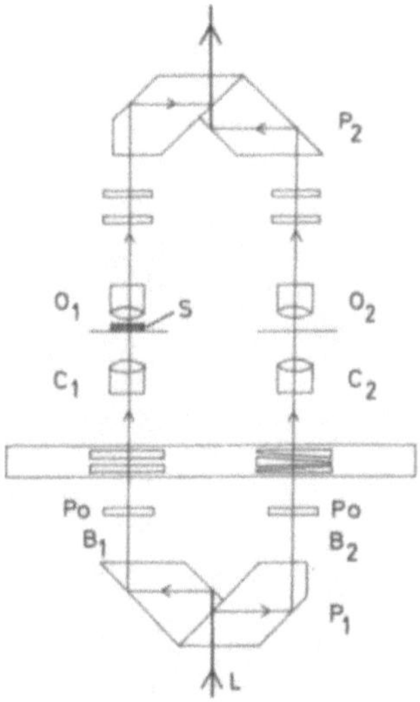

Interferons: Macromolecular antiviral agents produced by ↑ macrophages and various other cells.

Merigan, T.C., Friedman, R.M. (eds.): Interferons. New York: Academic Press 1982; Pestka, S. (ed.): Interferons. Vols. 78 and 79. New York: Academic Press 1982; Stringfellow, D.A. (ed.): Interferon and Interferon Inducers. Clinical Applications. New York: Marcel Dekker Inc. 1980; Kishida, T., Oku, T.: Yakugaku Zasshi, J. Pharm. Soc. J. *103*, 483 (1983)

Interglobular spaces, of Owen (spatium interglobulare*): Incompletely mineralized dentinal areas (IS) beneath ↑ dentinoenamel junction (arrowhead). I. are delimited by calcified ↑ dentinal globuli (DG) and contain only organic matrix of ↑ dentin (D). E = ↑ enamel (Fig. = human)

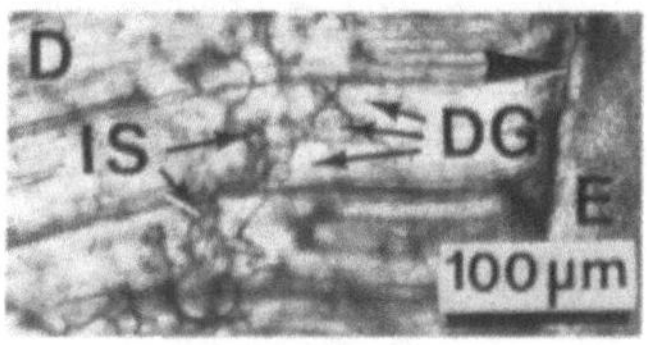

Interkinesis: ↑ Interphase

Interlobar columns: ↑ Renal columns

Interlobular arteries, of kidney (arteriae corticales radiatae; arteriae interlobulares*): ↑ Muscular arteries (IA), branches of ↑ arcuate arteries (AA), supplying ↑ renal lobules. I. leave arcuate arteries at right angles and traverse cortical parenchyma radially, each lying between two ↑ medullary rays (R). I. give off ↑ afferent arterioles (aa). (Fig., left = monkey; fig., right = human, India ink vascular ↑ injection)

↑ Kidney, vascularization of

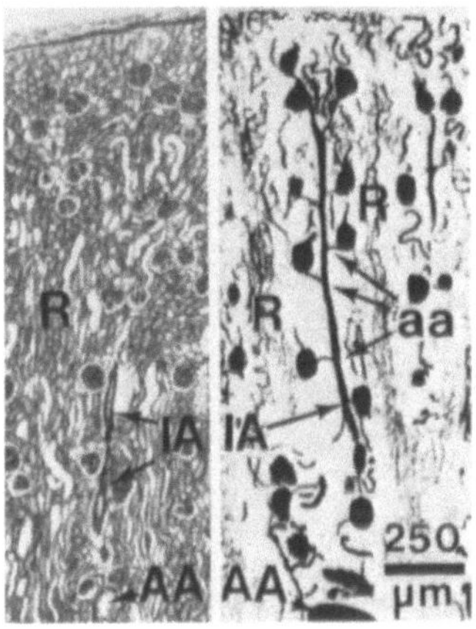

Interlobular arteries, of liver (arteria interlobularis*): ↑ Muscular arteries of small caliber, branches of hepatic artery, each situated in a ↑ portal canal of liver. I. assure nutrient circulation of liver by vascularizing ↑ stroma, extralobular bile ducts, and also delivering small volumes of arterial blood directly to hepatic ↑ sinusoids. Frequently, ↑ epithelioid cells are found in tunica media of I.; they are probably involved in regulation of blood flow.

↑ Liver, vascularization of

Interlobular ducts, of exocrine glands (ductus interlobularis*): Excretory tubes (ID) lined with ↑ simple cuboidal to columnar epithelium situated within septa (S) separating lobules (L) of an ↑ exocrine gland. I. empty into excretory duct (E). (Fig. = ↑ parotid gland, human)

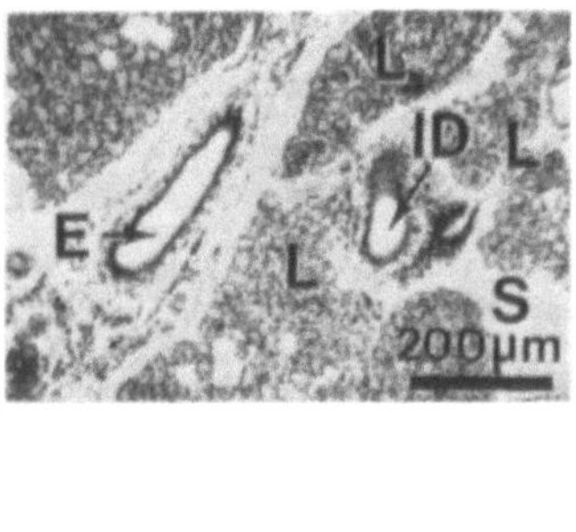

Interlobular veins (vena interlobularis*): Thin-walled veins each located in a ↑ portal canal of liver. I. are branches of portal vein; they give off ↑ inlet venules, which open into ↑ liver sinusoids.

↑ Liver, vascularization of

Intermediate body: ↑ Mid-body

Intermediate cells: ↑ Trachea, epithelium of

Intermediate cells, of stria vascularis: ↑ Stria vascularis

Intermediate erythroblast: ↑ Erythroblast, polychromatophilic

Intermediate filaments (cytokeratin filaments): Noncontractile microfilaments (C), about 10 nm in diameter, participating in formation of ↑ cytoskeleton. In ↑ absorptive cells, I. are concentrated in ↑ terminal web (TW) from which they extend as a three-dimensional meshwork up to cell base. They are particulary abundant in ↑ Sertoli's and ↑ myoepithelial cells. I. are also found around nucleus, fixing it within cell. I. consist of prekeratinlike protein and ↑ vimentin. (Fig. = absorptive cell, rat)

↑ Intermediate microfilaments

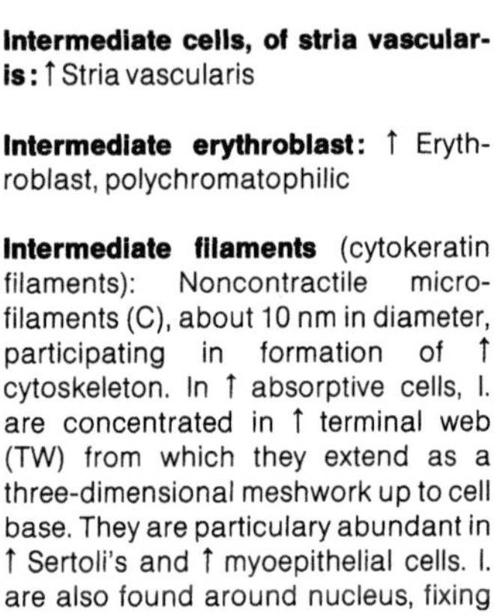

Henderson, D., Weber, K.: Exp. Cell Res. *129*, 441 (1980); Jahn, W.: Eur. J. Cell Biol. *26*, 259 (1982); Lazarides, E.: Nature *283*, 249 (1980); Osborn, M., Geisler, N., Shaw, G., Sharp, G., Weber, K.: Intermediate Filaments. Cold Spring Harbor Symposia on Quantitative Biology. Vol. 46, Part. 1. New York: Cold Spring Harbor Laboratory 1982

Intermediate glands, of stomach: Glands between ↑ gastric glands proper and ↑ pyloric glands occupying a zone about 1 cm wide. Structurally, I. display fewer ↑ parietal and ↑ chief cells than gastric glands proper, with more mucoid cells similar to ↑ neck mucoid cells.

Intermediate junctions: Specialized intercellular ↑ attachment devices comprising an intercellular space about 20 nm wide filled with ↑ glycoprotein material. In area of I., cell membrane is unchanged; toward cytoplasm, fine osmiophilic filaments irradiate. Two types of I.: I) ↑ Zonula adherens of ↑ terminal bar; 2) ↑ fascia adherens of ↑ intercalated disc.

Intermediate lobe, of hypophysis: ↑ Hypophysis, pars intermedia of

Intermediate microfilaments (intermediate-sized myofilaments, skeletin filaments, 10-nm-filaments): Noncontractile, cross-striated rigid ↑ microfilaments (10 nm in diameter, periodicity 2.5 nm) found in ↑ cardiac and ↑ smooth muscle cells, ↑ Purkinje cells of ↑ impulse-conducting system of heart, ↑ skeletal muscle fibers (↑ desmin microfilaments), and some other cells. I. consist of four subfibrils; they form bundles of 20–50 units and converge toward ↑ desmosomes and ↑ dense bodies. I. make part of cytoskeleton. According to recent data, it seems likely that ↑ intermediate filaments, although of different chemical nature, belong to the family of I.

↑ Smooth muscle cells, microfilaments of; ↑ Smooth muscle cells, molecular biology of contraction of

Eriksson, A., Thornell, L.-E.: J. Cell Biol. *80*, 231 (1979)

Intermediate neuroglial cells: Inconspicuous small round cells located in ↑ neutropil near blood capillaries. I. have a large oval nucleus rich in ↑ heterochromatin and poorly developed ↑ organelles. I. are considered precursors of ↑ neuroglial cells which may differentiate under certain conditions into both ↑ astrocytes and ↑ oligodendrocytes. Since I. may become phagocytic in areas of neuronal destruction, they are regarded as one of the sources of ↑ microglia.

Intermediate sinus (cortical sinus, perforating sinus, sinus corticalis perinodularis*): A lymphatic space binding ↑ subcapsular sinus with ↑ medullary sinus of a ↑ lymph node. I. accompanies trabeculae; structurally, it is identical to subcapsular sinus.

Intermediolateral nucleus (nucleus intermediolateralis): A group of small ↑ preganglionic neurons (Iml) located in ↑ intermediolateral gray column of ↑

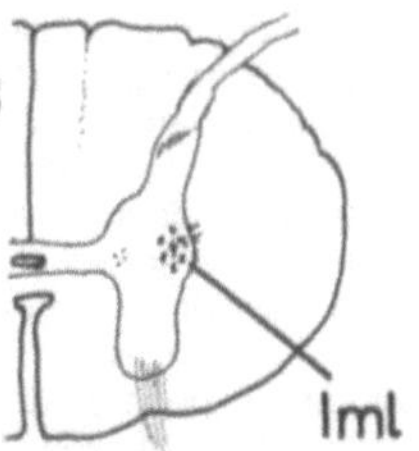

spinal cord. Between segments C_8 and L_3, cells of I. are sympathetic and prominent; between segments S_2 and S_4, they are parasympathetic and inconspicuous.

Intermediomedial nucleus (nucleus intermediomedialis): A group of small ↑ preganglionic neurons (Imm) located in interior or gray matter of ↑ spinal cord. Between segments C_8 and L_3, cells are rare and sympathetic; between segments S_2 and S_4, they are conspicuous and parasympathetic.

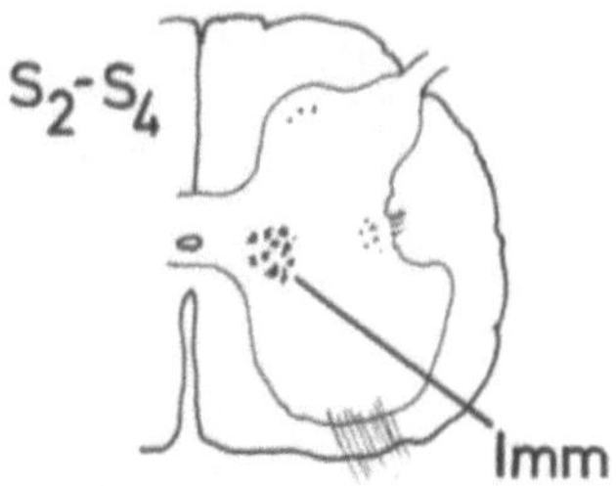

Internal elastic lamina: ↑ Elastic lamina, internal

Internal glial limiting membrane: ↑ Membrana limitans gliae perivascularis

Internal pyramidal layer: ↑ Brain cortex, homotypical isocortex of

Internal remodeling, of bone: ↑ Bone formation, secondary

Internal reticular apparatus: ↑ Golgi apparatus

Internal sphincter, of anus: ↑ Anal canal

Internal tongue process: ↑ Tongue process, of oligodendrocytes

Interneuronal synapse: ↑ Synapse, interneuronal

Interneurons: Nerve cells intercalated between two or more other ↑ neurons in a neuronal circuit. I. act as modulators of synaptic transmission by excitation or inhibition; they are usually ↑ Golgi cells type II.

Brazier, M.A.B. (ed.): The Interneuron. Berkeley, Calif.: University of California Press 1969; Tömböl, T., Babosa, M., Hajdù, F., Somogyi, Gy.: Acta Morphol. Acad. Sci. Hung. *27*, 297 (1979)

Interneurons, of cerebellar cortex: Nerve cells (↑ basket cells, ↑ granule cells, ↑ stellate cells) interposed between afferent (↑ mossy fibers, ↑ climbing fibers) and efferent fibers (↑ Purkinje cell axons) of cerebellar cortex. Basket cells, large granule cells, and stellate cells inhibit ↑ Purkinje cells (inhibitory neurons), whereas granule cells excite them.

↑ Cerebellum, neuronal circuits of

Internodal segment (internodes, segmentum internodale*): The portion (IS) of a ↑ nerve fiber (NF) between two ↑ nodes of Ranvier (NR). Length of I. is proportional to circumference of contained ↑ axon (A). In ↑ peripheral nervous system (PNS), I. are in general longer (up to 1.5 mm) than in ↑ central nervous system (CNS). A ↑ Schwann's cell (SC) furnishes ↑ myelin sheath (MS) for only one I., however, an ↑ oligodendrocyte (O) can give rise to as many as 40–50 I.

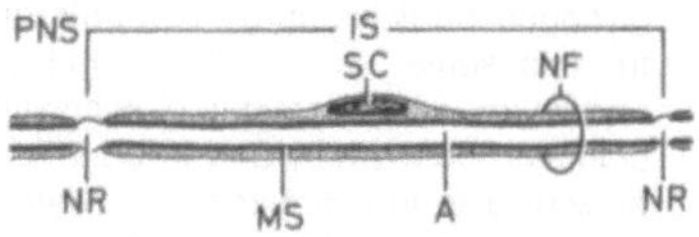

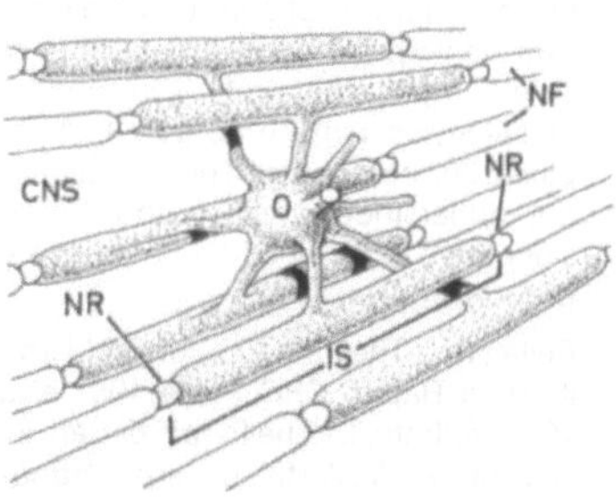

Internodes: ↑ Internodal segments

Internucleosomal DNA: Short segments of DNA molecule connecting ↑ nucleosomes to form a ↑ chromatin fiber.

Interoceptors: A group of sensory receptors receiving stimuli from internal organs (↑ aortic and ↑ carotid bodies, ↑ macula densa, etc.). ↑ Chemoreceptors and ↑ baroreceptors are I.

Interpapillary pegs: ↑ Epidermal ridges

Interperiod line: ↑ Intraperiod line; ↑ Myelin sheath

Interperiod membrane: ↑ Intraperiod line; ↑ Myelin sheath

Interphase (interkinesis): All phases of ↑ cell cycle between end of one division and beginning of the next. During I., a cell effects, without any visible morphological changes in nucleus, its genetically programed functions and also replicates its DNA content; it is, therefore, incorrect to designate I. as "resting" phase.

Interprismatic substance, of enamel: The calcified matrix (IS) between ↑ enamel prisms (EP). In I., ↑ hydroxyapatite crystals (H) are arranged perpendicular to those of the prisms. (Fig. = newborn rat)

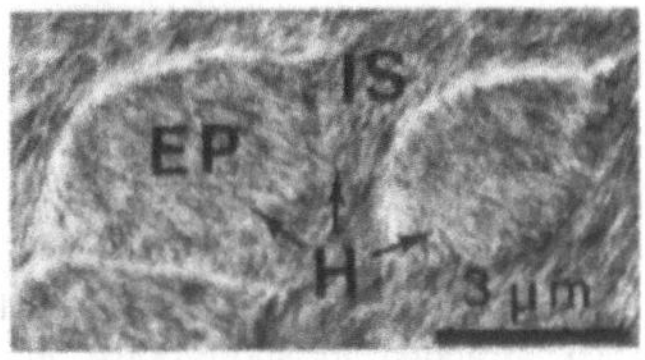

Termine, J.D., Torchia, D.A., Conn, K.M.: J. Dent. Res. *58*(B), 773 (1979)

Interstice (interstitium): Total spaces and holes within a ↑ tissue or ↑ organ filled with ↑ tissue fluid and/or a tissue not specific for the tissue or organ concerned (mostly ↑ loose connective tissue). In some organs, I. may contain groups of particular cells (↑ interstitial cells).

Interstitial cells, of Cajal: Oval or stellate cells (IC) randomly scattered among ↑ smooth muscle cells (SMC) of tunicae musculares of intestine. I. have an oval, deeply invaginated, ↑ heterochromatin-rich nucleus with a conspicuous nucleolus. Cytoplasm is clear and reduced to a narrow rim; it encloses a few mitochondria, an inconspicuous Golgi apparatus, some short rough endoplasmic cisternae, an abundance of free ribosomes, some ↑ dense bodies, and numerous ↑ micropinocytotic vesicles. I. are incompletely surrounded by a basal lamina (BL) and are in close contact both with autonomic nerve fibers (NF) originating from ↑ myenteric plexus and, via ↑ nexus (N), with adjacent smooth muscle cells. Function of I. is not known: It is thought that they represent a particular type of undifferentiated or immature smooth muscle cell acting as pacemaker for peristaltic contractions of tunicae musculares.

Rumessen, J.J., Thuneberg, L.: Anat. Rec. *203*, 115 (1982); Rumessen, J.J.,

Thuneberg, L., Mikkelsen, H.B.: Anat. Rec. *203*, 129 (1982); Thuneberg, L.: Adv. Anat. Embryol. Cell Biol. *71*, 1 (1982); Vajda, J., Fehér, E.: Acta Morphol. Acad. Sci. Hung. *28*, 251 (1980)

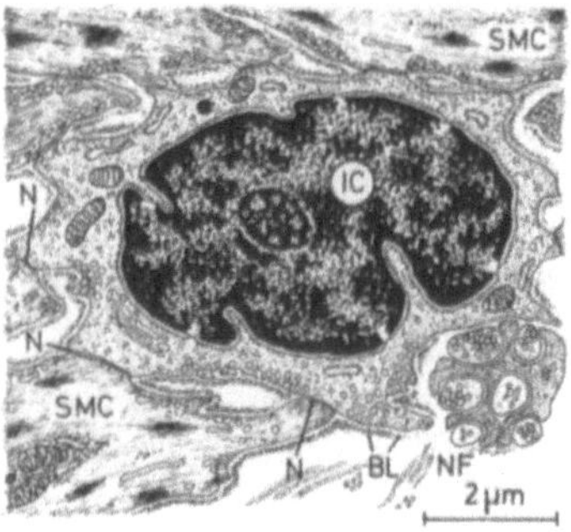

Interstitial cells, of liver: ↑ Perisinusoidal cells

Interstitial cells, of ovary (interstitiocytus ovarii*): Large oval clear cells (IC) arranged in short cords in ↑ stroma of ↑ ovary. Similar to ↑ lutein cells, I. have an elliptical nucleus, large mitochondria with both cristae and tubules, a small Golgi apparatus, short cisternae of rough endoplasmic reticulum, tubules of smooth endoplasmic reticulum, free ribosomes, and numerous ↑ lipid droplets (L). I. occur before birth and disappear after ↑ menopause. It is believed that they originate from hypertrophied cells of ↑ theca interna of atretic secondary ↑ ovarian follicles and that they produce ↑ estrogens for prepubertal development of secondary sexual characteristics. I. are well developed in rodents and poorly developed in humans. (Fig. = mongolian gerbil)

↑ Interstitial gland, of ovary

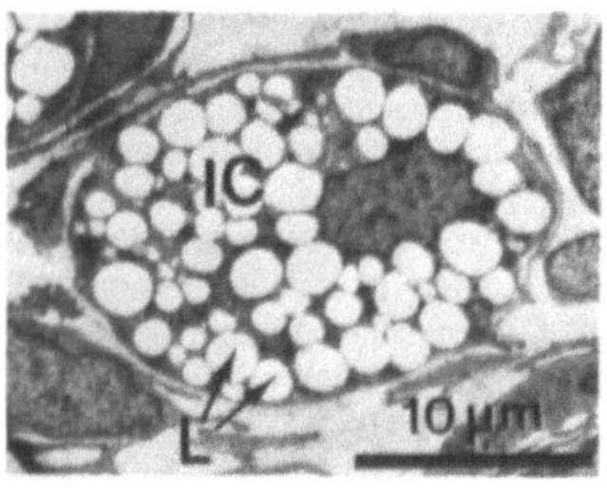

Guraya, S.S.: Int. Rev. Cytol. *55*, 171 (1978)

Interstitial cells, of pineal organ (dark cell, pineal neuroglial cell, supporting cell, cellula obscura*): Stellate cells with long processes incompletely sur-

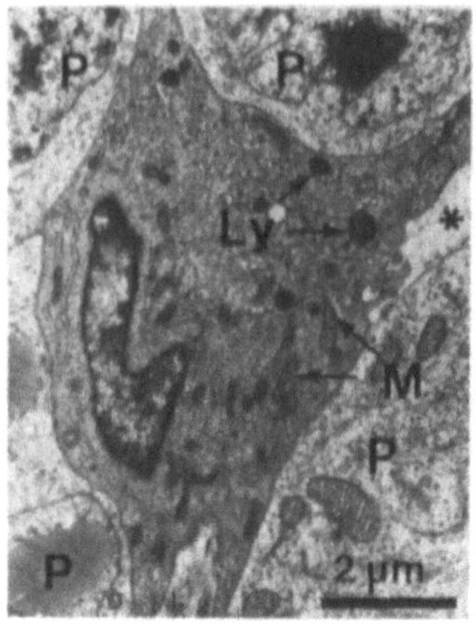

rounding ↑ pinealocytes (P), ↑ pericapillary spaces, and ↑ pineal canaliculi (asterisk). I. have an elongated nucleus rich in ↑ heterochromatin; cytoplasm contains small filiform mitochondria (M), a moderately to well-developed Golgi apparatus, elongated flattened rough endoplasmic cisternae, some small granules filled with osmiophilic material considered to be primary ↑ lysosomes, some secondary lysosomes (Ly), occasional ↑ glycogen particles, large amounts of free ribosomes, a few microtubules, and many microfilaments, 5–6 nm in diameter, both singly and in bundles, present also in cell extensions. Cytoplasmic matrix is of considerable electron density, although it may appear paler than that of surrounding pinealocytes. I. make up about 5% of pineal cell population. Their function is not understood; some authors consider them to be modified ↑ astrocytes. (Fig. = rat)

Reiter, R.J.: The Pineal Gland, Vol. I and II. Boca Raton: CRC Press 1981 and 1982; Vollrath, L.: The Pineal Organ. In: Oksche, A., Vollrath, L. (eds.) Handbuch der mikroskopischen Anatomie des Menschen, Vol. 6, Part 7. Berlin, Heidelberg, New York: Springer-Verlag 1980

Interstitial cells, of testis (Leydig's cells, endocrinocytus interstitialis*): Round, polygonal, or fusiform acidophilic cells (about 20 μm across) lying singly or in clusters between ↑ seminiferous tubules and adjacent to capillaries (Cap). Nucleus is eccentric and poor in ↑ heterochromatin; one or two nucleoli are prominent. Cytoplasm is abundant and contains numerous elongated mitochondria with cristae and tubules, a moderately developed Golgi apparatus (G), ↑ centrioles, and small stacks of short and flattened rough endoplasmic cisternae (rER). Smooth endoplasmic reticulum (sER) is extensive in the form of irregularly branching and anastomosing tubules. Besides a moderate number of free ribosomes and some ↑ glycogen particles, I. contain numerous ↑ peroxisomes, ↑ lysosomes (Ly), ↑ lipochrome (Lch) and ↑ lipofuscin (Lf) granules, ↑ lipid droplets (L), and, frequently, ↑ Reinke's crystals (C) associated at times with groups of short ↑ microtubules (Mt). I. produce ↑ testosterone after puberty and stay under control of hypothalamic ↑ gonadotropin-releasing factor and ↑ interstitial cell-stimulating hormone (ICSH). By a negative ↑ feedback mechanism, testosterone inhibits production of gonadotropin-releasing factor and ICSH.

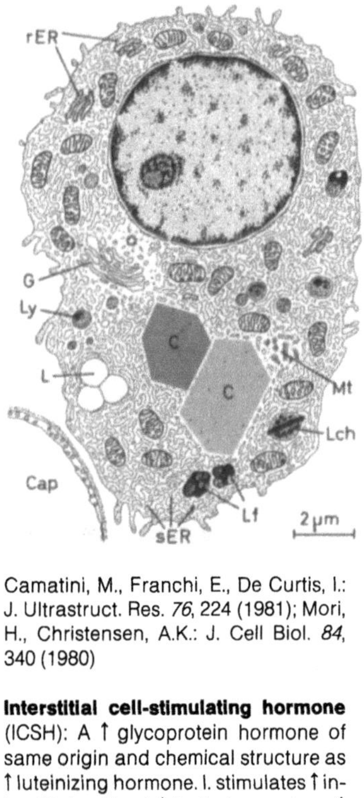

Camatini, M., Franchi, E., De Curtis, I.: J. Ultrastruct. Res. *76*, 224 (1981); Mori, H., Christensen, A.K.: J. Cell Biol. *84*, 340 (1980)

Interstitial cell-stimulating hormone (ICSH): A ↑ glycoprotein hormone of same origin and chemical structure as ↑ luteinizing hormone. I. stimulates ↑ interstitial cells of ↑ testis to produce ↑ testosterone.

Interstitial fluid: ↑ Tissue fluid

Interstitial gland, of ovary (interstitium ovarii*): Sum of ↑ interstitial cells (IG) of an ↑ ovary. Poorly developed in humans, I. has the form of short cellular strands predominantly located near atretic ↑ ovarian follicles (AF). CA = ↑ corpus albicans (Fig. = 44-year-old woman)

Mossman, M.H., Koering, M.J., Ferry, D. Jr.: Am. J. Anat. 115 (1964)

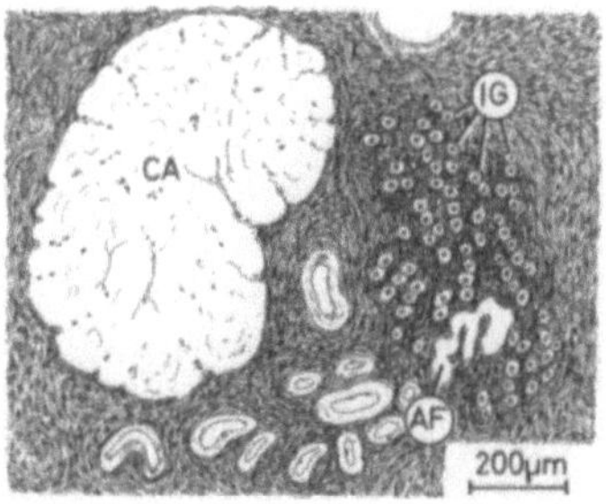

Interstitial growth, of cartilage: ↑ Cartilage, growth of

Interstitial lamellae, of bone (lamellae interstitiales*): Irregularly shaped fragments of lamellar bone (IS), in general devoid of ↑ Haversian canals, filling spaces between secondary ↑ osteons (O). From the latter, I. is demarcated by ↑ cement lines (arrowheads). In some cases, there are connections between ↑ canaliculi I. and neighboring osteons. I. represent remnants of earlier generations of secondary osteons removed during constant secondary ↑ bone formation. (Fig. = human)

↑ Bone, compact

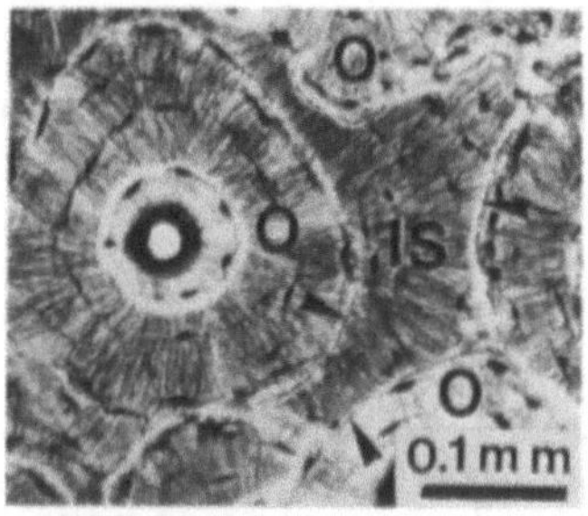

Interstitial segment, of oviduct: ↑ Oviduct

Interstitial tissue, of kidney (renal interstitium): Sum of ↑ loose connective tissue with blood and lymphatic capillaries and nerve fibers surrounding ↑ nephrons and excretory duct system. 1) In cortex, I. predominantly contains ↑ fibroblasts (F) and, under normal conditions, rare cells similar to ↑ histiocytes, as well as a small amount of reabsorbed fluid passing from tubules (T) to capillaries (C). 2) In medulla, I. is more abundant than in cortex and contains spindle-shaped cells with branching processes, ↑ lipid droplets, and a prominent, probably contractile, microfilamentous system (possibly ↑ myofibroblasts). Function of I. consists of production of ↑ ground substance of loose connective tissue, compression of tubules by contractile cells, ↑ phagocytosis, and synthesis of ↑ prostaglandins PHE_2, $PGF_{2\alpha}$, and prostacyclin. I. makes up about 5% of kidney mass. (Fig. = cortex of a perfused rat kidney with artificially enlarged intercellular spaces)

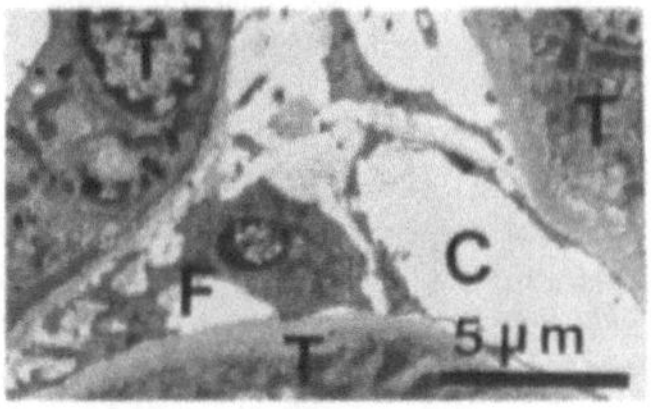

Interstitial tissue, of testis (interstitium testis*): Sum of Leydig's ↑ interstitial cells, blood and lymphatic capillaries, and nerve fibers embedded in ↑ loose connective tissue enveloping ↑ seminiferous tubules. Some cells with prominent microfilaments are thought to be ↑ myofibroblasts capable of contracting and provoking peristaltic contractions of seminiferous tubules, which push ↑ spermatozoa toward ↑ rete testis.

Connell, C.J., Connell, G.M.: The Interstitial Tissue of the Testis. In: Johnson, A.D., Gomes, W.R. (eds.): The Testis, Vol. 4. New York, San Francisco, London: Academic Press 1977

Interstitium: ↑ Interstice

Interterritorial matrix (matrix interterritorialis*): A feebly basophilic and metachromatic ↑ cartilage matrix between ↑ isogenic groups of ↑ chondrocytes containing smaller amounts of ↑ chondromucoprotein than ↑ capsule of cartilage cells.

Intervertebral discs (discus intervertebralis*): A specialized type of ↑ symphysis (ID) connecting bodies of two adjacent vertebrae (V). Structure: 1) Plates of partly calcified hyaline ↑ cartilage (CC) apposed to lamellar ↑ bone of vertebrae and continuous with a zone of fibrous ↑ cartilage (FC). 2) Fibrous cartilage is continuous with anuli fibrosi (AF), which are lamellar rings of very strong, densely packed, spirally arranged ↑ collagen fibers. In each lamella (L), collagen fibers are parallel, but they lie at right angles to those of neighboring lamellae. 3) Central space of I. is filled by ↑ nucleus pulposus (NP). Recently, ↑ elastic fibers have been observed in lamellae of anuli fibrosi.

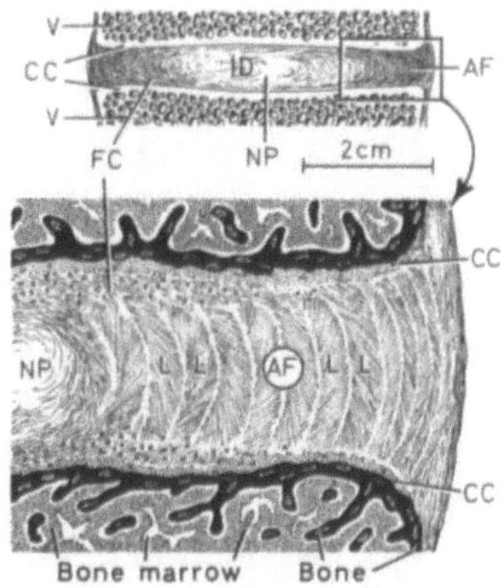

Buckwalter, J.A.: Fine Structural Studies of Human Intervertebral Disc. In: White, A., Gordon, S. (eds.): Idiopathic Low Back Pain. St. Louis: C.V. Mosby 1982; Inoue, H.: Spine *6*, 139 (1981); Johnson, E.F., Chetty, K., Moore, I.M., Stewart, A., Jones, W.: J. Anat. *135*, 301 (1982)

Intervillous spaces, of placenta (placental labyrinth, spatium intervillosum*): The enormous irregular system of interconnecting channels surrounding ↑ placental villi; on one side, it is limited by ↑ chorionic plate, on the other by the ↑ basal plate. I. contains maternal blood.

↑ Placenta, full-term, structure of

Intestinal glands: ↑ Lieberkühns' crypts

Intestinal juice: ↑ Succus entericus

Intestinal tract, endocrine cells of: ↑ Endocrine cells, of gastrointestinal tract

Intestinal villi (villus intestinalis*): Fingerlike and leaflike processes (0.5–1.5 mm in length) of ↑ tunica mucosa of ↑ small intestine. Structure: 1) Epithelium (E) = a ↑ simple columnar epithelium composed of ↑ absorptive cells and ↑ goblet cells. 2) Stroma or core (S) = a highly cellular ↑ loose connective tissue of ↑ lamina propria enclosing a central lymphatic vessel or ↑ lacteal (L) and a rich network of blood capillaries lying close to basal surface of epithelium. ↑ Smooth muscle cells (arrows) arising from lamina muscularis mucosae (arrowhead) anchor at tip and sides of I.; by their rhythmic contractions, I. effect a shortening and wavelike pumping motion, which empties lacteal into intestinal lymphatic vessels. There are about 40 I./mm^2 in ↑ duodenum and ↑ jejunum, and 10 I./

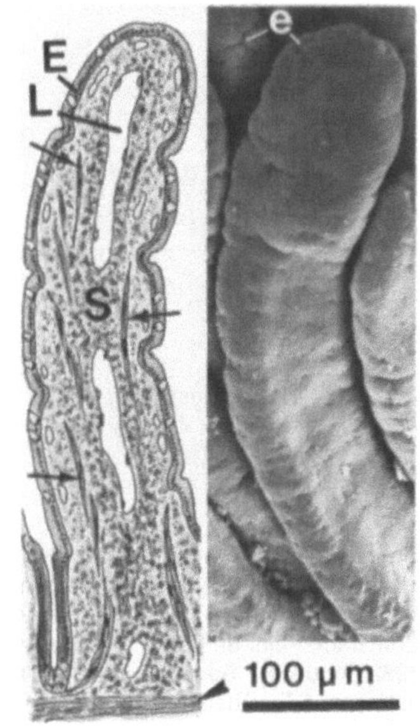

mm² in ↑ ileum. e = ↑ extrusion zone (Fig. = rat)

↑ Intestine, small, lymphatics of

Forrester, J.M.: J. Anat. *111*, 283 (1972)

Intestinal villi, blood circulation in: One to two arteries (A) run without branching through lamina propria up to tip of villus, then branch into a capillary network (Cap). One to two veins (V) collect capillaries and convey blood to submucosal veins. At tip of villus, there is a short direct arteriovenous ↑ anastomosis (AA) connecting arteries and veins. In absence of absorption, blood passes through arteriovenous anastomosis.

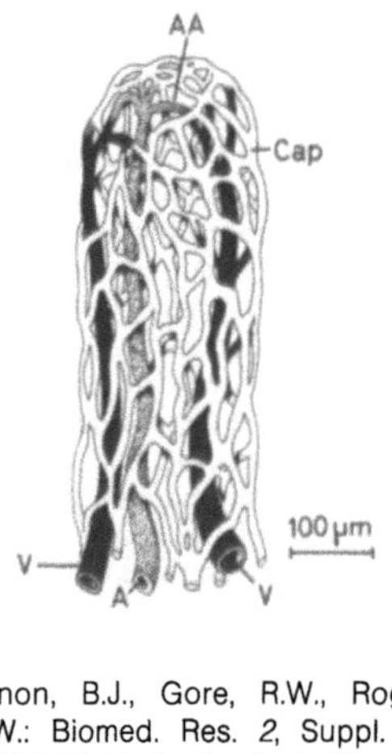

Gannon, B.J., Gore, R.W., Rogers, P.A.W.: Biomed. Res. *2*, Suppl. 235 (1981); Ohtani, O., Kikuta, A., Ohtsuka, A., Taguchi, T., Murakami, T.: Arch. Histol. Jpn. *46*, 1 (1983)

Intestine (intestinum*): A chain of tubular organs making up part of ↑ gastrointestinal tract. I. is situated between ↑ stomach and ↑ anal canal and comprises large and small ↑ I.

Intestine, absorption by: ↑ Absorption, intestinal

Intestine, innervation of: 1) Extrinsic innervation = preganglionic fibers of vagus and postganglionic fibers of ↑ sympathetic trunk, both probably efferent (sensory) in nature. 2) Intrinsic innervation = a) ↑ myenteric plexus (MP) for motor innervation of tunica muscularis (TMu); b) ↑ submucosal plexus (SP) for motor innervation of lamina muscularis mucosae (LMM) and ↑ smooth muscle cells of ↑ intestinal villi (V).

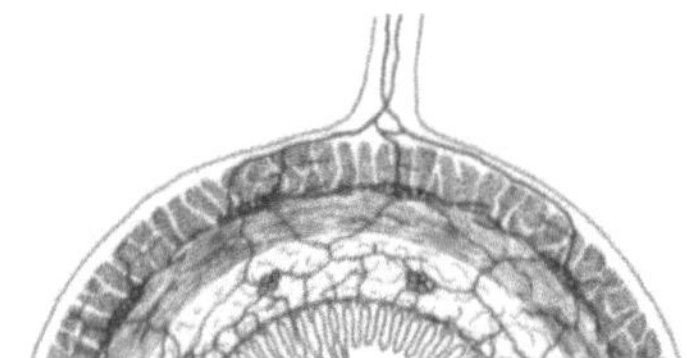

Gabella, G.: Int. Rev. Cytol. *59*, 129 (1979); Gershon, M.D.: Ann. Rev. Neurosci. *4*, 227 (1981)

Intestine, large (intestinum grassum*): A segment of ↑ gastrointestinal tract, about 1.5 m long, situated between ileocecal valve and ↑ anal canal. I. comprises ↑ appendix, ↑ cecum, ↑ colon (ascending, transverse, descending, and sigmoid), and ↑ rectum. In I., absorption of water and progressive solidification of intestinal contents occurs.

Intestine, small (intestinum tenue*): Part of ↑ gastrointestinal tract between ↑ stomach and large ↑ intestine. I. consists of ↑ duodenum, ↑ jejunum, and ↑ ileum; within it, absorption of nutritive substances takes place.

↑ Absorptive cells; ↑ Absorption, intestinal

Intestine, small, lymphatics of: At bases of ↑ intestinal villi (V), ↑ lacteals (L) anastomose with lymphatic capillaries between crypts of ↑ Lieberkühn (LC) and constitute a plexus on inner aspect of lamina muscularis mucosae (LMM). Branches of this mucosal plexus with valves penetrate muscularis

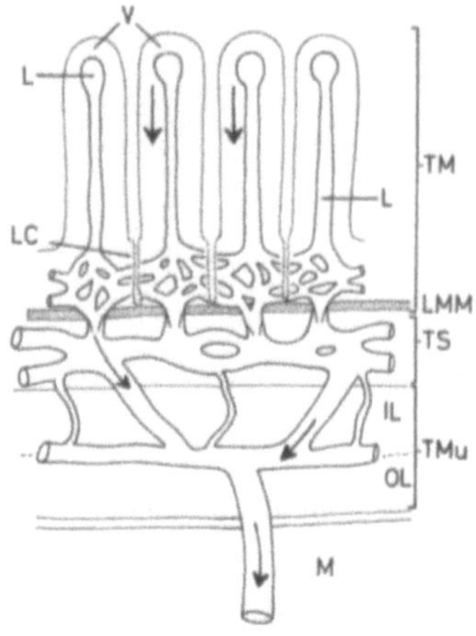

mucosae and form a plexus of large lymphatic vessels in tela submucosa (TS). ↑ Lymphatic capillaries surrounding ↑ lymphatic nodules of ↑ Peyer's patches (not shown) empty into this plexus. Lymphatics then perforate inner layer (IL) of tunica muscularis (TMu) and receive additional lymphatic capillaries situated between two muscular layers. Finally, larger lymphatic vessels penetrate outer muscular layer (OL) and continue through ↑ mesentery (M). TM = tunica mucosa

Intima (tunica intima*): The innermost layer of ↑ arteries, ↑ veins, and ↑ lymphatic vessels.

Intimal cushions (intra-arterial cushion, subendothelial cushion, pulvinar tunicae intimae*): Subendothelial aggregations of ↑ smooth muscle cells (IC) oriented parallel to axis of some ↑ muscular arteries and some veins. I. are present along arteries of ↑ carotid body, ↑ nasal mucosa, at branching sites of ↑ coronary, thyroid, renal, splenic, and intracranial arteries, as well as some veins, e.g., ↑ suprarenal vein (fig.). I. serve as valves for regulation of blood flow. (Fig. = human)

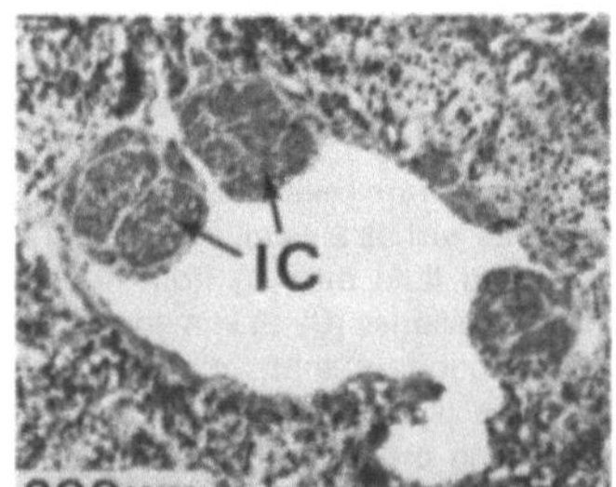

Hesse, M., Böck, P.: Z. mikrosk.-anat. Forsch. *94*, 471 (1980); Kardon, R.H., Farley, D.B., Heidger, P.M.Jr., Van Orden, D.E.: Anat. Rec. *203*, 19 (1982)

Intra-arterial cushions: ↑ Intimal cushions

Intra-articular cartilage: ↑ Disc, articular

Intracellular canaliculi: ↑ Canaliculi, intracellular

Intracleft lines: ↑ Intrasynaptic filaments

Intracristal space, of mitochondria: Extension (IS) of ↑ outer chamber (OC) between leaves of ↑ mitochondrial cristae (C); functionally identical to outer chamber.

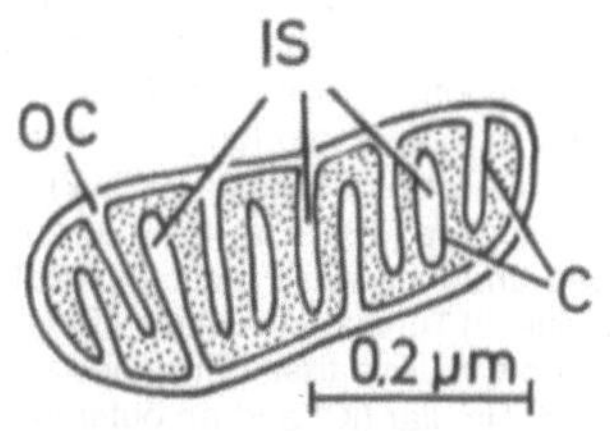

Intracytosis: Transport of substances in vesicles from one ↑ organelle to another (e.g., displacement of ↑ transfer vesicles from rough endoplasmic reticulum to ↑ forming face of ↑ Golgi apparatus, movement of primary ↑ lysosomes from Golgi apparatus toward ↑ phagolytic vacuole).

Intraepithelial glands: ↑ Endoepithelial glands

Intrafusal fibers: Specially differentiated ↑ skeletal muscle fibers situated within a ↑ neuromuscular spindle.

↑ Nuclear bag fibers; ↑ Nuclear chain fibers

Intraglomerular mesangial region: ↑ Mesangium

Intrahepatic bile ducts: ↑ Bile pathways

Intralobular ducts (ductus intralobularis*): Drainage pathways (ILD) of some ↑ exocrine glands situated within a ↑ lobule (L). I. are generally lined with ↑ simple cuboidal epithelium and are continuous with ↑ interlobular ducts. (Fig. = inactive ↑ mammary gland, human)

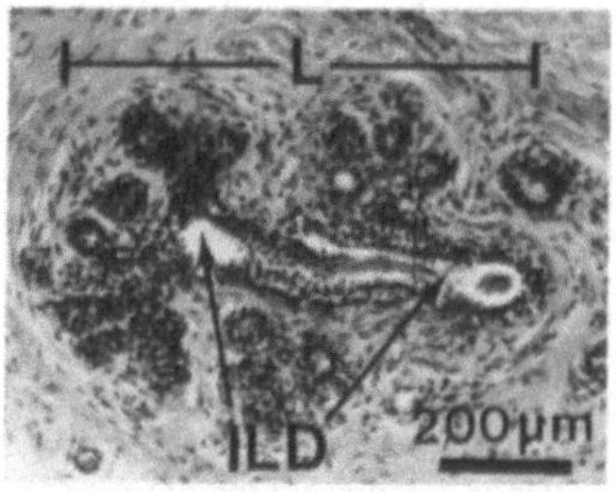

Intramembranous bones: ↑ Bones formed according to mechanism of direct ↑ bone formation.

Intramembranous particles: ↑ Membrane-associated particles

Intramitochondrial granules: ↑ Mitochondrial granules

Intramural pericytes: ↑ Mural cells, of retinal capillaries

Intranucleolar chromatin: ↑ Chromatin, intranucleolar

Intraperiod line (interperiod line, interperiod membrane): An electron-dense line, about 1.5 nm thick, situated between major dense lines of a ↑ myelin sheath.

Intrapulmonary bronchi: ↑ Bronchi, intrapulmonary

Intrarenal collecting ducts: ↑ Collecting tubules, of kidney

Intrasynaptic filaments (intracleft lines): Filaments, about 5 nm in width (arrows), of unknown significance extending transversally through ↑ synaptic cleft (SC) from presynaptic (Pre) to postsynaptic (Post) membrane. I. appear to be of a glycoprotein nature. (Fig. = rat)

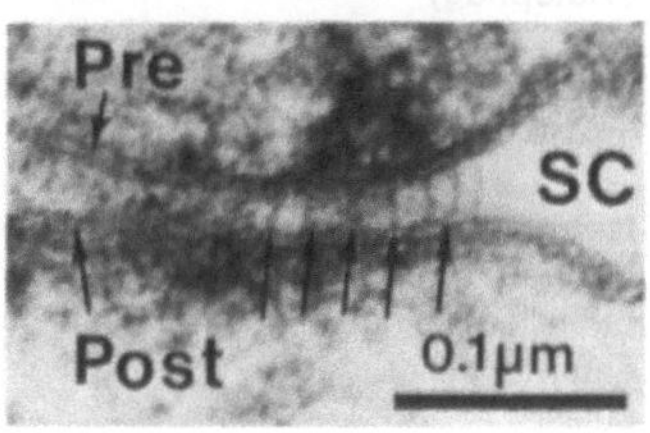

Intraventricular macrophages: ↑ Kolmer's cells

Intrinsic factor, gastric: ↑ Gastric intrinsic factor

Inulin: A metabolically inactive carbohydrate that rapidly passes unchanged through glomerular ↑ capillaries and reaches, without reabsorption in tubules, the definitive urine. Because of this property, I. is used for analysis of some renal functions.

Invaginating midget bipolar cells: ↑ Bipolar cells, of retina

Involution: 1) Decrease of parenchymal cell number in an organ (e.g., I. of ↑ thymus, I. of ↑ mammary gland after cessation of ↑ lactation, I. of ↑ corpus luteum). 2) Reduction of a hypertrophied organ to its normal size (e.g., I. of ↑ uterus to its usual size following childbirth).

↑ Atrophy; ↑ Hyperplasia; ↑ Hypertrophy; ↑ Thymus, accidental involution of; ↑ Thymus, involuted

Iodopsin: One of three cone cell pigments composed of ↑ retinene and ↑ photopsin. I. is contained in ↑ membranous discs of outer cone cell segment.

↑ Photoreceptors

Hurvich, L.M.: Color Vision. Oxford: Blackwell Scientific Publications 1981

Iridocorneal angle (angulus iridocornealis): Portion of ↑ anterior chamber delimited by ↑ limbus of eye and root of ↑ iris. ↑ Trabecular meshwork is located in I.

Iris*: A thin annular plate, 12 mm in diameter, with a central hole, the pupil. I. is most anterior part of ↑ uvea. I. extends from anterior termination of ↑ ciliary body (CB), ciliary margin (CM), to pupillary margin (PM). Structure: 1) Anterior surface lacks an epithelium; it is bordered by stromal ↑ fibroblasts and ↑ melanocytes. Deep invaginations of anterior surface form ↑ crypts (Cr). 2) Stroma (S) = a richly vascularized and innervated ↑ pigment loose connective tissue with ↑ clump cells (CC) and ↑ sphincter muscle (Sph). Number of melanocytes and ↑ melano-

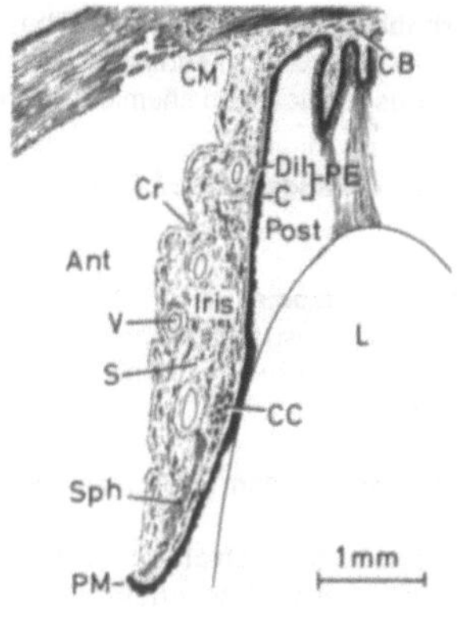

phores determines ↑ color of eye. Blood vessels (V) are ensheathed in a tubular envelope of collagen fibers, preventing interruption of circulation during movements of I. 3) Posterior pigment epithelium (PE) = a double layer of pigment epithelial cells (pars iridica retinae). Anterior layer forms ↑ dilator muscle (Dil); posterior cell layer consists of columnar cells (C) loosely attached to ↑ lens (L) by limiting membrane of I. Both layers are continuous with one another at pupillary margin. I. separates ↑ anterior chamber (Ant) from ↑ posterior chamber (Post) and acts as a diaphragm regulating quantity of light falling on ↑ photoreceptors.

Iron (Fe): A metallic element occurring in heme portion of ↑ hemoglobin and ↑ myoglobin, as well as in ↑ ferritin, ↑ transferrin, and ↑ hemosiderin. It is stored in ↑ siderosomes of various cells of ↑ bone marrow and ↑ spleen (↑ macrophages, ↑ reticular cells, ↑ erythroblasts). In a man weighing 70 kg, there are 2.3 g I. contained within heme, 0.56 g in hemosiderin, 0.45 in ferritin, and 0.14 g in myoglobin.

↑ Berlin blue reaction; ↑ Turnbull reaction

Bernat, I.: Iron Metabolism. New York: Plenum Press 1983; Richter, G.W.: Am. J. Pathol. *91*, 361 (1978)

Iron-hematoxylin staining: A combined staining technique using ↑ hematoxylin and iron alum which acts as a ↑ mordant and oxidizing agent, I. is employed for demonstration of ↑ mitochondria, ↑ myofibrils, ↑ intercalated discs, ↑ centrioles, etc. under light microscope.

Iron oxide: A substance in the form of heavily electron-dense particles of about 5 nm used for transmission electron microscopic demonstration of ↑ transendothelial transport.

Irritability, of nerve fibers: The capacity of nerve fiber endings to react to various physical and chemical agents.

Islets, of Langerhans: ↑ Pancreatic islets

Isochromosome: A ↑ chromosome anomaly consisting of transversal division of ↑ centromere rather than longitudinal division.

Isochronal rhythm: ↑ Ciliary motion

Isocortex, heterotypical: ↑ Brain cortex, heterotypical isocortex of

Isocortex, homotypical: ↑ Brain cortex, homotypical isocortex of

Isodesmosine: An amino acid involved, in addition to ↑ desmosine, in polymerization of ↑ elastic fibers.

Isogenic groups (chondrone, aggregatio chondrocytica*): Nests (IG) of two or more closely apposed ↑ chondrocytes (Ch) located deep inside ↑ cartilage. I. represent the progeny of a single ↑ chondrocyte as a result of its mitotic division during interstitial growth of cartilage. In hyaline ↑ cartilage (fig.), I. are surrounded by a common dense basophilic layer, the ↑ capsule (C), rich in ↑ chondromucoprotein; in elastic ↑ cartilage and fibrous ↑ cartilage, I. are composed of a few cartilage cells with a thin, inconspicuous capsule. ITM = ↑ interterritorial matrix. (Fig. = human)

↑ Cartilage, growth of

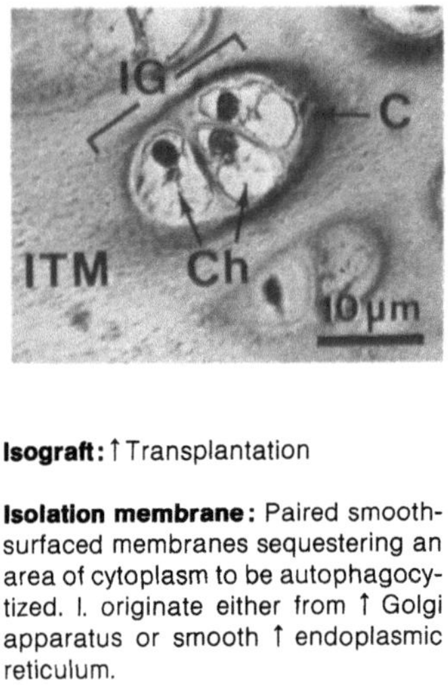

Isograft: ↑ Transplantation

Isolation membrane: Paired smooth-surfaced membranes sequestering an area of cytoplasm to be autophagocytized. I. originate either from ↑ Golgi apparatus or smooth ↑ endoplasmic reticulum.

↑ Autophagy

Isotopes, bone-seeking: ↑ Bone-seeking isotopes

Isotropy: The incapacity of inorganic or organic structures to cause a deviation in the plane of polarized light because of their randomly oriented molecules.

↑ Birefringence

Isotropic band, of striated muscle fibers: ↑ I-band

Isthmus, of oviduct: ↑ Oviduct

J

Jacobson's organ: ↑ Vomeronasal organ

Janus green B: A ↑ supravital and ↑ vital dye which stains ↑ mitochondria greenish-blue. Mitochondrial cytochrome oxidase system maintains the dye in its oxidized (colored) state.

Jejunum*: The segment of small ↑ intestine between ↑ duodenum and ↑ ileum. Structure: 1) Tunica mucosa (TM): a) epithelium (E) = a ↑ simple columnar epithelium formed of ↑ absorptive and ↑ goblet cells lining ↑ intestinal villi (IV) and ↑ Lieberkühn's crypts (C); b) lamina propria (LP) = a richly vascularized and innervated ↑ loose connective tissue; c) lamina muscularis mucosae (LMM) = a layer of ↑ smooth muscle cells. Tunica mucosa is thrown into long and numerous ↑ plicae circulares (PC). 2) Tela submucosa (TS) = a richly vascularized and innervated loose connective tissue. 3) Tunica muscularis (TMu) = smooth muscle tissue arranged in an inner circular (IC) and an outer longitudinal (OL) layer. 4) Tela subserosa (TSs) = a loose connective tissue. 5) Tunica serosa (arrowhead) = visceral ↑ peritoneum.

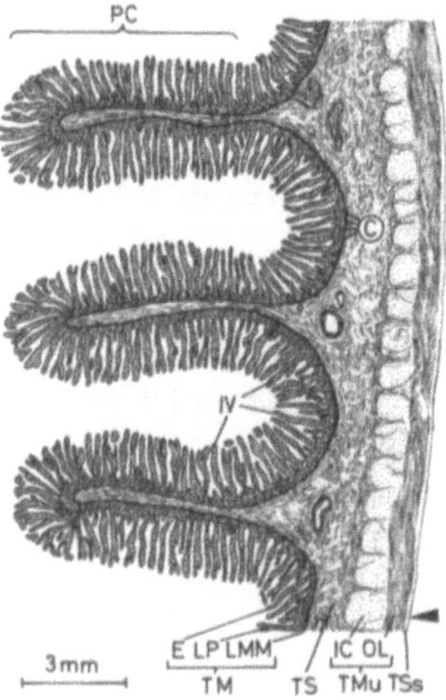

Shiner, M.: Ultrastructure of the Small Intestine Mucosa. Berlin, Heidelberg, New York: Springer-Verlag 1982

Joint cartilage: ↑ Articular disc

Joints (juncturae ossium*): Sites of contact between two or more bones permitting varying degrees of movement. In fig., linkage tissue is shown in brackets. (Asterisks indicate letters under which the terms are described.)

Sokoloff, L. (ed.): The Joints and Synovial Fluid. New York: Academic Press 1978; Wright, V., Dowson, D., Kerr. J.: The Structure of Joints. In: Hall, D.A., Jackson, D.S. (eds.): International Review of Connective Tissue Research. Vol. 6. New York: Academic Press 1973

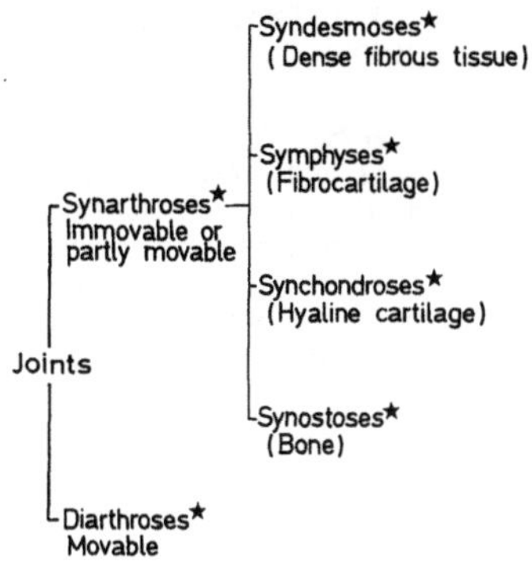

Junction, choledochoduodenal: ↑ Choledochoduodenal junction

Junction, dermoepidermal: ↑ Dermoepidermal junction

Junction, esophagogastric: ↑ Cardia

Junction, gap: ↑ Nexus

Junction, intermediate: ↑ Intermediate junctions

Junction, mucocutaneous: ↑ Mucocutaneous junction

Junction, myoendothelial: ↑ Myoendothelial junction

Junction, myoneural: ↑ Motor end plate

Junction, myotendinal: ↑ Myotendinal junction

Junction, sclerocorneal: ↑ Limbus, of eye

Junction, tight: ↑ Zonula occludens

Junctional complex, of epithelia (junctio intercellularis specialis*): A lateral beltlike attachment device (JC) of epithelial cells (↑ simple and ↑ pseudostratified epithelia) in area of ↑ apical pole. J. consists of ↑ zonula occludens (ZO) and ↑ zonula adherens (ZA). Some authors include ↑ desmosomes (D) in J. In light microscope, J. occurs as ↑ terminal bar.

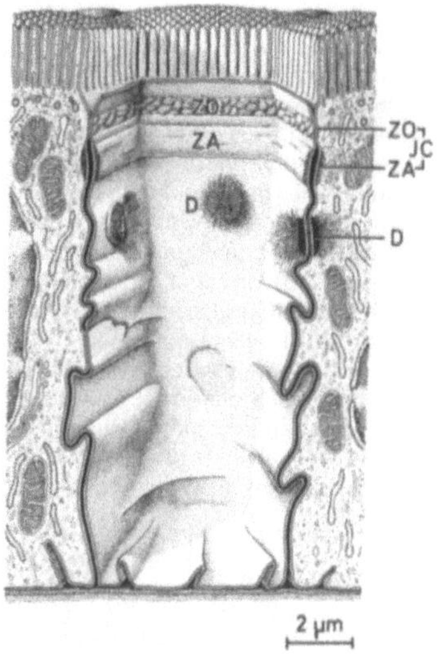

Pitts, J.D.: In Vitro *16*, 1049 (1980); Revel, J.P., Yancey, S.B., Meyer, D.J., Nicholson, B.: In Vitro 16, 1010 (1980); Staehelin, L.A.: Int. Rev. Cytol. *39*, 191 (1974); Staehelin, A., Hull, B.: Sci. Am. *238/5*, 140 (1978); Unvin, P.W.T., Zampighis, G.: Nature. *283*, 545 (1980)

Junctional folds: ↑ Subneural apparatus; ↑ Synaptic cleft, secondary

Junctions, intercellular (junctiones cellulares*): Specialized areas of cell membrane arranged as pairs on surface of opposed cells. J. contribute to cell cohesion and allow for communication between cells. J. include ↑ desmosome, ↑ fascia adherens, ↑ nexus, ↑ zonula occludens, ↑ zonula adherens, and ↑ interdigitations.

Juvenile cells: ↑ The term sometimes used to designate ↑ metamyelocytes.

Juxtaglomerular apparatus (juxtaglomerular complex, complexus juxtaglomerularis*): A common term for specially differentiated cell groups of ↑ vascular pole of ↑ renal corpuscle, consisting of: 1) ↑ Juxtaglomerular cells (JGC) of ↑ afferent arteriole (AA); 2) ↑ macula densa (MD) of ↑ distal tubule; 3) extraglomerular ↑ mesangium (EM). J. is involved in regulation of blood pressure and ↑ feedback control of glomerular filtration rate on an individual ↑ nephron level through ↑ renin-↑ angiotensin system. Some authors include ↑ afferent and efferent arterioles in J. (Fig. = mouse)

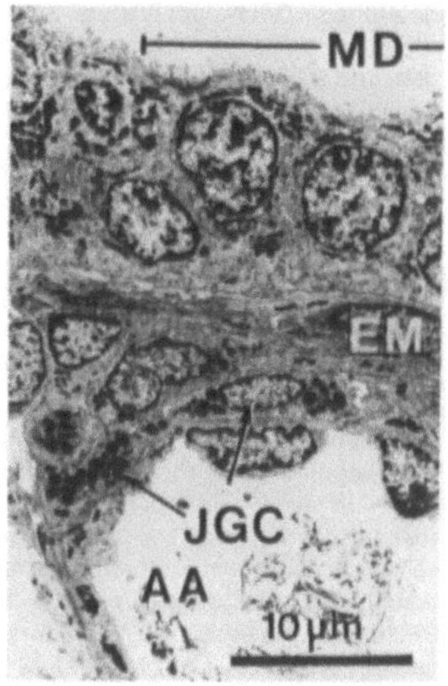

↑ Adrenal glands, cortex of; ↑ Adrenal glands, cortex, control of secretion; ↑ Aldosterone

Barajas, L.: Am. J. Physiol. *237*, F 333 (1979); Cantin, M., El-Khatib, E., Yunge, L.: Path. Biol. *27*, 261 (1979); Christensen, U.A., Bjaerke, H.A., Meyer, D.S., Bohle, A.: Acta Anat. *103*, 374 (1979); Gorgas, K.: Adv. Anat. Embryol. Cell Biol. *54/2*, 1 (1978)

Juxtaglomerular cells (↑ epithelioid cells, myoepithelioid cells, cellula juxtaglomerularis*): Modified smooth muscle cells (JC) constituting tunica media of ↑ afferent arteriole (A) at its entry into ↑ glomerulus. On one side, J. contact tunica intima (I) of arteriole, and on the other ↑ macula densa. J. are large and roughly spindle-shaped cells with an eccentric nucleus, numerous small mitochondria, a moderately

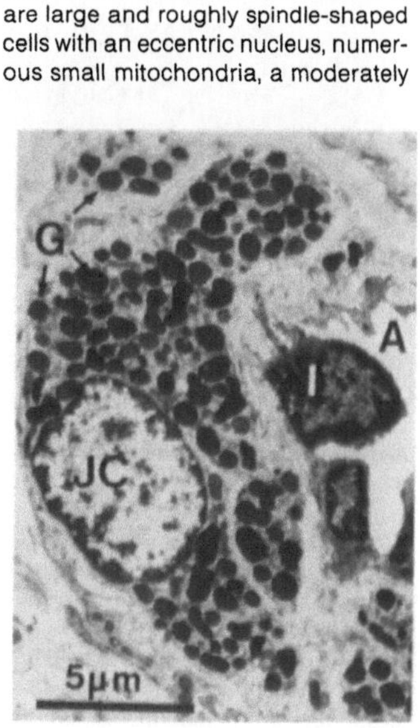

developed Golgi apparatus, short rough endoplasmic cisternae, ↑ glycogen, and some free ribosomes. An appreciable number of ↑ juxtaglomerular granules (G), which arise from Golgi apparatus, fill cytoplasm. Subjacent to plasmalemma are ↑ actin myofilaments and many ↑ caveolae. It is assumed that J. produce ↑ renin or its precursor. In renal ischemia and adrenalectomy, number of granules increases. J. are part of ↑ juxtaglomerular apparatus. (Fig. = mouse)

Bargmann, W.: Niere und ableitende Harnwege. In: Möllendorff v.W., Bargmann, W. (eds.): Handbuch der mikroskopischen Anatomie des Menschen. Vol. 7, Part 5. Berlin, Heidelberg, New York: Springer-Verlag 1978; Bucher, O., Kaissling, B.: Verh. Anat. Ges. *67*, 109 (1973); Cain, H., Boss, J.H., Egner, E.: Virchows Arch. A Pathol. Anat. *378*, 111 (1978); Désormeaux, Y., Ballak, M., Benchimol, S., Lacasse, J., Cantin, M., Genest, J.: Cell Tissue Res. *222*, 53 (1982); Zaki, F.G., Keim, G.R., Takii, Y., Inagami, T.: Ann. Clin. Lab. Sci. *12*, 200 (1982)

Juxtaglomerular granules (renin granules): Moderately osmiophilic, round or variously shaped ↑ unit membrane-limited granules (JG) found in ↑ juxtaglomerular cells. Precursors of J. are found within cisternae of Golgi apparatus; sometimes, granules have a crystalline interior. It is thought that J. contain ↑ renin. (Fig. = rat)

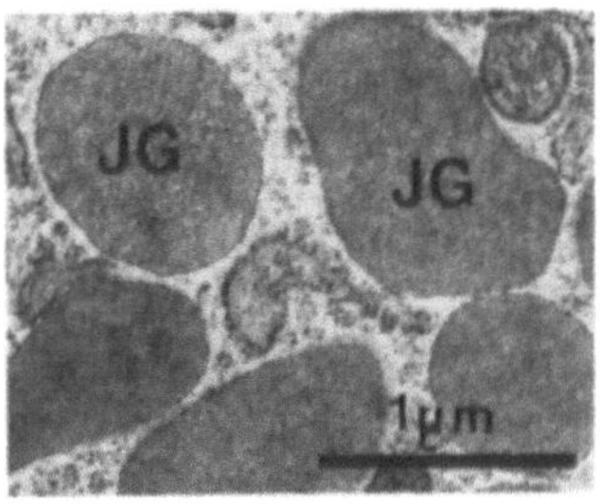

Taugner, Ch., Poulsen, K., Hackenthal, E., Taugner, R.: Histochemistry *62*, 19 (1979); Yun, J.C.H.: Nephron *23*, 72 (1979)

Juxtamedullary nephron: ↑ Nephron, juxtamedullary

Juxtaoral organs: Paired elongated structures located in muscular layer of each ↑ cheek in man and all other mammals. Structure: 1) ↑ Parenchyma (P) = a cellular cord, about 10 µm wide, composed of cells similar to ↑ keratinocytes of ↑ epidermis. Cells contain many ↑ tonofilaments and are held to one another by ↑ desmosomes and to basal lamina that surrounds parenchyma by numerous ↑ hemidesmosomes. 2) Capsule (C): a) outer layer = several concentric layers of flattened fibrocytes similar to ↑ perineurium; b) inner layer = one incomplete layer of fibrocytes (F). ↑ Collagen and ↑ elastic fibers are numerous in capsule; they run longitudinally. Between the two capsular layers there is a space (S) filled with ↑ tissue fluid. 3) Nerve fibers (NF) penetrate intracapsular portion of J., lose their ↑ myelin sheath, and contact outer aspect of basal lamina; some naked ↑ axons run among parenchymal cells. Function of J. is unknown: They could be ↑ mechanoreceptors, since they have a similarity to some ↑ corpuscles. Cap = capillary. (After Jeanneret-Gris, 1980)

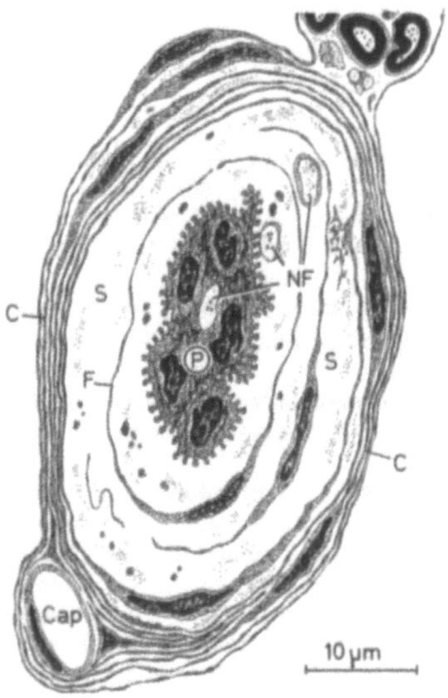

Jeanneret-Gris, B.: Arch. Anat. Microsc. *69*, 197 (1980); Müller, E., Zenker, W.: Histochemistry *71*, 279 (1981)

K

Kallikrein: A vasoconstrictive peptide produced by cells of ↑ striated ducts of ↑ parotid, ↑ submandibular, and ↑ sublingual glands, ↑ proximal convoluted tubules, initial segment of certain ↑ distal tubules of ↑ nephron, and in some ↑ papillary ducts. α-adrenergic stimulation increases discharge of K. into ↑ saliva.

Ørstavik, T.B.: J. Histochem. Cytochem. *28*, 881 (1980); Simson, J.A.V., Fenters, R., Chao, J.: J. Histochem. Cytochem. *31*, 301 (1983)

Kampmeier foci: Accumulations of ↑ lymphocytes and ↑ macrophages in the form of small white patches in ↑ mesothelium of mediastinal visceral ↑ pleura. It is believed that K. represent openings through which above-cited cells may pass onto pleural surface.

↑ Milky spots

Karyokinesis: Indirect nuclear division.

↑ Mitosis

Karyolymph (nuclear sap): The fluid component of ↑ karyoplasm in which ↑ chromatin and ↑ nucleolus bathe. K. consists of water with dissolved and suspended substances, such as soluble RNAs, salts, ↑ glycoproteins, and metabolites, which participate in an exchange between ↑ nucleus and ↑ cytoplasm. Most of K. is contained in area of ↑ euchromatin; in light microscope, K. appears completely unstained.

Karyolysis: ↑ Nucleus, degeneration and death of

Karyometry: A branch of ↑ stereology dealing with numerical expression of nuclear size and shape. Nuclear parameters (length, A; width, B; depth, C) can be measured directly in micro-

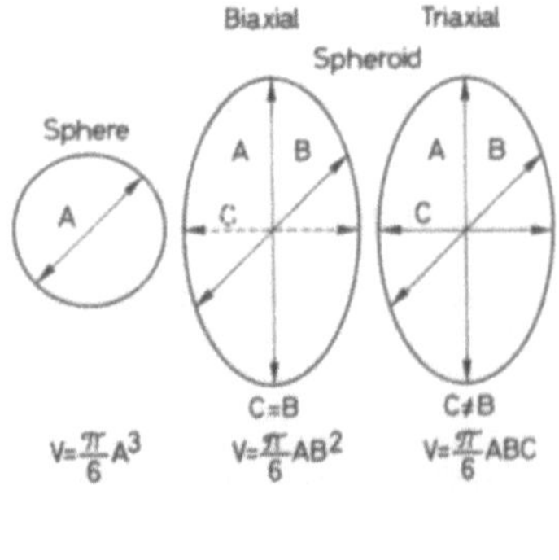

scope or by ↑ planimetry of projected images. Fig. gives simplest formulas for calculating nuclear volume (V) of directly measured nuclei.

Hildebrand, R.: Nuclear Volume and Cellular Metabolism. Adv. Anat. Embryol. Cell Biol. *60*, 1, Berlin, Heidelberg, New York: Springer-Verlag 1980; Palkovits, M., Fischer, J.: Karyometric Investigations. Budapest: Akademiai Kiado 1968

Karyoplasm (nucleoplasm, nucleoplasma*): A term designating entire ↑ nucleus when considered in relation to surrounding ↑ cytoplasm of cell. Together with cytoplasm, K. constitutes ↑ protoplasm of cell.

Karyopyknosis: ↑ Nucleus, degeneration and death of

Karyorrhexis: ↑ Nucleus, degeneration and death of

Karyosomes (chromatin particles, chromocenters): Individual membraneless clumps of ↑ heterochromatin, ranging from 0.1 to 2 µm in diameter. K. are scattered throughout ↑ karyolymph or marginated at inner aspect of ↑ nuclear envelope, leaving ↑ nuclear pores free.

Karyotheca: ↑ Nuclear envelope

Karyotype: The systematized array of metaphasic ↑ chromosomes of a cell or cell lineage obtained from photographs of disrupted cell nuclei. After cutting chromosomal pairs out from the photograph, they are ordered on the basis of position of ↑ centromere (Ce), size and shape, location of secondary constrictions and satellites (S). One member of a chromosomal pair originates from the mother (M) the other from the father (F). Chromosomal pairs are arranged in seven groups (A–G) whereby ↑ X- and ↑ Y-chromosomes are classified, according to their morphology, in groups C and G, respectively.

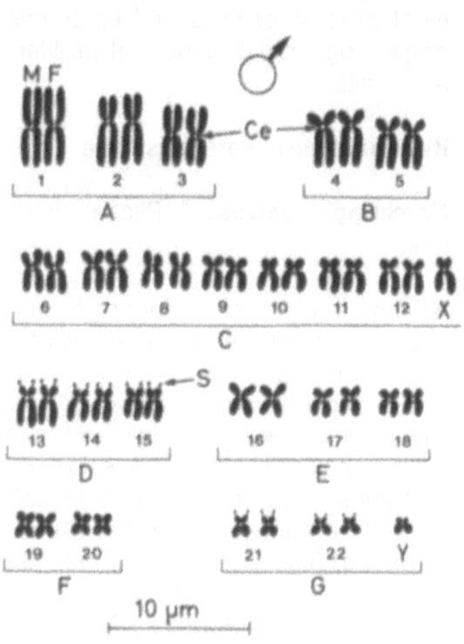

K-cells: A kind of ↑ endocrine cell of ↑ gastrointestinal tract that probably secrete ↑ gastric inhibitory peptide.

Keith and Flack node: ↑ Impulse-conducting system, of heart; ↑ Sinuatrial node

Keratan sulfate (keratosulfate): A ↑ glycosaminoglycan with N-acetylglycosamine-6-sulfate and galactose as repeating units. K. is present in ↑ ground substance of ↑ cartilage, ↑ nucleus pulposus, and particularly concentrated in ↑ corneal stroma, where it assures transparency.

Keratin: A family of water-insoluble elastic or hard proteins representing final product of ↑ epidermis, ↑ hair, and ↑ nail matrix. In the form of ↑ intermediate filaments, K. is present in almost all epithelia. K. has a cytoskeletal and protective role; it is divided into hard and soft ↑ K., although seven immunologically distinctive classes of K. have been described.

Tseng, C.G.S., Jarvinen, M.J., Nelson, W.G., Huang, J.-W., Woodcock-Mitchell, J., Sun, T.-T.: Cell *30*, 361 (1982)

Keratin, amorphous: A structureless, amorphous sulfur-containing protein that fills spaces between ↑ keratin filaments of both soft and hard ↑ keratins. K. of soft keratin is relatively insoluble in water and buffers between pH 2 and 11. In ↑ epidermis, K. originates from ↑ keratohyalin granules; in ↑ hairs, it originates from ↑ trichohyalin granules. Resistance of K. is due to oxidation of sulfhydryl groups into disulfide groups. In hard ↑ keratin, amorphous component contains more sulfur, making it insoluble in water. Origin of hard K. is not known.

Keratin filaments: ↑ Microfilaments (KF) measuring about 6–8 nm in diameter embedded in a mass of amorphous ↑ keratin. K. of soft ↑ keratin are presumably derived from ↑ tonofilaments (Tf), which are associated with ↑ keratohyalin granules in ↑ epidermis and with ↑ trichohyalin granules (T) in ↑ hair. Origin of K. of hard ↑ keratin is not known. K. contain a smaller amount of sulfur than amorphous keratin does; they are insoluble in buffers between pH 3 and 11 and anchor into cytoplasmic leaflet of plasmalemma of ↑ horny cells. K. are made up of fibrous α-protein, giving them flexibility and elasticity. (Fig., top = mongolian gerbil; fig., bottom = human)

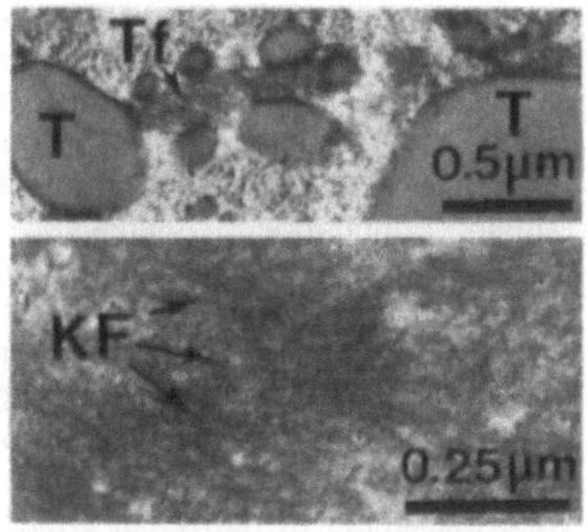

Steinert, P.M., Idler, W.W., Zimmermann, S.B.: J. Mol. Biol. *108*, 547 (1976)

Keratin, hard: A type of permanent and very solid keratin constituting bulk of ↑ nail plate, ↑ cuticle, and ↑ cortex of hairs. K. consists of very densely packed ↑ keratin filaments embedded in a highly electron-dense matrix of amorphous ↑ keratin. K. is insoluble and of higher sulfur content than soft ↑ keratin; it does not desquamate. K. is formed in ↑ nail matrix and cells of hair cortex without interposition either of ↑ keratohyalin or of ↑ trichohyalin granules; mechanism of its formation is not known.

Keratin, soft (horny matrix): A filamentary-amorphous substance filling ↑ horny cells of ↑ epidermis and cells of inner root sheath of ↑ hair follicle. K. consists of amorphous ↑ keratin (A) in which ↑ keratin filaments (KF) are embedded. (Fig. = horny cell, mongolian gerbil)

↑ Keratohyalin granules

Keratinization, of epidermis: A process of gradual ↑ differentiation of viable cells of ↑ epidermis into nonviable ↑ horny cells (HC). Phases are: 1) Synthetic phase (S) = formation of ↑ tonofilaments (Tf) by ↑ ribosome (R), formation of ↑ keratohyalin granules (KG) and ↑ membrane-coating granules (MCG) in stratum basale (SB), stratum spinosum (SS), and stratum granulosum (SG). 2) Degradative phase (D) = by activation of specific enzymes, cell nucleus and ↑ organelles are broken down and disappear, but tonofibrils and keratohyalin granules remain unchanged; immediately above stratum granulosum, they transform into a mass of soft ↑ keratin, which gradually fills cells of stratum corneum (SC). Membrane-coating granules, discharged from cells of stratum granulosum, probably provoke thickening of horny cell's plasmalemma and dissolve desmosomal contacts in stratum disjunctum (not shown). Time required for transformation of one basal cell into a horny, fully keratinized cell varies from 20 to 90 days. (Modified after Zelickson 1967)

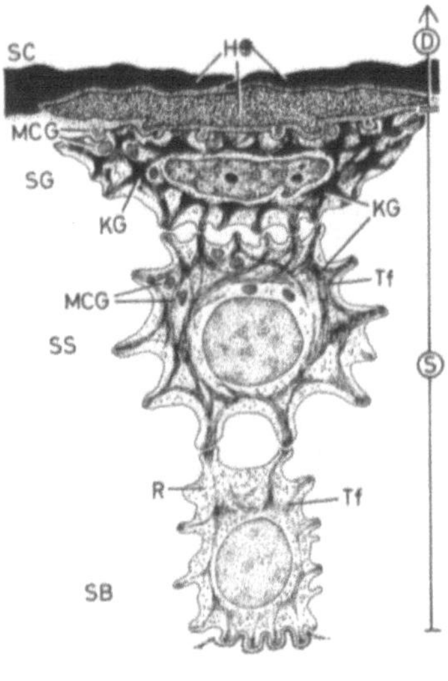

Matoltsy, A.G.: J. Invest. Dermatol. *65*, 20 (1976); Zelickson, A.S. (ed.): Ultrastructure of Normal and Abnormal Skin. Philadelphia: Lea & Febiger 1967

Keratinization, of nail: A process transforming proliferating cells in stratum germinativum of ↑ nail matrix into hard ↑ keratin of ↑ nail plate. During K., there is no presence of ↑ keratohyalin granules; cells to be keratinized flatten, amount of ↑ tonofilaments in their cytoplasm increases, and cell membrane thickens, probably due to discharge of ↑ membrane-coating granules into intercellular spaces.

Keratinized cells: ↑ Horny cells

Keratinizing system (Malpighian system): Main part of ↑ epidermis derived from cutaneous ↑ ectoblast. K. is constantly renewed from basal layer; final products of K. are dead ↑ horny cells.

↑ Pigmentary system

Keratinocytes: A term sometimes used to designate both ↑ basal and ↑ prickle cells of ↑ epidermis.

↑ Epidermis, ultrastructure of

Keratinosomes: ↑ Membrane-coating granules

Keratocytes (cornealocytes): Specialized ↑ fibroblasts (K) located in clefts between lamellae (L) of ↑ collagen fibers of corneal stroma. Stellate processes of K. spread parallel and perpendicular to collagen fibers. K. have an elongated, heterochromatin-rich nucleus with prominent nucleolus. Cytoplasm encloses threadlike mitochondria, a well-developed Golgi apparatus surrounded by numerous ↑ collagen secretory granules, extensive rough endoplasmic reticulum, a few lysosomes, and many free ribosomes. Beside collagen fibers, K. produce ↑ chondroitin and ↑ keratan sulfates, both responsible for the swelling property and transparency of ↑ cornea. B = ↑ Bowman's membrane, E = corneae epithelium, NF = nerve fibers

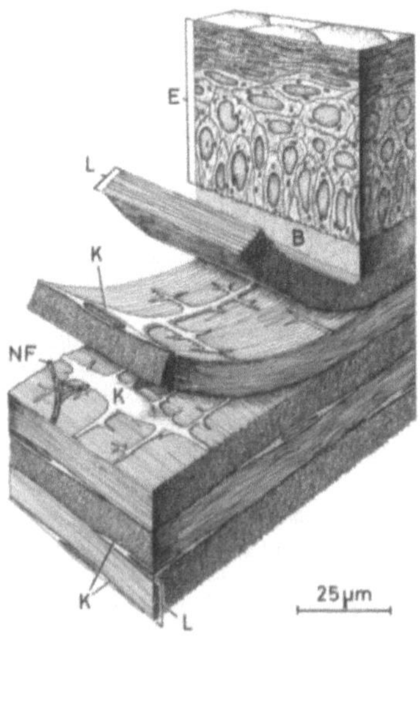

Keratogenous zone, of hair: The area (KZ) of keratinization of ↑ medulla, cortex, and ↑ cuticle of hair shaft, situated just above ↑ hair papilla (HP) tip.

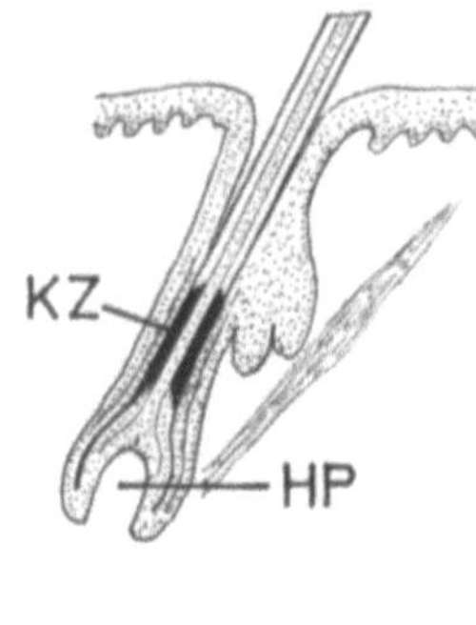

Keratohyalin granules (granulum keratohyalini*): Basophilic, membraneless, and irregularly shaped electron-dense structures (KG), ranging 0.2–5 µm in diameter, present mainly in cells of stratum granulosum of ↑ epidermis. K. consist of dense agglomerations of particles measuring 2 nm located among and around ↑ tonofilament bundles (Tf); K. may also appear within nucleus. K. contain proline and amino acids rich in sulfhydryl groups; they are involved in ↑ keratinization of ↑ epidermis, i.e., formation of soft ↑ keratin. Origin and exact chemical nature of K. are still unknown. (Fig. = human)

↑ Trichohyalin granules

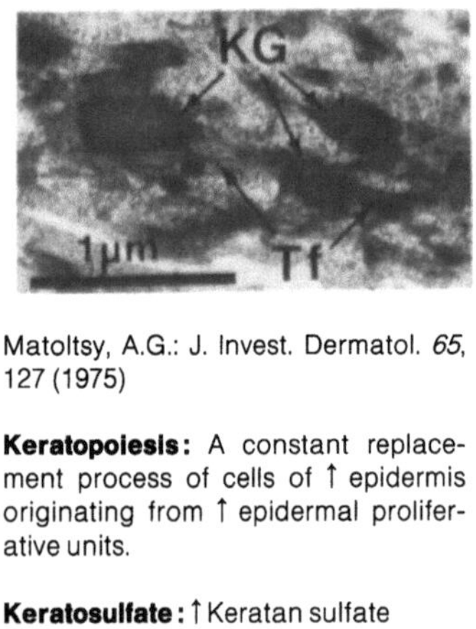

Matoltsy, A.G.: J. Invest. Dermatol. *65*, 127 (1975)

Keratopoiesis: A constant replacement process of cells of ↑ epidermis originating from ↑ epidermal proliferative units.

Keratosulfate: ↑ Keratan sulfate

Kerckring's valves: ↑ Plicae circulares

Kidney (ren*): Paired, bean-shaped organs responsible for excretion of ↑ urine. Structure: 1) Capsule (C): a) fibrous capsule (capsula fibrosa*) = a thin layer of ↑ dense connective tissue with occasional ↑ smooth muscle cells; b) adipose capsule (capsula adiposa*) = white ↑ adipose tissue. 2) ↑ Parenchyma: a) renal cortex (Co) = a 7 to 10-mm-thick granulated zone composed of ↑ renal corpuscles, ↑ proximal and ↑ distal convoluted tubules of ↑ nephrons, arched ↑ collecting tubules, and ↑ medullary rays (MR). Cortical substance penetrating between pyramids (Py) of medulla is referred to as ↑ renal columns (RCo); b) renal medulla = a finely striated zone formed by 8–18 ↑ renal pyramids (Py) penetrating minor ↑ calyces (MC) with their tips, and with their bases adjacent to cortex. Extending from bases of each pyramid are ↑ medullary rays, which reach into cortex. Medulla is composed of ↑ straight portions of proximal and distal tubules, ↑ loops of Henle, and ↑ collecting tubules. A pyramid with overlying cortex forms a ↑ renal lobe; a ↑ renal lobule is a medullary ray and its immediately adjacent cortical tissue. Cortical substance between medullary rays is referred to as renal labyrinth. 3) ↑ Renal sinus (S). 4) ↑ Renal hilus (H) = concave side of organ where vessels, nerves, and ↑ renal pelvis (P) enter and leave. About 5% of kidney parenchyma belongs to ↑ interstitial tissue. U = ↑ ureter.

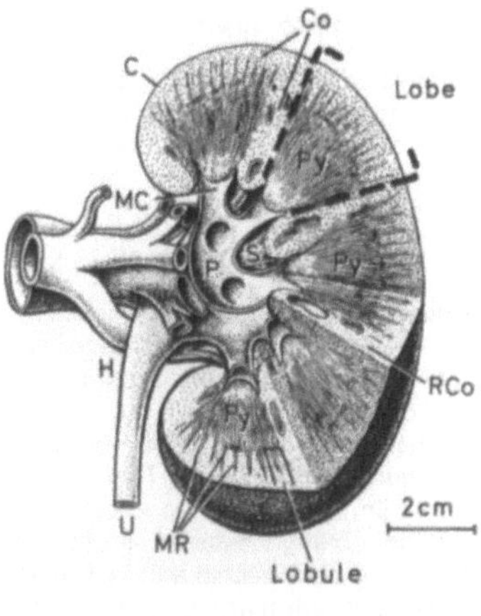

Andrews, P.M.: The Urinary System-Kidney. In: Hodges, G.M. and Hallowes, R.C. (eds.): Biomedical Research Applications of Scanning Electron Microscopy, Vol. 1. London, New York: Academic Press 1979; Bargmann, W.: Niere und ableitende Harnwege. In: Möllendorff, v.W., Bargmann, W.: Handbuch der mikroskopischen Anatomie des Menschen, Vol. 7, Part 5. Berlin, Heidelberg, New York: Springer-Verlag 1978; Maunsbach, A.B., Olsen, T.S., Christensen, E.I.: Functional Ultrastructure of the Kidney. London, New York: Academic Press 1980

Kidney: See also ↑ Renal

Kidney, collecting tubules of: ↑ Collecting tubules, of kidney

Kidney, columns of: ↑ Renal columns

Kidney, corpuscles of: ↑ Renal corpuscle

Kidney, cortex of: ↑ Renal cortex

Kidney, endocrine function of: The kidney produces ↑ renin, ↑ erythropoietin, ↑ prostaglandins, and ↑ kallikrein.

Peart, W.S., Lancet, September 10, 543 (1977)

Kidney, functions of: ↑ Kidney, histophysiology of; ↑ Kidney, endocrine function of

Kidney, glomerulus of: ↑ Renal glomerulus

Kidney, histophysiology of: By a complex process involving filtration, active absorption, passive absorption, and secretion, kidney liberates the body from waste products and foreign substances and assures osmotic stability of body fluids; it has also a pronounced endocrine function. 1) ↑ Renal glomerulus = about 10% of 1800 l blood passing daily through glomeruli forms ↑ glomerular filtrate due to ↑ filtration pressure and ↑ glomerular filtration barrier. About 99% of this ultrafiltrate is reabsorbed across tubular epithelium of ↑ nephron and ↑ collecting tubules and reenters blood vascular system. 2) ↑ Proximal tubule = complete reabsorption of glucose, 85% reabsorption of water, but only 40% reabsorption of ↑ urea, uric acid, and creatinine. Here, active reabsorption of amino acids, bicarbonate, proteins, Na^+, and vitamin C takes place. This portion of nephron actively secretes creatinine, para-aminohippuric acid, drugs, and some dyes used in clinical investigations. 3) ↑ Loop of Henle = site of urine concentration by passive diffusion of water in descending limb and active pumping of Cl^-, with Na^+ following passively, in ascending limb. 4) ↑ Distal convoluted tubule = active pumping of Na^+ and its partial replacement by secretion of K^+, hydrogen, and ammonia; urine becomes acidified here. 5) ↑ Collecting tubules = further passive water reabsorption. ↑ Antidiuretic hormone (ADH) and ↑ aldosterone control water and Na^+ reabsorption. In absence of ADH, distal and collecting tubules become impermeable to water, which results in production of a large amount of diluted urine. Aldosterone increases rate of Na^+ absorption in distal tubule. (See physiology texts for further information.) ↑ Kidney, endocrine function of

Bulger, R.E., Dobyan, D.C.: Anat. Rec. *205*, 1 (1983); Maunsbach, A.B., Olsen, T.S., Christensen, E.I. (eds.): Functional Ultrastructure of the Kidney. New York: Academic Press 1981

Kidney hormones: ↑ Kidney, endocrine function of

Kidney, innervation of: ↑ Myelinated and ↑ nonmyelinated ↑ sympathetic nerve fibers originate predominantly in celiac plexus and innervate renal artery and its branches. Some nerve endings have been described between cells of extraglomerular ↑ mesangium and in renal ↑ interstitial tissue. It is not definitely established if parasympathetic nerve fibers penetrate kidney. Since kidney functions after ↑ transplantation, nerve supply is not essential for its activity.

Kidney, interstitium of: ↑ Interstitial tissue, of kidney

Kidney, juxtaglomerular apparatus of: ↑ Juxtaglomerular apparatus

Kidney, lobes of: ↑ Renal lobe

Kidney, lobules of: ↑ Renal lobule

Kidney, lymphatics of: The lymphatics accompany larger blood vessels and leave kidney, together with renal vein, at ↑ hilus or pass through capsule. Lymphatics are absent in ↑ renal corpuscles and ↑ medullary rays.

Kidney, macula densa of: ↑ Juxtaglomerular apparatus; ↑ Macula densa

Kidney, medulla of (medulla renalis*): Sum of kidney ↑ pyramids. (For subdivision of K. into zones and stripes, see ↑ Nephron)

Kidney, nephrons, cortical, of: ↑ Nephron, cortical

Kidney, nephrons, juxtamedullary, of: ↑ Nephron, juxtamedullary

Kidney, nephrons of: ↑ Nephron

Kidney, pelvis of: ↑ Renal pelvis

Kidney, podocytes of: ↑ Podocytes

Kidney, pyramids of: ↑ Pyramid, of kidney

Kidney, sinus of: ↑ Renal sinus

Kidney, vascularization of: The renal artery divides within ↑ renal sinus into ventral and dorsal branches, which give off interlobar arteries (IA) situated in ↑ renal columns. At level of ↑ pyramid bases, interlobar arteries divide into arcuate arteries (AA), from which ↑ interlobular arteries (IlA) emerge radially and at fairly regular intervals. From them, at right angles, arise ↑ afferent arterioles (Aff), which break up into capillaries of ↑ renal glomerulus (G). Glomerular capillaries unite and form ↑ efferent arterioles (Eff). Efferent arterioles associated with cortical ↑ nephrons (CN) break up into a ↑ peritubular capillary network (PCN), which drains both into interlobular vein (IlV) and into radially oriented deep cortical veins (DCV), emptying into arcuate vein (AV). Efferent arterioles associated with juxtamedullary ↑ nephrons (JN) penetrate medulla at various levels as descending thin-walled nonbranched arterial vasa recta (AVR) (arteriolae rectae spuriae). These form a hairpin loop, turning upward toward cortex to empty as venous vasa recta (VVR) into interlobular vein. Interlobular arteries spread into richly anastomosing network of subcapsular area (SA) and finally form superficial cortical veins (SCV). These veins unite to form stellate veins (SV), which drain into interlobular vein. Some interlobular arteries perforate capsule and, as capsular arteries (CA), vascularize capsule (C); there are also anastomoses between capsular arteries and interlobular veins. Blood supply of the medulla is as follows: Most of medulla (90%) is vascularized by arterial vasa recta (AVR) of juxtamedullary nephrons; remaining 10% is supplied by branches of both arcuate and interlobular arteries, vasa recta vera (VRV) (arteriae medullares rectae verae*), which penetrate medulla, then turn as hairpin loops and drain, via venous vasa recta (VVR), into arcuate and interlobular veins (IV). Arcuate and interlobular arteries of a ↑ renal lobule are terminal ↑ arteries.

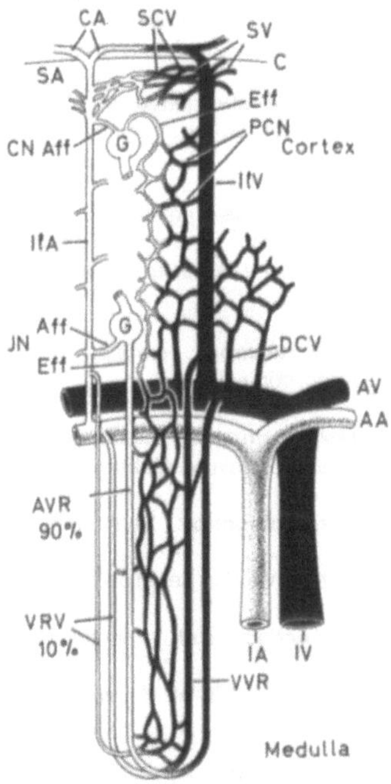

Hodson, J.C.: Current Problems in Diagnostic Radiology, Vol. 7, No 11, 1 (1978); Kriz, W.: Nephron *31*, 290 (1982); Kriz, W., Barret, J.M., Peter, S.: Int. Rev. Physiol. *2*, 1 (1976)

Kidney vein: ↑ Renal vein

Killer cells (cytotoxic lymphocytes, graft-rejection cells): ↑ T-lymphocytes responsible for cellular immunity because of their capacity to produce during ↑ immune response: a) cytotoxic substance directly destroying foreign cells (↑ lymphotoxin); b) ↑ lymphokines.

Berke, G.: Immunol. Rev. *72*, 5 (1983); Geiger, B., Rosen, D., Berke, G.: J. Cell Biol. *95*, 137 (1982)

Kinases: A group of intracellular enzymes activated by ↑ cyclic AMP. By acting on other enzymes, K. provoke a specific histophysiological response by a cell to a polypeptide hormone.

↑ Polypeptide hormones, action on the cell of

Kinetochoral microtubules: ↑ Microtubules, kinetochoral

Kinetochore (kinetochorus*): A specialized device of a ↑ centromere (Ce) in the form of a trilaminar disc (K) measuring about 300–500 nm in diameter. K. is situated at border between ↑ chromosome (Chr) and cytoplasm (C). Each K. consists of three distinct plates: Inner plate (IP) is osmiophilic and frequently indistinguishable from chromosome, it is about 40 nm thick; into outer plate (OP) of similar density and thickness anchor kinetochoral ↑ microtubules (Mt). A 20 to 30-nm-wide, electron-translucent zone separates both outer and inner plates. A strongly stained ↑ chromatin strand connects opposite K. and holds sister ↑ chromatids together. Interruption of this strand causes chromatids to separate. The term K. is frequently used as a synonym for a centromere. (Fig. = ovarian ↑ follicular cell, rat)

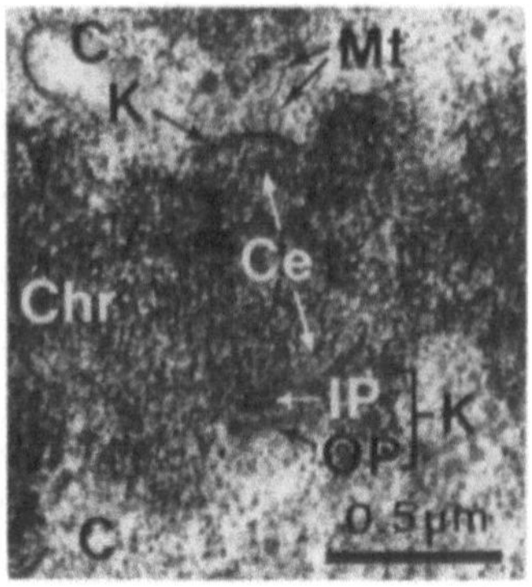

Goldstein, L.S.B.: Cell *25*, 591 (1981); Moens, P.B.: J. Cell Biol. *83*, 556 (1979); Pickett-Heaps, J.D., Spurck, T.P.: Eur. J. Cell Biol. *28*, 77 and 83 (1982); Rieder, L.C.: J. Ultrastruct. Res. *66*, 109 (1979); Roos, U.-P.: Cytobiologie *16*, 82 (1977)

Kinetosome: ↑ Cilium, basal body of

Kinocilia: ↑ Cilia

Kinocilia, formation of: ↑ Ciliogenesis

Klinefelter's syndrome: A group of signs and symptoms occurring in the case of an extra ↑ X-chromosome (44 + XXY) (undeveloped testicles, sterility, gynecomastia).

↑ Chromosomes, anomalies of

Knives: Special razors and specially fractured glass or diamond pieces for cutting ↑ sections for light and transmission electron microscopy. In light microscopy, most frequently used K. are plane-wedge K. (1) for ↑ sectioning ↑ paraffin blocks and plano-concave K. (2) for cutting ↑ celloidin blocks. For sectioning of ↑ semithin and ↑ ultrathin sections, K. are made of glass (G) or diamond (D) and equipped with a small metal or plastic trough (T) filled with water on which sections float.

↑ Microtome; ↑ Ultramicrotome

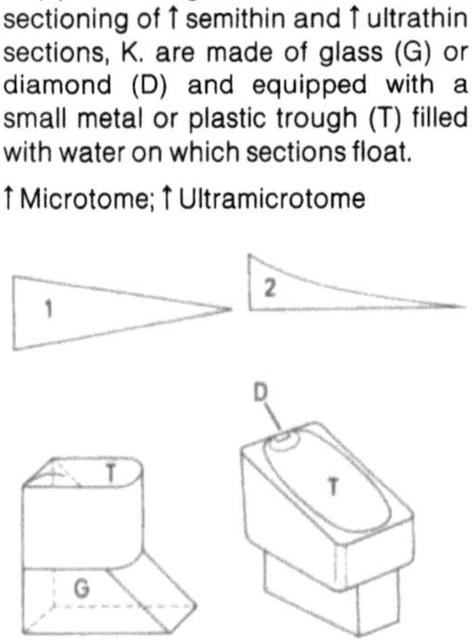

Kohn's pores: ↑ Alveolar pores

Kölliker's columns: ↑ Leydig-Kölliker columns

Kolmer's cells (epiplexus cells, supraependymal macrophages, intraventricular macrophages): Stellate cells (K) with numerous branched processes found freely wandering over epithelial cells of ↑ choroid plexus (C) and ↑ ependymal cells lining ↑ brain ventricles. K. have a round or elliptical nucleus and a prominent nucleolus. Their cytoplasm encloses scattered filiform mitochondria, a well-developed Golgi apparatus, a moderate number of short rough endoplasmic cisternae, and an appreciable number of ↑ lysosomes and ↑ phagolysosomes. Although exact function of K. remains unknown, it is believed that they represent a type of ↑ macrophage. (Fig. = rat)

Bleier, R., Albrecht, R.: J. Comp. Neurol. *192*, 489 (1980); Krstić, R.: Biomed. Res. *2*, Suppl. 129 (1981); Ling, E.A.: J. Anat. *133*, 555 (1981); Pietzsch-Rohrschneider, I.: Z. mikrosk.-anat. Forsch. *94*, 316 (1980); Sturrock, R.R.: J. Anat. *136*, 761 (1983)

Kolmer's inclusions: Occasional crystalloids found in ↑ horizontal cells of ↑ retina.

Korff's fibers: Argyrophilic strands of ↑ ground substance with ↑ reticular microfibrils (K) originating in ↑ pulp of tooth (P), passing between ↑ odontoblasts (O), and finally penetrating dentinal matrix (D). Some authors believe microfibrils to be of collagenous nature. A = ↑ ameloblasts

↑ Argyrophilia

Fox, A.G., Heeley, J.D.: Arch. Oral Biol. *25*, 103 (1980); Ten Cate, A.R., Melcher, A.H., Puddy, G., Wagner, D.: Anat. Rec. *168*, 491 (1970)

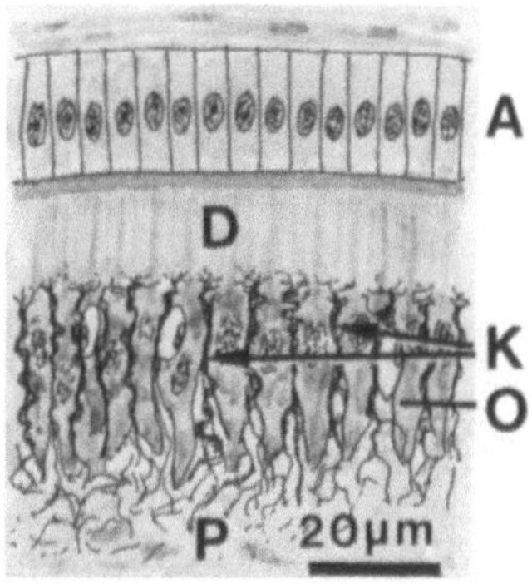

Krause's end bulb: ↑ Corpuscle, of Krause

Kulchitsky's cells: ↑ Enterochromaffin cells

Kupffer's cells (reticuloendotheliocytus stellatus*): Irregular, large stellate cells (K) located within ↑ liver sinusoids (S). Processes of K. extend, without any junctional devices, between ↑ endothelial cells (E) and frequently across sinusoid lumen. K. contain an oval nucleus, many mitochondria, a well-developed Golgi apparatus, short and peroxidase-reactive rough endoplasmic cisternae, numerous ↑ lysosomes (Ly), ↑ residual bodies, and occasional ↑ annulate

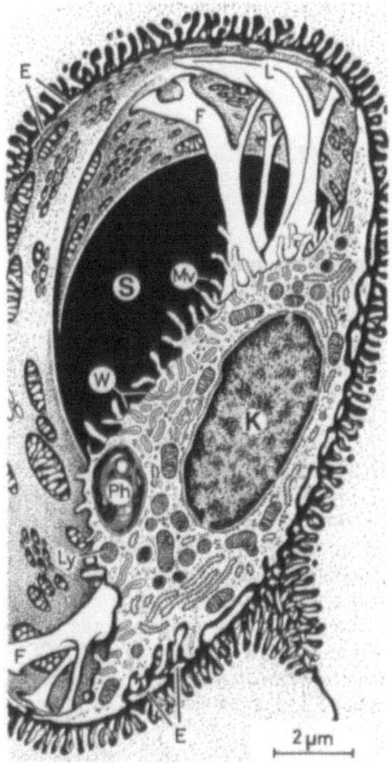

lamellae. K. also enclose large ↑ phagosomes (Ph), often containing engulfed aged ↑ erythrocytes and amorphous debris; deposits of ↑ iron and/or ↑ hemosiderin may also be found. Surface of K. displays irregular folds, ↑ filopodia (F), ↑ lamellipodia (L), ↑ microvilli (Mv), and invaginations; it is lined by thin ↑ glycocalyx. Plasmalemma forms ↑ wormlike bodies (W) with a dense central line. K. are ↑ macrophages representing an independent cell lineage. K. are considered activated endothelial cells by some authors. K. originate from other K. by mitotic divisions, but also from extrahepatic sources (bone marrow).

Motta, M., Muto, M., Fujita, T.: The Liver. An Atlas of Scanning Electron Microscopy. Tokyo, New York: Igaku-Shoin 1978; Sleyster, E.C., Knook, D.L.: J. Histochem. Cytochem. *47*, 484 (1982); Tamaru, E., Fujita, H.: Anat. Embryol. *154*, 125 (1978); Wisse, E.: Ultrastructure and Function of Kupffer Cells and Other Sinusoidal Cells in the Liver. In: Wisse E. and Knook, D.L. (eds.): Kupffer Cells and Other Liver Sinusoidal Cells. Amsterdam: Elsevier 1977

Kurloff's cells: Highly specific ↑ macrophages of blood and ↑ thymus of guinea pig.

Landemore, G., Quillec, M., Debout, C., Izard, J.: Cell Tissue Res. *231*, 457 (1983); Revell, P.A.: Int. Rev. Cytol. *51*, 276 (1977)

L

Labia majora*: A part of the external female genitalia in the form of large cutaneous folds with numerous ↑ sebaceous and ↑ sweat glands on either surface. ↑ Hypodermis contains a considerable amount of ↑ adipose tissue and a thin layer of ↑ smooth muscle cells. Outer surface is covered with ↑ hairs; inner surface is hairless. Present in the hypodermis are numerous sensory nerve endings and Meissner's, Pacini's, and genital ↑ corpuscles.

Labia minora*: Thin hairless cutaneous folds of the external female genitalia. L. consist of richly vascularized and innervated ↑ loose connective tissue (CT) covered by a thin, weakly keratinized ↑ stratified squamous epithelium (E) with deep, ramified ↑ epithelial ridges. Numerous large ↑ sebaceous glands (SG) are associated with the epithelium on both sides. Basal epithelial layer is slightly pigmented.

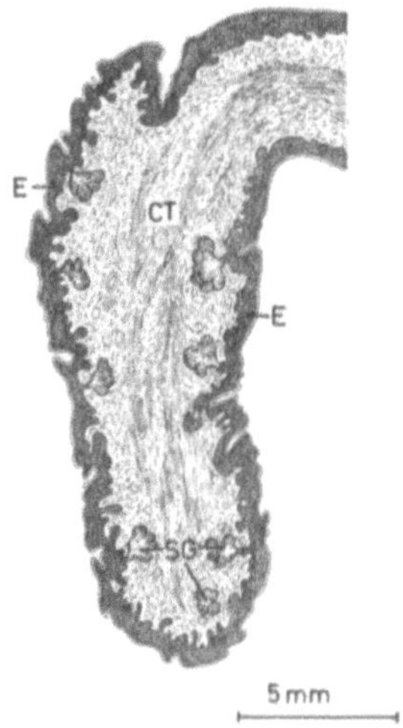

Labial glands (glandulae labiales*): ↑ Simple, small, ↑ mixed ↑ salivary glands (LG) scattered in the tunica mucosa of lower and upper ↑ lips. (Fig. = newborn, human)

Scott, J.: J. Biol. Buccale *8*, 187 (1980)

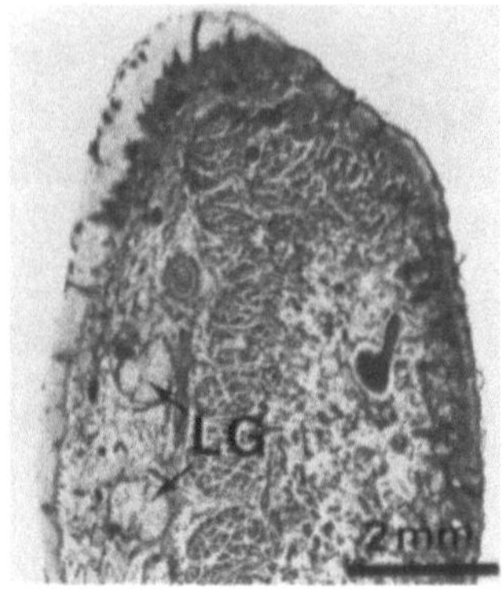

Labrocytes: A former term for ↑ mast cells.

Labyrinth, bony (labyrinthus osseus*): A bony capsule made up of compact bone in which membranous ↑ labyrinth is suspended. L. is filled with ↑ perilymph which flows through ↑ perilymphatic duct (PD) into ↑ subarachnoid space (SS).

↑ Semicircular canals

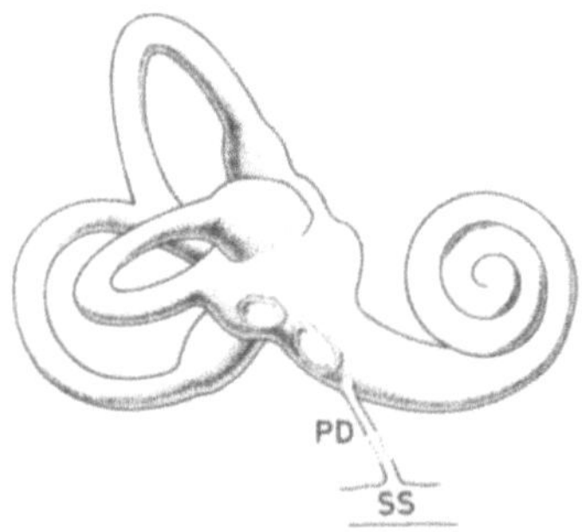

Labyrinth, membranous (labyrinthus vestibularis*): A complex system of fine tubes and sacs filled with ↑ endolymph and lined by a ↑ simple squamous epithelium overlying a very thin lamina propria. L. is composed of three ↑ semicircular ducts (SD) whose ↑ ampullae (A) communicate with ↑ utricle (U). Utricle communicates with ↑ saccule (S) by canalis utriculosaccularis (CUS). A fine tube, ↑ endolymphatic duct (ED), springs from canalis utriculosaccularis, runs through temporal bone, and ends as a blind enlargement, ↑ endolymphatic sac (ES). Saccule joins ↑ cochlear duct (CD) through ductus reuniens (DR). In ampullae are situated ↑ cristae ampullares (CA); in saccule and utricle, ↑ macula of saccule (MS) and utricle (MU); and in cochlear duct, ↑ Corti's organ (CO).

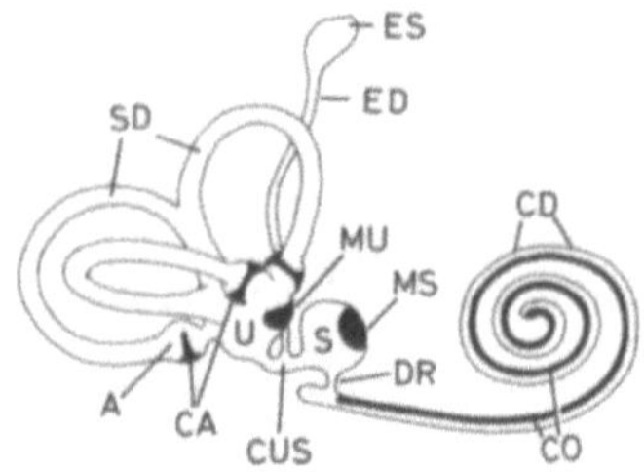

Labyrinth, membraneous, vascularization of: Blood supply of labyrinth is assured by arteria vestibularis and arteria cochlearis, both branches of arteria auditiva interna. Venous blood is drained by vena aqueductus vestibuli, vena aqueductus cochleae, and vena auditiva interna.

Labyrinth, membranous, wall of: L. consists of flattened ↑ simple squamous epithelium (Ep) resting on thin lamina propria (LP) composed of a framework of collagen fibers, ↑ fibrocytes, and pigment cells (PC). From lamina propria, thin connective strands (CS) spread toward ↑ endosteum (E) of ↑ semicircular canals (SC) and utricle to affix L. with bony ↑ labyrinth (BL). Cap = capillary, SD = ↑ semicircular duct

↑ Perilymphatic tissue

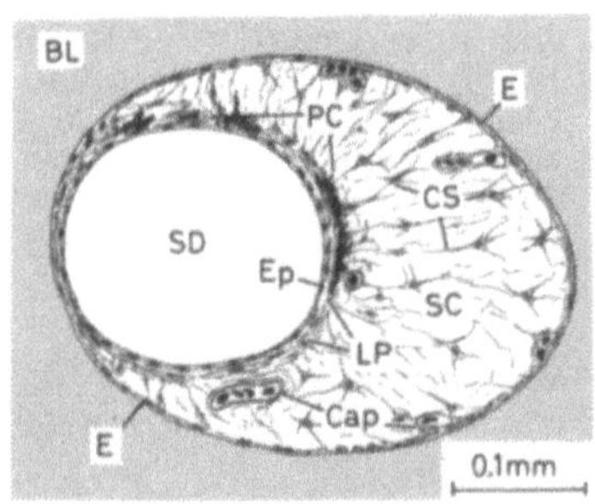

Lacis cells: ↑ Mesangium, extraglomerular

Lacrimal caruncle (caruncula lacrimalis*): A wartlike mucous fold (LC) located between two ↑ lacrimal ducts. L. is covered in some areas by a ↑ stratified squamous nonkeratinized epithelium (E), and in others with a ↑ stratified columnar epithelium containing ↑ goblet cells. In lamina propria

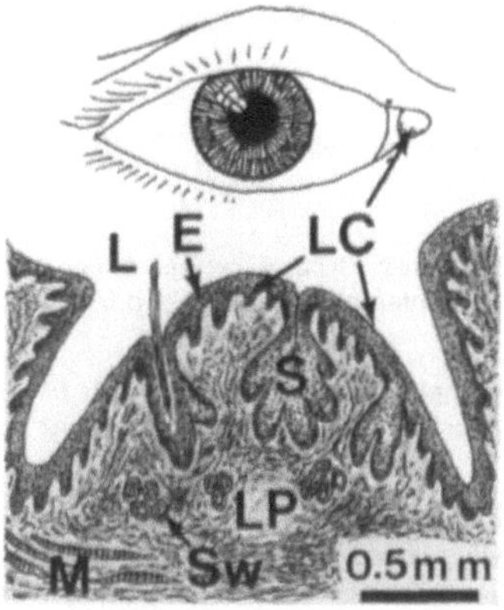

(LP) are scattered some ↑ skeletal muscle fibers (M) irradiating from musculus orbicularis oculi, ↑ lanugo hairs (L), and ↑ sebaceous (S), ↑ sweat (Sw), and accessory ↑ lacrimal glands. Their products form a whitish secretion visible in medial palpebral angle. L. is richly innervated.

Lacrimal ducts (ductuli lacrimales, lacrimal canaliculi, canaliculus lacrimalis*): Paired, short and narrow tubes (LD) beginning at lacrimal points (LPt) and reaching into lacrimal sac (LS). L. are lined with a nonkeratinized ↑ stratified squamous epithelium (E) resting on lamina propria (LP), which is rich in ↑ elastic fibers and ↑ plasma cells. L. conduct ↑ tears to lacrimal sac.

Adenis, J.P., Loubet, A., Leboutet, M.J.: Arch. Anat. Cytol. Pathol. *28*, 371 (1980)

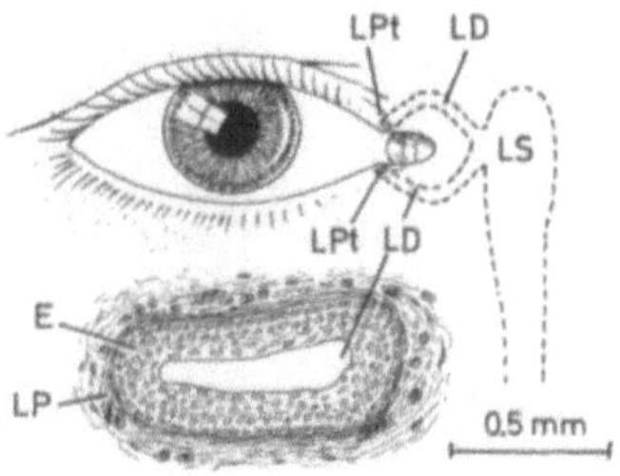

Lacrimal glands (glandula lacrimalis*): Paired compound ↑ tubuloacinar ↑ serous glands, one situated in the upper outer portion of each orbit. Structure: 1) Capsule = moderately ↑ dense connective tissue sending loose septa (S) into ↑ parenchyma. 2) Parenchyma is arranged in lobules (L), composed of elongated and ramified ↑ acini (A) surrounded by ↑ myoepithelial cells obviously invisible in routine

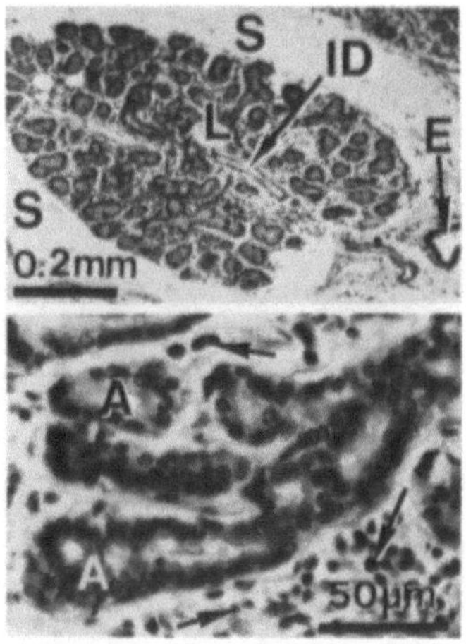

preparations. Acini empty into intralobular ducts (ID) lined with ↑ simple cuboidal epithelium. Interlobular ducts carry ↑ tears to 6–12 excretory ducts (E), also lined with simple cuboidal epithelium. Between acini are numerous ↑ lymphocytes and ↑ plasma cells (arrows); latter are probably involved in transfer of antibodies to tears. L. is an ↑ accessory organ of the eye. (Figs. = human)

Kühnel, W., Scheele, G.: Anat. Anz. *145*, 87 (1979)

Lacrimal glands, accessory (glandulae conjunctivales, glands of Krause, glandulae lacrimales accessoriae*): Small ↑ serous glands scattered throughout palpebral ↑ conjunctiva of upper ↑ eyelids and ↑ lacrimal caruncles. L. produce a watery secretion similar to that of ↑ lacrimal glands.

↑ Tears

Lacrimal glands, secretory cells of (lacrimocytus*): Cuboidal or low columnar cells with a predominantly basal nucleus, numerous slender mitochondria, well-developed ↑ ergastoplasm, an expanded Golgi apparatus, and a number of membrane-bound ↑ secretory granules (G) dis-

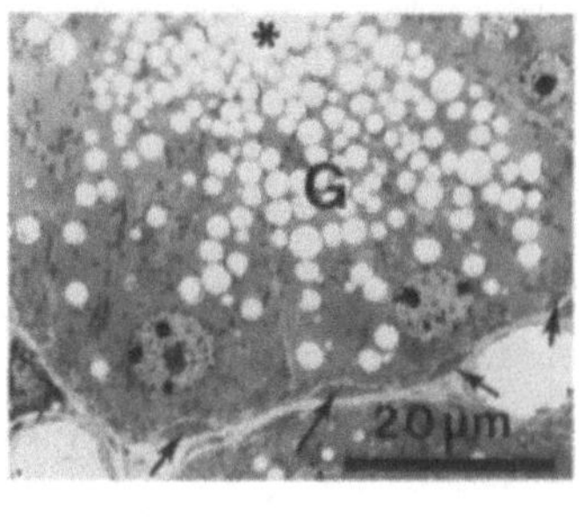

charging their contents into acinar lumen (asterisk). Secretory cells are surrounded by ↑ myoepithelial cells (arrows). (Fig. = mongolian gerbil)

↑ Tears

Lacrimal points (puncta lacrimalia*): Minute openings leading into ↑ lacrimal ducts.

Lactase: A digestive enzyme concentrated in ↑ striated border of ↑ absorptive cells; L. splits milk sugar into glucose and galactose.

Lactation (lactogenesis): 1) A process of synthesis and secretion of ↑ milk by glandular cells of ↑ mammary gland under influence of ↑ LTH-RF, ↑ prolactin, ↑ ACTH, ↑ corticosteroids, ↑ oxytocin and other hormones. a) ↑ Eccrine secretion of protein component, mostly ↑ casein granules (C): the first polypeptide chains are synthesized onto ribosomes of rough endoplasmic reticulum (ER) and transported to Golgi apparatus (G); in its vacuoles (V) occur the first casein granules, about 400 nm in diameter. By way of vacuoles, granules are transported to cell surface and discharged into lumen by ↑ exocytosis. b) ↑ Apocrine secretion of lipid component (L): ↑ lipid droplets (L) occur in ↑ cytoplasmic matrix, move toward ↑ apical pole, increase in volume, bulge into lumen, and finally pinch off from cell body, still surrounded by a narrow cytoplasmic rim (CR) that may include ↑ organelles such as mitochondria, etc. Expulsion of both milk components into excretory ducts is aided by contraction of ↑ myoepithelial cells (MC) in response to oxytocin secreted as consequence of ↑ suckling reflex. 2) Period following childbirth during which milk is formed in mammary gland.

↑ Mammary gland, endocrine control of

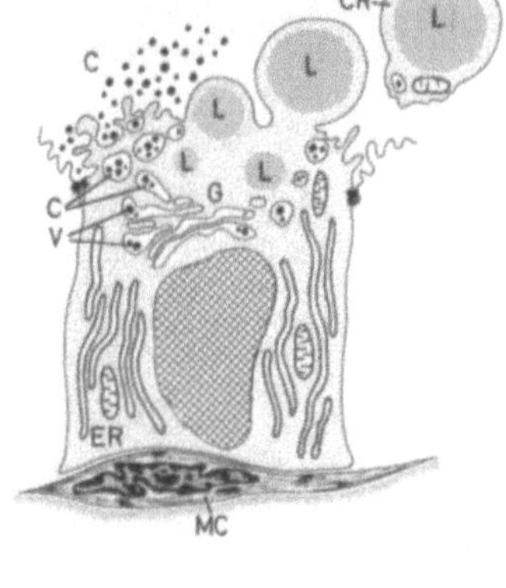

Cowie, A.T., Forsyth, I.A., Hart, I.C.: Hormonal Control of Lactation. Berlin, Heidelberg, New York: Springer-Verlag 1980; Larson, B.L., Smith, V.R. (eds.): Lactation. Vols. 1–4. New York: Academic Press 1974–1978; Ollivier-Bousquet, M.: Biol. Cell *39*, 21 (1980); Peixoto de Menzes, A., Pinto de Silva, P.: J. Cell Biol. *76*, 767 (1978); Tobon, H., Salazar, H.: J. Clin. Endocrinol. Metab. *40*, 834 (1975)

Lacteal (vas lymphaticum centrale*): A blind saclike ↑ lymphatic capillary within ↑ intestinal villus. L. receive absorbed fats (↑ chylomicrons) and become milky in appearance.

↑ Intestine, small, lymphatics of

Azzali, G., J. Submicrosc. Cytol. *14*, 45 (1982)

Lactiferous ducts (galactophores, ductus lactiferus colligens*): Fifteen to 25 canals (LD) extending from lobes of ↑ mammary gland to ↑ nipple, at tip of which they open as 0.4 to 0.7-mm-wide orifices. Here, L. are lined with keratinized ↑ stratified squamous epithelium (E), which is continuous with nipple ↑ epidermis. In ↑ lactiferous sinuses, epithelium is two- to three-layered, and in the remainder of L. it is two-layered cuboidal, the external layer being formed of ↑ myoepithelial cells. In distal segment, this layer becomes discontinuous. Cells of inner layer enlarge during pregnancy and ↑ lactation; they may be secretory active. L. are continous with interlobular ducts. A relatively dense lamina propria (LP) surrounds L. (Fig. = human)

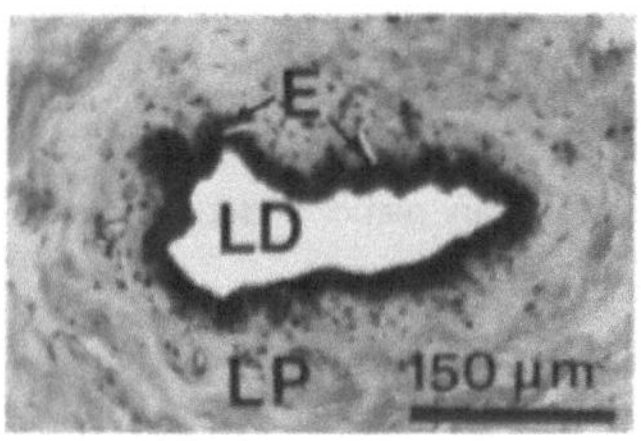

Lactiferous sinus (sinus lactifer*): A local expansion (LS) of ↑ lactiferous duct (LD) just beneath ↑ areola (A). The wall of L. is lined with two or three layers of cuboidal cells overlying a loose lamina propria (LP). Surrounding it is a layer of ↑ dense connective tissue (DCT) containing some ↑ smooth muscle cells (SMC). N = ↑ nipple

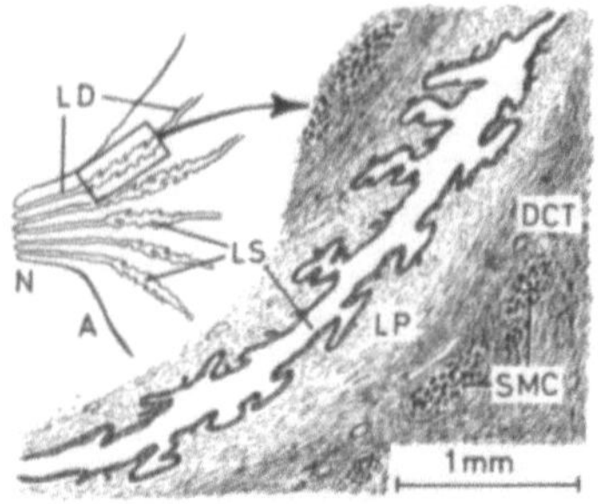

Lactogenesis: ↑ Lactation

Lactotropes: ↑ Mammotropes

Lacunae, of bone: (osteoplasts, lacuna ossea*): Almond-shaped cavities (L) located between ↑ lamellae and, more rarely, within lamellae themselves. Extending from L. throughout lamellae in all directions are ↑ canaliculi (C), which communicate with canaliculi of neighboring L. In each L. lies an ↑ osteocyte (O) which does not fill it completely; a narrow space averaging 0.2–2 µm containing nonmineralized bone matrix separates cell from wall of L. (Fig. = mongolian gerbil)

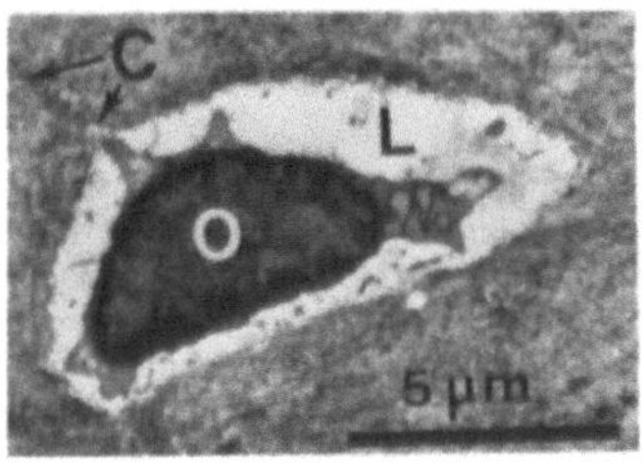

Wassermann, F., Yaeger, J.A.: Z. Zellforsch. *67*, 636 (1965)

Lacunae, of cartilage (lacuna cartilaginea*): Cavities in the ↑ cartilage matrix, each containing a ↑ chondrocyte.

Lacunae, of endometrium: Enlarged, thin-walled vascular spaces within ↑ endometrium. 1) Connecting L. or ectatic capillaries = arterial spaces formed by fusion of capillaries within functionalis; 2) collecting L. = venous enlargements carrying blood toward endometrial veins.

↑ Endometrium, vascularization of

Lacunae, of erectile tissue: ↑ Cavernous bodies, of penis; ↑ Erectile tissue

Lacunae, of Howship: ↑ Howship's lacunae

Lacunae, of Morgagni: ↑ Urethral lacunae

Lamellae, annulatae: ↑ Annulate lamellae

Lamellae, circumferential, of bone: ↑ Circumferential lamellae, of bone

Lamellae, concentric, of bone: ↑ Osteon, secondary

Lamellae, interstitial, of bone: ↑ Interstitial lamellae, of bone

Lamellae, of bone (lamella ossea*): Successive layers (L) of calcified ↑ bone matrix, about 3–7 µm thick, formed by ↑ osteoblasts as the result of successive waves of appositional growth. In adult bone, L. are maintained by ↑ osteocytes (Oc). In long compact ↑ bone, L. are organized in various patterns corresponding to physical forces acting on bone (outer circumferential L., ↑ osteons, ↑ interstitial L., and inner circumferential L.). ↑ Collagen fibrils (CF) of adjacent L. cross at different angles. In cancellous ↑ bone several L. arranged parallel form a ↑ trabecula.

↑ Bone, appositional growth of; ↑ Bone, long, growth in diameter; ↑ Bone formation, secondary; ↑ Circumferential lamellae, of bone

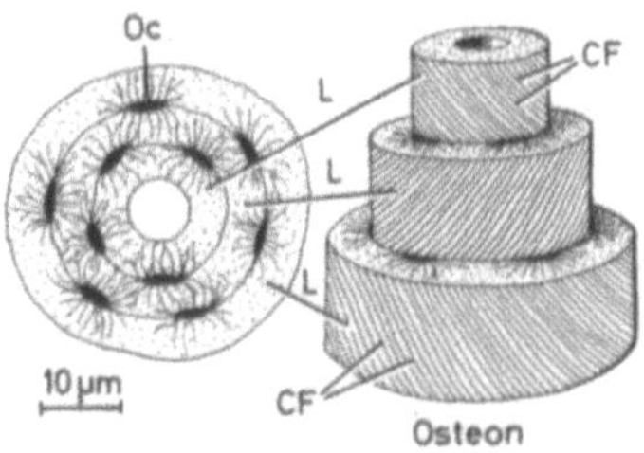

Knese, K.-H.: Gegenbaurs Morphol. Jahrb. *125*, 15 (1979)

Lamellae, of enamel: ↑ Enamel lamellae

Lamellae, of Golgi apparatus (cisterna, lamella*): ↑ Unit membrane-bound, flattened sacs (L) with somewhat expanded edges from which vesicles (V) and vacuoles, i.e., ↑ secretory granules (S), separate. L. are perforated by numerous fenestrae (F) permitting circulation of ↑ hyaloplasm between them.

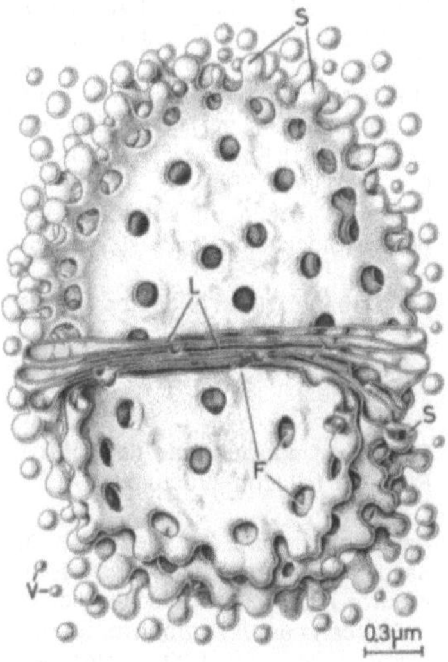

Lamellar bodies: ↑ Multilamellar bodies

Lamellar bone: ↑ Bone, mature

Lamellar phagosomes: ↑ Heterophagosomes (LPh), measuring 2–4 μm in diameter, found in supranuclear portion of ↑ pigment epithelial cells of ↑ retina. L. contain phagocytized external parts of outer segments of ↑ photoreceptors with ↑ membranous discs, which give them a lamellar aspect. Following breakdown, L. become transformed into ↑ residual bodies. (Fig. = mongolian gerbil)

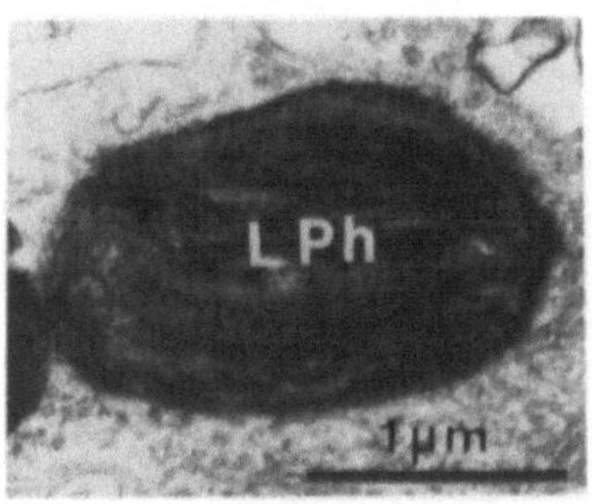

Lamellar structures, of oocytes: Proteinaceous filaments (F) arranged to form bands 30 nm thick, 0.12 μm wide, and 0.25–2.5 μm long. L. are grouped predominantly in the peripheral cytoplasm and do not contain RNA. Their function is unknown; however, it is believed that they could represent a protein reserve ready for use during segmentation. L. are less frequent in man than in rodents. (Figs. = rat)

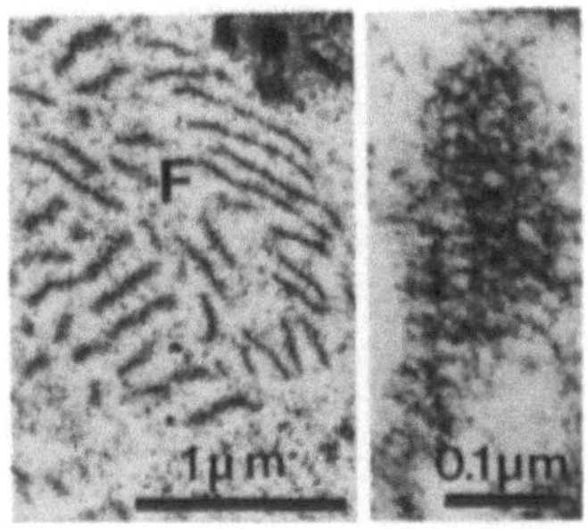

Baeckland, E., Heinen, E.: Bull. Assoc. Anat. (Paris) *64*, 173 (1980)

Lamellar tissue (textus connectivus fibrosus lamellaris*): A kind of ↑ loose connective tissue arranged in layers (L). L. is present in ↑ hypodermis, between galea aponeurotica and ↑ periosteum of skull bones, etc. L. permits free displacement of skin over bones and muscles.

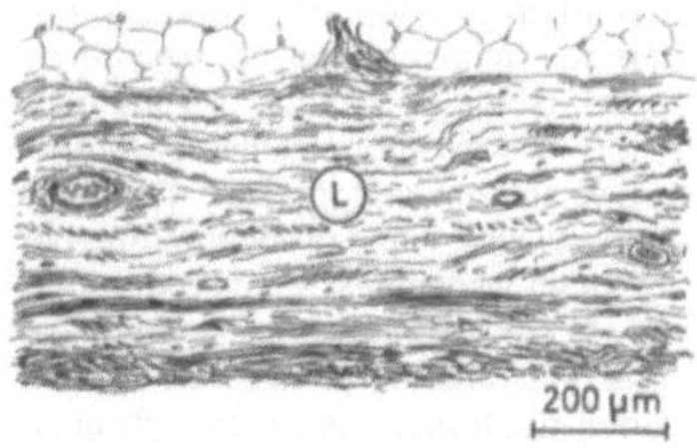

Lamellated dense bodies: ↑ Membrane-coating granules

Lamellated granules: ↑ Membrane-coating granules

Lamellipodia (undulating membrane, processus lamellosus*): Variously shaped, very attenuated cellular flaps (L), about 100 nm thick, extending from margins of ↑ wandering cells. L. effectuate waving movements and are involved in ↑ ameboidism and ↑ macropinocytosis. F = ↑ filopodia

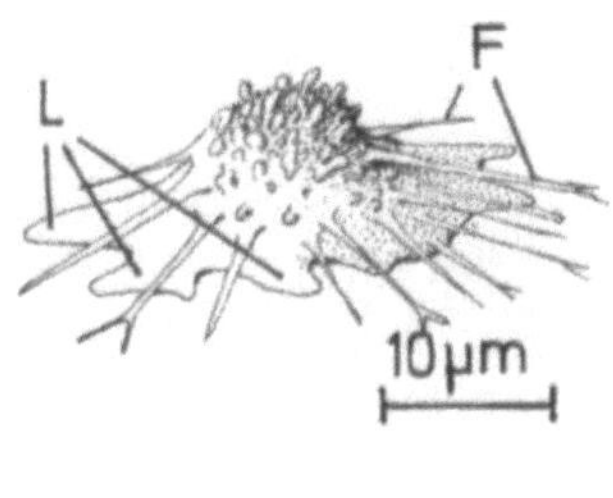

Lamina: 1) A more or less sharply delimited flat layer of nerve cells of the brain cortex; 2) a thin plate of any ↑ tissue.

Lamina, basal: ↑ Basal lamina

Lamina basalis, of endometrium: ↑ Endometrium

Lamina cribrosa*: A perforated layer of ↑ sclera, situated behind ↑ optic papilla, through which pass fibers of ↑ optic nerve, as well as retinal vessels.

Lamina densa: ↑ Glomerular basal lamina

Lamina, dental: ↑ Dental lamina

Lamina dura: Innermost bony lamella of ↑ alveolar bone adjacent to ↑ periodontal ligament. L. surrounds root of tooth and forms its socket.

Lamina dysfibrosa: ↑ An inconstant nerve cell layer of ↑ brain cortex.

Lamina epithelialis (epithelium*): Epithelial covering of a ↑ tunica mucosa.

Lamina functionalis, of endometrium: ↑ Endometrium

Lamina muscularis mucosae*: A layer of ↑ smooth muscle cells of tunica mucosa of ↑ gastrointestinal tract extending from ↑ esophagus to ↑ anal canal. Contraction of L. throws ↑ tunica mucosa into longitudinal folds. L. separates tunica mucosa from ↑ tela submucosa.

Lamina propria*: ↑ Loose connective tissue immediately underlying an ↑ epithelium. Papillary layer of ↑ skin, "stroma" of ↑ endometrium, and ↑ tela subserosa of various organs are examples of L.

Lamina rara externa: ↑ Glomerular basal lamina

Lamina rara interna: ↑ Glomerular basal lamina

Lamina spiralis* (bony spiral lamina): A thin bony projection (LS) of ↑ modiolus surrounding modiolus in a screwlike fashion. ↑ Spiral limbus (SL) rests on L.; through it pass ↑ dendrites of ↑ bipolar cells of ↑ spiral ganglion (G). ↑ Basilar membrane (B) is stretched between L. and ↑ spiral ligament (S). (Fig. = guinea pig)

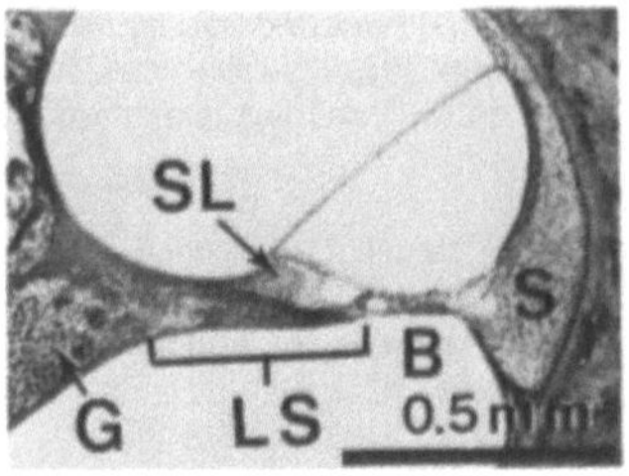

Lamina suprachoroidea*: Outermost layer of ↑ choroid, composed of flattened fibroblastlike cells, ↑ macrophages, and numerous ↑ melanophores.

Laminae, of brain cortex: ↑ Brain cortex; ↑ Brain cortex, heterotypical isocortex of; ↑ Brain cortex, homotypical isocortex of

Laminated plates: ↑ Membranous discs, of photoreceptors

Laminin: A widely distributed noncollagenous ↑ glycoprotein isolated from ↑ basal laminae of embryonic and adult animal and human tissues.

↑ Fibroblast, functions of

Hay, E.D.: J. Biol. *91*, 205s (1981); Martinez-Hernandez, A., Amenta, P.S.: Lab. Invest. *48*, 656 (1983); Laurie, G.W., Leblond, C.P., Martin, G.R., Silver, M.H.: J. Histochem. Cytochem. *30*, 991 (1982); Timpl, R., Rohde, H., Robey, P.G., Rennard, S.I., Foidart, J.-M., Martin, G.M.: J. Biol. Chem. *254*, 9933 (1979)

Lanceolate nerve endings: Specialized nerve endings participating in ↑ hair innervation.

Langendorff's cells: ↑ Colloid cells, of Langendorff

Langerhan's bodies (Birbeck's granules, Langerhan's granules, vermiform granules, X-bodies): Platelike structures (LB), about 0.1–0.5 µm long and 30 nm wide, found in ↑ Langerhans' cells. L. are enclosed by a ↑ unit membrane; through their center runs a striated lamella (SL) with periodicity of 9 nm; it is composed of two-dimensional paracrystalline arrays of cross-striated patterns visible on oblique sections (arrows). L. are formed by outpouching of Golgi cisternae (G); newly formed L. resemble a tennis racket. L. empty their contents into intercellular spaces (IS); chemical composition and function of L. are unknown.

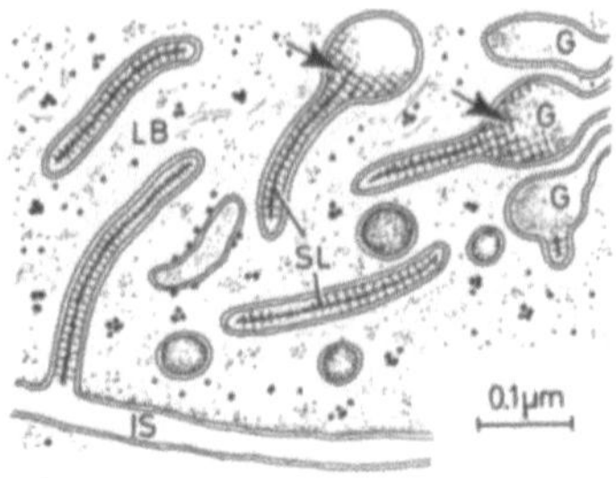

Breatnach, A.S.: Int. Rev. Cytol. *18*, 1 (1965); Katz, S., Tamaki, K., Sachs, D.H.: Nature *282*, 324 (1979); Zelickson, A.S.: Ultrastructure of Normal and Abnormal Skin. Philadelphia: Lea & Febiger 1967

Langerhans' cells, of epidermis (dendrocytus granularis nonpigmentosus*): Clear cells (L) found in all viable layers of ↑ epidermis or in contact with its basal lamina and not connected by ↑ desmosomes with surrounding cells. From the round cell body extend some slender dendrite-shaped processes. Nucleus is large and displays deep indentations. Cytoplasm contains numerous mitochondria, a well-developed Golgi apparatus, ↑ centrioles, short rough endoplasmic cisternae, many smooth endoplasmic ↑ tubules and vesicles, ↑ lipid droplets, a number of small lysosomes, specific granules (= ↑ Langerhans' bodies), and few free ribosomes. Granules bud off from Golgi apparatus, can fuse with plasmalemma, and empty their contents into extracellular space. The origin and function of L. is controversial and not understood: Recent studies have shown that L. are neither effete ↑ melanocytes nor ↑ macrophages. L. appear to be important in induction of cutaneous contact hypersensitivity; of all epidermal cells, only L. have ↑ Fc and C3 receptors. (Fig. = mongolian gerbil)

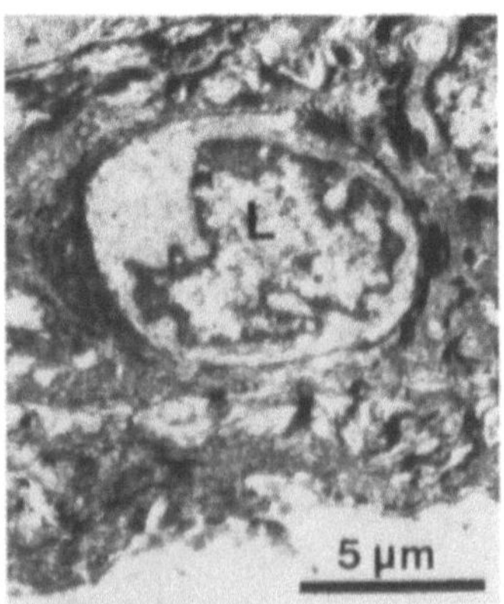

Figueroa, C.D., Caorsi, I.: J. Anat. *131*, 669 (1980); Gilette, T.E., Chandler, J.W., Greiner, J.V.: Ophthalmology *89*, 700 (1982); Murphy, G.F., Bhan, A.K., Sato, S., Harrist, T.J., Mihm, M.C.: Lab. Invest. *45*, 465 (1981); Rodriguez, E.M., Caorsi, I.: J. Ultrastruct. Res. *65*, 279 (1978); Shelley, W.B., Lennart: Acta Derm. Venereol. (Stockholm) Suppl. *79*, 7 (1978)

Langerhans' islets: ↑ Pancreatic islets

Langhans' cells: ↑ Cytotrophoblast; ↑ Trophoblast

Langhans' subchorionic fibrinoid: ↑ Placenta, full-term, structure of

Lanthanum (La): An element (La) used in its nitrate form in transmission electron microscopy for localization of ↑ glycocalyx and/or carbohydrate-rich extracellular material. Int = intercellular space, Mv = ↑ micropinocytotic vesicles (Fig. = ↑ subcommissural organ, rat)

↑ Alcian blue; ↑ Intercellular cement; ↑ Ruthenium red

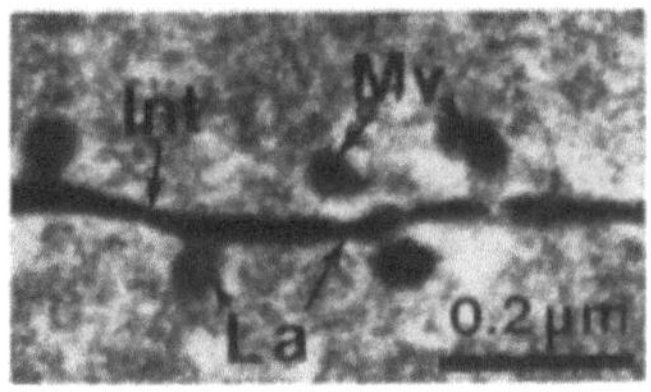

Shaklai, M., Tavassoli, M.: J. Histochem. Cytochem. *30*, 1325 (1982)

Lanugo hairs (lanugo*): Very fine, almost colorless fetal ↑ hairs.

↑ Vellus hairs

Large alveolar cells, of lung: ↑ Alveolar cells, type II

Large intestine: ↑ Intestine, large

Large veins: ↑ Veins, large

Laryngeal glands (glandulae laryngeae*): Simple ↑ mixed ↑ tubuloalveolar glands (LG) located in lamina propria of ↑ larynx. L. are simple invaginations of epithelium and possess neither ↑ intercalated nor ↑ striated ducts; ↑ mucous tubules (T) are frequently enlarged. L. include ↑ arytenoid glands, ↑ epiglottic glands, and glands situated in ventricular folds (VeF) and dorsal wall of the laryngeal ventricle. They are innervated by ↑ nonmyelinated nerve fibers and serve to moisten laryngeal mucosa and ↑ vocal cords (VC). (Fig. = human)

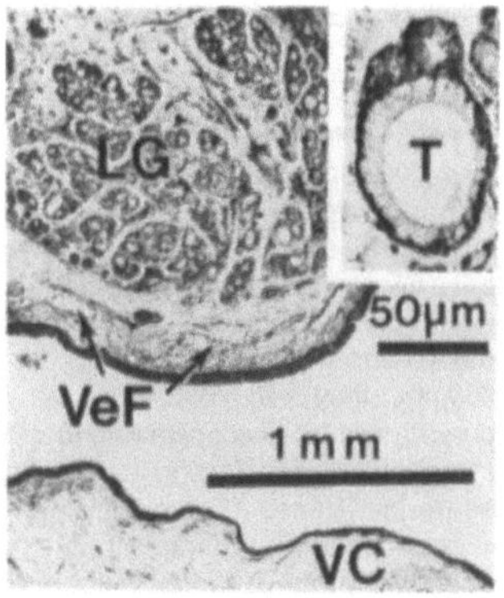

Laryngobronchial epithelium: ↑ Trachea, epithelium of

Laryngopharynx (hypopharynx, pars laryngea pharyngis*): Inferior segment of ↑ pharynx between upper edge of ↑ epiglottis and posterior aspect of ↑ cricoid cartilage. Structurally, L. corresponds to ↑ oropharynx.

Larynx*: A tubular musculocartilaginous organ, part of conducting portion of respiratory system situated between ↑ oropharynx above and ↑ trachea below. Structure: 1) Tunica mucosa (TM): a) epithelium (E) = tall ↑ pseudostratified ciliated epithelium, except for small area on ↑ vocal cords (VC) where it is ↑ stratified squamous (black). b) Lamina propria (LP) = ↑ loose connective tissue with ↑ lymphatic nodules (LN) around ventricular recess (VR). Tunica mucosa forms two sets of prominent folds: ventricular folds (VeF), with ↑ laryngeal glands (LG), and ↑ vocal cords (VC), with ↑ vocal ligament (VL) and vocal muscle (VM). Between the folds is the laryngeal ventricle (LV) with its narrow pouchlike prolongation, ventricular recess (VR). 2) Laryngeal cartilages = ↑ epiglottis, ↑ thyroid (TC), ↑ cricoid, ↑ arytenoid, ↑ corniculate, and ↑ cuneiform cartilages. Lamina propria is attached to ↑ perichondrium (P) of laryngeal cartilages. 3) Tunica adventitia = loose connective tissue (not shown). L. is most important part of the voice organ. Part of laryngeal wall limiting laryngeal opening is ↑ glottis.

↑ Conus elasticus

Fink, B.R.: The Human Larynx. A Functional Study. New York: Raven Press 1975; Hirano, M., Ito, T.: Biomed. Res. *2*, Suppl. 289 (1981); Lewis, D.J., Prentice, D.E.: J. Anat. *130*, 617 (1980); Ohyama, M., Ohno, I., Fujita, T., Adachi, K.: Biomed. Res. *2*, Suppl. 273 (1981)

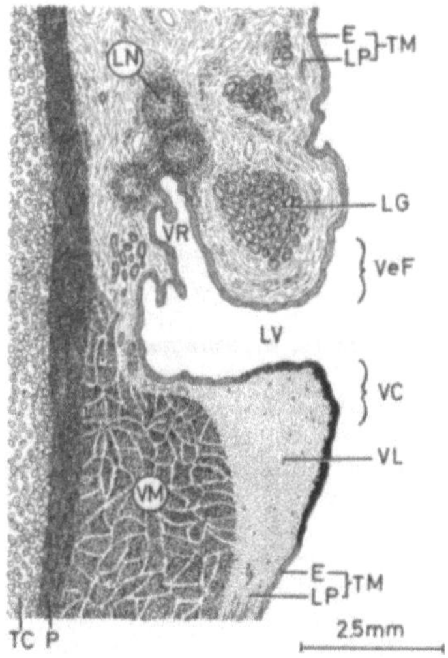

Laser (**l**ight **a**mplification by **s**timulated **e**mission of **r**adiation): An amplifier of light producing extremely intense coherent monochromatic light. By focusing L. light, one can produce a microbeam, 2.5 µm in diameter, for use as a ↑ radiation probe for selective destruction of entire living cells, cell ↑ organelles, special areas of ↑ chromosomes, etc. For this purpose, the target must be photosensitized by selective ↑ supravital staining (e.g., ↑ Janus green B for mitochondria, acridine orange for some parts of chromosomes), enabling destruction of only stained structure without damaging of nonphotosensitized parts of cell. Largely used in therapy, L. is also employed for ↑ optic diffraction.

Dinstl, K., Fischer, P.L. (eds.): Der Laser. Heidelberg, Berlin, New York: Springer-Verlag 1981; Hillenkamp, F., Pratesi, R., Sacchi, C.A. (eds.): Lasers in Biology and Medicine. New York: Plenum Press 1980

Late erythroblast: ↑ Erythroblast, orthochromatic

LATS: ↑ Long-acting thyroid stimulator

L-band, of striated muscle fibers: ↑ Pseudo H-band

Lead citrate: A substance used for ↑ contrasting ↑ ultrathin sections and for demonstrating ↑ glycogen in TEM.

↑ Uranyl acetate

Cattini, P.A., Davies, H.G.: Stain Technol. *58*, 29 (1983)

Lectins: Multivalent ↑ ligands of immune origin which agglutinate cells and/or precipitate ↑ glycoconjugates by strong affinity for oligosaccharide residues of plasmalemma. By forming bridges between neighboring cells, L. provoke formation of cell clumps. Membrane L., involved in cell recognition and ↑ micropinocytosis of proteins, have been found in a large number of cells (↑ endothelial cells of splenic sinusoids, ↑ Kupffer's cells, ↑ macrophages, etc.). L. also include plant L., such as phytohemagglutinin (PHA), and concanavalin A.

Barak Briles, E.: Int. Rev. Cytol. *75*, 101 (1982); Barondes, S.H.: Ann. Rev. Biochem. *50*, 207 (1981); Bartels, J.R., Frazier, A., Rosen, S.D.: Int. Rev. Cytol. *75*, 61 (1982); Franz, H., Ziska, P., Mohr, J.: Acta Histochem. *71*, 19 (1982); Roth, J.: Experimentelle Pathologie. Suppl. 3. The Lectins. Jena: Gustav Fischer 1978

Lemnocytes: ↑ Schwann's cells

Lens*: A transparent biconvex organ held by ↑ zonular fibers (F) of ↑ ciliary body immediately behind ↑ pupil. Anterior L. surface is spherically convex, whereas posterior is paraboloid. Structure: 1) ↑ Capsule (C) completely encloses L. and renders it impermeable to ↑ wandering cells. 2) Epithelium of L. (E) = a layer of cuboidal cells lining its anterior surface; at equator (Eq), cells divide by occasional mitoses, (so-called ↑ nuclear zone, NZ), elongate, and gradually differentiate into 3) ↑ lens fibers (LF), which form L. substance, subdivided into peripheral zone or cortex (Cx) and central zone or nucleus (N). Devoid of in-

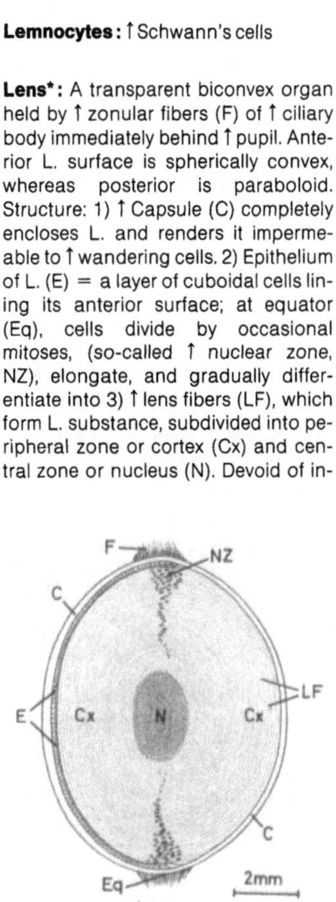

Lens

nervation and vascularization, L. is nourished by ↑ aqueous humor and ↑ vitreous body. L. is main refractive medium of eye (↑ refractive index 1.36–1.4); it is particularly rich in vitamin C and composed of about 70% water, 25% proteins, and 5% inorganic salts and fat. By traction, fibers of ↑ ciliary zonule prevent inherent tendency of L. to become spherical.

Bloemendal, H. (ed.): Molecular and Cellular Biology of the Eye Lens. New York: John Wiley & Sons 1981; Goodenough, D.A., Dick, J.S.B.II, Lyons, J.E.: J. Cell Biol. *86*, 576 (1980); Lo, W.K., Harding, C.V.: Invest. Ophthalmol. *24*, 396 (1983); McDevitt, D.S. (ed.): Cell Biology of the Eye. New York: Academic Press 1982

Lens epithelium: ↑ Lens

Lens fibers (fibrae lentis*): Elongated cells (LF) in form of ribbonlike six-sided prisms, 7–10 mm long, 5–12 µm wide, and 2–4 µm thick, large prism side being parallel with surface of ↑ lens. L. are arranged in concentric layers (about 1400 in newborn and 2000 in adult). In region of equator, each L. contains a central nucleus, occasional mitochondria, a few rough endoplasmic cisternae, and numerous ↑ polyribosomes. L. are closely apposed to one another and held together by numerous ↑ nexus and complicated "ball and socket" interdigitations (arrows), probably involved in change of shape of L. during accomodation movements. Toward interior of lens, L. lose their nuclei and ↑ organelles; their cytoplasm becomes highly condensed, giving rise to nucleus of lens. L. newly added to nucleus undergo such strong condensation that lens hardly grows during life. (Figs. = rat)

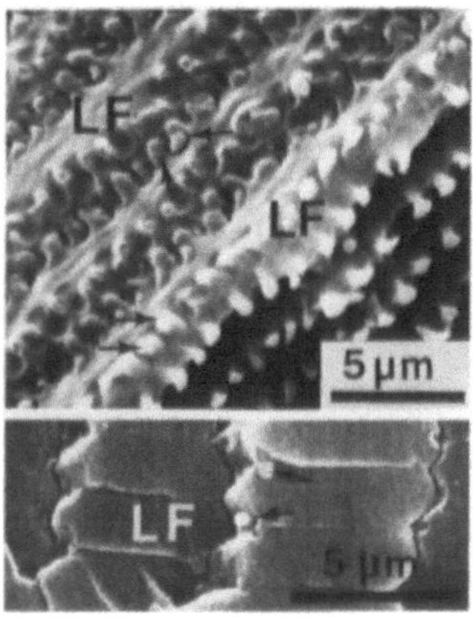

↑ Lens fibers, arrangement of; ↑ Nuclear zone, of lens

Dickson, D.H., Crock, G.W.: Invest. Ophthalmol. *11*, 809 (1972); Hollenberg, M.J., Wyse, J.P.H., Lewis, B.J.: Cell Tissue Res. *167*, 425 (1976); Hoyer, H. E.: Cell Tissue Res. *224*, 225 (1982); Kuszak, J.R., Rae, J.L., Pauli, B.U., Weinstein, R.S.: J. Ultrastruct. Res. *81*, 249 (1982); Zampighi, G., Simon, S.A., Robertson, J.D., McIntosh, T.J., Costello, J.: J. Cell Biol. *93*, 175 (1982)

Lens fibers, arrangement of: Lines where ↑ lens fibers (LF) meet one another end to end; the sutures (S) form two Y-like figures at anterior and posterior lens surfaces; they are at an angle of 60° to one another, so that an individual fiber has a complicated spiral course from suture to suture. (Modified after Vogt 1936. In: Möllendorff, W.v.)

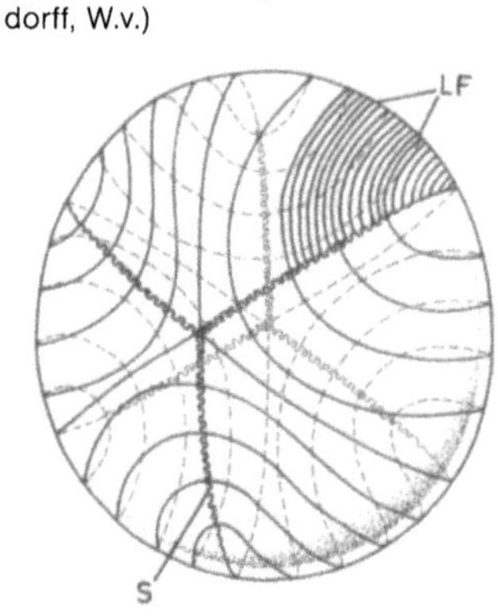

Möllendorff, W. v. (ed.): Handbuch der mikroskopischen Anatomie des Menschen. Haut und Sinnesorgane. Das Auge. Vol. 3, Part 2. Berlin: Julius Springer 1936

Lenticular papillae: Flattened ↑ lymphoreticular bulges (LP) measuring about 0.5–1 mm situated at free surface of root of ↑ tongue. L. enclose several ↑ lymphatic nodules (LN), but do not possess ↑ crypts as ↑ lingual follicles (LF) do. CP = ↑ circumvallate papillae.

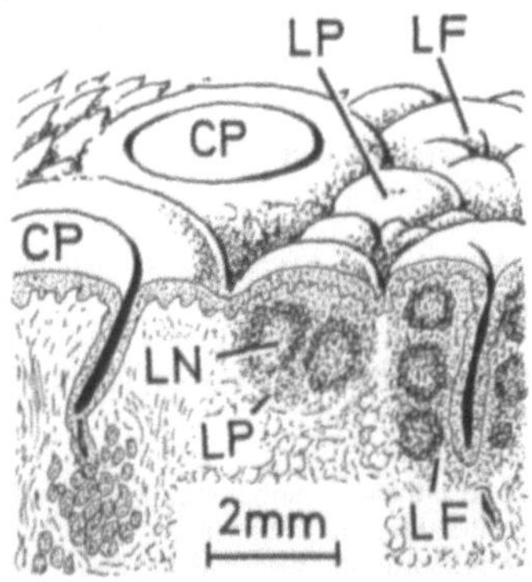

Leptomeninx: Collective term for ↑ arachnoid and ↑ pia mater.

Krisch, B., Leonhardt, H., Oksche, A.: Cell Tissue Res. *228*, 597 (1983)

Leptomeric fibrils (leptofibrils, leptomeric complexes, leptomeric organelles): Groups of parallel or irregularly arranged dense lines (L) found in ↑ striated muscle fibers within ↑ myofibrils, between these or in close relation with ↑ sarcolemma (S). Dense lines have periodicity ranging from 80 to 250 nm; they can bifurcate, be incurved, and intersect with one another at various angles, giving a zebra appearance. Dense lines are frequently in confluence with ↑ Z-lines. L. are predominantly found in muscle fibers with particular embryonal characteristics (↑ Purkinje fibers, ↑ intrafusal fibers). Chemical nature and function of L. are unknown. Because of connection between dense lines and Z-lines, it is thought that lines of L. are precursors of Z-lines. Some authors consider L. to be developing myofibrils; others, to be special devices involved in mechanical properties of muscle fibers. (Fig. = mongolian gerbil)

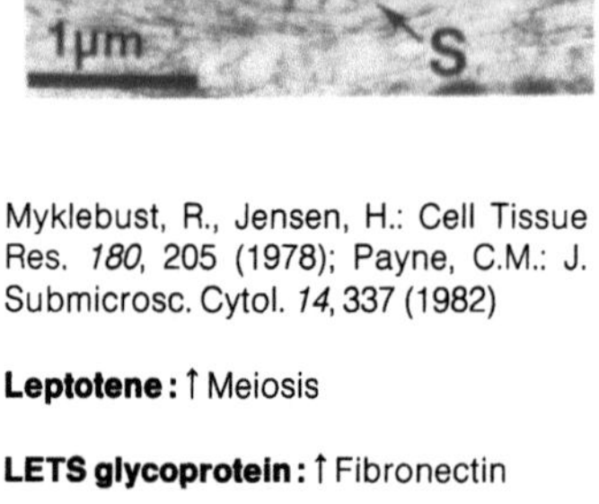

Myklebust, R., Jensen, H.: Cell Tissue Res. *180*, 205 (1978); Payne, C.M.: J. Submicrosc. Cytol. *14*, 337 (1982)

Leptotene: ↑ Meiosis

LETS glycoprotein: ↑ Fibronectin

Leucine aminopeptidase (LAP): An enzyme present in many kinds of cells with function of splitting leucyl residues.

Leukocytes (white blood cells, leucocytus*): A common term for all nucleated cells with ameboid properties circulating in blood. There are about 5000–10 000 L./mm^3 in human blood (or 5–10 × 10^9/l, ↑ SI).

↑ Granulocytes, basophilic; ↑ Granulocytes, eosinophilic; ↑ Granulocytes, neutrophilic; ↑ Lymphocytes; ↑ Monocytes

Badway, J.A., Karnowsky, M.L.: Ann. Rev. Biochem. *49*, 695 (1980); Keller, H.U., Till, G.O. (eds.): Leukocyte Locomotion and Chemotaxis. Basle, Boston, Stuttgart: Birkhäuser Verlag 1983; Hattori, A.: Biomed. Res. *2*, Suppl. 199 (1981); Newell, D.G.: The White Cell System. In: Hodges, G.M., Hallowes, R.C. (eds.): Biomedical Research Applications of Scanning Electron Microscopy, Vol. 2. London, New York: Academic Press 1980

Leukocytes, agranular: ↑ Lymphocytes; ↑ Monocytes

Leukocytes, basophilic: ↑ Granulocytes, basophilic

Leukocytes, circulating pools of: ↑ Granulocytes, pools of

Leukocytes, classification of: According to criterion of presence or absence of specific granulations in their cytoplasm, L. are classified as follows:

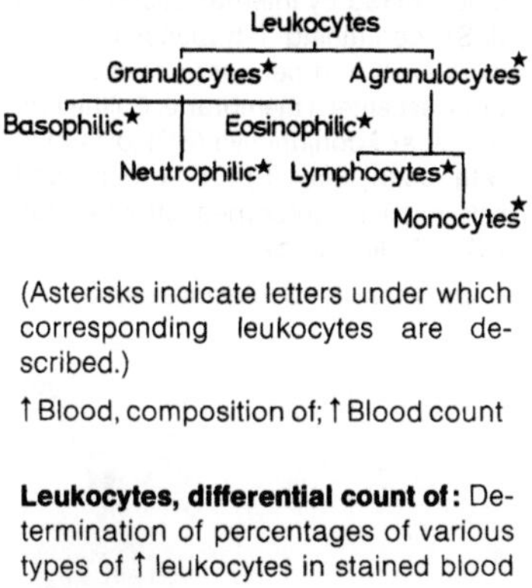

(Asterisks indicate letters under which corresponding leukocytes are described.)

↑ Blood, composition of; ↑ Blood count

Leukocytes, differential count of: Determination of percentages of various types of ↑ leukocytes in stained blood smears.

↑ Blood count

Leukocytes, eosinophilic: ↑ Granulocytes, eosinophilic

Leukocytes, granular: ↑ Granulocytes

Leukocytes, neutrophilic: ↑ Granulocytes, neutrophilic

Leukocytes, polymorphonuclear: ↑ Granulocytes whose circulating forms have a lobulated nucleus.

↑ Granulocytes, basophilic; ↑ Granulocytes, eosinophilic; ↑ Granulocytes, ↑ neutrophilic

Leukofuchsin: ↑ Schiff's reagent

Leukopoiesis: Process of formation and development of ↑ leukocytes.

↑ Granulopoiesis; ↑ Lymphopoiesis; ↑ Monopoiesis

Leukopoietin: A hypothetical ↑ tissue hormone acting on ↑ hematopoietic stem cells to cause ↑ differentiation, via granulopoietic ↑ colony-forming units, into ↑ myeloblasts.

Leukotrienes (slow-reacting substance of anaphylaxis, SRS-A): Ethanol-soluble thioesters produced in allergic reactions by ↑ leukocytes and possibly ↑ mast cells under influence of immunoglobulin E (IgE). L. provoke long-lasting strong contraction of tracheobronchial smooth muscle cells, vasodilatation, and increased vascular permeability.

↑ Immunoglobulins

Berti, F., Folco, G., Velo, G. (eds.): Leukotrienes and Prostacyclin. New York: Plenum Publ. Co. 1983; Buisseret, P.D.: Sci. Am. *247/2*, 82 (1982); Corey, E.J.: Experientia *38*, 1259 (1982)

Levator ani: ↑ Anal canal

Leydig's cells: ↑ Interstitial cells, of testis

Leydig-Kölliker columns (sarcostyles): Bundled ↑ myofibrils visible with light microscope in longitudinally sectioned ↑ cardiac and ↑ skeletal muscle fibers. In cross section, L. correspond to ↑ Cohnheim's fields. Some authors consider L. to be a shrinkage ↑ artifact.

LH-gonadotropes: ↑ Gonadotropes type II

LH-RF: ↑ Gonadotropin-releasing factor

Liberines: ↑ Releasing factors

Lids: ↑ Eyelids

Lieberkühn's crypts, of colon and rectum (crypta intestinalis*): Densely packed simple straight and parallel tubular infoldings (LC) of intestinal epithelium (E) reaching ↑ lamina muscularis mucosae (LMM). L. of both organs are composed mostly of tall ↑ goblet cells, small numbers of ↑ absorptive and ↑ undifferentiated cells, and occasional ↑ enterochromaffin cells located at bottom of L. In L. occurs replacement of the superficial cells. (Fig. = human.)

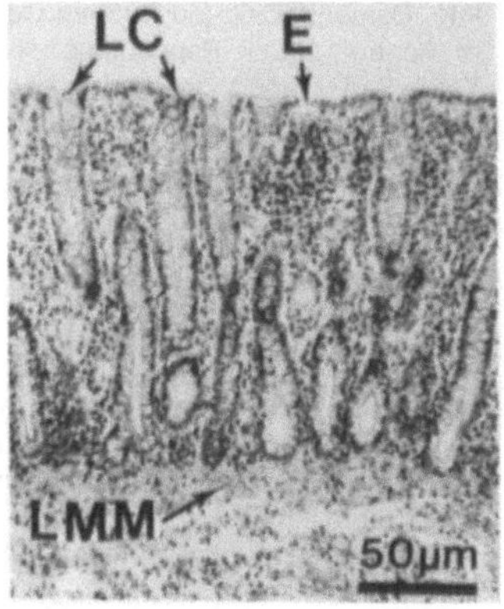

Lieberkühn's crypts, of small intestine (crypta intestinalis*): Tubular infoldings (L) of intestinal epithelium (E) beginning at porelike openings (arrows) between bases of ↑ intestinal villi (V) and reaching ↑ lamina muscularis mucosae (LMM). Structure: L. are lined with ↑ simple columnar epithelium composed of 1) ↑ undifferentiated cells, 2) ↑ absorptive cells, 3) ↑ goblet cells, and, at the bottom, 4) ↑ Paneth's cells (P) and 5) ↑ enterochromaffin cells. In L. occurs replacement of absorptive and goblet cells by mitotic activity (arrowheads) of undifferentiated cells and production of some digestive and bactericidal enzymes, as well as ↑ tissue hormones.

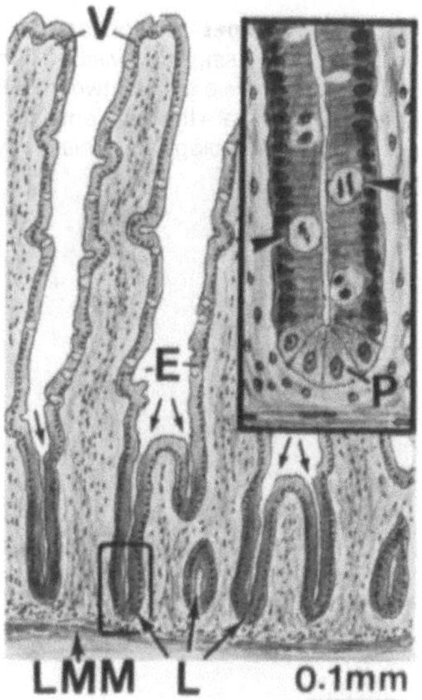

Appleton, D.R., Sunter, J.P., Watson, A.J. (eds.): Cell Proliferation in the Gastrointestinal Tract. London: Pitman Medical 1980; Cairnie, A.B., Lala,

P.K., Osmond, D.G. (eds.): Stem Cells of Renewing Cell Populations. New York, San Francisco, London: Academic Press 1976

Life cycle, of blood cells: ↑ Hematopoiesis

Ligament, spiral: ↑ Spiral ligament

Ligamenta lata: ↑ Broad ligaments

Ligaments (ligamentum*): More or less clearly delimited strong bundles of ↑ dense connective tissue connecting two or more bones. ↑ Collagen fibers of L. are continuous with those of ↑ periosteum.

Ligands: ↑ Mitogens with an affinity for chemical complexes on cell membrane. L., when combined with plasmalemma of ↑ lymphocytes, stimulate them to divide.

Light cells, of arched collecting tubules: ↑ Collecting tubules, of kidney

Light cells, of thyroid gland: ↑ C-cells, of thyroid gland

Light chief cells, of parathyroid glands: ↑ Chief cells, of parathyroid glands

Light meromyosin (LMM): A straight immobile part of ↑ myosin molecule (m.w. about 150000 daltons) participating in formation of ↑ myosin filament shaft.

Light microscopes: A large family of ↑ microscopes using the visible light spectrum (wavelength between 400 and 750 nm); L. are the main optical instruments in ↑ cytology and ↑ histology.

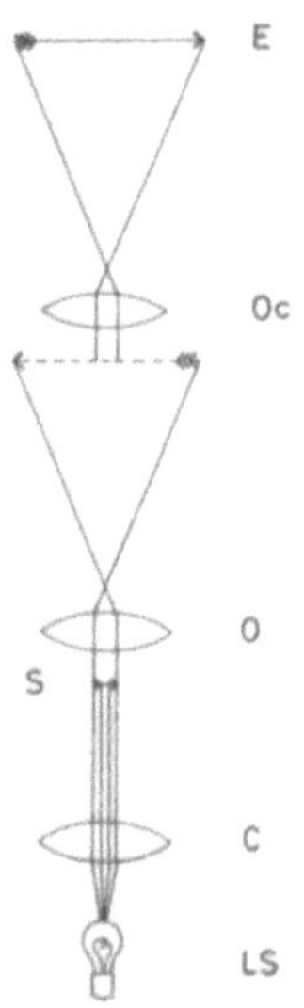

Principle: A beam of visible light is concentrated by ↑ condenser lens (C) from a light source (LS) onto plane of specimen (S). Light passing through fine structural details of specimen is then refracted by ↑ objective (O), forming a magnified image. A segment of image is then magnified by ↑ ocular (Oc) and presented to eye (E). ↑ Resolving power of the best L. is about 0.25 µm.

Turner, G.L.E.: Essays on the History of the Microscope. Oxford: Senecio Publ. Co. Ltd 1980

Light microscopes, categories of: Light microscopes are subdivided as follows:

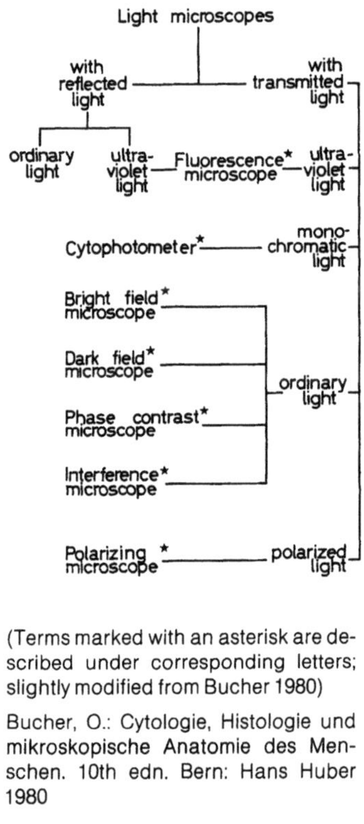

(Terms marked with an asterisk are described under corresponding letters; slightly modified from Bucher 1980)

Bucher, O.: Cytologie, Histologie und mikroskopische Anatomie des Menschen. 10th edn. Bern: Hans Huber 1980

Light microscopy: All techniques and instruments for obtaining and analyzing a ↑ microscopic preparation with ↑ light microscopes.

Limbic system: A collective term for a rim of unclearly delimited ↑ brain cortex surrounding hilus (H) of cerebral hemisphere (gyrus cinculi or limbicus, GC; area subcallosa, AS; gyrus dentatus, GD; gyrus fornicatus, GF; septal area, SA; hippocampus, Hip; uncus, U), functionally connected with deep nuclei (amygdala, thalamus, regio preoptica tuber cinereum, septal nuclei). Structurally, L. corresponds to ↑ allocortex. L. is responsible for affects and emotive expression. Through its connections with ↑ hypothalamus, L. considerably influences endocrine functions.

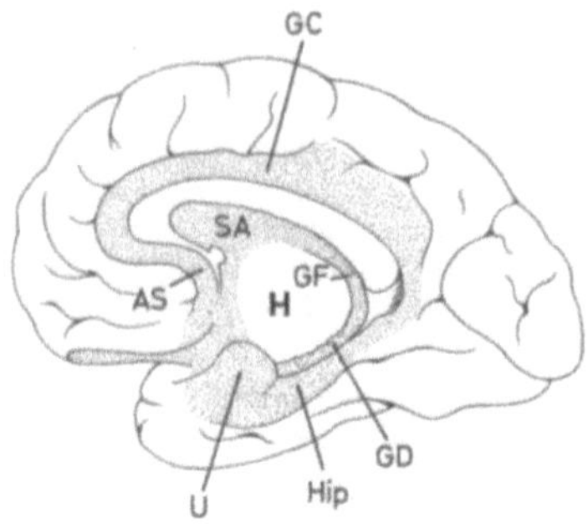

Isaacson, R.L.: The Limbic System. 2nd edn. New York: Plenum Publ. Co. 1982

Limbus, of eye (sclerocorneal junction, limbus corneae*): A circular boundary line (L) between ↑ sclera (S) and ↑ cornea (C). On outer side, L. is marked by external scleral sulcus (ESS) and peripheral termination of ↑ Bowman's membrane. On inner side, L. is marked by internal scleral sulcus (ISS), containing ↑ trabecular meshwork (TM) and peripheral termination of ↑ Descemet's membrane. Epithelium of bulbar ↑ conjunctiva (BC) joins at L. with epithelium (E) of cornea, and sclera joins with corneal stroma (CS). CB = ↑ ciliary body

↑ Arcus senilis; ↑ Schlemm's canal

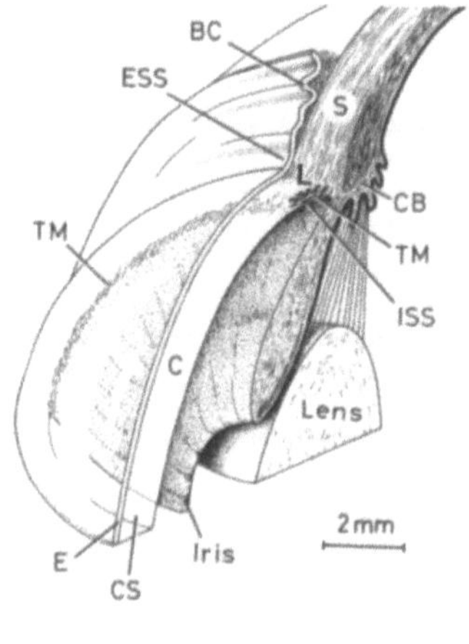

Limbus spiralis: ↑ Spiral limbus

Limiting plate, of liver (lamina hepatica limitans*): A nearly continuous lamina (LP) of ↑ liver parenchymal cells separating interior of a classic ↑ liver lobule (L) from ↑ portal canal (PC). Through openings of L. pass ↑ bile ductules, fine terminal branches of hepatic artery and of ↑ inlet venules, and ↑ lymph that drains into ↑ periportal tissue space (arrowheads). (Fig. = human)

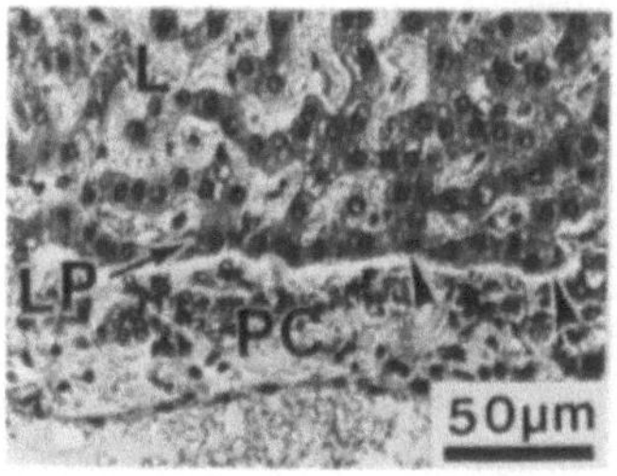

Linea anocutanea: ↑ Anal canal

Lines, of Hunter-Schreger (lamella enameli*): Alternating dark and clear lines (L) generally oriented perpendicular to surface of ↑ enamel (E). L. represent sections of neighboring zones of crossed ↑ enamel prisms. D = ↑ dentin, R = ↑ lines, of Retzius

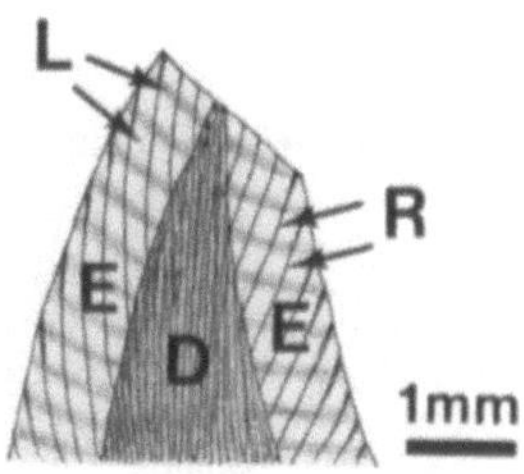

Lines, of Owen: ↑ Contour lines, of Owen

Lines, of Retzius (growth lines, of Retzius, incremental lines of Retzius, stripes of Retzius, lineae incrementales enameli*): Lines (L) running obliquely from surface of ↑ enamel (E) toward root of a tooth in a longitudinal section of the crown; in a cross section, L. appear as concentric lines oriented

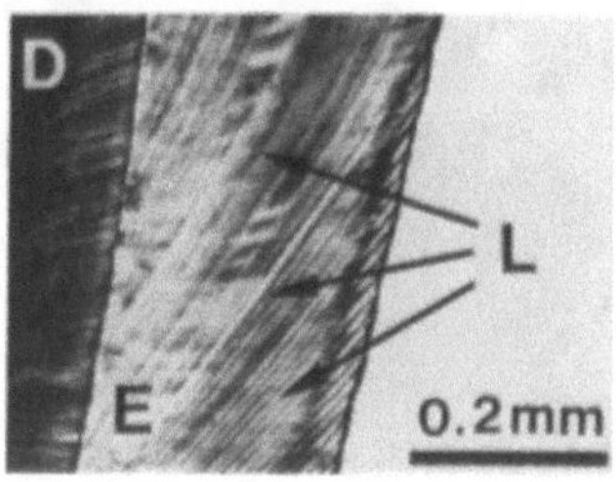

parallel to surface of enamel; they represent successive stages in mineralization of ↑ enamel prisms. D = ↑ dentin. (Fig. nondecalcified tooth, human)

Weber, D.F., Ashrafi, S.H.: Anat. Rec. *194*, 563 (1979)

Lingua: ↑ Tongue

Lingual corpuscles: ↑ Corpuscles, lingual

Lingual follicles: Small ↑ lymphoepithelial organs (LF) situated between sulcus terminalis of ↑ tongue and ↑ epiglottis. Each L. is composed of a short ↑ crypt (Cr) surrounded by ↑ lymphatic tissue (LT) with ↑ lymphatic nodules (L). Crypts are lined with nonkeratinized ↑ stratified squamous epithelium infiltrated with ↑ lymphocytes and ↑ macrophages. Each L. is enclosed in a thin connective capsule (C). Mucous lingual glands (MG) are situated outside this capsule and reach bottom of crypts with their excretory canals (EC). All L. together form ↑ lingual tonsil.

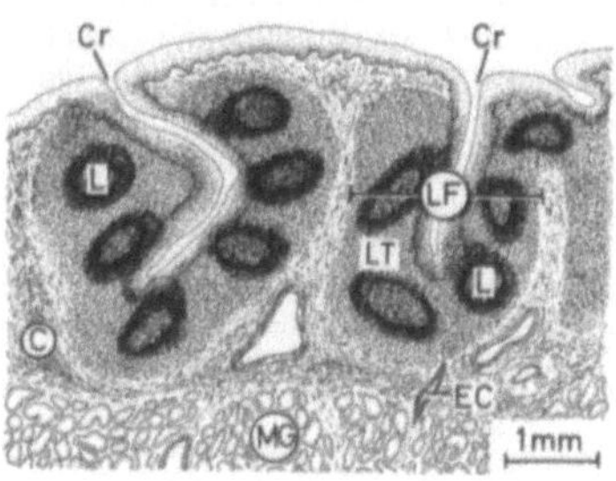

Lingual glands (glandulae linguales*): Small ↑ mixed, ↑ mucous, and/or ↑ serous glands located in ↑ tongue.

↑ Anterior lingual glands; ↑ Ebner's glands

Lingual papillae: ↑ Circumvallate papillae; ↑ Filiform papillae; ↑ Foliate papillae; ↑ Fungiform papillae; ↑ Lenticular papillae

Lingual thyroid: Remnant of thyroid tissue in ↑ tongue.

Lingual tonsil: ↑ Tonsil, lingual

Lining cells: ↑ Littoral cells

Lipid-containing Kupffer's cells: ↑ Perisinusoidal cells

Lipid droplets (gutta lipidis*): Predominantly regular round ↑ inclusions (L) with homogeneous and moderately

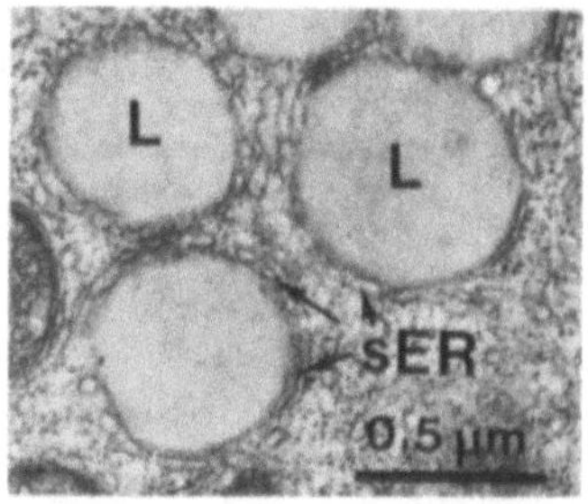

osmiophilic contents (after ↑ glutaraldehyde ↑ fixation). After fixation in OsO_4 L. become blackened. Only very small L. are bound by a ↑ unit membrane; larger L. are separated from ↑ cytoplasmic matrix by a peripheral condensation about 5 nm thick and reinforced by numerous parallel microfilaments. Normally, L. are surrounded by ↑ tubules of smooth endoplasmic reticulum (sER). Mechanism of L. production is not fully clear; there is an important relationship between synthesis of L. and smooth endoplasmic reticulum. L. serve as stored fuel and as material for production of ↑ steroid hormones. (Fig. = ↑ granulosa cells, rat)

↑ Adipose cells, white, histophysiology of; ↑ Fat; ↑ Lipid droplets, genesis of; ↑ Lipid droplets, microfilaments of

Imaizumi, M.: Arch. Histol. Jpn. *30*, 353 (1969)

Lipid droplets, genesis of: First ↑ lipid droplets (L) appear within ↑ Golgi apparatus (G); they are enclosed by a ↑ unit membrane (M). Once free in cytoplasm, only a fine osmiophilic condensation line, interface membrane (IM), separates droplet from ↑ cytoplasmic matrix. As droplets grow in size, they become surrounded by flattened profiles of smooth ↑ endoplasmic re-

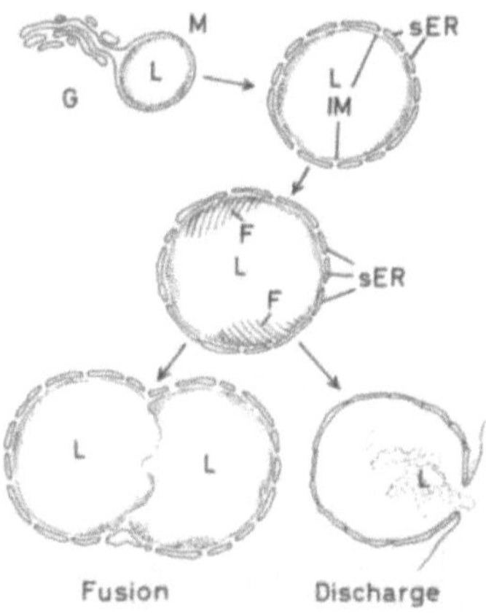

ticulum (sER), which are abundant in ↑ steroid hormone-producing cells, but less numerous in secretory cells of ↑ mammary gland and in ↑ adipose cells. Some droplets become surrounded by a basket of 5-nm-thick microfilaments (F). Droplets may remain separate, may fuse to form a large vacuole, or may be discharged from cell.

↑ Apocrine secretion; ↑ Holocrine secretion; ↑ Lipid droplets, microfilaments of

Lipid droplets, microfilaments of: Filiform structures (F), about 4–5 nm thick, and 30–50 nm apart, located at border between ↑ cytoplasmic matrix (C) and ↑ lipid droplets (L). Length of microfilaments is unknown; they are disposed in an orthogonal arrangement around lipid droplets. It is thought that microfilaments stabilize lipid droplets in cytoplasm and/or prevent their coalescence. BL = ↑ basal lamina, N = ↑ nucleus, NF = ↑ nerve fiber, RF = ↑ reticular fibers (Fig. = ↑ adipose cell, brown)

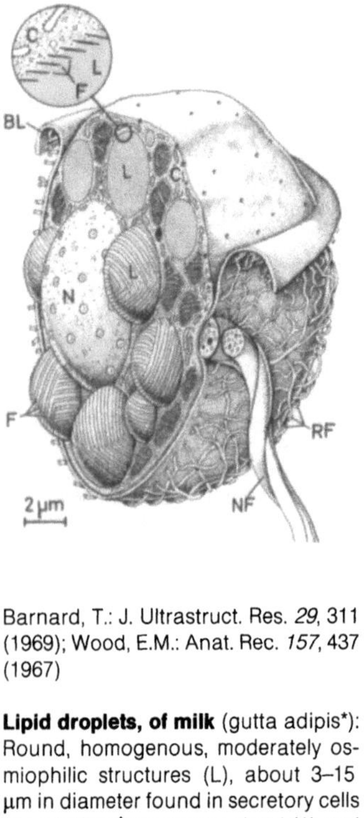

Barnard, T.: J. Ultrastruct. Res. *29*, 311 (1969); Wood, E.M.: Anat. Rec. *157*, 437 (1967)

Lipid droplets, of milk (gutta adipis*): Round, homogenous, moderately osmiophilic structures (L), about 3–15 µm in diameter found in secretory cells of lactating ↑ mammary gland (1) and in its ↑ tubuloalveoli (2). L. are separated from cell by ↑ apocrine secretion, but are always surrounded by a thin cytoplasmic rim (R) and its plasmalemma (inset, arrows); in some cases, larger amounts of cytoplasm with ↑ organelles (arrowheads = mitochondria) can be associated with L.C = ↑ casein granules. (Figs. = rat)

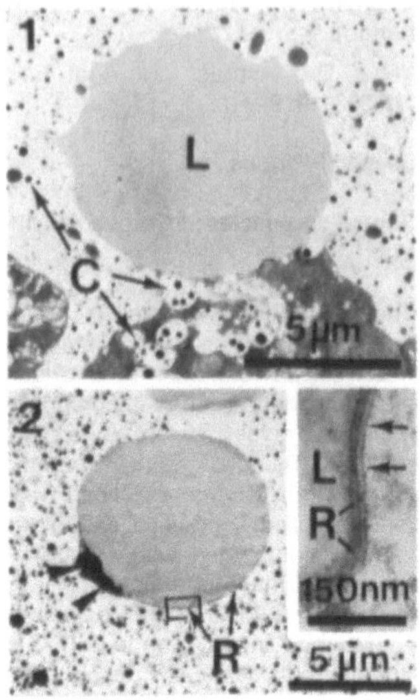

Lipids: Calorie-rich ↑ inclusions found in form of ↑ lipid droplets and vacuoles in many categories of cells, particularly those rich in smooth ↑ endoplasmic reticulum and ↑ mitochondria with tubules. L. are dissolved in ↑ paraffin sections; histochemical demonstration of L. under light microscope is possible using frozen sections and corresponding staining (↑ Sudan dyes). In ↑ transmission electron microscopy, after ↑ osmium tetroxide fixation, L. become resistant to dissolution.

Morii, S., Takigami, S., Kaneda, Y., Shikata, N.: Acta Histochem. Cytochem. *15*, 185 (1982)

Lipids absorption, in intestine: ↑ Absorption, intestinal, of lipids

Lipids, staining of: To preserve lipids for light-microscopic analysis, a tissue must be fixed in ↑ formalin and cut with a ↑ freezing microtome. Sections are then exposed to one of the ↑ Sudan dyes, which dissolve in ↑ lipid droplets and color them (= physical ↑ staining). Stained sections are mounted in a water-soluble medium to avoid lipid extraction. Fixed with OsO_4 for ↑ transmission electron microscopy, lipids stain black, probably due to formation of osmium esters. After preliminary aldehyde fixation, lipids do not blacken on treatment with OsO_4 (Ald/OsO_4), but appear pale gray.

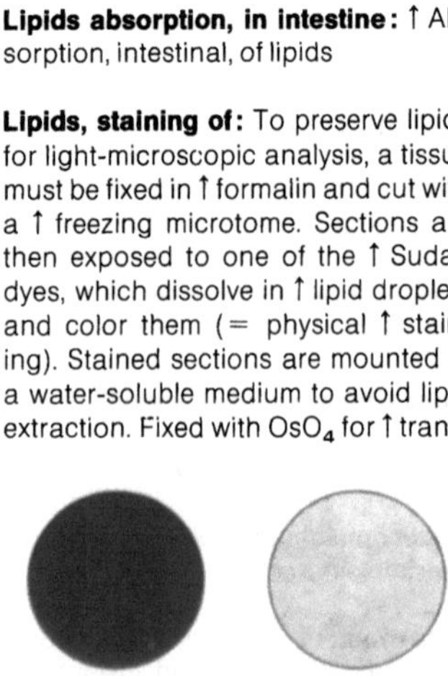

OsO_4 Ald/OsO_4

Lipoblast: The precursor cell of ↑ adipose tissues.

Lipochromes (granulum lipochromi*): Liposoluble exogenous ↑ pigments present in several vegetables (↑ carotenoids). Following absorption, L. penetrate some cells where they occur in the form of heterogeneous conglomerates of irregular clear lamellae and membraneless granules. L. are particularly abundant in ↑ interstitial cells of testis in elderly men; they cause the yellow color of white ↑ adipose tissue and ↑ corpus luteum. (Fig. = ↑ interstitial cells, of testis, human)

Lipocytes, of liver: ↑ Perisinusoidal cells

Lipofuscin (lipofuscin granules, granulum lipofuscini*): A brown endogenous ↑ pigment (Lf) present in many kinds of cells. L. is acid- and alkali-resistant, insoluble in fat solvents, does not undergo decoloration with hydrogen peroxide, and gives a primary golden-brown ↑ autofluorescence. L. granules, about 0.3–3 µm in diameter, are surrounded by a ↑ unit membrane (arrowheads); their contents include ↑ lipid droplets (L) and a variety of highly electron-dense particles and lamellae. L. repre-

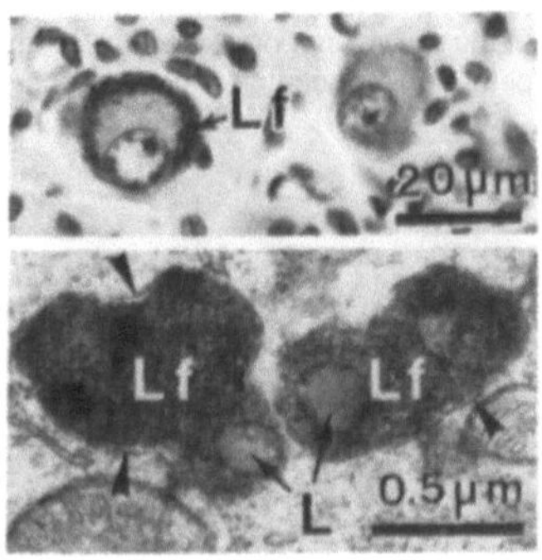

sents indigestible residues of auto- and heterophagocytized material; therefore it is considered a kind of secondary ↑ lysosome, equivalent to a ↑ residual body. Since L. accumulates in ↑ ageing cells (particularly in ↑ neurons and ↑ cardiac muscle cells), it is referred to as "wear and tear" pigment. (Fig., top = ↑ autonomic neurons, human; fig., bottom = ↑ bipolar neuron, rat)

Coleman, R., Silbermann, M., Gershon, D., Reznick, A.: Biol. Cell *46*, 207 (1982); Crichton, D.N., Busuttil, A., Ross, A.: J. Ultrastruct. Res. *72*, 130 (1980); Koobs, D.H., Schultz, R.L., Jutzy, R.V.: Arch. Pathol. Lab. Med. *102*, 66 (1978)

Lipoproteins: Proteins with an incorporated triglyceride core. On the basis of particle size of L., they are subdivided into: 1) Chylomicrons, 2) ↑ very low-density L. (VLDL), and 3) ↑ low-density L. (LDL) or beta lipoproteins. L. are form in which ↑ lipids enter bloodstream following absorption from small ↑ intestine.

Liposomes: Artificial particles, 0.5–2 µm in diameter, obtained after homogenization of white ↑ adipose cells. L. are enclosed by a single layer of peripheral osmiophilic condensation.

↑ Lipid droplets

Gregoridis, G.: N. Engl. J. Med. *295*, 765 (1976); Nicolau, C., Poste, G. (eds.): Liposomes in vivo. Biol. Cell *47*, 1 (1983); Ostro, J.M.: Liposomes. Basel: Marcel Dekker AG 1983

Lipotropins (adipokinetic hormones, LPH): Two polypeptide hormones (m.w. about 6000 daltons) synthesized and secreted together with ↑ adrenocorticotropic hormone by ↑ corticotropes of pars distalis of ↑ hypophysis. L. stimulate liberation of fatty acids from stored triglycerides, probably via ↑ cyclic AMP.

Schwandt, P.: Klin. Wochenschr. *52*, 153 (1974)

Lips (labium oris*): Mucocutaneous folds forming beginning of ↑ gastrointestinal tract. Structure: A) Outer surface = ↑ skin (S) with ↑ hairs, ↑ sebaceous glands, and ↑ sweat glands. B) ↑ Mucocutaneous junction or red portion of lip (arrow) = a transition of ↑ epidermis of skin to ↑ stratified nonkeratinized epithelium of posterior surface; richly developed blood vessels underlying this area are responsible for red color of L. C) Inner surface: 1) Tunica mucosa (TM) = a) stratified nonkeratinized epithelium (E) with

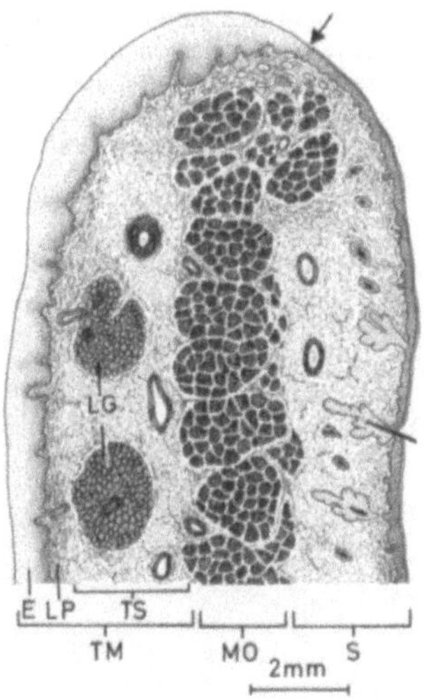

deep ↑ epithelial ridges, b) lamina propria (LP) = a thin layer of ↑ loose connective tissue forming numerous papillae; 2) tela submucosa (TS) = a loosely arranged ↑ fibroelastic tissue with ↑ mixed ↑ labial glands (LG). Midportion of L. is occupied by bundles of musculus orbicularis oris (MO.)

Liquor-contacting neurons: ↑ Cerebrospinal fluid-contacting neurons

Liquor folliculi*: A clear viscid liquid occurring in spaces between ↑ follicular cells of secondary vesicular and mature ↑ ovarian follicles; in further development of follicle, L. fills entire ↑ antrum folliculi. L. is produced, or at least modified, by ↑ follicular cells, and contains, besides ↑ hyaluronic acid, an appreciable concentration of ↑ estrogens.

Lithium carmine: A nontoxic red colloidal dye used for ↑ vital staining of ↑ macrophages.

Littoral cells, of bone marrow: Squamous endothelial-like cells (Li) lining sinuses (S) of ↑ bone marrow. L. have no basal lamina, their layer is incomplete (arrows) and constantly traversed by various blood cells. It is believed that L. represent flattened ↑ reticular cells, since they can phagocytize under certain conditions. (Fig. = rat)

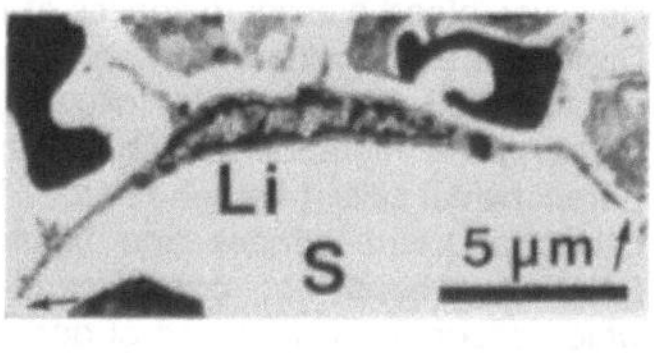

↑ Bone marrow, vascular sinuses of

Littoral cells, of lymph nodes (lining cells): Squamous endothelial-like cells (Li) lining all sinuses of ↑ lymph nodes. Whereas those separating ↑ capsule (C) from ↑ subcapsular sinus (S) lie on a ↑ basal lamina, those separating sinuses from ↑ lymphatic tissue (LT) possess no basal lamina. L. are continuous with ↑ endothelial cells of afferent lymphatic vessels. In addition to other ↑ organelles, L. contain a few lysosomes. Under certain conditions (excess of foreign material), L. can phagocytize. Through layer of L., ↑ lymphocytes (L) and other ameboid elements can enter sinuses. It is thought that L. are in fact flattened ↑ reticular cells. (Fig. = rat)

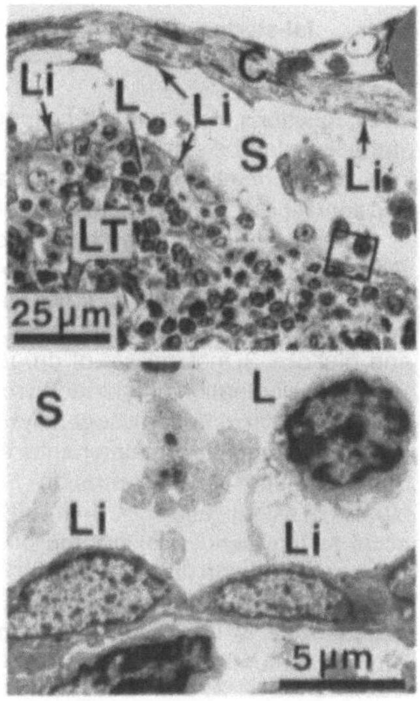

Littoral cells, of splenic sinuses: ↑ Endothelial cells, of splenic sinuses

Littre's glands: ↑ Urethral glands, of male urethra

Liver (hepar*): Largest gland of the body (about 1500 g in adult), situated in right hypogastrium. Structure: 1) ↑ Glisson's capsule; 2) ↑ stroma = a ↑ loose connective tissue separating the lobules and forming tissue of the ↑ portal canals; 3) ↑ parenchyma = ↑ liver parenchymal cells arranged in ↑ hepatic plates that form classic ↑ liver lobules. ↑ Liver sinusoids, situated between hepatic plates, take part in ↑ liver vascularization. ↑ Bile is discharged into ↑ bile pathways.

Elias, J., Sherrick, J.C.: Morphology of the Liver. New York: Academic Press 1969; Motta, P.: Arch. Histol. Jpn. *47*, 1

(1984); Tsuneki, K., Ichihara, K.: Arch. Histol. Jpn. *44*, 1 (1981)

Liver, acinus of: ↑ Acinus, of liver

Liver, bile capillaries of: ↑ Bile canaliculi

Liver, bile ducts of: ↑ Bile ducts; ↑ Bile pathways

Liver, blood supply of: ↑ Liver, vascularization of

Liver, capsule of: ↑ Glisson's capsule

Liver, cell plates of: ↑ Hepatic plates

Liver, central veins of: ↑ Central veins, of liver

Liver, fat-storing cells of: ↑ Perisinusoidal cells

Liver, functions of: The vital functions of the liver can be summarized as follows: 1) Exocrine function = synthesis and secretion of ↑ bile and cholesterol; 2) endocrine function = synthesis and secretion into blood of plasma proteins (albumins, globulins, ↑ fibrinogen, ↑ lipoproteins, ↑ prothrombin, etc.) and glucose; 3) metabolism of drugs, steroids, deiodination of ↑ triiodothyronine and ↑ thyroxine; 4) glycogenolysis and glyconeogenesis, maintenance of normal glucose concentration in blood; 5) esterification of free fatty acids to triglycerides; 6) storage of ↑ glycogen, fat, ↑ iron (in ↑ lysosomes); 7) detoxification of poisons and hydrogen peroxide; 8) ↑ hematopoiesis from 2nd to 8th month of intrauterine life.

Liver, innervation of: ↑ Adrenergic nerve fibers accompany hepatic artery and portal vein and follow their branching up to ↑ portal canals. From here, nerve fibers penetrate liver parenchyma where they form a randomly distributed intralobular network. ↑ Cholinergic nerve fibers originate from plexus vagalis; they also penetrate liver lobules.

Forssmann, W.G., Ito, S.: J. Cell Biol. *74*, 299 (1977); Moghimzadeh, E., Nobin, A., Rosengren, E.: Cell Tissue Res. *230*, 605 (1983)

Liver lobule, classic (lobulus hepaticus*): Smallest morphological unit (L) of liver ↑ parenchyma. Each L. has the form of a small polyhedral prism composed of ↑ hepatic plates (HP) converging toward ↑ central vein (CV). Transversally sectioned, an L. appears to be a more or less regular pentagon or hexagon demarcated by a very small amount of connective tissue (in human) with ↑ portal canals (PC) at corners. Each L. measures about 1–2 mm in diameter; the total number of L. in a human liver averages about one million.

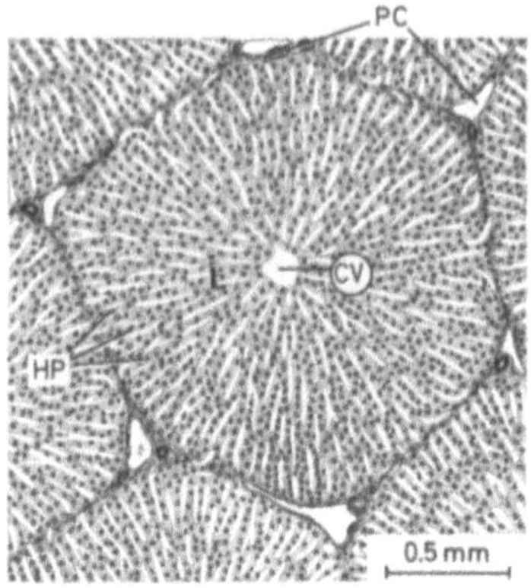

Vonnahme, F.-J., Müller, O.: Cell Tissue Res. *215*, 193 (1981)

Liver lobule, portal: Smallest functional unit of ↑ liver (interrupted lines), consisting of a roughly triangular zone of liver ↑ parenchyma with a ↑ portal canal (PC) in center and ↑ central veins (CV) at corners. L. includes segments of three or more contiguous classic ↑ liver lobules (L); blood flows from portal canal toward central veins, ↑ bile in opposite direction. A = ↑ acinus, of liver, InV = ↑ inlet venules, IlV = ↑ interlobular vein

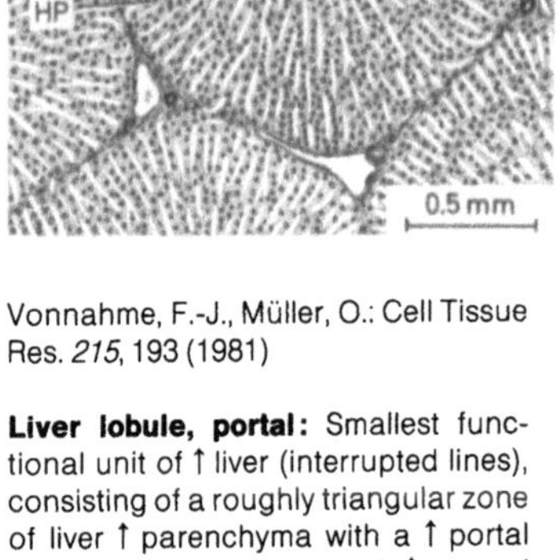

Liver, lymphatic circulation of: There are no intralobular lymphatic vessels. However, blood plasma in spaces of ↑ Disse can be considered lymph flowing toward ↑ portal canals, following exchange of metabolites with ↑ liver parenchymal cells. ↑ Lymph reaches ↑ periportal tissue spaces and finally lymphatic vessels of portal canal, which accompany branches of portal vein.

Liver parenchymal cells (hepatocyte, hepatocytus*): Polyhedral cells constituting ↑ hepatic plates. About 25% of them are binucleate; 70% of mononucleated L. are ↑ tetraploid, and about 2% ↑ octaploid. Each nucleus is round with one or more nucleoli. Cytoplasm encloses about 800 elliptical or oblong mitochondria (M) with transverse and longitudinal cristae; their matrix is of low density and often contains ↑ mitochondrial granules. The well-developed ↑ Golgi apparatus (G) is multiple (up to 50 complexes) and grouped near nucleus and ↑ bile canaliculi (BC); its vacuoles often contain 25–80 nm electron-dense particles, probably representing ↑ lipoproteins (↑ very low-density lipoproteins). Elongated rough endoplasmic cisternae (rER) are arranged as ↑ ergastoplasm and are frequently continuous with smooth endoplasmic tubules (sER). Total surface area of smooth endoplasmic reticulum ranges between 25% and 50% of that of rough endoplasmic reticulum. ↑ Lysosomes (Ly), ↑ peroxisomes (P), ↑ glycogen particles (Gly), ↑ lipid droplets (LD), and free ↑ ribosomes are numerous. Surfaces bordering ↑ liver sinusoids (S) bear many short and irregular microvilli (Mv); at their bases are numerous ↑ micropinocytotic and ↑ coated (CV) vesicles. Surfaces in contact with opposing L. display several ↑ zonulae occludentes and ↑ nexus (J) and also form half of a bile canaliculus (BC). D = ↑ Disse's space

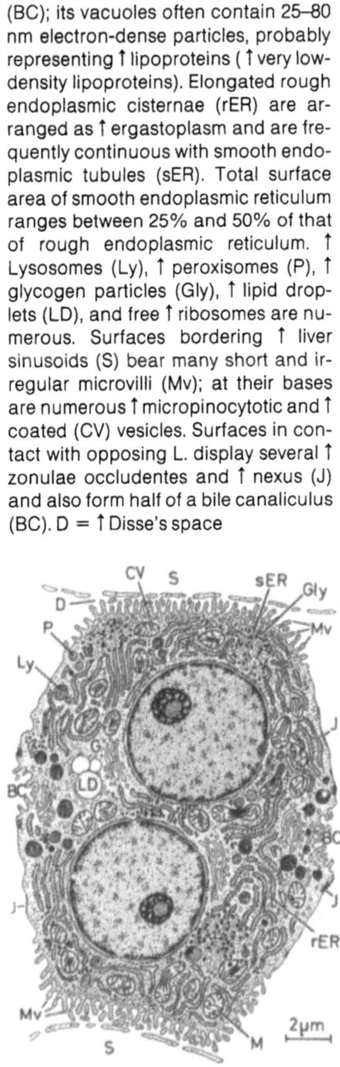

Babcock, M.B., Cardell, R.R.Jr.: Am. J. Anat. *140*, 299 (1974); Carriere, R.: Int. Rev. Cytol. *25*, 201 (1969); Lin, C., Chang, J.P.: Science *190*, 465 (1975); Morré, J. Eur. J. Cell Biol. *26*, 21 (1981); Motta, P., Muto, M., Fujita, T.: The Liver.

An Atlas of Scanning Electron Microscopy. Tokyo, New York: Igaku Shoin 1978

Liver plates: ↑ Hepatic plates

Liver, portal canals of: ↑ Portal canals

Liver sinusoids (hepatic sinusoids, vas sinusoideum*): Intralobular, irregularly shaped discontinuous capillaries (S) extending from ↑ inlet venules to ↑ central vein. L. form an anastomotic network between ↑ hepatic plates (P). Lining of L. consists of ↑ endothelial and ↑ Kupffer's cells. Through intercellular openings, as well as fenestrations (f) and ↑ sieve plates (sp) of endothelial cells (E), L. communicate with perisinusoidal space of ↑ Disse (D). Only a few scattered ↑ collagen and ↑ reticular microfibrils surround L. (Fig., top = human; fig., bottom = rat)

↑ Perisinusoidal cells

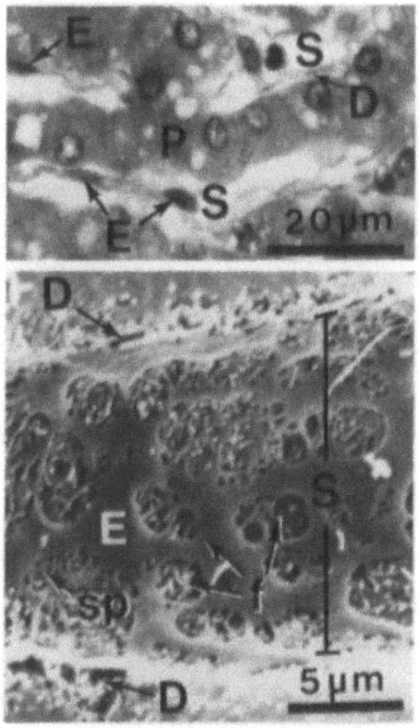

Bankston, P.W., Pino, R.M.: Am. J. Anat. *159*, 1 (1980); Gentrault, J.L., Montecinorodriguez, F., Cinqualbre, J.: Structure of the Normal Human Liver Sinusoid after Perfusion Fixation. In: Knook, D.L., Wisse, E. (eds.): Sinusoidal Liver Cells. Amsterdam: Elsevier Biomedical Press 1982; Lemoine, F., Hemet, J., Dubois, J.-P.: Biol. Cell *43*, 221 (1982); Tanuma, Y., Ohata, M., Ito, T.: Arch. Histol. Jpn. *46*, 401 (1983)

Liver, vascularization of: Rich, divided into: 1) Functional circulation (80% of volume) = portal vein, dividing into interlobar veins (IV); these branch and give off ↑ interlobular veins (IIV), situated in ↑ portal canals (PC). At regular intervals, interlobular veins produce short perpendicular branches, ↑ inlet venules (InV), which surround a segment of classic ↑ lobule (L). From inlet venules, blood passes into ↑ liver sinusoids (S) and circulates between ↑ hepatic plates (P) to be collected in ↑ central veins (CV). From here, it continues into ↑ intercalated vein (SV), then into ↑ collecting veins (CoV), and, finally, into ↑ hepatic veins, emptying into inferior ↑ vena cava. 2) Nutrient circulation (20% of volume) = branches of hepatic artery, interlobar arteries (IA), dividing into ↑ interlobular arteries (IIA) which supply, via their capillary network, ↑ stroma, portal canals, and ↑ bile ducts (BD). From here, blood passes into interlobular veins and sinusoids; only a small volume of blood enters sinusoids directly from interlobular arteries. Arrows indicate direction of blood flow.

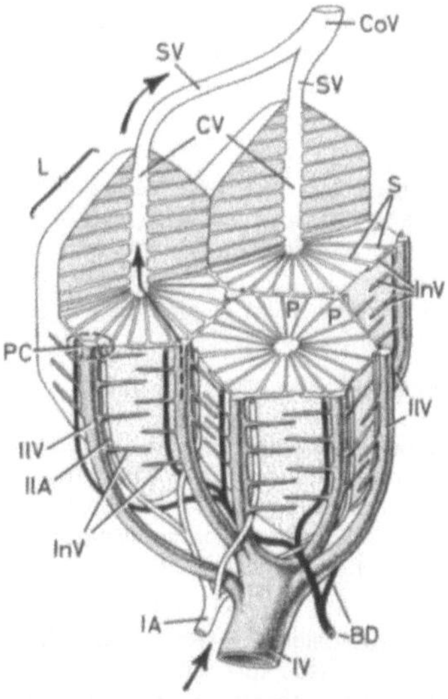

Brauer, R.W.: Physiol. Rev. *43*, 115 (1963); Burkel, W.E.: Anat. Rec. *167*, 329 (1970); Ohtani, O.: Biomed. Res. *2*, Suppl. 219 (1981)

LM: ↑ Light microscope(y)

Lobe (lobus*): A part of a parenchymatous organ with constant anatomical delimitation and constant form. Each L. is composed of various number of ↑ lobules.

Lobe, of kidney: ↑ Renal lobe

Lobule (lobulus*): A prismatic or pyramidal segment of a ↑ parenchyma more or less clearly delimited by connective septa. Several L. constitute a ↑ lobe.

Lobule, of auricle: ↑ Auricle

Lobule, of kidney: ↑ Renal lobule

Lobule, of liver: ↑ Liver lobule, classic; ↑ Liver lobule, portal

Lobule, of lung: ↑ Lung lobule

Lobule, of mammary gland (lobulus glandulae mammariae*): A group of ↑ tubuloalveoli emptying into one ↑ intralobular duct.

Lobule, of renal glomerulus: A segment of capillary network of ↑ renal glomerulus formed by one branch of ↑ afferent arteriole.

Lobule, of testis (lobulus testis*): One of about 250 pyramidal segments of testicular ↑ parenchyma delimited by ↑ septula testis. Tip of each L. is oriented toward ↑ rete testis, base toward ↑ tunica albuginea.

Lobule, of thymus (lobulus thymi*): A segment of ↑ thymus ↑ parenchyma well delimited by connective septa. Each L. consists of an outer cortex and an inner medulla; all are connected by the medulla.

Long-acting thyroid stimulator (LATS): An ↑ immunoglobulin of type IgG produced by cells of ↑ reticuloendothelial system. It has a stimulative action on ↑ thyroid follicular cells, as does ↑ thyrotropic hormone but with considerably longer effect. Production of L. is pathological.

Long-spacing collagen (LSC): A rare natural manner of side-to-side aggregation of ↑ tropocollagen molecules without longitudinal overlapping; resulting cross-striation has a periodicity of 280 nm instead of 64 nm. L. is found in ↑ subcommissural organ (fig.) of some species. Origin and functions unknown; probably identical to ↑ fibrous long-spacing collagen. (Fig. = rat)

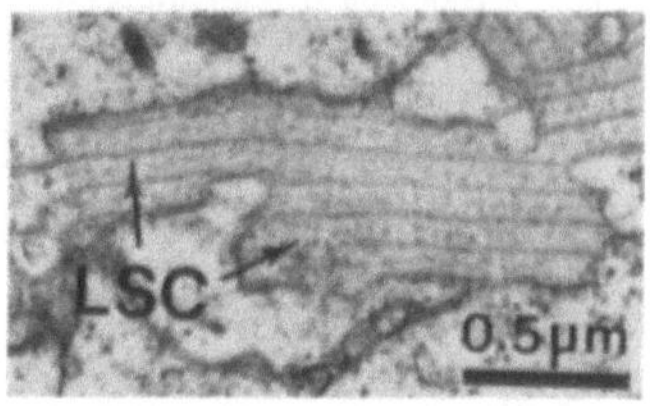

↑ Segment long-spacing collagen

Loop, of Henle (ansa nephroni*): The U-shaped portion of a ↑ nephron located partly in cortex and partly in medulla. L. consists of 1) ↑ straight portion of ↑ proximal tubule (PT), or thick descending limb, 2) ↑ thin segment (TS), and 3) straight portion of ↑ distal tubule (DT), or thick ascending limb. L. penetrates into medulla for varying depths, depending upon whether nephron is of cortical or juxtamedullary type. L. plays an important role in concentrating urine by serving as a countercurrent multiplier system. MR = ↑ medullary ray.

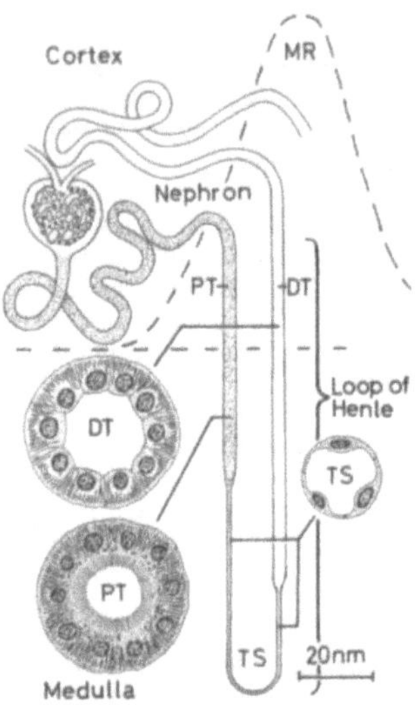

Neiss, W.F.: Anat. Embryol. *164*, 315 (1982)

Loose connective tissue (areolar tissue, textus connectivus fibrosus laxus*): A kind of ↑ fibrous connective tissue with relative predominance of cells over fibrous elements. L. is an irregular tridimensional meshwork formed of ↑ fibroblasts (Fb), ↑ fibrocytes (Fc), ↑ collagen (C), ↑ elastic (E), and ↑ reticular (R) fibers. The meshes or areolae are normally filled with small amounts of fluid ↑ ground substance and can greatly enlarge under certain conditions (allergic reaction, inflammation). They also enclose ↑ wandering cells: ↑ histiocytes (H), ↑ lymphocytes (L), ↑ monocytes (Mo), ↑ plasma cells (P), eosinophilic ↑ granulocytes (Gr), and ↑ mast cells (M). Along capillaries (Cap) there are ↑ pericytes (Pc) and ↑ nonmyelinated nerve fibers (NF). L. may also contain ↑ adipose cells; neutrophilic ↑ granulocytes are present under pathological conditions. L. is an interstitial tissue filling spaces between glandular lobules and lobes and between organs (tunica adventitia); it constitutes ↑ endomysium, ↑ endoneurium, and ↑ endotendineum; L. surrounds and accompanies vessels and nerves, forms ↑ stroma of ↑ testis, ↑ kidney, ↑ liver, etc., composes ↑ pia-

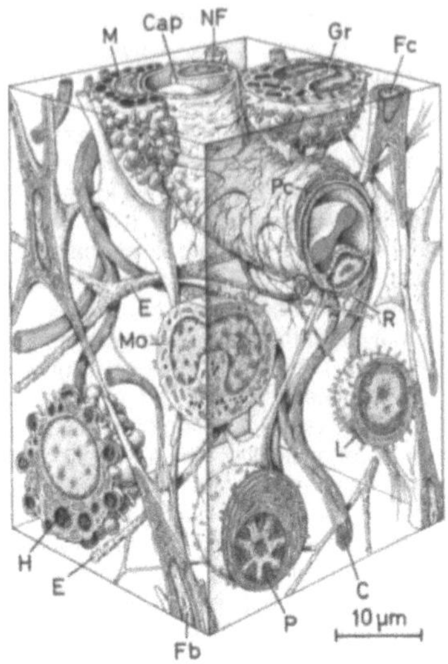

arachnoid, papillary layer of ↑ dermis, ↑ hypodermis, ↑ laminae propriae, ↑ telae submucosae, etc.

↑ Edema

Loral artery (arteria trabecularis*): A branch of superior hypophyseal arteries participating in formation of capillary loops in pars intermedia and posterior portion of pars distalis of ↑ hypophysis.

↑ Hypophysis, vascularization of

Low-density lipoproteins (LDL, beta lipoproteins): A kind of ↑ lipoproteins occurring in smooth ↑ endoplasmic reticulum of ↑ liver parenchymal cells and secreted into spaces of ↑ Disse.

Brown, M.S., Anderson, R.G.W., Basu, S.K., Goldstein, J.L.: Recycling of Cell-Surface Receptors: Observation from the LDL Receptor System. In: Cold Spring Harbor Symposia on Quantitative Biology, Vol. 46, Part 2. New York: Cold Spring Harbor Laboratory 1982

L-system, of striated muscle fibers (elementum tubulare*): Tubules of ↑ sarcoplasmic reticulum (L) running parallel or longitudinal to ↑ myofibrils (Mf). Tubules of L. open into ↑ terminal

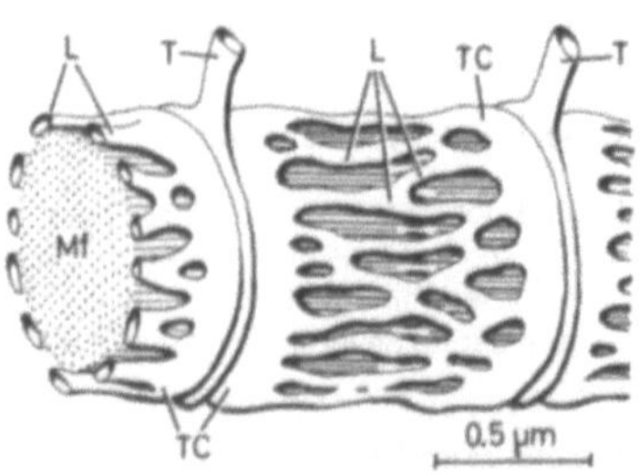

cisterna (TC). L. is more developed in ↑ skeletal muscle fibers than in ↑ cardiac cells. T = ↑ T-tubules

LTF: ↑ Lymphocyte-transforming factor

LTH: ↑ Prolactin

LTH-cells: ↑ Mammotropes

LTH-RF: ↑ Prolactin-releasing factor

Lubarsch's crystals: About 1-µm-long and 0.7-µm-wide bodies occasionally found in ↑ spermatogonia. L. consist of densely arranged microfibrils interspersed with dense granules; their function is unknown.

Lumen (pl. lumina): Space in interior of a tubular or hollow structure or organ (capillary L., intestinal L., etc.).

Luminosity horizontal cells, of retina: ↑ Horizontal cells

Lung circulation, functional: ↑ Lungs, vascularization of; ↑ Pulmonary arteries; ↑ Pulmonary veins

Lung circulation, nutrient: ↑ Bronchial arteries; ↑ Bronchial veins; ↑ Lungs, vascularization of

Lung lobule (pulmonary lobule, secondary lobule, lobulus pulmonis*): A roughly pyramidal segment of pulmonary ↑ parenchyma oriented with tip toward ↑ hilus and base, about 0.5–2 cm wide, toward visceral ↑ pleura (P). A ↑ bronchus (B) enters at tip of L.; once within L., it loses cartilage and becomes a ↑ bronchiole (Br), dividing

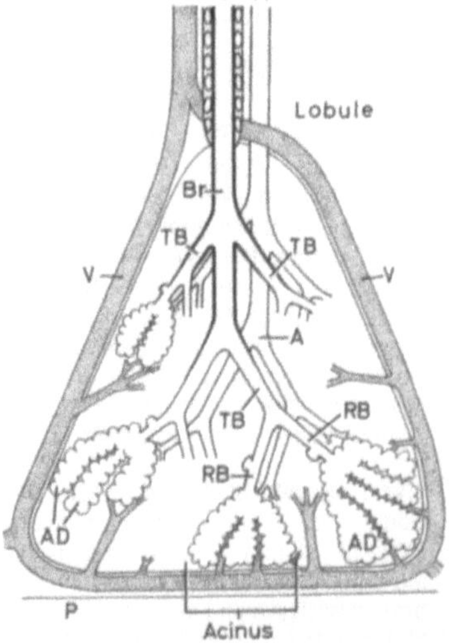

to give 50–80 terminal ↑ bronchioles (TB), 100–200 respiratory ↑ bronchioles (RB), and 600–1000 ↑ alveolar ducts (AD), corresponding to about 200–300 pulmonary ↑ acini. A branch of ↑ pulmonary artery (A) accompanies bronchiolar branching; blood is collected by a ↑ pulmonary vein (V) at periphery of L. Only at tip of L. does the vein join the bronchus.

Cummings, G. (ed.): Cellular Biology of the Lung. New York: Ettore Majorana International Science Service 1982

Lung lobule, primary: A term used by some authors to designate ↑ acinus of lung.

Lung lobule, secondary: A term used by some authors to designate ↑ lung lobule.

Lungs (pulmo*): Paired lobed organs situated in thoracic cage. Structure: 1) Visceral ↑ pleura; 2) ↑ stroma = thin connective septa dividing ↑ parenchyma into lobules; 3) parenchyma or respiratory structures = ↑ lung lobules composed of ↑ acini, ↑ alveolar ducts, ↑ alveoli; 4) bronchial tubes = extra- and intrapulmonary ↑ bronchi, ↑ bronchioles, terminal and respiratory ↑ bronchioles.

Gehr, P., Bachofen, M., Weibel, E.R.: Resp. Physiol. *32*, 121 (1978); Gould, V.E.: Lab. Invest. *48*, 507 (1983); Low, F.N.: Progr. Clin. Biol. Res. *59*, 289 (1981); Virdio, E.A., Galina, M.A. (eds.): 11th International Congress of Anatomy. New York: Alan R. Liss 1981; Weibel, E.R.: Oxygen Demand and the Size of Respiratory Structures in Mammals. In: Wood, S.C., Lenfant, C. (eds.): Evolution of Respiratory Processes. Vol. 13. New York, Basle: M. Dekker 1979

Lungs, airways of: ↑ Pulmonary airways

Lungs, airways of, branching of: ↑ Bronchi; ↑ Bronchioles

Lungs, alveolar cells of: ↑ Alveolar cells, type I and II

Lungs, alveolar ducts of: ↑ Alveolar ducts

Lungs, alveolar sac of: ↑ Alveolar sac

Lungs, alveoli of: ↑ Alveolus, of lung

Lungs, histophysiology of: The main function of the lung is to provide oxygen to the blood and to remove CO_2 through ↑ blood-air barrier. (See physiology texts for further information)

Lungs, lymphatics of: Two sets of lymphatics drain ↑ lymph from lungs to tracheobronchial ↑ lymph nodes. 1) Pleural lymphatics = relatively large superficial lymphatic vessels with valves draining lymph from ↑ pleura and interlobular septa. 2) Pulmonary lymphatics = lymphatic vessels accompanying ↑ bronchi, ↑ pulmonary artery, and vein. Finest vessels, paraalveolar lymphatics, closely approach ↑ alveoli, but ↑ alveolar septa are devoid of lymphatic vessels.

Grande, N.R., Ribeiro, J., Soares, M., Carvalho, E.: Acta Anat. *115*, 302 (1983); Lauweryns, J.M.: Prog. Clin. Biol. Res. *59*, 299 (1981)

Lungs, repair of: ↑ Regeneration of destroyed pulmonary tissue is impossible. All lost pulmonary ↑ parenchyma is replaced by fibrous scar.

Lungs, vascularization of: 1) Functional L. = branches of ↑ pulmonary artery (PA) accompanying bronchial branching and entering ↑ lung lobule at its tip. In lobule, a branch of pulmonary artery follows branching of ↑ bronchiole up to respiratory ↑ bronchiole (RB). Here, arterial branches form capillary network of alveoli (A). Arterialized blood (black) is then collected by short veins at periphery of lobule, veins of visceral ↑ pleura (P), and veins of interlobular septa (S). At tip of lobule, veins of interlobular septa join to form branch of ↑ pulmonary vein (PV), which follows bronchial tree. 2) Nutrient L. = ↑ bronchial arteries (BA) accompanying bronchi and bronchioli up to respiratory bronchioles, where they anastomose (arrowhead) with small branches of pulmonary artery. B = ↑ bronchus (Arrows indicate direction of blood circulation)

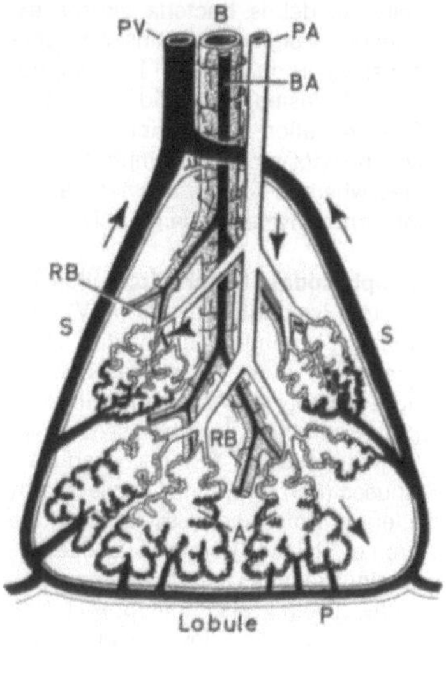

Kendall, M.W., Eissmann, E.: Anat. Rec. *196*, 275 (1980)

Lunule, of nail (lunula*): White crescent-shaped area (L) situated at root of ↑ nails. Partially covered by ↑ eponychium (E), L. is usually only visible on thumb and first finger. Although L. is not fully understood, it is believed that it represents distal extension of ↑ matrix (M) toward ↑ nail bed (B). Opacity of L. seems to be caused by considerable thickness of matrix in comparison with nail bed and by presence of large amounts of prekeratinous fibrils in proliferating matrix cells.

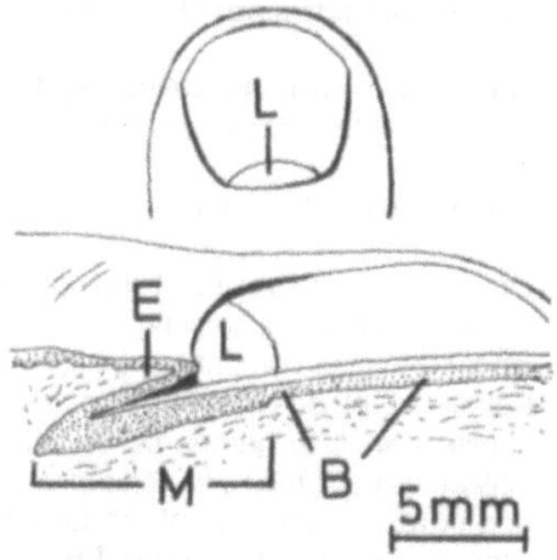

Luschka, ducts of: ↑ Ducts, of Luschka

Lutein cells (luteocytus*): Two kinds of polyhedral cells constituting ↑ corpus luteum: larger, clear, ↑ granulosa lutein cells and smaller, darker, peripheral ↑ theca lutein cells.

Luteinization: Process of postovulatory ↑ differentiation of ↑ granulosa cells of a ruptured secondary ↑ ovarian follicle and cells of its theca interna into ↑ granulosa and ↑ theca lutein cells. Sequence: 1) Enlargement of nucleus and loss of ↑ heterochromatin; 2) appearance of ↑ mitochondria with tubules; 3) multiplication of smooth endoplasmic ↑ tubules; 4) enlargement of Golgi apparatus; 5) increase in lysosomes and ↑ lipofuscin granules. L. is under control of ↑ luteinizing hormone.

↑ Corpus luteum; ↑ Ovulation

Crisp, T.M., Channing, C.: Biol. Reprod. *7*, 55 (1972)

Luteinizing hormone (LH in female = ICSH in male): A ↑ glycoprotein hormone (m.w. about 26000 daltons) secreted by ↑ gonadotropes type II of ↑ hypophysis. Together with ↑ follicle-stimulating hormone, L. provokes ↑ ovulation, converts ↑ follicular and ↑ theca interna cells into ↑ granulosa and ↑ theca lutein cells of ↑ corpus luteum, and stimulates secretion of ↑ progesterone. In the male, L. stimulates ↑ interstitial cells of testis to produce ↑ testosterone, probably acting by way of ↑ adenyl cyclase–cAMP system.

↑ Interstitial cell-stimulating hormone

Luteinizing hormone gonadotropes: ↑ Gonadotropes type II

Luteinizing hormone-releasing factor: ↑ Gonadotropin-releasing factor

Luteotropes: ↑ Mammotropes

Luteotropic hormone: ↑ Prolactin

Luteotropic hormone-releasing factor: ↑ Prolactin-releasing factor

Luteotropin: ↑ Prolactin

Lymph (lympha*): A clear, transparent, slightly opalescent, sometimes yellowish fluid formed by continuous ultrafiltration of blood plasma through ↑ capillary wall. L. bathes tissues; it is collected by lymphatic ↑ capillaries and flows through ↑ lymph nodes into ↑ thoracic duct. L. transports ↑ B- and ↑ T-lymphocytes, cell debris, antigens, resorbed fat (↑ chylomicrons), ↑ hormones, exogenous ↑ pigments, and waste products.

Lymph nodes (nodus lymphaticus*): Kidney-shaped encapsulated ↑ lymphatic organs (5–15 mm in longitudinal diameter) associated in relatively constant groups throughout body; absent in ↑ central nervous system. Structure: A) ↑ Stroma: 1) ↑ capsule (Ca) = thin layer of ↑ dense connective tissue perforated at convex aspect of L. by several afferent lymphatic vessels (ALV); 2) trabeculae (T) = thin extensions of capsule into parenchyma. B) ↑ Parenchyma: 1) ↑ cortex (C): a) outer cortex (OC) = ↑ lymphatic tissue with ↑ lymphatic nodules (LN), b) ↑ inner cortex (IC) = the diffuse ↑ lymphatic tissue that continues without clear demarcation into medullary cords; 2) medulla (M): ↑ medullary cords (MC) = cordlike branched extensions of lymphatic tissue of inner cortex. C) Sinuses: 1) ↑ subcapsular sinus (SS) = lymphatic space between capsule and outer cortex; 2) ↑ intermediate sinus (IS) = continuation of subcapsular sinus accompanying trabeculae; 3) ↑ medullary sinus (MS) = lymphatic spaces between medullary cords continuous with intermediate sinuses. D) Hilus (H) = concave side of L. where blood vessels (BV) enter and leave along with one or two efferent lymphatic vessels (ELV). In area of hilus, subcapsular and medullary sinuses communicate directly. L. are frequently embedded in adipose tissue (AT).

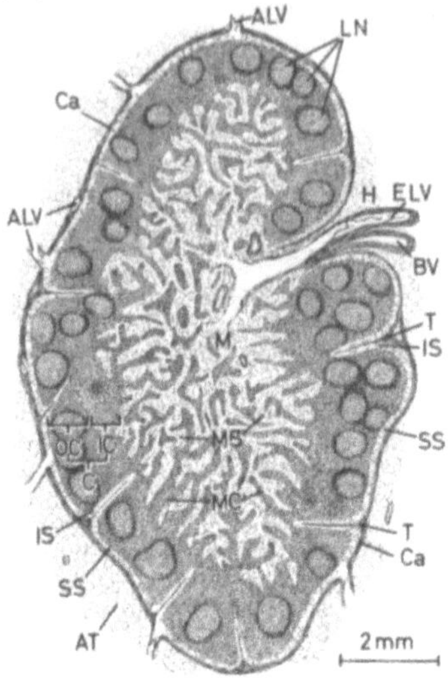

Belisle, C., Sainte-Marie, G.: Anat. Rec. *199*, 45 and 72 (1981) and Anat. Rec. *199*, 213 (1981); Fossum, S., Vaaland, J.L.: Anat. Embryol. *167*, 229 (1983)

Lymph nodes, capsule of: ↑ Capsule, of lymph node

Lymph nodes, cortex of: ↑ Cortex, of lymph node

Lymph nodes, histophysiology of: ↑ Lymph nodes represent active filters intercalated along the path of lymphatic vessels actively involved in: 1) Phagocytosis: ↑ reticular cells and free ↑ macrophages phagocytize foreign cells, cell debris, bacteria, viruses, exogenous ↑ pigments, toxins, and antigens; 2) production of ↑ B-lymphocytes, ↑ plasma cells, and antibodies; 3) recirculation of ↑ lymphocytes: B-lymphocytes move to ↑ lymphatic nodules, whereas ↑ T-lymphocytes wander into inner, thymus-dependent ↑ cortex.

Lymph nodes, lymph circulation in: Afferent lymphatic vessels (ALV) containing ↑ lymph from connective tissue spaces or more peripheral ↑ lymph nodes empty into subcapsular sinus (SS). From there, lymph passes via intermediate sinus (IS) into medullary sinuses (MS) to be finally collected by efferent lymphatic vessels (ELV). In area of ↑ hilus (H), lymph of subcapsular sinus reaches medullary sinuses directly. Lymph also infiltrates ↑ lymphatic tissue and flows slowly from ↑ cortex (C) to medulla (M). Direction of lymph circulation (arrows) is regulated by ↑ valves (V) that open toward ↑ capsule (arrowhead) in afferent lymphatic vessels and away from lymph node in efferent lymphatic vessels.

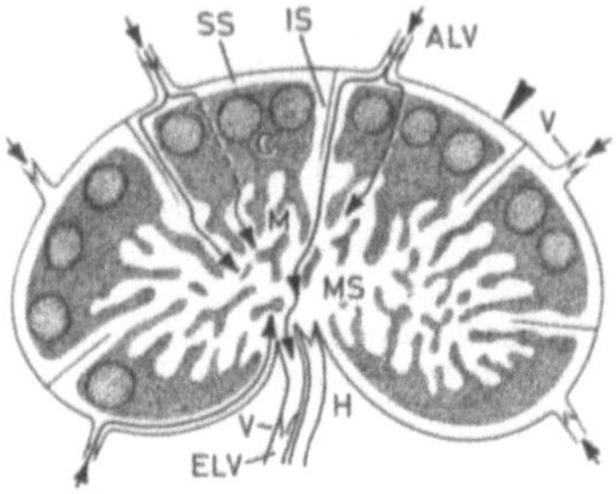

Ogata, T.: Biomed. Res. *2*, Suppl. 189 (1981); Ogata, T., Usui, T.: Biomed. Res. *2*, Suppl. 181 (1981)

Lymph nodes, medullary cords of: ↑ Medullary cords, of lymph node

Lymph nodes, vascularization of: A small ↑ muscular artery (A) penetrates ↑ lymph node through ↑ hilus (H), runs within trabeculae, branches, and enters ↑ medullary cords (MC). On reaching ↑ cortex, branches spread into capillary network around germinal centers of ↑ lymphatic nodules (LF). Blood is collected by ↑ postcapillary venules (PV) that course radially through inner cortex (IC), enter medullary cords, and unite into a larger vein (V) which leaves lymph node at hilus.

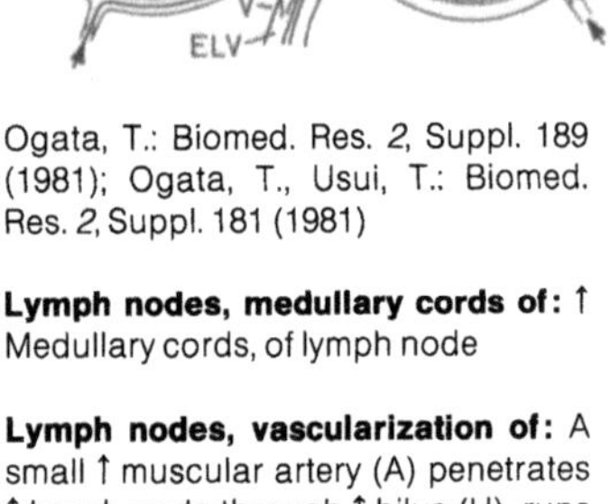

Anderson, A.O., Anderson, N.D.: Am. J. Pathol. *80*, 387 (1975); Irino, S., Takasugi, N., Murakami, T.: Scanning Electron Microscopy 1981/III, 89 (1981)

Lymphatic capillaries (vas lymphocapillare*): Largely anastomotic channels, 10–50 μm wide, lined by very thin ↑ endothelium. Only ↑ central nervous system and intralobular liver ↑ parenchyma are devoid of L. Structure: Very attenuated ↑ endothelial cells (E) connected to each other only by occasional ↑ zonulae occludentes and forming in general a continuous sheet.

In ↑ lacteals of ↑ intestinal villi, there are apertures between endothelial cells. Apposing cells overlap in flaplike fashion. Besides moderately developed ↑ organelles, cytoplasm of endothelial cells contains actin microfilaments (Mf) and sends out irregular luminal processes. L. have practically no basal lamina (BL). Firm connection between capillaries and surrounding connective tissue is maintained by numerous anchoring filaments (AF) originating from reinforced areas of outer endothelial surface and extending among adjoining ↑ collagen (CF), ↑ reticular, and ↑ elastic fibers. L. have neither adventitial cells nor ↑ pericytes. Function: Through semipermeable endothelium, L. reabsorb part of water containing electrolytes and some proteins that continually leaves blood capillaries. Shortening of actin microfilaments causes endothelial cells to separate from each other, permitting entrance of greater amounts of ↑ tissue fluid into L. (arrows). L. begin mostly as blind dilatations (lacteals).

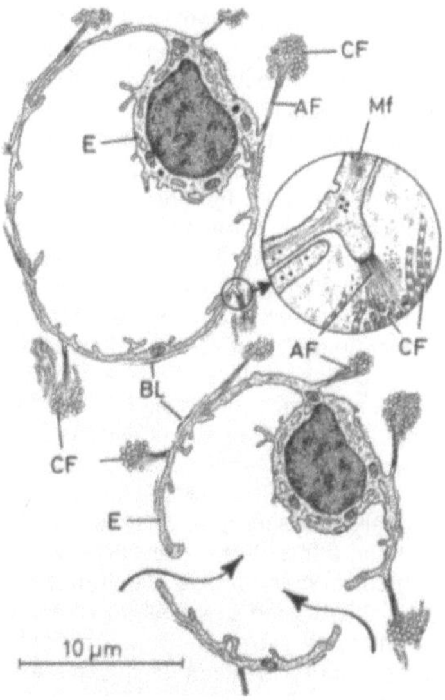

Huth, F., Bernhardt, D.: Lymphology *10*, 54 (1977); Leak, L.: The Fine Structure and Function of the Lymphatic Vessel System. In: Handbuch der allgemeinen Pathologie. Lymph Vessel System. Vol. 3/6. Berlin, Heidelberg, New York: Springer-Verlag 1972

Lymphatic circulatory system: Sum of all ↑ lymphatic vessels collecting ↑ lymph from tissues and draining it into venous circulation.

Lymphatic duct, right (ductus lymphaticus dexter*): Second lymphatic vessel of the body, emptying ↑ lymph into confluence of right jugular and subclavian veins. Structure: 1) Tunica intima = ↑ endothelium (E) and thin subendothelial layer composed of ↑ collagen and ↑ elastic fibers; 2) tunica media (M) = thin circular layer of ↑ smooth muscle cells; 3) tunica adventitia (A) = longitudinally oriented smooth muscle cells interspersed within ↑ loose connective tissue. L. is well innervated; blood supply is assured by ↑ vasa vasorum.

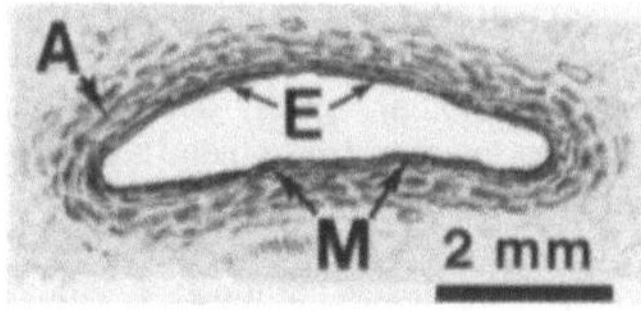

Altmann, H.-W. et al. (eds.): Handbuch der allgemeinen Pathologie, Vol. 3/6. Lymph Vessel System. Berlin, Heidelberg, New York: Springer-Verlag 1982

Lymphatic ducts (ductus lymphatici*): A collective term for the ↑ thoracic duct and right ↑ lymphatic duct. Structurally, they are identical.

Lymphatic nodules (nodulus lymphaticus*, folliculus lymphaticus*): Spherical aggregations of small ↑ lymphocytes found in various organs (↑ lymph nodes, ↑ spleen, ↑ tonsils, ↑ Peyer's patches, etc.). Morphofunctionally, L. may be divided into ↑ L. primary, ↑ L. secondary, and ↑ L. tertiary.

Lymphatic nodules, aggregated: ↑ Peyer's patches

Lymphatic nodules, capsule of: ↑ Capsule, of splenic nodules

Lymphatic nodules, of spleen: ↑ Spleen; ↑ White pulp

Lymphatic nodules, primary (lymphonodulus primarius*): Round or elliptical agglomerations of small ↑ lymphocytes found in ↑ lymph nodes, ↑ spleen, ↑ tonsils, etc. L. have no ↑ germinal center; they are present in the above-cited organs in the absence of exposure to antigens (intrauterine life; animals born and kept under sterile conditions).

Sakuma, H., Asano, S., Kojima, M.: Acta Pathol. Jpn. *31*, 473 (1981)

Lymphatic nodules, secondary (lymphatic follicle, folliculus lymphaticus*, lymphoid nodulus, lymphatic nodulus, nodulus lymphaticus*): Spherical accumulations (LN) of cells of lymphocyte series within ↑ stroma of diffuse ↑ lymphatic tissue (LT). In addition to ↑ cortex of ↑ lymph nodes, L. are present in ↑ spleen, ↑ tonsils, ↑ Peyer's patches, and ↑ appendix, and solitary L. are scattered along digestive, respiratory, and genital passages. Structure: 1) Capsule = thin layer of elongated ↑ reticular cells almost invisible in preparations because of heavy infiltration with small lymphocytes of cap. 2) Cap, corona, or mantle (C) = circular crescent-shaped dense peripheral aggregation of small ↑ lymphocytes. Cap is somewhat wider toward sinuses (S) in lymph nodes and toward nearest epithelium in other organs. 3) ↑ Germinal center (GC): a) light region (lr) = loosely arranged mixed population of ↑ immunoblasts, multiplying and differentiating ↑ B-lymphocytes, their precursors, and some ↑ plasma cells; b) dark region (dr) = closely packed immunoblasts, large and medium-sized lymphocytes, ↑ macrophages, and immunoblasts in the course of transformation into plasma cells. Latter are relatively few in number since they migrate into inner cortex (IC). L. represent most important sites of B-lymphocyte production and high-level antibody formation. They do not appear until after birth, in response to antigen exposure. Germinal center also contains small numbers of ↑ T-lymphocytes (↑ helper cells). (Fig. = human)

↑ Capsule, of splenic nodules; ↑ Lymphatic nodules, solitary

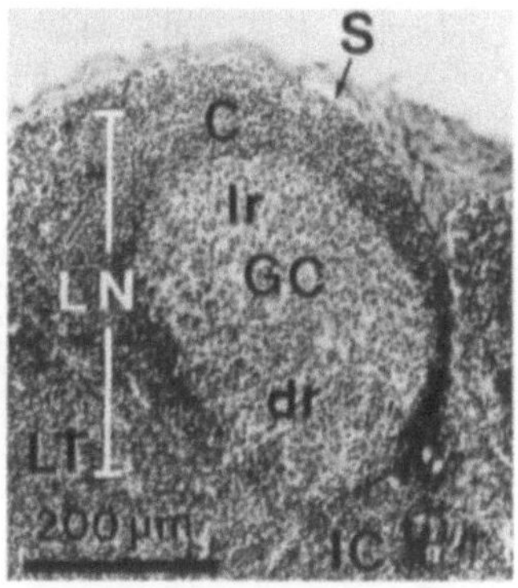

Lennert, K., Müller-Hermelink, K.H.: Verh. Anat. Ges. *69*, 19 (1975)

Lymphatic nodules, solitary (lymphonoduli solitarii*): Secondary ↑ lym-

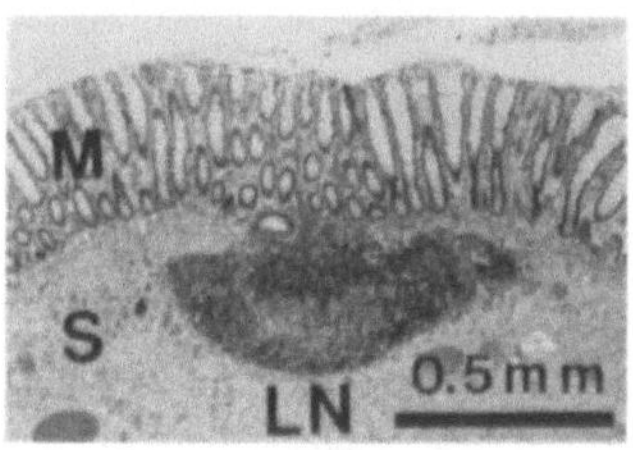

phatic nodules (LN) scattered in ↑ tunica mucosa (M) and ↑ tela submucosa (S) of various organs. (Fig. = ↑ colon, human)

Lymphatic nodules, tertiary: Agglomerations of ↑ T-lymphocytes (without ↑ germinal center) in inner ↑ cortex of ↑ lymph node.

Lymphatic organs (lymphoid organs, lymphoreticular organs): Organs formed completely of ↑ lymphatic tissue, such as ↑ lymph nodes and ↑ spleen.

Lymphatic organs, thymus-dependent zones of: Areas of ↑ lymphatic organs populated with ↑ T-lymphocytes = inner ↑ cortex of ↑ lymph nodes, central regions of ↑ periarterial lymphatic sheaths of ↑ spleen, and internodular region of ↑ Peyer's patches. A small number of T-lymphocytes are present in ↑ germinal centers of ↑ lymphatic nodules. After thymectomy, thymus-dependent zones are rapidly depleted of T-lymphocytes.

Lymphatic organs, thymus-independent zones of: Areas of ↑ lymphatic organs populated with ↑ B-lymphocytes = outer ↑ cortex and ↑ medullary cords of ↑ lymph nodes, ↑ marginal zone of ↑ white pulp of ↑ spleen, and cortex of ↑ lymphatic nodules.

Lymphatic ring, of pharynx (Waldeyer's throat ring, anulus lymphaticus pharyngis*): The ↑ lymphoepithelial organs surrounding beginning of ↑ gastrointestinal and respiratory tracts. To L. belong palatine (P), lingual (L), pharyngeal (Ph), and tubal (T) ↑ tonsils. L. is actively involved in early defense of organism.

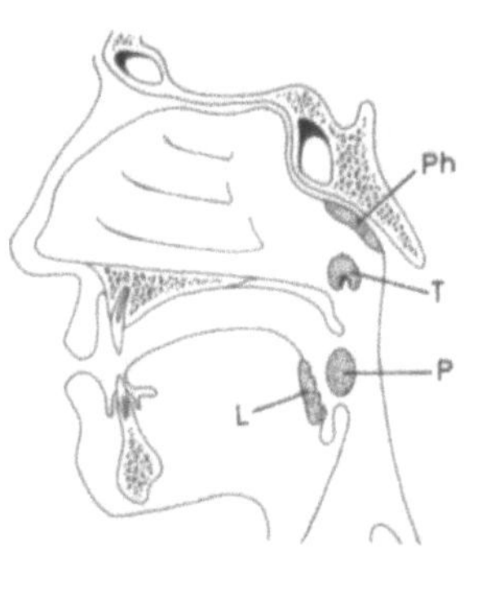

Lymphatic system (systema lymphaticum*): The tissues and organs involved in production of ↑ lymphocytes (↑ bone marrow, ↑ lymph nodes, ↑ tonsils, ↑ diffuse lymphatic tissue of ↑ gastrointestinal tract, ↑ spleen, ↑ thymus), conveying of ↑ lymph (↑ lymphatic vessels), and its filtration (lymph nodes, spleen). L. is defense system responsible for recognition and destruction of foreign material in course of ↑ immune response.

Lymphatic tissue (lymphoid tissue, lymphoreticular tissue): A category of ↑ reticular tissue specialized in production of ↑ B-lymphocytes. L. constitutes ↑ parenchyma of ↑ lymph nodes, ↑ spleen, and ↑ tonsils; it forms ↑ lymphatic nodules. L. is structurally identical to reticular tissue, but with an abundance of lymphopoietic cells and lymphocytes in its meshes.

Lymphatic tissue, diffuse: ↑ Diffuse lymphatic tissue, of gastrointestinal tract

Lymphatic tissue, of gastrointestinal tract: ↑ Diffuse lymphatic tissue, of gastrointestinal tract; ↑ Lymph nodules, solitary; ↑ Peyer's patches

Lymphatic vessels (vasa lymphatica*): A system of tubes carrying ↑ lymph from periphery of body to blood circulation. L. include: 1) ↑ Lymphatic capillaries; 2) collecting ↑ lymphatic vessels; 3) ↑ thoracic duct and right ↑ lymphatic duct.

Schwartz, C.J., Werthessen, N.T., Wolf, S. (eds.): Structure and Function of the Circulation. New York and London: Plenum Press 1981

Lymphatic vessels, collecting (vas lymphaticum myotypicum*): Musculoendothelial tubes (L) conveying ↑ lymph from ↑ lymphatic capillaries and draining it into ↑ thoracic duct. Structure: 1) L. below 0.2 mm in diameter have indistinctly delimited layers in their walls, in which predominate subendothelial, longitudinally oriented ↑ smooth muscle cells; ↑ valves are conspicuous. 2) L. over 0.2 mm in diameter display a tunica intima (I) with ↑ endothelium and subendothelial groups of longitudinal smooth muscle cells, a tunica media (M) with few circularly oriented smooth muscle cells, and a thin tunica adventitia (A). Valves are always present. (Figs. = human)

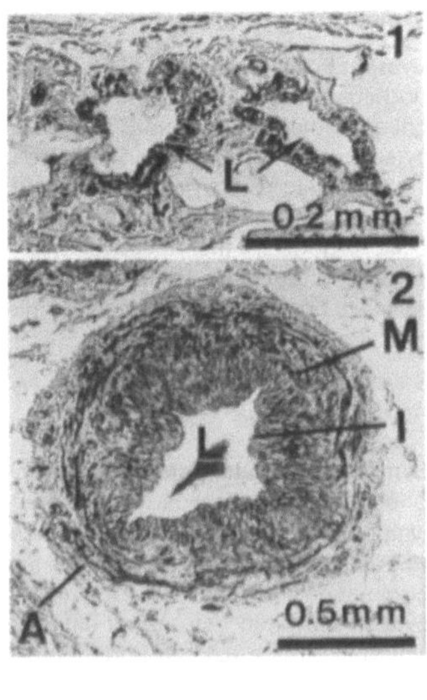

Lymphoblast: ↑ Immunoblast

Lymphocytelike cells, of bone marrow: ↑ Candidate stem cells

Lymphocrinia: Discharge of a ↑ hormone from ↑ endocrine cells or an ↑ endocrine gland into ↑ lymph.

Lymphocytes (lymphocytus*): Most common type of ↑ agranulocytes (25%–40% of circulating ↑ leukocytes). 1) Small L. = globular cells measuring 8 µm with a spherical nucleus containing very condensed ↑ heterochromatin and one to three nucleoli. Cytoplasmic rim, only 0.2–1 µm wide, surrounding nucleus contains occasional mitochondria, a small Golgi apparatus, ↑ centriole, some short rough endoplasmic cisternae, a few ↑ multivesicular bodies, and a considerable amount of free ribosomes, responsible for ↑ basophilia of L. In about 10% of small L., dense membrane-limited granules (0.2 µm in diameter) which appear azurophilic under ↑ light microscope are present; they correspond to primary ↑ lysosomes. Surface of small L. may be smooth or studded with numerous 1 to 2-µm-long microvilli. Small L. constitute majority of L. 2) Medium-sized L. = cells measuring up to 12 µm in diameter; a 2 to 3-µm-wide cytoplasmic rim surrounds nucleus. Ultrastructure of medium-sized L. is similar to that of small L. 3) Large L. = voluminous cells almost identical to ↑ immunoblasts; this type of L. does not reach the circulation under normal conditions. Although L. are mobile (migrating through epithelia, leaving bloodstream and ↑ lymph, wandering through connective tissues), they have only very weak phagocytotic activity. Small L. may transform into ↑ macrophages. Production sites: ↑ bone marrow, ↑ lymphatic organs, ↑ lymph nodes, ↑ spleen, solitary ↑ lymph nodules. According to their immunological capacity, L. are subdivided into ↑ B-lymphocytes and ↑ T-lymphocytes; no morphological differences exist between these two categories. Both are oxidase- and peroxidase-negative.

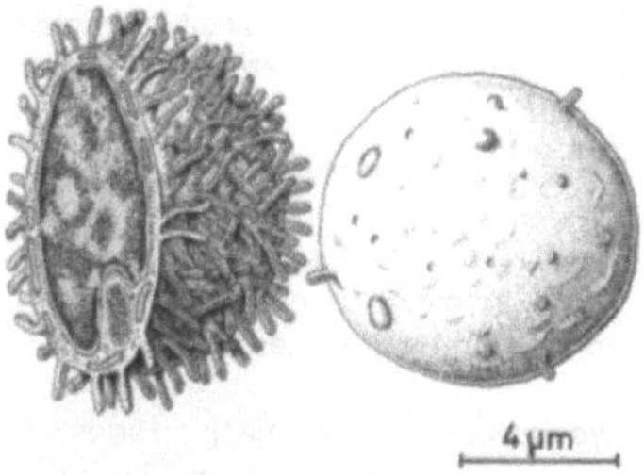

↑ Immune response

Bach, F., Bonavida, B., Vitetta, E., Fox, F.C.: T and B Lymphocytes: Recognition and Function. New York: Academic Press 1979; Bhalla, D.K., Braun, J., Karnovsky, M.J.: J. Cell Sci *39*, 137 (1979); Kataoka, K., Minowada, J.: Arch. Histol. Jpn. *42*, 355 (1979); Vos, J.G., Roholl, J.M., Leene, W.: Cell Tissue Res. *213*, 221 (1980)

Lymphocytes, activated: Immunocompetent ↑ lymphocytes which have recognized an antigen. Upon first contact with antigen, ↑ B-lymphocyte becomes activated under influence of ↑ T-lymphocyte (↑ helper cells), and T-lymphocyte under influence of ↑ macrophage. L. enlarge, chromatin loosens, ↑ ribosomes and ↑ polyribosomes abound in cytoplasm, and L. divide mitotically several times, forming ↑ clones and giving rise, via ↑ immunoblasts, to ↑ plasma cells and subpopulations of T-lymphocytes (helper cells, ↑ killer cells, ↑ suppressor cells). One kind of L. differentiates into recirculating ↑ memory cells ready to be directly activated by antigen.

Lymphocytes, antibody-secreting: ↑ B-lymphocytes; ↑ Lymphocytes

Lymphocytes B: ↑ B-lymphocytes

Lymphocytes, circulation of: Passage of lymphopoietic cells and ↑ lymphocytes between various ↑ lymphopoietic organs and tissues via blood and ↑ lymph. 1) Slow L. = movement of undifferentiated lymphocytes from ↑ bone marrow into bloodstream and from there into ↑ thymus and into ↑ bursa analogue to become mature immunocompetent ↑ B- and ↑ T-lymphocytes. After having reached the bloodstream again, B- and T-lymphocytes wander into tissues to be collected by afferent lymphatics and transported with lymph to ↑ lymph nodes. Via efferent lymphatics, B- and T-lymphocytes once again reach bloodstream, then tissues, and cycle repeats. Part of lymphocytes carried by blood are destined for lymphocyte recirculation through different organs. Slow L. requires several weeks. 2) Fast L. = ↑ lymphocytes, recirculation of. (Modified after Rogers 1983)

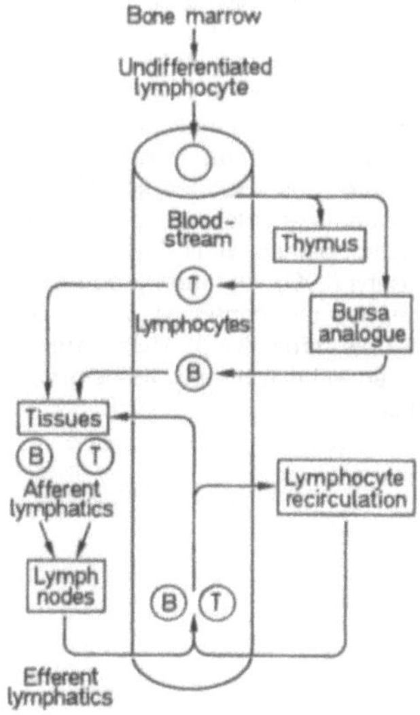

Rogers, A.W.: Cells and Tissues. London, New York: Academic Press 1983; Soussa, M. de: Lymphocyte Circulation. Experimental and Clinical Aspects. Chichester: John Wiley 1981

Lymphocytes, cytophysiological subdivision of: On the basis of their immunological capacities, lymphocytes are divided into ↑ B-lymphocytes and ↑ T-lymphocytes.

Lymphocytes, cytotoxic: ↑ Killer cells

Lymphocytes, immunocompetent: Mature, completely differentiated ↑ lymphocytes having matured either in ↑ bursa analogues (↑ B-lymphocytes) or in ↑ thymus (↑ T-lymphocytes).

Lymphocytes, killer: ↑ Killer cells

Lymphocytes, life cycle of: ↑ Lymphocyte recirculation; ↑ Lymphopoiesis

Lymphocytes, localization in bone marrow: In general, ↑ lymphocytes form groups around small radial arteries.

Lymphocytes, recirculation of: A fast migratory phenomenon of small ↑ B- and ↑ T-lymphocytes from blood to tissues and ↑ lymphopoietic organs and back into blood. Pathways: B- and T-lymphocytes circulating in tissues return to ↑ lymph nodes through afferent lymphatics (Aff), move through sinuses and ↑ parenchyma of nodes, and continue through efferent lymphatics (Eff) into ↑ thoracic duct, which discharges ↑ lymphocytes into blood. From blood, lymphocytes are distributed to tissues and lymphopoietic organs. In lymph nodes, ↑ lymphocytes cross wall of ↑ postcapillary venules (PcV) and enter ↑ lymphatic tissue, migrating through it to be again collected by efferent lymphatics. In ↑ spleen, lymphocytes also circulate through parenchyma and leave organ via its lymphatics and blood vessels. Although ↑ thymus produces T-lymphocytes, they do not recirculate through its parenchyma, but enter blood circulation directly. L. through blood takes about 0.6 h; transit through spleen requires about 6 h, and passage through lymph nodes 15–20 h. During this period, lymphocytes do not divide. Large lymphocytes (= ↑ immunoblasts) do not recirculate. The purpose of L. is constant patrolling of immunocompetent ↑ lymphocytes throughout organism and informing lymphopoietic organs about presence or absence of antigens in body. In presence of antigens, some lymphocytes settle down in lymphopoietic organs, begin to divide, and trigger ↑ immune response. Those that are lost during L. are compensated via slow lymphocyte circulation by means of production in bone marrow, thymus, and lymphatic tissue.

↑ Lymphocytes, circulation of

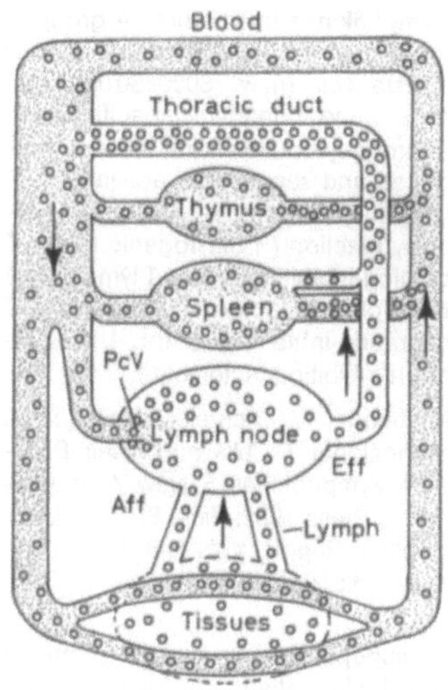

Ford, W., Smith, M.E.: Lymphocyte Recirculation Between the Spleen and Blood. In: Role of the Spleen in the Immunology of Parasitic Diseases. Tropical Diseases Research Series No 1. Basel: Schwabe & Co 1978; Mitchell, J.: Immunology *24*, 93 (1973)

Lymphocytes, suppressor: ↑ Immune response; ↑ Lymphocytes, activated; ↑ T-lymphocytes

Lymphocytes T: ↑ T-lymphocytes

Lymphocyte transforming factor (LTF): A ↑ lymphokine secreted by activated ↑ T-lymphocytes stimulating transformation and clonal multiplication of nonsensitized lymphocytes into ↑ immunoblasts.

Lymphoepithelial organs: A group of organs in which ↑ lymphocytes intimately contact or colonize an ↑ epithelium. L. include all ↑ tonsils, ↑ Peyer's patches, ↑ appendix, and ↑ thymus. L. are actively involved in immunological defense.

Lymphoepithelial tissue (lymphoepithelium): An epithelium normally associated with or colonized by ↑ lymphocytes.

↑ Diapedesis; ↑ Lymphoepithelial organs

Lymphogram: The X-ray of ↑ lymphatic vessels and ↑ lymph nodes injected with a nontoxic, radiopaque substance.

Lymphoid nodules: ↑ Lymphatic nodules

Lymphoid tissue: ↑ Lymphatic tissue

Lymphokines (mediators): A group of more than ten soluble proteinaceous substances (m.w. 8000–80000 daltons) without antigenic specificity and ↑ immunoglobulin characteristics, produced and secreted by activated ↑ T-lymphocytes at site of antigen-antibody reaction (↑ blastogenic factor, ↑ cloning-inhibiting factor, ↑ lymphocyte transforming factor, ↑ lymphotoxin, ↑ migration-inhibiting factor, ↑ proliferation-inhibiting factor, etc.).

Feldmann, M., Schreier, M.H. (eds.): Monoclonal T Cells and Their Products. Lymphokines, 5. New York: Academic Press 1982; Pick, E. (ed.): Lymphokine Report, Vols. 1–8. New York: Academic Press 1980–83

Lymphopoiesis: Process of formation and ↑ differentiation of ↑ lymphocytes. From ↑ totipotent hematopoietic stem cells (THSC) arise, by ↑ mitosis in ↑ bone marrow, B- and T-immunoblasts, which give immature large ↑ B- and ↑ T-lymphocytes by division. Further mitoses give rise to medium-sized lymphocytes and finally to small B- and T-lymphocytes. Very early in this phase lymphocytes are able to recognize antigens. However, to complete their differentiation and maturation, they must pass into blood and reach ↑ bursa analogues (BA) and ↑ thymus (T), respectively, where they become mature immunocompetent ↑ lymphocytes. Once more they pass into blood and wander into ↑ lymphatic tissue. Following contact with an antigen (Ag) during ↑ immune response, small lymphocytes differentiate into corresponding ↑ immunoblasts within ↑ lymphopoietic organs (LyO). Although production of lymphocytes continues within bone marrow, most lymphocytes are produced within lymphatic tissue and thymus.

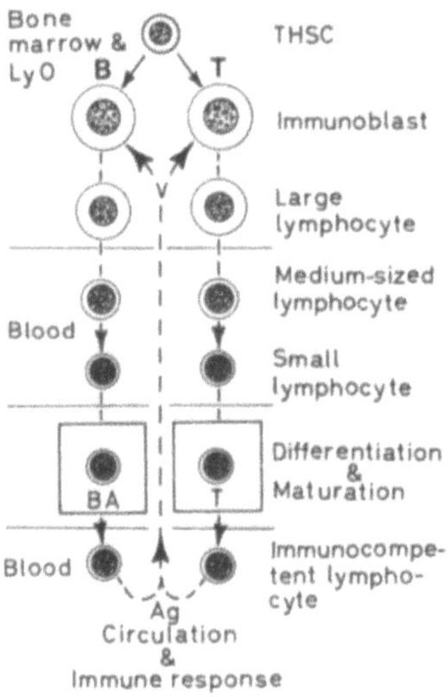

↑ Hematopoiesis, current view of; ↑ Lymphocytes, circulation of

Müller-Hermelink, H.K., Gaudecker, B. v.: Verh. Anat. Ges. *74*, 235 (1981)

Lymphopoietic organs: All tissues and organs in which ↑ lymphopoiesis takes place (↑ bone marrow, ↑ lymphatic tissue, ↑ lymphoepithelial organs, ↑ lymphatic organs).

Lymphoreticular organs: ↑ Lymphatic organs

Lymphoreticular tissue: ↑ Lymphatic tissue

Lymphotoxin (LT): A ↑ lymphokine secreted by ↑ killer cells causing cell ↑ lysis.

Lyoglycogen: A water-soluble and directly stainable fraction of ↑ glycogen, making up about 20% of its total amount.

Lysis: Liquefaction of a cell, generally provoked by various enzymes and/or toxins.

↑ Autolysis; ↑ Karyolysis; ↑ Lymphotoxin

Lysolecithin: A chemical substance found in membranes of ↑ catecholamine storage granules.

Lysosomes (lysosoma*): Digestive spherical ↑ organelles of cell containing large number of hydrolytic enzymes. Morphofunctionally, L. can be subdivided into primary lysosomes (1) and secondary lysosomes (2). Primary L. are lined by a ↑ unit membrane about 6 nm thick; they measure from 25 nm to 0.5 µm in diameter and contain a homogeneous, moderately osmiophilic, fine granular material. Primary L. are not involved in digestive activity; they transport hydrolytic enzymes to ↑ phagolytic vacuoles. Secondary L. measure from 0.5 to 1.5 µm in diameter and are enclosed by a unit membrane of same thickness as that of primary L. Secondary L. display heterogeneous contents composed of some moderately osmiophilic material, irregular, very dense particles, highly electron-dense lamellae, vesicles, and ↑ lipid droplets (arrows). Secondary L. have already been engaged in the digestive process: Their contents represent remnants of undigested material. (Fig. = ↑ interstitial cells, of testis, human)

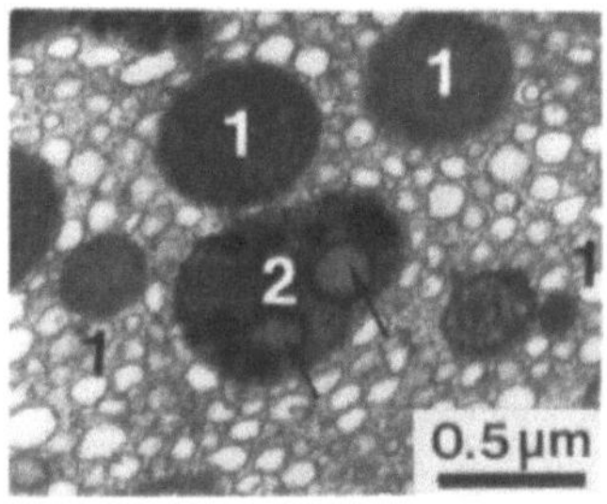

↑ Lipofuscin; ↑ Lysosomes, enzymes of; ↑ Residual bodies

Bainton, D.F.: J. Cell Biol. *91*, 66s (1981); Dingle, J.T. (ed.): Lysosomes in Biology and Pathology. Vols. 1–6. Amsterdam: Elsevier/North Holland 1979; Neufeld, E.F., Sando, N.G., Garvin, A.J., Rome, L.H.: J. Supramol. Struct. *6*, 95 (1977)

Lysosomes, autolytic: ↑ Autophagosomes

Lysosomes, enzymes of: Among numerous other enzymes (acid phosphatase, its phosphate esters including mononucleotides, etc.), ↑ lysosomes contain acid ribonuclease- and acid deoxyribonuclease-hydrolyzing nucleic acids, phosphoprotein phosphatase-hydrolyzing phosphoproteins, cathepsin- and collagenase-hydrolyzing proteins, α-glucosidase, β-N-acetylglucosaminidase, β-glucuronidase, β-galactosidase, α-mannosidase, and aryl-sulfatase hydrolyzing ↑ proteoglycans. The present list is valid for liver lysosomes; in other cells, enzymes of lysosomes may be different. Under the action of inadequate osmotic pressure, freezing and thawing, sonic vibrations, lecithinase, proteases, acid pH, high temperature, detergents, etc., the lysosomal limiting membrane becomes injured and enzymes penetrate cytoplasm, leading to ↑ autolysis. ↑ Corticoids stabilize lysosomal limiting membrane.

Gahan, P.B.: Int. Rev. Cytol. *21*, 1 (1967); Holtzmann, E.: Lysosomes. A Survey. Wien, New York: Springer-Verlag 1975

Lysosomes, function of: Polypeptide chains of lysosomal enzymes are synthesized at ↑ ribosomes of rough endoplasmic reticulum (ER), shed in its cisternae, and transferred in ↑ transport vesicles (TV) to ↑ Golgi apparatus (G). Here, synthesis is accomplished, and primary ↑ lysosomes (PLy) separate from Golgi apparatus. In the case of ↑ phagocytosis of foreign material (e.g., bacteria; B), primary lysosomes fuse (arrowheads) with ↑ phagolytic vacuole (PhV) and discharge their enzymes; the vacuole transforms into a ↑ heterophagosome (H) and its contents become degraded. Resulting structure is secondary lysosome (SLy). When areas of cytoplasm with ↑ organelles, ↑ inclusions, or ↑ secretory granules (SG) become sequestered, resulting autophagic vacuole (AV) transforms into ↑ autophagosome (A) after having been filled with enzymes of primary lysosomes. At end of digestion, autophagosome transforms into secondary lysosome. Contents of ↑ pinosomes or fused ↑ micropinocytotic vesicles (MV) can be digested, also leading to formation of a secondary lysosome. In further transformation, a secondary lysosome condenses and becomes ↑ residual body (RB), which can be stocked in cell as ↑ lipofuscin or discharged from cell by ↑ exocytosis (Ex).

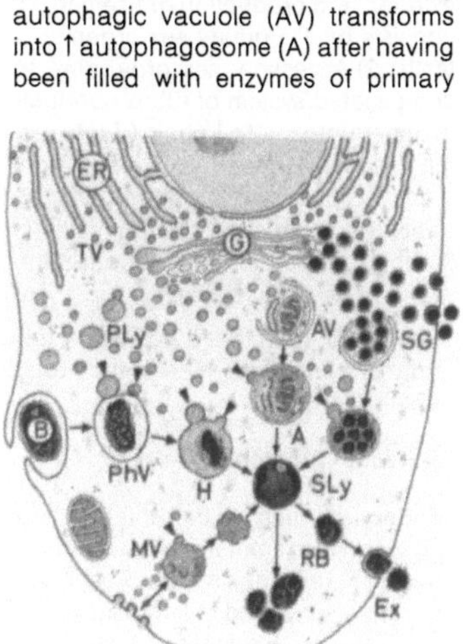

Aterman, K.: Histochem J. *11*, 503 (1979); De Duve, C.: Recherche *49*, 815 (1974); Pitt, D.: Lysosomes and Cell Function. London, New York: Longman 1975

Lysosomes, heterolytic: ↑ Heterophagosome

Lysosomes, primary: ↑ Lysosomes

Lysosomes, secondary: ↑ Lysosomes

Lysozyme: A bacteriolytic substance hydrolyzing glycosides in the cell wall of bacteria. L. is contained in specific granules of neutrophilic ↑ granulocytes, secretory granules of ↑ Paneth's cells, and in ↑ tears.

M

Macroblast: ↑ Erythroblast, polychromatophilic

Macrocytes: Pathological ↑ erythrocytes, larger than normal.

Macroglia: A collective term for all varieties of ↑ astrocytes and ↑ oligodendrocytes.

Macrometer: A mechanism for long-range movements of body tube and for coarse adjustment of focus in ↑ light microscopes.

Macrophage migration-inhibiting factor: ↑ Migration-inhibiting factor

Macrophage system: A historical term for ↑ reticuloendothelial system.

Macrophagelike cells: ↑ Synovial cells

Macrophages: A class of relatively voluminous cells capable of phagocytizing large particles, such as entire cells or cell debris. ↑ Alveolar macrophages, fixed and wandering ↑ histiocytes, ↑ Kupffer's cells, ↑ Kolmer's cells, ↑ microglia, ↑ monocytes, and ↑ reticular cells are all considered M. M. make up part of ↑ mononuclear phagocyte system. M. play an important role in regulation of ↑ immune response.

↑ Phagocytosis

Carr, I.: The Macrophage: A Review of Ultrastructure and Function. London: Academic Press 1973; Normann, J.S. (ed.): Macrophages and Natural Killer Cells. New York: Plenum Publ. Co. 1982; Polliack, A., Gordon, S.: Lab. Invest. *33*, 469 (1974); Vray, B., Saint-Guillain, M.L., Leloup, R., Hoebke, J.: J. Reticuloendothel. Soc. *29*, 307 (1981); Unanue, E.R., Rosenthal, A.S. (ed.): Macrophage Regulation of Immunity. New York: Academic Press 1980

Macrophages, alveolar: ↑ Alveolar macrophages

Macrophagocytosis: A process whereby whole cells, cell debris, or voluminous foreign particles are engulfed into ↑ macrophages by massive ↑ pseudopodia.

↑ Phagocytosis

Macropinocytosis: The interiorization by short cell processes of dissolved substances contained in ↑ pinosomes (P) into cell body. Pinosomes move toward ↑ Golgi apparatus (G) where their

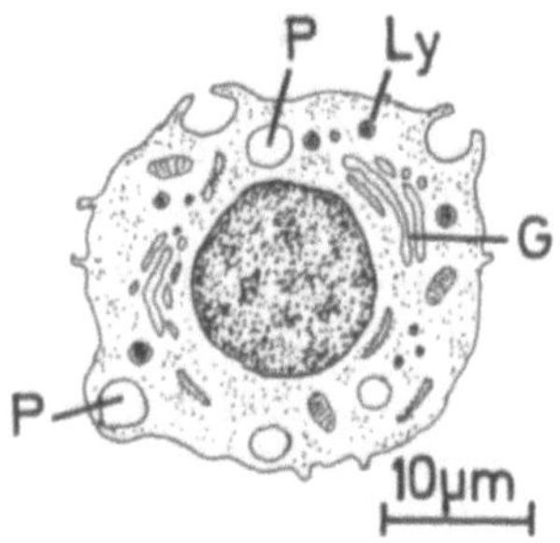

contents become digested by ↑ lysosomes (Ly). M. is visible with light microscope.

Macrovasculature: All blood vessels larger than 100 µm, i.e., visible with naked eye. ↑ Arteries and ↑ veins are considered M.

Macula adherens: ↑ Desmosomes

Macula cones: Rodlike elongated cones present in ↑ fovea centralis.

↑ Photoreceptors

Macula densa (pars maculata*): Part of ↑ juxtaglomerular apparatus consisting of a zone of 15–40 specialized epithelial cells (MD) in ↑ distal tubule wall of ↑ nephron lying over ↑ vascular pole of ↑ renal corpuscle. Cells of M. are cuboidal or columnar with many lateral interdigitations and short apical microvilli. Nuclei of M. cells are closer together than in neighboring nonspecific cells; they are relatively voluminous with prominent nucleoli. Mitochondria (M) are short and scattered in ↑ apical pole; Golgi apparatus is small and located mainly in upper two–thirds of cell body; rough endoplasmic cisternae are short and flattened. M. cells lie on an incomplete basal lamina; their basal poles are occupied by a moderately developed ↑ basal labyrinth. Some M. cells contact extraglomerular ↑ mesangial cells (MC); others contact ↑ juxtaglomerular cells (JC) through basal lamina. Function of M. is not fully understood: It is thought that M. serves as an ↑ osmoreceptor; all changes in sodium concentration in distal tubule are transmitted to juxtaglomerular cells, which increase or decrease ↑ renin production as a result. (Fig. = rat)

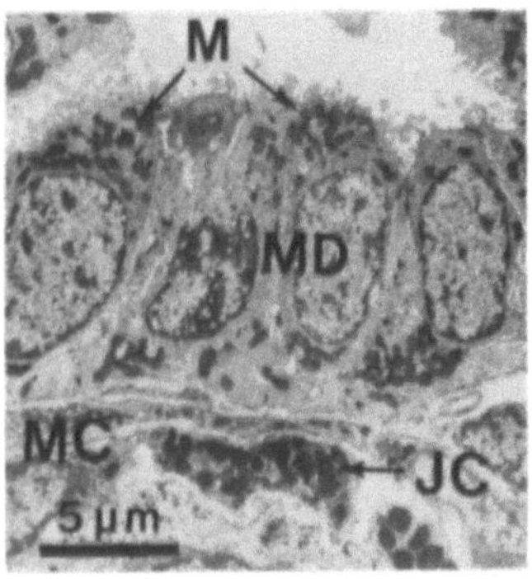

Isler, H., Krstić, R.: Arch. Histol. Jpn. *44*, 15 (1981); Schiller, A., Taugner, R.: Cell Tissue Res. *200*, 337 (1979); Sikri, K.L., Foster, C.L.: J. Anat. *132*, 57 (1981)

Macula lutea (macula*): A round, shallow depression (ML) of vitreal surface of ↑ retina (R), 1.5 mm in diameter, located about 4 mm temporally from ↑ optic papilla (OP). Macroscopically, M. appears as a yellow spot because of local accumulation of ↑ carotenoids. Margins of M. are thickened as a consequence of accumulation of ↑ ganglion cells (GC). Bottom of M. is ↑ fovea centralis (arrow).

↑ Ganglion cell layer, of retina

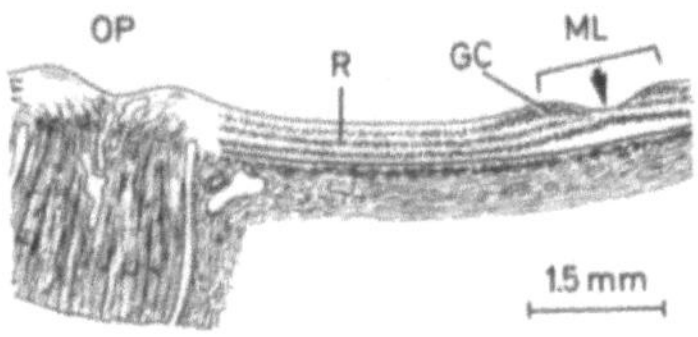

Macula occludens: An occasional circumscribed area of fused cell membranes, i.e., obliterated intercellular spaces, between two contiguous ↑ cardiac muscle cells at level of ↑ intercalated disc.

Macula, of saccule and utricle (macula sacculi et utriculi*): Well-delimited neuroepithelial areas, 4–6 mm^2, making up part of epithelium of ↑ saccule and ↑ utricle. Structure: 1) ↑ Neuroepithelium (Ne) = ↑ vestibular cells types I and II (VC I, VC II) and ↑ supporting cells (SC). Sensory ↑ hairs (H) of vestibular cells penetrate ↑ otolithic membrane (OM), which bears ↑ otoconia (O) on its surface. 2) Lamina propria (LP) = a thickened and well-vascularized ↑ endosteum of bony ↑ labyrinth transversed by many nerve fibers (NF) in contact with vestibular cells. Since otoconia have a specific gravity considerably greater than ↑ endolymph, every linear acceleration pulls hairs and alters activity of vestibular cells and nerve fibers.

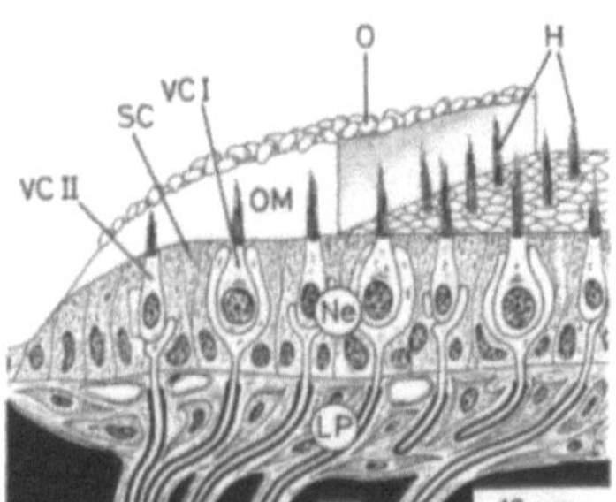

Harada, Y.: Biomed. Res. *2*, 391 (1981)

Macula pellucida: ↑ Stigma

MAG: ↑ Myelin-associated glycoprotein

Magnification, microscopic: The ratio of image size to object size. 1) ↑ Light microscope = a product of magnification (M) of the ↑ objective (MOb) and magnification of the ↑ ocular (MOc), i.e., M = MOb × MOc. 2) ↑ Transmission electron microscope = a product of magnification of the objective (MOb) and magnification of the ↑ projective (MPr), i.e., M = MOb × MPr. 3) ↑ Scanning electron microscope = a quotient between length of line that secondary electron beam (L) scans on screen of TV tube and length of line that primary electron beam (l) scans on surface of the object, i.e., M = L/l.

Major compartment, of spermatozoon tail: ↑ Spermatozoon, tail of

Major dense line: ↑ Myelin sheath

Malassez's rests: The former term for ↑ cementicles.

Male-determining chromosome: ↑ Chromosome, male-determining

Male reproductive system: A group of organs responsible for production of ↑ spermatozoa, their transport to the exterior, and production of ↑ semen. M. consists of: 1) Primary sex organ = ↑ testis; 2) Accessory sex organs = a) complicated system of tubes conducting spermatozoa to ↑ penis (↑ rete testis, ↑ ductuli efferentes, ↑ ejaculatory duct); b) copulatory organ = penis; c) a set of organs producing seminal fluid (↑ prostate, ↑ bulbourethral glands, ↑ seminal vesicles).

Brandes, D. (ed.): Male Accessory Sex Organs. New York: Academic Press 1974

Malleus: ↑ Auditory ossicles

Mallory staining: A trichrome ↑ staining technique combining acid fuchsin, orange G, and anilin blue. Result: nuclei are stained red, cytoplasm pink, ↑ collagen and ↑ reticular fibers and ↑ mucus blue.

Mall's space, of liver: ↑ Periportal tissue space

Malpighian corpuscles, of kidney: ↑ Renal corpuscles

Malpighian corpuscles, of spleen: ↑ Splenic nodules

Malpighian layer, of epidermis (stratum germinativum): A common term for stratum basale and stratum spinosum of ↑ epidermis.

Malpighian system: ↑ Keratinizing system

Mammary gland (breast, glandula mammaria*): A large compound ↑ tubuloalveolar gland composed of 15–25 irregular lobes. General structure: 1) ↑ Stroma: a) ↑ dense connective and adipose tissue enclose all ↑ lobes (L) and ↑ lactiferous ducts (LD), forming interlobar septa (S), b) ↑ loose connective tissue surrounds ↑ lobules (Lo) and ↑ intralobular ducts penetrating between ↑ tubuloalveoli. 2) ↑ Parenchyma = a) tubuloalveoli grouped in lobules; several lobules constitute a lobe, each provided by a lactiferous duct, which opens at tip of ↑ nipple (N); tubuloalveoli are surrounded by ↑ myoepithelial cells; b) excretory ducts = each lobule has an intralobular duct, which empties into an ↑ interlobular duct (ILD); several interlobular ducts merge to form a lactiferous duct. Beneath ↑ areola (A), lactiferous ducts become slightly enlarged to form ↑ lactiferous sinuses (LS). In the course of life, M. changes its histological structure; thus, one can distinguish following phases: 1) ↑ M. before puberty; 2) ↑ M. in puberty; 3) M. in an adult woman = a) ↑ M., resting, b) ↑ M. in pregnancy, c) ↑ M. in lactation, 4) ↑ M. in senile involution. Sk = ↑ skin, M = muscles

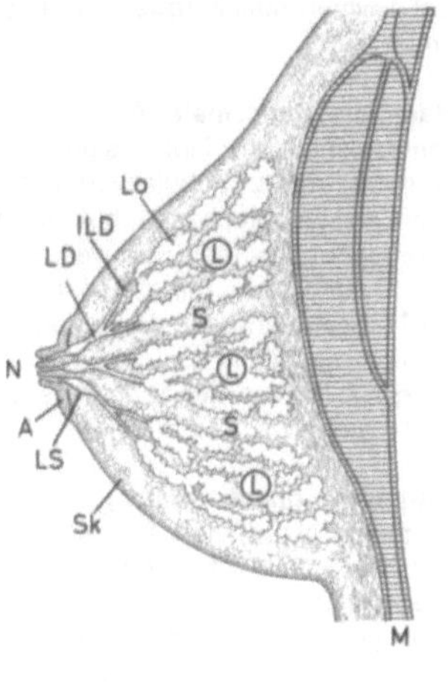

Cowie, A.T.: J. Invest. Dermatol. *63*, 2 (1974); Hallowes, R.C., Peachey, L.A.: The Mammary Gland and Human Breast. In: Hodges, G.M. and Hallowes, R.C. (eds.): Biomedical Research Applications of Scanning Electron Microscopy, Vol. 2. London, New York: Academic Press 1980

Mammary gland, areola of: ↑ Areola, of mammary gland

Mammary gland, before puberty: M. is composed of only a few ↑ lobules (L) consisting of short, poorly branched, canalized and/or solid epithelial strands (S). ↑ Lactiferous ducts (LD) are very short; ↑ stroma is relatively abundant and contains small groups of ↑ adipose cells. (Fig. = human newborn)

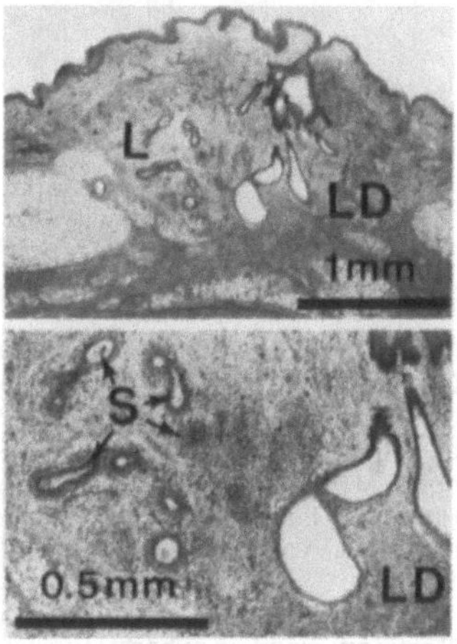

Mammary gland, endocrine control of: 1) ↑ Mammogenesis = during puberty, ↑ estrogens are responsible for growth of excretory ducts; ↑ progesterone stimulates development of ↑ tubuloalveoli. Small amounts of ↑ prolactin (LTH), ↑ growth hormone (STH), ↑ adrenocorticotropic hormone (ACTH), ↑ corticoids, ↑ insulin, ↑ thyroxin, and ↑ parathormone are also needed. 2) ↑ Lactation = all cited hormones are also needed in this phase. Estrogens and progesterone are predominantly of placental origin. Lactation begins at end of pregnancy, but ↑ milk starts to flow only after parturition, probably due to absence of ↑ placenta and a consecutive decrease in estrogens and progesterone levels, which renders mammary gland more sensitive to LTH. 3) ↑ Galactopoiesis = a predominant influence of LTH, thyroxin, and, especially, STH. ↑ Suckling reflex stimulates, via a neurosecretory pathway, production of ↑ prolactin-releasing factor, LTH, ACTH, and ↑ oxytocin; the latter provokes contraction of ↑ myoepithelial cells, which facilitates discharge of milk. There is practically no estrogens and progesterone influence.

Mammary gland, epithelium of: A simple epithelium composed of secretory cells (SC) and myoepithelial cells (MC). 1) Secretory or alveolar cells (lactocytus*) = poorly delimited asymmetrical cuboidal or prismatic cells with a round or elliptical nucleus and a conspicuous nucleolus. Cytoplasm contains numerous mitochondria, extensive ↑ ergastoplasm (E), and a large Golgi apparatus (G). During ↑ lactation, vacuoles separate along with ↑ casein granules (C) from Golgi apparatus. ↑ Lipid droplets (L) occur, bulging into lumen, and finally separate from cell, still surrounded by a thin cytoplasmic rim (R). Cells delimit deep intercellular ↑ canaliculi (arrow) in which are discharged the majority of casein granules. 2) ↑ Myoepithelial cells (myoepitheliocytus stellatus*) = spindle-shaped or stellate cells inserted between bases of secretory cells. Their contraction pushes ↑ milk toward excretory canalicular system.

↑ Suckling reflex

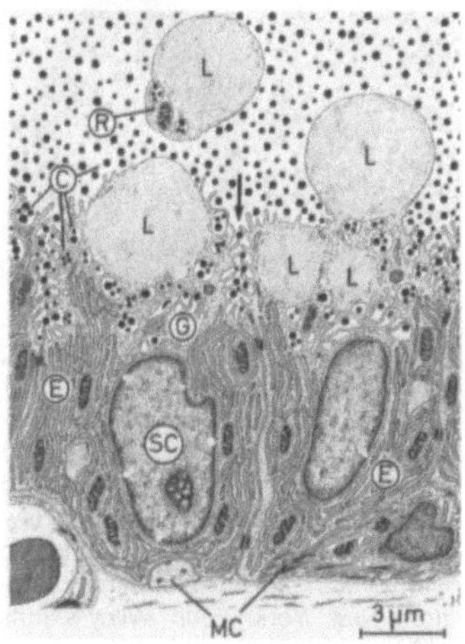

Brooker, B.E.: Cell Tissue Res. *229*, 639 (1983); Hollmann, K.H.: Cytology and Fine Structure of the Mammary Gland. In: Larson, B.L., Smith, V.R. (eds.): Lactation, Vol. 1. New York: Academic Press 1974; Villa-Porcille, E., Olivier, L.: Biol. Cell. *36*, 155 (1979)

Mammary gland, female: ↑ Mammary gland

Mammary gland, in lactation: A considerable increase in number and size of ↑ tubuloalveoli (T), gives a spongy appearance to ↑ lobules (L). In some tubuloalveoli, epithelium is low cuboidal, in others columnar; large ↑ lipid droplets (arrowheads) are visible in glandular cells. Cells display a marked increase in number and size of all ↑ organelles and RNA content. Increase in volume of lobules reduces septa (S) to narrow connective strands. ↑ Loose connective tissue between glandular elements is infiltrated with ↑ lymphocytes, ↑ plasma cells, and eosinophilic ↑ granulocytes. Some ↑ macrophages penetrate lumen of tubuloalveoli and transform into ↑ colostrum bodies. (Figs. = human)

↑ Lactation

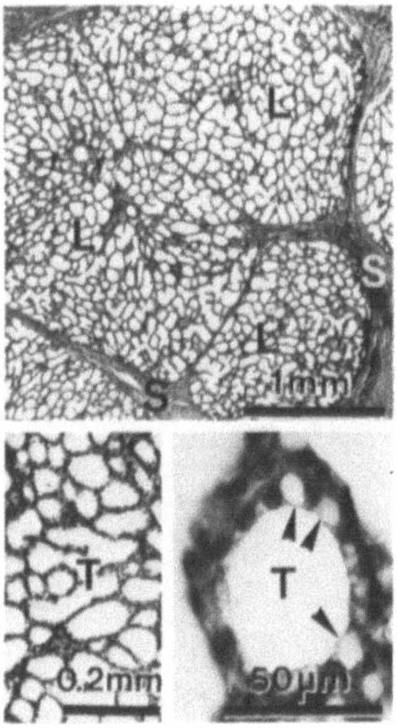

Murakami, M., Shimada, T., Nishida, T., Sakima, M.: Arch. Histol. Jpn. *40*, 421 (1977); Salazar, H., Tobon, H.: Morphologic Changes of the Mammary Gland During Development, Pregnancy and Lactation. In: Josimovitch, J.B., Reynolds, M., Cobo, E. (eds.): Lactogenic Hormones, Fetal Nutrition, and Lactation. New York: John Wiley & Sons 1974

Mammary gland, in pregnancy: Under influence of ↑ progesterone, ↑ estrogens, ↑ growth hormone, ↑ corticoids, and ↑ prolactin, multiplication and branching of ↑ tubuloalveoli begins, which initially remain small and collapsed. Gradually, ↑ lobules (L) increase in volume with a parallel reduction in ↑ stroma. Intralobular ↑ loose connective tissue becomes infiltrated with ↑ lymphocytes, eosinophilic ↑ granulocytes, and ↑ plasma cells; secretory epithelium is flattened. Toward end of pregnancy, tubuloalveoli (T) expand, their epithelium becomes cuboidal and starts to produce ↑ colostrum. Stroma is reduced to relatively narrow septa (S). Arrowheads = ↑ intralobular ducts (Figs. = human, last month of pregnancy)

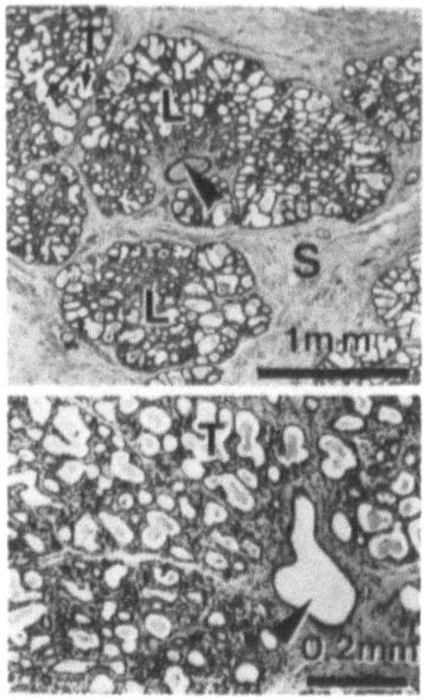

Salazar, H., Tobon, H.: Morphologic Changes of the Mammary Gland During Development, Pregnancy and Lactation. In: Josimovitch, J.B., Reynolds, M., Cobo, E. (eds.): Lactogenic Hormones, Fetal Nutrition, and Lactation. New York: John Wiley & Sons 1974

Mammary gland, in puberty: Proliferation, growth, and branching of excretory duct system with establishment of ↑ ovarian cycle. With successive cycles, ↑ lactiferous ducts elongate, mainly under influence of ↑ estrogens. ↑ Tubuloalveoli moderately proliferate under influence of ↑ progesterone, but regress in next follicular phase of cycle. Connective ↑ stroma, rich in adipose cells, increases considerably and furnishes bulk of M.

Graumann, W.: Z. mikrosk.-anat. Forsch. *59*, 523 (1953)

Mammary gland, in senile involution: Toward and after ↑ menopause, ↑ parenchyma gradually reduces, ↑ tubuloalveoli almost totally disappear, and only a small part of excretory system persists. Epithelium of remaining tubuloalveoli often become pluristratified, some excretory ducts enlarge cystically, and others become completely obliterated. Blood vessels and part of ↑ stroma display signs of hyaline degeneration in parallel with a certain numerical increase in ↑ elastic fibers and, frequently, ↑ adipose tissue.

Mammary gland, inactive: ↑ Mammary gland, resting

Mammary gland, innervation of: 1) Somatic sensory nerve fibers = many free nerve endings and Meissner's ↑ corpuscles in skin of ↑ nipple, end corpuscles resembling Merkel's and Krause's ↑ corpuscles beyond ↑ areola, Pacinian ↑ corpuscles in interior of ↑ parenchyma. Receptors for intramammary pressure probably exist. 2) Sympathetic nerve fibers = nonmyelinated nerve endings for innervation of ↑ smooth muscle cells of areola, nipple, and tunica media of arteries. Sensory innervation of M. is responsible for establishment of ↑ suckling reflex.

Mammary gland, involution of: With cessation of nursing, remaining ↑ milk particles are phagocytized by ↑ macrophages and transported into lymphatic vessels. ↑ Autophagy of ↑ casein granules, ↑ lipid droplets, and cell ↑ organelles occurs in epithelial cells; numerous cells degenerate and disintegrate. Cell debris is removed by macrophages. ↑ Tubuloalveoli gradually become narrower and rapidly diminish in number, so that structure of gland resembles resting ↑ mammary gland. However, excretory duct system and ↑ myoepithelial cells are not reduced in number and structure. Regressive changes in mammary gland are provoked by increasing level of ↑ estrogens and reestablishment of ↑ ovarian cycle.

↑ Involution

Mammary gland, lymph circulation of: There are no intralobular lymphatic vessels, they are arranged around ↑ lobules and run through septa and direct ↑ lymph mainly toward axillary ↑ lymph nodes.

Mammary gland, male: A small gland composed of a few ↑ lobules and short excretory ducts. Lobules consist of short, mostly collapsed ↑ tubuloalveoli. Between lobules there is a well-developed ↑ stroma and there is an interstitial ↑ loose connective tissue between tubuloalveoli. Only a quantitative difference exists between M. and the resting ↑ mammary gland of female.

Mammary gland, resting (inactive mammary gland): The ↑ parenchyma consists mainly of ↑ intralobular, ↑ interlobular, and ↑ lactiferous ducts (LD) with a small number of poorly branched, collapsed ↑ tubuloalveoli lined with ↑ simple cuboidal epithelium. ↑ Lobules (L) are small. ↑ Stroma largely predominates and consists of wide, ↑ dense connective tissue strands (S) rich in ↑ adipose tissue (A). A ↑ loose connective tissue fills interspaces between parenchymal elements of lobules. (Figs. = human)

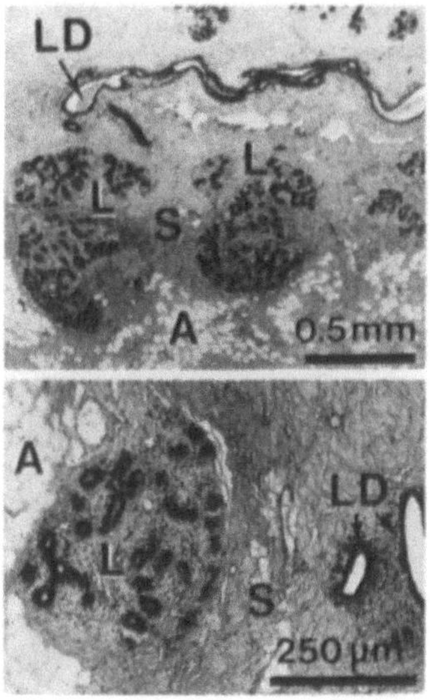

Mammary gland, vascularization of: The mammary gland is supplied with blood from intercostal, lateral thoracic, and internal mammary arteries. During pregnancy and ↑ lactation, blood supply increases with a parallel enlargement of intralobular capillary network. A system of subcutaneous veins becomes evident during pregnancy.

Mammary papilla: ↑ Nipple

Mammogenesis: The process of formation of ↑ mammary gland during puberty.

↑ Mammary gland, before puberty; ↑ Mammary gland, in puberty

Mammotropes (lactotropes, LTH-cells, prolactin cells, cellula mammotrophica*): Round, small acidophilic cells predominantly located in posterolateral parts of pars distalis of ↑ hypophysis. Nucleus is spherical with a distinct nucleolus. In inactive M., mitochondria (M) are few, Golgi apparatus (G) small, rough endoplasmic (rER) cisternae short and flattened, and ↑ lysosomes (Ly) occasionally occur. The ↑ unit membrane-bound ↑ secretory granules (SG), measuring 60–90 nm, arise from Golgi apparatus; they are relatively sparse in resting M. Number and size of M. greatly increase during pregnancy and ↑ lactation; Golgi apparatus enlarges, endoplasmic cisternae become longer, and secretory granules more numerous. M. produce ↑ prolactin; after suckling period, ↑ lysosomes increase in number and eliminate excess ↑ organelles and secretory granules by ↑ autophagy and ↑ crinophagy, respectively.

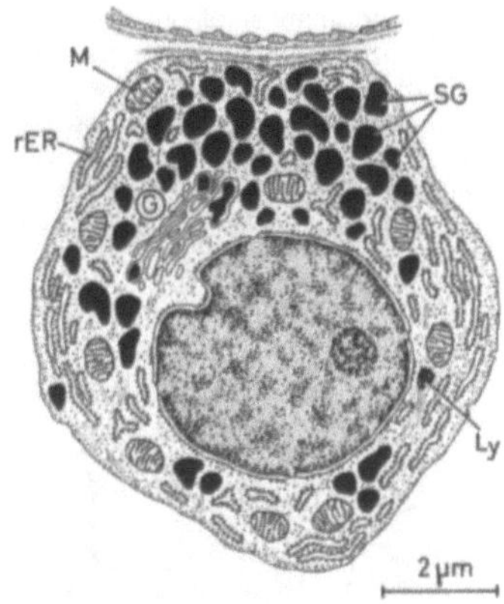

Dacheux, F.: Cell Tissue Res. *207*, 277 (1980); Harigaya, T., Kohmoto, K., Hoshino, K.: Acta Histochem. Cytochem. *16*, 51 (1983); Osamura, R.Y., Komatsu, N., Izumi, S., Yoshimura, S., Watanabe, K.: J. Histochem. Cytochem. *30*, 919 (1982); Smith, R., Farquhar, M.: J. Cell Biol. *31*, 319 (1966)

Mammotrophic hormone: ↑ Prolactin

Manchette, of spermatid: A conical aggregation of ↑ microtubules (M) around nucleus (N) of a ↑ spermatid occurring in the course of its elongation. M. is attached to posterior margin of ↑ acrosomal cap. In caudal portion of cell, M. has the form of a tube and is therefore termed caudal tube. Function of M. is unknown: It is thought to play a role in reshaping of spermatid nucleus. (Fig. = human)

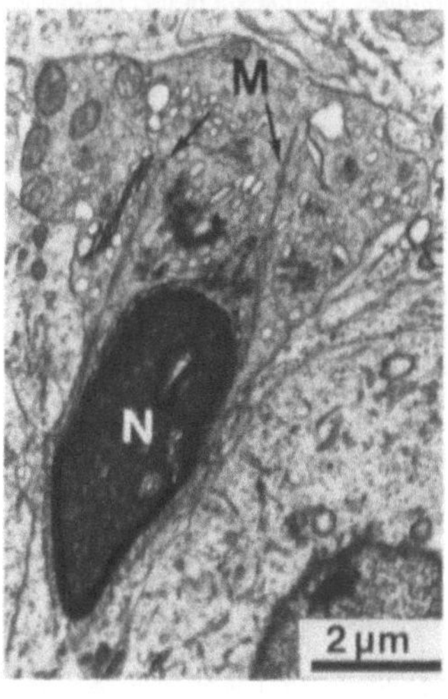

↑ Spermiogenesis

Mac Kinnon, E.A., Abraham, P.J., Svatek, A.: Z. Zellforsch. *136*, 447 (1973)

Mantle dentin: ↑ Dentin, cover

Mantle, of lymphatic nodules (cap, corona): The peripheral zone of densely aggregated small ↑ B-lymphocytes surrounding ↑ germinal center of secondary ↑ lymphatic nodules.

Marginal cells, of stria vascularis: Columnar cells (MC) with convex luminal surface and occasional short microvilli. Nucleus is located in apical half of cell. Golgi apparatus, some mitochondria, numerous ↑ micropinocytotic vesicles, short cisternae of rough endoplasmic reticulum, ↑ lysosomes, and ↑ multivesicular bodies are supranuclear. ↑ Lipofuscin granules and ↑ lipid droplets are also present. Most mitochondria are located in compartments of a well-developed ↑ basal labyrinth (BLt). It is believed that M. are involved in selective absorption of materials from ↑ endolymph and in secretion of potassium ions. M. contact basal cells (BC), intermediate cells (IC), and capillaries (Cap). BL = basal lamina

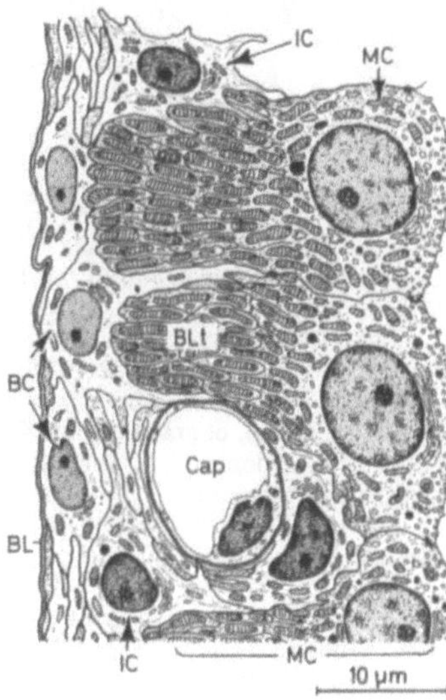

Santos-Sacchi, J.: Hear. Res. *6*, 7 (1982)

Marginal fold, of capillaries: A thin cytoplasmic flap (MF) overlapping lu-

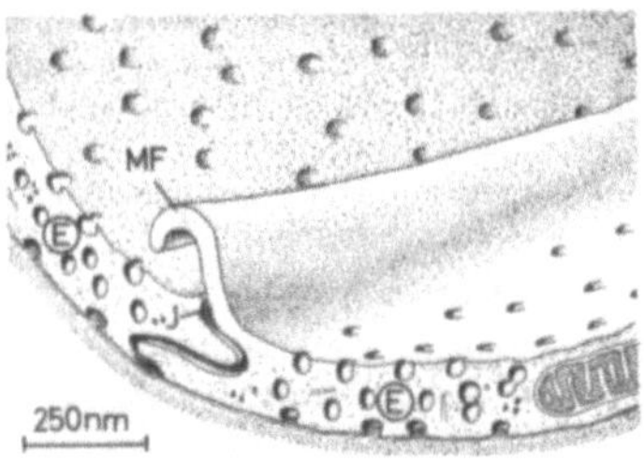

minal junction of two adjoining ↑ endothelial cells (E). Exact function of M. is unknown: It has been suggested that it may be involved in ↑ micropinocytotic activity and/or to retard blood flow near wall of vessels.

Marginal sinus: ↑ Subcapsular sinus

Marginal zone, of spleen: A condensation (MZ) of ↑ reticular cells, ↑ B-lymphocytes, and ↑ reticular fibers of ↑ red pulp around ↑ periarterial lymphatic sheaths (PALS) and ↑ splenic nodules (SN). M. can contain ↑ splenic sinusoids (S); it is continuous with ↑ splenic cords (SC). M. seems to represent major ↑ B-lymphocyte compartment of ↑ spleen. (Fig. modified after Weiss and Tavassoli 1970)

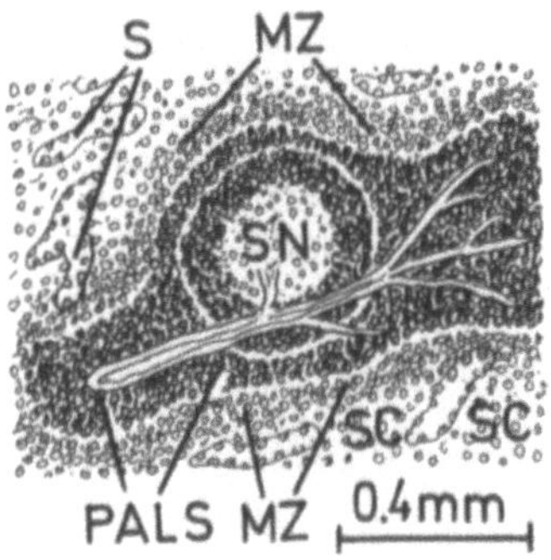

Blue, J., Weiss, L.: Am. J. Anat. *161*, 161 (1981); Pabst, R., Binns, R.M.: Cell Tissue Res. *226*, 319 (1982); Weiss, L., Tavassoli, M.: Sem. Hematol *7*, 372 (1970)

Marginating pool, of granulocytes: ↑ Granulocytes, pools of; ↑ Granulocytes, ready-reserve of

Marrow, bone: ↑ Bone marrow

Marrow cavity, primary: ↑ Bone formation, indirect

Martinotti's cells: Small ↑ neurons present in most layers of ↑ brain cortex. Their ↑ axons are directed toward surface and ramify in molecular layer. Structurally very similar to ↑ granule cells of CNS.

Mast cells (mastocyte, labrocyte, basophilus textus*): A type of large ↑ wandering cell (20–30 μm across) of ↑ connective tissue proper generally found along blood vessels. Cell body has an extremely variable form; it appears oval or spherical and covered by small folds (F), invaginations (I), and variously long microvilli (Mv). Nucleus is mostly elliptical and rich in ↑ heterochromatin; nucleolus is prominent. Mitochondria are small but numerous, Golgi apparatus (G) is well developed, cisternae of rough endoplasmic reticulum are sparse and short, free ribosomes are moderately numerous, and ↑ microtubules and microfilaments are present. Cytoplasm is loaded with

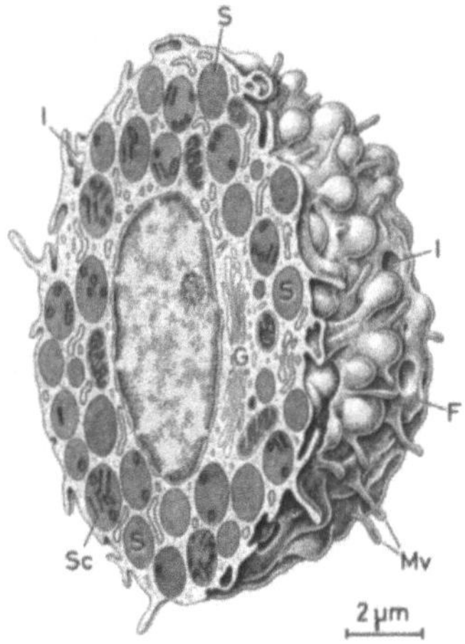

many metachromatic ↑ secretory granules (S) averaging 0.3–1.5 μm in diameter; they are surrounded by a ↑ unit membrane and filled with a heterogenous contents showing pronounced ↑ osmiophilia. In man, crystals and/or ↑ scrolls (Sc) may occur in granules. Granules contain ↑ histamine, ↑ heparin (in rodents, also ↑ serotonin), chimase, β-glucuronidase, and ↑ eosinophilic chemotactic factor of anaphylaxis (ECF-A). Outside secretory granules are found ↑ leukotrienes, ↑ prostaglandins, ↑ platelet-activating factor, and hydrogen peroxide. With secretion of all these substances, M. act on blood vessels, regulate composition and viscosity of ↑ ground substance, and assure an optimal relationship between blood vessels and connective tissue. M. are also involved in ↑ immediate sensitivity in cases of allergy, since a considerable amount of ↑ immunoglobulin E is bound to their plasmalemma; an antigen-IgE interaction provokes rapid discharge of all above-mentioned substances, resulting in consecutive ↑ edema and bronchospasm. Life span: 8–18 days.

Austen, K.F.: Harvey Lect. *73*, 93 (1979); Behrendt, H.: Verh. Anat. Ges. *74*, 261 (1980); Kiernan, J.A.: J. Anat. *128*, 225 (1979); Kitamura, Y., Yokoyama, M., Matsuda, H., Ohno, T.: Nature *291*, 159 (1981); Lawson, D., Raff, M.C., Gomperts, B., Fewtrell, C., Gilula, N.B.: J. Cell Biol. *72*, 242 (1977); Németh, A., Röhlich, P.: Eur. J. Cell Biol. *28*, 39 (1982)

Mastocytes: ↑ Mast cells

Mastoid "cells": ↑ "Cells," of mastoid process

Maternal placenta: ↑ Placenta, full-term, structure of

Matrix*: 1) The intercellular substance of ↑ connective and supporting tissues. 2) The formative portion of an ↑ organ. 3) A substance contained in ↑ inner chamber of ↑ mitochondria.

Matrix cells, of hair (cellulae matricis pili*): A collective term for four kinds of germinative cells (MaC) located in ↑ hair bulb. Undifferentiated M. are generally columnar with a large nucleus and a prominent nucleolus; their cytoplasm contains a moderate number of small mitochondria, a small Golgi apparatus, few rough endoplasmic cisternae, abundant free ribosomes, and some microfilaments. ↑ Hemidesmosomes (Hd) attach M. to basal lamina. M. divide mitotically and differenti-

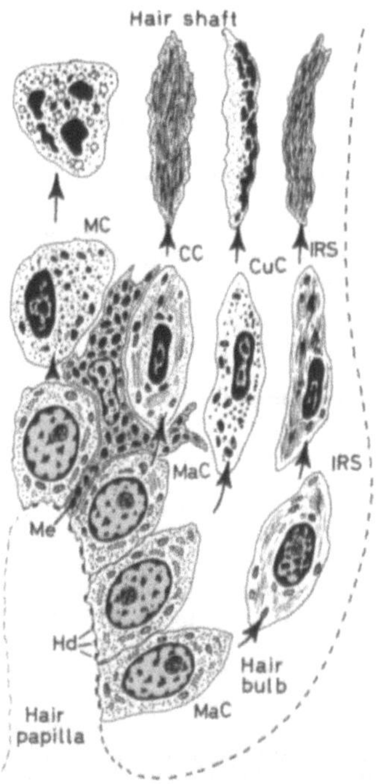

ate into: 1) ↑ Medullary cells (MC) (squamocytus polygonalis*) = a few cells located at tip of ↑ hair papilla; they constitute ↑ medulla of ↑ hair shaft. 2) Cortical cells (CC) = cuboidal cells, which gradually become elongated and filled with 6 to 8-nm-thick microfilaments; they transform into keratinized ↑ cortex of hair shaft. 3) Cuticular cells (CuC) (cellula cuticularis*) = spindle-shaped to flattened cells with 30 to 40-nm granules, which fuse together to form an amorphous condensation at cell periphery (= ↑ cuticle, of hair shaft). 4) Inner root sheath cells (IRS) = columnar cells with many 6 to 8-nm ↑ tonofilaments and ↑ trichohyalin granules; these cells differentiate into cells of ↑ Henle's and ↑ Huxley's layers. Toward hair shaft, they gradually become horny and disappear at opening level of ↑ sebaceous glands. Among M. are scattered ↑ melanocytes (Me), which transfer ↑ melanin granules to neighboring cells. (Fig. modified after Strauss and Matoltsy, in: Weiss, L., Greep, R.O.: Histology, 4th edn. New York: McGraw Hill Co. 1977)

Matrix granules, of cartilage: Irregular membraneless electron-dense structures (arrows), 10–15 nm in diameter, lying freely among or attached to ↑ collagen microfibrils. It is believed that M. represent agglomeration of macromolecules of ↑ chondromucoprotein. (Fig. = hyaline ↑ cartilage, rat)

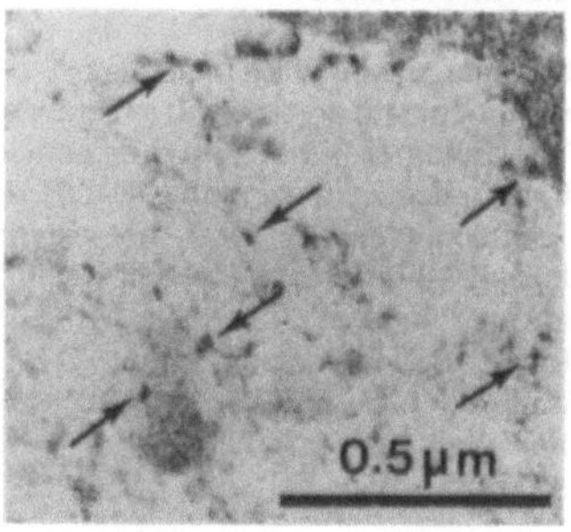

Matrix granules, of mitochondria: ↑ Mitochondrial granules

Matrix, mitochondrial: ↑ Mitochondrial matrix

Matrix, of bone: ↑ Bone matrix

Matrix, of cartilage: ↑ Cartilage matrix

Matrix, of connective and supporting tissues (substantia intercellularis*): A term designating ↑ fibers and ↑ ground substance of a ↑ connective and ↑ supporting tissue.

Matrix, of cytoplasm: ↑ Cytoplasmic matrix

Matrix, of hair (matrix pili*): The germinative portion of a ↑ hair constituting major part of ↑ hair bulb. M. is composed of ↑ matrix cells situated around ↑ hair papilla.

Matrix, of nail: ↑ Nail matrix

Matrix vesicles (calcification vesicles, calcium-accumulating vesicles, calcifying globules, spherulites): ↑ Unit membrane-bound, round or oval structures, 0.1–0.25 µm in diameter, found in ↑ capsule of chondrocytes, near ↑ chondrocytes of calcifying cartilage zone, and also near ↑ apical pole of ↑ odontoblasts and around ↑ osteoblasts. M. contain a fine granular dense material and often minute needles of ↑ hydroxyapatite. Concentrated within M. are acid and alkaline phosphatase, ATPase, and some lipids that bind calcium (phosphatidyl serine, sphingomyelin). It is generally accepted that M. serve as one of initial sites for hydroxyapatite deposition. (Fig. = rat)

↑ Bone, calcification of; ↑ Cartilage, calcification of; ↑ Dentin, calcification of; ↑ Dentinal matrix vesicles

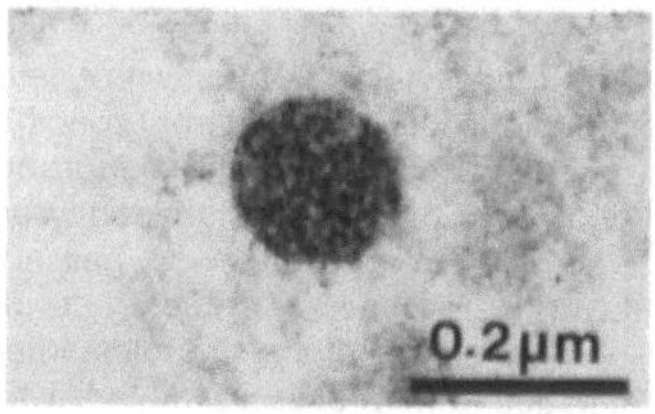

Ali, S.Y.: Fed. Proc. *35*, 135 (1976); Bab, I.A., Muhlrad, A., Sela, J.: Cell Tissue Res. *202*, 1 (1979); Ornoy, A., Atkin, I., Levy, J.: Acta Anat. *106*, 450 (1980)

Matrix vesicles, biogenesis of: The process of formation of ↑ matrix vesicles by ↑ chondrocytes, ↑ odontoblasts, and ↑ osteoblasts; not definitively elucidated. Hypotheses intending to explain M.: 1) Budding from cellular processes of above-mentioned cells; 2) extrusion of preformed structures; 3) cellular degeneration and disintegration; 4) secretion of subunits and their extracellular assemblage. Evidence shows that hypotheses 1 and 4 are most probable. (Fig. modified after Rabinovitch and Anderson 1976)

Oronoy, A., Atkin, I., Levy, I.: Acta Anat. *106*, 450 (1980); Rabinovitch, A.L., Anderson, C.H.: Fed. Proc. *35*, 112 (1976)

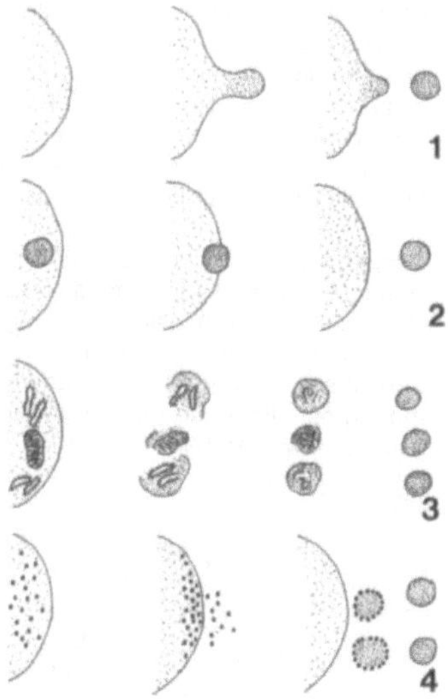

Matrixin: A proteinaceous material forming 2 to 3-nm-thick filaments of ↑ nuclear skeleton.

Maturing cartilage, zone of: ↑ Bone formation, indirect

Maturing face, of Golgi apparatus (distal face, trans face): The concave aspect (M) of ↑ Golgi apparatus, generally oriented away from nucleus. ↑ Glycoproteins and some lysosomal enzymes (particularly acid phosphatase) are concentrated within cisternae of M., which have thicker membranes than those of ↑ forming face (F). Presecretory granules (PG) separate from M.

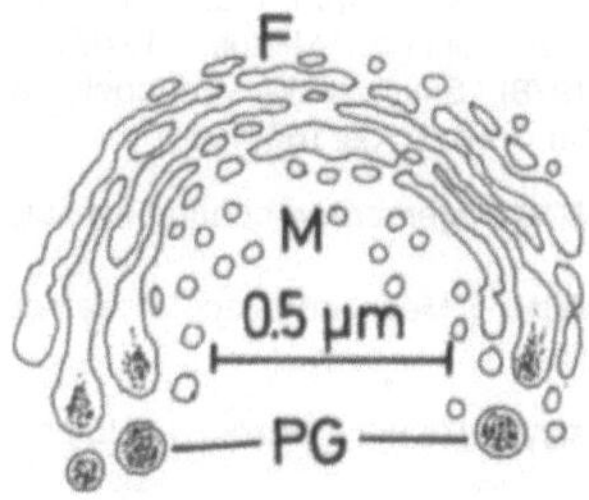

Maturing phase, of spermiogenesis: ↑ Spermiogenesis

May-Grünwald-Giemsa staining: A method for ↑ staining smears of ↑ blood and ↑ bone marrow.

M-cells: ↑ Synovial cells

M-cells (follicle-associated epithelial cells, FAE): Specially differentiated ↑ absorptive cells scattered in epithelium covering ↑ Peyer's patches. M. are characterized by loosely arranged ↑ microvilli (Mv), considerably much longer and less numerous than those composing ↑ striated border (SB) of neighboring ordinary absorptive cells (A). Apical surface of M. displays multiple pits (P). M. cell body and lateral surface are deeply invaginated by intraepithelial lymphocytes (L). Evidence is that M. transport intact foreign macromolecules and/or antigens (Ag) from intestinal lumen to adjacent intraepithelial lymphocytes or underlying ↑ lymphatic tissue (arrow). (Modified after Owen and Nemanic 1978)

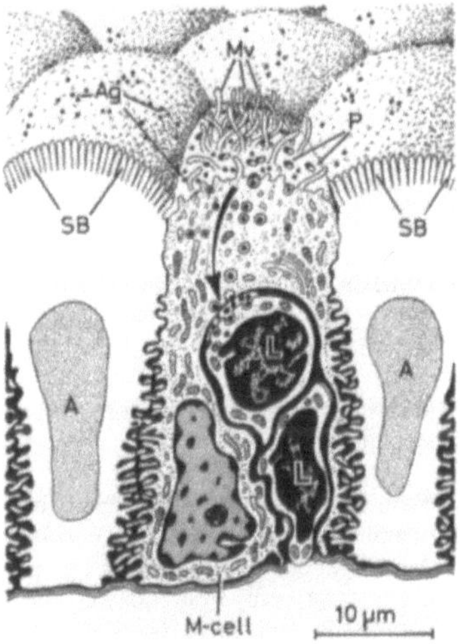

Bockman, D.E.: Arch. Histol. Jpn. *46*, 271 (1983); Kagnoff, M.F.: Immunology of the Digestive System. In: Johnson, L.R. (ed.) Physiology of the Gastrointestinal Tract. Vol. 2. New York: Raven Press 1981; Owen, R.L., Nemanic, P.: Scan. Electron Microsc. 1978/II, 367 (1978); Smith, M.W., Peacock, M.A.: Am. J. Anat. *159*, 167 (1980)

MCG: ↑ Membrane-coating granules

MCH: ↑ Mean corpuscular hemoglobin

MCV: ↑ Mean corpuscular volume

Mean corpuscular hemoglobin (MCH): The quantity of ↑ hemoglobin (Hb) contained in a single ↑ erythrocyte.

$$\text{MCH} = \frac{\text{Hb}}{\text{number of erythrocytes}}.$$

Normal values = 27–34 pg (↑ picogram).

Mean corpuscular volume (MCV): The mean volume of a single ↑ erythrocyte.

$$\text{MCV} = \frac{\uparrow \text{hematocrit}}{\text{number of erythrocytes}} = 81\text{–}99\ \mu m^3$$

or, according to ↑ SI, 81–99 fl (↑ femtoliter).

Meatus, auditory, external: ↑ Auditory meatus, external

Meatus, auditory, internal: ↑ Auditory meatus, internal

Mechanoreceptors: ↑ Exteroceptors responding to mechanical stimuli (touch, vibration, pressure). ↑ Corpuscles of Meissner, Vater-Pacini, Golgi-Mazzoni, Ruffini, and Krause, ↑ lanceolate nerve endings around ↑ hairs, ↑ free nerve endings in ↑ epidermis, and ↑ Merkel's cells are all considered M. In a larger sense of the term, ↑ hair and ↑ vestibular cells also belong to M.

Halata, Z.: The Mechanoreceptors of the Mammalian Skin. Adv. Anat. Embryol. Cell Biol., *50*/5,1 (1975)

Meconium corpuscles: Small supranuclear accumulations of ↑ bile pigments found in ↑ absorptive cells of ↑ intestinal villi. M. occur in about 4th month of intrauterine life as a consequence of ingestion of meconium.

Media: ↑ Tunica media

Median eminence (eminentia mediana*): A ↑ circumventricular organ making up part of floor of third ventricle between ↑ infundibulum of hypophysis and nuclei tuberales of ↑ hypothalamus. Structure: 1) Ependymal zone (E) = a layer of ↑ ependymal cells and ↑ tanycytes in contact with ↑ cerebrospinal fluid (F). 2) Internal zone (IZ) = ↑ neurosecretory axons of supraoptico- and paraventriculohypophyseal tract. 3) External zone or palisade layer (PL) = fine axons mostly devoid of ↑ neurosecretory granules intermingled with processes of tanycytes reaching outer surface. 4) Capillary layer (CL) = capillary loops of hypophyseal portal system. M. represents site of passage of ↑ releasing and ↑ inhibiting factors into hypophyseal circulation. Tanycytes are believed to transport biologically active molecules from ↑ cerebrospinal fluid to hypophyseal circulation.

↑ Hypophysis, vascularization of

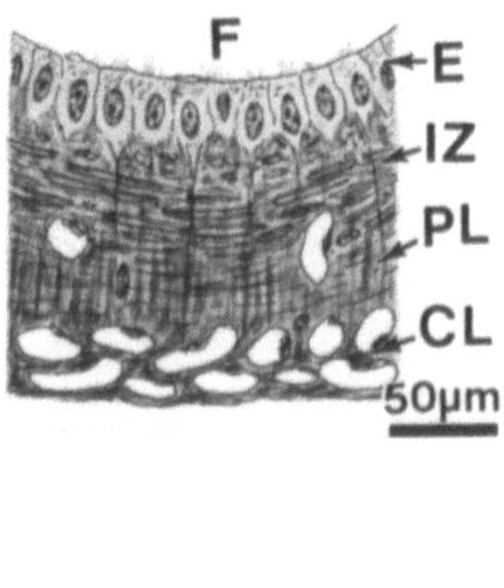

Ohtsuka, M., Yamamoto, Y., Daikoku, S.: Arch. Histol. Jpn. *46*, 203 (1983); Zaborsky, L., Schiebler, T.H.: Z. mikrosk.-anat. Forsch. *92*, 781 (1978); Zamora, A.J., Ramirez, V.D.: Cell Tissue Res. *226*, 27 (1982)

Mediastinum testis*: A thickening (M) of ↑ tunica albuginea (A) on posterior surface of ↑ testis. ↑ Rete testis (R) is situated in connective tissue of M. E = corpus epididymidis. (Fig. = human)

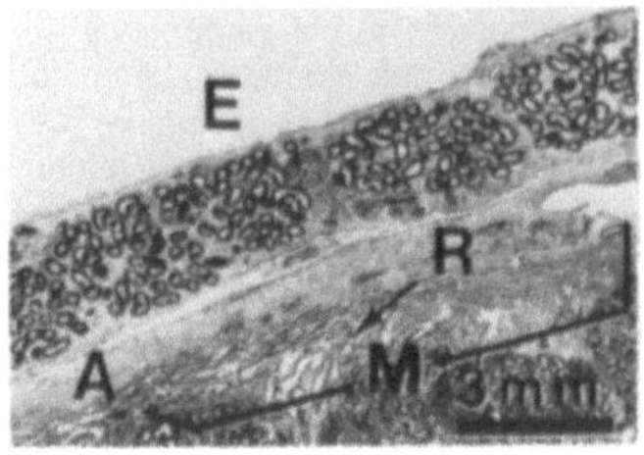

Mediators, of cell-mediated immune response: ↑ Immune response; ↑ Lymphokines

Medium-sized veins: ↑ Veins, medium-sized

Medulla: The inner portion of an ↑ organ, or its part, as distinguished from outer portion, or ↑ cortex.

Medulla, of adrenal glands: ↑ Adrenal glands, medulla of

Medulla, of hair shaft (medulla pili*): A row of ↑ medullary cells (M) in center of ↑ hair shaft. M. is not always present in human hairs; if present (mostly coarse ↑ terminal hairs), it does not extend through entire length of hair shaft. ↑ Lanugo hairs do not have M. M. is formed by ↑ matrix cells of hair. C = ↑ cortex, of hair shaft; Cu = ↑ cuticle, of hair shaft

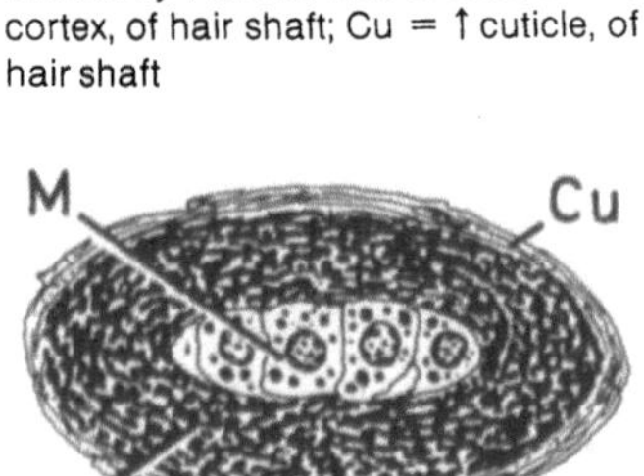

Medulla, of kidney: ↑ Kidney

Medulla, of lymph nodes: ↑ Lymph nodes

Medulla, of thymus: ↑ Thymus

Medullary artery (nutrient artery, arteria nutritia*): A small ↑ muscular artery usually accompanied by two veins. M. supplies blood to ↑ bone marrow, ↑ cancellous bone, and only innermost ↑ osteons of ↑ compact bone. M. penetrates ↑ medullary cavity via foramen nutritium.

Medullary cavity (cavitas medullaris*): The cavity within a long bone filled with ↑ bone marrow.

Medullary cells, of hair shaft (squamocytus polygonalis*): Cells (M) constituting ↑ medulla of some ↑ hairs. Near ↑ hair papilla, M. are cuboidal with a central round nucleus, a well-developed rough endoplasmic reticulum, occasional ↑ melanin granules, and numerous 30 to 50-nm ↑ trichohyalin granules. Toward surface of ↑ epidermis, M. become irregularly shaped, their nuclei become pyknotic (arrows), trichohyalin granules coalesce to fill entire cell body with hard ↑ keratin, in which small gaseous vacuoles may occur. M. arise from ↑ matrix cells of hair and generally do not extend through entire length of hair. (Fig. = ↑ eyelash, mongolian gerbil)

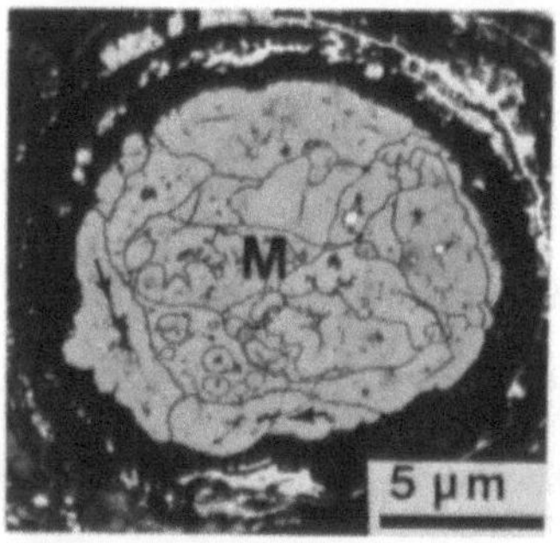

Clement, J.-L., Hagege, R., Le Pareaux, A., Carteaud, J.-P.: Scan. Electron Microsc. 1981/III, 377 (1981)

Medullary collecting tubules, of kidney: ↑ Collecting tubules, of kidney

Medullary cords, of lymph node (chordae medullares*): Irregularly branched strands (C) of ↑ lymphatic tissue forming major part of ↑ lymph node medulla. M. are continuous with inner ↑ cortex (IC) and lined by ↑ littoral cells (arrowheads). M. delimit ↑ medullary sinuses (S); through M. run arterial and venous blood vessels, capillaries (arrow), and ↑ nonmyelinated nerve fibers. M. are rich in ↑ B-lymphocytes, ↑ macrophages, and ↑ plasma cells. (Fig. = human)

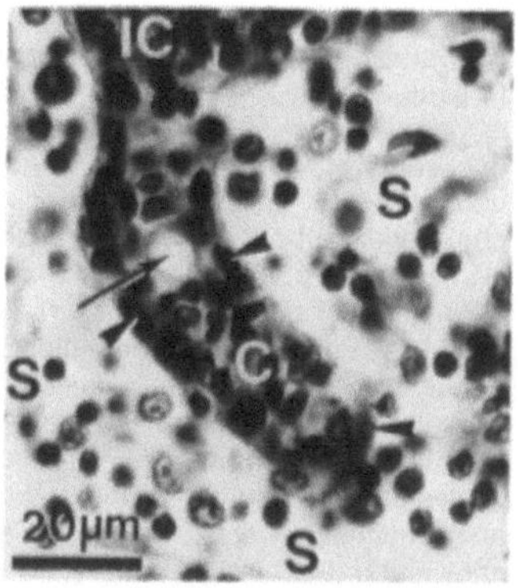

Medullary pyramids: ↑ Pyramids, of kidney

Medullary rays, of kidney (irradiatio medullaris): Finely striated extensions (MR) of base of a ↑ pyramid (Py) reaching into cortex (Co) and almost to renal capsule (Ca). M. are composed of straight portions of ↑ proximal (P) and ↑ distal (D) tubules of ↑ nephron and straight ↑ collecting tubules (C). Although structurally similar to medulla, M. are considered part of cortex. M. represent axis of a ↑ renal lobule. RC = ↑ renal corpuscle, IA = ↑ interlobular artery

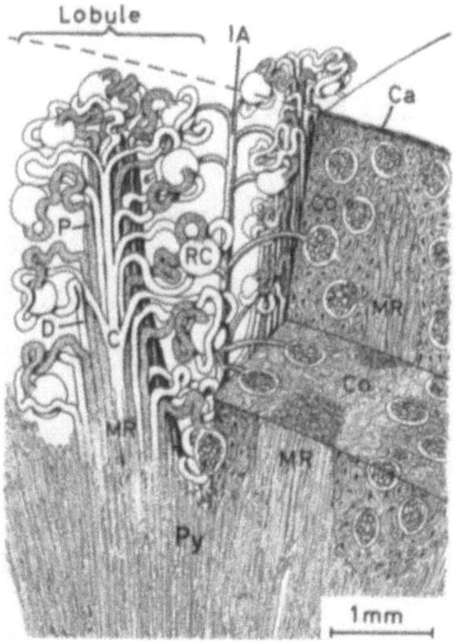

Medullary sinus (sinus medullaris*): An irregular lymphatic space surrounding ↑ medullary cords of a ↑ lymph node. In area of ↑ hilus, M. anastomoses directly with ↑ subcapsular sinus to which it corresponds structurally.

Megakaryoblast (megakaryoblastus*): The first distinguishable precursor cell of ↑ megakaryocyte. M. is a round of oval cell, 15–25 µm in diameter, with a large round or oval nucleus; frequently ↑ binucleated. Nucleus displays several deep invaginations; the masses of ↑ heterochromatin are denser than in a ↑ myeloblast. Nucleus encloses several nucleoli. In thin rim of agranular, slightly basophilic cytoplasm are situated small round mitochondria, a relatively well-developed Golgi apparatus, short profiles of rough endoplasmic reticulum, and a moderate number of free ribosomes. Some occasional electron-dense ↑ unit membrane-bound granules can also be present in cytoplasm. M. differentiates in the course of ↑ thrombopoiesis into a ↑ promegakaryocyte.

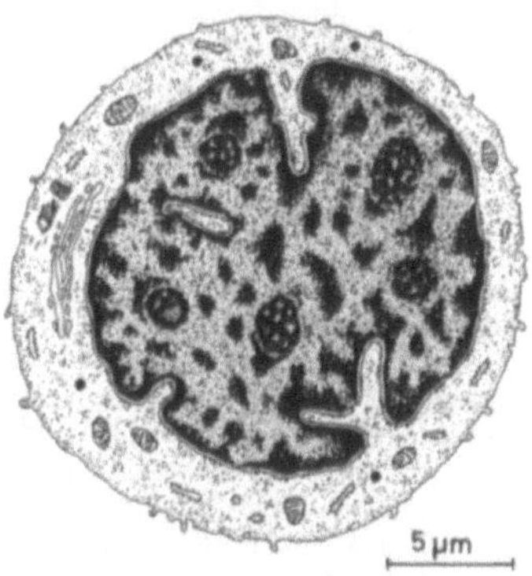

Freedman, M.H., McDonald, T.P., Saunders, E.F.: Cell Tissue Kinet. *14*, 53 (1981)

Megakaryocyte (megakaryocytus*): A giant cell, about 100 µm in diameter, with a very large ↑ polyploid lobulated nucleus (64 n chromosomes) containing several nucleoli. Enormous cytoplasm consists of a perinuclear zone enclosing a multiple Golgi apparatus, numerous ↑ centrioles, most of the mitochondria, rough endoplasmic reticulum, and a peripheral zone with abundant dense granules (Gr). This zone is subdivided by ↑ platelet demarcation channels (PDC) into cytoplasmic segments measuring 2–4 µm, each containing several dense granules. Platelet demarcation channels reach cell periphery, giving M. a highly irregular surface. M. are mostly found adjacent to sinusal wall (SW) of ↑ bone marrow. A massive protrusion of the cell penetrates sinus and fragments into ↑ platelets (P) or produces ↑ platelet ribbons (PR). From a platelet-producing M. (fig.) about 2500 platelets arise after fragmentation of cytoplasm;

nucleus and perinuclear cytoplasm are destroyed by ↑ macrophages. Resting M. is identical to platelet-producing M., but does not fragment into platelets.

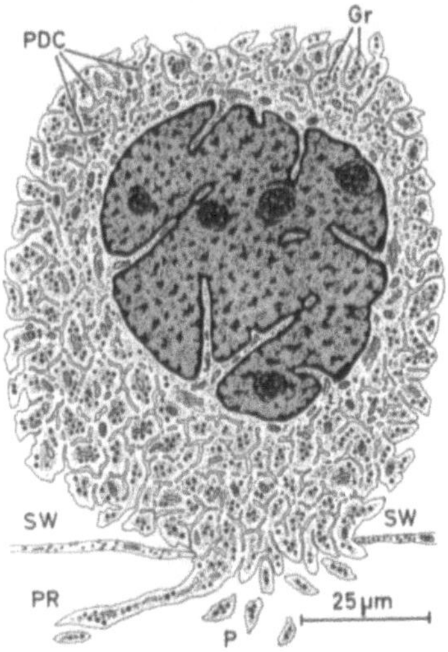

Evatt, B.L., Levine, R.F., Williams, N.T.: Megakaryocyte Biology and Precursors: In Vitro Cloning and Cellular Properties. New York: Elsevier/North Holland 1981; Fedorko, M.E.: Lab. Invest. *36*, 310 (1977); Ihzumi, T., Hattori, A., Sanada, M., Muto, M.: Arch. Histol. Jpn. *40*, 305 (1977); Penington, D.G.: Blood Cells *5*, 5 (1979)

Megakaryocytopoiesis: Formation and maturation of ↑ megakaryocytes.

↑ Hematopoiesis, current view of; ↑ Thrombopoiesis

Killmann, S.V.A., Cronkite, E.P., Mullerberat, C.N. (eds.): Haemopoietic Stem Cells. Copenhagen: Munksgaard 1983; Levin, J.: Blood *61*, 617 (1983); Williams, N., McDonald, T.P., Trabellino, E.M.: Blood Cells *5*, 43 (1979)

Meibomian glands: ↑ Tarsal glands

Meiosis*: A special type of nuclear division reserved for ↑ germ cells (↑ spermatocytes and ↑ oocytes). M. consists of reductional division followed by an equational division. **I) Reductional division:** Prior to prophase. DNA is replicated, so that chromosomal mass contains twice normal content of DNA (4 n DNA). Number of ↑ chromosomes is the same as in ↑ somatic cells (2 n Chr). Prophase I (prolonged): 1) Leptotene = chromosomes become visible as single, long, fine filaments. 2) Zygotene = homologous [paternal (p) and maternal (m)] chromosomes approach one another, form pairs, and come together in close lateral, ↑ gene to gene apposition (= ↑ synapsis). 3, 4) Pachytene = a contraction and thickening of chromosomes. Each chromosome splits into two ↑ chromatids without doubling of ↑ centromere. The four chromatids of each chromosomal pair constitute a ↑ tetrad. ↑ Crossing over of chromatids at points termed chiasmata occurs with exchange of segments of chromatids. Diplotene (not shown) = chromosomes coil and begin to separate. For a short time, they are held together at chiasmata, then fragmentation of ↑ nucleolus occurs. Diakinesis = chromosomes become broader, thicker, and more distinct; disappearance of nucleolus and ↑ nuclear envelope.

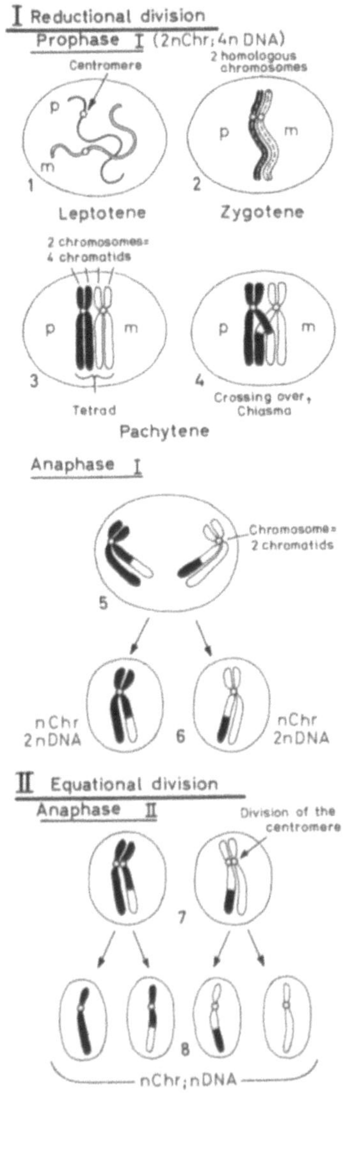

Metaphase I (not represented): Formation of ↑ spindle apparatus. Chromosomes form ↑ equatorial plate. 5) Anaphase I = centromeres of chromosomes do not divide as they do in ↑ mitosis, such that whole chromosomes, not chromatids, move to opposite cell poles. 6) Telophase I (not shown) = a short reconstruction of nucleus. ↑ Diploid number of chromosomes (2 n) is halved to ↑ haploid number (n) with 2 n DNA content. **II) Equational division:** Similar to mitotic division. Equational division soon follows reductional division, so that DNA cannot replicate during very short ↑ interphase (lack of S phase of ↑ cell cycle). 7) Anaphase II = centromeres divide as in mitosis, chromatids completely separate from each other. Each chromatid becomes a daughter chromosome of a new germ cell with a haploid number of chromosomes and only half the DNA content (n DNA) of somatic cells. Thus, M. results in formation of four genetically different germ cells (8). In the course of ↑ fertilization, diploid number of chromosomes and normal amount (2 n) of DNA are restored.

↑ Oogenesis, ↑ Spermatogenesis

Dodge, J.D., Vickerman, K.: Mitosis and Meiosis: Nuclear Division Mechanisms. In: Dooday, G.W., Lloyd, D., Trinci, A.P.J. (eds.): The Eukaryotic Microbial Cell. Soc. Gen. Microbiol. Symp. 30. Cambridge: Cambridge University Press 1980; Golubovskaya, I.N.: Int. Rev. Cytol. *58*, 247 (1979); Stern, H., Hotta, Y.: Ann. Rev. Genet. *9*, 37 (1972)

Meiotic spindle: ↑ Spindle apparatus

Meissner's corpuscle: ↑ Corpuscle, of Meissner

Meissner's plexus: ↑ Submucosal plexus

Melanin: An indole-containing reddish to dark brown endogenous ↑ pigment synthesized by ↑ melanocytes in the course of ↑ melanogenesis. M. is ↑ argentaffin, insoluble in water, organic solvents, dilute acids and bases, but decolored by hydrogen peroxide or potassium permanganate. M. is widely distributed: It is present in the form of ↑ melanin granules in stratum germinativum of ↑ epidermis, ↑ pigment epithelium, ↑ choroid, ↑ hairs, ↑ leptomeninges, ↑ iris, and locus niger of CNS. It is likely that reddish-brown M. is a type of specially polymerized M. It is synthesized under influence of ↑ melanocyte-stimulating hormone and ultraviolet light, against which it protects the body. ↑ Melatonin inhibits production of M. in lower vertebrates, but not in man.

Melanin granules (granulum melanini*): ↑ Unit membrane-bound ↑ organelles (M) with a highly osmiophilic content, ↑ melanin. M. develop from ↑ melanosomes in the course of ↑ melanogenesis. In ↑ pigment epithelial cells, M. are rod shaped (up to 3 µm long and 0.5 µm wide). In cells of stratum germinativum of ↑ epidermis and its ↑ melanocytes, M. are roughly round (0.5–1 µm in diameter) and, depending upon race, contain greater or lesser amounts of melanin. M. are transferred from ↑ melanocytes by an unknown mechanism to epidermal cells, this transfer increases under influence of ultraviolet light. M. function by scattering light and absorbing and protecting cells from ultraviolet light. (Fig. = pigment epithelial cell, mongolian gerbil)

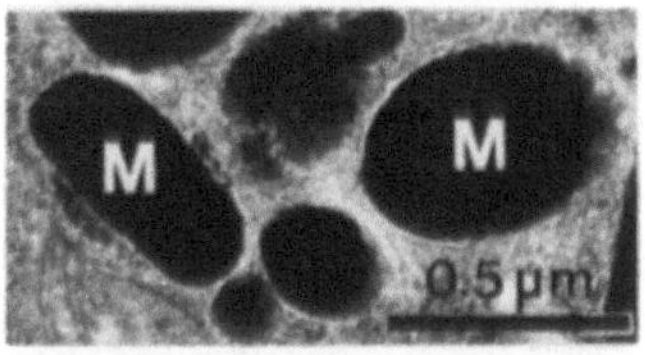

↑ Cytokrinia; ↑ Melanin granules, of Caucasoids; ↑ Melanophores

Melanin granules, of Caucasoids: ↑ Melanin granules of white race. Smaller (0.3–0.7 µm in length) than in Mongoloids and Negroids, these M. contain fragments of ↑ melanin (M) of various electron densities. (Fig. = ↑ basal cells of ↑ epidermis)

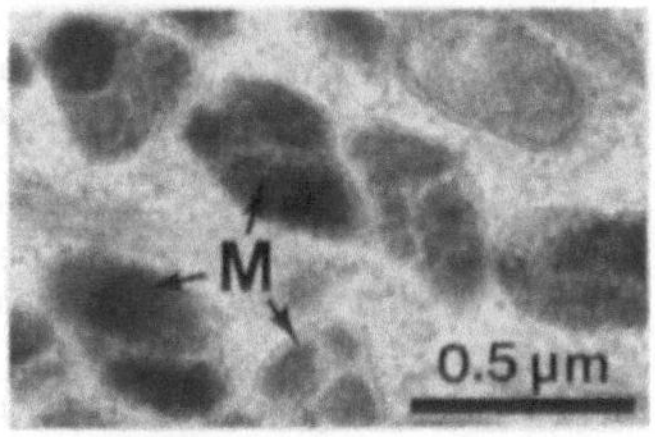

Szabo, G., Gerald, A.B., Pathak, M.A., Fitzpatrick, T.B.: Nature *222*, 1081 (1969)

Melanoblasts: Precursor cells of ↑ melanocytes originating from neural crest and migrating to ↑ epidermis. Without contact with epidermal cells, M. are unable to synthesize ↑ melanin.

Melanocyte-stimulating hormone (MSH): Two fractions of a polypeptide hormone (α- and β-MSH) produced by cells of pars intermedia of ↑ hypophysis and by ↑ corticotropes. M. stimulates synthesis of ↑ melanin. Production of M. is activated by M.-releasing factor; hypothalamic M.-inhibiting hormone and ↑ melatonin have antagonistic action on ↑ melanogenesis in some animals.

Tilders, F.J.H., Swaab, D.F., van Wimersma Greidanus, Tj. B. (eds.): Melanocyte Stimulating Hormone: Control, Chemistry and Effects. Basle: Karger 1977

Melanocyte-stimulating hormone-inhibiting factor (MIF): A hypothalamic polypeptide inhibiting synthesis and release of ↑ melanocyte-stimulating hormone.

Melanocyte-stimulating hormone-releasing factor (MSH-RF): A hypothalamic hypophyseotropic peptide of unknown nature acting on cells of pars intermedia of ↑ hypophysis to produce ↑ melanocyte-stimulating hormone.

↑ Releasing factors

Melanocyte system: All ↑ melanocytes of ↑ skin.

Fitzpatrick, T.B., Szabo, G., Seiji, M., Quevedo, W.C.Jr.: Biology of the Melanin Pigmentary System. In: Fitzpatrick, T.B. et al. (eds.): Dermatology in General Medicine. 2nd edn. New York. McGraw Hill 1979

Melanocytes (melanocytus*): Irregularly shaped ↑ argyrophilic cells (M) situated at ↑ dermoepidermal junction and between ↑ basal cells (BC) of ↑ epidermis. M. have a round or elliptical nucleus, which may be indented. Cytoplasm is clear and encloses numerous small mitochondria, a well-developed Golgi apparatus, short cisternae of rough endoplasmic reticulum, microfilamentous bundles, and some free ribosomes. Present in cell body and in cell processes (arrows) are numerous ↑ melanosomes (Me) and ↑ melanin granules (MG); by very long and branching cell processes, the latter are transferred to neighboring epidermal cells. M. have a common basal lamina with basal epidermal cells. There are neither ↑ desmosomes nor other junctional devices between M. and adjacent cells. Since M. contain an appreciable concentration of ↑ tyrosinase they give a positive ↑ DOPA reaction. M., which make up about 10%–25% of basal cell layer of epidermis in all human races, are responsible for ↑ melanogenesis. There are about 1200–1500 M./mm² skin surface. BP = ↑

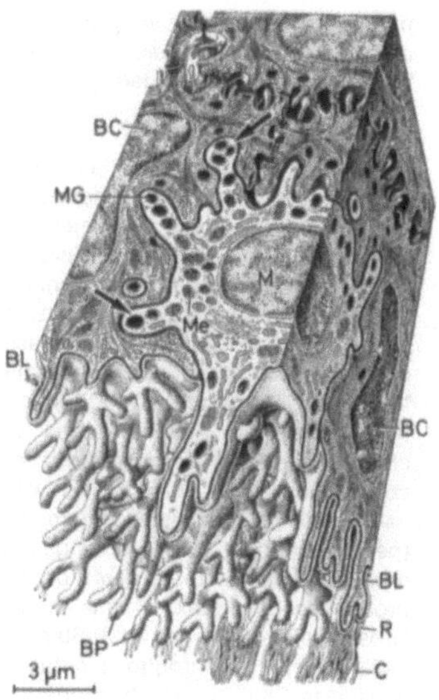

basal processes, C = ↑ collagen microfibrils, R = ↑ reticular microfibrils

↑ Cytocrinia; ↑ Melanoblasts

Hu, F.: J. Invest. Dermatol. *73*, 70 (1979); Quevedo, W.C.Jr., Fleischmann, R.D.: J. Invest. Dermatol. *75*, 116 (1980)

Melanocytes, of pigment connective tissue (cellula pigmentosa*): Stellate cells (Mc) with long processes (P), often in contact with ↑ collagen fibrils (C) and adjacent cells of ↑ pigment connective tissue. Nucleus is spindle-shaped; cytoplasm contains in addition to the usual ↑ organelles, a great number of ↑ melanin granules (M), among which there are also some ↑ melanosomes (Me). This supports the hypothesis that M. are not ↑ melano-

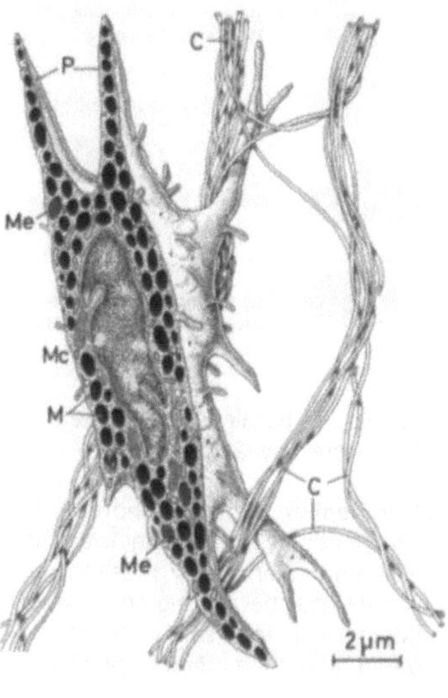

phores, but cells able to produce ↑ melanin. Thus, M. are cells of ectodermal origin having invaded tissue of mesodermal origin.

Melanogenesis: The very complex process, not completely understood, of ↑ melanin synthesis by ↑ melanocytes. Sequence: 1) Stage I melanosome = vesicles with fine granular amorphous proteinaceous material in center and short fibrillar ↑ tyrosinase molecules (TM) at periphery separate from ↑ Golgi apparatus (G). 2) Stage II melanosome = after a series of changes, vesicles elongate and become ovoid; their interior fills with many helicoidal tyrosinase filaments (periodicity about 10 nm), arranged in a parallel pattern, and an amorphous proteinaceous matrix. At this stage, tyrosinase becomes active. 3) Stage III melanosome = through increased tyrosinase activity, electron-opaque melanin progressively accumulates along filaments and on proteinaceous matrix; internal structure of melanosome becomes gradually obscure. 4) Stage IV melanosome = mature ↑ melanin granule – vesicles are completely filled with melanin; no more tyrosinase activity. A weak M., has been observed in ↑ pigment epithelial cells. rER = rough ↑ endoplasmic reticulum, TV = ↑ transport vesicles

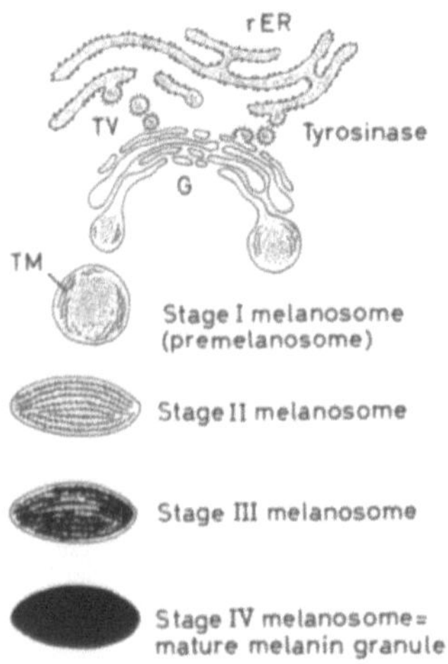

Stanka, P., Rethjen, P., Sahlmann, B.: Cell Tissue Res. *214*, 343 (1981)

Melanophores (melanophorus*): Stellate fibroblastlike connective tissue cells characterized by presence of a variable number of ↑ melanin granules (M) in their cytoplasm. Nucleus is elongated and in a central position; in cytoplasm are scattered rod-shaped mitochondria, a well-developed Golgi apparatus (G), and short cisternae of rough endoplasmic reticulum. M., together with ↑ melanocytes, form pigmented component of ↑ pigment connective tissue. M. are present in stroma of ↑ iris, ↑ ciliary body, ↑ choroid, ↑ leptomeninges, and ↑ dermis of genital skin. It is generally believed that M. are incapable of synthesizing melanin granules, but receive them from melanocytes; mechanism of transport of melanin granules from melanocytes to M. is not known.

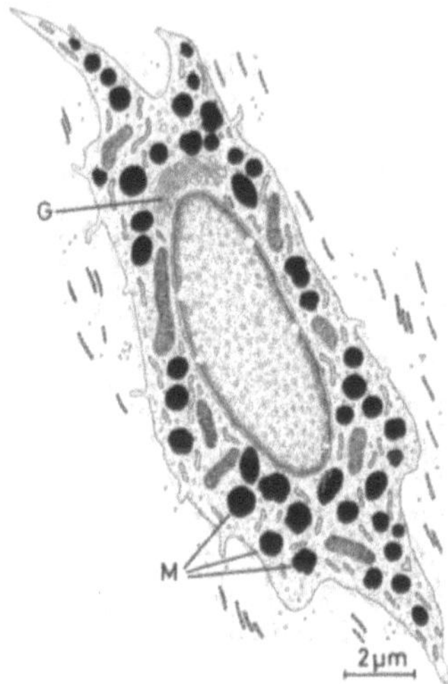

Melanosomes (melanosoma*): ↑ Unit membrane-bound, rounded or spherical ↑ organelles (Me), precursors of ↑ melanin granules (M, IV) found in ↑ melanocytes and cells of ↑ pigment epithelium. Depending on stage of ↑ melanogenesis, contents of M. may constitute only a fine granular proteinaceous matrix (P) with occasional peripheral ↑ tyrosinase filaments (F, I

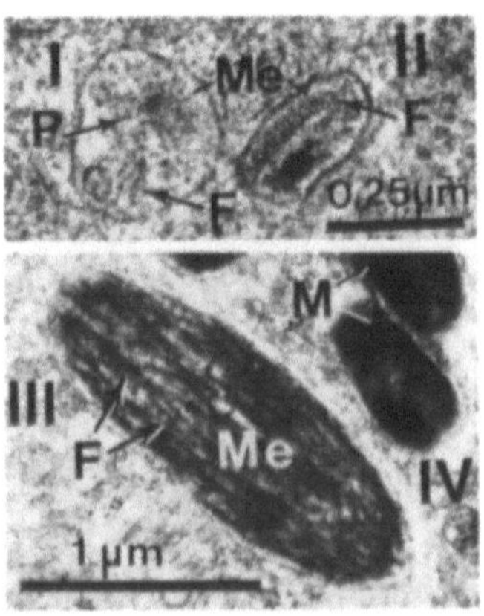

and II). In further M. stages number of tyrosinase filaments and highly osmiophilic ↑ melanin (II and III) increases. The same stages (II and III) of M. are characterized by high tyrosinase activity. (Fig., top = melanocyte, human; fig., bottom = ↑ pigment epithelium, mongolian gerbil)

Melanosomes, stage I (premelanosome, premelanosoma*): ↑ Unit membrane-bound elliptical cell ↑ organelles (Me), 0.2–0.5 µm long and 0.1–0.5 µm wide, found in ↑ melanocytes and ↑ pigment epithelium of ↑ retina. M. contain only short helicoidal ↑ tyrosinase filaments (F) and some amorphous proteinaceous material. Through successive stages of ↑ melanogenesis, M. gradually transform into ↑ melanin granules (MG). III = melanosome, stage III (Fig. = melanocyte, human)

↑ Melanosomes

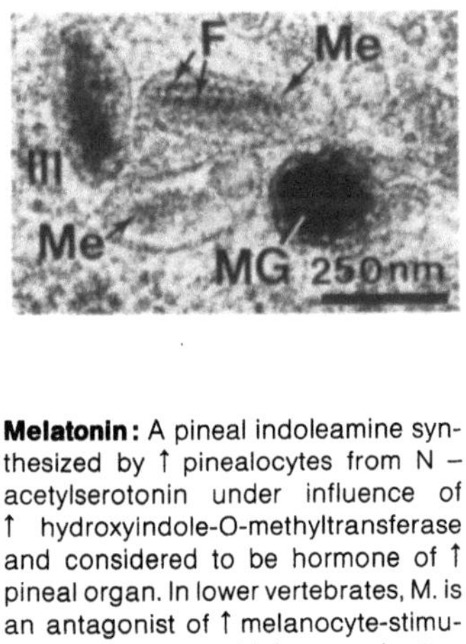

Melatonin: A pineal indoleamine synthesized by ↑ pinealocytes from N – acetylserotonin under influence of ↑ hydroxyindole-O-methyltransferase and considered to be hormone of ↑ pineal organ. In lower vertebrates, M. is an antagonist of ↑ melanocyte-stimulating hormone; in higher vertebrates, its effect on ↑ melanocytes is very weak. However, there is a regular rhythmicity of M. synthesis in man and some animals, with detectable nocturnal peak levels. M. has also an antigonadotropic effect by inhibiting secretion of hypophyseal ↑ gonadotropic hormones.

Cardinali, D.P.: Endocrine Reviews *2*, 327 (1981); Klein, C.D. (ed.): Melatonin Rhythm Generating System – Developmental Aspects. Basle: Karger 1982; Waldhauser, F., Wurtman, R.J.: The Secretion and Actions of Melatonin. In: Litwack, G. (ed.): Biochemical Action of Hormones, Vol. 10. New York: Academic Press 1983

Membrana fibroelastica infraglottica: ↑ Conus elasticus

Membrana fibroelastica laryngis: A layer of thick ↑ elastic and ↑ collagen fibers corresponding to tela submucosa of ↑ larynx. M. is subdivided into membrana fibroelastica infraglottica or ↑ conus elasticus and membrana fibroelastica supraglottica, which is considerably less developed; its fibers irradiate into ↑ ventricular folds.

Membrana fibroelastica supraglottica: ↑ Membrana fibroelastica laryngis

Membrana granulosa, of ovarian follicles (stratum granulosum*): Wall of secondary vesicular and mature ↑ ovarian follicles formed of several layers of ↑ granulosa cells.

Membrana limitans gliae perivascularis* (internal glial limiting membrane): An almost complete layer of astrocytic ↑ perivascular feet (PF) surrounding blood capillaries (Cap) of CNS. Feet are held together by ↑ nexus (N) and are separated from ↑ endothelial cells (E) only by a ↑ basal lamina (BL). M. represents less selective part of blood-brain ↑ barrier. M. lacks ↑ circumventricular organs.

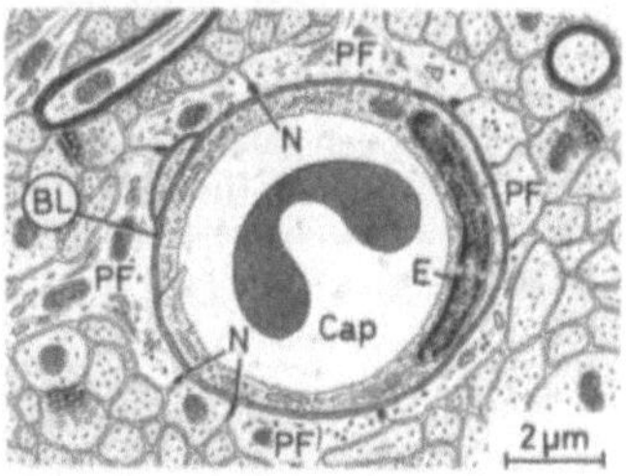

Reese, T.S., Karnovsky, M.J.: J. Cell Biol. *34*, 207 (1967)

Membrana limitans gliae superficialis* (external glial limiting membrane): A complete layer (M) of ↑ astrocytes (A) and their ↑ subpial feet separating ↑ central nervous system (CNS) from ↑ pia mater (PM). (Fig. = ↑ optic nerve, rat)

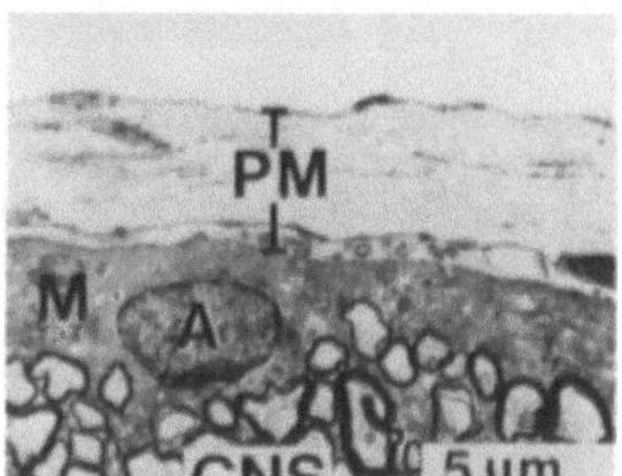

Bondareff, W., McLone, D.G.: Am. J. Anat. *136*, 277 (1973); Wagner, H.-J., Barthel, J., Pilgrim, C.: Anat. Embryol. *166*, 427 (1983)

Membrana pellucida: ↑ Zona pellucida

Membrana reticularis: ↑ Reticular membrane; ↑ Phalangeal cells, outer

Membrana synovialis: ↑ Synovial membrane

Membrana tectoria: ↑ Tectorial membrane

Membrana tympani: ↑ Tympanic membrane

Membrane, alveolocapillary: ↑ Blood-air barrier

Membrane, arachnoid: ↑ Arachnoid

Membrane-associated particles (intramembrane particles, intramembranous particles, MAP): Structures, about 9 nm in diameter (arrows), visible on ↑ EF and ↑ PF split surfaces of ↑ freeze-cleaved or ↑ freeze-etched ↑ cell membrane and other ↑ cytomembranes. M. probably represent cleaved molecules of proteins extending through entire phospholipid layers. ES = ES face of cell membrane. (Fig. = ↑ pinealocyte, rat)

↑ Cell membrane, models of; ↑ Freeze-cleaving, terminology of

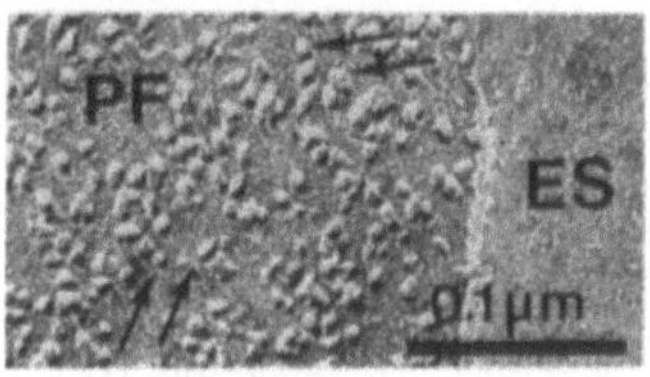

Robertson, J.D., Verage, J.A.: A Morphometric Freeze-Fracture-Etch (FFE) Analysis on the Nature of Intramembrane Particles. In: Bailey, G.W. (ed.): 38th Ann. Proc. Electron Microscopy Soc. Amer. San Francisco 1980

Membrane, basement: ↑ Basement membrane

Membrane, basilar: ↑ Basilar membrane

Membrane blisters: ↑ Vesicles, 0.2–0.6 µm in diameter, observed at cell surface; considered an artifact caused by combining ↑ glutaraldehyde and ↑ osmium tetroxide fixation.

Shelton, E., Mowczko, E.W.: Scanning *1*, 166 (1978)

Membrane bones: Bones which develop by direct ↑ bone formation (frontal, occipital, parietal, and temporal bones, mandibula, clavicula).

Membrane, cell: ↑ Cell membrane

Membrane, chemistry of: ↑ Cell membrane, chemical composition of

Membrane, choroid: ↑ Choroid

Membrane-coating granules (cementosome, dense body, keratinosome, MCG, lamellated dense body, Odland body, small dense body, transitory dense body, granulum lamellosum*): Round of elliptical ↑ unit membrane-bound structures (MCG), 0.1–0.3 µm in diameter, found in superficial cells (SC) of almost all ↑ stratified squamous epithelia independent of their degree of ↑ keratinization. M. contain a number of electron-dense lamellae (L), 3 nm wide, separated by a 5.5-nm electron-translucent band; the latter is divided by a thin intermediate band 1 nm thick. M. discharge their contents (arrow) by fusion with plasmalemma into intercellular spaces (IS), forming peculiar multilayered patterns (asterisks). M. are composed of ↑ glycoproteins, phospholipids, and ↑ lipoproteins; several lysosomal enzymes are concentrated in their interior. Biogenesis of M. is unknown: It has been postulated that M. are derived from degenerated mitochondria, Golgi apparatus, rough endoplasmic reticulum, and/or ↑ multivesicular bodies. M. are involved in production of ↑ glycocalyx which renders epidermis impervious to water; M. also induce breakdown of ↑ desmosomes and consecutive epithelial desquamation with their enzymes.

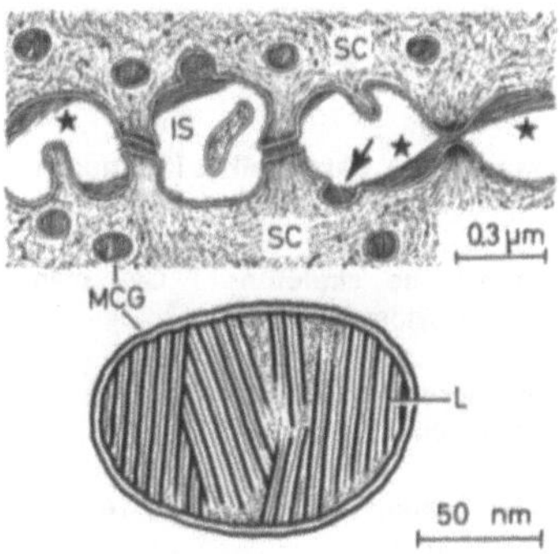

Hayward, A.F.: Int. Rev. Cytol. *59*, 97 (1979); Landmann, L.: Eur. J. Cell Biol. *33*, 258 (1984); Matoltsy, A.G.: J. Invest. Derm. *67*, 20 (1976)

Membrane, Descemet's: ↑ Descemet's membrane

Membrane, glassy, of atretic follicle: ↑ Glassy membrane, of atretic follicle

Membrane, glassy, of eye: ↑ Bruch's membrane

Membrane, glassy, of hair: ↑ Glassy membrane, of hair

Membrane, glassy, of ovarian follicles: ↑ Glassy membrane, of ovarian follicles

Membrane, hyaloid: ↑ Vitreous body

Membrane(s), mitochondrial: ↑ Mitochondrial membranes

Membrane, models of: ↑ Cell membrane, models of

Membrane, Nasmyth's: ↑ Cuticle, of enamel

Membrane, nuclear: ↑ Nuclear membrane

Membrane, of Bowman: ↑ Bowman's membrane; ↑ Cornea

Membrane(s), of brain: ↑ Meninges

Membrane, of Bruch: ↑ Bruch's membrane

Membrane(s), perforated: ↑ Elastic laminae

Membrane, periodontal: ↑ Periodontal ligament

Membrane, plasma: ↑ Cell membrane

Membrane(s), platelet demarcation: ↑ Platelet demarcation membranes

Membrane, serous: ↑ Serous membrane

Membrane, Shrapnell's: ↑ Shrapnell's membrane; ↑ Tympanic membrane

Membrane skeleton: ↑ Cell membrane, cytoskeleton of

Membrane space: ↑ Outer chamber, of mitochondria

Membrane, tectorial: ↑ Tectorial membrane

Membrane, vestibular: ↑ Reissner's membrane

Membranous bone: ↑ Bone, membranous

Membranous discs, of photoreceptors (laminated plates): Very flattened, ↑ unit membrane-limited, parallel sacs (D), about 2 µm in diameter and 14 nm thick, with an 8-nm-wide cavity. About 1300–1500 M. form outer segment of a ↑ photoreceptor. In rods (R), M. are enveloped by plasmalemma (P); in cones (C), they are often formed by deep plasmalemmal invaginations (arrows). M. consist of lipids and ↑ rhodopsin in rods and ↑ iodopsin in cones. In rods, protein is constantly synthesized on ↑ polyribosomes and in Golgi apparatus and assembled to M. at ciliary connection zone; approximately 90–100 newly synthesized M. move, in the course of about 10–13 days, toward ↑ pigment epithelium where they are phagocytized and destroyed in ↑ lamellar phagosomes. In cones, M. do not have such a turnover.

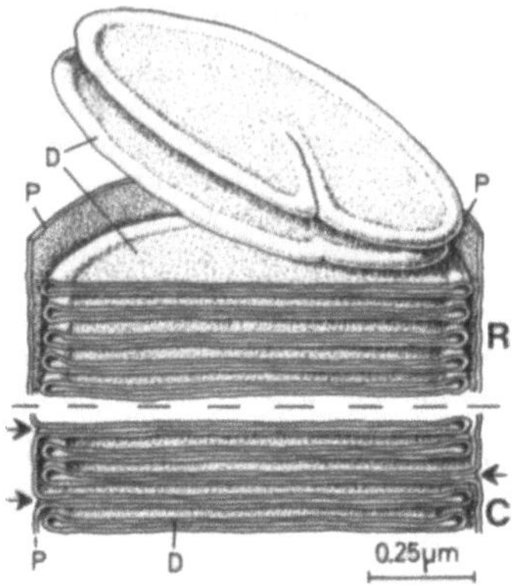

Olive, J.: Int. Rev. Cytol. *64*, 107 (1980); Roof, D.J., Heuser, J.E.: J. Cell Biol. *95*, 487 (1982); Roof, D.J., Korenbrot, J.I., Heuser, J.E.: J. Cell Biol. *95*, 501 (1982); Sjöstrand, F.J., Kreman, M.: J. Ultrastruct. Res. *65*, 195 (1978); Steinberg, R.H., Fisher, S.K., Anderson, D.H.: J. Comp. Neurol. *190*, 501 (1980)

Membranous labyrinth: ↑ Labyrinth, membranous

Membranous pneumonocytes: ↑ Alveolar cells, type I

Memory cells: Functional subcategories of ↑ B- and ↑ T-lymphocytes genetically programed to react rapidly to reintroduction of an antigen by ↑ differentiation into ↑ plasma cells and ↑ killer cells in the course of secondary ↑ immune response. Life span: several years.

Menarche: The first ↑ menstruation; occurs at an average age of 13 years.

Meningeal spaces: Spaces between ↑ meninges (↑ cisternae, of brain; ↑ subarachnoid space; ↑ subdural space).

Meninges*: Connective tissue sheets enveloping ↑ central nervous system. Composition: A) Pachymeninx = ↑ dura mater (DM). B) Leptomeninges: 1) ↑ Arachnoid (A); 2) ↑ pia mater (PM). Between dura mater and arachnoid is ↑ subdural space (SDS); between arachnoid and pia mater extends ↑ subarachnoid space (SAS).

↑ Brain, penetration of blood vessels into

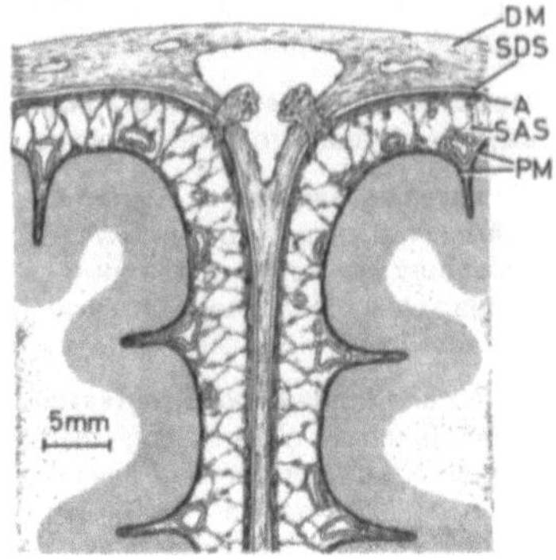

Meniscal tissue: A type of regular ↑ dense connective tissue (M) forming articular ↑ menisci. M. consists of very densely packed ↑ collagen fibers (CF) with a small number of ↑ fibrocytes (F) between them. Absence of a territorial matrix distinguishes M. from fibrous ↑ cartilage, although patches of the latter can be found in M. There are practically no ↑ wandering cells in M. (Fig. = human)

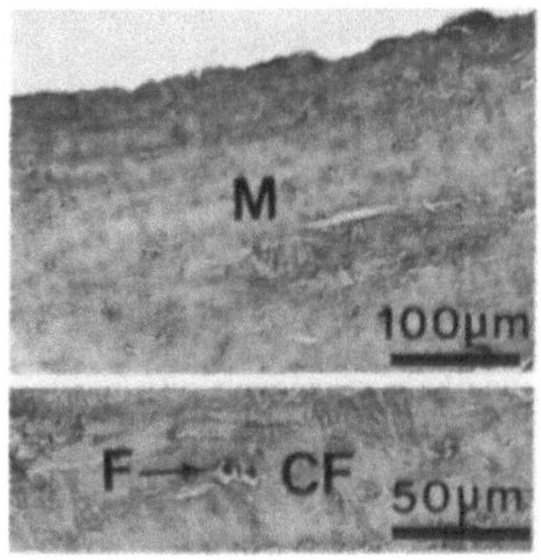

Ghadially, F.N., Lalonde, J.-M.A., Wedge, J.H.: J. Anat. *136*, 773 (1983); Somer, L., Somer, T.: Acta Anat. *116*, 234 (1983)

Meniscus (meniscus articularis*): A crescent-shaped strand (M) of ↑

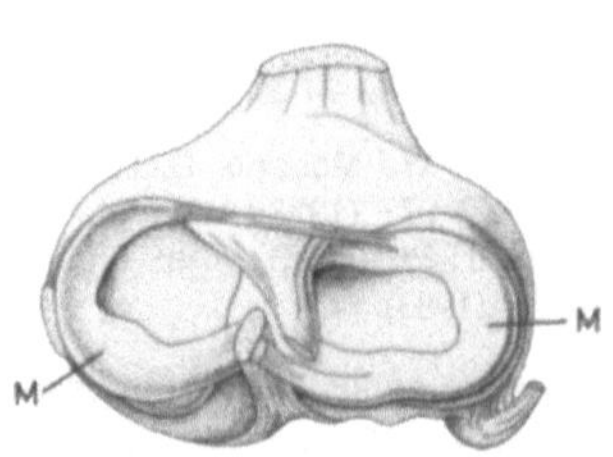

meniscal tissue found in some articulations, such as knee joint (fig.). The two ends of an M. are attached to bone; function of M. is to increase contact surface between two bones and to increase congruency between them.

Menopause (menopausa*): The gradual interruption of ↑ menstrual cycles provoked by exhaustion of ↑ oocyte stock in ↑ ovary; M. occurs at age of 45–50 years.

Perezpalacios, G., Cravioto, M.C., Medina, M., Ulloaaguirre, A.: The Menopause. In: Serra, G.B. (ed.): Ovary. Comprehensive Endocrinology. New York: Raven Press 1983

Menstrual cycle (cyclus menstrualis*): A sequence of morphological and functional changes of ↑ endometrium if secondary ↑ oocyte does not become fertilized and blastocyst implanted.

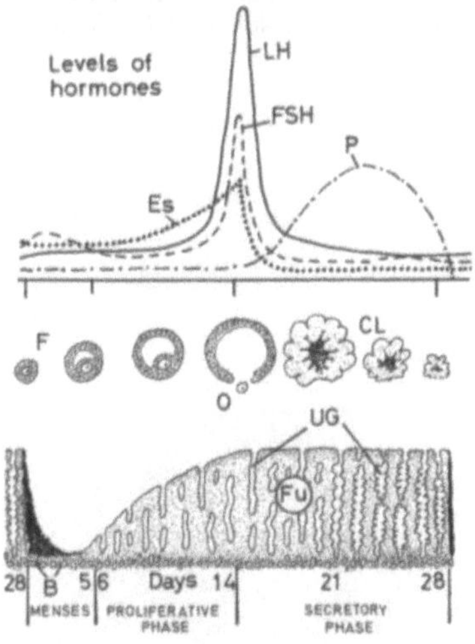

Phases: 1) Proliferative phase (days 5–14): Under influence of ↑ follicle-stimulating hormone (FSH), a new follicle (F) develops in ↑ ovary simultaneously with an increase in estrogen level (Es). Epithelium and stroma of endometrium regenerate by pronounced mitotic activity; in late proliferative phase, epithelial cells accumulate ↑ glycogen in their basal cytoplasm, endometrium becomes thicker, and ↑ coiled arteries become longer. 2) Secretory phase (days 15–28): FSH and ↑ luteinizing hormone (LH) together provoke ↑ ovulation (O); LH stimulates formation of ↑ corpus luteum (CL) and production of ↑ progesterone (P). The latter influences ↑ uterine glands (UG), which become tortuous; epithelial cells accumulate glycogen in their apical cytoplasm and start secreting a product rich in carbohydrate; ↑ fibroblasts of stroma turn into ↑ decidual cells, stroma becomes edematous and thickens to 5–6 mm. Spiral arteries lengthen and form dense coils. 3) Menstrual phase or menses (days 1–5): In absence of ↑ fertilization, i.e., in absence of ↑ human chorionic gonadotropin, LH level decreases, corpus luteum stops producing progesterone, FSH secretion is once again stimulated, and a new follicle develops. Simultaneously, functionalis (Fu) shrinks, coiled arteries are compressed, inducing an ischemia followed by extravasation of blood into stroma. These hematomas, together with liberation of lysosomal enzymes from cells of endometrium, cause endometrium to rupture and functionalis to be shed, with exception of basalis (B), and to be eliminated with menstrual blood.

↑ Ovarian follicles, growth of

Menstrual phase: The period of compression of endometrial blood vessels followed by hemorrhage and desquamation of functionalis during ↑ menstrual cycle.

Menstruation (menstrual flow, menses): A bloody vaginal discharge lasting roughly 4 days at end of each ↑ menstrual cycle.

Merkel's cells (cellula tactus*): Clear, round or elliptical cells located in basal layer of ↑ epidermis; particularly frequent where ↑ dermis is well vascularized and innervated. Nucleus is highly lobulated with a prominent nucleolus. Cytoplasm contains numerous rod-shaped mitochondria, a well-developed supranuclear Golgi apparatus (G), some short rough endoplasmic cisternae, a moderate number of free ribosomes, and bundles of peripheral microfilaments (Mf). From Golgi apparatus arise membrane-bound, electron-dense granules (SG), about

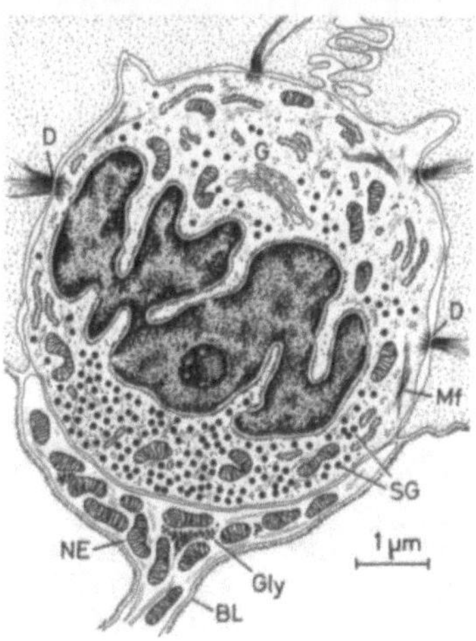

100 nm in diameter, which accumulate predominantly in basal cell portion. M. are held in contact with adjacent cells by small ↑ desmosomes (D). Basal cell surface is closely associated with a discoidal enlargement of a nonmyelinated afferent ↑ nerve ending (NE) containing many mitochondria and ↑ glycogen particles (Gly). Exact function of M. is unknown; they are thought to belong to ↑ paraneurons and represent potential neuroreceptor cells. BL = ↑ basal lamina

Garant, P.R., Feldman, J., Cho, M.I., Cullen, M.R.: Am. J. Anat. *157*, 155 (1980); Gottschaldt, K.M., Vahle-Hinz, C.: Science *214*, 183 (1981); Nurse, C.A., Mearow, K.M., Holmes, M., Visheau, B., Diamond, J.: Cell Tissue Res. *228*, 511 (1983); Tachibana, T., Tokio, N.: J. Anat. *131*, 145 (1980); Turner, D.F.: Anat. Rec. *205*, 197 (1983)

Merkel's corpuscle: ↑ Corpuscle, of Merkel

Merocrine secretion: ↑ Eccrine secretion

Meromyosins: ↑ Subunits of ↑ myosin molecule.

Mesangial cells (deep cells, mesangiocytus*): Roughly spindle-shaped or irregular cells situated between capillary loops (Cap) of ↑ renal glomerulus and embedded in ↑ mesangial matrix (MM), except at contact sites with other M. Nucleus is elongated, often deeply invaginated, and rich in ↑ heterochromatin; nucleolus is small. Cytoplasm is dark and contains a few small mitochondria, a moderately to well-developed Golgi appara-

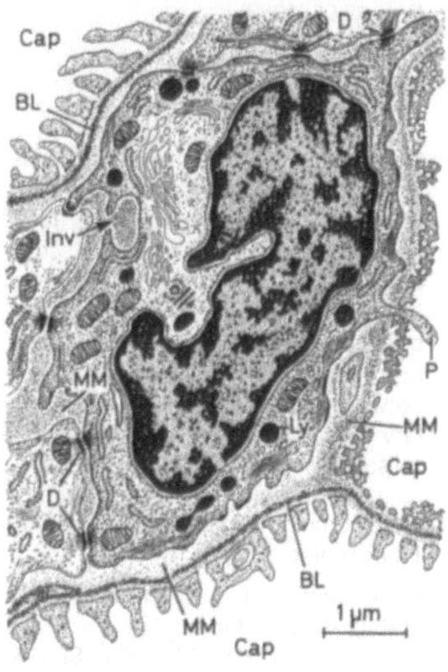

tus, ↑ centrioles, some short and flattened rough endoplasmic cisternae, a few lysosomes (Ly), subplasmalemmal actin microfilaments, and a considerable number of free ribosomes. Adjacent M. are held together by numerous ↑ desmosomes (D). Extensions (P) of M. penetrate between capillary loops and at times into capillaries. Deep invaginations (Inv) of cell surface are filled with mesangial matrix: For this reason, it is believed that M. phagocytize parts of ↑ basal laminae (BL), contributing in this way to clearing of large protein molecules retained here during filtration. M. are continuous with extraglomerular ↑ mesangium.

Ausiello, D.A., Kreisberg, J.I., Roy, C., Karnovsky, M.J.: Kidney Int. *16*, 804A (1979); Romen, W., Morath, R.: Virchows Arch. B Cell Pathol. *31*, 205 (1979)

Mesangial matrix (lamella hyalina*): An amorphous, basal lamina-like material (arrowheads) almost completely surrounding ↑ mesangial cells (MC). M. is of lesser electron density than lamina densa (arrows) and continuous with lamina rara interna of ↑ glomerular basal lamina. P = ↑ pedicels, of podocytes (Fig. = rat)

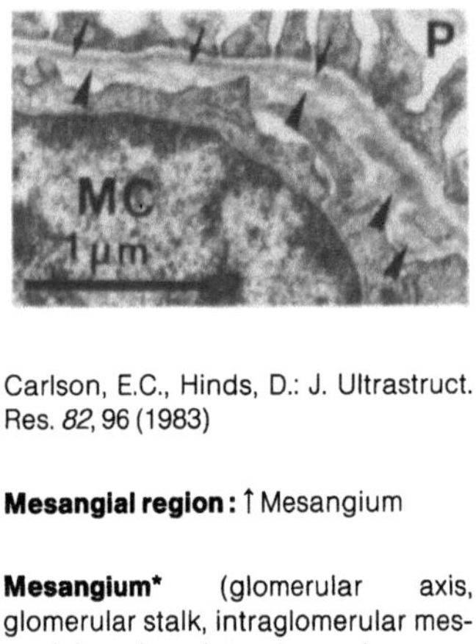

Carlson, E.C., Hinds, D.: J. Ultrastruct. Res. *82*, 96 (1983)

Mesangial region: ↑ Mesangium

Mesangium* (glomerular axis, glomerular stalk, intraglomerular mesangial region, intraglomerular mesangium, mesangial region): A conically shaped central supporting structure (M) for ↑ renal glomerulus (G). M. begins at ↑ vascular pole (VP) of a ↑ renal corpuscle and penetrates deeply between glomerular lobules and capillary loops (CL). M. is composed of ↑ mesangial cells (MC) and a ↑ mesangial matrix in which they are embedded. M. is continuous with extraglomerular ↑ mesangium (EM).

Burkholder, P.M.. Lab. Invest. *46*, 239 (1982); Roll, F., Madri, J.A., Furtmayr, H.: J. Cell Biol. *85*, 597 (1980); Steffes,

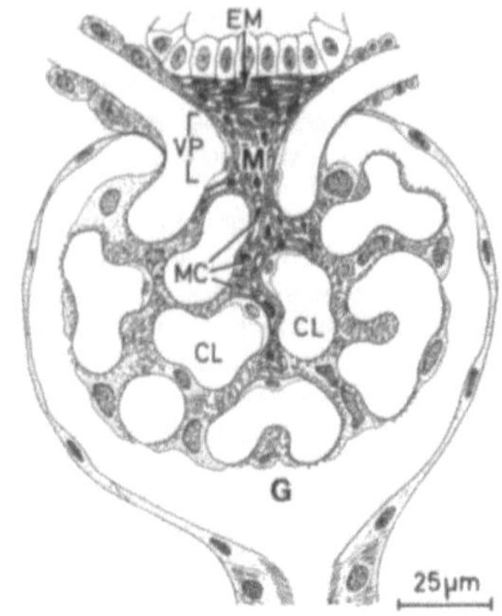

M.W., Barbosa, J., Basgen, J.M., Sutherland, D.E.R., Najarian, J.S., Mauer, M.S.: Lab. Invest. *49*, 82 (1983)

Mesangium, extraglomerular (Goormaghtigh's cells, lacis cells, polar cushion, Polkissen, insula juxtavascularis*): A group of conically arranged cells (EM) at ↑ vascular pole of a ↑ renal glomerulus immediately underlying ↑ macula densa (MD). Laterally, M. is delimited by ↑ afferent (Aff) and ↑ efferent arterioles (Eff) and continues into intraglomerular ↑ mesangium (M). M. is composed of flattened stellate cells with a central elliptical nucleus and condensed chromatin. Cytoplasm contains few ↑ organelles, some small lysosomes, and occasional secretory granules. Slender extensions of cells form an irregular network (lacis) devoid of blood and lymphatic vessels, but with occasional ↑ adrenergic nerve endings. Each cell is surrounded by a distinct basal lamina (not shown).

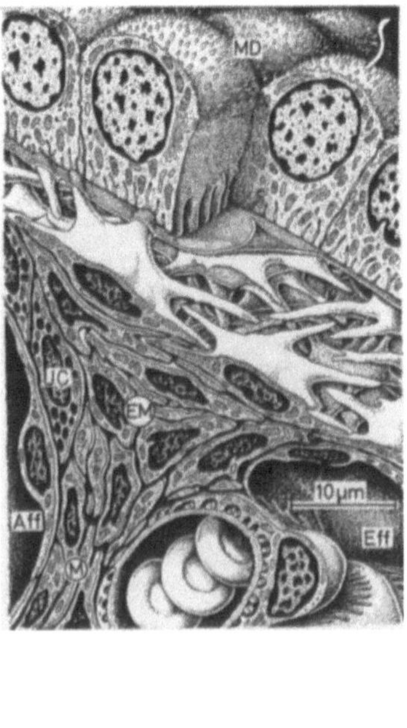

Some cells are in contact with ↑ juxtaglomerular ones (JC). Function of M. is unknown.

Spanidis, A., Wunsch, H., Kaissling, B., Kriz, W.: Anat. Embryol. *165*, 239 (1982)

Mesaxon*: A narrow cleft (M) formed by apposed segments of plasmalemma (P) of ↑ neuroglial cells extending from their surface to surround an ↑ axon (Ax). A) Transverse sections: I) Peripheral nervous system (PNS) = 1) ↑ Nonmyelinated nerve fibers have M. leading from basal lamina (BL) of ↑ Schwann's cell (Sch) to enclosed axons. 2) In ↑ myelinated nerve fibers, M. is subdivided into two parts in the course of wrapping of ↑ myelin lamellae around axon: Outer M. (MO) extends from surface of Schwann's cell to outermost myelin lamella (My), and inner M. (MI) extends between innermost lamella and axon. II) Central nervous system (CNS) = 3) Myelinated

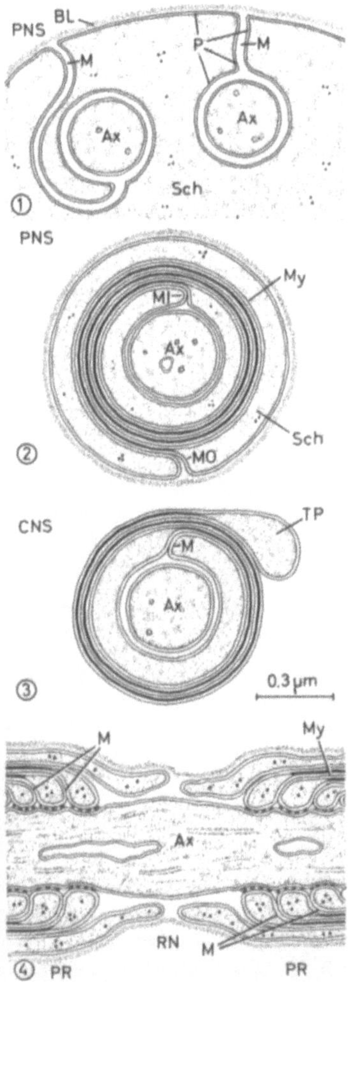

nerve fibers have only one M. leading from innermost myelin lamella to axon. B) Longitudinal section: 4) In ↑ paranodal regions (PR) of myelinated nerve fibers of both PNS and CNS, as many M. are present as myelin lamellae (My) enveloping axon. TP = ↑ tongue process of oligodendrocyte, RN = ↑ node, of Ranvier

Inokuchi, T., Higashi, R.: Arch. Histol. Jpn. *43*, 221 (1980)

Mesenchymal bone: ↑ Bone, membranous

Mesenchymal cells: Small stellate cells, 7–10 μm in diameter, with several extensions (P) of variable length, held in contact by means of ↑ nexus (N) with those of adjacent M. to form ↑ mesenchymal tissue. Nucleus is relatively large and rich in ↑ heterochromatin; it contains several nucleoli. In thin cytoplasmic rim are scattered some small mitochondria (M), an inconspicuous Golgi apparatus (G), a few short rough endoplasmic cisternae, and an abundance of free ribosomes. M. represent stem cells which give rise to a large number of tissues and organs by ↑ differentiation. It is believed that some M. also persist in adults to furnish stem cells of several tissues and organs, in case of need. M. have a high ↑ mitotic index; they possess ameboid and phagocytic capacities.

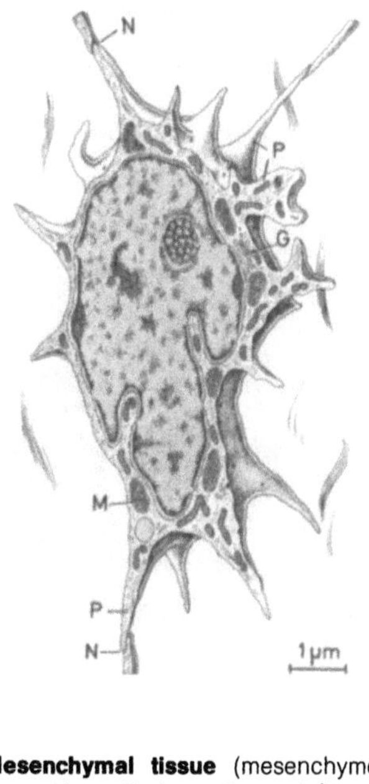

Mesenchymal tissue (mesenchyme, mesenchyma*): An embryonal ↑ connective tissue originating almost completely from ↑ mesoderm. Small part of M. originates from ↑ ectoderm. M. is composed of stellate or polymorphous ↑ mesenchymal cells (M) forming a tridimensional network. Intercellular spaces are large; they contain a still amorphous intercellular substance, occasional ↑ wandering cells (W), and capillaries (C). As an undifferentiated tissue, M. has a number of potentials and gives origin to: 1) All ↑ connective and supporting tissues; 2) ↑ myocardium and almost all ↑ smooth muscle cells (exceptions – iris muscles and ↑ myoepithelial cells); 3) ↑ endocardium, ↑ endothelium, and ↑ mesothelium; 4) ↑ lymph nodes, ↑ spleen, and ↑ lymphatic vessels; 5) ↑ blood cells; 6) ↑ synovial membranes of ↑ joints and bursae. (Fig. = human)

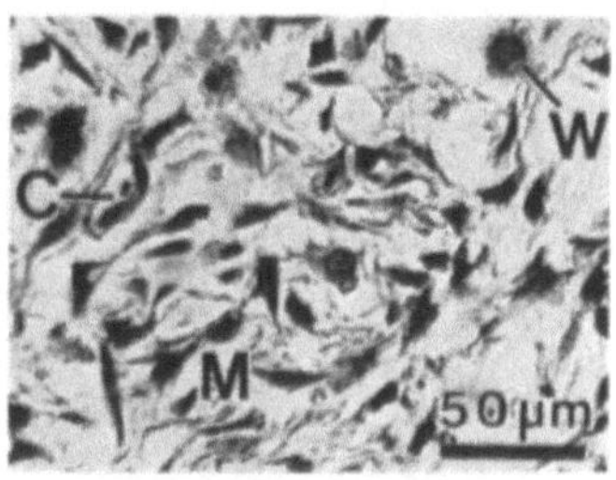

Mesenchyme: ↑ Mesenchymal tissue

Mesentery (mesenterium*): Two closely apposed sheets of ↑ peritoneum (P) attaching some segments of ↑ gastrointestinal tract to dorsal abdominal wall. Between these sheets lies ↑ loose connective tissue with ↑ fibroblasts (F), ↑ wandering cells, blood and lymphatic vessels, and nerve fibers; presence of ↑ milky spots. C = ↑ collagen fibers, LC = ↑ lymphatic capillary (Fig. = mongolian gerbil)

↑ Retiform tissue

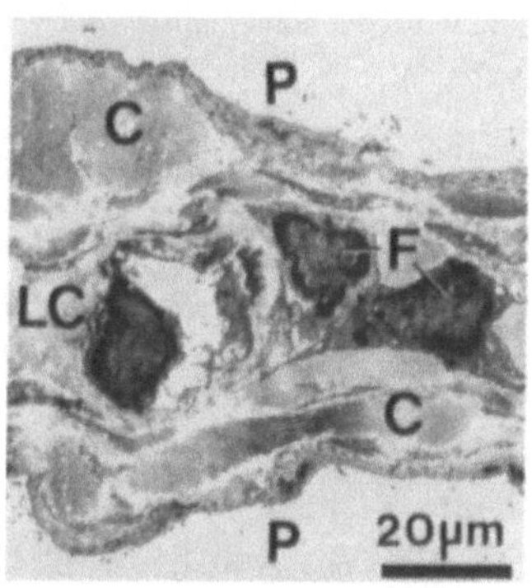

Gotloib, L., Digenis, G.E., Rabinovich, S., Medline, A., Oreopoulos, G.D.: Nephron *34*, 248 (1983)

Mesoblastic phase, of hematopoiesis: ↑ Hematopoiesis, prenatal

Mesoderm (mesoderma*): The middle primitive germ layer giving rise, via ↑ mesenchymal tissue, to ↑ connective and supporting tissues, ↑ cardiovascular system, ↑ lymphatic system, ↑ bone marrow, almost all ↑ smooth muscle cells, etc. (See embryology texts for further information.)

Mesoglia: ↑ Microglia

Mesosalpinx: A fold of ↑ peritoneum enclosing ↑ oviducts; structurally identical to ↑ mesentery.

Mesotendineum* (mesotenonium): A fold of a ↑ serous membrane (M) enclosing a segment of some ↑ tendons (T), connecting them to their fibrous vagina (V). M. consists of two closely apposed sheets of ↑ mesothelial cells between which run blood and lymphatic vessels and nerve fibers that enter and leave the tendon. M. is continuous with both sheets of ↑ synovial vagina (SV) of tendon. VC = vaginal cavity

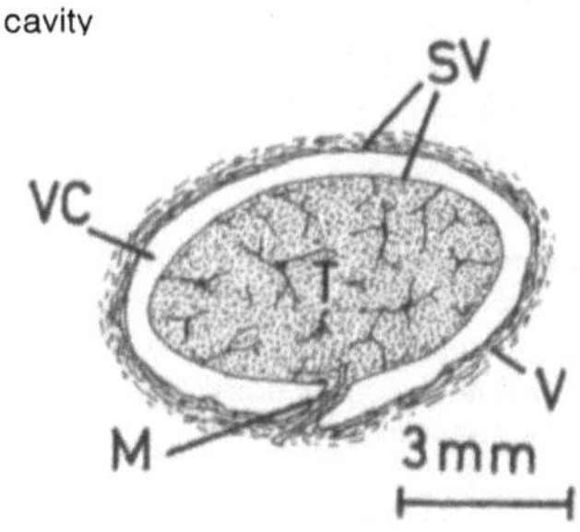

Mesotenonium: ↑ Mesotendineum

Mesothelial cells: Platelike cells forming ↑ mesothelium. M. have a flattened nucleus with a small nucleolus, a few small mitochondria, an inconspicuous Golgi apparatus (G), short rough endoplasmic cisternae, some ↑ lysosomes (Ly), and a moderate number of free ribosomes. Surface plasmalemma forms many ↑ micropinocytotic vesicles (Mp) and irregular microvilli (Mv) of variable length. Lateral cell borders are indented with adjacent M. and held together by junctional complexes (J) composed of ↑ zonulae occludentes and ↑ zonulae adherentes. M. lie on ↑ basal lamina (BL). They are of mesenchymal origin and contribute to intensive exchange of substances between ↑ serous cavities and underlying tissue; under certain conditions, M. may abandon their epithelial union and differentiate into ↑ macrophages. Cap = capillary

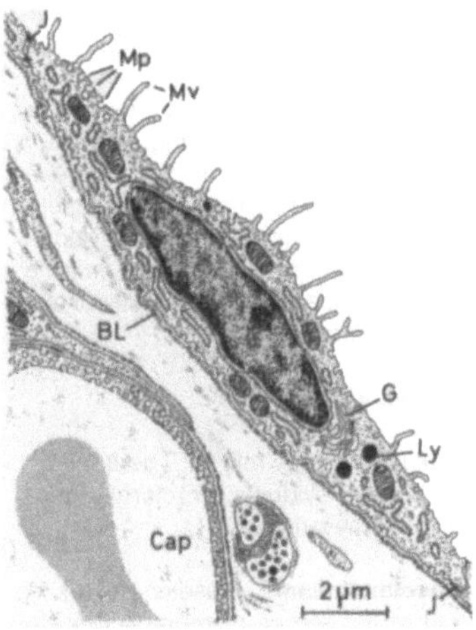

Baradi, A.F., Rao, S.N.: Tissue Cell *8*, 159 (1976)

Mesothelium*: A collective term for all epithelia originating from splanchnopleuric and somatopleuric ↑ mesenchyme. ↑ Peritoneum, ↑ pericardium and ↑ pleura are lined by M., which is also believed to line ↑ synovial vaginae of tendons.

↑ Mesothelial cells

Mesovarium*: A fold of ↑ peritoneum enclosing ↑ ovary; structurally identical to ↑ mesentery. Peritoneal ↑ mesothelium is continuous with ↑ germinal epithelium.

Messenger RNA (mRNA): A kind of ↑ ribonucleic acid engaged in transcription and transport into cytoplasm of genetic information from the nucleus. M. is synthesized in ↑ nucleolus under influence of RNA polymerase and passes into cytoplasm and links several ↑ ribosomes to ↑ polyribosomes. ↑ Transfer RNA uses its ↑ anticodons to read ↑ codons of M.; ribosomes move along M. molecule and slide off it after polypeptide chain has been synthesized. The released M. disappears rapidly, except in anuclear cells. Depending on length of message written on it, M. can measure 300–600 nm. (See molecular biology texts for further information)

Metachromasia: A term designating a staining reaction in which the stained material becomes colored differently (purple-red) from that of applied dye (blue). M. is due to presence of sulfated ↑ glycosaminoglycans such as ↑ heparin and ↑ chondromucoprotein. Metachromatic dyes are certain basic dyes (toluidin blue, azure II, thionin, methyl violet, etc.) Exact reason for change in absorption spectrum is unknown.

Metachronal rhythm: A component of ↑ ciliary motion.

Metamyelocyte, basophilic (metamyelocytus basophilicus*): A round cell, about 10–12 µm in diameter, gradually transformed within red ↑ bone marrow into basophilic ↑ granulocyte. Nucleus is horseshoe-shaped, rich in condensed ↑ heterochromatin, and without a nucleolus. Cytoplasm is relatively abundant and contains some mitochondria; a Golgi apparatus (G) and rough endoplasmic reticulum are poorly developed since synthesis of basophilic granules (BG) ceases in this stage of ↑ granulopoiesis. Cytoplasm also contains ↑ centrioles (C), a few free ribosomes, and small lysosomes (Ly).

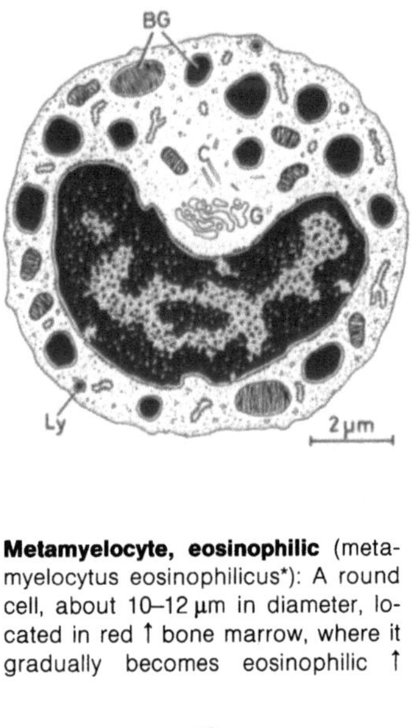

Metamyelocyte, eosinophilic (metamyelocytus eosinophilicus*): A round cell, about 10–12 µm in diameter, located in red ↑ bone marrow, where it gradually becomes eosinophilic ↑ granulocyte. Nucleus, horseshoe-shaped and filled with an appreciable amount of dense ↑ heterochromatin, contains no nucleolus. Since at this stage of ↑ granulopoiesis the production of ↑ eosinophilic granules has practically stopped, cell ↑ organelles, such as mitochondria, Golgi apparatus (G) and rough endoplasmic reticulum, are poorly developed. In addition to ↑ centrioles (C), cytoplasm contains several large eosinophilic granules with a crystalline core (EG) and some ↑ azurophilic granules (AG) considered to be primary ↑ lysosomes.

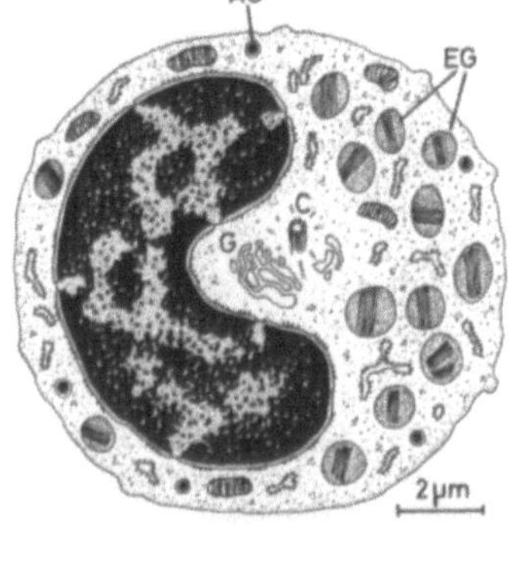

Metamyelocyte, neutrophilic (juvenile cell, metamyelocytus neutrophilicus*): A cell of granulopoietic series which gradually differentiates through ↑ band neutrophil into a neutrophilic ↑ granulocyte. Cell body is oval and measures about 12 µm in diameter. In early M., nucleus is reniform; in late M., it becomes horseshoe-shaped and rich in ↑ heterochromatin. Nucleoli are absent. Cytoplasm is abundant and contains sparse rod-shaped mitochondria, centrioles (C), a small number of free ribosomes, and some ↑ glycogen particles. Since M. do not produce specific granules, Golgi apparatus (G) is poorly developed, and the sporadic rough endoplasmic cisternae are short and flattened. Larger azurophilic granules (AG) and smaller specific granules (SG) are also present in cytoplasm in a ratio of 1 : 3 or 1 : 4. M. show ameboid movements and do not divide.

↑ Granulopoiesis; ↑ Granulocyte, neutrophilic, granules of

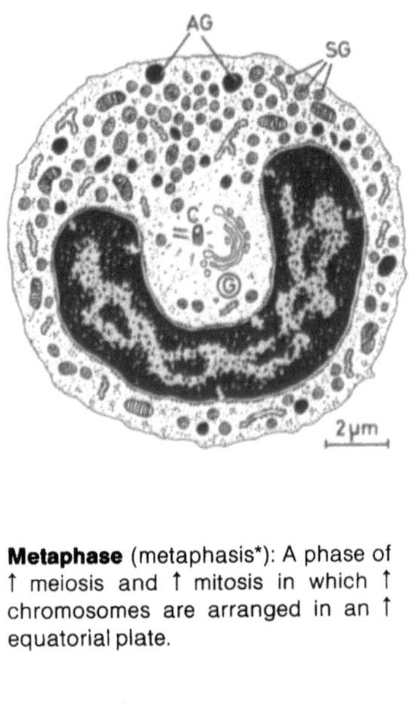

Metaphase (metaphasis*): A phase of ↑ meiosis and ↑ mitosis in which ↑ chromosomes are arranged in an ↑ equatorial plate.

Metaphysis*: A conical bone segment (M) connecting ↑ diaphysis (D) with ↑ epiphysis (E).

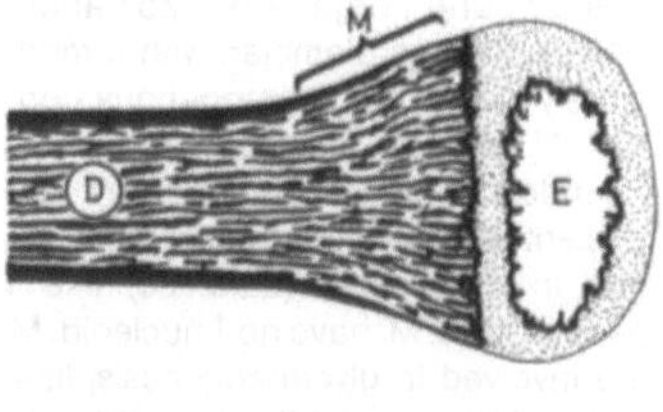

Metaplasia: The transformation of a mature, fully differentiated tissue of one class into a fully differentiated tissue of another. M. is possible only in epithelial tissues and, partially, in connective tissues. Examples: A) Experimental transformation of ↑ ductus deferens to ↑ ureter provokes transformation of ↑ pseudostratified epithelium of ductus (1) into ↑ transitional epithelium (2). B) Under pathological conditions, pseudostratified bronchial epithelium (1) changes into ↑ stratified squamous epithelium (2). C) Continuous stimulation of nonkeratinized stratified squamous epithelium (1) leads to its transformation into keratinized epithelium (2). D) Accumulation of fat in ↑ reticular cells (1) causes them to change into white ↑ adipose cells (2). The cited examples are in general reversible if the factors provoking M. disappear (interrupted lines).

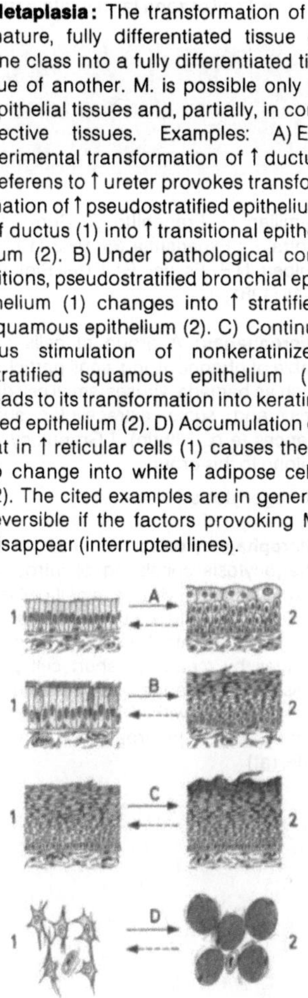

Woodworth, C.D., Mossman, B.T., Craighead, J.E.: Lab. Invest. *48*, 578 (1983)

Metaplasm: A former collective term designating all microfilamentous and microtubular structures occurring in cells during their ↑ differentiation (↑ gliofilaments, ↑ myofilaments, ↑ neurofilaments, ↑ tonofilaments, ↑ intermediate filaments, ↑ microtubules, ↑ neurotubules, etc.). M. is now classed among ↑ organelles according to their new definition.

↑ Cell, structure of

Metarterioles (precapillary arteriole, arteriola precapillaris*): Terminal conical segments of ↑ arterioles, 10–100 μm in length; caliber gradually diminishes to 5 μm. Structure: 1) Tunica intima = a) ↑ endothelial cells (EC) similar to those in arteriole; b) subendothelial layer = a few ↑ collagen and ↑ elastic fibers (EF), no internal ↑ elastic lamina. 2) Tunica media (TM) = one layer of helicoidally arranged ↑ smooth muscle cells, sometimes in contact with endothelial cells [↑ myoendothelial junctions (MJ)]; at end of M., some muscle cells are replaced by ↑ pericytes. 3) Tunica adventitia (TA) = a ↑ loose connective tissue with ↑ macrophages, ↑ mast cells (M), and ↑ nonmyelinated nerve fibers (NF). Intermittent contractions of M. at intervals of 2–8 s open and close arteriolar-capillary communication, so that new quanta of blood may be rhythmically distributed to ↑ capillaries. ↑ Sympathetic nervous system controls these contractions.

↑ Cardiovascular system; ↑ Microvascular bed; ↑ Microvasculature; ↑ Precapillary sphincter area

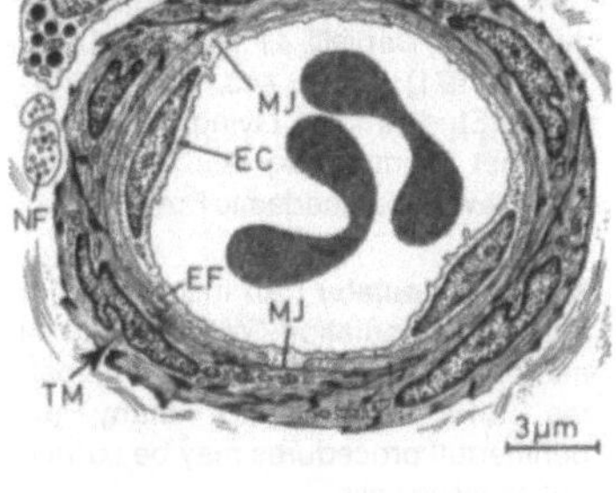

Metarubricyte: ↑ Erythroblast, orthochromatic

Methyl green–pyronin staining: A combined ↑ staining method for simultaneous demonstration of nucleic acids; DNA (desoxyribonucleotids of ↑ chromatin) is stained green and RNA (ribonucleotids of ↑ nucleolus and ↑ ergastoplasm) is stained red by pyronin.

Meynert's cells: ↑ Neurons found in ↑ striate area.

Microanalyzer, electron probe: ↑ Electron probe microanalyzer

Microbodies: ↑ Peroxisomes

Microdensitometer: An instrument for measuring optical density of photographic emulsions by analysis of light absorption. Some M. analyze secondary image produced in a microscope instead of photographic negative; in this case, M. is employed as a ↑ cytophotometer. Newer models of M. are equipped with an automatic scanning device to make serial scans of specimen; such instruments draw isodensity profiles, which can be converted into colors through a television system and/or analyzed by a computer. M. are used for light-microscopic quantitative analysis of products after cytochemical reactions, as well as for ↑ automatic cell recognition. In transmission electron microscope M. is employed for distinguishing ↑ cytomembranes, ↑ desmosomes, ↑ crystals, etc. For some authors, M. is synonymous with ↑ microspectrophotometer.

Altmann, F.P.: Microscopica Acta *79*, 327 (1977); Bloom, A.: Instruments and Techniques in Microdensitometry. In: G.L. Wied (ed.): Introduction to Quantitative Cytochemistry. New York and London, Academic Press 1966; Brugal, G., Chassery, J.-M.: Histochemistry *52*, 241 (1977)

Microdensitometry: A technique for measurement of concentration or mass of a substance in an optically defined region of a cell with a ↑ microdensitometer.

Bitensky, L.: Microdensitometry. In: Trends in Enzyme Histochemistry and Cytochemistry. Ciba Foundation symposium 73. Amsterdam Oxford New York, Excerpta Medica 1980

Microfibrils, collagen: ↑ Collagen microfibrils

Microfibrils, reticular: ↑ Reticular microfibrils

Microfilaments: Fine, proteinaceous threads, 5–12 nm in diameter, found singly and in bundles in cytoplasm of a great variety of cells. 1) Part of M. are contractile and responsible for cell motility, ↑ exocytosis, ↑ endocytosis, cytoplasmic streaming, etc. (↑ actin and ↑ actinlike M.). 2) A group of M. permit muscle contraction (↑ myofilaments). 3) A group of M. participate principally in dynamic organization of cytoplasm, i.e., formation of ↑ cytoskeleton and reinforcement of cell unions (↑ gliofilaments, ↑ intermediate microfilaments, ↑ neurofilaments, ↑ tonofilaments).

↑ Metaplasm

Elements of Organization: Microfilaments. In: Organization of the Cytoplasm, Part 2. New York: Cold Spring Harbor Symposia on Quantitative Biology, Vol. *46*, 1982

Microfluorometry: A technique for measurement of low concentrations of fluorescent products of histochemical reactions using a special ↑ fluorescence microscope.

Microfoliae: Smooth-surfaced tongue- or scalelike projections, 1–2 µm wide, 3–4 µm high, and roughly 0.5 µm thick, emerging from surface of squamous epidermal cells of sole. M. are thought to interdigitate with corresponding grooves of adjacent cells, thus reinforcing cell union.

Fujita, T., Tanaka, K., Tokunaga, J.: SEM Atlas of Cells and Tissues. Tokyo, New York: Igaku-Shoin 1981

Microglia* (Hortega's cells, mesoglia): Small stellate ↑ neuroglial cells located predominantly along capillaries (Cap) of CNS (inset). Slender extensions of M. are short with spikelike projections; oval nucleus (N) contains an appreciable amount of ↑ heterochromatin. In the scant ↑ perikaryon are scattered some mitochondria, a sometimes extensive Golgi apparatus, and a number of ↑ lysosomes (Ly), ↑ phagosomes (Ph), ↑ residual bodies (RB), and free ribosomes. ↑ Cytoplasmic matrix of M. is denser than that of other neuroglial cells. For visualization under light microscope, M. require specific metallic staining methods; identification under transmission electron microscope is difficult. M. are motile cells functioning as ↑ macrophages in CNS.

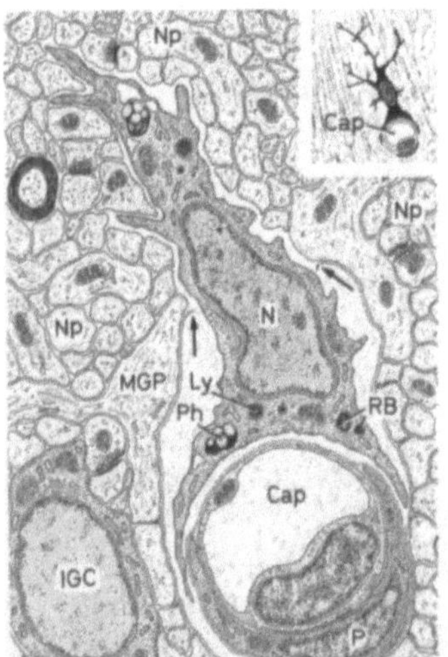

From capillaries, M. break ↑ basal lamina (arrows), cross ↑ membrana limitans gliae perivascularis (MGP), and wander into ↑ neuropil (Np) to phagocytize debris of ↑ myelin and dead cells, thereby forming, in some cases, ↑ compound granular corpuscles. It seems likely that M. originate from mesectoderm of neural crest and migrate along blood vessels into CNS during the period of its vascularization. M. are therefore thought to be differentiated ↑ pericytes (P) or ↑ intermediate glial cells (IGC). Another source for M. could be ↑ piarachnoid. Some authors believe that M. derive from ↑ monocytes.

Imamoto, K., Fujiwara, R., Nagai, T., Maeda, T.: Arch. Histol. Jpn. *45*, 505 (1982); Ling, E.A.: The Origin and Nature of Microglia. In: Fedoroff, S., Hertz, L. (eds.): Advances in Cellular Neurobiology, Vol. 2. New York; Academic Press 1981; Murabe, Y., Sano, Y.: Cell Tissue Res. *225*, 469 (1982); Tseng, C.Y., Ling, E.A., Wong, W.C.: J. Anat. 136, 251 and 837 (1983)

Micromanipulation: The finest mechanical handling procedure performed on living cells under light microscope using ↑ micromanipulators. M. includes dissection of a cell, extraction and replacement of its nucleus, grafting of one cell onto another, microinjection of substances, etc.

Jeon, K.W., Danielli, J.F.: Int. Rev. Cytol. *30*, 49 (1971); Kopac, M.J.: Microchirurgical Studies on Living Cells. In: Brachet, J. and Mirsky, A.E. (eds.): The Cell. New York: Academic Press 1959

Micromanipulator: An instrument for ↑ micromanipulation. With the aid of microknives, microhooks, micropipettes, and microneedles, many experimental procedures may be carried out on living cells.

Micrometer (µm): 10^{-4} cm or 10^3 nm; this term replaces former designation "micron."

Micrometer: A device for fine adjustment of focus in ↑ light microscope; can be used to measure thickness of ↑ sections

Micrometer, object: A glass disc with ruled scale (in general 1 mm, graduated in 0.01 mm), used for calibration of ocular ↑ micrometer.

Micrometer, ocular: A graduated disc mounted in ↑ ocular; when calibrated with object ↑ micrometer, it permits direct measurement of objects under light microscope.

Micron: A former term for 10^{-4} cm.

Microperoxisomes: ↑ Unit membrane-bound ↑ organelles (Mp), about 0.15–0.25 µm in diameter, with a moderately osmiophilic homogenous content. M. arise by bulging from smooth ↑ endoplasmic reticulum; they are present in a variety of cells. Although rich in peroxidase (catalase) like ↑ peroxisomes, M. have no ↑ nucleoid. M. are involved in glyconeogenesis, lipid metabolism, and synthesis of steroids. (Fig. = ↑ cell of zona fasciculata, rat)

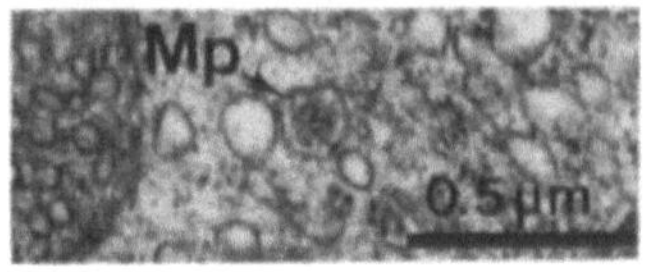

Familiari, G., Franchitto, G., Correr, S., Motta, P.: Experientia *35*, 1503 (1979); Tolbert, N.E., Essner, E.: J. Cell Biol. *91*, 271s (1981)

Microphages: A group of cells capable of phagocytizing small particles, such as bacteria, viruses, small cell debris, and some extracellular substances (e.g., ↑ fibrin). ↑ Granulocytes belong to this group.

↑ Phagocytosis

Microphagocytosis: A kind of ↑ phagocytosis consisting of introduction of bacteria, viruses, small cell debris, or extracellular substances, such as ↑ fibrin (F), into the body of ↑ microphages by means of short cell processes (P). Phagocytized material is broken down within ↑ heterophagosomes. (Fig. = neutrophilic ↑ granulocyte, rat)

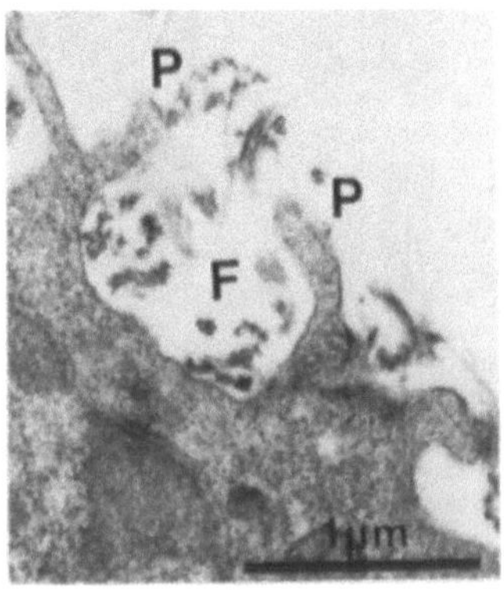

Microphotometer: ↑ Cytophotometer

Microphotometry: ↑ Cytophotometry

Micropinocytosis: ↑ Endocytosis of a fluid with dissolved substances from extracellular space in ↑ micropinocytotic vesicles.

Micropinocytosis vermiformis: ↑ Wormlike bodies

Micropinocytotic vesicles (plasmalemmal vesicles, vesicula pinocytotica*): Small spheres (M), measuring about 50–60 nm, found in a great variety of cells and are formed by invagination of their plasmalemma (P). Filled with dissolved substances, M. separate from plasmalemma and penetrate cell body. Here, they can fuse with one another and form larger vacuoles (V), whose contents are degraded by ↑ lysosomes (Ly); in some cases, limiting membrane of M. gradually disappears and their contents are liberated into ↑ cytoplasmic matrix. M. may occur uncoated or coated (Mc); the latter seem to contain predominantly dissolved proteinaceous material. In ↑ endothelial and ↑ smooth muscle cells, M. are known as ↑ caveolae. M. play an important role in ↑ transcytosis of macromolecules across capillary endothelium.

↑ Coated vesicles; ↑ Lysosomes, function of; ↑ Transendothelial channels

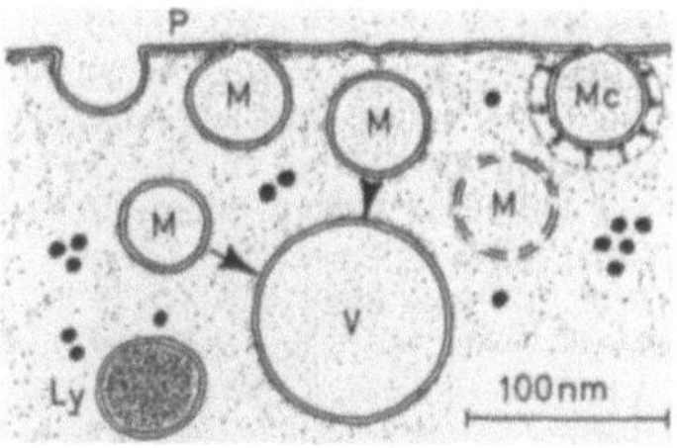

Inoué, T.: Scan. Electron Microsc. 1981/IV, 1 (1981); Inoué, T.: Biomed. Res. *2*, Suppl. 83 (1981); Frøkjaer-Jensen, L.: J. Ultrastruct. Res. *73*, 9 (1980); Singer, I.I.: Exp. Cell Res. *122*, 251 (1979)

Microplicae (microridges, microvillar ridges, vermiform ridges): Minute vermiform folds of free surface of various cells (gastric mucosa, ↑ facet cells, nephron tubules, ↑ corneal epithelium, etc.). M. are thought to increase cell surface area, to play a role in holding and spreading ↑ mucus and other secretion products, and to maintain moisture on free surface. (Fig. = facet cell, rat)

Andrews, P.M.: J. Cell Biol. *68*, 420 (1976); Sperry, D.G., Wassersug, R.J.: Anat. Rec. *185*, 253 (1976)

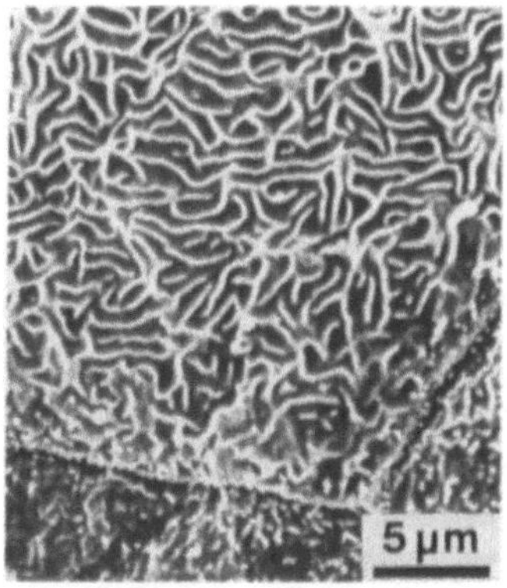

Microprobe, electron: ↑ Electron probe microanalysis

Microridges: ↑ Microplicae

Microscope: An optical or electronic instrument permitting observations of minute objects not visible with the naked eye.

Microscope, bright-field: ↑ Bright-field microscopy

Microscope, dark-field: ↑ Dark-field microscopy

Microscope, electron, high-voltage: ↑ Transmission electron microscope, high-voltage

Microscope, electron, scanning: ↑ Scanning electron microscope

Microscope, electron, scanning transmission: ↑ Scanning transmission electron microscope

Microscope, electron, transmission: ↑ Transmission electron microscope

Microscope, fluorescence: ↑ Fluorescence microscope

Microscope, interference: ↑ Interference microscope

Microscope, light: ↑ Light microscope

Microscope, numerical aperture of: ↑ Numerical aperture

Microscope, phase contrast: ↑ Phase contrast microscope

Microscope, polarizing: ↑ Polarizing microscope

Microscope, resolving power of: ↑ Resolving power

Microscope, ultraviolet: ↑ Ultraviolet microscope

Microscopic preparation, in electron microscopy: The succession of technical and technological procedures in preparing cells and tissues for study under an ↑ electron microscope. (Terms marked with an asterisk are explained under corresponding letters)

↑ Electron microscopy

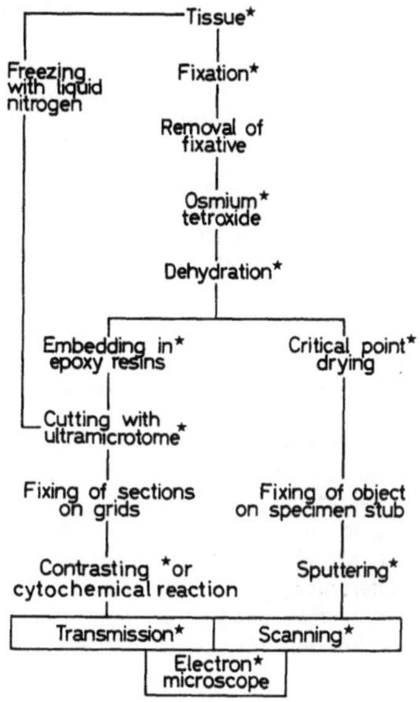

Hayat, M.A.: Principles and Techniques of Electron Microscopy. Vol. 1. 2nd edn. Baltimore: University Park Press 1981

Microscopic preparation, in light microscopy (histological preparation):

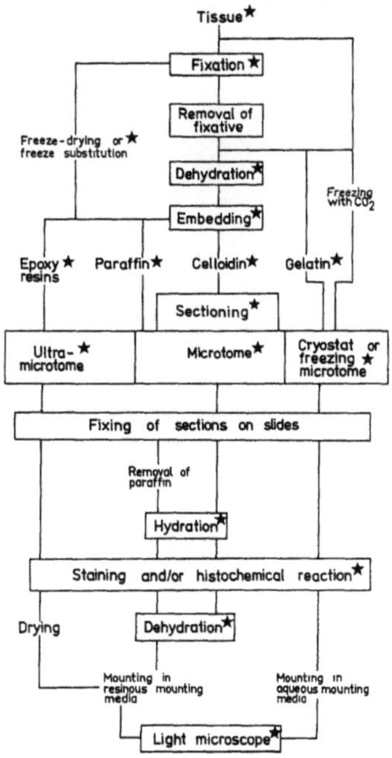

The succession of technical and technological procedures in preparing cells, tissues, or organs for analysis with the ordinary ↑ light microscope. For other types of light microscope, more sophisticated methods have been developed. (Terms marked with an asterisk are explained under corresponding letters) (Modified after Bucher 1980)

Bucher, O.: Cytologie, Histologie und mikroskopische Anatomie des Menschen. 10th edn. Bern: Hans Huber 1980; Burck, H.-C.: Histologische Technik. 5th edn. Stuttgart: Georg Thieme 1982; Locquin, M.V., Langeron, M.: Handbook of Microscopy. Borough Green: Butterworth 1982

Microsomes: Artificial structures (M) obtained as a consequence of fragmentation of rough ↑ endoplasmic reticulum (ER) during homogenization and ultracentrifugation of cells. M. appear as 0.1–0.3-µm vesicles with translucent contents; outer surface of their membrane is studded with ↑ ribosomes (R). Microsome fraction, separated by differential ↑ centifugation, is used for biochemical analysis. G = direction of centrifugal force

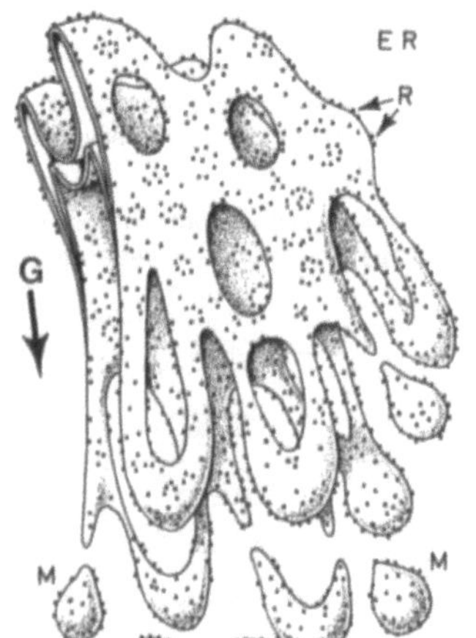

Lee, C.P., Schatz, G., Dallner, G.: Mitochondria and Microsomes. Amsterdam: Addison Wesley 1981

Microspectrophotometer: ↑ Cytospectrophotometer

Microspectrophotometry: A method for quantitative measurement of ↑ nucleic acids and certain other cytochemical substances using a special ↑ fluorescence microscope.

Ploem, J.S.: In: Trends in Enzyme Histochemistry and Cytochemistry. Ciba Foundation Symposium 73. Amsterdam, Oxford, New York: Excerpta Medica 1980

Microtome: An instrument for cutting ↑ sections for ↑ light microscopy. During rotation of wheel (W), a mechanical device pushes (horizontal arrow) block (B) toward ↑ knife (K) for a desired number of ↑ micrometers (1–40); a simultaneous vertical movement (vertical arrows) of block permits knife to cut a section (S). Repeated rotations of wheel furnish serial sections. Besides this rotary M., there are sliding M., some of which are specially adapted for sectioning of ↑ celloidin blocks.

↑ Ultramicrotome

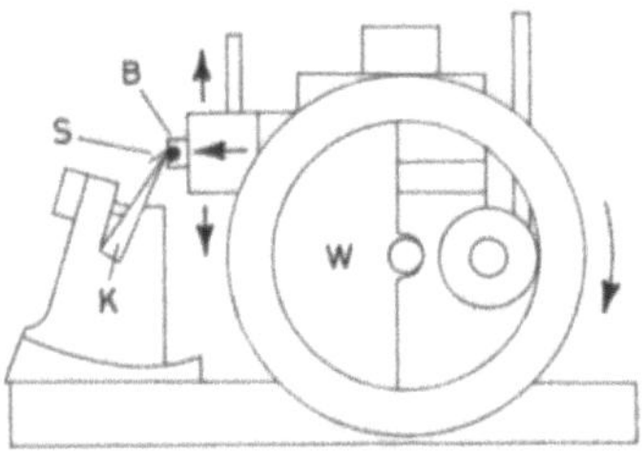

Microtrabeculae: Protein strands (M), 3–6 nm thick and of various length, forming an irregular tridimensional lattice within ↑ cytoplasmic matrix. On one side, M. are connected with ↑ microtubules (Mt), ↑ microfilaments, ↑ organelles, and ↑ ribosomes (R); on the other side, they are connected with ↑ ectoplasm (E) of plasmalemma (P). It is therefore believed that M. coordinate movement and orientation of these structures during cell motion. M. are visible only with high-voltage ↑ transmission electron microscope. M. belong to ↑ cytoskeleton. Mi = mitochondrion, rER = rough endoplasmic reticulum (Modified after Porter and Tucker 1981)

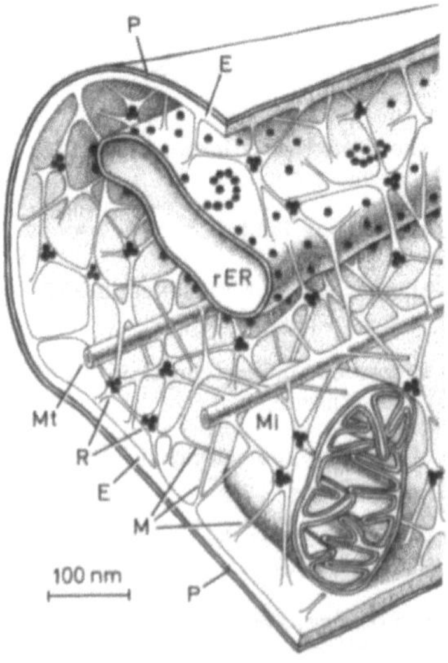

Porter, K.R., Tucker, J.B.: Sci. Am. *244*, 41 (1981); Porter, K.R., Anderson, K.L.: Eur. J. Cell Biol. *29*, 83 (1982)

Microtubule organizing center (MTOC, pericentriolar body, satellite): A globular accumulation of electron-dense material (MTOC), located at tip of ↑ basal foot (BF) and around ↑ centriole as ↑ satellites. M. is able to initiate assembly of ↑ microtubules (Mt) of centriole, ↑ basal bodies, and ↑ spindle apparatus.

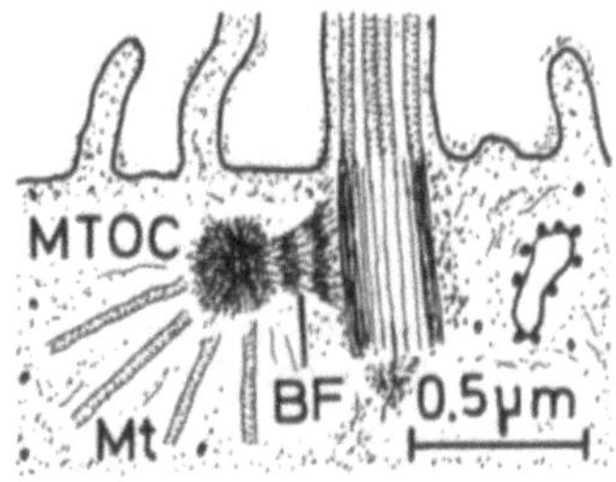

Brenner, S.L., Brinkley, B.R.: Tubulin Assembly Sites and the Organization of Microtubule Arrays in Mammalian Cells. Cold Spring Harbor Symposia on Quantitative Biology, Vol. 46, Part 1. New York: Cold Spring Harbor Laboratory, 1982; Pearson, P.J., Tucker, J.B.: Cell Tissue Res. *180*, 241 (1977); Raff, E.C.: Int. Rev. Cytol. *59*, 2 (1979); Watt, F.H., Harris, H.: J. Cell Sci. *44*, 103 (1980)

Microtubules (microtubulus*): Hollow, cylindrical, nonramified ↑ organelles (Mt), about 24 nm wide and several micrometers long, with a roughly 15-nm-wide electron-lucent hole and a 5-nm-thick wall. Wall consists of 13 parallel, helicoidally arranged protofilaments; each of these is 5 nm in diameter and made up of globular, 5-nm subunits of α- and β- ↑ tubulin, which unite to form tubulin dimers. M. are found in almost all cells: 1) Singly, scattered throughout cytoplasm; 2) in bundles, connected by fine bridges (kinetochoral M. of ↑ spindle apparatus, ↑ manchette, ↑ neurotubules); 3) fused, to form doublets (↑ axoneme of ↑ kinocilia and ↑ flagellum) and triplets (↑ basal body, ↑ centriole). M. are essential for cell division; they play an important role in maintaining cell shape (↑ cytoskeleton), cell movements and morphogenesis, cell polarity, ↑ endocytosis, ↑ exocytosis, and intracellular transport and distribution of organelles, ↑ secretory granules, ↑ synaptic vesicles, ↑ coated vesicles, and, probably, water and metabolites. M. are involved in distribution and displacement of cell-surface receptors and translation of information carried by ↑ hormones. Evidence suggests that M. transport information throughout the cell body, from cell surface to cell interior, and vice versa. M. are formed by polymerization of tubulin subunits by ↑ microtubule organizing center; shortening of M. is caused by depolymerization of subunits. The natural alkaloids colchicin, vinblastin, and vincristin block polymerization of M. (Fig. = ↑ follicular cell, of ovarian follicle)

↑ Microtubules, in morphogenesis; ↑ Microtubules, kinetochoral

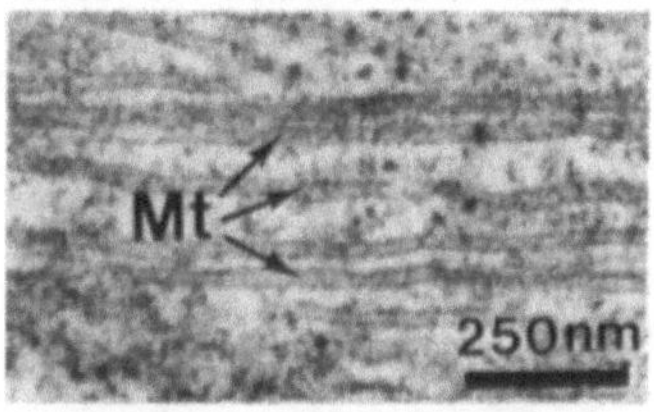

Brabander, M.J., de: La Recherche *145*, 810 (1983); Dustin, P.: Sci. Am. *243/2*, 58 (1980); Elements of Organization: Microtubules. In: Organization of the Cytoplasm, Part 1. New York: Cold Spring Harbor Symposia on Quantitative Biology, Vol. *46*, 1982; Heuser, J.E., Kirschner, M.W.: J. Cell Biol. *86*, 212 (1980); Roberts, K., Hyams, J.S., (eds.): Microtubules. New York: Academic Press 1980; Sakai, H., Mohri, H., Borisy, G.G., (eds.): Biological Functions of Microtubules and Related Structures. Tokyo: Academic Press 1982

Microtubules, assembly of: ↑ Tubulin

Microtubules, chromosomal: ↑ Microtubules, kinetochoral

Microtubules, continuous: ↑ Microtubules, polar

Microtubules, in morphogenesis: The action of ↑ microtubules on determination of cell shape. Examples: Development of ↑ axon depends upon parallel alignment of microtubules, i.e., ↑ neurotubules; parallel orientation of microtubules causes elongation of cells of lens vesicle to form ↑ lens fibers. A good example of morphogenetic role of microtubules is ↑ manchette of ↑ spermatid and microtubular bundles of inner and outer ↑ pillars.

Microtubules, kinetochoral (chromosomal microtubules, chromosomal "fibers"): ↑ Microtubules, about 20 nm in diameter, attached to ↑ kinetochores; they transmit or produce force necessary for ↑ chromosome movement during cell division. Large chromosomes have more M.

Moens, P.B.: J. Cell Biol. *83*, 556 (1979); Nicklas, R.B.: Chromosome Movements: Facts and Hypothesis. In: Little, M., Paveletz, N., Petzelt, H., Postingl, H., Schroeter, D., Zimmermann, H.P. (eds.): Mitosis, Facts and Questions. Berlin, Heidelberg, New York: Springer-Verlag 1977

Microtubules, of platelets: ↑ Platelets

Microtubules, polar (microtubules, continuous): A group of ↑ microtubules of ↑ spindle apparatus that continue through cell body from one ↑ centriole to the other without anchoring into ↑ kinetochores. During late telophase, central segments of M. form ↑ mid-body.

Microvascular bed: Part of ↑ cardiovascular system between arterial and venous circulations composed of ↑ arterioles, ↑ metarterioles, ↑ capillaries, and ↑ venules. M. is actively engaged in exchange of gases, nutrients, and waste products between tissues and blood.

↑ Microvasculature

Microvasculature: Blood vessels less than 100 µm in diameter. M. consists of: 1) Common sequence = ↑ arteriole (A) – ↑ metarterioles (Met) – ↑ capillaries (Cap) – ↑ postcapillary venule (PCV) – ↑ venule (V); 2) arteriovenous ↑ anastomoses (AA); 3) arterial ↑ portal system = arteriole (A) – capillaries (Cap) – arteriole (A) – capillaries (Cap) – venule (V) (↑ renal glomerulus, Gl); 4) venous portal system = venule (V) – capillaries (Cap) – venule (V) (↑ liver, hypophyseoportal system). M. forms ↑ microvascular bed. (Some authors consider M. to be all blood vessels less than 500 µm in diameter.) PSA = ↑ precapillary sphincter area

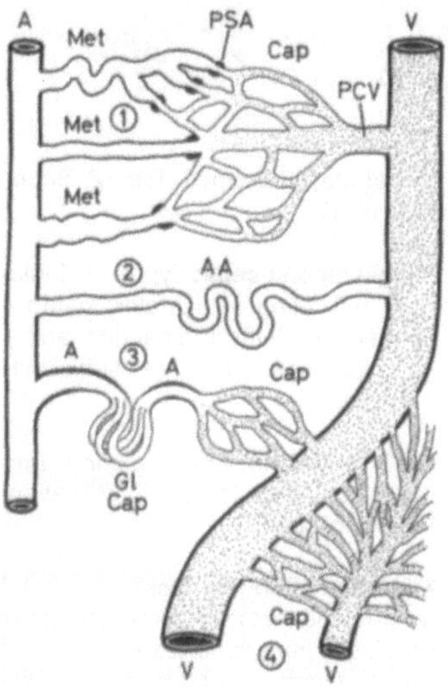

Anderson, B.G., Anderson, W.D.: Biomed. Res. *2*, Suppl. 209 (1981); Kaley, G., Altura, B.M., (eds.): Microcirculation. Vols. I–III. Baltimore: University Park Press 1977; Ohtani, O., Kikuta, A., Ohtsuka, A., Taguchi, T., Murakami, T.: Arch. Histol. Jpn. *46*, 1 (1983)

Microvillar contractile proteins: A common term for microfibrillar proteins (bundle of ↑ actin microfilaments; α-actinin), which combined with myosinlike microfilaments of ↑ terminal web induce shortening of ↑ microvilli.

Microvillar ridges: ↑ Microplicae

Microvilli*: Narrow (about 0.1 µm thick and 1–6 µm long), cylindrical cytoplasmic processes (Mv) projecting from free surface of various cells (↑ absorptive cells, ↑ proximal tubule cells of nephron, ↑ choroid plexus cells, ↑ syncytiotrophoblast, etc.), lined by plasmalemma (P) and covered by ↑ glycocalyx (G). Each M. encloses a compact bundle of 10–50 ↑ actin microfilaments (Ac) anchored in a dense, α-actinin (A)-containing zone at tip of M., running down M., and inserted into ↑ terminal web (TW) as rootlet filaments (RF). Together with myosinlike filaments (My) of terminal

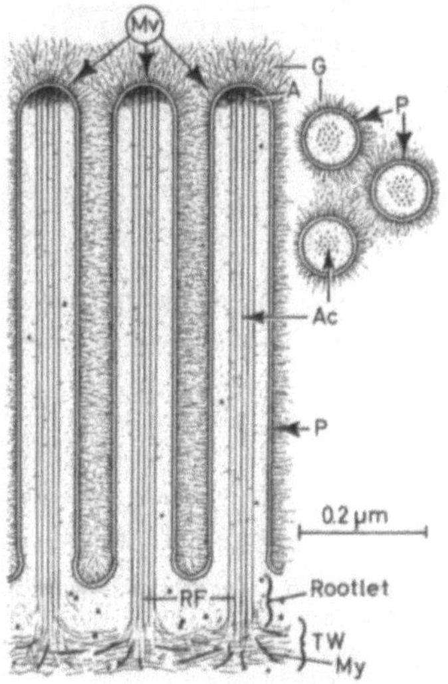

web, M. have ability to contract. Short M. (up to 2 µm) form ↑ striated border; longer M. (up to 6 µm) form ↑ brush border. Cells may also emit lateral M.; these are irregular, frequently branched, of various lengths, and devoid of a microfilamentous bundle. M. enormously increase absorptive cell surface area.

↑ Antennulae microvillares; ↑ Microvilli, shortening of

Bretscher, A.: Characterization and Ultrastructural Role of the Major Components of the Intestinal Microvillus Cytoskeleton. In: Cold Spring Harbor Symposia on Quantitative Biology, Vol. 46, Part 2. New York: Cold Spring Harbor Laboratory 1982; Drenckhahn, D., Gröschel-Stewart, U.: J. Cell Biol. *86*, 475 (1980); Loor, F.: Nature *269*, 272 (1976); Matsudaira, P.T., Burgess, D.R.: J. Cell Biol. *92*, 657 (1982)

Microvilli, shortening of: Several models to explain shortening of microvilli (Mv) have been proposed; all of them are based: 1) On presence of ↑ actin microfilaments (Ac) anchoring on tip of each microvillus on one side and in ↑ terminal web and ↑ zonula aderens (ZA) on the other; 2) on existence of myosinlike microfilaments (My) in terminal web. The interaction of these two kinds of microfilaments causes rootlet of a microvillus to slide deeper into cell, followed by a consecutive shortening of microvilli, probably with a pumping effect. (Fig. modified after Drenckhahn and Gröschel-Stewart 1980)

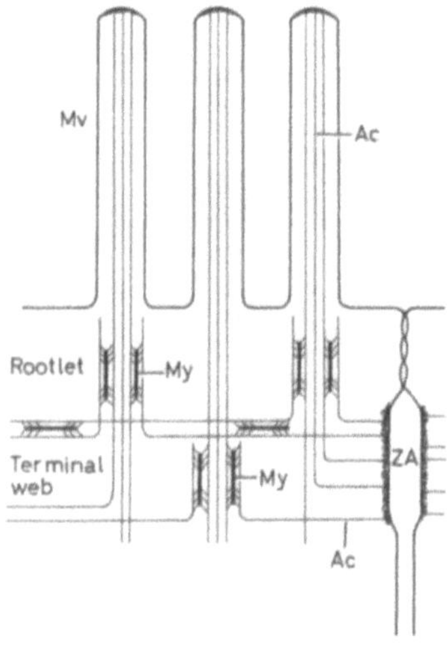

Drenckhahn, D., Gröschel-Stewart, U.: J. Cell Biol. *86*, 475 (1980); Hirokawa, N., Tilney, L.G., Fujiwara, K., Heuser, J.E.: J. Cell Biol. *94*, 425 (1982)

Mid-body (Flemming's body, intermediate body, relictum fusi*): A transitional, last link between two daughter cells before their definitive separation in the course of late telophase. M. contains densely associated polar ↑ microtubules (Mt) of ↑ spindle apparatus enclosed in an electron-dense material.

Mullins, J.M., McIntosh, R.J.: J. Cell Biol. *94*, 654 (1982)

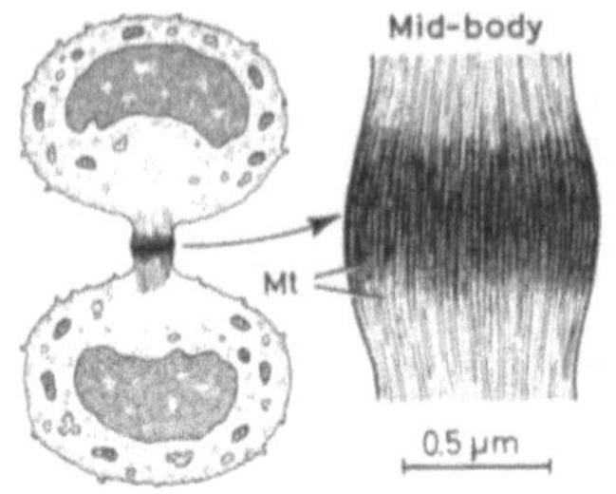

Mid-diaphyseal ring: ↑ Periosteal band

Middle ear (auris media*): Part of organ of hearing comprising ↑ tympanic membrane (TM), ↑ auditory tube (AT), ↑ auditory ossicles (AO), ↑ tympanic cavity (TC) with ↑ tensor tympani (TT) and ↑ stapedius muscle, and chorda tympani nerve.

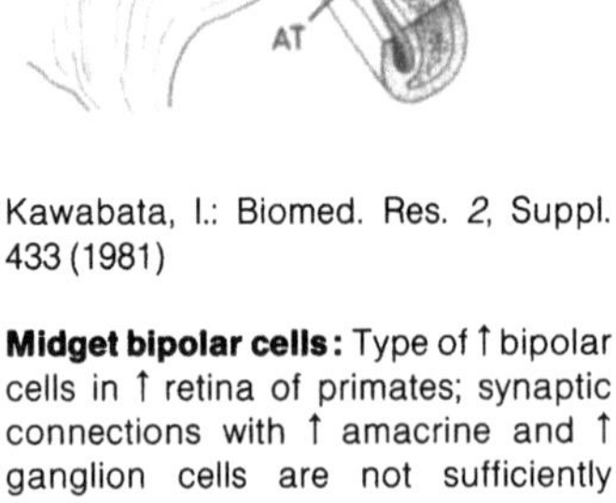

Kawabata, I.: Biomed. Res. *2*, Suppl. 433 (1981)

Midget bipolar cells: Type of ↑ bipolar cells in ↑ retina of primates; synaptic connections with ↑ amacrine and ↑ ganglion cells are not sufficiently understood.

MIF: ↑ Melanocyte-stimulating hormone-inhibiting factor; ↑ Migration-inhibiting factor

Migration-inhibiting factor (MIF): A ↑ lymphokine that immobilizes local and passing ↑ macrophages in vicinity of site of antigen- ↑ T- ↑ lymphocyte reaction, inducing them to function there instead of moving away.

Milk (lactus): The white to yellowish fluid secreted by lactating ↑ mammary gland under main influence of ↑ prolactin. Composition: 1) Fluid phase or milk plasma contains dissolved carbohydrates, vitamins, α-and β-lactalbumins, ↑ immunoglobulins, and many other substances in small quantities (sodium, potassium, ↑ iron, etc.). 2) Suspended particles: a) protein granules = ↑ casein granules (C) with bound calcium; b) lipid droplets (L) = mainly neutral triglycerides. 3) Desquamated epithelial cells and ↑ colostrum bodies. By transepithelial transport, ↑ immunoglobulin A, produced by ↑ plasma cells, is secreted in M. to protect newborn. Composition of M. varies between species: human M. consists of 88% water, 7% lactose, 4% fat, and 1% proteins. Color of M. originates from emulgated fat and content of ↑ carotenoids. (Fig. = rat) (See biochemistry texts for further information)

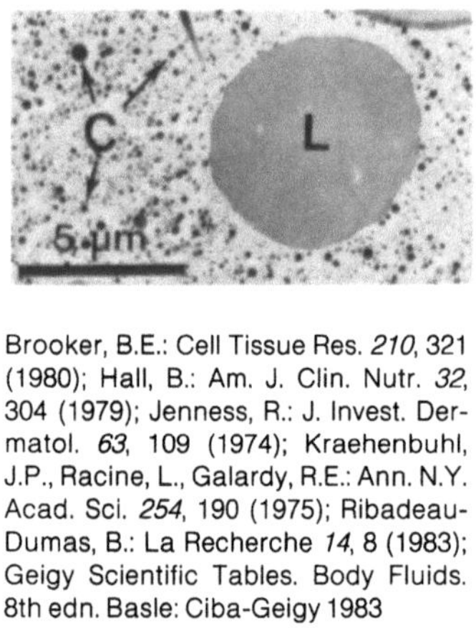

Brooker, B.E.: Cell Tissue Res. *210*, 321 (1980); Hall, B.: Am. J. Clin. Nutr. *32*, 304 (1979); Jenness, R.: J. Invest. Dermatol. *63*, 109 (1974); Kraehenbuhl, J.P., Racine, L., Galardy, R.E.: Ann. N.Y. Acad. Sci. *254*, 190 (1975); Ribadeau-Dumas, B.: La Recherche *14*, 8 (1983); Geigy Scientific Tables. Body Fluids. 8th edn. Basle: Ciba-Geigy 1983

Milk ejection reflex: ↑ Suckling reflex

Milk secretion: ↑ Lactation

Milky spots (macula lactea*): Opaque, well-vascularized aggregates of ↑ lymphocytes and ↑ macrophages located in ↑ mesentery and ↑ omenta, predominantly along blood vessels and covered by a discontinuous layer of ↑ mesothelial cells. M. also contain ↑ fibroblasts, ↑ granulocytes, ↑ plasma cells, and lymphatic ↑ capillaries. Exact function of M. is unknown; it is believed that they are involved in body defense system since their macrophages and lymphocytes can wander into peritoneal cavity and combat infections.

↑ Kampmeier foci

Hodel, Ch.: Eur. Surg. Res. *2*, 435 (1970); Mironov, V.A., Gusev, S.A., Barodi, A.F.: Cell Tissue Res. *201*, 327 (1979)

Millimicron (mµ): A former term for 10^{-7} cm; now replaced by ↑ nanometer (nm).

Mineralization: ↑ Calcification

Mineralocorticoids: A group of steroid hormones synthesized by zona glomerulosa of adrenal cortex; M. are involved in fluid and electrolyte balance. Among M. are ↑ aldosterone and deoxycorticosterone.

↑ Adrenal glands, cortex of

Minor compartment, of spermatozoon tail: ↑ Spermatozoon tail, principal piece of

MIT: ↑ Monoiodothyrosine

Mitochondria* (mitochondrium*): Membrane-bound, enzyme-loaded, semiautonomous ↑ organelles present in almost all cells. M. can be elliptical, ovoid, spherical, discoidal, rod-shaped, threadlike, etc., averaging 0.2–2 µm in width and 2–7 µm in length. Structure: 1) Outer membrane (OM) = ↑ unit membrane, about 5–7 nm thick; 2) ↑ outer chamber (OC) = an electron-lucent space, about 8–10 nm wide; 3) inner membrane (IM) = a unit membrane, about 5–7 nm thick, forming ↑ cristae (C), ↑ tubules, and vesicles; toward matrix this membrane encloses ↑ elementary particles (EP); 4) ↑ inner chamber (IC) with ↑ mitochondrial matrix, in which are concentrated many enzymes, ↑ mitochondrial DNA, ↑ mitochondrial granules (MG), ribosomes (R), and, sometimes, crystals, lipid droplets, and ↑ glycogen particles. Morphologically, M. can be divided into ↑ M. with cristae (fig.), ↑ M. with tubules, ↑ M. with longitudinal cristae, ↑ M. with prisms, and ↑ discochondria.

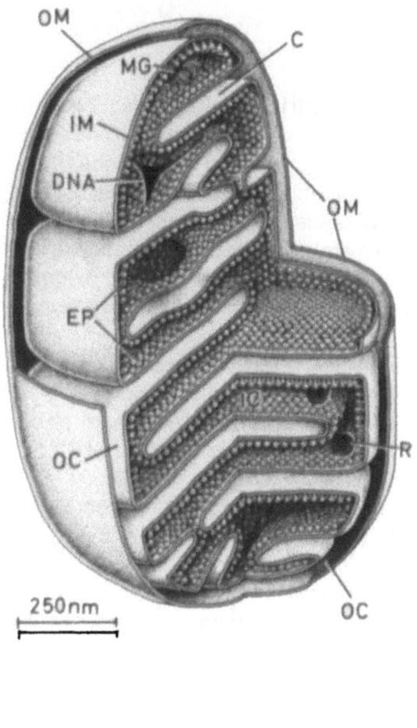

Bereiter-Hahn, J., Vöth, M.: Biol. Cell *47*, 309 (1983); Ernster, L., Schatz, G.: J. Cell Biol. *91*, 227s (1981); Lee, C.P., Schatz, G., Dallner, G.: Mitochondria and Microsomes. Amsterdam: Addison Wesley 1981; Tzagoloff, A.: Mitochondria. In: Sickevitz, P. (ed.): Cellular Organelle. New York: Plenum Press 1982; Whittaker, D.A.: Mitochondria: Structure, Function and Assembly. Harlow: Longman Group 1978

Mitochondria, biogenesis of: Although semiautonomous ↑ organelles, with their own DNA and RNA, mitochondria cannot reproduce without assistance of nuclear genetic program, since information content of ↑ mitochondrial DNA is insufficient for a completely autonomous biogenesis. 1) It is likely that nuclear genetic program controls synthesis of mitochondrial molecular components and that some essential parts of mitochondria are synthesized autonomously. 2) Radioautographic and other studies have shown that mitochondria divide (arrows) as shown in fig. (Fig. = skeletal muscle, rat)

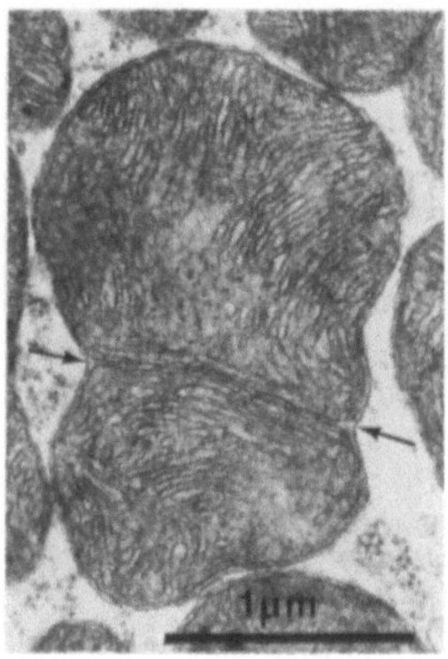

Tedeschi, H.: Mitochondria: Structure, Biogenesis and Transducing Function. Cell Biology Monographs, Vol. 4. Vienna: Springer-Verlag 1976

Mitochondria, elementary particles of: ↑ Elementary particles, of mitochondria

Mitochondria, functions of: Mitochondria are principal energy producers of cell. Energy is produced by Krebs tricarboxylic acid cycle and oxidative phosphorylation system, tightly coupled to respiratory chain. Besides oxygen, only fuel that mitochondria need is phosphate and adenosine diphosphate (ADP); principal final products are adenosine triphosphate (ATP), CO_2, and water. Mitochondria also oxidize and synthesize fatty acids, concentrate ↑ ferritin, proteins in crystalline form, accumulate some vital stains (↑ Janus green), transform themselves into ↑ yolk granules, and accumulate cations in ↑ mitochondrial granules. (See biochemistry texts for further information)

Mitochondria, life span of: After about 10 days, ↑ mitochondria are destroyed in ↑ autophagosomes.

Mitochondria, number of: M. varies from no mitochondria in ↑ erythrocytes to about 800 mitochondria in ↑ liver parenchymal cells.

Mitochondria, origin of: The so-called symbiont hypothesis postulates that ↑ mitochondria are intracellular parasites (endosymbiotic bacteria) that entered cytoplasm of higher cells in the course of evolution and established symbiotic relationship with them. Main proofs: 1) Outer mitochondrial membrane chemically resembles ↑ endoplasmic reticulum of higher cells; the penetrating parasite was thus enclosed within this membrane. 2) In inner mitochondrial membrane are located enzymes of respiratory chain as in membrane of bacteria. 3) Bacterial mesosomes – membranous projections comparable with mitochondrial ↑ cristae – also contain enzymes of respiratory chain like mitochondrial cristae. 4) Mitochondrial DNA is circular like bacterial DNA. 5) Protein synthesis in mitochondria and bacteria can be blocked by chloramphenicol, whereas extramitochondrial protein synthesis remains unaffected.

↑ Bioblast

Küntzel, H., Köchel, H.G.: Nature *293*, 751 (1981)

Mitochondria-rich cells: ↑ C-cells, of thyroid gland

Mitochondria, with cristae: The most frequent type of ↑ mitochondria, in which inner membrane (arrows) is thrown into numerous folds, ↑ cristae

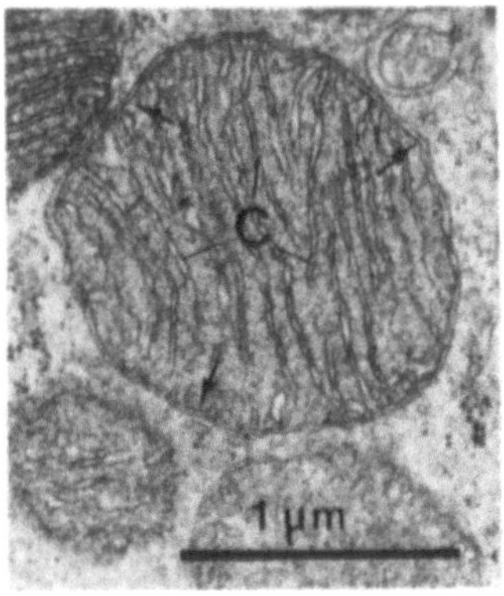

(C), in general perpendicularly oriented to longitudinal axis of ↑ organelle. (Fig. = brown ↑ adipose tissue, rat)

Mitochondria, with longitudinal cristae: A kind of mitochondria whose ↑ cristae (C) are parallel to long axis of ↑ organelle. Such M. are rare and found mostly in ↑ liver parenchymal cells. (Fig. = rat)

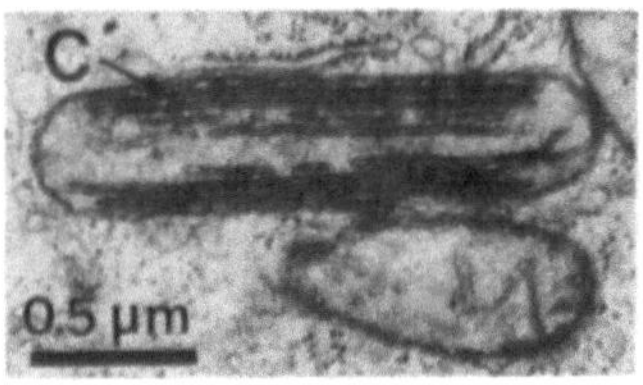

Mitochondria, with prisms: A kind of mitochondria, found in some ↑ neuroglial cells of ↑ brain, containing ↑ cristae (C) and a bundle of prism-shaped tubules (P). Function of prisms is unknown.

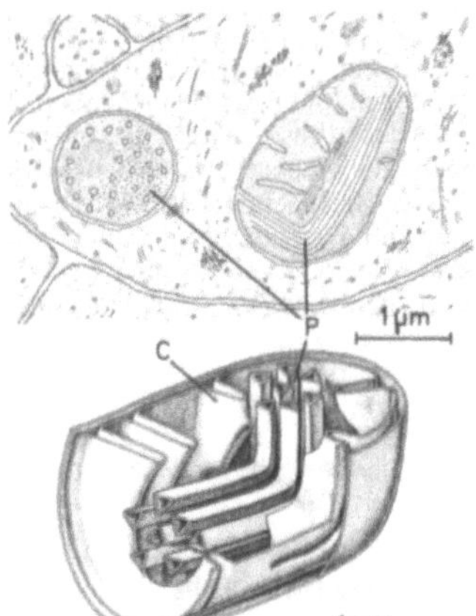

Duncan, D., Morales, R.: Anat. Rec. *175*, 519 (1973); Fernandez, B., Suarez, I., Gianonatti, C.: J. Anat. *137*, 483 (1983)

Mitochondria, with tubules: A type of mitochondria whose inner membrane makes numerous branched tubular projections (T), about 20 nm in diameter, toward ↑ mitochondrial matrix. Mitochondria of this type characterize cells that produce ↑ steroid hormones, such as adrenal cortex cells (fig.), ↑ follicular cells of ↑ ovarian follicles, ↑ interstitial cells of testis, etc. (Fig. = rat)

↑ Cells, of zona fasciculata; ↑ Cells, of zona glomerulosa; ↑ Discochondria; ↑ Tubules, of mitochondria

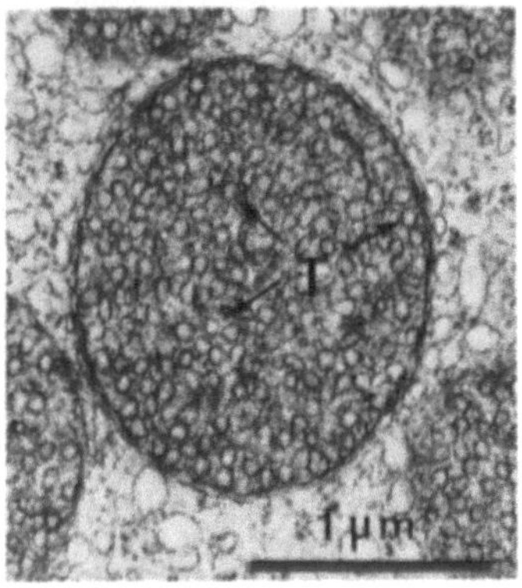

Munn, E.: The Structure of Mitochondria. New York: Academic Press 1975

Mitochondrial DNA (mtDNA): A highly twisted, double-stranded molecule having a circular shape. In mammalian cells, M. is about 5.5 µm long and 2 nm thick; a mitochondrion contains one or more DNA molecules. Amount of genetic information coded on M. is not sufficient to control synthesis of all mitochondrial proteins, but only about 5%–6% of them (probably insoluble proteins of inner membrane). Synthesis of all other proteins, as well as mitochondrial replication, is under control of nuclear DNA. M. has a higher guanine-cytosine content than the latter; M. resembles bacterial DNA, which is also circular and probably does not contain ↑ histones.

↑ Mitochondria, origin of

Grivell, L.A.: Sci. Am. *248/3*, 60 (1983); Slonimski, P., Borst, P., Attardi, G. (eds.): Mitochondrial Genes. Monograph 12. New York: Cold Spring Harbor Laboratory 1982

Mitochondrial genes: ↑ Mitochondrial DNA

Mitochondrial granules (intramitochondrial granules, granula mitochondrialia*): Highly osmiophilic membraneless particles (arrowheads), 20–50 nm in diameter, scattered in ↑ mitochondrial matrix. The bivalent cations Ca^{2+} and Mg^{2+} and others are thought to be concentrated in M. Under higher transmission electron microscopic magnification, M. show a fine granular structure, with an occasional lamellar pattern. (Fig. = ↑ liver parenchymal cell, rat)

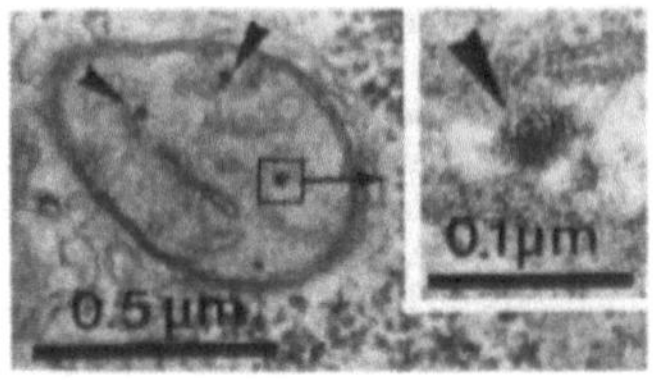

Barnard, T.: Scan. Electron Microsc. 1981/II, 419 (1981); Brdiczka, D., Barnard, T.: Exp. Cell Res. *126*, 127 (1980); Bullock, C.G., Gilmore, R.St.C., Wallace, W.F.M.: J. Anat. *135*, 835 (1982); Thomas, R.S., Greenawalt, J.W.: J. Cell Biol. *39*, 55 (1980)

Mitochondrial matrix (matrix mitochondrialis*): A homogeneous fine granular or filamentous substance (M) contained in ↑ inner chamber of ↑ mitochondria. M. is moderately osmiophilic and includes all soluble enzymes of citric acid cycle, those involved in oxidation of fatty acids, enzymes for protein synthesis, mitochondrial ribosomes, DNA, and ↑ mitochondrial granules (MG).

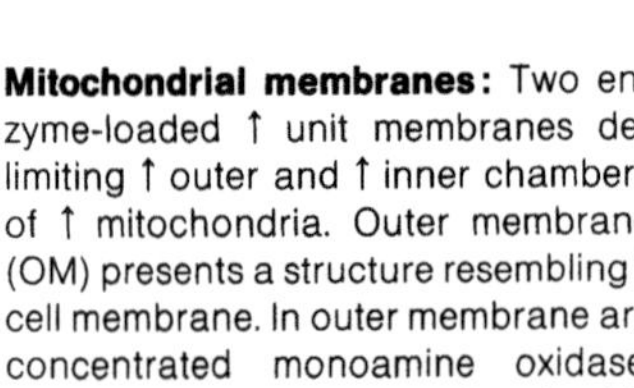

Mitochondrial membranes: Two enzyme-loaded ↑ unit membranes delimiting ↑ outer and ↑ inner chambers of ↑ mitochondria. Outer membrane (OM) presents a structure resembling ↑ cell membrane. In outer membrane are concentrated monoamine oxidase, nicotinamide-adenine dinucleotide cytochrome reductase, kynurenine hydroxylase, fatty acid coenzyme A ligase, adenosine triphosphate acyl coenzyme A synthetase, and phospholipases. Inner membrane (IM) contains ↑ elementary particles (EP), which become visible after special preparatory techniques when they protrude toward ↑ mitochondrial matrix (M). Inner membrane encloses respiratory chain enzymes (RC) localized along outer edge of ↑ cristae – adenosine triphosphate synthetase, succinate dehydrogenase, β-hydroxybutyrate dehydrogenase, carnitine fatty acid acyl transferase, and ferrochelatase. (Modified after De Roberts et al. 1975)

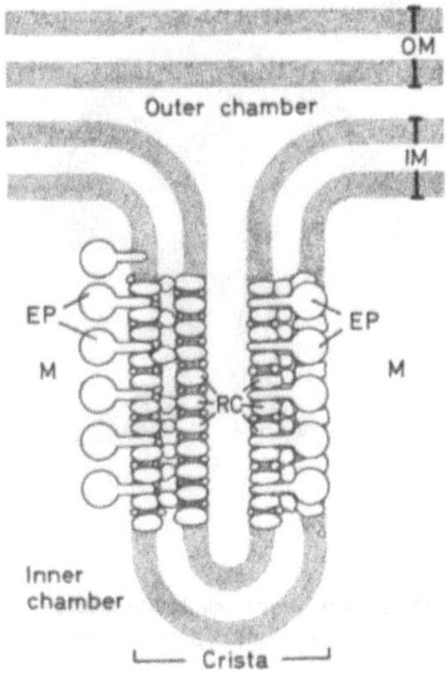

De Robertis, EPD, Saez, F.A., De Robertis, EMF, Jr.: Cell Biology, 5th edn. Philadelphia, London, Toronto: Saunders Co. 1975

Mitochondrial nucleus (nucleoid, genophore, mt-nucleus): DNA-containing area of ↑ mitochondrion, sometimes visible as an electron-transparent fine fibrillar zone in ↑ mitochondrial matrix.

Kuroiwa, T.: Int. Rev. Cytol. *75*, 1 (1982)

Mitochondrial sheath: Helicoidally arranged elongated ↑ mitochondria (M) joined end-to-end throughout middle piece (MP) of ↑ spermatozoon tail. Mitochondria surround outer ↑ dense fibers (ODF). To show M. clearly, mitochondria have been drawn at some distance from one another; in reality, they are tightly packed. M. produces energy for spermatozoon motility. M. extends to ↑ annulus (A). Ax = ↑ axoneme, P = plasmalemma (removed), PC = proximal centriole

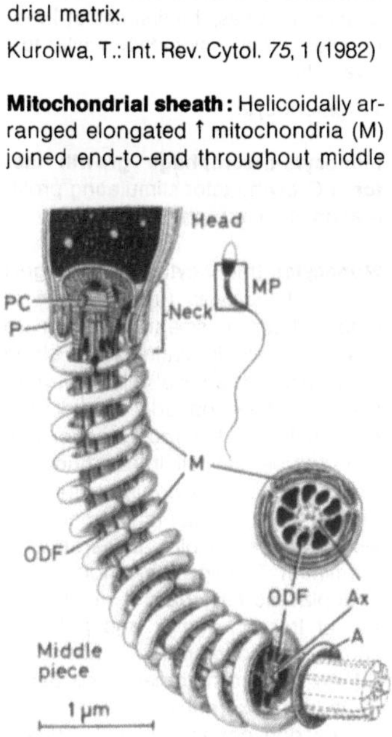

Mitogens: Nonantigenic substances that induce transformation of large lymphocyte populations into actively dividing ↑ immunoblasts. M. include ↑ lectins and ↑ ligands.

Mitosis (mitosis cellularis*): Indirect nuclear division (karyokinesis) followed by division of cell body (↑ cytokinesis); usual process of cell reproduction resulting in formation of two daughter cells with identical chromosome number (2 n) and deoxyribonucleic acid content (2 n DNA) to that of original cell. Before M., normal amount of genetic material in interphasic ↑ somatic cell nucleus (2 n DNA) is doubled (4 n DNA) during S phase of ↑ cell cycle. Sequence: Prophase: Gradual appearance of ↑ chromosomes (1), which become thicker, shorter, and coiled and form ↑ spireme (2). Division of centriole (2) and migration of two daughter ↑ centrioles, surrounded by ↑ achromatic apparatus (A, 2, 3), toward opposite pole of nucleus; fragmentation of rough endoplasmic reticulum and Golgi apparatus. Near end of prophase, both ↑ nuclear envelope and ↑ nucleolus (Nu) are lost from view. ↑ Karyoplasm mixes with cytoplasm, forming mixoplasm. At the same time, chromosomes split longitudinally into two identical halves (3), ↑ chromatids (4), held together by a ↑ centromere (4, C). Metaphase: Development of ↑ spindle apparatus. Regular arrangement of chromosomes about spindle equator, forming an ↑ equatorial plate (4, 5, EP). Viewed from one of poles, chromosomes form a starlike figure (↑ monaster). Anaphase: Begins by doubling and longitudinal division of centromeres. Identical sets of chromatids, now considered daughter chromosomes with their own centromeres, move toward opposite poles of spindle apparatus. Two chromosome groups are visible as a double star figure, called diaster (6). Divergent movement of daughter chromosomes is partially due to shortening (depolymerization) of ↑ microtubules of spindle apparatus. Bubbling of cytoplasm; appearance of numerous microvilli at cell surface. Anaphase is concluded when two chromosomal masses reach cell poles (7). Number of daughter chromosomes assembled at either cell pole at end of anaphase is exactly the same as number in cell at prophase (46 in man). Telophase: Daughter chromosomes grouped at cell poles begin to uncoil (8,9), become longer and indistinct, lose their stainability, and fuse to a single mass (10). With reconstruction of nuclear envelope and reappearance of nucleoli, karyokinesis is concluded. Mitochondria, lysosomes, free ribosomes, elements of Golgi apparatus, and cytomembranes are distributed in approximately equal parts to daughter cells. A deep constriction of cell body, beginning of cytokinesis, occurs midway between two newly formed nuclei. Remnant of spindle apparatus may persist for a brief time as a microtubular ↑ midbody in area of cleavage furrow. Finally, there is definitive separation of daughter cells, ending cytokinesis. Reconstruction phase: Daughter cells gradually become structured like those during ↑ interphase.

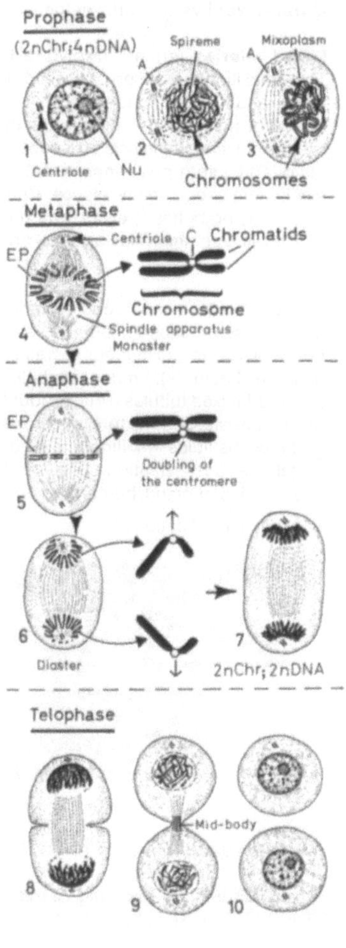

Bullough, W.S.: Biol. Rev. *50*, 99 (1975); Capco, D.G., Penman, S.: J. Cell Biol. *96*, 896 (1983); Dustin, P.: Bull. Ass. Anat. *63*, 121 (1979); Heneen, W.K.: Eur. J. Cell Biol. *25*, 242 (1981); Zimmermann, A.M., Forer, A. (eds.): Mitosis/Cytokinesis. In: Cell Biology: A Series of Monographs. New York, and London: Academic Press 1981

Mitotic index: Ratio of number of dividing cells in a tissue, regardless of phase, to number not undergoing ↑ mitosis; M. is expressed as a permillage.

Mitotic spindle: ↑ Spindle apparatus

Mitral cells: Large ↑ neurons with ↑ perikarya forming well-defined stratum mitrale of ↑ olfactory bulb. With their dendrites, M. participate in formation of ↑ olfactory glomeruli. Some ↑ axons of M. send impulses to granular layer, but majority of axons continue into lateral olfactory stria. Structurally, M. are identical to other neurons.

Mitral valve: ↑ Valves, atrioventricular

Mixed arteries (arteria mixotypica*): The transitional segment between ↑ elastic and ↑ muscular arteries characterized by presence in tunica media of islands of ↑ smooth muscle cells separating and interrupting fenestrated ↑ elastic laminae in many places. Certain aortic branches (axillary, external carotid, and common iliac arteries) are M.

Mixed glands (seromucous gland, glandula seromucosa*): Glands composed of serous and mucous secretory units, i.e., ↑ acini (A), ↑ mucous tubules (T), and ↑ mixed tubules (MT). Product of M. consists of a mixture of serous and mucous fluids resulting in a moderately viscous secretion. (Fig. = ↑ submandibular gland, human)

↑ Glands, classification of

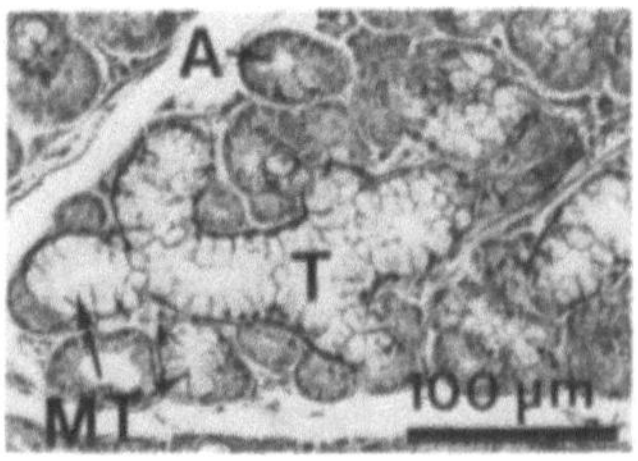

Mixed synapse: ↑ Synapse, mixed

Mixed tubules, of exocrine glands: Elements of ↑ mixed glands composed of a tubule of ↑ mucous cells (MC) with a ↑ demilune (D) of ↑ serous cells at the end. Secretory product of the latter passes through fine intercellular ↑ canaliculi (IC) to reach tubular lumen. Mucous part of M. is surrounded by stellate ↑ myoepithelial cells (My). ↑ Intercalated ducts (ID) of M. are very rare, since cells of duct transform themselves into mucous cells.

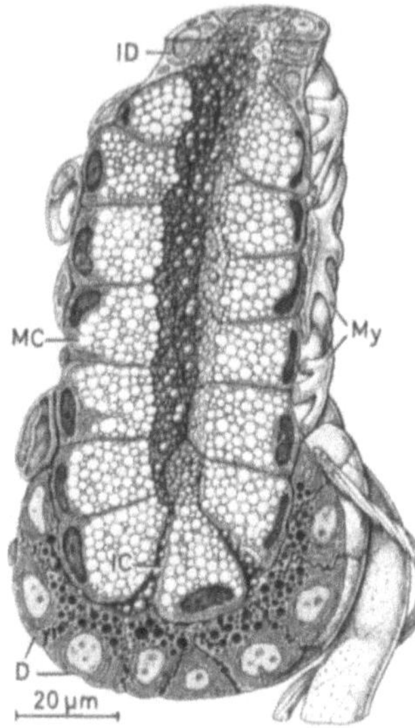

Mixoplasm: The ↑ karyoplasm mixed with ↑ cytoplasm after fragmentation of ↑ nuclear envelope at end of prophase of ↑ mitosis.

M-line (mesophragma*, linea M*): A dark line, about 85 nm wide, (M) situated in center of an ↑ A-band. M. is formed by central thickened segment of ↑ myosin myofilaments. In transversal sections, M. displays hexagonally arranged myosin myofilaments interconnected by six fine lateral extensions. M. of neighboring ↑ myofibrils are connected by transversally oriented ↑ intermediate ↑ desmin microfilaments. (Fig. = rat)

↑ Sarcomere, transversal sections of

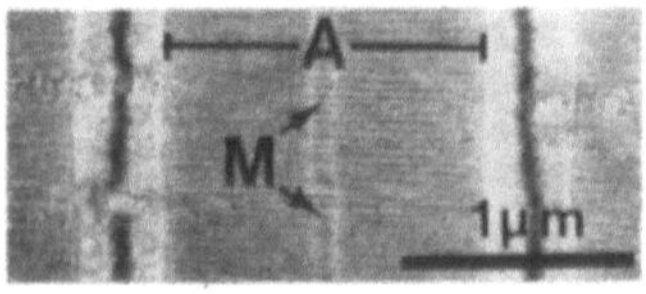

Luther, P., Squire, J.: J. Mol. Biol. *125*, 313 (1978); Sjöström, M., Kidman, S., Henrikson Larsen Ängquist, K.-A.: J. Histochem. Cytochem. *30*, 1 (1982); Wang, K., Ramirez-Mitchell, R.: J. Cell Biol. *96*, 562 (1983)

Modiolus*: A conical pillar (M) of cancellous ↑ bone forming bony axis of ↑ cochlea. In M. are situated ↑ spiral ganglion (SG) and ↑ axons of ↑ bipolar cells, which leave bone at base of M. ↑ Lamina spiralis ossea (LS) is a fine extension of M. IAM = internal ↑ auditory meatus

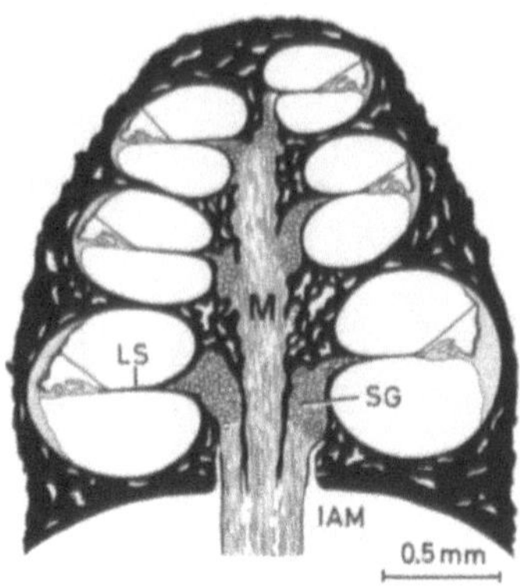

Molecular layer, of brain cortex: ↑ Brain cortex, homotypical isocortex of

Moll's glands: ↑ Ciliary glands

Monaster: A single star figure of ↑ chromosomes forming ↑ equatorial plate when viewed from one of cell poles. M. occurs in metaphase of ↑ mitosis.

Mongolism: ↑ Down's syndrome

Monoblast (monoblastus*): According to some theories, the first cell of ↑ monopoiesis, morphologically similar to ↑ myeloblast.

↑ Promonocyte

Monocyte-macrophage growth factor: ↑ Growth factor stimulating proliferation of ↑ macrophages.

Monocytes (monocytus*): The largest of all ↑ leukocytes (up to 17 µm in blood; 20 µm in smears) making up 2%–5% of all circulating white blood cells. An M. has an oval cell body, voluminous kidney-shaped eccentric nucleus rich in ↑ heterochromatin, and two to three nucleoli. It the abundant, weakly basophilic cytoplasm are situated numerous small mitochondria, a well-developed Golgi apparatus, a prominent ↑ diplosome, short rough endoplasmic cisternae, a small number of free ribosomes and polyribosomes, ↑ glycogen particles and 0.2 to 0.4-µm, ↑ unit membrane-limited, dense granules (azurophilic under light microscope) corresponding to primary

↑ lysosomes (Ly). M. have very strong ↑ ameboidism: Their surface may be smooth or studded with numerous protrusions, ↑ pseudopodia, ↑ filopodia, and ↑ microvilli depending on cell activity. M. remain in circulation about 40 h as blood ↑ macrophages; they then leave bloodstream, enter connective tissue, and transform via ↑ polyblasts into ↑ histiocytes. M. are involved in immunological reactions by processing of phagocytized antigens, secretion of some fractions of complement, and synthesis of ↑ interferon. Production site is ↑ bone marrow; life span is up to 60–90 days.

↑ Monocytopoiesis

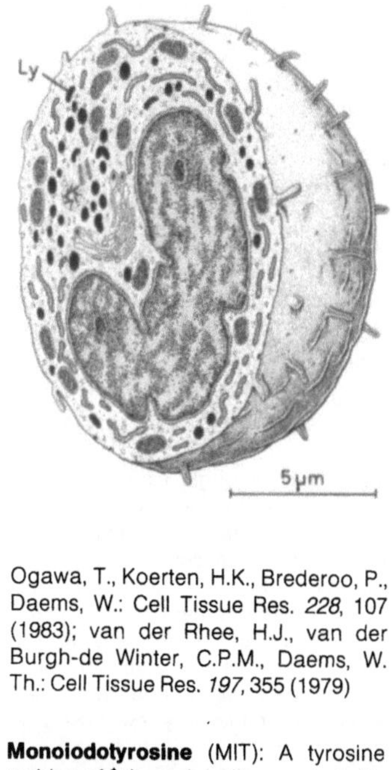

Ogawa, T., Koerten, H.K., Brederoo, P., Daems, W.: Cell Tissue Res. *228*, 107 (1983); van der Rhee, H.J., van der Burgh-de Winter, C.P.M., Daems, W. Th.: Cell Tissue Res. *197*, 355 (1979)

Monoiodotyrosine (MIT): A tyrosine residue of ↑ thyroglobulin coupled with one oxidized iodide ion.

↑ Thyroid follicular cells, biosynthesis of hormones in

Mononuclear phagocyte system (MPS): Concept comprising several classes of widely dispersed macrophagic cells characterized by: 1) Common origin from stem cell of ↑ bone marrow and ↑ monocytes (except ↑ microglia); 2) peculiar morphology; 3) presence of receptor sites for ↑ immunoglobulins on their plasmslemma; 4) high phagocytic activity induced by immumoglobulins and complement system. M. is primarily involved in defense of body by ↑ phagocytosis and antigen processing, but also in destruction of aged ↑ erythrocytes, extrahepatic bile production, and ↑ iron and fat metabolism. Concept of M. replaces earlier ↑ reticuloendothelial/reticulohistiocytic system. (Terms marked with an asterisk are described under corresponding letter. Modified after van Furth 1980)

Stem cell* (bone marrow)
↓
Monopoiesis*
↓
Monocytes* (blood)
↓
Macrophages* (tissues)
- alveolar*
- of bone marrow
- of lymph nodes
- of spleen
- of serous cavities
- histiocytes*
- Kupffer's cells*
- microglia*
- osteoclasts*
- synovial A-cells*(?)

Van Furth, R. (ed.): Mononuclear Phagocytes: Functional Aspects. 2 Vols. The Hague: Martinus Nijhoff Publ. 1980

Mononuclear wandering cells: All migratory cells of ↑ connective tissue proper.

Monophyletic theory, of hematopoiesis: ↑ Hematopoiesis, theories of

Monopoiesis (monocytopoiesis): The formation and development of ↑ monocytes – a process still awaiting full explanation. 1) According to one theory, ↑ totipotent hematopoietic stem cells (possibly the same as in ↑ granulopoiesis = CFU-G) differentiate into ↑ monoblasts (Mob), which divide and give rise to ↑ promonocytes (Pmoc). The latter divide and gradually differentiate during a period of 1–3 days into mature ↑ monocytes (Mc). 2) According to another theory, CFU-G cells generate ↑ promyelocytes (Pmyc) via ↑ myeloblasts (Myb); promyelocytes proliferate and give rise, after a series of divisions, to both ↑ myelocytes (Myc), belonging to granulopoiesis, and promonocytes (Pmoc), which finally transform into monocytes (Mc).

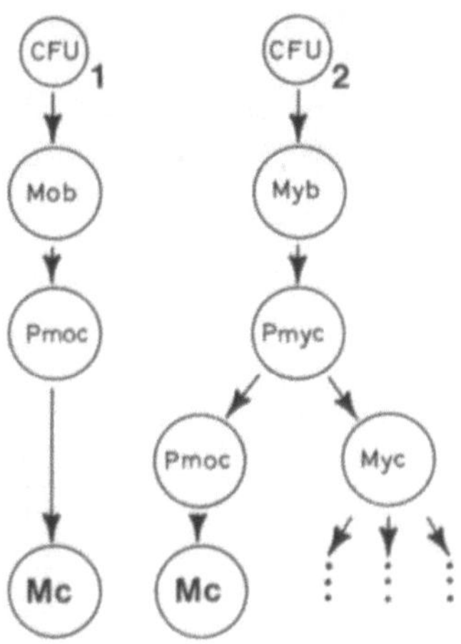

↑ Hematopoiesis, current view of

Leder, L.-O.: Der Blutmonozyt. Berlin, Heidelberg, New York: Springer-Verlag 1967; Metcalf, D., Moore, M.A.: Haemopoietic Cells. Amsterdam: North-Holland Publishing Co. 1971

Monoribosomes: Singly occurring ↑ ribosomes (R) either free or attached to membranes of ↑ endoplasmic reticulum. (Fig. = basal epidermal cell, rat)

↑ Polyribosomes

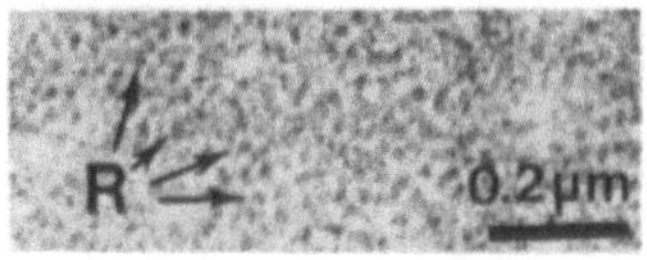

Monosomy: The case of ↑ chromosome anomaly when one ↑ autosome or one ↑ sex chromosome is lacking (2 n − 1). An example of M. is Turner's syndrome (44, X).

Montgomery's glands: ↑ Areolar glands

Mordant: A chemical substance (M) (usually salt, hydroxide, or alum of a bi- and trivalent metal) which modifies a substratum (S) that is difficult to stain and forms the link between it and the dye (D) during indirect ↑ staining.

↑ Hemalum

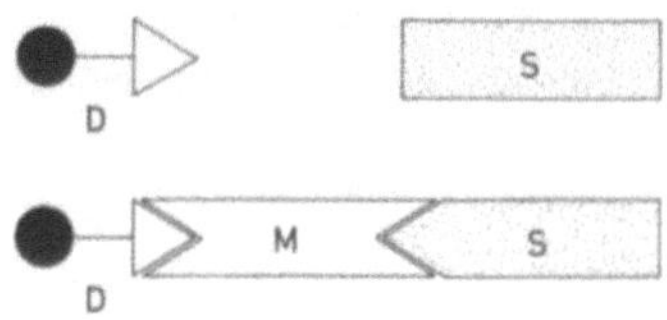

Morgagni, lacunae of: ↑ Urethral glands, endoepithelial

Morgagni, rectal columns of: ↑ Anal canal

Morphometry: ↑ Stereology

Mossy fibers: One of two kinds of afferent fibers (MF) reaching cerebellar cortex. M. are often branched and end in ↑ cerebellar glomeruli and establish synaptic contacts with dendrites of ↑ granule cells and axons of large granule cells. M. are characterized by very numerous, coreless, ↑ synaptic vesicles (SV) and a few dense-cored vesicles, 80–100 nm in diameter (large arrows). (Fig. = rat)

↑ Cerebellum, neuronal circuits of; ↑ Climbing fibers

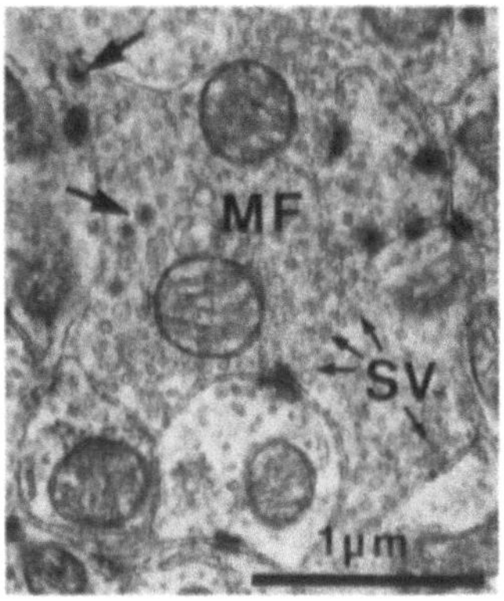

Motilin: A 22-amino acid polypeptide isolated from porcine duodenal extract and probably secreted by ↑ motilin cells. In the dog, M., increases motor activity of gut, thereby producing "housekeeper" contractions.

Seino, Y., Porte, D., Yanaihara, N., Smith, P.H.: Gen. Comp. Endocrinol. *38*, 234 (1979)

Motilin cells (EC_2-cells): Slender, ↑ argyrophilic epithelial cells scattered in ↑ intestinal villi and ↑ Lieberkühn's crypts of ↑ duodenum, ↑ jejunum, and ↑ ileum of dogs and humans. M. contain 200–400-nm ↑ unit membrane-bound granules and represent less than 10% of ↑ enterochromaffin cells. M. probably produce ↑ motilin.

↑ Endocrine cells, of gastrointestinal tract

Kobayashi, S., Iwanaga, T., Fujita, T., Yanihara, N.: Arch. Histol. Jpn. *43*, 85 (1980)

Motility: The capacity of living systems to exhibit motion and perform mechanical work with consumption of energy.

Allen, R.D.: J. Cell Biol. *91*, 148 s (1981); Goldman, R.D. (ed.): Cell Motility. New York: Cold Spring Harbor Symposia 1976

Motoneurons: ↑ Motor neurons

Motor cortex: ↑ Brain cortex, phylogenetic and structural subdivision of

Motor end plate (myoneural junction, neuromuscular junction, terminatio neuromuscularis*): The specialized site along ↑ skeletal muscle fiber at which a motor nerve fiber ends. M. is a flat, oval structure, 40–60 µm in diameter, situated midway along a muscle fiber and represents a synaptic contact between an efferent termination of a motor nerve fiber (NF) and a skeletal muscle fiber (MF). Approaching a muscle fiber, a motor axon (A) abruptly loses its ↑ myelin (MS) and ↑ Schwann's sheath (SS) and ramifies into several end branches (E), which are all covered by a plate of flattened ↑ teloglia cells (T). Teloglia cells are covered by a basal lamina (BL) continous with that of muscle fiber. Axon branches lie in corresponding shallow ↑ synaptic troughs of muscle fiber, primary synaptic clefts (SI). ↑ Sarcolemma of clefts invaginates into ↑ sarcoplasm and forms many secondary synaptic clefts (SC), collectively referred to

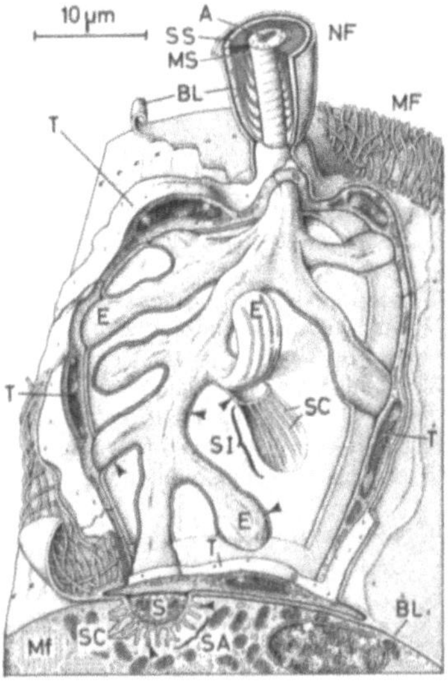

as ↑ subneural apparatus (SA). A flocculent substance continuous with basal lamina (BL) of muscle fiber fills subneural apparatus (arrowheads) and space between teloglia and muscle fiber, but is lacking between axon branches and teloglia cells. Axon branches contain numerous mitochondria and an abundance of ↑ acetylcholine-containing ↑ synaptic vesicles (S). Sarcolemma underlying subneural apparatus is devoid of ↑ myofibrils (Mf), but rich in nuclei, mitochondria, and rough endoplasmic reticulum, and is known as ↑ sole plate. An action potential reaching axon arborization provokes discharge of acetylcholine into subneural apparatus and a depolarization wave that initiates contraction of muscle fiber.

Couteaux, R.: Motor End-Plate Structure. In: Bourne, G.H. (ed.): The Structure and Function of Muscle. New York, London: Academic Press 1973; Gauthier, G.F.: The Motor End Plate. In: Landon, D.N. (ed.): The Peripheral Nerve. London: Chapman and Hall Ltd. 1976; Uehara, Y., Campbell, G.R., Burnstock, G.: Muscle and Its Innervation. London: Edward Arnold 1976

Motor neurons (motoneurons, somatic neurons): Large ↑ multipolar nerve cells (30–50 µm) of anterior column of ↑ spinal cord. M. have a voluminous nucleus, well-developed ↑ Nissl bodies, and a multiple ↑ Golgi apparatus. Their ↑ axons terminate in ↑ motor end plates of skeletal musculature. Axons of small M. innervate ↑ neuromuscular spindles. There are about 100 000 M. in man. M. innervating ↑ smooth muscle cells belong to ↑ autonomic nervous system.

↑ Spinal cord, nuclei of

Conradi, S., Kellerth, J.O., Berthold, C.H.: J. Comp. Neurol. *184*, 741 (1979)

Motor unit: A single ↑ motor neuron (N) together with one or a group of extrafusal ↑ skeletal muscle fibers (F) it innervates. In muscles with coarse movements, M. are large, i.e., one ↑ axon (A) innervates many muscle fibers; in muscles with fine movements (extrinsic muscles of eye, lumbricoid muscles), M. are small, i.e., one axon innervates only one or very few muscle fibers.

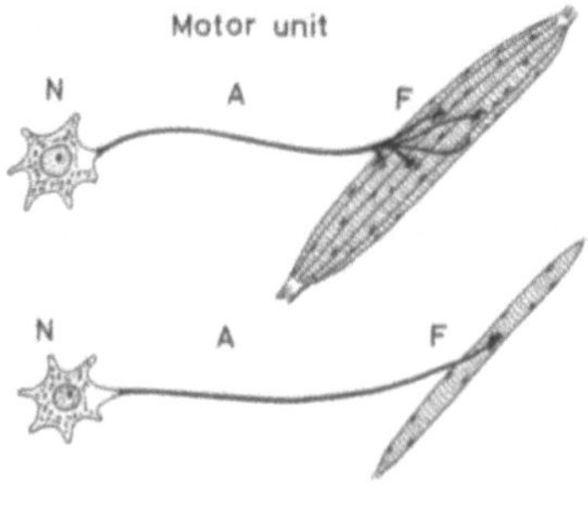

Mouth: ↑ Oral cavity

M-phase, of cell cycle: ↑ Cell cycle; ↑ Mitosis

mRNA: ↑ Messenger RNA

MSH: ↑ Melanocyte-stimulating hormone

MSH-RF: ↑ Melanocyte-stimulating hormone-releasing factor

mt nucleus: ↑ Mitochondrial nucleus

mtDNA: ↑ Mitochondrial DNA

MTOC: ↑ Microtubular organizing center

µm: ↑ Micrometer

Mucicarmine staining: A technique for light-microscopic demonstration of ↑ mucigen and ↑ mucin, which take on a red color.

Mucigen: A precursor of ↑ mucus.

Mucin: A protein-polysaccharide complex that is a constituent of ↑ mucus. There are several types of M. – sialomucins, sulfomucins, etc.

Leppi, T.J., Spicer, S.S.: Am. J. Anat. *118*, 833 (1966)

Mucocutaneous junction: The area of transition (MJ) between ↑ skin and a ↑ tunica mucosa. Toward tunica mucosa, keratinized layer (K) gradually decreases in thickness and disappears; ↑ epidermis (E) is replaced by a considerably thicker but nonkeratinized ↑ stratified squamous epithelium (SE) and, therefore, transparent. As a result, redness of underlying blood vessels (V) is visible, giving M. its red color. On mucous side, ↑ connective papillae (P) are considerably longer than ↑ dermal papillae in skin. (Fig. = M. of ↑ lip)

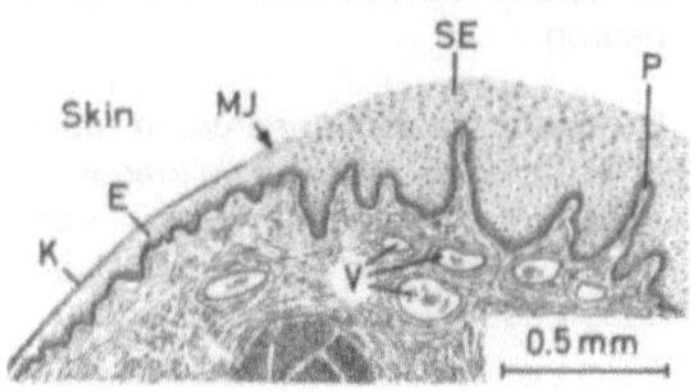

Mucogingival junction: The area of transition between nonkeratinized ↑ stratified squamous oral epithelium and keratinized stratified squamous epithelium of ↑ gingiva.

Schroeder, H.E., Amstad-Jossi, M.: Cell Tissue Res. *202*, 75 (1979)

Mucoid: A ↑ glycoprotein with more than 40% carbohydrates; similar to ↑ mucin. ↑ Pyloric, ↑ duodenal, and eccrine ↑ sweat glands produce M.

Mucoid cells, of sweat glands (dark cells, cellulae densae*): Pyramidal cells whose large side is directed toward glandular lumen. M. cover almost entire luminal surface. M. have a basal nucleus displaying deep indentations, condensed ↑ heterochromatin, and a small nucleolus. Cytoplasm contains rod-shaped mitochondria, a well-developed Golgi apparatus associated with numerous, ↑ unit membrane-bound, ↑ secretory granules (SG), elongated rough endoplasmic cisternae, and an abundance of free ribosomes. Secretory granules enclose a fine granular, electron-dense material composed of mucoprotein and ↑ glycosaminoglycans. By fusion with apical plasmalemma, granule contents are discharged into glandular lumen. Apical surface bears small, irregular microvilli (Mv); M. are held to adjacent cells by ↑ junctional complexes (J) and numerous lateral interdigitations (Int). M. rest on basal lamina (BL).

↑ Clear cells, of sweat glands; ↑ Sweat glands, eccrine

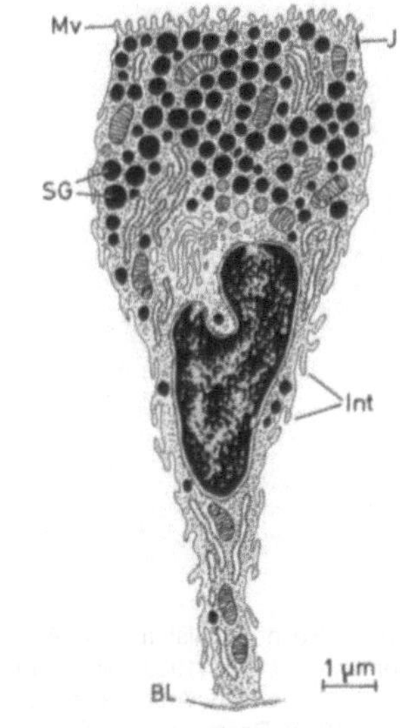

Mucoid oolemma: A term sometimes used to designate ↑ zona pellucida.

Mucopolysaccharides: ↑ Glycosaminoglycans

Mucosa: ↑ Tunica mucosa

Mucous cells: ↑ Goblet cells

Mucous cells, of bronchial glands: ↑ Bronchial glands

Mucous cells, of gastric glands proper: ↑ Neck mucous cells, of gastric glands proper

Mucous cells, of gastric surface: ↑ Surface mucous cells, of gastric mucosa

Mucous cells, of mucous and mixed salivary glands (mucocytus*): Large clear cells forming ↑ mucous tubules of ↑ mucous and ↑ mixed glands. Nucleus is situated in basal cell pole and contains condensed ↑ chromatin; nucleolus is small. Predominantly in basal cell portion are grouped small mitochondria, cisternae of rough endoplasmic reticulum, some lysosomes, and a considerable number of free ribosomes. The well-developed Golgi apparatus (G) is situated just above nucleus. Entire supranuclear and apical cell portions are crowded with ↑ unit membrane-bound, electron-lucent ↑ mucous droplets (MD) whose contents leave cell by ↑ exocytosis. Apical plasmalemma forms some short microvilli; M. lie on a ↑ basal lamina (BL).

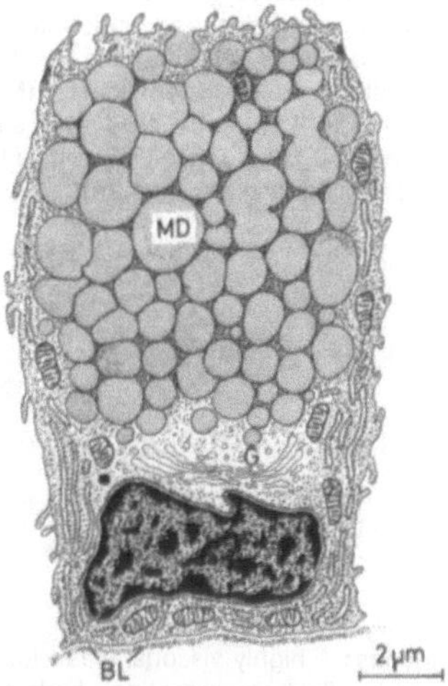

Mucous connective tissue: ↑ Gelatinous tissue

Mucous droplets (mucous granules): About 0.3–1-µm ↑ unit membrane-bound, pale structures (MD) present in

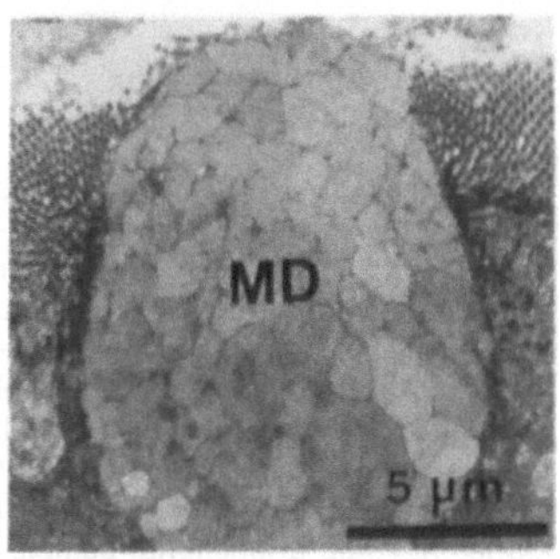

↑ goblet and other mucus-secreting cells. M. arise from Golgi apparatus and accumulate in ↑ apical pole. Droplets frequently fuse together; they are discharged from cell body by ↑ exocytosis. (Fig. = goblet cell, mongolian gerbil)

↑ Mucous cells, of mixed and mucous salivary glands

Mucous glands (glandula mucosa*): ↑ Exocrine glands producing ↑ mucus (e.g., ↑ mucous glands, of tongue, ↑ palatine glands, some ↑ pharyngeal glands, etc.).

↑ Glands, classification of

Mucous glands, of tongue: ↑ Mucous glands situated in root of ↑ tongue.

Mucous granules: ↑ Mucous droplets

Mucous membrane: ↑ Tunica mucosa

Mucous tubules (portio terminalis*): Secretory end pieces of ↑ mucous and ↑ mixed glands in the form of fine, frequently branched canalicules (MT) lined with ↑ mucous cells and surrounded by ↑ myoepithelial cells. (Fig. = ↑ sublingual gland, human)

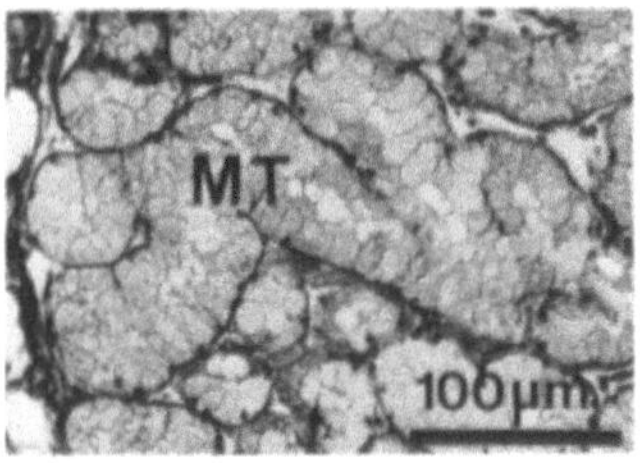

Mucus: A highly viscous, clear, lubricating fluid representing hydrated form of ↑ mucin. M. is also composed of other ↑ proteoglycans; it is produced by ↑ goblet cells, ↑ mixed, ↑ mucous, and other glands, and various kinds of ↑ mucous cells.

Flood, P.R.: Biomed. Res. *2*, Suppl. 49 (1981)

Müllerian hillock: ↑ Colliculus seminalis

Müller's cells (Müller's fiber, gliocytus radialis*): Tall, slender cells (MC) extending from ↑ outer (OLM) to ↑ inner limiting membranes (ILM) of ↑ retina. Both membranes are formed by M. Nucleus is elliptical and situated at level of inner nuclear layer (INL); apical cell portion (between nucleus and outer limiting membrane) is very narrow and carries at its tip a junctional complex (J) connecting M. with ↑ photoreceptors; some microvilli (Mv) penetrate between inner segments of photoreceptors. Most ↑ organelles are situated in conical basis of cell; mitochondria are rod-shaped, Golgi apparatus (G) is small and paranuclear, rough endoplasmic cisternae are short and sparse, in contrast to well-developed smooth endoplasmic (sER) tubules, and there are an appreciable number of ↑ glycogen particles. Cytoplasm encloses many bundles of ↑ microfilaments (Mf) spreading radially in cell basis. Lateral cell surfaces display some irregularly branched extensions, which penetrate between ↑ neurons. M. lie on a distinct basal lamina (BL). M. are main structural and metabolic supporting elements of retina and considered to be specially differentiated ↑ astrocytes.

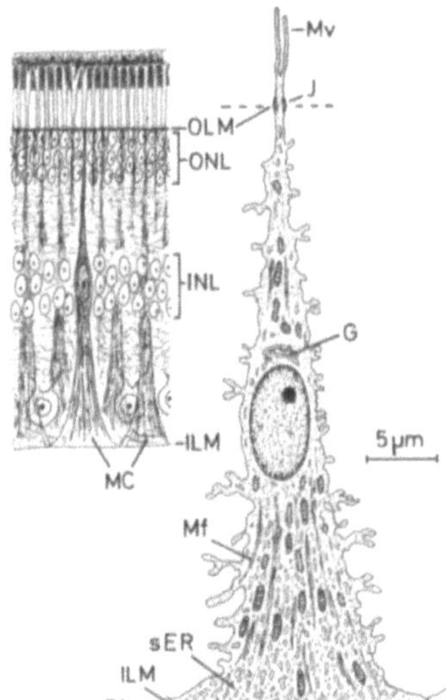

Ikeda, A., Yoshi, I., Mishima, N.: Arch. Histol. Jpn. *43*, 175 (1980); Masutani-Noda, T., Yamada, E.: Arch. Histol. Jpn. *46*, 393 (1983); Rasmussen, K.-E.: J. Ultrastruct. Res. *39*, 413 (1972)

Müller's muscle: A ↑ smooth muscle of ↑ ciliary body.

Müller's muscle, of eyelid: ↑ Tarsal muscle, superior

Multicellular glands: Glands consisting of many cells, as opposed to ↑ unicellular glands.

↑ Glands, classification of

Multilamellar bodies (cytosomes, lamellar bodies): ↑ Unit membrane-bound structures (MLB) found in cytoplasm of ↑ alveolar cells type II. M. measure 0.2–1.5 µm in diameter and contain a number of osmiophilic lamellae (L) composed of phospholipids, ↑ proteoglycans, and proteins. Besides acid phosphatase, a surface-active substance of the alveolar secretion, phosphatidylcholine is concentrated in M. Lamellar contents of M. are discharged from cell and spread over entire alveolar surface, giving rise to ↑ surfactant and ultimately to ↑ tubular myelin. M. develop from ↑ multivesicular bodies (MVB), which contain lamellae and vesicles (V). (Fig. = rat)

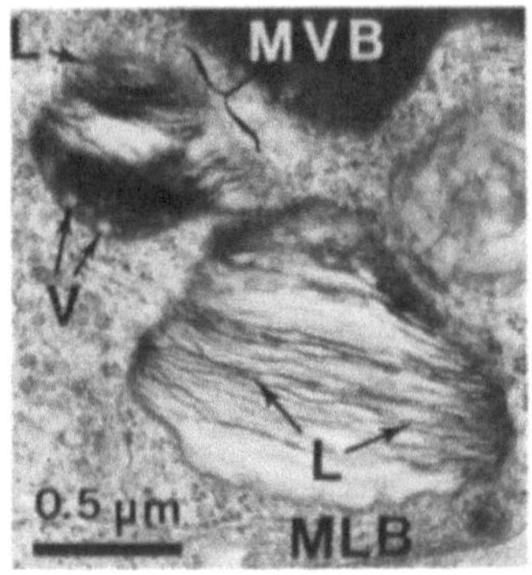

Sanders, R.L., Hassett, R.J., Vatter, A.E.: Anat. Rec. *198*, 485 (1980); Smith, D.S., Smith, U., Ryan, J.W.: Tissue Cell *4*, 457 (1972)

Multilaminar follicle: ↑ Ovarian follicle, secondary solid

Multiplication zone, of hyaline cartilage: ↑ Bone formation, indirect

Multipolar nerve cell: ↑ Multipolar neuron

Multipolar neuron (neuronum multipolare*): A ↑ neuron with an irregularly shaped ↑ perikaryon (P) and ↑ dendrites (D) oriented in various directions. (Fig. = ↑ spinal cord, human)

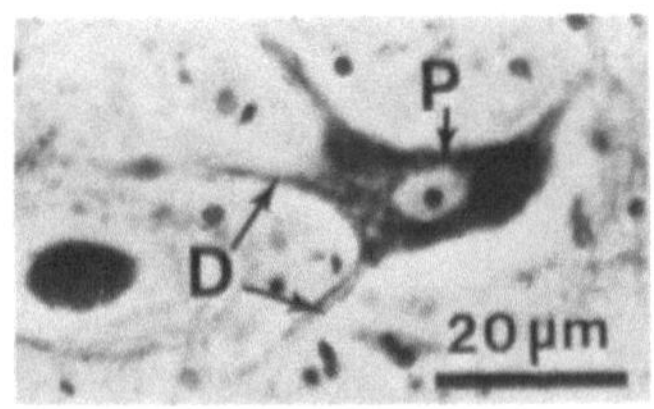

Multivesicular bodies (corpus multivesiculare*): ↑ Unit membrane-bound vacuoles (MVB), 0.3–2 µm in diameter, containing a variable number of

50-nm, seemingly empty vesicles (V). M. are frequent in ↑ liver parenchymal cells, epithelial cells of ↑ ductus epididymis (fig.), ↑ syncytiotrophoblast, and ↑ alveolar cells type II. Matrix of M. is very clear; it may sometimes be moderately osmiophilic. A fusion of vesicles, often coated (Golgi vesicles?), has also been observed with M. On the whole, origin, fate, and function of M. are still unknown. In alveolar cells type II, M. gradually transform into ↑ multilamellar bodies. M. are considered to be a kind of secondary ↑ lysosome, since their matrix (but not the vesicles) exibits an acid phosphatase activity. It seems that M. are also involved in formation of ↑ melanosomes and ↑ membrane-coating granules. (Fig. = rat)

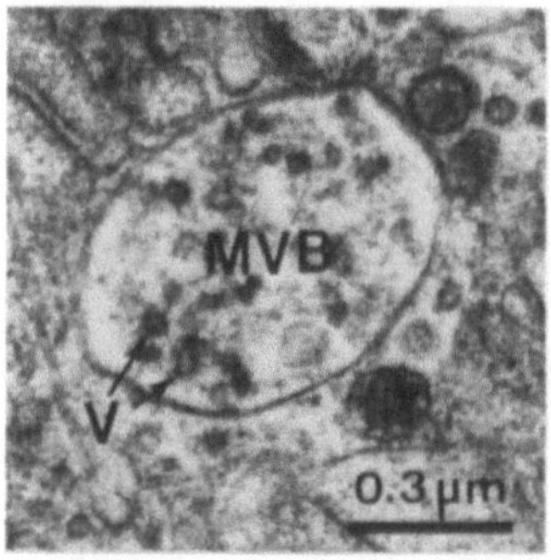

Abramowitz, J., Turner, W.A., Chavin, W., Taylor, J.D.: Cell Tissue Res. *182*, 409 (1977); Heine, H., Schaeg, G.: Z. mikrosk.-anat. Forsch. *94*, 160 (1980); Matoltsy, A.G.: J. Invest. Dermatol. *67*, 20 (1976)

Mural cells, of retinal capillaries (intramural pericytes): A type of retinal capillary cell (M) peripheral to ↑ endothelial cells (EC). M. have a large, elliptical ↑ heterochromatin-rich nucleus, some mitochondria, several cisternae of rough endoplasmic reticulum, an inconspicuous Golgi apparatus, and some ↑ glycogen particles. Cytoplasm contains ↑ actin microfilaments. M. are enclosed within a well-developed basal lamina (BL) continuous with that of endothelial cells. It is believed that M. represent a category of ↑ pericyte controlling retinal capillary flow. (Fig. = rat)

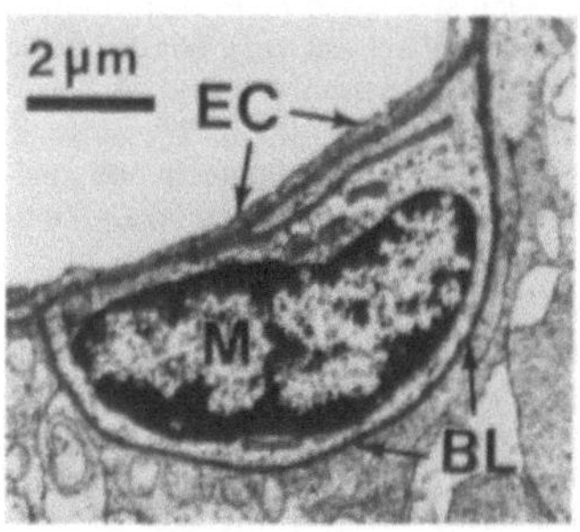

Muscle, arrector pili: ↑ Arrector pili muscle

Muscle, Brücke's: ↑ Ciliary body

Muscle, cardiac: ↑ Cardiac muscle

Muscle, cardiac, conducting system of: ↑ Impulse-conducting system, of heart

Muscle, cardiac, intercalated disc of: ↑ Intercalated disc

Muscle, ciliary: ↑ Ciliary body

Muscle, constrictor of pupil: ↑ Sphincter pupillae muscle

Muscle contraction: ↑ Smooth muscle cells, mechanism of contraction of; ↑ Striated muscle fibers, molecular biology of contraction of; ↑ Striated muscle fibers, morphology of contraction of

Muscle, dilator, of pupil: ↑ Dilator muscle

Muscle fascicle (primary bundle, fasciculus muscularis*): A variable number of parallel ↑ skeletal muscle fibers (MF) enveloped by ↑ perimysium (P), which permits each individual M. somewhat independent movements with regard to adjacent M. For this reason, an M. is considered smallest functional unit of a ↑ muscle, and is sometimes called a myon. Muscle fibers of an M. unite to compose the fine common tendon of an individual M.; these tendons together form main ↑ tendon of a muscle. ↑ Endomysium (E) surrounding each muscle fiber contains capillaries (Cap) and nerve fibers (NF). Many M. form a muscle. A = ↑ A-band, I = ↑ I-band, S = ↑ satellite cells, Z = ↑ Z-line

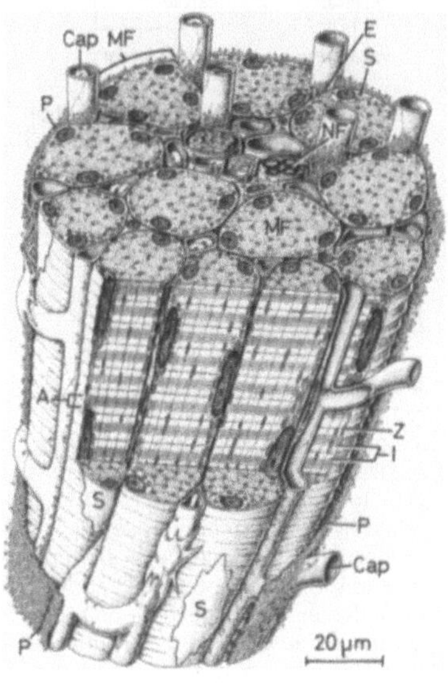

Muscle fiber: ↑ Striated muscle fibers

Muscle fiber, cardiac: ↑ Cardiac muscle fibers

Muscle fiber, sarcolemma of: ↑ Sarcolemma

Muscle fiber, skeletal: ↑ Skeletal muscle fibers

Muscle fiber, striated, myofibrils of: ↑ Myofibrils

Muscle fiber, striated, sarcoplasmic reticulum of: ↑ Cardiac muscle cells, sarcoplasmic reticulum of; ↑ Skeletal muscle fibers, sarcoplasmic reticulum of

Muscle, Müller's: ↑ Ciliary body

Muscle, myosin of: ↑ Myosin

Muscle, skeletal, motor end plate of: ↑ Motor end plate

Muscle, skeletal, neuromuscular junction of: ↑ Motor end plate

Muscle, smooth, mechanism of contraction of: ↑ Smooth muscle cells, mechanism of contraction of

Muscle, smooth, structure of: ↑ Smooth muscle cells

Muscle, sole plate of: ↑ Sole plate

Muscle spindle: ↑ Neuromuscular spindle

Muscle, striated, banding patterns of: ↑ Bands, of striated muscles

Muscle, striated, M-line of: ↑ M-line

Muscle, striated, myofilaments of: ↑ Myofilaments

Muscle, striated, sarcomere of: ↑ Sarcomere

Muscle-tendon junction: ↑ Myo tendinal junction

Muscular arteries (arteria myotypica*): ↑ Arteries with tunica media consisting almost completely of ↑ smooth muscle cells. All M. are divided into: 1) ↑M., large and 2) ↑M., small.

Muscular arteries, innervation of: ↑ Nonmyelinated ↑ adrenergic nerve endings terminate as fine knoblike endings between or close to ↑ smooth muscle cells. Discharge of ↑ neurotransmitter depolarizes muscle cells; impulses are then conducted to neighboring cells through ↑ nexus. Action of ↑ sympathetic nervous system constricts arteries, except ↑ coronary arteries, which are dilated upon sympathetic stimulation.

Muscular arteries, large (distributing arteries): Blood vessels, 2–10 mm in diameter, receiving blood from ↑ elastic arteries and distributing it to tissues and organs. Structure: 1) Tunica intima: a) ↑ endothelium (E) = a single continuous layer of squamous epithelial cells lying on basal lamina, b) subendothelial connective tissue layer or lamina propria intimae (LPI) = interlaced ↑ collagen and ↑ reticular fibrils with a few longitudinally oriented ↑ smooth muscle cells. In certain M., they form ↑ intimal cushions with a valve function. In some M., subendothelial layer may be absent, c) internal ↑ elastic lamina (IEL) = fenestrated elastic sheet (missing in umbilical artery). 2) Tunica media = 10–60 layers of helicoidally arranged smooth muscle cells. Between them runs a fine network of collagen and reticular fibrils and ↑ elastic fibers (EF). 3) Tunica adventitia or externa: a) external ↑ elastic lamina (EEL); b) an inner layer of ↑ dense connective tissue with longitudinally oriented elastic fibers (EF), gradually giving way to an outer layer of ↑ loose connective tissue, also with longitudinally oriented collagen fibrils and some scattered smooth muscle cells parallel to axis of M. ↑ Vasa vasorum (VV), lymphatic vessels, and nerve terminals (↑ nervi vasorum) penetrate most of peripheral layer of media. Through active constriction, M. regulate blood flow to various organs.

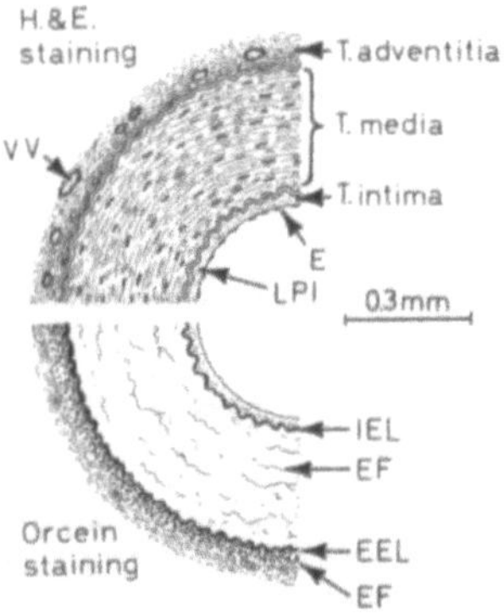

↑ Muscular arteries, innervation of

Muscular arteries, small: ↑ Muscular arteries 0.1–2 mm in diameter. Structure: 1) Tunica intima = ↑ endothelium (E) lies directly on internal ↑ elastic lamina (el); no subendothelial connective tissue layer. 2) Tunica media (TM) = only three to four ↑ smooth muscle cell layers with a few ↑ elastic fibers. 3) Tunica adventitia (TA) = an indistinct or absent external ↑ elastic lamina; some ↑ collagen fibrils, but neither smooth muscle cells nor ↑ vasa vasorum. (Fig. = dog)

↑ Arterioles

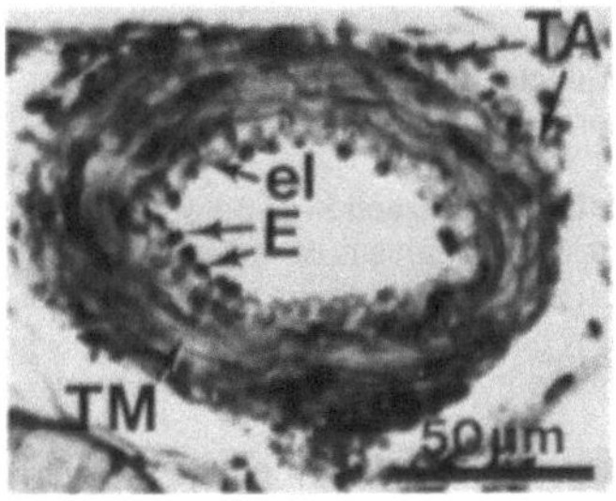

Muscular tissue (textus muscularis*): A group of ↑ tissues with a common and principal characteristic of contractility. According to its morphological properties and dependence on will, M. is divided as follows:

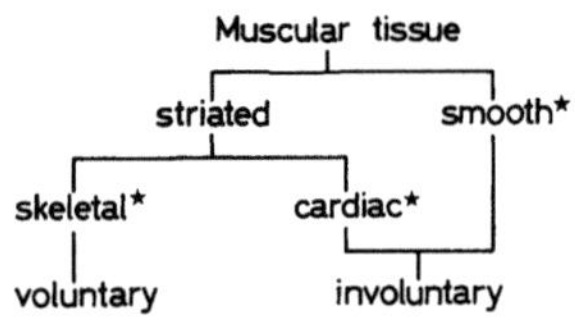

Main functions of M. are locomotion (↑ skeletal M.), heart contraction (↑ cardiac M.), and movements of various organs (↑ smooth M.). (Terms marked with an asterisk are described under corresponding letters)

Sawada, H.: Scanning Electron Microscopy 1981/IV, 7 (1981)

Muscular venules (venula muscularis*): Small blood vessels (MV), 50 µm – 1 mm in diameter, connecting ↑ postcapillary venules with small veins. Structure: 1) Tunica intima (TI) = a relatively thick continuous ↑ endothelium (E) whose cells are connected with one another by ↑ zonulae occludentes and ↑ nexus. Some processes of ↑ endothelial cells perforate basal lamina and form ↑ myoendothelial junctions. ↑ Collagen fibrils and ↑ elastic fibers may occur in varying numbers between endothelium and media. 2) Tunica media (TM) = one or two layers of flattened ↑ smooth muscle cells (M); in spleen and kidney, this layer is incomplete. 3) Tunica adventitia (TA) = a relatively thick, continuous layer of collagen fibrils (C), a few ↑ fibroblasts (F), their processes, and ↑ nonmyelinated nerve fibers. M. accompany ↑ arterioles and stay under control of ↑ autonomic nervous system; however, their contractibility is weak. (Figs. = rat)

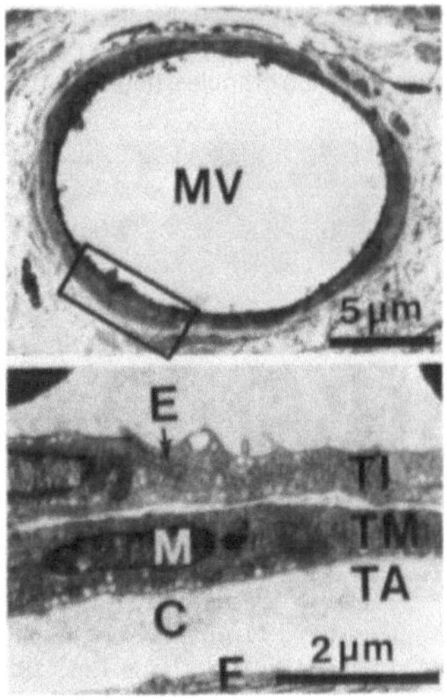

Muscularis: ↑ Tunica muscularis

Musculus cardiacus: ↑ Cardiac muscle

Muscularis mucosae: ↑ Lamina muscularis mucosae

Myelin: The insulating material forming ↑ myelin lamellae. M. is characterized by its high proportion of lipid in relation to protein. Lipid fraction of M. makes up about 70%–80% of its weight and consists of cholesterol, phospholipids, and glycolipid cerebroside specific to M. Protein fraction consists of at least two kinds of ↑ basic proteins, phospholipid protein, ↑ glycoproteins, and enzymes (carbonic anhydrase). Exact chemical composition of M. has not been completely elucidated.

Hollingshead, C.J., Caspar, D.L.D., Melchior, V., Kirschner, D.A.: J. Cell. Biol. *89*, 631 (1981); Norton, W.T.: Myelin: Structure and Biochemistry. In: Tower, D.B. (ed.): The Nervous System, Vol. 1. New York: Raven Press 1975

Myelin-associated glycoprotein (MAG): A ↑ glycoprotein (m.w. about 100000 daltons) synthesized by ↑ Schwann's cells. It is thought that M. plays a role in maintaining ↑ periaxonal space, ↑ Schmidt-Lanterman incisures, and outer ↑ mesaxons by preventing complete adherence of Schwann's cells and ↑ myelin lamellae.

Trapp, B.D., Quarles, R.H.: J. Cell Biol. *92*, 877 (1982)

Myelin bodies, of pigment epithelial cells: Phagocytized tips (MB) of outer segments of ↑ photoreceptors. M. are broken down within ↑ lamellar phagosomes. (Fig. = rat)

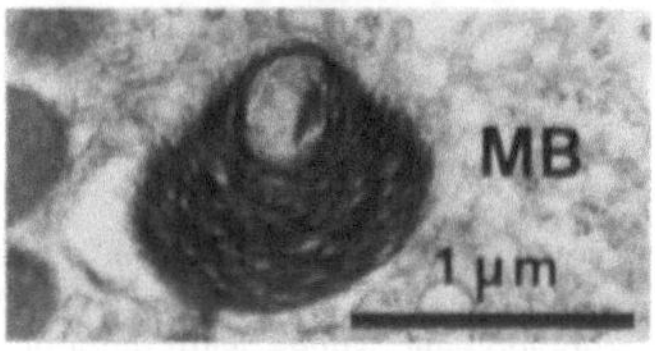

Myelin lamellae (lamella myelini*): The product of wrapping and fusion of ↑ cell membranes (M) of cytoplasmic sheets of ↑ oligodendrocytes or ↑ Schwann's cells around ↑ axon. Major dense line (MDL) contains a considerable number of molecules of ↑ basic proteins (B). The two adjacent lamellae forming ↑ intraperiod line (IL) contain molecules of proteolipid proteins (P) that seem to hold lamellae together and ↑ glycoproteins (GP) that may extend into lipid bilayer (LL). Major dense line and intraperiod line are connected by proteins (Pr) whose chemical composition is not completely known. All molecules have a radial arrangement. (Modified after Morell and Norton 1980)

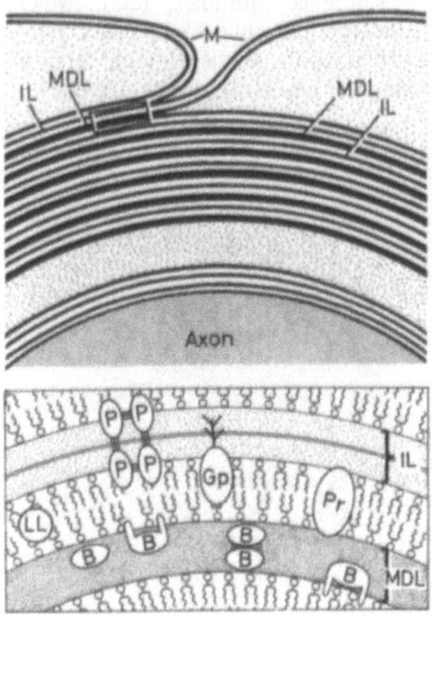

↑ Mesaxon; ↑ Myelinization in central nervous system; ↑ Myelinization in peripheral nervous system; ↑ Myelin sheath

Morell, P., Norton, W.T.: Sci. Am. *242/5*, 88 (1980)

Myelin sheath (stratum myelini*): A tubular lipoprotein envelope (MS) around an ↑ axon (A). M. is situated within ↑ Schwann's cells (Sch) in ↑ peripheral nervous system or formed by cytoplasmic sheets of ↑ oligodendrocytes in ↑ central nervous system. M. is composed of a variable number of concentrically wrapped series of ↑ myelin lamellae, appearing under transmission electron microscope as light and dark lines in a repeating pattern of about 12–15 nm. Structure: 1) Major dense lines (MDL), or major periods, are about 2.5–3 nm thick and represent apposition of cytoplasmic leaflets of plasmalemma in myelin-synthesizing cells. 2) Intraperiod lines (IL) are less dense, about 1.5 nm thick, and represent fused outer leaflets of plasmalemma in myelin-synthesizing cells. Depending upon preparation method and/or maturation degree of M., a 2-nm-wide clear gap, corresponding to extracellular space, can be seen in middle of intraperiod line (arrowheads). M. generally surrounds large axons of peripheral and central nervous systems; it is interrupted at ↑ nodes of Ranvier. There is no morphological difference between M. in peripheral and central nervous systems. (Fig. = peripheral nervous system, rat)

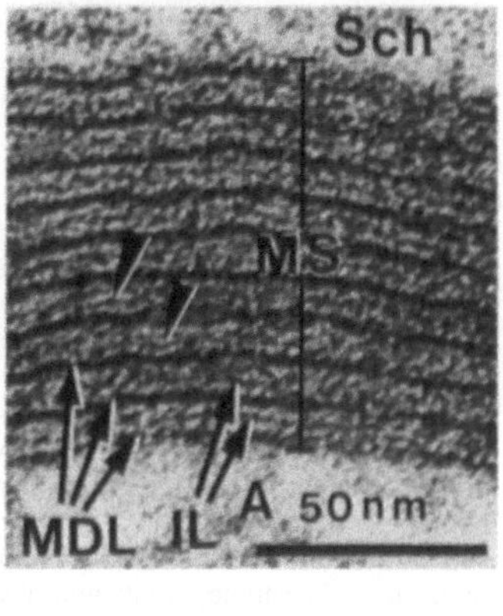

↑ Myelinated nerve fiber; ↑ Myelinization, in central nervous system; ↑ Myelinization, in peripheral nervous system

Myelin, tubular: ↑ Tubular myelin

Myelinated nerve fiber (neurofibra myelinata*): An ↑ axon (A) surrounded by a ↑ myelin sheath (M). In ↑ peripheral nervous system, myelin sheath is situated within ↑ Schwann's sheath, i.e., cytoplasm of ↑ Schwann's cell (Sch); in ↑ central nervous system, myelin sheath is furnished by ↑ oligodendrocyte. In addition, an M. of peripheral nervous system is surrounded by ↑ endoneurial sheath (ES). (Fig. = mongolian gerbil)

↑ Nerve fibers, of central nervous system; ↑ Nerve fibers, of peripheral nervous system

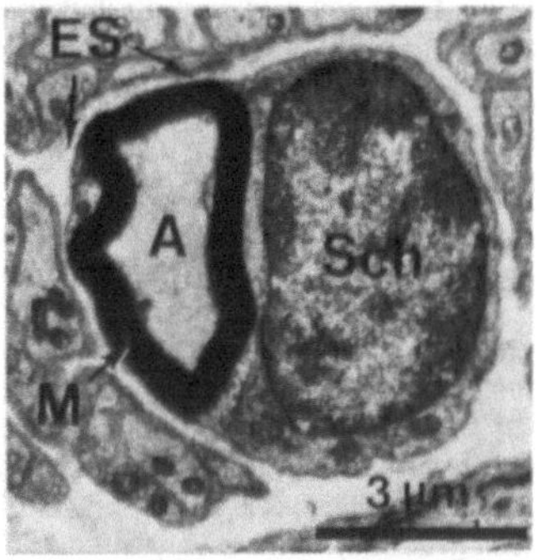

Myelinization (processus myelinopoieticus*): The process of synthesis and wrapping of ↑ myelin lamellae around ↑ axons of ↑ central and ↑ peripheral nervous systems. Morphologically, one can distinguish ↑ M. in central nervous system and ↑ M. in peripheral nervous system.

Myelinization, in central nervous system: The ↑ oligodendrocyte (O) forms 40–50 very thin ↑ roughly trapezoid

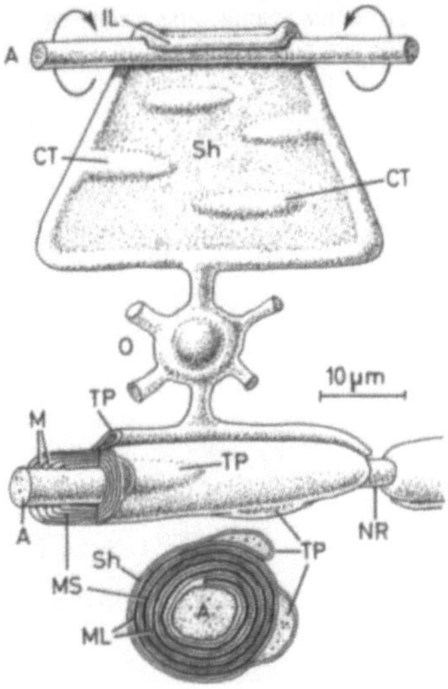

cytoplasmic sheets (Sh) with a thickened border and some transversally oriented cytoplasmic thickenings (CT). The sheets are affixed by their external edges (inner loops, IL) to neighboring ↑ axons (A) around which they are wrapped in a spiral fashion. During this process, some cytoplasm can remain in these sheets, constituting ↑ tongue processes (TP). In the course of wrapping, plasmalemma of both surfaces of sheets come into close contact with one another; in this way, ↑ meylin lamellae (ML) of ↑ myelin sheath (MS) are formed. The latter begins a few microns from ↑ axon hillock of a ↑ neuron and extends between two ↑ nodes of Ranvier (NR) along axons. Lateral thickened borders of sheets form ↑ mesaxons (M).

Myelinization, in peripheral nervous system: 1) An ↑ axon (A) invaginates cytoplasm of a ↑ Schwann's cell (Sch). 2) In ↑ paranodal region, axon affixes itself (arrowheads) to plasmalemma of Schwann's cell. Site of penetration of axon is marked by longitudinally oriented ↑ mesaxon (M). 3) Schwann's cell gradually forms a very attenuated cytoplasmic sheet (S) that begins to wrap spirally around axon (arrows). This is followed by elongation of mesaxon; outer aspects of its membranes fuse forming ↑ intraperiod line (IL) of ↑ myelin sheath (MS). 4) By constant elongation of mesaxon, further spiral wrapping of cytoplasmic sheet around axon, and fusion of inner leaflets of wrapped mesaxon, major dense lines (MDL) of myelin sheath are formed. Once formed, myelin sheath subdivides mesaxon into outer (OM) and inner (IM) mesaxon. Exact mechanism of spiral wrapping is unclear: It is thought that entire Schwann's cell wraps around axon. BL = ↑ basal lamina, ML = ↑ myelin lamellae

↑ Schwann's cell, unrolled

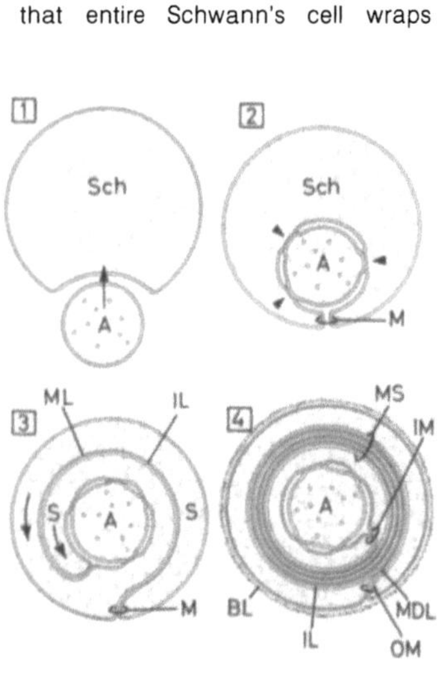

↑ **Myeloarchitectonics:** The spacial arrangement of ↑ myelinated nerve fibers in ↑ central nervous system.

↑ Fibroarchitectonics

Myeloblast (myeloblastus*): The first morphologically recognizable cell of granulopoietic series, measuring about 15 μm in diameter. Cell body is spherical; nucleus is voluminous and round with a small amount of peripheral ↑ heterochromatin and one or two nucleoli. Cytoplasm contains no granules, but numerous small mitochondria (M), an inconspicuous Golgi apparatus (G) with an adjacent ↑ centriole (C), flattened rough endoplasmic cisternae, occasional ↑ annulatae lamellae, and an abundance of free ↑ ribosomes (R) and ↑ polyribosomes (Pr) responsible for strong basophilia of M. under light microscope. M. make up 1% of all nucleated blood cells of red ↑ bone marrow.

↑ Granulopoiesis

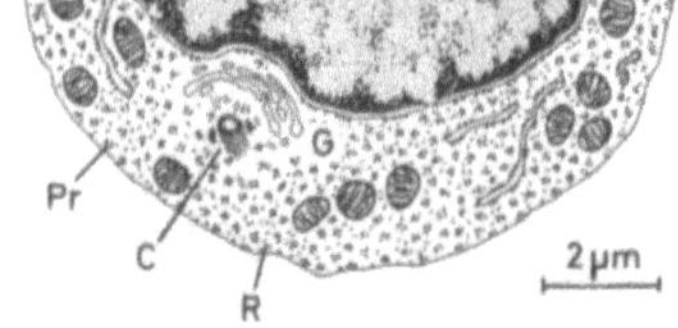

Myelocyte, basophilic (myleocytus basophilicus*): The first cell of granulopoietic series to have distinct basophilic granules. M. is a round, infrequent cell, up to 10–12 μm in diameter, with an elliptical nucleus relatively rich in ↑ heterochromatin. Cytoplasm contains a few small scattered mitochondria, a well-developed Golgi apparatus (G) with adjacent centrioles (C), some short, flattened rough endoplasmic cisternae, and a considerable number of free ↑ ribosomes and ↑ polyribosomes. From Golgi apparatus, vesicles about 300 nm in diameter arise with extremely dense contents; these fuse with each other to give rise to larger basophilic granules (BG). M. divide actively.

↑ Granulopoiesis

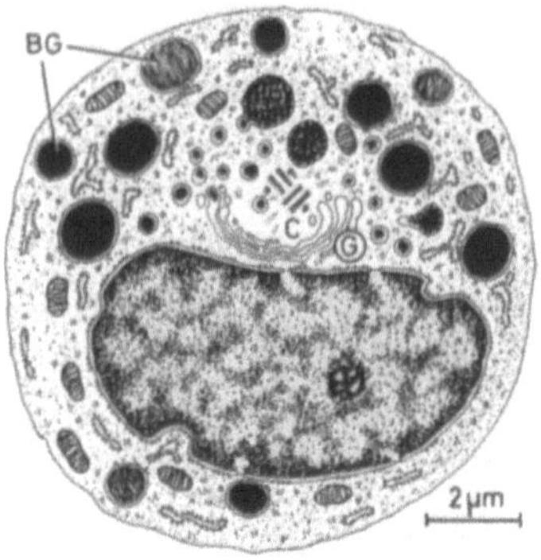

Myelocyte, eosinophilic (myelocytus eosinophilicus*): The first cell of granulopoietic series with recognizable ↑ eosinophilic granules. Cell body is round, 12–16 μm in diameter, with an eccentric elliptical nucleus and peripheral clumps of ↑ heterochromatin; nucleolus is small and frequently inconspicuous. Mitochondria are spherical and numerous; Golgi apparatus (G) has prominent centrioles (C); rough endoplasmic cisternae (rER) are wide, elongated, and richly branched. The latter contain a moderate amount of dense, peroxidase-reactive ↑ reticuloplasm. Besides occasional dense ↑ azurophilic granules (AG), whose production is stopped at this stage, cytoplasm contains a few less dense peroxidase-positive, immature eosinophilic granules, arising from convex Golgi saccules. These granules coalesce and give rise to large, mature eosinophilic granules (EG), which undergo maturation; a central crystalloid appears in their interior only at stage of late M. M. divide actively.

↑ Granulopoiesis

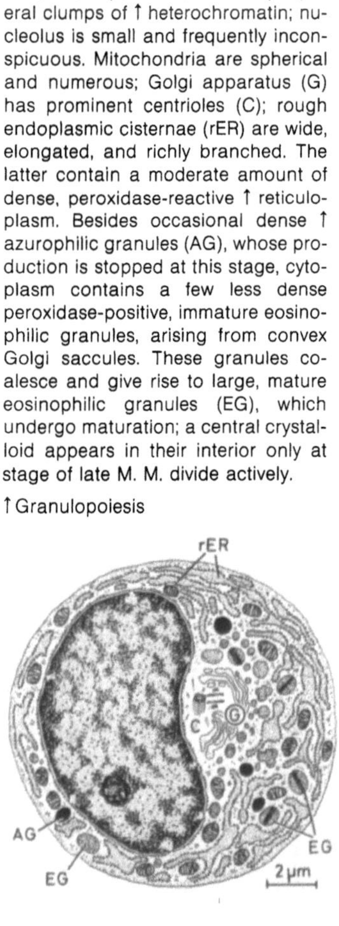

Walle, A.J., Parwaresch, M.R.: Cell Tissue Kinet. *12*, 249 (1979)

Myelocyte, neutrophilic (myelocytus neutrophilicus*): The first cell of granulopoietic series in which specific neutrophilic granules appear. Round or oval cell body measures about 12–16 µm in diameter. Nucleus is voluminous, eccentric, and elliptical, with frequent invaginations; it is moderately rich in ↑ heterochromation and contains one nucleolus. Cytoplasm encloses a small number of rod-shaped mitochondria, a well-developed Golgi apparatus (G) with accompanying ↑ centriole (C), a few short and flattened rough endoplasmic cisternae, a moderate number of ↑ azurophilic granules (AG), some free ↑ ribosomes, and ↑ glycogen particles. From convex aspect of Golgi apparatus arise approx. 100-nm peroxidase-negative specific granules loaded with low-density homogenous material. These immature granules fuse to form 200-nm mature specific granules (SG) with moderately osmiophilic contents. M. divide actively; they make up 5%–19% of all nucleated cells of red ↑ bone marrow.

↑ Granulopoiesis; ↑ Granulocytes, neutrophilic, granules of

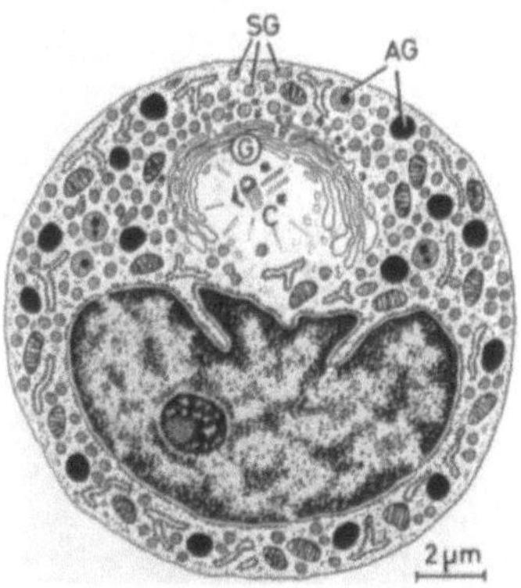

Myeloid bodies: Multilamellated, ↑ unit membrane-limited ↑ organelles (MB), 0.2–3 µm in diameter, produced predominantly by ↑ GERL complex, but also by ↑ endoplasmic reticulum and ↑ nuclear envelope (NE). Found in many kinds of cells, M. are eliminated by ↑ exocytosis; some of them show an acid

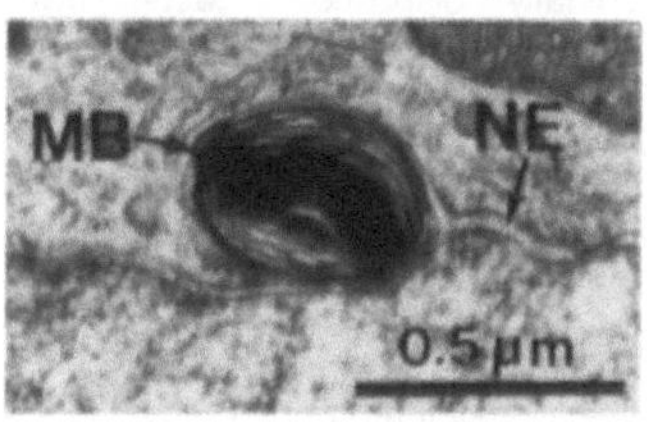

phosphatase activity. It is thought that M. contain phospholipids because of their similarity to ↑ multilamellar bodies. Function of M. is unknown. (Fig. = ↑ distal tubule cell, of nephron, rat)

Zwahlen, R., Richardson, B.P., Hauser, R.E.: J. Ultrastruct Res. *67*, 340 (1979)

Myeloid phase, of hematopoiesis: ↑ Hematopoiesis, prenatal

Myeloid tissue (textus myeloideus*): The tissue constituting ↑ bone marrow.

Myeloperoxidase: An enzyme (m.w. about 160 000 daltons) present in ↑ azurophilic granules of neutrophilic, basophilic, and eosinophilic ↑ granulocytes and ↑ monocytes. M. liberates from H_2O_2 activated oxygen that has a bacteriocidal effect. M. is also used in ultracytochemistry as a protein ↑ tracer; it is considerably heavier than ↑ horseradish peroxidase.

Myelopoiesis: ↑ Granulopoiesis

Myenteric plexus (Auerbach's plexus, plexus myentericus*): Groups of ↑ ganglia (arrowheads) and bundles of ↑ nerve fibers between circular (CL) and longitudinal layers (LL) of tunica muscularis of ↑ gastrointestinal tract.

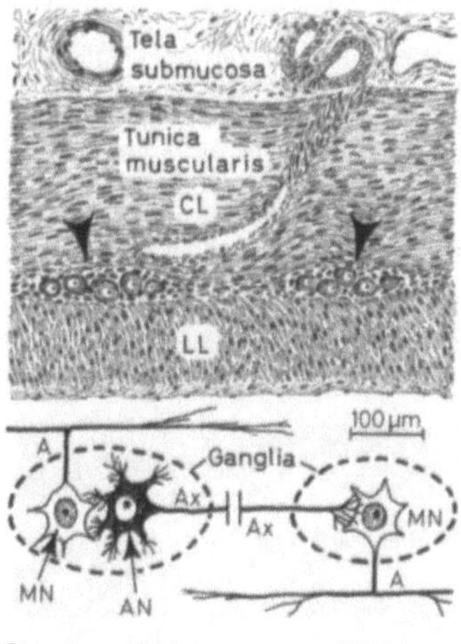

Structure: 1) Motor neurons (MN) have few short ↑ dendrites and ↑ axons (A) that penetrate muscular layers. 2) Association neurons (AN) send many short dendrites ending on bodies of motor neurons of same ganglion. Axons (Ax) of association neurons form connections with motor neurons of other ganglia. Together with ↑ submucosal plexus, M. is responsible for intrinsic innervation of gastrointestinal wall.

↑ Intestine, innervation of

Hoyes, A.D., Barber, P.: Cell Tissue Res. *209*, 329 (1980), Stach, W.: Z. mikrosk.-anat. Forsch. *95*, 161 (1981), *96*, 497 and 972 (1982)

Myoblast (myoblastus*): Elongated cells forming ↑ myotubes by fusion.

↑ Skeletal muscle fibers, histogenesis of

Myocardium*: The middle layer of ↑ heart wall composed of ↑ cardiac muscle with abundantly vascularized ↑ endomysium, a poorly-developed network of ↑ elastic fibers, ↑ autonomic nerve fibers, and abundant lymphatic vessels. Internal aspect of M. is modeled into anastomosing cords, trabeculae carneae, and conical muscular protrusions, papillary muscles.

↑ Cardiac muscle cells; ↑ Cardiac muscle fibers

McNutt, N.S., Fawcett, D.W.: Myocardial Ultrastructure. In: Langer, G.A., Brady, A.J. (eds.): The Mammalian Myocardium. New York: John Wiley & Sons, Inc. 1974; Sommer, J.R., Waugh, R.A.; Environ. Health Perspect. *26*, 159 (1978)

Myocardium, innervation of: A large number of fine ↑ nonmyelinated parasympathetic (vagus nerve) and sympathetic nerve fibers are present in myocardium, but there are no structures similar to ↑ motor end plates.

↑ Adrenergic nerve fibers; ↑ Cholinergic nerve fibers

Abraham, A.: Microscopic Innervation of the Heart and Blood Vessels in Vertebrates Including Man. New York: Pergamon Press 1969

Myocytus cardiacus: ↑ Cardiac muscle cells

Myoendothelial junction: Direct contact (MJ) in the form of a ↑ nexus between processes of ↑ endothelial cells (EC) and underlying ↑ smooth muscle cells (MC) established through fenestrae of internal ↑ elastic lamina (EL) or, in absence of the latter, through structures of subendothelial layer; frequently found in ↑ metarterioles.

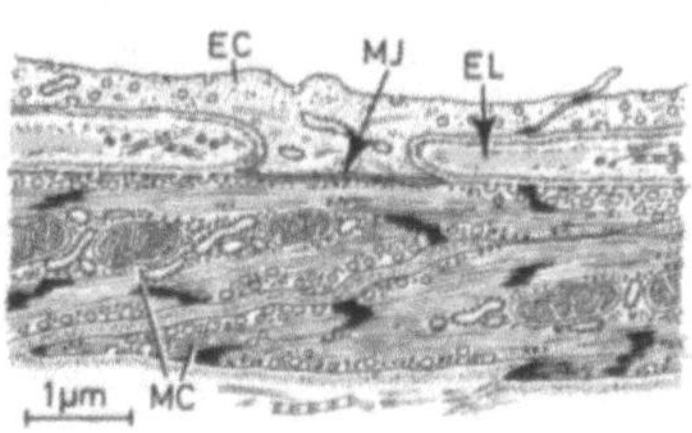

Myoepithelial cells (basket cells, myoepitheliocytus*): A class of spindle-shaped or stellate cells (arrows) surrounding secretory portions of ↑ salivary, ↑ sweat, ↑ lacrimal, and ↑ mammary glands. M. lie between epithelial cells and their ↑ basal lamina. Ultrastructurally, M. are very similar to ↑ smooth muscle cells, but of ectodermal origin. By means of their processes, M. are attached to ↑ collagen and ↑ reticular fibrils of underlying tissues. Through their contraction, M. accelerate expulsion of secretory product into canalicular system. (Fig., top = apocrine sweat gland, human; fig., bottom = lacrimal gland, mongolian gerbil)

↑ Basket cells, of mammary gland; ↑ Basket cells, of salivary glands; ↑ Mixed tubules; ↑ Myoepithelial cells, of sweat glands

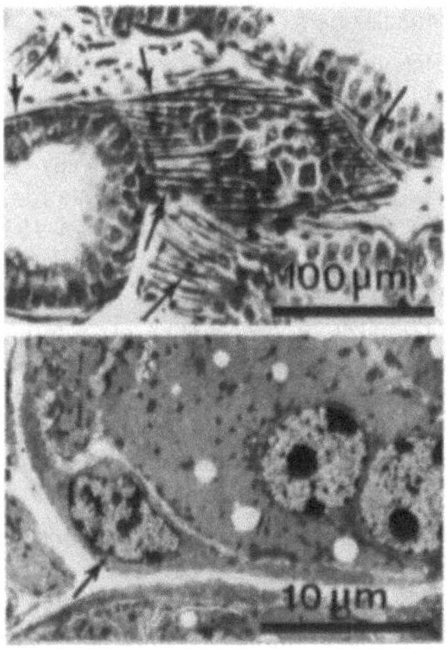

Abe, J., Atsuji, K., Tsunawaki, A.: Kurume, Med. J. *28*, 233 (1981); Murakami, M., Sugita, A., Abe, J., Hamasaki, M., Shimada, T.: Biomed. Res. *2*, Suppl. 99 (1981)

Myoepithelial cells, of sweat glands (myoepitheliocytus fusiformis*): Spindle-shaped cells (MC), 30–90 µm long, lying between basal lamina (BL) and secretory epithelial cells (EC). M. are oriented transversally or tangentially to secretory portions of ↑ sweat glands. M. have an elongated nucleus, cell ↑ organelles grouped in ↑ endoplasm (E), and many ↑ actin myofilaments (Mf) attached to ↑ dense bodies (D). With their numerous processes (P), M. are connected, by interposition of basal lamina, with ↑ collagen (C) and ↑ reticular (R) fibrils of ↑ dermis. By their contraction, M. aid in expulsion of ↑ sweat. M. are particularly numerous around axillary and perianal apocrine sweat glands.

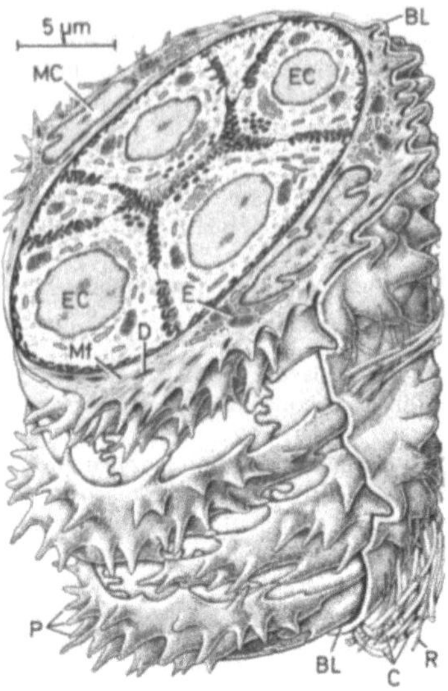

Nagato, T., Yoshida, H., Yoshida, A., Uehara, Y.: Cell Tissue Res. *209*, 1 (1980)

Myoepithelioid cells: ↑ Juxtaglomerular cells

Myoepithelium, of iris: ↑ Dilator muscle

Myofibrils (myofibrilla*): Bundles (M) of ↑ myofilaments, about 0.5–2 µm in diameter and of varying length, extending longitudinally through ↑ sarcoplasm of ↑ cardiac muscle cells and ↑ skeletal muscle fibers. M. show characteristic ↑ banding patterns (↑ A- and ↑ I-bands); in some cases, M. can be grouped to form ↑ Leydig-Kölliker columns. In cardiac muscle cells, M. attach with ↑ actin myofilaments to ↑ desmosomes of ↑ intercalated discs and to ↑ sarcolemma of digitiform processes

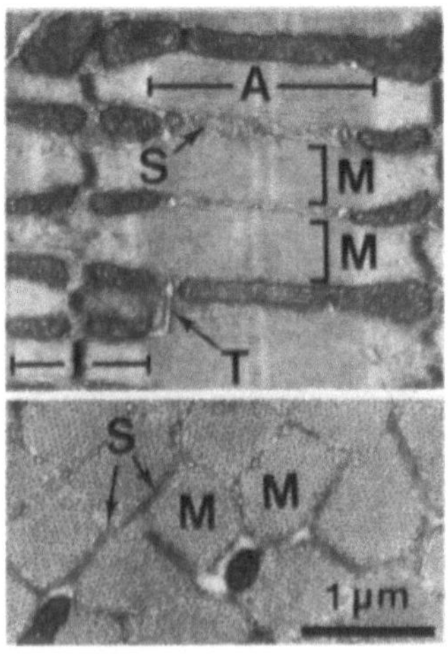

at end of skeletal muscle fibers. Each M. consists of a variable number of ↑ sarcomeres; ↑ sarcoplasmic reticulum (S) and ↑ T-tubules surround every M. M. in ↑ Purkinje cells of ↑ impulse-conducting system of heart have an irregular aspect and direction, but no T-tubules. (Figs. = rat)

↑ Myotendinal junction

Ishikawa, H.: Biomed. Res. *1*, 15 (1980)

Myofibrils, Cohnheim's field of: ↑ Cohnheim's fields

Myofibrils, Leydig-Kölliker columns of: ↑ Leydig-Kölliker columns

Myofibroblasts (myoid cells, contractile interstitial cells): Spindle-shaped cells (M) with long processes (P) resembling both ↑ fibroblasts and ↑ smooth muscle cells. M. have an elongated nucleus, rich in ↑ heterochromatin with a prominent nucleolus.

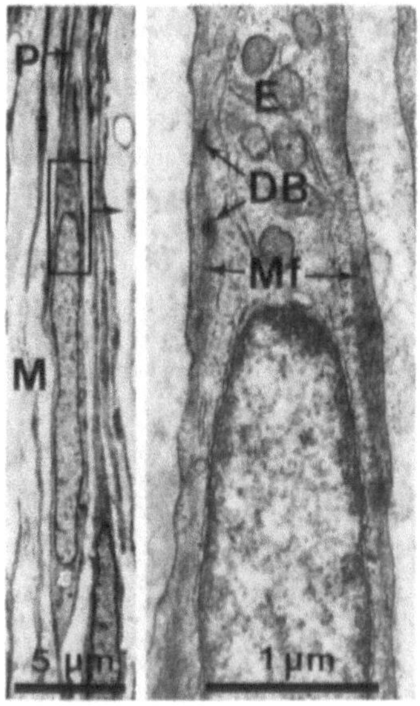

Numerous mitochondria, a conspicuous Golgi apparatus, well-developed rough endoplasmic cisternae, and an abundance of free ribosomes are predominantly located in conical cytoplasmic zone at pole of nucleus, similar to ↑ endoplasm (E). Cytoplasm also contains a large number of longitudinally disposed ↑ actin myofilaments (Mf), 4–8 nm thick, between which are scattered ↑ dense bodies (DB) like those found in smooth muscle cells. Processes of M. establish numerous ↑ nexus with one another, permitting intercellular communication. An incomplete ↑ basal lamina envelops M.; M. are present in three to six layers in lamina propria of ↑ seminiferous tubules (fig.) and are responsible for rhythmic contractions which force ↑ spermatozoa toward ↑ rete testis. M. have also been described in testicular ↑ tunica albuginea, in theca externa of ↑ ovarian follicles, and in adrenal capsule. M. play a very important role in wound contraction and may synthesize ↑ collagen fibrils. (Fig. = human)

Böck, P., Breitenecker, G., Lungmayr, G.: Z. Zellforsch. *133*, 519 (1972); Dym, M., Fawcett, D.W.: Biol. Reprod. *3*, 308 (1970); Gorgas, K., Böck, P.: Cell Tiss. Res. *154*, 533 (1974); Guber, S., Ross, R.: Surg. Gynecol. Obstet. *146*, 641 (1978)

Myofilaments (myofilamentum*): ↑ Microfilaments responsible for contraction of ↑ muscular tissue. Two kinds, ↑ actin and ↑ myosin M., together form ↑ myofibrils.

Myoglobin: An iron-containing oxygen-binding reddish endogenous ↑ pigment of red skeletal muscles. Oxygen-binding capacity of M. in man is small.

↑ Skeletal muscle fiber, red

Myoid cells, of seminiferous tubules: ↑ Myofibroblasts

Myoid cells, of thymus (Hammar's myoid cells): ↑ Skeletal muscle fibers of unknown significance occasionally found in medulla of ↑ thymus in mammals; M. are more frequent in birds and reptiles.

Myoid, of cone and rod cells: A term designating inner half of inner segment of ↑ photoreceptors.

Myometrium*: Middle layer (M) of ↑ uterus consisting of crossed and helicoidally arranged bundles of ↑ smooth muscle cells. M. is characterized by three vascular layers: 1) Stratum submucosum* or subvasculosum (Sm) = blood vessels of small caliber forming a network immediately below ↑ endometrium (E); 2) stratum vasculosum* (V) = thick-walled, muscular ↑ arcuate arteries and veins; 3) stratum supravasculosum* or subserosum (Ss) = blood vessels of small diameter in external zone of M. Through M. run numerous ↑ myelinated and ↑ nonmyelinated nerve fibers; lymphatics are particularly frequent between ↑ endometrium and M., and between M. and ↑ perimetrium (P). (Fig. = monkey)

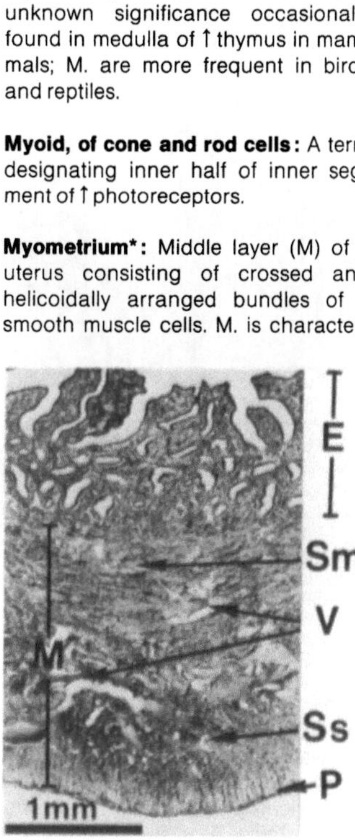

Myometrium, histophysiology of: ↑ Estrogens control ↑ differentiation of ↑ smooth muscle cells of myometrium at puberty and maintenance of their normal size and cytophysiological condition. In pregnancy, smooth muscle cells increase in size and from 25 μm may reach 5 mm; to existing cell are joined new muscle cells formed by differentiation from undifferentiated preexisting muscle cells. Contractions of M. during pregnancy are almost completely inhibited by hormone ↑ relaxin; it has been shown that ↑ progesterone has practically no effect on myometrial contractions. At end of pregnancy, ↑ oxytocin and ↑ prostaglandins initiate and maintain myometrial contractions. In ↑ menopause, smooth muscle cells of M. atrophy, since estrogens are lacking.

Myon: The smallest functional unit of a skeletal muscle, corresponding to a ↑ muscle fascicle.

Myoneural junction: ↑ Motor end plate

Myosin: A fibrillar macromolecule (m.w. roughly 450 000 daltons), about 150 nm long and 1.5–2 nm thick, found in ↑ muscular tissue and nonmuscle cells (epithelial cells, ↑ myoepithelial cells, brain tissue). M. molecule consists of subunits, meromyosins: 1) Light meromyosin (LMM) is straight and forms backbone of ↑ myosin filament with adjacent LMM subunits. 2) Heavy meromyosin (HMM) is composed of two parts – HMM S-2 fragment projected outward from backbone of myosin filament, and HMM S-1 fragment forming head or cross bridge of M. HMM S-1 represents active part of M., since it contains ATP and sites for binding ATP and ↑ actin myofilaments. During contraction, HMM S-2 spreads out (direction of arrows), and HMM S-1 establishes contact with actin myofilament, pulling it toward ↑ M-line. Several M. molecules form a ↑ myosin myofilament. (Modified after Huxley 1969) (See biochemistry texts for further information)

↑ Microvilli, shortening of; ↑ Striated muscle fibers, molecular biology of contraction of; ↑ Smooth muscle cells, molecular biology of contraction of

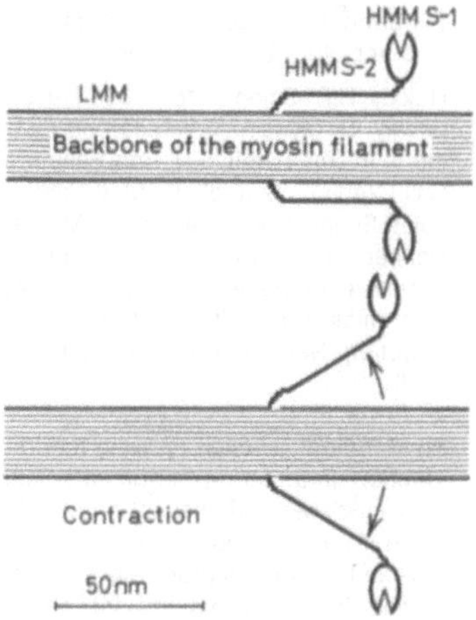

Gröschel-Stewart, U.: Int. Rev. Cytol. *65*, 194 (1980); Huxley, E.H.: Science *164*, 1356 (1969); Pepe, F.A.: Macromolecular Assembly of Myosin. In: Stracher, A. (ed.): Muscle and Nonmuscle Motility, Vol. 1. New York: Academic Press 1983; Toma, M., Berl, S.: Eur. J. Cell Biol. *28*, 122 (1982)

Myosin myofilament (A-band filament, myosin microfilament, myofilamentum crassum*): A 10-nm-thick microfilament, about 1.5 μm long, composed of several ↑ myosin molecules. Backbone or shaft of M. consists of light meromyosin (LMM). At regular intervals of 42.9 nm, six heads or bridges of heavy myosin molecules (HMM S-1) are arranged in a helical pattern around shaft of M. HMM S-1 subunit is connected to LMM by HMM

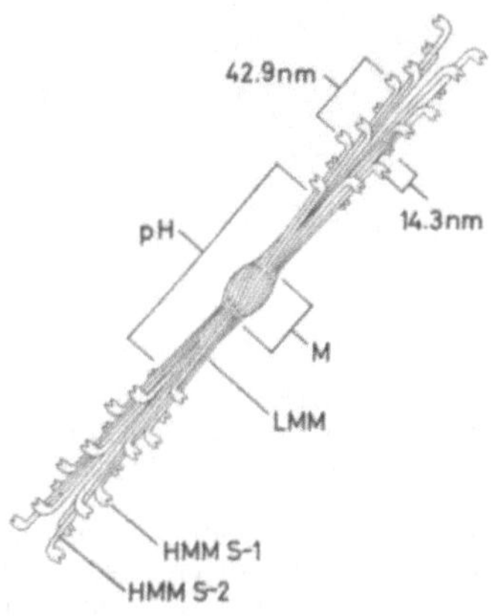

S-2 meromyosin. Central segment of M. is devoid of bridges and corresponds to ↑ pseudo-H band (pH). Interlacing LMM subunits of myosin molecule form a slight thickening in center of M., corresponding to ↑ M-line (M). In a cardiac and skeletal ↑ myofibril, M. compose main part of ↑ A-band.

↑ H-band; ↑ Striated muscle fibers, molecular biology of contraction of

Huxley, H.E.: Science *164*, 1356 (1969); Lowey, S.: Myosin: Molecule and Filament. In: Timasheff, S.N. and Fasman, G.D. (eds.): Biological Macromolecules Series. Vol. 5. New York: Marcel Dekker Inc. 1971; Myosin Filament Structure. In: Pepe, F.A., Sanger, J.W., Nachmias, V.T. (eds.): Motility in Cell Function. New York: Academic Press 1979

Myotendinal junction (muscle-tendon junction, junction myotendinea*): The contact between ↑ skeletal muscle fiber (MF) and ↑ tendon (T). At its end, muscle fiber tapers gradually and becomes deeply invaginated. Into these infoldings (F) enter ↑ collagen (CF) and ↑ reticular (RF) microfibrils; the latter penetrate ↑ basal lamina (BL), anchor to outer aspect of ↑ sarcolemma (S), form a loop, and interlace with ↑ collagen microfibrils, assuring a firm contact with tendon. A = ↑ actin myofilaments anchoring into internal aspect of sarcolemma

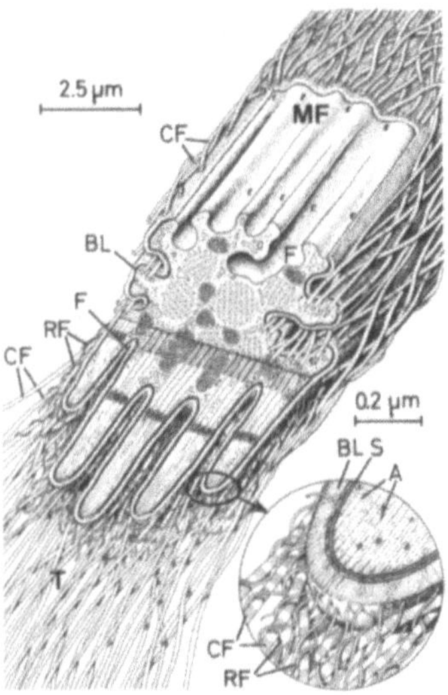

Hanak, H., Böck, P.: J. Ultrastruct. Res. *36*, 68 (1971); Trotter, J., Corbett, K., Avner, B.P.: Anat. Rec. *201*, 293 (1981)

Myotube (myotubus*): A cylindrical multinucleated cell, precursor of ↑ skeletal muscle fiber. Nuclei of an M. form a row in center of cell; ↑ myofibrils occupy peripheral ↑ sarcoplasm.

↑ Skeletal muscle fibers, histogenesis of

Myxedema: A group of symptoms accompanying thyroid hypofunction.

N

Nabothian cysts (Nabothian follicles, ovula Nabothi, vesicula cervicalis*): Occluded and dilated glands of uterine ↑ cervix, transformed into 5 to 6-mm-large vesicles.

Nail (unguis*): A platelike organ (N) situated on dorsal surface of distal end of phalanges (Ph) of fingers and toes. Structure: 1) Nail plate (NP) (stratum corneum*) = a slightly curved, semitransparent plate composed of many layers of fully keratinized, flattened cells firmly held together so that they do not desquamate. Cells are completely filled with hard ↑ keratin. Proximal edge of nail plate, nail root (NR), is embedded in ↑ epidermis (Ep), which also forms a thin fold, ↑ eponychium (E), covering nail root. Distally, nail plate ends in a free edge overlying ↑ hyponychium (H). Laterally, nail plate is delimited by two skin folds, nail walls (NW). Slit between nail walls and nail plate is nail groove (NG). 2) ↑ Nail bed (NB). 3) ↑ Nail matrix (M). 4) ↑ Dermis (D), underlying nail bed and nail matrix is connected by firm ↑ collagen and ↑ elastic fibers to ↑ periosteum (Pe) of phalange; dermis forms many longitudinally oriented ↑ dermal papillae (Pa).

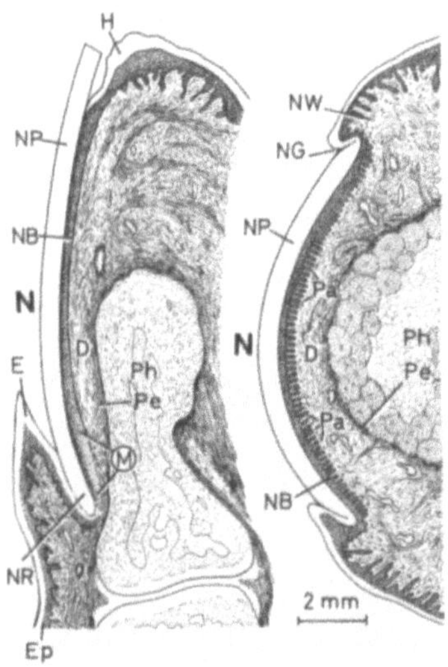

Zaias, N., Alvarez, J.: J. Invest. Dermatol. *51*, 120 (1968); Zaias, N., Baden, N.: Nails. In: Fitzpatrick, T.B. et al. (eds.): Dermatology in General Medicine. New York: McGraw-Hill Inc. 1979

Nail bed (lectulus unguis*): Several layers (NB) of epidermal cells, corresponding to stratum germinativum of ↑ epidermis, underlying nail plate (NP). Except in ↑ nail matrix, these cells do not participate in formation of nail plate. N. shows relatively regular longitudinal ↑ epidermal ridges (er) interdigitated with ↑ dermal papillae (dp). (Fig. = human, transversal section)

Nail keratin: ↑ Keratin, hard

Nail matrix (matrix unguis*): The thickened proximal part of ↑ nail bed (M) composed of 3–20 layers of actively dividing cells of stratum germinativum. New cells are constantly added to nail root and undergo direct keratinization without interposition of ↑ keratohyaline. By constant addition of new cells, nail plate (NP) is pushed forward slowly over nail bed, 0.5 mm/week. M. is visible on some fingers and toes as ↑ lunula. (Fig. = human, longitudinal section)

↑ Nail

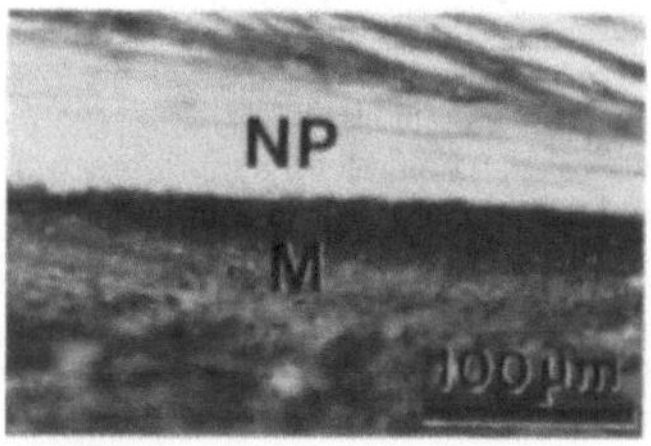

Nanometer (nm, millimicron, $m\mu$): 10^{-7} cm or one thousandth of a micron; N. replaces former Ångström (↑ Å) units.

Nasal cartilages (cartilagines nasi*): ↑ Cartilage, hyaline

Nasal cavities (cavitas nasi*): Part of conducting portion of ↑ respiratory system (NC) in the form of paired passages separated by ↑ nasal septum. N. begin at nostrils (N) and end at choanes (Ch). Each N. is subdivided into ↑ vestibulum nasi (VN) and nasal cavity proper. Vestibulum is lined with ↑ skin; nasal cavity proper is lined with ↑ respiratory (RM) and ↑ olfactory mucosa (OR), both attached to ↑ bone or hyaline ↑ cartilage. N. communicate with ↑ paranasal sinuses.

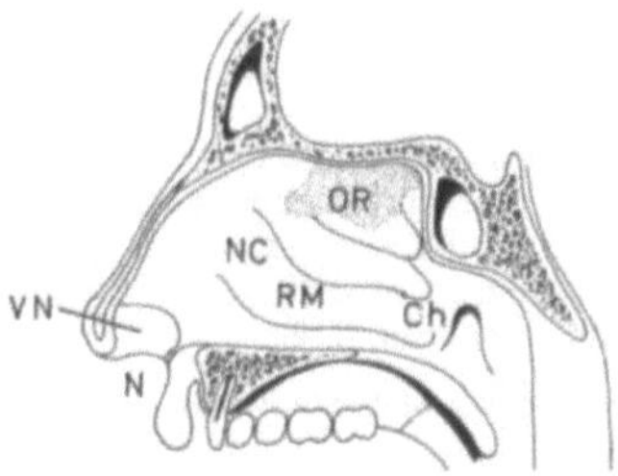

Nasal glands (glandula nasalis*): ↑ Mixed glands (G) located between veins of venous plexus (V) of ↑ respiratory mucosa, lining ↑ nasal cavities (Fig. = human)

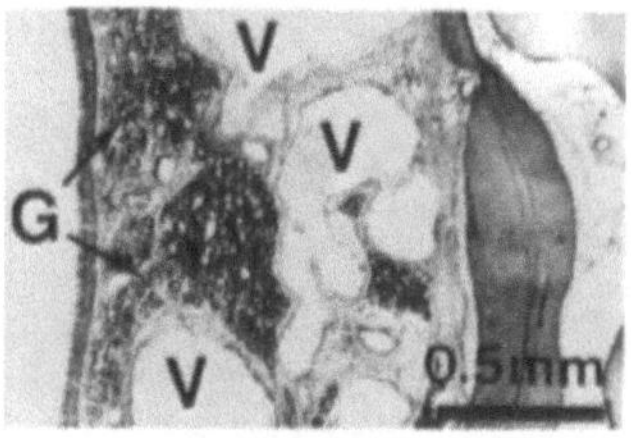

Nasal mucosa (tunica mucosa nasi*): The ↑ respiratory mucosa lining ↑ nasal cavities.

↑ Nasal glands; ↑ Pharyngeal glands

Boysen, M.: Virchows Arch. B Cell Pathol. *40*, 279 (1982); Schaeffer, J.P.: The Mucous Membrane of the Nasal Cavity and the Paranasal Sinuses. In: Cowdry, E.W. (ed.): Special Cytology, Vol. 1. New York: Paul B. Hoeber, Inc. 1982; Thaete, L.G., Spicer, S.S., Spock, A.: Am. J. Anat. *162*, 243 (1981)

Nasal septum (septum nasi*): The vertical mediosagittal separation (NS) dividing ↑ nasal cavity into halves. Structure: 1) Tunica mucosa (TM): a) epithelium (E) = ↑ pseudostratified ciliated epithelium with numerous ↑ goblet cells; b) lamina propria (LP) = a ↑ loose connective tissue with varying numbers of ↑ tubuloalveolar ↑ mixed glands (G) for moistening mucous surface, well-developed venous plexus (VP) for heating airstream, nerve fibers, and an appreciable number of ↑ macrophages and ↑ lymphocytes grouped in small follicles (F). Lamina propria is firmly attached to ↑ perichondrium and/or ↑ periosteum (P). 2) Thin plate of hyaline ↑ cartilage (C) and/or lamellar ↑ bone (B).

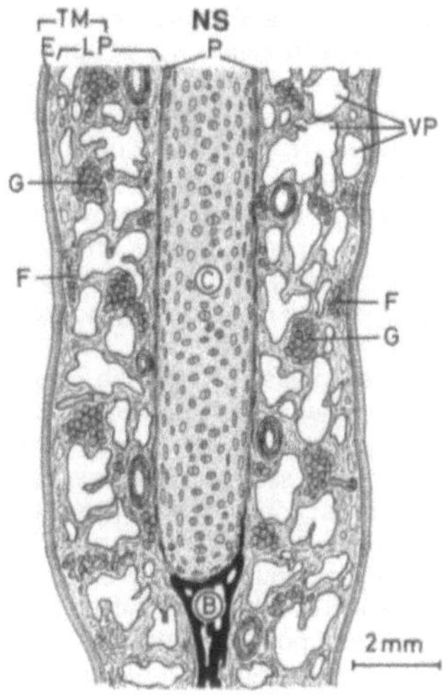

Nasmyth's membrane: ↑ Cuticle, of enamel

Nasolacrimal canal (canalis nasolacrimalis*): A channel about 2 cm long, conveying ↑ tears from ↑ lacrimal sac to middle meatus of nose. N. is lined by a ↑ pseudostratified ciliated epithelium (E) with two levels of nuclei; it overlies a loose, well-vascularized lamina propria (LP). In segment of N. passing through bone (B), lamina propria is attached to ↑ periosteum (asterisk). (Fig. = human)

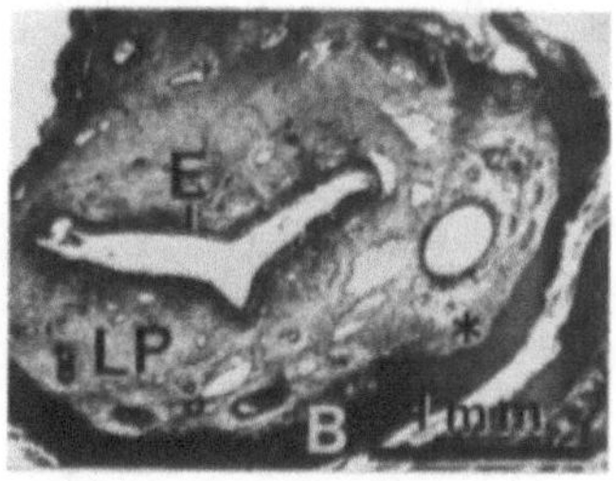

Nasopharynx (epipharynx, pars nasalis pharyngis*): A segment of ↑ pharynx situated between choanes, sphenoidal bone, posterior pharyngeal wall, and soft ↑ palate. ↑ Pseudostratified epithelium of N. changes to ↑ stratified squamous in area where ↑ uvula touches posterior and lateral walls of pharynx during act of swallowing. Tubal and pharyngeal ↑ tonsils belong to N. In upper part of N., musculature is lacking and replaced by fascia pharyngobasilaris.

Navicular fossa: ↑ Fossa navicularis

N-cells (neurotensin cells): Columnar cells found predominantly in epithelium of ↑ intestinal villi of small intestine in some animals; possibly existent also in humans. N. have an elliptical nucleus with finely dispersed ↑ chromatin and several invaginations of various depths. Cytoplasm is clear and contains numerous mitochondria, a well-developed and multiple Golgi apparatus (G), some short and flattened rough endoplasmic cisternae and a small number of free ribosomes. From Golgi apparatus arise ↑ unit membrane-bound secretory granules (SG) with dense contents; they measure 250–425 nm in diameter. The granules are predominantly located in basal cell pole, which overlies basal lamina (BL). ↑ Apical cell pole bears densely packed ↑ microvilli (Mv), like those of ↑ striated border. On the basis of immunocytochemical studies, it is believed that N. are a source of ↑ neurotensin. (Fig. after Helmstaedter et al. 1977)

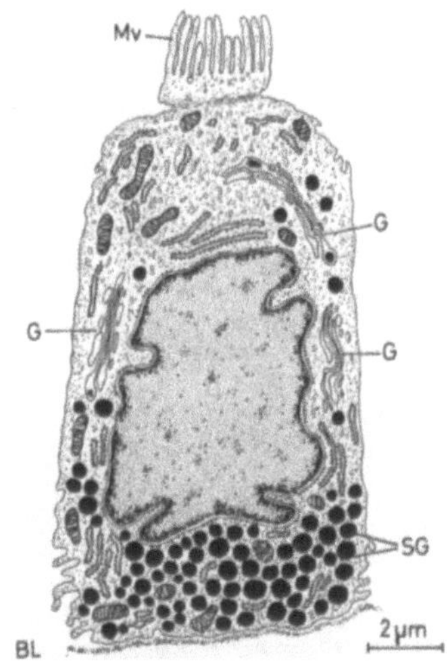

Helmstaedter, V., Feurle, G.E., Forssmann, W.G.: Cell Tissue Res. *184*, 445 (1977)

nDNA: ↑ Nucleosomal DNA

NEB: ↑ Neuroepithelial bodies

NEB-cells: Specialized columnar ↑ endocrine cells forming ↑ neuroepithelial bodies. Nucleus is prominent; cytoplasm contains small elongated mitochondria, a well-developed Golgi apparatus, relatively numerous but short cisternae of rough endoplasmic reticulum, a few lysosomes, and abundant free lysosomes. Predominantly in basal cell pole are located many ↑ unit membrane-bound granules (SG), measuring 100–300 nm, with contents of various densities. Apical plasmalemma forms some short microvilli (Mv) at air-contact surface. Basal cell surface is frequently in contact with ↑ cholinergic nerve fibers (NF). N. are believed to be endocrine elements storing and releasing ↑ catecholamines and/or polypeptide hormones. It is likely that N. and small granule cells of tracheal epithelium are the same type of cell. N. are classified among ↑ APUD cells.

↑ Trachea, epithelium of

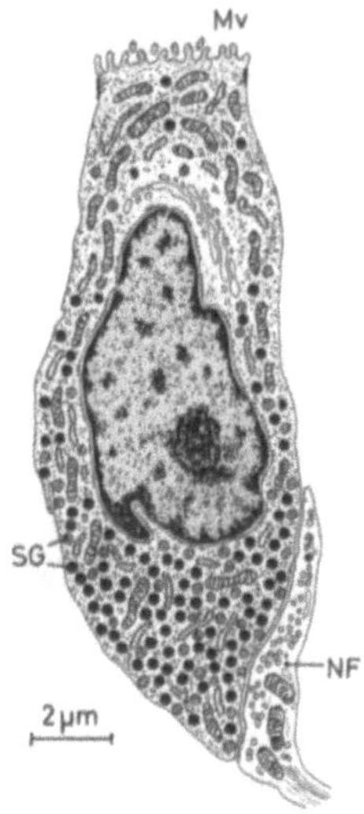

Neck mucous cells, of gastric glands proper (mucocytus cervicalis*): Pyramidal or pear-shaped cells lining necks of ↑ gastric glands proper. N. have an elliptical nucleus with a prominent nucleolus. Cytoplasm contains rod-shaped mitochondria, a supranuclear, well-developed Golgi apparatus (G), a small number of flat rough endoplasmic cisternae, occasional lysosomes, and a moderate number of free ribosomes. Supranuclear cell portion is occupied by large ↑ PAS-positive granules (SG) surrounded by a ↑ unit membrane. Moderately osmiophilic interior contains ↑ glycosaminoglycans. Luminal surface of N. is studded with short microvilli lined by thick ↑ glycocalyx (Glc); at lateral surfaces, N. display ↑ junctional complexes (J), microvillous interdigitations, and small ↑ desmosomes. N. may also be found deep in gastric glands; they are also present in cardiac and pyloric regions. Function of N. is not known; according to some opinions, they represent nondifferentiated replacement cells for ↑ surface mucous cells or precursor cells for both ↑ parietal and ↑ chief cells. Life span: about 6 days. BL = ↑ basal lamina

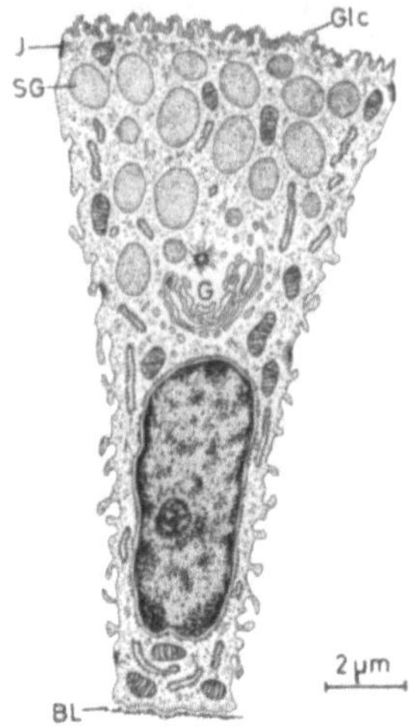

Helander, H.F.: Int. Rev. Cytol. *70*, 217 (1981); Suganuma, T., Katsuyama, T., Tsukahara, M., Tatematsu, M., Sakakura, Y., Murata, F.: Am. J. Anat. *161*, 219 (1981)

Neck, of spermatozoon: ↑ Spermatozoon, connecting piece of

Neck, of tooth (cervix dentis*): Region of tooth where ↑ crown meets root.

↑ Teeth

Neck region, of nephron: A short and inconstant segment of ↑ nephron connecting ↑ renal corpuscle with ↑ proximal tubule. N. is lined with simple squamous epithelium continuous with epithelium of ↑ Bowman's capsule.

↑ Urinary pole, of renal corpuscle

Necrobiosis: A progressive decrease of vital processes in a cell, tissue, or organ. Examples of physiological N. are loss of function of aged ↑ erythrocytes, cells of ↑ sebaceous glands, epidermal cells in the course of ↑ keratinization, etc.

Necrolysis: The postmortal dissolution of cells caused by liberated lysosomal enzymes.

Necrosis: A circumscribed local death in a tissue or organ. N. is accompanied by swelling of cell body, ↑ karyorrhexis, ↑ karyolysis, or ↑ karyopyknosis, vacuolization of mitochondria, Golgi apparatus, and endoplasmic reticulum, ↑ lysis of ↑ cytoplasmic matrix and ↑ cell components by liberated lysosomal enzymes, fragmentation of ↑

plasmalemma and other ↑ cytomembranes.

↑ Necrolysis

Wyllie, A.H., Kerr, J.F., Currie, A.R.: Int. Rev. Cytol. *68*, 251 (1980)

Nematosome: ↑ Nucleoluslike body

Neocortex: ↑ Brain cortex, heterotypical and homotypical isocortex of

Nephrocytes: A term designating both ↑ distal and ↑ proximal tubule cells of ↑ nephron.

Nephron (nephronum*): Fundamental morphofunctional unit of ↑ kidney. Parts: A) ↑ Renal corpuscle (RC) = 1) ↑ glomerulus (G); 2) ↑ Bowman's capsule (BC). B) Renal tubule = 1) ↑ neck region (N); 2) ↑ proximal tubule (PT): a) proximal convoluted tubule (PCT), b) straight portion (SP); 3) ↑ thin segment (TS); 4) ↑ distal tubule (DT): a) distal convoluted tubule (DCT), b) straight portion (SP); 5) connecting portion (CP). Straight portions of both proximal and distal tubules together with thin segment form ↑ loop of Henle. Renal corpuscle and convoluted tubules of both proximal and distal tubules are situated within cortex; their straight portions are situated within ↑ medullary rays and penetrate medulla for varying depths. Points of transition between straight portions of both proximal and distal tubules of a juxtamedullary N. into thin segment subdivide medulla into zones and stripes as indicated. There are about 2–2.5 million N. in the two kidneys. A completely entwined N. has, depending on length of loop of Henle, a length of 2–3 cm. Two types of N.: ↑ N. cortical and ↑ N. juxtaglomerular. CT = ↑ collecting tubule

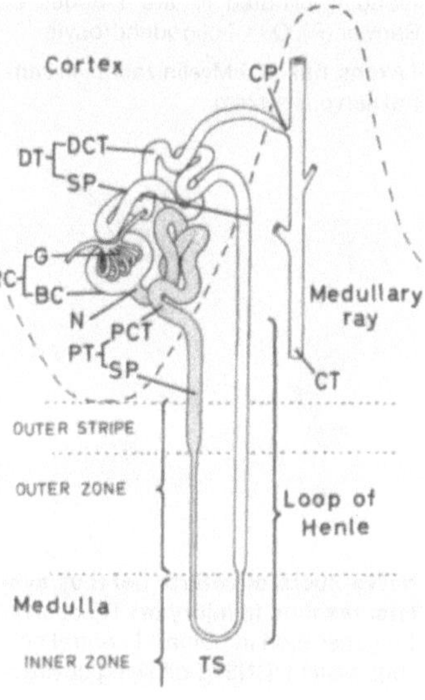

Nephron, cortical (nephronum corticale*): A kind of ↑ nephron with ↑ renal corpuscle (RC) located in peripheral region of cortex; N. are characterized by a short ↑ loop of Henle (LH) and a very short, often absent, ↑ thin segment (TS). N. make up about 80% of all nephrons.

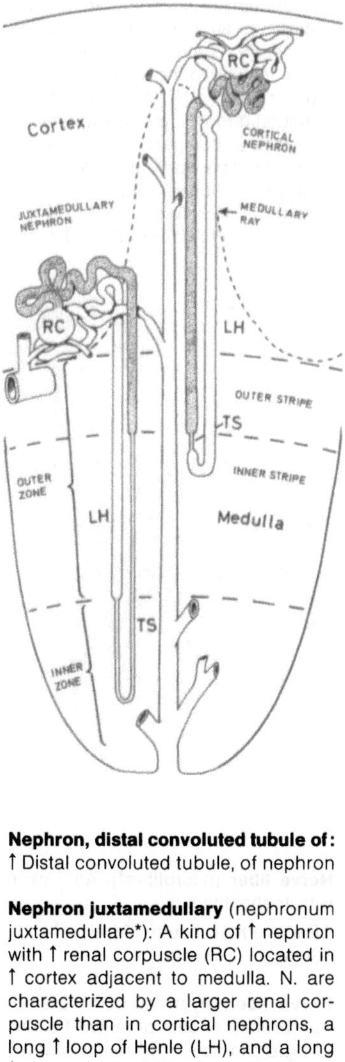

Nephron, distal convoluted tubule of: ↑ Distal convoluted tubule, of nephron

Nephron juxtamedullary (nephronum juxtamedullare*): A kind of ↑ nephron with ↑ renal corpuscle (RC) located in ↑ cortex adjacent to medulla. N. are characterized by a larger renal corpuscle than in cortical nephrons, a long ↑ loop of Henle (LH), and a long ↑ thin segment (TS). In man, N. constitute about 20% of all nephrons and are particularly active in urine concentration by water reabsorption. (See fig. under ↑ Nephron, cortical)

Nephron, loop of Henle: ↑ Loop of Henle

Nephron, proximal tubule of: ↑ Proximal tubule, of nephron

Nephron, renal corpuscle of: ↑ Renal corpuscle

Nephron, thin segment of: ↑ Thin segment, of loop of Henle

Nerve: ↑ Nerve, peripheral

Nerve calyx: A cup-shaped afferent nerve ending almost completely surrounding ↑ vestibular cells type I, either of ↑ cristae ampullares or ↑ macula of saccule and utricle.

↑ Synapse, calyceal

Nerve cell: ↑ Neuron

Nerve ending: A termination of an ↑ axon (A). According to their location, N. are said to be central or peripheral. 1) Central N. (CNS) = an axon forming ↑ synapses with other ↑ nerve cells (NC) inside ↑ central nervous system (CNS). 2) Peripheral N. (PNS) = an axon ending at a nerve cell (NC), nonnervous structure (↑ skeletal muscle fiber = ↑ motor end plate, MEP; ↑ glandular cell = ↑ neuroglandular synapse, NgS), or in a tissue outside CNS in the form of ↑ free N. (FN), ↑ expanded tip N. (ETN), and encapsulated N. or ↑ corpuscles (C).

↑ Nerve endings, peripheral, functional subdivision of

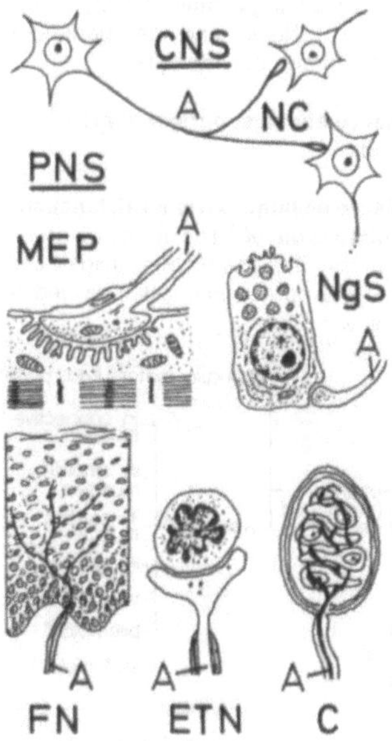

Nerve endings, autonomic: ↑ Adrenergic nerve endings; ↑ Cholinergic nerve endings

Nerve endings, encapsulated: ↑ Corpuscles

Nerve endings, free: ↑ Free nerve endings

Nerve endings, in connective tissues: ↑ Myelinated and ↑ nonmyelinated axon terminations present in all ↑ connective tissues. N. may be either ↑ free or encapsulated (= ↑ corpuscles).

↑ Nerve endings, peripheral, functional subdivision of

Nerve endings, in epithelia: Terminations of ↑ axons in surface, sensory, and glandular ↑ epithelia. ↑ Nerve endings in surface epithelia lose their ↑ myelin sheath as they enter ↑ epithelium; they become ↑ free nerve endings and have a sensory function (↑ epidermis, ↑ corneal epithelium, etc.). Another type of N. are ↑ expanded tip nerve endings like ↑ corpuscles of Merkel. In sensory epithelia, nerve endings are also ↑ nonmyelinated and make synapses with ↑ sensory cells; in glands, nerve endings may be both ↑ myelinated and ↑ nonmyelinated and have sensory and secretory functions (neuroglandular ↑ synapse).

↑ Nerve endings, peripheral, functional subdivision of

Nerve endings, intragemmal: ↑ Taste buds

Nerve endings, motor: ↑ Motor end plate; ↑ Nerve endings, peripheral, functional subdivision of

Nerve endings, perigemmal: ↑ Taste buds

Nerve endings, peripheral, functional subdivision of: The terminal arborization of ↑ axons (PNE) outside ↑ central nervous system is subdivided as follows:

PNE
- efferent* (motor and secretory)
  - somatic = motor end plate*
  - autonomic*
    - in connective tissues
    - in smooth muscles
    - in cardiac muscle
    - in glands
    - vasomotor*
    - pilomotor*
    - sudomotor*
- afferent* (sensory)
  - in epithelia } free*
  - in connective tissue } free*
  - in neurotendinal spindle*
  - in neuromuscular spindle*
  - in corpuscles*
  - baroreceptors*
  - chemoreceptors*
  - specialized (eye*, ear*)

(Terms marked with an asterisk are described under corresponding letters)

Nerve endings, vegetative: ↑ Adrenergic nerve endings; ↑ Cholinergic nerve endings

Nerve fascicle (fasciculus*): A bundle (NF) of parallel-running ↑ myelinated (M) and ↑ nonmyelinated (Nm) ↑ nerve fibers. Each N. is surrounded by ↑ perineurium (P); interior of an N. is occupied by ↑ endoneurium (E) with ↑ fibrocytes (F), ↑ vasa nervorum (Cap), etc. N. may be found singly in tissues; two or more N. enclosed by ↑ epineurium form a peripheral ↑ nerve. An N. can be composed of nonmyelinated nerve fibers only, like ↑ fila olfactoria. BL = ↑ basal laminae, CF = ↑ collagen microfibrils, PE = ↑ perineural endothelium, ZO = ↑ zonula occludens

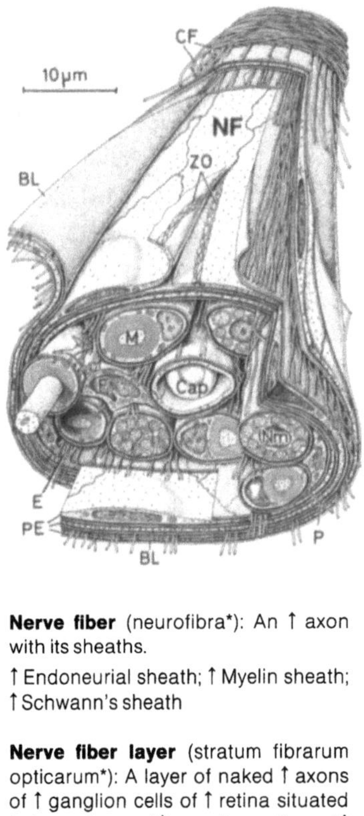

Nerve fiber (neurofibra*): An ↑ axon with its sheaths.

↑ Endoneurial sheath; ↑ Myelin sheath; ↑ Schwann's sheath

Nerve fiber layer (stratum fibrarum opticarum*): A layer of naked ↑ axons of ↑ ganglion cells of ↑ retina situated between layer of ↑ ganglion cells and ↑ inner limiting membrane.

Nerve fibers, adrenergic: ↑ Adrenergic nerve fibers

Nerve fibers, classification of: Main criteria for classification: 1) Morphology = a) ↑ myelinated, b) ↑ nonmyelinated; 2) location = a) central, b) peripheral; 3) direction of conduction = a) afferent (incoming), b) efferent (outgoing); 4) functional = a) motor, b) autonomic (sympathetic, parasympathetic), c) sensory, etc. (See physiology texts for further information)

↑ Nerve endings

Nerve fibers, cholinergic: ↑ Cholinergic nerve fibers

Nerve fibers, fusimotor (gamma fibers, gamma efferents, efferents II and III): Efferent ↑ myelinated intrafusal ↑ nerve fibers (NF), 2–8 µm in diameter, forming ↑ motor end plates with ↑ nuclear bag (NB) and ↑ nuclear chain (NC) muscle fibers of ↑ neuromuscular spindle (NS). PS = ↑ periaxial space

↑ Gamma motor neurons

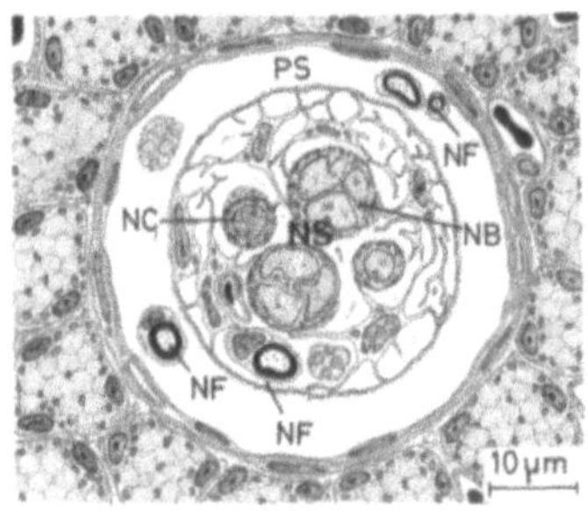

Nerve fibers, intrafusal: All ↑ nerve fibers situated within ↑ neuromuscular spindle.

Nerve fibers, nonmyelinated: ↑ Nonmyelinated nerve fibers

Nerve fibers, of central nervous system: Two kinds of N.: 1) ↑ Myelinated N. = an ↑ axon (A) surrounded by a ↑ myelin sheath (MS); 2) ↑ nonmyelinated N. = axons lacking any sheath. Along myelinated N. are ↑ nodes of Ranvier (R). O = ↑ oligodendrocyte

↑ Axons, naked; ↑ Myelinization, in central nervous system

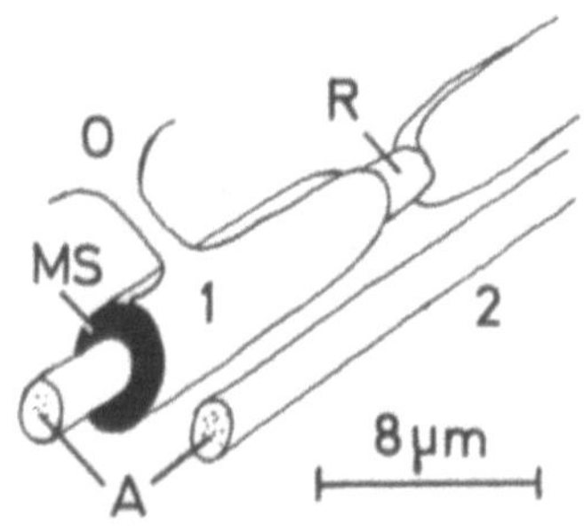

Nerve fibers of central nervous system, reaction to injury of: There is no ↑ regeneration in human ↑ central nervous system (CNS). Following destruction in CNS, corresponding nerve cells die, ↑ microglia and ↑ astrocytes phagocytize the debris, aided by blood-borne ↑ phagocytes. After ↑ phagocytosis of destroyed tissue, astrocytes proliferate and form a dense astrocytic scar ("glial scar") that fills damaged area. Oligodendrocytes delimit these areas by ↑ myelin lamellae, thus preventing growth of axonal sprouts. Despite above-mentioned restriction, recent investigations have shown that regeneration and remyelination can occur in CNS under favorable conditions. For both phenomena, astrocytes and oligodendrocytes seem to cooperate, aided by ↑ Schwann's cells migrating from ↑ peripheral nervous system through ↑ Redlich-Obersteiner zone.

↑ Compound granular corpuscles; ↑ Nerve fibers of peripheral nervous system, degeneration and regeneration of; ↑ Transneuronal degeneration

Allt, G.: Trends Neurosci. *2*, 226 (1979); Haber, B., Perez-Polo, J.R., Hashim, G.A. (eds.): Nervous System Regeneration. New York: Alan R. Liss Inc. 1983; Nicholls, J.G. (ed.): Repair and Regeneration of the Nervous System. Berlin, Heidelberg, New York: Springer-Verlag 1982

Nerve fibers, of peripheral nervous system: Two kinds of N.: 1) ↑ Myelinated N. = ↑ axon (A) surrounded by ↑ myelin sheath (MS), sheath of ↑ Schwann or neurilemma (Sch) = the cytoplasm of ↑ Schwann's cell (Sch), and ↑ endoneurial sheath (E) = the basal lamina (BL) of Schwann's cell and ↑ reticular and ↑ collagen micrifibrils (F). Myelinated N. show segments lacking myelin and

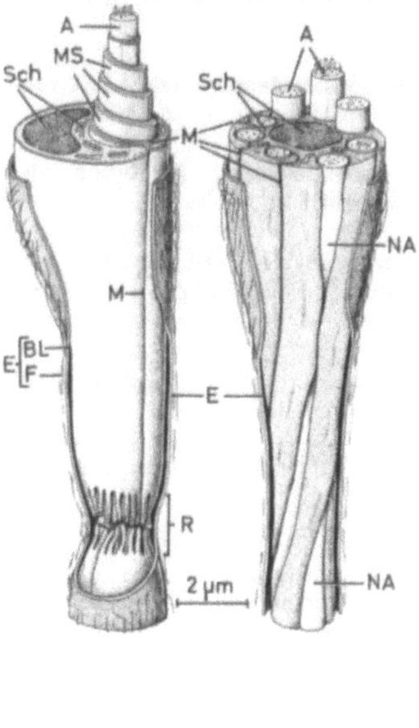

Schwann's sheaths, ↑ nodes of Ranvier (R), and with only one from outside visible ↑ mesaxon (M). 2) ↑ Nonmyelinated N. = several axons (A) surrounded by sheath of Schwann, and endoneurial sheath (E). Along nonmyelinated N., there are no nodes of Ranvier, but several mesaxons (M) and naked ↑ axons (NA) are present.

Nerve fibers of peripheral nervous system, degeneration and regeneration of: The succession of events after sectioning of a ↑ nerve fiber. A) Proximal stump (PS): 1) Primary or ascendent degeneration of ↑ axon (1st week) = destruction and degeneration of axon (Ax) extending two to three ↑ internodal segments in clean-cut and 2–3 cm in lacerated wounds; fusion of ↑ axolemma over severed ends and formation of ↑ retraction bulb (RB); disintegration of ↑ myelin sheath. 2) Axonal reaction = swelling of body of ↑ nerve cell, progressive dissolution of ↑ Nissl bodies (= retrograde chromatolysis),

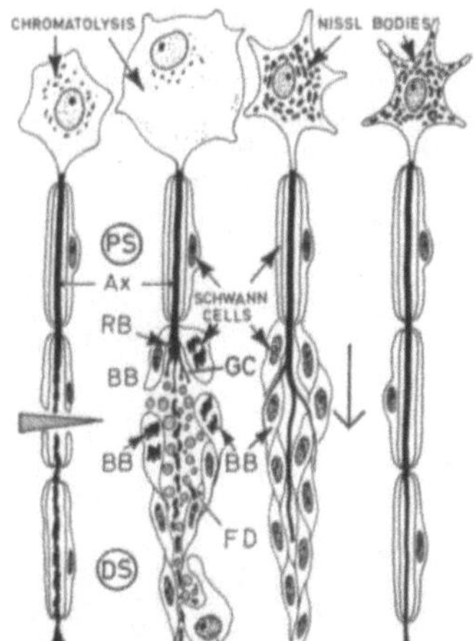

shifting of nucleus into eccentric position away from ↑ axon hillock ("fish-eye" cells). 3) After 4–6 weeks, recovery of nerve cell. 4) Axonal regeneration = fine axoplasmic sprouts [= growth cones (GC)] bulge from retraction bulb toward distal stump. 5) Proliferation of ↑ Schwann's cells in the form of bands of ↑ Büngner (BB) in the same direction. B) Distal stump (DS): 1) Secondary, descendent, or Wallerian degeneration = complete disintegration of axon, its collaterals and endings (= terminal degeneration); breakdown of myelin sheath into fatty droplets (FD); reaches maximum about 2 weeks after injury). 2) ↑ Phagocytosis of debris by ↑ macrophages. 3) Mitotic division of Schwann's cells, which also form bands of Büngner (BB). 4) The latter bridge the gap between stumps and join proliferated Schwann's cells of proximal stump. 5) Axon grows (arrow) toward distal stump (3–4 mm/day in mammals) using bands of Büngner as support and guide. 6) Still enveloped by Schwann's cells of the bands, which furnish a new myelin sheath, axon grows into distal stump and forms its collaterals and end arborization (several months). Conditions for regeneration: Body of nerve cell must be intact; connective tissue between two ends of cut nerve fiber must be absent; gap between two ends must not be very great.

↑ Nerve fibers of central nervous system, reaction to injury of; ↑ Transneuronal degeneration

Ide, C.: Arch. Histol. Jpn. *46*, 243 (1983); Kelly, J.P.: Reaction of Neurons to Injury. In: Kandel, E.R. and Schwartz, J.H. (eds.): Principle of Neural Science. New York, Amsterdam, Oxford: Elsevier/North Holland 1981; Patsalos, P.N., Bell, M.E., Wiggins, R.C.: J. Cell Biol. *87*, 1 (1980); Tessler, A., Autilio-Gambetti, L, Gambetti, P.: J. Cell Biol. *87*, 197 (1980); Wakefield, C., Shonnard, N.: Anat. Rec. *206*, 79 (1983)

Nerve fibers, parasympathetic: ↑ Nonmyelinated nerve fibers belonging to parasympathetic division of ↑ autonomic nervous system.

Nerve fibers, pilomotor: ↑ Pilomotor nerve fibers

Nerve fibers, sensory, encapsulated: ↑ Corpuscles

Nerve fibers, sudomotor: ↑ Sudomotor nerve fibers

Nerve fibers, sympathetic: ↑ Nonmyelinated nerve fibers belonging to sympathetic division of ↑ autonomic nervous system.

Nerve fibers, terminations of: ↑ Nerve endings

Nerve fibers, vasomotor: ↑ Vasomotor nerve fibers

"Nerve glue": ↑ Neuroglia

Nerve growth factor (NGF): A protein of crucial importance for development, growth, and maintenance of function of various types of nerve cells, particularly of peripheral sensory and sympathetic nerve cells. N. is mainly synthesized in ↑ submandibular gland and seems to be controlled by ↑ testosterone.

Brandshaw, R.A.: Ann. Rev. Biochem. *47*, 191 (1978); Edgar, D., Barde, Y.-A.: TINS *6*, 260 (1983); Schwab, M.E., Stöckel, K., Thoenen, H.: Cell Tissue Res. 169, 289 (1976); Slack, J.R., Hopkins, W.G., Pockett, S.: Muscle Nerve *6*, 243 (1983)

Nerve head, of retina: ↑ Optic papilla

Nerve, optic: ↑ Optic nerve

Nerve, peripheral (spinal nerve, nervus spinalis*): A group of several ↑ nerve fascicles (F) held together by ↑ epineurium (E). Each fascicle is surrounded by ↑ perineurium (P); the ↑ loose connective tissue separating ↑ nerve fibers forms ↑ endoneurium (En). Continuous with epineurium are loose connective and adipose tissues that surround nerve fascicles. ↑ Vasa nervorum (VN), ↑ nervi nervorum, and lymphatic vessels run within and between nerve fascicles. A loose connective sheath of various thickness, intermingled with groups of white ↑ adipose cells at outer aspect of epineurium, forms ↑ paraneurium (Pa).

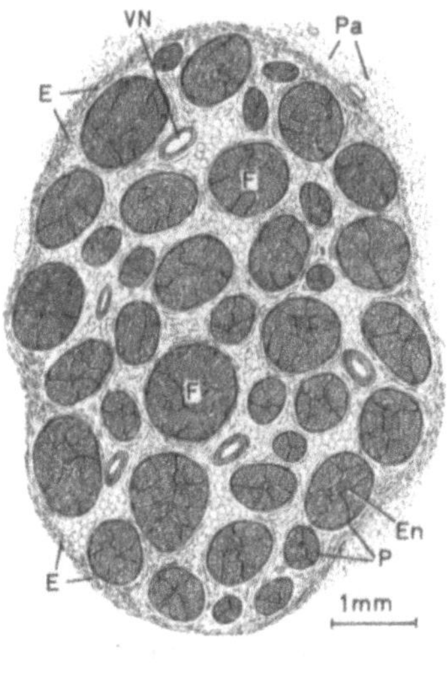

Nervi nervorum*: ↑ Nerve fibers innervating peripheral ↑ nerves

Nervi vasorum: Autonomic ↑ nonmyelinated nerve fibers innervating blood and lymphatic vessels.

Nervous system (systema nervosum*): The entire mass of ↑ nervous tissue in the body. N. is integrator of bodily functions and all of its parts work together to this end. Parts: 1) ↑ Central nervous system or CNS (↑ brain and ↑ spinal cord) is delimited from surrounding connective tissue and blood vessels (B) by ↑ membrana gliae limitans superficialis (MS) and perivascularis (MP), both formed by processes of ↑ astrocytes (As). In this particular milieu are situated ↑ nerve cells (N), whose ↑ axons (A) are myelinated by ↑ oligodendrocytes (O); ↑ microglia cells (M) also belong to CNS. 2) Peripheral nervous system or PNS is a continuation of CNS and comprises all nervous structures situated outside membrana gliae limitans superficialis. PNS consists of: a) ↑ Somatic nervous system = afferent, ↑ myelinated somatic nerve fibers (MNF) for voluntary innervation of skeletal muscles; b) ↑ autonomic nervous system = ↑ autonomic ganglia (AG) and ↑ nonmyelinated afferent nerve fibers (ANF) for autonomic innervation of interior organs; c) sensory efferent nerve fibers (SNF) = processes of ↑ pseudounipolar nerve cells of ↑ spinal ganglia (SG). All three kinds of fiber constitute peripheral ↑ nerve (PN). Besides neurons, PNS-includes special ↑ neuroglial cells [↑ Schwann's (Sch) and ↑ satellite cells (SC)]. Arrows indicate direction of nerve impulses. ROZ = ↑ Redlich-Obersteiner zone. (Modified after Rhodin 1974)

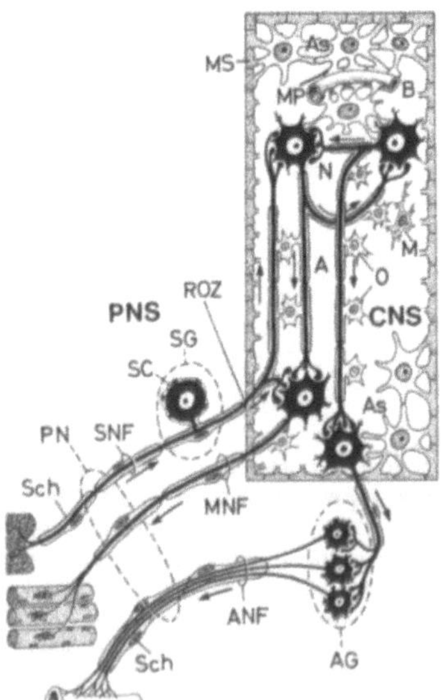

Hubbard, J.I. (ed.): Peripheral Nervous System. New York: Plenum Press 1974; Nathan, P.: The Nervous System. Oxford: Oxford University Press 1982; Rhodin, J.A.G.: Histology, A Text and Atlas. New York: Oxford University Press 1974; Sandri, C., Van Buren, J.M., Akert, K.: Membrane Morphology of the Vertebrate Nervous System. Progress in Brain Res. *46*, 2nd edn. Amsterdam: Elsevier Biomedical 1982; Seymour, R.M., Berry, M.: The Nervous System. In: Hodges, G.M. and Hallowes, R.C. (eds.): Biomedical Research Applications of Scanning Electron Microscopy. Vol. 1. London, New York: Academic Press 1979

Nervous system, autonomic: ↑ Autonomic nervous system

Nervous system, central: ↑ Central nervous system

Nervous system, craniosacral division of: ↑ Parasympathetic nervous system

Nervous system, functional parts of: On the basis of function, nervous system may be divided into: 1) ↑ Somatic nervous system and 2) ↑ autonomic nervous system.

↑ Nervous system

Nervous system, gray matter of: ↑ Gray matter, of CNS

Nervous system, interneurons of: ↑ Interneurons

Nervous system, limiting membranes of: ↑ Membrana limitans gliae superficialis; ↑ Membrana limitans gliae perivascularis

Nervous system, neurons of: ↑ Neurons

Nervous system, parasympathetic: ↑ Parasympathetic nervous system

Nervous system, peripheral: ↑ Nervous system

Nervous system, somatic: ↑ Nervous system; ↑ Somatic nervous system

Nervous system, supporting elements of: ↑ Neuroglia

Nervous system, sympathetic: ↑ Sympathetic trunk; ↑ Sympathetic nervous system

Nervous system, sympathetic trunk of: ↑ Sympathetic trunk

Nervous system, visceral: ↑ Autonomic nervous system

Nervous tissue (textus nervosus*): A highly specialized vascularized ↑ epithelium composed of ↑ neurons and ↑ neuroglia.

Nervous tissue, histogenesis of (neurogenesis*): In dorsal part of embryo, somatic ectoderm (SE) thickens and forms neural plate (NP), which invaginates and gives rise to neural groove (NG). Contact sites of somatic ectoderm with neural groove, – neurosomatic junctions (NSJ) – approach one another, fuse, and form neural tube (NT), initially lined with ↑ ventricular cells. From neurosomatic junctions, groups of cells proliferate and separate to form neural crest (NC). Epithelium or ↑ neuroepithelium (Ne) of neural tube soon transforms into a columnar pseudostratified which is mitotically very active at early stages of embryogenesis. Progressively, G_1 phase of some cells becomes longer, so that a considerable cell population becomes arrested in this phase of ↑ cell cycle. These cells form mantle zone (MZ) of neural tube (future ↑ gray matter). Expansive accumulation of postmitotic cells in G_0 phase provokes rapid growth of CNS. Outside mantle zone, up to basal lamina (BL), is the marginal zone (MaZ), which becomes invaded by processes of postmitotic cells (future ↑ white matter). Cells of mantle zone are customarily referred to as ↑ neuroblasts and ↑ glioblasts. By ↑ differentiation, neuroblasts give rise to mature ↑ neurons of CNS; by further divisions, glioblasts give rise to ↑ neuroglial cells. From epithelium (E) lining central canal arise: 1) ↑ Ependymal cells; 2) ↑ pituicytes; 3) ↑ pinealocytes, and epithelial cells of 4) ↑ choroid plexus. From neural crests derive: 1) Neuroblasts of PNS; 2) glioblasts of PNS (= ↑ satellite cells, of peripheral neurons and ↑ Schwann's cells); 3) ↑ sympathoblasts; 4) ↑ melanoblasts; 5) ↑ microglia; 6) mesectoderm, and, possibly; 7) ↑ C-cells of ↑ ultimobranchial bodies. (See embryology texts for further information)

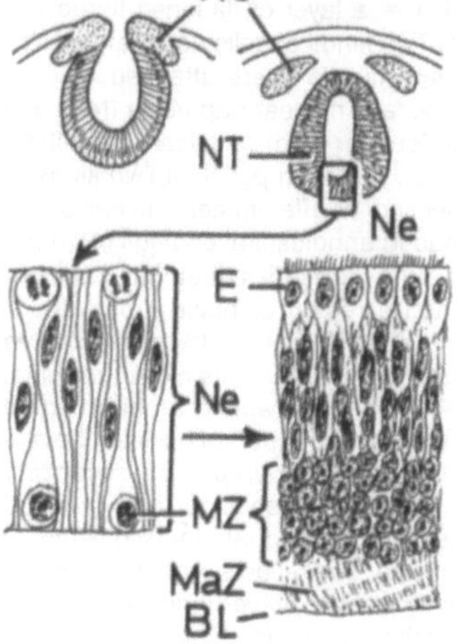

Neumann's sheath (peritubular dentin, dentinum peritubulare*): A layer (NS) of highly calcified ↑ dentin surrounding a ↑ dentinal tubule (DT).

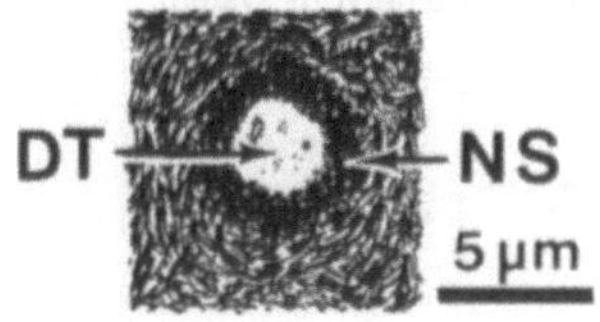

Neural tube: ↑ Nerve tissue, histogenesis of

Neuraxis: A term sometimes used to designate ↑ central nervous system.

Neurilemma: ↑ Schwann's sheath

Neurilemmal cells: ↑ Schwann's cells

Neurimidic acid: ↑ Sialic acid

Neuriminidase: An enzyme removing ↑ sialic acid from cell surface.

Neurite: ↑ Axon

Neurobiotaxis: The property of ↑ neurons to establish specific functional contacts (↑ synapses) in order to form neuronal circuits. N. is based on recognition capacity of neuronal ↑ glycocalyx and its adhesive faculty.

↑ Synaptic cleft

Neuroblast (neuroblastus*): The immature precursor of a ↑ neuron incapable of further division. N. gives rise to a mature neuron by ↑ differentiation. Since N. does not divide, the term "neuroblast" should be replaced by "young neuron."

Neurocrinia: The discharge of ↑ hormone(s) from ↑ endocrine cells into ↑ nervous tissue of ↑ central nervous system.

Neuroendocrine cells: ↑ Neurosecretory neurons

Neuroepithelial bodies (NEB): Specialized, richly innervated intraepithelial structures found within intrapulmonary airways of various mammalian species [mostly at bronchiolar bifurcations (BB)] composed of 10–80 argyrophilic ↑ NEB-cells. Nonmyelinated (cholinergic?) nerve endings (NE), filled with clear ↑ synaptic vesicles, are in contact with NEB-cells. It is believed that N. play a role as intrapulmonary ↑ chemoreceptors, reacting to changes in airway gas composition by release of ↑ catecholamines and/or polypeptide hormones into bronchiolar capillaries (BC), acting on airway smooth musculature. Mv = microvilli, SG = ↑ secretory granules

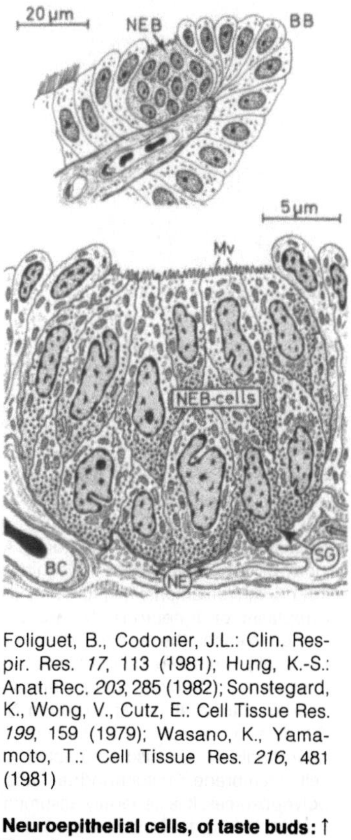

Foliguet, B., Codonier, J.L.: Clin. Respir. Res. *17*, 113 (1981); Hung, K.-S.: Anat. Rec. *203*, 285 (1982); Sonstegard, K., Wong, V., Cutz, E.: Cell Tissue Res. *199*, 159 (1979); Wasano, K., Yamamoto, T.: Cell Tissue Res. *216*, 481 (1981)

Neuroepithelial cells, of taste buds: ↑ Taste buds

Neuroepithelial portion of retina (pars optica retinae): Part of ↑ retina containing ↑ photoreceptors and situated posterior to ↑ ora serrata.

Neuroepithelium: 1) ↑ Pseudostratified epithelium of neural tube. 2) A surface epithelium containing ↑ neurons or ↑ sensory cells (↑ cristae ampullares, ↑ macula of saccule and utricle, ↑ olfactory epithelium, ↑ retina, ↑ taste buds, etc.).

↑ Nerve tissue, histogenesis of

Kessel, R.G., Kardon, R.H.: Biomed. Res. *2*, Suppl. 483 (1981)

Neurofibrils: Intertwined bundles (Nfb) of ↑ neurofilaments (Nfl), about

0.5–3 µm thick, running around nucleus (N), between mitochondria (M), Golgi apparatus (G), and other ↑ organelles to extend into ↑ dendrites (D) and ↑ axon (A). N. are stainable with metallic silver because of its precipitation around neurofilaments.

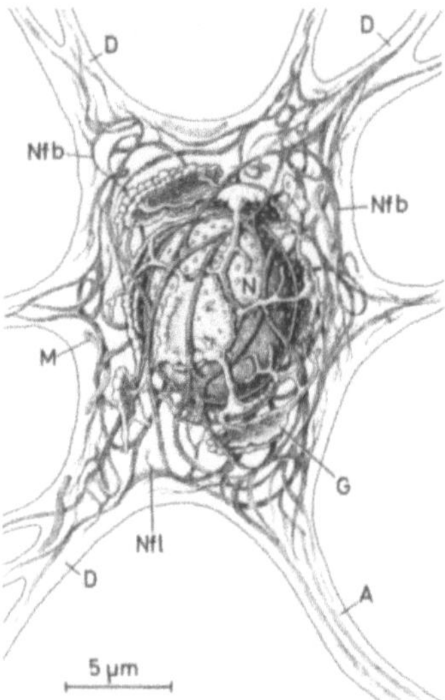

Wisniewski, H.M., Soifer, D.: Mech. Ageing Dev. *9*, 119 (1979)

Neurofilaments (neurofilamentum*): A kind of 7-nm-thick and indefinitely long ↑ microfilaments (N) present in cytoplasm of ↑ neurons. N. are attached to each other and to ↑ neurotubules (Nt) by cross-bridges about 5 nm thick and 10–20 nm long (arrowheads), spaced 40 nm apart along N. Bundles of cross-linked N. form ↑ neurofibrils. N. are also connected to cell membrane, mitochondria, and polyribosomes. It is generally assumed that N. belong to ↑ cytoskeleton and that they are involved in ↑ axoplasmic and ↑ dendritic flows; other functions of N. are unknown. (Fig. = ↑ axon in CNS, rat)

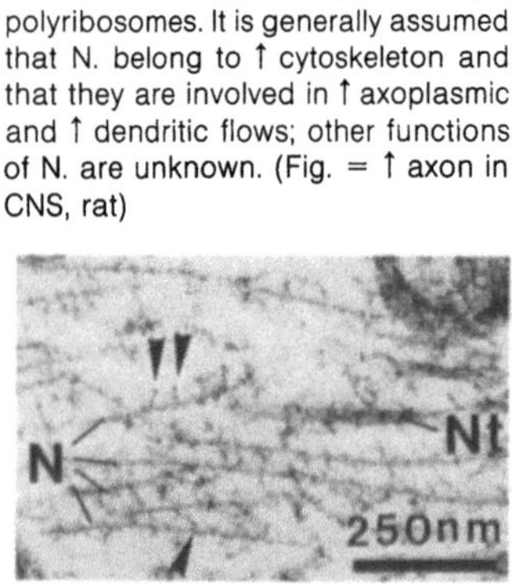

Letterier, J.-F., Liem, K.H., Shelanski, M.L.: J. Cell Biol. *95*, 982 (1982); Metuzals, J., Clapin, D.F., Chapman, G.D.: Cell Tissue Res. *223*, 507 (1982); Mori, H., Kurokawa, M.: Biomed. Res. *1*, 24 (1980); Rash, J.E.: TINS *6*, 209 (1983)

Neuroglandular synapse: ↑ Synapse, neuroglandular

Neuroglia* (glia cells, "nerve glue", neuroglial cells): A class of nonneuronal cells intimately surrounding ↑ perikarya, ↑ dendrites, and segments of ↑ axons of neurons, separating them from blood vessels and connective tissue. Two kinds of N. cells: 1) N. cells of ↑ central nervous system; 2) N. cells of ↑ peripheral nervous system.

Neuroglia
- Central
  - Ependymal cells*
  - Astrocytes* (Macroglia)
  - Oligodendrocytes* (Macroglia)
  - Microglia*
- Peripheral
  - Schwann's cells*
  - Satellite cells*

In addition to its mechanical supportive function, N. participates in nutritive processes of neurons, assures their chemical microenvironment, synthesizes ↑ myelin sheaths, insulates and protects neurons, removes degenerated neurons, and plays an important role in regeneration of ↑ nerve fibers of peripheral and central nervous systems. (Asterisks indicate letters under which corresponding terms are described)

Oksche, A.: Neuroglia I. In: Oksche, A., Vollrath, L. (eds.): Handbuch der mikroskopischen Anatomie des Menschen, Vol. 4, Part 10, Nervensystem. Berlin, Heidelberg, New York: Springer-Verlag 1980; Sears, T.A. (ed.): Neuronal-Glia Cell Interrelationships. Berlin, Heidelberg, New York: Springer-Verlag 1982

Neuroglial cells: ↑ Neuroglia

Neuroglial cells, of retina: ↑ Retina, neuroglial cells of

Neurohormones: Substances produced and released from ↑ neurosecretory neurons (e.g., ↑ antidiuretic hormone, ↑ oxytocin, ↑ relasing factors, etc.).

Neurohypophyseal hormones: ↑ Oxytocin and ↑ antidiuretic hormone are the only hypothalamic hormones stored in ↑ neurohypophysis. It is not yet known if ↑ pituicytes synthesize a hormone.

Diericks, K.: Int. Rev. Cytol. *62*, 120 (1980)

Neurohypophysis* (lobus posterior*): A common term for all parts of ↑ hypophysis derived from bottom of diencephalon: ↑ Median eminence, ↑ infundibulum, and lobus nervosus.

↑ Hypophysis, lobus nervosus of

Moses, A.M. (ed).: Neurohypophysis. Basle: Karger 1977; Seyama, S., Pearl, G.S., Takei, Y.: Cell Tissue Res. *206*, 291 (1980)

Neurohypophysis, Herring bodies of: ↑ Herring bodies

Neurokeratin: A proteinaceous, loosely arranged network (N) surrounding ↑ axons (A) of ↑ myelinated nerve fibers after extraction of lipid component of ↑ myelin.

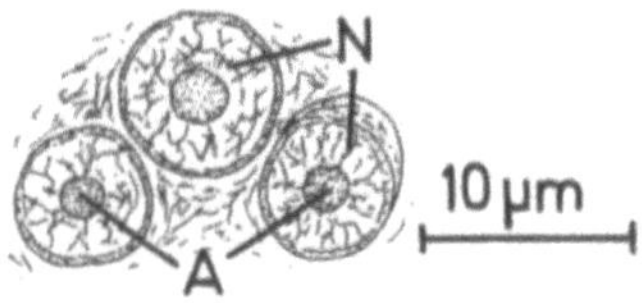

Neurolemma: ↑ Schwann's sheath

Neuromotor system: A morphofunctional connection of motor ↑ nerve fibers with effector organs, principally skeletal muscles.

Neuromuscular junction: ↑ Motor end plate

Neuromuscular spindles (muscle spindle, fusus neuromuscularis*): Fusiform, 100 to 200-µm-wide and 0.5 to 7-mm-long organs scattered within skeletal muscles. N. are parallel to ↑ extrafusual muscle fibers (EMF). Structure: 1) Extensible external connective tissue capsule (C) attached to ↑ perimysium (Pe) of neighboring extrafusal muscle fibers. Capsule is a continuation of ↑ perineurium (P). 2) ↑ Periaxial space (PS). 3) Inner capsule (IC) = a layer of flattened fibrocytes. 4) Two kinds of delicate intrafusal striated muscle fibers attached to capsule: a) ↑ nuclear bag fibers (NB), one to four per N.; b) ↑ nuclear chain fibers (NC), up to ten per N. 5) Two kinds of sensory or afferent nerve fibers; a) denuded annulospiral endings (Aff I) are wound around equatorial noncontractile region of nuclear bag fibers, basal lamina of the latter also encloses axon; small branches of annulospinal endings are wrapped around central noncontractile part of nuclear chain fibers; b) flower-spray, or secondary (Aff II) nerve endings are grouped around juxtaequatorial segments of nuclear chain fibers. Annulospiral and flower-spray nerve endings represent stretch receptors. 6) Motor or efferent nerve supply: Fine ↑ γ-motor nerve fibers (Eff II and III) terminate in small ↑ motor end plates (MP) or in grapelike endings on nonnucleated regions of both ↑ intrafusal muscle fibers. Much thicker ↑ α-fibers innervate extrafusal muscle fibers. Stretching of muscle provokes a lengthening of N.; excitatory impulses are generated in annulospiral endings and flow to ↑ spinal cord. From spinal cord, corresponding α- and γ-motor neurons send impulses for simultaneous contraction of both extrafusal and intrafusal muscle fibers via α- and γ-fibers, respectively. Simultaneous contraction of intrafusal fibers stretches noncontractile regions of nuclear bag and chain fibers where annulospiral endings are wound; in this way, γ-fibers increase sensitivity of N. to stretching. Thus, N. is an organ controlling degree of contraction of skeletal muscles.

↑ Tandem spindle

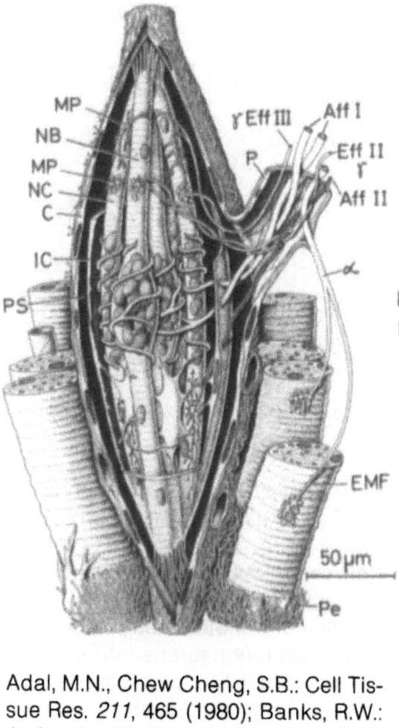

Adal, M.N., Chew Cheng, S.B.: Cell Tissue Res. *211*, 465 (1980); Banks, R.W.: J. Anat. *133*, 571 (1981); Dow, P.R., Shinn, S.L., Ovalle, W.K.: Am J. Anat. *157*, 375 (1980); Kucera, J.: Histochemistry *73*, 397 (1981); Taxi, J. (ed.): Ontogenesis and Functional Mechanisms of Peripheral Synapses. Amsterdam, New York, Oxford; Elsevier/North-Holland Biomedical Press 1980

Neuron (neuronum*, neurocytus*): A nerve cell with all its processes. N. consists of cell body with nucleus (N), soma or perikaryon (P), variable number of ↑ dendrites (D), and one ↑ axon (A), which often emits collateral(s) (C). Functionally, dendrites and cell body represent receptor portion of N.; axon with collateral(s) is conductile portion of N., and terminal branching is effector portion of N. N. is the biological, genetic, trophic, and morphological unit of ↑ nervous system. Together with ↑ neuroglia, N. constitute ↑ nervous tissue. CNS = ↑ central nervous system, O = ↑ oligodendrocyte, PNS = ↑ peripheral nervous system, ROZ = ↑ Redlich-Obersteiner zone, SC = ↑ Schwann's cells. (Fig. modified after Bloom and Fawcett 1975)

↑ Neurons, structure of

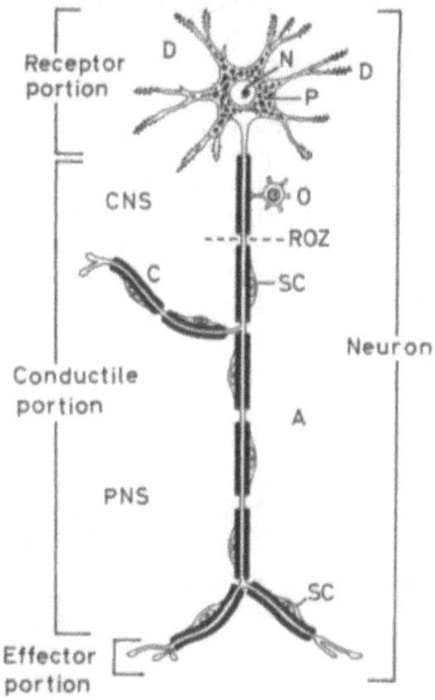

Bloom, W., Fawcett, D.W.: A Textbook of Histology. 10th edn. Philadelphia, London, Toronto, W.B. Saunders Co 1975; Bray, D., Gilbert, D.: Ann. Rev. Neurosci. *4*, 505 (1981); Peters, A., Palay, S., Webster, H. DeF.: The Fine Structure of the Nervous System. 2nd edn. Philadelphia: W.B. Saunders 1976; Stevens, C.F.: Sci. Am. *241*, 49 (1979)

Neurons, alpha motor: ↑ Alpha motor fiber

Neurons, association: ↑ Association neurons

Neurons, autonomic: ↑ Autonomic neurons

Neurons, axis cylinder of: ↑ Axon

Neurons, axon of: ↑ Axon

Neurons, bipolar: ↑ Bipolar neurons

Neurons, chromatophilic substance of: ↑ Nissl bodies

Neurons, classification of: On the basis of the number of processes and shape of cell body, ↑ neurons may be classified as: 1) Unipolar = from cell body emerges only one short ↑ axon (A); ↑ photoreceptors may be considered unipolar neurons. 2) ↑ Pseudounipolar neurons = during development, two processes approximate (arrows) and fuse together, forming a single process which bifurcates into a proximal process with function of an axon and a distal process with function of a ↑ dendrite. Both processes are, however, myelinated; to this type belong ↑ ganglion cells of ↑ spinal ganglia. 3) ↑ Bipolar neurons = relatively symmetrical fusiform neurons with two processes – one dendrite (D) and one axon (A), arising from opposite poles of cell body. To bipolar neurons belong ↑ bipolar neurons of ↑ retina, ↑ spiral and ↑ vestibular ganglia, and ↑ olfactory cells. 4) ↑ Multipolar neurons = ↑ soma gives rise to many dendrites (D) and one axon (A); this type is the most common in ↑ nervous system. Multipolar neurons may be basket-shaped, granular, stellate, pyramidal, or have a special form like ↑ Purkinje cells, ↑ Golgi cells type I and II, and cells of ↑ autonomic ganglia.

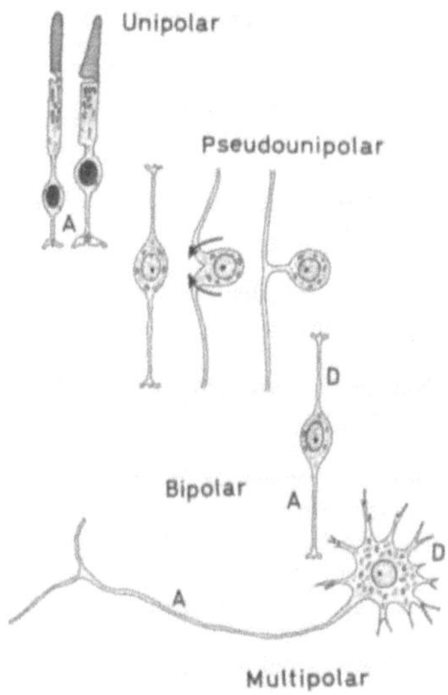

Neurons, dendrites of: ↑ Dendrites

Neurons, development of: ↑ Nerve tissue, histogenesis of

Neurons, gamma motor: ↑ Gamma motor neurons

Neurons, Golgi type I: ↑ Golgi type I neurons

Neurons, Golgi type II: ↑ Golgi type II neurons

Neurons, integrator: ↑ Interneurons

Neurons, motor: ↑ Motor neurons

Neurons, multipolar: ↑ Multipolar neurons

Neurons, myelinated: ↑ Bipolar neurons, of spiral and vestibular ganglia

Neurons, neurofibrils of: ↑ Neurofibrils

Neurons, neurofilaments of: ↑ Neurofilaments

Neurons, neurotubules of: ↑ Neurotubules

Neurons, Nissl bodies of: ↑ Nissl bodies

Neurons, of retina: ↑ Amacrine cells; ↑ Bipolar neurons, of retina; ↑ Ganglion cells, of retina; ↑ Horizontal cells

Neurons, olfactory: ↑ Olfactory epithelium

Neurons, postganglionic: ↑ Postganglionic neurons

Neurons, preganglionic: ↑ Preganglionic neurons

Neurons, processes of: ↑ Axon; ↑ Dendrites

Neurons, pseudounipolar: ↑ Pseudounipolar neurons

Neurons, pyramidal: ↑ Pyramidal neurons

Neurons, response to injury of: ↑ Nerve fibers of central nervous system, reaction to injury of; ↑ Nerve fibers of peripheral nervous system, degeneration and regeneration of

Neurons, satellite cells of: ↑ Satellite cells

Neurons, somatic: ↑ Somatic neurons

Neurons, structure of: In ↑ perikaryon (P), ↑ neuron generally contains one voluminous nucleus (N) with finely dispersed ↑ chromatin; binucleate neurons are found in some ↑ autonomic ganglia. Nucleus of some neurons (e.g., ↑ Purkinje cells) is ↑ tetraploid. ↑ Nucleolus (Nu) is large; in the female, ↑ Barr body (BB) is adjacent to nucleolus or ↑ nuclear membrane. ↑ Neuroplasm encloses numerous rod-shaped mitochondria, and well-developed and multiple ↑ Golgi apparatus (G) communicating through ↑ tubules of smooth ↑ endoplasmic reticulum with rough endoplasmic reticulum to form ↑ GERL complex. Cisternae of rough endoplasmic reticulum are aggregated in ↑ Nissl bodies (NB). ↑ Lysosomes (Ly) are numerous and close to Golgi apparatus; 60–80-nm granules are also adjacent to this ↑ organelle. ↑ Lipid droplets, ↑ multivesicular bodies, and a ↑ centriole are also present in neurons. Whereas almost every neuron contains ↑ lipofuscin (Lf), only a few neurons enclose ↑ melanin granules. In ↑ neurosecretory neurons, ↑ neurosecretory granules are produced in perikaryon. ↑ Neurofibrils (Nf) and ↑ neurotubules (Nt) run through perikaryon and penetrate into cell processes. From cell body extend several ↑ dendrites (D) and one ↑ axon (A), which begins at ↑ axon hillock (AH). On ↑ soma and extensions of a neuron end ↑ boutons terminaux (BT) of other neurons. Cap = capillary

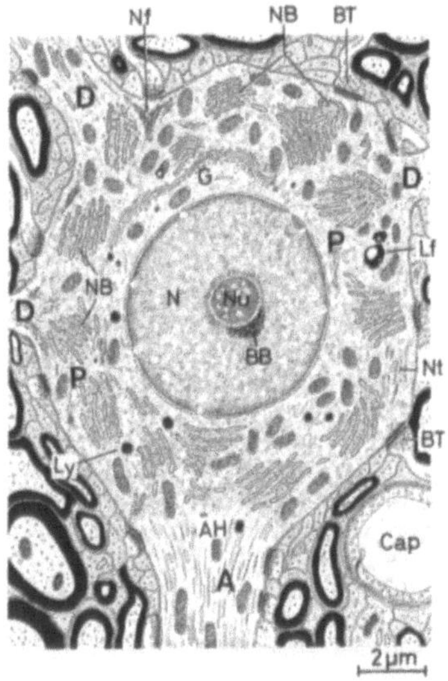

Pfenninger, K.H.: Ann. Rev. Neurosci. *1*, 445 (1978)

Neurons, synapses of: ↑ Synapses

Neurons, unipolar: ↑ Neurons, classification of; ↑ Photoreceptors

Neurophysins: Carrier proteins linked in 1:1 molar ratio by noncovalent bonds with hypothalamic hormones. To each hormone corresponds an appropriate N.; for example, in the rat, N. I is associated with ↑ antidiuretic hormone, and N. II (and III) with ↑ oxytocin. It seems that N. may crystallize within ↑ neurosecretory granules.

Bargmann, W., Gaudecker, B.V.: Z. Zellforsch. *96*, 495 (1969); De Mey, J., Vandesande F., Dierickx, K.: Cell Tissue Res. *153*, 531 (1974); Dierickx, K.: Int. Rev. Cytol. *62*, 120 (1980); Lauber, M., Camier, M., Masse, J.O., Cohen, P.: Biol. Cell. *36*, 111 (1979); Yulis, C.R., Rodriguez, E.M.: Cell Tissue Res. *127*, 93 (1982)

Neuropil: A densely packed conglomerate (N) of ↑ neuroglia, ↑ dendrites, and ↑ myelinated and ↑ nonmyelinated axons separating ↑ nerve cells (NC) of CNS from one another. Under light microscope, N. appears as a structureless substance containing some nuclei. About 17%–20% of N. is formed by intercellular spaces. (Fig., top = human ↑ spinal cord; fig., bottom = ↑ cerebellum, rat)

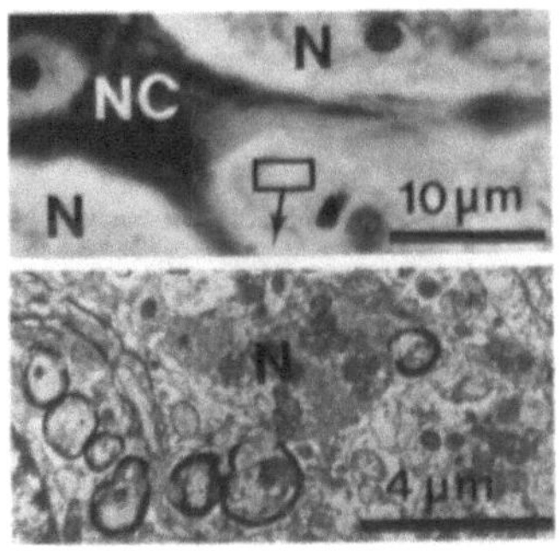

Wagner, H.-J., Barthel, J., Pilgram, C.: Anat. Embryol. *166*, 427 (1983)

Neuroplasm (cytoplasma neurocyti*): The ↑ cytoplasm of a ↑ neuron.

Neurosecretion: The phenomenon of production in ↑ perikaryon and release of ↑ neurohormones from ↑ axon of ↑ neurosecretory neurons into circulating body fluids (↑ blood, ↑ cerebrospinal fluid, ↑ nervous tissue).

↑ Hypothalamus

Farner, D.S., Lederis, K. (eds.): Neurosecretion. New York: Plenum Press 1981; Krieger, D., Hughes, J.C.: Neuroendocrinology. Oxford: Blackwell Scientific Publications 1980; Normann, T.C.: Int. Rev. Cytol. *46*, 1 (1976)

Neurosecretory axon: An ↑ axon (A) of a ↑ neurosecretory neuron transporting ↑ neurosecretory granules (Gr) to ↑ pericapillary spaces (PS) of ↑ hypothalamus or ↑ neurohypophysis. (Fig. = rat)

↑ Herring bodies

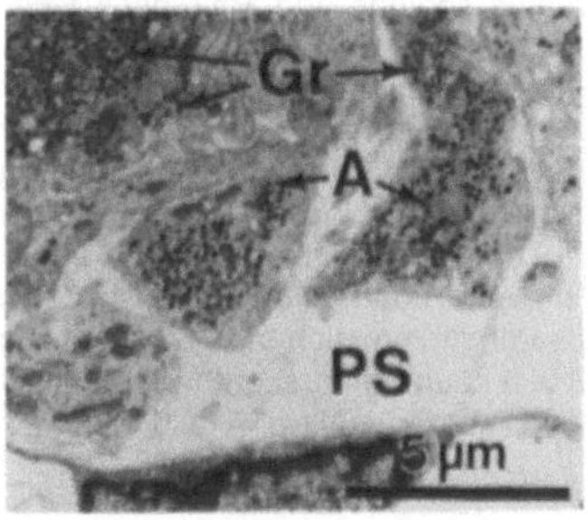

Loesch, A.: J. Ultrastruct. Res. *77*, 344 (1981)

Neurosecretory cells: ↑ Neurosecretory neurons

Neurosecretory granules (substantia neurosecretoria*): ↑ Unit membrane-bound ↑ secretory granules (NsG), 0.1–0.3 µm in diameter, filled with fine granular, strongly osmiophilic material. N. arise from Golgi apparatus of ↑ neurosecretory neurons and are transported by descendent ↑ axoplasmic flow along ↑ neurosecretory axons of ↑ hypothalamohypophyseal tract to some hypothalamic nuclei, ↑ neurohypophysis, and/or portal circulation of ↑ hypophysis, where they are released by ↑ exocytosis into ↑ pericapillary spaces. N. contain ↑ oxytocin, ↑ antidiuretic hormone, or other ↑ neurohormones of unknown composition bound to ↑ neurophysins. The content of disulfide groups in N. renders them selectively stainable with performic acid, alcian blue, or chrome ↑ alum hematoxylin. (Fig. = rat)

↑ Hypophysis, vascularization of

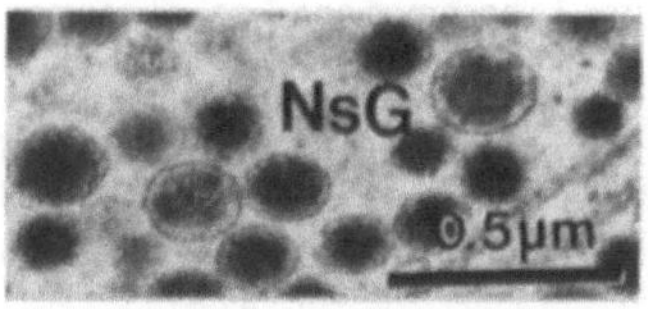

Nordmann, J.J., Labouesse, J.: Science *211*, 595 (1981)

Neurosecretory neurons (neuroendocrine cells, neurosecretory cells, neuronum secretorium*): Specialized ↑ multipolar neurons responding to stimuli by synthesis and release of ↑ neurohormones. ↑ Perikaryon of an N. is clearly pear-shaped; nucleus is round with finely dispersed ↑ heterochromatin and a voluminous nucleolus. Cytoplasm contains numerous mitochondria, a well-developed ↑ GERL complex (G), ↑ Nissl bodies (NB), and a moderate number of ↑ lysosomes and ↑ lipofuscin granules. From Golgi apparatus arise numerous ↑ neurosecretory granules (NsG), which penetrate axon (A). In the course of their movement through axon, accumulations of neurosecretory granules form thickenings, ↑ Herring bodies (HB). Nerve impulses are transmitted to N. by way of numerous ↑ synaptic contacts (SC) of their plasmalemma. N. form groups, nuclei, in ↑ hypothalamus. Cap = capillary, D = ↑ dendrites

↑ Hypophysis, vascularization of

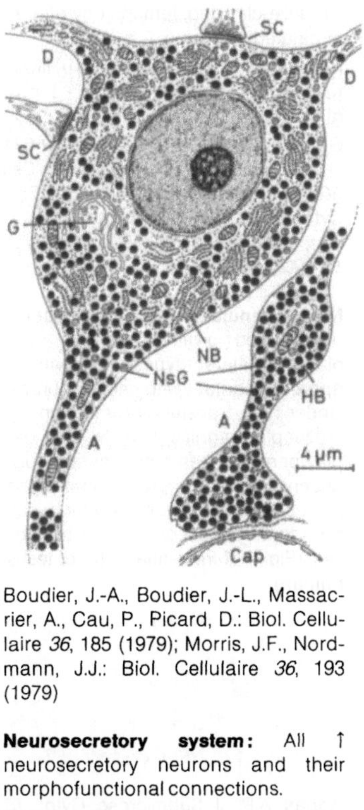

Boudier, J.-A., Boudier, J.-L., Massacrier, A., Cau, P., Picard, D.: Biol. Cellulaire *36*, 185 (1979); Morris, J.F., Nordmann, J.J.: Biol. Cellulaire *36*, 193 (1979)

Neurosecretory system: All ↑ neurosecretory neurons and their morphofunctional connections.

Neurosensory synapse: ↑ Synapse, neurosensory

Neurotendinal spindle: ↑ Golgi tendon organ

Neurotensin: An oligopeptide found in mammalian ↑ hypothalamus and ↑ N-cells of gut mucosa. N. stimulates contractility of ↑ smooth muscle cells and induces hyperglycemia and hyperglucagonemia.

↑ Neurotransmitter substances

Nemeroff, C.B., Prange, A.J. (eds.): Neurotensin, a Brain and Gastrointestinal Peptide. New York: Academy of Sciences, Vol. 400, 1982; Reinecke, M., Almasan, K., Carraway, R., Helmstaedter, V., Forssmann, W.G.: Cell Tissue Res. *205*, 383 (1980)

Neurothelium: A term sometimes used to designate ↑ mesothelium lining subdural aspect of ↑ dura mater.

Neurotransmitter substances (chemical mediators, transmitters): Chemical substances contained in ↑ synaptic vesicles and liberated by ↑ exocytosis into ↑ synaptic cleft, causing transmission of a nerve impulse to another nerve cell or to an effector organ. Terminations of ↑ autonomic nerve fibers release N. into ↑ tissue fluid. N. include: ↑ acetylcholine, ↑ norepinephrine, ↑ dopamine, ↑ substance P, ↑ histamine, ↑ serotonin, ↑ gamma-aminobutyric acid (GABA), some peptides (↑ enkephalins, ↑ endorphin, ↑ neurotensin, bombesin), etc.

↑ Synapse, "by distance"

Buijs, R.M., Pévet, P., Swaab, D.F. (eds.): Chemical Transmission in the Brain – The Role of Amines, Amino Acids and Peptides. Progress in Brain Research 55. Amsterdam: Elsevier Biomedical Press 1982; Kravitz, E.A., Treherne, J.E. (eds.): Neurotransmission, Neurotransmitters and Neuromodulators. In: J. Exp. Biol. *89*, Cambridge Univ. Press 1981; Mihaly, A.: Z. mikrosk.-anat. Forsch. *96*, 916 (1982); Neurotransmitters and Their Action. TINS *6*, 293, Special Issue 1983

Neurotubules (microtubulus*): Eight categories of ↑ microtubules (N), 24 nm in diameter and many micrometers long, present in ↑ perikaryon and extensions of a ↑ neuron. A threadlike dense core (arrows), about 4 nm thick, is situated within N. (endoluminal transport of some substances?). N. can be linked with ↑ neurofilaments (Nf) into ↑ neurofibrils to form ↑ cytoskeleton of nerve cell. It is likely that in the course of ↑ axoplasmic flow, ↑ synaptic vesicles or ↑ neurosecretory granules slide along N. (Fig. = nerve fiber of CNS, rat)

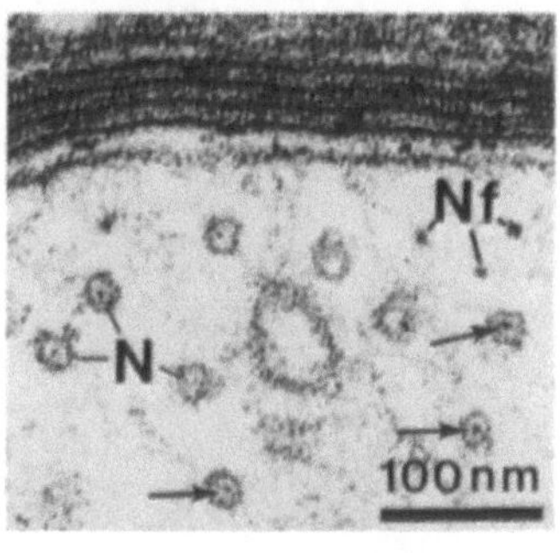

Gray, E.G., Westrum, L.E., Burgoyne, R.D., Barron, J.: Cell Tissue Res. *226*,

579 (1982); Livingstone, A.: Cell Tissue Res. *182*, 401 (1977); Wucker, R.B., Kirkpatrick, J.B.: Int. Rev. Cytol. *33*, 45 (1970)

Neutral red: An azin dye used for ↑ supravital staining of specific granules and ↑ phagosomes of polymorphonuclear ↑ granulocytes.

Neutrophilic granulocytes: ↑ Granulocytes, neutrophilic

Neutrophilic granulocytes, granules of: ↑ Granulocytes, neutrophilic, granules of

Neutrophilic metamyelocyte: ↑ Metamyelocyte, neutrophilic

Neutrophilic myelocyte: ↑ Myelocyte, neutrophilic

Neutrophils: ↑ Granulocytes, neutrophilic

Nexus* (gap junction, macula communicans*): A well-delimited heptalaminar ↑ junctional device (N) between numerous kinds of cells characterized by an intercellular gap, about 2 nm wide. N. consists of seven layers: Layers 1–3 and 5–7 represent cell membranes of two adjacent cells; layer 4 is intercellular space. Filled with ↑ lanthanum (La) and seen in oblique section or in ↑ freeze-cleaving replicas (R), N. reveals existence of polygonal, 9-nm subunits or ↑ connexons (Co) arranged in a hexagonal array. In some cases, a fine dark dot (d) can be observed in center of each subunit; it has been postulated that a fine canalicule (C) traverses subunits that could consist of a group of plasmalemmal proteins (P). N. is an area of low electrical resistance for cell-to-cell propagation of bioelectrical impulses and a preferential crossing-over point for exchange of molecules between two adjacent cells. For these reasons, N. is considered an electrical ↑ synapse. N. occur in ↑ epithelia, ↑ smooth and ↑ cardiac muscle cells, and ↑ connective and ↑ nervous tissues. A protein, ↑ connexin, has been isolated from N. EF, PF = surfaces of freeze-cleaved cell membrane

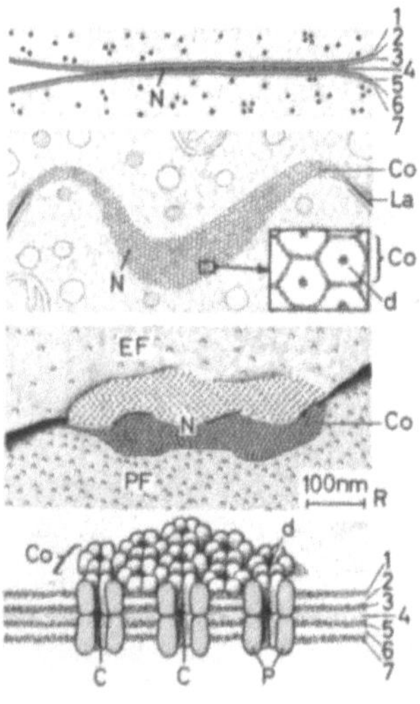

↑ Freeze-cleaving, terminology of; ↑ Intercalated disc

Baker, T.S., Caspar, D.L.D., Hollingshead, C.J., Goodenough, D.A.: J. Cell. Biol. *96*, 204 (1983); Fallon, R.F., Goodenough, D.A.: J. Cell Biol. *90*, 521 (1981); Hirokawa, N., Heuser, J.: Cell *30*, 395 (1982); Peracchia, C.: Int. Rev. Cytol. *66*, 81 (1980); Zampighi, G., Corless, J.M., Robertson, J.D.: J. Cell Biol. *86*, 190 (1980)

Nexus, annular (spherae occlusae): A ring-shaped profile found in cytoplasm of different types of cell without apparent contact with cell membrane. Under high transmission electron microscopic magnification, N. may display penta- or heptalaminar structure, depending upon preparation method. N. represents a ↑ nexus which surrounds a projection of neighboring cell. (Fig. = ↑ interstitial cells, of testis, human)

Aguas, A.P.: J. Submicrosc. Cytol. *13*, 85 (1981)

NGF: ↑ Nerve growth factor

nI-body (ν-body): ↑ Nucleosomes

Niche cells: ↑ Alveolar cells type II

Nile blue sulfate: An oxasin dye widely used for demonstration of fatty acids and phospho- and glycolipids.

Nipple (mammary papilla, teat, papilla mammae*): A musculocutaneous hairless process protruding over level of breast. N. is covered by ↑ epidermis (E) whose basal layer is heavily pigmented; numerous ↑ epidermal ridges are deeply interdigitated with ↑ dermal papillae. There are a ↑ number of ↑ sebaceous glands (SG) in skin of N. Free nerve endings and Meissner's ↑ corpuscles are particularly numerous at tip of N. Interior of N. is occupied by a ↑ dense connective tissue very rich in helicoidally arranged ↑ smooth muscle cells (M) around ↑ lactiferous ducts (LD). Touch, cold, and psychic stimuli provoke contraction of muscle cells leading to erection of N. Suckling acts via a neurohumoral pathway to establish ↑ suckling reflex.

↑ Areola, of mammary gland

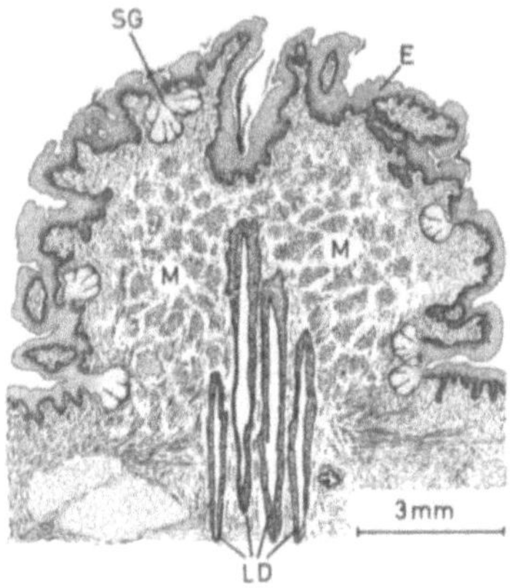

Nissl bodies (chromophilic substance, Nissl substance, tigroid substance, substantia chromatophilica*): Irregular basophilic granules (NB) found in ↑ perikaryon and ↑ dendrites of ↑ nerve cells. Under transmission electron microscope, an N. appears as a cluster of flattened, parallel ↑ cisternae (C) of rough endoplasmic reticulum (rER) (↑ ergastoplasm). Between cisternae, there are a considerable number of free ↑ ribosomes and ↑ polyribosomes. ↑ Axon hillock and ↑ axon are devoid of N., which represent sites of high ↑ protein synthesis. Fatigue and injury of nerve cell decrease content of N. in cell body. (Fig., top = ↑ multipolar neuron, spinal cord, human; fig., bottom = ↑ Purkinje cell, rat)

↑ Chromatolysis

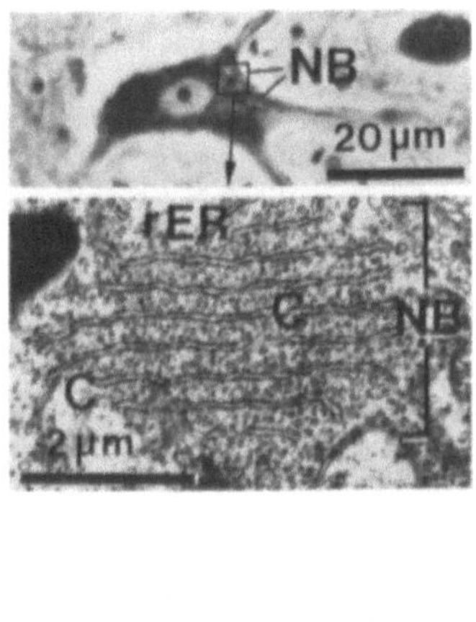

Broadwell, R.D., Cataldo, A.M.: J. Histochem. Cytochem. *31*, 1077 (1983)

Nissl substance: ↑ Nissl bodies

Nitabuch's fibrinoid (substantia fibrinoidea*): A layer of ↑ fibrinoid occurring in second half of pregnancy within ↑ basal plate. N. usually represents a split line between ↑ uterus and ↑ placenta during its expulsion.

↑ Decidua basalis; ↑ Placenta, full-term, structure of

Nitro B-T: A ↑ tetrazolium salt.

nm: ↑ Nanometer

Nociceptors: Pain receptors; mainly in the form of ↑ free nerve endings scattered throughout body and in epithelia. Some ↑ corpuscles are also considered to be N.

Nodal cells (pacemaker cells, P-cells, myocytus nodalis*): Cells, about 10 µm wide and up to 25 µm long, constituting ↑ atrioventricular and sinuatrial nodes. N. have a globular or elongated nucleus and a conspicuous nucleolus. The very clear cytoplasm contains several oval large mitochondria (M), a small Golgi apparatus (G), a few rough endoplasmic cisternae, short and scattered smooth endoplasmic ↑ tubules (sER), sparse ↑ myofibrils (Mf), relatively numerous subsarcolemmal vesicles (V), and a considerable amount of ↑ glycogen (Gly). There is neither a ↑ T-system nor ↑ intercalated discs in N. N. are held to neighboring cells by numerous ↑ zonulae adherentes (ZA). N. are enveloped in a basal lamina (BL) and establish contacts with both atrial muscle cells and ↑ transitional cells.

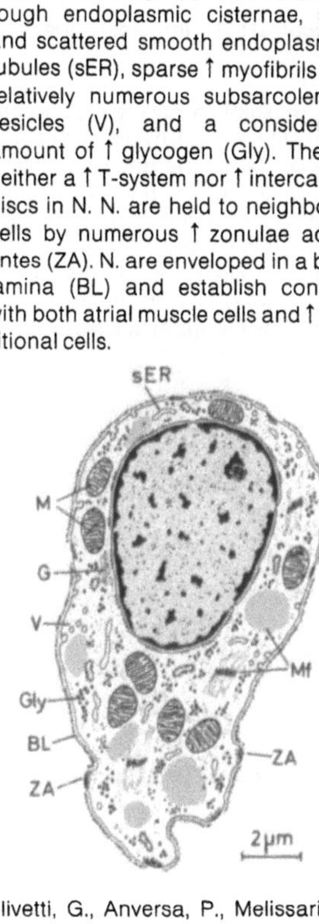

Olivetti, G., Anversa, P., Melissari, M., Loud, A.V.: Lab. Invest. *40*, 331 (1979)

Node, atrioventricular: ↑ Atrioventricular node

Node, sinuatrial: ↑ Sinuatrial node

Nodes, hemal: ↑ Hemal nodes

Nodes, lymph: ↑ Lymph nodes

Nodes, of Ranvier (nodus neurofibrae*): Short specialized interruptions (NR) in ↑ myelin sheath (MS) occurring along ↑ myelinated nerve fibers at meeting points of successive ↑ Schwann's cells (Sch) or ↑ oligodendrocyte processes. 1) ↑ Peripheral ner-

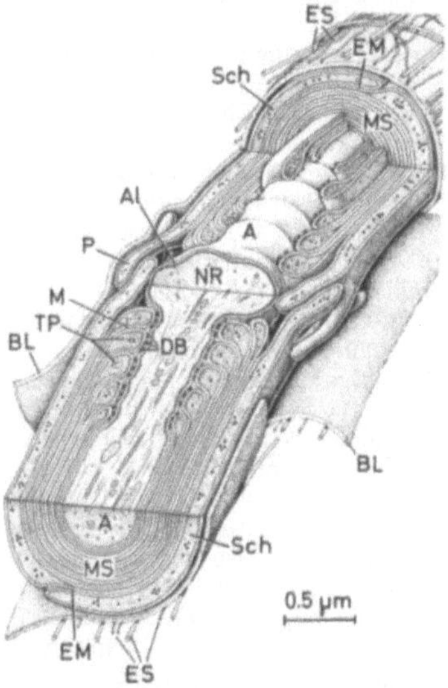

vous system: In the region of N., axon (A) is somewhat thickened and covered only by overlapping digitiform processes (P) of adjoining Schwann's cells, their common ↑ basal lamina (BL), and ↑ endoneurial sheath (ES). In ↑ paranodal region, ↑ myelin lamellae form successive ↑ mesaxons (M), separating a series of cytoplasmic ↑ tongue processes (TP) of Schwann's cell that are connected with ↑ axolemma (Al) by small dense bars (DB). Wrapping of Schwann's cells around axon leaves spiral impressions on its surface. Mesaxons represent longitudinal splitting of external mesaxon (EM). 2) ↑ Central nervous system: N. have identical ultrastructure as in 1), but lack basal lamina and endoneurial sheath. If an axon gives off collaterals or receives synaptic ↑ boutons, this takes place at N. Two N. delimit an ↑ internodal segment. N. are regions of high capacitance and low electric resistance responsible for saltatory conduction.

Berthold, C.-H., Rydmark, M.: J. Neurocytol. *12*, 475 (1983); Duchesne, P.Y.: C.R. Soc. Biol. *173*, 944 (1979); Ghabriel, M.N., Allt, G.: The Node of Ranvier. In: Harrison, R.J., and Navaratnam, V. (eds.): Progress in Anatomy, Vol. 2. Cambridge: Cambridge University Press 1982; Raine, C.S., Finch, H., Masone, A.: J. Neurocytol. *12*, 533 (1983); Peters, A.: Quart. J. Exp. Physiol. *51*, 229 (1966)

Nodules, lymphatic: ↑ Lymphatic nodules

Nodules, of Arantius: ↑ Arantius' nodules

Nomarski optics: A type of ↑ interference microscope.

Nonciliated cells, of bronchiolar epithelium: ↑ Bronchiolar epithelium

Nondisjunction: The failure of two ↑ chromatids (C) (in ↑ mitosis, fig.) or two homologous ↑ chromosomes (in ↑ meiosis) to separate and go to two daughter cells. N. of chromosomes 21 leads to ↑ Down's syndrome; N. of ↑ sex chromosomes leads to ↑ Klinefelter's, ↑ Turner's syndromes, etc. Ce = ↑ centromere

↑ Accidents, of cell division; ↑ Chromosomes, anomalies of

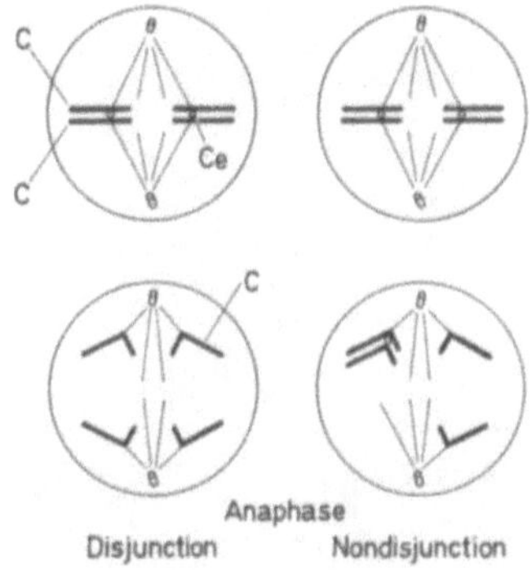

Nongranular leukocytes: ↑ Agranulocytes; ↑ Lymphocytes; ↑ Monocytes

Nonius periods: A shift in ↑ banding pattern (arrow) in ↑ skeletal muscle fi-

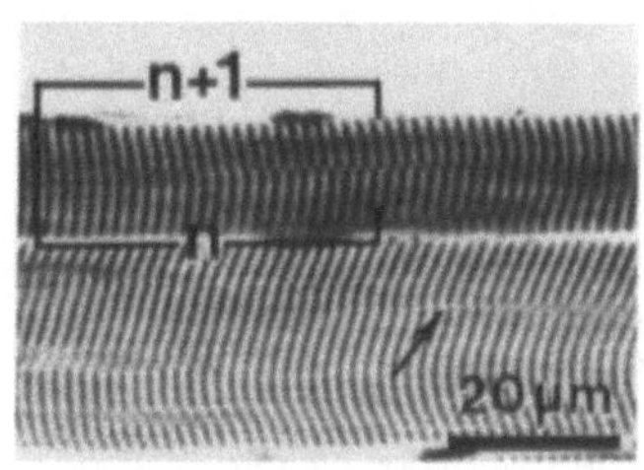

bers, so that to n A-bands in one ↑ myofibril correspond n + 1 in adjacent myofibril. It is thought that N. arise because of arrangement of ↑ Z-lines in the form of slightly spiral steps. (Fig. = human)

Häggquist, G.: Gewebe und Systeme der Muskulatur. In: Möllendorff, W.v. (ed.): Handbuch der mikroskopischen Anatomie des Menschen, Vol. II/4. Berlin: Springer-Verlag 1956

Nonmyelinated nerve fiber (unmyelinated nerve fiber, neurofibra nonmyelinata*): An ↑ axon without ↑ myelin sheath. 1) N. in ↑ peripheral nervous system = a ↑ Schwann's cell (S) with several ↑ axons (A) invaginated into its cytoplasm; Schwann's cell is lined by ↑ endoneurial sheath (E). In some N., several bundles of very fine axons are situated within a Schwann's cell (↑ fila olfactoria). There are no ↑ nodes of Ranvier along N. 2) N. in ↑ central nervous system = a completely naked ↑ axon. (Fig., top = two N. of rat; fig., middle = detail of a Schwann's cell with bundles of axons, ↑ olfactory mucosa, mongolian gerbil; fig., bottom = ↑ retina, rat)

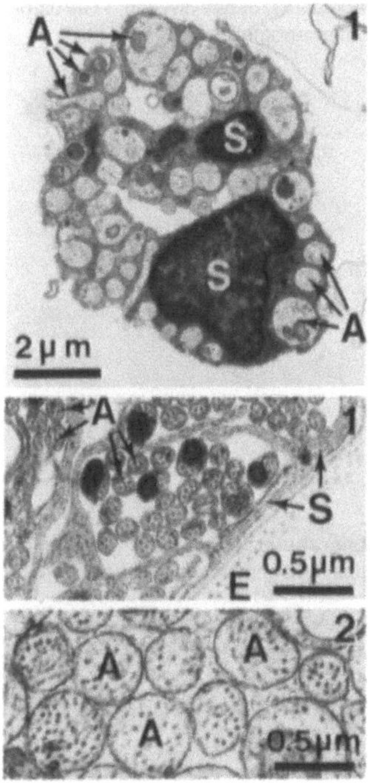

Nonpigmented cells, of ciliary body: ↑ Ciliary body

NOR: ↑ Nucleolus organizer region

Noradrenaline storage granules: ↑ Norepinephrine storage granules

Norepinephrine: A ↑ neurotransmitter produced and released by ↑ norepinephrine-producing cells of adrenal medulla, terminations of most ↑ adrenergic nerve fibers, some cells of ↑ hypothalamus, brain stem, ↑ cerebellum, ↑ spinal cord, etc. Besides its mediator role, N. elevates main arterial pressure, stimulates release of free fatty acids, stimulates CNS, etc. Chemically, N. is a ↑ catecholamine. (See physiology texts for further information)

↑ Dense-core vesicles

Norepinephrine-producing cells (norepinephrine-storing cells, norepinephrocytus*): ↑ Chromaffin cells making up part of ↑ parenchyma of medulla of ↑ adrenal glands. Cell body is polygonal with a large, eccentric, elongated nucleus and a conspicuous nucleolus. Cytoplasm encloses relatively voluminous mitochondria, short scattered cisternae of rough endoplasmic reticulum, a few lysosomes, occasional ↑ multivesicular bodies, and free ribosomes. From the well-developed Golgi apparatus (G) arise ↑ unit membrane-bound, ↑ norepinephrine storage granules (Gr) with dense contents; by ↑ exocytosis, contents of granules are discharged into ↑ pericapillary space (PS). ↑ Cholinergic preganglionic fibers (F) may form neuroglandular ↑ synapses (S) with N. Cap = capillary

↑ Epinephrine-producing cells

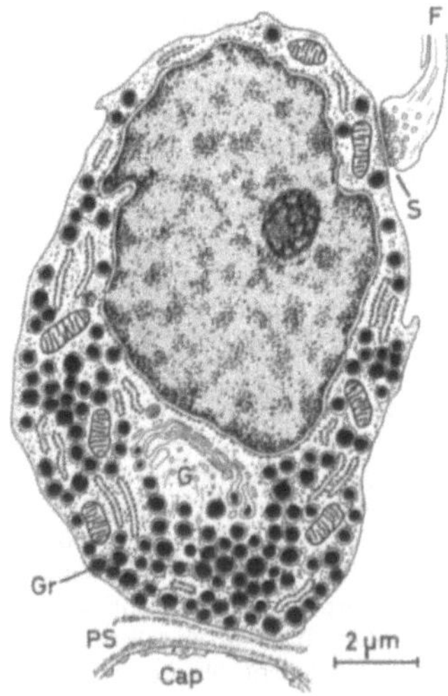

Norepinephrine storage granules (noradrenaline storage granules): ↑ Unit membrane-bound structures (NG), 15–30 nm in diameter found in cytoplasm of ↑ norepinephrine-producing cells of adrenal medulla (in man, N. and epinephrine storage granules are present in a single cell type). Unlike ↑ epinephrine storage granules (EG), contents of N. are of very high electron density. ↑ Norepinephrine is stored in N.; other ultracytochemical characteristics correspond to those of epinephrine storage granules. In preparations, N. are frequently deformed. (Figs. = rat)

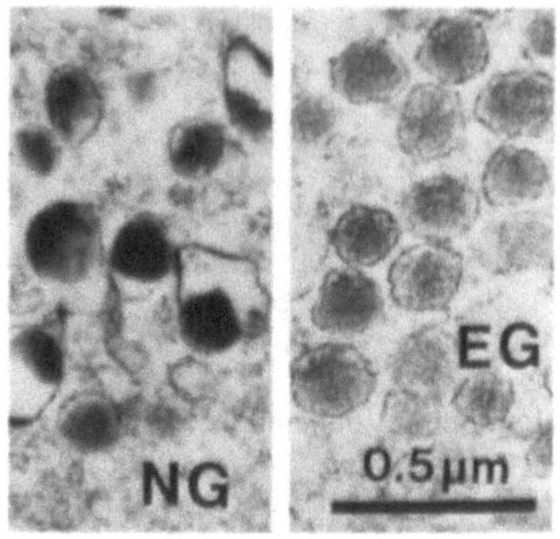

Kryvi, H. Flatmark, T., Terland, O.: Eur. J. Cell Biol. *20*, 76 (1979); Pollard, H.B., Pazoles, C.J., Creutz, C.E., Zinder, O. Int. Rev. Cytol. *58*, 195 (1979)

Normoblast: ↑ Erythroblast orthochromatic

Nose (nasus*): A hollow organ consisting of ↑ bone (B), hyaline ↑ cartilage (C), ↑ skeletal muscle, and ↑ connective tissue. Outer surface and ↑ vestibulum nasi (VN) are lined by ↑ skin (S); inner surface superior and posterior to vestibulum is lined by ↑ respiratory muscosa (RM). Nostrils (nares*, N) are delimited by ↑ wings of nostrils (W).

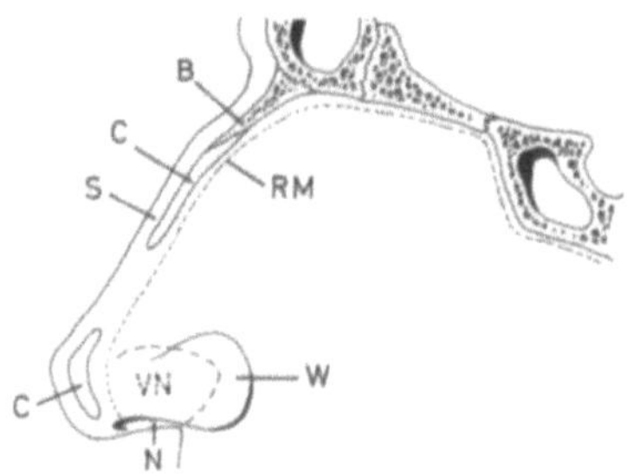

Nostril, wing of: ↑ Wing, of nostril

Notochord (chorda dorsalis, notochorda*): A primitive axial supporting structure originating from notochordal plate. N. is an important organizer for development of CNS and related structures. (See embryology texts for further information)

↑ Notochordal tissue

Jurand, A.: J. Embryol. Exp. Morphol. *32*, 1 (1974); Krause, W.J., Cutts, J.H.: Arch. Histol. Jpn. *45*, 155 (1982)

Notochordal tissue (chordoid tissue): A kind of primitive tissue (N) composing ↑ notochord. In mammalian embryos, N. is located within a cartilaginous mass (C) which will form, after indirect ↑ bone formation, bodies of the vertebrae (V); here, N. cells occupy a continuous tunnel (T) which becomes segmented during further development. N. consists of stellate cells with an elongated or globular nucleus and a conspicuous nucleolus. Cytoplasm contains only few mitochondria, some cisternae of rough endoplasmic reticulum, and a small Golgi apparatus. The relatively important microfilamentous bundles (Mf) anchor in ↑ attachment plaques of ↑ desmosomes (D). The finest ↑ collagen microfibrils (Cf) run near the cells. In man, N. persists until 7 years of age in ↑ nuclei pulposi of ↑ intervertebral discs.

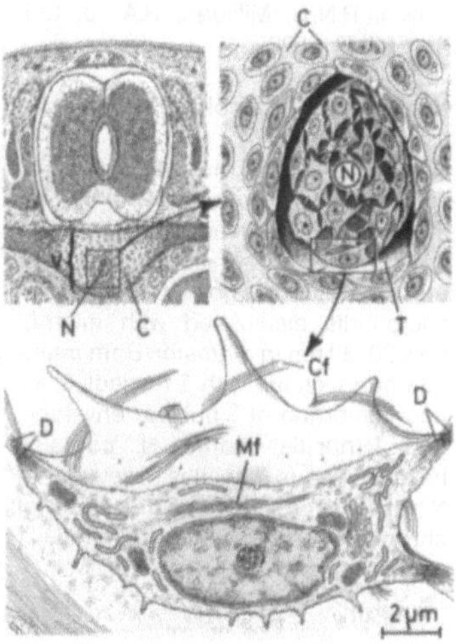

Bancroft, M., Bellairs, R.: J. Embryol. Exp. Morphol. *35*, 383 (1976); Kenney, C., Carlson, E.: Anat. Rec. *190*, 827 (1978); Trout, J.J., Buckwalter, J.A., Moore, K.C.: Anat. Rec. *204*, 307 (1982)

Nuage: An electron-dense, membraneless cell ↑ organelle (N) in the form of a spongelike agglomeration of a fine granular material. N. is present in cytoplasm of all ↑ germ cells throughout their development from the stage of primordial germ cell to the stage of secondary ↑ oocyte or ↑ spermatozoon. In the male, N. is frequently associated with ↑ chromatoid body. Function of N. is not known; it is though to play a role in ↑ differentiation of germ cells. (Fig. = ↑ spermatocyte, rat)

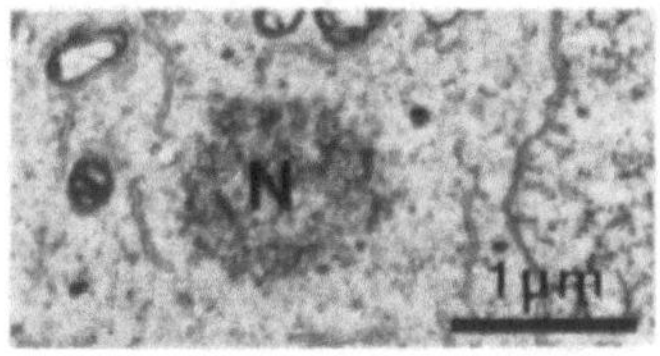

Russel, L., Frank, B.: Anat. Rec. *190*, 79 (1978); Söderström, K.-O.: Cell Tissue Res. *215*, 425 (1981)

Nuclear bag fibers (bursa nuclearis*): The longer kind of ↑ intrafusal muscle fibers of ↑ neuromuscular spindle. Nuclei (N) are grouped in middle portion of fiber, which lacks ↑ myofibrils (Mf) and is, therefore, noncontractile. Annulospiral endings (Aff I) wind around this portion. Extremities of N. are in contact with γ-motor fibers (Eff II). Like ↑ nuclear chain fibers, N. are connected by fine collagen fibrils (C) to capsule of neuromuscular spindle. There are about one to four N. per neuromuscular spindle. LO = ↑ leptomeric organelle

↑ Nuclear chain fibers

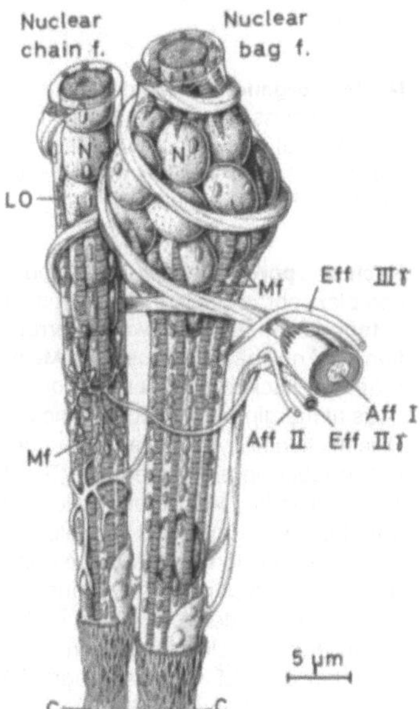

Kucera, J.: Histochemistry *75*, 113 (1982); Kucera, J., Hughes, R.: Cell Tissue Res. *228*, 535 (1983)

Nuclear bodies: Approximately 0.2–0.5-µm spherical structures (NB) surrounded by a thin, fibrillar, proteinaceous layer (FL) found in the vicinity of nucleolus (Nu) of some normal cells (adrenal cortex, ↑ plasma cells, ↑ prostate, ↑ thyroid). Core (C) of N. is composed of anastomosing rows of ribosomelike particles (10 nm in diameter) containing ↑ ribonucleoproteins. N. probably originate from nucleolus; their appearance is related to increased RNA synthesis.

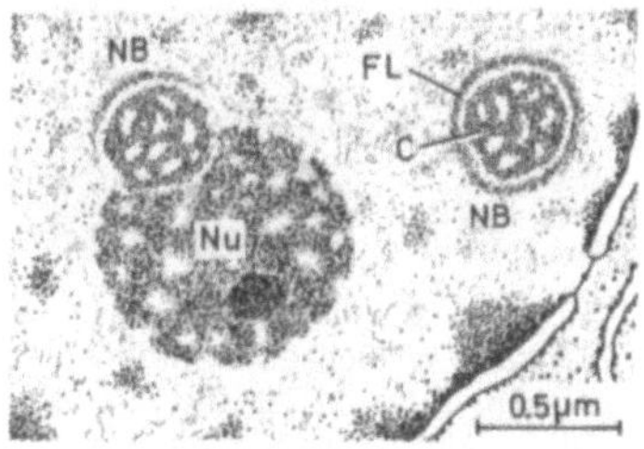

Doyle, D.G.: Am. J. Anat. *157*, 61 (1980); Padykula, H.A., Pockwinse, S.M.: Anat. Rec. *205*, 119 (1983); Palacios, G.S., Lafarga, M.: Z. mikrosk.-anat. Forsch. *93*, 951 (1979); Vagner-Capodano, A.M., Mauchamp, J., Stahl, A., Lissitzky, S.: J. Ultrastruct. Res. *70*, 37 (1980)

Nuclear chain fibers (vinculum nucleare*): The shorter kind of ↑ intrafusal muscle fibers of ↑ neuromuscular spindle. Nuclei form a row in the middle of the fiber, which is devoid of ↑ myofibrils (Mf) and, therefore, noncontractile; myofibrils display, as in ↑ nuclear bag fibers, many ↑ leptomeric organelles (LO). Central region of N. is in contact with coils of annulospiral endings (Aff I). Fine collagen fibrils connect N. and nuclear chain fibers to capsule of neuromuscular spindle. Flower spray endings (Aff II) branch around juxtaequatorial portions of N.; here γ-motor fibers (Eff II and III) form small ↑ motor end plates. There are up to ten N. per neuromuscular spindle. (See fig. under ↑ Nuclear bag fibers)

Kucera, J.: Histochemistry *74*, 183 (1982)

Nuclear envelope (karyotheca*): The boundary (NE) of a ↑ nucleus (N). Under light microsope N. appears as a single line; under transmission electron microscope, it is resolved into: 1) Outer nuclear membrane belonging to ↑ endoplasmic reticulum (R), 2) ↑ perinuclear cisterna (PC), and 3) inner nuclear membrane (M). N. is interrupted by ↑ nuclear pores (NP); in some cells, inner aspect of N. contacts ↑ fibrous lamina (FL). During cell division, N. fragments into vesicles, which reconstitute N. at end of anaphase and in telophase. N. represents a barrier to free diffusion of ions between nucleus and cytoplasm. (Fig., top = ↑ fibroblast, cell culture; fig., bottom = ↑ pinealocyte, rat)

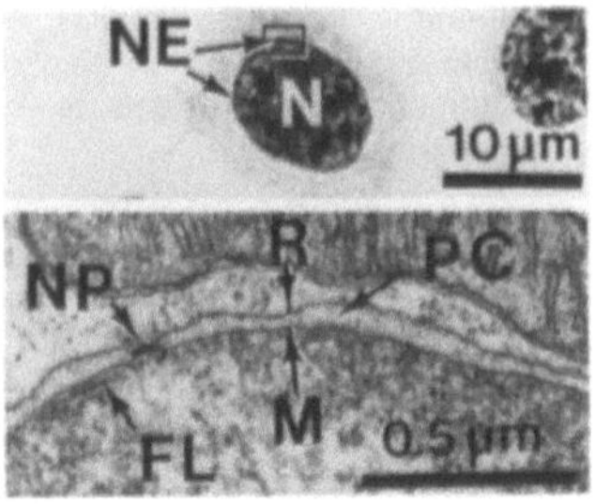

Franke, W.W., Scheer, U., Krohne, G., Jarasch, E.-D.: J. Cell Biol. *91*, 39 s (1981); Matsuura, S., Masuda, R., Sakai, O., Tashiro, Y.: Cell Struct. Funct. *8*, 1 (1983); Maul, G.G. (ed.): The Nuclear Envelope and the Nuclear Matrix. New York: A. R. Liss 1982; Zatsepina, O.V., Polyakov, V. Yu., Chentsov, Yu.S.: J. Cell Biol. *26*, 277 (1982); Zbarsky, I.B.: Int. Rev. Cytol. *54*, 295 (1978)

Nuclear growth: The increase in nuclear volume. 1) True N. = increase in number of ↑ chromosomes or their size leading to rhythmic ↑ N.; not to be confused with 2) false N. = a functional nuclear swelling provoked by entrance of water and metabolites into nucleus of cells exposed to high functional demand.

Hildebrand, R.: Nuclear Volume and Cellular Metabolism. Adv. Anat. Embryol. Cell Biology, Vol. 60. Berlin, Heidelberg, New York: Springer-Verlag 1980

Nuclear growth, rhythmic: The very regular increase in nuclear volume following a geometric progression (1:2:4:8, etc.) as a consequence of duplication of chromosomal number or volume by ↑ cryptoendomitosis or ↑ endomitosis.

↑ Polyploidy

Nuclear lamina: ↑ Fibrous lamina

Nuclear layers, of retina (stratum nucleare externum et internum*): Two layers: the outer one is formed by nuclei of ↑ photoreceptors, the inner one by nuclei of ↑ amacrine, ↑ bipolar, ↑ horizontal, and ↑ Müller's cells.

Nuclear limiting zone: ↑ Fibrous lamina

Nuclear matrix: ↑ Nuclear skeleton

Nuclear membrane: A term used in light microscopy to designate dense line surrounding ↑ nucleus. In transmission electron microscopy N. corresponds to ↑ nuclear envelope.

↑ Nuclear membranes, inner and outer

Nuclear membrane, inner (membrana nuclearis interna*): The leaflet (NM) of ↑ nuclear envelope (NE) in contact with ↑ chromatin (Chr) or ↑ fibrous lamina (FL) in nuclei that possess it. At border of ↑ nuclear pores (NP), N. is continuous with outer ↑ nuclear membrane (OM). Structurally, N. is a ↑ unit membrane. (Fig. = ovarian ↑ follicular cell, rat)

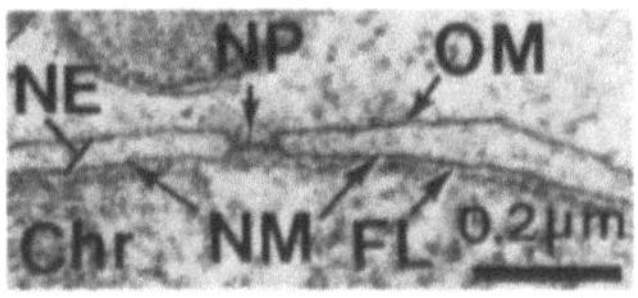

Nuclear membrane, outer (membrana nuclearis externa*): The outer leaflet (OM) of ↑ nuclear envelope (NE); N. belongs to rough ↑ endoplasmic reticulum. (See fig. under ↑ Nuclear membrane, inner)

Nuclear organelles: Membranous or membraneless structures present at periphery of and within a nucleus (↑ chromosomes, ↑ nuclear body, ↑ nuclear envelope, ↑ nucleolus).

Nuclear pores (nucleopore, pore complex, porus nuclearis*): Round or octagonal, 50 to 70-nm-wide interruptions of ↑ nuclear envelope (NE). Membranes of nuclear envelope are continuous at margin of N. Circumference of N. is thickened by a filamentous material grouped in eight granules (G) of 10-nm diameter, which form so-called annulus (A). Pore and annuli, together, are designated the pore complex. From the pores, filamentous annular material extends into both ↑ cytoplasm and ↑ nucleus. The pores are closed by a fine (5-nm) diaphragm (D) with a central granule (CG). Number of pores varies from 20–150/μm^2, so they may occupy 3%–35% of nuclear surface. Under certain cytophysiological conditions, number of N. can be increased or reduced. N. are considered to be a passageway between nucleus and cytoplasm. (Modified after Franke 1970)

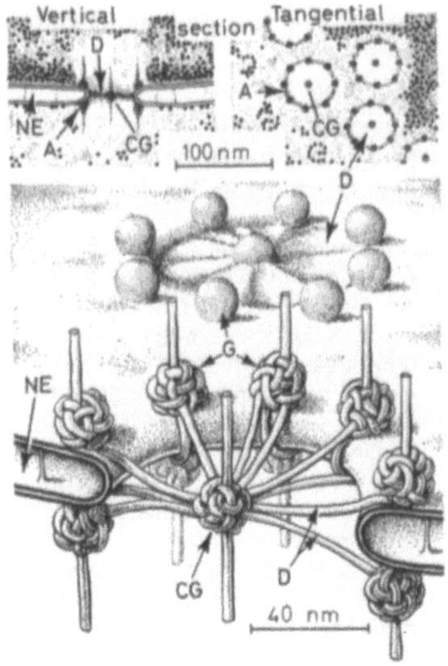

Franke, W.W.: Z. Zellforsch. *105*, 405 (1970); Gerace, L., Ottaviano, Y., Kondor-Koch, C.: J. Cell Biol. *95*, 826 (1982); Severs, N.J., Jordan, E.G.: Experientia *34*, 1007 (1978); Shatten, G., Thoman, M.: J. Cell Biol. *77*, 517 (1978); Unwin, P.N.T., Milligan, R.A.: J. Cell Biol. *93*, 63 (1982)

Nuclear sap: ↑ Karyolymph

Nuclear skeleton (nuclear matrix): A spongelike proteinaceous structural framework of nucleus visible after extraction of ↑ chromatin and ↑ karyoplasm. N. is formed by 2 to 3-nm-thick microfibrils associated with microfibrils 20–30 nm in diameter. Both maintain connections with ↑ nucleolus, inner membrane of ↑ nuclear envelope, and ↑ nuclear pores. N. contains heterogeneous RNA; it is believed that N. plays a role in arrangement of chromatin of meiotic and mitotic ↑ chromosomes.

↑ Matrixin

Berezney, R., Coffey, D.S.: J. Cell Biol. *73*, 616 (1977); Hancock, R.: Biol. Cell *46*, 105 (1982); Kuzmina, S., Buldyaeva, T., Troitskaya, L., Zbarsky, I.: Eur. J. Cell Biol. *25*, 225 (1981); Maul, G.G. (ed.): The Nuclear Envelope and the Nuclear Matrix. New York: A.R. Liss 1982; Miller, T.E., Huang, C.-Y., Pogo, O.: J. Cell Biol. *76*, 675 (1978)

"Nuclear vacuoles": Small, irregular clear spaces (NV) occurring between condensed ↑ chromatin particles (Chr) in ↑ karyoplasm of late ↑ spermatids (fig.) and ↑ spermatozoa. A fine filamentous substance of unknown significance fills N., which are more frequent in man than in animals. N. have no boundary membrane. (Fig. = human)

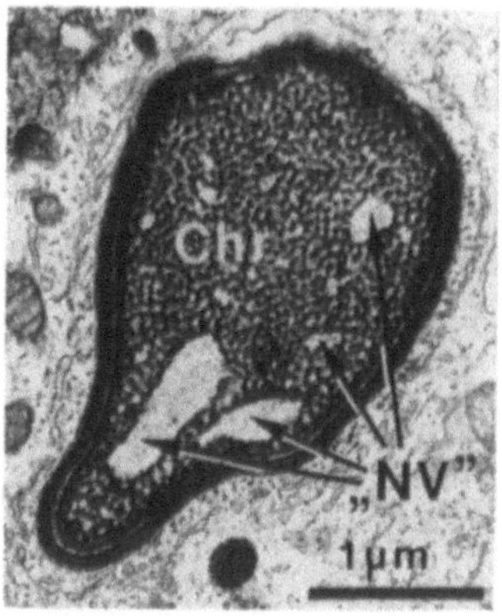

Nuclear zone, of lens: The ↑ area (NZ) at lens equator where cells divide to form ↑ lens fibers (LF). (Fig. = rat)

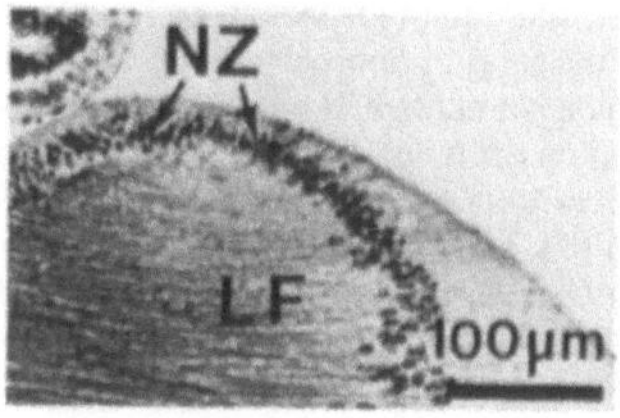

Nuclei, number of, in a cell: The large majority of cells are mononuclear; ↑ cardiac cells, ↑ fact cells, ↑ liver parenchymal cells, cells of ↑ autonomic ganglia are frequently ↑ binucleate. ↑ Chondroclasts, ↑ osteoclasts, ↑ skeletal muscle fibers, and ↑ syncytiotrophoblast are regularly multinuclear. ↑ Erythrocytes, some ↑ lens fibers, and keratinized cells of ↑ epidermis, ↑ nail, and ↑ hair are anuclear.

Nucleic acids: Collective term for ↑ deoxyribonucleic and ↑ ribonucleic acid.

Saenger, W.: Principles of Nucleic Acid Structure. Berlin, Heidelberg, New York: Springer-Verlag 1983

Nucleic acids, detection of: 1) Staining methods: a) ↑ Feulgen reaction, very specific for DNA; b) methyl-green pyronin, DNA = green, RNA = red; c) ↑ gallocyanin-chrome alum staining; d) azure B staining, etc. 2) ↑ Cytophotometry in UV light (maximum absorption about 260 nm; no possible distinction between DNA and RNA. 3) Quantitative measurement = ↑ cytospectrophotometry following Feulgen reaction.

Deitch, A.D.: Cytophotometry of Nucleic Acids. In: G.L. Wied (ed.): Introduction to Quantitative Cytochemistry. New York, London: Academic Press 1966; Pearse, A.G.E.: Histochemistry. Theoretical and Applied. 3rd edn. Edinburgh, London: Churchill Livingstone 1972

Nucleoid: A crystalline core (N) of a ↑ peroxisome (P) composed of very regularly arranged 9.5 to 11.5-nm-wide microtubules, each surrounded by ten smaller ones, 4.5 nm wide. In some species, N. is composed only of 4.5-nm microtubules in hexagonal array. It is believed that N. consists of urate oxidase, since human liver peroxisomes lack N. and show no urate oxidase activity. (Fig = ↑ liver parenchymal cell, rat)

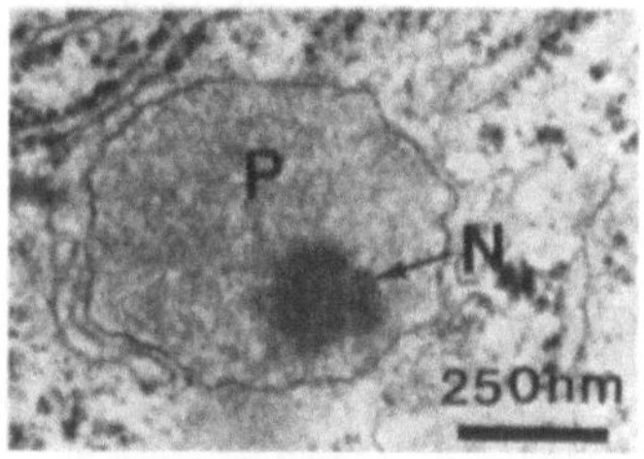

Goldman, B.M., Blobel, G.: Proc. Nat. Acad. Sci. *75*, 5066 (1978); Tsukada, T., Mochizuki, Y., Fujiwara, S.: J. Cell Biol. *28*, 449 (1966)

Nucleoid, mitochondrial: ↑ Mitochondrial nucleus

Nucleolar channel system (NCS): A labyrinthine intranucleolar system of membrane-bound tubes (T), about 50 nm in diameter, arranged in a double helix and surrounded by a dense amorphous material (DM); N. is found in cells of ↑ uterine glands during secretory phase of normal ↑ menstrual cycle (most prominent around day 19). It is believed that formation of N. depends on estrogen and progesterone levels; function unknown.

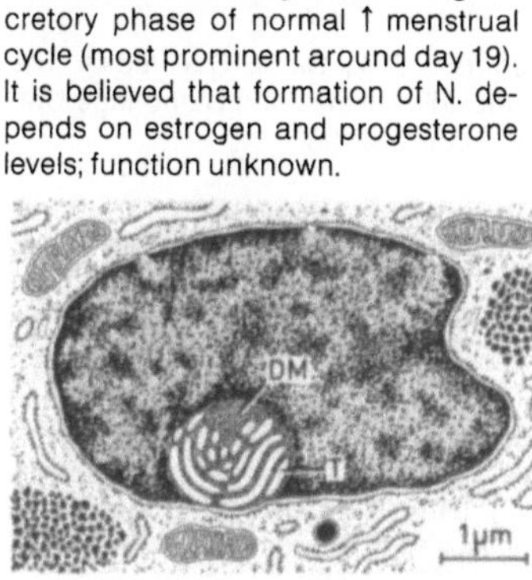

More, I.A.R., McSeveney, D.: J. Anat. *130*, 673 (1980); Pryse-Davies, T.A., Ryder, T.A., MacKenzie, M.L.: Cell Tissue Res. *203*, 493 (1979)

Nucleolar chromatin: ↑ Chromatin, intranucleolar

Nucleolar constriction: ↑ Nucleolus organizer region

Nucleolar fibrillar centers: ↑ Fibrillar centers

Nucleolar organizer: ↑ Nucleolus organizer region

Nucleolar zone: ↑ Nucleolus organizer region

Nucleolonema*: A spongelike structure (N) formed of irregularly convoluted and anastomosing strands, about 60–80 nm thick, of both pars granulosa and pars fibrosa of ↑ nucleolus. Intranucleolar ↑ chromatin (IC), continuous with perinucleolar ↑ chromatin (PC), penetrates meshes of N.

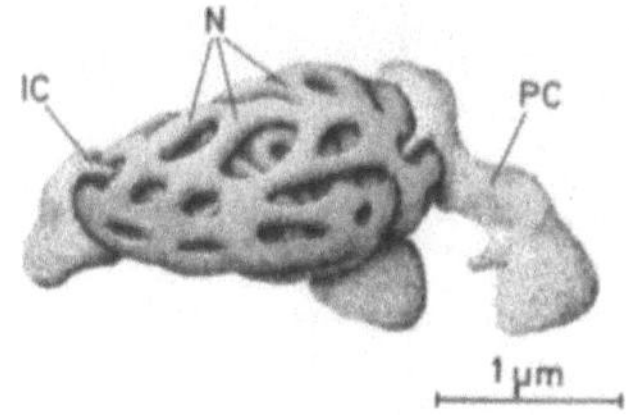

Nucleolus*: A membraneless, round or ovoid, strongly basophilic ↑ nuclear organelle visible during ↑ interphase in ↑ karyoplasm, free or attached to inner ↑ nuclear membrane. Structure: I) Light microscopy = a conspicuous refractile homogenous granule 1–3 µm in diameter (inset, Nu). II) Transmission electron microscopy = 1) Pars granulosa* (G) = predominantly peripheral agglomerations of 10–15-nm particles. 2) Pars fibrosa* (F) = dense, irregular accumulations of filaments, about 5 nm thick, predominantly situated in interior of N. Both pars granulosa and pars fibrosa contain RNA; they form ↑ nucleolonema. 3) Perinucleolar ↑ chromatin = patches of ↑ heterochromatin (Chr) attached to periphery of N. 4) Intranucleolar ↑ chromatin = single or twisted filaments of DNA that penetrate pars fibrosa and form, in some cases, well-delimited ↑ fibrillar centers (FC). 5) Matrix = moderately osmiophilic proteins containing described parts of N. Basophilia of N. is due to its ↑ ribonucleoprotein content. Some functionally very active and/or ↑ polyploid cells possess several N. ↑ Cistrons of ↑ nucleolus organizer region are responsible for synthesis of N. Depending upon kind of cell and its functional stage, N. may vary greatly in morphology; N. size has a conspicuous circadian rhythmicity. (Fig. = epithelial cell of ↑ lacrimal gland, rat; inset = ↑ autonomic nerve cell, human)

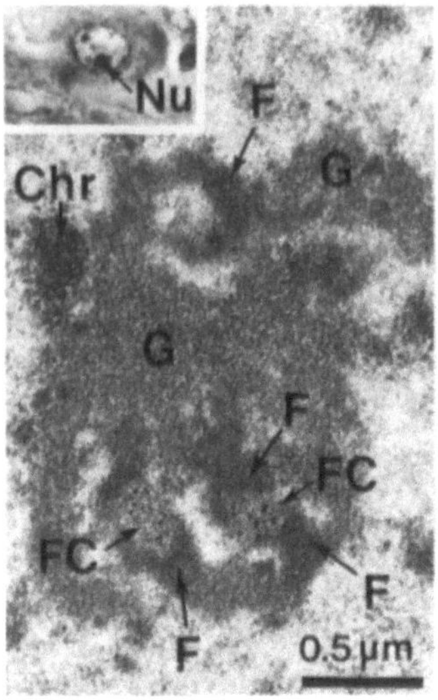

Fakan, S.: Int. Rev. Cytol. *65*, 225 (1980); Ghosh, S.: Int. Rev. Cytol. *44*, 1 (1976); Hernandez-Verdun, D., Bouteille, M., Ege, T., Ringertz, N.R.: Exp. Cell Res. *124*, 223 (1979); Miller, L.O. Jr.. J. Cell Biol. *91*, 15s (1981); Pébusque, M.-J., Seïte, R.: Ultrastruct. Res. *77*, 83 (1981); Recher, L., Sykes, J.A., Chan, H.: J. Ultrastruct. Res. *56*, 152 (1976)

Nucleolus-associated chromatin: ↑ Chromatin, perinucleolar

Nucleolus, chemical composition of: Major part of ↑ nucleolus consists of ↑ ribonucleic acid and ↑ ribonucleoproteins; therefore, nucleolus is ↑ Feulgen-negative and stains red with pyronin in ↑ methyl green-pyronin stained sections. A small amount of DNA in the form of perinucleolar and intranucleolar ↑ chromatin is also present in nucleolus, as well as an RNA polymerase and other enzymes involved in synthesis of ↑ ribosomal RNA. (See biochemistry and molecular biology texts for further information)

↑ Nucleolus, function of

Nucleolus, function of: The primary function of ↑ nucleolus is production of ribosomal subunits by repetitive rDNA ↑ cistrons of ↑ nucleolus organizer region. Mechanism: Along a strand of nucleolus organizer DNA, each cistron is separated from the next by a segment of inactive DNA (S). Nucleolus organizer appears as a series of trunks of a "Christmas tree" with branching parts corresponding to rDNA cistrons coding preribosomal 45 S rRNA, with "branches" of each tree corresponding to 45 S rRNA of varying lengths just synthesized. These branches form pars fibrosa of nucleolus. After a series of complicated transformations, 45 S rRNA is cleaved into two shorter RNA molecules (18 S and 32 S rRNA). The 18 S rRNA is rapidly released through ↑ nuclear pores (P) into cytoplasm, whereas 32 S rRNA associates with proteins and accumulates within pars granulosa. Here, 32 S rRNA is transformed into 28 S rRNA, which also enters cytoplasm and combines with 18 S rRNA to constitute ↑ ribosome. (See molecular biology texts for further information)

↑ Ribonucleic acids, synthesis of

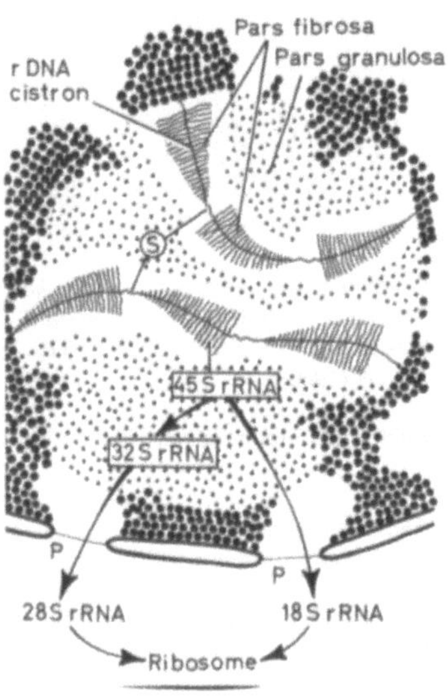

Jordan, E.G., Cullis, C.A. (eds.): The Nucleolus. New York: Cambridge University Press 1982

Nucleolus organizer region (NOR, nucleolar constriction, nucleolar organizer, nucleolar zone, SAT-zone): The secondary constriction (NOR) of certain ↑ chromosomes associated with formation of ↑ nucleolus. There are five N. per ↑ nucleus with a nucleolus. N. contains many rDNA ↑ cistrons for synthesis of 45 S ribosomal RNA. N. may be visible as ↑ fibrillar centers in nucleolus. C = ↑ centromere, CF = ↑ chromatin fiber, S = ↑ satellite

↑ "Christmas tree;" ↑ Nucleolus, function of

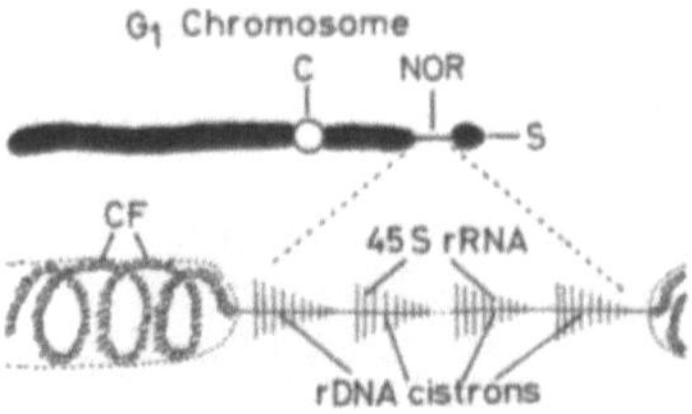

Clavaguera, A., Querol, E., Coll, D., Genesca, J., Egozcue, J.: Cell. Mol. Biol. *29*, 255 (1983); De Capoa, A., Ferraro, M., Lavia, P., Pellicia, F., Finanzi-Agro, A.: J. Histochem. Cytochem. *30*, 908 (1982); Schwarzacher, H.G., Wachtler, F.: Human Genetics *63*, 89 (1983); Vio-Cigna, M., Pébusque, M.-J., Seite, R.: Biol. Cell *44*, 329 (1982)

Nucleoluslike body (granular cytoplasmic body, granular fibrillar body, nematosome, terminal body): A round or spherical membraneless cytoplasmic structure (N), 0.7–1.5 µm in diameter, situated in the vicinity of ↑ nucleus, consisting of approximately 5 to 7-nm-thick filaments and a highly osmiophilic fine granular material. N. are found in various mammalian cells, but their origin and chemical nature are not known. It is believed that N. originate from nuclei or nucleoli. Recent investigations have shown that N. consist of ↑ ribonucleoproteins.

Heym, C., Addicks, K.: Anat. Embryol. *165*, 281 (1982); Katoh, Y., Shimizu, N.: Arch. Histol. Jpn. *45*, 325 (1982); Takeuchi, I.K., Takeuchi, Y.K., Cell Tissue Res. *226*, 257 (1982)

Nucleoplasm: ↑ Karyoplasm

Nucleoplasmic index: The proportion (NP) of nuclear to cytoplasmic volume. expressed as:

$$NP = \frac{V_n}{V_c - V_n}$$

(V_n = nuclear volume, V_c = volume of the cell). N. is constant for every kind of cell; large cells have voluminous nuclei, and small cells small nuclei. Decrease in N., i.e., increase in cytoplasm, seems to represent one triggering factor for cell division.

Nucleopores: ↑ Nuclear pores

Nucleoside diphosphatase: An enzyme found in high concentrations in membranes of ↑ Golgi apparatus.

Nucleosomal DNA (nDNA): Part of deoxyribonucleic acid molecule in direct contact with ↑ histone body to form a ↑ nucleosome.

Nucleosomes (ni body, ν-body): Particulate ↑ chromatin subunits (N), about 10 nm in diameter, composed of a central globular ↑ histone octamer (H) around which a DNA double helix winds in two full turns. DNA in direct contact with histone, so-called ↑ nucleosomal DNA (nDNA), consists of 166 nucleotide pairs. All N. are interconnected with 2-nm thick internucleosomal DNA (iDNA) in an elementary ↑ chromatin fiber (CF). N. are found in nuclei of all ↑ eucaryotic cells. H1 = histone H1. (Fig. = ↑ liver parenchymal cell)

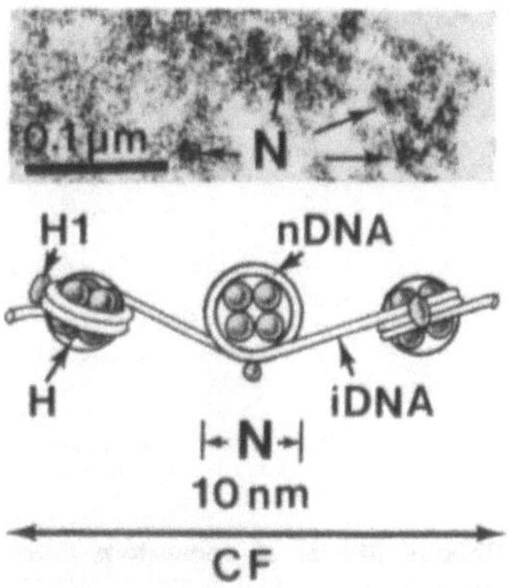

Kronberg, R.D., Klug, A.: Sci. Am. *244*, No. 2, 48 (1981); Noll, M.: Bull. Schweiz. Akad. Wiss. *34*, 321 (1978); Olins, D.E., Olins, A.L.: Am. Sci. *66*, 704 (1978); Rattner, J.B., Hamkalo, B.A.: J. Cell Biol. *81*, 453 (1979); Schwarzacher, H.G., Bielek, E., Ruzicka, F.: Hum. Genet. *35*, 125 (1977)

Nucleotide: A complex molecule constituted of phosphoric acid, pentose sugar (usually ribose or deoxyribose), and either purine or pyrimidine base. N. of deoxyribonucleic acid (DNA nucleotide) is composed of deoxyribose and one of the bases – adenine, thymine, cytosine, or guanine. N. of ribonucleic acid (RNA nucleotide) is made up of ribose and one of the bases – cytosine, uracil, adenine or guanine. ↑ Deoxyribonucleic acid and ↑ ribonucleic acid are linear polymers of many N. (See biochemistry texts for further information)

Nucleus*: The part (N) of a ↑ eucaryotic cell in which ↑ chromosomes are located. During ↑ cell cycle, N. passes through two phases: 1) Metabolic N. = N. during ↑ interphase; 2) mitotic N. = N. during ↑ mitosis. Metabolic N. consists of a ↑ nuclear envelope (NE), surrounding N. as a morphofunctional unit; ↑ karyolymph, ↑ chromatin (Chr), and ↑ nucleolus (Nu). For mitotic N., see ↑ mitosis. (Fig. = ↑ pancreatic acinar cell, rat)

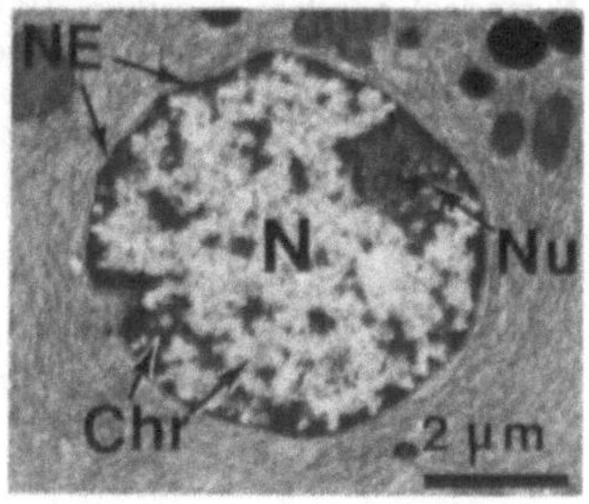

Busch, H. (ed.): The Cell Nucleus. Vols. 1–8. New York: Academic Press 1974–1982; Gilles, M.: Progr. Histochem. Cytochem. *13*, 1 (1980); McCready, J.S., Akrig, A., Cook, P.R.: J. Cell Sci. *39*, 53 (1979); Nucleus and Cytoplasm. In: Organization of the Cytoplasm, Vol. 46, Part 2. New York: Cold Spring Harbor Symposia on Quantitative Biology 1982

Nucleus, degeneration and death of: The process of gradual morphofunctional changes leading to final disappearance of ↑ nucleus. Morphological aspects: 1) Karyolysis = an enzymatic dissolving of entire nuclear material. 2) Karyopyknosis = shrinkage of ↑ nucleus accompanied by depolymerization of ↑ chromatin and formation of a small, heavily stained chromatin clump, which is finally enzymatically destroyed. 3) Karyorrhexis = a break up of ↑ nuclear envelope and chromatin into several fragments that gradually disappear in cytoplasm. N. parallels ↑ degeneration and death of entire cell.

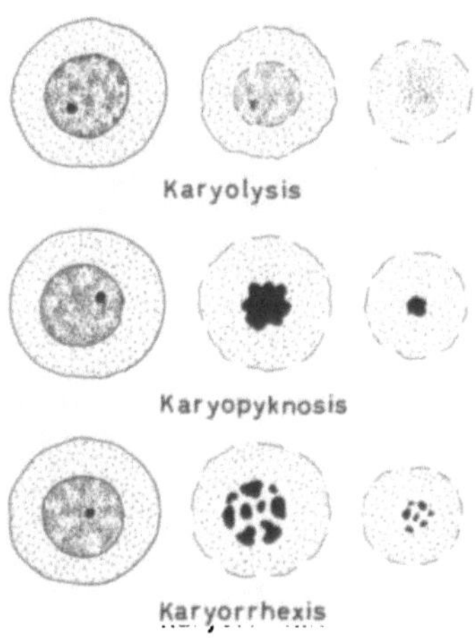

Nucleus, extrusion of: ↑ Erythroblast, orthochromatic, extrusion of nucleus from

Nucleus, forms of: Form of the ↑ nucleus obviously depends upon form of cell. In flattened cells (1), nucleus is also flattened; in cuboidal and round cells, nucleus is spherical (2); in prismatic (3) and spindle-shaped (4) cells, nucleus is elongated, etc. Special forms of nucleus, like lobated nucleus (5), nucleus of ↑ spermatozoa (6), reniform (7), mulberry-shaped (8), etc., are present in ↑ granulocytes, ↑ monocytes, ↑ megakaryocytes, etc.

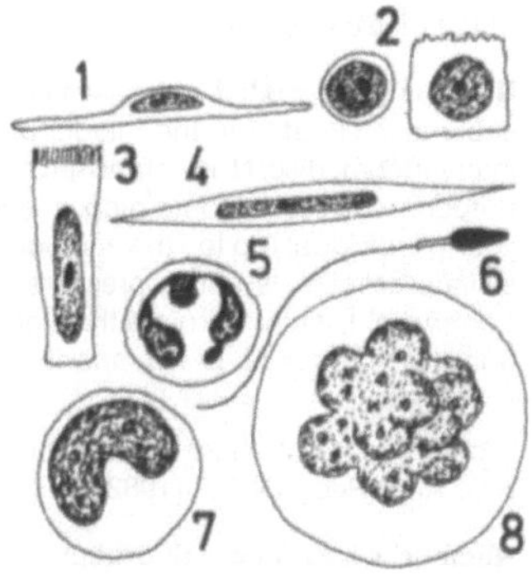

Nucleus, functions of: Nucleus is storage site of bulk of genetic information of a ↑ eucaryotic cell in the form of ↑ deoxyribonucleic acid (DNA). Therefore, nucleus is responsible for: 1) Regulation of cell ↑ differentiation, maturation, and function; 2) replication and transmission of genetic message to new cells, in spite of cell divisions; 3) synthesis of ↑ messenger, ↑ transfer, and ↑ ribosomal RNA and their transport into ↑ cytoplasm. Nucleus divides by ↑ mitosis, ↑ amitosis (↑ somatic cells), and ↑ meiosis (↑ germ cells).

↑ Chromatin; ↑ Nucleolus

Nucleus, in meiosis and mitosis: ↑ Meiosis; ↑ Mitosis

Nucleus intermediolateralis: ↑ Intermediolateral nucleus; ↑ Spinal cord

Nucleus intermediomedialis: ↑ Intermediomedial nucleus; ↑ Spinal cord

Nucleus, number of, in a cell: ↑ Nuclei, number of, in a cell

Nucleus, of central nervous system: An aggregate of ↑ perikarya of ↑ neurons having similar morphology and function (e.g., ↑ intermediolateral N., ↑ intermediomedial N., ↑ paraventricular N., ↑ supraoptic N., etc.).

↑ Hypothalamus; ↑ Spinal cord

Nucleus, of lens: Central zone of ↑ lens consisting of nonnuclear, tightly condensed ↑ lens fibers.

Nucleus, paraventricular: ↑ Paraventricular nucleus

Nucleus, position of, in a cell: In general, ↑ nucleus is central in cell body. Some cells (↑ erythroblasts, ↑ plasma cell) have an eccentric nucleus, whereas in other cells nucleus changes position depending on amount of product accumulated in cell body or direction in which cells discharge their secretory products (↑ mucous cells, cells of ↑ uterine glands, etc.).

Nucleus pulposus*: A pulpy, semifluid substance situated in the middle of ↑ intervertebral disc. N. is composed of ↑ notochordal tissue containing in the young individual (up to 7th year) some stellate cells. N. is under pressure of body weight; it may herniate through ↑ annulus fibrosus. N. is a remnant of ↑ notochord.

Trout, J.J., Buckwalter, J.A., Moore, K.C.: Anat. Rec. *204*, 307 (1982)

Nucleus, supraoptic: ↑ Supraoptic nucleus

Nude mouse (nu/nu): A practically hairless mouse mutant stem in which ↑ thymus fails to develop (aplasia thymi). N. is widely used in immunological research.

Fogh, J. Giovanella, B. (eds.): The Nude Mouse in Experimental and Clinical Research. New York: Academic Press 1978

Nuel's tunnel (Nuel's space, cuniculum medium*): A large space

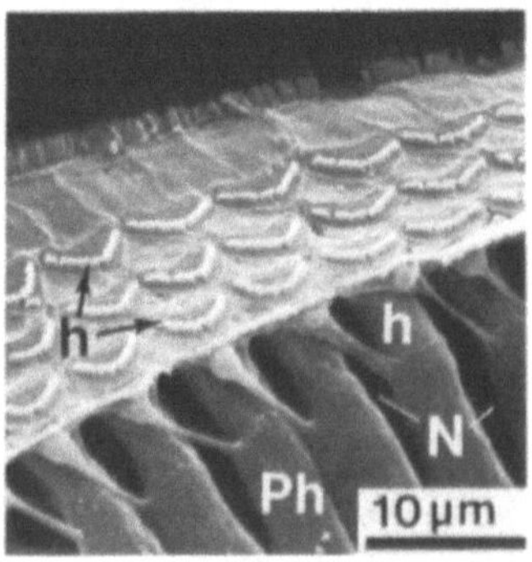

(N) limited by outer ↑ pillars, rows of outer ↑ phalangeal cells (Ph), and outer ↑ hair cells (h) of ↑ Corti's organ. N. is filled with a semigelatinous substance, which gives support to nerve fibers passing through it. (Fig. = mongolian gerbil)

↑ Corti's tunnel

Nuhn's glands: ↑ Anterior lingual glands

Numerical aperture (apertura, NA): Product of ↑ refractive index (n) of the medium between the object and ↑ objective and sine of angle between central and marginal rays entering the objective:

$$NA = n \cdot \sin \alpha$$

The maximal N. of a dry system is about 0.9, and of an ↑ oil-immersion objective, about 1.4.

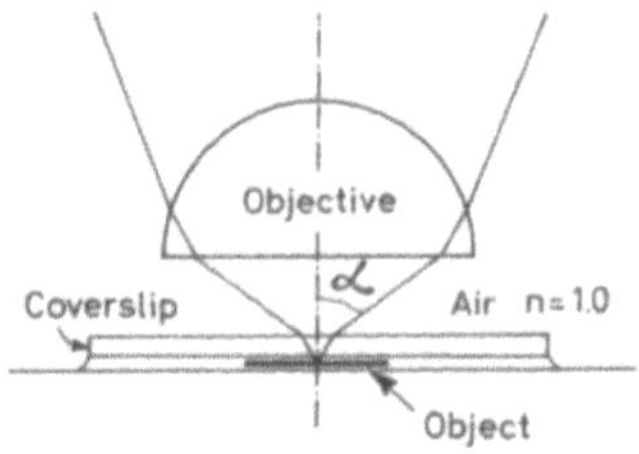

"Nurse cells": ↑ Sertoli's cells

Nutrient artery, of bone: ↑ Medullary artery

O

Objective: 1) ↑ Light microscopes: A system of glass lenses (L) providing initial magnification of object and projecting its image in direction of ↑ ocular. According to type of correction made to eliminate defects of lenses, following kinds of O. may be distinguished: a) Achromatic O. = a correction of chromatic aberration in middle part of visible spectrum only; b) apochromatic O. = a correction of chromatic aberration through whole visible spectrum; c) planar O. = a correction of field curvature, so that entire fields is in focus. The medium between O. and object may be dry (medium is air) or wet as in ↑ oil-immersion O. On each O. are engraved following characteristics: the kind of O., its ↑ magnification (M), its ↑ numerical aperture (NA), correction for length of body tube in millimeters (T), and correction for thickness of ↑ coverslip (C). 2) In electron microscopy, O. is an ↑ electromagnetic lens.

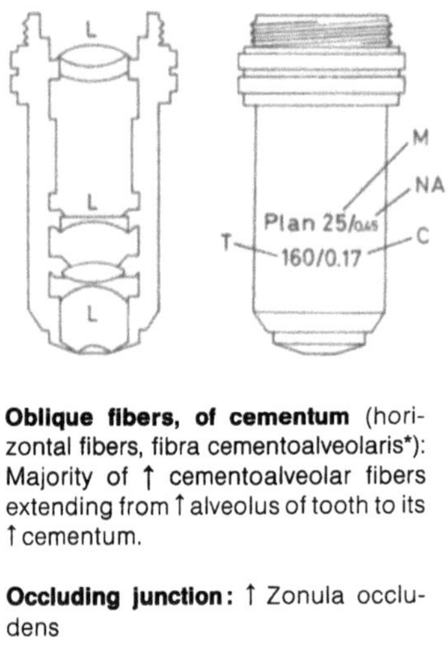

Oblique fibers, of cementum (horizontal fibers, fibra cementoalveolaris*): Majority of ↑ cementoalveolar fibers extending from ↑ alveolus of tooth to its ↑ cementum.

Occluding junction: ↑ Zonula occludens

Octaploidy: A cell ↑ nucleus with 8 n ↑ chromosomes.

Ocular (eyepiece): A system of glass lenses for further magnification and

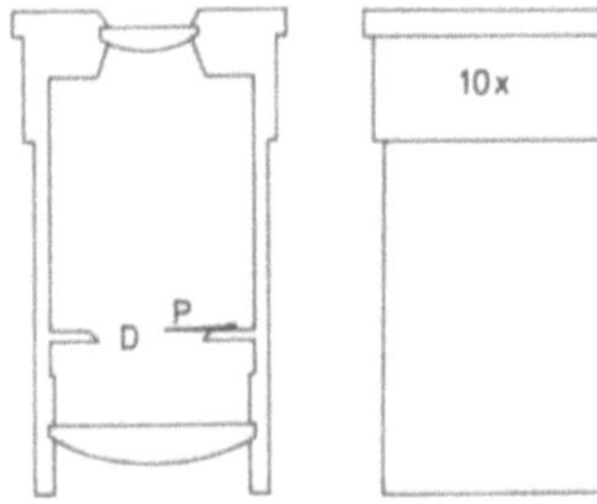

projection onto ↑ retina or photographic film of image created by ↑ objective. At the diaphragm level (D) of O., one can place a pointer (P), micrometric scale, or graticule for stereological analysis. Magnification of O. is marked on its upper or lateral side.

↑ Projective; ↑ Stereology

Oddi's sphincter: ↑ Sphincter, of Oddi

Odland bodies: ↑ Membrane-coating granules

Odontoblasts (odontoblastus*): Tall columnar cells (O) forming an ↑ epithelioid layer lining ↑ pulp cavity of ↑ teeth. Nucleus is basal and elongated, with a conspicuous nucleolus. Cytoplasm encloses many filiform mitochondria, extensive rough endoplasmic reticulum, and a very large Golgi apparatus (G), from which separate vacuoles (V) with striated contents. Toward ↑ apical pole this content gradually condenses, the vacuoles become smaller and turn into highly osmiophilic secretory granules (SG). Their contents, released by ↑ exocytosis, represent precollagen material and ↑ proteoglycans of ↑ predentin (P); some granules are thought to be extruded into predentin and transformed into ↑ dentinal globules (DG). A very long and sometimes branched ↑ Tomes' fiber (TF) extends from apical pole of O. Well-developed ↑ zonulae occludentes (J) assure firm cell contacts. O produce predentin throughout life of an individual, leading to a progressive narrowing of pulp cavity; between O. run numerous capillaries (Cap). D = ↑ dentin

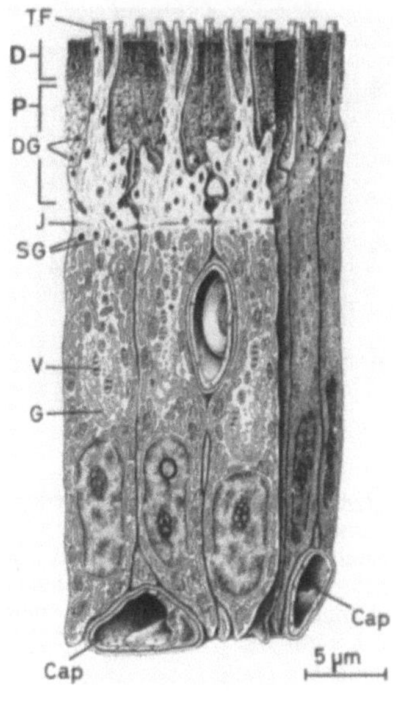

Goldberg, M., Escaig, F.: Biol. Cell *40*, 203 (1981); Sasaki, T., Ishida, I., Higashi, S.: J. Electron. Microsc. (Tokyo) *31*, 378 (1982); Weinstock, M.: Anat. Rec. *199*, 270A (1981)

Oil droplets: Single or grouped ↑ lipid droplets (OD) found in cells of ↑ pigment epithelium of ↑ retina of some animals. (Fig. = mongolian gerbil)

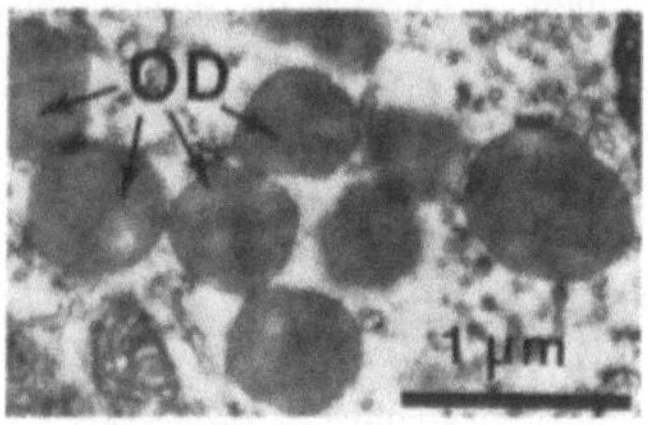

Oil-immersion objective: A special glass ↑ objective immersed in immersion oil dropped onto ↑ coverslip. Since immersion oil has same ↑ refractive index as glass (n = 1.515), O. permits the use of all light rays penetrating the frontal lens without any reflection (R), i.e., without any loss of luminosity, always present when there is air between lens and coverslip. O. are marked by the letters IMM and a horizontal black line above frontal lens.

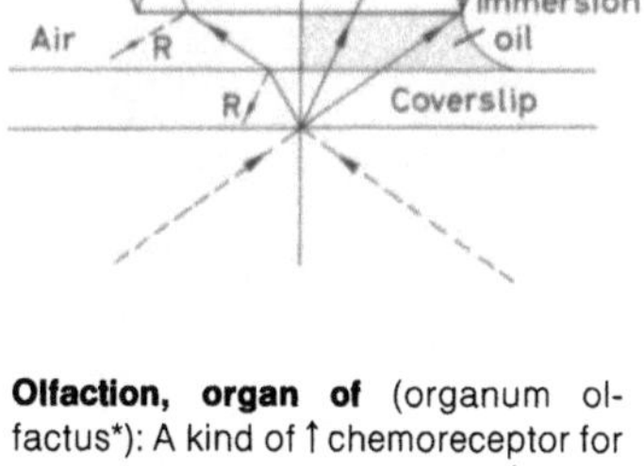

Olfaction, organ of (organum olfactus*): A kind of ↑ chemoreceptor for sense of smell, represented by ↑ olfactory mucosa.

↑ Teleceptors

Balboni, G.C.: Arch. Ital. Anat. Embriol. *86*, 1 (1981)

Olfactory bulb (bulbus olfactorius*): A bulbous protuberance of ↑ allocortex lying on lamina cribrosa of ethmoid bone. Structure: 1) Outer fibrous layer (stratum fibrosum externum) = ↑ nonmyelinated axons of ↑ olfactory cells. 2) Layer of ↑ olfactory glomeruli (Gl) (stratum glomerulosum). 3) Outer granular layer (stratum granulosum externum) = ↑ perikarya and initial segments of processes of large and small ↑ tufted cells (TC). 4) Outer plexiform layer (stratum plexiforme externum) = axons and dendrites of mitral (MC), tufted, and ↑ granule cells (GC). 5) Layer of ↑ mitral cells (stratum mitrale). 6) Inner plexiform layer (stratum plexiforme internum) = mainly collaterals of mitral cell axons. 7) Outer granular layer (stratum granulosum internum) = perikarya of granule cells. 8) White matter (substantia alba) = ↑ myelinated axons of mitral and tufted cells. O. is primary olfactory center.

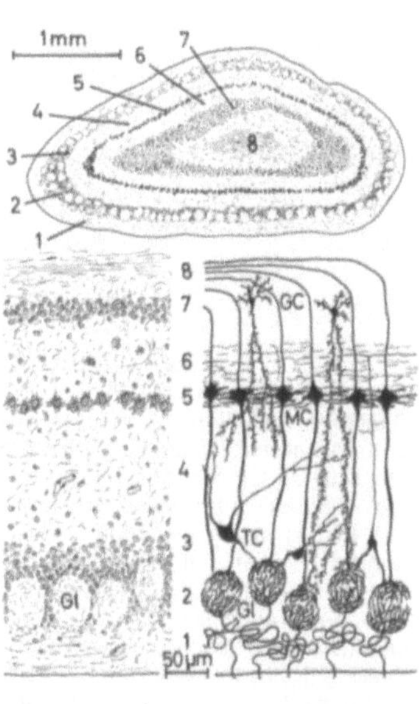

Hummel, G.: Anat, Histol. Embryol. *8*, 289 (1979); Pinching, A.J., Powell, T.P.S.: J. Cell Sci. *9*, 305 (1971); Rakic, P.: Neurosci. Res. Program Bull. *13*, 289 (1975); Shepherd, G.M.: The Olfactory Bulb: A Simple System in the Mammalian Brain. In: Brookhart, J.M. and Mountcastle, V.B. (eds.): Handbook of Physiology, Sect. 1.: The Nervous System. Bethesda, Md.: American Physiological Society 1977; Stephan, H.: Allocortex. In: Bargmann, W. (ed.): Handbuch der mikroskopischen Anatomie des Menschen. Vol. 4, Part 9, Nervensystem. Berlin, Heidelberg, New York: Springer-Verlag 1975

Olfactory cells (cellula neurosensoria olfactoria*): Specialized ↑ bipolar ↑ neurons scattered in ↑ olfactory epithelium.

Olfactory cilia (cilium*): Nonmotile ↑ cilia projected from ↑ olfactory vesicle (OV) almost parallel to surface of ↑ olfactory epithelium. Shaft (S) of O. consists of nine microtubular doublets and a central microtubular pair. In distal part (D), the doublets become singlets and the central pair gradually disappears; tip of O. contains only a few microtubules enclosed by ↑ cellmembrane. Shaft may measure 1–80 µm or more; distal part in some species may be four or five times longer than shaft. Each O. has a well-developed ↑ basal body (B). Some authors attribute limited motility to O. It is thought that O. carry receptors for odoriferous substances. (Fig. = mongolian gerbil)

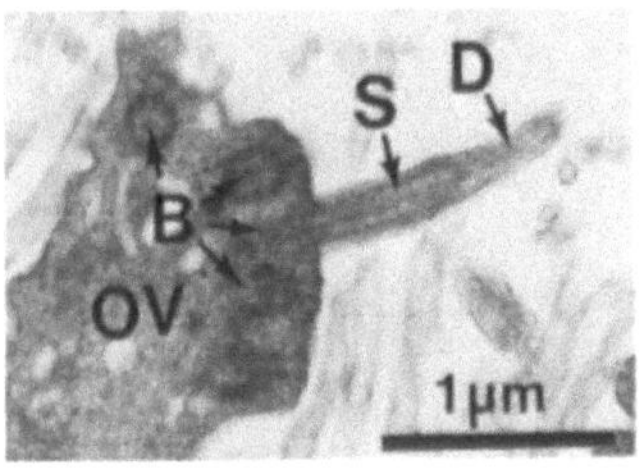

Olfactory epithelium: A yellowish ↑ pseudostratified columnar epithelium about 60 µm tall overlying ↑ olfactory region. Structure: 1) Olfactory cell (cellula neurosensoria olfactoria*) = a slender bipolar chemoreceptor ↑ neuron (OC) with nucleus situated in lower half of epithelium. Dendritic process (D) ends on surface as a bulblike swelling, ↑ olfactory vesicle (OV). Ten to twenty ↑ olfactory cilia (C) with typical ↑ basal bodies (BB) extend radially from olfactory vesicle, parallel to surface. Cilia are embedded within a thick layer of ↑ mucus. In ↑ perikaryon and ↑ dendrite, there is little ↑ ergastoplasm. Below nucleus, ↑ neuroplasm forms a 0.2-µm-thin ↑ axon (A), or central process, which perforates ↑ basal lamina (BL) and joins other axons to constitute nonmyelinated ↑ fila olfactoria (FO). 2) Supporting cell (sustentacular cell, cellula sustentacularis*) = a tall columnar cell (SC) with numerous microvilli (Mv) at its free surface. Apical cytoplasm contains a well-developed ↑ terminal web (TW) and, in some species, ↑ mucous granules (MG) produced by supranuclear Golgi apparatus (G). Nucleus is located in upper half of cell; it is poor in ↑ heterochromatin. Besides numerous mitochondria, supporting cells contain some cisternae of rough endoplasmic reticulum, abundant ↑ tubules of smooth endoplasmic reticulum, ↑ lysosomes, ↑ tonofilaments (Tf), ↑ lipid droplets, and ↑ lipofuscin granules. Supporting cells are connected to olfactory cells by many junctional contacts (no ↑ nexus) and partially ensheath dendrites and axons of the latter. From basal cell pole extend some slender fingerlike processes (FP). Supporting cells also lie on basal lamina (BL); they play a role comparable to ↑ neuroglial cells with regard to olfactory cells (metabolic exchanges, support). 3) Basal cell (cellula basalis*) = a small nonspecialized round or triangular cell (BC) adjacent to basal lamina (BL) without contact to free epithelial surface. Nucleus is round, frequently invaginated; ↑ organelles are poorly developed and rare. Basal surface forms some branching processes which may surround axons. Basal cells are capable of division and ↑ differentiation into supporting cells.

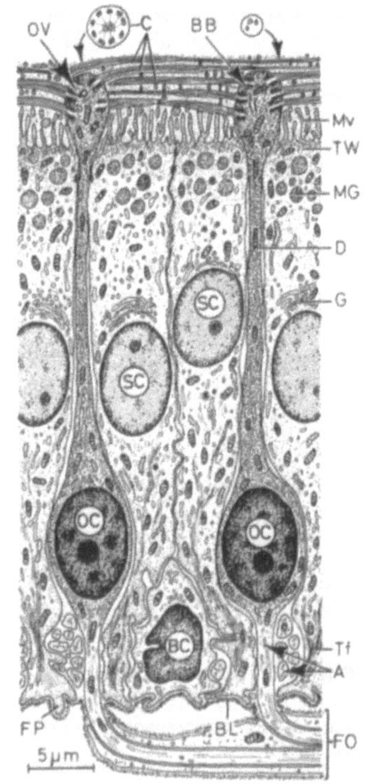

Arstila, A., Wersall, J.: Acta Otolaryngol. *64*, 187 (1967); Breipohl, W. (ed.): Olfaction and Endocrine Regulation. London: IRL Press Ltd. 1982; Breipohl, W., Ohyama, M.: Biomed. Res. *2*, Suppl. 437 (1981); Ohno, I., Ohyama, M., Hanamure, Y., Ogawa, K.: Biomed. Res. *2*, Suppl. 455 (1981)

Olfactory glands (Bowman's gland, glandula olfactoria*): Branched ↑ tubuloalveolar ↑ mixed glands (OG) located in lamina propria of ↑ olfactory mucosa (OM). Structure: 1) Light cells (serous cells) = pyramidal cells with a round basal nucleus, numerous elongated mitochondria, a small supranuclear Golgi apparatus, and an extensively developed rough endoplasmic reticulum. In ↑ apical pole and between rough endoplasmic ↑ cisternae are situated small electron-dense ↑ secretory granules, probably proteinaceous in nature. Small ↑ lysosomes and ↑ lipofuscin granules are present in same area and lend yellowish-brown color to olfactory mucosa. 2) Dark cells (mucous cells) = columnar cells with an oval nucleus, large mitochondria, and a predominantly supranuclear Golgi apparatus. In basal cell portion are scattered narrow rough endoplasmic cisternae, which are continuous with smooth endoplasmic ↑ tubules accumulated mostly in apical cytoplasm. Here are also large homogenous secretory granules with electron-lucent material containing a seemingly steroid component. Excretory canals (EC) of O. open at surface of ↑ olfactory epithelium (OE). Secretion of O. serves as a solvent for odoriferous substances, eliminates remnants of such substances, and moistens olfactory mucosa. (Fig. = human)

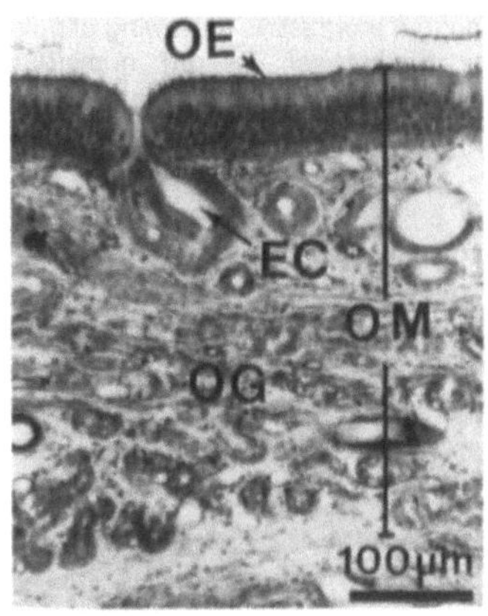

Olfactory glomeruli: Circumscribed areas (G) in ↑ olfactory bulb, where terminal ramifications of ↑ axons of ↑ olfactory cells form synaptic contacts with bushy endings of ↑ dendrites both of ↑ mitral and ↑ tufted cells. (Fig. = human)

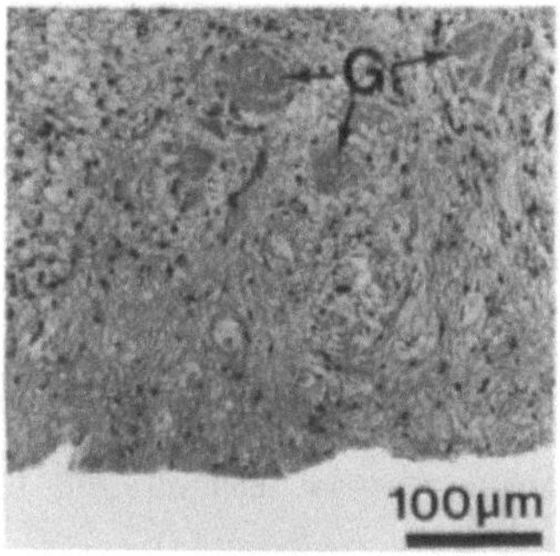

Pinching, A.J., Powell, T.P.S.: J. Cell Sci. *9*, 347 (1971); Willey, T.J.: J. Comp. Neurol. *152*, 211 (1973)

Olfactory knob: ↑ Olfactory vesicle

Olfactory mucosa: A yellowish-brown covering (OM) of ↑ olfactory region. Structure: 1) ↑ Olfactory epithelium (OE); 2) lamina propria (LP) = a ↑ loose connective tissue continuous with ↑ periosteum of underlying bone (B). Lamina propria contains numerous veins (arrowheads) belonging to venous plexus, ↑ muscular arteries (A), lymphatic capillaries, ↑ fila olfactoria (arrows), pigment cells, some lymphocytes, and ↑ olfactory glands (OG). (Fig. = dog)

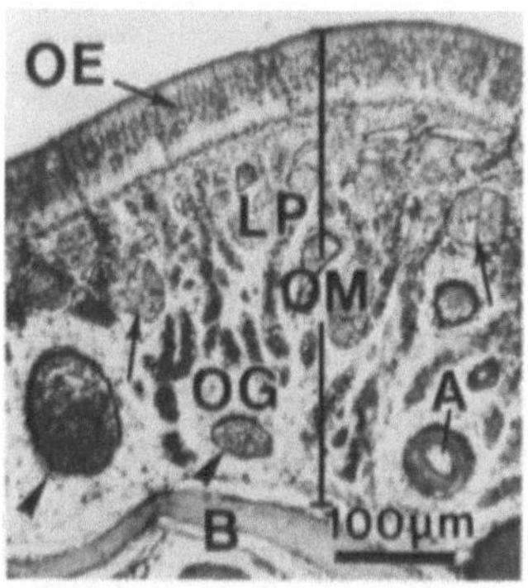

Yamamoto, M.: Arch. Histol. Jpn. *38*, 359 (1976)

Olfactory region (regio olfactoria tunicae mucosae nasi*): Area (OR) of each

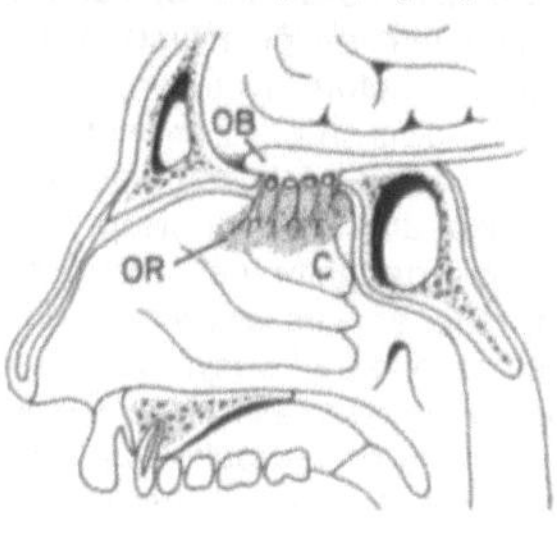

↑ nasal cavity lined with ↑ olfactory mucosa. O. extends in humans over part of roof of nasal cavity, plus small portions of upper third of ↑ nasal septum and superior nasal ↑ concha (C); total surface area of human O. is about 500 mm². OB = ↑ olfactory bulb

Olfactory vesicle (olfactory knob, bulbus dendriticus*): A knoblike dendritic terminal (V) of an ↑ olfactory cell protruding above surface of ↑ olfactory epithelium. Besides numerous mitochondria (M), smooth endoplasmic reticulum, ↑ neurofilaments, and ↑ neurotubules, O. contains basal bodies (arrows) of ↑ olfactory cilia (C). (Figs. = mongolian gerbil)

↑ Cilia, basal body of

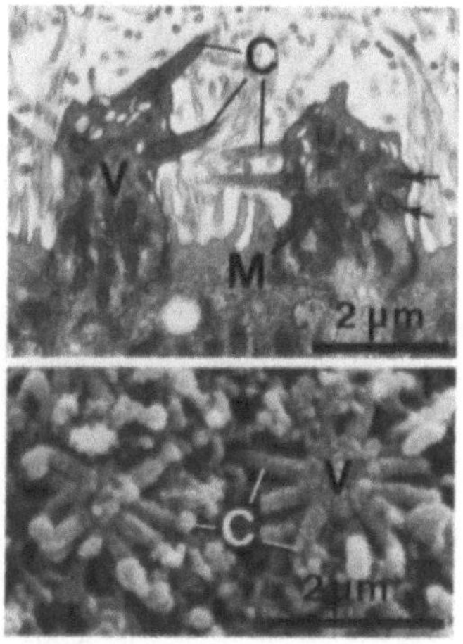

Oligodendroblasts: Precursors of ↑ oligodendrocytes.

Barbarese, E., Pfeiffer, S.E., Carson, J.H.: Dev. Biol. *96*, 84 (1983)

Oligodendrocytes (oligodendrocytus*): Small, dark ↑ neuroglial cells (O) present in ↑ gray and ↑ white matter of CNS. O. has a 6 to 8-µm-wide ↑ perikaryon with a relatively large nucleus rich in ↑ heterochromatin. In thin cytoplasmic rim, there are small mitochondria, a moderately developed Golgi apparatus, short and flattened rough endoplasmic cisternae, some bundles of ↑ gliofilaments, and a number of free ribosomes. The 10–50 roughly trapezoid processes (P) are in fact very attenuated cytoplasmic sheets, which in the course of ↑ myelinization in CNS furnish neighboring ↑ axons (A) with ↑ myelin sheath (MS). In this way, a single O. may myelinate 10–50 ↑ internodal segments. Two kinds of O. exist: 1) Interfascicular O. situated in white matter; 2) perineuronal O. situated near or associated with perikarya of ↑ neurons (N) of gray matter. Reasons for contractions of O. in ↑ tissue cultures are unknown: AH = ↑ axon hillock, arrows = ↑ nodes of Ranvier, ML = ↑ myelin lamellae, TP = ↑ tongue process

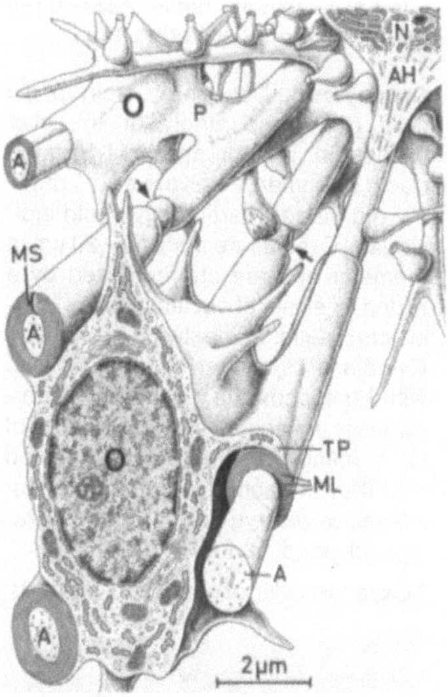

Gonatas, N.K., Hirayama, M., Stieber, A., Silberberg, D.H.: J. Neurocytol. *11*, 997 (1982); Hartman, B.K., Agrawal, H.C., Agrawal, D., Kalmbach, S.: Proc. Natl. Acad. Sci. USA *79*, 4217 (1982); Sturrock, R.R.: J. Anat. *132*, 429 (1981); Sturrock, R.R.: J. Anat. *134*, 771 (1982); Yoshioka, T., Inomata, K., Sugioka, K., Nakamura, K.: Brain Develop. *2*, 337 (1980)

Oligodendroglia cells: ↑ Oligodendrocytes

Oligomucous cells: Precursors of ↑ goblet cells with a few supranuclear ↑ mucous droplets and a basal nucleus. O. originate from ↑ undifferentiated cells of crypts of ↑ Lieberkühn.

Omentum*: A curtainlike peritoneal sheet extending from greater curvature of ↑ stomach downward over intestines. O. has structure of an irregular lacelike net, which is formed of ↑ retiform connective tissue covered by ↑

mesothelial cells. In some areas, ↑ macrophages form patches, ↑ milky spots. O. has a great defensive function in limiting infections in abdominal cavity (Fig. = cat)

Liebermann-Meffert, D., White, H.: The Greater Omentum. Berlin, Heidelberg, New York: Springer-Verlag 1983

Oncocytes: Columnar acidophilic cells (O) found singly or in small groups in ↑ acini and/or ↑ striated ducts of human ↑ salivary and ↑ parathyroid glands, particularly in old individuals. O. measure about 15–20 µm in diameter and are characterized by a round or elliptical nucleus rich in ↑ heterochromatin; nucleolus is prominent. Cytoplasm is almost completely filled with ↑ mitochondria, so that other ↑ organelles are absent or few. Function of O. is unknown; it has been suggested that they are active, degenerative, abnomal, or reserve cells. (Fig. = parathyroid gland, human)

↑ Oxyphilic cells, of parathyroid glands

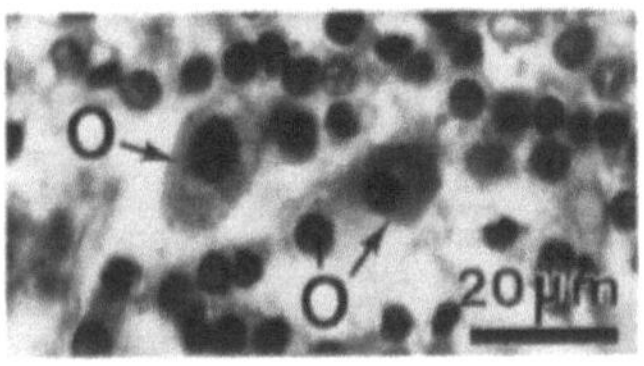

Schramm, U., Dahm, H.H.: Eur. J. Cell Biol. *19*, 227 (1979); Shimono, M., Yamamura, T.: J. Electron. Microsc. (Tokyo) *24*, 119 (1975)

Oocyte (ovocytus*): The female ↑ germ cell.

Oocyte, primary (ovocytus primarius*): A spherical cell (OP) surrounded by ↑ follicular cells (FC). Nucleus contains a ↑ diploid number of ↑ chromosomes in ↑ dictyotene stage; it is round and voluminous with a conspicuous nucleolus. ↑ Ooplasm contains a moderate number of small globular ↑ mitochondria, a single Golgi apparatus (G), short scattered cisternae of rough endoplasmic reticulum, occasional ↑ annulate lamellae (AL), and a few ↑ cortical granules. In the course of ↑ oogenesis, O. continuously becomes enlarged by an increase in volume of ooplasm. Mitochondria become more numerous; Golgi apparatus disperses into many multiple complexes, which produce an abundance of ↑ cortical granules. Annulate lamellae increase in number; ↑ oolemma forms many microvilli that penetrate ↑ zona pellucida occurring between O. and follicular cells. During ↑ ovulation, O. achieves first meiotic division, which results in first ↑ polar body and secondary ↑ oocyte. O. is found in primordial, primary, secondary, and mature ↑ ovarian follicles.

↑ Balbiani's vitelline body

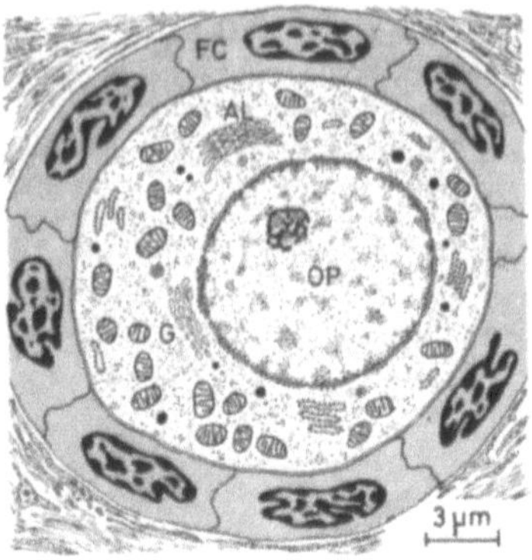

Masui, Y., Clarke, H.J.: Int. Rev. Cytol. *57*, 186 (1979)

Oocyte, secondary (egg, gamete, ovum, praeovum, ovocytus secundarius*): A giant spherical cell (OS) measuring about 100–150 µm in diameter, transported along ↑ oviduct, surrounded by ↑ zona pellucida (ZP) and ↑ granulosa cells (GC). O. is result of first meiotic division of primary ↑ oocyte; therefore, its nucleus contains a ↑ haploid number of ↑ chromosomes. Nucleus (N) is globular with finely dispersed chromatin; nucleolus (Nu) is voluminous. ↑ Ooplasm contains many small round mitochondria, a multiple Golgi apparatus (G), short parallel cisternae of rough endoplasmic reticulum, ↑ annulate lamellae (AL), ↑ Balbiani's body (BB), ↑ cortical (CG) and ↑ yolk granules, ↑ multivesicular bodies, and ↑ lamellar structures (LS). ↑ Oolemma forms many microvilli (Mv)

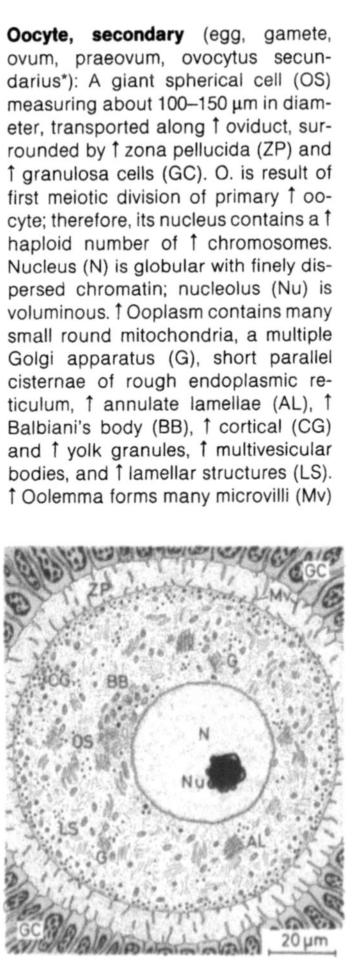

in contact with corresponding processes of granulosa cells. Soon after this stage, nucleus of O. enters second meiotic division, but chromosomal arrangement is arrested in metaphase until ↑ fertilization.

↑ Corona radiata; ↑ Oogenesis

Uebele-Kallhardt, B.M.: Human Oocytes and Their Chromosomes. Berlin, Heidelberg, New York: Springer-Verlag 1978; Weymarn, N.v., Guggenheim, R., Müller, H.: Anat. Embryol. *161*, 19 (1980)

Oocytes, lamellar structures of: ↑ Lamellar structures, of oocytes

Oogenesis (ovogenesis*): The process of development and maturation of ↑ oogonia to ↑ ovum. Phases: 1) Multiplication phase = mitotic divisions of oogonia up to birth; resulting primary

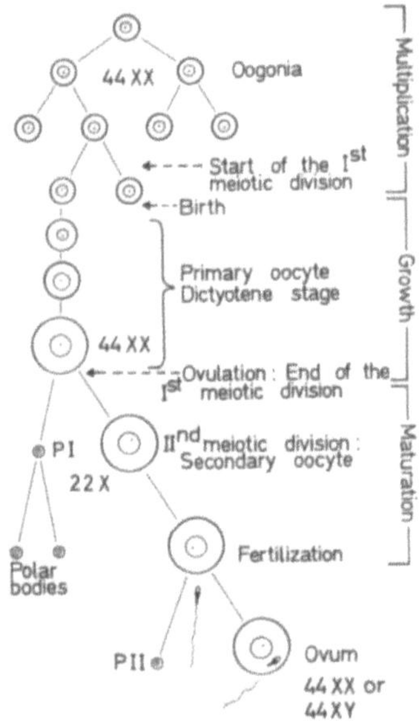

↑ oocytes are surrounded by ↑ follicular cells forming primordial ↑ ovarian follicles. As fetal life ends, primary oocytes have already entered long ↑ dictyotene stage of first (reductional) meiotic division. 2) Growth phase = an increase in volume of one primary oocyte up to ↑ ovulation. 3) Maturation phase = termination of first meiotic division just before ovulation, resulting in equal division of ↑ chromatin (= ↑ haploid number of ↑ chromosomes), but unequal division of cytoplasm, so that one daughter cell, first ↑ polar body (PI), remains very small and the other, secondary ↑ oocyte, receives almost all the cytoplasm. After expulsion of first polar body, nucleus of secondary oocyte starts second (equational)

meiotic division, but it just progresses to metaphase unless secondary oocyte is fertilized. Result of this division, taking place in ↑ oviduct, is second polar body (PII) and ↑ ovum. Since first polar body divides once, result of O. is one ovum and three polar bodies.

↑ Acrosome reaction; ↑ Follicles of ovary, growth of; ↑ Meiosis

Guraya, S.S.: Int. Rev. Cytol. *53*, 121 (1974); Kunz, W., Schäfer, U.: Oogenese und Spermatogenese. Jena: VEB Gustav Fischer 1978; Masui, Y., Clarke, H.J.: Int. Rev. Cytol. *57*, 185 (1979)

Oogonia: Primordial ↑ germ cell in the female, existing only up until birth.

↑ Oogenesis

Baker, T.G., Franchi, L.L.: J. Cell Sci. *2*, 213 (1967)

Oolemma: The ↑ cell membrane of ↑ oocytes and ↑ ovum.

Ooplasm: The ↑ cytoplasm of ↑ oocytes and ↑ ovum.

"Open circulation," of spleen: ↑ Spleen, vascularization of

Opsins: Hydrophobic proteins participating in formation of visual pigments and concentrated in ↑ membranous discs of ↑ photoreceptor outer segments. In rods, O. is termed ↑ scotopsin; in one kind of cone, O. is ↑ photopsin. Structure of O. in other categories of cones is yet not clearly defined.

Opsonins: Serum antibodies coating bacteria, cell debris, and dead cells, enhancing uptake and destruction by ↑ phagocytes.

Optic diffraction: A technique for analysis of transmission electron microscopic negatives using visible light in order to obtain more information about periodic structures than that obtained by simple visual examination. Principle: An HeNe ↑ laser (La) beam (b) is directed through a system of lenses (L) and diaphragms (D) onto periodic detail of negative (N); here, the beam diffracts and produces groups of more or less bright spots, due to interference of light, onto a screen or camera (C). Geometrical distribution of these spots is directly related to structural lattice of the crystalline or paracrystalline object.

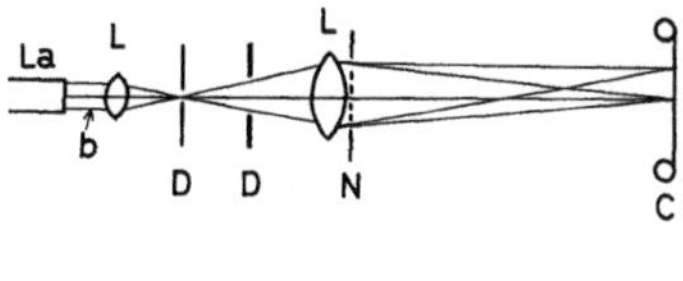

Glauert, A.M. (ed.): Practical Methods in Electron Microscopy. Vol. 1, Part 2, Electron Diffraction and Optical Diffraction Techniques. Amsterdam: Elsevier/North Holland 1974

Optic disc: ↑ Papilla nervi optici

Optic fibers (fibra optica): Partially naked and partially myelinated ↑ axons of retinal ↑ ganglion cells.

↑ *Optic nerve;* ↑ *Optic papilla*

Optic nerve (tractus opticus, nervus opticus*): A cylindrical strand of white matter connecting ↑ eyeball with ↑ brain. A) Intraocular part is very short and composed of naked ↑ axons of retinal ↑ ganglion cells; it extends to lamina cribrosa. B) Orbital part is about 3 cm long and begins behind lamina cribrosa; here, axons become myelinated by ↑ oligodendrocytes. Structure: 1) Meninges = a) ↑ dura mater (DM) continuous with ↑ sclera; b) ↑ arachnoid (A); c) ↑ pia mater (PM). ↑ Subarachnoid space is enlarged and termed vaginal space (VS). From pia mater extend numerous connective septa (S) into O., however, they are separated from ↑ optic fibers by ↑ subpial feet of fibrous ↑ astrocytes. 2) There are about 400 000 ↑ myelinated ↑ optic fibers in human O. Near eyeball, arteria and vena centralis retinae (arrowhead) enter and leave O. (Fig. = dog)

↑ Membrana limitans gliae perivascularis and superficialis

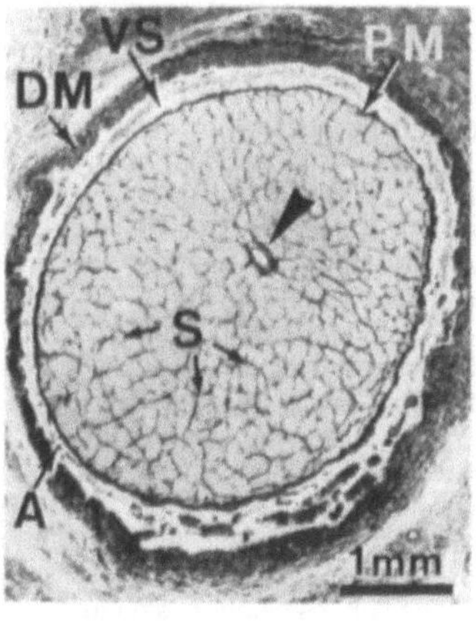

Optic papilla (nerve head, papilla nervi optici, optic disc, discus nervi optici*): A retinal area (OP), about 1.5 mm across, devoid of ↑ neurons, situated at posterior pole of ↑ eyeball. At O., ↑ optic fibers leave ↑ retina and retinal blood vessels (RV) enter and leave ("blind spot"). At margins of O., converging optic fibers pile up and form an annular ridge (AR) surrounding a slight excavation (E), also called physiological cup, where optic fibers are loosely arranged. Immediately behind O., optic fibers pierce ↑ sclera (S) in lamina cribrosa (LC) and become myelinated; here, ↑ optic nerve begins (ON). Increased intracranial pressure provokes a swelling of O; increased intraocular pressure produces a depression (cupping of disc). DM = ↑ dura mater, VS = vaginal space

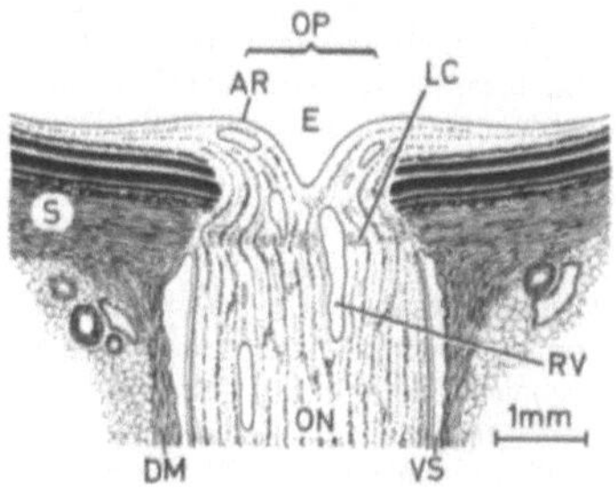

Ora serrata* (ora terminalis): A sharply delimited dentate line (white arrow), marking abrupt reduction in layers of optical portion of ↑ retina (R) to only two layers of ciliary portion of retina (CR). First, ↑ ganglion cells and ↑ optic fibers disappear, followed by ↑ photoreceptors. After disappearance of outer plexiform layer, the two nuclear layers fuse; finally, inner plexiform layer disappears. ↑ Pigment epithelium (P) continues into ciliary portion; ↑ Müller's cells become inner nonpigmented cell layer (M). (Fig. – mongolian gerbil)

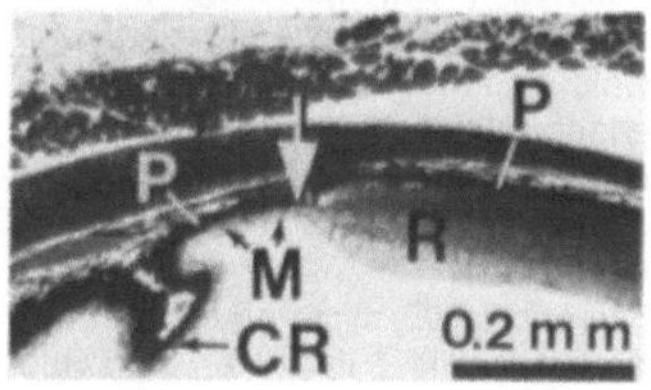

Ora terminalis: ↑ Ora serrata

Oral cavity (mouth, os*): Beginning of ↑ gastrointestinal tract. Topographically composed of: 1) Vestibulum oris (↑ lips, ↑ cheeks, ↑ gingiva); 2) ↑ teeth; 3) cavum oris proprium (↑ tongue, ↑ palate, ↑ palatine tonsils).

Ten Cate, A.R.: Oral Histology, Development, Structure and Function. St. Louis: C.V. Mosby 1980

Orange G: An anionic (acidic) dye used in several trichrome ↑ stainings (Mallory staining, Masson-Goldner staining, etc.).

Orbiculus ciliaris: ↑ Ciliary body

Orcein: A stain used for selective demonstration of elastic structures (↑ elastic fibers and ↑ elastic laminae).

Puchtler H., Meloan, S.N.: Histochemistry *64*, 119 (1979)

Organ: A part of the body clearly delimited from its surroundings and characterized by its form, position, structure, and function. An O. is a morphofunctional unit composed of two or more ↑ tissues.

↑ Parenchyma; ↑ Stroma

Organ culture: An experimental procedure consisting of removal of an embryonic or adult organ from the body and transferring it to a natural or artificial nutrient medium where it can be maintained for various periods under well-defined conditions.

Organ, of Corti: ↑ Corti's organ

Organ, pineal: ↑ Pineal organ

Organ system: A group of structurally and functionally similar formations widely spread throughout an organism (e.g., ↑ cardiovascular O., nervous O., etc.).

Organelles: Structures located in ↑ cytoplasmic matrix or ↑ nucleus serving to perform all individual functions within a cell under obligatory energy consumption. Most O. are membranous or ↑ unit membrane-enclosed structures, such as ↑ mitochondria, ↑ Golgi apparatus, rough and smooth ↑ endoplasmic reticulum, ↑ lysosomes, ↑ microbodies, ↑ multivesicular bodies, ↑ coated vesicles, etc. Nonmenbranous O. are ↑ nucleolus, ↑ ribosomes, ↑ microfibrils, ↑ microtubules, ↑ centrioles, ↑ cilia, ↑ flagellae, ↑ chromosomes, ↑ desmosomes, enzyme molecules, etc.

↑ Metaplasm

Frey-Wyssling, A.: Gegenbaurs morphol. Jahrb. (Leipzig) *124*, 455 (1978); Frey-Wyssling, A.: Experientia (Basel) *34*, 547 (1978); Kirschner, R.H.: Biomed. Res. *2*, Suppl. 71 (1978); Reid, R.A., Leech, R.M.: Biochemistry and Structure of Cell Organelles. Glasgow: The Blackie Publ. Group 1980; Tanaka, K., Naguro, T.: Biomed. Res. *2*, Suppl. 63 (1981)

Organelles, nuclear: ↑ Nuclear organelles

Organogenesis: A branch of ↑ embryology dealing with normal development of ↑ organs.

Organum vasculosum laminae terminalis (OVLT): A ↑ circumventricular organ in the form of a narrow vertical fold situated above optic chiasm in anterior part of third ventricle. O. is covered with one layer of polygonal ↑ ependymal cells overlying a richly branched capillary network. A system of supraependymal axons forms ↑ synapses with ependymal cells; ↑ neurosecretory axons whose perikarya do not make up part of O. end on the capillaries. Function of O. is unknown; it is presumed to be involved in exchange of biologically active molecules between blood and brain and in production of ↑ somatostatin and ↑ LH-releasing factor. (Fig. = monkey, horizontal section)

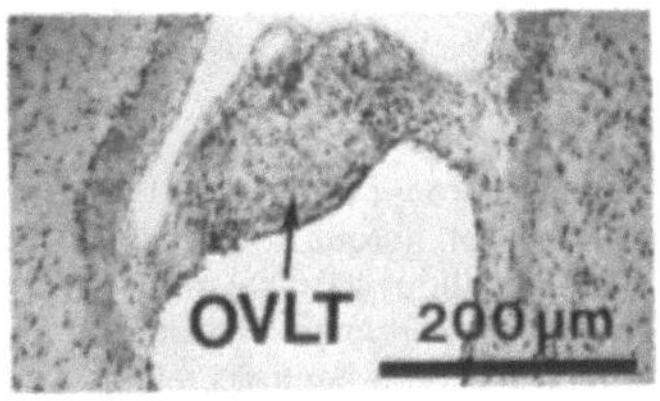

Bennet-Clarke, C., Joseph, S.A.: Cell Tissue Res. *221*, 493 (1982); Krstić, R.: Biomed. Res. *2*, Suppl. 129 (1981); Szabó, K.: Cell Tissue Res *233*, 579 (1983)

Oropharynx (mesopharynx, pars oralis pharyngis*): Portion of ↑ pharynx corresponding to projection of isthmus faucium and upper edge of ↑ epiglottis to posterior pharyngeal wall. O. is lined by nonkeratinized ↑ stratified squamous epithelium and contains ↑ mucous glands in tela submucosa.

Orthochromatic erythroblast: ↑ Erythroblast, orthochromatic

Osmiophilia: The capacity of certain cellular components to bind and reduce colorless osmium tetroxide to black osmium dioxide. In light microscope, an osmiophilic structure appears black; in transmission electron microscope, osmiophilic structures scatter electrons and, therefore, appear transparent on photographic negatives.

Osmoreceptors: More or less well-defined specialized structures scattered in viscera and in ↑ hypothalamus, responding to changes in osmotic pressure of internal environment (↑ cerebrospinal fluid-contacting neurons, ↑ macula densa, ↑ supraoptic nucleus).

Osmotic barrier: The luminal ↑ plasmalemma forming impermeable ↑ plaques of ↑ facet cells preventing dehydration of the latters by hypertonic urine. O. is constantly renewed by addition to surface plasmalemma of ↑ discoid vesicles.

Ossicles, auditory: ↑ Auditory ossicles

Ossification: ↑ Bone formation

Ossification center, primary: ↑ Bone formation, indirect

Ossification centers, secondary: ↑ Bone formation, indirect

Ossification, direct: ↑ Bone formation, direct

Ossification, ectopic: ↑ Bone formation, ectopic

Ossification, endochondral: ↑ Bone formation, indirect

Ossification, indirect: ↑ Bone formation, indirect

Ossification, intramembranous: ↑ Bone formation, direct

Ossification, of long bones: ↑ Bone formation, indirect

Ossification, secondary: ↑ Bone formation, secondary

Ossification zone: ↑ Bone formation, indirect

Osteoblasts (osteoblastus*): Roughly fusiform basophilic cells (Ob) with a large round nucleus (N), a prominent nucleolus, numerous mitochondria, and an extensive Golgi apparatus (G) surrounded by many vesicles. ↑ Cisternae of rough endoplasmic reticulum (rER) are large, long and branched, often parallel, and form an ↑ ergastoplasm; presence of small lysosomes, ↑ glycogen particles, occasional ↑ lipid droplets, and a considerable number of free ribosomes. Many vesicles (V) loaded with secretory material fuse with plasmalemma. O. form an ↑ epithelioid layer on surface of bone ↑ trabeculae (BT) and are responsible for production of ↑ osteoid (O). Its secretion is in general polarized toward border of newly mineralized ↑ bone matrix, but O. elaborate osteoid in all other directions, surrounding themselves with it and differentiating gradually into ↑ osteocytes (Oc). ↑ Matrix granules (MG) probably originate from microvillous processes (Mv) of O. O. arise by ↑ differentiation from ↑ osteoprogenitor cells.

↑ Appositional growth

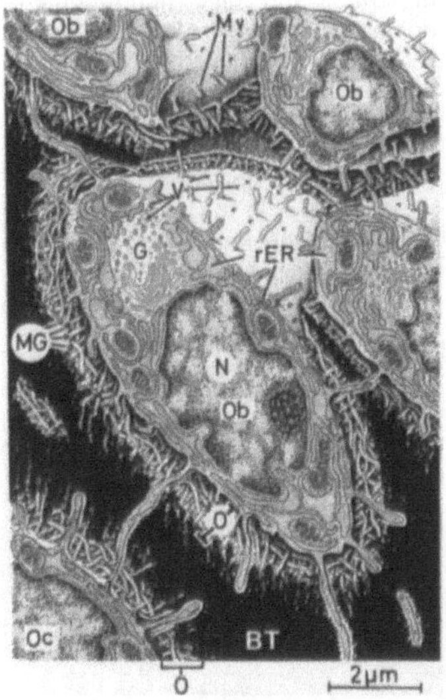

Knese, K.-H.: Stützgewebe und Skelettsystem. In: Möllendorff, W.v., Bargmann, W. (eds.): Handbuch der mikroskopischen Anatomie des Menschen, Vol. 2, Part 5. Berlin, Heidelberg, New York: Springer-Verlag 1979

Osteochondral complexes: ↑ Cartilage bones

Osteoclasia: ↑ The process of bone resorption by ↑ osteoclasts.

Osteoclasts (polykaryocyte, osteoclastus*): Giant multinucleated cells (Oc) measuring up to 100 µm in diameter. O. contain 5–50 nuclei (N) with prominent nucleoli and a corresponding number of ↑ centriole pairs. Nuclei have a life span of several days; they can multiply by ↑ amitosis. Mitochondria are numerous; Golgi complexes (G) are multiple, surrounded by a large number of ↑ coated vesicles; rough endoplasmic cisternae are short and sparse. Many vacuoles (V) with clear flocculent contents and an abundance of small dense granules (Gr), both with lysosomal properties, are scattered within cytoplasm. Activated O. develop a ↑ ruffled border (RB) toward resorbing ↑ bone matrix (BM). During this process, ↑ hydroxyapatite crystals (HC) are liberated and transported through infolding of ruffled border within vacuoles, where they are dissolved. At periphery of ruffled border, plasmalemma is very smooth (asterisks) and in close contact with underlying substratum; in this way, a specific acid microenvironment, created by glycogenolysis and consecutive accumulation of lactic acid, stimulates bone resorption. Other faces of cell are relatively smooth with some microvilli (Mv); the underlying cytoplasmic rim, ↑ ectoplasm (E), is clear and devoid of ↑ organelles. O. are located within ↑ Howship's lacunae. ↑ Parathormone stimulates O. to form ruffled border and resorb bone or calcified cartilage. ↑ Calcitonin inhibits O. and provokes a disappearance of ruffled border. ↑ Chondroclasts are morphologically identical to O.

↑ Osteoclasts, origin of

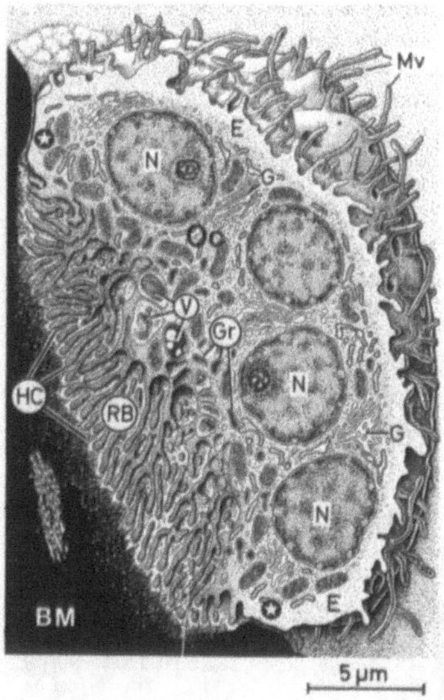

Addison, W.C.: Histochemical J. *11*, 719 (1979); Knese, K.-H.: Stützgewebe und Skelettsystem. In: Möllendorff, W.v., Bargmann, W. (eds.): Handbuch der mikroskopischen Anatomie des Menschen, Vol. 2, Part 5. Berlin, Heidelberg, New York: Springer-Verlag 1979; Rodinova, N.V.: Tsitologiya *25*, 655 (1983); Stanka, P., Bargsten, G., Herrmann, G.: Verh. Anat. Ges. *75*, 237 (1981); Takagi, M., Parmley, R.T., Toda, Y., Denys, F.: Lab. Invest. *46*, 288 (1982)

Osteoclasts, origin of: Osteoclasts arise by coalescence of mononuclear ↑ osteoprogenitor cells. It is possible that ↑ monocytes having left the circulation may also form osteoblasts by fusion. Finally, ↑ osteocytes liberated from ↑ lacunae may dedifferentiate and fuse into one osteoclast.

Hall, B.K.: Anat. Rec. *183*, 1 (1975); Jee, W.S., Nolan, P.N.: Nature *200*, 225 (1963); Hanaoka, H.: Clin. Orthop. *145*, 252 (1979); Stanka, P., Bargsten, G.: Cell Tissue Res. *233*, 125 (1983); Walker, D.G.: Science *180*, 875 (1973)

Osteocytes (bone cell, osteocytus*): Cells of mature ↑ bone. O. is a flattened cell lying within its own lenticular ↑ lacuna within calcified ↑ bone matrix (BM). O. is separated from lacuna by a 0.2 to 2-µm-wide space, containing collagen microfibrils (CF) and nonmineralized ↑ proteoglycan matrix. From cell body extend many fine processes (P) which penetrate ↑ canaliculi (C). Here, they are in contact, by ↑ zonulae occludentes and/or nexus (J), with similar processes of other O. Nucleus (N) is elliptical; depending upon phase of life cycle of O., mitochondria (M), Golgi apparatus (G), and rough endoplasmic reticulum (rER) may be well or moderately developed; lysosomes (Ly) may be more or less numerous. Fine microfilaments bundles (Mf) are grouped under plasmalemma; they extend into processes. There are 700–900 O./mm^3 of compact ↑ bone. Besides synthesizing organic bone matrix, O. may also resorb it (↑ osteocytic osteolysis), contributing to homeostasis of blood calcium. When liberated from lacunae during osteoclastic bone resorption, O. may revert to an ↑ osteoblast or fuse to form an ↑ osteoclast.

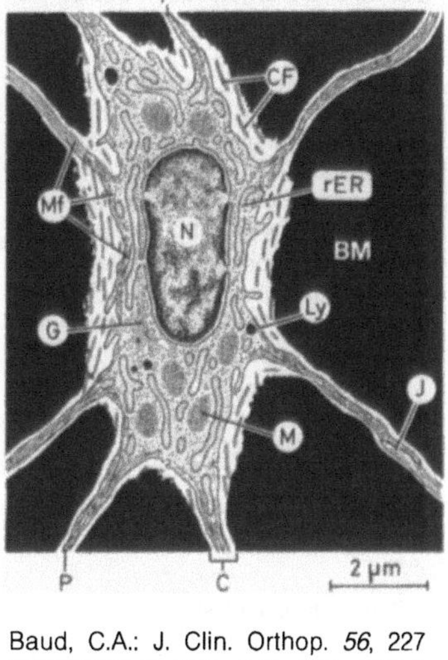

Baud, C.A.: J. Clin. Orthop. *56*, 227 (1968); Ejiri, S., Ozawa, H.: Arch. Histol. Jpn. *45*, 399 (1982); Jande, S.S., Bélanger, L.F.: Calcif. Tiss. Res. *6*, 280 (1971); Knese, K.-H.: Stützgewebe und Skelettsystem. In: Möllendorff, W.v,

Bargmann, W. (eds.): Handbuch der mikroskopischen Anatomie des Menschen, Vol. 2, Part 5. Berlin, Heidelberg, New York: Springer-Verlag 1979

Osteocytes, life cycle of: Three phases in life of an osteocyte: 1) Formative phase = expansion of cell ↑ organelles; synthesis of ↑ osteoid. 2) Resorptive phase = relative reduction in cell organelles; increase in number of ↑ lysosomes and destruction of osteoid by their proteolytic enzymes; partial demineralization of intercellular bone matrix (↑ osteocytic osteolysis). 3) Degenerative phase = vacuolization of nucleus and cytoplasm; swelling of cell organelles, senescence of osteocyte.

Jande, S.S., Bélanger, L.F.: Clin. Orthop. *94*, 281 (1973)

Osteocytic osteolysis: A process of bone resorption by ↑ osteocytes due to their proteolytic lysosomal enzymes. ↑ Parathormone stimulates and ↑ calcitonin inhibits O.

Bélanger, L.F.: Calcif. Tissue Res. *4*, 1 (1969)

Osteogenesis: Process of ↑ bone formation.

Osteogenic bud: ↑ Periosteal bud

Osteogenic cells: ↑ Osteoprogenitor cells

Osteogenin: A hypothetical ↑ bone inductor substance.

Lacroix, P.: L'organisation des os. Paris: Masson et Cie 1949

Osteoid (prebone, preosseous tissue, primary matrix): The organic substance of ↑ bone produced mainly by ↑ osteoblasts and, to a lesser extent, by ↑ osteocytes. O. consists of randomly oriented ↑ collagen fibrils masked by a ↑ ground substance containing ↑ chondroitin sulfate. O. later becomes fully calcified, giving rise to ↑ bone matrix of immature ↑ bone.

Osteoid tissue (textus osteoideus*): The noncalcified bone.

Osteolysis: The process of bone removal by ↑ osteoclasts and ↑ osteocytes.

↑ Osteocytic osteolysis

Osteon: ↑ Osteon, secondary

Osteon, primary (primitive Haversian system, osteonum*): A system of more or less concentric, not clearly demarcated bone ↑ lamellae with randomly oriented ↑ collagen fibrils, around one of vascular channels of the immature ↑ bone. O. are soon resorbed under action of ↑ osteoclasts and replaced during secondary ↑ bone formation by secondary ↑ osteons (definitive Haversian system) with highly oriented lamellae.

↑ Bone, mature

Osteon, secondary (secondary or definitive Haversian system, osteonum*): All concentrically arranged ↑ lamellae (LM) around a ↑ Haversian canal (HC) of ↑ compact bone. O. is functional unit composed of 3–25 lamellae, between which are situated ↑ lacunae (L) containing ↑ osteocytes (Oc). O. measure 100–500 µm in diameter and several centimeters in length; they are generally longitudinally oriented and have a complicated branched and anastomosing configuration. Since direction of collagen fibers is different in successive lamellae, one can see alternating light and dark layers even under an ordinary light microscope. O. are sharply demarcated by ↑ cement lines (CL) from each other and from ↑ interstitial lamellae. In some cases, ↑ canaliculi (C) of an O. may be connected with those of interstitial system or with other O. (arrow). In the course of secondary ↑ bone formation, O. are replaced by third, fourth, and higher orders of osteons.

↑ Bone, compact

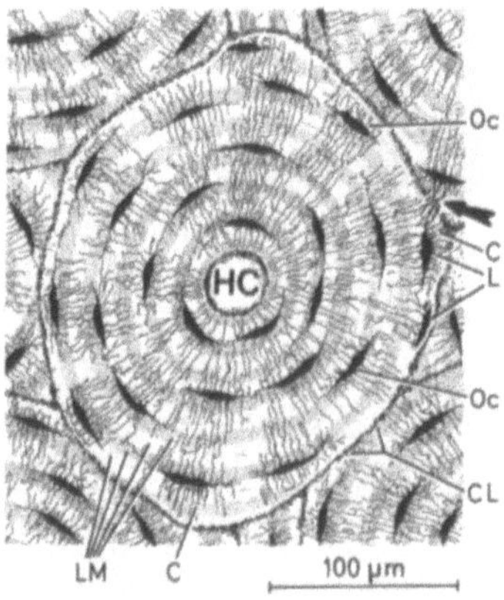

Cooper, R.R., Milgram, J.W., Robinson, R.A.: J. Bone Joint Surg. *48*, 1239 (1966); Frasca, P., Harper, R.A., Katz, J.L.: Scan. Electron Microsc. 1981/III, 339 (1981)

Osteonic canal: ↑ Haversian canal

Osteoplasts: ↑ Lacunae, of bone

Osteopoiesis: ↑ Bone formation

Osteoprogenitor cells (osteogenic cells, preosteoblasts): Flattened or spindle-shaped cells (Op) which differentiate from ↑ mesenchymal cells. O. have acidophilic or faintly basophilic cytoplasm, a flattened nucleus, a few mitochondria, an inconspicuous Golgi apparatus, and some short rough endoplasmic cisternae. Main function of O. is to divide (asterisk), forming a population of cells able to differentiate into ↑ osteoblasts (Ob). O. are particularly numerous around ↑ periosteal buds, along capillaries of primitive bone marrow, in the area of direct ↑ bone formation, and in ↑ cambium of ↑ periosteum (P). A number of O. persist in this layer of adult periosteum and in ↑ Haversian canals of mature bone. They represent reserve cells, which may be activated to differentiate into osteoblasts during secondary ↑ bone formation or in the case of ↑ bone repair. O. located in ↑ endosteum may fuse and give rise to ↑ osteoclasts.

↑ Differentiation

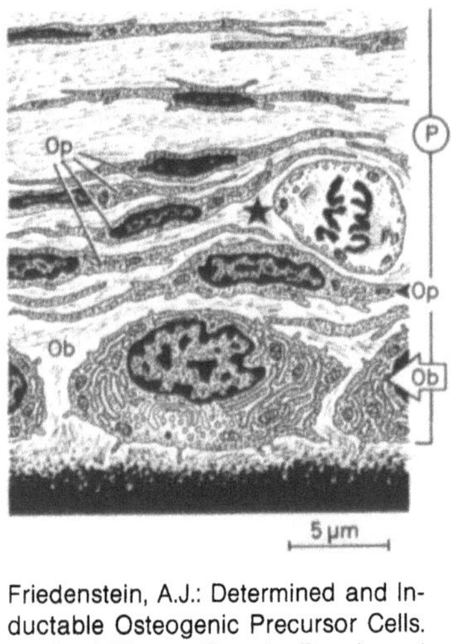

Friedenstein, A.J.: Determined and Inductable Osteogenic Precursor Cells. In: Hard Tissue, Growth, Repair and Remineralization. Ciba Foundation Symposium II. Amsterdam: Excerpta Medica 1973; Knese, K.-H.: Stützgewebe und Skelettsystem. In: Möllendorff, W.v., Bargmann, W. (eds.): Handbuch der mikroskopischen Anatomie des Menschen, Vol. 2, Part 5. Berlin, Heidelberg, New York: Springer-Verlag 1979; Owen, M.: Int. Rev. Cytol. *28*, 213 (1970); Owen, M.: Calcif. Tissue Res. *25*, 205 (1978)

Osteosynthesis: The surgical fixation of fragments of fractured bone with screws, wire, etc.

↑ Bone repair

Ostium, cardiacum: ↑ Cardia

Otoconia: ↑ Otoliths

Otolithic membrane (statolithic membrane, membrana statoconiorum*): A gelatinous substance (OM) covering ↑ neuroepithelium (N) of ↑ maculae of saccule and utricle. Although hairs of ↑ vestibular cells penetrate O., they are separated from it by a narrow space filled with ↑ endolymph. ↑ Otoliths (Ot) rest on O. (Fig. = guinea pig)

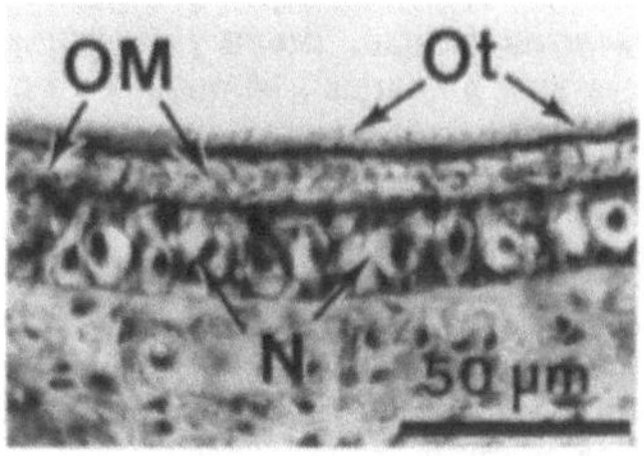

Harada, Y.: Biomed. Res. *2*, Suppl. 391 (1981)

Otoliths (otoconium, statolith, statoconium*): Crystals of calcium carbonate (calcite) and protein, about 3–5 µm long and 2 µm thick, overlying ↑ otolithic membrane of ↑ macula of saccule and utricle. Since O. have a higher specific weight than ↑ endolymph, linear acceleration provokes shearing motions of otolithic membrane, which displace hairs of ↑ vestibular cells and, thus, excite or inhibit them. (Fig. = mongolian gerbil)

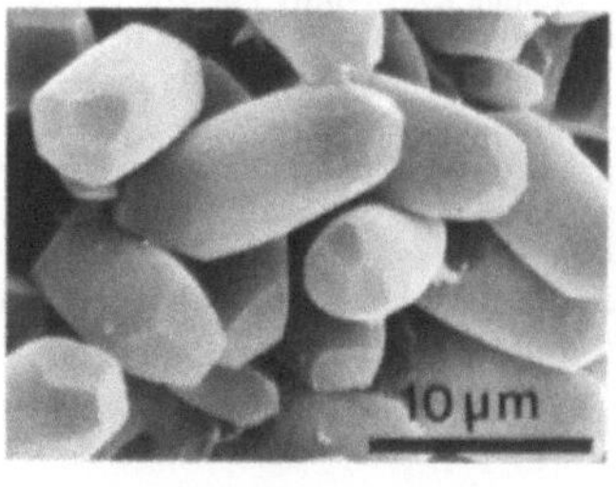

Ballarino, J., Howland, H.C.: Anat. Rec. *204*, 83 (1982); Harada, Y., Tagashira, N.: Biomed. Res. *2*, Suppl. 415 (1981); Lim, D.J.: Scann. Electron Microsc. *3*, 929 (1979); Marmo, F., Balsamo, G., Franco, E.: Cell Tissue Res. *233*, 35 (1983); Nakahara, H., Bevelander, G.: Anat. Rec. *197*, 377 (1980)

Outer band, of Baillarger: ↑ Brain cortex, homotypical isocortex of

Outer chamber, of mitochondria (membrane space, spatium intermembranosum*): A cleft (OC), about 10–20 nm wide, between outer (OM) and inner (IM) mitochondrial membrane. O. contains adenylate kinase and nucleoside diphosphokinase.

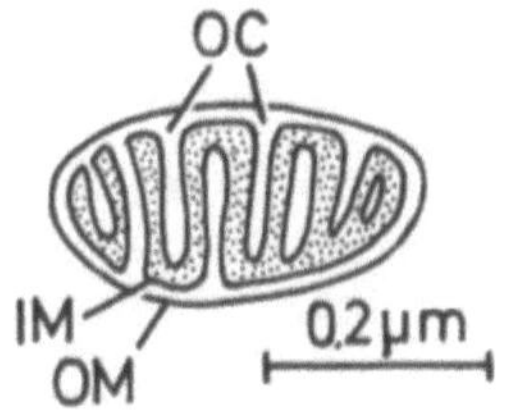

Outer circumferential lamellae: ↑ Bone, compact

Outer dense fibers: ↑ Dense fiber, outer, of spermatozoon tail

Outer granular layer, of brain cortex: ↑ Brain cortex, homotypical isocortex of

Outer limiting membrane, of retina (stratum limitans externum*): A dark line (OLM) separating bacillary layer (BL) from outer nuclear layer (ONL). In transmission electron microscope, O. corresponds to a sievelike row of junctions (↑ zonulae adherentes) between ↑ Müller's cells (MC) and ↑ photoreceptors (Ph) at outer border of outer nuclear layer. (Fig. top = human; fig., bottom = rat)

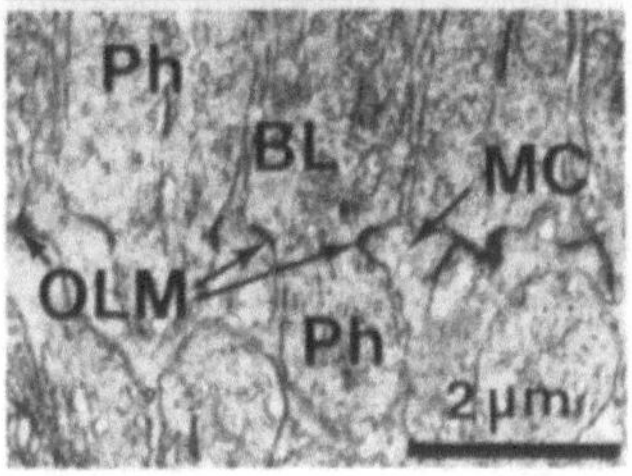

Outer mitochondrial membrane: ↑ Mitochondrial membranes

Outer plexiform layer, of retina (stratum plexiforme externum*): A site of synaptic contacts between ↑ axons of ↑ photoreceptors and ↑ dendrites of ↑ bipolar cells. Dendrites and axons of ↑ horizontal cells also participate in formation of O.

Outer root sheath, of hair: ↑ Hair follicle

Oval window (fenestra ovalis, fenestra vestibuli*): An oval opening (O) in bony ↑ labyrinth closed by an annular ligament (AL) holding footplate of stapes (S). Ligament forms a minute articulation around footplate of stapes. I = incus (Fig. = mongolian gerbil)

↑ Auditory ossicles

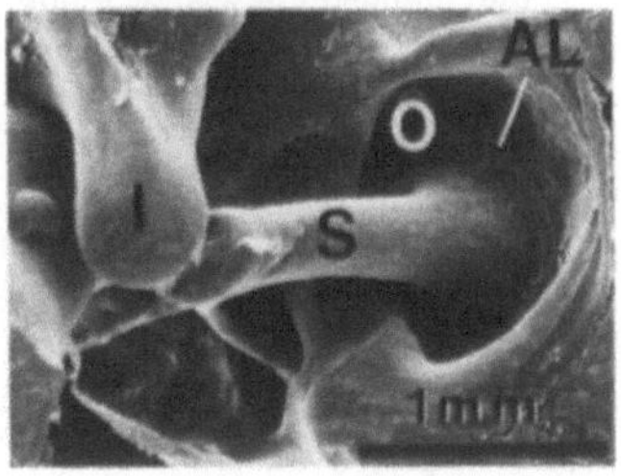

Révész, Gy., Lelkes, Gy., Aros, B.: Acta Morph. Hung. *31*, 327 (1983)

Ovarian bursa: A membranous sac enclosing ↑ ovary in some animals.

Martin, G.G., Sack, M., Talbot, P.: Anat. Rec. *201*, 485 (1981)

Ovarian cycle: The sequence of repetitive histophysiological events in ↑ ovary beginning at puberty and ending at ↑ menopause. O. consists of following phases: 1) Growth and maturation of ↑ ovarian follicles; 2) ↑ ovulation; 3) formation and secretion of ↑ corpus luteum. O. is controlled by ↑ hypothalamus and hypophysis, as well as by ↑ estrogens and ↑ progesterone.

↑ Ovary, control of activity

Ovarian follicle (folliculus ovaricus*): A spherical morphofunctional unit consisting of a female ↑ germ cell in primary ↑ oocyte stage surrounded by ↑ follicular cells. During growth of follicles, one can distinguish primordial, primary, secondary, and mature ↑ O.

Ovarian follicle, atretic (folliculus atreticus*): Any ↑ ovarian follicle during ↑ degeneration. 1) ↑ Atresia of primordial and primary ↑ ovarian follicles = ↑ pyknosis of nucleus and shrinkage of primary ↑ oocyte (O), dissolution of ↑ follicular cells (F). Cellular fragments are phagocytized by ↑ macrophages. Atresia of these follicles leaves no traces. 2) Atresia of secondary and mature ↑ ovarian follicles = ↑ pyknosis of nucleus, accumulation of ↑ lipid droplets and ↑ lysosomes in cytoplasm, shrinkage and breakdown of primary oocyte, penetration of vascularized strands of connective tissue into stratum granulosum and ↑ cumulus oophorus, degeneration of ↑ granulosa cells (GC), and hypertrophy of theca interna (TI), whose cells form temporary glandular tissue of ↑ interstitial gland. The follicle collapses, its antrum fills with ↑ fibroblasts and ↑ wandering cells. ↑ Zona pellucida may persist as an infolded membrane for some time. Thickness of ↑ glassy membrane (GM) increases. Atretic process ends with formation of a small irregular scar, ↑ corpus albicans.

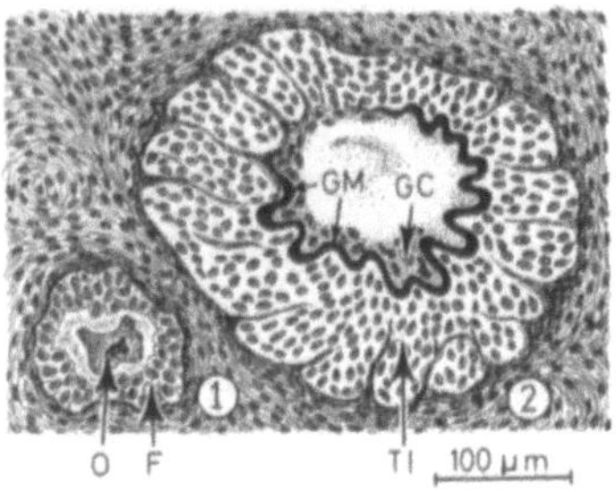

Peluso, J.J., England-Charlesworth, C., Bolender, D.L., Steger, R.W.: Cell Tissue Res. *211*, 105 (1980); Spanel-Borowski, K.: Cell Tissue Res. *214*, 155 (1981)

Ovarian follicle, control of growth: Before formation of follicular ↑ antrum, follicles are not under hormonal control. From secondary vesicular ↑ ovarian follicle up to ↑ ovulation, they develop under influence of ↑ gonadotropin-releasing factor and ↑ follicle-stimulating hormone (FSH). ↑ Luteinizing hormone is involved only in ↑ ovulation. Under control of FSH, ↑ follicular cells start to secrete ↑ estrogens.

↑ Ovarian follicles, growth of

Ovarian follicles, growth of: From a very large population of ↑ oogonia created during fetal life, only about 400000 primordial follicles develop. At birth, they are present in cortex of ↑ ovary. Between birth and puberty, there is practically no growth of follicles. At puberty, ↑ follicular cells and primary ↑ oocyte of a primordial follicle start to grow, giving rise successively to primary follicles, secondary solid and vesicular follicles, and mature follicles. At moment of ↑ ovulation, wall of the latter bursts and secondary ↑ oocyte, surrounded by ↑ zona pellucida (ZP) and ↑ corona radiata (CR), is extruded into ↑ oviduct. In woman, growth and maturation of a follicle

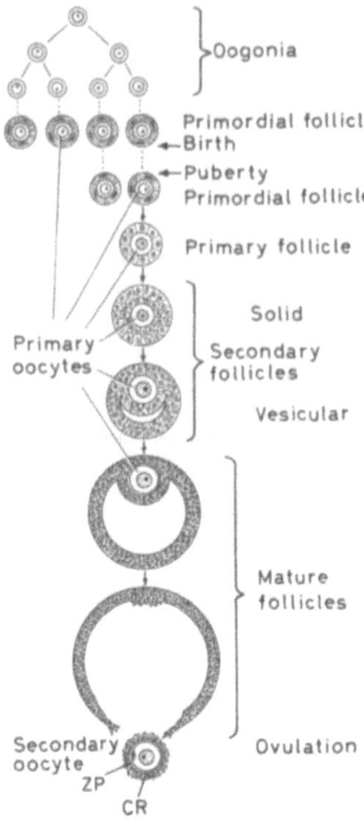

require about 10–14 days from beginning of ↑ menstrual cycle.

↑ Oogenesis; ↑ Ovarian cycle

Kanzaki, H., Okamura, H., Okuda, Y.: J. Anat. *134*, 697 (1982); Oakberg, E.F.: In Vitro *15*, 41 (1979)

Ovarian follicle, mature (Graafian follicle, tertiary follicle, folliculus ovaricus maturus*): A vesicular formation measuring about 10–20 mm bulging from ovarian surface. Most of O. occupies ↑ antrum folliculi (A) filled with ↑ liquor folliculi. Primary ↑ oocyte (O) is situated in a local eccentric thickening of ↑ stratum granulosum (SG), ↑ cumulus oophorus (CO). One or more

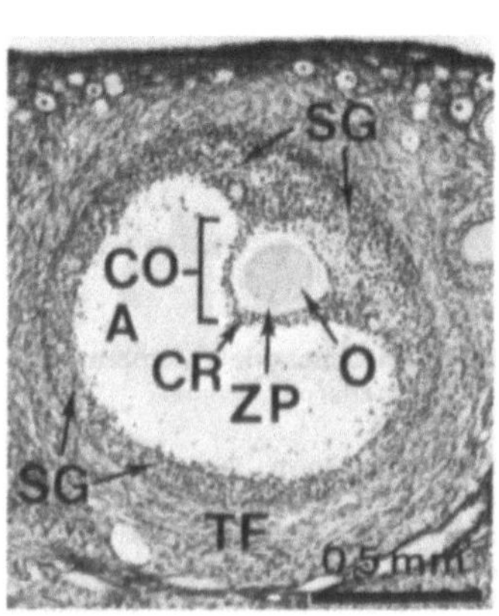

layers of columnar ↑ granulosa cells, ↑ corona radiata (CR), are attached to oocyte and accompany it after ↑ ovulation. ↑ Zona pellucida (ZP) is now 5–10 µm thick. Both ↑ theca interna and theca externa (TF) reach their maximal development. Only about 400 follicles reach the stage of O. during fertile period of a woman. (Fig. = rabbit)

Ovarian follicle, preovulatory: A very large mature ↑ ovarian follicle occupying full thickness of ovarian cortex and bulging above surface of ↑ ovary.

Ovarian follicle, primary (folliculus ovaricus primarius*): A primary ↑ oocyte (O) surrounded by a single layer of cuboidal or columnar ↑ follicular cells (F) separated from ovarian stroma (S) by a distinct basal lamina (BL). ↑ Zona pellucida (ZP) becomes visible under light microscope at the stage of O. G = ↑ germinal epithelium (Fig. = rabbit)

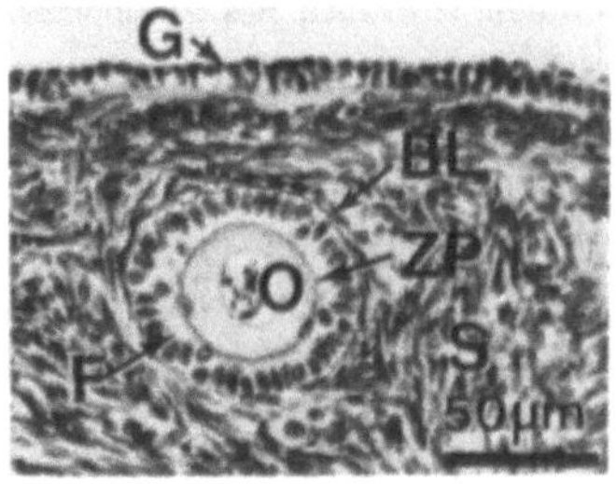

Ovarian follicle, primordial (unilaminar follicle, folliculus ovaricus primordialis*): A primary ↑ oocyte (O) surrounded by a single layer of flattened ↑ follicular cells (F). At birth, there are about 200000 O./ ↑ ovary. Through growth and multiplication of follicular cell and enlargement of primary oocyte, O. transforms into primary ↑ ovarian follicle. O. are situated in ovarian cortex. (Fig. = rabbit)

↑ Ovarian follicle, growth of; ↑ Oogenesis

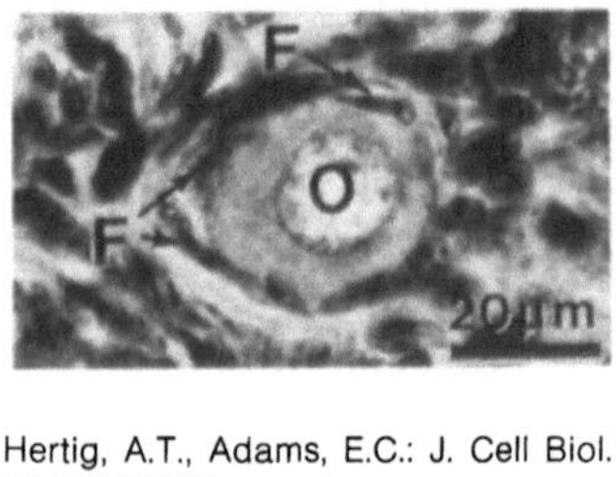

Hertig, A.T., Adams, E.C.: J. Cell Biol. *34*, 647 (1967)

Ovarian follicle, secondary solid (multilaminar primary follicle, folliculus ovaricus secondarius*): A primary ↑ oocyte (O) enclosed by a solid mass of stratified ↑ follicular cells; mitotic proliferation of these cells gives rise to ↑ granulosa cells (GC). Between oocyte and granulosa cells, ↑ zona pellucida (ZP) enlarges. A basal lamina (BL), known also as a ↑ glassy membrane, is interposed between granulosa and ovarian stromal cells (S), which begin to organize themselves into ↑ theca folliculi (TF). Plasmalemma of oocyte forms numerous microvilli that penetrate zona pellucida and contact projections of follicular cells. (Fig. = rabbit)

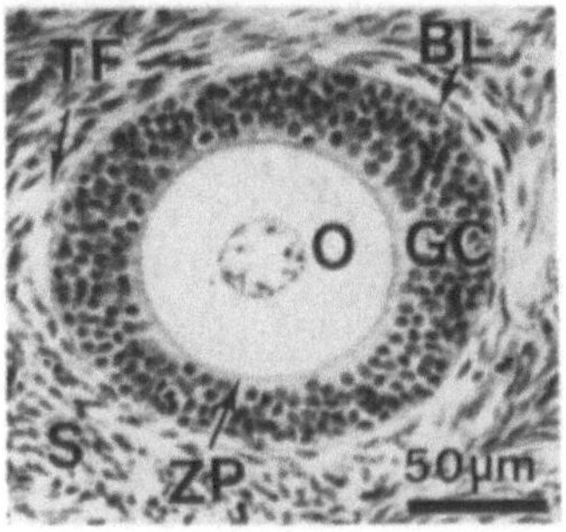

Apkarian, R., Curtis, J.C.: Scan. Electron Microsc. 1981/IV, 165 (1981); Kanzaki, H., Okamura, H., Okuda, Y., Takenaka, A.; Morimoto, K., Nishimura, T.: J. Anat. *134*, 697 (1982)

Ovarian follicle, secondary vesicular (antral follicle, folliculus ovaricus secundarius vesiculosus*): Stage of development of an ↑ ovarian follicle between secondary solid follicle and mature follicle. By accumulation of fluid, ↑ liquor folliculi, between ↑ follicular cells, intercellular spaces enlarge and fuse together to form a single crescent-shaped cavity, ↑ antrum folliculi (A). Some layers of ↑ granulosa cells (GC) lie on ↑ glassy membrane (GM), forming stratum granulosum (SG). ↑ Call-Exner bodies (CE) appear among these cells. Primary ↑ oocyte (O) no longer grows. ↑ Theca folliculi (TF) show an inner vascular and an outer fibrous layer. (Fig. = rabbit)

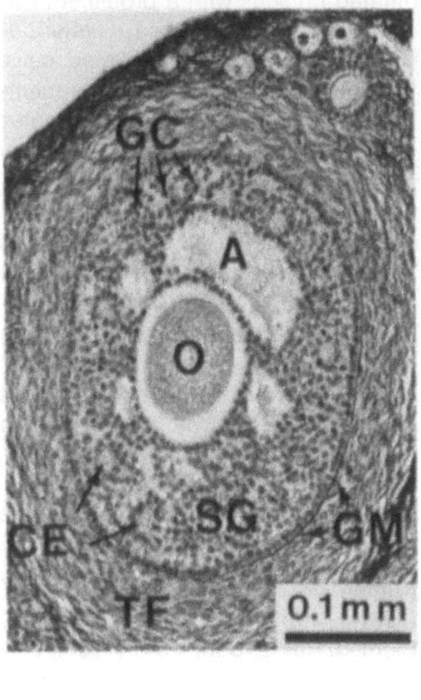

Ovarian follicle, unilaminar: ↑ Ovarian follicle, primordial

Ovarian follicular cells: ↑ Follicular cells, of ovarian follicles

Ovary (ovarium*): One of paired primary female sex organs (2.5–5 cm long, 1–3 cm wide, and 0.6–1.5 cm thick) for storage and development of ↑ oocytes prior to their expulsion and for production of steroid hormones stimulating development and function of secondary sex organs, ↑ placenta and ↑ mammary gland. Structure: A) Cortex (C): 1) ↑ Germinal epithelium (GE); 2) ↑ stroma = ↑ tunica albuginea (TA) continuous with ↑ "cellular" loose connective tissue, in which postnatal ↑ oogenesis takes place; 3) ↑ parenchyma = atretic (AF), primordial, primary (PF), secondary (SF), and tertiary (TF) ↑ ovarian follicles, ↑ interstitial cells, and, after puberty until ↑ menopause, ↑ corpus luteum (CL). ↑ Corpora albicantia (CA) persist after menopause. B) Medulla (M) or zona vasculosa = a loose connective tissue with many large blood vessels, ↑ lymphatic vessels, and nerves entering and leaving O. at ↑ hilus (H); all these elements run through ↑ mesovarium (Mes).

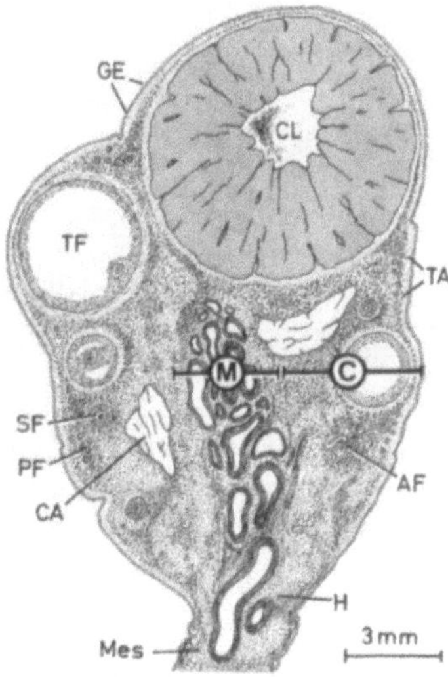

Channing, C.P., Marsh, I., Sadler, W.A. (eds.): Ovarian Follicular and Corpus Luteum Function. In: Advances in Experimental Medicine and Biology, Vol. 12. New York: Plenum Press 1979; Guraya, S.S.: Int. Rev. Cytol. *51*, 49 (1977); Motta, P.M.; Makabe, S.: Biomed. Res. *2*, 325 (1981); Zuckerman, L., Weir, B.J. (eds.): The Ovary, Vols 1–3. New York: Academic Press 1977

Ovary, control of activity: Under influence of ↑ follicle-stimulating hormone-releasing factor (FSH-RF) of ↑ hypothalamus, ↑ hypophysis (H) produces ↑ follicle-stimulating hormone (FSH), which stimulates secretion of ↑ estrogens by ↑ follicular cells. FSH and estrogens inhibit (−) production of FSH-RF and stimulate (+) secretion of ↑ luteinizing hormone-releasing factor (LH-RF), which, in turn, stimulates hypophysis to secrete ↑ luteinizing hormone (LH). Both FSH and LH induce maturation of ↑ ovarian follicles and ↑ ovulation. Under further influence of LH, ↑ corpus luteum, which produces ↑ progesterone, forms. It inhibits (−) production of LH-RF. When pregnancy does not occur, corpus luteum persists only during second half of ↑

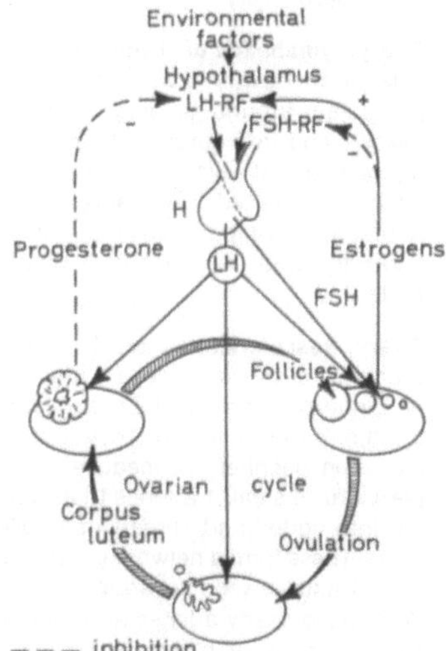

menstrual cycle. When pregnancy occurs, ↑ chorionic gonadotropin induces maintenance of function of corpus luteum and production of progesterone for about 4 months, after which corpus luteum degenerates. Production of progesterone is continued by ↑ placenta. Since progesterone inhibits ↑ prolactin secretion, levels of prolactin increase only after expulsion of ↑ placenta. Prolactin stimulates secretion of ↑ milk for a certain period, simultaneously inhibiting production of FSH/LH-RF. Gradually, concentration of prolactin decreases and ↑ ovarian cycle is reestablished. [In some animals (mouse, rat), prolactin is necessary to stimulate corpus luteum to produce progesterone.]

↑ Oogenesis; ↑ Ovarian follicles, growth of

Greenwald, G.S., Terranova, P.F. (eds.): Factors Regulating Ovarian Function. New York: Raven Press 1983; Serra, G.B. (ed.): Ovary, Comprehensive Endocrinology. New York: Raven Press 1983

Ovary, corpus luteum of: ↑ Corpus luteum

Ovary, follicles of: ↑ Ovarian follicles

Ovary, germinal epithelium of: ↑ Germinal epithelium

Ovary, innervation of: ↑ Nonmyelinated nerve fibers enter ↑ ovary through ↑ hilus, ramify around ↑ ovarian follicles, and innervate tunica media of arteries. A few sensory nerve endings have been found in ↑ stroma.

Ovary, interstitial tissue of: ↑ Interstitial tissue, of ovary

Ovary, lymphatics of: Lymphatic circulation in ↑ ovary is well developed, particularly around growing ↑ ovarian follicles and in ↑ corpus luteum. It is assmumed that ↑ lymphatic vessels of ovary are involved in transport of its hormones. ↑ Lymph drains into aortic, preaortic, and para-aortic ↑ lymph nodes.

Ovary, vascularization of: Tortuous branches of ovarian arteries, arteriae helicinae, penetrate at ↑ hilus and form a major part of medulla. They form a plexus in periphery of medulla; from plexus arise smaller arteries that radiate into cortex and spread as capillaries. These form a network in ↑ theca folliculi and between ↑ ovarian follicles. Veins accompany arteries and form a plexus in hilus. All blood vessels of O. have remarkable dynamics, since they follow changes during ↑ ovarian cycle.

Ovary, vestigial organs of: Embryonal remnants occurring in ↑ mesovarium – ↑ epoophoron, ↑ paraoophoron, and duct of Gärtner. (See embryology texts for further information)

Oviducts (Fallopian tube, uterine tube, tuba uterina*): Paired tubular organs, 10–12 cm long, for conducting ↑ ovum from ↑ ovary (Ov) to ↑ uterus (U). Four parts: A) Infundibulum with ↑ fimbriae (F); B) ampulla; C) isthmus; D) intramural or interstitial segment. Structure: 1) Tunica mucosa (TM) is highly folded, especially in ampulla = a) epithelium (E) = a ↑ simple columnar epithelium with ciliated (C) and secretory cells (S); b) lamina propria (LP) of mucosal folds is poorly developed. Submucosal loose connective tissue layer is minimal. 2) Tunica muscularis (TMu) = a) an inner circular (IC) or spiral, and b) an outer longitudinal (OL) layer of ↑ smooth muscle cells without a distinct boundary between the two. Thin in ampulla, tunica muscularis becomes thicker in isthmus. Peristaltic contractions of muscularis from infundibulum to uterus transport ovum from peritoneal opening (Op) of O. to uterine cavity. Some muscular bundles of outer muscular layer of ampullar segment lie immediately below tunica serosa (so-called subperitoneal musculature, arrowheads) and serve to move O., adapting its position to the protruding preovulatory ↑ ovarian follicle (arrow) of ovary. 3) Tela subserosa (TSs) = a loose connective tissue with blood and lymphatic vessels and nerve fibers. 4) Tunica serosa (TS) = squamous ↑ mesothelial cells of ↑ peritoneum. The serosa forms ↑ mesosalpinx.

↑ Oviduct, epithelium of

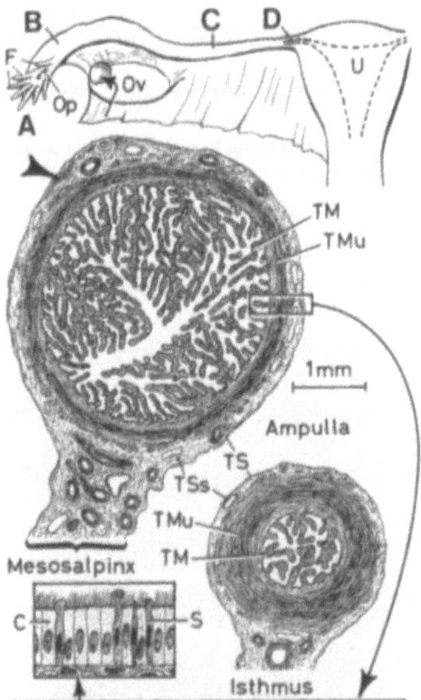

Hafez, E.S.E., Blandau, R.J.: The Mammalian Oviduct, Comparative Biology and Methodology. Chicago: The University of Chicago 1969

Oviducts, epithelium of: A simple columnar epithelium, tallest in ampulla and decreasing in height toward ↑ uterus. Two kinds of cells: 1) Ciliated cells (CC) (cellula ciliata*) are predominant and display a basal or central elliptical nucleus with a prominent nucleolus. Cytoplasm is clear with poorly developed ↑ organelles. Apical cell sur-

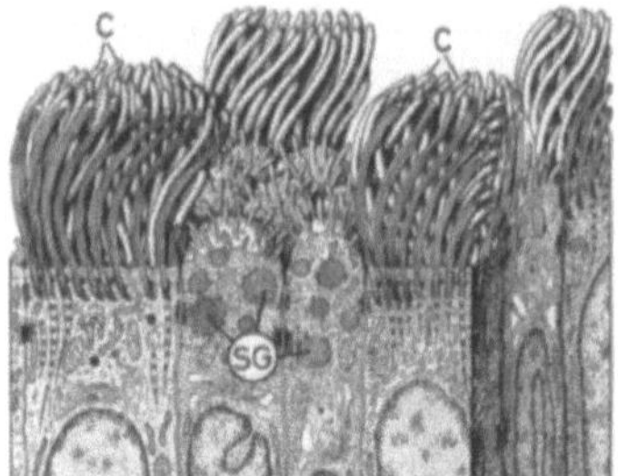

face bears numerous ↑ cilia (C), which beat toward uterus. It is likely that ciliary movements orient ↑ spermatozoa, rather than transport ↑ ovum; the latter seem to be swept by peristaltic contraction of ↑ oviduct. Ciliated cells increase in height and number during proliferative phase of ↑ menstrual cycle, reaching about 30 µm at ↑ ovulation; during secretory phase, they are fewer and measure about 15 µm in height. 2) Secretory cells or nonciliated cells (SC) (cellula nonciliata*) have a central elongated and deeply invaginated nucleus with a prominent nucleolus; cell ↑ organelles are better developed than those of ciliated cells; from Golgi apparatus separate membrane-enclosed ↑ secretory granules (SG), which are discharged at apical surface between ↑ microvilli. During secretory phase of menstrual cycle, number of cells increases; their product is a kind of nutrient material for ovum as well as a substance involved in ↑ capacitation of spermatozoa. According to some opinions, the two cell types of O. are only different functional states of one cell type. The existence of a third type of cells, basal, intercalated, or peg cells, is not generally accepted. After ↑ menopause, O. becomes simple columnar with fewer ciliated cells.

Kugler, P.: Histochemistry *73*, 137 (1981); Kühnel, W., Busch, L.C.: Biomed. Res. *2*, Suppl. 341 (1981); Lindenbaum, E.S., Beach, D., Peretz, B.A.: Anat. Rec. *203*, 67 (1982); Merchan, J.A., Gonzales-Gomez, F.: Int. J. Fertil. *25*, 293 (1980); Verhage, H.G., Bareither, M.L., Jaffe, R.C., Akbar, M.: Am. J. Anat. *156*, 505 (1979)

OVLT: ↑ Organum vasculosum laminae terminalis

Ovulation (ovulatio*): Rupture of preovulatory ↑ ovarian follicle and expulsion of secondary ↑ oocyte. Sequence: 1) Follicle (F) attains a diameter of 10–20 mm and bulges above surface of ↑ ovary; oocyte (O), surrounded by ↑ corona radiata (CR), is loosened from ↑ cumulus oophorus (CO). 2) Stratum granulosum (SG), theca folliculi (TF), and tunica albuginea (TA) become progressively thinner and weaker in a restricted area called ↑

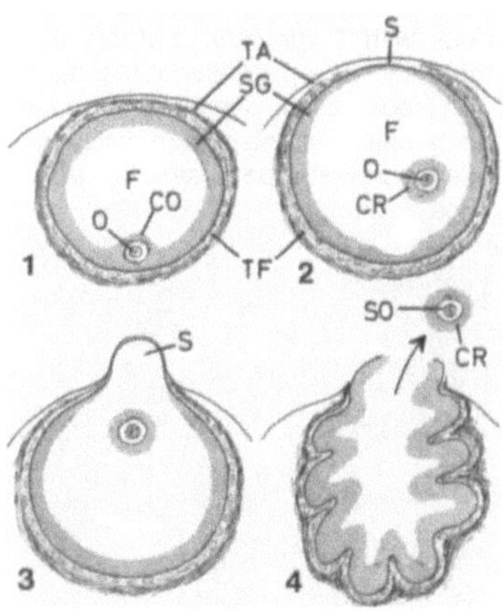

stigma (S), perhaps by action of ↑ collagenases acting on ↑ ground substance and ↑ collagen fibers. Blood circulation in this thinned area between follicular wall and surface of ↑ ovary is interrupted, leading to ↑ necrosis of cells in stigma; secretion of secondary ↑ follicular fluid. 3) Stigma protrudes as a clear cone and 4) ruptures; secondary oocyte (SO), attached to cells of corona radiata, flows out with follicular fluid into oviduct. Rupture of follicle is accompanied by a weak hemorrhage. Exact mechanism of O. is unknown. It is believed that bulging of follicle is partially caused by hyperemia of medullary blood vessels. ↑ FSH stimulates growth of follicle during first half of ↑ ovarian cycle; at about midcycle (13th–15th day), an additional surge of ↑ LH and ↑ prostaglandins seem to provoke rupture of stigma. O. occurs in middle of ↑ menstrual cycle.

Aron, C.: Bull. Ass. Anat. *63*, 345 (1979); Balboni, G.C.: Ovulation and Luteal Phase. In: Serra, G.B. (ed.): Ovary: Comprehensive Endocrinology. New York: Raven Press 1983; Downs, S.M., Longo, F.J.: Anat. Rec. *205*, 159 (1983)

Ovum* (egg): The secondary ↑ oocyte after ↑ fertilization, i.e., expulsion of secondary ↑ polar body. Some authors use the term O. as a synonym for secondary oocyte.

↑ Oogenesis

Hertig, A.T., Rock, J., Adams, E.C.: Am. J. Anat. *98*, 435 (1956), Teasrik, J., Dvořak, M.: Folia Morphol. *29*, 297 (1981)

Owen's contour lines: ↑ Contour lines, of Owen

Oxyntic cells: ↑ Parietal cells

Oxyphil cells, of parathyroid gland (Welsh's cells, cellula oxyphilica*): Large polygonal ↑ acidophilic cells found singly and in small groups in cellular cords of ↑ parathyroid glands, particularly in old individuals. Other than in humans, O. are found in monkeys and cattle. O. have a round central nucleus and an extensive number of ↑ mitochondria (M), which fill entire cytoplasm. Golgi apparatus and rough endoplasmic reticulum are sparse or absent. ↑ Lysosomes (Ly), ↑ glycogen particles (Gly), and ↑ lipid droplets (LD) can be present among mitochondria. Lateral cell borders are straight. Between chief cells and O., there are transitional forms. Although O. contain large amounts of oxidative enzymes, their function is unknown; some authors consider them to be ↑ oncocytes.

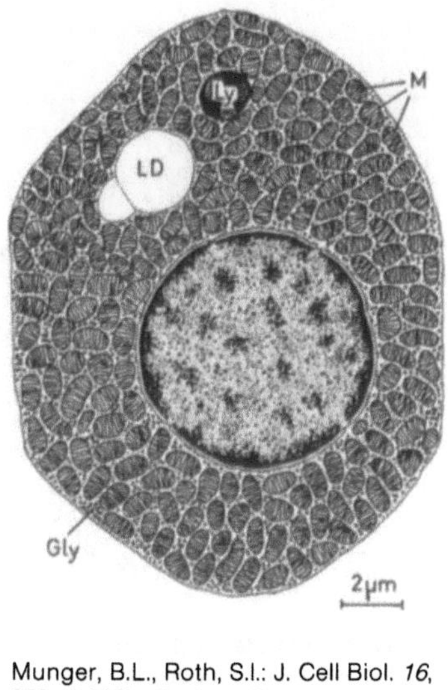

Munger, B.L., Roth, S.I.: J. Cell Biol. *16*, 379 (1963); Rother, P.: Z. mikrosk.-anat. Forsch. *79*, 533 (1969)

Oxytalan fibers (pre-elastic fibers): Ramified and anastomosing fibers up to 2 µm thick present in ↑ connective tissues and elastic ↑ cartilage. O. consist of bundles of ↑ microfilaments, 8–12 nm thick. They fail to stain with dyes commonly used for ↑ collagen, ↑ reticular, and ↑ elastic fibers, but stain with ↑ orcein, ↑ aldehyde fuchsin, and ↑ resorcin fuchsin only after previous oxidation. O. are also resistant to digestion with elastase and acid hydrolysis. Nature of O. is not fully understood; it is believed they represent precursors of ↑ elastic fibers.

Bradamante, Z., Švajger, A.: J. Anat. *123*, 735 (1977); Goldfischer, S., Coltoff-Schiller, B., Schwartz, E., Blumenfeld, O.O.: J. Histochem. Cytochem. *31*, 382 (1983); Jonas, I.E., Riede, U.N.: J. Histochem. Cytochem. *28*, 211 (1980)

Oxytocin: A hypothalamic hormone synthesized in ↑ neurosecretory neurons of ↑ supraoptic and ↑ paraventricular nuclei and anterior commissure nucleus. It is transported along ↑ hypothalamohypophyseal tract to ↑ neurohypophysis where it is stored. O. stimulates uterine contractions during coitus and delivery, and causes contraction of ↑ myoepitelial cells surrounding ↑ tubuloalveoli of ↑ mammary gland, allowing ↑ milk to be expelled.

↑ Suckling, reflex

Kawata, M., Sano, Y.: Anat. Embryol. *165*, 151 (1982); Rhodes, C.H., Morrell, J.I., Pfaff, D.W.: Cell Tissue Res. *216*, 47 (1981)

P

Pacchionian corpuscles: ↑ Arachnoid villi

Pacchionian granulations (arachnoid granulations, granulatio arachnoidalis*): Fungiform, fibrous avascular arachnoid organs (PG) that penetrate into ↑ dura mater (DM), sagittal sinus, and sometimes bone of skull near sinus. P. represent age-transformed ↑ arachnoid villi whose compact cellular mass is gradually replaced by ↑ reticular and ↑ collagen fibers. P. are lined by one or more layers of flattened cells and connected with ↑ arachnoid (A) by a thin stalk (S). P. serve as pathway for outflow of ↑ cerebrospinal fluid from ↑ subarachnoid space to venous circulation. Number and size of ↑ P. increase with age. (Fig. = human; preparation courtesy of Prof. W. Zaki, Lausanne)

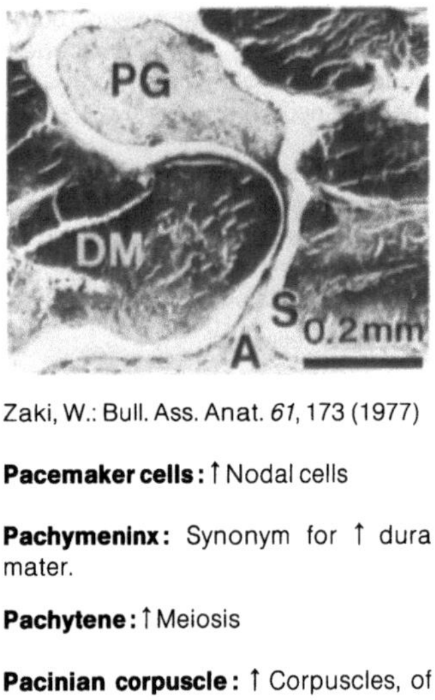

Zaki, W.: Bull. Ass. Anat. *61*, 173 (1977)

Pacemaker cells: ↑ Nodal cells

Pachymeninx: Synonym for ↑ dura mater.

Pachytene: ↑ Meiosis

Pacinian corpuscle: ↑ Corpuscles, of Vater-Pacini

Palate, hard (palatum durum*): Anterior part of ↑ oral cavity roof. Structure: Tunica mucosa (TM): a) epithelium (E) = nonkeratinized ↑ stratified squamous epithelium with thin keratinized patches, b) lamina propria (LP) = ↑ dense connective tissue firmly attached to ↑ periosteum (P) of bone (B). Several small mixed ↑ palatine glands (PG) are situated in lamina propria toward soft ↑ palate. (Fig. = frontal section)

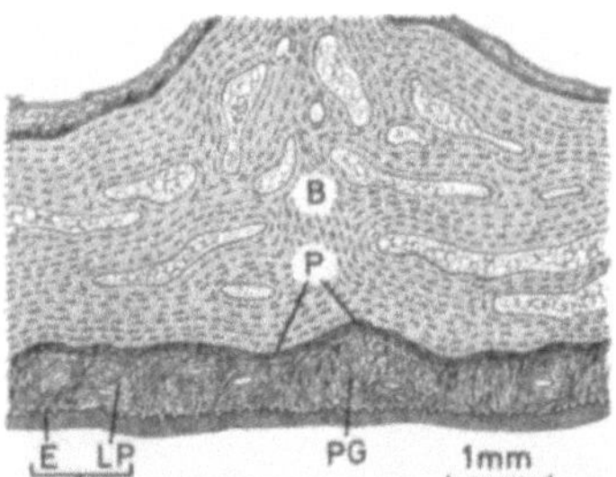

Palate, soft (palatum molle*, velum palatinum*): A mucous fold with a fibromuscular skeleton separating ↑ oral cavity from ↑ pharynx. Structure: 1) Tunica mucosa (TM): a) epithelium (E) = nonkeratinized ↑ stratified squamous epithelium with numerous ↑ connective papillae on oral side; deep toward choanes on pharyngeal side, this is replaced by columnar ↑ pseudostratified ciliated epithelium (arrow); b) lamina propria (LP) = ↑ loose connective tissue immediately underlying epithelium at both sides; on pharyngeal side, this layer contains mixed ↑ pharyngeal glands (PhG). 2) Elastic layer or stratum elasticum (EL) is composed of densely interwoven ↑ elastic fibers; this layer continues onto pharyngeal side, but lies between glands and muscles. 3) Tela submucosa or submucous layer (TS) only on the oral side = loose connective tissue with large ↑ palatine mucous glands (PG). 4) Muscular layer or lamina tendinomuscularis* (Mu) = ↑ skeletal muscle fibers and their fasciae that belong to musculi levator veli palatini, palatoglossus, and palatopharyngicus. Terminal part of P., the uvula (U), contains some muscle fibers of musculus uvulae (MU) and a group of ↑ mucous tubules (MT). (Fig. = mediosagittal section)

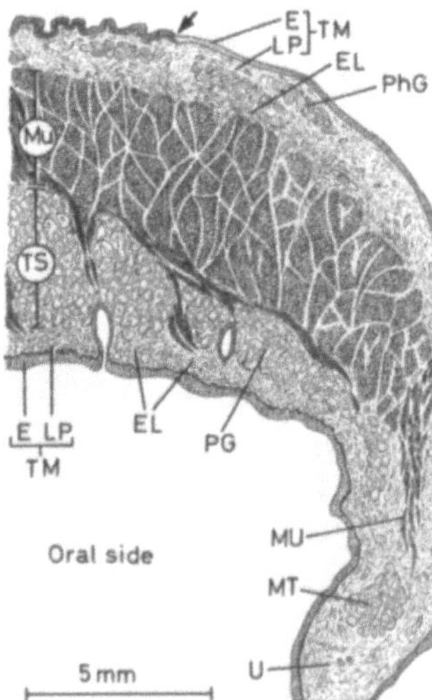

Palatine glands (glandulae palatinae*): Purely ↑ mucous glands situated in tela submucosa on oral side of soft ↑ palate.

Palatine tonsils: ↑ Tonsils, palatine

Paleocortex: ↑ Allocortex

Palladium shadowing: ↑ Shadowing

Palpebral conjunctiva (tunica conjunctiva palpebrarum*): Portion of ↑ conjunctiva lining posterior surface of ↑ eyelids.

Pancreas*: A lobulated retroperitoneal gland composed of an exocrine portion provided with an excretory duct system and an endocrine portion in the form of ↑ pancreatic islets. Structure: 1) Capsule (C) = thin layer of ↑ loose connective tissue. 2) ↑ Stroma = connective septa (S) dividing parenchyma into ↑ lobules (L). 3) ↑ Parenchyma: a) exocrine P. = compound, purely ↑ serous gland composed of ↑ pancreatic acini (A), which are continuous with ↑ intercalated ducts (ID), tributaries to larger ↑ interlobular ducts (IID) located in septa. Interlobular ducts empty into ↑ pancreatic ducts; b) endocrine P. = pancreatic islets (PI). (Figs. = human)

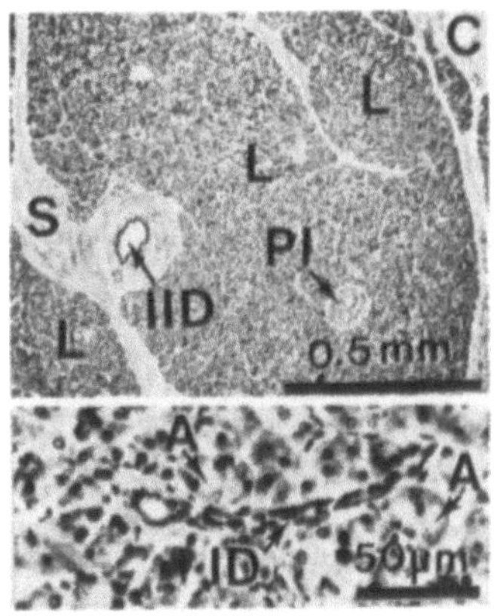

Ohtani, O.: Arch. Histol. Jpn. *46*, 315 (1983)

Pancreas, A-cells of: ↑ A-cells, of pancreatic islets

Pancreas, B-cells of: ↑ B-cells, of pancreatic islets

Pancreas, C-cells of: ↑ C-cells, of pancreatic islets

Pancreas, centroacinar cells of: ↑ Centroacinar cells

Pancreas, control of secretion: 1) Exocrine pancreas: a) ↑ nervous system = stimulation of vagus nerve provokes an increase of enzyme secretion; b) hormones = ↑ secretin and ↑ cholecystokinin-pancreozymin act to increase volume of enzyme-rich pancreatic juice with an alkaline pH (high bicarbonate content), which serves to neutralize acid stomach juice and provides a favorable pH for pancreatic enzyme activity. 2) Endocrine pancreas = ↑ pancreatic islets, control of secretion.

Pancreas, D-cells of: ↑ D-cells, of pancreatic islets

Pancreas, innervation of: Pancreas is innervated by predominantly ↑ nonmyelinated nerve fibers of celiac plexus and ↑ myelinated fibers of vagus nerve. Nerve fibers accompany arteries and capillaries; ↑ axons may form neuroglandular ↑ synapses with acinar and islet cells.

Pancreas, zymogen granules of: ↑ Zymogen granules

Pancreatic acinar cells (exocrine cells, of pancreas, cellula acinosa*): Basophilic pyramidal epithelial cells constituting major part of ↑ pancreatic acini. P. have globular nucleus (N) with one or two large nucleoli. Cytoplasm contains rod-shaped voluminous mitochondria, an extensive Golgi apparatus (G), and abundant ↑ ergastoplasm (E) which is predominant in basal and paranuclear cell portions. ↑ Unit membrane-bound vacuoles with moderate osmiophilic contents arise from Golgi apparatus (= ↑ condensing vacuoles, asterisk); they mature and transform into ↑ zymogen granules (ZG), whose contents are emptied into acinar lumen by ↑ exocytosis. Luminal cell surface bears some short microvilli. Neighboring cells are held together by apical ↑ junctional complexes and lateral interdigitations. P. overlie a ↑ basal lamina. P. synthesize trypsinogen, chymotrypsinogen, carboxypeptidase, part of pancreatic amylase, lipase, lecithinase, ribonuclease, and desoxyribonuclease, which are all contained in inactive form in zymogen granules. (Fig. = rat)

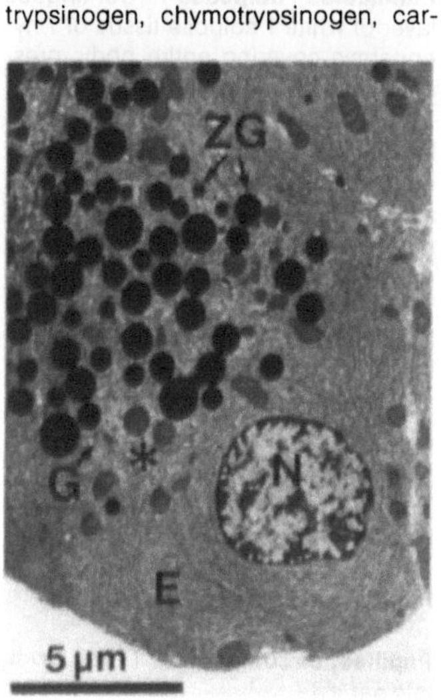

Bendayan, M., Roth, J., Perrelet, A., Orci, L.: J. Histochem. Cytochem. *28*, 149 (1980)

Pancreatic acinus (acinus pancreaticus*): Smallest secretory unit of exocrine ↑ pancreas, composed of ↑ pancreatic acinar cells (AC) and ↑ centroacinar cells (CC) which are continuous with cells of ↑ intercalated duct (ID). BL = ↑ basal lamina, Cap = fenestrated ↑ capillaries, NF = ↑ nonmyelinated nerve fibers

↑ Acinus, glandular

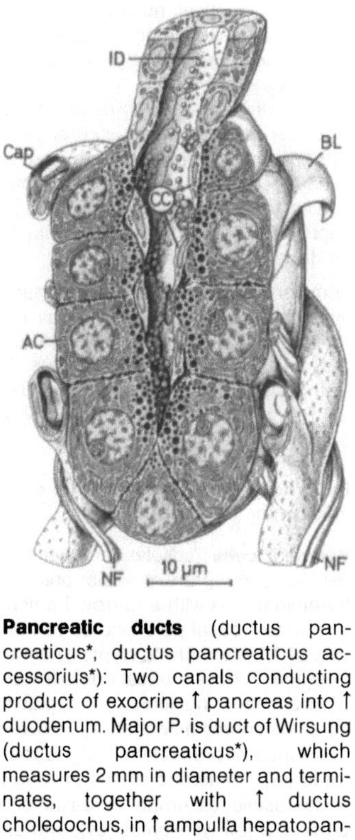

Pancreatic ducts (ductus pancreaticus*, ductus pancreaticus accessorius*): Two canals conducting product of exocrine ↑ pancreas into ↑ duodenum. Major P. is duct of Wirsung (ductus pancreaticus*), which measures 2 mm in diameter and terminates, together with ↑ ductus choledochus, in ↑ ampulla hepatopancreatica in three-fourths of the population. Minor P., called duct of Santorini (ductus pancreaticus accessorius*), is not always found; when present, it opens into duodenum independently. P. are lined by ↑ simple cuboidal or columnar epithelium (E), forming numerous glandlike outpocketings (G) in lamina propria (LP). Function of these structures is not clear; some authors think they represent true mucous glands, others relate them to formation of new ↑ pancreatic islets (PI). External to epithelium is a particularly thick lamina propria of ↑ dense connective tissue. (Fig. = human)

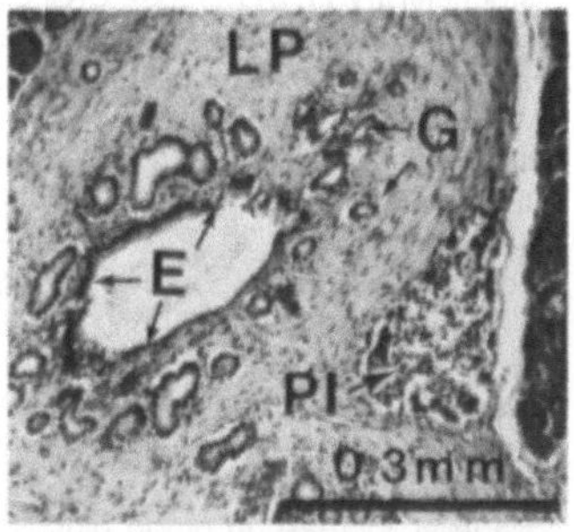

Fölsch, U.R., Creutzfeldt, W.: Gastroenterology *73*, 1053 (1977); Githens, S., Holmquist, D.R.G., Whelan, J.F., Ruby, J.R.: In Vitro *16*, 797 (1980)

Pancreatic islets (islets of Langerhans, insulae pancreaticae*): Multihormonal micro-organs (PI) in form of ovoid groups of clear ↑ endocrine cells scattered throughout exocrine pancreatic ↑ parenchyma (EP), predominantly occurring in tail of organ. Endocrine cells are arranged in irregular cords; P. are relatively well demarcated from exocrine tissue by fine ↑ reticular fibrils. P. measure about 0.1–0.2 mm in diameter and range in total number between 200 000 and 1 800 000 in human pancreas. They are highly vascularized by fenestrated ↑ capillaries and supplied by numerous autonomic nerve fibers. About 20% of cell population is ↑ A-cells, 75% ↑ B-cells, and 5% ↑ D-cells. ↑ C-cells are rare and ↑ E-cells probably do not exist in man. There are two categories of P.: juxtaduodenal, producing ↑ insulin and ↑ pancreatic polypeptide, and juxtasplenic, secreting both insulin and ↑ somatostatin. Recently, a periinsular ↑ Schwann cell sheath surrounding P. in different mammals has been observed. (Fig. = human)

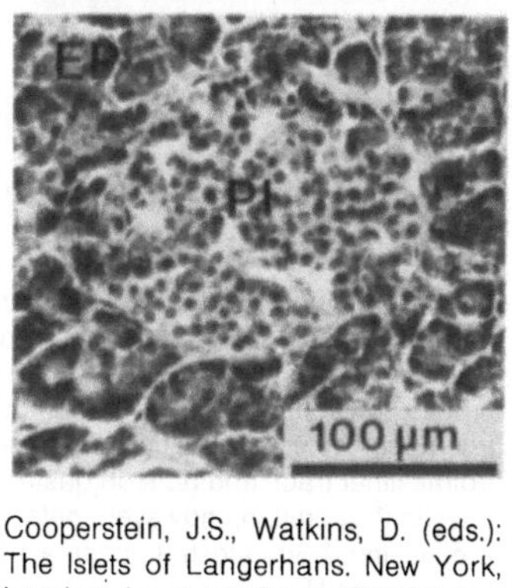

Cooperstein, J.S., Watkins, D. (eds.): The Islets of Langerhans. New York, London: Academic Press 1981; Donev, S.R.: Cell Tissue Res. *237*, 343 (1984);

Fiocca, R., Sessa, F., Tenti, P., Usellini, L., Capella, C., O'Hare, M.M.T., Socia, E., Histochemistry *77*, 511 (1983); Orci, L.: Perrelet, A.: Morphology of Membrane Systems in Pancreatic Islets. In: Volk, B.W., Wellmann, K.F. (eds.): The Diabetic Pancreas. New York: Plenum Publ. Co. 1977; Orci, L., Stefan, Y., Bonner-Weir, S., Perrelet, A., Unger, R.: Diabetologia *21*, 73 (1981); Orci, L., Unger, R.H.: Lancet *20*, 1243 (1975)

Pancreatic islets, control of secretion: Main factor determining activity of ↑ A-cells and ↑ B-cells is blood glucose level. Increase in blood glucose stimulates B-cells to produce ↑ insulin and inhibits A-cells from secreting ↑ glucagon. ↑ Sympathetic nervous system, ↑ cholecystokinin-pancreozymin (CCK-PZ), ↑ gastrin, ↑ cortisol, etc. stimulate glucagon secretion, which incites ↑ D-cells to produce ↑ somatostatin, which in turn inhibits both A-cells and B-cells. Insulin secretion is stimulated by ↑ parasympathetic nervous system, ↑ gastrin, glucagon, CCK-PZ, etc. Production of glucagon is also inhibited by ↑ secretin and insulin; secretion of insulin is inhibited by ↑ alloxan and ↑ sympathetic nervous system. (Partly after Faber and Haid, 1972)

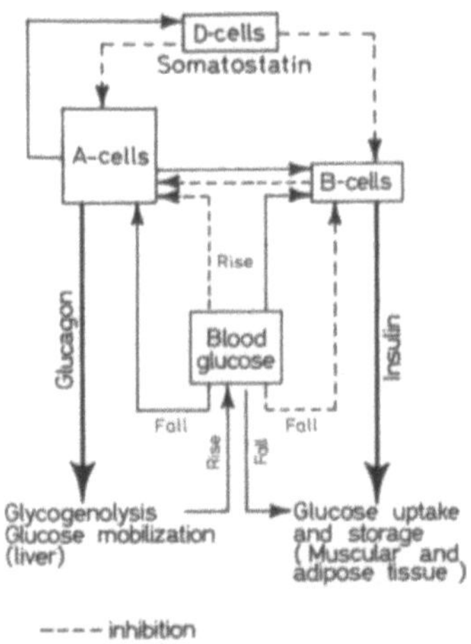

Faber, H.V., Haid, H.: Endokrinologie. Stuttgart: UTB 1972

Pancreatic polypeptide (PP): A linear polypeptide containing 36 amino acid residues. P. is produced by ↑ pancreatic polypeptide cells of ↑ pancreas and by some ↑ endocrine cells of ↑ gastrointestinal tract, and acts on gastrointestinal secretion; however, details are lacking. ↑ Somatostatin decreases production of P.

↑ Pancreatic islets, control of secretion

Johnson, D., Noe, B.D., Bauer, E.G.: Anat. Rec. *204*, 61 (1982); Lehy, T., Peranzi, G., Cristina, M.L.: Histochemistry *71*, 67 (1981); Orci, L., Baetens, D., Ravazzola, M., Stefan, Y., Malaisse-Lagae, F.: Life Sci. *19*, 1811 (1976); Schweistahl, M.R., Schweistahl, J.V., Frost, C.C.: Am. J. Anat. *152*, 257 (1978); Sessa, F., Fiocca, R., Tenti, P., Solcia, E., Tavani, E., Pliteri, S.: Virchows Archiv A *399*, 227 (1983)

Pancreatic polypeptide cells (F-cells, PP-cells): Kind of endocrine cells of pancreatic ↑ parenchyma not only found in juxtaduodenal ↑ pancreatic islets, but also associated with ↑ pancreatic acinar cells and found among epithelial cells lining small and medium-sized ↑ pancreatic ducts. P. have a round or elliptical nucleus; in cytoplasm are scattered filamentous mitochondria, short rough endoplasmic cisternae, and a considerable number of ↑ unit membrane-enclosed small dense-core ↑ secretory granules, 140–200 nm in diameter, with homogenous contents reacting with antibodies to human ↑ pancreatic polypeptide. Therefore, it is thought that P. produce this substance.

Bommer, G.: Die PP-Zelle des Endokrinen Pankreas. Stuttgart: Gustav Fischer 1981; Rassan, J., Girard, R., Marliss, E.B. (eds.): Diabetes Mellitus. A Pathophysiological Approach to Clinical Practice. New York: John Wiley & Sons 1980

Pancreozymin: ↑ Cholecytokinin-pancreozymin

Paneth's cells (cellula panethensis*, cellula cum granulo acidophilici*): Pyramidal cells with a narrow ↑ apical and wide ↑ basal pole located at bottom of ↑ Lieberkühn's crypts of small intestine. P. have a basal nucleus with a prominent nucleolus. Cytoplasm contains sparse mitochondria, a well-developed supranuclear Golgi apparatus (G), numerous stacks of rough endoplasmic cisternae in basal cell portion, a few lysosomes (Ly), and an appreciable amount of free ribosomes. Supranuclear cytoplasm encloses numerous large acidophilic ↑ secretory granules (S), which arise as ↑ unit membrane-bound clear vacuoles from Golgi apparatus. Contents of granules gradually concentrate and densify and small crystals may appear within. By ↑ exocytosis, contents of granules (corresponding to ↑ PAS-positive ↑ glycosaminoglycans) are discharged into lumen of Lieberkühn's crypts (Cr). ↑ Apical pole bears some short microvilli; P. rest on ↑ basal lamina (BL). Exact function of P. is unknown: there is evidence that they produce ↑ lysozyme, a bacteriolytic enzyme used for regulation of intestinal bacterial flora. It seems that P. do not produce peptidases needed for digestion of peptides. P. accumulate zinc. ↑ Mitoses of P. have not been observed. Since mitoses have only been observed in adjacent nondifferentiated cells, it is thought that these differentiate into P.; their renewal takes about 3 weeks.

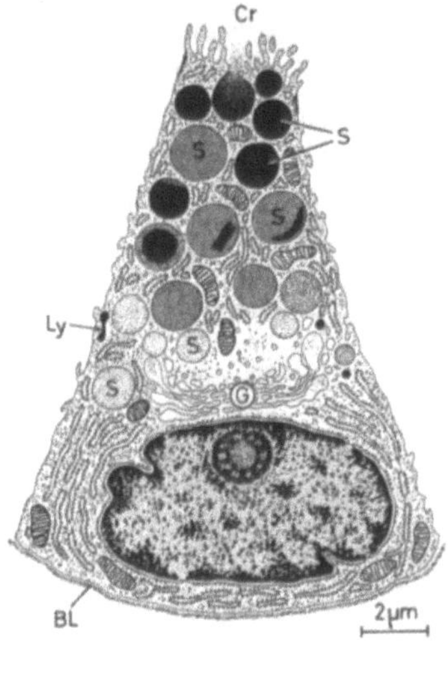

Elmes, M.E., Gwyn Jones, J.: Histochem. J. *13*, 338 (1981); Répassy, G., Lapis, K.: Acta Morph. Acad. Sci. Hung. *27*, 21 (1979); Rodning, C.B., Erlandse, S.L., Wilson, I.D., Carpenter, A.-M.: Anat. Rec. *204*, 33 (1982); Toth, D.M.: Cell Tissue Res. *211*, 293 (1980)

Panniculus adiposus*: Continuous layer of white ↑ adipose tissue of ↑ hypodermis covering entire body; present in infants and young children.

Panniculus carnosus (musculi cutanei*): Areas of ↑ skin where ↑ skeletal muscle fibers terminate in ↑ dermis. P. is vestigial in man, except in muscles of facial expression.

Papanicolaou's staining: Technique for ↑ staining smears of female genital tract for diagnostic purposes.

Marshall, P.N.: Microscopica Acta *87*, 233 (1983)

Papilla, of mammary gland: ↑ Nipple

Papilla nervi optici: ↑ Optic papilla

Papillae, circumvallate: ↑ Circumvallate papillae

Papillae, connective: ↑ Connective papillae

Papillae, dermal: ↑ Dermal papillae

Papillae, filiform: ↑ Filiform papillae

Papillae, foliate: ↑ Foliate papillae

Papillae, of kidney: ↑ Renal papilla

Papillary ducts (Bellini's ducts, ductus papillaris*): Terminal parts (P) of ↑ collecting tubules of ↑ kidney, P. are lined with ↑ simple columnar epithelium and open at ↑ renal papilla. (Fig. = monkey)

↑ Area cribrosa

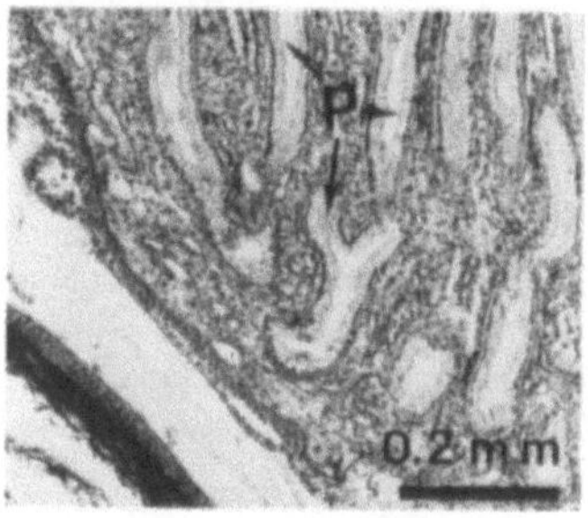

Papillary layer: ↑ Dermis; ↑ Skin

Paracortical zone, of lymph node: ↑ Cortex, of lymph node

Paraffin: Most commonly used ↑ embedding medium for ↑ light microscopy. Dehydrated piece of tissue is placed in intermediate organic solvent miscible with dehydrating agent and P. (toluene, methyl benzoate), then transferred into melted P. at about 60 °C. P. infiltrates tissue; specimen is then cooled and P. solidifies, forming a block ready to be cut with a ↑ microtome.

↑ Dehydration

Parafollicular cells: ↑ C-cells, of thyroid gland

Paraformaldehyde: ↑ Aldehyde-containing substance used for ↑ fixation in ↑ electron microscopy.

Paraganglia, achromaffin: ↑ Paraganglia, parasympathetic

Paraganglia, chromaffin: ↑ Paraganglia, sympathetic

Paraganglia*, parasympathetic (achromaffin paraganglia): Encapsulated groups of two cell types accompanying parasympathetic nerves (vagus, glossopharyngicus). Structure: 1) Type I cells (chief cells) = large, round or polygonal elements with pale cytoplasm, spherical finely structured nucleus, conspicuous nucleolus, moderately developed ↑ organelles, and many ↑ unit membrane-bound 50 to 200-nm electron-dense ↑ secretory granules (SG), which give no visible ↑ chromaffin reaction under light microscope. Recent studies have shown that every type I cell produces and stores ↑ catecholamines, but in amounts too small to permit light-microscopic identification by means of chromaffin reaction. Nerve endings (NE) with numerous clear ↑ synaptic vesicles make synaptic contacts (S) with type I cells. Function of type I cells is not clear; theory postulating their role in synthesis of ↑ acetylcholine should be revised, since they dispose of catecholamine-synthesizing enzymes and therefore belong to the ↑ sympathetic nervous system. 2) Type II cells (supporting cells), closely apposed to type I cells, are modified ↑ Schwann's cells with flattened bodies, dark cytoplasmic matrix, and ↑ heterochromatin-rich nucleus. P. are well vascularized by fenestrated ↑ capillaries (Cap). BL = basal lamina

↑ Autonomic ganglia

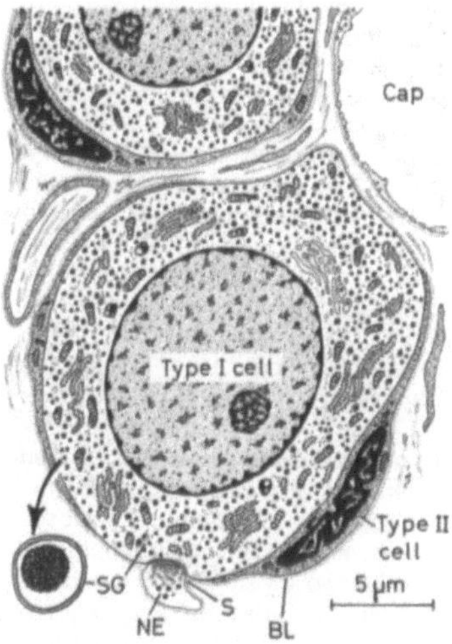

Böck, P.: The Paraganglia. In: Oksche, A., Vollrath, L. (eds.): Handbuch der mikroskopischen Anatomie des Menschen, Bd. 6, Teil 8. Berlin, Heidelberg, New York: Springer-Verlag 1982; Kummer, W., Addicks, K.: Cell Tissue Res. *224*, 455 (1982)

Paraganglia, sympathetic (chromaffin paraganglia, paraganglion sympathicum*): Small encapsulated agglomerations of ↑ chromaffin cells occurring in various areas throughout body, but mostly predominant in retroperitoneum near ↑ sympathetic trunk; some P. are found in heart, lungs, and kidneys. P. also include ↑ aortic and ↑ carotid bodies and are structurally identical to them.

↑ Zuckerkandl's organ

Böck, P.: The Paraganglia. In: Oksche, A., Vollrath, L. (eds.): Handbuch der mikroskopischen Anatomie des Menschen, Vol. 6, Part 8. Berlin, Heidelberg, New York: Springer-Verlag 1982

Paraganglion aorticum abdominale: ↑ Zuckerkandl's organ

Paraganglion suprarenale: Synonym for adrenal medulla.

↑ Adrenal glands, medulla of

Parallel fiber: T-branched ↑ axon of a cerebellar ↑ granule cell. Two branches together measure about 3.5 mm in length and pass through dendritic arborization of about 350 successive ↑ Purkinje cells. There are about 200 000 P. per dendritic tree of a single Purkinje cell. P. are excitatory to ↑ dendrites of Purkinje cells.

↑ Cerebellum, neuronal circuits of

Paralutein cells: ↑ Theca lutein cells

Parametrium (tunica adventitia uteri*): Subperitoneal ↑ loose connective tissue of lateral sides of ↑ corpus of uterus continuous with tissue of ↑ broad ligaments. P. is rich in blood and lymphatic vessels and nerve fibers.

Paranasal sinuses (sinus paranasalis*): Cavities in frontal, ethmoidal, sphenoidal bones and maxilla that communicate with ↑ nasal cavity. Structure: Tunica mucosa (TM): a) epithelium (E) = ↑ pseudostratified ciliated epithelium with numerous ↑ goblet cells, continuous with epithelium lining ↑ respiratory mucosa; b) lamina propria (LP) = thin layer of poorly vascularized ↑ loose connective tissue with rare small ↑ mixed glands. Lamina propria is firmly attached to ↑ perios-

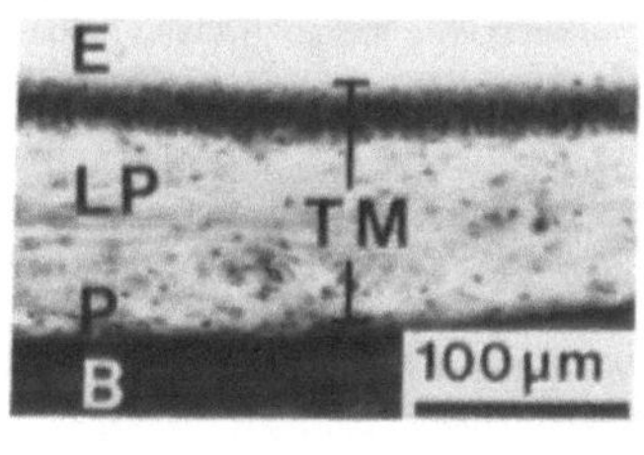

teum (P) of bone (B). (Fig. = sinus frontalis, human)

Schaeffer, J.P.: The Mucous Membrane of the Nasal Cavity and the Paranasal Sinuses. In: Cowdry, E.V. (ed.): Special Cytology, Vol. 1. New York: Paul B. Hoeber, Inc. 1982

Paraneurium: ↑ Loose connective and/or ↑ adipose tissue immediately surrounding a ↑ peripheral nerve.

↑ Epineurium

Paraneuron: A concept grouping categories of cells found outside the CNS which have some morphological, physiological, and metabolic properties in common with ↑ neurons. P. include cells which: 1) are able to produce substance(s) identical with or related to a ↑ neurotransmitter or suspected transmitter and protein/polypeptide substance(s) which may possess hormonal actions, 2) possess synaptic vesiclelike and/or neurosecretionlike granules, 3) release secretory product(s) in response to adequate stimulus of the receptor site of their cell membrane, and 4) originate in most cases in the same way as neurons. The following cells are classified as P.: ↑ chromaffin cells of adrenal medulla, ↑ SIF-cells of ↑ autonomic ganglia, type I ↑ glomus and ↑ paraganglion cells, ↑ C-cells of ↑ thyroid gland, ↑ pancreatic islet cells, ↑ endocrine cells of gastrointestinal, respiratory, and urogenital mucosa, sensory cells of ↑ taste buds, ↑ adenohypophyseal cells, ↑ parathyroid cells, ↑ hair cells of inner ear, ↑ photoreceptors, ↑ pinealocytes, ↑ Merkel's cells, ↑ melanocytes, and ↑ mast cells.

Fujita, T.: Biomed. Res. *4*, 239 (1983); Kanno, T. (ed.): Paraneurons, Their Features and Function. Amsterdam, Oxford, Princeton: Excerpta Medica International Congress Series 552, 1981; Kobayashi, S., Chiba, T. (eds.): Paraneurons: New Concept on Neuro-Endocrine Relatives. Arch. Histol. Jpn. *40*, suppl. (1977)

Paranodal region (paranode): Sum of cytoplasmic ↑ tongue processes (TP) and ↑ mesaxons (M) immediately bordering a ↑ node of Ranvier (NR). In P. each tongue process is connected to ↑ axolemma (A) by dense bars (arrowheads). Tongue processes correspond to thickened lateral edges of ↑ Schwann's cells (S) or ↑ oligodendrocyte processes. (Figs. = rat)

Wiley, C.A., Ellisman, M.H.: J. Cell Biol. *84*, 261 (1980)

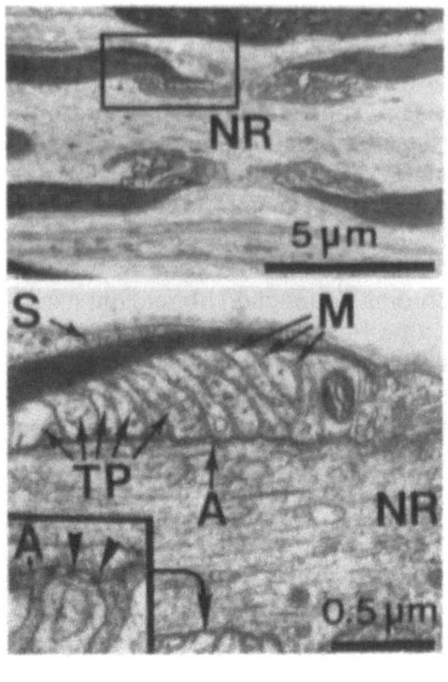

Paranode: ↑ Paranodal region

Paraoophoron*: Vestigial organ associated with ↑ ovary. P. is remnant of caudal part of mesonephros in form of irregularly arranged epithelial tubules within ↑ mesovarium. (See embryology texts for further information)

Paraplasm: ↑ Inclusions

Parasympathetic division: ↑ Parasympathetic nervous system

Parasympathetic ganglia: Ganglia belonging to ↑ parasympathetic nervous system. P. include: 1) Ganglia attached to cranial III, VII, IX, X, and XI and to 2nd, 3rd, and 4th sacral spinal nerves for parasympathetic innervation of neighboring organs; 2) numerous small peripheral or ↑ terminal ganglia scattered in all organs of neck, thorax, and abdomen. P. are structurally identical to ↑ autonomic ganglia.

Parasympathetic nervous system (craniosacral division, parasympathetic division, systema nervosum autonomicum, pars parasympathica*): One of two divisions of ↑ autonomic nervous system. Preganglionic neurons serving for parasympathetic innervation of head and trunk originate from nuclei of oculomotor and vagus nerves and nuclei salivatorii; those for pelvic organs merge from 2nd, 3rd, and 4th sacral segments of ↑ spinal cord. Innervation is effected via postganglionic neurons of ↑ terminal ganglia.

↑ Parasympathetic ganglia

Parasympathetic paraganglia: ↑ Paraganglia, parasympathetic

Paratendineum (paratenonium): Layer of ↑ loose connective tissue enveloping ↑ epitendineum. P. serves as sliding layer for ↑ tendons without ↑ synovial vaginae.

Paratenonium: ↑ Paratendineum

Parathyroid glands (glandulae parathyreoideae*): Four small ↑ endocrine glands intimately associated with ↑ thyroid gland. Structure: 1) Capsule (C) = thin layer of moderately ↑ dense connective tissue; 2) ↑ stroma = connective septa (S) extending into parenchyma from capsule; 3) ↑ parenchyma = epithelial cells arranged in irregular anastomosing cords (PC), dense cellular groups, and, less frequently, small follicles. Well-vascularized parenchyma consists of ↑ chief cells (CC) and ↑ oxyphil cells (OC). Third cell type, intermediate or transitional cells, are very probably only a functional modification of chief cells. P. produce ↑ parathyroid hormone and ↑ glutaurine. AC = ↑ adipose cells

↑ Parathyroid glands, control of activity

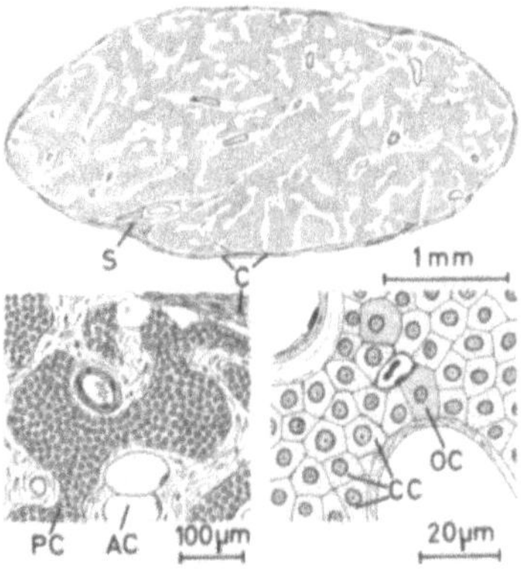

Aurbach, G.D. (ed.): Parathyroid Gland. Handbook of Physiology, Vol. 7. Am. Phys. Soc. Baltimore: Williams and Wilkins 1976; Coleman, R., Silberman, M., Bernheim, J.: Acta Anat. *106*, 424 (1980); Jones, C.T., Hunt, R.D. (eds.): Endocrine System. Berlin, Heidelberg, New York: Springer-Verlag 1983; Setoguti, T., Inoue, Y.: Cell Tissue Res. *228*, 219 (1983); Wild, P.: Acta Anat. *108*, 340 (1980)

Parathyroid glands, control of activity: Activity of ↑ parathyroid glands is controlled by level of Ca^{2+} in blood by a negative ↑ feedback mechanism. Decrease of Ca^{2+} in blood provokes increase in production and release of ↑ parathyroid hormone (PTH), which activates, via the ↑ adenyl cyclase-cAMP system, specific kinases of ↑ osteoclasts and a consecutive resorption of ↑ bone matrix. A high Ca^{2+} level in turn inhibits synthetic activity of parathyroid glands.

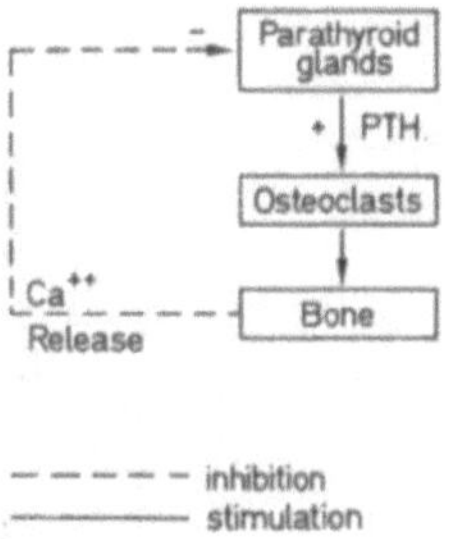

Dietel, M.: Funktionelle Morphologie und Pathologie der Nebenschilddrüsen. Stuttgart, New York: Gustav Fischer Verlag 1982

Parathyroid glands, principal cells of: ↑ Chief cells, of parathyroid glands

Parathyroid hormone (PTH): A polypeptide hormone (m.w. about 9000 daltons) synthesized by ↑ chief cells of ↑ parathyroid glands. P. raises level of blood calcium by: 1) Increasing absorption of Ca^{2+} in the intestine, 2) mobilizing of Ca^{2+} from bone by activation of ↑ osteoclasts, and 3) intensifying reabsorption of Ca^{2+} and excretion of phosphate by ↑ nephron. Action of P. is mediated by ↑ adenyl cyclase cAMP-system and probably by gene activation. (See physiology texts for further information)

↑ Glutaurine

Gray, T.K., Cooper, C.W., Munson, P.L.: Parathyroid Hormone, Thyrocalcitonin and the Control of Mineral Metabolism. In: McCann, S.M. (ed.): Endocrine Physiology. London: Butterworth 1974; Wong, E.T., Lindall, A.W.: Proc. Soc. Exp. Biol. Med. *148*, 387 (1975)

Paraventricular nucleus (nucleus paraventricularis*): Group of ↑ neurosecretory neurons (NPV) of ↑ hypothalamus situated immediately below ↑ ependyma of lateral walls of third ventricle (III). Axons of cells of P. end in ↑ neurohypophysis, carrying ↑ oxytocin by anterograde ↑ axoplasmic flow. By neural pathways, stimulation of ↑ nipple in suckling and dilatation of ↑ vagina stimulate neurons of P. to produce ↑ oxytocin. (Fig. = rat)

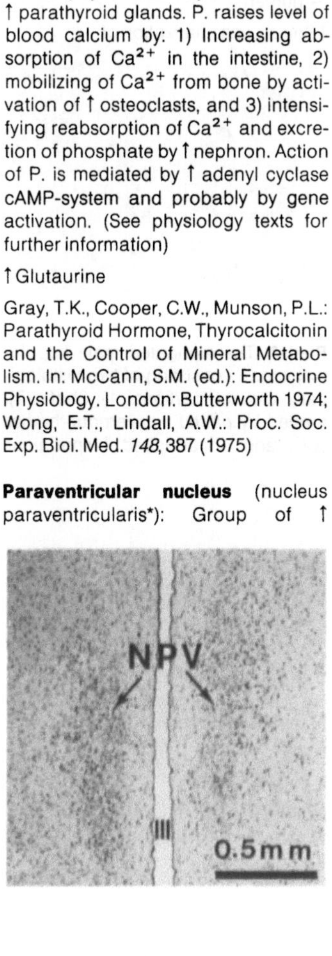

↑ Suckling reflex

Kiss, J.Z., Palkovits, M., Zaborsky, L., Tribollet, E., Szabo, D., Makara, G.B.: Brain Res. *265*, 11 (1983); Ono, T., Nishino, H., Sasaka, K., Muramoto, K., Yamo, I., Simpson, A.: Neurosci. Lett. *10*, 141 (1978); Panzica, G.C., Viglietti-Panzica, C., Contenti, E.: Cell Tissue Res. *227*, 79 (1982)

Paravertebral ganglia: ↑ Vertebral ganglia

Parenchyma: All elements making up specific morphological and functional part of an organ. P. is supported by a ↑ stroma which brings blood and lymphatic vessels as well as nerve fibers. Stroma divides P. into ↑ lobules and ↑ lobes. (See fig. under ↑ Stroma)

Parenchymal cells, of liver: ↑ Liver parenchymal cells

Parenchymal channels: ↑ Canaliculi, tissue

Parietal cells, of gastric glands proper (oxyntic cells, principal cells, cellula parietalis*): Large, well-delimited pyramidal or spheroidal acidophilic cells with bases bulging on outer surface of middle segment of ↑ gastric glands proper. Sometimes binucleate, P. contain numerous large elliptical ↑ mitochondria (M) with densely packed cristae and frequent ↑ mitochondrial granules, inconspicuous Golgi apparatus, few short rough endoplasmic cisternae, some smooth endoplasmic ↑ tubules, rare lysosomes, and free ribosomes. A ramified 1 to 2-µm-wide ↑ intracellular secretory canaliculus (SC), lined by numerous microvilli, extends as trenchlike invagination from apical cell surface, surrounds nucleus, and reaches almost to basal lamina (BL). A richly developed system of plasmalemmal invaginations forms in apical cytoplasm and around the canaliculus a network of smooth tubulovesicular profiles (T) with translucent contents. This system is abundant in resting cells; ↑ histamine, ↑ insulin, and/or direct vagal stimulation reduce its extent. Strong ↑ acidophilia of P. is based on abundance of mitochondria and smooth membranes. P. are connected by small ↑ desmosomes with ↑ chief cells (CC). P. synthesize hydrochloric acid by a not fully understood mechanism. It is thought that tubulovesicular profiles actively transport chloride ions across the cell. Liberted from reaction producing carbonic acid and catalyzed by carbonic anhydrase present in great concentration in P., H^+ ions cross plasmalemma by an active transport mechanism and unite with Cl^- ions to form 0.1N HCl.

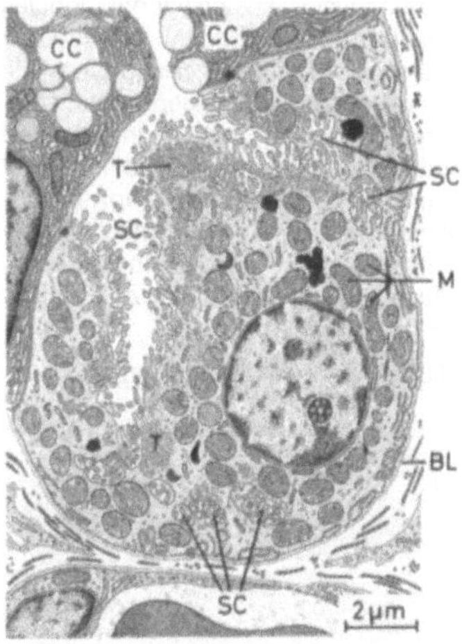

Dibona, R., Ito, S., Berglindh, T., Sachs, G.: Proc. Natl. Acad. Sci. USA *76*, 6689 (1979); Helander, H.F.: Int. Rev. Cytol. *70*, 217 (1981); Jacobs, D.M., Sturtevant, R.P.: Anat. Rec. *203*, 101 (1982); Osawa, W., Ogata, T.: Arch. Histol. Jpn. *41*, 141 (1978); Sato, A., Spicer, S.S.: Histochem. J. *13*, 495 (1981)

Parietal layer, of Bowman's capsule: ↑ Bowman's capsule

Parietal pleura: ↑ Pleura

Parotid acinar cells: ↑ Serous cells, of salivary glands

Parotid duct (Stenon's duct, ductus glandulae parotis*): Main excretory canal of ↑ parotid gland, about 4–6 cm long, opening into vestibulum of

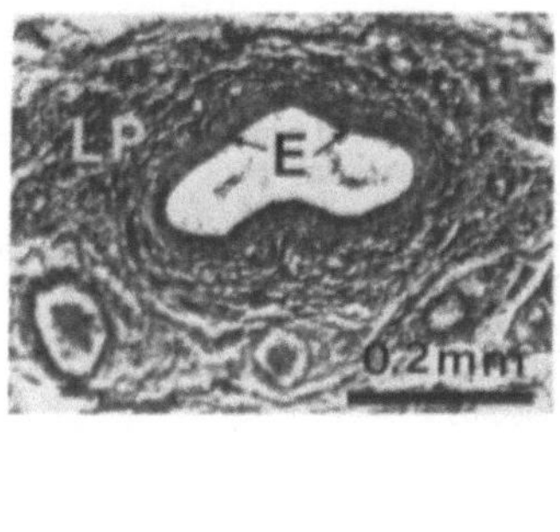

mouth. P. is lined by ↑ pseudostratified epithelium (E) overlying moderately dense and richly vascularized lamina propria (LP). (Fig. = human)

↑ Ducts, of salivary glands proper

Parotid glands (glandula parotis*): Compound ↑ tubuloacinar, purely serous ↑ salivary glands. Structure: 1) Connective capsule (C). 2) ↑ Stroma = connective septa penetrating into parenchyma dividing it into ↑ lobules (L). Through septa run blood and lympathic vessels and nerve fibers. 3) ↑ Parenchyma = polygonal lobules composed of ↑ acini (A) and variable number of ↑ adipose cells (AC). Acini, surrounded by ↑ myoepithelial cells, are connected with ↑ intercalated ducts (ID) that open into a ↑ striated duct (SD). Striated ducts empty into interlobular ducts (ILD) which finally unite to form ↑ parotid duct. P. produce watery component of ↑ saliva. (Figs. = human)

↑ Serous cell, of salivary glands

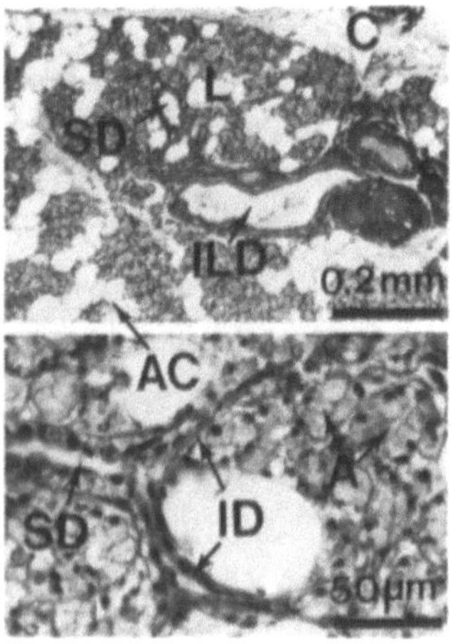

Ginsbach, G., Kühnel, W.: Anat. Embryol. *155*, 23 (1978); Mark, M.R., Sharawy, M., Pennington, C.: Scanning Electron Microscopy 1981/III, 137 (1981); Riva, A., Testa-Riva, F., Del Fiacco, M., Lantini, M.S.: J. Anat. *122*, 627 (1976)

Pars cardiaca: ↑ Cardiac area

Pars ciliaris retinae: ↑ Ciliary body

Pars distalis, of hypophysis: ↑ Hypophysis, pars distalis of

Pars fibrosa, of nucleolus: ↑ Nucleolus

Pars granulosa, of nucleolus: ↑ Nucleolus

Pars infundibularis, of hypophysis: ↑ Hypophysis, pars tuberalis of

Pars intermedia, of hypophysis: ↑ Hypophysis, pars intermedia of

Pars interstitialis, of oviduct: ↑ Oviduct

Pars maculata, of distal tubule: ↑ Macula densa

Pars membranacea, of urethra: ↑ Urethra, male

Pars nervosa, of hypophysis: ↑ Hypophysis, lobus nervosus of

Pars papillaris, of dermis: ↑ Dermis

Pars penile, of urethra: ↑ Urethra, male

Pars plana, of ciliary body: ↑ Ciliary body

Pars plicata, of ciliary body: ↑ Ciliary body

Pars prostatica, of urethra: ↑ Urethra, male

Pars reticularis, of dermis: ↑ Dermis

Pars spongiosa, of urethra: ↑ Urethra, male

Pars tuberalis, of hypophysis: ↑ Hypophysis, pars tuberalis of

Particles, intramembranous: ↑ Membrane-associated particles

PAS-reaction: ↑ Periodic acid-Schiff reaction

P-cells, of atrioventricular node: ↑ Nodal cells

Pectinate line: ↑ Anal canal

Pedicels (foot processes of podocytes, cytopodium*): Slender 1 to 3-µm-long and 0.2-µm-wide terminal extensions (P) of a ↑ podocyte adhering to external surface of ↑ glomerular basal lamina (BL). P. interdigitate in a complex manner with others from same podocyte and/or adjacent podocyte, resulting in very elaborate pattern of narrow clefts, ↑ slit pores (SP). All P. together form dense sieve around ↑ glomerular capillaries. E = ↑ endothelial cell (Figs. = rat)

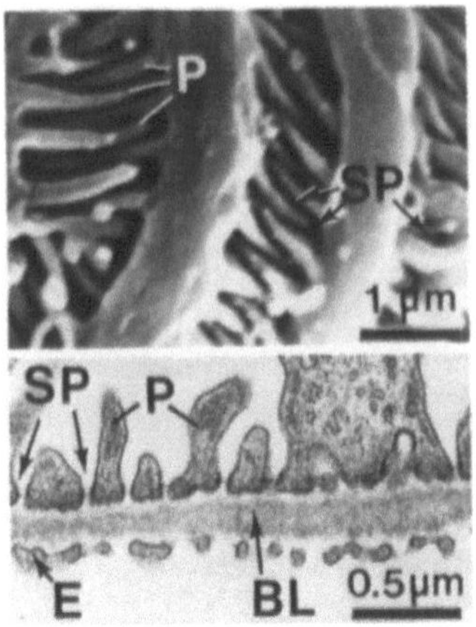

Penicillar arteries (arteries of red pulp, pulp arteries, penicilli, arteriolae penicillares*): Nonanastomosing slender branches (PA) of ↑ central artery (CA), about 25 µm in external diameter, located in ↑ red pulp (RP) of ↑ spleen. P. give off many sheathed ↑ capillaries. SN = ↑ splenic nodule (Fig. = human)

↑ Spleen, vascularization of

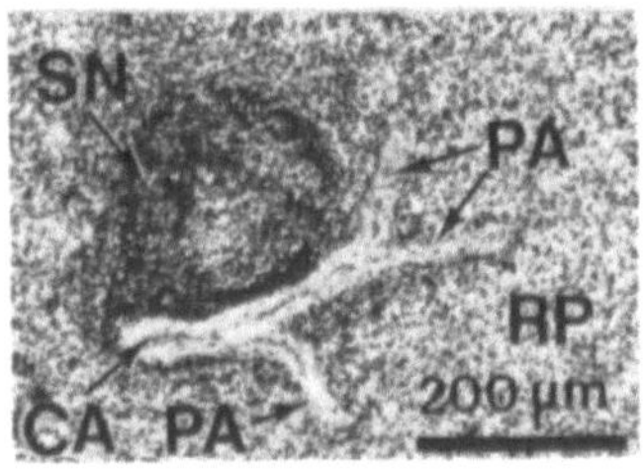

Penis*: The elongated male copulatory organ consisting of two histologically distinct portions: 1) Shaft (T), formed by paired ↑ cavernous bodies (CB) and an unpaired ↑ spongious

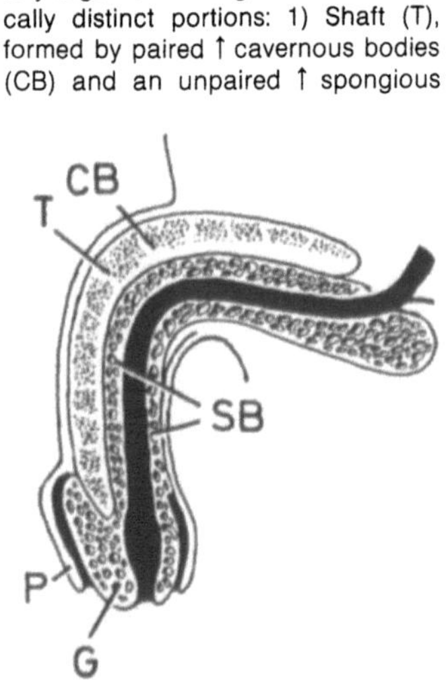

body (SB); 2) ↑ glans penis (G), formed by spongious body only. P = ↑ prepuce

↑ Penis, shaft of

Penis, cavernous bodies of: ↑ Cavernous bodies, of penis

Penis, erection of: ↑ Penis, vascularization of

Penis, innervation of: 1) Sympathetic innervation = nerve fibers from pelvic sympathetic system form nonmyelinated network among ↑ smooth muscle cells of trabeculae of ↑ corpora cavernosa and tunica media of ↑ helicine arteries; sympathetic vasoconstrictor impulses terminate ↑ erection. 2) Parasympathetic innervation = sacral plexus via pudendal nerve; nerve fibers originate from S_2–S_4 segment of ↑ spinal cord. Parasympathetic system is responsible for erection. 3) Sensory innervation = free nerve endings in epithelium of ↑ glans, prepuce, and urethra. ↑ Corpuscles of Meissner, Vater-Pacini, and genital ↑ corpuscles are present in ↑ dermis of glans.

Penis, lymphatics of: Present in ↑ skin of ↑ prepuce and shaft. ↑ Lymph of ↑ glans is collected in deep ↑ lymphatic capillaries that drain into dorsal superficial lymphatic vessels. Lymph flows to superficial inguinal ↑ lymph nodes.

Penis, shaft of (corpus penis*): P. consists of two ↑ cavernous bodies (CB) and one ↑ spongious body (SB) enclosed by ↑ tunica albuginea (TA) and surrounded by fascia penis (FP). Tunica albuginea enveloping spongious body is considerably thinner than that of cavernous bodies. Between tunica albuginea and fascia penis are situated arteria, vena and nervi dorsalis penis (AVNdp). Over fascia penis and loosely attached is almost hairless skin (S) characterized by pigmented basal cells of ↑ epidermis (E), large numbers

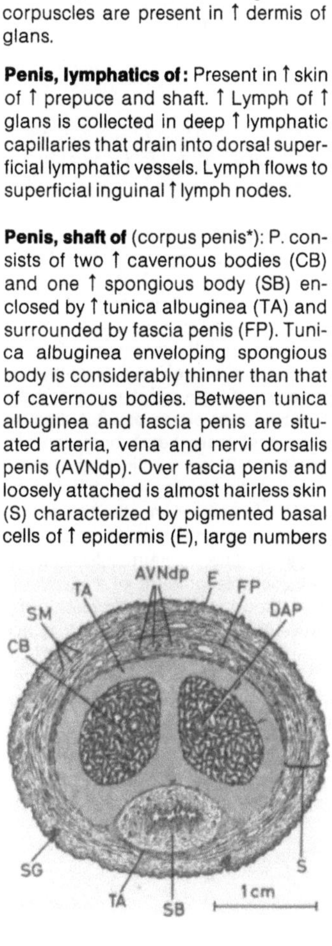

of ↑ corpuscles, scattered ↑ sebaceous (SG) and ↑ sweat glands, and abundance of ↑ smooth muscle cells (SM). DAP = ↑ deep artery, of penis

Penis, spongious body of: ↑ Spongious body, of penis

Penis, vascularization of: 1) ↑ Cavernous bodies. From ↑ deep artery of penis (A), blood flows in three directions: a) into arteriovenous ↑ anastomoses (An) which conduct blood directly into efferent veins (EV); b) into

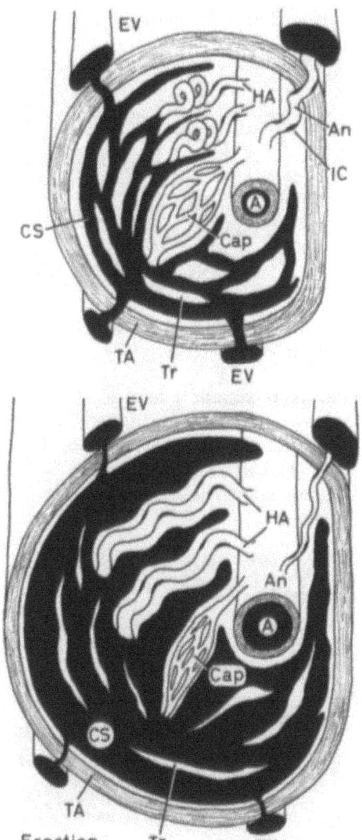

coiled ↑ helicine arteries (HA) which empty blood into ↑ cavernous sinuses (CS) (a and b = functional circulation); c) into short nutrient arteries which split into a capillary network (Cap) vascularizing trabeculae (Tr) and draining into sinuses. From here, blood is collected by efferent veins (EV) piercing ↑ tunica albuginea (TA). 2) ↑ Spongious body. Richly branched plexus of veins with ↑ intimal cushions. In flaccid state, blood flows almost exclusively through arteriovenous anastomoses into efferent veins, and only in small quantities into cavernous sinuses through nutrient capillary network, since helicine arteries are coiled and reversibly closed. In ↑ erection, under influence of parasympathetic vasodilatator nerves of sacral spinal cord (S_2–S_4), helicine arteries uncoil and dilate and blood fills sinuses under high pressure. Simultaneously, arteriovenous anastomoses and capillary network constrict and become compressed, leading to almost total interruption of blood outflow. Since tunica albuginea is only weakly extensible, cavernous bodies become enlarged and very rigid. During erection, spongious body is also filled with an increased amount of blood, but venous circulation continues freely and the spongious body remains less turgid, allowing ↑ semen to pass urethral lumen during ↑ ejaculation. According to recent research, intimal cushions (IC), if they exist in some vessels of cavernous bodies, do not play a role during erection.

Conti, G.: Acta Anat. *14*, 217 (1952); McConnel, J., Benson, G.S., Schmidt, W.A.: Anat. Rec. *203*, 475 (1982); Newman, H.F., Northrup, J.D., Devlin, J.: Invest. Urol. *1*, 350 (1964)

Pepsin: Proteolytic enzyme formed from ↑ pepsinogen under influence of gastric juice.

Pepsinogen (propepsin): A precursor of ↑ pepsin; secreted by ↑ chief cells of ↑ gastric glands proper.

Peptic cells: ↑ Chief cells, of gastric glands proper

Peptidases: Large group of enzymes which split peptide bond (–NH–CO–).

Gossrau, R.: Verh. Anat. Ges. *74*, 163 (1980)

Perforating fibers: ↑ Sharpey's fibers

Perforating sinus, of lymph node: ↑ Lymph node; ↑ Lymph node, sinuses of

Perfusion fixation: A technique of chemical ↑ fixation consisting of introduction of fixative solution into vascular system of experimental animal or isolated organ. Principle: A needle (N) is introduced into left ventricle (LV) or ↑ aorta (A) of anesthetized animal, followed by opening of right auricle (RA). At the same moment, washing of vascular system of blood with oxygenated Ringer's solution (R), is started. A few minutes later, this is replaced with a fixative solution (F). Rinsing and fixative fluids circulate in same direction as blood; both flow out from heart through opening (O) in right auricle (arrow). Well-perfused animal becomes stiff; small pieces are then cut and left for 1 h in same fixative. For P. of organs with important venous circulation (liver, lung), the needle is introduced into the arteria pulmonalis (AP), followed by opening of left auricle (LA). In the case of isolated organs, the needle is introduced into main arterial vessel. Advantages: Excellent preservation of cells, tissues, and organs with conservation of their exact anatomical position. Disadvantages: Applicable only in well-vascularized tissues and organs; more or less complicated apparatus. Since intracardiac P. requires opening of the chest and consecutive pneumothorax, it is useful to assure an artificial respiration to the experimental animal in order to prevent hypoxemia. Several techniques have been developed for P. of specific organs. CA = circulation in head and anterior members; CP = circulation in the abdomen and posterior members; L = lungs; T = three-way tap; BB = bubble trap

↑ Perfusion fixation, retrograde

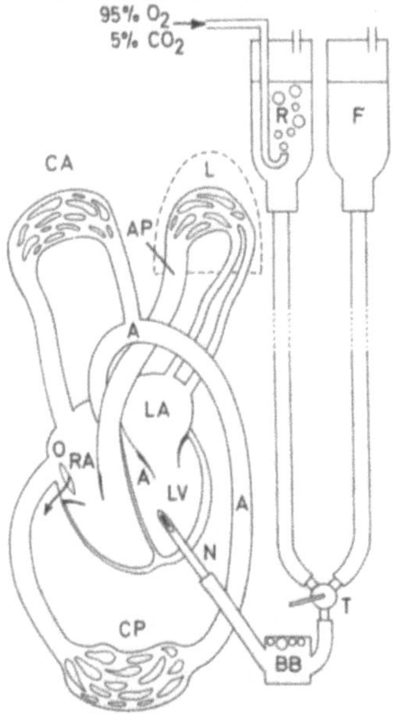

Coalson, J.J.: Anat. Rec. *205*, 233 (1983); Forssmann, W.G., Ito, S., Aoki, A., Dym, M. Fawcett, D.W.: Anat. Rec. *188*, 307 (1977)

Perfusion fixation, retrograde: A kind of ↑ perfusion fixation, consisting of introduction of a needle into abdominal ↑ aorta with consecutive opening of ↑ vena cava inferior. Rinsing and fixative solutions circulate in direction opposed to that of blood. P. avoids opening of thorax and prevents pneumothorax, so that the experimental animal respires up to moment of introduction of fixative.

Periarterial lymphatic sheath (PALS, vagina periarterialis lymphatica*): Cylindrical meshwork (P) of ↑ reticular cells around ↑ central artery (CA) of ↑ spleen. Small ↑ T-lymphocytes, located at periphery of P., predominate over ↑ macrophages, ↑ plasma cells, and other free cells lodged in meshwork. Immediately along central artery are located some ↑ B-lymphocytes. Scattered within P. are ↑ splenic nodules (SN). With decrease in caliber of central artery, P. attenuates. Peripheral to P. is ↑ marginal zone (MZ). Splenic nodules and P. form ↑ white pulp of ↑ spleen. (Fig. = human)

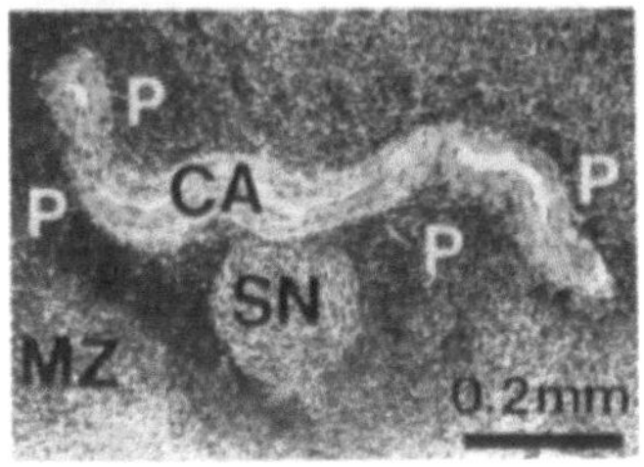

Janout, V., Weiss, L.: Anat. Rec. *172*, 197 (1972)

Periaxial space: Interstice (PS) between outer (OC) and inner capsule (IC) of ↑ neuromuscular spindle (NS). P. is filled with ↑ tissue fluid; intrafusal ↑ nerve fibers run through it. (Fig. = human)

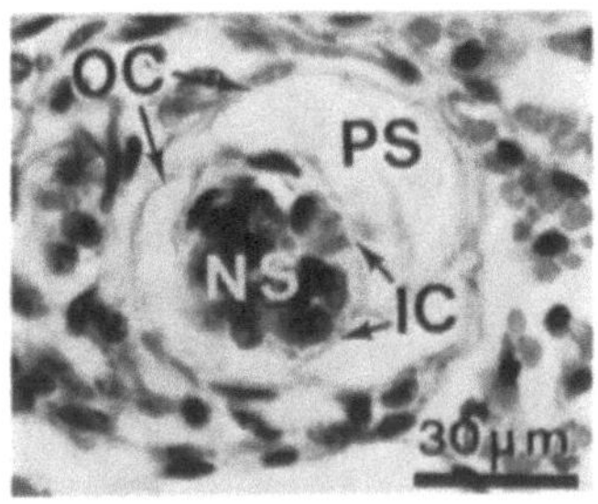

Periaxonal space: A gap (PS), about 20 nm wide, separating ↑ Schwann's cell (SC) or ↑ oligodendrocyte from ↑ axon (A). MS = ↑ myelin sheath (Fig. = rat)

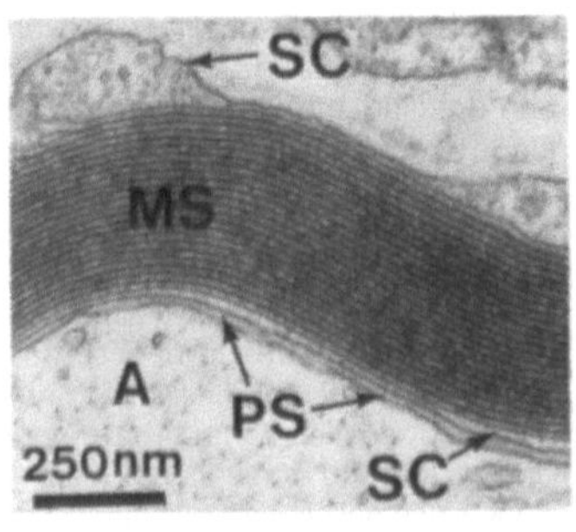

↑ Myelin-associated glycoprotein

Pericapillary space: Terminal extensions (P) of a ↑ perivascular space surrounding blood or lymphatic ↑ capillaries (Cap). Although ill-defined in ↑ loose connective tissue, P. in some tissues and glands is delimited by capillary basal lamina and basal lamina covering cells of respective tissue (↑ adipose tissue, ↑ muscular tissues, ↑ endocrine and ↑ exocrine glands). P. contain ↑ tissue fluid, ↑ reticular and ↑ collagen fibrils, ↑ fibrocytes, ↑ nonmyelinated nerve endings, and ↑ wandering cells. Width of P. is 0.5–15 µm; it is particularly large in stroma of ↑ iris and absent in area of ↑ blood-brain barrier and ↑ glomerular capillaries. P. can anastomose with tissue ↑ canaliculi. (Fig. = exocrine ↑ pancreas, rat)

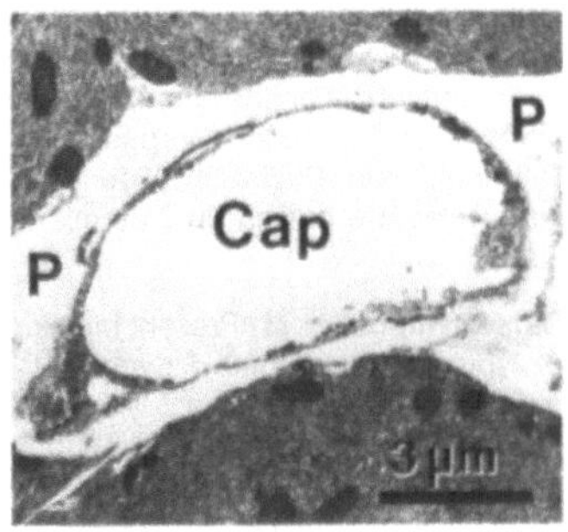

Pericardium*: Fibroserous sac around heart composed of two sheets: 1) Visceral sheet = ↑ epicardium*. 2) Parietal sheet = P. proper: a) Pericardium serosum* or tunica serosa* = ↑ mesothelium (M) overlying thin layer of ↑ loose connective tissue. Mesothelium is reflected onto great veins and arteries as they pierce parietal sheet to enter and leave heart; here, the two sheets of P. are continuous. b) Pericardium fibrosum* or tunica fibrosa (TF) = ↑ dense connective tissue consisting of several crossed strata of ↑ collagen fibers intermingled with a considerable amount of ↑ elastic fibers. Blood and lymphatic vessels and nerve fibers are also present. Space

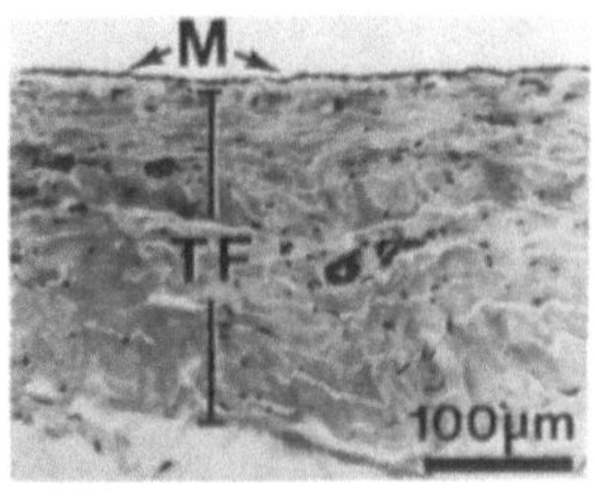

between two sheets of P., cavum pericardii, contains some serous fluid, liquor pericardii, produced by mesothelium to facilitate sliding of the sheets. (Fig. = human)

Pericentriolar material: ↑ Satellites, of centriole

Perichondrial bone formation: ↑ Bone formation, perichondrial

Perichondrium*: Layer of richly vascularized ↑ dense connective tissue (P) surrounding ↑ cartilage (C). It is absent over articular ↑ cartilage and fibrous ↑ cartilage. P. originates from ↑ mesenchyme surrounding area in which cartilage develops: Outer layer of ↑ mesenchymal cells differentiates into ↑ fibroblasts and ↑ fibrocytes, producing dense ↑ collagen fibers of fibrous layer. Mesenchymal cells of inner or chondrogenic layer differentiate into ↑ chondroblasts and ↑ chondrocytes which synthesize ↑ cartilage matrix. In adult, only fibrous layer persists. Cartilage is nourished by diffusion from vessels of P. (Fig. = ↑ trachea, human) ↑ Cartilage canals

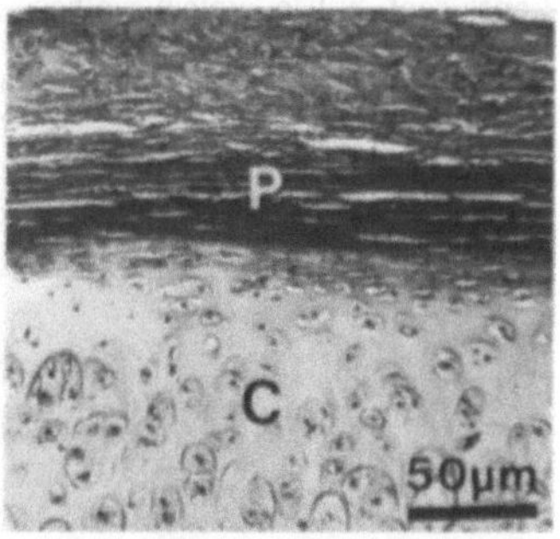

Perichoroidal space (spatium perichoroidale*): A virtual cleft between ↑ sclera and ↑ lamina suprachoroidea of ↑ choroid.

Perichromatin fibers: Filiform structures, 3 to 5-nm-thick, forming small bundles peripheral to ↑ heterochromatin masses. P. consist of ↑ ribonucleoproteins and ↑ RNA; they are thought to represent precursors of ↑ perichromatin and ↑ interchromatin granules.

Perichromatin granules: Round membraneless electron-dense structures (arrows), 40–45 nm in diameter, surrounded by a 25-nm-wide clear halo. P. are relatively rare and are found at the boundary between ↑ euchromatin (E) and ↑ heterochromatin (H). They consist of fibrils, about 3 nm thick, that connect them in some cases with neighboring P. Main substances of P. are ↑ ribonucleoproteins; it is believed that P. are linked to ↑ chromatin and involved in transmission of information from DNA. (Fig. = ↑ fibroblast, rat)

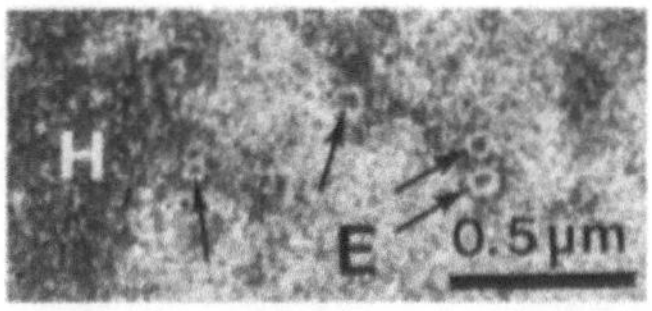

Paiement, J., Bendayan, M.: J. Ultrastruct. Res. *81*, 145 (1982); Puvion-Dutilleul, F., Puvion, E.: J. Ultrastruct. Res. *74*, 341 (1981)

Pericranium: ↑ Bone, flat, of skull

Pericryptal fibroblasts: Flattened connective cells (PF) immediately adjacent to external aspect of basal lamina of epithelial cells of ↑ Lieberkühn's crypts (LC). P. are thought to be involved in continuous upward migration of epithelium. (Figs. = ↑ colon, human)

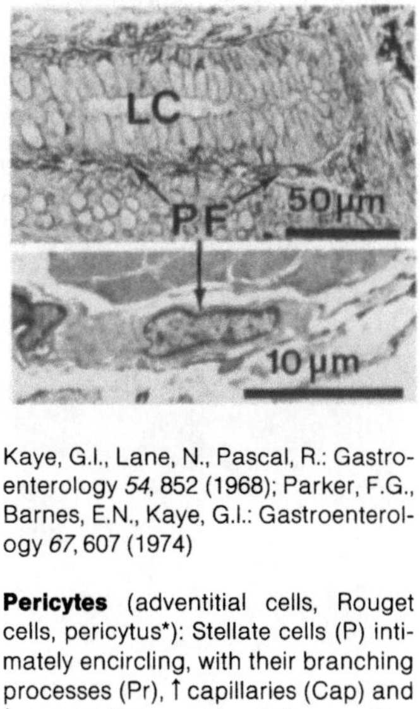

Kaye, G.I., Lane, N., Pascal, R.: Gastroenterology *54*, 852 (1968); Parker, F.G., Barnes, E.N., Kaye, G.I.: Gastroenterology *67*, 607 (1974)

Pericytes (adventitial cells, Rouget cells, pericytus*): Stellate cells (P) intimately encircling, with their branching processes (Pr), ↑ capillaries (Cap) and ↑ postcapillary venules. P. have a relatively large nucleus rich in ↑ heterochromatin, a few small mitochondria, an inconspicuous Golgi apparatus, several flattened rough endoplasmic cisternae, occasional lysosomes, microfilaments, and a moderate number of free ribosomes. P. are located between leaflets of capillary ↑ basal lamina; with fine foot processes, they make contact with endothelial cells through ↑ maculae occludentes. Role of P. in regulation of caliber of capillaries has not been accurately demonstrated; it is believed that they represent reserve cells capable of ↑ differentiating into ↑ macrophages. (Fig. = ↑ hypodermis, rat)

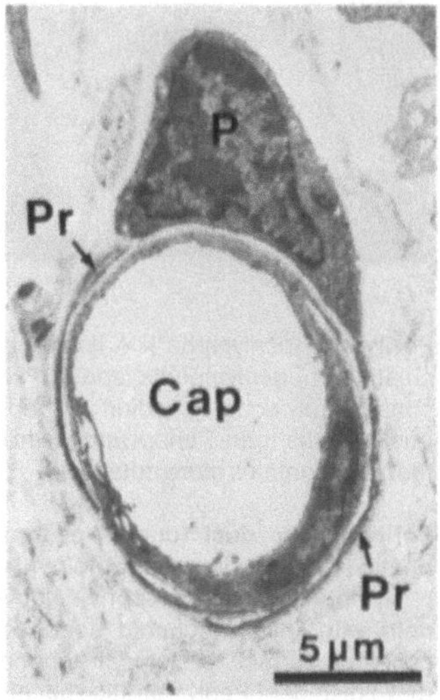

Murakami, M., Sugita, A., Shimada, T., Nakamura, K.: Arch. Histol. Jpn. *42*, 297 (1979)

Pericytic venule: ↑ Postcapillary venule

Peridental ligament: ↑ Periodontal ligament

Perigemmal cells, of taste buds: ↑ Taste buds, structure of

Perikaryon: Portion of ↑ cytoplasm with ↑ organelles surrounding ↑ nucleus (N). Term mostly used in connection with ↑ neurons.

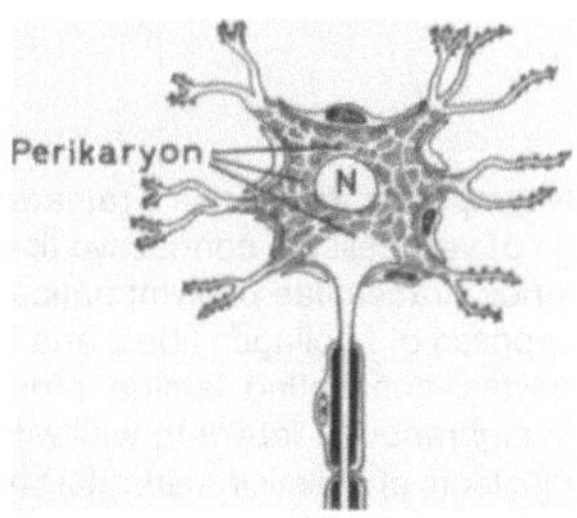

Perikymata: A succession of minute circular ridges and depressions (P) visible on external surface of ↑ enamel of young teeth. P. represent appositional growth patterns of enamel. (Fig. = human)

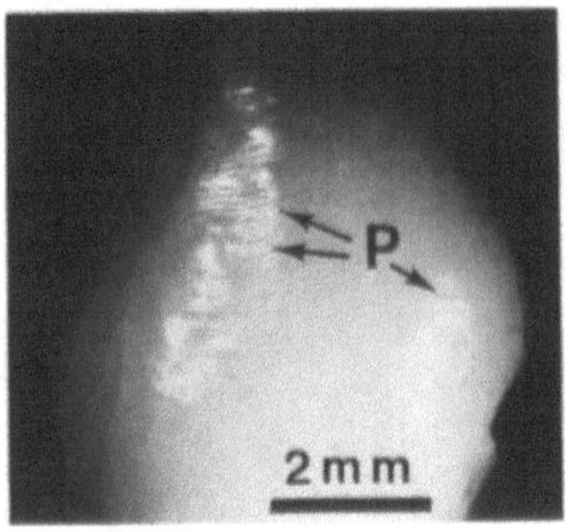

Perilymph (perilympha*): A liquid circulating in ↑ perilymphatic space. P. is considerably richer in sodium chloride and proteins than ↑ endolymph, while the latter contains more potassium.

Perilymphatic duct (ductus perilymphaticus*): Fine bony canal joining ↑ perilymphatic space of bony ↑ labyrinth with ↑ subarachnoid space of posterior surface of temporal bone. (See fig. under ↑ Perilymphatic space)

Perilymphatic space (spatium perilymphaticum*): Space (PS) situated between membranous ↑ labyrinth (ML) and bony ↑ labyrinth (BL) containing ↑ perilymph and, in area of ↑ semicircular ducts (SD) and utricle (U), ↑ perilymphatic tissue (PT). PD = ↑ perilymphatic duct

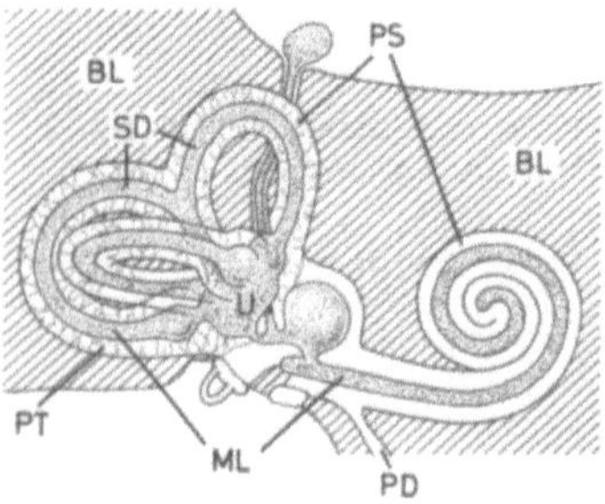

Perilymphatic tissue: ↑ Framework (PT) of very delicate connective tissue strands (trabeculae perilymphaticae*) composed of ↑ collagen fibers and ↑ fibrocytes, connecting lamina propria of membranous ↑ labyrinth wall with ↑ endosteum of ↑ semicircular ducts and ↑ utricle. ↑ Perilymph circulates through meshes of P. (See fig. under ↑ Perilymphatic space)

Perimetrium*: ↑ Tunica serosa of ↑ uterus structurally corresponding to ↑ peritoneum.

Perimysium*: Sheath of ↑ loose connective tissue (P) enveloping each ↑ muscle fascicle (F) of a ↑ skeletal muscle. P. consists of delicate collagen and elastic fibrils, some ↑ fibroblasts, and ↑ wandering cells. Through P. run ↑ arterioles, ↑ venules, ↑ lymphatic capillaries, and ↑ nerve fibers; within it are situated ↑ neuromuscular spindles. P. allows some independent movement to fascicles. (Fig. = human newborn)

↑ Endomysium

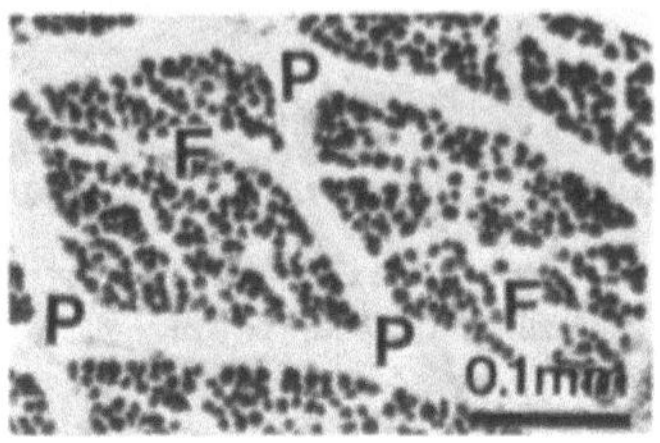

Perineurial endothelium (perineurial epithelium, pars epithelialis perineurii*): A layer of several concentrically arranged flattened cells forming part of ↑ perineurium. P. cells have elongated and attenuated nucleus; in addition to usual ↑ organelles, cytoplasm contains many ↑ micropinocytotic vesicles (MV). Cells are held together by numerous ↑ zonulae occludentes and ↑ nexus. Every cell layer is sandwiched between ↑ basal laminae (BL); in narrow spaces between cell layers run longitudinal ↑ collagen microfibrils (C). P. constitutes part of ↑ blood-nerve barrier. (Fig. = rat)

↑ Nerve fascicle

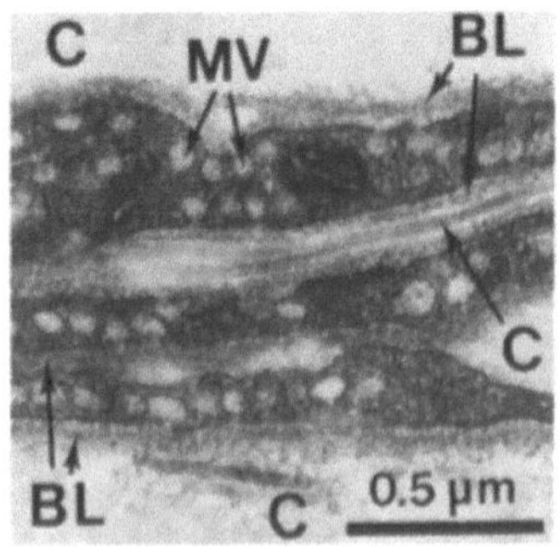

Akert, K., Sandri, C., Weibel, E.R., Peper, K., Moor, H.: Cell Tissue Res. *165*, 281 (1976)

Perineurium*: Concentric layer of ↑ dense connective tissue (P) enclosing each ↑ nerve fascicle (F). P. consists of ↑ perineurial endothelium and longitudinally oriented ↑ collagen fibrils. Toward end of nerve fascicle, P. has openings through which ↑ nerve fibers (NF) penetrate into tissue. E = ↑ endoneurium. (Fig. = ↑ hypodermis, rat)

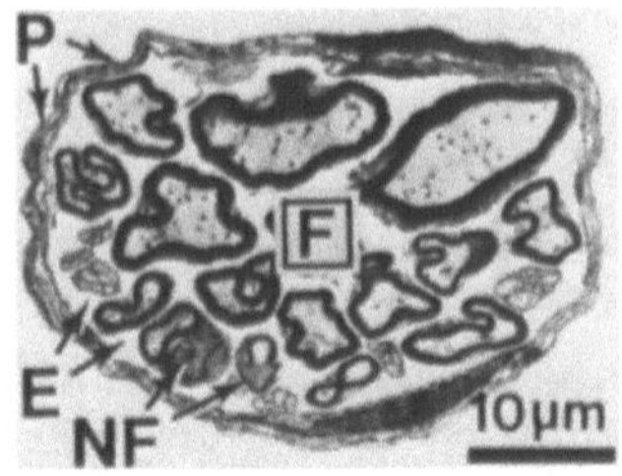

Shinowara, N.L., Michel, M. E., Rapoport, S.I.: Cell Tissue Res. *227*, 11 (1982)

Perineuronal satellite cells: ↑ Oligodendrocytes

Perinuclear cisterna (perinuclear space, cisterna karyothecae*): Cavity (PC) of ↑ nuclear envelope, about 15–25 nm wide. P. is delimited by inner ↑ nuclear membrane (M) and membrane of rough ↑ endoplasmic reticulum (R) in most cells. P. communicates with cisternae of endoplasmic reticulum (arrowheads). P. is generally translucent, but in protein-synthesizing cells it may contain a fine granular material identical to ↑ reticuloplasm (Rp). In antibody-producing cells (↑ plasma cells), antibodies start to accumulate within P., which is considered an integral part of endoplasmic reticulum. (Fig. = ↑ fibroblast, rat)

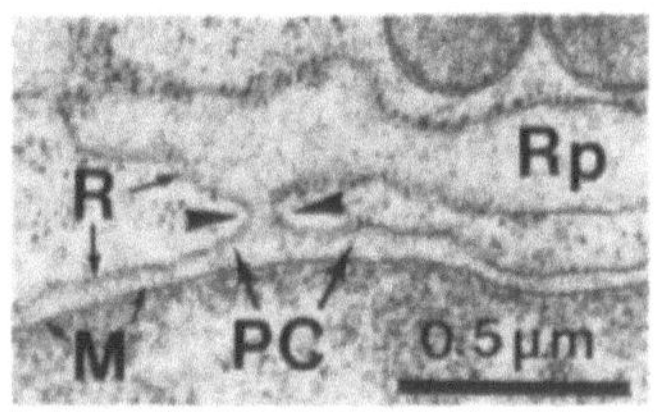

Perinuclear space: ↑ Perinuclear cisterna

Perinucleolar chromatin: ↑ Chromatin, perinucleolar

Periodate-leukofuchsin reaction: ↑ Periodic acid-Schiff reaction

Periodic acid–Schiff reaction (PAS-reaction, periodate-leukofuchsin reaction): Staining technique for histochemical demonstration of carbohydrate-rich macromolecules, such as ↑ glycogen, mucoproteins, ↑ glycoproteins, ↑ proteoglycans, etc. Periodic acid (HIO_4) oxidizes hydroxyl groups on two adjacent carbon atoms (arrows) and converts them to aldehydes. Upon contact with these aldehyde groups, ↑ Schiff's reagent produces stable red or magenta reaction product.

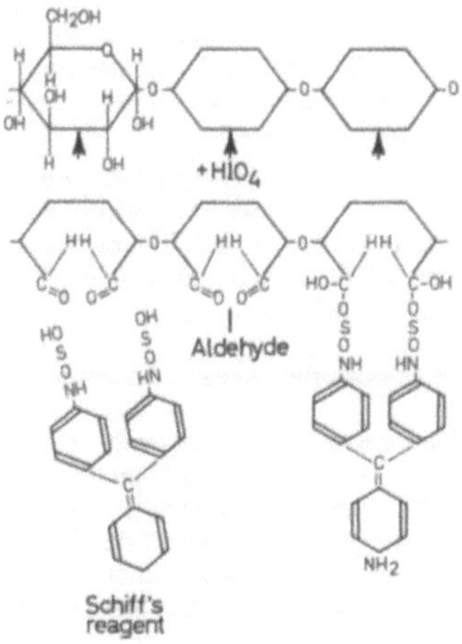

Periodontal ligament (periodontal membrane, peridental ligament, ligamentum periodontale*): ↑ Periosteum (PL) of ↑ alveolar bone, (AB) 0.1–0.3 mm thick, assuring firm contact between ↑ teeth and ↑ alveoli. P. is composed of densely packed ↑ collagen fibers and ↑ Sharpey's fibers (SF), which are embedded on one side in alveolar bone and on other side in ↑ cementum (C). Fibers of P. have slightly wavy course through ↑ periodontal space, permitting certain limited movements of the teeth; fibers also act as suspensory ligament, since most are oriented obliquely toward roots of teeth. Some ↑ collagen microfibrils do not have cross-striations; it has been shown that P. has fastest turnover of all collagen fibers of organism. There are no true ↑ elastic fibers in P., but rather a presence of ↑ oxytalan fibers. ↑ Fibroblasts and ↑ fibrocytes are common; occasional ↑ cementicles can be found. Blood and lymphatic vessels (V) are numerous and parallel to roots of teeth; they are accompanied by ↑ nerve fibers and are embedded in a

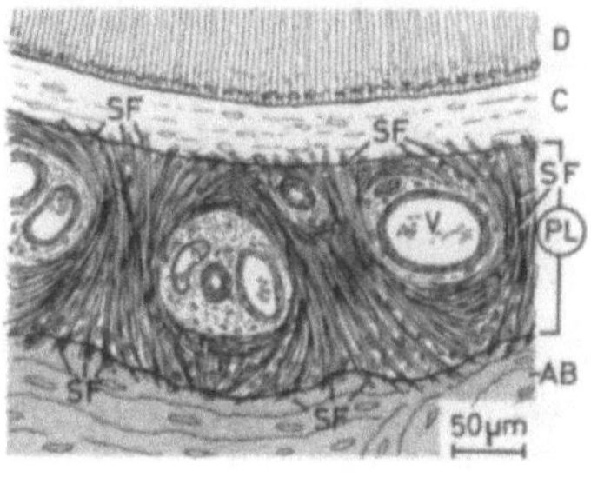

small amount of ↑ loose connective tissue. D = ↑ dentin (Fig. = cross section)

↑ Gomphosis

Beertsen, W., Brekelmans, M., Everts, V.: Anat. Rec. *192*, 305 (1978)

Periodontal membrane: ↑ Periodontal ligament

Periodontal space: ↑ Circular gap between ↑ alveoli of teeth and their roots, occupied by ↑ periodontal ligament.

↑ Periodontium

Periodontium*: A collective term for ↑ alveolar bone (AB), ↑ alveoli of teeth (A), ↑ periodontal ligament (PL), separations between alveoli and roots of teeth, ↑ Sharpey's fibers (= apical fibers, AF; oblique fibers, SF; alveolar crest fibers, ACF), and interdental fibers (not shown). C = ↑ cementum, Ce = ↑ cementicles, PS = ↑ periodontal space, V = blood vessels

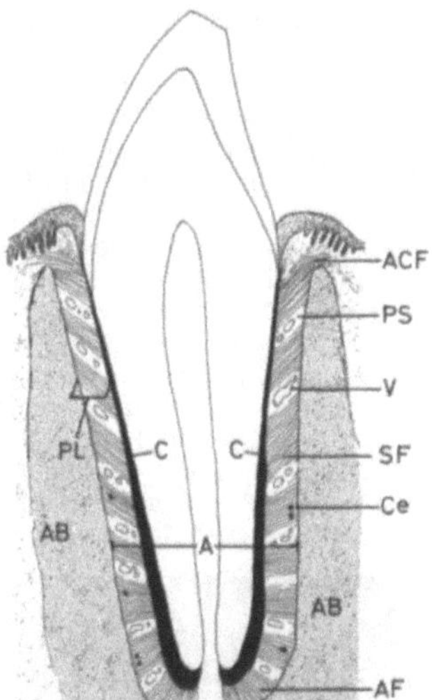

Halackova, Z., Ourdan, L., Kukletova, M.: Histochem. *67*, 173 (1980); Itoh, K. Wakita, M., Kobayashi, S.: Arch. Histol. Jpn. *44*, 453 (1981); Narayanan, A.S., Page, R.C.: Collagen Rel. Res. *3*, 33 (1983)

Periosteal bony band (bony collar, mid-diaphyseal ring, periosteal collar,

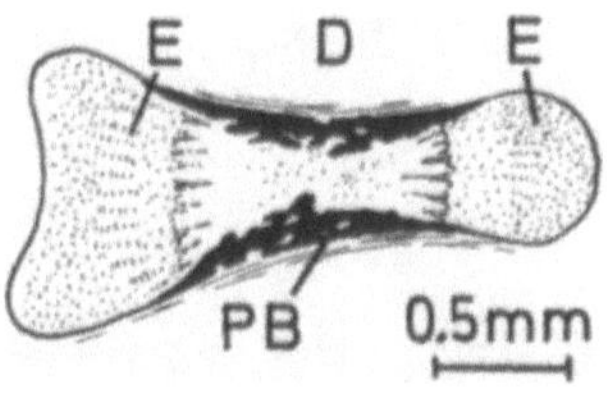

annulus osseus perichondralis*): Initial ring of membranous ↑ bone (PB) around midportion of ↑ diaphysis (D) of cartilaginous model. P. precedes indirect ↑ bone formation and has function of assuring firm connection between two ↑ epiphyses (E).

Periosteal bud (osteogenic bud, gemma osteogenica primaria*): Vascular sprout that perforates sleeve of ↑ periosteal bony band (PB). P. arises from inner layer of ↑ periosteum (Pe) and consists of a capillary loop (CL) accompanied by ↑ osteoprogenitor cells, ↑ osteoblasts (Ob), ↑ osteoclasts (Oc), primitive ↑ bone marrow cells (PMC), and ↑ macrophages. Through pulsation of capillary loop, as well as action of multinuclear endothelial structure (ES) with osteolytic properties on tip of loop and osteoclasts, P. perforates periosteal bony band and penetrates ↑ diaphysis of cartilaginous model at beginning of indirect ↑ bone formation.

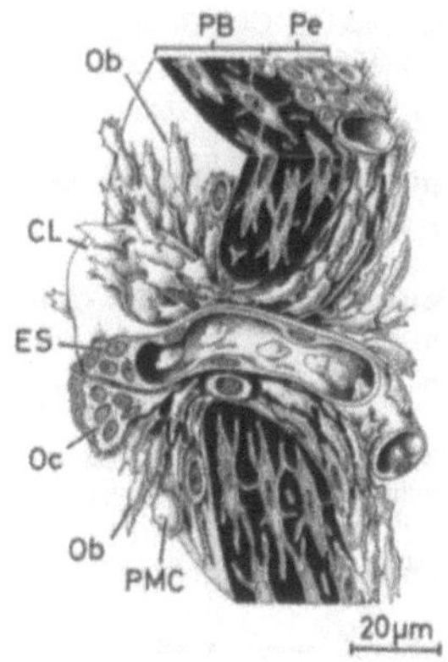

Periosteal collar: ↑ Periosteal bony band

Periosteum*: Thick layer of specialized ↑ dense connective tissue covering entire surface of bone, except ↑ articular cartilages and insertion sites of ↑ ligaments and ↑ tendons. Structure (in growing and young bones): 1) Inner layer, ↑ cambium or osteogenic layer (IL) = a loosely arranged and well-vascularized layer of ↑ osteoblasts (Ob); 2) middle layer (ML) = undifferentiated ↑ osteoprogenitor cells (Op); 3) outer or fibrous layer (OL) = irregularly arranged ↑ collagen and ↑ elastic fibers, some flattened ↑ fibroblasts, ↑ nerve fibers, and blood vessels. During ↑ bone formation, P. is responsible for ↑ bone's appositional growth. In adult bones, cambium is lacking, but some osteoprogenitor cells remain and differentiate into osteoblasts during ↑

bone repair. Fibrous layer in adult bones is thickened. P. is connected by ↑ Sharpey's fibers to bone. A well-developed sensory innervation is present, along with scattered tactile ↑ corpuscles. Through ↑ Volkmann's canals, P. provides blood supply to a great portion of cortex of long bones.

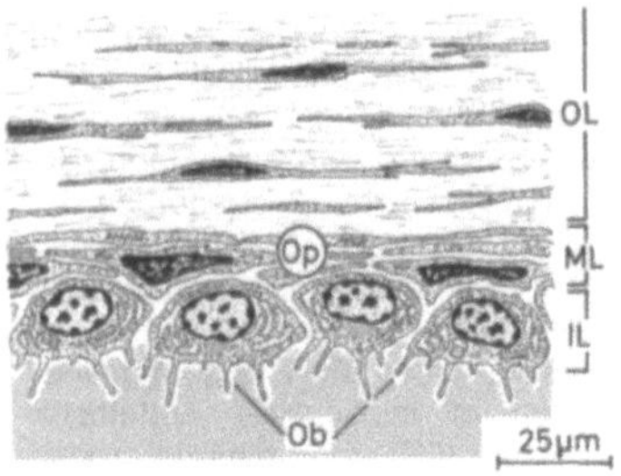

Knese, K.-H.: Stützgewebe und Skelettsystem. In: Möllendorff, W.v., Bargmann, W. (eds.): Handbuch der mikroskopischen Anatomie des Menschen, Vol. 2: Die Gewebe, Part 5. Berlin, Heidelberg, New York: Springer-Verlag 1979

Peripheral cells, of taste buds: ↑ Taste buds, structure of

Peripheral cells, of sweat glands: ↑ Sweat glands, eccrine, excretory duct cells of

Peripheral ganglia, satellite cells of: ↑ Satellite cells, of peripheral neurons

Peripheral nerve: ↑ Nerve, peripheral

Peripheral nerve endings: ↑ Nerve endings, peripheral

Peripheral nerve terminals, of autonomic nerve fibers: ↑ Adrenergic nerve fibers; ↑ Cholinergic nerve fibers; ↑ Myenteric plexus; ↑ Submucosal plexus; ↑ Sympathetic ground plexus

Peripheral nervous system (PNS, systema nervosum periphericum*): All elements of ↑ nervous system situated outside ↑ membrana gliae limitans superficialis.

Oksche, A., Vollrath, L. (eds.): Handbuch der mikroskopischen Anatomie des Menschen. Vol. 4, Part 1, Nervengewebe. Das peripherische Nervensystem. Das Zentralnervensystem. Berlin, Heidelberg, New York: Springer-Verlag 1978

Peripolesis: 1) Penetration of ↑ wandering cells between closely apposed fixed cells of a tissue. 2) Tendency of ↑ lymphocytes in ↑ cell culture to move away from each other.

↑ Emperiopolesis

Periportal tissue space (Mall's space): A lymphatic space (PS) between connective tissue of ↑ portal canal (PC) and ↑ limiting plate (LP). P. receives ↑ lymph from spaces of ↑ Disse (D) through occasional openings (arrowhead) in limiting plate. Lymph is drained from P. into lymphatic vessels of portal canals. (Fig. = human)

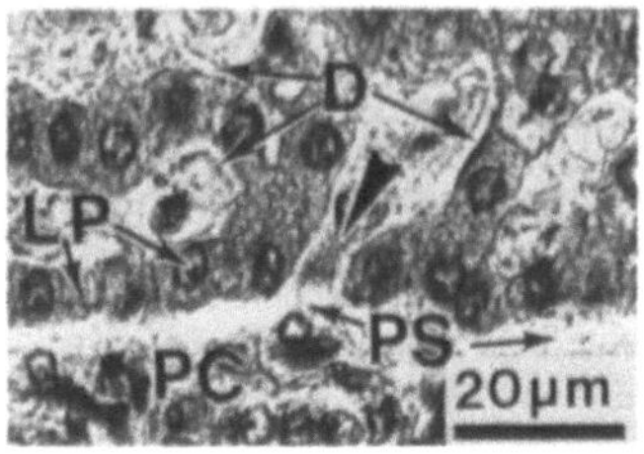

Perisinusoidal cells (cells of Ito, fat-storing cells, interstitial cells, lipocytes, lipid-containing Kupffer's cells, stellate cells, lipocytus perisinusoideus*): Stellate cells (PC) located within spaces of ↑ Disse (D). P. have several long processes contacting neighboring ↑ liver parenchymal cells (H), but without any ↑ junctional complex. Plasmalemma of P. forms irregular small microvilli which intermingle with those of liver parenchymal cells. Nucleus is rich in ↑ heterochromatin and deformed by large ↑ lipid droplets (L). ↑ Organelles are rather poorly developed; P. show weak endocytotic activity and possess no ↑ phagosomes. P. probably represent poorly differentiated ↑ mesenchymal cells, which may be considered to be resting hematopoietic stem cells; under pathological conditions, P. may be transformed into ↑ adipose cells, active blood stem cells, or ↑ fibroblasts. It is thought that P. are involved under normal conditions in accumulation of fat and vitamin A, as well as in production of intralobular ↑ reticular and ↑ collagen fibrils (CF). S = ↑ liver sinusoid (Fig. = rat)

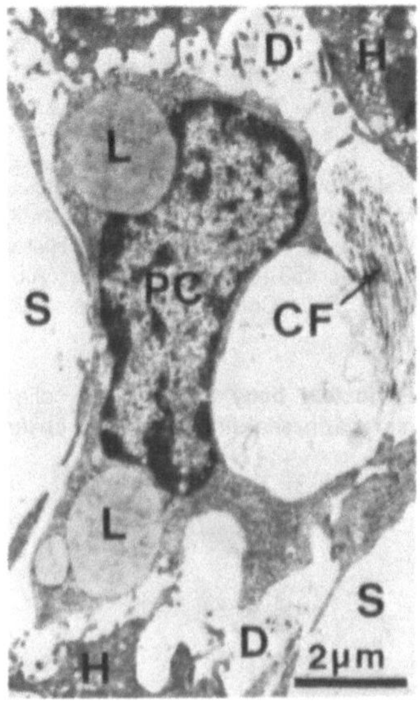

Bartok, I., Toth, J., Remenar, E., Viragh, Sz.: Anat. Rec. *194*, 571 (1979); Tanuma, Y., Ohata, M., Ito, T.: Arch. Histol. Jpn. *46*, 401 (1983); Wake, K.: Int. Rev. Cytol. *66*, 303 (1980)

Perisinusoidal space: ↑ Disse's space

Peristalsis: Involuntary rhythmic contractions of smooth muscular layers of ↑ gastrointestinal tract and some other tubular organs (↑ ductus deferens, ↑ ductus epididymidis, ↑ oviduct, ↑ ureter, etc.), initiated by ↑ interstitial cells of Cajal as pacemaker. P. propels contents of mentioned organs.

Peritendineum externum: ↑ Epitendineum

Peritendineum internum: ↑ Endotendineum

Peritenonium externum: ↑ Epitendineum

Peritenonium internum: ↑ Endotendineum

Peritoneum*: Serous membrane lining abdominal cavity (parietal P.) and enveloping most organs contained therein (visceral P.). Structure: 1) Epithelium = single layer of ↑ mesothelial cells (M); 2) lamina propria (tela subserosa) = thin layer of ↑ loose connective tissue (CT) with some ↑ fibroblasts, ↑ collagen fibrils, and ↑ wandering cells. P. constitutes tunica serosa of ↑ gastrointestinal tract and some other abdominal organs and forms ↑ mesentery. (Fig. = rat)

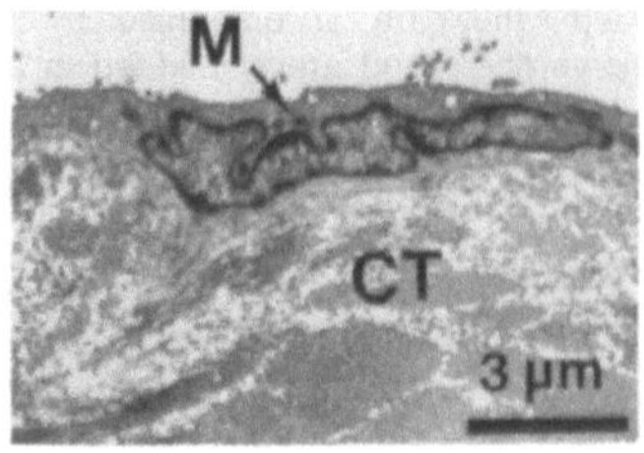

Peritubular capillary network, of kidney (rete capillare peritubulare, pars corticalis et medullaris*): Richly interconnected secondary capillary network (PCN) formed by branching of ↑ efferent arteriole (Eff) of cortical ↑ nephrons. P. supplies convoluted and straight segments of both ↑ proximal and ↑ distal tubules located in cortex. Structurally, P. is composed of ↑ capillaries of fenestrated type; it is believed that blood flows in opposite direction to that of fluid in tubular lumen. Some authors describe a deep P. formed by occasional branching of efferent arteriole of juxtamedullary ↑ nephrons; this P. participates in blood supply to deep cortex and medulla. Aff = ↑ afferent arteriole, G = ↑ renal glomerulus. (Fig. = human, India ink vascular ↑ injection)

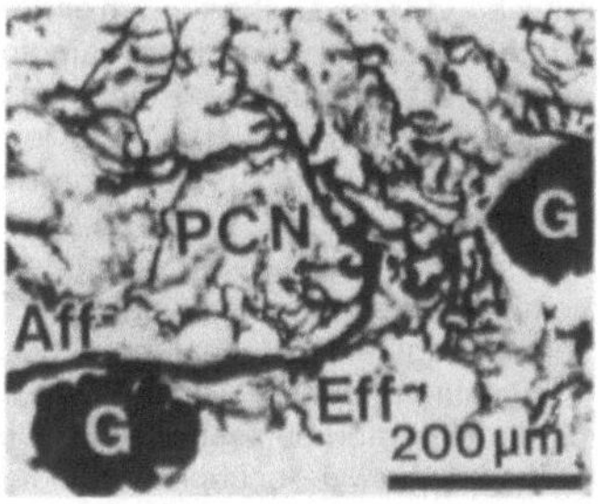

Steinhause, M., Eisenbach, G.-M., Galaske, A.: Pflüger's Arch. *318*, 244 (1970)

Peritubular contractile cells, of testis: ↑ Myofibroblasts

Peritubular dentin: ↑ Neumann's sheath

Perivascular feet (astroglial end feet, foot plate, processus vascularis*):

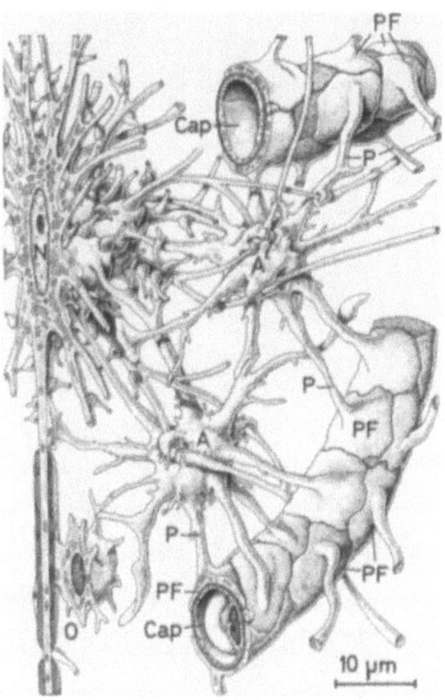

Flattened end expansions (PF) of astrocyte processes (P) applied to the capillary (Cap) wall in ↑ central nervous system and ↑ retina. P. cover about 80% of external capillary surface and form ↑ membrana limitans gliae perivascularis. P. may contain ↑ mitochondria, some short rough endoplasmic profiles, occasional lysosomes, free ribosomes, and ↑ gliofilaments. P. are held together by means of ↑ nexus. A ↑ basal lamina separates them from external aspect of capillary wall. A = ↑ astrocyte, N = ↑ neuron, O = ↑ oligodendrocyte

↑ Astrocyte, protoplasmic, fibrous, velate; ↑ Brain, penetration of blood vessels into; ↑ Subpial feet

Perivascular space: An ill-defined space of varying width surrounding blood and lymphatic vessels. P. contains, besides ↑ tissue fluid, ↑ reticular, ↑ collagen, and ↑ elastic fibrils, ↑ fibrocytes, ↑ wandering cells, and mostly ↑ nonmyelinated nerve fibers; it plays an important role in exchange between cell and circulation. P. is absent in general in CNS, where vessels are surrounded by a short conical ↑ Virchow-Robin space. Terminal extensions of P. are ↑ pericapillary spaces.

Perivitelline space: Narrow slit (PS) between ↑ oocyte (O) and ↑ zona pellucida (ZP). P. is filled by microvilli (Mv) of oocyte; ↑ polar bodies are also located there. (Fig. = rat)

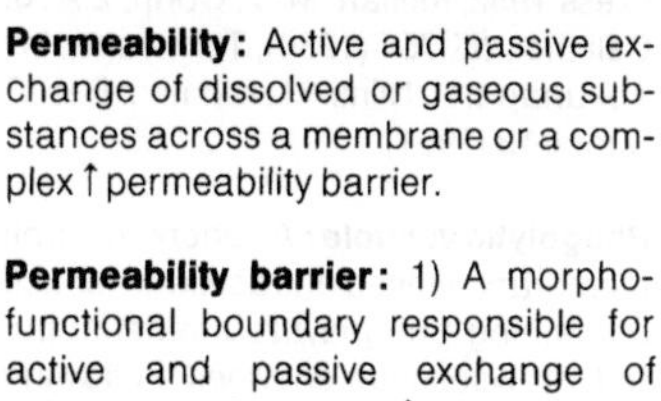

Permeability: Active and passive exchange of dissolved or gaseous substances across a membrane or a complex ↑ permeability barrier.

Permeability barrier: 1) A morphofunctional boundary responsible for active and passive exchange of substances between ↑ tissues and blood, composed of ↑ endothelial cells and their ↑ basal lamina. In a 75-kg adult, area of P. is about 6000 m^2. 2) ↑ Epidermis, representing a selectively permeable barrier against water loss and penetration of substances from the environment.

Elias, P.M., Friend, D.S.: J. Cell Biol. *65*, 180 (1975)

Peroxidase: An enzyme concentrated within ↑ peroxisomes reducing H_2O_2 into O_2 and water.

↑ Myeloperoxidase; ↑ Thyroperoxidase

Vacca, L.L., Abrahams, S.J., Naftchi, N.E.: J. Histochem. Cytochem. *28*, 297 (1980)

Peroxisomes (microbody, peroxysoma*): ↑ Unit membrane-bound spherical cell ↑ organelles (P) 0.2–0.8 µm in diameter. Human P. have moderately osmiophilic homogeneous contents; in other species, contents may be locally condensed or crystalline (= ↑ nucleoid, N). P. contain ↑ peroxidase, catalase, D-amino acid oxidase, and urate oxidase; they reduce H_2O_2 and are involved in α-keto acid production as well as in gluconeogenesis from lipids. Frequent in ↑ proximal tubule cells of kidney, ↑ liver parenchymal cells, ↑ macrophages, etc., P. are never discharged from cell. They play an important role in control of H_2O_2 metabolism. (Fig. = liver parenchymal cell, rat)

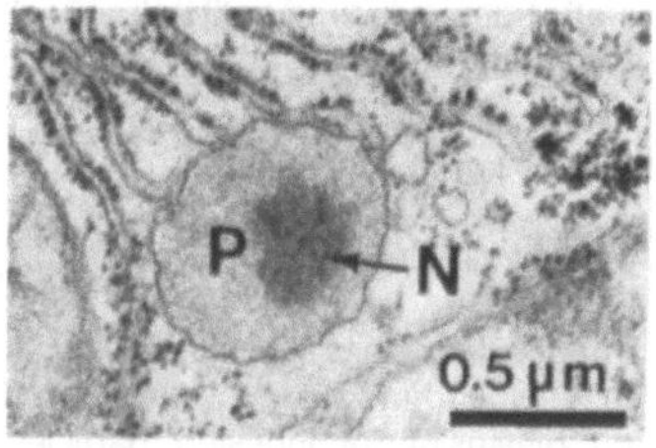

Böck, P., Kramar, R., Pavelka, M.: Peroxisomes and Related Particles in Animal Tissues. Cell Biology Monograph, Vol. 7. Berlin, Heidelberg, New York: Springer-Verlag 1980; De Duve, Ch.: Sci. Am. *248*/5, 52 (1983); Kindl, H., Lazarow, P.B.: Peroxisomes and Glyoxysomes. Annals of the New York Academy of Sciences, Vol. 386. New York Academy of Sciences 1982; Tolbert, N.E., Essner, E.: J. Cell Biol. *91*, 271s (1981)

Peyer's patches (folliculi lymphatici aggregati*): Well-delimited lymphatic organs (PP) composed of 200–400 ↑ lymphatic nodules (LN). P. are located in tela submucosa (TS) of ↑ ileum opposite insertion of ↑ mesentery. A P. is about 12–20 mm long and 8–12 mm wide; its longer diameter is parallel to axis of ileum. Nodules of P. interrupt lamina muscularis mucosae (LMM), penetrate lamina propria (LP), erase ↑ intestinal villi (V), and bulge into intestinal lumen. ↑ M-cells are scattered in epithelium covering P. There are about 300 P. during puberty; in adulthood, the total sinks to about 30–40. P. are involved in ↑ immune response and provide precursors of ↑ plasma cells. P. are thought to represent a ↑ bursa analogue in man.

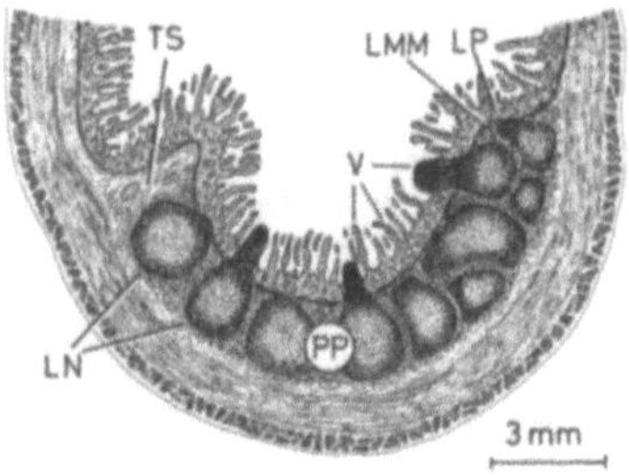

Abe, K., Ito, T.: Arch. Histol. Jpn. *41*, 195 (1978); Lause, D.B., Bockman, D.E.: Cell Tissue Res. *218*, 557 (1981); Owen, R.L., Jones, A.L.: Gastroenterology *66*, 189 (1974); Tomasi, T.B.Jr.: N. Engl. J. Med. *287*, 500 (1972); Yamaguchi, K., Schoefl, G.I.: Anat. Rec. *206*, 391, 403 and 419 (1983)

PF-face, in freeze-cleaving (plasmatic fracture face): A split half of ↑ cell membrane facing extracellular space.

↑ Cell membrane, freeze-cleaving of; ↑ Freeze-cleaving, terminology of

pg: ↑ Picogram

Phagocytes: Cells specially differentiated to perform ↑ phagocytosis. Based on dimension of ingested particles, one can distinguish ↑ macrophages and ↑ microphages.

Phagocytes, alveolar: ↑ Alveolar macrophages

Phagocytic recognition: ↑ Phagocytosis

Phagocytin: A very labile bactericidal substance contained in specific granules of neutrophilic ↑ granulocytes.

↑ Granulocytes, neutrophilic, granules of

Phagocytosis: Process of uptake and intracellular digestion of particles by ↑ phagocytes. Sequence: 1) ↑ Chemotaxis and locomotion = phagocyte approaches particles (P) to be engulfed, attracted by chemical substances they release and by some relatively nonspecific electrostatic or hydrophobic interaction between particles and phagocyte surface (particularly relevant for ↑ alveolar macrophages cleaning airways). 2) Phagocytic recognition and adherence = selection by phagocyte of particles to be phagocytized. Through ↑ immunoglobulin G (IgG) and ↑ complement (C) which coat surface of particles, as well as through presence of ↑ Fc and ↑ C3 receptors for IgG and complement on phagocyte membrane, particles are selected and adhere to plasmalemma of phagocyte. Prior to being phagocytized, pathogenic micro-organisms are coated with ↑ opsonins. 3) Ingestion = formation of ↑ pseudopodia (Ps) and introduction of particles into phagocyte. 4) Digestion = formation of ↑ phagolytic vacuoles (PV) into which primary ↑ lysosomes (Ly) discharge their enzymes; formation of ↑ phagosomes and, finally, ↑ residual bodies (RB).

↑ Lysosomes, functions of

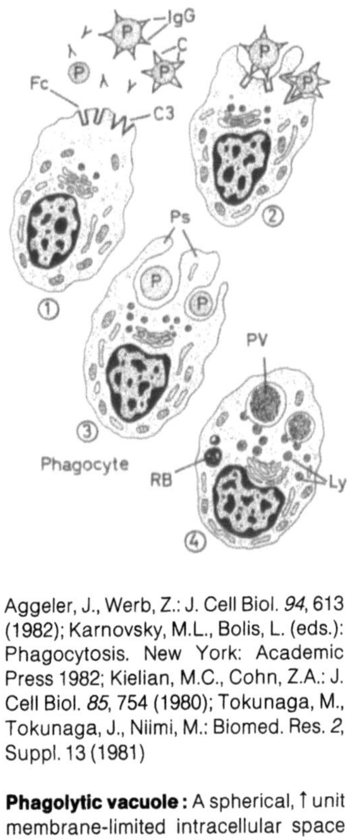

Aggeler, J., Werb, Z.: J. Cell Biol. *94*, 613 (1982); Karnovsky, M.L., Bolis, L. (eds.): Phagocytosis. New York: Academic Press 1982; Kielian, M.C., Cohn, Z.A.: J. Cell Biol. *85*, 754 (1980); Tokunaga, M., Tokunaga, J., Niimi, M.: Biomed. Res. *2*, Suppl. 13 (1981)

Phagolytic vacuole: A spherical, ↑ unit membrane-limited intracellular space containing phagocytized material but not yet lysosomal enzymes. Following discharge of enzymes into P., it becomes a ↑ phagosome.

↑ Lysosomes, functions of

Phagosome (phagosoma*): A ↑ phagolytic vacuole in which primary ↑ lysosomes have discharged their enzymes. With regard to origin of material contained in P., they can be subdivided into ↑ autophagosomes and ↑ heterophagosomes. Some authors use the term P. as synonym for heterophagosome.

Leung, K.-P., Allen, R.D.: Eur. J. Cell Biol. *29*, 1 (1982)

Phagosomes, lamellar: ↑ Lamellar phagosomes

Phalangeal cells, inner (Deiter's cells, cellula phalangea interna*): Slender tall cells (IPC) completely surrounding inner ↑ hair cells (IHC). P. are arranged in two rows between inner ↑ pillars (IP) and ↑ border cells (BC). P. have relatively small elliptical or round nucleus. Cytoplasm is of very low density and contains small number of filiform mitochondria, some short rough endoplasmic cisternae, occasional smooth endoplasmic ↑ tubules, lysosomes, and a few free ribosomes. Only internal row of P. reaches surface of ↑ Corti's organ with a small inner phalanx (IPh) bearing microvilli (Mv). With junctional complex (↑ zonula occludens and ↑ zonula adherens) at phalanx edges, P. maintains contact with hair cells, border cells, and adjacent P. A space about 2–6 µm wide, occupied by ↑ nonmyelinated nerve fibers (NF), remains between rows of P. P. rest on ↑ basilar membrane (BM) and serve as supporting elements for inner hair cells.

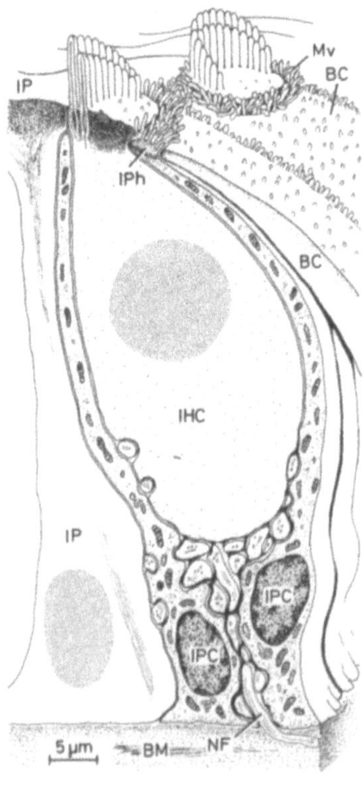

Babel, J., Bischoff, A., Spoendlin, H.: Ultrastructure of the Peripheral Nervous System and Sense Organs. Stuttgart: Georg Thieme 1970

Phalangeal cells, outer (Deiter's cells, cellula phalangea externa*): Tall asymmetrical cells (OPC) with a cylindrical body resting on ↑ basilar membrane (BM) and a long thin fingerlike stalk (S), which terminates in a rhomboid flat expansion or phalanx (Ph). There are 4–5 rows of P. in ↑ Corti's organ. Nucleus is round and located in basal cell pole; cytoplasm is pale and contains small scattered mitochondria, an inconspicuous Golgi apparatus, a few flattened and short rough endoplasmic cisternae, occasional lysosomes, moderate numbers of smooth tubules, and some free ribosomes. From cell base to phalanx extends a prominent bundle of ↑ microtubules (Mt) which serve to maintain cell shape. Well-developed junctional complex at phalanx edges, consisting of a ↑ zonula occludens (ZO) and a ↑ zonula adherens (ZA), maintains contacts with adjacent P. and outer ↑ hair cells (HC). From cytoplasmic side, junctional complex is surrounded by dense filamentous material (DM) in which microtubules end. Viewed from above, contours of phalanges form networklike pattern, ↑ reticular membrane. P. are supporting elements for outer hair cells, lower third of which rests in a shallow concavity of P. (arrow indicates position of ↑ stria vascularis).

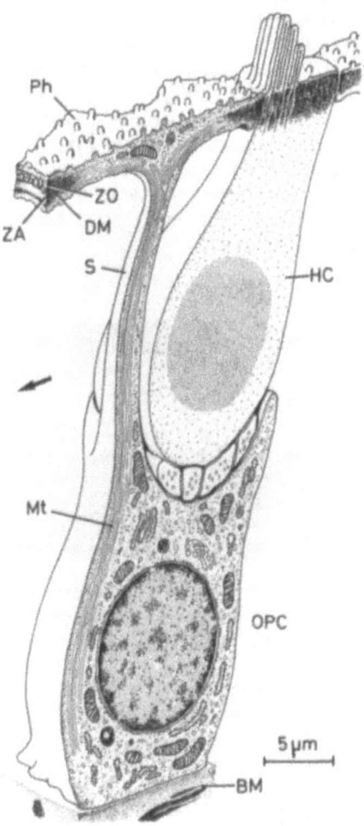

Iurato, S. (ed.): Submicroscopic Structure of the inner Ear. New York: Pergamon Press 1968; Wersall, J., Flock, A., Lundquist, P.-G.: Z. Hörgeräte Akustik *9*, 56 (1970)

Phalanx (caput cellulae phalangealis externae et internae*): Flattened rhomboidal extension of inner and outer ↑ phalangeal cells contacting inner and outer ↑ hair cells and ↑ pillar cells, thus participating in formation of ↑ reticular membrane.

Pharyngeal glands (glandula pharyngealis*): Submucosal glands of ↑ pharynx = ↑ mixed glands in ↑ nasopharynx and purely ↑ mucous glands in ↑ oropharynx and ↑ laryngopharynx. In soft ↑ palate, P. are situated in lamina propria.

Pharyngeal pituitary: A nest of adenohypophyseal tissue located in roof of ↑ nasopharynx; vestige of ↑ Rathke's pouch. (See embryology texts for further information)

Pharyngeal tonsil: ↑ Tonsil, pharyngeal

Pharyngoesophageal sphincter: Superior sphincter of ↑ esophagus.

Pharynx (throat): A musculofibrous organ, component of both ↑ gastrointestinal and respiratory tracts, in which passageways for gases and food merge and cross. P. is divided into three portions: ↑ nasopharynx, ↑ oropharynx, and ↑ laryngopharynx. Structure: 1) Tunica mucosa (TM): a) epithelium (E) = in nasopharynx, ↑ pseudostratified, ciliated with ↑ goblet cells; in oropharynx and laryngopharynx, nonkeratinized ↑ stratified squamous; b) lamina propria (LP) = ↑ loose connective tissue with solitary ↑ lymphatic nodules (LN) in nasopharynx and ↑ elastic fibers in oropharynx and laryngopharynx. 2) Tela submucosa (TS) = loose connective tissue with tubar (TT) and pharyngeal ↑ tonsils (TPh) in nasopharynx and richly branched elastic network in oropharynx and laryngopharynx. Submucosal ↑ pharyngeal glands in nasopharynx are ↑ mixed (MxG), those of oropharynx and laryngopharynx purely ↑ mucous (MG). 3) Tunica muscularis (TMu) = two layers of ↑ skeletal muscle fibers: inner longitudinal and outer oblique. 4) Tunica adventitia (TA) = loose connective tissue continuous with cephalopharyngeal ↑ aponeurosis (Ap).

↑ Respiratory mucosa

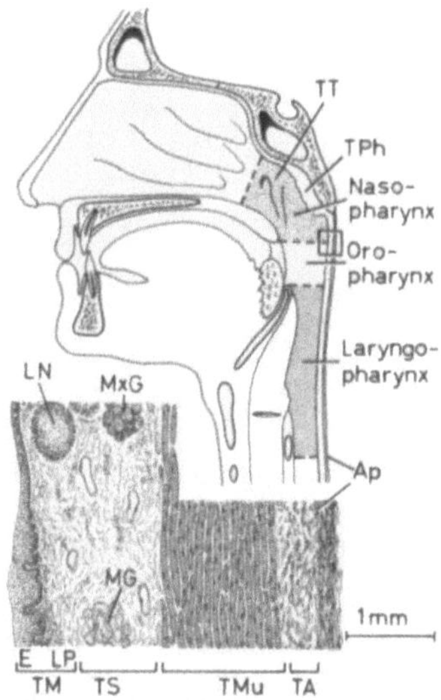

Phase contrast microscope: An optical instrument for study of living cells without any injurious intervention. Principle: 1) Amplitude (a) and phase of light traversing a nonabsorbent biological object (O) are same before and after object. 2) In the case of an absorbent object (stained section, S) of same thickness as in 1, light phase remains same before and after object, but its amplitude after object is diminished (a'). 3) An object of same optical properties as in 1, but thicker, does not affect amplitude, but provokes a delay of phase ($\lambda/4$). 4) in an object with a ↑ refractive index differing from that in cases 1, 2, and 3, amplitude is not affected, but decrease in light velocity implies phase change ($\lambda/4$). The object in cases 1, 3, and 4 is not visible, since ↑ photoreceptors react only to changes of amplitude. Different cellular components produce phase change, because they vary in thickness and refractive index. Thanks to phase optics, i.e., annular diaphragm (AD) between light source (L) and ↑ condenser (C) + annular diaphragm (AD) at back focal plane of ↑ objective (Ob), P. converts phase variations into amplitude variations, thereby enabling eye to detect contrast between different structures only by optical means. P. is basic instrument in ↑ cell and tissue culture. Sp = specimen stage with specimen, Oc = ↑ ocular

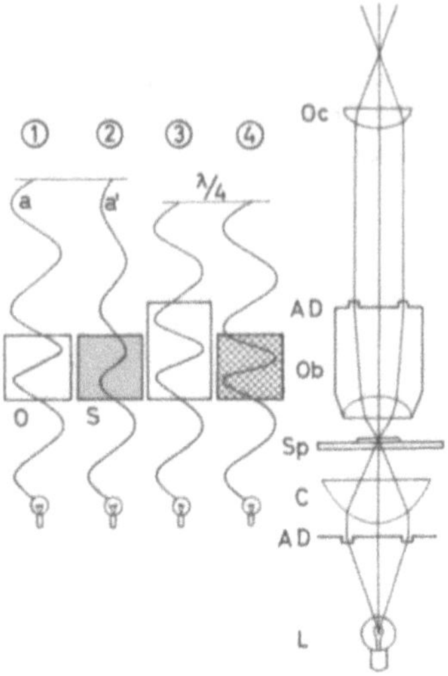

Ronchi, V., Mondal, P.K., Prabhakar, R.K., Subramanyam, S.: Atti Fond. G. Ronchi (Firenze) *35*, 527 (1980)

Pheochrome cells: ↑ Chromaffin cells

Pheochrome reaction: ↑ Chromaffin reaction

Phosphatases: Hydrolytic enzymes which split esters of natural organic phosphates (ATP, glucose-6-phosphate, glycerophosphate) and liberate phosphate ions as one of end products. Acid P. is present in ↑ lysosomes and is active at pH 4.5–6.0; alkaline P. is active at pH 9.0–9.6.

Photopsin: ↑ An ↑ opsin forming ↑ iodopsin with ↑ retinene in one category of cone cells.

↑ Photoreceptors

Photoreceptors (photoreceptor*): Two kinds of highly differentiated neuroepithelial cells in ↑ retina which are sensitive to light. 1) Rod cell (cellula optica bacilliformis*) = slender threadlike ↑ neuron (averaging 50 µm in length) with outer segment (OS) touching ↑ pigment epithelium (PE). Outer segment (rod proper) is enclosed by a plasmalemma and contains numerous closely packed ↑ membranous discs (MD). Outer segment is connected with inner segment (IS) by a modified ↑ cilium (C) lacking central microtubular pair, but having a well-developed ↑ basal body (BB) and a striated rootlet (R). Inner segment is subdivided into outer part or ellipsoid (E), containing numerous mitochondria and ↑ centriole (Ce), and an inner part, or myoid (M), with Golgi apparatus (G), some rough endoplasmic cisternae (rER), smooth endoplasmic ↑ tubules (sER), and ↑ microtubules. Inwardly, myoid attenuates abruptly and forms outer fiber (OF); ↑ zonula adherens (ZA) for junction with adjacent ↑ Müller cells (↑ outer limiting membrane) is situated at this level. Outer fiber is slender cytoplasmic cylinder which enlarges to contain the small elliptical nucleus (N), after which it again attenuates and is termed inner fiber (IF). In fact, this is ↑ axon of rod cell, which terminates in bulb-shaped ↑ rod spherule (RS). Rod cells are situated at periphery of optical retina; they are adapted for perception of light at dusk and for seeing black and white. There are about 120×10^6 rod cells in human retina. 2) Cone cell (cellula optica coniformis*) = neuron similar in size and structure to a rod cell. Outer segment (OS) is conical (the cone proper) and composed of a large number of flattened membranous discs continuous with plasmalemma. Inner segment (IS), outer fiber (OF), and inner fiber (IF) are wider than in rod cells. Nuclei are elliptical and situated in a single row immediately beneath outer limiting membrane. Inner fiber (IF), i.e., axon, enlarges in ↑ cone pedicle (P). Cone cells are situated in central part of optic retina (particularly in ↑ fovea centralis); they are responsible for color perception. Functionally, but not morphologically, cone cells are subdivided into three varieties: red-, green-, and blue-sensitive. Visual pigment of cone cells is ↑ iodopsin. There are about 6×10^6 cone cells in human retina. Both types of P. may vary in length depending on position in retina. SR = ↑ synaptic ribbons, SV = ↑ synaptic vesicles

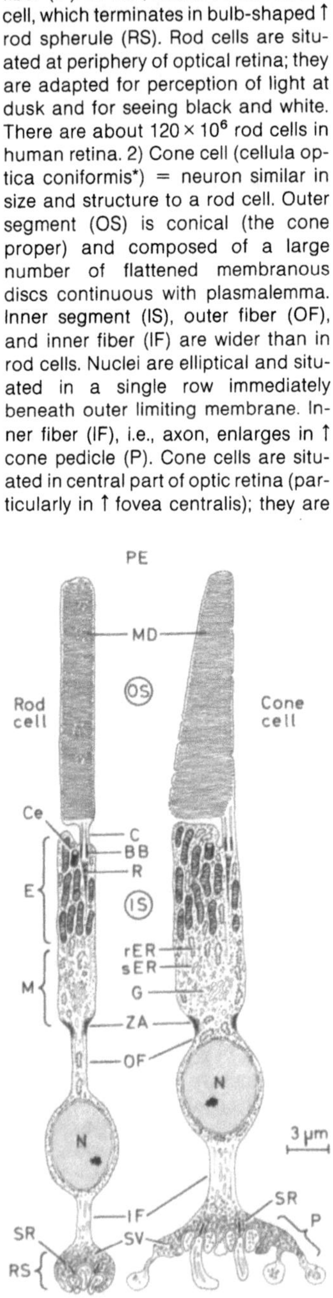

Anderson, D.H., Fischer, S.K.: J. Ultrastruct. Res. *67* 23 (1979); Borwein, B.: Anat. Rec. *205*, 363 (1983); Braekevelt, C.R.: Anat. Anz. *153*, 33 (1983); Murray, R.G., Jones, A.E., Murray, A.: Anat. Rec. *175*, 673 (1973); Young, R.W.: J. Cell Biol. *33*, 61 (1967)

Physiological cup: ↑ Optic papilla

Phytohemagglutinin (PHA): Plant ↑ lectin extracted from red kidney bean which can induce proliferation of ↑ T-lymphocytes in vitro.

Pia mater*: ↑ Leptomeninx adhering to tissue of ↑ central nervous system and initial segment of cerebrospinal nerves. P. invests all surfaces, including indentations, sulci, and fissures. Structure: Several layers of flattened ↑ fibroblasts (F) separated by fine ↑ collagen fibrils (Cf). P. is well vascularized and contains numerous nerve fibers, some for innervation of pial blood vessels and some with sensory function. ↑ Corpuscles of Krause are also present. Along blood vessels wander ↑ histiocytes and ↑ lymphocytes. In P. of base of brain, one can find ↑ melanocytes. P. is separated from ↑ nervous tissue by basal lamina of ↑ subpial feet (SF) of ↑ membrana limitans gliae superficialis. (Fig. = ↑ optic nerve, mongolian gerbil)

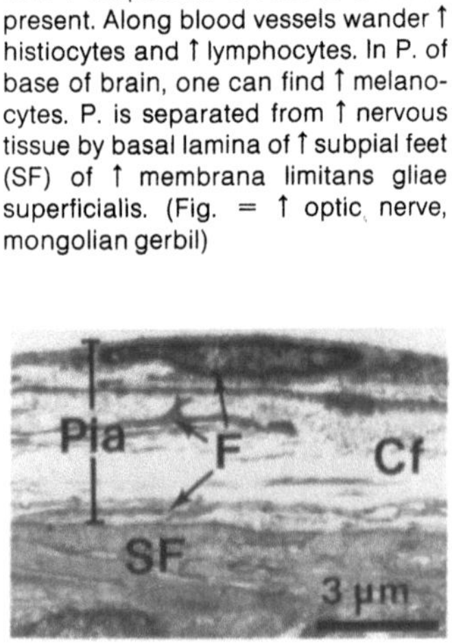

Sano, Y., Takeuchi, Y., Yamada, H., Ueda, S., Goto, M.: Histochemistry *76*, 277 (1982)

Piarachnoid (leptomeninx): Common term for ↑ arachnoid and ↑ pia mater because of their intimate structural relation.

Picogram (pg): 10^{-12} of a gram.

PIF: ↑ Proliferation-inhibiting factor

Pigment cells: Term including all cells synthesizing and transporting ↑ melanin (↑ melanocytes and ↑ melanophores).

Pigment cells, of ciliary body: ↑ Ciliary body

Pigment connective tissue (textus connectivus pigmentosus*): Kind of highly vascularized ↑ loose connective tissue constituting stroma of ↑ iris, ↑ ciliary body, and ↑ choroid. Besides ↑ fibroblasts and ↑ fibrocytes (Fc), P. is characterized by rare ↑ melanocytes (M), numerous ↑ melanophores (Me), ↑ mast cells, and ↑ macrophages. In iris, latter can be giant macrophages and contain phagocytized melanin granules (↑ clump cells). ↑ Collagen (CF) and ↑ elastic (EF) fibrils form delicate network; around capillaries (Cap), collagen fibrils have distinct circular arrangement. Number of melanocytes determines ↑ color of eye. NF = ↑ nerve fibers

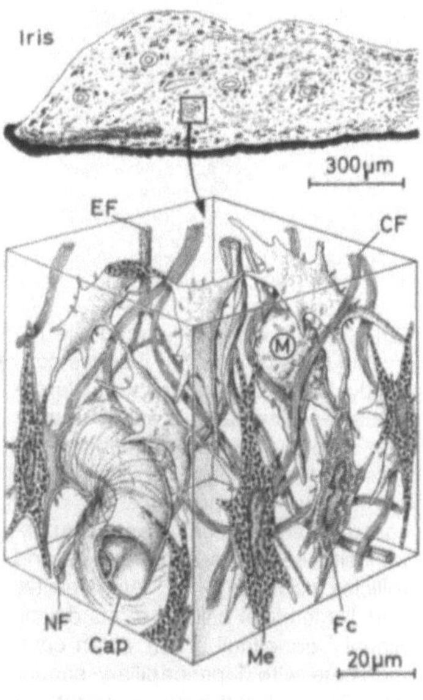

Pigment epithelial cells (cellula pigmentosa*): Cuboidal cells, about 14 μm wide and 10–14 μm tall (PC), constituting ↑ pigment epithelium. Viewed from above, P. appear as very regular hexagons (inset). Nucleus is round with prominent nucleolus. Cytoplasm contains numerous mitochondria, large supranuclear Golgi apparatus, scarce rough endoplasmic cisternae, and highly developed smooth ↑ endoplasmic reticulum (sER) in the form of a branched network of ↑ tubules. The supranuclear cytoplasm is occupied by large spindle-shaped ↑ melanin granules (MG), ↑ melanosomes (Ms), ↑ lysosomes (Ly), and ↑ residual bodies (RB) with lamellar debris (= ↑ lamellar phagosomes). Latter structures are transformed into ↑ lipofuscin granules. Small number of free ribosomes, as well as ↑ microperoxisomes, are also present in P. ↑ Apical cell pole is provided with short microvilli and longer, often branched, processes (Pr), which are intimately interdigitated with ↑ photoreceptor outer segments (Ph) but without any attachment devices. Melanin granules can penetrate into longer processes. Cells are held in contact with adjacent P. by ↑ junctional complexes. Lateral cell surfaces are slightly undulating. Basal plasmalemma forms a well-developed ↑ basal labyrinth (BLt). P. overlie a basal lamina (BL). P. are capable of weak ↑ melanogenesis. MD = ↑ membranous discs

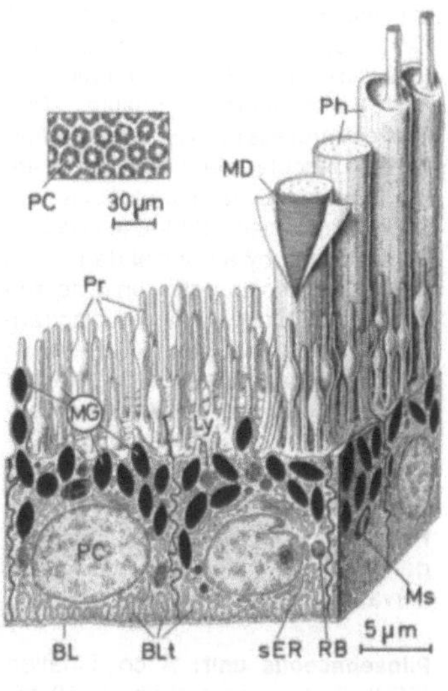

Novikoff, A.B., Leuenberger, P.M., Novikoff, P.M., Quintana, N.: Lab. Invest. *40*, 155 (1979)

Pigment epithelium, histophysiology of: In darkness, ↑ melanin granules (MG) accumulate in ↑ apical pole of ↑ pigment epithelial cells (PC); there are no melanin granules in cell processes (P). Sensitivity of eye increases, but ↑ resolving power diminishes. In light, melanin granules stream into cell processes which surround ↑ photoreceptor outer segments (Ph) like minute camerae obscurae. Resolving power of eye increases, but sensitivity diminishes. In human, P. is less developed than in birds and some other species.

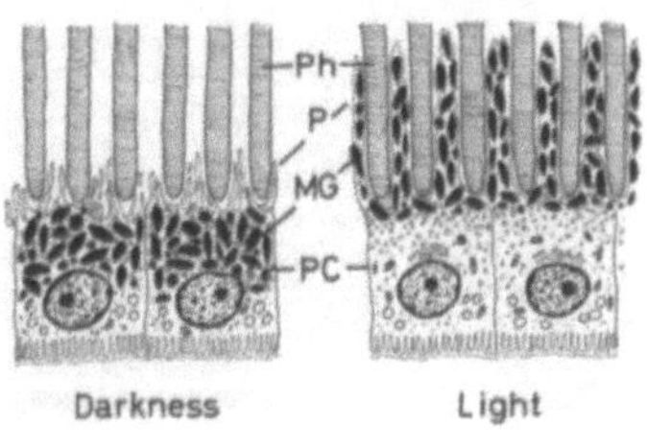

Pigment epithelium, of retina (pars pigmentosa retinae*): Outermost retinal stratum (PE) composed of one layer of cuboidal ↑ pigment epithelial cells (PC) situated between ↑ photoreceptors (Ph) and ↑ choroid (Ch), from which it is separated by ↑ Bruch's membrane (BM). ↑ Retina and P. are in firm contact only at ↑ optic papilla and at ↑ ora serrata. Functions of P.: 1) With ↑ junctional complexes between epithelial cells, P. selects metabolites diffusing into other retinal layers; 2) prevention of light reflection from external ocular tunics through absorption of light by ↑ melanin granules; 3) ↑ melanogenesis in restricted volume; 4) digestion and turnover of growing tips of photoreceptors; 5) regeneration of ↑ rhodopsin, if photoreceptors remain in contact with P.

↑ Lamellar phagosomes; ↑ Pigment epithelium, histophysiology of

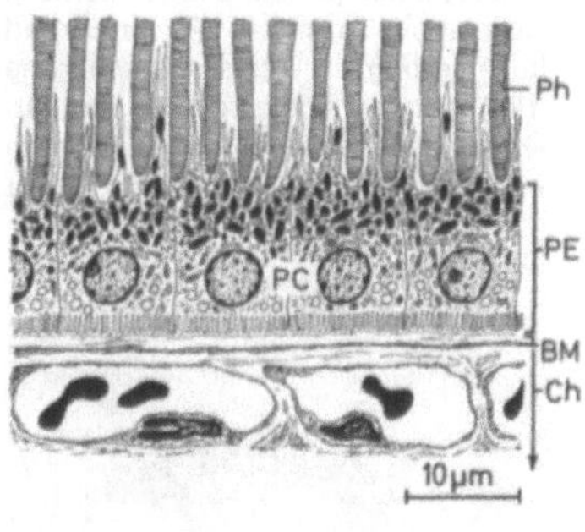

Goldman Herman, K., Steinberg, R.H.: Cell Tissue Res. *227*, 485 (1982); Lerche, W., Kaune, H.: Ophthalmologica *180*, 301 (1980)

Pigmentarchitectonics: Arrangement of pigmented ↑ neurons in ↑ central nervous system.

Pigmentary system: A collective term for ↑ melanocytes which originate from neural crest and colonize ↑ keratinizing system of ↑ epidermis during embryonal life.

↑ Nervous tissue, histogenesis of

Pigments: ↑ Inclusion substances (P) visible in living cells and tissues because of their color. On the basis of origin and chemical composition, they may be classified as follows (after Bucher, 1980):

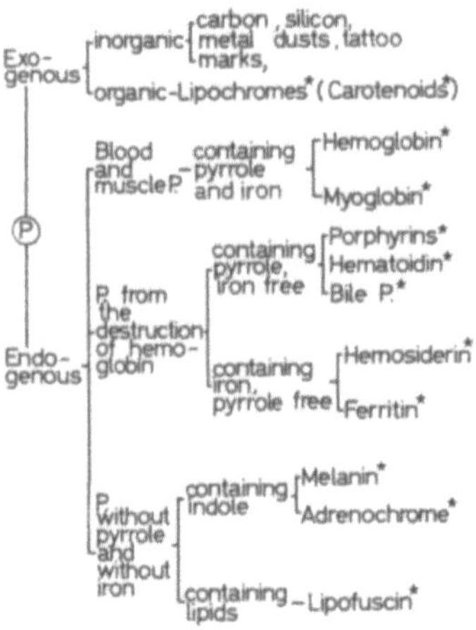

Exogenous P. are introduced into cells and tissues from exterior; endogenous P. are synthesized by cells of body. (P. marked with an asterisk are described under corresponding letters)

Bucher, O.: Cytologie, Histologie und mikroskopische Anatomie des Menschen, 10th edn., Bern: Hans Huber 1980

Pillars, inner (cellula pilaris interna*): Cells about 70 µm tall (IP) with broad base resting on ↑ basilar membrane (BM), slender middle portion, and flattened head with a concavity with which convex head of outer ↑ pillar (OP) articulates to form a ↑ Corti's tunnel (CT). Nucleus and cytoplasm are located basally and directed toward tunnel, mitochondria are small and scattered predominantly in basal cell portion, Golgi apparatus is inconspicuous, and rough endoplasmic reticulum is mainly tubular in form. From cell base to head extends a thick bundle (B) of straight, parallel, and regularly arranged ↑ microtubules (Mt), each surrounded by 6-nm-thick microfilaments (Mf, inset). Both end in a very osmiophilic ↑ terminal web (TW) in head of cell. P. touch both inner (IHC) and outer ↑ hair (OHC) cells and maintain, together with outer pillars, constant height of ↑ Corti's organ. There are about 5600 P.

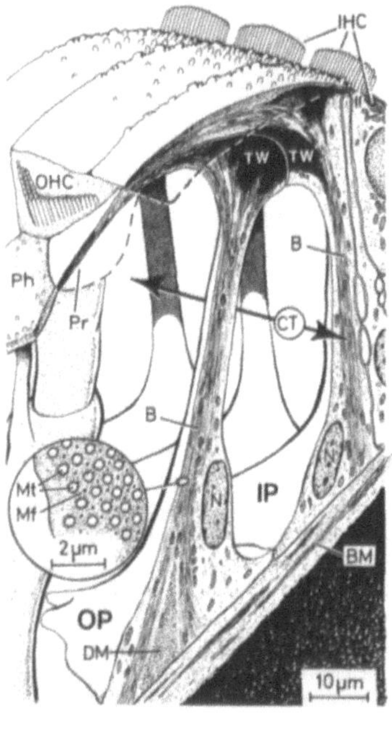

Pillars, outer (cellula pilaris externa*): Cells (OP) slightly longer than inner ↑ pillars (about 85µm) but of identical ultrastructure. Head of each P. fits into corresponding concavity of an inner pillar (IP). With a flattened tonguelike process (Pr), P. reach free surface of ↑ Corti's organ and make contact with innermost row of outer ↑ hair cells (OHC) and nearest ↑ phalanx (Ph). This cell process participates in formation of ↑ reticular membrane. There are two bundles (B) of microtubules (Mt) and microfilaments (Mf) at the base of P., separated by a conical dark mass (DM). The two bundles unite to run through cell body, ending in electron-dense ↑ terminal web (TW) within head. There are about 3800 P., so that about three inner pillars are connected with two P. (See fig. under ↑ Pillars, inner)

Pilomotor nerve fibers: ↑ Postganglionic sympathetic nerve fibers innervating ↑ arrector muscles.

Pilosebaceous unit: A combination (PSU) of one or several ↑ hairs (H), together with glands that discharge their secretory product into ↑ hair follicles (↑ apocrine sweat glands, ↑ sebaceous glands, S). (Fig. = human)

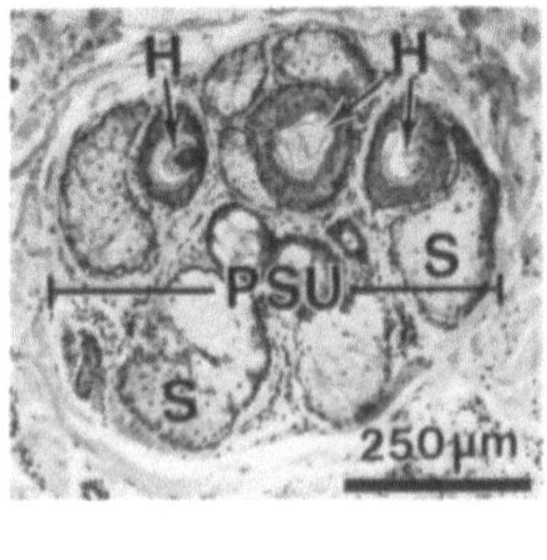

Pineal body: ↑ Pineal organ

Pineal canaliculi: ↑ Canaliculi, pineal

Pineal chief cells: ↑ Pinealocytes

Pineal gland: ↑ Pineal organ

Pineal interstitial cells: ↑ Interstitial cells, of pineal organ

Pineal neuroglial cells: ↑ Interstitital cells, of pineal organ

Pineal organ (epiphysis cerebri, pineal body, corpus pineale*): Conical organ 5–8 mm in length and 3–5 mm in width situated on roof of diencephalon, attached to brain by ↑ pineal stalk (S).

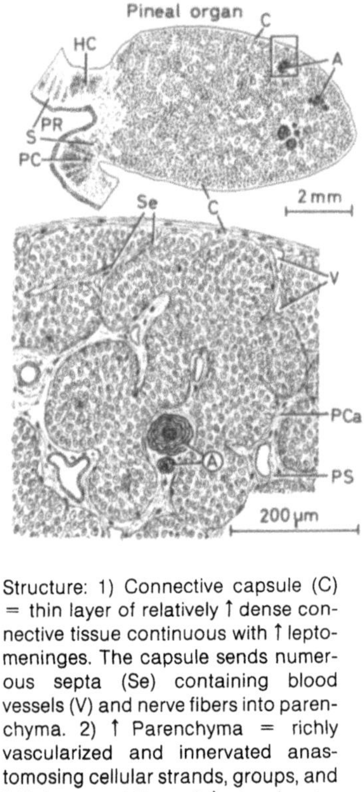

Structure: 1) Connective capsule (C) = thin layer of relatively ↑ dense connective tissue continuous with ↑ leptomeninges. The capsule sends numerous septa (Se) containing blood vessels (V) and nerve fibers into parenchyma. 2) ↑ Parenchyma = richly vascularized and innervated anastomosing cellular strands, groups, and follicles consisting of ↑ pinealocytes and ↑ interstitial cells. Strands delimit pineal ↑ canaliculi (PCa) which communicate with ↑ pericapillary spaces (PS). Through latter run fenestrated ↑ capillaries, some collagen fibrils, and ↑ adrenergic nerve fibers. ↑ Acervuli (A) are always present in P. of adult individuals. P. is innervated by postganglionic nonmyelinated nervi conarii originating from superior cervical ganglion. Some ↑ myelinated nerve fibers penetrate P. from both commissures. Vascularization of P. is assured by two posterior choroid arteries; venous blood is drained into great cerebral vein. HC = habenular commissure, PC = posterior commissure, PR = ↑ pineal recess

Karasek, M., Hansen, J.T.: Am J. Anat. *163*, 257 (1982); Møller, M.: Cell Tissue Res. *154*, 1 (1978); Reiter, J.R.: Am. J. Anat. *162*, 287 (1981); Schneider, T., Semm, P., Vollrath, L.: Cell Tissue Res. *200*, 41 (1981); Vollrath, L.: The Pineal Organ. Handbuch der mikroskopischen Anatomie des Menschen, Vol. 6, Part 7. Berlin, Heidelberg, New York: Springer-Verlag 1981

Pineal organ, histophysiology of: ↑ Melatonin is considered to be a true pineal hormone. Secretion follows 24-h rhythm, with maximum during night. ↑ Pineal organ responds to external stimuli, primarily visual, relayed to it via sympathetic nerve fibers, ↑ norepinephrine, and ↑ cyclic AMP, by changes in synthesis and release of melatonin, ↑ serotonin, and specific methylindoles. This rhythmically modifies function of some endocrine glands (gonads, hypophysis) in daily and/or seasonal rhythms. Therefore, pineal organ is referred to as neuroendocrine transducer. Very little is known about some chemically unclearly defined antigonadotrophic peptides opposed to maturation of gonads or about arginine vasopressin, a primitive neurohypophyseal hormone also found in pineal organ.

↑ Hydroxyindole-O-methyltransferase

Altar, A.: Dev. Neurosci. *25*, 166 (1982); Axelrod, J., Velo, C.P., Fraschini, F.: The Pineal Gland and Its Endocrine Role. New York: Plenum Publ. Co 1983; Cardinali, D.P., Vacas, I.M.: Pineal Function in Reproductive Physiology. In: Muldoon, T.G., Mahesh, V.B., Perez-Ballester, B. (eds.): Recent Advances in Fertility Research. Part A. New York: Alan R. Liss 1982; Reiter, R.J. (ed.): The Pineal and Reproduction. New York: Alan R. Liss 1978; Reiter, R.J. (ed.): The Pineal and Its Hormones. New York: Alan R. Liss 1982

Pineal organ, melatonin in: ↑ Melatonin

Pineal parenchymal cells: ↑ Pinealocytes

Pineal recess (recessus pinealis*): Extension of third ventricle (PR) delimited by habenular (HC) and posterior (PC) commissures. P. is lined by nonciliated ↑ ependymal cells among which occur islands of ciliated cells. Ependymal lining rests upon layer composed of filamentous glial processes and capillary network characterized by ↑ perivascular spaces. A = ↑ acervulus, PO = ↑ pineal organ (Fig. = human, courtesy Biomedical Research)

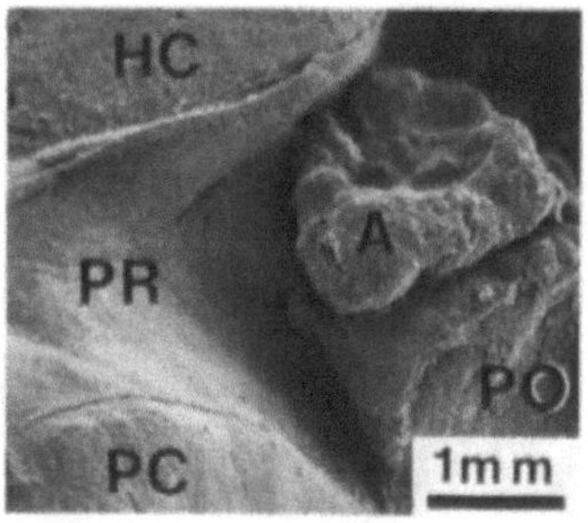

Pineal sand: ↑ Acervuli

Pineal stalk: Thin extension of brain tissue (PS) connecting ↑ pineal organ (P) with ↑ brain. P. contains nerve fibers from habenular (HC) and posterior commissures (PC), and represents floor of ↑ pineal recess (PR).

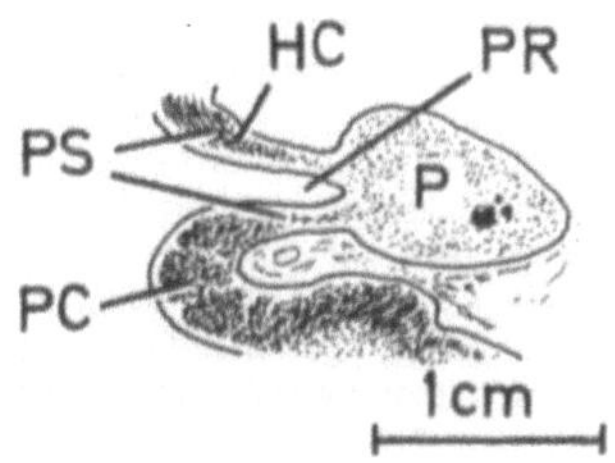

Pinealocytes (chief cell, clear pineal cell, pineal parenchymal cell, pinealocytus, cellula lucida*): Round or irregular stellate pale cells with two or more processes, some of which penetrate with clublike endings (= ↑ polar terminals, PT) into pineal ↑ canaliculi or ↑ pericapillary spaces (PS). Nucleus of a P. is generally round, sometimes with deep invaginations; nucleolus is very large. Cytoplasm contains voluminous polymorphous mitochondria, large and frequently multiple Golgi apparatus (G), centrioles (C), many short and flattened rough endoplasmic cisternae (rER), an abundance of atypical smooth endoplasmic reticulum (sER) (= intergrade reticulum), ↑ subsurface cisternae (SC), ↑ lipid droplets (L), ↑ lysosomes, free ribosomes, occasional ↑ cilia, and many ↑ microtubules (Mt). Adjacent to Golgi apparatus and in processes are ↑ unit membrane-bound, round or elongated vesicles, 0.2–1 μm in diameter, with fine granular and moderately dense contents, ↑ grumose bodies (GB). In some species (mouse), small vesicles with a dense core have been found. A characteristic feature of P. are ↑ synaptic bars (SB) found free in cytoplasm and/or in contact with plasmalemma. By unknown mechanism, P. secrete ↑ melatonin, ↑ serotonin, and some ill-defined pineal peptides. P. comprise about 90% of ↑ pineal organ cells. Cap = capillary, NE = ↑ adrenergic nerve ending

↑ Paraneuron; ↑ Pineal organ, histophysiology of

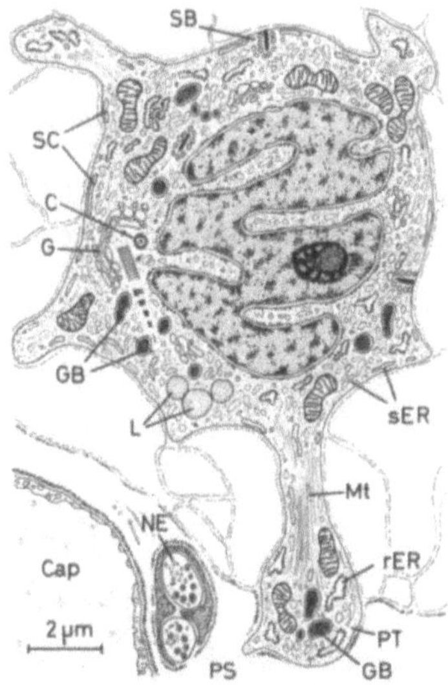

Altschule, M.D. (ed.): Frontiers of Pineal Physiology. Cambridge, Mass.: M.I.T. Press 1975; Ariens Kappers, J., Schadé, J.P. (eds.): Structure and Function of the Epiphysis Cerebri. Amsterdam, London, New York: Elsevier 1965; Ito, T., Matsushima, S.: Arch. Histol. Jpn. *30*, 1 (1968); Karasek, M., Endocrinol. Exp. *15*, 17 (1981); Pévet, P.: Neural Transm. *40*, 289 (1977)

Pinna: ↑ Auricle

Pinocytosis: ↑ Phenomenon of active intake of dissolved substances by cell. P. visible under light microscope is called ↑ macropinocytosis; P. visible only in electron microscope is referred to as ↑ micropinocytosis.

Pinosome: A vacuole in which a droplet of extracellular fluid is interiorized into cell body during ↑ macropinocytosis.

Pits, gastric: ↑ Foveolae gastricae

Pituicytes (pituicytus*): Elongated or very irregularly shaped cells with long and slender cytoplasmic processes (P). Nucleus is oval, frequently indented, and has small peripheral clumps of ↑ heterochromatin; nucleolus is prominent. Cytoplasm is abundant and clear; it contains a moderate number of mitochondria, a well-developed Golgi apparatus (G), ↑ centrioles (Ce), sparse short rough endoplasmic cisternae, frequent lysosomes, a few free ribosomes, bundles of ↑ microtubules (Mt) and 5–7-nm-thick microfilaments (Mf), varying numbers of ↑ lipid droplets (LD), and ↑ micropinocytotic vesicles. P. are in intimate contact with ↑ neurosecretory axons (NA) and often with capillaries. It is thought that they are specially differentiated ↑ neuroglia cells with a supportive function, but it is possible that they play some metabolic role in secretory process. P. represent an intrinsic cell population of ↑ neurohypophysis. Several classes of P. are described: major, dark, ependymal, oncocytic, granular.

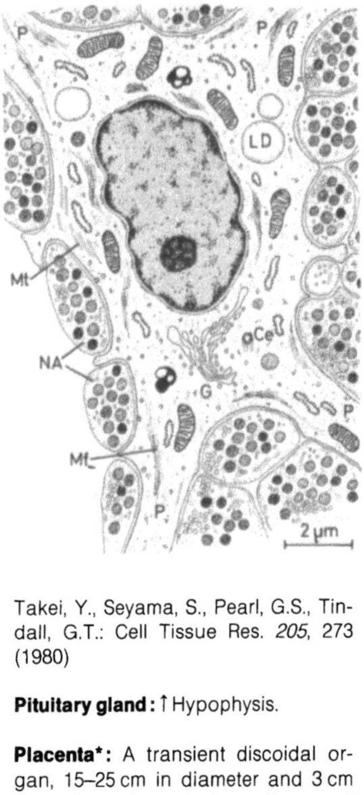

Takei, Y., Seyama, S., Pearl, G.S., Tindall, G.T.: Cell Tissue Res. *205*, 273 (1980)

Pituitary gland: ↑ Hypophysis.

Placenta*: A transient discoidal organ, 15–25 cm in diameter and 3 cm thick, formed partly by embryo and partly by ↑ uterus, where physiological exchanges take place between mother and developing embryo/fetus. Histologically, one can distinguish, on basis of its age, early ↑ P. and full-term ↑ P. (See embryology texts for further information)

Beaconsfield, P., Birdwood, G, Beaconsfield, R.: Sci. Am. *243/2*, 80 (1980); Boyd, J.D., Hamilton, W.J.: The Human Plancenta. Cambridge: W. Heffer & Sons, Ltd. 1970

Placenta, diffuse: Early stage of the placenta, when ↑ chorionic villi are uniformly distributed over outer surface of chorionic vesicle. (See embryology texts for further information)

Placenta, early: Placenta from formation up to 4th–5th month of pregnancy; characterized by ↑ placental villi with bilayered ↑ trophoblast. Number of ↑ cytotrophoblastic cells decreases with further placental maturation. (See embryology texts for further information)

Placenta, full-term, structure of: I) Fetal placenta. 1) Chorionic plate (CP) = a) ↑ amniotic epithelium (AE) lines amniotic cavity (AC); b) chorionic mesoderm (CM) continues as ↑ Wharton's jelly into ↑ umbilical cord; c) chorionic epithelium or ↑ trophoblast (Tr) covers all placental villi (PV) and forms ↑ trophoblastic shell (TrS); d) Langhans' subchorionic fibrinoid (LF). 2) ↑ Placental or chorionic villi (PV) = large stem villi branch many times and form small villi; some villi pierce trophoblastic shell and terminate on ↑ decidua basalis (DB) as ↑ anchoring villi (AV). Each stem villus with its branches has a separate fetal blood supply; this unit is referred to as fetal ↑ cotyledon. Villi are bathed by maternal blood circulating in ↑ intervillous spaces (IVS). 3) Trophoblastic shell (TrS) = ↑ cytotrophoblast of distal tip of stem villi which covers decidua basalis (DB) except at opening of maternal vessels. Trophoblastic shell is fetal part of decidua basalis. 4) Placental labyrinth = intervillous spaces (IVS) between fetal and maternal plancenta filled with maternal blood. II) Maternal placenta. 1) ↑ Decidua basalis (DB) forms, with ↑ basal plate (BP), decidual septa (DS) separating fetal cotyledons from one another. Decidua basalis attaches fetal placenta to ↑ myometrium. 2) Maternal blood vessels = ↑ uteroplacental arteries (UA) empty into intervillous spaces; maternal veins (V) collect maternal blood from intervillous spaces. 3) ↑ Nitabuch's basal fibrinoid layer (NF). 4) Rohr's fibrinoid layer (RF) below trophoblastic shell. 5) Enlarged ↑ uterine glands (UG) with loose interglandular stroma.

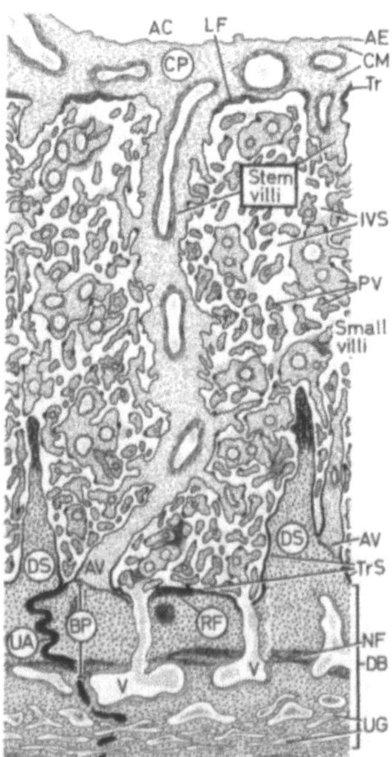

Becker, V., Schiebler, Th.H., Kubli, F. (eds.): Die Plazenta des Menschen. Stuttgart, New York: G. Thieme 1981; Göldner, H.-J.: Z. mikrosk.-anat. Forsch. *94*, 33 (1980); Kaufmann, P., King, B.F. (eds.): Structural and Functional Organization of the Placenta. Basel, München: Karger 1982; Ramsey, E.M.: The Placenta. Eastbourne: Praeger Scientific 1982

Placenta, function of: 1) Exchange of gases, electrolytes and metabolites between maternal and fetal blood, 2) excretion of fetal waste products into maternal blood, 3) production of ↑ placental hormones, 4) transmission of maternal antibodies to the fetus, 5) secretion of ↑ fibrinoid, etc.

Klopper, A. (ed.): Immunology of Human Placental Proteins. Eastbourne: Praeger Scientific 1982; Miller, R.K., Thiede, H. (eds.): Placenta: Receptors, Pathology and Toxicology. Eastbourne: Praeger Scientific 1981; Young, M., Boyd, E.D., Longo, L.D. Telegdy, G. (eds.): Placental Transfer. Eastbourne: Praeger Scientific 1981

Placenta, hemochorial: A type of ↑ placenta in which ↑ placental villi are directly exposed to maternal blood. Human placenta is of this type.

↑ Placenta, types of

Placenta, types of: Based on ↑ placental barrier composition, mammalian placentas are subdivided into: 1) Epitheliochorial, 2) syndesmochorial, 3) endotheliochorial, and 4) hemochorial. (See embryology texts for further information)

Steven, H.D. (ed.): Comparative Placentation, Essays in Structure and Function. New York: Academic Press 1975

Placenta, vascularization of: 1) Maternal placenta. Maternal blood penetrates ↑ intervillous spaces (IVS) via ↑ uteroplacental arteries (UA) and bathes ↑ placental villi (PV). From intervillous spaces, blood collects in maternal veins (V) and marginal sinus (MS) at border of placenta and from there reaches myometrial veins. 2) Fetal placenta. Oxygenation of fetal blood takes place in capillaries of placental villi through ↑ placental barrier. Arterial blood collects in umbilical vein (UV). Via umbilical arteries (UAr), venous blood is distributed to capillary network of placental villi. Each ↑ cotyledon has a separate blood supply. Under normal conditions, maternal blood and fetal blood are completely separated. Arrows indicate direction of blood flow.

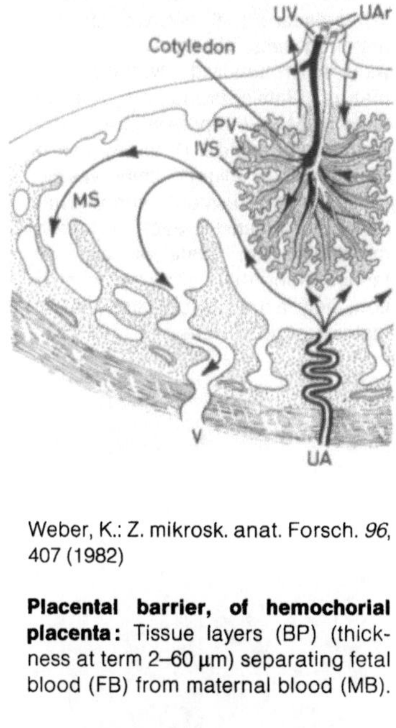

Weber, K.: Z. mikrosk. anat. Forsch. *96*, 407 (1982)

Placental barrier, of hemochorial placenta: Tissue layers (BP) (thickness at term 2–60 µm) separating fetal blood (FB) from maternal blood (MB).

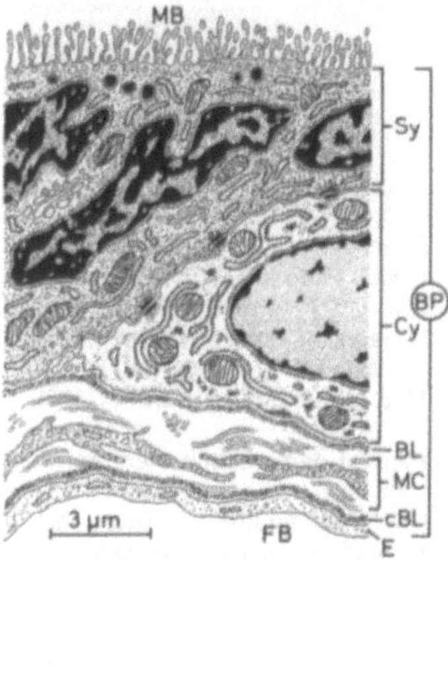

Structure: 1) ↑ Syncytiotrophoblast (Sy); 2) ↑ cytotrophoblast (Cy) (incomplete in second half of pregnancy); 3) their basal lamina (BL); 4) mesenchymal core (MC) of ↑ placental villi; 5) basal lamina of fetal capillaries (cBL), and 6) capillary ↑ endothelial cells (E). P., and especially its trophoblastic layers, acts as a membrane regulating selective transport of nutrients, wastes, and gases from maternal to fetal blood and vice versa. It also prevents penetration of some foreign materials (maternal blood, viruses, bacteria, toxins) into fetal circulation.

Friess, A.E., Sinowatz, F., Skolek-Winnish, R., Träutner, W.: Anat. Embryol. *158*, 179 (1980); Tedde, G., Delrio, A.N.: Arch. Ital. Anat. Embriol. *85*, 39 (1980); Wallenburg, H.C.S., Van Kreel, B.K., Van Dijk, H.P. (eds.): Transfer Across the Primate and Non-Primate Placenta. Eastbourne: Praeger Scientific 1981

Placental hormones: In addition to ↑ estrogens and ↑ progesterone, the ↑ placenta, particularly its ↑ syncytiotrophoplast, synthesizes ↑ human chorionic gonadotropin (HCG), ↑ human chorionic somatomammotropin (HCS), and ↑ human chorionic thyrotrophin (HCT).

Klopper, A., Genazzani, A., Crosignani, P.G. (eds.): The Human Placenta. New York, London: Academic Press 1981

Placental labyrinth: ↑ Intervillous spaces, of placenta; ↑ Placenta, full-term structure of

Placental villi, of early placenta (chorionic villi, villi chorioallantoici*): Large, plump, and poorly branched processes of chorionic plate covered by two layers of ↑ trophoblast. 1) Inner layer, ↑ cytotrophoblast (C) or Langhans' layer = layer composed of individual cells which decrease in number after 5th month of pregnancy. 2) Outer layer or ↑ syncytiotrophoblast (S) = continuous layer of multinucleated cytoplasm persisting until delivery. Fetal blood vessels (V) are of small caliber and in a central position; all fetal erythrocytes (E) are nucleated.

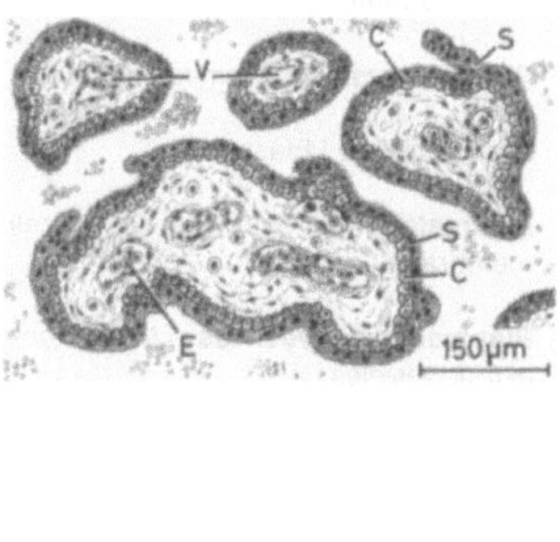

Kling, B.F., Mais, J.J.: Anat. Embryol. *165*, 361 (1982)

Placental villi, of full-term placenta (villi chorioallantoici*): Slender and richly branched processes of chorionic plate lined by ↑ syncytiotrophoblast (S) and only rare ↑ cytotrophoblastic cells. Fetal blood vessels (V) numerous and of great caliber, are situated just below ↑ trophoblast. P. are very numerous and divided into stem, intermediate, and terminal villi.

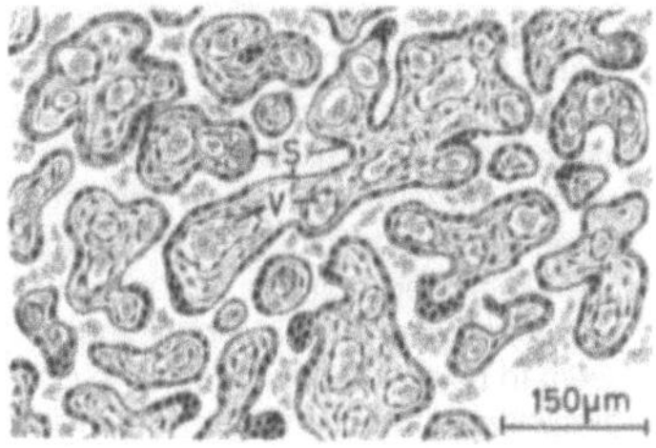

Kaufmann, R., Sen, D.K., Schweikhart, G.: Cell Tissue Res. *200*, 409 (1979); Sen, D.K., Kaufmann, R., Schweikhart, G.: Cell Tissue Res. *200*, 425 (1979)

Placentome: Term designating placental ↑ cotyledon.

Planar objective: ↑ Objective

Planimetry: One of basic methods of ↑ stereology for measuring surfaces of different cell and organ components in light and electron microscopy.

Planum semilunatum*: Area (PS) of tall hairless columnar cells at both ends of ↑ crista ampullaris (CA) where it joins wall of ↑ ampulla (A) of ↑ semicircular duct. C = ↑ cupula (somewhat shrunken). (Fig. = mongolian gerbil)

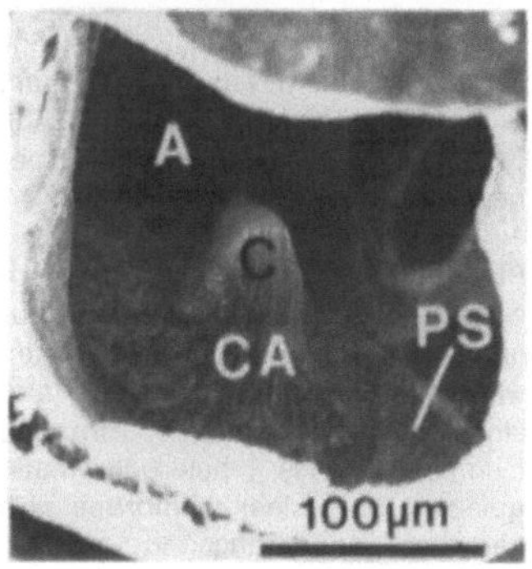

Plaques, of facet cells: Polygonal, well-delimited, slightly shallow and stiff areas (P) in apical ↑ cell membrane of

↑ facet cells. Arrangement of P. is similar to that of cobblestones. Surface area of individual P. varies from 0.05 to 0.25 μm^2; all P. together cover about 75% of total luminal surface. ↑ Unit membrane of P. is asymmetrical, as is that of ↑ discoid vesicles (DV); luminal leaflet measures 4.5 nm, cytoplasmic leaflet 3 nm. ↑ Freeze-cleaving reveals highly regular arrays of protein particles (MP) of about 12 nm in diameter with a center-to-center distance of 16 nm. Each particle consists of six 5-nm subunits (inset). P. probably play a role in impermeability of cell membrane to water. On the other hand, they serve as anchoring sites for microfilaments (Mf). Particle-free interplaque areas (Int) probably represent hinge areas between rigid P., allowing apical cell membrane to fold up in unstreched ↑ transitional epithelium. P. seem to be constantly renewed in Golgi apparatus and transported in form of discoid vesicles to apical plasmalemma where they fuse with it. (Modified after Staehelin et al., 1972)

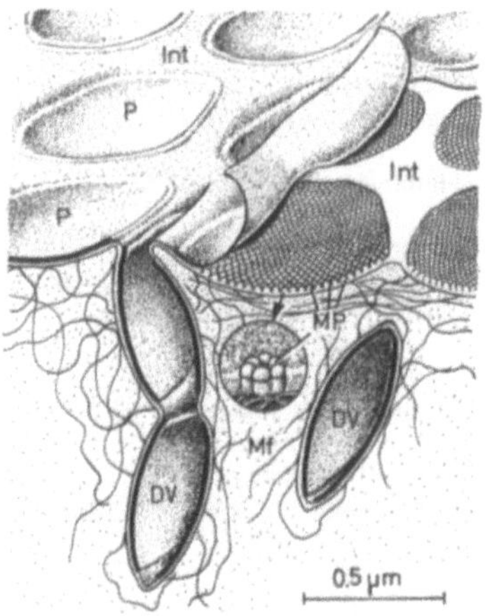

Knutton, S., Robertson, J.D.: J. Cell Sci. *22*, 355 (1976); Severs, N.J., Warren, R.C.: J. Ultrastruct. Res. *64*, 124 (1978); Staehelin, L.A., Chlapowski, F.J., Bonneville, M.A.: J. Cell Biol. *53*, 73 (1972)

Plasma cells (plasmacyte, plasmocytus*): Ovoid basophilic nongranulated cells, up to 20 µm in diameter, with round eccentric nucleus and small nucleolus. Radially arranged clumps of dense ↑ heterochromatin adjacent to nuclear membrane give cartwheel appearance to nucleus. Cytoplasm contains system of highly developed, frequently expanded rough endoplasmic (rER) cisternae in the form of ↑ ergastoplasm, responsible for ↑ basophilia of P. The moderately osmiophilic ↑ reticuloplasm consists of antibodies. In some normal P., and particularly during chronic infections, the rough endoplasmic cisternae are considerably distended, with masses of very dense ↑ acidophilic material, ↑ Russel fuchsinophilic bodies. Moderate numbers of mitochondria are situated between cisternae; a well-developed paranuclear Golgi apparatus (G), a prominent ↑ diplosome, scarce ↑ lysosomes (Ly), a few ↑ multivesicular bodies, and a great amount of free ribosomes are present. Cell surface is generally smooth. P. are constant elements of connective tissues, ↑ medullary cords of ↑ lymph nodes, and ↑ germinal centers of ↑ lymphatic nodules. They are weakly ameboid cells, with a life span of 10–30 days. P. synthesize and release ↑ immnoglobulins and are responsible for humoral immunity. Production sites: ↑ bone marrow, connective tissues, ↑ lymphatic tissue (in last case from ↑ B-lymphocytes).

↑ Immunoblasts; ↑ Immune response; ↑ Lymphocytes, cytophysiological subdivision of

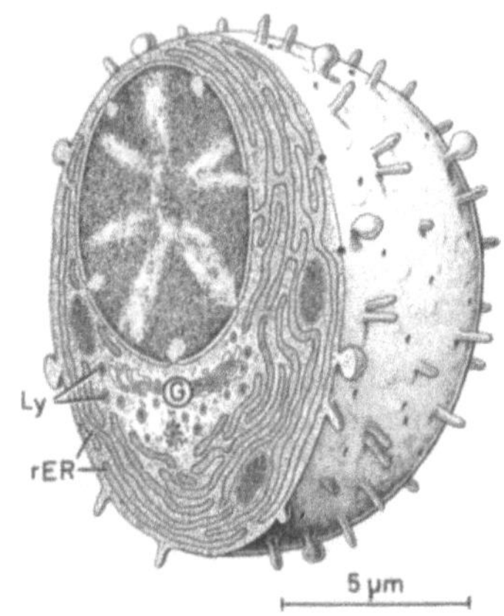

Plasma membrane: ↑ Cell membrane

Plasma, of blood*: Transparent, slightly yellowish fluid serving as transport vehicle for blood elements. (See biochemistry and physiology texts for further information)

Plasma, seminal: ↑ Seminal plasma

Plasmacytes: ↑ Plasma cells

Plasmalemma: Synonym for ↑ cell membrane.

Plasmalemmal vesicles: ↑ Micropinocytotic vesicles

Plasmoblast: Precursor of ↑ plasma cell similar to ↑ immunoblast.

Plasmodium: Protoplasmic mass containing many nuclei, resulting from successive amitotic divisions.

↑ Amitosis; ↑ Syncytium

Plasmolemma: ↑ Cell membrane

Plate, epiphyseal, of long bones: ↑ Epiphyseal plate

Plate, motor end: ↑ Motor end plate

Platelet-activating factor: A glyceryl ether phosphoglyceride secreted by ↑ mast cells provoking ↑ platelet aggregation and inducing release of ↑ serotonin.

Platelet demarcation channels: Three-dimensional system of meandering clefts (C), about 15–20 nm wide, which subdivide cytoplasm of a mature platelet-producing ↑ megakaryote (M) and represent boundaries of future ↑ platelets. P. are formed by confluence of aligned small vesicles (V) belonging to smooth endoplasmic reticulum. When P. reach surface of megakaryocyte, platelets are detached along them and enter circulation; only a thin cytoplasmic rim remains around nucleus.

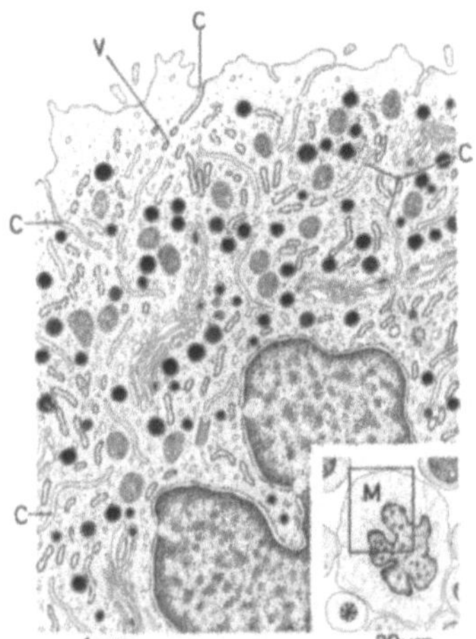

Behnke, O.: J. Ultrastruct. Res. *26*, 111 (1969); Fedorko, M.E.: Lab. Invest. *36*, 310 (1977); Ihzumi, T., Hattori, A., Sanada, M., Muto, M.: Arch. Histol. Jpn. *40*, 305 (1977); Shaklai, M., Tavassoli, M.: J. Ultrastruct. Res. *62*, 270 (1978)

Platelet demarcation membranes: ↑ Unit membranes of interconnected smooth endoplasmic vesicles lining ↑ platelet demarcation channels.

Platelet-producing megakaryocyte: ↑ Megakaryocyte

Platelet ribbons: Mode of platelet separation from platelet-producing ↑ megakaryocyte (M) in the form of long wormlike projections (PR) that emerge into lumen of vascular sinus (S) of ↑ bone marrow. P. fragment into single ↑ platelets in bloodstream.

↑ Bone marrow, vascular sinuses of

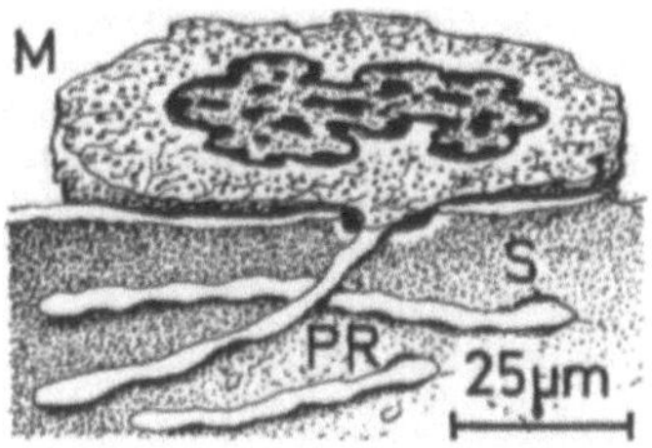

Platelet satellitism: Poorly understood phenomen of platelet agglomeration around nonlymphocytic ↑ leukocytes.

McGregor, D.H., Davis, J.W., Liu, P.I., Gates, B.A., Poindexter, A.R.: Lab. Invest. *42*, 343 (1980)

Platelet thrombus: An aggregation of ↑ platelets at site of damaged ↑ endothelium.

Platelets (thrombocyte, thrombocytus*): Round or oval biconvex anucleate corpuscle, 2–4 µm in diameter, involved in ↑ blood clotting. Cytoplasm is subdivided into two zones: 1) Central, granulomere (Gr) or chromomere, is highly refractile and contains several 0.2–0.3-µm purple ↑ α-granules (α) with a moderately dense contents, considered ↑ lysosomes, although they possibly contain ↑ thromboplastin, ↑ fibrinogen, and platelet factor 4. Another type of granules are very dense granules (VDG) which contain ADP, Ca^{2+} and in some species ↑ serotonin. Rare mitochondria, short rough endplasmic cisternae, ↑ glycogen particles, and ribosomes may also be found in the granulomere. A system of smooth interconnecting tubules (IT) filled with dense or light material communicates with external milieu. 2) Peripheral zone or hyalomere (Hm) is pale and homogeneous; it contains ↑ actin and ↑ myosin microfilaments (Mf) and a marginal ring of ↑ microtubules (Mt) which probably maintain the form of P. P. are slightly phagocytic. A cell membrane (about 8 nm thick) with an external 50-nm-thick ↑ glycocalyx (Gl; only partly represented), containing ↑ fibrinogen and thromboplastin, surrounds P. This coat is responsible for the agglutination of P. to one another and to ↑ endothelium. P. are produced by fragmentation of ↑ megakaryocyte cytoplasm. There are 250000–300000 P./mm³ of blood (250–300 × 10^9/l, ↑ SI). P. have life span of about 10 days; they are destroyed in ↑ spleen and ↑ lungs.

↑ Platelet demarcration channels; ↑ Thrombosthenin

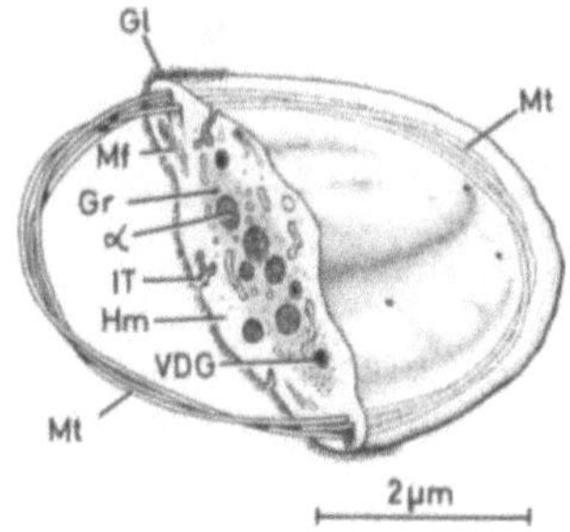

Debus, E., Weber, K., Osborn, M.: Eur. J. Cell Biol. *24*, 45 (1981); Hattori, A.: Biomed. Res. *2*, 199 (1981); Morgenstern, E.: Eur. J. Cell Biol. *26*, 315 (1982); Nachmias, V.T.: J. Cell Biol. *86*, 795 (1980); Pham, T.D., Kaplan, K.L., Butler, V.P.Jr.: J. Histochem. Cytochem. *31*, 905 (1983)

Platelets, release of: Process of liberation of ↑ platelets into blood by fragmentation of cytoplasm of platelet-producing ↑ megakaryocyte (PPM). 1) Within cytoplasm of such a megakaryocyte, generally adjacent to outer aspect of ↑ endothelial cells (EC) of ↑ bone marrow vascular sinuses (VS), azurophilic granules (AG) become clustered and aligned; beween them appear vesicles (V) which elongate, fuse, and form a more or less continuous three-dimensional system of ↑ platelet demarcation channels (C). 2) As soon as channels reach cell surface, megakaryocyte cytoplasm is fragmented; clusters of individual platelets (P) or ↑ platelet ribbons (R) are liberated and enter bloodstream through mural apertures of sinuses. In bloodstream, ribbons of platelets separate into individual platelets; around megakaryocyte nucleus only a thin cytoplasm zone remains. After having formed about 2000 platelets, megakaryocyte undergoes ↑ degeneration.

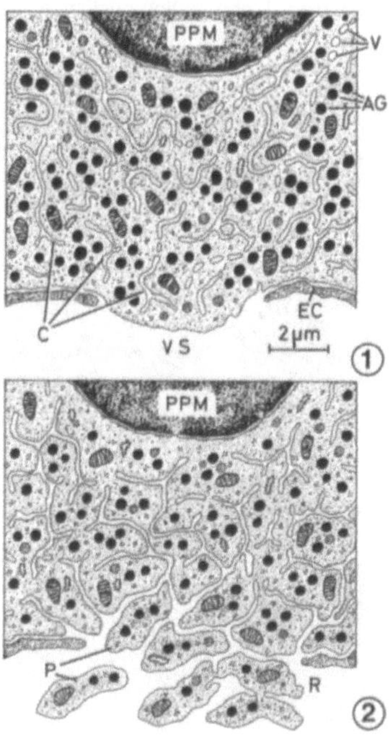

Becker, R.P., De Bruyn, P.P.: Am. J. Anat. *145*, 183 (1976)

Plates, of liver parenchymal cells: ↑ Hepatic plates

Platinum shadowing: ↑ Shadowing

Pleura (pleura pulmonaris*): Serous membrane lining ↑ lung (visceral P.) and interior surface of thoracic cavitiy (parietal P.): Structure: A) Visceral P.: 1) ↑ Mesothelium (M); 2. tela subserosa (S) = a ↑ dense connective tissue with numerous ↑ elastic fibers radiating into interlobular septa. Vascularization consists of ↑ bronchial arteries; ↑ lymph is drained into tracheobronchial ↑ lymph nodes. There are numerous sensitive nerve endings. B) Parietal P., is considerably thicker than visceral P., but with same stratification. In tela subserosa are some adipose cells. (Fig. = visceral P., human)

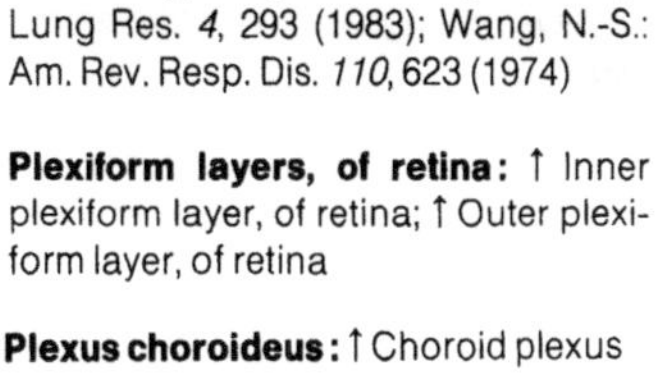

Mariassay, A.T., Wheeldon, E.B.: Exp. Lung Res. *4*, 293 (1983); Wang, N.-S.: Am. Rev. Resp. Dis. *110*, 623 (1974)

Plexiform layers, of retina: ↑ Inner plexiform layer, of retina; ↑ Outer plexiform layer, of retina

Plexus choroideus: ↑ Choroid plexus

Plexus, myenteric: ↑ Myenteric plexus

Plexus, of Auerbach: ↑ Myenteric plexus

Plexus, of Meissner: ↑ Submucosal plexus

Plexus, submucosal: ↑ Submucosal plexus

Plexus vesicalis: ↑ Urinary bladder

Plica secretoria utriculi: Mucosal fold projecting into membranous ↑ labyrinth in region of posterior opening of lateral ↑ semicircular duct into ↑ utricle. P. is lined by ↑ simple columnar epithelium composed of two cell types, both having highly specialized morphofunctional relationships with underlying ↑ loose connective tissue of lamina propria. It is believed that P. plays a role in composition of ↑ endolymph.

Hanak, H., Stockinger, L., Vyslonzil, E.: Cytobiologie *2*, 139 (1970)

Plicae circulares* (valves of Kerckring): Permanent transverse or spiral semilunar folds (PC) of both tunica mucosa and tela submucosa extending around one-half to two-thirds of circumference of small intestine. P. begin about 5 cm distal to ↑ pylorus, reaching their greatest development in distal portion of ↑ duodenum and proximal portion of ↑ jejunum; they gradually decrease in height and complexity in middle of ↑ ileum: P. provide an increase of intestinal absorptive surface.

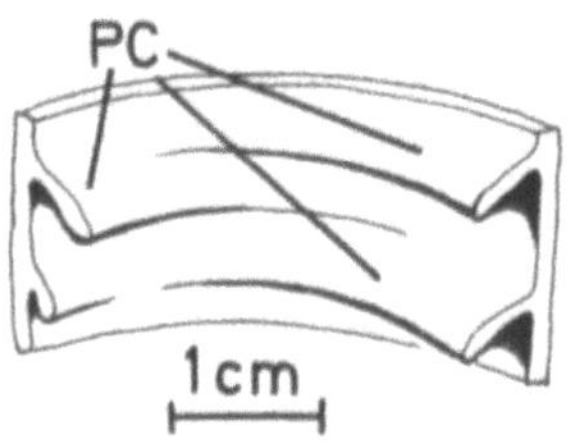

Plicae semilunares*: Folds of ↑ colon wall (PS) between two ↑ haustrae (H) formed by circular layer of tunica muscularis. P. may disappear, but never change their position along colon.

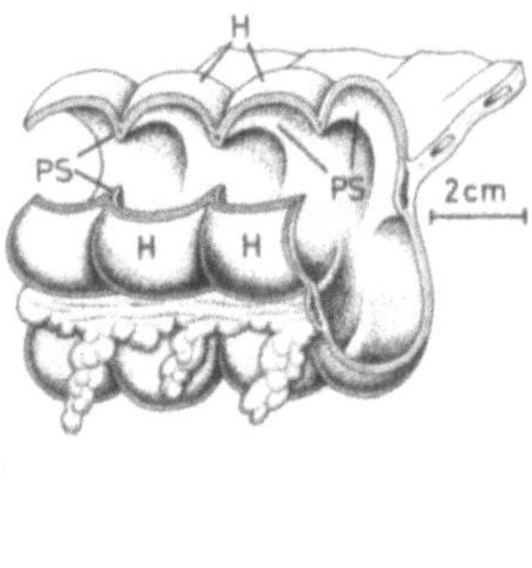

Plicae synoviales* (plicae alares): Flaplike protrusions of various length (PA) of ↑ synovial membrane into articular cavity (AC). ↑ Smooth muscle cells within P. probably serve to prevent them from being crushed between two articular surfaces. C = ↑ articular cartilage

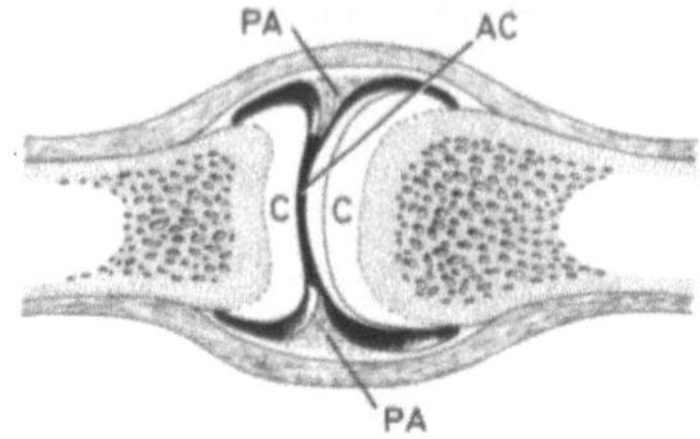

Pluricellular glands: ↑ Multicellular glands

Pluristratified glands: ↑ Exocrine glands composed of several layers of cells. P. are glands with a ↑ holocrine mechanism of secretion (↑ sebaceous glands, ↑ tarsal glands).

↑ Glands, classification of

Pneumonocytes, type I and II: ↑ Alveolar cells, type I and II

Podocytes (podocytus*): Large epithelial cells (P) surrounding external aspect of ↑ glomerular capillaries (GC). ↑ Perikaryon is round or elongated and bulges into ↑ Bowman's space (BS). Nucleus is oval with finely dispersed ↑ chromatin; nucleolus is prominent. Cytoplasm is clear, containing round or elongated mitochondria, well-developed and frequently multiple Golgi apparatus, short cisternae of rough endoplasmic reticulum, some ↑ lysosomes, ↑ microtubules, ↑ microfilaments, and a few free ribosomes. From cell body emerge long primary processes or trabeculae (PP) that ramify into secondary processes (SP) which then branch into tertiary processes (TP). All processes form ↑ pedicels (Pe) which are extremely interdigitated with pedicels of same or adjacent P. Pedicels rest on external surface of capillary ↑ basal lamina (BL). P. represent visceral layer of ↑ Bowman's capsule.

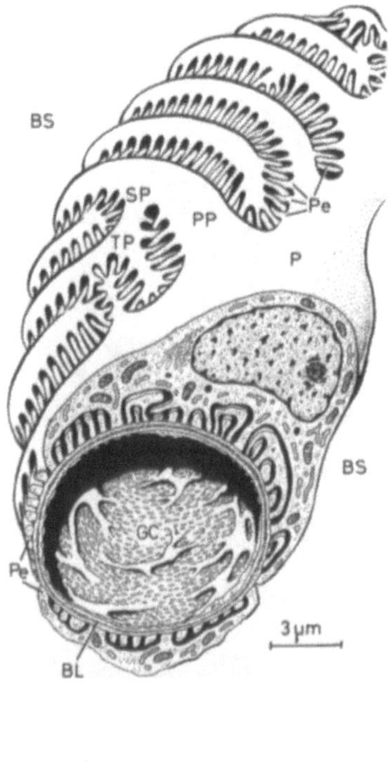

Fujita, T., Tokunaga, J., Edanaga, M.: Cell Tissue Res. *166*, 299 (1976); Fujita, T., Tanaka, K., Tokunaga, J.: SEM Atlas of Cells and Tissues. Tokyo, New York: Igaku-Shoin 1981; Zawistowski, S., Roszkievicz, J.: Ann. Acad. Med. Gedanensis *9*, 59 (1979)

Podocytic membranes: ↑ Podocytic plates

Podocytic plates (podocytic membranes): Attenuated cytoplasmic sheets extending from ↑ perikarya of ↑ podocytes and covering many podocytic processes and their ↑ pedicles.

Yoshinari, T., Fujita, T.: Arch. Histol. Jpn. *45*, 99 (1982)

Poietins: Partly hypothetical hormones stimulating transformation of ↑ hematopoietic stem cells into different lineages of blood cells (e.g., ↑ erythropoietin, ↑ leukopoietin, ↑ thrombopoietin).

Poikilocytosis: Deviation of ↑ erythrocytes from their normal shape.

Point counting: A basic method of ↑ stereology for calculation of volume density, surface density, and other parameters.

Polar bodies (polocytes, polocytus primarius et secundarius*). Minute cells occurring as a result of unequal cytoplasmic division during ↑ oogenesis. Frist (reductional) meiotic division (1) of primary ↑ oocyte (O) ends just before ↑ ovulation in ↑ haploid distribution of ↑ chromosomes between daughter cells; however, first P. (2, P I) receives only a minimum of cytoplasm. Its nucleus is composed of isolated ↑ chromosomes (Chr). It divides once more (3). After formation of first P., secondary ↑ oocyte enters into second (equational) meiotic division which arrests at metaphase (3). Only following

↑ fertilization (3), a second P. (4, P II) is expelled, like first, into ↑ perivitelline space (PVS). Nucleus of the second P. is lined by a regular ↑ nuclear envelope. All P. rapidly disintegrate and disappear. ZP = ↑ zona pellucida

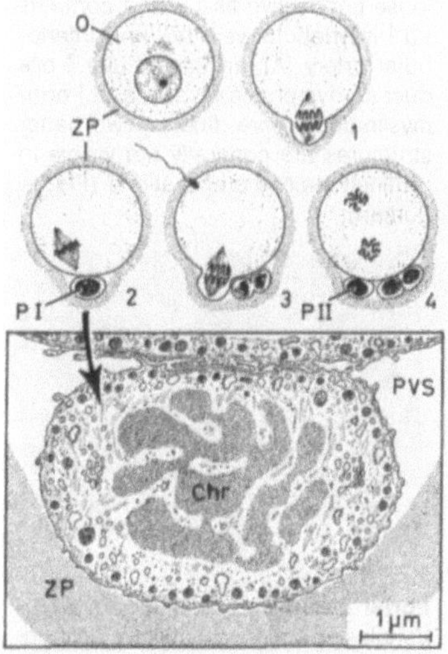

Hamaguchi, M.S., Hiramoto, Y.: Exp. Cell Res. *112*, 55 (1978); Zamboni, L., Mishell, D.S., Bell, L.H.Jr., Baca, M.: J. Cell Biol. *30*, 579 (1966)

Polar cushion: ↑ Mesangium, extraglomerular

Polar terminals (polar processes): Clublike endings of ↑ pinealocyte extensions penetrating into pineal ↑ canaliculi and ↑ perivascular spaces. P. contain mitochondria, ↑ grumose bodies, ↑ microtubules, rough endoplasmic cisternae, and smooth endoplasmic tubules. It is believed that ↑ pinealocytes release secretory product through P. and/or receive information concerning composition of surrounding ↑ cerebrospinal fluid.

Polarization, of cells: Structural and functional differences between ↑ apical pole and ↑ basal pole of same epithelial cell conditioned by its activity and/or transcellular transport. P. is also visible in ↑ wandering cells whose anterior pole has morphofunctional properties other than those of the posterior pole.

↑ Uropodium

Polarizer: Nicol prism or sheet of polaroid film placed between light source and ↑ condenser of a ↑ polarizing microscope to polarize light.

↑ Analyzer

Polarizing microscope: Optical instrument for indirect ultrastructural study of molecular arrangement of some tissue components by analysis of ↑ anisotropy and/or ↑ birefringence. Principle: From light source (L), ordinary light is polarized in ↑ polarizer (P), concentrated by ↑ condenser (C) on plane of specimen stage (S), refracted by ↑ objective (O), analyzed by ↑ analyzer (A), and observed through ↑ ocular (Oc). 1) With prisms in parallel position, polarized light passes through, and field of view (F) is bright. 2) With prisms in crossed position, polarized light is not transmitted and field of view (F) is dark. 3) If an anisotropic specimen (Sp) is placed on specimen stage when prisms are perpendicular, plane of polarization will deviate. By rotation of specimen stage or analyzer one can determine any change in character of polarization induced by object, and hence arrangement of molecules within it.

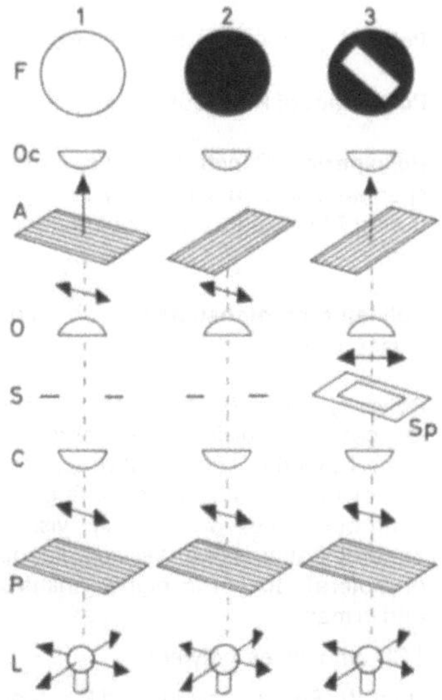

Polkissen: Mesangium, extraglomerular

Polocytes: ↑ Polar bodies

Polyblast: Intermediate stage of a ↑ monocyte during transformation into a ↑ histiocyte after having left circulation.

Polychromatophilic erythroblast: ↑ Erythroblast, polychromatophilic

Polychromatophilic normoblast: ↑ Erythroblast, polychromatophilic

Polykaryocyte: ↑ Osteoclast

Polymorphonuclear granulocyte: ↑ Granulocyte, basophilic; ↑ Granulocyte, eosinophilic; ↑ Granulocyte, neutrophilic

Polypeptide hormone-producing cells: ↑ Endocrine cells

Polypeptide hormones, action on the cell of: ↑ Hormone (H) as "first messenger" is recognized and bound to a specific receptor (SR), probably a complementary part of hormone molecule, located at external surface of ↑ cell membrane (CM). This hormone-receptor interaction, supported by a lipid transducer (TR), activates ↑ adenyl cyclase. Acting on ATP to form ↑ cyclic AMP (cAMP), adenyl cyclase amplifies the received signal to factor 10^9. Cyclic AMP, considered "second messenger," activates intracellular kinases which in turn activate other enzymes, leading to specific cellular response. Hormones acting this way include ↑ epinephrine, ↑ norepinephrine, ↑ adrenocorticotropic hormone, ↑ thyrotropic hormone, ↑ melanocyte-stimulating hormone, ↑ parathyroid hormone, ↑ luteinizing hormone, ↑ antidiuretic hormone, ↑ thyroxin, and probably some others.

↑ Steroid hormones, action on the cell of

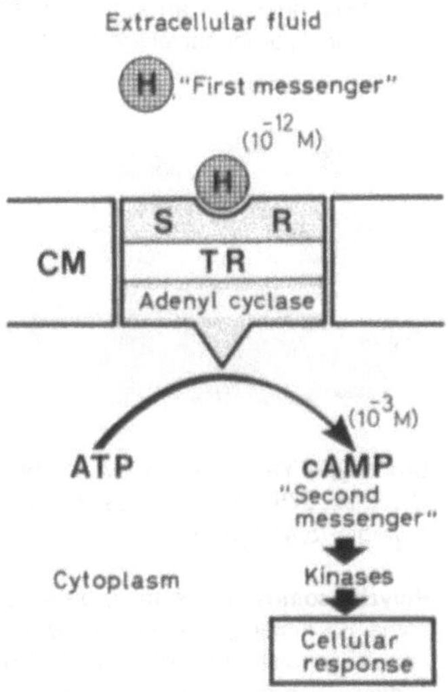

Kühnau, J.: Physiologie und Biochemie der endokrinen Regulationen und Korrelationen. In: Bargmann, W., Kühnau, J., Siebemann, R.E., Steiner, H., Uehlinger, E. (eds.): Handbuch der Allgemeinen Pathologie. Endokrine Regulations- und Korrelationsstörungen; Vol. 8/1. Berlin, Heidelberg, New York: Springer-Verlag 1971; Litwack, G. (ed.): Biochemical Actions of Hormones, 9.

New York: Academic Press 1982; Robinson, G.A., Sutherland, E.W.: Amer. N.Y. Acad. Sci. *185*, 5 (1971)

Polyphyletic theory, of hematopoiesis: ↑ Hematopoiesis, theories of

Polyploidization: Process of DNA replication with subsequent duplication of the chromosomal set, but without ↑ karyokinesis. P. can be a consequence of ↑ endomitosis, ↑ cryptoendomitosis, nuclear fusion, or formation of ↑ nuclear envelope after ↑ metaphase. Reasons for P. are poorly understood.

Polyploidy (cellula polyploidea*): State of cell ↑ nucleus containing various multiples of ↑ haploid number of ↑ chromosomes. Majority of human ↑ somatic cells are ↑ diploid (2n), having 23 pairs of chromosomes (23 + 23 = 46). Tetraploid nuclei (4n) contain 92 chromosomes (46 + 46 = 92), etc. Most frequently, P. is result of duplication of chromosomal set with a consecutive increase in nuclear volume. P. exists in highly differentiated cells without or with very limited mitotic capacity to adapt to high functional demand, e.g., ↑ liver parenchymal cells (8n and 16n), ↑ megakaryocytes (32n, 64n, or more). [Fig. = liver, monkey. A polyploid nucleus (arrow) contains two nucleoli; other nuclei are diploid]

↑ Polyploidization

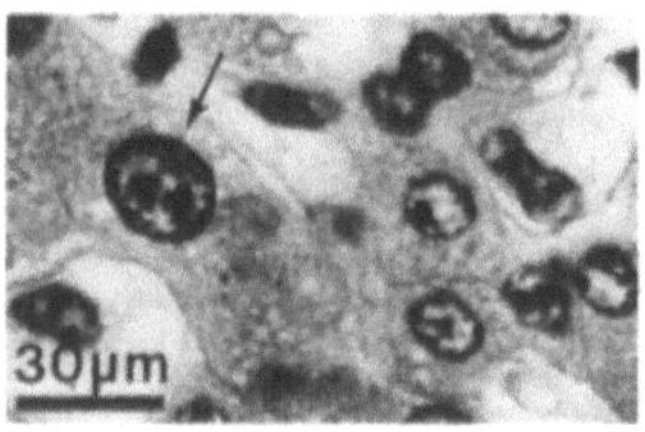

Brodsky, W.Ya., Uryvyeva, I.V.: Int. Rev. Cytol. *50*, 275 (1977); James, J.: Cytobiologie *15*, 410 (1977)

Polyribosomes (polysomes, polyribosoma*): Clusters, rosettes, helices, or spirals (arrowheads) of 3–30 ↑ ribosomes which may occur freely in cytoplasm or attached to membranes of rough endoplasmic reticulum. P. are held together by a 1.5-nm-thick filament of ↑ messenger RNA. (Fig. = ↑ liver parenchymal cell, rat)

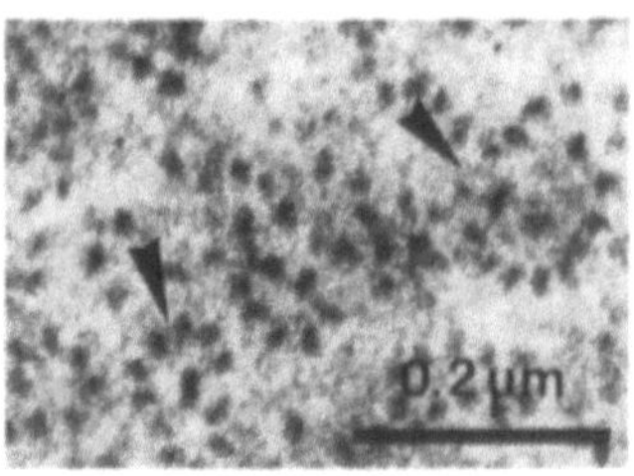

↑ Polyribosomes, helicoidal

Giddings, H.T., Staehelin, A.L.: J. Cell Biol. *85*, 147 (1980); Martin, K.A., Miller, O.L.Jr.: Dev. Biol. *98*, 338 (1983)

Polyribosomes, helicoidal: Corkscrew-shaped clusters of ↑ ribosomes (HP) involved in synthesis of microfibrillar structures, principally ↑ myofilaments (Mf). (Fig. = young ↑ skeletal muscle fiber, rat)

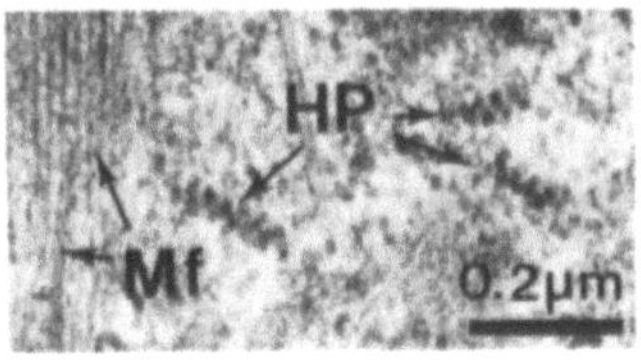

Polys: ↑ Granulocyte, neutrophilic

Polysomes: ↑ Polyribosomes

Polyspermy: Penetration by several ↑ spermatozoa of a secondary ↑ oocyte during ↑ fertilization; does not occur in man.

Polytenic chromosomes: ↑ Chromosomes, polytenic

Polyteny: Process of repeated ↑ DNA replication occurring in a ↑ chromosome, but without consecutive separation of ↑ chromatids. P. leads to giant ↑ chromosomes with easily visible banding patterns (e.g., salivary glands of diptera); absent in higher animals and in man.

↑ Chromosomes, polytenic

Nagl, W.: Endopolyploidy and Polyteny in Differentiation and Evolution. Amsterdam: Elsevier/North Holland 1978

Pore complex: ↑ Nuclear pores

Pores, of capillaries: ↑ Capillary pores, of fenestrated capillaries

Pores, of glomerular capillaries: ↑ Capillary pores, of glomerular capillaries

Pores, of Kohn: ↑ Alveolar pores

Porphyrins: A large group of endogenous noniron ↑ pigments consisting of four pyrrole molecules joined in a ring. One P., protoporphyrin III, forms the heme of ↑ hemoglobin when bound to an iron atom.

Portal canals (portal area, canalis portalis*): Small triangular or round spaces (PC) at corners of classic ↑ liver lobules (LL). Each P. consists of ↑ loose connective tissue and contains an ↑ interlobular vein (V), an ↑ interlobular artery (A), an interlobular ↑ bile duct (B), lymphatic vessels, and ↑ nonmyelinated nerve fibers. Two latter structures are generally not visible in routinely stained preparations. (Fig. = human)

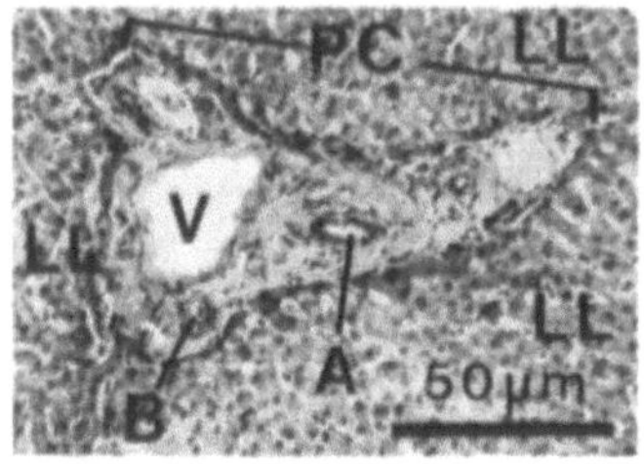

Portal circulation, of hypophysis: ↑ Venules connecting capillary network of pars tuberalis with that of pars distalis.

↑ Hypophysis, vascularization of

Portal lobule, of liver: ↑ Liver lobule, portal

Portal system (rete mirabile): A vascular network (capillaries, sinus, venules) interposed in course of an artery or vein. 1) Arterial P. = renal artery ... ↑ arteriola afferens (Aff) – ↑ glomerular capillaries (GC) – ↑ arteriola efferens (Eff) – ↑ peritubular capillary network (PCN) ... renal vein (RV). 2) Venous P. = vena portae ... ↑ interlobular vein (ILV) – ↑ inlet veins (InV) – ↑ liver sinusoids (S) – ↑ central vein (CV) ... hepatic vein (HV). 3) Hypophyseoportal system = superior hypophyseal artery – capillaries of the pars tuberalis – portal venules – capillaries of the pars distalis – hypophyseal veins.

↑ Hypophysis, vascularization of; ↑ Kidney, vascularization of; ↑ Liver, vascularization of; ↑ Microvascular bed

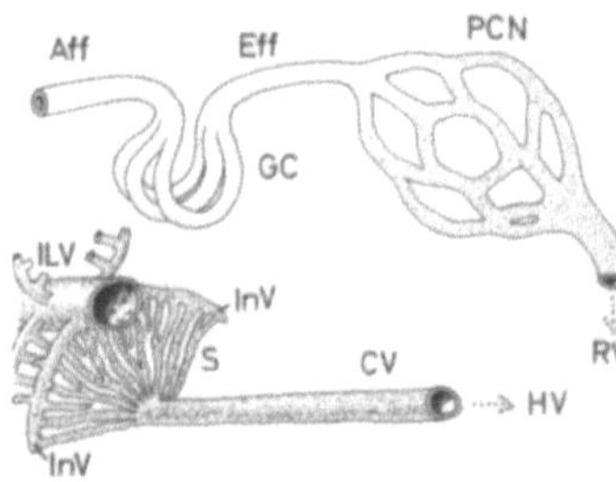

Portal triad (trias hepatica*): Term designating three most prominent elements of a ↑ portal canal: ↑ Interlobular vein, ↑ interlobular artery, and interlobular ↑ bile duct.

Portal vein (vena portae*): Large propulsive vein identical in structure to ↑ vena cava inferior.

Portio vaginalis, of uterus: ↑ Cervix, of uterus

Postacrosomal region, of spermatozoa: Part of ↑ spermatozoon head (PR) between caudal end of ↑ acrosome (A) and neck. Plasmalemma of P. is characterized by a slight thickening, postacrosomal sheath (postacrosomal dense lamina). P. is first region to fuse with plasmalemma of secondary ↑ oocyte during ↑ acrosome reaction.

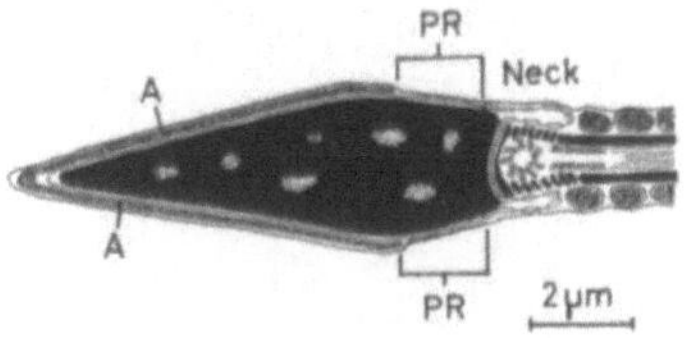

Postcapillary venules (pericytic venule, venula postcapillaris*): Blood vessels, 10–50 µm in diameter and 50–700 µm in length, receiving blood from venous ↑ capillaries. Structure: 1) Tunica intima = layer of ↑ endothelial cells (E), about 0.2–4 µm thick, held together by frequent overlapping and relatively limited ↑ zonulae occludentes. In endothelium, a few clusters of ↑ capillary pores may be present, as well as frequent ↑ myoendothelial junctions. 2) Tunica media = layer of flat ↑ pericytes (P). 3) Tunica adventitia (A) = layer of thin ↑ collagen fibers, threadlike fibrocytic processes, and some ↑ wandering cells. P. are believed to represent an important area of blood-interstitial fluid exchange. (Fig. = rat)

↑ Microvasculature

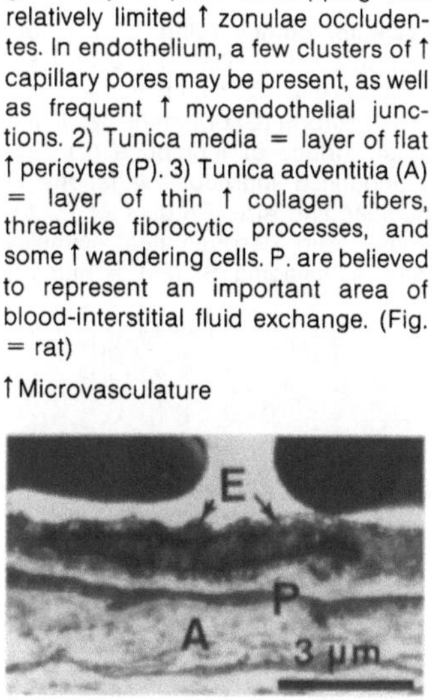

Cho, Y., De Bruyn, P.H.: J. Ultrastruct. Res. *69*, 13 (1979); Simionescu, N., Simionescu, M., Palade, G.E.: J. Cell Biol. *79*, 27 (1978)

Postcapillary venules, of lymph nodes (high-endothelium venules, venula postcapillaris*): Venous vessels (V), about 30–50 µm in diameter, situated in inner ↑ cortex of lymph nodes and ↑ lymphatic tissue of ↑ tonsils. ↑ Endothelium (E) is made up of stellate cuboidal to columnar cells with swollen bodies and several lateral processes interlaced with processes of adjacent cells. Wall is infiltrated by numerous ↑ lymphocytes (L) that pass from blood into perivascular thymus-dependent lymphatic tissue (LT). P. are a major pathway used by ↑ B-lymphocytes and ↑ T-lymphocytes to enter ↑ lymph nodes. (Fig. = rat)

↑ Lymphocytes, recirculation of

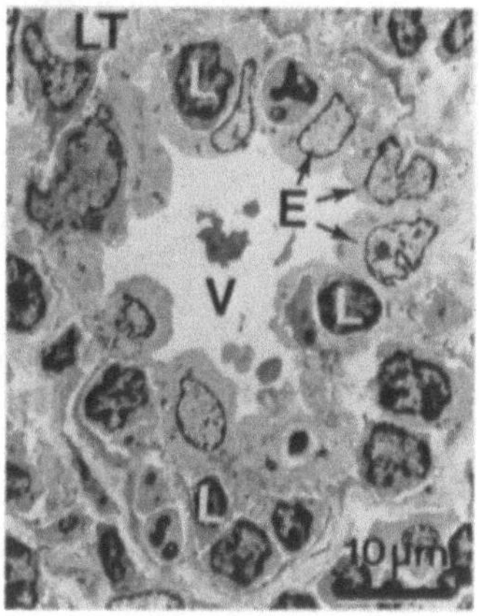

Anderson, A.O., Anderson., N.D.: Am. J. Pathol. *80*, 387 (1975); Andrews, P., Milsom, D.W., Ford, W.L.: J. Cell Sci. *57*, 277 (1982); Fujita, T., Tanaka, K., Tokunaga, J.: SEM Atlas of Cells and Tissues. Tokyo, New York: Igaku Shoin 1981; Irino, S,. Takasugi, N., Murakami, T.: Scanning Electron Microscopy 1981/III, 89 (1981); Yamaguchi, K., Schoefl, G.I.: Anat. Rec. *206*, 419 (1983)

Posterior chamber, of eye (camera posterior bulbi*): Narrow space (PC) filled with ↑ aqueous humor and limited by ↑ lens (L), posterior surface of ↑ iris (I), ↑ ciliary body (CB), and ↑ vitreous body (VB). P. communicates with ↑ anterior chamber (AC) through narrow slit between iris and lens.

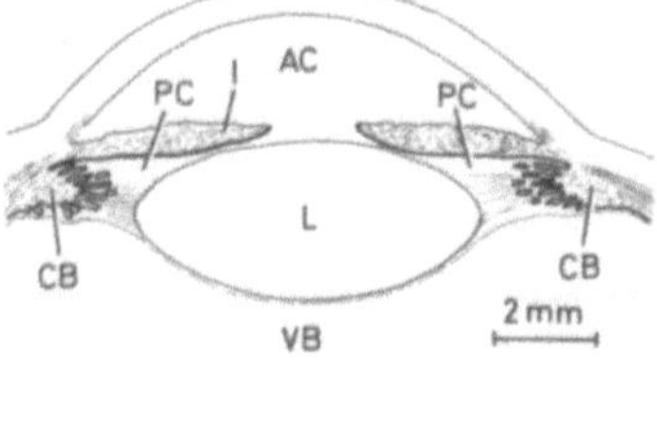

Posterior epithelium, of cornea: ↑ Cornea

Posterior glands, of tongue (glandulae linguales posteriores*): 1) ↑ Serous glands of Ebner associated with ↑ circumvallatae papillae. 2) ↑ Mucous glands of root of ↑ tongue.

Posterior lobe, of hypophysis: ↑ Hypophysis, lobus nervous of

Posterior pituitary: ↑ Hypophysis, lobus nervous of

Posterior ring, of spermatozoon head: ↑ Postacrosomal region; ↑ Spermatozoon, head of

Postganglionic nerve fibers: Mostly ↑ nonmyelinated axons of ↑ postganglionic neurons forming gray ramus communicans and visceral nerves. P. include ↑ pilomotor, ↑ sudomotor, and ↑ vasomotor nerve fibers, as well as ↑ sympathetic ground plexus for innervating glands and skeletal muscles. With exception of sudomotor nerve fibers, which are ↑ cholinergic, all P. are ↑ adrenergic.

Postganglionic neurons: ↑ Multipolar ↑ autonomic neurons situated in ↑ autonomic ganglia.

Postnatal hematopoiesis: ↑ Hematopoiesis, postnatal

Postsynaptic cisternae (subsynaptic cisternae): Particularly long and flattened ↑ subsurface cisternae (PC) parallel to basal plasmalemma of outer ↑ hair cells (HC) facing efferent nerve endings (Eff). Significance unknown; possibly involved in accumulation of Ca^{2+} ions. Aff = afferent nerve endings

↑ Corti's organ

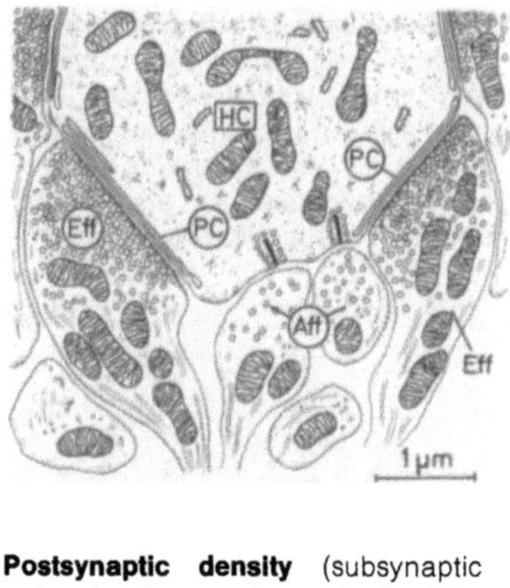

Postsynaptic density (subsynaptic web, densitas postsynaptica*): Fine dense filamentous material (PD)

spreading short distance from postsynaptic membrane (arrowheads) into cytoplasm of postsynaptic cell. It is believed that ↑ actin filaments associated with P. play role in synaptic plasticity. (Fig. = brain, rat)

↑ Presynaptic density; ↑ Synapse, chemical

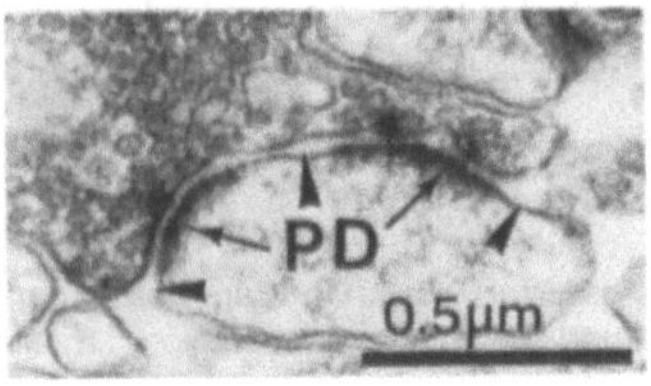

Fifková, E., Delay, R.J.: J. Cell Biol. *95*, 345 (1982); Matus, A. Pehling, G., Ackermann, M., Maeder, J.: J. Cell Biol. *87*, 346 (1980)

Potassium bichromate staining: ↑ Chromaffin reaction

PP-cells: ↑ Pancreatic polypeptide cells

Prebone: ↑ Osteoid

Precapillary arterioles: ↑ Metarterioles

Precapillary sphincter area (precapillary sphincter, sphincter precapillaris*): A slight thickening of tunica media of a ↑ metarteriole around site where it spreads into ↑ capillaries. P. is innervated by vasoconstrictor ↑ adrenergic nerve fibers; it is thought that P. play a role in regulation of blood flow throughout ↑ microvascular bed.

↑ Cardiovascular system; ↑ Microvasculature

Rhodin, J.A.G.: J. Ultrastruct. Res. *18*, 181 (1967)

Precursor cells, of adenohypophysis (reserve cells): Small chromophobic cells with polygonal cell body, little cytoplasm, and relatively voluminous nucleus. All cell ↑ organelles are very poorly developed; in transmission electron microscope, only very occasional electron-dense granules can be detected. It is thought that P. are resting elements which may transform into either acidophilic or basophilic cells; they may be degranulated chromophilic cells, especially ↑ corticotropes.

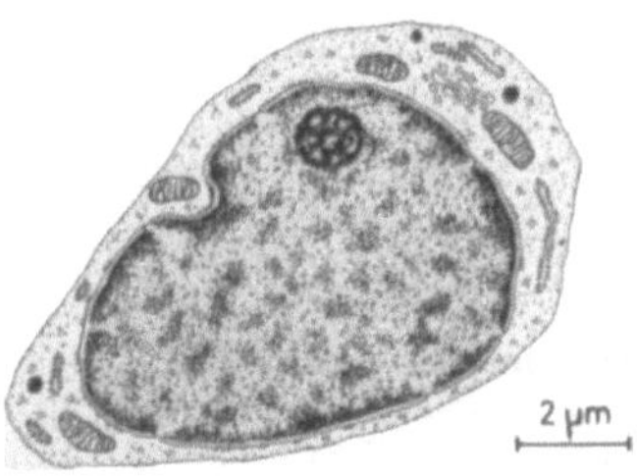

↑ Hypophysis, chromophobic cells of

Predentin (predentinum*): Nonmineralized organic material (P) produced by ↑ odontoblasts. P. consists of randomly oriented ↑ collagen microfibrils (C) of varied width embedded in an amorphous ↑ proteoglycan matrix. Through action of ↑ dentinal globules and collagen material, P. becomes calcified and transformed into ↑ dentin (D). (Fig. = newborn rat)

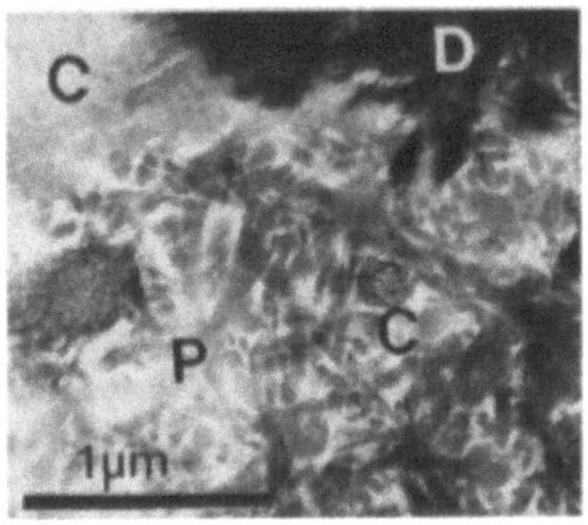

Preelastic fibers: ↑ Oxytalan fibers

Preenzyme granules: ↑ Zymogen granules

Preganglionic nerve fibers: Small ↑ myelinated axons of ↑ preganglionic neurons of ↑ spinal cord making synaptic contacts with ↑ postganglionic neurons of ↑ autonomic ganglia. All P. are ↑ cholinergic.

Preganglionic neurons: Small ↑ multipolar nerve cells of ↑ spinal cord whose ↑ axons form ↑ preganglionic nerve fibers and extend into ↑ peripheral nervous system to synapse with ↑ autonomic neurons of ↑ autonomic ganglia. P. of ↑ sympathetic nervous system form ↑ intermediolateral and ↑ intermediomedial nucleus between segments C_8 and L_3; those of ↑ parasympathetic nervous system form ↑ intermediomedial and ↑ intermediolateral nucleus between segments S_2 and S_4.

↑ Spinal cord, nuclei of

Premelanosomes (premelanosoma*): Former term for stage I melanosomes occurring during ↑ melanogenesis.

↑ Melanosomes

Prenatal hematopoiesis: ↑ Hematopoiesis, prenatal

Preosseous tissue: ↑ Osteoid

Preosteoblasts: ↑ Osteoprogenitor cells

Prepuce (preputium*): Circular cutaneous fold (P) overlying ↑ glans penis (G) attached to corona glandis and frenulum. Skin (OS) of outer surface is thin, elastic, without ↑ lanugo hairs, but with rare ↑ sweat and ↑ sebaceous glands that occur in absence of ↑ hair follicles. Inner skin (IS) is thinner than skin of outer surface and almost completely free of sweat and sebaceous glands. In common subcutaneous layer, there is no adipose tissue but considerable number of blood vessels and bundles of ↑ smooth muscle cells (F) and ↑ elastic fibers (El). (Fig. = adult human, ↑ orcein staining)

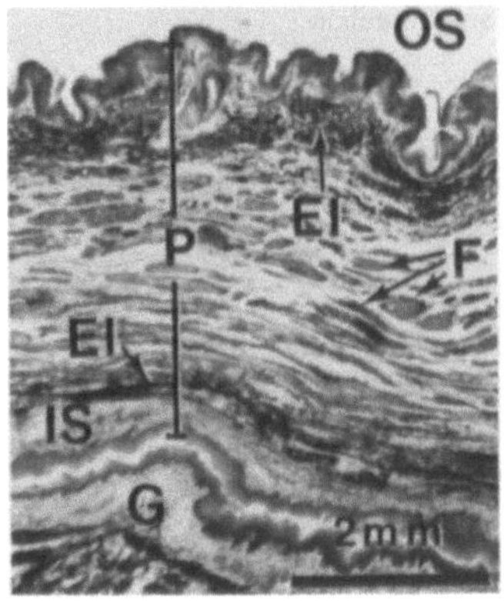

Preputial glands (glands of Tyson, glandula preputialis*): Peculiar ↑ sebaceous glands (PG) located in ↑ dermis of corona (C) of ↑ glans penis and inner surface of ↑ prepuce (P); very pronounced individual variations in number and size. Product of P., together with desquamated epithelial cells, is ↑ smegma (S). (Fig. = human)

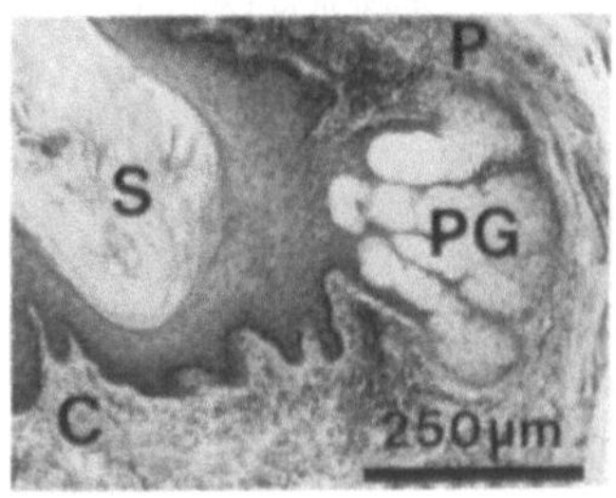

Presecretory granules: ↑ Zymogen granules

Prespermatid: ↑ Spermatocyte, secondary

Presumptive hematopoietic cells: ↑ Candidate stem cells

Presynaptic density (densitas presynaptica*): Agglomeration of proteinaceous material at presynaptic membrane (arrowheads) of a ↑ bouton terminal (BT). Cross-sectioned, P. shows short conical projections (p) toward interior of bouton terminal. ↑ Synaptic vesicles (sv) penetrate between these projections to reach presynaptic membrane. Oblique sections and ↑ freeze-etching preparations have shown that P. is arranged in a hexagonal lattice. PD = ↑ postsynaptic density (Fig. = brain, rat)

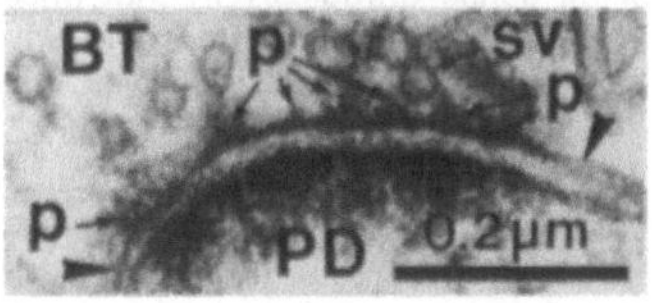

Westrum, L.E., Gray, E.G., Burgoyne, R.D., Barron, J.: Cell Tissue Res. *231*, 93 (1983)

Preterminal arborization: ↑ Telodendria

Preterminal bronchioles: ↑ Bronchioles, preterminal

Prevertebral ganglia (collateral ganglia, ganglia intermedia*): Category of sympathetic ↑ autonomic ganglia (PG) of various size and shape located some distance from ↑ spinal cord (SC) near ↑ aorta. By way of splanchnic nerves (SN), P. are connected with ↑ vertebral ganglia (VG); on nerve cells of P. end some of preganglionic visceral efferent fibers (1) originating from spinal cord. A portion of postganglionic visceral efferent fibers (2), as well as visceral afferent fibers (3), pass through P. Structure of P. corresponds to that of all ↑ autonomic ganglia. P. include aorticorenal, celiac, and mesenteric ganglia. TG = ↑ terminal ganglia

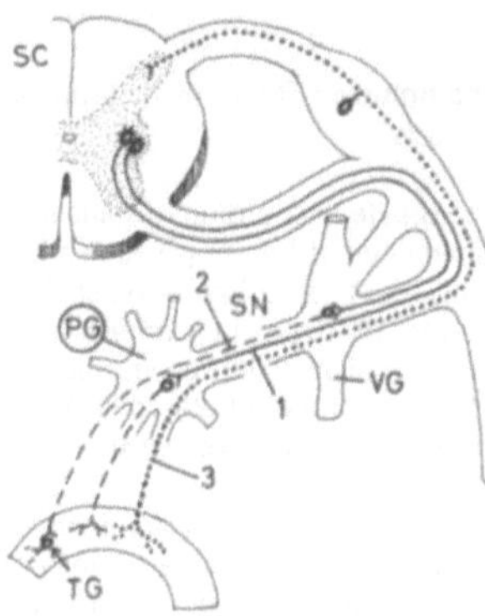

↑ Sympathetic nervous system

PRF: ↑ Prolactin-releasing factor

Prickle cells (keratinocyte, epidermocytus spinosus*): Voluminous polygonal cells (PC) forming several layers between stratum basale and stratum granulosum of ↑ epidermis. Nucleus is round; cell ↑ organelles are not particularly developed; an abundance of free ribosomes and ↑ tonofibrils (Tf) extends throughout cytoplasm from one ↑ desmosome to another. Cytoplasm encloses ↑ membrane-coating granules. P. are characterized by numerous short cytoplasmic processes, the spines (S) or prickles, extending radially from cell body. Spines of adjacent cells are attached, end to end or side by side, by well-developed desmosomes. P. are separated by wide ↑ interfacial canals (c). Mitotic activity of P. is not so high as in ↑ basal cells (BC). (Fig. = mongolian gerbil)

↑ Epidermis, ultrastructure of; ↑ Intercellular bridges, false

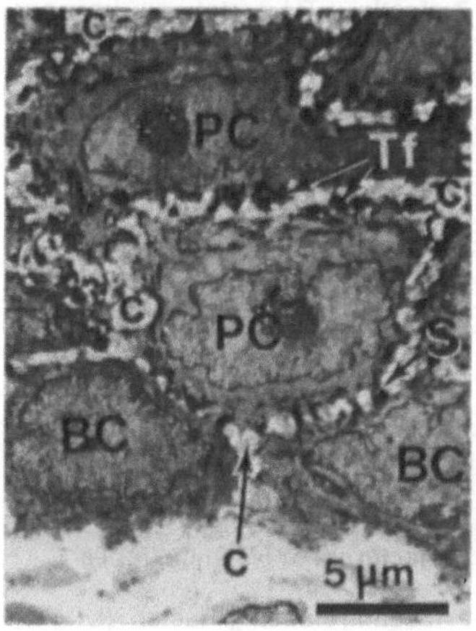

Merriman, J.A., Nieland, M.L., Wedmore, R.J.: J. Microsc. *116*, 243 (1979)

Primary constriction, of chromosomes: ↑ Chromosomes

Primary degeneration, of peripheral nerve fibers: ↑ Nerve fibers of peripheral nervous system, degeneration and regeneration of

Primary granules, of neutrophilic granulocyte: ↑ Granulocyte, neutrophilic, granules of

Primary lysosomes: ↑ Lysosomes

Primary marrow spaces: Spaces between ↑ trabeculae of membranous ↑ bone filled with hematopoietic red ↑ bone marrow.

Primary matrix, of bone: ↑ Osteoid

Primitive erythroblasts: Erythroblasts formed during mesoblastic phase of prenatal ↑ hematopoiesis. P. are somewhat larger than definitive ↑ erythroblasts.

Primitive marrow cavity: ↑ Bone formation, indirect

Primordial follicle: ↑ Ovarian follicle, primordial

Principal cells, of arched collecting tubules: ↑ Collecting tubules, of kidney

Principal cells, of ductus epididymidis: ↑ Ductus epididymidis, epithelium of

Principal cells, of gastric glands proper: ↑ Parietal cells, of gastric glands proper

Principal cells, of parathyroid glands: ↑ Chief cells, of parathyroid glands

Principal piece, of spermatozoon: Part of ↑ spermatozoon tail between middle piece and end piece.

Prism sheath: ↑ Enamel sheath

Prisms, enamel: ↑ Enamel prisms

Proacrosomal granules (granula proacrosomalia*): Round ↑ unit membrane-bound structures (arrows) found in Golgi apparatus (G) of ↑ spermatid. P. contain homogeneous fine granular material of moderate density; they are ↑ PAS-positive and rich in ↑ glycoprotein. With progression of ↑ spermiogenesis, P. unite and form single large globule, ↑ acrosomal granule. (Fig. = human)

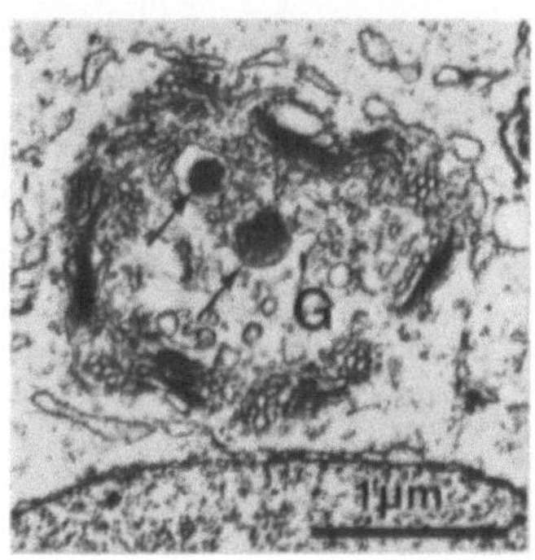

Procaryotes: Primitive cells with diffuse nuclear material in ↑ protoplasm. P. lack mitochondria; their ↑ genome is represented by a single large DNA molecule, and there is no mitotic replication (e.g., bacteria and cyanobacteria).

↑ Eucaryotes

Procentriolar organizer: ↑ Ciliogenesis; ↑ Deuterosome

Procentrioles: Precursors of ↑ centrioles occurring during ↑ ciliogenesis.

Procollagen: Precursor molecule of ↑ tropocollagen.

↑ Collagen, biosynthesis of

Bornstein, P., Erlich, H.P., Wyke, A.W.: Science *175*, 544 (1972); Fessler, J.H., Fessler, L.I.: Ann. Rev. Biochem. *47*, 129 (1978)

Proctodeal glands: ↑ Anal canal

Proelastin (tropoelastin): A presumably soluble precursor molecule of ↑ elastin (m.w. about 72000 daltons).

Proerythroblast (pronormoblast, rubriblast, proerythroblastus*): First cell of erythrocytic series. Cell body is oval or angular, up to 19 µm in diameter, with large nucleus poor in ↑ heterochromatin and one or two nucleoli. Cytoplasm contains several mitochondria, a small Golgi apparatus (G), some peripheral microtubular bundles (Mt), a ↑ centriole, and occasional short rough endoplasmic cisternae (rER). Moderate number of ↑ polyribosomes (Pr) and ribosomes (R) give cytoplasm mild ↑ basophilia visible under light microscope. P. theoretically generates 16 ↑ erythrocytes.

↑ Erythropoiesis; ↑ Erythropoiesis, cell divisions during

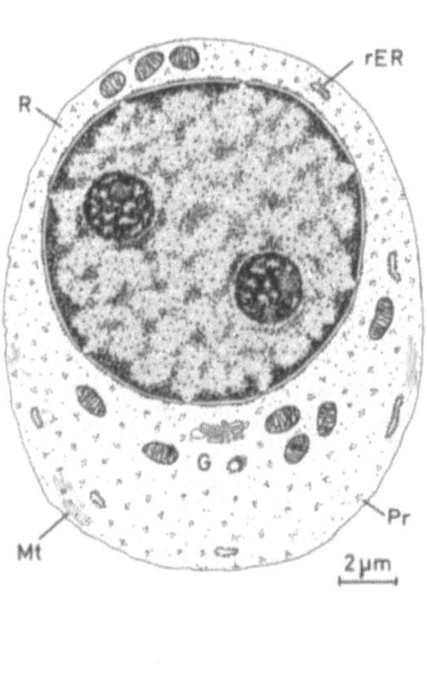

Progesterone: Female sexual hormone produced by ↑ corpus luteum and ↑ placenta, particularly after 4th month of pregnancy. P. is responsible for secretory phase of ↑ menstrual cycle, i.e., preparation of endometrial mucosa for ↑ implantation of blastocyst, as well as for development of embryo. P. stimulates proliferation of ↑ tubuloalveoli of ↑ mammary gland and inhibits further maturation of ↑ ovarian follicles. Mechanism of action of P. is via DNA receptors and gene stimulation. (See biochemistry and physiology texts for further information)

↑ Prolactin; ↑ Steroid hormones, action on the cell of

Progranulocyte: ↑ Promyelocyte

Proinsulin: Precursor of ↑ insulin, synthesized by ↑ B-cells of ↑ pancreatic islets.

Projective: Special ↑ ocular projecting image onto film. A P. in transmission electron microscopy is an ↑ electromagnetic lens with the function of projecting electron beam onto screen or film.

Prolactin (LTH, lactotropic hormone, luteotropin): Protein hormone (m.w. about 25000 daltons) synthesized and released by ↑ mammotropes of pars distalis of ↑ hypophysis. P. initiates and maintains secretion of ↑ corpus luteum in pregnancy, promotes development and secretory activity of ↑ mammary gland.

Flückiger, E., Del Pozo, E., Werder, K.v.: Prolactin. Heidelberg, Berlin, New York: Springer Verlag 1982; Giannattasio, G., Zanini, P., Rosa, J., Meldolesi, R., Margolis, R.K., Margolis, R.V.: J. Cell Biol. *86*, 273 (1980)

Prolactin cells: ↑ Mammotropes

Prolactin-releasing factor (LTH-RF, luteotropic hormone-releasing factor, luteotropic hormone-releasing hormone. LTH-RH, PRF): Hypothalamic hypophyseotropic peptide of unknown structure stimulating ↑ mammotropes to produce ↑ prolactin.

↑ Releasing factors

Proliferation-inhibiting factor (PIF): ↑ Lymphokine that inhibits ↑ mitosis in ↑ cell culture.

Proliferation zone, of hyaline cartilage: ↑ Bone formation, indirect

Proliferative phase, of endometrium: Period of ↑ regeneration from ↑ basalis of surface epithelium, ↑ uterine glands, and stroma of ↑ functionalis during ↑ menstrual cycle.

↑ Endometrium

Proliferative phase, of menstrual cycle: ↑ Menstrual cycle

Promegakaryocyte: Large smooth-surfaced round cell about 45 µm in diameter characterized by voluminous lobulated ↑ polyploid nucleus with several nucleoli. Cytoplasm is abundant and basophilic; it contains numerous small mitochondria, a multiple Golgi apparatus (G), many ↑ centrioles, extensive rough endoplasmic reticulum, an abundance of free ribosomes, and small dense granules (Gr) that arise from Golgi apparatus. (In some animals, other types of granules with very dense eccentric core can be found.) With the exception of a homogenous ↑ ectoplasm (E), cytoplasm encloses a very complex system of smooth vesicles, tubules, and flat cisternae that form ↑ platelet demarcation channels (PDC) by fusion. P. develops from ↑ megakaryoblast, and by further ↑ differentiation in the course of ↑ thrombopoiesis, it becomes either reserve or platelet-producing ↑ megakaryocyte.

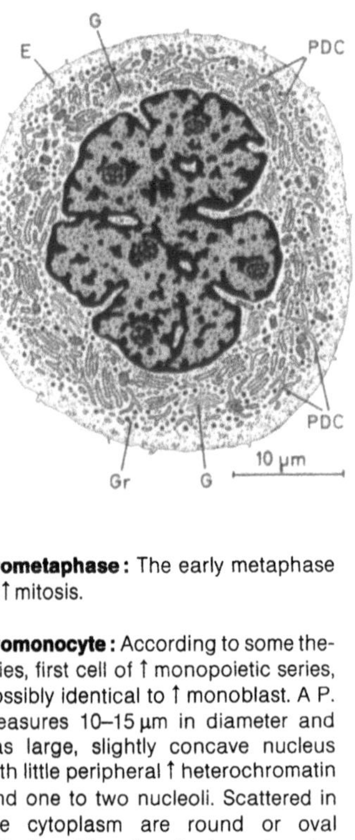

Prometaphase: The early metaphase of ↑ mitosis.

Promonocyte: According to some theories, first cell of ↑ monopoietic series, possibly identical to ↑ monoblast. A P. measures 10–15 µm in diameter and has large, slightly concave nucleus with little peripheral ↑ heterochromatin and one to two nucleoli. Scattered in the cytoplasm are round or oval mitochondria, a few short rough endoplasmic ↑ cisternae, and a prominent Golgi apparatus (G), from which arise immature ↑ azurophilic granules (IAG) in form of vesicles containing low-density material and a very osmiophilic core. Besides ↑ centrioles, numerous free ↑ polyribosomes (Pr) are present, giving pronounced ↑ basophilia to P. There are no mature ↑ azurophilic granules within P.

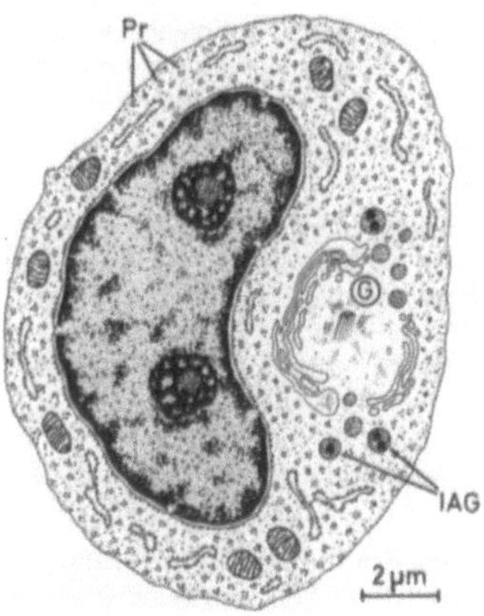

Nichols, B.A., Bainton, D.F., Farquhar, M.G.: J. Cell Biol. *50*, 489 (1971)

Promyelocyte (progranulocyte, promyelocytus*): Largest cell of ↑ granulopoiesis, measuring between 18 and 24 µm in diameter. Cell body is oval, nucleus round and large with small amount of peripheral ↑ heterochromatin and one or two nucleoli. In mature P., nucleus may become kidney-shaped. The cytoplasm is abundant and contains, besides small numerous mitochondria, an extensive Golgi apparatus (G) with adjacent centrioles (C), numerous expanded rough endoplasmic (rER) cisternae with a moderately dense, peroxidase-reactive ↑ reticuloplasm and a considerable amount of free ↑ ribosomes and ↑ polyribosomes. Together with rough endoplasmic reticulum, latter contribute to strong ↑ basophilia of P. In young P., a moderate number of immature azurophilic granules (IAG), 200–400 nm in diameter, arise from concave surface of Golgi apparatus. They are filled with a homogenous flocculent material and frequently contain a central dense core or crystalloid of 100–150 nm. Mature P. enlarges up to 24 µm; mitochondria decrease in number, whereas amount of mature ↑ azurophilic granules (AG) increases. These granules are formed by fusion of smaller granules with a dense core. P. actively divide and make up about 3% of all nucleated elements of red ↑ bone marrow. According to some authors, ↑ monoblast and ↑ monocyte can arise after several divisions of a P.

↑ Granulopoiesis; ↑ Monopoiesis

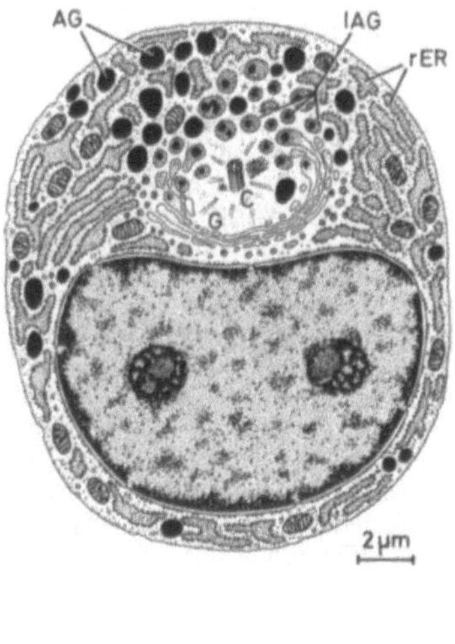

Pronormoblast: ↑ Proerythroblast

Propeptides: ↑ Registration peptides

Prophase, of meiosis: ↑ Meiosis

Prophase, of mitosis: ↑ Mitosis

Proplasmacytes: Cells intermediate in ↑ differentiation of ↑ plasmoblast to ↑ plasma cell.

Proprioceptors: Group of ↑ sensory receptors specialized in receiving information concerning position of body or its parts in space (↑ Golgi tendon organ, ↑ neuromuscular spindle, nerve terminals in joint ↑ capsule, ↑ vestibular apparatus).

Propulsive veins: Large ↑ veins of infracardial body region characterized by very thin tunica media and very thick tunica adventitia with longitudinally oriented bundles of ↑ smooth muscle cells. Their contractions help propel blood to heart. Inferior ↑ vena cava, azygos, external iliac, ↑ portal, ↑ renal, ↑ splenic, and superior mesenteric veins are all considered P.

Prorubricyte: ↑ Erythroblast, basophilic

Prostaglandins: Large group of hormones synthesized in most, possibly all, organs of body. Although of very similar chemical structure, P. often have opposite effects on platelet aggregation, blood pressure, etc. Some are involved in regulation of female reproductive cycle, others decrease gastric secretion, induce plasticity of erythrocytes, modify hypophyseal response to hypothalamic hormones, relax bronchial musculature, modulate release of ↑ neurotransmitters, etc. Most P. act via ↑ adenyl cyclase-cyclic AMP. (See biochemistry and physiology texts for further information)

Coceani, F.: Arch. Int. Med. *133*, 119 (1974); McGiff, J.C., Malik, K.U., Terragno, N.A.: Fed. Proc. *35*, 2382 (1976); Morley, J.: Sandorama *2*, 5 (1981); Sirgiu, P., Cossu, M., Perra, M.T.: Anat. Rec. *204*, (1982); Wilson, D.E.: Arch. Int. Med. *133*, 112 (1974)

Prostate (prostate gland, prostata*): Musculoglandular organ situated below ↑ urinary bladder and traversed by pars prostatica urethrae (U). Structure: 1) ↑ Capsule (C) = ↑ dense connective tissue rich in ↑ smooth muscle cells. 2) ↑ Stroma = broad fibromuscular septa (S) with abundant smooth muscle cells; septa are directed toward ↑ colliculus seminalis (CS). 3) ↑ Parenchyma: a) mucosal glands (MG) = short invaginations of urethral epithelium with somewhat enlarged ends; b) submucosal or periurethral glands (PG) = small ↑ tubuloalveolar glands situated beneath urethral epithelium and separated from main prostatic glands by mass of smooth muscle cells; c) main prostatic glands (MPG) = 30–50 ↑ tubuloalveolar glands opening into about 15–30 prostatic ducts (PD) entering ↑ urethra laterally from colliculus seminalis. P. is well vascularized and contains numerous lymphatic vessels, small sympathetic ganglia, and free intraepithelial nerve endings. Product of P. is a watery opalescent liquid (pH 6.5) containing acid phosphatase and several proteolytic enzymes, one of which is fibrinolysin. A part of ↑ prostaglandins is secreted by P.

↑ Prostatic glands, main

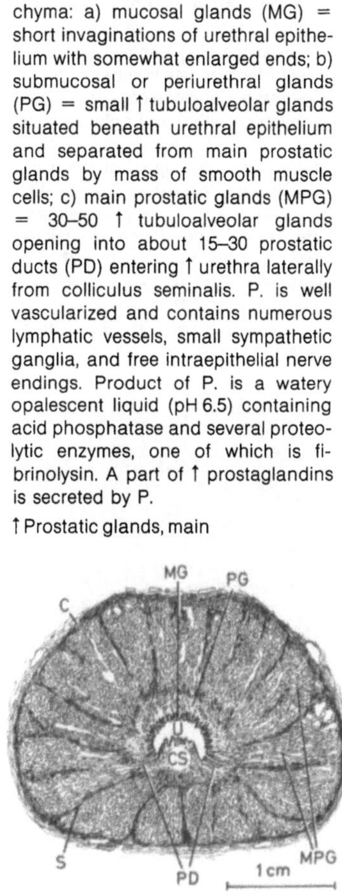

Aumüller, G.: Prostate Gland and Seminal Vesicles. In: Oksche, A., Vollrath, L.

(eds.); Handbuch der mikroskopischen Anatomie des Menschen, Vol. 7, Part 6. Berlin, Heidelberg, New York: Springer-Verlag 1979

Prostate, epithelium of: ↑ Simple cuboidal or columnar epithelium with patches of ↑ pseudostratified epithelium, composed of: 1) Principal cells = large columnar secretory cells measuring about 12 µm in height. Nucleus is situated in lower third of cell; it displays dispersed ↑ chromatin and a conspicuous nucleolus. Cytoplasm contains small scattered mitochondria, a moderately developed Golgi apparatus, long anastomosing cisternae of rough endoplasmic reticulum, some ↑ lysosomes (Ly), a few ↑ lipid droplets (L), and ↑ glycogen particles (Gly). Supranuclear and apical cell portions are filled with ↑ unit membrane-bound secretory vesicles (SV) of extreme polymorphism; some appear empty, and others display a fine granular content, but most contain a group of very osmiophilic granules. Contents of secretory vesicles empty by ↑ exocytosis. 2) Basal cells = trigonal or flattened cells situated below or between principal cells; they can reach lumen with their ↑ apical pole in some cases. Nucleus contains condensed chromatin and ↑ cytoplasmic matrix is very dense. Cell ↑ organelles are scarce and poorly developed; there is an abundance of microfilaments. It has been postulated that basal cells might represent precursor of principal cells. P. is remarkable for high concentration of acid phosphatase.

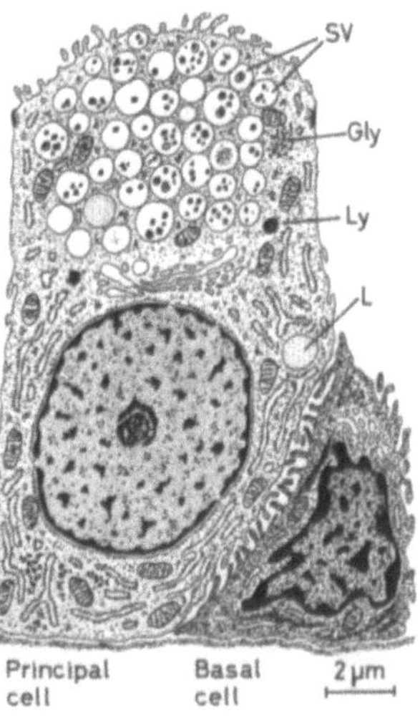

Aumüller, G.: Prostate Gland and Seminal Vesicles. In: Oksche, A., Vollrath, L. (eds.): Handbuch der mikroskopischen Anatomie. Vol. 7, Part 6. Berlin, Heidelberg, New York: Springer-Verlag 1979; Kachar, B., Pinto da Silva, P.: Anat. Rec. *198*, 549 (1980); Murphy, G.P., Sandberg, A.A., Karr, J.P. (eds.): Prostatic Cell: Structure and Function, Part A. New York: Alan R. Liss Inc. 1981

Prostatic concretions (corpus amylaceum, prostatolith, sympexion, concretio prostatica*): Concentric lamellated bodies (C) of variable dimensions (20 µm–2 mm) found in glandular cavities of ↑ prostate. P. are birefringent, may calcify, and may also be found in ↑ ejaculate. Number of P. increases with age. It is believed that P. originate through condensation of prostatic secretion around fragments of desquamated epithelial cells. P. consist of proteins, nucleic acids, cholesterol, and tertiary calcium phosphate. (Fig. = human)

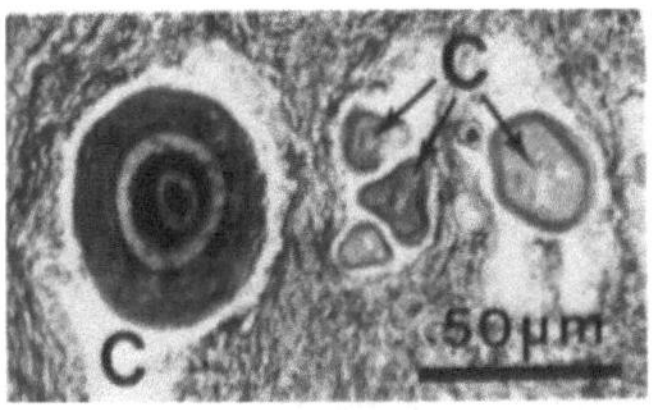

Prostatic ducts (ductuli prostatici*): About 15–30 short excretory ducts of ↑ prostate opening at base of ↑ colliculus seminalis. P. are lined by ↑ simple cuboidal to columnar epithelium.

↑ Urethral crest

Prostatic glands, main: Irregularly shaped ↑ tubuloalveolar glands (G) most frequently lined by ↑ simple columnar epithelium. In some ↑ tubuloalveoli (TA), epithelium can be simple cuboidal or ↑ pseudostratified; many epithelial folds characterize the glands. Thick septa (S) with numerous ↑ smooth muscle cells (M) separate tubuloalveoli; contraction of these muscle cells aids in ejaculatory discharge of prostatic fluid. Prostatic concretions (PC) are present within P. Tubuloalveoli unite to form ↑ prostatic ducts. C = ↑ capsule. (Figs. = human)

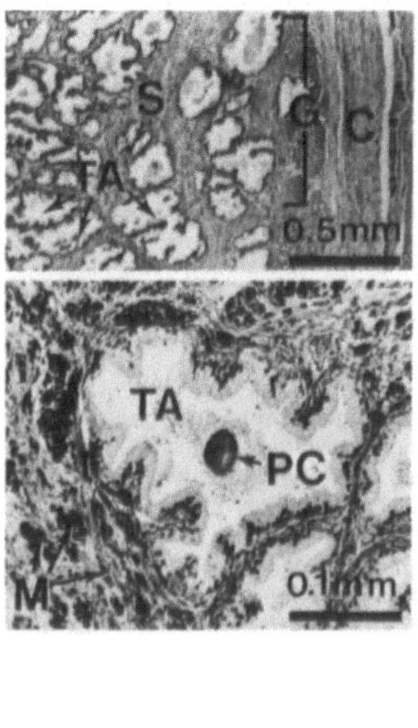

↑ Prostate, epithelium of

Prostatic muscle (stroma myoelasticum prostatae*): Muscle composed of all prostatic capsular and stromal ↑ smooth muscle cells whose contractions propel prostatic secretion into ↑ urethra during ↑ ejaculation.

Prostatolith: ↑ Prostatic concretions

Protective permeability barrier, of stomach: All ↑ surface mucous cells of stomach mucosa. By secretion of acid-resistant ↑ mucus, P. protects wall of stomach against action of hydrochloric acid and its reabsorption across mucosa. Alcochol and aspirin greatly diminish resistance of P.

Protein hormone-producing cells: ↑ Endocrine cells

Protein synthesis: Process of elaboration of complex nitrogenous organic compounds from amino acids under control of nuclear genetic program. Sequence: 1) ↑ Messenger RNA (mRNA) with transcribed information from uncoiled ↑ deoxyribonucleic acid (DNA) passes into cytoplasm through ↑ nuclear pores (P). 2) ↑ Ribosomal RNA (rRNA) enters cytoplasm and contributes to composition of two ribosomal subunits that unite into a ↑ ribosome when they contact mRNA. 3) ↑ Transfer RNA (tRNA) also traverses pores and binds enzymatically to an activated amino acid (A). The sequence of bases of the tRNA triplet (↑ anticodon, arrowhead) identifies its complementary sequence of bases (↑ codon) along mRNA molecule, links with it, and adds amino acid as specified by code triplet on mRNA. Amino acids are brought together in a polypeptide chain (PC) which enters ↑ cisterna of rough ↑ endoplasmic reticulum through a canal in larger ribosomal subunit. The tRNA, to which amino acid was previously attached, is liberated from it and mRNA and can thus be used again. (See molecular biology texts for further information)

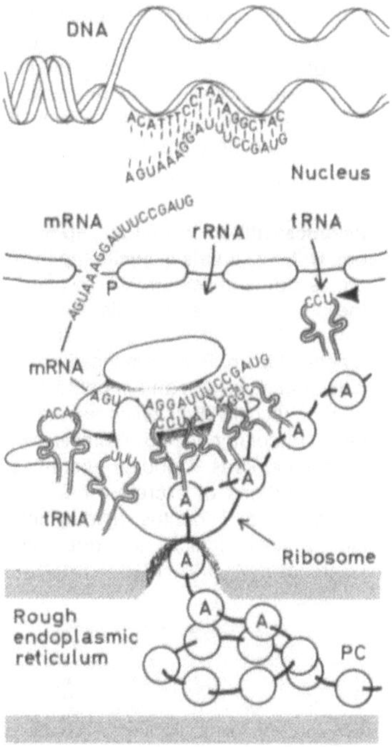

Bielka, H.: Acta Histochem. Suppl. Vol. XVII, 31 (1976); Kern, H.F.: Verh. Anat. Ges. *74*, 79 (1980); Lin, C.-T., Chang, J.P.: Science *190*, 465 (1975); Palade, G.: Science *189*, 347 (1975); Pérez-Bercoff, R. (ed.): Protein Biosynthesis in Eukaryotes. NATO Advanced Study Institutes Series, Series A: Life Sciences. New York: Plenum Press 1982

Proteoglycan subunits: ↑ Glycosaminoglycans

Proteoglycans: Very complex heterogenous macromolecules composed of sulfated ↑ glycosaminoglycans covalently bound at regular intervals along a carrier protein. P. are connected by link ↑ glycoproteins (LG) to long molecule of ↑ hyaluronic acid. Protein components of P. are synthesized on ↑ ribosomes of rough ↑ endoplasmic reticulum according to mechanism of ↑ protein synthesis. Sugar components are added in both rough endoplasmic reticulum and Golgi apparatus. P. are synthesized by ↑ chondroblasts, ↑ chondrocytes, ↑ fibroblasts, ↑ fibrocytes, ↑ mast cells, ↑ smooth muscle cells, type B ↑ synovial cells, ↑ osteoblasts, etc. P. represent main component of ↑ ground substance of ↑ connective and supporting tissues and blood vessels. They are responsible for diffusion of water-soluble substances throughout connective and supporting tissues and thus for their solidity. (Modified after Poirier et al. 1980)

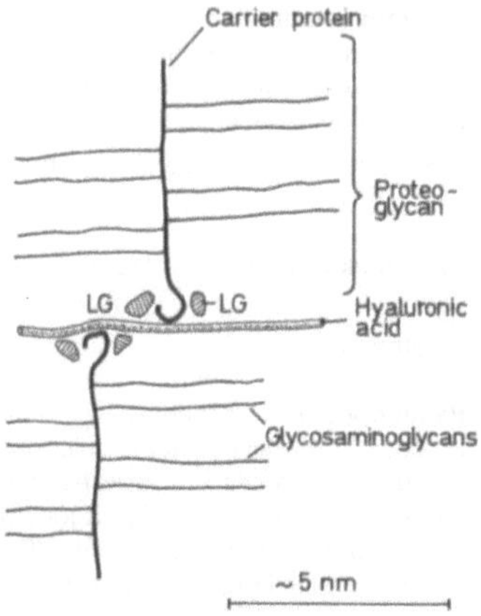

↑ Chondromucoprotein; ↑ Metachromasia

Greiling, H., Gressner, A.M., Stuhlsatz, H.W.: Verh. Anat. Ges. *75*, 115 (1981); Hay, E.D.: J. Cell Biol. *91* 205 s (1981); Lennarz, W.J. (ed.): The Biochemistry of Glycoproteins and Proteoglycans. New York, London: Plenum Press 1980; Poirier, J., Bernaudin, J.F., Robert, L.: Biologie de la cicatrisation. Pantin: Laboratoires Hoechst 1980; Varma, R.S., Varma, R. (eds.): Glycosaminoglycans and Proteoglycans. Basel, München: Karger 1982

Prothrombin: Precursor of ↑ thrombin. P. is converted into thrombin through action of ↑ thromboplastin.

Protochondral tissue (center of chondrification, scleroblastem, textus prechondralis*): Dense aggregation of ↑ mesenchymal cells that differentiate into ↑ chondroblasts at site of future formation of hyaline ↑ cartilage.

↑ Cartilage, histogenesis of

Protofibrils, collagen (protofibrilla*): First fibrillar linear polymer units formed as a result of polymerization of ↑ tropocollagen molecules. Diameter of P. is below 5 nm; several P. form a ↑ collagen microfibril. Periodicity of P. is not clearly visible.

↑ Collagen fibers, submicroscopic and microscopic organization of

Protofilaments: Thirteen threadlike subunits composing a ↑ microtubule.

↑ Tubulin

Protoplasm (protoplasma*): Living substance of plants and animals in form of a heterogenous aqueous phase containing biological components whose integrated functions exhibit properties of life. P. is subdivided into ↑ nucleus (karyoplasm) and ↑ cytoplasm. Smallest unit of P. is ↑ cell.

Proximal tubule, of nephron (pars proximalis tubuli nephroni*): Segment (P) of ↑ nephron, about 14 mm long and 30–60 μm wide, situated between ↑ renal corpuscle and beginning of ↑ thin segment of ↑ loop of Henle. P. consists of a proximal convoluted tubule (fig.) situated between ↑ medullary rays and a ↑ straight portion situated within medullary rays and partially within medulla. P. is lined by ↑ simple cuboidal epithelium with ↑ brush border (arrowheads). D = ↑ distal tubule (Fig. = monkey)

↑ Proximal tubule of nephron, cells of; ↑ Straight portion of proximal tubule, cells of

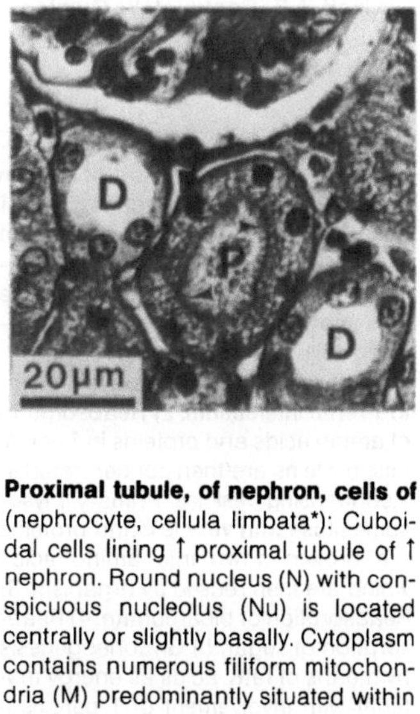

Proximal tubule, of nephron, cells of (nephrocyte, cellula limbata*): Cuboidal cells lining ↑ proximal tubule of ↑ nephron. Round nucleus (N) with conspicuous nucleolus (Nu) is located centrally or slightly basally. Cytoplasm contains numerous filiform mitochondria (M) predominantly situated within

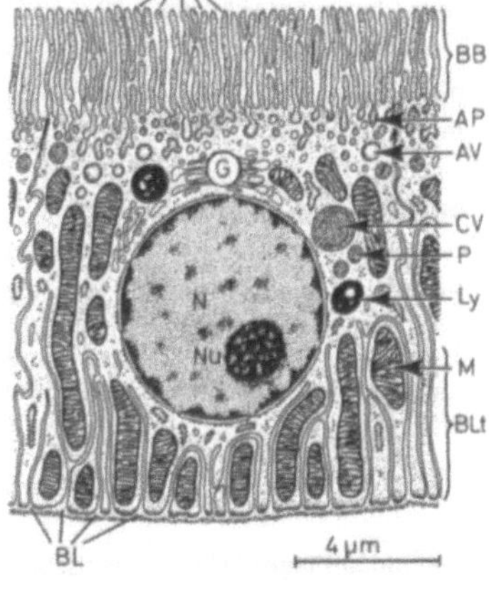

compartments of a well-developed ↑ basal labyrinth (BLt), a large supranuclear Golgi apparatus (G), short cisternae of rough ↑ endoplasmic reticulum, and ↑ tubules of smooth endoplasmic reticulum. ↑ Condensing vesicles (CV) with moderate osmiophilic contents, large secondary ↑ lysosomes (Ly) containing heterogenous material, small primary lysosomes, ↑ peroxisomes (P), and ↑ polyribosomes are also present. A very large number of ↑ apical pits (AP), as well as apical vesicles (AV), occur in apical cytoplasm. Numerous microvilli (Mv), 3–6 µm long, are present at ↑ apical pole, together forming ↑ brush border (BB). Lateral cell surfaces are thrown into simple and ramified folds. BL = ↑ basal lamina

↑ Distal tubule of nephron, cells of

Evan, P.A.Jr., Hay, D.A., Dail, W.G.: Anat. Rec. *191*, 397 (1978); Evan, P.A.: Biomed. Res. *2*, Suppl. 317 (1981); Novikoff, A.B., Spater, H.W., Quintana, N.: J. Histochem. Cytochem. *31*, 656 (1983)

Proximal tubule, of nephron, functions of: 1) Active transport of sodium ions into intercellular spaces and ↑ basal labyrinth by an Mg^{2+}-dependent Na^+, $-K^+$ activated ATPase pump located in the plasmalemma. Chloride ions move passively into basal labyrinth, followed by an osmotic movement of water through basal lamina into ↑ renal interstitium. 2) Reabsorption of amino acids and proteins in ↑ apical pits; proteins are then concentrated in ↑ condensing vesicles. Primary ↑ lysosomes fuse with vesicles, and proteins are broken down into amino acids which are then reused by organism. 3) Reabsorption of bicarbonate. 4) Reabsorption of sugar. 5) Glyconeogenesis by means of fatty acids as energy material with involvement of ↑ peroxisomes. 6) ↑ Excretion of many exogenous acids (such as penicillin) and organic bases into tubular lumen.

Prussian blue staining: ↑ Berlin blue reaction

Psammoma bodies: ↑ Acervuli

Pseudo H-bands (L-band): Narrow clear zones immediately adjacent to ↑ M-line. In P., cross-bridges connecting ↑ myosin myofilaments are lacking.

↑ H-band

Pseudocartilage: ↑ Chondroid tissue

Pseudoeosinophils: Neutrophilic ↑ granulocytes of some animals having conspicuous specific granules with predilection for eosin.

↑ Granulocytes, neutrophilic, granules of

Pseudoisocyanin: A metachromatic dye.

↑ Metachromasia

Pseudopodia (processus ameboideus*): Massive cytoplasmic processes (P) with cell ↑ organelles emitted by ↑ ameboid cells for ↑ phagocytosis and/or migratory movements. (Fig. = neutrophilic ↑ granulocyte, rat)

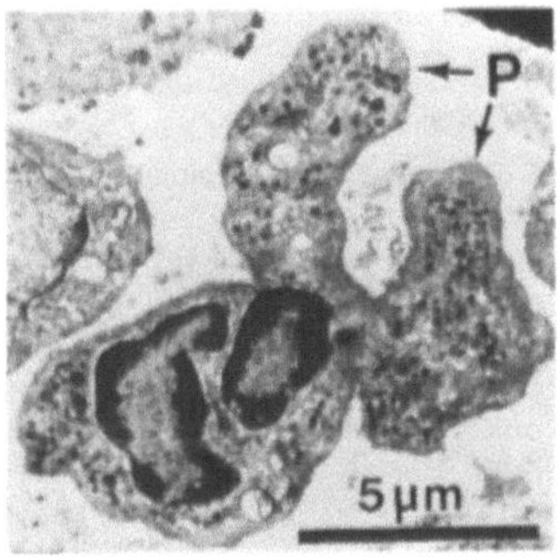

Pseudostratified columnar epithelium, with three and more levels of nuclei (epithelium pseudostratificatum*): Type of epithelium composed of tall columnar ciliated cells (CC), nondifferentiated basal cells (BC), and conical intercalated cells (IC) in course of ↑ differentiation into ciliated cells. Since all cells contact basal lamina (BL) but do not reach free surface, their nuclei lie on different levels. ↑ Goblet cells (GC)

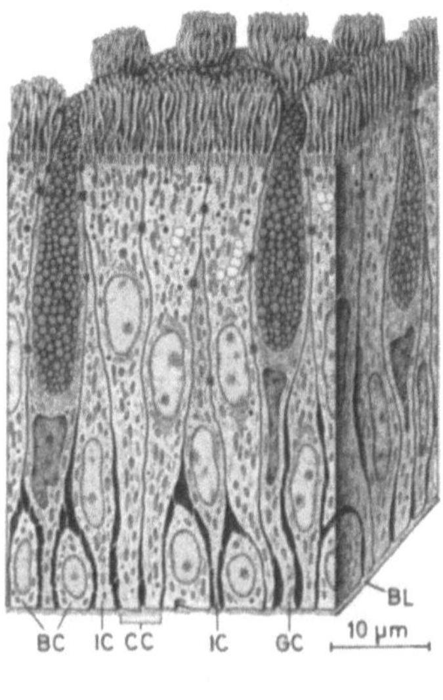

are usually present in P. Basal cells represent source for epithelial ↑ regeneration. P. lines ↑ respiratory mucosa of ↑ nasal cavities, ↑ nasopharynx, ↑ larynx (except ↑ vocal cords), ↑ trachea, and large ↑ bronchi.

↑ Trachea, epithelium of

Pseudostratified columnar epithelium, with two levels of nuclei (epithelium pseudostratificatum*): Type of epithelium in which all cells lie on basal lamina (BL), but only one cell type reaches lumen. P. is composed of tall columnar cells (CC) and nondifferentiated basal cells (BC) which assure renewal of columnar cells. Epithelia of ↑ ductus deferens, ↑ ductuli efferentes, some portions of excretory ducts of glands, and ↑ ductus epididymidis (fig.), where epithelium is studded with ↑ stereocilia (S), are P. (Fig. = rat)

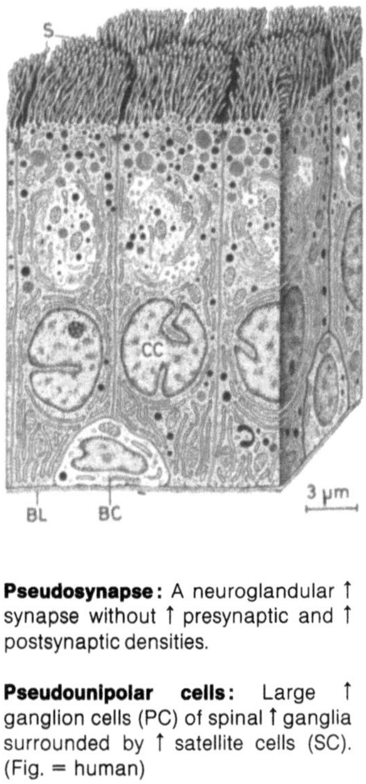

Pseudosynapse: A neuroglandular ↑ synapse without ↑ presynaptic and ↑ postsynaptic densities.

Pseudounipolar cells: Large ↑ ganglion cells (PC) of spinal ↑ ganglia surrounded by ↑ satellite cells (SC). (Fig. = human)

↑ Neurons, classifications of

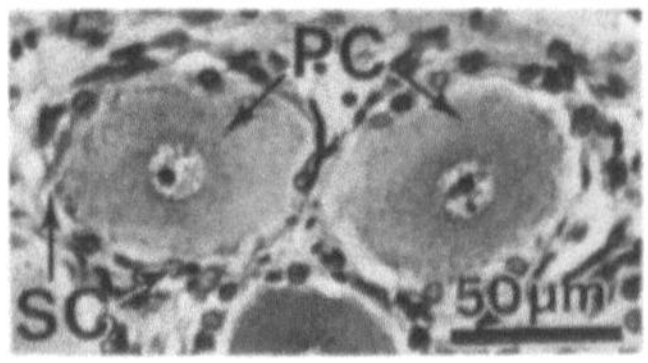

P-substance: ↑ Substance P

PS-surface, of cell membrane: ↑ Freeze-cleaving, terminology of

PTH: ↑ Parathyroid hormone

Ptyalin: An amylase of ↑ saliva; splits starch into smaller water-soluble particles.

Pulmonary airways: Part of conducting portion of ↑ respiratory system conveying air from primary ↑ bronchi up to ↑ alveolar ducts. P. consist of:

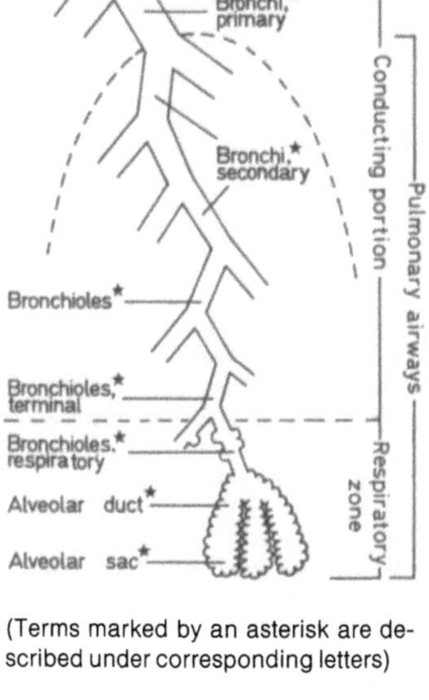

(Terms marked by an asterisk are described under corresponding letters)

Pulmonary alveolus: ↑ Alveolus, of lung

Pulmonary arteries (arteriae pulmonales*): Branches of pulmonary artery (PA) accompanying branching of secondary ↑ bronchi (B) as far as respiratory ↑ bronchioles. P. are of elastic type until their diameter decreases to less than 1 mm, at which point they gradually change to muscular type. Elastic P. have a thin intima and a relatively thin media composed of alternating layers of ↑ smooth muscle cells and ↑ elastic laminae. Adventitia is thinner than in pulmonary veins (PV)

of same caliber. P. of ↑ muscular type have a conspicuous internal ↑ elastic lamina and a considerable number of ↑ elastic fibers scattered among smooth muscle cells of media. P. belong to functional lung circulation. BA = ↑ bronchial artery (Fig. = calf, ↑ orcein staining)

↑ Lungs, vascularization of

Lane, B.P., Zeidler, M., Weinhold, C., Drummond, E.: Anat. Rec. *205*, 397 (1983)

Pulmonary capillaries: ↑ Blood-air barrier

Pulmonary epithelial cells: ↑ Alveolar cells type I and II

Pulmonary vascularization: ↑ Lungs, vascularization of

Pulmonary veins (venae pulmonales*): Thin-walled venous blood vessels (PV) collecting arterial blood from ↑ lung lobules and ↑ pleura. P. accompany secondary ↑ bronchi (B) up from tip of lobule. Tunica intima contains a meshwork of predominantly longitudinal ↑ elastic fibers underlying a thin ↑ endothelium. Tunica media, only present in P. larger than 100 µm, may contain a considerable number of ↑ smooth muscle cells and elastic fibers, causing P. to resemble ↑ pulmonary arteries of same caliber. Adventitia (A) is thicker than that of pulmonary arteries of same diameter. On sections, P. occur singly or together with bronchi and pulmonary artery (PA). P. belong to functional lung circulation. (Fig. = calf. ↑ orcein staining)

↑ Lungs, vascularization of

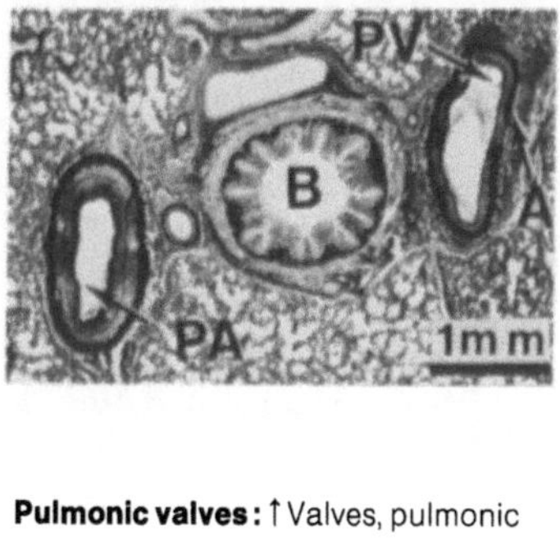

Pulmonic valves: ↑ Valves, pulmonic

Pulp arteries, of red pulp: ↑ Penicillar arteries

Pulp cavity (cavitas dentis*): Elongated chamber (PC) within tooth filled with ↑ pulp of teeth. P. communicates through an ↑ apical foramen (AF) with ↑ periodontal space (PS).

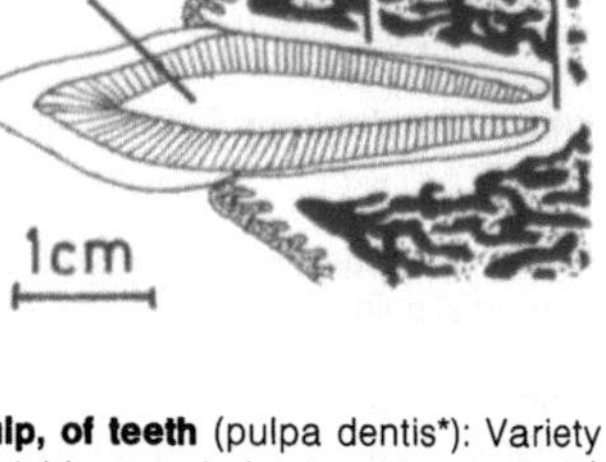

Pulp, of teeth (pulpa dentis*): Variety of richly vascularized and innervated ↑ loose connective tissue (P) filling ↑ pulp cavity of teeth. P. somewhat resembles ↑ mesenchyme; it is composed of a considerable number of fusiform and stellate ↑ fibroblasts and ↑ fibrocytes but has a relatively limited number of ↑ wandering cells. In metachromatic ↑ ground substance, ↑ collagen fibrils run in all directions and do not form bundles; ↑ elastic fibers exist only in wall of vessels. ↑ Capillaries (Cap) of P. are of fenestrated type; it is not yet known whether lymphatic vessels are present. ↑ Odontoblasts (O) make up part of P. NF = ↑ nerve fascicle, D = ↑ dentin (Fig. = human)

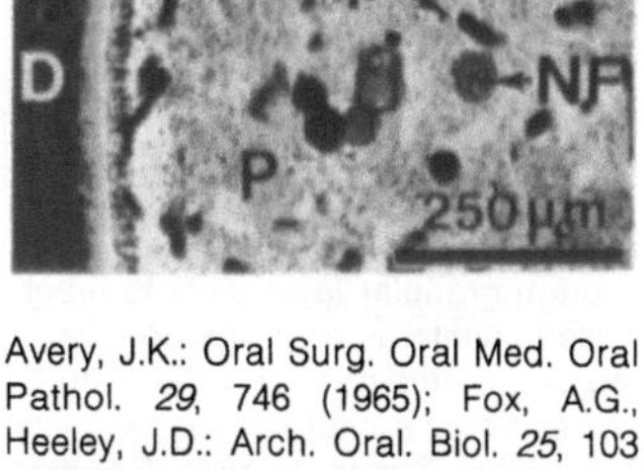

Avery, J.K.: Oral Surg. Oral Med. Oral Pathol. *29*, 746 (1965); Fox, A.G., Heeley, J.D.: Arch. Oral. Biol. *25*, 103 (1980); Johnsen, D.C., Harshbarger, J., Rymer, H.D.: Anat. Rec. *205*, 421 (1983)

Pulp of teeth, Korff's fibers of: ↑ Korff's fibers

Pulp, red, of spleen: ↑ Red pulp, of spleen

Pulp veins (vena pulparis*): Short veins (PV) in ↑ red pulp (RP) of ↑ spleen. P. collect blood from ↑ sinuses

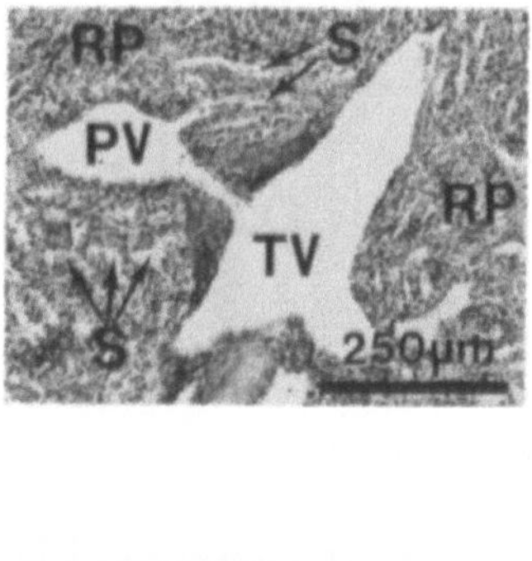

(S) and drain into ↑ trabecular veins (TV). Wall of P. is extremely thin and consists of ↑ endothelium and rare ↑ smooth muscle cells. (Fig. = human)

Pulp, white of spleen: ↑ White pulp, of spleen

Puncta lacrimalia*: ↑ Lacrimal ducts; ↑ Lacrimal points

Pupil (pupilla*): Circular opening in center of ↑ iris. Depending on degree of contraction and relaxation of ↑ dilator and ↑ sphincter muscles, diameter of P. changes constantly, thus regulating quantity of light that falls onto ↑ photoreceptors.

Purkinje cells, of cerebellum (neuronum piriformum*): Large ↑ neurons constituting ganglionic layer (GL) of cerebellar cortex. Ultrastructurally, P. correspond to other nerve cells. Two to three primary ↑ dendrites (PD) extend from voluminous and flattened ↑ perikaryon (P) and branch extensively into terminal dendrites (TD) studded with about 60000 ↑ dendritic spines (DS). Dendritic tree is situated within molecular layer (ML); ↑ dendritic field is very flat and always perpendicular to axis of ↑ folium (arrows). Due to this feature, P. are disposed one behind another, so that a ↑ parallel fiber (PF) may contact about 350 P. It has been estimated that over its entire surface, a P. receives as many as 200000 synaptic contacts. The single myelinated ↑ axon (A) of a P. runs through granular layer (GR) to reach nucleus dentatus as a unique type of cerebellar efferent fiber. ↑ Climbing fibers establish excitatory synapses with P., whereas ↑ granule cells, ↑ basket cells, and ↑ stellate cells form inhibitory ↑ synapses with their dendrites or bodies.

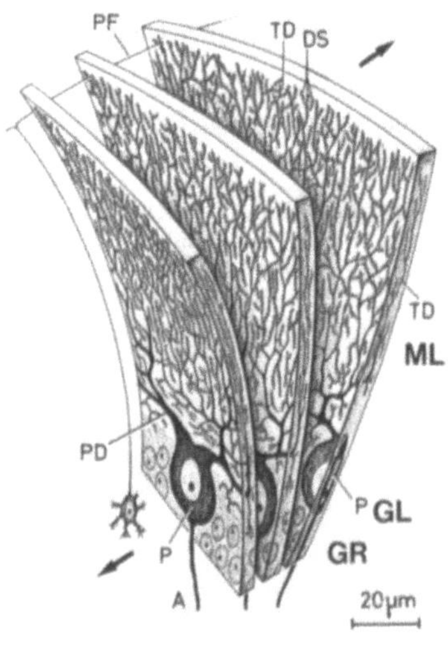

↑ Cerebellum, neuronal circuits of

Scherini, E., Bolchi, F., Biggiogera, M., Bernocchi, G.: J. Submicrosc. Cytol. *13*, 17 (1981)

Purkinje cells, of impulse-conducting system: Large, elongated, cylindrical cells (PC) arranged in an end-to-end and side-to-side fashion to form ↑ Purkinje fibers (PF). P. are about 100 µm long and 50 µm wide; they frequently contain two nuclei (Nc). ↑ Sarcoplasm is of low density; it encloses numerous small mitochondria (M), a multiple Golgi apparatus, some cisternae of rough and smooth ↑ endoplasmic reticulum, and a considerable amount of ↑ glycogen particles (Gly). ↑ Myofibrils (Mf) are predominantly grouped at cell periphery and run mostly longitudinally, but many of them have an oblique or spiral direction. ↑ T-system and ↑ intercalated discs are not present. P. are held together by numerous interdigitations studded with ↑ desmosomes (D), ↑ nexus (N), and ↑ fasciae adherentes (FA, see insets). There is no basal lamina (BL) between P.; it envelopes entire Purkinje fiber. P. transmit impulses to ↑ transitional cells, which transfer them to ↑ cardiac muscle cells (CMC), this leading to heart contraction.

↑ Impulse-conducting system

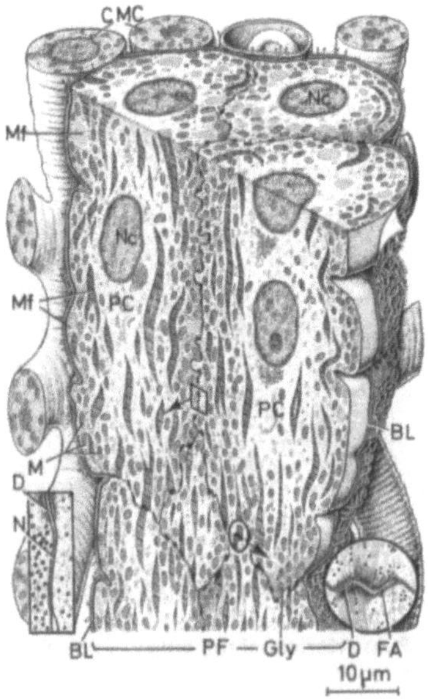

Bogusch, G.: Anat. Embryol. *155*, 254 (1979); Nùñez-Duràn, H.: Acta Anat. *107*, 177 (1980); Oculati, F., Franceschini, S., Cinti, S., Gazzanelli, G., Amati, S.: J. Submicrosc. Cytol. *12*, 73 (1980)

Purkinje fibers, of impulse-conducting system (myofibra conducens purkinjiensis*): Terminal ramifications (PF) of ↑ atrioventricular bundle composed of voluminous ↑ Purkinje cells (PC). P. run below ↑ endocardium (E) and in ↑ myocardium (M). In both cases, P. are separated from adjacent ↑ cardiac muscle cells by a connective tissue sheath (CS). P. conduct electrical impulses to ↑ transitional cells.

↑ Impulse conducting system

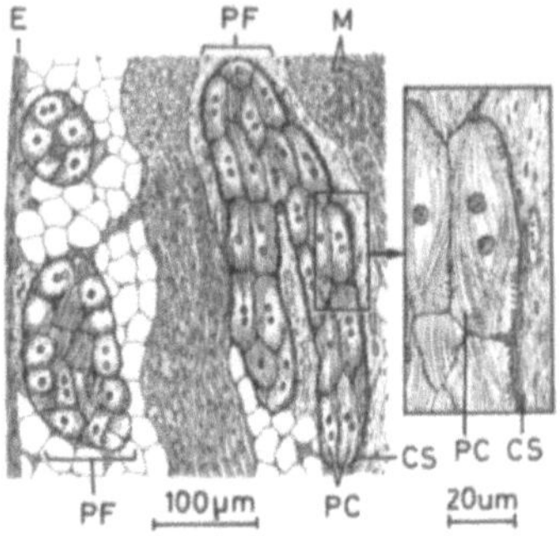

Canale, E., Campbell, G.R., Uehara, Y., Fujiwara, T., Smolich, J.J.: Cell Tissue Res. *232*, 97 (1983); Forsgren, S., Strehler, E., Thornell, L.-E.: Histochem. J. *14*, 929 (1982); Sommer, J.R., Johnson, E.A.: J. Cell Biol. *36*, 497 (1968)

Pus: An accumulation of massive numbers of dead and degenerating neutrophilic ↑ granulocytes and other ↑ leukocytes in abscesses and boils.

Pyknosis: ↑ Karyopyknosis; ↑ Nucleus, degeneration and death of

Pyloric antrum (antrum pyloricum*): Portion of ↑ stomach between body and ↑ pylorus. Structure: 1) Tunica mucosa (TM): a) epithelium = ↑ simple columnar epithelium invaginated into deep ↑ foveolae (F); b) lamina propria = ↑ loose connective tissue crowded with ↑ pyloric glands (PG); c) lamina muscularis mucosae (arrowhead) =

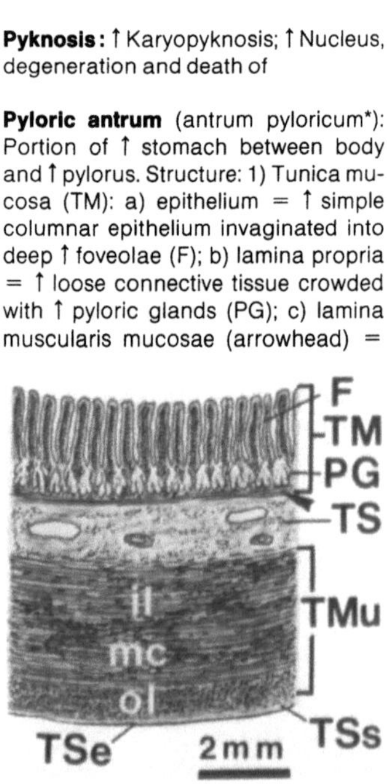

layer of ↑ smooth muscle cells. 2) Tela submucosa (TS) = loose connective tissue with blood and lymphatic vessels and ↑ submucosal plexus. 3) Tunica muscularis (TMu) = smooth muscle tissue organized into inner oblique layer (il), middle circular layer (mc), and outer longitudinal layer (ol). 4) Tela subserosa (TSs) = loose connective tissue. 5) Tunica serosa (TSe) = ↑ peritoneum.

Pyloric glands (glandula pylorica*): Branched and coiled tubular ↑ homocrine glands (PG) located in ↑ pylorus and ↑ pyloric antrum of ↑ stomach. P. are composed of clear cells resembling ↑ neck mucous cells; they produce a ↑ mucoid substance protecting mucosa from high acidity of gastric juice. ↑ Enterochromaffin and gastrin cells (↑ G-cells) are present in P. Two to three P. open into a ↑ foveola (F). Foveolae extend over up to half the tunica mucosa in these regions. (Fig. = monkey)

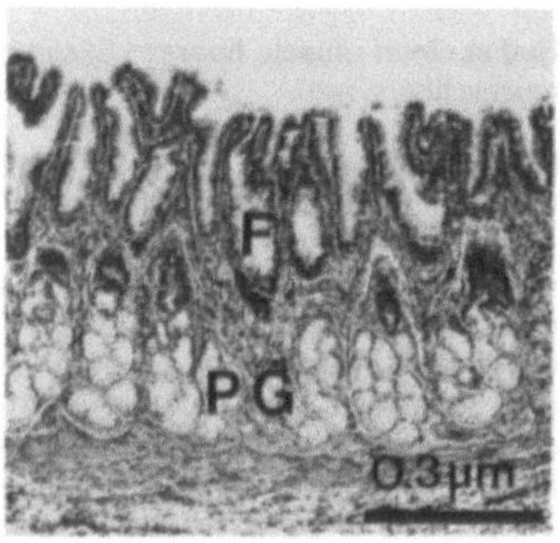

Pylorus* (pyloric channel): Short portion of ↑ stomach located between ↑ pyloric antrum and ↑ duodenum, structurally almost identical to pyloric antrum. Middle circular layer of the tunica muscularis is particularly well-developed and forms pyloric sphincter (musculus sphincter pylori*). In P. some ↑ pyloric glands may pierce lamina muscularis mucosae and penetrate tela submucosa.

Pyramid, of kidney: ↑ Renal pyramid

Pyramidal layer, of brain cortex: ↑ Brain cortex, heterotypical isocortex of; ↑ Brain cortex, homotypical isocortex of

Pyramidal neurons: Category of ↑ multipolar neurons (PC) with a pyramidal ↑ perikaryon. P. constitute pyramidal (III) and ganglionic layers (V) of ↑ brain cortex, although they are also present in layers II and VI. Ultrastructurally, P. correspond to other ↑ neurons showing a large and multiple Golgi apparatus and abundant ↑ Nissl bodies (NB). Ascending and lateral ↑ dendrites (D) emerge from tip and lateral sides of a P., and an ↑ axon (A) extends from its base. P. are completely embedded in ↑ neuropil (N), where processes of ↑ astrocytes (As) simultaneously contact both P. and capillaries (Cap). Several thousand ↑ boutons terminaux (B) make contact with a P.

↑ Perivascular feet

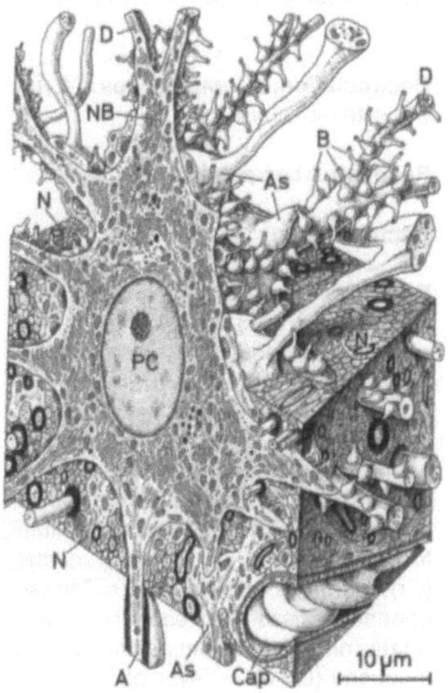

Pyroninophilic cells: ↑ Immunoblasts

Q

Q-cells: ↑ Candidate stem cells

Quenching: Term used to designate a technique of very rapid freezing of biological specimens in liquid air, nitrogen, helium, etc.

Quiescent hair: ↑ Club hair

Quinacrine mustards: ↑ Fluorochromes used to obtain banding patterns of ↑ chromosomes for more refined diagnosis of normal and abnormal chromosomes.

Caspersson, T., Farber, S., Foley, G.E., Kudynowski, J., Modest, E.J., Simonsson, E., Wagh, U., Zeck, L.: Exp. Cell Res. *49*, 219 (1968); Kovacs, M., Sellyei, M., Vass, L.: Folia Hered. Path. *23*, 81 (1974)

R

Racemose glands: Synonym for compound ↑ tubuloalveolar glands.

Radial branches of arcuate arteries, of endometrium: Relatively short muscular branches of ↑ arcuate arteries that enter ↑ endometrium and continue as ↑ coiled arteries.

↑ Endometrium, vascularization of

Radial columns, of brain: ↑ Columns, of brain

Radial layer, of tympanic membrane: ↑ Tympanic membrane

Radiation probe: Technique using microbeams of either protons, UV light, or visible ↑ laser light for experimental selective destruction of chosen cell components in living cells and the study of consequences that such destruction provokes on cell activity.

Radioautography: ↑ Autoradiography

Rami communicantes: ↑ Sympathetic nervous system

Ranvier's nodes: ↑ Nodes, of Ranvier

Rathke's cleft (lumen residuale sacculi hypophysialis*): Fine gap (Cl) in pars intermedia of ↑ hypophysis, a remnant of ↑ Rathke's pouch. R. is found in childhood; as a rule, it becomes discontinuous and transforms into ↑ Rathke's cysts in adult.

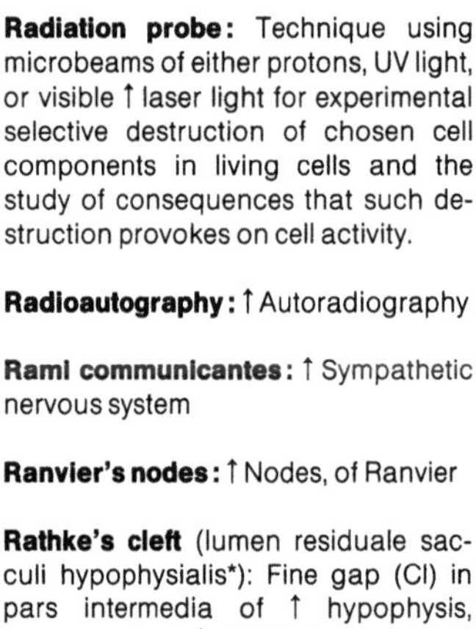

Correr, S., Motta, P.: Biomed. Res. *2*, Suppl. 109 (1981)

Rathke's cysts: Follicles (C) of pars intermedia of ↑ hypophysis lined with ↑ simple cuboidal epithelium which can also be ciliated. R. contain a colloid. It is thought that they are the result of fragmentation of ↑ Rathke's cleft (Cl) in the adult. (See fig. under ↑ Rathke's cleft)

Rathke's pouch (sacculus hypophysialis*): Ectodermal invagination of roof of primitive buccopharyngeal cavity. In course of further development, R. gives rise to pars distalis, pars tuberalis and pars intermedia of ↑ hypophysis. (See embryology texts for further information)

RBC: ↑ Erythrocyte

Receptors, sensory: ↑ Sensory receptors

Reciprocal synapse: ↑ Synapse, reciprocal

Recirculation, of lymphocytes: ↑ Lymphocyte recirculation

Reconstruction phase: ↑ Mitosis

Rectal columns: ↑ Anal canal

Rectum*: Terminal segment of ↑ gastrointestinal tract consisting of two parts: upper part, or R. proper, and lower part, or ↑ anal canal. Structure of R. proper: 1) Tunica mucosa (TM): a) epithelium (E) = ↑ simple columnar epithelium composed of occasional ↑ absorptive cells and an abundance of ↑ goblet cells (GC). The epithelium forms deep and numerous ↑ crypts (Cr); b) lamina propria (LP) = ↑ loose connective tissue with scattered ↑ lymphatic nodules; c) lamina muscularis mucosae (LMM) = layer of ↑ smooth muscle cells. Tunica mucosa of R. is thrown into three transversal folds (plicae transversales recti), one on right side (= Kohlrausch's valve) and two on left. 2) Tela submucosa (TS) = well-vascularized and innervated loose connective tissue. 3) Tunica muscularis (TMu) = smooth muscle cells disposed in an inner circular (IC) and an outer longitudinal (OL) layer. 4) Tela subserosa (TSs) = loose connective tissue. 5) Tunica serosa (TSe) = ↑ peritoneum. Instead of tunica serosa and tela subserosa, retroperitoneal area of R. is lined with ↑ tunica adventitia.

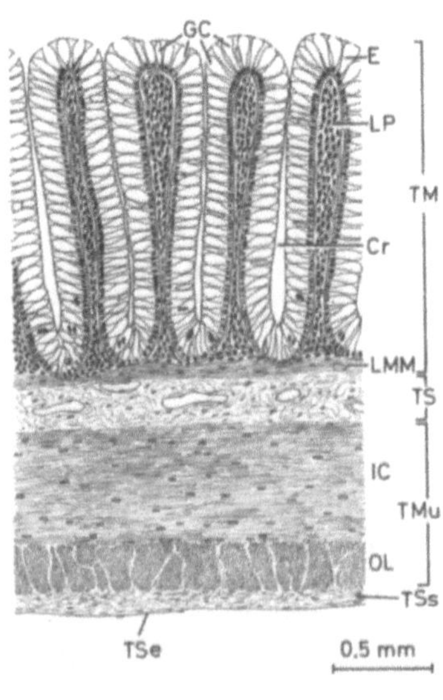

↑ Lieberkühn's crypts of colon and rectum

Neutra, M.R., Grand, R.J., Trier, J.S.: Lab. Invest. *36*, 535 (1977)

Red blood cell: ↑ Erythrocyte

Red blood corpuscle: ↑ Erythrocyte

Red pulp (pulpą rubra*): Collective term for ↑ splenic cords and ↑ splenic sinusoids filled with blood.

Blue, J., Weiss, L.: Am. J. Anat. *161*, 135 and 189 (1981); Chen, L.T., Weiss, L.: Am. J. Anat. *134*, 425 (1972); Grouls, V., Helpap, B.: Adv. Anat. Embryol. Cell Biol. *75*, 1 (1982)

Red skeletal muscle fibers: ↑ Skeletal muscle fibers, red

Redlich-Obersteiner zone: Limit (ROZ) between ↑ central (CNS) and ↑ peripheral (PNS) nervous system. For ↑ myelinated nerve fibers (MF), R. is characterized by abrupt cessation of ↑ oligodendrocytes (O) and appearance of ↑ Schwann's cells (SC) which ensheath, together with their basal lamina (BL), peripheral axons (A). R. is situated at level of first ↑ node of Ranvier (RN), dividing ↑ paranodal region belonging to an oligodendrocyte from paranodal region ensheathed by a Schwann's cell. For ↑ nonmyelinated fibers (NF), R. is the site where axons (A) become ensheathed by Schwann's cells. Basal lamina of ↑ nerve fibers is continuous with that lining ↑ subpial feet (SF) of ↑ astrocytes (BL).

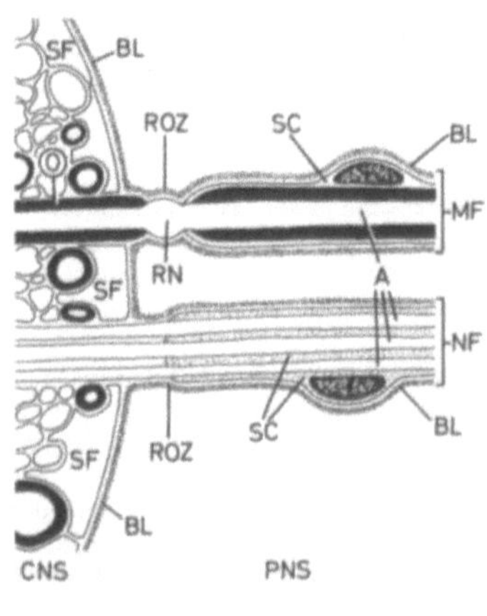

Kahle, W.: Nervensystem und Sinnesorgane. dtv-Atlas der Anatomie. Vol. 3. Stuttgart: Deutscher Taschenbuch Verlag 1976

Reductional division: Any division of a cell in which each daughter cell receives only half of ↑ chromosome set. First meiotic division is R.

↑ Equational division; ↑ Meiosis

Reflex, milk ejection: ↑ Suckling reflex

Refractive index (n): Relative velocity of light in a medium compared with velocity in air (n = 1). R. in air compared with water is n = 1.33; air compared with crown glass is n = 1.515, as for ↑ immersion oil.

↑ Oil-immersion objective

Refractive media, of eye: ↑ Eye, dioptric media of

Regeneration (repair, renewal): Capacity of ↑ tissues to replace by ↑ mitosis or ↑ meiosis lost cells by new cells which, optimally, should be morphofunctionally identical to those lost. Diagram shows varieties of R. In physiological R. loss of tissue is completely restored to normal level (L). In some cases, R. may be accompanied by an overproduction of tissue (e.g., ↑ callus). Physiological R. also comprises permanent R. of blood cells, surface and glandular epithelia, ↑ seminiferous epithelium, ↑ endometrium (= cyclic R.), etc. R. is possible only from tissue of same class; it depends upon degree of ↑ differentiation of tissue, age, blood supply, etc. The higher the differentiation degree, the less the capacity of R. In case of low regenerating capacity, tissue is replaced by a fibrous or glial scar. With regard to rapidity of R., tissues may be classified into: 1) Tissues with high R. rate (gastrointestinal epithelia, ↑ epidermis, hematopoietic cells, seminiferous epithelium, etc.); 2) tissues with low R. rate (liver parenchyma, kidney, thyroid, salivary glands, etc.); some of these organs may accelerate their R. rate if damaged, for example, liver parenchyma after partial hepatectomy (so-called conditional R.); 3) tissue without R. capacity (↑ neurons). (Modified after Bucher 1980)

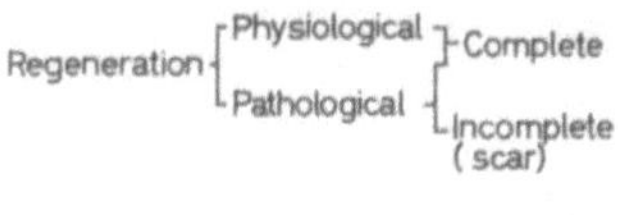

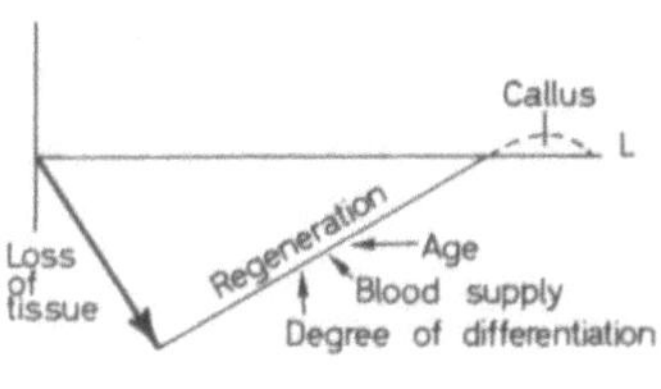

Bucher, O.: Cytologie, Histologie und mikroskopische Anatomie des Menschen. 10th edn. Bern: Hans Huber 1980

Regeneration, of nerve fibers: ↑ Nerve fibers of central nervous system, reaction to injury of; ↑ Nerve fibers of peripheral nervous system, degeneration and regeneration of

Registration peptides: Nonhelical peptide groups located at pro-α-chain ends of ↑ procollagen molecules. R. keep triple helical form of procollagen molecules in solution. Once in extracellular space, R. are cleaved by procollagen peptidase, so that procollagen molecules are transformed into ↑ tropocollagen molecules.

↑ Collagen, biosynthesis of

Regnaud's residual bodies: ↑ Residual bodies, of Regnaud

Reinke's crystals (Reinke's crystalloids, crystalloidum*): Intracytoplasmic proteinaceous ↑ inclusions (C) found in ↑ interstitial cells of human testis. R. are predominantly rectilinear, regular in geometric form, and with sharp angles; they may be up to 5 μm thick and 15 μm long. R. have little affinity for stains and appear transparent in light-microscopic preparations. In ↑ transmission electron microscope R. display highly ordered, 5-nm-thick filamentous macromolecules; depending on plane of section, crystalline lattice may appear linear (with periodicity of 18 nm), rhomboid, hexagonal (figs.), etc. It is thought that R. are formed by union of microfilamentous (arrow) and microtubular (arrowheads) structures present in their vicinity (inset). Although R. are biochemically similar to catalase, their function is unknown. (Figs. = human)

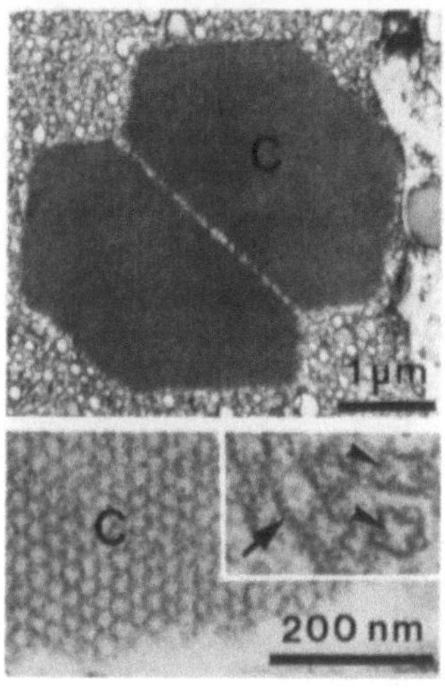

Nagano, T., Ohtsuki, I.: J. Cell Biol. *51*, 148 (1971); Payer, A.: Anat. Rec. *198*, 175 (1980)

Reissner's fiber: ↑ Glycosaminoglycan thread (R) produced by cells of ↑ subcommissural organ (SCO) and extending from it down into central canal of ↑ spinal cord. R. is formed of many thin filaments (F) that unite into a single fiber, 2–5 μm in diameter. Function of R. is unknown: it is believed to play a morphogenetic role for development of spinal cord, a role in agglutination of escaped erythrocytes (E), and a role in detoxification of biogenic amines circulating in ↑ cerebrospinal fluid. R. is lacking in humans. (Fig. = rat; courtesy Biomedical Research)

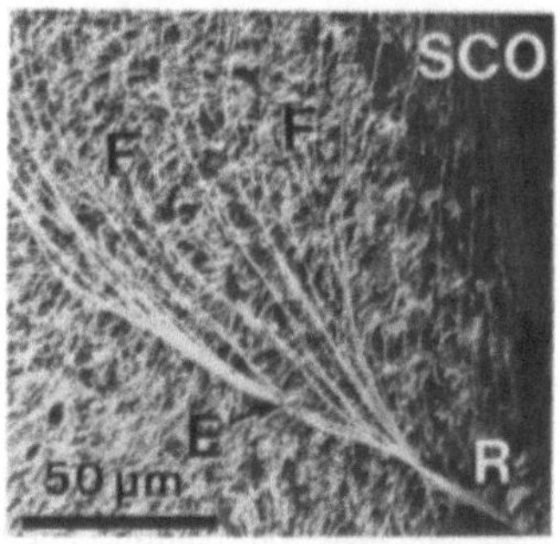

Hofer, H., Meinel, W., Erhardt, H.: Cell Tissue Res. *205*, 292 (1980); Krstić, R.: Biomed. Res. *2*, Suppl. 129 (1981); Tulsi, R.S.: J. Comp. Neurol. *211*, 11 (1982); Woollam, D.H.M., Collins, P.: J. Anat. *131*, 135 (1980)

Reissner's membrane (membrana vestibularis*, paries vestibularis*): Thin bilaminar sheet extending obliquely from internal edge of ↑ spiral limbus to ↑ spiral ligament and separating scala vestibuli (SV) from ↑ cochlear duct (CD). Aspect of R. oriented toward cochlear duct is lined by flattened epithelial cells (E) with nuclei thicker than neighboring cytoplasm. Apical surface of these cells bears some short microvilli; basal surface is considerably infolded and displays frequent vacuolar enlargements and numerous ↑ micropinocytotic vesicles. Epithelial cells are held together by ↑ zonulae occludentes (arrowhead). Epithelium rests on basal lamina. Toward scala vestibuli are situated occasional delicate ↑ collagen microfibrils (arrows) and a layer of very attenuated fibrocytelike cells (F) without a surrounding basal lamina. It is believed that R. plays a role in water and electrolyte transport. (Fig. = mongolian gerbil)

↑ Corti's organ

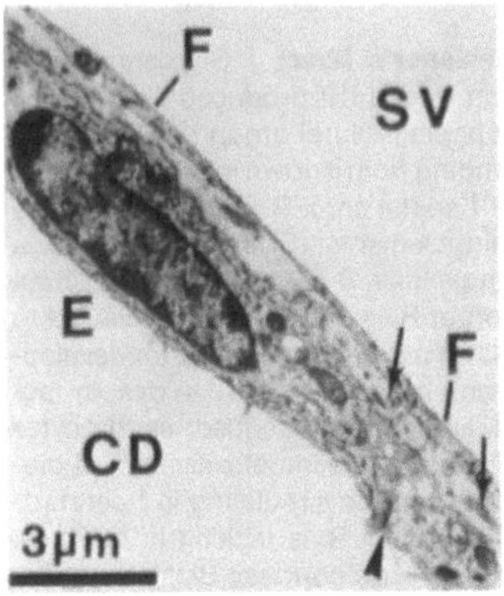

Relative volume: ↑ Stereology

Relaxin: Polypeptide hormone produced by ↑ granulosa lutein cells during late pregnancy and also in ↑ basal plate. R. seems to inhibit myometrial contraction in pregnancy and to stimulate dilatation of ↑ cervix.

Long, J.A.: Biol. Reprod. *8*, 87 (1973); Mathieu, Ph., Rahier, J., Thomas, K.: Cell Tissue Res. *219*, 213 (1981)

Releasing factors (hypophyseotropic hormones, liberines, releasing hormones, RF, RH, RH liberines): Polypeptides produced in perikarya of ↑ neurosecretory neurons of some ↑ hypothalamic nuclei under influence of brain cortex, sensory organs, and various internal stimuli. R. are discharged in ↑ pericapillary spaces of hypophyseoportal vessels, reach the adenohypophysis (Ah), and act on cells equipped with respective receptors inducing production of corresponding stimulating hormones, probably by ↑ adenyl cyclase-cAMP system. At present, following R. are known: CRF (↑ adrenocorticotropic hormone-RF, ACTH-RF), GRF, GH-RF or STH-RF (↑ growth or somatotropic hormone RF), FSH-RF (follicle-stimulating hormone-RF), LTH-RF or PRF (luteotropic hormone-RF or ↑ prolactin-RF), LH-RF (luteinizing hormone-RF), TSH-RF (thyrotropic hormone-RF), MSH-RF (melanocyte-stimulating hormone-RF). The common designation for FSH/LH-RF is Gn-RF (gonadotropin-RF). R. represent first step in neuroendocrine regulation of activity of endocrine glands; they constitute a potent amplification system of hormonal secretion in hypophyseal cells. Production of R. is controlled by a ↑ feedback mechanism (FB) and/or by direct nervous impulses from sensory organs through synaptic contacts to corresponding neurosecretory neurons of hypothalamic nuclei.

↑ Inhibiting factors; ↑ Hypophysis, vascularization of

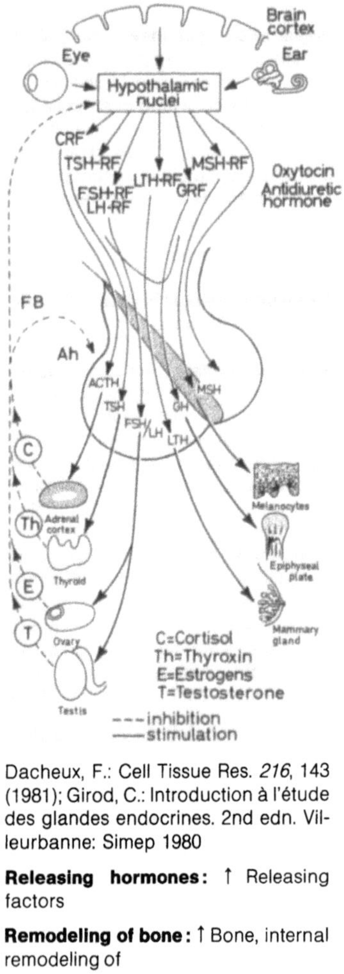

Dacheux, F.: Cell Tissue Res. *216*, 143 (1981); Girod, C.: Introduction à l'étude des glandes endocrines. 2nd edn. Villeurbanne: Simep 1980

Releasing hormones: ↑ Releasing factors

Remodeling of bone: ↑ Bone, internal remodeling of

Renal arteries: ↑ Muscular arteries, large

Renal calyces: ↑ Calyces, major; ↑ Calyces, minor

Renal columns (Bertin's columns, interlobar columns, columna renalis*): Cylindrical projections (RC) of cortical tissue (C) surrounding each ↑ pyramid (P) of ↑ kidney.

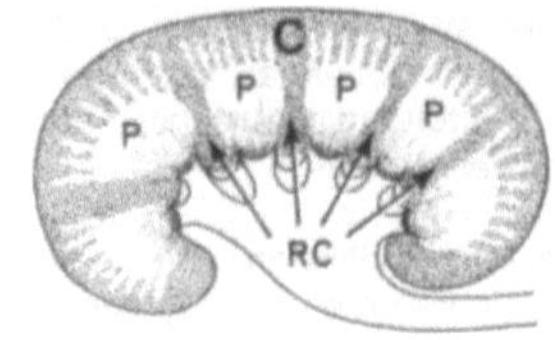

Renal corpuscle (Malpighian corpuscle, corpusculum renale*): The initial portion (C) of ↑ nephron, consisting of ↑ Bowman's capsule (B) and ↑ renal glomerulus (G). R. has a ↑ vascular pole (VP), at which ↑ afferent (Aff) and ↑ efferent (Eff) arterioles enter and leave, and a ↑ urinary pole (UP), at which ↑ proximal tubule (PT) begins.

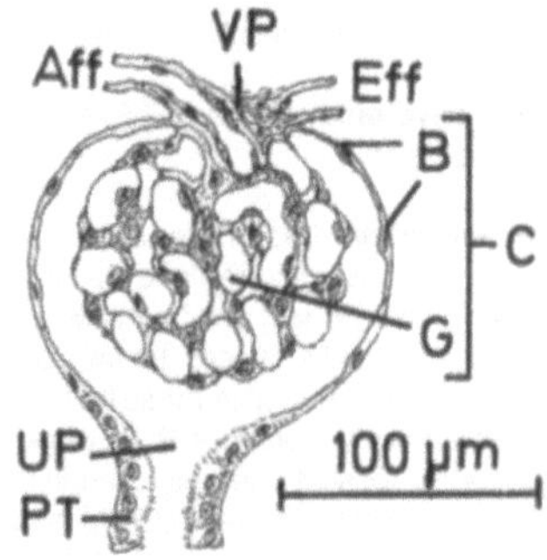

Renal cortex (cortex renalis*): Part of renal ↑ parenchyma, composed of ↑ renal corpuscles, ↑ proximal and ↑ distal convoluted tubules of ↑ nephrons, arched collecting tubules, and ↑ medullary rays.

↑ Collecting tubules, of kidney; ↑ Renal columns

Renal glomerulus (glomerulus corpusculi renalis*): Tuft of 10–50 capillary loops enclosed in a ↑ Bowman's capsule and covered by its visceral

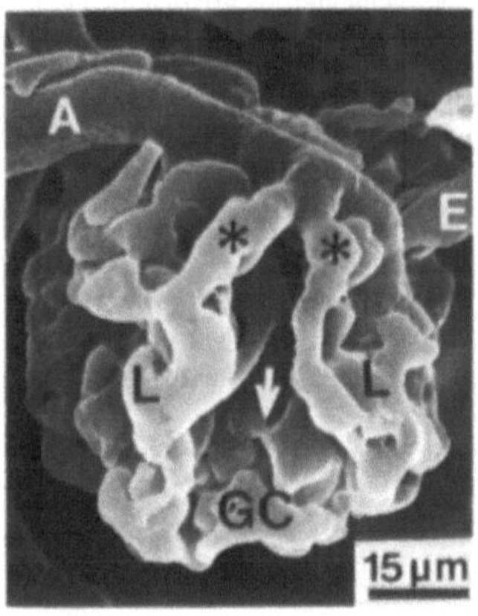

layer. Entering ↑ renal corpuscle, ↑ afferent arteriole (A) divides into 4–8 primary branches (asterisks), each of which spreads into a network of ↑ glomerular capillaries (GC). Capillaries originating form each primary branch form a glomerular lobule (L); there are anastomoses within lobules and between them (arrow). All capillaries reunite finally to form ↑ efferent arteriole (E), so that R. represents an arterioarterial anastomosis. With Bowman's capsule, R. composes a renal corpuscle. (Fig. = rat; microvascular cast and photo courtesy Prof. J. Kazimierczak, Lausanne)

↑ Portal system

Arakawa, M., Edanaga, M., Tokunaga, J.: Biomed. Res. *2*, Suppl. 307 (1981); Courtoy, P.J., Timpl, R., Farquhar, M.G.: J. Histochem. Cytochem. *30*, 874 (1982); Jones, D.: Lab. Invest. *37*, 569 (1977); Nizze, H., Csikos, A.: Int. Urol. Nephrol. *12*, 3 (1980); Rovenskà, E.: Acta Anat. *102*, 399 (1978)

Renal interstitium: ↑ Interstitial tissue, of kidney

Renal labyrinth (cortical labyrinth): Cortical tissue between ↑ medullary rays.

Renal lobe (lobus renalis*): Part of kidney ↑ parenchyma (L) composed of a ↑ pyramid (Py) and overlying ↑ renal cortex (C).

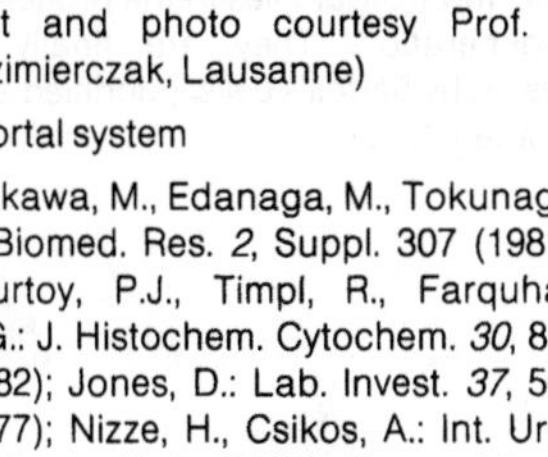

Renal lobule (lobulus renalis, lobulus corticalis*): Small portion of ↑ renal cortex (C) composed of a ↑ medullary ray (MR) with its immediately associated cortical tissue. M = medulla

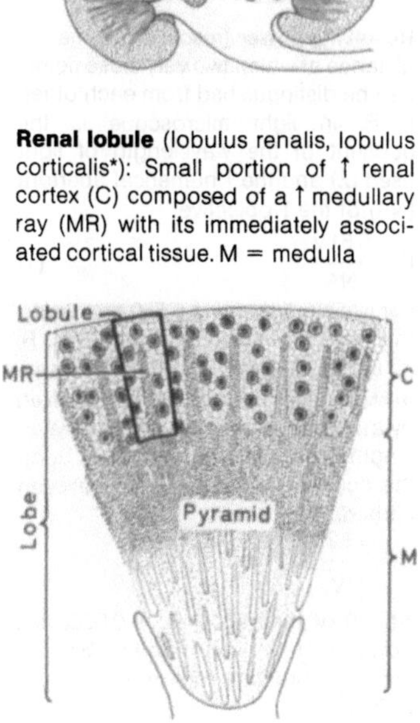

Renal medulla (medulla renalis*): Part of kidney ↑ parenchyma, consiting of straight portions of ↑ proximal and ↑ distal tubules of ↑ nephrons, ↑ loops of Henle, and ↑ collecting tubules.

Renal papilla (papilla renalis*): Tip (Pa) of a kidney ↑ pyramid (P). R. penetrates into a minor calyx (c); 200–700 ↑ papillary ducts (arrows) in form of ↑ area cribrosa open at R., which is lined by ↑ pseudostratified columnar epithelium continuous with ↑ transitional epithelium of ↑ calyces. (Fig. = monkey)

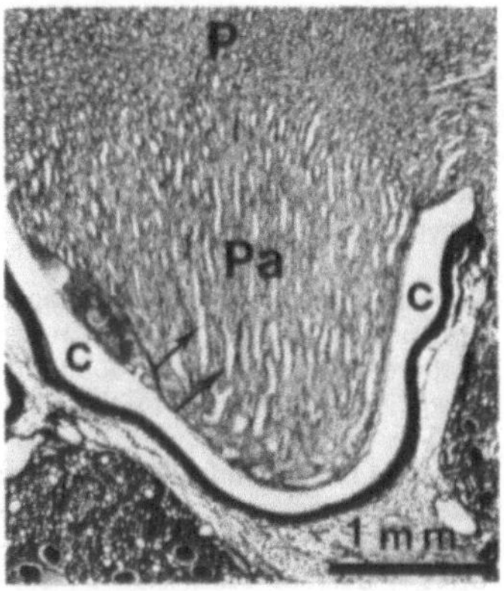

Renal pelvis (pelvis renalis*): Flattened, funnel-shaped enlargement (P) of extrarenal collecting system situated between major ↑ calyces (MC) and beginning of ↑ ureter (U). Structurally, R. is identical to major calyces; it is located in ↑ renal sinus.

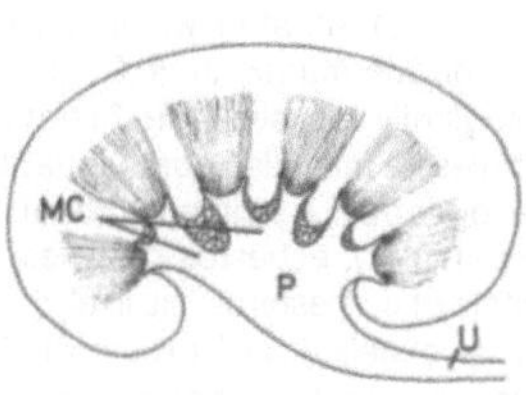

Verani, R., Bulger, R.E.: Am. J. Anat. *163*, 223 (1982)

Renal pyramid (medullary pyramid, pyramis renalis*): Conical mass of medullary substance (P) with slightly convex base adjacent to cortex (C), and an apex or papilla (p) that projects into a minor ↑ calyx. ↑ Medullary rays radiate from base of R. into cortex. Besides blood, lymphatic vessels, and ↑ nonmyelinated nerve fibers, R. consists of straight collecting tubules, ↑ papillary ducts, and ↑ loops of Henle. There are 8–18 R. in a human ↑ kidney. (Fig. = monkey)

↑ Calyces majores and minores; ↑ Collecting tubules, of kidney

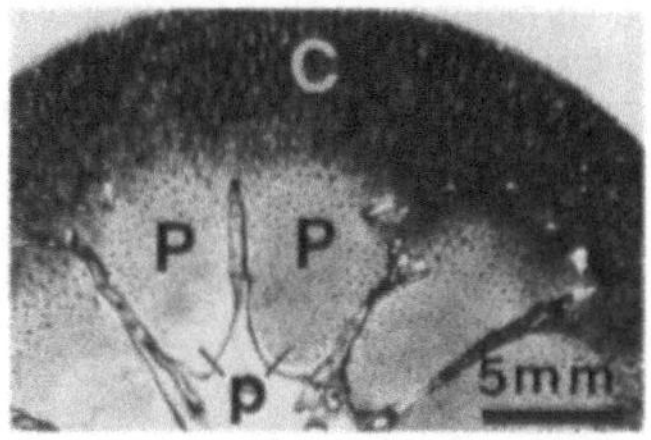

Renal sinus (sinus renalis*): Flattened cavity (RS) within ↑ kidney, through which pass ↑ renal artery, ↑ renal vein, lymphatics, and nerves; in R. are situated ↑ renal pelvis (P) and ↑ calyces. R. is filled mostly with ↑ adipose tissue and communicates with ↑ hilus (H).

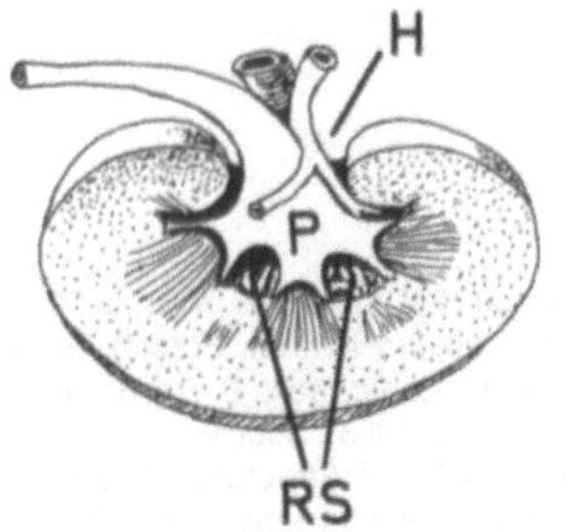

Renal tubules: ↑ Uriniferous tubules

Renal veins (vena renalis*): Large ↑ propulsive veins identical in structure to ↑ vena cava inferior.

Renewal: ↑ Regeneration

Renin: Proteolytic enzyme released mainly by ↑ juxtaglomerular cells, converting ↑ angiotensinogen to ↑ angiotensin I.

Faraggiana, T., Gresik, E., Tanaka, T., Inagami, T., Lupo, A.: J. Histochem. Cytochem. *30*, 459 (1982)

Renin granules: ↑ Juxtaglomerular granules

Rennin: Proteolytic enzyme involved in milk digestion, synthesized by ↑ chief cells of ↑ gastric glands proper of calves, but not of man.

↑ Milk

Săglam, M., Aşti, R.: Z. mikrosk.-anat. Forsch. *97*, 480 (1983)

Renshaw's cells: Small ↑ interneurons (R) lying in medial part of ventral horn of ↑ spinal cord and supplied by ↑ axon collaterals (C) of ↑ motor neurons (MN). ↑ Axons (A) of R. synapse with ↑ perikarya of other motor neurons. Not every motor neuron has a corresponding R. Inhibitory action of R. on motor neurons and their occurrence in the human have not been definitely confirmed. I = intercalated neurons

↑ Spinal cord, neurons of

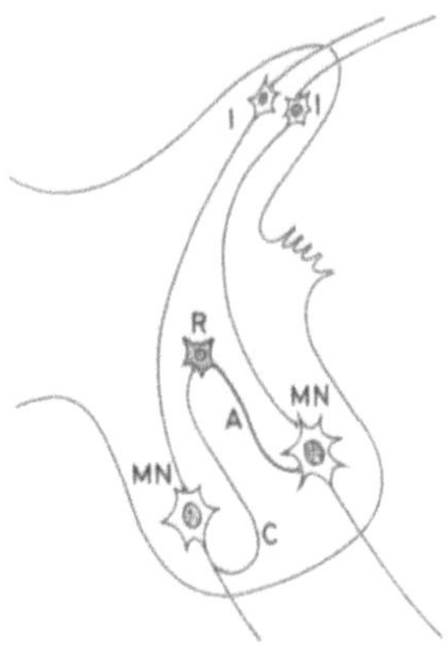

Kogai, Gy., Karcsu, S., Toth, L., Csillik, B.: Z. mikrosk.-anat. Forsch. *93*, 209 (1979); Lagerbäck, P.-A.: Brain Res. *264*, 215 (1983); Scheibel, M.E., Scheibel, A.B.: A Structural Analysis of Spinal Interneurons and Renshaw Cells. In: Brazier, M.A.B. (ed.): The Interneuron. Berkeley, Calif.: University of California Press 1969

Repair: ↑ Regeneration

Replica: An exact copy of a biological surface made of plastic (for light-microscopic and scanning electron microscopic observations) or of carbon under high vacuum after ↑ freeze-cleaving (for transmission electron microscopic observation).

Glauert, A.M. (ed.): Practical Methods in Electron Microscopy. Vol. 8. Amsterdam: Elsevier/North Holland 1980

Reproductive system, female: ↑ External genital organs, female; ↑ Ovary; ↑ Oviduct; ↑ Uterus; ↑ Vagina

Ferenczy, A.: The Female Reproductive System. In.: Hodges, G.M. and Hallowes, R.C.: Biomedical Research Applications of Scanning Electron Microscopy, Vol. 2. London, New York: Academic Press 1980; Möllendorff, W.v., Bargmann, W. (eds.): Weibliche Genitalorgane. In: Handbuch der mikroskopischen Anatomie des Menschen, Vol. 7, Parts 3 and 4/VI. Berlin, Heidelberg, New York: Springer-Verlag 1957 and 1966

Reproductive system, male: ↑ Bulbourethral glands; ↑ Ductus deferens; ↑ Epididymis; ↑ Penis; ↑ Prostate; ↑ Seminal vesicle; ↑ Testis

Möllendorff, W.v., Bargmann, W. (eds.): Männliche Genitalorgane. In: Handbuch der mikroskopischen Anatomie des Menschen. Vol. 7, Part 2/VIII. Berlin, Heidelberg, New York: Springer-Verlag 1930; Murakami, M., Sugita, A., Shimada, T., Hamasaki, M., Morizono, T., Nakamura, K.: Biomed. Res. *2*, Suppl. 355 (1981)

RES: ↑ Reticuloendothelial system

Reserve cells, of adenohypophysis: Term occasionally used to designate ↑ chromophobes.

Reserve megakaryocyte: ↑ Megakaryocyte

Reserve zone, of hyaline cartilage: Area of resting cartilage in ↑ epiphyses of the cartilaginous model during indirect ↑ bone formation.

Resident cells, of connective tissue: Fixed cells constantly present in connective tissues (↑ fibroblasts, ↑ fibrocytes, ↑ adipose cells, fixed ↑ histiocytes, ↑ reticular cells, ↑ tendon cells).

Residual bodies: Secondary ↑ lysosomes (R) containing indigestible remnants after any lysosomal action (↑ autophagy, ↑ crinophagy, ↑ heterophagy). R. are surrounded by a ↑ unit membrane. Their contents consist of dense granules of varying size, phospholipid myelinlike lamellae, and lipid particles (L) situated within an electron-opaque matrix; thus, R. are morphologically very similar to ↑ lipofuscin granules. R. may also contain particles of exogenous ↑ pigments (coal and dye particles), asbestos particles, and crystals of cholesterol. It is likely that R. may be expelled by ↑ exocytosis from some cells. As this is not possible from other cells (e.g., ↑ cardiac muscle cells, ↑ liver parenchymal cells, ↑ neurons,

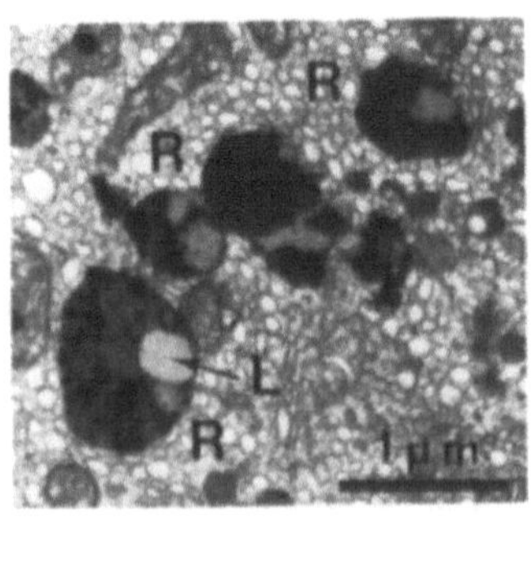

etc.), R. accumulate with age or wear and tear. In the latter situation, they are considered lipofuscin pigment. (Fig. = ↑ interstitial cell, of testis, human)

Residual bodies, of Regnaud (residual cytoplasm): Excesses of spermatozoon cytoplasms (C) which remain as ↑ unit membrane-limited anucleate masses (RB) within ↑ Sertoli's cells after freeing of ↑ spermatozoa in lumen of a ↑ seminiferous tubule. R. contain dense granules, ↑ lipid droplets, and ↑ organelles in the process of degeneration. They are finally digested by Sertoli's cells. (Modified after Fawcett 1973)

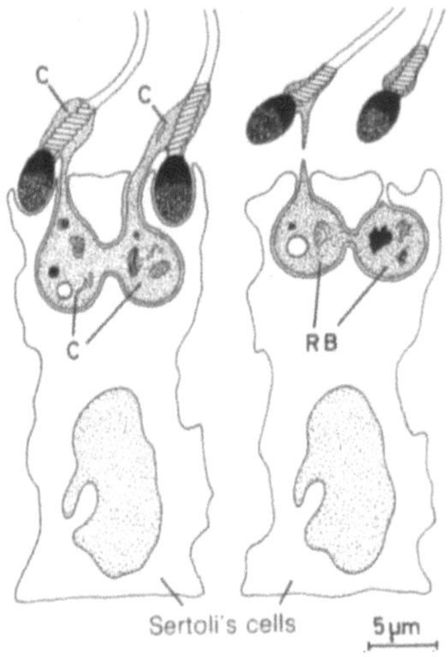

Fawcett, D.W., in: Segal, J.S. et al. (eds.): The Regulation of Mammalian Reproduction. Springfield, Ill.: Charles C. Thomas 1973

Resolving power (resolution): Smallest distance at which two very close points can be distinguished from each other. I) R. in light microscope is the quotient of the wavelength of light used (λ) and the ↑ numerical aperture (NA) of the ↑ objective:

$$R = \frac{0.61\,\lambda}{NA} \qquad (1)$$

For visible light of $\lambda = 550$ nm, using the best ↑ oil-immersion objective, R. is about 0.25 µm. II) R. in ↑ transmission electron microscope is given by the same expression (1). The wavelength (λ) of electrons depends upon the accelerating voltage (V) between cathode and anode:

$$\lambda = \frac{12.2}{\sqrt{V}} \cdot 0.1\ \text{nm}$$

At 100 kV theoretical R. is 0.0038 nm; however, this limit cannot be obtained in practice (actually best R. is 0.23 nm).

III) Theoretical R. in ↑ scanning electron microscope is inversely proportional to the diameter of the electron beam that scans the object (actually 2.5–7.5 nm).

Resolution: ↑ Resolving power

Resorcin fuchsin: A stain for selective demonstration of ↑ elastic fibers and ↑ elastic laminae.

Resorption cavity: ↑ Absorption cavity

Resorption lacunae: ↑ Howship's lacunae

Respiratory bronchioles: ↑ Bronchioles, respiratory

Respiratory mucosa (tunica mucosa nasi*): Tunica mucosa (RM) lining ↑ nasal cavities and ↑ nasopharynx (excepting ↑ olfactory mucosa). Structure: 1) Epithelium (E) = a ↑ pseudostratified ciliated epithelium with ↑ goblet cells; 2) lamina propria (LP) = a ↑ loose connective tissue with numerous ↑ mixed glands (G), free lymphocytes, and large veins with lacunar enlargements forming a richly ramified venous plexus (V), particularly well developed in ↑ nasal mucosa and ↑ conchae nasales. Lamina propria is firmly attached to ↑ periosteum (P) or ↑ perichondrium. R. moistens and warms inspired air. B = bone. (Fig. = human)

↑ Respiratory mucosa, vascularization of

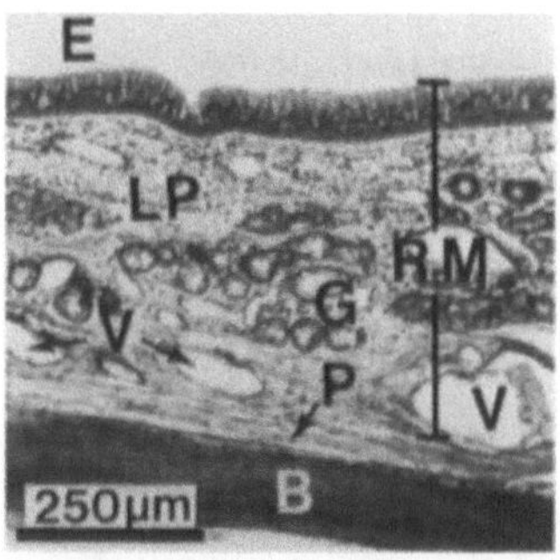

Andrews, P.: Biomed. Res. *2*, Suppl. 281 (1981); Jeffrey, P.K., Reid, L.M.: The Respiratory Mucous Membrane. In: Brain, J.D., Proctor, D.F., Reid, L. (eds.): Respiratory Defense Mechanisms. New York: Marcel Dekker Inc. 1977

Respiratory mucosa, vascularization of: Small ↑ muscular arteries (A) run vertically through mucosa, splitting just beneath epithelium (Ep) into superficial arcading branches, from which arise a dense network of fenestrated ↑ capillaries (Cap). They empty into venous lacunae (L) of venous plexus (VP). Musculature of wall of lacunae forms numerous sphincters (S). From lacunae, blood flows into thick-walled collecting veins situated in proximity to ↑ bone (B) or ↑ cartilage. Between arteries and veins there are numerous tortuous arteriovenous ↑ anastomoses (AVA) containing ↑ epithelioid cells (E). Contractions of lacunae sphincters and walls of collecting veins provoke swelling of venous plexus; its emptying is induced by relaxation of musculature of collecting veins. Arrows indicate direction of blood. (Modified after Körner 1937)

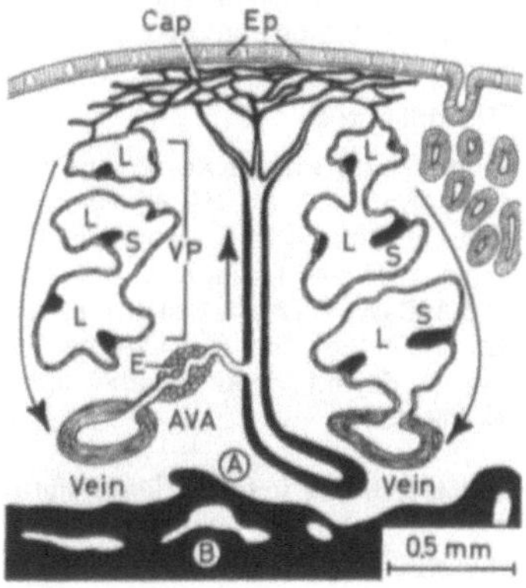

Cauna, N., Hinderer, K.H.: Ann. Otol. *78*, 865 (1969); Dawes, J.D.K., Prichard, M.M.L.: J. Anat. *87*, 311 (1953); Körner, F.: Z. mikrosk.-anat. Forsch. *41*, 131 (1937)

Respiratory region: ↑ Respiratory zone

Respiratory system (apparatus respiratorius*, systema respiratorium*): All organs primarily involved in intake of oxygen and elimination of carbon dioxide from blood. R. consists of: 1) Conducting portion with function of conducting, warming, moistening, and filtering inspired air = ↑ nasal cavity with associated sinuses, nasopharynx, ↑ larynx, ↑ trachea, ↑ bronchial tree; 2) ↑ respiratory zone responsible for exchange of gases; 3) ventilating portion with function of provoking inspiratory and expiratory movements = thoracic cage, intercostal muscles, and all ↑ elastic fibers of ↑ lungs.

Andrews, P.M.: The Respiratory System. In: Hodges, G.M., and Hallowes, R.C. (eds.): Biomedical Research Applications of Scanning Electron Microscopy, Vol. 1, London, New York: Academic Press 1979

Respiratory unit: Smallest morphofunctional component of ↑ lung. R. corresponds to a pulmonary ↑ acinus.

Respiratory zone: Collective term for all lung structures involved in gas exchange: ↑ alveoli, ↑ alveolar ducts, ↑ alveolar sacs, and respiratory ↑ bronchioles. Average surface of R. is about 150 m^2.

Response, immune: ↑ Immune response

Rete cutaneum: ↑ Skin, vascularization of

Rete mirabile: Historical term for ↑ portal system of vessels.

Rete ovarii: Vestigial system of imperfectly anastomosing mesonephrotic canaliculi situated in medulla of ↑ ovary. R. is lined with ↑ simple cuboidal epithelium. (See embyrology texts for further information)

Rete subpapillare (rete arteriosum subpapillare*): Flat vascular network (RS) at bases of ↑ dermal papillae. From R. arise hairpinlike capillary loops (CL) consisting of an ascending arterial and a descending venous limb. Only one loop penetrates each dermal papilla. (Fig. = human, India ink vascular ↑ injection)

↑ Skin, vascularization of

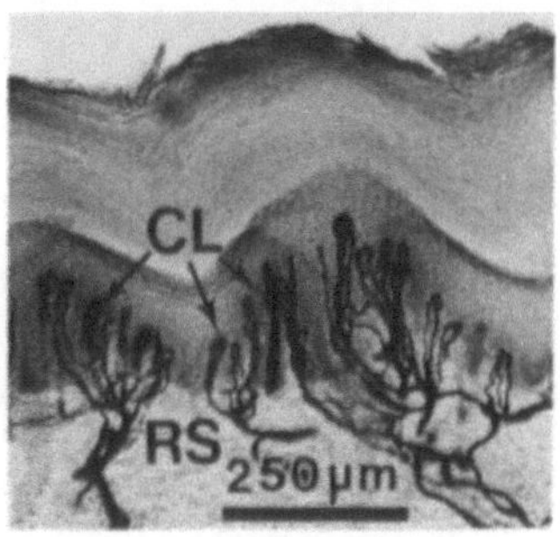

Rete testis*: Labyrinthine system of irregularly anastomosing canaliculi and chambers (RT) composing a complex spongelike network within ↑ mediastinum testis. R. connects ↑ tubuli recti to ↑ ductuli efferentes. Lining of R. consists of ↑ simple squamous to low columnar epithelium (E). Apical cell surfaces bear some microvilli and an occasional single ↑ cilium. It seems that the cells do not have a secretory function. Underlying connective tissue contains some ↑ fibroblasts and a considerable amount of collagen fibers (cf). (Figs. = human)

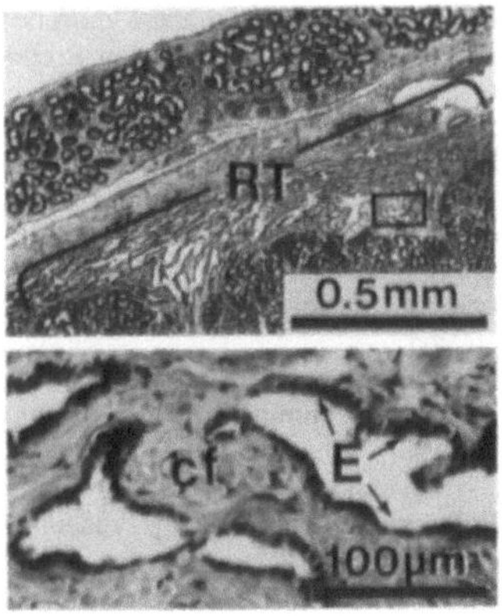

Dym, M.: Anat. Rec. *186*, 493 (1976); Kormano, M.: The Rete Testis. In: Johnson, A.D., Gomes, W.R. (eds.): The Testis. Vol. 4. New York, San Francisco, London: Academic Press 1977

Reticular cells (reticulum cell, cellula reticularis*): Stellate or irregularly shaped cells constituting fundamental

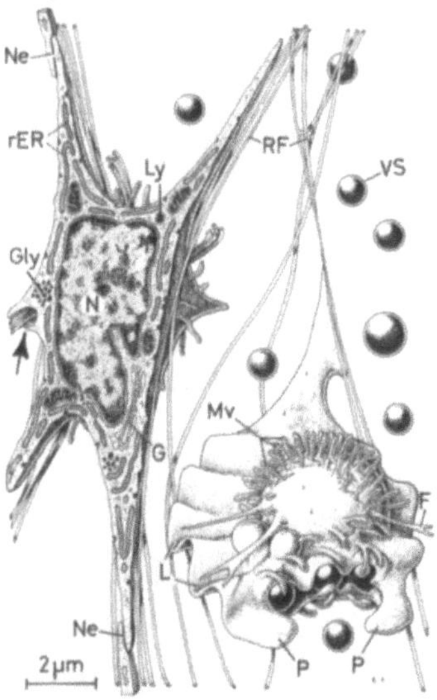

supporting network of ↑ bone marrow, ↑ lymph nodes, ↑ lymphatic nodules, ↑ spleen, and ↑ tonsils. R. have a central nucleus (N) with a prominent nucleolus; cytoplasm contains varying numbers of mitochondria, a small to well-developed Golgi apparatus (G), varying numbers of rough endoplasmic cisternae (rER) filled with moderately dense ↑ reticuloplasm, some lysosomes (Ly), an appreciable amount of free ribosomes, fine bundles of microfilaments, and ↑ glycogen particles (Gly). R. are adjacent to and/or enclose (arrow) bundles of ↑ reticular fibers (RF); they are interconnected by ↑ nexus (Ne). Several types of R. have recently been described. Functions: 1) Synthesis of reticular fibers (lesser collagen); 2) ↑ phagocytosis of dead cells, cell debris, infectious agents, and particles of inert foreign matter [e.g., ↑ vital stains, (VS)]; in such a state, R. can abandon reticular fibers and develop microvilli (Mv), ↑ lamellipodia (L), ↑ pseudopodia (P), and ↑ filopodia (F), accompanied by an increase in lysosomes and general development of ↑ organelles; 3) trapping of antigens on their surface and induction of adjacent immunocompetent ↑ B-lymphocytes to differentiate into ↑ plasma cells; 4) some R. with poorly developed organelles are thought to represent stem cells of ↑ lymphatic and ↑ myeloid tissues. (Not to be confused with ↑ reticulocytes)

↑ Dendritic cells

Biermann, A., Graf von Keyserlingk, D.: Acta Anat. *100*, 34 (1978); Heusermann, U., Zuborn, K.-H., Schroeder, L., Stutte, H.J.: Cell Tissue Res. *209*, 279 (1980); Saito, H.: Arch. Histol. Jpn. *40*, 333 (1977); Sakuma, H., Kasajima, T., Imai, Y., Kojima, M.: Acta Pathol. Jpn. *31*, 449 (1981)

Reticular fibers (reticular fibrils, fibra reticularis*): Kind of ↑ birefringent and ramified fibers (R), 0.1–1.5 µm in diameter, present in all connective tissues and constituting a three-dimensional supporting network of ↑ lymph nodes, ↑ spleen, and ↑ bone marrow. R. are present in interlobular septa of ↑ pancreas and ↑ salivary glands; they surround ↑ capillaries, ↑ adipose cells, ↑ smooth muscle cells, ↑ skeletal muscle fibers, ↑ peripheral nerve fibers, etc., and stay in close contact with ↑ reticular cells. R. are argyrophilic, ↑ PAS positive, and trypsin resistant; they contain about ten times more carbohydrate than ↑ collagen fibers, can be stretched considerably more than collagen fibers, do not swell in diluted acids, and yield no ↑ gelatin by boiling. One R. consists of several ↑ reticular microfibrils (RMf) enclosed by a ↑ glycoprotein and ↑ proteoglycan coating (C), probably responsible for ↑ argyrophilia and staining properties of R., which are different to those of collagen fibers. In lymph nodes, bone marrow, and spleen, R. are enclosed by long sheetlike processes (P) of ↑ reticular cells. E = ↑ endothelial cells of ↑ splenic sinusoids. (Fig., top = silverstained human lymph node; fig., middle = R. of rat lymph node; fig., bottom = R. of rat spleen)

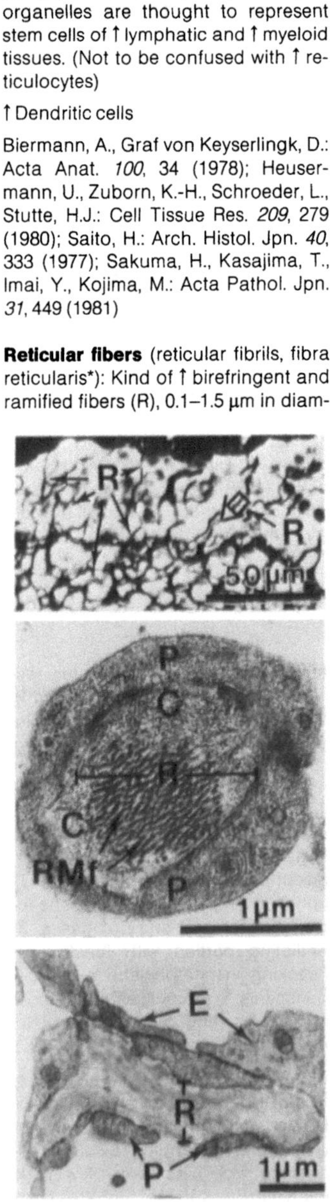

Montes, G.S., Krisztan, R.M., Shigihara, R., Tokoro, R., Mourao, P.A.S., Junqueira, L.C.U.: Histochemistry *65*, 131 (1980)

Reticular fibrils: ↑ Reticular fibers

Reticular lamina: Layer of ↑ basement membrane.

Reticular layer: ↑ Dermis; ↑ Skin

Reticular membrane, of Corti's organ (reticular lamina, membrana reticularis*): Network pattern (R) formed of contours of ↑ phalanges (P) of outer ↑ phalangeal cells and outer ↑ pillars (Op), interarticulating with outer ↑ hair

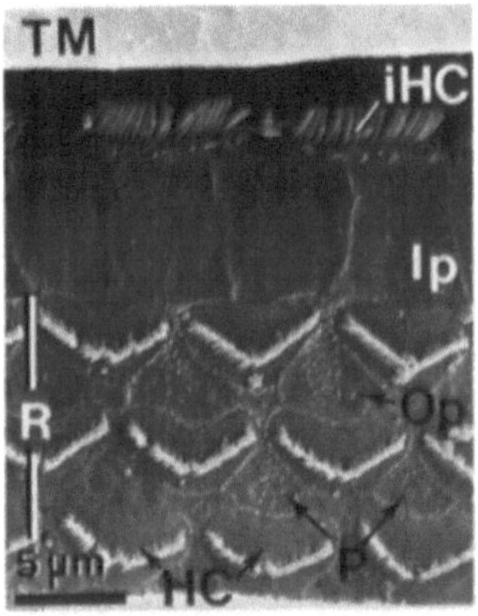

cells (HC). iHC = inner ↑ hair cells; Ip = inner ↑ pillars; TM = ↑ tectorial membrane. (Fig. = mongolian gerbil)

Gulley, R.L., Reese, T.S.: J. Neurocytol. *5*, 479 (1976); Hoshino, T.: Arch. Histol. Jpn. *37*, 25 (1974)

Reticular microfibrils: Threadlike structures (RMf), 9–12 nm in diameter, displaying delicate 15-nm cross-banding. R. are interposed between basal laminae (BL) of various epithelia and underlying ↑ collagen microfibrils (CMf). Similarly, R. assure contact between ↑ skeletal muscle fibers and ↑ tendon microfibrils; bundles of R. constitute ↑ reticular fibers (R). In general, R. are considered to consist of chemically modified ↑ tropocollagen, which can persist in this state or can transform into mature collagen microfibrils. Therefore, R. display considerable morphological differences depending on the organ in question. (Figs. = ↑ dermis, mongolian gerbil)

↑ Basement membrane

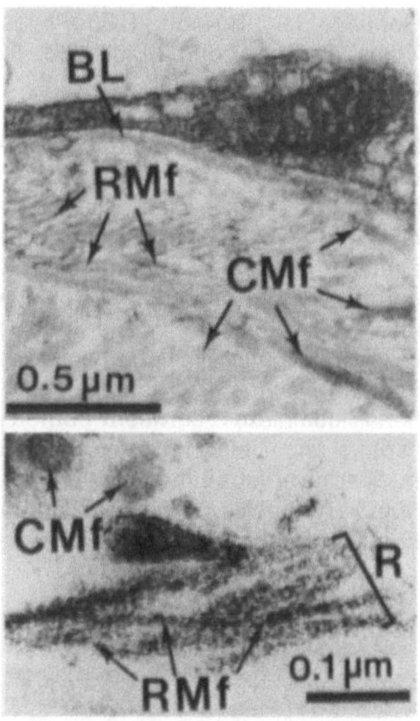

Fukuta, K., Mochizuki, K.: Arch. Histol. Jpn. *45*, 181 (1982)

Reticular tissue (textus connectivus reticularis*): A kind of ↑ fibrous connective tissue in form of a tridimensional meshwork of ↑ reticular cells (arrows) and ↑ reticular fibers. Within R. wander ↑ macrophages (M), ↑ lymphocytes (L), ↑ monocytes, ↑ plasma cells, and occasional ↑ eosinophils. R. is the basic tissue of ↑ lymphatic nodules, ↑ tonsils, ↑ lymph nodes, ↑ bone marrow, and ↑ spleen; it is actively involved in defense of organism and is therefore classified with ↑ reticuloendothelial system. (Fig. = ↑ subcapsular sinus, human)

↑ Lymphatic tissue

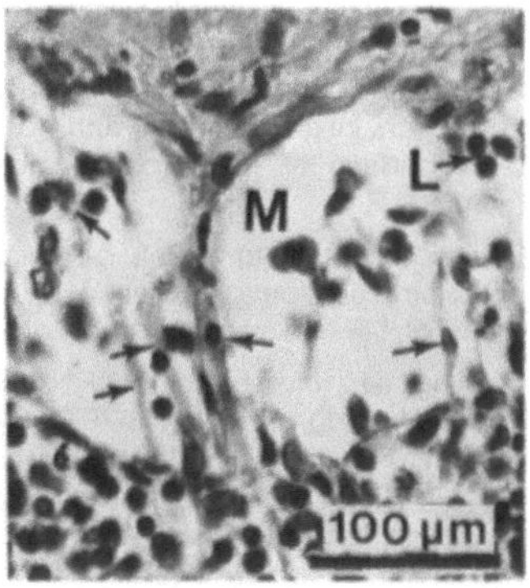

Clark, S.: Am. J. Anat. *110*, 217 (1962); Fujita, T.: Recent Adv. RES Res. *18*, 1 (1978)

Reticulocytes (polychromatophilic erythrocyte, reticulocytus*): Anuclear, round or polygonal cells (A), 7.2–9 µm in diameter, belonging to erythrocytic series, present within ↑ bone marrow and circulating blood (about 0.8%–1% of ↑ erythrocytes). Cytoplasm contains only a few mitochondria, some vesicles, several ↑ siderosomes (S), and a small number of residual ↑ polyribosomes (Pr). After usual staining of the blood smear (fig. A), R. appear bluish or greenish due to ↑ basophilia of residual polyribosomes and ↑ acidophilia of ↑ hemoglobin, hence the term polychromatophilic. In fresh blood, polyribosomes are precipitated with ↑ brilliant cresyl blue or 9-aminoacridin (fig. B) and form a weblike basophilic reticulum ["substantia granulo-filamentosa" (SGF), or "substantia reticularis"] giving these cells their name. Through ↑ micropinocytosis R. may take up ↑ ferritin and still synthesize hemoglobin. In the course of 24 h, R. lose their reticulum and become mature ↑ erythrocytes. (Not to be confused with ↑ reticular cells.)

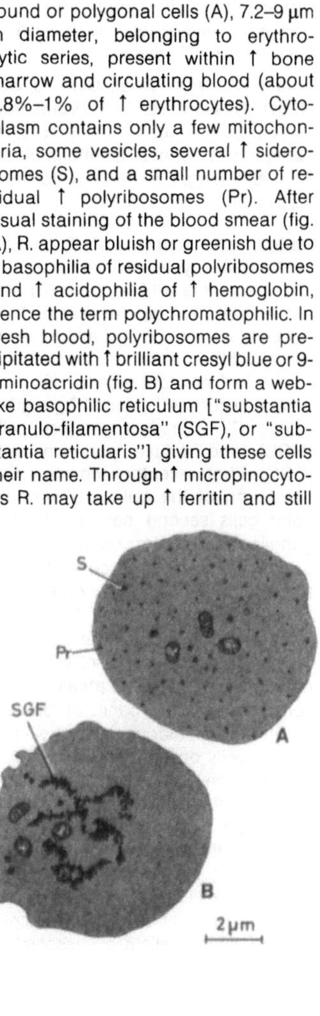

↑ Erythropoiesis; ↑ Supravital staining

Zweig, S.E., Tokuyasu, K.T., Singer, S.J.: Member-associated Changes during Erythropoiesis – On the Mechanism of Maturation of Reticulocytes to Erythrocytes. In: Marchesi, V.T., Gallo, R.C. (eds.): Differentiation and Function of Hematopoietic Cell Surfaces. Vol. 1. New York: Alan R. Liss Inc. 1982

Reticuloendothelial system (RES): Diffuse system of ↑ macrophages (↑ reticular cells) of ↑ lymphatic organs and specialized phagocytic endothelial cells lining ↑ blood sinuses of various organs. This classical RES (Aschoff 1924) includes:

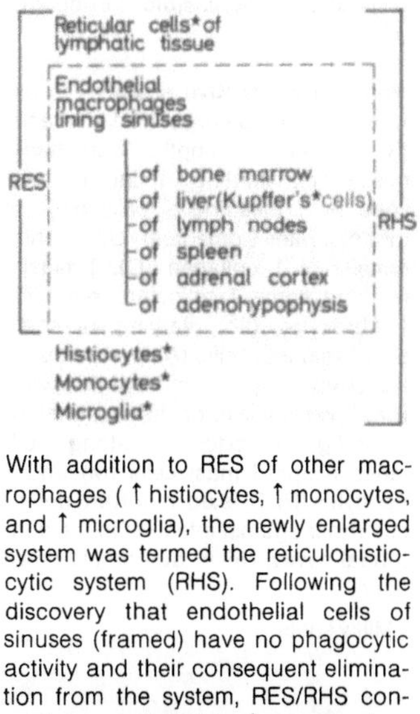

With addition to RES of other macrophages (↑ histiocytes, ↑ monocytes, and ↑ microglia), the newly enlarged system was termed the reticulohistiocytic system (RHS). Following the discovery that endothelial cells of sinuses (framed) have no phagocytic activity and their consequent elimination from the system, RES/RHS concept was replaced by ↑ mononuclear phagocyte system. (Terms marked with an asterisk are explained under corresponding letters)

Carr, I., Deames, W.T. (eds.): The Reticuloendothelial System. Vol. 1. Morphology. New York and London: Plenum Press 1980; Fujita, T.: Rec. Adv. RES Res. *18*, 1 (1978); Fujita, T., Kashimura, M.: Biomed. Res. *2*, Suppl. 159 (1981); Rose, N.R., Siegel, B.V. (eds.): The Reticuloendothelial System: A Comprehensive Treatise. Vol. 4. New York: Plenum Publ. Co. 1982; Sbarra, A.J., Strauss, R.R. (eds.): The Reticuloendothelial System. Vol. 2. Biochemistry and Metabolism. New York and London: Plenum Press 1980

Reticulohistiocytic system (RHS): Concept of a ↑ reticuloendothelial system extended by ↑ histiocytes, ↑ monocytes, and ↑ microglia.

Reticuloplasm: Fine granular contents of varying density (R) found in rough endoplasmic ↑ cisternae. In general, R. is composed of polypeptide chains. (Fig. = ↑ plasma cell, rat)

↑ Protein synthesis

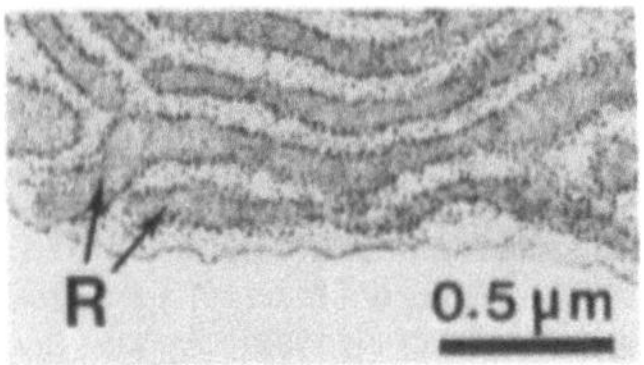

Reticulum cells: ↑ Reticular cells

Reticulum, endoplasmic: ↑ Endoplasmic reticulum

Retiform connective tissue: ↑ Loose connective tissue organized in strands (S) of various lengths and thicknesses constituting a lacelike net lined by ↑ mesothelial cells (MC); it forms greater ↑ omentum (O). Strands consist of ↑ collagen (C), ↑ elastic (E) and some ↑ reticular fibers (R), ↑ fibrocytes (Fc), varying numbers of ↑ adipose cells (A), numerous ↑ histiocytes (H), some ↑ lymphocytes, rare ↑ plasma cells, occasional eosinophilic ↑ granulocytes, and, along capillaries (Cap), ↑ mast cells (M) and ↑ pericytes (P). R. is an important defense barrier against infection in the peritoneal cavity. NF = autonomic nerve fiber

↑ Milky spots

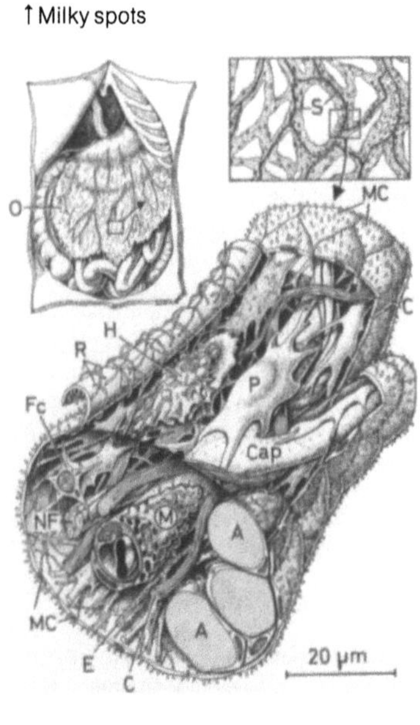

Retina*: Specially differentiated photosensitive part of ↑ central nervous system forming innermost coat of ↑ eye. R. is subdivided into a posterior or optic portion up to ↑ ora serrata (pars optica retinae*), a nonfunctional anterior portion covering ↑ ciliary body (pars ciliaris retinae*), and posterior surface of ↑ iris (pars iridica retinae*).

↑ Retina, optic, neurons of; ↑ Retina, optic, stratification of; ↑ Retina, pars ciliaris of; ↑ Retina, vascularization of

Dowling, J.E.: Invest. Ophthal. *9*, 655 (1970); Dowling, J.E., Werblin, F.S.: Vision Res. *3*, 1 (1971); McDevitt, D.S. (ed.): Cell Biology of the Eye. New York: Academic Press 1982

Retina, optic, neuroglial cells of: ↑ Müller's cells are largely predominant, some spindle-shaped ↑ neuroglial cells, mostly ↑ astrocytes, are present in inner nuclear layer, ganglion cell layer, and nerve fiber layer where their ↑ perivascular feet participate in formation of ↑ blood-retinal barrier. Occasional ↑ microglial cells have also been found in ↑ retina.

Böhme, G.: Verh. Anat. Ges. *75*, 961 (1981); Büssow, H.: Cell Tissue Res. *206*, 367 (1980); Kohno, T., Inomata, H., Taniguchi, Y.: Jpn. J. Ophthalmol. *26*, 53 (1982); Ling, E.A.: Arch. Histol. Jpn. *45*, 37 (1982)

Retina, optic, neurons of: The retina contains five types of ↑ neurons: 1) ↑ Photoreceptors (Ph) = cone (C) and rod (R) cells; 2) ↑ bipolar cells (BC) = flat bipolar cells (FBC), ↑ midget bipolar cells (MBC), and ↑ rod bipolar cells (RBC); 3) ↑ horizontal cells (HC); 4) ↑ amacrine cells (AC); and 5) ↑ ganglion cells (GC) = midget ganglion cells (MGC) and diffuse ganglion cells (DGC). Photoreceptors represent first (I) neuron of the optic pathway; bipolar cells second neuron (II), and ganglion cells third neuron (III). In ↑ fovae centralis one cone cell enters into contact with only one midget bipolar cell, which forms ↑ synapses with only one midget ganglion cell, thus assuring the best resolving power of vision. In other retinal areas, there is considerable convergence of rod cells on one rod bipolar cell and of several bipolar cells on one diffuse ganglion cell. Also, several cones converge on one flat bipolar cell and several flat bipolar cells make synaptic contacts with one diffuse ganglion cell. Overall convergence in retina is about 105:1 (126 million photoreceptors: 1.2 million ↑ optic fibers, OF). Amacrine and horizontal cells are ↑ association retinal neurons. (Modified after Dowling and Boycott 1969)

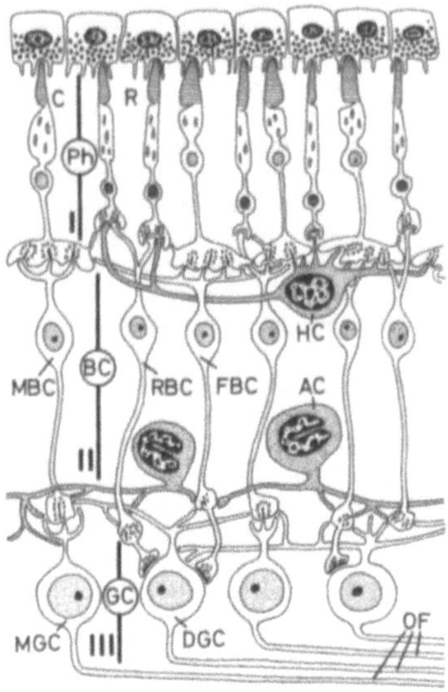

Dowling, J.E., Boycott, B.B.: Philos. Trans. R. Soc. Lond. [Biol] *255*, 80 (1969)

Retina, optic, stratification of: A routinely stained ↑ retina displays, from outside to inside, ten successive layers: 1) ↑ Pigment epithelium (pars pigmentosa*); 2) ↑ photoreceptors (stratum photosensorium*) = inner and outer segments of cone and rod cells; 3) ↑ outer limiting membrane (stratum limitans externum*) = junctional complexes between photoreceptors and ↑ Müller's cells; 4) outer nuclear layer (stratum nucleare externum*) = nuclei of photoreceptors; 5) outer plexiform layer (stratum plexi-

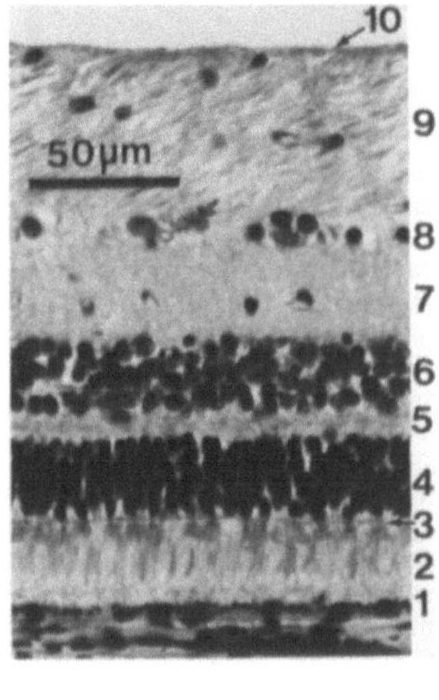

forme externum*) = ↑ synapses between photoreceptors, ↑ bipolar cells, and ↑ horizontal cells; 6) inner nuclear layer (stratum nucleare internum*) = nuclei of bipolar, ↑ amacrine, horizontal, and Müller's cells; 7) inner plexiform layer (stratum plexiforme internum*) = synapses between bipolar, ganglion, and amacrine cells; 8) ganglion cell layer (stratum ganglionare*) = ↑ ganglion cells of retina; 9) nerve fibers layer (stratum neurofibrarum*) = ↑ axons of ganglion cells (↑ optic fibers); and 10) inner limiting membrane (stratum limitans internum*) = feet of Müller's cells and their basal lamina. (Fig. = human)

Retina, pars ciliaris of: ↑ Ciliary epithelium

Retina, pars iridica of (pars iridica retinae*): A bilayered posterior epithelium of the ↑ iris.

Retina, photoreceptors of: ↑ Photoreceptors

Retina, pigment epithelium of: ↑ Pigment epithelium

Retina, vascularization of: ↑ Retina is supplied with blood from 1) ↑ choriocapillaris (Ch) which supplies ↑ pigment epithelium (PE), ↑ photoreceptors (Ph), and outer half of outer plexiform layer (OPL) by diffusion, and 2) arteria centralis retinae with branches (AC) running immediately below inner limiting membrane (ILM) and forming a capillary network (Cap) in inner nuclear layer (INL). Venous blood is collected by ↑ postcapillary venules which join at right angles with veins (V) in layer of nerve fibers (NF). The two principal branches of arteria centralis retinae are ↑ terminal arteries.

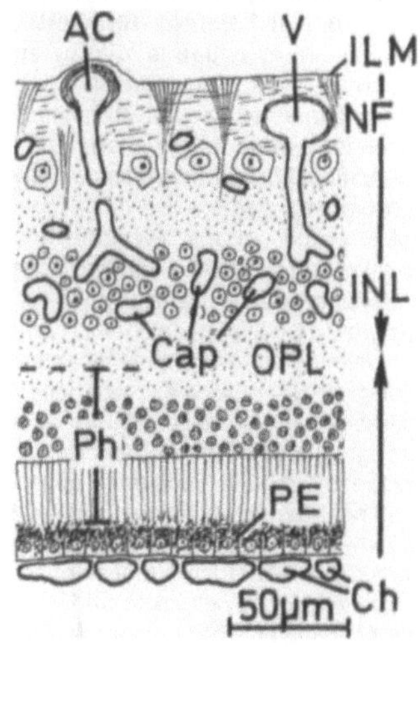

Retinal: ↑ Retinene

Retinal ganglion: All ↑ bipolar cells of the ↑ retina.

Retinal membranous discs: ↑ Membranous discs, of photoreceptors

Retinene (retinal): An aldehyde of vitamin A; with ↑ scotopsin it forms ↑ rhodopsin and with ↑ photopsin it constitutes ↑ iodopsin.

Retraction bulb: Dilatation of an ↑ axon at the disintegrated proximal stump end (PS) 48 h after ↑ axotomy as a consequence of local interruption of ↑ axoplasmic flow. R. contains mitochondria, vesicles, smooth cisternae, ↑ neurotubules, dense-cored granules, and ↑ acetylcholine. In case of ↑ regeneration of the cut axon, fine axonic processes emerge from R. – growth cones (GC) which extend between proliferated ↑ Schwann's cells of ↑ Büngner's bands (BB) toward the distal stump (DS); one of these cones, completely surrounded by Schwann's cells, will regenerate ↑ nerve fiber.

↑ Nerve fibers of peripheral nervous system, degeneration and regeneration of

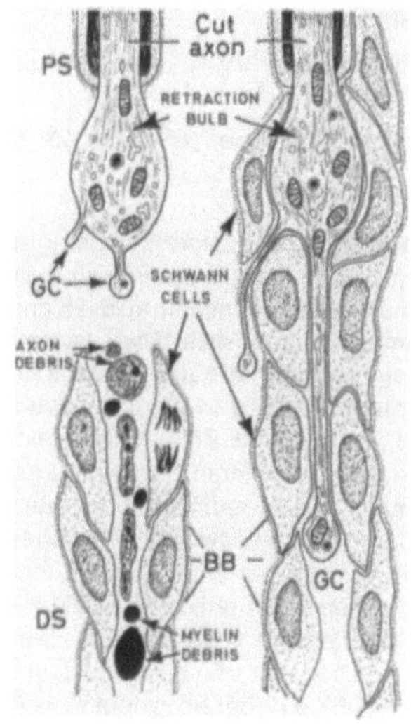

Retroannular recess: ↑ Annulus, of spermatozoon

Retrograde transneuronal degeneration: ↑ Transneuronal degeneration

Retzius' lines: ↑ Lines, of Retzius

RF: ↑ Releasing factors

RH: ↑ Releasing factors

Rhagiocrine cells: Term occasionally used to designate ↑ macrophages of ↑ loose connective tissue.

Rhodopsin (visual purple): Photosensitive pigment of rods, contained in their ↑ membranous discs. R. consists of ↑ scotopsin and ↑ retinene. Light bleaches R. by breaking retinene-scotopsin bond.

Blaustein, D.I., Dewey, M.M.: J. Histochem. Cytochem. *27*, 788 (1979); Hargrave, P.A.: Rhodopsin Chemistry, Structure and Topography. In: Osborne, N.N., Chader, G.J.: Progress in Retinal Research, Vol. 1. Oxford: Pergamon Press 1982

Rhopheocytosis: Type of ↑ micropinocytosis by which ↑ erythroblasts receive one-fourth of their ↑ iron in form of ↑ ferritin particles from ↑ reticular cells of ↑ bone marrow (↑ erythroblastic islands). Sequence: 1) Ferritin particles (Fe) extruded by ↑ exocytosis from reticular cell (R) attach to a ↑ glycocalyx (Gl) covering an approximately 0.1-µm-wide area of erythroblast plasmalemma (P); 2) this area invaginates within cell body to form a ↑ micropinocytotic vesicle (MpV); 3) vesicle with ferritin particles separates from cell membrane; 4) membrane of vesicle disappears; ferritin iron is available for ↑ hemoglobin synthesis. Remaining iron needed for hemoglobin synthesis is obtained in form of ↑ transferrin circulating in blood plasma; this process cannot be observed. R = ↑ ribosomes, Pr = ↑ polyribosomes, Hb = ↑ hemoglobin

↑ Erythroblast, basophilic

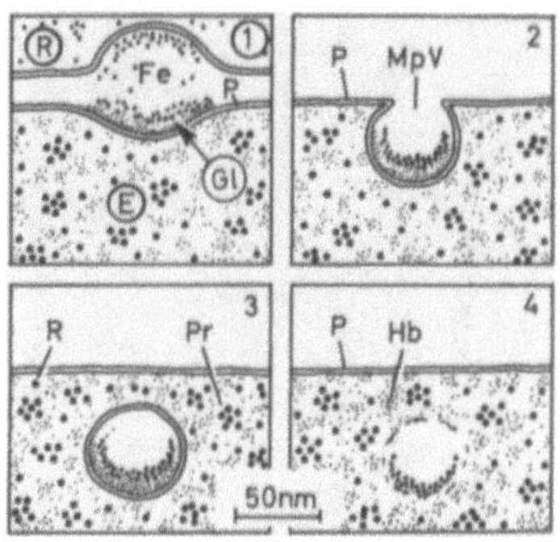

Policard, A., Bessis, M.: C.R. Acad. Sci. *246*, 3194 (1958)

Ribonucleic acid, messenger: ↑ Messenger RNA

Ribonucleic acid, ribosomal: ↑ Ribosomal RNA

Ribonucleic acid, transfer: ↑ Transfer RNA

Ribonucleic acids: Single-stranded ↑ nucleic acids with function of transporting transcribed information from DNA into cytoplasm and of synthesizing proteins. Pentose of R. is ribose; pyrimidine bases are cytosine and uracil, and the purines, adenine and guanine. There are three major classes of R.: ↑ Messenger R. (mRNA), ↑ ribosomal R. (rRNA), and ↑ transfer R. (tRNS). R., particularly rRNA, are responsible for ↑ basophilia of the cytoplasm in protein-producing cells. (See biochemistry and molecular biology texts for further information)

↑ Protein synthesis

Perry, R.P.: J. Cell Biol. *91*, 28s (1981)

Ribonucleic acids, detection of: ↑ Nucleic acids, detection of

Ribonucleic acids, synthesis of: RNAs are transcribed from corresponding ↑ cistrons of the DNA molecule. Ribosomal DNA cistron (rDNA) of ↑ nucleolar organizer region (NOR) is responsible for transcription of ↑ ribosomal RNA (rRNA) which forms, after a series of steps, 28S and 18S rRNA that participate in formation of 60S and 40S ribosomal subunits, respectively. ↑ Messenger RNA (mRNA) and ↑ transfer RNA (tRNA) are transcribed outside ↑ nucleolus and pass through ↑ nuclear pores (NP) into cytoplasm, where mRNA attaches itself to small ribosomal subunits. These, combined with large ribosomal subunits, form ↑ ribosomes. ↑ Protein synthesis begins following attachment to tRNA to these ribosomes. (See molecular biology texts for further information)

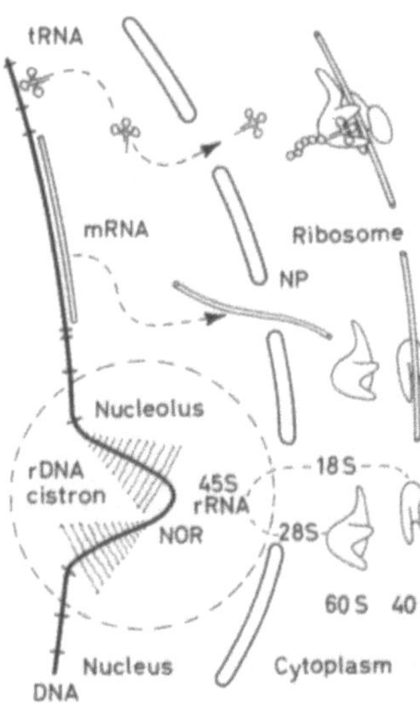

Fakan, S., Puvion, E.: Int. Rev. Cytol. *65*, 255 (1980)

Ribonucleoproteins: Proteins associated with some ↑ ribonucleic acids in ribonucleoprotein complexes. R. are visible in the nucleus as ↑ perichromatin granules, ↑ interchromatin granules, and ↑ perichromatin fibers. R. are also present in cytoplasm, since they are bound to ↑ messenger RNA as it leaves the nucleus. A special form of R. are ↑ informosomes.

Ribosomal RNA (rRNA): Large molecule of 45S with molecular weight of 4.5×10^6 daltons transcribed from ribosomal DNA ↑ cistron of ↑ nucleolus organizer region. This molecule cleaves into a 41S molecule which undergoes further degradation and yields one 20S rRNA molecule and one 32S rRNA molecule. The 20S rRNA is broken down into an 18S rRNA molecule that enters cytoplasm and joins into the small ribosomal subunit. 32S rRNA is transformed into 28S rRNA and passes into cytoplasm; here, two such molecules participate in formation of the large ribosomal subunit. (See molecular biology texts for further information)

↑ Ribonucleic acids, synthesis of; ↑ Ribosomes

Walker, A.T., Pace, N.R.: Cell *33*, 320 (1983)

Ribosomes (ribosoma*): Dense granular ↑ organelles (R) essential to ↑ protein synthesis, R. measure 15–25 nm in diameter and consists of two asymmetric subunits. Small subunit has a sedimentation rate of 40S and contains one ↑ ribosomal RNA (rRNA) molecule and 24 different proteins. Large subunit has a sedimentation rate of 60S and contains two rRNA molecules with 40 proteins. Small subunit includes a head (H), and a base (B) from which platform (P) protrudes. Large subunit consists of a central protuberance (CP), a ridge (Ri), and a stalk (S). ↑ Messenger RNA (mRNA) molecule passes through cleft (C) between head and platform of small subunit. It is likely that ↑ transfer RNA (tRNA) molecule and amino acids (AA) are situated between ridge and central protuberance of large subunit. R. may occur free in cytoplasm, in ↑ monoribosomes, or in clusters – ↑ polyribosomes attached to membranes of ↑ endoplasmic reticulum.

↑ Ribosomes, functions of

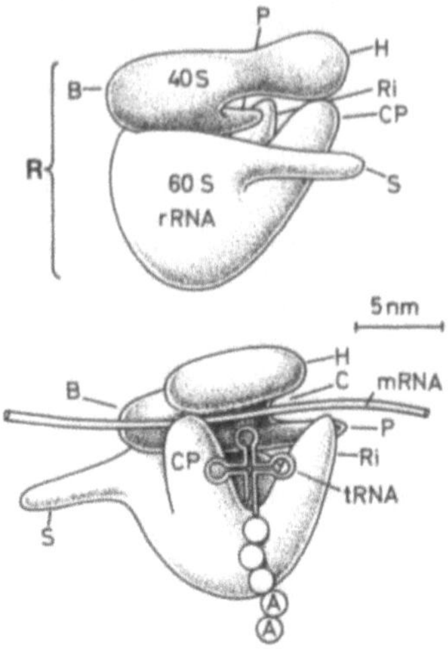

Bielka, H. (ed.): The Eukaryotic Ribosome. Berlin, Heidelberg, New York: Springer-Verlag 1981; Gross, B., Westermann, P., Bielka, H.: EMBO J. *2*, 255 (1983); Ivanov, I.E., Sabatini, D.D.: J. Ultrastruct. Res. *76*, 263 (1981); Lake, J.A.: Sci. Am. *245/2*, 56 (1981)

Ribosomes, bound: ↑ Ribosomes attached to membrane of ↑ endoplasmic reticulum.

↑ Ergastoplasm

Ribosomes, free: Ribosomes without contact with membrane of ↑ endoplasmic reticulum. R. are scattered in ↑ cytoplasmic matrix.

↑ Ribosomes, function of

Ribosomes, functions of: Small ribosomal subunit attaches to first ↑ codon of ↑ messenger RNA (mRNA) and forms part of ↑ transfer RNA (tRNA) binding site as codon is read by anticodon of tRNA. Large ribosomal subunit contains peptidyltransferase which is involved in formation of peptide bonds. This particle also contains two binding sites for two tRNA molecules. Large subunit therefore functions as part of tRNA binding sites, catalyzing peptidyl transfer and holding growing polypeptide chain. During ↑ protein synthesis, ↑ ribosomes move along mRNA molecule, read genetic message and receive amino acids brought there by tRNA, specified by codon of mRNA. Ribosomes provide binding sites for steric orientation of mRNA-tRNA interaction. Free ↑ ribosomes are involved in synthesis of proteins needed by cell itself; ribosomes bound to membranes of ↑ endoplasmic reticulum produce proteins destined to be secreted. (See molecular biology texts for further information)

Siekevitz, P., Zamecnik, P.C.: J. Cell Biol. *91*, 53s (1981)

Ridges, epidermal: ↑ Epidermal ridges

Riolan's ciliary muscle: ↑ Ciliary muscle, of Riolan

RNA: ↑ Ribonucleic acids

RNP: ↑ Ribonucleoproteins

Rod bipolar cells: ↑ Bipolar cells, of retina

Rod cells: ↑ Photoreceptors

Rod spherule (spherula terminalis bacilli*): Knoblike axon terminal of a rod cell (RS). R. generally contains one large mitochondrion (M), a few ↑ synaptic ribbons (arrowheads), and a large number of ↑ synaptic vesicles (SV). Dendrites of ↑ bipolar (B) and ↑ horizontal cells (H), and at times axons of the latter, penetrate an invagination of R. (Fig. = rat)

↑ Photoreceptors

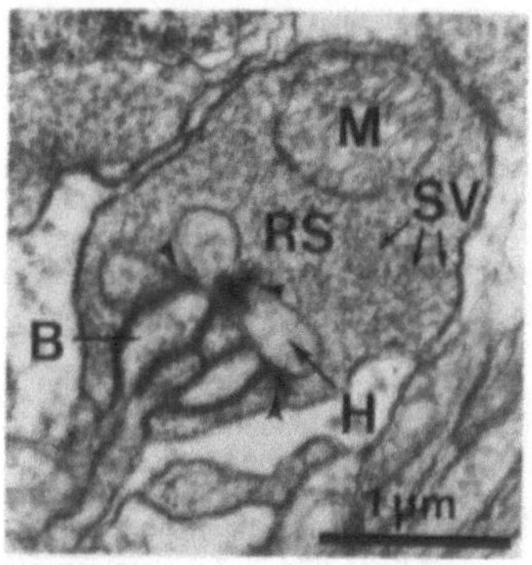

Rods, enamel: ↑ Enamel prisms

Rohr's fibrinoid: ↑ Placenta, full-term, structure of

Rokitansky-Aschoff crypts: ↑ Gallbladder

Roots, of teeth: ↑ Teeth

Rouget's cells: ↑ Pericytes

Rouleaux: Column-shaped aggregations of ↑ erythrocytes in stagnant blood or in thick smears of fresh blood. Formation of R. is probably result of

surface tension and of ↑ glycoproteins coating surface of erythrocytes.

rRNA: ↑ Ribosomal RNA

Rubriblast: ↑ Proerythroblast

Rubricyte: ↑ Erythroblast, polychromatophilic

Ruffini's corpuscle: ↑ Corpuscles, of Ruffini

Ruffled border (fluffy border): Numerous deep infoldings of plasmalemma of an active ↑ osteoclast, visible on surface which touches ↑ bone (B). Membrane of R. is studded on its cytoplasmic side with a myriad of bristlelike appendages about 15–20 nm long and 20 nm apart. Through narrow extracellular channels (EC) which delimit densely packed foliate (FP) and villuslike processes (P) of R., small free crystals of ↑ hydroxyapatite originating from calcified matrix are taken up. These are dissolved in cytoplasmic vacuoles (CV). R. is a very dynamic but inconstant cell differentiation which occurs only when osteoclast is active. R. is free of ↑ organelles. (Fig. = decalcified bone, rat)

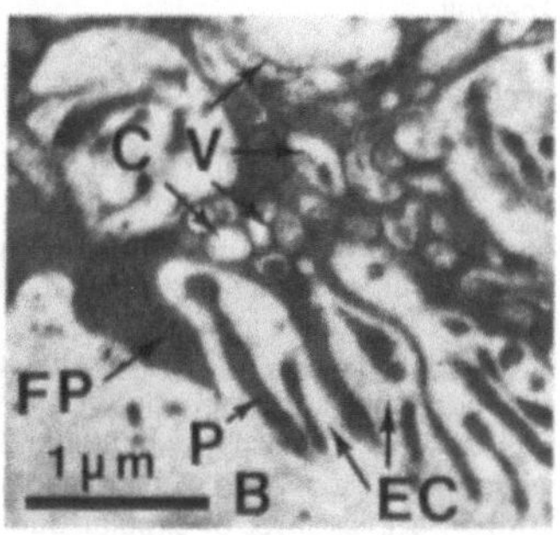

Bonucci, E.: Calcif. Tissue Res. *16*, 13 (1974); Kallio, D.M., Garant, P.R., Minkin, C.: J. Ultrastruct. Res. *37*, 169 (1971)

Rugae, of stomach: Longitudinal folds formed by contracted ↑ tunica mucosa of empty ↑ stomach.

Rugae, of vagina: Transverse folds of vaginal ↑ tunica mucosa.

↑ Vagina

Bussell's fuchsinophilic bodies: Accumulations of dense homogeneous material within distended rough endoplasmic ↑ cisternae in a small percentage of ↑ plasma cells. R. stain strongly with eosin or fuchsin and are considered an overproduction of incomplete ↑ immunoglobulin molecules. Frequency of plasma cells with R. increases in some chronic inflammatory processes; after death of plasma cell, R. may be found in surrounding tissue.

Ruthenium red: Stain for light-microscopic localization of some ↑ glycosaminoglycans and transmission electron microscopic demonstration of ↑ glycocalyx (G). Substances binding R. become electron opaque owing to strong dispersion of electrons by atoms of ruthenium. (Fig. = R. staining of ↑ proximal tubule of ↑ nephron, rat)

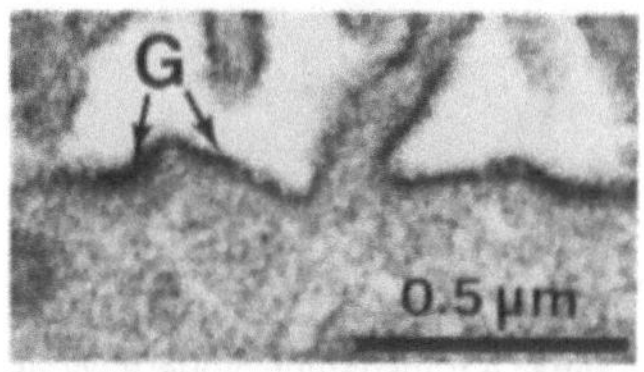

Dierichs, R.: Histochemistry *64*, 171 (1979)

S

Sac, alveolar: ↑ Alveolar sac

Sac, endolymphatic: ↑ Endolymphatic sac

Saccule, of ear (sacculus*): A vesicular enlargement of membraneous ↑ labyrinth containing ↑ macula of saccule.

Saccules, of endoplasmic reticulum (sacculus*): Spherically enlarged segments (S) of rough ↑ endoplasmic reticulum. (Fig. = ↑ ovarian follicular cell, rat)

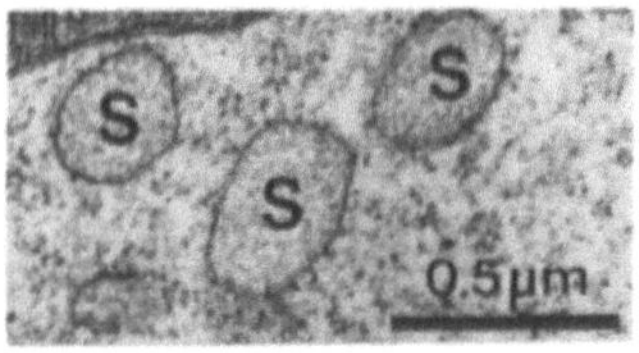

Saliva: A transparent, moderately viscous colorless fluid (pH about 7), product of various ↑ salivary glands. Besides water, S. contains ↑ proteoglycans, ↑ immunoglobulins, sodium, calcium, potassium, chloride, ↑ ptyalin, ↑ salivary corpuscles, and ↑ kallikrein. S. moistens mucosa of mouth, lubricating food, and thereby facilitates swallowing; its action is weakly bactericidal.

Geigy Scientific Tables. Body Fluids. 8th edn. Basle: Ciba-Geigy 1983

Salivary corpuscles (corpusculum salivare*): Desquamated epithelial cells contained in ↑ saliva.

Salivary ducts: ↑ Ducts, of salivary glands; ↑ Striated ducts

Salivary glands (glandulae salivares*): The oral cavity associated glands specialized in production of ↑ saliva. On the basis of size, location, and type of their secretory end pieces, S. may be classified as: 1) Major and minor ↑ S.; 2) ↑ buccal, ↑ labial, ↑ lingual, ↑ parotid, ↑ sublingual, ↑ submandibular S.; 3) ↑ serous, ↑ mucous, and ↑ mixed (seromucous) S.

↑ Mucous cells, of mucous and mixed salivary glands; ↑ Myoepithelial cells; ↑ Serous cells, of salivary glands

Pinkstaff, C.A.: Int. Rev. Cytol. *63*, 141 (1980); Young, J.A., Van Lennep, E.W.: The Morphology of Salivary Glands. London, New York: Academic Press 1978

Salivary glands, acinus of: ↑ Acinus, of glands

Salivary glands, basal cells of: ↑ Myoepithelial cells

Salivary glands, major (glandulae salivares majores): Large paired compound ↑ exocrine glands located outside oral cavity and provided with long excretory ducts opening into oral cavity. ↑ Parotid, ↑ sublingual, and ↑ submandibular glands belong to S.

Salivary glands, minor (glandulae salivares minores*): Small ↑ mucous or ↑ mixed ↑ exocrine glands scattered in wall of oral cavity, except in ↑ gingiva and hard ↑ palate. S. comprise: 1) ↑ Labial glands (↑ mixed); 2) ↑ buccal glands (mixed); 3) ↑ palatine glands (mucous); 4) ↑ glossopalatine glands (mixed); 5) ↑ anterior lingual glands (mixed) of ↑ tongue. The ↑ saliva of all these glands moistens oral cavity.

Salivary glands, mucous cells of: ↑ Mucous cells, of mucous and mixed salivary glands

Salivary glands, myoepithelial cells of: ↑ Myoepithelial cells

Salivary glands, serous cells of: ↑ Serous cells, of salivary glands

Sand granules, of pineal organ: ↑ Acervuli

SA-node: ↑ Sinuatrial node

Santorini's duct: ↑ Pancreatic ducts

Saphenous vein (vena saphena*): A special type of medium-sized ↑ vein. Structure: 1) Tunica intima (TI) = an ↑ endothelium overlying a very thick subendothelial connective tissue layer with numerous, mostly longitudinal ↑ smooth muscle cells. 2) Tunica media (TM) = a thick layer of circular bundles of smooth muscle cells intermingled with ↑ collagen and ↑ elastic fibers. 3) Tunica adventitia (TA) = a ↑ loose connective tissue with ↑ vasa vasorum, lymphatic vessels, nerve fibers, and longitudinal bundles of smooth muscle cells (arrowheads), as well as an abundance of elastic fibers parallel to axis of vessel.

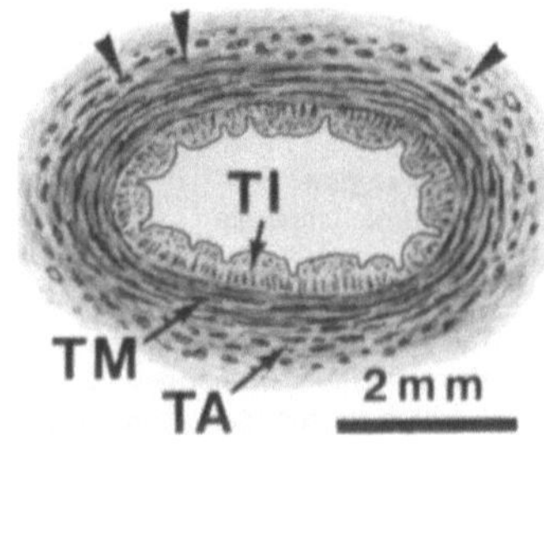

Crissman, R.S., Ross, J.N.Jr., Davis, T.: Anat. Rec. *198*, 581 (1980)

Sarcolemma*: The ↑ cell membrane (S) of ↑ cardiac and ↑ smooth muscle cells and ↑ skeletal muscle fibers. Under light microscope, S. appears as a delicate line, which is in reality composed of ↑ cell membrane, ↑ basal lamina (BL), and a fine network of ↑ reticular (R) and ↑ collagen (C) microfibrils. (Fig. = skeletal muscle fiber, rat)

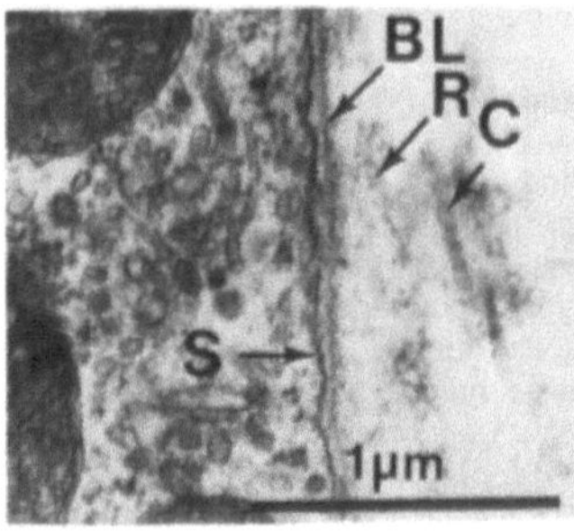

Sarcomere: A segment (S), about 2–3 μm long, between two successive ↑ Z-lines repeated along ↑ myofibril; S. is smallest contractile unit of ↑ striated muscle fibers. In skeletal myofibrils, S. contains two ↑ T-tubules, whereas in cardiac myofibrils, S. contains only one T-tubule. A system of transversally oriented ↑ intermediate filaments forms filamentous bridges between apposed Z-lines and ↑ M-lines of neighboring myofibrils, resulting in transversal alignment of sarcomeric striations. S. of same myofibril are held by a system of longitudinal intramyofibrillar intermediate filaments anchoring in Z-lines. (Fig. = ↑ skeletal muscle fiber, rat)

↑ A-band; ↑ Bands, of striated muscle fibers; ↑ I-band; ↑ Cardiac muscle cell

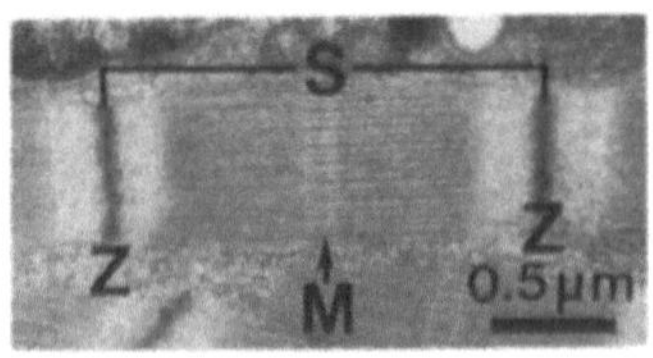

Wang, K., Ramirez-Mitchell, R.: J. Cell Biol. *96*, 562 (1983).

Sarcomere, transversal sections of: 1) At the level of ↑ I-band, ↑ actin myofilaments (Af) are arranged hexagonally. 2) At the level of ↑ A-band, where actin and ↑ myosin myofilaments (Mf) intermingle, each myosin myofilament is surrounded by six actin myofilaments. 3) At the level of ↑ M-line, hexagonally arranged myosin myofilaments are connected with conspicuous side extensions. 4) At the level of ↑ H-band, only myosin myofilaments are present. 5) At the level of ↑ Z-line, there occurs a complicated tetragonal lattice.

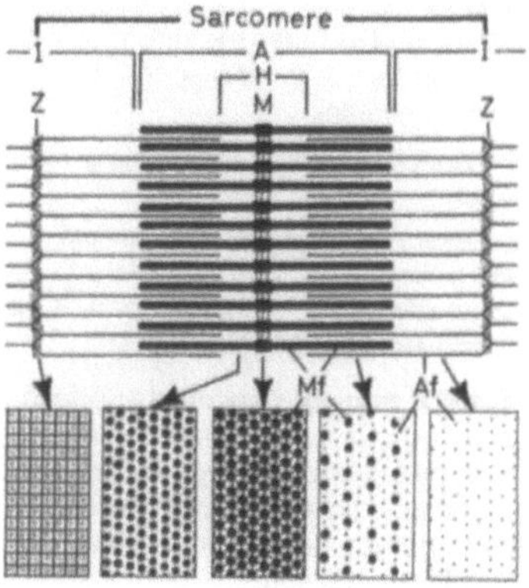

Sarcoplasm (sarcoplasma*): The ↑ cytoplasm of ↑ cardiac and ↑ smooth muscle cells and ↑ skeletal muscle fibers.

Sarcoplasmic reticulum (reticulum sarcoplasmaticum*): A kind of smooth ↑ endoplasmic reticulum present in ↑ cardiac muscle cells and ↑ skeletal muscle fibers.

↑ Cardiac muscle cells, sarcoplasmic reticulum of; ↑ Skeletal muscle fibers, sarcoplasmic reticulum of

Franzini-Armstrong, C.: Fed. Proc. *39*, 2403 (1980); Meis, L. de: The Sarcoplasmic Reticulum. Vols. 1 and 2. New York: John Wiley & Sons 1980/81; Sommer, J.R., Wallace, N.R., Junker, J.: J. Ultrastruct Res. *71*, 126 (1980); Van Winkle, W.B., Entman, M.L.: Life Sci. *25*, 1189 (1979)

Sarcosomes: A former term for ↑ mitochondria of ↑ skeletal muscle fibers.

Sarcostyles: ↑ Leydig-Kölliker columns

Sarcotubules: Tubules of ↑ L-system of ↑ sarcoplasmic reticulum.

SAT-chromosomes: A former term for ↑ chromosomes with a ↑ nucleolus organizer region.

SAT-zone: A former term for secondary constriction of ↑ chromosomes with a ↑ nucleolus organizer region. Since this constriction is weakly stained, it has been thought that it does not contain DNA or, formerly, thymonucleic acid (**s**ine **a**cido **t**hymonucleinico).

Satellite cells, of peripheral neurons (amphicytes, capsular cells, gliocytus ganglii*): Flattened peripheral ↑ neuroglial cells surrounding bodies of ↑ ganglion cells in ↑ spinal, cranial, and ↑ autonomic ganglia. In the latter, investment of individual ↑ neurons by S. is frequently incomplete. S. contain a small flattened to rounded nucleus (N) with an appreciable quantity of ↑ heterochromatin. Cytoplasm encloses numerous mitochondria (M), ↑ cisternae of rough endoplasmic reticulum (rER), many free ribosomes, a moderately developed Golgi apparatus (G), and some small lysosomes (Ly) in its vicinity. There are no junctional complexes between ganglion cells and S.; only a 15 to 20-nm-wide intercellular space separates these two elements. External surface of S. is covered by a basal lamina (BL). Some flattened ↑ fibrocytes (F) belonging to ↑ endoneurium surround S. S. play an important role in metabolic exchanges of ganglion cells.

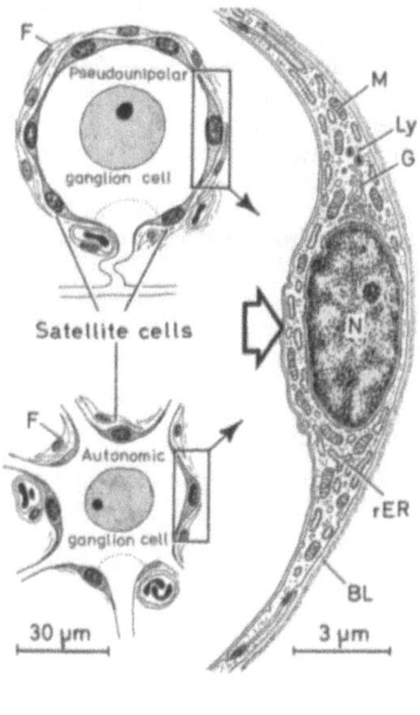

Pannese, E.: Adv. Anat. Embryol. Cell. Biol. *65*, 1 (1981)

Satellite cells, of skeletal muscle fibers (myosatellitocytus*): Inconspicuous cells (S) flattened in shallow depressions of ↑ skeletal muscle fibers (MF) and enclosed within common basal lamina (BL). The latter is lacking between muscle fiber and S. Nucleus occupies almost the whole cell body, and it is richer in ↑ heterochromatin than nuclei of muscle fiber; cell ↑ organelles are small and sparse. S. play an important role in ↑ regeneration of skeletal muscle fibers, providing them with nuclei. (Fig. = rat)

↑ Skeletal muscle fibers, regeneration of

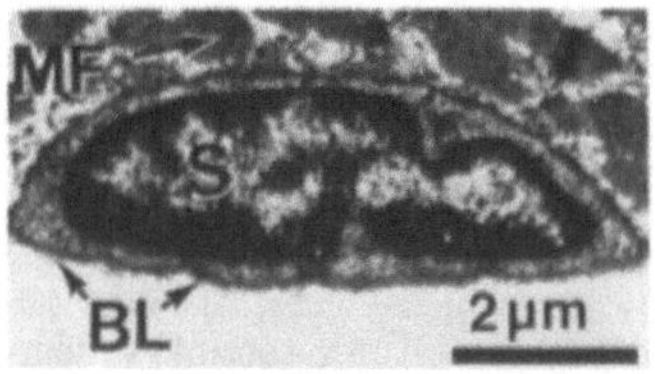

Gibson, M.C., Schultz, E.: Anat. Rec. *202*, 329 (1982); Kelly, A.M.: Anat. Rec. *190*, 891 (1978); Lipton, B.H., Schultz, E.: Science *205*, 1292 (1979); Schmalbruch, H., Hellhammer, U.: Anat. Rec. *185*, 279 (1976); Schulz, E.: Am. J. Anat. *147*, 49 (1976)

Satellites, of centriole (pericentriolar material): Globular aggregations of electron-dense fine granular material (S), about 75 nm in diameter, connected with triplets of a ↑ centriole (C). It is believed that S. can initiate assembly of ↑ microtubules during prophase and their depolymerization during anaphase, thus acting as a ↑ microtubule organizing center. (Fig. = oviduct ciliated epithelial cell, mongolian gerbil)

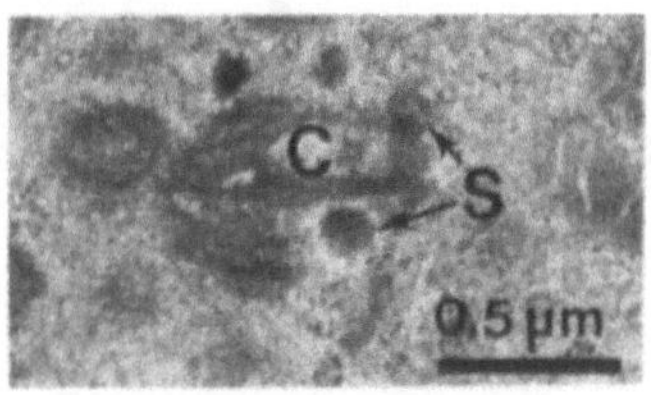

Berns, M.W., Ratner, J.B., Bremer, S., Meredith, S.: J. Cell Biol. *72*, 351 (1977); Gould, R.R., Borisy, G.G.: J. Cell Biol. *73*, 601 (1977)

Satellites, of chromosomes: ↑ Chromosomes, satellites of

Satellitism: ↑ Platelet satellitism

Scala media: ↑ Cochlear duct

Scala tympani: ↑ Cochlea

Scala vestibuli: ↑ Cochlea

Scanning electron microscope (SEM): An electro-optical instrument for direct visualization and three-dimensional rendition of surfaces. Principle: In a high vacuum (produced by pumps, P) an electron stream is produced by an incandescent cathode (C), accelerated between it and the anode (A), and concentrated by ↑ electromagnetic lenses (L) to a very fine coherent beam (B) that falls onto surface of a dehydrated and gold-metallized object (O). By means of impulses produced in a scanning generator (SG), electron beam scans surface of object in a regular scan pattern, synchronous with electron beams in two cathode ray tubes (CT), so that each luminous point of cathode ray tube corresponds to a point on the object. At impact site of electrons (a) on the object's surface, X-rays and secondary electrons (se) are generated; the latter are collected by a detector (D). Since each point of the object's surface emits a variable

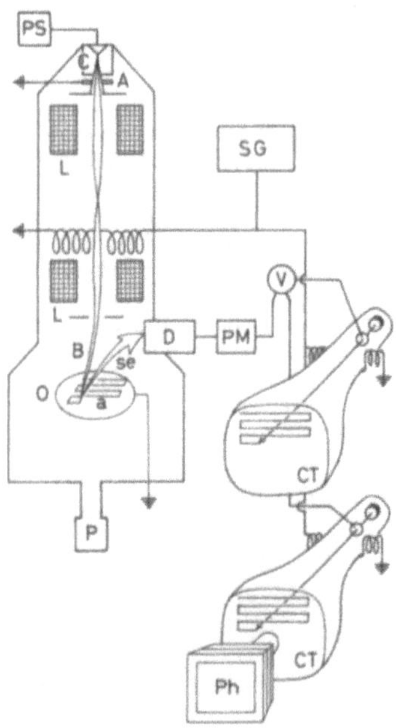

amount of secondary electrons, intensity of signal produced by the detector varies correspondingly and simultaneously modulates intensity of luminous points on screen of cathodic tubes. Thus, image of object's surface is obtained point by point along scan pattern and photographed with a camera (Ph). S. permits observation of very large surfaces; its depth of field is considerable; ↑ resolving power of a modern S. is about 3 nm. PM = photomultiplier, V = videoamplifier, PS = power supply, stabilizer of current, high-tension generator, etc. (Modified after Gautier 1980)

↑ Critical point drying; ↑ Magnification, microscopic; ↑ Sputtering

Fujita, T., Tanaka, K., Tokunaga, J.: SEM Atlas of Cells and Tissues. Tokyo, New York: Igaku-Shoin 1981; Gabriel, B.L.: Biological Scanning Electron Microscopy. New York: Van Nostrand Reinhold Co. 1982; Gautier, A.: Uni Lausanne *28*, 21 (1980); Goldstein, J.I., Newbury, D.E., Echlin, P., Joy, D.C., Fiori, Ch., Lifshin, E.: Scanning Electron Microscopy and X-Ray Microanalysis. New York, London: Plenum Press 1981; Hodges, G.M., Hallowes, R.C. (eds.): Biomedical Research Applications of Scanning Electron Microscopy. Vols. 1 and 2. London, New York, San Francisco: Academic Press 1979 und 1980

Scanning transmission electron microscope (STEAM): An ↑ electron microscope combining some properties of ↑ scanning and ↑ transmission electron microscopes. Principle: An electron beam (B) less than 1 nm in diameter scans a specimen (S) and passes through it. Passing through specimen, some electrons practically do not lose their speed [elastically scattered electrons (E e^-)], whereas others become retarded [inelastically scattered electrons (I e^-)]. Both kinds of electrons are separately analyzed according to their energy losses – elastically scattered, by a dark-field detector (DFD), and inelastically scattered, by a light-field detector (LFD) and energy analyzer (EA). Image is then displayed on a TV screen; it is possible to select either signals of separate electron beams or an appropriate combination of the two. Advantage of S. is in a considerably better extraction of information carried by electrons and in requiring much fewer electrons to produce usable micrographs.

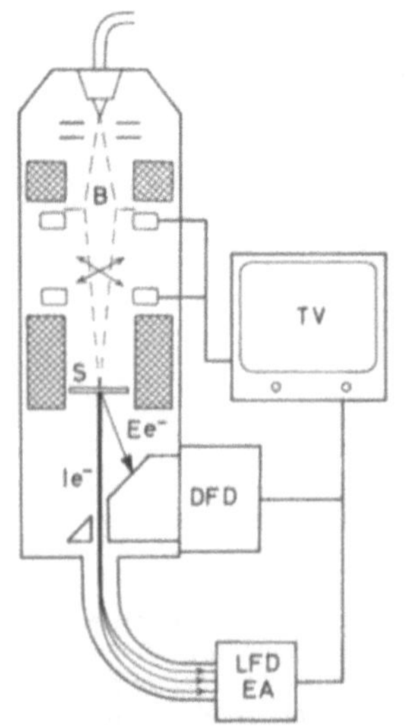

Crewe, A.V.: Science *221*, 325 (1983); Engel, A.: Ultramicroscopy *3*, 278 (1978); Kellenberger, E.: Trends in Biochemical Sciences *3*, N 135 (1978)

Scarpa's ganglion: ↑ Vestibular ganglion

S-cells (secretin cells): ↑ Endocrine cells found in ↑ Lieberkühn's crypts of ↑ duodenum. It is believed that S. produce ↑ secretin.

↑ Endocrine cells, of gastrointestinal tract

Larsson, L.-I., Sundler, F., Alumets, R., Håkanson, R., Schaffalitzky de Muckadell, O.B., Fahrenkrug, J.: Cell Tissue Res. *181*, 361 (1977)

Schiff's reagent (leukofuchsin, fuchsin-sulfurous acid): Colorless fuchsin-sulfurous acid giving a red or magenta-colored complex with aldehyde groups. S. is used in ↑ Feulgen and ↑ PAS-reaction.

Puchtler, H., Meloan, S.N.: Histochemistry *72*, 321 (1981)

Schlemm's canal (sinus venosus sclerae*): A flattened about 0.3-mm-broad, ringlike vein (Sch) without valves situated at anterior edge of ↑ sclera (S) in front of ↑ trabecular meshwork (TM). S. encircles periphery of ↑ cornea (C). Structure: 1) Very thin nonfenestrated ↑ endothelium (E); 2) discontinuous basal lamina (BL); 3) thin adventitial layer consisting of ↑ collagen fibrils (CF) and ↑ fibrocytes (F). S. receives ↑ aqueous humor from ↑

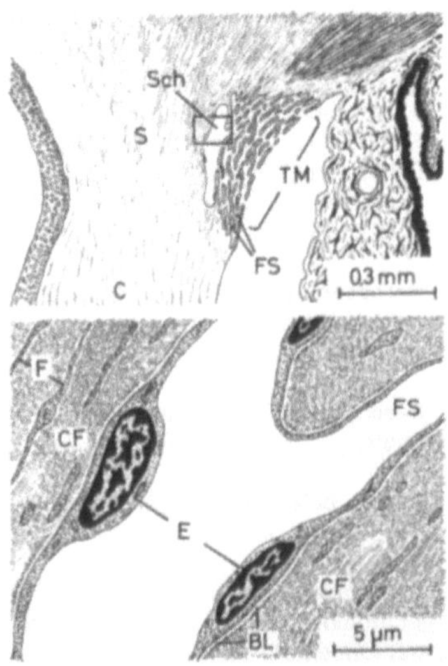

Fontana's spaces (FS) and drains it through 25–35 fine channels into episcleral veins.

Raviola, G.: Invest. Ophthalmol. *13*, 828 (1974)

Schmidt-Lanterman clefts (Schmidt-Lanterman incisure, incisura myelini*): Conical areas (SLC) in which lamellae (L) of ↑ myelin sheath (MS) are separated at the level of the major dense line. S. appear under light microscope after osmification as oblique funnel-shaped discontinuities in myelin sheath. S. are frequently found in ↑ myelinated peripheral nerve fibers and are almost completely absent in those of CNS; it is therefore thought that S. facilitate flexibility of nerve fibers. A = ↑ axon, M = outer ↑ mesaxon, BL = basal lamina of ↑ Schwann's cell, Sch = ↑ Schwann's cell

↑ Schwann's cells, unrolled

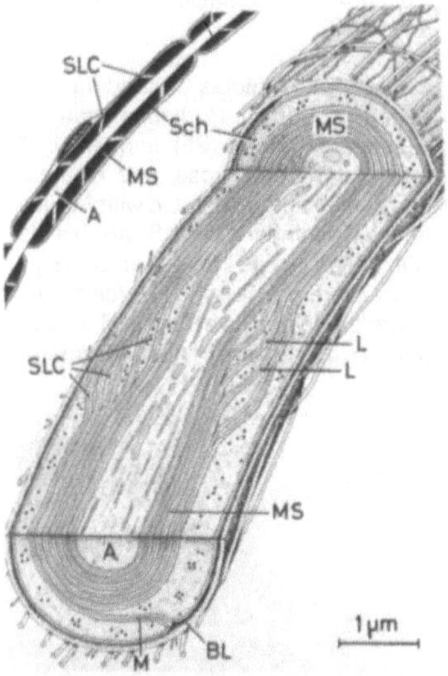

Celio, M.R.: Brain Res. *108*, 221 (1976); Friede, R.L., Samorajski, T.: Anat. Rec. *165*, 89 (1969)

Schmidt-Lanterman segments: Short portions (SLS) of a ↑ myelinated nerve fiber (NF) between two ↑ Schmidt-Lanterman clefts (SLC). (Fig. = human)

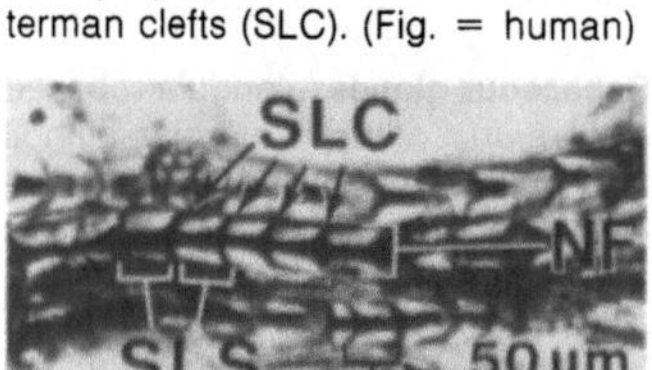

Schwalbe's line: A thickening of ↑ Descemet's membrane at periphery of ↑ cornea.

Schwann's cells (lemnocyte, neurolemnocytus, neurilemmal cells, sheath cells, neurolemmocytus*): Elongated ↑ neuroglial cells sheathing ↑ axons (A) in ↑ peripheral nervous system. Nucleus is located at midpoint of extension of each cell; it is peripheral, elongated, and crescent-shaped in ↑ myelinated nerve fibers (MF); round and central in ↑ nonmyelinated nerve fibers (NMF). Nucleus encloses masses of ↑ heterochromatin and an inconspicuous nucleolus. Cytoplasm contains mitochondria, a small Golgi apparatus (G), some short and flattened rough endoplasmic cisternae (rER), a moderate number of free ribosomes, occasional lysosomes (Ly), and a considerable number of ↑ micropinocytotic vesicles. S. are enveloped by a basal lamina (BL). In myelinated nerve fibers, each S. forms only a single ↑ internodal segment; borders between S. in nonmyelinated fibers are not easy to distinguish, since S. overlap. S. in myelinated nerve fibers furnish ↑ myelin sheath (MS) for a single axon; thus, they have only one inner (IM) and one outer ↑ mesaxon (OM). S. in nonmyelinated nerve fibers enclose 2–20 or more axons without furnishing them with a myelin sheath; thus, an S. displays a corresponding number of outer mesaxons (OM) but no inner mesaxons. S. are mechanical and metabolic supporting elements for axons, which also isolate the latter from ↑ endoneurium (E).

↑ Schwann's cells, unrolled; ↑ Schwann's sheath

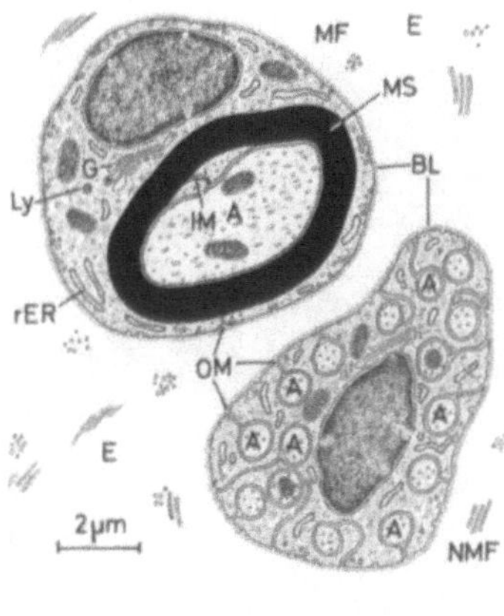

Bunge, B., Williams, A.K., Wood, P., Uitto, J., Jeffrey, J.J.: J. Cell Biol. *84*, 184 (1980); Carey, J.D., Eldridge, F.C., Cornbrooks, J.C., Timpl, R., Bunge, R.P.: J. Cell Biol. *97*, 473 (1983); Carlsen, F., Behse, F.: J. Anat. *130*, 545 (1980)

Schwann's cells, unrolled: S. have the form of a very thin trapezoidal cytoplasmic sheet (Sh) with elongated ↑ perikaryon (P), thickened borders (B), and cytoplasmic channels (C) obliquely oriented to ↑ axon (A). In the course of wrapping (arrows) around axon, lateral borders form successive ↑ mesaxons (M) of ↑ paranodal region, whereas cytoplasmic channels form ↑ Schmidt-Lanterman clefts (SLC).

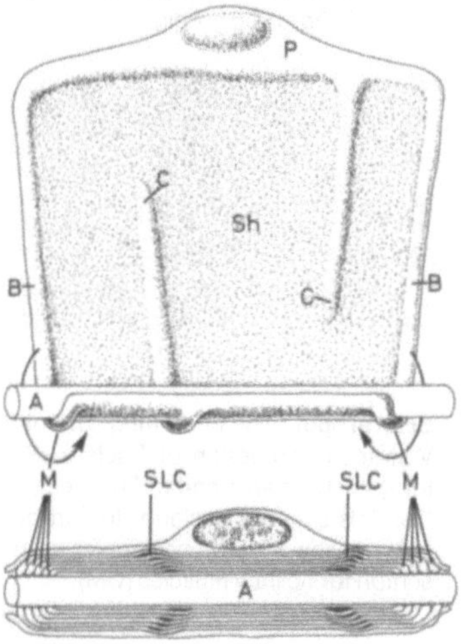

Schwann's sheath (sheath of Schwann, neurilemma, neurolemma*): The cytoplasm (SS) and nucleus (N) of a ↑ Schwann's cell (SC) surrounding ↑ myelin sheath (MS).

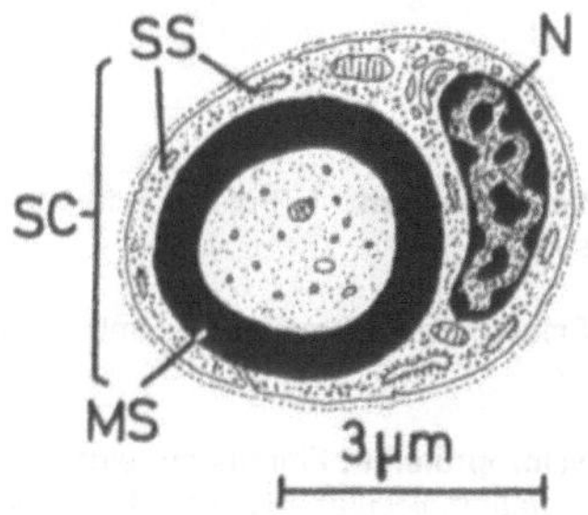

Schwann's tubes: ↑ Büngner's bands

Schweigger-Seidel ellipsoids: ↑ Capillaries, sheathed, of spleen; ↑ Ellipsoids, of Schweigger-Seidel

Sclera*: A layer, 0.4–0.6 mm thick, of ↑ dense connective tissue forming external coat of ↑ eyeball up to ↑ limbus of ↑ cornea. Structure: 1) Outer fibrous layer (OL) = a dense meshwork of crossed ↑ collagen fiber bundles parallel to surface; the bundles are intermingled with some scattered ↑ elastic fibers and stellate ↑ fibrocytes. This layer is vascularized, but ↑ glycosaminoglycans are lacking. 2) Lamina fusca (LF) = a thin inner layer of S. with slender bundles of collagen and some elastic fibers interspersed with ↑ melanocytes (m). In the area of ↑ optic papilla, S. is perforated by numerous small openings, ↑ lamina cribrosa. A narrow ↑ suprachoroid space (arrowhead) separates S. from ↑ choroid (c), and ↑ episcleral space (e) from ↑ capsule of Tenon. (Fig. = pig)

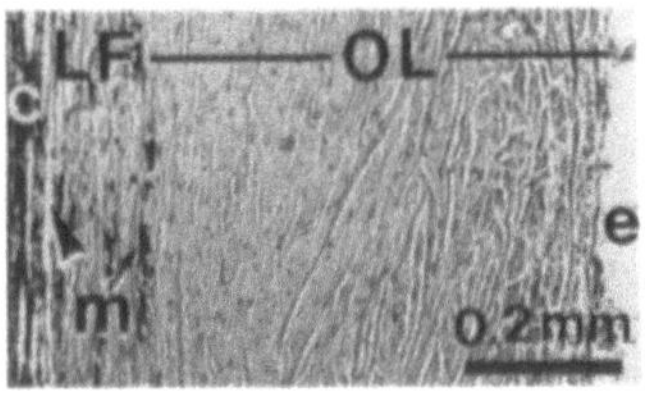

Scleral spur: The inner wing (SS) of a V-shaped depression of ↑ sclera (Sc) into which tapered end of ↑ cornea (C) fits. S. is situated posterior to ↑ trabecular meshwork (TM) and serves as insertion for ↑ ciliary muscles (CM).

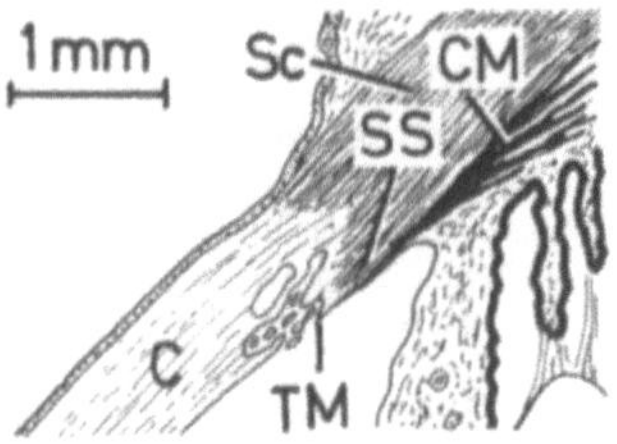

Scleroblastem: ↑ Protochondral tissue

Sclerocorneal junction: ↑ Limbus, of eye

Scleroproteins: Fibrous proteins with regular molecular structure. S. include ↑ collagen, ↑ elastin, and ↑ keratin. All S. are insoluble in water.

Scotopsin: An ↑ opsin that combines with ↑ retinene to form ↑ rhodopsin.

Scrolls: Cylinders (S), about 0.2 µm long and 50–80 nm wide, resembling enfolded sheaths, found in human ↑ mast cell granules (G). Function of S. is unknown.

Caulfield, J.P., Lewis, R.A., Hein, A., Austen, F.: J. Cell Biol. *85*, 299 (1980); Lagunoff, D.: J. Invest. Dermatol. *58*, 296 (1972)

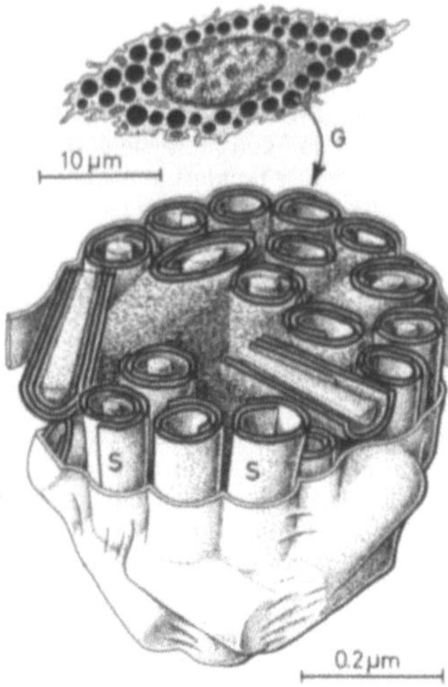

Scrotum*: A musculocutaneous sac containing testes. Histologically, the most characteristic structure of S. is ↑ skin with pigmented ↑ basal cells of ↑ epidermis and presence of ↑ tunica dartos.

Sebaceous cells: Large polygonal cells constituting ↑ sebaceous glands. S. lying on basal lamina (BL) are nondifferentiated (basal or stem cells) (BC) and display a round or flattened nucleus with a small nucleolus, a moderate number of round mitochondria, some rough endoplasmic ↑ cisternae and smooth endoplasmic ↑ tubules, a small Golgi apparatus, ↑ tonofibrils (Tf), considerable numbers of free ribosomes, occasional ↑ lipid droplets, and dark cytoplasmic matrix. By gradual ↑ differentiation, basal S. enlarge and become polygonal; their cytoplasm fills with lipid droplets (LD), which give a foamy aspect to S. Mitochondria are round, some of them with tubules; Golgi apparatus is inconspicuous; rough endoplasmic reticulum is reduced to only sparse cisternae, but smooth endoplasmic reticulum (sER) is abundant. Adjacent S. are held together by numerous small ↑ desmosomes (D). Toward centrum of gland, nucleus of S. becomes pyknotic (PN), mitochondria swell (SM), sER tubules and plasmalemma become disrupted, lipid droplets coalesce, and S. finally dies and is transformed into ↑ sebum (S). S. are constantly renewed by mitotic division of basal cells.

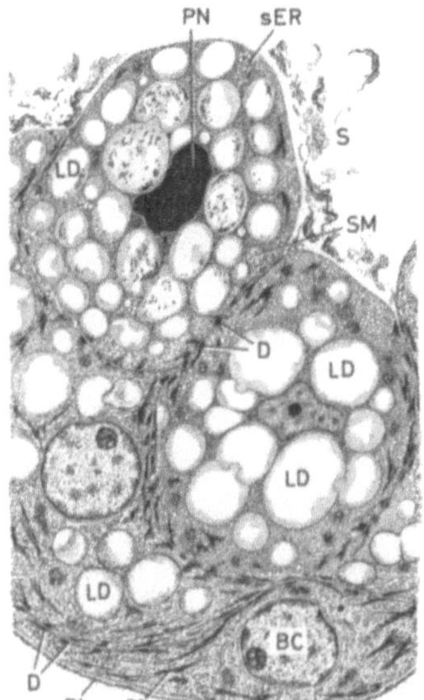

↑ Holocrine secretion

Mesquita-Guimarães, J., Pignatelli, D., Coimbra, A.: J. Submicrosc Cytol. *11*, 435 (1979)

Sebaceous follicles (glandula sebacea separata*): Particularly large ↑ sebaceous glands (SF) frequent in ↑ skin of forehead, nose, and external ear, sometimes associated with ↑ vellus hairs. Openings (O) of S. are visible with naked eye as "pores" on skin surface. There are about 800 S./cm² skin. S. are particularly sensitive to ↑ androgens. (Fig. = ↑ wing of nostril, human)

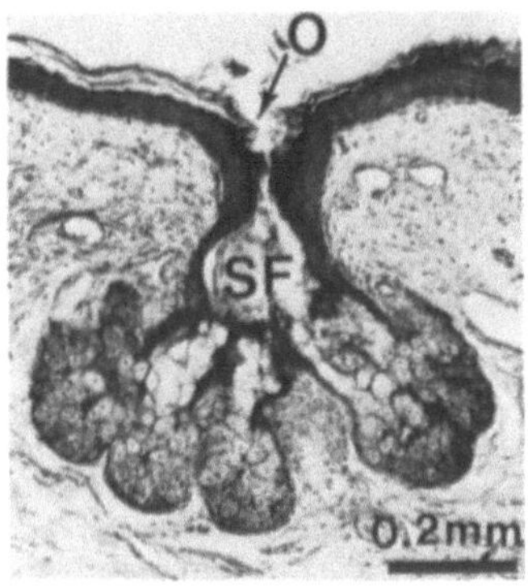

Sebaceous glands (glandula sebacea pili*): Pluristratified ↑ alveolar glands (0.2–2 mm across) with ↑ holocrine secretion. S. are scattered in ↑ dermis, but lacking in ↑ skin of palms and soles. S. consist of several alveoli (A) lined by a basal lamina and a thin layer of connective tissue (arrowhead). Alveoli unite and form a short, often keratinized duct (D), which opens into upper part of ↑ hair follicle (HF). S. of ↑ areola, ↑ nipple, ↑ eyelids, border of ↑ lips, ↑ labia minora, ↑ glans penis, and internal fold of ↑ prepuce open directly onto skin surface. Basal cells of S. are nondifferentiated; they divide and differentiate into ↑ sebaceous cells, which finally transform into ↑ sebum (S). There are about 100 S./cm² skin surface. S. are an integral part of ↑ pilosebaceous complex; they are stimulated by ↑ androgens. (Fig. = human)

↑ Sebaceous follicles

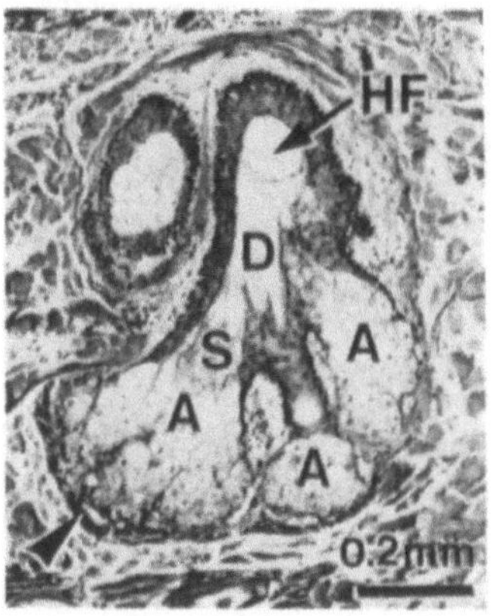

Strauss, J.S., Pochi, P.E.: Histology, Histochemistry and Electron Microscopy of Sebaceous Glands in Man. In: Gans, O., Steigleder, G.K. (eds.): Normale und pathologische Anatomie der Haut. Berlin, Heidelberg, New York: Springer-Verlag 1969; Strauss, J.S., Pochi, P.E., Downing, D.T.: J. Invest. Dermatol. *67*, 90 (1976)

Sebum: The product of ↑ sebaceous glands, result of ↑ holocrine disintegration of ↑ sebaceous cells. S. is composed of lipids, triglycerides, wax esters, squalene, and very small amount of sterol. S. lubricates ↑ skin and ↑ hairs, apparently keeping skin soft and supple. However, the role of S. is not fully understood.

Geigy Scientific Tables. Body Fluids. 8th edn. Basle: Ciba-Geigy 1983

"Second messenger": ↑ Polypeptide hormones, action on the cell of

Secondary constriction, of chromosome: ↑ Chromosomes; ↑ Nucleolus organizer region

Secondary degeneration, of peripheral nerve fibers: ↑ Nerve fibers of peripheral nervous system, degeneration and regeneration of

Secondary follicle, of ovary: ↑ Ovarian follicle, secondary solid; ↑ Ovarian follicle, secondary vesicular

Secondary granules, of neutrophilic granulocytes: ↑ Granulocytes, neutrophilic, granules of

Secondary lysosomes: ↑ Lysosomes

Secondary ossification: ↑ Bone formation, secondary

Secretin: A polypeptide ↑ tissue hormone probably released from ↑ S-cells of duodenal ↑ Lieberkühn's crypts. S. increases production of bicarbonate by cells of ↑ pancreatic ducts and causes secretion of watery pancreatic juice containing a few enzymes to neutralize acid gastric juice.

↑ Endocrine cells, of gastrointestinal tract

Secretin cells: ↑ S-cells

Secretion: An energy-consuming process of active uptake of small molecules by cells and their assembly into very complex products used by the organism in accomplishing certain functions. Sequence: 1) Uptake of small molecules in ↑ micropinocytotic vesicles (MV); mitochondria (M) furnish energy needed for all steps of secretory process; 2) synthesis of protein chains by ↑ polyribosomes of rough endoplasmic reticulum; 3) newly synthesized protein is transported to ↑ Golgi apparatus (G) in ↑ transport vesicles (TV), which fuse with cisternae of ↑ forming face; 4) incorporation of sugar into protein molecule within Golgi apparatus; 5) pinching off of immature secretory granules (ISG) from ↑ maturing face of Golgi apparatus; 6) formation of ↑ condensing vacuoles (CV); 7) maturation of secretory product and formation of ↑ secretory granules (SG); 8) release of secretory granule contents either by ↑ eccrine (E), ↑ apocrine, or ↑ holocrine secretion. Step 5) requires Ca^{2+}. Arrows indicate direction of S. Small molecules necessary for S. circulate in blood capillaries (Cap).

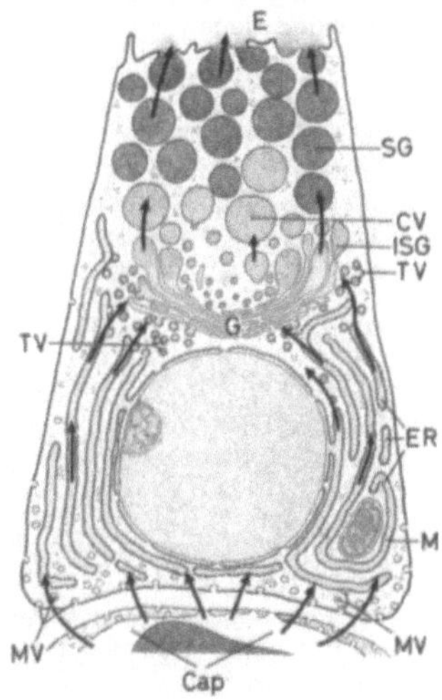

↑ GERL

Chailley, B.: Biol. Cell *35*, 55 (1979); Conn, M. (ed.): Cellular Regulation of Secretion and Release. New York: Academic Press 1982; Novikoff, A.B., Novikoff, P.M.: Biol. Cell *36*, 101 (1979); Orci, L.: Morphologic Events Underlying the Secretion of Peptide Hormones. In: James, V.H.T. (ed.): Excerpta Medica International Congress Series No 403. Amsterdam: Excerpta Medica 1976; Poisner, A.M., Trifaró, J.M. (eds.): The Secretory Granule. The Secretory Process. Vol. 1. Amsterdam: Elsevier Biomedical Press 1982

Secretion, apocrine: ↑ Apocrine secretion

Secretion, cytocrine: ↑ Cytocrinia

Secretion, eccrine: ↑ Eccrine secretion

Secretion, endocrine: ↑ Endocrine secretion

Secretion, exocrine: ↑ Exocrine secretion

Secretion, holocrine: ↑ Holocrine secretion

Secretion, merocrine: ↑ Eccrine secretion

Secretory capillaries: ↑ Canaliculi, intercellular

Secretory cells: All cells specializing in ↑ exocrine secretion.

↑ Mucous cells; ↑ Serous cells

Secretory cells, of oviduct: ↑ Oviduct, epithelium of

Secretory cells, of uterus: ↑ Uterus, epithelium of

Secretory complex, of IgA: ↑ Immunoglobulin A

Secretory ducts: ↑ Striated ducts

Secretory end piece: Characteristically arranged ↑ secretory cells at beginning of excretory ducts of an ↑ exocrine gland (↑ acinus, ↑ mucous tubule, ↑ mixed tubule).

Secretory granules (granulum secretorium*): Spherical, ↑ unit membrane-bound ↑ inclusions, 0.1–2 µm in diameter (S), with moderate to high osmiophilic contents present in ↑ glandular cells. S. pinch off from ↑ Golgi apparatus in the course of ↑ secretion. After a variously long process of maturation and condensation, contents of S. are discharged from cell by ↑ exocytosis. S. of ↑ endocrine cells (top) are in general smaller than those of ↑ exocrine cells (bottom). (Fig., top = ↑ A-cell, of pancreatic islet; fig., bottom = ↑ pancreatic acinar cell, mongolian gerbil)

↑ Condensing vacuoles

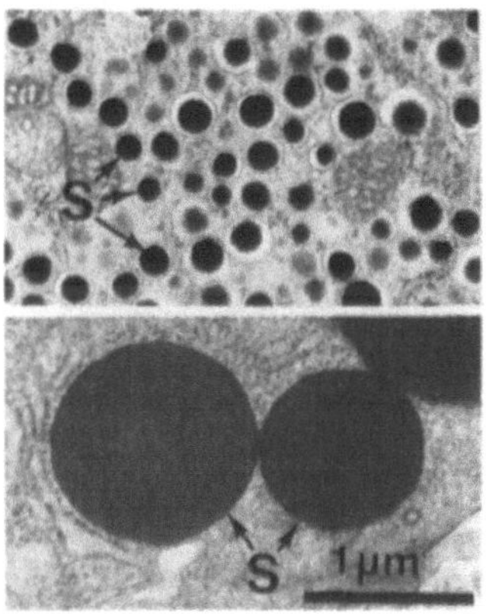

Poisner, A.M., Trifaró, J.M. (eds.): The Secretory Granule. The Secretory Process. Vol. 1. Amsterdam: Elsevier Biomedical Press 1982

Secretory immunoglobulin A: ↑ Immunoglobulin A

Secretory phase, of menstrual cycle: A period of synthesis and release of carbohydrate-rich products by ↑ uterine glands during ↑ menstrual cycle.

↑ Endometrium

Sectioning: A procedure of cutting ↑ sections with a ↑ microtome or ↑ ultramicrotome.

Sections (histological section): Thin tissue slices obtained in the course of ↑ sectioning. Thickness may be: 1) Normal (6–10 µm); 2) ↑ semithin (0.5–2 µm); or 3) ↑ ultrathin (20–60 nm). Occasionally, for special purposes, 40-µm-thick S. are cut (↑ alveolus of lung, elastic basket of).

Segi's cap: An agglomeration of ↑ argyrophilic cells (SC) on tip of ↑ intestinal villi in ↑ duodenum and ↑ jejunum of human fetus and fetuses of some animals. Majority of S. cells are ↑ enterochromaffin cells containing ↑ serotonin; rest of cap consists of cells giving a positive immunohistochemical reaction to ↑ gastrin, ↑ motilin, and ↑ somatostatin. Function of S. is not known.

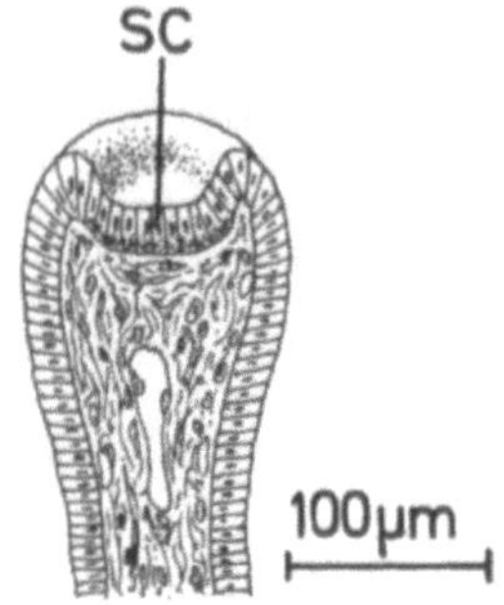

Iwanaga, T., Kobayashi, S., Fujita, T., Yanaihara, N.: Biomed. Res. *1*, 117 (1980); Kobayashi, S., Iwanaga, T., Fujita, T.: Arch. Histol. Jpn. *43*, 79 (1980); Segi, M.: Biomed. Res. *1*, Suppl. 1 (1980); Yamada, J., Kuramoto, H., Iwanaga, T.: Arch. Histol. Jpn. *44*, 193 (1981)

Segment long-spacing collagen (SLS-collagen): An artificial mode of side-to-side molecular packing of ↑ tropocollagen molecules (Tr), without overlapping, brought about by acidic collagen solution in presence of ATP. Resulting cross-striation has periodicity of 280 nm; vertical lines are formed by binding of a ↑ contrasting agent to thickenings (Th) along tropocollagen molecules.

↑ Fibrous long-spacing collagen; ↑ Long-spacing collagen

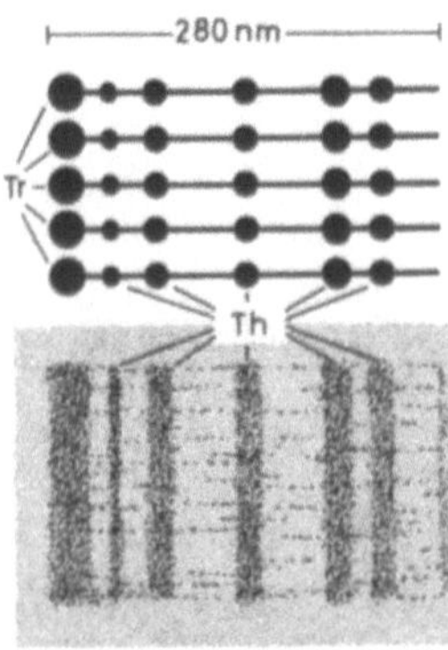

Bruns, R.R., Hulmes, D.J.S., Therrien, S.F., Gross, J.: Proc. Natl. Acad. Sci. USA *76*, 313 (1979)

Segmenta bronchopulmonalia: ↑ Bronchopulmonary segments

Segmented columns: ↑ Spermatozoon, connecting piece of

SEM: ↑ Scanning electron microscope(y)

Semen (sperm, sperma): A yellowish, slightly opalescent, viscous fluid representing a mixture of secretions of ↑ testis, ↑ epididymis, ↑ seminal vesicles, ↑ prostate, ↑ bulbourethral, and ↑ urethral glands. S. is composed of 10% ↑ spermatozoa (about 60000/mm^3), the rest being ↑ seminal plasma. S. also contains occasional epithelial cells, ↑ Sertoli's cells, ↑ spermatogonia, ↑ spermatocytes, ↑ wandering cells (predominantly ↑ lymphocytes), ↑ prostatic concretions, ↑ Böttcher's crystals, pigment granules, lipid droplets, amino acids, cholesterol, lactic acid, fructose, proteins, ↑ hyaluronidase, and ↑ prostaglandins.

↑ Ejaculate

Geigy Scientific Tables. Body Fluids. 8th edn. Basle: Ciba-Geigy 1983

Semen, crystals of Böttcher of: ↑ Böttcher's crystals

Semicircular canals (canales semicirculares*): Three bony tunnels lying in three planes perpendicular to one another. S. belong to bony ↑ labyrinth and are lined by thin ↑ endosteum, extensions of which reach lamina propria of ↑ semicircular ducts lodged within S. ↑ Perilymph circulates in S.

Semicircular ducts (ductus semicirculares*): Three membranous tubes (SD) excentrically lodged within ↑ semicircular canals (SC) and filled with ↑ endolymph. S. belong to membranous ↑ labyrinth. B = ↑ bone, E = ↑ endosteum, PS = ↑ perilymphatic space (Fig. = guinea pig)

↑ Labyrinth membranous, wall of

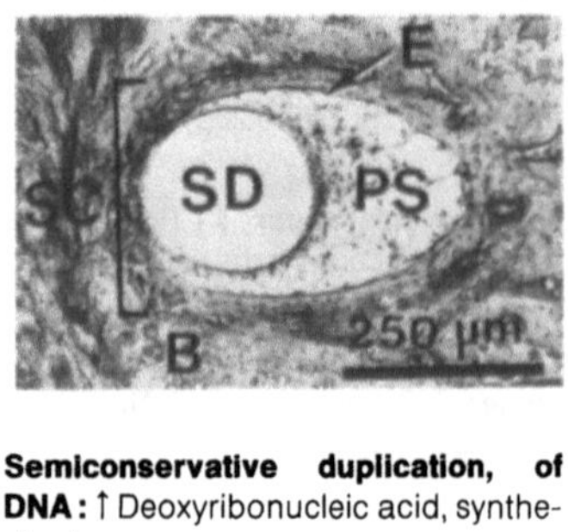

Semiconservative duplication, of DNA: ↑ Deoxyribonucleic acid, synthesis of

Semilunar valves: ↑ Aortic valves; ↑ Pulmonic valves

Seminal plasma: A weakly basic, milky, fructose-rich mixture of secretions of ↑ testis, ↑ epididymis, ↑ seminal vesicles, ↑ prostate, ↑ bulbourethral, and ↑ urethral glands representing about 90% of ↑ semen. P. serves as a nutritional and locomotive milieu for ↑ spermatozoa; it also neutralizes acidity of ↑ vagina.

Geigy Scientific Tables. Body Fluids. 8th edn. Basle: Ciba-Geigy 1983

Seminal vesicles (vesicula seminalis*): Irregularly shaped 3 to 5-cm-long paired accessory male sexual organs situated behind ↑ prostate and lateral to each ampulla of ↑ ductus deferens. S. are composed of several coils of a tube consisting of following layers: 1) Tunica mucosa (TM): a) epithelium = a ↑ simple columnar epithelium with patches of nonciliated ↑ pseudostratified columnar epithelium; epithelium is yellowish due to its content of ↑ lipofuscin; b) lamina propria = a ↑ loose connective tissue with vessels and ↑ elastic fibers. Mucosa forms numerous branching folds (F), ridges, and saclike evaginations (E) of various lengths. 2) Tunica muscularis (TMu) = ↑ smooth muscle cells arranged in an inner circular (IC) and an outer longitudinal layer (OL). 3) Tunica adventitia (TA) = a ↑ loose connective tissue with elastic network, representing capsule of S. Proximal extremity of S. joins end of ductus deferens to form an ↑ ejaculatory duct. Epithelium produces a yellowish alkaline secretion (pH 7.2) starting movement of ↑ spermatozoa and protecting them against vaginal secretion. Secretion contains fructose to nourish spermatozoa. A considerable concentration of ↑ prostaglandins can be found in secretion of S. Activity of S. is stimulated by ↑ testosterone.

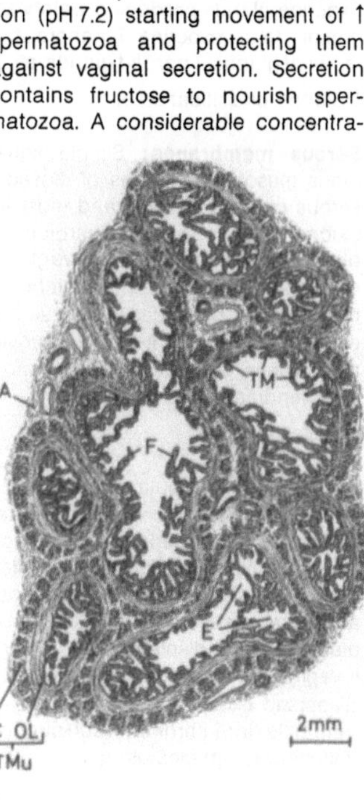

Aumüller, G.: Prostate Gland and Seminal Vesicles. In: Oksche, A., Vollrath, L. (eds.): Handbuch der mikroskopischen Anatomie des Menschen, Vol. 7, Part 6. Berlin, Heidelberg, New York: Springer-Verlag 1979; Tsukise, A., Yamada, K.: Acta Histochem. *70*, 276 (1982)

Seminiferous epithelium (epithelium spermatogenicum*): A highly specialized complex ↑ stratified columnar epithelium lining ↑ seminiferous tubules. S. consists of supporting cells [= ↑ Sertoli's cells (SC)] and spermatogenic cells responsible for ↑ spermatogenesis [= ↑ spermatogonia (Spg); primary ↑ spermatocytes (PSp); secondary ↑ spermatocytes (SSp); ↑ spermatids (St); ↑ spermatozoa (S)]. Only Sertoli's cells and ↑ spermatogonia have contact with basal lamina (BL). Sertoli's cells subdivide S. into basal and adluminal compartments. (Modified after Clermont 1963)

↑ Blood-testis barrier

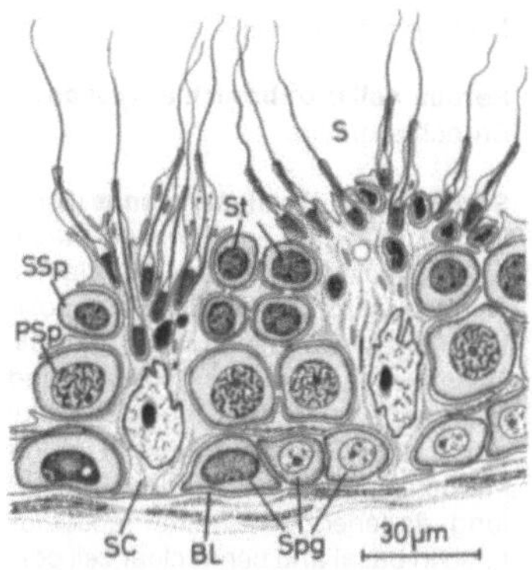

Clermont, Y.: Am. J. Anat. *112*, 35 (1963); Goto, K.: Biomed. Res. *2*, Suppl. 361 (1981)

Seminiferous epithelium, cycle of: A repeating sequence of characteristic grouping of ↑ germ cells around a ↑ Sertoli's cell (not represented) occurring in a given area of ↑ seminiferous epithelium. S. is a morphological expression of maturational changes in spermatogenic cells. In man, S. lasts 16±1 days; ↑ spermatogenesis requires four cycles or about 64±4.5 days. Each cycle in man passes through six stages. Ad = dark A ↑ spermatogonia, Ap = pale A spermatogonia, B = B spermatogonia, PSp = ↑ primary spermatocyte in various phases of meiotic prophase, RB = ↑ residual body, S = ↑ spermatozoa, Spg = spermatogonia, SSp = ↑ secondary spermatocyte, St = ↑ spermatids in various stages of ↑ spermiogenesis. (Modified after Clermont 1963)

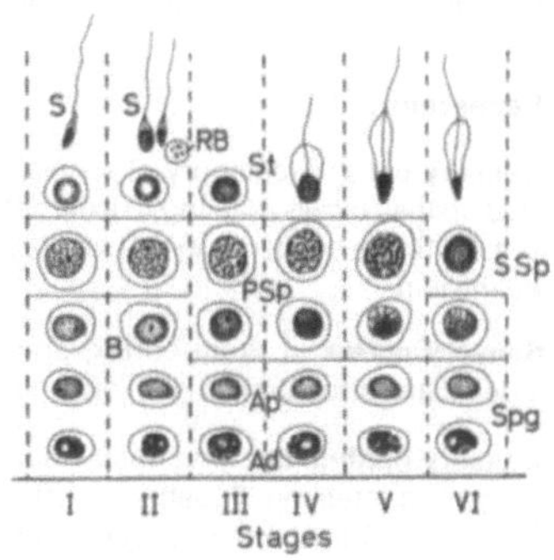

Clermont, Y.: Am J. Anat. *112*, 35 (1963); Clermont, Y.: Am J. Anat. *128*, 265 (1970)

Seminiferous tubules (tubulus seminifer convolutus*): Highly convoluted channels, 150–250 µm wide and 30–70 cm long (ST), forming excretory ↑ parenchyma of ↑ testis. S. are usually long loops which begin and end as short ↑ tubuli recti (TR) leading to ↑ rete testis (RT); they may also branch, anastomose, or begin blindly. One to four S. compose one of 250 lobules of testis (L), apex of which is oriented toward ↑ mediastinum testis (MT). S. are lined by ↑ seminiferous epithelium (SE) and surrounded by a basal lamina (BL) followed by thin lamina propria, composed of one or several layers of peritubular ↑ myofibroblasts (Mf) and some ↑ fibrocytes. Rhythmic contractions of myofibroblasts, independent of nerve impulses, gradually move ↑ spermatozoa toward rete testis. Total length of S. is around 250 m.

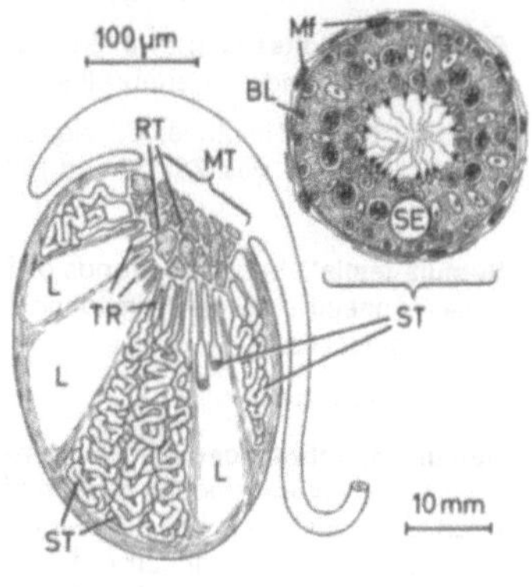

Semthin sections: Tissue slices ranging between 0.5 and 2 µm in thickness. In majority of cases, S. are cut with an

↑ ultramicrotome and analyzed, after appropriate ↑ staining or ↑ contrasting, under ↑ light microscope or conventional or high-voltage ↑ transmission electron microscope.

Böck, P.: Der Semidünnschnitt. Berlin, Heidelberg, New York: Springer-Verlag 1984

Senescene: ↑ Ageing

Sensory cells: A category of specialized epithelial cells reacting to external stimuli. S. are scattered in a ↑ sensory epithelium.

Sensory cortex: ↑ Brain cortex, phylogenic and structural division of

Sensory epithelium: A surface ↑ epithelium containing ↑ sensory cells. Epithelia of ↑ maculae of saccule and utricle, ↑ cristae ampullares, ↑ Corti's organ, ↑ olfactory epithelium, and ↑ taste buds are typical S.

Sensory ganglia: ↑ Spinal ganglia and ganglia associated with some cranial nerves. All S. have identical histological structure.

Sensory neurons: ↑ Neurons conducting nerve impulses from periphery of body (skin, internal and sensory organs) to ↑ central nervous system.

Sensory receptors: Specialized nerve endings or organs acting as transducers of various forms of energy into neural activity. According to source of impulses to which S. react, they can be classified into: 1) ↑ Exteroceptors; 2) ↑ interoceptors; 3) ↑ proprioceptors, and 4) ↑ teleceptors. Physiological classification divides S. according to the kind of energy to which they react: 1) ↑ Baroreceptors; 2) ↑ chemoreceptors; 3) ↑ mechanoreceptors; 4) thermoreceptors, and 5) ↑ photoreceptors.

Septal cells: ↑ Alveolar cells type II

Septate junctions: Kind of intercellular junctions found in invertebrates.

Graf, F., Noirot-Timothée, C., Noirot, Ch.: J. Ultrastruct. Res. *78*, 136 and 152 (1982)

Septula testis*: Very fine strands of ↑ loose connective tissue delimitating ↑ lobules of ↑ testis.

↑ Seminiferous tubules

Septum membranaceum*: Membranous part of interventricular septum of heart. S. consists of a 1 to 2-mm-thick layer of ↑ dense connective tissue covered on both sides by ↑ endocardium.

Seromucous glands: ↑ Mixed glands

Serosa: ↑ Tunica serosa

Serotonin (enteramine, 5-hydroxytryptamine, 5-HT): A vasoconstrictive substance produced by blood ↑ platelets, some ↑ endocrine cells of gastrointestinal tract (two-thirds of all body S.), some ↑ neurons in ↑ hypothalamus, ↑ pinealocytes, and in basal ganglia. S. provokes contraction of ↑ smooth muscle cells; through this it probably controls peristaltic movements of gut, together with ↑ interstitial cells of Cajal. S. also acts as a ↑ neutrotransmitter.

↑ Synapse, aminergic; ↑ Synapse, serotoninergic

Advances in Serotonin Methods. J. Histochem. Cytochem. *30*, 739–850 (1982); Azmitia, E.C.: J. Histochem. Cytochem. *30*, 739 (1982)

Serotoninergic synapse: ↑ Synapse, serotoninergic

Serous cavities: Closed spaces of body lined with ↑ mesothelium. Peritoneal, pleural and pericardial cavities, as well as cavity lined by ↑ tunica vaginalis of testis are S.

↑ Serous membranes

Serous cells, of bronchial glands: ↑ Bronchial glands

Serous cells, of salivary glands (serocytus*): Cuboidal cells forming ↑ acini and ↑ demilunes of ↑ salivary glands. Nucleus is round and in basal position, often containing two nucleoli. Cytoplasm encloses rod-shaped mitochondria, a supranuclear Golgi apparatus (G), and abundant rough endoplasmic reticulum (rER) in form of long flattened ↑ cisternae accumulated in basal and perinuclear cell portions. Rough endoplasmic reticulum is associated with large amount of free ribosomes, both structures being responsible for ↑ basophilia of basal half of S. From Golgi apparatus arise ↑ unit membrane-bound granules of low density (↑condensing vacuoles, CV), which transform into ↑ secretory granules (SG) with moderately osmiophilic material; their contents are discharged from cell by ↑ exocytosis. Apical plasmalemma forms short microvilli (Mv). S. are connected to adjacent cells by ↑ junctional complexes (J) and lateral ↑ interdigitations (Int). S. produce watery component of ↑ saliva, rich in amylase, ↑ peroxidase, ptyalin, ↑ glycosaminoglycans, and salts. According to a new wider classification, ↑ pancreatic acinar cells and gastric ↑ parietal cells also belong to S. BL = ↑ basal lamina

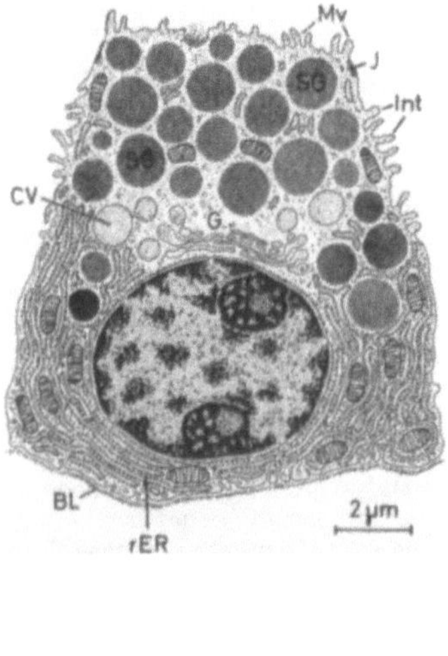

Pinkstaff, C.A.: Int. Rev. Cytol. *63*, 141 (1980); Riva, A., Riva-Testa, F.: Anat. Rec. *176*, 149 (1973)

Serous demilunes: ↑ Demilunes

Serous exudate: Watery fluid with low proportion of proteins; always present in small quantity in ↑ serous cavities where it is produced by ↑ mesothelium in order to facilitate sliding of organs. Under normal conditions, S. contains only a few free cells, of which the most frequent are free ↑ macrophages, small ↑ lymphocytes, desquamated mesothelial cells, eosinophilic ↑ granulocytes, and ↑ mast cells.

Serous glands (glandula serosa*): ↑ Exocrine glands producing watery secretion (e.g., exocrine ↑ pancreas, ↑ lacrimal glands, ↑ parotid glands, etc.).

↑ Glands, classification of

Serous membranes: Simple squamous mesothelial linings of closed ↑ serous cavities of body and most included organs (↑ pericardium, ↑ pleura, ↑ peritoneum, ↑ tunica vaginalis of testis). S. do not contain glands; ↑ mesothelium produces thin ↑ serous exudate similar to ↑ lymph. ↑ Synovial membranes, ↑ synovial vagina, and ↑ mesotendineum are kind of S.

↑ Tunica serosa

Sertoli's cells (cellula sustentacularis*): Highly irregular nondividing tall columnar cells extending from basal lamina (BL) to lumen (L) of ↑ seminiferous tubule. Nucleus (N) is voluminous and basal, measures about 9–12 μm in diameter, and displays several deep invaginations. Its ↑ chromatin is finely dispersed except for a dense ring of perinucleolar ↑ chromatin surrounding voluminous nucleolus (Nu). Cytoplasm is clear and contains numerous slender elongated mitochondria, a large but simple Golgi apparatus (G), some groups of short parallel rough endoplasmic cisternae (rER), and a considerable number of smooth endoplasmic tubules (sER), particularly around ↑ lipid droplets (LD) and ↑ acrosomal caps of each late ↑ spermatid (Sp). Cytoplasm also encloses ↑ Charcot-Böttcher crystals (Cr), numerous variously shaped ↑ lysosomes (Ly), ↑ lipofuscin granules (Lf), microfilamentous bundles (Mf), ↑ glycogen particles (Gly), and occasional ↑ annulate lamellae. At sites of contact between two S., there are long ↑ zonulae occludentes (ZO); ↑ subsurface cisternae (SC) of special structure are situated in this area. Identical cisternae surround heads of spermatids and ↑ spermatozoa (Spz). Zonulae occludentes between S. are responsible for formation of ↑ blood-testis barrier. Lateral surfaces of S. display deep recesses occupied by ↑ spermatogenic cells (SpC) during their multiplication and maturation, with long, frequently branched cell processes surrounding these cells. Functions: 1) Mechanical support for spermatogenic cells; 2) their protection; 3) participation in their metabolism; 4) providing of microenvironment to ↑ germ cells; 5) release of late ↑ spermatids into tubular lumen; 6) phagocytosis of degenerated germ cells and residual cytoplasm (RC) of late spermatids; 7) secretion of ↑ androgen-binding protein; 8) possible role in conversion of pregnenolone into ↑ testosterone; 9) formation of blood-testis barrier; 10) secretion of small amounts of ↑ estrogens.

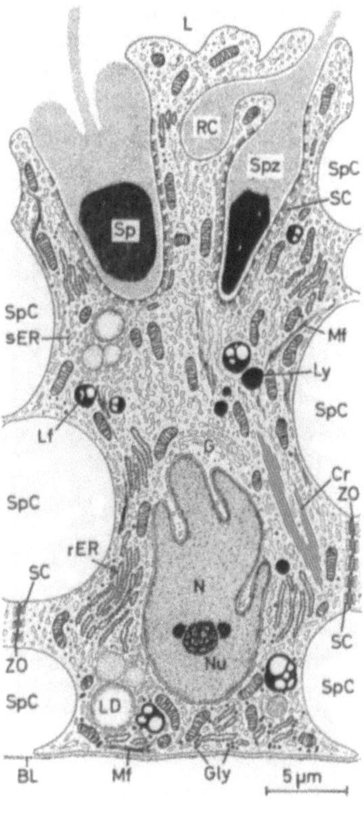

Camatini, M., De Curtis, I., Franchi, E.: J. Ultrastruct. Res. *79*, 314 (1982); Hatier, R., Grignon, G.: Anat. Embryol. *160*, 11 (1980); Morales, C., Clermont, Y.: Anat. Rec. *203*, 233 (1982); Suárez-Quian, C.A., Dym, M., Makris, A., Brumbaugh, J., Ryan, K.J., Canick, J.A.: Andrology *4*, 203 (1983); Russel, L.D.: Gamete Res. *3*, 179 (1980)

Sex chromatin: Inappropriate term for a condensed and inactived long arm of one of two ↑ X-chromosomes visible in female ↑ interphase nuclei as ↑ Barr body or ↑ "drumstick." Other X-chromosome is extended (having a ↑ euchromatin form and therefore being invisible), but very active in giving off its genetic information. S. is a kind of facultative ↑ heterochromatin.

Sex chromosomes (gonosome, heterochromosome, heterosome, gonosoma*): Pair of ↑ chromosomes that differ between the two sexes. Whereas two S. in female are morphologically identical (XX), those of male differ in size and location of ↑ centromere (XY). Diploid human cells have a chromosomal garniture of 44, XX in female and 44, XY in male. ↑ Y-chromosome contains genes for development of ↑ testis; ↑ X-chromosome in male is overpowered by Y-chromosome. Female sex is homozygous; male sex is heterozygous.

↑ Autosomes; ↑ Karyotype

Sex, determination of: Whereas all secondary ↑ oocytes are genetically identical, carrying 22 ↑ autosomes and one ↑ X-chromosome, there are two categories of ↑ spermatozoa: ↑ androspermatozoa and ↑ gynospermatozoa. ↑ Fertilization of a secondary oocyte with an androspermatozoon results in formation of ↑ zygote of a male (44, XY), and that with a gynospermatozoon results in zygote of a female (44, XX). Since gynospermatozoa have a nuclear surface larger than that of androspermatozoa and seem to be somewhat heavier (an X-chromosome is heavier than a small Y-chromosome), it is believed that androspermatozoa are somewhat faster than gynospermatozoa and therefore probably fertilize the secondary oocyte more frequently. This would explain the fact that for 100 newborn girls there are about 105 boys.

Sex hormones: Collective term for ↑ androgens and ↑ estrogens.

Sex organs, external, female: ↑ Clitoris; ↑ Hymen; ↑ Labia majora; ↑ Labia minora

Sex organs, external, male: ↑ Penis; ↑ Scrotum

Sex vesicle: Inappropriate term for agglomerated ↑ sex chromosomes in nucleus of primary ↑ spermatocyte. S. has no limiting membrane.

SGC-cells: ↑ Autonomic neurons, of sympathetic ganglia

Shadowing: Technique of depositing a fine layer of heavy metal (gold, chromium, palladium, platinum, uranium) on electron microscopic specimens in order to increase their contrast, give them a three-dimensional aspect, and/or produce ↑ replicas. 1) In ↑ transmission electron microscope, S. is effected in a high vacuum using an incandescent tungsten filament (TF) from wich metal evaporates in a given direction and at a given angle (α = 15°–45°). Metal (M) is deposited on one side of relief of specimen (Sp); on other side, shadows (S) form which permit calculation of height (h) of object ($h = S \cdot tg\alpha$). In ↑ freeze-cleaving, entire de-

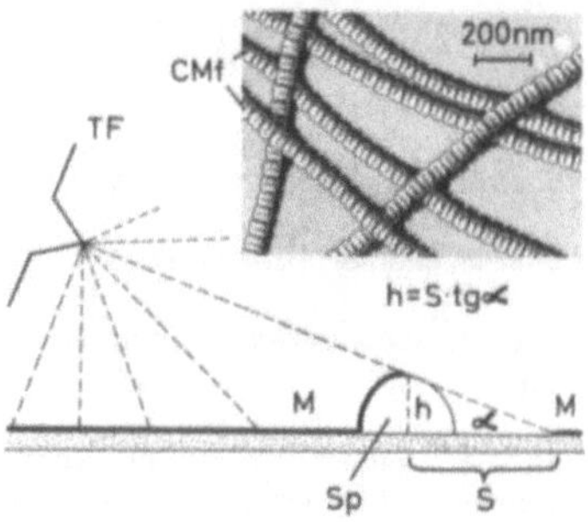

posited metallic layer with carbon support is freed from specimen surface and transferred onto specimen holder of transmission electron microscope. (The finest S. is obtained by simultaneous evaporation of platinum with carbon.) 2) In ↑ scanning electron microscope, S. gives contrast and electric conductibility to specimen. This kind of S. is effected in an atmosphere of argon by cathode ↑ sputtering of gold; S. is diffuse, and metallic particles penetrate deeply into slits and furrows of specimen. (Inset: chromium-shadowed ↑ collagen microfibrils, CMf)

Glauert, A.M. (ed.): Practical Methods in Electron Microscopy, Vol. 8. Amsterdam: Elsevier/North Holland 1980

Shaft, of bone: ↑ Diaphysis

Sharpey's fibers, of periosteum (perforating fibers, fibrae perforantes*): Coarse ↑ collagen fibers penetrating outer ↑ circumferential lamellae and even ↑ interstitial lamellae of bone from fibrous layer of ↑ periosteum. S. assure firm contact of periosteum with ↑ bone.

Sharpey's fibers, of tooth (cementoalveolar fibers, fibrae perforantes cementales*): Coarse ↑ collagen fibers (S) of ↑ periodontal ligament embedded at one end in ↑ alveolar bone (AB) and inserted at other end in ↑ cementum (C). S. extend obliquely across ↑ periodontal space (PS) toward ↑ apex of tooth, giving it support to withstand shock of biting and chewing. However, S. do permit a certain amount of motion. D = ↑ dentin. (Fig. = human)

↑ Gomphosis

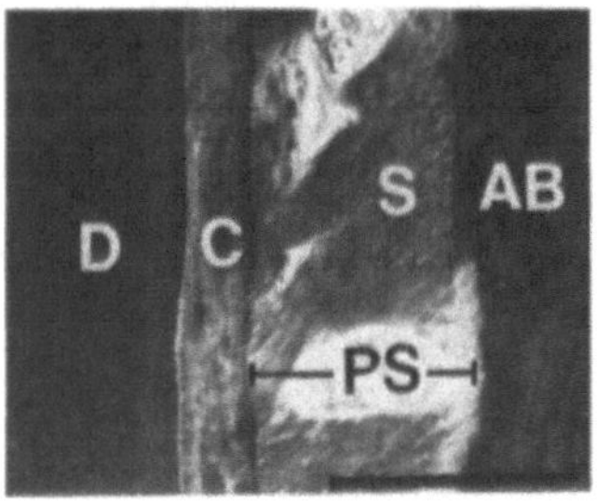

Sheath cells: ↑ Schwann's cells

Sheath, enamel: ↑ Enamel sheath

Sheath, endoneurial: ↑ Endoneurial sheath

Sheath, mitochondrial of spermatozoon: ↑ Mitochondrial sheath, of spermatozoon

Sheath, myelin: ↑ Myelin sheath

Sheath, neurolemmal: ↑ Schwann's sheath

Sheath, of Henle: ↑ Endoneurial sheath

Sheath, of Neumann: ↑ Neumann's sheath

Sheath, of Schwann: ↑ Schwann's sheath

Sheath, of Schweigger-Seidel: ↑ Ellipsoids, of Schweigger-Seidel

Sheath, periarterial, of spleen: ↑ Periarterial lymphatic sheath

Sheath, prismatic rod: ↑ Enamel sheath

Sheathed capillaries, of spleen: ↑ Capillaries, sheathed, of spleen

Shrapnell's membrane (pars flaccida membranae tympani*): Anterosuperior flaccid portion of ↑ tympanic membrane devoid of middle fibrous layer.

SI (**S**ystème **I**nternational d'Unités): New system of units recommended by World Health Organization for medicine and other sciences.

The SI for the Health Professions. World Health Organization: Geneva 1977; Plenert, W., Heine, W.: Normalwerte. 6th edn. Jena: VEB Gustav Fischer 1983; Wald, A.: Science *221*, 808 (1983)

Sialic acid (neurimidic acid): A carbohydrate linked to protein of plasmalemma in some cells (↑ lymphocytes); S. is responsible for distinctive capacities for homing and aggregating displayed by cells concerned. In absence of S., ↑ contact inhibition is lost, leading to uncontrolled cell movement over other cells. Enzyme ↑ neuriminidase removes S.

Schauer, R. (ed.): Sialic Acid. Cell Biology Monographs, Vol. 10. Wien, New York: Springer-Verlag 1982

Siderocyte: Young ↑ erythrocyte still containing some ↑ siderosomes in its cytoplasm.

Siderosomes: Highly electron-dense agglomerations (S) of ↑ iron in form of ↑ ferritin occurring in cytoplasm of ↑ erythroblasts, ↑ reticulocytes, and ↑ siderocytes. S. measure about 0.1–0.3 μm; in some cases, they may be surrounded by a ↑ unit membrane. (Fig. = rat)

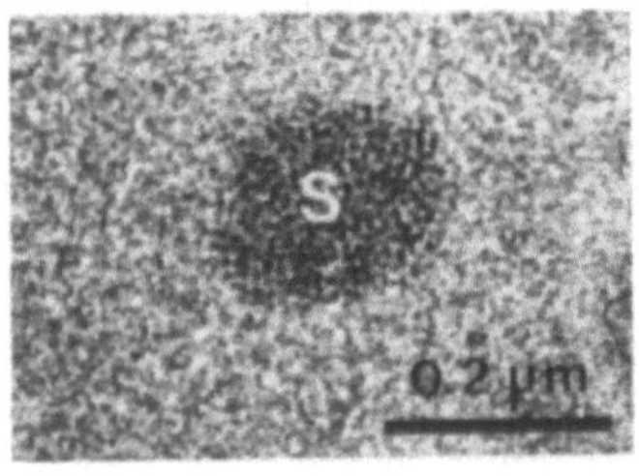

Ghadially, F.N., Lalonde, J.-M., Oryschak, A.F.: Virchows Arch. B. Cell Path. *22*, 135 (1976)

Sieve plates: Clearly delimited groups (SP) of fenestrations (F) in attenuated cytoplasm of ↑ endothelial cells (E), of fenestrated ↑ capillaries, and of ↑ liver sinusoids. (Fig. = adrenal cortex, rat)

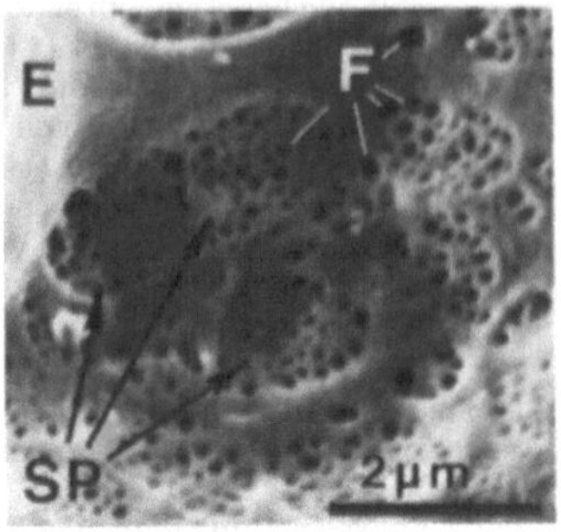

SIF-cells: ↑ Autonomic neurons, of sympathetic ganglia

Sigmoid colon: ↑ Colon

Silver staining (silver impregnation, silver nitrate staining): Use of silver nitrate for staining of some structures (intercellular clefts, ↑ neurofibrils, ↑ reticular fibers, ↑ neuroglia cells, ↑ neurons, ↑ A-cells of ↑ pancreatic islets, ↑ C-cells of ↑ thyroid gland, some ↑ endocrine cells of ↑ gastrointestinal tract, ↑ melanocytes, etc.). All these structures and cells are capable of reducing metallic silver from staining solution with or without previous treatment. Blackening of structures is based on deposition of silver precipitates.

↑ Argentaffinity; ↑ Argyrophilia

Simple columnar epithelium (unilayered prismatic epithelium, epithelium columnare simplex*): Single-

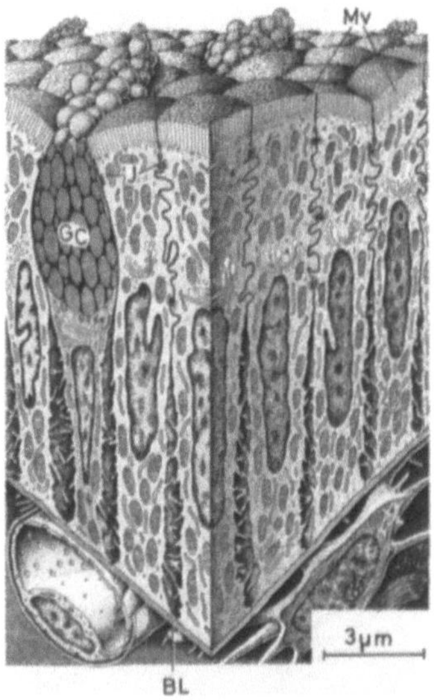

layered epithelium formed of tall and slender prismatic cells, height of which is several times greater than width. Cells have elongated nucleus and well-developed ↑ organelles and ↑ junctional complexes (J); they contact ↑ basal lamina (BL) with narrow base. Apical surface can be studded with ↑ microvilli (Mv), ↑ striated border, or ↑ cilia; basal pole can be formed in ↑ basal labyrinth. S. is found from ↑ cardia to ↑ anal canal, in ↑ oviduct, ↑ uterus, small ↑ bronchi, ↑ bronchioles up to terminal ↑ bronchioles, most secretory portions and ducts of ↑ exocrine glands, ↑ seminal vesicles, ↑ thyroid follicles, central canal of ↑ spinal cord, and some ↑ ovarian follicles. S. can be composed of several kinds of cells, among which ↑ goblet cells (GC) are most frequently represented.

Simple cuboidal epithelium (isoprismatic epithelium, epithelium cuboideum simplex*): Single layer of cells in form of small hexagonal prisms, height of which is approximately equal to width. Nuclei (N) are round, ↑ organelles and ↑ junctional complexes well developed. Cells of S. can have ↑ microvilli, ↑ brush border, ↑ cilia, ↑ basal labyrinth (BLt). S. includes epithelia of ↑ proximal and ↑ distal tubules (fig.) of ↑ nephron, ↑ ciliary body, ↑ choroid plexus, ↑ ependyma, ↑ pigment epithelium of ↑ retina, ↑ germinative epithelium of ↑ ovary, ↑ thyroid follicles, many glands, some excretory ducts, ↑ bile ducts, terminal ↑ bronchioles, and some ↑ ovarian follicles. Cells lie on a basal lamina (BL).

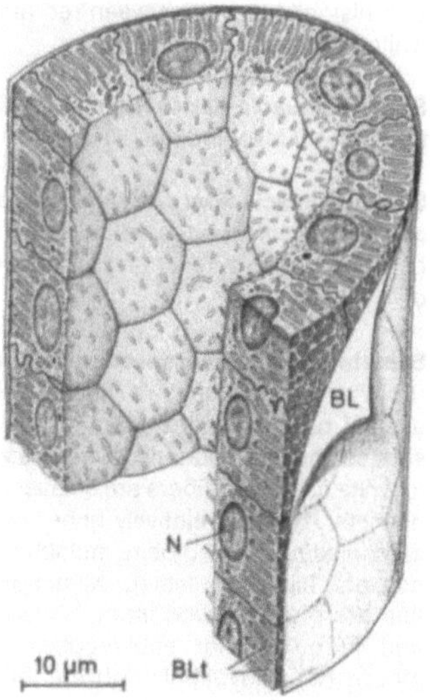

Simple glands: ↑ Exocrine glands with single nonramified excretory duct (↑ buccal glands, ↑ esophageal glands, ↑ labial glands, etc.).

↑ Glands, classification of

Simple squamous epithelium (epithelium squamosum simplex*): Unique layer of very attenuated polygonal cells held together by ↑ junctional complexes (JC) and intercellular ↑ interdigitations (Id). Nucleus (N) is generally thicker than surrounding cytoplasm and bulges toward free surface. In S. are included: ↑ endothelium, ↑ mesothelium, and epithelium of parietal sheath of ↑ Bowman's capsule, inner surface of ↑ tympanic membrane, nonspecialized area of ↑ membranous labyrinth, ↑ thin segment of ↑ Henle's loop, some segments of ↑ rete testis, initial segments of ↑ intercalated ducts, some ovarian ↑ follicular cells, and endothelium of ↑ cornea (fig.). D = ↑ Descemet's membrane

Sinuatrial node (node of Keith and Flack, nodus sinuatrialis*): Part of ↑ impulse-conducting system of mammalian heart, structurally identical to ↑ atrioventricular node.

Sinus caroticum: ↑ Carotid sinus

Sinus, intermediate: ↑ Intermediate sinus; ↑ Lymph nodes; ↑ Lymph nodes, sinuses of

Sinus, lactiferous: ↑ Lactiferous sinus

Sinus, marginal, of lymph nodes: ↑ Subcapsular sinus; ↑ Lymph nodes, sinuses of

Sinus, medullary, of lymph nodes: ↑ Lymph nodes; ↑ Lymph nodes, sinuses of

Sinus, of kidney: ↑ Renal sinus

Sinus, of valve: Circular space between ↑ valve and wall of a ↑ vein.

Sinus, perforating, of lymph nodes: ↑ Intermediate sinus; ↑ Lymph nodes; ↑ Lymph nodes, sinuses of

Sinus, subcapsular, of lymph nodes: ↑ Lymph nodes; ↑ Lymph nodes, sinuses of; ↑ Reticular tissue; ↑ Subcapsular sinus

Sinus venosus sclerae: ↑ Schlemm's canal

Sinuses, of bone marrow: ↑ Bone marrow, vascular sinuses of

Sinuses, of dura mater (sinus durae matris*): Large venous vessels (S) situated between two layers of ↑ dura mater (DM). Structure: 1) Tunica intima (TI) = ↑ endothelium (E) overlying a ↑ basal lamina; 2) fibrous layer (TF) = ↑ dense connective tissue connected in some areas with ↑ periosteum of skull bones. Since ↑ smooth muscle cells are completely lacking in wall of S., they cannot change their diameter. (Fig. = tentorium cerebelli, human)

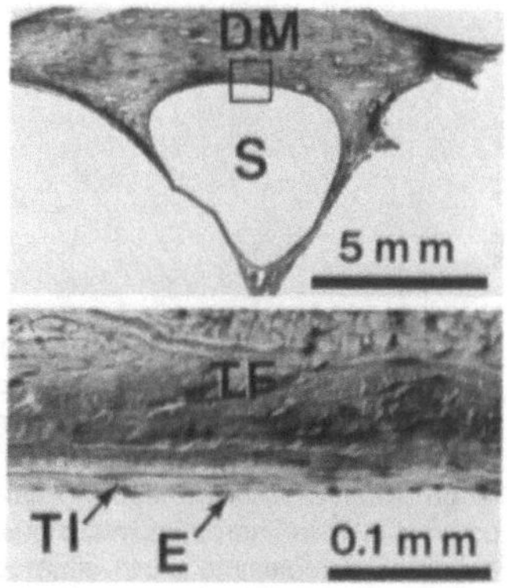

Sinuses, of lymph node: ↑ Lymph nodes, sinuses of

Sinuses, paranasal: ↑ Paranasal sinuses

Sinuses, vascular, of bone marrow: ↑ Bone marrow, vascular sinuses of

Sinusoids: ↑ Capillaries, sinusoidal

Sinusoids, hepatic: ↑ Liver sinusoids

Sinusoids, of liver: ↑ Liver sinusoids

Sinusoids, of spleen: ↑ Splenic sinusoids

Skeletal muscle (musculus*): Functional unit composed of large number of ↑ skeletal muscle fibers grouped in ↑ muscle fascicles (F) and enveloped in successive connective tissue sheaths. Structure: Each muscle fiber (MF) is invested by ↑ endomysium (E); several muscle fibers are associated into a muscle fascicle (F) enveloped by its own connective tissue sheath, ↑ perimysium (P). Variable numbers of fascicles compose entire S., which is ensheathed by ↑ epimysium (Ep) and then by ↑ fascia (Fa). B = ↑ bone, T = ↑ tendon

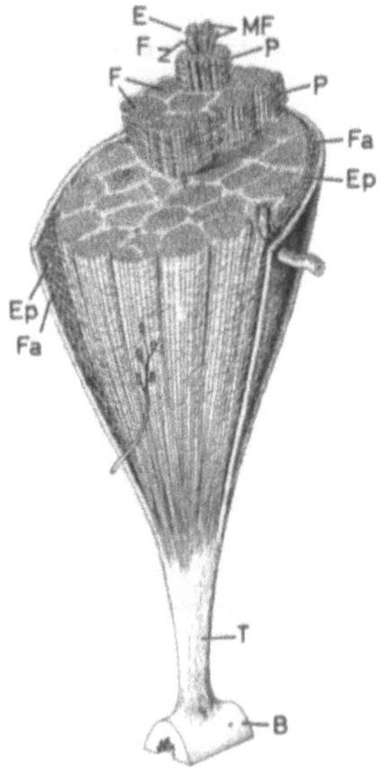

Skeletal muscle fibers (myocytus*): Very long cylindrical multinucleated cells (SMF) ranging from 10 to 40 µm in width and several centimeters in length. Each S. is surrounded by ↑ sarcolemma (S); in ↑ sarcoplasm (Sp) just beneath sarcolemma are situated hundreds or thousands of nuclei (N), Golgi apparatuses, and cisternae of rough endoplasmic reticulum. There are no ↑ centrioles in S. Main mass of sarcoplasm is occupied by ↑ myofibrils (Mf) separated from each other by threadlike mitochondria and ↑ sarcoplasmic reticulum. A considerable amount of ↑ glycogen, free ribosomes, and a variable number of ↑ lipid droplets can be found in sarcoplasm. On longitudinal sections (L), S. display longitudinal striation corresponding to myofibrils; their ↑ myofilaments form characteristic transversal ↑ banding patterns (↑ A-bands and ↑ I-bands). In transverse sections (T), myofibrils may be grouped into ↑ Cohnheim's fields. Several S. form a ↑ muscle fascicle of ↑ skeletal muscle. SC = ↑ satellite cell (Figs. = rat)

↑ Striated muscle fibers, molecular biology of contraction of; ↑ Striated muscle fibers, morphology of contraction of

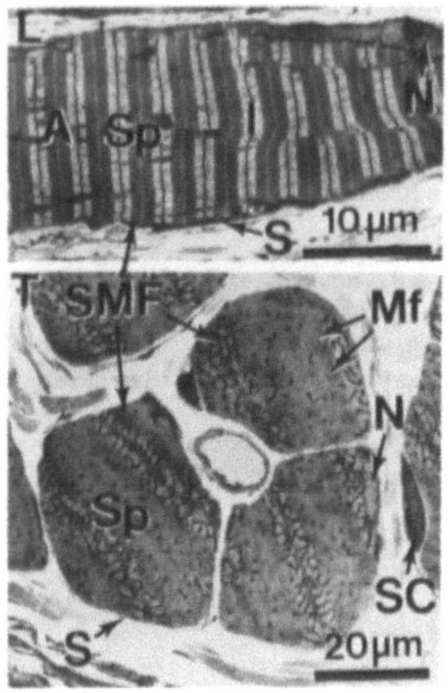

Bonilla, E.: J. Ultrastruct. Res. *82*, 341 (1983); Dowben, R.M., Shay, J.W. (eds.): Cell and Muscle Motility. Vol. 4. New York: Plenum Publ. Co. 1983

Skeletal muscle fibers, histogenesis of (striomyohistogenesis*): From myotomes (Myot) arise presumptive myoblasts (PMy) representing two cell populations. On one side, presumptive myoblasts differentiate into true ↑ myoblasts (TMy); on other, presumptive myoblasts remain undifferentiated and give rise to ↑ satellite cells (SC). True myoblasts range in rows, fuse together, and form ↑ myotube (Mt). To myotube adhere satellite cells, dividing there and furnishing nuclei (N) to myotubes. Certain number of satellite cells remain adjacent to myotubes as reserve cells for ↑ regeneration of ↑ skeletal muscle fibers. Myotube gradually differentiates into skeletal muscle fibers (MF). Insets: A) True myoblasts have vesicular rough endoplasmic reticulum (ER) from which bulge B) ↑ sarcotubules (St) which C) unite to constitute ↑ sarcoplasmic reticulum (SR). D) Between segments of sarcoplasmic reticulum, ↑ T-tubule gradually invaginates to form a ↑ triad (Tr). ↑ Myofilaments (Mf) are synthesized by helicoidal ↑ polyribosomes (hr). In many points S. is still not well understood.

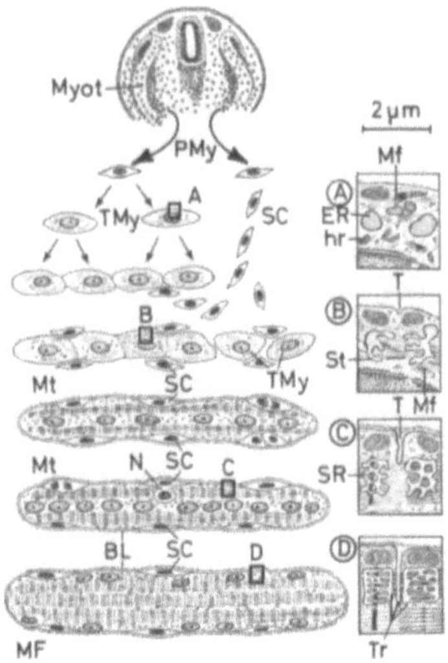

Goldspink, D.F. (ed.): Development and Specialization of Skeletal Muscle. Cambridge: Cambridge University Press 1980; Isobe, Y., Shimada, Y.: Cell Tissue Res. *231*, 481 (1983); Nag, A.C., Foster, J.D.: J. Anat. *132*, 1 (1981); Pearson, M.L., Epstein, H.F. (eds.): Muscle Development: Molecular and Cellular Control. New York: Cold Spring Harbor Laboratory 1982; Platzer, A.: Anat. Rec. *190*, 639 (1978); Russel, R.G., Oteruelo, F.T.: Anat. Embryol. *162*, 403 (1981)

Skeletal muscle fibers, innervation of: ↑ Motor end plate

Skeletal muscle fibers, intermediate: Type of muscle fibers morphologically and histochemically between red and white ↑ skeletal muscle fibers.

Skeletal muscle fibers, junction with tendon of: ↑ Myotendinal junction

Skeletal muscle fibers, mechanism of contraction of: ↑ Striated muscle fibers, molecular biology of contraction of

Skeletal muscle fibers, red (myofibra rubra*): Type of muscle fibers with usual ↑ banding patterns as in other striated muscle fibers. In comparison to white fibers, red fibers are smaller in diameter, but are relatively richer in ↑ sarcoplasm, ↑ myoglobin, mitochondria, and ↑ lipid droplets (L). Mitochondria are provided with many ↑ cristae and form irregular subsarcolemmal groups (G). ↑ Myofibrils (Mf) are thinner than in white fibers, ↑ Z-lines thicker than corresponding lines in white fibers. S. have a considerably richer blood supply than white fibers and show highest activity of succinic dehydrogenase of all types of skeletal muscle fibers. S. compose predominantly intercostal muscles, masticatory musculature, and muscle of eyeball, since red fibers have relatively slow contraction but are slower to fatigue than white fibers. Cap = capillaries, BL = ↑ basal lamina, T = ↑ T-tubules, SR = ↑ sarcoplasmic reticulum, arrowheads = openings of T-tubules

↑ Skeletal muscle fibers, white

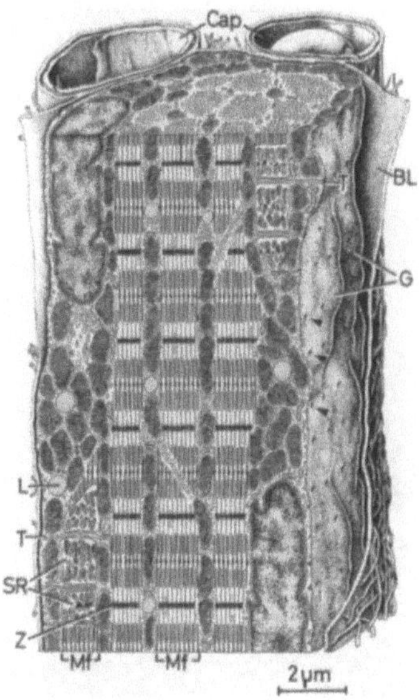

Skeletal muscle fibers, regeneration of: Repair of damaged muscle fibers (MF) is possible under two conditions:

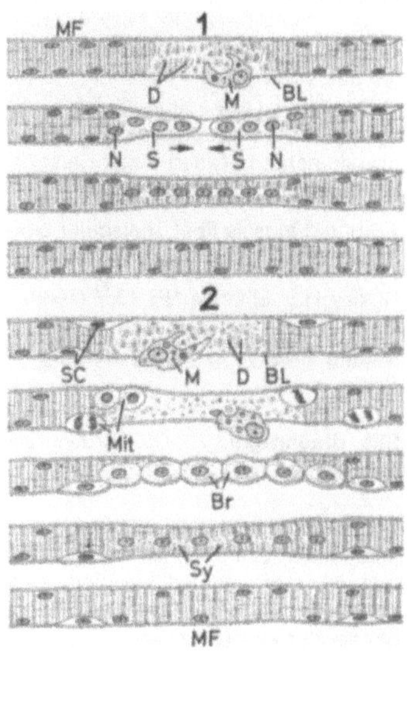

That ↑ basal lamina (BL) remains intact, and that destroyed segment is not too large. 1) Continuous regeneration = after ↑ macrophages (M) have destroyed debris (D), ends of two segments (S) bulge toward one another (arrows), fuse together, nuclei (N) multiply, and damaged segment becomes regenerated in about 2 weeks. 2) Discontinuous regeneration = after ↑ phagocytosis of debris by macrophages (M), ↑ satellite cells (SC) divide mitotically (Mit) and bridge (Br) damaged segment. By their fusion, there arises a ↑ syncytium (Sy) with nuclei in central position. By further ↑ differentiation of syncytium, morphofunctional continuity of skeletal muscle fiber is reestablished.

Allbrook, D.: Muscle Nerv. *4*, 234 (1981); Carlson, B.M., Faulkner, J.A.: Med. Sci. Sports Exerc. *15*, 187 (1983); Jakubiec-Puka, A., Kulesza-Lipka, D., Kordowska, J.: Cell Tissue Res. *227*, 641 (1982); Ontell, M., Hughes, D., Bourke, D.: Anat. Rec. *204*, 199 (1982)

Skeletal muscle fibers, sarcoplasmic reticulum of (reticulum sarcoplasmaticum*): Specialized kind of smooth ↑ endoplasmic reticulum (SR) surrounding each ↑ myofibril (Mf) of a ↑ skeletal muscle fiber (F). Sarcoplasmic reticulum extends from one A-I junction to another and consists of a longitudinal ↑ L-system of ↑ sarcotubules that communicate with transversally oriented ↑ terminal cisternae (TC), forming a ↑ triad (Tr) with an interposed ↑ T-tubule. During relaxation of muscle fiber within sarcoplasmic reticulum, calcium ions accumulate. An action potential spreading through T-tubules induces release of calcium

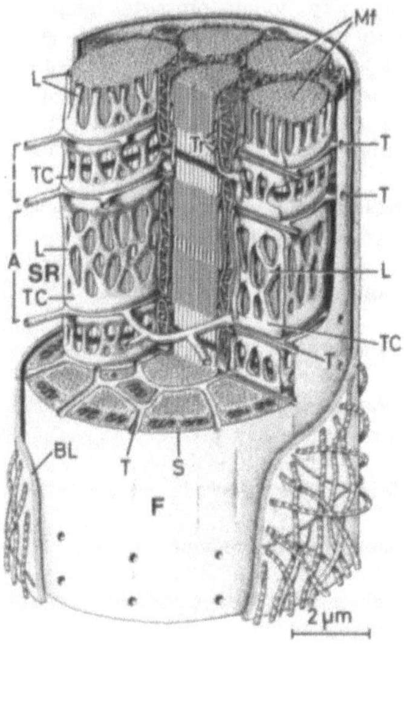

ions from sarcoplasmic reticulum into area of ↑ A-band, which in turn provokes contraction. BL = ↑ basal lamina, S = ↑ sarcolemma

↑ Cardiac muscle cells, sarcoplasmic reticulum of; ↑ Striated muscle fibers, molecular biology of contraction of

MacLennan, D.H., Campbell, K.P.: Trends Biochem. Sci. *4*, 148 (1979)

Skeletal muscle fibers, types of: On basis of morphological and histochemical properties, skeletal muscle fibers are subdivided into: 1) ↑ Skeletal muscle fibers, intermediate; 2) ↑ skeletal muscle fibers, red; 3) ↑ skeletal muscle fibers, white.

Gauthier, G.F.: J. Cell Biol. *82*, 391 (1979)

Skeletal muscle fibers, white (myofibra alba*): Type of muscle fibers with diameter considerably larger than that of red fibers, but containing relatively less ↑ sarcoplasm, ↑ myoglobin, and mitochondria. ↑ Myofibrils (Mf) are larger and stronger, but ↑ Z-lines are thinner than in red fibers. S. contract more rapidly than red fibers, but also fatigue more rapidly. S. compose predominantly muscles of extremities. BL = ↑ basal lamina, S = ↑ satellite cell

↑ Skeletal muscle fibers, red

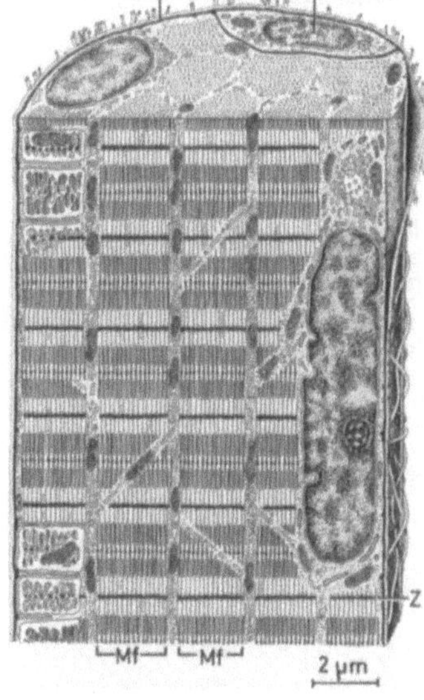

Skeletal muscle, vascularization of: Small arteries (A) run through ↑ perimysium parallel to ↑ muscle fascicles and turn at right angles, entering ↑ endomysium, where arteries bifurcate into two ↑ arterioles (Ar) running parallel to muscle fibers. Arterioles

form richly branched networks of capillary loops (C) crossing muscle fibers. S. is abundant; in adult man, total length of capillaries in skeletal muscles is about 40 000 km. (Fig. = guinea pig, vascular ↑ injection with lithium carmine)

Holley, J.A., Fahim, M.A.: Anat. Rec. *205*, 109 (1983)

Skeletal tissues: Tissues composing skeleton: ↑ bone, ↑ cartilages, and ↑ joints.

Skeletin: Protein (m.w. about 55 000 daltons) forming ↑ intermediate microfilaments.

Eriksson, A., Thornell, L.-E., Stigbrand, T.: J. Histochem. Cytochem. *27*, 1604 (1979)

Skeletin filaments: ↑ Intermediate microfilaments

Skeleton, cardiac: ↑ Cardiac skeleton

Skeleton, cellular: ↑ Cytoskeleton

Skene's glands: ↑ Urethral glands, of female urethra

Skin (cutis, integumentum commune*): ↑ Organ forming external surface of body. Structure: 1) ↑ Epidermis (E). 2) ↑ Dermis (D) = a) papillary layer or stratum papillare* (PL); b) reticular layer or stratum reticulare* (RL). ↑ Hairs (H), ↑ sebaceous glands (SG), ↑ sweat glands (SwG), and ↑ arrector muscles (A) are situated in dermis. 3) ↑ Hypodermis (Hyp). Based on thickness of epidermis and presence or absence of hairs, S. may be divided into: A) Thick S. = S. on palms of hands and soles of feet with very thick epidermis but no hairs or sebaceous glands; B) thin S. = S. covering remainder of body, with hairs and cutaneous glands. At about 16% of total body weight, S. is heaviest single organ of body.

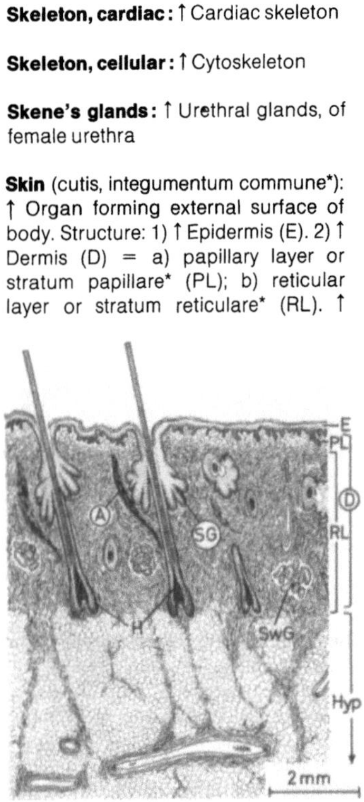

Brown, A.C.: The Integumentary System. In: Hodges, G.M., and Hollowes, R.C. (eds.): Biomedical Reseach Applications of Scanning Electron Microscopy, Vol. 2. New York, London: Academic Press 1980; Monatgna, W., Parakkal, P.F.: The Structure and Function of Skin, 3rd edn. New York: Academic Press 1974

Skin appendages: Collective term for structures developing from ↑ epidermis (↑ hairs, ↑ nails, ↑ mammary glands, ↑ sebaceous glands, and ↑ sweat glands).

Tsuji, T.: J. Microsc. *131*, 115 (1983)

Skin, dermis of: ↑ Dermis

Skin, epidermis of: ↑ Epidermis

Skin, functions of: 1) Protection of body against harmful physical and chemical agents, such as injury, ultraviolet irradiation, etc. (↑ keratin, ↑ melanin); 2) protection against desiccation (keratin); 3) thermoregulation (sweating, vascular supply); 4) maintenance of water balance (sweating); 5) ↑ excretion (↑ sweat glands); 6) sense organ for touch, pressure, temperature, and pain (↑ corpuscles, ↑ nerve endings); 7) metabolism of that (↑ hypodermis); 8) metabolism of vitamin D, and 9) production of odoriferous substances (apocrine ↑ sweat glands).

Skin, glands of: ↑ Sebaceous glands; ↑ Sweat glands

Skin, hair of: ↑ Hairs

Skin, hypodermis of: ↑ Hypodermis

Skin, Langerhans' cells of: ↑ Langerhans' cells

Skin, melanocytes of: ↑ Melanocytes

Skin, Merkel's cells of: ↑ Merkel's cells

Skin, pilosebaceous units of: ↑ Pilosebaceous units

Skin, sensory nerve endings in: ↑ Corpuscles; ↑ Nerve endings

Skin, vascularization of: Blood is distributed by ↑ muscular arteries forming network, rete cutaneum (RC), in ↑ hypodermis (Hyp) just below ↑ dermis (D). From one side of network, which is parallel to surface, there are branches supplying hypodermis, deep segment of ↑ hair follicles, and deeply situated ↑ sweat glands. From upper side of rete arise branches that ramify within dermis and form a superficial network, ↑ rete subpapillare (RS), which gives off capillary loops (CL) for ↑ dermal papillae (DP). Venous blood is collected in venous networks nearly parallel with arterial ones. Numerous arteriovenous ↑ anastomoses in skin form ↑ glomi.

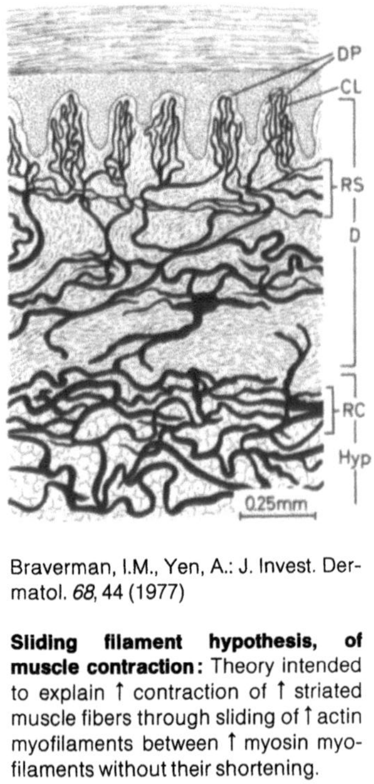

Braverman, I.M., Yen, A.: J. Invest. Dermatol. *68*, 44 (1977)

Sliding filament hypothesis, of muscle contraction: Theory intended to explain ↑ contraction of ↑ striated muscle fibers through sliding of ↑ actin myofilaments between ↑ myosin myofilaments without their shortening.

↑ Striated muscle fiber, molecular biology of contraction of

Huxley, H.E.: Science *164*, 1356 (1969)

Sliding theory, of ciliary beating: ↑ Cilium, motion of

Slit diaphragm: ↑ Glomerular slit membrane

Slit pores: Very complicated system of intercellular clefts formed between extensively interdigitated ↑ pedicels of ↑ podocytes. S. are bridged by ↑ glomerular slit membrane.

Slow-reacting substance (SRS): Lipid produced by basophilic ↑ granulocytes. S. is chemically similar to ↑ prostaglandins; it provokes prolonged vasodilatation and increased capillary permeability.

Slow-reacting substance, of anaphylaxis: ↑ Leukotrienes

SLS: ↑ Segment long-spacing collagen

Small alveolar cells: ↑ Alveolar cells, type I

Small dense bodies: ↑ Membrane-coating granules

Small-granule cells: ↑ Trachea, epithelium of

Small granule-containing cells: ↑ Autonomic neurons, of sympathetic ganglia

Small intensely fluorescent cells: ↑ Autonomic neurons, of sympathetic ganglia

Small intestine: ↑ Intestine, small

Small veins: ↑ Veins, small

Smegma: White oily substance found in corona of ↑ glans penis. S. consists mainly of desquamating epithelial cells in combination with secretion of ↑ preputial glands.

Smooth endoplasmic reticulum: ↑ Endoplasmic reticulum, smooth

Smooth muscle cells (smooth muscle fibers, myocytus nonstriatus*): Long spindle-shaped cells (20–500 μm in length, 5–8 μm in width) arranged so that thin end of one cell is adjacent to thickest portion of another cell. S. have central elongated rod-shaped nucleus with several deep invaginations, and one or two nucleoli. Bulk of ↑ sarcoplasm is occupied by longitudinally disposed microfilaments (↑ actin, ↑ myosin, ↑ intermediate) and ↑ dense bodies (DB). In ↑ endoplasm (E) are concentrated slender mitochondria, a small Golgi apparatus (G), short rough endoplasmic ↑ cisternae, few lysosomes, ↑ glycogen particles, and free ribosomes. Some smooth endoplasmic ↑ tubules (SR) are also present and may extend to cell periphery. ↑ Sarcolemma of S. forms many ↑ caveolae (C), lacking on areas of peripheral insertion of dense bodies. S. are enveloped in a basal lamina (BL), absent at ↑ nexus sites (N) where adjacent S. are held together. A fine network of ↑ collagen (CF) and ↑ reticular microfibrils (R), and ↑ elastic fibers (EF) surrounds every S.; reticular microfibrils penetrate into end invaginations of S. and reinforce cell union. In close contact with basal lamina or between S., there are ↑ autonomic nerve endings (NE). Besides their contractility, S. are capable of synthesizing collagen and elastic fibrils, as well as ↑ glycosaminoglycans of ↑ ground substance.

↑ Synapse, "by distance"

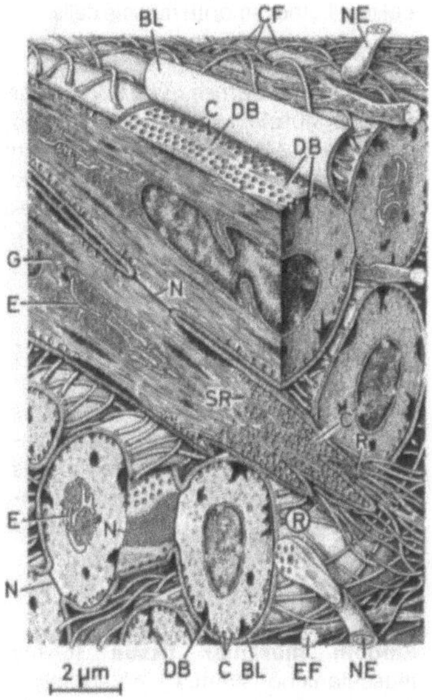

Bülbring, E., Branding, A.F., Jones, A.W., Tomita, T. (eds.): Smooth Muscle: An Assessment of Current Knowledge. London: Edward Arnold 1981; Gabella, G., Blundell, D.: J. Cell Biol. *82*, 239 (1979); Oakes, B.W., Batty, A.C., Handley, C.J., Sandberg, L.B.: Eur. J. Cell Biol. *27*, 34 (1982); Sawada, H.: Biomed. Res. *2*, Suppl. 153 (1981); Small, J.V., Sobieszek, A.: Int. Rev. Cytol. *64*, 241 (1980)

Smooth muscle cells, histogenesis of: ↑ Smooth muscle cells originate from ↑ mesenchymal cells by ↑ differentiation (except muscles of ↑ iris which are of ectodermal origin).

Smooth muscle cells, mechanism of contraction of: It is assumed that ↑ dense bodies (DB) are interconnected with one another by relatively rigid ↑ intermediate ↑ desmin microfilaments (IF). ↑ Actin myofilaments (A) also anchor into these bodies. In moment of contraction by an unknown mechanism, there appear in ↑ sarcoplasm ↑ myosin myofilaments (M) between which actin myofilaments slide, similar to the process in ↑ striated muscle fibers but considerably slower. This results in displacement of dense bodies due to forces transmitted along intermediate microfilaments; cell shortens, thickens, cell surface becomes deformed and nucleus (N) twisted. (Modified after Leblond, C.P., and Schulz, E. In: Ham, A.W.: Histology. 7th edn. Philadelphia: Lippincott 1974.)

↑ Smooth muscle cells, molecular biology of contraction of

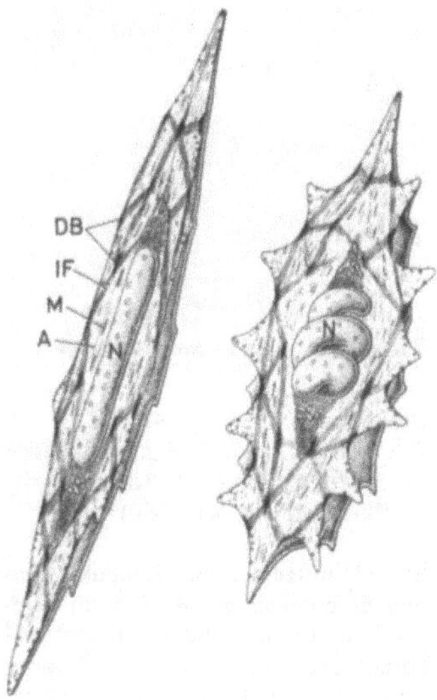

Ebashi, S., Maruyama, K., Endo, M. (eds.): Muscle Contraction. Its Regulatory Mechanisms. Berlin, Heidelberg, New York: Springer-Verlag 1980; Gabella, G.: Cell Tissue Res. *170*, 187 (1976); Smooth Muscle, British Medical Bulletin *35*, 209 (1979). London: British Council 1979

Smooth muscle cells, microfilaments of: 1) Myofilaments: a) ↑ Actin myofilaments (A) = 3 to 8-nm-thick contractile protein filaments more than 1 μm long anchoring into ↑ dense bodies (DB). More than 12 actin filaments surround a myosin myofilament in an irregular or rosettelike arrangement. b) ↑ Myosin myofilaments (M) = 12 to 15-nm-thick and about 0.5-μm-long contractile protein filaments. They are not always demonstrable, and it is therefore believed that they form under conditions favorable for contraction and dissolve afterwards. Myosin myofilaments are not attached to dense bodies. Both types of myofilamets are responsible for formation of actomyosin and contraction of smooth muscle cells. 2) ↑ Intermediate microfilaments (not visible) = 10-nm-thick ↑ desmin microfilaments extending from one dense body to next, forming ↑ cytoskeleton of S. They consist of proteins and are noncontractile, thus relatively rigid. During contraction, these fibrils assure simultaneous shortening of the whole smooth muscle cell. (Fig. = rat)

↑ Smooth muscle cells, mechanism of contraction of; ↑ Smooth muscle cells, molecular biology of contraction of

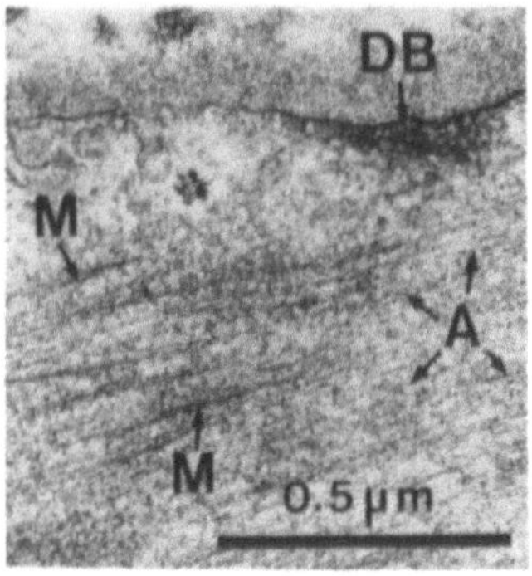

Brökelmann, J.: Verh. Anat. Ges. *75*, 603 (1981); Cooke, P.: Eur. J. Cell Biol. *27*, 55 (1982); Small, V., Sobieszek, A.: Int. Rev. Cytol. *64*, 241 (1980)

Smooth muscle cells, molecular biology of contraction of: ↑ Actin myofilaments (A) are always present in ↑ sarcoplasm and attached to ↑ dense bodies (DB). At beginning of contraction, ↑ myosin myofilaments (M) appear and actin myofilaments are pulled toward them. With progression of contraction, more and more myosin filaments are formed, so that actin filaments slide farther between them. Dense bodies approach each other and smooth muscle cell contracts (arrow). During relaxation, myosin myofilaments diminish in number and become difficult to observe. It is thought that they disaggregate reversibly in sarcoplasm between contractions. Many points of S. are still unexplained. (Modified after Panner and Honig 1970)

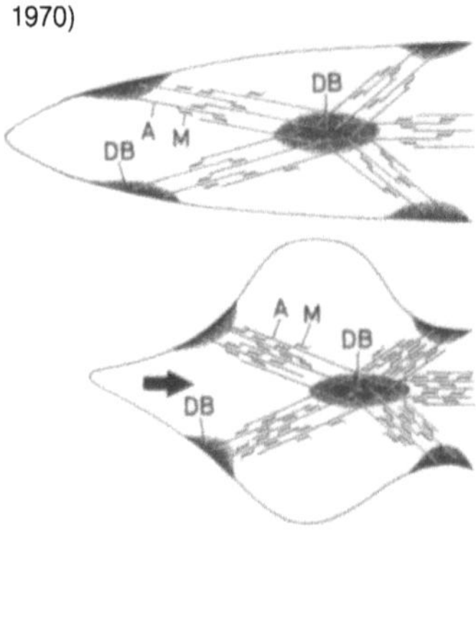

↑ Smooth muscle cells, mechanism of contraction of

Fay, F.S., Cooke, P.H.: J. Cell Biol. *56*, 399 (1973); Kelly, R.E., Rice, R.V.: J. Cell Biol. *42*, 683 (1969); Panner, B.J., Honig, C.R.: J. Cell Biol. *44*, 52 (1970)

Smooth muscle, vascular: All ↑ smooth muscle cells composing wall of blood or lymphatic vessels. Contraction of S. is regulated by impulses carried by ↑ autonomic nerve fibers; impulses do not spread between many cells. Structurally, cells of S. are identical to all other smooth muscle cells.

↑ Synapse, "by distance"

Crass, M.F., Barnes, C.D.: Vascular Smooth Muscle: Metabolic, Ionic, and Contractile Mechanism. Research in Topics and Physiology. New York: Academic Press 1982; Fujiwara, T., Uehara, Y.: Biomed. Res. *3*, 649 (1982); Fujiwara, T., Ikeuchi, M., Uehara, Y.: Biomed. Res. *4*, 225 (1983)

Smooth muscle, visceral: All ↑ smooth muscle cells forming muscular layers of wall of various organs. S. is characterized by autorhythmicity: impulses spread widely through ↑ nexus and provoke regular peristaltic contractions. Cells of S. are identical to other smooth muscle cells.

↑ Interstitial cells, of Cajal

Smooth muscular tissue (textus muscularis nonstriatus*): An aggregation of many ↑ smooth muscle cells forming the contractile portion of various organs (↑ gastrointestinal tract, ducts of some glands, ↑ bronchial tree, ↑ genitourinary system, ↑ arteries, ↑ veins, large ↑ lymphatic vessels, etc.). S. is also present in skin of ↑ penis, ↑ scrotum, ↑ areola of breast, ↑ nipple, and as ↑ arrector muscle in connection with ↑ hairs. ↑ Dilator and ↑ sphincter muscles are also S. In longitudinal sections (L) of S., nuclei (N) of smooth

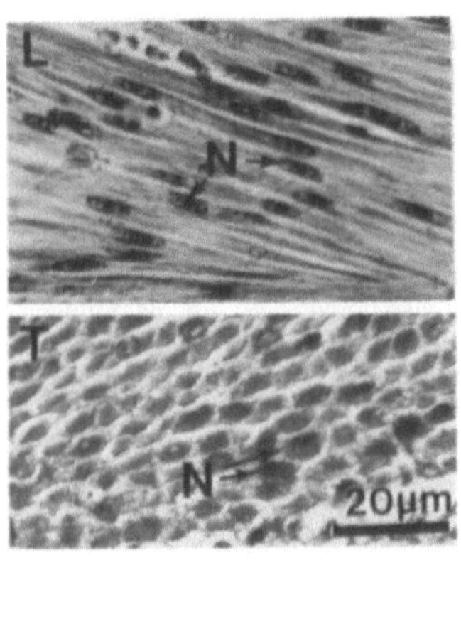

muscle cells are seen as rod-shaped; in transversal sections (T), they are round, but not seen in every cell, since in many cases they are not in path of section. In S., cells are arranged parallel, so that thick portion of one cell is juxtaposed to thin end of adjacent cells. (Figs. = ↑ jejunum, human)

Smudge cells: Dead ↑ leukocytes in blood smear with pyknotic nucleus, spongy cytoplasm, and weak stainability.

Soft palate: ↑ Palate, soft

Sole plate: Slightly elevated area of ↑ skeletal muscle fiber immediately underlying a ↑ motor end plate. ↑ Sarcoplasm of S. is particularly rich in nuclei and mitochondria.

Solitary lymphatic nodules: ↑ Lymphatic nodules, solitary

Soma (pl. somata): Term designating body of a ↑ cell; mostly used in connection with ↑ neurons.

Somatic cells: All cells of the body except ↑ germ cells. All S. have same DNA content in their nuclei (6.8×10^{-12} g per nucleus).

↑ Cells, types of

Somatic nervous system: Functional part of ↑ nervous system composed of appropriate part of ↑ central nervous system, cranial and ↑ spinal ganglia, ↑ sensory and ↑ motor neurons.

↑ Nervous system, functional parts of

Somatic neurons: ↑ Motor neurons

Somatodendritic synapse: ↑ Synapse, somatodendritic

Somatoliberin: ↑ Growth hormone-releasing factor

Somatomedins: Group of growth-promoting peptides probably synthesized in ↑ liver under action of ↑ growth hormone and possibly other factors of pars distalis of ↑ hypophysis. S. enhance sulfate incorporation into ↑ cartilage ground substance and play role in regulation of ↑ cell cycle.

Hall, K., Sara, V.: Period. Biol. *85*, Suppl. 1 (1983); Rothstein, H.: Int. Rev. Cytol. *78*, 127 (1982)

Somatosomatic synapse: ↑ Synapse, somatosomatic

Somatostatin (growth hormone-inhibiting factor, GIF, somatotropin-release inhibiting factor, SRIF): Peptide inhibiting growth hormone-releasing factor. S. is secreted by ↑ neurosecretory neurons of anterior ↑ hypothalamus, ↑ retina, ↑ C-cells of ↑ thyroid gland, ↑ D-cells of ↑ pancreatic islets, fetal large intestine, and ↑ D-cells of ↑ gastrointestinal tract.

Efendic, S., Uvnaswallensten, K.: Period. Biol. *85*, Suppl. 1 (1983); Fenoglio, C.M., King, D.W.: Human Pathol. *14*, 475 (1983); Krisch, B.: Immunocytochemistry of Neuroendocrine Systems. Prog. Histochem. Cytochem. *13*, 1 (1980); Pelletier, G.: Immunohistochemical Localization of Somatostatin, Prog. Histochem. Cytochem. *12*, 1 (1980)

Somatotropes, of hypophysis (alpha acidophil, STH-cell, cellula somatotropica*): Round or oval medium-sized acidophilic cell with spherical nucleus (N) and conspicuous nucleolus. Cytoplasm contains numerous oval mitochondria (M), well-developed ↑ Golgi apparatus (G), considerable number of long parallel cisternae of rough endoplasmic reticulum (rER) located predominantly in cell periphery, and abundance of free ribosomes. Very numerous electron-dense ↑ unit membrane-bound ↑ secretory granules (SG) of constant dimensions (300–350 nm) separate from Golgi apparatus. Their contents are discharged from cell by ↑ exocytosis (arrows). S. synthesize ↑ growth hormone; they are included in ↑ APUD-cells. BL = epithelial and capillary ↑ basal lamina, Cap = capillary

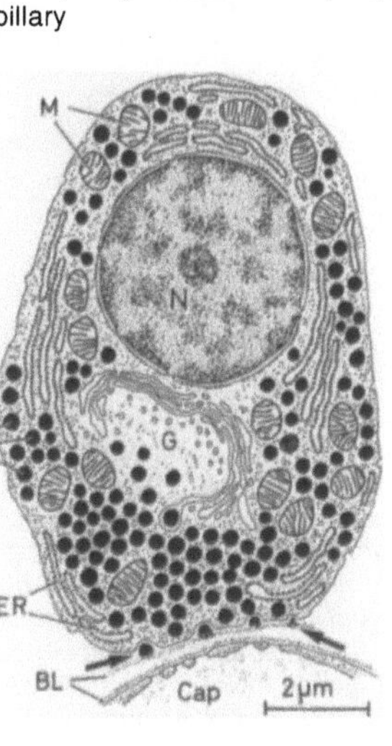

Dacheux, F.: Cell Tissue Res. *207*, 277 (1980); Takor, T.T., Pearse, A.G.E.: Histochemie *37*, 207 (1973)

Somatotropic hormone: ↑ Growth hormone

Somatotropin: ↑ Growth hormone

Somatotropin-release inhibiting factor: ↑ Somatostatin

Space, of Bowman: ↑ Bowman's space

Space, of Disse: ↑ Disse's space

Space, of Fontana: ↑ Fontana's space

Space, of Mall: ↑ Periportal tissue space

Space, of Nuel: ↑ Nuel's space

Space, of Tenon: ↑ Episcleral space

Space, pericapillary: ↑ Pericapillary space

Space, perisinusoidal: ↑ Disse's space

Space, perivascular: ↑ Perivascular space

Spatium perichoroidale: ↑ Perichoroidal space

Spatium perisinusoideum: ↑ Disse's space

Specializations, of cell surface: ↑ Epithelial cells, specializations of cell surface in

Specific endothelial organelles (Weibel-Palade bodies): Elongated, ↑ unit membrane-bound ↑ organelles (SEO), 0.3–0.6 µm long and about 0.15 µm in diameter, found in ↑ endothelium of ↑ endocardium, ↑ aorta, ↑ pulmonary arteries and veins, renal veins, dermal veins, and ↑ venules. In interior, S. contain moderately osmiophilic homogenous material, sometimes with a limited number of longitudinally oriented microtubules about 20 nm in diameter (arrowheads). Exact function of S. is unknown. They are not lysosomes, since they do not contain acid phosphatase; it is thought that S. may contain ↑ histamine and a procoagulative substance which is discharged into blood vessels under influence of ↑ epinephrine, Recently, von Willebrand protein has been demonstrated within S. (Fig. = dermal venule, mongolian gerbil)

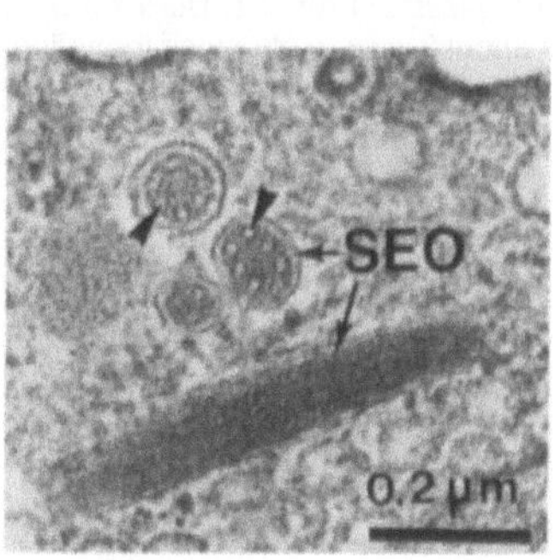

Berendsen, P.B., De Fouw, D.O.: Anat. Rec. *196*, 295 (1980); Cuevas, P., Gutierrez Diaz, J.A., Reimers, D.: Acta Anat. *114*, 303 (1982); Fujimoto, S.: Anat. Rec. *203*, 197 (1982); Heine, H., Schaeg, G., Henrich, H.: Z. mikrosk.-anat. Forsch. *95*, 617 (1981); Wagner, D.D., Olmsted, J.B., Marder, V.J.: J. Cell Biol. *95*, 355 (1982)

Specific granules, of neutrophilic granulocytes: ↑ Granulocytes, neutrophilic, granules of

Spectrin: An actin-containing filamentous polypeptide polymer attached to intramembranous particles of erythrocyte membrane.

Goodman, S.R., Shiffer, K.: Am. J. Physiol. *244*, C 121 (1983); Marchesi, V.T., Furthmayr, H., Tomita, M.: Ann. Rev. Biochem. *45*, 667 (1976); Scheven, C., Stibenz, D.: Gegenbaurs Morphol. Jahrb. *129*, 287 (1983); Williamson, P., Bateman, J., Kozarsky, K., Mattocks, K., Hermanowicz, N., Choe, H.-R., Schlegel, A.: Cell *30*, 725 (1982)

Spermatid transformation: ↑ Spermiogenesis

Spermatids (spermatidium*): Medium-sized round cells, daughter cells of secondary ↑ spermatocyte, which they closely resemble in their early stage. S. form a layer toward lumen of ↑ seminiferous tubule. Nucleus is somewhat smaller than in secondary spermatocytes; cytoplasm is pale and contains small mitochondria with vacuolized cristae, a well-developed Golgi apparatus (G) in which ↑ acrosomal vesicle (AV) may appear, appear, ↑ centrioles (Ce), some short rough endoplasmic

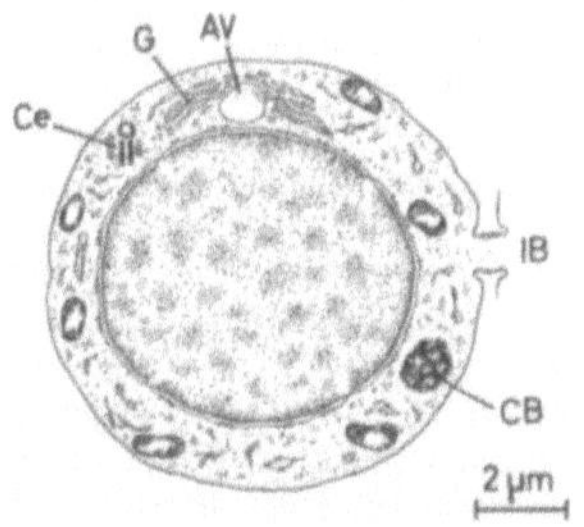

↑ cisternae, smooth endoplasmic ↑ tubules, and ↑ chromatoid body (CB). S. communicate through true ↑ intercellular bridges (IB). In course of ↑ spermiogenesis, S. differentiate into ↑ spermatozoa without any division.

↑ Spermatogenesis

Spermatic cord (funiculus spermaticus*): Long vascular strand extending from deep inguinal ring through inguinal canal into ↑ scrotum. S. includes, embedded in ↑ loose connective tissue, ↑ ductus deferens (D), arteria testicularis (At), arteria ductus deferentis (Ad), veins of plexus pampiniformis (P), musculus cremaster (C), lymphatics, and nerve plexuses.

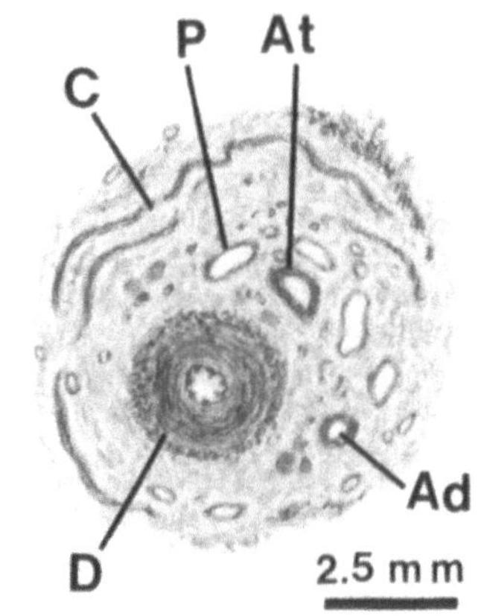

Spermatoblasts: ↑ Spermatogonia

Spermatocytes, primary (spermatocytus primarius*): Daughter cells of ↑ spermatogonia type B with initially similar structure (preleptotene S.). In the course of their ↑ differentiation, S. duplicate their DNA content (4nDNA, 2nChr), enlarge, move toward lumen of ↑ seminiferous tubule, and enter ↑ meiotic prophase. During leptotene (Lept) in nucleus, ↑ chromosomes in form of delicate filaments (F) occur. In zygotene (Zyg), chromosomes form ↑ tetrads; pairing of homologous ↑ chromosomes begins and ↑ synaptonemal complexes (SC) become visible. These complexes persist during pachytene (Pach), but in diplotene, chromosomes shorten and separate, so the complexes disappear. In ↑ sex vesicle (SV), short filament corresponds to core of ↑ Y-chromosome and longer one to ↑ X-chromosome. These two cores form a synaptonemal complex as well. Nucleus encloses a large main nucleolus (MNu) and several smaller nucleoli. Cytoplasm of S. is clear and contains, besides ↑ centrioles (Ce), peripheral mitochondria (M) with swollen cristae, a small Golgi apparatus (G) and short rough endoplasmic cisternae. S. are interconnected by true ↑ intercellular bridges (IB). Daughter cells of S., secondary ↑ spermatocytes, arise after reductional meiotic division and have a ↑ haploid number of chromosomes.

↑ Spermatogenesis

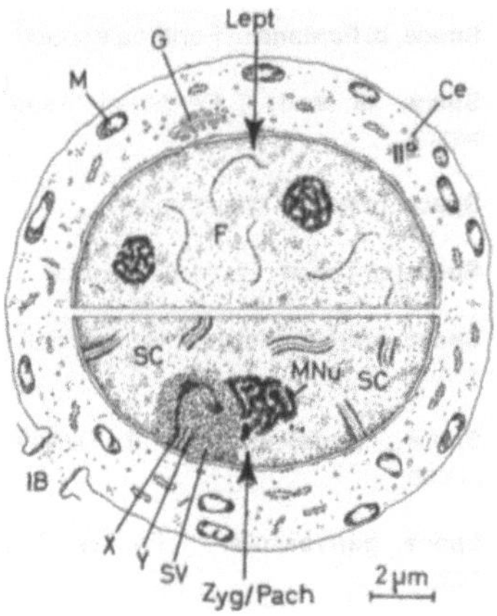

Moens, P.B.: Int. Rev. Cytol. *35*, 117 (1973)

Spermatocytes, secondary (prespermatid, spermatocytus secundarius*): Medium-sized round cells, daughter cells of primary ↑ spermatocyte situated more toward lumen of ↑ seminiferous tubule. Nucleus is round and contains 2nDNA concentrated in a ↑ haploid number of ↑ chromosomes (nChr). Chromosomes consist of two ↑ chromatids united by a ↑ centromere. In transmission electron microscope, chromosomes are visible as clumps of pale ↑ chromatin. Cytoplasm is clear and encloses small peripheral mitochondria with vacuolized cristae, an inconspicuous Golgi apparatus (G), ↑ centrioles (Ce) a few short rough endoplasmic ↑ cisternae, and smooth endoplasmic ↑ tubules. Duration of S. is about 8 h; soon after their formation,

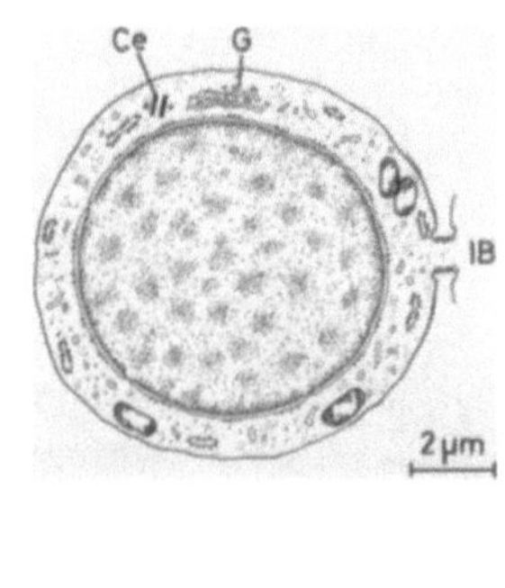

they enter into equational division without duplicating their DNA content (↑ S-phase of ↑ cell cycle does not occur). After division of centromeres, chromatids separate as in a regular ↑ mitosis. Daughter cells, ↑ spermatids, still have a haploid number of chromosomes and nDNA. S. are interconnected by true ↑ intercellular bridges (IB).

↑ Meiosis; ↑ Spermatogenesis

Spermatocytogenesis: Phase of ↑ spermatogenesis during which ↑ spermatogonia proliferate and give rise to primary ↑ spermatocytes.

Spermatogenesis: Process of formation and development of ↑ spermatozoa from ↑ spermatogonia. Phases: 1) Multiplication phase or ↑ spermatocytogenesis = proliferation of spermatogonia to primary ↑ spermatocytes. 2) Growth phase = increase in volume of primary spermatocytes and beginning of meiotic prophase. 3) Maturation phase = first (reductional) meiotic division and formation of secondary ↑ spermatocytes with

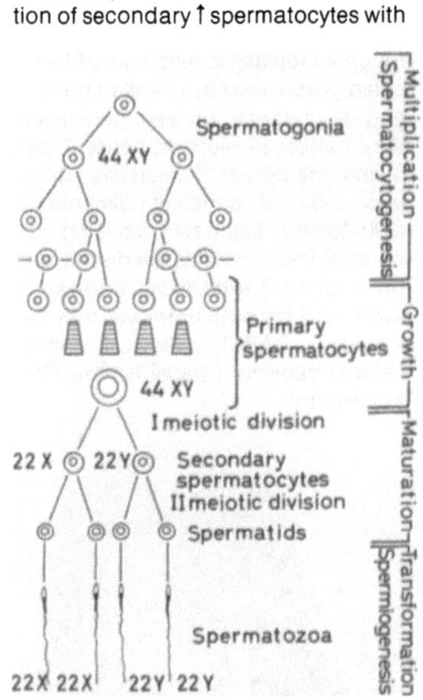

a ↑ haploid number of ↑ chromosomes, followed by second (equational) meiotic division of secondary spermatocytes into ↑ spermatids. 4) Transformation phase or ↑ spermiogenesis = gradual ↑ differentiation of spermatid into spermatozoa. Duration of S. is about 64 days, whereby pachytene stage of meiotic prophase lasts 16 days. S. begins at puberty and continues throughout life. It is temperature-dependent, a high body temperature inhibiting it; only in ↑ scrotum, where the ↑ tunica dartos maintains the temperature at 3 °C below body temperature, can ↑ testes produce spermatozoa. (Modified after Bucher 1980)

↑ Seminiferous epithelium, cycle of

Bucher, O.: Cytologie, Histologie und mikroskopische Anatomie des Menschen. 10th edn. Bern: Hans Huber 1980; Courot, M., Hochereau-de Reviers, M.-T., Ortavant, R.: Spermatogenesis. In: Johnson, A.D., Gomes, W.R. (eds.): The Testis, Vol. 1. New York, San Francisco, London: Academic Press 1970; Holstein, A.F., Roosen-Runge, E.C.: Atlas of Human Spermatogenesis. Berlin: Grosse Verlag 1983; Steinberger, E.: Physiol. Rev. *51*, 1 (1971)

Spermatogenesis, cycle of: ↑ Seminiferous epithelium, cycle of

Spermatogenic cells: Morphologically distinguishable evolutive forms of an ontogenetically unique cell type, male ↑ germ cell, from ↑ spermatogonia over ↑ spermatocytes and ↑ spermatids to ↑ spermatozoa.

Spermatogonia (spermatoblast, spermatogonium*): Two types of stem cells for ↑ spermatogenesis. 1) S. A* (A) = flattened cell (~12 µm in diameter) lying with its broad base on basal lamina (BL) of a ↑ seminiferous tubule (ST). Nucleus is ovoid or elliptical, 6–7 µm across, with finely dispersed ↑ chromatin and one or two nucleoli (Nu) applied to nuclear membrane. ↑ Karyoplasm and ↑ hyaloplasm of some S. A can be either pale (Ap) or dark (Ad). In addition, some S. Ad display a circumscribed chromatin rarefaction (R). Short mitochondria, often in small groups, are scattered in relatively sparse cytoplasm; Golgi apparatus is small; rough endoplasmic ↑ cisternae are few and short; ↑ centrioles are always present. ↑ Lubarsch crystals are occasionally found. S.A divide by ↑ mitosis; about half serve to replace themselves (stem cells), whereas the other half gradually differentiate in the course of four successive mitotic divisions, giving rise to S. B. It seems that S. Ad are stem cells, while S. Ap give rise to S. B. 2) S. B* (B) = pear-shaped cell with predominantly round nucleus containing peripherally massed chromatin; single nucleolus is in central position and accompanied by clumps of perinucleolar ↑ chromatin. Other structural features are identical to those of S.A. S.B contacts basal lamina with a narrow base. After mitotic division of S. B, daughter cells differentiate into primary ↑ spermatocytes. S. B remain in contact by means of true ↑ intercellular bridges (IB).

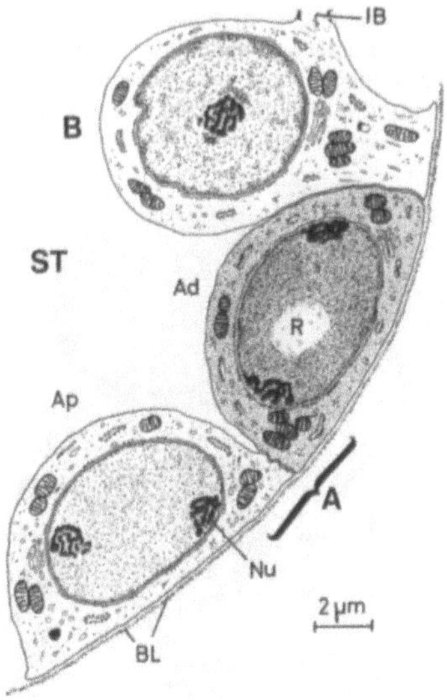

Bartmanska, J., Clermont, Y.: Cell Tissue Kinet. *16*, 135 (1983); Cairnie, A.B., Lala, P.K., Osmond, D.G. (eds.): Stem Cells of Renewing Cell Populations. New York, San Francisco, London: Academic Press 1976; Lok, D., Weenk, D.L., De Rooij, D.G.: Anat. Rec. *203*, 83 (1982)

Spermatozoon (spermatozoön*, spermium*): An approximately 60-µm-long actively motile male ↑ germ cell. S. is highly differentiated, composed of: 1) ↑ Spermatozoon head = nucleus, ↑ acrosome (A), and ↑ postacrosomal region (PR); 2) ↑ spermatozoon tail = a) ↑ spermatozoon neck, b) middle piece, c) ↑ principal piece, d) end piece.

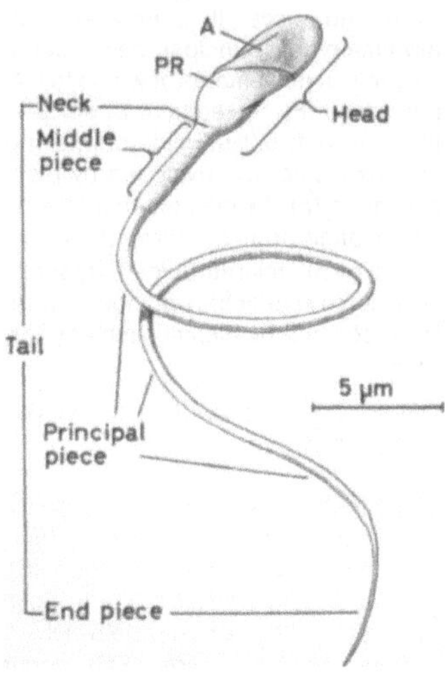

Breed, W.G.: Cell Tissue Res. *229*, 611 (1983); Fawcett, D.W., Bedford, M.J.: The Spermatozoon. München, Wien, Baltimore: Urban & Schwarzenberg 1979; Gould, K.G.: Int. Rev. Cytol. *63*, 323 (1980); Toyama, Y., Nagano, T.: Anat. Rec. *206*, 171 (1983)

Spermatozoon, connecting piece of: ↑ Spermatozoon, neck of

Spermatozoon, head of (caput*): An approximately 4 to 5-µm-long and 2 to 3.5-µm thick, anterior, hydrodynamically shaped portion of the spermatozoon. Viewed from face S. is ovoid, in profile it is pyriform. Structure: 1) Nucleus = bulk of S., containing very condensed and homogeneous ↑ chromatin except in areas of ↑ nuclear vacuoles (arrows). Nucleolus is not visible. In region covered by ↑ acrosome, nucleus is surrounded by a narrow ↑ nuclear envelope (NE) devoid of ↑ nuclear pores (NP). They exist only in postacrosomal region, where nuclear envelope becomes wider, separating from chromatin and entering ↑ spermatozoon neck. Immediately behind acrosome, outer leaflet of nuclear envelope fuses with plasmalemma (P), forming circular depression, posterior ring (R). 2) ↑ Acrosome = specialized organelle covering two-thirds of S. 3) ↑ Postacrosomal region (PR) = part of S. between acrosome and neck. S. is completed by ↑ implantation fossa (IF). Condensation of chromatin is thought to protect ↑ genome from external noxious influences. S. contains ↑ haploid number of chromosomes.

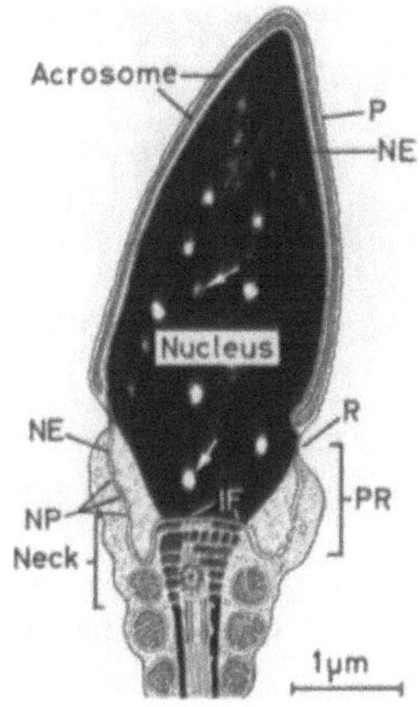

Balhorn, R.: J. Cell Biol. *93*, 298 (1982)

Spermatozoon, neck of (connecting piece, cervix*): An approximately 2-µm-long segment of ↑ spermatozoon tail immediately behind ↑ spermatozoon head (H) containing, besides some cytoplasm (Cy), capitulum

(C), segmented columns (SC), proximal ↑ centriole (PC), remnants of distal centriole (DC), beginning of ↑ axoneme (Ax), and one or two mitochondria (M). A circular fold of ↑ nuclear envelope (NE) with ↑ nuclear pores (P) may be situated within S. Structure: 1) Capitulum (capitellum) = slightly convex expanded anterior end of one of segmented columns of spermatozoon tail corresponding in shape to ↑ implantation fossa (IF), where capitulum is attached to ↑ basal plate (BP). Capitulum represents caudal articular surface between head and tail of spermatozoon. 2) Segmented columns (major columns) = cross-striated prolongations (about 1–2 µm) of outer ↑ dense fibers (ODF) toward spermatozoon head. Pattern of striation between dense bands (DB) is 66.5 nm; each dense band is composed of nine or ten transversal lines. A space roughly 15 nm wide separates dense bands. 3) Proximal centriole = nine microtubular triplets enclosed by material of segmented columns. 4) Distal centriole = only its remnant is present in mature spermatozoon. It is believed that distal centriole forms segmented columns and axoneme.

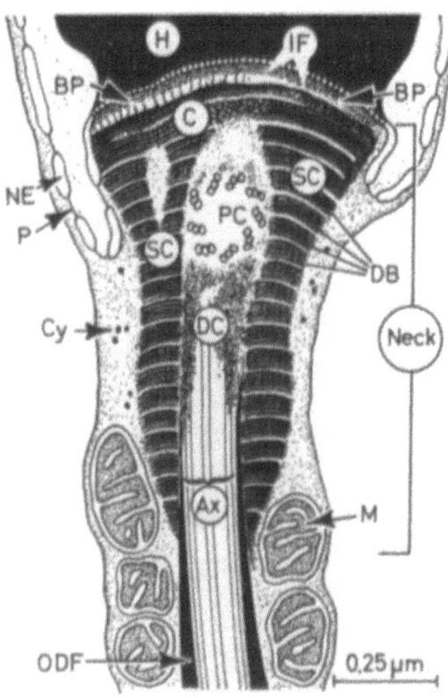

Irons, M.J.: J. Ultrastruct. Res. *82*, 27 (1983)

Spermatozoon, tail of (cauda*, flagellum*): An approximately 50 to 55-µm-long specially modified ↑ cilium, movements serving to propel ↑ spermatozoon. Structure: 1) ↑ Spermatozoon neck (N). 2) Middle piece (M) = roughly 5 to 7-µm-long by over 1-µm-broad segment following spermatozoon neck. Immediately beneath plasmalemma (p) there is a layer of helicoidally oriented mitochondria (m) arranged end to end to form ↑ mitochondrial sheath. This sheath sur-

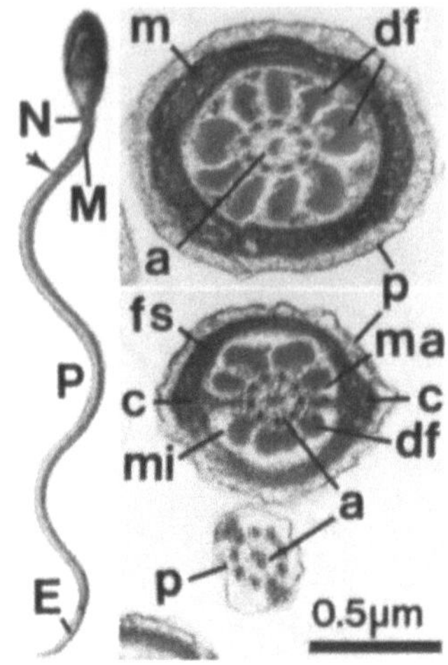

rounds nine outer ↑ dense fibers (df), concentrically disposed around ↑ axoneme (a). Mitochondria furnish energy for spermatozoon motility. Middle piece ends at level of ↑ annulus (arrow). 3) Principal piece (P) = an about 45-µm-long and 0.5-µm-thick segment located between middle and end piece. Beneath plasmalemma is ↑ fibrous sheath (fs) with columns (c) dividing principal piece into a major compartment (ma) containing four outer dense fibers (df) and a minor compartment (mi) containing three outer dense fibers concentrically arranged around axoneme (a). Principal piece tapers toward end piece. It is believed that asymmetry of principal piece could play role in movements of whole tail. 4) End piece (E) = segment about 5–7 µm long and 0.2 µm wide containing only axoneme (a) enveloped by plasmalemma (p). Axoneme terminates by dissociation of all doublets, thus consisting of 20 ↑ microtubules; they shortly become reduced in number but remain covered by plasmalemma. (Figs. = rat)

Omoto, C.K., Brokaw, C.J.: J. Cell Sci. *58*, 385 (1982)

Spermiogenesis: Four successive stages of ↑ differentiation of ↑ spermatid (Sp) into ↑ spermatozoon (S). Phases: I) Golgi phase (1,2) = in Golgi apparatus (G), glycoprotein-rich ↑ proacrosomal granules (PG) occur and fuse to give rise to an ↑ acrosomal granule (AG) within an ↑ acrosomal vesicle (AV), which becomes closely applied to outer leaflet of ↑ nuclear envelope. Newly synthesized proacrosomal granules fuse with acrosomal vesicle, contributing to its enlargement and to enlargement of acrosomal granule. Simultaneously, mitochondria (M) are displaced to cell periphery and ↑ centrioles (Ce) migrate to opposite side of acrosomal vesicle, where distal centriole (DCe) begins to form ↑ axoneme (A) of ↑ spermatozoon tail. Soon afterward, both centrioles approach cell nucleus. While distal centriole continues formation of axoneme, proximal one (PCe) attaches itself in ↑ implantation fossa (IF). ↑ Chromatoid body (CB) approaches centrioles, gradually transforming into ↑ annulus (An). II) Cap phase (3) = acrosomal vesicle spreads as thinly folded ↑ acrosomal cap (AC) over entire anterior of spermatozoon nucleus; in this area,

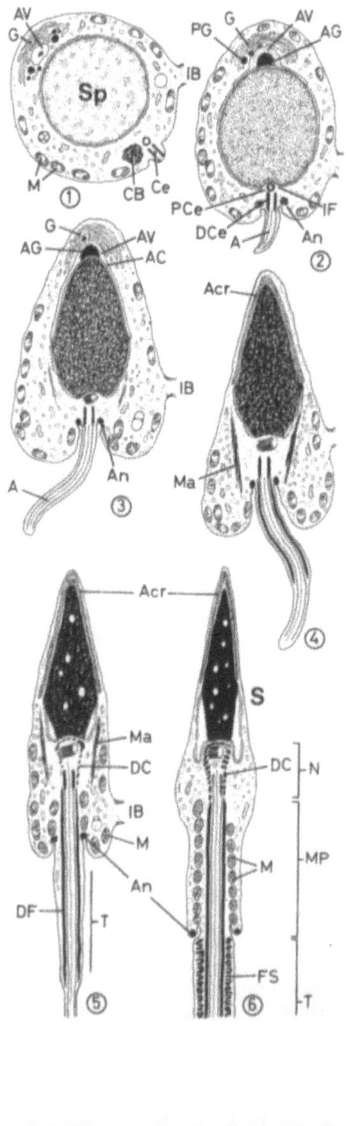

nuclear envelope becomes more dense and loses ↑ nuclear pores. III) Acrosome phase (4, 5) = substance of acrosomal vesicle gradually condenses over nucleus and constitutes ↑ acrosome (Acr); it is oriented toward base of ↑ seminiferous tubule. Nucleus elongates, ↑ chromatin condenses in small clumps. Nucleus becomes surrounded by microtubular ↑ manchette (Ma), dense columns (DC) form in ↑ spermatozoon neck (N), annulus migrates down ↑ spermatozoon tail (T), mitochondria (M) wrap around ↑ dense fibers (DF) up as far as annulus, thus forming ↑ mitochondrial sheath of middle piece (MP). Distal to annulus, fibrous sheath (FS) appears. IV) Maturation phase (6) = progression of condensation of chromatin, ↑ phagocytosis of residual cytoplasm by ↑ Sertoli's cells, release of spermatozoon into lumen of seminiferous tubule, and interruption of ↑ intercellular bridge (B).

↑ Spermatogenesis

Afzelius, B.A., Johnsonbaugh, R.E., Kim, J.-W., Plöen, L. Ritzen, E.M.: J. Submicrosc. Cytol. *14*, 627 (1982); Yasuzumi, G.: Int. Rev. Cytol. *37*, 53 (1974)

S-phase, of cell cycle: Replication phase of genetic material during ↑ cell cycle.

Spherulites: ↑ Matrix vesicles

Sphincter, inferior, of esophagus (esophagogastric sphincter): Reinforcement of circular layer of tunica muscularis at lower end of ↑ esophagus.

Sphincter, of Oddi (musculus sphincter ampullae hepatopancreaticae*): A smooth muscle investing terminal portions of ↑ ductus choledochus (C), ↑ pancreatic duct (P), and ↑ ampulla hepatopancreatica (A). Parts: 1) Musculus sphincter choledochus (SC) = ↑ smooth muscle cells surrounding choledochus prior to its junction with pancreatic duct; 2) musculus sphincter pancreaticus (SP) = smooth muscle cell layer surrounding intramural part of pancreatic duct; 3) fasciculi longitudinales (FL) = smooth muscle bundles extending from site of penetration of both choledochus and pancreatic duct to ampulla (A); 4) musculus sphincter ampullae (SA) = circular fibers surrounding ampulla. Smooth muscle cells of S. are continuous with tunica muscularis (TM) of ↑ duodenum. S. may vary greatly in form and structure among individuals. All sphincters stop flow of respective products into duodenum; generally, they relax after meals. Contraction of fasciculi longitudinales shortens ampulla and facilitates flow of ↑ bile into duodenum.

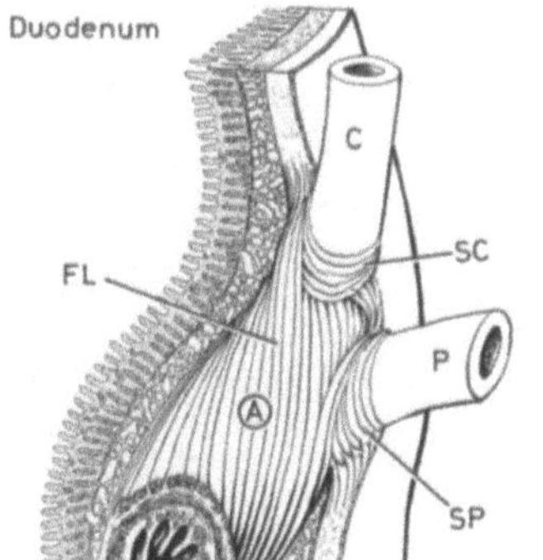

Kyösola, K., Rechard, L.: Z. mikr.-anat. Forsch. *91*, 287 (1978)

Sphincter, pharyngoesophageal: ↑ Sphincter, superior, of esophagus

Sphincter pupillae muscle (musculus sphincter pupillae*): Ring of ↑ smooth muscle cells (SP) surrounding pupillary border (arrowhead) of ↑ iris. Contraction of S. reduces pupillary diameter. S. is innervated by ↑ nonmyelinated parasympathetic fibers from ↑ postganglionic neurons of ciliary ganglion. (Fig. = pig)

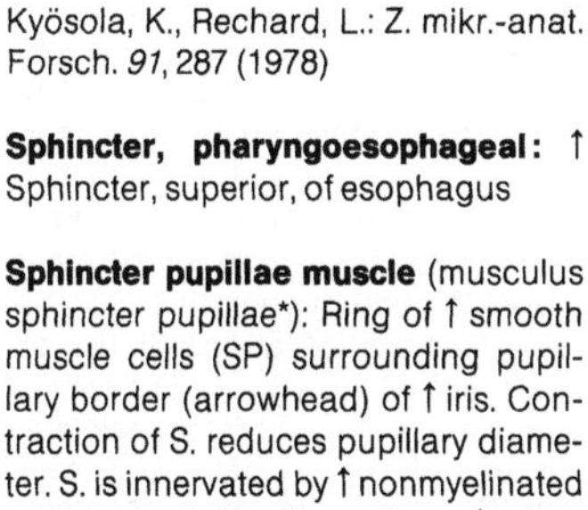

Sphincter, superior, of esophagus (pharyngoesophageal sphincter): Reinforcement of circular layer of tunica muscularis at beginning of ↑ esophagus.

Spicules, of bone: Small slivers or strands (S), of ↑ bone surrounded by single layer of ↑ osteoblasts (Ob). Adjoining S. fuse to form first ↑ trabeculae of a membranous ↑ bone. During indirect ↑ bone formation, first S. are constituted around calcified remnants of hyaline ↑ cartilage in the shape of pencil-like outgrowths.

↑ Bone formation, direct

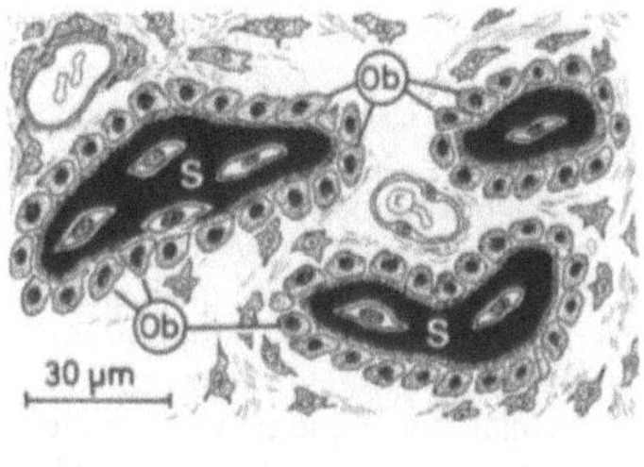

Spinal cord (medulla spinalis*): An approximately 40 to 45-cm-long and 0.7 to 1.5-cm-wide cylindrical part of ↑ central nervous system located in vertebral canal. Structure: 1) Gray matter or substantia grisea* = part of S. in which ↑ perikarya of ↑ neurons are situated. Gray matter is subdivided into anterior column (AC), posterior column (PC), and lateral column (LC); all three are united by central gray commissure (CGC) with a central canal (CC) in the middle. From basis of posterior column, reticular formation (RF), representing a zone where white and gray matters mix, arises. Dorsal extremity of posterior column displays two unclearly delimited areas; at tip of column, zona spongiosa (of Waldayer) (ZSp) overlies substantia gelatinosa of Rolando (SGel). Dorsolateral fasciculus of Lissauer (DF), a tract of white matter, is parallel to dorsal root fibers (DRF). 2) White matter or substantia alba* = ascending and descending nerve fibers grouped into anterior (AF), lateral (LF), and posterior (PF) funiculi. Posterior funiculi are separated from each other by posterior median septum (MS), anterior funiculi by anterior median fissure (AMF). Functionally, white matter is subdivided into numerous fiber tracts. ARF = anterior root fibers. (See neuroanatomy texts for further information)

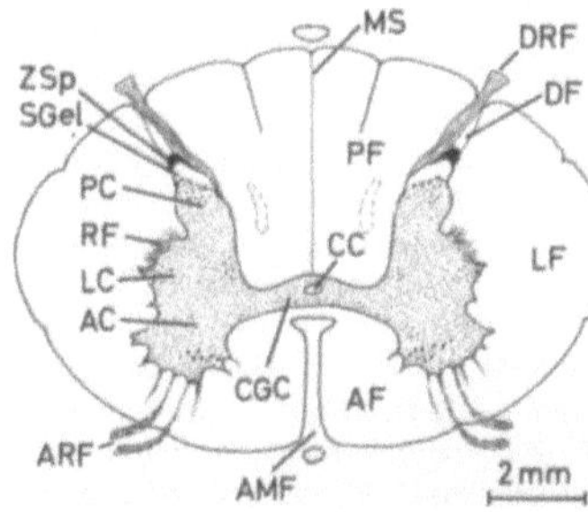

Carpenter, M.B.: Human Neuroanatomy. 7th edn. Baltimore: Williams & Wilkins Co. 1976

Spinal cord, morphology at various levels of: 1) Cervical segment = oval outline with greatest transverse diameter, large volume of gray matter, massive anterior gray columns (AC), relatively slender posterior gray columns (PC) with abundant white matter. 2) Thoracic segment = round to oval outline, small volume of gray matter, slender anterior and posterior gray columns, well-marked lateral columns (LC), relatively abundant white matter. 3) Lumbar segment = roughly circular outline with large volume of gray matter, massive anterior and posterior columns, relative predominance of gray matter over white matter. 4) Sacral segment = circular to quadrangular outline, abundance of gray matter, massive anterior and posterior columns, little white matter.

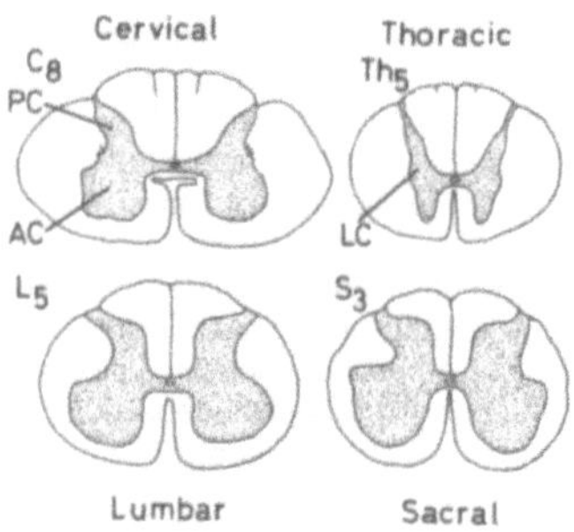

Spinal cord, neurons of: 1) Radicular neurons = ↑ neurons with axons leaving the spinal cord and forming anterior root (AR) of a peripheral ↑ nerve: a) ↑ Motor neurons (M); b) sympathetic neurons (S); c) parasympathetic neurons (P). 2) Endogenous neurons = nerve cells of which axons do not leave spinal cord: a) ↑ Association neurons (A) with ↑ perikarya situated within zona spongiosa (ZSp) and whose axons form part of fasciculi proprii (FP); b) intercalated neurons (I) with perikarya situated in substantia gelatinosa (SGel) and axons synapsing with motor neurons; c) commissural neurons (C) with axons crossing midline of gray commissure; d) neurons of nucleus dorsalis of Stilling-Clarke (ND) with axons forming tractus spinocerebellaris anterior and posterior (Tr); e) ↑ Renshaw's cells (RC). Axons of ↑ pseudounipolar neurons (PC) of ↑ spinal ganglion make complex synaptic contacts with all endogenous neurons, leading to establishment of various reflex arcs. PR = posterior root of a peripheral ↑ nerve. (See neuroanatomy and neurophysiology texts for further information)

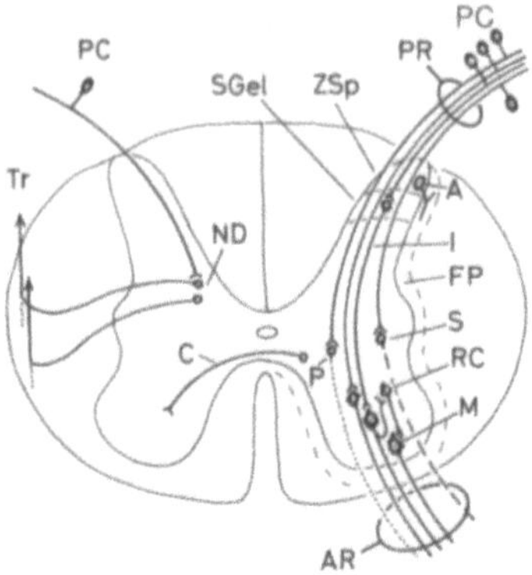

Brown, A.G.: Organization in the Spinal Cord. Berlin, Heidelberg, New York: Springer Verlag 1981

Spinal cord, nuclei of: 1) Anterior column (AC) = ↑ motor neurons associated in not always clearly delimited groups: nucleus ventromedialis, nucleus dorsomedialis, nucleus ventrolateralis, nucleus retrodorsolateralis. Extension of these nuclei is not constant in all segments of spinal cord. 2) Posterior column (PC) = ↑ interneurons forming nucleus dorsomarginalis in zona spongiosa (ZSp); nucleus dorsalis of Stilling-Clarke extends from C_8 to $L_{2/3}$. 3) Lateral column (LC) = nucleus intermediolateralis extends as sympathetic nucleus from C_8 to $L_{2/3}$ and parasympathetic nucleus from S_2 to S_4; nucleus intermediomedialis is sympathetic in thoracolumbal segment (C_8–$L_{2/3}$) and parasympathetic in sacral segment (S_2–S_4).

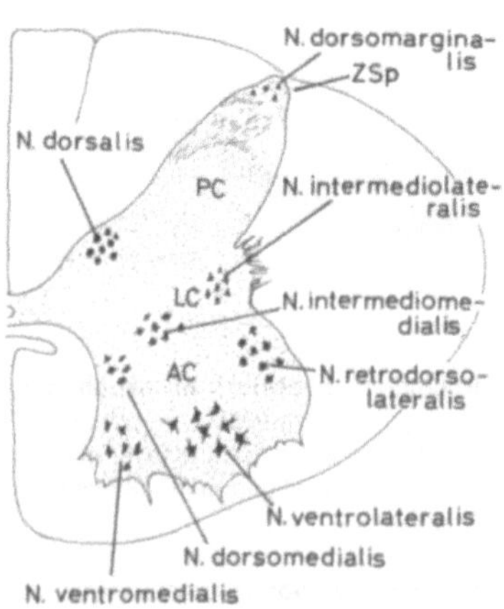

Spinal ganglion (sensory ganglion, posterior root ganglion, ganglion sensorium*): Fusiform organ (SG) about 1 cm long, incorporated within posterior root (PR) of a peripheral ↑ nerve (PN). Structure: 1) Capsule (C): a) Outer layer (OL) = ↑ dense connective tissue continuous with ↑ dura mater and ↑ epineurium; b) inner layer (IL) = concentric cellular sheaths continuous with ↑ arachnoid and ↑ perineurium. 2) Cells: a) ↑ Pseudounipolar ganglion cells (GC) largely predominate; b) small unipolar cells with nonmyelinated axons; c) association cells = unipolar ↑ interneurons connecting two ganglion cells (b+c not shown). 3) Nerve fibers (F) = bundles of ↑ myelinated and ↑ nonmyelinated nerve fibers embedded in well-vascularized ↑ endonurium (E). AR = anterior root

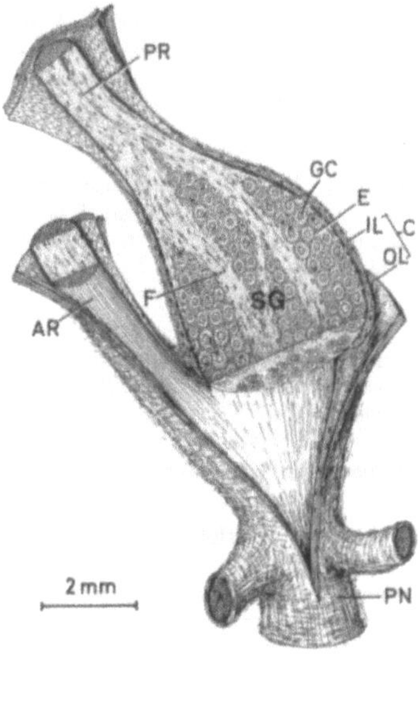

Spinal nerve: ↑ Nerve, peripheral

Spinal roots, of peripheral nerve (ramus anterior et posterior*): Two groups of ↑ nerve fiber bundles entering and leaving ↑ spinal cord at level of gray columns (GC); those from a given segment of the cord unite just beyond ↑ spinal ganglion (SG) to form peripheral ↑ nerve (PN). 1) Anterior root (AR) contains a) ↑ myelinated somatomotor nerve fibers for innervation of skeletal

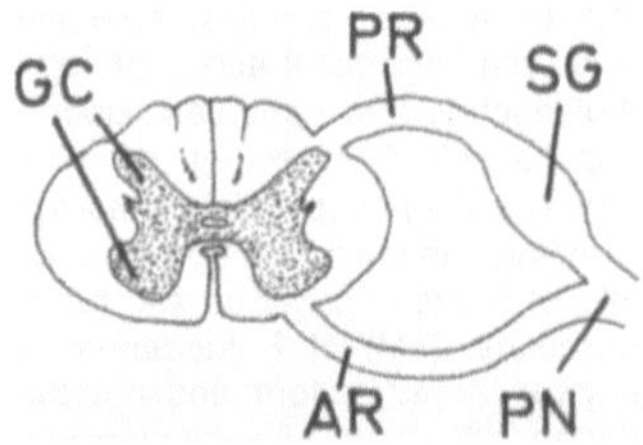

musculature; b) fine myelinated nerve fibers for motor innervation of ↑ intrafusal muscle fibers; c) thin, lightly myelinated nerve fibers of ↑ autonomic nerve system. 2) Posterior root (PR) contains a) sensory fibers emerging from ↑ pseudounipolar cells of ↑ spinal ganglia; b) afferent fibers from viscera.

Spindle apparatus (mitotic spindle, fusus mitoticus* apparatus fusalis*): Microtubular ↑ organelle formed during metaphase of dividing cells and responsible for regular distribution and precise separation of ↑ chromosomes (Ch) in ↑ meiosis or ↑ chromatids in ↑ mitosis. S. consists of: 1) Polar ↑ microtubules (PM), extending from one cell pole to other, i.e., one ↑ diplosome (D), to other; 2) kinetochoral ↑ microtubules (KM) growing from ↑ kinetochores (K), with task of grouping ↑ chromosomes at cell equator to constitute ↑ equatorial plate. During anaphase, kinetochoral microtubules slide along polar ones, due to presence of contractile actomyosin in S. and by simultaneous disassembly of ↑ tubulin dimers. Polar microtubules become still longer, elongating cell and pulling two sets of chromosomes or chromatids apart. At end of telophase, a bundle of polar microtubules persists for short time as ↑ mid-body (MB). (Slightly modified after Dustin 1980)

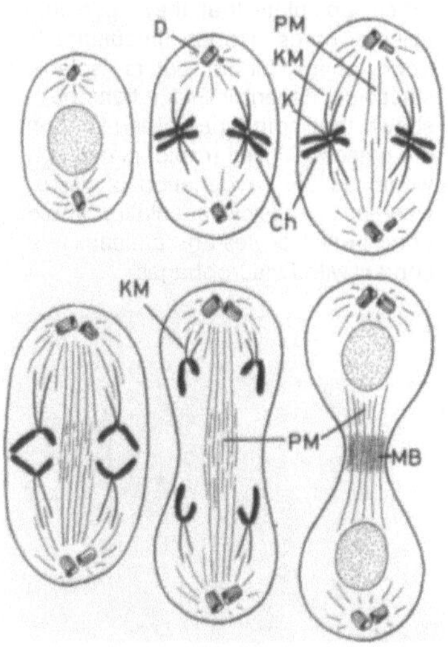

Dustin, P.: Sci. Am. *243/2*, 58 1980); Heneen, K.W., Czajkowski, J.: Biol. Cell *37*, 13 (1980); Inoué, S.: J. Cell Biol. *91*, 131s (1981); King, S.M., Hyams, J.S., Luba, A.: Eur. J. Cell Biol. *28*, 98 (1982); Paweletz, N., Finze, E.-M.: J. Ultrastruct. Res. *76*, 127 (1981); Petzelt, C.: Int. Rev. Cytol. *60*, 53 (1979)

Spindle, meiotic: ↑ Spindle apparatus

Spindle, mitotic: ↑ Spindle apparatus

Spindle, muscle: ↑ Neuromuscular spindle

Spindle, neuromuscular: ↑ Neuromuscular spindle

Spindle, neurotendinal: ↑ Golgi tendon organ

Spindles, enamel: ↑ Enamel spindles

Spine apparatus (spinula dendritica*): Special ↑ organelle (SA) situated within ↑ dendritic spine (DS). S. is composed of one or more smooth-walled flattened cisternae (C) separated from one another by dense amorphous material. Function of S. is unknown. (Fig. = ↑ cerebellum, rat)

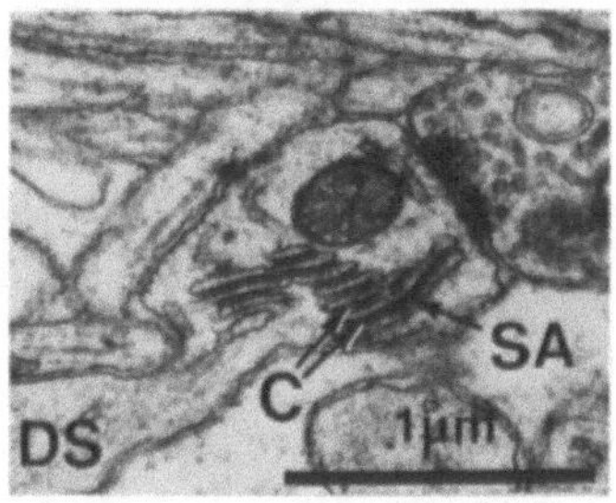

Manina, A.A.: Int. Rev. Cytol. *57*, 345 (1979); Westrum, L.E., Hugh Jones, D., Gray, E.G., Barron, J.: Cell Tissue Res. *208*, 171 (1980)

Spines, dendritic: ↑ Dendritic spines

Spinous cells, of epidermis: ↑ Epidermis; ↑ Prickle cells

Spiny bodies: ↑ Coated vesicles

Spiral arteries, of endometrium (coiled arteries, arteria spiralis*): Branches of uterine ↑ arcuate arteries for blood supply of ↑ endometrium.

↑ Endometrium, vascularization of

Spiral fibers: ↑ Corti's organ, innervation of

Spiral ganglion (cochlear ganglion, Corti's ganglion, ganglion spirale*): Spiral chain of ↑ neuron groups (SG) at base of bony spiral lamina (SL). S. is formed of true ↑ bipolar neurons, some of them enveloped in ↑ myelin sheath. Their dendrites (D) reach ↑ hair cells of ↑ Corti's organ, whereas their axons (A) leave ↑ modiolus and unite to form cochlear nerve. (Fig. = guinea pig)

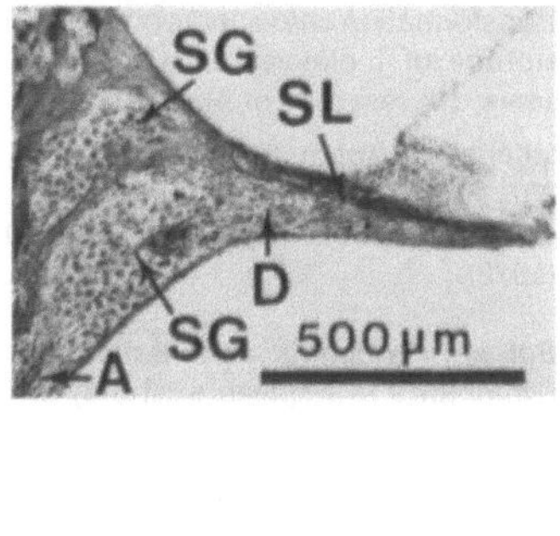

Duckert, L.G., Duvall, A.J.: Otolaryngology *86*, 434 (1978); Keithley, E.M., Feldman, M.L.: J. Comp. Neurol. *188*, 429 (1979); Merck, W., Riede, U.N., Loehler, E., Cuerten, I.: Arch. Otorhinolaryngol. *217*, 441 (1977)

Spiral ligament (ligamentum spirale*): Crescent-shaped, thickened and well-vascularized ↑ endosteum (SL) of outer cochlear limit (CL). S. begins on lateral wall of ↑ cochlear duct (CD), short distance above attachment of ↑ Reissner's membrane (RM), and extends short distance below level of ↑ basilar membrane (arrowhead), thus forming part of outer wall of both scala vestibuli (SV) and scala tympani (ST). Toward cochlear duct, S. is covered by ↑ stria vascularis (StV), below of which is situated ↑ spiral prominence (SP). S. represents outer site of attachment for basilar membrane. (Fig. = guinea pig)

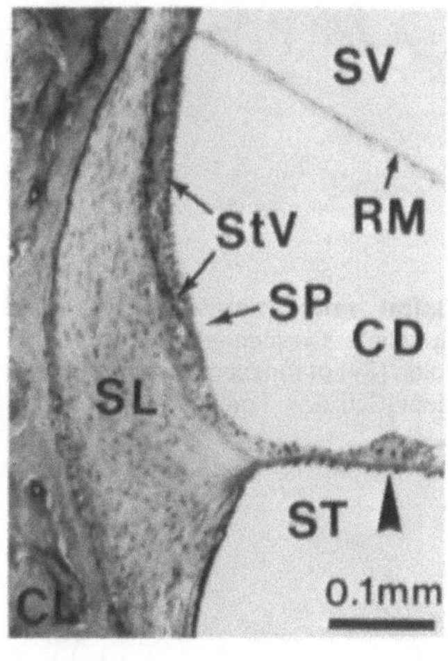

Morera, C., Dal Sasso, A., Iurato, S.: Rev. Laryngol. Otol. Rhinol. *101*, 73 (1980)

Spiral limbus (limbus laminae spiralis*): Thickened ↑ endosteum (SL) of ↑ lamina spiralis (L) protruding into ↑ cochlear duct (CD). S. is deliminated by vestibular (VL) and tympanic lips (TL), with inner spiral sulcus (asterisk) between. Majority of ↑ collagen fibers of S. are oriented vertically and form ↑ auditory teeth. Some fibers radiate from tympanic lip into pars arcuata of

↑ basilar membrane. In upper surface of S. are situated ↑ interdental cells (I), with their product, ↑ tectorial membrane (TM), extending from vestibular lip. ↑ Reissner's membrane (RM) is fixed to inner edge of S. (Fig. = guinea pig)

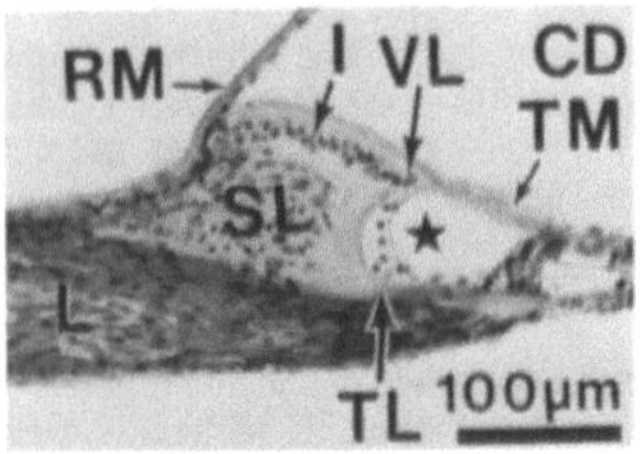

Spiral prominence (prominentia spiralis*): Highly vascularized area (SP) of ↑ spiral ligament (SL) protruding into ↑ cochlear duct (CD). S. is situated inferior to ↑ stria vascularis (SV), extending whole length of cochlear duct. S. is covered by ↑ simple squamous or cuboidal epithelium continuous with ↑ Claudius' cells (C). (Fig. = guinea pig)

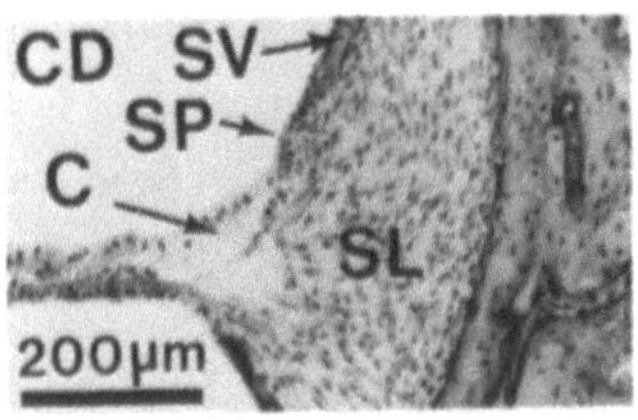

Spiral valve (Heister's valve, plica spiralis*): Crescent-shaped or spiral folds (SV) of tunica mucosa of ↑ cystic duct (CD). G = ↑ gallbladder

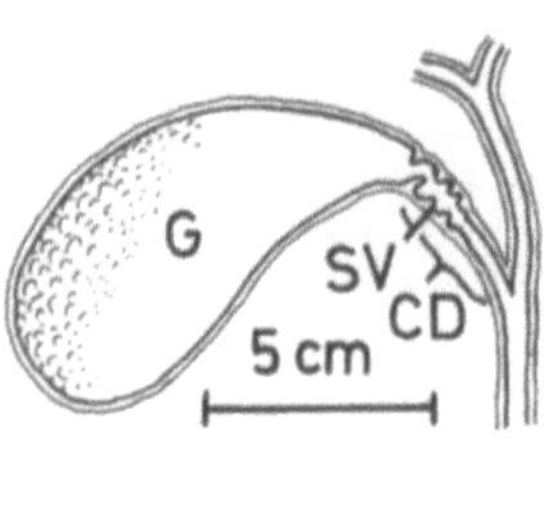

Spireme (glomus dispersum*): Coiled filamentous ↑ chromosomes occurring in ↑ nucleus during prophase of ↑ meiosis and ↑ mitosis.

Spleen (lien*, splen*): An elongated ↑ lymphatic organ weighing about 200 g situated in left hypogastrium and covered with ↑ peritoneum. Structure: 1) ↑ Capsule (C). 2) ↑ Stroma = ↑ trabeculae (Tr) emerging from ↑ hilus and inside capsule, ramifying and subdividing parenchyma into irregular communicating compartments. Trabeculae consist of moderately ↑ dense connective tissue with some ↑ smooth muscular cells, and contain ↑ trabecular arteries (TA) and veins (TV). 3) ↑ Parenchyma: a) White pulp = ↑ splenic nodules (SN) and ↑ periarterial lymphatic sheaths (PALS); b) red pulp (RP) = ↑ splenic sinusoids (SS), ↑ splenic cords (SC), and blood within both. CA = ↑ central arteries, PA = ↑ penicillar arteries

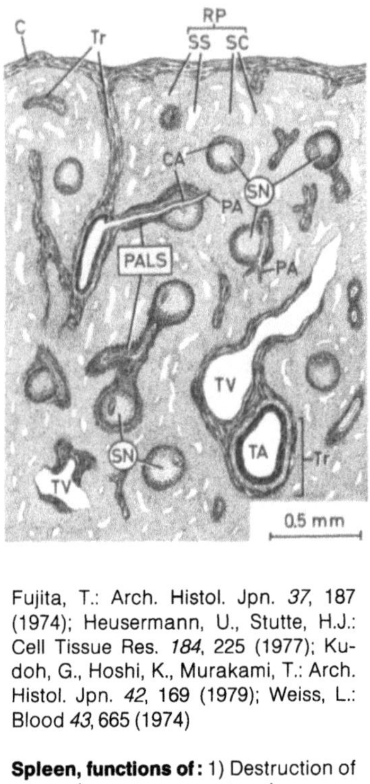

Fujita, T.: Arch. Histol. Jpn. *37*, 187 (1974); Heusermann, U., Stutte, H.J.: Cell Tissue Res. *184*, 225 (1977); Kudoh, G., Hoshi, K., Murakami, T.: Arch. Histol. Jpn. *42*, 169 (1979); Weiss, L.: Blood *43*, 665 (1974)

Spleen, functions of: 1) Destruction of aged ↑ erythrocytes by ↑ macrophages; 2) filtration of blood from particulate matter, damaged and aged cells; 3) production of ↑ B-lymphocytes; 4) trapping of blood-borne antigen; 5) production of antibodies; 6) immune defense by facilitating contact of ↑ B- and ↑ T-lymphocytes with macrophages and antigen; 7) storage of ↑ plasma cells; 8) sequestration of blood ↑ monocytes and stimulation of their transformation into macrophages; 9) storage of ↑ platelets in a ready reserve; 10) reservoir of blood.

Mitchell, J., Abbot, A.: Immunology *21*, 207 (1971); Song, S.H., Groom, A.C.: Can. J. Physiol. Pharmacol. *50*, 400 (1972)

Spleen, lymphatic circulation of: Lymphatics in ↑ spleen exist only in ↑ capsule, ↑ trabeculae, and ↑ white pulp; there are no lymphatic vessels within ↑ red pulp. At ↑ hilus, lymphatics unite to form deep lymphatics that drain into ↑ lymphatic ducts.

Spleen, vascularization of: Splenic artery branches and gives off ↑ trabecular arteries (TA) which leave ↑ trabeculae (Tr) and enter ↑ red pulp (RP), where they become surrounded by ↑ periarterial lymphatic sheath (PALS) up to ↑ splenic nodule (SN). Short arterial segment that perforates splenic nodule is ↑ central artery (CA) from which arise capillaries (C) for blood supply of nodule. At boundary between nodule and red pulp, central artery suddenly branches into two to six ↑ penicillar arteries (PA). Each of these branches into two or three sheathed ↑ capillaries (SC) that continue as simple arterial ↑ capillaries. Majority open into ↑ reticular tissue of ↑ splenic cords (SpC); rest open directly into ↑ splenic sinusoids (SS). Blood from splenic cords is collected within sinusoids. Several sinusoids join and form ↑ pulp veins (PV) that drain blood into ↑ trabecular veins (TV), from which it reaches splenic vein. Exact mode of termination of splenic arterial capillaries is still controversial: Some authors believe that they open into sinusoids ("closed circulation"); others postulate that they open into splenic cords ("open circulation"), from where blood enters splenic sinusoids. Recent investigations have shown that both types of termination are present: Closed (rapid) circulation would assure oxygenation of tissue, whereas open (slow) circulation would bring ↑ erythrocytes and antigens into contact with ↑ macrophages.

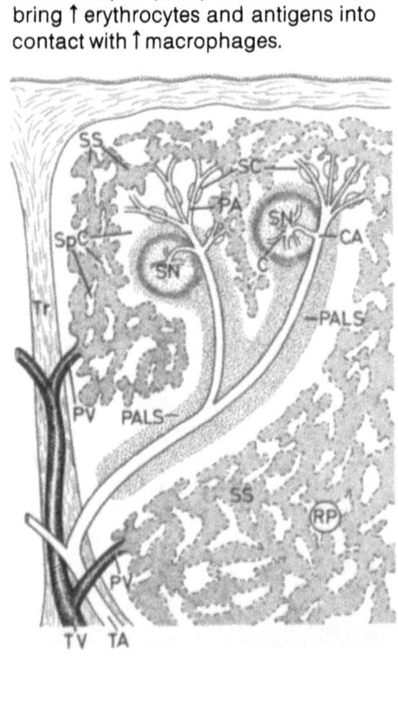

Chen, L.T.: Science *201*, 157 (1978); Irino, S., Murakami, T., Fujita, T.: Arch. Histol. Jpn. *40*, 297 (1977); Schmidt, E.E., MacDonald, I.C., Groom, A.C.: Cell Tissue Res. *225*, 543 (1982) and *228*, 33 (1983)

Splenic colonies: Small foci of clonal proliferation of a suspension of living hematopoietic cells from a nonirradiated donor injected into ↑ spleen of a lethally irradiated experimental animal. Since S. consist of all types of blood cell, they represent evidence that all types derive from one stem cell.

↑ Clone; ↑ Hematopoiesis, theories of

Splenic cords (Billroth's cords, chorda splenica*): Highly cellular strands (SC) of ↑ red pulp separating ↑ splenic sinusoids (S) from one another. S. are composed of ↑ reticular cells, ↑ macrophages, ↑ monocytes, ↑ lymphocytes, and all other mature blood cells, including numerous ↑ erythrocytes.

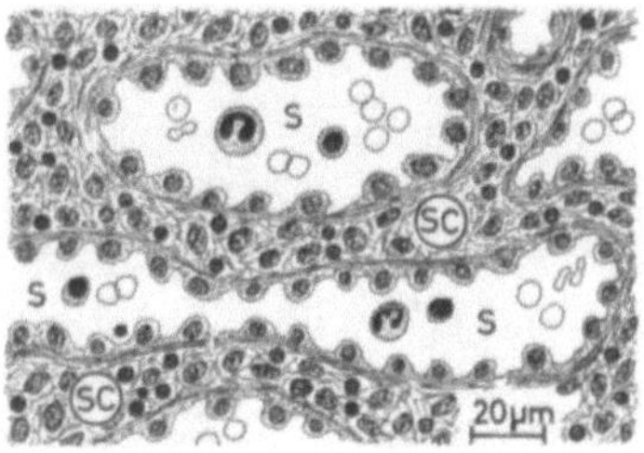

Splenic nodules (Malpighian corpuscle, lymphonodulus splenicus*): Small white spots, 0.3–0.5 mm in diameter, visible with naked eye on surface of freshly sectioned ↑ spleen. S. are spherical accumulations of ↑ lymphatic tissue around central artery (CA) scattered here and there within ↑ periarterial lymphatic sheath (PALS). Basic tissue of S. is a framework of ↑ reticular fibers and ↑ reticular cells in meshes of which lie other cellular elements. Structure: 1) ↑ Capsule (Cp). 2) Cap or mantle (C) = peripheral zone with great concentration of small ↑ B-lymphocytes; ↑ T-lymphocytes, ↑ plasma cells, and ↑ macrophages are sparse. 3) ↑ Germinal center (GC) = central clear zone composed predominantly of ↑ B-lymphocytes, B- ↑ immunoblasts, medium-sized to large lymphocytes, and cells differentiating into plasma cells. 4) ↑ Central artery (CA) = medium-sized ↑ muscular artery pushed into an eccentric position within S. Its adventitia is replaced by lymphatic tissue. Artery gives off many radial branches for blood supply of S. ↑ Erythrocytes do not leave vessels within S. but blood-borne antigens may diffuse through vessel ↑ endothelium. S. represent production sites of B-lymphocytes.

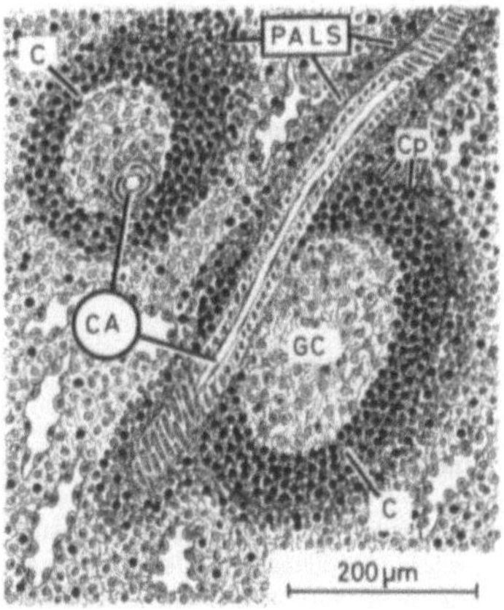

Weissman, I.L.: Transpl. Rev. *24*, 159 (1975)

Splenic nodules, capsule of: ↑ Capsule, of splenic nodules

Splenic sinusoids (sinus venosus*): Long, irregularly shaped, richly anastomosing channels (S), 12–50 μm in diameter, making up bulk of ↑ red pulp (RP). S. are lined by ↑ endothelial cells (E) that are supported outside by a system of circularly arranged ↑ reticular "ring" fibers (rf), 2–5 μm apart, and ↑ adventitial cells (A). Between endothelial cells there are 1 to 7-μm-wide interendothelial clefts (C) devoid of basal lamina. These clefts facilitate exchange of ↑ erythrocytes and ↑ wandering cells between S. and adjacent ↑ reticular tissue of red pulp. Confluence of several S. forms a ↑ pulp vein. (Fig., top = human, silver staining; fig., bottom = dog)

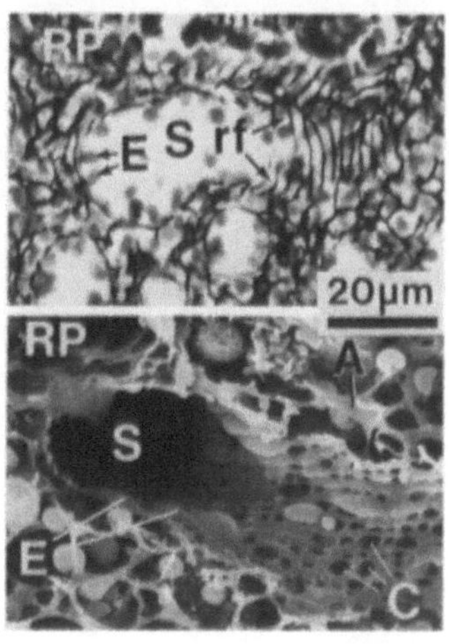

Splenic vein (vena lienalis*): Large ↑ propulsive vein with a structure identical to ↑ vena cava inferior.

Spongious body (corpus spongiosum*): Spongelike system (CS) of irregularly anastomosing veins (V) located within lamina propria (LP) of pars spongiosa of male ↑ urethra. S. also forms ↑ glans penis. Structure: Veins are almost round and increase in diameter toward the thin tunica albuginea (TA). They are characterized by presence of longitudinally oriented ↑ intimal cushions (C) of ↑ smooth muscle cells bulging into venous lumen. Trabeculae (T) between veins consist of ↑ loose connective tissue rich in ↑ elastic fibers, ↑ nerve fibers, and ↑ smooth muscle cells; arteriae helicinae (H) also run through trabeculae. In contrast to ↑ cavernous bodies of penis, blood also circulates through veins in absence of ↑ erection. During erection, venous drainage is not blocked as it is in corpora cavernosa, and S. is thus considerably less turgid, allowing ↑ semen to pass at moment of ↑ ejaculation.

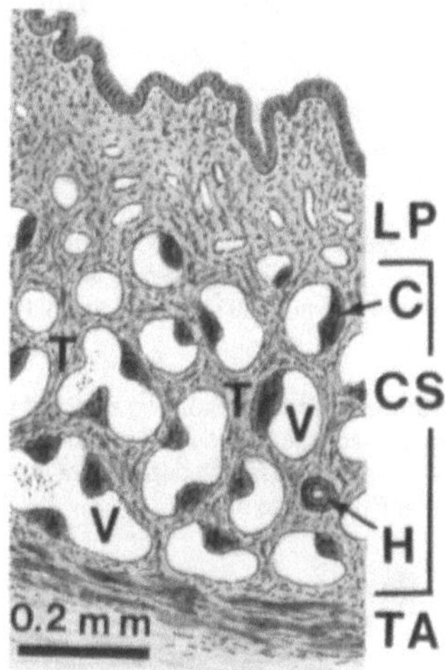

Millo, R., Fransi, A.T., Usai, E.: Arch. Ital. Anat. Embriol. *83*, 277 (1978)

Spongy bone: ↑ Bone, cancellous

Spongy portion, of urethra: ↑ Spongious body; ↑ Urethra, male

Sputtering: A technique of diffuse metallization of ↑ scanning electron microscope specimens in moderate vacuum and argon atmosphere in order to make them conductive and to increase production of secondary electrons, i.e., to increase contrast to image.

↑Shadowing

Echlin, P.: Scan. Electron Microsc. 1981/I, 79 (1981)

Squamous alveolar cells: ↑ Alveolar cells type I

Squamous epithelia: ↑ Simple squamous epithelium; ↑ Stratified squamous epithelium, keratinized; ↑ Stratified squamous epithelium, nonkeratinized

SRIF: ↑Somatostatin

SRS: ↑Slow-reacting substance

SRS-A: ↑Leukotrienes

Stab cells: ↑Band neutrophils

Stab neutrophils: ↑Band neutrophils

Staining: Procedure of conversion of image obtained by refraction of light into image obtained by absorption of light following treatment of cells and tissues with dye or combination of dyes.

Bancroft, J.D., Stevens, A. (eds.): Theory and Practice of Histological Technique, 2nd edn. London: Churchill Livingstone 1982

Staining, chemical: Any ↑ staining in which dye chemically binds with substratum. Almost all stainings are of this type.

Staining, direct: ↑Staining effected by direct contact of dye (D) with substratum (S) thanks to capacity of latter to link directly to dye.

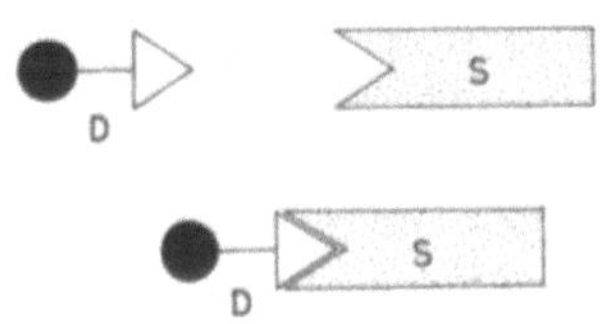

Staining, indirect: Technique of ↑ staining using a ↑ mordant.

"Staining," in transmission electron microscopy: A misnomer for ↑ contrasting. Since electrons in a ↑ transmission electron microscopy have only one wavelength, it is physically impossible to effect staining.

Staining, physical: ↑Staining in which dye colors substratum thanks to physical phenomena, e.g., solubility of dye in substratum (Sudan III, etc.). Not frequent.

Staining, progressive: Gradual treatment of substratum until desired nuance has been obtained.

Staining, regressive: Procedure during which a substratum is overstained with a dye which is then partially removed ("differentiated") with appropriate agents. This technique permits better control of ↑staining.

Staining, supravital: ↑ Supravital staining

Staining, vital: ↑Vital staining

Stapedius muscle (musculus stapedius*): Smallest ↑ striated muscle in mammals, stretching between wall of ↑ tympanic cavity and head of stapes. S. is innervated by facial nerve and influences mobility of stapes in ↑ oval window.

Burgener, J., Mayr, R.: Anat. Embryol. *161*, 65 (1980); Veggetti, A., Mascarello, F., Carpène, E.: J. Anat. *135*, 333 (1982)

Stapes: ↑Auditory ossicles

Statins: ↑Inhibiting factors

Statoliths: ↑Otoliths

Stellate cells, of adenohypophysis (type 5 cells): Cells (SC) with scanty cytoplasm around a deeply invaginated nucleus and long slender processes extending between adjacent granulated cells (GC). Cytoplasm contains moderately developed ↑ organelles and abundant microfilament bundles, but no secretory granules. Under experimental conditions provoking stimulation of granulated cells, S. become hypertrophied with well-developed organelles. They are therefore considered not only sustentacular elements, but also to be involved in supply of nutrients to granulated cells or disposal of cellular waste products. S. have been described in rabbits. (Fig. according to Shiotani 1980)

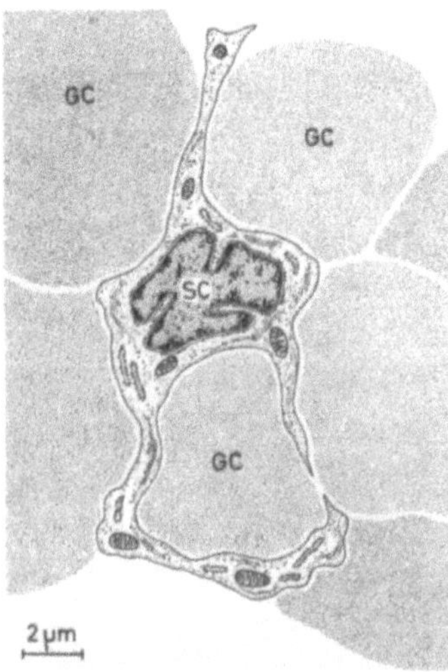

Shiotani, Y.: Cell Tissue Res. *213*, 237 (1980)

Stellate cells, of cerebellum: Small ↑ neurons (SC) included in ↑ Golgi type II neurons situated within molecular layer (ML) of ↑ cerebellar cortex; with their dendrites (D) they establish synaptic contacts with ↑ parallel fibers (PF), whereas their axons (A) branch onto dendrites of ↑ Purkinje cells (PC). Like ↑ granule cells, S. have an inhibitory action on Purkinje cells.

↑Cerebellum, neuronal circuits of

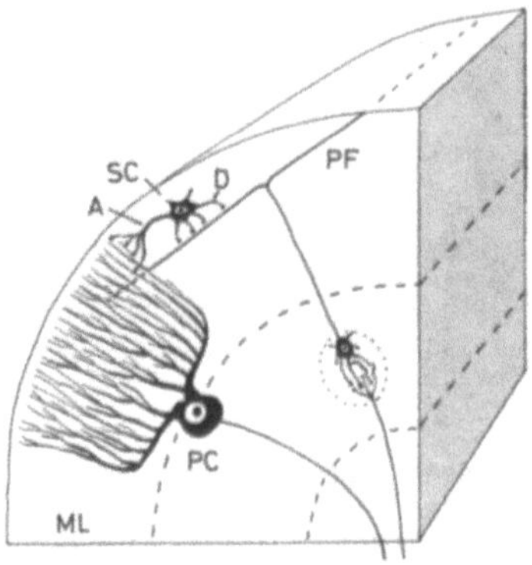

Lange, W.: Z. Zellforsch. *134*, 129 (1972)

Stellate cells, of liver: ↑ Perisinusoidal cells

Stellate cells, of pineal organ: Small cells stained with acid hematein found in rat and bovine pineal organs; it is likely that most S. are in fact ↑ interstitial cells.

Quay, W.B.: Acta Morph. Neer. Scand. *5*, 87 (1962); Wolfe, D.E.: The Epiphyseal Cell. In: Ariens Kappers, J. and Schadé, J.P. (eds.): Structure and Function of the Epiphysis Cerebri. Progress in Brain Research, Vol. 10. Amsterdam: Elsevier 1965

Stellate reticulum (enamel pulp, reticulum adamantinum*): An ↑ atypical epithelium filling ↑ enamel organ. (See embryology texts for further information)

Stem cells, hematopoietic: ↑ Hematopoietic stem cells

Stem cells, of spermatogenic epithelium: ↑ Spermatogonia

Stenon's duct: ↑ Parotid duct

Stereocilia (stereocilium*): Slender, nonmotile microvillous processes (S), 5–7 μm long and 0.1–0.2 μm thick, projected from apical surface of some epithelial cells. ↑ Hairs of ↑ hair cells are also S., with an internal filamentous core contributing to their stiffness. Epididymal S. (fig.) have no special internal structures and probably serve to increase cell surface.

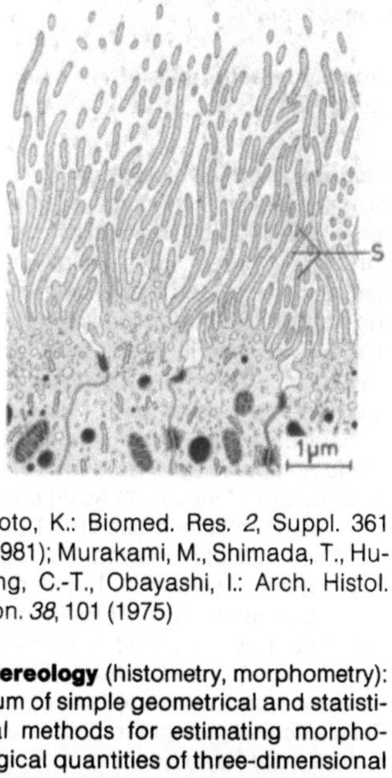

Goto, K.: Biomed. Res. *2*, Suppl. 361 (1981); Murakami, M., Shimada, T., Huang, C.-T., Obayashi, I.: Arch. Histol. Jpn. *38*, 101 (1975)

Stereology (histometry, morphometry): Sum of simple geometrical and statistical methods for estimating morphological quantities of three-dimensional structures from measurement of two-dimensional ↑ sections. Principle: An image of sectioned material is covered with or projected onto a suitable test lattice of points, lines, or areas, depending on parameters to be calculated. By point counting, one can obtain fundamental parameters: 1) Volume density or relative volume (V_V) = volume of ↑ component related to containing volume (e.g., nuclear or mitochondrial volume contained in unit volume of cytoplasm). Technique: Counting of points (P) falling on component A to be quantified. Volume density of component A expressed in percentage of unit volume is then:

$$V_{V_A} = \frac{P_A}{P_T} \cdot 100$$

where P_T is the total number of points falling on section and P_A the number of points lying component A. 2) Surface density (S_V) = surface area of component A per unit containing volume (e.g., area of endoplasmic reticulum contained in unit volume of cytoplasm). Technique: point counting of number of intersections (I) of a contour between component A and surrounding structure by lines of known length (L):

$$S_V = 2 \cdot I_L$$

3) Numerical density (N_V) = number of particles per unit of volume. For further details see references.

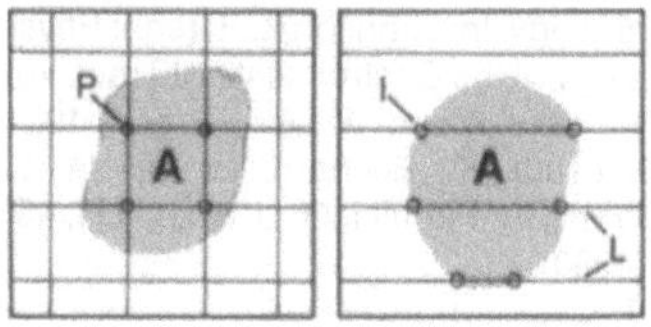

Aherne, W.A., Dunnill, M.S. Morphometry. London Edward Arnold 1982: Glauert, A.M. (ed.): Practical Methods in Electron Microscopy. Vol. 6, Part 2. Quantitative Methods in Biology. Amsterdam: Elsevier/North Holland 1980: Haug, H.: Microsc. Acta *78*, 197 (1976); Underwood, E.E.: Acta Stereol. *2*, 7 (1983); Weibel, E.R.: Stereological Methods. Vols. 1 and 2. London, New York: Academic Press 1979 and 1980; Weibel, E.R.: J. Histochem. Cytochem. *29*, 1043 (1981)

Stereoid hormone-producing cells: ↑ Endocrine cells

Steroid hormones: All ↑ hormones having a cyclopentanoperhydrophenantrene ring (steroid ring) as basic molecular structure. Precursor of S. is cholesterol, which can be incorporated by cell from blood or synthesized by it due to action of enzymes located in membranes of rough ↑ endoplasmic reticulum and in ↑ mitochondria. There is practically no S. stored in steroid hormone-producing cells, but there is normally a considerable amount of lipids, composed of triglycerides and cholesterol esters. To S. belong ↑ androgens, ↑ estrogens, and ↑ corticoids.

↑ Endocrine cells; ↑ Steroid hormones, action on the cell of

Steroid hormones, action on the cell of: A ↑ steroid hormone (H) penetrates a cell and binds with a corresponding cytoplasmic receptor (R) if one is present ("first step"); otherwise hormone leaves cell. Together with receptor, hormone enters nucleus, forming complex with a nuclear acceptor (A); this complex (C) has a great affinity for DNA, linking with it and changing mechanism of genetic transcription ("second step"). From this gene activation by the nuclear acceptor–hormone complex, there arises a new ↑ messenger RNA (mRNA) that enters cytoplasm, where it associates with ↑ ribosomes (r) and begins synthesis of enzymes which determine specific response of cells to particular steroid hormone. Evidence exists that "two step" mechanism, although not fully understood, represents most frequent manner of action of steroid hormones. Cap = capillary

↑ Polypeptide hormones, action on the cell of

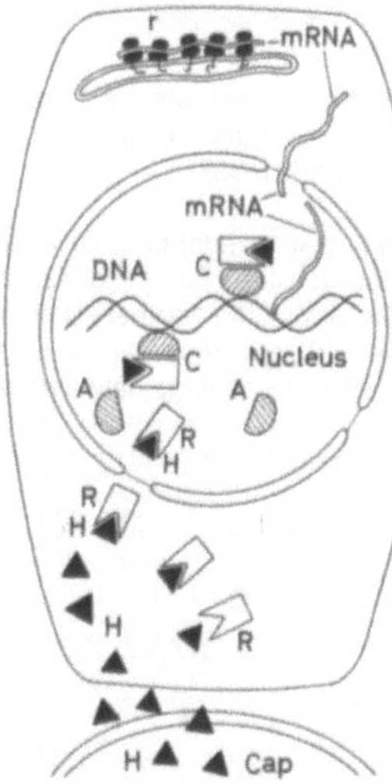

Barrack, E.R., Coffey, D.S.: The Role of the Nuclear Matrix in Steroid Hormone Action. In: Litwack, G. (ed.): Biochemical Actions of Hormones, Vol. 10. New York: Academic Press 1983; Frowein, J.: Dtsch. Med. Wochenschr. *101*, 589 (1976); Liao, S.: Int. Rev. Cytol. *41*, 87 (1975); Roy, A.K., Clark, J.H.: Gene Regulation by Steroid Hormones. Berlin, Heidelberg, New York: Springer-Verlag 1980; Stumpf, W.E., Sar, M.: Acta Histochem. Cytochem. *15*, 560 (1982)

STH: ↑ Growth hormone

STH-RF: ↑ Growth hormone-releasing factor

Stigma (macula pellucida*): Small restricted area (S) on surface of preovulatory ↑ ovarian follicle. Starting about 30 min before ↑ ovulation, blood circulation in S. slows, becoming progressively arrested; this hemostasis leads to discontinuity of ↑ germinal epithelium (E) and rapid thinning out of stratum granulosum (SG) and tunica albuginea. In last 5 min before ovulation, S. bulges as nipplelike cone above surface of follicle. (Fig. = rat)

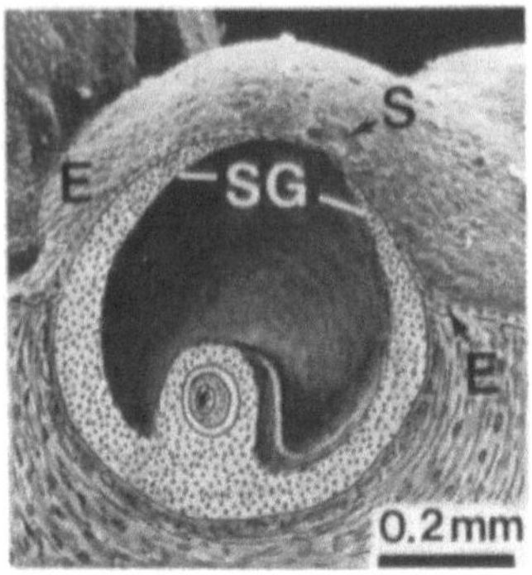

Tsujimoto, D.: Katayama, K., Tojo, S., Mizoguti, H.: Acta Obstet. Gynecol. Scand. *61*, 269 (1982)

"Stoma": An overlapping of a large cytoplasmic flap (S) at distal and proximal ends of ↑ endothelial cells (E). In silver-stained preparations, S. give impression of openings between cells.

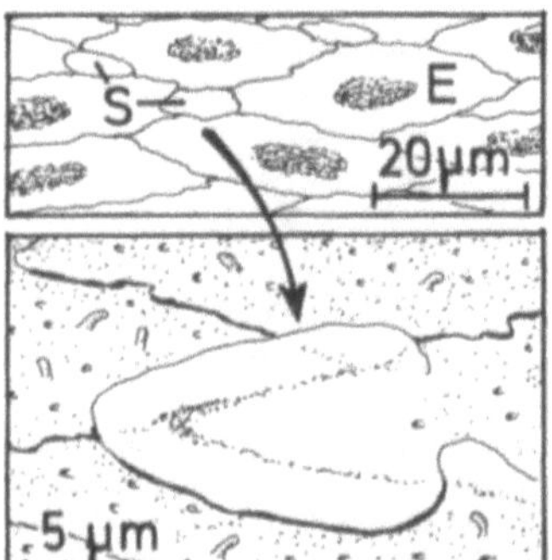

Reidy, M.A., Schwartz, S.M.: J. Ultrastruct. Res. *75*, 363 (1981)

Stomach (gaster, ventriculus*): Saccular dilatation of ↑ gastrointestinal tract situated between ↑ esophagus (E) and ↑ duodenum (D). Parts: 1) ↑ Cardia; 2) ↑fundus; 3) ↑ corpus; 4) ↑ pyloric antrum; 5) ↑ pylorus. S. is slightly curved, with lesser curvature, curvatura minor, (c) at right and greater curvature, curvatura major (C), at left. Mucosa of S. is thrown into longitudinal folds, rugae (R), which are particularly well developed along curvatures.

↑ Stomach, corpus of

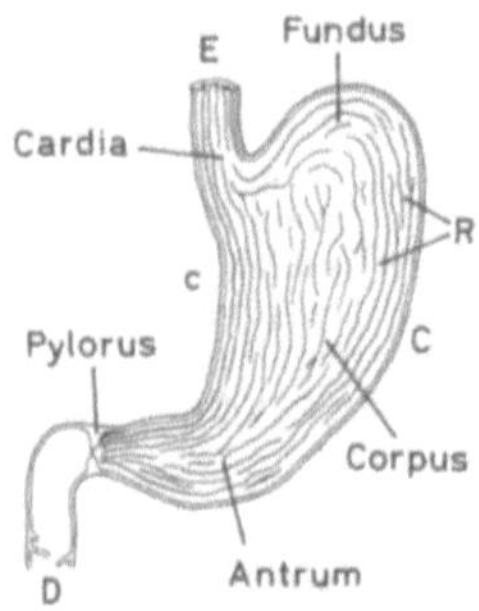

Helander, H.F.: Int. Rev. Cytol. *70*, 217 (1981); Ito, S.: Functional Gastric Morphology. In: Johnson, L.R. (ed.): Physiology of the Gastrointestinal Tract. Vol. 1. New York: Raven Press 1981; Kyösola, K., Rechard, L., Veijola, L., Waris, T., Penttilä, O.: J. Anat. *131*, 453 (1980)

Stomach, cardiac glands of: ↑ Cardiac glands

Stomach, corpus of (corpus ventriculi*): Central portion of ↑ stomach between ↑ fundus and ↑ pyloric antrum. Structure: 1) Tunica mucosa (TM) displays ↑ gastric areas (GA) and consists of: a) epithelium (E) = ↑ simple tall columnar epithelium forming ↑ foveolae gastricae (arrows) into which open ↑ gastric glands proper (GG); b) lamina propria (LP) = ↑ loose connective tissue with abundance of

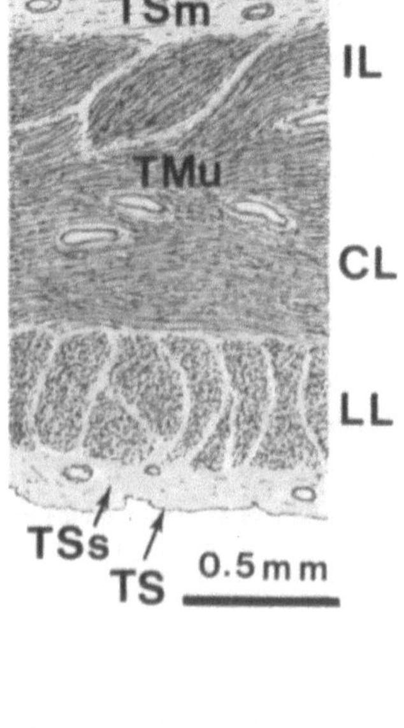

blood and ↑ lymphatic capillaries, ↑ wandering cells, and a few isolated ↑ lymphatic nodules; c) lamina muscularis mucosae (LM) = inner circular and outer longitudinal layer of ↑ smooth muscle cells. 2) Tela submucosa (TSm) = relatively dense connective tissue with numerous blood and lymphatic vessels as well as ↑ submucosal nerve plexus. 3) Tunica muscularis (TMu) = three layers of smooth muscle cells [inner layer (IL), circular layer (CL), longitudinal layer (LL)]. 4) Tela subserosa (TSs) = loose connective tissue. 5) Tunica serosa (TS) = peritoneal ↑ mesothelium.

↑ Stomach, tunica muscularis of

Stomach, fundic glands of: ↑ Gastric glands proper

Stomach, gastric pits of: ↑ Foveolae gastricae

Stomach, histophysiology of: 1) As an exocrine gland, ↑ stomach produces large volumes of hydrochloric acid (pH 2) and pepsin, both initiating digestion. It also produces ↑ gastric intrinsic factor which combines with vitamin B_{12} to permit maturation of ↑ erythrocytes. For protection of mucosa against acid gastric juice, ↑ surface mucous cells and ↑ neck mucous cells produce appreciable amount of an acid-resistant ↑ mucus. 2) As an endocrine gland, stomach secretes ↑ gastrin, ↑ somatostatin, ↑ vasoactive intestinal peptide, ↑ serotonin, ↑ histamine, and ↑ substance P. 3) Through peristaltic contractions, food filling stomach is compressed, churned, and mixed with gastric juice to form chyme. Peristaltic waves begin in fundus and move toward ↑ pylorus.

Stomach, intermediate glands of: Glands situated in a narrow zone between ↑ gastric glands proper and ↑ pyloric glands. Existence of S. in human is not generally accepted.

Stomach, protective permeability barrier of: ↑ Protective permeability barrier, of stomach

Stomach, pyloric glands of: ↑ Pyloric glands

Stomach, tunica muscularis of: 1) Outer layer = longitudinal ↑ smooth muscle cell layer arranged parallel to curvatures (c, C); this layer is continuous with longitudinal layers of ↑ esophagus (E) and ↑ duodenum (D). 2) Middle layer = thick layer of circularly arranged smooth muscle cells; this layer is continuous with corresponding layers of esophagus and duodenum. At level of ↑ cardia (Ca), middle layer is somewhat reinforced; at ↑ pylorus it forms sphincter pylori (SP). 3) Inner layer = oblique muscular bundles spreading from incisure between cardia and fundus (F) on anterior and posterior surfaces of ↑ stomach.

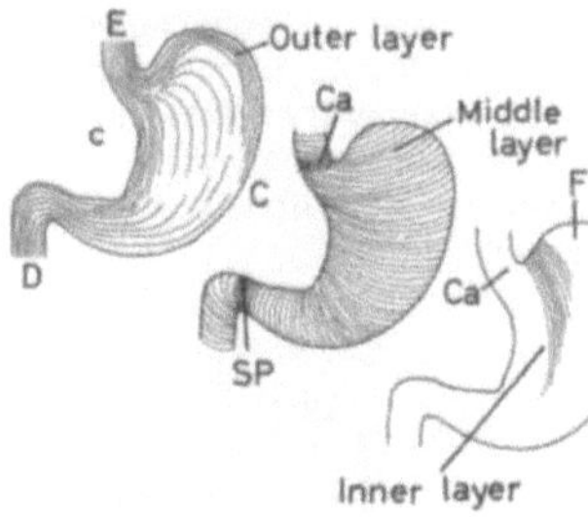

Stomach, vascularization of: Branches of gastric arteries perforate ↑ tunica serosa and ↑ tunica muscularis to form large submucosal plexus. From this plexus, small branches arise for blood supply of tunica muscularis, lamina muscularis mucosae, and tunica mucosa. Capillaries surround glands, especially ↑ foveolae. Blood is collected in fine veins that form plexus between bottoms of glands and lamina muscularis mucosae. Veins run from this plexus into ↑ tela submucosa to constitute submucosal venous plexus. Veins accompanying arteries leave submucosal plexus, pass through tunica muscularis, and leave wall of ↑ stomach.

Storage bodies: A term used occasionally to designate primary ↑ lysosomes.

Straight arteries, of endometrium: Muscular branches of ↑ coiled arteries of ↑ endometrium destined for blood supply of ↑ basalis. S. are not sensitized to cyclic action of hormones.

↑ Endometrium, vascularization of

Straight collecting tubules: ↑ Collecting tubules, of kidney

Straight portion, of proximal tubule (straight descending limb of loop of Henle, pars recta partis proximalis tubuli nephroni*): Segment of ↑ proximal tubule of ↑ nephron between proximal convoluted tubule and ↑ thin segment of ↑ loop of Henle. Cells of S. have less interdigitated borders than cells of proximal convoluted tubule; ↑ brush border (arrows) of S. is somewhat longer than that of proximal convoluted tubule, ↑ basal labyrinth is less well developed than in proximal convoluted tubule. Function of S. is same as in proximal tubule. (Fig. = rat)

↑ Proximal tubule of nephron, functions of

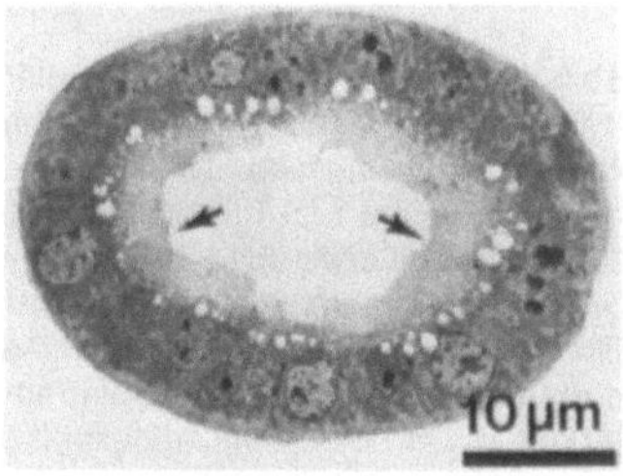

Straight portion, of proximal tubule, cells of: Cuboidal cells somewhat lower than cells of proximal convoluted tubules, and with simpler internal ultrastructural features. Lateral cell surface is thrown into fewer lateral folds than in cells of proximal convoluted tubule. Folds are infrequently branched and end as basal foot processes (BF) on ↑ basal lamina (BL). (Modified after Bulger 1965)

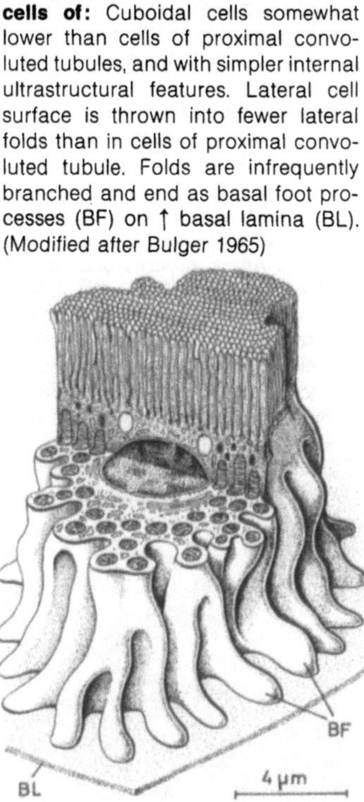

Bulger, R.E.: Am. J. Anat. *116*, 237 (1965)

Straight tubules, of testis: ↑ Tubuli recti

Strata, of epidermis: ↑ Epidermis

Stratified columnar epithelium (epithelium columnare stratificatum*): ↑ Epithelium composed of several layers of cells, among which only ↑ basal cells (BC) lie on ↑ basal lamina (BL). Superficial cells are cuboidal and/or columnar (CC). S. is rare in adult; it lines pars spongiosa of male ↑ urethra (fig.) and a segment of female ↑ urethra, ↑ stria vascularis, conjunctival ↑ fornices, and ducts of ↑ sweat glands. In general, S. is found in transitional zones between ↑ pseudostratified columnar epithelium and ↑ stratified squamous nonkeratinized epithelium. ↑ Seminiferous epithelium also belongs to S.

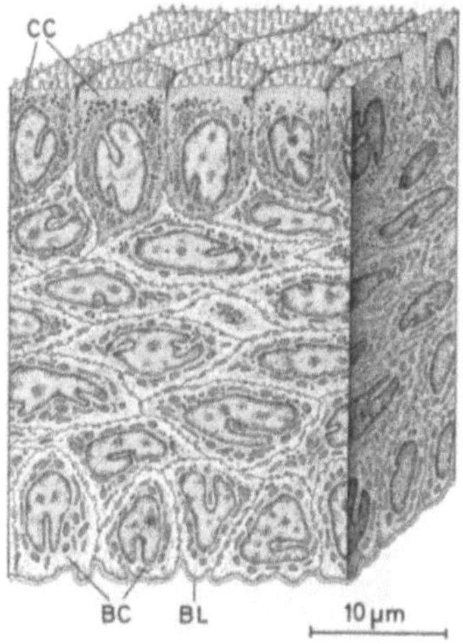

Stratified squamous epithelium, keratinized (epithelium squamosum stratificatum cornificatum*): Multilayered epithelium in which only ↑ basal cells (BC) contact ↑ basal lamina (BL). Cells overlying basal cell layer are transformed during ↑ keratinization into horny "scales" (S) which constantly exfoliate. Renewal of S. is assured by mitotic division of basal cells. Several other kinds of cells, such as ↑ melanocytes (M), ↑ Merkel's cells, ↑ Langerhans' cells, and ↑ nonmyelinated nerve fibers (NF), are scattered in S. To S. belong ↑ epidermis and epithelia lining ↑ gingiva, ↑ filiform papillae of tongue, and part of hard ↑ palate. Cap = capillary

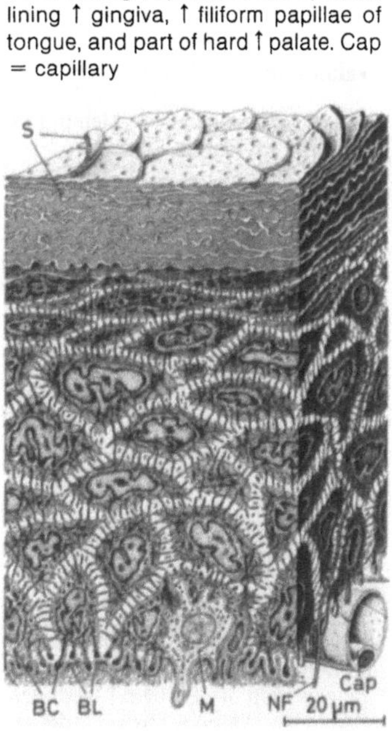

Stratified squamous epithelium, nonkeratinized (epithelium squamosum stratificatum noncornificatum*): ↑ Epithelium composed of several layers of cells in which only ↑ basal cells (BC) contact ↑ basal lamina (BL). All cells are held together by numerous ↑ desmosomes. Toward the free surface, cells are transformed into squamous "scales" (arrow) and shed; lost cells constantly being replaced by proliferating basal cells. S. is constantly moist by secretion of different glands. S. lines the major part of oral cavity, parts of ↑ pharynx and ↑ epiglottis, ↑ esophagus, ↑ vocal cords, intermediate zone of ↑ anal canal, ↑ vagina, terminal parts of male and female ↑ urethrae and anterior surface of ↑ cornea (fig.). BM = ↑ Bowman's membrane

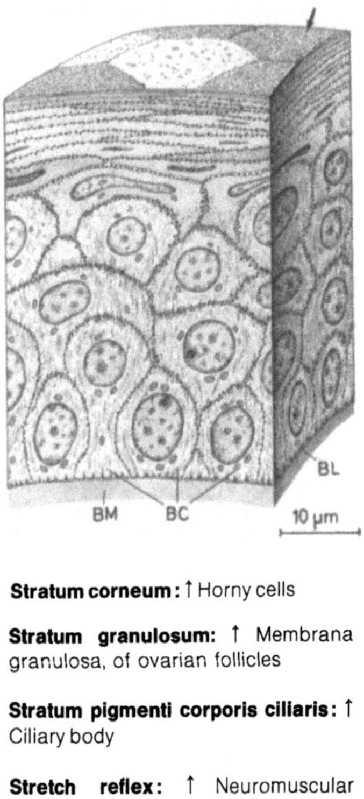

Stratum corneum: ↑ Horny cells

Stratum granulosum: ↑ Membrana granulosa, of ovarian follicles

Stratum pigmenti corporis ciliaris: ↑ Ciliary body

Stretch reflex: ↑ Neuromuscular spindle

Stress, action on neurosecretory system: A stressing nerve impulse reaches the brain where it triggers secretion of ↑ adrenocorticotropic hormone-releasing factor (CRF) which acts on ↑ corticotropes of ↑ adenohypophysis to produce ↑ ACTH. Carried by blood, ACTH acts on ↑ cells of zona fasciculata of ↑ adrenal cortex to produce ↑ corticosteroids which induce a corresponding response by tissues and organs. Corticosteroids in turn inhibit secretion of CRF by a ↑ feedback mechanism. During stress, adrenal cortex thickens and cells of zona fasciculata become hypertrophic.

↑ Adrenal glands, cortex, histophysiology of; ↑ Releasing factors

Stria vascularis*: Area of vascularized ↑ stratified columnar epithelium covering inner surface of ↑ spiral ligament. Structure: 1) ↑ Marginal cells (MC) with well-developed ↑ basal labyrinth. 2) Intermediate cells (IC) = starlike cells without contact with luminal surface. Central nucleus occupies major part of cell body. Golgi apparatus is near nucleus; mitochondria and lysosomes are scarce, as are short ↑ cisternae of rough endoplasmic reticulum and ↑ tubules of smooth endoplasmic reticulum. Branched processes of intermediate cells extend between marginal and basal cells; some processes end in vicinity of capillaries (Cap). Function unknown. 3) Basal cells (BC) = flattened irregularly shaped cells lying on basal lamina (BL) and comparable to other basal cells of various epithelia. Apical processes penetrate between marginal cells, the lateral ones intermingling with neighboring basal cells. Basal cells have a relatively voluminous nucleus, rare ↑ organelles, and some ↑ lipid droplets. Function unknown; possible stem cells. S. is only vascularized epithelium of the body; it is presumably involved in secretion of ↑ endolymph. P = ↑ pericyte

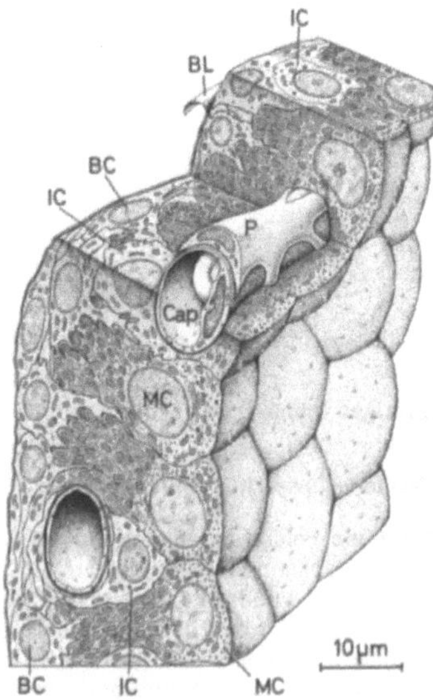

Forge, A.: Cell Tissue Res. *226*, 375 (1982); Fujimoto, S., Yamamoto, K., Hayabuchi, I., Yoshizuka, M.: Arch. Histol. Jpn. *44*, 223 (1981); Spector, G.J., Carr, C.: Laryngoscope *89*, Suppl. 1 (1979); Wright, A.: Arch. Otorhinolaryngol. *229*, 39 (1980)

Striate area (Brodmann's area 17, area striata*): Region (17) of visual brain cortex around sulcus calcarinus (SC) characterized by presence of Gennari's or Vicq d'Azyr's line (IVb) which subdivides inner granular layer (IV, arrow) into two strata (IVa) and IVc) composed of very numerous small ↑ granule cells. Some particularly large cells (Meynert's cells) lie in Gennari's line. S., as primary visual cortex, receives optic radiation from lateral geniculate body.

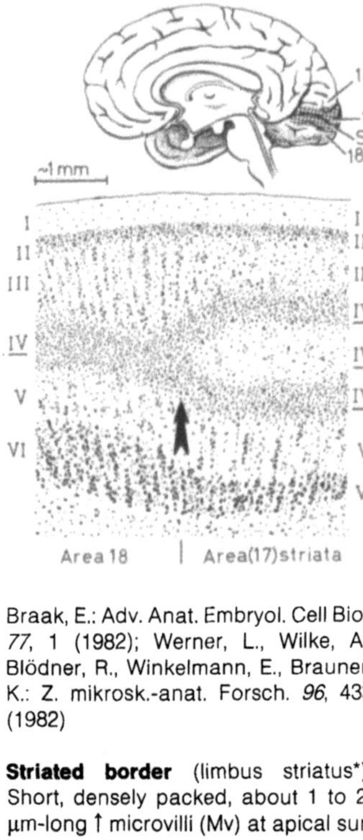

Braak, E.: Adv. Anat. Embryol. Cell Biol. *77*, 1 (1982); Werner, L., Wilke, A., Blödner, R., Winkelmann, E., Brauner, K.: Z. mikrosk.-anat. Forsch. *96*, 433 (1982)

Striated border (limbus striatus*): Short, densely packed, about 1 to 2-µm-long ↑ microvilli (Mv) at apical sur-

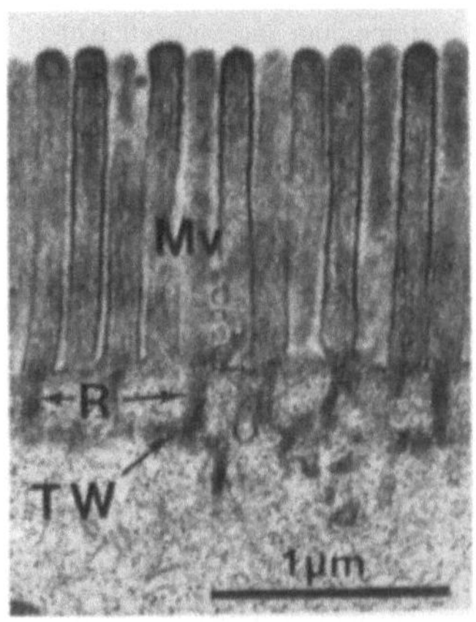

face of ↑ absorptive intestinal epithelial cells, visible as delicate striations at high light-microscopic magnification. S. greatly increases absorptive surface (in human: absorptive intestinal surface without S. = 2 m²; with S. = 200 m²). R = rootlets of microvilli, TW = ↑ terminal web (Fig. = absorptive cell, rat)

Burgess, D.R.: J. Cell Biol. *95*, 853 (1982); Chambers, C., Grey, R.: Cell Tissue Res. *204*, 387 (1979); Hirokawa, N., Tilney, L.G., Fujiwara, K., Heuser, J.E.: J. Cell Biol. *94*, 425 (1982); Matsudaira, P.T., Burgess, D.R.: Structure and Function of the Brush-Border Cytoskeleton. Cold Spring Harbor Symposia on Quantitative Biology, Vol. 46, Part 2. New York: Cold Spring Harbor Laboratory, 1982

Striated ducts (excretosecretory ducts, salivary ducts, secretory ducts, ductus striatus*): Intralobular channel system (SD) of ↑ salivary glands lined by acidophilic tall ↑ simple columnar epithelium with rod-shaped mitochondria (arrowheads) oriented vertically to base of cells, therefore "striated." S. receive ↑ intercalated ducts (IC), leave ↑ lobules, and connect with ↑ interlobular ducts. S. are most frequent in ↑ parotid gland, occurring only occasionally in ↑ sublingual gland. In adult male individuals, S. show richer ramifications than in female. (Fig. = parotid gland, human)

↑ Striated ducts, cells of

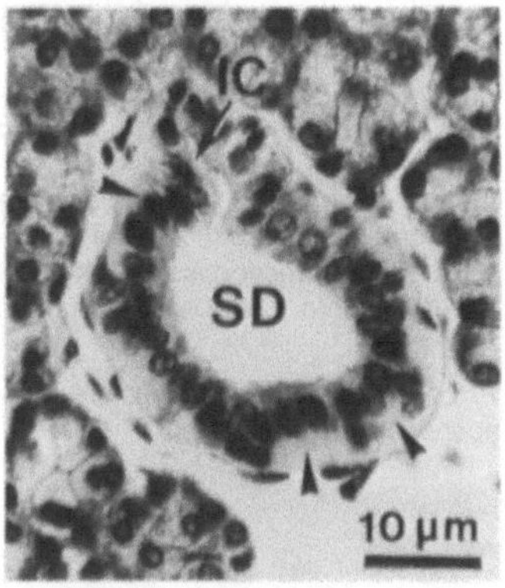

Hand, A.R.: Anat. Rec. *195*, 317 (1979)

Striated ducts, cells of: Tall columnar cells resting on a ↑ basal lamina. Nucleus (N) is round and generally central, mitochondria (M) are numerous and filamentous, arranged in columns parallel to vertical axis of cell. Between mitochondria, basal plasmalemma forms a well-developed ↑ basal labyrinth (arrows) involved in transport of water and reabsorption of sodium ions from ↑ saliva. Golgi apparatus and rough endoplasmic reticulum are inconspicuous. ↑ Apical pole is occupied by variable amount of dense osmiophilic ↑ secretory granules (S) containing appreciable quantity of ↑ kallikrein. Granules are discharged into lumen (L) of duct by ↑ exocytosis. (Fig. = mongolian gerbil)

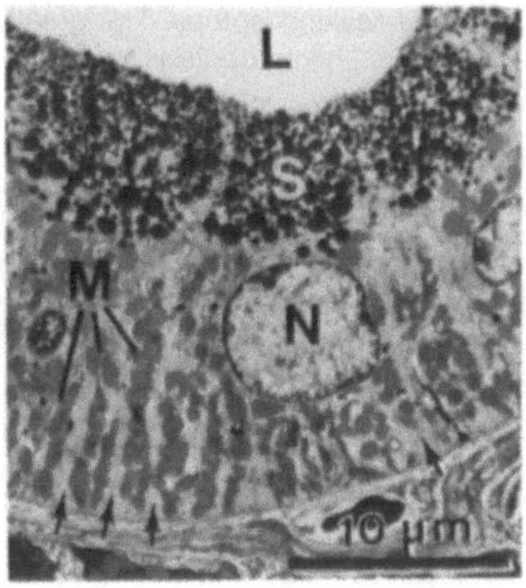

Striated muscle fibers: Morphofunctional units of ↑ striated muscles. In ↑ cardiac muscle, each S. is formed of many ↑ cardiac muscle cells, whereas ↑ skeletal muscles consist of ↑ skeletal muscle fibers.

Striated muscle fibers, molecular biology of contraction of: During relaxation, ↑ tropomyosin molecules (Tr) prevent physical contact between HMM S-1 "heads" (S-1) of ↑ myosin myofilament (My) and G monomers of ↑ actin myofilament (Ac). At start of contraction, under influence of Ca^{++}

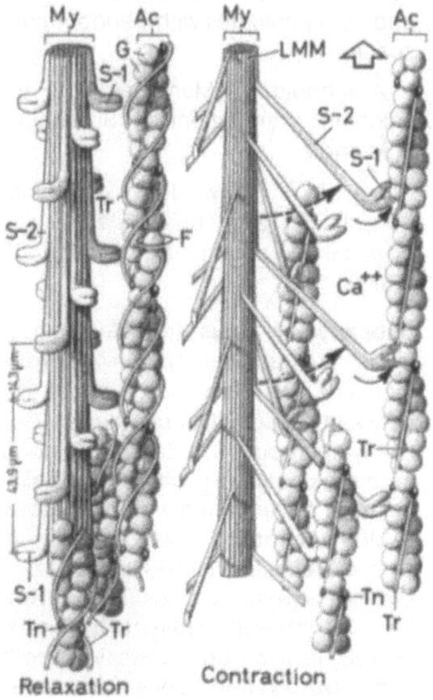

ions and ↑ troponin (Tn), tropomyosin occupies groove between twisted double strands of F-actin polymers, so that myosin "heads" contact G-actin monomers. Simultaneously, "rod" portions of myosin myofilaments, or HMM S-2 (S-2), swing out radially (large arrows) and pull actin myofilament in direction of hollow arrow. In addition, myosin "heads" may move (small arrows) over F-actin polymer like feet of caterpillar in motion, contributing to deeper sliding of actin myofilaments along myosin ones. This ↑ sliding filament hypothesis is currently widely accepted, but some points are still unresolved. LMM = ↑ light meromyosin

↑ Heavy meromyosin

Dowden, R.M., Shay, J.W. (eds.): Cell and Muscle Motility. New York: Plenum Press 1981; Ebashi, S., Maruyama, K., Endo, M. (eds.): Muscle Contraction, Its Regulatory Mechanisms. Berlin, Heidelberg, New York: Springer-Verlag 1980; Franzini-Armstrong, C., Peachey, L.D.: J. Cell Biol. *91*, 166s (1981); Huxley, H.E.: Molecular Basis of Contraction in Cross-Striated Muscles and Relevance to Motile Mechanisms in Other Cells. In: Stracher, A. (ed.): Muscle and Nonmuscle Motility, Vol. 1. New York: Academic Press (1983); Squire, J.: Structural Basis of Muscular Contraction. New York: Plenum Press 1981

Striated muscle fibers, morphology of contraction of: During contraction observed under light microscope, length of ↑ A-band remains constant, whereas ↑ I- and ↑ H-bands shorten. In transmission electron microscope, it is evident that neither ↑ myosin (My) nor ↑ actin (Ac) myofilaments shorten;

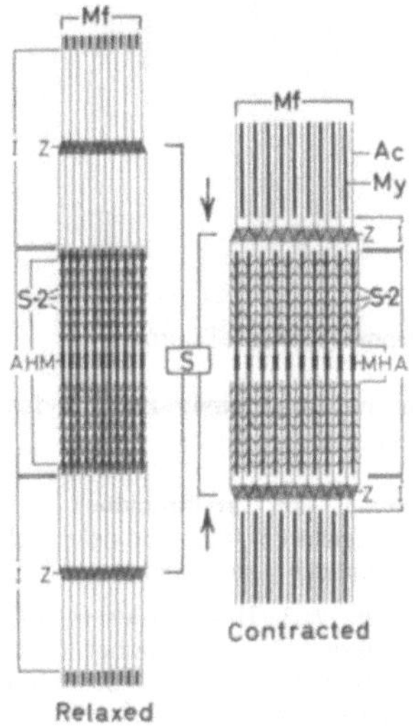

rather, actin myofilaments are pulled (arrows) toward ↑ M-band between myosin myofilaments, leading to shortening of ↑ sarcomeres (S). Through lateral pulling of myosin HMM S-2 portions (S-2), distance between myofilaments increases and ↑ myofibril (Mf) becomes wider. Strongly contracted striated muscle fibers display pronounced ↑ anisotropy under ↑ polarizing microscope. Z = ↑ Z-line

↑ Sliding filament hypothesis; ↑ Striated muscle fibers, molecular biology of contraction of

Bagshaw, C.R.: Muscle Contraction. New York: Chapman and Hall 1982

Striated muscle fibers, transversal section of: ↑ Cohnheim's fields; ↑ Sarcomere, transversal sections of

Striated muscles: Muscular tissues characterized by transverse ↑ bands. Cardiac and skeletal muscles are S.

Stripe, of Hensen: ↑ Tectorial membrane

Stroma: Network of ↑ loose or ↑ dense connective tissue serving as support for ↑ parenchyma (P) of various organs. It is organized in septa and/or ↑ trabeculae communicating with organ ↑ capsule or ↑ hilus. S. divides parenchyma into ↑ lobules, bringing blood (A, V) and lymphatic (L) vessels and nerve fibers (N) to it. (Exception: S. of ↑ iris, ↑ coliary body, and ↑ cornea does not support any parenchyma; in the latter it is avascular.)

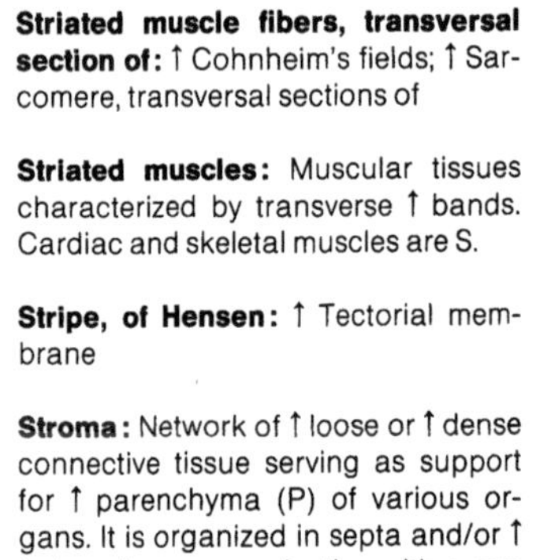

Stroma, corneal: ↑ Cornea

Stroma, of endometrium: ↑ Endometrium

Stroma, of erythrocytes: A ↑ cytoskeleton of red blood cells probably formed of ↑ stromatin.

Gratzer, W.B.: The Cytoskeleton of the Red Blood Cell. In: Stracher, A. (ed.): Muscle and Nonmuscle Motility, Vol. 2. New York: Academic Press 1983

Stromatin: Nonrespiratory protein isolated from ↑ erythrocytes. S. could form their stroma; however, no morphological evidence for this hypothesis exists.

↑ Stroma, of erythrocytes

Strontium (Sr): Chemical element accompanying calcium in ↑ hydroxyapatite; also found in pineal ↑ acervuli. As a ↑ bone-seeking isotope, ^{90}Sr greatly endangers ↑ hematopoiesis.

Subarachnoid angle: Lateral limit (SA) of ↑ subarachnoid space (SS), where ↑ spinal roots (SR) enter and leave. At level of S., arachnoid membrane (AM) is continuous with ↑ perineurium (P), and ↑ dura mater (D) with ↑ epineurium (E) of peripheral ↑ nerves (PN).

↑ Arachnoid

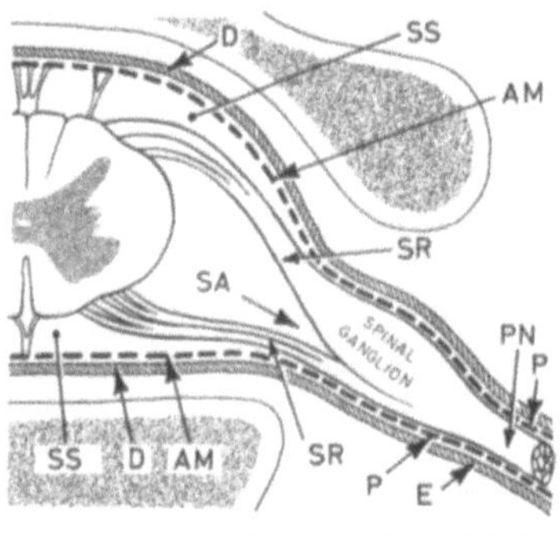

McCale, J., Low, F.N.: Anat. Rec. *164*, 15 (1969)

Subarachnoid space (cavitas subarachnoidealis*): Space between arachnoid membrane and ↑ pia mater filled with ↑ cerebrospinal fluid. S. contains some fixed and ↑ wandering cells and communicates with ↑ endoneurial space.

↑ Arachnoid; ↑ Meninges; ↑ Pacchionian granulations; ↑ Subarachnoid angle

Cloyd, M.W., Low, F.N.: J. Comp. Neurol. *153*, 325 (1974); Malloy, J.J., Low, F.N.: J. Comp. Neurol. *157*, 87 (1974)

Subcapsular sinus (marginal sinus, sinus subcapsularis*): An approximately 50 to 150-μm-large space (S) between capsule (C) and ↑ cortex (Co) of ↑ lymph node. S. is lined by ↑ littoral cells (Li) continuous with ↑ endothelial cells (E) of afferent lymphatic vessels (ALV). It contains ↑ reticular tissue, whose ↑ reticular cells (RC) and ↑ reticular fibers (R) cross sinus. ↑ Lymph enters S. through afferent lymphatic vessels (arrow) and flows into both cortex (arrow) and ↑ penetrating sinus to be filtered. ↑ Lymphocytes (L) can migrate between littoral cells into S. M = ↑ macrophage, Mo = ↑ monocyte, P = ↑ plasma cell

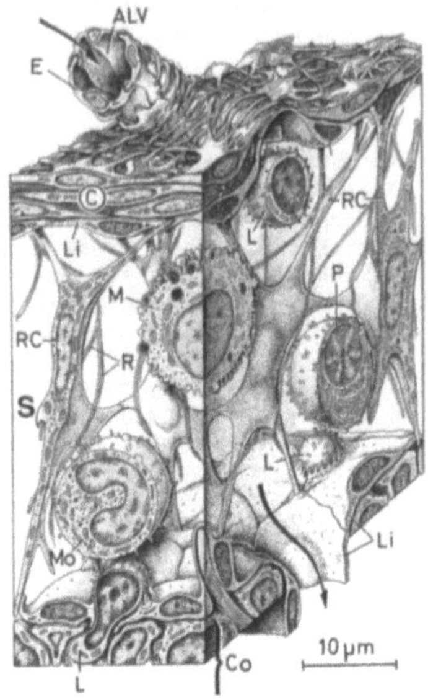

Forkert, P.-G., Thliveris, J.A., Bertalanffy, F.D.: Cell Tissue Res. *183*, 115 (1977)

Subcommissural organ (organum subcommissurale*): A unique completely ciliated ↑ circumventricular organ (SCO) situated below posterior commissure (PC) just at entrance to mesencephalic canal (arrowhead). S. is composed of tall, specially differentiated ↑ ependymal cells covering a layer of small subependymal cells. Cells of S. produce ↑ glycosaminoglycans which glide into mesencephalic canal to form ↑ Reissner's fiber. Although it is believed that S. plays role in control of glomerular zone of ↑ adrenal gland, its function is still unknown. S. is rudimentary or absent in human. (Fig. = rat)

↑ Long-spacing collagen; ↑ Subependymal layer

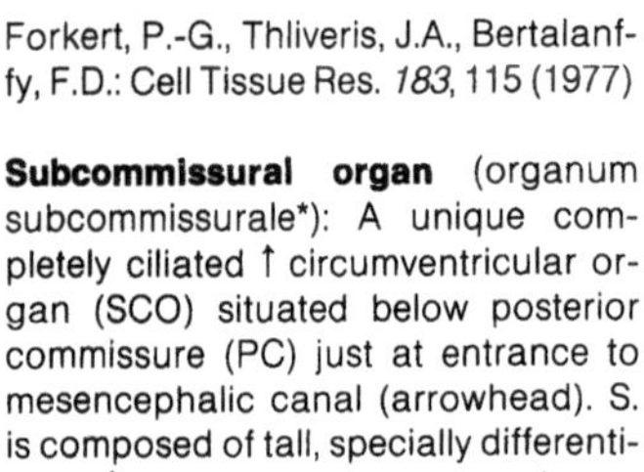

Marcinkiewicz, M., Bouchaud, C.: Biol. Cell *48*, 47 (1983); Sterba, G., Kiessig, C., Naumann, W., Petter, H., Kleim, I.: Cell Tissue Res. *226*, 427 (1982); Tulsi, R.S.: Cell Tissue Res. *232*, 637 (1983)

Subcutis: ↑ Hypodermis

Subdural space (spatium s. cavum subdurale*): Narrow serous space between ↑ dura mater and arachnoid membrane containing minimal amount of ↑ tissue fluid.

↑ Arachnoid; ↑ Meninges

Subendothelial cushions: ↑ Intimal cushions

Subependymal layer (hypendyma): Variously thick layer of small stellate ↑ neuroglial cells, ↑ myelinated and ↑ nonmyelinated nerve fibers, and occasional ↑ neurons immediately underlying ↑ ependyma. In some areas (↑ circumventricular organs), capillaries of S. are surrounded by ↑ pericapillary space; i.e., here ↑ blood-brain barrier is lacking.

Subfornical organ (SFO, organum subfornicale*): ↑ Circumventricular organ situated on roof of third ventricle near interventricular foramina. S. is lined by a ↑ symple squamous to cuboidal epithelium which is ciliated only at periphery of organ. Within S. are ↑ neurons, numerous ↑ neuroglial cells, and capillaries with large ↑ pericapillary spaces, representing areas free of ↑ blood-brain barrier. Function of S. is unknown; it seems to be involved in hydrodynamic homeostasis and in control and secretion of ↑ luteinizing hormone-releasing factor and ↑ somatostatin. (Fig. = mongolian gerbil; courtesy Biomedical Research)

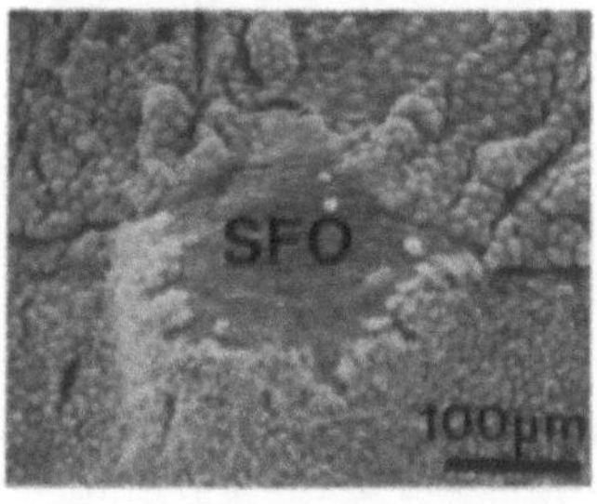

Dellmann, H.D., Simpson, J.B.: Int. Rev. Cytol. *58*, 333 (1979); Krstić, R.: Biomed. Res. *2*, Suppl. 129 (1981); Simpson, J.B.: Int. Rev. Cytol. *58*, 333 (1979)

Sublingual glands (glandula sublingualis*): Paired ↑ mixed ↑ tubuloacinar major ↑ salivary glands with a predominance of ↑ mucous tubules (MT). Structure: 1) Capsule (C); 2) ↑ stroma = connective septa (S) that divide parenchyma into lobules (L); 3) ↑ parenchyma = mucous tubules with and without serous ↑ demilunes; ↑ acini as well as ↑ intercalated and ↑ striated ducts (SD) are rare. ↑ Myoepithelial cells are present around secretory end pieces. After leaving lobule, striated ducts join to form interlobular ducts (ID); several interlobular ducts form an excretory duct. S. produce mucous component of ↑ saliva. (Figs. = human)

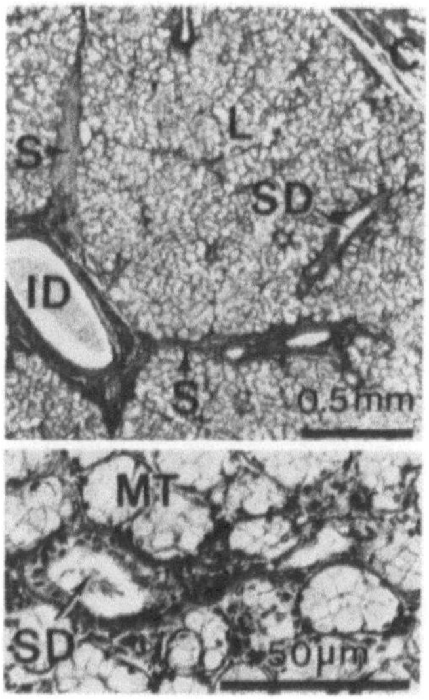

Sublobular veins: ↑ Intercalated veins

Submandibular duct (Wharton's duct, ductus glandulae submandibularis*): Main excretory canal of a ↑ submandibular gland, opening at floor of mouth. At its beginning, S. is lined with tall ↑ simple columnar epithelium which becomes ↑ pseudostratified in terminal portion. Moderately dense lamina propria with some ↑ elastic fibers surrounds epithelium.

↑ Ducts, of salivary glands proper

Sato, A.: Biol. Cell *39*, 237 (1980)

Submandibular glands (submaxillary gland, glandula submandibularis*): Paired ↑ mixed ↑ tubuloacinar major ↑ salivary glands with a predominance of serous elements. Structure: 1) Capsule (C); 2) ↑ stroma = connective septa (S) that divide parenchyma into ↑ lobules (L); 3) ↑ parenchyma = ↑ acini (A) predominate in parenchyma; they are continuous with ↑ intercalated ducts (ID); numerous ↑ striated ducts (SD). A minor part of the parenchyma is composed of ↑ mucous tubules (MT) with frequent serous ↑ demilunes; presence of ↑ myoepithelial cells around secretory end pieces. Striated ducts form interlobular ducts (ILD) which unite to constitute the principal excretory canal (↑ submandibular duct). S. produces mostly serous ↑ saliva and synthesizes over 20 biologically active peptides, among which is ↑ nerve growth factor. (Fig. = human)

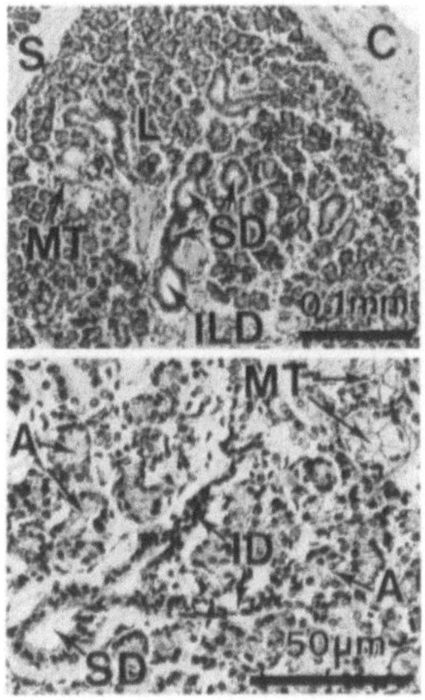

Barka, T.: J. Histochem. Cytochem. *28*, 836 (1980); Gibson, L.H.M.: Anat. Anz. *153*, 91 (1983); Murphy, R.A., Watson, A.Y., Metz, J., Forssmann, G.W.: J. Histochem. Cytochem. *28*, 890 (1980); Tandler, B.: J. Submicr. Cytol. *15*, 519 (1983); Testa Riva, F.: J. Submicr. Cytol. *9*, 251 (1977)

Submaxillary glands: ↑ Submandibular glands

Submucosa: ↑ Tela submucosa

Submucosal glands, of gastrointestinal tract: ↑ Esophageal glands proper; ↑ Duodenal glands

Submucosal plexus (Meissner's plexus, plexus submucosus*): Intrinsic, richly branched nervous network situated within ↑ tela submucosa of ↑ gastrointestinal tract. S. is composed of small ↑ multipolar neurons similar to type I ↑ autonomic neurons of sympathetic ganglia and a large numbers of ↑ nonmyelinated nerve fibers with bulbous endings filled with ↑ dense core ↑ synaptic vesicles. S. receives postganglionic sympathetic nerve fibers from ↑ sympathetic trunk. S. innervates ↑ lamina muscularis mucosae, ↑ lamina propria, epithelium, tela submucosa and its vessels, as well as glands and ↑ lymphatic tissue.

↑ Intestine, innervation of; ↑ Myenteric plexus

Gershon, M.D.: Ann. Rev. Neurosci. *4*, 227 (1981)

Submucosal vessels, of endometrium: Term sometimes used to designate radial branches of uterine ↑ arcuate arteries.

↑ Endometrium, vascularization of

Subneural apparatus (plicae postsynapticae*): Totality (S) secondary ↑ synaptic clefts (SC) formed by deep infoldings of ↑ sarcolemma of ↑ synaptic trough, increasing contact surface between ↑ axon (A) and ↑ skeletal muscle fiber (MF). S. is filled with a ↑ glycoprotein material continuous with basal lamina (BL) of muscle fiber. C = ↑ synaptic cleft, of motor end plate

↑ Motor end plate

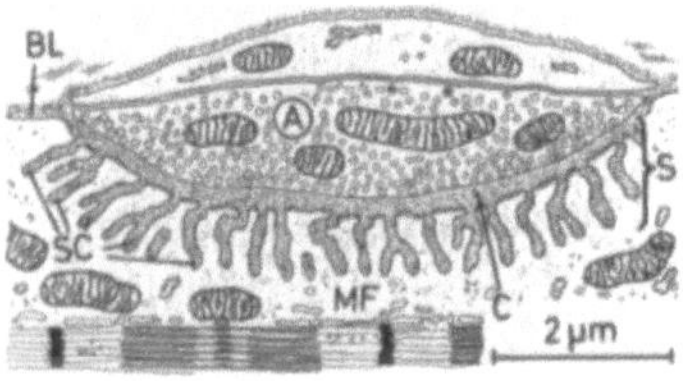

Subneural cleft: ↑ Subneural apparatus; ↑ Synaptic cleft, of motor end plate

Subperitoneal musculature, of oviduct: ↑ Oviduct

Subpial feet (processus pialis*): Flattened end expansions (SF) of ↑ astrocytes lying immediately below ↑ pia mater (PM). S. are held together by ↑ nexus (N) and are covered by a basal lamina (BL). All S. participate in formation of ↑ membrana limitans gliae superficialis (M).

↑ Brain, penetration of blood vessels into; ↑ Perivascular feet

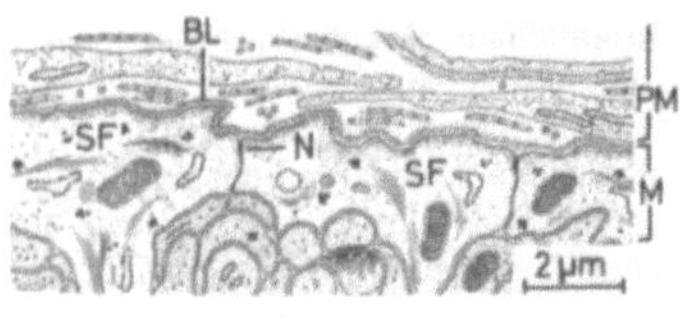

Wagner, H.-J., Barthel, J., Pilgrim, C.: Anat. Embryol. *166*, 427 (1983)

Subretinal space: ↑ Proteoglycan-containing cleft between ↑ pigment epithelium and ↑ photoreceptors. S. is filled with numerous processes of pigment cells that assure contact between pigment epithelium and outer segments of cone and rod cells.

Adler, A.J., Klucznik, K.M.: Curr. Eye Res. *1/10*, 579 (1982)

Subsarcolemmal cisternae: Cisternae (SC) of ↑ sarcoplasmic reticulum immediately underlying ↑ sarcolemma (S). S. represent a category of ↑ subsurface cisternae. (Fig. = rat)

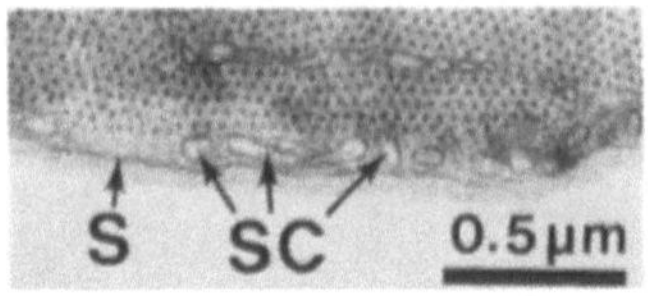

Substance P: Undecapeptide present in minute quantities in various regions of ↑ central nervous system (↑ hypothalamus, substantia nigra), ↑ retina, ↑ sensory ganglia, and intestine of man and some animals. S. is a powerful smooth muscle stimulating factor, provoking contraction of intestine; it is also a hypotensive factor for dilatation of blood vessels. S. is probably involved in transmission of pain and analgesic impulses. At level of ↑ synapses, S. influences replacement of ↑ neurotransmitter and amplifies its action on postsynaptic membrane by an increase in ↑ cyclic AMP. S. may also act as ↑ neurotransmitter and neuromodulator.

↑ Enterochromaffinlike cells

Chan-Palay, V.: Anat. Embryol. *156*, 225 (1979); Del Fiacco, M., Dessi, M.L., Atzori, G.M., Levanti, M.C.: Brain Res. *264*, 142 (1983); Gaudina, G., Fasolo, A.: Cell Tissue Res. *211*, 241 (1980); Lindner, G., Grosse, G.: Z. mikrosk.-anat. Forsch. *96*, 643 (1982); Porter, R., O'Connor, M. (eds.): Substance P in the Nervous System. Ciba Foundation Symposia, Vol. 91. London: Pitman 1982

Substantia alba: ↑ Central nervous system

Substantia compacta, of bone: ↑ Bone, compact

Substantia granulo-filamentosa: ↑ Reticulocyte

Substantia grisea: ↑ Brain cortex; ↑ Central nervous system

Substantia propria, of cornea: ↑ Cornea

Substantia reticularis: ↑ Reticulocyte

Substantia spongiosa, of bone: ↑ Bone, cancellous

Subsurface cisternae: Flattened ↑ unit membrane-limited spaces (SC) about 20 nm wide, lying parallel to ↑ plasmalemma (P) at distance of 10–15 nm. S. lack ribosomes (R) wherever they confront plasmalemma; often they are entirely devoid of them. It is therefore believed that S. belong to smooth ↑ endoplasmic reticulum. S. are found in excitable cells such as ↑ pinealocytes (1), muscle cells and fibers (2), ↑ hair cells (3), etc. In pinealocytes and ↑ smooth muscle cells they are frequently adjacent to ↑ autonomic nerve fibers; in smooth muscle cells they form couplings with sarcolemma (S). In hair cells, S. are arranged in several layers and are particularly abundant facing efferent nerve endings (↑ postsynaptic cisternae). A special kind of S. are found between ↑ Sertoli's cells (4) where they are connected with plasmalemma by periodic densities formed by aggregations of actin filaments (F) in a hexagonal array. The intercellular space in this zone is reduced to 7–9 nm. Similar S. surround heads of ↑ spermatids and ↑ spermatozoa.

↑ Subsarcolemmal cisternae

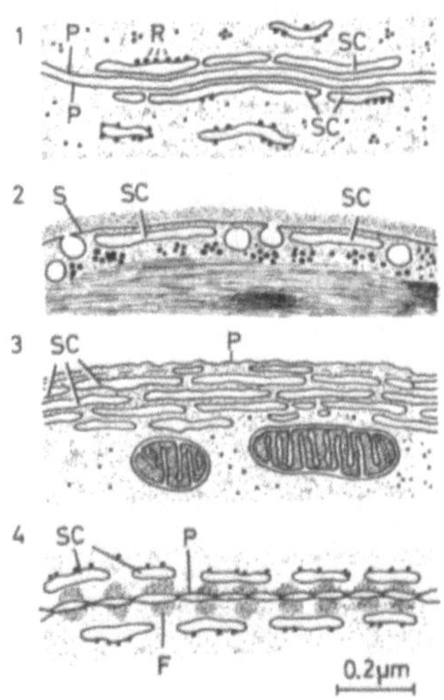

Lubitz, D.K.J., Ekström, von: Cell Tissue Res. *220*, 787 (1981); Dym, M., Fawcett, D.W.: Biol. Reprod. *3*, 308 (1970); Kolde, G., Themann, H.: Cell Tissue Res. *223*, 455 (1982); Møller, O.J., Østergaard Thomsen, O., Larsen, J.A.: Cell Tissue Res. *228*, 13 (1983); Saito, K.: Cell Tissue Res. *229*, 467 (1983)

Subsynaptic cisternae: ↑ Postsynaptic cisternae

Subsynaptic web: ↑ Postsynaptic density

Succus entericus (intestinal juice): Liquid produced mainly by ↑ Lieberkühn's crypts. S. consists of water, sodium chloride, bicarbonate, and a small amount of other components, but practically no enzymes.

Geigy Scientific Tables. Body Fluids. 8th edn. Basel: Ciba-Geigy 1983

Suckling reflex (milk ejection reflex): Neuroendocrine reflex consisting of contraction of ↑ myoepithelial cells of ↑ mammary gland and consecutive expression of ↑ milk in lactating female as response to suckling. Stimulation of ↑ nipple generates sensory impulses that are carried via ↑ spinal ganglia to ↑ brain cortex. From here, impulses are relayed to ↑ hypothalamus, where they stimulate ↑ neurosecretory neurons of ↑ paraventricular nucleus to produce and discharge ↑ oxytocin. This hormone provokes contraction of myoepithelial cells around ↑ alveoli and expulsion of milk into ↑ lactiferous ducts.

Tindal, J.S.: Neuroendocrine Control of Lactation. In: Larson, B.L. (ed.): Lactation, Vol. 4. New York: Academic Press 1978; Wakerley, J.B., Lincoln, D.W.: J. Endocr. *57*, 477 (1973)

Sudan dyes: Liposoluble weakly ionizable dyes widely used for histochemical demonstration of ↑ lipids (Sudan III, Sudan black).

↑ Staining, physical

Sudanophilia: Affinity of a cell or tissue component for ↑ Sudan dyes. S. is based on solubility of these dyes in cellular or tissue ↑ lipids.

↑ Staining, physical

Sudomotor nerve fibers: ↑ Postganglionic nerve fibers innervating ↑ sweat glands. Although sympathetic, S. release ↑ acetylcholine.

Jenkinson, D.M., Montgomery, I., Elder, H.Y.: J. Anat. *125*, 625 (1978)

Sulcus, external, of cochlea (sulcus spiralis externus*): Spiral space between vestibular and tympanic lips of ↑ spiral limbus.

Sulcus, internal, of cochlea (sulcus spiralis internus*): Small spiral sulcus at insertion line of ↑ basilar membrane to ↑ spiral ligament. S. is lined by ↑ Böttcher's and ↑ Claudius' cells.

Sulfomucin: Acid ↑ glycoprotein secreted by ↑ mucous glands.

Superficial cells: ↑ Sweat glands, eccrine, excretory duct cells of

Superficial cells, of transitional epithelium: ↑ Facet cells

Superior mesenteric vein (vena mesenterica superior*): Large ↑ propulsive vein with structure identical to that of ↑ vena cava inferior.

Supporting cells, of carotid body: ↑ Glomus cells, of carotid body

Supporting cells, of Corti's organ: ↑ Claudius' cells; ↑ Hensen's cells; ↑ Phalangeal cells

Supporting cells, of cristae ampullares and macula of saccule and utricle: Tall irregular cells interposed among ↑ vestibular cells (VC). Nucleus is oval, rich in ↑ heterochromatin, and located at cell base; it contains one nucleolus. Cytoplasm encloses some filiform mitochondria, relatively well-developed rough endoplasmic reticulum, and a small Golgi apparatus (G). An extensive filamentous ↑ terminal web (TW), an occasional ↑ cilium (C), and a ↑ centriole (Ce) are situated in ↑ apical pole. Conspicuous bundles of ↑ microtubules (Mt) run from cell base to terminal web. Some ↑ lysosomes (Ly) and small ↑ secretory granules (SG) are also scattered in apical cell portion. S. support vestibular cells; their secretory role has not been definitively proved. BL = ↑ basal lamina

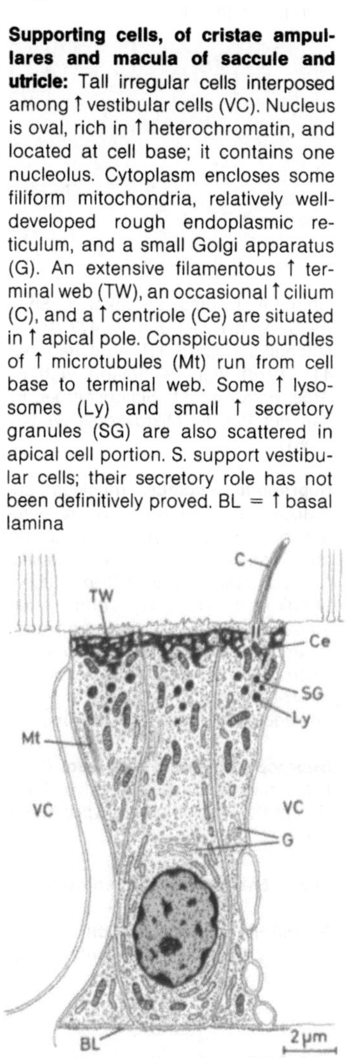

Supporting cells, of olfactory epithelium: ↑ Olfactory epithelium

Supporting cells, of parasympathetic paraganglia: ↑ Paraganglia, parasympathetic

Supporting cells, of pineal organ: ↑ Interstitial cells, of pineal organ

Supporting cells, of sympathetic paraganglia: ↑ Carotid body; ↑ Paraganglia, sympathetic

Supporting cells, of taste buds: ↑ Taste buds, structure of

Supporting tissues (textus sustentacularis*): Group of ↑ tissues characterized by a solid intercellular substance. S. comprise: 1) ↑ Bone, 2) ↑ cartilage, 3) ↑ cementum, and 4) ↑ dentin. In broad sense of term, S. belong to ↑ connective tissue.

Suppressor cells: Subpopulation of ↑ T-lymphocytes which regulate both humoral and cellular ↑ immune responses by their inhibitory action on ↑ helper and ↑ killer cells.

Suprachoroid space (spatium perichoroideale*): Potential cleft between ↑ sclera and ↑ choroid, crossed by numerous ↑ collagen fibers.

Supraependymal axon: Very thin, frequently branched process (A) of a ↑ supraependymal neuron overlying ↑ ependymal cells (E). S. may be over 100 µm long; they display many swellings (S) and end with a ↑ bouton terminal (arrowhead). Both swellings and bouton terminal contain ↑ synaptic vesicles with unknown ↑ neurotransmitter and form ↑ synapses with ependymal cells, as well as with ↑ supraependymal dendrites. Function of S. is not known. (Fig. = ↑ subfornical organ, rat)

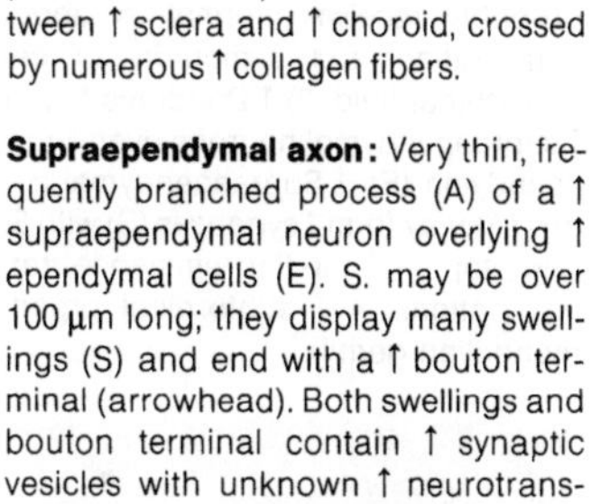

Supraependymal cells: Collective term for all single or clustered cells (SC) overlying ↑ ependyma (E). They include supraependymal macrophages (↑ Kolmer's cells), supraependymal neuroglial cells (fig.), and ↑ supraependymal neurons. Determination of the type of an S., based on SEM

micrographs, is difficult. The function of S. other than macrophages is not clear. (Fig. = ↑ subfornical organ, mongolian gerbil)

Supraependymal dendrites: 1) Short clublike processes (D) of ↑ cerebrospinal fluid-contacting neurons penetrating ↑ cerebrospinal fluid, S. emerge between ↑ ependymal cells (E) and probably receive information about composition and osmolarity of cerebrospinal fluid. 2) ↑ Dendrites (D) of ↑ supraependymal neurons overlying ↑ ependyma (E). ↑ Supraependymal axons (A) may form ↑ synapses (S) with S. (Fig., top = ↑ median eminence, rat; fig., bottom = ↑ subfornical organ, mongolian gerbil)

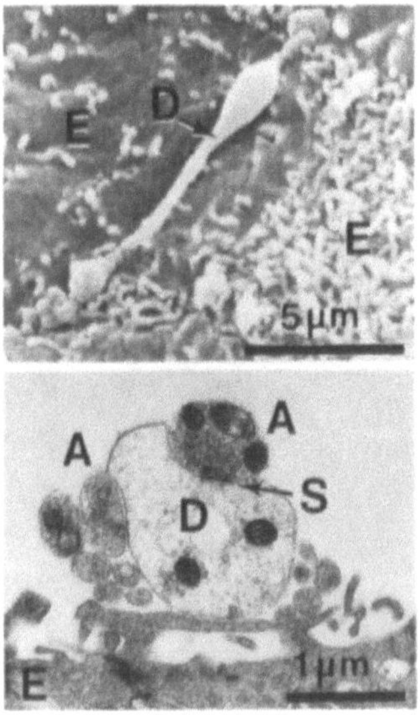

Derer, P.: J. Physiol. *77*, 211 (1981)

Supraependymal macrophages: ↑ Kolmer's cells

Supraependymal neurons: Nerve cells (SN) overlying ↑ ependyma of brain ventricles and ↑ circumventricular organs. S. have a long ↑ supraependymal axon (A) synapsing with ↑ ependymal cells (E) or penetrating between them to end in ↑ subependymal layer. Exact function of S. is unknown; it is believed that they may be involved in reception of physicochemical changes in ↑ cerebrospinal fluid, that they may produce and release bioactive substances into cerebrospinal fluid and/or regulate function of ↑ cerebrospinal fluid-contacting neurons. D = ↑ supraependymal dendrites. (Fig. = ↑ subfornical organ, mongolian gerbil)

Chiba, A., Honma, Y., Nakai, Y., Shioda, S.: Arch. Histol. Jpn. *44*, 477 (1981); Mitchell, J.A., Card, J.P.: Anat. Rec. *192*, 441 (1978); Vigh-Teichmann, Vigh, B., Aros, B., Kausz, M., Simonsberger, P., van der Pol, A.: Mikroskopie *38*, 337 (1981)

Supraoptic nucleus (nucleus supraopticus*): Group of ↑ neurosecretory neurons of ↑ hypothalamus situated immediately above chiasma fasciculi optici. Cells of S. produce ↑ antidiuretic hormone and small amounts of ↑ oxytocin, both transported and stocked in ↑ neurohypophysis. Axons of S. cells form ↑ supraopticohypophyseal tract. ↑ Neurons of S. are considered to be ↑ osmoreceptors.

Supraopticohypophyseal tract (tractus supraopticohypophysealis): Anterior part of ↑ hypothalamohypophyseal tract.

Suprarenal glands: ↑ Adrenal glands

Suprarenal paraganglion (paraganglion suprarenale): Synonym for adrenal medulla.

↑ Adrenal glands, medulla of

Suprarenal vein (vena suprarenalis*): Vessel collecting blood from ↑ adrenal gland. Branches (V) of S. run through medulla of gland and are characterized by agglomerations of longitudinally oriented ↑ smooth muscle cells forming well-developed ↑ intimal cushions (IC). Contraction of cushions provokes slowing down of blood flow with simultaneous enrichment of blood with adrenal hormones. (Fig. = human)

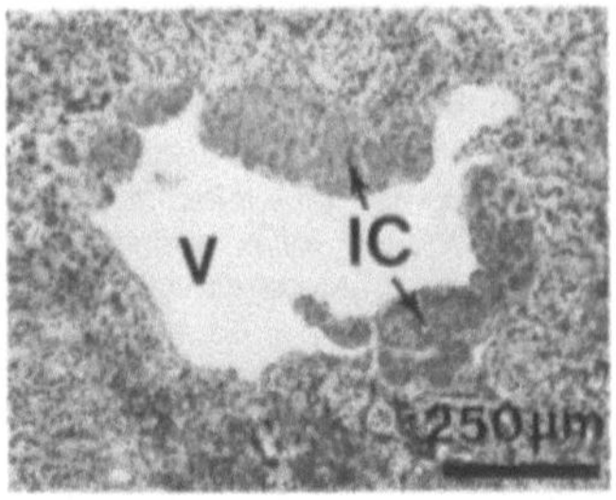

Supravital staining: ↑ Staining and observation of freshly obtained cells and their ↑ organelles in living state. Examples: S. of ↑ mitochondria with ↑ Janus green B; S. of granules and ↑ phagocytic vacuoles of neutrophilic ↑ granulocytes with neutral red; S. of ↑ reticulocytes by mixing of freshly drawn blood with a drop of ↑ brilliant cresyl blue or 9-aminoacridin.

Surface coat: ↑ Glycocalyx

Surface density: Parameter in ↑ stereology.

Surface mucous cells, of gastric mucosa (mucous cells, of gastric surface; cellula superficialis mucosa*): Tall columnar cells (20–40 µm in height) lining ↑ foveolae gastricae. Nucleus is elliptical, rich in ↑ heterochromatin, and located in a basal position; nucleolus is prominent. Cytoplasm contains rod-shaped mitochondria, a well-developed supranuclear Golgi apparatus, ↑ centrioles, sparse, flattened, rough endoplasmic ↑ cisternae, a few lysosomes, and a considerable amount of free ribosomes. In apical cell portion is found a large number of ↑ PAS-positive ↑ mucigen granules (SG) of various shapes. They are enclosed by a ↑ unit membrane and contain an electron-dense material consisting of ↑ glycosaminoglycans. Granules arise from Golgi apparatus (G); since ↑ exocytosis of their contents has not been observed, they are thought to leave the cell by diffusion. In foveolar lumen mucigen is converted into acid-resistant ↑ mucus, which lubricates mucosal surface and protects it against action of gastric juice. Apical surface bears short microvilli covered by ↑ glycocalyx (Glc). S. are held together by well-developed ↑ junctional complexes (J), lateral ↑ interdigitations (Int), and small ↑ desmosomes. Deeper in foveolae, S. become continuous with ↑ neck mucous cells. Life span about 3 days; ↑

mitosis takes place at bottom of foveolae. BL = ↑ basal lamina

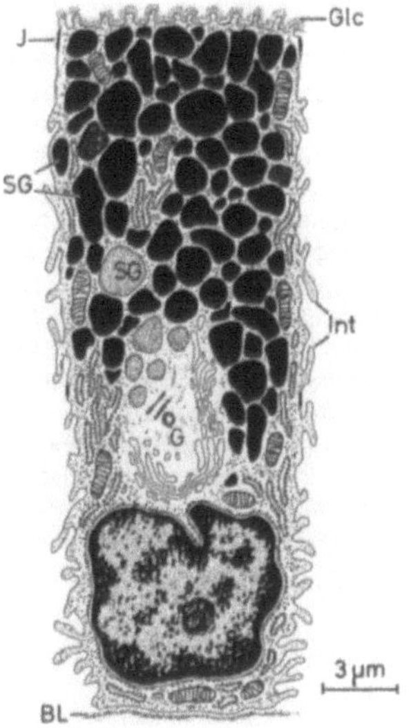

Kramer, M.F., Geuze, J.J.: J. Cell Biol. *73*, 533 (1977); Rubin, W., Ross, L.L., Sleisenger, M.H., Jeffries, G.H.: Lab. Invest. *19*, 598 (1968); Smits, H.L., Kramer, M.F.: Am. J. Anat. *161*, 365 (1981); van Huis, G.A., Kramer, M.F.: Am. J. Anat. *156*, 301 (1979)

Surfactant (antiatelectatic factor): Surface-active phospholipid coating (S) of alveolar surfaces in lung. S. resembles ↑ myelin lamellae; it consists of two or more highly osmiophilic lamellae, about 10 nm thick and separated from each other by a 10-nm-wide clear layer. From plasmalemma (P) of alveolar cells (A), S. is separated by a cleft (C) about 8–10 nm wide. S. is product of type II ↑ alveolar cells, i.e., their ↑ multilamellar bodies. Chemically, S. is related to phosphatidylcholine; it also contains a ↑ glycoprotein component and dipalmitoyil lecithin. S. reduces surface tension at air-fluid surface, thus preventing collapse (atelectasis) of ↑ alveoli. (Fig. = rat)

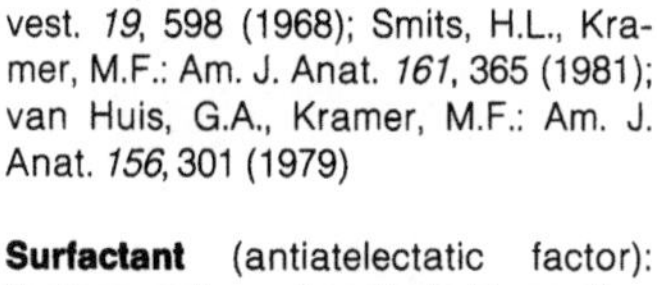

Dierichs, R., Lindner, E.: Histochemistry *60*, 335 (1979); Goerke, J.: Biochim. Biophys. Acta *344*, 241 (1979); Hulbert, W.C., Forster, B.B., Laird, W., Phil, C.E., Walker, D.C.: Lab. Invest. *47*, 354 (1982), Iwatsuki, H.: Acta Histochem. Cytochem. *13*, 646 (1980); Kikkawa, Y., Smith, F.: Lab. Invest. *49*, 122 (1983)

Suspensory apparatus, of lens: All structures contributing to mechanical stability of ↑ lens in ↑ eyeball (↑ ciliary processes, ↑ zonular fibers, and possibly ↑ vitreous body).

Curtis, R.: J. Anat. *136*, 69 (1983)

Suspensory folds, of gastrointestinal tract: ↑ Mesentery; ↑ Omentum

Suspensory ligament, of lens: ↑ Ciliary zonule

Sustentacular cells: ↑ Supporting cells

Sutura cranii*: ↑ Syndesmosis (S) connecting two skull bones (B). Space between them is filled with ↑ dense connective tissue (DCT), with ↑ collagen fibers penetrating into ↑ bone matrix. S. is bridged by ↑ Sharpey's fibers (SF) running through ↑ periosteum (P).

↑ Wormian bones

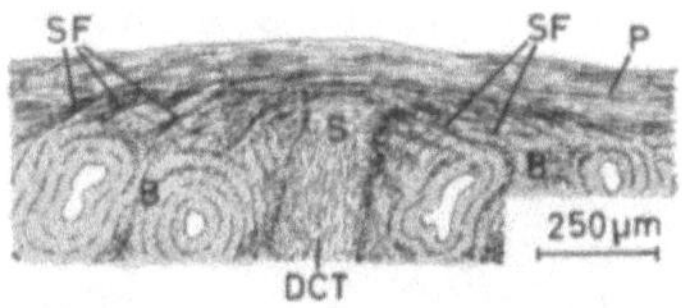

Sweat: Secretory product of ↑ sweat glands. S. consists of water, sodium, chloride, potassium, ↑ urea, and lactate. Whereas sodium ion is actively pumped between cells into excretory ducts, water is passively transported to restore isotonicity. Composition of S. depends upon environmental temperature. Apocrine ↑ sweat glands also produce odoriferous substances starting at puberty.

Geigy Scientific Tables. Body Fluids. 8th edn. Basel: Ciba-Geigy 1983

Sweat glands (glandulae sudoriferae*): Small coiled ↑ tubular and ↑ tubuloalveolar cutaneous glands divided into two groups: 1) ↑ S. apocrine and 2) ↑ s. eccrine.

McEvan Jenkinson, D., Montgomery, I., Elder, H.Y.: J. Anat. *129*, 117 (1979)

Sweat glands, apocrine (odoriferous sweat glands, glandula sudorifera apocrina*): Incorrect term for sweat glands (Ap) of axilla (fig.), pubic region, ↑ labia majora, ↑ scrotum, perineal and circumanal regions, and ↑ areola of breast. Electron microscopic investigations have shown that term "apocrine" is wrong, since these glands function by an ↑ eccrine secretion. Structure: 1) Body of gland is formed of numerous saclike ↑ tubuloalveoli (TA) lined with ↑ simple cuboidal epithelium. In comparison with eccrine ↑ sweat glands (Ecc), S. measure 1–5 mm, are more deeply situated in ↑ dermis, are innervated by ↑ adrenergic nerve endings, and secrete a more viscous product. Discharge is helped by numerous ↑ myoepithelial cells (MC). Secretion of S. is odorless; its decomposition by skin bacteria leads to formation of an acrid odor. At beginning of ↑ menstrual cycle and in pregnancy, axillar S. become larger and their cells more voluminous; during menstrual bleeding and after ↑ menopause, S. show regressive changes. 2) Excretory duct (ED) is lined by ↑ simple cuboidal epithelium lying on basal lamina. In contrast to eccrine ↑ sweat glands, excretory ducts of S. are connected with ↑ hair follicles (HF). S. also comprise sweat glands of nasal ↑ vestibulum, ↑ ciliary glands of ↑ eyelids, and ↑ ceruminous glands.

↑ Clear cells, of sweat glands; ↑ Mucoid cells, of sweat glands

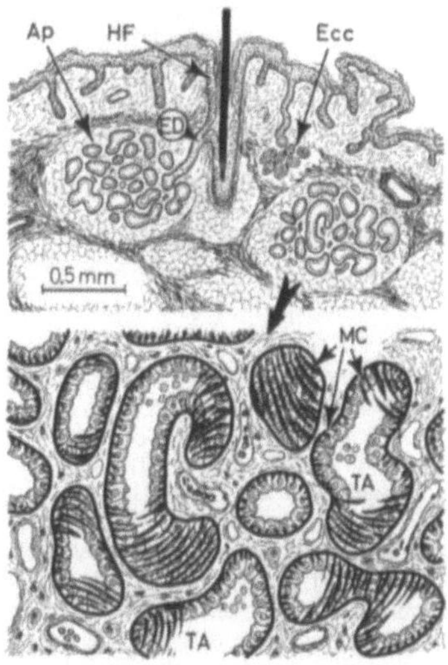

Biempica, L., Montes, L.F.: Am. J. Anat. *117*, 47 (1965); Hurley, H.J., Shelley, W.B.: The Human Apocrine Sweat Gland in Health and Disease. Springfield: C.C. Thomas, Publisher 1960

Sweat glands, eccrine (glandula sudorifera eccrina*): Simple tubular cutaneous glands composed of body (B), or secretory portion, and ↑ excretory duct (ED). Structure: 1) Body is a simple unbranched tube (T) coiled into a glomerulus 0.3–0.5 mm in diameter, situated in deep ↑ dermis and ↑ hypodermis. Epithelial lining is composed of a single layer of cuboidal or columnar secretory cells of two types: ↑ mucoid or dark cells and ↑ clear cells. Between bases of secretory cells lie numerous ↑ myoepithelial cells (MC). 2) Excretory duct is slightly twisted and lined with a stratified epithelium which is missing at keratinized layer (KL) of ↑ epidermis (E); here, duct becomes transformed into a corkscrewlike canal (C) opening onto epidermal surface. S. are present everywhere in skin, except for margins of ↑ lips, ↑ nipples, inner surface of ↑ prepuce, ↑ glans penis, ↑ nail bed, ↑ clitoris, and ↑ labia minora. There are 2–5 million S., producing a hypotonic watery fluid with some sodium chloride; thus, they are involved in thermoregulation and excretion. S. are innervated by ↑ sudomotor nerve fibers of ↑ sympathetic nervous system.

↑ Sweat; ↑ Sweat glands eccrine, excretory duct cells of

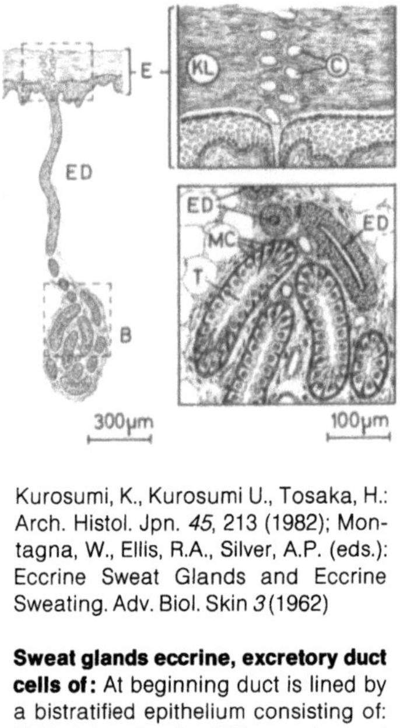

Kurosumi, K., Kurosumi U., Tosaka, H.: Arch. Histol. Jpn. *45*, 213 (1982); Montagna, W., Ellis, R.A., Silver, A.P. (eds.): Eccrine Sweat Glands and Eccrine Sweating. Adv. Biol. Skin *3* (1962)

Sweat glands eccrine, excretory duct cells of: At beginning duct is lined by a bistratified epithelium consisting of: 1) Peripheral cells (PC) = polygonal cells with a round and often deeply invaginated nucleus enclosing a prominent nucleolus. Cytoplasm contains numerous mitochondria (M) and ribosomes, whereas other ↑ organelles are sparse and poorly developed. 2) Superficial cells (SC) = irregular cells forming luminal surface of duct. They have mainly an irregular nucleus with a conspicuous nucleolus and poorly developed organelles. A well-developed ↑ terminal web (TW), interlaced with numerous ↑ tonofibrils (Tf), is located in apical cytoplasm immediately beneath apical plasmalemma. BL = ↑ basal lamina

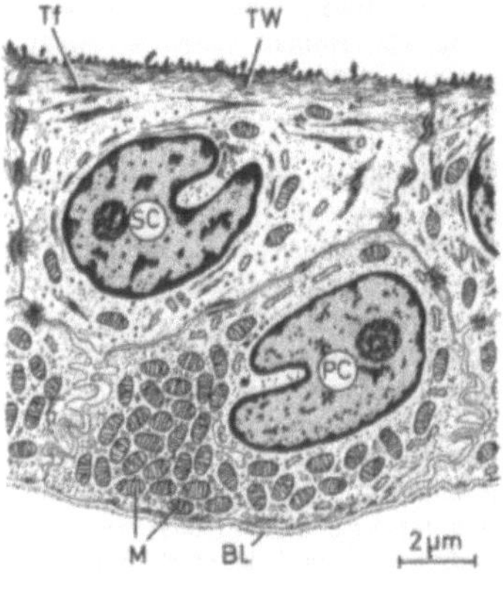

Montgomery, I., McEvan Jenkinson, D., Elder, H.Y.: J. Anat. *134*, 741 (1982)

Swell bodies: Specialized blood vessels (SB) with subendothelial sphincterlike ↑ intimal cushions (arrowheads) located in middle and superior ↑ concha nasalis. ↑ Adrenergic nerve endings are particularly abundant. S. regulate blood flow through venous plexus of conchae. (Fig. = human)

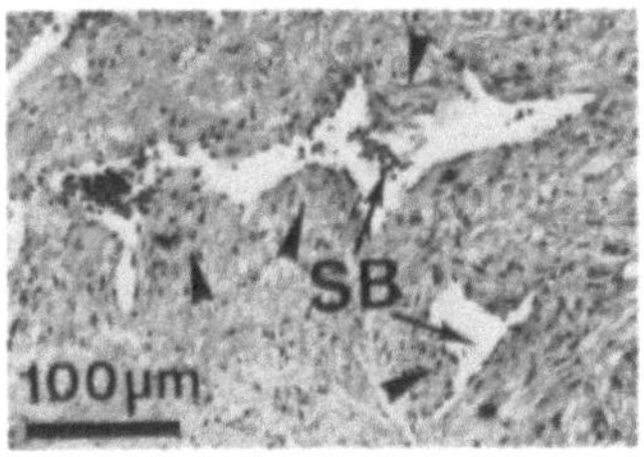

Temesrekasi, D.: Z. mikrosk.-anat. Forsch. *80*, 219 (1969)

Symmetrical synapse: ↑ Synapse, chemical, types of

Sympathetic chain ganglia: ↑ Vertebral ganglia

Sympathetic ganglia: ↑ Autonomic ganglia

Sympathetic ground plexus: Terminal ramifications of sympathetic ↑ postganglionic nerve fibers forming a network in various tissues and organs.

↑ Adrenergic nerve fibers; ↑ Sympathetic nervous system

Sympathetic nerve endings: ↑ Adrenergic nerve endings

Sympathetic nerve fibers: ↑ Adrenergic nerve fibers

Sympathetic nervous system (systema nervosum autonomicum, pars sympathica*): Part of ↑ autonomic nervous system composed of: 1) Sympathetic centers in ↑ spinal cord = ↑ intermediolateral and ↑ intermediomedial nucleus. 2) White rami communicantes = ↑ preganglionic ↑ myelinated axons of ↑ sympathetic neurons for innervation of viscera. 3) ↑ Sympathetic ganglia = a) ↑ vertebral ganglia or sympathetic chain ganglia forming two ganglionated ↑ sympathetic trunks of 21 or 22 pairs of ganglia along lateral sides of the vertebral column; b) ↑ prevertebral or collateral ganglia = unpaired, nonsegmental ganglia in front of ↑ aorta (celiac ganglion, aorticorenal ganglion, superior and inferior mesenteric ganglia); c) ↑ terminal ganglia in various organs. 4) Gray rami communicantes = ↑ postganglionic ↑ nonmyelinated nerve fibers for innervation of muscles and glands; these fibers join peripheral ↑ nerve (PN). SG = ↑ spinal ganglion

↑ Autonomic ganglia

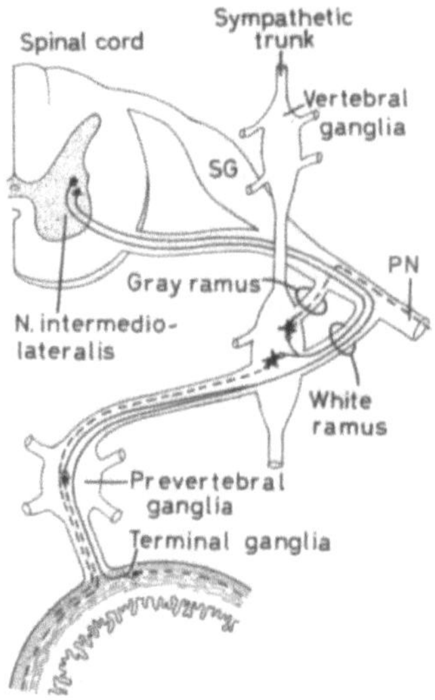

Sympathetic neurons: ↑ Autonomic neurons, of sympathetic ganglia

Sympathic paraganglia: ↑ Paraganglia, sympathetic

Sympathetic trunk (truncus sympathicus*): Paired chains of ↑ vertebral ganglia on either side of vertebral column. S. is connected with peripheral ↑ nerves and with ↑ prevertebral ganglia.

↑ Sympathetic nervous system

Sympathicotropic hilus gland: ↑ Hilus cells

Sympathoblasts (chromaffinoblasti*): Precursor cells originating from neural crest, giving rise to ↑ chromaffin cells of ↑ chromaffin ganglia and adrenal medulla.

↑ Adrenal glands, medulla of; ↑ Nervous tissue, histogenesis of

Sympexions: ↑ Prostatic concretions

Symphysis: Kind of ↑ synarthrosis in which fibrous ↑ cartilage connects two bones. There are two S. in the body: 1) ↑ symphysis pubica and 2) ↑ intervertebral discs.

Symphysis pubica*: ↑ Joint uniting two pubic bones. Structure: 1) Thin zone of hyaline ↑ cartilage (HC) continuous with 2) fibrocartilaginous discus interpubicus (DI) with a fine central cleft (C). Liggamenta pubicum superius (LPS) and arcuatum pubis (LAP) enclose S. and continue into ↑ periosteum of pubic bone.

↑ Symphysis; ↑ Synarthrosis

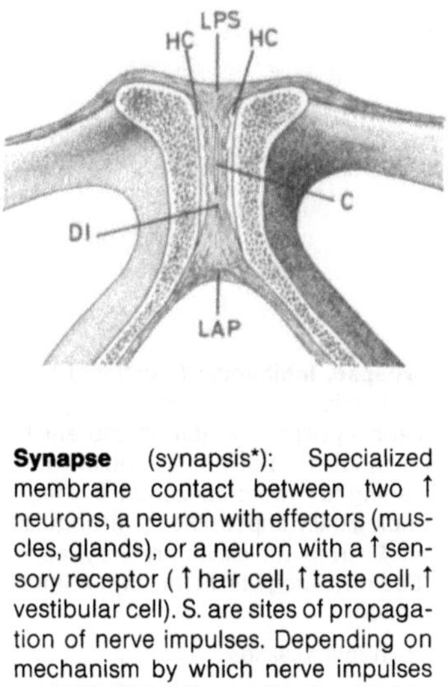

Synapse (synapsis*): Specialized membrane contact between two ↑ neurons, a neuron with effectors (muscles, glands), or a neuron with a ↑ sensory receptor (↑ hair cell, ↑ taste cell, ↑ vestibular cell). S. are sites of propagation of nerve impulses. Depending on mechanism by which nerve impulses are transmitted, S. may be classified as follows:

- Synapse
  - Chemical
    - Interneuronal
      - Axoaxonic
      - Axodendritic
      - Axosomatic
    - Neuromuscular
      - Motor end plate
    - Neuroglandular
    - Neurosensory
  - Electrical – Interneuronal
  - Mixed – Interneuronal

(Different categories of S. are described under ↑ Synapse,...)

↑ Bouton terminal, ↑ Synapse, chemical

Galbraith, S.: Acta. Anat. *107*, 46 (1980); Jones, G.D.: Adv. Anat. Embryol. Cell. Biol. *55/4*, 1 (1978); Kirsche, W.: Ergebn. Exp. Med. *25*, 9 (1977); Manina, A.A.: Int. Rev. Cytol. *57*, 345 (1979); Taxi, J. (ed.): Ontogenesis and Functional Mechanisms of Peripheral Synapses. Amsterdam: Elsevier/North Holland Biomedical Press 1980

Synapse, adrenergic: ↑ Bouton terminal with ↑ dense core ↑ synaptic vesicles containing ↑ epinephrine or ↑ norepinephrine.

↑ Synapse; ↑ Synapse, "by distance"; ↑ Synapse, G

Synapse, aminergic: ↑ Bouton terminal with ↑ synaptic vesicles containing one of the biogenic amines (↑ dopamine, ↑ gamma-aminobutyric acid, ↑ histamine, ↑ serotonin, etc.).

Synapse, asymmetrical: ↑ Synapse, chemical, types of

Synapse, axoaxonic (synapsis axoaxonica*): Synaptic contact between two ↑ axons.

↑ Bouton terminal; ↑ Nodes, of Ranvier; ↑ Synapse; ↑ Synapse, interneuronal

Zhu, C.G., Sandri, C., Akert, K.: Brain Res. *230*, 25 (1981)

Synapse, axodendritic (synapsis axodendritica*): Contact between a ↑ bouton terminal (BT) of one ↑ neuron with a ↑ dendrite (D) of another neuron. SV = ↑ synaptic vesicles (Fig. = ↑ cerebellum, rat)

↑ Synapse; ↑ Synapse, interneuronal

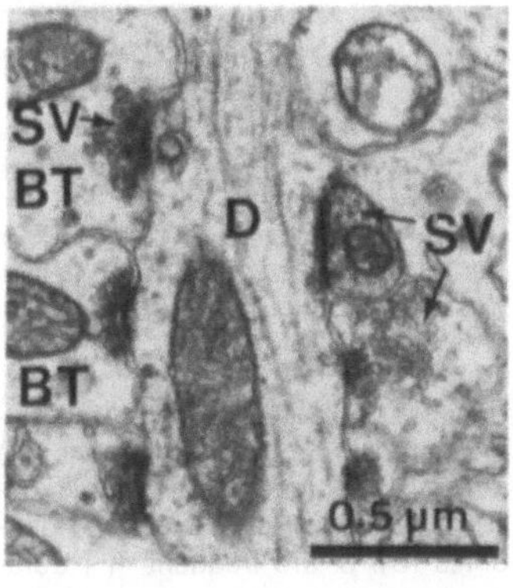

Tapia, R., Cotman, C.W. (eds.): Regulatory Mechanisms of Synaptic Transmission. New York: Plenum Press 1982

Synapse, axosomatic (synapsis axosomatica*): Synaptic contact between a ↑ bouton terminal (BT) and ↑ soma (S) of another ↑ neuron. (Fig. = brain, rat)

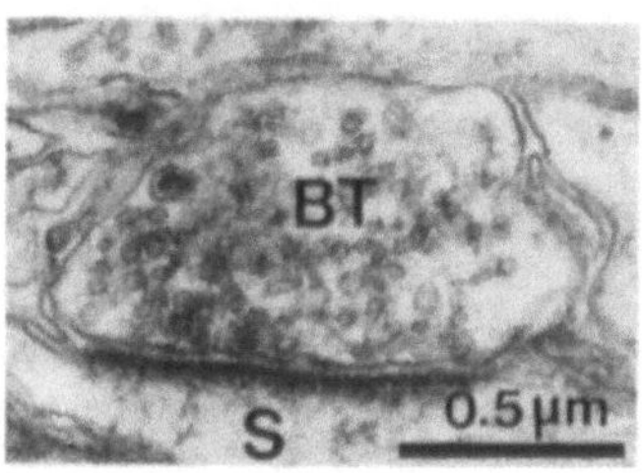

↑ Synapse; ↑ Synapse, interneuronal

Synapse, axospinous: ↑ Synapse betwen an ↑ axon and a ↑ dendritic spine.

↑ Synapse, axodendritic

Synapse, "by distance" (innervation by distance): A ↑ free nerve ending (E) is situated freely in a large tissue space without entering into contact with a target cell. ↑ Neurotransmitters contained in its ↑ synaptic vesicles are released into ↑ tissue fluid and act on many glandular or smooth muscle cells. S. are present in tunicae musculares of various organs, ↑ endocrine and ↑ exocrine glands, etc., most frequently located in ↑ pericapillary spaces (PS). In majority of cases, S. contain ↑ dense-cored vesicles. S. characterize ↑ autonomic nervous system.

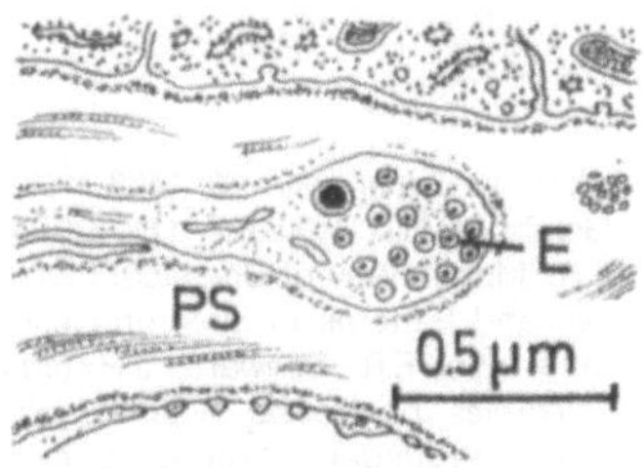

Synapse, calyceal: Intimate contact between a ↑ nerve calyx and a ↑ vestibular cell type I, either of ↑ cristae ampullares or ↑ macula of saccule and utricle. There are also some electrical ↑ synapses between nerve ending and cell.

Synapse, chemical: Most frequent type of ↑ synapse, in which propagation of nerve impulses is mediated by ↑ neurotransmitters released from ↑ synaptic vesicles into ↑ synaptic cleft. Structure: 1) Presynaptic element = a ↑ bouton terminal (BT) containing mitochondria (M), smooth endoplasmic reticulum (sER), ↑ neurofilaments (Nf), ↑ neurotubules (Nt), and ↑ synaptic vesicles (SV). Plasmalemma facing synaptic cleft is referred to as presynaptic membrane (PM); it is covered by ↑ presynaptic density (PD) on side of bouton. 2) ↑ Synaptic cleft (SC) = an interspace, about 20 nm wide, between bouton and postsynaptic element. Synaptic cleft contains some moderately dense material, sometimes disposed in fine transverse ↑ intrasynaptic filaments. 3) ↑ Postsynaptic element (PE) = postsynaptic membrane

(PsM) reinforced by ↑ postsynaptic density (PsD). Postsynaptic element may be a ↑ dendrite, ↑ soma, or ↑ axon of another ↑ neuron, a ↑ skeletal muscle fiber, a ↑ glandular or a ↑ sensory cell. A nerve impulse from bouton propagates only to postsynaptic element.

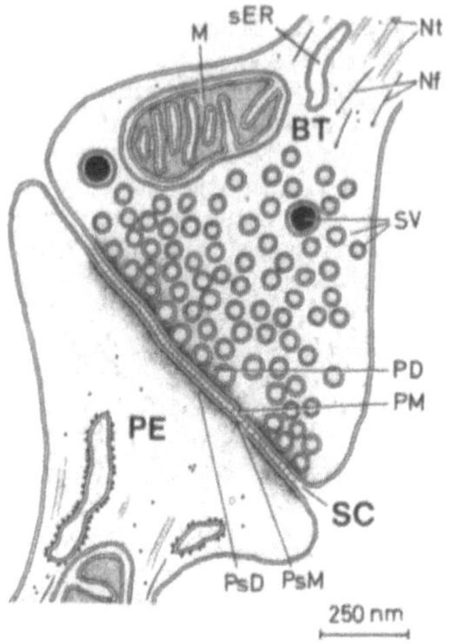

Akert, K.: Experientia *38*, 1408 (1982); Gershon, M.D., Schwartz, J.H., Kandel, E.R.: Morphology of Chemical Synapses and Patterns of Interconnection. In: Kandel, E.R., Schwartz, J.H. (eds.) Principles of Neural Science. New York, Amsterdam, Oxford: Elsevier/North Holland 1981; Manina, A.A.: Int. Rev. Cytol. *57*, 345 (1979)

Synapse, chemical, types of: On basis of differences in width of ↑ synaptic cleft (SC) and extent of ↑ postsynaptic density (PsD), chemical ↑ synapses can be subdivided into: 1) Type I, or Gray's type I synapse = synaptic cleft is about 30 nm wide and postsynaptic density is pronounced; because of asymmetry of this type it is also referred to as an asymmetrical synapse. 2) Type II, or Gray's type II synapse = synaptic cleft is approximately same width as intercellular gap (IG); postsynaptic density is not pronounced and is present as dots in ↑ bouton terminal (BT). This type of synapse is referred to as a symmetrical synapse. Between both types there are transitional forms.

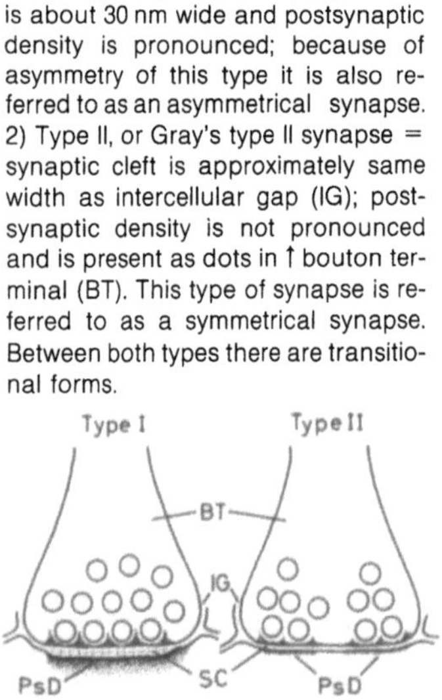

Synapse, cholinergic: ↑ Bouton terminal with ↑ synaptic vesicles containing ↑ acetylcholine as ↑ neurotransmitter.

Hanin, I., Goldberg, A.M. (eds.): Progress in Cholinergic Biology: Model Cholinergic Synapses. New York: Raven Press 1982

Synapse, dendrodendritic (synapsis dendrodendritica*): Rare type of synaptic contact between two ↑ dendrites in form of reciprocal ↑ synapse. Presynaptic dendrite contains ↑ synaptic vesicles morphologically identical to those in a ↑ bouton terminal.

Zhu, C.G., Sandri, C., Akert, K.: Brain Res. *230*, 25 (1981)

Synapse, electrical (electrotonic coupling area, electrotonic synapse, synapsis nonvesicularis*, synapsis electricalis*): Kind of intercellular contact (S) without ↑ synaptic vesicles or any ↑ neurotransmitter substance. Pre- and postsynaptic membranes are joined in form of a ↑ nexus. Electrical resistance in this area is very low, permitting electrical impulse to spread rapidly from one cell to the next. In some cases (lateral vestibular nuclei) an S. can be combined with a chemical ↑ synapse.

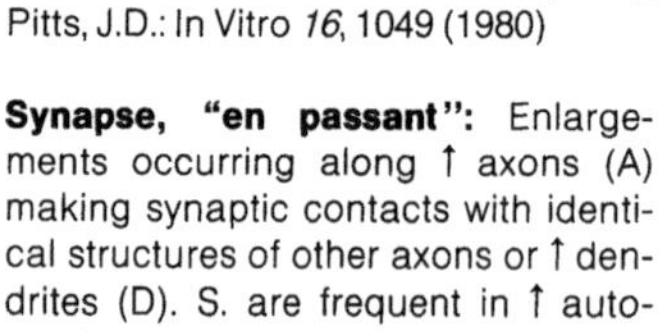

Akert, K.: Experientia *38*, 1408 (1982); Benett, M.V.L.: Fed. Proc. *32*, 65 (1973); Kosaka, T.: Brain Res. *271*, 157 (1983); Pitts, J.D.: In Vitro *16*, 1049 (1980)

Synapse, "en passant": Enlargements occurring along ↑ axons (A) making synaptic contacts with identical structures of other axons or ↑ dendrites (D). S. are frequent in ↑ autonomic nervous system, especially between nerve fibers innervating smooth musculature.

↑ Adrenergic nerve endings

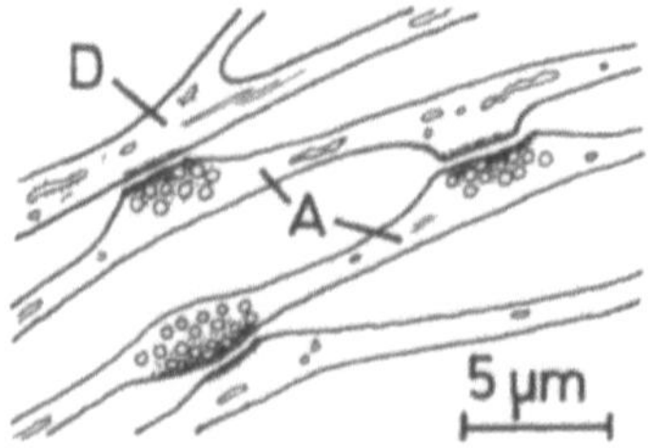

Synapse, excitatory: Functional type of chemical ↑ synapse depolarizing postsynaptic membrane by increasing its permeability to sodium ions. S. is morphologically identical to inhibitory ↑ synapse. (See physiology texts for further information)

Benshalom, G.: Cell Tissue Res. *200*, 291 (1979)

Synapse, G: Terminals (G) of certain classes of aminergic neurons in ↑ central nervous system characterized by 50-nm, ↑ dense-core ↑ synaptic vesicles (S) with a 25-nm electron-dense dot identical to those found in ↑ adrenergic nerve endings. ↑ Presynaptic density is lacking. (Fig. = ↑ pineal body, rat)

Synapse, "by distance"

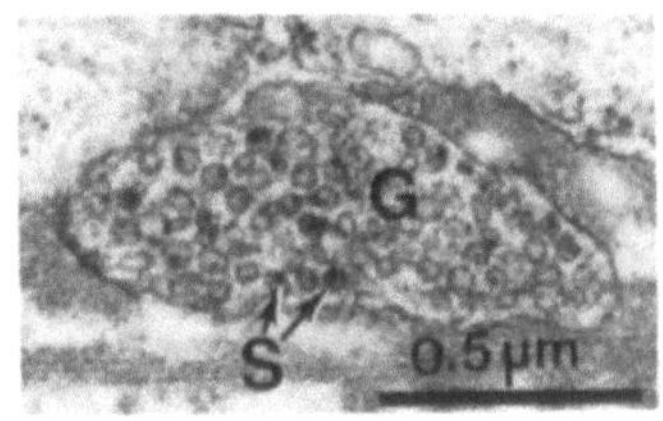

Synapse, inhibitory: Functional type of chemical ↑ synapse increasing permeability of postsynaptic membrane to chloride ion, but not to sodium ion, leading to hyperpolarization. It is morphologically impossible to distinguish an S. from an excitatory ↑ synapse. (See physiology texts for further information)

↑ Synaptic vesicles

Uchizono, K.: Nature *207*, 642 (1965)

Synapses, interneuronal: Depending on portion of a ↑ neuron reached by a ↑ bouton terminal (BT), following three principal combinations can be distinguished: 1) Axosomatic ↑ synapse = bouton terminal contacts ↑ soma (S) of a neuron to form any of following varieties of synapses: 1a) simple axosomatic synapse, 1b) invaginated axosomatic synapse, 1c) axosomatic spine. 2) Axodentritic ↑ synapse = bouton terminal contacts a ↑ dendrite (D) and may form: 2a) simple axodendritic synapse, 2b) axodendritic spine synapse, 2c) crest synapse, 2d) branched spine synapse, 2e) "en passant" ↑ synapse, 2f) reciprocal ↑ synapse, 2g) polysynaptic endings, 2h) interdigitated spine synapse. 3) Axoaxonic ↑ synapse = bouton terminal and axon (A) may form: 3a) simple axoaxonic synapse, 3b) axoaxonic synapse, 3c) "en passant" synapse. (Modified after Andres 1975)

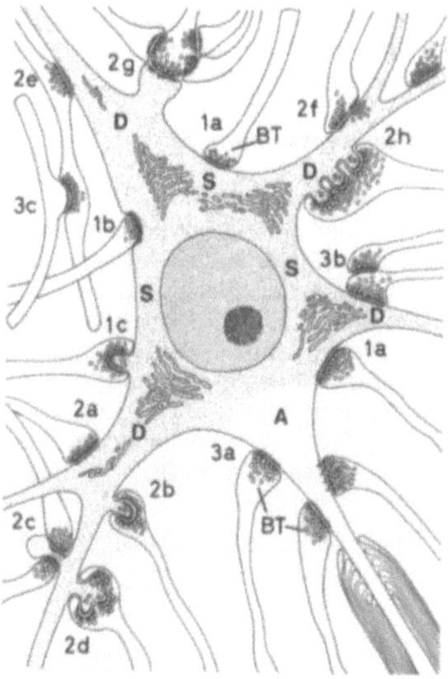

Andres, K.H.: J. Neural Transm. Suppl. *12*, 1 (1975)

Synapse, mixed: ↑ Bouton terminal (BT) simultaneously containing chemical (Ch) and electrical (El) presynaptic elements in contact with corresponding postsynaptic elements. S. are found in vestibular and mesencephalic nuclei.

↑ Synapse; ↑ Synapse, chemical; ↑ Synapse, electrical

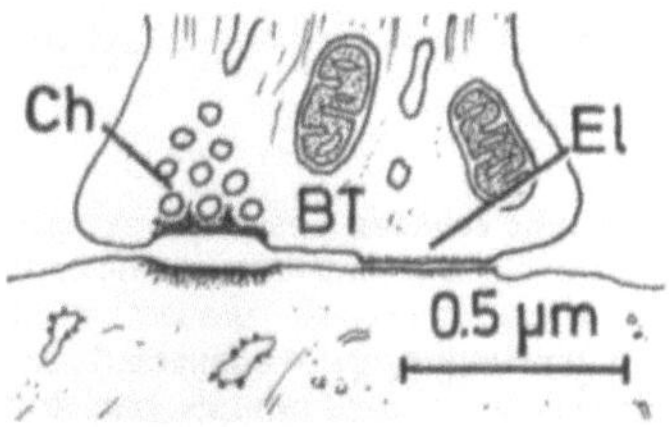

Synapse, neuroglandular: Contact between a ↑ bouton terminal (BT) and an endocrine or exocrine ↑ glandular cell (GC) and/or a brown ↑ adipose tissue cell. In general, bouton invaginates into cell body. Basal lamina (BL) accompanying axon is continuous with basal lamina of epithelium. In

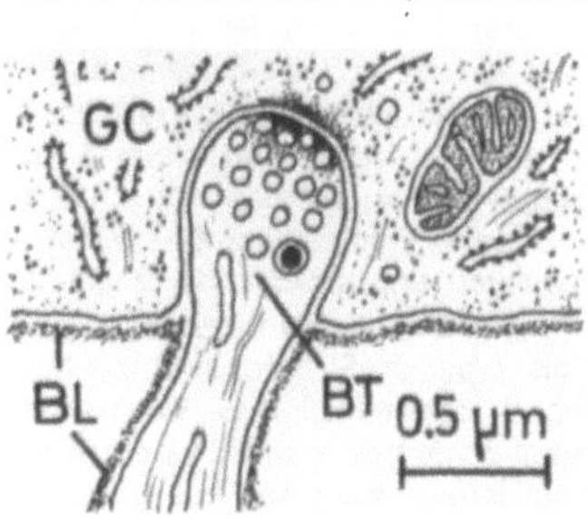

some S. ↑ pre- and ↑ postsynaptic densities may be absent; such S. are called pseudosynapses.

Andres, H.H.: J. Neural Transm. Suppl. *12*, 1 (1975)

Synapse, neuromuscular: ↑ Motor end plate

Synapse, neurosensory: Synaptic contact between a ↑ neuron and a ↑ sensory cell (↑ hair cell, ↑ taste cell, ↑ vestibular cell, etc.).

↑ Synapse

Synapse, reciprocal: Special dendrodendritic or somatosomatic ↑ synapse between two adjacent ↑ dendrites (D) found in thalamus, in ↑ olfactory bulb between ↑ mitral cells and ↑ granule cells, and between ↑ somata of type I glomus cells of ↑ carotid body. S. is characterized by a group of spherical ↑ synaptic vesicles (S) and a ↑ presynaptic density in one dendrite and same structures in opposed dendrite, but at a short distance from one another. It is assumed that impulses follow direction of arrows. S. is a rare form of ↑ synapse.

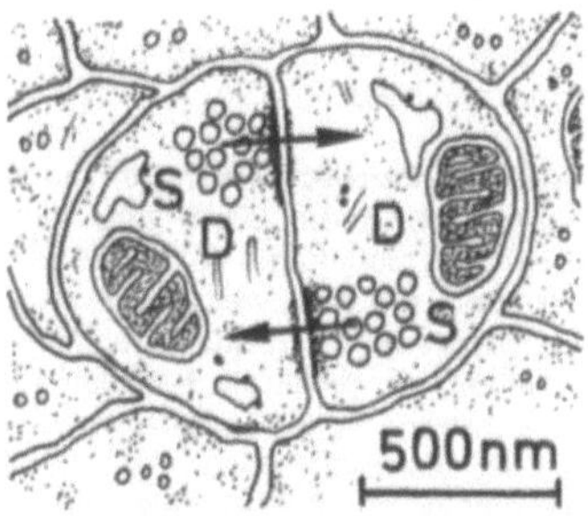

Andres, K.H.: Anatomy and Ultrastructure of the Olfactory Bulb in Fish, Amphibia, Reptiles, Birds and Mammals. In: Wolstenholme, G.E.W., Knight, J. (eds.): Taste and Smell in Vertebrates. London: J. & A. Churchill 1970; Dunn, R.F.: J. Comp. Neurol. *193*, 255 (1980); Matsumoto, S., Uchida, T., Nakajima, T., Ozawa, H.: Arch. Histol. Jpn. *43*, 275 (1980)

Synapse, serotoninergic: ↑ Bouton terminal with ↑ synaptic vesicles containing ↑ serotonin as ↑ neurotransmitter.

Synapse, somatodendritic (synapsis somatodendritica*): Rare type of ↑ synapse in which body of a ↑ neuron represents presynaptic element and a ↑ dendrite of another neuron, postsynaptic element.

Synapse, somatosomatic (synapsis somatosomatica*): Occasional type of ↑ synapse in which a limited area of body of a ↑ neuron makes synaptic contact (chemical or electrical) with body of another neuron. S. have been found in ↑ sympathetic ganglia and between type I ↑ glomus cells.

↑ Synapse, reciprocal

Synapse, symmetrical: ↑ Synapse, chemical, types of

Synapse triad: ↑ Triad, of retinal neurons

Synapsins: Two almost identical phosphoproteins present in mammalian ↑ nervous tissue. One seems to be specifically associated with ↑ synaptic vesicles.

De Camill, P., Cameron, R., Greengard, P.: J. Cell Biol. *96*, 1337 (1983)

Synapsis, in meiosis: Point-for-point pairing of homologous ↑ chromosomes (Chr) during prophase I of ↑ meiosis. Ultrastructurally, S. correspond to ↑ synaptonemal complexes (SC). A = axial filament of chromosome, CE = central element, LE = lateral element

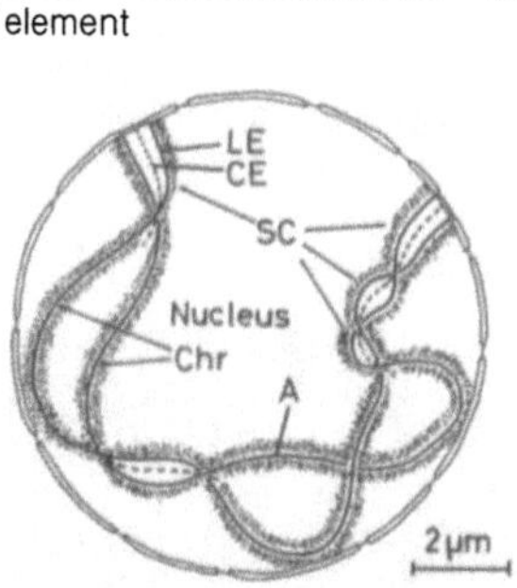

Synaptic bars (presynaptic body, synaptic body, synaptic rod): Cylindrical structures (SB), about 180–320 nm long and 35 nm thick, with a central thread (arrowhead), surrounded by 20–50 apparently empty ↑ synaptic vesicles (SV), containing an unknown ↑ neurotransmitter (↑ gamma-aminobutyric acid?). S. are found in ↑ hair cells, ↑ vestibular cells, and ↑ pinealo-

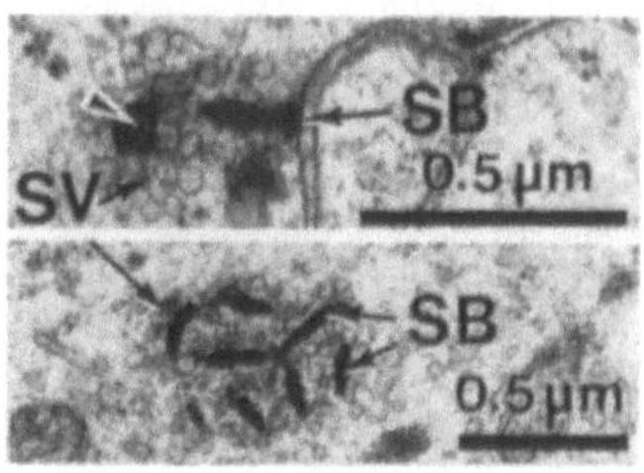

cytes. S. may be single or grouped, in contact with plasmalemma (fig., top), or free in cytoplasm (fig., bottom). It is thought that S. serve, at least in pinealocytes, in intercellular communications as an element directing synaptic vesicles to cell membrane. (Figs. = pinealocyte, rat)

↑ Synaptic ribbon

Bodian, D.: Anat. Rec. *197*, 379 (1980); Fry, K.R., Spira, A.W.: J. Histochem. Cytochem. *28*, 142 (1980); Krstić, R.: Cell Tissue Res. *166*, 135 (1976); Theron, J.J., Biagio, R., Meyer, A.C.: Cell Tissue Res. *217*, 405 (1981)

Synaptic bodies: ↑ Synaptic bars

Synaptic cleft, of motor end plate: Space (C) 80–100 nm in width separating presynaptic membrane (arrowheads) of ↑ axon (A) from ↑ sarcolemma (S). S. is filled with a fine granular and moderately osmiophilic material, similar to that of ↑ basal lamina. From this primary S. arise, by infoldings of sarcolemma, single and branched furrows called secondary ↑ synaptic clefts (SC). Same material filling primary synaptic cleft is present in secondary synaptic cleft. All secondary synaptic clefts constitute ↑ subneural apparatus. ↑ Synaptic vesicles (SV) discharge ↑ acetylcholine into S. following arrival of nerve impulse. (Fig. = rat)

↑ Motor end plate

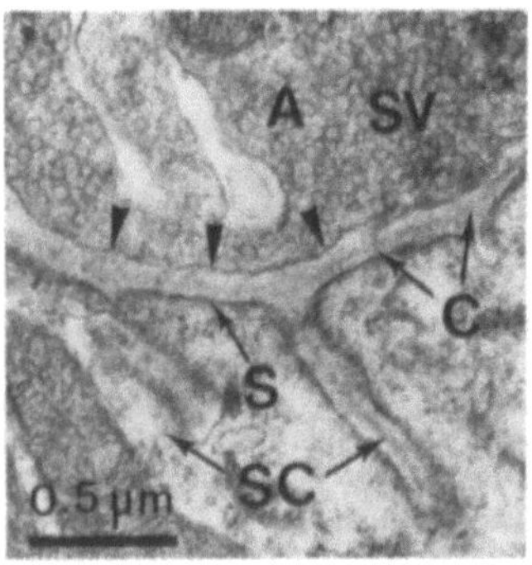

Synaptic cleft, of neuronal synapse (fissura synaptica*): Space (arrowhead) 15–30 nm in width, separating presynaptic membrane (Pre) of ↑ bouton terminal (BT) from postsynaptic membrane (Post). A certain amount of fine granular and moderately osmiophilic material (substantia intrafissuralis*) fills S., frequently forming 5-nm filaments extending perpendicularly across S. (↑ intrasynaptic filaments). Chemically, this material is a ↑ glycoprotein and represents a specialized kind of ↑ glycocalyx. It is responsible not only for adhesiveness between neurons, but also for their recognition in order to form specific functional connections (↑ neurobiotaxis). ↑ Synaptic vesicles (SV) empty into S. when nerve impulse arrives. (Fig. = rat)

↑ Diffusion channels

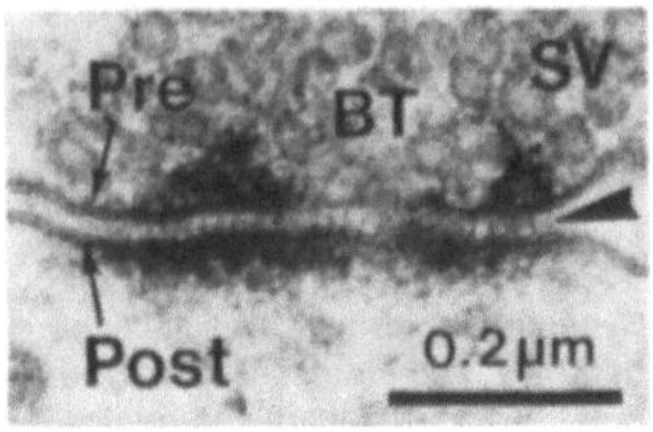

Pfenninger, K.H.: Progr. Histochem. Cytochem. *5*, 1 (1973)

Synaptic cleft, primary: ↑ Synaptic trough

Synaptic clefts, secondary (junctional folds, subneural clefts): Simple or ramified infoldings of ↑ sarcolemma of ↑ synaptic trough, penetrating 1–2 µm into ↑ sarcoplasm of ↑ skeletal muscle fiber. Together S. form ↑ subneural apparatus.

↑ Motor end plate; ↑ Synaptic cleft, of motor end plate

Synaptic ribbons: Arciform to round plates (SR), about 30 nm thick and 0.2–0.5 µm high, found in ↑ cone pedicles and ↑ rod spherules at site where ↑ dendrites of ↑ horizontal (H) and ↑ bipolar cells (B) penetrate. A leaflet, approximately 3 nm thick, is visible within S. Each S. is surrounded by a halo of apparently empty ↑ synaptic vesicles (SV) containing an unknown transmitter (possibly ↑ gamma-aminobutyric acid). S. are also present in ↑ pinealocytes of some animals. (Fig. = rat)

↑ Triad, of retinal neurons

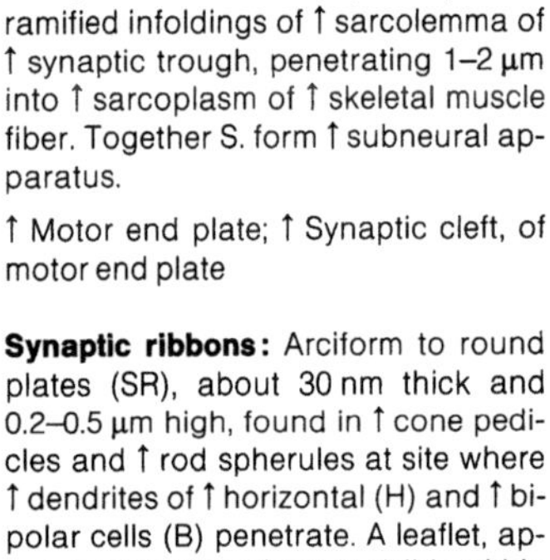

Karasek, M., King, T.S., Brokaw, J., Hansen, J.T., Petterborg, L.J., Reiter, R.J.: Anat. Rec. *205*, 93 (1983); Kosaras, B., Welker, H.A., Vollrath, L.: Anat. Embryol. *166*, 219 (1983)

Synaptic rods: ↑ Synaptic bars

Synaptic trough (primary synaptic cleft): Shallow single or branched depression (ST) on surface of ↑ skeletal muscle fiber occupied by ↑ axon (A). ↑ Sarcolemma (S) of S. forms numerous secondary ↑ synaptic clefts (SSC) constituting ↑ subneural apparatus (SA). A glycoprotein layer lining S. and penetrating into subneural apparatus has not been represented. BL = ↑ basal lamina, T = ↑ teloglia

↑ Motor end plate; ↑ Synaptic cleft, of motor end plate

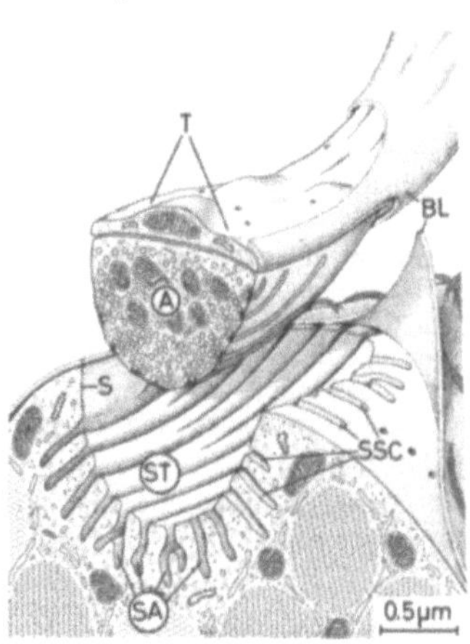

Synaptic vesicles (vesicula presynaptica*): ↑ Unit membrane-enclosed spherical structures present in ↑ boutons terminaux of chemical ↑ synapses and around ↑ synaptic bars and ↑ synaptic ribbons. Types of S.: ↑ Spherical or S-vesicles = S. with electron-lucent center measuring 40–60 nm in diameter and containing in most cases ↑ acetylcholine as a ↑ neurotransmitter, but also ↑ serotonin, ↑ gamma-aminobutyric acid, etc. 2) Flattened or F-vesicles = S. in the form of discs 40–60 nm in diameter and about 30 nm thick; their content is unknown. 3) Granular, G-vesicles or ↑ dense-core vesicles = S. 40–60 nm in diameter, with an electron-dense dot 25 nm in diameter containing either ↑ epinephrine or ↑ norepinephrine. 4) Large vesicles = S. measuring 80–100 nm in diameter and containing a 50-nm, fine granular, highly osmiophilic core. This type of S., probably containing ↑ dopamine, may be present among vesicles of other types and in cholinergic ↑ synapses. It is not possible to discern the functional capacity of S. on the basis of their morphology. It was thought that F-vesicles were characteristic of inhibitory ↑ synapses, but this opinion was not generally accepted, since flattening of S. is possibly a ↑ fixation ↑ artifact.

↑ Synapse, G

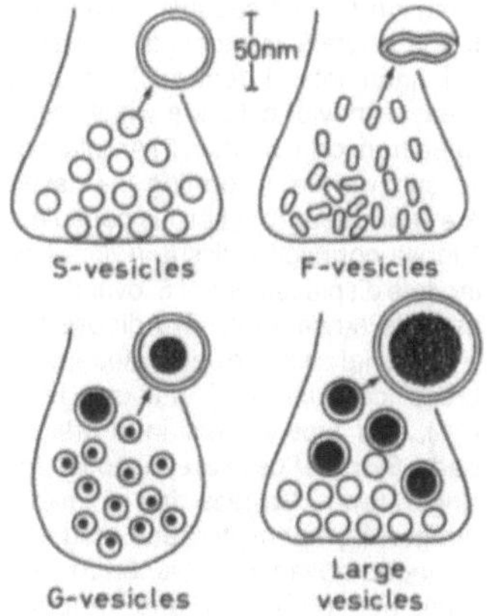

Bodian, D.: J. Cell Biol. *44*, 115 (1976); Paula-Barbosa, M.: Cell Tissue Res. *164*, 63 (1975); Rockel, A.J., Jones, E.G.: J. Comp. Neurol. *147*, 61 (1973); Uchizono, K.: Nature *207*, 642 (1965)

Synaptic vesicles, origin and recycling of: It is assumed that ↑ synaptic vesicles (SV) originate by pinching off from ↑ tubules of smooth endoplasmic reticulum (sER). Some of them are produced in ↑ perikaryon and transported to ↑ bouton terminal by ↑ axoplasmic flow. Others pinch off from smooth endoplasmic reticulum within boutons. Upon arrival of nerve impulse, vesicles move toward presynaptic membrane (PM), fuse with it, and discharge their ↑ neurotransmitter through ↑ diffusion channels (branched arrow) into ↑ synaptic cleft (SC). Incorporated membrane of former synaptic vesicle moves away from synaptic region (arrows) to be gradually transformed into a ↑ coated vesicle (CV). It is then taken up by ↑ micropinocytosis and moves toward smooth endoplasmic reticulum to fuse with its tubules. Simultaneously, another synaptic vesicle pinches off from smooth reticulum.

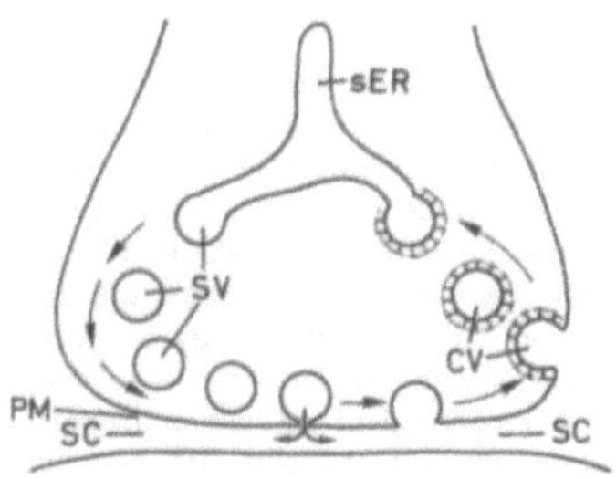

Heuser, J.E., Reese, T.S.: Structure of the Synapse. In: Kandel, E.R. (ed.): Handbook of Physiology. The Nervous System. Vol. 1, Part I. Bethesda: American Physiological Society 1977; Turner, P.T., Harris, A.B.: Nature *242*, 57 (1973)

Synaptoid vesicles: Spherical structures (SV), 30–50 nm in diameter, forming groups among ↑ neurosecretory granules (G) in endings of ↑ neurosecretory axons. S. do not stay in contact with any presynaptic membrane. They have electron-lucent interiors and contain a ↑ zinc iodide-osmium tetroxide-positive lipid substance. Function of S. is unknown. (Fig. = ↑ neurohypophysis, rat)

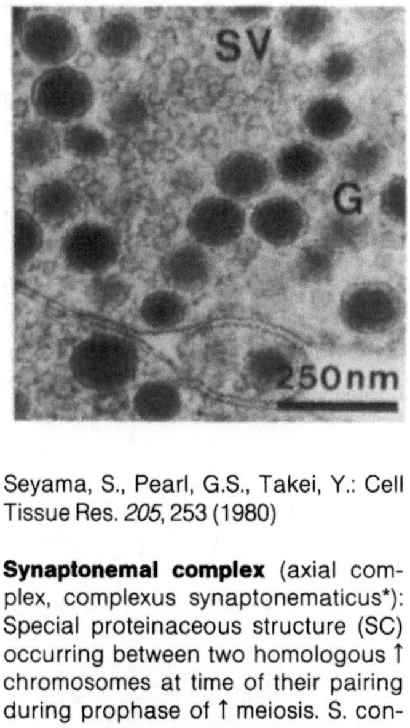

Seyama, S., Pearl, G.S., Takei, Y.: Cell Tissue Res. *205*, 253 (1980)

Synaptonemal complex (axial complex, complexus synaptonematicus*): Special proteinaceous structure (SC) occurring between two homologous ↑ chromosomes at time of their pairing during prophase of ↑ meiosis. S. consists of two dense lateral elements (LE), 20–80 nm wide (lateral lines or arms), formed of electron-dense fibrillar material representing axial filament of chromosome. Between two lateral lines is a central or medial element (CE) connected to lateral elements by fine transverse filaments (arrowheads). It is thought that pairing needed for recombination of DNA molecules of homologous ↑ chromatids takes place inside central element. S. is attached to inner aspect of ↑ nuclear membrane. Chemically, it is proteinaceous in nature and contains ↑ histones; transverse filaments contain DNA. (Fig. = primary ↑ spermatocyte, human)

↑ Synapsis, in meiosis

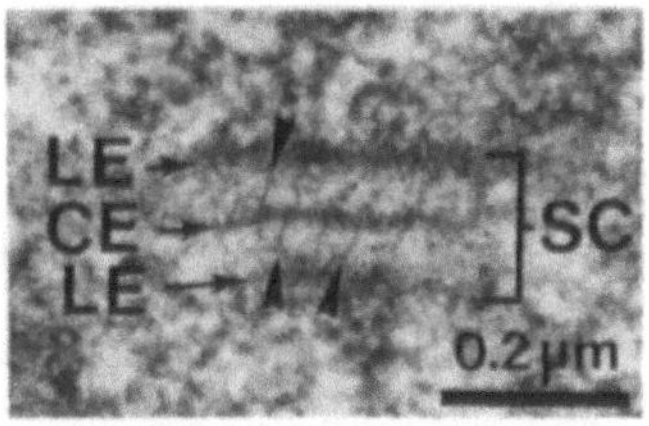

Maguire, M.P.: Exp. Cell. Res. *112*, 297 (1978); Moens, P.B.: Int. Rev. Cytol. *35*, 117 (1973); Solari, A.J.: Int. Rev. Cytol. *38*, 273 (1974); Westergaard, M., von Wettstein, D.: Ann. Rev. Genet. *6*, 77 (1972)

Synaptosomes: Detached ↑ boutons terminaux obtained by density gradient ↑ centrifugation of ↑ nervous tissue.

Meller, K.: Cell Tissue Res. *231*, 347 (1983); Whittaker, V.P.: The Subcellular Fractionation of Nervous Tissue. In: Bourne, G. (ed.): The Structure and Function of Nervous Tissue. III. New York, London: Academic Press 1969

Synarthrosis: Immovable or partly movable ↑ joint without an articular cavity. Depending on tissue that unites two bones (↑ dense connective tissue, fibrous ↑ cartilage, hyaline ↑ cartilage, ↑ bone), one can distinguish: 1) ↑ Syndesmosis, 2) ↑ symphysis, 3) ↑ synchondrosis, and 4) ↑ synostosis.

Synchondrosis*: Kind of ↑ synarthrosis in which hyaline ↑ cartilage temporarily connects two bones. At birth, an S. persists between os sphenoidale and os occipitale. All ↑ epiphyseal plates are S.

Syncytiotrophoblast (syncytiotrophoblastus*): External layer of ↑ trophoblast.

Syncytium: Multinuclear protoplasmic mass formed by fusion of many originally separate cells, and accompanied by consecutive disappearance of their cell membranes (e.g., ↑ skeletal muscle fibers, ↑ syncytiotrophoblast).

↑ Plasmodium

Syndesmosis*: Kind of ↑ joint in which ↑ dense connective tissue (DCT) connects two bones. ↑ Suturae (S) of skull bones are an example of a S.; another is ↑ gomphosis.

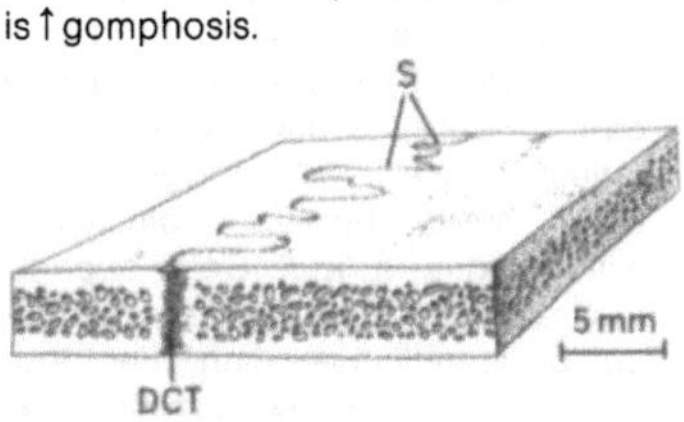

Synostosis: Kind of ↑ synarthrosis in which parts of skeleton become secondarily connected by ↑ bone. Examples: until puberty, osilium, pubis, and ischii are connected by hyaline ↑ cartilage (allowing for growth), which is replaced by bone during indirect ↑ bone formation to form os coxae. After having been connected by bone, sacral vertebrae form sacrum.

Synovial cells (synoviocyte, cellula synovialis*): Cells grouped into 1–6 epithelioid layers along inner border of ↑ synovial membrane (SM). S. are not epithelial cells, but fibroblastlike elements of mesenchymal origin lying among, rather than on, ↑ collagen fibers (CF) of fibrous stroma. There are two types of S.: 1) A-cells (macrophagelike cells, M-cells, synoviocytus phagocyticus*) = cells with elongated cell body, an oval nucleus, voluminous mitochondria, a moderately-developed Golgi apparatus, and some cisternae of rough endoplasmic reticulum. Cell membrane forms numerous irregular and ramified microvilli (Mv), many peripheral vacuoles, and ↑ micropinocytotic vesicles. A-cells do not overlap and are not held together by ↑ junctional complexes, as are epithelial cells. Their basal lamina is incomplete or nonexistent. ↑ Lysosomes and ↑ phagosomes (Ph) are present in each cell. A-cells are thought to be ↑ phagocytes. 2) B-cells (fibroblastlike cells, F-cells, synoviocytus secretorius*) = cells with polygonal cell body, round nucleus, voluminous mitochondria, a large Golgi apparatus, numerous cisternae of rough endoplasmic reticulum, many free ribosomes, and dense ↑ secretory granules (SG) 280–320 nm in diameter. Coarse cell processes (CP) containing secretory granules penetrate into synovial cavity. There are no junctional complexes between cells. Basal lamina exists; however, it is frequently interrupted on cell surfaces in contact with collagen fibers. Evidence is that B-cells secrete ↑ proteoglycans and ↑ hyaluronic acid, playing an important role in metabolism of synovial membrane and composition of ↑ synovial fluid. In some areas both A- and B-cells are missing; here, underlying tissue is directly exposed to synovial fluid. Cap = capillary

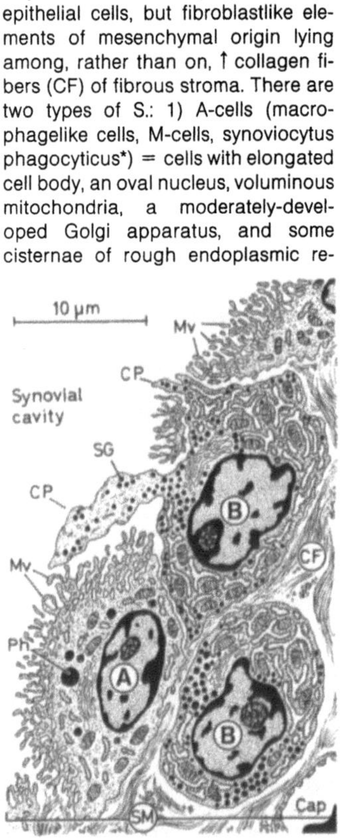

Graabeck, P.M.: J. Ultrastruct. Res. *78*, 321 (1982); Linck, G., Porte, A.: Biol. Cellulaire *42*, 147 (1981); Okada, Y., Nakanishi, I., Kajakawa, K.: Cell Tissue Res. *216*, 131 (1981)

Synovial fluid (synovia): Viscous transparent liquid present in small amounts in each articular cavity. S. is about 95% water; it is a protein-containing ultrafiltrate of blood circulating in vessels of ↑ synovial membrane in which ↑ synovial B-cells add ↑ proteoglycans and ↑ hyaluronic acid. S. lubricates articular surfaces and nourishes ↑ articular cartilage by diffusion.

Geigy Scientific Tables. Body Fluids. 8th edn. Basel: Ciba Geigy 1983

Synovial membrane (membrana synovialis, stratum synoviale*): Layer (SM) lining inner surface of articular capsule, except for ↑ articular cartilage. S. may be thrown into folds (SF) and processes [↑ articular villi, (AV)]. Synovial cells (SC) contained in S. tend to group along inner border of S.; in their relation to fibrous capsule (FC) three types of S. can be distinguished: 1) Fibrous type = synovial cells lie directly on fibrous capsule. This type is found where S. is exposed to pressure (↑ ligaments, ↑ tendons); in some areas, synovial cells are widely separated from one another. 2) Areolar type = synovial cells, forming 3–6 layers, are separated from fibrous capsule by ↑ loose connective tissue (CT), allowing free displacement of S. over fibrous capsule (knee joint). 3) Adipose type = synovial cells are separated from fibrous capsule by ↑ adipose tissue (AT); this type covers intra-articular pads. Synovial cells are disposed in a single layer. Besides few ↑ macrophages, ↑ lymphocytes, and irregularly arranged collagen fibers, S. contains numerous blood (V) and lymphatic vessels, as well as ↑ myelinated and ↑ nonmylinated nerve fibers.

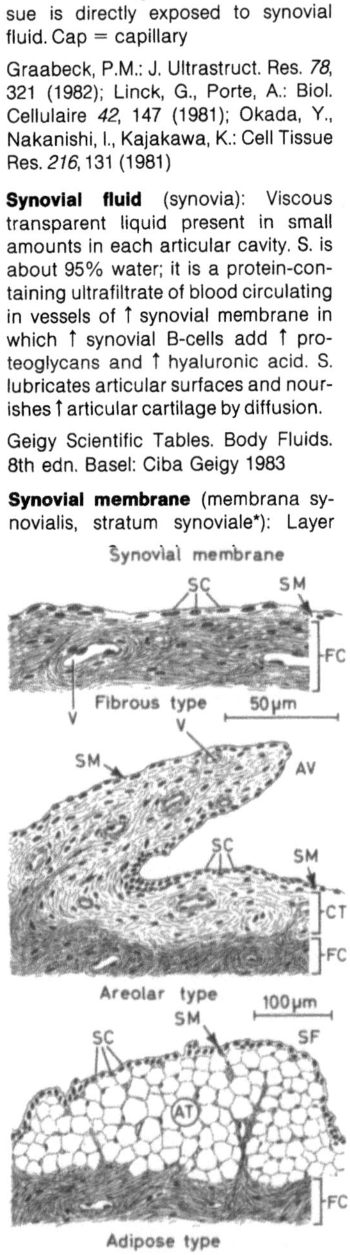

Date, K.: Arch. Histol. Jpn. *42*, 517 (1979); Franchimont, P. (ed.): Articular Synovium. International Symposium on Articular Synovium, Bruges, October 1982, Basel: Karger 1982; Hasselbacher, P.: Clin. Rheum. Dis. *7*, 57 (1981); Wassilev, W.: Verh. Anat. Ges. *75*, 221 (1981)

Synovial vagina, of tendon (vagina synovialis tendinis*): Double-walled serous tube (SV) in which segments of some ↑ tendons (T) slide. S. consists of an inner ↑ serous membrane (IS) connected by ↑ loose connective tissue of ↑ epitendineum (E) to tendon, and an outer serous membrane (OS) lining interior surface of ↑ fibrous vagina (FV) of tendon. In area of ↑ mesotendineum (M) these two membranes are apposed; at ends of S. they are continuous with one another. S. is lined with ↑ mesothelial cells which release a small amount of ↑ synovial fluid into

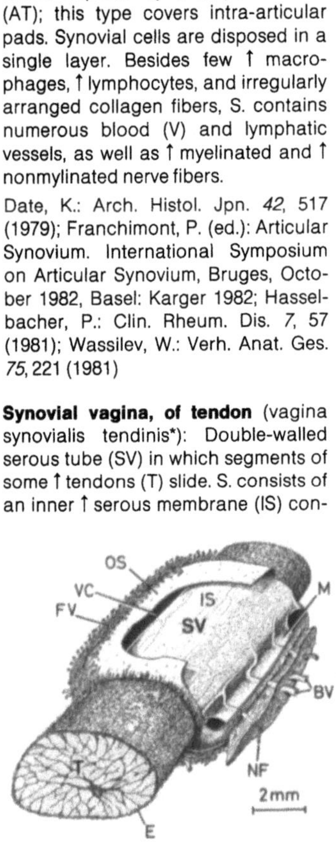

vaginal cavity (VC), facilitating sliding of tendons. BV = blood vessels, NF = ↑ nerve fiber

Synoviocytes: ↑ Synovial cells

Systemic circulation: Total of blood vessels distributing and collecting blood from body, with exception of ↑ lungs.

System, of organs: ↑ Organ system

T

T_3: ↑ Triiodothyronine

T_4: ↑ Thyroxin

Tactile cells: Modified ↑ Schwann's cells present in ↑ corpuscle of Meissner.

Taeniae, of colon (taeniae coli*): Three thickenings of longitudinal layer of ↑ tunica muscularis of ↑ colon.

Lineback, P.E.: Am. J. Anat. *36*, 357 (1925)

Tandem spindle: Two ↑ neuromuscular spindles arranged in line; ↑ nuclear bag fibers continue from first spindle to second.

Tannic acid staining: Technique used in ↑ light and ↑ transmission electron microscopy for demonstration of amorphous ↑ elastin, microfibrillar structures, and ↑ proteoglycans within ↑ basal lamina, ↑ basal labyrinth, ↑ terminal web, etc.

Cotta-Pereira, G., Guerra-Rodrigo, F., David-Ferreira, J.F.: Stain Technol. *51*, 7 (1976)

Tanning, of skin: Histophysiological response of ↑ melanocytes and ↑ melanosomes on exposure to sunlight. As an immediate response, melanosomes in melanocytes and cells of germinal layer of ↑ epidermis transform into ↑ melanin granules; after several days melanocytes synthesize new melanosomes, i.e., melanin granules.

Tanycytes: Specialized ↑ ependymal cells (T) with a variously long basal process (BP), the end of which is directly apposed onto a capillary (Cap) of underlying ↑ neuropil. A T. is a cuboidal or columnar cell, 10–15 µm in diameter, with an elongated nucleus and one or two distinct nucleoli. Cytoplasm encloses numerous mitochondria predominantly grouped in ↑ basal pole, a prominent supra- or infranuclear Golgi apparatus, cisternae of rough endoplasmic reticulum, some lysosomes, ↑ multivesicular bodies, ↑ microtubules, numerous free ↑ ribosomes, and ↑ glycogen particles. A smooth, unbranched basal process extends from basal pole, ending in a terminal foot (TF) on capillary ↑ basal lamina. ↑ Apical pole bears numerous microvillilike extensions and rare ↑ cilia (C). T. occur in some areas of lateral wall of third ventricles, in infundibular recess, in ↑ median eminence, where they establish a morphofunctional link between ventricular lumen and vasculature of median eminence, i.e., with ↑ portal system of ↑ hypophysis. They are also present in ↑ organum vasculosum laminae terminalis with a function similar to that of median eminence, i.e., integration of two separate and distinct fluid components – ↑ cerebrospinal fluid and blood.

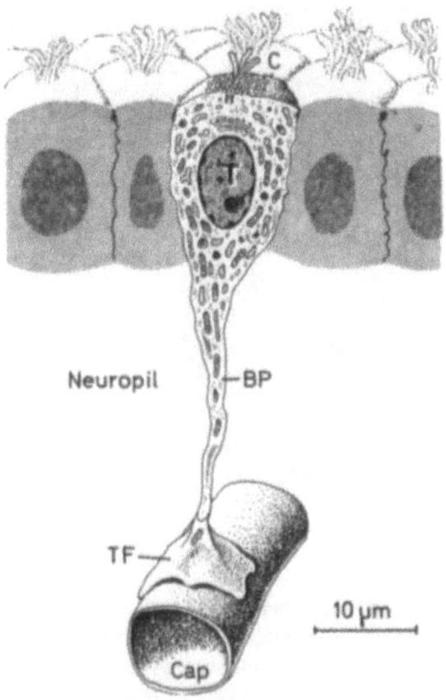

Bruni, J.E., Clattenburg, R.E., Millar, E.: Anat. Anz. *153*, 53 (1983); Burnett, B.T., Felten, D.L.: Anat. Rec. *200*, 337 (1981); Coates, P.W., Davis, S.L.: Anat. Rec. *203*, 179 (1982); Güldner, F.-H., Wolff, J.R.: Brain Res. *61*, 217 (1973); Scott, D.E., Pauli, W.K.: Cell Tissue Res. *200*, 329 (1979)

Tapetum lucidum*: Layer of highly differentiated cells interposed between stratum vasculosum and ↑ choriocapillaris of ↑ choroid of some nocturnal animals. Tapetal cells are characterized by parallel rod-shaped inclusions originating from ↑ endoplasmic reticulum. Recent data support hypothesis that tapetal cells are highly specialized choroidal ↑ melanocytes which reflect light toward ↑ photoreceptors.

Braekevelt, C.R.: Anat. Embryol. *163*, 201 (1981); Lesiuk, T.P., Braekevelt, C.R.: J. Anat. *136*, 157 (1983)

Tarsal glands (Meibomian gland, glandula tarsalis*): Branched ↑ sebaceous glands (T) arranged in layer within ↑ tarsal plate (TP) of ↑ eyelids. T. are parallel to one another and perpendicular to tarsal plate. Orifices of T. form a row at free edge of lids posterior to ↑ eyelashes (E). Transversally sectioned, T. display flowerlike figure (inset). There are about 30 T. in upper lid and 25 in lower lid; oily product of T. prevents overflow of ↑ tears and seals lid margins when eyes are closed.

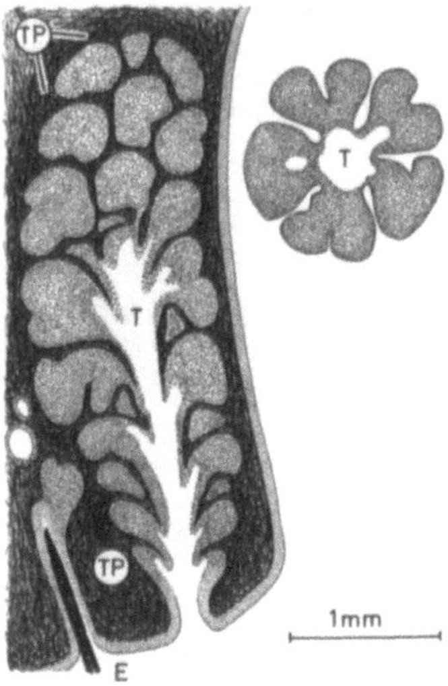

Tarsal muscles (musculus tarsalis superior et inferior*): 1) Superior T. = small ↑ smooth muscle whose cells intermingle with ↑ skeletal muscle fibers of musculus levator palpebrae superioris. 2) Inferior T. = minute smooth muscle located between inferior ↑ tarsal plate and fornix conjunctivae inferior. Tonus of both muscles regulates lid slit (rima palpebrarum).

↑ Eyelids

Tarsal plates (tarsus*): Crescentlike plaques (TP) of ↑ dense connective tissue making up skeleton of ↑ eyelids. Upper T. is about 10 mm in breadth, and lower, about 5 mm. ↑ Tarsal glands (TG) are lodged in T. Rigidity of eyelids in due to T.

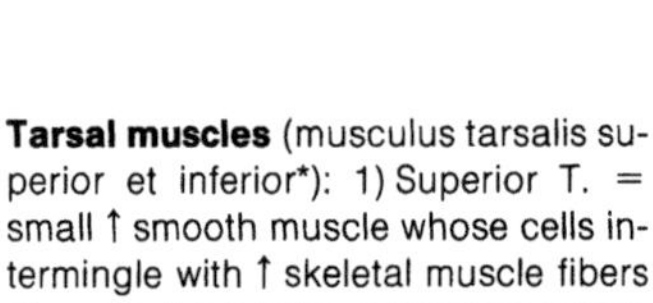

Tarsus: ↑ Tarsal plate

Taste buds (caliculus gustatorius*): Barrel-shaped ↑ chemoreceptors (TB) present primarily in epithelium (E) of lateral sides of ↑ circumvallate papillae, and less frequently in epithelium of adjacent wall of circular furrow. T. occur rarely in ↑ foliate and ↑ fungiform papillae, soft ↑ palate, and laryngeal surface of ↑ epiglottis. There are about 2000 T. in man, half of them situated in circumvallate papillae. T. consist of five cell types; cell divisions lead to division of whole T. In such cases, a fine epithelial septum separates two T. (Fig. = rabbit)

↑ Taste buds, structure of

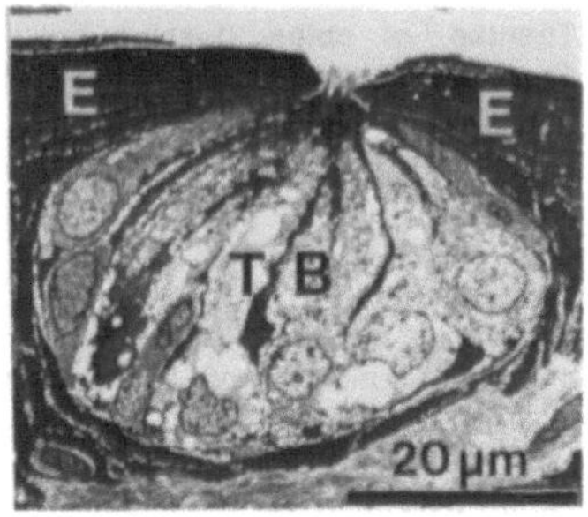

Farbman, A.I.: Cell Tissue Kinet. *13*, 349 (1980); Fujimoto, S.: Arch. Histol. Jpn. *45*, 365 (1982); Toyoshima, K., Shimamura, A.: Biomed. Res. *2*, Suppl. 459 (1981)

Taste buds, structure of: Onionlike groups 20–30 fusiform cells located within ↑ stratified squamous epithelium (E) of some lingual papillae, palate, etc. T. extend from ↑ basal lamina (BL) to ↑ taste chamber (TC), which communicates with oral cavity by means of ↑ taste pore (TP). T. consist of five cell types. 1) Type I cells (dark cells, supporting cells, cellula sustentacularis*) = tall columnar cells with ↑ microvilli (Mv) at ↑ apical pole. Elongated heterochromatin-rich nucleus is located in basal cell portion. Cytoplasm contains numerous mitochondria, cisternae of rough endoplasmic reticulum, many free ribosomes, and a dark ↑ cytoplasmic matrix. In vicinity of supranuclear Golgi apparatus are dense ↑ unit membrane-bound granules (Gr) whose ↑ glycosaminoglycan (Gly) content is secreted into taste chamber. 2) Type II cells (clear cells, taste cells, cellula gustatoria*) = narrow cells extending from basal lamina to taste chamber. Their apical plasmalemma forms a bouquet of coarse microvilli (formerly called taste hairs) immersed in glycosaminoglycan material within taste chamber. Large, clear round nucleus is located in basal cell half. Mitochondria and rough and smooth endoplasmic reticulum are well developed; several primary and secondary ↑ lysosomes, free ribosomes, as well as some microfilaments are also present. In contact with basal cell part are nonmyelinated intragemmal nerve endings (IGE). Absorption of substances onto microvillar plasmalemma is converted into electric impulses, transmitted to nerve endings, and perceived as taste. 3) Type III cells (clear cells, taste cells, cellula gustatoria*) = columnar cells similar to those of type II. At basal plasmalemma there are densities (D) surrounded by synapselike vesicles. Club-shaped nerve endings remain in contact with these densities. Cells of both types II and III are considered sensory elements, i.e., principal gustatory transducers. 4) Type IV cells (basal cells, cellula basalis*) = small undifferentiated cells located at base of taste buds. They have poorly developed ↑ organelles and a relatively large nucleus. Basal cells divide and differentiate into other taste bud cells; their processes may ensheath nerve endings. It is believed that type IV cells arise from adjacent perigemmal cells. 5) Type V (perigemmal cells, peripheral cells) = crescentlike cells located at periphery of taste buds. They have a central elongated nucleus, scarce organelles, many microfibrillar bundles, and some ribosomes. Cells enclose perigemmal nerve endings (PGE). Cells of taste buds have a life span of about 10 days. According to recent studies, it seems that all five types of taste cells are evolutionary stages of same cell type (basal cells).

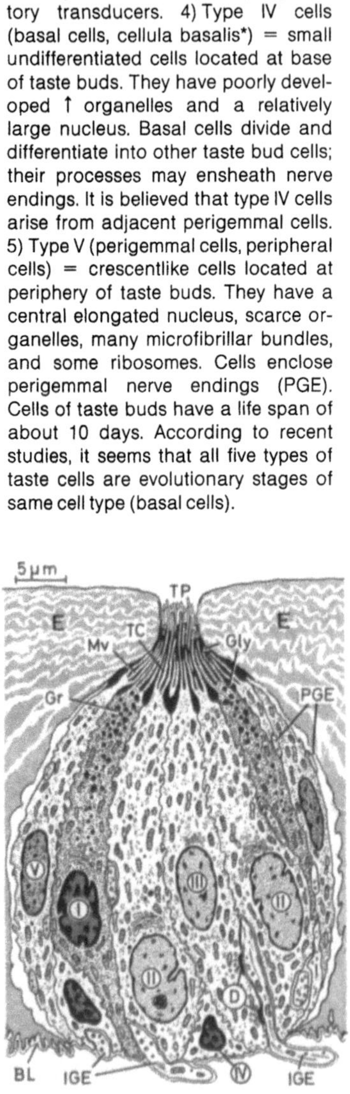

Arvidson, K., Cottler-Fox, M., Friberg, U.: J. Anat. *133*, 271 (1981); Farbman, A.I.: Cell Tissue Kinet. *13*, 349 (1980); Takeda, M., Hoshino, T.: Arch. Histol. Jpn. *37*, 395 (1975); Takeda, M.: Arch. Histol. Jpn. *40*, 243 (1977); Wolstenholme, G.E.W., Knight, J. (eds.): Taste and Smell in Vertebrates. London: J. & A. Churchill 1970

Taste cells: ↑ Taste buds, structure of

Taste chamber: Small dilatation (TC) at bottom of ↑ taste pore (TP), partially filled with osmiophilic ↑ glycosaminoglycan substance (Gly) into which ↑ microvilli (Mv) of various taste cells penetrate. E = lingual epithelium (Fig. = rabbit)

↑ Taste buds, structure of

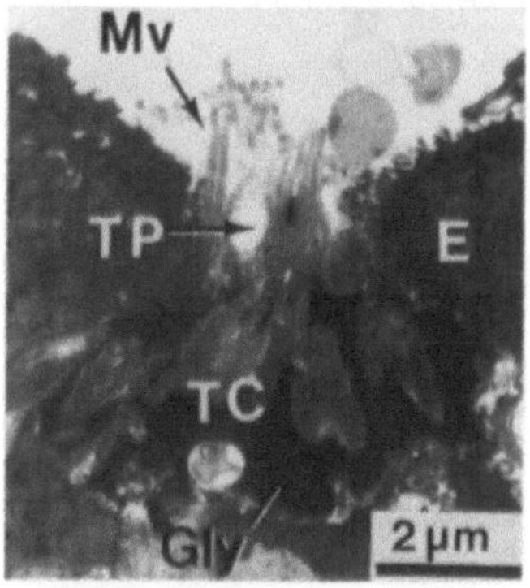

Taste hairs: ↑ Taste buds, structure of

Taste pit: ↑ Taste pore

Taste pore (taste pit, porus gustatorius*): Small opening in epithelium covering a ↑ taste bud connecting ↑ taste chamber with oral cavity.

T-cells: ↑ Transitional cells

Tears: Sterile, colorless, watery liquid, product of ↑ lacrimal glands. T. contain sodium chloride, traces of proteins, and some bactericide enzymes (↑ lysozyme). In awake state lacrimal glands produce about 0.5 g T./day.

Geigy Scientific Tables. Body Fluids. 8th edn. Basel: Ciba-Geigy 1983

Tectorial membrane (membrana tectoria*): Jellylike cuticular structure (TM) extending from vestibular lip (VL) of ↑ spiral limbus (SL) to slightly beyond external row of outer ↑ hair cells (HC) of organ of ↑ Corti (C). T. has a striated appearance and consists of radial microfibrils, 9 nm thick, embedded in a homogeneous substance. A nonstriated dark band on under surface of T. is called stripe of Hensen (HS); tips of tallest ↑ hairs of hair cells are affixed to this layer. Gelatinous substance of T. is secreted by ↑ interdental cells (IC); it is composed of mucoproteins and ↑ glycosaminoglycans. RM = ↑ Reissner's membrane

↑ Cuticle

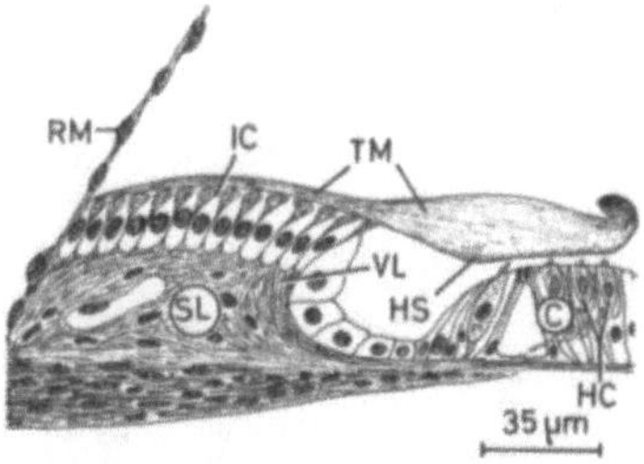

Kawabata, I., Nomura, Y.: Acta Otolaryngol. *91*, 29 (1981); Kronester-Frei, A.: Arch. Otorhinolaryngol. *224*, 3 (1979); Steel, K.P.: Hear. Res. *9*, 327 (1983); Tanaka, T., Takiguchi, T., Aoki, T.: Auris Nasus Larynx *3*, 109 (1976)

Teeth (dentes*): Hard conical organs with a central soft tissue, ↑ pulp (P). T. are implanted in corresponding ↑ alveoli of jaws. Parts: 1) Crown (corona dentis*) projects above ↑ gingiva (G). 2) Root (radix dentis*) is inserted into socket of ↑ alveolar bone (AB) and ends as apex. 3) Neck (cervix dentis*) is a narrow zone between crown and root. 4) Pulp chamber (cavitas dentis*) (PC), communicates with ↑ periodontal space (PS) through apical foramen (AF) at apex of tooth. Crown is covered by ↑ enamel (E); ↑ dentin (D) surrounds entire pulp chamber and ↑ cementum (C) surrounds root.

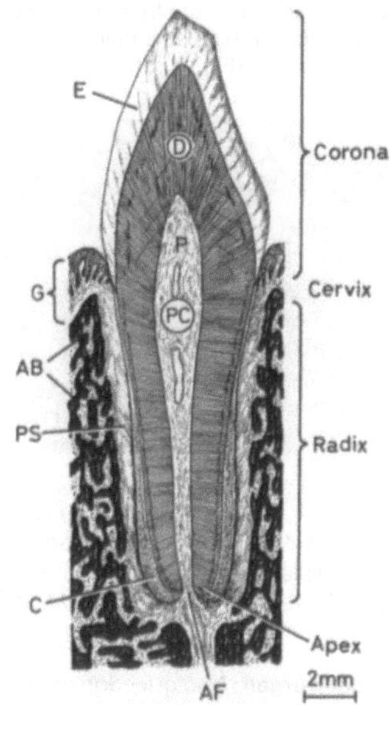

Kurten, B.: Teeth New York: Columbia 1982; Schumacher, G.-H., Schmidt, H.: Anatomie und Biochemie der Zähne. Stuttgart: Gustav Fischer 1983

Teeth, auditory: ↑ Auditory teeth

Teeth, primary (teeth deciduous): Set of first 20 teeth occurring after birth. T. are shed between 6th and 13th year and replaced by secondary teeth with identical histological structure.

↑ Teeth, replacement of

Teeth, replacement of: Up to 6th month of life, a primary and a secondary tooth occupy same ↑ alveolus. Later, they become separated by a thin bony lamella of ↑ alveolar bone. T. begins with resorption by ↑ osteoclasts (O) of this bony lamella, after which they resorb ↑ cementum (C) and ↑ dentin (D) of root of tooth, attacking even ↑ enamel. During resorption, osteoclasts form conspicuous ↑ Howship's lacunae (HL). Osteoclasts originate from ↑ periodontium (P) and ↑ pulp of primary tooth. Once devoid of its root, a tooth is shed. Development of secondary tooth is accompanied by formation of new alveolar bone and a new ↑ periodontal ligament.

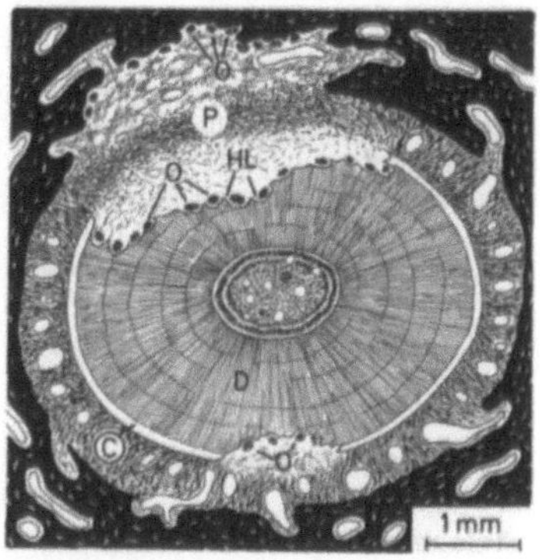

Teeth, secondary (teeth succedaneous, teeth permanent): Set of 32 teeth which gradually replace primary ↑ teeth.

Tela: More or less distinctly delimited layer of a ↑ tissue.

Tela choroidea* (choroidal tela): Very vascular ↑ lamina propria of ↑ choroid plexus. T. derives from ↑ pia mater and occurs at sites where it directly adheres to epithelial lining of ↑ brain ventricles. Here, pia mater protrudes into ventricles to form villi of choroid plexus.

Tela submucosa* (tunica submucosa): Layer of richly vascularized and innervated ↑ loose connective tissue situated between ↑ tunica mucosa and ↑ tunica muscularis of various tubular organs.

↑ Submucosal plexus

Tela subserosa* (lamina propria serosae, tunica subserosa): Layer of ↑ loose connective tissue (TS) situated between ↑ tunica muscularis (TM) and ↑ tunica serosa (TSe). (Fig. = ↑ colon, pig)

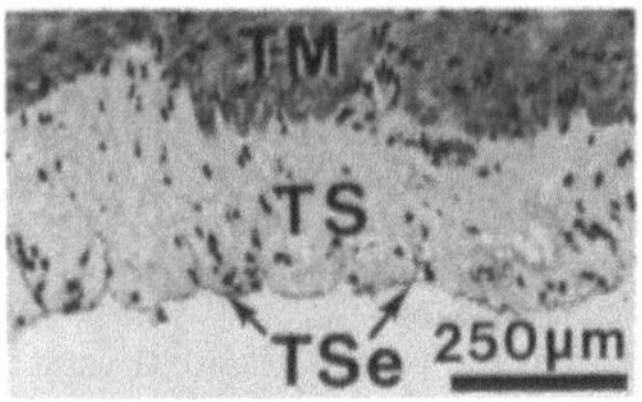

Teleceptors: Group of ↑ sensory receptors reacting to distant stimuli (↑ ear, ↑ eye, ↑ olfactory mucosa).

Telodendria (preterminal arborization): Many fine branches of an ↑ axon near its terminal structures.

Telogen hair: ↑ Club hair

Teloglia: Specially differentiated flattened ↑ Schwann's cells covering an ↑ axon of a ↑ motor end plate. T. are covered with ↑ basal lamina continuous with that of ↑ skeletal muscle fiber.

Telomere: ↑ Chromosomes, telomere of

Telophase, in meiosis: ↑ Meiosis

Telophase, in mitosis: ↑ Mitosis

TEM: ↑ Transmission electron microscope(y)

Tendinous cords: ↑ Chordae tendineae

Tendon cells (cellula tendinea*): Elongated fibroblastlike cells (T) with central elliptical nucleus and prominent nucleolus. Cytoplasm contains usual ↑ organelles; rough endoplasmic reticulum is well developed. With their long sheethlike processes (P), T. embrace ↑ tendon fibers (TF) and reach adjacent T. T. are responsible for synthesis and turnover of all kinds of fibers and ↑ ground substance of tendons. E = ↑ elastic fibers

↑ Tendon microfibrils

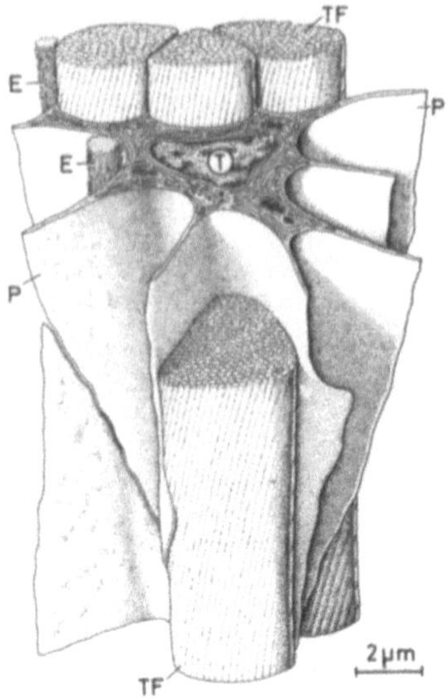

Trelstad, R.L., Hayashi, K.: Dev. Biol. *71*, 228 (1979)

Tendon fibers (fibreae tendineae*): Strong bundles of ↑ tendon microfibrils, among which there are some ↑ elastic fibers.

Tendon microfibrils: ↑ Collagen microfibrils (C) forming ↑ tendon fibers. T. range from 40 to 250 nm in diameter; longitudinally sectioned, they display identical morphological patterns and periodicity to collagen microfibrils elsewhere in organism. Transversally sectioned, T. are frequently interconnected by thin osmiophilic bridges (arrowheads). (Fig. = rat)

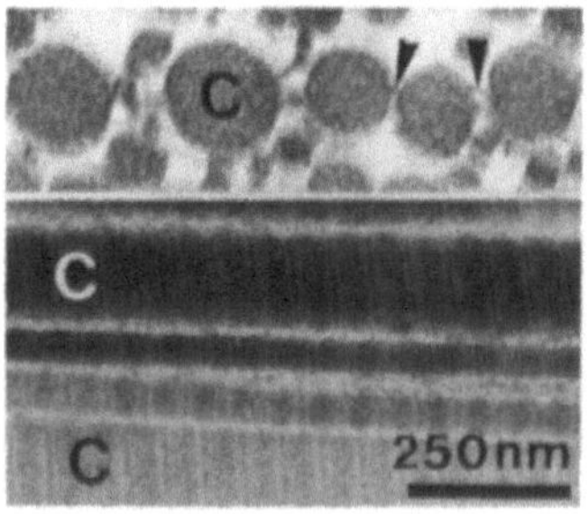

Dyer, F.R., Enna, C.D.: Cell Tissue Res. *168*, 247 (1976); Marchini, M., Morocutti, Castellani, P.P., Leonardi, L., Ruggeri, A.: Connect. Tissue Res. *11*, 103 (1983)

Tendon organ: ↑ Golgi tendon organ

Tendons (tendo*): ↑ Organs connecting ↑ skeletal muscles to bones. T. consist of regular ↑ dense connective tissue, i.e., ↑ tendon fibers arranged parallel to axis of T. Among fibers there are ↑ tendon cells. Tendon fibers are grouped into primary fascicles (PF) enclosed by an ↑ endotendineum; each group of primary fascicles forms a secondary fascicle (SF) surrounded by a ↑ peritendineum. All secondary fascicles are held together by ↑ epitendineum continuous with a layer of ↑ loose connective tissue, ↑ paratendineum. Some T. are enclosed by a ↑ synovial vagina and possess a ↑ mesotendineum. Blood vessels, lymphatics, and nerve fibers are situated in central region of T.

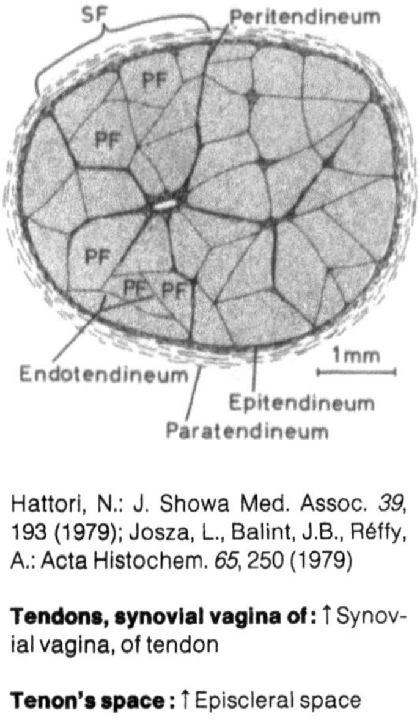

Hattori, N.: J. Showa Med. Assoc. *39*, 193 (1979); Josza, L., Balint, J.B., Réffy, A.: Acta Histochem. *65*, 250 (1979)

Tendons, synovial vagina of: ↑ Synovial vagina, of tendon

Tenon's space: ↑ Episcleral space

Tensor tympani muscle (musculus tensor tympani*): Small ↑ skeletal muscle attached by a fine ↑ tendon to manubrium of malleus. ↑ Endomysium of T. is rich in capillaries, nerve fibers, and ↑ adipose cells. Characteristic of T. are numerous ↑ neuromuscular spindles.

↑ Auditory ossicles; ↑ Tympanic cavity

Brzezinski, D.K.: Arch. Ohren Nasen Kehlkopfheilkd. *179*, 550 (1962); Veggetti, A., Mascarello, F., Carpène, E.: J. Anat. *135*, 333 (1982)

Terminal arteries: ↑ Muscular arteries with branches having no or very few ↑ anastomoses with neighboring arteries. Branches of ↑ coronary arteries and arteria centralis retinae are T.

Terminal bar: Dark dot or short dense thread visible with ↑ light microscope (LM) between adjacent cuboidal or columnar epithelial cells, immediately below their free surface. Horizontally sectioned, T. are visible as a beltlike, dense, continuous line surrounding perimeter of each cell. In ↑ transmission electron microscope (TEM), T. correspond to ↑ zonula occludens (ZO) and ↑ zonula adherens (ZA) of ↑ junctional complex.

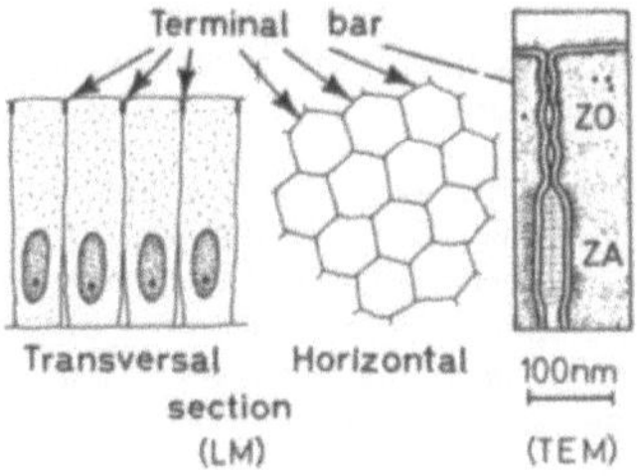

Terminal bile ductules: ↑ Bile ductules

Terminal body: ↑ Nucleoluslike body

Terminal bronchioles: ↑ Bronchioles, terminal

Terminal cisternae (cisterna terminalis*): Enlargements of ↑ sarcoplasmic reticulum (TC) into which tubules of ↑ L-system open. In ↑ skeletal muscle fibers, T. are continuous and completely surround each ↑ myofibril; two adjacent cisternae do not communicate, since between them a ↑ T-tubule is interposed to form a ↑ triad. In ↑ cardiac muscle cells, T. are flattened and discontinuous, leading to formation of ↑ dyads.

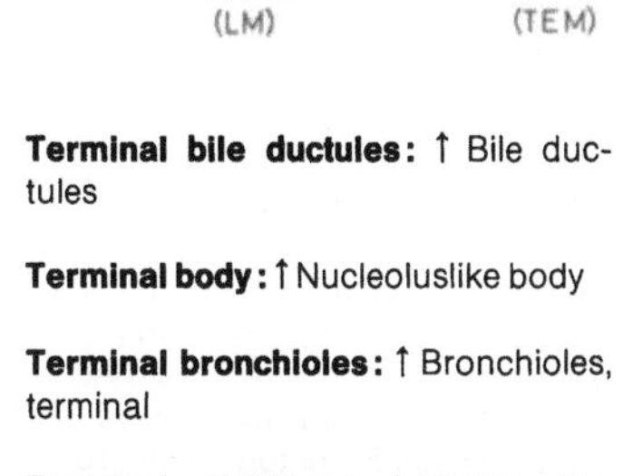

Terminal degeneration: Earliest morphological changes occurring during Wallerian degeneration which begins at nerve terminals: Swelling and loss of synaptic vesicles (12–24 h after ↑ axotomy); increase of ↑ neurofilaments with consequent increase in ↑ argyrophilia; ballooning of ↑ boutons terminaux, and increase in density of ↑ axoplasm ("dark reaction").

↑ Nerve fibers of peripheral nervous system, degeneration and regeneration of

Terminal ganglia: Category of small ↑ autonomic ganglia (mostly sympathetic) located close to or within organs they innervate. Structure of T. corresponds to that of other autonomic ganglia; it is likely that ↑ neurons of some T. are incompletely surrounded by ↑ satellite cells. T. include ciliary and cardiac ganglia, and ganglia of ↑ myenteric and ↑ submucosal plexus. Axons of ↑ prevertebral ganglia end in T.

↑ Sympathetic nervous system

Terminal hairs: Long, coarse, pigmented ↑ hairs covering scalp, pubis, axilla, beard, arms, and legs. ↑ Eyelashes, ↑ tragi, and ↑ vibrissae are also T.

Terminal hepatic venules: ↑ Central veins

Terminal portal venules: ↑ Inlet venules

Terminal villi: ↑ Placental villi extending freely into ↑ intervillous space.

Terminal web: Network of interwoven microfilaments (TW) immediately beneath free surface of some epithelial cells (in general, those with a ↑ brush border or ↑ striated border). Structure: 1) Apical zone (Az), or microvillar rootlets (R), consists of ↑ actin microfilaments closely associated with sealing elements of ↑ zonula occludens (ZO) and of core actin microfilaments (AM). 2) Adherens zone (Adz) is composed of 6 to 8-nm-thick actin microfilaments reinforced with 10-nm ↑ tonofilaments and ↑ intermediate filaments. All these microfilaments attach laterally in ↑ zonula adherens (ZA). 3) Basal zone (Bz) is formed by tonofilaments ending in ↑ desmosomes (D). Intermediate filaments extend deeper into cytoplasm. T. also contains α- ↑ actinin, short ↑ myosin

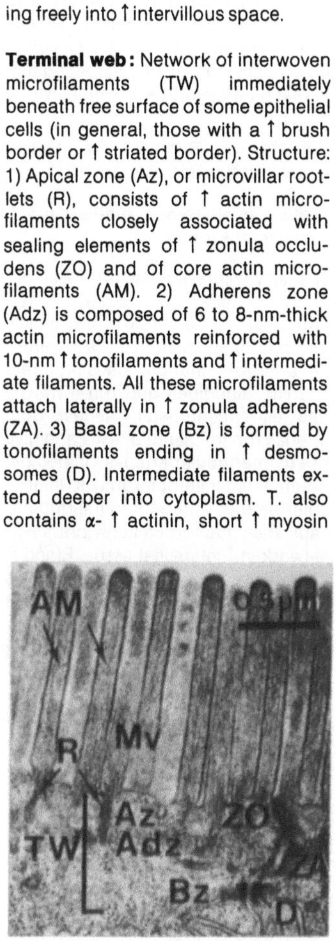

microfilaments, and ↑ tropomyosin, and contributes, as part of ↑ cytoskeleton, to mechanical stability of epithelial cells and contractility of ↑ microvilli (Mv). (Fig. = rat)

↑ Cuticular plate; ↑ Microvilli, shortening of

Horokawa, N., Tilney, L.G., Fujiwara, K., Heuser, J.E.: J. Cell Biol. *94*, 425 (1982); Hull, B.E., Staehelin, L.A.: J. Cell Biol. *81*, 67 (1979); Leeson, T.S.: J. Anat. *134*, 653 (1982)

Territorial matrix, of cartilage: ↑ Capsule, of chondrocytes

Tertiary cortex, of lymph node: ↑ Cortex, of lymph node

Tertiary follicle: ↑ Ovarian follicle, mature

Tertiary lymphatic nodules: ↑ Lymphatic nodules, tertiary

Testis*: Primary male sex organ, structured as a compound tubular ↑ exocrine gland for production of ↑ spermatozoa and an ↑ endocrine gland for production of ↑ testosterone. T. has an ovoid form; it measures 4.5 cm in length, 2.5 cm in breadth, and 3 cm in anteroposterior diameter; it weighs about 15 g. Structure: 1) ↑ Tunica albuginea (TA); at posterior side of T. it forms ↑ mediastinum testis (M). 2) ↑ Stroma = ↑ septula testis (S), delimiting about 250 ↑ lobuli testis (LT); septula converge toward ↑ mediastinum testis where blood, lymph vessels, and nerve fibers enter and leave. 3) ↑ Parenchyma = a) 500–1000 ↑ seminiferous tubules (ST), emptying into the ↑ rete testis (RT), and b) ↑ Leydig's cells. T. is situated in ↑ scrotum and partially enveloped by ↑ tunica vaginalis.

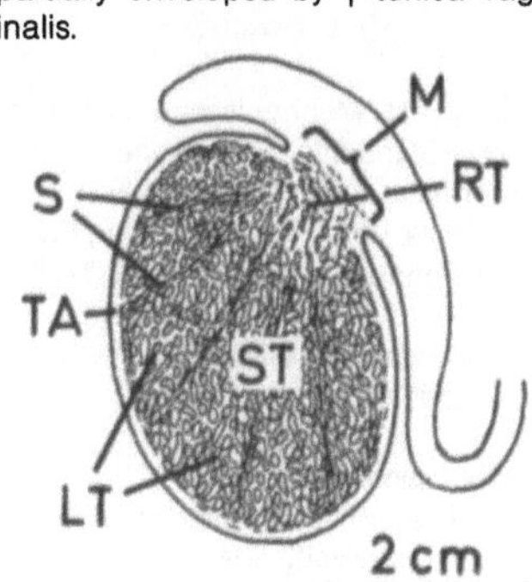

Burgos, M.H., Vitale-Calpe, R., Aoki, A.: Fine Structure of the Testis and Its Functional Significance. In: Johnson, A.D., Gomes, W.R. (eds.): The Testis, Vol. 1. New York, San Francisco, London: Academic Press 1970

Testis, control of secretion: ↑ Spermatogenesis is controlled by ↑ hypothalamohypophyseal complex, i.e., ↑ gonadotropin-releasing factor (LH/FSH-RF), which stimulates ↑ gonadotropes I of hypophysis to release ↑ follicle-stimulating hormone (FSH). It stimulates growth of ↑ seminiferous tubules and, via ↑ adenyl cyclase-cyclic AMP system, secretion of ↑ androgen-binding protein (ABP) by ↑ Sertoli's cells. Gonadotropin-releasing factor also acts on ↑ gonadotropes II to produce ↑ luteinizing hormone (LH/ICSH), which in turn stimulates ↑ Leydig's cells to produce ↑ testosterone, necessary for maturation of ↑ spermatozoa. By negative ↑ feedback (interrupted lines), testosterone inhibits LH/FSH-RF. Action of brain cortex and sense organs on hypothalamohypophyseal complex plays an important role in T.

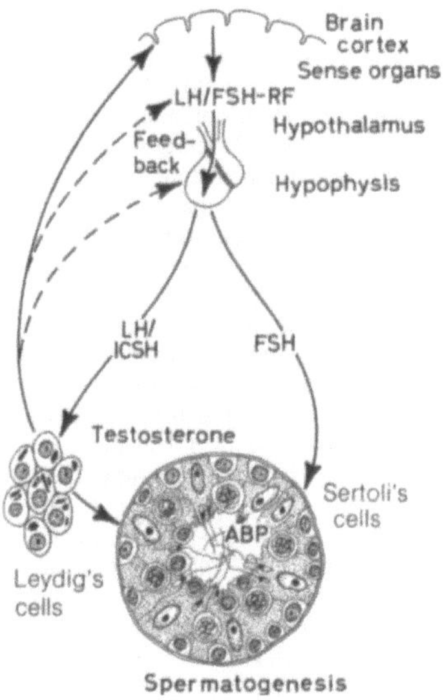

Means, A.R.: Mechanisms of Action of Follicle-Stimulating Hormone (FSH). In: Johnson, A.D., Gomes, W.R. (eds.): The Testis, Vol. 4. New York, San Francisco, London: Academic Press 1977

Testis, excretory ducts of: ↑ Ducts, excretory, of testis

Testis, histophysiology of: 1) Exocrine secretion = a) ↑ holocrine secretion of ↑ spermatozoa during ↑ spermatogenesis, beginning at puberty and lasting throughout life; b) ↑ eccrine secretion of ↑ androgen-binding protein by ↑ Sertoli's cells. 2) Endocrine function = production of ↑ testosterone by ↑ interstitial cells starting at puberty.

Ewing, L., Brown, B.L.: Testicular Steroidogenesis. In: Johnson, A.D., Gomes, W.R. (eds.): The Testis. Vol. 4. New York, San Francisco, London: Academic Press 1977; Hall, P.F.: Endocrinology of the Testis. In: Johnson, A.D., Gomes, W.R. (eds.): The Testis. Vol. 2. New York, San Francisco, London: Academic Press 1970

Testis, interstitial tissue of: Totality of tissue between ↑ seminiferous tubules. T. consists of ↑ loose connective tissue with numerous ↑ fibroblasts, ↑ fibrocytes, ↑ mast cells, ↑ lymphocytes, ↑ macrophages, ↑ myofibroblasts, blood and lymphatic vessels, ↑ nerve fibers, and strands or groups of ↑ interstitial cells.

Hooker, C.W.: The Intertubular Tissue of the Testis. In: Johnson, A.D., Gomes, W.R. (eds.): The Testis. Vol. 1. New York, San Francisco, London: Academic Press 1970

Testis, lobules of: ↑ Lobule, of testis

Testis, lymphatics of: ↑ Lymph of testis is collected by numerous thin-walled intertubular lymphatics and flows into larger lymphatic vessels along ↑ septula testis toward ↑ mediastinum testis and/or ↑ tunica albuginea to reach para-aortic and renal ↑ lymph nodes.

Holstein, A.F., Orlandini, G.E., Möller, R.: Cell Tissue Res. *200*, 15 (1979)

Testis, seminiferous tubules of: ↑ Seminiferous tubules

Testis, vascularization of: Testicular artery divides into several main branches that penetrate organ, mostly through ↑ mediastinum testis (centrifugal arteries); other branches perforate ↑ tunica albuginea and run toward ↑ rete testis (centripetal arteries). Both arteries run in ↑ septula testis and give rise to intertubular arteries located between ↑ seminiferous tubules. Interlobular arteries spread into intertubular capillaries that form a richly branched network in ↑ interstitial tissue. Blood is drained by ↑ postcapillary venules whose confluences form centrifugal veins directed toward tunica albuginea, and centripetal veins running toward rete testis. Blood is finally carried out from testis into pampiniform plexus.

Kormano, M., Suoranta, H.: Anat. Rec. *170*, 31 (1971)

Testosterone: Principal male sex hormone, synthesized by ↑ interstitial cells of ↑ testis. T. stimulates ↑ spermatogenesis, development and maintenance of male secondary sexual characteristics; it acts on brain to maintain male sexual behavior and inhibits, by a negative ↑ feedback mechanism, secretion of ↑ gonadotropin-releasing factor, as well as hypophyseal gonadotropins. T. has an anabolic effect, stimulating ↑ protein synthesis.

Tetrachrome staining: Staining technique combining four dyes for distinct demonstration of various organ constituents, (e.g., Herlant staining for ↑ hypophysis, etc.)

Tetrad (bivalent, chromosoma quadrivalens*): Four ↑ chromatids, but only two ↑ centromeres of a homologous ↑ chromosome pair (= two ↑ dyads) occurring during pachytene stage of meiotic prophase I.

↑ Meiosis

Tetraiodothyronine: ↑ Thyroxin

Tetraploidy: ↑ Polyploidy

Tetrazolium salts: Soluble colorless substances leading to ↑ formazan upon reduction. T. are widely used for cytochemical localization of various dehydrogenases, diaphorases, succinate oxidase system, and monoamine oxidase. They include nitro-blue T. (nitro B-T) and tetranitro-blue T.

Altman, F.P.: Prog. Histochem. Cytochem. *9*, 1 (1976)

TF: ↑ Transfer factor

Theca cone: Wedge-shaped proliferation (TC) of ↑ theca folliculi (TF), oriented toward ovarian surface. Growth of T. displaces primordial ↑ ovarian follicles (PF) laterally, so that they are not unduly compressed by expanding follicle.

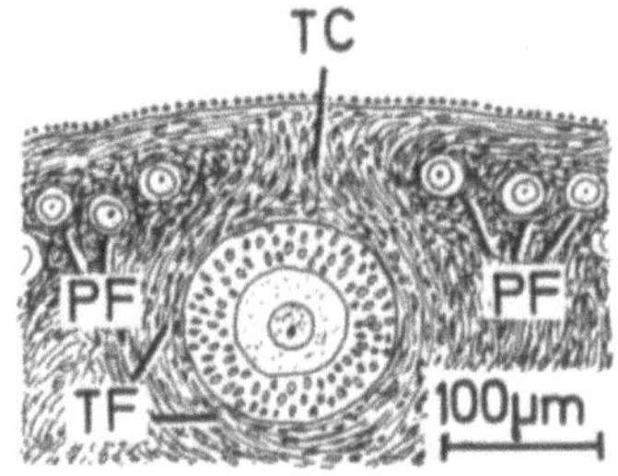

Theca externa: ↑ Theca folliculi; ↑ Thecal cells

Theca folliculi (thecae folliculi*): Bilayered concentric stromal sheath (TF) around growing and mature ↑ ovarian follicles (F). Structure: 1) Theca interna (TI) = a highly vascular and cellular ↑ loose connective tissue layer separated by ↑ glassy membrane

(GM) from ↑ membrana granulosa (MG). ↑ Thecal cells of theca interna become ↑ theca lutein cells after ↑ ovulation; simultaneously, loose connective tissue and capillaries penetrate ↑ corpus luteum and form its ↑ stroma. 2) Theca externa (TE) = a layer of concentrically arranged ↑ collagen fibers and a few fibroblastlike cells. In degenerating corpus luteum, connective tissue of this layer replaces its epithelium and forms ↑ corpus albicans.

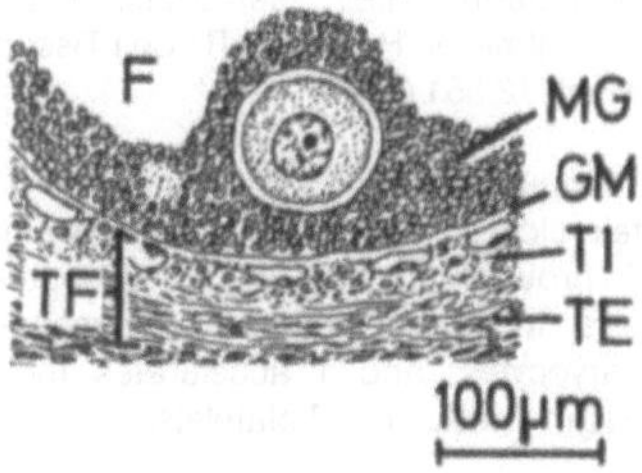

Theca interna: ↑ Theca folliculi; ↑ Thecal cells

Theca lutein cells (paralutein cells, thecoluteocytus*): Polygonal cells measuring up to 20 µm, forming minor, peripheral part of ↑ corpus luteum. T. have a round and ↑ heterochromatin-rich nucleus with a prominent nucleolus. Cytoplasm is somewhat darker than that of ↑ granulosa lutein cells and encloses rod-shaped and spherical mitochondria (M), some with tubules, a well-developed and multiple Golgi apparatus (G), flattened rough endoplasmic ↑ cisternae (rER), often arranged in whorls, an appreciable amount of smooth endoplasmic ↑ tubules (sER), frequently continuous with rER, ↑ lipid droplets (LD), a few lysosomes, and a considerable numer of free ribosomes. T. are derived from theca interna cells during ↑ luteinization; under control of ↑ prolactin, they produce ↑ progesterone.

↑ Lutein cells

Crisp, T.M., Derousky, D.A., Denys, F.R.: Am. J. Anat. *127*, 37 (1970)

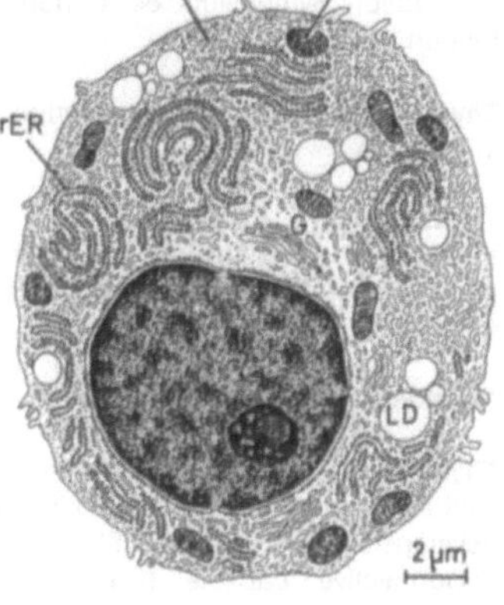

Thecal cells (cellulae thecales*): Two kinds of cells forming ↑ theca folliculi around antral and mature ↑ ovarian follicles. 1) Theca interna cells (TIC) = ovoid or spindle-shaped cells with an elliptical or round nucleus, numerous small mitochondria, and a moderately developed Golgi apparatus. Some have an extensively developed system of rough endoplasmic cisternae; others, particularly frequent before ↑ ovulation, have a considerable number of branched smooth endoplasmic ↑ tubules and ↑ lipid droplets. Cells contain a variable number of free ribosomes. They are thought to produce ↑ estrogens. After ↑ ovulation, some of them degenerate; others differentiate into ↑ theca lutein cells. 2) Theca externa cells (TEC) = fibroblastlike, elongated cells with a long, narrow nucleus, some round mitochondria, long rough endoplasmic ↑ cisternae, a small Golgi apparatus, ↑ glycogen particles, and a considerable number of contractile microfibrils. For this reason, some authors consider these cells ↑ myofibroblasts. Theca externa cells have no secretory activity; however, there is evidence that they play a role in postovulatory collapse of the follicle. Cap = capillary, F = ↑ follicular cells, of ovarian follicles (Fig. = rat)

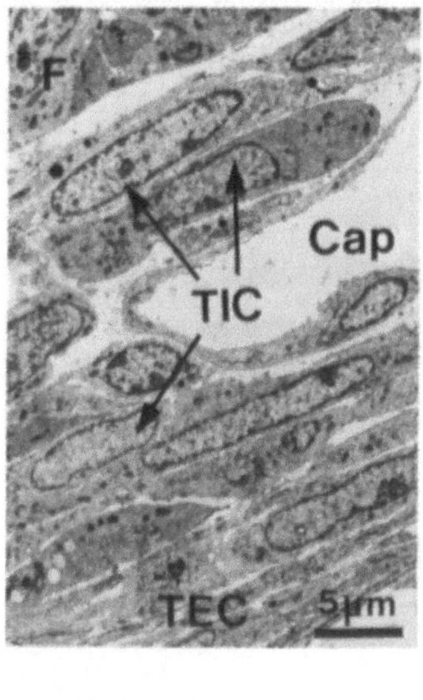

Hiura, M., Fujita, H.: Arch. Histol. Jpn. *40*, 95 (1977); Nunez Duran, H.: Acta Anat. *98*, 24 (1977)

Thermoreceptors: ↑ Sensory receptors reacting to heat and cold. Only cold receptors have been morphologically identified as nonmyelinated ↑ axons (A) invaginated a few microns into ↑ basal cells (BC) of ↑ epidermis. Axons are accompanied up to basal cell layer by ↑ Schwann's cells (Sch), ↑ basal lamina (BL) of which is continuous with that of epidermis. (Modified after Hensel et al. 1974)

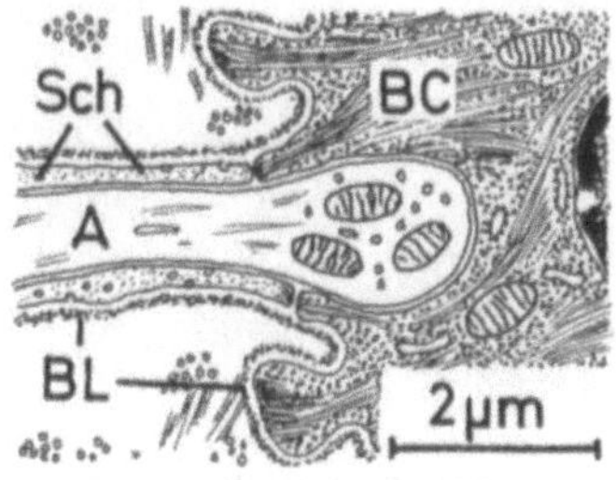

Halata, Z.: Diagnostik *11*, 39 (1978); Hensel, H., Andres, K.H., Düring, M.V.: Pflügers Arch. Ges. Physiol. *352*, 1 (1974)

Theta antigen (θ): Surface antigen characteristic of ↑ T-lymphocytes of mice.

Thiamine pyrophosphatase: Enzyme present in high concentrations in one to two cisternae of ↑ maturing face of ↑ Golgi apparatus. Role of T. in Golgi apparatus is not completely understood.

Thick filaments, of striated muscle: ↑ Myosin myofilament

Thin filaments, of striated muscle: ↑ Actin myofilament

Thin limb: ↑ Thin segment, of loop of Henle

Thin segment, of loop of Henle (thin limb): U-shaped tubule, 15–30 µm in diameter (TS), lined with ↑ simple squamous epithelium 1–5 µm thick, so that nuclei of some epithelial cells (E) bulge into lumen. T. is situated between straight portions of ↑ proximal and ↑ distal tubules of ↑ nephron. It is particularly long in juxtamedullary nephrons, short or absent in cortical nephrons. In both cases, T. lies within ↑ renal medulla. Two types of T. have been described: type I with numerous intercellular ↑ interdigitations and short ↑ zonulae occludentes, and type II without interdigitations and long zonulae occludentes. T. are permeable to active and passive transport of sodium and water, and apparently serve to concentrate urine. (Fig. = rat)

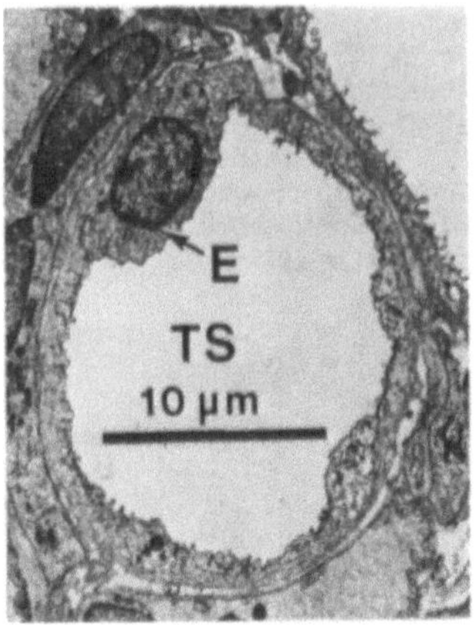

Bachmann, S., Kriz, W.: Cell Tissue Res. *225*, 111 (1982); Kokko, J.P.: Kidney Int. *22*, 449 (1982); Nagle, R.B., Altschuler, E.M., Dobyan, D., Dong, S., Bulger, R.E.: Am. J. Anat. *161*, 33 (1981); Schiller, A., Taugner, R., Kriz, W.: Cell Tissue Res. *207*, 249 (1980); Schwartz, M.M., Venkatachalam, M.A.: Kidney Int. *6*, 193 (1974)

Thionin: Basic dye used for staining of ↑ Nissl bodies and for ↑ metachromatic staining of ↑ mast cells and ↑ ground substance.

Thoracic duct (ductus thoracicus*): Main ↑ lymphatic vessel of body, communicating with venous circulation at angle between left vena jugularis interna and left vena subclavia. Structure:

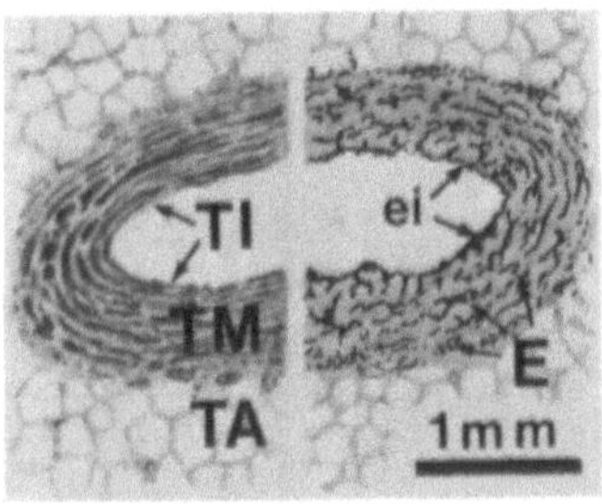

1) Tunica intima (TI) = a) ↑ endothelium and b) fine subendothelial ↑ loose connective tissue with internal ↑ elastic lamina (ei). 2) Tunica media (TM) = circularly and helicoidally arranged smooth muscle bundles intermingled with a richly developed elastic network (E). 3) Tunica adventitia (TA) = loose connective tissue with ↑ vasa and ↑ nervi vasorum. (Fig., left = conventional staining; fig., right = ↑ orcein staining)

Thrombin: Proteinase converting fibrinogen of ↑ fibrin.

Thrombocytes (thrombocytus*): Nucleate cells in blood of birds, homologous to ↑ platelets in mammals. Term "T." is largely used as a synonym for platelets.

Thrombocytopenia: Deficiency in number of ↑ platelets in the circulating blood (below 250×10^9/l, ↑ SI).

Thrombocytopoiesis: ↑ Thrombopoiesis

Thromboplastin, of platelets: Phospholipid, probably contained in α-granules of ↑ platelets; it supports ↑ blood clotting initiated by tissue ↑ thromboplastin.

Thromboplastin, of tissues: Factor 3 of ↑ blood clotting released from injured areas of blood vessels. T. starts a series of complex transformations of prothrombin to ↑ thrombin.

Thrombopoiesis (thrombocytopoesis*): The process of formation of ↑ platelets. Sequence: From a stem cell (↑ CFU-M) arises ↑ megakaryoblast, whose nucleus divides incompletely, the two sets of daughter ↑ chromosomes not separating, but forming a larger bilobulate nucleus with 4n chromosomes. After an interval another division of daughter sets of chromosomes follows, leading to lobulated ↑ polyploid nucleus (8n) of a ↑ promegakaryocyte. Number of chromosomes in nucleus continues to increase according to mechanism described up to 32n, and cell becomes a

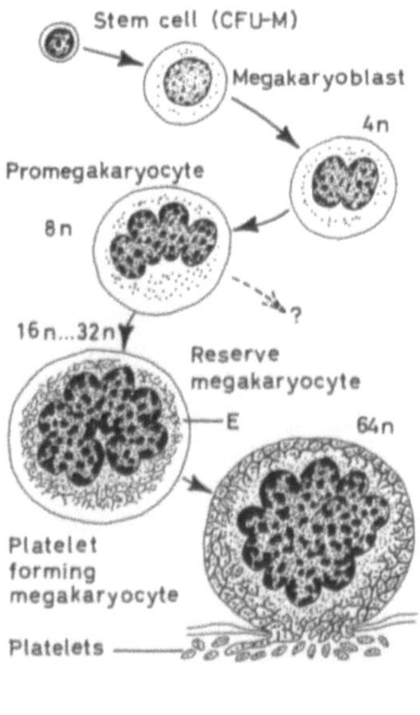

reserve megakaryocyte measuring 50–70 µm. From reserve megakaryocyte, characterized by a clear ↑ ectoplasm (E), differentiates the platelet-forming ↑ megakaryocyte (100 µm in diameter, 64n), whose cytoplasm becomes fragmentated by ↑ platelet demarcation channels into platelets. Transformation from stem cell to platelets requires about 10 days.

↑ Megakaryocytopoiesis; ↑ Platelets, release of

Wichmann, H.E., Gerhardts, M.D., Spechtmeyer, H., Gross, R.: Cell Tissue Kinet. *12*, 551 (1979)

Thrombopoietin: Serum ↑ glycoprotein closely related to ↑ erythropoietin. T. probably stimulates ↑ thrombopoiesis, increases number of ↑ megakaryocytes, and ↑ accelerates their fragmentation into ↑ platelets.

Thrombosthenin: Contractile protein (actomyosinlike complex?) associated with microfilaments of ↑ platelets. Contraction of T. provokes a change in shape of platelets and, together with released T., clot retraction. Some authors consider microfilaments of platelets to be made up of T.

↑ Blood clotting

Thrombus: Mixture of ↑ fibrin, ↑ platelets, and ↑ erythrocytes formed in a blood vessel of a living organism.

Thymic corpuscles: ↑ Hassall's bodies

Thymic humoral factor (thymosin): Protein containing 108 amino acid residues, probably produced by thymic ↑ epithelial-reticular cells, and stimulating ↑ differentiation and maturation of ↑ T-lymphocytes.

Bach, J.-F., Papiernik, M.: Cellular and Molecular Signals in T-Cell Differentiation. In: Porter, R., Whelan, I. (eds.): Microenvironments in Haemopoietic and Lymphoid Differentiation. Ciba Foundation Symposium 84. London: Pitman 1981

Thymocytes: ↑ T-lymphocytes lodged in ↑ thymus.

Thymopoietin: Presumably thymic hormone inducing formation of T-lymphocyte surface markers on precursors of ↑ T-lymphocytes.

Thymosin: ↑ Thymic humoral factor

Thymus*: ↑ Lymphoepithelial organ located in anterior mediastinum; most developed before puberty. Structure: 1) Connective capsule (Ca); 2) ↑

stroma = ramified septa (S) penetrating from capsule into parenchyma; 3) ↑ parenchyma = ↑ lymphoepithelial tissue subdivided into polyhedrallob--ules (L) 0.2–3 mm in diameter. Labules display a darker peripheral zone rich in ↑ lymphocytes, cortex (C), and a central zone rich in ↑ epithelial-reticular cells, medulla (M), with ↑ Hassall's corpuscles (HC). Lobules are continuous with one another via narrow medullar bridges.

↑ Thymus, cortex of; ↑ Thymus, medulla of

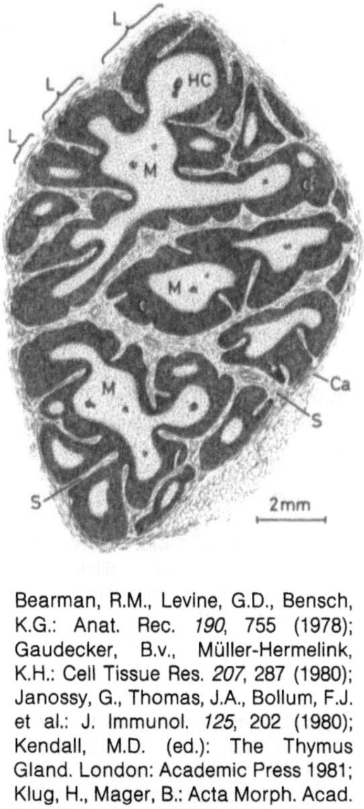

Bearman, R.M., Levine, G.D., Bensch, K.G.: Anat. Rec. *190*, 755 (1978); Gaudecker, B.v., Müller-Hermelink, K.H.: Cell Tissue Res. *207*, 287 (1980); Janossy, G., Thomas, J.A., Bollum, F.J. et al.: J. Immunol. *125*, 202 (1980); Kendall, M.D. (ed.): The Thymus Gland. London: Academic Press 1981; Klug, H., Mager, B.: Acta Morph. Acad. Sci. Hung. *27*, 11 (1979)

Thymus, accidental involution of: Acceleration of normal, gradual, age-dependent ↑ involution of ↑ thymus in response to a large variety of stress factors (acute infections, chronic diseases, poisoning, dietary deficiencies, ionizing radiation) and hormonal influences (adrenocorticotropic hormone, adrenal and gonadal steroids; T. during gravidity). Because of massive death of small cortical ↑ lymphocytes and their destruction by ↑ macrophages, thymus diminishes rapidly in size while fat accumulates in ↑ reticular-epithelial cells. In dietary deficiencies ↑ Hassall's bodies are reduced in number; in acute infection they are increased. During stronger T. there is ↑ atrophy of reticular-epithelial cells, reduction of thymic medulla, and disappearance of lobular structure and stromal connective tissue. (Fig., bottom modified after Hammar 1906)

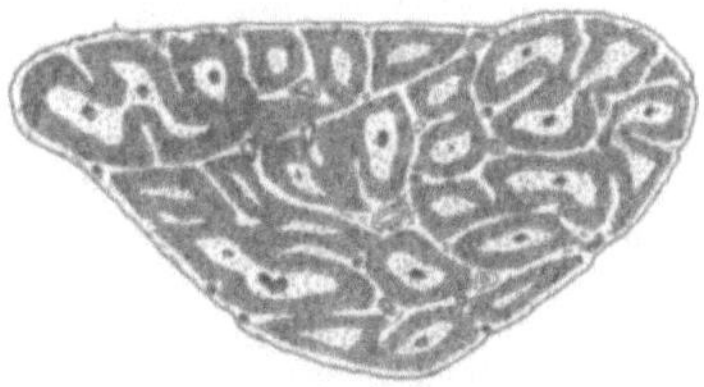

Thymus, 3-year-old boy

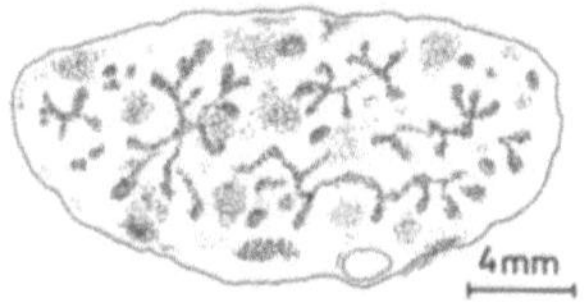

Thymus, 17-year-old boy: Accidental involution

Hammar, J.A.: Arch. Anat. Physiol. Suppl. 91 (1906)

Thymus, aged: ↑ Thymus, involuted

Thymus, blood barrier of: ↑ Blood-thymus barrier

Thymus, cortex of (cortex thymi*): Dark peripheral zone of thymic lobule separated from connective septa (S) by flattened ↑ epithelial-reticular cells (ERC). Between relatively few epithelial-reticular cells there is an extensive population of small ↑ T-lymphocytes (sL). Medium-sized (mL) and large lymphocytes (lL) are considerably less numerous. A few ↑ macrophages (M) and ↑ plasma cells are also present in T. (Fig. = rat)

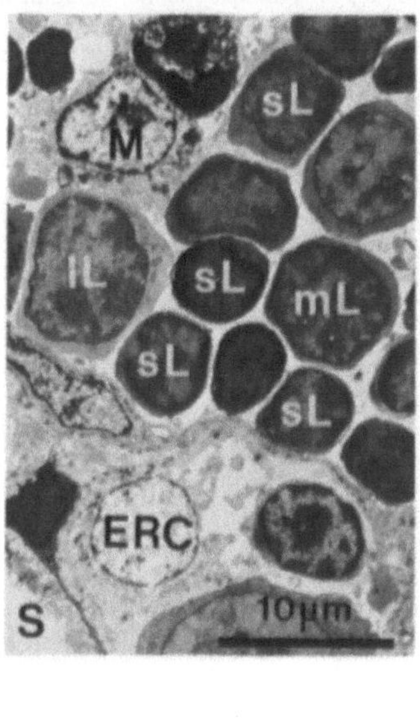

Thymus-dependent zones, of lymphatic organs: ↑ Lymphatic organs, thymus-dependent zones of

Thymus, functions of: As central immunological organ of the body, thymus is responsible for: 1) Production of ↑ T-lymphocytes, 2) providing T-lymphocytes with specific receptors capable of recognizing foreign proteins or antigens, and 3) production of ↑ thymic humoral factor.

Thymus, Hammar's myoid cells of: ↑ Hammar's myoid cells, of thymus

Thymus, Hassall's bodies of: ↑ Hassall's bodies, of thymus

Thymus, histophysiology of: In newborns and children up to puberty, ↑ thymus is well developed. From puberty on the thymus shows age ↑ involution changes and becomes gradually replaced by ↑ adipose tissue of anterior mediastinum; however, its function remains. Diseases, severe stress, toxins, ↑ ACTH, ↑ cortisol and gonadal steroids, etc. can provoke accidental involution of the thymus.

↑ Thymus, accidental involution of; ↑ Thymus, involution of

Haar, J.L.: Anat. Rec. *179*, 463 (1974)

Thymus-independent zones, of lymphatic organs: ↑ Lymphatic organs, thymus-independent zones of

Thymus, involution of: Gradual transformation of prepubertal thymus into corpus adiposum thymi. After puberty thymus undergoes a series of changes consisting in decrease in weight, de-

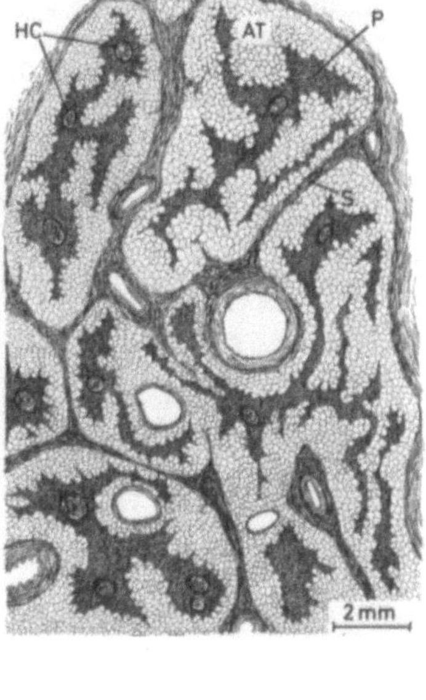

crease of lymphocyte production, and replacement of cortex by adipose tissue (AT). ↑ Parenchyma (P) shrinks, limit between the cortex and medulla gradually disappears, septa (S) enlarge, and ↑ Hassall's corpuscles (HC) increase in number and size. In adults thymus finally is transformed into a mass of ↑ adipose tissue (corpus adiposum thymi) containing scattered islands of parenchyma composed of enlarged ↑ reticular-epithelial cells. However, even with this change, thymus remains functional in the adult.

↑ Involution

Bartel, H.: Z. mikrosk.-anat. Forsch. *93*, 537 (1979); Gaudecker, B.v.: Cell Tissue Res. *186*, 507 (1978)

Thymus, lymphocyte circulation in: In fetal life thymic epithelium becomes colonized with stem cells (immature ↑ T-lymphocytes) from liver. Later, these immature T-lymphocytes originate from ↑ bone marrow. Lymphocytes divide in thymus and rapidly migrate into thymodependent zones of lymphatic organs; therefore ↑ lymphatic nodules are lacking in thymus. No ↑ lymph drains into thymus.

↑ Lymphatic organs, thymus-dependent zones of

Thymus, medulla of (medulla thymi*): Clear central zone of thymic lobules connecting each of them to a common treelike structure. ↑ Epithelial-reticular cells (ERC) are large, stellate and cover considerable areas. ↑ Lymphocytes are much less abundant; large and medium-sized lymphocytes (mL) predominate over small ↑ T-lymphocytes (sL); ↑ macrophages, ↑ plasma cells, and eosinophilic ↑ granulocytes (eG) are rare; occasional ↑ myelocytes can be found. ↑ Postcapillary venules are prominent. ↑ Hassall's corpuscles are characteristic for T. (Fig. = rat)

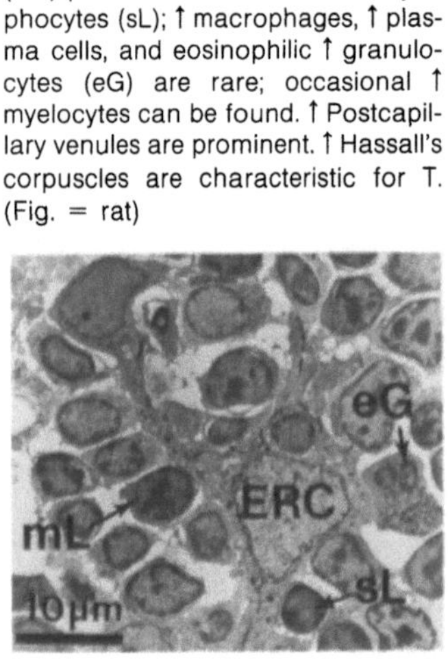

Thymus, vascularization of: Arteries (A) supplying thymus run through septa (S) and give off lateral branches situated in secondary septa. At corticomedullary boundary arteries divide into ↑ arterioles (Art) that follow corticomedullary boundary. Arterioles give off capillaries (Cap) for blood supply of cortex, whereas only a few capillaries are destined for blood supply of medulla. The capillaries form subcapsular arcades, turn toward medulla, and join to form ↑ postcapillary venules (Post). Venules empty into veins (V) also situated at corticomedullary boundary; veins drain into interlobular veins (IV) following arterial branching. A small portion of cortical capillaries drain directly into interlobular veins.

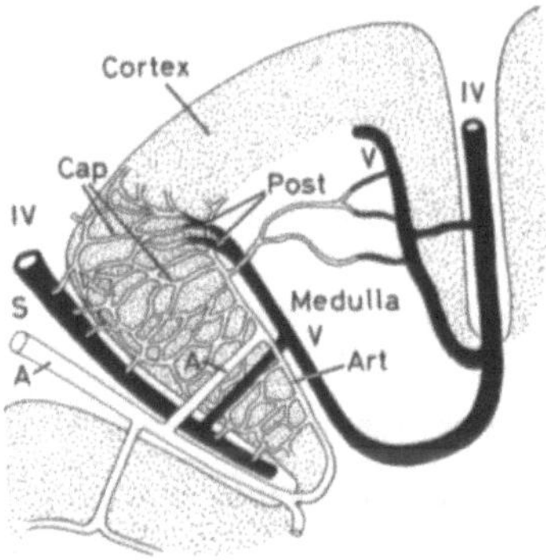

Bearman, R.M., Bensch, K.G., Levine, C.D.: Anat. Rec. *183*, 485 (1975); Irino, S., Takasugi, N., Murakami, T.: Scan. Electron Microsc. 1981/III, 89 (1981); Kardon, R.H., Kessel, R.G.: Biomed. Res. *2*, Suppl. 173 (1981)

Thyrocalcitonin: ↑ Calcitonin

Thyroglobulin: ↑ Glycoprotein synthesized and secreted into follicular lumen by ↑ thyroid follicular cells. T. contains 2%–4% hexosamine, galactose, mannose, fucose, other carbohydrates, and various iodinated amino acids (↑ triiodothyronine and tetraiodothyronine = ↑ thyroxin).

Marriq, C., Rolland, M., Lissitzky, S.: EMBO Journal *1*, 397 (1982)

Thyroid: ↑ Thyroid gland

Thyroid cartilage (cartilago thyroidea*): A ↑ hyaline cartilage of ↑ larynx.

Thyroid follicles (folliculus*): Basic structural units (F) of thyroid ↑ parenchyma in form of 50 to 500 µm-wide spheres. A T. is lined with ↑ thyroid follicular cells (FC) and filled with ↑ colloid. Each T. is surrounded by a dense network of fenestrated ↑ capillaries (Cap) and several nerve fibers (NF). In resting ↑ thyroid gland T. are large, with a flattened epithelium; in active gland they are small and lined with columnar epithelium. As a rule, ↑ C-cells (C) do not reach lumen of follicle.

↑ Ultimobranchial follicle

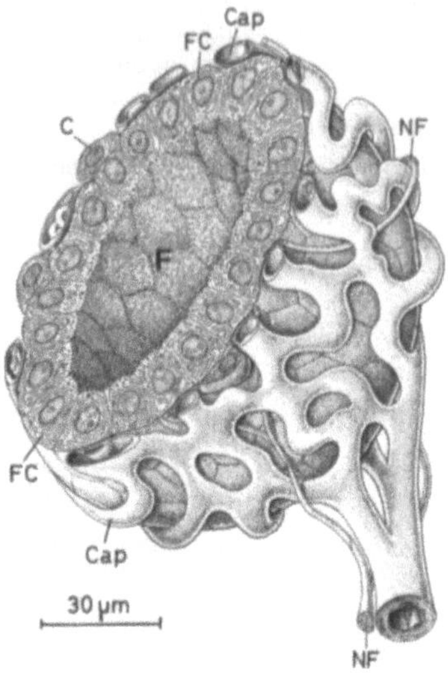

Chan, A.S.: Cell Tissue Res. *233*, 693 (1983); Sobrinho-Simoes, M., Johannessen, J.V.: Submicrosc. Cytol. *13*, 209 (1981); Young, B.A., Baker, T.G.: J. Anat. *135*, 407 (1982)

Thyroid follicles, "cold:" ↑ Thyrotropic hormone-unreactive follicles with extremely flattened epithelium; found in older individuals.

Tamura, S., Fujita, H.: Arch. Histol. Jpn. *44*, 177 (1981)

Thyroid follicular cells (cellula follicularis*): Cuboidal to columnar basophilic cells with a round central nucleus and a prominent nucleolus. Cytoplasm contains filamentous mitochondria, a number of branching and anastomosing rough endoplasmic ↑ cisternae, many free ribosomes, and medium-sized primary ↑ lysosomes (Ly). Well-developed Golgi apparatus is associated with ↑ unit membrane-enclosed ↑ apical vesicles (AV) loaded with colloid, each 50–200 nm in diameter, which fuse with apical plasmalemma and release their contents into follicle. Larger ↑ colloid vacuoles (CV) may also appear in ↑ apical pole. Free cell surface is studded with short microvilli (Mv). Adjacent cells are held together by ↑ interdigitations and ↑ junctional complexes. T. lie on a ↑ basal lamina (BL). Active T. (during synthesis of ↑ thyroglobulin and hormones, fig., top) become columnar;

Golgi apparatus enlarges, rough endoplasmic reticulum expands and fills with moderately dense ↑ reticuloplasm, and apical vesicles increase in number. At cell surface flaplike ↑ pseudopodia (P) appear, and microvilli become longer and more frequent. During release of hormones number of colloid vacuoles increases and T. become cuboidal, with extensive rough endoplasmic reticulum. Resting T. (fig., bottom) flatten, as do their nuclei and rough endoplasmic cisternae (rER). Apical vesicles become rare, microvilli decrease in number and shorten.

↑ Thyroid follicular cells, biosynthesis of hormones in

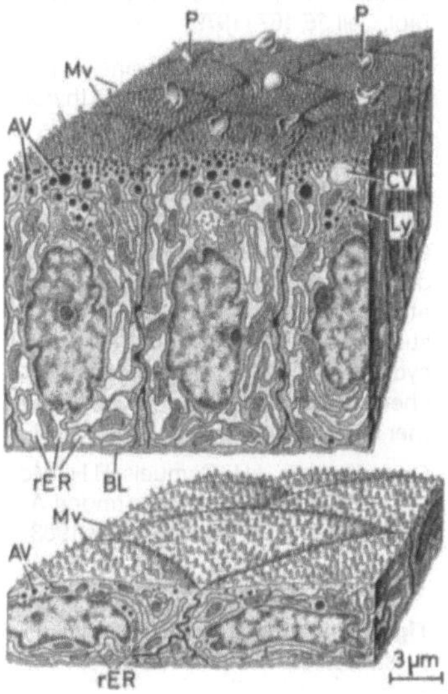

Herzog, V., Miller, F.: Biol. Cell *36*, 163 (1979); Garcia, J.G., Romero, A.R., Alonso, J.L.L.: An. Anat. *32*, 67 (1983); Klink, G.H., Certel, J.E., Winship, T.: Lab. Invest. *22*, 2 (1970); Nunez, E.A., Becker, D.: Am. J. Anat. *129*, 369 (1970); Miyagawa, J., Ishimura, K., Fujita, H.: Cell Tissue Res. *223*, 519 (1982)

Thyroid follicular cells, biosynthesis of hormones in: Follicular cells produce, store, and release ↑ triiodothyronine (T_3) and ↑ thyroxin (T_4) in following steps: 1) Synthesis and storage = a) uptake from capillaries (Cap) of amino acids (A), monosaccharides (M), and iodide ($2\,I^-$); b) synthesis of polypeptide chains in rough endoplasmic reticulum (ER); c) their transport to Golgi apparatus (G) where synthesis of ↑ thyroglubulin is achieved; d) ↑ exocytosis of thyroglobulin granules (TG) in form of ↑ apical vesicles into lumen of follicle; e) oxidation of iodide into iodine (I_2) by ↑ thyroperoxidase (H_2O_2); f) iodination of tyrosine residue of thyroglobulin at level of apical plasmalemma and successive formation of mono- (MIT) and diiodotyrosine (DIT). One MIT and one DIT molecule give one T_3 molecule, and two DIT molecules form one T_4 molecule. Material is stored as ↑ colloid (C). 2) Release: g) ↑ endocytosis of colloid (C); h) formation of a ↑ colloid vacuole (CV); i) fusion of lysosomes (Ly) with vacuole and formation of a ↑ phagosome (Ph) containing colloid; j) hydrolysis of thyroglobulin into globulin, T_3, and T_4; k) diffusion of both hormones into capillaries. Step f) is controversial: Some authors believe that iodination takes place in rough endoplasmic reticulum. (Modified after Fawcett et al. 1969)

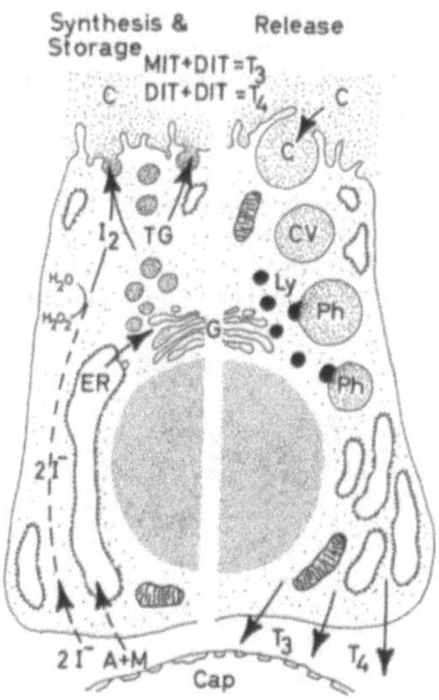

Fawcett, D.W., Long, J.A., Jones, A.L.: Progr. Hormone, Res. *25*, 315 (1969); Fujita, H.: Acta Histochem. Cytochem. *14*, 391 (1981)

Thyroid gland (glandula thyroidea*): ↑ Endocrine gland situated at anterior surface of ↑ trachea, just below ↑ larynx. Structure: 1) Capsule = dense connective tissue surrounding T.; 2) ↑ stroma = connective septa (S) with blood and lymphatic vessels and nerve fibers; 3) ↑ parenchyma = ↑ thyroid follicles (F) lined with a single layer of ↑ thyroid follicular cells (arrows). Peripheral to epithelium are ↑ C-cells (C), in contact with common ↑ basal lamina but not with lumen of follicles. T. is well vascularized; each follicle is surrounded by a dense pericapillary network; lymphatic vessels are numerous. Autonomic ↑ nonmyelinated nerve fibers innervate blood vessels. ↑ Regeneration of epithelium is effectuated by ↑ mitosis; follicles can multiply by budding. (Fig., top = resting human T.; fig., bottom = active rat T.)

↑ Ultimobranchial body

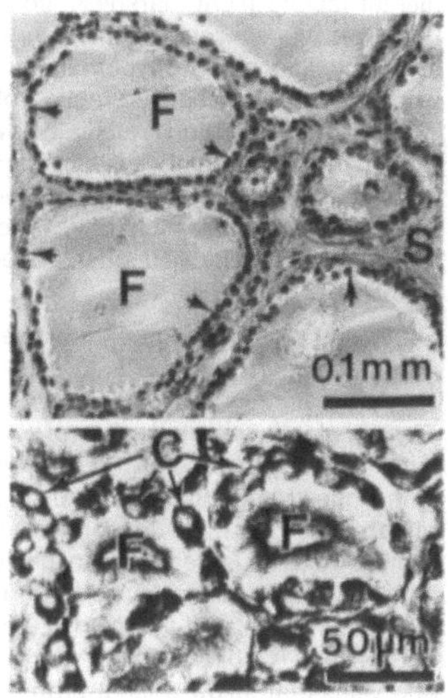

Chambard, M., Gabrion, J., Mauchamp, J.: J. Cell. Biol. *91*, 157 (1981); Garcia, G.J., Reina, L.J., Valls, G.B.: An. Anat. *32*, 47 (1983); Girod, C.: Introduction à l'étude des glandes endocrines. Villeurbanne: Simep 1980; Shimada, T.: Biomed. Res. *2*, Suppl. 243 (1981); Sobrinho-Simoes, M., Johannessen, J.V.: J. Submicr. Cytol. *13*, 209 (1981)

Thyroid gland, C-cells of: ↑ C-cells

Thyroid gland, colloid of: ↑ Colloid, of thyroid follicles

Thyroid gland, control of secretion: Under influence of ↑ thyrotropin-releasing factor (TSH-RF), ↑ thyrotropes of adenohypophysis release ↑ thyrotropic hormone (TSH), which acts on ↑ thyroid follicular cells to stimulate release of ↑ triiodothyronine (T_3) and ↑ thyroxin (T_4). Liberated thyroid hor-

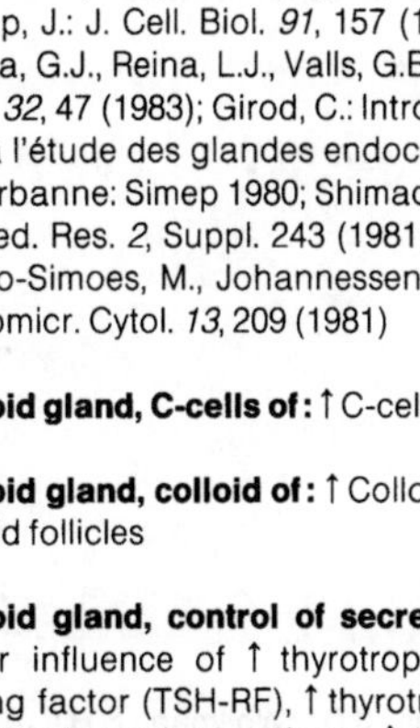

mones inhibit production of TSH, and perhaps TSH-RF, by negative ↑ feedback mechanism. Activity of thyroid gland depends upon concentration of iodine in blood, sexual cycle, environmental factors (light, darkness, cold, heat), etc. Via sensory organs, neural stimuli influence ↑ hypothalamus where TSH-RF is produced. Action of TSH on thyroid function is mediated by ↑ adenyl cyclase-cAMP system after binding of TSH to corresponding receptors on plasmalemma of thyroid follicular cells. (Modified after Faber and Haid 1972)

Faber, H.v., Haid, H.: Endokrinologie. Stuttgart: Eugen Ulmer 1972

Thyroid gland, vascularization of: Arteries for blood supply of ↑ thyroid gland penetrate its capsule and give off branches that run in connective septa. Arterioles spread into a richly anastomosing capillary network (Cap) surrounding ↑ thyroid follicles (F). Venous blood is collected into venules and veins that follow arterial branching. (Fig. = microcast, rat)

↑ Capillaries, fenestrated

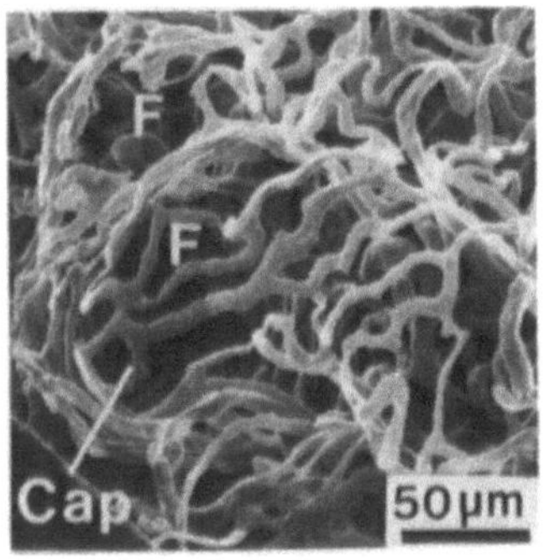

Zeligs, J.D., Wollman, S.H.: J. Ultrastruct. Res. *75*, 291 (1981)

Thyroid hormones: ↑ Calcitonin; ↑ Thyroxin; ↑ Triiodothyronine

Thyroid-stimulating hormone: ↑ Thyrotropic hormone

Thyroidectomy cells: Hypertrophic ↑ thyrotropes of pars distalis of ↑ hypophysis, occurring following ablation of ↑ thyroid gland. T. are characterized by an expanded Golgi apparatus, increased number and width of rough endoplasmic ↑ cisternae, and decreased number of ↑ secretory granules. Administration of thyrostatica provokes a similar effect.

Thyroliberin: ↑ Thyrotropin-releasing factor

Thyroperoxidase: ↑ Peroxidase present in ↑ thyroid follicular cells, responsible for iodination of ↑ thyroglobulin.

Thyrotropes (beta$_2$ basophils, beta cells, thyrotrophs, TSH-cells, cellula thyrotropica*): Irregular, polygonal, elongated or stellate cells grouped in center of pars distalis of ↑ hypophysis making up 1%–2% of its population. T. have a large elliptical nucleus, a well-developed, frequently multiple Golgi apparatus (G), numerous small mitochondria, some flattened rough endoplasmic ↑ cisternae arranged in parallel, a few ribosomes, and a clear cytoplasmic matrix. From Golgi apparatus arise numerous ↑ unit membrane-bound ↑ secretory granules (SG), 100–160 nm in diameter, with a moderately osmiophilic content. These are grouped in periphery of cell body and discharged into ↑ pericapillary (PS) and intercellular spaces by ↑ exocytosis (arrows). In light microscope, granules are strongly basophilic; they contain ↑ thyrotropic hormone. After thyroidectomy and administration of thyrostatica, T. transform into ↑ thyroidectomy cells. Cap = capillary

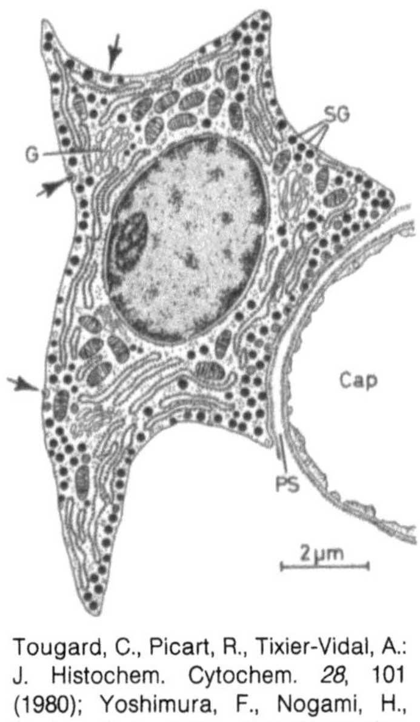

Tougard, C., Picart, R., Tixier-Vidal, A.: J. Histochem. Cytochem. *28*, 101 (1980); Yoshimura, F., Nogami, H., Yashiro, T.: Anat. Rec. *204*, 255 (1982)

Thyrotropic hormone (thyroid-stimulating hormone, TSH, thyrotropin, thyrostimuline): ↑ Glycoprotein (m. w. about 25000 daltons) secreted by ↑ thyrotropes of ↑ adenohypophysis. T. stimulates iodine uptake by ↑ thyroid follicular cells, and synthesis and release of iodine-containing thyroid hormones. Action of T. is mediated by ↑ adenyl cyclase-cyclic AMP system. Secretion of T. is controlled by ↑ thyrotropin-releasing factor.

Thyrotropin-releasing factor (thyroliberin, TRF, thyrotropin-releasing hormone, TRH, TSH-RF): Hypothalamic polypeptide stimulating ↑ thyrotropes to produce and release ↑ thyrotropic hormone. Release of T. is controlled by levels of ↑ thyroxin and ↑ triiodothyronine via a negative ↑ feedback mechanism. T. also seems to stimulate ↑ prolactin secretion.

↑ Releasing factors; ↑ Thyroid gland, control of secretion

Griffiths, E.C., Bennetti, G.W.: Thyrotropin-Releasing Hormone. New York: Raven Press 1983; Tixier-Vidal, A., Moreau, M.F., Picart, R., Gourdji, D.: Biol. Cell *36*, 167 (1979)

Thyroxine (tetraiodothyronine, T$_4$): Iodine-containing hormone of ↑ thyroid follicular cells. Main function of T. is to control basal metabolic rate; it stimulates catabolism, glycogenolysis, mobilization of fat from storage sites, etc. and is indispensable for normal growth. Action of T. seems to be mediated by gene activation in some ↑ tissues, and in other tissues via ↑ adenyl cyclase-cyclic AMP system. (See biochemistry and physiology texts for further information)

Oppenheimer, J.H., Samuels, H.H.: Molecular Basis of Thyroid Hormone Action. New York: Academic Press 1983

Tight junction: ↑ Zonula occludens

Tigroid substance: A former term for ↑ Nissl bodies.

Tissue (textus*): Coherent association of cells differentiated in same direction and having same function. There are four basic T. in body: 1) Epithelial tissue or ↑ epithelium (textus epithelialis*); 2) ↑ connective and supporting tissue (textus connectivus*), including ↑ blood (sanguis*); 3) ↑ muscular tissue (textus muscularis*); 4) ↑ nervous tissue (textus nervous*).

↑ Differentiation

Tissue, adipose, brown: ↑ Adipose tissue, brown

Tissue, adipose, white: ↑ Adipose tissue, white

Tissue, areolar: ↑ Loose connective tissue

Tissue, bone: ↑ Bone

Tissue, cartilaginous: ↑ Cartilage

Tissue, "cellular": ↑ "Cellular" connective tissue

Tissue, chondroid: ↑ Chondroid tissue

Tissue, chordoid: ↑ Notochordal tissue

Tissue, connective and supporting: ↑ Connective and supporting tissues

Tissue culture: Experimental procedure in which group of cells is removed from body under sterile conditions and transferred to a natural or artificial nutritive medium. Under favorable and well-controlled conditions cells can be kept for long periods. T. is used in a very large number of investigations (e.g., cell ↑ differentiation, cell growth, cell interactions, cell nutrition, cell multiplication).

Nuzzolo, L., Vellucci, A.: Tissue Culture Technique. St. Louis: Warren H. Green 1983

Tissue, dense connective: ↑ Dense connective tissue

Tissue, elastic: ↑ Elastic connective tissue

Tissue, epithelial: ↑ Epithelium

Tissue fluid (extracellular fluid, interstitial fluid): Filtrate of blood plasma occurring at arterial ends of ↑ capillaries under influence of blood pressure. T. represents a water-rich phase of ↑ ground substance of ↑ connective tissue, and consists of a simple solution of nutrients and gases. Most T. is absorbed at venous ends of capillaries; part of it drains into ↑ lymphatic capillaries. (See physiology texts for further information)

Tissue, gelatinous: ↑ Gelatinous tissue

Tissue hormones: ↑ Hormones produced by ↑ tissue without morphological characteristics of an ↑ endocrine gland, or, at least, by tissue whose main function is not hormone synthesis. To T. belong ↑ gastrointestinal hormones, ↑ kidney hormones, and ↑ placental hormones.

Tissue, loose connective: ↑ Loose connective tissue

Tissue, lymphatic: ↑ Lymphatic tissue

Tissue, lymphoid: ↑ Lymphoid tissue

Tissue, lymphoreticular: ↑ Lymphatic tissue

Tissue macrophages: ↑ Histiocytes, fixed; ↑ Histiocytes, wandering

Tissue, mesenchymal: ↑ Mesenchymal tissue

Tissue, mucous connective: ↑ Gelatinous tissue

Tissue, muscular: ↑ Muscular tissue

Tissue, nervous: ↑ Nervous tissue

Tissue, notochordal: ↑ Notochordal tissue

Tissue, pigment: ↑ Pigment connective tissue

Tissue, reticular: ↑ Reticular tissue

Tissue, retiform: ↑ Retiform tissue

Tissue, skeletal: ↑ Skeletal tissues

Tissue, smooth muscle: ↑ Smooth muscle tissue

Tissue, supporting: ↑ Supporting tissues

TL-antigen: Surface antigen of mouse ↑ T-lymphocytes.

T-lymphocytes (thymus-dependent lymphocytes): Functional category of ↑ lymphocytes (35% of circulating lymphocytes) responsible for cellular immunity. T. are produced in ↑ bone marrow, from which they migrate to colonize ↑ thymus, where they divide, differentiate, and mature. T. are then released into blood and wander into thymus-dependent zones of ↑ lymph nodes and ↑ spleen. Surface of a T. bears only about 700 molecules of specific immunoglobulins. In contact with a corresponding antigen, T. proliferate and differentiate into T-immunoblasts which generate ↑ helper, ↑ killer, ↑ suppressor, and ↑ memory cells. Killer cells may surround bacteria, grafted cells, etc., adhere to them, and provoke their ↑ lysis by an irreversible alteration of plasma membrane. As helper cells, T. regulate transformation of ↑ B-lymphocytes into ↑ plasma cells, i.e., control of humoral immunity. Life span of killer cells is a few days; that of memory cells, several years. Lost T. are renewed, chiefly by thymus and partly by bone marrow.

↑ Immune response, ↑ Lymphatic organs, thymus-dependent zones of

Cohn, M.: Cell *33*, 657 (1983); Pabst, R., Trepel, F.: Anat. Rec. *195*, 341 (1979); Petrzilka, G.E., Schroeder, H.E.: Cell Tissue Res. *201*, 101 (1979); Scollay, R., Jacobs, S., Jerabek, L., Butcher, E., Weissman, I.: J. Immunol. *124*, 2845 (1980)

Toluidine blue: Basic dye giving ↑ metachromasia with corresponding substrate.

Tomes' fiber (odontoblast's process, processus odontoblasti dentinalis*): Long apical process (TF) of an ↑ odontoblast lying in a ↑ dentinal tubule. T. extends from ↑ pulp cavity to ↑ dentinoenamel and dentinocemental junction (fig.), where it becomes ramified, still lying in dentinal tubule. T. and dentinal tubules assure metabolite transport in ↑ dentin. Asterisk = ↑ Tomes' granular layer (Fig. = human)

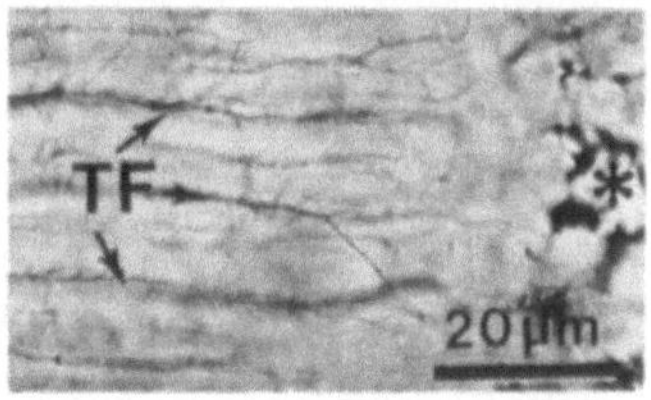

Gunji, T., Kobayashi, S.: Arch. Histol. Jpn. *46*, 213 (1983)

Tomes' granular layer (stratum granulosum dentini radicis*): Zone of small noncalcified dentinal spaces (TGL) immediately under dentinocemental junction.

↑ Dentin

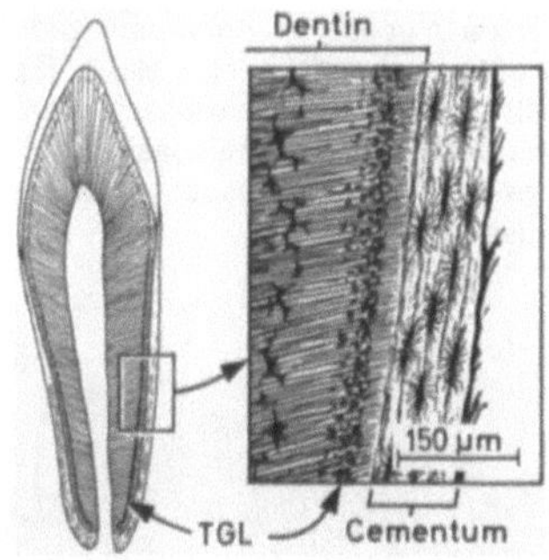

Ten Cate, A.R.: Anat. Rec. *172*, 137 (1972)

Tomes' process (ameloblastic process): Apical protrusion (TP) of an ↑

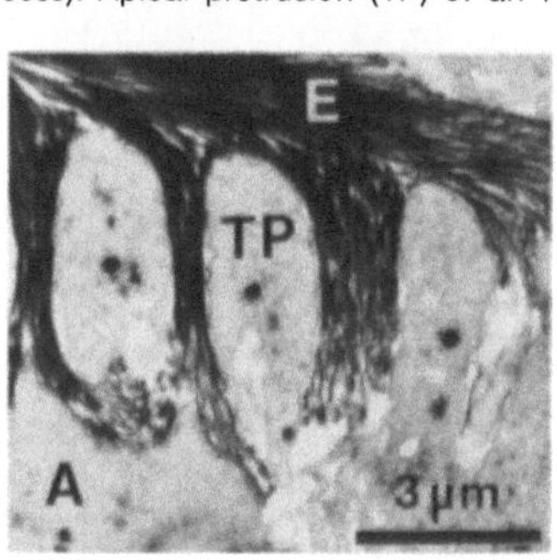

ameloblast (A), roughly 5 µm long, involved in orientation of one ↑ enamel prism (E). (Fig. = rat)

↑ Amelogenesis

Kallenbach, E.: Tissue Cell *5*, 501 (1973); Wakita, M., Tsuchiya, H., Gunji, T., Kobayashi, S.: Arch. Histol. Jpn. *44*, 285 (1981)

Tongue (lingua*): Muscular organ covered by mucosa of oral cavity. A) Structure of dorsal surface: 1) Tunica mucosa = partially keratinized ↑ stratified squamous epithelium (Ep), strongly indented with lamina propria (LP). Both mucosal layers form various kinds of papillae with special distribution – ↑ circumvallate papillae, (Circ); ↑ filiform papillae, (Fil); ↑ foliate papillae, (Fol); ↑ fungiform papillae, (Fung); ↑ lenticular papillae, (Le). 2) ↑ Aponeurosis linguae (ApL) = no tela submucosa on dorsum linguae. 3) Muscles (Musc) = interlacing bundles of crossed ↑ skeletal muscle fibers running in three planes (vertical, transversal, longitudinal). 4) Lingual glands = ↑ anterior lingual glands (ALG) are ↑ mixed; purely ↑ serous glands of ↑ Ebner open into furrows of ↑ circumvallate papillae; purely mucous posterior lingual glands lie between muscle fibers of root of T. 5) Lingual ↑ tonsil (LT). B) Structure of ventral surface: 1) Tunica mucosa = nonkeratinized ↑ stratified squamous epithelium (Ep) with low indentation; loose lamina propria (LP). 2) Tela submucosa (TS) = ↑ loose connective tissue. 3) Lingual muscles (Musc).

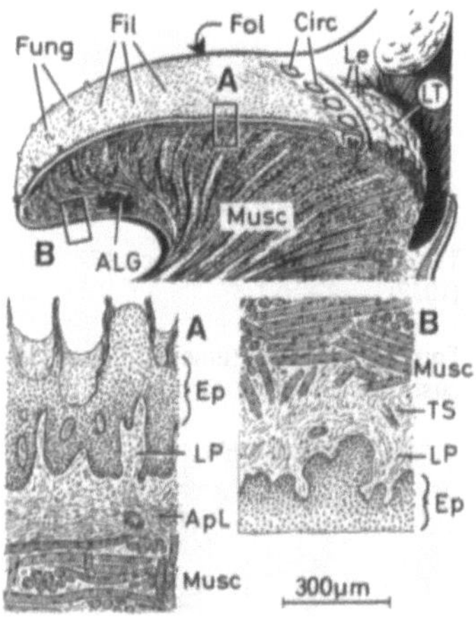

Tongue, follicles of: ↑ Lingual follicles; ↑ Tonsil, lingual

Tongue, papillae of: Differentiations of tunica mucosa of dorsal and lateral ↑ tongue surfaces.

↑ Circumvallate papillae; ↑ Filiform papillae; ↑ Foliate papillae; ↑ Fungiform papillae; ↑ Lenticular papillae

Tongue processes, of oligodendrocyte: Knifelike areas of cytoplasm remaining between two ↑ cell membranes (P) of an oligodendrocyte sheath (OS) in course of its wrapping (arrow) around an ↑ axon (A). T. are parallel to axon and contain ↑ microtubules and sometimes cell ↑ organelles. Shorter side of trapezoidal sheath directly contacting axon is internal T. (IT); longer side of sheath is external T. (ET). There are also some intermediate T. (T) found between ↑ myelin lamellae (ML). Lateral edges (E) of sheath furnish ↑ mesaxons of ↑ paranodal region.

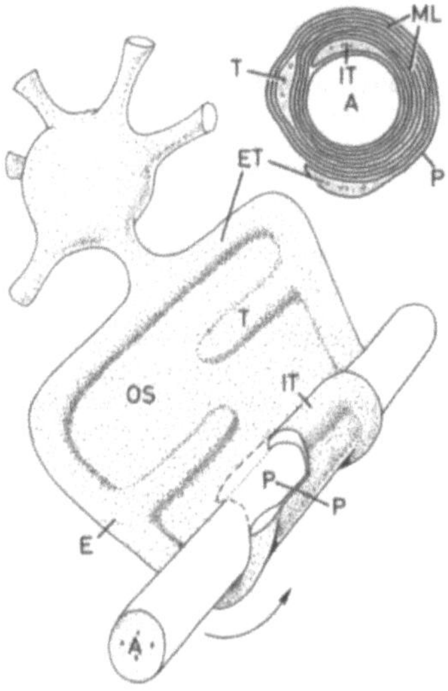

Tongue, taste buds of: ↑ Taste buds; ↑ Taste buds, structure of

Tonofibrils (tonofibrilla*): Threadlike cytoplasmic ↑ organelles (Tfb) composed of bundles of ↑ tonofilaments, particularly abundant in germinal layer of ↑ epidermis (fig.). T. traverse cell body in various directions and extend into cell expansions, to anchor themselves in ↑ attachment plaques of ↑ desmosomes (D) and ↑ hemidesmosomes. T. act as a supportive cytoplasmic element reinforcing intercellular cohesion. (Fig. = rat)

↑ Metaplasm

Brody, I.: J. Ultrastruct. Res. *30*, 209 and 601 (1970)

Tonofilaments (tonofilamentum*): Noncontractile cytoplasmic microfilaments (Tfl), 5–9 nm thick, obviously inserted into ↑ attachment plaques (AP) of ↑ desmosomes (D). Bundles of T. constitute ↑ tonofibrils (Tfb) visible in light microscope. Material of T. is birefringent fibrous protein ↑ keratin, rich in sulfhydryl groups. T. participate in formation of ↑ cytoskeleton. (Fig. = rat).

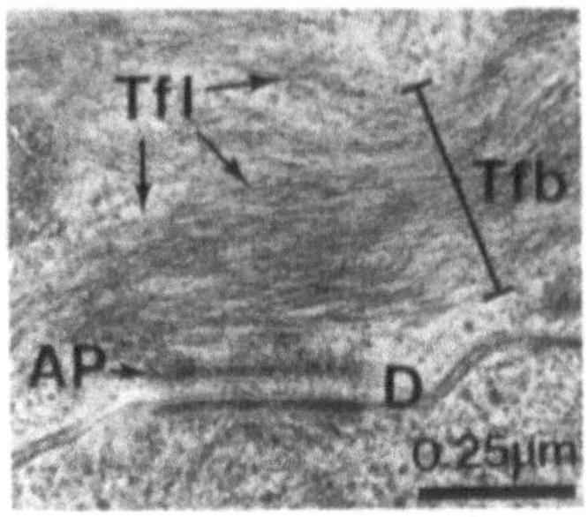

Lane, E.B., Klymowsky, M.W.: Epithelial Tonofilaments: Investigating Their Form and Function Using Monoclonal Antibodies. Cold Spring Harbor Symposia on Quantitative Biology, Vol. 46, Part 1. New York: Cold Spring Harbor Laboratory 1982; Yen, S.-H., Liem, R.K.H., Jenq, L.-T., Shelanski, M.L.: Exp. Cell Res. *129*, 313 (1980)

"Tonsil, abdominal" (tonsilla abdominalis*): Term sometimes used to designate ↑ appendix because of its richness in ↑ lymphatic tissue.

Tonsil, lingual (tonsilla lingualis*): All ↑ lingual follicles (LF) grouped at root of ↑ tongue. T. belongs to ↑ lymphatic ring of ↑ pharynx.

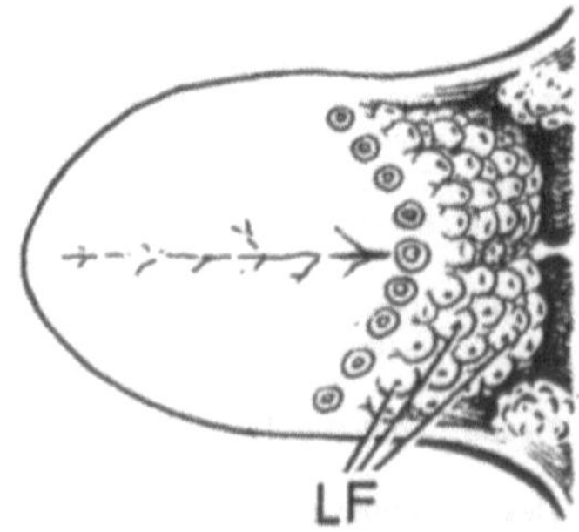

Tonsil, pharyngeal (tonsilla nasopharyngealis* s. adenoidea*): ↑ Lymphoepithelial organ situated in roof of ↑ nasopharynx. T. is lined by ↑ pseudostratified columnar ciliated epithelium (E), containing many ↑ goblet cells and invaded by ↑ lymphocytes and ↑ macrophages. Epithelium does not form ↑ crypts, but rather folds (F). A roughly 2-mm-thick ↑ lymphatic tissue (LT) with numerous ↑ lymphatic nodules (L) lies under epithelium. T. is surrounded by a thin connective capsule (C) which sends some septa into lymphatic tissue. Outside capsule are small ↑ mixed glands (MG) with excretory ducts (ED) opening into epithelial folds. T. belongs to ↑ lymphatic ring of pharynx.

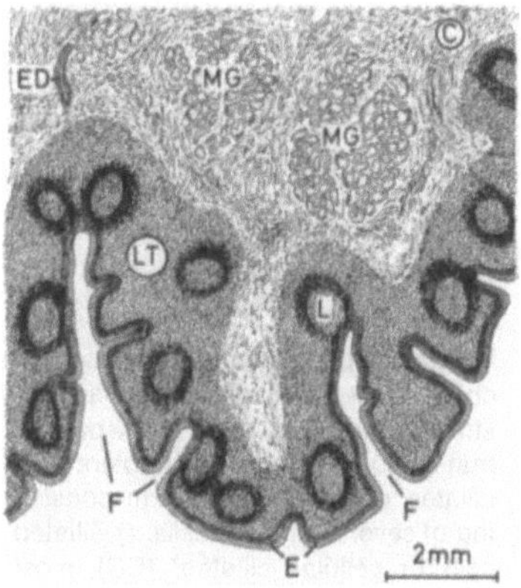

Tonsils (tonsillae*): ↑ Lymphoepithelial organs forming a ↑ lymphatic ring of pharynx (Waldeyer's ring). T. include: 1) Palatine ↑ T., 2) lingual ↑ T., 3) pharyngeal ↑ T., and 4) tubal ↑ T. Function: Defense of organism against early viral and/or bacterial infections by direct action of ↑ lymphocytes and production of antibodies.

↑ Immunoglobulin A

Antoni, F, Staub, M. (eds.): Tonsils: Structure, Immunobiology and Biochemistry. Budapest: Akademiai Kiado 1978

Tonsils, crypts of: ↑ Crypts, of tonsils

Tonsils, palatine (faucis, tonsilla palatina*): Paired ↑ lymphoepithelial organs situated between glossopharyngeal and pharyngopalatine arches. T. consist of 10–20 deep ↑ crypts (Cr) lined by ↑ stratified squamous nonkeratinized epithelium, heavily invaded by ↑ lymphocytes. Lamina propria around crypts is occupied by ↑ lymphatic tissue (LT) with numerous ↑ lymphatic nodules (L). T. are surrounded by a dense connective capsule (C) that sends a few septa (S) into lymphatic tissue. Outside capsule one can found occasional ↑ mucous glands (MG) with excretory ducts opening at surface of T. and some ↑ skeletal muscle fibers (MF) belonging to pharyngeal musculature. T. is a part of ↑ lymphatic ring of pharynx.

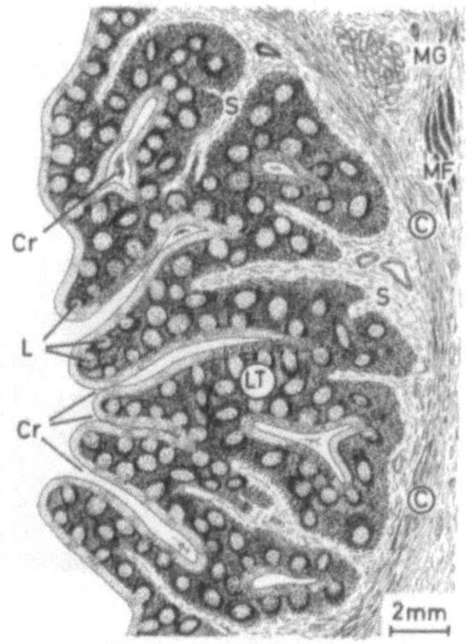

Gaudecker, B.v., Müller-Hermelink, H.K.: Cell Tissue Res. *224*, 579 (1982); Hoffmann-Fezer, G., Löhrs, U., Rodt, H.V., Thierfelder, S.: Cell Tissue Res. *216*, 361 (1981); Howie, A.J.: J. Pathol. *130*, 91 (1980); Yamamoto, M., Ohyama, M., Hanamure, Y., Ogawa, K.: Biomed. Res. *2*, Suppl. (1981)

Tonsils, tubal (Gerlach's tonsil; tonsilla tubaria*): Small collections of ↑ lymphoreticular tissue near opening of ↑ auditory tubes. T. is covered with ↑ pseudostratified ciliated epithelium and is structurally very similar to pharyngeal ↑ tonsil.

↑ Lymphatic ring, of pharynx

Tooth: ↑ Teeth

Totipotent hematopoietic stem cell (THSC): Presumably, a stem cell for all hematopoietic cell lineages whose existence has been proved only indirectly. In absence of morphological identification, some authors prefer term ↑ candidate stem cell.

↑ Hematopoiesis, current view of; ↑ Hematopoietic stem cells

Trabeculae, arachnoid (trabecula arachnoidalis*): Fine threadlike or ribbonlike strands forming a three-dimensional spiderweb network between arachnoid membrane and ↑ pia mater. T. consist of processes of ↑ fibroblasts accompanied by bundles of ↑ collagen microfibrils.

↑ Arachnoid

Trabeculae carneae*: Anastomosing bundles of ↑ cardiac muscle fibers forming a networklike relief on inner aspect of heart ventricular cavities.

Trabeculae, of bone (trabecula ossea*): Thin anastomosing ↑ spicules, beams, or threads (T) of ↑ bone devoid of ↑ Haversian canals. T. appear in course of direct ↑ bone formation by fusion of little islands of immature ↑ bone; during indirect ↑ bone formation, first T. are formed around remnants of calcified cartilage. During secondary ↑ bone formation T. are replaced by mature ↑ bone. T. constitute a three-dimensional framework of a cancellous ↑ bone. (Fig. = skull, human fetus)

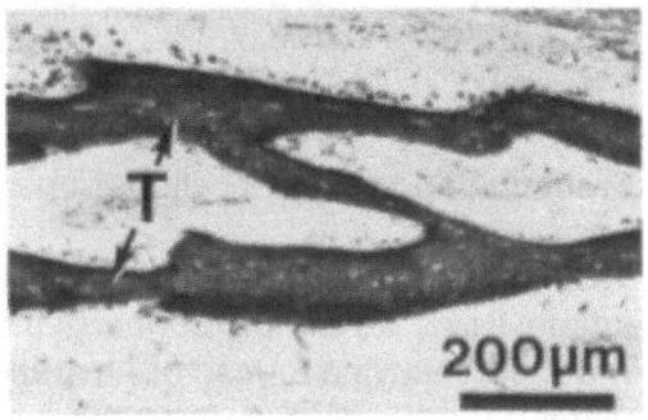

Trabeculae, of cavernous bodies: ↑ Cavernous bodies, of penis

Trabeculae, of lymph node: ↑ Lymph node

Trabeculae, of podocyte (cytotrabecula*): Long primary processes of a ↑ podocyte.

Trabeculae, of spleen (trabecula splenica*): Extensions of moderately ↑ dense connective tissue (Tr) emerging from ↑ hilus and inside of ↑ capsule. Some ↑ smooth muscle cells may also be found in T. Through a T. run ↑ trabecular artery (TA) and vein (TV). (Fig. = human)

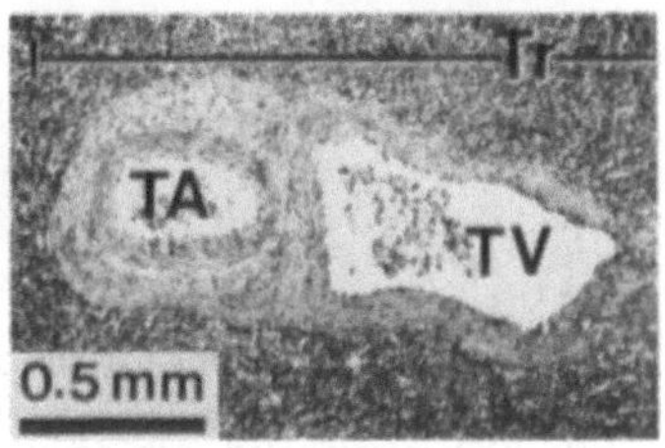

Trabeculae, perilymphaticae: ↑ Perilymphatic tissue

Trabecular artery (arteria trabecularis*): ↑ Muscular artery running through a ↑ trabecula (Tr) of ↑ spleen. All layers of T. are well developed; tunica adventitia (A) clearly delimitates T. from tissue of trabecula. T. are branches of splenic artery; branches of T. are ↑ central arteries. I = tunica intima, M = tunica media (Fig. = human)

↑ Spleen, vascularization of

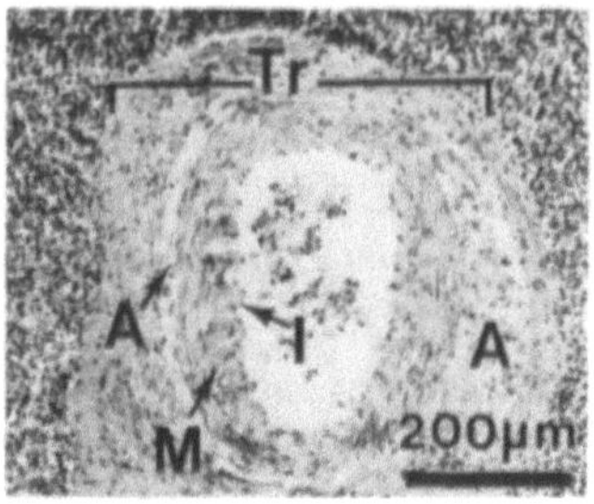

Trabecular bone: ↑ Cancellous bone

Trabecular gland: Term used to designate ↑ liver as an ↑ exocrine gland with ↑ parenchyma formed of ↑ hepatic plates.

Trabecular meshwork (trabeculum, reticulum trabeculare*): Spongy network (T) of fine connective strands (S) situated deep in stroma of ↑ limbus in angle of ↑ anterior chamber between

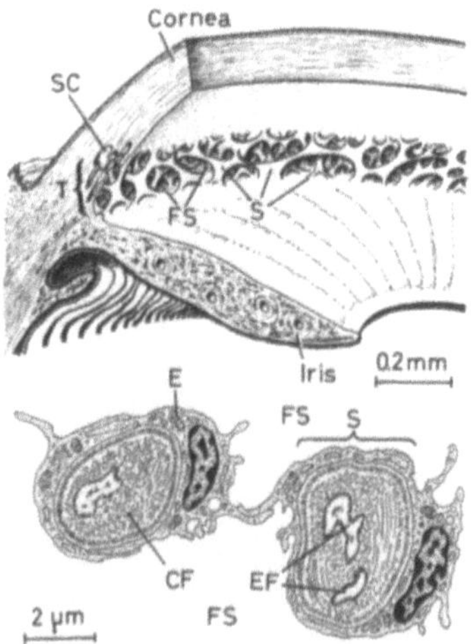

root of ↑ iris and ↑ cornea. Core of strands is composed of ↑ collagen fibrils (CF) intermingled with some ↑ elastic fibers (EF) and ensheathed by an endothelium (E) continuous with that of cornea. Strands delimit a labyrinthine system of narrow passages – intertrabecular spaces or ↑ Fontana spaces (FS) – establishing contact between anterior chamber and ↑ Schlemm's canal (SC) toward which ↑ aqueous humor streams.

Futa, R.: Folia Ophthalmol. Jpn. *31*, 188 (1980); Hansson, H.-A.: Biomed. Res. *2*, Suppl. 465 (1981); Raviola, G.: Invest. Ophthalmol. *13*, 828 (1974)

Trabecular veins (vena trabecularis*): Branches (TV) of ↑ splenic vein running through ↑ trabeculae (Tr) of ↑ spleen. Except for an ↑ endothelium (E) lining their lumen, T. have no wall: their wall is composed of tissue of trabeculae. PV = ↑ pulp vein (Fig. = human)

↑ Spleen, vascularization of

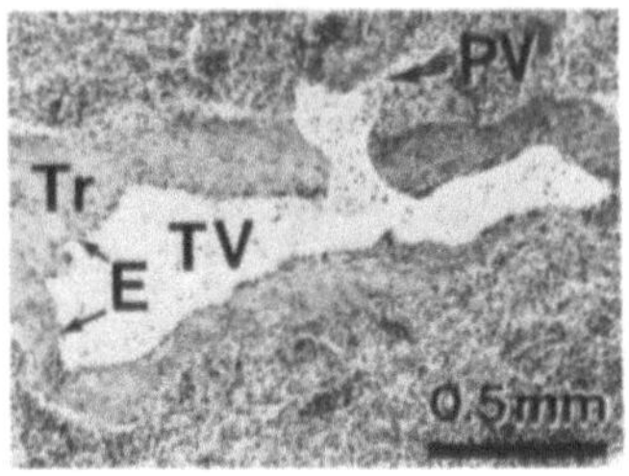

Tracers: Molecules of known dimensions used for visualization of their passage through and/or between cells. 1) Particulate T. = with the transmission electron microscope, directly visible molecules such as ↑ ferritin, dextran, ↑ glycogen, colloidal thorium, iron oxide, etc. 2) Mass T. = heavy osmiophilic products of ultrahistochemical reactions, e.g., ↑ horseradish peroxidase with diaminobenzidine (indirect visualization).

Trachea*: Airway tube, about 20 cm long, situated between base of ↑ larynx and ↑ carina. Structure: 1) Tunica mucosa (TM): a) epithelium (E) = a ↑ pseudostratified epithelium resting upon an unusually thick basal lamina;

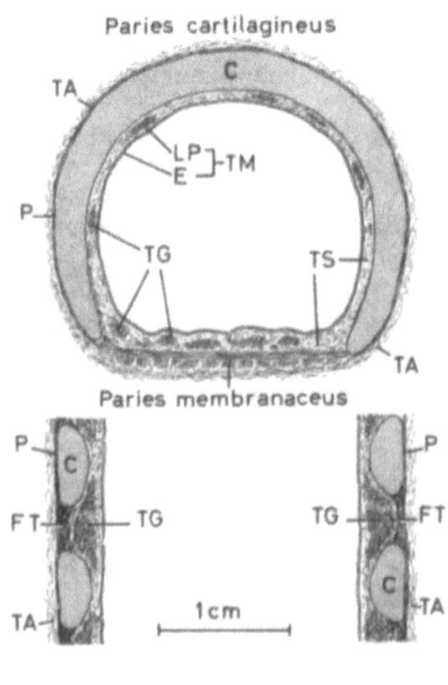

b) lamina propria (LP) = a thin layer of ↑ loose connective tissue rich in ↑ elastic fibers. 2) Tela submucosa (TS) = some loose connective tissue with numerous ↑ wandering cells. ↑ Tracheal glands (TG) are situated in this layer. 3) Tunica fibromusculocartilaginea: a) paries cartilagineus or cartilaginous part = 16–20 successive C-shaped tracheal ↑ cartilages (C) bound together by a vertically arranged layer of dense ↑ fibroelastic tissue (FT) inserted in ↑ perichondrium (P) of rings. Space between cartilages is occupied by tracheal glands; b) paries membranaceus or membranous part = a layer of ↑ smooth muscle cells stretched between tips of cartilaginous rings. In tela submucosa of this part numerous tracheal glands are also present. 4) Tunica adventitia (TA) = a loose connective tissue. Primary ↑ bronchi have an identical structure.

↑ Trachea, epithelium of

Trachea, epithelium of (laryngobronchial epithelium, epithelium pseudostratificatum ciliatum*): An approximately 30-µm-high, ↑ pseudostratified ciliated columnar epithelium consisting of several kinds of cells. 1) Ciliated cells or cellulae ciliatae* (CC), most numerous = tall columnar cells with a nucleus at various levels, many mitochondria, a large Golgi apparatus, some cisternae of rough endoplasmic reticulum, lysosomes, and free ribosomes. Adjacent cells are connected at ↑ apical pole with ↑ junctional complexes, and along their lateral surfaces with small ↑ desmosomes. From apical pole emerge 100–200 ↑ cilia (C) averaging 5–7 µm in length and, between them, numerous ↑ cytofila (Cf). Cilia beat with a frequency of 25 bps in direction of ↑ larynx to move mucus and inspired dust particles to tussigenic zones. 2) ↑ Goblet cells (GC). 3) Basal cells (BC) = small polygonal cells adjacent to ↑ basal lamina (BL). Basal cells have a relatively voluminous nucleus, poorly developed ↑ organelles, some free ribosomes, and numerous ↑ tonofibrils. Basal cells differentiate into either ciliated or goblet cells via intermediate cells. 4) Intermediate cells (IC) = pyramidal elements lying on basal lamina. Their apical poles do not reach free epithelial surface. Nuclei of intermediate cells are situated below nuclei of ciliated cells. Organelles are moderately developed; cytoplasm contains some ↑ tonofilaments. 5) Brush cells, nonciliated cells or cellulae penicillatae* (BrC) = columnar cells resting on basal lamina and reaching luminal surface with apical pole. Instead of cilia, brush cells bear ↑ microvilli (Mv). Organelles are moderately developed; some lysosomes, free ribosomes, glycogen particles and sometimes precursors of basal bodies can also be found. Since sensory nerve endings (NE) make ↑ synapses with basal cell pole, it is believed that brush cells are sensory receptors. 6) Small-granule cells or tracheal endocrine cells (SGC) = two kinds of solitary columnar cells with predominantly infranuclear heterochromatin-rich nucleus, a small Golgi apparatus, few mitochondria, and short and narrow cisternae of rough endoplasmic reticulum. Basal cytoplasm and cellular process along basal lamina contain 100 to 300-nm dense-core granules. ↑ Cholinergic nerve endings (CNE) enter into contact with small granule cells which are believed to be endocrine elements; one class of small granule cells are neurosecretory cells producing ↑ catecholamines; others secrete polypeptide hormones.

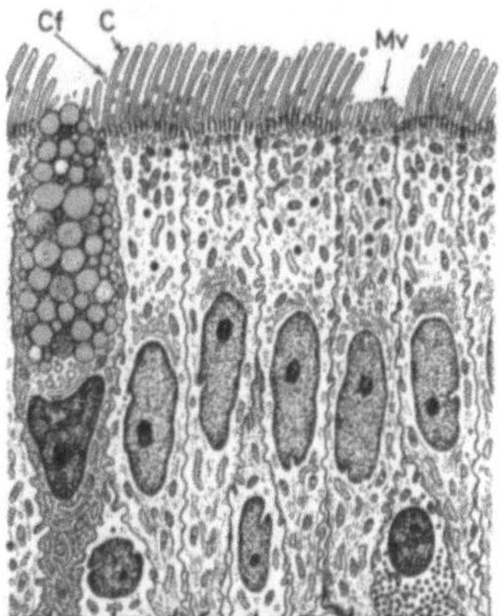

Dalen, H.: J. Anat. *136*, 47 (1983); Hoyt, R.R.Jr., Sorokin, S.P., Feldman, H.; Exp. Lung Res. *3*, 273 (1982); Kawamata, S., Fujita, H.: Arch. Histol. Jpn. *46*, 355 (1983); McDowell, E., Combs, J.W., Newkirk, C.: Exp. Lung Res. *4*, 205 (1983); Wu, R., Smith, D.: In Vitro *18*, 800 (1982)

Tracheal endocrine cells: Synonym for small-granule cells in epithelium of ↑ trachea.

Dey, R.D., Shannon, J.R., Hagler, H.K., Said, S.I.: J. Histochem. Cytochem. *31*, 501 (1983)

Tracheal glands (glandulae tracheales*): Small ↑ mixed glands (TG) situated in lamina propria of both cartilaginous (C) and membranous parts; they are particularly abundant in latter and between cartilages. T. moisten luminal surface of ↑ epithelium (E). (Fig. = human)

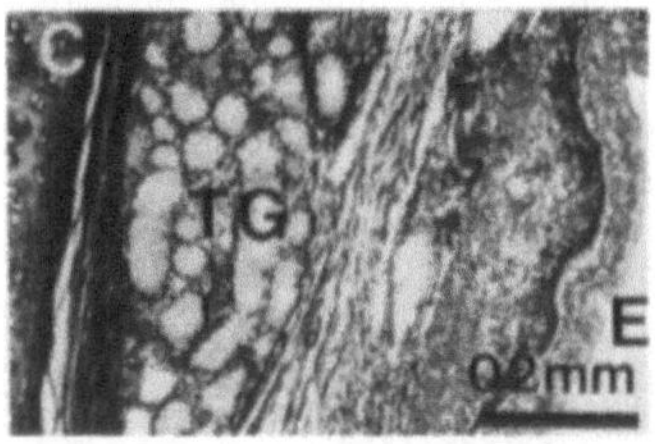

Tragi*: Coarse ↑ terminal hairs growing at entrance of external ↑ auditory meatus; number increases with age. T. have no ↑ arrector pili muscles.

Transcapillary channels: ↑ Transendothelial channels

Transcription: Process of synthesis of ↑ messenger, ↑ transfer, and ↑ ribosomal RNA from ↑ deoxyribonucleic acid.

↑ Ribonucleic acids, synthesis of

Transcytosis (cytopempsis, diacytosis): Process of transcellular and/or ↑ transendothelial transport of substances within ↑ micropinocytotic vesicles (MV) without these substances being utilized by the cells. T. may be followed in transmission electron microscope after injection of colloidal metallic solutions, with particles (P) passing, inside vesicles, from capillary lumen (CL) into ↑ pericapillary space (PS), crossing ↑ endothelial cell (E) and its ↑ basal lamina (BL).

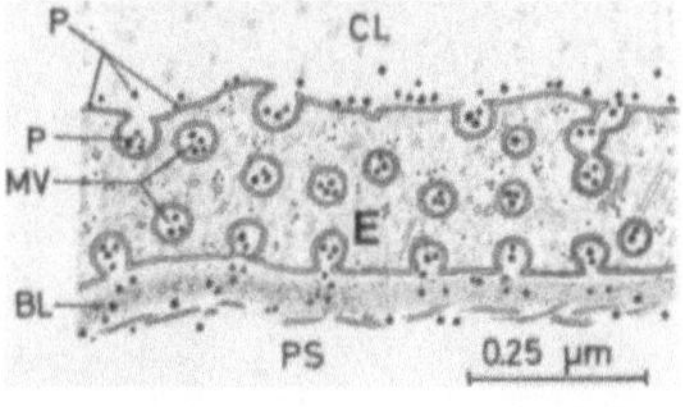

Herzog, V.: J. Cell Biol. *97*, 607 (1983); London, M.F., Michel, C.C., White, I.F.: J. Physiol. (Lond.) *296*, 97 (1979); Simionescu, N.: Transcytosis and Traffic of Membranes in the Endothelial Cell. In: Schweiger, H.G. (ed.): International Cell Biology 1980–1981. Berlin, Heidelberg, New York: Springer-Verlag 1981

Transendothelial channels (transcapillary channels): Fine canals (TC) running transversally through thickness of ↑ endothelial cells (E) of continuous ↑ capillaries. T. are formed by a single ↑ micropinocytotic vesicle (MV, 1), by two fused vesicles (2), or by a chain of three or more merged vesicles (3). Plasmalemma (P) and vesicle membrane (VM) are continuous with both sides of T. Passage of some macromolecules is greatly facilitated through T.

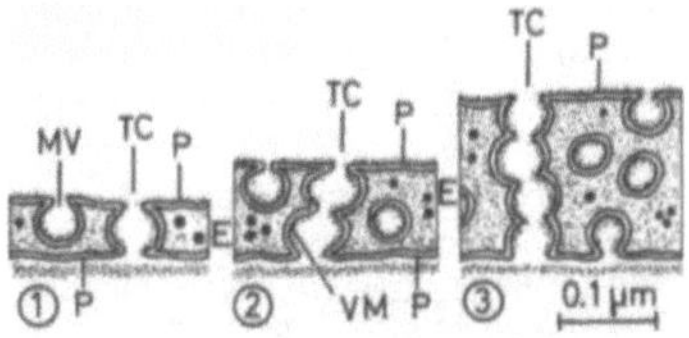

Simionescu, V., Simionescu, M., Palade, G.E.: J. Cell Biol. *64*, 586 (1975)

Transendothelial transport: Passage of substances across ↑ capillary wall. This occurs through ↑ transendothelial channels, by way of ↑ micropinocytotic vesicles, and by diffusion.

↑ Transcytosis

Simionescu, N.: Physiol. Rev. *63*, 1536 (1983)

Trans face: ↑ Maturing face, of Golgi apparatus

Transfer factor (TF): An antigen-specific protein produced by ↑ lymphocytes immediately after contact with antigen. T. is capable of conferring antigen specificity to quiescent circulating lymphocytes.

Transfer RNA (tRNA): Type of ↑ ribonucleic acid reading ↑ codons of messenger RNA (mRNA) with its ↑ anticodon and recognizing corresponding amino acid that codons specify. tRNA is synthesized outside ↑ nucleolus and has clover-leaf form. Most specific segment of tRNA carries anticodon which links with codon of mRNA. Amino acid acceptor end (AA) of molecule recognizes and attaches a specific amino acid. One of two lateral segments (R) recognizes and links with ↑ ribosome, and other segment (E) carries site for recognition and binding of specific amino acid-activating en-

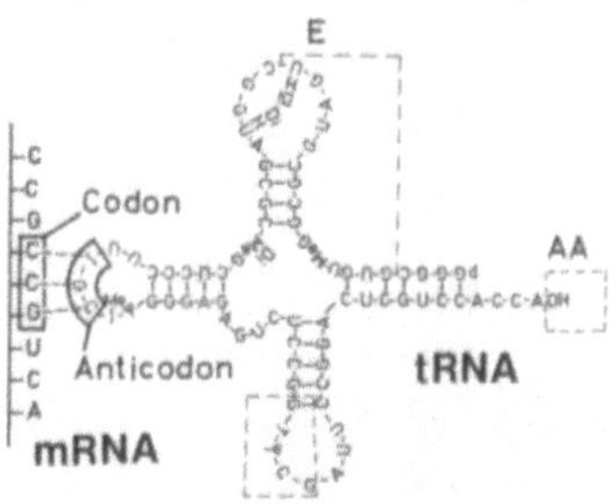

zyme (tRNA–aminoacyl synthetase). (See molecular biology texts for further information) (Slightly modified after De Robertis et al. 1975)

↑ Ribonucleic acids, synthesis of

De Robertis, E.P.D., Saez, F.A., De Robertis, E.M.F. Jr.: Cell Biology, 6th edn. Toronto, Philadelphia, London: Saunders 1975; Rich, A., Soung Hou Kim: Sci. Am. *238*/1, 52 (1978); Schimmel, P.R., Söll, D., Abelson, J.N. (eds.): Transfer RNA: Structure, Properties, and Recognition. Monograph 9A. New York: Cold Spring Harbour Laboratory 1979); Söll, D., Abelson, J.N., Schimmel, P.R. (eds.): Transfer RNA: Biological Aspects: Monograph 9B. New York: Cold Spring Harbour Laboratory 1980

Transfer vesicles: ↑ Transport vesicles

Transferrin: An iron-binding plasma globulin involved in transport of ↑ iron.

↑ Ferritin

Iacopetta, B.J., Morgan, E.H. Yeoh, G.C.T.: J. Histochem. Cytochem. *31*, 336 (1983)

Transillumination: Passage of light through thin ↑ tissues or ↑ organs. T. is used for light-microscopic study of blood dynamics and lymph circulation in various living organs.

Transitional cells (T-cells): A kind of elongated cells constituting, together with ↑ nodal cells, ↑ atrioventricular and ↑ sinuatrial nodes, and also connecting ↑ Purkinje cells with ↑ cardiac muscle cells (CMC). T. have elongated nucleus (N) with condensed ↑ chromatin; cytoplasm encloses numerous small mitochondria (M), a poorly developed Golgi apparatus, some rough and smooth endoplasmic profiles, a few ↑ glycogen particles, and no ↑ T-system. Through cytoplasm run numerous longitudinal and spiral ↑ myofibrils (Mf) which are attached to ↑ junctional complexes (JC), mostly ↑ zonulae adherentes. There are no ↑ intercalated discs, neither between different T. nor between T. and cardiac muscle cells.

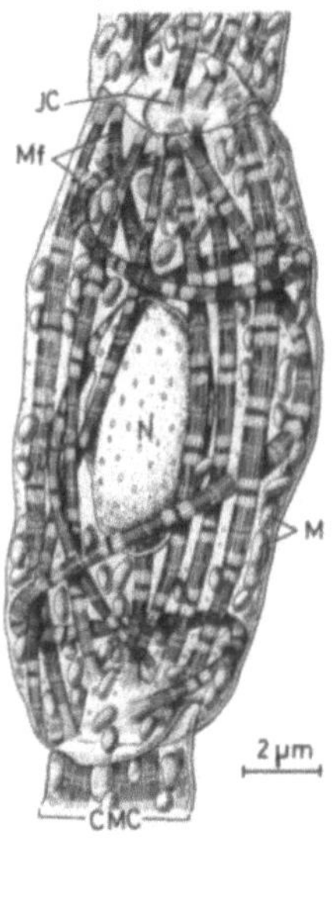

Akester, A.R.: J. Anat. *133*, 161 (1981); Roberts, N.K., Castelman, K.R.: Anat. Rec. *195*, 699 (1979)

Transitional endoplasmic reticulum: ↑ Endoplasmic reticulum, transitional

Transitional epithelium (urothelium, epithelium transitionale*): Multilayered epithelium lining renal ↑ calyces, ↑ renal pelvis, ↑ ureter, ↑ urinary bladder, and initial part of male and female ↑ urethra. T. is composed of layer of basal cells (BC), or stratum basale*, in contact with ↑ basal lamina (BL), a few layers of elongated intermediate cells (IC), or stratum intermedium*, also contacting basal lamina, and large ↑ facet cells (FC), or stratum superficiale*, containing ↑ discoid vesicles (DV). Controversy persists concerning classification of T.: According to several authors, it belongs to ↑ pseudostratified epithelium, since fine processes(?) (whose existence is not universally accepted) of facet cells penetrate up to basal lamina.

↑ Transitional epithelium, histophysiology of

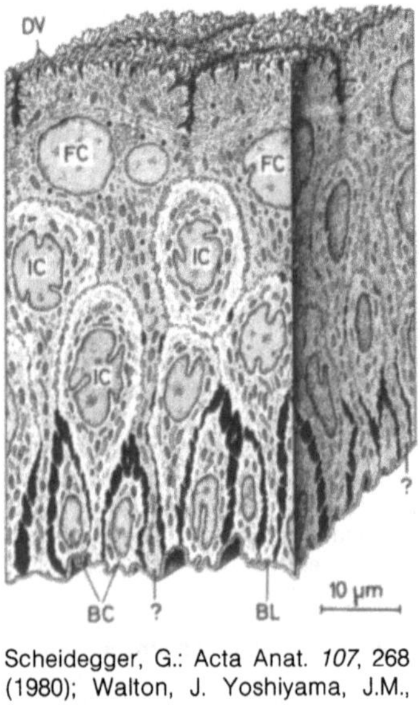

Scheidegger, G.: Acta Anat. *107*, 268 (1980); Walton, J. Yoshiyama, J.M., Vanderlaan, M.: J. Submicrosc. Cytol. *14*, 1 (1982)

Transitional epithelium, histophysiology of: In relaxed (empty) ↑ urinary bladder, ↑ transitional epithelium is high and ↑ facet cells (FC) bulge into lumen; their apical surfaces are highly folded. In distended (full) urinary bladder, transitional epithelium stretches and becomes flattened displaying only two cell layers, and facet cells become cuboidal or squamous; their apical cell membrane becomes smooth and covers a considerably wider area. This process is supported by existence of ↑ discoid vesicles in facet cells. BC = basal cells, BL = basal lamina, IC = intermediate cells

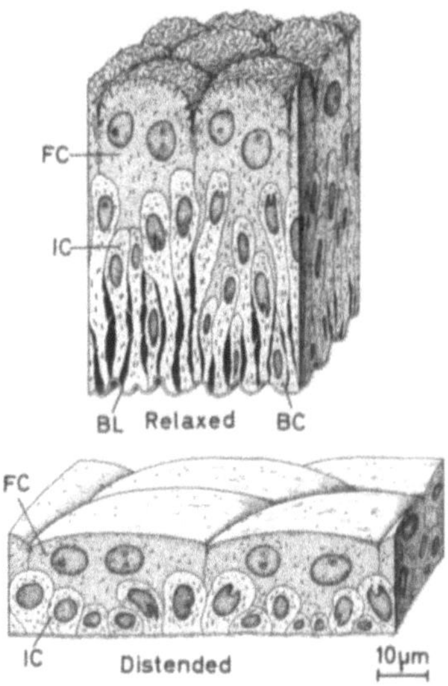

Minsky, B.D., Chlapowski, F.J.: J. Cell Biol. *77*, 685 (1978)

Transitional hematopoietic cells: ↑ Candidate stem cells

Transitional vesicles: ↑ Transport vesicles

Transitory dense bodies: ↑ Membrane-coating granules

Translation: Decoding of genetic information contained in ↑ messenger RNA molecule and assembly of proteins according to sequence of its nucleotides. (See genetic and molecular biology texts for further information)

Translocation, chromosomal: Chromosome anomaly consisting in abnormal transfer of a segment of one ↑ chromosome to a nonhomologous chromosome during prophase I of ↑ meiosis. T. results in two chromosomes with abnormal morphology.

↑ Chromosomes, anomalies of

Transmission electron microscope (TEM): Observation and photographic instrument using electrons in place of light and ↑ electromagnetic lenses in place of glass lenses. Principle: In a high vacuum (produced by pumps, P), an incandescent tungsten cathode (C) emits electrons, which are then accelerated in an electric field between cathode and anode (A) and focused onto object (O) by condenser lens (CL). Electrons are transmitted through object and scattered; objective lens (OL) forms magnified image of object. A segment of this image is finally projected by projective lens (PL) onto fluorescent screen (S) or photographic plate (Ph). Image in T. is due to scattering of electrons on atomic nuclei of elements constituting object; dispersed electrons are arrested by a fine objective diaphragm (OD), and image results from absence of those electrons captured by diaphragm. The higher the atomic number, the higher the electron dispersion, i.e., the greater the contrast of image. Since energy of electrons is very low, T. requires very thin objects or ↑ sections. ↑ Resolving power of modern T. is about 0.25 nm; practical resolving power for biological objects is about 1.0 nm. PS = power supply, stabilizer of current, high tension generator, etc.

↑ Contrasting; ↑ Ultramicrotomy

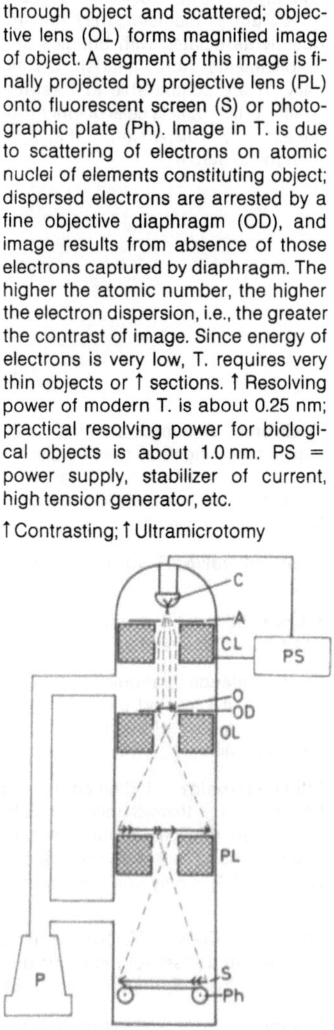

Hagege, R., Hagege, J.: La pratique du microscope électronique conventionnel. Paris: Masson et Cie 1981; Lange, R.H., Blödorn, J.: Das Elektronenmikroskop TEM + REM. Leitfaden für Biologen und Mediziner. Stuttgart, New York: Georg Thieme 1981; Weakley, B.K.: A Beginner's Handbook in Biological Transmission Electronmicroscopy. 2nd edn. Edinburgh: Churchill Livingstone Ltd. 1981

Transmission electron microscope, high-voltage: ↑ Transmission electron microscope operating with accelerating voltage of 1000 kV or more in order to obtain greater ↑ resolving power (0.15–0.2 nm) and greater penetration of electrons. T. permits observation of tissue sections 0.2–3 µm thick and of certain kinds of living cells and organism such as bacteria in special chambers.

Griffith, J.D. (ed.): Electron Microscopy in Biology. Vol. 1. Chichester: John Wiley & Sons, Ltd. 1981; Hirabayashi, M., Hiraga, K., Shindo, D.: Ultramicroscopy *9*, 197 (1982)

Transmitters: ↑ Neurotransmitter substances

Transneuronal degeneration (transsynaptic degeneration): Extension of degenerative changes when synaptic contact from ↑ neuron (N) to neuron has been interrupted (Int). On basis of direction of this degenerative effect, one can distinguish: 1) Anterograde T. = propagation of degenerative changes in direction (arrow) of synaptic transmission; 2) retrograde T. = spreading of degenerative changes contrary to direction (arrow) of synaptic transmission.

↑ Nerve fibers, of peripheral nervous system, degeneration and regeneration of

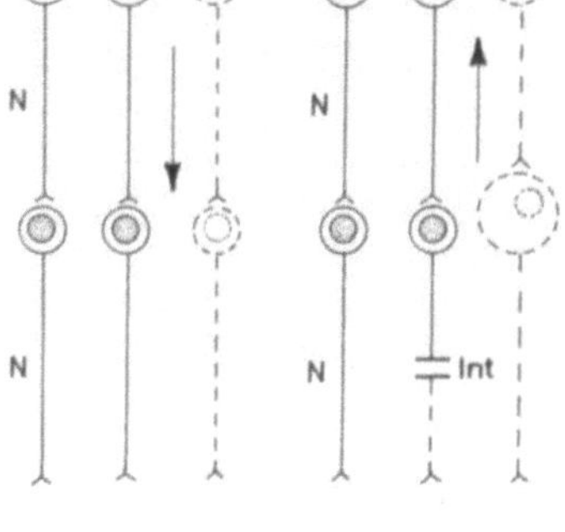

Fuentes, C., Marty, R.: Z. mikr.-anat. Forsch. *95*, 905 (1981)

Transparent chamber: Technique for light-microscopic study of living cells and ↑ tissues by installing transparent windows (W) in ↑ skin of laboratory animals. Most frequently, T. is mounted in ↑ auricle (A) of a rabbit; external skin (ES) and subcutaneous tissue lie between two windows, held by metallic rings (R), whereas corresponding pieces of cartilage (C) and internal skin (IS) are cut off. T. also permits observation of autografts thanks to their vascularization from neighboring tissue. T. can be implanted into dorsal skin of rat and mouse, cheek pouch of hamster, etc. A natural T. is ↑ anterior chamber of eye, frequently used for observation of denervated living tissues and ↑ organs.

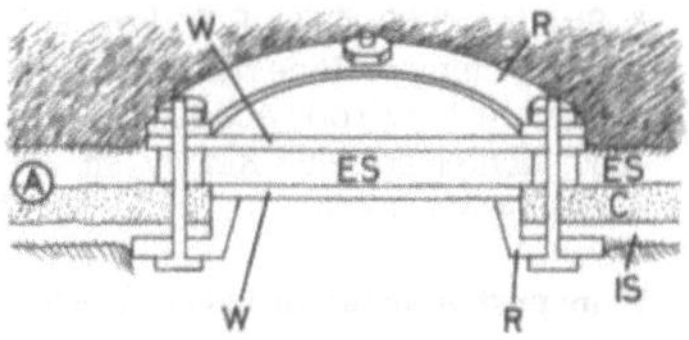

Endrich, B., Asaichi, K., Götz, A., Messmer, K.: Res. Exp. Med. *177*, 125 (1980); Sewell, I.A.: J. Anat. *100*, 839 (1966); Williams, R.G.: Int. Rev. Cytol. *3*, 359 (1954)

Transplantation (grafting): Transfer of ↑ tissue or ↑ organ from its natural environment into another position in same body or into another individual. On basis of degree of genetic parentage shared by donor and host, one can distinguish: 1) Autograft (autoplastic T., autologous T., autochthonous T.) = transfer of tissues or organs into new position in same individual. 2) Isograft (isoplastic T., isologous T., syngeneic T.) = transfer of tissues or organs between members of same homozygote twin pair. 3) Homograft (allograft, homoplastic T., homologous T.) = transfer of tissues or or-

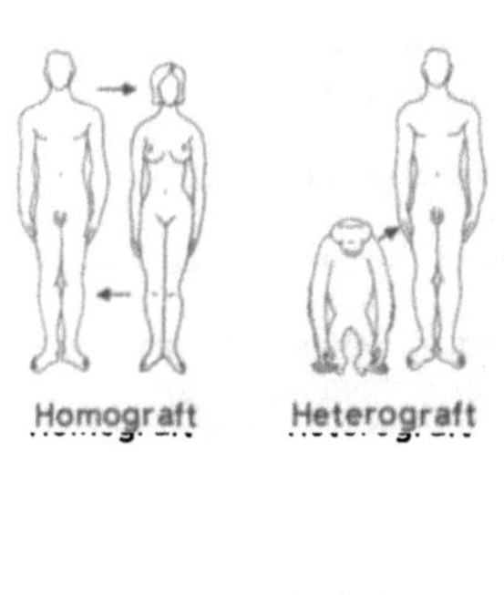

gans between two genetically differing members of same species (most frequent type of T. – blood transfusion, ↑ bone marrow, ↑ cornea, ↑ cardiac muscle, ↑ kidney, ↑ skin, etc.). 4) Heterograft (heteroplastic T., heterologous T.) = transfer of tissues or organs between members of different species (monkey-human, pig-human, dog-human, etc.). In both homograft and heterograft, ↑ graft rejection reaction occurs.

↑ Histocompatibility

Brent, L., Fabre, J.W., Elves, M.W., Sells, R.A., Rappaport, F.T.: Transplantation Today. Vol. VII. San Francisco: Grune & Stratton 1983; Gale, R.P., Fox, F.C. (eds.): Biology of Bone Marrow Transplantation. New York: Academic Press 1980; Schreiner, G.E.: Kidney Int. *23*, Suppl. 14, S-4 (1983)

Transport vesicles (transfer vesicles, transitional vesicles): ↑ Coated vesicles, 50–60 nm across (arrowheads), involved in transportation of products from rough ↑ endoplasmic reticulum (ER) to ↑ Golgi apparatus (G), and in some cells from there to ↑ lysosomes. T. are formed by budding off from membranes of both rough and transitional ↑ ER. It is thought that lamellae of Golgi apparatus are formed by fusion of T. (Fig. = epithelial cell, rat)

↑ Golgi apparatus, origin of

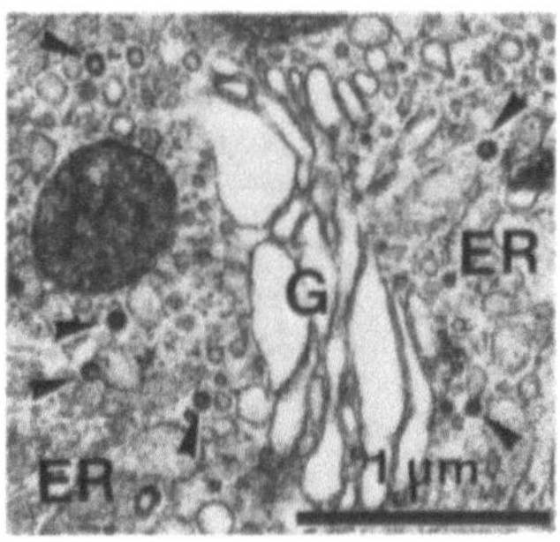

Transporting epithelium: ↑ Epithelium, transporting

Transsynaptic degeneration: ↑ Transneuronal degeneration

Transversal division, of centromere: ↑ Accidents, of cell division

Transverse colon: ↑ Colon

TRH: ↑ Thyrotropin-releasing factor

Triad, of retinal neurons (triad synapse): A characteristic symmetrical figure formed of a ↑ cone pedicle (CP), two ↑ horizontal cell ↑ dendrites (H) and, between them, an inserted central dendrite (B) of a midget ↑ bipolar neuron. F = dendrites of flat bipolar neurons, SR = ↑ synaptic ribbon. (Modified after Raviola and Gilula, 1975)

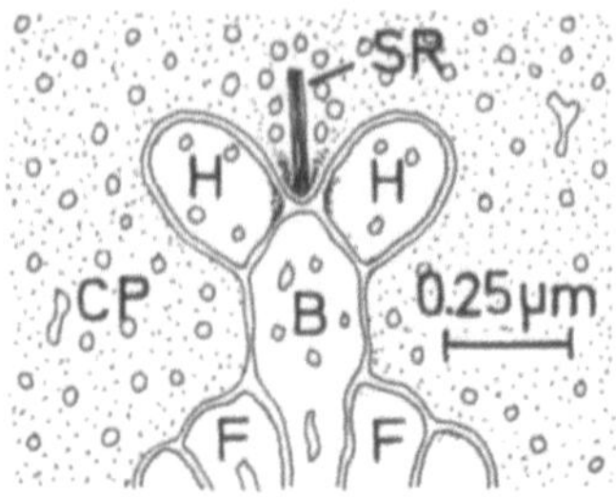

Raviola, A., Gilula, N.B.: J. Cell Biol. *65*, 192 (1975)

Triad, of striated muscular tissue (trias*, triades*): A characteristic structure (Tr) composed of two adjacent ↑ terminal cisternae (TC) of ↑ sarcoplasmic reticulum plus an interposed ↑ T-tubule. A slit, roughly 12–14 nm wide, separates T-tubule from terminal cisternae; membranes of the latter are connected with that of T-tubule by small "units" (arrowheads) which seem to allow passage of ions and small molecules. In mammalian ↑ skeletal muscle fibers (top), T. are situated at level of A-I junction, thus there are two T. per ↑ sarcomere. In ↑ cardiac muscle cells (bottom), T. lie at level of ↑ Z-line, thus there is one T. per sarcomere. T. are less frequent in cardiac muscle cells than in skeletal muscle fibers; ↑ dyads are most common in cardiac muscle cells. (Figs. = rat)

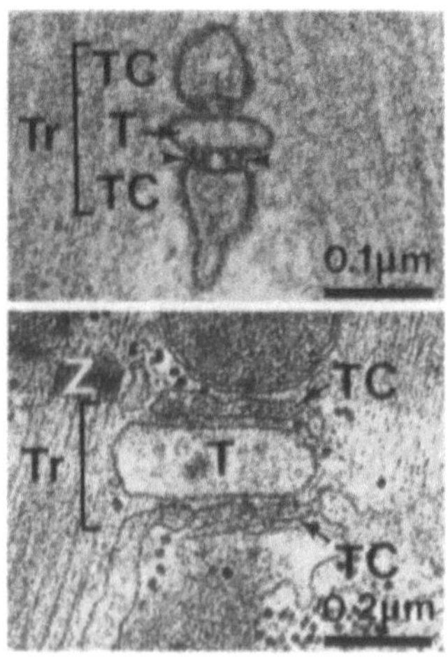

Mitchell, R.D., Saito, A., Palade, P., Fleischer, S.: J. Cell Biol. *96*, 1017 (1983)

Triad, portal (trias hepatica*): Three most conspicuous elements of a ↑ portal canal: ↑ interlobular artery, ↑ interlobular vein, and interlobular ↑ bile duct.

Triad synapse: ↑ Triad, of retinal neurons

Trichohyalin granules (granulum trichohyalini*): Membraneless amorphous ↑ inclusion granules (T) of varying density present in cells of ↑ Huxley's layer of ↑ hair follicle. T. average 0.1–0.2 µm in diameter; they are frequently associated with bundles of ↑ tonofilaments (arrowheads). Chemically, T. are comparable to ↑ keratohyalin. T. fill ↑ cuticle cells of inner root sheath by dispersion. (Fig. = mongolian gerbil)

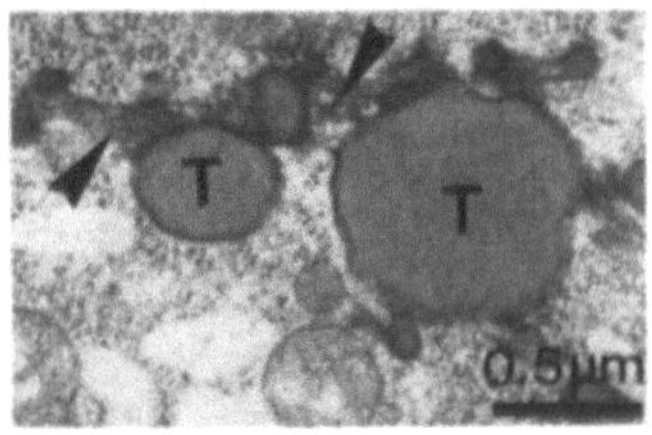

Trichrome staining: Staining technique combining three dyes for distinct demonstration of various organ constituents (e.g., Azan, Mallory staining, Masson staining, etc.).

Tricuspid valve: ↑ Valves, atrioventricular

Trigona fibrosa: ↑ Annuli fibrosi, of heart; ↑ Cardiac skeleton

Trigonal glands (glandula trigoni vesicae*): Small ↑ mucoid glands located in tunica mucosa of trigonum of ↑ urinary bladder.

Triiodothyronine: Iodine-containing hormone of ↑ thyroid follicular cells with action identical to, but more pronounced than ↑ that of thyroxin. Secretion rate of T. is only one-tenth that of thyroxin.

Chopra, I.J., Cody, V.: Triiodothyronine in Health and Disease. Berlin, Heidelberg, New York: Springer-Verlag 1981

Trisomy (trisomia*): Case of chromosome anomaly when there is one ↑ autosome or one ↑ sex chromosome too many. Cause of T. is ↑ nondisjunction.

↑ Chromosomes, anomalies of; ↑ Down's syndrome

Triticeal cartilage (cartilagines triticeae*): Small hyaline and/or fibrocartilaginous pieces of ↑ larynx.

↑ Cartilage, hyaline; ↑ Cartilage, fibrous

tRNA: ↑ Transfer RNA

Trophoblast (chorionic epithelium, trophoblastus*): Bistratified epithelial lining of ↑ chorion composed of superficial syncytiotrophoblast and internal cytotrophoblast. T. covers internal aspect of chorionic plate, ↑ placental villi, and ↑ decidua basalis bordering intervillous spaces. Structure: 1) Syncytiotrophoblast (S) = continuous layer of multinucleated cytoplasm in which nuclei have a tendency to clump together. Superficial cytoplasm is ↑ acidophilic, because it contains a great number of ↑ micropinocytotic vesicles, smooth surfaced vacuoles, large, irregularly shaped vesicles, and ↑ tubules of smooth endoplasmic reticulum. Basal cytoplasm is ↑ basophilic due to its large amount of rough endoplasmic ↑ cisternae and many free ribosomes; it also presents a moderately developed ↑ basal labyrinth. Throughout cytoplasm are scattered ↑ Golgi complexes surrounded by small membrane-bound granules, ↑ multivesicular bodies, and occasional ↑ lipid droplets. ↑ Mitochondria are small and slender, with both tubules and cristae. A myriad of pleomorphic microvilli, sometimes forming ↑ brush border, project into ↑ intervillous spaces (IS). Syncytiothrophoblast is connected to cytotrophoblast cells by small ↑ desmosomes and ↑ zonulae occludentes. Syncytiotrophoblast persists up to birth, but as number of cytotrophoblastic cells decreases, T. becomes pseudostratified and comes into contact (arrowheads) with fetal capillaries (FC). Syncytiotrophoblast frequently forms ↑ trophoblastic islands. 2) Cytotrophoblast or Langhans' cells = complete layer of individual cuboidal or flattened cells (C) which gradually decrease in number after 4th month of pregnancy; however, some cytotrophoblastic cells persist in placenta at term (fig.). Nucleus of cytotrophoblastic cells is round or elliptical, the cytoplasm contains rod-shaped mitochondria, elongated narrow rough endoplasmic cisternae, a Golgi apparatus surrounded by small vesicles and dense granules, lysosomes, smooth endoplasmic ↑ tubules, microfilaments, some ↑ glycogen particles, a few free ribosomes, and, in some cases, intracellular ↑ canaliculi. All cytotrophoblastic cells lie on ↑ basal lamina and are interconnected by a few desmosomes and zonulae occludentes. Cytotrophoblastic cells proliferate by ↑ mitosis; their fusion leads to formation of syncytiotrophoblast. (Fig. = human)

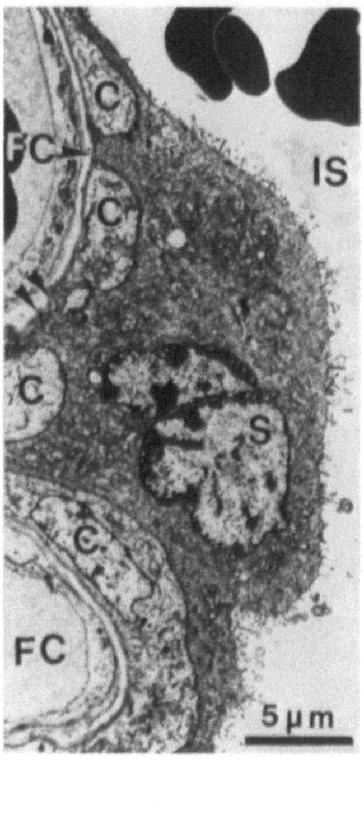

↑ Chorionic villi; ↑ Placental barrier; ↑ Placenta, early; ↑ Placenta, full-term, structure of; ↑ Trophoblast, histophysiology of

Enders, A.C.: Obstet. Gynecol. *25*, 378 (1965); Martin, B.J., Spicer, S.: Anat. Rec. *175*, 15 (1973); Metz, J., Weihe, E.: Anat. Embryol. *158*, 167 (1980); Metz, J., Weihe, E., Heinrich, D.: Anat. Embryol. *158*, 41 (1979); Sibley, C.P., Bauman, K.F., Firth, J.A.: Cell Tissue Res. *229*, 365 (1983)

Trophoplast, histophysiology of: Trophoblast forms principal part of ↑ placental barrier. Syncytiotrophoblast synthesizes placental protein hormones, such as ↑ human chorionic somatomammotropin and ↑ human chorionic thyrotropin, and ↑ steroid hormones (↑ estrogens and ↑ progesterone). Cytotrophoblast probably produces human chorionic gonadotropin and possibly ↑ fibrinoid.

Nishishira, M., Yagishashi, S.: Acta Histochem. Cytochem. *12*, 434 (1979)

Trophoblastic giant cells: ↑ Giant cells, of trophoblast

Trophoblastic islands (proliferation islands): Sprouts (TI) of proliferated ↑ syncytiotrophoblast found at external surface of ↑ placental villi (V). T. may separate from villi and enter maternal pulmonary circulation via ↑ intervillous spaces (IV). As ameboid elements, T. may pinch off from ↑ anchoring villi and invade ↑ endometrium as ↑ giant cells of ↑ trophoblast. (Fig. = early ↑ placenta, human)

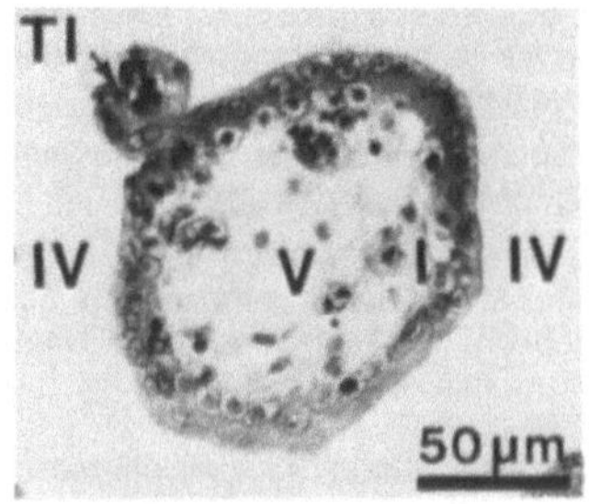

Trophoblastic shell: Layer of ↑ trophoblast covering ↑ decidua basalis.

↑ Placenta, full-term, structure of

Tropocollagen: Macromolecular monomer (T), fundamental element of ↑ collagen, 280 nm long and 1.5 nm thick. T. is composed of three polypeptide α-chains (α) coiled around central axis to form right-handed helix. Each chain is composed of repetitive sequence of amino acids (AA), every third amino acid being glycine (Gly). Chains are held together by hydrogen bonds between proline (Pro) and glycine. The sugar components glucose (Glu) and galactose (Gal) are fixed onto hydroxylysine (Hyl). T. molecule ends with nonhelical aminoterminus (AT) and carboxyterminus (CT) having different electrical charges (so-called tail and head of T. molecule). During alignment of T., aminoterminus of one T. molecule comes into contact with carboxyterminus of another. Chemically, T. consists of 30% glycine, 25% proline/hydroxyproline, remainder representing other amino acids. Molecular weight of T. is about 300000 daltons; every chain is composed of about 1000 amino acids. T. is ↑ birefringent.

↑ Bone collagen; ↑ Collagen, fibrillogenesis of

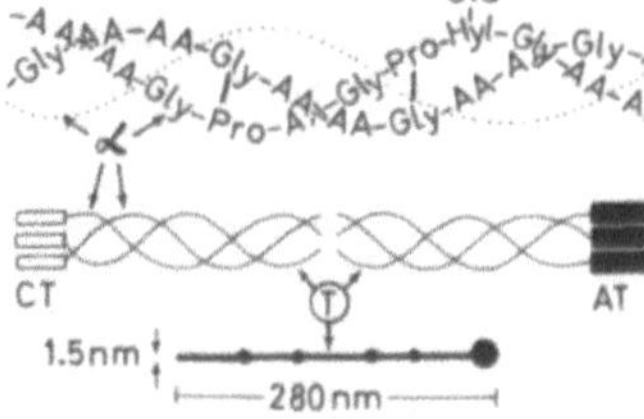

Tropoelastin: ↑ Proelastin; ↑ Elastin

Tropomyosin: An approximately 40-nm-long muscle protein providing sensitivity of actomyosin to calcium ions and triggering muscle contractions. T. is situated in groove of actin helix and serves to bind ↑ troponin to ↑ actin myofilament.

↑ Striated muscle fiber, molecular biology of contraction of

Schloss, J.A., Goldmann, R.D.: J. Cell Biol. *87*, 633 (1980); Yamaguchi, M., Robson, R.M., Stromer, M.H.: Ultrastruct. Res. *80*, 111 (1982)

Troponin: Structural calcium-receptive protein (m.w. about 50 000 daltons) distributed at regular intervals of about 40 nm along ↑ actin myofilament and attached to it through ↑ tropomyosin. When Ca^{2+} concentration is low, T. inhibits contact of S-1 portion of ↑ myosin myofilament with actin myofilament by interposition; an increase of Ca^{2+} leads to displacement of T., formation of actomyosin, and muscle contraction.

↑ Striated muscle fiber, molecular biology of contraction of

Cohen, C.: Sci. Am. *233*, 36 (1975)

Trypan blue: Vital dye principally used for selective demonstration of ↑ macrophages. After intravenous or intraperitoneal injection, particles of T. are phagocytized by macrophages, coloring them blue. Some T. particles pass ↑ glomerular filtration membrane and are taken up by ↑ athrocytosis into ↑ proximal tubule cells of ↑ nephron. Similarly, epithelial cells of ↑ choroid plexus may also store T.

Trypsin: Proteolytic enzyme secreted by ↑ pancreatic acinar cells.

TSH: ↑ Thyrotropic hormone

TSH-cells: ↑ Thyrotropes

TSH-RF: ↑ Thyrotropin-releasing factor

T-system: All ↑ T-tubules of a ↑ skeletal muscle fiber or a ↑ cardiac muscle cell.

T-tubules (tubulus transversus*): Transversally oriented canalicular invaginations (T) of ↑ sarcolemma (S) of ↑ skeletal muscle fibers (Sk) and ↑ cardiac muscle cells (C) into ↑ sarcoplasm. T. are richly ramified, surrounding each ↑ myofibril transversally and separating two adjacent ↑ terminal cisternae. In skeletal muscle fibers, T. are situated at junction of ↑ A-bands and ↑ I-bands; in cardiac muscle cells, T. lie at level of ↑ Z-line. T. of latter are considerably wider and filled with an amorphous material continuous with basal lamina (BL). T. participate in a very rapid spreading of action potential from cell surface into interior of sarcoplasm to each ↑ myofibril. (Figs. = mongolian gerbil)

↑ Dyad; ↑ Triad, of striated muscular tissue

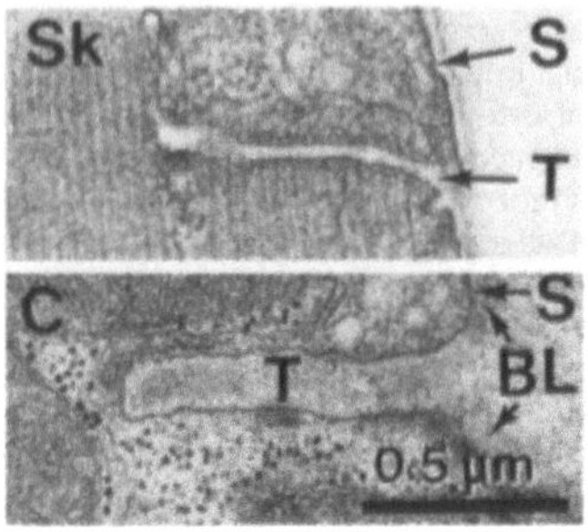

Leeson, T.S.: Acta Anat. *108*, 374 (1980); Moses, R.L., Kasten, F.K.: Cell Tissue Res. *203*, 173 (1979)

Tubal tonsil: ↑ Tonsil, tubal

Tuberoinfundibular tract (tuberohypophyseal tract, tractus tuberoinfundibularis): Posterior part of ↑ hypothalamohypophyseal tract.

Tubes, auditory: ↑ Auditory tubes

Tubes, Fallopian: ↑ Oviducts

Tubular glands (glandula tubulosa*): Simple glands in form of tubes of various lengths (e.g., ↑ gastric glands proper). T. with branched end piece is called branched T. (e.g., ↑ pyloric glands); T. with coiled terminal piece is termed coiled T. (e.g., eccrine ↑ sweat glands).

↑ Glands, classification of

Tubular myelin (fingerprint figures): Tubular lipid crystalloids about 45 nm in diameter, densely packed in quadratic lattice; found in ↑ alveoli of lung. T. originates from ↑ multilamellar bodies; it is considered to be a degradation product of ↑ surfactant. T. may be phagocytized by ↑ alveolar macrophages. (Fig. = rat)

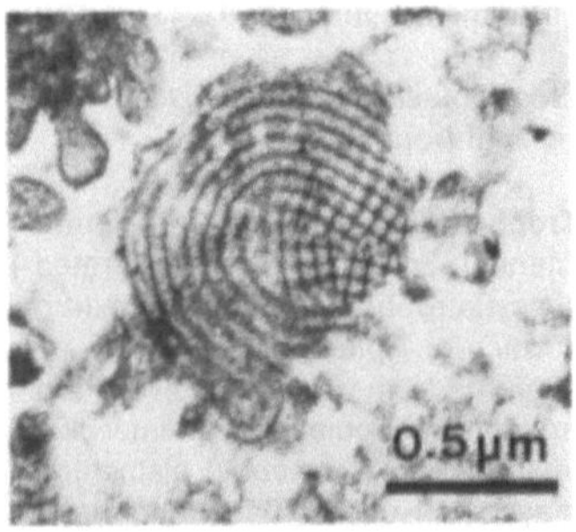

Hassett, R.J., Engleman, W., Kuhn, C. III: J. Ultrastruct. Res. *71*, 60 (1980); Stratton, C.J.: Tissue Cell *9*, 301 (1977); Weibel, E.R., Kistler, G.S., Töndury, G.: Z. Zellforsch. *69*, 418 (1966); Williams, M.C.: Exp. Lung Res. *4*, 37 (1982)

Tubules, dentinal: ↑ Dentinal tubules

Tubules, mucous: ↑ Mucous tubules

Tubules, of mitochondria (tubulus mitochondrialis*): Fingerlike invaginations (T) of inner mitochondrial membrane (IM) into ↑ mitochondrial matrix. T. frequently end with a bulbous enlargement (B). OC = outer chamber, OM = outer mitochondrial membrane

↑ Mitochondria; ↑ Mitochondria, with tubules; ↑ Mitochondrial membranes

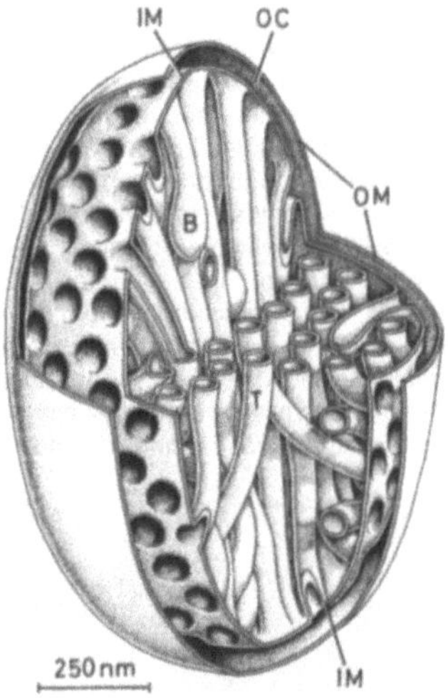

Tubules, of smooth endoplasmic reticulum (tubules*): Three-dimensional system of irregularly branching and anastomosing channels (T), 30–60 nm in diameter, often continuous with ↑ cisternae of rough ↑ endoplasmic reticulum. (Fig. = ↑ pigment epithelial cell, rat)

↑ Endoplasmic reticulum, smooth

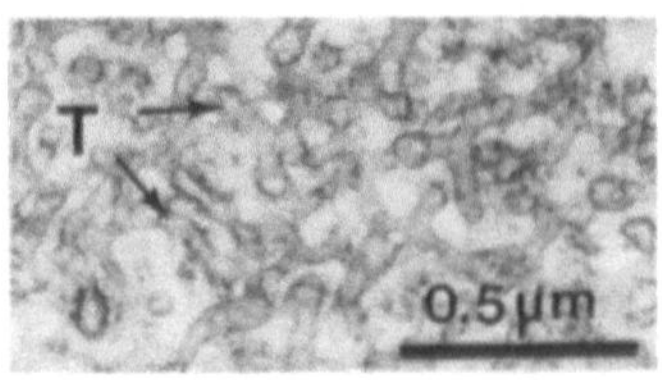

Tubules, seminiferous: ↑ Seminiferous tubules

Tubules, transverse: ↑ T-tubules

Tubules, uriniferous: ↑ Uriniferous tubules

Tubuli recti (straight tubules, tubulus seminifer rectus*): Approximately 1-mm-long and 0.1 to 0.25-mm-wide channels (TR) connecting ↑ seminiferous tubules with ↑ rete testis (RT). Initial portion of T. is lined with modified ↑ Sertoli's cells; rest is lined by ↑ simple cuboidal epithelium with cells bearing irregular microvilli and occasional ↑ cilia. T. are surrounded by ↑ dense connective tissue in which occasional ↑ smooth muscle cells can be present. (Fig. = human)

Dym, M.: Am. J. Anat. *140*, 1 (1974); Lindner, S.G.: Andrologia *14*, 253 (1982); Osman, D.I., Plöen, L.: Anat. Rec. *192*, 1 (1978)

Tubulin: ↑ Basic protein of ↑ microtubules composed of two subunits, α and β monomers, measuring 5 nm in diameter. The monomers associate to form double molecules or dimers. Through a linear assembly of dimers, there occur strands, protofilaments, arranged side by side in a sheet (not represented). By curling, sheet forms a tube, microtubule. Assembly of dimers is controlled by ↑ microtubule organizing center. (Modified after Dustin 1980)

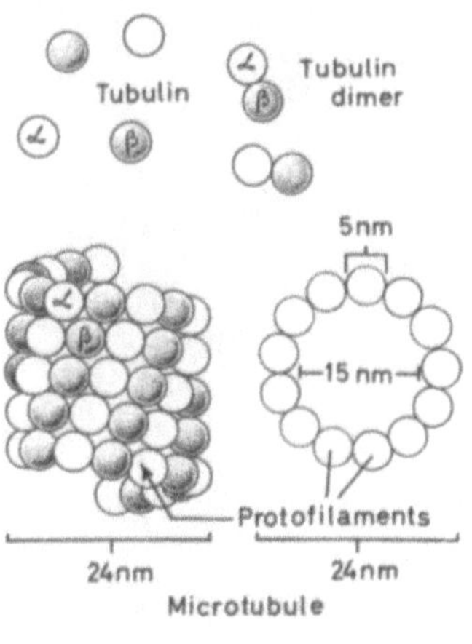

Dustin, P.: Sci. Am. *243/2*, 58 (1980); Ludeana, R.F., Little, M.: Bio Systems *14*, 231 (1981); Sakai, H., Mohri, H., Borisy, G.G. (eds.): Biological Functions of Microtubules and Related Structures. New York: Academic Press 1983; Willingham, M.C., Yamada, S.S., Pastan, I.: J. Histochem. Cytochem. *28*, 453 (1980)

Tubuloacinar glands (glandula acinosa*): Glands with branched tubular excretory ducts (ED) ending in ↑ acini (A). To pure T. belong ↑ serous glands of organism (exocrine ↑ pancreas, ↑ parotid glands, ↑ Ebner's glands, etc.). Since in ↑ mixed glands, ↑ mucous tubules (MT) are in fact transformed ↑ intercalated ducts (ID) and ↑ demilunes (D) are transformed acini, to T. also belong mixed ↑ salivary glands, glands of respiratory system, etc.

↑ Glands, classification of

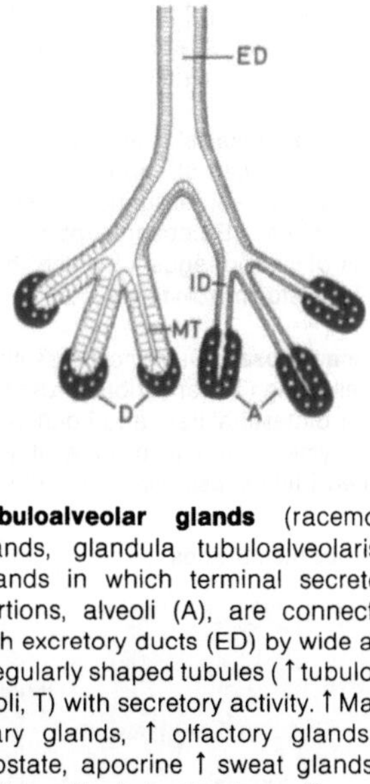

Tubuloalveolar glands (racemose glands, glandula tubuloalveolaris*): Glands in which terminal secretory portions, alveoli (A), are connected with excretory ducts (ED) by wide and irregularly shaped tubules (↑ tubuloalveoli, T) with secretory activity. ↑ Mammary glands, ↑ olfactory glands, ↑ prostate, apocrine ↑ sweat glands, ↑ bulbourethral glands, etc. are T.

↑ Alveolus, of mammary gland; ↑ Glands, classification of

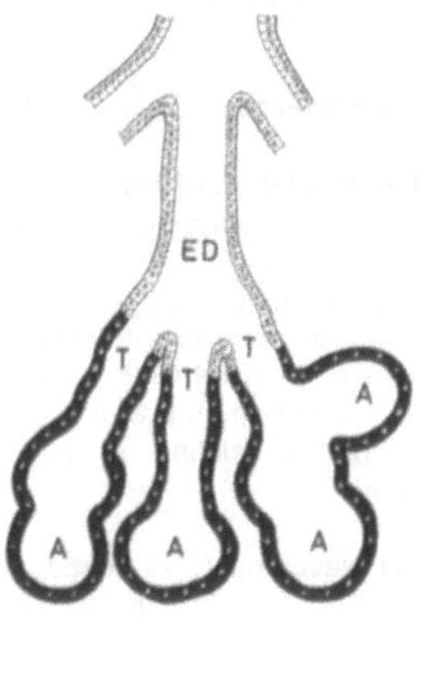

Tubuloalveoli: Morphofunctional units of ↑ tubuloalveolar glands.

Tufted cells: Two varietes of ↑ neurons with ↑ perikarya situated in outer granular layer of ↑ olfactory bulb. Their dendrites break up into many branches which form compact, rounded, bushy terminals which interlace with ↑ axons of ↑ olfactory cells to form ↑ olfactory glomeruli. Axon of a T. enters white matter and continues via olfactory tract into lateral olfactory stria. Structurally, T. correspond to other ↑ neurons.

Tufts, enamel: ↑ Enamel tufts

Tunica* (tela, tunic): One of enveloping layers of an ↑ organ, particularly that with a tubular form.

Tunica adventitia* (tunica fibrosa*): Outermost layer of various organs, made up of ↑ loose connective tissue which connects ↑ organ to surrounding parts. T. contains blood and lymphatic vessels, and ↑ nerve fascicles.

Tunica albuginea*: Very thick whitish ↑ capsule of some organs (↑ penis, ↑ ovary, ↑ testis).

Tunica albuginea*, of ovary: Layer of ↑ dense connective tissue (TA) between ↑ germinal epithelium (GE) and primordial ↑ ovarian follicles (PF). (Fig. = cat)

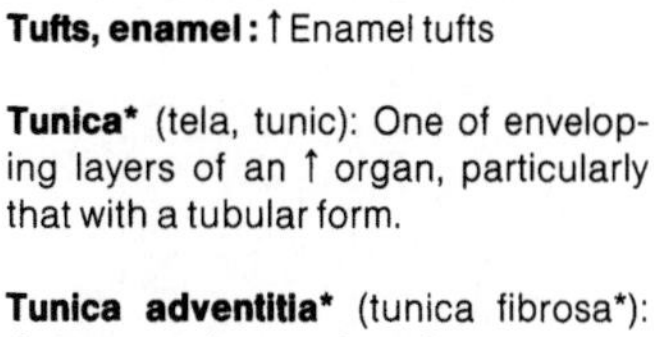

Tunica albuginea*, of penis: Thick layer of ↑ dense connective tissue (A) surrounding ↑ cavernous bodies (CB) and ↑ spongious body of ↑ penis. T. of spongious body is considerably thinner than those of cavernous bodies. (Fig. = human newborn)

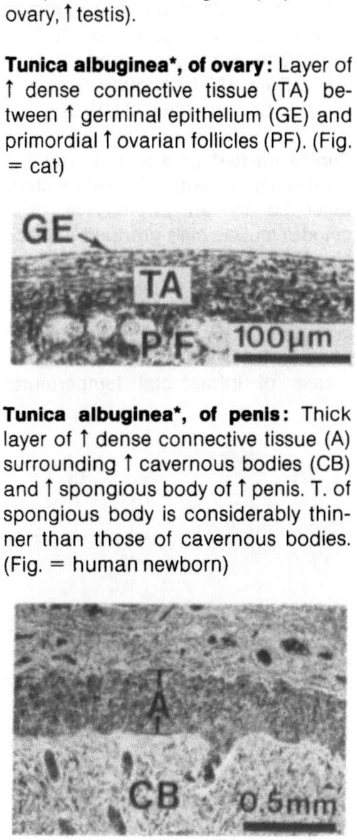

Tunica albuginea*, of testis (capsule): Thick layer of ↑ dense connective tissue (TA) surrounding ↑ testis. Inner surface of T., ↑ tunica vasculosa (TV), contains large blood vessels (arrowheads) of testis. Major part of outer aspect of T. is covered with ↑ mesothelium (arrow) of ↑ tunica vaginalis. At posterior margin of testis, T. thickens and forms ↑ mediastinum testis. Besides elements of connective tissue, T. contains ↑ myofibroblasts and some ↑ smooth muscle cells. T. exerts a certain pressure on ↑ seminiferous tubules (ST), pushing immobile ↑ spermatozoa toward ↑ epididymis. Rhythmic contraction of smooth muscle cells in man has not been proved.

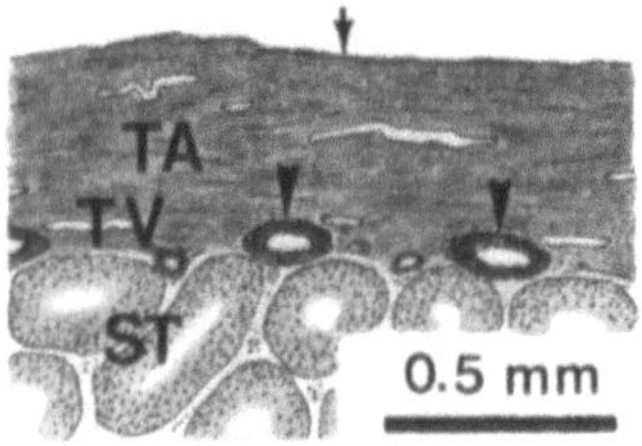

Chacon-Arellano, J.-T., Wolley, D.M.: J. Anat. *131*, 263 (1980); Davis, J.R., Lagford, G.A., Kirby, P.J.: The Testicular Capsule. In: Johnson, A.D., Gomes, W.R. (eds.): The Testis. Vol. 1. New York, San Francisco, London: Academic Press 1970; Langford, G.A., Heller, C.G.: Science *179*, 573 (1973)

Tunica dartos* (musculus dartos*): ↑ Dermis (D) of ↑ skin of ↑ scrotum characterized by an abundance of ↑ smooth muscle cells grouped in fascicles (F) which attach to papillary layer and to adventitia of blood vessels. T. acts as thermoregulator for ↑ spermatogenesis: Contraction of T. diminishes surface of scrotum, causing increase of intrascrotal temperature; relaxation has opposite effect. (Fig. = human)

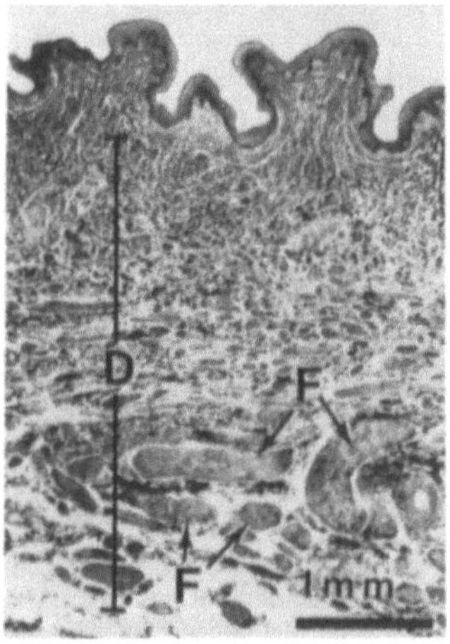

Tunica intima*: Innermost layer of blood and ↑ lymphatic vessels. T. consists of ↑ endothelium and subendothelial layer, thickness and structure depending on diameter of vessel in question.

Tunica media* (media): Middle layer of ↑ macrovasculature, ↑ microvasculature (except ↑ capillaries), ↑ lymphatic ducts, and ↑ thoracic duct.

Tunica mucosa* (mucous membrane, mucosa): Innermost mucous layer of all cavities and canals of body which connect with exterior (↑ gastrointestinal tract, ↑ respiratory system, and ↑ genitourinary system). T. is made up of: a) ↑ epithelium, b) ↑ lamina propria, and sometimes c) ↑ lamina muscularis mucosae. Glands are present except in T. of most of ↑ genitourinary system.

Tunica muscularis*: Smooth muscular tissue in wall of various ↑ organs with function of assuring peristaltic contractions. (Exception: upper two-thirds of ↑ esophagus, in which there are ↑ skeletal muscle fibers.)

Tunica serosa*: Outermost mesothelial cell lining (TS) of various organs of ↑ gastrointestinal tract and ↑ genitourinary system. Lamina propria of T. is termed ↑ tela subserosa (TSs). (Fig. = rat)

↑ Serous membranes

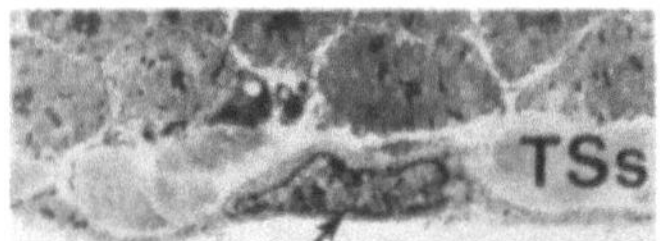

Tunica submucosa: ↑ Tela submucosa

Tunica subserosa: ↑ Tela subserosa

Tunica vaginalis*, of testis: ↑ Serous membrane surrounding anterior and lateral surfaces of ↑ testis. Visceral leaf adheres to ↑ tunica albuginea; parietal leaf lines inner surface of ↑ scrotum. T. consists of ↑ mesothelium which produces a small amount of serous fluid enabling free gliding of testes in scrotum.

Tunica vasculosa*, of testis: Well-vascularized layer of ↑ loose connective tissue on inner aspect of ↑ tunica albuginea.

Tunnel, of Corti: ↑ Corti's tunnel

Tunnel outer, of Corti's organ: ↑ Nuel's tunnel

Turbinate bones: ↑ Conchae nasales

Turnbull reaction: Histochemical method for demonstration of ↑ ferrous salts (Fe^{2+}) with potassium ferricyanide.

↑ Berlin blue reaction

Turner's syndrome: ↑ Chromosomes, anomalies of; ↑ Monosomy

Tympanic cavity (tympanum, cavitas tympanica*): Irregular air-filled space (TC) in temporal bone (TB) belonging to middle ↑ ear. Laterally T. is limited by ↑ tympanic membrane (TM), anteromedially it communicates with ↑ auditory tube (AT) and posteriorly with mastoid ↑ "cells" (MC). Structure: 1) Tunica mucosa is connected to ↑ periosteum (P): a) epithelium (E) = ↑ simple squamous to cuboidal epithelium with some ↑ pseudostratified patches around opening of auditory tube and attachment line of tympanic membrane. In children, epithelium is predominantly ciliated (C) with ↑ goblet cells; b) lamina propria (LP) = richly vascularized and innervated ↑ loose connective tissue with few ↑ mucous glands grouped in anterior part of T. ↑ Lymphocytes and ↑ lymphatic capillaries are present. In T. are located ↑ auditory ossicles (AO), ↑ tendons of ↑ tensor tympani and ↑ stapedius muscles, and chorda tympani nerve.

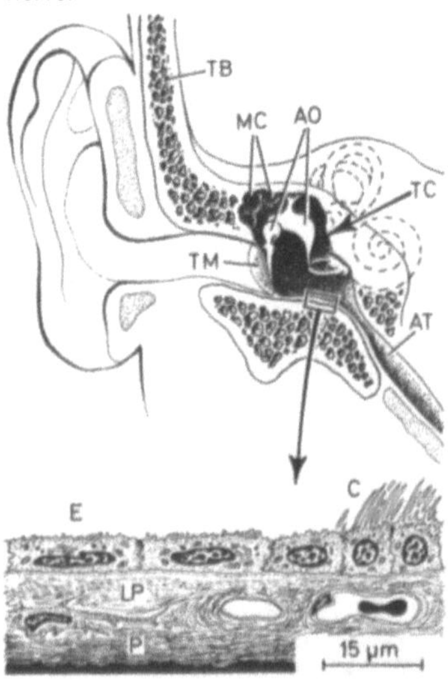

Tympanic membrane (ear drum, membrana tympani*): A thin membrane (T) separating external ↑ auditory meatus from ↑ tympanic cavity (tc). Major portion of T. is tightly stretched [= pars tensa* (pt)]; remainder is loosely arranged [= pars flaccida* or membrane of Shrapnell (pf)]. Structure: 1) Cutaneous layer or stratum cutaneum* (CL): a) extremely thin ↑ epidermis (Ep) continuous with that of auditory meatus; b) very poorly developed ↑ dermis blends with underlying fibrous layer. 2) Fibrous layer (FL) = ↑ collagen fibers (cf) disposed in an inner circular layer or stratum circulare* (ci) and an outer radial layer or stratum radiatum* (r). Fibers combine with hyaline ↑ cartilage to form a thickened ring at periphery of T., annulus fibrocartilagineus* (AF). Numerous ↑ elastic fibers (ef) and capillaries (Cap) run through this layer. Pars flaccida is devoid of a fibrous layer. 3) Mucous layer or stratum mucosum* (ML) = a very flattened epithelium of tympanic cavity. T. is vascularized by branches of arteries of auditory meatus and tympanic cavity; auriculotemporal nerve, auricular branch of vagus and tympanic branch of the glossopharyngeal nerves innervate T. (Fig., top = human; fig., bottom = rat)

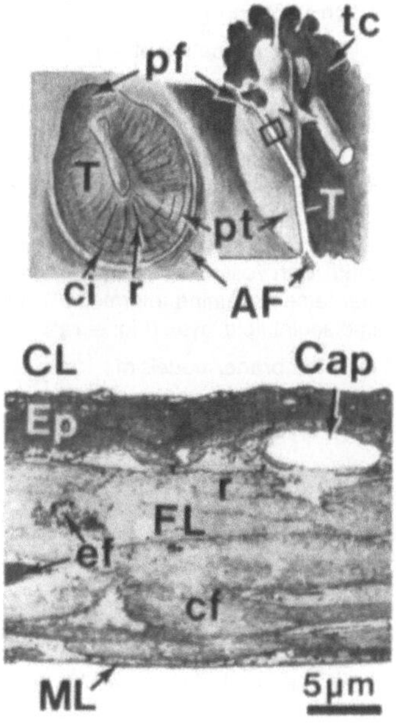

Hattori, Y., Yuge, K.: Otologia Fukuoka *25*, Suppl. 1115 (1979)

Type I cell, of carotid body: ↑ Carotid body

Type I cell, of lung: ↑ Alveolar cells, type I

Type I cell, of macula of saccule and utricle: ↑ Vestibular cells

Type I cell, of parasympathetic paraganglion: ↑ Paraganglia, parasympathetic

Type I cell, of semicircular canal: ↑ Vestibular cells

Type I cell, of sympathetic paraganglion: ↑ Carotid body

Type I cell, of taste bud: ↑ Taste buds, structure of

Type I gonadotrope: ↑ Gonadotropes, type I

Type II cell, of carotid body: ↑ Carotid body

Type II cell, of lung: ↑ Alveolar cells, type II

Type II cell, of macula of saccule and utricle: ↑ Vestibular cells

Type II cell, of parasympathetic paraganglion: ↑ Paraganglia, parasympathetic

Type II cell, of semicircular canal: ↑ Vestibular cells

Type II cell, of sympathetic paraganglion: ↑ Carotid body

Type II cell, of taste bud: ↑ Taste buds, structure of

Type II cell, of gonadotrope: ↑ Gonadotropes, type II

Type III cell, of taste bud: ↑ Taste buds, structure of

Type IV cell, of taste bud: ↑ Taste buds, structure of

Type V cell, of taste bud: ↑ Taste buds, structure of

Tyrosinase: A copper-containing enzyme converting tyrosine into ↑ DOPA and then into dopaquinone, which is converted, after a series of steps, into ↑ melanin. T. is synthesized on ↑ ribosomes of rough ↑ endoplasmic reticulum of ↑ melanocytes, then reaches ↑ reticuloplasm, to be transported in ↑ transport vesicles to ↑ Golgi apparatus, from which T.-containing vesicles separate (stage I of ↑ melanogenesis). T. molecules are thin, very coiled filaments, about 1 nm thick, 20–500 nm long, and with a periodicity of about 10 nm.

Tyson's glands: ↑ Preputial glands

U

U-cells: Cells of ↑ ultimobranchial follicles.

Ultimobranchial bodies (corpus ultimobranchiale*): Small cell groups and/or follicles originating from 5th pharyngeal pouch and embedded within developing mammalian ↑ thyroid. In adult thyroids, U. are not obvious, since they gradually regress during postnatal life; however, their remnants produce ↑ ultimobranchial follicles. U. of mammals provide ↑ C-cells, which originate from neural crest and migrate into U. during development. In lower vertebrates, U. are separate organs containing a considerable amount of ↑ calcitonin.

LiVolsi, A.V., Merino, M.J.: Am. J. Surg. Path. *2*, 133 (1978)

Ultimobranchial cells: A term sometimes used to designate ↑ C-cells of ↑ thyroid gland.

Ultimobranchial follicles: Kind of ↑ thyroid follicles (F) originating from ↑ ultimobranchial bodies. U. consist of two or more layers of flattened cells, ↑ U-cells (U), surrounding a cavity containing cell debris (D). Basal cells have developed ↑ hemidesmosomes on their basal plasmalemma, abundant free ribosomes, and a considerable number of ↑ tonofilaments bound in ↑ tonofibrils (Tf). Cell surfaces of all other

cells are studded with numerous microvillous projections bearing an ↑ attachment plaque on their tips. The innermost cells degenerate and transform into debris. Epithelium of U. resembles a thin ↑ stratified squamous nonkeratinized epithelium. Significance of U. is still unknown. (Figs. = rat)

Fraser, B.A., Duckworth, J.W.A.: Acta Anat. *105*, 269 (1979); Ketelbant-Balasse, P., Nève, P.: Cell Tiss. Res. *166*, 49 (1976)

Ultracryotomy: Cutting of ↑ ultrathin sections from deeply frozen unfixed blocks to demonstrate various enzymatic and other fixation-labile substances in ↑ transmission electron microscope. U. is performed with an ↑ ultramicrotome equipped with a special freezing device.

Ultramicrons: Particles ranging from 1 to 100 nm in diameter taken up by cells in the course of ↑ athrocytosis.

Ultramicroscope: A former synonym for ↑ ultraviolet microscope; now, U. means ↑ transmission electron microscope.

Ultramicrotome: An instrument for ↑ sectioning ↑ ultrathin sections. Principle: A metallic bar (B), fixed on one side, carries the object (O) on its other side. Through controlled heating (H), the bar expands toward a glass or diamond ↑ knife (K), which cuts a section (S). Along with this expansion, the bar makes rotatory movements (interrupted line) permitting production of ribbons of sections. Besides this most frequently used thermodynamic U., there are also mechanical and piezoelectric U.

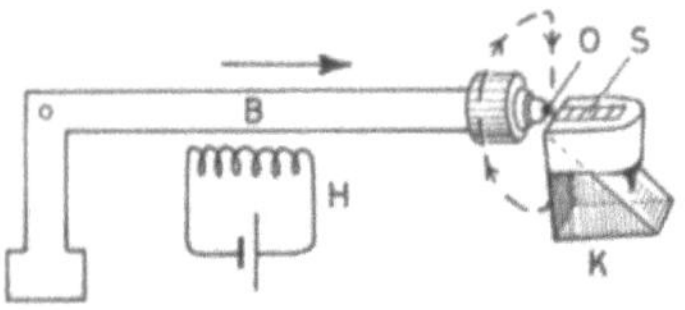

Ultramicrotomy: All preparatory, technical, and instrumental procedures for ↑ sectioning with an ↑ ultramicrotome.

Glauert, A.M. (ed.): Practial Methods in Electron Microscopy. Vol. 3, Part 2, Ultramicrotomy. Amsterdam: Elsevier/North Holland 1975; Pease, D.C., Porter, R.K.: J. Cell Biol. *91*, 287 s (1981)

Ultrathin sections: ↑ Sections with thickness ranging between 20 and 60 nm. U. are cut with an ↑ ultramicrotome and observed under ↑ transmission electron microscope.

Sakai, T.: J. Electron Microsc. (Tokyo) *29*, 369 (1980)

Ultraviolet microscope (ultramicroscope): A light microscope using ultraviolet light of 250 nm wavelength in order to increase ↑ resolving power. In practice, this is never done, and U. is employed as a basic instrument of the ↑ cytophotometer and ↑ fluorescence microscope.

Ultraviolet microspectrophotometry: ↑ Microspectrophotometry

Umbilical cord (funiculus umbilicalis*): A roughly 50-cm-long white strand connecting umbilicus of fetus to ↑ placenta. Structure: 1) ↑ Amniotic epithelium (A) = a ↑ simple cuboidal to ↑ stratified squamous epithelium continuous with ↑ epidermis of abdominal ↑ skin. 2) ↑ Gelatinous tissue (GT) enclosing two umbilical arteries (UA) and one umbilical vein (UV). U. has neither nerve fibers nor lymphatic vessels. In some U., remnants of allantois (Al) may be found. (See embryology texts for further information)

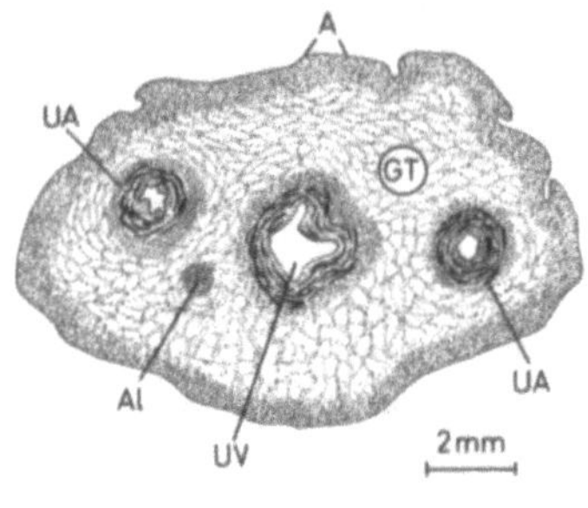

Parry, E.W.: J. Anat. *107*, 505 (1970)

Undifferentiated cells, of Lieberkühn's crypts: Narrow columnar cells (UC), each with a relatively large basal nucleus, prominent nucleolus, poorly developed ↑ organelles, and a large number of free ribosomes. By mitotic division (M), U. replace ↑ absorptive (A) and ↑ goblet cells (G), which migrate from ↑ Lieberkühn's crypt toward apex of an ↑ intestinal villus with simultaneous further ↑ differentiation. An intermediate step between U. and goblet cells is ↑ oligomucous cell (O).

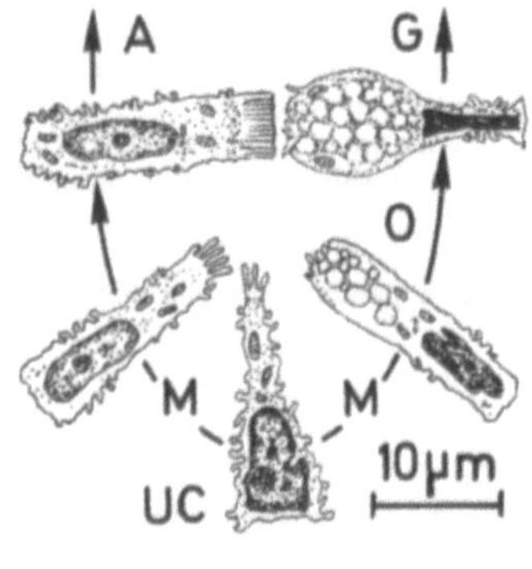

Cairnie, A.B., Lala, P.K., Osmond, D.G. (eds.): Stem Cells of Renewing Cell Populations. New York, San Francisco, London: Academic Press 1976

Undulating membranes: ↑ Lamellipodia

Unicellular glands: ↑ Glands consisting of only one cell (e.g., ↑ goblet cell).

↑ Endoepithelial glands; ↑ Glands, classification of

Unilaminar follicle: ↑ Ovarian follicle, primordial

Unipolar neurons: ↑ Neurons, classification of; ↑ Photoreceptors

Unistratified glands: ↑ Exocrine glands composed of only one layer of ↑ secretory cells. Except for ↑ pluristratified glands (↑ sebaceous and ↑ tarsal glands), all other glands are U. with an ↑ eccrine and/or ↑ apocrine secretion mechanism.

↑ Glands, classification of

Unit membrane (elementary membrane): A trilamellar sheath (U), about 7.5–11 nm thick, forming ↑ cell membrane and ↑ cytomembranes. Dark lamellae (lamina externa* et interna*) correspond to protein layers impregnated with osmium atoms; central clear lamella (lamina intermedia*) is a bimolecular lipid layer. (Fig. = rat)

↑ Cell membrane, models of

Luftig, R., Wehrli, E., McMillan, P.: Life Sci. *21*, 285 (1977); Pelttari, A., Heminen, J.H.: Biol. Cell *47*, 343 (1983)

Unmyelinated nerve fiber: ↑ Nonmyelinated nerve fiber

Uranyl acetate: A radioactive substance widely used for ↑ contrasting ↑ ultrathin sections; treatment of sections with U. is usually followed by treatment with ↑ lead citrate.

Urate oxidase: An enzyme concentrated in ↑ peroxisomes.

Urea: One of end products of metabolism, excreted in ↑ urine.

↑ Excretion

Ureter*: A tube conducting ↑ urine from ↑ renal pelvis to ↑ urinary bladder. Structure: 1) Tunica mucosa (TM): a) epithelium (E) = ↑ transitional epithelium; b) lamina propria (LP) = a well-vascularized and innervated ↑ loose connective tissue. Mucosa is folded when U. is empty. 2) Tunica muscularis (TMu) = helicoidally arranged anastomosing smooth muscular strands separated by abundant loose connective tissue. Although not clearly delimited, one can distinguish an inner longitudinal (IL) and a middle circular (MC) layer; in lower part of U. (pars pelvina*), an outer longitudinal (OL) layer is added. Tunica muscularis is responsible for descending peristaltic contractions of U. 3) Tunica adventitia (TA) = a loose connective tissue rich in adipose cells, blood and lymphatic vessels, and nerve fibers.

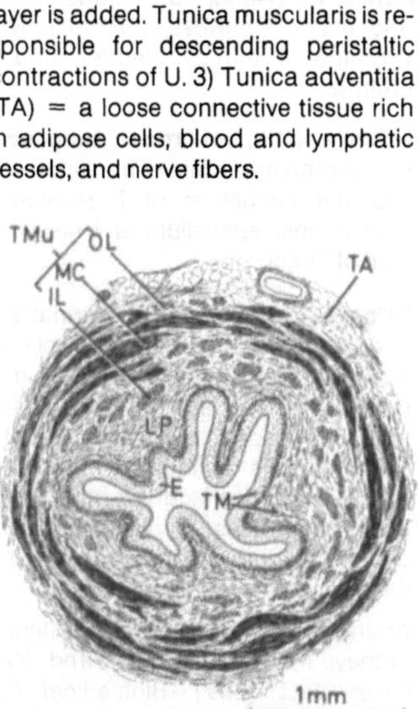

Bergman, H. (ed.): The Ureter. 2nd edn. Berlin, Heidelberg, New York: Springer-Verlag 1981; Liu, H.C.: Am J. Anat. *111*, 1 (1962)

Urethra, female (urethra feminina*): A tube, roughly 8 mm wide and 2–4 cm long, conducting ↑ urine from ↑ urinary bladder to exterior: Structure: 1) Tunica mucosa: a) epithelium (E) = ↑ transitional epithelium near urinary bladder, followed by an intermediate zone covered by ↑ stratified columnar epithelium, ending as nonkeratinized ↑ stratified squamous epithelium; epithelium forms glandular outpocketings, ↑ urethral glands (G); b) lamina propria = a ↑ loose connective tissue containing venous plexuses (VP) similar to ↑ spongious body in male urethra. 2) Tunica muscularis (TM) consists of a circular and a longitudinal layer. Circular ↑ smooth muscle cells unite with those of ↑ urinary bladder to form a sphincter muscle around internal opening of U. In its intermediate portion, U. is surrounded by ↑ skeletal muscle fibers of sphincter urethrae; end portion of U. is enclosed by skeletal muscle fibers of vaginal sphincter muscle. 3) Tunica adventitia (TA) = a thin layer of ↑ loose connective tissue.

↑ Urethral chromaffin cells

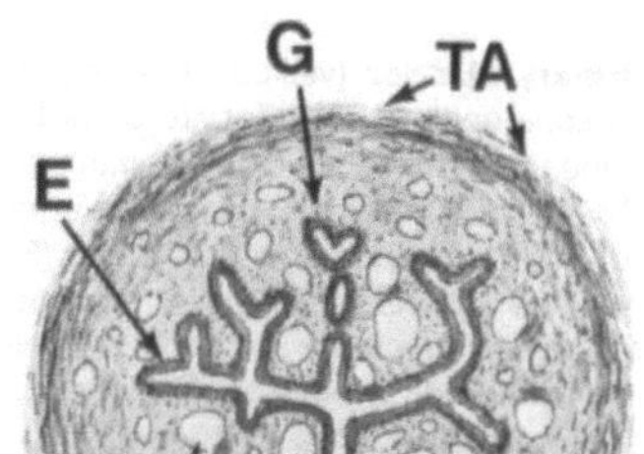

Cullen, W.C., Fletcher, T.F., Bradley, W.E.: Anat. Rec. *199*, 177 (1981)

Urethra, male (urethra masculina*): A tube, roughly 8 mm wide, conducting ↑ urine from ↑ urinary bladder to exterior; also serves for passage of ↑ semen during ↑ ejaculation. U. has three portions. A) Pars prostatica* (2 cm): 1) Tunica mucosa: a) epithelium = ↑ transitional epithelium with patches of ↑ stratified columnar epithelium; b) lamina propria = a ↑ loose connective tissue continuous with prostatic stroma and containing many ↑ smooth muscle cells. B) Pars membranacea* (18 mm): 1) Tunica mucosa: a) epithelium = a ↑ stratified columnar epithelium; b) lamina propria = a loose connective tissue. 2) Tunica muscularis = a thin inner longitudinal and a thick outer circular layer of ↑ smooth muscle cells. Since this portion corresponds to urogenital and pelvic diaphragmata, some ↑ skeletal muscle fibers can be found. C) Pars spongiosa* (15–18 cm, penile portion, spongy portion, fig.): 1) Tunica mucosa: a) epithelium (E) = stratified columnar epithelium changing into nonkeratinized ↑ stratified squamous in ↑ fossa navicularis. Epithelium forms deep branching tubules, ↑ urethral glands (UG), epithelial invaginations and ↑ urethral lacunae; b) lamina propria (arrowheads) = a thin layer of loose connective tissue continuous with ↑ spongious body (SB). 2) ↑ Tunica albuginea (TA) contains ↑ collagen and ↑ elastic fibers, as well as smooth muscle cells; it is considerably thinner than that of ↑ cavernous bodies (C) of ↑ penis. (Fig. = human newborn)

↑ Urethral chromaffin cells

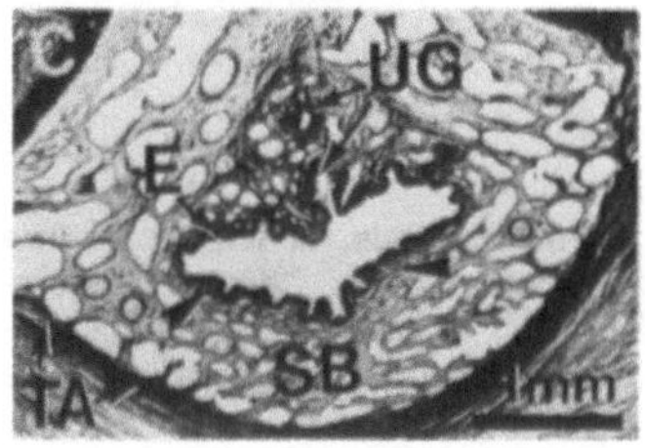

Alm, P., Colleen, S.: Acta Pathol. Microbiol. Scand. Sect. A Pathol. *90*, 103 (1982); Cullen, W.C., Fletcher, T.F., Bradley, W.E.: Anat. Rec. *199*, 187 (1981)

Urethral chromaffin cells: ↑ Chromaffin cells scattered in epithelia of male and female ↑ urethra. U. are closely associated with sensory axons; functional significance of U. is unknown.

Dixon, J.S., Gosling, J.A., Ramsdale, D.R.: Z. Zellforsch. *138*, 397 (1973)

Urethral crest (crista urethralis*): A roughly 1.5-cm-long longitudinal fold (UC) of posterior wall of prostatic (P) portion of ↑ urethra (U). Central part of U. is ↑ colliculus seminalis (CS), onto which ↑ utriculus prostaticus (UP) and ↑ ejaculatory ducts (ED) open. ↑ Prostatic ducts (PD) open lateral to U.

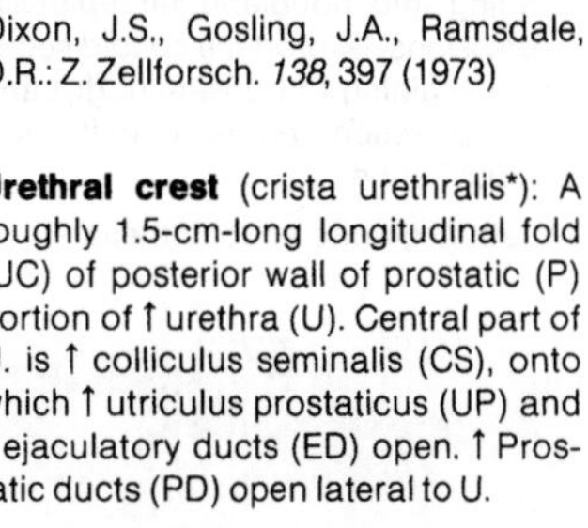

Urethral glands, endoepithelial (glandulae urethrales*): Epithelial outpocketings (UG) in female and male ↑ urethra lined with clear mucous cells

(MC). Secretion of U. protects epithelium against ↑ urine.

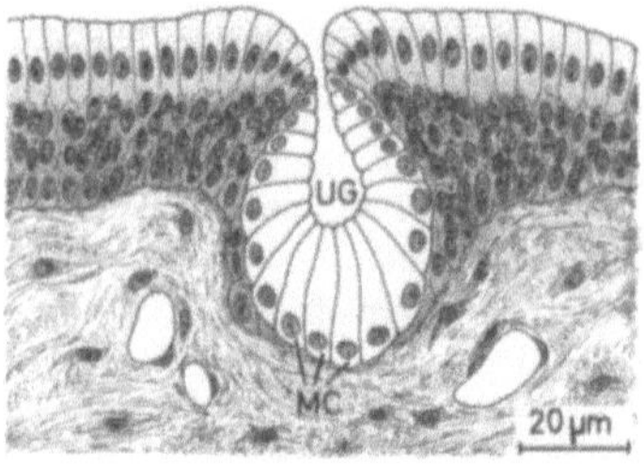

Urethral glands, of female urethra (Skene's glands, glandulae urethrales*): Small ↑ mucous glands situated in lamina propria of wall of female ↑ urethra; morphofunctionally, very similar to ↑ urethral glands of male urethra.

Urethral glands, of male urethra (Littre's glands, glandulae urethrales*): Epithelial invaginations (U) located predominantly at dorsal wall of ↑ urethra. Two kinds of U.: 1) Short U. = ↑ endoepithelial glands composed of clear mucous cells (inset); 2) long and branched U., penetrating between veins (V) of ↑ spongious body. Long U. are provided with ducts (D), sometimes opening into nonglandular epithelial depressions. U. produce colloid secretion containing ↑ glycosaminoglycans; their secretion protects epithelium against ↑ urine.

↑ Urethral glands, endoepithelial

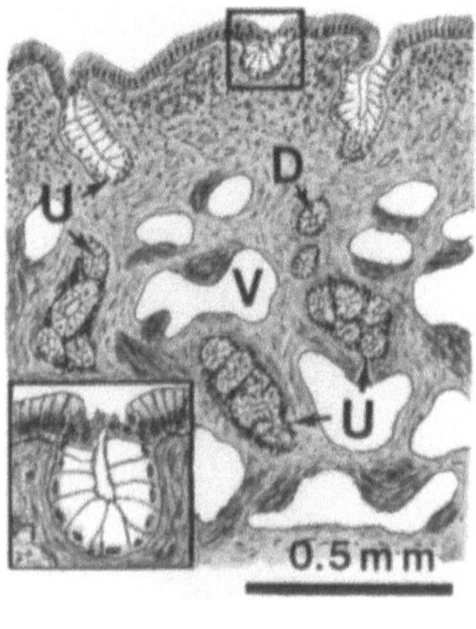

Urethral lacunae (lacunae urethrales*): Nonglandular epithelial depressions (UL) of female and male urethral wall into which ducts of ↑ urethral glands sometimes open. SB = ↑ spongious body. (Fig. = human)

↑ Urethra, female; ↑ Urethra, male

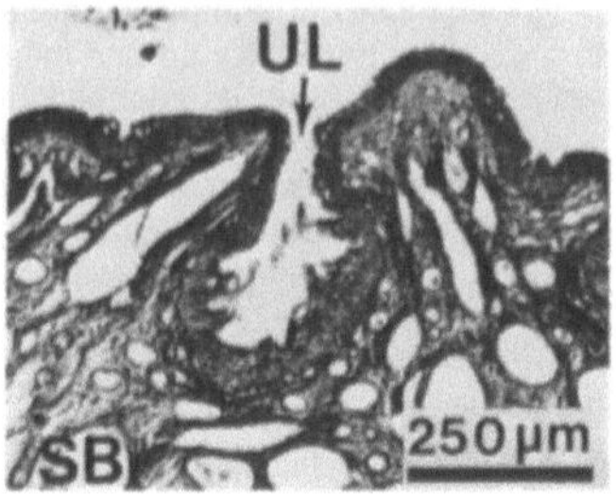

Urinary bladder (vesica urinaria*): A musculoepithelial sac for storage of ↑ urine. Structure: 1) Tunica mucosa (TM), folded in empty U., except in area of trigonum vesicae: a) epithelium (E) = ↑ transitional epithelium (eight to ten layers in empty U., two to three layers in filled U.), some mucoid glands are present in area of trigonum; b) lamina propria (LP) = a ↑ loose connective tissue. 2) Tela submucosa (TS) = an abundant, loose connective tissue rich in ↑ elastic fibers, blood and lymphatic vessels, and some occasional ↑ lymphatic nodules. 3) Tunica muscularis (TMu) = three intermingled and not clearly defined smooth muscular layers: Outer (O) and inner (I) layers are predominantly longitudinal, the middle (M) one circular. In trigonum, muscularis forms internal sphincter surrounding internal opening of urethra; presence of numerous ↑ elastic fibers, autonomic nerves, and small ganglia. 4) Tela subserosa (TSs) = a loose connective tissue with blood and lymphatic vessels and numerous nerve fibers. 5) Tunica serosa (TSe) = ↑ peritoneum, present only on superior and lateral surfaces of U. Tunica adventitia (TA) = a richly vascularized, loose connective tissue with a well-developed network of autonomic nerve fibers (plexus vesicalis) whose sensory terminations penetrate between epithelial cells. Recently, lamina muscularis mucosae of U. has been described.

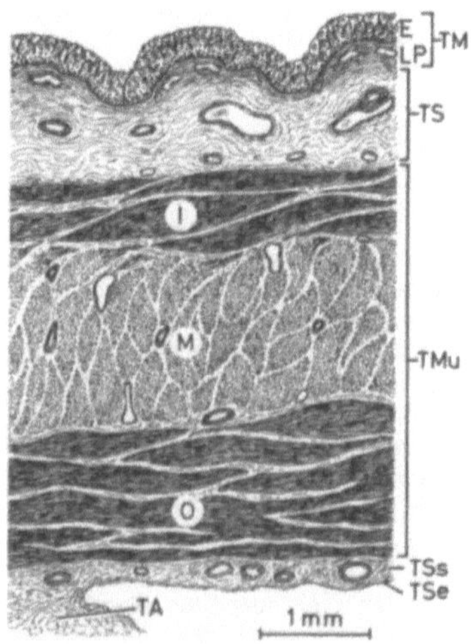

↑ Trigonal glands

Dixon, J.S., Gosling, J.A.: J. Anat. *134*, 617 (1982); Dixon, J.S., Gosling, J.A.: J. Anat. *136*, 265 (1983); Hicks, R.M.: Biol. Rev. *50*, 215 (1975); Hodges, G.M.: The Urinary System – Bladder. In: Hodges, G.M., and Hallowes, R.C. (eds.): Biomedical Research Applications of Scanning Electron Microscopy, Vol. 1. London, New York: Academic Press 1979

Urinary passages: All ↑ organs conducting ↑ urine from kidney ↑ parenchyma to exterior. U. include minor and major ↑ calyces, ↑ renal pelvis, ↑ ureters, ↑ urinary bladder; and ↑ urethra.

Urinary pole, of renal corpuscle (polus tubularis*): Site of transition of ↑ capsular epithelium of ↑ Bowman's capsule into epithelium of ↑ neck region of ↑ nephron.

Urinary system (organa urinaria*): The group of ↑ organs responsible for ↑ excretion of waste products of metabolism, maintenance of osmotic balance of the body, and production of ↑ urine and its conduction from the body. U. includes ↑ kidneys, major and minor ↑ calyces, ↑ renal pelvis, ↑ ureters, ↑ urinary bladder, and ↑ urethra.

Andrews, P.M.: The Urinary System – Kidney. In: Hodges, G.M. and Hallowes, R.C. (eds.): Biomedical Research Applications of Scanning Electron Microscopy, Vol. 1. London, New York: Academic Press 1979

Urine: Slightly yellowish, transparent, watery excretory product of ↑ kidney. 1) Primary U. = an ultrafiltrate of blood filling ↑ Bowman's space. 2) Secondary U. (definitive U.) = primary U. after reabsorption of water, electrolytes, glucose, albumins, etc., by cells of ↑ nephron and ↑ collecting tubules. (See physiology texts for further information)

Geigy Scientific Tables. Body Fluids. 8th edn. Basel: Ciba-Geigy 1983

Uriniferous tubules: The entire tubular excretory system of ↑ kidney. Each U. consists of two parts: 1) ↑ Nephron and 2) ↑ collecting tubules.

Urogastrone (hEGF): A gastric antisecretory hormone isolated from human ↑ urine, probably identical with ↑ epidermal growth factor. U. inhibits gastric hydrochloric acid secretion; it seems to be produced by ↑ duodenal glands.

Carpenter, G., Cohen, S.: Ann. Rev. Biochem. *48*, 193 (1979)

Uropodium: A small cytoplasmic tail (U), particularly visible when ↑ wandering cells move over a flat surface in cell culture.

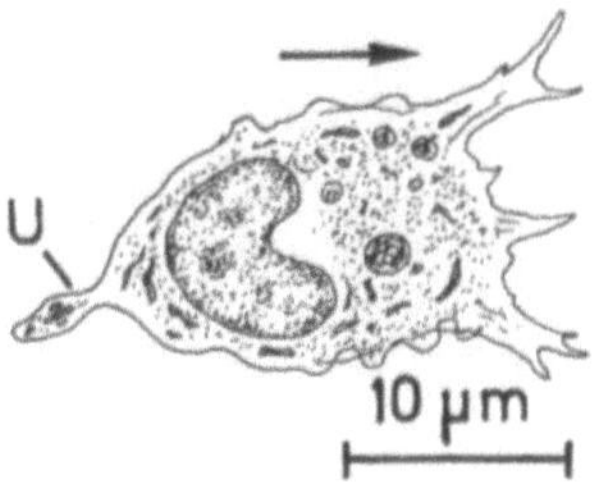

Urothelium: ↑ Transitional epithelium

Uterine glands (glandulae uterinae*): Simple ↑ tubular glands (UG) situated in ↑ endometrium with bases reaching ↑ myometrium. Here, some U. may branch. U. are surrounded by endometrial stroma (S) and lined with a ↑ simple columnar epithelium, with occasional ciliated cells. (Fig. = human)

↑ Uterus, epithelium of

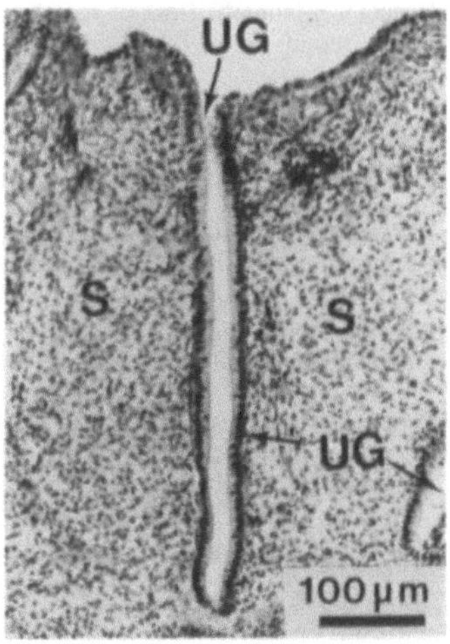

Perry, J.S., Crombie, P.R.: J. Anat. *134*, 399 (1982)

Uterine glands, before puberty: Short tubular invaginations (UG) of uterine epithelium into endometrial stroma (S). M = ↑ myometrium (Fig. = 6-year-old girl)

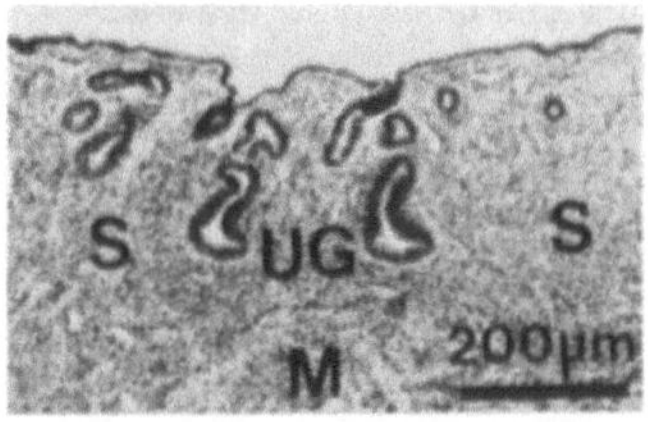

Uterine glands, cyclic changes of: 1) Proliferative phase of ↑ menstrual cycle = under influence of ↑ estrogens, ↑ uterine glands regenerate from glandular remnants in ↑ basalis (B). As a consequence of high mitotic activity, uterine glands rapidly elongate, but preserve their straight shape. At the end of this phase, uterine glands become slightly sinuous. 2) Secretory phase of menstrual cycle = under influence of ↑ progesterone, uterine glands become quite tortuous with numerous saccular enlargements, giving a spongy aspect to middle stratum of ↑ endometrium (spongiosa, S). Epithelium of uterine glands becomes scalloped and locally pseudostratified. Short straight segments traversing compacta (C) connect enlarged bodies of uterine glands with uterine lumen.

↑ Uterine glands, cyclic changes of epithelium of

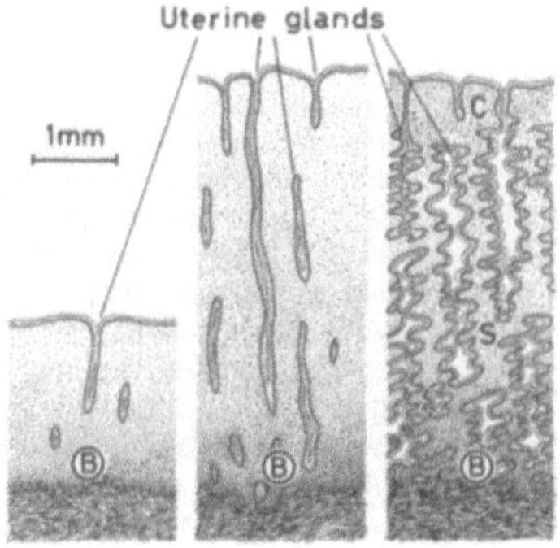

Uterine glands, cyclic changes of epithelium of: 1) Proliferative phase of ↑ menstrual cycle (5th–14th day): Secretory cells are narrow, with a nucleus adapted to cell shape and, in basal pole, poor development of cell ↑ organelles. 2) Secretory phase of menstrual cycle: a) First half (10th–21st day) = cells enlarge, each nucleus becomes more voluminous and nucleoli more prominent, mitochondria increase in size and volume, Golgi apparatus ex-

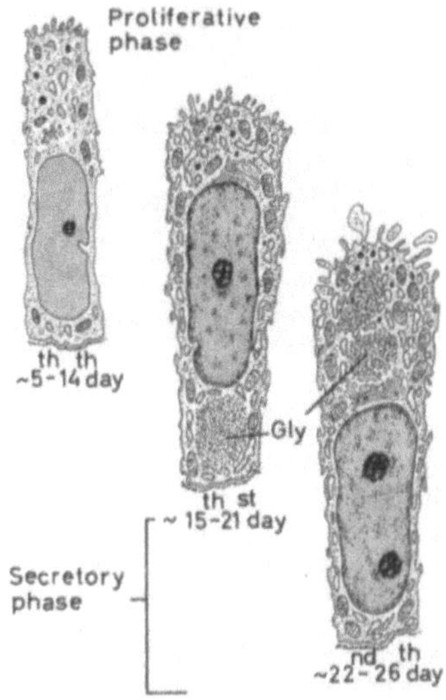

pands, in basal cell pole ↑ glycogen (Gly) accumulates and pushes nucleus toward middle of cell body. Cisternae of rough endoplasmic reticulum enlarge and increase in number; ↑ polyribosomes are abundant. b) Second half (22nd–26th day) = cells continue to enlarge; whereas cell organelles display approximately same degree of development as in first half of secretory phase, with glycogen accumulating in ↑ apical pole, so that nucleus becomes displaced again toward basal cell pole. Apical plasmalemma forms numerous microvilli and broad cell protrusions loaded with glycogen. Protrusions may separate from ↑ apical cell pole and enter glandular lumen. c) Premenstrual phase (27th–28th day) = mitochondria shrink, in cytoplasm myelinlike figures occur, Golgi apparatus becomes smaller, and apical plasmalemma is disrupted, liberating accumulated glycogen into lumen.

↑ Uterus, epithelium of

Uterine glands, in pregnancy: During pregnancy, ↑ uterine glands (UG) become irregularly shaped and flattened, giving a ladderlike appearance to the epithelium. U. form a spongelike layer between ↑ decidua basalis (DB) and ↑ myometrium (M). After expulsion of ↑ placenta, U. regenerate epithelium of ↑ uterus. CA = ↑ coiled arteries. (Fig. = human)

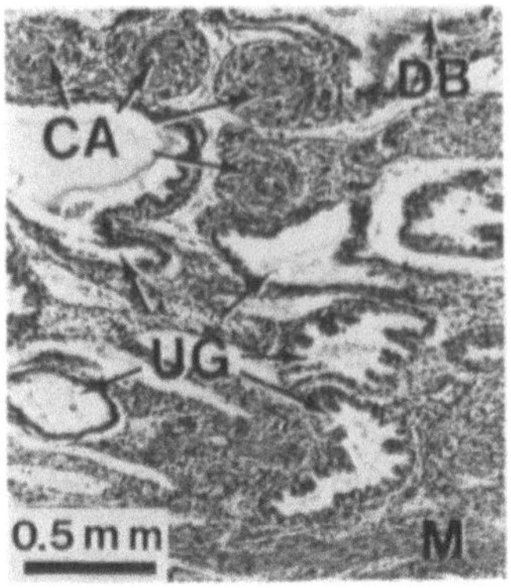

Uterine secretion: A mucoid fluid, product of secretory cells of epithelium of ↑ uterus. Pinching off of cell protrusions, loaded with ↑ glycogen, contributes to formation of U. In some animals, a low-molecular-weight, progesterone-dependent protein, uteroglobin, has been insolated from U.

Beyer, H.M.: Uterine Secretion Protein Patterns Under Hormonal Influences: In: Current Topics in Pathology, Vol. 62, Berlin, Heidelberg, New York: Springer-Verlag 1976; Beier, H.M.: Endometrial Secretion Proteins – Biochemistry and Biological Significance. In: Beller, F.K., Schumacher, G.F.B. (eds.): The Biology of the Fluids of the Female Genital Tract. Amsterdam: Elsevier/North Holland 1979; Savouret, J.-F., Milgrom, E.: DNA *2*, 99 (1983)

Uterine tubes: ↑ Oviducts

Uteroglobin: ↑ Uterine secretion

Uteroplacental arteries: ↑ Coiled arteries of ↑ endometrium altered during pregnancy. Tunica media of U. is partly invaded by ↑ cytotrophoblast.

↑ Endometrium, vascularization of

Uteroplacental veins: Veins of ↑ endometrium during pregnancy. ↑ Trophoblast does not invade wall of U.

Uterus*: An unpaired, hollow, pear-shaped, thick-walled muscular organ in which ↑ ovum develops into child. In nonpregnant women, U. is about 7.5 cm long, 3–4 cm across, and 2.5 cm thick, consisting of an upper expanded portion, ↑ corpus of uterus (C), which continues into ↑ cervix of uterus (Ce). Part of corpus between ↑ oviducts (O) is referred to as ↑ fundus (F), whereas a narrow segment between corpus and cervix is termed the isthmus (I). Portion of U. penetrating vagina (V) is ↑ portio vaginalis (PV).

↑ Endometrium; ↑ Myometrium; ↑ Perimetrium

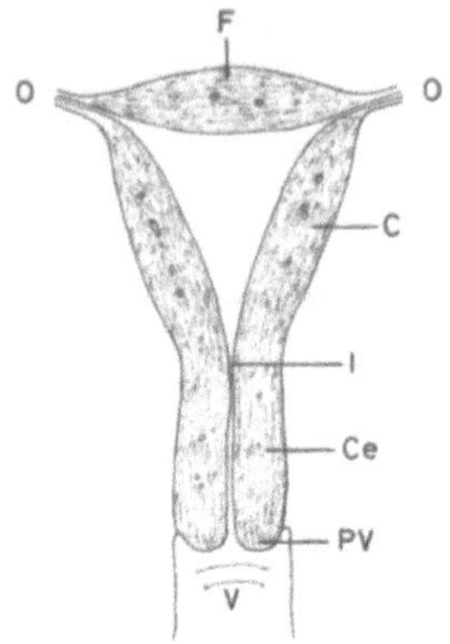

Uterus, cervix of: ↑ Cervix, of uterus

Uterus, corpus of: ↑ Corpus, of uterus

Uterus, endocervix of: ↑ Endocervix

Uterus, endometrium of: ↑ Endometrium

Uterus, epithelium of: A ↑ simple columnar epithelium lining uterine cavity and ↑ uterine glands, consisting of two kinds of cells: 1) Secretory cells (SC) or cellulae nonciliatae* = columnar cells each with an elongated nucleus containing one or two nucleoli. Cytoplasm encloses numerous mitochondria and a well-developed supranuclear Golgi apparatus from which arise some ↑ unit membrane-bound dense secretory granules (SG). Cisternae of rough ↑ endoplasmic reticulum are frequent, as are ↑ polyribosomes. In basal and ↑ apical poles, ↑ glycogen (Gly) accumulates during secretory phase of ↑ menstrual cycle. Apical plasmalemma forms irregular microvilli and coarse protrusions, which, loaded with glycogen, pinch off from cell. Despite conspicuous secretory granules, exact mechanism of secretion of this cell type is not fully understood. 2) Ciliated cells (CC) or cellulae ciliatae* = columnar cells with an oval nucleus and an inconspicuous nucleolus. Cytoplasm contains some mitochondria, a small Golgi apparatus, some short cisternae of rough endoplasmic reticulum and free ribosomes. Apical cell pole bears ↑ cilia (C) with rootlets penetrating deep into cytoplasm. The cilia beat toward ↑ vagina and in uterine glands toward their opening. There are numerous ciliated cells in U. lining uterine cavity, but they are rare uterine glands. BL = ↑ basal lamina

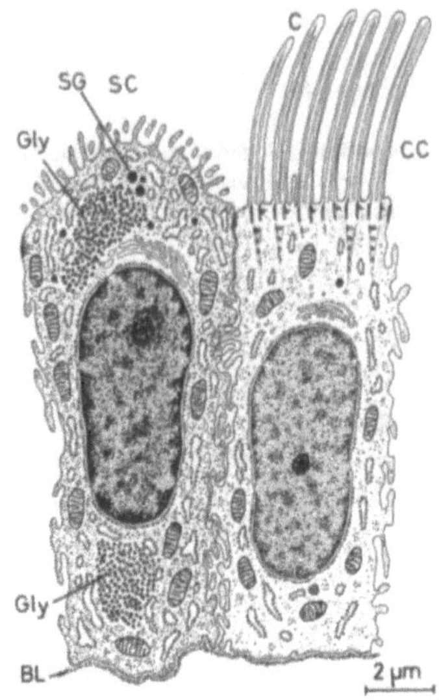

Uterus, isthmus of: ↑ Uterus

Uterus, myometrium of: ↑ Myometrium

Uterus, vascularization of: ↑ Endometrium, vascularization of; ↑ Myometrium

Utricle (utriculus*): Part of membranous ↑ labyrinth in which ↑ macula of utricle is situated.

Utriculus masculinus: ↑ Utriculus prostaticus

Utriculus prostaticus* (uterus masculinus, utriculus masculinus, vagina masculina): A short blind pouch (U) opening onto tip of ↑ colliculus seminalis (CS) and lined with ↑ stratified

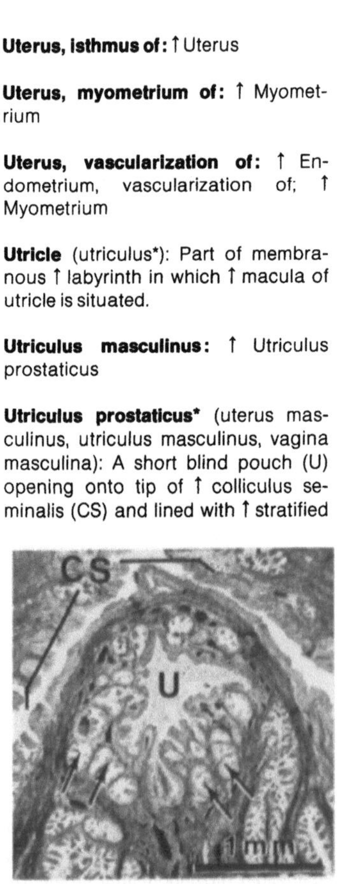

columnar epithelium. Within colliculus, U. branches into several glandlike formations (arrows) covered by ↑ pseudostratified epithelium. There is evidence concerning functional engagement of U. as an accessory male sexual gland; U. is a remnant of fused caudal ends of Müllerian ducts; it is homologous to ↑ vagina. (Fig. = human)

Uvea (tunica vasculosa bulbi*): Richly vascularized middle tunic of ↑ eyeball subdivided into: ↑ Choroid, ↑ ciliary body, and ↑ iris. U. is responsible for vascularization of almost entire eyeball, accomodation of eye, and regulation of quantity of light penetrating eyeball.

↑ Eye, vascularization of

Uvula*: A small unpaired tonguelike process of soft ↑ palate hanging down into mouth. U. is lined with a nonkeratinized ↑ stratified squamous epithelium overlying a ↑ loose connective tissue of lamina propria, rich in ↑ elastic fibers. A few groups of ↑ mucous tubules and ↑ skeletal muscle fibers of musculus uvulae are situated within interior of U.

V

Vacuole (vacuola*): An intracellular ↑ unit membrane-limited spherical space filled either with a fluid and/or a more or less solid content.

Vacuoles, condensing: ↑ Condensing vacuoles

"Vacuoles, nuclear": ↑ "Nuclear vacuoles"

Vagina*: A tubular organ situated between ↑ cervix of ↑ uterus and vestibulum vaginae; lower limit of V. is marked by ↑ hymen or carunculae hymenales. Structure: 1) Tunica mucosa (TM): a) epithelium (E) = a nonkeratinized ↑ stratified squamous epithelium about 0.2 mm thick, with numerous ↑ epithelial ridges; b) lamina propria (LP) = a ↑ dense connective tissue with an abundance of ↑ elastic fibers and an extensive plexus (P) of small veins. Variable numbers of ↑ wandering cells, ↑ lymphocytes, and ↑ macrophages are present in lamina propria during ↑ menstrual cycle. 2) Tunica muscularis (TMu) = an inner circular and an outer longitudinal layer of ↑ smooth muscle cells; some ↑ skeletal muscle fibers of bulbocavernous muscle surround entrance of V. 3) Tunica adventitia (TA) = a ↑ loose connective tissue; fornix vaginae posterior is lined with ↑ peritoneum of excavatio rectouterina. Wall of V. contains numerous lymphatic vessels and nerve fibers. Lubrification of V. is assured by glands of cervix. Reaction of vaginal fluid is acid; it is a consequence of fermentation of ↑ glycogen by bacterial flora (lactobacilli vaginales).

↑ Vaginal cycle

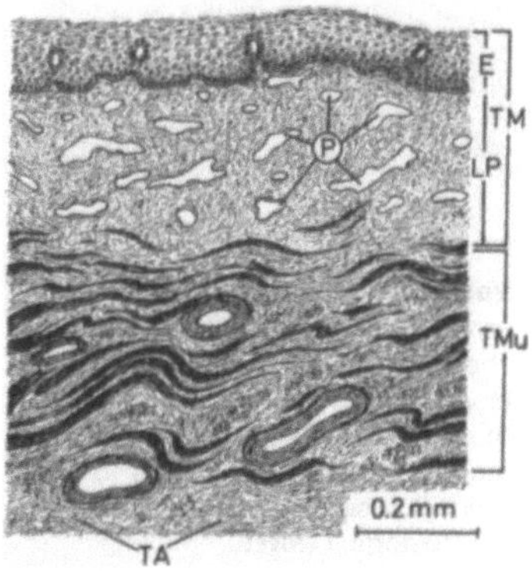

King, B.F.: J. Ultrastruct. Res. *83*, 99 (1983); Parakkal, P.F.: Anat. Rec. *178*, 529 (1974)

Vagina masculina: ↑ Utriculus prostaticus

Vagina, synovial, of tendon: ↑ Synovial vagina, of tendon

Vaginal cycle: Sequence of changes occurring in vaginal epithelium throughout 28-day period of a woman. 1) A few days before ↑ ovulation, under influence of ↑ estrogens, epithelium is thickest (about 250 μm); some ↑ keratohyalin granules occur in the superficial cells with condensed nuclei. In smears, cells are polygonal and both acidophilic and basophilic. 2) At time of ovulation, cells of superficial and spinocellular layers accumulate considerable amounts of ↑ glycogen and begin to desquamate. The liberated glycogen is fermented under action of lactobacilli vaginales to lactic acid (pH 4), protecting vagina from other bacteria. In smears, all cells are acidophilic, with pyknotic nuclei, and very swollen. 3) Under influence of ↑ progesterone, during secretory phase of ↑ menstrual cycle, desquamation continues, epithelium becomes thinner (about 150 μm), amount of glycogen and acidity of vagina decrease. In smears, cells are basophilic; they occur in clumps and have folded borders.

↑ Papanicolaou's staining

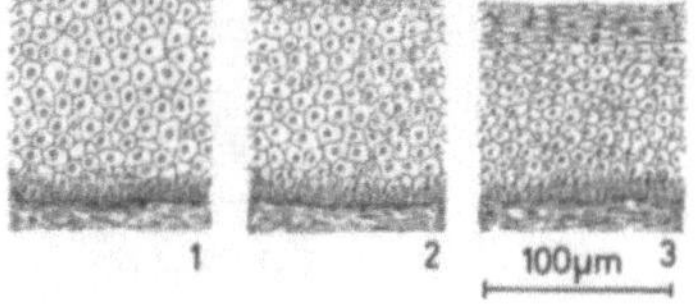

Vaginal space, of optic nerve: ↑ Optic nerve; ↑ Optic papilla

Vallatae papillae: ↑ Circumvallate papillae

Valve, mitral: ↑ Valves, atrioventricular

Valve, tricuspid: ↑ Valves, atrioventricular

Valves, aortic (semilunar valves, valva aortica*): Three pocketlike sheets of ↑ dense connective tissue arranged around root of ↑ aorta. Structurally, V. are very similar to atrioventricular ↑ valves. V. have no ↑ chordae tendineae, but a rounded connective tissue accumulation at the middle of their free edges, nodulus of ↑ Arantius. There is also no extension of ↑ cardiac muscle fibers into V. In some cases, ↑ smooth muscle cells can be found in V.

Bairati, A., DeBiasi, S.: Anat. Embryol. *161*, 329 (1981)

Valves, atrioventricular (valvae atrioventriculares*): Membranous supple flaps or cusps attached to ↑ annuli fibrosi (AF) and extending from border of atrioventricular opening into ventricles. Structure: 1) Both atrial and ventricular surfaces are covered by ↑ endocardium (Ec) common to the heart cavities. 2) Subendocardial layer is thickened on atrial side (A) and contains a dense network of ↑ elastic fibers and scattered ↑ smooth muscle cells. Ventricular side (V) of V. is thinner and more loosely arranged. 3) Central flat sheet of very ↑ dense connective tissue (DT) is continuous with annulus fibrosus. Near annulus fibrosus, some ↑ cardiac muscle fibers may penetrate V. From free edges of V., numerous ↑ chordae tendineae (CT) extend to papillary muscles. There are no blood vessels in V. in man; their vascularization is achieved by circulating blood. Mitral and tricuspid valves are V.

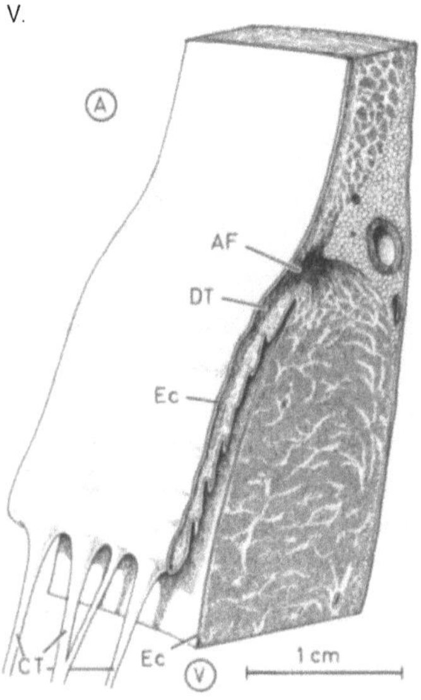

Lim, K.O.: Jpn. J. Physiol. *30*, 455 (1980)

Valves, of Kerkring: ↑ Plicae circulares

Valves, of lymphatic vessels (valvula lymphatica*): Paired opposed folds (V) of ↑ endothelium of ↑ lymphatic vessels permitting ↑ lymph to circulate only toward heart. (Fig. = human)

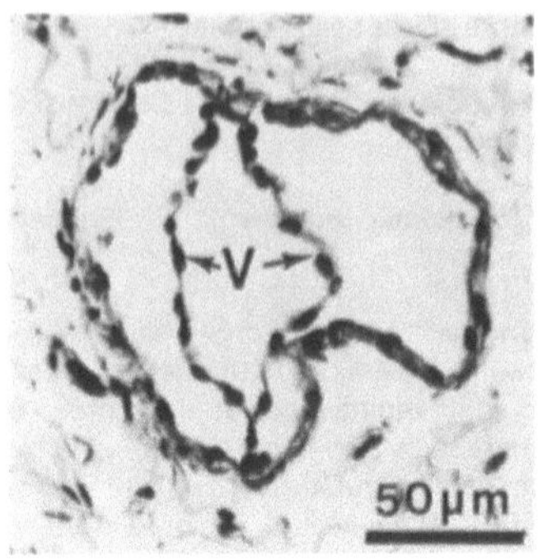

Albertine, K.H., Fox, L.M., O'Morchoe, C.C.C.: Anat. Rec. *202*, 453 (1982)

Valves, of vein (valvula venosa*): Slender folds (V) of various lengths of ↑ tunica intima protruding into lumen of a ↑ vein. Free edges of V. point in direction of blood flow. V. are paired structures situated opposite one another; thus, they permit blood to flow only toward heart. The space between wall of vein and V. is sinus (S). V. primarily occur in ↑ postcapillary venules; they are present in veins conducting blood against gravity, thus V. are particularly numerous in veins of legs. Cerebral veins, venous sinuses of skull, jugular veins, ↑ vena cava superior, and veins of viscera and ↑ bone marrow have no V. (Fig. = human)

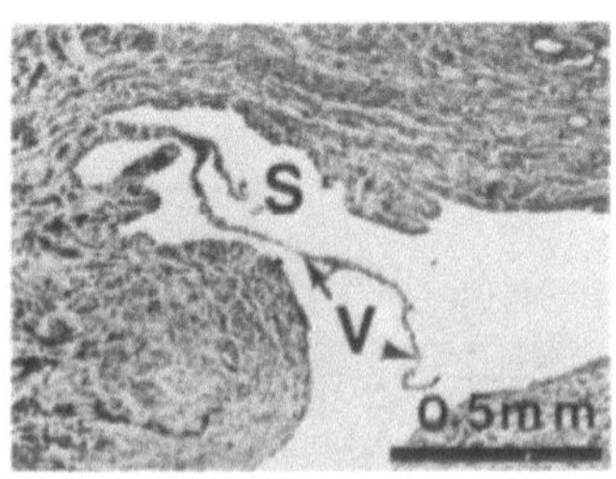

Valves, pulmonic (valvulae semilunares*): Flaplike, avascular sheets of ↑ dense connective tissue covered with ↑ endothelium. V. are situated at beginning of truncus pulmonalis and are connected to ↑ annulus fibrosus. V. are structurally identical to aortic ↑ valves.

Valves, semilunar: ↑ Aortic valves

Valves, sinus of: ↑ Sinus, of valve

Valves, spiral: ↑ Spiral valve

Vas deferens: ↑ Ductus deferens

Vasa nervorum: ↑ Small vessels (VN) supplying blood to peripheral ↑ nerves. V. run in ↑ epineurium, penetrate ↑ perineurium (P), and branch in ↑ endoneurium (E). (Fig. = human)

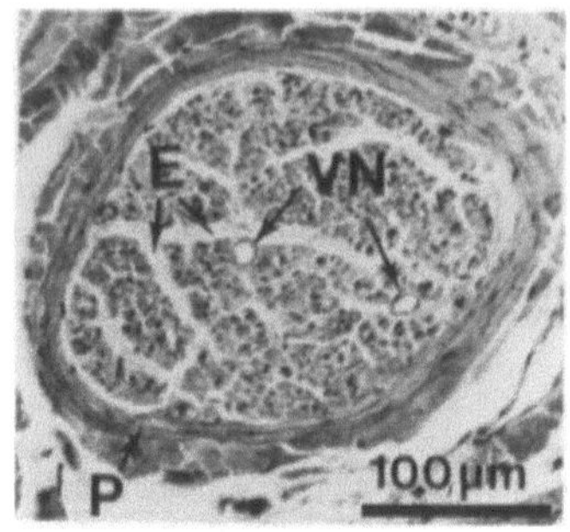

Vasa recta* (fasciculus vascularis*): A common term designating thin-walled blood vessels (VR) of small caliber descending at various levels from ↑ renal cortex (C) into ↑ renal medulla (M) and turning back toward cortex. Whereas arterial V. have a continuous ↑ endothelium, venous V. have fenestrated endothelium. V. serve as countercurrent exchangers for diffusable substances. A = ↑ arteria arcuata, G = ↑ renal glomerulus, PCN = ↑ peritubular capillary network. (Fig. = human; vascular ↑ injection with India ink)

↑ Kidney, vascularization of

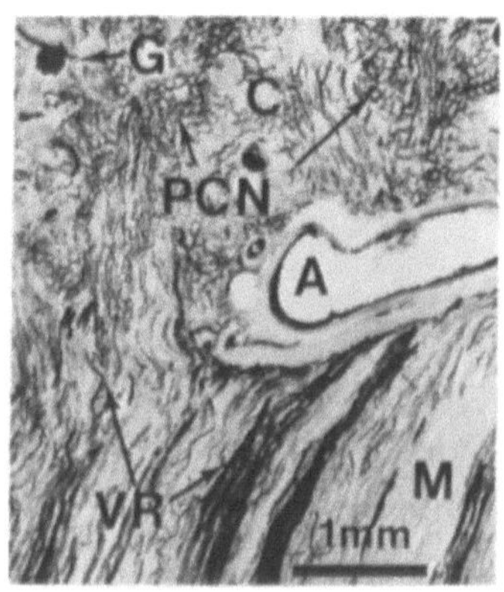

Vasa vasorum*: Small blood vessels (VV) vascularizing tunica adventitia and external part of tunica media of larger ↑ arteries, ↑ veins, and ↑ lymphatic ducts (vasa lymphaticorum*). (Fig. = ↑ vena cava inferior, human)

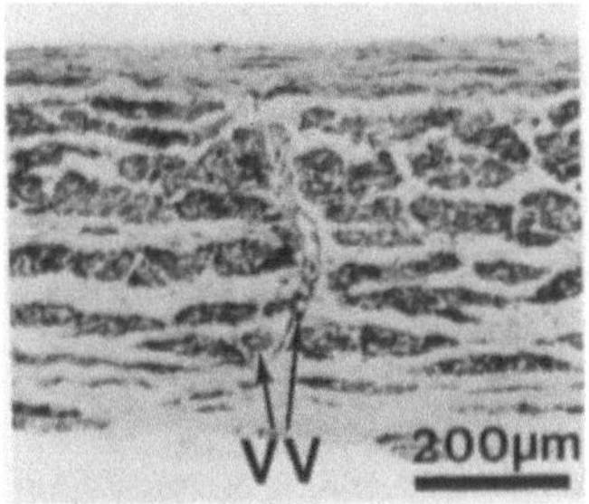

Vascular component, of bone marrow: Total vascular sinuses and small arteries and veins of ↑ bone marrow.

↑ Bone marrow, vascular sinuses of

Vascular injection: ↑ Injection, vascular

Vascular smooth muscle: ↑ Smooth muscle, vascular

Vascular papillae: ↑ Connective and ↑ dermal papillae (VP) containing numerous capillary loops (CL) for vascularization of the thick ↑ stratified squamous epithelium (E). LP = ↑ lamina propria. (Fig. = ↑ esophagus, human)

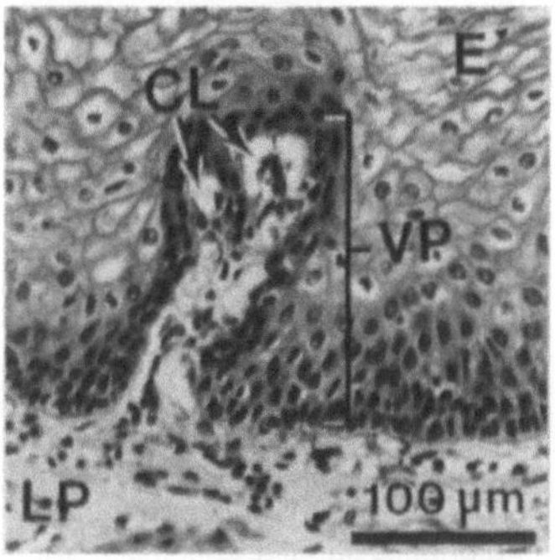

Vascular pole, of cell: Part of cell body facing blood or lymphatic ↑ capillary. ↑ Secretory granules are concentrated in V. of ↑ endocrine cells.

Vascular pole, of renal corpuscle (polus vascularis*): An area (VP) where ↑ afferent and ↑ efferent arterioles enter and leave ↑ renal corpuscle (RC). At V., the two walls of ↑ Bowman's capsule are continous. (Fig. = monkey)

Frank, M., Kriz, W.: Anat. Rec. *204*, 149 (1982)

Vascular sinuses: ↑ Capillaries, sinusoidal

Vascularization: Supplying of a ↑ tissue or an ↑ organ with ↑ blood vessels.

Vas spirale*: A small vein running in scala tympani just under ↑ Corti's organ. Some authors consider V. to be a small artery.

Vasoactive intestinal polypeptide (VIP): A peptide containing 28 amino acids found in peripheral ↑ nerves which supply some ↑ exocrine glands, ↑ genitourinary system, surface epithelia, ↑ blood vessels, and nonvascular smooth muscles. V. is also secreted by ↑ D_1-cells of ↑ gastrointestinal tract. V. provokes vasodilatation of coronary, pulmonary, and abdominal vessels and inhibition of HCl and pepsin production in ↑ gastric glands proper.

↑ Endocrine cells, of gastrointestinal tract

Alm, P., Alumets, J., Hakanson, R., Owman, Ch., Sjöberg, N.-O., Sundler, F., Baecker, B., Yanaihara, N., Forssmann, W.G.: Anat. Embryol. *167*, 173 (1983); Fuji, S., Kobayashi, S., Fujita, T., Yanaihara, N.: Biomed. Res. *1*, 180 (1980); Walles, B.: Cell Tissue Res. *205*, 337 (1980)

Vasomotor nerve fibers: ↑ Postganglionic ↑ nonmyelinated nerve fibers (V) innervating ↑ smooth muscle cells (S) of blood vessels. (Fig. = ↑ arteriole, rat)

↑ Synapse, "by distance"

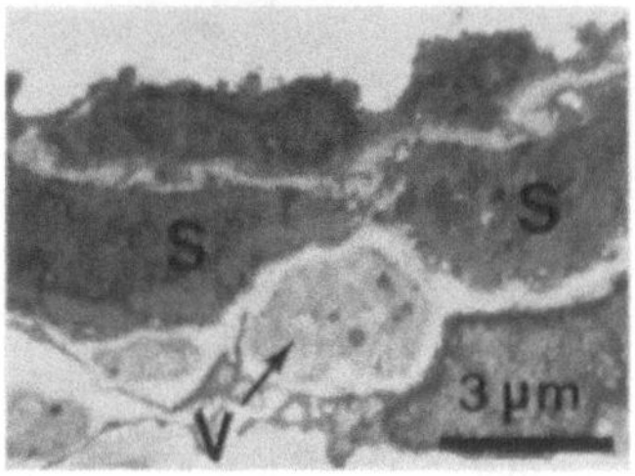

Vasopressin: ↑ Antidiuretic hormone

Vater-Pacini corpuscles: ↑ Corpuscles, of Vater-Pacini

Vater's ampulla: ↑ Ampulla hepatopancreatica

Vegetative nerve endings: ↑ Adrenergic nerve endings; ↑ Cholinergic nerve endings

Vegetative nervous system: ↑ Autonomic nervous system

Vein (vena*): A blood vessel conveying blood to ↑ heart.

Vein, central longitudinal: ↑ Bone marrow, vascularization of; ↑ Central longitudinal vein, of bone marrow

Vein, portal: ↑ Portal vein

Vein, saphenous: ↑ Saphenous vein

Vein, splenic: ↑ Splenic vein

Vein, suprarenal: ↑ Suprarenal vein

Veins, bronchial: ↑ Bronchial veins

Veins, central, of liver: ↑ Central veins, of liver

Veins, collecting: ↑ Collecting veins, of liver

Veins, intercalated: ↑ Intercalated veins

Veins, interlobular: ↑ Interlobular veins

Veins, large: Blood vessels with diameter greater than 10 mm formed by confluence of several medium-sized ↑ veins. Structure of majority of V. corresponds to structure of ↑ vena cava inferior. Smooth muscle tissue of ↑ tunica adventitia is poorly represented in V. of supracardiac portion of body. ↑ Sinuses of ↑ dura mater with a wall consisting of ↑ dense connective tissue completely lacking in smooth muscle cells are also considered V.

Veins, medium-sized: Blood vessel 1–10 mm in diameter situated between small ↑ veins and large ↑ veins. Structure: 1) Tunica intima = ↑ endothelium overlying a basal lamina and thin layer of subendothelial bundles of ↑ collagen and ↑ reticular fibrils; ↑ elastic fibers may also be present. 2) Tunica media = several layers of ↑ smooth muscle cells intermingled with bundles of collagen fibrils and elastic fibers. In some V., an internal ↑ elastic lamina may separate tunica intima from tunica media. V. of legs have more layers of smooth muscle cells. 3) Tunica adventitia = a ↑ loose connective tissue containing predominantly longitudinally oriented collagen bundles, ↑

vasa vasorum, lymphatic vessels, and nerve fibers.

↑ Saphenous vein

Veins, pulp: ↑ Pulp veins

Veins, renal: ↑ Renal veins

Veins, small: Blood vessels (V) 0.2–1 mm in diameter situated between ↑ muscular venules and medium-sized ↑ veins. Structure: 1) Tunica intima = a nonfenestrated ↑ endothelium (E) overlying a ↑ basal lamina. 2) Tunica media (M) = two or three complete layers of ↑ smooth muscle cells. 3) Tunica adventitia (A) = some ↑ fibroblasts and ↑ fibrocytes, ↑ collagen bundles,- and occasional ↑ elastic fibers. Presence of ↑ valves.

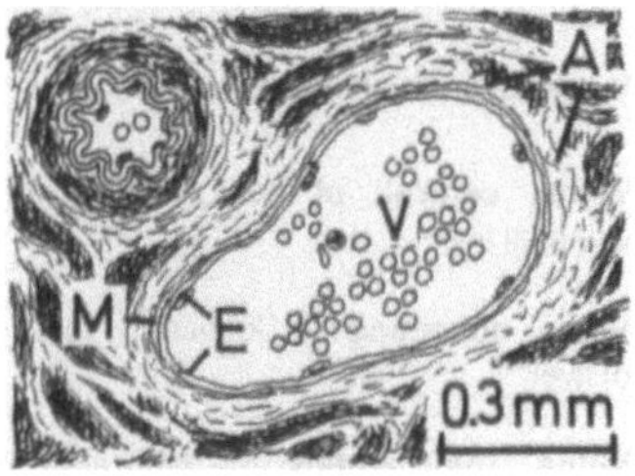

Veins, trabecular: ↑ Trabecular veins

Veins, uteroplacental: ↑ Uteroplacental veins

Vellus hairs: Small, fine, usually colorless ↑ hairs almost invisible to naked eye covering much of the body in females and, to a lesser degree, in males. Hairs of forehead and back are typical V.

Vena cava inferior: Main venous blood vessel of infracardial region of body. Structure: 1) Tunica intima (TI): a) ↑ endothelium = ↑ simple squamous epithelium with cells held together by numerous ↑ zonulae occludentes and ↑ nexus. ↑ Basal lamina and internal ↑ elastic lamina are largely fenestrated, allowing endothelial cells to establish some ↑ myoendothelial junctions. 2) Tunica media (TM) is very thin and consists only of some interspersed ↑ smooth muscle cells arranged in a circular fashion and embedded in a mass of ↑ collagen and ↑ elastic fibers; presence of ↑ fibrocytes. External ↑ elastic lamina is absent. 3) Tunica adventitia (TA) = thickest of all layers; composed of ↑ loose connective tissue with an abundance of longitudinally oriented smooth muscle cells grouped in bundles (B) and intermingled with numerous elastic and collagen fibers. ↑ Vasa vasorum, lymphatic vessels, and nervous plexus are well developed. Contraction of muscle cells actively propels blood to heart, thus V. is a propulsive vein. Thickness of adventitia resists hemostatic blood pressure. Azygos, external iliac, ↑ portal, ↑ renal, ↑ splenic, and superior mesenteric veins have the same structure. (Fig. = human)

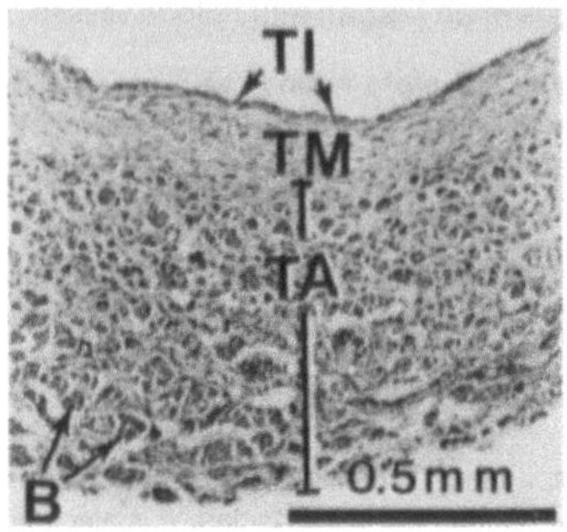

Vena cava superior*: A blood vessel collecting blood from supracardiac region of body. Structure: 1) Tunica intima (TI) = a thin ↑ endothelium overlying an inconspicuous layer of ↑ collagen fibrils intermingled with some ↑ elastic fibers. 2) Tunica media (TM) = a layer with occasional circular ↑ smooth muscle cells. 3) Tunica adventitia (TA) = a ↑ loose connective tissue with some longitudinal bundles (B) of smooth muscle cells. Wall of V. is considerably thinner than that of ↑ vena cava inferior. (Fig. = human)

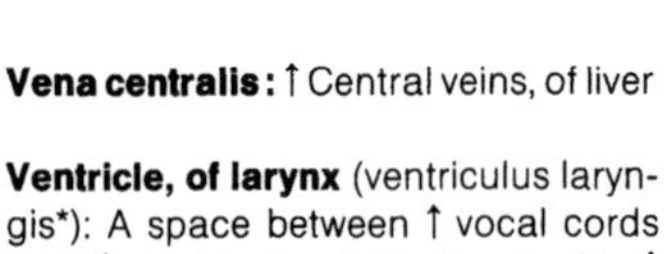

Vena centralis: ↑ Central veins, of liver

Ventricle, of larynx (ventriculus laryngis*): A space between ↑ vocal cords and ↑ ventricular folds lined with ↑ pseudostratified ciliated epithelium. Excretory ducts of ↑ laryngeal glands open into V.

Ventricles, of brain: ↑ Brain ventricles

Ventricular cells: Tall cells extending from lumen to basal lamina of early neural tube during its histogenesis. V. participate in formation of ↑ neuroepithelium. (See embryology texts for further information)

↑ Nervous tissue, histogenesis of

Ventricular folds (false vocal cords, plicae vestibulares*): Superior lateral pleats of laryngeal mucosa covered by ↑ pseudostratified ciliated epithelium. Lamina propria contains some ↑ mixed ↑ laryngeal glands and ↑ lymphatic nodules.

↑ Membrana fibroelastica laryngis

Venule (venula*): A thin-walled blood vessel 15–50 μm in diameter and 50–700 μm in length formed by confluence of several ↑ capillaries. Structurally, V. are subdivided into ↑ postcapillary V. and ↑ muscular V.

Venules, muscular: ↑ Muscular venules

Venules, pericytic: ↑ Postcapillary venules

Venules, postcapillary: ↑ Postcapillary venules

Venules, postcapillary, of lymph node: ↑ Postcapillary venules, of lymph node

Vermiform appendix: ↑ Appendix

Vermiform granules: ↑ Langerhans' bodies

Vermiform ridges: ↑ Microplicae

Vertebral ganglia (paravertebral ganglia, sympathetic chain ganglia, ganglia trunci sympathici*): 21–22 pairs of ↑ autonomic ganglia forming two chains on either side of vertebral column; structurally corresponding to ↑ autonomic ganglia.

↑ Sympathetic nervous system

Verumontanum: ↑ Utriculus prostaticus

Very low-density lipoproteins (VLDL): ↑ Lipoproteins composed of an apoprotein and a mixture of cholesterol and phospholipids. V. occur within ↑ tubules of smooth endoplasmic reticulum in ↑ liver cells as amorphous membraneless granules 30–80 nm in diameter. From here, granules (VLDL) reach ↑ Golgi apparatus (G) and accumulate in its saccules after a carbohydrate component has been added. V. are transported in vesicles of Golgi apparatus to plasmalemma and discharged by ↑ exocytosis into spaces of ↑ Disse. (Fig. = rat)

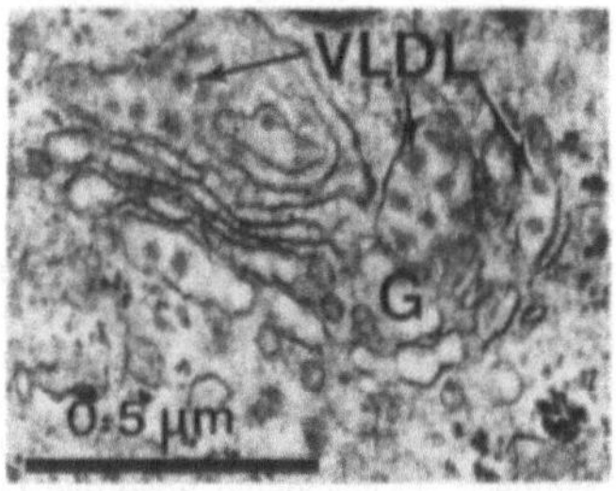

Vesica urinaria: ↑ Urinary bladder

Vesicle, acrosomal: ↑ Acrosomal vesicle

Vesicles, coated: ↑ Coated vesicles

Vesicles, condensing: ↑ Condensing vesicles

Vesicles, intermediate: ↑ Transport vesicles

Vesicles matrix: ↑ Matrix vesicles

Vesicles, micropinocytotic: ↑ Micropinocytotic vesicles

Vesicles, olfactory: ↑ Olfactory vesicles

Vesicles, plasmalemmal: ↑ Caveolae

Vesicles, seminal: ↑ Seminal vesicles

Vesicles, transfer: ↑ Transport vesicles

Vesicles, transitional: ↑ Transport vesicles

Vesicles, transport: ↑ Transport vesicles

Vesicular supporting tissue: ↑ Chondroid tissue

Vesiculation: A process of intensive formation of ↑ micropinocytotic vesicles observed in some ↑ endothelial cells and ↑ smooth muscle cells.

↑ Caveolae

Vessels, blood: ↑ Blood vessels

Vessels, lymphatic: ↑ Lymphatic vessels, collecting

Vestibular apparatus (labyrinthus vestibularis*): A group of sensory organs belonging to membranous ↑ labyrinth reacting to changes in position of head in space. V. includes three ↑ cristae ampullares and ↑ macula of saccule and utricle.

Vestibular cells, type I (hair cell, cellula cum synapse caliciformi*): Piriform sensory ↑ mechanoreceptor cells scattered in ↑ sensory epithelium of ↑ cristae ampullares (CA) and ↑ maculae of saccule and utricle (M). V. have a round central nucleus with finely dispersed chromatin, small mitochondria, a moderately developed Golgi apparatus (G), some short rough endoplasmic ↑ cisternae, numerous smooth vesicles and tubules, occasional lysosomes (Ly), and some free ribosomes. ↑ Apical pole carries a tuft of 40–100 ↑ hairs (H) and a single nonmotile ↑ cilium (C) with an expanded tip. Central fibrillar cores (CF) of the hairs penetrate ↑ cuticular plate (CP), which is interrupted in the area of ↑ basal body (BB) of cilium. V. lie in a chalicelike enlargment of an afferent nerve ending (AF); ↑ synaptic bars (SB) are adjacent to basal plasmalemma facing nerve ending. Electrical ↑ synapses (ES) may be found between V. and nerve chalice. In crista ampullaris, hairs and cilium are considerably longer (about 50 µm) than in maculae.

↑ Hair cells; ↑ Synapse, calyceal

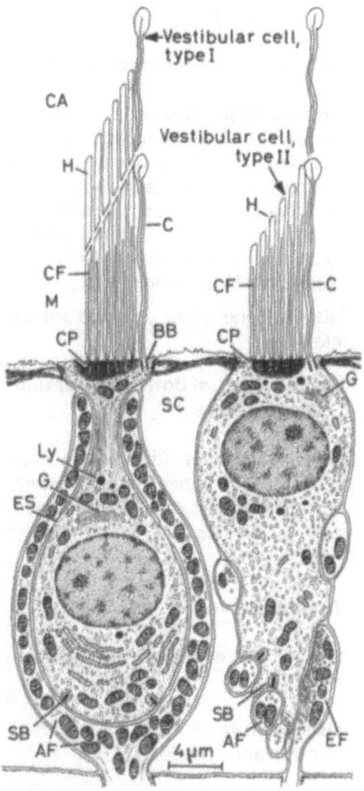

Hamilton, D.W.: J. Ultrastruct. Res. *23*, 98 (1968); Kessel, R.G., Kardon, R.H.: Scan. Electron Microsc. *3*, 967 (1979); Lindeman, H.H.: Ergeb. Anat. Entwickl. Gesch. *42*, 1 (1969); Wersäll, J., Flock, A.: Acta Otolaryngol., Suppl. *192*, 85 (1964); Zahm, D.S.: Am J. Anat. *158*, 263 (1980)

Vestibular cells, type II (hair cell, cellula cum synapse disseminata*): Columnar sensory ↑ mechanoreceptor cells scattered in ↑ sensory epithelium of ↑ cristae ampullares (CA) and ↑ maculae of saccule and utricle (M). V. have a round and predominantly apical nucleus with a nucleolus. Cytoplasm contains some short mitochondria, a small supranuclear Golgi apparatus (G), a few short rough endoplasmic ↑ cisternae, a large number of smooth endoplasmic profiles, occasional lysosomes, and a few free ribosomes. ↑ Apical pole bears 40–100 ↑ hairs (H), which increase in height toward single nonmotile ↑ cilium (C) with an expanded tip. Central filamentous core (CF) of hairs penetrates and at times runs through ↑ cuticular plate (CP). ↑ Synaptic bars (SB) are adjacent to basal plasmalemma facing individual endings of afferent nerve fibers (AF). V. are also in contact with efferent endings (EF), characterized by a large number of ↑ synaptic vesicles. In crista ampullaris only the cilium is very long (about 50 µm), whereas hairs have approximately same length as in maculae. SC = ↑ supporting cells (See fig. under ↑ Vestibular cells, type I)

Bagger-Sjoback, D., Gulley, R.L.: Acta Oto-Laryngol. *88*, 401 (1979); Flock, Å.: J. Cell Biol. *22*, 413 (1964); Gualtierotti, T. (ed.): The Vestibular System: Function and Morphology. Berlin, Heidelberg, New York: Springer-Verlag 1981; Hudspeth, A.J.: TINS *6/9*, 366 (1983); Kelly, J.P.: Vestibular System. In: Kandel, E.R. and Schwartz, J.H.: (eds.): Principle of Neural Science. New York, Amsterdam, Oxford: Elsevier/North Holland 1981

Vestibular ganglion (ganglion of Scarpa, ganglion vestibulare*): A group of myelinated ↑ bipolar neurons located in internal ↑ auditory meatus. Their ↑ dendrites reach ↑ cristae ampullares and ↑ maculae of saccule and utricule, whereas ↑ axons form vestibular division of vestibuloacoustic nerve.

Vestibular glands (Bartholin's glands, vulvovaginal glands, glandulae vestibulares majores et minores*): Compound paired ↑ tubuloalveolar glands (about 5–6 mm in diameter) located in lateral wall of vaginal vestibule. Structure: Several fibromuscular septa (S) separate ↑ parenchyma into small lobules. Intralobular ducts (arrows) unite to form a single excretory canal (EC) opening into angle between ↑ labia minora and ↑ hymen (or carunculae hymenales). Excretory canal is lined by ↑ simple columnar epithelium, which transforms at surface into ↑ stratified columnar epithelium (SC). ↑ Tubuloalveoli (T) are lined by large clear columnar cells producing ↑ mucoid lubricating fluid. After age of 30 years, V. begin to undergo gradual ↑ involution. V. are homologous to ↑ bulbourethral glands in male.

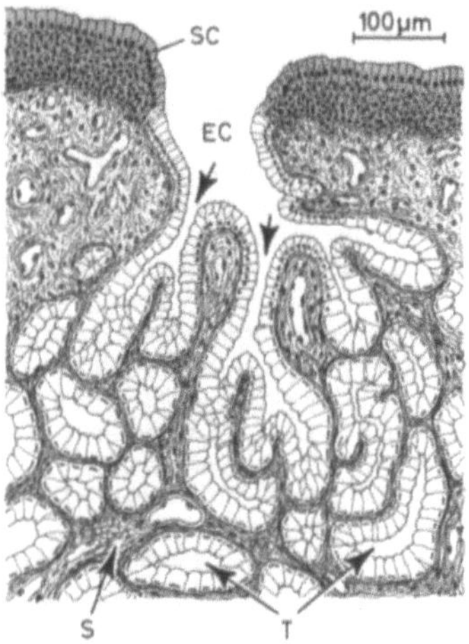

Vestibular lip: ↑ Auditory teeth; ↑ Limbus spiralis

Vestibular membrane: ↑ Reissner's membrane

Vestibule, of nose (vestibulum nasi*): The anterior part of each ↑ nasal cavity delimited laterally by ↑ wing of nostril and medially by ↑ nasal septum. V. is divided into two parts, both lined by ↑ epidermis and continuous with that of face: 1) External part, with ↑ vibrissae associated with apocrine ↑ sweat glands (↑ glandulae vestibulares nasi); 2) internal part, without hairs and without any cutaneous glands, continuous with ↑ respiratory mucosa.

Vibrissae*: ↑ Hairs in ↑ vestibulum nasi. ↑ Glandulae vestibulares nasi open into ↑ hair follicles of V. V. have no ↑ arrector pili muscles.

Vicq d'Azyr's line: ↑ Striate area

Villi, arachnoid: ↑ Arachnoid villi

Villi, chorionic: ↑ Chorionic villi

Villi, intestinal: ↑ Intestinal villi

Villi, placental: ↑ Placental villi, of early placenta; ↑ Placental villi, of full-term placenta

Vimentin: A protein (m.w. about 57000 daltons) found in some ↑ intermediate filaments.

Henderson, D., Geisler, N., Weber, K.: J. Mol. Biol. *155*, 173 (1982)

Vinculin: An intracellular protein (m.w. 130000 daltons) localized at specialized sites where microfilaments (mostly actin) terminate at ↑ cell membranes in various cells (e.g., ↑ zonula adherens of epithelial cells, at sarcolemma-associated ↑ dense bodies of ↑ smooth muscle cells, at ↑ fascia adherens of ↑ intercalated disc membranes, at ↑ costameres, etc.). It is thought that V. participate in anchoring microfilaments to specific membrane sites of various cells. It has also been postulated that V. binds to ↑ actin myofilament ends and inhibits actin polymerization. V. does not comprise part of cell membrane protein.

Geiger, B., Tokuyasu, K.T., Dutton, A., Singer, S.L.: Proc. Natl. Acad. Sci. USA *77*, 4127 (1980); Otto, J.J.: J. Cell Biol. *97*, 1283 (1983)

VIP: ↑ Vasoactive intestinal peptide

Virchow-Robin spaces: Conicial extensions of ↑ subarachnoid space surrounding blood vessels entering and leaving brain; delimited toward brain by ↑ pia mater. Under normal conditions, V. contain ↑ cerebrospinal fluid, ↑ fibrocytes, ↑ collagen fibers, and occasional ↑ wandering cells.

↑ Brain, penetration of blood vessels into

Visceral layer, of Bowman's capsule: ↑ Bowman's capsule

Visceral muscles (involuntary muscles): Muscles innervated by ↑ autonomic nervous system and independent of will. Cardiac and ↑ smooth muscles are V.

Visceral pleura: ↑ Pleura

Visual cells, of retina: ↑ Photoreceptors

Visual purple: A common term for ↑ iodopsin and ↑ rhodopsin.

Vital staining: A procedure in which a nontoxic dye (lithium carmin, ↑ trypan blue) is injected into a living organism to study vital processes of cells capable of accumulating it, either by ↑ athrocytosis or ↑ phagocytosis (↑ proximal tubule cells of ↑ nephron, ↑ macrophages).

↑ Supravital staining

Vitellogenesis: A process of formation of ↑ yolk granules in ↑ oocytes. It is assumed that protein moiety of ↑ yolk is synthesized by rough ↑ endoplasmic reticulum and carbohydrate moiety by Golgi apparatus. In the course of this synthesis, ↑ vitellogenin absorbed by oocyte is added to yolk. According to some authors, yolk granules arise from mitochondria (M) by their gradual loss of ↑ cristae and transformation of ↑ mitochondrial matrix in yolk (Y). (Fig. = rat)

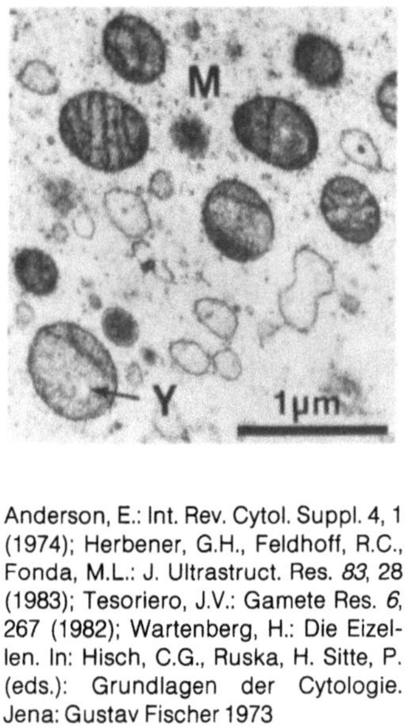

Anderson, E.: Int. Rev. Cytol. Suppl. 4, 1 (1974); Herbener, G.H., Feldhoff, R.C., Fonda, M.L.: J. Ultrastruct. Res. *83*, 28 (1983); Tesoriero, J.V.: Gamete Res. *6*, 267 (1982); Wartenberg, H.: Die Eizellen. In: Hisch, C.G., Ruska, H. Sitte, P. (eds.): Grundlagen der Cytologie. Jena: Gustav Fischer 1973

Vitellogenin: A ↑ yolk protein synthesized in ↑ liver and absorbed in ↑ micropinocytotic vesicles by vitellogenic ↑ oocytes.

Vitreous: ↑ Vitreous body

Vitreous body (vitreous humor, vitreous, corpus vitreum*): A colorless, amorphous, transparent, and gelatinous mass filling cavity between ↑ lens (L) and ↑ retina (R). As a kind of extremely thin connective tissue, V. consists of nearly 99.9% water with a ↑ refractive index of 1.334. Structure: 1) Solid phase = a) ↑ hyalocytes; b) a very small number of microfibrils about 10 nm thick, probably collagenous (periodicity 12 nm). Microfibrils are attached to ↑ inner limiting membrane of retina. Hyalocytes, occasional ↑ macrophages, and microfibrils are predominantly located at periphery, capsule (C) of V. Capsule is slightly thickened at anterior limit of V. forming ↑ hyaloid membrane (HM). 2) Liquid phase = long chains of polymerized ↑ hyaluronic acid binding large amounts of water. In certain V., a roughly 1-mm-long ↑ hyaloid canal extends between lens and ↑ optic papilla (OP). Functions of V.: Transmission of light rays, maintenance of lens in its position, maintenance of inner retinal layers in contact with ↑ pigment epithelium, and participation in retinal metabolism.

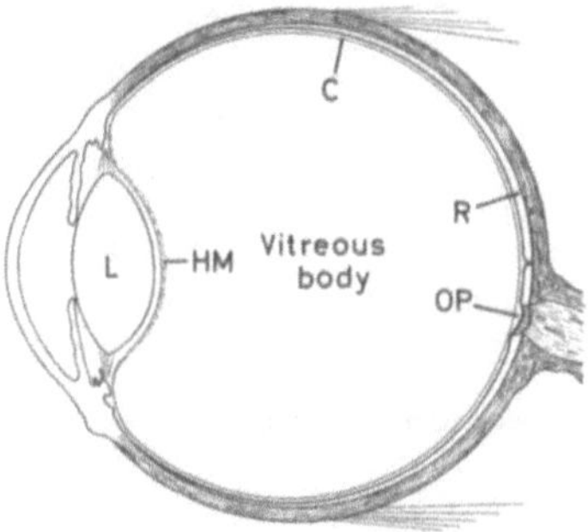

Hansson, H.-A.: Biomed. Res. *2*, Suppl. 465 (1981); Rhodes, R.H.: Histochemistry *78*, 125 (1983); Smith, G.N.Jr., Newsome, D.A.: Dev. Biol. *62*, 65 (1978)

Vitreous cells: ↑ Hyalocytes

Vitreous humor: ↑ Vitreous body

VLDL: ↑ Very low-density lipoprotein

Vocal cords (vocal folds, true vocal cords, plica vocalis*): Inferior pleats (VC) of laryngeal tunica mucosa. Structure: 1) Epithelium covering V. is ↑ stratified squamous nonkeratinized epithelium (region between arrows), passing gradually into ↑ pseudostratified ciliated epithelium. 2) Lamina propria contains a dense accumulation of ↑ elastic fibers composing ↑ vocal ligament (VL) and a striated ↑ vocal muscle (VM) at its lateral side. V. are involved in voice production.

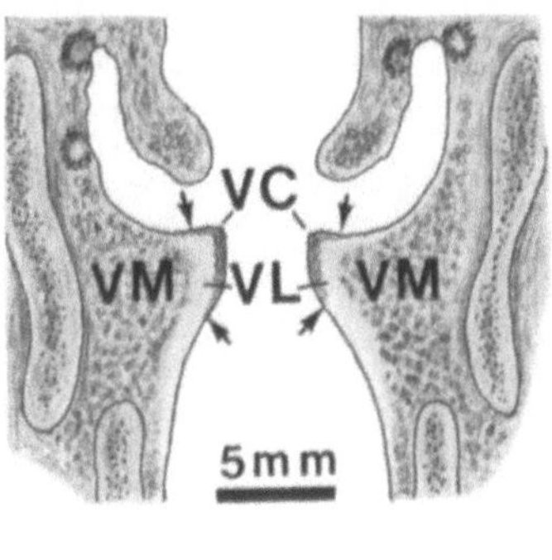

↑ Larynx

Vocal folds: ↑ Vocal cords

Vocal ligament (ligamentum vocale*): A band of ↑ elastic fibers stretched between ↑ arythenoid carilage and midline point of ↑ thyroid cartilage. V. is part of ↑ vocal cord.

Vocal muscle (musculus vocalis*): Intrinsic skeletal laryngeal muscle stretched between processus vocalis of ↑ arythenoid cartilage and midline point of ↑ thyroid cartilage. By contraction, V. confers different widths to rima glottidis and thus plays an important role in phonation.

↑ Glottis

Volkmann's canals (canalis perforans*): Channels (VC), about 20–50 μm in diameter, connecting ↑ Haversian canals (HC) with one another and with bone surfaces (periosteal, P; endosteal, E). V. are oblique or transverse to axis of a long bone. Through V., blood and lymphatic vessels and nerve fibers, predominantly from ↑ periosteum (P), reach Haversian canals; a minor part of these elements originates from medullar cavity (MC). V. are not surrounded by concentric ↑ lamellae of bone.

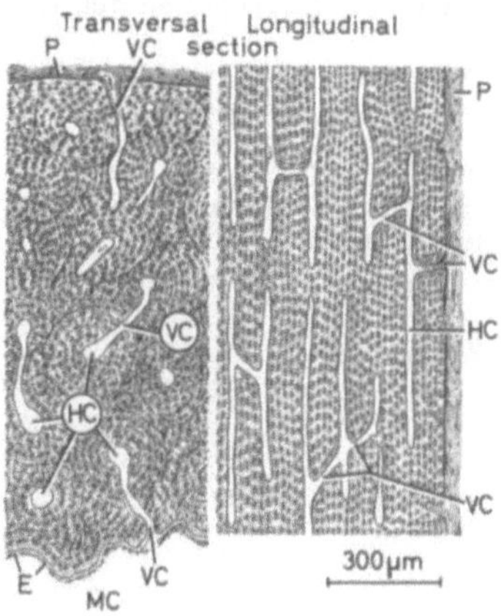

Volume density: A parameter in ↑ stereology.

Voluntary muscles: All ↑ skeletal muscles controlled by ↑ somatic nervous system.

↑ Motor end plate

Vomeronasal organ (Jacobson's organ, organum vomeronasale*): A complex tubular structure composed of a ↑ sensory epithelium, supporting hyaline ↑ cartilage, blood vessels, and nerves. V. is a sensory organ probably involved in olfaction in animals, but present in humans only during embryonal life.

Vaccarezza, O.L., Sepich, L.N., Tramezzani, J.H.: J. Anat. *132*, 167 (1981)

W

Wall, of cell: ↑ Cell wall

Wallerian degeneration: ↑ Nerve fibers of peripheral nervous system, degeneration and regeneration of

Wandering cells, of connective tissue proper: Mobile cells present in ↑ connective tissue proper with the function of: 1) Defense of organism, either through ↑ phagocytosis or synthesis of antibodies (↑ granulocytes, eosinophilic and neutrophilic, ↑ histiocytes, ↑ lymphocytes, ↑ monocytes, and ↑ plasma cells); 2) production of part of ↑ ground substance of connective tissue proper (↑ mast cells).

"Wear and tear" bodies: A term sometimes used to designate ↑ residual bodies.

↑ Lipofuscin

Web, terminal: ↑ Terminal web

Weibel-Palade bodies: ↑ Specific endothelial organelles

Wharton's duct: ↑ Submandibular duct

Wharton's jelly: ↑ Gelatinous tissue

White adipose tissue: ↑ Adipose tissue, white

White blood cells: ↑ Leukocytes

White matter, of central nervous system (substantia alba*): The mass of CNS, consisting of ↑ myelinated and ↑ nonmyelinated axons, ↑ oligodendrocytes, fibrous ↑ astrocytes, ↑ microglia, and blood vessels. Bodies of ↑ neurons lack W.

White pulp, of spleen (pulpa alba*): ↑ Total ↑ periarterial lymphatic sheaths and ↑ splenic nodules.

↑ Marginal zone, of spleen; ↑ Red pulp

Dijkstra, C.D., Döpp, E.A.: Cell Tissue Res. *229*, 351 (1983); Veerman, A.J.O., van Ewijk, W.: Cell Tissue Res. *156*, 417 (1975); Weiss, L.: Bull. Johns Hopkins Hosp. *115*, 99 (1974)

White skeletal muscle fibers: ↑ Skeletal muscle fibers, white

Wing, of nostril (ala nasi*): A cutaneocartilaginous fold comprising lateral wall of each nostril. In ↑ skin of outer surface, there is an abundance of ↑ sebaceous follicles (SF), generally not associated with hairs, a few eccrine ↑ sweat glands (SwG), and occasional hairs (H). Number of sebaceous glands and hairs decreases gradually toward inner surface of W., where ↑ glandulae vestibulares nasi (GV) and ↑ vibrissae (V) can be found. In interior of W., a piece of hyaline ↑ cartilage (C) is situated; ↑ striated muscle fibers of musculus nasalis (MN) are attached to ↑ periochondrium (P).

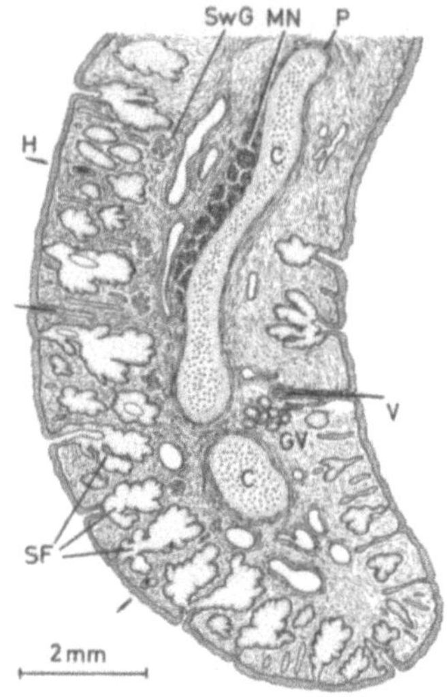

Wirsrung's duct: ↑ Pancreatic ducts

Witch's milk: A tiny colostrumlike milk secretion of ↑ mammary glands of some newborns of either sex for 3–4 days after birth. Secretion of W. is presumably due to increased levels of maternal ↑ prolactin.

↑ Colostrum

Wormian bones: Isolated ossicles (W) found within ↑ dense connective tissue of a ↑ syndesmosis (S). W. arise by detachment of groups of ↑ osteoblasts from margins of adjoining bones (B). (Fig. = ↑ sutura cranii, rat)

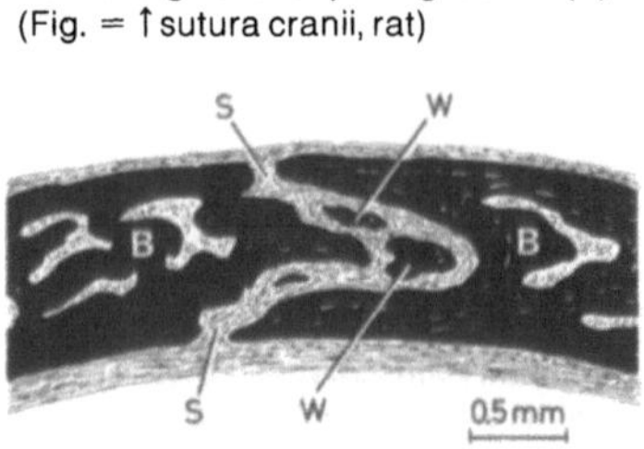

Wormlike bodies (micropinocytosis vermiformis*): Tortuous, sometimes branched, tubular invaginations (W) of cell membrane of ↑ Kupffer's cells, 0.5 μm or more long and of constant diameter (100–150 nm), found in some animals and probably also in man. W. display a median dense line (L), which possibly represents condensed ↑ glycocalyx. Exact function of W. is not known; it is thought that they might be an expression of initial stage of particle engulfment or a membrane reservoir permitting Kupffer's cells to enlarge their surface area.

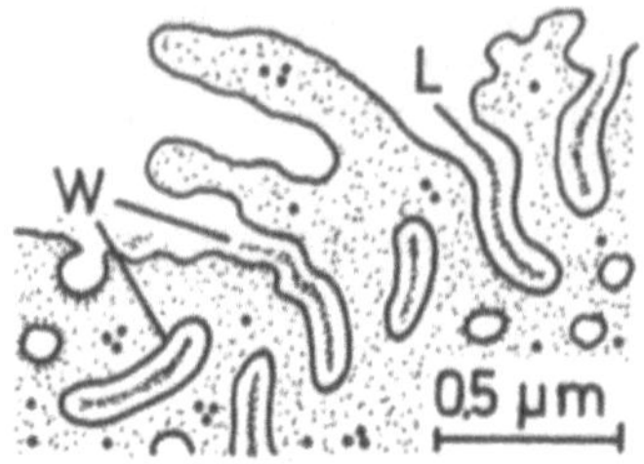

Tanuma, Y.: Arch. Histol. Jpn. *41*, 113 (1978)

Woven bone: ↑ Bone, immature

X

X-bodies: ↑ Langerhans' bodies

X-cells: ↑ Candidate stem cells

X-chromosome (X-chromosoma*): A medium-sized submetacentric ↑ heterochromosome determining the female sex when X. occurs in pairs. It seems in this case, only one X. is genetically active, this active X. also carries a great many genes of more general importance for the organism. Part of other X. becomes inactivated and transforms into ↑ Barr body and ↑ "drumstick."

↑ Chromosomes, submetacentric; ↑ Heterochromatin, facultative

Epstein, H.F., Wolf, S. (eds.): Genetic Analysis of the X Chromosome. Studies of Duchenne Muscular Dystrophy and Related Disorders. New York: Plenum Press 1982; Lyon, M.F.: Biol. Rev. *47*, 1 (1972)

X-ray diffraction: A technique based upon diffraction of X-rays on biological crystals or biological objects having a regular internal structure in at least one direction (↑ collagen, ↑ hemoglobin, ↑ myelin, ↑ myoglobin, ↑ nucleic acids, etc.). Principle: A narrow X-ray beam (X) of known wavelength is directed onto object (O) to be analyzed; it diffracts X-ray on the planes of

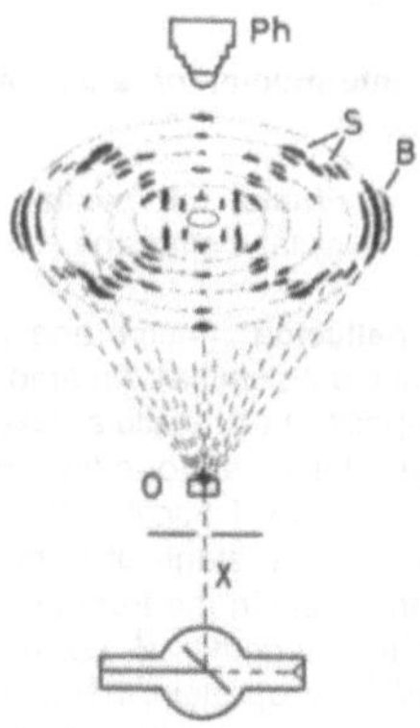

its lattice and produces a series of concentric spots (S) or bands (B) on a screen or film (Ph). The spots and bands result from interference between different diffracted rays. Distances between planes of lattice can be calculated from angle of a defined spot in diffraction pattern and wavelength of X-ray used. Since wavelength of an X-ray at 10 kV is 0.123 nm, X. permits three-dimensional analysis of atomic organization of a molecule.

Huskins, D.W.L. (ed.): X-Ray Diffraction by Disordered and Ordered Systems. Oxford. Pergamon Press 1982; Kendrew, J.C.: Science *139*, 1259 (1963); Perutz, M.F.: Angew. Chemie, *75*, 589 (1963); Watson, J.D., Crick, F.H.C.: Nature *171*, 737 (1953)

X-ray microanalysis: ↑ Electron probe microanalysis

X-ray microanalyzer: ↑ Electron probe microanalyzer

Y

Y-chromosome (Y-chromosoma*): A small acrocentric ↑ heterochromosome without ↑ satellite carrying genes for development of ↑ testes in all mammals. Y. can be visualized in interphasic nuclei after staining with ↑ quinacrine mustard.

↑ Chromosomes, acrocentric

Yellow bone marrow: ↑ Bone marrow, yellow

Yolk (vitellus*): A nutritive ↑ inclusion material found within ↑ oocytes, utilized subsequent to ↑ fertilization for support of earliest morphogenesis. Y. is concentrated within ↑ yolk granules and consists of proteins (↑ vitellogenin), lipids, and polysaccharides. In mammals, Y. seems to be partially synthesized within oocytes and partially by cells other than oocytes (↑ liver parenchymal cells) and interiorized by oocytes in ↑ micropinocytotic vesicles.

↑ Vitellogenesis

Yolk granules: Approximately 0.5-µm ↑ unit membrane-bound spherical ↑ inclusions (YG) of ↑ oocytes with predominantly homogenous fine granular contents. Y. are formed during ↑ vitellogenesis and contain ↑ yolk. (Fig. = rat)

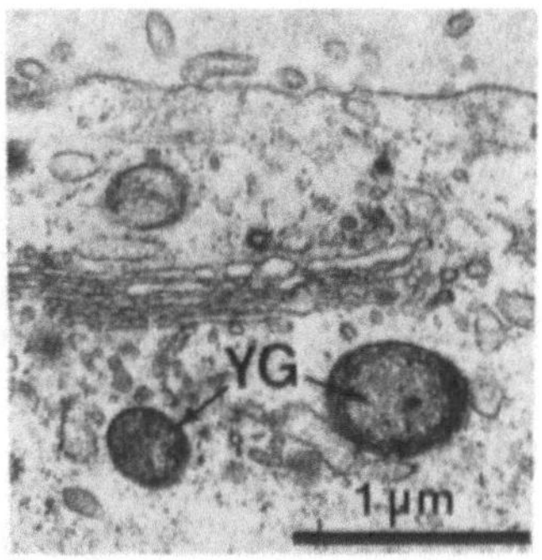

Korfmeister, K.-H.: Verh. Anat. Ges. *75*, 621 (1981); Pasteels, J.J.: Yolk and Lysosomes. In: Dingle, J.T. (ed.): Lysosomes in Biology and Pathology. Vol. 3. Amsterdam, London: North Holland Publishing Co. 1973

Z

Z-disc: ↑ Z-line

Zeis' glands: ↑ Glands, of Zeis

Zenker's fixative: A fixative mixture composed of potassium bichromate, sodium sulfate, mercuric chloride, and acetic acid.

↑ Fixation

Z-filaments: ↑ Z-line; ↑ Z-line, possible model of

Zinc iodide-osmium tetroxide (ZIO): A mixture of zinc iodide with OsO_4 reacting with some lipid-containing substances resulting in high contrast in transmission electron microscope. With Z., some ↑ dense-core synaptic vesicles (SV) of ↑ adrenergic nerve endings (fig.), ↑ synaptoid vesicles, some Golgi cisternae and their vesicles, ↑ cytosomes, etc. are impregnated. (Fig. = ↑ pineal organ, rat)

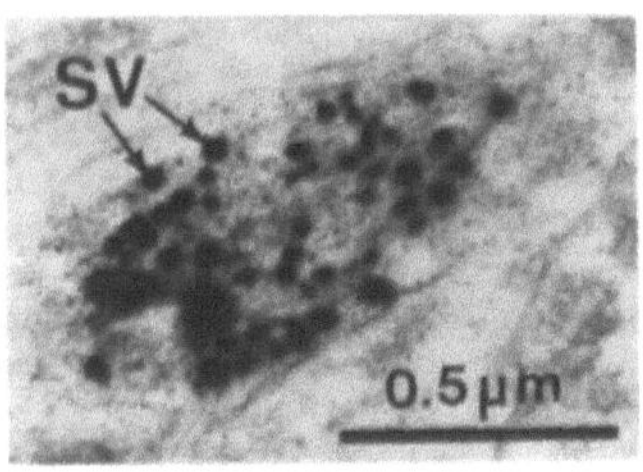

Blümcke, S., Kessler, W.D., Niedorf, H.R., Becker, N.H., Veith, F.J.: J. Ultrastruct. Res. *42*, 417 (1973); Osborne, M.P., Thornhill, R.A.: J. Neurocytol. *3*, 459 (1974); Pellegrino de Iraldi, A.: Experientia *29*, 844 (1973); Rodriguez, E.M., Gimenez, A.R.: Z. mikrosk.-anat. Forsch. *95*, 257 (1981); Vrensen, G., De Groodt, D.: Brain Res. *74*, 131 (1974)

ZIO: ↑ Zinc iodide-osmium tetroxide

Z-line (Z-disc, linea Z*, telophrama*): A roughly 0.1-μm-wide dark line (Z) situated in the middle of ↑ I-band. Structure: 1) In longitudinal section (L), one can observe that ↑ actin myofilaments (Ac) approaching Z. split into two ↑ Z-filaments (Zf) that intermesh with the two Z-filaments on opposite side of Z., creating a zigzag appearance. 2) In transversal section (T), Z. displays a regular square or basket weave pattern (arrows). All structures of Z. are embedded in a ↑ Z-matrix (Zm). Exact organization and chemical composition of Z. are very complex and not completely understood. Z. of adjacent ↑ myofibrils are connected with one another and with ↑ sarcolemma by a system of transversally oriented ↑ intermediate ↑ desmin microfilaments. (Fig. = rat)

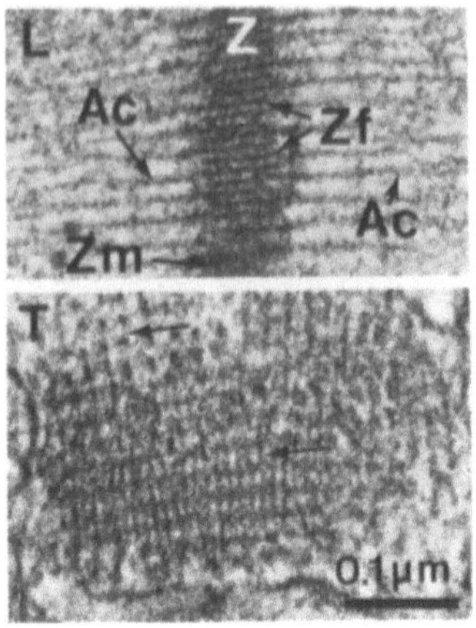

Chowrashi, P.K., Pepe, F.A.: J. Cell Biol. *94*, 565 (1982); Saetersdal, T., Engedal, H., Lie, R., Mykelbust, R.: Cell Tissue Res. *207*, 21 (1980); Sjöström, M., Kidmen, S., Larsen, H., Ängquist, K.-A.: J. Histochem. Cytochem. *30*, 1 (1982); Wang, K., Ramirez-Mitchell, R.: J. Cell Biol. *96*, 562 (1983); Yamaguchi, M., Robson, R.M., Stromer, M.H., Cholvin, N.R., Izumimoto, M.: Anat. Rec. *206*, 345 (1983)

Z-line, possible model of: Approaching ↑ Z-line (Z), ↑ actin myofilaments (A), having a hexagonal arrangement around ↑ myosin myofilaments, (M) are displaced laterally about 10 nm and become tetragonally arranged. The termination of each actin myofilament splits into four ↑ Z-filaments (Zf) arranged in the form of a small pyramid. Each Z-filament of one side of Z-line is connected with Z-filaments of opposite side, so that in longitudinal sections one can see a characteristic zigzag pattern, and in transversal sections a regular square pattern. Presence of ↑ Z-matrix (Zm) permits better observation of square lattice of transversally sectioned Z-line. Recent investigations have shown that Z-line has a considerably more complicated structure than in this model, since more than four Z-filaments arranged in series have been discovered.

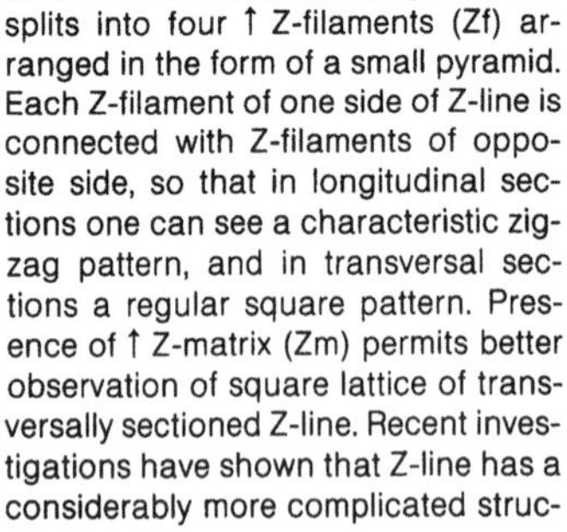

Goldstein, M.A., Schroeter, J.P., Sass, R.L.: J. Cell Biol. *83*, 187 (1979); Kelly, D.E., Cahill, M.A.: Anat. Rec. *172*, 623 (1972)

Z-matrix: A fine granular electron-dense material found in ↑ Z-line. Although ↑ tropomyosin and ↑ actinin are known to be present in Z.; its exakt chemical composition is still unknown.

Zona arcuata, of basilar membrane: ↑ Basilar membrane

Zona basalis, of decidua basalis: ↑ Decidua basalis

Zona columnaris, of anal canal: ↑ Anal canal

Zona compacta, of decidua basalis: ↑ Decidua basalis

Zona cutanea, of anal canal: ↑ Anal canal

Zona fasciculata: ↑ Adrenal glands, cortex of

Zona glomerulosa: ↑ Adrenal glands, cortex of

Zona intermedia, of adrenal cortex: An inconstant narrow cell layer between zona glomerulosa and zona fasciculata.

Zona intermedia, of anal canal: ↑ Anal canal

Zona pectinata, of basilar membrane: ↑ Basilar membrane

Zona pellucida* (membrana pellucida, mucoid oolemma): An amorphous ↑ PAS-positive extracellular layer (ZP), 5–10 μm thick, surrounding primary and secondary ↑ oocytes (O); Z. first appears at the stage of primordial ↑ ovarian follicle in the form of plaques which fuse together. Z. consists of a gell-like ↑ glycoprotein; in transmission electron microscope it is well delimited and composed of fine granular and moderately osmiophilic material. Numerous microvilli (arrows) of both oocyte and ↑ follicular cells (F) penetrate Z. and make contact. Some authors believe that Z. is produced by follicular cells; others are of the opinion that oocyte also participates in its formation. During ↑ acrosome reaction, Z. is dissolved by acrosomal ↑ hyaluronidase. (Fig. = rat)

↑ Cuticle

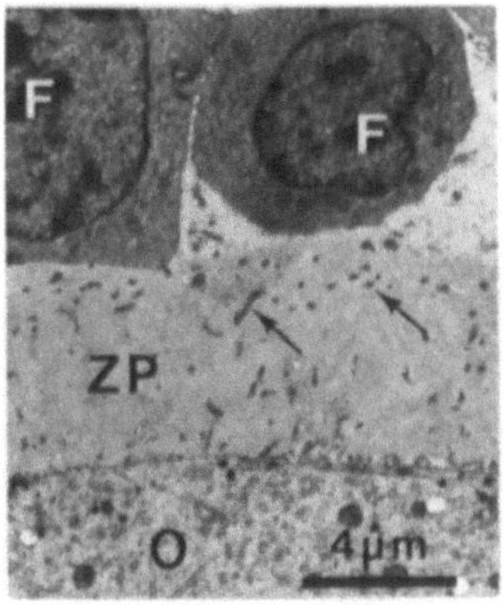

Bleil, J.D., Wassarman, P.M.: Dev. Biol. *76*, 185 (1980); Greve, J.M., Salzmann, G.S., Roller, R.J., Wassarman, P.M.: Cell *31*, 749 (1982); Philips, D.M., Shalgi, R.: J. Ultrastruct. Res. *72*, 1 (1980); Weymarn, N.V., Guggenheim, R., Müller, H.J.: Anat. Embryol. *161*, 19 (1980)

Zona reticularis: ↑ Adrenal glands, cortex of

Zona spongiosa, of decidua basalis: ↑ Decidua basalis

Zone, extrusion: ↑ Extrusion zone

Zone, of Weil: A relatively cell-poor zone (W) of tooth ↑ pulp (P), immediately underlying layer of ↑ odontoblasts (O). D = ↑ dentin. (Fig. = human)

Fox, A.G., Heeley, J.D.: Arch. Oral Biol. *25*, 103 (1980)

Zones, of cytoplasm: Three more or less delimited concentric areas in cell body: 1) ↑ Centrosphere in the middle of the cell surrounded by ↑ Golgi apparatus (G); 2) ↑ endoplasm containing ↑ nucleus (N) and cell ↑ orgenelles [except for ↑ centriole (C)]; 3) ↑ ectoplasm free of organelles.

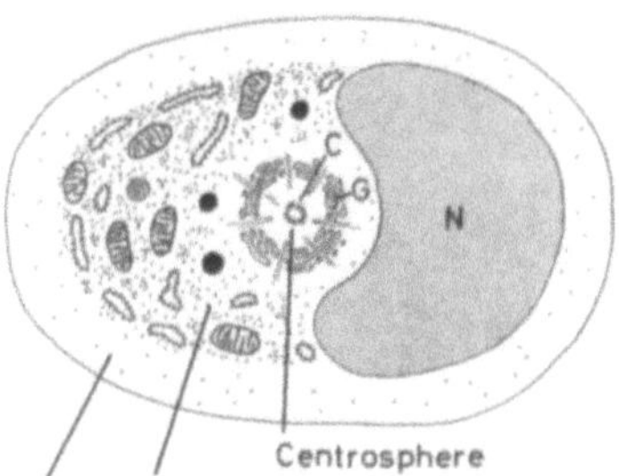

Zonula adherens* (intermediate junction, adhering junction): Part (ZA) of ↑ terminal bar (TB) situated just below ↑ zonula occludens (ZO). In the area of Z., intercellular gap (G) measures 15–20 nm and contains a moderately dense osmiophilic amorphous material of an unknown nature. Cytoplasmic leaflets of ↑ cell membranes facing Z. are slightly thickened; fine 5 to 7-nm-thick ↑ actin microfilaments (Mf) are anchored there, radiating into ↑ terminal web (TW). Like zonula occuludens, Z. runs as a circumferential belt around ↑ apical pole and contributes to maintenance of cell cohesion. Z. represents an insertion area for terminal web. Extensive Z. are present in ↑ intercalated discs; there, Z. have a slightly modified structure and are referred to as ↑ fasciae adherentes.

↑ Microvilli, shortening of

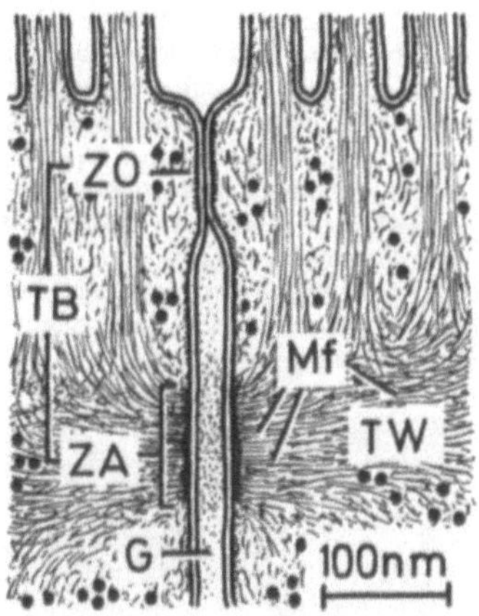

Zonula ciliaris: ↑ Ciliary zonule

Zonula occludens* (occluding junctions, tight junction): A roughly 0.1 to 0.5-µm-wide beltlike zone (ZO) of fusion of outer leaflets of ↑ cell membranes to two adjacent epithelial cells. High transmission electron microscopic magnification shows that fusion is effected only at points (arrowheads). There is no ↑ glycocalyx within Z. Z. represents outermost component of ↑ terminal bar and runs around ↑ apical pole. Z. seals off underlying intercellular cleft from contact with outside environment; it contributes to cell adhesion and participates in ionic and metabolic cell interactions. (Fig. = ↑ gallbladder epithelium, mongolian gerbil)

↑ Zonula occludens, freeze-cleaving of; ↑ Zonula occludens, possible molecular structure of

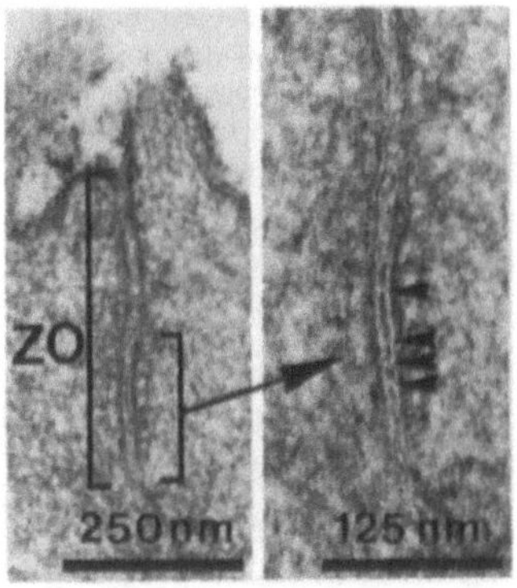

Briggman, J.V., Bank, H.L., Bigelow, J.B., Graves, J.S., Spicer, S.S.: Am. J. Anat. *162*, 357 (1981); Kachar, B., Reese, T.S.: Nature 296, 464 (1982); Shimono, M., Nishihara, K., Yamamura, T.: J. Electron Microsc. *30*, 29 (1981); Wade, J.B.: J. Cell Biol. *60*, 168 (1974)

Zonula occludens, freeze-cleaving of: In ↑ replicas, ↑ zonula occludens (ZO) appears as a characteristic weblike array of anastomosing ridges (R) at ↑ PF face of ↑ cell membrane and as corresponding shallow depressions (D) at ↑ EF face of cell membrane. ↑ Freeze-cleaving has enabled some hypotheses to be developed concerning molecular structure of zonula occludens. (Fig. = ↑ surface mucous cell of stomach, dog)

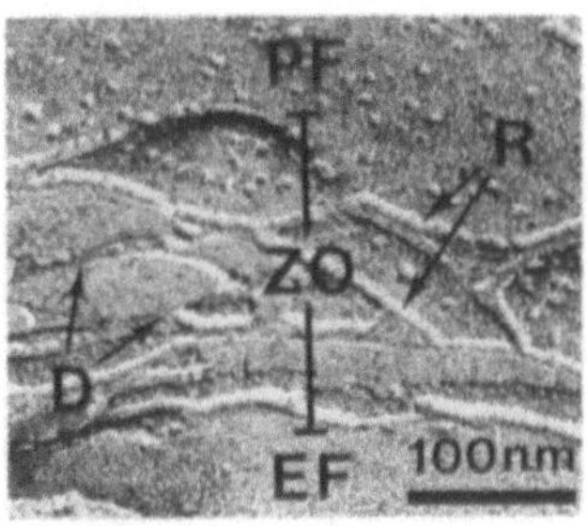

↑ Zonula occludens, possible molecular structure of

Hirokawa, N.: J. Ultrastruct Res. *80*, 288 (1982); Van Deurs, B., Koehler, J.K.: Cell Biol. *80*, 662 (1979)

Zonula occludens, possible molecular structure of: The punctate contacts (PC) appear to correspond to protein particles (P) traversing apposed ↑ cell membranes (CM) and fusing by their tips with inner leaflets (IL)

of cell membranes. When inner leaflet of cell membrane is moved out by ↑ freeze-cleaving denuded protein particles forming ridges (R) are visible; when inner leaflet remains intact, one can observe shallow depressions (D) corresponding to sites where tips of particles fuse with inner plasmalemmal leaflet. (Modified after Staehelin 1973)

↑ Zonula occludens; ↑ Zonula occludens, freeze-cleaving of

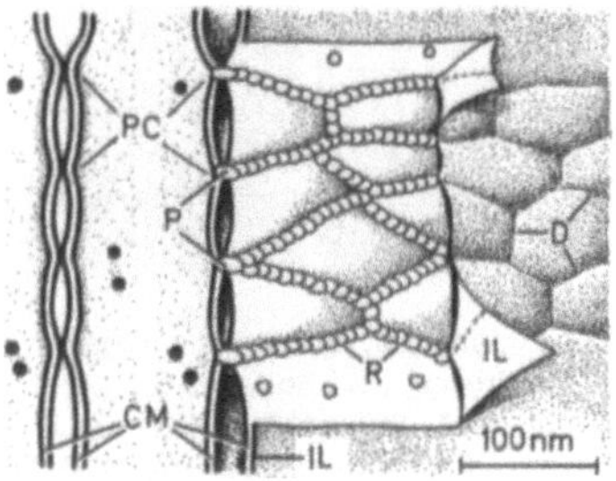

Staehelin, L.A.: J. Cell Sci. *13*, 763 (1973)

Zonular fibers (zonular filaments, zonules, fibrae zonulares*): Bundled microfilaments (ZF) forming ↑ ciliary zonule. Z. are stretched between ↑ ciliary epithelium (CE) and equator of ↑ lens. Z. measure about 10 nm in thickness and display a periodicity of approximately 12.5 nm (fig. bottom); in transversal sections, Z. have a tubular appearance. They are attached to ↑ basal lamina (BL) of ↑ ciliary processes and lens ↑ capsule. Near lens equator, Z. fuse in thicker fibers and form about 140 bundles. Their chemical composi-

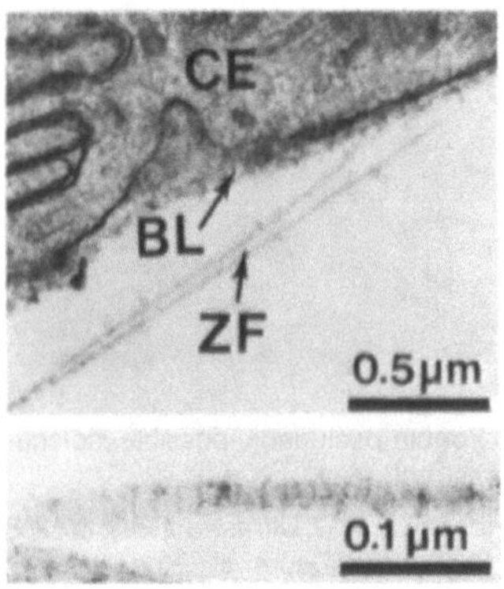

tion appears to correspond to that of ↑ elastic fibers, since they are digested by ↑ elastase and not by ↑ collagenases. Tension of Z. is controlled by contraction and relaxation of ↑ ciliary muscles: Contraction relaxes Z. and lens springs into a more convex shape, thus increasing its refractive power. (Figs. = mongolian gerbil)

Curtis, R.: J. Anat. *136*, 69 (1983)

Zuckerkandl's organ (paraganglion aorticum abdominale): The largest of sympathetic ↑ paraganglia situated at beginning of arteria mesenterica inferior.

Zygote (zygota*): The first ↑ diploid cell of a new organism formed by fusion of two ↑ haploid ↑ germ cells of opposite sexes.

↑ Fertilization

Zygotene: ↑ Meiosis

Zymogen granules (presecretory granules, granulum zymogeni*): Spherical, ↑ unit membrane-bound ↑ inclusions (ZG), 0.3–1.5 μm in diameter, with fine granular contents present in ↑ apical pole of ↑ chief cells of ↑ gastric glands proper (fig., left) and ↑ pancreatic acinar cells (fig., right). In the first cells, Z. are weakly osmiophilic; in the second cells, they are very electron dense. Z. arise from Golgi apparatus (G) and contain, apart from numerous other inactive enzymes, propepsin in chief cells and trypsinogen in pancreatic acinar cells. Contents of Z. are released by ↑ exocytosis into glandular or acinar lumen; activation of enzymes takes place only in lumen of ↑ stomach or in ↑ duodenum. (Figs. = rat)

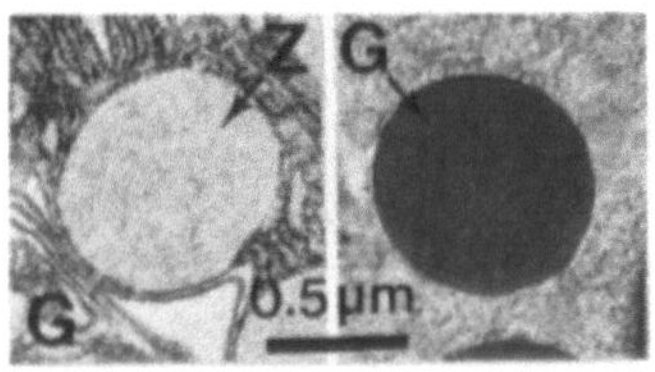

Pâquet, M.R., St.-Jean, P., Roberge, M., Beadouin, A.R.: Eur. J. Cell Biol. *28*, 20 (1982); Reggio, H.A., Palade, G.E.: J. Cell Biol. *77*, 288 (1978)

Zymogenic cells: ↑ Chief cells, of gastric glands proper

MIX
Papier aus verantwortungsvollen Quellen
Paper from responsible sources
FSC® C105338

If you have any concerns about our products,
you can contact us on
ProductSafety@springernature.com

In case Publisher is established outside the EU,
the EU authorized representative is:
Springer Nature Customer Service Center GmbH
Europaplatz 3, 69115 Heidelberg, Germany

Printed by Libri Plureos GmbH
in Hamburg, Germany